Commonly Used
Ornamental Plants

The Editors

Dr. R. L. Misra served the cause of floricultural science for more than 16 years in temperate regions of Himachal Pradesh and about 24 years in sub-tropical conditions of New Delhi. After his joining at IARI, New Delhi in 1984, he was associated with all the 7-8 floricultural courses being offered to the M.Sc. and Ph.D. students, apart from looking his mandated research programmes. Out of his whole official career, as a chairman, he guided 16 Ph.D. and 6 M.Sc. students apart from being members of the advisory committees of many more students, and developed 30 gladiolus varieties which have high demand countrywide. Now all these students are serving various institutions as Professors and Associate Professors and in other departments holding a very high position. His contributions are acclaimed worldwide. As a teacher as well as a researcher, his contribution is well known as he has in all some 500 publications out of which some 170 are research publications, 80 symposium papers, 80 book chapters, a few review articles including 6 books. He is the founder member of Indian Society of Ornamental Horticulture from 1993 to 2004 (12 years continuously) he edited its regular publication Journal of Ornamental Horticulture. He is Fellow of Indian Society of Genetics and Plant Breeding since 80's, and Indian Society of Ornamental Horticulture since 2010. For his most significant contribution in the field of floriculture, he was honoured with 'Gold Medal' of Horticultural Society of India in 2013. He is life member in 11 professional societies. Once he was Officer-in-Charge, Indo-Israel Project of Research and Development on Farms, IARI, New Delhi which confines to the protected cultivation of vegetables and flowers; and joined as a regular Project Coordinator of All India Coordinated Research Project on Floriculture of ICAR at Division of Floriculture and Landscaping, IARI, New Delhi where he coordinated some 21 centres across 17 states. He served IARI for 40 years and retired as Project Coordinator *cum* Principal Scientist in May 2008. Only because of his highly significant contributions in the field of teaching and research, after his retirement he was inducted as adjunct Faculty of IARI in Floriculture.

Dr. Sanyat Misra is a graduate from UAS, Bangalore; postgraduate from GBP University of Agriculture and Technology, Pantnagar; and Ph.D. from Dr. YSP University of Horticulture and Forestry, Solan (H.P.), and then after completing his Ph.D. degree served for 3 years as Research Associate at NBPGR, New Delhi and Shimla, and in 2004 joined as regular Scientist *cum* Assistant Professor (Floriculture) in the Department of Horticulture, Birsa Agricultural University, Kanke, Ranchi-834006 (Jharkhand). So far he has guided three M.Sc. students for their thesis work. He is a very active member of the horticulture faculty looking after research, teaching and extension. He has to his credit about 20 publications, two bulletins and three books of international repute.

Commonly Used
Ornamental Plants

– Editors –

R. L. Misra

Sanyat Misra

2023

Daya Publishing House®

A Division of

Astral International Pvt. Ltd.

New Delhi – 110002

ISBN: 9789359191041 (HB)

Published by : **Daya Publishing House®**
A Division of
Astral International Pvt. Ltd.
– ISO 9001:2015 Certified Company –
4736/23, Ansari Road, Darya Ganj
New Delhi-110 002
Ph. 011-43549197, 23278134
E-mail: info@astralint.com
Website: www.astralint.com

Prof. R.B. Singh
Chancellor
Central Agricultural University, Imphal

Formerly
President, National Academy of Agricultural Sciences, New Delhi
Member, National Commission on Farmers, New Delhi
Assistant Director General & Regional Representative for Asia & the Pacific, FAO of the United Nations, Bangkok
Chairman, Agricultural Scientists Recruitment Board, New Delhi
Director & Vice-Chancellor, Indian Agricultural Research Institute, New Delhi

Foreword

Ornamental plants have been an integral part of the rich Indian culture and tradition since the time immemorial. Giving its proverbial agro-ecological and geographical diversity, India is one of the mega centers of origin and diversity of ornamental plants. As the civilization evolved, a good number of these plants were domesticated not only for their aesthetic value but also for medicinal usages and religious purposes. Couched in the rich background of the wild forms, the production and distribution (marketing) of ornamental plants has emerged as an important socio-economic enterprise.

Floriculture in India is a fast expanding industry producing economically rewarding and intellectually stimulating opportunity particularly for the youth in agriculture. Yet, India's share in the global floricultural trade is less than one percent. The research, academic, extension,, and training pursuits in this field must be commensurate with the value-chain management implications of the sub-sector. Starting in the late 1950s organized research and UG and PG education in floriculture/landscaping/ornamental crops in the country was initiated during the past 40 years in horticultural and agricultural universities and in related ICAR institutes, culminating in the All India Coordinated Research Project on Floriculture and the Directorate of Floriculture, Non-ICAR institutions, such as CSIR-National Botanical Research Institute, have also been pursuing floriculture research. These initiatives and future opportunities must be adequately reflected in the university curricula and teaching programmes towards creating quality human resources and enriching research and extension activities.

With the above backdrop, I must compliment Drs. R.L. Misra and Sanyat Misra, leading experts in floriculture, for bringing out the book **"Commonly Used Ornamental Plants".** The book contains 41 chapters separately dealing with 30 various common ornamental crops such as *Aglaonema, Alstroemeria, Araucaria, Begonia, Bellis, Bougainvillea, Canna, Catharanthus, Codiaeum, Columnea, Dieffenbachia, Euphorbia, Ficus, Fuchsia, Geranium & Pelargonium, Hedera, Holmskioldia, Impatiens, Ixora, Lathyrus, Monstera, Mussaenda,* Ornamental Banana, *Petunia, Philodendron, Primula, Saintpaulia, Spathiphyllum, Streptocarpus* and *Viola;* and 11 groups of ornamentals such as Annuals with 126 genera, **Bromeliads** with 34 genera, **Cacti** with 105 genera, **Carnivorous Ornamentals** with 12 genera, **Ferns and Allied Plants** with 53 genera, **Flowering Indoor Plants** with 188 genera, **Foliage Plants** with 382 genera, **Lawn** with 37 genera, **Ornamental Gingers** with 7 genera, **Proteaceous Ornamentals** with 12 genera and **Succulents Other Than Cacti** with 192 genera. Details on each crop and group include: nomenclature, origin, brief history and botany; means of propagation including micropropagation; classification, species and varieties; production technology; manipulation of growth and development; plant protection; and post harvest technology in very lucid terms. Each chapter gives a succinct account of significant scientific works carried out worldwide.

This invaluable book will be an asset to the students, researchers and other stakeholders and fills a long-felt gap of an authentic scientific compilation on ornamental plants. I understand that authors have brought out two other complementary publications *viz.* (i) **Commercial Ornamental Crops: Cut Flowers** and (ii) **Commercial Ornamental Crops: Traditional and Loose Flowers,** first one already published through Kruger Brentt Publishers, England, and the second one is in the press. I congratulate the authors for their efforts for bringing out such a unique publication, and I wish this outstanding father-son team all success in its future endeavours.

(R. B. Singh)

Preface

Since the time immemorial, the flowers in Indian system are being used traditionally irrespective of the caste, creed and religions. There are many plants and trees which are therefore attached with our day-to-day life and many a plants have been associated with gods and goddesses. All over the country, people from different walks of life, irrespective of their sex, faith, creed and caste grow some ornamentals near their homes to be used in needs. Also in the villages, mostly there used to be one patch where local flowers were allowed to grow on their own from where villagers used to pluck them for ceremonial occasions and for offering to gods and goddesses. The trees or shrubs emitting pleasant fragrances especially by the sunset used to be planted in villages. Then there was no flower market, no flower business or any other thing alike and the people used to collect them unhesitantly from anywhere, irrespective of the fact that who is growing these. For making the look of the house and its surrounding beautiful, people were and are using ornamentals by planting around. Even in the mango orchard one can see such plantings and there had been no restrictions that who will use them. These are *Amaranthus, Amaryllis, Barleria, Bougainvillea, Calotropis, Canna, Catharanthus, Celosia, Clerodendron, Codiaeum, Crossandra, Dahlia, Dendranthema, Eranthemum, Ervatamia, Gaillardia, Helianthus, Heliconia, Hibiscus, Impatiens, Ixora, Jasminum, Lawsonia, Narcissus, Nerium, Nyctanthes, Pandanus, Polianthes, Rosa, Tagetes, Thevetia* and so on, which people used to grow and enjoy, and out of which a few were being used even for extraction of essential oils. These flowers were being plucked or harvested to be used for certain arrangements, decorations, for making garlands or for offering at the religious ceremonies or meant for social celebrations. In Thailand no marriage is solemnized without exchange of floral garlands of *Calotropis* between brides and bridegrooms. Loose flowers are, in fact, not really the loose ones as they may be quite compact but the people pluck them from the parent plant for net instant use. Since there was no concept in Asian countries for enjoying them indoors by arranging them in flower vases so such flowers were without a part of the stem as in the cut flowers but now cut flowers have also achieved an important place in the country. The cut flower concept we borrowed from western cultures. However, Asians residing in western countries, for their traditional use on certain occasions, import loose flowers from this country. We have dealt with '**Traditional and Loose Flowers**' separately in a book and the 'Commercial Ornamental Crops - Cut Flowers' separately but here it is something different to those. This country has several recognized markets in different cities now. Here in this book, we have put our all efforts to include almost all the important ornamentals which not only in India are being grown but world over. We have tried to furnish here their 'A to Z' proper cultivation practices in a truely comprehensive and scientific manner.

ICAR has formulated an uniform course programme in Floriculture for Master's and Doctoral programmes, therefore, this book has been written to avoid any confusion that what are important loose, cut or most commonly used ornamentals. Also one more query comes to the mind that what is the difference between a loose flower and a cut flower? ***Loose flowers*** may be short-lived, may not last long and have no long stalks attached with them, though ***cut flowers*** last long and have floral stalks attached with them which help the flowers sucking the water and nutrient solution from the vases. We assure that not only to the students and their teachers and professors of horticulture this book will be useful but will also serve an useful guide to the scientific community of floriculture, the landcapers and horticulturists, the growers and other garden enthusiasts. The authors have tried their best to make it an ever-published academic book of its kind, looking into its coverage and contents. We claim that all these publications [(i)'**Commercial Ornamental Crops - Cut Flowers.** (ii) **Commercial Ornamental Crops - Traditional and Loose Flowers**, (iii) **Commonly Used Ornamental Plants** and (iv) **Soilless Crop Production**] together will be quite helpful and a complete guide in commercial floricultural crops. However, almost those crops cover under commercial ornamental crops, cut flowers, traditional and loose flowers are not in cluded in this book.

R. L. Misra

Sanyat Misra

Contents

List of Contributors

Ajit Kumar

(Bromeliads, Cacti, Carnivorous Ornamentals)

Department of Horticulture, G.B. Pant University of Agriculture & Technology, Pantnagar - 263 145, Distt. Udham Singh Nagar (Uttarakhand)

Bag, M.K.

(*Araucaria*)

Germplasm Evaluation Division (Old Building), N.B.P.G.R., Pusa Campus, New Delhi-110 012.

Banerji, B.K.

(*Bougainvillea*)

B-4/108, Shrinathji Vihar, 538, Sitapur Road, Lucknow-226020 (U.P.). E-mail: banerjibk@yahoo.co.in

(Retd. from: Floriculture section, National Botanical Research Institute, Rana Pratap Marg, Lucknow-226001).

Bhavya Bhargava

(Carnivorous Ornamentals)

Institute of Himalayan Bioresource Technology, CSIR, Palampur-176061 (H.P.).

Chandrashekar, S.Y.

(*Codiaeum variegatum, Euphorbia, Ficus, Holmskioldia sanguinea*)

Department of Floriculture and Landscape Architecture, College of Horticulture, Mudigere-577132, Chikmangalur dist. (Karnataka).

Dhatt, K.K.

(*Viola*)

Department of Floriculture and Landscaping, Punjab Agricultural University, Ludhiana - 141 004.

Dhiman, S.R.

(Proteaceous Ornamentals)

Department of Floriculture and Landscaping, Dr. YSP Univ. of Hort. and Forestry, Nauni, Solan-173230 (H.P.)

Dubey, R.K.

(***Aglaonema*, *Bellis*,** Bromeliads, **Ferns and Allied Plants, *Lathyrus***)

Department of Floriculture and Landscaping, P.A.U., Ludhiana-141004 (Punjab).

Ganga, M.

(***Spathiphyllum***)

Horticultural College and Research Institute, T.N.A.U., Coimbatore-641003 (T.N.).

Gupta, Y.C.

(Carnivorous Ornamentals, Proteaceous Ornamentals)

Department of Floriculture and Landscaping, Dr Y S Parmar University of Horticulture and Forestry, Nauni-173230, Distt.- Solan (H.P.).

Indrajit Sarkar

(**Bromeliads**)

Department of Floriculture, Medicinal and aromatic Plants, Faculty of Horticulture, Uttar Banga Krishi Vishwavidyalaya, Pundibari, Coochbehar-736 165 (W.B.).

Jagadeeswari, V.

(*Spathiphyllum*)

Horticultural College and Research Institute, T.N.A.U., Coimbatore-641003 (T.N.).

Janakiram, T.

(*Impatiens*, ***Monstera*, *Mussaenda***, Ornamental Banana, ***Philodendron***)

Division of Floriculture and Landscaping, Indian Agricultural Research Institute, New Delhi-110012.

Jayoti Majumder

(***Ixora***)

Department of Floriculture and Landscaping, Faculty of Horticulture, B.C.K.V., Mohanpur-741252, Nadia (W.B.).

Kalkame Ch. Momin

(**Proteaceous Ornamentals**)

Department of Floriculture and Landscaping, Dr. YSP Univ. of Hort. and Forestry, Nauni-173230, Solan (H.P.)

Latha, S.

(*Euphorbia*, *Holmskioldia sanguinea*)

Department of Floriculture and Landscape Architecture, College of Horticulture, Mudigere-577132, Chikmangalur dist. (Karnataka).

Mekala, P.

(*Spathiphyllum*)

Horticultural College and Research Institute, T.N.A.U., Coimbatore-641003 (T.N.).

Meenu Kumari

(Columnea & Petunia × hybrida)

Division of Floriculture and Landscaping, I.A.R.I., New Delhi-110 012.

Misra, R.L.

(*Aglaonema,* Annuals, *Araucaria,* ***Bagonia***, *Bougainvillea,* Bromeliads, **Cacti**, Canna, Carnivorous Ornamentals, *Catharanthus, Euphorbia,* Ferns and Allied Plants, *Ficus,* Flowering Indoor Plants, *Fuchsia × hybrida, Geranium* & *Pelargonium, Ixora,* **Lawn**, *Primula,* Proteaceous Ornamentals, *Streptocarpus,* Succulents Other Than Cacti, *Viola*)

C-4, Brahma Apartments, Plot 7, Sector 7, Dwarka, New Delhi-110 075 (Mob.: 9968287841).

(Retired from Division of Floriculture and Landscaping, IARI, New Delhi-110 012).

Munikrishnappa, P.M.

(*Codiaeum variegatum, Euphorbia, Ficus*)

Department of Floriculture and Landscape Architecture, College of Horticulture, Mudigere-577132, Chikmangalur dist. (Karnataka).

Namita

(***Columnea***,*Fuchsia × hybrida, Hedera, Monstera,* ***Petunia***, *Streptocarpus*)

Division of Floriculture and Landscaping, I.A.R.I., New Delhi-110 012.

Nataraj, S.K.

(*Codiaeum variegatum, Euphorbia, Ficus, Holmskioldia sanguinea*)

Department of Floriculture and Landscape Architecture, College of Horticulture, Mudigere-577132, Chikmangalur dist. (Karnataka).

Padmadevi, K.

(*Spathiphyllum*)

Horticultural College and Research Institute, T.N.A.U., Coimbatore-641003 (T.N.).

Palai, S.K.

(Cacti)

College of Horticulture, O.U.A.T., Chiplima, Sambalpur-768025 (spalai@hotmail.com).

Parminder Singh

(Ferns and Allied Plants)

Department of Floriculture and Landscaping, P.A.U., Ludhiana-141004 (Punjab)

Poonam Kumari

(*Columnea, Fuchsia × hybrida, Primula*)

Division of Floriculture and Landscaping, I.A.R.I., New Delhi-110012.

Pragya Ranjan

(***Araucaria***, Bromeliads, ***Dieffenbachia***)

Indian Institute of Vegetable Research, P.O. Jakhani, Shahanshahpur, Varanasi-221 305 (U.P.).

Preeti Hatibarua

(Bromeliads)

Regional Research Station, Assam Agricultural University, Kahikuchi, P.O. Azara, Guwahati – (Assam)

Priyanka Sharma

(Carnivorous Ornamentals)

Department of Floriculture and Landscaping, Dr Y S Parmar University of Horticulture and Forestry, Nauni-173230, Distt.- Solan (H.P.).

Raju, D.V.S.

(*Viola*)

Division of Floriculture and Landscaping, I.A.R.I., New Delhi – 110 012.

Ranjan, J.K.

(*Araucaria, Dieffenbachia*)

Indian Institute of Vegetable Research, Jakhini, Shahanshahpur, Varanasi-321 005 (U.P.)

Ritu Jain

(*Mussaenda,* **Ornamental Banana**)

Division of Floriculture and Landscaping, I.A.R.I., New Delhi-110 012.

Sanjay Kumar

(*Alstroemeria, Catharanthus, Impatiens*)

CSIR-Institute of Himalayan Bioresource Technology, Palampur-176 061 (H.P.)

Sanyat Misra

(*Aglaonema,* **Annuals**, *Araucaria, Begonia, Bougainvillea,* Bromeliads, Cacti, ***Canna***, Carnivorous Ornamentals, *Columnea, Dieffenbachia, Euphorbia,* Ferns and Allied Plants, *Ficus,* **Flowering Indoor Plants, Foliage Plants**, *Geranium* & *Pelargonium, Hedera, Impatiens, Ixora, Lathyrus,* Lawn, *Petunia*, Proteaceous Ornamentals, *Saintpaulia, Streptocarpus,* **Succuleants Other Than Cacti**, *Viola*)

Department of Horticulture, Birsa Agricultural University, Kanke, Ranchi-834 006 (Jharkhand).

Sapna Panwar

(***Fuchsia* × *hybrida, Geranium* & *Pelargonium*, *Hedera*,** *Monstera, Petunia,* ***Primula, Saintpaulia, Streptocarpus***)

Division of Floriculture and Landscaping, I.A.R.I., New Delhi-110 012.

Sathyanarayana Reddy, B.

(***Codiaeum variegatum, Euphorbia, Ficus, Holmskioldia sanguinea***)

Department of Floriculture and Landscape Architecture, College of Horticulture, Mudigere-577132, Chikmangalur dist. (Karnataka). dr.reddybs@gmsil.com.

Sellam Perinban

(*Ixora*)

Division of Agricultural Engineering, I.A.R.I., New Delhi-110 012.

Sheela, V.L.

(**Ornamental Ginger**)

Department of Pomology and Floriculture, College of Agriculture, Vellayani, Thiruvananthapuram-695522 (Kerala)

Sheena, A.

(Ornamental Ginger)

Department of Pomology and Floriculture, College of Agriculture, Vellayani, Thiruvananthapuram-695522 (Kerala)

Simrat Singh

(*Aglaonema, Bellis,* Ferns and Allied Plants, *Lathyrus*)

Department of Floriculture and Landscaping, Punjab Agricultural University, Ludhiana-141 004.

Singh, D.R.

(*Araucaria*)

NRC on Orchids, Pakyong-737 106, East Sikkiim (Sikkim).

Singh, M.K.

(***Alstroemeria***, ***Catharanthus***, ***Impatiens,*** *Viola*)

Division of Floriculture and Landscaping, I.A.R.I., New Delhi-110 012.

Soudamini Karjee

(*Columnea*)

Division of Floriculture and Landscaping, I.A.R.I., New Delhi-110 012.

Thaneswari

(*Primula*)

Division of Ornamental Crops, I.I.H.R., Bengaluru-560089 (Karnataka).

Usha Sonkhle

(*Philodendron*)

Division of Floriculture and Landscaping, I.A.R.I., New Delhi-110 012.

Zahor Ahmed

(*Bellis*)

Department of Floriculture and Landscaping, Punjab Agricultural University, Ludhiana-141 004.

1

Aglaonema (Family: Araceae)

R.K. Dubey, Simrat Singh, Sanyat Misra and R.L. Misra

[**Common names**: Aglaonema, Cast iron plant, Chinese evergreen (*Aglaonema modestum*)]

Introduction and Origin

All the genera under Araceae are called aroids and *Aglaonema* is one of them. It derives its name from the Greek *aglaos* meaning 'bright' and *nema* meaning 'thread', referring to the stamens. It is native to the tropical and sub-tropical swamps, as well as rainforests of southeastern Asia, extending from Bangladesh east to the Philippines, and north to southern China. *Aglaonema* is a genus of about 50 species of evergreen perennial herbaceous foliage plants suitable for growing indoors and in the greenhouses. One of the reasons for the Chinese Evergreen's popularity and long life as an interior plant is its ability to cope with lower light levels. In its natural habitat it thrives on the dark tropical forest floor where it receives a bright sunlight only occasionally. Apart from their foliage beauty, they are also cultivated for their arum-like spathes and clusters of long-lasting large red berries as being produced by certain species such as *A. commutatum*. The Chinese were the first to cultivate the species *Aglaonema modestum*. *Aglaonema* species are also considered to be one of the most resistant indoor plants to formaldehyde pollution (Junhui Zhou, 2011).

The sap of this plant is poisonous. It causes skin irritation and if ingested, the sap causes irritation of the mouth, lips, throat and tongue. The agalonema plant also possesses certain medicinal properties. *Agalonema treubii* is a valuable source for glycosidase inhibitors that are antidiabetic, antimetastatic, antiviral and acts as immunomodulatory agent. In particular, α-glycosidase inhibitors such as α-homonojirimycin and β-homonojirimycin isolated from *Aglaonema treubii* have been shown to be potentially therapeutic for diabetes type 2 and HIV-1 infection (Chen *et al.*, 2007).

Botany

Aglaonema is an erect and tufted herbaceous perennial growing from 20 to 150 cm in height with thick and shallow roots that produce basal shoots with short stems. The leaves are more or less in tight spirals with long sheathed petiole, dark to medium green but often beautifully variegated, alternate, oblong-lanceolate to narrowly ovate, 10-45 cm long and 4-16 cm broad, with a thick costa and few lateral nerves and often patterned. The flowers are petalless, relatively inconspicuous and are borne on sessile or short-stalked spadix which give rise to red berries only in certain species, with white or greenish-white arum-like straight spathes that are convolute below and open above. Peduncles are borne in clusters and are shorter than petioles.

Species and Varieties

Out of some 50 *Aglaonema* species, only 9-13 are grown indoors, some of which are described below:

Aglaonema brevispathum hospitum (syn. *A. hospitum*): A native of Thailand, the original species bears plain green leaves but of *A. b. hospitum* the leaves are 20 cm long, leathery, deep glossy-green, ovate with creamy white spots and are borne on stalks as long as or longer, the spathe is pale-green normally with a pointed tip and 4 cm long. A form of *A. brevispathum* also bears white midrib. It has creeping stem similar to *A. costatum* to which it is closely related.

Aglaonema commutatum: Native of SE Asia (Philippines, Malucca Islands), this species is highly variable. It grows up to 15 cm high with spread of 20-30 cm, bearing thick-textured oblong-lanceolate leaves which are deep green marked with silvery-grey and some 25 cm long. It produces green-white spathes some 5 cm long on some 8 cm long stalk in July, followed by heads of dark red berries. *A. c. pseudobracteatum* bears green marbled white stems and yellow & green mottled narrower leaves with distinct white veins. 'Malay Beauty' is quite robust. In fact, *A. treubii* is also its smaller and distinct form correctly known as *A. c.* 'Treubii', which was recorded from Indonesia, is a very compact house plant growing up to 22 cm high with about 30 cm spread, bearing silvery-grey marbled bluish-green, leathery and narrow leaves. *A. c.* 'Tricolor' bears variegated and broadly elliptical leaves with pinkish-white leaf stalks. *A. commutatum* has given most of the commercial hybrids which are available in the *Aglaonema* trade.

Aglaonema costatum: Native of Malaya. A very attractive dwarf species growing up to 15 cm in height and some 25 cm in spread, bearing stiff and broadly ovate dark green leaves with ivory-coloured or white midribs and white spots. Spathes are pale-green, some 2.5 cm long and appear in July on a 5 cm long stalk.

Aglaonema modestum (syn. *A. acutispathum*): It is a hardiest species, withstanding the temperature of 7 °C and is native of Asia (south of China). Its leaves are glossy dark green, leathery, waxy, 15-20 cm long, pendent and ovate-acuminate. Spathes are pale-green and 7.5 cm high. *A. m.* 'Variegatum' bears irregular creamy-white variegation.

Aglaonema nitidum (syn. *A. oblongifolium*): It is a native to Malaya, growing some 1 m tall with a spread of some 30-40 cm. The leaves are stiffly erect, elliptic, dark glossy green and more than 20 cm long. Small spathes are some 5-10 cm long, green with white margins and appear in july. *A. n. curtisii* is the largest cultivated species with 22.5 cm long leaves which are dark green with neat silver markings, and the spathes which appear in July are green-white, and 5-10 cm long.

Aglaonema pictum: A native of Malaya and Sumatra. It grows 30-60 cm in height with 22 cm spread, bearing stippled leaves in three shades of green, *i.e.* from

***Aglaonema* 'Abidjan'**

***Aglaonema* 'Amelia'**

***Aglaonema* 'B.J. Freeman' (syn.'Cecelia' or 'Gabrielle')**

***Aglaonema* 'Black Lance'**

***Aglaonema* 'Brilliant'**

***Aglaonema* 'Cory'**

Aglaonema 'Deborah'

Aglaonema 'Emerald Star'

Aglaonema 'Emerald Bay'

dark to silvery, 20 cm long and broadly ovate. In August it produces 5 cm long yellow-green spathes. *A. p. tricolor* bears pinkish leaf stalks having leaves with two shades of green, mottled with silver-grey.

Aglaonema rotundum: : A native to Malaya and Sumatra is a most beautiful short compact plant. This dwarf species bears short-stalked broadly ovate leaves which are more than 15 cm long, leathery, dark metallic glossy-green with pink midrib and lateral veins, and where reverse side is wine red with pink veins.

Aglaonema simplex: It is native of Java and is very similar to *A. modestum* except the leaves which are thinner and larger. The leaves in *A. s.* 'Angustifolium' are little corrugated, quite narrow, slender and pointed.

There have been some hybrids that are of commercial importance. Over the last few years, 15-20 new varieties have been introduced. These introductions have come from three main sources, *viz.* University of Florida, Partha & Mukundan of India, Sunshine Foliage World and Zolfo Springs, Florida. The University of Florida through its breeding programme has introduced *Aglaonema* 'Stripes' and 'Silver Bay'. These new varieties have been bred for indoor use in view of their colourful stems and/or leaves, suckering potential, and resistance to chilling. Partha & Mukundan of India with their 'Stars of India' collection has introduced two new hybrids which include 'Jewel of India' and 'Emerald Star'. The beautiful combination of leaf colours in *Aglaonema*, makes it well known in Asia, especially South East Asia. Variety of *Aglaonema* could be evolved by employing conventional and non-conventional methods. Through particle bombardment technique, a new *Aglaonema* variety with a combination of red and white leaf colour has also been developed (Totik *et al.*, 2011a).

Certain outstanding *Aglaonema* varieties are 'Abidjan' growing compact with longer and broader leaves and is an introduction from Ivory Coast. 'Amelia' grows stocky with good suckering, bearing large dark green leaves with random light green patches all over the leaf, and is an introduction from Sunshine Foliage World to the elite *Aglaonema* series. 'B.J. Freeman' (syn. 'Cecilia' or 'Gabrielle') is quite large-growing as a specimen plant in 25-35 cm pot, bearing large grayish-green leaves. 'Black Lance' is a good suckering variety with dense canopy of lance-shaped narrow leaves having broad silver lining in the centre. 'Brilliant' has white stems with bright foliage and requires low maintenance, surviving well under poor light conditions. 'Copry' is a good suckering variety bearing white or cream stems with a feathery silver-green striping on a medium green leaf and stands well to below freezing temperatures and low light conditions indoors. 'Deborah' is a sport of 'Queen of Siam' in the elite group having white to cream stems & petioles, and golden-green leaves at the edges & the veins, and gray-green centre. 'Emerald Bay', an introduction from University of Florida, USA and being produced through tissue culture by Agri-Starts, Inc. in Apopka, Florida, tolerates near to freezing temperature.

The 'Emerald Star' developed by Partha & Mukundan of India is an upright growing variety with *Dieffenbachia* look having excellent suckering and bearing wide and dark glossy three-shaded green leaves with a very prominent mid-vein, and bears lower temperatures to a great extent.

Aglaonema **"Golden Bay"**

Aglaonema **"Green Lady"**

Aglaonema **"Illumination"**

'Golden Bay' is a product of University of Florida but being multiplied through tissue culture by Agri-Starts, Inc., Apopka, Florida. It suckers heavily, lasts longer as indoor plant and has good colouration. 'Green Lady' from elite series is a stocky large plant with good suckering ability, bearing dark green leaves with light green blotches. 'Illumination', a new introduction to the elite series, is a vigorous but short-statured hardy plant, having white to cream stems with silvery-green leaves blotched and variegated golden along the central vein. 'Jewel of India', developed by Partha & Mukundan of India suckers well, tolerates freezing temperatures and bears silvery leaves. 'Silver Frost' from elite series is a medium growing plant bearing broad leaves with bold silvery patch in the centre. 'Silver Queen', a very popular aglaonema for the last 40 years and can be grown as a specimen plant. 'Stripes' is the first aglaonema variety evolved by University of Florida. Its leaves have narrow silvery-white bands along the veins that overlay a fainter and more diffuse pattern in the same area. 'White Lance' is quite new in the elite series with dense growth and good suckering ability, bearing white to cream stem and light green leaves having narrow leaf blades. 'White Rain' from the elite series was developed by Sunshine Foliage World which has good and dense suckering ability. The plant is medium, stems white to cream and the leaves are quite broad.

Aglaonema **'Jewel of India'**

Aglaonema **'Silver Queen'**

Aglaonema **'Silver Frost'**

Aglaonema **'White Rain'**

Aglaonema **'Stripes'**

Aglaonema **'White Lance'**

Propagation

Aglaonema is usually propagated through seeds, separation of suckers and through stem cuttings. For most cultivars, stem cuttings are still considered the most efficient method. **Seeds** can be used to propagate *Aglaonema commutatum* and its cultivars. Other species and their cultivars may also be multiplied through seeds if these set the berries with viable seeds. Barring the pure species and their botanical varieties which generally breed true, most of the other varieties through seeds may not come true. Fruits are harvested when their colour turns bright red, cleared of pulp and the fresh seeds are sown in the seed compost, *i.e.* sphagnum peat or sphagnum moss, at a depth of 1.25 cm and kept at a temperature of 21-27 °C, and these germinate within two weeks, though slightly stored ones from 1 to 3 months. At lower temperatures, fewer seeds may germinate and may take longer time. Its seeds lose viability rapidly, therefore, it is advised to sow them just after harvest. **Tip cuttings** are most commonly used method for propagation in *Aglaonema*. Cuttings should be taken with 3-6 leaves on them, more the leaves better is success, and usually the 3-leaf cuttings do not root well after potting as compared to those with 5 or 6 leaves. Cuttings are taken in February in sub-tropical regions, after mid-October to January in the tropical regions and in April to May in the temperate regions of India, and are planted in pots or directly in the propagating chamber in the growing compost at a temperature of 18-21 °C for better rooting. Kurniawan Budiarto (2011) reported better rooting in cuttings taken from medial part of stem. Cuttings root well in peat, peat-perlite or peat-styrofoam medium with a pH of 5.5 to 6.0. Liquid fertilizer of NPK in 20-20-20 ratio should be applied at the rate of 25 g per 13 litres of water per square meter when roots start to emerge. The temperature of the rooting medium must be kept at 21-27 °C for better root initiation within 4 to 6 weeks. At low rooting medium temperature, the rooting is delayed, and it may take even up to 12 weeks depending upon the extent of low temperature given. The misting cycle during rooting should be set for 15 seconds per 30 minutes during daylight hours. The cuttings should be planted usually at 8 cm apart for most cultivars. Although not used commonly, **cane cuttings** are a means to propagate many plants from a few stock plants. Each cane is divided into as many sections as possible leaving 1-2 nodes per piece. Cane sections are placed some 2.5 cm apart on a propagation bench having the sterilized medium as for tip cuttings. Once the roots have formed and buds start to grow, apply 20-20-20 NPK liquid fertilizer as for tip cuttings, once a month until plants are large enough to transplant, which usually takes 4 to 6 months. **Division** is commonly suited for the cultivars that produce large numbers of suckers such as 'Fransher Evergreen' and 'Silver Queen'. Clumps are divided, each basal shoot or sucker having a few roots, and these are planted in 20-25 cm pots filled with suitable mixture, *i.e.* leaf mould, garden loam and fresh coarse sand. Even unrooted suckers with several leaves attached, root in pots of the growing compost in a propagating cell at 18-21 °C.

Micropropagation of *Aglaonema* var. 'Cochin' was performed using axillary shoots as explants on MS medium supplemented with 1.5 mg/l TDZ and 3 mg/l benzylamino purine (BAP) to avoid callus growth. After the 5th subculturing (10 weeks) 1,000 shoots were obtained from two initial axillary shoots (Totik *et al.*, 2011b). Chen and Yeh (2007) carried out *in-vitro* shoot multiplication from the axillary buds on MS medium containing BAP. Application of IBA at 9.8 or 19.7 mM to the base of the microcuttings resulted in 100 per cent *ex vitro* rooting with longest roots. A total of thirteen different bacterial species were identified as contaminants from micropropagated *Aglaonema* cultures. Three antibiotics, including gentamicin, tetracycline and chloramphenicol were found to be effective at low concentrations of 4-32 mg/l to inhibit bacterial growth under *in vitro* conditions (Jong and Yu, 2012).

Cultural Practices

Aglaonema thrives well between **light intensity** of 1,000 and 2,500 fc, depending upon temperature. Light intensities of 1,500 to 2,500 fc will produce excellent plants. Unlike most foliage plants, aglaonema positions its foliage according to the light intensity. These are the plants which can be grown from shade to filtered sun, minimum being the 50 fc for 12 hours daily (Virginie and Elbert, 1989), especially when the leaves are variegated, though all the species develop large, healthy and brilliant leaves at 100-400 fc. Light intensity is satisfactory when leaves tilt 45 ° or more from the plant but light intensity is high when the tilt is less than 45 °. Plants growing under excessive light usually show symptoms of bleached foliage and slight tip-burn. Most foliage plants have their origins in the tropics and require relatively high night **temperatures** to sustain rapid growth. Any significant change in growth rates would be the effect of environmental seasonal influences (Benedetto *et al.*, 2006). Best growth is obtained when soil and air temperatures are relatively high. Soil temperatures between 21.1- 26.6 °C are most desirable for root growth, while air temperature during day-time should be at least 21.1 °C and during nights between 15.5 to 18.3 °C, provided root temperatures are maintained above 15.5 °C. If temperatures drop below 10 °C, chilling injury occurs characterized by loss of the silvery colour in 'Silver Queen Evergreen' aglaonema. The minimum winter temperature needed is 10 °C, however, 16 °C is ideal but *A. costatum* and *A. pictum* need a little more heat than the other species (Hay and Beckett, 1971). During summer when temperature reaches 21 °C, ventilation is provided.

No single **medium** may prove best for aglaonema cultivation. *Aglaonema* requires a growing medium which has excellent aeration and has high water-holding capacity. Aglaonema grows well in peat moss:sharp coarse sand (3:1). The recommended best medium for aglaonema is 2 parts peat moss, 1 part vermiculite and 1 part perlite. Peat moss or the sphagnum peat is the product prepared through decomposition of sphagnum moss which is largest and fastest growing of all the mosses. After decomposition it becomes brown, is fibrous and absorbs or holds large quantities of water. Its nutritive value is very low but pH value is acidic. Peat is also prepared through sedge, a member of Cyperaceae but this is more acidic and is considered inferior for aglaonema. Vermiculite is prepared of mica (silica crystal) by subjecting to high temperatures so that its books exfoliate. It is crisp and brittle, not absorbing the water but its many surfaced structures hold moisture by surface tension. Perlite is prepared by subjecting the obsidian volcanic stone to intense heat so that it transforms into white granules which are very light, firm and gritty. It is used in the mixes on the place of sand for better aeration and for absorbing more moisture. Water is given once or twice a week and care should be taken to avoid the plants being exposed directly to rain as it may cause water-logging which may cause rotting of the roots. In fact, any medium which is having non-capillary pore space of 8-12 per cent and capillary pore space of 55-65 per cent, each by volume, is suitable for aglaonema cultivation. In fact, the potting media should always be moist but never soggy. However, the pH of medium should be brought down in the range of 5.5-6.0 by incorporating sufficient dolomite into the potting medium. Sandra *et al.* (2011) reported media formulations of peat-bark-starlite-rice hulls-coir (2:2:3:1:2) and peat-bark-perlite-rice hulls-coir (4:1.5:2.5:1:1) giving 8.9 to 9.5 per cent taller plants after eight weeks of their growth, when compared with standard peat-vermiculite-perlite (5:2:3) medium. Aglaonemas are normally initiated in a 12-15 cm pot, filled with damp peat, moss or vermiculite. They need only shallow potting with enough room to permit the roots to spread and produce offsets. These are **repotted** every second or third year for growing larger specimens. However, repotting can be substituted by applying liquid feed once a month during the growing season, but even then before becoming pot-bound these are required to be repotted.

Virginie and Elbert (1989) suggest monthly **feeding** of aglaonema with ¼ strength of a 20-20-20 NPK fertilizer. It would be wise to apply 50 per cent nitrogen from ammonium source and 50 per cent from nitrate source for healthy growth of aglaonema. Chaves *et al.* (2006) reported that plant K+ uptake increased with application of K+ or basic nutrient solution. The uptake and transport of calcium (Ca) were enhanced by the use of NO_3—N and inhibited by the presence of other cations in the medium (NH_4+, K+, Na+) and under alkaline pH. Magnesium uptake increased with NO_3—N application. NPK ratios widely acceptable for aglaonema production include 1:1:1 and 3:1:2. Excellent quality aglaonemas can be grown by application of 1,360 kg of nitrogen per hectare per year in greenhouses. For a successful aglaonema crop the suggested optimum ranges of elements are N 2.5 to 3.5 per cent, P_2O_5 0.4 to 0.7 per cent, K_2O 1.5 to 2.5 per cent, Ca 0.5 to 1.0 per cent, Mg 0.4 to 0.8 per cent, boron 10 to 50 ppm, copper 10 to 60 ppm, iron 50 to 300 ppm, manganese 50 to 300 ppm and zinc 25 to 200 ppm. Aglaonemas flower within 23 to 48 weeks of planting, depending on the cultivar, following a single spray with 250 ppm **GA_3**. *Aglaonema* responds well to frequent **watering** (2 to 3 times per week) when grown in a well- drained and highly aerated medium. Pot-watering systems such as surface emitters are suggested to reduce foliage wetting. Plants grown in poorly drained and aerated potting media often grow slowly but in such media the growth can be improved if plants are watered less frequently, but it is much easier to use the correct potting medium.

Insect-Pests, Diseases and Physiological Disorders

Several genera of **nematodes** especially species of lesion nematodes, *Pratylenchus,* and root-knot, *Meloidogyne* have been reported to limit growth of aglaonema. Plants infested with lesion nematodes show symptoms of stunting, and the basal leaves turn chlorotic and droop. Roots may rot making the base completely devoid of roots. Infested roots are galled though in severe infestation these rot. Nematodes need not become serious problems in production when the propagules are taken from nematode-free stock. To ward off nematode infestation, nematode-free media should be used, nematode-infested plants should not be introduced by following proper quarantine measures, and the plants should be grown on raised benches. In nematode infested soil, Furadan or carbofuran may be mixed in the soil, followed by irrigation.

Aglaonema does not appear to be seriously affected by insects, mites, or related pest problems with the possible exception of periodic infestations by lepidopterous **caterpillars**, root-infesting mealy bugs as well as aglaonema and latania scales. Dipel O (*Bacillus thuringiensis*) 50WP is labelled for greenhouse use against lepidopterous larvae. Spreader-sticker should be added to improve coverage and persistence of foliar residues.

Aglaonema is almost disease-free provided it is grown properly. The best way to grow this plant is to use correct potting media, light, fertilizer and provide optimum water at the surface of the potting medium. Numerous foliar diseases have been reported on aglaonema, but the only serious problems arise due to bacterial pathogens, *Erwinia and Xanthomonas*. **Bacterial blights** (leaf spots) and **stem rots** (leaf spots) are caused by *Erwinia carotovora* and *E. chrysanthemi*. Watery leaf spots mostly with disintegration of the centres occur. Bacterial **stem rots** are generally first noticed with the cuttings when these show stickiness. At this time, the cut end of the stem becomes mushy with foul smelling, cuttings turn yellow, and the rooting process is delayed if not altogether halted. ***Xanthomonas* leaf spot** (*Xanthomonas campestris* pv. *dieffenbachiae*) causes **r**eddish-brown areas on edges of leaves with bright yellow margins. Under wet and warm conditions, bacteria also spread into leaf centres and lesions expand until they reach to the leaf veins. To ward off bacterial problem, the stock material should be pathogen-free and there should be minimal foliage wetting in the field. Foliar applications of copper or antibiotic compounds on a weekly basis provide adequate control. Both the diseases (*Erwinia* and *Xanthomonas*) can be controlled by keeping the foliage dry by avoiding overhead watering. *Erwinia* is prevalent on cuttings imported from Central and South America with symptom expression occurring during the propagation period. However, once cuttings are rooted and foliage is allowed to dry, these pathogens do not create problems on new foliage.

***Fusarium* stem rot** (*Fusarium* spp.) typically appears as a soft rot at the base of a cutting or rooted plant. The rotten area frequently has a purplish to reddish margin. *Fusarium* sometimes forms tiny, bright red and globular structures (fruiting bodies) at the stem base of severely infected plants. Before planting, the cuttings should be dipped in anti-fungal solution. In case of stem rot, the infected plants or cuttings should be removed from stock nursery as soon as the symptom is detected. ***Myrothecium* leaf spot** (*Myrothecium roridum*) is one of the most common foliage diseases in aglaonema with easy diagnosis, and the tan to brown spots some 2.5 cm large with a bright yellow border generally appear on wounds. Examination of the lower leaf surface shows the black and white fruiting bodies of the pathogen in concentric rings near the outer edge of the spot. Avoiding excessive wetting of the foliage and soil will minimize its incidence. **Root rot** (*Pythium* spp., *Phytophthora* spp.) is symptomised by yellowing of lower leaves and wilting of plants. The roots themselves are brown to black and the outer portion of infected roots can easily be pulled away from the inner core. Use of sterilized pots and potting media, growing of plants on raised benches and use of pathogen-free planting material will help controlling this problem. Soil drenching with fungicides will help controlling infection of both the pathogens, *Pythium* or *Phytophthora* root rot.

Symptoms of **bent tip** generally appear on terminal leaves where leaves have a 'fishhook' appearance, together with some older leaves becoming terminally 'hooked'. In fact, the new leaf tip appears to be obstructed and caught by the succeeding leaf, resulting in the 'fishhook' appearance. Excessive light and water stress are associated to increase severity in susceptible cultivars. Therefore, it is suggested that susceptible plants should be placed at favourable light, temperature and humidity so that such situation is avoided. When leaves assume more or less **vertical** or **low angle position** instead of the normal 45 to 90 ° from the stem, leaves become lighter in colour or exhibit washed-out (bleached) appearance, and in extreme cases leaf tips become palish-white, it is a sure case of excessive light and/or temperature conditions. Optimum light and temperature conditions will ensure normal growth in such cases though leaf-bleaching may not recover fully. **Chilling injury** generally occurs in middle to lower leaves by developing gray and chlorotic splotches in most of the species and cultivars, more prominent being the `Silver Queen'. Though most cultivars may tolerate 7.2-10.0 °C temperatures but `Silver Queen' can not tolerate 12.7 °C or below. Ethylene treatment hastens **leaf and bract abscission** or **senescence** in most of the foliage plants. Treating *Aglaonema* 'Mary Ann' with 0.9 µl 1-MCP (1-methylcyclopropene), a gaseous ethylene-binding inhibitor, for 4 to 5 h at 21 °C reduces the deleterious effects of ethylene during shipping (Andrew *et al.*, 2011). Terminal leaves are more prone to **copper deficiency** as they become chlorotic and sometimes even dwarfed and deformed with serrated edges. Older leaves become lighter green and in severe cases terminals and lower breaks abort. Cultivar `Fransher' is especially susceptible to copper deficiency. Application of copper sulphate at a rate equivalent to 750 g per 100 sq. m. of the soil or copper spraying to foliage will solve this problem. It is preferable to include copper in the potting medium or use a periodic micronutrient application of copper. Soil temperatures of 18.5°C or below will contribute to copper deficiency as roots become poorly efficient to remove copper from cold soils. Therefore it is suggested either to raise soil temperature or foliar spraying of copper should be done at such time.

References

Andrew J. Macnish, Ria T. Leonard and Terril A. Nell, 2011. Sensitivity of potted foliage plant genotypes to ethylene and 1-methylcyclopropene. *HortSci.*, **46**(8): 1127–1131.

Benedetto, A. D., J. Molinari, C. Boschi, D. Benedicto, M. Cerrotta and G. Cerrotta, 2006. Estimating crop productivity for five ornamental foliage plants. *Intern. J. agric. Res.*, **1**(6): 522-533.

Chaves L. A., J. Garcia, S. Jimenez and M.T. Lao, 2006. Influence of the modification of nutritional parameters in *Aglaonema commutatum*: K, Ca_2, Mg_2 and Na. *Communications in Soil Sci. & Pl. Analysis.*, **37**(15-20): 2927-2937.

Chen, J, R.J. Henny and F. Liao, 2007. Aroids are important medicinal plants. *Acta Hort.*, No. 756, pp. 272-275.

Chen, W. L. and D.M. Yeh, 2007. Elimination of *in vitro* contamination, shoot multiplication, and *ex- vitro* rooting of *Aglaonema. HortSci.*, **42**(3): 629-632.

Hay, R. and K.A. Beckett, 1971. *Reader's Digest Encyclopaedia of Garden Plants and Flowers*, pp. 26-27. The Reader's Digest Association Limited, London, England.

Jong, Y. F. and R.H. Yu, 2012. Molecular identification and antibiotic control of endophytic bacterial contaminants from micropropagated aglaonema cultures. *Pl. Cell, Tissue & Org. Cult.*, **110**(1): 53-62.

Junhui Zhou (2011). Purification of formaldehyde-polluted air by indoor plants of Araceae, Agavaceae and Liliaceae. *J. Food, Agric. & Environ.*, **9** (3 & 4): 1012-1018.

Kurniawan Budiarto, 2011. Conventional propagation of several aglaonema accessions using split single bud stem cutting. *J. agri. Vigor.*, **10**(2): 99-104.

Sandra, B., Keona Wilson, L. Muller, Chris J. Wilson, Regina A. Incer, Peter Stoffella and Donald Graetz, 2011. Evaluation of new container media for *Aglaonema* production. *Communications in Soil Sci. & Pl.* Analysis, **40**(17-18): 2673-2687.

Totik S.M., Any Fritiani, W. Adhityo and C. Tet-fatt, 2011a. NMU-induced mutation in aglaonema by particle bombardment. *Intern. J. Basic & Appl. Sci., IJBAS-IJENS*, **11**(03): 59-67.

Totik Sri Mariani, Any Fitriani, A. Jaime and Teixeira da Silva, 2011b. Micropropagation of *Aglaonema* using axillary shoot explants. *Intern. J. Basic & Appl. Sci.*, **11**(01): 46-53.

Virginie, F. and G.A. Elbert, 1989. Foliage Plants *for Decorating Indoors*, pp. 133-135. Timber Press, Portland, Oregon, USA.

2

Alstroemeria (Family: Alstroemeriaceae)

M.K. Singh and Sanjay Kumar

[**Common names**: Brazilian parrot lily, Chilian lily (*A. chiliensis*), Golden Peruvian lily/Peruvian lily (*A. aurea* or *A. aurantiaca*), Herb lily/Inca lily (*A. pelegrina*), Lily of Peru, Lily of the Incas, Parrot lily (*A. pulchella*), St. Martin's flower (*A. ligtu*)]

Introduction

The name *Alstroemeria* is given by Linneaus after the Swedish naturalist Klas von Alstroemer (1736-94). The centre of origin of *Alstroemeria* is found in Latin America, especially countries like Chile, Peru and Brazil. Its habitat has been recorded from dry, warm to moist and cools to temperate regions and from 26 ° to 40 °C south latitude (Bridgen, 1997). Recently, it has become popular as a garden flower, as a potted flowering plant, and more so as a cut flower crop. The plants produce beautiful large inflorescence in many different colours, *viz.* lavender, orange, yellow to dark yellow, white, pink, red, purple and bicolours. Flowers are characterized by black dots at the base of the petals and throats. The cut flowers have a long post harvest life, *i.e.* up to two weeks. These valuable characteristics have made *Alstroemeria* one of the top cut flowers at the Dutch Flower Auction Centres. The plants are high yielder and possess an ever blooming habit after flower initiation has once occurred but in general most cultivars flower best during spring and early summer but only in sub-temperate areas (neither in the plains nor in the subtropical regions). The straight, un-branched shoots emerge from underground rhizomes which also produce enlarged storage roots. It is gaining popularity in Indian flower market due to its long floral stem and prolonged vase life and the growers have started venturing its commercial cultivation recently in the sub-temperate areas. Under the All India Coordinated Research Project of ICAR on Floriculture running at various centres of the country, in general, and at Institute of Himalayan Bioresource Technology (CSIR), Palampur (H.P.), in particular for Himachal Pradesh and other hill states, the improvement of agro-techniques for production of its cut flower and planting material are being standardized.

History, Botany and Breeding

It is an herbaceous perennial consisting of multi-stemmed rhizome from which aerial shoots and fibrous roots develop. Later on these fibrous roots become thickened storage roots which are called 'radices medullosae'. These roots are white, fleshy, very brittle and dense. Leaves have parallel veins which are grey to dark green and mostly hairless on both the sides. Leaf petioles twist 180 ° on the stems so that original underside becomes the top side (Bayer, 1989). Stomata on leaves are present on both the sides, *i.e.* upper and lower (Healy and Wilkins, 1985; Heins and Wilkins, 1979). Inflorescence is cyme and each cyme is sympodially branched with up to four florets per cyme that open one after another. The flowers of *Alstroemeria* are zygomorphic, terminal, simple or compound, rarely solitary, tepals 6, free, clawed, the 3 inner tepals narrower and longer than the outer;

stamens 6, declinate, attached to base of tepals, anthers basifixed, style slender, stigma 3-fid, ovary inferior, 3-celled. Stamens open and shed their pollens before the stigma becomes receptive (Traub, 1943). Fruits are many seeded capsule.

The chromosomes in *Alstroemeria* are very large and the basic number is n=8.The somatic chromosome numbers of wild *Alstroemeria* species have been determined by karyotype analysis and it is 2n=2x=16 (Srasburger 1882; Taylor 1926; Hang and Tsuchiya 1988; Stephens *et al.,* 1993).

Primary objectives in breeding programmes for *Alstroemeria* are to develop the hybrids that possess large and well-formed buds showing complete colour, attractive flower colours with large and durable flowers on long stems, good yield, fragrance and compact growth habit, *vis-a-vis* disease resistance. A lot of cultivars exist which have been developed through interspecific and intraspecific hybridization. Presently, biotechnological approaches are also being employed to make further improvement in the interspecific hybrids through ovule cultures, especially in those cases where sexual incompatability occurs. Plant regeneration through callus culture has been reported. Particle bombardment and *Agrobacterium*-mediated procedures are applied for genetic transformation and some transformed plants with marker genes have been produced. In *Alstroemeria*, isolation of egg cells and zygotes from ovules has been attempted in order to develop an *in vitro* fertilization technique (Hoshino *et al.,* 2006).

Species and Varieties

The genus *Alstroemeria* comprises of more than 60 species. Hofreiter and Rodriguez (2006) reported 39 species from Brazil, 33 from Chile, 10 from Argentina, 2 from Peru and 1 from Bolivia. Van Scheepen (1991) described some of the most common *Alstroemeria* species such as ***A. angustifolia*** Herbert, a native of Chile growing to 60 cm bearing narrowly lanceolate leaves, and pale-pinkish 7-8 cm across flowers where upper laterals finely spotted and with a yellow band; ***A. aurea*** Graham (*A. aurantiaca* D.Don) from southern Chile, suitable for growing in warm porous soils and susceptible to winter, roots fleshy, growing to 1 m in height, subsessile and lanceolate leaves 7-10 cm long and often twisted to expose their grey-green undersides, umbels 3-7 rayed and each ray with 1-3 flowers, tepals 4-5 cm long, outer broadly ovate, obtuse, tipped green and inner acute with upper part spotted and flecked-red though lowermost occasionally without spots, flower colour bright orange or yellow with purple stripes, and its popular varieties are 'Angustifolia', 'Dover Orange', 'Flava', 'Lutea', 'Major', 'Moerheim', 'Orange King', 'Rubra', 'Splendors', *etc.*; ***A. brasiliensis*** Spring., a native of Brazil, growing to 120 cm, leaves are erecto-patent, linear, glabrous and 5-10 cm long, though those of sterile stems 7.5-10.0 cm and lanceolate, with silver-grey midrib, umbels normally 5-rayed each with 1-3 red-yellow flowers, tepals 4 cm, inner ones flecked-brown, and stamens shorter than tepals; ***A. caryophyllacea*** is native to Brazil with fragrant flowers where sepals and lower tepals are rose while upper with white centres, and grows to 45 cm; ***A. chilensis*** hort. from Termas de Chillen, growing to 75 cm, leaves scattered, narrowly obovate and margins minutely ciliate, umbels 5-6 rayed, each with 2 flowers, and tepals pale pink to blood red, upper inner tepals yellow striped red-purple; ***A. haemantha*** Ruize & Pav. from Chile, growing to 90 cm, subsessile leaves narrowly lanceolate, 7-15 cm long, grey-green underside and margin hairy, umbel 3-15 rayed each with 1-4 flowers, tepals 4-5 cm, oblanceolate-spathulate, narrow, orange to deep vermilion, outer ones tipped green, upper inner ones yellow-orange striped purple while lowermost one orange to orange-red and striped dark red, its most important variety is 'Rosea'; ***A. hookeri*** Lodd., a native to Peru, resembles to *A. pelegrina* but with pink tepals, upper inner ones blotched yellow and flecked red-purple; ***A. ligtu*** L. from Chile to Argentina, grows some 45-60 cm long with 5-8 cm narrowly lanceolate to linear-lanceolate and erecto-patent leaves, umbels 3-8 rayed each with 2-3flowers having 5.0-7.5 cm pedicel length, tepals white to pale lilac to pink-red, obovate, upper inner ones usually yellow, spotted and streaked white or purple or yellow and red, and stamens shorter than tepals; ***A. pelegrina*** L. from Chile, a species used extensively in breeding programme, growing to 30-60 cm, glabrous and lanceolate leaves 5-8 cm, flowers solitary or in 2-3 rayed umbels, tepals 5 cm, off-white flushed mauve or pink with a darker central zone, inner tepals yellow at base, flecked brown or maroon, and the prominent varieties are 'Alba', 'Rosea', *etc.*; ***A. psittacina*** Lehm. from Brazil, very similar to *A. pulchella*, growing to 90 cm, stems spotted mauve, glabrous and lanceolate leaves up to 7.5 cm, umbels 4-6 rayed, tepals 4.0-4.5 cm, green overlaid with dark wine red, spotted and streaked red or maroon, and stamens as long as tepals; ***A. pulchra*** Anon. grows to 45 cm with white to light grayish-pink or soft lilac flowers spotted yellow, red or purple; ***A. pygmaea*** Herb. from Argentina is very similar to *A. aurea*, growing only to 20 cm as the stems are subterranean which produce a group of grey-green leaves of 2.5 cm size, flowers 5.0 cm long, yellow but inner tepals spotted red; ***A. versicolor*** Ruiz & Pav is native to Chile and resembles *A. psittacina* though smaller, growing to 60 cm with linear leaves of 2.5 cm size, umbels 2-4 rayed, tepals 2.5 cm, pale yellow or orange with purple spots but inner segments spotted and streaked; ***A. violacea*** Philippi., a Chilean species growing to 100 cm with glabrous, ovate-oblong, 6-9 cm long and spreading & scattered leaves, umbels 3-6 rayed each with 3-5 flowers, tepals 3.5-5.5 cm, bright violet, outer ones obovate, blunt, shortly cuspidate, inner tepals oblong-acute, white at base, spotted purple, sometimes flushed orange, and stamens shorter than tepals; ***A. zoellneri*** E. Bayer from Aconagua and Santiago provinces of Chile, stems 30-60 cm, leaves linear and around 2.5 cm, flowers 5-6 cm across, the outer segment ovate and acuminate with a dark spot near the apex while the inner laterals recurved having a yellow band with purple streaks; *etc.*

Some most promising varieties are 'Amber' (orange), 'Aladdin' (lemon yellow), 'Bambi' (orange, a Brasiliensis variety), 'Bombay' (bright pink), 'Bounty' (dark pink), 'Claudia' (pink), 'Disco' (lilac), 'Gold Finger' (yellow), 'Jesssica' (deep pink), 'Mandarino' (orange), 'Petit Rouge' (red, a Brasiliensis variety), 'Pluto' (orange), 'Sylvia' (creamy-white), 'Symphony' (red), 'Visa' (red, a Brasiliensis variety), *etc.* At IHBT, Palampur the varieties performing better for commercial cultivation have been identified as 'Aladdin', 'Amor', 'Capri', 'Cinderella', 'Pluto', 'Rosita', 'Serena', and 'Tiara' (Singh, 2006).

Propagation

Alstroemeria is vegetatively propagated by **rhizome division** or through **micro-propagation**. Asexual propagation allows plants to grow true to type. Bed-growing plants of *Alstroemeria* are divided at least every second and third year, depending on the variety and growth characteristics. About one to two weeks prior to division, plants should be severely pruned, leaving only the youngest 15 to 20 cm long shoots. Roots of the rhizomes grow 30-40 cm deep so digging of the roots should be done properly to get the feeding roots along with growing points for making division. The rhizomes with roots and shoots should be planted separately. The crop is also propagated by **seed** for cut flower production and to develop new varieties. Rhizomes of *Alstroemeria* cv. 'Pluto' were when planted under 35 per cent shade-nets under Srinagar (J&K) conditions with various number of storage roots on the rhizomes and nitrogen fertigation, Singh *et al.* (2010a) recorded maximum values in terms of per cent plant establishment (61.39 per cent), vegetative shoot number (4.78), rhizome cluster weight (13.11 g), number of developed rhizomes (2.38), longest rhizome's length (5.50 cm), new storage root number, as well as fibrous roots (5.57 and 6.31) and the propagation coefficient (31.18) with highest level of storage roots, *i.e.* 2-4 whereas among the nitrogen levels tested, 200 ppm recorded highest values in terms of per cent established plants (63.34 per cent), vegetative shoot number (4.45), rhizome cluster weight (11.61 g), new storage roots as well as fibrous root numbers (4.82 and 6.14) and the propagation coefficient (24.32 per cent) over lower levels of nitrogen. In an experiment to study the effect of storage root number and BA on growth and rhizome production, Singh *et al.* (2010b) recorded maximum sprouting and plant establishment percentages, vegetative shoot number, rhizome cluster weight per plant, developed rhizome number, longest rhizome length, new storage root number, new fibrous root number and propagation coefficient with highest level of storage roots, *i.e.* 2-4 whereas 60ppm BA recorded highest per cent sprouting, developed rhizome number, longest rhizome length, new storage root number, fibrous roots per plant and propagation coefficient over other BA levels (0, 20 and 60 ppm).

Cultural Practices

Alstroemeria can flower round the year through selection of proper cultivars and further management of the crop. When *Alstroemeria* is planted in greenhouse during October-November, it flowers in the following March to August. However, in the second year same plants produce floral stems from March to July though 30 to 35 per cent flowers are produced in May only (Singh, 2005). *Alstroemeria* prefers location having cool temperature, free from water-logging and strong winds. It does not prefer direct sunlight and can be grown successfully in cool place under polyhouse conditions. Cooling system is required in the place having higher temperature. The crop is most comfortably suited to a relative humidity between 65-85 per cent and **CO_2** concentration at 900 ppm which provide earlier flowering, and increased yield with better flower quality. Sufficient **light** (5,000 ft candles) is important to prevent bud abortion and to produce strong and heavy quality floral stalks. Long day speeds up the development of the flower, but decreases the production of shoot. Short days decrease flower production, and suppress rhizome formation & growth. The optimum photoperiod appears to be 12-16 hours, although day length more than 16 hours induces earlier flowering but decreases total flowering shoots, so it is not recommended for commercial production. **Temperature** has also a very important role to play as flower induction occurs at a soil temperature of 14-20 °C. Prolonged temperatures at or higher than 24 °C should be avoided as it decreases and/or inhibits the flower induction completely, however, it can tolerate as low as 5 °C in winter. *Alstroemeria* shows differential response to temperature. Optimum day/night air temperatures for its cultivation should be 18-20/12-16 °C.

In general, **soil** with a pH reaction of 5.5 to 7.0 and EC less than 1.0 are optimum for *Alstroemeria* growing. For best results, the most important factors are soil structure and its water retention capacity but with proper drainage facilities as water-logging may invite damping off or root rot diseases. *Alstroemeria* has luxuriant and quick growth in peat and sandy-loam soil than in clay. Clay soil may be improved by mixing sand and well decomposed FYM to make it light, porous and fertile. The soil of the plots should be deeply ploughed with at least 30-40 cm depth to allow full growth of its roots for a 2-year production cycle, firmly pulverized through frequent ploughings and planking after mixing some 30 metric tonnes of well-decomposed compost or FYM but at the last ploughings the plot should have sufficient moisture. Now proper lay out is made, bunds and channels are prepared and the beds of 1.2 m width should be **prepared** with 50-60 cm clearance between two beds for walking and cultural operations. Under the even climatic conditions, *Alstroemeria* can be **planted** year round but planting time can differ because of the required flowering time and to avoid high temperatures. It is planted in 2-rows system parallel to each other, first row to other row at 30 cm and then distance between one 2-row to other 2-row some 45 cm, and the plant to plant distance some 30 cm though distances may differ depending upon vigorosity of the variety and thus some 6-9 plants may be accommodated per square metre area. Rhizomes some 20 cm deep by spreading the roots properly are planted in the row. *Alstroemeria* can also be grown in larger pots for decoration. One plant per 13 cm pot or 3 plants per 25 cm pot are recommended. Light intensity and temperature influences its flowering but after planting normally it takes 3-5 months for the first flower to appear. October–November polyhouse planting in sub-temperate conditions starts flowering from March to June and produces greatest number of generative shoots. In open cultivation, **shading** is necessary to protect the crop from direct sun and therefore a 25 per cent green shading net is recommended, however, in case of even polyhouse cultivation the roof-tops may be required to be shaded. It is important to find a good balance between shading for cooling and to keep optimum light intensity in the polyhouse. *Alstroemeria* plants grow 50 to 150 cm tall depending upon the cultivars so taller plants require **staking** to hold them straight and for this 2 to 3 layers of wide nets should be used. It should be spread in the beds as soon as the rhizomes start sprouting. The height of supporting net should be raised as the plants grow in height. The lower most net should be fixed at 30 cm above the ground level. **Thinning** of undesirable shoots promotes the production of new shoots and improves the quality of flowers. It is advisable to remove weak and blind (non-flowering) shoots on a monthly basis in late autumn and winter from the production plot. One year old plants of *Alstroemeria* cv. 'Serena' under polyhouse conditions were when thinned at 0, 30, 60 and 90 per cent of vegetative shoots on monthly basis during flower production period (March-July), Singh *et al.* (2006) recorded maximum stems, *i.e.* 75.00 stems per plant with maximum 'A' grade flowers per year and maximum length of flowering stem, *i.e.* 91.37 cm under 30 per cent thinning, followed by 61.50 stems and 87.5 cm stalk length under 60 per cent thinning as has also been reported by Healy and Wilkins (1991) that stem removal by pulling technique stimulates bud break and formation of new aerial shoot. Without thinning of vegetative shoots, plants are likely to produce less number of floral stems. Before planting soil analysis should be done to ascertain the fertility status of the soil. It requires fertile soil with high nutrient levels once the plants are established. Regular **fertilization** with 300 ppm N_2 and 300 ppm K_2O through $Ca(NO_3)_2$ and KNO_3 is recommended per plant per week for good growth and quality flower production. Nitrogen in the form of only nitrates should be provided as it is a crop of cool growing conditions. It is sensitive to salt; therefore, only good quality water should be used in its growing. Rain water is highly suitable for its cultivation. The salt content in the **irrigation** water should be less than 10 micro mol per litre. Top 30 cm of soil should be continuously kept moist because of the root development taking place at this depth. *Alstroemeria* crop can evaporate 3-6 litres of water per square metre area per day, so if sufficient water is not given to the crop the flower production may hamper. Newly planted crop should not be over-irrigated.

Yield of floral stems depends chiefly upon the cultivars, plant spacing, growing conditions and other cultural practices; however, an *Alstroemeria* plant can produce 50-75 cut stems per year under polyhouse conditions.

Growth and Development

Wazir *et al.* (2010) tried various concentrations of Alar (Daminozide), Chlormequat (CCC) and Paclobutrazol (Cultar) as spray, drench and combination of both on pot-grown *Alstroemeria* cvs 'Selection No.14' and 'Riana' under glasshouse conditions at Mashobra (H.P.) for two successive flushes revealed that both the cultivars were favourably affected as a maximum of 31.5 per cent height reduction in the first flush was recorded with Alar 1,500 ppm spray whereas a combination of drench and spray of Alar 1,500 ppm resulted in maximum 28 per cent height reduction in the second flush, 'Riana' recording highest reduction but with maximum spread in the second flush, though 'Selection No.14' recording minimum days to bud formation, producing greater number of cymes per inflorescence in second flush and with a better pot presentability than cv. 'Riana' which remained in flowering for maximum duration in both the flushes. Significant reduction in days to bud formation was recorded with PP_{333} 50 ppm drench and spray in the first flush. A marked positive effect of PP_{333} 50 ppm drench and spray along with Alar 1,500 ppm drench and spray was observed on number of cymes per inflorescence in the first flush whereas in the second flush, CCC 1,500ppm drench and spray resulted in maximum number of cymes per inflorescence. All the treatments significantly increased duration of flowering and pot presentability in both the flushes, with spray application of Alar 1,500 ppm coupled with drench and spray application of CCC 1,500 ppm and PP_{333} 50 ppm producing the best results.

Gupta and Gupta (2009) when soaked *Alstroemeria* cv. 'Alladin' rhizomes in thidiazuron (TDZ) at 0, 15, 30 and 45 ppm for 12 hours, followed by low temperature storage for 0, 4, 6 and 8 weeks at 2-4 °C and then planting in the first week of September as well as in March in polythene bags filled with sand+soil+FYM, recorded maximum per cent sprouting *cum* establishment of plants at 6 weeks of cold storage treatment under 15 ppm TDZ. Though March planting without cold storage treatment was found at par to those cold-treated, however, in September planting the shoot number produced after 40, 60 and 80 days was found significantly increased by storage durations, best for shoot production being 6 weeks of storage and those rhizomes produced maximum number of shoots when TDZ treatment was not given. The shoot production was higher in spring as compared to autumn.

Post Harvest

In *Alstroemeria*, floral shoot emerges after 5-6 months of planting depending upon the cultivars and it further takes 25-35 days to flower. Flowering shoots should be pulled when the first floret are fully coloured and the majority are showing colour in the inflorescence. For long distance market, shoots may be harvested when the first floret is swollen and about to open. Flowers should be kept immediately in the fresh water after cutting the lower portion of shoots. It should be stored at 2-4 °C temperature. Flowers and leaves are sensitive to ethylene and turn yellow after harvest. Sometimes cut flowers of *Alstroemeria* exude a sap that can cause skin irritation, so it is advisable to wear rubber gloves while handling the flowers. Dhiman (2009) placed the cut stems of *Alstroemeria* cv. 'Alladin' in 0, 25, 50 and 75 ppm GA_3, benzylaminopurine (BAP) or BAP + GA_3 solutions and recorded delayed senescence as measured by days to 50 per cent leaf yellowing which was 14 days with GA_3 (75 ppm) + BAP (75ppm) which indicate that GA_3 + BAP each at 75 ppm has the potential to be used commercially as cut flower preservative. *Alstroemeria* cut flowers held in pulsing treatment with sucrose (10 per cent) + 8-HQC (300 ppm) + BA (25ppm) solution for 8 hours resulted in maximum vase life, however, after storage for 24 hours at 4 °C in sucrose (4 per cent) + 8-HQC (200 ppm) + BA (10 ppm) solution and then held in fresh solution of same composition resulted in maximum vase life with best appearance of flower (Manikrao, 2007).

Insect-Pests, Diseases and Physiological Disorders

The main insects attacking *Alstroemeria* plants are aphids, caterpillars, grasshoppers and whiteflies. The nymphs and adult **aphids** are usually found on young leaves and flower buds sucking their sap and their infestation retards the plant growth and hamper the flower quality, *vis-a-vis* act as vector for viral diseases. Two sprays of Malathion or Rogor @ 1.0-1.5 ml per litre of water at 15-20 days interval can control it. The green **caterpillars** are particularly active in summer whose presence can be felt through rolling leaves and flower buds and damage the foliage, buds and flowers. These can be controlled with Malathion or Metasystox spray @ 1.0-1.5 ml per litre of water. **Grasshoppers** are particularly active in summer and rainy season and these generally feed on leaves, buds and flowers through their biting and chewing mouth parts. Regular spray of Monocil or Metasystox @ 0.2 per cent will control these pests. **Whiteflies** are mostly found on young foliage, especially underside of the leaves and the buds sucking their sap which retards the growth of the plant so flower quality becomes poor. These also act as vector for certain viral diseases. Its multiplication rate is very high, therefore should be controlled early to check their menace. Regular spray of Acetamiprid @ 3 g per 10 litres of water on the plant and beneath the surface of leaves can control their attack.

Nematode spp. (*Pratylenchus penetrans* and *P. bolivianus*) generally feed the plants underground so aerial growth is hampered, flower colours become dull and poor and serious attack may kill the plants. Infected plant shows light brown strains on the underground stem. Control measures include application of *neem* -based products, crop rotation with marigold, soil sterilization by steam or formaldehyde drenching at 2 per cent before planting, and incorporation of Furadan in the soil.

Alstroemerias are susceptible to number of fungal pathogens such as *Botrytis*, *Pythium* and *Rhizoctinia* spp., which attack the plants and flowers. One of the methods to prevent these diseases is soil disinfection. This is done after incorporation of organic materials in the ground and before planting. Disinfection of soil by steam controls ground fungi and nematodes. At the same time it also keeps the soil free of weeds for a longer period. It is also essential to provide adequate ventilation and spacing for plants. In ***Botrytis*** infected plant, brown spots on flower petals develop when the weather is quite humid. It can be controlled by providing adequate ventilation and keeping the crop dry during rainy season and spray of mancozeb @ 2.0 g per litre of water at an interval of 10 to 12 days. Root-rot is caused by the fungus ***Pythium*** in water-logged conditions or in the heavy and compact soils in moist conditions. It can be controlled by sterilization of soil before planting, better air circulation and decreasing the moisture content of growing media. ***Rhizoctonia*** is the main causal organism for foot rot, where infected plant shows rotting of shoots just at ground level. Watering in mid-day during warm weather should be avoided and growing temperatures should be controlled. *Alstroemeria* wilt caused by ***Fusarium oxysporum*** in India was first time reported by Shanmugam *et al.* (2007). Symptoms appear as leaf chlorosis and slight vein clearing on outer leaflets, followed by leaf yellowing and abscission, discolouration of stem vascular tissue and ultimate plant death. *Fusarium oxysporum* causing vascular wilt or basal rot inflict heavy crop damage and rhizome rot. Shanmugam *et al.* (2011) screened seven *Alstroemeria* cultivars, *viz.* 'Alladin', 'Amor', 'Butterscotch', 'Capri', 'Pluto', 'Rosita' and 'Tiara' against *Fusarium oxysporum* where they observed 'Tiara' as more resistant than other cultivars in terms of mean per cent disease index. Though *Fusarium* infected plants cannot be saved but treating the plantations with Captan at 0.2 per cent alternate with Bavistin 0.1 per cent at fortnightly intervals in the standing crop and dipping of the rhizomes before planting and after lifting in Captan will check further spread.

Since it is propagated vegetatively through rhizome division so the **virus** infected plants perpetuate if not controlled in time. There are some 10 viruses infecting *Alstroemeria* plants such as Alstroemeria mosaic potyvirus, Alstroemeria streak potyvirus, Alstroemeria carlavirus, Cucumber mosaic cucumovirus, Tomato spotted wilt tospovirus/Impatiens necrotic spot tospovirus, Alstroemeria ilarvirus, Tobacco rattle tobravirus, Arabis mosaic nepovirus, Freesia cucumber potyvirus and Rhabdovirus. Potyvirus infection can be easily detected through serological (ELISA, ISEM, RIA, Immunodiffusion and Immunoblottiing) and nucleic acid based techniques. Tissue culture accompanied by chemotherapy and thermotherapy are useful for quality planting material from *Alstroemeria* infected with potyvirus. Besides these, control of aphids, thrips, weeds, nematodes, *etc.* and use of resistant cultivars are the measures to arrest potyviruses (Mehra *et al.*, 2005). Infection of *Alstroemeria* hybrids in India with a subgroup 1 strain of *Cucumber mosaic virus* (CMV) is reported for the first time in India. The virus was identified on the basis of host range, transmission by *Aphis gossypi* and *Myzus persicae* in the non-persistent manner, ELISA, electron microscopy and RT-PCR using CMV-specific primers. CMV was detected in nine hybrids and 61 per cent of plants by dot-blot hybridization (Verma *et al.*, 2005).

*Alstroemeria*s are sensitive to **fluoride**. Irrigation water containing fluoride can cause leaf scorch. Therefore fluoride containing water and fertilizers such as superphosphate should be avoided. Leaf scorch can also be induced by widely fluctuating polyhouse temperatures, which needs to be maintained properly. Flower bud abortion/**blasting** can be caused by very low light intensity.

References

Bayer, E. 1989. Die gattung *Alstroemeria* in Chile. In: *Mitteilungen der Botnischen Staatssammlung Munchen,* Band 24. Botnische Staatssammlung Munchen, Munchen, p. 3.

Bridgen, M.P. 1997. *Alstroemeria*. In: *The Physiology of Flowering Bulbs* (eds Hertogh, A. De and M. Le Nard), pp. 201-209. Elsevier, Amsterdam.

Dhiman, M.R. 2009. Effect of GA_3 and benzylaminopurine on foliar chlorosis of cut *Alstroemeria* stems. *National conference on Floriculture for Livelihood and Profitability*, IARI, New Delhi, held on 16-19 March, p. 2.3.

Gupta, R., and Y.C.Gupta, 2009. Effect of low temperature storage and thidiazuron application on establishment of plant and shoot production in *Alstroemeria. Nat. Conf. for Livelihood and Profitability,* IARI, New Delhi, held on 16-19 March, p. 2.2.

Hang, A. and T.Tsuchiya,1988. Chromosome studies in the genus *Alstroemeria*. II. Chromosome constitutions of eleven additional cultivars. *Plant Breeding*, **100**: 273-279.

Healy, W.E. and H.F.Wilkins 1985. *Alstroemeria*. In: *Handbook of Flowering,* vol. I (ed. Halevy, A. H.), pp. 415-424. CRC Press, Boca Raton, Florida.

Healy, W.E. and H.F.Wilkins,1991. *Alstroemeria*: The Cut flower Crop. In: *Ball Red Book*, 15th ed (ed. Ball, V.), pp. 311-316. Geo. J. Ball Publishing, USA.

Heins, R.D.and H.F.Wilkins, 1979. Effect of soil temperature and photoperiod on vegetative and reproductive growth of *Alstroemeria* 'Regina'. *J. Amer. Soc. hort. sci.*, **104**: 359-365.

Hofreiter, A. and E.F. Rodriguez, 2006. The Alstroemeriaceae in Peru and neighbouring areas. *Revista Peruana Biologia*, **13**: 5-69.

Hoshino,Y., N. Murata and K. Shinoda, 2006. Isolation of individual egg cells and zygotes in *Alstroemeria* followed by manual selection with a microcapillary-connected micropump. *Ann. Bot.*, **97**:1139-1144.

Manikrao, D. R. 2007. Studies on Possible Alternatives of Silver Thiosulphate in the Postharvest Handling of *Alstroemeria* and Carnation Cut Flowers. Ph.D. Thesis, Dr. Y.S. Parmar University of Horticulture and Forestry, Nauni, Solan (H.P.).

Mehra, A., V. Chandel, M. Singh, N. Verma, M.K. Singh and A.A. Zaidi, 2005. Viruses of *Alstroemeria*: Investigation, Control and Production of Virus free Plants. *Bhartiya Vaigyanik evam Audyogik Anusandhan Patrika*, **13**(1): 30-37.

Shanmugam,V., S. Kumar, M.K. Singh, R. Verma, V. Sharma and N.S. Ajit, 2007. First report of *Alstroemeria* wilt caused by *Fusarium oxysporum* in India. *Plant Pathol.*, p.1587.

Shanmugam,V., K. Atri and M.K. Singh, 2011. Screening *Alstroemeria* hybrids for resistance to *Fusarium oxysporum*. *Indian Phytopath.*, **64**(4): 388-389.

Singh, M.K. 2005. Production technology of *Alstroemeria* as a cut flower –a review. *Bhartiya Vaigyanik evam Audyogik Anusandhan Patrika*, **12**(1): 34-38.

Singh, M.K. 2006a. Performance of *Alstroemeria* cultivars under polyhouse conditions. *Indian J. Hort.*, **63**(2): 195-198.

Singh, M.K., R. Ram and R.Prasad, 2006. Effect of thinning of vegetative shoots on cut flower production of 'Serena' *Alstroemeria* hybrids under polyhouse conditions. *Indian J. agric. Sci.*, **76**(3): 181-182.

Singh, A., I.T. Nazki, Z.A. Qadri, Z. Ahmed and Sofi Inam-ul-Rehman, 2010a. Effect of storage on root number and nitrogen fertigation on clonal multiplication of *Alstroemeria* cv. 'Pluto'. *Natl. Symp. on Lifestyle Floriculture: Challenges and Opportunities*. Dr. Y.S. Parmar University of Horticulture and Forestry, Nauni, Solan, (H.P.), held on 19-21 March, p. 3.12.

Singh, A., I.T. Nazki, Z.A. Qadri, Z. Ahmed and Sofi Inam-ul-Rehman, 2010b. Effect of storage on root number and benzyladenine on clonal multiplication of *Alstroemeria* cv. 'Pluto'. *Natl. Symposium on Lifestyle Ffloriculture: Challenges and Opportunities*. Dr Y.S. Parmar University of Horticulture and Forestry, Nauni, Solan, (H.P.), held on 19-21 March, p. 3.12.

Stephens, J.L. T. Tsuchiya and T. H. Hughes, 1993. Chromosome studies in *Alstroemeria pelegrina* L. *Intern. J. Pl. Sci.*, **154**: 565-571.

Srasburger, E. 1882. Über den Teilunsvorgang der Zellkeme und das Verhältnis der Kemteilung zur Zeleteilung. *Archiv für Mikroskopische Antomie*, **21**: 476-590.

Taylor, W.R. 1926. Chromosome morphology in *Fritillaria*, *Alstroemeria*, *Silphium* and other genera. *Amer. J. Bot.*, **13**: 179-193.

Traub, H.P. 1943. Dichogomy and interspecific sterility. *Herbertia*, **10**: 131-132.

Van Scheepen, J. (ed.), 1991. International Checklist for Hyacinths and Miscellaneous Bulbs. *Royal General Bulb Grower's Association* (*KAVB*), Hillegom, The Netherlands, p. 409.

Verma, N., A.K. Singh, L. Singh, G. Raikhy, S. Kulshrestha, M.K. Singh, V. Hallan, Rajaram and A.A. Zaidi, 2005. Cucumber mosaic virus (CMV) infecting *Alstroemeria* hybrids in India. *Australasian Pl. Path.*, **34**: 119-120.

Wazir, J.S., Y.D. Sharma and S.R. Dhiman, 2010. Response of potted *Alstroemeria* to different growth retardants under protected conditions. *National Symposium on Lifestyle Floriculture: Challenges and Opportunities*. Dr. Y.S. Parmar University of Horticulture and Forestry, Nauni, Solan, (H.P.), held on 19-21 March, pp. 5.9.

3

Annuals

Sanyat Misra and R.L. Misra

Introduction

Annual ornamentals are a group of herbaceous plants which grow from seeds and complete their entire growth cycle, *i.e.* seed to seed in one season or in a maximum of one year. In their brief journey of life, they display a spectacular range of colours and are floriferous for a long period, especially when prevented from setting seeds. Annuals are known for their easy cultivation and that from seeds. They are meant for instant landscaping when planted in pans or pots, and also beautify the gardens in beds and borders with or without perennials and bulbous ornamentals, in sunken gardens, in window boxes, in hanging baskets, on walls and trellises and as ground covers. Apart from open garden beauty, there are a few which make good indoor display. Flowers of certain annuals can be used as cut flowers, as loose flowers, in pharmaceutical uses and for extraction of pigments. There are annuals whose flowers or seedheads are excellent for drying while some more are suitable for their sweet fragrance. There are but a few annuals those display only foliage beauty.

Classification of Annuals

Due to variation in the season of growth and flowering, several species of annuals thrive well in summer or rainy season, while majority of annuals grow well in the winter season. Generally annuals are classified into three broad types based upon the seasons. These three types are winter, summer and rainy season annuals. **Winter season** annuals display a greatest variety of flower colours. Various annuals for winter season are *Acroclinium roseum* (*Helipterum roseum*), *Agrostemma githago, Althaea rosea, Alyssum maritimum, Ammobium alatum, Antirrhinum majus, Arctotis grandis, Arenaria balearica, A. grandiflora, A. montana, A. purpurascens, Arnebia* spp., *Asperula orientalis* (*A. azurea-setosa*), *Bellis perennis, Brachycome iberidifolia, Brassica oleracea* var. *acephala, Briza maxima, Bromus briziformis, Browallia speciosa, B.* × *major, B. viscosa, Calceolaria herbeohybrida, Calendula officinalis, Callirhoe pedata, Callistephus chinensis, Campanula medium, Centaurea cyanus, C. moschata, Cheiranthus cheiri, Chrysanthemum carinatum* (syn. *C. atrococcineum, C. tricolor*), *C. coronarium* (syn. *Glebionis coronarium, G. coronaria*), *C. majus, C. maximum, C. multicaule, C. parthenium* (syn. *Matricaria eximia, Tanacetum parthenium*; though perennial but treated as an annual chrysanthemum), *C. segetum, C. viscosum* (*C. viscidi-hirtum*), *Cladanthus arabicus, Clarkia elegans, Delphinium hybridum, Dianthus barbatus, D. caryophyllus* (spray), *D. chinensis, Diascia barberae, Dimorphotheca annua* (*D. pluvialis*), *D. sinuata, Echium plantagineum, E. rubrum, Eschscholzia californica, Gamolepis tagetes* (syn. *G. annua*), *Gazania splendens, Godetia grandiflora, Gypsophila elegans, Helichrysum bracteatum, Iberis amara, I. umbellata, Ionopsidium acaule, Lagurus ovatus, Lathyrus odoratus, Legousia speculum-veneris* (*Specularia speculum-veneris*), *Limnanthes douglasii, Limonium sinuatum, L. suworowii, Linaria maroccana, Linum grandiflorum* var. *rubrum, Lupinus hartwegii, Malva sylvestris, Myosotis alpestris, Matthiola incana, Mesembryanthemum criniflorum,*

Mimulus tigrinus, Molucella laevis, Nemesia strumosa, Nemophila maculata, N. menziesii (N. insignis), Nicotiana alata grandiflora, N. suaveolens, Nigella damascena, Papaver nudicaule, P. rhoeas, P. somniferum, Perilla frutescens (syn. *P. nankinensis*, grown for its red-purple leaves), *Petunia hybrida, Phacelia campanularia, P. tanacetifolia, Phlox drummondii, Pimpinella monoica, Platystemon californicus, Polygonum affine, Rudbeckia bicolor, Salvia splendens, S, orminum, Saponaria calabrica, Scabiosa atropurpurea, Schizanthus hybridus grandiflorus* (syn. *S. pinnatus*), *Senecio arenarius, S. cruentus* (syn. *Cineraria cruenta*), *S. elegans, S. maritima (S. cineraria, Cineraria maritima), Silene armeria, S. coeli-rosa (Lychnis coeli-rosa, Viscaria elegans), S. compacta, S. pendula, Silyburn marianum, Tagetes erecta, T. patula, Thunbergia alata, Tropaeolum majus, T. peltophorum (T. lobbianum), T. peregrinum (T. canariense), Ursinia versicolor* (syn. *U. pulchra*), *Venidium fastuosum, Verbena hortensis (V. × hybrida), Viola cornuta, Viola tricolor* var. *hortensis, Viola × wittrockiana, etc.* Summers are generally associated with dry spell of wind and hot sunny days, so it is often felt that such seasons suppress the beauty to a great extent by giving a bare look to the garden, but with proper selection of annuals, the beauty and charm of the garden to some extent can be maintained even in this season similar to winter. To get colour in garden during **summers**, different annuals which can be grown are *Ageratum conyzoides, A. houstonianum* (syn. *A. mexicanum*), *Amaranthus caudatus, Anchusa capensis, Calandrinia discolor, C. umbellata, Catananche* spp., *Celosia argentea, Cleome spinosa (C. pungens), Collinsia bicolor, Coreopsis drummondii, C. tinctoria (C. bicolor), Cosmos bipinnatus, C. sulphureus, Crepis rubra, Didiscus caerulea, Felicia bergeriana, Gaillardia pulchella, Gomphrena globosa, Helianthus annuus, Impatiens balsamina, Ipomoea purpurea (Convolvulus major), I. tricolor (I. rubro-caerulea), Kochia scoparia* var. *trichophila, Lunaria annua (L. biennis), Malcomia maritima, Mina lobata (Ipomoea lobata, I. versicolor), Myosotis alpestris, Nicandra physaloides, Portulaca grandiflora, Quamoclit pennata (Ipomoea pennata), Reseda odorata, Saponaria calabrica, S. vaccaria, Stachys lanata, Tagetes erecta, T. patula, Tithonia rotundifolia, Torenia fournieri, Ursinia anthemoides, U. cakilifolia, U. versicolor (U. pulchra), Zinnia elegans, etc.* For getting display of flowers in **rainy season**, option for several annuals is there for growing in the garden. The beauty of almost all of summer annuals can be enjoyed during rainy season by adjusting their sowing and planting times. Such annuals are *Ageratum conyzoides, A. houstonianum (A. mexicanum), Amaranthus caudatus, Catananche* spp., *Celosia argentea, Cleome spinosa, Coreopsis tinctoria, Cosmos bipinnatus, C. sulphureus, Exacum affine, Felicia bergeriana, Gaillardia pulchella, Gilia lutea (Leptosiphon hybridus), Gomphrena globosa, Helipterum bulboldtianum, H. manglesii (Rhodanthe manglesii), H. roseum (Acroclinium roseum), Impatiens balsamina, Ipomoea purpurea (Covolvulus major), I. tricolor (I. rubro-caerulea), Helianthus annuus, Kochia scoparia, Lunaria annua (L. biennis), Mina lobata (Ipomoea lobata, I. versicolor), Portulaca grandiflora, Quamoclit pennata (Ipomoea pennata), Reseda odorata, Sanvitalia procumbens, Saponaria calabrica, S. vaccaria, Sedum caeruleum, Tagetes erecta, T. patula, Tithonia rotundifolia (T. speciosa), Torenia fournieri, Ursinia anthemoides, U. cakilifolia, U. versicolor (U. pulchra), etc.* Also there are **all season** annuals which can be grown throughout the year such as *Ageratum conyzoides, A. houstonianum* (syn. *A. mexicanum*), *Coreopsis drummondii, C. tinctoria, Cosmos bipinnatus, C. sulphureus, Gaillardia pulchella, Helianthus annuus, Hymenantherum* spp., *Tagetes erecta, T. patula, Tithonia rotundifolia (T. speciosa), Torenia fournieri, Ursinia anthemoides, U. cakilifolia, U. versicolor (U. pulchra), etc.*

Most annuals tolerate any soil type but certain others can be grown in specialized soils. The flowers are massed irregularly with respect to their height but grouped in suitable colour combinations to produce natural, harmonious and pleasing effects. If properly selected and timely-planted, they are the most beautiful plants for landscaping. Regular geometrical designs for beds and borders of annuals look more attractive. For quick seasonal colour in the garden, the fast growing annuals with a variety of flower colours in every hue imaginable are unsurpassed. Within a short time, in some cases as little as 6 to 8 weeks from seed to flower, the annuals produce a riot of colour. While working out colour scheme with annuals in a garden the same basic principles of colour are used as those adopted by an artist or interior decorator. Combining annuals of different colours to produce pleasing and restful effect is most rewarding. While grouping annuals particularly in herbaceous border, plant height, flower colour, the plant architecture, time of flowering and flower size are important considerations. Tall plants like *Althaea, Antirrhinum, Chrysanthemum, Clarkia, Coreopsis, Cosmos, Dahlia, Delphinium, Helianthus, Tithonia, etc.*, should be planted at the back while semi-tall annuals like *Calendula, Callistephus, Cheiranthus, Dianthus, Godetia, Matthiola, Phlox, Schizanthus, Tagetes erecta, Verbena, etc.* in the centre and dwarf annuals like *Alyssum, Iberia, Mesembryanthemum, Petunia, Portulaca, Viola, etc.* are most useful for edging in the foreground.

Ornamental Uses

Cut Flowers

Acroclinum roseum (*Helipterum roseum*; acroclinum, everlasting flower, immortelle, paper flower), *Amaranthus caudatus, Ammobium alatum, Antirrhinum majus, Arctotis hybrida* syn. ×*Venidio-arctotis, Bellis perennis, Calendula officinalis, Callistephus chinensis, Catananche caerulea* (blue cupidone), *Centaurea cyanus* (cornflower), *Centaurea moschata* (sweet sultan), *Cheiranthus cheiri* (wallflower), *Chrysanthemum carinatum* syn. *C. tricolor, C. coronarium, C. frutescens, C. multicaule, C. parthenium* syn. *Matricaria eximia, C. segetum, C.* × *spectabile* (annual chrysanthemums), *Delphinium ajacis* (rocket larkspur), *D. consolida* (larkspur), *Dianthus barbatus* (sweet william), *Dianthus chinensis* (Indian pink), *Dianthus caryophyllus* (annual carnation), *Didiscus caerulea* (blue lace flower), *Digitalis ferruginea* (foxglobe), *D. purpurea, Dimorphotheca annuaa* (*D. pluvialis*), *D. calendulacea, Gaillardia pulchella* (gaillardia), *Gamolepis tagetes* syn. *G. annua, Gazania* × *hybrida, Godetia amoena* (*Oenothera lindleyi*; farewell to spring), *G. bottae, G. grandiflora* (*Oenothera whitneyi*), *Gomphrena globosa, Gypsophila elegans, Helianthus annuus* (sunflower), *Iberis amara* (candytuft), *Layia elegans*(tidy tips), *Legousia speculum-veneris*, syn. *Specularia speculum-veneris* (venus's looking glass), *Limonium bounduellii, L. sinuatum, L. suworowii* (sea lavender, statice), *Lupinus hartwegii* (lupin, lupine), *Matthiola incana* (stock), *Molucella laevis* (bells of Ireland), *Nemesia strumosa, Nigella damascena* (love-in-a-mist), *Pimpinella monoica* (lady's lace, pimpinella), *Reseda odorata* (mignonette), *Rudbeckia bicolor* (coneflower), *Salvia horminum, Saponaria calabrica, S. vaccaria* (creeping zinnia), *Scabious atropurpurea* (scabious), *Schizanthus pinnatus* (butterfly flower), *Scirpus* (bulrush), *Tagetes erecta* (African marigold), *T. patula* (French marigold), *Ursinia anthemoides, U. cakilifolia, U. versicolor* (*U. pulchra*), *Venedium fastuosum, V. hirsutum* (monarch of the veldt), *Zinnia elegans* (zinnia), *etc.* are quite suitable as cut flowers.

Loose or Traditional Flowers

Arctotis × *hybrida, Amaranthus caudatus* (amaranth), *Bellis perennis* (daisy), *Calendula officinalis* (pot marigold), *Callistephus chinensis* (China aster), *Celosia argentea* (cockscomb), *Centaurea cyanus* (corn flower), *C. moschata* (sweet sultan), *Chrysanthemum carinatum* syn. *C. tricolor, C. coronarium, C. frutescens, C. multicaule, C. parthenium* syn. *Matricaria eximia, C. segetum, C.* × *spectabile* (annual chrysanthemums), *Coreopsis drummondii, C. tinctoria, Cosmos bipinnatus, C. sulphureus, Dahlia variabilis* (annual or bedding dahlias), *Datura ceratocaula, S. metel, D. meteloides, Dianthus barbatus* (sweet william), *D. caryophyllus* (spray type annual carnation), *D. chinensis* (Indian pink), *Impatiens balsamina* (balsam), *Helianthus annuus* (sunflower), *Rudbeckia bicolor* (coneflower), *Tagetes erecta* (African marigold), *T. patula* (French marigold), *etc.*

Fragrance

Some annuals are favoured for their attractive appearance and some others for highly fragrant flowers. These types of annuals possess both attractive flowers and also good odour. These types of annuals not only beautify the garden, but also fill the adjoining area or house with delicious fragrance. Various annuals having fragrant flowers are *Alyssum, Centaurea cyanus, Cheiranthus cherei, Cladanthus arabicus* (leaves and flowers fragrant), *Cleome spinosa* (*C. pungens*), *Dianthus barbatus, D. caryophyllus* (annual), *Helipterum bulboldtianum, Lathyrus odoratus, Lunaria annua* (*L. biennis*), *L. rediviva, Nicotiana, Nasturtium, Myosotis, Petunia, Phlox, Reseda odorata, Tagetes, Verbena hortensis* (*V.* × *hybrida*), *etc.*

Rockeries

There are some annuals which can thrive well in rocky areas and can be used for adding colour in different seasons there such as *Ageratum, Alyssum, Arctotis, Arenaria, Arnebia, Asperula, Brachycome,Calandrinia, Calceolaria, Callirhoe, Campanula, Cerastium, Cheiranthus, Crepis rubra, Dimorphotheca, Echium, Erodium, Eschscholzia, Gilia lutea, Godetia, Gypsophila, Ionopsidium acaule* (violet cress), *Limonium, Linaria alpina* (toadflax), *Matthiola, Myosotis, Nemesia, Phlox, Platystemon californicus* (cream cups), *Portulaca grandiflora* (sun plant), *Sanvitalia procumbens, Saponaria calabrica, S. vaccaria, Sedum caeruleum, Tropaeolum, Verbena, etc.*

Shady Locations

Most annuals like to grow in full sun and produce their best flowers only if they have plenty of light. To landscape semi-shady locations, there are *Ageratum, Alyssum, Browallia speciosa, Calceolaria, Delphinium, Exacum affine, Godetia, Iberis amara, Impatiens, Lobelia, Linaria, Lupinus, Myosotis, Nicotiana, Phlox, Salvia, Senecio arenarius* (cineraria), *S cruentus* (syn. *Cineraria cruenta*), *S. elegans, S. maritima* (*S. cineraria, Ceneraria maritima*), *Torenia, Verbena, Viola, etc.*

Edging and other Formal Works

There are short statured annuals which are used for formal work in the garden including edging of paths and other features of the garden. Annuals which can be utilised for this purpose are dwarf varieties of *Agertatum, Antirrhinum, Alyssum, Bellis perennis, Brachycome, Browallia, Iberis, Impatiens, Limnanthes douglasii* (poached egg flower), *Lobelia, Malcomia maritima* (Virginian stock), *Matthiola incana, Mesembryanthemum, Nemophila maculata, N. menziesii, Petunia hybrida, Phacelia, Phlox, Platystemon californicus* (cream cups), *Portulaca, Saponaria calabrica, S. vaccaria, Tagetes patula, Verbena, Viola, etc.*

Screens

For temporary screening, quickly covering a small space of trellis or a bare wall, arches, pergolas, low fences or partitions in a garden, some annual climbers can be recommended like canary creeper, *Ipomoea purpurea* (*Covolvulus major*), *I. tricolor* (*I. rubro-caerulea*), *Lathyrus odoratus, Mina lobata* (*Ipomoea lobata, I. versicolor*), *Quamoclit pennata* (*Ipomoea pennata*), *etc.*

Pots and Bowls

Annuals are also used in the garden as pot plants. Pot plants are utilised for getting instant colour in the garden, as by putting pots for display in the garden having annuals which are in bloom. Colour in the garden can be added by using pot with annual flowers. Various annuals suitable for pots are *Antirrhinum* (dwarf ones), *Calceolaria, Callistephus, Celosia, Collinsia bicolor, Dianthus, Diascia barberae, Didiscus caerulea, Dimorphotheca annua* (*D. pluvialis*), *Gazania splendens, Godetia, Kochia scoparia* (summer cypress), *Legousia speculum-veneris* (*Specularia speculum-veneris*; venus's looking glass), *Linum grandiflorum, Nemesia strumosa, Nemophila maculata, N. menziesii* (*N. insignis*), *Petunia hybrida, Phacelia, Sanvitalia procumbens, Senecio arenarius* (cineraria), *S. cruentus* (syn. *Cineraria cruenta*), *S. elegans, S. maritima* (*S. cineraria, Ceneraria maritima*), *Tagetes, Thunbergia alata* (black-eyed susan), *Ursinia anthemoides, U. cakilifolia, U. versicolor* (*U. pulchra*), *Venedium fastuosum, V. hirsutum, Viola, etc.*

Hanging Baskets and Window Boxes

Annuals can also be utilised for hanging baskets, as these make wonderful display in the garden or patio area of the house. Annuals suitable for hanging baskets are *Alyssum, Felicia bergeriana* (kingfisher daisy), *Iberis amara, Nasturtium, Petunia hybrida, Portulaca, Sanvitalia procumbens, Tagetes patula, Verbena hortensis* (*V. × hybrida*), *Viola wittrockiana, etc.*

Long-lasting

There are variety of annuals with long lasting flowers and can be added in garden for longer duration of flowering. Some of the annuals with long lasting flowers are *Acroclinium, Amaranthus caudatus, Ammobium alatum, Gomphrena, Gamolepis, Helichrysum (Bracteantha), Helipterum, Limonium, etc.*

Drying

The flowers of many of the annuals may be dried for winter decoration indoors and these are *Acroclinium, Amaranthus, Ammobium alatum, Briza maxima* (pearl grass), *Bromus briziformis* (brome grass), *Catananche, Eryngium, Gomphrena, Helichrysum, Helipterum, Lagurus ovatus* (hare's tail grass), *Limonium, Lunaria annua* (*L. biennis*), *L. rediviva, Nicandra physaloides* (fruiting branches), *Salvia horminum, etc.*

Quick Flowering

There are annuals which take three to four months to come into flowering, while some will bloom in five to six weeks. Sometimes there is need to plant quick flowering annuals for getting colour in the garden for some important occasions. These types of annuals provide colour in the garden in a very short interval of time. They are *Ageratum, Alyssum, Balsam, Browallia,* Candytuft, *Celosia, Dianthus, Gomphrena, Portulaca,* marigold and *Zinnia.*

Propagation

All the annuals are propagated through seeds. Though regular removal of dead flowers encourages more flowers to appear as energy consumed for seed formation is diverted towards flower production, but in case when after the flowering the aim is to produce seeds, the dead flowers are allowed to mature into seed heads, and after ripening these are collected, processed, cured, packed and stored as per requirement of the individual species. And when next season appears, these seed packets are taken out from storage and sown at right time in seed pans, trays, pots or on raised beds for nursery raising. Seeds of the winter-flowering annuals in the northern plains and in rest of the country with tropical or sub-tropical climatic conditions are sown during autumn, *i.e.* September-October, and in the temperate regions during September-October as well as in March-April. Seeds of the summer-season annuals are sown during February in the sub-tropical regions, in January-February in the tropical conditions and in February-March in temperate areas. However, rainy-season flowering annuals are required to be sown during May-June throughout the country. The bold-seeded crops such as *Althaea rosea, Ipomoea* spp., *Lathyrus odoratus, Lupinus* spp., *Tropaeolum* spp., *etc.* can be sown directly at their permanent positions and thinned out afterwards if required. The mixture for sowing of seeds for nursey raising should consist of one part each of garden soil, coarse sand, farmyard manure and leaf mould, all by volume. The same mixture can be used when seeds are raised in some containers. The nursery beds should be some 15 cm above the soil level and some 1 metre in width for proper manual operations just by sitting in the trenches made to both the sides of the beds. At the time of bed preparation, the soil should thoroughly be mixed with farmyard manure, sand and leaf mould to a depth of some 15 cm and at the time of sowing the nursery beds should have sufficient moisture. It would be better to treat the soil with 2 per cent formalin and then covered with black polythene sheets for 2 days and then polythene should be removed and the soil should be worked out but sowing of seeds should be done only after a week of treatment. Seeds may also be treated with 0.2 per cent Captafol, dried in shade and then sown. Seeds of winter-flowering annuals are normally sown at temperatures from 10-18 °C, some requiring as low as 10 ° though certain others even 18 °C. While sowing the seeds, some 2-3 cm deep furrows should be made with fingers or wooden sticks or pencils some 5-6 cm apart. After sowing the seeds thinly and evenly, these are covered some 1.0 cm with the same soil mixture or sand and then the beds are covered with straw to conserve the moisture, to facilitate watering so that seeds do not get disturbed from their original place of sowing, and to save them from damage by birds, *etc.* Daily in the morning it should be checked for germination and if surface of the soil has started splitting the seeds have started germinating so now immediately the straw mulch should be removed otherwise the seeds will become crooked and lanky. After germination, light watering should be carried out daily with water can having fine rose. The nursery beds should be kept weed-free throughout. Normally the seedlings are planted in the afternoon when these have attained 3 pairs of leaves or 8-12 cm height, *i.e.* 30-45 days of seed sowing, followed by immediate watering to settle the roots and to overcome the lifting shock. Seeds of *Nigella* and *Senecio* germinate only in the dark, of *Echium, Lobelia* and flowering *Nicotiana* germinate only when first exposed to sunlight (Raghava, 2001), and there are certain others which though immediately after harvesting germinate but after sometime acquire some sort of rest within which can be broken through some chemical treatment.

There are certain annuals such as *Arenaria, Arnebia, Bellis, Briza, Campanula, Cerastium, Stachys, etc.* which are also propagated through division of their clumps, especially in the temperate regions where these become biennial due to highly favourable agroclimatic conditions. Division of clumps is carried out only during planting season of individual crops but it should be cool temperature so that no pathogen may harm the injured parts. Divisions can be planted in equal parts of peat, garden loam and sand mixture (v/v).

Certain annuals are also capable of coming up well through cuttings when they are in vegetative phase. Lateral (*Cheiranthus cheiri* 'Harpur Crewe', *Datura meteloides* with 1 year old and 7.5-10.0 cm long cuttings, *Dianthus chinensis, Gazania, Portulaca, Tagetes, etc.*), basal (*Antirrhinum, Arnebia, Campanula, Catananche, Iberis, Lupinus, Viola, etc.*) or root (*Anchusa, etc.*) cuttings 5.0-7.5 cm long should be taken and inserted in equal parts (v/v) of peat-sand medium and when these root these are potted singly in 7.5 cm pots in mixture containing equal parts (v/v) of garden loam, coarse sand, leaf mould and farmyard manure and kept in cold frame. These can be planted out when proper planting season appears, as per requirement of the individual species.

Genera, Species and Varieties

The elaborations given here pertaining to various genera, their species and concerned varieties are as per details furnished on various crops by Bailey (1942), Desai (1962), Pizzetti and Cocker (1968), Hay and Beckett (1971), Helleyer (1983), Beckett (1985, 1987), Brickell (1994), Dole and Wilkins (1999), and Raghava (2001).

***Adonis* spp.** (Ranunculaceae). *Adonis* takes its name from the god Adonis, worshipped as the fertility god in the Middle East, and in the mythology known as the beloved of both Aphrodite and Persephone. The genus comprises of some 20 species of heady annuals and herbaceous perennials from Europe and Asia. They are suitable for borders, for rock gardens and for cutting. They are winter-flowering. They are propagated through seeds sown in September. The flowers at bud stage cut with long stalks in case of both the species described here, when are placed in little soil-filled pots with copious watering, the buds continue to expand for long period of time.

A. annua (*A. autumnalis*; pheasant's eye) from Mediterranean and South Europe is 20-40 cm tall annual with thread-like leaf segements. Flowers are cup-shaped, 1.5-2.5 cm across and with 5-8 bright scarlet petals having dark basal spot.

A. flammeus from Central Europe to Caucasus and Asia Minor is an branching annual growing erect up to 30 cm tall with very succulent and leafy stems. The leaves are much-divided similar to those of *Anthemis.* Solitary terminal and flame-coloured flowers appear during winter to spring.

Ageratum spp. (Family: Asteraceae, common names: ageratum, floss flower). A genus of some 60 annual species from tropical America, the ones of ornamental values are *A. conyzoides* and *A. houstonianum* (syn. *A. mexicanum*). *Ageratum* in Greek takes its name from *a* for 'no' and *geras* for 'old', alluding to the comparative longevity of the flowers. *A. houstonianum* is native to Mexico, growing up to 60 cm high, bearing heart-shaped leaves in opposite pairs, and the shaving-brush type clustered flowerheads some 10 cm across, appearing during summer and rainy season and also during winters in terminal clusters, are fluffy, pompon-like, and in shades of white, pink, rose and blue. Its seeds are raised at 15-18 °C temperature. Varieties are 'Blue Bedder', 'Blue Blazer', 'Blue Denube', 'Blue Marie', 'Blue Monk', 'Blue Perfection', 'Fairy Pink', 'F_1 hybrids', 'Midget Blue', 'Summer Sky', 'Summer Snow', *etc.*

Agrostemma githago (Family: Caryophyllaceae, common name: corn cockle). It takes its name from the Greek *agros* meaning 'field' and *stemma* meaning 'a crown or garland'. There are three summer-flowering hardy annual species from the Mediterranean and Southern Russia, *e.g. A. coeli-rosa* (syn. *Silene coeli-rosa*), *A. coronaria* (syn. *Lychnis coronaria*) and *A. githago* from Mediterranean. *A. githago* grows up to 1 m with thin stems, bearing mid-green lanceolate leaves and open trumpet-shaped red-purple flowers some 3-5 cm wide with 5 petals. Its seeds are tiny, dark brown, rounded and poisonous. Seeds are raised during spring. The varieties are 'Milas' (lilac-pink, 6 cm), 'Purple Queen' (rose-purple, 8 cm), *etc.*

Althaea (Alcea) rosea (syn. *A. chinensis*, family: Malvaceae, common name: hollyhock). In Greek, *althaia* means 'to cure' due to medicinal properties of some species. It is a winter-flowering annual to biennial from China, growing from 1.5 to 3.0 m tall and is suitable for back-side planting in the borders. It bears light green, rough, hairy, palmate, long-stalked and 5-6-lobed leaves. The flowers in the varieties are single (5-petalled), semi-double or double and more than 10 cm across in shades of white, yellow, pink, red, vermillion and purple, and appear in terminal racemes. Varieties are 'Begonia-Flowered Crested Mixed', 'Carter's Double', 'Carter's Double hybrids', 'Chater's Double' (double paeony-shaped flowers), 'Deepika', 'Double Triumph Mixed', 'Dulhan', 'Fordhook Giant', 'Gauri', 'Giant Double Powderpuff', 'Pusa Apricot Supreme' (F_1 hybrid), 'Pusa Hollyhock Gulabi', 'Pusa Hollyhock Lalima', 'Pusa Hollyhock Krishna', 'Pusa Hollyhock Sweta', 'Pusa Pastel Pink Supreme' (F_1 hybrid), 'Pusa Pink Beauty' (F_1 hybrid), 'Pusa Yellow Beauty' (F_1 hybrid), 'Semperflorens', 'Single Mixed', 'Summer Carnival', *etc.* The seeds are raised at 13-16 °C temperature.

Alyssum maritimum (syn. *Lobularia maritima*, family: Brassicaceae, common names: madwort, sweet alyssum). It is a fast-growing hardy annual from South Europe and Western Asia. It takes its name from the Greek *a* meaning 'no' and *lyssa* meaning 'rage or madness', alluding to supposed medicinal properties against rabies. *Lobularia* in Latin derives from *lobules* 'a small pod' as the seed pods are quite small, rounded and slightly inflated. This species is mat- or hummock-forming with densely-branched, slender and spreading stems that grow up to 15 cm or more tall and bear greyish-green, usually hoary, up to 6 cm long and linear to oblanceolate leaves, and about 6 mm wide, 4-petalled white, lilac or purple and fragrant flowers in clusters appearing in terminal, rounded and some 5 cm long racemes in profusion during winter to spring. Seeds are sown at 10-13 °C temperature. Popular varieties are 'Lilac Queen' (deep lilac), 'Little Dorrit' (syn. 'Little Gem', pure white), 'Minimum' (syn. 'Snow Carpet', quite dwarf, pure white), 'Oriental Night', 'Pink Heather', 'Rosie O'Day' (rose-red), 'Royal Carpet' (violet-purple), 'Snow Carpet', 'Snow Cloth', 'Tetra Snowdrift', 'Violet Queen' (violet-purple), 'Wonderland' (deep purplish-pink), *etc.*

Amaranthus caudatus (Family: Amaranthaceae, common name: love-lies-bleeding). In Greek *amaranthos* means 'unfading' as some species retain their colours similar to everlasting flowers. This is native to tropical, sub-tropical and temperate regions of the world with about 60 annual species, where some are weedy but a few make handsome subjects for summer growing either for their boldly coloured foliage or unusual tail-like clusters of densely-produced tiny flowers. The plants are erect bearing alternate and lanceolate to ovate leaves of various colours.

In *A. caudatus*, the leaves are green and ovate, flowers are red, catkin-like tailed, clustered, pendant and some 40 cm or more long which appear during summer to autumn. Its varieties are 'Flame' (firey), 'Flaming Fountain' (flame), 'Green Balls' (ball-like clusters of green flowers), 'Pygmy Torch' (yellow), 'Red Fox' (red), 'Viridis' (pale-green), *etc.*

A. hypochondriacus (syn. *A. hybridus*, common name 'prince's feather') bears up to 15 cm long leaves which are green and oblong-lanceolate, and erect panicles bearing red flowers though individual clusters arch at the tips, and its normally grown form is *A. h. erythrostachys* bearing reddish-purple leaves.

A. tricolor (syn. *A. gangeticus*) from India has fascinated stems, bearing ovate and some 20 cm long leaves which are flushed with yellow, red or bronze though flowers are inconspicuous and hidden in the foliage. Its hybrids are 'Joseph's Coat' (leaves have dark green, yellow, golden-orange and red patterning), 'Molten Fire' (leaves lance-shaped, drooping and patterned with orange-red and bronze), *etc.* They should be grown through seeds at 15 °C temperature. The popular varieties of amaranths are 'Blue Bird', 'Flame', 'Flaming Fountain', 'Pygmy Torch', 'Red Fox', *etc.*

Amberboa moschata (syn. *Centaurea moschata*, Sweet sultan; Asteraceae). A hardy annual from SW Asia (Turkey, Caucasus) grows up to 90 cm tall with thin stems, bearing greyish-green, entire, narrow, lanceolate to pinnately-lobed or lyrate leaves. Solitary and fragrant flowers some 7.5 cm across in shades of white, yellow, pink and purple appear during winter to spring. These make excellent cut flowers, and are also suitable for garden display as bedding plants and in borders. These are propagated through seeds by sowing these in autumn or spring in the nursery beds or at their permanent site and then thinning out afterwards. Popular varieties of sweet sultan are 'Alba', 'Favourita', 'Graceosa', 'Imperialis', 'Odorata', 'Splendens', 'Suaveolens', *etc.*

Ammi majus (Lady's lace; Apiaceae): A native of South Europe, this white-flowering annual may grow up to 1.5 m tall or even more though usually it is smaller than stated, and is excellent for planting in the beds, borders and as cut flowers. Its leaves are mid-green, linear to spathulate or obovate, and the corymbs or racemes of white flowers. They are propagated through seeds. First these are raised in the nursery and then after four weeks the seedlings are transplanted in the prepared beds in the field.

Ammobium alatum (Family: Asteraceae, common name: winged everlasting). In Greek *ammos* means 'sand' and *bio* means 'to live', referring to native habitat. There are only two annual species from Eastern Australia, and *A. alatum* is the only grown species from New South Wales which grows erect some 60 cm in height with winged stems. The flowers are everlasting and appear in flowerheads 2-3 cm across with chaffy but petal-like bracts. Flowers are silvery-white with yellow central florets. The flowerhead of *A. a.* 'Grandiflorum' is 4-5 cm wide. They are propagated through seeds sown during autumn or in spring.

Anastatica hierochuntica (Family Cruciferae, common names: Resurrection plant, Rose of Jericho). In Greek, *anastasis* is for 'resurrection'. It has only one species native of Morocco to South Iran. It is more curious plant than being ornamental and is a desert plant. It is an annual growing up to 15 cm high with much-branched stem. Leaves at lower ranks are entire, the upper ones slightly toothed, are obovate-spathulate and stellate-pubescent. Flowers are small white in spikes along the branches, sepals 4, petals 4 and obovate, stamens 6 (4 longer), fruit a silique, small and dehiscent. After flowering the leaves drop and the radiating branches curve inward in the form of a skeletal ball and dry out but when rains remoisten, the branches of these plants uncurl to resurrect these as the live plants.

Anchusa capensis (Anchusa; Boraginaceae). Greek for *ankousa* means 'alkanet', as some species yield rouge. It is an annual or biennial species native to South Africa, and grows 60 cm in height. Leaves are mid-green hairy, narrowly lance-shaped, acuminate and 10 cm or more long. Blue flowers are some 6 mm across with red mouth and white throat and appear during rainy season in elongated panicles. It is most suitable as a pot plant or in the mixed border. Its variety 'Blue Bird' produces sprays of flowers similar to *Myosotis*. These can be raised by sowing the seeds at temperatures of 13-16 °C. Its popular varieties are 'Bedding Bright Blue', 'Blue Bird', 'Dropmore', 'Morning Glory', 'Pride of Dover', 'Royal Blue', *etc.*

Antirrhinum majus (Family: Scrophulariaceae, common names: antirrhinum, bunny mouth, bunny rabbit, dog flower, snapdragon). In Greek *anti* means 'like' and *rhin* means 'nose or snout', alluding to the appearance of the flowers. It is semi-hardy annual to perennial, native to Southwest Europe and Mediterranean, growing straight up to 1.2 m in height, bearing glossy, mid-green and lanceolate to ovate leaves up to 7 cm long. Flowers are 3.0-4.5 cm long, pink and spurred but the varieties may be white, yellow, orange, pink, red, bronze and purple. It is grown throw seeds and stem cuttings. These are grouped according to size and floral types, *i.e.* taller ones growing from 60 cm to 1 m, intermediates 45 cm and dwarf ones 20-30 cm in height. It is grown in beds, borders and pots. Tall types are most suitable as cut flowers. Further these are classified into preloric (taller ones with regular tubular-shaped flowers) such as 'Coronette' (wide range of colours); penstemon (trumpet-shaped flowers), double, and penstemon-type irregular tubular-shaped flowers such as 'His Excellency' (intermediate, scarlet). The tall, large-flowered sorts such as 'Tetra Snaps' are excellent for cutting, and the smallest 'Tom Thunb', 'Magic Carpet', *etc.* which grow hardly 15 cm in height are suitable for pots. 'Taffs White' with cream-margined leaves is a beautiful foliage plant. Other popular varieties are 'Cavalier', 'Commander', 'Glacier', 'Guardsman', 'High Noon', 'Juliana', 'Maximum Orchid', 'Maximum Scarlet', 'Maximum Yellow', 'Sunlight', 'Temple', 'Tinker Bell', 'Volcano', 'White Spire', *etc.* F_1 hybrids are 'Golden Dragon', 'High Life', 'Princess Crimson', 'Princess Scarlet', 'Princess Yellow' (red eye), 'Rocket Bronze', 'Rocket Lemon', 'Rocket Pink', 'Rocket Red', 'Rocket White', 'Rose Dragon', 'Royal Carpet Crimson', 'Royal Carpet Orange', 'Super Jet', 'Supreme Vanguard', 'Venus', 'White Dragon', *etc.*

Arctotis breviscapa (syn. *A. speciosa,*family: Asteraceae, common names: African daisy, Blue-eyed daisy, Transvaal daisy). In Greek *Arktos* means 'a bear' and *otosi* means 'an ear' as the pappus scales on the seed appear as bear's ear. This annual is from South Africa, where stems and leaves are woolly in silver-green, growing more than 15 cm tall and bears pinnately lobed oblong-lanceolate leaves, and the flowerheads have daisy-like orange-yellow flowers, outer petals being long, tubular and light-coloured though central disc brown, purple or violet. These bloom throughout the year, *i.e.* winter, summer to rainy season and during autumn. These are excellent for border, for pots and for cutting though last only a few days. Its flowers have tendency to close during dull weather and by evening. Other species are perennials but are grown as annual.

A. acaulis (*A. scapigera*) from Africa grows up to 25 cm high and is suitable for edging, for front row border and for window boxes. Lobed leaves are white-haired beneath and the flowers some 8.5-9.0 cm across appear from orange to deep red in the varieties.

A. × *hybrida* grows up to 60 cm tall and is a hybrid with many interspecific crosses and whose some 10 cm wide flowers last longer in vase. The colours of the flowers in the hybrids may be white, cream, yellow, orange, apricot, carmine and red, many with attractive zoning.

A. stoechadifolia from Africa grows up to 90 cm in height and is suitable for planting in the back of the border. The stems are ribbed and thick, the leaves are grey-green, narrow and toothed, and the flowers are 7.5 cm across, and pearl-white with blue central disc surrounded by a golden zone. *A. s. grandis* bears larger flowers in shades of white to primrose. These are propagated through seeds sown in autumn or during March-April at 18 °C temperature. 'Sutton's Triumph' is its most popular strain.

Arenaria spp. (Family: Caryophyllaceae, common name: sandwort). In Latin, *arena* means 'sand' as these were recorded from the sandy-loam soil of the northern temperate zone. These are hardy to half-hardy low-growing annuals to seed-growing perennials suitable for rock gardens and alpine houses. They are mostly small, creeping or mat-forming tufted plants, bearing filiform to linear-lanceolate or ovate leaves in opposite pairs. Solitary or in panicles, the white (rarely pink) and 5-petalled flowers appear during winter to spring so seed-sowing or division-planting (*A. balearica* and *A. purpurascens*) or 2.5-5.0 cm long basal shoot cutting-planting (*A. grandiflora* and *A. montana*) is carried out during autumn to late winter.

A. balearica is from Balearic Islands and Corsica which grows to 3 cm in height but about 50 cm in spread, bearing mid-green, ovate and minute leaves, and some 5 mm across, white and star-shaped flowers. It is suitable for planting in the shady side of rockery.

A. grandiflora is native to European Alps and North Africa and grows up to 8 cm in height and some 30 cm in spread, bearing densely produced green and pointed leaves and funnel-shaped white flowers some 1.8 cm across.

A. montana from Spain and France grows up to 15 cm in height and some 45 cm in spread, bearing green and narrowly lanceolate leaves. White saucer-shaped flowers some 1.2-1.8 cm across are produced on branched stems in spring to May.

A. purpurascens from Pyrenees and Spain grows prostrate up to 8 cm in height and 30 cm in spread, bearing mid-green, lanceolate and narrowly pointed leaves. Pink to purple star-shaped flowers, some 1.2-1.8 cm across appear from winter to rains in the cluster of 2-3.

Arnebia spp. (Family: Boraginaceae, common name: prophet flower). Its Latin name comes from the Arabic *sagaret el arneb*. The members of the genus are tender to hardy annuals to perennials from Tropical Africa, Mediterranean regions and Himalayas, but only one perennial species (*A. echioides*, syn. *Macrotomia echioides, Echioides longiflorum* from Armenia and NE Turkey) is in general cultivation as an annual, which is suitable for planting on large rock gardens and at front in the borders. It grows up to 30 cm tall, bearing rough, hairy and narrowly oblong leaves. The primrose-yellow and funnel-shaped flowers appear in spike-like cymes where all the 5 lobes are rounded with a basal black spot which fades with the aging of the flowers. It is propagated through seeds during autumn, and thick root cuttings some 3-4 cm long or stem cuttings with a heel during spring at 13-16 °C temperature.

Asperula orientalis (syn. *A. azurea-setosa*, family: Rubiaceae). In Latin *asper* means 'rough', because of its rough stems. An annual from Syria, suitable as spring-flowering pot plants and as cut flowers, attains height of some 30 cm, bearing mid-green, hairy and narrowly lanceolate leaves. Its tubular, pale-blue and fragrant flowers some 1.8 cm across appear in terminal clusters. Seeds are broadcast in the prepared field on the flowering site in autumn in the plains and in April on temperate regions and then thinned out at required spacing when have attained 6-10 cm height.

Bassia scoparia (Kochia, Summer cypress; syn. *Kochia scoparia, K. scoparia* var. *trichophylla*;

Chenopodiaceae). Genus *Kochia* was named for Wilhelm Daniel Josef Koch (1771-1849), professor of botany at Erlangen, West Germany. Out of 80-90 species under this genus, only *K. scoparia* is cultivated one which is native of South Europe to Japan. It is a half-hardy summer to autumn annual growing pyramidal or columnar to a height of some 1m, and is suitable as a specimen plant in the bedding schemes, for edging and as a pot plant. It has decorative foliage with a profusion of pale-green, 3-6 cm long, narrowly lanceolate to linear and pointed leaves, and barely noticeable tiny green and petalless flowers are insignificant. At the end of the autumn, the plants are blushed with coppery shade. *K. s. trichophylla* 'Childsii' strain is more uniform and neater than the former. It can be propagated through seeds sown in mid-spring in the beds or at 16-18 °C temperature in the glasshouse.

Bellis perennis (Asteraceae; common names: Common daisy, English daisy, Lawn daisy). It takes its name from Latin *bellus* meaning 'pretty'. This hardy perennial though in cultivation is treated as a winter annual, is native to Europe, the Mediterranean region and Asia Minor. It is suitable for planting in border edging, carpet bedding, in rockery and in pots. These are evergreen, clump-forming and low-growing plants (hardly growing up to 15-20 cm height), bearing 3-8 cm long, mid-green and orbicular-ovate leaves. Solitary flowerhead some 2- 3 cm across, consists of white ray-like florets, rarely tinged pink, and disc yellow. The non-seed producing cultivars such as 'Dresden China' (pink) and 'Rob Roy' (red) are propagated through division though other cultivars are raised through seeds sown in September-October. The cultivars are in two groups: 'Monstrosa' which bears very large, usually double or semi-double blooms more than 5 cm across, and 'Miniature' with 1-2 cm across pom-pom type double heads with quilled florets, both the groups bearing white, pink or crimson blooms that appear true from seeds. 'Mammoth' strains have double flowers on long stalks. Its most common varieties are 'Aetna', 'Chevreuse', 'Liliput', 'Longfellow', 'Monstrosa', 'Pomponett', 'Ruby', 'Snowball', 'Super enorma', 'Vesuv', *etc.*

Brachycome iberidifolia (Family: Asteraceae, common names: brachycome, swan river daisy). In Greek *brachys* means 'short' and *kome* means 'hair', due to bristle-like pappus crowning the fruit. A native of Australia, this is a half-hardy annual growing up to 45 cm tall, bearing pinnatifid leaves with linear segments, and blue-purple or white ray flowers on a 2-3 cm across flowerhead. This is suitable for planting in borders, in edgings and as pot plants. Seeds are sown during autumn or in late winter at 15-18 °C temperature.

Bracteantha bracteatum (syn. *Helichrysum bracteatum*; Family: Asteraceae; common names: Everlasting flower, Helichrysum, Immortelle, Strawflower). In Greek, *helios* is for 'sun' and *chrysos* for 'golden', named so after the flowerheads of some species. The annual species are suitable for the borders, for cutting and their showy daisy-like flowers with persisting chaffy bracts are everlasting so are dried for dry arrangements. These are propagated through seeds sown in greenhouses at a temperature of 18 °C. *B. bracteatum* (*B. macranthum*) from Australia is a half-hardy, robust and branching annual growing up to 1.2 m tall. Its leaves are bright green, large, entire, oblanceolate and acuminate. Yellow or orange flowers are terminal and some 3-4 cm across. Its prominent varieties are 'Album' (white), 'Atrococcineum' (deep scarlet), 'Atrosanguineum' (blood-red), 'Bicolor' (flowers red, marked dark red and pink), 'Borussorum Rex', 'Dwarf Double', 'Dwarf Spangle Mixed' (up to 30 cm so suitable for bedding also, mixed colours), 'Fireball' (orange), 'Luteum' (yellow), 'Ferrugineum Monstrosum' (heads large, red), 'Macranthum' (blooms up to 6.25 cm across), 'Monstrosum' (larger heads, more bracts), 'Monstrosum Double Mixed' (tall, flowers double and mixed colours), 'Nanum' (dwarf form), 'Purpureum' (purple), 'Roseum' (rosy), 'Tall Double', *etc.*

Brassica oleracea (Flowering kale/Ornamental cabbage; Brassicaceae). A native of the Mediterranean to temperate Asia, these plants in their improved forms are being grown in the tropical to sub-tropical ornamental gardens for their loosely rosette, round and variously coloured (white, green, pink, mauve and purple) leaves. They make excellent pot plants and are good for borders and beddings. Its attraction is superimposed when temperature falls up to 10 °C though it does well up to 20 °C temperature. Ornamental cabbage feather-leaved types are 'Coral Queen' (centre red), 'Coral Prince' (centre white), 'Tokyo Series' (pink and white cultivars), *etc.*, round-leaved types being 'Osaka' and 'Pigeon Series' (white, pink, red and purple leaves), and fringed-leaved types such as 'Kamome' and 'Nagoya Series' (leaves being white, pink, rose and red), while flowering kales are 'Red Peacock' (red centre), 'White Peacock' (white centre), 'Feather Red', 'Feather White', *etc.*

Briza maxima (Graminae, common names: pearl grass, quaking grass). It is an ancient Greek name for 'a grass', probably rye, which Linnaeus adopted for this genus. A native to Mediterranean, is a hardy annual grass growing up to 60 cm in height, forming small tufts of erect stems, bearing bright green, 5-20 cm long and 3-8 mm wide, narrow and pointed leaves. They are grown for their pendent and loose ornamental panicles with some

6-8 mm long and ovate spikelets which are pale-silvery-green or rarely purple-flushed, appearing on 10 cm upper length of stalk. They are useful in floral arrangements or as dry flowers. It can be propagated *in situ* in the flowering site and thinned out to required spacing when sufficiently grown. Another annual species, *B. minor* (lesser quaking grass) from Mediterranean grows more than 40 cm in height, bearing 3-14 cm long and 3-9 mm wide leaves, and 4-20 cm long panicles where usually shining green spikelets are triangular-ovoid to rounded and 3-5 mm wide.

Bromus brizaeformis (Graminae, common name: brome grass). *Bromus* is ancient Greek name for oat. A hardy annual grass from Europe, SW and Central Asia, grows up to 60 cm in height with 5-15 cm long, one-sided and nodding panicles where spikelets are strongly flattened, oblong-ovate and about 2.5 cm long. They are suitable for drying and in floral arrangements. It is propagated through seeds sown *in situ* directly in the prepared field and thinned out when sufficiently grown.

Browallia americana (syn. *B. elata*; Solanaceae). A genus of six half-hardy annual species from NS America and West Indies. *Browallia* was named for John Browell (1707-1755), a champion of Linnaeus, Bishop of Abo and a Swedish botanist. These are grown in the open in the mild climate, in the pots, in the hanging basket and for cutting. Leaves are loose, alternate or opposite and the racemes are normally terminal though sometimes with solitary flowers. The flowers appear during winter in colour range of blue, violet or white, tubular with 5-lobed mouth where upper one is broader. Seeds are sown by autumn. These are propagated through seeds at 18 °C temperature.

B. speciosa (bush violet) from Colombia grows up to 1.2 m tall bearing bright green, 10 cm long, slender-pointed and ovate leaves. Flowers are 5 cm across at mouth, purple above and paler beneath. To keep plant in shape, it requires pruning. Its varieties are 'Blue Troll' (dwarf), 'Silver Bells' (white), 'White Troll' (dwarf, white), *etc*.

B. viscosa from Colombia grows up to 60 cm tall. It is a compact plant and is suitable for bedding and as a pot plant. The leaves are mid-green, 2-4 cm long, sticky-hairy and ovate. The violet-blue flowers have a white eye, 2.5 cm across and the petals are deeply notched. Its var. 'Sapphire' is more compact and grows up to 25 cm.

Calandrinia spp. (Portulacaceae). *Calandrinia* was named for Jean Louis Calandrini (1703-1758), professor of mathematics and philosophy at Geneva. Though there are some 130 annual and perennial species under this genus but only two are in general cultivation, both being perennial but cultivated as annuals as both are propagated through seeds every year during spring and planted out during September or sown directly to their site of flowering during autumn and thinned out afterwards but require protection from frost. Both the species are semi-prostrate plants bearing a profusion of short-lived attractive flowers. They are excellent for borders, for rock gardens and as pot plants.

C. discolor from Chile grows up to 50 cm high, bearing spathulate leaves that are grey-green above and purple beneath, and bowl-shaped pale-purple flowers some 5 cm across with contrasting yellow stamens appear in summer.

C. umbellata from Peru is best among the species which grows up to only 15 cm in height. The plants are very compact, do well under dry situations and are suitable for edging paths, in the rock garden, and as pot plants at cold places. It bears grey-green, hairy, narrow, linear leaves and forms a bushy mat of ground cover. Cup-shaped crimson and some 1.8 cm across flowers appear in clusters during summers and monsoon.

Calceoraria × herbeohybrida (syn. *C. hybrida*, family: Scrophulariaceae, common names: calceolaria, pocket book plant, pouch flower, slipper flower, slipper-wort). In Latin, *calceolus* means 'a little shoe or slipper', due to its flower shape. Most of the species are native from Mexico to South America.

Calceoraria × herbeohybrida, a giant-flowered strain is a half-hardy biennial hybrid of garden origin though grown as annual, suitable for pots for keeping temporarily indoors. It is a giant-flowered strain developed mainly from *C. amplexicaulis, C. corymbosa, C. crenatiflora, C. integrifolia* (*C. rugosa*) and *C. purpurea*. These grow up to 45 cm high, bearing opposite pairs or whorls of mid-green, soft, slightly hairy and broadly ovate to linear leaves. The 2-lipped flowers, lower lip greatly inflated and pouched similar to the toe of a fancy slipper, more than 6 cm wide, in shades of yellow, orange, bronze, scarlet and red, and blotched and spotted in various combinations and are borne in large terminal clusters from late spring to summer. These can be propagated through tiny seeds sown during autumn or spring at a minimum teperature of 18 °C. During winter, the temperature should not go below 10 °C as plants will be damaged. Commercially available horticultural varieties are 'Giant-Flowered' (orange, bronze and scarlet marked with contrasting tints), 'Grandiflora Tigrata' (large inflorescences in various colour combinations or self-coloured), 'Intermedia' (large inflorescences in shades of yellow, orange, pink and bronze, marked with red), 'Monarch strain' (range of colours marked exquisitely), 'Multiflora

Nana' (dwarf, compact and smaller inflorescences with a wide range of bizarre shades and combinations of yellow, orange, pink, scarlet and bronze), *etc.*

C. gracilis from Ecuador is a half-hardy annual that grows up to 45 cm high, bearing mid-green, sticky, deeply cut and hairy leaves, and some 1.2 cm long and lemon-yellow flowers come up in profusion during summer to monsoon.

C. mexicana from Mexico is a half-hardy bushy annual suitable for bedding, that grows up to 25 cm high, bearing laciniated mid-green leaves, and about 1.2 cm long and pale-yellow flowers during late monsoon.

C. scabiosifolia from Peru, a half-hardy annual suitable for bedding, grows 60 cm in height, bearing, mid-green, hairy and deeply divided leaves, and some 1.2 cm long pale-yellow flowers in dense clusters from spring to autumn.

Calendula officinalis (Family: Asteraceae, common names: calendula, pot marigold). *Calendula* originated from the South Europe, and takes its name from the Latin *calendae* meaning 'the first day of the month', and *kalendarium* meaning 'account'; the almost perpetual flowering of this plant in its native haunts was a constant reminder that interest was owed on the first day of each month. *Calendula officinalis* plant is a hardy annual in bushy-shape, growing up to 60 cm tall with several branches, leaves are light green, alternate, oblanceolate, 7-14 cm long and narrow, and the flowers produced freely are some 10 cm across, yellow and orange with daisy-like flowerheads. Achenes (seeds) are bow-shaped to almost circular, cylindrical with 2-3 forms being present in each head. They are suitable for bedding, as pot plants, for borders and for cutting in winter to spring. These are grown as winter annuals by sowing the seeds in September-October. Important varieties are 'Apricot Beauty' (apricot-yellow), 'Art Shades' (45 cm tall with semi-double flowers in cream, apricot and gold), 'Baby Gold' (up to 40 cm tall, fully-double, golden-yellow), 'Baby Orange' (up to 40 cm tall, semi-double to double, bright orange), 'Chrysantha' (chrysanthemum-like yellow flowers), 'Cream Beauty' (cream-white), 'Crested Mixed' (two outer rows of plain petals surrounding a quilled centre), 'Geisha Girl' (fully double, red-orange with incurving petals), 'Kelmscott Giant Orange' (fully double, bright orange), 'Lemon Beauty' (lemon-yellow with a brown centre), 'Meteora' (large orange and yellow), 'Nova' (single, orange with a maroon centre), 'Orange Ball' (highly double flowers), 'Orange King' (30 cm tall, large double orange), 'Pacific Beauty' (apricot, pink or delicate pink), 'Radio' (deep orange and golden yellow, petals quilled, curled and pointed), 'Sunnyside' (golden, centre chocolate and surrounded by a collar of tubular petals), *etc.* Some other important varieties are 'Golden Emperor', 'Golden Gem', 'Gold Fink', 'Gold Star', 'Lemon Coronet', 'Lemon Gem', 'Muraji', 'Orange Coronet', 'Orange Gem', 'Yashima', 'Yellow Colossal', *etc.*

Callirhoe pedata (Malvaceae, common name: poppy mallow). It is a Greek mythological name for the daughter of Achelous, a minor Greek god. A native to Texas (USA), though it is a perennial but cultivated as annual as from seeds it comes well and flowers in the first season. It is suitable for planting on the banks, in the borders, in the rock gardens and thrive well in dry soils, and its roots survive mild winter in sandy soils. It grows up to 60 cm tall with erect stems carrying lobed or dissected green leaves and bowl-shaped crimson flowers some 5 cm across during spring to summer. It is propagated through seeds sown at flowering site in late winter, and thinning out when sufficiently grown, or through cuttings taken in May in temperate areas and during September in the plains.

Callistephus chinensis (Family: Asteraceae, common names: aster, China aster). In Greek, *kallos* means 'beauty'and *stephanos* for 'crown', alluding to the large colourful flowerheads. It is only species in the genus which is half-hardy annual native to China and Japan. It grows to more than 60 cm in height with strong, woody and branched stems, branching generally only in the upper side and not basal. They are suitable for cutting, as loose flowers, for borders and beddings and as pot plants. Its basal leaves are stalked, upper ones sessile, they are mid-green, coarsely toothed and ovate. Flowers are daisy-like, single or a few together surrounded by leafy bracts, ray florets deep purple, pink or white and disc yellow. Through spontaneous mutation and crossing many varieties are listed including the dwarf ones, and these flower from summer to autumn in the temperate regions of the country though during late winter to early summer in the plains. The varieties are in shades of purple, red, pink and white. There are various strains such as '**Ball Type strain**' which grows up to 75 cm tall, is mid-season and wilt resistant, flowers are recurved and fully double on long and clean stems and is ideal for cutting. '**Giants of California strain**' grows up to 75 cm high with spreading habit, flowering mid- to late-season with large ruffled flowers. '**Bouquet Powder Puffs strain**' grows up to 60 cm in height, is early to mid-season, resistant to wilt and weather and produces fully double and large flowers with quilled centres in various colours, 'Miss Europe' being a clear pink form. '**Duchess strain**' grows up to 60 cm tall, is wilt-resistant and late-flowering, bearing heavy flowerheads so requires support, and the petals are thick and incurving. Its form 'Fire Devil' bears bright red flowers. '**Paeony-Flowered strain**' grows up to 60 cm tall bearing pink flowers with broad and spreading petals. '**Single Sinensis strain**' is wilt-resistant, grows up to 60 cm tall, and flowering in the mid-season with clear yellow disc surrounded by long and graceful petals. '**Unicum Mixed** or **Spider Aster**' is a mid-season, growing up to 60 cm high and bears large double flowers with slender quilled petals. '**Ostrich-Plume strain**' grows up to 38 cm tall with spreading branches, is wilt-resistant and the flowers are large ostrich-feather type (the best known double form) in various colour range and is suitable for bedding. '**Pompon strain**' grows up to 38 cm tall, is weather-resistant and mid-season and bears clear button-like flowers, some with contrasting discs. These are ideal for cutting. 'Pirette' bears scarlet flowers with a white centre. '**Liliput strain**' grows up to 38 cm tall in pyramidal shape, is weather-resistant, flowering in mid-season and bears a mass of button-like quilled flowers in wide range of colours. It is ideal for pot-growing and cutting. '**Queen of the Market strain**' grows compact up to 38 cm tall and bears large and fully double flowers suitable for pot-growing and for cutting. '**Waldersee Mixed**' is early to mid-season growing upright up to 30 cm, bearing a profusion of small star-shaped flowers. '**Milady series**' plants are erect bushy annuals, growing 25-30 cm high, bearing double rose-pink to purplish-blue flowerheads in summer and early autumn. '**Chrysanthemum-Flowered strain**' grows compact up to 25 cm tall bearing tightly packed double flowers, suitable for bedding. 'Chater's Erfurt' is its rose-pink form. '**Thousand Wonders series**' grows erect and bushy with 20 cm height and bears double flowerheads in blue, purple, red, pink and white colours. Cultivars growing 60 cm and above are considered tall and these include 'Giant Perfection' (flowers double and largest with incurved petals, excellent for cutting), 'Madeleine' (most graceful among all the asters, flowers large and single, ideal for cutting), 'Princess' (flowers double with narrow petals on long erect stems), *etc.*; 45-59 cm as intermediates and these include 'Bouquet' (some 20 double flowers on one plant), 'Pompon' (plant compact, masses of button-shaped flowers), 'Ostrich Plume' (stems much branched, flowers many, large and double with leathery and recurved petals), 'Unicum or Needle-Flowered' (many large double flowers with narrow or almost needle-shaped petals); 25-38 cm as dwarfs and these include 'Liliput' (plant-form pyramidal, flowers numerous, small, tubular and double), 'King of the Dwarfs' (compact growth with double flowers); and 24 cm and below as very dwarf such as 'carpet-forming' (almost prostrate growing, flowers double and in profusion); *etc.* The types are still evolving. Various other varieties are 'American Beauty', 'Ariake Pompon', 'Azure Blue', 'Benihanabi', 'Blue Wonder', 'Bouquet Mid Blue', 'Bouquet Powderpuff', 'Bouquet White', 'Comet', 'Crego', 'Crego Azure', 'Early Bird', 'Giant Massagno', 'Invincible', 'Kamini', 'Mid Blue', 'Phule Ganesh Pink', 'Phule Ganesh Purple', 'Phule Ganesh Violet', 'Phule Ganesh White', 'Poornima', 'Pot'n Patio Blue', 'Pot'n Patio Pink', 'Queen of the Market', 'Shashank', 'Totem Pole', 'Violet Cushion', 'White Kurenai', *etc.* These all are propagated through seeds sown in September-October in the plains and in March in temperate regions at 16 °C temperature.

Campanula medium (Family: Campanulaceae, common names: bellflower, bluebell, canterury bell, harebell). *Campanula* is diminutive of Latin *campana* meaning 'a bell' as in many species the flowers are bell-shaped.

C. medium is native to North and Central Italy and Southeast France. This biennial is commonly used as bedding plant, potted flowering plant, cut flower and the smaller species in hanging baskets (Dole and Wilkins, 1999). It is a fully hardy, erect, hairy, evergreen and clump-forming biennial growing from 60-100 cm, bearing rosetted fresh green leaves, basal some 30 cm wide and others, all are obovate to lanceolate and toothed, and single or double lilac-blue or white flowers some 5 cm long which appear during spring in the plains and in summer on the hills. The bell-shaped flowers are 5-lobed. 'Bells of Holland' is its very popular cultivar.

C. pyramidalis (chimney bellflower) from South Europe is a half-hardy, erect and branching biennial growing to 2 m in height so needs staking. The leaves are heart-shaped and the racemes are long bearing star-shaped blue or white flowers in summer. These are propagated through seeds sown in October or March-April and also by division of clumps.

Celosia argentea (Amaranthaceae, cockscomb). In Grrek, *kelos* stands for 'burning', alluding to the flame-like shape and colour of some of the species. A compact-growing summer annual (in fact it is a perennial but cultivated as an annual) native to Tropical Asia, growing up to 1 m tall, bearing 5-7 cm long, pale-green and lanceolate to ovate leaves, and erect or feathery plume-like, sometimes somewhat drooping flowers which including the bracts are silvery-white. *C. a.* var. *cristata* is a dwarf compact plant which bears mid-green and oval leaves and feathery *cum* fan-like and cristate flowerheads in yellow, orange, pink, apricot and red shades. Its varieties include 'Fire Chief', 'Jewel Box', *etc. C. a pyramidalis* (syn. *C. plumosa*) produces erect, pyramidal and plume-like panicles in shades of yellow, orange, pink and red. These bear mid-green and ovate leaves. They are used for cutting, as dry flowers, in bedding and as pot plants. The varieties are 'Apricot

Brandy', 'Argentea Plumosa', 'Empress', 'Fairy Fountains', 'Fancy Plume', 'Fiery Feather', 'Fire Glow', 'Floradale', 'Forest Fire Improved', 'Geisha Series', 'Golden Feather', 'Liliput Mixed', 'Magnifica', 'Pampas Plume', 'Red Fox', 'Silver Feather', 'Thompsonii Magnifica Mixed', *etc.* These all are propagated through seeds sown in February-March at 18 °C temperature.

Centaurea spp. (Asteraceae). In Greek mythology, *centauros* or *centaur* stands for 'half-man, half-horse'. The flowers are composed of a globular base wherefrom appear numerous slender and tubular florets. *C. cyanus* (cornflower), a well-branched, erect and hardy annual is native to SE Europe (Great Britain), and grows up to 90 cm tall, bearing grey-green, remotely toothed, sometimes narrowly lobed and narrowly lanceolate leaves. It is winter- to summer-flowering with branched sprays of white, pink, red, purple and blue flowers 2.5-5.0 cm across. Tall varieties, such as 'Ball strain', 'Blue Diadem', *etc.* are suitable for cutting, though dwarf ones growing up to 30 cm tall are ideal for growing in pots and for bedding, and these varieties include 'Blue Boy', 'Dwarf Rose Gem' (rose-red), 'Dwarf Blue' (blue), 'Emperor William', 'Frosted Queen', 'Jubilee Gem' (blue), 'Little Boy', 'Pinkie', 'Polka Dot strain' (range of colours), 'Red Boy', 'Snowman', *etc. C. dealbata* (American sweet sultan) grows up to 1 m tall, producing large thistle-like lilac-blue flowers some 10 cm across, is suitable for growing in the beds and borders, and for cut flowers.

Chrysanthemum spp. (Asteraceae, annual chrysanthemum). *Chrysanthemum* derives from the Greek *chrysos* meaning 'gold' and *anthos* meaning 'a flower'. Annual chrysanthemums are meant for bedding, borders, loose flowers and also for cut flowers. Botanically there are several species being counted under annual chrysanthemum such as *C. carinatum, C. coronarium, C. frutescens, C. multicaule, C. parthenium, C. segetum, C. viscosum.*

C. carinatum (syn. *C. atrococcineum, C. tricolor*; tricolor chrysanthemum, tricolor daisy) is a hardy annual native of North Africa (Morocco) which was introduced in the European gardens in 1796. It grows up to 60 cm tall with erect, rigid and branched stems, bearing bright green, a little leathery, 5-8 cm or even more longer and bipinnatifid leaves. The single flowers some 6.25 cm across are produced freely on long stems from winter to summer, having purple discs surrounded by ray florets with banding of different colours, predominantly white with yellow base though in garden cultivars these may be banded with red, maroon or purple. Its crossing with other species has given good results. Also there are cultivars with double flowers. The varieties are 'Atrococcineum', 'Burridgeanum', 'Dunnett's Mixed' (double, range of colours), 'Eclipse', 'Flammenspiel', 'John Bright', 'Merry', 'Merry Mixed' (single, range of colours), 'Monarch Court Jesters' (single, range of colours), 'Northern Star' (single, pure white zoned yellow), 'Pole Star' (flowers white, zoned primrose-yellow), 'Rainbow Mixed' (flowers in combinations of white, yellow, lavender, scarlet, red, bronze, *etc.*), 'Torch' (flowers bronze and red, zoned crimson and purple), *etc.*

C. coronarium (syn. *Glebionis coronarium, G. coronaria*; crown daisy, garland chrysanthemum) from Mediterranean areas and Portugal is a hardy annual with erect and much-branched stems growing from 40 to 120 cm tall, bearing feathery, pale-green, much-divided (bipinnatisect), deeply cut and stem clasping leaves. Its flowerheads are 4-5 cm across with yellow discs surrounded by broad golden-yellow to white ray florets which may be single or double. The varieties are 'Albo', 'Cecilia' (90 cm tall with strong stem, flowers pure white and some 10 cm across, superb for cutting), 'Floreplenum' (75 cm tall, flowers semi- or fully-double in shades of golden-yellow and whte), 'Golden Crown' (120 cm tall, flowers single and deep yellow), 'Golden Gem' (up to 20 cm tall, flowers double, button-like and lemon-yellow), 'Golden Glory' (up to 90 cm tall, flowers single canary-yellow, superb for cutting), 'Luteum', 'Nanum Compactum' (a dwarf form), 'Nivea', 'Oreon', 'Primrose Gem', 'Sunray' (1 m tall, plant very vigorous, flowers single and canary-yellow), 'Tetra Comet', 'Tom Thumb', *etc.*

C. frutescens (*Argyranthemum frutescens*) from Canary Isles is though perennial but usually grown as an annual and is suitable for bedding and as a pot plant. Its evergreen and usually bipinnatisect leaves are pale to glaucous-green and some 4-8 cm long, and the ray florets are white with pale-yellow disc, 2-5 cm across which continue appearing intermittently all through the year. This sub-shrubby species grows from 30-45 cm tall. It has certain garden hybrids with certain other Canary Isles species such as *C. coronopifolium* (Paris daisies) so there had been a lot of colour variation now. All these varieties are suitable for planting in window boxes and in conservatory.

C. multicaule (yellow daisy) from Algeria grows compact and up to 30 cm tall, and is an excellent plant for the rock gardens, for edging and in the window boxes. Its leaves are bluish-grey, glaucous, little leathery, 5-10 cm long, spathulate, trifid or pinnatifid and coarsely toothed. Its golden-yellow flowers some 2.5 cm across appear during summer and rains.

C. parthenium (syn. *Matricaria eximia, Tanacetum parthenium*; bachelor's button or feverfew) from Great Britain is though hardy perennial but usually grown as an annual as blooms the first season through the seeds, and may attain the height of 20-45 cm. Its light green leaves are soft, hairy and deeply cut and have pungent aroma. Flowers are small, white and some 2 cm across. It is suitable for edging, in the small beds, in the window boxes and in the rock gardens. Its some of the best varieties are 'Aureum' (foliage golden-yellow, flowers white and suitable for edging), 'Ball's Double White' (flowers small double and white, ideal for cutting), 'Golden Ball' (compact growing, masses of small button-like, double and yellow flowers), 'Gold Star' (20 cm long, a row of white ray petals surround the yellow centre), 'Snowball' (30 cm tall, flowers pompon-like and ivory-white), 'Snow Puffs' (dwarf for pot culture, masses of small button-like white flowers, blooms long-lasting), 'White Bonnet' (height 75 cm, flowers double and pure white), *etc.*

C. segetum (corn marigold) from Great Britain grows from 30-60 cm tall is much-branched bushy annual, bearing stalkless, dark green, deeply incised and oblong leaves. Flowers are single, solitary, some 6.25 cm across, rich yellow with conspicuous central disc, terminal and appear during rains. Its varieties mentioned below are roughly 45 cm tall and these are 'Blanca', 'Eastern Star' (primrose-yellow, disc brown), 'Eldorado' (flowers canary-yellow with a black centre), 'Evening Star' (flowers golden-yellow), 'Gloria', 'Golden Glow' (double and golden-yellow), 'Isabel', 'Morning Star' (yellow), 'Romeo', 'Yellow Stone', *etc.*

C. viscosum (*C. viscidi-hirtum*; viscous chrysanthemum) from North Africa and Spain is a bush-like vigorous annual growing up to 45 cm tall, bearing thick and sticky hairs on the surface of deeply cut foliage. The coarse-textured yellow flowers appear during rainy season. It is ideal for growing in the wild garden at sunny site in poor soils.

C. × *spectabile* is a hybrid between *C. carinatum* and *C. coronarium* suitable for border cultivation and cutting. These can grow up to 120 cm high, bearing grey-green and deeply toothed leaves. Variously coloured single flowers some 7.5-10.0 cm across are produced during rains. Its var. 'Cecilia' is tetraploid, long-stemmed, white with bright yellow disc, ray florets banded yellow. All these can be propagated through seeds at a temperature of 13 °C.

Cladanthus arabicus (Asteraceae). In Greek, *klados* means 'a branch' and *anthos* 'a flower', owing to the flowerheads occurring at the end of the branches. There are only four annual species in this genus native to South Spain and North Africa, but only one is in general cultivation. It is an aromatic bushy annual growing up to 90 cm tall with slender stems, that branches repeatedly just below each flowerhead. Leaves are alternate, pinnatisect with linear lobes, and the lobes being further lobed. The flowerheads are golden-yellow some 4 cm across and appear during rains. This is propagated through seeds sown in spring.

***Clarkia* spp.** (Onagraceae). It was named for Captain William Clark (1770-1838) who was co-leader with Capt. Meriwether Lewis of the first expedition to cross the Rockies from east to west to the Columbia River (1804-06). There are some 36 species of annuals from WN America and Chile including all the species once in the past classified as *Godetia*. They are erect, slender, free-branching with alternate and linear to lanceolate leaves and 4-petalled flowers, which are sometimes lobed, in terminal clusters. They make good pot plants, are suitable for borders, *vis-a-vis* cutting during winter and spring.

C. amoena (*Godetia amoena, Oenothera amoena*) from California is a bushy annual growing up to 60 cm tall, bearing 1-6 cm long and lanceolate leaves, and 4-6 cm across and erect flowers in shades of white, pink, lilac or red on a slender stem. *C. a. whitneyi* (*C. grandiflora* of gardens) bears larger flowers (8 cm across) with a central red blotch. This has both dwarf and tall strains in single and double forms.

C. elegans (*C. unguiculata, Oenothera elegans*: common names: clarkia, mountain garland) from California grows erect up to 60 cm tall with reddish stems, and green, 1-6 cm long and ovate leaves. Flowers are double, up to 5 cm across and in a wide variety of shades from salmon to lavender and red-purple on up to 30 cm tall spikes.

C. pulchella from WN America grows up to 45 cm tall, bearing mid-green, linear to lanceolate leaves and 4 cm across petals which are strongly 3-lobed and in pink to white on up to 23 cm long spikes. Many of the strains are available, mostly in semi-double form in the shades of white, pink and purple. They are propagated through seed by sowing in the nursery beds or directly at the site of flowering in autumn or late winter. *Clarkia* has many varieties such as 'Alba', 'Apple Blossom', 'Brilliant', 'Brilliant Rose', 'Chamois', 'Enchantress', 'Fire Brand', 'Lady Satin Rose', 'Lilac', 'Royal Bouquet', 'Salmon Queen', 'Scarlet Queen', *etc.*

Cleome spinosa [syn. *C. pungens, C. hasslerana*; family: Cleomaceae (Capparidaceae); Common names: bee plant, spider flower, stinking clover]. Its generic name was given by Theophrastus for a mustard-like plant but afterwards adopted by Linnaeus for this genus. There are many species in the genus but normally only one is cultivated and that is *Cleome spinosa* which is native to Tropical America. It grows erect up to 1.2 m tall with hairy, sticky-glandular and aromatic stems, and stipules spine-tipped on the base of each leaf stalk. The leaves are mid-green, digitate and divided into 5-7 oblong-lanceolate leaflets. The flowers which appear during rains are 4-10 cm across, white, flushed pink, petals 4 and narrow and stamens long. It is propagated through seeds sown in March at 18 °C temperature. Its varieties are available in various colours including white, yellow and pink to rose-purple such as 'Colour Fountain' (scented deep purplish-pink), 'Golden Sparkler' (golden), 'Helen Campbell' (white), 'Pink Queen' (pink), 'Rose Queen' (rose), *etc.*

Collinsia bicolor (syn. *C. heterophylla*; family: Scrophulariaceae). *Collinsia* was named for Zaccheus Collins (1764-1831), Vice President of the Philadelphia Academy of Natural Sciences, and an American botanist. It is a genus of 20 annuals from North America, mainly West Coast, and only one species, *i.e. C. bicolor* from California is in general cultivation. This grows erect up to 60 cm height with diffused branching. The leaves are mid-green, 2-7 cm long, in opposite pairs, oblanceolate and usually toothed. Clustered flowers are tubular, 1.5-2.5 cm long, 2-lipped, upper lip white to pale-lilac and the lower violet to rose-purple which appear in the upper axils of the leafy spikes from summer to autumn. It is propagated through seeds either sown *in situ* in spring or during autumn. Its var. 'Salmon Beauty' though has salmon-rose flowers but there are many other varieties in shades of white, lilac, rose and purple.

Consolida ambigua (syn. *Delpinium ajacis*; family: Ranunculaceae; common names: annual larkspur, rocket larkspur). The Greek *delphinion* comes from *delphis* 'a dolphin', alluding to the shape of flower of certain annual species. *Delphinium* is a large genus comprising of 250 species of hardy and half-hardy annuals and herbaceous perennials with spurred flowers, though now all the annual forms have constituted a separate genus *Consolida*. These all are excellent border plants and useful for cutting. These have showy spires of white, purple and true blue flowers though pure red and yellow are not yet available.

Consolida ambiguum (or *C. ambigua*) is a hardy annual from Mediterranean region, growing upright up to 1 m tall with sparsely branching stems. The leaves are mid-green, fern-like and finely cut, basal being cut into oblong segments. Racemes are spire-like, loose, some 60 cm long with blue or violet flowers some 2.0-3.5 cm wide and spurs some 1.3-1.8 cm long, and follicles some 1.5-2.0 cm long, and these flowers appear during late winter and spring in the Indian plains and from summer to rains in the temperate regions. It has produced hyacinth-flowered garden group ranging from 30 cm to 60 cm in height, suitable for pot-growing, and these bear double flowers on blunt-ended spikes having a poor branching-tendency than the type species, apart from its other cultivars or strains in white, pink and purple colours.

Delphinium regale (*D. consolida*, common larkspur) is a hardy annual from Europe including Great Britain, grows erect up to 1.2 m tall, well-branched from the base and much like *C. ambiguum* but basal leaf segments are linear and ferny, racemes are 22-38 cm long, branched and panicle-like, flowers are densely set in white, pink, red, purple and blue colours, floral spurs 1.2-2.5 cm long and the ripe follicles some 0.8-1.5 cm long. This species has given rise to 'Giant Imperial strain' which is a base-branching type and is available in various colours such as 'Blue Spire' aand 'Dazzler' (carmine-scarlet); the 'stock-flowered group' which flowers earlier than the former strain and the 'Rosamund' variety under this group bears bright rose flowers. Both the strains are ideal for border planting and for cutting. They are propagated by sowing the seeds either *in situ* at their place of flowering or in the nursery beds during autumn or in late winter. However, as the seeds are short-lived, so until sowing these should be stored at 2-5 °C.

Coreopsis spp. (syn. *Calliopsis*; family: Asteraceae; common name: tickseed). In Greek, *koris* means 'a bug' and *opsis* is for 'seeds', which in this case is bug-like. The genus contains 120 species of annuals and perennials from America and Africa, and among the annuals only two are generally cultivated such as *C. drummondii* (*C. basalis*) from Texas and *C. tinctoria* (*C. bicolor*) from N. America. Leaves are in opposite pairs and usually pinnately lobed. Flowers are terminal, daisy-like with small disc and ray florets broad. Cypselas (fruits) are narrow and flattened with or without a pappus of scales or bristles and often black. These are summer-flowering annuals propagated through seeds sown during autumn, spring and summer. These are ideal for border planting and for cutting.

C. drummondii grows erect up to 80 cm tall and *C. tinctoria* to 1 m. Stems are stiff and branched bearing mid- to dark-green leaves. The flowers some 5 cm across appear in profusion, bright yellow with a deep purple disc and a red-brown blotch at each ray floret base in case of *C. drummondii* and bright yellow in case of *C. tinctoria*. Single colour strains in yellow, crimson and crimson-scarlet are available, the 'Golden Crown' being deep golden-yellow with brown centre in *C. drummondii* though *C. tinctoria* varieties are 'All Double Mixed' (double and semi-double yellow to chestnut-colour combinations), 'Dwarf Mixed' (single, yellow or chestnut-colour with blotches of maroon or dark crimson), *etc.* Other noteworthy varieties are 'Early Sunrise', 'Pusa Tara', 'Sunbeam', 'Sunburst', *etc.*

Cosmos **spp.** (Family: Asteraceae; common name: cosmea)**.** In Greek, *kosmos* means 'beautiful'. Both the species being described here, *viz. C. bipinnatus* and *C. sulphureus* which are half-hardy annuals from Mexico, former growing up to 90 cm tall with slender stems bearing mid-green pinnatifid leaves having linear segments and racemes generally with 8 ray florets and the flowerheads some 5-10 cm across in the shades of white, pink, rose or crimson; and the latter growing only 60 cm tall, bearing darker green bipinnatipartite leaves with lanceolate lobes, and flowerheads 4-5 cm across with yellow ray florets. They are mainly erect plants with leaves in opposite pairs and the flowerheads daisy-like, terminal and with very broad ray florets. They are excellent for growing in borders, in pots and for cutting and display their beauty during summer to autumn. These can be propagated through seeds by sowing these at temperature of 16 °C. The varieties bear larger and semi-double flowerheads in shades of gold to orange-red. *C. bipinnatus* varieties are 'Alipore Beauty', 'Dazzler', 'Gloria', 'Pink Sensation', 'Purity', 'Radiance' (flowers deep rose and crimson bocoloured), 'Sensation Mixed' (early and large-flowering strains), 'Versailles', *etc. C. sulphureus* varieties are 'Fiesta', 'Klondyke' (double, golden-orange), 'Mandarin', 'Orange Flare', 'Orange Ruffle', 'Sunset' (semi-double, vermilion-red), *etc.*

Crepis rubra (Asteraceae; common name: dandelion)**.** In Greek, *krepis* means 'boot or foowear' though meaning is obscure. Out of the annual types, only *C. rubra* from South Italy, Greece, Balkan Peninsula and Crete is in general cultivation which grows up to 30 cm tall, is fully hardy, tolerating sun or shade, preferring well-drained soil, suitable for front of the border and in the rock gardens, and the propagation is through seeds. The member of this genus, in general, have long tap root, flat rosettes of pale-green and toothed leaves and many-petalled dandelion-like flowerheads. *C. rubra* is a clump-forming species where stems are with black and white hairs and bear a basal cluster of light green and oblanceolate, dentate to pinnatifid leaves some 15 cm long and these surround the floral stems, and solitarily (rarely 2) orange to rose or white flowerheads some 2.5-3.5 cm across during summer to autumn.

Datura **spp.** (Solanaceae; Jimson weed)**.** *Datura* is the Latinized version of the Indian *dhatura* or possibly of Arabic *tatorah*. It is a genus of 24 species of annuals, shrubs and trees, shrubby ones sometimes included in *Brugmansia* though this genus now does not exist any more. They are native to warm temperature regions, mainly in Central America. There is a myth that 'Lakshmi or Lakmè', the beauteous daughter of a Brahmin priest in the late 19th century and 'Gerard', a British officer, fell in love but when she realized that such marriage is impossible, she ate *Datura stramonium* flowers and died. In fact, it is a poisonous plant in all its parts as it contains scopolamine, hyoscyamine, and atropine (daturine) poisonous alkaloids, atropine producing irregularity in the heart beat coupled with rapid breathing with ultimate psychological reactions and even assuming maniacal proportions. Very little dose of plant extract makes the human beings good-humoured, little dose becomes delirious and a little more will make a human insane for whole life or will die. Shrubby species are very susceptible to cold. In tropical climate, these can be planted in the garden for their prolonged flowering period but it should always be borne in the mind that these are poisonous plants well known to the human beings in India so should be planted away from children and with a written caution. In India, its only a few seeds are mixed with 'bhang' (*Cannabis sativa*) dry leaves, ground with almond, fennel, peeled watermelon seeds, various other ingredients and water, diluted and served to the people during 'Holy Festival' to make the people good-humoured. Its leaves and flowers are also offered by Hindus to lord Shiva during 'Mahashivaratri Worshippings'. These are summer- to autumn-flowering, mostly with hanging flowers but in a few cases the flowers remain erect. These bear green, very large, alternate, ovate and acuminate leaves with 2-3 clear angles on each side of the margins. Flowers are large, trumpet-shaped and open to 5-spreading lobes. These are free-flowering and suitable for conservatory. These are propagated through seeds sown in February-March.

Datura ceratocaula (*D. cornigera, Solandra herbacea*) from Mexico and tropical Americas is a tender annual found growing up to 90 cm tall in damp and swampy areas. Stems are thick, fleshy but strong, hollow, little branched and usually forked, and reddish covered with grey powder. Alternate leaves are mid-green, stalked, oval-lanceolate with undulated and deeply cut margins, and underside silvery-green. Very large (15 cm long), solitary and fragrant-white flowers suffused with violet interior are axillary and erect or obliquely borne, which open in the late afternoon and close in the morning. The fruits are smooth.

Datura metel (*D. fastuosa, Stramonium fastuosum*) from India is though shrubby perennial but grown as an annual with beautiful drooping flowers. Its specific name *metal* derives from the Arabic *jawz māthil* which means 'narcotic nut'. It grows up to 1.2 m tall, stems being red with thick hairs and poor-smelling. Its leaves are ovate, grey-green at upper surface while silvery-green below. Highly fragrant, solitary and trumpet-shaped flowers some 15-20 cm long and erect are white with greenish-white at the base. The fruits are quite large, the size of an apple, globose and pendulous and covered with similar prickles throughout. It grows comfortably in sandy soils so are found along side the seaside gardens. Its popular varieties are 'Chlorantha' (double, yellowish-green), 'Fastuosa' (stems dark purple, flowers with double to triple corollas, exterior purple-violet, interior cream-white), 'Huberiana' (vigorous growth and larger flowers than the type, flower colour variable), *etc.*

D. stramonium (Jimson weed, thorn apple) from tropical to warm temperate zones of India is the most common Indian *Datura*. It is a tender annual growing up to 1.2 m tall with an erect cylindrical main stem which is much-branched with smooth branches. Leaves are dark green though undersurface little silvery, some 20 cm long and up to 13 cm wide, long-stalked, ovate-acuminate, with deeply and widely dentate margins. Flowers white, solitary, erect, and terminal & axillary but at the end of the stem or branches some 10 cm long are short-stalked, narrowly trumpet-shaped but mouth bell-shaped. Fruit is large to the size of the small apple covered with equal size and uniformly distributed small prickles. Its varieties 'Tatula' bears pale-violet corolla and dark violet anthers, 'Inermis' with smooth fruits, *etc.*

Delphinium grandiflorum (syn. *D. chinensis*; Delphinium; Ranunculaceae). A close relative of annual larkspur (*Consolida ajacis*), this is a perennial plant but is cultivated as an annual. There are some 259 species from northern temperate zone and mountains farther south. It prefers cool climates, in the temperate regions it remains as a perennial, and its flowers are more attractive than larkspur. It bears about 20 cm long, alternate, palmately or digitately lobed basal leaves, and tall, thick and slender stalk up to 90 cm tall with terminal racemes of spurred flowers that have 5 petaloid sepals and 2-4 quite small and sometimes with contrast colouration. They are suitable for pots, borders and for cutting. They are available in colour range of white, creamish-yellow, bluish-white, red, mauve, blue and purple. They are propagated usually through division or cuttings.

Dianthus spp. (Caryophyllaceae). In Greek, *di* is for 'Jove or Zeus', and *anthos* for 'flower' ~ the flower of the gods. Though there are many half-hardy to hardy annual to biennial and perennial species under the genus but *D. barbatus* (sweet william), *D. caryophyllus* (annual/spray carnation, clove pink, gillyflower) and *D. chinensis* (syn. *D. sinensis*; dianthus, Indian pink) are evergreen to semi-evergreen annual to biennial and are only of common interest. Their native haunts are Northern Hemisphere, rarely the Southern Hemisphere and are quite common in southern Europe. *D. barbatus* is from East Europe, Russia and China; *D. crayophyllus* (annual type) from WS France and *D. chinensis* from E. Asia. All are excellent for cutting, for borders and beddings, and for pot-planting and are propagated through seeds sown during autumn or in late winter at a temperature of 13 °C. *D. barbatus* and *D. caryophyllus* are also excellent in fragrant gardens.

D. barbatus, a hardy biennial growing from 30-60 cm tall was introduced to the European gardens in 1573. Its leaves are dark green, shining, wide and form a dense basal mass. Floral stalks are thick, robust and leafy. Its flattened flowerheads are densely packed, some 7.5-12.5 cm across in singles or doubles in shades of white, pink, scarlet, red, crimson and violet, overlaid with various other colours even in concentric zones. Its hybrids with *D. woodii* has yielded 'Sweet Wivefield' strain which has flowers 17.5-20 cm across in loose heads but in greater range of colours than the type species. Its most common varieties are 'All Double' (white, pink or crimson), 'Auricula Eye' (red and maroon with white eye), 'Bright Eyes' (plant small, mixture of colours), 'Giant White' (large white), 'Indian Carpet' (dwarf, mixture of colours) 'Pink Beauty' (giant flowers), 'Red Monarch' (plant small, flowers scarlet), 'Scarlet Beauty' (salmon-scarlet), 'Wee Willie' (plants 10-15 cm tall, flowers in mixed colours), *etc.* Apart from these there are many more good varieties such as 'Albus', 'Black Beauty', 'Copper Red', 'Crimson Beauty', 'Diadem', 'Giant White', 'Holborn Glory', 'Homeland', 'Ishii Red', 'Kurukawa', 'Marginatus', 'Nigrescens', 'Oculatus', 'Pink Beauty', 'Scarlet Glory', 'Scented Beauty', 'White Giant', *etc.*

The original annual species *D. caryophyllus* grows up to 60 cm tall and is a progenitor of the border and perpetual-flowering perennial carnations, bearing a typical bright grey-green foliage, and 2.5-8.0 cm across dull-purple and single flowers with dentate petals, and these flowers have strong but sweet clove-fragrance. Varieties are 'Apricot Bizarre', 'Aurora', 'Chabaud', 'Cherry Flake', 'Enfant de Nice', 'King Cup', 'Madonna', 'Malmaison', 'Margaret', 'Marguerite', 'Nero', 'Orange Sherbat', 'Picotee Fascination', 'Pixie Delight', 'Princess Alice', 'Riviera Giants', 'Scarlet Flake', 'Scarlet Luminette', 'Snow Clove', *etc.*

D. chinensis is tufted species, growing up to 45 cm in height, bearing mid- to dark-green, and lanceolate to linear leaves. Clustered flowers in a bunch of 4-6 from red to white often with darker centre, and some 3.0-6.0 cm across appear on the top of the branched inflorescence. In its varieties the flowers may be single or double though type species is single. Its beautiful forms are 'Baby Doll' (dwarf up to 15 cm high, mixed colours), 'Bravo' (bright red), 'Fireball' (double scarlet), 'Gaiety' (mixed colours), 'Gaiety Double' (mixed colours), 'Heddewigii' (single or double mixed), 'Snowball' (large double white), *etc.* Apart from these, there are many other beautiful varieties such as 'Black Prince', 'Blue Peter', 'Bravo', 'Carpet', 'Cyclops', 'Diadem', 'Fire Storm', 'Magic Charm Hybrid','Miss Miwako', 'New Mikado', 'Pink Beauty', 'Pink Flesh', 'Red Bedder', 'Rich Crimson', 'Salmon King', 'Salmon Queen', 'Scarlet Queen', 'Snowball', 'Snowdrift', 'Snow Storm', 'Sunfire Hybrid', *etc.*

Diascia barberae (Scrophulariaceae). *Diascia* derives from the Greek *di* which means 'two', and *askos* 'a sac', alluding to the two nectary spurs. *D. barberae* from South Africa is a half-hardy and well-branched annual growing 30 cm tall with slender stem, bearing dark green, glossy, ovate and toothed leaves, and shell-like 2-spurred rose-pink flowers some 1.5-2.0 cm across that appear during summer and rains. Flower throat is spotted green. It is an excellent plant for edging border and paths and as a summer-flowering pot plant. It is propagated by seeds sown in February-March at 16 °C temperature.

Dimorphotheca spp. (Family: Asteraceae; common names: African daisy, dimorphotheca). In Greek, *dis* is for 'twice', *morpho* for 'form or shape' and *theka* for 'fruit', referring to the 2 different cypsela (seed) shapes in each head. They are of spreading growth with narrowly lanceolate to oblong leaves and daisy-like solitary flowerheads with white or brightly coloured flowers. These are beautiful winter- to spring-flowering annuals suitable for cuttings and for borders. They are propagated by seeds sown in September-October at 16-18 °C.

D. annua (*D. pluvialis*), a branching annual from South Africa grows up to 30 cm tall with spreading or erect habit and bears hairy, 2.5-8.0 cm long and narrowly obovate leaves with blunt teeth. It bears golden-brown floral disc which is surrounded by ray florets which on the upper surface are white or creamy-white while purple beneath.

D. sinuata (*D. aurantiaca* of gardens, *D. calendulacea*; star of the veldt), an attractive and branching annual from South Africa grows up to 30 cm tall with slender stems. Its leaves are dark green, oblong to spathulate, some 8 cm long, wavy margins and coarsely toothed. Pale-yellow to white flowers with greying-blue centre and measuring some 6.25 cm across appear during late winter to spring. Its commonly grown varieties are 'Buff Beauty', 'Goliath', 'Orange Glory', 'Orange Improved', 'Salmon Beauty', 'White Beauty', *etc.*

Dorotheanthus bellidiformis (syn. *Mesembryanthemum criniflorum*; Ice plant, Mesembryanthemum; Aizoaceae). The genus was named for Dorothea, mother of professor C. Schwantes, specialist in the study of *Mesembryanthemum* and its allies, and Greek *anthos* is for 'a flower'. A South African (Cape region, Angola and Namibia) genus of 10 succulent annuals, formerly classified as *Mesembryanthemum*, but only two are in general cultivation. Stalked leaves are cylindrical to spathulate or linear with conspicuous water cells, and the flowers are daisy-like opening only in sun. Flowering occurs during transitional period of winter and the spring in the plains, however, in the temperate regions flowering occurs from May to September. These can be grown in the pots elsewhere under sunshine and before flowering may be brought indoors but indoors their closed flowers will also give enchantment. These are also suitable for edging. These are propagated through seeds sown in September or in February. *M. crystallinum* is highly invasive and naturalizes well.

D. bellidiformis (syn. *Mesembryanthemum criniflorum*; common names: fig marigold, ice plant, living stone daisy, mesembryanthemum) from SW Cape is a hardy, prostrate-growing and mat-forming annual some 30 cm across and 8-15 cm tall with light-green, some 8 cm long, obovate to almost cylindrical leaves covered with tiny glistening spots appearing as sugar grains. A mass of solitary and brightly coloured flowers some 2.5-4.0 cm across in shades of white, orange, pink, red or bicoloured and zoned with darker centres and blackish-purple anthers appear from winter to spring.

D. tricolor (*D. gramineus, Mesembryanthemum tricolor*) from South Africa is a hardy mat-forming *cum* spreading (30 cm across) annual some 8 cm tall bearing dark green, 5-8 cm long, narrow and cylindrical leaves. During spring solitary mass of flowers some 4 cm across appear in shades of white, pink or red, often with contrasting centre and black-purple stamens.

Echium spp. (Family: Boraginaceae; common name: viper's bugloss). *Echium* derives from the Greek word *echion*, a name for viper's bugloss. *E. plantagineum* (*E. lycopsis*; purple viver's bugloss) is a base-branching bushy annual (sometimes biennial) native to Mediterranean and Great Britain which grows up to 1 m tall. Its leaves are mid-green, 5-14 cm long, tough but softly white-hairy, ovate but narrowing up to the stalks so that these become oblong-lanceolate, and flowers some 2.5 cm long are tubular blue, pale-purple or pink which appear on one side of the long branching spikes from June to August in the temperate areas and in late winter to spring in the plains. These are suitable for borders but the dwarf ones are suitable for pot planting. It is propagated through seeds sown during autumn or during late winter at 13-16 °C temperature. *E. rubrum* (*E. russicum*) from East Europe to Central USSR is closely allied biennial to *E. planatagineum*. It is hardy and bushy plant growing to 1 m tall suitable for planting in borders. Leaves are mid-green, 5-10 cm long, narrower, lanceolate and quite pointed. Clustered bright red flowers are tubular, some 1.2 cm long with protruding yellow stamens and are formed on a 50 cm tall spike-like inflorescence during winter to spring in the plains and during May to July on the hills. 'Burgundy' variety bears dark red flowers on long spikes.

Emilia coccinea (syn. *Cacalia coccinea;* Tassel flower; *Asteraceae*). It is a rosette-forming leafy-stemmed annual native of Tropical Africa and India, which grows up to 60 cm tall, bearing yellow, orange or red flowers about 2 cm across. It is suitable for growing in the border and for cut flowers. It is propagated through seeds, which are sown first in the nursery in September-October and then transplanted in October-November in the prepared beds at 30 ×40 cm distances.

Erysimum cheiri (*syn. Cheiranthus cheiri*; Wallflower; Brassicaceae). The name *Cheiranthus* derives from the Greek *cheir* which stands for hand, alluding to the custom of carrying its bouquet at certain occasions. A hardy biennial from Europe though grown as annual, growing erect to 20-60 cm tall, bearing dull-green and up to 10 cm long oblong-lanceolate leaves, and orangish-yellow mustard-like flowers 1.25-4-0 cm across in dense spikes. Selections have brought about a wide range of colours such as cream, yellow, orange, scarlet, crimson and purple in single and double forms. It is ideal for beds, borders and rock gardens. Except the var. 'Harpur Crewe' which is propagated through lateral heel cuttings some 3.5-5.0 cm long, the wallflowers are propagated through seeds sown during autumn. Its noteworthy varieties include 'Blood Red' (deep red), 'Carmine King' (carmine), 'Cloth of Gold' (golden-yellow), 'Compacta' (height 60 cm and spread 1.2 m), 'Dresden Forcing' (orange), 'Eastern Queen' (salmon-red), 'Feltham Early' (mahogany), 'Fire King' (rich orange-red), 'Giant Brown' (coppery), 'Giant Fire King' (deep orange), 'Golden Bedder' (orange), 'Goliath Forcing' (deep orange), 'Hamlet', 'Harpur Yellow' (25 cm tall, sterile, fully double, yellow), 'Othello', 'Paris', 'Persian Carpet' (mixed), 'Primrose Bedder' (primrose-yellow), 'Primrose Monarch' (primrose-yellow), 'Purple Queen' (purple), 'Ruby Gem' (ruby red), 'Scarlet Bedder' (red), 'Scarlet Emperor' (red), 'Tom Thumb Mixed' (range of colour), 'Vulcan' (deep velvet crimson), *etc*.

Eschscholzia californica (Family: Papaveraceae; common name: Californian poppy). The genus originates to commemorate Dr. Johann Friedrich Eschscholz (1793-1831), a Russian doctor and naturalist of German extraction who accompanied Russian scientific expeditions including one of the Pacific coast of North America. Only two annual species of general garden interest are cultivated, *E. caespitosa* (*E. tenuifolia*) from California and *C. californica* from California to Washington. They are tufted with alternate, several times ternately and deeply dissected leaves, and poppy-like 4-petalled flowers, and afterwards long pod-like seed capsules. They are propaged through seeds sown during September or in February-March *in situ* or in nursery beds.

E. caespitosa, a hardy annual grows 10-20 cm tall and is suitable for edging and in the rock gardens. Leaves are blue-green and divided finely in thread-like forms. Flowers are yellow, some 2.5-4.0 cm across and appear in profusion during winter and spring. 'Miniature Primrose' (primrose-yellow), 'Sundew' (primrose-yellow), 'Sunglow' (lemon-yellow) are most common varieties.

E. californica grows up to 20-40 cm tall and is suitable for edging, borders and rock gardens. Leaves are glaucous to grey-green and dissected into many narrow segments, and the flowers are lustrous orange-yellow, 4-6 cm wide and appear during winter and spring. Many mixed colour strains in single and double forms are available such as 'Alba' (creamy-white), 'Aurantiaca' syn. 'Orange King' (orange), 'Aurora', 'Ballerina' (semi-double, petals fluted in shades of yellow, orange, pink, carmine and crimson-scarlet), 'Carmine King' (deep rose-pink), 'Carmine Queen', 'Crocea', 'Dazzler', 'Fireglow', 'Flame', 'Gleaming', 'Golden West', 'Ivory White', 'Mandarin', 'Mikado' (mahogany-red), 'Monarch Art Shades' (semi-double in shades of yellow and orange to pink, purple and white) 'Red Chief', 'Sutton's Flame', *etc*.

Exacum affine (Gentianaceae). *Exacum* derives from the Gaelic name *exacon*, a vernacular for *Centaurium* but Linnaeus used it for this genus. It is an annual to biennial from Socotra Island (Indian Ocean) growing compact up to 25 cm tall, bearing shiny green, 1.5-4.0 cm long, opposite, entire and broadly ovate to elliptic leaves. Solitary or clustered, rounded and fragrant flowers some 1.5-2.0 cm across, 5-petalled, rotate, lavender-blue to purplish-blue with darker petal bases and yellow stamens, emerge from the leaf axils during winter to spring. It is most suitable for growing indoors. It is propagated through seeds sown in September or in spring at 18 °C temperature.

Felicia bergeriana (Asteraceae; kingfisher daisy). *Felicia* was named for Herr Felix (*d.* 1846), a German official of Rogensburg. *Felicia bergeriana* (*Aster bergeriana*) from South Africa is a half-hardy bushy annual growing some 15 cm tall, bearing a dense mat of grey-hairy, entire, lanceolate to narrowly oblong, some 2.5-4.0 cm long and toothed leaves. Profusion of bright blue with yellow to almost black centred flowers some 2-3 cm across are borne during winter to spring on hairy stalks. The petals curl back during dull weather and open when it becomes sunny. It is suitable for bedding out, for edging the borders, in pots and for growing in window boxes. It is propagated through seeds sown at 16 °C temperature.

Gaillardia pulchella (Asteraceae; blanket flower). *Gaillardia* genus was named for Gaillard de Charentonneau, French magistrate and patron of botany. *G. pulchella*, a fully hardy annual from Central and South USA and Mexico grows up to 45 cm tall. Its leaves are grey-green, hairy and lanceolate to spathulate. Single flowers some 5 cm across in shades of yellow, pink or red appear throughout the year. It is suitable for herbaceous border and cutting. It is propagated through seeds sown at 15 °C temperature. Garden cultivars vary in forms, having enlarged ray florets with quilled central disc or ball-shaped flowerheads. 'Lollipops' grows some 30 cm tall with red or yellow flowers, is free-flowering and weather-resistant. 'Picta Lorenziana' is taller up to 45 cm and bears double flowers in shades of bright colours of yellow through bronze to red. Other popular varieties are 'Bremen', 'Burgundy', 'Dazzler', 'Double Tetra Fiesta', 'Gaiety Double', 'Goblin', 'Indian Chief', 'Lollypop Orange', 'Lollypop Yellow', 'Monarch Strain', 'Regalis', 'Sanguinea', 'Sunshine Strain', *etc.*

Gamolepis tagetes (syn. *G. annua*; family: Asteraceae; common names: gamolepis, sunshine daisy). In Greek, *Gamolepis* is for united scales, referring to the involucres. It is a hardy to half-hardy annual of wiry growth from South Africa suitable for borders and mass effects. It grows up to 30 cm tall with much-branched stem, bearing alternate and pinnatifid or pinnately-parted leaves with 5-7 lobes or leaflets on either side of rachis, and leaflets entire or lobed. Involucre nearly urn-shaped, the flowerheads bright yellow or orange and about 2 cm across, which being free-flowering continue appearing from winter to rainy season. It is propagated through seeds sown in late winter.

Gazania splendens (Family: Asteraceae; Common names: African daisy, gazania). *Gazania* was named for Theodore of Gaza (1398-1478), translator into Latin of the botanical works of Theophrastus. *Gazania splendens* is though a perennial species but grown as an annual and flowers from the seeds the same season during winter to spring. Its flowers close in the evening or after pollination. It is suitable as pot plants, for bedding out and for cutting. It is native to South Africa and grows about 45 cm tall. Leaves are dark green at upper side and white on the underside and spathulate. Bright orange daisy-like ray florets with black and white spots at petal bases, disc yellow but briefly surrounded by black colouration and flowerheads some 7.5 cm across are borne on hairy stalk. *G.* × *hybrida* growing nearly of the same height was developed by crossing *G. splendens* with other species. Leaves are dark green, grey on the underside, lanceolate and toothed. The flowers appear in a range of colours, predominantly orange. 'Monarch Mixed' produces flowers in shades of yellow, orange, pink, brown, red and ruby, all strikingly zoned. Other varieties are 'Bridget' (orange-yellow, zoned maroon), 'Carnival Hybrids', 'Hazel' (mahogany-crimson, tipped orange-yellow and zoned purple-brown), 'Minister', 'Snuggle Bonnie' (glowing yellow but paling towards the tips), 'Sunshine', *etc.* They are propagated through seeds sown during autumn or in January-February at a temperature of 16 °C.

Gilia spp. (Polemoniaceae). *Gilia* was named for Fillipo Luigi Gilii (1756-1821), an Italian astronomer. Though there are some 120 species in the genus but only a few are hardy annuals which are ornamental, such as *G. achilleifolia*, *G. capitata* (blue thimble flower), and *G. tricolor* (bird's eyes), all hardy annuals from California. They all have attractive foliage and very showy flowers, flowering in late winter to spring, and make quite suitable pot plants, though dwarf ones can be grown in the rock gardens. For propagation, the seeds are sown either in September or in February-March in nursery beds or at their permanent sites *in situ*.

G. achilleifolia grows bushy up to 60 cm tall. Its leaves are mid-green, covered with soft sticky hairs and are finely dissected. Funnel-shaped, bluish and clustered flowers are borne on the top of stalks in the globular flowerheads some 2.5 cm across.

G. capitata is a slender-stemmed plant growing up to 50 cm tall or more. Its leaves are mid-green, some 10 cm long, ferny and bi- or tripinnatifid with linear segments. Blue to lavender flowers some 6-8 mm long, and some 50 or more such flowers together crowd to make a rounded pincushion-type terminal head some 2.5 cm across. This species is most suitable for border-massing and for cutting.

G. tricolor an erect and bushy plant grows up to 60 cm tall with slender but leafy stems that carry pale-violet and clustered bell-shaped flowers some 2 cm across. Flowers are ringed maroon at the base, and the yellow tubes are marked with purple spots. Leaves are mid-green, and bipinnate with narrow segments.

Godetia grandiflora (Family: Onagraceae: common names: farewell-to-spring, godetia, satin flower). The genus *Godetia* was erected for G.H. Godet, a Swiss botanist. Though this genus once existed but now all of its species have been included under *Clarkia*. These are suitable for pot planting, in the borders and for cutting. The genus comprises of 20 hardy annuals, all having bushy growth with brightly coloured single or double flowers. *G. amoena* from North America and *G.*

grandiflora (*G. amoena whitneyi*) from California are the two most cultivated species having bushy growth.

G. amoena has slender stems and grows to 60 cm, bearing mid-green lanceolate leaves. Inflorescence is a loose spike with funnel-shaped lilac or red-pink flowers some 5 cm across that appear in winter to spring.

G. grandiflora is the most popular species growing compact up to 40 cm tall, bearing mid-green, oblong and pointed leaves. Clustered funnel-shaped rose-purple flowers some 5 cm across appear from June to August in temperate conditions and from late winter to spring in the plains. This has produced many garden varieties in single and double forms and in the shades of red, lilac and white, such as 'Azalea-flowered Mixed' (30 cm tall, flowers semi-double and petals frilled or wavy), 'Crimson Glow' (20-25 cm tall with intense colouring), 'Dwarf Mixed' (30-40 cm, best mixture of named dwarf varieties such as white, pink, salmon, crimson and/or red), 'Kelvedon Glory' (35-40 cm tall, deep salmon-orange), 'Sybil Sherwood' (45 cm tall, salmon-pink), *etc.*

Gomphrena globosa (Amaranthaceae; globe amaranth). *Gomphrena* is a Latin name for an unknown sort of amaranth. It is native to Tropical Asia. There are numerous species under *Gomphrena*, but generally only one is in cultivation and that is *G. globosa*. It is a half-hardy, erect, much-branched annual growing up to 45 cm tall (dwarf ones only up to 15 cm). Leaves are light-green, hairy, in opposite pairs, up to 10 cm long and oblong to elliptic and taper towards the base. Flowers are tiny each with a 5-lobed perianth, hidden within the chaffy and papery bracts, all forming terminal ovoid heads some 2.5-4.0 cm in diameter from summer to autumn. The bracts may be in shades of white, yellow, orange, purple or pink. 'Cissy', 'Dwarf Buddy Purple' (compact plant with 15 cm height, flowers vivid purple), 'Dwarf White', 'Liliput Buddy', 'Nana Compacta Rubra' (15 cm high, flowers purple) are its dwarf forms. Its other varieties are 'Alba' (white), 'Aurea superba' (syn. *Gomphrena haageana*; golden-yellow), 'Rosea' (pink), 'Rubra' (purple-red), *etc.* *Gomphrena* is suitable for bedding, and the everlasting flowerheads can be severed along with the stalk and dried for floral arrangement. These can be propagated through seeds sown in March-April at the temperatures of 15-18 °C.

Gypsophila spp. (Caryophyllaceae). In Greek, *gypsos* is for 'gypsum' and *philos* for 'loving', referring to various species loving the gypsum rocks or limestone in the wild. Though the genus comprises 125 species but among annuals only two are generally cultivated, *viz.* G. *elegans* (baby's breath, chalk plant, gypsophila) from Iran, Turkey and Caucasus growing to 50 cm tall; and *G. muralis* from Europe, Asia Minor, the Caucasus and Siberia growing up to 15 cm tall.

G. elegans is erect-growing much-branched plant with graceful flower stalk, bearing opposite leaves which are stalkless, glaucous-green, lower ones lanceolate-spathulate and the upper ones elongated-lanceolate to linear. Its profuse white or pink flowers some 1 cm across, axillary or terminal are arranged in racemes and borne solitarily on highly fragile leafless stems during summer to autumn in the temperate areas while during winter to spring in the plains. Corolla is star-like, 5-petalled and the white ones sometimes are externally marked with 3 violet or reddish-green longitudinal stripes. It is used in the borders, as cut flower and as a filler. Its various important varieties are 'Covent Garden' (large white), 'Double White' (white), 'Early Snowball' (white), 'Grandiflora Alba' (large white), 'Pacifica', 'Paris Market' (white, profuse flowering), 'Repens Rosea' (bright rose), 'Single White' (white), *etc.*

G. muralis is quite dwarf species bearing much-branched and widespread wiry stems with bright green, minute and linear, smooth and opposite leaves. Profuse flesh-pink tiny flowers appear along *G. elegans*. This little plant is suitable for planting on dry walls. These are propagated through seeds sown during autumn or late winter.

***Helenium* spp.** (Asteraceae; sneezeweed). *Helenium* derives from the Greek *helenion*, named for a plant after Helen of Troy and used by Linnaeus for this genus. Also its derivation may be from the Greek *helios* for 'sun' because of the association of flowering in the direction of the sun or *Helenus* the son of Priam. A genus of about 40 species of hardy annuals and herbaceous perennials from North and South America. They are suitable for naturalizing effect, in group of irregular plantings in the garden here and there, in herbaceous borders and for cutting. These can be propagated through seeds sown in the autumn and during spring.

H. setigerum (*Amblyolepis setigera*) from Texas is an above-branching tender but attractive annual growing up to 45 cm tall, bearing 5.0-7.5 cm long, alternate and oblong leaves with erect stalks. Golden-yellow, solitary and terminal flowers some 6.25-7.5 cm across appear from spring to autumn depending upon the time of seed sowing.

H. tenuifolium from SE United States is an aromatic annual of graceful habit, growing up to 45 cm tall and is commonly known as sneezeweed as Indians in USA use its dried flowers for snuffing to relieve stuffed noses. It bears minute and narrow heather-like pale-green leaves. Profusion of rich yellow flowers appearing on a much-branched slender stem are 3.75 cm across with dentate petal tips and one stem with flowers giving appearance of a bouquet.

Helianthus annuus (Asteraceae; sunflower). In Greek, *helios* is for 'sun' and *anthos* 'flower', as flowers usually are sun-facing. Some 150 species are recorded with a few of annuals but only *H. annuus* (sunflower) from United States and Mexico which grows up to 3 m tall, and *H. debilis* (*H. cucumerifolius*) from North America which grows to a height of up to 45 cm, are in common cultivation during summer for ornamental use, both being suitable for border planting, the former in the back while latter in the middle. These can also be used as cut or loose flowers. These are propagated through seeds sown *in situ* at the place of flowering or in the glasshouse at 16 °C.

In *H. annuus* (*H. lenticularis*) the hairy stems are though strong enough but hollow, bearing up to 35 cm long, stalked, alternate, hairy on both the surfaces, ovate (heart-shaped) and irregularly dentate leaves. Flowerheads may be 5-45 cm in diameter with variously coloured dark brown to black and velvety disc while ray florets are yellow to orange. Ligules (petals) are numerous, oval-lanceolate and radiate around the central disc. The large central flowerhead is solitary but later in the leaf axils develop several small flowers. The varieties under this species are 'Abendsonne' (petals purplish but changing to chestnut-red or yellow and disc maroon, flowers 15 cm in diameter), 'Autumn Beauty' (sulphur-yellow), 'Bicolor' (petals reddish, tipped yellow), 'Californicus' (semi-double, disc shiny yellow), 'Chrysanthaeflorus' (flowers large and double, petals lacerated), 'Citrinus' (primrose-yellow), 'Daisetsuzan', 'Gelber Knirps' (dwarf, flowers double), 'Globe of Gold' (dwarf, double), 'Globosus Fistulosus' (enormous globose flowers), 'Goldener Neger' (much-branched, disc small and dark, ray florets golden), 'Indianerin' (much-branched, disc small, centre dark, ray florets golden), 'Intermedius' (syn. *H. hybridus*, similar to original species but flowers smaller), 'Italian White' (pale-primrose), 'Mammoth', 'Mars' (yellow, disc dark brown), 'Nanus Florepleno' (flowers double), 'Purpureus' (flowers pink shading to purple), 'Pygmy Dwarf', 'Red' (red), 'Russian Giant' (tallest, normally used for seed production), 'Sunnengold' (Sungold, double), 'Sutton's Autumn Beauty', 'Sutton's Double Orange', 'Sutton's Red', 'Tall Sungold', 'The Sun', 'Variegatus' (leaves variegated), *etc.*

H. debilis (*H. cucumerifolius*) from southern United States is a bushy, hairy and much-branched species with reddish-blue stems spotted with yellow, where leaves are glossy green, cordate or triangular, some 10 cm long and marginally toothed. Terminal and axillary flowers are solitary, bright yellow and some 7.5 cm across or more. Its varieties are 'Dazzler' (maroon-spotted orange), 'Diadem' (pale-lemon-yellow, disc almost black), 'Excelsior' (mixed strain), 'Herbstschönheit' (yellow-bronze with a dark maroon zone around the disc), 'Orion' (golden-yellow, with a little twisted petals), 'Perkeo' (blooms small in various shades of yellow), 'Plumosus' (disc pale to intense yellow), 'Purpureus' (petals pale-pink to intense purple), 'Rotstern' (reddish-brown), 'Stella' (intense golden-yellow, petals curled and disc dark violet-maroon), 'Sunburst' (a multi-hued strain), *etc.*

Heliopsis buphthalmoides (Asteraceae; orange sunflower). In Greek, *helios* is for 'sun' and *opsis* for 'like', referring to the likeness to sunflower. A genus of 12 species of annuals and herbaceous perennials from Americas, including Peru. *H. buphthalmoides* is an annual which can be propagated through seeds sown from spring to summer to have floral display from summer to autumn. It is suitable for borders, for informal bedding, for naturalizing and for cutting. It is native to North America and grows up to 75 cm tall. Leaves are scabrous, lanceolate and 5.0-7.5 cm long. Flowers are single yellow alike single sunflowers, solitary and terminal.

Helipterum roseum (syn. *Acroclinium roseum*, family: Asteraceae, common names: acroclinum, everlasting flower, immorteles, paper flower). In Greek *helios* means 'sun' and *pteron* means 'a wing'~ the pappus on the fruit~ seed ~ is composed of radiating plumose bristles like a diagramatic sun. It is a hardy annual from Western Australia growing more than 40 cm high. Leaves are grey-green, linear to lanceolate and about 6 cm long, and flowerheads solitary, semi-double, papery, pink (rosy) to white and some 2.5 cm across though varieties have larger flowers than the type. Varieties are 'Albo' (white), 'Blue Hawaii', 'F_1 hybrids', 'Goliath', 'Large Flowered Mixed' (syn. 'Grandiflorum', rose-pink or white), 'Rose', 'Rose and White', 'Royal Hawaii', 'White Hawaii', *etc.*

Helipterum bulboldtianum, a hardy annual from Australia grows up to 50 cm tall. This has narrow pointed leaves and woolly-textured near-white fragrant flowers some 7.5 cm across. They are suitable for edging the borders. The flowers turn green after drying.

H. manglesii (*Rhodanthe manglesii*) from Australia is also a fast-growing annual with greyish-green and pointed-oval leaves, and papery, red, pink or white flowerheads. They are good for planting in borders. The flowers for drying are cut before the petals are fully expanded, and hang them upside-down in an airy and cool place. They are winter-flowering and the seeds are raised at 16 °C temperature.

***Hibiscus* spp.** (Malvaceae; mallow). In Greek, *Hibiscus* is for a 'mallow', applied by Linnaeus to this genus. The genus comprises some 300 species of annuals, perennials, shrubs and trees with alternate leaves that are often palmately-lobed. Flowers are very showy and 5-petalled. All the annuals being described below are hardy suitable for mixed borders and for loose flowers, and are propagated through seeds sown in March at 13-16 °C.

H. abelmoschus (*Abelmoschus moschatus*) from India is an annual growing up to 1.8 m tall with erect habit. Leaves are dark green, palmate with 5-7 lobes, dentate and about 15 cm wide. Flowers are beautiful yellow with a red centre, solitary and 10 cm in diameter. Its seeds are amicably aromatic.

H. sabdariffa (*H. rosella*; common names: Jamaica sorrel, red sorrel, roselle or southern cranberry) from Tropical Africa is an annual growing up to 2.1 m tall. It is a very vigorous, fast-growing, stems being reddish and has robust branching. Seeds should be sown in January-February for flowering during July-August. Its basal leaves are dark green, entire and some 20 cm wide but stem leaves are digitate, each lobe lanceolate and dentate. Bright yellow and some 15 cm across flowers have red bracts and calyx, where calyx becomes enlarged and fleshy which is used for preparing jams and jellies and as a flavouring agent in cold drinks.

H. trionum (*H. africanus, H. versicarius*; flower-of-an-hour or goodnight at noon) from North Africa and Mediterranean regions is a hardy annual bearing dark green, ovate and coarsely-toothed (deeply lobed) leaves with three conspicuous segments which grows up to 75 cm tall. Cream-white to pale-yellow flowers with bright maroon-chocolate centre and some 5.0-7.5 cm across continue appearing throughout summer and autumn, each flower being followed by an inflated bladder-like calyx bearing seed-capsule. Though now varieties are available where flowers remain open for whole of the day but the original species open in the morning only for a few hours.

Hunnemannia fumariifolia (Papaveraceae; Mexican tulip poppy). This generic name was coined to honour John Hunnemann, English botanist who died in 1839. A native of highlands of Mexico, it is closely allied to *Eschscholzia*. The half-hardy genus comprises only one species with only one variety. It is a herbaceous perennial with slender stems but always grown as an annual through seeds sown in September, and it grows to a height of 60 cm. Its leaves are glaucous silver-grey, triternately divided and fern-like, and the flowers are large (7.5 cm across) brilliant yellow and poppy-like with numerous orange-yellow stamens though sepals are two and caduous and petals four and spreading. It is suitable for borders, for beddings and for cutting.

Iberis spp. (Family: Brassicaceae; common name: candytuft). *Iberis* derives from Iberia which is ancient name of Spain, since most species were discovered in Spain. The specific name of *Iberis amara* comes from the Latin *amarus* which refers to its bitter flavour. Candytuft is made of two words, *i.e.* 'candy' (deriving from Candia the former name of Iraklion on the Island of Crete) and 'tuft' (originating from the clustered flowers or tufted plant growth). They are suitable for growing in borders and for cutting while the dwarf ones are suitable for rock gardens or dry walls. All the annual species are propagated through seeds by sowing either in autumn or during late winter. They bear deep green, alternate, linear to spathulate or obovate leaves and corymbs or racemes of 4-petalled flowers with 2 being larger.

I. amara (syn. *I. coronaria*; rocket candytuft) is the most common garden ornamental native to Central and Southern Europe including Great Britain. It is a hardy annual growing up to 40 cm tall with erect and branched stems, bearing 7.5-10.0 cm long, alternate, widely pinnatifid or toothed leaves that are spathulate to oblanceolate and wider at the tip. White to purple short-stalked flowers appear in large compact corymbs during late winter to spring. Its taller varieties are 'Hyazinthenblutige Riesen', 'Imperialis', both being hyacinth-flowered. Its other taller varieties are 'Empress', 'Giant Snowflake', 'Iceberg', 'Improved White Spiral', 'Sempervirens White', 'Spital White', 'White Pinnacle', *etc.* Dwarf ones are 'Little Prince', 'Tom Thumb', 'White Pinnacle', *etc.*

I. umbellata from Southern Europe is a hardy annual with erect branching habit where lower branches are generally larger than those at the summit. It grows 15-40 cm tall, bearing mid-green alternate leaves where basal ones are lanceolate and the rest linear. Violet-lilac suffused pink flowers appear in compact umbels and when in full bloom the plants expand to 40 cm in diameter with a flattish umbrella-shape. Its popular varieties are 'Alba' (white), 'Dwarf Fairy Mixed' (up to 25 cm tall, flower colours a mixture of white, lavender, rose-pink and red), 'Formosa Purpurea' (purple-red), 'Königin von Italien' (lilac), 'Red Flash' (up to 30 cm tall, vivid carmine), 'Rose Cardinal' (up to 25 cm tall, rose-scarlet), 'Vulcan' (up to 40 cm tall, tetraploid with large carmine flowers), *etc.*

Impatiens spp. (Balsaminaceae; balsam). The genus comprises some 700 species, mostly of hardy and half-hardy annuals, but only a few are of ornamental importance and in common cultivation. *Impatiens* derives from the Latin for impatient, as ripe seed capsules burst open explosively. They are suitable for bedding, in the borders and as loose flowers. They are propagated through seeds sown during March-April to flower during summer to autumn.

I. balsamina (rose balsam, touch-me-not), a half-hardy compact plant from India, China and Malaysia that grows compact up to 75 cm tall and is branched sparingly, bears pale-green, 7-15 cm long and narrowly to broadly lanceolate, acuminate and toothed leaves. The flowers in shades of white, yellow, pink, crimson, red and purple and some 4 cm wide appear in summer in the upper leaf axils. The forms are 'Camellia-Flowered Mixed' (45 cm tall) and 'Tom Thumb Mixed' (25-30 cm tall), both with large double flowers in various colours. Other types are 'Dwarf Bush Flowered', 'Rose Flowered', 'Royal Balsam', 'Tall Double', *etc.* 'Accent', 'Blitz', 'Minette' and 'Super Elfin' are the F_1 hybrids.

I. glandulifera (*I. roylei*; Himalayan touch-me-not,Indian balsam, policeman's helmet), a hardy annual from Himalaya grows up to 1.5 m high with thick, succulent and branching stems, bearing light green, up to 15 cm long, paired or whorled in 3, ovate and sharply toothed leaves. Pale to deep rose-purple or white and spurred flowers some 3-5 cm long are produced in terminal and axillary panicles from summer to autumn.

I. noli-tangere (touch-me-not), a very hardy annual from C and N Europe eastwards to Siberia, is brilliantly green plant with main stem being succulent and the nodes quite swollen. Leaves are light green, dentate and oval-lanceolate. Brilliant yellow flowers with red spots in the throat appear during summer and autumn under filtered situations.

I. repens, a procumbent or creeping annual to perennial with rooting stems from India and Sri Lanka grows only 5 cm tall, bearing small and oval to rounded leaves, and turmeric-yellow flowers with large hairy spurs.

Ionopsidium acaule (Family: Brassicaceae; common names: diamond flower, violet cress). In Greek, *ion* stands for 'violet' and *opsis* for 'like', due to its flower colour. This tufted genus comprises only 1 hardy annual species, *I. acaule* from Portugal, suitable for pot-growing in the shaded greenhouse or indoors, in the rock gardens and in the crevices of pavings. The plant is almost stemless, the leaves are mid-green, long-stalked, 5.0-7.5 cm long, dense, glabrous, the blades are rounded and 0.8-1.2 cm wide, and lilac to white 4-petalled solitary flowers tinted violet and some 6 mm to 1 cm wide are produced freely from the leaf axils above the foliage level. This is propagated through seeds sown from spring to autumn for whole the year flowering

Ipomoea spp. (Convolvulaceae). *Ipomoea* derives from the Greek *ips* for 'worm' and *homoios* for 'similar to', probably due to the twining stem tips or coiling tendrils. There are some 500 species in the genus but ornamental annuals are only a few in cultivation such as *I. hederacea* (*Convolvulus hederaceus, Pharbitis hederacea*), an annual climber from Tropical America growing up to 2.5 m long; *I. purpurea* (*Covolvulus major, C. purpureus, Pharbitis purpurea*; common morning glory) from Tropical America, an half-hardy annual climber growing up to 3 m long with hairy stems; and *I. tricolor* (*I. rubro-caerulea, Pharbitis rubro-caerulea*) from Tropical America is though perennial but cultivated as an annual that grows 3 m long. These all make fine pot plants, however, these can also be trained on arcs and walls.

I. hederacea bears 15 cm long and broadly ovate-cordate leaves which may be superficially 3-lobed. Purple, blue, red and pink flowers with broad bristly-sepals that contract suddenly from a linear to spreading recurved tip are produced from summer to autumn.

I. lobata (syn. *I. versicolor*) from Mexico is a half-hardy perennial species though grown as a tender annual, now correctly known as *Mina lobata*. It grows 15-25 cm tall with climbing stems and 3-lobed dark green leaves. Spikes are slender and carry up to 12 crimson flowers though changing to orange, yellow and white with age, each measuring some 5 cm long, and appear during summer to autumn. These are suitable for cutting.

I. purpurea bears up to 13 cm wide and broadly heart-shaped or 3-lobed leaves, white, pink, magenta and purple-blue funnel-shaped flowers some 7 cm long having white throats and oblong and short pointed bristly sepals. Its variety 'Violacea' bears double flowers. *I. tricolor* bears almost heart-shaped, 15-25 cm wide and slender-tipped leaves. Purplish-blue and white-tubed funnel-shaped large flowers some 10 cm long and wide are produced during summer to nearly autumn. Its two noteworthy varieties are 'Heavenly Blue' (rich sky blue) and 'Flying Saucers' (striped blue and white).

Lagurus ovatus (Graminae; hare's tail grass). In Greek, *logos* is for 'a hare' and *oura* for 'a tail', alluding to the flower spikes that resemble so. The genus *Lagurus* comprises only one species (*L. ovatus*) of hardy annual grass from the Mediterranean, and is used in borders and as dry material for flower arrangement. It is winter-growing, tufted and grows up to 45 cm in height, bearing grey-green, downy and narrowly linear leaves. Stems are slender and terminate in floral panicles which are white, dense, ovoid, 2.5-4.0 cm long and woolly. This is propagated through seeds sown *in situ* during September-October.

Lathyrus odoratus (Family: Leguminosae; common names: everlasting pea, sweet pea). In Greek, *lathyrus* is for 'a pea'. Though there are about 130 species of hardy annuals, sub-shrubs and herbaceous perennials, mostly tendril-bearing climbers, but out of all the annuals, only two ornamentals are in common cultivation and these are *L. odoratus* from Italy, growing up to 3 m long, and *L. tingitanus* from the Mediterranean growing to a length of 1.8 m. They are excellent as cut flowers, for making colourful screens and as pot plant. These all have the characteristic pea-flower shape with large wing petals (standard) and central keel petals, and the leaves are pinnatifid.

L. odoratus bears single pairs of mid-green, smooth and ovate leaflets which terminate into tendrils. Flowers are fragrant and in shades of white, pink, red and purple some 2.5 cm across. There are many varieties which are divided into groups as per their height and flowering time. 'Cuthbertson Floribunda' varieties grow up to 2.4 m long and bear 5-7 flowers on each stem and this includes the varieties 'Jenny' (white) and 'Robert' (mid-blue). 'Multiflora Gigantea' also grows 2.4 m long, and is earlier in flowering, each stem bearing 6-10 flowers some 5 cm in size, and the varieties are 'Colorama' (mixture of colours) and 'Ramona' (orange). The 'Spencer' varieties are the most popular ones, growing up to 3 m long and flowering late with 4-5 blooms on each long stem. The varieties under this group are 'Air Warden' (scarlet), 'Carlotta' (carmine), 'Early Flowering Frilled', 'Gertrude Tingay' (deep lavender), 'Countess Spencer', 'Giant Frilled', 'Late Flowering Spencer', 'Leamington' (deep lilac), 'Noel Sutton' (blue-purple), 'Princess Elizabeth' (salmon-pink), 'Royal Flush' (cream-pink), 'Sonata' (salmon-pink), 'Spotlight' (ivory, flushed pink), 'Stylish' (mid-blue), 'Swan Lake' (white), 'Tell Tale' (white, edged pink), *etc.* There are three main groups in dwarf sweet peas such as 'Bijou' (bush-type), 'Knee-hi' (vigorous than "Bijou" and growing 1.2 m), and 'Colour Carpet' (grows up to 25 cm).

L. tingitanus from Mediterranean regions is a half-hardy annual growing up to 1.8 m long with narrow, lanceolate and acuminate leaves, long slender flower stems bearing clusters of large, red and purple blooms. Its var. 'Roseus' bears pink flowers.

Lavatera trimestris (Malvaceae; lavater or mallow). *Lavatera* was named for J.R. Lavater, a 16th century doctor and naturalist in Zurich. The genus contains some 25 species of annuals, perennials and shrubs from Canaries, Mediterranean, coastal Europe, California, Asia and Australia. The annual species, *L. trimestris* (*L. rosea*) from South Europe grows erect and is suitable for planting in borders and for cutting and is the most beautiful in the genus. Leaves are pale-green, alternate and ovate to palmately-lobed. Racemes are terminal and leafy, the clayx-like bracts are 3-6 and sometimes even 9 and all are united at the base, and rosy flowers some 10 cm across are 5-petalled with obovate and often notched petals, which appear in temperate regions from July to September. Its most popular varieties are 'Loveliness' (deep rose), 'Splendens Alba' (large white), *etc.* They are propagated through seeds sown *in situ* in September.

***Legousia* spp.** (Campanulaceae; venus's looking glass). *Legousia* was erected for Legous de Gerland, founder of the Dijon Botanic Garden in 1773. There are about 15 species of annuals but the two *L. hybrida* (*Specularia hybrida*) and *L. speculum-veneris* (*Specularia speculum-veneris*; poached egg flower) are generally in cultivation. They are basally branched with erect or shortly decumbent stems, bearing oblong to obovate or spathulate leaves. Five-petalled violet flowers with tubular bases appear from the upper axils from winter to spring. They are suitable for bedding, for planting in the rock gardens, *vis-à-vis* window boxes and on dry walls and can be propagated through seeds sown in the nursery beds during September or *in situ* on their permanent flowering places which may be thinned out later on.

L. hybrida from SW Europe grows about 20 cm tall, bearing undulate, crenate and 2 cm long leaves and lilac-blue to reddish-purple flowers in clustered corymbs. Calyx lobes are clearly awl-shaped. Flowers appear during winter and spring.

L. speculum-veneris from SW & SC Europe, north of Holland grows taller, *i.e.* up to 40 cm tall bearing 4 cm long leaves and violet-blue or white and 1.0-2.5 cm long flowers appearing often in paniculate clusters in winter or spring. Calyx lobes are clearly five.

***Limonium* spp.** (Family: Plumbaginaceae; common names: sea lavender, statice). In Greek, *leimon* is for 'a meadow', owing to their sea-marsh habitat. There may be some 300 species of annuals, perennials and sub-shrubs in the genus but only three, *L. bonduellii* from Algeria, *L. sinuatum* from Mediterranean areas and *L. suworowii* from Caucasus and Iran to Central Asia (W. Turkestan) are in general cultivation for flowering during winter to spring in the plains and summer to autumn in the temperate areas. They are tufted plants having branching at the base and are suitable for planting in the borders, in pots, for cutting and for drying as everlasting flowers. *L. suworowii* is particularly excellent even for bedding. These may be propagated through seeds sown during autumn or in February-March, if in glasshouse, at a temperature of 13-16 °C.

L. bonduellii is strictly a perennial but is best treated as a half-hardy annual, grows to about 30 cm tall with erect stems and is suitable for cutting and drying. Basal rosette pale-green leaves are ovate and deeply lobed. Yellow and loose-clustered flowers some 7.5 cm across are borne during spring and summer.

L. sinuatum (notch leaf, winged statice) is a most popular half-hardy perennial species though treated as an annual, growing to about 45 cm tall with broadly winged stems, bearing mid- to dark-green oblanceolate and waving leaves some 8-10 cm long with deep pinnately lobed margins. Inflorescences are with much-branched dense panicles. Blue and cream flower clusters some 7.5-10.0 cm long, calyces funnel-shaped, purple and 1.0-1.2 cm long, flowers being surrounded by green bracts, are borne during Spring to summer. The varieties in various shades of calyces (yellow, orange, salmon, rose-pink, red, carmine, blue and lavender) are available, such as 'Apricot Beauty', 'Atrocoerulea', 'Art Shades', 'Candissima', 'Chamois Rose', 'Fast Blue', 'Heavenly Blue', 'Iceberg', 'Market Grower's Blue', 'Market Rose', 'Midnight Blue', 'Pastel Shades', 'Purple Monarch', 'Rosea', 'Snow Queen', *etc.*

L. suworowii (syn. *Psyllostachys suworowii*; common statice, rat's tail statice, Russian statice) is a half-hardy annual growing to about 45cm high bearing basally rosette mid-green, oblanceolate, lobed and waving leaves some 15-25 cm long. Panicles of fingered spikes some 45 cm long are borne erect along with radiating ones from the base with dense arrangement of rose-pink flowers some 4 mm long appear during spring and summer. It is most suitable pot plant and the inflorescences as cut flowers.

Linanthus androsaceus (syn. *L. lutea, Leptosiphon hybridus*; family: Polemoniaceae). The generic name derives from the Greek *linon* which means 'flax' and *anthos* for 'flowers'. Genus *Linanthus* though comprises of 15 annual species but only *L. androsaceus* is generally in cultivation. It comes from California and is most suitable for edging the borders, and in rock gardens, especially in the crevices of the pavings as it grows erect up to 15 cm high with base-branched stems. Its stem-clustering leaves some 1-3 cm long are mid-green and deeply divided into pointed 5-9 linear to oblanceolate segments. Its slender *cum* long-tubed, star-shaped flowers some 1.5 cm long and 1.25 cm across with 5 rounded lobes in shades of yellow, orange, pink, lilac, red and white continue appearing from early spring to rainy season.

Linaria maroccana (Family: Scrophulariaceae; common names: linaria, toadflax). *Linaria* derives from the Greek *linon* meaning 'a flax', as its leaves are similar to those of flax. It is a genus of some 150 species of hardy and half-hardy annuals, herbaceous perennials and sub-shrubs, and the three (*L. bipartita, L. maroccana, L. reticulata*) described here are hardy annuals. There are many other annual-type perennial species wshich enrich *Linaria* collection. These are effective in the borders and beddings, dwarf ones for rock gardens and for edging the paths and *L. maroccana* is also effective as pot plant. These are winter- to spring-flowering annuals so seeds are sown in September or in February-March.

L. bipartita from Portugal and North Africa is an erect growing compact and bushy annual growing to 38 cm tall with small, narrow and pointed leaves. The inflorescence is a 15-cm long spike bearing purple-violet flowers marked orange on corolla-edges. Its varieties 'Alba' (white), 'Splendida' (purple) are quite popular.

L. maroccana from Morocco is a most beautiful hardy annual with bushy and branched habit, growing erect to about 38 cm tall. Its leaves are light glaucous-green, small, narrowly linear and pointed. Its violet-purple flowers with yellow throat are 2-3 cm long, snapdragon-like with a basal and pointed spur, *vis-à-vis* a white or yellow blotch on the lower lip. There are numerous varieties such as 'Alba' (pure white), 'Excelsior Mixed' (30 cm tall, flowers violet, blue, crimson, pink and yellow), 'Fairy Bouquet' (20 cm tall and about 2 cm long flowers in different shades), 'Rosea' (pink), *etc.*

L. reticulata from Portugal is a hardy annual growing to a height of 90 cm, bearing pale-green, narrow and pointed leaves. Purple flowers with an orange or yellow blotch on the lower lip and some 2 cm long flowers with basal spur appear in winter and spring.

Linum grandiflorum var. *rubrum* (Family: Linaceae; common names: flax, linum, scarlet flax). *Linum* derives from the ancient Latin name flax. It is a genus of 230 species of hardy and tender annuals, perennials and sub-shrubs of cosmopolitan distribution, but especially the Mediterranean region. *L. grandiflorum* from Algeria (North Africa) is a hardy annual growing up to 45 cm tall with slender, smooth and much-branched stem, and many pale-green, narrow, alternate, elongated-lanceolate and pointed leaves. Single saucer-shaped rose-coloured flowers some 4 cm across are carried in profusion in thread-like terminal branched clusters during winter and spring. It is most suitable for pot-growing and is propagated through seeds sown in September. Its several varieties are found in the garden such as 'Bright Eyes' (38 cm tall, flowers 5 cm across, white with carmine eye), 'Caeruleum' (sky-blue), 'Coccineum' (scarlet), 'Kermesinum' (crimson), 'Rubrum' (30 cm tall, brilliant crimson), 'Venetian Red' (large carmine-red), *etc.*

***Lobelia* spp.** (Family: Campanulaceae/Lobeliaceae). Named so to honour Matthias de l'Obel (1538-1616), botanist and physician to James 1. There are some 370 species of annual to perennial lobelias with some aquatics, shrubs and tree-like, often with acrid milky sap which may irritate skin. These naturally occur in tropical and temperate regions worldwide, especially in North, Central and South America from marshes, wet meadows, and riverbanks, to woodland, well-drained hilly and mountainous slopes, and deserts. They are fully hardy (*L. tupa* up to -10 °C) to frost tender. All the lobelias bear simple, alternate, often stalkless leaves and terminal solitary, paniculate or racemose 2-lipped tubular flowers each with 5 lobes, upper 2 being often erect but lower 3 often spreading and fan-like. Perennials look effective either in a mixed border or beside water; annuals for edging, hanging baskets, in the window box or in the pots; and aquatic species look exquisite in a wildlife pool. Shrubby and tree-like species are usually not cultivated. Plants may be infested by slugs. *L. splendens* is more susceptible to viruses, *L. siphilitica* to leaf blotch where crowns may rot during damp and mild conditions. These are propagated through seeds sown in September-October on nursery beds at 13-18 °C in case of annuals, and as soon as the seeds are ripe in case of perennials. Perennial plants for multiplication are divided in spring while aquatics in summer. *L. cardinalis* is propagated through bud cuttings taken in midsummer. Lobelias present a great diversity of forms from montane tree species such as *L. aberdarica*, through a range of handsome perennials of varying hardiness to small species grown as annuals. The flowers in case of *L. cardinalis* (cardinal flower) are brilliant scarlet-red. The perennial lobelias such as *L. dortmanna* (Water/Aquatic lobelia) and *L. paludosa* (Swamp lobelia) bear pale-blue to pale-violet, and pale-blue flowers, respectively. *L. erinus*, a low-growing perennial though grown as an annual through seeds bears blue or violet flowers while its cultivars in different colours. *L. laxiflora*, a perennial lobelia bears red and yellow flowers, *L. richardsonii*, an evergreen perennial is a trailer and bears white-throated bright lilac-blue flowers, *L. siphilitica* (Blue cardinal flower), a clump-forming perennial bears long-lasting bright blue flowers, *L. splendens* (syn. *L. fulgens*; Scarlet lobelia) is a perennial with scarlet flowers, *L. tupa* is an upright perennial bearing brick-red to orange-red flowers, *etc.* The species such as *L. fenestralis, L. gracilis, L. heterophylla, L. inflata, L. subnuda, L. tenuior, etc.* are annual (occasionally perennial). Its noteworthy varieties are 'Bees' Flame', 'Brightness', 'Cherry Ripe', 'Dark Crusader', 'Erinus cultivars', 'Cascade Series', 'Cobalt Blue', 'Compliment Series', 'Crystal Palace', 'Fan Series', 'Illumination', 'Lilac Fountain', 'Moon Series', 'Mrs. Clibran', 'Palace Series', 'Queen Victoria', 'Regatta Series', 'Riviera Series', 'Rosamund', 'Sapphire', 'Speciosa Cultivars', 'Will Scarlet', *etc.*

Lunaria annua (syn. *L. biennis*; family: Brassicaceae; common names: honesty, Moonwort, satin flower, silver shilling, St. Peter's Pence). In Latin, *luna* or *lunar* is for 'moon', alluding to the silvery membranous septum that divides each seed pod. It is a genus of 3 species, only one is commonly grown. *Lunaria annua* (*L. biennis*) from Europe is cultivated as an annual though it is a biennial and sometimes living even for three years. The plant grows from 30 to 75 cm tall bearing dark green, slightly coarse-textured, cordate and coarsely dentate leaves. Reddish-violet 4-petalled fragrant flowers are cross-shaped and some 2.5 cm across, and the disc-like flat silicles are some 4 cm across so when cut and dried provide excellent display at home. Also a white form is available. 'Variegata' has variegated leaves and crimson flowers. It is winter- to spring-flowering plant propagated through seeds sown in early September.

Lupinus hartwegii (Legiminoceae; lupin). In Latin, *lupus* is for 'the wolf', as it was believed that lupins robbed the soil of fertility. There are some 200 species of hardy and half-hardy annuals, herbaceous perennials and sub-shrubs. They are excellent for group-plantings in beds and borders and for cutting. Their flowers are basically shaped like those of peas but standard (the upper petal) is folded back while lower petals are laterally compressed to form a keel. Their spire-like raceme is slender and densely arranged with flowers. These are propagated through seeds sown directly at the place of flowering in September to flower by winter and spring. All the species described below are annuals.

L. hartwegii from Mexico is a delightful annual with bushy and branched growth, growing up to 90 cm and bearing soft green and hairy leaves divided into 7-9 leaflets, and about 30 cm tall spike with bright blue flowers flushed pink on the standard. This has numerous varieties in shades of dark blue, sky blue and white and even the dwarf forms such as 'Nanus' (45 cm tall and flowers in mixture), 'Pixie Delight' (45 cm tall, flowers in mixture of red, pink, purple and blue), *etc.*

L. luteus from Europe is a hardy annual growing up to 30 cm tall, bearing small dark green and digitate leaves divided into 8-9 lanceolate and hairy segments. Inflorescence is stout, erect, some 25 cm long and bears dense masses of strongly fragrant yellow flowers.

L. mutabilis from South America is a much-branched tall (140 cm tall) annual bearing glaucous-green long-stalked leaves that are digitate and divided into long and thin segments. Inflorescence is an erect and quite decorative spike some 45 cm long comprising of numerous large and widely-spaced large white and sweetly-scented flowers with erect standard petal often flushed blue with yellow markings. It contains numerous varieties such as 'Cruckshanksii' (flowers vivid-blue suffused white, yellow or purple), 'Roseus' (pink), 'Versicolor' (blue), *etc.*

L. nanus from California is an attractive free-flowering growing up to 30 cm tall with much-branched stems, and bearing light-green, hairy and 5-7 segmented leaves. Inflorescence is a short spike bearing large blue flowers though standards with white spot which is dotted purple. Very effective in group planting.

Malcolmia maritima (Brassicaceae; Virginian stock)**.** *Malcolmia* is named for William Malcolm (*d.* 1820) and his son also with the same name (1769-1835), the London nurserymen. It is a genus of about 30 species of annual and perennials from Mediterranean to Central Asia and Afghnistan, one of which is in common cultivation which is suitable for planting in drifts in borders and for edging. *M. maritima* from SW Greece and S. Albania is a hardy annual growing 20 cm tall with grey-green, oblong to obovate and blunt leaves. The cross-shaped red, lilac, rose or white and sweetly fragrant flowers some 1.25 cm across are carried on slender stems. Sowing of seeds should be done in September and March at their permanent site which will start giving blooms after one month of sowing and will continue up to one month more.

Malva sylvestris (Malvaceae; common mallow)**.** *Malva* is ancient in vernacular for mallow. The genus comprises some 40 species of annuals and perennials, closely allied to *Lavatera. Malva sylvestris* from Europe, temperate Asia and North America is cultivated as an annual though it is a biennial and sometimes even perennial. However, it is propagated through seeds sown in September to flower by spring to summer. It is suitable for planting in the borders and in tubs and pots. It grows pyramidally erect up to 90 cm tall, is rough-hairy and much-branched laterally, branches also erect and highly leafy. Leaves 5-7 lobed, rough and cordate and the flowers are large purple-rose. Petals are thrice as long as calyx.

Matthiola incana (Family: Brassicaceae; common names: gillyflower, stock)**.** *Matthiola* is named for Pierandrea Mattioli (1500-1577), an Italian doctor and botanist. It is a genus of 55 species of annuals, perennials and sub-shrubs from Mediterranean, Canary Islands to Azores and Europe to Central Asia. They have rosette to alternate, linear to oblong and sometimes pinnatifid leaves and erect racemes with 4-petalled heavily scented flowers in pastel shades or rich colours being massed in spikes. They are suitable for beds and borders, for pot-cultivation and for cutting. *M. incana* from South Europe grows up to 60 cm tall and is a hardy biennial though grown as an annual. The leaves are grey-green, felty in texture, long and narrow. The spikes are compact, up to 25 cm long and densely packed with pale-purple flowers which appear during winter and spring. It has numerous varieties, a few of them being 'Giant Column', 'Giant Excelsior Column', 'Giant Rocket', 'Miracle Column', 'Snow Wonder', 'White Wonder', 'X-mas Ocean', 'X-mas Pink', 'X-mas Rouge', 'X-mas Snow', *etc.* under non-branching (column type); and 'Beauty of Nice', 'Dwarf Ten Week', 'Early Giant Imperial', 'Giant Perfectioin Ten Week', 'Trisomic Seven Week', *etc.* under branching type; both in singles and doubles and in shades of white, yellow, gold, pink, crimson, lavender, purple, copper, bicolours, *etc.*

Melampodium paludosum (Melampodium; Asteraceae)**.** It is a hardy annual from Mexico which may grow up to 60 cm tall, and is suitable for growing in the pots, in herbaceous borders and in beds at a sunny situation. It is a free-flowering annual bearing star-shaped yellow to golden-yellow flowers in profusion. Its popular varieties are 'Medallion Golden Yellow', 'Golden Globe', 'Million Gold', 'Show Stars', 'Derby', *etc.*

Mentzelia lindleyi (syn. *Bartonia aurea*; Blazing star; Loasaceae)**.** This annual is native of California and has been named for Christian Mentzel (1622-1701), a German physian and botanist. This genus comprises of some 60 species of annuals, biennials, perennials and shrubs from North to South America, but only the one described above is in cultivation though is not common in India. It is an hardy annual growing up to 45 cm tall, bearing sessile, hispid, alternate, pinnatifid and 2-17 cm long leaves. Its night-blooming, fragrant and glistening golden-yellow flowers with orange-red eye are axillary, coming up from the upper leaves or in small terminal clusters, 5.0-6.5 cm across and 5-petalled. It is most suitable for bedding. It is propagated through seeds.

***Mimulus* spp.** (Family: Scrophulariaceae; common names: mimulus, monkey flower)**.** *Mimulus* derives its name from the Latin *mimus* which means 'to mimic', as flowers of the first known species fancifully resembling the face of a monkey. It is native of temperate regions of the world, mainly in the Americas (not in Europe) with some 100 species of hardy annuals and herbaceous perennials. In their natural habitat the plants grow in bog conditions, though in the garden these do well in ordinary conditions. They are suitable as pot plants, in poolside planting, in the rock gardens, and in the borders. They have showy ovate to linear-lanceolate leaves in opposite pairs, flowers tubular but opening to

5-lobed mouth. Many of the perennial forms as described below are being grown as annuals from seeds sown in September or under glass in March at a temperature of 11-16 °C.

M. brevipes from California is a showy viscid-pubescent large-flowered annual growing up to 38 cm tall, bearing bright green, lanceolate to linear and some 10 cm long leaves. The masses of canary-yellow flowers some 3.5-5.0 cm long, with broad limbs, rounded lobes and unequal *cum* acuminate calyx-teeth, appear during winter to spring in the plains of India and during rainy season in the temperate regions. This requires abundance of moisture and sun during its growing though requires protection when grown in summer in temperate areas. It is propagated through seeds at 13-16 °C temperature sown in spring as well as through cuttings.

M. cupreus from South Chile is a tender herbaceous perennial growing up to 38 cm tall and is parent of many of the beautiful perennial hybrids, though is grown as an annual. The leaves are mid-green, 2.5-3.0 cm long and oblong-ovate. Flowers some 2.5-4.0 cm long opening during winter to spring in the plains and during summer on the hills, while opening these are yellow but later change to coppery-orange spotted-brown. It requires partial shade for its growing. Though propagated through seeds but off-springs do not appear true to type so for choice of the colour these are propagated through division and cuttings. Its popular varieties are 'Bee's Dazzler' (scarlet), 'Leopold' (yellow, marked orange), 'Plumtree' (lovely pink), 'Queen's Prize' (pink and yellow), 'Red Emperor' (crimson), 'Scarlet Bee' (15 cm tall, deep scarlet), 'Whitecroft Scarlet' (10 cm tall, bright orange-scarlet), *etc.*

M. luteus (*M. hybridus tigrinus*; monkey flower or monkey musk) from Alaska to New Mexico and Chile is an herbaceous perennial growing up to 30 cm tall with sprawling to semi-erect stems, bearing stem leaves smaller, the basal sometimes laciniate and up to 7.0 cm long, parallel-veined, ovate to roundish to subcordate and sharply toothed though teeth only a few but regular. Normally, it is grown as an annual through seeds. Calyx up to 1.2 cm long, flowers yellow with red spots inside the throat and on the lobes, and corolla 2.5-5.0 cm long. The species requires more moisture for its growing and sometimes survive well even in shallow water. It is free-flowering for longer period, *i.e.* from January to May in the plains and from May to September on the hills. It is propagated through seeds sown during September or in spring at 13-16 °C or through division and cuttings. With *M. guttatus* (*M. longsdorfi*), it has produced many popular hybrids such as 'Alpinus' (mat-forming, good for rock garden, flowers large marked brown-purple), 'Duplex' (it is named 'hose-in-hose' as its calyx being petal-like which appears that one flower is inside the other, flowers yellow marked bronze), *M. l. guttatus* is quite dwarf but bears large flowers which are clearly blotched brown-purple, *etc. M. l. variegatus* from Chile is quite similar to *M. luteus* but with large flowers up to 5.0 cm long and with varying mixture of yellow and purple, and where two yellow longitudinal lines are dotted with brown on the middle lobe of the lower lip, and all the lobes are bright crimson-purple with a violet exterior. Though colour may not appear true to the type but these are propagated as an annual through seeds either sown in September or during early spring at 13-16 °C temperature. By division and through cuttings also these can be propagated.

Molucella laevis (Family: Labiatae; common names: bells of Ireland, shell flower, molucella). Though its derivation is uncertain but reputedly its generic name derives from Molucca Island (Indonesia) as one species is from there, though in the genus there are only four half-hardy species from Mediterranean to NW India. *M. laevis* (Molucca balm) from Turkey to Syria is an annual growing more than 50 cm tall with stout stems, which bear light-green and long-stalked leaves some 5.0 cm long in opposite pairs which are boldly crenate and rounded to triangular-ovate. Flowers are white, 1.0 cm long, tubular and 2-lipped, calyces green, 2-3 cm wide and bowl-shaped (shell-like) in terminal spikes and these appear during winter and spring. It is propagated through seeds sown in September or during March on the hills at 13-15 °C under glass. It is suitable for planting in the borders, for cutting and for drying.

***Myosotis* spp.** (Family: Boraginaceae; common names: forget-me-not, scorpion grass). It derives from the Greek *mus* 'a mouse' and *ous* or *otos* 'ear', due to leaves looking-like mouse-ears. *Myosotis* is the ancient vernacular name for many plants with short, pointed and hairy leaves but Linnaeus adopted this as a genus only for this plant. It is a genus of more than 50 species of annuals and perennials from temperate parts of the world, particularly in Europe, Asia and Australia. All the species described below being alpine in nature can be propagated as annuals by sowing the seeds during autumn or during April in the cold frame in boxes or pans for use in rock gardens in the temperate areas.

M. alpestris (*M. rupicola, M. sylvatica alpestris*) from Europe is an alpine, short-lived, and clump-forming hardy perennial growing up to 15 cm tall with the tufts of green and hairy leaves, just above which dense clusters of tiny bright blue and fragrant flowers about 8 mm across appear with distinctive creamy-yellow eyes during spring to early summer in temperate areas. It prefers gritty soil and is most suitable for growing in rock gardens as a scree plant.

M. arvensis from Europe and Asia is an annual or biennial, growing from 20-30 cm tall with highly branched and hairy stems. Its leaves are stalkless, oblong and obtuse. Though it is quite similar to *M. alpestris* but on racemes the individual flowers are more widely spaced, the calyx are deeply divided in regular 5 lobes, and the flowers are blue or white which appear from March to July. It thrives well in poor soils, in sun or half-shade and is propagated through seeds sown during autumn.

M. caespitosa (*M. rehsteineri*) from the swamps and inundated places of Europe is a appressed-hairy annual to short-lived perennial having short and clustered rootstocks, stems erect some 12 cm tall, terete or with decurrent lines below the leaves, and the leaves are dark green, lanceolate and leathery. It is a subject of temperate regions and produces sprays of rounded bright blue flowers in summer.

M. sylvatica (*M. alpina, M. oblonga* or *M. oblongata*) from Europe including Great Britain is a hardy and compact biennial to perennial species though generally grown as an annual in the temperate regions, growing from 30 to 45 cm tall with mid-green, somewhat woolly-surfaced, soft and pointed leaves. It is suitable as a border plant, for naturalizing in the woodlands and for cut flower production. It is the major parent of most of the garden cultivars. Inflorescence comprises of tiny groups of small, scented and vivid blue flowers some 8 mm across which continue opening from spring to early summer. It can be grown in full sun or partial shade, even in dry soils though its varieties require moist and rich soils. A few of its choicest varieties are 'Alba' (20-25 cm tall, white), 'Blue Ball' (15-20 cm tall, indigo-blue), 'Blue Bird' (30 cm tall, deep blue), 'Rose Pink' (15-18 cm tall, pink), 'Ultramarine' (15 cm tall, flowers dark blue), *etc.*

Nemesia strumosa (Scrophulariaceae). *Nemesia* derives from the Greek vernacular name for a plant of similar appearance. It is a genus of 50 species of annual and herbaceous perennial plants and sub-shrubs. Their stems are 4-angled bearing pairs of lanceolate leaves and 2-lipped tubular flowers with a short spur or pouch, the lower lips may be entire or notched though upper one is divided into four. One among the annual species, *viz. N. strumosa* from South Africa is half-hardy highly ornamental annual, growing 30-60 cm tall with erect and branching stems, bearing pale-green leaves in oppsosite pairs, lanceolate, coarsely toothed and some 8 cm long. Its pouched but unspurred and shortly tubular and obliquely funnel-shaped flowers with pouched bases and some 2 cm across in shades of white, yellow or purple and often with spotted and bearded throats appear in terminal racemes during winter and spring. Its most common form in the garden is 'Suttoni' (20-30 cm tall in shades of white, yellow, orange, pink, red, blue and purple). Other important varieties are 'Blue Gem' (20 cm tall, flowers smaller and sky-blue), 'Carnival Mixed' (plants dwarf up to 23 cm tall and compact), 'Fire Ball', 'Hybrid Aurora', 'Hybrid Blue', 'Hybrid Mixed', 'Orange Prince', 'Sutton's Mixed' (30 cm tall, flowers large in wide range of colours), 'Triumph', *etc.* The species and its varieties are suitable for formal bedding and for mixed borders, as well as in pots. Its flowers are also useful for cutting. It is propagated through seeds sown in September.

***Nemophila* spp.** (Hydrophyllaceae; California bluebell). In Greek, *nemos* is for 'a wooded pasture or grove', and *phileo* 'to love', as some species grow in open pine forest. It is a genus of 13 annual species from North America. These are slender-stemmed decumbent annuals, bearing narrow, toothed or pinnately lobed leaves which may be alternate or opposite and cup-shaped or rotate, 5-petalled flowers in blue, purple or white and sometimes with clear spots that appear solitarily from the upper leaf axils on long stalks in spring. They are suitable for growing in front borders, in beds, for edging the paths and in pots. They are propagated through seeds sown in September or March.

N. aurita (*Pholistoma auritum*; Fiesta flower) from Pacific Coast of United States is a much-branched attractive annual climber with slender prickly stems growing up to 2.0 metres. Leaves are small, alternate, hairy, dentate and highly dissected into lanceolate segments. Clustered lavender-blue or violet flowers some 3.2 cm across appear in loose, open and terminal racemes during spring.

N. maculata (five spots) from Sierra Nevada Mountains in California is a dwarf (25 cm tall), branching and spreading plant with a little succulent and erect lateral growths. Leaves are light green, opposite and 5-9 lobed. Flowers are flattish bell-shaped (saucer-shaped), white but slightly veined purple and with a dark purple spot on each of the rounded petal-tips, so it is commonly named 'five spots'. Variety 'Albida' is white, 'Grandiflora' is large-flowered and 'Purpurea' bears purple petal-marbling.

N. menziesii (syn. *N. atomaria, N. insignis, N. modesta, N. pedunculata*; common names: baby blue eyes, California bluebell) from California and Oregon is a little (15-20 cm tall) annual most popular in gardens. The plant is quite dense and erect bearing feathery, light green though sometimes marbled white, opposite or alternate and pinnatifid with 6- 9 segments. Flowering stalk is axillary, longer than leaves, sky-blue flowers shading to white towards the centre, solitary, broad, 3.0-3.8 cm across, flattish and saucer-shaped with 5 slightly overlapping petals. This species has been botanically recognized with 19 different subspecies. Important varieties are 'Alba' (white), 'Argentea' (white, striped violet-mauve), 'Discoidalis' (large purple eye), 'Elegans' (white, centre chocolate), 'Marginata' (ligaht blue, edged white), 'Oculata' (white, centre purple), 'Purpurea-Rubra' (wine), 'Vittata' (almost black, petals edged white), *etc.*

Nicandra physaloides (Family: Solanaceae; common names: apple of Peru, shoofly plant). The genus *Nicandra* was erected for Nikander (about 150 A.D.), a poet of ancient Greek city, Colophon, in Turkey, who wrote about plants and their uses. The genus has only one unusual and attractive hardy annual species, *N. physaloides* fom Peru, suitable for border with attractive foliage, flowers and fruits. It is thought to repel flies. It grows about 90 cm tall with spreading branches, bearing mid-green, ovate, waving and toothed leaves. The pale-violet flowers with white throats and some 4 cm across, open only for a few hours daily at mid-day, and appear during summer to autumn on the hills and during spring to Summer in the plains if seeds are sown in September in the plains or in March-April under glass at a temperature of 15 °C. Its fruits are globular, 3.75-54.0 cm across and are enclosed by bright green and purple calyces, which with the branches can be removed for dry flower display or with flower arrangement.

***Nicotiana* spp.** (Solanaceae). *Nicotiana* was named for Jean Nicot (1530-1600), ambassador to Portugal who introduced tobacco to France. There in the genus are 66 species of half-hardy or tender annual and herbaceous perennial plants, and those described here are grown as annuals in borders or pots whose flowers open mostly by the evening and emit a heavy fragrance. These are propagated through seeds sown under glass in February or March at 18 °C temperature.

N. alata (*N. affinis*; flowering tobacco, nicotiana) from South America (SE Brazil, Uruguay, Paraguay and Argentina) is a sticky-hairy perennial though grown as an annual in its various hybrid forms. It grows up to 75-165 cm tall with much-branched stems, having stalkless, wide and large (10 cm long) basal leaves while those at stems are lanceolate. Floral stalks bear loose and open racemes of fragrant and tubular pinkish-white flowers with some 5 cm wide mouth. The hybrids can be white, cream, pink, crimson, lilac or greenish-white. The flowering is from spring to summer or even up to autumn. Its notable varieties are 'Crimson Bedder' (crimson), 'Dwarf White Bedder' (white, remains open during daytime), 'Lime Green' (greenish-yellow), 'Sensation Mixed' (mixed colours), *etc.*

N. suaveolens from Australia is an annual or biennial species with an erect, branched and shrubby habit growing up to 60 cm tall. Leaves are robust, sticky-hairy and lanceolate. Inflorescence is a terminal raceme bearing 5 cm long drooping and fragrant flowers with mouth being 2.5 cm wide, pale-green exterior and reddish-yellow interior. Its var. 'Variegata' has marbled foliage with white-edging, 'Macrantha' (syn. *N. fragrans*) with large white flowers, 'Undulata' with large undulated foliage, *etc.*

N. sylvestris from Argentina is a half-hardy biennial growing up to 1.5 m tall with stout and branched stems. The basal leaves are rosette-forming, mid-green, rough-surfaced and oblong while stem leaves are mid-green, stalkless, widely oval or lyre-shaped. It is day-flowering, and the flowers are fragrant white and borne in terminal corymbs some 8.75 cm long with 3.75 cm wide mouth.

Nigella damascena (Ranunculaceae). *Nigella* in Latin is diminutive form of *niger* 'black', as its seeds are black. There are some 20 species of annuals, out of which only two, *N. damascena*, a hardy annual from South Europe and *N. bispanica* from Spain are in general cultivation. These bear feathery foliage, saucer-shaped flowers and striking seed pods, and are excellent for borders and as cut flowers. These are propagated through seeds sown during autumn or in early spring for winter to spring flowering.

N. damascena (devil in a bush, fennel flower, Jack-in-the-green, Jack-in-prison, lady-in-the-bower, love-in-a-mist, nigella) grows erect up to 60 cm tall, bearing bright-green finely cut foliage. Beautiful blue or white flowers some 4 cm across from among the leafy crown of thread-like bracts are produced during winter to spring. Fruit, an inflated seed-pod, after ripening becomes pale-brown, 2.5 cm long with red vertical linings and similar leafy bracts which can be used as dry flower if harvested a little earlier. Its popular varieties are 'Altpreussen' (indigo-blue), 'Miss Jekyll' (large semi-double blue flowers), 'Oxford Blue' (blue), 'Persian Jewels Mixed' (flower colour mixtures of white, rose-pink, mauve, blue, *etc.*), *etc.*

N. bispanica grows to a height of up to 60 cm. The plants are vigorous with dark-green deeply divided leaves. Deep blue and faintly scented flowers some 6.25 cm across with a cluster of bright red stamens appear during winter and spring, followed by slightly inflated seed pods crowned with horn-like projections.

***Omphalodes* spp.** (Boraginaceae). In Greek, *omphalos* is for 'a navel', referring to the cupped seed (nutlet) which appears as a human navel. It is a genus of some 28 species of hardy annuals and herbaceous perennials from Europe, Asia and Mexico. *O. linifolia* (*Cynoglossum linifolia*; Venus's navelwort) from SW Europe is an erect annual growing up to 30 cm tall, bearing 1-5 cm long, glaucous and spathulate to lanceolate leaves. Its white or blue-tinted 5-petalled flowers being similar to *Myosotis* are 1.0 cm across and appear during winter to spring. It is propagated through seeds sown *in situ* during autumn or spring. It is suitable for beddings, borders and in the rock gardens.

Papaver spp. (Papaveraceae). *Papaver* in Latin vernacular is for 'poppy', probably for *P. somniferum*. More than 50 species in this genus are present but only five are in common cultivation, and these are *P. alpinum* (alpine poppy) from European mountains to North Arctic, *P. glaucum* (tulip poppy) from Syria, *P. nudicaule* (iceland poppy) from Arctic regions, south to USA and Asia, *P. rhoeas* (common corn poppy, field poppy) from Europe (Great Britain) and Asia and *P. somniferum* (opium poppy) from SE Europe and West Asia.

P. alpinum is a tufted perennial species though usually raised annually from seeds, growing from 10 to 25 cm tall with slender and leafless stalk having basal leaves which are grey-green, glaucous, hairy, and pinnate to bi- or tripinnatisect. Solitary white, yellow, orange or red flowers some 2.5-5.0 cm across appear during spring.

P. glaucum, a hardy species grows to a height of 45 cm with erect stalks. Leaves are grey-green, smooth, entire or lobed. Crimson-scarlet flowers are tulip-like and some 10 cm across and appear in spring.

P. nudicaule is a tufted half-hardy perennial though grown annually from seeds. It grows erect with 45-75 cm leafless stalks. Leaves are basally rosetted, soft green, 10-15 cm long, hairy and pinnatifid (deeply lobed). Fragrant white, yellow, orange or red flowers some 4.5-6.5 cm across appear during spring to summer, petals with tissue-paper texture. This is the only species which is suitable as cut flowers. There are forms with large double flowers. Its garden varieties are 'Kelmscott Strain' (a long-stemmed mixture of pink, salmon, apricot, orange, golden-yellow and scarlet), 'Champagne Bubbles' (F_1 hybrid with large flowers in various colours), *etc.*

P. rhoeas is a hardy and hairy annual some 60-90 cm tall having basal rosette of leaves which are pale-green and deeply lobed. Red flowers with black centre are 7.5-10.0 cm across and are produced from spring to summer. In gardens, it is represented by Shirley poppies, a mixed strain in shades of red, pink and white, often with picotee margins, and even the large double forms.

P. somniferum is a hardy, robust and glaucous annual growing up to 1 m tall. Leaves are pale grey-green, smooth, 7-12 cm long, stem-clasping, deeply lobed and ovate-oblong. The flowers in shades of white, pink, red and purple with blackish spots and some 10 cm across are produced during spring to summer, followed by capsules from which opium may be extracted. Its strains 'Paeony-Flowered Mixed' with fully double flowers and 'Pink Chiffon' with clear pink double flowers are quite popular.

Perilla frutescens (syn. *P. nankinensis*; family: Labiatae). Its exact derivation is unknown, probably it originates from the Latin *pera* meaning 'bag', alluding to the shape of the fruiting calyx. It is a genus of 5-6 species of annuals from India to Japan, out of which one (*P. frutescens* syn. *P. ocymoides* from China) is in general cultivation. It is a bushy and erect half-hardy annual growing up to 90 cm in height. Its stems are square and leaves undivided, in opposite pairs some 11 cm long, broadly ovate and pointed with marked veining, and deeply toothed. This species is grown for its red-purple ornamental leaves which when crushed emit a spicy smell. Insignificant white flowers some 6 mm across are small and tubular with 5 lobes and appear in July and August in Himalayas on a 10 cm long spike. Variety 'Atropurpurea' bears dark-purple leaves, 'Foliis Atropurpurea Laciniata' bears dark purple, crumpled and deeply cut leaves and 'Nankinensis' (*P. laciniata*) has deeply dissected bronze to purple foliage. It is propagated through seeds sown in February or March at a temperature of 18 °C.

Pericallis* × *hybrida (syn. *Cineraria cruentus*; Cineraria; Asteraceae). A native to Canary Islands, this annual grows up to 60 cm tall, bearing rounded-cordate, downy beneath, 10-20 cm long with shallow lobes. Flowerheads are with colour range of pink, red, purple, blue and bicolours even in the white, 3-8 cm long. Its Multiflora strain grows up to 50 cm high with dense rounded heads of flowers; Multiflora Nana (Grandiflora Nana) grows compact up to 30 cm high with dense rounded heads; and Stellata grows up to 75 cm tall bearing starry flowers in loose clusters.

Petunia* × *hybrida (Solanaceae; petunia). In Latin, *petun* in Brazil is for 'tobacco', and the genus name was coined by Antoine Laurent Jussieu due to similarily of the plant to tobacco. The genus comprises 40 species of tufted annuals or herbaceous perennials with stickily glandular-hairy native to the eastern part of South America. The petunias which we grow in the garden today are perennials though grown as half-hardy annual as it flowers the same season through seeds. They are suitable for bedding, as pot plants, for hanging and window boxes and for edging beds. Only three are of any horticultural interest, *P. axillaris, P. inflata* and *P. violacea*, as because of the crossing of these three species together we find the modern *Petunia* cultivars in the garden which we now call *Petunia* × *hybrida*.

P. axillaris (*P. nyctaginiflora*) from Argentina grows up to 60 cm tall with highly fragrant flowers. The stems are erect, leaves large and oval covered with sticky hairs. Strongly night-scented flowers are dull white some 3.75-5.0 cm across. It is one of the parents of modern garden petunia hybrids.

P. inflata from Argentina and Paraguay grows up to 45 cm tall and appears similar to *P. violacea* but the leaves are smaller and more linear and corolla tube is inflated and swollen. This is also one of the parents in the development of the modern garden petunia hybrids.

P. violacea (*P. integrifolia*) from Argentina grows up to 30 cm tall with slender, prostrate and sticky stems. The leaves are short-stalked, oval and sticky. Flowers when massed are very attractive, are 3.75 cm long, pinkish-red or reddish-violet. This is one of the parents for developing modern garden cultivars.

P. × *hybrida* is of garden origin which contains a group of perennial cultivars developed by crossing among *P. axillaris, P. inflata* and *P. violacea*. Most cultivars appear true to type when raised from seed. Since varieties vary in size (plants compact to elongated from 12 to 40 cm tall and flowers 5 cm to 13 cm across), flower form (single, double, frilled to heavily crested) and colour (white, yellow, red, purple, bi- and multi-coloured), these are grouped into 'Multiflora' which has 15-30 cm tall and bushy plants with single or double flowers some 5 cm across and the varieties being 'Apple Blossom' (F_1 hybrid, pale-pink), 'Brass Band' (F_1 hybrid, deep cream), 'Cherry Tart' (double F_1 hybrid, cherry pink and white, and carnation-like flowers), 'Dream Girl' (deep rose to pink), 'Plum Blue' (soft clear blue), 'Plum Crazy' (F_1 hybrid, pink, lavender, purple and yellow mixture), 'Polaris' (F_1 hybrid, deep blue with white star), 'Red Satin' (scarlet), 'Snowdrift' (pure white), 'Sugar Plum' (F_1 hybrid, orchid-pink with mauve veining), 'Summer Sun' (yellow), *etc.*; 'Grandiflora' with 15-30 cm tall bushy plants with fewer but larger single or double flowers some 10 cm across and the varieties being 'Cascade' (F_1 hybrid, mixture of red, pink, white and blue), 'Fluffy Ruffles' (ruffled and frilled in shades of crimson, pink, lavender, purple and white, and veined in darker shade), 'Happiness' (rose-pink), 'Mariner' (deep blue), 'Miss Blanche' (white), 'Pan American All Double Mixed' (flowers ruffled), 'Polynesia' (coral-salmon), 'Superbissima Mixed' (petals ruffled and throat veined), *etc.*; and 'Nana Compacta' with low-growing plants (not exceeding 15 cm in height) though with large flowers often with waved petals and most suitable for pot cultivation and the varieties being 'Alderman' (indigo-violet), 'Blue Bedder' (blue), 'Dwarf Resisto' (blue, red and pink), 'Dwarf Giants of California' (mixed colours), 'Fire Chief' (scarlet), 'Rose of Heaven' (pink), *etc.*; and 'Pendula' with low-growing plants having trailing habit so are most suitable for window and hanging baskets and the varieties being 'Avalanche' (mixed colours), 'Balcony Blended Mixed' (very popular strain with mixed colours), *etc.*

Phacelia spp. (Hydrophyllaceae). In Greek, *phaselos* is for 'bundle or bunch' as the first described species had bunched flowers. The genus comprises some 200 species of mostly annuals or a few herbaceous perennials native to the SW USA, particularly Texas, New Mexico and California. Though these plants as such are not so attractive but their blue flowers are much appreciated. They are tufted, erect to decumbent and the leaves are alternate, linear to ovate and sometimes 5-lobed. The flowers are tubular and rotate in scorpiod cymes often in panicles. These are suitable for borders, for edging beds and for pot-growing. These are propagated through seeds sown in September or in the temperate areas in March and April.

P. campanularia from California is a hardy and dwarf bushy annual which grows up to 25 cm tall with basally-branched erect stems which are reddish-brown, bearing small, dark green tinted reddish, foetid on crushing, ovate and irregularly toothed leaves. The flowers are campanulate (bell-shaped), upturned, 2.5 cm across and brilliant gentian-blue which appear in spring in the plants but during summer and autumn in temperate areas.

P. tanacetifolia (*P. tripinnata*) from California is a hardy annual which grows with erect stems to 60 cm or more and where whole plant is hairy. The leaves on the stems are alternate, dark green, finely dissected and lobed. The flowers in crowded spikes are lavender and bell-shaped and most liked by the bees, some 8 mm long and appear in July on the hills though during late-spring in the plains.

P. viscida (*Eutoca viscida*) from California is a hardy annual of upright habit, growing over 60 cm tall, bearing pale-green, almost heart-shaped to oval, acuminate and dentate leaves covered with viscous hairs, and the leaves on crushing emit formalin-smell. Spikes are 10-20 cm long, bearing some 2.5 cm across bell-shaped vivid-blue flowers with white centre which is speckled blue and the stamens are contrasting white. This also has a white form known as 'Alba'.

P. whitlavia (*P. grandiflora, Whitlavia grandiflora*) from southern California is a much-branched hardy annual growing up to 45 cm tall with erect stems, bearing small, oval, dentate and acuminate leaves. Purple or purple-blue bell-shaped racemose flowers some 2.5 cm long with spreading mouth appear during rains in the hills and in Spring in the plains. The species has many beautiful varieties such as 'Alba' (white), 'Gloxinioides' (white with blue centre), *etc.*

Phlox drummondii (Polemoniaceae; phlox). *Phlox* is a Greek name originally coined by Theophrastus for another flame-coloured flower as then probably he did not know the present-day phlox, and now this name applies to the present-day phlox, used by Linnaeus. It is a genus of some 66 species of hardy herbaceous perennials, half-hardy and hardy sub-shrubs and annuals. The annuals are suitable for borders, beddings, pot plants and for cutting. The flowers appearing in dense clusters are usually salver-shaped, sometimes the petals are cleft or notched and in colours of pink, red, purple, lavender and white. Among the annual species, it is only *P. drummondii* from Texas which is in general cultivation. It is a half-hardy clump-forming annual with erect to straggling stems which are slightly hairy and grow 30-50 cm tall, leaves are light-green, sessile but partially stem-clasping, and lanceolate (linear to ovate), and some 7 cm long in opposite pairs. Flowers in colour from purple-pink to pinkish-red are borne in dense heads some 7.5 cm across, individual flowers being 2.5 cm across, 5-lobed where petals are widely ovate, tubular and in terminal cymes or panicles. It can be propagated by sowing of seeds in September for winter to spring flowering but for late-flowering in March or april under glass at 13-16 °C temperature. The species has many cultivars which are grouped under three different heads: (i) 'Grandiflora' with large flowers (30 cm tall and available in mixture of colours such as cream, rose, red, crimson and violet), (ii) 'Cuspidata, Star or Stellaris' having star-shaped flowers with slender-pointed petals and sometimes with fringes, such as the strains 'Twinkle' (15 cm tall) which is compact with variously coloured flowers, and (iii) 'Nana Compacta' which includes dwarf and compact types growing up to 23 cm tall and these are available in mixtures and in separate colours. Its various named varieties are 'Alba', 'Art Shades', 'Atropurpurea', 'Brilliant', 'Coccinea', 'Fordhook Mixed', 'Giant Tetra Mixed', 'Glamour', 'Globe Beauty', 'Globe Mixed', 'Snowball', 'Tetra Red', 'Twinkle', 'Vermilion', 'Violacea', *etc.*

Pimpinella monoica (Family: Umbelliferae; common names: lady's lace, pimpinella). It takes its name probably from Latin *bipinnula* meaning 'bipinnate', alluding to its leaf structure. Though there are some 75 species of perennial herbs native to India but only *P. monoica* is in general cultivation as an annual for winter-flowering though season can be extended even up to autumn, growing up to 1.5 m tall and is suitable for planting in the borders, in the beds and for cutting. Its leaves are deeply dissected, the tiny flowers are white and formed in a mass of terminal umbels. Its another form *P. major* var. *rosea* bears light or pale-pink flowers though not much in cultivation. It is propagaterd through seeds sown in September and the seedlings are transplanted when some 4 cm tall.

Platystemon californicus (Papaveraceae; cream cups). In Greek, *platys* is for 'a broad' and *stemon* for 'stamen', alluding to the extended stamen filaments. It is a genus of one hardy annual species, *P. californicus* from California, which branches at base and is suitable for edging borders and in the drifts of a rock garden. It is a compact plant growing up to 30 cm tall, lower stems decumbent, bearing low-clustered, grey-green, hairy, 2-7 cm long and narrowly oblong (linear) leaves and a mass of cream or pale-yellow saucer-shaped and 6-petalled flowers some 2.5 cm across from upper leaf axils in July under temperate conditions. It is propagated through seeds sown either in September for early spring flowering or in March-April for July flowering.

Portulaca grandiflora (Family: Portulacaceae; common namaes: rose moss, sun moss, sun plant wax pink). *Portulaca* is the Latin vernacular name for purselane. The genus comprises some 200 species of annuals and perennials from tropics and warm temperate regions, particularly the Americas. They are dwarf and often with prostrate spreading habit where leaves and stems both are usually succulent. Their leaves are bright green, alternate or almost opposite and ovate to linear. These bloom throughout the year except in very cold climates. The flowers appear in profusion, are saucer-shaped, and 2.5 cm across in the true species with a central boss of bright yellow stamens. The flowers open only in clear weathers and close down in dull weathers and by evening. The species in general cultivation is *P. grandiflora* from Argentina and Brazil which is though perennial but cultivated as annual through seeds or through cuttings. It is tufted annual-type perennial growing up to 25 cm tall with a spread of 40 cm. Its flowers are single orange or purple though its varieties have shades of white, yellow, orange, apricot, pink, salmon, scarlet, red and purple, *vis-à-vis* mottled or striped, single or double and in various sizes including the giants. Though there are numerous named varieties but it is not easy to get seeds which may breed true as varieties are highly intercrossable so to maintain the character of a variety, these should be propagated only through cuttings of the new growth. The named varieties are 'Albiflora' (pure white), 'Bedmannii' (white striped purple), 'Caryophyllus' (red striped white), 'Double Mixed' (various colours but flowers double), 'Splendens' (purple-red), 'Sulphurea' (deep yellow), 'Tellusonii' (scarlet), 'Thorburnii' (deep yellow), *etc.*

***Quamoclit* spp.** (Family: Convolvulaceae)**.** In Greek, *quamoclit* stands for 'a dwarf kidney bean'. *Quamoclit coccinea* (*Ipomoea coccinea*; red morning glory, star ipomoea) from SE USA is a climbing annual growing up to 3 m long. This bears arrow to heart-shaped long-pointed leaves some 15 cm wide on slender stalks, often with toothed margins. Fragrant, tubular and bright scarlet flowers with yellow throat, 2 or more together on long stalks and some 4 cm wide appear during summer to autumn. The seed capsules are reflexed. *Q. pinnata* (syn. *Ipomoea pinnata, I. quamoclit*; common names: cypress vine, Indian pink) from Tropical America. It is a half-hardy climbing annual from Tropical America which grows up to 1.8 m long. It bears bright green, ovate and pinnately cut leaves, cuts many up to the midrib and in the form of well-spaced linear-ferny lobes which make the leaves in palmate-shape. Solitary or paired, scarlet or orange, 2.5 cm long and wide and tubular flowers to most of their length are formed on long peduncles and then expand at the top into 5 ovate lobes. Flowering occurs from summer to autumn. These make fine pot plants, and can also be trained on arcs and walls. These are propagated through seeds sown in March.

Reseda odorata (Resedaceae; mignonette)**.** In Latin, *resedo* is 'to heal', as in the past certain species were used to treat bruises. A genus of 60 species of annuals, biennials and perennials from the Mediterranean to Central Asia, and where only one, *Reseda odorata* from North Africa is in general cultivation. It is a short-lived hardy perennial or annual grows up to 38 cm tall with branching at the base. The leaves are mid-green, smooth, alternate, oblong, entire or 3-lobed and some 7.5 cm long. Yellowish-white strongly sweet-scented flowers some 6 mm across with many orange-yellow stamens are borne during summer to autumn on terminal racemes with 6-lobed petals. It is suitable for planting in borders, for cutting and also for late-winter flowering. It is propagated through seeds in March, April or September on the hills at the site of flowering or under glass at a temperature of 13 °C. Its varieties are 'Bismark' (red), 'Cloth of Gold' (yellow), 'Crimson Giant' (red), 'Golden Queen' (yellow), 'Goliath' (large spikes of red flowers), 'Golden Goliath' (golden-yellow), 'Machet' (red-tinged flowers), 'Parson's White' (white), *etc.*

Rudbeckia bicolor (Family: Asteraceae; common names: coneflower, gloriosa daisy)**.** The genus *Rudbeckia* was erected for Olof Rudbeck (1630-1702), a Swedish anatomist, botanist and antiquarian, and his son of the same name (1660-1740), both professors at Uppsala University. The son befriended Linnaeus who later named the genus in their honour. It is a genus of 25 species of hardy annuals and herbaceous perennials. Among the annuals, only *R. bicolor* from Texas is in general cultivation which is suitable for planting in the borders and for cutting. It is an upright branching annual where stems and leaves are bristly. It grows some 60 cm tall, bearing mid-green and oblong leaves. The flowers are yellow with a black central cone and some 5 cm across appear from July to August in the temperate areas. It can't tolerate harsh summer temperatures in the plains. The varieties are numerous such as 'Autumn Frost', 'Double Daisy', 'Giant Gloriosa Daisy', 'Golden Daisy', 'Goldflame', 'Kelvedon Star' (flowers gold-yellow with mahogany zoning, cone dark), 'Mein Freund', 'Pink Wheel', 'Starlight', 'Superba' (maroon on petal back), *etc.* These are propagated through seeds sown in pots in March and April.

Salpiglossis sinuata (Solanaceae; painted tongue)**.** *Salpiglossis* derives from the Greek word *salpinx* 'a trumpet' and *glõssa* 'a tongue', alluding to the form of corolla and style. *S. sinuata* from Chile is a beautiful tender annual with elongated sticky foliage and tapering funnel-shaped flowers suitable for planting in the borders and in the pots. Though flowers are sticky but with care can be handled even as cut flowers. It grows up to 60 cm tall with branched stems that carry light green, alternate, broadly lanceolate and wavy-edged leaves. The funnel-shaped flowers are produced from the leaf axils in terminal clusters some 5 cm long and across at mouth, and are multi-coloured in shades of yellow, orange, lavender, crimson and scarlet, petals are velvety in texture and the flowers are veined in deeper or contrasting colours and are produced over a long period in spring in the plains while from July to September in temperate areas. It is propagated from seeds sown under glass during February at 18 °C temperature, or in September. Its most popular varieties include 'Grandiflora' (60 cm tall, and flowers large in compact heads with wide range of colours), 'Grandiflora Nana' (40 cm tall, flowers in mixed colours), 'Splash' (F_1 hybrid in mixed colours), 'Superbissima' (syn. 'Imperiale'; 75 cm tall, flowers very large which appear solitary).

Salvia spp. (Labiatae)**.** In Latin, *salvus* means 'safe or well', because of the reputed medicinal properties of several species. There are some 750 hardy, half-hardy and tender species of annuals, biennials, perennials and mainly evergreen sub-shrubs from world-wide distribution. The annual species are generally half-hardy and are used in mixed borders, in beddings and for drying. Their flowers are tubular and 2-lipped, mostly hooded and appear in terminal clusters.

S. horminum from South Europe is a true annual species growing up to 45 cm tall with erect and branched stems, bearing mid-green and ovate leaves. It produces some 1.25 cm long and insignificant pale-pink or pale-purple flowers during winter and spring. The terminal tufts of the coloured bracts are some 3.75 cm long in various colours which make attraction of the flowers. The varieties are 'Blue Beard' (bracts blue-purple), 'Monarch Bouquet' (mixture containing various colours such as white, rose, red, blue and purple), 'Oxford Blue' (bracts blue), 'Pink Sundae' (bracts rose-carmine and red), *etc.*

S, patens from Mexico is a perennial species but grown as an annual. It grows erect with slender and branched stem to a height of 60 cm. The leaves are mid-green, ovate and pointed and the flowers are clear blue, some 5 cm long and distantly spaced on the slender stems.

S. splendens (scarlet sage) from Brazil is though half-hardy perennial but grown as annual up to 38 cm tall, bearing bright green, 9 cm long, ovate, acuminate and serrated leaves. Its scarlet flowers some 3.75-5.0 cm long are surrounded by same colour of bracts some 2.5 cm long and the flowers arise in densely packed terminal spikes. A few of its varieties are 'America', 'Blaze of Fire' (25 cm tall, flowers scarlet), 'Carabiniere Scarlet', 'Early Bonfire', 'Fireball' (38 cm tall, velvety-scarlet), 'Fire Dwarf', 'Gaiety', 'Harbinger', 'Hot Jazz', 'Hussar', 'Panorama', 'Pirate', 'Purple Blaze' (30 cm tall, violet-purple to slate-colour), 'Red Pillar', 'Salmon Pygmy' (23 cm tall, salmon-pink), 'Scarlet Picola', 'Scarlet Pygmy', 'Scarlet Queen', 'St. John's Fire', 'Sutton's Fireball', 'Violet Queen', *etc.*

Sanvitalia procumbens (Asteraceae; creeping zinnia). The genus *Sanvitalia* was erected for Prof. Fedserico Sanvitali (1704-1761). It is a genus of 7 hardy species of annuals, perennials and shrubs from SW USA, Mexico, C. America and Bolivia to Argentina. Only one annual species, *S. procumbens* native of Mexico is in general cultivation as border plants, in rock gardens, as pot plants and in the hanging baskets. It grows to 15 cm tall, has prostrate habit with trailing stems. Its leaves are mid-green, ovate and pointed. The flowers are daisy-like, bright yellow with black disc and 2.5 cm across which appear in winter and spring. There are forms ('Flore Pleno') even with double flowers. It is propagated through seeds sown in September at their flowering places or during March-April in temperate areas in glasshouse at 13-16 °C temperature.

***Saponaria* spp.** (Caryophyllaceae). In Latin, *sapo* is for 'soap' as once the roots and leaves of *S. officinalis* were used as a soap substitute. It is a genus of some 30 hardy to half-hardy annual, biennial and perennial species from Europe (mainly the Mediterranean) and Asia. They are tufted and erect with terminal cymes of salver-shaped 5-petalled flowers where calyces are tubular. Each flower in its throat has a ring of 5 scales forming a small corona. They are excellent in the borders, for rock gardens, edging paths and as cut flowers or used as filler. The annual species are propagated through seeds sown in September for winter and spring flowering.

S. calabrica from Italy is a hardy annual with compact growth and growing up to 15 cm tall. The stem clasping leaves are light green, smooth and narrowly lanceolate. Flowers are deep rose, some 1.2 cm across and tubular and are produced in profusion during winter to spring.

S. vaccaria (*Vaccaria pyramidata, V. segetalis*) from Europe is a hardy annual growing up to 75 cm tall with smooth branches. Quite smooth leaves are light green, narrow and lanceolate. Flowers are star-like, deep pink, some 1.25 cm across and are borne in large sprays. It has a number of varieties including 'Alba' and 'Pink Beauty'.

Scabiosa atropurpurea (Family: Dipsacaceae; common names: pincushion flower, scabious). In Latin, *scabies* is for 'a disease' which certain members of the genus were once supposed to cure. The genus comprises some 100 annual and herbaceous perennial species which are tufted to clump-forming with opposite pairs of entire or pinnatifid leaves and daisy-like cushion-forming flowerheads where each floret is 4-5-lobed and tubular to funnel-shaped. Outer florets are much longer. *S. atropurpurea* being described here is suitable for bedding, in borders and for cutting, and the dried seedheads in flower arrangements. It can be propagated by sowing the seeds in September in the plains for flowering during winter and spring or in March-April only in the temperate areas where these flower during summer. *S. atropurpurea* (sweet scabious) from South Europe to Turkey is a hardy annual to biennial and short-lived perennial, growing up to 90 cm tall with branched stems, having a compact rosette of mid-green, narrow and dissected leaves. It is grown as an annual. Stems are slender and bear dark crimson flowers some 5 cm across from July to September in temperate areas though during winter to spring in the plains. The varieties are in shades of white, salmon, lavender, pink, cherry red, blue and dark purple. Most popular varieties are 'Black Night', 'Blue Moon', 'Bridesmain', 'Cockade Mixed', 'Coral Moon', 'Giant Imperial Hybrids', 'Heavenly Blue', 'Loveliness', 'Oxford Blue', 'Peace', 'Silver Moon', 'Tom Thumb', *etc.*

Schizanthus pinnatus (syn. *S. grandiflorus, S. hybridus grandiflorus*; family: Solanaceae; common names: butterfly flower, poor man's orchid). In Greek, *schizo* means 'to divide or split' and *anthos* for 'a flower', as the corolla is deeply cleft. These are half-hardy annuals consisting of some 15 species, suitable for annual borders, as pot plants and for cutting from winter to spring. These are propagated through seeds sown during autumn and late winter at 16 °C temperature. Only one species is in general cultivation. *S. pinnatus*, a bushy plant from Chile grows from 45 cm to 1.2 m tall with erect stems. The leaves are pale-green, deeply divided (pinnate to bipinnate) and ferny. Showy orchid-like single flowers some 2.5-4.0 cm wide in shades of rose, purple, yellow and white are produced during winter to spring in the plains and from June to October in the temperate regions. The flowers are tubular, two lips broadly lobed and spreading, the upper with two lobes and the colour contrasting at the centre (often yellow) with purple spots and streaking, and the lower with three lobes. A garden hybrid, *S.* × *wisetonensis* evolved by crossing *S. pinnatus* and *S. retusus grahamii* produces many coloured strains in the garden, some dwarf ones quite suitable for pot planting. 'Dwarf Bouquet' is a compact plant growing up to 30 cm tall and producing flowers in shades of amber to soft pink, rose, salmon and crimson. 'Pansy Flowered' is a self-coloured strain growing up to 45 cm tall and producing flowers in a variety of colours. Other popular varieties are 'Angel Wings', 'Brilliance', 'Butterfly Giants', 'Cattleya Orchid', 'Crimson Cardinal', 'Dr. Badger's Mixed', 'Dwarf Banquet', 'Excelsior', 'Giant Hybrids', 'Monarch', *etc. S. r. grahami* is now not found in cultivation.

Sedum caeruleum (Crassulaceae). In Latin, *sedo* is 'to sit', alluding to the way some species grow on walls and rocks. It is a large genus of 600 species of succulent perennials and sub-shrubs, including a few annuals, biennials and monocarps. *S. caeruleum* from islands of Western Mediterranean (Southern Europe) is a branched annual growing to 10-15 cm tall and wide, and is suitable for growing in borders or on rock gardens and dry walls. It is propagated by sowing its seeds *in situ* in March and April. The leaves of this species are pale-green but reddish in dry sunny sites or when flowering commences and then even stems turn red, minute and 1-2 cm long, fleshy, cylindrical, ovoid to oblongoid and forms a dense covering over the soil. Sky-blue flowers that are white at base in the centre are 0.6 cm across and 7-9 petalled, and appear during July in the temperate regions while during spring to summer in the plains.

Senecio spp. (Asteraceae). In Latin, *senex* is for 'an old man', an oblique reference to the white and hoary pappus. This genus comprises some 3,000 species but annuals are only a few such as *S. arenarius* (cineraria), *S. cruentus* (*Cineraria cruenta*), *S. elegans, S. maritima* (*S. cineraria, Cineraria maritima*). All the species bear alternate leaves and daisy-like flowers. These are tufted pot plants and are propagated from seeds.

S. arenarius from South Africa grows to 60 cm tall bearing mid-green oblong leaves often covered with sticky hairs. Pale-purple flowers some 2 cm across with yellow centre appear in loose clusters during summer.

S. cruentus from Canary Islands is a half-hardy perennial but grown as annual. It grows to 45 cm tall bearing mid- to dark green ovate or palmate leaves toothed marginally. White, pink, rose, violet and in various other colours some 2.0-7.5 cm across flowers are produced in compact masses during winter and spring. The hybrids include 'Hybrida Grandiflora' with varieties such as 'Brilliant', 'Exhibition Mixed' and 'Monarch'; 'Double Flowered' with varieties as 'Double Duplex'and 'Gubler Mixed'; 'Multiflora Nana' with 30-40 cm height such as 'Berlin Market', 'Dwarf Large Flowered Mixed', 'Gaytime Mixture', 'Gem Mixed'; and 'Stellata' growing 45-75 cm tall with varieties as 'Feltham Beauty', 'Mixed Star', *etc.*

S. elegans from South Africa grows up to 45 cm tall, bearing dark green, oblong-ovate, usually deeply pinnately lobed and purple flowers some 2.5 cm across that appear during spring. Its varieties being single or double in shades of white, pink, lavender and mauve.

S. maritima (*S. cineraria, Cineraria maritima*) from Mediterranean regions is a hardy evergreen perennial though treated as a half-hardy annual, and grows to a height of 60 cm with entire plant (stem and leaves) white-felted giving showy silvery appearance. Leaves are oblong-ovate and deeply lobed. Yellow flowers some 2.5 cm wide are borne during late spring though foliage being more attractive than the flowers. Its strains are 'Candicans' with deeply lobed leaves, 'Diamond' with almost white and deeply dissected leaves and 'Silver Dust' which is densely white-felted with dissected ferny leaves.

Silene spp. (Caryophyllaceae; common names: campion, catchfly). The genus name is an old Greek name for 'sticky or German catchfly'. Though there are some 500 species under this genus but the cultivated annuals, or perennials treated as annuals, are only a few. They are tufted to clump-forming, prostrate to erect with opposite pairs of narrow leaves and terminal cymes of flowers. Flowers are 5-petalled and their calyces are tubular to inflated. They are suitable for borders and bedding out. These are propagated through the seeds sown during autumn and in late winter *in situ*. In greenhouses, these require 15 °C temperature until germination.

S. armeria is a hardy species from South Europe growing erect about 30-60 cm tall, carrying elliptic grey-green leaves and where upper part is sticky. Purple rose or white flowers are borne some 2 cm across in dense clusters of single flowers during spring.

S. coeli-rosa (*Lychnis coeli-rosa, Viscaria elegans*; rose of heaven, viscaria) from Mediterranean areas is a hardy annual, growing 45 cm tall, bearing grey-green and oblong leaves, and rose-purple flowers having white centre and some 2.5 cm across during spring to early summer. Its noteworthy varieties are 'Candida' (white), 'Cardinalis' (crimson), 'Blue Pearl' (lavender-blue), 'Fire King' (scarlet), 'Rose Beauty' (deep pink), 'Oculata Nana Compacta' strain (seldom exceeding 15 cm), 'Loyalty' (blue), 'Love' (rose-carmine), *etc.*

S. compacta (*S. orientalis*) from Asia Minor is a hardy biennial though grown as an annual, growing up to 45 cm tall with slender stems and grey-green, smooth, elliptic and pointed leaves. Flowers are bright pink some 2.5 cm across and appear in losse clusters during spring.

S. pendula from Mediterranean regions is a hardy and most popular annual, growing 15-23 cm tall with erect stems. The leaves on the stems are mid-green, hairy, oblong and pointed, and pale-pink axillary flowers some 1.25 cm across appear in loose clusters during spring. Its 'Compacta Mixed' strain grows only 15 cm high bearing salmon-pink or crimson double flowers and dark green foliage which is suitable for edging paths and rock gardens, and 'Triumph Mixed' also dwarf is available in shades of orange, pink and salmon.

Silybum marianum (Family: Asteraceae; common names: blessed thistle, holy thistle, milk thistle, our lady's thistle, St. Mary's thistle). In Greek, *silybon* refers to 'a thistle-like plant'. A genus of two biennial species from the Mediterranean, though both are grown as hardy annuals through seeds, however, only one is under general cultivation. It is a thistle-like plant growing from 60-120 cm tall with robust stems which may be simple or lax-branched. Leaves are basal in flat rosette, glossy dark green, 30-60 cm long, ovate, lobes sinuate to pinnatifid with spiny margins, and marbled with white veins. Flowerheads erect or nodding, some 5 cm wide, flowers are red-purple, tubular and faintly fragrant. The involucral bracts are spine-tipped. This species has more attraction in its foliage than its flowers. Flowers appear during winter and spring. This is most suitable plant for the wild garden and for the back of the border.

***Tagetes* spp.** (Asteraceae). *Tagetes* was probably named for the Etruscan god Tages. It is a genus of some 50 annuals and a few perennial species from southern USA to Argentina. Those under general cultivation are annuals such as *T. erecta, T. patula* and *T. signata*. These all grow erect with free-branching habit bearing highly foetid and pinnately lobed leaves, and solitary or clustered flowerheads which are terminal and have ray florets and discs. They are suitable for bedding, in the borders, for pot cultivation and the dwarf ones for edging, however, the flowers serve most of the loose flower needs apart from being the good cut flowers. These are propagated through seeds and also cuttings.

T. erecta (African marigold) from Mexico to Central America is a half-hardy strong-growing annual growing up to 1 m high with glossy dark green and pinnatifid leaves. The flowerheads are yellow to orange, 5-13 cm across, broad-petalled single but cultivated forms mostly semi- to fully double and in various shapes such as crimped, quilled or 2-lobed florets. The species and its forms are predominantly winter- to early summer-flowering, however, there are a few varieties which produce flowers even during summer. Its varieties are innumerable and some of most promising are 'Burpee's First White', 'Burpee's Giant Fluffy', 'Crackerjack' (mixture of colours) growing up to 75 cm tall, 'Cupid', 'Golden Age', 'Guys and Dolls', 'Mr. Moonlight', 'Odourless', 'Orange Hawaii', 'Pineapple Crush Improved', 'Snowbird', 'Spun Gold' (30 cm high, chrysanthemum-flowered type, flowers 7.5 cm across), 'Spun Yellow', 'Sugar and Spice', 'Sunset Giant', 'Sutton's Double Orange', 'Sutton's Double Yellow', 'Sweet 'n' Gold', 'Sweet 'n' Yellow', 'Yellow Fluffy', 'Yellow Stone', *etc.* 'F_1 hybrids' such as 'Climax Series' (fully double flowers in separate and mixed colours), under which the 'Yellow Climax' (best exhibition variety of the series including the primrose, golden-yellow and orange) is outstanding; 'Gold Coin Series' (60 cm tall, flowers 15 cm across), the best one being 'Double Eagle' (golden-orange with frilled and ruffled petals); dwarf ones such as 'First Lady' (F_1, a profusion of primrose-yellow flowers some 9.75 cm across), *etc.* Certain other F_1 hybrids are 'Apollo Moonshot', 'Beauty Gold', 'Beauty Orange', 'Beauty Yellow', 'Deep Orange Lady', 'Diamond Jubilee', 'First Lady', 'Gold Lady', 'Golden Climax Improved', 'Golden Jubilee', 'Gold Galore', 'Inca Gold', 'Inca Orange', 'Inca Yellow', 'Primrose Lady', 'Royal Gold', 'Royal Orange', 'Royal Yellow', 'Toreador', 'Tresbien Orange', 'Tresbien Yellow', 'Yellow Climax', 'Yellow Galore', *etc.*

T. patula (French marigold) from Mexico and Guatemala is a half-hardy bushy annual which can grow from 15 cm to 45 cm high, the dwarf types being most compact and floriferous. The leaves are dark green, pinnatifid, and some 5 cm across flowerheads are yellow to orange marked with darker orange to brown. It has so many varieties and is easily crossable with *T. erecta*. Its many of the varieties and forms flower even during summer months. Its choicest varieties are 'Bolero', 'Bonanza Flame', 'Bonanza Orange', 'Boy Scout',

'Butterscotch' (up to 25 cm high, single mahogany flowers with golden-crested centres), 'Carmen', 'Dainty Marietta', 'Fiesta', 'Giant Crested', 'Golden Boy', 'Gold Finch', 'Goldie', 'Harmony' (22.5 cm high, flowers single with mahogany collar and disc golden), 'Harmony Boy', 'Happy Orange', 'Happy Yellow', 'Honeycomb', 'Honey Sophia', 'King Tut', 'Lemon Drop', 'Monarch Mixed' (22.5 cm tall, flowers double ball-shaped and in shades of yellow to deep mahogany), 'Naughty Marietta' (30 cm tall, flowers rich golden with maroon blotch), 'Orange Boy', 'Orange Sophia', 'Petite' (a strain growing 15 cm tall, flowers in singles and doubles in full colour range), 'Petite Gold', 'Petite Spry', 'Petite Yellow', 'Queen Sophia', 'Red Cherry', 'Red Glow', 'Red Pygmy', 'Scarlet Sophia', 'Spanish Bricade' (30 cm high, mahogany and gold double flowers), 'Spry Boy', 'Star Dust', 'Sunbeam' (15.0-22.5 cm tall, flowers single yellow with crested centre), 'Sparky', 'Yellow Boy', 'Yellow Nugget', *etc.* 'Pusa Narangi Gainda' and 'Pusa Basanti Gainda' have been developed in IARI, New Delhi.

T. signata (*T. tenuifolia*; signet marigold) from Mexico and Central America is a half-hardy slender and branched annual, growing up to 60 cm with light green, pinnatifid and sweet-smelling leaves. It is free-flowering with 2.5 cm across and bright yellow flowerheads. Its varieties 'Golden Gem' (20 cm tall, flowers bright orange with deeper orange marks at the bases of ray florets) and 'Lemon Gem' (22.5 cm tall, flowers clear lemon-yellow) are very popular.

Thunbergia alata (Family: Acanthaceae; common names: black-eyed susan, clock vine). *Thunbergia* was named for Carl Peter Thunberg (1743-1828), a Dutch professor of botany at Uppsala University, and doctor who worked in South Africa and Japan. It is a genus of some 200 species of annuals and perennials, many of which have climbing nature. The leaves are entire in opposite pairs and irregularly 5-lobed. Flowers may be solitary from the leaf axils or in racemes. Those growing below up to 3 m tall make good pot plants when its branches are trained against a globular structure.

T. alata is though an herbaceous perennial climber but is generally grown as annual. It is native to SE Africa and was first introduced into England in 1823 from Zanzibar. The growth of this climber is quite vigorous, some 3 m long and its leaves which are some 8 cm long resemble to that of ivy in ovately-triangular shape with winged leaf stalks. Its calyx is surrounded by 2 green bracts, the flowers which are borne in the leaf axils comprise of 5 petals, flower diameter 3.75-5.0 cm and the colour is orange-yellow with a chocolate-brown centre, though now there are forms with cream and yellow-maroon flowers. Flowering occurs from winter to spring. It is grown for its handsome foliage and beautiful flowers. Its varieties are 'Alba' (white), 'Aurantiaca' (orange, centre reddish-black), 'Bakeri' (white), 'Dodsii' (foliage 7.5 cm long and variegated green and white, flowers brownish-orange with a chestnut centre), 'Fryeri' (pale-yellow with white eye), 'Lutea' (yellow), *etc.*

T. fragrans (mountain creeper) from India is a tender evergreen and perennial climber though grown as annual, growing up to 3 m long, with 5.0-7.5 cm long, triangular and more pointed leaves. The flowers are fragrant white, axillary, tubular, flat-wide mouthed and some 3.75 cm across. Flowering occurs during winter to spring.

T. gibsonii (*T. gregorii*) from Tropical Africa is a tender perennial climber but generally grown as annual, which grows to 4.5 m long, bearing 5.0-7.5 cm long and ovate leaves. Flowers are tubular with wide mouth, 3.5-4.5 cm across and brilliant orange with a darker central eye and appear from spring to summer. It can be trained against a wall.

T. grandiflora (blue trumpet vine, sky flower) from India is a perennial climber but grown as annual and is found growing some 3-6 m long with flowers appearing during late summer to early autumn in West Bengal climate. Its stems are woody, rough-textured leaf surfaces and the leaves are ovate with angular serration. Blue flowers which are some 7 cm long are carried in pendent racemes. It is excellent for training against a wall or on the trees.

Tithonia rotundifolia (syn. *T. speciosa, T. tagetiflora, Helianthus speciosus*; family: Asteraceae; common name: Mexican sunflower). *Tithonia* has mythological derivation as Tithonus was favourite of Aurora. A native of Mexico which in the British gardens was introduced before 1713, is a perennial though grown as an annual through seeds and grows bushy, stout and erect up to 2.40 m tall having hairy and reddish stems bearing alternate, cordate or widely ovate, 3-lobed, dentate, up to 25 cm long and glabrous leaves. Solitary scarlet-orange sunflower-like blooms some 8-9 cm across are produced from summer to autumn on long stems. Flowers are most suitable for cutting. One more species, *T. diversifolia* is also in little cultivation which is native to Mexico and Guatemala. It is similar to *T. rotundifolia* in height and other characters except that stems are covered with fine hairs, leaves are entire or 3-5 lobed, rounded-ovate, 15-20 cm long, and long-stalked solitary flowers are orange-yellow and some 7.5 cm across.

Torenia fournieri (Scrophulariaceae; wishbone flower). The genus *Torenia* was erected for Reverend Olaf Toren (1718-1753), Chaplain to the Swedish East

India Company in India and China. A genus of 50 species of annual and herbaceous perennials, and the one described here is a half-hardy annual suitable for bedding out and for pot planting for summer flowering. *T. fournieri* from South Vietnam grows from 30-38 cm tall with semi-pendent habit. Its leaves are light green, in opposite pairs, 3.5-5.0 cm long and half as wide, oval, dentate and acuminate. The curious-looking flowers are attractive and like the small flowers of snapdragon, tubular, 4-lobed, upper petals pale sky-blue, the three lower violet to dark purple and centre yellow, and appear solitary or in small racemes during summer in such profusion that entire foliage is completely hidden. The varieties are 'Alba' (white), 'Bicolor Compacta' (blue-white), 'Grandiflora', 'Nana Compacta Gefion' 10 cm tall, flowers pale-blue marked darker and throat white, 'Speciosa', *etc.*

Trachymene caerulea [syn. *Didiscus caeruleus*; family: Hydrocotylaceae (Umbelliferae); common name: blue lace flower]**.** This is a half-hardy annual, which was formerly the only member under the genus *Didiscus*, but now there are 12-40 species under *Trachymene*. In Greek, *trachys* means 'rough' and *menix* for 'a membrane', alluding to the fruit of some species. It is native to Western Australia which grows bushy up to 60 cm tall with erect and branched stem. The stem and leaves, both are sticky. The narrow leaves are pale-green and have toothed lobes through ternate or biternate divisions. Slightly fragrant, tiny, lavender-blue and unequally 5-lobed flowers appear in summer on long-stalked umbels, 2.5-7.0 cm wide with spherical heads which look like the small annual scabious in appearance. The flowers are excellent for cutting. It is propagated by seeds sown in March at 15 °C temperature.

Tropaeolum spp. (Tropaeolaceae; nasturtium)**.** The generic name derives from the Latin *tropaeum* 'a trophy' as a plant of *T. majus* in flower resembles a trophy pillar hung with shields (leaves) and bloody helmets (flowers) of the defeated army. These are annual to perennial climbers, comprising of up to 90 species, often with tuberous roots and alternate leaves which may be rounded, shield-shaped or digitate. Flowers comprise of 5 sepals where uppermost possesses long nectary spur, and 5 broad petals. They are propagated through seeds sown during autumn or through offsets, divisions or basal cuttings. They make beautiful pot plants and are suitable for window sills.

T. majus (nasturtium) from Peru is a strong-growing and most widely grown annual climber which can grow up to 3 m in length. Its leaves are mid- to bright green, smooth, almost circular and 5-15 cm wide with waved edges, and peltate. On crushing, both the stems and leaves emit a pungent smell. Faintly scented yellow, orange or red flowers some 5-6 cm across with long spurs appear during late winter to spring. It has numerous varieties and forms. There is one of purple shade throughout the stem and foliage. This when hybridized with *T. minus* has produced many colours and forms. 'Tom Thumb' is dwarf strain growing hardly up to 25 cm with single or double flowers, the varieties being 'King of Tom Thumb' (flowers scarlet, foliage purple-tinted) and 'Golden King of Tom Thumb' (flowers deep yellow). 'Jewel Mixed' (semi-double in shades of yellow, pink, salmon, scarlet and crimson) grows up to 30 cm. The 'Gleam' strain such as 'Golden Gleam' (semi-double, golden-yellow), 'Indian Chief' (scarlet semi-double, foliage purple-tinted) and 'Orange Gleam' (bright orange, semi-double) has trailing habit and most suitable for hanging baskets. There are fully double varieties with trailing habit such as 'Burpeei' and 'Hermine Grasshof' which can be propagated only through cuttings.

T. peltophorum (*T. lobbianum*) from Colombia and Ecuador is an annual very similar to *T. majus*, however, the undersurface of leaves is downy, the orange-red flowers are 2.0-2.5 cm across and the lower petals serrated.

T. peregrinum (*T. canariensis*; canary creeper) from Peru and Ecuador is an annual or short-lived perennial with climbing stems up to 4 m long. Leaves are round with 5-7 deep divisions. Canary-yellow flowers with a small red spot at petal bases, the 2 upper larger petals are erect and dissected almost to half way into many narrow segments so that this looks as if fringed, appear during winter to spring.

Ursinia **spp.** (Asteraceae)**.** The genus was erected for Johannes Heinrich Ursinus (1608-67), German botanist and author. The genus comprises some 80 species of annuals, herbaceous perennials and shrubs, and the annuals and the annual-type which are being described here are suitable for growing in the beds, borders, pots and for cutting. Its flowers have tendency of closing by the evening and in dull weather conditions. Flowers are produced during winter to spring in the plains of India and from June to August in the temperate regions.

U. anethoides from South Africa is though a half-hardy perennial but is grown very widely as an annual in the gardens. It grows to 45 cm tall with light green and finely dissected leaves. Bright orange-yellow daisy-like flowers with a purple central disc are produced some 5 cm across. Garden varieties are more compact than the type species and are available in a range of orange shades. 'Sunstar' bears orange-scarlet flowers with a claret-coloured zone.

U. anthemoides from S. Africa is small growing (30 cm tall) half-hardy annual quite similar to *U. anethoides* except that the leaves are flatter undersides and the ray florets are purple.

U. cakilifolia from S. Africa is small growing (30 cm tall) hardy annual bearing light green, smooth and deeply dissected leaves and bright orange-yellow flowers some 5 cm across.

U. versicolor (*U. pulchra*) from S. Africa growing 30 cm tall is a half-hardy annual of bushy habit. The scented leaves are dark green and finely dissected and the flowers are 5 cm across and yellow or orange with a purple zone.

Venidium fastuosum (Family: Asteraceae; common names: Monarch of the veldt, Namaqualand daisy)**.** In Latin, *vena* is for 'a vein', alluding to the ribbed fruit or seed. The genus contains some 30 species of half-hardy annuals and perennials from South Africa but only *V. fastuosum* is usually available which is an annual growing from 60-90 cm tall with branching from the base. Leaves are woolly silver-white, irregularly and deeply lobed, and 10-15 cm long. Solitary flowers with 10-15 cm wide flowerheads, where rays are rich orange with purple-brown at the base which suround the black central disc. Flowers open only in sunshine during winter to spring in the plains and from June to October in the temperate regions. Its one of the strains with variously coloured flowers with the name of 'Hybrida' is available where colours are white to lemon. *V. fastuosum* with *Arctotis grandius* and *A. breviscapa* bear profusion of flowers in various forms and some 6.0-7.5 cm wide flowerheads. It is excellent for bedding out and in the borders, as well as for cutting. *V. hirsutum* is also a half-hardy annual growing to a height of 30 cm and is most suitable for pot-planting. Its leaves are grey, soft and deeply lobed. A profusion of bright orange flowers with dark central disc and 5 cm across appear during winter to spring. These are propagated through seeds sown during autumn in the nursery beds or during winter-end in greenhouses at 16 °C temperature.

Verbena hortensis (syn. *V.* × *hybrida*; family: Verbenaceae; common name: florist's or garden verbena, vervain)**.** *Verbena* originates from the old Latin name vervain (*V. officinalis*), a herb with medicinal properties. Some 250 species are in the genus, mostly perennials, shrubs and sub-shrubs and a few annuals. *V. hortensis* has probably evolved by crossing *V. peruviana* with *V. incisa, V. phlogiflora* and *V. platensis*. It is a tender perennial race usually grown as a half-hardy annual, annually through seeds sown during autumn. Its stems are procumbent to ascending, 20-30 cm tall, leaves mid- to dark green, ovate and serrated, and 5-10 cm long, and 1-2 cm across fragrant flowers are formed during winter to spring in dense and compact head-like spikes (more than 7.5 cm across) in shades of scarlet, purple, lavender, creamy-yellow and white. These are suitable for bedding and in hanging baskets. Its varieties are divided into two groups: (i) 'Grandiflora' ~ large-flowering plants, 30-40 cm high which includes the varieties such as 'Ellen Willmott' (salmon-pink, eye white), 'Royal Bouquet Mixed' and 'Mammoth Mixed'; and (ii) 'Compacta' ~ grows 22-30 cm high with more spread and its varieties are 'Amethyst' (mid-blue), 'Blaze' (bright scarlet), 'Delight' (coral-pink), 'Sparkle' (scarlet, eye white), 'Sparkle Hybrids' (bright and varied colours), 'White Ball' (pure white), *etc*. Some of the other most valued varieties are 'Hybrida Nanissima' (10-17 cm tall, flowers quite compact), 'Rainbow Mixture' (25-30 cm tall, compact in shades of apricot, lavender, mauve, pink, salmon, crimson and scarlet), 'Rose Vervain' (30-35 cm tall, vivid pink), 'Sweet Lavender' (38 cm tall, inflorescences globe-shaped, soft lavender), *etc.*

Viola spp. (Violaceae). *Viola* is a Latin name for 'a violet', perhaps originally from the Greek. It is a genus of 500 hardy species of annuals and perennials, out of which the three mentioned here below are though perennials but treated just as annuals in the garden and are therefore propagated through seeds by sowing these in nursery beds, or in the glasshouse at 16 °C, during autumn and then transplanting. They are suitable for bedding, as ground covers, in the pots, as edging, in the rock gardens and alpine houses, and in the hanging baskets. They are mainly clump-forming, sometimes stoloniferous with cordate, linear to orbicular leaves. They are dwarf to tall, flowers small to large and highly colourful in the shades of white, yellow, orange, pink, scarlet, red, purple and violet, in various colour combinations and provide enchantment for a long period from winter to early summer in the plains and for whole of summer to autumn in the temperate areas. The flowers consist of five petals which in the pansy types are rounded while violet petals are strap-shaped. Upper two petals are often longer and erect, and the lower one usually broader and with a nectary spur. Gardener's perception is this that those species and hybrids with flat-faced flowers and conspicuous leafy and lobed stipules are pansies. The seed capsules are ovoid and on ripening split into three sections of valves with steel-grey to black seeds.

V. cornuta (horned violet, tufted pansy, viola, violet) from Pyrenees is though perennial but grows as annual up to 30 cm tall with a spread of 30-38 cm. Leaves are mid-green, ovate to oval with rounded teeth. Deep lavender flowers similar to those of angular violets, 2.5 cm across and with a slender spur appearing in winter and spring to early summer. It has various varieties such as 'Admiration' (violet-blue), 'Alba' (white), 'Arkwright Ruby' (crimson-shaded and suffused terracotta), 'Campanula Blue' (quite sky-blue), 'Chantreyland' (suffused apricot-orange), 'Jersey Gem' (rich blue-purple and broader petals), 'Minor' (height 5 cm, flowers small lavender, foliage making a close mat), 'Purple Perfection' (wine-red), 'White Lady' (white), 'White Perfection' (cream-white), 'Yellow Bedder' (golden-yellow), and a hybrid race 'Funny Face' (15-18 cm tall, flowers resemble little faces, a mixture of colours), *etc.*

Viola tricolor (European wild pansy, field pansy, Johnny-jump-up, miniature pansy, viola) from Europe including Great Britain is though hardy perennial but grows as an annual up to 15 cm tall with the spread of 30 cm or more, having square and spreading stems, bearing mid-green ovate to lanceolate (heart-shaped) leaves with rounded teeth. Flowers are tricoloured, pansy-like and solitary in shades of white, yellow, dark blue and purple-black or bi- to tri-coloured and appear in profusion. Its varieties are 'Early Flowering Dutch Giants' (early-flowering, available in mixture or self colours), 'Swiss Giants' (plant compact with erect stems, unlimited colour combinations, flowers enormous and 7.5 cm across, flowering early spring to mid-sumemr), 'Gay Jester' (winter-flowering, flowers very large and in a wide range of colours), 'Jumbo Giants' (an early-flowering American strain with flowering duration of six months, colour shades apricot, bronze, pink, purple and red), and 'Majestic Giants' (flowers enormous with 10 cm across, stems long so good for cutting, colours many). All the varieries and hybrid strains mentioned under *V.* × *wittrockiana* naturally come under *V. tricolor*.

Viola × *wittrockiana* (syn. *Viola hortensis, V. tricolor* var. *hortensis, V. t. maxima*; garden pansy, hearetsease, pansy) is a group of strains arising out by crossing *V. lutea, V. tricolor* and *V. altaica*, out of all these the contribution of *V. tricolor* is immense, so the hybrids and varieties are also the short-lived perennials though are grown only as annuals. It may grow vigorous up to 25 cm high with soft, and erect to decumbent stems bearing mid-green ovate to lanceolate leaves with rounded teeth. The flowers are basically similar to *V. tricolor* but larger (varying from 2.5-10 cm across).

The varieties under *V.* × *wittrockiana* and *V. tricolor* are 'Arkwright Ruby' (15 cm tall, flowers fragrant bright crimson), 'Blue Heaven' (15 cm tall, self-coloured in white, golden-yellow, red, violet and blue), 'Celestial Queen' (light blue), 'Helios' (golden-yellow), 'Ice King' (white), 'Jackanapes' (15 cm tall, dark blue-purple and yellow bicolour), 'Swiss Giants strain' (15 cm tall, flowers with velvety texture) with varieties such as 'Super Swiss Giants', 'Rhinegold' (deep golden-yellow), 'Ullswater Blue' (flowers blue) and 'Westland Giants' (flowers 15 cm across, the largest), *etc.* Among the named varieties are 'Admiration' (dark purple), 'Apenglow', 'Berna', 'Black Prince', 'Color Festival Mixed', 'Coronation Gold', 'Elli's Oregon Giants', 'Engelma's Giant', 'Irish Molly' (copper yellow), 'Lake of Thun', 'Lily' (lemon-yellow), 'Maggie Mott' (light mauve), 'Masterpiece', 'Mont Blanc', 'Norah Leigh' (lavender-blue), 'Primrose Dame' (primrose-yellow), 'Purple Queen', 'Rhine Gold', 'Super Beacon', 'Super French Giant', 'Thor Giants', 'Tromardeau Mixed', 'Tropez Mixed', 'White Swan' (white), 'Yellow Queen', *etc.*

F_1 hybrids of *V. tricolor* and *V.* × *wittrockiana* are 'Clear Crystal', 'Crystal Bowl Mixed', 'Deep Blue Clean', 'Felix', 'Gold Bedder', 'Imperial Blue', 'Imperial Gold', 'Imperial Ocean', 'Imperial Orange', 'Imperial pink Shades', 'Imperial Purple', 'Imperial Red', 'Imperial Silver Blue', 'Imperial Yellow', 'Light Blue Clean', 'Majestic Giant' (18 cm tall, flowers 10 cm across in shades of white, yellow, scarlet, crimson, purple and light blue), 'Orange Clean', 'Orange Prince', 'Purple Bedder', 'Queen Alexandrine', 'Red Bedder', 'Rose Bedder', 'Scarlet Clean', 'White Bedder', 'White Clean', 'Yellow Clean', *etc.*

***Zinnia* spp.** (Asteraceae). The genus *Zinnia* was erected for Johann Gottfried Zinn (1727-1759), German professor of botany at Gottingen. The genus comprises 20 species of annuals and perennials. They are erect plants with opposite pairs of linear to ovate leaves and solitary flowerheads with tubular to bell-shaped involucres and 1 to several rows of ray florets. Though zinnias are non-fragrant even then are special as these grow and flower in the plains of India in summer when temperature is in between 40-47 °C persistently for months and size and colours being enormous. They are useful for planting as edge plants, in the borders and beddings and for cutting. The annual zinnias are propagated through seeds sown in March on the nursery beds or in the greenhouses at 16-18 °C temperatures until germination. Only two of its annual species are in general cultivation.

Z. angustifolia (*Z. aurea, Z. ghiesbreghtii, Z. haageana, Z. mexicana, Sanvitalia mexicana*) from Mexico is a half-hardy and small-flowered annual species which grows up to 30-75 cm tall with upright and much-branched hairy stems marked with red and light green, stalkless, opposite, 7-10 cm long and half as wide, basal leaves oval-lanceolate but others much narrower and oblong and slightly covered with bristly hairs, and up to 4 cm across, single and bright orange flowers from May to October in the plains and from July to September in the temperate climates. Its var. 'Persian Carpet' (38 cm tall and compact, flowers small and fully double and bicoloured in circular zones in wide range of colours) is most popular.

Z. elegans (youth and age) from Mexico, the most common half-hardy robust annual species growing erect to 60-75 cm tall with solitary purple flower per stem, is suitable for beds, borders and cutting. The leaves are light to mid-green, oval to lanceolate and pointed. Stems are much-branched, and leaves and stems both are rough-surfaced. Flowers are dahlia-like, over 5 cm across, composed of external ray florets (ligules or petals) with bright colouring and yellow or orange disc florets in the centre, appearing during summer to autumn. Its varieties have almost limitless range of colours and are generally double-flowered.

Most popular classes of zinnia are 'Double' (75 cm tall, flowers 7.5-10.0 cm across), 'Liliput or Pompon' (45 cm tall, flowers 4-5 cm across), 'Dahlia Flowered' (75-90 cm tall, flowers 10-15 cm across with outward curving petals), 'Californan Giant' (more than 90 cm tall, flowers largest *i.e.* >15 cm across), 'Scabiosa Flowered' (about 80 cm tall, flowers scabiosa-type and around 8 cm across), 'Cactus Flowered' (75-90 cm tall, flowers 10 cm across with little recurved and reflexed petals), 'Pumila' (45 cm tall and compact, flowers highly double and about 7.0 cm across), 'Thumbelina' (10-15 cm tall, flowers semi-double to double and 3-4 cm across), 'Sombrero' (38-45 cm tall and excellent for cutting, flowers single, 6.25 cm across and crimson-scarlet tipped bright yellow), 'Belvedere Dwarfs' (large double orange), 'Border Beauty Rose' (double deep pink), 'Burpee Hybrids' (double flowerhead in range of colours), 'Envy' (double green flowerheads), 'Fantastic Series' (20 cm tall, flowerheads double and in various colour range), 'Peppermint Stick Series' (flowerheads pompon-dahlia type and are variously striped, blotched and stippled), 'Peter Pan Series' (20 cm tall, flowerheads double in a range of colours), 'Pulcino Series' (30 cm tall, flowerheads double and in a range of colours), 'Ruffles Series' (flowers double, pompon and ruffled in a mixture of colours), 'Sunshine Series' (pompon double in a range of colours), *etc.* Its popular varieties are 'Burpeana Giant', 'Burpee's New Gigantea', 'Burpee's Tetra', 'Button Box', 'Cactus Mixed', 'Canary Bird', 'Candy Cane', 'Dream', 'Envy', 'Exquisite', 'Golden Ball', 'Kumamoto Scarlet', 'Liliput', 'New Gigantea', 'Polar Bear', 'Pulcino', 'Ruffled Jumbo', 'Salmon Beauty', 'Scarlet Flame', 'State Fair', 'White Ball', 'Whirligig', *etc.*

Z. linearis from Mexico is though a perennial species but is grown as an annual. Its height is 20-25 cm, flowering from June to November in the temperate regions and from May to October in the plains. It has not gained that popularity which it deserved due to its flower colouring and remarkably long flowering period. It is a wide-spreading dwarf species with many slender and semi-trailing stems. Profusion of solitary flowering on slender erect stems occurs in a manner that the flowers hide the complete plant. Flower size is 5 cm across. These are excellent for bedding and borders, in the edging, as a temporary ground cover, in the rock gardens and on dry walls. There is also a rare white flower form.

Cultural Practices

The seedlings from the nursery are lifted when the field for planting is ready and the seedlings have attained 4-leaf stage, *i.e.* normally 4-6 weeks after sowing, depending upon the weather conditions and type of species. If the seedlings have been raised under glass, these require to be hardened off gradually by exposing them to sunlight. One day prior to planting the container in which seedlings have been raised or the nursery beds, should be watered so that there may be minimum root damage while pulling the seedlings out, and together the field should be ready for transplanting akin to the requirement of individual species or varieties. Site should be as per requirement of individual annuals as some require full sun, some require half-shade, certain others partial shade and a few even marshy places. Though annuals are short-season ornamentals, soil pH is of little value if the range is 5.5 to 8.5, but definitely most of the annuals are very rampant and comfortable if the range is from 6.5-8.0, with exception of only a very few, and the soil type sandy-loam. While **preparing the land**, the soil should be fortified with farmyard manure (FYM) @ 3-5 kg/m^2. In case FYM can not be managed, this should be substituted with compost, dung or stable manure, poultry or sludge manure or through green manuring, but green manuring will not work in totally exhausted soil. Generally, annuals are not provided with potassic and phosphatic fertilizers since these grow, flower and seed in one season or a year and their action is very slow. In a very exhausted soil, at the time of soil preparation, the soil should thoroughly be mixed with 30-60 g muriate of potash and 60-120 g superphosphate per square metre area. Normally three deep ploughings followed by plankings every time is carried out to make the soil completely pulverized and then beds are prepared after taking out the rootstocks of perennial weeds and other unwanted material which may hinder the growth of the crop. The soil is thoroughly leveled and the beds are prepared of any length but with 1.6 m width for convenience of intercultural and other operations. Between two beds, there should be 50 cm clearance for ease of watering and for walking so these paths should be slightly lower than the beds.

Planting for winter- and spring-flowering annuals is done during October while for summer to rains in March-April and for autumn-flowering during June but only either during cloudy weathers or by evening. Summer crops require high density of planting than the winter crop. Dwarf or less-spreading annuals such as *Ageratum, Alyssum, Arctotis, Bellis, Brachycome, Calceolaria, Cheiranthus, Dianthus, Eschscholzia, Felicia, Godetia, Gomphrena, Iberis, Linaria, Linum, Lobelia, Matricaria, Matthiola, Mesembryanthemum, Mimulus, Nemesia, Phacelia, Phlox, Venidium, Verbena, Viola, etc.* are planted quite close, *i.e.* 25 × 25 cm or 30 × 30 cm spacing; medium ones such as *Amaranthus, Antirrhinum, Calendula, Celosia, Centaurea,Cineraria, Clarkia, Coreopsis, Cosmos, Delphinium, Dianthus caryophyllus, Dimorphotheca,Gypsophila, Impatiens, Kochia, Limonium, Lupinus, Nigella, Papaver, Rudbeckia, Salvia, Schizanthus, Tagetes erecta, etc.* are planted at 45 × 45 or 45 × 30 cm spacing; while the tallest ones such as *Althaea, Chrysanthemum, Helianthus, Helichrysum, Heliotrope, Molucella, Pimpinella, etc.* are planted 60 × 45 or 60 × 60 cm apart. The planting should immediately be followed by watering to settle the roots and to negate the planting shock. Until new growth starts, the seedlings should be regularly watered at alternate days if the weather is dry.

Mostly the annuals have horizontal branching so after sometime whole gap between the plants are covered by plant's growth and no room is left for **weed** growth and development. However, for up to one month and sometimes even up to two months of planting, since there is poor growth so weeds develop very fast which should be pulled out or knocked down immediately in the soil so that these may not be able to rob the main crop of its nutrients, moisture and sunlight and may not be able to act as host for various insect-pests and diseases. It would be better if the beds are **hoed** for twice at fortnightly intervals for better aeration in the soil and also for earthing up the plants to make their bases strong enough to support whole growth of the plant. Weeding and hoeing are simultaneous processes. In case of those taller plants which are unable to support their whole growth, require to be **staked** with some sticks or through strings stretched around the beds and in between the lines, on strong pegs, and this will keep plants intact at their places even when there is strong winds or heavy rains. A thick layer of organic **mulch** (3-5 cm thickness) in the beds will also prevent growth of weeds, conserve the moisture, balance the temperature and in turn provide nutrient to the soil after decomposition.

In case the growth of the plants is poor, **urea** at 2 per cent should be sprayed on the plants once or twice during growth period or broadcasting of urea @ 60 g/m^2 followed by watering should be done to encourage growth. Muriate of potsash or superphosphate should never be applied once the crop has been planted, however, it is only nitrogenous fertilizer and that too urea which is effective at this stage but its frequent spraying or application may encourage the growth on the cost of flowering. In those cases where the crop is meant for pot or cut-flower production, **liquid manure** (slurry) may be started after one month of planting until flowering at fortnightly intervals. For making liquid manure, fresh dung should be diluted to its four times of its volume and kept in field in an earthen pitcher with mouth of pitcher covered with some earthenware or thick cloth. Daily it requires stirring and after one week in summer though after two weeks in winter these are fully fermented and decomposed, it is then strained in another pitcher and diluted again to four times of its volume, and thus one litre of dung finishes into 16 times of its original volume. In this liquid, some 160 grammes of urea may be dissolved to quicken the growth process. For the summer crops, the **watering** should be once in five days though for winter crops at 10-12 days, however, this may vary from crop to crop and weather conditions. Heavy soils should be lightly and less frequently irrigated to avoid water stagnation at any time though light sandy-loam soils require frequent and to some extent deep irrigations. For display in the pots or in shows and also for production of loose flowers in certain crops, the plants may be **disbudded** and **pinched** once to thrice depending upon the crop's requirement so that more lateral buds are encouraged to grow. There are plants where pinching or disbudding should never be practiced, such as *Althaea, Antirrhinum, Delphinium, Lathyrus odoratus, Linaria, Linum, Lupinus, Matthiola, Quamoclit, Tropaeolum, etc.* Inclement weathers or late planting causes emergence of floral buds before time so such buds should be removed otherwise vegetative growth will be hampered. In case where larger flowers are required, some lateral buds should be disbudded.

Growth and Flowering

Growth and flowering in most of the annuals are affected by various cultural conditions and environmental factors. *Dianthus barbatus* and *Nigella damascena* are typical long-day plants, *Callistephus, Cosmos* and ornamental *Nicotiana* are typical short-day plants, both for vegetative growth and flowering; though there are annuals that require long-days for flower initiation followed by short-days for flower development. Imbalance of N nutrition is also instrumental as its deficiency in the soil causes premature flowering in certain annuals such as *Clarkia, Iberis* and *Salvia* though its excess in the soil causes premature flowering in *Helianthus, Lupinus,* ornamental *Nicotiana* and *Tagetes erecta* (Raghava, 2001).

Application of growth regulating chemicals at vegetative phase of plant growth, especially the retardants such as CCC (1,000-2,000 ppm), B-Nine (2,000-5,000 ppm) and SADH (1,000-3,000 ppm) have been found retarding plant height, increasing number of lateral branches, *vis-à-vis* improving flowering in *Althaea, Arctotis, Centaurea moschata, Coreopsis, Cosmos, Phlox* and *Viola*, whereas 100-400 ppm GA_3 gives encouraging results with respect to growth and flowering in *Antirrhinum majus, Callistephus chinensis* and *Tagetes erecta* (Raghava, 2001).

Post Harvest Management

Most annuals are grown for garden display, *viz.* beddings, borders, rock gardens and screes, edging, hanging baskets, window sills and container gardening for instant landscaping but only a few are cultivated for cut-flowers such as *Amaranthus, Antirrhinum, Callistephus, Centaurea cyanus, C. moschata, Chrysanthemum* (annual), *Delphinium, Dianthus caryophyllus* (annual), *Gaillardia, Gomphrena, Gypsophila, Helianthus, Limonium, Matthiola, Molucella, Rudbeckia, Scabiosa, Tagetes erecta, Zinnia, etc.*; and for loose-flowers such as *Calendula, Callistephus, Catharanthus, Chrysanthemum* (annual), *Dianthus barbatus, D. caryophyllus* (annual), *D. chinensis, Gaillardia, Helianthus, Impatiens, Lathyrus, Rudbeckia, Tagetes, etc.* When in flower, these are collected for local marketing. They are harvested either early in the morning or by evening when a portion of flowers have started showing colours, or just opening or when just fully opened, and kept with their cut ends in a bucket of water. At this time proper post harvest management practices should be applied akin to requirement of individual flowers so that maximum enchantment can be availed while putting indoors. These are graded as per floral size and stem lengths, flower shape, colour and freshness. One should be very careful about *Gypsophila, Limonium, Matthiola, Molucella* and *Scabiosa* when preparing for export to international markets. For loose flowers, normally they are harvested when just fully opened and the yield of such flowers may be from 5 tonnes per hectare in case of *Catharanthus, Impatiens, Lathyrus, etc.*, 10-12 t/ha in case of *Helianthus, Calendula, Callistephus,Chrysanthemum* (annual), *Tagetes patula, etc.*, and 20-22 t/ha in case of *Tagetes erecta*.

To avoid glut in the market, these are cold-stored akin to requirement of individual species such as *Antirrhinum* at 0.6-1.7 °C, *Dianthus caryophyllus* (annual) 0-2 °C, *Gypsophila* at 4 °C for 1-2 days, *Limonium* for 2-3 weeks at 2 °C, and *Matthiola* at 4 °C. For prolonging the life of cut flowers, various formulations are recommended such as 8-HQC 300 ppm and sucrose 1.5 per cent for *Antirrhinum*; 8-HQS 250 mg/l + sucrose 60 g/l + CCC 70 mg/l + $AgNO_3$ 50 mg/l for *Callistephus*; 8-HQC 200 ppm + sucrose 10 per cent for cut *Dianthus caryophyllus* (annual); $AgNO_3$ 25 ppm + sucrose 5-10 per cent for *Gypsophila*; 8-HQC 0.3 g + CCC 0.05 g + sucrose 50 g/l for *Matthiola*; *etc.*

Insect-Pests and Diseases

Since annuals are from various families, genera and species so collectively the insect-pests and disease problems are in plenty. *Aphis gossypii* feed on the underside of the leaves and on the tender parts in gardens and in greenhouses, and the **peach aphids** (*Myzus persicae*) discolour and distort the leaves similar to virus infection. *Macrosiphum solanifolii* suck the plant sap from aerial parts of the plants and weaken them. These may be controlled by spraying with 0.1-0.2 per cent Malathion, Lindane, Parathion, oxydemeton methyl. *Macrosiphum* spp. and *Anuraphis maidi-radicis* are **root aphids** infesting especially on *Callistephus* and their infestation causes yellowing and plant death. Chlordane soil application before planting or Lindane spraying on the crop controls these pests. **Lacewing** (*Chrysopa carnea*) effectively predates on the aphids if introduced into the greenhouses. **Tarnished plant bug** (*Lygus lineolaris*) feeds on the plants by puncturing the terminal shoots below the flower buds causing the buds or flowers to droop. Spraying with 0.2 per cent Chlorpyriphos or methyl parathion will kill this pest. **Leaf hoppers** (*Macrosteles fascifrons*) are vectors for viral diseases which may be controlled by spraying with 0.2 per cent methyl parathion. **Leaf miners** (*Liriomyza compositell*) feed inside the leaves and calyx through tunnelling and in its serious infestation leaves and calyx turn brown and die. Adult fly lays its eggs inside the leaves through puncturing, and after hatching its yellow maggots start feeding inside the leaves by making the mines. Spraying of Triazophos or methyl-o-demeton at 0.05 per cent controls this pest. Adult **thrips** (branded greenhouse thrips and western flower thrips) and its nymphs feed by piercing on the tender parts of the plant and its serious infestation deforms the plants. These are controlled with 0.2 per cent Metasystox spraying. **Cottony cushion scale** (*Iceria purchasei*) may become a problem in temperate regions on China aster, and sometimes even **soft scales** also attack this crop. Various species of **whiteflies** (greenhouse whitefly, iris whitefly, silver leaf whitefly and sweet potato whitefly) have been observed attacking this crop. The infestation of **spider mite** (*Tetranychus telarius*) makes the foliage webby, discoloured and deformed which may be controlled through spraying with Kelthane at 0.1 per cent. Red spider mite (*Tetranychus telarius*) becomes serious during dry weathers. Cyclamen mite (*Steneotarsonemus pallidus*) and a 'red and black stink bug' have been found infesting China aster. These are controlled through the use of Aramite, Dimite, Kelthane or Ovex. Looper larvae are controlled through foliar spraying of Trichlorfon or methyl parathion. *Helicoverpa armigera* caterpillars are serious pests of antirrhinum plants including the seed capsules. It can be controlled by Thiodan 0.3 per cent or Ripcord 0.45 per cent spraying. **Aster beetle** or black blister beetle (*Epicauta pennsylvanica*) feeds on foliage and flowers and destroys the whole plant. Weekly spraying of Methoxychlor during infestation period will control this pest. **Asiatic beetle** (*Autoserica castanea*) feeds on the foliage during night hours and hides itself in the soil at the base of the plant or in the vicinity. Soil application of Chlordane or Dieldrin will eliminate the pest at its larval stage. Lead arsenate spraying on the plants once or twice during night will kill the feeding adults as well as larvae. **Chrysomelid beetle** (*Aulacophora foveicollis*) attacks the crop at initial stage by cutting holes in the leaves and tender shoots while its grubs feed on the underground parts including the roots which results into plant death. Carbofuran or Phorate at 1kg/ha soil application will keep this pest under control. **Leaf and flower eating caterpillars** (*Helicoverpa armigera*) attack the crop at the onset of winter and feed on leaves and flowers while ***Phycita* green caterpillars** with brown heads feed on ovaries and stamens of the buds and flowers by grooving through the central receptacle. Metasystox or methyl parathion 0.3 per cent spraying will kill these larvae. Female moth of the **stem borer** (*Platyptilia molopias*) lays its eggs on the tips of the plants and the black-headed creamy larvae coming out after hatching of these eggs feed inside by boring the shoots, stems and the laterals thus affected shoots become hollow and wilt. Carbofuran or Phorate at 1 kg/ha soil application will control this pest. **Semilooper**, a caterpillar (*Clenoplusia albostriata*) which is blue-green in colour feeds on the leaves and also on the flowers. Sometimes the **cutworms**, **corn earworms** and **cabbage looper** attack these crops. Spraying Chlorpyriphos or Quinalphos at 0.05 per cent or Carbaryl at 0.1 per cent will control this pest.

Root-knot nematode (*Meloidogyne incognita*) infest the plants through galling of the roots which stunts the plants, and **foliage nematode** (*Aphelenchoides ritzema-bosi*) sucks plant juice causing their stunting, malformation, cupping and yellowing of leaves and ultimate plant death in its serious infestation. Foliar nematodes cause leaf blight and leaf fall and its serious infestation may kill the plants. Nemacur soil application and Demeton foliar spray will control both types of nematodes. Moreover, use of Carbofuran in the soil will also control their infestation.

Diseases which infest annuals are damping off (*Pythium splendens*), wilt (*Fusarium oxysporum* f. sp. *callistephi*, *F. conglutinans* var. *callistephi*, *Sclerotium rolfsii*, *Verticillium albo-atrum*, *Acrostalagmus vilmorinii*, *Phialophora fastigiata*), collar and root rot (*Phytophthora cryptogea*), stem rot (*Pellicularia filamentosa*), grey mould (*Botrytis cinerea*), rust (*Coleosporium solidaginis*), canker (*Phomopsis callistephi*), leaf spot (*Alternaria*, *Stemphylium callistephi*, *Septoria callistephi*, *Ascochyta asteris*) and viruses. Parrini and Rumine (1989) in Italy described the diseases such as *Botrytis cinerea*, *Peronospora antirrhini*, *Puccinia antirrhini*, *Pythium*, *Rhizoctonia solani*, *Sclerotinia sclerotiorum* which were recorded infecting snapdragon. From Italy, Pasini *et al.* (1996) isolated ***Rhizoctonia solani*** from China aster and antirrhinum causing collar or foot rot. *Fusarium* infection affects the vascular system which is expressed first as yellowing of the plants, followed by death. It has been found occurring along with *Acrostalagmus* and *Verticillium* causing wilt or with *Botrytis cinerea* or *Rhizoctonia* causing wilt. To control ***Pythium splendens*** on China aster seedlings, when Niebisch and Kelling (1986) used Previcur N (propamocarb) and Ridomil (metalaxyl) + zineb, they recorded Ridomil superior over Previcur N, and application of Captan 80 or Thiram FW on seeds and seedlings, they recorded increased yield of cut flowers. ***Phialophora*** also causes vascular wilt, first recorded from New Zealand. Against all these pathogens, soil sterilization with formalin or chloropicrin is a must before taking the next crop of China aster from the same field, however, resistant cultivars such as 'Bouquet Scharbach', 'Balls Blue', *etc.* are also available. Before sowing, the seeds should also be soaked in a 0.1 per cent solution of mercuric chloride for 30 minutes. Once disease occurs on the plants, there is no control so in disease-prone areas, benomyl 0.1 per cent alternate with Thiride 0.2 per cent should be sprayed to the standing crop fortnightly. **Collar and root rot** becomes serious under moist conditions. Its infection is symptomized by water-soaked areas at the stem base, rotting of the stem at base, root rot and wilting of the plants. Captan 0.2 per cent spraying is very effective against the infection of this pathogen. Restricted irrigation and use of resistant cultivars are suggested. **Stem rot** can be controlled through spraying with 0.2 per cent Captan or Thiride. **Grey mould** spreads more rapidly during cool and humid weathers. This disease generally occurs 3-4 days after sowing or 10-12 days after pricking out, and its serious infection causes plant blight. Spraying of such plants with Dithane M-45 or Dithane Z-78 at every 5-7 days during inclement weathers will keep this disease under check. Attack of **rust** to China aster plants causes bright yellowish-orange spots on underside of leaves, more serious being on the young plants. Rust (*Puccinia antirrhini*) is the most serious pest of snapdragons

(Horst, 1990) which affects the crop in field as well as in greenhouses. In its infection brown pustules are observed on stems, leaves and sepals. The severity of the disease increases during dry weathers. Thiram or a mixture of sulphur and zineb at 0.2 per cent sprayings frequently will control this problem. There are now many varieties available which are genetically resistant to this disease. **Canker** affects the lower part of the stem, and distal portion completely collapses, though roots are not affected. Spraying with 0.1 per cent Bavistin alternate with 0.2 per cent Captan will control this problem. **Leaf spots** are first yellowish, turning brown and then black and increasing in size, lower leaves being infected first. Maneb or zineb sprayings at 0.2 per cent will keep these diseases under check.

Damping off is a serious problem at seedling stage. This is caused either by ammonical fertilizer application or due to *Pythium* infection (Horst, 1990). *Rhizoctonia solani* and *Pellicularia praticola* cause havoc at high temperatures coupled with high soil moisture. To some extent these problems can be avoided through proper sanitation, thin sowing or distant planting and by regulating proper water supply. Application of Brassicol at 0.3 per cent or sterilized soil can also check these diseases. ***Antirrhinum* wilt** is caused due to infection with *Fusarium solani* and *F. solani* var. *merittii* which are favoured by high temperatures (29-35 °C) and high soil moisture. Seed dressing with 0.2 per cent Captan is very effective in keeping this problem under check. **Stem rot** is caused by *Sclerotinia sclerotiorum. Alternaria* and *Helminthosporium* also attack snapdragon. All these pathogens can be kept under check by regulating the watering and through regular spraying of the crop fortnightly with Bavistin 0.1 per cent alternate with Captan 0.2 per cent. **Snapdragon blight** is caused by *Phyllosticta antirrhini* which produces brownish spots on the foliage and ashy-grey spots on stems, and the attack of this pathogen becomes severe during high temperatures. Spraying the crop with 0.2 per cent Dithane M-45 or Dithane Z-78 controls this pathogen. ***Botrytis*** (Horst, 1990) also infects this crop during windy, chilly and humid conditions. Its control is the same as to that of snapdragon blight. **Powdery mildew** [*Oidium* sp. (Horst, 1990), *Podosphaera leucotricha*] is symptomized by formation of a white powder-coating on young stems, both the leaf surfaces and on the sepals, and its severity is more under high humid conditions and dry soil. Through spraying with wettable sulphur or Karathane at 0.2 per cent or Bavistin 0.1 per cent controls these pathogens. **Downy mildew** [*Peronospora antirrhini* (Horst, 1990)] forms mealy-white patches on the lower side of leaves, and in severe infection the plants become crippled. Seedlings are badly affected with this disease. Bordeaux mixture at 0.1 per cent or Dithane M-45 at 0.2 per cent sprayings at fortnightly intervals will keep this pathogen under check. Metalaxyl and Manzate are also quite effective. **Snapdragon anthracnose** is a common disease of greenhouse snapdragons caused by *Colletotrichum antirrhini* and *C. fuscum.* It is symptomized by pale-yellowish-green to grey spots with a narrow border on the stems and leaves, being sunken and oblong on the old stems. It is controlled by destroying the infected old debris and by spraying with Dithane M-45, Dithane Z-78 or Captan 0.2 per cent, or Bavistin 0.1 per cent. ***Septoria antirrhini* leaf spot** causes small circular spots with a pale centre bordered with purplish-brown margins. ***Cercospora antirrhini*** also causes leaf spot. Dried infected leaves should be collected and destroyed. The control measures adopted in case of anthracnose will control these pathogens also.

'**Aster yellows virus**' causes stunting of the plants, formation of numerous adventitious shoots, pale-yellowish tinge on the leaves, off-colouring of the blooms and the ray florets of China aster usually yellowish-green, and the virus is transmitted through *Macrosteles fascifrons* leaf hopper. Its Californian strain is also reported. Control of the leaf hopper and destroying the infected plants will check further spread of this virus. A seed-borne viral disease '**chrysanthemum mosaic virus**' causes mild dwarfing and serious bloom distortion. '**Spotted wilt virus**' and '**curly top virus**' also infect China aster so the affected plants should be uprooted and burnt. A strain of **CMV** which in nursery is transmitted through the aphids causes plant stunting and little leaf. **Tomato spotted wilt virus** and **impatiens necrotic spot virus** cause brown or black stem lesions which often do not appear until just before flowering (Laughner and Corr, 1996). Such plants should be rogued out and the aphids should be controlled regularly. Several researchers (Dimock, 1958; Forsberg, 1958; Nelson, 1962; Williamson, 1962; Porter and Aycock, 1967; Engelhard, 1971) have reviewed the control of antirrhinum diseases.

Since antirrhinum plants continue absorbing high levels of nutrients even up to the time of harvest, Hood *et al.* (1993) objected the practice of reducing the mineral nutrition during the final stages of production prior to harvest as this causes tip-breaking problem. Excessive grassy growth can occur due to improper selection of cultivars and excessive use of nitrogen.

References

Bailey, L.H. 1942. *The Standard Cyclopedia of Horticulture* (3 vol.). The Macmillan Company, New York.

the Mapuche people and native Australians. The dried seeds are made into a flour which the Araucanos then use to make a fermented beverage (muday). The seeds are also fed to livestock, especially during winter. Local people use to cut trees for fuelwood and construction but large-scale logging of *Araucaria* forests followed the arrival of Europeans in the region in the 19th century. Once the monkey puzzle was being considered to be the most valuable timber in the southern Andes for making railway sleepers, pit props in mines, ship masts, and for paper pulp. In the 1940s, it was even reported as being used in the construction of aeroplanes.

Botany

Araucaria, an evergreen coniferous genus comprises usually the large trees with a massive erect stem reaching to a height of 30–80 metres and horizontally spreading branches growing in whorls. The leaves are leathery, spirally-arranged and overlapping, ovate or awl-shaped, or needle-like. The banana-like male flowering cones or catkins (male strobili) being produced in terminal clusters of 2-6, are cylindrical and shed pollen in June and then turn dark brown, and female trees produce ovoid to globular cones, 10-18 cm long and 7.5-13 cm wide on the upper side of some top branches. *Araucaria araucana* has pyramid-shaped tree, up to 50 m high with a trunk circumference of up to 2.5 m; leaves ovate, hard, glossy dark green, spine-tipped, up to 5 cm long with a life of 10-15 years; strobili in both sexes on usually different trees, male some 7.5-12.5 cm long, ripe cones globular, 10-18 cm long and hardy. Araucarias are mostly dioecious, with even some being monoecious or changing sex with time. Female cones are globose and larger in size (7-25 cm) than male cones (4-10 cm) producing wind-dispersed pollen. Female cones may produce upto 80-100 large edible seeds similar to pine nuts which disperse due to gravity owing to their large size (2–4 cm long, 1–2 cm wide) and heavy weight (3.5–5.0 g) (González *et al.*, 2006). Araucaria seeds can also be dispersed over greater distances by birds, rodents and other animals (Veblen, 1982; González *et al.*, 2006). The chromosome number has been recorded 2n=26 in *Araucaria angustifolia, A. araucana, A. bidwillii, A. brasiliensis, A. cunninghami* and *A. excelsa*.

Classification, Species and Varieties

Under the family Araucariaceae, earlier there were only two genera ~ *Araucaria* de Jussieu and *Agathis* Salisbury but in 1994 a third genus *Wollemia* Jones was recorded from Australia (Zonneveld, 2012). A total of 19 species are reported in *Araucaria* while there are 15-17 in *Agathis* and only 1 in *Wollemia* (Eckenwalder, 2009; Farjon, 2010). *Araucaria* has a disjunct distribution with two species in South America and the other 17 in Australasia. They are usually divided in four sections ~ Section Eutacta is distributed in New Caledonia (13 spp.), Norfolk Island (1 sp.) and Australia/New Guinea (1 sp.); Section Araucaria is distributed in Brazil (1 spp.) and Chile (1 spp.); section Bunya in Australia (1 sp.); and section Intermedia in New Guinea (1 spp.). The important species in Eutacta are A. *cunninghamii*, A. *heterophylla*, A. *columnaris*, A. *montana*, *etc. Araucaria heterophylla*, a slow-growing tree, attaining up to 1.5 m height, is the most suitable species sold as house plant whose growth is restricted making it pot-bound for indoor use. Its leaves are spirally arranged and are usually broadly triangular to needle-like; male and female cones are green, maturing to brown and are normally produced on separate trees, male being conical or cylindrical while females spherical, ovoid or ellipsoid. A. *cunninghamii* is a large, unbuttressed, symmetrical tree, 50-70 m high; bole straight, cylindrical, self-pruning, clean to 30 m or more; mature trees 1.2-1.7 m in diameter; trunk internodes variable, 1-4 m; crown pyramidal to flat, monoecious; male strobili usually borne on lower and mid-crown branches, terminal, green, yellow at anthesis, red-brown later, elongated, about 90 × 10 mm; cone green, ovoid, 70-100 × 60-80 mm, covered with short spines, 9-10 mm long, deflexed; seed in the form of ovulate cone scales, more or less flat, woody, triangular, with 2 thin wings, indehiscent scale terminating in a sharp spine and is reddish-brown.

Section Araucaria contains two species, *viz.* A. *angustifolia* (Bertol.) Kuntze and *A. araucana* (Molina) K. Koch distributed in Brazil and Chile, respectively. *Araucaria araucana* is an impressively large, hardy and long-lived conifer, reaching to 50 m height, 2.5 m in diameter and up to 1,300 years in age (Montaldo, 1974), thus sometimes described as a living fossil. It is regarded as the national tree of Chile. *A. angustifolia* is native to Brazil and Argentina and is used worldwide as house plant. Both these important timber species can cross easily with each other and are the only araucarias that are mostly dioecious. The large seeds of both, produced abundantly, have been an important food source for local people and wildlife. *A. bidwillii* is a member of section Bunya while *A. hunsteinii* is from Itermedia.

Propagation

Primary method of propagation is through **seeds**. Germination is favoured at 25-30°C and seed starts germinating within 15-25 days of sowing. Seedlings can be raised by pre-germination techniques or by sowing into beds. *A. heterophylla* can also be propagated by **stem tip cutting**. Tip cuttings of 7-10 cm length should be taken from vertical shoot tips at midsummer and rooted in a cold frame. Cuttings should not be taken from horizontal side branches as they never form an erect tree. A. *cunninghamii* can be successfully **grafted** by using scion budding with material taken from the apical leader of the main stem, or by side-approach grafting and bottle grafting using the apical shoot of the main stem. Grafted branch produces plagiotropic grafts and has little use other than for pollen production. Asexual reproduction by **root suckering** has been reported on the Andes and the Coastal Range (Schilling and Donoso, 1976; Cortés, 2003), particularly under severe disturbance regimes (Cortés, 2003; González *et al.,* 2006).

Cultivation

Araucaria can be successfully grown under a range of **soils** including clay, loam or sandy which may be slightly acidic or alkaline but it prefers well-drained clay soil. It can tolerate some drought and salt. Although they provide some shade, they are not suitable for patios or terraces because they are too large and their large surface roots being so common. In addition, columnar-formed trees generally cast limited shade due to the narrow crown. They often have an attractive pyramidal form when they are small, but they quickly grow too tall for most residential sites. They can live as a house plant for a long time if not overwatered. These keep themselves in shape as if pruned. However, appearance of multiple trunks or leaders mars its charm forever and then these are pruned to grow only one central leader. Rigorous **weeding** is necessary until the canopy closes, as *Araucaria* species suffer from grass competition, grow only slowly and often turn chlorotic. Young plants should be **watered** well, especially during periods of drought. Continuous **piercing sun** during May-June in the subtropical areas of the country burns its top and exposed foliage, and a persiting **low winter temperature** when it goes down 10 °C, its tops and the tender parts show frosty-rot in the subtropical regions of India and then before such conditions occur the araucarias require protection ~ only thatching at the top during fierce summer, especially during day hours, and thatching the whole plant during chilly winters, especially in the nights. In nature it is a tropical plant.

Seed Collection and Storage

In *A. cunnighamii*, female flowering commences when it is about 12 years old while male flowering does not occur until it is 22-27 years. However, in seed orchards of this species the age of production of male cones has been reduced to only 5 years and female cones 2-3 years using physiologically mature grafting material but ripening in 2-3 years then falling apart. On maturity, the cones are covered in golden spikes. The seed production is often unreliable and low, a time lapse of 8-10 years between good seed crops is common (Schmidt, 2000). Seeds are orthodox in nature. The viability of seeds that have been dried to moisture contents in equilibrium with ambient environment has been found maintaining for 8 years at -9 to -15 °C. Similarly, 50 per cent germination after 50 weeks of air-dry storage at -12 °C has been reported. If mature seeds are dried to 5 per cent moisture, long-term storage is possible in sealed containers at 3 °C or lower. There are approximately 2,400-4,000 seeds pr kilogramme.

Diseases and Insect-Pests

Araucaria in nature is very rarely attacked by diseases and pests. Very few diseases such as honey fungi, root rot, anthracnose or needle necrosis, and insect-pests such as scale insect and mealy bugs have sometimes been recorded. Out of the *Armillaria cepistipes, A. gallica, A. mellea* and *A. ostoyae* **honey fungi**, the attack of first and last species is utterly rare, though if once infected with *A. mellea* and *A. ostoyae* the damage is heavy whereas the damage by *A. gallica* and *A. cepistipes* is negligible. Infection of *Armillaria* is characterized by presence of white fungal growth between the bark and wood usually at ground level. Clumps of honey coloured toad-stools sometimes appear briefly on infected stumps during autumn. As a result, the upper parts of the plant may die, sometimes the death is sudden during hot and dry weathers, indicating failure of the root system, and sometimes the branches start gradual dying back over several years. Leaves of such plants become smaller and paler, and the plants fail to flower or unusually heavy flowering followed by an unusually heavy crop of fruit (usually just before death). Cracking and bleeding of the bark at the base of the stem, and mushrooms are produced in autumn from infected plant material if suitable environment for the fungi prevails. This fungus spreads from an infected plant to healthy plant through the soil using black or brown root-like cords called rhizomorphs, which are the origin of the name 'bootlace fungus'. These develop mostly 2.5–20.0 cm below the soil surface in moist soil but may be found deeper in dry soils. When the growing tips come into contact with the roots of susceptible living plants, they are able to penetrate the tissues and grow through to the inner layers of the bark. To diagnose whether the plant is infected with honey fungus, peeling away the bark at the base shows the white or creamy white paper-thin layer of fungal tissues (mycelium). Till date no effective control measure is available. If honey fungus is confirmed, the only effective remedy is to excavate whole root along with stumps and destroy these by burning or through landfill. This will destroy the food base on which the rhizomorphs feed, and they are unable to grow in the soil when detached from infected material. Spread of fungi can be prevented to unaffected areas through using a physical barrier such as a 45 cm deep vertical strip of butyl rubber (pond lining) or heavy duty plastic sheet buried in the soil that actually blocks the rhizomorphs. It should protrude 2-3cm above the soil level. Regular deep cultivation will also break up rhizomorphs and limit the spread. **Root rot** (*Cylindrocladium* and *Pythium* spp.) is comparatively more common disease of garden or nursery where plant is generally grown in pots. Over-watering of plants in case of root rot leads to loss of vigour and eventual death of whole plant. Ideally, plant should be kept barely moist. If plant suffers from root rot, plants are unpotted, whole lot of soil is knocked down and washed away, all the infected roots are trimmed off and then is followed by treatment with an effective fungicide and then the plants are potted into a pot filled with the sterilized soil medium. The pot should have good drainage. After repotting, the pots are sparingly watered to keep the soil barely moist. It is better to let the plant dry out a bit before watering than to over-water. **Anthracnose** (needle necrosis) caused by*Colletotrichum derridis* mostly due to overhead watering, starts as small dead areas on the needles, thereby branches turn brown and needles fall. To check the menace of this disease, the plants are only surface-watered, and affected parts of the plants are pruned followed by fungicide application. **Blight** (*Cryptospora longispora*) symptom first appears in the lower branches and then gradually move upward when plants are some 5-6 years old. Infected limbs die and their tip-ends break off. Infected branches should be pruned off and burnt. Quarantine measures should be taken before release to the nurserymen.

Mealybugs (*Pseudococcus aurilanatus, Planococcus citri* and *Pseudococcus ryani*) are small, wingless, dull-white and soft-bodied insects producing a waxy powdery-covering. They often look like small pieces of cotton and tend to congregate where leaves and stems branch. They suck the sap out of plant tissues. The youngs tend to move around until they find a suitable feeding spot, and they hang out in colonies and feed. They weaken the plant leading to yellow foliage and leaf drop. To control the pest, Acephate 2 g or Prophenophos 2 ml per litre of water should be sprayed. Natural enemies such as lady beetles in the garden help reducing its population levels. **Scales** (*Eriococcus araucariae*) are problem on a wide variety of plants. Young scales crawl until they find a good feeding site, but after attaining the adulthood, the females lose their legs and remain static on a spot protected by its hard shell layer. They appear as bumps, often on the lower sides of leaves. Scales can weaken a plant leading to yellowing and falling of foliage and through their honeydews attract ants which may lead to unsightly spot and the attack of sooty mould. Once established they are hard to control. Infested plants should be isolated and sprayed with Malathion or Acephate @ 2g or Orophenophos 2ml per litre of water, especially during crawling stage. Other pests include **larvae** of *Septromorpha rutella* which are known to infest *A. cunninghamii* seed, **weevils** (*Vanapa oberthueri*) that are associated with damage on trees following poor pruning or thinning, termite (*Coptotermes elisae*) and a leaf-footed bug (*Amblypelta cocophaga*). These are automatically controlled while controlling mealy bugs or lepidopterous larvae.

References

Aagesen, D.L. 1998a. Indigenous resource rights and conservation of the Monkey-Puzzle tree (*Araucaria araucana*, Araucariaceae): a case study from southern Chile. *Econ. Bot.*, **52:** 146–160.

Aagesen, D.L. 1998b. On the northern fringe of the South American temperate forest: The history and conservation of the Monkey-Puzzle Tree. *Environ. Hist.*, **3**: 64–85.

Burns, B. 1993. Fire-induced dynamics of *Araucaria araucana–* Nothofagus Antarctica forest in the Southern Andes. *J. Biogeogr.*, **20**: 669–685.

Cortés, M. 2003. Dinámica y Conservación de *Araucaria araucana* (Mol.) Koch. en la Cordillera de la Costa de Chile. Tesis de Magíster en Ciencias, Mención Recursos Forestales. Facultad de Ciencias Forestales, Universidad Austral de Chile, Valdivia. Chile.

Eckenwalder, J.E. 2009. *Conifers of the World*. Timber Press, Portland, OR, USA, 720 pp.

Farjon, A. 2010. A Handbook of the World's Conifers, vol. *1* (ed. Brill, E.J.). Leiden & Boston, U. K.

Farjon, A. and C.N. Page, 1999. Conifers. Status Survey and Conservation Action Plan. IUCN/SSC Conifer Specialist Group. IUCN, Gland, Switzerland and Cambridge, UK.

Gallo, L., F. Izquierdo, L.J. Sanguinetti, A. Pinna, G. Siffredi, J. Ayesa, C. Lopez, A. Pelliza, N. Strizier, M. Gonzales Peñalba, L. Maresca and L. Chauchard, 2004. *Araucaria araucana* forest genetic resources in Argentina. In: Challenges in Managing Forest Genetic Resources for Livelihoods: Examples from Argentina and Brazil (eds Vinceti, B., W. Amaral and B. Meilleur), pp. 115-143. International Plant Genetic Resources Institute, Rome.

González, M. and T.T. Veblen, 2007. Incendios en bosques de *Araucaria araucana* y consideraciones ecológicas al madereo de aprovechamiento en areas recientemente quemadas. *Rev. Chil. Hist. Nat.*, **80**: 243–253.

González, M., M. Cortés, F. Izquierdo, L. Gallo, C. Echeverría, S. Bekessy and P. Montaldo, 2006. *Araucaria araucana*

(Molina) K. Koch. Araucaria, Pehuén, Pino piñonero, Pino de Neuquén, Monkey Puzzle Tree. In: *Las Especies Arbóreas de los Bosques Templados de Chile y Argentina* (ed. Donoso, C.). *Autoecología*, pp. 36-53. Marisa Cuneo Ediciones, Valdivia, Chile.

Kunzmann, L. 2007. Araucariaceae (Pinopsida): aspects in palaeobiogeography and palaeobiodiversity in the Mesozoic. *Zoologisher Anzeiger*, No. 246, pp. 257-277.

Montaldo, P. 1974. La Bioecología de *Araucaria araucana* (Mol.) Koch. Boletí Técnico No. 46. Instituto Forestal Latinoamericano de Investigación y Capacitación. Mérida, Venezuela.

Sanguinetti, J. and T. Kitzberger, 2009. Factors controlling seed predation by rodents and non-native Sus scrofa in *Araucaria araucana* forests: potential effects on seedling establishment. *Biol. Invasions*, **12**: 689–706.

Schilling, G. and C. Donoso, 1976. *Reproducción vegetativa natural de Araucaria araucana* (Mol.) Koch. *Invest. Agrícola*, **2**: 121–122.

Schmidt L., 2000. Guide to Handling of Tropical and Subtropical Forest Seed. DFSC.

Shepherd, J.D., R.S. Ditgen, 2005. Human use and small mammal communities of *Araucaria* forests in Neuquén, Argentina. *Mastozool. Neotrop.*, **12**: 217–226.

Veblen, T.T. 1982. Regeneration patterns in *Araucaria araucana* forests in Chile. *J. Biogeogr.*, **9**: 11–28.

Zonneveld, B. J. M. 2012. Genome sizes of all 19 *Araucaria* species are correlated with their geographical distribution. *Plant Syst Evol*, 298:1249–1255.

5

Begonia (Family: Begoniaceae)

R.L. Misra and Sanyat Misra

[**Common names**: Beefsteak geranium, Begonia, Blooming-fool begonia/Christmas begonia/Lorraine begonia {*B.* x *cheimantha* (*B. dregei* x *B. socotrana*)}, Davis begonia (*B. davisii*), Dewdrop begonia (*B. carrierei*), Elephant's ear, Fire king begonia (*B. goegoensis*), Grape-leaf begonia/Maple-leaf begonia (*B. dregei,* syn. *B. parvifolia*), Grape-vine/Maple-leaf begonia {*B.* x *weltoniensis* (*B. dregei* x *B. sutherlandii*)}, Hardy begonia (*B. grandis* ssp. *evansiana*), Hollyhock begonia (*B. gracilis*), Hybrid tuberous begonias (*B. tuberhybrida*, a garden race of tuberous begonias), Winter-flowering begonias [*B.* x *hiemalis* syn. *B.* x *elatior* {*B. socotrana* x (*B.* x *tuberhybrida*)}], *etc.*]

Introduction and Origin

The genus *Begonia* was named after Michel Begon (1638-1710), a French patron of botany, and for a time Governor of French Canada. Begonias are indigenous to Mexico (*B. monophylla,* tuberous, Barkley and Boghdan, 1972; *B. alice-clarkae*, shrubby canelike belonging to section Liebmannia, Ziesenhenne, 1976; *B. lyniceorum, B. multistaminea* and *B. sousae*, all rhizomatous, Burt-Utley, 1983; *B. roseibractea*, having large pea-green bracts overlaying one another, Ziesenhenne, 1983; *B. lachaoensis*, white flowered some 15 cm tall species belonging to the section Huzia A.DC., Ziesenhenne, 1985; *B. manicata*, Ziesenhenne, 1988), Central and South America [Argentina (*B. descoleana* Smith et Schubert nov. sp., Smith and Schubert, 1950), Bolivia (*B. leathermaniae,* a shrubby and canelike species with pale-pink flowers, O'Reilly and Karegeannes, 1983), Brazil (*B. epipsila,* shrub-like most suitable for hanging basket, Thompson, 1978b; *B. odeteiantha*, a shrubby species with pendent stems, Thompson, 1978c; *B. edmundoi,* Thompson, 1979; *B. vitifolia*, growing to about 3.6 metres high, Doorenbos, 1979; *B. macduffieana*, a species having unspotted and straight leaves, Smith and Schubert, 1985; *B. grisea*, previously coded as U001, Ziesenhenne, 1986; *B. solimutata*, previously coded as U003, its leaf colour varying with light intensity, Smith and Wasshausen, 1990), Colombia, Ecuador, Guatemala, Peru, West Indies, *etc.*], Asia [Bhutan, China, India, Japan, Java (*B. robusta, B. multangula* both rhizomatous; Doorenbos, 1980), Malaya (Kiew, 1989; from Trengganu, a beautiful species), Nepal, Pakistan, Philippines (*B. fenicis, a* large-leaved rhizomatous species belonging to Diploclinium section; and *B. cumingii,* a shrub-like species; Thompson, 1978a, 1986), Sri Lanka, Sumatra, Taiwan (Peng *et al.,* 1988; *B. ravenii,* both tuberous and stoloniferous, showy pendant inflorescence, chromosome no. n = 18), *etc.*] and South Africa [Congo, Guinea, Jamaica, Madagascar (Ziesenhenne, 1973; *B. bogneri,* a grass-like species with long narrow strap-like leaves and tuberous stem base), South Africa (Cape of Good Hope, Natal), Tanzania, Zambia, *etc.*]. Torode (1984) recorded *B. chlorosticta* in 1967 from Sarawak. O'Reilly (1991) has given an account of the unidentified species as coded (Nos. U178-U186) by the American Begonia Society, 2 collected from Guatemala, 2 from Panama, 1 from Solomon Islands, 3 from Philippines and 1 from Venezuela. Lee (1979) has described *B. pavonina* which is native to SE Asia, and possesses blue foliage colour due to light interference and not due to pigmentation. Sands (1990) has described six species collected in 1984 and has put them in three SE Asian sections, *viz.* Petermannia (*B. malachosticta, B. cauliflora* and *B. erythrogyna*), Bracteibegonia (*B. imbricata* and *B. kinabaluensis*) and Platycentrum (*B. amphioxus*). All the begonias are suitable for greenhouse culture, and some can be used for outdoor beddings and also as an effective indoor- and basket-plant. Maier and sattler (1977) and Sattler and Maier (1977) reported the structure of spiphyllous appendages [hair-like, and leaf-like (similar to main leaf) appendages] on the upper side of the leaves in addition to trichomes in *B. hispida* var. *cucullifera* and stated that trichomes lack vascular and other tissues such as chlorenchyma, a hypodermis and typical epidermis though hair-like and leaf-like appendages have an upper epidermis with a large-celled hypodermis, palisade and spongy tissue and a lower hypodermis and epidermis with stomata. They are most suitable as single pot-specimens and as window-garden subjects. *Begonia rhopalocarpa* is grown in the greenhouse mainly for its club-shaped yellow and red fruits (Doorenbos, 1980a). The leaf-stalks of some of the *Begonia* species are used as the leaf-stalks of rhubarb (Doorenbos, 1981). The rhizomes of many species, especially native to South America are bitter and astringent and these are employed by local people there against certain fevers and for syphilis. Some species also contain purgative principles. The sour soup of one of the Asiatic species is used for cleaning the weapons. Shomer-Ilan *et al.* (1973) extracted an endogenous gonadotrophin-like proteinaceous plant factor from fresh leaves of *B. semperflorens* which has growth-promoting activity in plants as well as in animals. Doskotch *et al.* (1969) recorded cucurbitacin B, dihydrocucurbitacin B, cucurbitacin D and an unidentified compound of possible terpenoid structure, the cytotoxic principles from the tubers of *B. tuberhybrida* var. *alba.*

Cyanidin 3-*0*-β-(2^G-xylosylrutinoside) and cyanidin and pelargonidin 3-*0*-β-(2^G-glucosylrutinoside) are reported to occur in begonia flowers and leaves as rutin, kaempferol 3-glycoside, quercetin 3-glucoside and quercetin 3-xyloside (Harborne and Hall, 1964), apart from 1-Caffeoyl- and 1-feruloyl-glucose, though cyanidin 3-*0*-β (2^G-xylosylrutinoside) is major pigment component found in majority of the species (Langhammer and Grandet, 1974). Ensemeyer *et al.* (1980) isolated a dimeric proanthocyanidin and a by-product (+)- catechin from *B. glabra.* From the leaves of *B. erythrophylla*, Vereskovskii *et al.* (1987) for the first time isolated flavonoids 3-*0*-methylquercetin, 3-*0*-methylkaempferol, quercet, luteolin, quercetin-3-*0*-rutinoside, quercetin-3-*0*-rhamnoside and luteolin-7-0-glucoside. Perpetual flowering begonia, *Begonia semperflorens* cvs 'Rosea' and 'Carmen' leaves change their pigmentation, light shade reduces the red colour intensity in 'Rosea' leaves but deeper shade brightens the leaf colour in 'Carmen' (Silis and Stanko, 1972). Jangoux (1990) in Brazil also reported that leaf colour of the Brazilian begonia species, *B. soli-mutata* changes with shading, and the change is pronounced within 10 minutes and is complete within 30 minutes, probably due to reorientation of the chloroplasts in the cell cytoplasm. Jürgensmeier (1961) stated that oxalic acid oxidase-peroxidase system affects the decolourization of anthocyanin *in vitro.* Jürgensmeier and Bopp (1961) stated that the enzyme system which decolourizes red anthocyanin in begonia leaves was found to possess water soluble (peroxidase system) and glycerine soluble (oxalodehydrogenase or oxalic acid oxidase) fractions. Pynot and Martin (1969) reported that in *B. gracilis* var. 'Carmen', higher the temperature lower was the content of anthocyanins due to its poor production. Lee *et al.* (1979) while working with *B. pavonia* reported that abaxial anthocyanin layer in leaves is enhancer of light capture in deep shade.

Botany

Begonias are ususlly succulent herbs, shrubs or climbers grown either for their elegant flowers or for their beautiful foliage. Roots are tuberous, rhizomatous or fibrous and the tubers remain dormant in winters. Leaves form a rosette at the apex of rhizome, and are mostly coloured in red, purple, brown and/or silvery. First inflorescence emerges usually from 5th or 6th leaf axil in case of tuberous begonias. Flowers are unisexual, both male (more showy than female) and female (having prominent winged ovaries) together in the inflorescence, and red, pink, white, yellow or orange in colour, sometimes even bicoloured, and often double. Corolla segments are 2 + 2 (ovate-oblong, and 2 shorter) in male flowers and 2-6 (almost of equal size) in females. Styles are forked; three in number with papillose stigmas and fruit is a loculicidal bony capsule. Charpentier *et al.* (1989a) reported that in *Begonia horticola* (a South African Species in the section Teraphila) the placentation of the inferior ovary of the female flower is mostly axile with upper part being parietal. Anatomically they recorded that at early stages of development the inferior ovary grows underneath the perianth primordia, where axile part develops from an axial meristem and parietal placentas by the development of the septa towards the inside of the cavity. The anatomy of the mature female flower is similar to that of other *Begonia* species. In *B. dregei*, Charpentier *et al.* (1989b) found that inferior ovary develops below the perianth and the placental tissues have been recorded of mixed type with structurally intermediate ovary wall between axial and appendicular organs. Lecocq (1977) in *B. tuberhybrida* recorded inferior position of its three ovaries which are united themselves and with the four or five parts of the perianth, and most frequently there had been variations in the number of carpels (giving rise to meiomerous or pleiomerous gynoecia) and in the position of the ovaries.

With exception of the 'pendula group' which has 2n = 28 chromosomes, all other hybrids are tetraploids (2n = 56). Legro and Doorenbos (1969) studied somatic chromosome numbers in about 100 *Begonia* species out of which 14 were found with 2n = 22 chromosomes, 35 species with 2n = 28 and 17 species with 2n = 56 chromosomes, and concluded that the sections Pritzelia and Begoniastrum encountered with a great diversity in the chromosome numbers. In a further studies, Legro and Doorenbos (1971) worked out somatic chromosome numbers in 90 *Begonia* species originating in Africa, Asia or America and recorded 16 different numbers ranging from 16 to 76, most common being 28 (26 species), 38 (10 species) and 56 chromosomes (15 species). Arends (1970) studied 23 Elatior begonias (*Begonia* x *hiemalis*) and recorded that 21 were triploids each with a diploid number of 26, 27 or 28 long chromosomes in addition to a consistent number of 14 short chromosomes, and stated that triploid Elatior begonias resulted from crossing tetraploid tuberous hybrids having various numbers such as 52, 54 and 56 of long chromosomes with the diploid, 2n = 28, *i.e. B. socotrana* which has short chromosomes, however, he further stated that the pentaploid variety 'Flambeau' (2n = 66) would have resulted from fusion of a tetraploid gamete which was produced by a tetraploid tuberous hybrid, with a normal haploid gamete of *B. socotrana*. Seedling progeny from the original Lorraine parents, *B. dregei* and *B. socotrana*, were treated with colchicine and the tetraploid shoots were perpetuated vegetatively and then these fertile Lorraines were crossed with selected *B. socotana* diploids to obtain triploid Elatior types (Horn, 1971). Horn *et al.* (1976) obtained allotetraploid Lorraine begonias through colchicine treatment of seeds and seedlings and by crossing induced autotetraploids of *B. socotrana* and *B. dregei*, and through backcrossing these with diploid *B. socotrana* and by crossing tetraploid *B. socotrana* and diploid *B. dregei* they produced several triploid progenies, out of which several outstanding selections were made. Seitner (1977ab, 1978) described about a compact and much branched yellow-flowered hybrid with peltate leaves with numerous staminate flowers named as 'Gold Coast' derived from *B. prismatocarpa* x *B. staudtii dispersipilosa* and the other foliage rhizomatous F_1 hybrid named as 'Rajkumari' with red leaves and petioles derived from *B. sudjanae* and *B. rajah*. Due to differing ploidy levels in the hybrids, breeding among them is highly complicated. Leçoço and Dumas (1975) described teratological floral parts corresponding to petals and stamens capped by stigmatoid lobes in double male flowers. The stigmatoid formations show a glandular activity and release essentially hydrophobic exudates. Zeilinga (1962) reported that diploid and tetraploid varieties of *B. semperflorens* were more sensitive to adverse growing conditions than its triploids. Matouš (1965) reported cytoplasmic male sterility in *B. semperflorens* by crossing it with various species and also by collecting spontaneous pollen-sterile mutants. The studies on the hybridization between *B. semperflorens* and other species (Matouš, 1969) for obtaining male sterile lines, as well as spontaneous occurrence of male-sterile mutants such as abscission of male flower buds before opening, reduced pollen viability or complete pollen sterility without bud drop are presented. Reimann-Philipp and Lorenz (1971) described about the 'cinderella' factor by which a particular type of anther deformity results and make the plants male-sterile. Under this, through hybridization the plants homozygous for this factor are produced and then these are used in producing triploid F_1 hybrids as in case of *B. semperflorens* x *B. gracilis*. Reimann-Philipp and Lorenz (1978) picked up a single diploid plant with brown foliage (a character normally appears only in tetraploid material) to study the inheritance of this character. They found brown leaf character dominating the green leaf though sib-matings or the backcrosses of the offsprings did not produce expected ratio of monohybrid segregation deficient with brown leaf but there were populations showing green leaf deficiency also which were picked up to be used as pollinators for tetraploid brown leaf inbred lines for the production of triploid brown F_1 hybrids. Karper (1973) reported two new Dutch hybrids, *viz.* 'Nocturne' (cv. Sarabande x *B. daedalea*) and 'Andante' (cv. Marble Arch x *B. daedalea*) which are propagated through leaf cuttings, each plant yielding some 30 cuttings which become saleable plants in 4-5 months. Doorenbos (1973a) involved many diploid, triploid and tetraploid cultivars in breeding programme using *Begonia socotrana* as male parent and got good success with tetraploid ones. He reported that yellow and white flower colours were recessive. Doorenbos (1973b) reported a triploid red cv. 'Turo' in Elatior begonia. Doorenbos and Legro (1968) studied the cytology of numerous clones of the original Gloire de Lorraine type and the Konkurrant subgroup and reported that Konkurrant subgroup members of begonia are triploid. They crossed *B. socotrana* and *B. dregei* and through colchicine they introduced tetraploidy in the hybrids which were used afterwards as seed parents through back-crossing with *B. socotrana* and then the Konkurrent type plants were produced. Preil (1974) described transformation of anthers into stigmas as a trait inherited independently of that of exposed ovules and the both are inherited by recessive genes with participation of maternal factors. Begonia var. 'Fireflush', a pentaploid hybrid derived by crossing *B. robusta* var. *rubra* (octaploid) with *B. annulata* (Doorenbos, 1975) is described. Hahn (1978, 1980) crossed rhizomatous Mexican *Begonia boweri* (introduced from south Mexico to Germany in 1948) with *B. mazae, B. smaragdina* and *B. heracleifolia* to produce Mexicross begonias resistant to mildew (*Erysiphe cichoracearum*) in winter and such hybrids require partial or full shade, most suitable for cultivation for miniature indoor gardens with colourful leaf patterns, but not too much humidity and remain evergreen throughout the year without any rest. These now have many cultivars. Preil and Lorenz (1983) made several reciprocal crosses involving a pink-flowered clone of *B. socotrana* (2n = 28, 33) and two lines (carmine-red male-sterile and white male-fertile) of the hybrid *B. semperflorens* (2n = 30-35) and recorded mainly uniform matromorphic, patromorphic or intermediate progenies with descriptions of their possible genetic origins such as parthenogenesis, gynogenesis and androgenesis.

Knuth (1962) through dip treatment of tip cuttings with colchicine obtained polyploid tissues but this reverted back soon. Lindermann (1968) recorded spontaneous mutants, *viz.* 'Schwabenland Orange' from 'Schwabenland' and 'Riegers Goldlachs Typ Fünfhausen I' & 'Riegers Goldlachs Typ Fünfhausen II' from 'Goldlachs'. Mikkelsen *et al.* (1975) used leaf cuttings of the cv. 'Schwabenland Red' and stem cuttings of the cv. 'Aphrodite Rose' (both sterile triploids) for γ- and fast-neutron irradiation and obtained two outstanding mutants, *viz.* No. 8173 and No. 8291 from 'Schwabenland Red', both producing numerous adventitious shoots though No. 8291 is dwarf but profuse in flowering. Also one orange-flowered mutant from the same cultivar was obtained where 40 per cent or more of the terminal flowers are female. Sandved (1975) obtained 'Aida', a mutant with red and white flowers from the cv. 'Liebesfeuer' of *B.* x *hiemalis*, which is easily propagated from tip cuttings. Shigematsu and Matsubara (1972) cultured *in vitro* the small pieces of mature leaves of *Begonia rex* cv. 'Winter Queen' having silver-white leaves on White's agar medium and their adventitious buds arising from them were γ-irradiated which produced some 7 per cent plants with sectorial chimaeric variegation in their leaves. The pieces of leaves of such plants were again cultured which produced young mutant plants after about 5 months having green leaves with silver spots. A completely mutant plant from the previous lot was obtained through frequent cutting-back, layering and through division of such plants (Matsubara *et al.,* 1974). Matsubara (1982) irradiated small explants of *Begonia rex* cv. 'Winter Queen' and *B. masoniana* cv. 'Iron Cross' *in vitro* on agar or Kanuma soil and the plants so obtained were subjected to frequent pruning to isolate sectorial mutations. This way he obtained non-chimaeric induced mutants 'Gin-Sei' and 'Ryoku-Ha' in *B. rex* and 'Big Cross' & 'Kaede-Iron' in *B. masoniana* though in *B. masoniana* he recorded two chimaeric mutants, 'Mini-Mini-Iron' and 'Orange Iron'. Soedjono (1988) when irradiated the plants of *B. cucullata* at 0-100 Gy in single and repeated doses, obtained a flower mutant from red to pink under single dose of 5 Gy, and the shoot production from the centre of the plant when the plant was given first 15 Gy initially and then 75 Gy 24 h later. Benetka (1987) irradiated the leaf cuttings of *B.* x *hiemalis* cv. 'Schwabenland' with γ-rays at 1.5, 2.0 or 2.5 krad, in vM_3 generation 14 mutants with shortened internodes were obtained, so the plants were shorter with about 50 per cent plants producing small leaves excepting the two mutants which had enlarged leaves. He also recorded two flower colour mutations. X-ray irradiation studies were also carried out on Rieger Elatior begonia, *Begonia* x *hiemalis* (Doorenbos and Karper, 1975; Molnar, 1976; Roest *et al.,* 1981; Lina and Molnar, 1983). Doorenbos and Karper (1975) stated to have obtained 18-35 per cent spontaneous mutants of flower colours, growth habit, colour, shape and size of the leaves and flowers though through leaf cutting irradiation with 0, 1500, 2000 or 2500 rad this percentage increased to 80 per cent in all the four clones tested. Roest *et al.* (1981) obtained almost cent per cent solid (non-chimaeric) mutants out of 30 per cent recorded for colours, form and size of leaves and flowers through detached leaf irradiation in two genotypes when grown *in vitro*. Through X-ray treatment at wageningen of a red but sterile progeny of a yellow tuberous begonia, a new cultivar 'Tiara' with double golden-yellow flower colour has been obtained (Anon., 1975a). Molnar (1976) and Lin and Molnar (1983) when working with the cv. 'Renaissance' (scarlet single-flowered), produced 'Northern Sunset', a semi-double fairly powdery mildew (*Erysiphe cichoracearum*) resistant mutant with profusion of flowers after subjecting three weeks of short days not exceeding 10 hours and which is suitable for year-round pot plant production or for bedding out, and another 'Saanred' compact mutant with attractive double bright red flowers having strong stems and finely serrated leaves, most suitable as potted plant for year-round greenhouse production, through X-ray treatment

Classification, Species and Varieties

Hoover (1989) estimated that only about 25 per cent of the total number of species described (up to 1981) are in cultivation, *i.e.* 64 out of 539 from Asia, 33 out of 165 from Africa and 210 of 610 from Latin America. There are more than 900 perennial species under the genus *Begonia* and these are classed into three major groups as under.

I. Tuberous (Tuberhybrida hybrids, the most striking flowering types with wide range of colours except blue and mauve) *e.g. Begonia biserrata* (leaves 30 cm tall, flowers in dense axillary cymes not exceeding leaves, tepals white and serrate); *B. bogneri* (leaves 12-15 cm tall, male flowers with 4 tepals, outer pair pale-pink, inner white, female flowers 6 tepals); *B. x cheimantha* (*B. dregei x B. socotrana*, semi-tuberous, flowers large generally pink, varieties are 'Love Me', lilac pink, 'Marjorie Gibbs', pink, 'White Marina', white edged pink); *B. cinnabarina* (stems 60 cm, flowers fragrant, orange-red); *B. davisii* (leaves 10 cm, flowers bright red); *B. dregei* (syn. *B. parvifolia*, 'grape-leaf' or 'maple-leaf begonia', tuberous or semi-tuberous, leaves pale-green with grey spots and purple veins above, dull red beneath, flowers white, *B. suffruticosa* is probably a variety of this species with white to pale-pink flowers); *B. froebelii* (underside of leaf purple, with fleshy purple hairs above and beneath, flowers pendant and crimson to scarlet); *B. gracilis* (hollyhock begonia, syn. *B. bicolor or B. diversifolia*, stems 60-100 cm, bulbils present in leaf axils, flowers pink, its var. *martiana* has fragrant flowers); *B. grandis* ssp. *evansiana* (syn. *B. evansiana*, a very hardy begonia, stems 60-100 cm, leaves copper green above with red veins, red beneath, flowers pendant cymes, fragrant, pink or white, varieties are 'Alba' white, 'Claret Jug' pink, 'Simsli' with large flowers); *B. homonyma* (syn. *B. caffra*, very close to *B. dregei, flowers white); B. x intermedia* (*B. boliviensis x B. veitchii*, its var. 'Bertinii' evolved by crossing *B. boliviensis* with *B. x intermedia*, stems 30 cm, flowers vermillion); *B. josephi* (flowers pink); *B. micranthera* (stems 30 cm, flowers white or pale-pink); *B. octopetala* (stemless, flowers ivory-white); *B. partita* (stem strongly swollen and rather tuberous or caudiciform at base, flowers small white); *B. picta* (stems 38 cm, leaves green mottled with white and purple-bronze, flowers pink and fragrant); *B. sikkimensis* (tuber woody, stems 45 cm, flowers bright red, its forms are 'Gigantea' and 'Variegata'); *B. socotrana* (rarely in cultivation but superseded by the hybrid strain known as 'Lorraine begonias', semi-tuberous, stems 30 cm, flowers rose-pink, a parent of many important winter-flowering hybrids, *viz.*, 'Ege's Favourite', 'Gloire de Lorraine', 'Marina' and 'Mrs. Lionel de Rothschild', all bearing pink flowers); *B. sutherlandii* (stems 10-80 cm, trailing, veins and margins of leaves red, flowers orange to orange-red); *B. x tuberhybrida* (first named so in 1896, hybrid tuberous begonias, also known as *B. x tuberosa* hort. and not *B. tuberosa* Lam., this by crossing with *B. boliviensis, B. clarkei, B. davisii, B. froebeli, B. gracilis, B. pearcei* and *B. veitchii* has given evolution of many beautiful cultivars, stems absent or present, and erect to pendulous, about 60 cm high, flowers single or double in axillary clusters of various colours, *viz.*, white, yellow, orange, pink to red, bicoloured and fragrant); *B. veitchii* (syn. *B. rosiflora*, stems 30-75 cm, flowers bright scarlet and fragrant); *B. x weltoniensis* (*B. dregei x B. sutherlandii*, commonly known as 'maple-leaf' or 'grapevine begonia', tuberous or semi-tuberous, stems up to 1 m, leaves veined purple above and pale-green beneath, flowers white or pink, its var. 'Alba' is white); *B. wollnyi* (semi-tuberous, stem 30 cm, flowers green-white), *etc. B. x tuberhybrida* varieties under 'Hiemalis' group {*B. socotrana* x (*B.* x *tuberhybrida*), the most commonly grown begonia as a flowering potted plant} *i.e.* large flowered group includes 'Bolero' (double, orange), 'De Ridder's Yellow' (double yellow), 'Eveleen's Orange' (deep orange), 'Exquisite' (pink), 'Fantasy' (vivid pink, everblooming), 'Man's Favourite' (single, white), 'Renaissance' (double, frilled, coral red), 'Schwabenland Orange' (bright apricot), 'Thought of Christmas' (double white), 'Van de Meer's Glory' (salmon-orange), *etc.* The varieties under 'Pendula group' are 'Dawn' (pale yellow), 'Golden Shower' (golden-yellow), 'Lou-Anne' (rose-pink), 'Red Cascade' (scarlet), 'Rose Cascade' (rose-pink), *etc.* Other varieties include 'Allan Langdon' (red), 'Bali Hi' (pale cream edged red), 'Bernat Klein' (double, pure white), 'Bonfire' (red), 'Buttermilk' (cream flushed apricot pink), 'Diana Wynyard' (large white), 'Elaine Tartelin' (rose pink), 'Fairylight' (milky white edged pink), 'Festiva' (yellow), 'Falstaff' (rose pink), 'Harlequin' (white edged pink), 'Jamboree' (ruffled yellow and orange-red), 'Jean Blair' (yellow edged red), 'Jo Rene' (crimson to pink), 'Kupfergold' (copper gold), 'Lulandii' (large, pink), 'Madame Richard Galle' (orange), 'Mary Heatley' (orange), 'Olympia' (crimson), 'Orange Cascade' (yellow flushed burnt apricot), 'Pink Princess' (pale pink), 'Queen Fabiola' (red), 'Rhapsody' (rose-pink), 'Richard Robinson' (semi-tuberous, white), 'Roy Hartley' (salmon rose), 'Santa Barbara' (yellow), 'Santa Teresa' (white, edge ruffled and pink), 'Saturn' (salmon-orange), 'Seville' (yellow edged pink), 'Speckled Roundabout' (semi-tuberous, pink and white), 'Sweet Dreams' (pink), 'Switzerland' (red), 'Tahiti' (coral orange), 'Thelma' (pink, double, edges frilled), 'Torsa' (pink), 'Trisha' (picotee yellow edged apricot), *etc.*

Further they are divided in 13 groups; i) **Single**-flowers large, tepals 4 and usually flat, ii) **Crispa** or

Frilled group- single, large with frilled and ruffled tepals, iii) **Cristata** or **Crested** group- same as ii) but outgrowth in centre of the flowers, iv) **Narcissiflora** or **Daffodil**-flowered group- double, large, central tepals erect, resembling as narcissus corona, v) **Camellia** or **Camelliflora (Large-flowered double)** group- double, large, tepals regular and flower camellia type but not ruffled, vi) **Ruffled Camellia** group, same as v) but with ruffled tepals, vii) **Rosiflora** or **Rosebud** group- large with flower centre resembling rose bud, viii) **Fimbriata Plena** or **Carnation** group- flowers large and double, tepals fimbriate, ix) **Picotee** group- flowers large, usually double, camellia shaped, tepals edged with various shades or colours, x) **Marginata** group- as in ix) but edged with distinct colour, xi) **Marmorata** group- flowers as in v), pink marbled with white, xii) **Pendula** or **Hanging-basket** group- stems pendulous or trailing, flowers many, single or double, and small to large, and xiii) **Multiflora** group- plants low, bushy and compact, flowers many, small and single or double.

II. Rhizomatous (Rex Cultorum hybrids and related species with colourful foliage and inconspicuous flowers) *e.g. B. boweri* (height 15-25 cm, a plant of foliage beauty, emerald-green leaves with a broken chocolate-brown margin, flowers pale-pink or white), *B. daedalea* (height up to 30 cm, good as foliage plant, mid-green leaves with a red network turning brown, flowers white tinged pink), *B. x feastii* (height 25 cm, a foliage plant with mid-green leaves but red beneath, flowers pale-pink), *B. manicata* (height 45 cm, a winter flowering begonia with mid-green leaves having narrow red margins, flowers pink), *B. masoniana* (commonly known as 'iron cross', height 25 cm, an outstanding foliage plant flowering seldom, 4-5 deep bronze purple bars radiate from mid-green hairy leaf centre forming a cross), *B. rex* (height 30 cm, the major parent of the foliage begonias known as Rex begonias, leaves dark green and wrinkled and a silvery zone close to the margin, and flowers pale pink), etc.

III. Fibrous-rooted (Cultorum hybrids, the wax begonia), *e.g. B. albo-picta* (grown for its bright green narrow leaves covered with silver spots, flowers green-white), *B. coccinea* (height 180 cm, leaves red margined, flowers bright coral-red, a hybrid variety 'President Carnot' is more vigorous with paler flowers), *B. corallina* (height 210 cm, leaves speckled white above and flushed red beneath, flowers bright coral-pink, var. 'Lucerna' has spotted silver leaves flushed red beneath), *B. fuchsioides* (height 125 cm, bright red or pink flowers), *B. haageana* (height 125 cm, leaves deep green above and purple-red beneath, flowers are pink tinged white), *B.* x *hiemalis* [syn. *B.* x *elatior*, parents being *B. socotrana* x (*B.* x *tuberhybrida*)], commonly known as 'winter flowering begonias', tubers absent but tending to die back to swollen bases, flowers single or double, white to pink, yellow, orange or red and flowering in winter or year-round], *B. ludwigii* (stems 1 m, swollen at base but roots fibrous, leaves white hispid above and beneath green with veins red at sinus above, flowers many, cream-white with green and pink marks externally), *B. metallica* (height 90 cm, flushed-pink white flowers), *B. semperflorens* (height up to 15-25 cm, dark purple or bright green leaves, red, pink or white flowers, the hybrids of this species are most popular for summer bedding where green foliage forms include 'Flamingo' with white flowers edged with deep pink, 'Organdy' with a mixture of red, pink and white, 'Pandy' with blood-red flowers, and purple-brown foliage forms include 'Carmen' with rose-pink flowers and 'Rosea' with red foliage, 'Coffee and Cream' pure white, 'Indian Maid' with deep scarlet flowers), etc.

Bakker (1968) stated that two new forms of leaf begonias, *viz.* 'Hyde Park' and 'Marble Arch' were developed at Wageningen by crossing ***Begonia bowery*** and ***B.*** *mazae*, leaves having darker markings along the veins. Doorenbos (1972) described of a var. 'Fireflush' which is probably a tetraploid hybrid between *B. annulata* and *B. robusta* and its correct name is *B.* x *leopoldii* cv. 'Bettina Rothschild'. Dipner (1975) provided some additional names of Elatior begonia cultivars such as 'Riegers Nixe', 'Riegers Ballerina', 'Aphrodite-Twinkles', 'Weisse Siegfried Merholz', 'Limelight', 'Turo' and 'Agnita'. O'Reilly (1978) described about a rhizomatous cultivar 'Lospe' derived from *Begonia boweri* 'Bowtique' x *B. carrieae*, with typical leaf margins and white flowers. O'Reilly (1980) described about a creeping rhizomatous cultivar with unknown parentage which has large metallic silver flushed rose-pink leaves when young with a blackish-burgundy centre and border but when mature the rose-pink deepens to red, and there appears no flower. Certain other popular brgonia varieties are 'Alba', 'Blue Boy', 'Carpet', 'Gustav Knaak', 'Leuchtfeuer', 'Mardi Gras', 'Marble Arch', 'Medora', 'Nocturne', 'Radio', 'Sarabande', 'Shirt Sleeves', 'Onnenschein', *etc.*

The species and varieties in any crop are evaluated for their plant shape, size, height and yield, for their presentability of decorative parts such as foliage or flowers and their colours, the season of growing and the period of presentation whether for certain period or year round (evergreen) including the durability of inflorescence or foliage, for their general appearance and floriferousness (in case of foliage beauty the compactness of the plant and foliage colour variation during different seasons), for their shade, light and temperature requirements, for their multiplicability, for resistance to adverse weather conditions, certain accountable diseases and other pests, *etc.* Loeser (1979a) laid out an experiment on Elatior begonias, in one case, taking four varieties, *viz.* 'Nixe', 'Baluga', 'Balalaika' and 'Aphrodite' at three different locations in Germany, *viz.* Heidelberg, Munster or Hamberg, and in another case, 21 varieties only at Heidelberg conditions, both under full sun or semi-shade, and found that in each locality 'Balalaika' and 'Balega' were outstanding though under Heidelberg conditions seven varieties performed well under full sun and three varieties under semi-shade. Westerhof (1980a) in Holland evaluated Elatior-type pot begonias for their year-round culture through two trials, one consisting of 17 cultivars and the other of 35 cultivars. He found various 'Schwabenland' types, 'Aida', 'Trudy', 'Tamara', 'Toran' and 'Lorina' quite satisfactory for winter-growing though other cultivars developed bud drop when grown in autumn or winter though cultivars 'Nixe', 'Elfe', 'Symphe', 'Sirene' and 'Elfride' were found quite suitable for summer growing. Loeser (1988a) studied 46 potted Elatior begonia hybrid cultivars for their space requirements and their response to short-day treatments on staggered plantings, and observed that short-day treatment had little effect on selling dates when compared to control but without short-day treatment the space requirement per 1000 plants was larger due to having larger plant diameter. Loeser (1988b) had grown 34 Elatior begonia hybrids outdoors in beds with 61-71 per cent shading or in balcony boxes and found that 'Fiorella', 'Ilona', 'Heidi' and 'Nelly' grew well under both the circumstances though 'Ariane', 'Lorina' and 'Korona' also performed well in beds. With 14 *Begonia tuberhybrida* hybrid cultivar trial, Loeser (1987b) recorded strongest and tallest plants in 'Stargazer' and 'Happy End'; largest flowers in 'Imperial Orange' and 'Memory'; earliest flowering in 'Champion', 'Clips', 'Nonstop' and 'Kamilla'; and by the end of September the severe mildew susceptible ones were 'Kamilla', 'Royal', 'Rokoko' and 'Stargazer'. Nelson and Cover (1976) studied 26 fibrous-rooted potted begonia (*B. multiflora*) cultivars in USA through seed being sown on July 17, January 21 and June 30 and recorded 'Amba', 'Derby', 'Pink Comet', 'Vodka', 'Whisky', 'Bella 7' and 'Blushing Baby' quite satisfactory. Meylan and Tripod (1980) in Switzerland tried 67 fibrous-rooted begonia (*B. semperflorens*) as bedding plants and reported that 'Fantastic Compacta' (scarlet), 'Fortuna Compacta' (white, large flowers), 'Ambra' (Salmon-red edged white), 'Verdo' (pink), 'Supernova' (scarlet) and 'Gladiator Red' (red) were outstanding with regard to plant vigour and height, colour and quality of foliage and flowers, size of the flowers and resistance to pests and diseases. With 32 *Begonia semperflorens* cultivar trial, Loeser (1987a) recorded strongest growth in 'Danica', 'Party', 'Stara', 'Hybris' and 'Avalanche'; the tallest growth in 'Danica', 'Stara', 'Hybris', 'Avalanche' and 'Fantastica'; the largest flowers in 'Hybris', 'Fantastica', 'Glamour' and 'Forto'; the earliest ones were 'Athena', 'Ecrin', 'Ascot', 'Avalanche', 'Ersa' and 'Forto'; and the most outstanding cultivars were 'Danica', 'Ambra', 'Vision', 'Athena', 'Stara', 'Hybris', 'Ecrin', 'Ascot', 'Olympia', 'Eureka', 'Rapid' and 'Juwel'.

Propagation

Usually single begonias and species are fertile and produce **seeds** but it takes two years from seed to seed. In double begonias since stamens are petaloid so they lack pollen. In case when potting compost does not have sufficient nutrients, the flowers become semi-double or single. Seeds are used for propagation only when new varieties are to be developed. Tuberous begonia hybrids are commercially propagated through seeds. The seeds are sown in February on the surface of a thoroughly moistened peat medium without covering. The medium is adjusted to a pH of 5.5 to 6.5. The growing temperature is provided from 23-26° C until germination which takes about one fortnight. They remain in the greenhouse for 4-6 weeks and then they are planted finally outdoor in the beds, usually in mid-May. Afterwards they are placed at 18° C night and 22° C day temperature for proper growth. Temperature at 20° C delays the emergence but supplying 23-25° C temperature for 5-7 days induces earlier germination when humidity is high. This way the tubers are harvested by mid-October. Sowing seeds for tuber production in *B. tuberhybrida* cv. 'Fimbriata Scharlaken' in Belgium on March 1 instead of January 15 saved the time and labour in pricking out the seedlings in the glasshouse bench though number of tubers produced was slightly lower than usual sowing time owing to the losses after planting out seedlings in the open but there was not much difference in the percentage of large (>4 cm diameter) tubers harvested (Anon., 1979a). Haegeman (1967) collected seeds of large-flowered double begonia (*B. tuberhybrida*) from eight seed producers as selections of yellow, scarlet and pink and had sown them in Holland in late January, and early- & mid-November, and the resultant tubers after lifting in October were planted in the following season and flowered in August and then their blooms were assessed for shape and colour. In case of fibrous-rooted and rhizomatous begonias, seeds are sown in pans or boxes containing seed compost as a medium in February-March at 16° C greenhouse temperature where after germination they are pricked off after first true leaf is visible, and then the seedlings are set in boxes of the growing compost, hardened and are finally set at their flowering sites from late May. These are discarded after flowering. Under rhizomatous types, a few seed-propagated cultivars are identified which breed true. In 'Hiemalis' begonias, 'Charisma' is known as seed propagated series.

Seeds of *Begonia semperflorens* varieties stored at room temperature in a brown paper seed envelope, and in a waxed paper within a polythene envelope lost viability drastically in second and third year when sown in the first week of December, however, those stored in the envelope of very thin polythene inside a paper seed envelope which was then wrapped in aluminium foil and sealed with adhesive plastic tape gave highest percentage of germination in the first year as well as in the second or even the third year (Bisaillon, 1968). Bass (1980) suggested storage of seeds of *B. semperflorens* at 4-5°C with 35-40 per cent RH. Haegeman and Van Onsem (1966) compared six media for the seedling growth of tuberous begonia cv. 'Dubbel Kopergeel', *viz.* (i) fertilized Finnish peat, (ii) Finnish peat alone, (iii) a 1:1 mixture of pine needle litter and Finnish peat, (iv) a 1:1 mixture of pulverized peat and pine needle litter, (v) a 1:1 mixture of leaf mould and 1-year-old pine needle litter, and (vi) pure peat added with salts. The seeds were sown on February 2 in peat with and without fertilizers, germination appeared first in the unfertilized peat, and all the seedlings were pricked out on February 21 and finally planted out in the open on March 25 and 28. Fertilized Finnish peat with or without pine litter with a pH of 5.2 was found best for growth and development though peat without fertilizers having pH of 2.9 gave poorest results throughout. *Begonia semperflorens* 'Dwarf Carmen' seeds germinated best (71 per cent) at 25°C when were exposed for short duration to room daylight every other day, though in light the germination was 62 to 100 per cent after 8 days, however, compared to 4 per cent germination in darkness, it could be found from 39 per cent to 80 per cent in dark when the seeds were treated with GA though potassium salt had no any effect and thiourea inhibited it (Pollard and Roberts, 1968). The hybrid seeds between the two clones of *B. socotrana* were when subjected to 9 to 15 h at 1200 to 2000 lx daylengths, gave maximum germination at 20°C, poor at 15°C and nil at 10°C, the short days (9 h) accelerated flower initiation and formed flowering shoots at the leaf bases whereas with long days (15 h) most plants flowered but with a few remaining vegetative for over one year and those flowered the flowers appeared on elongated leaf-bearing shoots, however, no bulbil was found formed above 12.5 h daylength (Zimmer, 1972).

In tuberous begonias, the **tubers** can be **divided** by cutting them into pieces, number of pieces depending on the size of the tuber. It would be better if only 2-4 pieces in a tuber is made, but if the tuber is quite large even up to eight pieces can be obtained, but it would be better if each piece has one healthy shoot. *B. evansiana* (*B. grandis* ssp. *evansiana*) can also be propagated through axillary bulbils collected in the autumn when the stem is dying, and if planted in the spring will produce the flowers the same year. *B. gracilis* produces minute bulbils in large clusters enclosed in papery bracts in the leaf axils, require one more season for developing flowering size bulbs. Tubers of *B. rex* can also be divided and propagated as with other rhizomatous or tuberous begonias. These will flower the same season when planted in the spring. *B. sutherlandii* also forms bulbils in the leaf axils after flowering which may be stored in winter and planted in spring. When growth starts in the spring, the bulbils are planted individually on the surface of the container containing damp potting medium and then the containers are kept at 7-10° C temperature for further growth. Rhizomatous begonias are also propagated through division of their rhizomes, each division having a healthy growing point. These are also planted in March-April. The growth through tuber is very slow because of their tendency to grow slowly as these tubers may be as old as 10 to 20 years or even older as there is no annual renewal.

Sub-tuberous begonias which are dwarf and shrubby, shed their leaves and even stems by late autumn, leaving a thickened tapering **caudex** partly above the soil level so at this time watering should be reduced. At complete falling of leaves, the pots are stored dry over winter at a minimum temperature of 7° C but are repotted in the spring when growth has initiated.

Tuberous begonias, the 'hiemalis' begonias and the fibrous-rooted begonias are propagated by **stem cuttings**, and the 'hiemalis' and many species and varieties, particularly the large-leaved rhizomatous begonias such as *B. masoniana* and *B. rex* through leaf cuttings also. A rooted terminal vegetative stem cutting having one fully expanded leaf is most suitable for propagation. The stock plants from where the cuttings are to be taken are kept vegetative by providing 16 h long days and 18-20° C growing temperature. Cuttings taken from such plants under long day conditions, root within 3-4 weeks at a medium temperature of 21-22° C. Cuttings are rooted better directly in the pots in which they are to be sold. Cuttings can be raised under polyhouse with intermittent mist in the summer but no mist in the winter. Plants produced from terminal cuttings save the production time. Grappelli *et al.* (1985) when used earthworm castings alone, or mixed with peat and sand (1:1) 3 parts to earthworm castings 1 part, recorded better results in mixed one. In fact, earthworm castings contain growth regulators such as GA_3, cytokinins (IPA) and auxins (IAA). Van Onsem (1955) reported that cuttings of the varieties of *Begonia multiflora* root well between 0.2 per cent and 0.4 per cent IBA, and 0.2 per cent of NAA, though there had been varietal differences, and sometimes even their combined treatments in woody cuttings, when used with charcoal. Chitra (1998) while working with *B. rex* also recorded success with stem cuttings and with the use of 10 and 100 ppm IBA the cuttings established and sprouted well in the medium sand + leaf mould in 1-1 ratio where plants reached to their maximum height, leaf area and flowering.

Besides terminal cuttings, they can also be multiplied through leaf discs, petioles, leaf blade, flower stalk, apex explants, anther and petals. **Leaf cuttings** are grown at 18° C soil temperature and supplemental light to encourage more branches and adventitious shoots but the process takes more time, *i.e.*10-13 weeks to form a plantlet (with 3-5 shoots and good root system as their sink areas are double), roots and shoots. Within 17 to 22 weeks these plants become saleable, and this way some 30 cuttings may be obtained from each mother plant. The stock plants from which leaf cuttings are to be taken should be given SD (12-13 h light) treatment for 4 weeks to promote shoot production. Djurhuus (1984) when had grown stock plants of *B.* x *tuberhybrida* cv. 'Karelsk Jomfru' at 2 CO_2 levels (330 or 1100 µl/litre) and 7 or 15 W/m^2 in 10-h SD or 24-h LD treatments, recorded 23 per cent more cutting production under CO_2 alone and 41 per cent under CO_2 + LD, and in rooted cuttings the LD or 7 W/m^2 encouraged growth and lateral shoot formation though CO_2 alone had no any such effect. In a trial with begonia 'Schwabenland', it was found that during summer and autumn the top grade cuttings are produced under pH 4.6, in winter at pH 5.1 and in spring at pH 4.0 (Anon., 1975b). Heide (1967a) noted the endogenous auxin levels of main growth promoting substances such as IAA and a gibberellin-like factor in the leaves of *B.* x *cheimantha* in the acidic fraction of ether extracts, being higher in the plants grown under long days than those grown in short days. They observed increased level of auxins at high temperatures, *i.e.* 20 and 25°C though gibberellin activity was not much affected; however, several other growth promoting substances were also observed in the non-acidic fraction. They attributed the reduction in auxin levels associated with low temperature and a greater number of short days which strongly suggests that the corresponding enhanced budding ability and depressed rooting ability result from a lowering of the ratio of endogenous auxins to endogenous cytokinins in the leaves. Furnish (1947) rooted leaf cuttings successfully in a simple nutrient solution consisting of one level teaspoonful of balanced garden fertilizer mixture dissolved in 1.14 litre of water and when two months old these were successfully potted in compost consisting of 2:2:$^1/_2$ (v/v) of leafmould, sand and loam. Fearon (1965) found faster growth in *B. rex* and stated that Elatior or large flowered ones become dormant during winter so giving them long days through tungsten light from November onwards prevents them going dormant and further flowering so cuttings from such plants could be taken from January. Bigot (1967) cultured calibrated leaf fragments of 14 *Begonia rex* varieties at 16-h days and 20-22°C temperatures and obtained rooting within five weeks while shoot formation capacity was found more on the upper than on the lower surface of the leaf fragment. Rooting hormone should be used for better results. Chlyah-Arnason and TranThanh (1968) through weekly spraying, on attached leaves of *Begonia rex* cv. 'President' kept at high R.H., of the aqueous solutions (10^{-4}M) of a natural cytokinin 6-γ, γ-dimethylallyl amino purine extracted from *Corynebacterium fascians*, obtained tiny shoot meristems over the veins of entire leaf surface after 5-7 weeks of treatment. Chlyah and Van (1971) when sprayed the intact (attached) leaves with 6•DMAAP or BA, both at 10^{-4}, buds developed particularly at the junction of the main leaf veins and produced leaves but no roots, in former case leaves remained immature while in latter case leaves were mature, however, low temperature (12-22°C) alone accelerated the bud formation on intact leaves or when cut across the main vein and sprayed with water while high temperature (24-22°C) alone inhibited this on intact leaves but when incised and water-sprayed a few roots developed on the distal side of the cut and on attached leaves treated with IAA at 10^{-4}M no roots were formed but on cutting the main vein roots were formed. Chitra (1998) with *B. rex* found leaf cuttings as most reliable and successful propagules than stem cuttings and with IBA treatment only 100 ppm dip for 6 hours provided encouraging results in terms of success in sprouting percentage, time taken for sprouting, growth rate of sprouts, total number of sprouts emerging per leaf, and the total number of leaves being produced from new plantlets otherwise control proved better than IBA when the treatment was for lesser period. Wildern and Criley (1975) with leaf cuttings of *B. nulambafolia* reported that N6BA at 100 ppm and PBA at 200 ppm increase shoot development while IBA has little effect on root numbers only. Prevot (1968) when used 2- and 6-cm long pieces of leaves of the begonia vars 'Raureiff' and 'President Carnot', and the removal of the adventitious buds as they appeared at the proximal end of the leaf piece promoted the appearance of the buds, but not the roots, at the distal end. When leaf pieces from 1 to 20 cm long were taken it was noted that smaller the piece the greater were the numbers of buds and roots in proportion to the length, and the treatment with BA showed this difference quite marked though completely suppressed the root formation in all the cases. Further it was noted that the veins isolated from the laminae with epidermises intact formed buds without polarity. Guirfanova and Tokin (1942) through sieve-like punching (3-5 mm circular holes in diameter) in one or two of the leaves attached with the plants of *Begonia rex* and then keeping the plants at 20-25°C temperatures in propagation boxes in a humid atmosphere, noted that about two-third of such leaves had fallen but in 16 of such leaves they obtained rooting on the edges of the circles after a month followed by appearance of new leaves a week later on the upper side of 10 of the leaves. Gisløord (1974) found 8000 lx light intensity as optimum for shoot and root initiation in leaf cuttings of *B.* x *hiemalis* cv. 'Schwabenland' and from 4500-8000 for 'Liebesfeuer'. Hilding (1974) found 18°C as the best temperature for adventitious bud formation in *B.* x *hiemalis* 'Schwabenland' leaf cuttings as well as mother plants though growth of the adventitious shoot and root formation was found better even in 'Liefesfeuer', and the cv. 'Nelly Visser' was not found suitable for leaf cuttings as there only a few adventitious buds were found formed after 60 days of the planting of the cuttings. However, keeping the mother plants at short day conditions stimulated adventitious bud formation and caused more vigorous rooting and cuttings of 'Liebesfeuer' in three weeks at short days flowered whereas 'Schwabenland' cuttings remained vegetative. Hentig (1978) treated the mother plants in 'Schwabenland' group of begonia cultivars, *viz.* 'Ballerina', 'Balalaika', 'Baluga', 'Elfe' and 'Nixe' with daylengths of 12-16 h with a night break of 4 h before midnight in the winter which resulted in the production of more leaf cuttings than those kept under 10 h daylengths though no difference was found whether the plants were given 17 or 21°C temperature. However, the 'Aphrodite' group were best produced from tip cuttings from the mother plants kept under 14-16 h days with a similar night break in winter. Hendriks and Ludolph (1990) had grown the Elatior begonia mother cultivars at 20/22°C or 16/18°C (day/night) temperature and at 40, 80 or 120 klx h/day illumination and the cuttings were taken from December to March, maximum number of cuttings being recorded at 80 klx h/day though at 40 klx h/day the results were at par to control, however, illuminated plants formed shorter internodes and smaller leaves. *Begonia socotrana* which normally flowers by Christmas, should suitably be propagated in April and May and then should be kept vegetative through temperature and daylength treatments as Dutch varieties when kept at 26.7°C remain vegetative irrespective of daylength (SD or LD) treatments albeit the varieties such as 'Lady Mac', 'Marjorie Gibbs', 'Melior' and 'Tove' when kept at 26.7°C flower during normal short days but no buds initiate with a daylength of 15 hours, however, at 15.6°C all the varieties flower whether it is long or short days, though flowering appears later under long days (Horton, 1953). Rünger (1957a, 1957b) subjected the leaf cuttings of Lorraine begonias first to short-day conditions for varying durations, then transferred these to long-days, and recorded that the number of flowering shoots increased with increasing periods of short-day treatments and after 60 short days some adventitious shoots produced the flowers, though those kept only on short days recorded less adventitious shoot formation. In a further experiment, he recorded that plants become generative in $12^1/_2$ hours photoperiods but remain vegetative with 13 hours. BA at 1 and 0.5 per cent solution or with talc was very effective in giving high percentages of adventitious buds and roots when applied to the cut surfaces of leaf cuttings of cv. 'Peterson' and the cuttings taken between August and September were the best as compared to those taken from June to September, and those taken in November and after differentiated well with NAA 1 per cent and talc though 1 per cent TIBA resulted in root formation and not the bud (Nagamura and Urabe, 1976). Rünger (1959, 1960) in Lorraine begonia leaf cuttings recommended 15-20°C temperatures for first 20-30 days after planting, for root as well as adventitious shoot formation, best being in summer if the plants are kept cool in cellar and given artificial illumination, and by taking the cuttings in May or in June the saleable plants may be prepared in December. For root formation, he recommended 25-30°C for first 40 days, then 20-25°C, however, at 15° the process was slow and at 10°C no rooting occurred. At 15-30°C but at short day conditions the shoots formed the flowers though at long day the plants became vegetative. In further trials, Rünger (1970, 1978b) obtained initiation of adventitious shoots under short days, earlier being in cuttings from larger leaves (6-8 cm diameter) than those from smaller leaves (2.5-3.5 or 4-6 cm diameter) and the adventitious shoots which grow vegetatively in short days produce smaller leaves with shorter nodes and remain vegetative for longer in short days than those with large leaves in long days. The cuttings with larger leaf blades (up to 7 cm diameter) have been observed producing greater number of roots and shoots in *B.* x *cheimantha*. In *B.* x *cheimantha* mother plants, Heide (1964, 1965) obtained better rooting of cuttings when the plants were grown in continuous light during winter and kept at 16-18° or 12-14°C than at 20-22°C, however, 18°C being the optimum though for optimum growing it required higher temperature. Heide (1967) obtained promoted shoot formation but with inhibited rooting when applied cytokinins under low temperatures and short days though at high temperatures and long days it was reverse. Cohl and Moser (1976) with Rieger begonia cultivars reported best results in terms of percentage of cuttings with shoots, and their rate of development when mother plants were grown at 15.6°C with SD treatment while their cuttings being given LD treatment. Supplementary lighting of 25 or 75 lx to the mother plants of Rieger begonia cultivars resulted into satisfactory number of leaves, and with leaf cuttings in a greenhouse the assured formation of adventitious shoots and the root development were obtained when the shelves were provided with 2500 lx lighting (Hentig, 1980). Horváth and Horváth (1969) obtained rooting in petioles of *B. semperflorens* with 10^3 mg/l IAA in 8 days compared to 14 days with 2.2 mg/l kinetin and 11 days in control. Mikkelsen and Sink (1978) reported that epidermal and subepidermal cells formed a callus at the basal portion of petiole in the leaf petiole cuttings of Rieger begonia cv. 'Aphrodite Peach' and rooting occurred from parenchymal cells of the petiole and from the cells of the internal portion of the callus. Lagerstedt (1967) through punching of the begonia leaf discs by a cork borer, each leaf yielding some 40-50 discs, these discs were sterilized with 10 per cent Clorox for 3-4 minutes and then were cultured under continuous fluorescent light on filter paper with 3 ml of distilled water as the culture medium, and this way each disc yielded a new plant resulting in a 10-fold increase in plant production. Rooting was encouraged with 50-100 mg/l IBA for 30 minutes or IBA 50 mg/l + kinetin up to 10 mg/l. Creighton (1968) propagated *B. phyllomaniaca* 'Templini' which has variegated leaves through partially mature epidermal cells of leaves, stems and peduncles. The plants were stimulated to subdivide by injury or by partial removal of the root system. This way he noted development of two stipulate leaves on leaves, and stems, and male flower primordia on peduncles, and the flower primodia though was found abscissed after 1-2 months but the leafy outgrowths could remain alive for as long as two years. None of the outgrowths on leaves and less than 1 per cent of those on stems developed a stem meristem, and where these occurred the leaves of the mature types developed where through induction of the roots these could be used as a separate plantlet with bulbous mass of xylem forming at the base of the outgrowth. No vascular connection between the parent plant and the outgrowth was found.

B. x tuberhybrida and haemalis can also be multiplied through **leaf explants**. Leaf sections (2 x 2 cm) having a part of the central vein is cultured in MS medium with vitamins, sucrose, 1 mg/l NAA and 5 mg/l IBA. Cultures are maintained in a 16 h photoperiod and at 20° C. After 8-10 weeks, the explants start bud differentiation when they are divided and sub-cultured to the liquid MS medium with all the initial supplements but 1 mg/l IBA instead of 5 mg for proliferation and development of propagules. These propagules are now placed *in vitro* in a 2 mg IBA for 10 days to induce rooting. This whole process takes about five months. Thakur (1973) obtained shoot buds through leaf explants of *B. semperflorens* grown on Nitsch basal medium supplemented with a cytokinin (6-furfurylaminopurine, BAP or SD 8339) at 10^{-5} or 10^{-6}M. Addition of IAA, IBA or NAA at these concentrations or GA at lower rate induced rooting. With *Begonia coccinea* leaf explants on MS medium with or without 1 mg BA + 0.1 mg NAA/litre, there had been 100 per cent plantlet formation (Gui *et al.*, 1985). With *B. fimbristipula* leaf explants cultured in liquid SH (Schenk and Hildebrandt) medium supplemented with 0.125 mg/litre 2, 4-D, 0.25-0.5 mg/litre BA and 10 per cent (v/v) CM or 0.2 per cent (w/v) CH produced embryoids after 50 day's incubation (Zhang and Guo, 1988). Direct bud formation (no callusing) occurred when 1-cm^2 young leaf explants of *B. masoniana* were cultured on MS medium amended with 2 mg/litre BA + 0.2 mg/litre NAA and after 60 days these plantlets (0.5-1.0 cm in height) were further subcultured for next 75 days on MS medium with 0.5 mg/litre BA + 0.2 mg/litre IAA and so some 9-fold increase in the number of plantlets was noted (Mei and Ai, 1987). Through leaf blade explants on modified MS medium with 1 mg NAA and 1 mg BA/litre and under 14 hour days at 1000 lx light intensity, 20°C and 70-80 per cent RH of Elatior begonia cvs 'Schwabenland Red' and clone 'SO1', the success was obtained within 4 months of culturing (Roest and Bokelmann, 1980). With *B. x tuberhybrida*, leaf sections (2 x 2 cm) including a major vein were cultured in semi-solid medium containing MS salts, vitamins, sucrose, 1.0 mg/l NAA and 5.0 mg/l BA at 16 h photoperiod and 20°C temperature, and in 8-10 weeks when explants had begun bud differentiation, were divided and subcultured to a liquid medium containing MS salts, vitamins, sucrose, 1 mg/l NAA and 1 mg/l BA and then proliferating propagules were placed *in vitro* in an IBA (2 mg/l) solution for 10 days to induce rooting, and this way it took some 5 months from explant culturing to placing of rooted propagules in the pots (Peck and Cumming, 1984). Iida *et al.* (1986) cultured leaf segment explants of *B. tuberhybrida* on MS medium containing 1 mg/l NAA and found good rooting when transferred to half-strength MS medium containing 1 mg/l NAA and after 30 days these could be transferred to soil, and this way 7 x 7 mm^2 of leaf segment could give 10^5 plantlets in a year.

Immersing the petioles of detached leaves of *Begonia* x *cheimantha* in ABA for 24 hours before planting formed adventitious buds, but reduced the number of roots produced only at higher concentrations (Heide, 1968). Fonnesbech (1974) with **petiole** segments of *B.* x *cheimantha* obtained normal plants on media containing 0.01 mg/l NAA and 0.5-1.0 mg/l BA and at 15-20°C temperatures, having no shoots in lower concentration of BA though at higher concentration shoot formation was promoted but the shoots were abnormal with malformed leaves. Thakur (1975) found that to petiole segment explants of *B. picta* cytokinins accelerated the differentiation of shoot buds and auxins the roots, recording further that low concentrations of auxins stimulated shoot bud differentiation while BAP and IBA for formation of the forked roots. Sehgal (1975) by culturing leaf and petiole segments of *B. semperflorens* on White's basal medium supplemented with kinetin (0.1-1.0 ppm) and casein hydrochloride/yeast extract, obtained adventitious buds and the embryo-like structures. Welander (1981) cultured leaf petiole explants of *Begonia elatior* hybrids on agar medium supplemented with 0.1 mg/l NAA and 0.05 mg/l BA and obtained shoot and root initiation. Imelda (1983) with petiole segments of Rieger begonia cultivars cultured on MS medium and obtained best shoot and root production at 0.1 mg NAA + BAP + 0.2 or 0.4 mg/l of 2iP. Viseur and Lievens (1987) found flowering in *in vitro* cultured plants of *B.* x *tuberhybrida* within 6-7 months by planting in peat containers. After flowering, tuberization occurred in normal coarse with average diameter of 4.0 cm. Khoder *et al.* (1981) cultured Elatior-Rieger begonia explants from various sources on modified MS medium and found induced shoot growth in medium low in ammonium nitrate and containing NAA at 0.1 mg/l + kinetin at 2 mg/l, *vis-à-vis* accelerated root growth after 50 days in similar medium with reduced level of kinetin, *i.e.* 0.4 mg/l, and thus after 7 months they obtained some 1,15,000 plants from one plant from an apical explant. Bigot (1981) cultured leaf, petiole, stem and inflorescence explants of Elatior begonia cultivars on MS medium supplemented with vitamins, 0.1-0.5 mg/litre NAA and 1-3 mg 2iP (dimethylallyladenine) and got success in producing the buds, as well as roots when same medium was fortified with 2 g activated charcoal and a mixture of 0.5 mg 2iP and either 0.2 mg IBA or 2 mg IAA. After 7 subcultures the ability of tissues declined to produce buds rapidly.

Reuther and Bhandari (1981) cultured the **shoot tips** of various Elatior hybrid begonia cultivars on Linsmaier/Skoog medium supplemented with 1 ppm IBA + 0.5 ppm BA to raise *Xanthomonas begoniae* (*X. campestris* pv. *begoniae*)-free stock plants. Leaf lamina segments (5 x 5 mm) from *in vitrto* plantlets were cut and cultured on the same basal medium containing IBA in combination with different cytokinins to induce adventitious organogenesis for mass propagation. BA at 0.5 ppm stimulated the formation of adventitious leaves whereas kinetin and 2iP promoted shoot initials on the wounded periphery of the explant. Each new plant was differentiated entirely from a single cell.

Pierik and Tetteroo (1987) cultured young **flower bud** explant of the Brazilian species *B.* venosa, a little known pot plant which is priced for its grey, hairy and succulent leaves. The callus was induced by BA and NAA at 0.5 mg/l, at 21°C and at low irradiance, its subculturing was found best on a medium containing 0.1 mg/l BA and 2 per cent glucose, the shoot development was found enhanced by lowering the BA and glucose concentrations to 0.01 mg/l and 0.5 per cent, respectively, and optimal rooting was recorded on a growth regulator-free medium with 0.5 per cent glucose. Sucessfully, the plantlets were transferred to the soil with 95 per cent success. Zang *et al.* (1982) cultured *B. semperflorens* **anthers** containing microspores at the late uninucleate stage which formed calli best on B_5 medium supplemented with kinetin at 1 mg/l, 2, 4-D at 2 mg/l and 5 per cent sucrose. Callus differentiation was induced on MS medium supplemented with zeatin, BA, kinetin, NAA and IAA, and shoot formation was found best on MS medium + zeatin at 10 mg/l, IAA at 0.5 mg/l and sucrose at 3 per cent. These shoots after separation from calli were transplanted into vermiculite and watered with half-strength MS nutrient solution containing 0.1 mg/l IAA where these rooted within 2 weeks. These plantlets had originated from pollen grains. Khoder *et al.* (1984) cultured anthers of an Elatior begonia (3n = 42) on Margara N45K medium where after 5 weeks callus had developed, out of which some calli produced shoots and thereafter plantlets while others only roots, and cytological studies revealed that four of the plantlets were monoploid with 21 chromosomes. Margara and Phelouzat (1984) cultured **petals** of *B.* x *elatior* cv. 'Schwabenland Rot' on media containing 10^{-6}M BA and (a) 10^{-6}M NAA, (b) 10^{-5}M 2, 4-D, (c) 5 x 10^{-8}M 2, 4-D or (d) no auxin. When both BA and auxin were present the petals explants produced roots, leaves and petaloid structures though with medium (b) organogenesis was temporarily inhibited. On subculturing, neoformation of abnormal structures was observed in (d). All the structures originated from epidermal cells except the roots which appeared from deeper parenchymatous cells. Numerous embroids both normal and atypical were formed and the external cells of neoformations accumulated anthocyanins and tannins in their vacuoles.

In a further studies, Margara and Piollat (1984) when cultured petals on modified Margara N5K nutrient agar medium containing both auxin and a cytokinin, obtained small petals which were transferred to various nutrient media (N5Ca, N5K and N30K) two months later with or without growth substances. In absence of growth substances, little or no rooting occurred but leaves and in most cases petals regenerated, and in the presence of NAA, kinetin or BA the explants formed roots, leaves and petals, N5K giving best results especially with BA or kinetin. In the presence of NAA, in all the media rooting was copious, and petals subculture resulted into intensified organogenesis.

Soils, Land Preparation, Planting and Other Cultural Operations

The growing medium should contain high level of organic matter. Soil mix should be moist but not wet, and do not allow it to dry completely at any time. Irrespective of the groups they belong to, begonias perform well in a medium which has a **pH** reaction of 5.2 to 7.0. Acidic medium should be amended by addition of calcium carbonate and the medium having high pH is amended by applying gypsum or double superphosphate. Bik (1967) found optimal pH of 5.2 for raising begonia cuttings and 4.7-5.2 for seed germination. Regarding proper **CO_2** concentration in the glasshouse, Lindemann (1973) tried 0.03 per cent (control, as already existing in the environment), 0.06, 0.1, 0.2 and 0.3 per cent atmospheric CO_2 concentrations and recorded 1.0 per cent, followed by 0.06 per cent giving improved growth and quality, stronger branching, increased number of leaves, buds and flowers and reducing the length of the production period in Elatior begonias; Anderson (1990b) reported 600 ppm as optimum in case of *Begonia* x *hiemalis* cv. 'Schwabenland'; Ingenillem *et al.* (1984) stated 800 ppm as optimum which stimulated the plant growth by a fifth, *i.e.* 14 days; while Mortensen and Ulsaker (1985) reported 900 µl litre^{-1} concentration as the optimum for increasing the dry weight, numbers of leaves and flowers and for reducing the time for flowering in hybrid Elatior begonia cv. 'Schwabenland' at a light level range of 45, 130, 270 and 390 µmol m^{-2}s^{-1} though varying temperature levels (16, 20, 24 and 28°C) did not affect it.

A light **soil mixture** is essential for the development of fine root system of begonias. They perform well in a soil mixture consisting of 7 parts good garden loam, 3 parts peat, 2 parts sand and fertilizer mixture (2 parts ammonium nitrate, 2 parts lime superphosphate and 1 part potassium sulphate, 6 g of this fertilizer to 1 kg of medium). Almost the same mixture is John Innes Potting Compost No. 2 where it is 'hoof and horn meal' instead of ammonium nitrate. Bugbee and Frink (1986) stated 1:1 (v/v) peat-vermiculite potting medium as the best one where they suggested that aeration may be adjusted by altering the particle size of the two constituents. They mentioned that plants grow best in the aeration range of 11.3 to 20.0 per cent by volume. Wasscher (1953a) compared the soil mixtures comprising peat, peaty material, and dredgings with farmyard manure or fertilizers on various varieties of begonia, and reported that var. 'Exquisite' produced more flowers, branches and leaves in peat and peaty material mixtures, and var. 'Nelly Visser' produced more flowers in dredging mixture though leaf and branch numbers were recorded greatest in peaty material. Kamińska (1967) raised *Begonia semperflorens* seedlings in leaf mould of pH 7.1 and sphagnum moss peat limed to 6.1 and enriched with micro-elements, and the fertilizers of 1N:1P_2O_5:1K_2O:0.3 MgO or 2N:0.8 P_2O_5:1.5K_2O:0.3 MgO were also applied at 1, 2, 3 or 4 g/l of substrate and reported that peat applied with 3-4 g/l of 1N:1P_2O_5:1K_2O:0.3 MgO produced heaviest seedlings. Schwemmer (1970) raised Lorraine begonia cv. 'Marina' in Germany on two peat substrates (Flora-Torf, pH 3.5, and Alpentorf, pH 5.1) supplied with three liquid fertilizers (Wuchsal, pH 6.9, Gabi plus, pH 7.4, and Kamasol grün, pH 6.4), and recorded best results with Alpentorf + Gabi plus because of the most suitable pH range in this medium. Witte and Sheehan (1974) obtained vigorous Rieger begonia plants with highest number of flower clusters in 6:4 bark/sand medium with Osmocote (18N, 9P_2O_5, 9K) at 7 kg/m^3 when compared to 4:6 bark/sand mixture or 1:1:1 bark/sand/peat mixture. Excellent results in begonia cv. 'Schwabenland' were obtained when it was grown in rockwool blocks measuring 10 x 10 x 10 cm fertilized at every watering with 2 g Nutriflora (9 per cent N + 5 per cent P_2O_5 + 19.5 per cent K_2O + 5 per cent MgO + Ca and trace elements) per litre of water (Anon., 1977). Nagamura (1980) used plastic or clay pots filled with soil:sawdust:rice husk (40:30:30) as control, sawdust:rice husk (75:25) or 100 per cent sawdust and fertilized once to thrice per week with N:P:K at 100:53:93 ppm, and found good results when fertilized thrice irrespective of media and pot type used. Marahrens and Toop (1986) found peat-vermiculite and peat-perlite substrates excellent when amended with superphosphate and calcium nitrate where cuttings of *B.* x *lucerna* developed strong main roots and very extensive fibrous branching *vis-à-vis* excellent and compact shoot growth compared with vermiculite or perlite alone.

Begonia **tubers** grouped in 4 sizes, with diameter ranging from 1.9-2.5 cm (2.3-5.9 g) to 4.1-5.0 cm (19.35-41.35 g) when were planted in Germany, best results were obtained with the largest tubers as these sprouted $9^1/_2$ days earlier and flowered 12 days earlier than the smallest group (Maatsch, 1956). Tubers of *B. x tuberhybrida* are started into growth by planting in March-April barely covered (only lightly) at 18/13° C (day/night) temperatures at a bright filtered location, spaced 6-8 cm apart or in 10.0-12.5-cm pots containing growing compost, and finishing off in 15-20-cm pots. Roots come up from different points of the tuber surface so light covering of the medium is must. Since tubers take 12-15 weeks to come into flower so for early flowering early planting is necessary. When nights are warm and danger of frost is over, the pots may be taken out. Skalská (1964a) planted the tubers of five sizes of *B. tuberhybrida* at four **spacings**, *viz.* 15 x 15 cm, 20 x 20 cm, 25 x 25 cm and 30 x 30 cm for tuber production and reported that 15 x 15 cm spacing was too close as this encouraged the development of *Oidium begoniae,* 20 x 20 cm was found sufficient for manual cultivation while remaining two spacings were found suitable for mechanical cultivation, and he recommended to use wider spacings between the rows and closer within the rows. Pot spacings on the bench with Rieger begonias in 15-cm pots have been tried at 22.5, 25, 27.5 or 30 cm apart (centre to centre) under 14-h days for the first 7 weeks from potting on November 16 in USA and under natural short days thereafter and these plants were ready for sale by January 29, and it was found that 22.5 cm spacing has been adequate for proper growth under short-day conditions (Nelson and Bost, 1975). Skvortsova (1972) raised the seedlings of *Begonia semperflorens* in the spring and planted out in early June at 20 x 20 cm, 10 x 10 cm, 5 x 5cm and 1 x 1 cm spacings, and he reported that only 20 x 20 spacing was proper for plant growth, branching and flowering while 1 x 1 cm spacing had been absurd and there was no branching. For early summer flowering, the tubers should be started in January and for mid-summer flowering in March. When shoots are 5-8 cm, they are removed gently from the containers where they had started and are planted in 10-cm individual pots, preferably the earthen pots but for larger tubers it may be 15-cm pot. It is now time to start with weak liquid fertilizer. While taking out the plants from the containers, care should be taken that no root of the fellow plants are disturbed. When they attain 15-cm height, they are moved into final pots or in the garden. For one tuber, a 20-cm pot is sufficient.

For bedding outside, after hardening the tuber begonias in cold frame, they are planted in prepared beds from April to May and the **tubers** are lifted in October, *i.e.* before the danger of frost occurrence. Earlier lifted plants may be planted in moist peat inside the greenhouse and allowed to grow until natural leaf senescence occurs. When leaf yellowing starts, watering is withheld to dry off the plants, and their tubers are over-wintered at 7° C in moist peat or vermiculite. Regularly they are checked that due to excessive dryness the tubers are not shrivelling or due to excess moisture they are not rotting. They are stored so until planting. Okagami *et al.* (1977) through applications of daminozide (B-9), chlorphonium (AMO-1618) or chlormequat (CCC) to the cuttings of *Begonia evansiana* delayed the onset of dormancy in aerial tubers but not in the detached ones. Tubers of *Begonia tuberhybrida* stored after lifting at 3-4°C until late August in Holland and then brought into the glasshouse at 18°C normally produce shoots more quickly than those cold-stored throughout, however, breaking of the first-formed 2-3 mm shoots delays shoot growth but more shoots are formed, and dipping of tubers in daminozide (Alar) at 1.5 mg/l for 10 minutes prevents one-fifth of the plants to go without flowering (Anon., 1979b). Price and Cunningham (1987) when stored the tubers for 8 weeks at 2.5°C of *B.* x *tuberhybrida* cvs 'Ruffled Pink' and 'Roseform Salmon', these were found respiring more rapidly during the next 24 hours at 20-21°C than those stored at 5 or 7.5°C, and storage in a 2-3 per cent O_2 atmosphere for up to 20 weeks at 5°C increased the sprouting percentage than those stored in air, however, no further growth or otherwise differences were observed in the tubers stored in the two atmospheres.

Half-hardy *B. evansiana* tubers are planted 5.0-7.5 cm deep in the beds outside but at a sheltered position though they are left as such in the field during winter where climate is mild but requires **mulching** to generate more heat. In cool greenhouse they are started in March-April by setting one tuber in 8-10 cm pot or three tubers in a 15 cm pot and grown at 10° C temperature with proper ventilation. Morning and late afternoon sun is alright but direct sunlight by mid-day is unfavourable from which the crop should be protected through shading. After sprouting till flowering, they are given with liquid manure at fortnightly interval. When leaves turn yellow, water is withheld gradually and the tubers are pot-dried and over-wintered at 2-4° C.

The pruned plants of 'Lorraine' and other winter-flowering begonias are started in March-April at a raised temperature regime of 10-13° C and when they grow up to 5.0-7.5 cm long, the shoots are removed from the base and planted singly in 6-cm pots containing equal parts of peat and sand (v/v) for rooting at 18° C, and then shifted to larger pots and finally in 15-cm pots. From this time now they are fed with liquid manure fortnightly till flowering. Initially the plants are **pinched** out 2-3 times for making the plants bushy. The greenhouses during summer are ventilated and plants are sprayed with water to give cooling effect. Just to save the energy for plant development, first flowers are nipped off in September-October so that these may give flowers profusely from November onwards. After the flowering is over and leaves start yellowing, the plants are cut back at 10-15 cm height, water is withheld but medium is kept moist (never dry) and pots are stored at 7° C.

Rhizomatous and fibrous-rooted begonias are grown as house plants or in the greenhouses in 15-20 cm pots (larger pots or tubs for *B. coccinia* and *B. corallina*, or planted out in the greenhouse border) at 10° C winter **temperature** though the species *B. boweri, B. masoniana* and *B. rex* require 13° C winter growing temperature. During winter the pots require to be kept moist not wet and the plants may not require water spraying but when temperature increases to 18° C in summer, they are water-sprayed to keep the environment cool. It also requires proper aeration, preferably in summer. During summer direct sunlight is avoided. They are repotted in March-April, followed by liquid feeding from May to September fortnightly. Bedding or pot plants of *B. semperflorens* are started through seeds in trays or pans having seed compost in Februaary-March at 16° C and when first true leaf appears these are pricked off and the seedlings are set in pots having growing compost. After hardening they are planted outdoors at their final sites from late May but in containers they are planted in 8-9 cm pots initially but finally in a 12.5-cm pot. Plants are discarded after flowering. Bowes (1990) recorded sporadic epiphyllous (adventitious buds as leaflet-like outgrowths) buds on intact plants of *Begonia rex*, and these buds develop into large leafy shoots which root rapidly. Certain rhizomatous begonias having been derived from crosses involving *Begonia manicata* var. *aureo-maculata* cv. Crispa, *B. heracleifolia* and many other Mexican species as are known for genetic growth anomalies, the varieties derived from them may also show such anomalies as has been reported by Boardman (1991) in Texas that single plant in two such varieties 'Essie Hunt' and 'Silver Surf' showed direct flower formation from leaves.

Lorraine and certain other tall growing or pendulous begonias require **staking**. Bamboo sticks or *Salix* canes should be inserted at each potting stage to support the slender growths. The insertion should not damage the growing tubers. Wasscher and Scholten (1954) reported that flower removal increases the size of the tubers. Van Onsem and Haegeman (1962) in France when **deblossomed** the plants of *Begonia tuberhybrida* in August, the tuber size was found increased. With var. 'Maxima', the deblossoming in mid-August and further repeated at short intervals, and in the large flowered varieties deblossoming thrice (to a maximum), increased tuber size in Holland (Van Onsem *et al.*, 1962). Maatsch and Zimmer (1964) promoted tuber growth in begonia seedlings through deblossoming as well as through SD treatment and the deblossming starting on August 20 had the greatest effect, more pronounced being under SD treatment. Skalská (1965) in tuberous begonia reported that bud and flower removal increases the tuber size significantly when buds are removed by the end of July to whole August due to days becoming shorter these days and this practice provides about 25 per cent more profit due to tuber sales. Hallig *et al.* (1966) reported flowering after 12 days of short day treatment before September 1, followed by pinching either at the start of SD treatment or 10 days before or only 8 day's SD treatment after September 1 and the optimum temperature being 18-20°C to Christmas begonia (*Begonia* x *cheimantha*) cv. 'Mørk Marina'. A growing temperature range of 11-14°C though produces better quality flowers but flowering is delayed. Sandved (1971c) while working with *Begonia* x *cheimantha* cvs 'Regent', 'Trine' and 'Storblomstret Astrid', reported that non-pinched plants flowered the earliest while pinching delayed it, more delayed flowering being under severe pinching. White *et al.* (1973b) pruned manually between 1 to 13 November retaining 2-3 of the shortest shoots, *vis-à-vis* other plants of the three Elatior begonia cvs 'Schwabenland Red', 'Goldlachs' and 'Krefield Orange' potted on August 30 were treated with chemical pruning agents, and reported that manual pruning was the best for increasing the branch numbers by 87, 95 and 100 per cent, respectively, in the three varieties but the flowering was delayed while synthetic cytokinins, especially SD 4901 (a formulation of benzyladenine) showed some promise. Agnew and Campbell (1983) with *B.* x *hiemalis* cv. 'Northern Sunset' reported that dikegulac spraying from 0.05 – 0.16 per cent in combination with hand pinching or double spraying of dikegulac (0.1 and 0.16 per cent) reduced the flower count significantly. Chlormequat (0.15 per cent) and hand-pinching reduced internode length but not the overall plant height, diameter, visual quality or dry weight.

Weeds from container-grown begonias should be taken out manually. Just to conserve the moisture if mulches like sawdust, perlite, straw, grasses, *etc.* or polyethylene are used, the weeds will automatically be controlled. Through soil fumigation with metam-sodium or methyl bromide the weeds will automatically be controlled because of their weed-killing properties but after fumigation no weedicide is applied. Benfluralin followed by Simazine as pre-plant or Simazine 5-6 days after planting followed by Bentazon 14-18 days after planting have been found effective in keeping the weeds under check. Van Onsem and Haegeman (1962) obtained slight increase in tuber size by mulching *Begonia tuberhybrida* field with black plastic film. Baranowski (1974) recommended Alipur (cycluron + chlorobufam) 1.8 kg/ha as weedicide in *Begonia semperflorens* field though Nexoval at 2.7 kg/ha was found phytotoxic as pre- or post-planting. Bing (1983) found napropamide 4.5 kg/ha as a best selective pre-emergence weedicide in begonia field though there was unacceptable injury. Himme *et al.* (1986) reported that 1.5-2.0 kg/ha 3 days post-planting ± 2.5-7.5 kg bentazone 3 weeks later in *Begonia maxima* field caused unacceptable injury, however, a single dose with 0.5 kg/ha bentazone 24 days after planting selectively controlled *Cardamine hirsute* and *Capsella bursa-pastoris.*

Water and Nutrient Management

Begonia is resistant up to 1.5 ppm chlorine for 3 hours but persisting treatment and/or higher concentrations cause necrosis and bleaching of old and middle-aged leaves, however, if chlorine is exposed to B-Nine treated plants or if affected plants are treated with B-Nine, there is little toxicity to the plants and also when after exposure there is lack of sunshine for long or under darkness the damage is found reduced (Brennan *et al.,* 1965) though there is no build up of chlorine in the plant tissues even if its high concentration is present in the atmosphere. Leone and Brennan (1969) in New Jersey (USA) while experimenting with 14 *Begonia semperflorens* cultivars stated that certain varieties were sensitive to ozone and SO_2, and the severity increases with increase in R.H. Gardner and Ormrod (1976) when exposed the plants of Rieger begonia cv. 'Schwabenland Red' to O_3 and SO_2 either separately or together at up to 30 pphm in former case and up to 180 pphm in latter case, for 4th-day for 5 days, at the early vegetative, pre-flowering or flowering stages, recorded the first two stages of plants having more leaf damage than exposure to single gas when exposed to 15 pphm of O_3 and 60 pphm of SO_2. Reinert and Nelson (1979) exposed the 12 Elatior begonia cultivars, *viz.* 'Schwabenland' cultivars (Pink, Red and Gold), 'Whisper O' Pink' and 'Improved Krefeld Orange' all being most sensitive, 'Renaissance', 'Heirloom', 'Nike', and 'Fantasy' all intermediate, and 'Ballerina', 'Mikkell Limelight' and 'Turo' all being least sensitive, and recorded poor dry weight of aerial parts in 9 of the 12 cultivars tested, and at 25 ppm O_3 in 4 varieties and at 50 ppm O_3 exposure in 8 of the 12 varieties there was inhibition of flower growth and development. Five Elatior begonia cultivars, *viz.* 'Schwabenland Red' (sensitive), 'Whisper O' Pink' (sensitive), 'Renaissance' (moderately sensitive), 'Fantasy' (moderately sensitive), and 'Turo' (insensitive) were when exposed to charcoal filtered air (control), O_3 (0.25 ppm), SO_2 (0.5 ppm) or O_3 (0.25 ppm) + SO_2 (0.5 ppm), injury occurred as expected in different cultivars as per their sensitivity but damage in all the plants was found further increased with combined exposure of both the gases (Reinert and Nelson, 1980).

The repeated spray or drench applications of zinc through Dithane Z-78 for controlling diseases was found beneficial even for dry matter production and for increasing flower numbers, and so was the case even with 10 or 100 ppm of inorganic Zn compounds (oxide or sulphate; Kahl and Wittmann, 1969). The foliar analyses of youngest canopy leaves 5 cm or wider in Rieger Elatior begonia cv. 'Schwabenland Red' indicated that the minimum critical levels for K were in the ranges of 0.93-0.95 per cent, for Mg it was 0.22-0.25 per cent and for B it was 13.0-14.0 ppm (Nelson *et al.,* 1979). Elliott and Nelson (1981) when applied 1.0 mM H_3BO_3 at progressive stages of growth to the potting compost in *Begonia* x *hiemalis* cv. 'Schwabenland Red', the B injury occurred initially on mature leaves, the chlorosis progressing from leaf margins to an eventual necrosis extending inwards for 3-5 mm from the margin and such leaves showed 125 to 258 ppm of B concentrations, however, older plants showed injury even at lower B concentrations in the leaves.

Humic acid and fulvic acid alcoholic extracts obtained through garden compost soil were when sprayed on the foliage of *Begonia semperflorens* in Czechoslovakia, these increased oxygen consumption by the leaves, the chlorophyll content and the growth of roots (longer growth with more branched laterals) and tops (Sladký, 1959).

A 12:20:26 **fertilizer** at the rate of 8-10 kg/100 m^2 in France had been reported quite encouraging with *Begonia tuberhybrida* (Cortvrieendt and De Groote, 1952). Skalská (1964) obtained larger tubers with 600 kg/ha than 400 kg/ha of Citramfoska (7:11:17.3) and with an additional application of 400 kg potassium sulphate, however, the combination mixture of 11:18:22 NPK at 365 kg/ha had given best results out of the three combinations used. Skalská (1966) stated that when two weeks before planting of *Begonia tuberhybrida* in the field, the soil was applied with, in addition to 80 kg N and 100 kg P_2O_5, 160 kg/ha K_2O either in the form of potassium sulphate (0 per cent Cl) or muriate of potash (12 per cent Cl) or 40 per cent potassium salt (42-48 per cent Cl), on analysis varying degrees of P_2O_5, K_2O, CaO, MgO, Na_2O and Cl were observed in the aerial parts during the growing season though the contents, especially P_2O_5, Na_2O and Cl were always observed lower in the tubers, and the Cl present in different fertilizers was in no case found giving any adverse effect on the plants or tubers. He further recorded that weight of aerial parts was increased from planting to blooming stage and in the tubers from blooming to harvesting. It was also stated that annual nutrient uptake with potassium sulphate application per 100 square metre, *i.e.* 2,500 plants spaced 20 x 20 cm, was about 0.89 kg N, 0.56 kg P_2O_5, 1.36 kg K_2O, 0.93 kg CaO, 0.43 kg MgO, 0.14 kg Na_2O and 0.40 kg Cl. Gugenhan and Deiser (1969) in Germany used 'Super Manural' (a peat-mineral fertilizer with standard formulation of 1 per cent N in mineral form of which $^1/_3{}^{rd}$ was for slow release, 1 per cent P_2O_5 and 1.5 per cent K) in *B. semperflorens* and found increasing yields with increasing applications, however, the new formulation with 2 per cent N and 2 per cent K was found even better than 'Super Manural'. Pivot (1975) found quite appropriate 152 mg N/litre of peat, 128 mg P_2O_5/litre of peat and 248 mg K_2O/litre of peat in pots with Gloire de Lorraine begonias. Bik (1978) in a pot trial with begonia cv. 'Schwabenland' found reasonably good pot plants when used 6-8 g/l of potting soil 14:14:14 NPK slow-release fertilizer Osmocote, and best results when combined with weekly top-dressings with a complete fertrilizer solution. Shakhova *et al.* (1979) with one-year and 2-3-year old plants of *B. rex, B. masoniana* and *B. diadema* when used 3:1:3 NPK top dressing from spring to autumn, obtained many decorative leaves, 0.2 per cent application rate proving most appropriate for first half season while 0.4 per cent thereafter. Schmidt and Tretner (1984) compared 400, 600 or 800 mg N/pot in the form of five slow-release fertilizers (Triabon, Osmocote, Plantosan 4D, Plantocote 4M and Nutricote) and recorded Osmocote and Plantosan 4D at 600 mg N/pot while Plantocote 4M at 800 mg N/pot giving best results with Elatior begonia cvs 'Najade' and 'Elfe'. Chase and Poole (1987) with fibrous-rooted begonia cv.'Prelude Red' grown in 10-cm square plastic pots found best results with Osmocote (19:26:10 NPK) at 0.5-3.5 g/pot when applied as top dressing. Will (1986) in Germany had grown *Begonia semperflorens* cvs 'Scharlach Juwel' and 'Rosa Juwel' in 10.0/22.5 cm Kultipack pots filled with peat compost and supplied with various amounts of Triabon, a slow-release fertilizer containing NPK + trace elements at 40 g/m^2 but found no difference in growth whether applied by spreading on the top of the substrate or mixed in. Four different types of fertilizers (a granular fertilizer, a liquid fertilizer, a slow-release coated fertilizer and an urea-based fertilizer, all applied at the rates of 0.75 and 1.00 g N/litre substrate) containing N, P, K and Mg were applied on Elatior begonia hybrid cv. 'Schwabenland Rot' but only small differences in flower development were noted among the treatments, though use of liquid fertilizer applied thrice per week over 8 weeks yielded highest plants (36-40 cm) with a total of 130 flowers per plant (Burghardt, 1988), followed by slow-release coated fertilizer. He recommended 0.75 g N/litre as sufficient. Hendriks and Scharpf (1984) rercommended 0.8 or 1.0 g/litre of 15:11:15 N-P-K fertilization in commercial compost or white peat applied 0-2 times daily from potting to flower development stage or 1.6 g/litre in the first 3-4 weeks after potting followed by 0.8 g/litre until sale. The effects of differential N and K nutrition either through calcium nitrate or ammonium nitrate at 6 levels and potassium sulphate at 7 levels have been given by Pawlowski (1967) on begonias. He reported 50 or 100 ppm N equally good for dry matter production, and stated that there was little difference between 100 and 200 ppm, and from 50-400 ppm the difference between the two nitrates had no or little difference, though at 800 ppm the results were utterly poor especially with calcium nitrate. The K from 0 to 1600 ppm applied, optimum was recorded at 800 ppm at optimum N levels and the highest level showed a very poor response even when compared to 50 ppm. For flowering there was no difference between the two sources of nitrogen and the response from 50 to 100 ppm was good and almost the same though higher doses showed poor response. In case of K for flowering, there was marked difference between 0 and 50 ppm though very little between 50 and 800 ppm especially at optimum N levels, and at 600 ppm there were fewer flowers than at 800 ppm.

For growing of seeds, the **medium** should contain 2 parts medium loam, 1 part peat and 1 part sand + fertilizer (lime superphosphate and finely ground chalk or limestone in 2:1, 1.5 g of this to 1 kg of medium). For pot culture, 7 parts of loam, 3 parts peat and 2 parts sand + fertilizer mixture (2 parts ammonium nitrate, 2 parts lime superphosphate and 1 part potassium sulphate, 6 g of this to 1 kg of medium) are most suitable. This medium is almost the John Innes Potting Compost where on the place of hoof and horn meal, ammonium nitrate is used. The basic principle of the medium is that this should be high in organic matter and most media should consist of sphagnum peat moss 45-60 per cent amended with soil, coarse perlite and vermiculite or coarse sand along with sufficient quantity of superphosphate to avoid phosphorus deficiencies which are most common. After one to $1^1/_2$ months of their planting in the final pots, they are fed at fortnightly intervals with liquid manure. It would be better if it is slurry prepared by dissolving 1 kg of fresh cow dung manure to 3 litres of water and 320 g of urea, filled in earthen pitcher covered with earthen lid and kept in sun for decomposition for about 10 days in summer and 15 days in winter, and then the volume is made to its 4 times, filtered and then sprayed in the medium. Finally if through evaporation some water is lost, that is added to make it 16 litres. While filling the pots, 1 kg of substrate should contain 95 mg N, 40 mg P_2O_5, 120 mg K_2O, 800 mg Ca and 400 mg Mg for optimum growth. If crop growth is still poor, despite regular application of liquid manure, nitrogen at the rate of 25-30 g/m^2 can be applied once but before appearance of first flowering buds. Hiemalis begonias are not heavy feeder, and a fortnightly application of liquid fertilizer of 100-125 ppm N is sufficient but ammonium levels in the winter should be reduced. Bosmans (1971) recommended spraying of urea at 10 g/l once or more times for better results.

In most of the *Begonia* species and varieties, stomata occurs singly on the cotyledons and on the primary leaves about half of these are in groups of 2-5, each group having a single respiratory chamber (Neubauer and Beissler, 1971) though on stipules this grouping occurs only half as frequently and on the mature leaves the majority of the stomata are closely grouped, being as many as 20 to a respiratory chamber in some cases. Hoover (1986) also reported stomatal clusters in many of the species as per the studies from Mexico forests on six population samples in *B. heracleifolia* and nine in *B. nelumbifolia*, an unusual character. There was an increase in the frequency of single stomata, the stomatal length showing a parabolic response, and the number of stomata per square milimetre decreasing with increase of the altitude in case of *B. nelumbifolia* and increasing with increase in the altitude in case of *B. heracleifolia*. In fact the stomatal clusters in begonia assist in water conservation. Medium should always be kept moist but not wet, and it should never be allowed to dry completely. At every change of container, they are watered. Afterwards also they are regularly **watered** lightly and before repeating watering every time, the pots should look dry. During cool weathers, the watering should be at the bottom of the plant and not above the crown. Therefore peripheral watering is preferred. The water requirement in tuberous begonias is more compared to other groups. Usually, 5 litres of water is required per m^2/day, and for a 5-month duration crop it comes to 750 litres/m^2. Wax begonias (*B. semperflorens*) require high humidity or fogging during germination or sprouting, and at any level these should not be water-stressed. Rex begonias require regular moist medium for proper root development but they do not tolerate over-watering. To avoid appearance of leaf spot diseases, the watering is carried out through drip or sub-irrigation and not on the crown. Hiemalis begonias which have fibrous root system can not bear either water stress or over-watering for long periods. For glasshouse crop, frequent water spraying is required in warm weathers to provide cooling effect to the crop, *i.e.* when temperature rises above 18° C but in winter when temperature is around 13° C no such spraying is required. Gislerød and Mortensen (1990) studied the effect of relative humidity (60±5 per cent or 90±5 per cent) and nutrient concentrations (1, 2 and 4 mS/cm) with *Begonia* x *hiemalis* in Norway and obtained 56 per cent lower transpiration rate under high RH and 10-20 per cent lower water consumption at higher nutrient solution. They recorded plant dry weight, height, width and leaf size higher under 2 mS/cm nutrient level than in 1 mS/cm solution at high RH but not at low, though highest nutrient solution either had no effect or proved detrimental. The number of flowers and percentage of flowering plants were found higher at 90 than at 60 per cent RH and at higher nutrient solutions. Djurhuus and Gislerød (1985) found *Begonia* x *hiemalis* tolerable to a wide range of moisture conditions whether watered once a day or several times but for substrate they found that when the compost contained more clay the growth became poor. Potting substrate containing some 20 per cent perlite is sufficient for proper aeration, and irrigation through automatic flooding system caused browning of Elatior begonia roots at the base due to water stagnation, low pH (4.0-4.5), O_2 deficiency and high ammonium concentrations (Scharpf and Grantzau, 1985). Haas and Röber (1988) found water requirement for Elatior begonias as 106-113 ml per plant daily and stated that through a closed ebb-and-flow irrigation system (which records the total water used) the pots should be watered when the substrate water tension reaches 60-90 hPa. Kwast *et al.* (1989) with Elatior begonia cv. 'Aphrodite Radiant' grown in peat with or without 25 per cent (v/v) perlite, watered the pots through ebb-and-flow system by flooding the benches for 20 minutes or 3 h waterlogging treatment using a nutrient solution with or without 20 per cent of the N supplied as nitrite and with or without aeration of the solution, and recorded that fresh weights were almost unaffected, albeit the same was found decreased when nitrite was added in the nutrient solution though at the base of the pot the root growth was found increased due to nitrite and waterlogging, and further increased due to inclusion of perlite in the medium. Hammer (1973) tried 10 cultivars of Rieger begonia in 12.5 cm plastic pots planted on April 9 and watered through capillary watering while half of the pots were hand watered. Recording of the data for plant heights and fresh & dry weights after 78 days of transplanting showed that capillary watering produced larger plants though there were considerable varietal differences.

Growth, Development, Flowering and Flower Forcing

Tuberous begonia hybrids are not sufficiently hardy; therefore, they are grown as an annual crop. Its ancestors originated in the Andean mountains where no marked fluctuations in temperatures occur though there from the falls to early spring drought period occurs when tubers go dormant. Long days of 14 h promote vegetative growth and subsequently flowering shoots, and if this condition is provided from March onwards, they will flower whole summer. Under short days the tuber formation is stimulated but within 20-30 days apical meristem stops developing and the buds in the lower part of the stem become dormant resulting into complete stoppage of flowering. Tuberization is favourably affected due to low temperature of 4-5° C.

Horton (1951) reported that tuberous rooted begonias remain in flower throughout the year at **day lengths** of about 14 hours, and the plants in flower in August when subjected to 9-h days for 1-8 weeks, then cut back and given 141/2 h days, followed by two weeks of short photoperiods which causes partial dormancy while three weeks of treatment causes complete dormancy. Matouš (1967) stated that begonia tubers grow most rapidly in short days and a 14 h day is considered to be critical for tuber formation. Short days induce dormancy in tuberous begonias, that is why the tubers immediately after harvest do not sprout and this state of **dormancy** remains for a long period. Even otherwise, the tubers harvested in late October remain dormant up to January, i.e. 3 month's resting. Though dormancy can be broken by subjecting the tubers for 60-90 days at 1-5° C temperature but this effort is of no significance as by this time the tubers will become non-dormant even under natural conditions. Esashi and Nagao (1973) described two types of dormancy in the aerial tubers of *Begonia evansiana,* one which occurs in immature tubers, i.e. photo-sprouting stage which requires blue or far-red light, and the other in mature tubers which requires chilling. Oxygen is required not only for tuber sprouting but also during the chilling process at 2-5°C to break tuber dormancy, and if the mature tubers are exposed to blue light during chilling period their dormancy is broken even when O_2 concentration is as low as 3 per cent. Blue light pre-treatment promotes sprouting of immature tubers only when given under low O_2 concentrations, while red light pre-treatment becomes effective in inducing dormancy in the immature tubers and in prolonging dormancy in mature tubers as O_2 tension is increased. Esashi (1969) found blue or far-red light inducing sprouting in immature tubers (not in matured tubers at room temperature) at high temperatures, optimum being 29°C in *Begonia evansiana*, and promotes dormancy release caused by chilling (2-5°C) in mature tubers but low temperatures of 15-17°C induce dormany in immature tubers, though red light inhibits both, i.e. sprouting in immature tubers and the dormancy release in mature tubers at low temperatures. Nagao and Okagami (1966) stated that CCC, applied either to the buds or to the nutrient solution in which the basal parts of rooted cuttings were immersed, accelerated tuber initiation in *B. evansiana* grown first under short-day conditions and then under long days, and prevented dormancy. Okagami and Esashi (1972) obtained promotion of sprouting in mature and immature aerial tubers but not stimulating tuber enlargement in *Begonia evansiana* by treatment with morphactin which overcomes sprout inhibiting action of GA. Esashi and Leopold (1967) stated that a DNA- and RNA- dependent protein synthesis is an essential part of the dormancy induction in begonias. They studied the immature *Begonia evansiana* tubers treated with inhibitor solutions for a period of three weeks at low temperature of 15°C, followed by washing and placing in light at 25°C to favour sprouting, a condition under which a non-dormant tubers will readily sprout. Dormancy induction during the period of low temperature was prevented by a wide range of inhibitors such as 8-azaguanine, ethionine, cyclohexamide, chloramphenicol, p-fluorophenylalanine, thiouracil, 5-fluorouracil, 5-fluorodeoxyuridine and 5-iododeoxyuridine at 34 x 10^{-5} to 10^{-3} M. The first two inhibitors were tested for their effects on the light regulation of dormancy, they again prevented induction by red light and enhanced the blue and far-red effect. In a further study, Esashi and Leopold (1969) induced the dormancy in *B. evansiana* tubers kept on filter paper moistened with solutions of inhibitors of nucleic acid and protein synthesis, by exposing them to red light and 15°C. They stated that all the inhibitors prevented dormancy induction (and also depressed the protein content of the tubers by 10-60 per cent) except actinomycin D. Okagami (1972) found gibberellins$_{1-4}$ inhibiting sprouting in aerial tubers of *B. evansiana*, and this action was favoured by high O_2 and inhibited by low O_2 levels, p-nitrophenol, resorcinol, salicylaldoxine, 2,4-dintrophenol, sodium azide and cycloheximide. Okagami and Nagao (1973) reported that with the progress of natural dormancy in *B. evansiana* and also with GA treatment, acidic, neutral and basic ethyl acetate-soluble and n-butamol-soluble growth inhibitors increase which suppress the sprouting of non-dormant aerial tubers which can be overcome by incubating the tubers in a low-O_2 atmosphere. In early dormant-stage-tubers the inhibitors increase markedly during incubation at 28°C in light.

For **forcing**, the tubers are started prior to more than one year. The freshly harvested tubers in October are placed in November in the greenhouse where these are subjected to long days treatment through high light intensities, *i.e.* 100 W/m^2 at 16° C for growth stimulation, after development of the first leaves so that flowering is obtained in February-March for spring shows. Gossens (1963) stated that large begonia tubers which were forced in November, sprouted earlier, flowered earlier and produced longer stems than smaller tubers. Beuzenberg *et al.* (1977) forced cv. 'Schwabenland' to flower for Christmas by treating the soil or substrate with fungicides such as AAterra (etridiazole) and GA 38/40, latter alone or mixed with Polyram Maneb as both do not inhibit growth but are suitable for controlling root rot, potting in 38th week (September 15), and spraying of CCC weekly from 43rd or 44th (depending upon plant growth and size) week to 48th week.

Through seeds, these are grown (sown) in November-December by providing long days with high light intensity, *i.e.* 100 W/m^2 during winter months till March, and now by May these may be offered for sale on commercial basis. If leaves are being formed continuously, flowering being dependent on leaf formation will continue as usual because flowering buds are formed in the leaf axils and the floral buds become apparent when there had been at least three nodes. Leaves and flowers, both are initiated under a 12 h LD, but highly promotive being 14-16 h. Less than 12 h photoperiod will stimulate tuber formation but leaf and flower initiation will stop. After 20-30 short days, apical meristem activity stops, adventitious buds at the stem base develop and tubers become dormant.

Tuberous-rooted begonias respond well to **photoperiod**. Lewis (1951a) when provided 14 or 16 hour days or interrupted dark period providing a total of 12 hours of light, the camelliaflora and multiflora begonias successfully flowered in winter, however, at 12 hours or less light plants stopped growing and flowering, and resulted into tuberization. Lewis (1951b) stated that when leaves reach to the right stage and the temperature is maintained at 15.6-22.2°C, even during short winter days the plants do not die and instead the combination of excess carbohydrates with short days encourage the tuber production, however, long days favour vegetative growth and flowering. Lewis (1953) noted that under short photoperiods the plant remains quiescent in growth but after a certain stage of leaf growth tuberization occurs if temperature is favourable though under long photoperiods there is no tuberization, albeit under short days and at 15.6°C or higher temperatures tuber weight is increased. Wasscher (1953b) through photoperiodic adjustments forced Lorraine begonia cultivars by providing short-day treatment for 3-4 weeks in July or August, and the plants treated in the beginning of July were in full bloom after two months with higher number of flowers as compared to late July and August treatment though in former case the plants were smaller. In large-flowered begonias, 3-4 weeks short day treatment either in second fortnight of July or in the beginning of August increases tuber size. In Germany, Maatsch and Rünger (1955a) stated that for tuberous-rooted begonias the short-day (6-12 hours) treatments during summer months inhibited the both, the vegetative growth and the reproduction though its action reverts back if the treatment is only for 10-25 days where tuber weight also increases from $1^1/_2$-$7^1/_2$-fold, but the longer duration causes first death of the growing point and then of the leaves and stems, however, 13 hours treatment causes long-day reactions, and a 17-day treatment starting by July end provides greater increase in tuber weight than treatment starting in the beginning of July or in August or September. Van Onsem (1961b) subjected the Ghent tuberous-rooted begonias to 9 hours short-day treatment for 20 or 30 days at six different periods between August 10 and November 10 and recorded increased tuber weight (18.3 g in 20-day treatment from 10-31 August in cv. Maxima Rose) as compared to control (8.4 g) though the treatment duration and the suitable time for treatment depended on the cultivars. Short-day treatment for three weeks in July in *B.* x *cheimantha* reduces the flowering time by half though August and September treated plants remain at par to control, however, lower night temperatures (11-14° and 15-17°C) delay flowering (Anon., 1965), and at early short-day treatment a night temperature of 18-20° but later in the season a night temperature of 15-17°C give best results. It is further recorded that pinching in July, 10 days before or after or at the start of the short-day treatment, does not have any effect on the start of flowering albeit flower quality is good though later pinching in the plants given short-day treatment in August-September slightly delays the flowering. As little as two short days are sufficient to stimulate flowering, however, 12 days treatment gives the best results. The cuttings taken from November 13 to March 8 from the flowering mother plants of *B. cheimantha* 'Karelske Jomfru' subjected to short days and deblossoming, resulted into rooting but not into bud-breaking in the first two batches of cuttings due to plants still being in reproductive phase (Anon., 1966). Through regulation of day length for flowering of begonias during Christmas has been scheduled by Heide (1969a) in Norway that under natural short day conditions the temperature should be kept at 20-22°C until mid-October, then 18°C until 15-20 November, and finally at 15°C until marketing of plants. This way though flower development was slower but quality was better. Zimmer (1969a) stated that *B. boweri* 'Cleopatra' bearing flowers and ornamental foliage flowered in short days and remained vegetative in long days, critical day length being slightly over 12 hours at 20°C. Zimmer (1971, 1975) stated that daily interruption of a 16-h darkness by fluorescent light breaks of 2 or 3 hours during the 4 to 7 and 11 to 14 hours of the dark period prevented flowering but did not happen so if it was applied in the middle of the night, i.e 8th to 10th hours in the short day plants, *B. boweri* and *B.* ***boweri*** var. ***nigra-marga.*** The bulblet formation in the two clones of *B. socotrana* was found suppressed by exceeding daylength by 12.5 hours in a 16-h periods interruption by a 3-h night break but this effect was found negated when the night break was made between the 5th and 7th or between 12-14th hour of the dark period (Zimmer, 1976). Peters (1974a) reported critical daylength for tuber formation in *B.* x *tuberhybrida* as 12-13 h, while above this no tuber was found formed in any of the treatments. Growing Rieger-elatior begonia (*B.* x *hiemalis*) cv. 'Rieger' under 16-h long days in the early part of their growing season, followed by three weeks of 10-h short-day conditions brought the plants in full bloom after three weeks from the termination of short day treatments (Molnar, 1974). Krebs and Zimmer (1983) stated that the inhibition of flowering by a red light break during the first light sensitive phase (5th hour of a 14-16 hour dark period) in *B. boweri* 'Cleopatra' could only be overcome by far red irradiation if the red light break was less than 5 minutes long and far red irradiation did not exceed 10 minutes. They recorded more than 5 minutes of red light inhibiting flower formation and more than 10 minutes of far red irradiation reducing the flowering. Sandved (1967) recorded differing requirements to day lengths and temperatures for different varieties of *B. hiemalis*, but in general he found that 7 days of SD treatment were sufficient to induce flowering but the flowering was more uniform if the SD treatment was extended to 14 or 21 days. Sandved (1971a) noted *B.* x *hiemalis* cvs 'Schwabenland' and 'Liebesfeuer' as SD plant with a critical day length of less than 13 hours so growing them at 21-24°C higher temperatures than at 15 or 18°C, the growth was rampant. Powell and Bunt (1978) reported that plants of *B.* x *hiemalis* 'Schwabenland Red' grown in summer and winter under both long and short days in England resulted into leaf size increase at maturity during short days but actively growing leaves were found unaffected due to change in environment from long to short days, however, the rate of leaf appearance under long days in winter as well as in summer was similar but decreased rapidly under short days, greater being in short days in winter than in summer. With *B.* x *hiemalis* cvs 'Romeo', 'Julie' and 'Nadia', 10-h short day treatment resulted in rapid and satisfactory flower initiation but at 13- or 16-h long days the flower initials formed were fewer and their development was very slow (Sandved, 1978). Loeser (1979) tried 26 Elatior begonia cultivars against short-day treatment and recorded that a few cultivars were day neutral, while most cultivars were benefited when a 3-week short-day treatment was given, however, under short day treatment 'Radiant' required 52 days and 'Turo' 77 days for flowering while without short day treatment the earliest was 'Ninon' requiring 58 days and the latest 'Turo' requiring only 85 days. While working with Elatior-type pot begonias, Westerhof (1980b) recommended blacking out for a week, 1-2 weeks after potting up to shorten the crop season by 2-4 weeks in case of 'Trudy Rood', and shortening of crop season by one week in case of 'Schwabenland', blacking out too early or for two weeks in case of 'Tango' flowering was found severely delayed, and blacking out for more than two weeks was detrimental to plant quality in all the five cultivars under study, though blacking out of the cultivars 'Tocora' and 'Toran' was of little value. In winter, growing of two Elatior-type varieties (Schwabenland and Baluga) at 6.8-142.8 mW/m^2 light intensities during winter, Westerhof (1982) found 90 mW/m^2 (about 21 lx) as the oiptimum for proper growth and for formation of side-shoots on the main stem. Jungbauer (1981) at 18°C found three weeks of SD for flower bud induction followed by 13-h days to stimulate growth and flowering in 'Aphrodite' and 'Schwabenland' begonias. Oloomi and Payne (1982) with *B.* x *tuberhybrida* F_1 hybrid seedlings observed that short days greatly enhanced tuber development though long days hampered it, and pinching though delayed flowering but did not affect tuber development. When the plants of *B.* x *cheimantha* cv. 'Dunkle Marina' (a short-day plant requiring 8-h days to flower) were grown with a night break of different durations and at various times of the dark period or different times after the start of the main light period, at 13-14 h of night break after the start of the main light period maximum effect was seen in preventing the flowering (Rünger *et al.,* 1984). Tonecki (1986) found the tuber size in *B.* x *tuberhybrida* directly proportional to the number of short days applied.

Rex begonias are grown exclusively for foliage though they flower during winter and spring. At short days and low temperatures they become dormant. They grow at 29/17° C day/night temperatures, average being 23° C. For maintaining proper colouration of the leaves,

2000-2500 fc through fluorescent lights, especially the red ones are most suitable.

Hiemalis begonias are first grown as LD for proper foliage growth and then shifted to SD for flower initiation though both are temperature as well as variety dependent. *B. x hiemalis* at 24° C though does not flower at 16 h photoperiod but flowers profusely at 10 h photoperiod. The decrease in temperature at 12° C encourages flowering even at LD of 16 hours. At 24° C the requirement of critical photoperiod is 12-13 hours. However, in winter, these may be provided with 14 h photoperiod at temperature greater than 18° C otherwise the plants will become dormant and tuberize. For commercial production, three weeks of SD is initially provided and then it is followed by LD for proper plant development and flowering. SD longer than three weeks will drastically reduce the number of flowers. The use of supplemental **CO_2** from 300-3000 ppm results in superior plants but after this treatment the plants are immediately shifted to SD environment. CO_2 at 600 ml/m^3 under optimal (2000 lx) supplemenatary light and 20/18^0 or 20/16^{0}C day/night temperature conditions yielded best quality plants in shortest culture period in Elatior begonias (Schmidt, 1985). Andersson (1990a) advocated 900 ppm CO_2 for 6 hours in the night when supplementary lighting was provided, and in one of the glasshouses which had two ventilators the air was circulated fully at 0.2-0.4 m/s about 4 times/h while in the second glassdhouse which had no ventilator the air was forced at 0.6-7.0 m/s, and the forced air circulation shortened the consumption time for CO_2 by about 20 per cent and reduced the RH. Gundersen (1959) applied GA at 10 or 100 ppm to *Begonia semperflorens* x *B. fuchsioides* and obtained stem elongation partly by cell wall elongation and partly by acceleration of cell division, *vis-à-vis* acceleration of cambial activity though lateral branch development was depressed and flowering was inhibited.

During and after SD, the plants of *Begonia tuberhybrida* var. *gigantea* were grown at **temperatures** of 10, 15, 20 and 25°C where low temperatures were found aggravating further the inhibiting effect on growth, and after 20 days at 20-25°C combined with short photoperiods when were changed to 10-15°C of low temperatures death of the apical buds occurred but 20°C and 25°C high temperatures following SD treatment reversed the growth inhibiting effect (Maatsch and Rünger, 1955b). Aerial tubers in *B. evansiana* began to sprout when isolated and exposed to continuous light while actively thickening but after the completion of thickening the tubers stopped sprouting in response to light but were made to sprout through low temperature treatment and whether they sprouted or not, they developed roots (Esashi and Nagao, 1959). Heide (1962) found high night temperature accelerating bud appearance in short days and delaying in long days though in both the cases a night temperature of 21°C rapidly accelerated the flower development. It was found that 9 hours of day length for two weeks was sufficient to induce abundant flowering. Rünger (1963, 1968, 1978a) stated that in Lorraine begonias including *B. limmingheimiana* when plants having only one or a few shoots were kept at different temperatures and day lengths, highest production of new shoots was found at 25°C though at 30° the growth was retarded and at 10°C there appeared no growth, however, the flower formation was optimum at short days and around 15°C temperature. In a further studies in cv. 'Marina' under LD (14 hours to continuous lighting), temperatures at 25°C or above inhibited the flowering drastically while at 10° or 15°C flowering was found initiated irrespective of the day length, and the response at 20°C of the long days have been quite unpredictable, sometimes delaying the flowering and other times preventing it completely. In the third trial with the rooted cuttings of cv. 'Dunkle Marina' being grown under 16-h long days and at 24°C, followed by 8-h short days for 6 or 10 days under 15-30°C night/day temperature range in all the combinations, and then again bringing back to long days at 24°C, flower buds were found formed 23.6 days after the end of the 10-day short day treatment at day/night temperatures of 25/20°C though those given 6-day SD treatment at low night temperatures resulted into late flower but formation by up to 50 days. Haegeman and van Onsem (1964-65) had grown tuberous begonias in 10°, 14°, 18°, 22° and 26°C heated soil at natural day lengths from July 24 to November 4, and recorded largest tubers in unheated soil, suggesting critical temperature at 10°C, below which tuber formation is stimulated. Sandved (1968) found rising temperature up to 18°C reducing the period from bud stage to full-bloom stage in *B.* x *cheimantha* as only 17 days though the flowers were smaller and even some buds failed to open, as compared to 12°C which took to attain this stage in 49 days. A temperature of 18°C for 2-3 weeks from Oct. 21, then 15°C improved the quality with optimum extension growth having 1.0-1.5 weeks advanced flowering, however, temperatures higher than 18°C yielded smaller flowers (Sandved, 1971b). Waines (1972) with *B. dregei* observed that 4-h night break in the middle of the dark period than complete 16-h dark period caused cent per cent flowering, and those receiving temperatures below 18.3°C with no night break developed abnormal inflorescences. White *et al.* (1973a) obtained best growth in Rieger-Elatior begonias with a minimum night temperature of 18.3-20.0°C during long day cycle of three weeks (minimum) in 12.5-15-cm pots, and weekly fertilization with 200 ppm N + 4-h of supplementary incandescent light or every 2-3 days subirrigation + Osmocote (14-14-14) + 4 (or 8) hours of light. The long days raised plants at 20°C were treated for 20, 40 or 60 days at 10, 15, 20 or 25°C for long (16 h) or short (8 h) photoperiods, and recorded optimum temperature range for tuber formation as 15-20°C though slightly higher for aerial growth, in the lowest and in the highest temperatures only small tubers were being formed, and 40-60 short days were found necessary for tuberization (Peters, 1974b). Jungbauer (1975) for getting flowering plants in December in Lorraine begonias recommended 20°C temperature until the appearance of the flower buds and then reducing it to 16°C.

Wax begonias (*B. semperflorens*-Cultorum hybrids) take 6-7 weeks from seed sowing to transplanting and the plugs require 8-9 weeks. **Transplanting** is carried out when seedlings have developed four true leaves. They are day neutral; therefore do not require any specific photoperiod or temperature. However, at 830 fc supplemental lighting for two weeks the flowering is more rapid.

Growth regulators such as GA_3, IAA, IBA, kinetin, ethrel (ethephon) and a morphactin [(CF 125), chlorflurecolmethyl], *etc.* affect growth and development in begonias. Certain growth promoting chemicals have also been isolated from begonia leaves and certain other parts. Campbell and Bakir (1985) in the leaves of shaded-greenhouse growing *B.* x *masoniana* recorded increased oncentrations of free IAA under long days whereas highest bound IAA during short days, and apart from this they recorded six cytokinin-like substances, *viz.* adenine, adenosine, zeatin, zeatin riboside, isopentyl-adenine and isopentyl-adenosine. Heide and Skoog (1967) in the leaves of *B.* x *cheimantha* recorded 30-300 μg kinetin per kg of fresh leaves, and the content was favoured by SD conditions. From the leaf extracts of *B.* x *cheimantha* cv. 'Nova', Odén and Heide (1988, 1989) recorded gibberellins IAA, GA_1, GA_4, GA_9, GA_{19} and GA_{20} and the quantification of IAA, A_4, A_9, A_{19} and A_2 grown under different temperature and daylength conditions, however, GA_1 contents could not be quantified due to its presence in a very low quantity. Hansen *et al.* (1988) extracted three unidentified glucosides and four cytokinins, *viz.* trans-zeatin, trans-zeatin riboside, isopentenyladenine and isopentenyladenosine, the quantity being several times higher at 15° than at 24°C, and at 15°C there was no clear effect of photoperiod, though at 24°C short days increased the activity many times as compared to long day conditions. Sano and Nagao (1970) from *Begonia evansiana* leaves extracted IAA oxidase when SD treatment had been given to form aerial tubers, being maximum after 6 days of SD treatment with interrupted nights. Hänisch Ten Cate *et al.* (1975) recorded 1 per cent IAA from female flowers in the monoecious *B. fuchsioides* hybrid where male flower buds were found abscising, and in *B. davisii* hybrid where there was rise of male bud drop in winter due to fall in the IAA content of the buds. Wasscher (1947) reported bud or flower drop in winter-flowering begonias, sometimes a serious problem, for which he recommended spraying of 5.0-12.5 mg/l NAA which controlled up to 65 per cent of the problem, and reported that higher concentration causes leaf and stalk curvature and closing of the corolla. Hänisch Ten cate and Bruinsma (1973) applied various growth substances to entire plants and also to pedicel explants with or without buds and recorded retarded pedicel abscission by IAA application or when flower buds were present though ethylene to a greater extent and ABA to a lesser extent accelerated abscission which could not be checked by application of GA and kinetin. Nagao and Mitsui (1959) with *Begonia evansiana* described that GA at 0.1-100 ppm could not break dormancy though 10-100 ppm inhibited the sprouting of aerial tubers and the dormancy was found prolonged and this dormancy could be broken by low-temperature treatment. Van Onsem and Haegeman (1961) recorded GA spraying at 0.2-1.96 g/plant in the concentrations of 10-50 ppm increasing stem length and tuber weight in certain varieties of *B. tuberhybrida.* Berghoef and Bruinsma (1979) with *B. franconis* reported that gibberellins promoted organ initiation in the buds and removal of the first bud of the inflorescence primordium strongly reduced organ initiation in the remaining buds when gibberellins were present.

Prevot (1969) induced the formation of buds and roots at the distal ends by dipping the proximal ends of *Begonia rex* leaf pieces in GA solutions. Okagami *et al.* (1977) when applied GA_3 from beginning to the 10th day of SD treatment to the buds of *B. evansiana* stem cuttings, found increased sprouting of developing tubers but inhibited aerial tuber formation though it did not inhibit sprouting in aerial tubers detached from CCC-treated cuttings, and when applied after the 10th day of SD treatment, it hastened the completion of dormancy and prolonged it, and in case of detached aerial tubers it caused increased maturation. Tonecki (1986) reported that GA_3 (25-250 ppm) or its combination with SD promoted growth, tuber weight and increased number of flowers in *B.* x *tuberhybrida* while IAA, kinetin and ethephon increased tuber weight in plants grown under natural daylength though the increase was considerably lower than that under short days. Spraying of triiodobenzoic acid (TIBA) either only on September 7 or two sprays, *viz.* September 7 and 22 on the plants of *B. tuberhybrida* cultivars at 25, 50 and 100 mg/l resulted into slightly reduced weight of tubers though percentage of marketable tubers was greater (Van Onsem, 1961a). ABA at 3 x 10^{-6} - 10^{-5} M inhibited sprouting in vernalized aerial tubers of *B. evansiana* but sometimes it caused sprouting at 10^{-6} M though under long day conditions it was ineffective for forming aerial tubers (Hashimota and Tamura, 1969). Berghoef and Bruisma (1979) found BA at 10^{-6}M initiating all the flower buds and their further development up to anthesis and low sucrose levels inhibiting female differentiation which in turn increases number of male flowers, and concluded that sexual differentiation is regulated endogenously by the central region of the inflorescence primordium, the carbohydrate level being a limiting factor for female differentiation. Ethrel 0.5 per cent at 2000 litres per hectare sprayings either on September 2 and 18 or separately on 2nd or 18th on various varieties of tuberous begonias resulted in increase of the tuber numbers as well as tuber weight by 8 grammes, though varying due to different varieties, weather conditions and the plant growth rate (Schelstraete, 1982). In a further studies, Schelstraete (1983) when sprayed ethephon at 500-1200 g in 750 or 1500 litres of water per hectare on September 2 and 23 on various cultivars of tuberous begonia and assessed the effects on tuber production on October 21 in Holland, recorded increased number as well as weight of tubers though varying with different dilutions and cultivars.

Height Control

To dwarf the plants in 'hiemalis' begonias, Cycocel (chlormequat) at 500-3000 ppm depending on the cultivars is sprayed to reduce the internodal length but flowering is slightly delayed. In tuberous begonias grown through seeds, CCC at 500 ppm may be used at the end of production to hold the plants if sales are slow. Its 250 ppm spraying once when the plants are in plugs and the next just after they are planted, results in 10-cm plants. In *B. semperflorens,* CCC 500 ppm on white-flowered cultivars and 1000 ppm spraying on others after development of the fourth leaf gives significant response. In rex begonia, height control is not required. Zimmer (1969b) controlled the plant growth and intensified the leaf colouring through three sprayings of CCC, B_9 or 1244 at 10-day intervals about two months after propagation. Spraying of *B. tuberhybrida* plants with 0.1, 0.3, 0.5, 0.7 or 1.0 per cent CCC, Patáková (1969) recorded restricted late summer growth, optimum being 0.5 per cent though higher concentrations injured the plants and reduced the tuber yield considerably. At 18°C but not at 21 or 24°C under 2 x 10^{-2} M, CCC drenching reduced stem and leaf growth in *B.* x *cheimantha* at all the concentrations with promotion of flowering in long days as well as at near critical temperature, particularly under low light intensity in winter (Heide, 1969). CCC at 5 x 10^{-3} and 2 x 10^{-2} M stimulated the growth and formation of adventitious buds in the leaves at 21 and 24°C whereas 8 x 10^{-2} M was inhibitory, and at 18°C bud formation was inhibited by all the CCC concentrations, though root formation at high concentrations was found stimulated but with reduced elongation. In detached leaves planted at 18 and 21°C, CCC at 10^{-4}-10^{-2} M stimulated bud formation though those at 24°C the budding was inhibited under 10^{-2}M but lower concentrations were stimulatory even at this temperature and rootings were stimulated. Phosfon drenching at 4 x 10^{-4} M was not effective but in detached leaves it stimulated bud formation with an optimum concentration of 10^{-6} M though it at all the temperatures stimulated root formation, optimum being 10^{-5} M. Spraying of CCC at 0.08-0.1 per cent to the plants of *B.* x *hiemalis* following the short day treatment resulted into good quality plants with restricted growth (Sandved, 1972). Will (1977) through plant spraying or drenching in Elatior begonias through 0.05-0.2 per cent CCC (chlormequat) obtained 15-20 per cent plant height reduction without affecting the plant diameter, however, daminozide (B-9) was not found as effective. Schenk and Brundert (1980) when sprayed 0.5 or 0.75 per cent CCC on Elatior begonias, recorded 22-45 per cent height reductions and 8-22 per cent diameter reductions but with necrosis on the leaf margins, and suggested that 0.3 per cent CCC after four weeks of potting may be more effective. Hentig and Knösel (1984) recommended 0.3 per cent CCC treatment in the 7th week of growth followed by two weeks SD from 8th week to improve plant quality in summer in vigorous growing Elatior begonia cultivars though weaker cultivars like 'Elfe' and 'Nixe' require only 0.15-0.3 per cent CCC spraying. Hilding (1975) for selling potted plants of *B. hiemalis* varieties recommended growing the plants at a good light intensity and at a constant 18°C temperature, pinching any time after one week of SD (SD starting from mid-February and the latest by mid-September) treatment, reduction of growing temperature from 5th week after the start of SD treatment, and spraying 0.1 per cent a.i. of CCC after three weeks of SD treatment will make the plants ready for sale in 6-7 weeks after the beginning of SD treatment. Elatior begonia plants being grown in winter in Germany were sprayed on 21 January with CCC, Bonzi 1 (PP333; 0.0625 per cent) and Bonzi 2 (0.125 per cent) and after one month again half the plants received repeat spray and nine weeks later since the first spray the plant height and diameter but not the floral characters were found reduced in all the treatments, and in Bonzi 2 double spray floral size and stalk length were also slightly reduced though number of days to flowering was slightly increased in the double spray of CCC and in both the sprays of Bonzi 2 (Horn and Wischer, 1987). Krauskopf and Nelson (1976) found reduced height, being most effective in winter when Rieger-Elatior begonia plants were sprayed with 0.3 per cent CCC though drenching of 0.3 per cent under low light conditions was injurious and drenching 0.3 and 0.6 per cent under high light conditions were ineffective, daminozide sprays up to 0.5 per cent and ancymidol sprays at 16.5 or 33 ppm were found ineffective but as a drench ancymidol at 0.125 mg per 15-cm pot under both high and low light conditions was quite effective. Paclobutrazol spray at 0.15, 0.3 and 0. 45 mg a.i. per plant or soil drench 0.15 mg a.i. per pot (pot size 15 x 12.5 cm) and uniconazole spray at 0.025, 0.05 or 0.75 mg a.i. per plant or soil drench 0.025 mg a.i. per pot applied to the seedlings of *B. semperflorens* (*B. cucullata*) controlled the height effectively but daminozide spray at 5000 ppm was not effective (Banko and Stefani, 1988). The plug-grown *Begonia semperflorens*-cultorum 'Cocktail Mix' were transplanted into jumbo six packs and when the seedlings had attained ~ 5 cm in height or width, Kuehny *et al.* (2001) sprayed them with 1000/800, 1250/1250 or 1500/5000 mg l^{-1} of chlormequat/daminozide solutions which reduced the plant height significantly.

In tuberous begonias, for getting large flowers for shows, the plants in largest pots (20-25 cm) are pinched 2-3 weeks after transplanting, to retain only 2-4 side shoots (2-4 nodes with leaves on the plant) and pinched wastes may be used for multiplication. Compact types are pinched only when plants become leggy. Pendulous types are pinched to control its apical growth and to make it bushy. For mass display in the summer, the first few buds are removed as in bedding compactness is more important than height. In rex begonias which are rhizomatous, pinching is not required.

Post Harvest

Plants in wax begonia (*B. semperflorens*) can be retained at 14-17° C for cool transportation. Hiemalis begonias are sensitive to ethylene gas. As low as 0.1 ppm for 24-72 h exposure results in flower bud abscission while marketing the potted flowering plants but its effects are negated through spraying with 0.5-1.0 mM silver thiosulphate. These require 2-6°C optimum transit temperature. Rex begonias are not easy to transport due to its brittle leaves and reflexed stems so these are sent only to the local markets. During transit these are kept at 5° C temperature. For interiorscaping the plants perform well at 150-500 fc light and more than 30 per cent humidity. Tuberous begonias yield some 20-25 flowers per plant and these flowers are normally used as corsages. Before transportation, the potted plants sprayed with silver thiosulphate at 0.3 mM, maintain properly up to 15-20 days and flower abscission is also reduced. Transportation of flowering potted plants is done at 5° C and of the plugs at 6-8° C. Haegeman (1967) in seed-grown *B. tuberhybrida* plants defined three main flower shapes, *viz.* "rose-bud" (the most desirable one), "camellia" and the "cylindrical". He further stated that pink selections produced fewer "rose-bud" flowers than scarlet and yellow selections. During six days of simulated transit, the plants of *Begonia* cv. 'Medora' were when subjected to 0.02 gravitational force vibrations of 4.1 cycles per second, leaf abscission, necrotic leaf areas and severed stem tips were caused and when STS was applied as pre-vibration spray (0, 0.28, 0.56 or 1.12 mM Ag) leaf abscission was found increased as ethylene emanation at stem tips rose with the increase of the STS concentrations irrespective of the vibration though during four weeks of simulated retail holding there was a decrease in leaf abscission due to STS (Auer and McConnell, 1984). Rystedt (1982) when stored potted *Begonia* 'Nixe' plants in the dark for 0-14 days at 20°C, though chlorotic and necrotic areas appeared on leaves even in 7 days stored plants but there was no bud abscission. With Elatior begonia cv. 'Serene', Høyer (1984) when kept the plants under conditions simulating transport for 72 hours in an: (a) ethylene-free atmosphere, (b) ethylene-free atmosphere for 48 hours followed by ethylene concentrations of 0.1-5.0 ppm, and (c) at these ethylene concentrations for the whole period, and then all these plants were kept for 12 days at ordinary conditions to compare with those kept already at room or greenhouse conditions, and recorded that higher ethylene concentrations affected the plants badly causing bud and flower drop, the length of exposure surpassing the concentrations, and thus suggested that during storage and transport there should be good ventilation free of ethylene where the plants will remain fresh for 72 hours in the dark. Fjeld (1989) gave exposure of 0.15 μl/litre ethylene in growth chambers of *B.* x *cheimantha* for 2, 6 or 11 days at various photon flux densities (14, 50 or 140 μmol m^{-2} s^{-1}) and temperatures (15, 18 or 21°C) and then the plants were held in a simulated interior climate (50-60 per cent RH, 20-21°C and 14 μmol m^{-2} s^{-1} photon flux density) for two weeks where he recorded stimulated abscission of flower buds and flowers under ethylene, high temperature and long exposure period irrespective of the light intensity though high light intensity during ethylene exposure period negates this problem markedly. Spraying *Begonia* x *cheimantha* plants with 100-2000 ppm ethephon increases flower bud abscission, *vis-à-*

vis flower malformation, more damaging being at later stages, *i.e.* stages III and IV than at earlier stages, *i.e.* stages I and II, and a high ethephon concentration at stages I and II inhibits the bud opening completely but this problem can be overcome completely by treating the plants with 1.05 mM STS (Moe and Smith-Eriksen, 1986). Fjeld (1986) by keeping the *B.* x *cheimantha* 'Nova' plants for the last three weeks of cultivation at (a) 15, 18 and 21°C and light conditions (natural with or without shading), and (b) different light intensities (2500, 5000 and 10000 lx) and 50 or 90 per cent RH, reported that increasing light intensities resulted in better keeping quality of plants developing more flower buds and flowers, decreasing the temperature produced more compact plants with more number of flower buds and flowers, and high RH promoted stem elongation. Fjeld (1990) by growing *B.* x *cheimantha* 'Nova' pot plants at: (a) 15, 18 or 21°C by exposing them to natural daylight conditions varying from 2.3-5.0 mol m^{-2} day^{-1} PAR or (b) natural daylight conditions + supplementary irradiation with either 40 or 80 μmol m^{-2} s^{-1} PAR through fluorescent light for 18 h/day for assessing the keeping quality by holding the plants for four weeks in a room with 20°C temperature, fluorescent light of 13.5 μmol m^{-2} s^{-1} at plant level for 12 h/day, and 47-68 per cent relative humidity (except for about 4 hours/day after watering when it rose to 80-90 per cent), and reported improved plant quality at marketing stage and keeping quality with increased number of buds and flowers in treatment (b) where supplementary irradiation of 80 μmol m^{-2} s^{-1} was given, however, increased temperature though increased the length of the flower stem but also increased flower bud abscission.

For sale of tubers, these are packed in boxes lined with expanded polystyrene which insulate the boxes against outside temperatures below freezing (up to -7°C) and even at -19°C temperature the tubers freeze only after 40 hours. Meeasurement of the electrical resistance of tubers shows with 95 per cent certainty whether or not the tubers have been frozen. Expanded polystyrene lining is better than four layers of corrugated cardboard (Van Onsem *et al.,* 1963a,b).

Lifting of Tubers, Storage and Dormancy

Haegeman (1967) stated that lowest yield of tubers is associated with yellow begonias, and rounded tubers with concave centre are associated with scarlet and pink flower colours, in *B. tuberhybrida.* After flowering when leaves have started yellowing, and after harvesting of the bulbils wherever formed, the tubers are stored dry in their pots with a minimum winter temperature of 7° C and repotted in the spring with the new signs of growth. *B. socotrana* is a true bulbous begonia which starts growing in September and continues flowering in the winter, *i.e.* November to March and then after flowering, its bulbs along with the bulblets formed at the base of the stem become dormant, restarting the growth in September again. Its 'Gloire the Lorraine' hybrids are the best and can be grown in hanging baskets for flowering all through the year. Sub-tuberous begonias which are dwarf and shrubby, shed their leaves and even stems by late autumn, leaving a thickened tapering caudex partly above the soil level so at this time watering should be reduced. At complete falling of leaves, the pots are stored dry over winter at a minimum temperature of 7-10° C but repotted in the spring when growth has initiated. Stocks of *B. semperflorens* are discarded after flowering. In cases where tubers have been lifted in autumn as in *B. x tuberhybrida* hybrids, they are dried in a thin layer at <20° C temperature for about 36-48 hours, cleaned and are stored in bags containing peat at 10-12° C. Hiemalis begonias are switched from long days to short days when there had been sufficient vegetative growth though these are dependent on temperature. It is an obligatory SD plant at temperatures above 24° C though at lower temperatures slowly flower initiation occurs under LD. At 16 h photoperiod and 24° C there is no flowering though at 10 h photoperiod and 24° C temperature all the varieties flower. Decrease in temperature (18, 16, 12° C) the absolute SD requirement decreases and most cultivars flower at 12° C and 16 h photoperiod. Rex begonias which are rhizomatous may go dormant under SD and low temperature. Tuberous begonias go dormant in nature under water stress and slight day length shift but not due to temperature. Under SD and decreasing temperatures, initiation of leaves and flowers and their development stops, shoots senesce and abscission layer between stem and tuber is formed. Autumn harvested tubers do not sprout until February so for breaking this dormancy they are stored at 1-5° C for 60-90 days. Leaves and flowers initiate from 12-16 h photoperiod, 14-16 h being more promotive, though less than 12 h photoperiod accelerates tuber formation and discourages leaf and flower initiation. Wax begonias (*B. semperflorens*) are day neutral.

Tubers of *Begonia tuberhybrida* (4-9 g, and 10-21 g) were stored for various periods at different temperatures before being planted at 23-25°C in Germany. Storage of tubers at higher temperatures (25-35°C) cause them to sprout more rapidly but their long storage causes rotting either stored dry in polythene bags or in moist peat, though storage for 90 days there is much the same sprouting in all the temperature treatments (Maatsch *et al.,* 1965).

Insect-Pests, Diseases and Physiological Disorders

Tuberous begonias are attacked by aphids, mealy bugs, thrips and whiteflies but are not of significant importance. The larvae of the vine weevil damage the tubers seriously. In greenhouses and under dry hot conditions, mites (cyclamen or tarsonemid or broad mites, *Polyphagotarsonemus* (*Tarsonemus*) *laetus*; and privet or false spider mites, *Brevipalpus obovatus*) attack the growing points of the shoots and produce brown blotches along the veins on the underside of the young leaves. Petterson (1979) for the control of the mite recommended dicofol or dimethoate. Zangheri (1957-58) recorded muscid (*Pegomyia bicolor*) attacking begonias in Italy, larvae mining the leaves, larval development taking 15-20 days and pupation in the soil. It produces four generations in a season. Nicotine or phosphoric acid esters control this pest. Wax begonias (*B. semperflorens*) are attacked by aphids, mealy bugs, fungus gnats (*Bradysia* species maggots infesting the roots which are controlled by *Bacillus thuringiensis israelensis* and soil sterilization) and shore flies. Rex begonias are infested with aphids, mealy bugs, mites, thrips, whiteflies, fungus gnats and shore flies. Hiemalis begonias are attacked by aphids, mites, thrips and foliar nematodes. Green peach aphids (*Myzus persicae*) crinkle and deform the foliage which can be controlled through nicotine sulphate or tobacco water. Thrips (*Frankliniella lilivora*) attack is characterized by silvery or brownish streaks on buds and flowers and sometimes even on younger leaves. Thrips are also vector for tomato spotted wilt viruses. These pests can be controlled through 0.2 per cent Malathion sprayings. Mealy bugs (*Pseudococcus calceolariae* and *Rhizoecus pritchardi*) also feed on begonia but their control is not easy. Smith *et al.* (1966) recommended 0.125 per cent malathion as drench to the pots, however, exposure of the infested pots to freezing temperature kills this pest. Whiteflies (*Trialeurodes vaporariorum*) resemble a whitish-yellow tiny moth which lays its eggs on the underside of leaves. After hatching these new crawlers start feeding on the leaves and its attack makes the plants chlorotic. These along with fungus gnats and shore flies can be controlled through regular spraying with 0.2 per cent Rogor or methyl parathion. Black vine weevil (*Otiorhynchus sulcatus*) larvae feed on tubers by boring into them and the adults feed at night by notching foliage and flowers. Foliage application of pyrethroids is effective against adults and soil drenches or soil sterilization will control this pest. Moorhouse *et al.* (1990) recommended introduction of entomogenous fungus, *Metarhizium anisopliae* to begonia plants in the glasshouse immediately before egg laying by the *Otiorhynchus sulcatus*, however, later application was effective but to a lesser extent.

Aphelenchoides fragariae foliage nematodes also infest 'hiemalis' begonias and cause irregular brown or blackish blotches on leaves. It causes reddening along the veins in one week and of the whole blade in three weeks of its infestation (Riedel and Larsen, 1974). Powell and Riedel (1973b) recommended Temik 10G (aldicarb) granules, Nemacur granules or spray, Vydate granules or spray, Systox (demeton) spray and Metasystox-R (oxydemeton-methyl) spray effective in lowering down the population of *Aphelenchoides fragariae* drastically but Nemacur granules and Temik were found phytotoxic at rates necessary to give practical control, Vydate granules were though phytotoxic but 1 per cent spray proved quite effective when was applied every two weeks for 24 weeks. They also suggested HWT above 49°C. Van Den Brande *et al.* (1955) also recommended 60 minute treatment with 0.5 per cent formaldehyde heated to 43°C, and apart from this they also stated 0.05 per cent and 0.025 per cent CBP (chlorobromopropene) emulsion (20, 40 or 60 minutes, and 40 or 60 minutes, respectively). Pieroh *et al.* (1959) stated that Trapex (methylisothiocyanate) kills not only free living eelworms and root-knot nematodes but also cysts of *Heterodora* species, together with *Rhizoctonia* and *Pythium*, exhibits herbicidal properties and kills soil insects. The cutting back of the foliar nematode infested Rieger begonia crop in mid-January and then application of Temik 10G (a 10 per cent aldicarb granules) after 15 days at 200 to 600 mg/15-cm pot, and again after $1^1/_2$ months application to half the plants had full control of the nematodes but in the repeat application, especially with 600 mg the plants were found stunted but leaf or flower size remained the same (Fisher, 1973). Powell and Riedel (1973a) applied $^1/_{16}$ or $^1/_8$ teaspoon of Temik 10G/15-cm pot to Rieger begonia plants, one week before leaf nematode inoculation or as an eradicant at the same rates to previously inoculated plants, and recorded severe deformation and stunting in standing crop, especially at higher concentration, and those that were cut back recovered and produced normal growth 4-5 weeks after treatment. Strider (1973) got eradicated the foliar nematodes from begonia plantings by applying oxamyl as soil drench followed in 10 days with a top application of the same, or its single drench or foliar application + aldicarb 10G applied one week before application or three applications of parathion. Rasmussen (1970, 1971) stated that *A. fragariae* and *A. ritzemabosi* can be controlled effectively in winter-flowering begonias through parathion (0.03 per cent), mevinphos (0.05 and 0.1 per cent) and thionazin (0.05 per cent) sprayings twice with a week's interval, as well as through HWT at 46°C for 10 minute dipping of the planting material. Riedel and Powell (1974) and Riedel *et al.* (1973) advocated 0.025-0.2 per cent oxamyl as foliar application fortnightly or twice at 4-week interval, or granular applications of Bay 68/38 [ethyl 4-(methylthio)-m-tolyl isopropylphosphoroamidate] and oxamyl at 5.625 kg and 13.5 kg/ha, respectively. Begonia leaf blight nematode (*Aphelenchoides olesistus*) causes leaf blight and kills the leaves rapidly, sometimes even the whole plant, and this may be controlled through Systox soil drench twice, 2-3 weeks apart (Anon., 1956) or through submersion of plants in hot water for one minute at 49.5-49.0°C, two minutes at 48.3-47.2°C, or for three minutes at 48.0-46.1°C, treatment being given at least three months before the marketing season. Pratylenchus *penetrans* and *Meloidogyne* spp. nematodes feed on the roots. Allen and Raski (1952) stated that although root-lesion nematode (*Pratylenchus penetrans*) may satisfactorily be controlled through fumigation with DD or CBP, the tubers may still carry a fairly high percentage of nematode infection. D'Herde *et al.* (1961) stated that cause of poor plant growth and the root rot in begonia field is due to occurrence of *P. penetrans* which could be controlled through steaming (soil sterilization) and treatment with various nematicides. Coolen and D'Herde (1968) suggested treating the soil every year before planting with DD at 6.75 kg/ha which gave constant tuber yield of the same quality and quantity year after year. Gillard and Van Den Brande (1955) obtained satisfactory control of *Meloidogyne arenaria* through HWT at 45°C for 60 minutes or at 48°C for 30 minutes. However, Vozna *et al.* (1984) in glasshouse trials in Russia when sprayed ammonium nitrate or urea dissolved in water (0.2 per cent) reduced nematode population, while spraying of potassium nitrate (0.3 per cent) or Rastvorin (0.4 per cent) increased resistance in plants against *Meloidogyne incognita* nematodes. Heungens (1968-69) stated that DD and dichloropropene either no or little affected the populations of beneficial organisms in begonia fields, and at certain times their populations were found even increased whereas nematode populations were greatly reduced. The attack of slugs and snails can be checked through repeat applications of metaldehyde baits. Methiocarb is also quite effective against these pests.

Bosmans and Kamoen (1970) made survey of begonia nurseries in Holland for three years to record incidence of various diseases and found *Xanthomonas begoniae, Agrobacterium tumefaciens, Corynebacterium fascians, "damping-off"* caused by various species, *Oidium begoniae, Botrytis cinerea, Rhizoctonia solani, Verticillium dahliae,* and various other fungi causing tuber rots. Botrytis blight is caused by *Botrytis cinerea* as water-soaked lesions to tan and necrotic spots on leaves, petals and petioles, and developing flower stalks rot and a part or whole plant collapses. In humid and cold weathers with restricted ventilation, and when plants are crowded and overwatered the white sclerotia can be seen on the affected part. It can be managed through proper sanitation of the planted area, excising all infected areas and painting them as well as leaf scars, wounds and growth cracks with Ziram paste (Zerlate, zinc dimethyldithiocarbamate), minimizing humidity, proper aeration even through adjustment of plant spacing, peripheral irrigation and proper canopy structuring, *vis-à-vis* 0.2 per cent spraying of Dithane M-45 (mancozeb) at 10-days interval. Kamoen (1972) reported that *B. cinerea* secretes citric acid and oxalic acids and these aggravate the situation where plant cells are rich in sugars. Dalchow (1968) reported that *Verticillium dahliae* affected plants were smaller with fewer leaves, and leaves were soft and flaccid with dull surfaces, so suggested that such plants should immediately be destroyed. *Fusarium oxysporum* f. sp. *begoniae* causes plant yellowing and tuber & root rot as this fungus affects the vascular system of the tubers. *F. solani* also attacks tuberous begonia. Regular fortnightly spraying with 0.1 per cent benomyl or Bavistin alternate with 0.2 per cent Captan throughout the standing crop will keep these diseases under check. Collar rot is caused by *Rhizoctonia solani* which affects roots and tubers and damping off by rotting the base of the stem, especially on poorly drained soils. Tuber rots are caused by *Pythium debaryanum, Cylindrocarpon radicicola* as a secondary parasite, and *Thielaviopsis basicola*. The schedule applied in controlling *Fusarium* will control all these pathogens also. *Verticillium dahliae* is also a vascular disease which will also be controlled together with *Fusarium. Oidium begoniae* and *Erysiphe chicoracearum* cause small mealy spots (powdery mildew) on the surface of the leaves which afterwards spread to entire foliage and the flowers. Any sulphur fungicide will control these diseases. Wasscher and Scholten (1954) used "Witex", an emulsifiable material containing polysulphides which proved very effective in controlling *O. begoniae* (mildew) if applied to flowering plants in winter. Zobrist (1946) recommended 0.1 per cent Cupromaag (copper carbonate) against mildew and 0.3-0.4 per cent Déryl against insects; Prota (1964) recommended a single spraying with 0.09 per cent Karathane; Tunblad (1947) in Switzerland and Andrén (1952) in Norway suggested Ultramare (F.D. White Oil) 1 per cent spraying; Viennot-Bourgin (1951) suggested 0.2 per cent copper oxychloride just at appearance of the disease; Schmidt (1952) stated that 2 per cent of a sulphur preparation (Ultraschwefel Geigy) + 0.1 per cent Etilon spreader against mildew and thrips both; Vasil'evskij and

Kareva (1958) suggested a copper-soap mixture (150 g green soap + 15 g $CuSO_4$/10 litres of water) application followed by another spraying after three weeks with 150 g green soap + 50 g colloidal sulphur/10 litres; Besemer (1961) suggested 0.2 per cent Wepsyn [amino phenyl bis (dimethylamido) phosphoryltriazole]; Valášková (1964) though tried many formulations which controlled *Oidium begoniae* from 63- 86 per cent, but recommended zineb 0.2 per cent and Phaltan 0.2 per cent before and Karathane 0.05 per cent after appearance of the symptom; Strider (1974) found sulphur dust, dinocap, triforine, and cupric hydrochloride + sulphur, all effective for up to four weeks but first two being more effective; Ríordáin (1979) found weekly sprayings of pyrazophos at 0.01 per cent or triforine at 0.014 per cent quite effective; Wohanka (1982) recommended Afugan at 0.05 per cent and Baymat (bitertamol) at 0.075 per cent and Ann *et al.* (1984) found bupirimate and triadimefon extremely good for controlling *Oidium begoniae* in begonia. Thompson (1961) obtained complete eradication of *Erysiphe* through 2 per cent Karathane dust, 1 and 2 ppm Actidione, or three applications of 0.05-0.06 per cent Karathane WD, at 3- or 7-day intervals. Qvarnström (1976) recommended Afugan (pyrazophos) at 0.04 per cent for best control against *Erysiphe polyphaga*, followed by Karathane (dinocap) at 0.03 per cent, by 5 times sprayings at 10-12 days intervals. Hemer (1978) stated that against powdery mildew control, Nimrod (bupirimate) at 0.04 per cent is the best formulation and the second best being Afugan (pyrazophos) at 0.05 per cent. Krebs (1988) tried Afugan (pyrazophos), Baymat (bitertanol), Nimrod (bupirimate), Saprol/Tarsol (triforine) and Bayfidan (tariadimenol) against Elatior begonia mildews (*Microsphaera begoniae*) and found best control with Bayfidan (0.01 per cent in the irrigation water). Bosmans (1971) suggested four sprays of benomyl (0.03 per cent) or thiophanate (0.06 per cent) for effective control of *Botrytis cinerea, Oidium begoniae* and *Rhizoctonia solani* on standing crop of begonia. Early application of Captan 80 to begonia cv. 'Gloire de Lorraine' was found controlling *Botrytis cinerea* (Niebisch and Kelling, 1986).

The bacterium *Xanthomonas campestris* pv. *begoniae* causes leaf spot (bacterial blight) which may be controlled by the use of $HgCl_2$. Riedel and Larsen (1974) detailed the interrelationship of *Xanthomonas begoniae* bacterium and *Aphelenchoides fragariae* nematode, former causes small water-soaked lesions at the leaf margin in 3-4 weeks, followed by marginal necrosis though in company of the latter it causes water-soaked patches over the whole leaf in a week and death of the leaf in 10-14 days. Jodon and Nichols (1974) reported that *Xanthomonas begoniae* survives on pelargoniums and *X. pelargonii* can survive on begonias without showing any symptom. The pathogen after its infection to begonias moves in the system rapidly within the plant so removal of diseased leaves will not make the plant healthy, and infection reaches to the soil through falling of the diseased leaves. This bacterium tolerates as high as 65°C temperature. Use of various chemicals proved futile though Kocide 101 and streptomycin sulphate had good control of this bacterium. Strider (1975b) stated that *X. begoniae* remained viable and virulent for 17 months in naturally dried and diseased begonia leaves stored in glass tubes at room temperature. Harri *et al.* (1977) reported that whatever cultivars of fibrous-rooted, tuberous and elatior begonias were tested proved susceptible and all the rex begonia cultivars were found resistant through spray inoculation of *X. begoniae*. Rattink (1979) stated that *X. begoniae* spreads to healthy plants from plant to plant in drops of water, by direct contact and through propagating knife and when the plants are grown in a warm, moist and shaded glasshouse than in a cool, dry and lighted house the symptoms are more pronounced. Rattink and Vruggink (1979) compared three methods (a normal isolation technique, immunodiffusion and immemofluorescence) in establishing the presence of *X. begoniae* even in small numbers in apparently healthy tissue of Elatior type begonias and found that immunofluorescence technique was the most reliable and thus *Xanthomonas*-free begonia plants can be chosen for taking leaf cuttings. Lovrekovich and Klement (1962) for its control suggested dipping of cuttings into 4:4:50 Bordeaux mixture. Knösel (1969) recommended buffer solutions of streptomycin and oxytetracycline + glycerol or mineral oil for good control of *X. begoniae*. Strider (1975a) reported that cupric hydroxide 0.2 per cent and streptomycin sulphate at 200 ppm applied bi-weekly provided complete protection against the natural spread of *X. begoniae* from infected plants through sub-irrigation. He further recommended methyl bromide (900 g/100 ft^3 of soil) fumigation which eliminates the bacterium from infected begonia tissues buried in soil. Hiemalis begonias are affected by *Erwinia carotovora* which causes soft rot on stems and leaves. Kamerman (1971), and Saaltink & Kamerman (1971) described the symptoms of *Erwinia chrysanthemi* infection on *Begonia bertinii*. They reported that infection spreads when tubers are divided for propagation or through contaminated tools used in dividing infected dahlia tubers and in its infection plants are stunted having unusually dark green leaves and the tubers have soft-brown areas along with decay-free areas showing brown streaks. They suggested selection of plants without discoloured xylem vessels. Stark (1964) reported leafy gall of begonia in Germany caused by *Corynebacterium* species. Dowson *et al.* (1938) reported a yellow bacterium which they provisionally named *Phytomonas begoniae* which in its infection produces minute glassy spots on the undersides of the leaves and when these spots increase in size these become clearly visible as water-soaked pale areas enveloping entire leaf afterwards and the affected tissues are flabby but later on turning brown and die. Its spread can be checked by burning such plants and through maintaining field hygiene. Streptomycin will control bacterial diseases. *Sclerotinia sclerotiorum* and *Sclerotium rolfsii* also infect 'hiemalis' begonias which can be controlled while controlling *Botrytis* blight. Rex begonias are also infected with *Myrothecium roridum* which causes leaf spots. Its control is also like to that of *Botrytis* blight. Apart from many others described above, *B. semperflorens* is infected with *Phytophthora* spp. which can be controlled when controlling *Fusarium*. To control various soil-borne pathogens, Gjaerum (1963) mixed 150 g of captan or 300-400 g of maneb or zineb per cubic metre of soil before potting begonia, which did not record any ill-effect on plants, however, ferbam was found phytotoxic, and PCNB and thiram made the flower colours paler. Kiplinger *et al.* (1973) applied several chemical formulations 7 times at 2-week intervals as soil drenches, foliar sprays or fumigants to begonia plants and recorded severe leaf scorch with dimethoate (Cygon) sprays or drenches, malathion sprays and parathion sprays. Sprays of benomyl, Agrimycin-17, Metasystox-R (oxydemeton-methyl), and lindane, drenches of lindane and Systox (demeton) or cyanide fumigation caused slight scorch; while diazinon 50 w.p., malathion 25 w.p. and lindane 25 w.p. caused foliar dicolouration due to spray residues. Wohlmuth (1975) through fungicide drenching at 2.5-3.0 l/m^2 one week after pricking out and thrice thereafter at 3-week intervals to 17 varieties of *B. semperflorens*, recorded that chinosol at 0.05 or 0.1 per cent, Albisal (oxyquinoline) at 0.1 or 0.2 per cent and Dithane Z-78 (zineb) at 0.2 or 0.3 per cent were tolerated well by the plants though increased frequency of drenching or increased volume of solutions inhibited growth to varying degrees.

Tomato spotted wilt virus, impatiens necrotic spot virus, Arabic mosaic virus and cucumber mosaic virus infect tuber begonias, the latter two being sometimes serious to 'multiflora' group. The main viruses encountered in double diffusion agar plate tests were tobacco mosaic, potato virus A and tobacco ringspot (Keller, 1987). Infected plants should be destroyed. Welvaert and Samyn (1975) obtained the highest percentage of healthy plants with a thermotherapy of 36-38°C for about 60 days. Heat treatment of begonia tubers at 38°C for 28 days resulted in a high percentage of healthy plants (Welvaert *et al.*, 1980).

Short day conditions may cause poor seedling development in the flat or plug tray. In tuberous begonias, too low humidity will cause poor and slow development of seedlings or death after germination. In 'hiemalis' begonias, a high night temperature (24° C or above) delays flowering. Edema (rupturing and corking of epidermal cells at the underside of leaves) is thought to be caused by high medium moisture content and excessively high day and a very low night temperatures or when the medium is persistently warm at night but air temperature is low. These situations can be avoided by reducing moisture, maintaining constant temperatures and reducing humidity. In rex begonia, chilling damage can occur at a low temperature *i.e.* 2° C. Winter short days can induce dormancy. Sudden temperature shifts may injure the leaves and may induce dormancy. In tuberous begonia, short days encourage tuber formation and abnormal shoot growth.

References

Agnew, N.H. and R.W. Campbell, 1983. Growth of *Begonia* x *hiemalis* as influenced by hand-pinching, dikegulac, and chlormequat. *HortSci.*, **18**(2): 201-202.

Allen, M.W. and D.J. Raski, 1952. Soil fumigation to control root-lesion nematode, *Pratylenchus* sp., in tuberous begonia. *Plant Dis. Reptr*, **36**:201-202.

Andersson, N.E. 1990. Mekanisk luftfordeling til blomstrende planter. *Gartner Tidende*, **106**(22): 609-610.

Anderson, N.E. 1990b. CO_2 til blomstrende potteplanter. *Gartner Tidende*, **106**(22): 606-607.

Andrén, F. 1952. Besprutningsförsök mot begoniamjöldagg. *Växtskyddsnotiser*, No. 4, pp. 57-59.

Ann, D.M., D.L. Steward and M.A. Scott, 1984. Evaluation of fungicides against powdery mildew of Rieger begonia. *Tests of Agrochemicals and Cultivars*, No. 5 (*Annals of Applied Biology* **104**, Supplement), pp. 30-31.

Anon. 1956. Foliar nematode disease of ferns and begonias. *Agric. Gaz. N.S.W.*, **67**: 258-259.

Anon. 1965. Short-day treatment of winter-flowering begonia (*Begonia* x *cheimantha* Everett) (Danish). *Tidsskr. Planteavl*, **68**: 881-884.

Anon. 1966. From the lighting trials at Rå. *Begonia cheimantha* Karelske Jomfru (Danish). *Gartneryrket*, **56**: 908.

Anon. 1975a. Nieuw *Begonia - ras : 'Tiara'. Vakblad Bloemist.*, **30**(27): 25.

Anon. 1975b. pH van stekgrond voor begonia 'Schwabenland'. *Vakblad Bloemist.*, **30**(51/52): 18-19.

Anon. 1977. Potplanten in steenwol. *Vakblad Bloemist.*, **32**(8): 27.

Anon. 1979a. Proef : besparing van produkttekosten in de begoniateelt. *Verbondsnieuws voor de Belgische Sierteelt*, **23**(3): 91.

Anon. 1979b. Scheutvorming bij *Begonia tuberhybrida*. *Verbondsnieuws voor de Belgische Sierteelt*, **23**(5): 167-168.

Arends, J.C. 1970. Somatic chromosome numbers in 'Elatior'-begonias (Dutch). *Meded. LandbHogesch, Wageningen*, No. 70-20, pp. 18.

Ascmidt, K. and G.R. Tretner, 1984. Sommerkultur von Begonia – Elatior – Hybriden. Vergleich von Langzeitdüngern. *Deutscher Gartenbau*, **38**(42): 1848-1850.

Auer, C.A. and D.B. McConnell, 1984. Simulated transit vibration and silver thiosulfate applications affect ethylene production and leaf abscission of begonia and schefflera. *HortSci.*, **19**(4): 517-519.

Bakker, J. 1968. Two new forms in the range of ornamental leaved begonias. *Vakblad Bloemist.*, **23**: 1749.

Banko, T.J. and M.A. Stefani, 1988. Growth response of selected container-grown bedding plants to paclobutrazol, uniconazole and daminozide. *J. environ. Hort.*, **6**(4): 124-129.

Baranowski, T. 1974. The phytotoxic effect of Nexoval and Alipur on some ornamental plants. *Roczniki Akademii Rolmiczei w Poznaniu, Ogrodnictwo*, **69**(5): 19-25.

Barkley, F.A. and K.S. Boghdan, 1972. *Begonia monophylla* Pavon. *Begonian*, **39**(11): 240-241.

Bass, L.N. 1980. Flower seed storage. *Seed Science & Techn.*, **8**(4): 591-599.

Benetka, V. 1987. Indukce kompaktních mutantù u *Begonia* x *hiemalis* Fotch odrùdy 'Schwabenland'. *Sbornik UVTIZ, Zahradnictvi*, **14**(1): 75-80.

Berghoef, J. and J. Bruinsma, 1979. Flower development of *Begonia franconis* Liebm. III. Effects of growth-regulating substances on organ initiation in flower buds *in vitro*. *Zeitschrift für Pflanzenphysiologie*, **93**(5): 377-386.

Besemer, A.F.H. 1961. Experiences with new materials for mildew control on various crops. *Meded. LandbHogesch. Gent*, **26**: 1343-1357.

Beuzenberg, M.P., J. Westerhof and C.V. Noordegraaf, 1977. Bloeiende begonia voor kerst. *Vakblad Bloemist.*, **32**(31): 20-21.

Bigot, M.C. 1967. Natural aptitudes of some *Begonia rex* varieties to form buds and roots. *A.R. Acad. Agric. Fr.*, **53**: 1005-1010.

Bigot, M.C. 1981. Multiplication vegetative *in vitro* de *Begonia* x *hiemalis* ("Rieger" et "Schwabenland"). I. Méthodologie. *Agronomie*, **1**(6): 433-440.

Bing, A. 1983. 1982 studies on tolerance of field grown annuals to postplant preemergence herbicides. In: *Proc. 37th ann. Meet. Northeastern Weed Sci. Soc.*, 1983, pp. 343-346.

Bik, R.A. 1967. The importance of a good pH in sowing and rooting media (Dutch). *Vakblad Bloemist.*, **22**: 517, 519.

Bik, R.A. 1974. Osmocote bij *Begonia* 'Schwabenland'. *Vakblad Bloemist.*, **29**(43): 13.

Bisaillon, A. 1968. Experiment on preserving seed vitality of *Begonia semperflorens*. *Plant Propagator*, **14**(1): 3-4.

Boardman, T. 1991. Flowering leaves. *Begonian*, **58**(May-June): 100-103.

Bosmans, P. 1971. Disease control and manurial trials in begonia (Dutch). *Mededelingen van de Faculteit Landbouwwetenschappen, Rijksuniversiteit Gent*, **36**(3): 1042-1048.

Bosmans, P. and O. Kamoen, 1970. Experiments on the occurrence, nature and control of begonia diseases. *Verhandelingen over Plantenziekten*, No. 22, 58 pp.

Bowes, B.G. 1990. Epiphyllous buds on intact plants of *Begonia rex* 'President'. *Begonian*, **57**(Sept-Oct): 178-182.

Brennan, E., I.A. Leone and R.H. Daines, 1965. Find that chlorine in air can damage plants. *N. J. Agric.*, **47**(3): 10-12.

Bugbee, G.J. and C.R. Frink, 1986. Aeration of potting media and plant growth. *Soil Sci.*, **141**(6): 438-441.

Burghardt, H. 1988. Im Vergleich : verschiedene Düngungsmethoden. Alle Düngungsmethoden bei Begonia – Elatior – Hybriden bringen gute Ergebnisse. *Gb + Gw, Gärtenbörse und Gartenwelt*, **88**(45): 1968-1970.

Burt-Utley, K. 1983. Three new species of *Begonia* (Begoniaceae) from Mexico. *Brittonia*, **35**(2): 115-119.

Campbell, R.W. and M.A. Bakir, 1985. Determination of seasonal levels of auxin and cytokinins in *Begonia* x *masoniana* Irmsch 'Iron Cross' leaves. In: *Procs., twelfth annual meeting, Pl. Growth Regulator Soc. Amer., Boulder, Colorado*, pp. 194-202.

Charpentier, A., L. Brouillet and D. Barabé, 1089a. Organogenèse de la fleur pistillée du *Begonia horticola* (Begoniaceae). *Canadian J. Bot.*, **67**(2): 559-572.

Charpentier, A., L. Brouillet and D. Barabé, 1089b. Organogenèse de la fleur pistillée du *Begonia Dregei et de l'Hillebrandia sandwicensis* (Begoniaceae). *Canadian J. Bot.*, **67**(12): 3625-3639.

Chase, A.R. and R.T. Poole, 1987. Effect of fertilizer rate on growth of fibrous-rooted *Begonia*. *HortSci.*, **22**(1): 162,

Chitra, D.V. 1998. Standardisation of Propagation Technique and Growing Media in Rex Begonia {*Begonia rex* (Putz.) Inimitable}. M.Sc. Thesis. College of Agriculture, Kerala agricultural University, Vellayani, Kerala.

Chlyah-Arnason, A. and M. van TranThanh, 1968. Budding capacity of undetached *Begonia rex* leaves. *Nature*, **218**:493.

Chlyah, A. and M.T.T. Van, 1971. Comparison of the localization of nucleic acid synthesis during bud formation on leaf fragments and on intact undetached leaves of *begonia rex*. *Biologia Plantarum*, **13**(3): 184-188.

Cohl, H.A. and B.C. Moser, 1976. Environmental control of shoot initiation by Rieger begonia leaf cuttings. *HortSci.*, **11**(4): 378-379.

Coolen, W.A. and C.J. D'Herde, 1968. Some aspects of the chemical disinfection of the soil in begonia cultivation. *Meded. Rijksfac. LandbWetensch., Gent*, **13**: 739-750.

Cortvrieendt, S.F. and R. De Groote, 1952. Essais de fumure sur gloxinia et begonia. *Riv. Agric. Brux.*, **5**: 1311-1318.

Creighton, H.B. 1968. Epidermal outgrowths in *Begonia phyllomaniaca* 'Templini'. *Amer. J. Bot.*, **55**: 705 (abstr.).

Dalchow, J. 1968. A hitherto unreported disease of Lorraine begonias : *Verticillium dahliae, Gartenwelt*, **68**: 264, 266.

D'Herde, J., J. De Maeseneer and J. Van Den Brande, 1961. Soil sickness in begonia growing. *Meded. LandbHogesch. Gent*, **26**: 1133-1143.

Dipner, H. 1975. Erweitertes Elatior – Begonien – Sortiment. *Schweizerische Gärtnerzeitung*, **78**(51): 594-595.

Djurhuus, R. 1984. The effect of CO_2, daylength and light on the production and subsequent growth of *Begonia* x *tuberhybrida* cuttings. *Acta Hort.*, No. 162, pp. 65-74.

Djurhuus, R. and H.R. Gislerød, 1985. Vanningsfrekvens og dyrkingemedia til *Aphelandra* og *Begonia* x *hiemalis*. *Gartneryrket*, **75**(6): 132-133.

Doorenbos, J. 1972. What is *Begonia* 'Fireflush'? *Begonian*, **39**(10): 230-231.

Doorenbos, J. 1973a. Breeding Elatior begonias (*B.* x *hiemalis* Fotsch). *Begonian*, **40**(12): 275-277, 290-291.

Doorenbos, J. 1973b. 'Turo', a new 'Elatior' begonia (Dutch). *Vakblad Bloemisterij*, **28**(42): 25.

Doorenbos, J. 1975. What is *Begonia* 'Fireflush'? (2). *Begonian*, **42**(2): 41, 49.

Doorenbos, J. 1979. *B.* (*Begonia*) *vitifolia* and other elusive tree-like begonias. *Begonian*, **46**: 234-240.

Doorenbos, J. 1980a. *B. rhopalocarpa*, a begonia with colorful fruit. *Begonian*, **47**: 102-103.

Doorenbos, J. 1980b. Two species from java new to cultivation. *Begonian*, **47**: 213-215.

Doorenbos, J. 1981. A begonia people grow as a vegetable. *Begonian*, **49**: 14, 17.

Doorenbos, J. and R.A.H. Legro, 1968. Breeding Gloire de Lorraine begonias. *Meded. LandbHogesch, Wageningen*, **68**(19): 14.

Doorenbos, J. and J.J. Karper, 1975. X-ray induced mutations in *Begonia* x *hiemalis*. *Euphytica*, **24**(1): 13-19.

Doskotch, R.W., M.Y. Malik and J.L. Beal, 1969. Cucurbitacin B, the cytotoxic principle of *Begonia tuberhybrida* var. *alba*. *Lloydia*, **32**: 115-122.

Dowson, W.J., W.C. Moore and L. Ogilvie, 1938. A bacterial disease of begonia. *J. roy. Hort. Soc.*, **63**: 286-290.

Elliott, G.C. and P.V. Nelson, 1981. Acute boron toxicity in *Begonia* x *hiemalis* 'Schwabenland Red'. *Communications in Soil Science and Plant Analysis*, **12**(8): 775-783.

Ensemeyer, M., L. Langhammer and H.W. Rauwald, 1980. Isolierung und Konstitutionsaufklärung eines dimeren Proaanthocyanidins in *Begonia glabra* Aubl. *Archiv dert Pharmazie*, **313**(1): 61-71.

Esashi, Y. 1969. The relation between light and temperature effects in the induction and release of dormancy in the aerial tuber of *Begonia evansiana*. *Plant Cell Physiol.*, **10**: 583-595.

Esashi, Y. and M. Nagao, 1959. Studies on the formation and sprouting of aerial tubers in *Begonia evansiana* Andre. II. Effects of light and temperature on the sprouting of aerial tubers. *Sci. Reps Tohoku Univ., Ser. 4, Biol.*, **25**: 191-197.

Esashi, Y. and M. Nagao, 1973. Effects of oxygen and respiratory inhibitors on induction and release of dormancy in aerial tubers of *Begonia evansiana*. *Plant Physiol.*, **51**(3): 504-507.

Esashi, Y. and A.C. Leopold, 1967. Regulation of dormancy induction in *Begonia evansiana* tubers by nucleic acid and protein synthesis. *Plant Physiol.* (suppl.), **42**: 54 (abstr.).

Esashi, Y. and A.C. Leopold, 1969. Regulation of the onset of dormancy in tubers of *Begonia evansiana*. *Plant Physiol.*, **44**: 1200-1202.

Fearon, J.R. 1965. The effect of artificial illumination on crops for early marketing. *N.A.S.S. Quart. Rev.*, No. 70, pp. 74-79.

Fischer, S. 1973. Temik trials on Rieger begonia. *Maryland Florist*, No. 186, pp. 6-7.

Fjeld, T. 1989. Effect of ethylene exposure, temperature and light intensity on keeping quality of *Begonia* x *cheimantha*. *Acta Hort.*, No. 261, pp. 373-376.

Fjeld, T. 1986. The effect of relative humidity, light intensity and temperature on keeping quality of *Begonia* x *cheimantha* Everett. *Acta Hort.*, No. 181, pp. 251-255.

Fjeld, T. 1990. Effects of temperature and irrandiance level on plant quality at marketing stage and the subsequent keeping quality of Christmas begonia (*Begonia* x *cheimantha Everett*). *Norwegian J. agric. Sci.*, **4**(3): 217-223.

Fonnesbech, M. 1974. The influence of NAA, BA and temperature on shoot and root development from *Begonia* x *cheimantha* petiole segments grown *in vitro*. *Physiologia Plantarum*, **32**(1): 49-54.

Furnish, G.B. 1947. Rooting rex begonia cuttings by hydroponics. *Nat. hort. Mag.*, **26**: 35-40.

Gardner, J.O. and D.P. Ormrod, 1976. Response of the Rieger begonia to ozone and sulphur dioxide. *Scientia Hort.*, **5**(2): 171-181.

Gillard, A. and J. Van Den Brande, 1955. Quelques problems concernant les nematodes des raciness (*Meloidogyne* spp.) en Belgique, particulièrement la disinfection des tubercules à l'eau chaude. *Parasitica*, **11**: 74-80.

Gislerød, H.R. 1974. Forsøk med bladstiklinger av hiemalisbegonia. *Gartneryrket*, **64**(22/23): 499-502.

Gislerød, H.R. and L.M. Mortensen, 1990. Relative humidity and nutrient concentration effect nutrient uptake and growth of *Begonia* x *hiemalis*. *HortSci.*, **25**(5): 524-526.

Gjaerum, H.B. 1963. The effect of some fungicides on root formation, growth and flower colour of begonias. *Gartneryrket*, **53**: 997-999.

Goossens, A. 1963. Tuber size and selection of tuberous begonias (Flemish). *Tuinbouwberichten*, **27**: 18.

Grappelli, A., U. Tomati, E. Galli and B. Vergari, 1985. Earthworm casting in plant propagation. *HortSci.*, **20**(5): 874-876.

Guba, E.F. and C.J. Gilgut, 1938. Control of the begonia leaf blight nematode. *Bull. Mass. Agric. Exp. Sta.*, No. 348, pp. 12.

Gugenhan, E. and F. Deiser, 1969. Cultivation experiments with Super Manural (German). *Erwerbsgärtner*, **23**:1334-1337.

Guirfanova, K. and B. Tokin, 1942. Shoots sent forth by a leaf of *Begonia rex* not severed from parent plant (Russian). *C.R. Acad. Sci. U.R.S.S.*, **35**: 122-124.

Gui, Y.L., S.R. Gu and T.Y. Xu, 1985. Organogenesis in leaf explants of *Begonia coccinea* (Chinese). *Acta Botanica Sinica*, **27**(5): 550-552.

Gundersen, K. 1959. Some experiments with gibberellic acid. *Acta Hort. Gotoburg.*, **22**: 87-110.

Haas, H.P. and R. Röber, 1988. Begonia – Elatior – Hybriden im geschlossenen System. *Gb + Gw, Gärtenbörse und Gartenwelt*, **88**(45): 1971-1973.

Haegeman, J. and J.G. van Onsem, 1964-65. The influence of soil temperature on tuber formation in hybrid begonias (Belgian). *Land-en Tuinb. Jaarb., Gent*, **19**: 181-183.

Haegeman, J. 1967. Comparative trials with tuberous begonias (*Begonia tuberhybrida* Voss) from various sources (Dutch). *Meded. Rijksstat. SierplVered. Melle*, **11**: 36.

Haegeman, J. and J.G. Van Onsem, 1966. An investigation into the value of a number of substrates for raising seedlings of tuberous begonias (Dutch). *Meded. Rijksstat. Sierplvered. Melle*, No. 10, pp. 11-22.

Hahn, E. 1978. Mexicross – Begonien. *Gb + Gw*, **78**(27): 645-646.

Hahn, E. 1980. Mexicross – Begonien. *Gb + Gw*, **80**(15): 336, 339.

Hallig, V.A., O.V. Christensen and F. Rehnstrom, 1966. Control of flowering in Christmas begonias (*Begonia* x *cheimantha* Everette) (Danish). *Tidsskr. Planteavl*, **70**: 170-178.

Hammer, A. 1973. Capillary watering of Rieger begonias. *Focus on Floriculture*, **1**(2): 14-15.

Hänisch Ten Cate, C.H., J. Berghoef, A.M.H. van der Hoorn and J. Bruinsma, 1975. Hormonal regulation of pedicel abscission in *Begonia* flower buds. *Physiologia Plantarum*, **33**(4): 280-284.

Hänisch Ten Cate, C.H. and J. Bruinsma, 1973. Abscission of flower bud pedicels in *Begonia*. I. Effects of plant growth regulating substances on the abscission with intact plants and with explants. *Acta Botanica Neerlandica*, **22**(6) 666-674.

Hansen, C.E., C. Kopperud and O.M. Heide, 1988. Identity of cytokinins in *begonia* leaves and their variation in relation to photoperiod and temperature. *Physiologia Plantarum*, **73**(3): 387-391.

Harborne, J.B. and E. Hall, 1964. Plant polyphenols. XIII. The systematic distribution and origin of anthocyanins containing branched trisaccharides. *Phytochem.*, **3**: 453-463.

Harri, J.A., P.O. Larsen and C.C. Powell, Jr. 1975. Bacterial leaf spot and blight of Rieger begonia cv. Aphrodite Rose. *Ohio Florists' Assocn. Bull.*, No. 552, pp. 4-6.

Harri, J.A., P.O. Larsen and C.C. Powell, Jr. 1977. Bacterial leaf spot and blight of Rieger elatior begonia : systemic movement of the pathogen, host range, and chemical control trials. *Plant Dis. Reptr*, **61**(8): 649-653.

Hashimoto, T. and S. Tamura, 1969. Effects of abscisic acid on the sprouting of aerial tubers of *Begonia evansiana* and *Dioscorea batatas*. *Bot. Mag., Tokyo*, **82**: 69-75.

Heide, O.M. 1962. Interaction of night temperatures and day-length in flowering of *Beegonia* x *cheimantha* Everett. *Physiol. Plant.*, **15**: 729-735.

Heide, O.M. 1964. Temperature and day length for mother plants and leaf cuttings of begonias (*Begonia* x *cheimantha*) (Norwegian). *Gartneryrket*, **54**: 1314-1316.

Heide, O.M. 1965. The optimum temperature for cuttings of winter-flowering begonias (Norwegian). *Gartneryket*, **55**: 1134, 1136.

Heide, O.M. 1967. The auxin level of *Begonia* leaves in relation to their regeneration ability. *Physiol. Plant.*, **20**: 886-902.

Heide, O.M. 1967b. The regulation of the regenerating ability of plant tissue (Norwegian). *Nord. JordbrForskn.*, **49**:278 (abstr.).

Heide, O.M. 1968. Stimulation of adventitious bud formation in begonia leaves by abscisic acid. *Nature*, **219**: 960-961.

Heide, O.M. 1969a. Christmas flowering of begonias regulated by day length (Norwegian). *Gartneryrket*, **59**: 694-697.

Heide, O.M. 1969b. Interaction of growth retardants and temperature in growth, flowering, regeneration, and auxin activity of *Begonia* x *cheimantha* Everett. *Physiol. Plant.*, **22**: 1001-1012.

Heide, O.M. and F. Skoog, 1967. Cytokinin activity in *Begonia* and *Bryophyllum*. *Physiol. Plant.*, **20**: 771-780.

Hemer, M. 1978. Echter Mehltau. Bekämpfung an blühenden Zierpflanzen. *Gb + Gw*, **78**(31): 737.

Hendriks, L. and D. Luydolph, 1990. Mutterpflanzenbelichtung verbessert Stecklingsproduktion. *Deutscher Gartenbau*, **44**(19): 1275.

Hendriks, L. and H.C. Scharpf, 1984. Bewässerungsdüngung von Elatior - Begonien. *Deutscher Gartenbau*, **338**(25) 1082-1084.

Hentig, W.U. von, 1978. Zur Vermehrung von Elatiorbegonien. *Gb+Gw*, **78**(9): 193-195.

Hentig, W.U. von, 1980. Elatiorbegonien. Einfluss von Bleuchtungsstärke und Blattstiellänge bei der Vermehrung von Rieger - Sorten. *Gb+Gw*, **80**(6): 125-126.

Hentig, W.U. von and K. Knösel, 1984. So behalten Kleinpflanzen im Sommer ihren character. Weitere Versuchsergebnisse zu Begonia - Elatior - Hybriden aus Geisenheim. *Gb + Gw*, **84**(43): 1023-1026.

Heungens, A. 1968-69. The effect of some nematicides on the beneficial soil fauna in begonia culture (Belgian). *Land- en Tuinb. Jaarb., Gent*, **23**: 227-228.

Hilding, A. 1974. Effects of daylength and temperature on the propagation of *begonia* x *hiemalis* by leaf cuttings. *Lantbrukshögskolans Meddelanden, A*, No. 209, 15 pp.

Hilding, A. 1975. Inverkan av temperature, toppning och retarderande medef på tillväxt och utveckling av höstbegonia, *Begonia* x *hiemalis* (Sv=Swede). *Lantbrukshögskolans Meddelanden, A*, No. 252, 26 pp.

Himme, M. van, R. Bulcke and J. Stryckers, 1986. Bloementeelt. *Mededeling van het Centrum voor Onkruidonderzoek van de Rijksuniversiteit Gent*, No. 44, pp. 121-125.

Hoover, W.S. 1986. Stomata and stomatal clusters in *Begonia* : ecological response in two Mexican species. *Biotropica*, **18**(1): 16-21.

Hoover, W.S. 1989. Species of *Begonia* in cultivation. *Begonian*, **56**: 149-150.

Horn, W. 1971. New methods in breeding winter-flowering begonias. *Gartenwelt*, **71**(5): 101-102.

Horn, W. and M. Wischer, 1987. Bonzi, ein never Hemmstoff bei Elatior-Begonien. *Gb + Gw*, **87**(51): 1894-1895.

Horn, W., H. Bundies and K. Zimmer, 1976. Untersuchungen zur Züchtung triploider F_1-Hybriden bei Lorraine - Begonien. *Zeitschrift für Pflanzenzüchtung*, **76**(3): 177-189.

Horton, F.F. 1951. Short day causes dormancy in tuberous rooted begonias. *Bull. N.Y. St. Flower Grs*, No. 76, p.3.

Horton, F.F. 1953. Keep begonias vegetative. *Bull. N.Y. St. Flower Grs*, No. 92, pp. 2.

Horváth, M. and G. Horváth, 1969. The petiolate rooting of *Begonia semperflorens*. *Acta Agron. Hung.*, **18**(1/2): 251-254.

Høyer, L. 1984. Begonia – elatior 'Serene'. Forringet holdbarhed efter påvirkning af aetylen og mørke (Danish). *Gartner Tidende*, **100**(37): 1170-1171.

Iida, T., K. Yabe, S. Wasida and Y. Sakurai, 1986. Mass propagation of *begonia tuberhybrida* Voss plantlets using tissue culture (Japanese). *Res. Bull. Aichi-ken agric. Res. Cent., Japan*, No. 18, pp. 186-190.

Imelda, M. 1983. Tissue culture of Rieger begonia petiole segments. *Annales Bogorienses*, **8**(1): 1-11.

Ingenillem, U., W. Mass and W. Schindler, 1984. Ein neuer Weg zur CO_2 - Düngung. *Deutscher Gartenbau*, **38**(38): 1624-1627.

Jangoux, J. 1990. The 'instant suntan' begonia. *Begonian*, **57**: 219-220.

Jodon, M.H. and L.P. Nichols, 1974. Bacterial leaf spot of begonia. *Bull. Pennsylvania Flower Grs*, No. 272, pp. 1, 8-9.

Jungbauer, J. 1975. Vorsicht mit Temperaturabsenkung bei Lorrainebegonien. *Gartenwelt*, **75**(17): 370-371.

Jungbauer, J. 1981. Elatiorbegonien. Die Reaktion auf die Tageslänge. *Gb + Gw*, **81**(18): 416-417.

Jürgensmeier, H.L. 1961. The co-operation of different factors in the decolourization of anthocyanin in begonias. *Planta*, **56**: 233-236.

Jürgensmeier, H.L. and M. Bopp, 1961. Enzymatic breakdown of anthocyanin in begonias (German). *Naturwissenschaften*, **48**: 80-81.

Kahl, E. and W. Wittmann, 1969. Investigations on the problem of the phytotoxic effect of plant protection materials containing zinc (German). I. *Tätigkeitsber. 1966-1969, Bundesant. PflSch., Wien.*, pp. 303-308.

Kamerman, W. 1971. A bacterial rot in *Begonia bartinii*. *Bloembollencultuur*, **81**(38): 1005-1006.

Kamiñska, M. 1967. The effect of different substrates and nutrient levels on seedling growth in *Salvia splendens* and *Begonia semperflorens* (Polish). *ZesŸ. Nauk. Szk. glów. Gosp. Wiejsk., Ser. Ogrodnictwo*, No. 4, pp. 153-165.

Kamoen, O. 1972. Pathogenesis of *Botrytis cinerea* on tuberous begonias. Thesis, *Rijksuniversiteit, Gent, Belgium*, 106 pp.

Karper, J.J. 1973. The new foliage begonias Nocturne and Andante (Dutch). *Vakblad Bloemisterij*, **28**(3): 17.

Keller, J.RE. 1987. Virus detection in cultivated begonias. *Begonian*, **54** (Jan.-Feb.): 4-6.

Khoder, M., P. Villemur and R. Jonard, 1981. La multiplication végétative de l'espèce florale *Begonia elatior* (cultivar Rieger) á partir de différents organes cultivés *in vitro.* Comptes Rendus des Séances de l'Académie des Science, III, Sciences de la Vie, **293**(7): 403-408.

Khoder, M., P. Villemur and R. Jonard, 1984. Obtention de plantes monoplödes et triplödes par androgenèse *in vitro* chez le *Begonia* x *hiemalis* Fotsch cv. (A). *Bulletin de la Société Botanique de France, Lettres Botaniques*, **131**(1) 43-48.

Kiew, R. 1989. *Begonia rajah* : refound! *Begonian*, **56**: 53-54.

Kiplinger, D.C., H.K. Tayama and G. Staby, 1973. Tests and observations on Rieger begonias. *Ohio Florists' Assocn Bull.*, No. 520, pp. 2-3.

Knösel, D. 1969. The control of *Xanthomonas begoniae* by foliar spraying with antibiotics and other preparations. *NachrBl. Dtsch. PflSchDienst., Braunschweig*, **21**: 100-102.

Knuth, M. 1962. Studies on colchicines – treated Gloire de Lorraine begonias. *Arch. Gartenb.*, **10**: 42-53.

Krauskopf, D.M. and P.V. Nelson, 1976. Chemical height control of Rieger Elatior begonia. *J. Amer. Soc. hort. Sci.*, **101**(5): 618-619.

Krebs, E.K. 1988. Mehltau und Elatior – Begonien. *Gb + GW (Gärtenbörse und Gartenwelt)*, **88**(45): 1982-1984.

Krebs, O. and K. Zimmer, 1983. Blütenbildung bei *Begonia boweri* Ziesenh. Und einem Abkömmling von Begonia 'Cleopatra'. X. Reversibilität des HR-Licht-Effektes durch anschliessendes DR-Licht. *Gartenbauwissenschaft*, **48**(4): 167-171.

Kuehny, J.S., A. Painter and P.C. Branch, 2001. Plug source and growth retardants affect finish size of bedding plants. *HortSci.*, **35**(2): 321-323.

Kwast, A., H.J. Luck, E. Grantzau and H.C. Scharpf, 1989. Bewässerungsdauer und Nitrit. Probleme bei der Anstaubewässerung. *Gb+Gw, Gärtnerbörse und Gartenwelt*, **89**(15): 720-723.

Lagerstedt, H.B. 1967. Propagation of begonia from leaf discs. *HortSci.*, **2**: 202.

Langhammer, L. and M. Grandet, 1974. Über die Verbreitung eines seltenen Anthocyans in der Familie der Begoniaceae. *Planta Medica*, **26**(3): 260-268.

Lecoço, M. and C. Dumas, 1975. Histophysiologie des stigmates normaux et des formations stigmatoïdes chez *Begonia tuberhybrida*. I. Observations preliminaries. *Canadian J. Bot.*, **53**(12): 1252-1258.

Lecocq, M. 1977. Le gynécée du *Begonia tuberhybrida* et ses variations. *Canadian J. Bot.*, **55**(5): 525-541.

Lee, D.W. 1979. *B. (Begonia) pavonina*, the begonia with blue leaves. *Begonian*, **46**: 210-213.

Lee, D.W., J.B. Lowry and B.C. Stone, 1979. Abaxial anthocyanin layer in leaves of tropical rain forest plants : enhancer of light capture in deep shade. *Biotropica*, **11**(1): 70-77.

Legro, R.A.H. and J. Doorenbos, 1969. Chromosome numbers in *Begonia. Neth. J. agric. Sci.*, **17**: 189-202.

Legro, R.A.H. and J. Doorenbos, 1971. Chromosome numbers in *Begonia*. 2. *Neth. J. Agric. Sci.*, **19**(3): 176-183.

Leone, I.A. and E. Brennan, 1969. Sensitivity to begonias to air pollution. *Hort. Res.*, **9**: 112-116.

Lewis, C.A. 1951a. Daylength controls flowering of tuberous-rooted begonias. *Bull. N.Y. St. Flower Grs*, No. 67, pp. 2-3, 8.

Lewis, C.A. 1951b. Some effects of daylength on tuberization, flowering and vegetative growth of tuberous-rooted begonias. *Proc. Amer. Soc. hort. Sci.*, **57**: 376-378.

Lewis, C.A. 1953. Further studies on the effects of photoperiod and temperature on growth, flowering and tuberization of tuberous-rooted begonias. *Proc. Amer. Soc. hort. Sci.*, **61**: 559-568.

Lindemann, A. 1968. Mutations in the Rieger strain of Elatior begonias (German). *Gartenwelt*, **68**: 266-267.

Lindemann, A. 1973. Effects on growth of enriching greenhouse air with carbon dioxide at different rates (German). *Zierplanzenbau*, **13**(19): 778-779.

Lin, W.C. and J.M. Molnar, 1983. Rieger Elatior begonia Saanred. *Canadian J. Pl. Sci.*, **63**(2): 563-564.

Loeser, H. 1979a. Begonia – Elatior – Hybriden als Beetpflanzen. *Deutscher Gartenbau*, **33**(47): 1977-1979.

Loeser, H. 1979b. Kurztagbehandlung bei Elatior-Begonien. *Deutscher Gartenbau*, **33**(47): 1985-1987.

Loeser, H. 1987a. *Begonia semperflorens* in einer mehrjährigen Sortenprüfung in Heidelberg 1984-86. *Zierpflanzenbau*, **27**(6): 198-199.

Loeser, H. 1987b. *Begonia tuberhybrida* in eine Sortenprüfung in Heidelberg 86. *Zierpflanzenbau*, **27**(6): 200-203.

Loeser, H. 1988a. Mit richtiger Sorte – schnelle Kultur. *Gb + Gw, Gärtenbörse und Gartenwelt*, **88**(45): 1976-1981.

Loeser, H. 1988b. Warum nicht auch im Balkonkasten? *Gb + Gw, Gärtenbörse und Gartenwelt*, **88**(45): 1985-1987.

Lovrekovich, L. and Z. Klement, 1962. The occurrence in Hungary of *Xanthomonas begoniae* (Takimoto) Dowson the causal agent of wilt and leaf spot of begonias. Növénytermelés, **11**: 191-194.

Maatsch, R. 1956. The effect of tuber size on the development of tuberous begonias (German). *Gartenwelt*, **56**: 343-344.

Maatsch, R. and A. Herklotz, 1965. The effect of storage temperature on the dormancy of tuberous begonias (German). *Gartenbauwiss.*, **30**: 69-74.

Maatsch, R. and W. Rünger, 1955a. Über das oberirdische Wachstum und die Knollenbildung von Knollenbegonien nach kurzfristiger Kurztagsbehandlung. *Gartenbauwiss.*, **1**(n.s.): 457-464.

Maatsch, R. and W. Rünger, 1955b. The effect of temperature on the photoperiodic response of tuberous begonias. *Gartenbauwiss.*, **2**: 478-484.

Maatsch, R. and K. Zimmer, 1964. Tuber formation by begonias (German). *Gartenwelt*, **64**: 8-9.

Maier, U. and R. Sattler, 1977. The structure of the epiphyllous appendages of *Begonia hispida* var. *cucullifera*. *Canadian J. Bot.*, **55**(3): 264-280.

Margara, J. and R. Phelouzat, 1984. Structure et ontogenèse des néoformations observées *in vitro* sur le pétale de *Begonia* x *elatior*. *Canadian J. Bot.*, **62**(12): 2798-2803.

Margara, J. and M.T. Piollat, 1984. L'aptitude à l'organogenèse des pétales de *Begonia* x *elatior* néoformés *in vitro*. *Comptes Rendus des Séances de l'Académie des Sciences, III* (Sciences de la Vie), **298**(20): 583-586.

Marahrens, E. and E. Toop, 1986. The effect of substrate on the rooting of *Begonia* x *lucerna* cuttings. *Acta Hort.*, No. 178, pp. 231-236.

Matouš, J. 1965. Male sterility in cultivated plants and its use in breeding hybrid varieties (Czech). *Vd Pr. Výzk. Úst. Okrasn. Zahrad. V Prùhonicích*, **3**: 13-38.

Matouš, J. 1967. Selection of begonia (*Begonia tuberhybrida* Voss) for tuber size (Czech). *Vd Pr. Výzk. Úst. okrasn. Zahrad. v Prùhonicích*, **4**: 13-24.

Matouš, J. 1969. The cytoplasmic male sterility in *Begonia semperflorens* hort. and possibilities of its use in breeding (Czech). *Acta prùhon.*, N. 19, pp. 1-45.

Matsubara, H. 1982. Mutation breeding in ornamental plants – technique used for radiation induced mutant in begonia, chrysanthemum, abelia and winter daphne. *Gamma Field Symposia,* No. 2, pp. 55-67.

Matsubara, H., K. Shigematsu and H. Suda, 1974. The isolation and fixation of a completely mutant plant of *Begonia rex* from a sectorial chimaera induced by gamma irradiation (Japanese). *J. Jap. Soc. hort. Sci.*, **43**(1): 63-68.

Mei, B.J. and H. Ai, 1987. The rapid propagation of *Begonia masoniana in vitro* (Chinese). *Plant Physiol. Comm.*, No. 2, pp. 27-30.

Meylan, G. and R. Tripod, 1980. Contribution à l'étude comparative de 67 cultivars de *Begonia semperflorens. Revue Horticole Suisse,* **53**(12): 407-409.

Mikkelsen, E.P. and K.C. Sink, Jr. 1978. Histology of adventitious shoot and root formation on leaf-petiole cuttings of *Begonia* x *hiemalis* Fotsch ';Aphrodite Peach'. *Scientia Hort.*, **8**(2): 179-192.

Mikkelsen, J.C., J. Ryan and M.J. Constantin, 1975. Mutation breeding of Rieger's Elatior begonias. *Amer. Horticulturist,* **54**(3): 18-21.

Moe, R. and A. Smith-Eriksen, 1986. The effect of ethephon and STS treatment on flower malformation and flower bud abscission in *Begonia* x *chiemantha* Everett., *Acta Hort.*, No. 181, pp. 155-160.

Molnar, J.M. 1974. Photoperiodic response of *Begonia* x *hiemalis* cv. Rieger. *Canadian J.Pl. Sci.*, **54**(2): 277-280.

Molnar, J.M. 1976. Rieger Elatior begonia cv. Northern Sunset. *Canadian J. Pl. Sci.*, **56**(4): 1003.

Moorhouse, E.R., A.T. Gillespie and A.K. Chornley, 1990. The progress and prospects for the control of the black vine weevil, *Otiorhynchus sulcatus* by entomogenous fungi. In: Proceedings and Abstracts, Vth intern. Colloquium on Invertebrate Pathology and *Micarobial Control, Adelaide, Australia,* August 20-24, 1990.

Mortensen, L.M. and R. Ulsaker, 1985. Effect of CO_2 concentration and light levels on growth, flowering and photosynthesis of *Begonia* x *hiemalis* Fotsch. *Scientia Hort.*, **27**(1/2): 133-141.

Nagamura, S. 1980. Studies on standard composts for potted flowers. 5. The relationship between the growth of cyclamen and begonia and the physical properties of several media and pot materials (Japanese). *Bull. Nara agric. Exp. St.*, No. 11, pp. 31-40.

Nagamura, S. and S. Urabe, 1976. The effects of 6-benzylaminopurine and auxin applied with talc on begonia leaf cuttings. *Bull. Nara agric. Exp. St.*, No. 7, pp. 24-30.

Nagao, M. and E. Mitsui, 1959. Studies on the formation and sprouting of aerial tubers in *Begonia evansiana* Andre. III. Effect of gebberellin on the dormancy of aerial tubers. *Sci. Reps Tohoku Univ., Ser. 4, Biol.*, **25**: 199-205.

Nagao, M. and N. Okagami, 1966. Effect of (2-chlorethyl) trimethylammonium chloride on the formation and dormancy of aerial tubers of *Begonia evansiana. Bot. Mag. Tokyo,* **79**: 687-692.

Nelson, P.V. and T. Bost, 1975. Proper spacing for Rieger Elatior begonia. *North Carolina Fl. Grs' Bull.*, **19**(1): 2-3.

Nelson, P.V. and D. Cover, 1976. Varieties of fibrous begonia for a flowering plant crop. *North Carolina Flower Grs' Bull.*, **20**(1): 1-3.

Nelson, P.V., D.M. Krauskopf and N.C. Mingis, 1979. Minimum critical foliar levels of K, Mg, and B in Rieger elatior begonia. *J. Amer. Soc. hort. Sci.*, **104**(6): 793-796.

Neubauer, H.F. and I. Beissler, 1971. Stomata and stomatal groups in different types of begonia leaves (German). *Gartenbauwissenschaft*, **36**: 45-50.

Niebisch, R.M. and K. Kelling, 1986. Results of chemical control of fungal diseases in ornamental plant production (German). *Gartenbau*, **33**(7): 215-218.

Odén, P.C. and O.M. Heide, 1988. Detection and identification of gibberellins in extracts of *Begonia* leaves by bioassay, radioimmunoassay and gas chromatography – mass spectrometry. *Physiologia Plantarum,* **73**(4): 445-450.

Odén, P.C. and O.M. Heide, 1989. Quantitation of gibberellins and indoleacetic acid in *Begonia* leaves: relationship with environment, regeneration and flowering. *Physiologia Plantarum,* **76**(4): 500-506.

Okagami, N. 1972. The nature of gibberellin – induced dormancy in aerial tubers of *begonia evansiana. Plant Cell Physiol.*, **13**(5): 763-771.

Okagami, N. and Y. Esashi, 1972. Dormancy regulation by morphactin in aerial tubers of *Begonia evansiana. Planta,* **104**(3): 195-200.

Okagami, N. and M. Nagao, 1973. Gibberellin-induced dormancy in *Begonia* aerial tubers. Increase of the growth inhibitor content by gibberellin treatment. *Plant and Cell Physiol.,* **14**(6): 1063-1072.

Okagami, N., Y. Esashi and M. Nagao, 1977. Gibberellin – induced inhibition and promotion of sprouting in aerial tubers of *Begonia evansiana* Andre. in relation to photoperiodic treatment and tuber stage. *Planta*, **136**(1): 1-6.

Oloomi, H. and R.N. Payne, 1982. Effects of photoperiod and pinching on development of *Begonia* x *tuberhybrida*. *HortSci.*, **17**(3): 337-338.

O'Reilly, T. 1978. Begonia profiles. *Begonia* 'Lospe'. *Begonian*, **45**: 67, 86.

O'Reilly, T. 1980. [*Begonia*] "Bewitched", a Halloween mustery. *Begonian*, **47**: 271-272.

O'Reilly, T. 1991. Unidentified *Begonia* species list. *Begonian*, **58**: 170-171.

O'Reilly, T. and C. Karegeannes, 1983. *Begonia leathermaniae*, a new Bolivian species. *Begonian*, **50**(6): 144-147.

Patáková, S. 1969. The effect of CCC on the growth of tubers of *Begonia tuberhybrida* Voss. *Acta prùhon.*, No. 18, pp. 47-56.

Pawlowski, H.E. 1967. The effects of calcium nitrate and ammonium nitrate on the growth of petunias, chrysanthemums and begonias. I. yield. *Gartenbauwiss.*, **32**: 193-211.

Peck, D.E. and B.G. Cumming, 1984. *In vitro* propagation of *Begonia* x *tuberhybrida* from leaf sections. *HortSci.*, **19**(3): 395-397.

Peng, C.I., Y.K. Chen and H.F. Yen, 1988. *Begonia ravenii* (Begoniaceae), a new species from Taiwan. *Botanical Bull. Academia Sinica, Taiwan*, **29**(3): 217-222.

Peters, J. 1974a. Der Einfluss von Wachstumsregulatoren auf das Knollenwachstum von *Begonia* x *tuberhybrida*. *Gartenbauwissenschaft*, **39**(2): 151-154.

Peters, J. 1974b. Einfluss der Temperatur auf das oberirdische Wachstum und die Knollenbildung bei *Begonia* x *tuberhybrida* (Voss.). *Gartenbauwiss.*, **39**(3): 301-308.

Pettersson, M.L. 1979. Angrepp av dvärgkvalster, *Tarsonemus pallidus* Banks, på utplanterings växter. *Växtskyddsnotiser*, **43**(3): 38-40.

Pierik, R.L.M. and F.A.A. Tetteroo, 1987. Vegetative propagation of *Begonia venosa* Skan *in vitro* from inflorescence explants. *Plant Cell, Tissue Organ Cult.*, **10**(2): 135-142.

Pieroh, E.A., H. Werres and K. Raschke, 1959. Trapex – a new nematicide for soil sterilization. *Anz. Schädlingsk.*, **32**: 183-189.

Pivot, D. 1975. Recherche de femure sur begonia "Gloire de Lorraine". *Pépiniéristes, Horticulteurs, Maraîchers*, No. 156, pp. 35-44.

Pollard, J. and E.H. Roberts, 1968. Successful germination of begonia and nemesia seeds. *Grower*, **70**: 374.

Powell, C.C. and R.M. Riedel, 1973a. Temik phytotoxic on Rieger begonia. *Ohio Florists Assocn Bull.*, No. 521, pp. 6.

Powell, C.C. and R.M. Riedel, 1973b. Foliar nematode control on Rieger begonia. *Ohio Florists Assocn Bull.*, No. 530, pp. 4-5.

Powell, M.C. and A.C. Bunt, 1978. Leaf production and growth in *Begonia* x *hiemalis* under long and short days. *Scientia Hort.*, **8**(3): 289-296.

Preil, W. 1974. Übert die Verweiblichung männlicher Blüten bei *Begonia semperflorens* – Untersuchungen zur Vererbung und Physiologie von zu Narben umgebildeten Antheren und freiliegenden Samenanlagen. *Zeitschrift für Pflanzenzüchtung*, **72**(2): 132-151.

Preil, W. and A. Lorenz, 1983. Uber das Auftreten matromorpher, patromorpher und intermediärer Nachkommen nach Kreuzungen zwischen *Begonia socotrana* (Hook.) und *Begonia* x *semperflorens-cultorum*. *Zeitschrift für Pflanzenzüchtung*, **91**(3): 253-260.

Prevot, P. 1968. Some aspects of the ability to form new buds in *Begonia rex*. *C.R. Acad. Agric. Fr.*, **54**: 545-549.

Prevot, P. 1969. Gibberellic acid, applied to the proximal end, induces bud formation at the distal end of *Begonia rex* leaves (French). *C.R. Acad. Agric. Fr.*, **55**: 1186-1191.

Prince, T.A. and M.S. Cunningham, 1987. Response of tubers of *Begonia* x *tuberhybrida* to cold temperatures, ethylene, and low-oxygen storage. *HortSci.*, **22**(2): 252-254.

Prota, U. 1964. An interesting case of mildew on begonia (*B.* x *argenteo-guttata* hort.). *Riv. Ortoflorofruttic. Ital.*, **48**: 50-55.

Pynot, M. and C. Martin, 1969. Biosynthesis of anthocyanins in *Begonia gracilis* var. Carmen in relation to temperature (French). *Bull. Soc. franç. Physiol.*, **15**: 47-53.

Qvarnström, K. 1976. Besprutningsförsök mot mjöldagg på begonia 1976. *Växtskyddsnotiser*, **40**(5): 166-167.

Rasmussen, A.N. 1970. Chemical control of *Aphelenchoides fragariae* in begonias (Danish). *Månedsoversigt Statens Plantepatologiske Forsøg*, No. 455, pp. 110-113.

Rasmussen, A.N. 1971. Control of leaf nematodes in Lorraine begonia. *Gartner Tidende*, **87**(23): 303-304.

Rattink, H. 1979. Onderzoek naar bacterieziekte bij begonia (1). *Vakblad Bloemist.*, **34**(12): 30-31, 33, 35.

Rattink, H. and Vruggink, H. 1979. A method to obtain *Xanthomonas*-free begonia plants. *Mededelingen van de Faculteit Landbouwwetenschappen, Rijksuniversiteit Gent*, **44**[1(1)]: 439-443.

Reimann-Philipp, R. and A. Lorenz, 1971. The breeding of triploid F1 hybrids of *Begonia semperflorens*. Elimination of stamen removal by the use of the "Cinderella" characteristic.*Gartenwelt*, **71**(5): 99-101.

Reimann-Phillips, R. and A. Lorenz, 1978. Zur Vererbung des Merkmals braune Laubfarbe bei *Begonia semperflorens* Link und Otto. *Zeitschrift für Pflanzenzüchtung*, **81**(2): 166-175.

Reinert, R.A. and P.V. Nelson, 1979. Sensitivity and growth of twelve Elatior begonia cultivars to ozone. *HortSci.*, **14**(6): 747-748.

Reinert, R.A. and P.V. Nelson, 1980. Sensitivity and growth of five Elatior begonia cultivars to SO_2 and O_3, alone and in combination. *J. Amer. Soc. hort. Sci.*, **105**(5): 721-723.

Reuther, G. and N.N. Bhandari, 1981. Organogenesis and histogenesis of adventitious organs induced on leaf blade segments of begonia-Elatior-hybrids (*Begonia* x *hiemalis*) in tissue culture. *Gartenbauwissenschaft*, **46**(6): 241-249.

Riedel, R.M., D.Q. Peirson and C.C. Powell, 1973. Chemical control of foliar nematodes (*Aphelenchoides fragariae*) on Rieger begonia. *Plant Dis. Reptr*, **57**(7): 603-605.

Riedel, R.M. and P.O. Larsen, 1974. Interrelationship of *Aphelenchoides fragariae* and *Xanthomonas begoniae* on Rieger begonia. *J. Nematol.*, **6**(4): 215-216.

Riedel, R.M. and C.C. Powell, 1974. Control of *Aphelenchoides fragariae* on Rieger begonia with oxamyl. *Plant Dis. Reptr*, **58**(10): 911-913.

Ríordáin, F.Ó. 1979. Powdery mildew, caused by *Oidium begoniae*, of Elatior begonia – fungicide control and cultivar reaction. *Plant Dis. Reptr*, **63**(11): 919-922.

Roest, S., M.A.E. van Berkel, G.S. Bokelmann and C. Broertjes, 1981. The use of an adventitious bud technique for mutation breeding of *Begonia* x *hiemalis*. *Euphytica*, **30**(2): 381-388.

Roest, S. and G.S. Bokelmann, 1980. Vermeerdering van begonia in kweekbuizen. *Vakblad Bloemist.*, **35**(5): 116-117.

Rünger, W. 1957a. A study on the effect of different durations of short-day treatment after taking leaf cuttings on the development of adventitious shoots by the begonia varieties Konkurrent and Marina (German). *Gartenbauwiss.*, **22**: 352-357.

Rünger, W. 1957b. Shoot development of leaf cuttings of the begonia varieties Konkurrent and Marina in different day lengths (German). *Gartenbauwiss.*, **22**: 358-359.

Rünger, W. 1959. The effect of temperature and day length on the formation and development of adventitious roots and shoots on leaf cuttings of the begonia varieties Konkurrent and Marina (German). *Gartenbauwiss.*, **24**: 472-487.

Rünger, W. 1960. The effects of temperature on adventitious shoot and root formation in leaf cuttings of Lorraine begonias (German). *Gartenwelt*, **60**: 344-345.

Rünger, W. 1963. The effect of temperature and day length on growth and flower formation of *Begonia limmingheiana*. *Gartenbauwiss.*, **28**: 571-574.

Rünger, W. 1968. The effect of temperature and day length on flower initiation in Lorraine begonias. *Gartenbauwiss.*, **33**: 469-475.

Rünger, W. 1970. The development of adventitious shoots on Lorraine begonia leaf cuttings with leaves of different sizes (German). *Gartenbauwissenschaft*, **35**: 121-126.

Rünger, W. 1978a. Einfluss von Tages – und Nachttemperatur während Kurztag – und Langtag perioden auf die Blütenbildung von Lorraine-Begonien. *Gartenbauwissenschaft*, **43**(2): 83-87.

Rünger, W. 1978b. Stecklingsblattgrösse und Adventivwurzel – und Adventivtriebbildung bei Lorrainebegonien. *Gartenbauwissenschaft*, **43**(3): 117-120.

Rünger, W., E. Preller and E. West, 1984. Wirkung der Nachtunterbrechung nach verschiedenen Hauptlichtperioden bei Lorrainebegonien. *Gartenbauwissenschaft*, **49**(2): 85-87.

Rystedt, J. 1982. Holdbarheden hos *Hibiscus rosa-sinensis* og *Begonia* 'Nixe' efter uphold i mørke. *Tidsskrift for Planteavl*, **86**(1): 37-46.

Saaltink, G.J. and W. Kamerman, 1971. *Begonia bertinii*, a new host of *Erwinia chrysanthemi*. *Neth. J. Plant Path.*, **77**: 25-29.

Sands, M.J.S. 1990. Six new begonias from Sabah. *Kew Mag.*, **7**(2): 57-85.

Sandved, G. 1967. Control of flowering in *Begonia* x *hiemalis* (Norwegian). *Nord. JordbrForskn.*, **49**: 284 (abstr.).

Sandved, G. 1968. The effect of temperature on the rate of development from the appearance of the bud to full bloom in winter-flowering *Begonia* x *cheimantha* Everett (Norwegian). *Garneryrket*, **58**: 148-150.

Sandved, G. 1971a. The effect of daylength and temperature on growth and flowering of *Begonia* x *hiemalis* 'Schwabenland' and 'Liebesfeuer'. *Gartneryrket*, **61**(19): 378-379.

Sandved, G. 1971b. Careful regulation of flowering in Christmas begonia, *Begonia* x *Cheimantha* Everett (Norwegian). *Gartneryrket*, **61**(37): 678, 680, 681.

Sandved, G. 1971c. Light and severe pinching of winter-flowering begonia (*Begonia* x *cheimantha* Everett) (Norwegian). *Gartneryrket*, **61**(38): 698-700.

Sandved, G. 1972. The effect of CCC on *Begonia* x *hiemalis*. *Gartneryrket*, **62**(24/25): 508.

Sandved, G. 1975. 'Aida' – en ny interessant sort av hiemalis-begonia (*Begonia* x *hiemalis* Fotch). *Garteneryrket*, **65**(34): 549, 562.

Sandved, G. 1978. Effekt av daglengde og temperature på vekst og blomstring hos *Begonia* x *hiemalis* 'Romeo', 'Julie' og 'Nadia'. *Gartneryrket*, **68**(32): 914-915.

Sano, H. and H. Nagao, 1970. Changes in the indole-3-acetic acid oxidase level in leaves of *Begonia evansiana* cuttings at the time of aerial tuber formation under short day conditions. *Plant and Cell Physiol.*, **11**: 849-856.

Sattler, R. and U. Maier, 1977. Development of the epiphyllous appendages of *Begonia hispida* var. *cucullifera* : implications for comparative morphology. *Canadian J. Bot.*, **55**(4): 411-425.

Scharpf, H.C. and E. Grantzau, 1985. Sauerstoff - Mangel bei Ebbe und Flut setzt Schäden. Abhilfe durch Belüftung der Nährlösung im Behälter ist einfach möglich. *Gb+Gw*, **85**(10): 410-411.

Schelstraete, A. 1982. Is chemisch bloemplukken bij knolbegonia's verantwoord? *Berbondsnieuws voor de Belgische Sierteelt*, **226**(13): 599-603.

Schelstraete, A. 1983. Onderzoek naar de optimale dosis Ethefon bij chemische bloemplukken begonia's. *Verbondsnieuws voor de Belgische Sierteelt*, **27**(11): 585, 587-589.

Schenk, M. and W. Brundert, 1980. Elatiorbegonien. CCC - Behandlung in den lichtarmen Monaten. *Gb + Gw*, **80**(47): 1061-1062.

Schmidt, E. 1952. Die Mehltaubekämpfung auf Lorraine - Begonien. *Schweiz. Gärtnerztg*, **55**(3):1.

Schmidt, K. 1985. Die kultur von Elatiorbegonien im Winterhalbjahr. *Deutscher Gartenbau*, **39**(26): 1264-1266.

Schwemmer, E. 1970. Lorraine begonias in peat without basic fertilization. *Erwerbsgärtner*, **24**: 1207.

Sehgal, C.B. 1975. *In vitro* differentiation of foliar embryos and adventitious buds from the leaves of *Begonia semperflorens* Link and Otto. *Indian J. exp. Biol.*, **13**(5): 486-488.

Seitner, P.G. 1977a. *Begonia* 'Gold Coast'. *Begonian*, **44**: 264-269.

Seitner, P.G. 1977b. Daughter of a rajah : *Begonia* 'Rajkumari'. *Begonian*, **44**: 180-185.

Seitner, P.G. 1978. Comparing of two begonia hybrids. *Begonian*, **45**: 15-17.

Shakhova, G.I., L.I. Vozna and V.N. Pogodina, 1979. Some characteristics of the mineral nutrition of begonias grown for their foliage (Russian). *Byulleten' Glavnogo Botanicheskogo Sada*, No. 111, pp. 66-72.

Shigematsu, K. and H. Matsubara, 1972. The isolation and propagation of the mutant plant from a sectorial chimaera induced by irradiation in *Beegonia rex*. *J. Japanese Soc. hort. Sci.*, **41**(2): 196-200.

Shomer-Ilan, A., R.R. Avtalion and Y. Leshem, 1973. Further evidence for the presence of an endogenous gonadotrophin-like plant factor, "phytotrophin": isolation and mechanism of action of the active principle. *Australian J. boil. Sci.*, **26**(1): 105-112.

Silis, D. Ya and S.A. Stanko, 1972. The influence of growing conditions on the red leaf colour in certain ornamentals (Russian). *Sbornik Nauchnych Rabot NI Zonal'n. Inst. Sadovodstva Nechernozem. Polosy*, **4**: 348-353.

Skalská, E. 1964a. The effect of plant spacing in *Begonia tuberhybrida* Voss. On the size of the tubers (Czech). *Acta prùhoniciana*, No. 8, pp. 21-31.

Skalská, E. 1964b. A contribution to the manuring of tuberous begonias (Czech). *Acta prùhoniciana*, No. 9, pp. 53-61.

Skalská, E. 1965. The effect of pinching out the flowers and buds on tuber size in tuberous begonias (Czech). *Rostlinná Výroba*, **11**: 111-118.

Skalská, E. 1966. A contribution to the nutrition problem of tuberous begonias (*Begonia tuberhybrida* Voss.). Part II (Czech). *Acta prùhoniciana*, No. 12, pp. 69-92.

Skvortsova, N.K. 1972. The effect of the nutrition area on the development and morphological characteristics in *Begonia semperflorens*. *Izvestiya Timiryazevskoi Sel'skokhozyaistvennoi Akademii*, No. 2, pp. 68-76.

Sladký, Z. 1959. The application of extracted humus substances to overground parts of plants. *Biol. Plant., Prague*, **1**: 199-204.

Smith, F.F., A.L. Boswell and G.S. Langford, 1966. Controlling mealybug infestations on African violet roots. *Flor. Nursery Exch.*, **146**: 14-15.

Smith, L.B. and B.G. Schubert, 1950. Una nueva begonia Argentina. *Lilloa*, **23**: 143-146.

Smith, L.B. and B.G. Schubert, 1985. A new cane begonia from Amazonian Brazil. *Beegonian*, **52**: 135-136.

Smith, L.B. and D.C. Wasshausen, 1990. *Begonia solimutata*, a new Brazilian species whose leaf color varies with light intensity. *Begonian*, **57**: 217-218.

Soedjono, S. 1988. Teknik radiasi sinar gamma terhadap keragaman tanaman *Begonia semperflorens*. *Bull. Penelitian Hortikultura*, **16**(1): 8-15.

Stark, C. 1964. Leafy galls on Lorraine begonias. *Gartenwelt*, **64**: 516-517.

Strider, D.L. 1973. Control of *Aphelenchoides fragariae* of Rieger begonias. *Plant Dis. Reptr*, **57**(12): 1015-1019.

Strider, D.L. 1974. Resistance of Rieger Elatior begonias to powdery mildew and efficacy of fungicides for control of the disease. *Plant Dis. Reptr*, **58**(10): 875-878.

Strider, D.L. 1975a. Chemical control of bacterial blight of Rieger Elatior begonias caused by *Xanthomonas begoniae*. *Plant Dis. Reptr*, **59**(1): 66-70.

Strider, D.L. 1975b. Susceptibility of Rieger Elatior begonia cultivars to bacterial blight caused by *Xanthomonas begoniae*. *Plant Dis. Reptr*, **59**(1): 70-73.

Thakur, S. 1973. *In vitro* foliar shoot bud formation in *Begonia semperflorens*. *Curr. Sci.*, **42**(12): 430-432.

Thakur, S. 1975. Differentiation of shoot buds and roots in petioler segments of *Begonia picta* Smith. *Indian J. exp. Biol.*, **13**(5): 517-520.

Thompson, H.S. 1961. Control of powdery mildew on tuberous begonia in Canada. *Canad. J. Plant Sci.*, **41**: 227-230.

Thompson, M.L. 1978a. *B. fenicis* Merrill. *Begonian*, **45**: 41-43.

Thompson, M.L. 1978b. *Begonia epipsila* Brade. *Bigonian*, **45**: 145-148.

Thompson, M.L. 1978c. *Begonia odeteiantha* Handro. *Begonian*, **45**: 91-94.

Thompson, M.L. 1979. Species survey : *Begonia edmundoi* Brade. *Begonian*, **46**: 31-33.

Thompson, M.L. 1986. *Begonia cumingii. Begonian*, **53**: 64-65.

Tonecki, J. 1986. Effect of short photoperiod and growth regulators on growth, flowering and tuberization of *Begonia* x *tuberhybrida. Acta Hort.*, No. 177, vol. I, pp. 147-156.

Torode, S.J. 1984. *Begonia chlorosticta. Plantsman*, **5**(4): 243-245.

Tunblad, B. 1947. Ett bekämpningsförsök mot mjöldagg på begonia. *Växtskyddsnotiser*, No. 2, pp. 24-27.

Valášková, E. 1964. The control of powdery mildew on begonias. *Acta prùhoniciana*, No. 9, pp. 63-75.

Van Den Brande, J., R.H. Kips and J.D. D'Hende, 1955. The control of *Heterodera rostochiensis* cysts on begonia and gloxinia corms (Belgian). *Meded. LandbHogesch. Gent*, **20**: 271-278.

Van Onsem, J.G. 1955. Le bouturage de variétés de *Bégonia multiflora au moyen d'hormones* végétales. *Rev. Agric. Brux.*, **8**: 1331-1336.

Van Onsem, J.G. 1961a. Effect of triiodobenzoic acid on tuber formation in *Begonia tuberhybrida* (French). *Rev. agric., Brux.*, **14**: 551-553.

Van Onsem, J.G. 1961b. Short-day treatment of tuberous-rooted begonias (French). *Rev. agric., Brux.*, **14**: 554-556.

Van Onsem, J. and J. Haegeman, 1961. Trials on the effect of gibberellins on growth and tuber weight in *Begonia tuberhybrida. Rev. Agric., Brux*, **14**: 263-272.

Van Onsem, J. and J. Haegeman, 1962. The influence of certain cultural techniques on tuber formation in *Begonia tuberhybrida* Voss. *Proc. 16th int. hort. Congr., Brussels*, vol. 1, p. 359.

Van Onsem, J.G., J. Haegemann and P. Torfs, 1963a. Studies on the packing and insulation material used in the transport of begonia tubers (Dutch). *Meded. Rijksstat. SierplVered. Melle*, **3**:1-9.

Van Onsem, J.G., J. Haegemann and P. Torfs, 1963b. Electrical conductivity of begonia tubers in relation to freezing (Dutch). *Meded. Rijksstat. SierplVered. Melle*, **3**:10-13.

Van Onsem, J. and P. Torfs, 1962. A study of the correlation between the frequency and the start of flower removal and tuber formation of tuberous begonias (Dutch). *Publ. Rijksstat. SierplVered. Melle*, **2**: 7-16.

Vasil'evskij, A.P. and V.M. Kareva, 1958. Powdery mildew on begonias. *Bjull. Gav. Bot. sada*, No. 31, p. 100.

Vereskovskii, V.V., S.V. Gorlenko, Z.P. Kuznetsova and T.V. Dovnar, 1987. Flavonoids of *Begonia erythrophylla* leaves (Russian). *Khimiya Prirodnykh Soedinenii*, No. 4, pp. 599-600.

Viennot-Bourgin, G. 1951. *Oidium begoniae* Puttemans. Maladie nouvelle pour la France. *Ann. Épiphyt.*, **2**(n.s.): 381-387.

Viseur, J. and C. Lievens, 1987. *In vitro* propagation and regeneration of plants from calluses of *Begonia* x *tuberhybrida. Acta Hort.*, No. 212, vol. II, pp. 705-709.

Vozna, L.I., G.I. Shakhova and M.A. Matveeva, 1984. Effect of top dressings on the development and gall nematode resistance of leaf begonias (Russian). *Byulleten' Glavnogo Botanicheskogo Sada*, No. 130, pp. 49-54.

Waines, J.G. 1972. Interaction of night temperature and day-length on floral initiation in *Begonia dregei. Hort. Res.*, **12**(1): 1-4.

Wasscher, J. 1947. Het voorkomen van knopval en bloemval bij begonia's door bespuiting met groeistof - oplosssingen. *Meded. Direct. Tuinb.*, **10**: 547-555.

Wasscher, J. 1953a. Begonia. Vergelijking molm, molming goed en baggerf. *Jversl. Proefst. Bloem. Aalsmeer*, p. 25.

Wasscher, J. 1953b. Begonia. Bloeivervroeging door kortedagbehandeling bij Lorraine-begonias. *Jversl. Proefst. Bloem. Aalsmeer*, pp. 24-25.

Wasscher, J. 1954. Begonia. *Jversl. Proefst. Bloem. Aalsmeer*, pp. 36-40.

Welander, T. 1981. Effect of polarity on and origin of *in vitro* formed organs in explants of *Begonia elatior* hybr. *Swedish J. agric. Res.*, **11**(2): 77-83.

Welvaert, W. and G. Samyn, 1975. Thermotherapy of cucumber mosaic virus infected begonias. *Meded. Faculteit Landbouwwetenschappen, Rijksuniversiteit Gent*, **40**(1): 185-196.

Welvaert, W., G. Samyn and E. van Wymersch, 1980. On the production of virus free *Begonia tuberhybrida* Voss. cv. *multiflora* varieties. *Fifth int. Symposium on virus Diseases of Ornam. Plants. Acta Hort.*, No. 110, pp. 253-257.

Westerhof, J. 1980a. Begonia geschikt voor jaarondteelt? *Vakblad Bloemist.*, **35**(43): 74-75, 77.

Westerhof, J. 1980a. Proef met verduistering begonia. *Vakblad Bloemist.*, **35**(8): 42-43, 45.

Westerhof, J. 1982. Dagverlenging bij *Begonia*, Hoeveel kunstlicht is voldoende? *Vakblad Bloemist.*, **37**(38): 38-39.

White, J.W., J. Holcomb and T. Maczko, 1973a. Research at Penn State, Rieger begonias, Progress report I. *Bull. Pennsylvania Flower Growers*, No. 263, pp. 6-9.

White, J.W., H. Guthrie and B. Watt, 1973b. Rieger Elatior begonias. Research at Penn State. *Progress report IV. Bull. Pennsylvania Flowers Grs*, No. 264, pp. 8-10.

Wildern, J.A. and R.A. Criley, 1975. Cytokinins increase shoot production from leg cuttings of begonia. *Plant Propagator*, **20-21**(4, 1): 7-9.

Will, H. 1977. Wachstumsregulatoren bei Elatiorbegonien. *Gb + Gw*, **77**(35):840-841.

Will, H. 1986. Langzeitdünger erleichtern Kultur in Verkaufseinheïten. *Gb + Gw*, **86**(2): 54-55.

Witte, W.T. and T.J. Sheehan, 1974. Effects of media and fertility on growth and flowering of Rieger begonia. *Proc. Fla St. hort. Soc.*, **87**: 508-512.

Wohanka, W. 1982. Mehltaubekämpflung bei Elatiorbegonien in Hydrokultur. Vorsicht beim Zusatz von Fungizidm zur Nährlösung. *Deutscher Gartenbau*, **36**(49) 2039, 2042.

Wohlmuth, N. 1975. Verträglichkeit von *Begonia semperflorens* gegenüber Fungiziden. *Gartenwelt*, **75**(5): 106.

Zangheri, S. 1957-58. Observations on *Pegomyia bicolor* Wied. With notes on the larval morphology of other species of the genus *Pegomyia. Oll. Zool. Agr. Bachic. Serie II*, **1**: 151-176.

Zang, S., Y. Fan and J. Zhao, 1982. Studies on anther culture of *Begonia semperflorens* Link & Otto. *Acta Hort. Sinica*, **9**(4): 61-64.

Zeilinga, A.E. 1962. Cytological investigation of hybrid varieties of *Begonia simperflorens*. *Euphytica*, **11**: 126-136.

Zeisenhenne, R. 1973. *Begonia bogneri* : a new species from the Malagasy Republic. *Begonian*, **40**(4): 76-80.

Zeisenhenne, R. 1976. *Begonia alice-clarkae* Ziesenh. *Begonian*, **43**: 63-67.

Zeisenhenne, R. 1983. *Begonia roseibractea*, a new species discovered in Mexico. *Begonian*, **50**: 18-21.

Zeisenhenne, R. 1985. *Begonia lachaoensis*. *Begonian*, **52**: 85-88.

Zeisenhenne, R. 1986. *Begonia* U001 is *Begonia grisea*. *Begonian*, **53**: 73-77.

Zeisenhenne, R. 1988. *Begonia manicata* Brongniart and its varieties. *Begonian*, **55**: 85-89.

Zhang, L.Y., G.G. Li and J.Y. Guo, 1988. Study on somatic embryogenesis in leaves of *Begonia fimbristipula* Hance *in vitro*. *Acta Botanica Sinica*, **30**(2): 134-139.

Zimmer, K. 1969a. *Begonia* cv. 'Cleopatra'. 1. Flower initiation (German). *Gartenwelt*, **69**: 519-520.

Zimmer, K. 1969b. *Begonia* cv. 'Cleopatra'. 2. Treatment with growth retardants (German). *Gartenwelt*, **69**: 520-521.

Zimmer, K. 1971. Further studies on the effect of light break on flower formation in begonias (German). *Gartenbauwissenschaft*, **36**(4): 303-308.

Zimmer, K. 1972. Studies with *Begonia socotrana* Hook. *Gartenbauwissenschaft*, **37**(2): 889-895.

Zimmer, K. 1975. Zur Störlichtwirkung bei einigen Abkömmlingen von *Begonia* 'Cleopatra' – Kurze Mitteilung. *Gartenbauwissenschaft*, **40**(2): 69-70.

Zimmer, K. 1976. Weitere Untersuchungen an *Begonia socotrana*. *Gartenbauwissenschaft*, **41**(3): 140-142.

Zobrist, L. 1946. L'oïdium des begonias. *Rev. hort. Suisse*, **19**: 109-114.

6

Bellis (Family: Asteraceae)

R.K. Dubey, Simrat Singh and Zahor Ahmed

[**Common names**: Annual daisy (*Bellis annua*), Bruisewort/Common daisy/Daisy/Day's eye/English daisy/Eye of the day/Lawn daisy/Mary's rose/Woundwort (*B. perennis*)]

Introduction and Origin

Bellis derives from the Latin *bellus* meaning 'pretty'. It is a genus of 15 species of perennials from Europe and Mediterreanean, often grown as annuals or biennials. *Bellis* is found in all types of mown, trampled or grazed calcareous and neutral grassland. It thrives best in areas that become fairly wet for part of the year. The daisy which is corruption of 'day's eye', owing to its flowers opening by the dawn and closing by the dusk, is known chiefly as a weed of lawns, pasture and roadside verges, but it also occurs on riverbanks, dune-slacks, and lake margins. Many other plants of Asteraceae family are also called 'daisies' owing to the utter similarity of their flowers to the main 'daisy'. Its historical name 'woundwort' is now more closely associated with *Betony* or *Stachys*. Chaucer called it 'eye of the day'. In Medieval times, *Bellis perennis* or the English daisy was commonly known as 'Mary's rose', and is also considered to be a flower of children and innocence. Daisy is used as a girl's name and as a nickname for girls named Margaret, after the French name for the oxeye daisy, marguerite. The genus has an important role as a source of medicinal plant. In folk medicine, an infusion made of the flowers was used in curing pleurisy and upper-respiratory ailments, and at present a poultice is prepared from the boiled leaves for the treatment of abscesses, bruises, sores, boils, *etc.* At the same time this species has commercial importance as an excellent ornamental plant (Desevedavy *et al.*, 1989; Jin-Bo *et al.*, 1995; Baytop 1996; Walthelm *et al.*, 2001; Siatka and Kašparová, 2001; Genç and Özhatay, 2006; Cakilcioðlu and Türkoglu, 2009) for edging borders, for carpet bedding, in window boxes and in pots for winter flowering and for cut flowers. In traditional medicine, *B. perennis* is used to treat rheumatism and as an expectorant (Avato and Tava, 1995).

Botany

The *Bellis* is a hardy or half-hardy, low-growing and clump-forming annual to evergreen perennial, bearing green, small, hairy and spoon-shaped leaves, making clear and flat rosettes. The upturned flower heads look like single flowers, but actually consist of a number of small, tightly packed individual flowers or 'florets' on inflorescence heads which are known as capitulum. The flower heads have bright golden-yellow central discs, composed of disc florets, which are surrounded by petal-like white ray-florets that often have deep pink or reddish flushes on the underside. There are several reports on the chromosome number of *Bellis* species (Inceer *et al.*, 2007; Vogt and Oberprieler 2008) with some variation occurring though diploid number ($2n = 18$) is most common (Negodi, 1936; Fernandes and Queiros, 1971). The species having diploid number with $2n = 18$ are *Bellis annua* and *B. perennis* while tetraploid number with $2n = 36$ is reported in *B. sylvestris*.

Species and Varieties

The *Bellis* group has been included in subtribe Asterinae (tribe Astereae) along with another 117 genera representing more than 3,000 species (Bremer, 1994). *Bellis* species are annuals or perennials. The varieties may be kept in three groups, (i) the **Pompons** (**pomponette** with miniature flowers) having all the smaller flowering forms with height and spread of 10-15 cm, (ii) the **Monstrosa** (**carpet series** or **goliath**) with exceptionally large-flowered types having height and spread of 15-20 cm, and (iii) the **Cockscomb** types which are in fact Monstrosa forms but consisting of several smaller inflorescences fused together in the form of crest. Sometimes one more type is mentioned such as **White carpet** with the same growing habits as the carpet series but the plants are compact, large and white-flowered and the height and spread some 15 cm. These all the forms are found only in *Bellis perennis* (Pizzetti and Cocker, 1968; Brickell, 1989). Some of the noteworthy *Bellis* species with good ornamental values are:

Bellis annua: *Bellis annua*, commonly known as Annual Daisy, belongs to the group of annual and biennial plants. This species occurs in Morocco, Algeria and Tunisia. It is a decidious plant with simple leaves. They are spoon-shaped with entire margins. The plants grow to a height of approximately 12 cm. The flowers are many-petaled. The plants bloom from February to June. The flowers are arranged solitary. It prefers a sunny to half shady site. It grows best in pebbly, loamy soil that is moderately moist.

Bellis longifolia: It is a perennial species growing from 5 to 20 cm in height. The roots are fleshy and stalk is leafless. Leaves are bright green, oblong-lanceolate and narrowing at the base gradually. Flowering period extends from April to July. Fruit is more or less bare. The species grows on rocky soils at altitudes from 1,200 to 2,000 meters.

Bellis perennis : A native throughout the meadows and pasture lands of Europe and into western Asia but is naturalized in California. Its species name *perennis* in Latin means 'everlasting'. This is a slow-growing and clump-forming evergreen species having spathulate or oblanceolate and 3-8 cm long leaves, campanulate involucrum, 1-2 seriate phyllaries, conical receptaculum, 1seriate ray flowers, white or pink ligules, yellow disc flowers and obovate achenes (Davis *et al.,* 1988). It is a hardy herbaceous perennial which is the parent of many of the fine horticultural types having white, pink or red flowers in singles, semi-doubles and doubles, and sometimes with spots, streaks, marbling or bicolouration. However, 'Hen-and-Chickens-daisy' is an unusual form with large white or pale-pink flowers, surrounded by many smaller flowers, which makes it curious and extends its flowering period. Another curious variety bears large spathulate leaves brightly marbled with yellow veins on a white base. Some of the other varieties with attractive flowers are *Bellis perennis* var. *alba* (pure white), *B. p.* 'Alice' (pale-pink), *B. p.* 'Brilliant' (large double red), *B. p.* 'China Pink' (smaller bright pink), *B. p.* 'Dresden China' (miniature, double, pink), *B. p.* 'Etna' (brilliant red), *B. p.* 'Giant Double' (mixed colours), *B. p.* 'Liliput' (miniature, double, crimson), *B. p.* 'Longfellow' (deep pink), *B. p.* var. *lutea* (cream), *B. p.* 'Pomponette' (miniature, double, white to crimson with small quilled petals), *B. p.* 'Quilled Mixed' (larger than miniature, double, mixed colours), *B. p.* 'Red Buttons' (miniature, double, carmine-red), *B. p.* 'Rob Roy' (miniature, double, red), *etc.* Many of its forms and varieties are sterile. Its obovate-spathulate leaves narrowing towards the stalk and having one central vein each, form a basal rosette. In the temperate regions of India it flowers perpetually from February to November under sheltered positions though in the Indian tropics and sub-tropics from December to March, the peak being during spring season when the field or meadows are full with solitary single white or pinkish flowers of the diameter of 18 to 31 mm though of the improved varieties it may range from 30-40 mm, borne on finely hairy stalk growing hardly 15 cm in height. Ligules (outer petals) are generally white with mostly pinkish reverse while the central yellow disc is composed of ray florets. The flowers close by the dusk and reopen at the dawn. It grows either prostrate or straight but almost stemless and spreads rapidly. It is propagated through seeds or division. *Bellis perennis* varieties having inherited the long and even winter-flowering habit, are most suitable for pot cultivation in the cool greenhouses.

Bellis rotundifolia: A native to southwest Spain, Morocco and Algeria, This perennial species grows hardly 30 cm in height, normally attaining only 10-15 cm of height and is being grown usually as a non-hardy annual since it can not tolerate frost. It has simple leaves arranged in basal rosettes with ovate margins, are more or less hairy, serrated, 1.8 cm to 3.7 cm in length and with 5 cm long leaf stalks. Flower stalks which appear during March to August, rise well above the leaves and bear many-petalled single flowers some 3.7 cm in diameter on slender stems, followed by achenes. Ligules (outer petals) are white and lilac and the central disc florets yellow. *Bellis rotundifolia* prefers a sunny to half shady site and can withstand temperatures down to -6.6 °C. It grows best in pebbly and loamy soil that is moderately moist.

Bellis sylvestris: A native of S. Europe is similar to *B. perennis* but grows from 10 to 45 cm in height, bearing narrowly obovate to oblong-spathulate leaves with winged stalks, floral heads 2-4 cm wide, ray florets white but tinged red-purple or entirely that colour. Flowers appear during spring to summer.

Propagation

Named varieties of all the perennial species are divided immediately after flowering. *Bellis perennis* can be divided in early spring or after they have finished flowering. In mild winter areas, the seeds are sown in the garden in late summer. Daisy seeds are very small, some 4,900-6,600 seeds per gramme of achenes, and some 3-5 seeds per plug are sown. Seeds in the plugs or in the media start with the radicle breaking through the testa and the roots touching the medium. It ends with fully developed cotyledons. In colder regions, seeds should be sown in September or early spring. *B. perennis* seeds can also be sown indoors by maintaining 21.1 °C temperature within the growing medium until germination, which takes 10-15 days. Seeds are sown in boxes or in the nursery beds in the open in late spring and when the seedlings are ready in 4-5 weeks for transplanting, these are planted during autumn for flowering the following spring.

Cultural Practices

The daisy is usually considered a weed of lawns but in an informal setting it can create a pretty addition to short grass areas. Most cultivation advice for this species concentrates on how to remove it from a lawn - by spraying with herbicide or digging up plants, with their roots, and removing them. The daisy will usually appear in a lawn without any care and mowing the grass every two or three weeks will keep it short enough for the daisy to survive and flower. It survives the cut of the lawn mower by having compact, ground-hugging rosettes of leaves. *B. perennis* generally blooms in the temperate regions from early to mid summer, although when grown under ideal conditions, they have a very long flowering season and will even produce a few flowers in the middle of mild winters

A well-drained, growing **substrate** having 5.5 to 6.2 pH range and 0.5 to 0.75 EC is most suitable for its growing provided the medium is fortified with 15-30 per cent clay, 0-15 per cent organic parts (*e.g.* compost), 0.2 kg/m^3 slow release fertilizer (3-6 months), iron-chelate and certain micronutrients. It can be grown at 10-12 °C outdoors. In mid-December the plants start to grow for 5-7 weeks at 10-12 °C. However, freezing temperatures should be avoided. Temperatures above 12 °C will result in big foliage growth, thin flower stems and a poor flowering initiation. Temperatures below 6 °C delay the bud development. At warm outdoors temperatures, if these are planted in the greenhouse, the vents require to be opened. Full light can promote plant growth with perfectly green leaves and number of flowers also increase. Its **fertilizer** requirement is moderate. Weekly application of 150-200 ppm nitrogen along with the complete balanced fertilizers will give encouraging results. At 5 °C, plants take the minerals properly. High nitrogen levels should be avoided, *vis-à-vis* ammonical nitrogen should also be to the minimum possible. Fertlizer application should be stopped after October begins, however, during spring 150-200 ppm nitrogen along with K_2O at 1:1.5 (N: K_2O) should be applied. Its roots are sensitive to high salt levels in the substrate. **Weeds** are a problem in daisy cultivation and manual weeding is cumbersome and very expensive. Lenacil at 2.2-4.4. kg/ha applied one week after planting *Bellis perennis* in autumn, was safe and controlled autumn-germinating annual broad-leaved weeds (Anon., 1972).

No **insect-pests** have been recorded infesting on daisies. However, sometimes seedling damping-off, gray mould, leaf spots, anthracnose, downy mildews have been found infecting the crop but are of insignificant importance.

References

Anonymous, 1972. Weed control in fall-planted annual flowers. Research Branch Report 1971, Canada Department of Agriculture, p. 337.

Avato, P. and A. Tava, 1995. Acetylenes and terpenoids of *Bellis perennis*. *Phytochemistry*, **40**(1):141-147.

Baytop, A. 1996. Farmasötik Botanik Ders Kitabý. Ýstanbul Üniversitesi Eczacýlýk Fakültesi Yayýnlarý.

Bremer, K. 1994. Asteraceae, Cladistics and Classification. Timber Press, Portland.

Brickell, C. 1989. *The Royal Horticultural Society Gardeners' Encyclopedia of Plants and Flowers* (ed-in-chief), pp. 444-445. Dorling Kindersley Limited, London, UK.

Cakilcioglu, U. and I. Türkoglu, 2009. Elazýðmerkez bölgesinde hemoroid sorunlarý için bitkilerinkullanýmý. *Acta Hort.*, No. 826, pp. 89-96.

Davis P.H., R.R. Mill and K. Tan, 1988. Flora of Turkey and the East Aegean Islands (Suppl.). EdinburghUniv. Press, Edinburgh, vol. **10**: 201-210.

Desevedavy, C., M. Amoros, L. Girre, C. Lavaud and G. Massiot, 1989. Antifungal agents: *in vitro* and *in vivo* antifungal extracts from the common daisy, *Bellis perennis*. *J. Nat. Prod.*, **52**(1): 184-185.

Fernandes, A. and M. Queiros, 1971. Contributiona la connaissance cytotaxonomique des Spermatophytadu Portugal (ser. 2). *Bol. Soc. Brot.*, **45**: 5-121.

Genç, G.E. and N. Özhatay, 2006. An ethnobotanical study in Çatalca (European part of Istanbul). II. *Turkish J. Pharm. Sci.*, **3**(2): 73-89.

Inceer, H., S. Hayirlioglu and M. Ozcan, 2007. Chromosome numbers of the twenty-two Turkish plants. *Caryologia*, **60**(4): 349-357.

Jin-Bo, Wang Yue Xin and Liu Chun, 1995. Breeding and selection of annual and biennial ornamental plants. *Acta Hort. Sinica*, **22**(1): 97-98.

Negodi, G. 1936. Cariologia delle specie italiane delgenere *Bellis* e contributo all appressamento dellostato energetico dei genomi di specie omoploidi. *Biol. Gen.*, **12**: 546-558.

Pizzetti, I. and H. Cocker, 1968. *Flowers ~ A Guide for Your Garden*, pp. 112-117. Harry N. Abrams, Inc., Publishers, New York, USA.

Siatka, T. and M. Kašparová, 2001. Growth and flavonoid production in *Bellis perennis* L. callus cultures. *Herba Polonica*, **47**(1): 17-21.

Vogt, R. and C. Oberprieler, 2008. Chromosome numbers of North African hanerogams. VIII. More counts in Compositae *Willdenowia*, **38**: 497-519.

Walthelm, U., K. Dittrich, G. Gelbrich and T. Schöpke, 2001. Effects of saponins on the water solubility of different model compounds. *Planta Med.*, **67**: 49-54.

7

Bougainvillea (Family: Nyctaginaceae)

B.K. Banerji, R.L. Misra and Sanyat Misra

[**Common names**: Bougainvillea, Glory of the garden, Paper flower (*B. glabra*)]

Introduction

Bougainvillea, the most popular garden plant with free-flowering habit and a very low maintenance requirement, is a genus first discovered by Philibert Commerson of France from Rio de Jeneiro (Brazil) and erected in 1766 to honour Louis Antoine de Bougainville (1729-1811), a French navigator, sailor, explorer and commander of the French ship 'Bordeuse' through which they sailed to Brazil from France while sailing around the world in 1767-1769. Bougainvillea was though initially recorded from Brazil but its primary original centres are South America from Brazil west to Peru and south to southern Argentina (Chubut Province). Though the genus was erected as *Buginvillea* in 1766, and the first authentic publication under this generic name was made out in 1789 by A.L. de Jussieu in *Genera Plantarum* but soon after, this became *Bougainvillea*. First species, *Bougainvillea spectabilis* was described and brought to Europe in 1829 (Holttum, 1970), second species described was *B. glabra* which was exhibited in Britain in 1861 and the third *cum* last was *B. peruviana* which was brought to Britain in 1930. *Bougainvillea spectabilis* was introduced into India in the year 1860 at the Royal Horticultural Society of India, Calcutta. Morphologically, *B. glabra* and *B. spectabilis* are closer to each other, differing only with respect to villous character of the latter, while *B. peruviana* shows distinct features in leaf outline, division of floral cyme, shape of flower tube and unbranched main shoot (Zadoo *et al.*, 1975a). The colour display in bougainvillea is because of the bracts, which occur in groups of three to which flowers are attached on the midrib. Though the flowers are insignificant but being surrounded by brilliantly coloured three large papery bracts persisting on the plants longer make it so spectacular. A range of colours is present in modern cultivars ranging from white, yellow, gold, orange, pink, lavender, magenta, violet, red, bicolours and mixed ones besides the bracts being in single and double forms, and certain varieties having even variegated leaves. Many varieties change colours from initial emergence to maturity thus giving an effect of being bicoloured while some do produce bracts of two colours like `Mary Palmer'. In India, *Bougainvillea peruviana* grows very well throughout the plains, *B. spectabilis* can tolerate even sub-temperate areas while *B. glabra* tolerates even more cooler climates, *i.e.* up to 1,500 metres a.m.s.l. or even more. These three species were taken by the colonial powers to various countries like Australia, Philippines, Malaysia, India, East and South Africa. Now this is bring grown in Bangladesh, France, Germany, India, Italy, Jamaica, Kenya, Malaysia, Mauritius, Pakistan, Philippines, Singapore, South Africa, South America (Brazil, Colombia, Ecuador, Peru; all the native haunts), Spain, Sri Lanka, Trinidad, Uganda, United Kingdom, United States of America (California, Florida, Hawaii, Texas, *etc.*) and in the tropical and subtropical regions of many other countries. Initaially, when a few varieties were introduced in R.H.S. Garden, Calcutta in 20th

Various Variegated and Non-variegated Bougainvillea Varieties.

century, *B. spectabilis* being the first in 1860 and many others subsequently, the breeding work was initiated though its popularity was achieved in 1920's with introduction of cv. 'Mrs. Butt' in Calcutta from R.B.G., Kew (England). Cv. 'Scarlet Queen' by Percy Lancaster in 1920 is the first variety developed in India, cv. 'Mary Palmer' again in Calcutta in 1949, and then in 1960 the Lalbagh Botanical Garden, Bangalore released 'Thimma'. Cv. 'Million Dollar' developed in Philippines in 1967 is still very popular. Dr. B.P. Pal initiated improvement work at IARI, and six varieties, *v iz.* 'Dr. R.R. Pal', 'Sonnet', 'Spring Festival', 'Stanza', 'Summer Time' and 'Vishakha' were developed. Later, a large number of varieties in India were developed from N.B.R.I., Lucknow, B.A.R.C., Mumbai and I.I.H.R., Bangalore. A var. 'Dr. H.B. Singh' developed by IIHR, Bangalore got patented in Australia with the name of 'Krishna'. International Society for Horticultural Science, Belgium has appointed IARI, New Delhi as the 'International Crop Registration Authority for *Bougainvillea*' for registration of its cultivars developed worldwide, and IARI has a good collection at the moment. *Bougainvillea* is a very drought tolerant genus and is ideal for growing in climates with long dry seasons as long as it gets water occasionally.

Uses

Bougainvilleas are so versatile that these provide brilliant colours to the garden even when the glamour of winter season annuals is over, *i.e.* up to October. These are the best landscape plants for training over porches, arbours, pergola and trees, as brilliantly coloured standards, in rounded or rectangular forms and as bonsai, as an espalier on walls and boundaries, as hedges, on road side and dividers, as a ground cover, as specimen plant in the corner of a lawn, for hanging baskets, as a bush, as a climber, in large containers, on factory sites and on gentle slopes for cascading effect. Even on the road dividers of National Highways where pollution level is high due to automobile emission, growth of bougainvillea is luxuriant. Bougainvilleas are relatively free from diseases and pests if grown in full sun though shade-growing plants may cause plants becoming unhealthy. 'Blondie', 'Begum Ali Yawar Jung', 'Dr. R.R. Pal', 'Mahara', 'Mary palmer', 'Sonnet', *etc.* are most suitable as **specimen** plants. Suitable cultivars for training on **trees** are 'Dr. R.R. Pal', 'Elizabeth', 'Formosa', 'Glory', 'Gopal', 'Louise Wathen', 'Meera', 'Pink Beauty', 'Purple Star', 'Spring Festival', *etc.* The most suitable bougainvilleas for **group plantings** are 'Arjuna', 'Begum Sikander', 'Cherry Blossom', 'Chitra', 'Flame', 'Gopal', 'Mahara Variegata', 'Mary Palmer', 'Padma', 'Purple Star', 'Tomato Red', *etc.* 'Begum Sikander', 'Blondie', 'Cherry Blossom', 'Chitra', 'Dr. R.R. Pal', 'Flame', 'Formosa', 'Happiness', 'Jayalakshmi', 'Krumbiegal', 'Lady Mary Baring', 'Lady Richard', 'Los Banos Variegata', 'Mahara', 'Mary Palmer', 'Mary Palmer Special', 'Mrs. H.C. Buck', 'Palekar', 'Parthasarthy', 'Poultoni Special', 'Roseville's Delight', 'Scarlet Queen Variegata', 'Shubhra', 'Sonnet', 'Splendens', 'Spring Festival', 'Summer Time', 'Stanza', 'Thimma', 'Tomato Red', 'Trinidad', 'Vishakha', 'Wazid Ali Shah', 'Zakiriana', 'Zulu Queen', *etc.* are most suitable for **pot planting**. Cultivars suitable for **topiary** are 'Cherry Blossom', 'Formosa', 'Happiness', 'Lady Richard', 'New Red', 'Parthasarthy', 'Perfection', 'Purple Star', *etc.*; and for **hedges** are 'Dr. R.R. Pal', 'Campbell', 'Dream', 'Killy Campbell', 'Lady Mary Baring', 'Mary Palmer', 'Mary Palmer Special', 'Mrs. H.C. Buck', 'Mrs. McClean', 'Partha', 'Parthasarthy', 'Sanderiana', 'Scarlet Queen Variegata', 'Shubhra', 'Thimma', *etc.* For making of **bush** the suitable cultivars are 'Asia', 'Blondie', 'Dream', 'Dr. H.B. Singh', 'Dr. R.R. Pal', 'Glabra', 'Los Banos Variegata', 'Louise Wathen', 'Mahara', 'Mary Palmer', 'Mary Palmer Special', 'Mrs. H.C. Buck', 'Parthasarthy', 'Poultoni Special', 'Red Triangle', 'Roseville's Delight', 'Shubhra', 'Splendens', 'Tetra Mrs. McClean', 'Thimma', 'Trinidad', 'Versicolor', 'Zakiriana', 'Zulu Queen', *etc.*; suitable as **shrubs** are 'Flame', 'Dr. R.R. Pal', 'Mary Palmer', 'Sonnet', 'Spring Festival', 'Summer Time', 'Thimma', 'Tomato Red', *etc.*; while as **climbers**, **espalier** and for training on **arches, pergola** or as **bonsai** suitable bougainvilleas are 'Begum Sikander', 'Cherry Blossom', 'Chitra', 'Dr. H.B. Singh', 'Dr. R.R. Pal', 'Dream', 'Glabra', 'Glabra Sanderiana', 'Golden Glow', 'Golden Glory', 'Lady Mary Baring', 'Los Banos Beauty', 'Los Banos Variegata', 'Louise Wathen', 'Mahara', 'Mary Palmer', 'Mrs. H.C. Buck', 'Palekar', 'Partha', 'President', 'Poultoni Special', 'Red Triangle', 'Rooseville's Delight', 'Scarlet Queen', 'Scarlet Queen Variegata', 'Shubhra', 'Singapore Dark Red', 'Splendens', 'Shubhra', 'Shweta', 'Thimma', 'Tomato Red', 'Walker', 'Yellow Queen', 'Zakiriana', 'Zulu Queen', *etc.* Suitable for **hanging baskets** are 'Blondie', 'Dream', 'Dr. H.B. Singh', 'Dr. R.R. Pal', 'Glabra', 'Glabra Variegata', 'Jawaharlal Nehru', 'Los Banos Variegata', 'Mary Palmer', 'Mrs. H.C. Buck', 'Palekar', 'Partha', 'Sanderiana', 'Scarlet Queen Variegata', 'Shubhra', 'Tomato Red', 'Zinna Barat', 'Zulu Queen', *etc.* and as **ground covers** are 'Dr. H.B. Singh', 'Dr. R.R. Pal', 'Dream','Glabra', 'Glabra Variegata', 'Los Banos Variegata', 'Mary Palmer', 'Mrs. H.C. Buck', 'Palekar','Scarlet Queen Variegata', 'Shubhra', 'Splendens', 'Thimma', 'Zinna Barat', 'Zulu Queen', *etc.* Bougainvilleas for **cascading** effects are 'Asia', 'Begum Ali Jawar Jang', 'Blondie', 'Cherry blossom', 'Dog Star', 'Dream', 'Dr. H.B. Singh', 'Dr. R.R. Pal', 'Enid Lancaster', 'Glabra', 'Isabel Green Smith', 'Lilac Perfection', 'Mary Palmer', 'Palekar', 'Shubhra', 'Splendens', *etc.* Bougainvilleas for making **standards** are 'Asia', 'Begum Sikander', 'Chitra', 'Formosa', 'Glabra', 'Gokul', 'Isabel Green Smith', 'Jayalakshmi', 'Lady

Mary Baring', 'Lord Willingdon', 'Los Banos Variegata', 'Mahatma Gandhi', 'Mary Palmer Special', 'Mrs. Butt', 'Palekar', 'Poultoni Special', 'Refulgens', 'Rose Queen', 'Scarlet Queen Variegata', 'Sensation', 'Shubhra', 'Soeciosa', 'Spring Festival', 'Tetra Mrs. McClean', 'Thimma', 'Zulu Queen', *etc.* and for **umbrella canopy** are 'Archana', 'Dr. R.R. Pal', 'Enid Lancaster', 'Meera', 'Parthasarthy', 'Summer Time', 'Trinidad', 'Zakiriana', *etc.* The suitable bougainvilleas for **road side** plantings and for **inner verges** are 'Blondie', 'Dream', 'Dr. R.R. Pal', 'Glabra', 'Mary Palmer', 'Mrs. H.C. Buck', 'Palekar', 'Red Triangle', 'Shubhra', 'Thimma', 'Zulu Queen', *etc.*

Plant extracts of *Bougainvillea spectabilis* have antifungal (Lakshmanan and Mohan, 1989; Lakshmanan *et al.*, 1990; Molina and Sanchez, 1991; Sindhan *et al.*, 1999), insecticidal (Stein and Klingauf, 1990; Rao *et al.*, 1992; Bhatnagar and Sharma, 1999), nematicidal (Sundarababu *et al.*, 1990) and antiviral (Nagarajan *et al.*, 1990; Sadasivam *et al.*, 1991; Cheema *et al.*, 1993; Balasaraswathi *et al.*, 1998; Jayashree *et al.*, 1998; Bharathi, 1999; Pun *et al.*, 1999) properties. In laboratory tests, the leaf extract of *Bougainvillea spectabilis* inhibited 'tomato spotted wilt virus' on *Capsicum annum* (Balasarawasthi *et al.*, 1998) and reduced 'okra yellow vein mosaic virus' infecting okra (Pun *et al.*, 1999). Antivirus protein was already characterized by Balasarawasthi *et al.* (1998). Diabetes mellitus is a metabolic syndrome characterized by an increase in the blood glucose level. Methanolic leaf extract of *Bougainvillea spectabilis* containing D-pinitol (3-O-methylchiroinositol) which is believed to exert insulin-like effect, significantly reduces intestinal glucosidase activity with regeneration of insulin-producing cells and increase in plasma insulin (Narayanan *et al.*, 1987) and its aqueous and methanolic leaf extracts increase the glucose-6-phosphate dehydrogenase activity, *vis-à-vis* hepatic and skeletal muscle glycogen content after 21 days of treatment. Bougainvillea is also used in treatment of cough and sore throat. A decoction of 10 g dried stem in 4 glasses of water in case of hepatitis, and the decoction of 10 g dried flowers in 4 glasses of water against leucorrhoea, *vis-à-vis* blood vessel disorder are recommended. In Mandsaur, the bougainvillea leaves are used by traditional practitioners for treatment of stomach acidity, diarrhoea and a variety of certain other body ailments. In Panama, local people use an infusion of *Bougainvillea glabra* flowers for treating low blood pressure. Bougainvilleas have unique capacity to absorb various types of pollutants.

Betalains which comprise a class of nitrogen-containing plant pigments in the plant cell sap are edible being antioxidant and having free radical scavenging propertiers. Prativa Lakhotia *et al.* (2014) used various biotic [methyl jasmonate (MJ) and ß-glucan] and abiotic elicitors calcium chloride, FeEDTA and copper sulphate) on biosynthesis of betalain pigments in bougainvillea callus cultures on MS medium. MJ at 0.5 μM was found most effective in inducing betalain biosynthesis in the callus. Among all the concentrations of ß-glucan, 0.5 mg/l was most effective in increasing the betacyanin and betaxanthin contents, though among abiotic elicitors, calcium chloride at 5 g/l showed maximum response coefficient (78.75 per cent), followed by FeEDTA at 100 μM and CuSO4 at 20 μM recording higher betazanthin and betacyanin content.

Botany

Bougainvillea is a woody but tender shrub or scrambling climber with hooked thorns on stems and is native to tropical and subtropical South America, with about 20 species, though Saifuddin *et al.* (2009, 2010) mention only 18 species of *Bougainvillea*, however, Zadu *et al.* (1975a) mention that only three species (*Bougainvillea glabra, B. peruviana* and *B. spectabilis*) possess colourful bracts and are of ornamental value, though Khoshoo (1998) mentions three hybrid groups, *viz. B.* × *buttiana* (*glabra* × *peruviana*), *B.* × *specto-peruviana* and *B.* × *specto-glabra*. The plant height varies from a metre to several, even beyond 12 m. These are planted in the frost-free areas, in tropical world and in warmer parts of America (California, Florida, Hawaii, Texas, *etc.*), especially on porches. Leaves are alternate, glabrous or pubescent, simple, entire, ovate to elliptic-lanceolate or ovate-acuminate, 4-13 cm long and 2-6 cm wide and petioled. Flowers 3 almost insignificant and generally white, corollaless and minute, each consisting of 3 tubular tepals and the whole surrounded by 3 large papery and persistent or non-persistent bracts which are colourful and very attractive, and such flowers surrounded by bracts are arranged in panicle-like clusters in profusion, stamens 6-8, basifixed and on filaments of differing heights, and ovary stipitate. The species are differentiated from one another on the basis of certain characters such as hair on the plants, the tip and colour of the bracts (Bor and Raizada, 1954), floral tube and leaf hairiness (Pancho and Bardenas, 1959; Holttum, 1970) and leaf shape and bract colour (Pancho and Bardenas, 1959). Swarup and Singh (1964) and Nair (1965) suggested pollen morphology, especially number and size of brochi in pollen grains as a more reliable source for identification of species, hybrids and varieties.

Breeding and Cytology

When these species were grown together in nurseries in their countries of introduction, these started hybridizing naturally through butterflies. As *Bougainvillea* is self-incompatible therefore selfing is not possible and even the inter-plant crossing within a cultivar is also out of question as these plants are clonally propagated and therefore share the same S-alleles. Therefore, any seed which was collected by the nurserymen, was essentially of interspecific origin resulting from any two of the three species (Zadoo *et al.*, 1976). The seeds were sown and any novelty thus produced was perpetuated by vegetative multiplication, however, the records of their parentage were seldom maintained. The interspecific hybridization resulted in great variation in growth habit, colour, periodicity of flowering and various other morphological features generated by recombination, transgressive segregation and release of latent mutations, interaction between various colour alleles and their intensifiers. Two typical examples of heterosis being brought about by interspecific hybridization are 'Mrs. Butt' and 'Mrs. H.C. Buck'. They recombine the floriferous nature of *B. spectabilis* with recurrent blooming habit of *B. glabra* and *B. peruviana*, respectively (Ohri, 1995; Ohri and Zadoo, 1979; 1980). Cytogenetics of cultivated bougainvillea including morphological variations, pollen mechanism and breeding system, bud sports, variation in meiotic system, tetraploid and restoration of fertility in steriles, hybridization studies and origin and evolution of ornamental taxa has already been worked out by Zadoo *et al.* (1975a-f; 1976) and Ohri and Zadu (1986). A case of interchange heterozygosity and nature of self- incompatibility in cultivated bougainvillea has been studied in detail by Zadoo and Khoshoo (1968, 1975). Cross-compatibility studies in some bougainvillea cultivars, cross-compatibility relations and origin of bougainvillea, precocious centromere division and origin of polyploidy taxa have been explained by Ohri and Zadoo (1975, 1979). Nuclear DNA content in different bougainvillea cultivars has been determined by micro-densitometer (Ohri and Khoshoo, 1982). Ohri (1979, 1995) made a detailed study on the cytogenetics of bougainvillea cultivars and their breeding mechanisms.

In fact, 'Mrs. Butt' was so different from any of these species that it was first thought to be a different hybrid species *B.* × *buttiana*. The three group hybrids can be easily distinguished by the intermediate characters between their respective parents (Zadoo *et al.*, 1975a). The cultivars belonging to *B. glabra-peruviana* have ovate leaf outline, unbranched flowering shoot and division of cyme like to that of *B. peruviana* while shape of leaf tip, pubescence of young leaves and shoots and shape of flowers are similar to those of *B. glabra*. The cultivars of *B.* × *specto-peruviana* combine the characters like branched main shoot (as in *B. spectabilis*) and division of flowering cyme more than twice and recurrent blooming habit (as in *B. peruviana*). Similarly the third group *B.*× *specto-glabra* is intermediate between parental characters with respect to degree of hairiness and shape of flower tube while being recurrent bloomers like *B. glabra* (Zadoo *et al.*, 1975a). This classification has been evolved taking into consideration the principles and procedures involved in horticultural taxonomy. A detailed description of 122 cultivars along with their parentage (wherever known), and salient features have been provided by Iredell (1990).

Further colour diversification and leaf variegation have been produced by somatic mutations. Two bud sport families are well recognized, such as *B.* × *buttiana* family and 'Mrs. H.C. Buck' family (Zadoo *et al.*, 1975c). Bud sports representing eight cultivars of *B.*× *buttiana* family have been selected because of three types of changes, *viz.* change in bract colour, imperfect flower tube, and leaf variegation. In the other family cultivar 'Mary Palmer' which arose from 'Mrs. H.C. Buck' is very peculiar in having either magenta or parchment white bracts on the same plant. It is therefore an intricate chimera as clonal propagation of branches with magenta bracts produces only magenta coloured plants, while white and semi-white branches vegetatively produce chimeral plants with both types of bracts. 'Thimma' differs from 'Mary Palmer' only in having variegated leaves while 'Shubhra' is a pure parchment white sport from white sector of 'Mary Palmer'. An electrophoretic study in the family has shown that white pigments of 'H.C. Buck' and magenta sector of 'Mary Palmer' are separable into three bands of betacyanins and two of betaxanthins, those of 'Shubhra' show only one band of betaxanthin and complete absence of any betacyanin band (Ohri, 1979, 1995; Kochhar and Ohri, 1979). A definite conclusion, which has emerged from these studies is that bud sports invariably occur in the direction of partial or complete loss of betacyanins with concurrent qualitative and/or quantitative gain of betaxanthins. It is now obvious that natural intercrossing and isolation of bud sports by the gardeners has produced an immense range of variation in bougainvilleas, which can be further exploited in planned breeding programmes with definite objectives and strategies based on a thorough understanding of genetic system.

The basic chromosome number in bougainvillea is x=17. Cooper (1931) recorded triploidy (2n=51) and tetraploidy (2n=68) in bougainvillea cultivars. Banerji and Banda (1967) recorded 'Poultoni Special' being tetraploid. Sharma and Bhattacharya (1960) when tested 40 cultivars, they found their diploid nature in all the cultivars except three which were observed triploids. Khoshoo and Zadoo (1969) studied 60 bougainvillea cultivars and recorded 58 cultivars to be diploid with 2n=34 and the two, *viz.* 'Perfection' and 'Poultoni Special' triploid with 2n=51. They induced tetraploidy in cvs 'Mary Palmer', 'Mrs. McClean' and 'President Roosevelt', and further by crossing tetraploid 'Mary Palmer' with diploid 'Lady Mary Baring' they found triploid, and by crossing triploid with diploid they produced aneuploids with 2n=36 (2x+2) and 2n=40 (2x+6). A further cytological investigation of 70 cultivars belonging to the three species (*B. glabra, B. peruviana* and *B. spectabilis*) and their hybrid groups has shown that all the taxa are diploid (2n=34) except the cultivar 'Poultoni Special' and 'Perfection' which are triploid (2n=51; Zadoo *et al.*, 1975d). These two triploid cultivars show typical alloploid behaviour with the formation of bivalent and univalent. Pancho and Capinpin (1961) reported a natural haploid in cv. 'Pequinito'. Ninan *et al.* (1959) also reported 2n=34, however, they recorded 2n=20 in buds of the red sector of 'Mary Palmer'. Pancho *et al.* (1960) reported 'Cypheri', 'Lateritia' and 'Temple Fire' as triploids. Tetraploidy (2n=68) has been recorded in a number of cultivars, *viz.* 'Crimson King', 'Magnifica', 'Mahara', 'Princess', 'Shubhra' and 'Thomasii' (Pancho *et al.*, 1960); and 'Lady Mary Baring', 'Mrs. McClean', 'President Roosevelt', 'Zakiriana' and 'Thimma' (Khoshoo and Zadoo, 1969; Zadoo *et al.*, 1975e). Many bougainvillea cultivars including all the multibracted ones are sterile, and therefore, are propagated only vegetatively, *i.e.* cuttings and air layerings.

Species

Bougainvillea glabra, a native of Brazil was first identified by Choicy about 1850, and this grows some 4 m wide and high, a vigorous spiny shrubby climber with small and curved thorns at the tips, leaves bright green, glabrous but with some puberulence sometimes, and ovate, acuminate or elliptical with glossy sheen, and bracts long and differing in sizes but triangular, cordate-ovate, bright rosy-red (cyclamen-purple or mauve, and sometimes white), distinctly veined, flowers swollen at base and hairy. Seed grown plants of this species show a lot of variation in size, bract shape and bract colours though predominantly in the mauve-purple shades or white. It is amenable to be trained to any shape, *vis-à-vis* standards. It should be noted that *B. glabra* and *B. spectabilis* are very much alike in general appearance, the main differences are the bloom cycle and *B. glabra* is hairless while *B. spectabilis* is hairy. Both these species may show wide structural variation when grown from seeds. Its varieties are 'Alexandra', 'Cypheri' (red), 'Magnifica' (syn. 'Traillii'), 'Mrs. Leano', 'Sanderiana', 'Snow White', 'Variegata', *etc.* ***B. peruviana***, first identified by Humbolt around 1810 is a native of Pacific Coast area of Northern South America, a climbing evergreen species having green coloured bark with short and fairly straight thorns, most stable species when grown from seeds, growth lanky with long branches and sometimes leafless at juvenile stage, and is similar to *B. spectabilis* in many aspects, except that it lacks hairs and bears slender flowers, and smaller bright purplish-pink (magenta to pink) bracts having crimped margins with floral tune constricted in the middle. This blooms frequently during dry periods. Leaves are green, usually long and thin, strongly ovate and hairless. Important varieties are 'Ecuador Pink', 'Lady Hudson', 'Princess Margaret Rose', *etc.* ***B. spectabilis***, a native of Brazil, was the first species to be identified in 1798 by Willdenow, and is very similar to *B. glabra*, growth dense, taller and hairy stems with stout hooked spines, leaves large, broadly elliptic to ovate, some rippling along the edges, hairy underneath and thicker, flowers in large panicles, flowers cream coloured, bracts quite large (4 cm or more), floral tube swollen at base, and flower colour deep rose-red varying to purple and greenish. Its blooming cycle is seasonal being triggered due to dry or cold spells. Its most popular varieties are 'Lateritia', 'Mary Palmer' bearing both carmine and white bracts, 'Thomasii', *etc.* *B.* × *buttiana* is a hybrid of *B. glabra* and the rarely cultivated species *B. peruviana*.

Hybrids and Varieties

Garden bougainvilleas have originated from three elemental species, *viz. B. spectabilis*, *B. glabra* and *B. peruviana* by natural interspecific hybridization under nursery conditions. These three group hybrids, *viz. B.* × *specto-glabra, B.* × *specto-peruviana* and *B.* × *glabra-peruviana* can be identified by their morpholoigical features. A great variation in growth habit, colour, and periodicity of flowering could be generated by recombination, transgressive segregation, release of latent mutations, *etc.* Further variation was created by isolating bud sports with a range of colours and also by chlorophyll mutations. The hybrids between *B. glabra* and *B. spectabilis* are partially fertile while those of *B. peruviana* with *B. glabra* and *B. spectabilis* are fully sterile. The flowers are usually pollinated by butterflies and the breeding system is characterized by sporophytically controlled self-incompatibility. Thus no selfing is possible on a single plant or with other plants of the same variety as these have been clonally multiplied and share the same S-alleles. Further breeding and improvement was constrained by narrow germplasm base. Therefore, colchiploidy was induced to restore fertility in a number of varieties, which show normal bivalent pairing and fertile pollen. Hybridization was now possible at both intra (2x × 2x, 4x × 4x) and interploidal (2x × 4x) levels giving rise to 2x, 3x, 4x and aneuploid hybrids. Any attractive hybrids thus produced have been perpetuated by vegetative multiplication. *Bougainvillea* breeding, therefore, has a tremendous future as entirely new variability can be introduced through aneuploid and polyploidy breeding by judicious use of polyploidy and hybridization. NBRI, Lucknow has a omprehensive collection of bougainvillea germplasm and its details has been explained by Roy *et al.* (2007). Indian Agricultural Research Institute, New Delhi-110 012, at its Division of Floriculture & Landscaping, is the International Registration Authority for registration of bougainvillea germplasm, appointed by the International Society for Horticultural Science, Belgium since 1966, and at present the repository maintains more than 90 varieties such as 'Abraham Kavoor', 'Alick Lancaster', 'Arjuna', 'B.T. Red', 'Begum Sikander', 'Blondie', 'Cascade', 'Chandrabieri', 'Cherry Blossom', 'Chitra', 'Cleopatra', 'Dr. B.P. Pal', 'Dr. Bhabha', 'Dr. H.B. Singh', 'Dr. Hado', 'Dr. R.R. Pal', 'Dr. Rao', 'Dream', 'Elizabeth Angus', 'Enid Lancaster', 'Fantasy', 'Flame', 'Flomen', 'Formosa', 'Gangaswamy', 'Garnet Glory', 'Glady's Heburn', 'Glorious', 'Gokul', 'Gold', 'Gopal', 'Hawaiian White', 'Isabel Greensmith', 'Jayalakshmi', 'Killie Campbell', 'Krumbiegal', 'Lady Hope', 'Lady Hudson', 'Lady Mary Baring', 'Lady Richards', 'Lakshmi', 'Los Banos Beauty', 'Los Banos Variegata', 'Los Banos Variegata Jayanti', 'Louise Wathen', 'Mahara', 'Mahatma Gandhi', 'Manohar Chandra', 'Mary Palmer Special', 'Mataji Agnihotri', 'Meera', 'Mrs. Bakery', 'Mrs. Butt', 'Mrs. Fraser', 'Mrs. H.C. Buck', 'Padmi', 'Pallavi', 'Partha', 'Pink Beauty', 'Poultoni', 'Poultoni Special', 'Profusion', 'R.S. Bhatt', 'Radha', 'Red September', 'Refulgens', 'Rosea Fuchsia', 'Roseville Delight', 'Sanderiana', 'Sensation', 'Shubhra', 'Shweta', 'Singapore Red', 'Sofia Mutant', 'Sonnet', 'Souva', 'Spectabilis', 'Splendens', 'Spring Festival', 'Stanza', 'Summer Time', 'Superba', 'Sweetheart', 'Tetra Mrs. McLean', 'Thimma', 'Tomato Red', 'Torch Glow', 'Versicolor', 'Vishakha' and 'Zakiriana' (Janakiram *et al.*, 2013).

Interspecific hybrids of bougainvillea are *B.* × *buttiana* (*B. glabra* × *B. peruviana*), first discovered by Mrs. R. Butt in a Trinidad garden, so it was named first as 'Mrs. Butt' in U.K. but in USA it was known as 'Crimson Lake' and now it has a dozen other names as well. It has straight and short thorns, possessing large and ovate (sometimes cordate) leaves which may be slightly hairy, flowers cream with some pink shades, bracts round and in shades of red, dark pink, repeat bloomer and requires pruning to keep it bushy and in proper shape. Its popular cultivars are 'Batik Yellow' (bract yellow and foliage variegated), 'Brilliant' (coppery-orange), 'Brilliant Variegata' (leaves variegated, centre deep green surrounded by white, bract red), 'Easter Parade' (pink), 'Golden Glow' (yellow), 'Jamaica White' (white), 'Killie Campbell' (orange), 'Mrs Butt' (crimson to magenta), 'Orange King' (orange), 'Pigeon Blood' (deep red), 'Scarlet O'Hara' (Orange-red), 'Surprise' (a chimera with rose-purple, white or mixture of the two), *etc. B.* × *spectoperuviana* is large growing and spreading, bears straight thorns and is usually a hairless hybrid with dark green, large and ovate leaves, flowers cream, bracts initially coppery-red but with age turning to shades of pink or magenta. It is a repeat bloomer. *B.* × *spectoglabra* is a well-branched hybrid with thick stems having numerous curved thorns and is a repeat bloomer, leaves small and dark green, flowers almost white and the bracts in shades of mauve or purple. Its most popular variety is 'Sanderiana'.

Most popular bougainvillea varieties are:

White

'Alba', 'Double Pink', 'Double White', 'Dr. B.P. Pal', 'Hawaiian White', 'Himani', 'Jennifer Fernie', 'Moonlight Madonna', 'Shubhra', 'Shweta', 'Snow Queen', 'Snow White', 'Sova', *etc.*

Yellow

'Enid Lancaster', 'Golden Glow', 'Lady Mary Baring', 'Miggi Ruser', 'Suverna', 'Yellow Queen', *etc.*

Orange

'Golden Glow', 'Lateritia' (syn. Brasiliensis, Dar-es-Salam), 'Louise Wathen', 'Margery Lloyd', 'Mrs. McClean', 'Orange King', 'Roseville's Delight', 'Tetra Mrs. McClean', 'Zakiriana', *etc.*

Pink

'Alick Lancaster' (syn. Lilac Queen), 'Boise de Rose' (biscuit colour), 'Cascade', 'Dwarf Gem', 'Grüss aus Budenweiler', 'Kalyani', 'Kayata', 'Los Banos Beauty' (syn. Pink Champagne), 'Los Banos Variegata', 'Maharaja of Mysore', 'Mahatma Gandhi' (syn. Mrs. H.C. Buck), 'Nawab Ali Yavur Jung', 'Pixie', 'Poultoni Special', 'Princess Margaret Rose' (syn. Lady Hudson), 'Spring Festival', *etc.*

Red

'Aida','Begum Ali Yavur Jung', 'Camarilo Fiesta', 'Crimson Jewel', 'Dr. R.R. Pal', 'Flame', 'Isabel Greensmith', 'Mahara', 'Meera', 'Mrs. Butt' (syn. Crimson Lake, Ruby), 'Mrs. R.B. Carrick', 'Palekar', 'Rosa Catalina', 'Scarlet Glory', 'Scarlet O'Hara', 'Scarlet Queen', 'Sri Durga', 'Stanza', 'Tomato Red', 'Versicolor', *etc.*

Mauve

'Dr. Harbhajan Singh', 'Dream', 'Enid Walker', 'Formosa', 'Gopal', 'President', 'Sachidananda', 'Trinidad', *etc.*

Magenta

'Asia', 'Gopal', 'Magenta Queen' (syn. Mrs. Butt Magenta, Purple Queen), 'Manohar Chandra', 'Mrs. H.C. Buck', 'Mrs. lancaster', 'Ruarka', 'Sonnet', *etc.*

Purple

'Alexandra', 'Columbina', 'Dream', 'Eclipse', 'Formosa', 'Gagarin', 'Glabra', 'Glabra Magnifica', 'Glabra Sanderiana', 'Jubilee', 'Magnifica', 'Mahara' (syn. Manila Magic Red, Manila Red, Million Dollar), 'Partha', 'Refulgens', 'Sanderiana', 'Speciosa', 'Splendens', 'Zulu Queen', *etc.*

Bi- and Multi-coloured

'Archana' (young leaves coppery variegated though irregular green centre is surrounded by pale-yellow margin in mature ones), 'Arjuna' (young in coppery variegation but mature ones blotched grey-green in the centre and creamish-white margin), 'Begum Sikander' (2n+3x-2=49); bicolour, purple to magenta margin and centre white), 'Carmencita' (lavender and mauve), 'Cherry Blossom' (multibracted, bicoloured, white with pink + purple at the top), 'Chitra' (bicolour, rose and white), 'Dr. Bhabha' (leaves yellow-green variegated), 'Dr. Rao' (leaves have white variegation in the margins), 'Fantasy' (bicolour, rose-purple and sulphur-purple), 'Hawaiian Beauty' (leaves blotched irregularly with pale-yellow), 'Los Banos Variegata' (ovate leaves having yellow variegation), 'Louise Wathen Variegated' (young foliage tinged orange but mature ones with creamish-white margin), 'Mahara Variegata' (leaves yellow-green variegated), 'Marietta' (young leaves margined pink but mature ones margined creamish-white), 'Mary Palmer' (syn. 'Surprise' in USA; bicolour, magenta, parchment white + magenta, and white), 'Mary Palmer Special' (2n=3x=51; bicolour, magenta and white), 'Mata Ji Agnihotri' (younger leaves with coppery variegation but mature ones margined pale-yellow), 'Odisee' (rose, white and purple with blotches), 'Parthasarthy' (young foliage cream with pinkish tinge, mature ones irregularly variegated with mosaic pattern of grey-green and cream), 'Scarlet Queen Variegated' (young foliage tinged pinkish but matured ones with creamish-white margin), 'Thimma' (bicolour, rose-magenta and white), 'Vishakha' (younger leaves with coppery variegation but matured ones margined white, 'Wajid Ali Shah' (2n=3x=51; bicolour, initially blotched with white and purple but changing to purple), *etc.*

Multibracted

In such cases generally flowers are absent. The varieties under this group are 'Archana', 'Carmencita' (syn. Gladalan Red, Philippines Red), 'Cherry Blossom', 'Double Pink', 'Double White', 'Godrej Cherry Blossom', 'Los Banos Beauty', 'Los Banos Variegata', 'Los Banos Variegata Silver Margin', 'Mahara', 'Mahara Variegata','Marietta', 'Pallavi', 'Roseville's Delight', *etc.*

Varieties with Variegated Foliage

Khoshoo and Zadoo (1969) strengthened the concept in bougainvillea breeding through picking up natural bud sports and through induced physical and chemical mutations as has also been supported by Nath *et al.* (1983), Banerji *et al.* (1983, 1987), Banerji and Datta (1987a,b), Banerji (2002), Jayanthi and Datta (2006). Leaf chlorophyll variegation adds new dimension to bougainvillea germplasm by exhibiting their multicoloured leaves, especially during the off season when the plant is not in bloom. Datta *et al.* (1995) induced chlorophyll variegation in bougainvillea and Datta and Banerji (1994a,b, 1997) even on double or multibracted bougainvilleas. The work on the line has also been investigated by Banerji and Datta (1996). Banerji (2007, 2008) studied ^{60}Co gamma ray induced mutations and analysis of multibracted bougainvilleas. Many of the chlorophyll variegation mutants are described here with, *viz.* 'Abhimanyu' (central green leaf lamina encircled by white margin), 'Angus Supreme'(bract purple and foliage edged gold), 'Archana' (leaf centre with green patch and margins light yellow), 'Arjuna' (leaf centre light and margins creamy-white), 'Batik Yellow' (bract yellow and foliage variegated), 'Bhabha' (leaves cream and green), 'Brilliant Variegata' (leaves variegated, centre deep green surrounded by white, bract red), 'Cinderella' (bract red and double and leaves variegated in green and cream), 'Coconut Ice' (leaves variegated and bracts white splashed pink), 'Danger Ivy' (leaf lamina cream and green and bracts pink), 'Daya' (leaves variegated green margined light yellow, bract faint pink), 'Danger Ivy' flowering only during extremely dry season, leaf lamina variegated with cream and green, and bracts pink), 'Eva's Ice-Cream' (free-flowering, bicolour bracts light mauve and white), 'Eva's Ice-Cream Highlight' (leaves cupped and bicolour, bracts dark mauve and white), 'Eva's Heart' (variegated bracts white blended pink), 'Eva's Wonder' (bracts purple and leaves variegated with green and yellow at early stage), 'Gangamma' (centre green and margins white), 'Gangaswamy' (centre green and margin white), 'Gautama Batik" (leaves variegated with green and white), 'Glabra Variegata' (variegation of cream or white), 'Golden Lady' (golden leaf variegation), 'Godrej Cherry Blossom' (leaves variegated with broadly ovate base, and bracts cream-yellow), 'Harissi' (foliage small with variegation of green and white, and bract purple), 'Hati Gadis' (bracts large and bright purple, and leaves variegated with green and white), 'Hati Gadis II' (leaf variegation in yellow and green), 'Hawaiian Beauty' (irregular grey-green and pale-yellow blotches on entire leaf), 'Hujan Panas II- Pink' (bract pink, leaves variegated with golden dots in the centre of leaf lamina), 'Jarum' (leaf yellow splashed green all over the surface and leaves and bracts crimpled), 'Jayalaxmi Variegata' (leaves variegated, younger ones coppery-red and the mature ones creamish-yellow, and bracts brick red), 'Jawahar Lal Nehru' (leaves variegated, younger ones coppery and the mature ones with irregular central patches and yellow margin), 'Jayalaxmi Variegata' (leaves variegated, younger ones coppery-red and the mature ones creamish-yellow, and bracts brick red), 'Kuvempu' (young shoots and leaves reddish and mature ones variegated with creamy-white and green while bract is red), 'Lady Hudson of Ceylon Variegata' (leaves variegated in green and pale-yellow), 'L.N. Birla' (green and creamish-yellow variegation), 'Laxminarayan' (cream and green variegation), 'Los Banos Variegata' (centre irregularly green and margin white and slightly wavy), ' Los Banos Variegata Jayanthi' (green and yellow variegation), 'Los Banos Variegata Silver Margin' (leaves variegated with green margined silver and the bracts purple; Banerji, 2002), 'Louise Wathen Variegata' (grey-green variegation in the centre and creamy-white margin), 'Magnificent Barry' (green and yellow leaf variegation), 'Mahara Variegata' (leaves irregulary variegated with combination of creamish-yellow and various shades of green overlapping patches, younger coppery while older ones dull yellow and faded green, and the bract purple), 'Mahara Variegata Abnormal Leaves' (leaves irregularly variegated in shades of green and light green with undulated margins), 'Mahatma Gandhi Variegated' (leaves variegated in green and white), 'Malaysia Indah' (leaves green contrasting gold), 'Manohar Chandra Variegata' (young leaves coppery but old ones with green centre and yellow margin), 'Marietta' (foliage variegated in white and green, and bract purple), 'Mata Ji Agnihotri' (centre green and light yellow margin), 'Mini White' (foliage small and variegated, and bracts white), 'Ms. Alice's Coat' (green leaves and lamina variegated with yellow, and bract white), 'Mrs. Butt Variegata' (leaves variegated with irregular patches of yellow and green), 'Mrs.Eusnia Rajasinghe' (central leaf portion green and the margin ash), 'Mrs. Eva Variegata' (leaves variegated, bracts green and white), 'Mrs. Eva II' (similar to 'Mrs. Eva Variegata' but leaves differently variegated in green and yellow), 'Mrs. Eva Variegata Purple' (leaf variegated and bracts purple), 'Mrs. Eva Variegata White' (leaf variegated and bract white), 'Mrs. Eva Variegata White II' (variegation on leaves though green and yellow but differently than its parent), 'Mrs. McClean Nirmal' (young leaves green to coppery white and the old green ones variegated with white patches), ' Muni Venkatappa' (green leaves variegated with white margin), 'Pallavi' (various shades of light and pale-green), 'Parthasarthy' (irregular variegation of green and pale-white), 'P.V. Sane' (initially centre greenish and outer pinkish-red but changing to yellow centre on yellow-green base), 'Phoenix' (leaf centre green and surrounded by golden and cream colouration, bracts dark red; Jan Iredell, 1986), 'Pixie Variegata' (evolved through 0.02 per cent EMS in cultivar 'Pixie', leaf variegation in the mutant by cream-white margin, and the bracts pink; Banerjee, 2009), 'Poultoni Variegata' (leaves variegated with green and light yellow patches all over), 'Preeti' (leaf centre dark green and margin light yellow, and the bracts deep pink), 'Proton Saga' (leaf gold dotted green and drooping bract panicles are palest-lavender-pink), 'Purple Gem' (plant shy bloomer, leaves variegated and small bracts bright purple),'Queen Margaret' (irregular variegation of cream and green shades on entire leaf lamina), 'Rao' (leaf centre green but periphery white), 'Refulgens Variegata' (green leaves streaked white), 'Royal Daupline' (centre grey-green with greenish-yellow periphery in irregular shape), 'Scarlet Queen Variegata' (leaf variegation in cream and green all over and bracts deep carmine-scarlet), 'Sharma' (a bud sport of 'Mrs. Fraser' with green and cream leaf variegation along the irregular red patches, and bract terracotta), 'Silver Top' (boat-shaped leaves are crinkled and variegated, older ones with narrow white border and adaxial surface silver mosaic), 'Singhe' (green in the centre of leaf surrounded by ash colour), 'Soundarya' (leaf variegated bud sport from 'Mrs. H.C. Buck'), 'Spectabilis Variegata' (leaves with cream-white variegation and bract purple), 'Surekha' (a pale-yellow patched variegated bud sport of 'Scarlet Queen Variegata' with carmine to purple bracts), 'Suvarna' (a leaf variegated sport with bract colour like its parent 'Lady Hudson of Ceylon'), 'Thimma' (a bud sport of 'Mary Palmer' with same bract colour and leaf centre yellow), 'Vellayani' (leaves in variegation of green and cream and bracts crimson red), 'Vellayani Variegata' (leaves irregularly variegated with green and cream and bracts crimson), 'Vishakha' (leaves variegated with white margin and bract red), 'White Proton Saga' (leaves variegated and bracts white), *etc.*

A few of the variegated varieties under *Bougainvillea* × *spectoperuviana* group are 'Magic Makris' (syn. 'Magic Icecream'), a 'Makris' bud sport having variegation of gold splashed in the central portion of leaf lamina and bract softer pink and white; 'Makris' (syn. 'Icecream'), a bud sport of cv. ' Mary Palmer' with pink and white colouring of bracts; 'Thimma' (syn. 'Vicky'), a bud sport of 'Mary Palmer' possessing bicoloured (pink and white) bracts and large splash of gold colouring in the centre of leaves.

The interspecific varieties are 'Baby Rose' (leaves dark green and bract white dotted mauve), 'Batik Orange' (a bud sport of 'Orange King' with variegated leaves), 'Batik Pink' (leaves variegated and bracts pink), 'Batik Red' (a bud sport of 'Mrs. Butt' with green and white leaves and red bracts), 'Blue Moon' (leaves variegated and bracts deep red), 'Chili Delight' (foliage green and distorted and orange bracts are also twisted and distorted), 'Chili Red Batik' (leaves variegated and bracts red), 'China Beauty' (foliage variegated and bracts soft rosy-pink turning to deep red with rise of temperature), 'Hong Kong Beauty' (a bud sport of 'Juanita Hatten' with central green patch of leaf lamina surrounded by golden colouring and bracts red), 'Hujan Panas' (syn. 'Red Fantasy'), a bud sport of ' Juanita Hatteri' has variegated leaves dotted gold and red bracts), 'Hujan Panas Orange'(foliage variegated and bracts orange), 'Iguana Variegata' (variegation prominent at younger stage but at ageing it turns green and the bracts orange but ageing to dusty pink), 'Ikan Bilis' (syn. Ray Fish', 'Puteri Emas') with distorted foliage and bracts, bract yellow), 'Indian Beauty' (syn. 'Miss India'), plants tall and looking like a plastic plant with shining and fresh foliage and the bract colour is red), 'Lady Bird' (leaves greenish and little cupped when young and bracts white with red dots), 'Lady Bird Batik' (a bud sport of 'Lady Bird' with green and white variegated foliage but bracts comparatively smaller), 'Lady Pink' (variegated leaves comprise of gold splashed with green colouration and bract pink), 'Mona Lisa' (this dwarf variety looks like 'Indian Beauty', leaves crispy and bracts red), 'Mona Lisa Yellow' (a bud sport of 'Mona Lisa' with yellow bracts), 'Mahsuri Reflex' (a bud sport of 'Puteri Mahsuri'), 'Mahsuri Reflex Pink' (a bud sport of 'Puteri Mahsuri', bracts pink), 'Marietta' (syn.'Cinderella') has variegated leaves and multibracted bracts of red colour), 'Mini Marbles' (syn. 'Tang Long', 'Chinese Lantern', has green leaves and orange bracts curling inward making impression of hanging lanterns), 'Muar's Butterfly' (bract red), 'New Pink' (leaves variegated and pink bract), 'Ninja Turtle' (syn.'Sirih Junjung Batik', has small leathery and variegated leaves, and magenta-red bracts crowded on closely-spaced nodes), 'Orange Batik II' (leaf variegation less clear and bract orange), 'Orange Puteri Mahsuri' (a bud sport of 'Puteri Mahsuri', bearing orange bracts), 'Poultoni Batik' (foliage variegated with yellow and green, and bracts red), 'Poultolni Orange Variegata' (leaves variegated and bracts orange), 'Puteri Mahsuri' (leaves variegated, and foliage and bracts distorted and twisted on inflorescence), 'Queen Marbel' (leaves and bracts distorted, and leaves with striped gold-lines and bracts orange), 'Queen Marbel Red' (leaves striped and bracts red), 'Ratna Mauve' (a free-flowering bud sport from 'Mrs. Eva', foliage grayish and distorted and drooping, bracts leafy, mauve and distorted), 'Ratna Orange' (syn. 'Orange Butterfly', an orange-bracted variety with characters of 'Ratna Red'), 'Ratna Pink' (syn. 'Pink Butterfly, characters of 'Ratna Red' but bracts pink), 'Ratna Purple' (syn. 'Purple Butterfly', a bud sport of 'Elizabeth Angus' with purple bracts and leaves highly wrinkled), 'Ratna Rainbow' (a bud sport of 'Sakuri', bracts white but pink at tips and leaves and bracts both are distorted), 'Ratna Red' (syn. 'Red Butterfly', a bud sport of 'Ratna's Butterfly', similar to 'Ratna Rainbow'), 'Ratna's Butterfly' (looking like 'Ratna Rainbow'), 'Ratna White' (syn. 'White Butterfly', foliage grayish due to intermixing of white and yellow tissues in different layers and bracts distorted but pure white), 'Ratna Yellow' (syn. 'Yellow Butterfly', though bract is yellow but other characters just to 'Ratna Red'), 'Red Batik' (leaves and bracts distorted, bracts red and so attractive as a number of butterflies are resting), 'Red Lotus' (a bud sport from 'Mini Marble' bearing cupped up and variegated leaves), 'Red Ribbons' (leaves variegated and bract red but both have distortion though bract texture fine), 'Red September' (leaves as 'Hong Kong Beauty' but variegation less prominent, bract red and pink), 'Sakura' (syn. 'Flamingo Pink', leaves dark green and bracts white with pink-shading towards tips), 'Sakura Batik' (a bud sport of 'Sakura' where prominent green and white variegation spreads over entire leaf lamina), 'Senjakala' (a bud sport from'Glabra' with purple bracts), 'Snow White' (a free-flowering bud sport of 'Mrs. Eva White' having green-marked cream foliage), 'Strawberry' (a bud sport of 'Queen Marbel' with variegated foliage where leaf lamina is multicoloured and bract contrasting red), 'Strawberry Delight' (a bud spoprt of 'Strawberry', leaves and bracts distorted and bract colour red), 'Sweet Dream' (a thornless variety, having golden and pale-lavender bracts), 'Tembikai' (syn. 'Watermelon', a rare beauty, foliage variegated and bracts soft pink and colour deepening with age), 'Tiger' (leaf lamina with gold marks splashed with green shades and bract red), 'Yellow Batik II' (foliage variegated and bracts yellow), 'Yellow Wonder' (leaves variegated but becoming only green with ageing and bracts yellow), *etc.*

Propagation

Bougainvillea is though grown through **seeds** but only for evolution of new varieties and there exist certain varieties which set seeds, some of which are 'Formosa', 'Lateritia', 'Maharaja of Mysore', 'Mrs. Fraser', 'Thomasii', 'Trinidad', *etc.* (Pal and Swarup, 1974). Seed setting in *Bougainvillea* is influenced due to environment which may vary from varietiy to variety, age of the plant as seed set is obtained only after two years of age, in many varieties even after fourth year and the setting varies from place to place as under Bangalore conditions setting is good compared to Delhi or Lucknow where the same variety may show sterility or self-incompatibility. Through induction of tetraploidy, the sterile varieties such as 'Lady Mary Baring', 'Mrs. McClean', 'President Roosevelt', 'Shubhra', 'Thimma' and 'Zakeriana' were restored of their fertility (Khoshoo and Zadu, 1969; Zadu *et al.*, 1975e) which proved a boon for breeders to improve those sterile cultivars which previously did not have any access for their improvement. Many varieties are self sterile. Seed set is common in the varieties with persisting bracts.

Difficult to root varieties such as 'Cypheri', 'Lateritia', 'Louise Wathen', 'MaryPalmer', 'Rosa Catalina', 'Tomato Red', *etc.* (Pal and Swarup, 1974) are multiplied preferably through budding or sometimes through layering (gooties). For **layering**, new mature shoots are selected and prepared with the help of various moist media such as leaf mould, compost, farmyard manure, soil and vermiculite and then wrapped around the cuts with the help of black polythene films or hessian cloth and when proper rooting has occurred, these are detached and transplanted in pots or directly in the prepared beds when weather is persistently humid and cloudy. These are watered immediately after planting. Virupaksha (1961) reported leaf mould the best medium for air layers. This propagation method provides us quite large size of plants. **Budding** is used in case of the varieties which are successful neither through cuttings nor through layering and in this case easy to root varieties such as 'Dr. R.R. Pal' is used as rootstock and 'T' or inverted 'T' budding is the usual practice.

Commercial way of multiplication in bougainvillea is only through half-ripe or mature green stem **cuttings** measuring 9-10 cm having five nodes (two for burying in the medium for root formation and three exposed above for shoot formation) with lower leaves removed, treated with some effective rooting hormone and rooted around 24 °C (Dole and Wilkins, 1999), however, for maintaining leaf turgidity mist or fog can be used (Bailey, 1933; Hackett and Sachs, 1967; Tse *et al.*, 1974; Bailey and Bailey, 1976; Kamp-Glass and Ogden, 1991). Singh and Rathore (1977) recommended semi-hardwood cuttings better than others as they obtained 90 per cent success under shaded polythene tents though better success was obtained in cvs 'Mary Palmer' and 'Thimma' with softwood cuttings than the hardwood, however, Singh and Motial (1979) reported that cv. 'Thimma' performs better with softwood cuttings than the semi-hardwood. Bhattacharjee and Balakrishna (1983a) found better success with apical cuttings than those of semi-harwood or hardwood cuttings. Misra (1971) studied 64 bougainvillea varieties for stage of cuttings for better rooting and found profuse rooting and maximum survival in hardwood cuttings of 'Blondie', 'Elizabeth', 'Pixie', 'Rosa Catalina' and 'Sundari' though rooting in white coloured and variegated types did not occur at all as has also been reported by Bhattacharjee and Balakrishna (1983b). Mahros (2000) recorded better rooting in 'Sanderiana' and 'Snow White' when cuttings were taken from middle portion though *Glabra variegata* performed better with basal cuttings. The cutting should be roughly of pencil thickness, and should be retained with the leaves as much as possible because Bhattacharjee and Balakrishna (1983a) recorded 100 per cent rooting as well as survival in cv. 'Usha'with four leaves, though no rooting in leafless cuttings even when treated with IBA (Mukhopadhyay and Bose, 1966). Aishabi (1985) recorded maximum rooting when basal ends of the cuttings in vars 'Jayalakshmy' and 'Cherry Blossom' were soaked in 500 ppm IBA solution. However, weather has greater role to play as cuttings planted in sand in February root much better (Mukherjee *et al.*, 1976; Singh and Motial, 1979; Bhattacharjee and Balakrishna, 1983a), though Yadav *et al.* (1978) recorded mid-August as the best planting time under West Bengal conditions. Generally, within a week sprouting takes place but some cultivars are late in sprouting where it may take about 6-8 week's time. Rooting has been found better in *Bougainvillea glabra* and *B.* × *buttiana* varieties compared to others (Mishra and Singh, 1984). Rooting in mist chamber is quite encouraging. The study on the rooting and sprouting behaviour of the stem cuttings of bougainvillea cv. 'Los Banos Variegata Silver Margin' planted in 15 cm pots filled with coarse sand and kept under glass-roofed mist chamber after treating these with auxin (IBA) at 1, 000, 2, 000, 3,000, 4,000 and 5,000 ppm by quick dip method (10 seconds) or 250, 500 and 1,000 ppm through long dip method (24 h) revealed that 1,000 ppm IBA induced cent per cent rooting with higher number of roots (32.0) and further survival (90 per cent) in the long dip method (Gupta *et al.*, 2002). Bougainvillea cuttings treated with IAA (Pillai, 1963; Rao, 1967; Lal *et al.*, 1971; Joshi *et al.*, 1989) and IBA root better (Pillai, 1963), 100 ppm IBA followed by 100 ppm NAA in case of cvs 'Akola', 'Formosa', 'Magnifica', 'Refulgens' and 'Shubhra' (Harris and Singh, 1991), 100 ppm IBA in case of 'Glabra Variegata', 'Sanderiana' and 'Snow White', wounding and subsequent dipping in IBA in case of 'Mrs. Butt' (Hosni *et al.*, 2000) as per a study in Egypt, hardwood cuttings of 'Mary Palmer' treated with 500 ppm IBA (Kale and Bhujbal, 1972) or 500 ppm of IAA/NAA (Maurya *et al.*, 1974), contrarily, Mukhopadhyay and Bose (1966) reporting 'Mary Palmer' not responding to either of IBA, NAA and Seradix B treatments though getting success in cv. 'Partha' with all the three chemicals and in cv. 'H.C. Buck' with IBA and NAA both, 2000 ppm of IBA in case of 'Dr. R.R. Pal', 'Mrs H.C. Buck' (Mishra and Sharma, 1995) and in hardwood cuttings of cvs 'Alok' and 'Thimma' (Panwar *et al.*, 1994, 1999), 3,000-4,000 ppm best in case of cv. 'Thimma' irrespective of season (Singh, 1993) and 10,000 ppm IBA in case of 'Mahara' (Kanamadi *et al.*, 1997) have been reported enhancing better rooting. Shoot tip cuttings treated with 4,000 ppm IBA resulted into best rooting in cv. 'Garnet Glory' (Gupta and Kher, 1991). Verma *et al.* (1992) found higher rooting in hardwood cuttings of cvs 'Mahara' and 'Dr. R.R. Pal' treated with IBA, ascorbic acid and sucrose. Bhattacharjee and Balakrishna (1983a) when studying 25 bougainvillea cultivars for rooting, recorded 4,000 ppm NAA responding better in case of 'Glabra Sanderiana', 'Golden Glow', 'Isabell Green Smith' and 'Roseville's Delight', 4, 000 ppm IBA in the other 20 cultivars though cv. 'Usha' responded positively in both, *i.e.* 4,000 NAA, *vis-à-vis* 4,000 and 6,000 ppm IBA.

Micropropagation is the best way to multiply shy-rooting varieties or those which are in great demand. Through shoot tip proliferation, the cultivars 'Magnifica' (Chaturvedi *et al.*, 1978; Sharma *et al.*, 1981) and 'Scarlet Queen Variegated' (Sharma and Chaturvedi, 1988), and through direct and indirect adventitious shoot development in cv. 'Shubhra' (Rajan *et al.*, 1998), having good rooting in the medium containing 2.5 mg/l NAA + 1.0 mg/l kinetin. Aishabi (1985) reported shoot apices and immature axillary stem segments as the most potent sources for callus formation in MS medium supplemented with NAA 1.0 mg + BA 0.5 mg l^{-1} (KIN was less effective than BA). In case of var. 'Mahara', there was maximum number of callused cultures as compared to vars 'Cherry Blossom' and 'Spring Festival'. He recorded MS medium containing BA 2.0 mg + IAA 1.0 mg l^{-1} as optimum with regard to proliferation of axillary buds whereas rooting was recorded maximum in half strength MS medium supplemented with 1.6 mg l^{-1} IBA.

Cultural Practices

This being a versatile landscape ornamental plant, adapts well to various agroclimatic regions of the country from tropical to subtemperate climatic conditions, from arid to high rainfall areas and to varying **soil** types, *i.e.* acid to alkaline, gritty and stony land to marshy sites via light, sandy-loam, loam, clayey and high polluted sites. However, it prefers 5.5 to 6.0 **pH** of soil reaction though at 5.5-7.5 pH reaction almost all the required nutrients dissolved in the soil are easily absorbed by the root hairs of the plants but this crop does not face any crisis even at the pH range of 5.6 to 8.5, however, Criley (11997) reported that pH of medium above 6.5 may lead to iron deficiency. The pH bar of the soil is very important factor for the availability of the soil nutrients to bougainvillea plants. The wider the bar, more available are the nutrients. Calcium, magnesium and potassium—the exchangeable bases—are most available at high pH and unavailable at low pH. Nitrogen and sulphur are also available at similar pH ranges. Iron, manganese, zinc, and copper are less available at high pH values. Phosphorus and boron are unavailable at both the levels, *i.e.* low as well as high pH. Once it is established, it does not require any further maintenance and goes on growing on its own, though regular watering or sufficient rains results into luxuriant vegetative growth but poor flowering while dry conditions cause them blooming continuously and profusely for long periods. However, plants prefer outdoor conditions, well-drained **loamy soil**, with low rainfall and intense hot conditions having at least five hours of direct sunlight per day. Bract colours become deeper and elegantly bright during hot and dry climate though under shade the colour remains dull and the blooming shyly or nil. In fact, bougainvilleas are facultative SD plant (Allard, 1935), requiring 80 days for flowering under 12 h daylength though 54 days under 10 h daylength while plants usually flower under LD but only fewer buds are formed (Allard, 1935). Flowering is accelerated at high photointensity (Hacket and Sachs, 1966, 1968, 1985). They also stated that older plants with more stem diameter flower earlier than newly rooted plants with poor stem diameter. Criley (1977) stated high irradiance and inductive 8 h SD promoting flowering, though under inductive SD cycles and optimum temperatures, the flowering was delayed due to low irradiance (Hackett and Sachs, 1966). High light intensity of 4,000-5,000 fc encourages rapid compact growth with shining flowers (Dole and Wilkins, 1999). Hackett and Sachs (1966) stated that a light/dark temperature regime of 24/21 °C is optimum for flowering under SD, however, cool night temperatures (15 °C) promote flowering under LD conditions though a night temperature of 26 °C inhibits the flowering under SD conditions. Spells of rainy and dry seasons cause them blooming with various flushes, however, winter of subtropical, subtemperate and temperate areas cause hinderance in their blooming and during low temperature or frost plants require protection.

Though it can be **planted** throughout the year but planting in hot summer, especially in subtropical regions may result in poor establishment or drying out of the plants, cuttings and their roots and the rigours of the winter in subtropical and temperate areas may kill the plants outdoors so it would be advisable to plant them preferably at the onset of monsoon for their proper establishment. Spacing depends on the vigour of the species or varieties, however, for hedging a closer spacing, for specimen plants wider and in the corner of the lawn or along roadside proper distance, *i.e.* 2.0-2.5 m should be maintained. The pits of 90 cm diameter and 60 cm depth should be dug out and some 20-25 kg well decomposed stable/cow dung/farmyard manure or compost and some 10 kg of coarse sand should be mixed in the soil along with some 250 g bone meal and 250 g BHC. After filling of the mixture, the pits are thoroughly drenched to settle down the soil mixture. At monsoon, the bougainvilleas are planted akin to the level of previous marking on the plants. After planting, immediately these are watered thoroughly for establishement of roots. Until proper root establishment these are regularly watered.

Normally, bougainvilleas require large **pots** for its growing so it may be cement pots or the half cut drums or the specially prepared earthen pots. Looking into the size of the plants and the pots, a plant may be retained in the same pot for 3-10 years. Pots should be filled with loamy soil 3 parts, FYM 1 part, coarse sand ½ part and one tablespoon of bone meal, all by volume (Pal and Swarup, 1974). In case of ordinary soil there should be equal volume of FYM and in case of clayey soil the mixture should contain coarse sand half the volume of the soil. Planting time may be as per the need but preferably during July-August to avoid any casualty. Potted plants require frequent and regular watering but its regulation will cause inducement in flowering at intervals. Through pruning their compact shape is maintained and growth is controlled. In case of poor growth, a little of fertilizers may be added to the potting mix. Looking into the size of the plants commensurate to the containers, these are first pruned to maintain its shape and then repotted, and that too preferably during rainy season. Potted bougainvilleas and their bonsais adorn the frontal gate, the balconies and terraces. Through CCC and B-Nine (once at 4,000 or 5,000 ppm, and twice with 8,000 ppm) foliar sprayings on potted plants, the growth is made compact and dwarf, and together produces a large number of axillary branches and flowering shoots (Bhattacharjee, 1972). Soil drench with CCC at 2,000 and 4,000 ppm once, and twice with 4,000 ppm is also quite effective for dwarfing the plants. Hackett and Sachs (1967) reported that ancymidol (A-Rest), chlormequat (cycocel) and daminozide (B-Nine) hasten flowering in bougainvillea plants if grown under SD though not as effective in case of LD showing this to be environmental and temperature responsive as neither CCC nor A-Rest caused height reduction in 'Raspberry Ice' nor in 'San Diego Red' in South Carolina but both the retardants were found effective reducing vegetative growth of 'San Diego Red' in California (Dole and Wilkins, 1999) and they suggested that A-Rest should be applied as a drench at 50 ppm while Bonzi drenching at 25-100 ppm, however, Leapold (1971) reported that GA_3 sprayings inhibit floral initiation as well as delays or inhibits even bract initiation. Norcini *et al.* (1992) observed that Dikegulac (Atrimmec) sprayings on bougainvilleas give variety and temperature specific responses as in cv. 'Rainbow Gold' it enhanced flowering while there was no response in case of cvs 'Raspberry Ice' and 'San Diego Red'. It increased axillary branching in 'Rainbow Gold', 'Raspberry Ice' and 'San Diego Red' though stem elongation was found reduced in case of 'Raspberry Ice' and 'San Diego Red'. Dikegulac at 400-1,600 ppm reduces even bracteole size during summer but not when temperature starts falling and when the bracteoles were already smaller due to falling atemperatures (Norcini *et al.*, 1994).

Pruning and Pinching

A bougainvillea has a bloom cycle followed by a rest period whether it is pruned or not. In fact, the bloom initiation does not depend upon pruning. However, to keep the plants bushy, in shape and compactness, these are regularly **pruned** and pinched. These have a definite bloom cycle though it is influenced highly by prevailing temperature, relative humidity and soil moisture. These practices are best attempted after the flowering is almost over. Only one stem is allowed initially, however, at the crotch 5-7 branches may be retained. Pruning is highly related to the utility of the plant and as per specific requirement of the variety. A standard is pruned to conform the shape of an umbrella, specimen plants in globular form, ramblers as for the purpose these are desired and for climbing up the object, hedges drastically, a potted plant to retain its properly specified shape and compact form, for cascading effect these are pruned and trained through bending on steeps downward, *etc.* Pruning is also required to stop the bougainvillea vines growing outward and for making them quite bushy. Shy blooming varieties are scarsely thinned. After pruning, if desired, it may be manured. For a flush of profuse blooms all around the plants, **pinching** will be required to initiate more stem breaks. First soft pinching is carried out when new growth is 7.5 cm which is approximately 10 days after propagation, and every 4-weeks the plants can be pinched (Kamp-Glass and Ogden, 1991). New stems will usually grow from each axillary bud and less pinching is required to obtain a nicely branched and well spread bougainvillea plant. 'Soft pinch' is removal of the shoot tips of new growth to promote branching of the lateral shoots. Normally 2-3 shoots from the leaf buds just below the pinched buds start coming out. The 'hard pinch' is removal of already grown out branches, normally once a year, to keep the plant in shape. This is usually practiced in bougainvilleas grown for hanging baskets. BA sprayings twice at 50-100 ppm, once 24 h after the first pinch and the second application after second pinch will increase the branching (Dole and Wilkins, 1999).

Feeding and Watering

Though bougainvillea is a very care-free and self-sustaining crop of negligible maintenance but by nature it is a heavy feeder for best vegetative growth. Ideal fertilizing dose for bougainvillea is 250 g NPK in the ratio of 1-1-1 or 1-3-2 (Pal and Swarup, 1974) per plant in an exhausted soil which can easily be met by application of some 20 kg well rotten farmyard manure or compost per pit at the time of planting or after a long spell of many years when the plant is showing excessive weak growth, otherwise bougainvilleas planted in the ground are seldom fed, however, fertilizer application if is to be made that should be preferably done after pruning. Dole and Wilkins (1999) stated 150-200 ppm nitrogen but in the cold seasons they suggested to avoid its ammonical forms. There are reports from Florida (USA) that fertigation causes profuse flowering (Popenoe, 1961; Hackett and Sachs, 1965). Application of manure or FYM even improves the texture of the soil so it is preferable over chemical fertilizers. A few of the varieties are sensitive to iron, manganese and magnesium deficiencies and sometimes show cholorosis symptoms in leaves though it is a very rare and not a common phenomenon (Kamp-Glass and Ogden, 1991). Though it is a very low-maintenance crop not requiring even watering once properly established but in exceptional cases if acute deficiency of any element occurs then it shows the symtoms. Nitrogen deficiency in bougainvillea appears as yellowish-green old foliage, sometimes with reddish veins. Its severe deficiency causes young shoots becoming slender and hard with smaller leaves. Phosphorus deficiency causes petiole and veins on leaf lamina turning red to purple and lamina becoming mottled yellow to blotchy-brown. In case of potassium deficiency, initially the edges of the older leaves start turning purplish and tip burning and then afterwards affecting younger leaves. In case of magnesium deficiency the leaf lamina of older leaves is blotched yellow or tan with veins remaining green. Zinc deficient plant of bougainvillea exhibits symptoms which are almost similar to magnesium but in this case leaves become small, crinkled and crooked. Sulphur deficiency shows turning of young leaves from green to complete yellow and vein colour always brighter compared to lamina. Molybdenum deficiency causes distorted stem and entire lamina mottling with little pigmentation. In case of manganese deficiency, lamina turns yellow and afterwards entire lamina is covered with brown blotches. Calcium is normally not available to the plants above 7.8 soil pH and its deficiency causes death of tips and margins of young leaves, and terminal buds becoming dead, smaller or shrivelled. Boron deficiency symptoms are similar in buds as in case of calcium but stems and leaves become brittle and leaf base turns yellow. Iron as well as aluminium deficiency symptoms are observed when soil is too alkaline. The deficiency symptoms of both these elements appear as leaf lamina and smaller veins turn pale or yellow but at later stage lamina becomes brown though larger veins remain unaffecterd. Chlorosis or yellowing of the leaf is commonly caused by iron, magnesium and nitrogen deficiencies. Many of the bougainvillea cultivars with chlorosis respond to watering with iron chelates together with magnesium at the recommended rate at 15 days interval until the leaves become usual green. Standard fertilizer application is recommended for overcoming N, P, and K deficiencies in soil. If minor elements are missing from the soil, the formulations of soluble trace elements should be used. However, the ample use of organic nutrients may compensate the need of all the needed elements. Foliar spray of fertilizers along with 1 per cent commonly used home detergent at 2 to 4 week's intervals and with alternate sprayings of iron chelates, potassium nitrate and magnesium sulphate (1 tsp to 4 litres of water) prevents chlorosis and gives better results.

In potted bougainvilleas if the potting mixture is porous, they require regular **watering**, *i.e.* twice a week during hot and dry periods. Thus, it is a very drought tolerant genus and is ideal for growing in climates with long dry seasons as long as it gets water occasionally. Peaty soil in the mixture if once dries out, it does not imbibe water as usual, therefore, either such conditions should be avoided or regular watering should be carried out. In case of ground planting, once established it does not demand manual watering and survives well under rainfed conditions.

Growth and Flowering

In case of bougainvilleas, growth regulators for flowering are used to control plant growth so that floriferousness is maintained throughout the blooming period, therefore, Cycocel 800 ppm for plant retardation and 4,000 ppm for increase in flowering is used (Krishan Swarup *et al.*, 2010).

Insect-Pests and Diseases

Bougainvilleas if grown at an open and sunny situation are relatively pest-free provided its surroundings are clean, however, growing these under shade may invite some pests to infest on them. Though not common but the pests observed on bougainvilleas are nematodes, mites, aphids, caterpillars, leaf miners, scale insects, termites, thrips, whiteflies and snails & slugs. Pathak *et al.* (1995) reported the association of phytophagous **nematodes** with rhizosphere soil and roots of bougainvillea for the first time from Bihar (India) which can be controlled through use of Nemacur or Carbofuran. Marginal leaf curling, cupping and distortion are normally the symptoms caused by **mites** (*Phylcoptes bougainvilleae*, from Brazil by Costa and Carvalho, 1960; *Vittacus bougainvilleae* on *Bougainvillea glabra* var. 'Sanderiana' from Kenya by Awad and Elhanhawy, 1991; and *Ornithonyssus sylviarum*) including spider mites which harbour on the plants making their colonies there. These secrete the hormone which interferes with plant growth. Mites cause damage by sucking cell contents from leaves, especially on the undersides so sometimes bronze colouration on the leaves is obsereved due to their infestation and in severe attack leaves yellow and fall off. Neem oil spray is the best treatment to control the mites. In severe cases, fortnightly spraying with systemic acaricide regularly and thereafter at the onset of warm weather will control these pests. Spray with 0.2 per cent Confidor also controls this problem. **Scales** (*Coccus hesperidium* from USA, Pirone *et al.*, 1960) which appear under the cover of wax suck the sap from bougainvillea plants, especially from the leaf undersides and around leaf joints. Its infestation causes withering of the stems. The pesticide sprays are effective when its juveniles are in crawling stage. Use of organic pesticide made from pyrethrin will also control this pest. **Aphids** (*Myzus persicae* Sulzer) are sap-sucking insects infesting young tender leaves, new growth and flowering shoots. Cool showery nights and late afternoon watering in cool weather often results in its infestation. The aphid infested plant shows curled and twisted young leaves and shoots. Aphids in most of the cases leave behind secretion that attracts ants and promotes mould growth. Lady bird beetle (syn. lady bug, lady beetle or lacewing) are its natural enemies. Yellow sticky traps near the infected plants kill these aphids. Forced jet water spraying will also kill these insects. Boiled tobacco solution is also quite effective. The size of bougainvillea **looper** [inchworm or measuringworm (*Melachacka jeseri*)] is approximately 3cm and its colour is green to brown. The larva mimics stem and it is difficult to recognize it even from a close distance. It feeds on young branches and foliage from the edges during night. Spraying Sevin during evening hour will save the plants from the menace of this caterpillar. Neem based biological insecticides which are ecofriendly will also control the infestation of this pest. The larvae of majority of **leaf-mining insects** (*Bedellia somnulentella* Zeller) tunnelling the lamina belong to Lepidoptera and a few from Diptera, Coleoptera and Hymenoptera and their type of damage is similar irrespective of their families to which these belong. Sanitation in the plant surrounding will keep these pests at bay. Infested leaves should be collected and burnt. Spraying with Metasystox at 0.2 per cent will also control these pests. **Mealy bugs** which also suck its sap can be controlled by spraying aqueous solution of Confidor at 0.2 per cent fortnightly. **Thrips** (thunder flies, thunder bugs, storm flies or corm lice; *Frankliniella occidentalis*) are also sucking insects whose infestation may cause discolouration and deformation of bougainvillea foliage. Thrips are tiny, slender insects with fringed wings. Two families of parasitoid hymenoptera, *i.e.* Eulophidae and Trichogrammatidae are known to haunt them. **Whiteflies** (*Aleyrodidae* sp.) normally feed on leaf undersides by tapping phloem portion and these together also release toxic saliva which decreases the plant turgor pressure and excrete honeydew which promotes mould growth. These also carry viruses from different host plants. Their control is a cumbersome process as these develop rapid resistance to most of the chemical pesticides. Use of yellow sticky traps is very effective to control whiteflies. Though infestation is rare but attack of **white ants** (termite) is very rapid and difficult to control. These move in armies eating away everything from the roots and coming upwards judiciously. The white ant feeds on the bark of the plant and digest all the tissues even some part of the hard wood (Banerji, 2007). Bougainvillea beds should be treated with Thimet. Use of neem cake is recommended as a prophylactic measure. Solar treatment also cuts down the chances of termite infestation in the field.

The causal organism of leaf spot is **bacterium** *Pseudomonas andropogonis* (reported in USA by Walker, 1991; Walker and Hodge, 1991; and in Darussalam-Brunei by Sivapalan and Hamdan, 1997) and *P. stizilobii*. Persisting wet weather conditions are favourable for its infection and in severe cases up to 80 per cent leaves fall down, especially in younger plants. Symptom appears initially on leaf surface in the form of small, rounded and reddish spots usually surrounded by pale green haloes which merge into the normal green pigment of the leaf lamina. The spots gradually expand and become irregular dark patches, and later on becoming black. Infected foliage remaining on the plant becomes the main source of further infection in the following year when the atmospheric conditions are favourable. The disease is transmitted by water and can be prevented by removing infected dead leaves from the plant and maintaining proper sanitation. Spraying copper oxychloride at 3 g/l at 10-15 day's interval controls these pathogens effectively.

Leaf spots caused by **fungi** have also been found due to infection of *Cercosporidium bougainvilleae* (reported in USA by Sobers and Martinez, 1966; Sobers and Seymour, 1968; in Indonesia by Kobayashi and Oniki, 1994; in Colombo by Pardo-Cardona, 1998), *Cercospora bougainvilleae, Colletotrichum dematum bougainvilleae* and *Gladosporium arthinioides.* Spraying with fungicides such as Captan, maneb and mancozeb at 1 g/l at 10-15 day's interval will control these infections. *Bougainvillea* blight caused by *Phytophthora nicotianae* var. *parasitica* affecting twigs, leaves and flowers has been reported from Port Blair (Sinha and Salam, 1988) and Jabalpur (Nema *et al.*, 1999) of India, and Taiwan (Ann and Ann, 2000). Here irregular ashy green hydrotic spots first appear at the tip and margin of youngest leaves, then affecting to middle leaves, petioles and stems which results in drooping of blackened branches. Spraying of copper chloride at 3 g/l at 10-15 day's intervals will control this disease.

Post Harvest

Though bougainvilleas are only landscape plants, its flowers are not meant for cuttings but it is human nature to find out even such possibilities and therefore Mancarelli and Hugo (1991) treated small apical branches of *Bougainvillea glabra* with ethanol at 4.5, 6.5 and 9.0 per cent and found complete inhibition of flowers. Silver thiosulphate at 0.5mM and NAA at 25 and 50 ppm significantly prevented bract drop and prolonged its life at certain stages in potted plants of cvs 'Purple Flowers' and 'Taipei Red' (Chang and Chen, 2001) while shipping. Gago *et al.* (1997) suggested that one application of 500 ppm spray of NAA at full bract formation prolongs the postharvest life. The vase life in bougainvillea was found increased when Kochhar *et al.* (1992) kept the cut shoots in solution of citric acid and succinic acid supplemented with NAA to prevent bract shedding. While transporting the potted bougainvilleas, care should be taken that a temperature of 3 °C is maintained as at this temperature plants feel comfortable but only up to six days. Higher temperature at this time may be harmful. Ethylene causes leaf and bract abscission (Cameron and Reid, 1983; Moe and Fjeld, 1987; Nowak and Rudnicki, 1990; Nell, 1993). In Denmark, growers use Argyline (a silver containing compound) spray at 'run off' at 2.5-3.0 g/l per week or every 2-3 weeks before shipment (Dole and Wilkins, 1999).

References

Aishabi, K.A. 1985. Standardization of Macro and Micro Propagation Techniques in Bougainvillea (M. Sc. Thesis). Kerala Agricultural University, Vellanikkara, Thrissur (Kerala).

Allard, H.A. 1935. Response of the woody plants *Hibiscus syriacus, Malvaviscus conzatti* and *Bougainvillea glabra* to day length. *J. agric. Res.*, **51**: 27-34.

Ann Pao Jen and P.J. Ann, 2000. New diseases and records of flowering potted plants caused by *Phytophthora* species in Taiwan. *Plant Path. Bull.*, **9**(1): 1-10.

Awad, Abou, B.A. and E.M. Elhanhawy, 1991. New mites of the family Eriophydiae from Kenya (Acari: Eriophyoidea). *Acarologia*, **32**(4): 329-333.

Bailey, L.H. 1033. *Standard Cyclopedia of Horticulture*, pp. 533-534. Macmillan, New York.

Bailey, L.H. and E.Z. Bailey, 1976. *Hortus Third ~ A Concise Dictionary of Plants Cultivated in the United States and Canada*, p. 174. Macmillan Publishing, New York.

Balasaraswathi, R., S. Sadasivam, M. Ward and J.M. Walker, 1998. An antiviral protein from *Bougainvillea spectabilis* roots: purification and characterization. *Phytochem.*, **47**(8): 1561-1565.

Banerji, B.K. 2002. Induction of mutation in multibracted *Bougainvillea* cultivars 'Mahara' and 'Los Banos Beauty'. Paper presented at '*National Symposium on Indian Floriculture in New Millennium*', held at Lal Baugh Botanic Garden, Bangalore, during Feb 25-26, 2002.

Banerji, B.K.2007. Multibracted Bougainvillea. In : *Hi-Tech Floriculture* (eds Sathyanarayana Reddy, B.S., T. Janakiram, S. Balaji Kulkarni and R.L. Misra), pp. 190-192. ISOH, IARI, New Delhi.

Banerji, B.K. 2008. Mutation and mysterious world of multibracted bougainvillea. *Floriculture Today*, January 8.

Banerji, B.K. 2009. 'Pixie Variegata' – A new *Bougainvillea* cultivar developed by NBRI, Lucknow. *Indian Bougainvillea Ann.*, **22**: 18-20.

Banerji, B.K. and G.K. Banda, 1967. A natural triploid in *Bougainvillea. Indian J. Hort.*, **24**: 106-108.

Banerji, B.K. and S.K. Datta, 1987a. Gamma ray induced chlorophyll variegated mutants in *Bougainvillea* cv. 'Los Banos Beauty'. *J. Nuclear Agric. & Biol.*, **16**(1): 48-50.

Banerji, B.K. and S.K. Datta, 1987b. Gamma ray induced genetic improvement of double bracted *Bougainvillea.* Paper presented at '*All India Seminar on Advances in Botanical Research in India during the last 10 years*', held on Nov. 1-3, at Dept. of Botany, Dungar College, Bikaner.

Banerji, B.K. and Datta, S.K. 1996. Bougainvillea mein utparivartan prajanan. *Vigyan Vani*, **2**: 51-53.

Banerji, B.K., P. Nath and M.N. Gupta, 1983. Gamma ray induced somatic mutation in *Bougainvillea* cv. 'Roseville's Delight'. I. Induction of chlorophyll mutation. In: '*National Symposium on Recent Development in Nuclear and Allied Techniques and Their Application in Agriculture, Biology & Animal Science*', held on May 5-7, at G.B. Pant Univ. of Agric. & Tech., Pantnagar, p. 41.

Banerji, B.K., P. Nath. and S.K. Datta, 1987. Mutation breeding in double bracted *Bougainvillea* cv. 'Roseville's Delight'. *J. Nuclear Agric. & Biol.,* **16**(1): 45-47.

Bharathi, M. 1999. Effect of plant extract and chemical inhibitors on cucumber mosaic virus of brinjal. *J. Mycol. Pl. Path.*, **29**(1): 57-60.

Bhatnagar, A. and V.K. Sharma, 1999. Relative toxicity and persistence of plant products against maize stem borer on maize. *Ann. Pl. Prot. Sci.*, **7**(2): 144-149.

Bhattacharjee, S.K. 1972. Effect of growth retardants of several varieties of *bougainvillea. Indian Agric.*, **16**(1): 85-90.

Bhattacharjee, S.K. and M. Balakrishna, 1983a. Propagation of *Bougainvillea* from stem cuttings. I. Effect of growth regulators, rooting media, leaf number, length and woodiness of cuttings. *Haryana J. hort. Sci.*, **12**: 7-12.

Bhattacharjee, S.K. and M. Balakrishna, 1983b. Propagation of *Bougainvillea* from stem cuttings. II. Effect of growth regulators. *Haryana J. hort. Sci.*, **12**: 13-18.

Bor, N.L. and M.B. Raizada 1954. Some beautiful Indian climbers and shrubs. *The Bombay Natural History Society*, Mumbai, pp. 267-274.

Cameron, A.C. and M.S. Reid, 1983. Use of silver thiosulfate to prevent flower abscission from potted plants. *Scientia Hort.*, **19**: 373-378.

Chang, Y.S.and H.C. Chen, 2001. Variability between silver thiosulfate and 1-naphthaleneacetic acid applications in prolonging bract longevity of potted *Bougainvillea. Scientia Hort.*, **87**(3): 217-224.

Chaturvedi, H.C., A.K. Sharma and R.N. Prasad, 1978. Shoot apex culture of *Bougainvillea glabra* 'Magnifica'. *HortSci.*, **13**: 36.

Cheema, S.S., S. Kapila and S. Arunakumar, 1993. Efficacy of various bio-products and chemicals against tobacco mosaic virus in tomato and cucumber mosaic virus in bottle gourd. *Pl. Dis. Res.*, **8**(2): 110-114.

Cooper, D.C. 1931. Microporogenesis in *Bougainvillea glabra. Amer. J. Bot.*, **18**: 337-358.

Costa, A.S. and A.M.B. Carvalho, 1960. Marginal leaf curl in *Bougainvillea* caused by mites. *Bragantia*, **19** (Supplement): 134-140.

Criley, R.A. 1977. Year round flowering of double bougainvillea : Effect of daylength and growth retardants. *J. Amer. Soc. hort. Sci.*, **102**: 775-778.

Criley, R.A. 1997. *Bougainvillea.* In: *Tips of Growing specialty Potted crops* (eds Gaston, M.L., S.A. Carver, C.A. Irwin and R.A. Larson), pp. 28-31. Ohio Florists' Association, Columbus, Ohio, USA.

Datta, S.K. and B.K. Banerji, 1994a. Improvement of double bracted bougainvillea through induced mutation. '*National Symposium on Frontiers in Plant Science Research*', held at N.S.A. Khan Center of P.G. Studies and Research, A.U. College, Mallepally, Hyderabad, on Feb. 13-14.

Datta, S.K. and B.K. Banerji, 1994b. 'Mahara Variegata'- a new mutant of *Bougainvillea. J. Nuclear Agric.& Biol.*, **23**(2): 114-116.

Datta, S.K. and B.K. Banerji, 1997. Improvement of double bracted *Bougainvillea* through gamma ray induced mutation. *Frontiers in Plant Science* (ed. Khan, I.A.), **64**: 395-400.

Datta, S.K., A.K. Dwivedi and B.K. Banerji, 1995. Investigations on gamma ray induced chlorophyll variegated mutants. *J. Nuclear Agric. & Biol.*, **24** (4): 237-247.

Dole, J.M. and H.F. Wilkins, 1999. *Bougainvillea*. In: *Floriculture Principles and Species*, pp. 235-238. Prentice Hall, New Jersey, USA.

Gago, C.M., J.A. Monteiro and H.M. Rodrigues, 1997. Extending potted bougainvillea post-production: NAA, STS, and ethanol (abstr.). *HortSci.*, **32**: 517.

Gupta, V.N. and M.A. Kher, 1991. A note on the influence of auxins in regeneration of roots in the tip cuttings of *Bougainvillea* sp. var. 'Garnet Glory' under intermittent mist. *Haryana J. hort. Sci.*, **20**(1-2): 85-87.

Gupta, V.N., B.K. Banerji and S.K. Datta, 2002. Effects of auxin on the rooting and sprouting behaviour of stem cuttings of bougainvillea cv. 'Los Banos Variegata Silver Margin' under mist. *Indian J. hort. Sci.*, **31** (1&2): 42-44.

Hackett, W.P. and R.M. Sachs, 1965. Factors affecting flowering in *Bougainvillea. Calif. Agric.*, **19**(9): 13.

Hacket, W.P. and R.M. Sachs, 1966. Flowering in *Bougainvillea* 'San Diego Red'. *Proc. Amer. Soc. hort. Sci.*, **88**: 606-612.

Hackett, W.P. and R.M. Sachs, 1967. Chemical control of flowering in *Bougainvillea* 'San Diego Red'. *Proc. Amer. Soc. hort. Sci.*, **90**: 361-364.

Hackett, W.P. and R.M. Sachs, 1968. Experimental separation of inflorescence development from initiation in *Bougainvillea. Proc. Amer. Soc. hort. Sci.*, **92**: 615-621.

Hackett, W.P. and R.M. Sachs, 1985. *Bougainvillea*. In: *Handbook of Flowering* (vol. II, ed. Halevy, A.H.), pp. 38-47. CRC Press, Boca Raton, Florida, USA.

Harris, C.V. and D.B. Singh, 1991. Role of auxin on rooting of cuttings of *Bougainvillea* cultivars during rainy and spring season. *New Agric.*, **2**(1): 12-19.

Holttum, R.E. 1970. *Bougainvillea*. In: *Flowering Vines of the World ~ An Encyclopaedia of Climbing Plants* (ed. Menninger, E.A.), Hearthside Press Inc., New York.

Hosni, A.M.E.L., S.A. Gendy, M.R. Shedeed and A.K. Ebrahim, 2000. Improvement of rooting in *Bougainvillea × buttiana* 'Mrs. Butt' by and/or IBA application(s) to cutting basal ends. *Ann. agric. Res.*, **45**(2): 659-678.

Iredell, J. 1990. *The Bougainvillea Growers Handbook*. Simon and Schuster's, Australia.

Janakiram, T., Ritu Jain, Kishan Swaroop, Kamlesh Kumar and Pavan Kumar P. 2013. *Bougainvillea Repository*. A Leaflet by Division of Floriculture and Landscaping, IARI, New Delhi, 4 p.

Jayanthi, R. and S.K. Datta, 2006. Improvement of *Bougainvillea* varieties using chemical mutagens. In: '*National Conference on Bougainvillea*', p. 17, held on April 12-13, at NBRI, Lucknow.

Jayashree, K., K.B. Pun and Sabitha Doraiswamy, 1998. Management of yellow vein mosaic virus disease of pumpkin. *Madras agric. J.*, **85**(2): 127-129.

Joshi, A.R., V.K. Mahrkar and K.T. Sadawarte, 1989. Studies on rooting of cuttings in some *bougainvillea* varieties as influenced by plant growth regulators. *PKV Res. J.*, **13**: 166-167.

Kale, P.N. and B.G. Bhujbal, 1972. Use of growth regulators in rooting of cuttings of *Bougainvillea* var. 'Mary Palmer'. *Indian J. Hort.*, **29**: 307-309.

Kamp-Glass, M. and M.A.H. Ogden, 1991. *Bougainvillea. GrowerTalks*, **55**(8): 17.

Kanamadi, V.C., S. Patil, Y.H. Ryagi, R.M. Shirol and Vijayakumar, 1997. Effect of growth regulators on the rooting of cuttings of *Bougainvillea* variety'Mahara'. *Adv. in agric. Res. in India*, **7**: 43-45.

Kishan Swarup, T. Janakiram, D.V.S. Raju, K.V. Prasad, Kanwar Pal Singh, Namita and Ritu Jain, 2010. *Practical Manual on Bougainvillea*, 30 p. Division of Floriculture and Landscaping, IARI, New Delhi-110 012.

Kobayashi, T. and M. Oniki, 1994. Circular leaf spot of *Bougainvillea* caused by *Circosporidium bougainvilleae* in Indonesia. *Ann. phytopath. Soc. Japan*, **60**(2): 221‘-224.

Khoshoo, T.N. 1998. Perspectives in bougainvillea breeding. *Bougainvillea N.L.*, **6**(2): 7-10.

Khoshoo, T.N. and S.N. Zadoo, 1969. New perspectives in *Bougainvillea* breeding. *J. Hered.*, **60**(2): 357-360.

Kochhar, V.K. and D. Ohri, 1979. Biochemical analysis of bract mutations in bougainvilleas. I. 'H.C. Buck' family. *Z. Pflanzenzuchtg.*, **79**: 47-51.

Kochhar, V.K., R. Shukla, A.S. Murthy and M.A. Kher, 1992. Studies on prolonging vase life of *Bougainvillea. Indian J. agric. Res.*, **26**(1): 15-24.

Lakshmanan, P. and S. Mohan, 1989. Anti-fungal properties of some plant extracts against collar rot of *Phaseolus aureus. Madras agric. J.*, **76**(5): 266-270.

Lakshmanan, S. Mohan and R. Jeyarajan, 1990. Anti-fungal properties of some plant extracts against *Thanatephorus cucumeris*, the causal agent of collar rot disease of *Phaseolus aureus. Madras agric. J.*, **77**(1): 1-4.

Lal, N., S.N. Yadav and L.S. Srivastava, 1971. Effect of some plant growth regulators on rooting behaviour of cuttings of *Bougainvillea* 'Thimma'. *Punjab hort. J.*, **1**: 278-279.

Leapold, A.C. 1971. Antagonism of some gibberellin actions by substituted pyrimidine. *Plant Physiol.*, **48**: 537-540.

Mahros, O.M. 2000. Rootability and growth of some types of *Bougainvillea* cutting under IBA stimulation. *Assiut. J. agric. sci.*, **31**(1): 19-37.

Maurya, A.N., S.P. Singh and S. Lal, 1974. Effect of plant growth regulators on the rooting of stem cuttings in *Bougainvillea. Punj. Hort. J.*, **14**: 84-86.

Mencarelli, F. and L. Hugo, 1991. Control of flower and bract abscission of *Bougainvillea* branches by ethanol solutions. *Agricoltura Mediterranea*, **121**(3): 282-286.

Mishra, S.M. and C.P. Sharma, 1995. Effect of plant growth regulators on rooting of stem cuttings in *Bougainvillea. Progressive Hort.*, **27**(1): 33-38.

Mishra, H.P. and K.P. Singh, 1984. Varietal differences in rooting of *Bougainvillea* stem cuttings. *South Indian Hort.*, **32**(2): 113-114.

Misra, A.K. 1971. Rooting behaviour of cuttings of *Bougainvillea* with the aid of Seradix B_3. *Indian J. Hort.*, **28**: 68-69.

Moe, R. and Fjeld, T. 1987. Keeping quality of potted plants as influenced by ethylene. *Gartner Tidende*, **101**: 1580-1583.

Molina, M.M. and L.M. Sanchez, 1991. Foliar inhibitors of *Bougainvillea spectabilis* chemical nature and anti-viral activity on sugarcane mosaic virus (SCMV). *Revista de ProtecionVegetal.*, **6**(1): 58-66.

Mukherjee, T.P., T. Roy and T.K. Bose, 1976. Standardisation of propagation from cuttings under mist. II. Effect of rooting media on root formation in cuttings of ornamental plants. *Punjab hort. J.*, **16**: 153-156.

Mukhopadhyay, D.P. and T.K. Bose, 1966. Improvement in the method of vegetative propagation in some varieties of bougainvilleas. *Indian J. Hort.*, **23**: 185-186.

Nagarajan, K., N.S. Murthy and T.S.N. Reddy, 1990. Utilisation of botanicals possessing antiviral principle against tobacco mosaic virus. *Botanical pesticides in integrated pest management*, pp. 407-412. *Proc. nat. Symp.*, held on January 21-22 at Rajahmundry, India.

Nair, P.K.K. 1965. Significance of pollen morphology in the study of cultivated plants. *Indian Agric.*, **9**:53-87.

Narayanan, C.R., D.D. Joshi, A.M. Majumdar and V.V. Dhekne, 1987. Pinitol, a new antidiabetic compound from the leaves of *Bougainvillea spectabilis. Curr. Sci.*, **56**: 139-141.

Nath, P., B.K. Banerji and M.N. Gupta, 1983. Spontaneous and induced mutations in *Bougainvillea. News Letter, Bougainvillea Festival*, Bougainvillea Society, Lucknow.

Nell, T. 1993. *Bougainvillea*. In: *Floewering Potted Plants, Prolonging Shelf Performance*, pp. 33-34. Ball Publishing, Batavia, Illinois, USA.

Nema, S., N.D. Sharma and S. Nema, 1999. *Phytophthora nicotianae*: incitant of blight of *Bougainvillea. J. Mycol. & Pl. Path.*, **29**(2): 262.

Ninan, T., M.P. Singh and M.S. Swaminathan, 1959. Meiotic behavior and pollen fertility in some varieties of *Bougainvillea. J. Indian bot. Soc.*, **38**: 140-145.

Norcini, J.G., J.M. McDowell and J.H. Aldrich, 1992. Effect of dikegulac on flowering and growth of *Bougainvillea* 'Rainbow Gold'. *HortSci.*, **27**: 35-36.

Norcini, J.G., J.H. Aldrich and J.M. McDowell, 1994. Flowering response of *Bougainvillea* cultivars to dikegulac. *HortSci.*, **29**: 282-284.

Nowak, J. and R.M. Rudnicki, 1990. *Posharvest Handling and Storage of Cut Flowers, Florist Greens and Potted Plants.* Timber Press, Portland, Oregon, USA.

Ohri, D. 1979. Cytogenetics of garden gladiolus and bougainvillea. Ph.D. Thesis. Panjab Univ., Chandigarh.

Ohri, D. 1995. *Bougainvillea* breeding. In: *Advances in Horticulture* (vol.12): *Ornamental Plants* (eds Chadha, K.L. and S.K. Bhattacharjee), pp. 363-376. Malhotra Publishing House, New Delhi.

Ohri, D. and T.N. Khoshoo, 1982. Cytogenetics of cultivated bougainvilleas. X. Nuclear DNA content. *Z. Pflanzenzuchtg*, **88**: 168-173.

Ohri, D. and S.N. Zadoo, S.N. 1975. Cross-compatibility studies in some *Bougainvillea* cultivars. *Incompatibility New Letter*, **6**: 35-39.

Ohri, D. and S.N. Zadoo, 1979. Cytogenetics of cultivated bougainvilleas. VIII. Cross-compatibility relationships and origin of *Bougainvillea* 'H.C. Buck' family. *Z. Pflanzenzuchtg.*, **82**: 182-186.

Ohri, D. and S.N. Zadoo, 1986. Cytogenetics of cultivated bougainvilleas IX. Precocious centromere division and origin of polyploid taxa. *Plant Breeding*, **97**: 227-231.

Pal, B.P. and V. Swarup, 1974. *Bougainvilleas.* ICAR, New Delhi.

Pancho, J.B. and A.E. Bardenas 1959. Bougainvilleas in the Philippines. *Baileya*, **7**(3): 81-100.

Pancho, J.B. and J.M. Capinpin, 1961. Haploidy in *Bougainvillea. Philippines Agric.*, **45**: 88-94.

Pancho, J.B., J.M. Capinpin and A.E. Bardenas, 1960. Chromosome number and fertility tests in *Bougainvillea. Phillipines Agric.*, **45**: 11-18.

Panwar, R.D., A.K. Gupta, J.R. Sharma and Rakesh, 1994. Effect of growth regulators on rooting in *Bougainvillea* var. 'Alok'. *Intern. J. Trop. Agric.*, **12**(3-4): 255-261.

Panwar, R.D., A.K. Gupta, R. Yamdagni and R.S. Saini, 1999. Effect of growth regulators on the rooting of cuttings of *Bougainvillea* cv. 'Thimma'. *Haryana agric. Uni. J. Res.*, **29**(1-2): 11-17.

Pardo-Cardona, V.M. 1998. *Cercosporidium bougainvilleae* (Montanola) Sobers and Seymour (Hyphmycetes, Dematiaceae) new plant pathogen of *Bougainvillea* sp. *Revista Facultad National de Agronomia Medellin*, Colombia, **51**(1): 253-255.

Pathak, K.N., Pranay Kumar, H.P. Mishra and P. Kumar, 1995. Association of phytophagous nematodes with rhizosphere soil and roots of some ornamental plants. *J. Ornam. Hort.*, **3**(1-2): 55-58.

Pillai, P.K. 1963. A new technique on rooting of plant cuttings using growth regulators. *Madras agric. J.*, **50**: 29-30.

Pirone, P.P., B.O. Dodge and H.W. Rickett, 1960. *Diseases and Pests of Ornamental Plants.* Constable & Co. Ltd, U.K.

Popenoe, J. 1961. *Bougainvillea* culture. *Ann. Hort. Mag.*, **40**(4): 319-324.

Prativa Lakhotia, Krishan P. Singh, S.K. Singh, M.C. Singh, K.V. Prasad and Kishan Swaroop, 2014. Influence of biotic and abiotic elicitors on production of betalaim pigments in bougainvillea callus cultures. *Indian J. Hort.*, **71**(3): 373-378.

Pun, K.B., D. Sabitha and R. Jeyarajan, 1999. Screening of plant species for the presence of antiviral principles against 'okra yellow vein mosaic virus'. *Indian J. Phytopath.*, **52**(3):221-223.

Rajan, G.B., G.J. Rani and G. Balakrishnamurthy, 1998. Plant regeneration in *Bougainvillea* cv. Shubhra through direct and indirect adventitious shoot development. *South Indian Hort.*, **46**(3-6): 214-215.

Rao, L.M. 1967. Effect of certain plant growth regulation substances on the rooting of *Bougainvillea spectabilis.* I. Stem cuttings. *J. Japanese Soc. hort. Sci.*, **36**: 445-448.

Rao, S.J., K.C. Chitra, P.K. Rao and K.S. Reddy, 1992. Antifeedant and insecticidal properties of certain plant extracts against brinjal spotted leaf beetle, *Henosepilachna vigintioetopunctata. J. Insect Sci.*, **5**(2): 163-164.

Roy, R.K., B.K. Banerji and A.K. Goel, 2007. Bougainvillea germplasm collections at NBRI. *Indian Bougainvillea Annual* (April 2007), **20**: 10-13.

Sadasivam, S., S. Rajamaheswari and R. Jeyarajan, 1991. Inhibition of certain plant viruses by plant extracts. *J. Ecobiol.*, **3**(1): 53-57.

Saifuddin, M., A.B.M.S. Hossain, O. Normaniza and K.M. Moneruzzaman, 2009. Bract size enlargement and longevity of *Bougainvillea spectabilis* as affected by GA_3 and phloemic stress. *Asian J. Pl. Sci.*, **8**(3): 212-217.

Saifuddin, M., A.B.M.S. Hossain and O. Normaniza, 2010. Impacts of shading on flower formation and longevity, leaf chlorophyll and growth of *Bougainvillea glabra. Asian J. Pl. Sci.*, **9**(1): 20-27.

Sharma, A.K. and H.C. Chaturvedi, 1988. Micropropagation of *Bougainvillea* × *buttiana* 'Scarlet Queen Variegated' by shoot tip culture. *Indian J. exp. Biol.*, **26**: 285-288.

Sharma, A.K., R.N. Prasad and H.C. Chaturvedi, 1981. Clonal propagation of *Bougainvillea glabra* 'Magnifica' through shoot apex culture. *Plant Cell Tiss. Org. Cult.*, **1**: 33-38.

Sharma, A.K. and U.C. Bhattacharya, 1960. Cytological investigations on *Bougainvillea* as an aid in interpreting the evolution and affinities of different species and varieties. *The Nucleus*, **3**: 19-76.

Sindhan, G.S., Indra Hooda and R.D. Parashar, 1999. Effect of some plant extracts on the vegetative growth of root rot causing fungi. *J. Mycol. Pl. Path.*, **29**(1): 110-111.

Singh, S.P. 1993. Effect of auxins and planting time on crarbohydrate and nitrogen fractions in semi-hard wood cuttings of *Bougainvillea* cv. 'Thimma' under intermittent mist III. *Advances in Horticulture & Forestry*, **3**: 157-163.

Singh, I.P. and S.V.S. Rathore, 1977. Rooting and survival of *Bougainvillea* cuttings as affected by maturity of wood and planting aenvironment. *Haryana J. hort. Sci.*, **6**: 201-203.

Singh, S.P. and V.S. Motial, 1979. Propagation of *Bougainvillea* cv. 'Thimma' under intermittent mist. *Plant Sci.*, **11**: 53-59.

Sinha, A.R.P. and M.A. Salam, 1988. Pathogen fungi of Andaman-I. *Adv. Plant Sci.*, **1**(2): 214-218.

Sivapalan, A. and F.H. Hamdan, 1997. Bacterial leaf spot of *Bougainvillea* caused by *Pseudomonas andropogonis* in Brunei Darussalam. *Bull.OEPP*, **27**(2): 273-275.

Sobers, E.K. and A.P. Martinez, 1966. A leaf spot of bougainvillea caused by *Cercospora bougainvilleae*. *Phytopath.*, **56**: 128-130.

Sobers, and Seymour, 1968. Distribution of pathogenicity and taxonomy of *Cercosporidium bougainvilleae*. *Proc. Fla St. hort. Soc.*, **81**: 397-401.

Stein, U. and F. Klingauf, 1990. Insecticidal effect of plant extracts from tropical and subtropical species. Traditional methods are good as long as they are effective. *J. App. Entom.*, **110**(2): 160-166.

Sundarababu, R., C. Sankaranarayana and S. Vedivelu, 1990. Nematode management with plant products. *Indian J. Nematol.*, **20**(2): 177-178.

Swarup, V. and B. Singh, 1964. Pollen morphology and hairs in classification of *Bougainvillea*. *Indian J. Hort.*, **21**: 155-164.

Tse, A.T.Y., A. Ramina, W.P. Hackett and R.M. Sachs, 1974. Enhanced inflorescence development in *Bougainvillea* 'San Diego Red' by removal of young leaves and cytokinin treatments. *Plant Physiol.*, **54**: 404-407.

Verma, S.C., M.M. Haider and A.S. Murty, 1992. Note on effect of chemicals on rooting in *Bougainvillea*. *Indian J. Hort.*, **49**(3): 284-286.

Virupaksha, M. 1961. Effect of different rooting media on air layers of croton, hibiscus and *Bougainvillea*. *Indian J. Hort.*, **6**: 4-8.

Walker, S.E. 1991. Bacterial leaf spot of *Bougainvillea*. *Plant Path. Circ.* – Gainesville, **342**: 2.

Walker, S.E. and N.C. Hodge, 1991. Bacterial leaf spot of *bougainvillea* in Florida caused by *Pseudomonas andropogonis*. *Plant Dis.*, **75**(9): 968.

Yadav, L.P., A.P. Bhattacharya and H.S. Pandey, 1978. Effect of season on the rooting of *Bougainvillea* cuttings. *Progreesive Hort.*, **9**: 72-73.

Zadoo, S.N. and T.N. Khoshoo, 1968. Cytogenetical studies in *Bougainvillea*. A case of interchange heterozygosity. *Genetica*, **39**: 353-360.

Zadoo, S.N. and T.N. Khoshoo, 1975. Nature of self-incompatibility in cultivated bougainvilleas. *Incompatibility Newletter*, **5**: 73-75.

Zadoo, S.N., R.P. Roy and T.N. Khoshoo, 1975a. Cytogenetics of cultivated bougainvilleas. I. Morphological variation. *Proc. Indian Natn. Sci. Acad.*, **41B**: 121-132.

Zadoo, S.N., R.P. Roy and T.N. Khoshoo, 1975b. Cytogenetics of cultivated bougainvilleas. II. Pollination mechanism and breeding system. *Proc. Indian Natn. Sci. Acad.*, **41B**: 498-502.

Zadoo, S.N., R.P. Roy and T.N. Khoshoo, 1975c. Cytogenetics of cultivated bougainvilleas. III. Bud sports. *Z. Pflanzenzuchtg*, **74**: 223-239.

Zadoo, S.N., R.P. Roy and T.N. Khoshoo, 1975d. Cytogenetics of cultivated bougainvilleas. IV. Variation in meiotic system. *La Cellule*, **73**: 311-322.

Zadoo, S.N., R.P. Roy and T.N. Khoshoo, 1975e. Cytogenetics of cultivated bougainvilleas. V. Tetraploidy and restoration of fertility in sterile cultivars. *Euphytica*, **24**: 517-524.

Zadoo, S.N., R.P. Roy and T.N. Khoshoo, 1975f. Cytogenetics of cultivated bougainvilleas. IV. Hybridization. *Z. Pflanzenzuchtg*, **75**: 114-137.

Zadoo, S.N., R.P. Roy and T.N. Khoshoo, 1976. Cytogenetics of cultivated bougainvilleas. VII. Origin and evolution of ornamental taxa. *Indian J. Hort.*, **33**: 278-288.

8

Bromeliads (Family: Bromeliaceae)

Indrajit Sarkar, Sanyat Misra, R.L. Misra, Pragya Ranjan, Preeti Hatibarua, Ajit Kumar and R.K. Dubey

[**Common names**: Air plant, Angel's/Friendship plant/Queen's tears (*Billbergia nutans*), Banded urn plant (*Billbergia* × 'Santa Barbara'), Bird's nest bromeliad (*Nidularium innocentii,* syn. *N. amazonicum*), Blushing bromeliad (*Neoregelia carolinae* 'Tricolor', *Nidularium fulgens* syn. *N. pictum, Guzmania picta*), Caroa/Caroa of Brazil (*Neoglaziovia variegata*), Coralberry (*Aechmea fulgens*), Earth star/starfish (*Cryptanthus*), Fingernail plant (*Neoregelia spectabilis*), Flaming sword (*Vriesea splendens*), Flaming torch (*Guzmania berteroana*), Friendship plant/Queen's tears (*Billbergia nutans*), Green earth star (*Cryptanthus acaulis*), Heart of fame (*Bromelia balansae*),Ivory pineapple (*Ananas comosus* 'Variegatus'), King of the bromeliads (*Vriesea hieroglyphica*), Lacquered wine-cup (*Aechmea victoriana* × *A. racinae* = 'Foster's Favorite'), Lobster claws (*Vriesea carinata*), Makimbeira of Brazil (*Neoglaziovia concolor*), Manda's urn plant (*Billbergia* × 'Bob Manda'), Marble plant (*Neoregelia marmorata*), Old man's beard/Spanish moss (*Tillandsia usneioides*), Painted feather (*Vriesea psittacina, V.* × *mariae*), Pineapple (*A comosus* syn. *A. sativus*), Pinguin of Jamaica/ Wild pine (*Bromelia pinguin*), Red feather (*V.* × *erecta*), Red pineapple/Wild pineapple (*Ananas bracteatus*), Scarlet star (*Guzmania lingulata*), Silver vase/Urn plant (*Aechmea fasciata*, syn. *Billbergia rhodocyanea*), Striped urn plant (*Billbergia pyramidalis* 'Striata'), Wax torch (*Aechmea bromeliifolia*)]

Introduction

Bromeliads are among the more recent plant groups to have emerged. Bromeliads belong to the family Bromeliaceae and are members of the class Liliopsida (monocots). *Bromeliaceae* was named so by the French botanist Jaume St. Hilaire in 1805 by combining the genus name *Bromelia* with ending of 'ceae', the name *Bromelia* was formally established in 1754 by Dr. Carl von Linne (Linneaus) to honour Olaf Bromel, a Swedish botanist who was well known in Europe at that time, though original name for this generic name was proposed by Charles Plumier, an early French explorer of the West Indies (Palmer, 1964;Smith, 1951). If a plant has strap-shaped leaves arranged in a rosette, leaves adorned with some type of scale or scurf, and the flowers arranged in threes with dissimilar sepals and petals, then the plant is almost certainly a bromeliad. Traditionally the family Bromeliaceae has been divided into three subfamilies: Pitcairnioideae, Bromelioideae and Tillandsioideae. The neotropical family Bromeliaceae, growing virtually exclusively in the New World tropics and subtropics, comprises 58 genera with over 3,170 species, all being almost entirely restricted mainly to the tropical Americas with a few species found in the American sub-tropics and one in tropical West Africa (Mabberley, 1997). The greatest number of primitive species inhabit the Andean highlands of South America, where they originated in the tepuis of the Guyana Shield (Givnish *et al.,* 2004). The most basal genus, *Brocchinia* is endemic to these tepuis, is placed as the sister group to the remaining genera in the family (Barfuss *et al.,* 2005). The tropical

West African species *Pitcairnia feliciana* is the only bromeliad not endemic to the Americas, and is thought to have reached Africa via long-distance dispersal about 12 million years ago (Givnish *et al.,* 2004). They can be found at altitudes from sea level to 4,200 m above mean sea level, and from rainforests to deserts. Approximately half the species are epiphytes, a few are lithophytes and others are terrestrial. These can therefore be found in the Andean highlands, from northern Chile to Colombia, in the Sechura desert of coastal Peru, in the cloud-forests of CS America, in southern U.S. from southern Virginia to Florida to Texas, and in far southern Arizona. In South America, the greatest number is found in Brazil, ranging from Chile and Argentina in South America through Central America and the Caribbean reaching their northern limit around Virginia in the SE United States in diverse habitats from hot and dry deserts to moist rainforests to cool mountainous regions, most living as **epiphytes** in forests where they cling tightly to the branches and trunks of the trees, shrubs or cactus but sometimes they cling to the telephone poles or even on the wires (lines) often fully exposed to the elements. A study in a forest revealed 175,000 bromeliads growing naturally in one hectare of a forest area, and collecting some 50,000 liters of water. This capability to take their nutrition and moisture from the atmosphere has earned these bromeliads the name 'air plants'. **Terrestrial** species are found growing in the ground as many other plants grow. They may be found growing in bright sun along sandy beaches to the shady understorey of a forest among the leaf litter and debris. **Saxicolous** species are found growing on rocks. They may grow on hard rocky outcrops where their roots may penetrate cracks and fissures to locate moisture or organic nutrients or sometimes they are found growing tenuously on sheer cliff faces. Usually they are inexpensive with brilliant and long-lasting blooms and ornamental foliage, and require very little care to grow. There are wide range of plants in the family having wide range of sizes from tiny miniatures to giants, and in various forms such as un-pineapple-type Spanish moss (*Tillandsia usneioides*) which is neither Spanish nor a moss, some resemble aloes or yuccas while still others look green leafy-grasses. They are available in astonishing array of colours and textures, and even when

their showy flower displays are discounted, bromeliads are beautiful foliage plants with strappy leaves in yellow, green, orange, red or purple colours with banded, striped, spotted or other combinations. In fact, they are attractive foliage plants and their blooms are a bonus rather than an essential. Some blooms, notably *Billbergia*, last only a few days, while others stay attractive for many months. Most sorts are tropical but stand a wider range of temperatures and are generally much more tolerant of fluctuating cultural conditions. Indoors, these like cooler climates though outdoors where temperatures are near freezing, these require protection.

Bromeliads entered recorded history some 500 years ago when Columbus introduced the pineapple (*Ananas comosus*) to Spain upon return from his second voyage to the New World in 1493. On that voyage he found it being cultivated by the Carib-Indians in the West Indies. Within 50 years, this tropical fruit was being cultivated in India and other Old World countries. Afterwards, in 1776 *Guzmania lingulata,* in 1828 *Aechmea fasciata*, and in 1840 *Vriesea splendens* were introduced into Europe. In the last hundred years, bromeliads became fully established as ornamental plants. Still the new species are recorded and the breeders are evolving new hybrids.

Although the pineapple is the only member of the family cultivated for food, several species including Caroa (*Neoglaziovia variegata*) are cultivated as a source of fibre. Humans have been using bromeliads for thousands of years where Incas, Aztecs, Mayas and others used them for food, protection, fibre and ceremony, just as these are being still used today. Spanish moss (*Tillandsia usneoides*) contains a tough, wiry core which was once used as a material for stuffing upholstry. Pineapple stems are a source of the protein-digesting enzyme **bromelain** used as a meat tenderizer. Because fresh pineapple also contains bromelain, it cannot be used in gelatin moulds since the enzyme breaks down the congealing proteins. Bromelain, a group of enzymes belonging to the cystein category of proteases, which breaks down or helps in digesting the proteins, was first observed in pineapple juice. In addition to the pineapple, proteases are reported for two species of *Bromelia* (*B. pinguin* and *B. hieronyni*). The Pineapple Research Institute of Hawaii, stated that stems, leaves and fruits of all pineapple varieties contain bromelain and suggested that all species of Bromeliaceae probably contained similar proteases. In pineapple, the concentration of bromelain is reported to increase as the plant matures. The bromelain obtained from fruits has different characteristics than that obtained from stems (the commercial source). Three or perhaps four distinct proteases are present in stem extracts while in fruits only two. Also certain species of bromeliads remove air pollutants such as xylene.

Botany

Bromeliaceae is among the basal families within the Poales and is unique because it is the only family within the order that has septal nectarines and inferior ovaries (Judd, 2007) that characterize the subfamily Bromelioideae (Sajo, 2004). The family is diverse enough to include those having developed certain special appendages from leaf structures which are called **trichomes** in the form of scurfy, stalked and peltate scales or hairs, that may be found sometimes even on other plant surfaces (Anon., 2009), to capture water in cloud-forests and to aid in reflecting sunlight in desert environments as in case of grey-leaved epiphytic *Tillandsia* species, some others have developed tightly overlapped leaves at the base to make a tank to capture rain water and nutrients in the absence of a well-developed root system while a large number of species have developed succulence nature for desert-dwelling (Katharina *et al.*, 2009). All bromeliads are composed of alternate, a spirally arranged, parallel-veined, usually lorate or strap-shaped and troughlike leaves with a sheathing base which sometimes make a rosette and commonly have spiny margins (Anon., 2009). The number of degrees between successive leaves varies from species to species with a few having a 180° separation between leaves. This causes the plant to grow in a flattened configuration with its leaves lined up in a single plane. The bases of the leaves in the rosette may overlap tightly to form a water reservoir and such plants are called 'tank bromeliads'. Certain such bromeliads are shelter for a wide variety of organisms taking advantage of the pools of water trapped by bromeliads. A study of 209 plants from the Equadorian lowlands identified 11,219 animals, representing more than 300 distinct species, many of which are found only on bromeliads such as ostracod crustacean species, about 2.5 cm long small salamanders, tree frogs and various other animals including insects, scorpions and snakes. Jamaican bromeliads harbour a 2 cm across reddish-brown crab (*Metopaulias depressus*), which has evolved social behaviour to protect its youngs from predation by the larvae of a damselfly (*Diceratobasis macrogaster*) living in the bromeliad. Some bromeliads even form homes for other species of bromeliads. These central cups also collect leaf litters and frog & insect wastes as their nutrition. The more ancestral terrestrial bromeliads do not have water storage capability and rely primarily on their roots for water and nutrient absorption. Tank bromeliads rely less heavily on their roots for nourishment and are more often found as epiphytes. The roots of epiphytic species harden off after growing and become quite strong as wire to hold plants securely to their hosts and except this support, these epiphytes do not take any sustenance from them, though in Spanish speaking countries these are called as *parasitos* meaning 'parasites'. In some species, the bases of the leaves form small chambers as they overlap and these protected spaces are often home to ants whose wastes may provide the bromeliad with extra fertilizer. In species found in desert regions where the air is hot and dry and the sun beats down relentlessly, the tiny trichomes (scales) also help the plant to reduce water loss and shield the plants from the solar radiation. These plants are so covered with scales that they appear silvery-white and feel fuzzy. On many species, especially in more humid areas, the scales are smaller and less noticeable. Sometimes the scales can form patterns and banding on the leaves that add further to plant beauty. From the centre of the rosettes, the inflorescence appears with few exceptions. The floral scape is normally long with erect or pendent and solitary to many flowers appearing quite away from the plant or on short scapes in rosette form. Conspicuous colourful scape bracts in the form of leaf-like appendages are quite attractive and entice pollinators. Most cultivated bromeliads bloom only once in their lifetime. Bromeliads mature and bloom at different ages, depending on the species and growing conditions. Their blooms are sometimes insignificant but are surrounded by showy bracts that may mostly last longer, sometimes even for months, which are liked by people. Blooms of *Billbergia* last but a week or so while many others last for many months. Flowers appear when plants stop producing the leaves and then the leaves are not formed afterwards though the plants produce offsets called 'pups' near the base of the plant inside the sheath of a leaf, deriving their nourishment from the mother plant until are fully grown and established to set out their own roots to grow independently. The mother plant may survive a generation or two. Since new leaves appear from the centre so after emergence of the inflorescence the new leaf growth is blocked though there are species (*Dyckia* and *Hechtia*) that produce lateral inflorescences than from the centre of the plant so such plants continue producing leaves from the centre while producing pups freely. A few uncommon species such as *Deuterocohnia* reblooms on an existing inflorescence, and so some can continue blooming even up to six years on one of these perennial inflorescences. Pups are sometimes produced either on long stolons or atop the inflorescence as in pineapple where green leafy top may be removed and planted to start a new plant.

Variegation in Bromeliads

Over 60 per cent of cultivated bromeliads are variegated having bands, dots, lines and streaks with different pigmentations though in the wild this is a rare phenomenon as such plants are less hardy and in nature are unable to compete with their normal counterparts hence are gradually lost. Variegation is rare in most of the members of the subfamilies Pitcairnoideae and Tillandsioideae though it has been observed occurring

in the genera *Alcantarea*, *Guzmania* and *Vriesea* and in a few species of *Tillandsia*. In the subfamily Bromelioideae, variegation is quite common, especially in the genera *Aechmea, Ananas, Billbergia, Cryptanthus, Neoregelia* and *Nidularium*. Anthocyanin pigments are present in many bromeliads in the epidermal cells hiding both, chlorophyll-pigmented and albino tissues. In *Aechmea orlandiana* var. 'Ensign', anthocyanin produces a quite spectacular red or rose patterning in the albino tissue. Reddish-brown stripes and bands appear in several hybrids such as *Aechmea* 'Red Ribbion' and *Neoregelia* 'Amazing Grace', *vis-a-vis* sometimes variegation in the inflorescenses and floral bracts are also found as with some *Guzmania* hybrids.

Variegated plants usually have slower growth with poor root formation as in case of *Vriesea* and *Guzmania*. In such cases, the newly formed inflorescences should be removed so that energy is diverted towards formation of new lateral pups with faster growth. The variegated pups should be allowed to grow half the size of mother plant before removal.

Of the many forms of variegation occurring in bromeliads, perhaps the most familiar is the white- and green-variegation of *Ananas comosus* var. *variegatus* (pineapple). This arises from a fusion, or chimera of two very different cell types, with white cells arising from a genetic mutation in the growing point of the leaf. Green sectors contain abundant chlorophyll, but white or pale- yellow sectors contain much less pigment. The chimeral type of variegation though is rare in nature, and results in slower growth compared to non-variegated varieties, as only the greener parts of the leaf can photosynthesize sufficiently to provide the sugars and energy for growth and the white sectors consume these resources. Other forms of variegation include the more functional 'developmental' variegation, or the appearence of colouration over time. Many *Guzmania* and *Neoregelia* have developmental variegation with younger leaves becoming red immediately before flowering. Variegation may also form bands of leaf hairs traversing the leaf or from variation in the structure of the leaf hairs over the leaf surface, although the function if any of these bands is not known. However, the 'discolour variegation', in which the leaf underneath is red and the upper surface is green, directly benefits photosynthesis and plant growth. Red underneath to the leaf is thought to act as 'red mirrors' reflecting light back up into the leaf and increasing photosynthesis and growth in shaded conditions. *Guzmania musaica* var. *discolor* (variegated) grows terrestrially on the shaded forest floor in the cloud forest at Cerro Jefe (central Panama), along with *Guzmania musaica* var, *concolor* (unvariegated) as epiphyte, though on the lower hill slopes these forms are replaced by a variety with a different form of variegation. This pattern is made up of darker and lighter green portions, often also mirrored by the red anthocyanin pigments on the underside of the leaf. As the principal function of leaves is to capture light and CO_2 gas, producing energy and structural materials for the plant, any variation in leaf colouration will affect the absorbance of light and the photosynthetic process.

Classification, Genera, Species and Varieties

DNA analysis to examine the relationship between the genera of Bromeliaceae in the University of Wisconsin revealed that Tillandsioideae and Bromelioideae subfamilies are monophyletic, *i.e.* derived from a single ancestral form though Pitcairnioideae was found to be paraphyletic and required five new groups to be split from it. This new reorganization of the genera within the most ancestral subfamily Pitcairnioideae hopefully now comes closer to representing their true evolutionary (phylogenetic) relationships. As part of the DNA sequencing, it was determined that the monospecific genus *Ayensua* which contains only *Ayensua uaipanensis*, actually belongs to the genus *Brocchinia*, though *Brocchinia serrata* was found to have no relation to the other members of its genus so was separated into a newly created genus *Sequencia*. So while *Ayensua* was lost in the new organization the number of genera remains at 58 with the addition of *Sequencia* (the name a tribute to its origin as a result of DNA sequencing). The subfamilywise genera are listed below (Luther, 2000):

Pitcairnioideae

This subfamily contains the most ancestral bromeliads where many resemble belonging to the grass family from which Bromeliaceae evolved. Almost all are terrestrial and rely on an extrnsive root system for their moisture and nutrients. The members are generally spiny-leaved and have dry capsules with small wingless seeds. It consists of 16 genera and 1,030 species, with six further divisions such as (i) **Brocchinioideae** (*Brocchinia*); (ii) **Lindmanioideae** (*Connellia, Lindmania*); (iii) **Hechtioideae** (*Hechtia*); (iv) **Navioideae** (*Brewcaria, Cottendorfia, Navia, Sequentia* (formerly *Ayensua*), *Steyerbromelia*); (v) **Pitcairnioideae** (*Deuterocohnia, Dyckia, Encholirium, Fosterella, Pepinia, Pitcairnia*); and (vi) **Pitcairnioideae** (*Puya*).

Bromelioideae

The members of this subfamily are highly diverse containing greatest number of genera (33) though least number of species (861). Here most of the species are epiphytic so obtaining most of their water and nutrients through their leaves rather than roots and are characterized with rosette-like form, many forming a water holding tank, the leaves edged with spines of varying sizes and the fruits are berry-like containing wet seeds which are generally distributed by birds and animals consuming the fruits. Its genera are *Acanthostachys, Aechmea, Ananas, Androlepis, Araeococcus, Billbergia, Bromelia, Canistropsis, Canistrum, Cryptanthus, Deinacanthon, Disteganthus, Edmundoa, Eduandrea, Fascicularia, Fernseea, Greigia, Hohenbergia, Hohenbergiopsis, Lymania, Neoglaziovia, Neoregelia, Nidularium, Ochagavia, Orthophytum, Portea, Pseudaechmea, Pseudananas, Quesnelia, Ronnbergia, Streptocalyx, Ursulaea* and *Wittrockia.*

Tillandsioideae

This subfamily contains only nine genera though a large number of species (1,277), mostly epiphytes with spineless leaves, the dry capsules containing feathery-plumose seeds, where dispersal is by wind and the plumes help the seeds adhering to a suitable epiphytic surface for germination. The members of this family with grey leaves have special adaptations for survival in very dry (xerophytic) conditions (Foster, 1951). This includes nine genera such as *Alcantarea, Catopsis, Glomeropitcairnia, Guzmania, Mezobromelia, Racinaea, Tillandsia, Vriesea* and *Werauhia.*

Thus in all, these become 58 genera, *viz. Acanthostachys, Aechmea, Alcantarea, Ananas, Androlepis, Araeococcus, Billbergia, Brewcaria, Brocchinia, Bromelia, Canistropsis, Canistrum, Catopsis, Connellia, Cottendorfia, Cryptanthus, Deinacanthon, Deuterocohnia, Disteganthus, Dyckia, Edmundoa, Eduandrea, Encholirium, Fascicularia, Fernseea, Fosterella, Glomeropitcairnea, Greigia, Guzmania, Hechtia, Hohenbergia, Hohenbergiopsis, Lindmania, Lymania, Mezobromelia, Navia, Neoglaziovia, Neoregelia, Nidularium, Ochagavia, Orthophytum, Pepinia, Pitcairnea, Portea, Pseudaechmea, Pseudananas, Puya, Quesnelia, Racinaea, Ronnbergia, Sequentia, Streptocalyx, Steyerbromelia, Tillandsia, Ursulaea, Vriesea, Werauhia* and *Wittrockia.*

Hybrid Genera

- ×*Aechopsis* (*Aechmea* × *Canistropsis*) (Butcher, D. 1999. *J. Brom. Soc.*, **49**: 14).
- ×*Anagelia* (*Ananas* × *Neoregelia*) (Smith, E.L. 1983. *J. Brom. Soc.*, **33**: 72).
- ×*Anamea* (*Ananas* × *Aechmea*) (Anon., 1979. *Int. Checklist Brom. Hyb.*, p. 22).
- ×*Anananas* (*Ananas* × *Pseudananas*) (Beadle, D.A. 1991. *Prelim. List. Cult. Grex Brom.*, p. 20).
- ×*Androlaechmea* (*Androlepis* × *Aechmea*) (Anon., 1951. *Brom. Soc. Bull.*, **1**: 24).
- ×*Billmea* (*Billbergia* × *Aechmea*) (Williams, K. 1974. *J. Brom. Soc.*, **24**:26).
- ×*Billnelia* (*Billbergia* × *Quesnelia*) (Hawkes, A.D. 1959. *Brom. Pap.*, **1**(6): 52).
- ×*Canegelia* (*Canistrum* × *Neoregelia*) (Butcher, D. 1991. *Hybridists Handbook Ed.*, **3**: 2).
- ×*Canmea* (*Canistrum* × *Aechmea*) (Foster & Foster, 1973. *J. Brom. Soc.*, **23**: 175).
- ×*Cryptananas* (*Cryptanthus* × *Ananas*) (Beadle, D.A. 1991. *Prelim. List. Cult. Grex Brom.*, p. 36).
- ×*Cryptbergia* (*Cryptanthus* × *Billbergia*) (Anon., 1952. *Brom. Soc. Bull.*, **2**: 72).
- ×*Cryptmea* (*Cryptanthus* × *Aechmea*) (Smith, E.L. 1983. *J. Brom. Soc.*, **33**: 72).
- ×*Deuterocairnia* (*Deuterocohnia* × *Pitcairnia*) (Butcher, D. 2002. *J. Brom. Soc.*, **52**: 51).
- ×*Dyckcohnia* (*Dyckia* × *Deuterocohnia*) (Anderson, G.H. ex Grant, 1998. *Selbyana*, **19**:116).
- ×*Dycktia* (*Dyckia* × *Hechtia*) (Beadle, D.A. 1991. *Prelim. List. Cult. Grex Brom.*, p. 82).
- ×*Guzlandsia* (*Guzmania* × *Tillandsia*) (Anon., 1979. *Int. Checklist Brom. Hyb.*, p. 35).
- ×*Guzvriesea* (*Guzmania* × *Vriesea*) (Hawkes, A.D. 1959. *Brom. Pap.*, **1**(5): 45).
- ×*Hechcohnia* (*Hechtia* × *Deuterocohnia*) (Anderson, G.H. ex Grant, 1998. *Selbyana*, **19**:117).
- ×*Hohenelia* (*Hohenbergia* × *Quesnelia*) (Butcher, D. 2002. *J. Brom. Soc.*, **52**: 51).
- ×*Hohenmea* (*Hohenbergia* × *Aechmea*) (Sousa, Silva, & Sousa, 2003. *J. Brom. Soc.*, **53**: 71).
- ×*Hohentea* (*Hohenbergia* × *Portea*) (Beadle, D.A. 1991. *Prelim. List. Cult. Grex Brom.*, p. 89).
- ×*Neobergia* (*Neoregelia* × *Billbergia*) (Smith, E.L. 1983. *J. Brom. Soc.*, **33**:73).
- ×*Neobergiopsis* (*Neoregelia* × *Hohenbergiopsis*) (Butcher, D. 2001. *J. Brom. Soc.*, **51**: 7).
- ×*Neomea* (*Neoregelia* × *Aechmea*) (Foster, M.B. 1958. *Brom. Soc. Bull.*, **8**: 75).
- ×*Neophytum* (*Neoregelia* × *Orthophytum*) (Foster, M.B. 1958. *Brom. Soc. Bull.*, **8**: 73).
- ×*Neorockia* (*Neoregelia* × *Wittrockia*) (Butcher, D. 1999. *J. Brom. Soc.*, **49**: 14).

☆ ×*Neostropsis* (*Neoregelia* × *Canistropsis*) (Butcher, D. 1999. *J. Brom. Soc.*, **49**: 14).

☆ ×*Neotanthus* (*Neoregelia* × *Cryptanthus*) (Anon., 1979. *Int. Checklist Brom. Hyb.*, p. 47).

☆ ×*Nidbergia* (*Nidularium* × *Billbergia*) (Butcher, D. 1982. *Checklist Brom. Hybrids Australia*, p. 28).

☆ ×*Nidulistrum* (*Nidularium* × *Canistrum*) (Hawkes, A.D. 1963. *Brom. Pap.*, **3**(9): 85).

☆ ×*Nidumea* (*Nidularium* × *Aechmea*) (Smith, L.B. 1968. *Brom.Soc Bull.*, **18**: 93).

☆ ×*Niduregelia* (*Nidularium* × *Neoregelia*) (Hawkes, A.D. 1963. *Brom. Pap.*, **3**(9): 85).

☆ ×*Ortholarium* (*Orthophytum* × *Nidularium*) (Foster & Foster, 1973. *J. Brom. Soc.*, **23**: 175).

☆ ×*Orthomea* (*Orthophytum* × *Aechmea*) (Smith, E.L. 1983. *J. Brom. Soc.*, **33**:75).

☆ ×*Orthotanthus* (*Orthophytum* × *Cryptanthus*) (Anon., 1974. *J. Brom Soc.*, **24**: 26).

☆ ×*Portemea* (*Portea* × *Aechmea*) (Ariza-Julia, 1978. *J. Brom. Soc.*, **28**: 21).

☆ ×*Pseudanamea* (*Pseudananas* × *Aechmea*) (Baensch & Baensch, 1994. *Bluh. Brom.*, p. 249).

☆ ×*Puckia* (*Dyckia* × *Puya*) (Butcher, D., 2002. *J. Brom. Soc.*, **52**: 52).

☆ ×*Pucohnia* (*Puya* × *Deuterocohnia*) (Anderson, G.H. ex D.A. Beadle, 1991. *Prelim. List. Cult. Grex Brom.*, p. 200).

☆ ×*Quesistrum* (*Quesnelia* × *Canistrum*) (Graf, H. ex D.A. Beadle, 1998. *Brom. Cult. Reg.*, p. 350).

☆ ×*Quesmea* (*Quesnelia* × *Aechmea*) (Knobloch, 1972. *J. Brom. Soc.*, **22**: 58).

☆ ×*Quesregelia* (*Quesnelia* × *Neoregelia*) (Carrone, J. 1983. *J. Brom.Soc.*, **33**: 207).

☆ ×*Streptolarium* (*Streptocalyx* × *Nidularium*) (this is obsolete as *Streptocalyx* is now *Aechmea*).

☆ ×*Ursumea* (*Ursulaea* × *Aechmea*) (Butcher, D. 2005. *J Brom. Soc.*, **55**(2): 56).

☆ ×*Vriecantarea* (*Vriesea* × *Alcantarea*) (Grant, J.R. 1996. *Phytologia*, **79**: 256).

☆ ×*Vrierauhia* (*Vriesea* × *Werauhia*) (Shiigi, D. ex D.A. Beadle, 1998. *Brom. Cult. Reg.*, p. 363).

☆ ×*Vrieslandsia* (*Vriesea* × *Tillandsia*) (Chevalier, C. 1931. *Bull. Soc. Natl. Hort. France* **V, 4**: 213-215).

The ones now in common cultivation are *Acanthostachys, Aechmea, Ananas, Androlepis, Billbergia, Brocchinia, Bromelia, Canistrum, Catopsis, Cottendorfia, Cryptanthus, Deuterocohnia, Dyckia, Encholirium, Fascicularia, Fosterella, Greigia, Guzmania, Hechtia, Hohenbergia, Navia, Neoglaziovia, Neoregelia, Nidularium, Ochagavia, Orthophytum, Pitcairnea, Portea, Puya, Quesnelia, Streptocalyx, Tillandsia, Vriesea* and *Wittrockia* which are compiled (Bailey, 1960; Everard and Morley, 1970; Hays and Beckett, 1971; Hellyer, 1983; Beckett, 1983, 1987; Huxley *et al.*, 1992; Brickell, 1994) and being described here below.

Acanthostachys

This genus from Brazil, Paraguay and N. Argentina consists of only one species ***A. strobilacea*** closely allied to *Ananas* though is epiphytic, and derives from the Greek *akantha* which means 'a thorn' and *stachys* for 'a spike' as the flower spikes have prickle-tipped bracts. Its leaves are quite slender, 30-90 cm long with tapered point, strongly chanelled, being serrated and teeth spiny-tipped, arching outwards and form the loose and open rosettes. Scape is little shorter than leaves and is topped by 1-2 long leaf-like bracts and up to 5 cm long dense and cone-like spike, which bears 3-petalled, up to 2.5 cm long yellow flowers coming up from the axil of a red or orange bract. The berries resemble a tiny pineapple, mature to yellow and are sweet and edible. It is most suitable for hanging basket and is propagated through division or through cuttings of mature but non-flowered rosette.

Aechmea

This derives from the Greek *aichme* 'a point' as the sepals have strong points or spines in the species *A. paniculata*, the first species described in the genus. It is native to Brazil (South America) and comprises some 150 species, mostly epiphytic. An *Aechmea* bears largest leaves in the family, some 2.7 m long. Most of these make attractive pot plants. Its sword-shaped, erect to arching, toothed or spiny leaves form a close rosette to hold water in the centre. Three-petalled flowers appear usually on long spike-like panicles which normally bear coloured bracts, and develop into brightly coloured berries and many offsets at the base before plant dies. These offsets are means of further propagation. ***A. bracteata*** from Mexico to Cental America grows more than 1.5 m tall, bearing grey-scaly, pale-green and up to 90 cm long leaves in dense rosettes. Scape appears during autumn some 60 cm long in pyramidal panicles bearing pale-yellow flowers with red bracts and afterwards blue berries. ***A. bromeliifolia*** from Honduras to Argentina grows up to 90 cm high, bearing dark grey-green *cum* waxy-white-scaly, up to 90 cm long, strap-shaped, tapering and spine- tipped leaves. Interior of the flowers yellowish-green, and bracts greenish-white and the flowers appear in dense clusters. ***A. calyculata*** from Brazil and Argentina grows up to 60 cm, bearing green leaves with grey scales, linear, up to 50 cm long with rounded ends which terminate into a prick. Spike is dense with yellow flowers and red bracts. ***A. caudata*** from Brazil grows up to 1.0 m tall with more than 60 cm long linear leaves which are dark grey with greenish scales and rounded ends, and terminating into a short spine. Spike is short up to 5 cm long with yellow flowers and brownish-pink bracts which turn yellow with age. The leaves of *A. c. variegata* (*Billbergia forgetii*) is prominently banded pale-yellow. ***A. chantinii*** (*Billbergia chantinii*) from Amazon region of Brazil to Peru grows up to 1.0 m in height with funnel-shaped rosette, bearing green, 40 cm long leaves cross-banded with pinkish to grey-white and terminating into short spine. Flower spike is loose and oval and some 10-15 cm long and flowers are yellow with red bracts. The plant spreads by stolons. ***A. coelestis*** from Brazil grows up to 70 cm tall. Leaves are grey-scaly with rounded tips though terminate into a short spine. Sky-blue flowers appear during winter on a 40 cm erect stem to a top 20 cm length in a dense pyramidal cluster, and the fruits are green with pink tints. ***A. comata*** (*A. lindenii*) from Brazil grows up to 1.0 m tall with rosetted, almost erect and some 60 cm long leaves. It is usually a winter-blooming species with scape length of 60 cm on which dense spike of 15-20 cm originates, bearing yellowish-red sepals and yellow petals. ***A. cylindrata*** from Brazil grows up to 60 cm in height, bearing 30-40 cm long leaves which are dark green above and paler below, spiny-serrated and the tips rounded and terminating with a short spine. Its scape some 40 cm tall with 20 cm long but dense rachis appears during winter with pink sepals and pale-blue petals. ***A. dealbata*** from Brazil is closely related to *A. fasciata* and grows up to 80 cm tall with up to 70 cm long leaves and 60 cm long scape. Leaves form an erect cylindrical rosette, are banded silvery-grey, lanceolate and with pointed tips. Flowers are violet to red and bracts red with grey to white felts where flowers and bracts make dense cone-like spikes. ***A. distichantha*** from Brazil to Argentina grows some 70 cm tall, bearing rigid and arching leaves up to 1.5 m long and scape some 60 cm long. Leaves are grey-scaly, somewhat channelled, margins brown and prickly and tips pointed. Panicle is rigid and appears on up to 60 cm long scape with purple to blue flowers from the axils of pink bracts which have white felts. *A. d. albiflora* bears white flowers and *A. d. glaziovii* (*A. glaziovii*) bears shorter spikes with fewer flowers. ***A. fasciata*** (*Billbergia rhodocyanea*) which was in 1828 introduced to Europe is from Brazil, grows some 60 cm tall and is liked for its beautiful purple-blue flowers and pink bracts. Green or grey-green leaves form an open vase-shaped rosette. Leaves are wide having thickly placed grey-scales with silver-white banding and black spines. Flowers are tubular, blue at first but later turning rose and make a dense pyramidal panicle some 15 cm long. The bracts are pale-pink and spined. *A. f. variegata* bears longitudinal stripes of creamy-yellow colour. *In vitro* propagated *A. fasciata* shoot tips gave 20 per cent variant somaclones having several smaller, grass-like, non-thorny, without anthocyanin pigmentation and non-waxy leaves on the elongated stem. Use of chlormequat and paclobutrazol on normal and variant plants retarded their growth and paclobutrazol induced waxy leaf formation, reduced leaf number and promoted a massive development of adventitious buds on variant somaclones (Ziv *et al.*, 1986). ***A. fulgens*** from Brazil is liked for its violet-purple flowers which turn red. It grows up to 50 cm tall with 45 cm long leaves which are broad, minutely spined and grey-green with cross-banding, *vis-à-vis* form an open rosette. Scape is some 20 cm long and carries numerous stalked, branched and waxen blue flowers. It bears vermilion berries which last long. The lower side of leaves is red in case of *A. f. discolor* (*A. discolor*). ***A. gamosepala*** from Brazil and Argentina is also a flowering plant growing some 50 cm in height with up to 35 cm long leaves. The leaves are green with grey-scaly back,and strap-shaped with spined tip. Rachis is loose with pale-blue flowers and brownish-red bracts. ***A. luddemanniana*** from Guatemala and Honduras is a flowering plant up to 70 cm tall, bearing strap-shaped some 50 cm long leaves and only up to 4.0 cm long spikes bearing blue flowers which turn red. Leaves are end-rounded, and green to reddish-brown bearing grey scales. The fruits are white and purple. ***A. mariac-reginae*** from Costa Rica is a summer-flowering bromeliad with some 80 cm long leaves, more than 60 cm long stalk and up to 20 cm long cone-like spike. Leaves are green, have densely toothed margins and rounded tip bearing short spine. White flowers with purple tips and large, lanceolate and bright red bracts in the form of inverted shuttlecock are densely arranged on the spike, especially upward. ***A. mexicana*** from Mexico to Ecuador is a winter-flowering species, bearing up to 90 cm long, pale-green and darker blotched leaves. The leaves are pointed or round-tipped and spiny serrated. Flowers are red but sometimes lilac in a pyramid-like stiff panicle up to 60 cm long but on a short stalk and the fruits are white. ***A. miniata*** from Brazil bears some 45 cm long leaves and up to 10 cm long panicle with red flowers. Leaf points are stiff and margins prickly- toothed. Panicle is broad and the fruits are red. *A. m. discolor* bears leaves flushed red-purple beneath. ***A. nudicaulis*** from Mexico and from West Indies to Brazil is a flowering bromeliad up to 70 cm tall and leaves up to 90 cm long. Leaves are green with cross-banding and grey beneath. Floral spike is loose, flowers are yellow and the bracts are small and bright red. *A. n. cuspidata* bears yellow flowers and bracts. ***A. orlandiana*** from

Brazil is a flowering bromeliad with 50 cm long stalk, bearing yellow to white flowers and long red bracts on dense and some 10-cm long spike. Some 35 cm long, linear and light green leaves with mottling and dark purplish cross-banding end into a spine-tipped structure. ***A. penduliflora*** (*A. paniculata, Billbergia paniculata*) from Costa Rica to Peru and Brazil bears leaves up to 60 cm long which are wider towards the tip and densely spiny- toothed though upper part may be smooth and grey-scaly. Each spiral panicle some 6 cm long comes up on a 30-45 cm long stalk which is partly sheathed by red bracts. Flowers are orange and the fruits are yellow to red. ***A. pineliana*** from Brazil to Peru bears up to 60 cm long, silvery grey-scaly leaf surfaces ornamented often with cross-banding and the leaves are semi-erect where end is rounded and terminates into a spine. Teeth on leaf margins are reddish-brown, the flowers are yellow arranged in a dense cone-like spike some 7 cm long on the top of some 45 cm long stalk. The spike is enclosed by waxy-red, long and overlapping bracts. ***A. racinae*** from Brazil is a flowering plant growing up to 50 cm tall and bears some 30 cm long bright green leaves and some 45 cm long arching stems so is most suitable for hanging basket. Leaf margins are prickly-toothed or smooth and the racemes are pendent-tipped and emerge from among or beneath the leaves. Sepals are red and some 12 cm long. In *A. r. erecta* the stem is short and erect. ***A. ramosa*** from Beazil is a flowering bromeliad some 1.0 m tall, bearing green with greyish-or pinkish-scaly, strap-shaped leaves which may grow more than 60 cm long and yellow flowers. The fruits appear on red or orange stalks. ***A. recurvata*** from Brazil has closely overlapping leaves as at the base these are quite expanded so form a rosette with a bulbous base. Leaves are narrow (1 cm wide), tapering, recurved, margins spiny and becoming red when exposed to sun. Flowers have red sepals and lilac-red to red petals which appear densely on the red-bracted spike some 6 cm long. *A. r. ortgiesii* forms quite short or no floral stalk though *A. r. benrathii* has purple-spotted leaf bases with a very few or no marginal teeth. ***A. tillandsioides*** from Mexico to Colombia bears up to 90 cm long and 2-5 cm wide leaves having brown spiny teeth on the margins. Panicle is stiff, 10-30 cm long, open and with yellow flowers on a 30 cm long stalk. ***A. veitchii*** from Costa Rica, Panama and Colombia grows to 1.0 m tall, bearing up to 90 cm long and tongue-shaped leaves where upper surface is pale-green and grey-scaly below. Sepals are pink-tipped, flowers white and bracts are red. ***A. weilbachii*** from Brazil grows up to 70 cm tall, bearing more than 45 cm long leaves which are dark green, shining and tipped spiny. Spike is loose, some 15 cm long and bears lilac-purple flowers with red bracts. The genus has produced hybrids, noteworthy are 'Foster's Favorite' (*A. victoriana* × *A. racinae*) and its mutant 'Foster's Favorite'.

Ananas

Ananas is the Tupi (South American) name for these plants. In fact, Guarani tribe of Brazil named it so as its derivation is from South American Indian languages (Smith, 1952). It is native of the tropics of Central and South America, with particular reference to Brazil. European interest began in bromeliads when Spanish conquistadors returned with pineapple, which became so popular as an exotic food that the image of the pineapple was adapted into European art and sculpture. The genus comprises some five perennial, evergreen and terrestrial species. Their leaves are sword-shaped, spiny-edged and clump-forming with erect rosettes. The flowers are 3-petalled, red or blue which appear in a spike, followed by syncarpous fruits being formed by fusion of fleshy fruits. They make a nice pot plant. Their propagation is through suckers or by their leafy shoots which top the fruits. These shoots can be sliced off and planted at a temperature of 21-26 °C in a propagating chamber for rooting. Slips or shoots forming at the base of the plants may be detached after winter is over. ***A. bracteata*** from Brazil and Paraguay bears dark green leaves some 1.2 m long with well-spaced marginal spines. Bracts on the stalks and beneath the florets are pink-red and flowers are lavender. Though it is not the common pineapple but produces brownish-red edible fruits. Whole plant is ornamental. *A. b.* 'Striatus' ('Tricolor') bears bronze-green centre, yellow margins and red spines. The common and true pineapple is ***A. comosus*** (*A. sativus*) from Brazil and Colombia which grows up to 90 cm tall, bearing grey-green, strap-shaped, channelled, pointed and up to 90 cm long arching leaves with closely set small marginal spines. The leaves form a dense rosette 0.9-1.5 m across. Flower stem is short and stout, the bracts inconspicuous but pink and the flowerhead some 7.5 cm long and ovoid. Flowers are 3-petalled and purple-blue with green bracts. The fruits are fleshy, succulent and coalesce to form the typical pineapple. In Britain it was introduced at Richmond (Surrey) where it produced edible fruits first in 1715. Its variegated forms are cultivated for ornamentation and the improved ones for fruits. *A. c.* 'Porteanus' is green with central yellow striping, and *A. c.* 'Variegatus' has broad creamy margins on the leaves, and whole centre of the plant turns bright rose at flowering.

Androlepis

In Greek, *aner* is for 'a man', and *lepis* for 'scale', referring to the two apical projections on its filaments. It is native to Guatemala to Costa Rica with only one species, ***A. skinneri***. It is stolonferous perennial growing up to 2.0 m when in flower. Leaves are ligulate, acute or acuminate, numbering 20 per plant, 1.0-1.8 m long, upper glabrous and below scurfy, full exposure to sun turns these to deep pink, margins closely dentate with 3 mm long spines, and leaf-sheaths are large, brown-scaly and suborbicular. Scape is strong, erect, white-mealy with overlapping straw-yellow papery bracts which are lanceolate-elliptic and pungent, the compound inflorescence is narrowly pyramidal to cylindrical, some 1.0 m long, inflorescence bracts subulate, flowers spicate, unisexual, sessile and under minute bracts, female flowers subglobose and 1-5 per spike, male flowers more densely arranged, *i.e.* 4-5 per spike, sepals are green, thick, free, broadly triangular, convex, winged, mucronate and 5-6 mm long, petals are pale-yellow, a little fused, elliptic and 1.0 cm long, filaments are fused and short, and anthers are with 2 apical appendages.

Billbergia

Billbergia was named for J.G. Billberg (1772-1844), a Swedish botanist. Its native haunts are in tropical and sub-tropical America (Argentina, Bolivia, Brazil, Peru, Uruguay, *etc.*). It is a genus of 50 evergreen perennial species and countless hybrids, most being epiphytic and a few terrestrial. The plants are easy to grow indoors or in the greenhouse. Its leaves are wide, strap-shaped, leathery and usually in a rosette which is tubular and holds water at the base. Rosettes are generally in clusters. Flowers are 3-petalled, often colourful, tubular and are borne within the showy bracts. Stalks may be erect but is mostly arching. These are propagated through division or by offsets. ***B. amoena*** from Brazil bears green cross-banded-silvery, strap-shaped, spiny and 20-60 cm long leaves. Racemes are 10-15 cm long on arching stalks, bracts rosy and flowers green. It has many forms with differing cross-banding. *B. a. penduliflora* bears highly pendent stalks and orange bracts; *B. a. rubra* bears pinkish-red leaves striped and spotted white; *B. a. viridis* bears white spots and striping on the leaves; *B.* × 'Albertii' (*B. distacya* × *B. nutans*) bears grey scaling on leaves, bracts rosy-red and flowers green and blue; *etc.* ***B. decora*** from Brazil, Peru and Bolivia produces short stems having dark green, grey-mottled, cross-banded, strap-shaped, pointed, up to 60 cm long leaves with margins minutely toothed and in erect and tubular rosettes. Racemes pendent with large carmine bracts and flowers green with rolling petals. ***B. distachya*** from Brazil is similar to *B. nutans* though differs by having grey-white scurfy and sometimes leaves are purple-tinted. *B. d. maculata* bears yellow-blotched leaves and *B. d. concolor* with green flowers. 'Albertii' (*B. distachya* × *B. nutans*) bears linear-oblong, some 30 cm long and dark green leaves in tubular rosettes with grey scaling, and the bracts are rosy-red while the flowers green and blue. 'Santa Barbara' (a hybrid involving probably *B. distachya* × *B. nutans* × *B. pyramidalis*) bears grey-green leaves striped with yellow and pink and make the narrow and longer rosettes. ***B. morelii*** from Brazil bears stiff, green, some 60 cm long and sword-shaped leaves with spiny margins. Floral stalk is about 30 cm long topped with 5 cm long and pendent spike with bright red bracts and green, blue and red flowers. ***B. nutans*** from Brazil, Uruguay, Paraguay and Argentina is clump-forming plant growing up to 45 cm tall, bearing dark green, some 40 cm long, erect, narrow and serrated leaves in tubular rosettes. Floral stalk is erect but arching at the end with 7.5-10.0 cm long

Aechmea fasciata
Photo: Blossom World Bromeliads

Billbergia nutans
Photo: Michael Andreas

Billbergia pyramidalis
Photo: Dorothy Berg

drooping cluster, flowers are 2.5-3.75 cm long, tubular and greenish-yellow but edged purple-blue, the calyx is pink and the bracts are pink or red. The petals reflex to show golden-yellow stamens. The species can tolerate as low as 2 °C temperature for a short period. 'Windii' (*B. decora* × *B. nutans*) grows 45 cm in height with stiff grey-green leaves covered with moisture-holding grey scales. The floral stalk is partly ensheathed in large pink bracts, further extending to a cluster of pendent and tubular flowers some 2.5-3.75 cm long where sepals are pink and the curled-petal margins are green-blue. ***B. pyramidalis*** from Brazil grows up to 40 cm tall forming vase-shaped rosette of broad, green, strap-shaped and finely toothed leaves which surround the narrow, erect and slightly taller panicle. Floral stalk is felted and ensheathed with pink bracts from which closely packed some 10-15 cm long globular flower cluster appears. Flowers are red with blue tips. *B. p.* 'Striata' bears cream-striped dark green leaves. *B.* × 'Bob Manda' is its quite compact hybrid with coppery green basic colour, sometimes with a purple tint and blotched-cream. ***B. saundersii*** (*B. rubro-cyanea*) from Brazil bears strap-shaped some 40 cm long leaves which form narrowly tubular rosettes. Leaf's upper surface is dark green and purplish to bronze beneath and both sides spotted white and sometimes even red. Floral stalk is arching bearing red bracts and pendent yellow-green flowers edged blue. 'Fantasia' (*B. saundersii* × *B. pyramidalis*) is though similar to *B. saundersii* but larger. Leaves are coppery-green but with bold blotches of cream and pink. Raceme is rather more compact than *B. saundersii* but with pink bracts and blue flowers. ***B. vittata*** from Brazil grows smaller than leaves bearing dark grey-green and scaly leaves at upper surface while prominently grey cross-banded beneath, and the leaves are more than 90 cm long, strap-shaped to lanceolate and forms an erect and broadly tubular rosette. Short stalks are topped by a pendent spike which bears blue, red and greenish-white flowers with petals rolling back and the red bracts are some 20 cm long. ***B. zebrina*** from Brazil has leaves about 90 cm long, strap-shaped, dark green with grey-white cross-banding and form tubular rosettes. Stems are some 40 cm long, have felted scales and bear large pink bracts with terminal spikes, each greenish-yellow flower some 8 cm long in bud and when open the petals coil as spring.

Brocchinia

Generic name *Brocchinia* is for G.B. Brocchi (1772-1826), an Italian naturalist. There are some 18 terrestrial or epiphytic perennial species native to South America (Brazil, Guyana, Colombia and Guyana Highlands of Venezuela). It is a rosette-forming herb growing up to 8.0 m tall, where basal sheaths are large, leaves linear or narrowly triangular, glabrous and pungent, scape much-branched and inflorescence is erect. Flowers are small and numerous, sepals green and free, white or green and free petals are larger than sepals, and fruit is a capsule. ***B. micrantha*** from E. Guyana and Venezuela bears woody and columnar stem. Leaves are 1.2 m long and the sheath is dark brown. Scape is strong, the inflorescence on scape is about 2.6 m tall and glabrous, bearing white flowers some 5 mm long. The floral bracts are spreading and acute, the sepals obtuse, petals emarginated, almost circular and little longer than sepals, and the capsules have 2 mm long seeds which have wings elongated to both the ends. ***B. paniculata*** from SE Colombia and Venezuela in scale-covered inflorescence, ovate floral bracts, 15 mm long flowers, and oblong sepals and petals. ***B. reducta*** from Tepuis of Guyana Highlands and Venezuela is terrestrial and is found in wetlands. Leaves are a few, bright yellow-green, lamina round-apiculate, cylindrical, clasping, venation arcolate, absorbing trichomes present, sheaths brown, and the leaves form a cylindrical tank emitting sweet odour to attract small animals as it is carnivorous, and the internal surface of the tank is laced with fine waxy powder so that insects are not able to make a grip. Inflorescence is tripinnate, 50-70 cm long, branches some 13 cm long, bracts 3.5 cm, flowers are short-pedicelled, 5 mm long and erect, sepals are glabrous, ovate-elliptic, apiculate and equalling the petals.

Bromelia

The genus was named so for Bromel, a Swedish botanist and is grown for the stiff form so most suitable as tropical hedge, and clustered flowers. This terrestrial, rosette-forming and evergreen genus comprises some two dozens of hothouse herbaceous species from tropical America. It can tolerate a minimum of 5-7 °C temperature and prefers full sun. Its leaves are stiff and pineapple-like. Flowers appear in heads or in panicles, calyx has 3 ovate-oblong sepals and the corolla is 3-parted. These are propagated through suckers. Important cultivated species are *B. balansae*, *B. binotii* and *B. pinguin*. ***B. balansae*** is an evergreen, clump-forming and basal-rosetted perennial growing some 1 m high with about 1.5 m spread, bearing mid- to grey-green, narrowly strap-shaped and arching leaves with large hooked spines. Panicles are club-shaped bearing tubular red- or violet-purple flowers with long bright red bracts. ***B. binotii*** from Brazil has spreading habit with lax panicle and sepals top-rounded. ***B. pinguin*** from West Indies grows 90-120 cm high, bearing bright green leaves which turn pink to red with age and are spiny-toothed. Panicle is densely packed with reddish and pubescent flowers with acute sepals though rachis is mealy. Its fruits are acidic and as large as the plums.

Canistrum

Canistrum in Greek is for 'a little basket', referring to the inflorescence in a basket of bracts. The genus comprises of some 10 epiphytic or terrestrial tropical species from Brazil, and is similar to *Nidularium*. The leaves are densely tufted, acute, spiny-margined and in cup-shaped rosette where from the centre the compound inflorescence emerges on a quite short stalk similar to *Nidularium* or on a longer exserted stalk and bears normally green or sometimes golden or blue flowers. ***C. amazonicum*** (*Aechmea amazonica, Karatas amazonica, Nidularium amazonicum*) produces some 15-20 leaves, greenish-brown above and light brown beneath and about 25-50 cm long which are wider in the middle. The bract-leaves are greenish-brown, and white flowers with a green tube are borne in a dense head. ***C. aurantiacum*** (*Aechmea aurantiaca*) grows vigorous with leaves expanded in the middle and the yellow flowers are some 5 cm long. ***C. lindenii*** (*Guzmania fragrans, Nidularium lindenii*) produces a dense rosette with about 20 tomentose and green-spotted leaves, bract-leaves creamy-white and the flowers white or greenish.

Catopsis

Catopsis in Greek means 'compound' or 'view', which has no known application. There may be more than 15 species found wild in tropical America with mostly rosulate leaves which may be strap-shaped or lanceolate. The racemes or spikes bear white or yellow flowers where petals and sepals are separate at the base, stamens are shorter than sepals and stigma sub-sessile. Their cultivation requirement is similar to erect *Tillandsia*. ***C. nitida*** (*Tillandsia nitida*) from West Indies grows 15-45 cm tall with rosettes comprising of shiny-green oblong-mucronate leaves and slender spikes with white flowers. ***C. penduliflora*** from Peru possesses some 15 cm long oblong-elliptic rosette leaves having thin denticulate margins, and branched racemose scape some 45 cm long bearing short-stalked white pendent flowers.

Cottendorfia

The generic name derives to honour Count Cotta von Cottendorf (1796-1863), a German botanist. The genus comprises 17 rosette-forming species from Guyana, Venezuela and NE Brazil. It grows some 3.0 m in height, bearing usually unarmed stiff leaves with white or grey-scales. Inflorescence is simple bearing small flowers and floral bracts. Flowers are white-green, sepals and petals are free and elliptic, and fruit is a small ovoid capsule. ***C. florida*** from NE Brazil which grows 1-3 m tall with 1.0 m long leaves has small sheaths, blades are grasslike, unarmed, pungent and beneath the leaves the white scales are quite densely placed. Inflorescence is up to 40 cm long formed on a tall scape with spreading branches having many white flowers up to 7 mm long, the sepals being obovate and petals elliptic.

Cryptanthus

The generic name derives from the Greek *krypto* meaning 'to hide' and *anthos* 'a flower' as the floral buds are hidden by bracts. This dwarf terrestrial genus originates in Brazil with 22 species of rosette-forming evergreen perennials that are excellent indoor foliage plants and for bottle garden. The narrow and leathery leaves, which in certain species are succulent, do not have watertight bases as others have and are highly *cum* variously coloured. They normally have toothed edges. Flowers though insignificant but are white and 3-petalled, and are borne in spikes. They are propagated through suckers, and offsets appearing either from the leaves or from the base. ***C. acaulis*** grows about 7.5 cm in height, forming 15-30 cm wide and flattened rosettes of leaves, comprising of green, white-scaly, narrowly triangular to lanceolate with wavy and spiny edges. It is propagated through offshoots (plantlets) that are formed several inches high, which may fall off from the plant if knocked down. Its var. 'Apple Blossom' grows up to 12.5 cm high having mauve leaf stalks and the leaves are basally cream-white, edged and mottled bright green and edges are smooth. 'Argenteus' grows 10 cm high with silvery scales on leaves. 'Rubra' bears smaller leaves which are coloured purple-brown overlying green. ***C. bivittatus*** grows up to 7.5 cm high where narrow toothed leaves lie flat to form a small wavy rosette some 15-35 cm across with two leaves growing up to 23 cm long. Leaves are strap-shaped, edges wavy and somewhat fleshy. The green leaves are striped darker and tinted rosy-red, especially at the base. Here too the offshoots (plantlets) may fall off when plant is shaken. ***C. bromelioides*** also forms a little flat irregularly arranged rosette spreading up to 35 cm across and in height as well. Leaves borne on slender stalks, are mid-green, strap-shaped, erect to arching, wavy and minutely toothed. The var. 'Tricolor' ('Rainbow Star') is more spreading with longitudinal stripes of white, cream-yellow, green and rose. ***C. fosterianus*** has flattened rosette with more than 60 cm spread but only up to 8.0 cm high. Leaves lie flat, are stiff, thick, wavy and toothed, coppery-green to purple-brown and cross-banded irregularly with grey to buff. ***C. zonatus*** grows up to 10 cm tall, bearing green but cross-banded grey-buff, thin, strap-shaped, wavy and finely toothed leaves that form a flat rosette some 50 cm across. *C. z. fuscus* bears red-brown leaves while *C. z.* 'Zebrinus' bears bold silvery banding. ×*Cryptbergia* (*Cryptanthus* × *Billbergia*) is a durable house plant having characteristics of both the parents. ×*C. meadii* (*Billbergia nutans* × *Cryptanthus beuckeri*) is a clump-forming with almost erect and 20-30 cm long leaves which are green mottled with pink. ×*C. rubra* (*Billbergia nutans* × *Cryptanthus bahianus*) is clump-forming having erect, a little broad and bronze-red leaves whereas flowers which appear seldom are small and white.

Deuterocohnia (formerly *Abromeitiella*)

It is a terrestrial bromeliad native to temperate regions of south America with five species. It was named so for the German botanist J. Abromeit (1857-1946). The species are perennial evergreens with attractive and closely packed cushion-type mat-forming (hummock-forming making some 70 cm high cushions) many small rosette where individual plant normally does not exceed 10 cm. Leaves are stiff, fleshy with densely overlapping bases, densely scaly triangular blades that are grey-tinted, sparsely spiny and apex sharply pungent. The 3-petalled yellowish-green flowers appear close to the leaves in groups of three in terminal bundles on short branches, sepals and petals are twisted with latter being much longer. Fruit is a pear-shaped capsule. They are succulent and match well with other succulents in the conservatory in large pans or at the sunny window indoors. They are propagated through division or by cuttings of single rosette. ***A. brevifolia***, a native to Andean slopes of South Bolivia and NW Argentina is a slow-growing plant with the spread of up to 30 cm. Its leaves are grey & scaly, 2-3 cm long, toothed at base and ovate-tringular, and 1-3-clustered greenish flowers some 3 cm long appearing only when plant is fully mature. Sepals are 1.3 cm long, and petals are intense green with a white-green scale. ***A. chlorantha*** from Andean slopes of Argentina is similar in appearance to former species but has serrated and some 2 cm long leaves in smaller rosettes. ***A. lorentziana*** from NW Argentina bears spiny, robust and some 15 cm long leaves, 1.6 cm long sepals and green petals with a larger fringed basal scale.

Dyckia

The genus is named for Prince Joseph Salm-Reifferscheld-Dyck (1773-1861), a German botanist who wrote several books on succulents. They are native to South America comprising of 80 clump-forming arid terrestrial species which are evergreen perennials with rigid, usually semi-succulent and narrow leaves in rosettes having edged with sharp spiny teeth which sometimes may be hooked. The shaded yellow 3-petalled flowers appear on long spikes well above the leaves. They are suitable for growing indoors and in conservatory. They are propagated through seeds or offsets. ***D. brevifolia*** (*D. princeps, D. sulphurea*) from Brazil bears arching, white-scaly green leaves being 20-30 cm long with 2 mm long spines on the edges. Spike is about 30 cm long with yellow and green flowers some 2.0 cm long. ***D. fosteriana*** from Brazil is a charming miniature species with rosettes spreading up to 13 cm across. Leaves are green to silvery-grey, strongly recurved and with hooked spines at the edges. Racemes are some 20 cm long, spiralling, and loose, bearing some 1.2 cm long and deep yellow to orange flowers. ***D. rariflora*** from Brazil bears green and grey-scaly, recurved and some 15 cm long leaves adorned with soft and blackish spines on the edges. Stem is about 60 cm in height bearing well-spaced some 12 orange flowers some 1.2 cm long. ***D. remotiflora*** from Southern Brazil and Uruguay and North Argentina bears strongly recurved linear leaves some 30 cm long where margins are adorned with small, straight, hard and pale spines at the edges. Floral stalk is 30-45 cm long bearing 8-12 orange-yellow flowers some 2.0-2.5 cm long.

Encholirium

Encholirium derives from the Greek *enchos* meaning 'sword' and *leirion* for 'a lily', referring to the shape and sharpness of the leaves. It contains some 12 short-stemmed perennial species native to NE Brazil. When in flower, it attains the height of up to 5.0 m. Leaves are narrowly triangular with coarse-spiny margins, sheaths are large, and forming a dense rosette. Inflorescence is simple or little-branched and scapose, bearing many cream to yellow or yellow-green and pedicellate flowers where sepals are overlapping and free, the petals narrower than sepals, rolling and free, and the capsules contain seeds with sickle-shaped appendages. ***E. spectabile*** attains up to 5.0 m height when in flower, leaf-sheath is large, the leaves up to 60 cm, upper leaf surface glabrous and lower scaly, marginal teeth some 1.0 cm long, inflorescence is stout, glabrous and densely racemose with a few branches, floral bracts entire and linear, yellow-green pedicellate flowers some 2.5 cm long, sepals are ovate and blunt, and the petals are lanceolate and some 2.0 cm long.

Fascicularia

The generic name derives from the Latin *fasciculus* meaning 'a bundle' on account of its densely clustered leaf rosettes. There are some five evergreen epiphytic or terrestrial species, making either mounds of narrow-leaved rosettes or forming clumps, native to Chile. From the centre of rosettes the clustered, tubular and 3-petalled flowers appear on mature plants, making an unusual specimen indoors or on the border of conservatory. They are propagated through offsets. ***F. bicolor*** bears stuffly leathery, arching, sharply pointed, inner leaves of the rosette bright red, and some 45 cm long and 1.5 cm wide leaves which are spiny-edged. Clustered flowers are pale-blue snd some 4.0 cm long in dense and rounded racemes. Bracts surrounding the flowers exceed them. ***F. pitcairnifolia*** is though identical to the former but with smaller floral bracts and blue flowers.

Fosterella

Named so for M.B. Foster, an American bromeliad specialist. It is native to S. Mexico to Peru, Paraguay and N. Argentina and contains some 13 stemless perennial species. It grows up to 80 cm in height. Leaves are thin and soft, a few, entire, or little toothed and form the rosette. Inflorescence simple or little panicled and subglabrous, floral bracts inconspicuous, flowers small and pedicellate, sepals rolling and free, petals white and longer than sepals, petal bases stuck to filaments, and capsules are spherical to compound pyramidal. ***F. micrantha*** from S. Mexico, Guatemala and El Salvador produces stems some 70 cm tall when in flower. Leaves are lanceolate, entire, acute, glabrous above and densely scaly beneath, and some 30 cm long. Inflorescence is complex, much-branched and with a cobweb-like coating, floral bracts ovate and acute, flowers pendent and bell-shaped, sepals ovate-triangular and blunt, and the petals are narrow-elliptic and blunt. ***F. penduliflora*** from C. Peru to NW Argentina bears linear-lanceolate leaves with soft and narrow apices. Inflorescence is short, paniculate and glabrous, sepals are lanceolate and the petals are 2 mm wide and lanceolate-oblong.

Greigia

The genus was erected in 1865 for Greig, major-general and a Russian horticulturist. It is a rock-loving large terrestrial bromeliad with 26 species native to S. Mexico to C. Peru to S. Chile, usually without distinct stems. It is allied to *Cryptanthus* but it has simple inflorescence instead of panicled. Sometimes it is also confused with *Billbergia* and *Bromelia*. It has perfect flowers with sepals either free or slightly joining at the base, and are linear, ovate-lanceolate or nearly subulate. The perianth-segments are free, elliptical, apex-rounded and rose-coloured or white suffused with rose which turn brownish and the stamens are generally shorter. ***G. sphacelata*** (*Billbergia sphacelata, Bromelia sphacelata*) from Chile is a very showy and stout pineapple-like plant some 90 cm high, bearing spiny-edged strong, recurving or spreading leaves. Rosy flowers in the axils of the leaves appear in dense heads where outer bract is spiny and acute.

Guzmania

The genus *Guzmania* commemorates Anastasio Guzman, an 18th century Spanish naturalist. It is native to tropical America, especially West Indies, comprises of 110 evergreen perennials which are rosette-forming and leaf-bases overlap so to form water-holding reservoirs, and majority of them are epiphytic. The diameter of the species may vary from 18-20 cm to 90-120 cm from one to another. The leaves are strap-shaped, glossy and smooth-edged. Yellow to flame-red *via* orange floral bracts are usually brilliant and last for several months. Yellow to white, 3-petalled and tubular flowers which are short-lived appear on panicled spike either on a stalk or sunk in the leaf rosette and are often held within showy bracts. They prefer humid locations and are excellent indoors or in conservatory. They are propagated through rooted offshoots produced by mother plant. ***G. berteroana*** from Puerto Rica produces erect, robust and some 40 cm long floral stalks which bear mid-green but flushed red at sun-exposed sites, broadly strap-shaped to elliptic glossy-red, and some 30 cm long leaves, broadly oval and overlapping bracts, and deep yellow flowers some 5-6 cm long with flared petal-tips. ***G. lingulata*** was when introduced to Europs in 1776, caused a sensation among gardeners unfamiliar with such a plant. This species from West Indies, Guyana, Colombia, Brazil, Bolivia and Ecuador produces floral stalks some 18-30 cm long bearing some 4 cm long clustered yellow-white flowers enclosed with crimson-red bracts some 6.25 cm long which give star-shaped floral shape. Flowers are short-lived but bracts remain for weeks. The leaves are metallic-green, narrow and ridged, some 45 cm long and 3 cm wide. ***G. monostachya*** (*G. tricolor*) from Florida and West Indies to Brazil produces cylindrical floral stalk some 30-40 cm long, stalk clasped by small salmon-red to green bracts striped brownish-red and bear 2.5 cm long white flowers, and on the top of spike the bracts are brightest vermilion. The leaves are green, up to 45 cm long and some 2 cm wide, slightly arched, and form a rosette up to 75 cm across. ***G. musaica*** from Panama and Colombia produces pale-green 60-75 cm long leaves, broadly strap-shaped and with round tips, and are adorned with zoned patterns of dark green to red-brown lines. Floral stalk is erect and 60-75 cm long with short head-like spike enclosed in overlapping, elliptic and pink bracts, though spikes bear glossy, cupped and pink bracts and bright yellow flowers some 5 cm long. ***G. sanguinea*** from Costa Rica to Colombia and Ecuador produces some 30 cm long, outcurving, red-tinged and lanceolaate leaves and in the rosettes the outer ones are green though inner ones bright red with yellow-green centres. Flowers are nearly stemless, tubular, pale-yellow or white and some 6.25 cm across. ***G. vittata*** from Colombia to Brazil produces some 60 cm long and dark glossy-green leaves cross-banded with silvery scales, and these form an erect rosette with vase-like outline. Floral stalks are erect, some

50 cm long, covered with tapered lanceolate bracts which overlap partially, and this terminates by dense panicle of bracts and 2 cm long white flowers. ***G. zahnii*** from Costa Rica produces pale to green, some 60 cm long, strap-shaped and pointed leaves lined longitudinally deep red. The lower part of some 50 cm long floral stalk is concealed by overlapping leaf-like reddish bracts, topped by smaller redder bracts from which some 15 cm long ovoid panicle emerges which comprises some 3 cm long yellow flowers and smaller yellow bracts. *Guzmania* has produced numerous hybrids, some with glossy leaves and deep red flowers.

Hechtia

The generic name was erected to honour Julius Gottfried Konrad Hecht (*d.* 1837), a councillor to the King of Prussia (Germany). The genus comprises of 15-45 species of terrestrial and dioecious bromeliads related to *Dyckia* which is native to arid regions of Mexico. Leaves are slender and succulent which form solitary or densely clustered rosettes, and the margins have sharp and hooked teeth. The unattractive panicles are large, diffuse and 3-petalled. ***H. argentina*** from central Mexico produces linear, low-arching and some 45 cm long leaves with silvery-white patina on both the surfaces. It is perhaps the most decorative species but is very slow-growing and rare. Its panicles are some 45 cm long, bearing greenish or white clustered flowers some 1 cm long. ***H. epigyna*** from Mexico produces 30-45 cm long leaves which are bright green but grey-scaly, linear, little channelled and with white hooked teeth. Floral stalk is more than 45 cm long bearing many 5 mm long lilac flowers. ***H. marnier-lapostollei*** from Mexico forms clustered but loose hummocks of rosettes. The leaves are narrowly triangular and recurved, up to 13 cm long, edged and tipped with spines, and are reddish-tinted when exposed to bright sun. The panicles bearing small white flowers rank well above the leaves.

Hohenbergia

It was named for Prince Hohenberg, Prince of Württenberg (*d.* 1830), a patron of botany. It is closely allied to *Aechmea*, is a tropical plant and performs well in the pots. The genus contains some 40 stemless terrestrial or epiphytic herbs from W. Indies and SE Brazil, growing up to 2.0 m in height. Leaves form a dense rosette, edged spiny and terminate into a stout spine. Scape is quite tall and bears the bi- or tripartite panicle with dense, short, sessile or stipitate spikes. White or blue sessile flowers are borne in the axils of large bracts. ***H. augusta*** (*Aechmea augusta, Hoplophytum augustum*) from Brazil produces some 90 cm long and 7.5 cm broad leaves which are spiny with pale scales on both the surfaces. The panicles are pyramidal and much longer to the leaves, covered with soft hairs, and bearing some 1.0-1.2 cm long blue flowers. ***H. legrelliana*** (*Aechmea legrelliana, Guzmania legrelliana*) from Uruguay is strong-growing *Billbergia*-like plant, bearing 7-12 entire, stiff and brown-scaly leaves. Scape bears 1.2-2.1 m tall and simple spike with serrated floral bracts and red flowers. ***H. stellata*** (*Aechmea glomerata*) from Trinidad and Brazil produces leaves some 90 cm long and 7.5 cm broad with long marginal spines. Panicles are either to the rank of leaves or slightly exceeding and the blue flowers are about 2.5 cm long.

Navia

Named so for Bernard Sebastian von Nau (1766-1845), a German naturalist and physicist. There are some 74 perennial species from SE Colombia, N. Brazil, E. Venezuela and Guyana. The plants are much-branched, cushion-forming and often lie flat on the ground but when in flower, these can grow up to 4.0 m high though many species are short-stemmed. Leaves either form a rosette or densely grow spirally along the stem, entire or toothed and narrow with large sheaths. Inflorescence is terminal and sometimes in compound panicles and flowers usually aggregate into a compound head. Sessile to sub-sessile flowers are actinomorphic, bearing overlapping, coiled and free or fused sepals, and fused petals forming a slender tube. Fruit is a capsule and the seeds are nearly bare. ***N. acaulis*** from S. Colombia is short-stemmed with occasional branching, bearing glabrous leaves some 12 cm long which are densely toothed with sharply acute tips. Rosette-sunk inflorescence is sessile and globose, floral bracts lanceolate to acuminate and 5 mm long, sepals broadly ellipsoid, slightly fused, 4-6 mm long and the posterior winged and keeled, stamens are exserted, and the fruit is woody to leathery and broadly ellipsoid. ***N. arida*** from S. Colombia has either unbranched short erect stem or stemless, bearing some 20 rosetted leaves some 38 cm long, green above and white-scaly below, caudate-attenuate, linear, spreading, and slightly toothed with 2 mm spines. Simple and centrally-located sessile inflorescence bears densely arranged small flowers with strong pedicels some 5 mm long, where outer bract is leaflike and tinged red, floral bracts are triangular, 2.5 cm long, membranous, red and white-scaly, sepals are keelless, thin, quite narrowly triangular and some 2.5-4.0 cm long, the petals are yellow with pink tips, and the fruit is about 8 mm and ovoid.

Neoglaziovia

It was named for A. Glaziou (1828-1906), French plant collector and landscape architect, where in Greek the prefix *neo* is for 'new'. It is native to Brazil, with only two species allied to *Billbergia* but having only two ovules than many in each locule. These are terrestrial rocky plants spreading *via* rhizomes. Its stems are short, leaves are narrow and spiny, longer than inflorescence, in a bundle-like rosette, the blades are linear and the leaves are used for making good fibre. Inflorescence is simple, loosely racemose, bracteate, erect and terminal, lower floral bracts are linear while upper triangular, sepals and petals free, almost symmetrical and ligulate. ***N. concolor*** grows up to 1.0 m when in flower. Leaves are 5-8 in number, 40-60 cm long, long-acuminate, densely white-scaly, scarcely serrated with 4 mm long and curved spines. Inflorescence is some 30 cm long, scape slender, canescent, lower floral bracts longer than flowers, pedicels slender and only 5-7 mm, sepals are red, broadly ovate and obtuse and petals are some 2 cm long and bright purple with 2 fimbriate basal scales. ***N. variegata*** (*Billbergia variegata, Bromelia variegata, Dyckia glaziovii*) grows up to or more than 90 cm tall with glabrous or minutely dotted leaves above, lighter broad white cross-banding beneath, pungent, and the margins incurved. Inflorescence some 25 cm long where lower floral bracts equal the scarlet flowers. ***N. concolor*** bears quite short stem with leaves being uniformly white-lepidote and having felts on the younger leaves. The calyx is scarlet and petals are violet.

Neoregelia

The generic name derives from the Greek *neo* for 'new' and *regelia*, an allied genus named for E. Albert von Regel (1815-1892), Director, Imperial Botanic Garden in St. Peterburg. An evergreen genus native to tropical South America, mainly from Brazil, comprising of 52 perennial epiphytic and terrestrial species but only about 20 species, their varieties and hybrids are in cultivation. Its broadly strap-shaped leaves with widened and overlapped bases form a perfect cup-shaped water-holding rosette. Erect flowers in colour range of blue, violet or white appear on erect stalks. When the flowering is over the rosettes die and the plant survives with its offsets. ***N. ampullacea*** from Brazil suckers profusely and only a few leaves form small tubular-rosette. Leaves are linear, some 13 cm long, red-tipped and banded red-brown beneath, and finely toothed. Blue flowers appearing 1.2-2.5 cm across, rise just above the neck of the rosette-tube. ***N. carolinae*** (*Aregelia marechalii, N. marechalii, Nidularium meyendorfii*) from Brazil, Colombia and Peru produces glossy-green (green above and darker green beneath) with purplish to orange-red inner base, linear, strap-shaped, stiff, 40 cm long and finely spine-toothed leaves which form the rosette. White to blue-purple flowers, sometimes edged white appear throughout the year from bright red bracts on the spike-like inflorescences. Being sensitive to chilling, the plants can be stored for up to two weeks at 12.2 °C. It is watered in the cup where leaves attach to the stem. Being sensitive to copper, no copper-based pesticides should be used. *N. c.* 'Tricolor' forms white longitudinal stripes which age to pink tints. ***N. marmorata*** (*Aregalia marmorata*) from Brazil produces soft-textured light green leaves blotched informally red-brown, wide-spreading, strap-shaped, and 40 cm long with red spines on the tips. Flowers are pale-purple to lavender some 2 cm across. Most of its commercial hybrids with *N. spectabilis* having green leaves with maroon blotches and a red tip are often available as *N. marmorata*. ***N. spectabilis*** from Brazil produces dull green leathery strap-shaped leaves some 40 cm long having sharp teeth and bright red foliage tips. Blue flowers emerge from purple banded green bracts.

Nidularium

Nidus in Latin is for 'a nest', referring to the cluster of leaves around the flowers. It is a genus of 22 rosette-forming epiphytic perennial species native to South America, mainly Brazil in their rain forests on rotting tree stumps though only 12 species with their varieties and hybrids are in cultivation. It is closely allied to *Neoregelia* but differs due to possessing erect and hooded tepals. The leaves are strap-shaped and prickle-toothed with overlapped bases forming water-holding cups. Bracts at flowering are pink, scarlet or red-purple and occur as a raised inflorescence with the floral branches between the bracts. Small tubular and clustered flowers close to the centre of the rosette are borne in white, blue, violet or rarely yellow colour, sometimes on longer stalks. ***N. billbergioides*** from Brazil produces bright green, sword-shaped, almost erect, up to 40 cm long and serration shortly toothed. Floral stalk erect and some 25 cm long with irregularly distributed sheathing bracts, which are topped by a cluster of triangular-ovate and yellow floral bracts with flaring greenish tips. The flowers partially concealed in the bracts are white and 2 cm long. ***N. fulgens*** (*N. pictum, Guzmania picta*) from Brazil produces sword-shaped 30 cm long and 4.5 cm wide shining pale-green leaves with darker mottling and coarse marginal spines. Blue flowers are enclosed with bright scarlet bract-leaves in the rosette's centre. *N. × chantrieri* (*N. fulgens × N. innocentii*) is white-flowered and appears as the finer version of *N. fulgens*. ***N. innocentii*** (*N. amazonicum*) from Brazil produces green, and finely spine-toothed linear leaves measuring some 30 cm long and 5 cm wide, upper surface bright deep purple with a metallic sheen though below red-purple. White flower appears in a dense head enclosed by pink-to brick-red bracts. ***N. i.*** var. ***innocentii*** grows 45 cm in height and the leaves similar to the parent species form the compact rosettes. Orange-red bracts are some 10 cm across with occasional green tips ensheathing the 1.9 cm wide white flowers. Under the species there are only six varieties, most showy being 'Striatum' where bracts are rose-red and leaves are striped yellow, followed by 'Lineatum' where bracts are red-tipped and the green leaves are striped with many white lines.

Ochagavia

It is a Chilean genus comprising 4-5 species. It was named so in 1853 for Sylvestris Ochagavia, Minister of Education in Chile. It is a stiff, shrubby, often erect-stemmed plant with many linear, toothed and spinose leaves making utricular rosette. Inflorescence on a short scape, globose or in a subracemose head with many flowers and is sunk into the centre of the rosette, sepals free, petals rose-pink or yellow, narrowly elliptic or suboblong, stamens exserted, and fruits swollen with persisting sepals. ***O. carnea*** grows up to 60 cm with 20-50 cm leaf length. Leaves are deflexed and the dense rosette is formed with 30-50 toothed (5 mm long spines) leaves which are white or grey-scaly below and shiny above, and their sheaths are triangular or ovate. Inflorescence is many-flowered, globose on a short peduncle, the outer bracts are bright rose-pink with laciniate margins, floral bracts lanceolate and white or pale-brown scaly. Flowers are narrowly elliptic, petals are pink to lavender and up to 3.0 cm long, stamens are bright yellow, exserted and the ellipsoid fruits are up to 1.8 cm across. ***O. elegans*** occurs only on the island of Juan Fernandez. It makes a vase-shaped rosette with unequal size of green leaves above though grey beneath, that arch at the tips and the margins are spiny. Dense, compact and highly showy purplish-pink flowers having yellow protruding stamens with much lower rank of the leaves fill in the whole rosette. ***O. lindleyana*** was introduced to Europe around 1851 where afterwards it became common in cultivation.

Orthophytum

Orthophytum is derived from Greek *ortho* meaning 'straight' and *phyton* is for 'a plant', referring to its straight inflorescence. This short-stemmed to stemless perennial genus from E. Brazil comprises of 17 semi-succulent species. The green or copper-green rosette-forming, narrowly triangular, serrate with soft spines, long attenuate and clasping leaves are many in many rows with large sheaths. Centrally-formed and sometimes sessile inflorescence formed on an erect scape is bipinnate with numerous dense heads. Inflorescence bracts are leaflike and spreading, large floral bracts are pungent, and sessile or short-pedicelled flowers are perfect with sepals erect to sub-erect and free, petals white, free with two scales and the inner stamina whorl is fused to petals. The fruits are berry-like and the seeds are without appendages. ***O. navoides*** produces spreading and stolon-forming stems some 40 cm long and when the plant matures for flowering the plant looks brilliant red. Some 30 cm long and sparsely scaly leaves form a dense rosette with dense tooths having 1.0 cm long and inward curving spines. Densely capitate inflorescence is sunk into the centre of the rosette. Floral bracts are narrowly triangular with small teeth, flowers are sessile, petals white and the sepals are some 3.0 cm long, free, straight and narrowly triangular. ***O. saxicola*** is stoloniferous and some 13 cm high when in flower. Its pale-green, erect, leathery and fleshy leaves which are 3-6 cm long and suberect, are little toothed with 2-3 mm long hooked spines. Its compact inflorescence bears only a few almost sessile flowers. Floral bracts are leafy, green and white sepals some 14 cm long are with papery margin, and white spreading petals are free, some 1.4 cm long, blades oblong and above the base these have two lacerated scales. The stamens are exserted. ***O. vagans*** has branched rhizomes which forms extensive mats. The leaf-sheaths almost encircle the rootstock, the leaves are thin, deeply channelled, suborbicular, some 12 cm long, toothed with 2 mm spines, blades green and are scaly below. Inflorescence is terminal and branched each with 8-15 densely packed subsessile flowers at apex and its bracts are leaf-like bright red to orange with size-reduction (2 cm only) at the apex though floral bracts are erect, narrowly lanceolate. Triangular and tomentose sepals are narrowly triangular and 1.2-1.5 cm long, whereas palish-green petals are linear, some 2.1 cm long and above the base bear 2 lacerate scales some 2 mm long.

Pitcairnea

It was named for William Pitcairn (1711-1791), a London physician who had a private botanic garden at Islington. It is a genus of some 250 species of evergreen, usually terrestrial bromeliads, native to tropical America with one species, *Pitcairnea feliciana* in West Africa. The leaves are strap-shaped to elliptic or linear, often toothed spiny and make the rosettes. Flowers panicled or racemose, 3-petalled, asymmetrical and tubular. ***P. corallina*** is notable for the lobes of its petals which never open widely though are edged with white margin. The leaves and spikes are quite large and arching, with greenish-red stalk and floral bracts of the spike, though leaves are green above grey beneath and broad with fine marginal serration, and the flower size larger at the base but gradually becoming smaller towards the top and there are more than 15 flowers per spike. ***P. feliciana*** from Africa has been found growing in rock crevices in Guinea and was once to comprise a new genus in the Liliaceae but in fact it was *Pitcairnia*. ***P. integrifolia*** from Venezuela and Trinidad produces leaves matt-green above, grey-white-scaly beneath, 60-90 cm long, linear and tapering to a thread-like point. Panicled inflorescence is pyramidal in shape, 45 cm or more long and bears 3-4 cm long bright red flowers. ***P. maidifolia*** from Costa Rica to Colombia, Guyana and Surinam produces green, lanceolate, 45-90 cm long leaves tapering to a stalk. Racemose inflorescence is spike-like some 30-45 cm long, bearing green and yellow or white flowers some 5 cm in length being formed individually in the axil of a broadly oval, green or yellow bract some 3 cm long.

Portea

Portea, a native of E. Brazil is named for Dr. M. Porte, a French tropical plant collector. It comprises 7 species of terrestrial perennials and when in flower it attains up to 2.0 m height. The leaves are dark green, stiff, scaly, spiny and form the rosette. Compound inflorescence formed on a slender central scape is erect and much-branched with brightly coloured bracts. Pedicellate blue or violet flowers are perfect, sepals are asymmetrical, fused and pungent, petals strongly fused to inner whorl and fruit is a fleshy berry. ***P. petropolitana*** is stemless and attains up to 1.0 m height when in flower. Leaf sheaths are large and some 80 cm long leaves are sub-elliptic, thickly brown-scaled, margins and apices toothed bearing some 4 mm long erect or curving and black spines with apex-spine coarse and brown. Scape is stout, inflorescence is sub-cylindrical, 3-pinnate, branched and attains some 50 cm length, branches some 12 cm long, red-brown, elliptic with overlapping bracts ~ lower inflorescence-bracts broadly lanceolate but other reduced, however, narrowly triangular floral-bracts some 5 mm long are formed on 1.0-4.0 cm pedicel, sepals are pink-orange with a lateral wing and some 1.5 cm long which are fused to half their length, and lavender-blue petals are some 3.0 cm long and blades narrowly elliptic with 2 fimbriate basal scales.

Puya

Puya is for 'a point' taken from the Mapuche Indians of Chile (Smith, 1952). It comprises of about 120 terrestrial xerophytic evergreen species often with woody-base, native to drier parts of South America, especially Chile. Some of the species in this genus are succulent. The plants produce thick rosettes of stiff, linear-tapered and sharply spiny-edged, often hooked leaves. The spikes, racemes or panicles are formed well above the leaves and nearly bell-shaped flowers are formed. In some species the lower half of the inflorescence-branches are crowded with flowers while upper half is sterile and bare where birds alight to pollinate the flowers. This likes sunny and well-ventilated situation and tolerates a minimum temperature of 5-7 °C. These are best grown in pots and borders. Propagation is from seed sown at 18-21 °C or by offsets. Some bromeliads are faintly scented, while others are heavily perfumed. ***P. alpestris*** produces leaves bright green above, densely pale-scaly beneath, 60 cm long and 1.0-2.5 cm basally wide. Metallic blue-green flowers some 5 cm long are produced on branched panicles up to 1.5 m tall. ***P. berteroniana*** produces some 1 m long leaves and some 5 cm wide at base, thinly greyish-scaly above and thickly beneath. Panicles are oblong, some 3-4 m tall, and bear 5-6 cm long blue-green flowers. The tips of the panicle branches bear only a few small bracts which help humming birds pollinating the flowers. ***P. caerulea*** (*Pitcairnia caerulea*) produces linear, very acute pineapple-like leaves some 60 cm long which are almost glabrous and edged spiny. Peduncle is 90-120 cm tall, bracts membranous, the inflorescence a little branched, flowers narrow and tubular, petals blue and oblong-obtuse, and sepals quite short and green. ***P. chilensis*** though is similar to *P. berteroniana* but leaves are glaucous and the flowers soft yellow. ***P. gigas*** from Colombia is the tallest member of the Bromeliaceae, growing up to 9 m in height. ***Puya raimondii*** is the largest bromeliad reaching 3–4 m tall when is vegetative with a flower spike 9–10 m tall, though the smallest is *Tillandsia usneioides* (Spanish moss). It is a succulent having caudex at the base. ***P. venusta*** produces silvery-grey-scaly and some 60-90 cm long leaves, and deep violet-red flowers in panicles some 1.5 m tall.

Quesnelia

It was christened so in honour of M. Quesnel, French Consul at Cayenne. There are dozen of caulescent or acaulescent herbaceous species native to Guyana and Brazil. The leaves are densely tufted and spiny. Scape is erect, arching or pendent, inflorescence is spicate, the bracts are entire or serrate, membranous or coriaceous and obtuse or acuminate. Flowers are sessile, perfect with usually free sepals, petals are free and convolute, and stamens 6 equal to or shorter than petals. ***Q. arvensis*** (*Q. cayennensis, Billbergia quesneliana*) from East Brazil produces some 75 cm long and 3.75 cm wide leaves armed with stout hooked spines. Spike is some 20 cm long with densely set flowers some 5 cm long. The bracts are entire and apex-rounded, sepals below webby, petals blue above and stamens quite shorter than petals.

Streptocalyx

It derives its name from the Greek *streptos* meaning 'twisted' and calyx. It is a native of E. Brazil, French Guinea, Peru and Ecuador and consists of 14 terrestrial or epiphytic perennial species growing up to 3.0 m high. Leaves are toothed and form a dense rosette.

Inflorescence is an extension of scape, and densely subglobose to slightly paniculate, bearing sessile flowers. Sepals are free to fused, asymmetrical and with a broad lateral wing, petals are free and narrow, the fruit is a fleshy berry and the seeds are ovoid or ellipsoid. ***S. longifolium*** from Colombia to Amazonian Bolivia and Brazil attains up to 50 cm height when in flower. Leaves are many, linear, 40-120 cm long, become maroon when fully exposed to sun, densely scaly, but below pale-scaly, slightly toothed with some 2.5 mm long and flat spines, while sheaths are small, broad and dark brown. Inflorescence is sessile or only with a short scape, some 7-15 cm long, ovoid or ellipsoid and with pale red-brown scales, the bracts are ovate or elliptic, pink to rusty-red, up to 2.5 cm long and toothed. Floral bracts are broadly ovate and 2.5-3.0 cm long, sepals are free and some 1.4-2.0 cm long with teeth and a short point, and the petals are white with dark margins and some 2.5-3.0 cm size. ***S. poeppigii*** from Colombia to Amazonian Bolivia and Brazil, attains some 50-80 cm height when in flower. Leaf sheaths are large and the leaves are many, some 1.6 m long, dark brown covered with dense buff-scales, elliptic, thick, leathery, linear, glabrous above and white-scaly below, slightly toothed with 4 mm long, dark and curved spines, and the leaves taper to a pungent pointed tip. Cylindrical or pyramidal inflorescence some 20-40 cm long is formed on a curved red scape where bracts are overlapping, bright pink, linear and toothed and the inflorescence is densely bipinnate and white-scurfy, floral bracts reniform with a small point, sepals sharp-pointed, apex purple and free with a 1.6-1.9 cm wing, the petals are bright purple to violet, and the fruit is a white long-lasting berry.

Tillandsia

Tillandsia was named for Elias Til-Landz (*d.* 1693), a Swedish botanist and professor of medicine. It is a genus of some 500 tender evergreen perennial species native to southern USA to South America but one species in West Africa most of which forming rosette, and a majority of them being epiphytic though only a few are in cultivation. A few species have slender and pendent stems with small and well-spaced leaves. Nearly all the species bear silvery-scaled hairs which absorb and retain moisture. Small flowers mostly emerge from the colourful bracts, mostly being solitary but remain always in spikes or panicles. Most of these require warm greenhouse conditions and a high degree of humidity. These can be grown on tree branches and in pots. These are propagated by removing the offsets in rosette-forming species and the pendent ones by division. ***T. cyanea*** from Ecuador is an epiphytic species growing about 23 cm high and some 30 cm wide. The leaves are spineless, red-brown-based and with brown longitudinal lines which form the narrow rosettes. Partially concealed floral stalks are robust, the inflorescence is elliptic, some 5 cm across and 8.75 cm high, comprising of formally overlapping bracts which have blunt tips. Violet-blue 3-petalled flowers some 5 cm across appear almost from alternate bracts which at flowering become pastel rose to red tinged with green. ***T. lindeniana*** from Peru is an epiphytic species growing up to 50 cm in height and 30-40 cm in spread, comprising of narrow linear leaves whose upper surface is dark green and lower purple. It forms a large rosette of leaves. Stalk is some 30 cm tall bearing some 20 cm long and flattened spike bearing showy coral-pink or carmine bracts from where large white-lipped deep blue flowers emerge in succession with 1-2 at a time. ***T. plumosa*** from Mexico forms leaf-rosettes of large clumps, bearing linear, channelled, dense, covered with white scales, the bases are expanded and swollen containing water-storage tissues and collectively create a bulky base in the rosette. Inflorescence is a simple or clustered head-like spike, rising above the leaves, bearing green flowers some 2 cm long and being subtended by pink-tipped bracts. ***T. purpurea*** from Peru is an extensive colony-forming terrestrial species in the coastal desert zone where it binds the sand, producing mostly a branching stem about 30 cm long or sometimes forming a rosette, bearing spirally arranged, crowded and up to 30 cm long leaves which are slender, arching, triangular and densely grey-scaly. The spike is branched and bears white and violet-blue fragrant flowers some 2 cm long, subtended by lavender to purple bracts. ***T. usneoides***, smallest plant in the family (less than 5 cm high) from sub-tropical and tropical America (from SE USA to Chile and central Argentina) is remarkable in the genus for lacking the usual cup-shaped water-storing rosette of leaves, though a covering of absorbent scales on plant surface occurs where between the scales, moisture fills the air spaces, the several feet long grey shoots hanging in festoons from trees and telegraph wires take on a greenish tinge and bear tiny blue or pale-green flowers. These shoots become the nesting material for birds and thus plant spreads vegetatively (Everard and Morley, 1970).

Vriesea

Vriesea is named for the Dutch botanist Willem Hendrick de Vriese (1806-1862), professor at Leiden and Amsterdam. It is native of tropical America (Cuba and southern Mexico to northern Argentina and the West Indies and Caribbean Islands), comprising of 190-246 species of mostly epiphytic perennials. The evergreen leaves are glossy-green or purple-red at bases, smooth-edged, overlapping at the base, arching stiffly outwards, leathery, strap-shaped, rosette-forming from a few centimeters to 1.8 m long and often netted, striped or marbled darker and the rosette diameter also from a few inches to 1.8 m. The rosettes make the cups for storing water. The spike is simple or branched, and slanting or horizontal which bears flattened, boat-shaped, parallel ranked, coloured and long-lasting bracts from which small, short-lived, tubular and 3-petalled flowers emerge out, yellow-flowered ones blooming during daytime while white-flowered ones during night. They are grown in pots or on tree-branches and are propagated through rooted offsets. It is chill-sensitive and can be stored at a minimum temperature of 13 °C. There had been maximum hybridization in this genus and the first hybrids were produced some 125 years back, and then again some 70 years back hybridizers started taking keen interest. The noteworthy hybrids are ***V.* × *erecta*** (*V.* × *poelmannii* × *V. rex*) bearing glossy-green foliage, deep red waxy bracts and yellow flowers; ***V.* × 'Favourite'** (*V. ensiformis* hybrid) bearing lustrous dark green leaves, maroon floral bracts and yellow flowers; ***V.* × *mariae*** (*V. carinata* × *V. barilletii*) bearing light green leaves tinted pink, salmon-red and yellow bracts and yellow flowers; ***V.* × *poelmannii*** (*V. gloriosa* × *V. vangeertii*) bearing light green leaves and flattened spikes of crimson bracts and yellow flowers; and ***V.* × *retroflexa*** (*V. psittacina* × *V. simplex*) bearing waxy pale-green leaves forming rosette and the bracts and flowers yellow in pendent spikes. The species in general cultivation are described here with. ***V. carinata*** from Brazil produces pale-green and some 20 cm long leaves, stalks some 30 cm tall topped with a broad and flattened spike bearing bracts which are scarlet below and yellow above and from which yellow flowers some 5 cm long appear. ***V. fenestralis*** from Brazil is known for its foliage beauty which grows up to 75 cm tall with spiraling rosettes comprising of yellow-green, broad and recurving leaves with networks of dark green veins and sometimes with circles of purple colouration undersides. Sulphur-yellow flowers some 6.25 cm long and tubular are borne on a spike which rises some 45 cm above the foliage ***V. fosteriana*** from Brazil produces numerous arching, dark to yellow-green leaves with irregular maroon cross-banding, broadly strap- or tongue-shaped, and some 45-60 cm long. Floral stalk is erect and about 90 cm tall topped with 25 cm long distichous spike which bears broad and red-brown spotted green sepals and greenish-yellow tipped brownish-red flowers some 4 cm long. Flowers open at night with rotten fruit smell and are pollinated by bats. ***V. gigantea*** (*V. tessellata*) from Brazil produces numerous dark green with a yellowish checkered pattern leaves which are strap-shaped and some 60 cm long and form a 90 cm across funnel-shaped rosette. Well-branched inflorescence is some 90-150 cm long which bears green bracts and some 4 cm long greenish-yellow flowers. It is a very attractive foliage plant looking like *V. fenestralis*. ***V. hieroglyphica*** from Brazil grows up to 60 cm high making rosettes of broad (10 cm) yellow-green leaves ornamented with dark purple irregular hieroglyphics and therefore it is grown for its foliage beauty. Floral stalk is about 75 cm high where upper third is branched bearing pale-green bracts and yellow tubular flowers some 5 cm long. ***V. incurvata*** from Brazil is very showy species which produces numerous lustrous green, strap-shaped, arching, some 40 cm long and pointed leaves, forming a rosette some 40-60 cm across. Floral stalk is short, topped with some 25 cm long blade-like spike ensheathed with keeled and overlapping red bracts which are margined yellow and from where some 5 cm long and yellow flowers protrude with green tips. ***V. psittacina*** from Brazil and Paraguay produces 40-60 mm long stalk topped with 20-30 cm long distichous spike, bearing red and green bracts and yellow flowers with green tips. Leaves are some 40 cm long, sword-shaped and form a loose rosette some 50-60 cm wide. ***V. rodigasiana*** from Brazil produces some 30 cm long leaves where upper surface is green, lower is brown-spotted while the base has brown suffusion. Stalk which is some 45 cm long is topped with branched spike, bearing about 2.5 cm long and waxy-yellow bracts and pale-green flowers about 3.5 cm long. ***V. saundersii*** (*V. botafogensis*) from Brazil produces numerous grey-green, strap-shaped, arching and about 45 cm long leaves adorned with irregularly placed red-brown spots, especially at the undersurface. Yellow flowers about 5 cm long are subtended by well-spaced greenish-yellow bracts on stout panicle some 20-30 cm long which come up well above the leaves. ***V. splendens*** from Guyana, Trinidad and Venezuela was introduced into Europe in 1840. It grows up to 45 cm high with rosettes composed of dull green, slender, mostly linear, stiff, spiny, parallel-veined, often coloured and attractive leaves patterned with striking purple-black cross-banding. Flower stalk some 60 cm long is sword-shaped, half of which is spike protruding high above the foliage, bearing brilliant red bracts with yellow, green or sometimes white long-lasting flowers some 5.0-7.5 cm long. This species has tendency to grow for years without producing flowers and therefore is classified as a foliage plant. Its var. 'Major' is more robust and with brighter colour.

Wittrockia

Wittrockia was named for V.B. Wittrock (1839-1914), a Swedish botanist. There are some 7 epiphytic, terrestrial or saxicolous perennial species from SE Brazil growing up to 40 cm tall. It is a rosette-forming, toothed and the blades are strap-shaped. Scape is topped by compound and dense inflorescence having densely overlapping leaflike bracts, and surrounded by a dense involucre of bracts. Flowers are sessile, sepals are asymmetrical, petals fused or joined at base, and the fruit is small and berry-like. ***W. superba*** from Brazil produces a dense rosette with 80-100 cm long leaves having large sheaths. Leaves have 4 mm long spines marginally, blades are yellow-green, shiny, broadly obovate, acute, densely brown-scaly and apices blood-red. Scape is stout and short, the bracts are ovate and red, and the upper toothless bracts aggregate below many-flowered almost hemispherical inflorescence, floral bracts are entire, lanceolate and membranous, and the apices thick and pungent. The length of ovate sepals is some 2.3 cm which are fused up to 0.5 cm, attenuate to a pungent apex and are longer than petals. Petals are acute, white with blue apices, fused up to 5-6 mm, and bear 2 fimbriate basal scales.

Propagation

Some kinds of bromeliads survive for several years after blooming and form huge clumps of offspring (as shown below), and become more attractive with age, but in most species (about 99 per cent), the bromeliads die slowly after blooming within one or two years and replace themself with one to several but generally three **offshoots or offsets** (offsets called **'pups'**) at the side of the base while declining. When inflorescence ceases to be ornamental, it is just chopped off as it will soon start producing offsets (pups) by redirecting the energy into growing the plants of the next generation. Bromeliads can start forming pups at any time, but most start pupping after they bloom. These pups are ready to be separated when they reach about $^1/_3$ to $^1/_2$ the size of the parent plant as the pups start forming the roots by this time so the plants can survive on their own and start forming their own central cup. Pups may be removed by cutting with a sharp clippers as close to the mother plant as possible. The longer you leave the pups on the mother plant the quicker they will reach maturity (taking nourishment from mother). The leaves of the mother plant should be trimmed back if they start interfering with pup's growth. Alternatively, taking the pups a bit smaller will encourage the mother plant to throw more pups sooner. It depends on whether you want a bunch of plants or only a few that will mature quicker. In temperate regions one should wait until spring to remove pups that would otherwise be ready to remove in the winter as pups don't usually root well when it is cold. However, if a number of evenly spaced pups around the mother are being formed, the mother plant may be cut away to let the pups form a clump.

Some bromeliads (*Alcantarea*, *Tillandsia edithae*, *T. grandis*, *T. rauhii*, *T. viridiflora* and *Vriesea glutinosa*) produce several grass-like **'adventitious pups'** usually well before the plant matures. It is best to leave them on the plant until they are 15-20 cm tall and then immediately removed otherwise the mother plant will loose its ability to reproduce the pups further. The method of removing the offsets depends on the species and its growing habit. Some are easily popped off by grabbing the pup at its base and twisting or pulling it away from mother plant. Some are quite firmly attached and need a more forceful means but always these should be carefully cut off with sterile snippers as close to the mother plant as possible and potted individually. These should not be pulled or twisted very hard since there is risk of crushing or breaking the meristem near the base and the pup may die. Some bromeliads (several *Pitcairnia* species) produce shoots from the underground **rhizomes** which are removed with shovel in case of large plants and knife when the plants are small. Other *Pitcairnia* species form **bulbous-like offsets**, which can be broken apart to make new plants. Tightly attached pups should be cut through the offset with a sharp knife or pruning clippers and pulled away from the parent

plant. If the pup is on a long **woody stolon**, cut through it with a large part of wood with pruning shears or even by loopers and then are set in empty pot to let the cut harden off by making callus which prevents infection of the pathogens. Some pups are difficult to root so in such cases the rootless offsets are placed in an empty plastic pot after putting a small amount of peat moss and nothing else, at the bottom, and then it should be watered similar to other plants. New roots will appear soon. Potted pups will usually mature and bloom in one to three years depending upon the genus and growing conditions. Pups are potted not more than 2.5 cm deep, braced with some rocks or stakes until the roots can hold it in place, potting mixture is kept dry, and such pots are kept at a shady and more humid place for a few weeks until the cut heals and the new roots begin to grow. In case the new plant is too wet and it has no stimulus to grow new roots, the cut may start rotting due to infection by some fungi. Until the plant is firmly rooted, these do not require any feeding. Offsets of some bromeliads are enclosed in dry, hard leaf-like scales. The new roots can emerge more easily if these scales are pulled off. In cool or wet climate, it is good to treat the cut with sulphur dust or some other effective fungicide. If the offset bears a few or no roots at all, it should be potted deep to keep it from falling over but during humid weather, this may cause rotting of the plant.

Tillandsia sommianus produces pups at each plant's base, and along, as well as at the end of its inflorescence. Some bromeliads such as *Cryptanthus, Orthophytum* and some *Tillandsia* produce **plantlets** right on the flower spike with a highly fragile attachment that even a very slow stroke can cause these to fall. These can be twisted off easily when they are ready to survive on their own, the brown or dead leaf if any should be removed, then cut ends should be dipped in a fungicide dissolved with some rooting hormone such as Rootone or Seradix. A variation to this approach occurs when pups are formed on the stem of the plants as in case of *Cryptanthus* and *Guzmania sanguinea.*

Some bromeliads such as common pineapple, *Tillandsia edithae, T. flexuosa, T. latifolia, T. lymanii, T. secunda* and *T. sommianus* are **viviparous** and form the pups on their inflorescence. Usually only one, but sometimes two are produced initially, which should be removed carefully without destroying the mother plant when these are at least 10-15 cm or one-third of the size of the mother plant. If the pup is removed successfully, a further batch of one or two pups may be produced.

Some populations of one species at a particular location may have particular reproductive strategies not shared with other populations of that species as in case of *Tillandsia latifolia*, where some of the populations may be viviparous while others not. A normal monocarpic species sometimes also produces pups such as *Tillandsia complanata* and *T. utriculata*. Benzing (1980) stated that 'resource availability plays a major role in monocarpy' as in case of *Tillandsia dasylirifolia*. In parts of its range in Mexico, this large epiphyte produces occasional or often weakly growing offshoots, but only when fruit set is poor. This kind of behaviour suggests that the seed crop is favoured when resources are allocated during the growth process. Asexual activity is held in abeyance as a fail-safe option in case the sink [young (seed) capsules] is too small to store enough of the available nutrients (Benzing, 1980).

There are **monocarpic** bromeliads producing no pups at all (*Tillandsia prodigiosa* and certain other species, and *Puya raimondii*) therefore these are reproduced only through **seeds**. The seeds in bromeliads are contained in succulent and juicy berries or in dry capsules which form after flowering. Bromeliads with dry capsules possess tiny seeds which through natural dispersal by wind get established and take 3-30 years to attain the blooming size (Hay and Beckett, 1971). Seeds lose vigour soon, therefore after harvesting these should be cleaned, dried and sown immediately at right conditions in pans containing a moisture-retentive mixture of equal parts of

leaf mould, sand, grit and pumice granules (by volume) or two parts moss-peat and one part coarse sand, and kept at 27 °C temperature at a place having continuous humid atmosphere. These after germination and when have attained four leaves the covers are removed for increasing periods for acclimatization, and when these attain 6-leaf stage the seedlings are pricked out some 2.5 cm apart into well-drained pots or pans (Hays and Beckett, 1971).

Cultural Practices

As a general rule of thumb, bromeliads will thrive in the same conditions as epiphytic orchids. However, they are considerably more tolerant than orchids of fluctuations in temperature, drought and careless feeding.

Requirement of **light** is based on the type of bromeliad grown. Different genera of bromeliads are tolerant to different levels of lights as some can withstand full tropical sun, while others quickly scorch. *Nidularium* requires the least amount of light and grey-leafed *Tillandsia* the most. The intense translucent red sheen in many *Neoregelia* species usually cannot be retained, if grown solely in the house. The other genera mentioned fall somewhere in between these two in light requirements. Adequate light is the most critical factor for growing healthy and colourful bromelaids. If plant leaves are soft and flexible and especially if they are spineless such as *Guzmania* and *Vriesea*, they probably grow in shady understorey in the wild and would do best in a low-light area. Those plants with stiffer leaves (usually spiny) such as *Aechmea* and *Neoregelia* or *Tillandsia*, these enjoy bright, filtered light. There are some plants that will tolerate full sun but most like a little protection. In general, the plants prefer well-lit, bright windowsills, but not direct sunlight. Plants that are yellowish may be receiving too much light which should be shifted to filtered situations, while dull dark green, leggy and elongated plants may be receiving too little light so such plants should be moved gradually to a brighter area where these will respond by regaining their original colours and tightening up to form a dense rosette with shorter leaves. Those requiring low light levels are easy to bring into flowering at home than those requiring higher light levels. Most will grow satisfactorily if are provided with 2,000 foot candles of light for 12 h a day. They will show better form and more intense colour with 3,000 foot candles, however, high humidity is necessary at these higher light levels. Within the acceptable range for each species, higher light intensities produce more compact growth and better foliage and inflorescence colour. The rule is to give a plant as much light as it can take without burning or bleaching. However, too much light may in many cases fade or bleach the brightness. Sometimes a plant when kept for longer in dim light conditions, sudden change to bright light may cause their fading or burning so such plants should be taken to sunny conditions in stages to acclimatize them. Bromeliads can be grown indoors under some types of artifical light. The standard incandescent bulbs do not produce the adequate colour balance for good plant growth though fluorescent tubes provide a broader spectrum of light rays with more intense light and less heat, *vis-a-vis* uses less electricity. There may be special tubes for plants though a combination of a warm white and a cool white is quite effective.

Temperature is not very important within a broad range. Bromeliads are highly tolerant of temperature variations. Most commonly available bromeliads will tolerate nights near freezing in winter, and days near or above 38 °C in summer. Their appearance may suffer a little after several weeks in such extremes, but they will recover as soon as the temperatures moderate. However, plants in hotter conditions will need more humidity. For the best performance, the nights should be 10 to 27 °C and the days 21-32 °C during most of the year, though many can tolerate even higher if there is good air circulation. A day-night differential of at least 10 °C is best. Most bromeliads have a special kind of photosynthesis called CAM (crassulacean acid metabolism) which creates sugars that allows these plants in hot and dry climates to open their stomates at night rather than during the day, which reduces water loss (Cowan, 1967). The *Guzmania* and *Vriesea* tend to be less tolerant of temperature extremes, especially cold nights, whereas *Guzmania* and soft-leaved *Tillandsia* are least tolerant to hot temperatures and make excellent plants for cooler conditions.

There is a relationship between air circulation, humidity and heat. **Air circulation** is mostly not important as long as the air is not stuffy. Mounted bromeliads require good air circulation. Good air movement is essential at very high temperatures or at humidity levels near saturation. Plants that do well in the sun outdoors, may burn in an unshaded window, because the still air fails to conduct the heat away. Stagnant or saturated air such as occurs in a tight greenhouse or very humid climates, encourages diseases. A fan can be used to move the air in an enclosed space. At the other extreme, homes with air conditioning or forced heated air that has humidity levels below 20 per cent, the plants will suffer. In such dry air, strong air circulation may actually cause leaf tips to dry out and die. Dead leaf tips are also a symptom of alkaline or salty water. **Relative**

humidity should be 50-70 per cent. There should be higher humidity levels if the temperature is higher with more vigorous air movement. *Nidularium*, *Vriesea* and *Guzmania* require higher humidity. When growing bromeliads indoors, one can provide extra humidity with wet trays of gravel placed under the plants, misting the plants frequently, grouping the plants together, setting the pots in a plastic saucer filled with water and by use of the humidifiers. Very high humidity seems to reduce the density of leaf scales. Since the beautiful leaf patterns of many bromeliads are created by banding of scales, plants may become less attractive under very humid conditions. Brighter light should help with this problem.

Potting and Mounting

Though in nature most bromeliads are epiphytic requiring a light and well-draining mix, standard recipes in use being equal parts of mulch/pine bark nuggets, perlite and composted peat/professional **potting mix** (a soil-less mix) and this medium moistens easily and drains well. Perlite, coarse builder's sand, tree-fern fibre, small gravel, red wood, pine, cypress, or fir or orchid bark are also suitable as potting mix. They can also easily be grown in a fast-draining potting soil. Terrestrial bromeliads where most do not bear rosetted leaves to hold water to fulfil their water needs (succulent *Dyckia, Hechtia,* grass-like *Pitcairnea* and pineapple) do better in a mix that retains a bit more moisture. Any potting mix is acceptable as long as it is acidic, sufficiently firm to hold the plants steady, retains moisture, is coarse-textured to drain away quickly and provides aeration so the roots can breathe. Bromeliad roots quickly suffocate in a tight, packed or saturated mix. There are almost many different potting mixes. What works at one location may not work very well for other locations because of the varying growing conditions. The pups should be removed in spring to summer for establishment, better rooting and active growth. *Tillandsia* is mounted with nut-sedge, cedar or grapewood, with branches that have dried out completley. However, for potting these, it should immediately be done as soon as these have been received. Too wet or soggy potting mixture will cause rotting of the plants whether too dry mixture will affect the growth adversely so such pots should be watered immediately after potting and afterwards regularly and appropriately. The one and only point that everyone

agrees on is 'potting mixes are not the same as soil'. Soils into the pots should never be added as it never works. Orchid bark/very coarse perlite and humus (3:1) works good for almost all bromeliads that can be potted. Some other ingredients that can be used are pumice, lava rock, charcoal, coarse peat or sphagnum moss and tree fern fibres. A mixture of 2/3rd peat-based soil mix and 1/3rd sand is a good mix. Whatever mix it may be, a different layer of other materials should be added to help the drainage, such as gravel or clay pot shards on the bottom of the pot. **Pot type** depends on individual choice. In a very heavy humid climate, it is better to use clay pots or baskets though in dry climates, plastic or ceramic pots should be used as they will retain moisture much longer than the porus clay pots. Over-potted bromeliads stay too wet, and encourage rot. Most bromeliads have very small root system, they should be grown in pots much smaller than a regular plant of the same size. A 10-15 cm pot is appropriate for a single plant of most species. In case the plant is in the incorrect pot size, transfer it into a larger pot.

Many bromeliads can be **mounted** on pieces of woods, wood boards, logs, roots, tree-fern slabs, rocks, corks or any non-toxic substrate that will last for several years. Wood, not the pressure-treated ones, that is naturally resistant to decay, such as cork bark, manzanita, cedar or grape are also good. Driftwood of the sea-water should be thoroughly soaked in non-salty water for several days to leach out salts before mounting the plants. Plastic coated wire, staples, various glues or narrow strips of nylon stockings should be used to tie or secure plants but in the process the basal portion of the plants should not be damaged. Bromeliads such as *Aechmea*, *Billbergia* and *Tillandsia* can be mounted to grow as epiphytes. Grey-leaved *Tillandsia* is drier-growing and prefers mounting. Some *Tillandsia* species will never show root attachment even after years of growth. For most species, a layer of moss or other water-holding medium is not necessary. It can harm some of the more arid species. The moss commonly seen on *Tillandsia* in the stores is cosmetic. It conceals all the glue. *Tillandsia* should be mounted onto wooden or cork mounts well before the blooming of the plant so that its roots penetrate the mounts because when in flower, normally these do not produce roots and instead divert their energy towards flowering, seed development and then forming the pups. *Neoregelia* is benefited from moss. Smaller plants can result without it, even in very humid climates. *Tillandsia* and *Aechmea* are especially recommended for mounting. Mounted bromeliads require a higher level of humidity and a lot more watering than the potted ones, consistently throughout the year. In dry air, the mounted plants may fail to develop sufficient roots to attach themselves to the substrate and may suffer from extreme dehydration of the foliage. There are at least three good ways to attach bromeliads to their supports. The most time consuming but permanent method is to tie the plant with nylon monofilament fishing line or strips of nylon stocking. Tie it so firmly that the plant can not wiggle, or the new roots will be broken off before they can attach themselves to the wood. This method is difficult with single plants that lack any basal stem. Bromeliads can be fastened to the substrate with non-toxic waterproof glues, hot glue and liquid nails. However, these tend to lose their grip after a couple of years, especially if the plants are soaked in water. Adhesives containing the ingredient Toluene, Shoe Goo, and Plumbers Goop are completely waterproof so these should be used just enough to fasten the plant, not burying the base of the stem in the glue as the roots will not be able to grow through it. Bromeliads with thick woody stolons can be stapled or nailed onto the wood. Nails can be driven through the thick stolons into the wood and, the staples straddle the thinner ones.

Feeding and Watering

They require the same nutrients as all other plants do. Macronutrients are required in large quantity though micronutrients in small quantity. Magnesium is critical for the production of chlorophyll and flowers. It helps strengthen cell walls and improves uptake of nitrogen, phosphorus and sulphur. Therefore, a small addition of epsom salts (magnesium sulphate) may help promote growth and initiate blooming in slow-growing and premature plants.

Epiphytic bromeliads differ from plants that grow in soil in that they absorb the **nutrients** mostly through their leaves instead of roots. In nature, the rosettes collect the dust, fallen leaves, bird droppings and rain, all of which contain nutrient minerals. In addition, the water in bromeliad tanks often contain blue-green algae (*Cyanobacteria*) which converts gaseous, elemental nitrogen from the air into nitrogen compounds that the plants can use. In cultivation, most of these sources of nutrition are absent so bromeliads should be fed at regular intervals. Before applying liquid fertilizer to bromeliads, the leaves are throughly wet with water, which ensures that the leaves are in their best condition to absorb the nutrients in the liquid fertilizer. Bromeliads are not heavy feeders and during the growing season, liquid fertilizer at ½ to ¼ strength will suffice, however, in case of slow-release fertilizer pellets, one pellet per plant would be sufficient for a season if mixed with soil compost. However, frequent feeding with dilute water-soluble fertilizer, about $^1/_8{}^{th}$ to $^1/_4{}^{th}$ the rate specified on the label would be sufficient for a month for good display. A complete water-soluble fertilizer (30:8:8 NPK) should be used through drenching of potting mix, foliage and central reservoirs. The fertilizer should be acidic. It is also very important that most of the nitrogen should be in the form of ammonium or nitrate and not the urea. Urea is not absorbed by leaves of bromeliads and instead sometimes damages bromeliads as the soil bacteria is necessary to convert urea into a usable form of nitrogen. Fertilizers with dye can stain the grey-leaved plants, and slow-release fertilizers can be added to the potting medium in addition to/or instead of foliar feeding. Slow release pellets such as Osmocote 14-14-14 or Magamp 7-40-6 can be added on top of the soil at a dose of about $^1/_4$ teaspoon for a 12.5 cm pot every 3 to 4 months, but fertilizer pellets should never be used in the cup. Most plants do well if sprayed with a dilute solution of fertilizer from time to time.Tank-bromeliads can be fed exclusively through the roots. Over-fertilizing causes loss of foliage colour. *Neoregelia* and its various variagated cultivars are very susceptible to such greening and so are members of some other genera. Many growers do not fertilize *Neoregelia*, or stiff-leaved *Aechmea* as they look better when grown in a slightly stressed condition. Most growers refrain from fertilizing during periods of very slow growth, often caused by low light levels and lower temperatures present in winter. *Guzmania* and *Vriesea* require fertilization to obtain large colourful spikes. Too much fertilizer will produce overgrown rosettes with very poor form. If a bromeliad gets leggy or loses foliage colour, it can be corrected by providing brighter light and less feeding. Excessive foliar feeding promotes growth of unsightly algae on the leaves, especially on the scaly species.

Bromeliads are highly tolerant of drought conditions but excess is always bad as plants show tip-browning. Under normal conditions indoors, it is not necessary to keep the central cup filled with **water** if light levels, the temperature and humidity are sufficiently high. In case central cup is filled with water, make sure to flush the central cup every so often to remove any built-up salts. However, it would be better if the bromeliad plants are watered through the soil every week during growth period but frequency is drastically reduced during winter months when these are resting as standing water during this period is dangerous. Mounted plants depend on their leaves to absorb needed moisture and food. Such plants should entirely be watered at least twice a week, and should be doused thoroughly in a tub or sink weekly. If water is allowed to be there in the leaf axils at least in case of fuzzy or grey-leaved *Tillandsia*, this may cause rotting. **Water quality** is important, and most of the domestic water is satisfactory, however, too salty or alkaline water is not fit for bromeliads. The use of a good acidic (pH 4.0-7.0, the optimum being 6.5) and low-salt water promotes the development of show quality plants. Hard water contains calcium and/ or magnesium salts or alkaline deposits (pH above 7) as unsightly spots on the foliage as also on bath room mirrors and dishes but does no serious damage to most non-tank terrestrial species though *Cryptanthus* is exception, however, high salt concentrations make essential elements less available to the plants. Species from higher elevations and the miniature ones are highly sensitive to salts. Many bromeliads grow well even up to water pH of 8.0. Bromeliads with shiny foliage may become unattractive but, the species with very scaly, grey foliage as *Aechmea fasciata* and most *Tillandsia* have a tendency not to show the hard water spots. Sodium and some other minerals, especially boron, copper and zinc are very harmful to bromeliads. **Leaf tip die-back** is usually one of the first symptoms. Evaporation of water from the central water reservoirs can concentrate salts enough to kill bromeliads. Artificially softened water is very salty due to high concentration of sodium so this should never be used on bromeliads. Water with a low to medium mineral content will prevent leaf tip die-back and reduce spotting if central water reservoir is flushed out weekly but if still the problems persist, it would be better to use rainwater or deionized water. However, the use of distilled water will pull out the nutrients from the plant tissues so when using it a little quantity of fertilizer should also be added. Collection and storing rainwater should be done in plastic containers, not the metal ones as copper is always dangerous to bromeliads.

A strong change in growing conditions, such as light or excessive dryness, may trigger a mature plant to bloom. Most bromeliads flower only once but their new off-shoots, premature or even mature plants not likely to bloom soon can be induced to flower by treating the plants with ethylene gas. The simplest method for doing this is to enclose the plant in a plastic bag with 1-2 ripe apples for a week under filtered light at room temperature which will cause stopping of leaf formation and will trigger flower initiation but this may take several weeks after the treatment to express the results. There are chemicals such as ethylene gas, calcium carbide and napthalene acetic acid that force bromeliads to bloom, but these are usually tricky to regulate, and often interfere with pup's development. It is usually best to be patient, and allow nature to take its course.

Insect-Pests and Diseases

Since tank bromeliads are shelter for frogs and various insects including **mosquitoes**, where mosquitoes

become a nuisance for human beings as they transmit various diseases though these are not as such harmful to bromeliads. Adult mosquitoes are winged, which lay their eggs at soggy places, and their larvae and pupae are strictly aquatic. As of 2004, the species of native mosquitoes recorded in Florida were 78 belonging to 12 genera, out of which a few have been found associated with tank-bromeliads. Mosquito-borne diseases are caused by protozoa (malaria transmitted by *Anopheles*), nematodes (filarial and dog heartworm, transmitted by *Aedes* and *Culex*) and viruses (yellow fever and dengue, transmitted by *Aedes* though encephalitis by *Culex*) in human beings. *Bacillus thuringiensis israelenis* (BTI) granules once in 1-2 months can be broadcast or put in the tank to control these. Liquid dish-washing soap is also effective as this breaks down the surface tension of the water so that mosquito egss do not hatch.

Bromeliads are relatively pest free. The main pests of bromeliads are scales and mealy bugs. **Scales** will look like little round dots covering both the surfaces of the leaves. **Mealy bugs** look like the white cottony patches. For small outbreaks, these can both be treated by wiping over them with a cotton swab dipped in rubbing alcohol. For more wide spread infestations, try mixing a little mild dishwashing detergent or baby shampoo with water and spraying the infected plants. The soap will coat and suffocate the insects. Orthene (Acephate), Cygon 2-E (Dimethoate) or Malathion at 0.2 per cent spraying or plant dipping (oil-based insecticides should be avoided) without any spreader will control these pests but no insecticide should be used in the cup. The plant should be thoroughly watered the day before treatment. It is recommended to rinse the plants off with clean water to make sure the pores on the leaves are open so the plant can breathe. Heavy oil-based insecticides should be avoided as these are likely to choke the plants as well. **Lygaeid bugs,** *Ozophora hohenbergia* has been recorded feeding on *Hohenbergia penduliflora, H. polycephala* and *H. urbaniana* seeds in Jamaica (Slater and Baranowski, 1978); and *Acroleucus bromelicola* on *Tillandsia dasyliriifolia, A. nexus on Tillandsia oaxacana,* and *T. violacea & A. tensus* on *Hechtia podantha,* all from Mexico (Brailovsky and Cervantes, 2008); though *Lygofuscanellus alboannulatus* was initially collected on an bromeliad in Costa Rica. Adults and nymphs of **lace bugs** (*Megalocysta gibbifera)* were recorded feeding on *Aechmea* leaves at Costa Rica, where sometimes the nymphs were found parasitized by larvae of small wasps possibly of Braconidae (Picado, 1913), often found stuck firmly in the gelatinous gum exuded by the host plant. A moderate infestation of rice root **aphid** (*Rhopalosiphum rufiabdominalis*) was recorded on roots and flowers of *Aechmea poeppigii* and certain other bromeliads, normally on roots and in hydroponic systems (Halbert, 1996). Woodley and Janzen (1995) recorded larvae of *Melanagromyza rosales* fly mining the leaves of *Bromelia pinguin* in Costa Rica. Larvae of an hymenopterous wasp (*Eurytoma* sp.) have been found feeding and destroying floral buds of *Vriesea friburgensis* and *Warauhia gladioliflora* in Brazil (Gates and Cascarite-Marin, 2004; Grohme *et al.,* 2007).

Some 576 weevil species are known from Florida, out of which 526 are native where only a few are pests. Among the 50 species of foreign origin, 5 were introduced as biological control of weeds such as hydrilla, water hyacinth, water lettuce and water milfoil and one to control *Melaleuca*. Each of them is a specialized feeder on one of those weed species. *Metamasius hemipterus* (cane weevil) attacks pineapple and certain other plants while *M. callizona* (Mexican bromeliad weevil, the black one) infests bromeliads. Florida's rarest native bromeliads *Catopsis nutans, C. berteroniana, Guzmania monostachia, Tillandsia fasciculata, T. flexuosa, T. pruinosa* and *T. utriculata* are in imminent danger as populations of a non-indigenous weevil (*Metamasius callizona* native of southern Mexico and Central America which was detected in Florida in 1989) encroaches whole of the Florida bromeliad populalation (Frank and Thomas, 2000). Weevil larvae mine the meristematic tissues of bromeliads. A fly has been observed in Central America parasitizing their larvae. *M. flavopictus* and *Cactophagus validirostris* were also recorded feeding on bromeliads in Mexico. *Metamasius mosieri* (the orange-black Florida bromeliad weevil) was also recorded feeding on bromeliads in Florida. The weevils have been found most commonly on *Tillandsia bulbisiana, T. utriculata, T. variabilis* and *T. setacea*, although it also occurs on several additional species of native bromeliads, including *T. fasciculata, T. simulata, T. paucifolia, T. flexuosa, T. bartramii* and the natural hybrid *T.* × *smalliana*. When it occurs on larger species such as *Tillandsia utriculata* and *T. fasciculata*, it appears limited to seedlings and immature individuals of those plants. In shade- houses, these have been found feeding on small species such as *Tillandsia bergeri, T. bulbosa, T. concolor, T. germiniflora, T. gardneri, T. hondurensis, T. ionantha, T. jucunda, T. rhomboidae, T. streptophylla, T. stricta* and *T. vernicosa*. From egg to adult, *Metamasius mosieri* takes from 18 to 22 weeks, in winter it takes more time though during summer less, and these survive 4-6 h even at freezing temperatures, *i.e.* -1 °C, though remains active at 13-32 °C. Adults feed externally on leaves, flowerstalks and inflorescences as do adults of other weevils that are specialists on bromeliads. Females of the Florida bromeliad weevil deposit eggs singly and not more per plant (to avoid their cannibalism nature) within the leaf slits made by adults and on the leaves in the centre. Upon hatching, the larvae start feeding on the outer layer of leaves and the larger ones chew the leaves in the centre of the plant, sometimes even tunneling into base of the plant and rarely even the floral stalks. These can be controlled by spraying 0.2 per cent Metacid-50 every 2-3 months. A few of the leaf beetles chew or mine the bromeliad leaves such as *Acentroptera pulchella* in southern South America (Mantovani *et al.*, 2005), and *Calliaspis rubra* on *Aechmea nallyi* in eastern Peru (Burgess *et al.*, 2003).

Picado (1913) reported larvae of *Acrolophus pallidus* and *Holcocera bromeliae* infesting on bromeliads in Costa Rica. *Acrolophus vigia* larvae producing silk were recorded from water present in the leaf axils of *Aechmea mexicana* and *Vriesea chiapensis* in Mexico feeding on leaves and its bases, and the large larvae of *Acrolophus* sp. mining the leaves of *Warauhia werckleana* bromeliad in Panama, where one of them was also found parasitized by hymenopterous larvae of *Bracon* sp. (Beutelspacher, 1969). Banana moth (*Opogona sacchari)* larvae were recorded on stems of various bromeliads and crowns of pineapple (Davis and Peña, 1990; Vorsino *et al.*, 2005). Lepidopterous larvae of *Epimorius testacellus* have been recorded feeding inside the flower pods of *Tillandsia fasciculata* in Florida (Heppner and Frank 2007) which there was being parasitized by a tiny hymenopterous parasitoid *Eurytoma sp.* (Bugbee, 1975). *Castnia boisduvalii (Castniidae)* larvae were noted feeding on *Tillandsia aeranthos* in Brazil (Biezanko, 1961). Its eight other species were also recorded from Brazil and Venezuela, most of whom feeding on pineapple and in large epiphytic bromeliads (*Tillandsia, Vriesia* and *Werauhia sp.*) in natural areas in Mexico, Guatemala, Honduras, and Panama. The fully grown pale-larvae are large (5cm) and their mining activities surely cause death of the plant. Albertoni *et al.* (2012) found larvae of *Geyeria decussata* feeding on the leaf bases of *Aechmea nudicaulis, Vriesea phillippocoburgii* and *Wittrockia superba* in southern Brazil. The larvae of a butterfly *(Strymonziba spp.)* have been noted infesting the fruits of *Aechmea bracteata* in Mexico, *Aechmea nudicaulis* in Brazil, and *Ananas comosus* in Mexico, Honduras, Costa Rica, Venezuela, Trinidad and Peru (Robbins, 2010). *Strymon megarus* larvae have been reported infesting leaves and flowers of *Bromelia pinguin* in Costa Rica, and pineapple in Brazil and Argentina. Six of other species such as *S. azuba, S. gabatha, S. lucena, S. oreola and S. serapio* are also reported feeding on various bromeliads (Robbins, 2010). Grishin and Durden (2012) described larvae of *Strymon solitario* as a new species detected in Texas (USA) and Tamaulipas (Mexico) feeding on flowerbuds, flowers and fruits of *Hechtia texensis*. The caterpillar of *Napaea eucharilla* was recorded damaging the leaves of *Werauhia sanguinolenta* in Panama (Schmidt and Zotz, 2000) and leaves of *Aechmea bracteata* and *A. nudicaulis* in Mexico (Beutelspacher, 1972), while *Caria ino* caterpillars on the leaves of *Tillandsia caput-medusae* in Mexico (Beutelspacher, 1972). Caterpillars of three South American species of *Dynastor* were observed feeding on bromeliad leaves (Penz *et al.*, 1999; Romero *et al.*, 2005).

Fungal diseases are not a common problem, but if any, can be controlled by using Banrot or Benomyl at 0.15 per cent, or Captan at 0.2 per cent by taking the plants outside. Viral infection is not a real problem as a disease, however, it is exploited to induce attractive variegations.

References

Albertoni, F.F., S.S. Moraes, J. Steiner and A. Zillikens, 2012. Description of the pupa and redescription of the imagines of *Geyeria decussata* and their association with bromeliads in southern Brazil (Lepidoptera: Castniidae). *Entomologia Generalis,* **34**: 61-74.

Anon., 2009. An update of the Angiosperm Phylogeny Group classification for the orders and families of flowering plants: APG III. *Bot. J. Linnean Soc.,* **161**(2): 105-121 (doi:10.1111/j.1095-8339.2009.00996.x Check date values in: |accessdate= ().

Bailey, L.H. 1960. *The Standard Cyclopedia of Horticulture* (3 vols.). The Macmillan Company, New York.

Barfuss, M.H., Rosabelle Samuel, Walter Till, and Todd F. Stuessy, 2005. Phylogenetic relationships in subfamily Tillandsioideae (Bromeliaceae) based on DNA sequence data from seven plastid regions. *Amer. J. Bot.,* **92**(2): 337-351.

Beckett, K.A. 1983. The Concise Encyclopedia of Garden Plants. Orbis Publishing Ltd., London.

Beckett, K.A. 1987. The RHS Encyclopaedia of House Plants including Greenhouse Plants. Salem House Publishers, Massachusetts, USA.

Benzing, D.H. 1980. The Biology of Bromeliads. Mad River Press, Eureka, California.

Beutelspacher, B.C.R. 1969. Una especie nueva de *Acrolophus* Poey, 1832, de bromeliáceas (Lepidoptera: Acrolophidae). *Anales del Instituto de Biología, Universidad Nacional Autónoma de México, Serie Zoología,* **40**: 43-48.

Beutelspacher, B.C.R. 1972. Some observations on the Lepidoptera of bromeliads. *J. Lepidopterists' Soc.,* **26:** 133-137.

Biezanko, C.M. 1961. Castniidae, Zygaenidae, Dalceridae, Eucleidae, Megalopygidae, Cossidae et Hepialidae da Zona Missioneira do Rio Grande do Sul (Contribução ao conhocimento da fisiografia do Rio Grande do Sul). *Arquivos de Entomologia Série B, Escola de Agronomia 'Eliseu Maciel',* **14**: 1-12.

Brailovsky, H. and L. Cervantes, 2008. Two new species and distribution records of the genus *Acroleucus* in Mexico (Hemiptera: Heteroptera: Lygaeidae: Lygaeinae). *Florida Entomologist,* **91**: 49-93.

Brickell, C. 1994. *The Royal Horticultural Society Gardeners' Encyclopedia of Plants and Flowers.* Dorling Kindersely Ltd., London.

Bugbee, R.E. 1975. A new species of the genus *Eurytoma* (Hymenoptera: Eurytomidae) from a pyralid occurring in the flower pods of *Tillandsia fasciculata. J. Georgia entom. Soc.,* **10**: 91-93.

Burgess, J., E. Burgess and M. Lowman, 2003. Observations of a beetle herbivore on a bromeliad in Peru. *J. Brom. Soc.,* **53**:221-224.

Cowan, R.S. 1967. Bromeliaceae of the Guyana Highland. *Memoirs of the New York bot. Grd.,* **14**(3), 214 p.

Davis, D.R. and J.E. Peña, 1990. Biology and morphology of the banana moth, *Opogona sacchari* (Bojer) and its introduction into Florida (Lepidoptera: Tineidae). *Proc. entom. Soc. Washington,* **92**: 593-618.

Everard, B. and B.D. Morley, 1970. *Wild Flowers of the World.* Peerage Books, London.

Foster, M.B. 1951. How well do you know your bromeliads? *The Bromeliad Society Bulletin,* v 1(5), September-October, pp49-50.

Frank, J.H. and M.C. Thomas. 2000. Weevils that eat bromeliads. Published on WWW at http://bromeliadbiota.ifas.ufl.edu/wvbrom.htm or http://entnem.ifas.ufl.edu/frank/bromeliadbiota/wvbrom.htm.

Gates, M.W. and A. Cascante-Marín, 2004. A new phytophagous species of *Eurytoma* (Hymenoptera: Eurytomidae) attacking *Werauhia gladioliflora* (Bromeliales: Bromeliaceae). *Zootaxa,* No. 512, pp. 1-10.

Givnish, T.J., Kendra C. Millam, Timothy M. Evans, Jocelyn C. Hall, J. C. Pires, Paul E. Berry, and Kenneth J. Sytsma, 2004. Ancient vicariance or recent long-distance dispersal? Inferences about phylogeny and South American-African disjunctions in Raptaceae and Bromeliaceae based on ndhf sequence data. *Intern. J. Pl. Sci.,* **165**(4): 35-54.

Grishin, N.V. and C.J. Durden, 2012. New bromeliad-feeding *Strymon* species from Big Bend National Park, Texas, USA and its vicinity. *J. Lepidopterists' Soc.,* **66**: 81-110.

Grohme, S., J. Steiner and A. Zillikens, 2007. Destruction of floral buds in the bromeliad *Vriesea friburgensis* by the phytophagous larvae of the wasp *Eurytoma* sp. in southern Brazil (Hymenoptera: Eurytomidae). *Entomologia Generalis* **30**: 167-172.

Halbert, S.E. 1996. *Rhopalosiphum rufiabdominalis* (Sasaki), rice root aphid, on roots of a potted ornamental bromeliad in Florida, *Aechmea poeppigii* (an Amazonian bromeliad). *Tri-Ology,* **35**(3) (online Journal of the Florida Department of Agriculture and Consumer Services).

Hays, R. and K.A. Beckett, 1971. In: *Reader's Digest Encyclopaedia of Garden Plants and Flowers.* The Reader's Digest Association Ltd., London, Great Britain.

Hellyer, A. 1983. *The Collingridge Illustrated Encyclopedia of Gardening.* Collingridge Books. Collingridge Press, England.

Heppner, J.B. and J.H. Frank, 2007. Bromeliad pod borer. University of Florida, IFAS, USA (http://creatures.ifas.ufl.edu/orn/bromeliad_pod_borer.htm).

Huxley, A. (Ed-in-Chief), M. Griffiths (Ed.) and M. Levy (Manag. Ed.), 1992. *The New Royal Horticultural Society Dictionary of Gardening* (4 vols.). The Macmillan Press Ltd., London.

Judd, W.S. 2007. *Plant systematics a phylogenetic approach* (3rd ed.). Sinauer Associates, Inc., Sunderland, NE England.

Katharina Schulte, Michael H. Barfuss and George Zizka, 2009. Phylogeny of Bromelioideae (Bromeliaceae) inferred from nuclear plastid DNA loci reveals the evolution of the tank habit within the subfamily. *Molec. Phylogenet. Evol.*, **51**: 327-339.

Luther, H.E. 2000. *An Alphabetical Listing of Bromeliad Binomials* (Seventh Edition). The Bromeliad Society International, California, USA.

Mabberley, D.J. 1997. *The Plant Book*. Cambridge University Press, Cambridge, U.K.

Mantovani, A., N. Magalhães, M.L. Teixeira, G. Leitão, C.L. Staines and B. Resende, 2005. First report on host plants and feeding habits of the leaf beetle *Acentroptera pulchella* Guérin-Méneville (Chrysomelidae, Hispinae). In: *Contributions to Systematics and Biology of Beetles* (Pensoft Series Faunistica), Sofia, Pensoft, No 43, pp. 153-157 (papers celebrating the 80th birthday of Igor Konstantinovich Lopatin).

Palmer, E.H. 1964. Nomenclature. *The Bromeliad Soc. Bull.*, v XIV (3), May-June, pp. 59-61.

Penz, C.M., A. Aiello and R.B. Styrgley, 1999. Early stages of *Caligo illioneus* and *C. idomeneus* (Nymphalidae: Brassolinae) from Panama, with remarks on the food plants for the subfamily. *J. Lepidopterists' Soc.*, **53**: 142-152.

Picado, C. 1913. Les broméliacées épiphytes considerées comme milieu biologique. *Bulletin des Sciences de la France et de la Belgique, No.* 47, pp. 215-360.

Robbins, R.K. 2010. The "upside down" systematics of hairstreak butterflies (Lycaenidae) that eat pineapple and other Bromeliaceae. *Studies on Neotrop. Fauna and Envirvon.*, **45**:21-37.

Romero M.F., P. Bermúdez and J.M. González, 2005. Notes on the life history of *Dynastor darius* (Fabricius) (Nymphalidae: Brassolinae) from Venezuela. *J. Lepidopterists' Soc.*, **59**: 37-39.

Sajo, M.G. 2004. Floral anatomy of Bromeliaceae, with particular reference to the epigyny and septal nectaries in commelinid monocots. *Plant Syst. Evol.*, **247**: 215-231.

Schmidt, G. and G. Zotz, 2000. Herbivory in the epiphyte *Vriesia sanguinolenta* Cogn. & Marchal (Bromeliaceae). *J. Trop. Ecol.*, **16**: 829-839.

Slater, J.A. and R.M. Baranowski, 1978. A new species of bromeliad lygaeid from Jamaica. *Florida Entomologist*, **61**: 83-88.

Smith, L.B. 1951. Why Bromeliad? *The Bromeliad Soc. Bull.*, v 1(2), March-April, p.11.

Smith, L.B. 1952. More Bromeliad Derivations. *The Bromeliad Society Bulletin*, v 2(1), January-February, pp. 60-61.

Vorsino, A.E., G.Y. Taniguchi and M.G. Wright, 2005. *Opogona sacchari* (Lepidoptera: Tineidae), a new pest of pineapple in Hawaii. *Proc. Hawaiian entomol. Soc.*, **37**: 97-98.

Woodley, N.E. and D.H. Janzen, 1995. A new species of *Melanagromyza* (Diptera: Agromyzidae) mining leaves of *Bromelia pinguin* (Bromeliaceae) in a dry forest in Costa Rica. *J. Nat. Hist.*, **29**: 1329-1337.

Ziv, M., T. Yogev and O. Krebs, 1986. Effects of paclobutrazol and chlormequat on growth pattern and shoot proliferation of normal and variant *Aechmea fasciata* Baker plants regenerated *in vitro*. *Israel J. Bot.*, **35**(3/4): 175-182.

9

Cacti (Family: Cactaceae)

R.L. Misra, Sanyat Misra, S.K. Palai and Ajit Kumar

[**Common names**: Agave cactus (*Leuchtenbergia principis*), Arizona organ pipe cactus (*Stenocereus thurberi*), Ball cacti (*Notocactus*), Barbados gooseberry/Lemon vine (*Pereskia aculeata*), Beaver tail cactus (*Opuntia basilaris*), Bird nest cactus (*Mammillaria camptotricha*), Bishop's cap cactus (*Asrrophytum myriostigma*), Bishop's mitre cactus (*A. m. quadricostatum*), Blue candle/Blue myrtle cactus (*Myrtillocactus geometrizans,* syn. *Cereus geometrizans*), Brain cactus (*Echinofossulocactus zacatecasensis*), Bunny ears (*Opuntia microdasys*, syn. *O. pulvinata*), Button cactus (*Epithelantha micromeris*), Cactus (pl. cacti), Christmas cactus (*Schlumbergera × buckleyi*), Colombian ball cactus (*Parodia vorwerkiana*), Coral cactus (*Rhipsalis cereuscula*), Cotton ball cactus/Peruvian old-man cactus (*Espostoa lanata*), Crab cactus/Lobster cactus (*Schlumbergera truncata*), Dahlia cactus (*Wilcoxia poselgeri, Echinocereus poselgeri, E. tuberosus*), Dancing bones/Drunkard's dream/Spice cactus (*Rhipsalis salicorioides,* syn. *Hariota salicornioides, Hatiora salicornioides*), Dumpling cactus/Mescal button/Peyote (*Lophophora williamsii*), Easter cactus (*Rhipsalidopsis gaertneri,* syn. *Schlumbergera gaertneri*), Feather cactus (*Mammillaria plumosa*), Fire crown cactus (*Rebutia sinilis*), Fish bone cactus (*Epiphyllum anguliger*), Fish hook cactus (*Ferocactus latispinus*), Giant torch thistles (*Echinocactus*), Glory of Texas (*Thelocactus bicolor*), Goat's horn cactus (*Astrophytum capricorne*), Golden ball cactus (*Parodia leninghausii*), Golden ball/Golden barrel/Mother-in-law's seat (*Echinocactus grusonii*), Golden ball cactus (*Notocactus leninghousii*), Golden lily cactus (*Lobivia aurea,* syn. *Echinopsis aurea*), Golden Tom Thumb cactus (*Parodia aureispina*), Grizzly bear cactus (*Opuntia erinacea ursina*), Hedgehog cactus (*Echinocereus pectinatus*), Indian fig (*Opuntia ficus-indica*), King/Princess of the night (*Selenicereus pteranthus*), Lace cactus (*Echinocereus reichenbachii,* syn. *E. caespitosus* and *Mammillaria elongata*), Living rock cacti (*Ariocarpus*), Melon cacti/Turk' head (*Melocactus communis*), Mexican sunball/Red crown cactus (*Rebutia minuscula*), Miniature show-bug (*Pelecyphora aselliformis*), Mistletoe cactus (*Rhipsalis baccifera*), Myrtle cacti (*Myrtillocactus*), Night blooming cereus (*Selenicereus*), Old man cactus/The old man of the desert (*Cephalocereus senilis*), Old man of the Andes (*Borzicactus celsianus,* syn. *Cleistocactus celsianus, Oreocereus celsianus*, and *B. trollii*), Old man of the mountains (*Borzicactus celsianus*, syn. *Oreocereus neocelsianus, Pilocereus celsianus*), Old woman (lady) cactus (*Mammillaria hahniana*), Orchid cactus (*Epiphyllum ackermannii,* syn. *Nopalxochia ackermannii*), Organ cactus (*Myrtillocactus geometrizans*), Organ-pipe cactus (*Lemaireocereus marginatus,* syn. *Marginatocereus marginatus*), Peanut cactus (*Chamaecereus silvestrii*), Powder-puff cactus (*Mammillaria bocasana*), Queen of the night (*Hylocereus undatus, Selenicereus grandiflorus*), Pitahaya dulce (*Lemaireocereus thruberi*),Prickly pears (*Opuntia*), Rainbow cactus (*Echinocereus rigidissimus,* syn. *E. pectinatus rigidissimus*), Rat's tail cactus (*Aporocactus flagelliformis*), Red spike (*Cephalophyllum alstonii*), Rose cactus (*Pereskia grandifolia,* syn. *Rhodocactus grandifolius*), Rose pincushion (*Mammillaria zeilmanniana*), Ruby dumpling (*Mammillaria tetracantha*), Saguaro/ Sahuaro (*Carnegiea gigantea,* syn. *Cereus giganteus*), Sand dollar cactus/Sea urchin cactus (*Astrophytum asterias*), San Pedro cactus (*Echinopsis pachanoi*), Scarlet bugler (*Cleistocactus baumannii,* syn. *Cereus baumannii*), Scarlet ball cactus (*Parodia haselbergii*, syn. *Notocactus haselbergii*), Sea fig (*Carnegiea chilensis*), Sea urchin cacti (*Echinopsis* spp., *Astrophytum asterias*), Seven stars (*Ariocarpus retusus*), Silver torch (*Cleistocactus strausii*), Snowball cactus (*Espostoa lanata*), Snowdrop cactus (*Rhipsalis houlletiana*), Spider cactus (*Gymnocalycium denudatum*), Star cactus (*Ariocarpus fissuratus*, syn. *Roseocactus fissuratus*), Strawberry cactus (*Echinocereus enneacanthus*, and *Thelocactus setispinus* syn. *Echinocactus setispinus, Hamatocactus setispinus, Ferocactus setispinus*), Tetejo fig orchid (*Pachycereus columnatrajani*), The southern head-bearing cactus (*Austrocephalocereus*), Torch cactus (*Trichocereus candicans*, syn. *Cereus candicans*; and *Trichocereus spachianus*), Totem pole cactus (*Lophocereus schottii* 'Monstrosus', syn. *Pilocereus schottii*), Vegetable sheep (*Tephrocactus*), *etc.*]

Introduction and Origin

It is thought that the first trace of the cactus family appeared on the earth at the end of the Mesozoic and the beginning of the Tertiary periods, a time marked by the great development of angiosperms. It is believed that all the present day cacti species already existed as far back as the early Quaternary period. Due to continuous deterioration of the climatic conditions, especially repeated periodic crisis in the amount of rainfall, the plants throughout tropical South America growing in areas that were greatly affected to exposure of the sun, began to develop adaptations that enabled them to withstand temporary periods of drought by storing water in their stems, which caused them to become more fleshy. Initially, these predecessors of cacti did not differ much from other leafy plants but gradually these perfected the changes in their body organs that were necessary for succulence to survive the adversities (Slaba, 1992).

The common name cactus (pl. cacti or cactuses) is derived from the ancient Greek word *kaktos*, meaning 'cardoon', a tall thistle-like plant with edible leaves and roots, but then it referred to spiny plants. Generally, all these are the plants with thick fleshy stems bearing spines but no leaves. Most cacti are stem succulents. In 1753, Carl Linnaeus knew only about 24 species of cacti which he placed under one genus ~ *Cactus* by shortening the name *Melocactus*. Since afterwards it was discovered that the genus *Cactus* runs in thousands of species of highly varied morphological forms, so the genus *Cactus* was adopted as a common name, and all the species having similar morphological characters were kept under one wrap by assigning its distinct generic name, and so there had been genera of cacti running into more than one hundared (approximately 134) with collectively more than 1,650 species. However, from time to time, frequent revisions have made especially the generic names very confusing, and sometimes there are up to half a dozen of synonyms of one genus. Cacti are mostly stem succulents, some even epiphytic, though a few growing as trees. They are native to America, ranging from Patagonia in the south to parts of western Canada in the north, except for *Rhipsalis baccifera* which grows in Africa and Sri Lanka. The classical habitat of the cactus is hot semi-desert (xeromorphic) though species extend north and south of Central America to British Columbia and Patagonia, and ascend to over 3,050 m in the Andes. Where no rain falls and the region is quite arid so no cacti can be found, though deserts having infrequent rainfalls are home for many of cacti and other succulents. *Epiphyllum* and their relatives come from rain-forest areas, and these are chiefly epiphytes which take support for their growing of certain other plants, especially trees but not depending on them for nourishment. Cacti have several features not possessed by members of other plant families, they all are succulents except *Pereskia*, and their flowers though are very attractive in majority of the cases but are stalkless so can not be used as cut flowers. Their roots are shallow and wide-spreading, shoots often photosynthetic, stems (plant body) simple and are usually globose (globular or spherical) or cylindrical having ribs (pleats or ridges) or tubercles (swollen parts of the plant body usually arranged spirally though most *Opuntia* and all *Epiphyllum* have flattened stems that are composed of jointed segments, *Opuntia* segments are technically known as pads), branched or cushion-forming, cuticle usually thick, stems usually with tufts of short barbed irritant hairs (spines, technically known as **glochids** though *Lopophora* is spineless) coming out from spiralled hairy pin-cushion-type structure called **areole** that acts as regenerating organ as this contains axillary buds or complete microscopic shoots in itself and produces, from or near a group of spines, usually single, multipetal and tubular flowers. Areoles have bud-scales and leaves modified into spines. Certain cacti do exhibit very small leaves initially which soon fall off, though a few primitive cacti still exist with fully developed leaves. Cacti are modified in such a form (condensed form, thick cuticle, leaflessness, and presence of areoles and glochids on the place of leaves and shoots) to withstand dry conditions by having a tough-skinned fleshy stem varying in shape which even after great drought have ability to regenerate. Spines conserve a layer of air around the stem which helps reducing the water loss, is an area where dew condenses, have protection to the plants from enemies, and they help dispersion of the species far and wide just by clinging to the furry-coating of the birds, animals and human beings. The leaves are usually very small, alternate, thick or fleshy and ephemeral or quite absent. Flowers at areoles, though rarely at branch-tips or in terminal cymes such as *Pereskia*, usually solitary, sessile, bisexual, more or less regular, often large and conspicuous, perianth with spirally arranged numerous segments, gradually transitioning from sepaloidy to petaloidy, united at base in an hypanthium, stamens numerous and developing centrifugally spirally or in groups from hypanthium and within this is nectary in a ring form, ovary of 3-numerous fused carpels, inferior (though superior and poorly fused in *Pereskia*), unilocular, style 1 with as many prongs as ovary, ovules many with parietal placentas though basal in *Pereskia*, fruit a berry (rarely dry dehiscent) containing seeds with curved embryo but no endosperm though perisperm starchy, if present. Pollination in their flowers is performed by hummingbirds, bats, bees or hawkmoths.

Cacti have a peculiar beauty and attraction of their own imparted by strange morphological characteristics of the plants and often by unique magnificence of the flowers hardly paralleled by any other in the plant kingdom. The infinite and unique variation in shape, size and colour of areoles and spines make the plants look curious and interesting. Therefore, they are grown as ornamental plants, and most suitable as indoor plants.

Various species of cacti are cultivated in more than 20 countries primarily for their fruits. Cactus pear (*Opuntia ficus-indica*) is a Crassulacean Acid Metabolism (CAM) species that is cultivated worldwide for fruit, forage, and vegetable (*nopalitos*) production (Barbera *et al.*, 1995). Several cultivars of cactus pear are used in NW Argentina as a source of forage, fruits, vegetables and medicines (Ochoa *et al.*, 1992). Cactus pear cultivation is important in subsistence agriculture in many countries and is an alternative succulent crop that is able to reduce the consequences of droughts (Pimienta-Barrios, 1993). Fruits of special flat-stemmed cactus cultivars of the genus *Opuntia* are sweet and highly valued for eating. The young tender pads of *Opuntia* and *Nopalea* are extensively used as a green vegetable in Mexico and Southern Texas. In Egypt, prickly pears have been grown for many years, especially in arid areas. The *Opuntia* plants are grown not only for fruit production but also as defensive hedges or for erosion control in reclaimed areas. A number of species of cacti have been shown to contain psychoactive agents, chemical compounds that can cause changes in mood, perception and cognition through their effects on the brain. Two species have a long history of use by the native peoples of the Americas ~ *Lophophora williamsii* in North America, and the San Pedro cactus *Echinopsis pachanoi* in South America, both containing mescaline which is concentrated in the photosynthetic portion of the stem above ground. Cochineal is a red dye which is produced by scale insects that live on *Opuntia*.

Botany, Breeding and Varieties

Cacti live in extremely dry environments, even being found in the Atacama Desert, one of the driest places on earth. Cacti show many adaptations to conserve water. Most species of cacti have lost true leaves which are modified to spines. Spines help prevent water loss by reducing air flow close to the cactus and providing some shade. Cactus spines (glochids) are produced from specialized cushion-like structure called areole which is an identifying feature for cacti. Areole is a kind of highly reduced shoot which is a regenerating organ, giving rise to usually tubular and multi-petaled flowers. The flowers have inferior ovaries, that lie below the sepals and petals, often deeply sunken into a fleshy receptacle. In the absence of leaves, enlarged stems carry out photosynthesis, unlike many other succulents. Cactus stems also store water, and are often ribbed or fluted, which allow them to expand and contract easily. Cacti occur in a wide range of shapes and sizes. The tallest free-standing cactus is *Pachycereus pringlei*, with a maximum recorded height of 19.2 m, and the smallest is *Blossfeldia liliputiana*, only about 1 cm in diameter at maturity. The smaller cacti usually have globe-shaped stems, combining the highest possible volume with the lowest possible surface area. Like other succulent plants, most cacti employ a special mechanism called 'crassulacean acid metabolism' (CAM) as part of photosynthesis. Transpiration, during which carbon dioxide enters the plant and water escapes, does not take place during the day at the same time as photosynthesis, but instead occurs at night. The plant stores the carbon dioxide which it takes in as malic acid, retaining it until daylight returns, and only then using it in photosynthesis. Because transpiration takes place during the cooler and more humid night hours, water loss is significantly reduced.

The basic chromosome number in most Cactaceae is 11. A species such as *Rhipsalis baccifera* may be diploid with 2n = 22 at one area of orign (Bolivia, Brazil and Paraguay), tetraploid with 2n =44 in another area of origin (Cameroon, Costa Rica, Kenya, Madagascar and Mexico) while still the octaploid with 2n = 88 in third area of origin or distribution (Madagascar). The flowers of cacti are distinguished according to the manner of pollination and fertilization and may be either autogamous (self-pollinated) or allogamous (cross-pollinated). Cross pollination is the prevalent method among the whole cactus family, and thus practically in all the instances at least two genetically different individuals are necessary for fertilization. Usually, the flowers are pollinated by insects and birds. Self pollination is common in *Melocactus* and *Frailea*, and occasional in *Notocactus* and *Rebutia*. Over the long history of domestication of cacti in several parts of the world, in search of novel and promising horticultural varieties, nurserymen, researchers and specialist collectors have attempted to hybridize intra-specific categories or even involving two or more different species and genera and thousands of attractive hybrids of commercial importance have been evolved such as ×*Aporberocereus* (*Aporocactus* × *Weberocereus*), ×*Aporechinopsis* (*Aporocactus* × *Echinopsis*), ×*Aporepiphyllum* (*Aporocactus* × *Epiphyllum*), ×*Borzimoza* (*Borzicactus* × *Denmoza*), ×*Borzipostoa* (*Borzicactus* × *Espostoa*), ×*Cephalepiphyllum* (*Cephalocereus* × *Epiphyllum*), ×*Chamaelobivia* (*Chamaecereus* × *Lobivia*), ×*Espostingia* (*Espostoa* × *Browningia*), ×*Espostocactus* (*Espostoa* × *Cleistocactus*), ×*Ferobergia* (*Ferocactus* × *Leuchtenbergia*), ×*Heliaporus smithii* syn. 'Mallisonii' (*Aporocactus flagelliformis* × *Heliocereus speciosus*), ×*Hylocalycium* (the first bigeneric cactus chimera), ×*Hylocalycium* (first bigeneric chimera between *Hylocereus* and *Gymnocalycium*), ×*Hylocalycium triangulare* (a graft chimera between *Gymnocalycium mihanovichii* var. *friedrichii* 'Hibotan' × *Hylocereus undatus*; Rowley, 1989), ×*Meierara* (*Aporocactus* × *Cryptocereus* × *Heliocereus*), ×*Pacherocactus* (*Pachycereus* × *Bergerocactus*), ×*Meierara* (*Aporocactus* × *Cryptocereus* × *Heliocereus*), *etc.* Panda and Das (1995) stated record of sectorial chimeras from *Echinopsis* stock carrying scions of *Ariocarpus kotschubeyanus, Epithelantha micromeris, Gymnocalycium denudatum, G. mihanovichii* and *Gymnocalycium* 'Hibotan' but these all are short-lived and too unstable. In fact, generic names of most of the species hybridised have undergone name changes due to division and merger of genera and species so frequently.

Some rare and unusual freak vegetative forms, *i.e.* monstrous, crested and variegated may occur in a large number of species of cacti. These monstrous and crested forms are usually mutations which can either appear as a seedling or as a branch on an otherwise normal plant. A crested plant is one where a head or stem is initially fanned out whereby new cells continue to form along a continuous undulating line along the fan. At maturity they are very contorted and become quite large. A variegation is caused by absence of chlorophyll on part or entire surface of a plant. If a part of the plant is affected it can survive as it is, but when entire plant is variegated it can't survive and grafting is necessary to perpetuate it. Some of these freak forms are quite attractive like the totally red variegated form of *Gymnocalycium mihanovichii var. friedrichii* cv. 'Hibotan'. This plant is propagated on a vast scale in Europe and the Far East and is usually grafted on to *Hylocereus* stock. All these freak forms can also be developed artificially by careful growth manipulation.

Well grown plants of species belonging to genera *Astrophytum, Ferocactus, Gymnocalycium mihanovichii var. mihanovichii, Melocactus* and *Rebutia* were selected as female parents and *Gymnocalycium mihanovichii var. mihanovichii cv. Fredrichii* as the male for crossing, and after fruit maturity the seeds were collected and sown which germinated in a week. From the seedling populations, plants having some definite colour, pattern, variegation, spination or form were selected, grafted on to suitable rootstock and propagated in large number by vegetative method for perpetuation. For induction of cristateness, the apical portion of the normal plant was removed allowing several off-sets to develop on the areoles. These off-sets were repeatedly removed until a large number of closely-spaced areoles were produced to align along a continuous undulating line from where new growths are allowed to form crest-like structures. A large number of new forms of cacti resulting due to inter-generic hybridization and growth manipulation at the Regional Plant Resource Centre, Bhubaneswar are being maintained (Das *et al.*, 1999). Some of these varieties are ×*Astrocalycium* 'David Hunt' (*Astrophytum* × *Gymnocalycium mihanovichii var. friedrichii*), a short globular black-green marveled cactus having white flakes on whole body and areoles bearing 7 spines with central one longer; ×*Astrocalycium* 'Dhruba' (*Astrophytum myriostigma* × *Gymnocalycium mithanovichii* var. *friedrichii*), a globular deep green cactus blotched cream to yellow combined with silvery scales, having 5 sharp and bent ribs, areoles woolly and sunken having 3-5 spines with central spine being stout and larger; ×*Astrocalycium* var. 'Uttam' (*Astrophytum myriostigma* × *Gymnocalycium mihanovichii* var. *friedrichii*), a globular deep green cactus blended with white and yellow mixed throughout, areoles sunken and woolly with 3-4 spines where central one is stout and larger; ×*Chamaelobivia* 'Arunima' (*Lobivia famatimensis* × *Chamaecereus silvestrii*), a cylindric cactus with massive light yellow colouration having pale-green blotches though apex pure green, ribs 13 and areoles deep brown with 13 spines out of which one large, 6 big and 6 small; ×*Ferocalycium* 'Madhuban' (*Ferocactus horridus* × *Gymnocalycium mihanovichii* var. *friedrichii*), a globular cactus with 13 ribs, 14 spines (7 upper ones whitish and hair-like, 6 in second whorl and 1 central reddish-brown) and yellow flowers; ×*Ferocalycium* 'Chilika' (*Ferocactus horridus* × *Gymnocalycium mihanovichii*), a green and globular cactus having 13 ribs, 10 spines, and areoles raised and wooly; ×*Ferocalycium* 'Sarala' (*Ferocactus horridus* × *Gymnocalycium mihanovichii*), a globular yellow cactus with 13 deep ribs and flat top, areoles raised, wooly and with 10 spines, 4 upper hair-like, 1 central deep brown and bent downwards and remaining 5 sharp and straight; *Gymnocalycium mihanovichii* cv. 'Ananta', a subglobose green cactus with pinkish blotches, ribs 12, areoles raised and wooly and spines 7, 6 large and 1 small; *G. mihanovichii* cv. 'Anindita', a globular cactus where green colouration is masked by orange, ribs 10 and spines 5, 2 long, 2 medium and 1 small, and tepals pink; *G. mihanovichii* cv. 'Ashoka', a globular cactus with lower part green and upper brick-red, ribs 9, areoles slightly raised and wooly; *G. mihanovichii* cv. 'Dr. B.B. Sundareshan', a columnar cactus vatiegated in orange, scarlet and green, yellowish-orange towards apex, ribs 12, areoles raised and wooly, and spines 4, 2 larger and 2 smaller; *G. mihanovichii* cv. 'Dr. J.B. Pattnaik', a globular and orange cactus with 14 ribs, sunken areoles and 4-5 deep brown spines; *G. mihanovichii* cv. 'Dr. R.S. Paroda', a pinkish-violet globular cactus with 9 ribs, areoles raised and wooly, and spines 4-5, 1 long, 1 small, and flowers pale-green; *G. mihanovichii* 'Dr. M.S. Swaminathan', a globular orange cactus with basal portion rusty and upper half garnet lake, top depressed, ribs 14 and spines straight and 5-6; *G. mihanovichii* cv. 'Dr. H.H. Haines', a globose brick-red cactus with 8 ribs and raised areoles bearing 3 pale-brown spines where lower one is longer; *G. mihanovichii* cv. 'H.F. Mooney', a reddish-orange globose cactus with 8 ribs, areoles raised and 3 erect spines where upper one is smaller; *G. mihanovichii* cv. 'Kalinga', a wine-red globose cactus with greenish-red base, areoles wooly and raised, ribs 11 and spines 5 where basal 1 is smaller, and flowers light pink; *G. mihanovichii* cv. 'Khandagiri', an orange cylindrical cactus with 13 ribs, sunken areoles, 4-5 spines with 2 longer and orange flowers some 6 cm long; *G. mihanovichii* cv. 'Krishna', a olive green globular cactus with pink bands across the ribs, ribs 13, areoles raised, spines 7 where central one larger than others and pink flowers some 4.5 cm long; *G. mihanovichii* cv. 'Lingaraj', a brick-red globular cactus with top paris green, ribs 9, areoles raised and wooly and spines 5 and straight where central one is large; *G. mihanovichii* cv. 'Mangal', a brick-red globular cactus tapering towards apex, ribs 10 with grey across banding, areoles slightly sunken and 4-5 dissimilar spines; *G. mihanovichii* cv. 'Mother Teresa', a red (lower half blood-red and upper half flesh-red) globular cactus with deep violet furrows, ribs 10, areoles wooly and raised, spines 3-4 where 2 are larger, and flowers pinkish with green stripes; *G. mihanovichii* cv. 'Mr. Thongler', an upper half orange and greenish below sub-globose to cylindrical cactus bearing 9-12 ribs, raised and wooly areoles, spines 5 and flowers pinkish-white; *G. mihanovichii* cv. 'Prof. H.Y. Mohanram', a crimson-red globular to sub-globose cactus, ribs 8 with prominent white bandings, areoles wooly and raised, and spines 4 and straight; *G. mihanovichii* cv. 'Sambhu', an olive-green sub-globose to columnar cactus narrowing at base, ribs 8, areoles raised and wooly, spines 3 with 1 shorter and flowers pink with green linings and 6 cm long; *G. mihanovichii* cv. 'Biju Patnaik', an orange and globular cactus with 10 ribs, areoles raised and 5 equal spines; *G. mihanovichii* cv. 'Vishwanath', a crimson-red cylindrical cactus with olive-green and orange blotches, ribs 11 with green banding, areoles raised and wooly, and brownish-black spines 5 with 1 being smaller; *G. mihanovichii* cv. 'Ananta' fa. caristata, a scarlet-red cristate cactus with greenish-yellow in the centre and greenish in furrows, areoles slightly raised, and light brown spines 3-4; *G. mihanovichii* cv. 'Dr. M.S. Swaminathan' fa. cristata, a crested cactus having violet in the middle and orange-violet outside, areoles slightly raised and wooly, and 3-5 pale-brown spines; *G. mihanovichii* cv. 'Duet' fa. cristata, a vertical half ivy-green crested cactus, other half scarlet-red with yellow towards centre, areoles raised and wooly, and 5 deep brown spines; *G. mihanovichii* cv. 'Mayuri' fa. cristata, a vertically half crimson-red cactus with orange furrows, other half olive-green and centre yellowish-green, areoles raised and wooly, and 3 upward bending spines where 1 is smaller; *G. mihanovichii* cv. 'Vichitra' fa. cristata, a multicoloured variegated (yellow, green, scarlet and orange) crested cactus, areoles wooly and slightly raised, and dark brown 5-6 spines which curve downward; *G. mihanovichii* var. *friedrichii* cv. 'Arun', a cherry-red cylindrical cactus with 10 ribs having prominent bands and wavy furrows, areoles raised and wooly, and 6-7 spines where lower 1 is longer; *G. mihanovichii* var. *friedrichii* cv. 'Balabhadra', a pale-orange globular cactus with scarlet on angles of ribs below, ribs 11, areoles raised and wooly, and deep brown 5-6 spines where 1 at left curving inward; *G. mihanovichii* var. *friedrichii* cv. 'Bhargavi', an orange and globular cactus with green furrows, ribs 9, areoles raised and woolly, and brown-tipped 6 spines with 2 being longer; *G. mihanovichii* var. *friedrichii* cv. 'Bimala', a crimson-red globular cactus with orange-yellow in the centre while furrows green, ribs 10, areoles slightly raised and wooly, and 6 pale-brown unequal spines where lower 2 curving inward; *G. mihanovichii* var. *friedrichii* cv. 'Britton', a carrot-red globular cactus with apricot yellow above and green centre, areoles raised and wooly, and brown-tipped 1 is small and curving downward; *G. mihanovichii* var. *friedrichii* cv. 'Chitrotpala', a tapering globular cactus having pale-yellowish-orange at upper half and violet on ribs and furrows, areoles pointed and raised, and spines

3-4 curving downwards; *G. mihanovichii* var. *friedrichii* cv. 'Daya', a globular cactus with 12 ridges, bottom crimson-red, middle orange and green in furrows, areoles raised and wooly, and curving spines 7-8 with central 1 longer; *G. mihanovichii* var. *friedrichii* cv. 'Dhauli', an yellowish-green globular cactus with pinkish-violet centre and furrows green, areoles slightly raised, and downward projecting 5-7 spines with 1 upper one smaller; *G. mihanovichii* var. *friedrichii* cv. 'Ekamra', a globular pale-green cactus having orange lower half, apex depressed, ribs 8-10 with inconspicuous banding, areoles raised and wooly, spines 5-6 and curving towards the furrows where central 1 is larger, and flowers pink; *G. mihanovichii* var. *friedrichii* cv. 'Gordon Rowley', a globular scarlet-red cactus with depressed apex apricot, areoles raised and wooly, and brown-tipped irregularly bent spines 6-8 where 1 is larger; *G. mihanovichii* var. *friedrichii* cv. 'Jagannath', a crimson-red globular cactus with black furrows, tapering upward, areoles raised and wooly, and dissimilar 4-5 spines with 1 or 2 longer; *G. mihanovichii* var. *friedrichii* cv. 'Madhura', a globular and upward tapering pale-yellow cactus with pinkish top, ribs 9, areoles raised, and brown to grey radiating 11-12 spines where 1 central erect, 5 upward curving small and 2-3 upper ones straight; *G. mihanovichii* var. *friedrichii* cv. 'Meera', a globular cactus having carmine-red 12 ribs with orange-green towards furrows though body's lower part scarlet, rib banding clear creamish-yellow, areoles less wooly and slightly raised, and brown-tipped radiating 8-9 spines which curve downward though 1 central curving outward; *G. mihanovichii* var. *friedrichii* cv. 'Meghana', an orange globular cactus with upper part yellow and furrows deep green, areoles less wooly and slightly raised, and downward curving 5-7 spines where upper 2 large and straight, middle 2 big and lowest 1 small; *G. mihanovichii* var. *friedrichii* cv. 'Mrs. Thongler', a scarlet-red upward tapering globular cactus with orange upper half and green furrows, ribs 9, areoles raised and wooly, and deep brown 5-6 spines with upper 2 larger and the lowest 1 smallest; *G. mihanovichii* var. *friedrichii* cv. 'Prachi', a crimson-red globular cactus with orange-red centre and green furrows, ribs 7 with creamish-white bands, areoles raised and wooly, and brown spines 4 with lowest 1 curving; *G. mihanovichii* var. *friedrichii* cv. 'John M. Poehlman', a flat apex globular cactus of scarlet-red colouration but lower part of furrows orange-red, ribs 10 with rosy-white bands, areoles raised and wooly, and brownish-black 3 spines; *G. mihanovichii* var. *friedrichii* cv. 'Prof. P.K. Sen', a crimson-red sub-globose cactus with deep red at bottom, orangish at apex and green at top centre, areoles closely set, raised and wooly, and apex curving 5-6 spines; *G. mihanovichii* var. *friedrichii* cv. 'Prof. P. Maheswari', an apex depressed crimson-red globular cactus with 9 ribs, areoles raised and wooly, and brown spines 3, all equal; *G. mihanovichii* var. *friedrichii* cv. 'Radha', a globular pinkish-red cactus with upper half in furrows orange-green, areoles raised and wooly, and spines 6 equal and radiating; *G. mihanovichii* var. *friedrichii* cv. 'Shanti', a scarlet-red globular cactus with top orange-yellow, ribs 8, areoles raised and wooly, and deep brown spines 4-5 curving downward with upper 3 larger and lower 1 smallest; *G. mihanovichii* var. *friedrichii* cv. 'Sneha', a deep orange globular cactus with 11 ribs having prominent whitish banding, wavy furrows, raised and wooly areoles, and 6-8 spines; *G. mihanovichii* var. *friedrichii* cv. 'Suvarna', an apex-depressed yellow and globular cactus having pinkish colouration on areole's top, ribs 9, areoles raised and wooly, and deep brown 11 spines where upper 2 larger; *G. mihanovichii* var. *friedrichii* cv. 'Soumya', an apex-depressed globose cactus with 10 ribs having prominent deep green banding, rib's angles maroon and furrows green, areroles raised, upward projecting 4-5 spines with 3 larger, and pale-pink flowers veined deep maroon and 5.5 cm long; *G. mihanovichii* var. *friedrichii* cv. 'Sudha', a sub-globose cactus with broader apex, ribs 8 where lower half at ribs scarlet-red and green at upper ribs and furrows, areoles slightly raised and wooly, spines 3 where upper 2 larger and curving upward, and flowers pink with green markings; *G. mihanovichii* var. *friedrichii* cv. 'Tarun', a globular cactus with narrower base and broader apex, burnt orange but furrows green, ribs 12, areoles normal and wooly, and deep brown spines 5 where 3 are larger; *G. mihanovichii* var. *friedrichii* cv. 'Udayagiri', a crimson-red globular cactus with a bit constricted in the centre, ribs 9 with white banding and with green furrows and green-blotched angles, and brown-tipped downward curving 3-4 spines where lowest 1 smaller; *G. mihanovichii* var. *friedrichii* cv. 'Utkal', a scarlet-red globular cactus with broader and pale-orange centre and green furrows, areoles raised and wooly, and spines 4 where lowest 1 smallest and curving downwards; *G. mihanovichii* var. *friedrichii* cv. 'Vijay', a crimson-red at bottom and above on rib's angles, and globose cactus having orange centre and green furrows, ribs 10-12 with prominent yellow banding, areoles raised and wooly, and dissimilar and brown 5-6 spines curving towards the furrows; *G. mihanovichii* var. *friedrichii* cv. 'A.B. Lau' fa. cristata, a cristate cactus with carmine-red throughout the rib's angles, greenish-yellow on the two sides, centre orange-yellow while furrows greenish-yellow, areoles raised, and spines 4 pointing towards the central furrow; *G. mihanovichii* var. *friedrichii* cv. 'Amitabh' fa. cristata, a pinkish-orange crested cactus blotched greenish in the centre and both sides of the crest though furrows green, areoles slightly raised and wooly, and brown-tipped 6 spines with central one smaller; *G. mihanovichii* var. *friedrichii* cv. 'Biren' fa. cristata, a deep scarlet-red crested cactus having orange central furrow, areoles slightly raised and woolly, and spines 3-4 with sometimes lowest 1 curved; *G. mihanovichii* var. *friedrichii* cv. 'Hillary' fa. cristata, a crested cactus having carmine-red ribs, yellowish-green furrows, green centre, lower ridges and furrows pale, areoles raised, and spines 4, straight and brown; *G. mihanovichii* var. *friedrichii* cv. 'Prof. J.C. Borg' fa. cristata, a crested cactus where lower half is olive-green to pinkish-brown, ribs cherry-red, furrows green to yellow, areoles raised and woolly, and radiating spines 3 where 2 are larger; *G. mihanovichii* var. *friedrichii* cv. 'Prof. P. Chandra' fa. cristata, a crested cactus in variegation of cherry red and amaranth purple on both the sides and central furrow orange-green, areoles slightly raised, and spines 3 where 2 upper ones larger and curving towards the centre though 3rd smaller one curving against the centre; *G. mihanovichii* var. *friedrichii* cv. 'Prof. P. Parija' fa. cristata, a scarlet-red crested cactus blotched with regular green, rib's angles yellow on both the sides, areoles slightly raised, and grey spines 3-5 tipped brown and mostly curved; ×*Gymnorebutia* 'Prashanna' (*Rebutia senilis* × *Gymnocalycium mihanovichii* var. *friedrichii*), a light orange globular cactus with depressed apex and centre greenish-yellow, white hair covered tubercles arranged spirally, areoles closely arranged, white radiating spines 10, and flowers red; *Mammillaria herrerae* cv. 'Lady Diana', an elongated to sub-globose, tubercles creamish-yellow, soft, white, radial and dissimilar spines more than 100 and flowers small and pink; *Melocactus ernestii* cv. 'Madhubabu', a sub-globose and deep green cactus having irregular pale-yellow blotches, ribs 10, areoles wooly and 8 × 6 mm, and spines 13-14 where central ones red, lower most radial and brown and 1 red and smallest; and *M. peruvianus* cv. 'Srinibash' fa. *Variegata*, a depressed globose and throughout deep green cactus blotched pale-yellow, areoles sunken, and reddish-brown spines 10-12 where central one stout while radials curved.

Classification and Species

Classification and naming of cacti has been difficult and controversial since they were discovered. Carl Linnaeus in1737 placed the cacti in two genera, *Cactus* and *Pereskia* though in ***Species Plantarum*** (1753) he placed all cacti under one genus, *Cactus*. Later, Philip Miller in 1754, divided cacti into several genera. Antoine Laurent de Jussieu (1789) placed all cactii in his newly created family Cactaceae. By the early 20th century, botanists felt that the genus *Cactus* named by Linnaeus is confusing, hence, it should not be used as a genus name. The 1905 Vienna Botanical Congress rejected the genus name *Cactus* and instead declared *Mammillaria* as the type genus of the family Cactaceae. However, the name of the family 'Cactaceae' was conserved and all cacti were placed under this family. In 1984, it was decided that the Cactaceae Section of the International Organization for Succulent Plant Study should set up a working group to work out consensus classification down to the level of genera. This group is now called as the International Cactaceae Systematics Group (ICSG). Their system has been used as the basis of subsequent classification. This group (ICGS) through discussion concluded that the family includes only around 130 genera. Herwig (1986) mentions *Arequipa paucicostata*, a cactus globular at first then cylindrical with only a few ribs and carmine flowers covered with hairs; *Austrocephalocereus dybowskii*, a cactus completely covered with silky white hairs and central spines some 3 cm long, flowers white, 4 cm long and appear only when overwintered in a cool dry environment; *Pyrrhocactus reconditus* (syn. *Neochilenia recondita*), a grey-green globular cactus with little sunken areoles, grey spines and pale-pink flowers; and *P. scoparius*, a spherical cactus with black-green colour bearing 12-16 well-marked ribs having tufted areoles with stout and black to grey long spines and creamy flowers appearing from the crown. Panda and Das (1995) mention that in India, the family Cactaceae is represented by 509 taxa belonging to 449 species and 73 accepted genera. They have tabulated 70 genera with number of species at the world scenario and the ones available in India. Here, with each genus the number of species found globally and the ones being cultivated in India are furnished below in brackets, *i.e. Acanthocalycium* (1/1), *Acanthocereus* (1/6), *Aporocactus* (2/2), *Ariocarpus* (5/6), *Astrophytum* (4/4), *Aztekium* (1/1), *Blossfeldia* (1/1), *Browningia* (2/7), *Carnegiea* (1/12), *Cephalocereus* (1/2), *Cereus* (10/30), *Cleistocactus* (9/64), *Coleocephalocereus* (2/10), *Copiapoa* (5/23), *Correocactus* (1/20), *Coryphantha* (9/45), *Denmoza* (1/9), *Discocactus* (3/9), *Echinocactus* (3/6), *Echinocereus* (22/47), *Echinopsis* (41/60), *Epiphyllum* (2/15), *Epithelantha* (1/1), *Escobaria* (4/16), *Espostova* (5/7), *Espostoopsis* (1/1), *Ferocactus* (12/23), *Frailea* (2/15), *Gymnocalycium* (18/50), *Haageocereus* (6/25), *Harrisia* (9/20), *Hatiora* (1/5), *Horridocactus* (1/1), *Neoporteria* (1/66), *Hylocereus* (4/20), *Leuchtenbergia* (1/11), *Lophophora* (1/2), *Maihuenia* (1/2), *Mammillaria* (62/150), *Melocactus* (15/30), *Micranthocereus* (2/9), *Mila* (1/1), *Myrtillocactus* (4/7), *Neobuxbaumia* (2/8), *Neolloydia* (7/15), *Neoporteria* (13/25), *Obregonia* (1/1), *Opuntia* (30/220), *Oreocereus* (18/30), *Ortegocactus* (1/1), *Pachycereus* (4/13), *Parodia* (27/50), *Pediocactus* (1/6), *Pelecyphora* (2/2), *Peniocereus* (2/20), *Pereskia* (4/16), *Pereskiopsis* (1/14), *Pilosocereus* (4/42), *Polaskia* (1/2),

Rebutia (22/30), *Rhipsalis* (2/40), *Schlumbergera* (3/6), *Sclerocactus* (1/19), *Selenicereus* (6/20), *Stenocactus* (3/10), *Stenocereus* (6/25), *Strombocactus* (1/1), *Thelocactus* (7/11), *Uebelmannia* (1/5) and *Weberocereus* (3/9). After scanning various literature sources (Bailey, 1942; Šubik, 1968; Everard and Morley, 1970; Hay and Beckett, 1971; Beckett, 1987; Brickell, 1989), the orchid genera found are *Acanthocalycium, Ancistrocactus, Aporocactus, Arequipa, Ariocarpus, Armatocereus, Arrojadoa, Arthrocereus, Astrophytum, Austrocactus, Austrocephalocereus, Aylostera, Aztekium, Bergerocactus, Blossfeldia, Borzicactus, Brachycereus, Brasilicactus, Browningia, Calymmanthium, Carnegiea, Cephalocereus, Cephalophyllum, Cereus, Chilenia, Cipocereus, Cleistocactus, Coleocephalocereus, Coloradoa, Copiapoa, Corryocactus, Coryphantha, Cylindrorebutia, Denmoza, Discocactus, Disocactus, Dolichothele, Echinocactus, Echinocereus, Echinofossulocactus, Echinopsis, Epiphyllum* (including *Epicactus*), *Epithelantha, Eriocactus, Eriocereus, Eriosyce, Escobaria, Escontria, Espostoa, Espostoopsis, Eulychnia, Facheiroa, Ferocactus, Frailea, Glandulicactus, Gymnocalycium, Haageocereus, Hamatocactus, Hatiora, Harrisia, Heliocereus, Hylocereus, Islaya, Jaminocereus, Lemaireocereus, Leocereus, Lepispium, Leuchtenbergia, Lobivia* (including *Chamaecereus*), *Lophophora, Loxanthocereus, Maihuenia, Malacocarpus, Mammillaria, Matucana, Mediolobivia, Melocactus, Micranthocereus, Mila, Myrtillocactus, Neobuxbaumia, Neochilenia, Neolloydia, Neoporteria, Neoraimondia, Neowardermannia, Nopalea, Notocactus, Obregonia, Ophthalmophyllum, Opuntia, Oreocereus, Oroya, Orridocactus, Ortegocactus, Pachycereus, Parodia, Pediocactus, Pelecyphora, Pereskia, Pereskiopsis, Pfeiffera, Pilosocereus, Phyllocactus, Polaskia, Pseudolobivia, Pseudorhipsalis, Pterocactus, Pygmaeocereus, Pyrrhocactus, Quiabentia, Rathbunia, Rebutia, Rhipsalidopsis, Rhipsalis, Samaipaticereus, Schlumbergera, Sclerocactus, Selenicereus, Sulcorebutia, Stenocactus, Stenocereus, Stephanocereus, Stetsonia, Strombocactus, Sucorebutia, Tacinga, Thelocactus, Trichocereus, Trichodiadema, Turbinicarpus, Uebelmannia, Weberocereus, Weingartia, Wigginsia,* and *Wilcoxia*; and 1,400–1,500 species (though certain other authentic literature sources mention even more than 1,650 species) which were then arranged into a number of tribes and sub-families. The ICSG classification of the cactus family recognizes four subfamilies, the largest of which is divided into nine tribes. The subfamilies are:

Subfamily Pereskioideae (Pereskieae) K. Schumann

One genus, *Pereskia* is placed under this sub-family. The features of this genus is considered closest to the ancestors of the Cactaceae. Plants of this genus are trees or shrubs with leaves; stems are smoothly round in cross section, rather than being ribbed or having tubercles (Anderson,2001).

Sub-family Opuntioideae (Opuntieae) K. Schumann

Nearly 15 genera are included in this subfamily. They may have leaves when they are young, but these are lost later. Their stems are usually divided into distinct 'joints' or 'pads' known as cladodes. Plants vary in size from the small cushions of *Maihueniopsis* to tree like species of *Opuntia*, rising to 10 m (33 ft) or more(Anderson, 2001). Certain others being *Grusonia, Nopalea, Pereskiopsis, Pterocactus*, etc,

Sub-family Maihuenioideae P. Fearn

This sub-family includes only one genus *i.e. Maihuenia*, with two species, both of which form low-growing mats. It has some features that are primitive within the cacti. Plants have leaves, and crassulean acid metabolism is wholly absent (Anderson, 2001).

Sub-family Cactoideae (Cereae)

This is the largest subfamily, including typical cacti. Members of this sub-family are highly variable in habit, varying from tree-like to epiphytic. Leaves are normally absent, although sometimes very reduced leaves are found in young plants. Stems are usually not divided into segments, and are ribbed or tuberculate. Two of the tribes, *Hylocereae* and *Rhipsalideae* contain climbing or epiphytic forms with a rather different appearance, their stems are flattened and may be divided into segments (Anderson, 2001). This is further divided into 8-9 tribes.

Tribe 1. *Cereanae*

The members of this tribe are bushy, erect, terrestrial or sometimes diffused, slender or stout, stems and branches more or less jointed and generally very spiny. Day or night blooming flowers are usually one but sometimes many and only on the upper parts of the mature areoles. The important genera are *Cephalocereus, Cereus, Cleistocactus, Espostoa, Haageocereus, Myrtillocactus, Oreocereus, Pachycereus, Pseudoespostoa, Trichocereus, etc.*

Tribe 2. *Hylocereanae*

These are often vine-like epiphytic, elongated, pendent or trailing, stems and branches ribbed or angled though rarely flat, and mostly night-blooming with large flowers such as *Aporocactus, Hylocereus, Selenicereus, etc.*

Tribe 3. *Echinocereanae*

Commonly known as 'hedgehogs', these low-growing, solitary or clump-forming cacti have ribbed and spiny stems with flowers emerging from the old lateral areoles such as *Chamaecereus, Echinocereus, Echinopsis, Lobivia, Rebutia, etc.*

Tribe 4. *Echinocactaneae*

Terrestrial, small and low-growing though sometimes several meters tall, solitary or clump-forming, stems spherical or shortly cylindrical, ribbed, flowers very showy and are borne from areoles on the crown of the plant such as *Ariocarpus, Astrophytum, Echinocactus, Ferocactus, Gymnocalycium, Hamatocactus, Lophophora, Notocactus, Parodia, etc.*

Tribe 5. *Cactanae*

Commonly known as 'melong cactus' or 'Turk's cap', these terrestrial cacti form wooly or bristle-like crowding structure on the crown of the plants due to closely set areoles such as *Melocactus, Discocactus, etc.*

Tribe 6. *Coryphanthanae*

Terrestrial low-growing, globose though sometimes cylindrical and rarely elongated, solitary or seldom clustering, stems tuberculate with spiralling tubercles, and flowers solitary either at lateral or apical areoles but never at the spiny aeroles such as *Coryphantha, Mammillaria, Thelocactus, etc.*

Tribe 7. *Epiphyllanae*

Tropical epiphytic cacti generally growing on trees, rocks also on the ground bearing flattened leaf-like, multi-branched and usually spineless stems with large and showy flowers,usually night-blooming such as *Epiphyllanthus, Epiphyllum* (syn. *Phyllocactus*), *Zygocactus, etc.*

Tribe 8. *Rhipsalidanae*

These epiphytic 'chain cacti' generally grow on trees, sometimes even on rocks bearing cylindric or flat, slender, much-branched and spineless stems, though there are exceptions as two genera bear small and tubeless flowers, such as *Erythrorhipsalis, Pseudorhipsalis, Rhipsalidopsis, Rhipsalis, etc.*

The cactus family includes about $1/5^{th}$ of the estimated 10,000 known succulents distributed among many unrelated botanical families and all richly rewarding to the collector in search of the unusual. The Cactaceae is of special interest to botanists for the combination of primitive, specialized flowers with highly advanced vegetative organs. The number of genera under the family is probably inflated by over-familiarization in horticulture and over-stressing of trivial features conspicuous there in. Many of the genera and species created under pressure from collectors and commercial growers are nearly equivalent to sub-genera, sub-species or varieties. According to Mabberley (1987) the family cactaceae comprises of 130 genera and 1,650 species but Heywood (1987) puts these numbers at 87 and 2,000, respectively. International Organization for Succulent Plants Study (IOS) has accepted 93 genera under the family Cactaceae (Hunt and Taylor, 1990). Hunt (1992) has brought out an account of 2,508 accepted and provisionally accepted species of cacti in the world, of which only some 1,208 names have been firmly accepted by all.

Brief details about some of the most popular cacti are being furnished here:

Acanthocalycium

This small genus of 10 species from Argentina is closely related to *Echinopsis.* These tolerate some frost and fairly hot sunlight. These are mostly slow-growing and dwarf, usually unbranched, initially spherical but with age becoming columnar, and are fairly rot-prone. Flowers in the shades of white, yellow, pink, orange or red appear during daytime. *A. violaceum* is the largest and most common species in this genus growing up to 60 cm in height and with a diameter of 15 cm. It can tolerate up to -10 °C temperature and its propagation is only by seeds.

Acanthocereus

This is native to Southern Mexico, Central America, NS America, the Caribbean and Florida. This comprises some six species after merger of *Monvillea* into the genus, and out of these *A. tetragonus* is most common. These are shrubby to sprawling plants, often with sprawling stems. Angular stems have 2-7 ribs, mostly 3-5 and may be segmented. These produce highly fragrant large white flowers in most species during night. Spines are small and stout and are present even on the floral tubes though fruits may or may not be. Globular fruits mature to red, the pulp is also red and the seeds black.

Ancistrocactus

The cacti stems in this genus are short globose to spherical or columnar, each stem bearing 10-20 tuberculate ribs. The spine arrangement on the body is quite symmetrical and charming whose density and structure sometimes hinder the flower from opening. These are propagated by seeds. *A. scheerii* (syn. *A. megarhizus*) grows some 10 cm in height with globose to columnar stems. The ribs at and near the tops of the body are cushioned with very small bristles though other areoles that are cushioned on the ribs on the sides bear 20-25 sharp yellowish, gently curved and radial spines out of which the lowest and longest one is stout, straight, darker and hooked.Funnel-shaped straw-coloured flowers appear in spring. *A. uncinatus* grows some 20 cm in height and bears blue-green, globose to columnar stems. Areoles produce one very long, reddish and hooked spine and 15-18 straight ones. Cup-shaped brown-green or reddish flowers some 2 cm across appear in spring.

Aporocactus (Rat's tail cactus)

Six species, all are suitable for hanging baskets and for sunny window ledges. These are native of Mexico near S. Jose del Oro, S. Bartolo on the Rio Grande. These pendent epiphytic cacti of pencil thickness are grown for their slender, fleshy stems and bright purple-red zygomorphic flowers resembling to those of *Epiphyllum*. They prefer partial shade for their growing, and are propagated through stem cuttings and through grafting on *Opuntia*, and its specific name *A. flagelliformis* means whip-like, referring to the whip-like long slender stems.

Ariocarpus

A genus of 6-7 highly desert species with most part of the body buried in the ground, one species from USA while remainder from Central Mexico. Only top is broadly-exposed bearing spineless angular tubercles from the centre of which large white or pink flowers appear. *A. fissuratus* (syn. *Anhalonium engelmannii*), a native of Texas (USA) to Mexico in the limestone hills, and has 5-13 cm across flat tops covered with 1.25-2.5 cm across, imbricated and triangular tubercles which are as wide at base as the length.Flowers shading from whitish to rose appear in the centre of the top some 2.5 cm long and wide. *A. lloydii* from Central Mexico has 10 cm or more wide rounded top with tubercles 20 mm base-wide, but upper portion rounded, whole surface irregularly fissured and imbricated, and bears 3 cm long purple flowers. *A. kotschubeyanus* (syn. *A. sulcatus*) from Central Mexico has its whole body concealed by the ground, up to 2.5 cm broader at top and bears 3 cm long rose-pink flowers. *A. retusus* (syn. *Anhalonium prismaticum*) from Mexico mountains also possesses some 7.5-20 cm across flat tops, initially bearing imbricate and triangular tubercles with acute, and cartilaginous tips but with age disappearing and making tubercles quite blunt. Old tubercles are 1.8-2.5 cm long with almost the same width at base, and the flowers appear from the centre of tops of rose colour. Other species are *A. furfuraceus, A. mcdowellii* very similar to *A. kotschubeyanus*, and *A. trigonus*.

Arrojadoa

A genus of less than 10 species from Brazil consists of long rambling stems with usually greenish-pink flowers being produced on the tips during whole of the summer which come out from among the long bristles, and the next season new stems start forming from the flowering area and then flowering in the same fashion. *A. dinae* is less leggy but even produces 1 m long stems having diameter of 2 cm, covered throughout with tan spines but at the flowering time the bristles at the tips become up to 1 cm long. It produces light pink flowers with light yellow centre and some 3 cm long. It is propagated through seeds and cuttings. Though it is easily grown but can not tolerate frost. *A. pennicillata* produces sparsely-spined stems of pencil-thickness, and bears short brown spines though stem ends become bristly from where clustererd pink flowers some 2 cm across appear. Though stems can grow several metres long but are usually unbranched. It also does not tolerate frost and is propagated through cuttings or seeds.

Astrophytum (Bishop's Hood)

A Mexican genus of six glasshouse species with several hybrids and varieties. It derives its name from the Greek *astron*, 'a star' and *phyton* 'a plant'. These are globular, ribbed, have white-scaled bodies and produce magnificent flowers from the top of the body which last for 4-5 days. Though these grow solitary but when old, occasionally produce offsets but not in case of *A. asterias*. Generally raised through seeds. Its most popular species is *A. asterias* which is slow in growth and grows up to 6 cm high and some 10-25 cm wide in the form of a flattened sphere with 8 ribs marked with white scales. Though lacks spines but white wooly tufts are found in the widely spaced rows of areoles. This flowers freely and produces 3 cm across yellow flowers. *A. capricorne* is a globular cactus but becoming cylindrical with age, body is light green marked with white wooly scales, possesses 7-9 sharply crested ribs with depressed areoles from where several (about 10) curved and brown-black spines appear, growing up to 25 cm high and some 13 cm wide. It produces star-shaped fragrant yellow 7-10 lobed flowers with red centre some 5-7 cm across. *A. c. senile* which is black-green bears 8 ribs thickly wool-tufted with long curved spines resembling horns. The flowers are yellow throated-red, 7 cm across and emerge from the centre of the crown. *A. myriostigma* (syn. *Echinocactus myriostigma*) is a pentagonal globular cactus but turning columnar-cylindrical with age and grows up to 60 cm. Grey body is covered with dense white scales, ribs 5-8 (only with 4 ribs in *A. m. quadricostatum*), though spineless but sometimes produces a few spines. Pale-yellow flowers some 4-6 cm across are produced during summer. *A. ornatum*, a slow-growing species grows up to 30 cm with the spread of 15 cm. Initially it is globular but changing to cylindrical with age. The body is grey with darker markings and bands of white scales, *vis-à-vis* 8 prominent ribs either straight or spiral. Areoles bear 5-11 stout, 5 cm long, first yellow but changing to black afterwards. The flowers are pale-yellow and 5-9 cm across.

Austrocephalocereus

This small genus from Brazil is clustered due to much basal-branching and erect-growing columnar cactus growing up to 2 m tall and 10 cm in diameter having cylindrical stems bearing true flowering cephalia (densely spined crowns) on the sunny side of the plant where flowers appear during nighttime and close with the sunrise. It is somewhat rot-prone and requires warmth during winter as long period of frost damages this. When flowering starts the orientation of the plant should be changed so that next part of the stem faces sun to form new cephalia. *A. dybowskii* is a clustering species with columnar stems some 2 m tall and 10 cm thick, covered with soft white hairs and short tan-brittle spines. It produces woolly lateral cephalium 20-40 cm long from where 3 cm wide whitish flowers appear when plants are mature though in cultivation it rarely matures. These are most suitable for pot-planting. It is propagated through cuttings or seeds. *A. lehgmannianus* is an attractive species growing about a metre tall bearing light blue 2-3 cm thick stems producing thin, brittle and tan spines and brownish *cum* woolly lateral cephalium when attains 30-50 cm height and where 3 cm wide off-white flowers appear during night hours and close by early morning.

Aylostera

A native of Bolivia, these are popular small cacti cultivated like *Rebutia*, therefore Bödecker in 1932 described *A. kupperiana* as *Rebutia kupperiana* in honour of Prof. Kupper of the Munich Botanical Garden though initially in 1931 this species was discovered in Bolivia by F. Ritter. These are 3 cm across globose cacti, later turning to columnar. Their body is green splashed bronze with a purplish tinge, 10-12 spiral ribs, and 16-20 spines, radial spines 15 with 5-8 mm in length and central spines 1-4 with 4.2 cm long and dark brown. Flowers in abundance from vermilion to bright orange-red and 10 cm across. *A. firebrigii* bears white and bristly spines and orange flowers. *A. pseudominuscula* a native of Argentina bears round to columnar stem which afterwards forms clusters and bears numerous small bright red flowers with long slender tubes. *A. deminuta* is also native of Argentina. All these are multiplied through seeds and from shoots.

Aztekium ritteri

It is native to Nuevo Leon, Mexico. It takes its name from Aztec people who founded the Mexican empire and the stem of this plant resembles the sculptures made by Aztec people. Its specific name *ritteri* is in honour of the well known cactus collector F. Ritter. It is a handsome monotypic genus where stem is flat from top which is covered with grey-white wool, the plant is 3-4 cm high and 6 cm wide, the stem is dull grey-green with 9-11 ribs and closely set areoles which are also covered with grey-white wool and 1-3 grey spines. Several pale-pink flowers some 2 mm long appear at the crown at a time. It is generally grafted on *Eriocereus jusbertii, E. bonplandii* or on *E. tortuosus*.

Blossfeldia

It is native to steep mountain slopes in the borders of Argentina and Bolivia, and possibly has 2-3 spineless species. It was named for Harry Blossfeld, a German nurseryman and collector of cacti. Their storage roots are turnip-shaped and stems globular having thin cuticle to absorb atmospheric moisture (rain, mist and dew) while in the wild these adapt well to drought without any harm, by becoming completely dry and papery, however, in no case these should be over-watered, and even whenever the weather is actually very hot, instead of usual watering their body should be sprayed with water once or twice a week. They are propagated through seeds, by offsets and by root cuttings. *B. liliputana*, the main species, forms a colony with globular to disc-shaped individual stems bearing dot-like white wooly areoles. It grows up to 2.5 cm and its pale-yellow flowers appearing at the crown are 5 mm wide.

Borzicactus

It is native to Ecuador to N. Argentina, and was named so for the Italian botanist Prof. Antonio Borzi (1832-1921), the Director of the Palermo Botanic Garden. This genus of usually globose to columnar cacti contains some 10 species of dissimilar sizes, some of which were previously known as even *Matucana, Oreocactus, Pilocereus* and *Submatucana*. The important species are *B. aurantiacus* (syn. *Submatucana aurantiaca*), a native of Peru, more or less globose and clump-forming dark green cactus growing 15 cm tall and producing angular tubercles. Areoles are wooly with 20-30 spines some 2.5-4.0 cm long and red to orange-yellow flowers some 9 cm long. *B. celsianus* (syn. *Oreocereus neocelsianus, Pilocereus celsianus*), a native of Bolivia is clump-forming, slow-growing, erect and sylindrical cactus some 1.2 m tall having 10-18 rounded and notched ribs, and large oval areoles full of whitish matted hairs covering whole of the body especially when young, radial spines almost hidden though 1-4, stout and reddish central spines are 5-8 cm long. Usually only mature plants produce 9 cm long red flowers. *B. haynei* (syn. *Matucana blanckii, M. cereoides, M. elongata, M. haynei,* etc.), a native of Peru has globose to oblongoid, 60 cm high and 10 cm or more thick stem, growing slow. Ribs 25-30 but inconspicuous and bear round tubercles. Areoles oval and full of yellow felts, and white and spreading spines altogether 28-38 and 4 cm or more long and with a complete net-working on entire stem when young. Orange-red 8 cm long flowers appear only on mature plants. *B. samaipatanus* grows often prostrate with branched cylindrical stems, having 8-12 ribs, small white spines and purple-red flowers which appear laterally. *B. trollii* (syn. *Oreocereus trollii*), a native of S. Bolivia to N. Argentina grows some 60 cm or more in height and 13 cm thick with erect and cylindrical clumpy stems though sometimes solitary, and is slow-growing. Ribs 10-25 with large oval areoles having sufficient white hairs up to 5 cm long, and these cover entire body. Spines are 8-12, reddish, yellowish or white, radials some 2 cm, and 1 of the 3 centrals some 5 cm long. Carmine-red 9 cm long flowers appear only on mature plants.

Brasilicactus

It is native to Rio Grande do Sul (Brazil), and the genus was established by C. Backeberg when he splitted the genus *Notocactus*, and then two species, *B. graessneri* and *B. haselbergii* were created as these had entirely different floral structure and seed shape, and the specific name is after the well known cactus grower Graessner of Perleberg. *B. haselbergii* which was initially described in 1885 as *Echinocactus haselbergii* by F. Haage of Erfurt is spherical, 12-15 cm across, crown with thick wooly felt and whole body with silvery-white spines having yellow tips. The ribs are around 30, areoles round and about 2 mm across and filled with white felt, 20 or more sharp and bristly radial spines which are initially yellow but later on becoming whitish and obliquely erect and the 4 central spines are longer, initially yellowish but later on changing to white. The fiery-red to orange, some 1.5 cm funnel-form flowers appear on the crown. It is propagated through seeds. *B. graessneri* is a very handsome and completely spherical (10 cm high and 10 cm across) light green cactus with depressed top. Whole body is thickly covered with yellow-brown spines. The tuberculate ribs are spirally arranged and more than 60, only 2-3 mm high and areoles are round with yellowish felt containing 65-66 spines some 2 cm long (about 60, thin, pale-yellow and glassy radial spines and 5-6 stout brownish-yellow central spines). The pale-green flowers appear about 2.5 cm in length. Its var. *albisetus* is similar to its parent but the spines are pale yellow-white. These plants are raised on *Eriocereus jusbertii*, however if seedlings are grafted on hybrid *Echinopsis*, the effect is more charming. However, its propagation through seeds is quite encouraging.

Browningia

A genus of columnar tree-like slow-growing spiny cactus with silvery- or green-blue stems possessing 20 or more ribs, is crowned by stout branches, can be propagated easily through seeds, and its most popular species is ***B. hertlingiana*** (syn. *Azureocereus hertlingianus*) with golden spines and tufted areoles.

Carnegiea gigantea

With the local name of Saguaro (an Indian origin name pronounced as *sa-wha-ro*, it is the State Flower of Arizona (USA), it has been found growing to an elevation of 1,370 m. A native of SW USA (Arizona and SE California) and Mexico (Sonora district), named for Andrew Carnegie (1835-1919), the Scottish born American industrialist and philanthropist, and has only 1, a very slow—growing species, the seedlings growing only 7 to 14 mm annually for several years reaching to a height of some 0.6 m in 20 years, 2.7 m in 35 years and 7.6 m in 80 years, in total it grows to 18 m in height. It is a green, columnar, cylindrical, branching and spiny species with 12-24 ribs. This bears short, funnel-shaped and fleshy white flowers at stem tips only though grows up to 18 m high, and can be multiplied through seeds. Pollination is carried out during the daytime by insects and Western white-winged doves, and at nights by bats. Areoles bear several spines, central one up to 8 cm and radials up to 2 cm, and flowering only on old specimens when plants are some 25 years old and have attained at least 4 m height. Flowers are white and 11 cm across. The fruits are ovoid, 4-chambered and contain thousands of shiny black seeds embedded in crimson pulp, the pulp is highly sweet and is liked by birds and Indians who prepare jam and jelly.

Cephalocereus

Though it is reported with 50 species but according to taxonomist Backeberg there is only one, *C. senilis* (syn. *Cereus senilis, Pilocereus senilis*). It derives its generic name from the Greek *kephalo* - head and *Cereus* – a genus of cacti, a native to southern USA to Brazil, a giant (15 m with 30 cm diameter) columnar cactus though grows not more than 2 cm a year, and where whole body even of small seedlings is covered by long white hairs as those of old men, hence its specific name *senilis*. Young plants have 12-15 ribs, and closely set areoles which produce 1-5 yellow spines together with 20-30 soft hair-like waved bristles which hang down to a length of 12 cm. Stems usually unbranched but sometimes branched and produce funnel-shaped flowers of white, yellow, pink or red on a cephalium-like structure on the sides or top of the stem where some 20-30 ribs change into spirally arranged tubercles only on the large plants when these attain 6-8 m height. *C. chrysacanthus* (syn. *Pilocereus chrysacanthus*), a native of Mexico grows straight up to 3 m with 10-12 ribs and basal branching. Closely set areoles are small and bear some 12 golden-yellow or brownish 1-2 cm long spines and a number of hair-like yellow bristles. Pink flowers are 8 cm long where tube is densely covered with pale-yellow or whitish hairs. *C. royenii* (syn. *Pilocereus floccosus, P. royenii*), a native of West Indies, branched but erect growing more than 2 m and of 4-8 cm thickness bearing 6 clear ribs, 14-15 spines, 4-5 central spines 2 cm or more long and 10 radial spines, together with numerous white hairs. Cephalium spiny and wooly and from where red flower apears.

Cereus

A native of South America and West Indies, it is the Latin name for a wax taper due to shape of some of the columnar species. The taxonomists have put 25-50 species under the genus. These are very fast-growing and cylindrical cacti becoming tree-like (9 m), and only after maturity these branch out from the upper part. The 5.0-7.5 cm white trumpet to funnel-shaped and often fragrant flowers open in the night and are followed by red, juicy and edible fruits. Their various species have now been included in other genera such as *Cereus acranthus, C. decumbens* and *C. versicolor* under *Haageocereus acranthus, H. decumbens* and *H. versicolor*, respectively; *Cereus baumannii, C. smaragdiflorus* and *C. strausii* under *Cleistocactus baumannii, C. smaragdiflorus* and *C. strausii*, respectively; *Cereus candicans, C. chiloensis* and *C. spachianus* into *Trichocereus candicans, T. chiloensis* and *T. spachianus*, respectively; *Cereus geometrizans* into *Myrtillocactus geometrizans; Cereus giganteus* into *Carnegiea gigantea;, Cereus senilis* into *Cephalocereus senilis; Cereus serratus* and *C. speciosus* into *Heliocereus serratus* and *H. speciosus, respectively; Cereus silvestri* into *Chamaecereus silvestri;* etc. Important *Cereus* species are *C. coerulescens* from Argentina growing up to 1 m or more and 3-4 cm thick, blue-green when young but becoming green with age and possesses 8 ribs with almost black areoles some 1.25 cm apart. Spines in the areoles strong, black, 13-16, 4 central ones 2 cm long and 9-12 radial ones, and the areoles are covered with white wool. Mature plants produce 20 cm long funnel-shaped white or red flowers which are scaly outside. *C. forbesii* from Argentina has branched stems, blue-green when young but afterwards darker, and 4-7 deep wing-like ribs. Large and yellowish to brownish areoles bear spines with bulbous bases, 5 cm long 1-2 central spines and 5-7 awl-shaped radials. Flowers are quite large, 25 cm long, inside white and outside green and purplish-brown. *C. hankeanus* from Argentina has columnar unbranched and 3-4 m high and 8 cm thick stems which when young are bluish-green but afterwards darker with 4-5 deep wing-like but notched ribs, and in these notches areoles are formed with one 3 cm long central spine and 3-4 yellowish-brown radials. Flowers greenish exterior, pinkish-white interior and 12 cm long. *C. hexagonus* from northern region of S. America is slow-growing and has branched but erect 10 m long and some 60 cm wide stems bearing 4-5 winged and notched ribs. Large areoles are initially yellowish but becoming grayish with age and bear 10 or more dissimilar spines, the longest being 6 cm. Flowers of pure white colour and 14 cm long are produced on quite large plants. *C. jamacaru* from S. America is quite similar to *C. forbesii* except that it has 5-8 more deeply notched ribs on the green stem bearing 8 yellow spines on each areole where 1 central spine is 12 cm long. Flowers are whitish-green and 20-30 cm long. *C. peruvianus* from SE Brazil and W. Argentina is a 10 m long and 10-20 cm wide, columnar and much branched cactus with 5-8 thicker and blunt ribs and brown spines, 1 central awl-shaped and 12 mm long and some 7 radials. Flowers 16 cm long and brownish-green. *C. p. monstrosa* has a number of crested mutations with flattened and convoluted stems. *C. serpentinus* (syn. *Nyctocereus serpentinus*) is a clump-forming, erect at first then clambering against walls or bushes, creeping without support or hanging with up to 3 m long jointed stems, bearing wooly areoles, white to brownish spines, and night-blooming large funnel-shaped white flowers. *C. tetragonus* from E. Brazil is a slow-growing erect cactus with free branching habit and grows some 3 m or more. It bears slightly notched and wing-like 4-5 ribs. Areoles whitish, large and bear 7-8 brown to black spines, six 1 cm long radials and 1-2 longer central spines. Only large specimens flower, and the flower colour is red and length some 12 cm.

Chamaecereus

Chamaecereus derives from the Greek *chamai* 'dwarf' and *Cereus*, a genus of cacti and comprises only one species, ***C. silvestrii***, a desert species from Argentina. Some botanists put it under *Lobivia*. Its stem segments root easily for propagation and requires little attention for its growing. It is a small (main stem 15 cm long and 1.3 cm in diameter) and mat-forming cactus spreading up to 30 cm, producing 8-10-shallow-ribbed pale-green, cylindrical, very soft and finger-like stems with numerous small and white areoles bearing tiny bristle-like some 10-15 short and whitish spines. It produces oblong-ovoid (peanut-shaped) branches some 6 cm long and 1.5 cm wide. Funnel-shaped some 5-7 cm long, 2.5 cm wide and orange-scarlet flowers appear during spring. It has produced hybrids with *Lobivia* which have flower colours in yellow, orange, red and purple.

Cintia knizei

With only one species, this genus was discovered in 1996 and the *Cintia* was named after the town of Cinti near Potosi in Bolivia. It is quite similar to *Copiapoa* or *Eriosyce*. Its roots are large and tuberous. Its ribs are irregularly tuberculate. Its new areoles on the top of the stem bear yellow flowers which are similar to *Copiapoa*.

Cleistocactus

It takes its name from the Greek *kleistos* meaning closed as the flowers do not open fully. A native to S. America with 30 slow-growing species having columnar, slender erect or rambling stems which bear many shallow ribs. Areoles closely set, round and possess numerous spines. Solitary flowers which are narrowly tubular and curved to almost straight appear from upper areoles and last 4-5 days. It is multiplied through seeds or by stem-tip cuttings. *C. baumannii* (syn. *Cereus baumannii*), a native of Argentina, Paraguay, Uruguay and SE Bolivia grows up to 2 m long and 2-3.5 cm across, bearing 12-16 rounded and shallow ribs divided by deep grooves, and closely set elliptic and 2-3 mm long areoles covered with short wool and yellow-brown hairs, each areole with 15-20 stiff and white-brown needle-like spines some 4 cm long though lowest ones being shortest. Flowers are tubular, curving at tips (zygomorphic), orange-scarlet to fiery-red in colour and some 6-7 cm long appear during summer in profusion. *C. b. flavispinus* bears pale-yellow areoles and spines. *C. dependens*, a native of Bolivia is a prostrate or pendulous cactus with 10-12 shallow ribs. Ribs bear white-tipped black and hairy areoles, each areole with 11-17 spines, and pink tipped pale-green flowers some 4.5 cm long. *C. smaragdiflorus* (syn. *Cereus smaragdiflorus*), a native of Argentina is so similar to *C. baumannii* that sometimes it is classified as one of its varieties. It grows up to 1.2 m bearing sylindrical and unbranched stem with 12-14 low ribs and numerous (22-28) spines, radial thin and central spines dark brown, stout and some 18 mm long in the areole. Flowers 3.5 to 6.25 cm long, tubular with orange exterior or reddish-yellow to scarlet tipped. *C. strausii* (syn. *Borzicactus strausii, Cereus strausii, Pilocereus strausii*), a native of Argentina and Bolivia grows erect with stems up to 1.2 m height and 7.5 cm diameter, sometimes branching from the base and bears up to 25 shallow ribs and white-felted closely-set areoles with 30 or more bristly white spines, sometimes being so densely formed that plant body is hardly seen, however, the older plants may produce a few (3-4) pale-yellow to white central spines measuring up to 5 cm. Flowers are tubular, dark red and 7.50-9.0 cm long which appear near top of the stem. *C. s. jujuyensis* bears dark red-brown central spines. *C. tominensis*, a native of Bolivia grows 1.2 m high and bears unbranched erect and light green stem with 18-20 ribs. Areoles give rise to 1.25 cm long 8-9 radial spines and 2.5 cm 3 long central spines. Flowers of tubular and light red colour are borne some 2.5 cm long during summer.

Copiapoa

It is a genus of 15-30 species (as per classifier) from Taltal (Chile). It takes its name from Copiaco, a province in Chile where these are found in abundance. Rich green stems tinged lightly brown under the sun, stems solitary, globular to somewhat conical or columnar, globose when young but afterwards columnar, throwing side shoots freely at the base, growing 5-20 cm high and 5-10 cm across and usually wooly at the top. Flowers are funnel- to bell-shaped and yellow flushed red. Most suitable for window sils and conservatory. These are multiplied through seeds, by offsets and through grafting. *Copiapoa* has many members known to cactus growers for more than 165 years as F. Ritter through Winter firm maintained many of the best species such as *C. alticostata, C. columna alba, C. dealbata, C. dura, C. hypogaea, C. longistaminea, etc. C. cinerea* (syn. *Echinocereus cinerea*), initially it starts as solitary but later on making clumps of glistening grey-white and cylindrical stems which grow more than 60 cm and have 18 broad and rounded ribs bearing 5-6 spines at younger age but with age these fall off leaving a 2 cm sharp spine. The flowers are yellow with 3-5 cm in length. *C. coquimbana* was collected from Coquimbo area of Chile hence is so its specific name, its stem usually solitary when cultivating though its old specimens make a clump at their natural haunts, is 13-20 cm tall, globular to conical and bears blunt and warty 10-17 ribs with large round and wooly areoles which contain black but graying with age 1 cm long 8-10 erect to curved radial spines and 2.5 cm long 1-2 central spines. From the wooly top of the stems bell-shaped, 3 cm long and bright yellow flowers appear during summer. *C. echinoides* (syn. *Echinocactus echinoides*) also produces usually solitary, globular to cylindrical stems up to 13 cm tall with some 10 well defined ribs which bear properly spaced large areoles. From areoles rise 5-7 radial spines, one 3 cm long erect central spine and 4 cm long greenish-yellow flowers. *C. gigantea* produces clustered stems which are globular when young but later on columnar and 90 cm tall with 14-22 ribs. Areoles are projected on the ribs and bear 9-14 mm long 8-9 yellow spines tipped darker, and from the crown which is felted brownish-red, emerge yellow flowers. *C. humilis* is clump-forming, 2.5-5 cm tall shortly cylindrical stems bearing inconspicuous ribs divided into clear tubercles, each one topped by a felted areole and 10 or more tiny radial spines. Flowers are yellow and up to 2.5 cm in diameter. *C. marginata* (syn. *Echinocactus marginata*) produces globular to cylindrical stems topped with yellow felt. Ribs rounded and 12-15. Large areoles are felty with two 2.5 cm long central spines and 7-9 radial spines. Flowers more than 2 cm in length and yellow. *C. montana* bears 10-17 prominent ribs which are divided into large (7 mm high) tubercles. White or brown felted areoles produce brownish-black to black, straight or slightly curved, 2 cm long 5 to 9 spines, radials 4-6 and the central stronger and 1-3. The faintly scented glossy and pale-yellow flowers, often many at a time, and 5.5 cm in diameter appear only on the crown though hidden under thick wool. Flowers are borne on 2-3 year's old plants. The word *montana in C. montana* means mountainous.

Coryphantha

Their natural haunts are from N. America to Cuba. This genus was formerly included under *Mammillaria*. This genus contains 64 species though some botanists consider it of 20-30 species only. It derives its name from the Greek *korypha* meaning 'summit' and *anthos* meaning 'a flower', and quite funnel-shaped flowers are borne on the crown of the stem. The stems are rounded to cylindrical, solitary or clumping widely with age, 2.5-15 cm long and bear conspicuous tubercles which on the upper side have a marked groove tipped with areole bearing spines. Propagation is through seeds or offsets. The cultivated species are being described here under. *C. andreae* is a glossy deep green globular to elongated cactus growing some 10 cm tall bearing 2 cm long rounded tubercles with felted grooves. Spines 15-17, central ones 5-7 stout and 2.6 cm long and the radials normally 10. Flowers are pale-yellow and 5-6 cm wide. *C. bergeriana* is a club-shaped cactus some 13 cm tall and with solitary stems which bear conical and grooved (1-1.5 cm long) tubercles. Four yellowish central spines are recurved and 2 cm long, while radial spines are often recurved and 18-20 in an areole, and the yellow flowers are 7 cm wide. *C. bumamma* by some botanists is considered to be a form of *C. elephantidens* which bears red to pink flowers. It bears solitary, globular and 12 cm long stems bearing rounded but flattened-top tubercles with fluted grooves. Central spines absent but radial spines are 6-8. Flowers are yellow to reddish and 8-10 cm wide. *C. clava* produces solitary bluish to grey-green cylindrical stem some 30 cm tall. Pointed tubercles are more than 1.5 cm long and obliquely conical and the areoles are small and wooly with about 12 mm long, 6-11 and yellowish radial spines while 1-4 radial spines are 2.5 cm long and thicker. Flowers are pale-yellow and 8 cm wide. *C. cornifera* (syn. *C. radians*) produces solitary, 10 cm or more long globular to elongated grey-green stems bearing obliquely conical tubercles wooly at the base. Though central spines in the areoles are absent but some 3 cm long 20 or more attractive cushion-forming radial spines are there. Flowers yellow, red at the base and some 7 cm wide. *C. c. echinus* (syn. *C. radians pectinata*) bears smaller spines and flower base green. *C. dasyacantha* (syn. *Escobaria dasyacantha*) produces 15 cm high, usually solitary (sometimes in small groups) and globular stems which shortly become cylindrical, *vis-à-vis* cylindrical and pointed 8 cm long tubercles. Areoles round, small and when young felted white though disappear later. Bristle-like white radial spines 25-35 and up to 1.5 cm long though brown to black and thicker central spines 7-13 and 2 cm long. Flowers are pink and white and 4.5 cm wide. *C. erecta* though has clustering of the stems in the wild but under cultivation these are solitary and up to 30 cm in height. Tubercles are 8 mm long, conical and blunt-tipped, and bear wooly areoles at younger stage. Spines 10-18, radials 8-14, 12 mm long and yellow to brown or grey while centrals 2-4, the lowest one reflexed and up to 2 cm long. Flowers yellow, 6-8 cm wide and stamens red-tipped. *C. hesterii* with oblong body and offset-producing cactus, bears wooly white head and purple flowers. *C. macromeris* is a green clump-forming cylindrical cactus some 15 cm tall. Cylindrical tubercles are 1.5-3 cm long and at young age the areoles are wooly. Awl-shaped reddish to whitish radial spines are 10-17 and 1-4 cm long while thicker brown to black central spines are 2-4 and 5 cm long. Pink to red-purple 8 cm wide flowers appear during summer. *C. m. runyonii* (syn. *C. runyonii*) in all its parts is grey-green and smaller. *C. minima* is only 5 cm tall with solitary or only a few basal branches, bearing about 3 mm long tubercles. White bristle-like radial spines are 13-18 though central spines only 2-4, thicker and slightly longer. Deep pink to purple flowers are 2.8 cm wide. *C. pallida* produces bluish-green, 13 cm tall, more or less spherical and solitary stems though sometimes form small clumps. Closely set tubercles bear 8 mm long about 20 white radial spines and black or black-tipped awl-shaped 4 cm long 3 central spines with lower one often down curved. Flowers are pale-yellow and 7 cm wide. *C. poselgeriana* (syn. *Echinocactus poselgerianus*) bears 20 cm high bluish-green, globular and solitary stems with thromboid 2 cm long tubercles and areoles with no wool. Radial spines 5-7, bulbous at bases, upper ones erect and fascicled, yellow to brown and up to 5 cm long 1 straight or arching central spine. Flowers pink and 4-5 cm wide. *C. sulcata* is grey-green, 8 cm tall and clustered stems which are globular to slightly elongated. About 1 cm long tubercles are rounded and flattened. White radial spines are 15 and central spines 1-3. Flowers yellow banded red and 5 cm wide. *C. sulcolanata* produces dark green and white-felted solitary globular stems depressed at the crown, 5 cm high and about 10 cm wide and after full establishment forms small clumps. Tubercles are 2 cm long, conical but somewhat angular and blunt-tipped. At younger phase the areoles are wooly. Fairly robust, the longest one being 4 cm long and 8-10 spines (all radial) are produced having yellow colouring tipped red to brown. Flowers are bright yellow and 8 cm wide. *C. vivipara* is a variable species with 15 cm tall, grey-green, globular to elongated clustered stems. Tubercles are cylindrical. Usually white radial spines number 12-20 or sometimes even more and brownish central spines number 4-6, 2.5 cm long and somewhat bulbous at the base. Flowers greenish outside and reddish inside open for a short time each day and measure 5 cm long.

Denmoza rhodacantha

The genus only with one small species is native to Mendoza and Salta provinces of Argentina, having globose or short columnar shape, bearing coloured yellow to red, short to long, and stout or sometimes even thread-like radial spines, and the spines may be either erect or recurved. Flowers are red, bilaterally symmetrical and resemble to those of *Cleistocactus* or *Oreocereus* as there also these are tubular with anthers sticking out the end.

Discocactus

A native to the eastern South America, this ultra-slow-growing, single to clump-forming genus with about 7 species, having compact growth habit, growing only up to 7.5 cm high and 25 cm across, produces enchanting white flowers on the top of the body. Its body is flattened to globose, ribbed and sometimes tuberculate. Areoles are filled either with fuzzy fibres or dense spines. Highly fragrant and long-tubed white flowers appear from the terminal woolly cephalium during night. To accelerate its growth, it is frequently grafted on quick-growing cactus rootstock.

Disocactus

It takes its name from the Greek *dis* meaning 'twice or double', *isos* meaning 'equal' and *cactus*, referring to the equal numbered rings of petals. It is native of northern S. America, Central America, Mexico and West Indies. This comprises some seven epiphytic species allied to *Epiphyllum*, most suitable for hanging baskets. The main stem is slender and cylindrical bearing flattened leaf-like branches which on the notches of the edges produce trumpet-shaped flowers. Propagation is through seeds, by cuttings, through entire lateral branches or their tips. The cultivated species are *D. macranthus* (syn. *Pseudorhipsalis macrantha*), a native of Mexico, grows up to 90 cm in length with arching to pendent stems while flattened side branches grow up to 5 cm, and yellow flowers 5 cm long and 8 cm wide appear during autumn or winter; and *D. nelsonii* (syn. *Chiapasia nelsonii, Epiphyllum nelsonii*), a native of Guatemala and South Mexico, initially grows erect but afterwards arching to pendent and up to 120 cm long, flattened side branches are up to 25 cm long and 3-4 cm wide, and the beautiful rose-purple flowers 10 cm long and 6 cm wide are violet-scented and free-flowering.

Dolichothele

The genus *Dolichothele* was erected in 1923 by Britton and Rose from the genus *Mammillaria* as in *Dolichothele* the seeds are not corky, the large flowers are long-tubed and the tubercles are long. It derives from the Greek *dolichos* meaning long and *thele* meaning a nipple due to very prominent tubercles. All the 12 species of this genus originate in Texas (USA) and Mexico and are closely related to *Mammillaria* having globular and clustering stems with large cylindrical tubercles and large funnel-shaped flowers. These are propagated through seeds and offsets and sometimes also through tubercles. *D. camptotricha* (syn. *Pseudomammillaria camptotricha*) is a clustering and globular cactus with prominent tubercles and entangled yellow spiny hairs and fragrant flowers. *D. longimamma* (syn. *Mammillaria longimamma, M. magnimamma*) is from Hidalgo (Central Mexico) where *longimamma* means long nippled. Spherical stems grow from 8 to 15 cm high and wide and through clustering make a clump, measuring 2-7 cm long and 1.0-1.5 cm across at base. Axils may be bare or with hairs but areoles have white felts, 3-12 white, erect and radiating radial spines measuring 0.5-2.0 cm though one erect central spine (sometimes 3) is smaller and pale-brown tipped black, and the 5-6 cm across and 6 cm long canary-yellow flowers normally three at a time appear on the crown. *D. sphaerica* (syn. *Mammillaria sphaerica*) from S. Texas and North Mexico bears 5 cm high and wide clustering stems, pale-green tubercles up to 1.5 cm long, and a rosette of 9-15, slender, 1 cm long yellow to white spines, and 6 cm or more wide pale-yellow flowers. *D. uberiformis* bears cylindrical, dark green, flabby and blunt tubercles, and four radial but no central spines.

Echinocactus

In Greek *echinos* means a hedgehog. A slow-growing genus of 10 species, native to southern North America, having globular when young to cylindrical stems which may grow in the wild 90 cm in diameter. Though in cultivation most species do not flower but their spines are very attractive. Clearly ribbed stems bear large areoles wooly at the top. Usually yellow, funnel-shaped and 3-7 cm long flowers are produced only on large plants. Many of its former species have been merged with other genera such as *E. cinerea, E. echinoides* and *E. marginata* as *Copiapoa cinerea, C. echinoides* and *C. marginata*, respectively; *E. eyriesii* as *Echinopsis eyriesii*; *E. poselgerianus* as *Coryphantha poselgeriana*; *E. pulcherrima* as *Frailea pulcherrima*; *E. rinconensis* and *E. setispinus* as *Thelocactus rinconensis* and *T. setispinus*, respectively. Genus *Homalocephala* is merged with *Echinocactus*. However, two American botanists ~ Britton and Rose placed the genus *Echinocactus* in the new genus *Gymnocalycium*. *E. grusonii*, a globular cactus native to San Luis Potosi to Hidalgo (Central Mexico), is a beautiful golden-spiked cactus growing some 1.3 m tall and 0.8 m across with wooly crown, bearing about 20-37 sharp ribs having deep fissures dotted with large and yellow to grey-wooly areoles which comprise of 8-10 erect or a bit curved and 2-3 cm long radial spines, and the 4-5 cm long and thicker 3-5 golden-yellow central spines arranged in form of a cross. Bell-shaped yellow flowers some 4-6 cm long and 5 cm across appear near the crown in several rows, forming a circle. *E. horizonthalonius* is from Mexico and southern USA, growing slow and grows up to 25 cm high and even more in diameter with solitary and blue-grey globular stem flattened at top. Stems comprise of 7-13 broad ribs, bearing many wooly areoles, 1 yellowish to pink or reddish thick central spine some 5 cm long which turns grey with age, and 6-9 erect or curved and 2-4 cm long radial spines. Pink flowers some 5-7 cm long and 5 cm wide appear even on young plants. *E. ingens* from Mexico has globular and depressed stems which grow up to 1 m and cylindrical with age. Ribs are 5-8 at younger stage but afterwards may go up to 50, and the areoles are well-spaced and yellowish wooly bearing one stiff and brown central spine some 2-3 cm long and 8 radial spines. Flowers are reddish-yellow. *E. platycanthus*, a native of Mexico is slow-growing, up to 60 cm high and wide, globular and depressed. Ribs are 30 and densely wooly at upper portion, spines grey-brown with 3-4 central ones which are some 2.5 cm long while 4 radial spines some 1.5 cm long. *E. texensis* (syn. *Homalocephala texensis*), a globular and flattened cactus native of Texas and Mexico grows up to 15 cm high and 30 cm across with tops having dense brown felt and the body with 13-27 acute ribs which bear well-spaced wooly areoles. Areoles bear reddish radial spines, 2 spines of 4 cm length and 4-5 remaining ones about 2 cm, though central spine is one and curving downward. Bell-shaped fragrant flowers which are pink with red in throat are some 6 cm long.

Echinocereus

This genus originates in southern part of North America [Mexico and USA (New Mexico, Texas, Oklahoma, Arizona and Colorado)]. It derives its name from the Greek word *echinos* meaning hedgehog and *cereus* a separate genus with which it is closely allied. A genus of generally 75 clump-forming species having oval to cylindrical growing stems up to 40 cm long, mostly with highly spiny areoles though fruits are spineless. These can be divided into two types: fast-growing ones forming sprawing or pendent branches from the base, and slow-growing straight growing single-stemmed species with beautiful spines. The flowers are long-lasting with reflexed petal tips, 5 cm wide, showy and funnel-shaped. These are propagated through seeds and stem cuttings. *E. baileyi* (*E. reichenbachii albispinus*) from Oklahoma (USA) bears usually solitary and spherical first but cylindrical later and erect but only 10 cm high with base-branching, possessing 16 white and erect radial spines about 5 cm long and no central spine. Its pink flowers are 5 cm long with toothed petals. *E. blanckii* grows up to 35 cm with blue-green cylindrical stems some 3 cm thick, at first erect but afterwards becoming prostrate and bearing 5-6 ribs with brownish areoles. Central spines are black while radial spines white and usually 9. Flowers are 9 cm long and reddish-violet, sometimes with pale centre. *E. chloranthus* bears 20 cm long erect and cylindrical stem which occasionally branches from near the base, having 18-20 low and almost rounded ribs and closely formed areoles. Areoles bear 1 cm long 12-20 radial spines and 3-6 central spines where 1 is 2.5 cm long. Flowers of yellow-green suffused brown which are funnel-shaped and some 2.5 cm long are borne along the stems. *E. cinerascens* is a clump-forming 30 cm long and 7 cm wide cylindrical cactus freely branching from the base, each stem having 5-12 ribs with whitish or yellow areoles, and each areole with 8-15 yellowish-white to reddish and 2 cm long spines, radials 8-11 and central spines 1-4. Floral buds in masses appear only on fully mature plants, initially brownish-violet but open to bright pink or purple in trumpet-shape and the flowers are 7-12 cm long. *E. dubius* has 10-20 cm long yellowish-green cylindrical stems, initially erect but soon becoming semi-prostrate with free-branching from the base. Ribs are 7-9 with rounded and yellowish-green areoles, each areole bearing 6-12 spines, radials 5-8, spreading and 1-3 cm long while centrals 1-4, 7 cm long and sometimes curving. Flowers are pink, 6 cm long and funnel-shaped. *E. enneacanthus* is 30 cm long cylindrical cactus more or less erect, soft in texture and freely branching from the base. Ribs are 8-13 with round and whitish areoles, each areole bearing 9-19 spines, radials white, 8-15 and some 1.5 cm long while centrals 1-4, brownish and up to 4 cm long. Red flowers some 5-6 cm wide are freely produced. *E. knippelianus* produces dark green 5 cm tall rounded to shortly columnar solitary stems bearing 5 shallow and spaced ribs and small rounded areoles. Young plants normally spineless but later 1-3, 1.5 cm long, white recurved and bristle-like. Flowers are purplish-violet, funnel-shaped and 4 cm wide. *E. pectinatus* is a 20 cm or more high and 3-6 cm across cactus having oval to cylindrical stem with branching from the base. Ribs low, straight and 20-23 bearing white and closely set areoles which at younger stage are hairy. Areoles are elliptic, felted white, 3 mm in length, curving and equipped with 22-31 spines, radials first pinkish but afterwards white, 22-25 and 9 mm long while central ones 2-6 in a single row and quite small. Bright pink funnel-shaped flowers some 6-8 cm long, white-hairy and spiny outside are produced freely. *E. pentalopus* has cylindrical, up to 13 cm long, prostrate and light green stems which branch out from the base and comprise of usually 5 spiralling ribs bearing closely set white areoles with only 3-5 whitish or yellowish radial spines initially but turning grey later on. Lilac to pinkish-violet 13 cm long flowers open almost flat. *E. procumbens* is of sprawling habit, growing up to 10-15 cm and is very similar to *E. pentalopus* but with darker central spines and purple flowers. *E. pulchellus* has simple joints to produce shoots. Individual joints are spherical to short cylindrical, up to 10 cm high and some 5 cm across with 5 mm wide, nearly 15 straight or spiral ribs which bear 8 mm apart spaced and 2 mm across round or oval areoles covered with white felt when young but later on shedding. Spines 1 cm long are 3-5, whitish or pale-brown. The pale-pink flowers are 4 cm long and wide. *E. reichenbachii* (syn. *E. caespitosus*) produces solitary and spherical but sooner changing to cylindrical stems, growing some 20 cm tall, bearing 10-19 ribs with dense areoles. Radial spines comb-like, white or yellowish with darker tips and 12-36 while central spines though often absent, sometimes 1-2 and rarely 7. Flowers are bright pink and 7 cm long. *E. rigidissimus* (syn. *E. pectinatus rigidissimus*) is an erect growing cylindrical cactus up to 30 cm long with 16-23 ribs and densely set areoles. Spreading and comb-like 1.5 cm long radial spines in various colours (whitish, pale-pink, brownish or reddish) are arranged in distinct bands of rainbow-like separate colours. Flowers are 6-7 cm long and of pink colour, exterior with white felt and spines. *E. triglochidiatus* (syn. *E. paucispinus triglochidiatus*) grows up to 20 cm tall with oval to shortly columnar stems branching from the base. Stems are erect as well as prostrate with 5-10 ribs bearing white-felted areoles. Central spine 1 which is 4 cm long though radial spines 3-6, 2.5 cm long and grey. *E. scopulorum* is a columnar cactus densely covered with small white spines. It produces large purple flowers.

Echinofossulocactus

It is a genus from Mexico which includes 32 species of variable cacti though some botanists put this under *Stenocactus*. This derives its name from the Greek word *echinos* meaning a hedgehog, from the Latin *fossulo* meaning 'ditch' and *cactus* due to the hollows between the spiny ribs being highly marked. The members of this genus comprise of globular to shortly cylindrical stems with graceful spination and highly attractive and almost free-flowering blooms. *E. albatus* produces some 13 cm wide bluish-green, solitary and globular stem with flattened top but turning cylindrical with age. Ribs are undulating and many with white-felting above. Areoles with rounded white felting. Spines yellowing to white, radials 1.5 cm long, bristle-like and nearly 10 though centrals 4, thicker and up to 11 cm long. The set up of areoles, their spines and colouration become very charming. Yellowish-white 2 cm long flowers appear from the top. *E. coptonogonus* produces some 10 cm wide spherical stems flattened at top, bearing up to 14 wide ribs with well spaced areoles. Spines 3-5, curving upward, longest one 3 cm, soft and red initially but becoming brown to grey and horny. It is free-flowering, with diameter of the floral mouth 4 cm and flower colour white with a central pink to purplish lining. *E. zacatecasensis* is named so as it was recorded from near Zacatecas (Mexico) and is the handsomest of all in the genus. It produces pale-green 10 cm wide and globular stems bearing some 55 thin and undulating ribs. Upper areoles felted-white, each bearing white, slender and spreading 10-12 radial spines up to 1 cm long, and 3 central spines where middle one is up to 4 cm long, straight, thick and flattened though outer 2 smaller and hooked. Flowers are white tipped pink and 3-4 cm across.

Echinopsis

It is a genus of 35 hardy cacti with globular or cylindrical, solitary or clump-forming stems ribbed prominently, grown for their trumpet-shaped sweetly scented pink or white flowers opening by evening and lasting for two days and where tube length is 10-20 cm. These are propagated through offsets or seeds. It takes its name from the Greek *echinus* meaning a hedgehog and *opsis* meaning similar. *E. eyriesii* is named so in honour of A. Eyries, cactus collector, who brought it to Le Havre in 1830. It bears globular base-branching stem which elongates with age and grows up to 15 cm in diameter. Ribs are 11-18 bearing rounded well-spaced felted areoles from where arise 10 radial and 4-8 central spines, all similar and dark brown. Pure white flowers 12 cm across are borne on 22-25 cm long tube. *E. intricatissima* is a free-branching, globular at first then columnar cactus which produces large white and fragrant flowers opening fully during open weathers. *E. multiplex* (syn. *E. oxygona*) is a native of Brazil, growing to a height of 15 cm. The plant body is pale to yellow-green and globular but elongating with age and branches freely from any part of the body and base thus forming a mounding clump. This possesses some 12-14 sharp ribs, bearing well-spaced areoles with stout and brown spines (radial awl-shaped and 8-9 while central 2-5, 4 cm in length and all brown tipped black). The flower tube is some 20 cm long and the fragrant pink flowers being produced freely are 15 cm across. *E. rhodacantha* (syn. *Denmoza rhodacantha*) from Argentina grows up to 15 cm or more in height, with initially globular stems which later on turn cylindrical. Body is dark green with about 15 prominent ribs. Spines are initially blood-red changing to grey via rust-red. Flowers are red, some 7.5 cm long and appear at the crown. *E. rhodotricha* (syn. *E. forbesii*) is a native of Argentina and Paraguay, growing to 30-80 cm tall with oval to cylindrical stems branching from the base. Stems bear 8-13 ribs with well-spaced yellowish to grey areoles. All yellowish to brown spines are 4-8, thick radials 4-7 and some 2 cm long while central spine nil to 1 and 3.5 cm long. Flowers along with tube some 15 cm long, 8 cm across and white. *E. werdermanniana* (syn. *Trichocereus werdermanniana*) is a native of Bolivia, attaining base diameter of 60 cm with green to grey cylindrical stems. It is highly spectacular when young and bears up to 6 ribs but later to 12. Closely set areoles are white with up to 4 cm long 12-18 yellow turning grey radial spines, and horny 6-9 reddish-brown central spines. Flowers appear at the top, pink when young but turning red with age.

Epiphyllum

This genus also includes *Epicactus* which has strap-shaped flattened green stems with notched edges from where flowers appear, and is propagated through stem cuttings.The members of *Epiphyllum* belong to Tropical South America to Mexico. It derives its name from the Greek *epi* meaning 'upon' and *phyllon* meaning 'a leaf' as flowers appear from leaf-like stem edges. It is a genus of 21 species of shade-loving cacti having woody base, growing up to 90 cm, and bearing leaf-like green and flattened or triangular fleshy stems in sections with notched edges from where large and trumpet-shaped flowers of white, yellow, red, scarlet, violet or their intermediate colours appear through their hundreds of varieties, all listed as *E.* × *ackermannii* though original species is seldom observed in flowering. Some species produce the flowers during day time while the fragrant ones mostly by evening. Flowers are showy, bell-shaped and up to 15 cm across in singles and doubles. In fact, the hybrids were produced involving *E. ackermannii* and *E. crenatum* with species of *Heliocereus* and *Selenicereus* and such hybrids are still called 'phyllocacti' or 'orchid cacti' but are listed as *E. ackermannii*. The choicest varieties are 'Autumn' (pink), 'Cambodia' (tepals ruffled purplish-red), 'King Midas' (golden yellow, centre darker), 'Pacesetter' (inner tepal's edge serrated paler, outer tangerine edged purplish and centre orange), 'Professor Ebert' (pink), 'Reward' (yellow but inner tepals paler-cream striped yellow), 'Sun Burst' (burnt orange but inner tepal's eye red-bronze), 'Truce' (inner tepals white and outer pale-green), *etc.* Though red type varieties such as 'London Glory' are free-flowering but white and yellow hybrids are mostly fragrant producing flowers from the base of the stems and are not free-flowering, however, the plants are vigorous such as 'London Sunshine' (yellow with white centre). These are very suitable for hanging baskets. *E. ackermannii* (syn. *Nopalxochia ackermannii*) grows up to 60 cm with crenate margins which bear tiny areoles having only a few spines, and crimson flowers some 15 cm long. *E. anguliger* is an ererct and bushy plant with leaf-like lanceolate branches having deep rounded lobes and 15 cm long and 10-13 cm wide fragrant flowers which are white within and yellowish-brown outside. *E. crenatum* grows up to 3 m, first erect and then pendent flattened leaf-like stems which bear 20 cm across white and slightly fragrant flowers in summer. *E. oxypetalum* also grows 3 m with flattened and frequently branched 12 cm across stems, first erect and then pendent, and produce tubular white flowers some 25 cm long by the evening.

Epithelantha

It originates in the SW USA and Mexico and comprises quite appealing three species which superficially appear as *Mammillaria* though flowering only at tips instead of from their bases, and out of these *E. micromeris* is widely cultivated as house plant. They are clump-forming and quite charming even without flowers. It takes its name from the Greek *epi* meaning upon and *thele* to the mode of its flowering. Their propagation is through seeds and offsets, however, young seedlings are grafted on hybrid *Echinopsis* soon after sprouting. In *E. micromeris* the specific name means made up of tiny parts and it originates to Mexico and New Mexico, Arizona and Texas, first producing solitary stems which sometimes later becoming more or less spherical, globe-shaped or columnar and clump-forming of the size of 2.5-5.0 cm long and wide with slightly depressed top though in cultivation these are much more, tapering below and flattened above, and the body is covered with a short tuft of thick white felt. Many closely set spirally arranged ribs are divided into numerous 1 mm high tubercles which bear some 20 white radiating radial spines some 2 mm long, 2-4 white (sometimes black-tipped) central spines some 1-6 mm long. White or pale-pink flowers some 6 mm in diameter appear on the crown from newest areoles.

Eriocactus

A genus of columnar cacti growing to more than 1 m in height and 10 cm acros in its natural haunts of Rio Grande do Sul (Brazil) producing side shoots at the base with age freely. The stem is green with brown felts on the crown. These are propagated through seeds and from side shoots. Its most popular species is *E. leninghausii* which was introduced into Europe in 1895 and was named as *Pilocereus leninghausii* by F. Haage though in the same year it was further named as *Echinocactus leninghausii* by Prof. K. Schumann, and further in the coming years A. Berger classified it under the genus *Notocactus* and from which it was separated in 1942 by C. Backeberg and classified in the genus *Eriocactus* due to this being columnar and globular bearing yellow stigma though other 'notocacti' have purplish-violet. *Eriocactus leninghausii* is broadly globular but later branching with golden-yellow cylindrical stems covered with spines, has oblique crown covered with spines and wool which shed later on, and with up to 30 or more, 5-7 mm high, slightly wavy and narrow ribs.Radiating radial spines are 0.5-1.0 cm long and 15-20 in number though darker, flexible, stiffer, erect or a bit curved central spines are 3-4 in number and 4 cm long. Yellow flowers several in number at a time and 6 cm across appear on the crown. *E. manifucus* var. *nigrispinus* is a clump-forming globular cactus having about 10 ribs with areoles, each areole with white or yellow spines and yellowish-cream flowers.

Eriocereus

It is native to northern Chaco (Argentina) and derives its name from *erion* meaning wool and *cereus* referring to the wool in the axils of the bracts on the ovary. The slender plants are usually erect initially but later on creep often forming large clumps. Flowers are large and the ovary is covered with bracts bearing felts and spines in their axils and fruits are also spiny. *E. jusbertii* is thought to be a hybrid between *Echinopsis eyriesii* and a *Cereus* species and has more erect branches up to 2 m long and usually with 6 well-spaced low ribs bearing very short spines and funnel-form white fragrant flowers some 20 cm long. This cactus is a very desirable stock with cactus lovers. *E. martini* grows straight up to 2 m with basal branching producing dark green 2.0-2.5 cm thick shoots bearing 5-6 ribs which merge with age. Areoles 3 mm across are round felted grey, each bearing 5-6 red radial spines and 2-3 cm long pale-brown or white centrals with darker tip. Some 20 cm or more longer and 17-18 cm across night-blooming white flowers are funnel-form. The other prominent species are *E. pomanensis* and *E. tortuosus.*

Eriosyce

It is native of Santiago de Chile and is found there to an elevation of 180-2,250 m. Its main species is *E. ceratistes* which is known to cactus growers since 1837. Investigations in Chile of this particular species revealed that there are many of its varieties/ecotypes such as *E. c.* var. *combarbalensis, E. c.* var. *coquibensis, E. c.* var. *jorgensis, E. c.* var. *mollesensis, E.c. vallenarensis* and *E. c.* var. *zorilaensis* as named by Backerberg after the locations these were recorded. The most popular specimens among cactus growers from Ritter's collections as classed by Krainz and Ritter as species are *E. ausseliana, E. lampampaensis* and *E. lhotzkyanae. E. ceratistes* produces some 50 cm across and even taller spherical stems which are dull pale-green bearing felted crowns and with up to 30 or even more 2-3 cm high and acute ribs, *vis-à-vis* large white-wooly areoles bearing 18-20 spines where both types of spines are not easily distinguishable, however, the bulbous-base brownish-yellow to brownish-black spines are erect or slightly curved and 3 cm long and the carmine flowers are 4-5 cm long, 3 cm across and bell- or funnel-shaped.

Escobaria

This genus of small plants from SW USA (Texas) to Mexico is closely related to *Coryphantha.* Its pink to yellow to tan and golden-brown flowers appear usually 1 cm across, sometimes even up to 6 cm in some species. *E. hesteri* from Texas appears in clusters some 10 cm across, producing 4 cm wide stems though plant stature is dwarf and its dark pink flowers are about 3 cm across which continue appearing throughout the summer. Under dry conditions it can tolerate a temperature of -10 °C and is somewhat rot-prone. It is propagated from seeds or cuttings. *E. leei* from Texas produces 1 cm across stems in clusters spreading up to 15 cm. Under dry conditions it can tolerate up to -7 °C temperature. Its brownish-pink flowers appear in early summer. It is propagated through offsets, seeds and sometimes even through grafting. *E. minima from* Texas is close to *E. leei* in appearance and is a protected species. Its 20 cm wide clusters comprise of several 2 cm wide stems. It produces several flushes of showy pink flowers about 2 cm across. Under dry conditions it tolerates up to -7 °C temperature. It is propagated through seeds, cuttings and occasionally through grafting. *E. roseana* from Mexico produces densely spined stems in 12 cm wide clusters, each stem being some 3.5 cm thick. It is rot-prone and survives up to -7 °C temperature if kept dry. It is propagated from seeds. *E. vivipara* from SW USA is found in a quite wide area with several varieties where certain survive prolonged cold temperature of up to -23 °C. It throws out offsets with stems about 6 cm thick in a cluster of about 20 cm diameter. Its pink flowers about 4 cm in diameter appear during spring. It is propagated through seeds.

Espostoa

The name has derived to commemorate Nicholas E. Esposto, a botanist of Lima (Peru). It is a genus of 6 to 11 species scattered in South America, especially Peru with interesting columnar bushy or large tree-like and branching cacti bearing 10-30 ribs. Most species have thick white wooly-covering, being more at the sunny side but only after 30 years of age, masking the short and sharp spines but some other members have normal armature and resemble *Cereus*. Plant flowers on lateral pseudocephalia (the elongated section of the stem tips modified from several ribs) only but on the upper part of the stem when fully mature. Propagation is through seeds and tip-cuttings of stem-branches. *Espostoa blossfeldiorum* (syn. *Thrixanthocereus blossfeldiorum*) from Peru is wrapped in fine hairs and grows solitary or with a few branches up to 1 m or so and 10-15 cm in diameter bearing some 18-25 low ribs felted dense-white in areoles which soon fall off. The radials are 20-25, glossy and 1 cm long while central 1, awl-shaped, 3 cm long and black-tipped. Flowers are white, funnel-shaped and 5 cm long. Also from Peru and Ecuador is *E. lanata*, a most spectacular but slow-growing arborescent cactus up to 5 m high and branches up to 1 m long and 15 cm thick with 20 low and rounded ribs which are covered with dense white wools. Areoles are full of persistent white or yellowish hairs and acicular variously coloured spines, radials 12 or more and yellowish or reddish and 6 mm long. Flowers emerging from pseudocephalium are white with pink interior, 6 cm across and appear only on mature plants on the upper part of the stem. The cristate forms of this species are among the most popular.

Eulychnia

With seven species in the genus, being native to the deserts along the coast of Chile and Peru, these large plants grow up to 7 m in height having many branches. With little or no rainfall, these cacti survive on the moisture through condensation brought about by the heavy and frequent fog of the area. Large spines are patterned on tuberculate ribs of the stems. The floral tubes are covered with spines, wool, scales or all the three and when the flowers open look like a ball popping open to one side.

Ferocactus

It takes its name from the Latin *ferox* meaning fierce and *cactus*. A slow-growing genus from North America (USA and Mexico) comprises some 35 unbranched species of barrel-shaped or spherical which become columnar after many years, sharply ribbed and coloured, usually curved and highly spiny cacti with apical funnel- to bell-shaped flowers which are not produced very often in cultivated ones, but these are cultivated especially for their interesting spines. These are sun-loving so are not suitable as house plants. *F. acanthodes* from California though produces glaucous-green, solitary and oval stem which with age becomes cylindrical and up to 3 m tall, having 13-23 blunt ribs and sunken and grey wooly areoles along the ridges. Usually red or pink and sometimes even white or yellowish curved radial spines, 9-20 in number and 4-6 cm long mesh up on the body to form a basket-like covering though somewhat flattened, flexible and with hooks at tips, 1 red central spine is up to 12 cm long. Flowers are 5 cm long and yellow to orange. *F. chrysacanthus* has green and spherical stem, growing up to 1 m high and 60 cm wide and with 15-20 ribs which are thickly covered with curved and yellow-white spines. Funnel-shaped 5 cm across flowers of yellow colour (rarely red) appear not before the plants attaining 25 cm of diameter. *F. haematacanthus* (syn. *F. stainesii haematacanthus, Hamatocactus haematacanthus*) from Mexico grows up to 50 cm. The stems are bright green, globular to shortly columnar, crown felted and has 12-20 wide, erect and notched ribs, and in the areole bearing 4 cm long 6 hooked and dark red coloured spines tipped yellow, while 6-12 cm long, thick and some 4 central spines. Flowers are purplish-red and 6 cm long. *F. horridus* from California produces globular and top flattened stems which later become oval and finally cylindrical some 2 m long. Ribs are 13, wide and blunt with well spaced areoles. Areoles bear 8-12, spreading, 3-4 cm long white radial spines, and 6-8 stout, spreading, 5-7 cm long central spines where 1 is flattened, hooked and 10 cm long but on large specimens. *F. latispinus* (syn. *Echinocactus corniger*) from Mexico is a green, flattened and spherical cactus growing up to 30 cm long with 40 cm spread. Stems bear 8-20 narrow ribs with thick, hooked and yellow or red spines, radials 6-12 in number and up to 2.5 cm long though centrals about 4 stout and 3.5 cm long whereas 1 hooked and quite flattened, 7 cm long, reddish, and bright when young. Plants only after attaining the 10 cm width, produce flowers of creamy or red colour having 3.5 cm length. *F. melocactiformis* from Mexico grows up to 60 cm in height. In cultivation the stem is generally solitary and blue-green though in wild it produces offsets. It comprises of 25 ribs on grown up plants. Areoles bear 9 recurved radial spines and 3-4 centrals, all of yellow colour and up to 6.25 cm long. Pale-yellow flowers are produced during summer with 5 cm length. *F. recurvus* is a bright green globular plant bearing about 13 ribs with 8 yellowish spines though central spine is flat, red and hooked. Though flowers are carmine but appear quite rarely. *F. wislizenii* is native from Texas to Arizona and produces up to 2 m stems which are first globular later on becoming oval and finally cylindrical. Stems bear 15-25 acute ribs with large and spaced areoles. From areoles arise some 24 spines, laterals 20, bristle- or awl-like and 5 cm long, while tip-hooked centrals 4, quite thick and flattened, and 5-6 cm long. Reddish-yellow flowers with green exterior are borne some 5-6 cm in length.

Frailea

The name of this genus commemorates Manuel Frail (b. 1850), a Spanish gardener who remained for many years Superintendent of the U.S. Department of Agriculture's Cactus Collection in Washington D.C. This genus belongs to Andes and sub-tropical South America. There are some one dozen species in this genus which comprise of small, spherical to columnar cacti with tuberculate ribs and large flowers which appear in profusion. The spines are short bristle-like. These are propagated through seeds and by offsets. *F. castanea* from Brazil and Paraguay grows 2-4 cm wide bearing globular with flat top and 10-15 brown to green wide but shallow ribs. Areoles are minute, each one having some 5 short and reflexed black spines. Flowers yellow and as large as the plant. During cloudy weathers though flowers do not open but through self-pollination there is proper seed setting. *F. columbina* from Colombia bears some 4 cm wide sub-globose stems with top depressed and 15-18 shallow ribs. Areoles are very small though give rise to 20-25 clustered yellowish spines, largest one being only 6 mm. Flowers are yellow and some 2.5 cm long. *F. gracillima* from Paraguay grows up to 10 cm tall, 2.5 cm wide and clustered, initially bearing globular stems but turning cylindrical later on and bearing 12-14 low and rounded ribs. Small areoles give rise to 18 clustered and bristle-like white spines some 6 mm long. Flowers are produced some 3 cm long of yellow colour with reddish throat. *F. pulcherrima* (syn. *Echinocactus pulcherrima*) from Uruguay forms dark green stem clusters which are shortly ovoid, 5 cm tall and 2.5 cm wide, and bear 18-21 low and rounded ribs. Each areole gives rise to 10-14 minute white to pale-brown spines. Flowers are yellow and up to 2.5 cm long. *F. pygmaea* from Argentina and Uruguay bears dark green, globular and 3 cm wide stems with flattened top and forming a cluster. Ribs are low, rounded and 12-21. Areoles are minute with 6-9 recurved to adpressed bristle-like spines. Flowers are yellow and up to 2.5 cm long.

Geohintonia mexicana

A quite unlike of any other cactus, this distinct genus discovered in 1991 has the only one species with globose to columnar body after attaining the maturity, and is found only in Nuevo Leon in the cliffs of pure gypsum. Some 20 flat and quite prominent ribs resembling the cooling fins on a circular radiator are formed on the frosty-blue-grey body. The rib edges are dotted with white areole bearing flexible short spines on the top of the plant. Dark pink funnel-shaped flowers appear from the woolly apex bearing the yellow stamens and a white stigma. It is propagated through seeds and grafting.

Gymnocalycium

It takes its name from the Greek *gymnos* meaning naked and *kalyx* meaning a bud as the flower buds here lack the spines, spine-like bristles or hair formation. However, the two American botanists ~ Britton and Rose put *Echinocactus* also under this genus. A native of South America, especially, Argentina, Brazil and Paraguay, this genus contains 60 species of globular to shortly cylindrical cacti with clear ribs bearing areoles above 'chin-like' projections and a few stout and generally recurved spines. The crowns generally bear smooth and scaly buds and the masses of trumpet-shaped flowers. Propagation is through seeds and offsets. *G. andreae* from Argentina produces somewhat flattened, bluish-green and globular stems up to 4 cm across, bearing 8 ribs and whitish areoles, each producing some 1 cm long 7 white radial spines and 1-3 upward curved central spines. Flowers are yellow, funnel-shaped and 3 cm across. *G. baldianum* (syn. *G. venturianum*) from Argentina was named so in 1905 by Spegazzini, an Argentinian botanist. It produces solitary dark grey-green spherical stems growing to 4.5 cm high and 7 cm wide with depressed top, bearing 9-11 rounded and broad ribs which are divided through deep grooves into tubercles, small sunken areoles with 5-7 yellowish to grey and red-tinged at base, thin and some 6 mm long, and straight or curved radial spines though central spines are absent. Red and funnel-shaped flowers some 4 cm long appear on the crown. *G. bicolor* (syn. *Echinocactus bicolor*) is globular and no offset-forming cactus with flattened top. Flowers are buff-white throated-carmine and appear 1-3 on the crown at a time. *G. denudatum* from Brazil and Argentina was sent from southern Brazil or Uruguay in 1825 by Sellow to Berlin where three years later it was first named as *Echinocactus denudatus* but in 1845,

Dr. Pfeiffer put this under genus *Gymnocalycium*. The specific name *denudatum* means naked. Its two forms exist, both are quite attractive with profuse flowering habit though in one case the flower size is small. The stem is pale to rich green, spherical but with age becoming slightly columnar and base-branching, top-depressed, 20 cm tall and 5-15 cm wide and with 5-8 broad and shallow ribs divided by faint cross grooves and lightly felted large areoles. Yellowish-brown turning grey and up-curving only radial spines, some 5-8 and 10 mm long are present in each areole. Flowers arise near the crown and are white or pale-pink, some 5-7 cm long and funnel-shaped. *G. gibbosum* from Argentina produces blue-green and spherical (when young) to grey-green and columnar (when old) stems growing to 20 cm tall and 9-15 cm wide. The meaning of specific name *gibbosum* is lumpy. Stems bear 12-19 tuberculated ribs, tubercles some 1.5 cm protruding. Large round to elliptic areoles are wooly and grey-brown with clear 'chins'. Spines 8-12, straight or slightly curved, and are pale-brown turning grey. Radial spines are 3.5 cm long, 7-10 and obliquely erect and 1-2 central spines are awl-shaped. The flowers are funnel-form, 6 cm long and white to reddish. *G. friedrichii* from Gran Chaco (Argentina) produces rough-textured and brownish-red or in some species even dark purple spherical stems growing only 5-6 cm high and wide and bearing 8 acute ribs. Younger ones have pale stripes on the sides of the ribs, shining and wet-like skin and white felted areoles bearing fragile yellowish spines so that it looks like a precious marble stone. Pale-pink flowers appear with slender blue-green tubes through scaly flower buds from the depressed top. *G. lafaldense* (syn. *G. bruchii*) from Argentina produces globular and clump-forming stems some 4 cm high and bearing some 12 low ribs and elongated areoles that are wooly to some extent. Areoles bear 12-18 spines, slender radials 12-15, some 5-10 mm long and at least with white tips while centrals 0-3 and slightly longer. Flowers are purplish-pink and 3-5 cm long. *G. mihanovichii* (syn. *Echinocactus mihanovichii*) from Paraguay produces greyish-green, globular with somewhat flattened top and 5 cm wide stems bearing broadly triangular some 8 ribs, with closely set small areoles. Only 7-8 yellowish to pale-brown radial spines some 4 cm long are present. Flowers are yellow-green to reddish and some 4-5 cm long. Many of its varieties and mutants such as 'Black Cap', 'Blondie' (yellow), 'Hibotan' (yellow), 'Optima Rubra' (pink), 'Red Cap', 'Red Head', 'Rosea', 'Yellow Cap', *etc.* in yellow and red colours have arisen which for their photosynthetic functioning are grafted on some green cacti, especially *Hylocereus*. *G. multiflorum* from South Brazil to Argentina produces solitary or clustering pale-green and globose stems some 9 cm high and 13 cm across bearing 5-15 tuberculate ribs, each large tubercle with 1 cm long elliptic areole from where appear 7-10 stout, awl-shaped, comb-forming and yellowish to pinkish radial spines, the largest spine being up to 3 cm, and up to 4 cm long, short-tubed and campanulate white to pinkish flowers. This species is reported having many forms at its place of origin. *G. pileomayensis* (syn. *Weingartia pileomayensis*) is a bluish-green tuberculated spherical, each tubercle with felted-white areole, bearing 6-11 white and stout spines. Profusion of yellow flowers appear on the crown. *G. platense is* from Argentina. *G. quehlianum* from Argentina grows globose up to 5 cm high and 7 cm wide when young and has grey-blue to brown, flattened and cylindrical stems up to 15 cm high with age, bearing some 8-13 knobby-rounded and tuberculate ribs. Each areole which is oval and wooly-white and about 2 mm across produces some 5 mm long 5 curved spines which are grey-brown to brownish-red at the base, and several 6-7 cm long and 5 cm across funnel-shaped white flowers with red throat appearing on the crown. *G. saglione* from Argentina produces fast-growing solitary rounded stems about 30 cm high and wide, bearing 30 or more, grey to blue-green and rounded ribs with large, oval and wooly areoles. Central spines are straight and shorter than radials. Awl-shaped and recurved radial spines are 7-10 or even more, 2.5-4.0 cm long and off-white or greyish to reddish. Flowers are pale-pink or white and 4 cm long. *G. schickendantzii* from Argentina grows up to 10 cm high and across with dark green spherical stems bearing 7-14 deeply notched ribs, long grey-brown tipped red radial spines and greenish-white to pale-pink flowers some 5 cm across. *G. spegazzinii* (syn. *G. loricatum*) from Argentina produces usually solitary globose or shortly cylindrical stems some 15 cm high bearing 10-13 flattish ribs and round and brownish areoles. Only 5-7 brown to grey, awl-shaped, recurved and sturdy radial spines some 3 cm or more long are present. Flowers are pink to white and some 8 cm long.

Haageocereus

The genus name is combination of two words, *Haage* (a nursery firm in Germany) and *cereus*. A genus which originated in Peru, comprises of 40, mostly erect or some completely prostrate growing species, of cylindrical, thickly and small- spiny cacti that produce funnel-shaped nocturnal flowers, the prostrate-growing ones are most suitable for hanging baskets. These are propagated through seeds and cuttings. *H. acranthus* (syn. *Cereus acranthus*) is a clump-forming as branches from the base, and grows erect some 90 cm in length and 8 cm wide and the stems bear 10-14 clear ribs that are notched on the large areoles, each areole giving rise to 20-30 yellow radial spines some 1 cm long while 2 awl-shaped 2 cm long central spines. Flowers are white with green splashing and some 6-9 cm long though a pink form of this species has also been recorded. *H. chosicensis* is a green stemmed upright species, growing 1.5 m tall and 1 m wide in clumps, individual stem being some 10 cm across with 19 or so ribs and the areoles bear dense bristle-like white radial spines and white, golden or red central spines. Tubular flowers some 7 cm long appear near the crown in white, lilac-white or pinkish-red colour. *H. decumbens* (syn. *Cereus decumbens, Borzicactus decumbens*) is clump-forming and prostrate growing species with 30-90 cm tall and 1 m width of the plant though individual green stem is only 10 cm across bearing 20 or so inconspicuous ribs which are notched between small and yellowish-felted areoles, each areole bearing about 30 some 5 mm long radial spines (white when young), and black to dark brown, 2-5 in number and some 2-4 cm long central spines. Flowers are fragrant white and some 7 cm long. *H. versicolor* (syn. *Cereus versicolor*) is a base-branched clump-forming species, erect growing with 2 m height and some 1 m spread, individual green stem being only 8 cm across and bearing 12-22 inconspicuous ribs and small wooly areoles which are closely set, each areole bearing 25-30 yellow to golden, red, brown or bicoloured some 5 mm long radial spines, and 1-2 with a few pointing downwards central spines which measure some 4 cm in length. *H. v.* 'Aureospinus' has complete yellow spines. The long-tubed flowers appearing near the crown are white and 10 cm long.

Hamatocactus

A native of Texas and N. Mexico, the genus has one very attractive species, *H. setispinus* which means bristly, owing to the species being full of attractive spines. The species produces dark green, globose at first then cylindrical, and 15 cm high and 10 cm or more across stems bearing 12-15 sharp and pronounced ribs with oval areoles having short white felt. Radiating, thin and oblique radial spines are dark brown or white and 12-15, the bottom ones being some 5 mm long, and the central spine is 1, erect and hooked, some 2-4 cm long, and dark brown with pale tip. Flowers having broad limb, yellow interior and carmine throat, and some 7 cm long appearing near the crown. The species has many varieties, all being hardy and rewarding.

Harrisia

In the wild, it is recorded from Florida through the Caribbean islands and in South America from its eastern part to Argentina. *Harrisia* comprises of some 20 species but only a few are in cultivation in the warmer regions. This bears thin stems with only a few ribs. Sufficiently spaced areoles are present on the stem filled with several needle-like spines, and the large white flowers with bristles or woolly hairs in the tube appearing from the areoles are nocturnal and fragrant. Fruits are sphere-shaped, yellow or red at maturity, and fleshy with sparse bristles on the body. *H. jusbertii*, probably an intergeneric hybrid of *H. pomanensis* × *Echinopsis eyriesii*, is only the species which is in cultivation.

Hatiora

It is a genus of epiphytic cacti with short, cylindrical and jointed stems, each being swollen at one end like a bottle. These require partial shade and properly drained soil for their growing. These are propagated through stem cuttings and seeds. *H. clavata* (syn. *Rhipsalis clavata*) is a pendent epiphytic cactus growing 60 cm in height and 1 m in spread with multi-branched dark green cylindrical stems, each widening towards the tips. Bell-shaped 1.5 cm wide terminal flowers mass the plants in late winters when the plants are 30 cm high. *H. cylindrica* is a highly branched cactus with red-spotted round joints. Flowers are red and orange. *H. salicornioides* is a free-branching bushy epiphytic cactus growing some 30 cm high and wide, where jointed stems have joints with expanded tips and bell-shaped terminal golden flowers in spring.

Heliocereus

It derives its name from the Greek *helios* meaning the sun and *cereus*, a genus under which once these cacti were classified. In fact, the flowers in *Cereus* open in the night though those of *Heliocereus* open in the sun. A native of Mexico and Central America, the genus includes 3-7 slender, branched and jointed, and angled or wing-stemmed columnar cacti with crenate ribs, some being more or less self-supporting erect while others possess scrambling stems and require support. Ribs bear bristly spines. Flowers are large, showy, funnel-shaped and are borne only one at a time at any point on the stem, each flower lasting for 2-3 days. These cacti are most suitable as indoor plants but at a sunny side. They are propagated through seeds or cuttings. *H. amecaensis* from Mexico with its light green 5 cm thick stems grows prostrate or clambering some 90 cm in height and spread, stems bearing 3-5 ribs with white bristle-like short spines. Flowers appear in summer in funnel-form with green exterior petals and white interior petals, and are 10-13 cm in diameter. *H. cinnabarimus* from Guatemala branches at the base, stems being erect, spreading, with aerial rooting and some 2.2 cm thick though plant height and spread is about 90 cm. Stem ribs 3-4 with each areole having 10 white bristle-like spines. The funnel-shaped flowers of the diameter of 15.0-17.5 cm appear during summer with outer petals green, inner petals white and rose-pink style. *H. serratus* (syn. *Cereus serratus*) from

Guatemala is an erect growing, some 30 cm in height, and branching cactus where stems are 4-angled with toothed angle-margins and bear some small yellowish spines. *H. speciosus* (*Cereus speciosus*) from Central America and Mexico is often hybridised with *Epiphyllum*. It has erect to trailing 4-angled and base-branching stems growing up to 90 cm or more in height and 5 cm in breadth and bearing numerous small needle-like yellowish to brownish spines some 12-19 mm long. Initially at younger stage the stems are red but later turn bright green. The ribs are 3-5 and undulating with felted areoles some 3.75 cm apart. Scarlet funnel-shaped and some 20 cm long and 15 cm across flowers with a bluish sheen open during summer and last for several days.

Hylocereus

It takes its name from the Greek *hyle* meaning the wood and *cereus* a genus from which it was created, owing to its native haunts in the tropical woodland. A native of West Indies and Mexico to Peru, this fast-growing handsome genus comprises some 20 climbing semi-epiphytic or epiphytic cacti jointed into sections, having slender, erect, climbing, angled and winged rooting stems and large trumpt-shaped flowers opening by evening. These are propagated by cuttings. These are also used as under-stock for grafting. *H. lemairei* from Trinidad and Tobago bears triangular deep green stems some 3 m or more in height with notched edges having well spaced areoles, each areole with 1-2 small spines. Flowers have white exterior, reddish-flushed interior and some 25 cm long. Its 8 cm long and ovoid fruits are edible. *H. triangularis* from Tropical America grows up to 20 cm height with softly angled stems, stem wings horny and flowers white some 20 cm long. *H. undatus* from Tropical America is a fast-growing, free-branching and dark green climbing cactus growing up to 3-5 m in height, stems being 7 cm wide, jointed into sections, 3 (sometimes even 2)-angled, notches round at margins, and singly spined or in a tuft of 2-5. Flowers are flattish white and some 30-35 cm across but lasts only for a night. Fruits are ovoid, red, 10 cm or more long and edible.

Islaya

It is native of southern Peru. Todate its detail information is not available though it's one of the species, *I. flavida* (syn. *Islaya grandiflorens*) was discovered by F. Ritter who mentioned that it has pale-yellow spines with darker tips (*flavidus* meaning yellowish). Backeberg thought it to be identical to *I. grandiflorens*. These have been found growing in the arid deserts of Pacific coast of southern Peru to northern Chile where there is no rainfall even for years and these cacti lie alive in the sandy or stony deserts without any roots. It is very slow-growing and grows some 10 cm in height. These cacti have a rich coat of wool on the crown and typical spines. Yellow-red (reddish outside and yellow inside) flowers of 4 cm width are produced by this species even at seedling stage. These cacti growing on the Pacific coast have shown that their fruits are red, thin-skinned and hollow though black seeds are filled only in the upper part of the fruit encased in a special sac, and these seeds have ability to remain dormant with a long periods of drought which is apparently connected to the plant's habitat.

Lemaireocereus

It is a genus of columnar cacti with dark green, ribbed and spiny stems that grow up to 3 m in height, so slow-growing that attaining 1 m height in 5-10 years and produces flowers only when attains at least 2 m of height. Requires full sun for its growing and is propagated through seeds as well as stem cuttings. *L. euphorbioides* (syn. *Rookshya euphorbioides*) is an columnar cactus growing up to 3 m in height and 1 m in spread and stems 10 cm across bearing 8-10 deep ribs, areoles on the rib-edges, and each areole with 1-2 black spines. Flowers are red and funnel-shaped, appearing in summer. *L. marginatus* (syn. *Marginatocereus marginatus, Stenocereus marginatus*) is a branching columnar cactus growing up to 7 m in height and 3 m in spread, stems shiny, some 30 cm across and 5-6 ribbed, ribs with closely set small areoles bearing minute spines. Flowers are white and funnel-shaped and produced in summer. *L. thurberi* is a columnar cactus with branching at low levels, growing some 7 m in height and 3 m in spread, stems dark green, glossy, 5-6 ribbed having closely set areoles with short spines on rib-edges. Flowers are white, funnel-shaped and appear during summer. The genus is now merged with stenocursus.

Lepismium

Lepismium is closely allied to *Acanthorhipsalis, Lymanbensonia* and *Pfeiffera*, and even to *Rhipsalis*. Some 15 species are recorded from Bolivia, along with Brazil and Argentina. They are either epiphytic or lithophytic with hanging branches and mesotonic, *i.e.* multiple branching in the middle of the centre. The stems may be flat to angled or rounded. Small white to red flowers (larger than *Rhipsalis*) arise along the line of the stem to the entire length and at both the sides, followed by bright coloured fruits on maturity. For their cultivation, these require more tropical and wetter conditions. They are propagated through stem segments.

Leuchtenbergia

A native of San Luis Potosi and Hidalgo (Mexico), there is only one species, *i.e. L. principis* in this genus.

The genus name *Leuchtenbergia* is in honour of Prince Maximilian E.Y.N. von Beauharnais (1817-1852) of Leuchtenberg (Germnay) and specific name *principis* means 'princely'. The body is simple like to that of small agave, roots parsnip-like, stems slightly ovoid, base branching with age, can attain up to 70 cm of height and covered with slender tubercles. This is an unique un-cactus-like cactus having curiously leaf-like quite elongated (up to 12 cm in length) and sharply triangular tubercles in cross-section bearing long papery and wavy spines on the tip of each tubercle. The bluish-green, erect and spreading tubercles gradually turn grey, then brown and finally die and break off from the body leaving prominent scars. The papery spines are yellow-brown and soft with tapering tips, irregular in length and flattened, radiating radials 6-14, up to 10 cm long and variously reflexed, and 1-2 centrals are some 15 cm long and straight or slightly curved. Glossy, fragrant and funnel-shaped yellow-green flowers tipped brown, and some 8 cm long and 5-6 cm wide arise on the top from inner side of the areole of the youngest tubercles only when plants are 5-6 years old. Sometimes the petals terminate in a small and somewhat broader spine. It does not tolerate direct sunlight during growing period. It is propagated through seeds.

Lobivia

Lobivia is an anagram of Bolivia. It is a large genus of 75 species from Andean South America, of small to medium-sized, spherical (round) to cylindrically columnar cacti which forms the clumps with age. The ribs are divided into tubercles and bear conspicuous spines and bristly hairs. Flowers are large, funnel- to bell-shaped, colourful usually red or yellow, freely-borne and opening by the day and closing at night, or lasting only for two days. These are propagated through seeds or offsets. *L. allegraiana* from Peru grows up to 15 cm high, body being bright green, clump-forming with 6-11 ribs bearing straw-coloured spines, 7 radials, and some 2 cm long 1 central. Flowers are pink to red and 3.75-5.0 cm across. *L. aurea* (syn. *Echinopsis aurea*) from Argentina grows up to 10 cm high with globose to elongated stems which bear quite prominent 12-15 ribs. Ribs bear brownish areoles with 8-10 and yellow-tipped brown radial spines, and 7.5 cm long yellow-tipped 1 central spine. Flowers funnel-shaped, 23 cm long, yellow and freely produced. *L. famatimensis* from South America has underground inverted egg-shaped tuberous cylindrical root, grows with solitary sylindrical stems some 3.5 cm high and 2.8 cm across with almost oblique but rounded crown having depressed centre on the top. Stems bear 24 longitudinal and perpendicular (sometimes spirally arranged) soft green ribs some 3-4 mm across and with depressed tip. The areoles are white-felted and pectinate, some 1-2 mm long, thin and white spines arranged in 2 rows, each with 6. Orange and some 3.2 cm across long flowers with rich yellow base appear about $^{1}/_{3}$rd distance below on the stem. It has many varieties and forms ranging from yellow to red with a wide range of shades. *L. hetrichiana* from Peru bears up to 10 cm high, solitary or free-clustering glossy bright green globular stems with some 11 ribs notched above the round felted areoles. Spines 7-9, radials 1.5 cm long, 6-8 and yellow-brown while those of central 1 longer and paler. Profuse bright scarlet flowers appear some 5-6 cm long. *L. jojoiana* is recorded from Salta Province of Argentina at 2,700 m elevation. It grows 5-7 cm across, solitary or in clumps with pale-green spherical stems which turn columnar with age, and from seedlings it takes 3-4 years to mature. Stems bear 12-20 sharply angled prominent tubercles. Areoles some 3 mm across are equipped with grey-white felt and the spines are pale and blackish-brown when young but later on turn grey. Central spine 1-3, deep brown to black and double the length of radials and 1 upper one 3 cm and hooked, while 1 cm long radial spines are 10 and reddish to white. Flowers are 6.5 cm across and 6 cm long, dark purple-red with even darker throat, and with a bluish sheen. *L. nealeana* is a low cylindrical cactus with 14 ribs bearing short needle-shaped spines and large bright red flowers. *L. pentlandii* (syn. *Echinopsis pentlandii*), a native of Bolivia and Peru grows some 10 cm high and is clump-forming. Dark to greyish-green stems are globular to shortly cylindrical with 10-20 prominent ribs bearing large oblong tubercles and 6-20 spined areoles, radials 5-10, brown, recurved and up to 3 cm long, though central from nil to 1, longer and straight or curved. The flowers are funnel-shaped, 5-8 cm long and 6-10 cm wide, and dark purple-red with paler throats and bluish sheen. This species has many varieties of various stem size and shape and flower colours in white, pink, purple or orange. *L. silvestri* (syn. *Chamaecereus silvestri*, the genus *Chamaecereus* being composed of two words, the *chamai* which in Greek means dwarf, and *cereus* a separate genus to which it is closer, having only one species *C. silvestri* but now merged with *Lobivia*). It is a native of NW Argentina (Tucuman province). It is a small mat-forming prostrate (caespitose) cactus 30 cm across with cylindrical and branched oblong-ovoid (pea-shaped) stem segments bearing pale-green joints and segments some 6-10 cm long and 1.5 cm wide having 8-10 shallow ribs and several small white, tuberculate and felt-dotted areoles. Areoles bear quite small bristle-like spines, 1-2 mm long and thin radial 10-15 and a grayish-white central spine which is sometimes absent. Flowers are funnel-shaped, orange-scarlet to vermilion, and 3-7 cm long and wide. It has certain hybrids with *Lobivia*.

L. wrightiana from Mantanaro River Valley in Central Peru is not suitable indoors. It has large turnip-like root and small ash-green body with thin partly hooked spines which with age are often variously curved. The body bears 15-17 spirally arranged ribs separated by shallow grooves. Radial spines 10 and hardly 1 cm long though afterwards these become antenna-like and up to 7 cm long. Flowers are beautiful lilac-pink with long tube.

Lophocereus

It is native of Mexico and SW States and has only two species with *L. schottii* (syn. *Pilocereus schottii*) being commonly cultivated one. *Lophocereus* derives its name from the Greek *lophos* meaning a crest, and *cereus* due to its similarity to the genus *Cereus*. It is a tall cactus with good accent and is suitable for indoor gardening or outside. It is propagated through seeds and cuttings. *L. schottii* is native of W. Mexico and S. Arizona. Its stems are erect, columnar, base-branching, 3 m or more in height, ribs prominent and 6-7 though sometimes may be less or more and areoles are large. Areoles at the upper part of the flowering stem bear tufted, twisted and bristle-like grey spines, though at the lower part as well as in younger stems they bear 4-10 black to grey, thick and conical spines. Floral tube is short, and the flowers are nocturnal, 4 cm across and flat, and open to pink. *L. s.* 'Monstrous' is spineless with irregularly swollen ribs and curious-looking.

Lophophora

A native of southern Texas and Mexico is highly succulent composed of 2-3 spineless species bearing globular bodies which can be multiplied through seeds. *Lophophora* derives its name from the Greek *lophos* meaning a crest and *phoreo* meaning to bear, referring to each areole hair-tufts. Flowers are composed of numerous narrow petals.Its species *L. williamsii* bears 10-15 cm long turnip-shaped roots, is a slow-growing cactus with round stem, branching at base but with age, soft, flexible and grey-green, initially solitary but with age offset-forming, whose height is up to 7.5 cm and the breadth up to 15.0 cm, and bear 5-15 broad and flattened ribs which are divided in some forms of tubercles, having well-spaced areoles without spines (though seedlings have very small ones) though an erect tufting of white wool on the crown. Flowers appear singly from the crown of the plant, some 2.5 cm wide, pale-violet having darker central stripe (sometimes white) and lasting for 2-3 days. This plant contains poisonous alkaloids.

Maihuenia

A genus of slow-growing fully frost-hardy alpine cacti with cylindrical stems which forms clumps with age. For growing it likes full sun, fertile and well-drained soil. It is propagated through seeds and stem cuttings. *M. poeppigii* is a clump-forming, 3 cm high and 30 cm wide, and a slow-growing cactus with green-brown, cylindrical and branching stems where most branches bear the spikes of cylindrical green leaves at the tips and with funnel-shaped yellow flowers on the crown.

Maihueniopsis

A native of Peru to Argentina and Bolivia through Chile at high elevations where there is sudden drastic temperature fluctuation the same day, *i.e.* freezing to well above freezing. Though some 19 species under the genus are listed but new cactus lexicon lists only 7 as two are placed under *Tephrocactus* and the rest as synonyms of *M. glomerata*. Its tightly-packed round to globular stem-segments form the ground-cushion. Those forming taproots below have only little growth above. When new growth ages, its leaves fall away. The members of the genus may have small or long, and dense or tight spines on the body. Though flowers may be white or red, but they are mostly yellow and in case of *M. clavarioides* the flowers are pure light yellow and appear at the apex of the top stem segment. Its many of the species are in cultivation. These are propagated through stem segments as well as through seeds.

Mammillaria

This genus is native mostly to Mexico though only a few to SW North America to northern South America, especially West Indies, Colombia and Venezuela. *Mammillaria* takes its name from the Latin *mammilla* meaning a teat (nipple), on account of teat-like protuberances of tubercles with which the small globe-shaped solitary or clustered stems are covered. These stems do not have ribs but only tubercles which sometimes spiral around the plant. The areoles are felted and from where small funnel- to bell-shaped flowers appear freely. These are propagated by offsets or by seeds. Under this genus there are 250-300 species, most of them are widely and easily cultivated. The major cultivated species of *Mammillaria* are being described here briefly. *M. albicans* is clump-forming spherical at first but later becoming cylindrical with wooly axils. It produces large pink flowers in ring form slightly below the crown. *M. applanata* (syn. *M. texensis*) from Texas has small height (2.5-5.0 cm) but large diameter (10 cm) with club-shape or cylindrical forms. *Applanata* means flattened owing to crown being flattish along with the flattened tubercles, and the crown is covered with white wool. Areoles on the depressed but angular tubercles on the crown bear 5-12 mm long 15-20 whitish radial spines and 1 erect brown tipped black and stronger central spine. Cream to pinkish flowers some 2 cm long and 3.5 cm wide appear in profusion in a ring around the crown. *M. bocasana* (Mexico) is globose to shortly cylindrical, 15 cm high and 4-5 cm wide, and clump-forming some 13-15 cm across. Body is 5 cm across, blue-green and covered with fine white spines and silky hairs from the centre of which up to 2 cm long 1-4 needle-like spines hooked yellow or red project out beyond the felt and are surrounded by numerous white and bristle-like radial spines some 2 cm long that end in silky hairs. Stems bear 1 cm long and cylindrical tubercles. Flowers are cream (pale-yellow) with red mid-veins and 1.5 cm long. *M. bachmannii* is a dark green semi-spherical cactus with sunken top and full of white wool in the areoles. It bears tuberculated and spirally arranged some 20-25 ribs. Central thorns are black and longer than others and flowers are pink. *M. bellacantha* is a dark green, ribbed and tuberculated globular cactus bearing white radial and reddish-brown central spines. Axils are wooly and flowers are pink. *M. bombycina* from Mexico was described prior to World War I when De Laet firm of Belgium imported its first specimens. The specific name *bombycina* means silky on account of its silky and shining spines. It is a fairly hardy and robust cactus, mostly grown on its own roots though if grafted on robust stock, it forms a huge clump of strong stems. It grows usually in clusters though sometimes solitary some 20 cm high and 6 cm across, globose at first and then becoming shortly cylindrical and bears pale-green stems covered with conical to cylindric spiralling nipple-like tubercles from the axils of which thick white felts emerge. Radial spines are white, pectinate, 30-40 and 2-10 mm long while fairly weak, yellowish tipped brownish-red and usually 2-4 central spines are 1 cm long though lower one is up to 2 cm and hooked. Profusion of pale-carmine centred darker flowers appear on the crown which are 1.5 cm in length and breadth. *M. camtotricha* from Mexico grows some 3-5 cm high with clustering, dark green and globose stems which bear conical to cylindrical tubercles. From the axils of tubercles tufts of spirally arranged white bristles emerge. Areoles bear only radial spines numbering 4-8 which are wavy, pale-yellow and the longest 1 up to 3 cm long. Flowers are white and green and some 1.2 cm long. *M. candida* is a slow-growing, clump-forming, 15 cm high and across, green-stemmed and columnar cactus covered with short and stiff white spines. Flowers are cream to rose and appear some 1-2 cm across. *M. carmenae* from central Tamulipas of Mexico grows up to 10 cm, globular to broadly obovoid, first with solitary stems and then clustering and bearing spirally arranged conical tubercles which are pinkish but with upper parts green. Yellow-white hair-like radial spines are more than 100 but no central spine. White or pink-tinted flowers are some 8 cm long. *M. centricirrha* from Mexico grows up to 30 cm high and 10 cm or more across with dark green, globose stems turning later cylindrical. Stems branch freely from the base. The specific name *centricirrha* means central. Tubercles are nipple-shaped, 2 cm long, angular and arranged in spiralling rows. The axils of tubercles and the areoles bear thick whitish felts and 1 central spine while there are about 2 cm long 4-5 radial spines which are erect or slightly curved, all pale in colour and tipped dark. The pale-pink flowers with darker central stripe and some 2.5 cm across appear in the form of a wreath on the top of the plant. Even its small plants of 3-4 years old flower year after year. *M. centricirrha* has sometimes been considered as synonymous to *M. magnimamma* but the latter bears yellowish flowers. Craig mentioned some 116 synonyms, out of which more than half of them as its varieties, and K. Schmann mentions the best ones as *M. c. bockii, M. c. divergens, M. c. recurva* and *M. c. krameri. M. densispina* from Mexico is slow-growing, clump-forming and grows up to 20 cm high with dark green, beautifully round or cylindrical stems bearing cone-shaped tubercles. Areoles bear white spines, 2 large central and some 25 yellow and smaller radials covering whole body but being quite dense on the upper surface and on the crown. This cactus with spines looks quite charming. Purple-red small flowers emerge around the crown in summer. *M. elegans* from Mexico is a slow-growing cylindrical cactus bearing pale-green, 20-30 cm high and 20 cm wide solitary stems when young but clustering with age having closely set ovoid to conical tubercles angular at base. At younger stage the areoles are felted white and bear 20-30 needle-like short white some 5 mm long radial spines which initially cover the body, and twice as long, 1-4 brown-tipped central spines which project from each areole beyond the white covering. Carmine to purple-red flowers are some 1.5 cm long. *M. elongata* (syn. *Leptocladodia elongata*) from Mexico, most abundantly in Hidalgo state, is erect when young but afterwards with age reclining and clump-forming, 15 cm tall and 30 cm wide clumps, bearing columnar stems some 3 cm across, is most suitable house plant requiring very little watering. The whitish tips to the spirally spreading small and conical aggregates (tubercles) of cylindrical green stems take their colour from the short-lived covering of wool clothing the young areoles. Each areole bears 15-20 needle-shaped, 1.2 cm long yellow, golden or brown curved radial spines, and sometimes a little longer 1 white, yellow or brown central spine, and it is spine colour which alters the appearance of entire plant. The flowers are 1.5 cm long, bell-shaped and white to yellowish, sometimes with red-zoning on tepals. *M. e.* 'Cristata' bears brain-like monstrous stems.

M. erythrosperma from Mexico grows only 5 cm high. It is a free-clustering cactus, soon forming a cushion with globular stems. Spines are glossy with 3-4 yellow centrals out of which one is hooked. Deep pink flowers appear in summer. *M. fragilis* (syn. *gracilis*), a shallow-rooted but quick-rooting clustering cactus from Mexico grows some 10 cm high and 20 cm wide with freely top-branching, green cylindrical stems which form dense mounds. Numerous small side branches are produced which detach with a mild touch hence its specific name *fragilis*, and these branches are best source for its multiplication. Conical tubercles are arranged spirally and their axils are wooly. The stems have closely set areoles bearing thickly covered, short (5 mm or more longer), arching, 12-16, white radial spines, and longer, 2-4, brownish-tipped central spines, the longest ones being 2.5 cm. Flowers 1-2 cm across and 1.2 cm long appear in white to cream-yellow. *M. geminispina* from Mexico is a green cylindrical cactus up to 25 cm in height and 50 cm spread, with stems first solitary but later dense clump-forming, bearing spirally arranged cylindrical to conical tubercles where axils are wooly and bristly. Areoles bear about 5 mm long, white and needle-shaped some 15-20 radial spines, and brown-tipped 2-4 central spines where longest ones are 2.5 cm. It is highly spectacular due to its spines. Flowers are red to carmine and 2 cm long. *M. hahniana* from Guanajuato (Mexico) grows some 40 cm high with 15 cm diameter. This cactus is solitary and spherical at first but with age branches out forming large clumps of heads. Green stems bear 5 mm long spherical to triangular and green tubercles with narrow and blunt-pointed tips, and from their angles short white felts and tufts of up to 4 cm long bristles are borne. Small areoles are elliptic with short white shedding felt. The 20-30 white radial spines are hair-like, wavy and flexible and 5-15 mm long, though single, erect and acicular, and brown-tipped central spine is up to 4 mm long. Funnel-shaped some 2 cm long and 1.2-1.5 cm wide flowers of wine-red with greenish-white throat in profusion appear near the crown in a ring form. It is easily grown through seeds.*M. magnimamma* (syn. *M. longimamma, Dolichothele longimamma*) is sometimes considered even as *M. centricirrha* though latter bears pale-pink flowers while former yellowish. It is already described under the genus *Dolichothele* as *D. longimamma*. *M. microhelia* from Mexico grows some 20 cm high with 40 cm spread. It is a columnar, with age clump-forming and branching green cactus having 5 cm stem diameter. Spines are cream or brown and fading with age. Flowers yellow or pink and some 1.5 cm wide appear during spring. *M. multiceps* (syn. *M. prolifera multiceps*) from Texas and Mexico, most suitable for growing in pans or shallow pots, grows 6 cm high and 15 cm or more across, is mound-forming with shortly cylindrical and branched stems bearing conical tubercles. Radial spines are numerous, hair-like and white while those of central spines are thicker and red-tipped at younger stage. Yellow to whitish-yellow flowers are funnel-shaped and 1.5 cm long. *M. parkinsonii* is highly wooly and clump-forming cylindrical cactus with flattened top white, long and stout spines coming up in the areoles set on tubercles. Flowers are yellow and 2 cm across. M. *pectinifera* (syn. *Solisia pectinifera*) from Mexico grows some 6 cm high with clustering and shortly cylindrical stems. Tubercles are spirally arranged, small and conical. Radial spines are white, about 4 mm long, appressed to tubercle and 20-40. Flowers are small and yellow. *M. pennispinosa* is clump-forming highly wooly, cylindrical cactus with light marginal and feathery spines. Central spines are red and hooked. The flowers are white with a pink central stripe. *M. plumosa* from Mexico grows some 12 cm high and with 40 cm spread producing clump-forming green and globose stems covered completely with feathery white spines, radials 40 or more, slender, branched and minute feather-like and interlocking the others to make an umbrella-like full covering. Tubercles up to 1.2 cm long and cylindrical with wooly axils. For its growing it requires calcium rich soil. Flowers are white to cream with red or green markings, 2 cm long and emerge during winter. *M. prolifera* (syn. *M. pusilla*) from West Indies grows up to 10 cm high with 30 cm spread, is clump-forming with globose to shortly cylindrical green stems bearing 5-7 mm long and dark green tubercles. Radial spines up to 40, white and bristle-like while central spines 5-9, yellowish and some 8 mm long. Flowers are produced in masses of yellow flushed green and are 1.5 cm long and 1-2 cm wide. Berries are red with taste and flavour of strawberry. *M. rhodantha* from Mexico grows some 60 cm high and with almost the same spread, bearing spherical to columnar green stems which branch with age from crown. The stem is densely covered with abundance of brown to yellow spines. The bright red flowers are some 1-2 cm across. *M. schiedeana* from Mexico grows 10 cm high with 30 cm spread, and is a clump-forming cactus with green stems covered with short, feathery and yellow spines which later turn white. Flowers are cream and merge with the colour of the spines. *M. sempervivi* from Mexico is a slow-growing spherical cactus with dark green stems and short white spines. It grows 7-8 cm high and wide. Plants above 4 cm height bear white wool between short, spiralling and angular tubercles. Arrangement of closely set tubercles is highly ornate. Flowers in the form of a wreath of deep vivid pinkish-red and some one dozen at a time appear on the crown. *M. sheldonii* from Mexico is a clump-forming, cylindrical, tuberculate cactus with bluish-green stems, growing 10-25 cm high and 15 cm across. Stems are tinted brownish-red when in the sun. Tubercles are fairly short, cylindrical and topped with slightly wavy and round areoles with a good network of spines. Areole bears white, 6-9 mm long and 10-15 radial spines while 1-3 central spines where 1 is stout, 1.2 cm long, reddish-brown and hooked. Pale-pink flowers edged pale and some 3 cm wide are produced in the upper part of the stem. It can be cultivated with ease only after grafting. *M. spinosissima* is a cylindrical cactus with wooly caps and axils. Areoles bear 20-30 radial and 7-10 thin, white and wooly spines covering the body. Pinkish funnel-shaped flowers with flat top appear aroung the crown in profusion. *M. tetracantha* from Mexico grows globose to cylindrical up to 30 cm tall, usually with solitary stems though in the wild this clusters. Tubercles almost 4-sided and their axils are with a few felts. Radial spines normally absent or in the form of deciduous bristles. Central spines are usually 4 and in cross-form, erect or curving, yellow-brown to reddish or grayish and 1.0-2.5 cm long. Flowers are carmine-red and 2 cm long. *M. umbrina* from Mexico grows up to 10 cm high and blooms carmine-red in summer when quite young. It produces solitary stems at first but with age forms clusters. Dark green cylindrical stems have depression at the crown and bear long and conical tubercles with 25 white slender radial spines and about 4 stout and deep red central spines. *M. vagaspina* bears grey-green globular stems with white tufted crown. Stem is tuberculate bearing up to 6 cm long grey-brown central spines. Creamy-white or yellowish flowers in profusion appear around the crown. *M. vaupelii* 'Cristata' is a misshapen monstrous variety of the original cylindrical species. Such crested types sometimes emerge directly as seedlings when growing the seeds. *M. winteriae* is a bluish-green globular cactus with square tubercles, wooly axils and with 3 cm long and stout radial spines. Flowers are creamy with a yellow central stripe and some 3 cm across. *M. zeilmanniana* from Mexico bears glossy, pale-green, globular and up to 15 cm tall and 30 cm spreading bodies with clump-forming 4.5 cm across stems having closely set 6 mm long and 3-4 mm thick, ovoid or short cylindrical tubercles. Areole bears white deciduous felt, 18 white marginal and thin radial spines some 1 cm long, and 4 reddish-brown central spines some 8 mm long with 1 being hooked. Flowers near the crown appear in a ring during spring-summer in large number in deep violet–red colour and some 1.5-2.0 cm across. This cactus was discovered in 1931 by E. Georgio of Saltilla and its specific name *zeilmanniana* is after H. Zeilman, a member of the German Cactus Society.

Matucana

A native of Peru, the flowers of this cactus is so similar to *Borzicactus* that sometimes it is included with that. It may be solitary to clustered and spherical to columnar. These are not cold-hardy but some originating at high elevations can tolerate some cold, however, these should be protected during cold and their roots rot due to cold and wet medium. White, yellow, orange, pink or red flowers that are bilaterally symmetrical (zygomorphic) are produced during spring. *M. aureiflora* has less-spined and flattened stems some 12 cm across. This species is an exception to the genus as bears radially symmetrical (actinomorphic) bright yellow flowers which appear during spring in several flushes. This requires protection against cold and intense warmth. It is propagated from seed. *M. formosa* grows with solitary barrel-shaped tall and wide body up to 17 cm wide. Several flushes of its orange-red flowers some 5 cm long and 3 cm wide appear during summer. This requires to be protected from intense sunlight and heavy frost. It is propagated through seeds. *M. intertexta* produces bright green solitary stem some 30 cm tall and 15 cm wide. Its light orange flowers some 5 cm long and 3 cm wide are produced during summer where petal-edges are darker. It can not tolerate intense sunlight or heavy frost. *M. madisoniorum* is heavily spined at seedling stage, at lower elevation the body becomes spineless after maturity while at higher elevations these retain their 4 cm long and curving spines even after maturity. Afterwards spination is quite variable on grey-green and rough body of the plant. Several flushes of orange flowers which are normally 5 cm long and 3.5 cm wide appear during summer. It is propagated through seeds and requires protection against frost.

Mediolobivia

This genus has been recorded from Argentina and Bolivia. It is a genus where body at younger stage is simple and globose, round or oval but later on clump-forming. Flowers are funnel-shaped and appear from upper part. Its most cultivated species are *M. aureiflora* from Argentina from rocky hills, and *M. ritteri* from Bolivia at 3,000 m elevation. *M. aureiflora* grows some 5 cm across, when young the body is simple, dull green but in sun tinged brown and globose but later on, by producing side shoots freely, it becomes clump-forming with a bit depressed top, *vis-a-vis* whole body covered with spines. Tubercles are 15-17, spirally arranged and some 3-4 mm high. Areoles 1-2 mm long are covered with thin and white felts, each areole bearing about 6 mm long 13-20 spines, out of which some 3-4 are central ones 1 cm or even longer. Flowers are orange-yellow with whitish throat, 4.0-4.5 cm long and 4.0 cm across. This species has many botanical varieties such as *M. a.* var. *albiseta*, *M. a.* var. *boedeckeriana*, *M. a.* var. *rubelliflora*, *M. a.* var. *sarothroides*, *M. a.* var. *duursmaniana*, *etc.* The species can easily be grown through seeds or by grafting on higher stock of *Piptanthocereus peruvianus*, *P. dayami*, and certain others for getting sooner and profusion of blooms. *M. ritteri* bears dark green, small, round or oval bodied clumps. Stems have reddish-purple tinge. The ribs are around 15 bearing low tubercles. Areoles are elliptic with yellowish felt. Spines are 8-10, weak and pale-brown, some 1 cm long and radiating. The flowers are funnel-shaped, glossy and vermilion with purplish throat and about 4.5 cm across.

Melocactus

Melocactus takes its name from the Latin *melopepo* meaning an apple-shaped melon as certain species are shaped so. It is a genus of 30 solitary, globose to shortly columnar species native to Tropical America and West Indies. Similar to *Cereus*, these have ribs (prominent) and spiny stems with white tufted crown when attaining flowering size, and these crowns are surrounded by brown young spines at the top and grey-white spreading spines on the rest of the body. After producing wooly crowns the stems stop growing but wooly crowns develop into columns. The body shape, the ribbing and arrangement of spines and the crown shape and spine arrangement there make these cacti quite charming. Body is grey-green with equi-distant, quite raised and spacious ribs. Body in summer produces funnel-shaped small flowers on terminal cephaliums covered with felts and bristler spines. These are propagated through seeds. *M. bahaiensis* from Brazil bears dull green spherical body growing 10 cm high and 15 cm across bearing 10-12 prominent ribs, each rib with up to 7 areoles and 14 stout, slightly curved and dark brown spines which become paler with age, 10 radial about 2.5 cm long, and 4 straight, and brown central about 3.75 cm long. Flowers 1-2 cm across are pink and appear from the cephalium at the crown. *M. cunispinus* (syn. *M. oaxacensis*) bears green, spherical to columnar stems some 20 cm high and 15 cm wide and about 15 rounded ribs on which each areole has curved radial spines and straight central spine. Deep pink some 1 cm across flowers appear on flat wooly crown. *M. intorus* (syn. *M. communis*) grows up to 20 cm in height and 25 cm in width, is a flattened spherical cactus bearing 18-20 with yellow-brown spines. On maturity the crown bears white column with fine and brown spines. Flowers are pink and are produced in summer. *M. matanzanus* from Mexico grows about 9 cm high and 10 cm wide with 8-9 thick ribs, and each rib with 5 areoles bearing 9-10 spines (8-9 radial and 1 central), each about 2 cm long, curved, awl-shaped and reddish to yellowish. Flowers are 2 cm long and pink. *M. maxonii* is a globular cactus with dark green stems bearing 12 prominent ribs and tufted spiny top, the body covered entirely with some 15-20 sharp reddish spines where radial and centrals are not distinguishable. When mature, it produces cephalium at the crown.

Myrtillocactus

The name derives from the Latin *myrtillus* meaning a small myrtle as berry fruits resembling those of a myrtle. A native of Mexico and Guatemala constitute 4 species of branching columnar (tree) cacti with blue-green, ribbed and spiny stems bearing star-shaped flowers opening during nights. These like sunny position and are propagated through seeds and cuttings. Its outstanding species is *M. geometrizans* (syn. *Cereus geometrizans*) from Mexico which grows up to 4 m high and with about 2 m spread, and whose stem is blue-green, jointed, vertically much-branched and cylindrical and columnar having 6 ribs bearing areoles about 2.5 cm apart with only a few short and black spines over 30 cm stem height and white flowers some 4 cm wide in clusters of 4-9 appearing during night. Radial spines are 5-9 and 1.2 cm or more long while central spines 1, flattened and about 8 cm long.

Neobuxbaumia

A native to the hills of E&S Mexico where these form the cactus-forest, this genus comprises of 8-9 large tree-like species attaining the height of about 16 m. It produces columnar stems with numerous ribs, and these stems branch out when are aged and have attained good girth and height. Younger plants are more spined as lower part of adult or older plants shed their spines or size of the stem becomes so large that spines look dwarfed. Ribs are a little tuberculate whose visibility is not marred due to spines present over the stem. Bell-shaped white or pink flowers which are bat-pollinated, appear during night. Round fruits have spines over them and the fruit pulp is white. *N. polylopha* is the most cultivated species and its form *cristata* looks very interesting. It is propagated through stem segments and seeds.

Neochilenia

This genus comprises of some 53 species, the most charming cultivated species among the Chilean ones being *N. jussieuii* whose specific name *jussieuii* is in honour of Prof. A.L. Jussieu of Paris. Other charming species are *N. fobeana* bearing thick covering of stout spines and orange-red flowers, *N. fusca* with dark green stems bearing 13 radial spines and 1-2 central spines some 3 cm long and yellow flowers, and *N. occulata* which is brown in colour and bears 1-4 cm long irregularly-spaced spines and yellow flowers with reddish stripe. These all are propagated through seeds and small seedlings are grafted onto *Echinopsis* or on short stock of *Eriocereus jusbertii. N. jussieuii* is a species with brownish-green stem having dark purple tinge, initially bearing globose body which after some time becomes cylindrical and some 8-10 cm across. It bears 13-16 ribs divided by cross-grooves, large areoles felted yellow, pale spines which with age turn brown tipped darker, radial spines 7-14 though central spines 1-2, up to 2.5 cm large (longer than radials) and little curved. Glossy and funnel-shaped flowers measuring up to 4 cm are pale-orange with darker stripe. *N. recondite* (syn. *Pyrrhocactus reconditus*) is a grey-green globular cactus with slightly sunken areoles, grey spines and pale-pink flowers.

Neolloydia

It is a genus of spherical to columnar cactus bearing spirally arranged short tubercles and dense spines. It requires full sun for its cultivation and is propagated through seeds. Most of its species are very difficult to cultivate unless grafted on some suitable stock. *N. conoidea* grows some 10 cm in height and some 15 cm wide, is a clump-forming columnar cactus with blue-green stem bearing abundance of white radial spines and longer black spines. The flowers appear funnel-shaped and purple-violet in summer.

Neoporteria

It derives its name from the Greek *neo* meaning new and the genus *Porteria* named for Carlos Porter, a Chilean entomologist. A genus from Peru, Chile and Argentina comprises 66 species of globose to shortly cylindrical cacti, though some scientists put some 53 species of this under *Neochilenia*. Here stems are ribbed and areoles are wooly, spiny, and sometimes the crowns are also wooly. The flowers of this genus are bell- to funnel-shaped. It is propagated through seeds or offsets. *N. chilensis* (syn. *Neochilenia chilensis, Nichelia chilensis*) from Chile grows up to 30 cm in height and 10 cm in width. When young the pale-green stems are globular though afterwards cylindrical and are densely covered with stout and golden spines of varying lengths. There are thick and green 20 ribs which are little notched. Radial spines are 20, yellow to whitish and some 1.2 cm long while those of 6-8 central spines are twice as long and yellow to brownish. Funnel-shaped flattish bright pink to carmine-red or sometimes even white flowers some 5 cm long and wide appear during summer from the crown. *N. napina* (syn. *Neochilenia napina*) from Chile has tuber-like swollen leaves with stems 3-9 cm tall and 5 cm wide, globose to elongated and reddish to brownish-green bearing deeply notched 14 spiralling ribs. Areoles bear 3-9 flat-pressed black or brown spines some 3 mm long. Funnel-shaped 3.0-3.5 cm long and 5.0 cm across, and flattish yellow flowers are borne on the crown during summer. *N. nidus* is spherical to columnar and grows 10 cm long bearing long, soft and grey spines which encircle the dark green-brown stem completely. Pink to cerise tubular flowers some 3-5 cm long having paler bases are borne on the crown with opening at the tips in summer or autumn. *N. paucicostata* is light green globular cactus with off-white areoles. Spines are sturdy black at first later lighter. Rose-white flowers appear in spring. *N. senilis* (syn. *N. gerocephala*) is easy to grow when grafted on suitable rootstock. It has globular body bearing irregularly curved and interwined spines across the entire body. Flowers are red. *N. subgibbosa* (syn. *Chilenia acutissima*) from Chile produces green to greysih-green globose stems at first but afterwards turning cylindrical and 8-30 cm or more tall and 10 cm wide, and bearing 14-16 deeply clefted ribs. Areoles are rounded and wooly with 3 cm long brownish and stout spines. Broadly funnel-shaped flattish pink to light red flowers appear 4-5 cm long and across. *N. villosa* is a clump-forming branched cactus growing to 15 cm high and 10 cm wide, having green to dark grey-green stems full of thickly spread 3 cm long spines which are sometimes curved. Tubular pink or white flowers appear in spring or autumn. *N. wagenknechtii* is globular at first then cylindrical with green stems bearing 15-20 prominent ribs which are tuberculated with areoles. Areole bears stout, long and grayish-brown numerous spines. Many flowers appear together at the crown.

Nopalxochia

It is a genus of epiphytic cacti having flattened and strap-shaped stems, so is closely related to *Epiphyllum* with which it hybridizes easily.In this genus the spines are not significant. Stems may die after the flowering. For its cultivation it requires partial shade and is propagated through stem cuttings. *N. ackermannii* is already described under *Epiphyllum*. *N. phyllanthoides* var. 'Deutsche Kaiserin' is pendant cactus growing some 60 cm long with about 1 m spread, having glossy, green, flattened and toothed stems, each stem 5 cm across. Pink flowers appear from stem margins some 10 cm across in spring. The flattened stems of 'Grandiflorum' bear 10 cm long red-purple flowers.

Notocactus

This genus originates in southern South America, especially Brazil and Uruguay and comprises of 15-25 species, formerly included under the genus *Echinocactus*. Its scientific name derives from the Greek *noto* meaning southern, and *cactus*. Stem is spherical and solitary to clusterd bearing attractively coloured spines and widely trumpet-shaped beautiful flowers with narrow petals opening flat at the top of the crown. On the body the ribs are prominent and the tubercles are divided by deep notches. Areoles are densely wooly. Flowers appear either solitary or 2-3 at a time from near the crown and individual flower lasts for about one week. They all are propagated through seeds though clustered ones through offsets also. *N. apricus* (syn. *Parodia apricus*) from Uruguay is a clump-forming species growing up to 8 cm high and with about 6 cm spread. The stems are light green, globular, bearing 15-29 almost flattenend ribs. Areoles bear 18-20 grey, bristly and curved radial spines, and red-yellow 4 larger central spines. Yellow flowers some 10 cm long with tinged red exterior of outer petals appear during summer. *N. concinnus* from S. Brazil and Uruguay grows some 6-8 cm high and with 10 cm spread having glossy and light green broadly globular stem with slight depression at the crown. Specific name *concinnus* means gentle. Ribs are notched, low, blunt, 16-20 and divided into tubercles, each tubercle bearing white felted areole the felt shedding afterwards, and each areole with some 10-12 thin and bristly yellow radial spines some 5-7 mm long, and the 4 stout central spines with crosswise arrangement are yellow and brown at the base, lowest being quite stout and some 2 cm long though other 3 only up to 1.5 cm. Funnel-shaped glossy flowers appearing freely several at a time, sulphur-yellow inside while red outside and of some 7-10 cm length during spring and summer on the crown. It is propagated easily from seed, and seedlings bear the flowers usually the third year of growing. This species was discovered by Sellow in Uruguay and reached Paris (France) in 1838 but one year later it was described by Monville as *Echinocactus concennus*. *N. graessneri* (syn. *Parodia graessneri*) from Brazil attains the height and breadth of 10-13 cm, having pale-green globular and flattened stem with depression on the top and whole body covered with yellow spines. There are some 50-60 notched ribs which bear areoles with yellow felts, various bright yellow radial spines, and 3-6 stout central spines. Green-yellow flowers of 19 mm across appear from June to September. *N. haselbergii* (syn. *P. haselbergii*) from Brazil grows 15 cm high and 10-13 cm wide. The stems are bright green, globular or cylindrical, flattened and little depressed at the apex and are covered with white-yellow spines. Ribs some 30 or more are low and spiraling. Areoles are felted white, each with 20 white radial spines and 3-5 pale-yellow central spines. Bright orange-red flowers, some 3.75-5.0 cm across appear in spring and early summer. *N. horstii* (syn. *Malacocarpus horstii*) is a palish-green globular cactus topped with white wool, having 15-20 deep and blunt ribs adorned with areoles, each areole with 15-20 sharp and thin yellow spines, and more than one large yellow flowers on the crown. *N. leninghausii* (syn. *Parodia leninghausii*) from Brazil is highly slow-growing and grows about 1 m with 10-20 cm breadth taking many years in attaining this dimension. The stems are light green, cylindrical and solitary at first but on maturity branching out from the base, bearing about 30 or even more low ribs, and crown clothed with white wool. Areoles are closely set and white wooly bearing about 20 white radial spines and 3-5 pale-yellow some 4 cm long central spines reflexed at the tips. It is shy in flowering with 4.0 cm long and 2.5 cm across lemon-yellow flowers having green exterior and appearing during June to August on one side of the stem. *N. l.* 'Cristata' has fascinated stems with crest formation. *N. magnificus* from Brazil is clump-forming, bearing broadly ovoid to shortly cylindrical stems having flattened top, with 12 or more ribs which are deep and sharply edged. Areoles are small and in a continuous row with short felts, bearing longer, hair-like and yellowish spines and yellow flowers. *N. mammulosus* (syn. *Parodia mammulosus*) from Argentina and Uruguay grows up to 13 cm with up to 10 cm spread. The stem is dark green, globular and with flattish top. Deeply notched ribs are 18-20 with small protuberance in each areole. Each areole bears 10-13 yellow-brown radial spines, and 2-4 brown-tipped central spines, 1 erect while others pointing downwards. The top centre of the stem is spineless. Flowers are yellow with outer petals having 1 red stripe, which appear from June to August. *N. ottonis* (syn. *Malacocarpus ottonis, Parodia ottonis*) from Brazil, Argentina, Paraguay and Uruguay grows about 10 cm high and with 7.5-18.0 cm spread. It may be solitary or clump-forming, bearing bright green globular or cylindrical stem which is flattened at the top and bears 10-13 broadly rounded ribs. Areole is felted-white bearing 10-18 yellow-brown radial spines and 3-4 red-brown central spines. Up to 10 cm long and yellow flowers appear on the crown from May to July. *N. purpureus* (syn. *Malacocarpus purpureus*) is globular at first then cylindrical, top sunken and thickly wooly. Stems bear high networking of yellow to brown spines coming up from the tufted areoles. Flowers are salmon and appear on the crown. *N. scopa* (syn. *Parodia scopa*) from Brazil and Uruguay grows about 18 cm high with 7.5 cm spread. Stems are solitary, pale-green, appearing globular but becoming cylindrical or club-shaped later on, and bear spine-concealed 30-40 low and notched ribs. Areoles when young are felted white but later on produce 40 or more 5 mm long, white and bristle-like radial spines, and 3-4 stout red-brown central spines. The yellow and 4 cm long and 5 cm wide flowers, some four in number at a time appear on the crown in April-May. *N. tubularis* from Brazil and Uruguay grows about 10 cm in height with 10-15 cm spread. Stems are glaucous-green, globular or semi-cylindrical with 16-23 rounded and notched ribs. Areoles are white-felted, each bearing 16-18 white radial spines and 4 red central spines. Flowers are yellow, tinged red at the petal bases and appear during summer.

Obregonia

This cactus is of Mexican origin (calciferous desert in Lanos del Joumave in Taumalipas state) was discovered by A.V. Friè in 1923, having only one species, *O. denegrii* (syn. *Strombocactus denegrii*), allied to *Leuchtenbergia* because of its non-cactus (artichoke-like) appearance though looks quite different from *Leuchtenbergia* due to its shorter tubercles. It was named so to honour Don Alvaro Obregon (1880-1928), then President of Mexico, and the specific name in the honour of Denegri, Minister of Agriculture in Mexico. This monotypic genus is a transition between *Ariocarpus, Strombocactus* and *Leuchtenbergia*. It is not easy to cultivate this cactus until it is grafted on hybrid *Echinopsis* where it feels comfortable and flowers within three years if kept at filtered situation after planting in porous clay soil having proper drainage. The dark green body is generally solitary though clustering casually and grows up to 6 cm high and 10-13 cm across with base terminating in a turnip-like strong root. Those grown under bright sun have grayish-green body tinged pinkish. The stem is covered with overlapping, flattened, triangular and spirally arranged leaf-like tubercles that appear as a *Sempervivum* rosette. Tubercles are thick, 1.5 cm long, and 2.0-2.5 cm across at the base, bearing areoles at the tip with 2-4 thin and pale-yellow deciduous spines tipped darker which measure some 1.5 cm long. Whitish or cream-coloured flowers some 2 cm across arise from the youngest areoles on the crown with almost naked floral tube and ovary. When flower dries, the ovary recedes into the plant and after ripening a soft white berry as to that of *Ariocarpus* or *Melocactus* arise in the thick white felt.

Opuntia

A genus from North and south America (Argentina, Bolivia, Chile, Guatemala, Mexico, Peru, USA, *etc.*) comprising of some 250-300 species of usually branched and shrubby cacti having distinctive jointed stems. *Opuntia* was named so originally for a Greek plant now not known, and was used for this genus for no known reason. These cacti vary in size from coarse tree-like plants to small species growing some 10-15 cm in height. In between the joints of the stem, the section called as 'pad' is either cylindrical, globular or flattened into round pads. These cacti may also bear small, cylindrical or conical and usually fleeting leaves. All the species bear tufts of glochids and spines, mostly hooked, on each areole though number and size of spines may differ. The short-tubed flowers are produced from areoles on the edges of the joints, usually red or yellow, broadly bell-shaped and roughly 7.5 cm across during summer, followed by fleshy, usually edible fruits such as *O. ficus-indica*. These like sun and plenty of space around the root zone. Though when small they can be used as house plants but when large enough, they should be kept outside. Certain species are also suitable as hedges. They are propagated by seed or by cuttings of mature pads. Most ornamental species are furnished here under. *O. basilaris* from SW USA is a bushy cactus growing up to 90 cm high with bluish-green sometimes suffused red, 10-20 cm long and broadly oval to obovate pads. Red-brown areoles are many but lacking the spines or sometimes 1. Flowers are 6-8 cm wide and pale-pink to carmine. *O. bergeriana* from unknown origin is more or less an erect bush growing up to 3 m bearing 25 cm long pale to greyish-green and narrowly oval to oblong pads. Areoles are though well spaced but unequal and bear 2-3 irregularly-sized awl-shaped spines, longest one being up to 4 cm. Freely produced flowers are 5 cm wide, usually deep red but sometimes orange. *O. brasiliensis* is a tree-like cactus growing up to 5.5 m high and 3 m in spread. The green stem is cylindrical having bright green branches of oval, flattened and spiny segments though 2-3-year old side branches fall off. The plants when attain 60 m height only then these bear masses of shallowly saucer-shaped yellow flowers some 4 cm across during spring and summer. *O. clavarioides* (*Austrocylindropuntia clavarioides*) is often in the shape of a hand or fan with brownish joints and does well when grafted on to a vigorous *Opuntia*. Flowers yellow splashed brown. *O. c.* var. *cristata* 'Minima' is a crested variety with elongated and nearly spherical joints. This variety also requires to be grafted on to a vigorous *Opuntia* stock. *O. cylindrica* is a bushy cactus with a height of up to 6 m and spread some 1 m with cylindrical stems some 4-5 cm across, bearing dark green, deciduous and cylindrical leaves some 2 cm long on new growth. Areoles are without spines or with 2-3 barbed ones. A mass of shallowly saucer-shaped pink-red flowers appear on plants over 2 m tall. *O. decumbens* from Mexico to Guatemala is a spreading shrub growing to 30 cm or more high and three times as wide. Pads are dark green, 10-20 cm long and broadly oval to elliptic-oblong. Areoles felted, often with reddish or purplish halo, spines though absent but sometimes 1-2 of up to 4 cm length and yellowish. Flowers some 5 cm wide are pale-yellow and normally ageing to reddish. *O. engelmannii* (syn. *O. ficus-indica, O. megacantha*) from Tropical America grows up to 5 m in length and spread and is used as a hedge because with age it becomes shrubby in tropical countries. Its pads are blue-green, oblong, flattened and up to 45 cm long, and the areole bears 1-5 spines, each measuring some 2.5 cm long though there are even its spineless forms. Flowers are 10 cm across and bright yellow which are followed by 5-10cm long yellow or red-streaked edible fruits. *O. erinacea* from SW USA is a spreading to semi-prostrate bushy cactus growing up to 80 cm high and with 2 m spread, having bluish-green stem with 8-15 cm long, oval and flattened segments. Areoles bear 5-20 cm long hair-like, 6-15 flattened spines and masses of saucer-shaped pink, yellow or white and 6-8 cm across flowers in summer. *O. humifusa* (syn. *O. compressa, O. opuntia*) from USA is a prostrate growing loosely mat-forming cactus with up to 15 cm length and 1 m spread. Stem segments (pads) are dark green, 7-18 cm long, flat, rounded to oval, and purple-tinged, with well-spaced areoles. Each marginal areole bears 3 cm long 0-3 spines, and 8 cm wide and yellow flowers which appear during spring-summer. Its var. *O. h. austrina* (syn. *O. compressa* of gardens) grows up to 90 cm high erect stems with more spiny pads. *O. imbricata* (syn. *O. arborescens*) bears cylindrical dense branching-joints instead of flattenend stem-joints. It grows to a height of 1.8 m and to a trunk breadth of 5.6 cm bearing wooly areoles with papery spine sheaths, 6-20 barbed spines with the longest one measuring up to 3.1 cm. Flowers are purple to pink, followed by dry yellowish fruits which remain attached with the plants for many months. In the following summer the new stem-joints bear quite large leaves. *O. lindheimeri* is found growing wild in USA (Luisiana and Texas) and NE Mexico and grows some 3.7 m high with pronounced trunk but has also been recorded with a prostrate habit. The plants of this species are highly variable in shape and size of stem-joint and fruit, spininess and flower colour. Each young bluish-green stem joint bears numerous narrow fleshy leaves which shed soon exposing the areoles with glochids including 1-6 large and stout spines as the central and others smaller and spreading. Flowers are from red to yellow followed by pear-shaped fruits some 5.6 cm in length. A species, *O. leptocarpa* with low bushy habit and elongated fruits from Texas is probably the hybrid between *O. lindheimeri* and *O. macrorhiza*. *O. macrorhiza* from USA (dry prairies from Misssouri and Kansas to Texas via Massachusetts to Georgia) to some extent appears as *O. compressa* (syn. *O. humifusa*), is a low spreading cactus having woody roots, and the areole with nil to 3 spines. The flowers are yellow and tinged red. *O. microdasys* (syn. *O. pulvinata*) from Mexico is a slow-growing much-branched bushy cactus with height of 60-100 cm and spread of 60 cm. This bears pale-green, oval and flattened stem-segments 8-18 cm long which develop brown marks under low temperatures. The stems bear oval, 1-2 mm across spineless areoles but with white, yellow, brown or red glochids set closely in diagonal rows. Masses of funnel-shaped yellow flowers some 5 cm across rise in summer on plants over 15 cm tall though under cultivation it flowers seldom. There are several varieties of this cactus all with yellow flowers. Its var. *albispina* is a bushy cactus growing up to 60 cm high having 30 cm spread, with green, oval and flattened segments. Areoles are though spineless but slender barbed white hairs are set in diagonal rows. Its some of the other varieties are *O. m. minima*, a dwarf variety with 5 cm long pads and dark yellow glochids, *O. m. pallida* with pale-yellow glochids, *O. m. rufida* with thick grey-green pads and dark red-brown glochids, *O. m. undulata* grows crested-type having flattenend joints fairly densely covered with golden-yellow tufts bearing minute bristles which with slightest contact come out and itch the skin, etc. *O. ovata* from Andes (Argentina), a subject of high altitude, is hummock-forming and 20-30 cm tall cactus with pale-green, thick and oval to broadly cylindrical pads some 4-7 cm long, and pale-yellow glochids with 5-9 awl-shaped, grey and 4 cm long spines. Though it hardly flowers under cultivation but when flowers appear they are 4-5 cm wide and yellow flushed red. *O. pentlandii* (syn. *Tephrocactus pentlandii*) from Peru, Bolivia and Argentina (Andes region) is a densely clump-forming cactus spreading up to 60 cm or even more. Pads are some 5 cm or even longer, bright green, obovoid to cylindrical or quite narrowly ovoid and with broad and low tubercles. Spines are often absent but upper areoles may bear 2-10 yellowish or brownish spines to the maximum of 6 cm length. Yellow flowers ageing to red are borne some 6 cm wide. *O. phaeacantha* from USA (Texas to California) and Mexico is a variable prostrate and mat-forming species. Pads are broadly obovate to orbicular and 15-30 cm long. Areoles are well-spaced, each bearing 1-4 down-pointing spines with longest one being 6 cm. Flowers are bright yellow though sometimes red-throated and some 8 cm wide. *O. robusta* from Mexico grows up to 60 cm tall under cultivation though in the wild even more than 5 m in height and spread. It is bushy with silvery-blue stems which are rounded to oval. Stem segments (pads) are flattened, some 30 cm across and oval to almost circular. Each areole has brown glochids with or without spines, but if present these spines are white to yellow, 8-12 and some 5 cm long. Saucer-shaped yellow flowers appear in spring some 7 cm across, followed by rounded and red fruits some 8 cm long. In Mexico, this species is grown for its edible fruits. *O. salmiana* from Brazil, Paraguay and Argentina grows some 50 cm in height (in wild up to 1 m) with branching and sprawling habits and requires staking usually. The stems are glossy dark green and cylindrical with about 1.25 cm in diameter, and bear yellow glochids and short spines. Several white flowers some 2.5 cm across appear freely during summer, followed by dark red spiny fruits which drop off and root. *O. scheeri* from Mexico is very slow-growing, and in the wild it reaches tree-like proportions though under cultivation only up to 90 cm. It is spreading species having grey to bluish-green, 15-20 cm long and rounded to oblong pads (joints) which are covered with golden spines and hairs, and with yellow-brown glochids. Areoles bear 1 cm long 8-12 spines. Flowers are 10 cm across and yellow ageing to red though under cultivation ordinarily it does not flower. It performs well under temperate conditions. *O. tunicata* (syn. *O. furiosa*) from Mexico to Texas (USA) and Chile is a handsome mounded dwarf and bushy shrub up to 60 cm tall and 1 m spread with cylindrical and whorled stems. Pads are blue-green, 15 cm or more long and oblongoid with prominent tubercles. Golden spines are in loose papery sheaths, 6-10, barbed, 4-5 cm long and densely covering the whole body. Flowers are yellow splashed greenish and some 3-5 cm wide. *O. verschaffeltii* from Bolivia is clump-forming cactus with dark green, spineless, cylindrical stems hardly growing 30 cm long though its real height is about 15 cm and spread some 1-2 m. Length of the pad is 15 cm or more in cultivation though in wild it seldom exceeds 4 cm. Persistent cylindrical leaves are borne on stem tips during spring to autumn which shed after a long time. Sometimes 1-3 thread-like some 6 cm long spines are noted. The flowers are orange-red, 4 cm across and appear during spring. *O. vulgaris* (syn. *O. monacantha*) has bright green oval joints bearing usually only 1 spine in each areole and creamy flowers tinged red on tips of the pads.

Oreocereus

A native to southern Bolivia to northern Argentina, and its species *O. neocelsianus* was first brought from Bolivia by Bridges which he described in 1850 as *Pilocereus celsianus*. In wild it is clustering with columnar bodies covered with long white hairs, and from areoles acicular, stout and whitish to yellowish or red spines emerge. It is either grown on its own roots or grafted on strong stock, and is propagated through seeds. Stems grow erect attaining the height of up to 1 m. It branches out at later stage. Its glossy-green shoots some 8-12 cm thick turn darker at a later stage bearing about 8 mm raised 10-17 ribs which are divided by deep grooves. Areoles of the 10-18 mm dimension are oval, bearing deciduous thick and yellow felt, each areole bearing some 5 cm or more long, white and wavy hairs later turning grey. All the spines are pale-yellow, orange or sometimes red, nine radial spines are 2 cm long and radiating while 1-4 central spines are up to 8 cm long, Flowers are narrow up to 10 cm long, red to pale-brown and emerge near the crown. *O. celsianus* is green and globular covered with white to brown long wool covering the body entirely along with stout and brown spines tipped black, ribs prominent, 10-20 and flowers rose-red.

Oroya

It is a genus of spherical cacti which require sunny situation for their growing and are propagated through seeds. Its one of the species, *O. peruviana* (syn. *O. neoperuviana*) grows 25 cm tall with 20 cm width, bearing dark green, highly ribbed and tuberculate stems covered with 1.5 cm long yellow spines with darker bases which arise from elliptically elongated areoles. Flowers are pink with yellow bases which appear around the crown during spring-summer.

Ortegocactus

This monotypic Mexican genus is though uncommon in cultivation but is quite handsome. It is a 10-cm across clustered cactus with 3 cm wide globular yellowish-green body bearing 6-8 long radiating spines of tan shade from the centre of each bold and projected tubercle. This genus is said to be a link between *Mammillaria* and *Coryphantha*. It is multiplied through grafting, cutting and seeds. *O. macdougallii* produces bright yellow flowers on the tips of its stems with orange to reddish base-shading which sometimes envelops the entire plant though well aerated and brightly lighted plants do not face this problem. Its mixture should be shallow and coarse to avoid getting it rotted. It can tolerate -7 °C temperature.

Pachycereus

In Greek *pachy* means thick and *cereus* a genus of cacti. These are usually branched columnar trees with woody trunks requiring full sun for their cultivation. Flowers are funnel-shaped and generally appear in the wild where growth exceeds at least 3 m. This genus includes some 10 slow-growing species, all native to Mexico, and is closely related to *Lemaireocereus*. They are propagated through seeds. Its important cultivated species are being described here with. *P. chrysomallus* (syn. *Pilocereus chrysomallus*) from Mexico is a tree-like with erect branches and grows to a height of 9 m bearing 13 ribs and areoles with long hairs and amber yellow spines, radial spines 11-13 with upper ones 1.25 cm long, the lower ones 2.5 cm, the 4 centrals still longer, and all the spines turn brown with age. Cephalium terminal or sometimes unilateral, 30 cm long and setose. *P. columna-trajani* (syn. *Pilocereus columna-trajani*) is tree-like growing up to 15 m in height and 60 cm in diameter, areoles are large and elliptic bearing 10-12 radial spines, upper ones very short though lower ones some 2.5 cm long, whereas central spines 2, upper one 2.5 cm long and the lower one 10-13 cm. Flowers some 5 cm long and yellow. *P. marginatus* (syn. *Cereus marginatus, C. gemmatus*) from Mexico has simple or apex-branching stems having 5.0-7.5 cm diameter and 5-6 entire wooly obtuse ribs. Spines are 7-9, almost equal in length and stout. Flowers are brownish-purple and about 3.75 cm long. *P. pectin-aboriginum* (syn. *Cereus pectin-aboriginum*) grows 11 m in height and about 3 m in spread with dark green columnar stems which bear 9-11 deep ribs. Areoles bear 1 cm long, dark brown with red bases but fading to grey 8 radial spines and longer central spines. *P. pringlei* (syn. *Cereus pringlei*) from N. Mexico grows up to 11 m in height and with 3 m spread in columnar form branching at base. It possesses some 10-15 ribs which may have furcation above, bearing closely set large elliptical areoles on the edges. Each areole bears 15-25 black-tipped white spines projecting downward. In cultivation usually it does not flower. *P. schottii* (syn. *Lophocereus schottii*) is dark green columnar cactus branching with age and grows up to 7 m with spread of 2 m. Stem bears 4-15 highly raised ribs and areoles on the edges of the ribs, each areole with small white spines facing upward. Flowers are funnel-shaped, pink and are produced during night. *P. s.* 'Monstrosus' is an olive to dark green columnar cactus with stems having 4-15 ribs but no spines. Funnel-shaped flowers are pink and some 3 cm wide opening in the night.

Parodia

The first *Parodia* was discovered by Schickendantz in 1886 near Catamarca in the mountains of Tucuman Province of Argentina. This specimen was named as *Echinocactus microspermus* by Dr. Weber due to its microscopically small brown seeds. In 1892, O. Kunze discovered another species from N. Argentina with erect spines and quite different from the above species which was named by Prof. Schumann in 1898 as *Echinocactus chrysacanthion* due to golden spines. Dr. C. Spegazzini in 1926 introduced the generic name *Parodia* for these species, in honour of the Argentine botanist Parodi. It is a genus of 30-35 species of South American origin, are small globular to shortly cylindrical cacti which may grow solitary or may form clump. Often the ribs are spiral around the green stem and bear tubercles. Crown bears wooly buds. The flowers are broadly funnel-shaped. They are most suitable for growing in pots. They are propagated through seeds or by separating offsets. Out of important cultivated species, *P. apricus* (syn. *Notocactus apricus*), *P. graessneri* (syn. *Notocactus graessneri*), *P. haselbergii* (syn. *Notocactus haselbergii*), *P. leninghausii* (syn. *Notocactus leninghausii*), *P. mammulosus* (syn. *Notocactus mammulosus*), *P. ottonis* (syn. *Parodia ottonis*) and *P. scopa* (syn. *Notocactus scopa*) are described under *Notocactus*. Others are being described here under. *P. aureispina* from Argentina grows solitary, globular and about 6 cm high. Stems bluish-green with numerous slightly spiral and closely set ribs which are divided into tubercles. Some 40 white bristle-like radial spines about 6 mm long, and bright yellow 6-7 central spines some 1.2 cm long with 1 hooked at tip are borne in an areole. Flowers are yellow and 3 cm wide. *P. chlorocarpa* is a small cylindrical cactus with seedlings growing very slowly, hence, it requires grafting on suitable rootstock. Large yellow flowers appear many together on the crown. *P. chrysacanthion* from Argentina bears globular to slightly elongated, solitary or clump-forming stems up to 6 cm in length and 8-10 cm across. Whole plant is densely covered with erect, yellow to honey-brown spines up to 4 cm long. Stems bear about 24 spiralled ribs which bear small yellow and felted areoles, each areole with 30-40 spines, radials bristle-like, quite pale-yellow and 5 mm long though centrals 19 mm long and golden-yellow to brown. Flowers of rich yellow colour and 2 cm long appear near the crown in early spring and continue appearing for several months. *P. commutans* is a globular cactus later becoming elongated with wooly top. Central spines are browny-yellow and 2-5 cm long and yellow flowers appear on the crown. *P. gracilis* is spherical at first then ovoid with wooly top having several yellow flowers on the top. *P. microsperma* from Argentina is clustering cactus growing up to 10 cm in height and bearing globular stems with about 20 spiralling ribs. Areoles bear 10-20 and glossy-white radial spines about 5 mm long while 3-4 red-brown centrals about 2 cm long with one longest one hooked at tip. Flowers are pale-red outside and orange inside, and 3.5-5.0 cm across. *P. minima* is a small cactus which grows well when grafted on a rootstock. The stem is deep green with globular and tuberculate body. Tubercles are closely set and slightly in spiralling manner. Spines are stout, small, white and 10-13. Large flowers more than one of yellow colour appear on the flat crown. *P. mutabilis* from the mountains of Argentina (2,520 m.a.s.l.) grows 8 cm across and up to 10 cm high with blue-green and globular stems, with the crowns being slightly wooly. Spirally arranged inconspicuous ribs are divided into low tubercles bearing small white-felted areoles. Highly slender bristle-like, white, radiating and about 50 radial spines some 1 cm long, and darker tipped, reddish, brown or white and usually 4 longer centrals (1.0-1.5 cm long) arranged in the form of a cross with one tip-hooked are borne in each areole. Flowers are golden-yellow (rarely red-throated) and 3-5 cm across. *P. nivosa* from Province of Salta (Argentina), discovered by A.V. Friè, is an ovoid and much –ribbed miniature alpine cactus growing up to 15 cm high and with 10 cm across having pale-green stem with olive tinge, the ribs being composed of spirally arranged conical tubercles that bear white-felted areoles, each areole bearing white, stout and straight about 18 radial spines each 1-2 cm long while 4 strong and crosswise-arranged central spines more than 2 cm long, the central one being sometimes brownish. It bears felted crown and bright red flowers, often several at a time, and 3-5 cm across. Though it does well on its own roots but the seedlings can also be grafted on hybrid *Echinopsis*. It is propagated only through seeds. *P. rutilans* (syn. *Notocactus rutilans*) is a columnar cactus growing up to 10 cm high and 5 cm in spread bearing green stem with areoles, each areole with about 15 small and white radial spines, and quite large, stout and mustard-brown 2 upward or downward pointing central spines. In summer, cream-centred pink flowers arise from around the crown. *P. sanguiniflora* (specific name means red flowers) discovered in 1928 in the mountains of Salta (Argentina) by A.V. Friè who named it initially *Microspermia sanguiniflora* but Dr. Spegazzini transferred these cacti to the genus *Parodia*. Prior to World War I there had been only three known *Parodia* species and today all the species collected from their mountainous habitat of NW Argentina, Bolivia, Paraguay and Brazil total more than one hundred, some being placed under certain other genera. It is a clump-forming cactus with dark green stem growing up to 10 cm or more high and the same across. The much-ribbed stem spherical but with age becoming columnar, bearing spirally arranged conical tubercles with white-felted areoles. Each areole bears 15 thin white to brown radial spines some 6-8 mm long, and 4 brown central spines arranged across, some being hooked especially the lowest one which is 2 cm long. The blood-red (rarely yellow) flowers some 4 cm across appear near the crown in spring. *Parodia vorwerkiana* (*Wigginsia vorwerkiana*) is a slow-growing spherical cactus that forms clump with age, and with flattened top growing some 8 cm in height and 9 cm in width. Stems are glossy green with 20 conspicuous, wart-like and pyramidal ribs which are shallowly tuberculate, each tubercle with one areole. The top of the body, especially at the crown, is tufted with thick white wool. Each areole bears 7-10 stout, yellow-white and radiating spines facing downward, the 1or 3 in the centre are longer. Yellow flowers appear in summer.

Pediocactus

This small genus from USA (Arizona, New Mexico, Colorado, Utah, Oregon and Washington) is most difficult to cultivate on its own roots, therefore, its offsets or seedlings are generally grafted for rapid growth. In their native habitats these are accustomed to hot and dry summers and extended periods below freezing. *P. simpsonii* from USA is easier to grow than other species of the genus. It can be grafted but comes up well with seeds or cuttings on its own roots. It has numerous hybrids and *P. s.* var. *minor* is the most popular among the varieties. Normally it is a clump-forming but low-growing with about 8 cm thick stems in a 20 cm clump. Flowering buds start forming in winter though 2 cm wide light pink flowers open during spring. It can tolerate up to -18 °C temperature.

Pelecyphora

This genus of attractive, little and more or less globular cacti comprises two endangered species both of these of Mexican origin, very close to *Ariocarpus* and are quite easy to grow indoors. It takes its name from the Greek *pelekys* meaning a hatchet and *phoreo* meaning to bear, referring to the shape of the flattened tubercles. These clearly bear the rhombic tubercles with flattened top, and elongated areoles having small spines with comb-like arrangement. These are propagated by seeds. The flowers are terminal, short-tubed and large for the size of the plant. *P. asseliformis* bears solitary or clustered, globose and greyish-green stem when young but later turning ovoid and up to 10 cm high. Hatchet-shaped tubercles are spirally arranged bearing narrowly elliptic areoles on the flattened tips. Spines some 8-60 are blunt and some 5 mm long. It produces rose-purple and 3 cm wide flowers. *P. strobiliformis* (syn. *Ariocarpus strobiliformis, Encephalocarpus strobiliformis*) bears grey-green, 4-6 cm across globular stems with somewhat cone-like, triangular and scaly tubercles which overlap and incurve. Flowers are violet-pink to magenta and 3 cm wide.

Peniocereus

A native to SW USA and Mexico, the genus comprises of 18 species where most have large underground caudex. Its stems are obscure and nondescript, flowers are showy white (rarely red), fragrant and nocturnal, and the tubes are slender and long. Areoles are quite bold along with bristles or spines. The fruits are large, juicy with red pulp and black seeds. These are propagated through stem cuttings and seeds. *P. greggii* var. *transmontanus* is in maximum cultivation.

Pereskia

These 20 or so non-succulent tree-like cacti (also shrubby and climbing forms) produce well-developed true leaves which supports the view that areoles in cacti represent a condensed shoot system. These have been recorded from Mexico and the West Indies to South America. It is thought that true cacti of today evolved from these ancestral cacti. The generic name *Pereskia* is in the honour of French naturalist, Nicholas Claude Fabre de Peiresc (1580-1637). Areoles are present in the leaf axils. Most species have many-petalled showy flowers and fleshy edible fruits. These are propagated by cuttings or seeds. *P. aculeata*, the species which is only in cultivation, is from Mexico to Argentina and was cultivated at Hampton Court in 1696 and at Kew (England) since 1760 where regularly it is bearing its fragrant light yellow flowers. Though it is an erect, semi-evergreen woody shrub some 9 m high, densely armed with wicked spines but older plants throw out climbing branches some 9 m long which climb by means of the hooked spines arising from the areoles. The young branches bear 1-3 recurved spines beneath each leaf whereas older branches bear numerous straight spines from the areoles in the leaf axils. Mid-green fleshy leaves some 5-10 cm long are lanceolate to elliptic with prominent midribs. Flowers are saucer-shaped, 2.5-4.0 cm across, fragrant and pale-pink, yellow or white which appear during autumn. The edible fruits are pale-yellow, spiny at first though smooth when ripe. In pots the height of this plant can be maintained up to 1.5 m and enjoyed indoors at a sunny situation.

Pereskiopsis

A genus of 8 small species from the group of opuntiad, surprisingly the genus resembles a deciduous shrub which produces non-segmented, frequently branched and round stems with included in the areoles, bearing deciduous, flat and often persisting leaves, as well as flowers appear from yellow to pink to red similar to opuntia-shape, and the fruits are colourful, rarely juicy with glochids present on the body. These are propagated through cuttings. The flowers in *P. velutina* are yellow. *P. spathulata* is most popular among the growers. Its round stems are small and as it is fast-growing, seedings are grafted over it.

Pfeiffera ianthothele (P. cereiformis, Rhipsalis ianthothele, R. cereiformis; Jungle cactus)

Pfeiffera was named after Ludwig Pfeiffer, a distinguished student of cactus. It is an epiphytic cactus closely allied to *Rhipsalis* and is native to Argentina. First it grows erect some 30-60 cm long with branching at the base, 1.9-2.5 cm in diameter and then become pendent. Its stems are mostly 4-angled (sometimes even 3) with tuberculate ribs, the areoles are at the summit of tubercles, short-woolly and bear 6-7 small (6 mm long) acicular spines, and the flowers are short-tubed, bell-shaped, small (2.5 cm long and little less wide), regular and purple-red but interior white, ovary and fruits are spiny, the rose-red fruits are berries some 1.25 cm in diameter and seeds are black.

Phyllocactus

This generic name is of old Greek origin meaning leafy cactus, and now this name relates to the hybrids (*P. hybridus*) of original botanical species growing as epiphytes in Mexico, Honduras, Guatemala and Brazil and the large-flowered cerei. The form is that of leaf-like shrubby cactus with expanded shoots which hang down from the trees having aerial roots at their terminal ends for absorbing nourishment and as also these help to climb them against the objects. These produce large flowers during spring to summer in various colours such as red, pink, cream, yellow or in the combinations of multicolours in single and double forms with up to 20 cm diameter. These grow well under filtered sunlight. They are propagated from the mature leaf-like shoots after drying the cut ends.

Pilosocereus

A native of Tamaulipas (Mexico) is allied to *Cephalocereus* as some of its species have been merged with this. It grows erect as a shrub or as an arborescent up to 6 m height and 1 m spread with about 20 branches, individual stems being dark or pale-brown and some 8 cm across, bearing wool-like spines in flowering zones at crown. Its most important species being *Pilocereus palmeri* (*P. leucocephalus*) with columnar growth. It bears some 7-12 broad and blunt ribs having white-haired crown and with 8-12 weak and 2-3 cm long radial spines while only 1 central some 3 cm long, all the spines being either brown or grey. The bell-shaped tubular flowers are pink and about 6 cm across opening in the evening with an unpleasant odour and remain open till the morning but emerge on the plants which have attained 1.5 m height. It is propagated from seed. *P. palmeri* var. *victoriensis* bears white petals with a violet tinge.

Pseudolobivia

This genus is thought to be a transition between the thermophilic and columnar plants of the genus *Echinopsis* which bear white nocturnal flowers with long tubes, and the small mountain *Lobivia* which blooms during day time with small bright flowers having shorter tubes. *P. aurea* was recorded from Cordoba (Argentina) whose body is dull green, spherical and grows 10 cm across with perpendicular ribs. The plants appear as small *Echinopsis* due to its stout, erect and 1-3 cm long spines, about 9 white radials and 4 brown centrals. It produces golden-yellow flowers more than 10 cm in length and 8 cm across, almost the same dimension as to that of its body. This cactus has many varieties in various forms of flowers but of the same colour. *P. kermesiana* was discovered in 1942 bearing carmine-red flowers though prior to this only white pseudolobivias were known except *P. aurea* bearing yellow flowers. In fact, *kermesiana* means carmine-red. Its green body is semi-spherical to spherical, attaining a diameter of even 15 cm or more. The low ribs whose number may increase with age are generally 15-23 on an old plant with 1.5 cm width, these being divided by grooves in tuberculate forms. Areoles bear rusty-yellow brown-tipped 11-16 thin radial spines up to 1.2 cm long, and 4-6 centrals some 2.5 cm long and these spines turn grey later. Flowers are some 18-20 cm long and 9 cm wide, green on the outside, the petals some 6 cm long and 2 cm wide along with stamens carmine-red, style pink and stigma yellow, and the tube is covered with greenish scales and grey wool. Collectively for the stem shape and flower size and colour, this species is outstanding.

Pterocactus

A small and endemic genus to Patagonia (Argentina) belonging to the subfamily Opuntioideae where glochids remain present, some species having in abundance while certain others are lax. Their large roots are tuberous and continue to grow even when stems are gone and subsequently replaced by new one. Its flowers are embedded in the stem in the manner that entire stem appears a long floral tube. Flowers appearing at the end of the stems, terminate the stem growth and then the stems branch out from the side and continue to grow until it also bears the flower at the apex. After successful fertilization of the flowers, the stem-ends develop into the dry fruits which after maturity split and release wide-winged seeds, an unique feature within Cactaceae in this case. It is propagated through stem cuttings and seeds. Most cultivated species is *P. tuberosus*.

Pygmaeocereus

An endemic genus to the fog-zone area along the coast of Peru, this genus comprises three clump-forming species bearing small cylindrical stems with shallow and tuberculate ribs which bear many spreading spines. Floral tubes are long, scaly and hairy, and the flowers are large, white and nocturnal. These are propagated through stem cuttings and seeds. Most cultivated species is *P. bylesianus*.

Quiabentia

A native from Bolivia to Argentina, this is allied to *Opuntia* hence belongs to the same subfamily as to *Opuntia*. They range from a shrub some 2 m tall to trees about 15 m high. The shrubby plants bear large fleshy leaves during the growing period. They have spines as long as as the leaves. They are columnar cacti with fleshy, lanceolate, pointed and glossy green leaves having light green colouration similar to the stem. *Q. chacoensis* from Argentina is a shrubby plant some 3 m tall where 5 cm long barbed spines exceed the length of the leaves. It is a quite fast growing species susceptible to frost and is grown through cuttings.

Rebutia

Being allied to *Aylostera*, a few of *Rebutia* species are included in *Aylostera*. A genus with 27 species of small green cacti from Bolivia and Argentina, bears globular or somewhat elongated, solitary or clustered stems, where flowers appear in profusion from plant bases, usually 2-3 years after raising from seeds. The stems are much-ribbed and tuberculate with short spines. *R. arenacea* (syn. *Sulcorebutia arenacea*) grows some 5 cm in height and 6 cm in width and is a brown-green spherical cactus with minute white spines on quite symmetrical and spirally arranged low tubercles which themselves make it very attractive, and white felt in the centre of depressed crown. A profusion of golden-yellow blooms about 3 cm across appear from plant bases in spring. *R. aureiflora* (syn. *Mediolobivia aureiflora*) from Argentina grows some 10 cm high and with 20 cm spread, the width of individual stem being about 5 cm. Stem is dark green but often tinged violet-red, spherical to shortly cylindrical, and clustering, bearing spiralling rows of ribs divided into 6 mm tubercles. Each areole bears creamy or brownish, bristle-like and 6-9 mm long spines, radials a bit stout, and centrals some 15-20 soft and longer. Masses of golden-yellow flowers appear from the crown some 4-5 cm long and wide in late spring, *i.e.* April and May. *R. chrysantha* from Argentina has clustered, globular but becoming shortly cylindrical and 12-15 cm tall stems which around them bear spiralling tubercles. Areoles bear 25-30 bristle-like, white-yellow and 1-2 cm long radial spines. Brick-red flowers are some 3 cm long. *R. deminuta* (syn. *Aylostera deminuta*) from Argentina grows 7.5-10 cm in height and width. A clump-forming cactus with globular body has depression at the crown. Stems are divided into 11-13 tuberculate ribs which have spiral arrangement. The areole bears 10-12 white tipped brown and 0.6-1.0 cm long radial spines. Deep orange-red flowers some 7.5 cm long and 1.9 cm wide are produced during summer. *R. d. grandiflora* bears larger flowers. *R. fiebrigii* from Bolivia is clustering and globose cactus growing up to 6 cm tall. Stems have tubercles in 18 rows, and bristle-like whitish and 2 cm long radial spines are 25-40. Bright yellow-red flowers are produced to 4 cm long in summer. *R. gracilis* is a cluster-forming spherical cactus with deep green tuberculate body where areoles have white tufts and 8-13 white, sharp and stout spines and large orange-red flowers appearing on the crown. *R. krainziana* from Bolivia is clustering cactus with globular and 5 cm tall stems. Stems bear small tubercles and areoles with 8-12 small bristle-like white radial spines. Orange-red flowers some 3.5 cm long are produced in summer. *R. kupperiana* from Argentina and Bolivia grows up to 5 cm in height and 10 cm in spread. It is a branching and clump-forming species with red-green spherical stems having up to 7.5 cm diameter. Stems bear small tubercles, and areoles with 12-18 needle-like copper-tipped white radial spines some 5 mm long, and 1 central spine. Red flowers some 2.5-4 cm across emerge during summer. *R. marsoneri* from Argentina was discovered by H. blossfeld and O. Marsoner in 1935. It grows best on its own roots and also when grafted on *Piptanthocereus peruvianus*. It is propagated by seed. This species grows 6 cm in height and 10 cm in diameter. The body with single stem is dark green but becoming grey at the base later, flattened, spherical and the crown depressed. Rarely the stems send out the shoots. The ribs are divided often into small and spirally arranged 2 mm high tubercles which are set 2.5-4.0 cm apart. The areoles are on small tubercles and bear bristly hairs and 25-35 yellow radial spines tipped brown and some 2 cm long, and from among them the centrals can not be differentiated, however, there are some of its forms which bear glassy white to yellow spines. The golden-yellow flowers some 4-5 cm long and 4 cm across appear generally at the base of the stem. *R. minuscula* from Argentina is a clump-forming globular cactus having green stem with somewhat flattened and sunken at the top, having 5 cm height and 10-12 cm width. Stems bear pale-green tubercles in spiralling fashion arranged in about 20 rows. Radial spines are about 30, pale-brown, 3-6 mm long and cover entire body uniformly. Masses of trumpet-shaped red-crimson flowers some 4 cm wide emerge in spring. *R. muscula* bears dark green stem and is clump-forming, growing up to 10 cm high with 15 cm width. The stem is densely covered with soft and white spines some 0.5 cm long. Bright orange flowers some 2-3 cm across appear from the stem base in profusion in late spring. *R. neocumingii* (syn. *Weingartia neocumingii*) grows to 10 cm in height and spread bearing green, spherical and tuberculate stem. Areoles bear dense clusters of yellow spines some 1.5 cm long, some being thicker than others. Masses of dark yellow cup-shaped flowers some 3 cm long appear in spring. *R. pseudodeminuta* from Argentina possesses height and spread of 7.5-10 cm. Stems are bright green, clump-forming, globular and depressed at the crown. Tubercles are prominent bearing small areoles, each areole bearing white tipped brown spines, 11 radials and 2-3 centrals. The golden-yellow and 2.5 cm across flowers appear in summer. *R. pygmaea* (syn. *Lobivia pygmaea*) from Bolivia and Argentina grows some 12 cm high and 10 cm wide, bearing grey- to purple-green, ovoid to columnar and clump-forming stems. Tubercles are spirally arranged and bear comb-like minute 9-11 radial spines, white at first but turning grey and are pressed against the body. Trumpet-shaped pink or rose-purple flowers some 2.5 cm across appear in summer. *R. rauschii* (syn. *Sulcorebutia rauschii*) grows some 5 cm high and 10 cm in spread bearing grey-green flattened and spherical stems. Stems bear minute comb-like golden to black spines. Flattish deep purple flowers some 3 cm wide emerge in spring. The performance of grafted plants is better. *R. senilis* from Argentina is a pale-green clump-forming and spherical cactus having greyish base with top-depressed, growing 8 cm high, 10 cm wide and 15 cm in spread, branching freely at base, bearing spirally arranged tubercles with soft, straight and 25-40 glassy-white tipped-brown bristly spines some 1.3 cm long matted around the entire body It may bear trumpet-shaped 5 cm across bright red flowers in spring. It is propagated through seeds or through shoots grafted on piptanthocerei. *R. spegazziniana* from Argentina is a light green clump-forming spherical cactus becoming columnar with age, and growing up to 4.0-6.25 cm across, 13 cm high and 20 cm spread. The stems are spirally tuberculate, each areole bearing about 14 white radial spines pressed back against the stem, and 2 yellow tipped brown central spines. Masses of slender-tube deep orange-red flowers some 10 cm across appear from stem bases in late spring. *R. spinosissima* from Argentina grows up to 7.5 cm high and with 5 cm spread. It is a clump-forming with pale-green globular stems which are some 5 cm across with slight depression on the top. Closely set tubercles are spirally arranged and bear white hairy areoles from where 1 cm long numerous white radial spines arise along with 5-6 thicker and yellow tipped brown central spines. Deep pink to brick-red flowers some 4 cm long and wide arise during June to August. *R. tiraquensis* (syn. *Sulcorebutia tiraquensis*) is a variable species growing 15 cm high and with 10 cm spread having green stem. Areoles are elongated and bear gold or red-white spines. Flowers of pink to orange-red appear during spring in masses from the base of the stem. *R. violaciflora* from Argentina is a clump-forming cactus with olive-green globular stems, growing 5 cm high and 15 cm wide (spread) with depressed top. Ttubercles in 20-25 rows on the stems are spirally arranged. Each areole bears yellow-grey wool, 15-25 and some 0.5-1.0 cm long white to yellowish radial spines, and 5-10 stout and longer central spines that are first white then changing to yellow. Lilac-rose long-tubed flowers some 4 cm long and wide emerge in summer. *R. xanthocarpa* from Argentina grows up to 4 cm high and with 10-20 cm spread. The plant is clump-forming and possesses pale-green globular stems some 10 cm in diameter, covered with numerous slender bristle-like glassy-white spines up to 6 mm long. Red flowers some 2 cm across appear during summer.

Rhipsalidopsis

Rhipsalidopsis derives from the genus name *Rhipsalis* and the Greek *opsis* meaning 'like', referring to the close resemblances between two genera. This genus comprises two Brazilian species of woody-based epiphytic shrub-like cacti with leafless but branched and jointed stems. Between the joints the stems are flattened as to those of leathery leaves. Pendulous flowers occur singly at the end of the branches with tubular bases but with many narrow tepals opening as funnel-like, containing narrow and spreading-lobed stigmas and make a beautiful specimen in hanging baskets. These flowers last for 3-4 days. They are propagated through seeds and stem cuttings having 1-3 joints. *R. gaertneri* (syn. *Epiphyllopsis gaertneri, Schlumbergera gaertneri*) from Brazil grows 45 cm long with 90 cm spread, the jointed stems are dull green with purple and notched margins, spreading and pendulous, the joints are fleshy and usually flat, stem segments oblong and glossy, sometimes 3-6 angled, and each segment end bears areoles from where new joints and orange-red flowers some 6.25 cm across emerge in spring. *R. g. makoyana* lacks purple markings on the joints and has stiffer joints with numerous stiff bristles on the areoles. The areoles bear short white wool and a few bristles. *R. rosea* from Brazil grows erect at first but later becomes pendent with about 15 cm height and 30 cm spread, free-branching and is a very charming dwarf bushy plant having fleshy leaf-like stem segments. Stems are green but usually tinged purple, slender, drooping, flat or 3-5 angled, bristly and segments some 5 cm long, crenate and green but sometimes margins flushed with red. Areoles on top of the stems have short bristles. Slightly scented flowers appearing freely in masses at stem tips are purple-pink, bell-shaped and 4 cm across which appear in late spring to early summer. The hybrid *A. × graeseri* evolved by crossing both the species is intermediate in habit producing 7.5-10.0 cm wide red flowers similar to *R. rosea*.

Rhipsalis

This genus with about 60 species has its origin in Brazil with centre of distribution to southern Brazil but one species, *R. baccifera* extends in America from Florida to Brazil and Peru, and also occurring in tropical Africa and Sri Lanka, probably as a centre of distribution as there it would have been taken over by man or birds. Even otherwise, except *Rhipsalis*, all other cacti are native to Americas. It is a genus of epiphytic pendent cacti having variously formed stems. It prefers 80 per cent relative humidity, higher than most cacti. It is propagated by seed or stem cuttings. These cacti are leafless, somewhat shrubby, epiphytic, pendent and free-branching with cylindrical or flattened leaf-like stem segments. The flowers with few tepals are small and funnel-shaped, followed by rounded, white, semi-transparent and slightly sticky berries. The plants are most suitable for hanging baskets or in pots and pans. Its many species are cultivated. *R. anceps* (syn. *Lepismium anceps*) grows about 45 cm in length with arching to pendent free-branching stems which are flattened as fleshy leaves or have 3-angles, notched and mostly purplish-margined. Flowers are pendulous, small, violet or white. *R. baccifera* (syn. *R. cassutha, R. cassytha*) a native to Florida to Brazil and Peru, Africa and Sri Lanka grows about 2 m long or even more with light green, pendent and cylindrical stems. Greenish flowers some 6 mm across are followed by whitish-translucent berries some 4-5 mm wide. *R. capilliformis* grows up to 1 m long having some 50 cm spread, free-branching, cylindrical and its pale-green stems hang down vertically, main stem joints 10-15 cm or more and about 3 mm thick though lateral joints much smaller, little thinner, and often in whorls of 3-7. Lustrous greenish-white and funnel-shaped flowers some 1 cm wide and about 6 mm long with recurved tips, are followed by 6 mm wide globular and white fruits. *R. cereuscula* grows 60 cm long with 50 cm spread, having green, pendent, branching, cylindrical or 5-angled stems, stem branches 3 cm long and whorled. Each stem tip bears bell-shaped white flowers during winter-spring. *R. crispata* grows up to 1 m long with indefinite spread, pale to yellowish-green stems are more or less erect, first bushy then pendent, stem joints leaf-like, elliptic to oblong, 8-13 cm long and with undulating edges from where funnel-shaped cream or pale-yellow flowers appear in clusters of 1-4 and 1.2 cm wide with recurved tips in winter-spring, and the fruits are 6 mm wide, globular and white. *R. cruciformis* (syn. *Lepismium cruciforme*) bears trailing to pendent stems some 60 cm or more long and glossy green, stem segments 15-30 cm long, winged or 3-5 ribbed, shallowly notched bearing therein areoles with some white bristles, and the segments taper towards the tips. White flowers in clusters of 1-5 and some 1.2 cm width are followed by rich purplish-red globular fruits which appear of the same breadth as the flowers. *R. houlletiana* with pendent or scrambling stems grows some 1 m or more long and up to 5 cm wide with flattened, and toothed or sharp tooth-like lobes on the margins bearing areoles and bell-shaped flowers some 6 mm wide and white in colour, which are followed by red fruits. *R. micrantha* from Ecuador and Peru grows 1m or more in length and bears flattened or 3-4 angled stems. White flowers some 6 mm across are followed by white or reddish fruits some 7 mm wide. *R. pachyptera* grows up to 60 cm or more in length having erect to spreading stems. Stem joints thick, flattened, broadly elliptic to oblong, reddish, margins notched and 8-20 cm long. Bright yellow flowers with red-tipped tepals are followed by red and globular fruit. *R. paradoxa* grows bushy at first then pendent and up to 1 m in length and with indefinite spread, the stems are green and triangular bearing alternately set segments at different angles. The flowers appearing from stem edges during spring-winter are white, funnel-shaped, 2 cm across and with recurved tips, followed by red berries. *R. pilocarpa* (syn. *Erythrorhipsalis pilocarpa*) grows up to 1 m long with cylindrical erect stems when young but becoming pendent. Stem joints grey-white, bristly-hairy, 5-12 cm in length and mostly whorled. White or pale-pink flowers are 2 cm wide, followed by purple-red broadly ovoid and some 1.2 cm long fruits bearing tufts of white bristles. *R. salicornioides* (syn. *Hatiora salicornioides, Hariota salicornioides*) grows up to 40 cm long with free-branching habit, stems erect at first but spreading and arching with age, stem segmenats clavate, 1-3 cm long and 0.4-0.8 cm thick and in whorls of 3-5, and appear as tiny green bottles are joined end to end. Bright yellow flowers some 1.2 cm long are followed by white, obovoid and translucent fruits. *R. shaferi* from Paraguay and Argentina grows up to 1 m or more in length with cylindrical and arching or pendent stems. White flowers some 9 mm across are followed by white or pink some 3 mm wide fruits. *R. tucumanensis* grows about 1 m long and with about 50 cm spread bearing less-branched, green, pendent and cylindrical stems 1 cm across. Pale-pink flowers in masses appear during summer, followed by pinkish-white berries. *R. warmingiana* grows 1 m long and having about 50 cm spread with stems growing erect first and then pendent bearing green, slender, cylindrical 2-4 angled stem branches, sometimes tinged red or brown. Greenish-white some 1 cm long flowers in masses appear in winter-spring, followed by violet berries.

Schlumbergera

This genus was erected to honour Frederick Schlumberger, a Belgian horticulturist of the late nineteenth century. A Brazilian genus of two epiphytic cacti allied to *Rhipsalidopsis* and sometimes still known as *Zygocactus* as all the latter ones have been merged with *Schlumbergera*. These cacti are epiphytic wth thin green stems flattened into clear jointed leaf-like segments having toothed margins, bearing small bristly areoles. Drooping tubular flowers appear at the end of the stem singly or in pairs with a little reflexed petals and protruding stamens. These are suitable for hanging baskets and pots. These can be propagated through seeds or stem cuttings with two stem segments or grafted on to *Pereskia aculeata*. *S.* × *buckleyi* (syn. *S.* × *bridgesii*) was developed by W. Buckley probably in 1852 by crossing the two species, *S. truncata* and *S. russelliana* (then called as *Epiphyllum russellianum*). It grows 30 cm high with a spread of 1 m, erect at first then pendent, having arching branched-stems bearing glossy mid-green flat joints up to 5 cm long and 2.5 cm wide with untoothed round indentations on the margins and ends. This produces in winter the rose to magenta flowers some 5.5-7.5 cm long with mouth-reflexed petals. *S. russelliana* grows about 45 cm long having about 90 cm spread with arching and spreading brown branches which are composed of some 4-6 cm long and 3 cm wide, oval or elongated joints, bearing on their margins 1-2 sharp teeth (notches) on their margins and one each of the upper corners, and bristles. Bright violet-pink 5.25-7.5 cm long flowers, paler inside and with highly reflexed petals appear freely in the late autumn and winter. *S. truncata* (syn. *Zygocactus truncata*) from Rio de Janeiro and Organ Mountains (Brazil) was introduced into England in 1818 and since then its many cultivars such as 'Bicolor' (shades of rose-red), 'Joanne' (red), 'Noris' (red and purple), 'Weinachtesfreude' (paler and dark red, throat purple), 'Westland' (shades of rose-red), *etc.* have become the popular house plants bearing masses of white, orange, salmon, rose, purple or blue flowers. It grows up to 30 cm high and with some 40 cm spread bearing arching branches. Stem joints are bright green turning red, free-branching with flat joints some 5.0-7.5 cm long and 3 cm wide, each with 2-4 deeply cut notches on each side, upper one more conspicuous than lower ones. Flowers of pink to deep red colour and some 5.0-7.5 cm long, the lower ones reflexing, appear during winter.

Sclerocactus

A native to southern USA and northern Mexico, this genus is difficult to grow as it hates organic material and survives purely on mineral substrate. The genus comprises 14 species but taxonomists strongly feel that this genus should be splitted out in four other genera such as *Ancistrocactus, Echinomastus, Glandulicactus* and *Toumeya* so the process is going on. These normally grow in small globular or cylindrical form, patterned with tuberculate ribs. Each tubercle bears an areole filled with pad-like woolly bristles and more than nine radiating medium to large white to brown spines in a form that whole body is densely covered and where central spine is prominently hooked. From the apex of the body erect, funnel-shaped and in the shades of white to greenish to yellow to pink and brown flowers appear, followed by *Ferocactus*--like dehiscent fruits. They are propagated through offsets and seeds. Most cultivated species are *S. blainei, S. contortus, S. mesae-verdae, S. parviflorus* and *S. polyancistrus*.

Selenicereus

It is a genus of 20 summer-flowering epiphytic cactus species originating in the region starting from southern USA, through Central America and West Indies into northern South America. Since it is night bloomer hence its generic name comes from *séléné* meaning 'moon' and *cereus* the name of the genus to which this relates. Stems are 2 cm across, green, trailing or climbing, elongated, long ribbed (4-10 ribs) or angled and are found clambering in the rocks or along the trunks and branches of trees producing aerial roots. Most species have spine-bearing areoles. Flowers are usually very large, sweetly scented, funnel-shaped, mostly white, and open flat in the night. It is propagated through seeds and 10-15 cm long stem cuttings. *S. grandiflorus* from Jamaica and Cuba is distinguished by other species of the genus by having tawny or whitish hairs appearing from the areoles on the ovary and floral-tube. These are night-blooming bearing funnel-shaped white flowers 18-30 cm across in summer. It is up to 5 m tall and with indefinite spread though stem width being 1-2 cm, stem fast growing, branching and rooting, having 5-8 low ribs and white-hairy areoles with yellow spines up to 1.5 cm long. Fragrant flowers opening in the summer-nights being white inside and buff-yellow exterior and are 18-25 cm long. *S. hamatus* from Mexico grows up to 5 m in height and is fast growing. The stem is slender with 4 ribs. Ribs bear short and down-pointing spurs at intervals of about 5 cm, each spur having a brown areole and 4-6 whitish bristle-like spines. Fragrant white to yellow flowers some 20-25 cm long and the same across are borne in summer and autumn. *S. pteranthus* from Mexico grows up to 2-3 m long bearing glaucous-green and often purple-flushed stems which are thick, stiff, freely-producing aerial roots, and are 4- to 6-ribbed. Fragrant white to cream flowers some 30 cm long are produced in winter during nights.

Stenocactus

Recorded from Mexico, this genus is closely related to *Ferocactus* and *Thelocactus* and is sold even with the name of *Echinofossulocactus*. It rarely produces individual stems larger than 12 cm and the stems have many thin and wavy ribs. Stems often cluster with age. These are propagated through seeds. *S. captonogonus* resembles more closely to *Ferocactus* than to *Stenocactus* and usually grows solitary some 10 cm tall and wide, producing 3 cm across pink and white flowers. It tolerates up to -10 °C temperature

Stenocereus

A native from S. Arizona (USA) to Mexico, the species of this genus are either low, bushy and rambling or much-branched columnar. This genus now includes the genera *Hertichocereus, Isolatocereus, Lemaireocereus, Mariginatocereus, Marshallocereus, Pachycereus* and *Ritterocereus*. The white to red flowers are either nocturnal or diurnal in opening. Most of the species are tolerant to intense light and heat. *S. thurberi* is a clump-forming species >3 m wide with individual stems being about 3 m tall and 12 cm wide, bearing red spines. Its pink flowers some 7 cm wide appear in summer. Frost kills the young plants and seedlings but in case of mature plants only tip is damaged. It is propagated through seeds.

Stetsonia

It is a NW Argentinian and S. Balivian genus of only one tree-like species, *S. coryne*, with short, stout and swollen trunk up to 8 m tall with spread of 4 m, bearing 8- or 9-ribbed blue-green stems with black spines fading to white tipped-black with age. Funnel-shaped white flowers appearing during summer nights are 15 cm long. It is propagated through seeds.

Strombocactus

This originated in Mexico. It is a genus of highly slow-growing cactus. It takes 5 years to attain 12 cm height when grown from seeds. This cactus is very difficult to grow, and is highly susceptible to overwatering. It is propagated through seeds or through grafting on hybrid *Echinopsis* or on short stock of *Eriocereus jusbertii*. Flowers are funnel-shaped and some 4 cm across. With the transfer of many of its species under the genus *Turbinicarpus*, *Strombocactus disciformis* remains the only representative of the genus. From *Turbinicarpus* it differs by its body structure, flowers and size of the seeds. Its first specimen was collected from Mexico by Coulter in 1829 and then afterwards by Ehrenberg in 1836. Its body is blue-green with grayish tinge, hemispherical to cylindrical, flattened, reaching up to 20 cm high and 9 cm across. Body below terminates in a strong turnip-like root. At an age the base is covered with brown corky spots. The crown is slightly depressed with felts. The ribs are divided into 1.0-1.8 cm high and hard rhomboid tubercles which are spirally arranged. Areoles bear 4-5 deciduous pale spines tipped-dark and some 1.2-2.0 cm long. Cream to pale-yellow flowers variously spotted purple at the tips and in throat arise on the crown and are 3.5 cm long and wide, and these appear in spring and remain open for several days.

Tephrocactus

From Argentina and Peru, this genus is often included in *Opuntia*. This has typical jointed growth and barbed glochids similar to *Opuntia*. Several species are found in the Andes at an altitude of more than 3,500 m and therefore its cushions are flattened and covered with white hairs. They require to be saved from extreme weather conditions. *T. articulatus* from Argentina is a variable species with length of joints ranging from 4 cm to 15 cm. These usually bear glochids but not the spines or bear white to brown flattened spines. Though it tolerates -10 °C temperature but if this condition persists the joints detach at the union and fall to the ground which can be used for multiplication as these root easily if planted during spring. Mostly it is propagated from cuttings and sometimes even through seeds.

Thelocactus

In Greek *thele* means 'a nipple' and cactus the common name. A native of southern USA (especially Texas) and Mexico, this genus comprises 40 or more species of mostly very small cacti resembling *Coryphantha*, and now also including members of *Ancistrocactus, Hamatocactus* and certain other allied genera. These are spherical to columnar with tuberculate or ribbed stems. Funnel-shaped flowers emerge from the elongated areoles on the crown. These flower regularly throughout the year. These are propagated by seeds, and offsets where possible. *T. bicolor* from Texas and Mexico bears pale-green and globose to conical stems, up to 25 cm high and almost of the same spread and 8-13 slightly spiralling ribs divided into square-based tubercles. White-felted areoles bear numerous (usually 9-18) needle-like red and yellow, and up to 3 cm long radial spines, and 4 spreading to erect, usually flattened and yellow with red base, and 5 cm long central spines. Flowers are glossy, purplish-pink, bell-shaped and up to 6 cm long and wide which open after sunrise. *T. leucacanthus* is a dark green clump-forming, spherical to columnar, 10 cm high and with 30 cm spread cactus bearing 8-13 tuberculate ribs which contain areoles, each areole bearing up to 20 short golden spines and yellow flowers some 5 cm across in summer. *T. medowellii* (syn. *Echinomastus medowellii*) is sometimes included in *Neolloydia*. It grows 15 cm in height and spread. The stem is dark green, spherical and tuberculate, and densely covered with short and white spines some 3 cm long. Violet-red flowers some 4 cm across appear in spring. *T. rinconensis* (syn. *Echinocactus rinconensis*) from Mexico reaches to 8 cm high and 13 cm thick and bears oblong stems with about 13 ribs divided into quite clear laterally flattened tubercles with somewhat diamond-shaped bases. One to four undifferentiated central and radial spines are awl-shaped with longest one up to 1.3 cm long. White flowers with yellow throat and some 4 cm wide appear during summer. *T. setispinus* (syn. *Echinocactus setispinus, Ferocactus setispinus, Hamatocactus setispinus*) from Texas and Mexico reaches to a height of up to 30 cm with globose to ovoid stems, and the offsets are produced only by old plants. Thirteen slightly spiralling ribs are divided into large areoles, each areole bears 4 cm long needle-like 10-15 radial spines, and a bit longer and thicker 1-3 central spines, 1 with hooked tips. Funnel-shaped yellow flowers with crimson throat appearing during summer are 8 cm long.

Trichocereus

Trichocereus derives from the Greek *trix* meaning 'hair' and the allied genus *Cereus*. It resembles *Cereus* but differs in ovary and flower's tubular bases being highly hairy but lacking spines or bristles. It is a genus of 25 South American species of largely columnar cacti. These are propagated by seeds and stem cuttings. *T. bridgesii* (syn. *Echinopsis bridgesii*) is a columnar cactus growing up to 5 m tall with 1 m spread. It is base-branching cactus with 4-8 ribs bearing areoles and from each areole up to 6 spines appear, along with scented, funnel-shaped white flowers in the night of summer. *T. candicans* (syn. *Cereus candicans, Echinopsis candicans*) from Argentina is a clump-forming and base-branching cactus growing up to 1 m with indefinite spread, bearing up to 11 broad and rounded ribs with large white-felted areoles, each areole with yellow or white 10-15 radial spines some 4 cm long, and up to 10 cm long 1-4 darker central spines. Fragrant white flowers some 25 cm long are funnel-shaped and open at night in summer only on large specimens. *T. chiloense* (syn. *Cereus chiloensis*) from Chile is a fairly slow-growing species with plant height of up to 8 m, branching above, and the old specimens becoming tree-like. Slightly grey-green stems have 10-15 rounded and notched ribs bearing large whitish areoles. Brownish-yellow to grey, thick and awl-shaped radial spines are 8-12 and 2-4 cm long, while central spines are 1-4 and 6.5 cm long. *T. pasacana*, a dark green cactus growing quite large and globular bearing 10-15 prominent ribs and white to brown, sharp, and 15-20 very long spines. *T. spachianus* (syn. *Cereus spachianus, Echinopsis spachianus*) is a clump-forming and base-branching cactus growing up to 2 m in height and spread and bears glossy-green stems with 10-15 ribs and yellowish to grey areoles and pale-golden spines. Radial spines are bristle- or needle-like and 1 cm long while central spines 1-2, thick and long. White flowers some 20 cm long opening at night of summers are fragrant and funnel-shaped, and appear only on the old specimens. *T. werdermanniana* (syn. *Echinopsis werdermanniana*) is already described under *Echinopsis*.

Turbinicarpus

From Mexico (San Luis Potosi state), the species *T. turbinicarpus* was collected by Sauer in 1934 at the elevation of 1,080 m. All the members of this genus are rare gems of Mexico's flora and the pride of every collector, these are so lovely and beautiful. The cultivation of *T. tunbinicarpus* becomes easy when grafted on hybrid *Echinopsis* or on short stock of *Eriocereus jusbertii*. It is propagated from seeds though seedlings require to be grafted immediately after germination. The body of *T. turbinicarpus* is globose at first when young but later becoming columnar and reaches up to 8 cm in height and 5 cm across. The root is turnip-like and the crown is covered with dense white wool. Stem is dull grey-green bearing white-topped ribs with 4-6 cornered flat tubercles which bear usually 3-4 pale and short radial spines and one 1 cm long central spine. Pink-tinged white flowers which are glossy and silky, and some 3.5 cm in diameter appear in summer.

Uebelmannia

A genus of six species from Minas Geraes (Brazil) which was first described in 1973. This genus appreciates the tropical warmth, *vis-à-vis* some humidity as too hot and dry climates are not congenial for its growing, and mostly these after raising the seedlings are grafted on some suitable rootstocks to save their roots from winter injury. This is a quite handsome cactus but often develops round purple blotches on the sides which scar the plants. *U. buiningii* is the smallest species in the genus whose purple body expands to 8 cm wide while the height is around 10 cm. It flowers for several weeks during spring producing 1.5 cm wide flowers, and afterwards it flowers intermittently during whole summer. It does not tolerate frost and is quite rot-prone. *U. flavispina* is often thought as an ecotype of *U. pectinifera* as mature plants in both the species are quite the same though at juvenile stage the body of *U. flavispina* is dark green with yellow spines white in case of *U. pectinifera* the body is dark purple with black spines. *U. flavispina* at maturity bears golden spines on its olive green body, grows to about 30 cm tall and 12 cm wide, and does not tolerate any frost. The height of *U. pectinifera* is about 50 cm and breadth some 15 cm with woolly crown from where about 5 cm across flowers appear sporadically from spring to summer. It can not tolerate frost.

Wigginsia

This genus originated in Tropical South America (Argentina, Brazil, Uruguay, *etc.*) with 13 species of globose to ovoid stemmed cacti closely related to *Notocactus*. It is named so in honour of Dr. Ira L. Wiggins (*b.* 1899), an American taxonomist. These are propagated through seeds. *W. corynodes* (syn. *Malacocarpus corynodes*) from Argentina, Brazil and Uruguay grows some 20 cm tall with top-felted. Stems are globular at first, then shortly cylindrical with 13-16 narrow and notched ribs, bearing round and white-wooly areoles. Awl-shaped some 2 cm long radial spines which are yellow with darker patches are 7-12 whereas central spine is no to 1 and longer. Flowers are canary-yellow, 5 cm wide and funnel-shaped. *W. erinaceus* from Brazil and Uruguay grows up to 15 cm tall with top-wooly. The stems are globular to shortly cylindrical with 15-20 notched, blunt and spiralling ribs and white wooly areoles. Brownish 6-8 radial spines 1-2 cm long and 1 central spine is 2.5 cm long. Bright yellow and funnel-shaped flowers are some 7 cm wide. W. *vorwerkiana* (syn. *Parodia vorwerkiana*) is described under *Parodia*.

Wilcoxia

A genus from SW USA and Mexico comprises some 6-8 species of tuberous rooted cacti with somewhat slender trailing stems bearing small bristle-like spines and large funnel-shaped flowers. *Wilcoxia* is named so to honour Brigadier General Timothy E. Wilcox (*b.* 1932), an American enthusiast of succulents.These are propagated through seeds and stem cuttings. *W. albiflora* (syn. *Echinocereus leucanthus*) from Mexico is a clump-forming prostrate and tuberous cactus reaching to a height of up to 20 cm and spread 30 cm, with stem thickness of 0.6 cm, bearing several (6-7) grey-green shallow ribs. Spines per areole are small, 10-12 and whitish. Terminal white with dark centre and softly streaked-purple flowers some 4-6 cm wide appear during summer. *W. schmollii* (syn. *Echinocereus schmollii*) is an erect to prostrate and tuberous cactus growing up to 30 cm in height and width bearing purplish-green 8-10 ribbed stems with white spines. Pinkish-purple flowers appear during spring and summer. *W. poselgeri* (syn. *Echinocereus poselgeri, E. tuberosus*) from Texas and Mexico grows up to 90 cm long and some 1.5 cm thick with dark green stems bearing about 8 low ribs. White to greyish spines are about 10-13, measuring 3-6 mm long. Deep pink and 5 cm wide flowers appear during summer and last for several days.

Zygocactus

An epiphytic cactus having flattened, jointed, dichotomously branched stems with serrated margins, bearing 2 prominent teeth at apex. Its species *Z. truncatus* has already been described under *Schlumbergera* as *S. truncata*.

Propagation

Cacti can be successfully grown from seeds, cuttings, offsets and also through grafting. It takes years for the seedlings to attain proper size specimen because of slow growth rate. Small seedlings require special care for their nourishment. Therefore, very few growers attempt to grow cacti from seeds. As many cacti cannot be propagated from cuttings or offsets because of their growth habit, raising them from **seeds** has become necessary. As many species do not flower or set seeds under our conditions, people raise cacti from imported seeds. Moreover, it is compulsory to raise the hybrids resulting from inter-generic or inter-specific crosses from seeds. Raising plants from seeds is, however, not difficult but cumbersome. *Mammillaria* and many other species produce seeds in large quantities which germinate easily. Seeds are preferably germinated in greenhouses in a medium commonly known as seed compost. A good seed-compost is composed of equal parts of well decomposed leaf mould and fine sand. A little methyl parathion powder may be mixed with the compost to keep off the ants which may take away the seeds after sowing. The compost should be sterilized before sowing. Seed is sown in a moist growing medium and then kept in a covered environment, until 7–10 days after germination, to avoid drying out. A very wet growing medium causes to rot the seeds and seedlings. A temperature range of 18–30 °C is considered ideal for germination, soil temperature of around 22 °C promotes best root growth. Low light levels are adequate during germination, but afterwards semi-desert cacti need higher light levels to produce strong growth.

Reproduction by **cuttings** makes use of parts of a plant that can grow roots. Some cacti produce 'pads' or 'joints' that can be detached or cleanly cut off. Other cacti produce offsets that can be detached and planted. Stem cuttings can be made ideally from relatively new growth. It is recommended that any cut surface be allowed to dry for several days to several weeks until a callus is formed over the cut surface. Rooting can then take place in an appropriate growing medium at a temperature of around 22 °C. This method has some limitations also. The most suitable time to take cuttings is from March to September. The plant from which cuttings are to be taken should not be watered for few days and thus the plant is forced to utilize its excess water reserve if any. This operation helps the cutting to be ready for use quickly and chances of rotting are minimized. Cuttings are taken from the mother plants with a sharp knife so that the cut is clean and horizontal. Immediately after cutting, the cut ends should be applied with fine dry sterilized sand and kept in shade for 4-5 days depending upon the size of the cutting. Then the cutting should be planted in a compost as suggested earlier under glass or polythene cover. August and September are the most ideal months for Delhi growers. Generally, roots strike within 12-15 days of planting. In some cases more time may be required. This is a very useful method because it helps in large scale multiplication of rootstocks to be used for grafting.

Grafting is not only an interesting method of propagation but also very useful one. This helps many species of cacti attaining proper specimen size rapidly. Seedlings of some delicate species which do not perform well on their own roots do better being grafted on rootstocks which are much hardier and can grow on varied soil and climatic conditions. Healthy part of a diseased or rotted plant can be utilized as scion and thus rare and uncommon species can be saved. Species with large globular or solitary stems are multiplied through grafting. Tall growing genera like *Trichocereus, Cepahalocereus, etc.* and some other genera having cristate, monstrous or irregular forms and those without chlorophyll can be multiplied through grafting with advantage. *Mamillaria angularis* var *'cristata'*, *Opuntia microdasys 'cristata' etc.* can be propagated through grafting. *Chamaecereus silvestrii*, *Aporocactus flagelliformis*, *Echinocereus pentalobous* and *Selenicereus* spp. can be grown as standards to improve their display. Sometimes abnormally rapid growth prevents proper spine formation and the grafted plants in such cases may not resemble the mother. In general, grafted plants grow better and flower profusely. *Cephalocereus* species for *Cleistocactus strausii* 'Cristata' and *Haageocereus bicolor*; *Cephalocereus albispinus* for *Cephalocereus senilis, C. s.* 'Cristatus', *Cleistocactus strausii* 'Cristata', *Espostoa plumosa, E. plumosa* 'Cristata', *Haageocereus bicolor, Mammillaria angularis* 'Cristata', *Notocactus leninghausii, Opuntia tunicata, etc.*; *Cereus* for *Lobivia famatimensis*; *Cleistocactus baumannii* for *Aporocactus flagelliformis, Chamaecereus silvestrii, Mammillaria elongata, M. gracilis* (syn. *M. fragilis*), *etc.*; hybrid *Echinopsis* for *Astrophytum asterias, A. capricornes, Epithelantha micromeris, Eriosyce ceratistes, Neochilenia jussieii, Obregonia denegrii, etc.*; hybrid *Echinopsis* and *Eriocereus jusbertii* for *Brasilicactus graessneri*; short stocks of *Eriocereus jusbertii* for *Neochilenia jussieii* & *Eriosyce ceratistes*; higher rootstocks of *Eriocereus jusbertii, E. bonplandii* or *E. tortuosus* for *Aztekium*; *Myrtillocactus geometrizans* for *Cephalocereus senilis cristatus,Gymnocalycium mihanovichii rubra* and its various other hybrids; *Opuntia* species for *Aporocactus flagelliformis* & *O. tomentosa* for *O. microdasys*; strong stock of *Piptanthocereus peruvianus, P. jamacaru* or *P.*

dayamii for *Espostoa lanata* & *Mediolobivia aureiflora*; *Rebutia senilis* on *Piptanthocereus*; *Trichocereus spachianus* for *Echinocereus pulchellus*; etc, Apart from these there are many other rootstocks such as *Hylocereus triangularis, Opuntia ficus-indica, Pereskiopsis velutina, Selenicereus grandiflorus, etc.* Das *et al.*(1982) have studied the graft compatibility of these rootstocks with large number of cacti species. Grafting methods employed are (i) **flat grafting** where success is great. For most of the species of cacti flat grafting method has been found to be very common and is the easiest and most successful one. These involve the process of bringing together two flat surfaces, that of the beheaded stock plant and top of the scion required to be grafted. For that grafting, both the stock and scion should be in active state of growth, just enough portion of stock and scion should be removed to reveal clearly distinguishable vascular bundle and the differently coloured ring towards the centre of the plant. No time should be wasted in getting the two freshly cut surfaces together, ensuring that their respective vascular bundles are in contact with each other; a slight of setting of the two shoot achieve this end. Light pressure is needed for few days to facilitate the union of the tissues, which can best be done with a band of elastic/plastic film loop over the top of the scion and beneath the base of the pot. The bang should not be thin or too tightly placed to cut through the plant tissue. Narrow strip of a polythene sheet tied in the groove tightly at the end of two cylindrical sticks pointing at the lower end for easy penetration to the growing medium make the most effective clamps for grafting cacti; the clamps can be used for repeated operations. Though union takes place in three days, the clamps need to be removed after about a week when the scion shows signs of new growth. (Das, 1973); (ii) **cleft grafting** is done with the pendent species such as *Aporocactus flagelliformis, A.mallisoni, Chamaecereus silvestrii, Hildwintera aureispina, Echinocereus pentalophus*, etc. In this process, a V-shaped notch is made at the growing point of the rootstock and a wedge-shaped similar cut is made to the basal end of the scion. Then the scion is inserted in the cleft of the stock and held in position by pinning two long cactus spines through the stock and the scion. In addition, it may be necessary to tie them together with ordinary jute thread or any other material used for tying. Thread and strings must be removed carefully once the union is complete (Bhattacharjee, 1975); and (iii) **splice grafting** where one side each of the scion and the stock is spliced at a 45° angle to match the cut surfaces of each other, both these are held in position by wrapping with the thread which should neither be too thin nor tied too tightly so as to injure the stock. In general, flat graft should be preferred for grafting in cacti unless otherwise there is a need for any of the other methods (Bhattacharjee, 1975).

Cultural Practices

Many of the larger cactus collections are housed under **protective structures** of polythene and glass cover. Although expensive, these structures provide favourable growing conditions in areas where maximum and minimum temperatures vary widely in summer and winter months. The conditions in a polyhouse or glasshouse also favour better germination of cactus seeds and growth of young seedling and sensitive types such as species of *Melocactus* and others that could not be grown outdoors. Plants are commonly grown in pots arranged on shelves or benches. Some growers prefer to raise plants in groups by planting them directly on raised beds or in the soil on the glasshouse/polyhouse floor. It is often observed that a cactus grown under semi-shade if suddenly removed to a more sunny situation develops burns due to scorching. It is for such reasons that the domesticated cacti should be housed in suitable locations. Cacti love **sunshine**, exposure to the morning sun should be preferred. In regions where it becomes very hot during summer, it is advisable to avoid exposure to mid-day sun for a few hours. Cacti grown under shade become lanky and often fail to flower, or flower sparsely. Excess of water is harmful for most of the species and hence all cacti should be grown under glass or ploythene roof, although a few hardy species of *Cereus, Echinops*is and *Opuntia* can be grown in the open under our climatic conditions. For production of flowers in large numbers and for development of healthy plants, abundant circulation of fresh air is essential. Hence, a cactus house should have only the roof made of glass, fibre-glass or polythene, supported on pillars, while the sides should remain open. Such houses should run in the north-south direction (Das and Panda, 1995). Under outdoor conditions most cacti prefer a sunny situation. Many genera are well adapted to thrive in conditions of great **light intensity** and summer heat. Fine wool, dense spines and deeply carved ribs act as protection against sun-rays and excessive light. Plants grown indoors or in glasshouses will grow faster and have an appealing appearance, but they can be more sensitive to sunburn during the hot periods of mid-summer if exposed to outside conditions. An excessive buildup of heat in glass or polythene covered areas can be avoided by shading with arrangements for adequate **ventilation**. The brown scarring of the surface-tissues of cacti display pale-colouration and result in an elongation and tapering of the growing tips with poor flowering.

Cacti are grown under varied **temperature** regimes in different parts of India. In Delhi, Kalimpong and Darjeeling where winter temperature goes down to 2 °C, the cacti enjoy a winter rest and to a large extent suspend growth. However, best growth has been recorded in most species when the temperature is between 15-30 °C. The period from March to April under our condition, when average temperature variation is from 25-35°C, is the best time for most species of cacti to flower and bear fruits. Therefore, under glass or polythene covers most cacti can be grown throughout the year in different climatic zones of India.

Cacti grown on porous soil mixture require generous **watering** at 3-4 days intervals during summer months. However, during winter the frequency of watering is drastically reduced and it is normally once in a week or in a fortnight depending on the species grown. In principle, watering to 'run-off-stage' once or twice a week is better than more frequent watering. Soil should be given time to dry out between watering, rather than being kept wet. Freshly potted plants should not be watered until rooting, usually a week or two after planting. Dripping water is fatal to cacti, therefore, any crack or hole, if detected in the roof of the cactus house need to be repaired immediately. Water should ideally have a slightly acidic reaction, be close to air temperature and have an abundance of dissolved oxygen.

For balanced growth and flowering of cacti, in general, a good porous **compost** is quite helpful. One part of sandy-loam soil, one part of well-decomposed cowdung manure, two parts of screened leaf mould, one part of fine sand, small quantity of charcoal powder and a little of bone meal make a good compost for cacti. With sufficient precautions while handling the plants, **potting** or **planting** of cacti in pot or ground can be done any time of the year. It is best to transfer the plants when they are dormant, *i.e.* during winter months. New growth starts from March and continues up to October. However, potting and planting should be avoided during rainy season. Porous earthen **containers** are always superior to glazed plastic or metallic types. Exchange of air takes place through pores which facilitates healthy root growth. The pots should be clean and dry while potting. Size of the pots should be just large enough to accommodate the plants, too big pots with enough compost keep extra water which may damage the plants. In small pots, plants become pot-bound quickly and growth ceases. Two to three layers of clean crocks should be put with concave side towards the bottom so as to facilitate easy drainage. Careful examination of root of the plants to be repotted or planted is essential. The dead or diseased roots and roots with tubercles caused by nematodes should be cut off. The wounds may be dressed with sulphur dust. Root mealy bugs which cling to the roots very often, if present, may be washed with methylated spirit in decoction of nicotine. Crystals of para-dichloro-benzene (PDB) may be put on the bottom layer of the compost above the crocks to keep mealy bugs away from roots. The roots of the treated plants should not be moist at the time of planting.

Insect-Pests and Diseases

The ample porosity in the medium and excessive dryness of the compost are the reasons for colonization of the medium by the ants. **Ants** (Formicoides, *Formica* spp.) tunnel the root zone for their movement which causes exposure of the roots from the soil so old fibrous roots dry up and new ones do not come in contact of the soil, *vis-à-vis* movement of the ants also disturbs the root system badly so plants look sickly. With them the ants may also carry mealy bugs, aphids and other insects for their food which may also attack cacti. It is advisable to treat the medium with methyl parathion dust while potting. However, the affected plants should be repotted by soil treatment. **Mealy bugs** (*Pseudococcus* spp. on stem and *Rhizoecus falcifer* on roots) are slow-moving creatures with thick powdery covering and can be found clustering at a point whether on the roots or on the aerial parts, sucking the plant sap. **Stem mealy bugs** (*Pseudococcus*) are not common excepting on *Opuntia*, so its infestation may be controlled by swabbing the stems with strongly soapy water, plain rubbing alcohol, malathion or pirimiphos methyl sprayings frequently fortnightly. The thick body coating protects the mealy bugs from insecticides. Unless the plants are examined of their roots for poor or sickly appearance, it is difficult to detect the incidence of root mealy bugs (*Rhizoecus falcifer*). Roots of plants affected by mealy bugs should be washed in decoction of nicotine containing methylated spirit. The excess water on roots should be allowed to dry off before planting. If the old pots and crocks are to be used, those should be cleaned in boiling water for 15 minutes before use. Crystals of para-dichloro-benzene (PDB) should be spread sparingly over the lowermost of compost on crocks and then the pots should be filled and planted with the treated plants. The fumes emitted by this chemical will keep off mealy bugs from roots. Introduction of ladybird (*Cryptolaemus montrouzieri*) will also control mealy bugs. **Root mites** are almost unknown to the growers as are red spiders but they cause a lot of damage. The infected plant may appear to be bleached and become spotted. Plants growing under shade and in drier parts are observed to be more susceptible to mites. Decoction of nicotine may be drenched at fortnightly intervals to get rid of this pest. With infestation by **scale insects** (*Cactoblastics* sp., *Dactylopius* sp.), the plant body becomes sticky with excrement (honeydew) and blacken with sooty moulds. Plant growth is also adversely affected. They are yellow, brown, dark grey or white and up to 6 mm long and are found sticking to the shady portion of the body or on the joints. Tar oil wash in winter or spraying with malathion or pirimiphos methyl during summer and autumn proves very effective in controlling these when newly hatched nymphs are present. **Red spider mites** (*Tetranychus urticae*) appear to be tightly fixed to the plants and are difficult to eradicate. Many species of *Trichocereus* and *Oleocereus* are commonly affected by scales. There is no satisfactory remedy. Only way to save the plant is to use the healthy part of affected plant for propagation. The infected parts should be destroyed. The female mite overwinter in the crevices of dead parts or in the soil. Malathion, dimethoate or formothion will protect the plants from its attack. Regular wetting of plant body with water will help checking the pest. Fumigants in the greenhouse will be very effective. **Woodlice** also known as slaters or pill bugs are grey or brownish-grey, sometimes with white or yellow markings, and up to 1 cm long with hard segmented bodies feeding at night though hide away in dark shelters during daytime. In fact, these feed on the decaying material and are often found on plants already damaged by other pests and diseases. These may make holes in the buds and flowers and seedling plants, and sometimes on the tender parts of the plant body, especially the growing tips, though in fact, they are not real pests. These are controlled by cleaning the plant debris and the dead plant parts. Scattering the methiocarb slug pellets will also control these pests. Lepidopterous **caterpillars** (*Agrotis segetum* and other species, *Helicoverpa armigera, Mamestra brassicae, Pieris brassicae, P. rapae, Plusia orichalcea*) feed on tender growing shoots, buds and flowers except the *Agrotis* which feed normally underground during daytime though at night above soil level. They can be picked and killed. Spraying pirimiphos methyl or fenitrothion in the soil for cutworms and spraying on aerial parts for other caterpillars will be effective. Alternatively, the soil treatment with chlorpyriphos + diazinon is effective for cutworms. *Bacillus thuringiensis* bacterium can also be introduced which will parasitize these. **Grasshoppers** (*Hieroglyphus* spp.) feed on the plants by chewing the tender parts. All the chemicals used for controlling caterpillars will also control grasshoppers. **Whiteflies** (*Trialeurodes vaporariorum, Aleyrodes proletella*) are very active tiny creatures some 2 mm long with white wings. These rest beneath the young leaves. The immobile nymphs are scale-like whitish-green. These feed on the tender leaves and flowers covering the affected areas with sticky white excrement (honedew) and a form of sooty mould. These can be controlled through spraying with permethrin, pyrethrum, pirimiphos methyl or insecticidal soaps thrice or four times at 5-days intervals. Parasitic wasp (*Encarsia formosa*) is an effective biological control in the polyhouse. **Slugs** and **snails** feed on seedlings, stem parts and flowers when the cacti are grown at humid and darker places. These normally do not ascend on those cacti where spines are small, closely set, bristle-like and easily detachable even with a soft touch. While feeding these batter the plants completely leaving a trail of silvery slime on the surface as well as on the earth surface. They are attracted to organic fertilizers and mulches. Soil should be cultivated regularly to expose their eggs, and slug pellets prepared of metaldehyde or methiocarb should be scattered around the affected plants. Regular sanitation and polyhouse fumigation for control of any insect will together control all these pests also.

Nematodes (*Melodogyne incognita, M. hapla, M. javanica*), the microscopic creatures are highly devastating to cacti as these interfere with the growth of plants by reducing the size of root system, its rotting and galling. These restrict supply of water and nutrients to the plants from the soil. No satisfactory permanent remedy to nematode infestation is known as these invade the plant tissues and feed inside the plant body. Temporary relief can be obtained by cutting away the affected roots from the plants and planting them again after dressing the cut ends of roots with sulphur dust against possible fungus attack through the wounds. Treating the potting compost before potting or repotting with Furadan or Thimet G-3 @ 2.5 g per cubic feet will effectively control root-knot nematodes.

Fungi, bacteria and viruses attack cacti, the first two particularly when plants are over-watered. **Brown spot** or **rot** (*Botryodiplodia theobromae*) is found on opuntias and cereus with irregular, elongated, blackish-brown to grayish spots on the phylloclades (Sohi, 1992). **Blossom blight** (*Choanephora* spp.) of *Dendrocereus* occurs only when there is rain and high humidity. In its infection the flowers are covered with white fungal growth later turning brown to purple with a definite metallic luster. It can be kept under check by growing plants on well-drained land. **Anthracnose** (*Mycosphaerella opuntiae*), **leaf scorch** (*Hendersonia opuntiae*), **leaf spot** (*Septoria cacticola*) and **stem rot** (*Aspergillus alliaceus, Helminthosporium* spp.) are also sometimes observed, especially on *Cereus serpentinus* (Sohi, 1992). Infected parts should immediately be removed and burnt and use of zineb or ziram (0.2 per cent) will control all these diseases. Moulds cause 'damping off" or young seedling collapse due to rotting of the base in saline medium or through wounds inflicted on plant body. Therefore low pH value peat-mix and crushed bricks in the medium will prevent this problem. The drenching of seedling pots or pans with benomyl, captan or bavistin will help preventing the mould. The plants or plant organs should be planted only after getting dried the wounded parts under sun for a few days and by treating with sulphur dust. The main fungi which attack cacti are *Phytophthora*, *Rhizoctina* and *Fusarium*. **Fusarium** can gain entry through a wound and causes rotting accompanied by red-violet mould. ***Helminthosporium* rot** is caused by *Bipolaris cactivora* (syn. *Helminthosporium cactivorum*). *Phytophthora cactorum* causes similar rotting in cacti. Fungicides may be of limited value in combating these diseases. Unless the attack is detected at an early stage it is difficult to save the affected plants and in most of the cases they are lost. Isolation of the diseased plant should be done immediately and if required, the plant has to be destroyed in order to save other plants from infection. Prevention is better than cure holds good in tackling disease problem in cacti. Correct watering and sterilized compost help in bringing down disease incidence by fungus considerably. Excess watering is harmful because aeration of roots is impaired. Drenching the soil and spraying the plants with Thiride or Blitox once every month may prove useful.

Several **viruses** have been found in cacti, including 'cactus virus X'. These appear to cause only limited visible symptoms, such as chlorotic (pale green) spots and mosaic effects (streaks and patches of paler colour). However, in an *Agave* species, 'cactus virus X' has been shown to reduce growth, particularly when the roots are dry. There are no treatments for virus diseases except destroying the plants.

Cacti **splitting**, as well as **ringed growth** and **deformities** especially in columnar cacti can be avoided through frequent shallow watering, especially during summer months. **Off-colouration** and **deformed** winter growth can be prevented by providing proper exsposure of sunshine and temperature akin to their requirement, *vis-à-vis* reduction in the frequency of watering during winter. Excessive heat and strong sunlight may cause **brown scarring** of plants so placing the plants at correct positions with proper ventilation will prevent this problem.

References

Anderson, E.F. 2001. *The Cactus Family*. Timber Press, Portland, Oregon, USA.

Bailey, L.H. 1942. *The Standard Cyclopedia of Horticulture* (vol. I-III). The Macmillan Co., New York.

Barbera, G., P. Inglese and E. Pimienta Barrios (eds), 1995. Agroecology cultivation and uses of cactus-pear. FAO, Plant Production and Protection paper 132.

Beckett, K.A. 1987. *The RHS Encyclopaedia of House Plants Including Greenhouse Plants*. Salem House Publishers, Massachusetts, USA.

Bhattacharjee, S.K. 1975. Cacti-a mysterious group of plants. *J. Sci. Club*, **29**:111-121.

Brickell, C. 1989. *The Royal Horticultural Society Gardener's Encyclopedia of Plants and Flowers*. Dorling Kindersley Limited, London.

Das, P. 1973. Grafting cacti for pleasure and profit. *Indian Hort.*, **18**(2): 25-26.

Das, P., S. Panda and R.C. Dash, 1982. A noted cactus rootstock. *Indian Hort.*, **26**(4):17-18.

Das, P. and P.C. Panda, 1995. Protected cultivation of cacti and other succulents. *Advances in Horticulture*, Vol – 12. *Ornamental Plants* (eds Chadha, K.L. and S.K. Bhattacharjee), pp. 819-851. Malhotra Publishing House, New Delhi.

Das, P., P.C. Panda and A.B. Das, 1999. *New Cactus Cultivars*. Regional Plant Resource Centre, Bhubaneswar, 5 p.

Everard, B. and B.D. Morley, 1970. *Wild Flowers of the World*. Peerage Books, London.

Hay, R. and K.A. Beckett, 1971. *Reader's Digest Encyclopaedia of Garden plants and Flowers*. The Reader's Digest Association Ltd., London.

Herwig, R. (1986). *2850 House & Garden Plants*. Crescent Books, New York, USA.

Heywood, V.H. 1987. *Flowering Plants of the World*. Croom Helm, London.

Hunt, D. 1992. *CITES-Cactaceae Checklist*. Royal Botanical Garden, Kew, U.K.

Hunt, D. and N. Taylor, 1990. The genera of Cactaceae : Progress towards concensus. *Bradleya*, **8**: 85-107.

Mabberley, D.J. 1987. *The Plant Book*. Cambridge University Press, Cambridge, U.K.

Ochoa, J., G. Ayrault, C. Degano and M.E. Alonso, 1992. El cultivo de la Tuna (*Opuntia ficus-indica*) en la provincia de Santiago del Estero, Argentina – II. Congreso Internacional de Tunay Cochinilla, held on September 22-25, 1992, in Santiago, Chile.

Pimienta-Barrios, E. 1993. El Nopal (*Opuntia* spp.): Una alternativa ecológica productiva para laszonas ridas y semi ridas. *Ciencia*, **44**: 339-350.

Rowley, G.D. 1989. *The Illustrated Encyclopaedia of Succulents.* Leisure Books, England.

Slaba, R. 1992. *The Illustrated Guide to Cacti.* Chancellor Press, London, U.K.

Sohi, H.S. 1992. *Diseases of Ornamental Plant in India.* ICAR, New Delhi.

Šubik, R. 1968. A Concise Guide in Colour ~ Cacti and Succulents. The Hamlyn Publishing Group Ltd., England.

10

Canna (Family: Cannaceae)

Sanyat Misra and R.L. Misra

[**Common names**:Achira, Arrowroot/Edible canna (*C. edulis*), Canna, Canna lily/Crozy canna/Gladiolus canna/ French dwarf canna (*C. generalis*), Common garden canna, Indian Shot (common English name for *C. indica* and many garden hybrids due to the seeds being very hard which resemble the shot or pellets in shotgun cartridges), Giant-flowered canna/Iris-flowered canna/Italian canna/Orchid-flowered canna (*C.* x *orchioides*), common Hindi names 'keli' due to banana-like leaves and '*vijayanti*']

Introduction and Origin

Its definite derivation is not known as it is of oriental origin but it is presumed that the word *Canna* derives from the Greek word *kanna* meaning a 'reed' as some species are tall with reed-like stems. All the species under the genus *Canna* are herbaceous perennials with rhizomatous rootstock and its native haunts are tropical and subtropical South America and Asia with one species native to USA (Florida).

Almost all the available cannas are clump-forming hybrids with fleshy rhizomes. The rhizomes of certain species such as *C. edulis* and *C. tuerckheimii* (syn. *C. gigantea*) yield edible starch. In West Indies the starch prepared from the indigenous species (*C. edulis*) is known as *tous-les-mois* meaning every month. The mature seeds in some countries are used as beads. Based on the vars 'The President' (red), 'Dorothy' (pink) and 'Copper Giant' (orange), the major anthocyanin pigment present in the flowers are cyanidin 3-rutinoside and its concentration differences account for the variation in flower colours (Ashtakala and Maloney, 1971).

Cannas are ideal as garden plants, for formal beds and borders, mass planting of single colour against a wall, amongst shrubbery, evergreen backdrop or at the waterside, and dwarf ones are most suitable for container planting even in the temperate regions but with winter protection to highlight deck, patio or entrance to buildings and homes. Its stately habit and majestic flower colours provide a very bold effect in the garden. The flower colour in canna is pale-yellow, yellow-green, pale-green, pale-flesh, yellow-red, sulphur-yellow, greenish-sulphur, greenish-red, orange, pale-scarlet, rose, rose-crimson, red-yellow, red, pale-purple, purple and may be variously tinged, striped, spotted and splashed with rose, red and other colours. True blue and blacks are the missing colours in canna. Up to the last quarter of 19th century, cannas were also being grown for their foliage beauty only but due to researches carried out from 1848-92 transformed these into attractive flowering ornamental. Even otherwise, its usually green (light- to blue-green, and in various shades of bronze & variegated) foliage with attractive shape and size also make a pleasing display even when not in flowers, and its single leaf is also suitable as backdrop for flower arranging and vase decoration. This ornamental has become quite homely to this country and no garden can be found in the country without a bed of canna. It is so versatile in nature that it can beautify any neglected nook or corner, wasteland or the rugged part of the garden *vis-à-vis* a situation which remains permanently moist or damp. Albeit, it is not suitable as a cut flower but its ultra dwarf types with bronze foliage or bold flower colours with green or variegated foliage planted in small containers give utmost enchantment when kept indoors in full bloom. Those bulbs suitable either for cut flowers or for loose flowers commercially qualify as commercial bulbous ornamentals but in this case though it does not qualify the level but it is an easy and rampant clump-forming grower with a very high multiplication rate and its easy cultivation and sustainability even in poor, and dampy soils has made it most prominent and important in this country and in the remaining tropical and subtropical world hence is being described here.

Botany

The rhizomes are horizontal, fleshy, stocky, and often stubby with many projections where leading buds are rounded, bearing numerous thick and branched roots, stems erect, unbranched and up to 3 metres in height depending upon the species or the cultivars, leaves usually green and sometimes variegated or stained purple or bronze having clear midrib and numerous lateral veins, are spirally arranged with lamina up to 1 metre long, inflorescence a terminal raceme or panicle, bracteate, upper bract subtending the solitary or paired, sessile or shortly pedicellate flowers, flowers hermaphrodite, tubular, showy, cream, yellow, orange, pink, rose, red, purple, rarely white, and oftenly tinged, spotted, striped or flushed with other colours, sepals 3, persisting, equal and free, overlapping, petals 3 (quill-like objects that form the cover to the coloured portion are the real petals and what usually we call as petals are, in fact, staminodia), coloured, unequal, long but not very wide and jointed in basal tube, stamens 5 out of which 1 with solitary, petaloid, and marginal anther, emerging from the base of the fleshy style and staminodes, staminodes petaloid, 1-4, the small and often recurved innermost one is termed as 'labellum' (forming lower lip) but outer staminodes exceed petals, normally only one stamen being fertile, ovary inferior, 3-loculed and with numerous ovules, style single, long and petaloid, fruit a loculicidal capsule, and seeds numerous, round to ellipsoid, black to brown and hard. Highly ornamental canna cultivars are usually of low fertility, the cultivars 'America' and 'President' have a maximum of 19.4 per cent pollen fertility (Feofilova, 1975). Shevchenko and Fiofilova (1981) worked out pollen viability in *Canna glauca,* the Crozy (Crosie) canna cvs 'The President', 'Louise von Ratibor' and 'Louis Cayeux', the orchid-flowered canna cvs 'Andenken an Wilhelm Ffitzer' and 'Feuer Vögel' and the hybrids between the cultivars and wild species, and reported that pollen viability ranged from 0 to 93 per cent, hybrids resembling the cultivars had low viability than those resembling the species and *C. glauca* triploid hybrids also had low viability. The basic chromosome number in canna is n = 9.

Tiwari and Gunning (1986a, 1986b) stated that in a *Canna indica* x *C.* sp. hybrid, prior to meiosis, tapetal nuclei divide without cytokinesis, and between mid- and late-prophase the tapetal cell walls break down releasing protoplasts into the locular cavity. The protoplasm does not fuse to form a periplasmodium, they are initiatlly more or less spherical, but may produce amoeboid processes in late stages of microsporogenesis. Meiosis is normal and cytokinensis is successive. They described that anther of *C. indica* x *c.* sp. hybrid contains a non-syncytial, invasive tapetum, considered to be intermediate between the secretory and invasive types since with the onset of prophase I the tapetal walls dissolve and the protoplasts move into the locular cavity, but there is no periplasmodium formation, and with the wall dissolution, the tapetal protoplasts develop a 17 mm thick extracellular granulofibrillar cell coat. Gupta (1966) through acute exposure of ^{60}Co gamma rays at 1000-2000 r induced colour mutations in canna. ^{60}Co irradiation of canna rhizomes beyond 2000 r suppressed or retarded sprouting, retarded plant development, delayed flowering considerably and produced mutations of changed flower colour and shape (Chemarin *et al.,* 1970). Chemarin *et al.* (1973) recorded changed flower colour mutants by treating the rhizomes of the three canna varieties 'Nadezhda, 'President' and 'Louise von Rattibor' from 0.5 to 6.0 krad ^{60}Co λ-rays. They found critical dose depressing plant growth and development as 2 krad for cv. 'Nadezhda' and 4 & 6 krad for the two triploid cvs 'President' and 'Louise von Rattibor', respectively.

Classification, Species and Varieties

There are 50-55 species under the genus *Canna.* The most noteworthy ones are *C. discolor*, *C. edulis, C. flaccida, C. glauca* var. *rubra-lutea, C. indica, C. iridiflora, C. lagunensis, C. langunose, C. liliflora, C. limbata, C. lutea, C. nepalensis* (syn. *C. chinensis*), *C. pedunculata, C. speciosa* and *C. warscewiczii* (syn. *C. sanguinea*). *C. discolor* (syn. *C. rotundifolia*) is native to C. America, height up to 180 cm, leaves 90 x 30 cm, green with purple margin and underside purple, and flowers red, yellow and purple; *C. edulis* (syn. *C. esculenta*) is native to West Indies, South America and South Africa, height 240-300 cm, leaves 60 x 15-20 cm, and green with purplish-red underside, and flowers red and yellow; *C. flaccida* (*C. angustifolia*) is native to Florida, height 180 cm, leaves 25 x 10-15 cm, inflorescence simple, flowers paired, iris-like and yellow, and a principal parent of modern garden hybrids; *C. glauca* (*C. schlechtendaliana, C. mexicana, C. stolonifera, C. lanceolata*) is native to West Indies and S. America, height 150 cm, leaves 30-50 x 3-15 cm and white-margined, raceme simple or forked, flowers paired and yellow-green, and its var. *rubra-lutea* has deep yellow flowers tinted red or rarely deep purple; *C. indica* (syn. *C. patens,* now *C. orientalis*) is native to West Indies, CS America and South Africa, one of the oldest species in cultivation, height 90-150 cm, leaves 20-50 x 10-18 cm and green but rarely stained red-purple, inflorescence simple or forked, flowers solitary or paired and colour red to pink-red or orange, and its var. 'Purpurea' has purple-bronze leaves and small red flowers, and 'Thompson and Morgan hybrids' have bronze leaves, and shades of cream, yellow and red flowers; *C. iridiflora* (syn. *C. gigantea, C. latifolia*) is native to Peru, height up to 250 cm, leaves 120 x 45 cm, inflorescence simple or branched, flowers solitary, pendant, rose-coloured, petals forming slender tube and petaloid stamens are notched at tip, and one of the earlier parents for developing pendant flowers in France; *C. lagunensis* is native to Mexico and C. America, medium size, leaves pale-margined, petals red-spotted; *C. liliflora* is native to Bolivia, Colombia and Panama, height up to 3 metres, leaves 90-120 x 45 cm and subhorizontal, inflorescence simple or branched, flowers lonicera-scented, solitary, in corymbose panicle and colour white which finally become tinged with brown; *C. limbata* (syn. *C. aureo-vittata, C. floribunda, C. laeta, C. patens, C. recurvata, C. variegata, C. ventricosa*) is native to Brazil, largest staminodia 2-lobed, the medium one emarginated and the other entire, and colour red with yellow margins; *C. lutea* (syn. *C. commutata, C. densifolia, C. floribunda, C. maculata, C. sulphurea*) is native to Mexico to Brazil, height 75-120 cm, leaves oblong or broad-lanceolate, raceme simple or rarely forked, flowers (staminodia) pale-yellowish-white; *C. nepalensis* (syn. *C. chinensis*) is native to Himalayas, height up to 180 cm, leaves broad-oblong, flowers elongated raceme or rarely paniculate, petals linear-lanceolate, reflexed and pale-purple, staminodia bright red and lip yellow, and it is one of the parents of earliest hybrids; *C. pedunculata* (syn. *C. buekii, C. reflexa*) native to West Indies and S. America, stems 150-180 cm, leaves oblong-lanceolate and 30-60 x 7.5-10 cm, inflorescence a many-flowered long raceme with hairy rachis, petals reflexed and greenish-yellow, staminodia emarginated and pale-yellow, and lip oblanceolate and yellow; *C. speciosa* (syn. *C. leptochila, C. saturate-rubra*) is native to Himalayas, height 150-180 cm, leaves broad-oblong, flowers in elongated raceme or rarely paniculate, petals linear-lanceolate, erect and pale-purple, staminodia emarginated and bright red, and lip emarginated and yellow; and *C. warscewiczii* (syn. *C. sanguinea*) is native to Costa Rica and Brazil, stems 90-120 cm and claret-purple, leaves oblong, some 15 cm long, 8 cm broad and more or less claret- or bronze-tinged, raceme simple and dense, sepals lanceolate and glaucous purple, petals lanceolate, reddish and glaucous, staminodia and lip oblanceolate and bright scarlet, and it is one of the earlier parents for developing hybrids with showy foliage and drooping flowers; *etc.* But now species are known only in herbaria and little known outside. These are only hybrid varieties which are available in the gardens or elsewhere.

M. Année of France is probably the first man to grow cannas through seeds sown in 1848 of the real species *Canna nepalensis* which with all probability was pollinated with *C. glauca*. He named the resultant race of tall cannas as *C. annaei*. Further in 1863, a new race with the name of *C. ehemanni* (syn. *C. iridiflora hybrida*) was developed through *C. iridiflora* and *C. warscewiczii* which was intermediate in stature and had showy foliage and more pleasing drooping flowers. Though *C. ehemanni* race is still in the trade but the original one is almost extinct. This race was further crossed with other species and races and a selection of dwarf but large-flowered type race was developed in France from a large population, and therefore this race is known as French cannas or Crozy (their one of the renowned breeder) cannas. In the middle of 20th century, in Italy, another race Italian or orchid-flowered cannas was developed by using *C. flaccida* with garden forms and with *C. iridiflora* and the resultant hybrids had iris-like outlines but the flowers were short-lived. The varieties under this race are 'America', 'Austria', 'Bavaria', 'Burbank', 'Burgundia', 'Italia', 'Pandora', *etc.* Bosse (1968) in Germany laid out a comparative trial of the new cultivars and the old ones and found that the standard old cultivars such as 'President', 'Gartenfeuer' (Liebesglut), 'Felix Ragout' and 'Garteninspektor Nessler' still stand better than the new ones, however, the yellow-flowered new cultivar 'Schwabenland' though less attractive but was more vigorous and highly tolerant to weather conditions. There during the unusually hot summer of 1967, all the new cultivars tested, *viz.* the dwarf growing 'Dondoscharlachfeuer', 'Granat', 'Charme', 'Claiiudia', and 'Citrus' and the standard-height cultivars 'Fanal', 'Favourite', 'Frohsinn', 'Edelgard', 'Dondoblutrot', 'Beatrix', and 'Aphrodite' grew and flowered well. Gubanov and Khalaburdin (1987) with a performance trial of canna cultivars in Turkmenia reported that lowest propagation coefficient was shown by cvs 'Auzhurnyi', 'Rosenkranz' and 'Kishinev' (4-5 plants/stock plant), and the highest by ''Lunnyi Svet and 'Kron' (18-19 plants/stock plant).

Canna x *generalis* (*C. glauca* x *C. indica* x *C. iridiflora* x *C. warscewiczii*) and *C.* x *orchioides* (*C. glauca* x *C.*

indica x *C. iridiflora* x *C. warscewiczii* x *C. flaccida*) are horticultural species with a range of plant height from 50 to 160 cm under which all the ornamental cultivars of hybrid origin are included whereas the height of the elemental species (*C. flaccida, C. glauca, C. indica, C. iridiflora* and *C. warscewiczii*) ranges from 89 cm to 500 cm. The characteristic features of *C.* x *generalis* hybrids are short to tall and slender, leaves from glaucous grey and leathery to dark chocolate-red and thin, flower shape and colour from small narrow segments to large and ruffled, colour being from pale-yellow to orange or scarlet, and of *C.* x *orchioides* hybrids the flowers are very large, tubular at base, petals reflexed, usually splashed or mottled and 3 broad wavy staminodes exceed by the lip, but now both are so much interbred that these are now referred to only as *C.* x *generalis.* The varieties under this group are 'Ambrosa' (dwarf, pink), 'America' (leaves copper-purple, flowers red), 'Angel Pink' (foliage green, flowers creamy-white speckled elongated rose-pink), 'Apricot Dream' (foliage ovoid and green with white margin, flowers apricot and salmon), 'Assault' (leaves purple-brown, flowers orange-scarlet), 'Bangkok' (syn. 'Trinacria Variegata', plants dwarf, leaves with yellow lines parallel to the veins, flowers small and light yellow streaked off-white inside), 'Black Knight'(syn. 'Ambassador', leaves brown, flowers red), 'Burgundy Blush' (tall, leaves large, round and deep green with burgundy blush, flowers small and cherry red), 'Cattleya' (foliage green, flowers golden-yellow blotched orange-red with ruffled petals), 'Cherry Red' (leaves dark, flower bright red), 'City of Portland' (height 100 cm, flowers rosy-pink), 'Cleopatra' (leaves splashed black, flowers red and yellow combined and bicoloured), 'Confetti' (leaves large and green, flowers large and creamy-white speckled elongated rose-pink), 'Crimson Beauty' (bright crimson), 'Cupid' (pink in large clusters), 'Dazzler' (height 120 cm, flowers burgundy), 'Di Bartolo' (leaves brown tinged purple, flowers pink), 'Elizabeth Hoss' (tall with green foliage, flowers clear yellow blotched irregular scarlet), 'Eureka' (dwarf, flowers cream tinged yellow), 'Feuerzauber' (syn. 'Fire Magic', leaves copper-purple, flowers orange-red), 'Golden Girl' (dwarf, leaves green, flowers bright yellow with irregular scarlet spots), 'Grumpy' (dwarf to 50 cm, flowers brilliant red), 'Halloween' (golden with red throat), 'J.B. van der Schoot' (yellow speckled red), 'King City Gold' (yellow), 'Kanchan' (dwarf with small and yellow flowers splashed red), 'King Humbert' (tall, leaves bronze-red, flowers orange), 'King Midas' (120 cm, flowers gold), 'Liebesglut' (leaves copper-purple, flowers orange-red), 'Lucifer' (crimson-red with yellow edging), 'Nirvana' (leaves striped yellow, flowers deep yellow, buds red), 'Orange Perfection' (pale-orange), 'Orchid' (pink), 'Park Princess' (pale-pink, many-flowered), 'Pfitzer's Chinese Coral' (pink), 'Pfitzer's Primrose Yellow' (cluster large, pale-yellow), 'Pfitzer's Salmon Pink' (pink), 'Pfitzer's Scarlet Beauty' (bright scarlet), 'Picasso' (both yellow and red), 'Pink Sunburst' (dwarf but with giant variegated leaves striped green, yellow, pink and bronze and the flowers salmon-pink), 'President' (leaves large and green, free-flowering with scarlet flowers), 'Rosamund Cole' (height 120 cm, flowers scarlet edged yellow, reverse golden with red overlay), 'Red Dazzler' (tall, leaves bold bright green, flowers lily-like and red), 'Red King Humbert' (height 210 cm, leaves bronze-red, flowers orange-scarlet), 'Richard Wallace' (height 120 cm, flowers golden-yellow), 'Rosemond Coles' (tall, foliage green, flowers orange-red margined yellow), 'Striatus' (syn. 'Bengal Tiger', tall, foliage distinct green and yellow striped, flowers orange-red), 'Seven Dwarfs' (flowers yellow, salmon, pink, red, crimson), 'Stadt Fellbach' (orange), 'Striped Beauty' (leaves veined cream and pale-gold, flowers scarlet in clusters), 'Tropical Sunrise' (foliage deep green, recurrent bloomer with large flowers of apricot and pink colour, cut flowers standing well in water), 'Wyoming' (leaves bronze, flowers orange), 'Yellow King Humbert' (height 135 cm, foliage bright green, free-flowering, flowers yellow dotted crimson), *etc.* Agri-Horticultural Society, Calcutta introduced 51 canna cultivars from Italy in between 1895 to 1904, and 10 from USA after 1904, *viz.*, 'Africa', 'Alemannia', 'America', 'Aphrodite', 'Asia', 'Atlanta', 'Attika', 'Australia', 'Austria', 'Bavaria', 'Borussia', 'Britannia', 'Burbank', 'Burgundia', 'Campania', 'Charles Naudin', 'Crown Prince of Italy', 'Edouard Andre', 'Emelia', 'Hellas', 'Heinrich Siedel', 'H. Wendland', 'Iberia', 'Indiana', 'Ischia', 'Italia', 'King Herbert', 'Kronos', 'La France', 'Mrs. Kate Grey', 'Oceanus', 'Pandora', 'Partenope', 'Pennsylvania', 'Pereus', 'Philadelphia', 'Phoebe', 'Pluto', 'Prof. Traub', 'Queen of Italy', 'Rhea', 'Roma', 'Romagna', 'Rossi', 'Sicilia', 'Suevia', 'Trinacria', 'Umbria', 'Wintzer's Colossal', 'Wm Beck', and 'Wyoming'. Other introduced or indigenously developed varieties in NBRI, Lucknow are 'Aida', 'After Glow', 'Aga Khan', 'Ailsen', 'Ali Petzi', 'Alison', 'Anarkali', 'Angel's Robe', 'Aristocrat', 'Ariel', 'Arjun', 'Atom Bomb', 'Bardara', 'Bharat', 'Black Knight', 'Bo Peep', 'Bridal Veil', 'Brocade', 'Carmine King', 'Charmion', 'Cherub', 'City of Portland', 'Claire', 'Colette', 'Daphne', 'Dainty Maid', 'Diana', 'Doris', 'Dragon's Tongue', 'Dream', 'Edith', 'Electra', 'Enchantress', 'Ethel', 'Eureka', 'Excelsior', 'Fair Maid', 'Flaccida-type', 'Florence', 'Gladiator', 'Gloria', 'Golden Standard', 'Goldilocks', 'Goliath', 'Heart's Desire', 'Imperator', 'Indiana', 'Isobel', 'Janet', 'Jehangir', 'Julia', 'King Alfred', 'Lord Buddha', 'Lord Reading', 'Lorelei', 'Louise', 'Louis Cayeux', 'Mamie', 'Masterpiece', 'Matchless', 'Morning Glow', 'Nerissa', 'New Red', 'Olive', 'Orange King', 'Oriole', 'Percy Lancaster', 'Perfection', 'Pink Satin', 'Plume', 'President', 'Primary Hybrid I', 'Primary Hybrid

II', 'Primary Hybrid III', 'Primary Hybrid IV', 'Prince Philip', 'Prof. Thacker', 'Queen Elizabeth', 'Queen Mab', 'Queen of Italy', 'Rajaji', 'Raj Mahal', 'Rosamund Coles', 'Rose Queen', 'Sangrila', 'Sirius', 'Sir John Anderson', 'Sans Souci', 'Soldier Boy', 'Star of India', 'Striped Queen', 'Stromboli', 'Sun Set', 'Sweet Heart', 'The Queen', 'Trinacria Variegata', 'Wintzer's Colossal', 'Yellow Gal', *etc.*

Propagation

The canna rhizomes are produced for commerce in the Holland, Japan and USA. Certain other countries such as Italy and France also produce it but to a limited scale. Whether it is tropical, subtropical or the subtemperate areas but these are planted in spring when danger of frost is over. For their multiplication and for maintaining varietal identity, cannas are propagated only through their rhizomes. Though it is not more demanding but the growth as well as multiplication are luxurious if the soil is rich in organic matter. It should be planted at a sunny situation in a sandy-loam soil with a pH reaction of 6-7. However, it does well even in heavy soil if it is not water-logged. The land where the cannas are to be propagated should be incorporated with compost or cow dung manure @ 5 kg/m^2. Soil should be at least 30 cm deep so for its planting the beds should be prepared by mixing the organic manure and sand, by taking out all the stones, pebbles and grits and by removing rootstocks of all the perennial weeds. The soil is made thoroughly pulverized by ploughing or digging thrice. At the time of planting there should be sufficiaent moisture in the soil. The pieces of rhizomes having atleast two nodes or one prominent bud are planted at 30 cm apart. The rhizomes may better be started in store by moistening the soil regularly and when sprouts are visible or become of good size these can be planted outdoors. They are well cared throughout the full growing season by providing timely irrigation but the portion of stems having flower buds are nipped when just emerged so that energy is not wasted in flower formation and instead is diverted for the development of rhizomes. The rhizomes should be lifted just before the onset of winter (in sub-temperate regions) when still the plant is green or in March (in the tropical and sub-tropical regions) when the sprouts are to emerge. Early lifted rhizomes, after drying of surface moisture, can be stored at 5-10° C, preferably at 7° C till marketing but one should be careful that these do not dry out in the store. The multiplication rate in canna is very high and within one year these form clumps. Porozov (1953) reports that 4-5°C winter storage of canna rhizomes results into 70-80 per cent losses but if left as such in the field these produce 6-8 new plants which may be lifted and divided in autumn, heeled up in the glasshouse, watered as per requirement, and kept at 12-15°C with their leaves removed as new ones appear in the spring. These plants are now potted and placed in the frames for hardening and then planted out.

Through meristem tips in *Canna indica*, Kromer and Kukulczanka (1985) found success with half strength MS medium supplemented with auxins and cytokinins. Buds developed from isolated meristems, obtained after a 6-month culture, were kept at 6°C for four winter months and in early springs the buds quickly developed shoots and roots, and nutrient solutions containing kinetin (2 mg/dm^3), adenine sulphate (100 mg/dm^3) and NAA (0.2 mg/dm^3) gave good results. Plants thus raised were propagated by axillary buds, detached and transferred to fresh nutrient solution where they grew and developed naturally.

For developing new varieties, they are grown through seeds which are brown to black, almost round and very hard. These should be soaked for 24 hours in warm water before sowing or seed coat is chipped and sown in late winter into pots filled with soilless medium or in sandy-peat mix, at 18-22° C temperatures until these germinate. Seeds soaked in sulphuric acid for 40 minutes causes germination of up to 80 per cent of seeds. After the germination, these are repotted gently without breaking or damaging their brittle roots, in the rich soil mix *i.e.* half good garden soil and half peat + sand, and the pots are kept at 16° C for further growing and hardening. Bouman and Grootjen (1987) and Grootjen and Bouman (1988) studied the structure of canna seeds and reported that seeds of canna lack a silicified endotesta, a micropylar collar and aril and do not germinate via an operculum, contrary to the other families of Zingiberales. They are pachychalazal in structure and only a small part of the testa is of tegumentary origin. Most of the storage tissue is derived from the chalaza. The outer layer comprises of raphe, chalaza and outer integument differentiates into a continuous exotesta of Malpighian cells. The intact testa is impermeable where water uptake and germination can occur only after shedding of the inhibition lid, a preformed part of the raphal side of the seed. Jadhav (1960) when harvested the seeds at the brown soft stage found 89 per cent germination though after removing the radicle cap it reached to 94 per cent, however, no untreated hard black seeds germinated which could be germinated to 72 per cent only when treated with sulphuric acid for three hours or 82 per cent after removing the radicle caps. Narayansamy *et al.* (1986) scarified the mature seeds of canna with concentrated sulphuric acid for 0-60 minutes, and found 80 per cent germination in six days in those treated for 40 minutes, however, 60 minute treatment gave only 2.68 per cent germination, untreated ones did not germinate and 20 minute treatment germinated in 18 days.

Cultural Practices

Though cannas grow well in heavy soils, dampy soils, in dry (to some extent) as well as water-logged soils, and in rugged poor and even in infertile soils and under semi-shady conditions but the charm is most astounding when the soil is light (sandy-loam) and water-retentive but not water-logged and at a sunny situation. Under complete shade the plants become taller and etiolated with the marred elegance of the foliage, *vis-a-vis* no flowering at all, and under semi-shady conditions the growth is rampant but the plants are shy in blooming. Preferably, its soil should have a pH reaction of 6-7. Therefore it is advisable to plant it only at sunny situations. In the sub-tropical conditions there is either no blooming during May-June when the weather is too hot with little relative humidity or the blooming is sporadic and of poor quality, and during winter too there is no flowering as the aerial part is unable to sustain itself the chills of the winter. In sub-temperate conditions, these are not winter-hardy and the aerial part of the plant burns during winter so requires winter protection though there is no any ill-effect on the rhizomes. However, under tropical conditions with sufficient atmospheric humidity these continue flowering throughout the year. In the temperate regions, the rhizomes may also rot due to frost injury during winter months so a thick layer of straw or some other organic material like dried leaves, saw-dust, manures, chopped sugarcane bagasse, paddy straw, wood shavings, vermiculite, ashes, *etc.* is covered over whole planting before the danger of frost. Due to frost when aerial parts of the plants burn, another layer of covering material should again be put so that frost may not penetrate through the frost-burnt parts of the plants. Protection may also be provided by erecting triangular polyhouses. In the congenial atmosphere and in the proper soil these form the clumps and give enchantment through their beautiful foliage and majestically attractive flowers. Once planted these are not disturbed for years until these become soil-bound so the soil should be quite fertile for continuous and profuse blooming.

For its planting, some 250-300 quintals of farmyard/ cow dung/stable manure or compost or some 100-150 quintals of pig or poultry manure per hectare of land should be incorporated in the soil along with 150 kg of P_2O_5 at the time of first ploughing to a depth of at least 30 cm so that it is properly mixed in whole of the soil and then the land is planked after taking out the rootstocks of the perennial weeds such as *Cynodon, Cyperus, Oxalis, Trifolium, etc.* with hand forks or manually. Brick pieces, stones, pebbles and grits of any kind should also be taken out. The rhizomes of *Cyperus* and the bulbs of *Oxalis* are very difficult to remove so one should be careful that these are thoroughly removed otherwise these will rob the cannas of their nutrients and if once established will be very difficult to remove from among the standing crop. Second ploughing is carried out immediately to pulverize the soil, for taking out the left out foreign material and for knocking down the germinating weeds. Just before planting, third and final deep ploughing is carried out followed by planking and then the beds of convenient sizes are prepared by drawing bunds and irrigation channels for walking and irrigations. Ailinci (1969a) in Italy recommended application of NPK fertilizers at 3-leaf stage of growth and at first appearance of the flower.

Under all the climatic conditions (sub-tropical, sub-temperate and temperate), the rhizomes are planted in March or April. However, in case of emergencies, the rhizomes may be lifted at any stage of plant growth and immediately planted in the prepared field which will give a new plant very soon because canna rhizomes either do not have absolute dormancy or the rhizomes are never juvenile and physiologically immature. Lifting of rhizomes and subsequent planting even when the clump is in flowering is also quite effective for cropping. Albeit, before planting the rhizomes are moistened regularly in a well-ventilated and aerated room for initiation of sprouts, *vis-à-vis* roots. When the sprouts are well developed in the store, these may be taken out for outdoor planting orienting the sprouts up at 90° angles, *i.e.* perpendicular to earth. Each piece of rhizome should have 2-3 nodes with at least one developed bud. The planting is done at 45 cm apart and 8-15 cm deep (setting with 8-15 cm of soil over the rhizomes) and watering is done sparingly until growth begins. Erushkevich (1978) reported that growing canna plants loose 369.9 to 518.2 $g/m^2/h$ of water through transpiration, being low in the beginning of summer, then increased and finally decreased. In the hot weather but under low relative humidity, transpiration was found decreased and the daily maximum occurred in the morning, however, increased soil moisture increased transpiration, and the plants stored at 17-25°C had a lower transpiration rate than those stored at 4-7°C. After planting, the minimum temperature of the field should not be less than 16°C in any case. For their proper growing, cannas need a constant temperature, a moist atmosphere and plenty of water. With the growth of plants the frequency and quantity of water goes on increasing substantially but where winter is severe the watering is stopped after the flowering is over so that crop is saved from the rhizome rot. Again in spring the watering should be carried out to facilitate sprouting. Where severe winter conditions are experienced, the stems should be allowed to blacken at the end of the season, then rhizomes are taken out, surface-dried for a day or two and then are stored in vermiculite, wood shavings, saw dust or in coarse sand. In store these are checked regularly and if the medium is too dry and affecting the shape of the rhizomes, it is moistened a little. Storage temperature should be around 7° C. Ailinci (1969b) when stored the rhizomes of two *Canna indica* varieties at 6°, 9° or 20°C and 75-80 per cent relative humidity, recorded respiration rate relatively high at all the temperatures and rhizomes starting sprouting after 60 days, however, the rate after 30 days of storage was related to the temperature, being highest at 20°C. He further stated that sprouting response was related to the cultivar rather than to the storage temperature.

Flowering starts within $2^1/_2$-3 months of planting if planted at optimum time. Late planting induces early flowering and early planting delays the duration. Faded flowers should be removed regularly without damaging the fruit set as at this stage also these plants look attracative because before maturation the fruits are green with special and dense protrusion all around for a long period. Moreover, the seeds after maturation may be collected for developing new varieties and for making beads of high quality. Weeds may be a problem only at initial stage of planting but afterwards its growth is so rampant that weeds do not get space to survive. Mulching is very effective for plant growth and flowering at early stage when still the temperature is low but in summer when the temperature is high it can simply inhibit the sprouting and growth of the weeds. Kwack *et al.* (1990) when planted *Canna* cv. 'American Red Cross' in Korea on April 25 with or without a vinyl film mulch, recorded 15 days earlier emergence, *vis-à-vis* flowering with increased number of suckers as well as flowers because of an increase in the soil temperature though the plots mulched on May 25 did not show any encouraging effect as the soil temperature had already been high at this time. For controlling weeds in the canna field, Vivekanandan and Granam (1975) used aminotriazole which induced complete chlorosis in *Canna edulis* leaves.

Insect-Pests and Diseases

Though Japanese beetles, slugs and snails attack this crop off and on but none of them causes any serious damage. Mallea and Suárez (1969) reported that young larvae of *Cobalus cannae* butterfly which severely damages the leaves of canna can effectively be controlled through several sprayings with carbaryl at 150 g in 100 litres of water or by 20 per cent hexochloroepoxy-octahydro-endo-dimethane naphthalene at 120 ml/100 litres of water. Reinert *et al.* (1983) in Florida evaluated 39 canna cultivars for resistance to *Calpodes ethulias* (skipper butterfly) which there causes extensive leaf damage in canna, and reported that for egg laying the butterfly preferred the varieties of red leaf colour or those which bear red, orange or scarlet flowers while it had least preference for varieties of other leaf colours or those bearing yellow, rose red, yellow-red or pink flowers. Andersson (1967) found the nematode, *Ditylenchus destructor* causing rhizome rot in *C. indica*, and on examination the nematodes were found bordering the healthy tissues. He suggested HWT for its possible control. Mottle (virus) disease of *Canna indica* is transmitted through *Myzus persicae, Rhopalosiphum maidis, Aphis gossypii* and *A. rumicis* (Gupta and Roychaudhuri, 1975). 'Cucumber mosaic virus' (CMV) and 'canna mosaic virus' (CaMV) transmitted through aphids and 'aster yellows' though leafhoppers are sometimes observed in isolated cases so such plants should be uprooted with the rhizomes along with roots and destroyed so that it may not get any chance to spread further. Chemical control of aphids and leafhoppers will also check the spread of these diseases. Brierly and Smith (1948) reported that the mosaic disease found on *Canna indica* also attacks the varieties of *C. glauca* and *C. generalis* but stated that the var. 'The President' has been immune. Broschat *et al.* (1983) in USA tested 52 cultivars of *C.* × *generalis* and *C. indica* and found that red-leaved cultivars, *viz.* 'Wyoming' and 'Ambassador' were highly resistant to 'hippeastrum mosaic' (HM) while out of 38 cultivars tested against canna rust (*Puccinia thaliae*), none was found completely resistant, however, 'Halloween' and 'Yellow King Humbert' were found highly susceptible. This rust in India had been the first record in India *on Canna indica* from West Bengal (Maji, 2003). *Botrytis cinerea* causes blight on leaves, stems and flowers and in its severe attack may affect rhizomes and roots causing grey mould. The damping off problem in the seedlings may also occur. This is favoured by mild, wet and humid environment. Excessive watering, especially during winter season should be avoided. Two weekly sprays with Dithane M-45 or Z-78 will control this disease. Petal blight in canna is caused by *Alternaria tenuis* (Thompson, 1961). Crisan and Szenyei (1987) isolated *Alternaria alternata* from *C. indica* and reported that cultivars 'Pictor Grigorescu', 'Agnes', 'Extra', 'America', 'Cardinal' and 'Signal' were resistant. For its *in vitro* control, they suggested Mycodifol, Euparen (Dichlofluanid) and Vitavax (carboxin). Narayanasamy and Durairaj (1972) recorded a new leaf spot disease (*Curvularia uncinata*) on *C. indica* from India. The bacterium *Xanthomonas cannae* causes water-soaked lesions first on leaves then spreading to the stem in advanced cases and finally attacking vascular system which may result in collapse of entire plant but its incidence on canna is very sporadic. Proper sanitation of the field and use of copper fungicides will control the disease to some extent.

References

Ailinci, N. 1969a. The influence of inorganic NPK fertilizers on some physiological processes in *Canna indica* L. plants (Italian). *Lucr. Sti. Inst. Agron. Iasi, Agron.-Hort.*, pp. 429-443.

Ailinci, N. 1969b. The influence of storage temperature on the respiration intensity of tubers of *Canna indica* L. and *Dahlia variabilis* Des. (Italian). *Lucr. Sti. Inst. Agron. Iasi, Agron.-Hort.*, pp. 419-427.

Andersson, S. 1967. *Ditylenchus destructor* in *Canna indica. Nematologia*, **13**:479.

Ashtakala, S.S. and R.J. Maloney, 1971. Characterization of anthocyanin pigments in three cultivars of garden *Canna. J. Amer. Soc. hort. Sci.*, **96**(6): 755-757.

Bosse, G. 1968. Trials with new varieties of *Canna indica* (German). *Neue Landschaft*, **13**: 99-102.

Bouman, F. and C.J. Grootjen, 1987. Canna seeds, structure and germination (abstr.). *Acta Botanica Neerlandica*, **36**(2): 149-150.

Brierley, P. and F.F. Smith, 1948. Canna mosaic in the United States. *Phytopath.*, **38** 230-234.

Broschat, T.K., J.A. Reinert and H.M. Donselman, 1983. Resistance of canna cultivars to canna rust and hippeastrum mosaic. *HortSci.*, **18**(4): 451-452.

Chemarin, N.G., I.A. Zabelin and A.N. Glazurina, 1970. The effect of ionizing radiation on cannas (Russian). *Trudy Gosudarstvennogo Nikitskogo Botanicheskogo Sada*, **46**: 194-198.

Chemarin, N.G., A.N. Glazurina and I.A. Zabelin, 1973. The uses of ionizing radiation in canna breeding (Russian). *Byulleten' Gosudarstvennogo Nikitskogo Botanicheskogo Sada*, No. 1(20), pp. 52-57.

Crisan, A. and A. Szenyei, 1987. Aspecte de patogenitate si combtere a unor specii de *Alternaria* de pe plante ornamentale. In: *Contributii Botanice Gradina Botanica Universitatea din Cluj-Napoca*, pp. 257-262.

Erushkevich, S.V. 1978. Summer transpiration of canna leaves (Russian). In: *Introduktsiya, Akhlimatiz I Selektsiya Tsvetoch.-Dekor. Rast. V Kirgizi. Frunze, Kirgiz SSR*, pp. 78-86.

Feofilova, G.F. 1975. Evaluation of some results of garden canna pollen analysis (Russian). *Byulleten' Gosudarstvennogo Nikitskogo Botanicheskogo Sada*, No. 3 (28), pp. 21-23.

Grootjen, C.J. and F. Bouman, 1988. Seed structure in Cannaceae : taxonomic and ecological implications. *Ann. Bot.*, **61**(3): 363-371.

Gubanov, V.N. and A.P. Khalaburdin, 1987. Promising canna cultivars for planting in Ashkhabad and their propagation coefficient (Russian). *Izvestiya Akademii nauk Turkmenskoi SSR, Biolgicheskikh Nauk*, No. 6, pp. 26-32.

Gupta, M.D. and S.P. Roychaudhuri, 1975. Studies on the mottle disease of *Canna indica* L. *Indian J. Hort.*, **32**(1/2): 106-109.

Gupta, M.N. 1966. Induction of somatic mutations in some ornamental plants. Reprint from *Proc. All India Symp. Hort.*, pp. 107-114.

Jadhav, A.S. 1960. Canna seeds can be germinated. *Poona agric. Coll. Mag.*, **51**: 28-29.

Kromer, K. and K. Kukulczanka, 1985. *In vitro* cultures of meristem tips of *Canna indica* (L.). *Acta Hort.*, No. 167, pp. 279-285.

Kwack, B.H., H.K. Kim and K.M. Lee, 1990. The effect of polyethylene film mulching in early spring on the growth and flowering of field-planted *Canna hybrida* L.(Korean). *J. Korean Soc. hort. Sci.*, **31**(2): 162-168.

Maji, M.D. 2003. A new rust disease of *Canna indica* in India. *Indian Phytopath.*, **56**(3): 302.

Narayanasamy, P. and P. Durairaj, 1972. A new leaf spot disease of *Canna indica* L. *Sci. Cult.*, **38**(2): 96-97.

Narayanasamy, P., G.N. Mohankumar, U.G. Nalawadi and H.S. Surendra, 1986. Acid scarification for the improvement of germination of canna (*Canna indica* Linn.) seeds. *South Indian Hort.*, **34**(2): 121-122.

Mallea, A.R. and J.H. Suárez, 1969. The biology and control of *Cobalus cannae* in Mendoza. *Idia*, No. 256, pp. 6-10.

Porozov, A.K. 1953. Something new in the cultivation of canna (Russian). *Sad I Ogorod*, No. 11, pp. 76-78.

Reinert, J.A., T.K. Broschat and H.M. Donselman, 1983. Resistance of *Canna* spp. to the skipper butterfly, *Calpodes ethlius* (Lepidoptera : Hespeiidae). *Environmental Ent.*, **12**(6): 1829-1832.

Shevchenko, S.V. and G.F. Feofilova, 1981. Pollen viability in canna hybrids and the original species and cultivars (Russian). *Byulleten' Gosudarstvennogo Nikitskogo Botanicheskogo Sada*, **3**(46): 94-98.

Thompson, H.S. 1961. Petal blight, a new disease of the canna Pfitzer's Dwarf. *Canad. J. Plant Sci.*, **41**: 503-506.

Tiwari, S.C. and B.E.S. Gunning, 1986a. Development of tapetum and microspores in *Canna* L. : and example of an invasive but non-syncytial tapetum. *Ann. Bot.*, **57**(4): 557-563.

Tiwari, S.C. and B.E.S. Gunning, 1986b. Development and cell surface of non-syncytial invasive tapetum in *Canna*: ultrastructural, freeze-substitution, cytochemical and immunofluorescence study. *Protoplasma*, **134**(1): 1-16.

Vivekanandan, M. and A. Granam, 1975. Studies on the mode of action of aminotriazole in the induction of chlorosis. *Plant Physiol.*, **55**(3): 526-531.

11

Carnivorous Ornamentals

Priyanka Sharma, Y.C. Gupta, Bhavya Bhargava, Sanyat Misra, R.L. Misra and Ajit Kumar

Introduction

Carnivorous plants derive some or most of their nutrients from trapping and consuming small animals such as insects and spiders. They are more widely called carnivorous plants because some use small animals like snails and frogs as their food (Schnell, 2002). Insect capturing by plants is believed to be an adaptation to life in nitrogen-poor soils such as acidic peat bogs and rock outcroppings. There, many of the essential nutrients, including nitrogen compounds, are leached away by water or consumed by anaerobic bacteria. As a result, oxygen is in short supply as well. Unlike carnivorous animals, however, such plants do not use their prey as an energy source but only make use of their nitrogen and phosphorus (Wallace *et al.,* 1990; Kamarainen *et al.,* 2003). Carnivorous plants are a fascinating group of plants having specialised leaves that are specifically designed to capture insects and certain other creatures as a source of food. There are a large variety of carnivorous plants, each variety having their unique structural characteristics. These insectivorous plants comprise of three groups, *viz.* (i) '**fly traps**' with spiny-edged leaves which are hinged in the middle, (ii) the '**sticky-leaved plants**' with hairs which secrete insect-catching fluid, and (iii) the '**pitcher plants**' with leaves which are water-filled funnels. While some varieties of carnivorous plants are more commonly available, others are very rare and therefore difficult to obtain. The paucity of soil nutrients in these sites is offset by the availability of high numbers of insects or other potential prey items. Supplementing low soil nutrients with prey capture thereby allows carnivorous species to out-compete species without such nutritional supplements. Examples of ecosystems that meet most or all of these specifications include bogs, wet pine savannas, marshes, swamps, and fens. Some carnivorous species extend into less characteristic habitats such as shady epiphytic perches, dry deserts, limestone cliffs, serpentine soils, and standing freshwater. Again, these habitats are generally low in nutrients but may be shady, aquatic, or arid–abiotic conditions not generally thought to be ideal for carnivorous plant success. Acidic soil seems to be another prerequisite for the occurrence of carnivorous plants. A good indicator of favourable carnivorous plant habitat, in fact, is the occurrence of *Sphagnum* moss, a species indicative of acidic soils. *Sphagnum* moss increases soil acidity as it slowly decomposes; this acid-generating decay, in turn, further limits nutrient availability and thereby favours carnivorous plants. Essential plant nutrients may be tied up or simply lacking in the soil and, therefore, unavailable for plant absorption. Securing limited nutrients by way of carnivory allows carnivorous plants a much needed mechanism to increase the uptake of nitrogen and other elements. They share three attributes that operate together and separate them from other plants, and these are capturing and killing the prey, having a mechanism to facilitate digestion of the prey, and to derive a significant benefit to facilitate digestion of the prey. Carnivorous plants, about 600 species worldwide, grow mainly in tropical or semi-tropical areas (Table 1) (Lee, 2008).

Indian native insectivorous plants are *Aldrovanda vesiculosa* (Droseraceae) which is found in the marshes of Sunderbans; three species of *Drosera* (Droseraceae), *viz. D. burmanni* in eastern and central India, *D. indica* in the Western Coast, and *D. peltata* in the plains all over India; *Nepenthes khasiana* (Nepenthaceae) in Khasi, Jaintia and Garo hills of Meghalaya; *Pinguicula alpina* (Lentibulariaceae) from Kashmir to Sikkim; and *Utricularia brachiata* (Lentibulariaceae) from Sikkim, Arunachal Pradesh and West Bengal; *U. caerulae* from Western Coast and Western Ghats; and *U. inflexa* var. *stellaris* from Khasi, Jaintia and Garo hills of Meghalaya.

Uses

Carnivorous plants are a fascinating group of plants, and have long been the subject of popular interest. Due to the various shapes and sizes of the leaves of carnivorous plants there are wide ranges of choice for landscaping purposes. Furthermore, their leaf colour variation as well as their beautiful flowers make these plants decorative with a high commercial potential. Large and colourful *Nepenthes* and *Sarracenia* are recommended as garden plants. *Sarracenia* is a cold-tolerant garden plant and many can be grown from parts of southern Canada to Florida, across most of the United Kingdom, much of Europe, and through most of non-tropical Australia and New Zealand. It is one of the easiest plants to grow in temperate climates, thrives best in open gardens, and usually shows the richest colour in full sun (Romanowski, 2001). However, most carnivorous plants are not recommended for outside landscaping because they require nutrient-poor and acidic soils. Accordingly, carnivorous plants can be used only for special outdoor landscaping such as swamps and bogs. Thus, carnivorous plants are utilized more for interior landscaping purposes. Their use as pot plants is most popular, and growing them in hanging baskets or terrariums is also gaining popularity. This is especially true because a terrarium is an excellent container for carnivorous plants.

Recently, attempts on the improvement of carnivorous plants by hybridization have been carried out mainly by many nurseries and growers. Presently, breeders use chemical mutagens to induce novel characters in many carnivorous plants. It was once considered that carnivorous plants could be used as one way of insect control because of their ability to eat insects. However, the amount of a catch is too low and it takes too long for the plants to digest insects. Consequently, the idea of pest control was questioned and consumption of carnivorous plants decreased.

Recently, a volatile compound called plumbagin, which has antifeedant activities and insecticidal effects, has been discovered, boosting interest in this group of plants. Many carnivorous plants produce volatile compounds called naphthoquinones similar to plumbagin that have anticancer and antimicrobial

Table 1: Worldwide Distribution of Carnivorous Ornamentals

Taxon		*No. of Species*	*Global*
Genus	Family		
Darlingtonia	Sarraceniaceae	1	NW USA
Heliamphora		5	NC South America
Sarracenia		10	SE USA, East Canada
Nepenthes	Nepenthaceae	70	Indonesia to Austrailia, Madagascar
Aldrovanda	Droseraceae	1	Eurasia
Dionaea		1	USA (North Carolina)
Drosera		141	Global
Drosophyllum	Drosophyllaceae	1	Portugal, W. Spain
Triphyophyllum	Dioncophyllaceae	1	W. Africa, Ivory Coast
Byblis	Byblidaceae	5	NW Austrailia
Cephalotus	Cephalotaceae	1	SW Austrailia
Genlisia	Lentibulariaceae	19	South America, Africa
Pinguicula		69	N. America, Europe, Asia
Utricularia		219	Global
Ibicella	Martyniaceae	1	South America
Brochinnia	Bromiliaceae	2	South America
Catopsis		1	USA (Florida), South America

effects. Therefore, growing carnivorous plants that yield a high amount of volatile compounds inside houses or in any closed area might help to increase anticancer and antimicrobial functions of people who inhale those volatiles (Didry *et al.*, 1994; Parimara and Sachidanandam, 1993). In *Dionaea muscipula* and *Drosera capensis*, two major groups of pharmaceutically important substances include: naphthoquinones, plumbagin, and ramentaceone; and (2) flavonoids: myricetin, and quercetin. These materials are known to have anticancer, antimicrobial and antispasmodic activities (Krolicka *et al.*, 2008), anticancer (Parimara and Sachidanandam, 1993), antimicrobial (Didry *et al.*, 1994), antimalaria (Likhitwitayawuid *et al.*, 1998) and antifungal (Shin *et al.*, 2007). Among the various medicinal properties of the plant, the digestive juice of the unopened pitcher plant *Nepenthes khasiana* Hook. is used as an eyedrop for cataract and night blindness (Behera *et al.*, 2007). Extracts from several species of *Pinguicula* are currently used in pharmaceutical formulations as cough suppresant and mucolytic. Indeed, the chemical characterization of *P. lusitanica* extracts revealed the presence of compounds with pharmacological interest (Grevenstuk *et al.*, 2010).

Trapping Mechanisms

Carnivorous plants grow on swampy or stony poor soils where nitrogen seems to be scarse. Carnivorous plants trap their prey in one of two basic ways, *i.e.* active or passive. **Active trapping** is used by the most famous of all carnivorous plants, the 'venus flytrap' (*Dionaea*). It uses it's well known 'snap trap' to catch its food. Each side of the trap has 2-3 trigger hairs. The 'meal' must touch a trigger on each side before the trap will close. The trap, which is a part of the leaf, will only work for about four times (give or take) after which the leaf dies and the plant must grow a new leaf. If the traps are closed repeatedly with no food, the plant will suffer and could die. **Passive trapping** is normally divided into two types: (i) sticky traps and pitfall traps, where sticky traps are used by the 'sundews' (*Drosera*) and 'butterworts' (*Pinguicula*). In sundews, the leaf tip or sometimes the whole leaf is covered with glands with a drop of sticky 'dew'. The prey gets mired down in the dew and dies. The plant then secretes digestive enzymes from small pores along the leaf and digests the prey. Butterworts work the same way, but on a much smaller scale catching only very tiny insects. Pitfall traps are the mechanisms used by 'American pitcher plants' (*Sarracenia*) which have a reservoir of fluid into which the prey falls and drowns. The plant then secretes digestive enzymes and digests the insects. Sometimes the pitcher plant gets indigestion by catching more than it can digest. The decaying bugs then leach through the side of the pitcher causing a brown blotch to appear.

Classes and the Species

There are five types of carnivorous plants based on their trapping mechanism, *viz.* pitcher, flypaper, snap, suction and lobster pot traps. **Pitcher** plants trap prey in a rolled leaf that contains a pool of digestive enzymes or bacteria. Usually insects are attracted by bright flower-like anthocyanin patterns in the leaves or by nectar bribes secreted by peristomes. They then fall into the pitcher due to slippery wax lining inside leaves. The plants that produce attractants, like those in Nepenthaceae are called pitcher traps, and plants with no attractants like those in Sarraceniaceae are called pitfall traps. **Flypaper** traps capture prey by using sticky mucilage. Sundews (*Drosera*), butterworts (*Pinguicula*) and *Drosophyllum* belong to this type. **Snap or steel traps** utilize rapid leaf movement to capture prey. Only two species, the *Aldrovanda vesiculosa* and 'venus flytrap' (*Dionaea muscipula*) belong to this type. These snap traps close rapidly when triggered to trap prey between two lobes. At the beginning of capture, the insect is held just strong enough to prevent its escape, but once plants detect protein, the hold becomes stronger. In this way, unnecessary snapping triggered by materials other than insects can be prevented. The prey inside the tightly closed leaves is digested over a period of one to two weeks. Leaves can be reused 3-4 times before they become unresponsive to stimulation but usually leaves wither after one capture. **Suction (bladder) traps** are exclusive to the genus *Utricularia*. Suction traps suck in prey with a bladder that generates an internal vacuum by pumping ions out of the interior and allowing water to enter by osmosis. The bladder has a small opening, sealed by a hinged door. In aquatic species, the door has a pair of long trigger hairs. Aquatic invertebrates such as *Daphnia* touch these hairs and deform the door by lever action, releasing the vacuum. Then the invertebrates are sucked into the bladder, where they are digested. **Lobster pots** are the trapping mechanism in *Genlisea* (corkscrew plant). Lobster-pot traps force prey to move towards a digestive organ with inwardly pointing hairs. Some of the most popular insectivorous genera are described below.

Aldrovanda (Droseraceae)

A. vesiculosa (waterwheel plant) is the submerged aquatic carnivore distributed over East Asia to Japan, India, Europe, Tropical and South Africa. In the cold areas the growing tip of the plants forms a dormant bud which starts growing when weather becomes warm. Its central stem produces the leaves in whorl-form (in the form of wheel spokes), and these leaves have traps similar to venus fly. Flowers appear above the water short stalks.

Cephalotus (Cephalotaceae)

Cephalotus is 'Albany pitcher or fly-catcher plant' and in Greek it means head-shaped, owing to knob-like swelling behind each anther. It has only one species *C. follicularis* which was first discovered by La Billardiere in 1792 from SW Australia, growing in swamps, often among dense scrub and reeds. It forms a thick underground rhizomatous taproot and rosetted small pitchers on slender stalks around the top of taproot, and these stalks rest on the ground at 45° angle having three ribs with ladder-like row of hairs to help insects climbing up, and the pitcher has a lid over this similar to *Nepenthes*. Pitcher leaves appear first in June or July, grow about 5.0 cm in length and 1.9 cm in diameter, and mature until December or January, though foliage leaves appear in July and August and mature by September or October, developing up to 13 cm in length. There are a number of ordinary flat leaves mainly during winter, and at short inflorescence at the end of a long (60 cm) stalk sweet-scented small white flowers appear. Also this has a few translucent window-like structures to pass on light inside the pitcher so that insects may not be afraid entering it. The pitcher inside at the rim has below-facing teeth-like spikes to make impossible for insects to bail out once inside. Insects are enticed inside by nectar glands so after reaching the slippery zone and falling in, they are digested in the liquid at the bottom of the pitcher. It can be cultivated in peaty soils in light shade.

Darlingtonia (Sarraceniaceae)

Darlingtonia from California and Oregon (USA) was named so for Dr. William Darlington (1782-1863), an American botanist. This genus has only species, *D. californica* (California pitcher or cobra plant). It is a rosette-forming herbaceous perennial, bearing 10-45 cm long, erect, yellow-green and net-veined leaves. Leaves individually in the rosette fold and fuse to form a slender container and the top a 2-lobed hood which is green and brown and appears as the head of a cobra. The pitcher contains a fluid in which when any insect falls, the enzymatic fluid helps in digesting the protein and other nitrogenous ingredients. Its flowers are solitary, bell-shaped and nodding with 5 greenish sepals and 5 purple petals. It is an intriguing pot plant for a cool house. The mixture for its growing should constitute of equal parts of peat and sphagnum moss and the pot or pan should be kept in a saucer filled with water as it likes damp conditions. They are propagated by division or through seeds in spring.

Dionaea (Droseraceae)

The generic name derives from *Dione*, the Greek name for Venus. *Dionaea muscipula* from damp mossy spots in the south-eastern states of the U.S.A. (North and South Carolina), commonly known as 'venus flay-trap', is the sole member of the genus. It was discovered about 1760 by Arthur Dobbs, then governer of North Carolina. It is a insectivorous herbaceous perennial. It makes a fascinating pot plant in a 50:50 mix of peat and sphagnum moss, pot stood in a saucer filled with water. It is not difficult to cultivate but during summer it should be protected from direct sun and the potting mix should always be kept damp. These are propagated by division of clumps or through seeds in the spring. Its rosettes have some eight, 8-15 cm long and spreading leaves. The leaves have a winged stalk, becoming broader upward, then prominently constricted and then a 2-toothed jaw with each leaf is formed, the jaws normally opening at 40° to 50° angle and forming the trap for small insects. The two sets of teeth with which the jaws are edged, interlock when closed to prevent the escape of larger prey. When an insect touches 3 or more irritable and hinged prongs, closure of the jaws is actuated. In the inner surface of each jaw are scattered many small digestive and alluring glands, most being just inside the marginal teeth. Once inside, the insect is killed and enzymatic juices secreted by digestive glands break down the substance which is then absorbed by the leaf.

Drosera (Droseraceae)

In Greek, *droseros* is for 'dewy', the glandular hairs on the leaves bearing dew-like droplets. *Drosera* is the second largest genus of carnivorous plants having 141 species with global distribution but maximum being confined to south-west corner of Western Australia and South Africa. It is a deciduous and evergreen perennial from the tropics to temperate regions. They are rosette-forming, and solitary or in tufts. First leaves of the seedlings are glandless but number of glands increases on successive leaves. Leaves vary in shape from almost circular to spathulate in European species, and linear or divided and multiforked in Australian species. In mature plants the leaves form rosettes which survive the cold season mostly as tight winter-buds and expand only when congenial season is arrived. The leaves are orbicular to linear, adorned with prominent red-stalked tentacles (hairs) that secrete a glistening and sticky dew-drop-like fluid, where this fluid is mistaken by insects and other small creatures for nectar and get stuck to it. The insects when start struggling, other tentacles bend towards the prey, smothering and pushing it down to the leaf surface. At this time more fluid is excreted and the enzyme breaks down the animal protein ready to be absorbed by the plant. On completion of the digestion process the leaf unfolds to blow away the dry shell of the insects. Flowers, which are 5-8 petalled are carried in racemes well above the leaves. They are propagated through division of clumps and seeds in spring. There are two groups under *Drosera*, the tuberous group and the fibrous-rooted. **Tuberous group** grows in the soil which remains damp during growing season but most of these dry out with their aerial plants during hot summer leaving the tubers in the soil for the next growing season. More than 40 tuberous species are recorded from the southwest corner of Western Australia, four or more from Victoria and South Africa and two to East Coast. The tubers are mostly of red ink colour. Some of the species produce a clear rosette of more or less rounded leaves. The members of this group may produce single stalks (sometimes branched one) one metre tall, however, *D. gigantea* grows even taller. *D. hamiltonii* of this group is found growing in the swamps in the extreme south of Western Australia by changing its tuberous habit into fibrous-root system. They are difficult to propagate outside their nativity but *D. hamiltonii* is easily cultivated through root cuttings propagated in damp peat moss and filtered situation. Pygmy *Drosera* grows only a few centimeters across, bearing usually round or spoon-shaped leaves, and the plant centre is protected by white bract-like stipules. Their roots are long **fibrous** which help growing in the damp soil. Most of such species are confined to Western Australia but with a few elsewhere. Their flowers are very attractive white, pink, orange or red and in many species they are large for the plant. The white or pale-flowered species are *D. dichrosepala, D. nitidula* and *D. paleacea*, and orange-flowered ones are *D. drummondii* and *D. platystigma*, mostly having black centres which makes these very attractive. *D. pygmaea* bearing small white flowers is the smallest species recorded from the eastern states of Australia in sandy soil underlaid by damp soil. Though these produce seeds but pygmy types also produce small green structures called 'gemmae' which splash around due to rain water and form new plants, with exception of *D. glanduligera* (Western Australia to New South Wales) which bears larger leaves some 1-2 cm across, and bright orange flowers with black centres but no gemmae and the plants die in late spring to early summer and the new seedlings appear after the autumn rain. However, *D. petiolaris* is a large species growing 10 cm across and is found in northern Australia. Pygmy sundews are not easy to grow. These are grown in pots topped with coarse sand and underlaid with damp soil with no leaf touching the mixture, and these require protection from direct rain water. *D. binata* (*D. dichotoma*) is the largest east Australian and New Zealand species growing in the swampy soil which never dries out fully, leaves are green to reddish, long-stalked, erect, narrow and blades with two linear lobes (sometimes multiforked) 7-15 cm long and resembling the prongs of a hayfork, and the flowers are numerous in pure white to pinkish shade, 1.5-2.0 cm wide, that appear in corymbose racemes some 30-50 cm tall in summer. Its roots are black and fleshy. It is easy to grow in peat or sphagnum moss which remains damp throughout. *D. arcturi* is a wet-alpine species found in New South Wales, Victoria and Tasmania and also in New Zealand. Its leaves are 1-10 cm long and up to 1 cm wide and the flowers are single white. It bears small fibrous roots. *D. spathulata*, an evergreen herbaceous perennial, which propagates through seeds and leaf cuttings, forms tufted to clump-forming clear rosettes of spreading leaves, flat to semi-erect, nearly spoon-shaped and tapering gradually into a broad, reddish stalk. The alternate leaves are 5 cm long though the blade only 5 mm, and red if at sunny situation otherwise green. The flowers are 5-8 petalled, 1 cm wide, white or pink similar to *D. burmanii* which has wider leaves, and the both are easy to cultivate. *D. adelae* bears lanceolate leaves and white or red flowers. *D. prolifer* has slender stalks with broad and rounded leaves though leaves of *D. schizandra* are broader, blunt and often notched. *D. indica* is an annual species bearing narrow leaves with wide global distribution. *D. anglica* (great sundew), a native of Europe to North Asia, bears more or less erect and long-stalked leaves with 3 cm long blades, and the flowers some 5 mm wide, appear in summer, and are white with 5-8 petals. *D. rotundifolia* (common sundew) is a deciduous perennial from temperate areas, with flattened rosettes which may be solitary or tufted, leaf blades orbicular and up to 1 cm wide, and the flowers which are some 5 mm wide, appear during summer bearing usually 6 white petals. Its hybrid with *D. anglica* is known as *D.* × *obovata*. *D. menziesii* (pink rainbow), a deciduous species from Western Australia grows from a small tuber with erect stems some 10-30 cm tall. Leaves with a width of 1.2 cm appear in small scattered clusters, adorned with spreading and red-flushed tentacles (hairs), and the pink to red flowers some 2.5 cm wide appear in summer. *D. capensis* from South Africa, an evergreen and clump-forming species bears narrowly oblong and 7 cm or more longer leaves, and purple 5-petalled flowers some 2.0 cm wide appear in summer.

Drosophyllum (Droseraceae)

Drosophyllum means 'dew-leaved', with only one species, *D. lusitanicum* from South Spain, Portugal and Morocco. Here the leaves are revolute than being involute as in *Drosera*. It is a sub-shrubby small plant growing with simple stem some 5-15 cm high where on the top are long, linear and glandular insectivorous leaves. Bright yellow flowers some 3.75 cm across are borne on a 30 cm tall stalk with 5 obovate, thin and twisted petals, 10-20 stamens alternating in length and with short anthers, 5 filiform styles and a narrow capsuler fruit having 5 valves.

Heliamphora **(Sarraceniaceae)**

Heliamphora in Greek means 'sun pitcher'. There is only one species, *H. nutans* from British Guiana which is rhizomatous, therefore it is propagated through sigle-crown rhizomes. It is a perennial growing some 60 cm in height, bearing radical and pitcher-like leaves, pitchers tubular and enlarging with erect, oblique and open mouth which has covering of a rudimentary lid which is hairy inside. Lid forms by terminating the midrib. Pitcher is veined red. White or pale-rose several nodding flowers appear on a slender scape. The corolla is 4-6-parted and ovate-pointed. The plant is grown in a mixture of peat, sphagnum moss and sand, and then topped with living sphagnum moss and then pot is kept in a saucer filled with water.

Nepenthes **(Nepenthaceae)**

Nepenthes is the Greek name for another dissimilar plant used for this genus by Linnaeus, supposedly due to its similar medicinal properties. *Nepenthes mirabilis* (tropical pitcher plant or monkey cup as monkeys drink rain water from their cups) with large global distribution (Indonesis to Australia, Madagascar, Philippines, India), is the scrambling climber which rambles over the tropical swamps or stony grounds some 10 m long (largest species) with leaves more or less lanceolate and thin where a tendril-like structure by prolongation of the midrib is formed in the end which wraps around anything that supports it and this prolonged midrib swells out into a pitcher-like pouch complete with a neat-looking lid. The pitcher-shaped leaf structure is cylindrical, upright, swollen at the base, and constricted above. Its flowers are inconspicuous, and male and female flowers are borne on separate plants. It requires warm humid conditions for its growing and does not appreciate temperatures below 10 °C. It has a fixed lid which protects the pitcher from excess rain. Near the top of the inside of the pitcher and on the underside of the lid it secretes out sweetness to attract insects, insects fall into the pitcher and drown in its digestive liquid at its base. There are some 70 species of *Nepenthes*, a genus with its centre of distribution in Borneo, but extending to Madagascar in the west and New Caledonia in the east, occurring from sea level to an altitude of 3,350 m and likes nutrient-deficient soils with high acidity such as peat swamps. The thickened rim of the pitcher-like organ which exudes a sweet juice which is eagerly sought by small running insects. The liquid in the pitcher-bottom is an enzymatic fluid that contains substances similar to those of the stomach with digestive properties corresponding to pepsin. Slimy walls of the pitcher inside and the fluid have a narcotic substance which makes the insects motionless, and in the process when several insects are caught the lid closes and then pitcher starts the process to digest them. Enzymatic liquid and special glands near the bottom of the pitcher decompose the insects by producing pepsin-like fluid and the plant absorbs the all important nitrogen of the insects leaving only the chitinous armour. After sometime the pitcher opens to start its business again. *Nepenthes khasiana* is the only pitcher plant indigenous to India in the high rainfall region of north-east region with plenty of sunlight, endemic to Meghalaya from West to East Khasi, Jaintia and Garo Hills from 1,000 to 1,500 m altitude (http://en.wikipedia. org/wiki/ Nepenthes_khasiana), which is vigorous, dioecious, rare, endangered and protected species, therefore, included in the Appendix- I of CITES and Negative List of Exports of the Government of India (Sharief and Murthy, 2010; Venugopal and Devi, 2003). It is an endangered medicinal plant of eastern India and is included in the Botanical Garden List of Rare and Threatened Species of India compiled by the International Union for Conservation of Nature and Natural Resources. Its plant is vigorous, the leaves are lanceolate, the pitchers are rounded to tubular, some 10 cm long and green flushed red. Multiplication of this ecologically and economically important wild plant through seeds and cuttings (macropropagation) is difficult though cultivation is not difficult. In spite of this, an effort has been made for its *ex situ* conservation. *N. khasiana* from Shillong has been grown, maintained and conserved for more than past 40 years in the National Orchidarium and Experimental Garden of Botanical Survey of India, Yercaud. Too rich and too dry soil is not suitable for their cultivation, however, *N. nortiana* is found on limestone in Sarawak, and *N. phyllamphora* in dry scrub vegetation of Sri Lanka, which are dry places. The pitchers formed on ground are different than those formed on climbing parts of *Nepenthes* that may have even three different shapes of pitcher. On the ground they form rosettes of pitcher-forming leaves. In *N. rafflesiana*, the pitchers on the ground are broad though upper aerial pitchers are more tapered, and the size of pitchers may range from 12.5-30.0 cm long. *N. hookeriana* from Borneo is thought to be a natural hybrid between *N. rafflesiana* and *N. ampullaria*, in leaves and stems it is similar to *N. ampullaria* but pale-green pitchers which are boldly spotted and purple-blotched are broadly oval some 13 cm long and 8 cm across. *N. ampullaria* from Malaysia grows up to 3.0 m or more bearing 30 cm long and oblong leaves and 5.0 cm long, ovoid and green pitchers, sometimes blotched red.

Pinguicula (Lentibulariaceae)

In Latin, *pinguis* means 'fat' ~ alluding to the slippery and glossy leaves having a greasy appearance. The *Pinguicula* (butterwort) has some 69 species of carnivorous perennials with wide distribution mainly in temperate climates. They bear linear to broadly ovate, fleshy, pale- to yellow-green leaves rising from the fleshy rootstock to form a rosette, where upper surface bears glands secreting sticky and mucilaginous fluids to entice and trap tiny insects. When an insect is trapped, leaf margins roll inward, the glands secrete enzyme to digest the soft parts and absorb nitrogenous and other substances. The violet-type flowers appear singly with long slender spurs. Persistently a minimum temperature of 10 °C or below is injurious. These like sunshine and damp site. These are cultivated or propagated by division, leaf cuttings or through seeds in spring and summer in all-peat mixture with 2 parts sphagnum moss, *vis-a-vis* topping of living sphagnum moss. *P. caudata* (*P. bakeriana*, common name bog violet) from Mexico is a tender species having loose rosettes, bearing obovate to rounded and 10 cm long leaves, though in winter rosettes become dense and less than 5 cm wide with narrow and fleshy leaves, and the flowers some 2.5 cm across, long-spurred and violet-purple to rich carmine which appear on a 7.5-18.0 cm tall stem during autumn. *P. grandiflora* (large-flowered or greater butterwort) is a hardy species from West Europe with loose rosettes. The yellow-green leaves are ovate-oblong in a flat rosette and 2-8 cm long, the flowers that appear in summer on 8-20 cm tall stems are violet-purple, 2.5-3.75 cm wide and with long spur, and it overwinters as a rootless bud. *P. gypsicola* from Mexico is frost-tender, bears dense sempervivum-like rosettes with linear-spathulate and some 5-6 cm long leaves which widen at the base, flowers that appear rosy-violet in winter on a 10 cm or more slender and tall stem are 2.5 cm wide and long-spurred with white colour. *P. vulgaris* (common butterwort), a native of northern arctic and temperate zones is hardy with loose rosettes, leaves ovate-oblong and flowers 1.0-1.5 cm wide which appear on 5-15 cm long stems.

Polypompholyx (Lentibulariaceae)

The flesh-flowered *Polypompholyx* species are annuals but related to *Utricularia*. *P. multifida* is much larger than *P. tenella*, former is confined to Western Australia and is distinctly lobed while *P. tenella*, a much smaller species is found in Victoria, Western and South Australia. Both the species dry in summer producing seeds.

Sarracenia (Sarraceniaceae)

Sarracenia is a genus of 10 species of hardy and half-hardy, evergreen, stemless, herbaceous and carnivorous plants that are tufted and rosette-forming. They are found in boggy areas throughout North and East America and are suitable for growing in a cool greenhouse. It commemorates Michel Sarrasin de l'Etang (1659-1734), a French physician at Quebec, who sent its first specimens to Joseph Pitton de Tournefort, the celebrated 17th century botanist. The pitcher-shaped variously coloured leaves appear in shades of red and purple, and the pitcher contains a watery fluid in which insects are drowned and digested. Solitary, pendulous and bowl-shaped flowers appear some 5.0-10.0 cm across and these are composed of five prominent sepals, *vis-à-vis* petals and a umbrella-shaped stigma. These are planted in spring in 20-25 cm pans in a mixture containing equal parts (v/v) peat and chopped sphagnum, topped with living sphagnum moss. The mixture is kept regularly moist by putting the pans in trays of water. The glasshouse temperature is maintained not below 5 °C in winter with maximum lighting, and 10 °C temperature in other seasons, and in case the temperature increases the glasshouse should have free ventilation. These can be repotted every third year. These are propagated in spring through division or by seed- sowing in compost at a temperature of 13-16 °C. *S. drummondii* (*S. leucophylla*), a half-hardy species from North America is most colourful of the genus and grows to a height of 30-75 cm with a spread of 23.0 cm. Its pitchers are long, slender and erect, green below and white upper including the pitcher-lid and veined purple. Yellow to red-purple flowers some 5.0-7.5 cm across appear in April and May. *S. flava*, a half-hardy species from North America grows up to 60 cm in height and 23.0 cm in spread. Its pitchers are long, erect, yellow- to yellow-green with a veining of crimson to purple in the throat. Yellow flowers have pungent flavour, appear during April and May and are 7.5-10.0 cm across. *S. purpurea* is hardier than *S. drummondii* and *S. flava*, growing up to 23 cm high and 30 cm spread and is native to North America. The semi-erect purple and green pitchers are formed in neat rosettes. The base of the pitchers is narrow and stem-like, then widen to about 5 cm towards the top and then gradually narrowing to the mouth, and the lids are green, veined purple. It bears some 5.0-7.5 cm across green-purple flowers in April. *S. purpurea* (sidesaddle plant, devil's boots) bears rosette of long leaves, each leaf as a whole forming a pitcher, and solitary large carmine flowers with undulated petals. These leaves appear from a short rhizome buried in the sodden moss. At outer surface of the pitcher are warts and nectar glands which lure the insects to its mouth. The inner surface of the pitcher is divided into one upper glandular zone with downward-pointing hairs, an alluring and more densely glandular zone with a velvety texture, a glassy glandular zone; and lowest of all a zone of downward pointing hairs that serve to trap the lured insects. The water in pitcher drowns the creature and then digestion takes place partly by the juices secreted by pitcher walls and partly by action of bacteria.

Utricularia (Lentibulariaceae)

In Latin *utriculus* means a 'little bottle' ~ alluding to the bladders, hence common name of *Utricularia* is bladderwort. It is the largest genus of 219 species of carnivorous plants, found globally and their traps (only a few mm long) either are underground, underwater or epiphytic of worldwide distribution. The leaves are linear to rounded, those of the aquatic species divided into thread-like segments, the land species are usually found in moist moss, and the epiphytes have swollen water-storing branches. Leaves being in the form of a tiny 2-valved bladder equipped with tripping mechanism and a spring that only opens inwards and closes the door, all within the fraction of a second. These catch the mosquito larvae, water insects, tadpoles, and small young fishes. When any tiny creature comes close enough to touch the sensitive trigger mechanism, it is sucked inside the bladder and just then the bladder closes, enzymes break down the body of the prey and digests. *Utricularia* constitutes the two groups: one that floats on the water and the other that grows in wet or damp soil often along the *Drosera*, generally having minute leaves ~ only a few mm long so are normally overlooked until these come to flowering when become quite attractive in white, yellow, blue or purple colour where flowers have two different parts, the smaller upper one, and other larger apron-like or fan-shaped below. Flowers are either solitary or in racemes and are strongly 2-lipped and spurred. *U. dichotoma* and *U. hookeri* are purple-flowered. *U. menziesii* is an outstanding scarlet-flowered species from Western Australia, which though dries up during summer but for further perennation, it leaves behind small corms. *U. aurea* and *U. exoleta* with yellow flowers grow in fish ponds but are not suited in aquarium indoors. *U. intermedia* from N. Hemisphere grows with stems some 10-25 cm long, some with green leaves with few or no bladders though others colourless, buried in the mud and bear bladders, leaves are bristly, palmately lobed and 4-12 mm long, and bladders 3 mm long, though flowers are bright yellow with red-brown lines and 0.8-1.2 cm long, that appear in short racemes above water in summer. This likes acidic water most. *U. minor* (lesser bladderwort) from N. Hemisphere is though similar to *U. intermedia* but smaller and grows with stems some 7-25 cm long, bearing smooth and branched leaf segments, and the pale-yellow flowers appearing in summer are 6-8 mm long. *U. vulgaris* (greater or common bladderwort) also from N. Hemisphere is a free-floating species with stems growing 15-45 cm long, all bearing 2.0-2.5 cm long, pinnatifid green leaves, segments with bristly teeth, with many bladders some 3 mm long, and bright yellow flowers 1.2-1.8 cm long.

Propagation

Carnivorous plants are propagated theough seeds but it is a very slow method. Seeds are sown in damp medium, *i.e.* peat:sphagnum moss in equal parts (v/v). Many species such as *Nepenthes* are propagated through cuttings which take one season to come into bloom. Venus flytrap, pitcher plants and many others are propagated through division of clumps as well as by dividing the underground organs such as corms, tubers and rhizomes which they produce. *Darlingtonia* produces stolons for multiplication. In certain cases, propagation is carried out through leaf pullng, such as Venus flytrap and pitcher plants. *Cephalotus* can be propagated through leaf or pitcher cuttings. Explants used in micropropagation are leaves, leaf segments, seedlings and shoot tips.

Cultural Practices

The **soils** of carnivorous plant habitats are characterized by very low nutrients such as nitrogen, phosphorus, and alkali ions, as well as high acidity. Since they obtain nutrients by consuming animal prey rather than absorbing *via* roots, the nutrient absorbing ability of roots is very limited. As a result, the roots of carnivorous plants will not tolerate nutrient-rich commercial horticultural mix. Nutrient-poor, acidic sphagnum peat moss or peat moss to perlite 3:1 or 2:1 are recommended as **growing medium** (Lee, 2007). They can be grown in pure peat moss or sphagnum moss; mixture of peat and sphagnum moss; or a mixture of peat moss, perlite and washed coarse river sand. Three parts peat moss, 1 part perlite and 1 part washed coarse river sand (v/v) works very well. *Drosera adelae* can be grown successfully in sphagnum moss, butterwort hybrid *P. emarginata* × 'Weser' in both, *i.e.* pure peat moss and pure sphagnum moss, and *P. moranensis* grows best in pure sphagnum moss. Generally, the habitats of carnivorous plants are warm, sunny and constantly moist so the plants experience relatively little competition from other low growing plants. For **repotting**, the size of the plant should be relative to the size of the pot. Right size of the plants to the right size of the pots will look aesthetic and will save the grower of his/her labour and maintenance. When repotting, the use of a pot that is little larger than it should have been is advantageous as it will allow some extra room for the plant to grow. Greenhouse facility is ideal for commercial cultivation of such plants. As a general rule, majority of carnivorous plants, including 'cobra lily' (*Darlingtonia californica*), 'venus flytrap', 'sundews' (not the tropical ones), 'pitcher plants' (*Sarracenia*) and 'bladderworts' (*Utricularia*) require at least four hours of direct **sunlight**. The 'lance-leaved sundew' (*Drosera adelae*) is an example of a tropical sundew that is native to northern Queensland. This as well as *Nepenthes* can grow in semi-shaded as well as under full-sunny conditions. Tropical *Drosera* can benefit from being kept humid. This can be accomplished by growing it in a terrarium or by placing an inverted jar over the pot. The bottle is translucent in colour and therefore allows only filtered light to enter and will keep the plant warm and humid. However, plants those are kept this way should be placed in an area that receives filtered light as full sunlight can cause overheating inside the bottle. *Pinguicula* requires a shaded position. Bright sunlight is highly beneficial for a number of species for producing strong and healthy plants full of colour. 'Thread-leaved sundew' (*Drosera filiformis* ssp. *filiformis*) can be grown in shade as well as in direct sunlight but those grown in full sunlight produce healthy green leaves with bright red colouration on the digestive glands, though plants grown under shade produce only colourless sticky glands (Amoroso Steve, 2008). Likewise, growing of *Sarracenia leucophylla* in full sunlight produces pitchers with white tops while remainder of the leaf remains green, however, in shade the whiteness of the top, *vis-à-vis* fenestrations are gone. So when growing inside, a fluorescent light supplement, 15-30 cm above the plants is recommended.

Most carnivorous plants grow in bogs, so almost all are quite resistant to drying due to low humidity. For plants growing on bogs, undersurface **watering** with a complete change of water once every 3-5 days is recommended. To increase air humidity, frequent water spray is needed, except for flypaper trap type plants in which frequent spray will wash out digestive enzymes in the leaves, resulting in retarded plant growth. A drip tray is essential always with 1 cm of water in the bottom applying an overhead watering every day during summer. In the winter months watering can be lessened but still making sure that the sphagnum is nice and damp. At this time, most species go into a stage of hibernation, therefore, the plants are facilitated to have a rest, so these should be left in cooler conditions but make sure that the temperature stays above -5 °C. Carnivorous plants like a humid atmosphere so misting is a must. In case of *Darlingtonia californica*, if there are hot days, cooling of the roots should be provided. This can be done using a nice thick clay pot and watering with ice water during the hottest part of the day. Even placing ice cubes on top of the sphagnum in the morning to melt throughout the day will prove advantageous. *Sarracenia psittacina* and *Drosera arcturi* pots during winter require to be placed into a container of water that covers all but the top 1-2 cm of sphagnum. This method applies all the year round and water can be allowed to freeze completely during the frost. Species such as *Drosera capensis* and *D. spatulata* do not require any hibernation time at all and will grow quite happily in a warm environment all the year around. This also goes for *Nepenthes* and *Cephalotus*. *Nepenthes* species require a large amount of misting to help in pitcher development, unlike *Cephalotus* that does not require misting. *Nepenthes* which is most suitable as a basket plant, does not require as much water as certain others but growing medium should still be kept moist. Some butterworts can sit in a very shallow tray of water, but watering from the top of the pot should be avoided as leaves could potentially rot. With exception of the tropical pitcher plant and some butterworts, most carnivorous plants should sit in shallow water trays. Rainwater is recommended for use, however tap water may be used provided plants are watered from the top of the pot to prevent the salts from building up in the growing medium. Tuberous sundews should be kept moist during cooler winter months and dry after they go dormant, usually at the end of October or early November to February. In the end of summer tuberous sundews should be returned to the water tray so the plant can grow out of the tuber, ready for the next growing season. At the end of a growing season, *Nepenthes* requires being removed from its old potting medium so by flushing away whole of the old season's medium from the root mass, this is repotted in fresh growing medium. This is due to the fact that *Nepenthes* breaks down its growing medium quite faster and produces a high level of root exhaustion. Sometimes this operation is necessary with even other carnivores. Any dead or damaged leaf or pitcher is **pruned off**. As carnivorous plants grow, they will produce new leaves replacing the older ones. As older leaves begin to accumulate on plants, it is advisable that these are removed to allow room for new growth and to prevent plants from being attacked by pests. A build up of a large number of old leaves restricts air movement and provides congenial environment for various insect-pests and diseases.

Very old withered leaves from the larger varieties of non-tuberous sundews such as *Drosera capensis* and *D. filiformis* and from *Sarracenia* can be pulled off the plants easily. Alternatively, in *Sarracenia*, old browning pitchers can be cut off from the base of the plant. However, when this practice is used, a small section of the pitcher remains attached to the rhizome. As these sections turn brown and accumulate, these should be pulled off the rhizome to prevent harbouring of insect-pests and pathogens in these sections. Generally **fertilization** of carnivorous plants is not recommended. In rare cases, when mineral deficiencies do really occur, foliar spray or undersurface watering supplemented with Hyponex is recommended (Lee *et al.*, 2003a,b; Joe *et al.*, 2003; Kim *et al.*, 2003).

Most carnivorous plants grow in habitats that are inhospitable during certain seasons. To survive these seasons, plants produce seeds and die, or enter a dormant period. Popular ornamental plants, such as Venus flytrap and *Sarracenia,* have a cool **dormancy** period of about three months. Plants should be kept at temperatures below 4 °C during this period, and watered once or twice a month on warm days. 'Well-rested' plants grow more vigorously than those that do not undergo a dormant period. Most carnivorous plants are high **frost-tolerant**. In fact, most species will grow better being exposed to a cold winter as in their natural habitat the frost is of common occurrence. *Darlingtonia* requires cold temperatures around its roots so that it is able to thrive as it is used to live in areas with freezing water flowing its root system. It should be ensured that water does not get too hot during the summer months, as water heating is a common cause of fatalities among cobra lilies. As much as possible, the people should try to give agroclimatic conditions to the plants akin to their natural habitat.

Insect-Pests and Diseases

While carnivorous plants are generally free of insect-pests or diseases, sometimes a few of problems are encountered which are being described here with. **Aphids** infest the plants and cause distortion, especially during spring. These may be killed through pressing with thumb and fingers on large leaves such as sarracenias though in small plants like to those of Venus fly traps and sundews these overwinter and cause extensive damage when new leaves appear during spring.

A systemic insecticide is the best way to deal with them. Use of soft soaps is the best organic way to control these. Nymphs and adults of the **mealy bugs** in clusters are found feeding in the leaf axils or on the ribs of the leaves by sucking the plant sap so plants are weaken. These insects are a serious problem on *Darlingtonia, Nepenthes* and *Sarracenia*. These should be rubbed with a brush dipped in methylated spirit or alcohol. Alternatively, a systemic insecticide should be sprayed. **Scale insects** are small brown blister-like insects some 1-2 mm long, and these become very serious on *Sarracenia psittacina*. These are controlled by dabbing each insect with a paintbrush dipped in methylated spirit or alcohol and spraying systemic insecticide.

Though little green **caterpillar** is not a big problem, but just one can create quite a few holes in *Sarracenia* and many other plants. Just keep hunting this is the best way to get rid of. **Slugs** and **snails** can occasionally feed on *Sarracenia purpurea* ssp. *purpurea* and many other carnivores that are grown in damp and shaded sites and these can be controlled through metaldehyde poison baits.

Botrytis is the fluffy grey mould starting on dead growth, so good husbandry is a way to prevent it. Botrytis is something to watch for particularly in spring and autumn on *Sarracenia* and Venus flytraps. Some species and hybrids are more prone to it than others. Removal of dead foliage regularly provides plenty of air circulation around the plants, especially in still and damp weather when it is cold. Spraying the plants every week with 0.2 per cent Dithane M-45 or Dithane Z-78 is quite effective in keeping this problem at bay. However, proper field sanitation should be maintained. As the rhizome of *Sarracenia* increases in length and becomes older, the ends become brown and disintegrate due to **fungal rot**, and ultimately whole plant collapses. Plants suspected of fungal rot should be removed from the pots and burnt.

References

Amoroso Steve, 2008. *Drosera filiformis* ~ varieties, cultivation and propagation. *Carniflora Australis, N.L. Australasian Carnivorous Plant Society*, No. 11, pp. 22–26.

Behera, K. K., S. Sahoo and P.N. Mohapatra, 2007. *Ethnobot. Leaflet*, No. 11, pp. 106–112.

Didry, N., L. Dubrevil and M. Pinkas, 1994. Activity of anthraquinonic and naphthoquinonic compounds on oral bacteria. *Die Pharmazie*, **49**:681-683.

Grevenstuk, T., N. Coelho, S. Gonçalves and A. Romano, 2010. *In vitro* propagation of *Drosera intermedia* in a single step. *Biologia plantarum*, **54**(2): 391-394.

Joe, H.T., J.K. Hwang and C.H. Lee, 2003.Effect of media, shading, watering and liquid fertilizer on growth of *Dionaea muscipula* (abstr.). *J. Korean Soc. hort. Sci. Technol.*, **21**(2):91.

Kamarainen, T., J. Usitaro, J. Jalonen, K. Laine and A. Hohtola, 2003. Regional and habitat differences in 7-methyjuglone content of Finnish *Drosera rotundifolia*. *Phytochem.*, **63**: 309-314.

Kim, Y.J., J.K. Kim, J.K. Hwang and C.H. Lee, 2003. Effect of media, shading, watering and liquid fertilizer on growth of *Drosera rotundifolia* (abstr.). *J. Korean Soc. hort. Sci. Technol.*, **21**(2): 92.

Krolicka, A., A. Szpitter, E. Gilgenast and A. Romanik, 2008. Stimulation of antibacterial naphthoquinones and flavonoids accumulation in carnivorous plants grown *in vitro* by addition of elicitors. *Enzyme Microbiol Tech.*, **42**:216-221.

Lee, C.H. 2007. Insectivorous plants. In: *Horticulture in Korea* (eds Lee, J.M., G.W. Choi and J. Janick). Publication of the Korean Society of Horticultural Science, Korea.

Lee, C. H. 2008. Carnivorous plants ~ new ornamentals. *Chron. Hort.*, **48** (4): 11-14.

Lee, J.Y., J.K. Hwang and C.H. Lee, 2003a. Effect of media, shading, watering and liquid fertilizer on growth of *Drosera aliciae* (abstr). *J. Korean Soc. hort. Sci. Technol.*, **21**(2): 93.

Lee, J.Y., J.K. Hwang and C.H. Lee, 2003b. Effect of media, shading, watering and liquid fertilizer on growth of *Sarracenia purpurea* (abstr.). *J. Korean Soc. hort. Sci. Technol.*, **21**(2): 92.

Likhitwitayawuid, K., R. Kaewamatawong, N. Ruangrungsi and J. Krungkrai, 1998.Antimalarial naphthoquinones from *Nepenthes thorelii. Planta Med.*, **64**:237-241.

Parimara, R. and P. Sachidanandam, 1993. Effect of plumbagin on some glucose metabolizing enzymes studied in rats in experimental hepatoma. *Mol. Cell Biochem.*, **12**: 59-63.

Romanowski, N. 2001. Gardening with Carnivores ~ *Sarracenia* Pitcher Plants in Cultivation and in the Wild. UNSW Press, Inc., Australia.

Schnell, D.E. 2002. Carnivorous Plants of the United State and Canada. Timber Press, Inc., Portland, Oregon, USA.

Sharief, M. U. and G.V.S. Murthy, 2010. *ENVIS Newsletter*, **15**(1): 3–4.

Shin, K.S., S.K. Lee and B.J. Cha, 2007. Antifungal activity of plumbagin purified from leaves of *Nepenthes ventricosa maxima* against phytopathogenic fungi. *Plant Pathol. J.*, **23**:113-115.

Venugopal, N. R. and Devi, 2003. *Feddes Report* (India), **114**(1–2): 69–73.

Wallace, R.A., G.P. Sanders and R.J. Ferl, 1990. Biology ~ The Science of Life (3th ed.). Haroer Collins Publishers Inc., New York.

12

Catharanthus (Family: Apocyanaceae)

M.K. Singh, Sanjay Kumar and R.L. Misra

[**Common names**: Greater periwinkle (*Catharanthus major*), Lesser periwinkle/Common periwinkle/Creeping myrtle/Running myrtle (*C. minor*), Cape periwinkle/Madagascar periwinkle/Old maid/Rose periwinkle (*C. roseus*, syn. *Ammocallis rosea, Lochnera rosea, Vinca rosea*), Periwinkle; Hindi- Sadabahar, Nayantara; and Tamil-Sudukadu mallikai]

Introduction and Origin

Catharanthus (syn. *Lochnera*) derives from the Greek *katharos* meaning 'pure, clean or without blemish' and *anthos* meaning 'a flower'. Its native haunts are tropical and sub-tropical West Indies and Malagasy but has become naturalized among the native flora in the West Indies and throughout other tropical areas, with one species in India. Possibly about 1750, it was brought to Paris by the French and from there it got its way to England where Philip Miller grew it at Chelsea in 1757 (Everard and Morley, 1970). The popularity of *Catharanthus* has been increased due to its low maintenance, long blooming period even under adverse conditions and resistance to insect-pests and diseases (Corley, 1981). In trade, it is known as *Vinca* and it shot into prominence due to discovery of anti-neoplastic activities of leaf alkaloids (Noble *et al.*, 1958).

Genetic improvement has resulted in improved self-branching habit combined with production of rose, dark pink, purple to white coloured flowers with red centered eye. Round the year flower production and lush green foliage make it an excellent bedding as well as hanging basket plant. It has become popular as a garden flower and also as a potted flowering plant for giving enchantment in the balconies and terraces. It can also be planted in the herbaceous border, in the interspaces of the shrubbery, on the roof garden, at a shady place between two buildings, in the rock and paved gardens, inside the houses when in bloom, and in the pots, tubs or urns. It has immunity to animal browsing due to strong acridity of all plant parts

Previously its alkaloids were being used as an adulterant in *Rauvolfia* roots but now it has occupied an independent position as an alternate source for raubasin and serpentine alkaloids used in the preparation of hypertension-relieving drugs. Due to presence of a small amount of ricinoleic acid, roots and leaves of this species are used in the treatment of diabetes (Garg *et al.*, 1987). It possesses various alkaloids (more than 60) which are used in pharmaceutical industries. It is a good source of a new anti-tumour drug for use in traditional medicine (Arcamone *et al.*, 1980). It is also the source of anticancer drugs vincristin and vinblastin. Three new dimeric alkaloids (leurocolombine, vinamidine and pseudovincaleukoblastine) have been found in *Catharanthus* (Tafur *et al.*, 1975). Germination and growth inhibitors present in the plants are used in folk medicine as a natural alternative to synthetic herbicides and for veterinary purposes (Dionello Basta and Basta, 1984). Its leaf extracts also possess some fungitoxic properties (Shivpuri *et al.*, 1997). It is effective against *Fusarium solani* (Vimala *et al.*, 1993). Extract of *C.*

roseus showed maximum inhibition of mycelial growth and spore germination against *Pyricularia oryzae* (*Magnaporthe grisea*) and *Helminthosporium oryzae* (*Cochliobolus miyabeanus*) *in vitro* (Ganguly, 1994). Plant extracts show moderate inhibition against the white muscardine fungus (*Beauveria bassiana*) (Raghavaiah and Jayaramaiah, 1987). Laboratory studies revealed that leaf and flower extracts of *Catharanthus roseus* had the greatest repellent effect against adult beetles (*Maladera insanabilis*) (Dwivedi and Awasthi, 1994). Water extracts of 100 g leaves of *Catharanthus roseus* significantly reduced the population and reproductive potential of *Meloidogyne incognita*, with maximum reduction in root and soil population occurring in the highest concentration (Alagumalai *et al.*, 1991). Ethanolic extracts of the flowers, leaves and stems showed insecticidal activity against *Culex quinquefasciatus* (Mohsen *et al.*, 1989). Plant extracts is also effective against *Plutella xylostella* larvae (Sreenivasan *et al.*, 2003). Its ovicidal activity against *Corcyra cephalonica* under laboratory conditions was evaluated and leaf extracts at S/100 dose level exhibited inhibition of egg hatching (Dwivedi and Awasthi., 1999). Its aqueous extract of leaves showed antifertility effect on male albino mice (Murugavel *et al.*, 1989). It is less susceptible to sulphur dioxide exposure (Howe and Woltz, 1981). In India, it is commercially grown in Karnataka, Tamilnadu and Maharashtra for domestic consumption as well as for export.

Botany and Genetics

The plants of *Catharanthus roseus* are free flowering and bloom round the year. It grows like weeds in many parts of the country. Plants are 30-60 cm tall with numerous erect branches. Stem is soft, angular and purple and contains usually a colourless sap. Plant is covered with bright green, smooth oval or oblong leaves which are cuneate at the base, membranous, upto 7.5 cm long, 2.5 cm broad, glabrous, shining green above, paler below, petiole up to 1.25 cm long. The flowers are bracteolate, pedicellate, hermaphrodite, actinomorphic, complete, hypogynous and typically pentamerous, solitary in the axils of the upper leaves, rosy-purple (*V. rosea*) or pure white (*V. alba*), the latter with or without a very small redish eye, very fragrant, seated on very short pedicels; calyx-tube very short, divided into five sepals, which are linear, acute in shape, hairy on the back and about 0.5 cm long, aestivation quincuncial; corolla tube 2.5-3.0 cm long, hairy outside inflated above, with a narrow throat, having a hairy ring inside below the stamens and with hairy rugosities above, lobes five, oblong-rounded, about 2.5 cm long, spreading; stamens five, free, alternate with the petals, situated on the swollen portion of the tube, anthers introse, often sagittate, bicelled, free or adherent (by viscid exudates) to the stigma or 'clavuncle'; filaments very short; disc present, higher than the ovary, consisting of two narrow obtuse fleshy scales; ovary of two separate and distinct carpels, hairy at the top, each ovary superior to half-inferior, unilocular with parietal placentation when syncarpous, bilocular with axile placentation; single style and stigma, style simple, about 2.5 cm long; stigma thick or dumble-shaped; and ovules few to many. The fruit of two separate-like follicles, seldom drupaceous and seeds with fleshy endosperm and a straight embryo.

The full chromosome complement of 2n=16 in *Catharanthus roseus* includes two metacentric, four subtelocentric and two telocentric pairs with the presence of a single nucleolus organizer region (Guimaraes *et al.*, 2012). Presently more than 100 cultivars with flowers of different colours and shapes are available (van der Heijden *et al.*, 2004). The new ornamental varieties developed through classical breeding have a wider range of flower colours and sizes. At present, the most commonly observed colours in *C. roseus* are a pink corolla with red eye, white corolla with red eye and orange-red corolla with white eye (Flory, 1944; Kulkarni *et al*, 1999). The determination of these flower colours in *C. roseus* involved at least six major interacting genes (Sreevalli *et al.*, 2002). These flower colours are the result of the accumulation of different members of a class of secondary metabolites, the anthocyanin pigments (Filippini *et al.*, 2003). Recently the first transgenic *C. roseus* plants, which were obtained following regeneration from hairy roots, have been reported (Choi *et al.*, 2004).

Periwinkle is one of the most important anticancerous drug-producing medicinal plants. Two dimeric alkaloids, vincristine and vinblastine obtained from its leaves find extensive use in the treatment of human neoplasms. Recent advances in biotechnology and genetic engineering and metabolic flux enhancement in secondary metabolites pathway(s) of periwinkle have been elaborated. The focus has been shifted towards the identification of alternative input routes for desired secondary metabolites, and related isolation of unique gene(s) or their cDNAs and DNA sequence(s) for marked activities leading to active metabolites productivity. The enzyme 'strictosidine synthase' has been considered a key point for limiting flux in the pathway. Full length cDNA coding for this enzyme has been obtained. The veracity of the clone got confirmed as it produced tryptophan decarboxylase activity when expressed in *Escherichia coli*. The production of novel alkaloids with higher biological activities and potential versatile uses as nematicide and phagodeterrent in addition to its already recognized medicinal values promise a greater and diverse role of *C.roseus* in future (Srivastava, 2003).

Species and Varieties

There are about eight deciduous or evergreen species in the genus, seven species are endemic to Madagascar, *e.g. Catharanthus coreaceus, C. lanceus, C. longifolius, C. ovalis* ssp. *grandiflorus* forma *ovalis, C. roseus, C. scitulus,* and *C. trichophylus*, all from Madagascar, and one *C. pusillus* is confined to India and Sri Lanka (Stearn, 1975). Out of these only *Catharanthus roseus* is grown for ornamental as well as for medicinal purposes, hence it is being described here with. It is an evergreen perennial and spreading shrubby plant but is grown as an annual through seeds year after year though to maintain a particular type or colour, propagation through cutting is the best option. In western countries it is normally a glasshouse crop due to being tender. This is the only species native to the tropics (Madagascar) which grows from 20-60 cm in height and up to 75 cm diameter in spread. Its leaves are opposite, dark glossy geen, oblong-ovate to lanceolate and from 3-10 cm long. Its flowers at species level are white to rose-pink, 3-4 cm in diameter and blooms throughout the year under tropical conditions of the country but in sub-tropical conditions it remains unflowered from December to February. *C. roseus ocellatus* bears white flowers with a red or pink eye. It is frost tender and can tolerate the temperature to a minimum of 5-7 °C. For its cultivation, it requires full sun and well-drained soil. Potted specimens should be watered moderately and even less when temperatures are low.

Important popular cultivars of *Catharanthus roseus* are'Bright Eyes', 'Creeping Carpet' (prostrate), 'Grape Cooler', 'Little Blanche', 'Little Bright', 'Little Bright Eye', 'Little Delicata', 'Little Linda', 'Little Pinkie', 'Little Pretty', 'Orchid Cooler', 'Peppermint Cooler', and 'Pretty in Pink'. CSIR-CIMAP, Lucknow also released 'Dhawal', 'Nirmal' and 'Prabal' cultivars of periwinkle where 'Prabal' has been developed through pure line selection. 'Nirmal' is resistant to dieback, collar rot and root rot diseases under field conditions.

Propagation

Periwinkle is propagated from seeds or stem cuttings including tip cuttings. It is very readily raised from seeds which it produces abundantly though there may be variation in plant type and colour of flowers. Seedlings can be raised either by direct sowing in the bed or through transplanting. Seeds are sown during March in pans or boxes of seed compost (2:1:1 by volume the loose bulk of medium loam:loose bulk of peat:loose bulk of sand and to one kiligramme of this mixture some one gramme of superphosphate of lime + half gramme of ground chalk) at a temperature of 15-18 °C. For direct sowing in the field a high seed rate of 2.5 kg/ha by drilling these in rows 45 cm apart or broadcasted but when transplanting, only 500 g seeds are required for use in the nursery for planting one hectare of land. After sowing the seeds are lightly covered with compost for proper setting and germination, and it takes 1-2 weeks (normally 10 days) for germination and 60 days for transplanting when these have attained a height of 15-20 cm, planted at 30 cm distance in rows spaced 45 cm which requires 75,000 plants per hactare. Fresh to 1-year old seeds should be taken for sowing. The seed germination vigour was obtained 93 per cent on the 4th day and it was 96 per cent on the 6th day at 20-25 °C, and the germination rate of fresh seed was 98 per cent and after 1, 2, 3 and 4 year's storage it dropped to 95, 53, 28 and 0 per cent, respectively (Mitrev, 1976). Seedlings are required to be pricked off when large enough to handle, first into boxes and then transferred into 8-10 cm pots containing compost as elaborated in the first paragraph of Cultural Practices somewhere further. This, in fact, is known as John Innes Potting Compost No. I. The seedlings are pinched once or twice to encourage bushy growth. If these are required to be planted in pots further for use in the greenhouses or elsewhere for decoration, these are potted on as necessary otherwise those are hardened off for outdoor bedding in a cold frame before planting them out in April in the plains of sub-tropical regions of the country and in late May in the temperate and sub-temperate regions.

Cuttings some 5.0-7.5 cm long are taken from plants that have been cut back in February in the sub-tropical regions, in March in the sub-temperate or temperate regions of the country, and in April of glasshouse plants. Cuttings are inserted in potting compost consisting of peat and sand, both by volume in equal parts, and placed in a chamber with 15-18 °C temperature. These cuttings root within 7-10 days and then these are transferred in 7.5-10.0 cm diameter pots filled with John Innes Compost No. I as defined under seed in the previous paragraph. The growing points should be pinched off to encourage lateral growth. These, after the season is gone, are overwintered at 10-13 °C, and then again when season starts these may be potted on as necessary or set out for bedding. In an experiment comparing cultivars 'Grape Cooler' and 'Orchid Cooler', plants propagated vegetatively flowered and finished 17-25 days earlier than those propagated from seeds (Qing and Kobza, 1997).

For ***in vitro*** **seed germination**, *Catharanthus roseus* seeds can be cultured in petri dishes containing MS medium supplemented with BA (0, 0.2, 0.5 and 1.0 mg/l) and NAA (0, 0.05, 0.1 and 0.2 mg/l) kept under continuous darkness where one can find higher percentage of germination. This first initiates the callus whose diameter increases with increasing rates of growth regulators. **Shoot tip explants** when were also cultured on MS medium containing different concentrations of BA and NAA, all the treatments induced shoot multiplication, highest number of proliferated shoots being in medium supplemented with BA at 1.0 mg/l and NAA at 0.1 mg/l, though rooting occurred on half-strength MS medium without growth regulators, and the plants grew well in small pots containing sand, soil and peat moss in a 1:1:1 ratio with 100 per cent survival (Atta Alla, 1997).

Cultural Practices

Periwinkle is a crop of tropical climate, and being frost-tender it can bear a minimum temperature of 7 °C though can tolerate high temperature and humidity, and in the climatic conditions of India, it feels quite comfortable from March to November in the sub-tropical regions, April to October in the sub-temperate and temperate regions and thoughout the year under tropical conditions. In the summer to rainy season, these have luxurious growth and profuse flowering, though during spring and autumn also their growth is rampant but in spring there is no flowering and during autumn the flowering is lessened. It flowers almost all the year round in mild climates and in summer and rainy seasons in other climates. In cold climate, it is cultivated as an indoor or glasshouse plant. *Catharanthus* should be grown at fully exposed **site** as for its good growing it requires full sun or slightly filtered (partial shade) when grown outdoors but indoors in 12-15 cm pots at a well-lit corner. The effects of daylengths of 9, 12 or 15 h on seeds and seedlings were investigated at 24/20 °C (day/night) but it did not express any effect on germination though seedling colour was found best at 15 h (Qing *et al*, 1997) daylength. A **temperature** range of 20-25 °C is optimum for plant growth and flowering. In perennial crops, it may continuously be grown at minimum winter temperature of 13 °C which will ensure a long flowering season (Hay and Beckett, 1971). In case when plants are not required for flowering, they may be kept at a winter temperature of 7 °C to go in rest but during this time the mixture is kept just moist and watering is drastically reduced. In glasshouse crop, if temperature exceeds 18 °C, it requires to be ventilated freely. Young seedlings are highly susceptible to cold as compared to grown up plants, and at a minimum temperature of 10 °C seedling growth is restricted and at 1°C these die though growth of grown up plants ceases at or below 7 °C. Plants of 'Grape Cooler' and 'Orchid Cooler' cultivars were when raised from seeds at day/night temperatures of 18/14, 24/20, 29/25 or 34/30 °C, at 18/14 °C (day/night) tremperatures the germination had been very poor with deformed leaves, though with every increase in temperature the growth was faster but uneven and the seedlings were spindly (Qing *et al*, 1997). The plants are not so particular about **soil** types though they do not like water-logged or highly alkaline soils. Any soil which is porous with good water-holding capacity but is not waterlogged and is reasonably fertile is optimum for its growing though for potting the compost should by volume have 7:3:2 loose bulk loam:peat:sharp coarse sand and to each kilogramme of this mixture some 3 g of 2:1:1 superphosphate of lime:bone meal:potassium sulphate and less than one gramme of finely ground chalk. A good light potting compost is always preferable. Heavy clay soils are generally not chosen for its cultivation, however, in such soils, it is advised to mix sand and well decomposed farmyard manure to make it porous. Good drainage is vital to avoid root rot and other fungal diseases. Anwar *et al.* (1988) reported that there had been 61.3 per cent leaf and 51.9 per cent root yield reduction at a salinity of EC 12 ds/M over control.

Preparation of land for large scale cultivation, the field should be excavated deeply through cultivator, at least three times to a depth of 30 cm, and after each ploughing it should be properly planked so that the soil is completely pulverized, all the weeds and their rootstocks should be taken out along with the polythene remains, bricks, stones or any other hard object, if any there, and then the land is properly levelled and beds of convenient sizes are prepared, keeping proper provision of water channels at least 75 cm wide. At the time of second ploughing, some 150-200 quintals of well-rotten farmyard manure should be mixed thoroughly in the soil. Conveniently, the width of the beds should be not more than 1.6 m with some 60-75 cm clearance in between the beds for carrying out cultural operations and watering. Bonemeal 25 g/m^2 and *neem* cake 20 g/m^2 should also be incorporated in the soil at the time of bed preparation. The recommended spacing is 30-45 cm apart depending on the variety and its spread. About 0.25 g seeds/m^2 is required when it is sown directly in the bed and 0.05 g/m^2 when it is transplanted at a **spacing** of 45 × 20 cm. Seeds are sown in nursery or in pans or pots in March-April and the seedlings are **transplanted** at an age of 30-45 days in April-May in the sub-tropical and temperate conditions though seed sowing in September – October and transplanting in October-November in the plains.

Seedlings take about two months for flowering after planting. Perennial plants are potted or repotted annually in March but the plants are seldom worth keeping for more than two years. New shoots are encouraged to appear with good quality of flowers if undesirable weak and staggered shoots are thinned out regularly. The feasibility of producing *Catharanthus roseus* under Akola conditions (altitude 281 m a.s.l., mean annual rainfall 843.5 mm, RH 22-87 per cent, temperature 10.9-41.4 °C) when was evaluated by sowing the seeds on raised beds and transplanting them at 45 × 30 cm, the growth was luxuriant and it yielded dried roots at 660 kg/ha and its total alkaloid content was 2.604 per cent (Vitkare and Phadnawis, 1988). The plants can be **headed back** every six months and replaced every second year when it becomes straggly or flower badly. Periwinkle plantings require four **weedings**, the first at 45 days after direct sowing or transplanting, second at 30 days after the first weeding, third after one month of second weeding and the last at 30-40 days of the third weeding. Pareek *et al.* (1991) reported effective control of weeds in periwinkle field by pre-plant use of 0.75 kg a.i./ha of fluchoraline or pre-emergence use of alachlor at 1.0 kg a.i./ha.

Pot Culture

Though recommended pot size for pot culture of periwinkle is 12-15 cm but this can also be grown in larger pots for decoration. The right dimension of the pots should be determined based on the size of the plants and their expected spread. These need to be in the right proportion. One plant per 13cm pot or 3 plants per 30 cm pot are planted. Generally, standard pots or extra tall pots are most suitable. A good potting mix may contain soil, leaf moulds, vermicompost or farmyard manure, sand and other coarse materials. It should be well-draining. Media containing a lot of fine particles should be avoided as this may lead to water-logging in the pot. Media should be kept evenly moist but not waterlogged. Wet soil can lead to the fungal rot, first affecting the roots and then afterwards whole plant. It is important to allow the water to drain out from the bottom of the pot. After a few years, the potted plants tend to become too heavy for the original container, so they are repotted in a new and larger pot, *i.e.* 15 cm in diameter, which is filled with fertile potting medium and then is watered copiously. Growth of *Catharanthus roseus* 'Grape Cooler' was investigated with two types of media (a peatlite mix, and another mix containing 25 per cent pine bark) and five types of nutrient charges in the peatlite media (sulphated micros, chelated micros, sulphated or chelated micros with pH adjustment to 5.5, and no charge). Effect of various sources of nitrogen on growth was investigated taking five different ratios of nitrate to ammonium. Greatest growth was measured on shoot length and shoots dry weight when grown in the peatlite medium either at sulphated micro or chelated micros adjusted to pH 5.5, and at the highest ratios of nitrate to ammonium. Root dry weight and growth were negatively affected by high levels of ammonium in the fertilizer solution (Thomas and Latimer, 1995).

Watering and Fertilization

Periwinkles require **watering** regularly but sparingly. Well distributed annual rainfall of 100 cm or more is ideal for cultivation. It can also be grown as a rainfed crop. In field trials with *C. roseus*, the plants were irrigated initially before being subjected for 5 months to 4 levels of soil moisture deficit (SMD), *viz.* 25, 50, 75 or 95 per cent at the root zone. With increasing SMD there was a marked reduction in fresh and dry weights as well as in the moisture contents of the various plant parts. The leaves contained the highest alkaloid concentrations, determined colorimetrically as vincaleukoblastine, in comparison with stems and roots at all SMD levels (Talha *et al.*, 1975). In a field experiment to study the response of *Catharanthus roseus* to different irrigation levels, the highest dry foliage and root yields were obtained with the scheduling of irrigation at 50 mm CPE accounting to 18 irrigations (Khode *et al.*, 2000).

It is advisable to apply 15 tonnes of farmyard manure per hectare at the time of field preparation. The optimum values of **nutrients** recommended for *C. roseus* in tissue nutrient analysis are nitrogen 4.9-5.4 per cent, phosphorus 0.4-.0.5 per cent, potash 2.9-3.6 per cent, calcium 1.4-1.6 per cent, magnesium 0.4-0.5 per cent, iron 95-150 ppm, manganese 165-300 ppm, zinc 40-45 ppm, copper 5-10 ppm and boron 25-40 ppm (Dole and Wilkins, 1988; Joiner *et al.*, 1981; Peterson, 1982). It also responds well with the application of inorganic fertilizer of 2.5 g nitrogen, 5 g phosphorus and 7.5 g potash per square metre area. Entire phosphorus and potash are applied as a basal dose before planting but nitrogen is applied @ 5 g/m^2 as top-dressing after two months of transplanting. The highest foliage and root yields of periwinkle were obtained with 20:30:30 kg NPK/ha (recommended dose) though significant increase in yield was obtained even with 10:15:15 kg NPK/ha, *i.e.* half recommended dose (Khode *et al.*, 2000). For three rainy seasons, transplanted seedlings of *C. roseus* were treated with N or P at 0, 20 or 40 kg/ha and harvested at 11 months. N fertilization did not increase foliage, root or alkaloid yield but P at 40 kg/ha increased the concentration of alkaloid (0.13-0.15 per cent) in roots over two seasons (Dahatonde and Joshi, 1982). Pareek *et al.* (1985) on average soils of Delhi reported 80 kg of N/ha as ideal for irrigated lands and 40 kg/ha for rainfed crops when applied in three split doses, *i.e.* at planting and at 45 and 90 days of transplanting. Hedge (1988) obtained optimum root yield (800 kg/ha) with 100 kg N+40 kg P_2O_5+50 kg K_2O and leaf yield (1,800 kg/ha) with 80 kg N+30 kg P_2O_5+60 kg K_2O. Field experiments were conducted to control lime-induced Fe-chlorosis, Zn-dwarfness and N-deficiency in a chlorosis susceptible ornamental plant, *C. roseus*. FeEDDHA, ZnEDTA and urea were added to the soil or to the foliage. Soil application of fertilizers was superior to the foliar application in correcting chlorosis, increasing chlorophyll content and improving foliage growth and flowering. Effect of FeEDDHA applied to the soil started to diminish after 6-8 weeks, therefore, frequent application of Fe-chelate was advised. Fe application was found to enhance the uptake of Zn, Cu and Mn by plants. The use of total Fe contents as an index for Fe status in plants was reliable when these contents were based on fresh weight and was misleading when was based on dry weight. The use of dry weight also led to erroneous results regarding the interrelationships between micronutrients. The determination of active iron (HCl extractable Fe) improved the diagnosis of Fe status in plants. Chlorotic leaves had less active Fe than green leaves (AboRady, 1988).

Growth and Flowering

Periwinkle flowers round the year depending on the care and attention of the crop. *Catharanthus roseus* seeds were soaked prior to sowing in 10, 20 or 50 per cent aqueous extracts of *Ocimum sanctum* leaves, and the same concentrations were when applied thrice to plants from treated seeds after 30, 60 and 270 days of growth, flowering was found induced earlier by all the treatments with shortened internodes and increased number of leaves. Extracts at 10 and 20 per cent increased the average number of flowers per plant throughout the year (Bhandari *et al.*, 1970-1971).

Effect of various growth retardants for dwarfing the periwinkle plants exhibited A-Rest (amcymidol) and cycocel (chlormequat chloride) being more effective than Floral (ethephon) but less effective than Sumagic (uniconazole) whereas B-nine did not have any effect (Adriansen, 1985).

In an experiment *C. roseus* seeds were irradiated with 10, 20 or 30 krad before planting. Survival, plant height and the number of fruits/plant were greatest in plants from untreated seeds. Flowering was earliest and pollen sterility greatest in plants from the 30-krad treatment and the number of seeds/fruit was greatest in the 10-krad treatment. Irradiation had no marked effects on plant morphology (Bose *et al.*, 1972). The plug-grown *Catharanthus roseus* 'Cooler Pink' were transplanted into jumbo six packs and when the seedlings had attained ~ 5 cm in height or width, Kuehny *et al.* (2001) sprayed them with 1000/800, 1250/1250 or 1500/5000 mg l^{-1} of chlormequat/daminozide solutions which reduced the plant height significantly.

Post Harvest

Since the **loose flowers** are used only for traditional use, therefore these are plucked early in the morning when opening but the flowers should not be moist with dew, and filled in the baskets loosely for marketing. Tight packing will damage the flowers and will make these unfit for marketing. If need be, these can be stored at 7-10 °C for 3-4 days but before sending to the market these will have to be brought to room temperature so that these may not desiccate. Flowers of *Catharanthus roseus* were when collected at four different stages of development (from tight bud to senescing flower) and studied for changes in total soluble carbohydrates, starch and reducing sugar contents, and amylase and invertase activities in the petals, it was found that flower development, total soluble carbohydrate and reducing sugar contents showed initial increases, followed by decreases at the 4th and 3rd stages, respectively though starch content and enzyme activities gradually declined to the 4th stage (Mehta *et al.*, 1998). Pot plants of *Catharanthus roseus* were found most susceptible to ethylene when were exposed to 3 ppm at 20 °C for one day (Woltering, 1984). Effects of plant growth regulators and mineral salts on the postharvest life of white flowers of *Catharanthus roseus* were investigated by placing 12.5 cm long twigs bearing mature flowers in 2 ppm GA_3 + 2 per cent sucrose, 2 ppm kinetin + 2 per cent sucrose or 120 ppm calcium nitrate + 150 ppm cobalt nitrate + 2 per cent sucrose which affected the fresh weight, protein and amino acid contents of flowers, and promoted vase life, however, the combinations containing plant growth regulators were more effective than those containing only mineral salts (Joseph *et al.*, 1996).

The crop becomes ready for harvest of **roots** after one year when planted for medicinal purpose. Two leaf strippings can be taken, first one after six months and the second after nine months of sowing. Aerial parts are cut and the soil is dug up for recovery of roots. Fruits for the purpose of further sowing should be collected without damage.

Insect-Pests and Diseases

Insect-pests which infest on periwinkles are aphids, whiteflies, caterpillars, grasshoppers and nematodes. The

aphids are usually found on young leaves and flower buds. These suck the sap of foliage and buds and cause retarded growth of the plants with poor quality flowers. It also acts as a vector for viral diseases. Two spray of Malathion or Metasystox @ 1.5-2.0 ml per litre of water at 15-20 days interval can control it. The **caterpillars** and **semiloopers** are particularly active in summer. Semiloopers feed on flowers, floral buds and leaves and roll the leaves to live in and feed. Its control is the same as for aphids. **Grasshoppers** are particularly active in summer and rainy seasons. They generally feed on leaves, floral buds and flowers through their biting and chewing mouth parts. These are mostly found in closed enclosures like polyhouses. Monocil spray or the treatments given to aphids will control these pests also. **Whiteflies** mostly feed on young foliage and underside of the leaves. These retard plant growth and cause poor quality of flowers by sucking the sap of foliage and buds. They may also act as a vector for viral diseases. Its spread is very fast if not checked early. Regular spray of acetamiprid @ 3 g/l on the plant and beneath the surface of leaves will control the whitefly infestation. Placing of yellow sticky strips will also trap the flies. **Root-knot nematodes** (*Meloidogyne incognita* and *M. javanica*) have been found infesting the roots of periwinkle in India (Alam, 1981). Their severe root-invasion causes severe plant stunting and leaf deformities. Control measures include sterilization of soil by steam or drenching by formaldehyde @ 2 per cent before planting, but afterwards incorporation of Furadan granules in the soil. Application of *neem*-based products and crop rotation with marigold can also minimize the nematode infestation.

Periwinkle is susceptible to a number of fungal **diseases** such as leaf spot, foliage blight, top rot, stem rot, die-back, mycoplasma-like organisms (MLOs), and other viral diseases such as 'cucumber mosaic virus' (CMV) and 'bean yellow mosaic virus' (BYMV).

In *Catharanthus roseus* plantings, Shanta Kumari and Chandrasekharan Nair (1981) recorded *Colletotrichum gloeosporioides* causing **leaf spots**. Symptoms first appear as dark brown to black girdling lesions on the twigs. When lesions occur at the ground level, entire runners die. Where healthy twigs touch the soil or infected plant parts, new lesions are developed. Within a few weeks, the disease spreads to stems and leaves and causes large sections of the bed to die. As long as cool and damp conditions remain in the field, the disease is found spreading in the planting. The spores of the fungus disseminate primarily by splashing and flowing water. Acervuli containing masses of spores and dark setae are observed within lesions (Sharma *et al.*, 2013). *Colletotrichum capsici* has been observed causing leaf and petiole **anthracnose** and **twig blight** in periwinkle (Mhaeskar, 1967). *Colletotrichum dematium* which causes **twig blight** was isolated from *Catharanthus roseus* 'Pretty in Pink'. Symptoms consist of wilting of the shoot tips followed by chlorosis and ultimate necrosis of the shoot tips. Necrotic tissues are typically covered with masses of acervuli with setae. The isolated fungus produces falcate conidia as well as abundant sclerotia on the host and culture, which is typical of *C. dematium*. Dithane M-45 at 0.2 per cent applied as a protective spray provides significant disease control (McMillan and Graves, 1996; Khatua *et al.*, 1981). Manoharachary (1975) reported *Alternaria longipes* ('EII' & 'EV'), and Singh *et al.* (1967) *Myrothecium roridum*, both causing **leaf spot** diseases in periwinkle. *Leveillula taurica* causes **powdery mildew** (Butler *et al.*, 1931; McRae, 1924; Mundkur, 1938) in this crop which can be controlled through use of sulphur. *Macrophomina phaseolina* has been recorded casusing **foot and root rot** (Khatua *et al.*, 1981), *Ophiobolus catharanthicola* **shoot blight** (Rao and Pande, 1980), and *Phyllosticta vincae* **twig blight** and **leaf spot** diseases (Tiwari *et al.*, 1978). **Foliage blight** (*Phytophthora parasitica* var. *nicotianae*) is favoured by high humidity, surface moisture and temperature (Gill *et al.*, 1977) and infects *Catharanthus roseus* plants in the nursery. Symptoms of the disease are grayish-brown discoloration of shoot tips and foliage followed by wilting, necrosis and rapid plant death. One shoot or the entire plant may be initially infected. Shoot pruning increases its incidence (Keim, 1977). The resistance of *Catharanthus roseus* cvs 'Little Blanche', 'Little Bright', 'Little Linda', 'Little Pinkie', 'Little Pretty' and 'Cooler' was when evaluated to *P. parasitiaca*, 'Little Linda' exhibited highest resistance to the pathogen (Rivera *et al.*, 2000). The systemic fungicide metalaxyl (Ridomil) is effective against this disease. Soil solarization also significantly reduces the relative area under the disease progress curve and final blight incidence (McGovern nd Garibaldi, 1996). Symptoms of **web blight** (*Rhizoctonia solani* 'AG-1'), which together causes even **root** and **stem rot**, were observed on the leaves of many plants of *C. roseus* as confirmed on the basis of characteristics of mycelium cultivated *in vitro* (Bertetti nd Garibaldi, 2006), and the stem rot is effectively controlled by Benlate or Terraguard drench (Chase, 1993). *Rhizopus stolonifer* fungus causes water-soaked lesions on stems and leaf tissues, rapidly followed by a **soft rot** spreading from the injured site, resulting into collapse and death of leaves and unthickened stems. Nearly 80 per cent of the periwinkle plants in experimental plots grown for the production of anti-cancer alkaloids were affected by a severe **die-back** caused by*Pythium butleri*. Affected leaves and twigs yielded an isolate of *P. butleri*, the pathogenicity of which was confirmed experimentally and was associated with the production of toxic metabolites (Janardhanan *et al.*, 1977). Periwinkles suffer with **top rot** (*Pythium debaryanum*) and **foot rot** (*Sclerotium rolfsii*) diseases during cultivation. Spraying such plants with oxytetracycline at 125 ppm thrice at fortnightly intervals results in their recovery (Chakravarty *et al.*, 1976). *Phytophthora colocasiae* causing **root rot** (Butler and Bisby, 1931; Dastur, 1916; Thomas and Ramakrishnan, 1948) has also been recorded on *C. roseus*. When the seeds of *Catharanthus* were inoculated with *Pseudomonas fluorescens* strain E6, Yuen and Schroth (1986) recorded increased growth and fresh weight (18-41 per cent greater) after 3-4 weeks of treatment.

Plants infected with '**mycoplasma-like organisms**' (MLOs) show symptoms of wilting, progressive dwarfing of flowers, crinkling and yellowing of the leaves and eventual complete inhibition of growth and flower production. Examination of affected periwinkle plants under the electron microscope reveal presence of MLOs in the phloem which are oval, mycelial or thread-like, all surrounded by a 3-layer membrane. In serological tests the pathogen of 'vinca yellows' shows affinities with 'clover phyllody' and 'potato round leaf'. *Catharanthus roseus* is the highly sensitive indicator for *Spiroplasma citri*, and shows the symptoms of wilting, stunting, marginal chlorosis, *vis-a-vis* a few and small flowers. This is transmitted by the leafhoppers (*Scaphytopius nitrides* and *Neoaliturus tenellus*). Periwinkle plants inoculated with 'eggplant little leaf disease' (a yellowing disease) were when heat-treated at 40 °C for 3-15 days or at 44 °C for 2-10 days and the effects assessed six months later showed that 44 °C for 10 days eliminated the pathogen completely from the stems and roots while 40 °C for 9-15 days or at 44 °C for 6-8 days was only partially effective (Anjaneyulu and Ramakrishnan, 1973). *Catharanthus roseus* was found exhibiting little leaf symptoms such as phyllody. These pathogens were present in the phloem elements of leaves, phyllodes, stems and roots of infected plants. Nymphs of *Cicadellidae* spp., *Hishimonus phycitis* and *Orosius albicinctus* (*O. orientalis*) are the vectors which transmit the 'little leaf disease' from diseased to healthy plants (Raj and Lakshmanan, 1982). MLOs were readily found in the sieve tubes of affected tissues of periwinkle. Symptoms after graft transmission included virescence and phyllody followed by proliferation of branches and dwarfing of foliage. The initial 'witches broom' effect comes from branches arising from the modified stamens of phylloid flowers. At high temperature the 'witches broom' did not develop, and leaves and flowers were normal in size and colour although flower proliferation may occur (McCoy and Thomas, 1980). The development of virescent flowers due to infection of different strains of mycoplasmas could be correlated with marked changes in the levels of endogenous cytokinins. In all cases the levels in the roots and mature leaves decreased while that in the flowers increased. Three peaks of cytokinin activity which co-eluted with glucosylzeatin, ribosylzeatin and zeatin were detected in these organs. No qualitative differences in the cytokinin complement of the infected plants were detected. At present it is not known whether or not the mycoplasmas used synthesized cytokinins which upset the cytokinin balance of *C. roseus* plants or cause changes in the synthetic capacity and/or translocation pattern of the infected plants (Davey *et al.*, 1981).

The **viruses** also infect periwinkle crop. The most common one is '**cucumber mosaic virus**' (CMV). It is also affected by '**phyllody'** and '**little leaf'** diseasees (Jhon, 1957). It is a natural host of BBMV serotype II (Shukla *et al.*, 1980). A common mosaic disease of periwinkle was shown to be related to CMV on the basis of physical properties, host range, particle morphology and serology (Nariani *et al.*, 1978). Effect of herbicides on the morphology of CMV particles was studied in periwinkle plants. Particles of this virus were reduced in size when treated with Monuron (100 ppm), Simazine (100 ppm), Nitrofen (10 ppm) and Treflan (10 ppm), though actual size was unaffected by 2, 4-D and Alachlor. The structure of the virions was deformed in treated host plants. When Monuron, Simazine, Nitrofen or Treflan was mixed with purified CMV, they caused a reduction in size and change in structure of the virus particles (Rao and Roychaudhuri, 1978). Viruses can be controlled through proper sanitation with appropriate weed control in the field, exclusion of the vector from infected plants, and control of thrips with insecticides. Rouging-off infected plants after the flowering is over, and removal of weeds which harbour these vectors will significantly minimize the problem. Procurement of the plants should be ensured only from certified virus-free mother stocks. Thrips can be monitored in the growing area with the use of yellow sticky cards. Cards should be checked on weekly basis and insecticides applied if needed. To prevent resistance of insecticides on thrips population, its application should be rotated with pyrethroids, carbamates, chlorinated hydrocarbons, organophosphates and soaps.

References

AboRady, M.D.K. 1988. Effect of iron deficiency on growth, micronutrient status and chlorophyll content of *Vinca rosea* grown in calcareous soils. *Arid Soil Res. and Rehabil.*, **2**(4): 275-283.

Adriansen, E. 1985. Kemisk Vaekstregulering, In: *Potteplanter I-Produktion*, Metoder, Midler, pp. 142-162.

Alagumalai, K., C.S.S. Hepsyba. and P. Ramaraj, 1991. Effect of leaf extract of *Vinca rosea* on the host plant cowpea infected with *Meloidogyne incognita. Current Nemat.*, **2**(2): 109-112.

Alam, M. M. 1981. Some additions to the hosts of root-knot nematodes. *Indian Phytopath.*, **34**(4): 243.

Anjaneyulu, A. and K. Ramakrishnan, 1973. Thermotherapy of eggplant little leaf disease. *Mysore J. agric. Sci*, **7**(3): 423-428.

Anwar, M., P.V. Singh and K. Subramanyam, 1988. Safe limits of soil salinity for three important medicinal plants. *Intern. J. Trop. Agric.*, **6**: 125-128.

Arcamone, F., G. Cassinelli and A.M. Casazza, 1980. New antitumor drugs from plants. *J. Ethnopharmac.*,**2**(2): 149-160.

Atta Alla, H.K., 1997. *In vitro* seed germination, callus formation and shoot proliferation of *Vinca rosea* L. *Ann. agric. Sci.*, **35**(3): 1555-1566.

Bertetti, D. and A. Garibaldi, 2006. Presence of *Rhizoctonia solani* Kuhn on *Vinca rosea* L. (*Catharanthus roseus* (L.) G. Don) in a public garden of Turin (Northern Italy). *Informatore Fitopatologico*, **56**(9): 33-35.

Bhandari, M.M., J.R. Mathur and H.C. Sharma, 1970-1971. Holy basil promotes flowering in periwinkle. *Rajasthan Agric.*, **10/11**: 11-14.

Bose, S., S. Gupta and B. Banerjee, 1972. Effect of treatment of X-rays on plant growth, flowering and fruiting in *Catharanthus roseus* (L) G. Don syn. *Vinca rosea* L. *Sci. and Cult.*, **38**(4): 196-198.

Butler, E.J. and G.R. Bisby, 1931. The fungi of India. Imperial Council of Agricultural Research. *Indian Science Monograph* 1, XVIII, Kolkota.

Chakravarty, D.K., D.C. Khatua and T.K. Das, 1976. Pathological problems in the cultivation of *Vinca rosea. Indian Agric.*, **20**(4): 289-295.

Chase, A.R. 1993. Fungicides for control of *Rhizoctonia* on potted ornamentals. *CFREC-Apopka Research Report.* http://ufdc.ufl.edu/UF00065835/00001.

Choi, P.S., Y.D. Kim, K.M. Choi, H.J. Chung, D.W. Choi and J.R. Liu, 2004. Plant regeneration hairy-root cultures transformed by infection with *Agrobacterium rhizogenes* in *Catharanthus roseus. Plant Cell Reports*, **22**:828-831.

Corley, W.L. 1981. Low maintenance annual flower evaluations. *Research Report*, College of Agriculture Experiment Stations, Georgia University, No. 368, p. 15.

Dahatonde, B.N. and B.G. Joshi, 1982. Studies on the effect of nitrogen and phosphorus on *Vinca rosea* syn. *Catharanthus roseus. Proc. Nat. Sem. on Medicinal and Aromatic Plants*, Maharashtra, pp. 11-13.

Dastur, J.F. 1916.*Phytophthora* on *Vinca rosea. Mem. Dept. Agric. Indian Bot. Ser.*, **8**: 233-242.

Davey, J.E., J. van Staden and G.T.N. de Leeuw, 1981. Endogenous cytokinin levels and development of flower virescence in *Catharanthus roseus* infected with mycoplasmas. *Physiol. Pl. Path.*, **19**(2): 193-200.

Dionello Basta, S.B. and F. Basta, 1984. Germination and growth inhibitors in plants used in folk medicine. *Ciencia e Cultura*, **36**(9): 1602-1606.

Dole, J.M. and H.F. Wilkins, 1988. Tissue Analysis Standards. *Minnesota State Florists Bulletin* (Univ. of Minnesota), **37**(6):10-13.

Dwivedi, S.C. and C.J. Awasthi, 1994. Relative efficacy of plant extracts on beetles *Maladera insanabilis* (Coleoptera: Scarabaeidae). *J. Ecotoxicol. and environ. Monitoring*, **4**(1): 77-79.

Dwivedi, S.C. and A. Kumar, 1999. Ovicidal activity of six plant leaf extracts on the eggs of *Corcyra cephalonica* Stainton (Lepidoptera: Pyralidae). *U. P. J. Zool.*, **19**(3): 175-178.

Everard, B. and B.D. Morley, 1970. *Wild Flowers of the World*, p.71. Peerage Books, London, UK.

Filippini, R., R. Caniato, A. Piovan and E.M. Cappelleti, 2003. Production of anthocynins by *Catharanthus roseus. Fitoterapia*, **74**: 62-74.

Flory, W.J. 1944. Inheritance studies of flower colour in periwinkle. *Proc. Amer. Soc. Hort. Sci.*, **44**:525-526.

Ganguly, L.K. 1994. Fungitoxic effect of certain plant extracts against rice blast and brown spot pathogen. *Environ. and Ecol.*, **12**(3): 731-733.

Garg, S.P., M.R.K. Sherwani, A. Arora, R. Agarwal and, M. Ahmad, 1987. Ricinoleic acid in *Vinca rosea* seed oil. *J. Oil Technol.Assoc. India*, **19**(3): 63-64.

Gill. H.S., O.K. Ribeiro and G.A. Zentmeyer, 1977. *Phytophthora* blight of periwinkles in the Coachella Valley of California. *Pl. Dis. Rep.*, **61**: 560-561.

Guimarães, G., L. Cardoso, H. Oliveira, C. Santos, P. Duarte and M. Sottomayor, 2012. Cytogenetic characterization and genome size of the medicinal plant *Catharanthus roseus* (L.) G. Don**.** *AoB Plants*. Published online. doi: 10.1093/aobpla/pls**002**.

Hay, R. and K.A. Beckett, 1971. *Vinca*. In: *Reader,s Digest Encyclopaedia of Garden Plants and Flowers*, pp. 732-733. The Reader's Digest association Limited, London, Great Britain.

Hedge, D.M. 1988. Response of periwinkle [*C. roseus* (L.) G. Don] to nitrogen, phosphorus and potassium fertilization. *Agric. Res. J. Kerala*,**26**: 227-233.

Howe, T.K.and S.S. Woltz, 1981. Symptomology and relative susceptibility of various ornamental plants to acute airborne sulfur dioxide exposure. *Proc Fla St. hort. Soc.*, **94**: 121-123.

Janardhanan, K.K., M.L. Gupta and A. Husain, 1977. Pythium die-back, a new disease of *Catharanthus roseus. Indian Phytopath.*, **30**(3): 427-428.

John, V.T., 1957. A review of plant viruses in India. *The Madras Univ. J.*, **27**(B): 373-450.

Joiner, J.N., C.A. Conover and R.T. Poole, 1981. Nutrition and fertilization. In: *Foliage Plant Production* (ed. Joiner, J.N.), pp. 229-268. Prentice-Hall, Englewood Cliffs, New Jersey, USA.

Joseph, L., K.V.S. Kumar and P.M. Mehta, 1996. Effect of growth regulators and chemical treatments on post-harvest life of *Ixora coccinia* and *Vinca rosea* flowers. *Adv. Pl. Sci.*, **9**(2): 187-193.

Keim, R. 1977. Foliage blight of periwinkle in southern California. *Pl. Dis. Repr*, **61**(3): 182-184.

Khatua, D.C., D.K. Chakravarty and C. Sen, 1981. Some new diseases of vegetable, ornamental and plantation crops. *Indian Phytopath.*, **34**: 231-232.

Khode, P.P., P.U. Ghatol, S.A. Bhuyar, D.D. Deo and V.M. Dhumal, 2000. Performance of *Vinca rosea* (*Catharanthus* spp.) to different fertility and irrigation levels. *Agric. Sci. Digest*, **20**(2): 135-136.

Kuehny, J.S., A. Painter and P.C. Branch, 2001. Plug source and growth retardants affect finish size of bedding plants. *HortSci.*, **35**(2): 321-323.

Kulkarni, R., K. Baskaran and N.Suresh, 1999. Inheritance in periwinkle; leaf pubescence and corolla colour. *J. Herbs, Spices and Medic. Pl.*, **6**: 85-88.

Manoharachary, C. 1975. New host records of some interesting fungi from India. *Sci. & Cult.*, **41**: 237-238.

McCoy, R.E. and D.L. Thomas, 1980. Periwinkle witches' broom disease in South Florida. *Proc. Fla St. hort. Soc.*, **93**: 179-181.

McGovern, R.J. and J.P. Begeman, 1996. Reduction of *Phytophthora* blight of Madagaskar periwinkle in the landscape by soil solarization. *Proc.Fla St. hort. Soc.*, **108**: 58-60.

McMillan, R.T., Jr. and W.R. Graves, 1996. Periwinkle twig blight caused by *Colletotrichum dematium* on *Catharanthus roseus* L. *Proc.Fla St. hort. Soc.*, **109**: 19-20

McRae, W. 1924. Economic Botany Part III. Mycology, *Annual Report of Board of Scientific Advice,* India, pp. 31-35.

Mehta, P.M., K.V.S. Kumar and Joseph Lucyamma, 1998. Some biochemical changes during development and senescence of *Vinca rosea* and *Ixora coccinea* flowers. *Adv. in Pl. Sci.*, **11**(2): 309-312.

Mhaeskar, V.G. 1967. A new twig blight disease of *Vinca rosea* from India. *Pl. Dis. Rep.*, **51**: 480-481.

Mitrev, A. 1976. Laboratory investigations on the biological properties of *Vinca rosea* seeds produced in Bulgaria. *Rasteniev"dni Nauki*, **13**(1): 85-90.

Mohsen, Z.H., M.J. Abdul Latif, B.M. Al-Chalabi and A. Al-Naib, 1989. Insecticidal activity of *Vinca rosea* against *Culex quinquefasciatus*. *J. Biolog. Sci. Res.*, **20**(3): 437-446.

Mundkur, B.B. 1938. *Fungi of India, Supplement I. ICAR Science Monograph*, **12**:1-54.

Murugavel, T., A. Ruknudin, S. Thangavelu and M.A. Akbarsha, 1989. Antifertility effect of *Vinca rosea* (Linn.) leaf extract on male albino mice - a sperm parametric study. *Curr. Sci.*, **58**(19): 1102-1103.

Nariani, T.K., S.P. Raychaudhuri, D.R. Rao, S.M. Viswanath and N. Prakash, 1978. Purification and serology of a virus causing mosaic in *Vinca rosea* L. *Curr. Sci.*, **47**(7): 232-233.

Noble, R.L., C.T. Beer and J.H. Cuts, 1958. Role of chance observations in chemotherapy: *Vinca rosea*. *Ann. N.Y. Acad. Sci.*, **76**: 882-894.

Pareek, S.K., M.L. Maheshwari and R. Gupta, 1985. Cultivation of periwinkle in North India. *Indian Hort.*, **30**(3): 9-12.

Pareek, S.K., M.L. Maheshwari and R. Gupta, 1991. *Chemical Weed Control and Water Management in Periwinkle, Vetiver and Palmarosa Oil Grass* (ed. Ray Choudhuri, S.P.), pp. 167-170. Today and Tomorrow Printers and Publishers, New Delhi.

Peterson, J.C. 1982. Monitoring and managing nutrition. Part IV: Foliar analysis. *Ohio Florists Assocn Bull.*, No. 632, pp. 14-16.

Qing, Li and F. Kobza, 1997. Further studies of technology in *Catharanthus roseus* (L.) G. Don (*Vinca rosea* L.). Roczniki Akademii Rolniczej w Poznaniu, Ogrodnictwo. Faculty of Horticulture, Mendel Univ. of Agriculture and Forestry, Brno, Czech Republic, **25**: 65-72.

Raghavaiah, G. and M. Jayaramaiah, 1987. Antifungal activity of selected plant extracts against the white muscardine fungus. *Curr. Res.* (Univ. of Agric. Sci., Bangalore), **17**(5): 62-64.

Raj, S.P. and M. Lakshmanan, 1982. Vectors and vector-borne diseases. *Proc. All India Symp.*, South India, pp. 143-150.

Rao, D.R. and S.P. Raychaudhuri, 1978. Effect of herbicides on the morphology of 'cucumber mosaic virus' particles when studied under electron miscroscope. *Nat. Acad. Sci. Letters*, **1**(4): 125-126.

Rao, V.G. and A. Pande, 1980. An undescribed species of *Ophiobolus* Riess on periwinkle. *Sci. & Cult.*, **46**(8): 295-296.

Rivera, M.C., O.S.F. Delfino, E.R. Wright and A. Rivera Gonzalez, 2000. Screening of cultivars of *Catharanthus roseus* in response to *Phytophthora parasitiaca*. *Horticultura Argentina*, **19**(47): 52-55.

Shanta Kumari, P. and M. Chandresekharan Nair, 1981. Some new host records for *Colletotrichum gloeosporioides* in India. *Indian Phytopath.*, **34**: 402-403.

Sharma, P., P.D. Meena and Y. P. Singh, 2013. New record of twig blight on *Catharanthus roseus* in India. *African J. Microbiol. Res.*, **7**(38):4680-4682.

Shivpuri, A., O.P. Sharma.and S.L. Jhamaria, 1997. Fungitoxic properties of plant extracts against pathogenic fungi. *J. Mycol. and Pl. Path.* **27**(1): 29-31.

Shukla, D.D., D.S. Teakle and K.H. Gough, 1980. Periwinkle, a latent host for broad bean wilt and cucumber mosaic viruses in Australia. *Pl. Dis.*, **64**(8): 802-803.

Singh, D.V., R.S. Singh and H.S. Srivastava, 1967. *Myrothecium* leaf spot of *Vinca rosea* L. *Sci. & Cult.*, **33**: 70-71.

Sreenivasan, N., R.P. Sridhar and P. Gnanamurthy, 2003. Efficacy of some plant extracts against *Plutella xylostella* L. on cabbage. *South Indian Hort.*, **51**(1/6): 183-185.

Sreevalli, Y., R.N. Kulkarni and K.Baskaran, 2002. Inheritance of flower colour in periwinkle: Orange-red corolla and white eye. *J. Hered.*, **93**: 55-58.

Srivastava, H.K. 2003. Genetic and biotechnological approaches for enhanced indole alkaloid production in periwinkle (*Catharanthus roseus*). *Recent Progress in Medicinal Plants.* Vol. 4: *Biotechnology and Genetic Engineering* (eds Govil, J.N. *et al.*), pp. 399-413. *Studium Press, LLC, USA.*

Stearn, W.T. 1975. A synopsis of the genus *Catharanthus* (Apocynaceae). In: *The Catharanthus Alkaloids* (eds Taylor, W. and N.R. Farnsworth), pp. 9-44.New York, USA.

Tafur, S., W.E. Jones, D.E. Dorman, E.E. Logsdon and G.H. Svoboda, 1975. Alkaloids of *Vinca rosea* L. (*Catharanthus roseus* G. Don). XXXVI: Isolation and characterization of new dimeric alkaloids. *J. Pharmaceut. Sci.*, **64**(12): 1953-1957.

Talha, M., A.S. Radwan and S. Negm, 1975. The effect of soil moisture deficit on growth and alkaloidal content of *Catharanthus roseus* G. Don. *Curr. Sci.*, **44**(17): 614-616.

Thomas, K.M. and T.S. Ramakrishnan, 1948. Studies in the genus *Phytophthroa* II. *Proc. Indian Acad. Sci.*, **27B**: 55-73.

Thomas, Paul A. and Joyce Latimer, 1995. Effects of soil media composition, nutrient charge and nitrogen source on growth of *Catharanthus roseus*. *J. Pl. Nutr.*, **18**(10): 2127-2134.

Tiwari, D.P., A.R. Lodh and A.I. Jacob, 1978. Diseases of *Catharanthus roseus* (= *Vinca rosea*) in Jabalpur, M.P. *Nat. Acad. Sci. Letters*, **1**(5): 163-164.

van der Heijden, R., D.I. Jacobs, W. Snoeijer, D. Hallard and R.Verpoorte, 2004. The *Catharanthus* alkaloids: Pharmacology and biotechnology. *Curr. Medicinal Chem.*, **11:** 1241-1253.

Vimala, R., K. Sivaprakasam and K. Seetharaman, 1993. Effect of plant extracts on the incidence of brinjal wilt. *Madras agr. J.*, **80**(7): 409-412.

Vitkare, D.G. and B.N. Phadnawis, 1988. Performance of *Vinca rosea* and *Solanum viarum* under Akola conditions. *Ann. Pl. Physiol.*, **2**(1): 104-106.

Woltering, E.J. 1984. Influence on ornamentals tested. Many species susceptible to ethylene. *Vakblad voor de Bloemisterij*, **39**(4): 112-113.

Yuen, G.Y. and M.N. Schroth, 1986. Interactions of *Pseudomonas fluorescens* strain E6 with ornamental plants and its effect on the composition of root-colonizing microflora. *Phytopath.*, **76**(2): 176-180.

13

Codiaeum variegatum (Family: Euphorbiaceae)

B. Sathyanarayana Reddy, S.Y. Chandrashekar, S.K. Nataraj and P.M. Munikrishnappa

(**Common names**: Croton, Joseph's coat, Rushfoil, South sea laurel)

Introduction and Origin

The former generic name comes from the Greek '*kroton*' meaning 'tick' which refers to the shape of the seeds of certain species, and its present name *Codiaeum* derives from the Indonesian (Moluccan) vernacular name *kodiha*. It originates from Malaysia to the Pacific and northern Australia *via* India, Sri Lanka and the Western Pacific Ocean Islands. Croton, a popular garden plant, is an evergreen tropical shrub most suitable for garden adoration and as potted plant, *vis-à-vis* for making hedges and in group plantings. It is valued for its exquisite foliage colours varying from green to variegation (dots, streaks and patches in various colours and patterns) and irregular leaf forms including twisting. In big parks, it can also be planted in continuous patches to make a croton garden. It comes up very well both in tropical and subtropical climates. In tropical climates, it enjoys humid warm conditions and exhibits its beautiful colours to the fullest extent though in sub-tropical climates these can survive only when planted in groups or at sheltered situation otherwise these will not be able to survive in India due to the harsh dry winds of summer months (May-June) and biting cold of winter (December to mid-February).

The plant is also well reputed for the production of valuable secondary metabolites of alkaloids, terpenes

Codiaeum variegatum

***C.v.* var. 'Blume'**

Red Croton

Croton Planting

and flavanoids in nature (Nasib *et al.*, 2008). Several species of croton produce red sap, while some species produce proanthocyanidins and/or alkaloids. Contact with its foliage latex may aggravate skin allergies. The leaf extracts have purgative, sedative, antifungal, anti-amoebic and anti-cancerous activities. The plant extracts are used to treat absence of menstruation (amenorrhea). San Blas Indians used to treat body aches and eye diseases. It is being used to treat jaundice, various liver ailments and as amoebicidal. Bark is used as a pink dye in Bangladesh. Some of the other common medicinal uses include treatment of cancer, constipation, diabetes, digestive problems, dysentery, external wounds, fever, hypercholesterolemia, hypertension, inflammation, intestinal worms, malaria, pain, ulcers and weight-loss. Several other effects of croton substances have been registered, including anti-hypertensive, anti-inflammatory, antimalarial, antimicrobial, antispasmodic, antiulcer, antiviral and myorelaxant (Antonio *et al.*, 2007).

Botany

It is an evergreen shrub growing up to 4 metres tall. Its leaves which diversify both in shapes and colours are ornamentally thick, shiny, 10-30 cm long, glabrous, ovate to narrowly linear to oak- or fiddle-shaped, sometimes waved, twisted and constricted, entire or lobed or spirally green or tinted and variegated. Leaf colour is basically green, spotted, blotched or suffused to various shades of green, yellow and often with shades of red, white, purple, orange or pink, being usually strongest along the veins. The inflorescence length ranges from small to long racemes with male and female flowers on separate inflorescences. The male flowers are white with scale-like petals ranging from 3-6 (mostly five) and 20-30 stamens. The female flowers are yellowish having 5 sepals but without petals. The fruit is a capsule measuring 9 mm in diameter and the ovary is 3-celled with one ovule in each cell.

The study by Deng *et al.* (2010) showed that the chromosome numbers of seven cultivars of *Codiaeum variegatum* (L.) ranged from 2n = 24, 30, 60, 66, 70, 72, 76, 80, 82, 84 and 96, which shows that the cultivars are polyploid and hyperaneuploid. Olusola Ogunwenmo *et al.* (2007) reported that cultivars 'Spirale', 'Royal', 'Broad Spotted Guinea' and 'Punctatum' had diploid chromosome number of 2n=60, while 'Sunray' and 'Royal-like' possessed 2n= 30 and 24, respectively. Genetic mutation and chromosome variability could account for the wide morphological vitiation and morpho-types in this plant. Crotons often produce spontaneous shoots that are different from parent plant and these can be isolated and multiplied.

Species and Varieties

There are some 15 species, *viz. Codiaeum affine* Merr., *C. bractiferum* (Roxb.) Merr., *C. ciliatum* Merr., *C. finisterrae* Pax & K.Hoffm., *C. hirsutum* Merr., *C. ludovicianum* Airy Shaw, *C. luzonicum* Merr., *C. megalanthum* Merr., *C. oligogynum* McPherson, *C. peltatum* (Labill.) P.S.Green, *C. stellingianum* Warb., *C. trichocalyx* Merr., *C. variegatum* (L.) Rumph., *etc.* but only *Codiaeum variegatum* is in general cultivation and that too the variety *pictum* is very common with a greater number of forms though many exist with Latin names as if these are themselves species. It is native of Malaysia and Polynesia, grows up to 3 metres with leathery, glossy and basically ovate leaves having spots, blotches or suffused with shades of yellow, orange, pink and red, being stongest along the veins. There are several hundred cultivars selected and bred for their foliage. Among them the selected varieties are 'Andreanum', 'Ann Rutherford', 'Apple Leaf', 'Acubifolium', 'Bangalore Beauty', 'Bangkok Queen', 'Bravo', 'Bruxellense', 'Carnival', 'Carrierei', 'Craigii', 'Corazon Deoro', 'Crispum', 'Delaware', 'Disraeli', 'Dr. B.P. Pal', 'Dread Locks', 'Elite', 'Excellent', 'Fantasy', 'Fascination', 'Fire', 'Fred Sander', 'Gandhiji', 'Gloriosum', 'Gloriosum Superbum', 'Glory', 'Golden Ring', 'Happiness', 'Harvest Moon', 'H.D. Maity', 'Imperialis', 'Indira Gandhi', 'King George', 'Magesticum', 'Nevada', 'Norma', 'Norwood Rubrum', 'Otto van Oerstedt', 'President', 'Pride of India', 'Pride of Mysore', 'Princess', 'Punctatum Aureum', 'Rainbow', 'Reidii', 'Show Girl', 'Spirale', 'Vivekananda', 'Vishnu', 'Warrenii', 'Wild Beauty', *etc.*

Propagation

Crotons are propagated through stem cuttings, air layering, grafting, seeds and also through tissue culture. When grown through **cuttings** in temperate conditions, bottom heat improves the rooting percentage. Whether it is hardwood (20-30 cm), semi-hardwood (8-15 cm) and tip cuttings (7.5 cm), all from the strong shoots of lateral growth can be used for multiplication. The cuttings of 20-30 cm length will take about 30-40 days for rooting and initiation of buds and thus the plants will be ready for planting within a span of 6-8 months. Percentage of rooting increases under humid (60-70 per cent or even 100 per cent) and warm conditions (24 °C) with porous and well drained soil as medium. In heavy rainfall areas and open situation, rooting performance will be better on raised beds, however, better these should be rooted in pots, polybags or trays. Incorporation of sand and farmyard manure will enhance the performance of rooting. The ideal temperature for better rooting is 24 °C though up to 30 °C these may be rooted easily. Growth regulators, especially IBA at 500-1,000 ppm is quite useful as dipping for rooting of cuttings. Cuttings planted under mist root better with enhanced rooting percentage. The rooted cuttings can be lifted and planted either in the polythene bags or in the pots as per the requirements. Lorenzo *et al.* (1982) reported better number, volume, length, fresh and dry weights of roots in media-E (100 per cent composted pine bark) which has an optimum intersection point of the air-water curve on pine bark in 8 cm high pots thus providing proper aeration and draining out of excess water. Sandy medium is always better for rooting. Beds prepared with addition of sand and farmyard manure (1:1) facilitates quick rooting and helps in the growth of easy-rooted plantlets. To get bigger plants and for immediate planting, plants are multiplied through air layering. Though cuttings root easily under mist, **air layering** provides well shaped plant in a short time. The process of air layering involves making a diagonal cut through one-third to one-half the diameter of the stem. Damp sphagnum moss is packed around the area, with the moss carefully covered with plastic wrap. When the stem roots, it is planted in a container filled with light-weight potting soil or a mixture of ingredients such as sand and peat moss.

Codiaeum variegatum (L) Blume cv. 'Corazon Deoro' and 'Norma' have successfully **micropropagated** by taking explants from newly sprouted shoots. The establishment and multiplication were possible at 1 mg/l BA + 3 mg/l 2iP in former case and 1 mg/l IAA + 3 mg/l 2iP in latter case in MS medium (Radice Silvia, 2010).

Cultural Practices

Though it adapts well to a wide range of **soils** but performance in medium fertile red and lateritic soils has been found best. Croton needs well-drained soils having pH range of 6.0-8.0 though adopts well where soil pH is 6.5-7.0. Slightly alkaline and acidic soils do not hamper its growth if it is rich in organic matter. However, a very low or quite high pH can be adjusted with dolomite or sulphur.

When grown outdoors, crotons benefit from bright direct **sunlight** at least 6-8 hours per day, especially during winter months for proper colouration though during summer such light is not preferred as this creates dullness in the leaves due to high temperature and therefore the plants are shaded. However, excessive shading also hampers the colour development in the foliage. High light intensity reduces leaf size though high shading results into larger leaves. While crotons are still attractive when the plant is grown in partial shade, full sunlight brings out the most intense colour variations and contrasting shades of the foliage under moderate temperature ranges. Its natural habitat is warm and humid with plenty of water. In tropical dry conditions, crotons have to be moved to the partial shaded areas. These can be grown indoors but will have to be taken out twice a week for retaining proper colour and shade. While growing indoors, these require 3,000-5,000 fc (50-70 per cent shading) illumination. Likewise, these also do not survive in temperate climatic conditions, though with utmost care these can be successfully grown in sub-temperate zones. For full display of colour and proper growth, these can be grown under 20-30 °C **temperature** range. They are susceptible to draughts and sudden change in temperature which may cause leaves to fall. During winters it remains comfortable at 16 °C temperature but in no case the temperature should drop below 13 °C, however, for short periods it can even be grown at 10 °C. For planting as hedge, the plants should be **spaced** 30-40 cm across while in shrubbery the spacing may be maintained at 1.5-2.0 m × 1.0 m. For planting, 30-40 cm pits are prepared and filled with excavated top soil and FYM, half and half, and in case the soil is heavy and not porous, 1/5th of the total volume the coarse sand is also added. After filling the pits are thoroughly watered to settle the mixture properly. One week after, planting has to be done by keeping the plant intact with ball of earth, maintaining the same depth as had been in the nursery. While planting in the pots, the same mixture should be filled in the pots and one plant per pot of 12-18 cm size should initially be planted in March-April, followed by watering. In later stages, the plants should be repotted in larger size of pots commensurate to the size of the plants

Croton requires regular **pruning** before onset of monsoon in order to keep the plants in shape, to maintain need-based spread and height, and to encourage the new shoots to form. Usually for large bushy specimens no pruning is required for providing the shape. Dead-wood, leggy, sparse-leaved branches and diseased parts are hard-pruned. While pruning, one-third of the longest branches should be sheared off close to the base of the plant but in a symmetrical way. Regular pruning is done in March to encourage new growth.

Weeds pose a problem in the shrubberies initially, therefore these should be removed regularly manually or through machines. When the croton plants have sufficient horizontal growth, the shade of their branches will not allow weeds to grow further.

Crotons need sufficient **water** for their luxurious growth and further development so that colouration of the foliage is expressed to its fullest strength. Depending upon the weather conditions and soils, crotons are watered evey 10-15 days, but in harsh summers the watering should be done at every 5 days coupled with

basal shoots, 'Rebecca', a sport of 'Camille'with shining yellow-green to near-white foliage having green edges and blotches, 'Roehrsii' with pale-yellow-green leaves and the lateral veins ivory, 'Rudolph Roehrs' has entire leaf colouring to pale-yellow-green except the green colouring on the narrow margins and the veins, *etc.*; ***D. oerstedii***, a native to Mexico and Costa Rica grows robust with more than 90 cm stems, leaves deep green, ovate to oblong-lanceolate and mostly asymmetrical, and its form *D. o.* 'Variegata' bears leaves with ivory-white midribs; ***D. seguine***, a native of Tropical America and one of the largest species growing more than 2 m in height bearing glossy deep green and slightly fleshy leaves, ovate to lanceolate or elliptic, 45 cm long leaves, and its varieties are 'Irrorata' bearing thin yellow-green leaves with deep green edges and blotches and whitish petioles, 'Lineata' with green leaves but leaf stalks white striped, 'Liturata' having cream zone to either side of the midrib, 'Nobilis' with spotted shining pale-green leaves, *etc.*

Certain other important hybrids are 'Gold Rush' with white mid-vein and the margin with a thin stripe of green, 'Parachute', a sport from hybrid 'Paradise' having leopard spots of green on yellow background, 'Snow Flakes', a selection from 'Tiki' with almost entire foliage white and grey and only a little green, 'Sparkles' having deep yellow leaves outlined with random green blotches around a white midvein, 'Sterling' having deep green leaves with simple white branching midvein, 'Tiki', a sport of *Dieffenbachia* × *memoria-corsii* with narrow and wavy leaves having green and grey variegation spotted white, 'Tropic Honey', a compact plant though similar to 'Tropic Marianne' but its yellow-green leaves have bright white midrib and dark green margins, 'Tropic Marianne' having almost solid yellow leaves with a thin margin of green, *etc.* Henny (1977) also reported that though there are about 100 varieties in *Dieffenbachia* since release of its first variety in 1870 but commercially only 20 are being cultivated (Chen *et al.*, 2002). Henny (1977) also reported one more hybrid, *i.e. D.* × *splendens* by crossing *D. leopoldii* with *D. maculata*.

Propagation

Dieffenbachias are propagated sexually through seeds, vegetatively through tip or cane (stem) cuttings and through micropropagation. Seed growing is carried out only in very special cases when new varieties are to be evolved. Seed production is very limited and seed germination is also very poor, hence, it is propagated only vegetatively. The plants are stripped of its lower leaves from April to June in the temperate climates and in September in the sub-tropical regions of the country, the tips or growing points are detached with a portion of stem about 7-8 cm in length containing three nodes (Marlatt, 1969), *vis-à-vis* the rest of the stem is also cut into pieces of the same size and rooted as the cuttings in the medium containing equal parts peat and sand or in a mixture of peat and leaf moulds or compost, peat, loam and sand and kept at a temperature of 21-24 °C. Morioka *et al.* (1990) advocated 16 °C as the best temperature for rooting. McConnell *et al.* (1981) in a study recorded that fresh root weight of *Dieffenbachia* 'Exotica Perfection' cuttings initially chilled at 7 °C for six days recorded only 59 per cent than that of control. Rooting takes place from 3-8 weeks at 25 °C media temperature. After rooting these are hardened off on greenhouse benches and then potted into 10 cm pots containing porous medium. Parent plant usually breaks again from the base but normally these are discarded as these become misshapen. Normally the cuttings are taken from the lower leafless stems for good success and such cuttings are inserted horizontally into the rooting medium. Usually soilless media such as sphagnum moss, sand, perlite *etc.* are better for rooting.

Through indexing procedure for getting disease-free plants, Knauss (1976) used shoot tip explants for development of protocols for direct and indirect mode of propagation. Shen *et al.* (2007) and Chen (2008) used leaf explants and established the protocols for indirect regeneration in four varieties, *viz.* 'Camouflage', 'Camille', 'Octopus' and 'Star Bright'. Tissue-cultured liners can be planted in 15-cm, 20-cm or 25-cm pots depending on the requirement of the cultivar but since peat is a product of wetland ecosystem having some detrimental effects (Barber, 1993; Buckland, 1993) so scientists are trying to find out some better alternative to it. Now throughout the world, the Govt. and private agencies are coming up for producing quality planting materials. Govil and Gupta (1997) stated that Commercial Tissue Culture firms in India export some 40 million tissue culture plantlets to the States and other countries.

Cultural Practices

Though it is basically a tropical plant, being the most chilling-sensitive, but it can be grown indoors in almost all the climates but under temperate conditions it requires either greenhouse cultivation or under certain other protection during inclement weathers and winters, and especially during nights. However, under sub-tropical conditions of North India as in Delhi too, this requires winter protection during nights, and summer (April to July) protection during bright sunny days. It does well at a temperature range of 16 to 33 °C (ideal being 21-27 °C) and at a relative humidity of 60-100 per cent without any quality deterioration. In adequate warmth the plants remain evergreen and full of display all year round without any resting period. However, direct sunlight should be avoided at temperatures above 27 °C to save the foliage of the plants from sun-burning so under such conditions commensurate shading is required though draughts are dangerous. Persisting temperatures below 16 °C and above 27 °C are not congenial for its proper growing. These survive even at 13 °C but with improper growth. A temperature as low as 7 °C can also be tolerated for a short period but the plants lose some of their lower leaves but this mars the beauty of the plant (Hay and Beckett, 1971). Semeniuk *et al.* (1986) reported severe leaf injury to *D.* 'Rudolph Roehrs' after subjecting the plants to two days at 5 °C. In a further study, McMahon *et al.* (1994) recorded visible foliar injury at 27 hours (older leaves showing water-soaking in 75 hours) of chilling at 1.2 °C to var. 'Rudolph Roehrs'. Li *et al.* (2008) reported that the var. 'Panthera' is chilling-tolerant though 'Tropic Honey' highly sensitive. The light intensity requirement of dieffenbachia is almost moderate though there is genotypic variation as 'Snow Flake' and 'Star Bright' tolerate as low as 50 foot candles light levels to look attractive while 'Camille' requires 150-250 fc though plants raised in commercial greenhouses are provided 80 per cent shade, *i.e.* 2,500 fc. Low light levels than required cause reduction in the level of variegation and stem elongation.

These are potted on initially in 15-cm pots in March in the sub-temperate regions of the country, in April-May in the temperate regions of the country while in September in the subtropical and by October end in the tropical (eastern and southern plains) regions of the country. However, large specimens are planted in 30-cm pots filled with 7 parts loam, 3 parts peat and 2 parts coarse and fresh sand. To each litre of this mixture some 6 g of 2:2:1 bone meal, lime superphosphate and potassium sulphate and 1.25 g of ground chalk are incorporated when these are planted in 10-20 cm pots and 12 g of 2:2:1 mixture + 2.5 g ground chalk when planted in 20-cm and larger pots. The pH of the media should be 6.0-6.5 with soluble salt levels not more than 1dS/m, therefore, growers for monitoring the nutrient status of the media should use an EC meter for checking soluble salts biweekly. However, the fertilization should be stopped before one month of shipment. Usually, the plants do not require fertilizing up to two months of their purchase. Organic wastes likes coconut coir dust (Stams and Evans, 1997), and tree bark and tea waste in 4:1 to 2:1 by volume (Hatamzadeh, 2004) may prove good alternative to peat-based media. Potting media are kept moist but not wet when controlled-release fertilizers are used. It is fertigation at 100 ppm water-soluble N per month used for *Dieffenbachia* plants, however, watering with simple water is usually twice a week in summer while sparingly during winter season.

For home growing, a humid atmosphere is necessary and for this the growing media should be highly porous and mixed with absorbent material which is kept moist though watering is carried out only when necessary. In cool conditions these are seldom watered to avoid rotting of the main stems. However, in hot weather (at and above 21 °C temperatures) the greenhouse crop requires foliage-syringing twice a day with proper aeration.

Indoor Care and Transportation

Every part of the plant being poisonous, these are handled after bearing the gloves to prevent irritation caused due to contact with the sap. In the interiorscaping these are being used for over 100 years but since the plant is highly fragile, especially the leaves being wide, a bit leathery and long, it's a challenge to maintain its aesthetic appearance indoors. Due to excessive shade the plants may become tall and leggy, the foliage may lose variegation, the leaf edges may dry and may turn brown and lower leaves may die or due to chilling injury the stem along with the entire foliage may become water-soaked and flaccid. Therefore, the plants should be protected during weathers when temperature may fall below 15 °C and selection of the varieties should be made as per the availability of the light levels indoors. During shipping, the sleeves should be used around entire foliage which are highly fragile and should be shipped around 15-16 °C temperature as lower temperature may cause chilling injury to the foliage. Ethylene in the storage temperature is often harmful to postproduction quality of almost all the foliage plants including *Dieffenbachia*. Exposure to 4.7 micro l^{-1}ethylene during storage caused leaf chlorosis in var. 'Tropic Marianne', however, these plants pretreated with 1-MCP for six hours reduced the number of chlorotic leaves when compared with control plants when stored in air containing 1 micro l^{-1} ethylene (ShuTing and DerMing, 2010).

Insect-Pests and diseases

Two-spotted spider mites (*Tetranychus urticae*) cause foliar webbing, yellowing and speckling and in their severe attack even the entire plant dies. Hamlen (1978) suggested use of predator mite, *Phytoseiulus macropilis* which controls the population within two weeks, though biological control programmes are effective only at small scale but not in the large commercial greenhouses. **Aphid** (*Aphis* spp.) infestation causes plant distortion at the site of new growths, and its attack causes honeydews and sooty moulds. **Rice root aphid** or reddish aphid (*Rhophalosiphum rufiabdominalis*) infests on the roots which can be controlled by drenching the plants with 0.1 per cent Malathion. **Mealybug** (*Planococcus* spp.) eggs appear as white cottony masses in leaf axils, lower leaf surfaces and on roots and its infestation stunts the plants, *vis-à-vis* causes honeydew and sooty moulds. Regular application of systemic insecticides at crawler stage may control this pest though at mature stage it is difficult to control except the hand-picking and killing.

Dieffenbachia **soft rot** or blight is caused by *Erwinia carotovora* pv. *carotovora* and *E. chrysanthemi*. Bazzi *et al.* (1987) reported the bacterium *Erwinia chrysanthemi* pv. *dieffenbachiae* infecting *Dieffenbachia* in Italy. The disease is widespread during summer in excessive humid greenhouses and during winters when production facilities limit external air exchange which again causes excess humidity levels inside. The bacteria are systemic and enter the stem of the plants that become active when rooting the cuttings. As soon as any plant shows symptoms of leaf yellowing and browning coupled with mushy spots emitting rotten and fishy smell, and the centres of spots falling out leaving the holes, especially during summer months, the copper compounds and antibiotics though are effective but to a limited extent, however, keeping the foliage dry keeps away the infections. Moreover, the plants should have wide spacing. *Xanthomonas axonopodis* pv. *dieffenbachiae* infects most of the plants of Araceae family including *Dieffenbachia* by causing quite dark (almost black) leaf spots surrounded by a yellow halo, spots starting from the leaf edges and then enlarging. Mode of spread is through soil, splashing water, cutting tools and insects. There is no chemical control to this disease, however, regular spray of copper containing bactericides will minimize the disease and cross-contamination from other genera should be avoided, field should be kept clean, if possible the pots should be watered through drip system instead of overhead (Norman and Henny, 1996), use only disinfected clippers and tools, and the crop should be properly fertilized as poor fertilization may favour spread of the disease (Chase, 1992).

Fusarium* stem rot** (*Fusaarium oxysporum* and *F. solani*) and *Erwinia* soft rot symptoms are quite similar in appearance but in *Fusarium* rot the stems turn mushy at the base with reddish or purple discolouration and the leaves show papery spots with light and dark rings similar to bull's eye. In this case, disease-free planting material is recommended. **Anthracnose** (*Colletotrichum gloeosporioides*) favours cool temperatures and its infection intensifies after pesticide or fertilizer damage. Its infection appears as tan water-soaked spots with concentric rings and the situation aggravates due to condensation dripping from the glass or plastic roof which splashes the spores spreading the disease. Proper distance among the plants, watering early in the day so that foliage dries out quickly, and spraying with 0.2 per cent mancozeb and Dithane M-45 will control the disease. ***Myrothecium* leaf spot** (*Myrothecium roridum*) infects the plants through wounds inflicted on the plants and breaks in the main vein. The tissue-cultured newly planted explants are highly susceptible to this pathogen where petiole rot occurs in the older leaves coupled with leaf spots. The spots caused by this pathogen are watery and contain usually black and white fungal fruiting bodies in concentric rings near the outer edge of the spot which are better seen from the undersides of the leaves. Care should be taken not to injure the plants at any level and use of chlorothalonil, mancozeb or thiophanate methyl + mancozeb provides the best control. Lorenzini *et al.* (1983) reported ***Colletotrichum gloeosporioides (*Glomerella cingulata*) infecting *Dieffenbachia amoena*. This problem can be controlled by applying thiophanate methyl + mancozeb.

Dasheen mosaic virus (DMV) which spreads through aphids and human beings, is very serious on dieffenbachia cultivars 'Bausei', 'Memoria-corsii', 'Perfection', *etc.* and its infection causes mosaic, leaf distortion and stunting. As the symptoms are not easily noticeable and the chemical control is not always effective so the planting materials should be taken only from the indexed plants, however, the hosts such as *Aglaonema, Philodendron, Spathiphyllum, etc.* should be properly monitored if planted nearby.

References

Arditti, J. and E. Rodriguez, 1982. *Dieffenbachia* ~ uses, abuses and toxic constituents ~ a review. *J. Ethnopharnmacology*, **5**(3): 293-302.

Bailey, L.H. and E.Z. Bailey, 1976. *Hortus Third ~ A Concise Dictionary of Plants Cultivated in the United States and Canada*. Macmillan Publishing Co., Inc., New York.

Barber, K.E. 1993. Peatlands as scientific archives of past biodiversity. *Biodiv. Conserv.*, **2**: 474-489.

Barnes, B.A. and L.E. Fox, 1955. Poisoning with *Dieffenbachia*. *J. History Med. Allied Sci.*, **10**: 173-181.

Bazzi, C., P. Minardi and U. Mazzucchi, 1987. Bacterial diseases of flower and ornamental plants in Italy (Italian). *Informatore Fitopatologico*, **37**(6): 15-24.

Buckland, P. 1993. Peatland archaeology ~ a conservation resource on the edge of exhibition. *Biodiv. Conserv.*, **2**: 556-566.

Chase, A.P. 1992. Common diseases of dieffenbachia. *Southern Nursery Digest*, **26**(12): 20-21.

Chen, J. 2008. Effects of genotype, explant source and plant growth regulators on indirect shoot organogenesis in

Dieffenbachia cultivars. *In vitro Cell Devel. Bio. Plant*, **44**(4): 282-288.

Chen, J. and R.J. Henny, 2006. Somaclonal Variation ~ An Important Source for Cultivar Development of Floriculture Crops. In: *Floriculture, Ornamental and Plant Biotechnology* (ed. Teixeira da Silva), pp. 244-253. Global Science Books, London.

Chen, J., R.J. Henny and D.B. McConnell, 2002. Development of New Foliage Plant Cultivars. In: *Trends in New Crops and New Uses* (eds Janick, J. and A. Whipkey), pp. 466-472. American Society of Horticultural Science Press, Alexandra, USA.

Cote, G.G. 2009. Diversity and distribution of idioblasts producing calcium oxalate crystals in *Dieffenbachia seguine* (Araceae). *Am. J. Bot.*, **96**(7): 1245-1254.

Croat, T.B. 2004. Revision of *Dieffenbachia* (Araceae) of Mexico, Central America, and the West Indies. *Ann. Missouri bot. Grdn*, **91**(4): 668-772.

Fochtman, F.W., J.E. Manno, C.L. Winek and J.A. Cooper, 1969. Toxicity of the genus *Dieffenbachia.Toxicol. Appl. Pharmac.*, **15**: 38-45.

Genua, J.M. and C.J. Hillson, 1985. The occurrence, type and location of calcium oxalate crystals in the leaves of fourteen species of Araceae. *Ann. Bot.*, **56**: 351-361.

Govil, S. and S.C. Gupta, 1997. Commercialization of plant tissue culture in India. *Plant Cell Tiss. Org. Cult.*, **51**:65-73.

Hamlen, R.A. 1978. Biological control of spider mites on greenhouse ornamentals using predaceous mites. *Proc. Fla St. hort. Soc.*, **91**: 247-249.

Hatamzadeh, A., V. Karimi, S. Vasseh-Mosalla and M.N. Padasht-Dehkaee, 2004. Evaluation of tree bark, tea waste compost and rice hull for preparing suitable substrates in training of indoor plant, dieffenbachia (*Dieffenbachia amoena*). *J. agric. Sci.*, **1**(2): 29-38.

Hay, R. and K.A. Beckett, 1971. Dieffenbachia. Reader's Digest Encyclopaedia of Garden Plants and Flowers, pp. 229-230. Reader's Digest Association, London.

Henny, R.J. 1977. Breeding,growing and observing *Dieffenbachia* species and seedlings. *Proc. Fla St. hort. Soc.*, **90**: 94-96.

Henny, R.J. 1995. 'Star Bright' *Dieffenbachia*. *HortSci.*, **30**: 164.

Henny R.J. and J. Chen, 2003. Foliage Plant Cultivar Development. *Plant Br. Rev.*, **45**: 245-290.

Jones, G.E. 1957. Chromosome numbers and phylogenetic relationship in the Araceae. Ph. D. Diss., University of Virginia, Charlotesville, USA.

Knauss, F. 1976. A tissue culture method for producing *Dieffenbachia picta* cv. 'Perfection'free of fungi and bacteria. *Proc. Fla St. hort. Soc.*, **89**: 293-296.

KuanHong, U. and Y. DerMing, 2010. Assessment of formaldehyde removal efficiency of shoots and medium-root zones of potted *Dieffenbachia maculata*. *J. Taiwan Soc. hort. Sci.*, **56**(3): 171- 181.

Li, Q.S., J.J. Chen, R.H. Stamps and L.R. Parsons, 2008. Variation in chilling sensitivity among eight *Dieffenbachia* cultivars. *HortSci.*, **43**(6): 1742-1745.

Lorenzini, G., E. Triolo and G. Scaramuzzi, 1983. Further contribution to the knowledge of new or little known diseases of ornamental plants in Tuscany (Italian). *Rivista della Ortoflorofrutticoltura Italiana*, **67**(1): 23-34.

Marlatt, R.B. 1969. Propagation of dieffenbachia. *Econ. Bot.*, **23**(4): 385-388.

McConnell, D.B., D.L. Ingram, C. Croda-Bada and J.T. Sheehan, 1981. Chilling effects on propagation of *Dieffenbachia* 'Exotica Perfection'. *Foliage Digest*, **4**: 3-4.

McMahon, M.J., A.J. Pertuit and J.E. Arnold, 1994. Effects of chilling on *Episcia* and *Dieffenbachia*. *J. Amer. Soc. hort. Sci.*, **119**: 80-83.

Molano-Flores, B. 2001. Herbivory and calcium concentrations affect calcium oxalate crystal formation in leaves of *Sida* (Malvaceae). *Ann. Bot.*, **88**(3): 387-391.

Morioka, K., Y. Sugiyama and K. Yonemura, 1990. Rooting cuttings of foliage plants in water. *Res. Bull. Aichi-ken agric. Res. Cent.*, No. 22, pp. 199-203.

Norman, D.J. and R.J. Henny, 1996. *Xanthomonas* and *Erwinia* resistance in twenty *Dieffenbachia* cultivars. Univ. Fla/ IFAS Central Fla Res. and Edn Cent Res. Rep. RH-96-9.

Pamies R.J., R. Powell, A.H. Herold and J. Martinez, 1992. The dieffenbachia plant-case history. *J. Fla Med. Assocn*, **79**(11): 760-761.

Semeniuk, P., H.E. Moline and J.A. Abbott, 1986. A comparison of the effects of ABA and an antitranspirant on chilling injury of coleus, cucumbers and dieffenbachia. *J. Amer. Soc. hort. Sci.*, **111**: 866-868.

Shen, X., J. Chen and M.E. Kane, 2007. Indirect shoot organogenesis from leaves of *Dieffenbachia* cv. Camouflage. *Plant Cell Tissue Organ Cult.*, **89**: 83-90.

ShuTing, F.F. and Y. DerMing, 2010. Postproduction quality affected by ethylene during storage and 1-MCP protection of *Dieffenbachia*. *J. Taiwan Soc. hort. Sci.*, **56**(1): 1-10.

Spoerke, D.G., Jr. and S.C. Smolinske, 1990. *Dieffenbachia* species. In: *Toxicity of Houseplants*, pp. 119-122. CRC Press, Boca Raton, USA.

Stams, R.H. and M.R. Evans, 1997. Growth of *Dieffenbachia maculata* 'Camille' in growing media containing sphagnum peat or coconut coir dust. *HortSci.*, **32**(5): 844-847.

Sweet, R. 1839. *Hortus Britanicus*. Ridgeway, London.

16

Euphorbia (Family: Euphorbiaceae)

B. Sathyanarayana Reddy, Sanyat Misra, R.L. Misra, S.Y. Chandrashekar, P.M. Munikrishnappa, S.K. Nataraja and S. Latha

[**Common names**: Annual poinsettia/Fire on the mountain/Hypocrite plant/Mexican flower plant/Painted leaf/Painted spurge (*Euphorbia heterophylla*); Cactus spurge (*E. pseudocactus*); Candelillo (*E. antisyphilitica*); Caper spurge/Mole plant (*E. lathyris*); Christmas flower/Christmas star/Crown of the Andes/Easter flower/Lobster flower/Lobster plant/Mexican flame-leaf/Painted leaf/Poinsettia/Spurge (*E. pulcherrima*); Cow's horn (*E. grandicornis*); Crown of thorns (*E. milii*); Cypress spurge (*E. cyparissias*); Finger tree/Rubber spurge (*E. tirucalli*); Flowering spurge (*E. corollata*); Gingham golf ball (*E. obesa*); Ghost weed/Snow on the mountain (*E. marginata*); Indian tree spruce/Milk bush (*E. tirucallii*); Ipecac spurge (*E. ipecacuanhae*); Little Christmas flower/Snowflake euphorbia/Snow bush/Snows of Kilimanjaro/White-laced euphorbia (*E. leucocephala*); Medusa's head (*E. caput-medusae*); Melon spruce (*E. meloformis*); Palo Amarillo (*E. fulva*); Scarlet plume (*E. fulgens*); Wood spurge (*E. amygdaloides*)]

Introduction and Origin

Euphorbia is the largest and most diverse genera in the family Euphorbiaceae which consists of over 2,000 species of leafy annuals, herbaceous biennials, perennials and sub-shrubs including cactus-like succulents, and deciduous and evergreen shrubs, many of them being succulents, of cosmopolitan distribution. They differ enormously in structures and forms but have recognition only through floral characters. Generic as well as family names derive from Euphorbus, the doctor to the king of Maurentania. In 1,200 B.C., the King Juba of Maurentania (Greece) named this plant after his physician Euphorbus in response to Augustus Caesar dedicating a statue to Antonius Musa, his own personal physician. Carl

Euphorbia leucocephala

E. pulcherrima

Poinsettia (Red)

Potted Euphorbia

Linnaeus adopted the name *Euphorbia* to the entire genus as well as to its family Euphorbiaceae, in physician's honour. Even otherwise the word itself comes from two Greek words *eu* means 'well' and *pherbo* means 'feed', 'nourish' or 'fat'. *Poinsettia*, the former generic name of *E. pulcherrima* was christened after Joel Roberts Poinsett, who introduced the plant to the USA. The common name spurge for all the euphorbias derives from the Middle English as the sap of most species is purgative. Though the species of *Euphorbia* of horticultural importance which are in cultivation shall be briefly described under the **species** column but further cultivation details shall be furnished only of two commercial species ~ *Euphorbia pulcherrima* and *E. leucocephala*.

Euphorbia pulcherrima (syn. *Poinsettia pulcherrima*) is native to tropical areas of Central America and Mexico. Among many species of *Euphorbia*, *E. pulcherrima* and *E. leucocephala* are of commercial important. *E. pulcherrima* is the number one potted flowering plant in the United States and is sold for Christmas holiday for decoration during Christmas hence it is also known as Christmas plant. It is also one of the major crops in Europe and Australia. These are predominantly the landscape plants. These are grown as specimen plant, as hedge, for mass plantings and along the paths and roads, and the dwarfer ones in the hanging baskets and as pot plants. Sometimes the bracts are also used as cut flowers. Basically these possess two types of leaves, crinkled usually with short stature and plain with tall stature. These continue giving spectacular coloured look throughout the year, however, on the hills from 600 to 1,200 m elevations, especially in the NEH regions and hilly terrain of West Bengal, these continue producing the large clusters of their coloured bracts all the year round except in January-February. The centre of the bracts contains the small true flowers known as cyathia. The specific name of ***E. leucocephala*** is from the Greek where *leucos* means 'white' and *kephale* means 'head', *i.e.* white head on account of the colour of its bracts.

Poinsettia latex has been used in hair removal cream in Mexico and Guatemala. A red dye has been obtained from the bracts. In Guatemala, the latex has been used as a remedy for toothache and to cause vomiting. Poultices of leaves have been applied to treat aches and pains. Although many species in the genus *Euphorbia* are highly toxic, *Poinsettia* is not among them. The toxic compounds are di-terpene esters found in the milky latex, but they have very low toxicity. If ingested may induce occasional abdominal pain with vomiting and diarrhea. Frequent contact with skin may cause redness, swelling and blisters lasting only for a few minutes.

Botany

In Euphorbiaceae where there are some 2,000 species of annuals, perennials, shrubs or succulents of deciduous or evergreen types of cosmopolitan distribution but mainly from subtropic to warm temperate zones. In the family the leaves are alternate, the actual flower is the small yellow structure found in the centre of each leaf bunch called cyathia or individual flower cyathium which comprises a tiny cup-shaped whorl of fused bracts with many male flowers reduced to 1 stamen each and a single female reducing to a 3-lobed ovary where the tips of the cyathical bracts bear crescent-shaped nector-bearing glands or petal-like structures. Many of the species have arrangement of cyathia in diachasial cymes where stalks are called as rays and grouped into umbel-like pseudumbels (umbel-like clusters) at the stem tips. Pairs of bracts called raylet leaves are at the base of cyathia and at the base of each pseudumbel there is a ring of larger bracts known as pseudumbel leaves which may sometimes be coloured. The fruit is a 3-lobed capsule.

E. pulcherrima is a shrub to small tree, varying in height up to 0.6 to 4 m. The plant bears medium to dark green and oval leaves that are toothed on the margins, pointed at the tips and measure 7 to 16 cm in length. The top leaves known as bracts (modified leaf), are flaming red, pink, white, marbled or combinations are often mistaken for flower petals because of their clusterings and colouration, but are actually leaves. The basic chromosome number of the family Euphorbiaceae is x=6-12. The meiotic chromosome of *Euphorbia pulcherrima* is 2n=42, with bi- and trivalent being most prevalent having multivalent associates in typical ring, chain, Y and cross type configurations. The chromosome number of *Euphorbia pulcherrima* is 2n=28 (Bempong and Sink, 1968). Out of nine *E. pulcherrima* cultivars studied 5 had 28 somatic chromosomes and 4 had 56. Progenies with 42 chromosomes could not be obtained from 28 × 56 crosses. It is suggested that the cultivars are polyploids, the basic number for the species being 7 (Ewart and walker, 1960).

Classification, Species and Cultivars

The species in Euphorbiaceae differ greatly in their forms but it is the floral characters which bind them into one family ~ Euphorbiaceae. As defined by Bailey (1960), Hay and Beckett (1971), and Beckett (1985, 1987) the important cultivated species of *Euphorbia* are:

E. abyssinica from Ethiopia, a succulent with erect and robust stems with almost wing-like and wavy 8 angled, each with a brownish-grey horny margin, leaves some 5 cm long, linear-lanceolate and only on the tips of young stems but soon falling off, and thorns paired and long, reaching up to 3 m so suitable only for large containers.

E. amygdaloides (wood spurge) from Europe, SW Asia, N. Africa, a perennial with seldom branched but erect stem of 30-80 cm, leaves oblanceolate, dark green, 3-8 cm long, pseudumbels 12 cm or more wide, raylet leaves yellow-green, fused to cup-shaped pairs.

E. antisyphilitica (*Tricherostigma antisyphilitica*) from Mexico, growing erect to 1 m with rod-like branches, almost leafless, and yielding a useful wax.

E. aphylla from Canary Islands, a bushy mound-forming 45 cm or more long shrub, stems leafless, erect, fleshy and shining grayish to yellowish-green.

E. atropurpurea from Tenerife (Canarife Islands), an erect shrub growing to 1 m or more with stems somewhat succulent, leaves glaucous, oblanceolate, up to 10 cm long in dense rosettes or tufts at the stem tips, and raylet leaves and cyathia emerging in a panicle-like cluster from the red-purple leaf rosettes from winter to spring.

E. biglandulosa (syn. *E. rigida*) from S. Europe and is quite similar to *E. myrsinites* but with ascending stems and lanceolate leaves.

E. bubalina from S. Africa growing up to 30 cm and appearance to that of a miniature palm, stem marked with leaf scars, 19 cm thick, green and lanceolate leaves on the upper portion which fall off during winter, and flowers minute, green and bell-shaped that appear during July and set seeds.

E. bupleurifolia from South Africa bearing 20 cm high and 8 cm thick and unbranched stem which is erect, succulent, ovoid to shortly cylindrical and prominently tubercled by the persisting leaf stalk bases, leaves up to 15 cm long at the top, light green, lanceolate and deciduous in autumn, and solitary cyathia rising only at a 5 cm stalk in summer with two green to red raylet leaves.

E. canariensis from Canary Islands is a highly decorative, succulent but very slow-growing shrub growing up to 1 m in containers though three times more in the wild where it forms wide mound-like clumps, branching at base but forking seldom from above, green with waxy-white patina, stems 4-5 angled each angle with a horny margin with a row of 12.5 cm long and paired thorns, and solitary and reddish-green cyathia at the tip of stems in the wild ones but rarely in containerized plants.

E. caput-medusae (syn. *E. commelinii*) from Cape region (S.A.) is curiously interesting plant, stems obconical and short having numerous 15-30 cm long and 2.5-5.0 cm thick branches, leaves about 2.5 cm long and linear-lanceolate, cyathia solitary and at short and thick peduncle, and glands erect and white.

E. capitulata from Balkan Peninsula, a hardy mat-forming rhizomatous perennial, stems only 10 cm long, leaves obovate, 1 cm, dense and grey-green, cyathium 1 and purplish.

E. cereiformis (syn. *E. erosa, E. leviana*) from South Africa is an erect and succulent shrub growing up to 1 m with basally branched dark green stems which above have 9-15 furrows where ridges are spiny and tubercled, leaves minute and short-lived, cyathia 1 at stem tip, and the raylet leaves purple.

E. characias from Portugal and Mediterranean, a hardy sub-shrubby but erect evergreen perennial having 60-180 cm stem length, grayish-green leaves linear to oblanceolate, 3-13 cm long, pseudumbels tiered and forming 15 cm or longer broad columns, raylet leaves yellow-green, cyathical nector glands reddish-brown, and good for cutting. ***E. characias wulfenii*** (syn. *E. veneta, E. wulfesii*) from E. Mediterranean, more robust than the type, leaves brighter grey-green, raylet leaves shining greenish-yellow and good for cutting.

E. corollata (*syn. Tithymalopsis corollata*) from E. USA, a hardy herbaceous perennial growing to 1 m, stems glabrous, slender and diffusely branched above, leaves 2.5-5.0 cm long though in the inflorescence much smaller and opposite, ovate-oblong to lanceolate, involucral glands 5 with clear white structures, and used for cutting as gypsophila.

E. cyparissias from Europe, an invasive weed-like species, rhizomatous, hardy herbaceous perennial 20-50 cm tall, grey-green leaves which turn reddish in autumn to 1-4 cm long, and raylet leaves greenish-yellow.

E. echinus from Morocco, a succulent species growing to 1 m with heavily branched, leafless and green stems, 6-7 sided with wavy ridges bearing paired grey spines and small green bell-shaped bisexual flowers in July.

E. epithymoides (syn. *E. polychroma*) from CS Europe, a clump-forming hardy herbaceous perennial, stems 30-60 cm, softly hairy 2-5 cm long leaves oblong to lanceolate, pseudumbels 8 cm or more and raylet leaves chrome-yellow.

E. fulgens from Mexico grows up to 1.2 m, spread 30-40 cm, a deciduous greenhouse shrub with thin branches bearing mid-green and elliptic-lanceolate leaves, and clear scarlet and up to 30 cm long bracts appear from November to February in recurved wand-like terminal sprays.

E. fulva (syn. *E. elastica*) from Mexico is a small tree used for extraction of rubber, leaves lanceolate, acute, pubescent and clustered at the tips, cyathia few, bracts small, glands ovate and capsule conical and 2.5 cm long.

E. gorgonis from Cape region of South Africa, having 10 cm high and wide globose, fleshy and tubercled swollen leaf bases, a crowd of succulent branches some 5 cm long which are green or sometimes red-tinted radiating from the top, leaves very small and soon falling off, and cyathia 1 only from the main stem and with brownish to bright crimson bands.

E. grandicornis from Natal to Kenya in Southern Africa, a 2 m high succulent shrub, stems highly branched and constricted into rounded or ovoid segments with 3 wing-like wavy and horny-margined angles bearing wide-spreading pairs of spines some 2-5 cm long.

E. grandidens from Cape region of South Africa, a succulent tree some 15 m tall in its natural habitat but is tamed to 2 m or below in containers, stems erect or in whorled branches having 3-4 spirally twisted horny angles with toothed margin bearing pairs of 4-6 mm long spines, and leaves short-lived and very small.

E. griffithii from Himalaya, a hardy clump or colony-forming rhizomatous herbaceous perennial, stems 1 m, leaves lanceolate with reddish mid-ribs, raylet leaves light red.

E. heterophylla from E. USA, an hardy annual with 30-90 cm stem, leaves sometimes variegated and ovate to fiddle-shaped, and pseudumbel and raylet leaves partly to full scarlet.

E. horrida from Cape to Karroo in South Africa, a clump-forming cactoid succulent shrub growing to 1 m with basal branching, stems erect, up to 14 furrows where ridges are toothed bearing either single or hardened flowering peduncle-derived clustered spines 4 cm long, and cyathia 1 and green.

E. ingens from Southern Africa, north to Zimbabwe and Zambia, a leafless succulent tree growing up to 10 m in the wild but is tamed to 1.5 m or below in the containers, young plants and branches of old trees are dark green on maturity and constricted in the end of each growing season, and have 3-5 wing-like ribbed and wavy angles bearing sometimes tiny spines.

E. ipecacuanhae (syn. *Tithymalopsis ipecacuanhae*) from E. USA, a perennial herb with only a 7.5-15 cm long forking inflorescence above ground (stem usually subterranean and scale-like bearing alternate leaves) bearing oval to linear (on different plants) glabrous and opposite leaves in red or green colours, cyathia long peduncled and gland appendages rudimentary.

E. lathyris from Europe, 60-180 cm tall hardy annual, leaves at lower stem opposite, towards bluish-green, slightly oblong and 7-10 cm long, and raylet leaves green to yellow-green.

E. leucocephala, a spreading shrub growing up to 2 m with elliptic to lance-shaped leaves, 3-8 cm long and appearing in whorls of 4-12, flowering during winter in the tropical India with a mass of creamy-white bracts as to that of white poinsettias, small greenish-yellow and fragrant flowers coming up at the branch end in trusses covering entire plant, and like many others in the family, the plant secretes white milky latex which is skin irritant.

E. mammillaris from Cape region of South Africa is a small (20 cm tall though up to 45 cm wide) cactoid succulent suckering profusely, cylindrical (up to 5 cm thick), semi-erect to prostrate stems having 7-17 tuberculate angles, leaves scale-like and short-lived, flowering peduncle woody, persistent and with 1.2 cm long spines, and cyathia 1 with yellowish or purple glands.

E. marginata from N. America, hardy annual 30-100 cm tall, leaves oblong to ovate, pale green and 8 cm long, and upper pseudumbel and raylet leaves broadly white-margined.

E. mellifera from Madeira and Canary Islands, a half-hardy evergreen shrub growing up to 15 m in wild, dark green leaves narrowly lanceolate, raylet leaves reddish-brown and cyathia honey-scented.

E. meloformis from South Africa, a rare but curious succulent often mistaken for cactus, spineless, globose or pyriform, 7.5-13 cm thick body with 8-10 ribs that are tuberculated at almost acute angles though sides darker but striped light green which wrinkle with age, leaves small and few, and flowers at flattened apex.

E. milii (syn. *E. splendens*) from Madagascar, a very popular flowering pot plant with height and spread of 30-60 cm, a semi-prostrate, semi-succulent, and spiny shrub with woody base, mid-green and lanceolate-obovate leaves at the tips, inflorescences 5.0 to 7.5 cm wide with branching cymes which are borne on long stalks with two kidney-shaped crimson bracts extending beneath the flowers, and flowering throughout but the peak being in winter.

E. myrsinites from S. Europe to W. Turkey, an evergreen hardy and tufted perennial with more than 40 cm long unbranched prostrate stems, leaves obovate to rounded, mucronate, crowded and brightly glaucous, pseudumbels up to 10 cm wide, and raynet leaves chrome-yellow.

E. nicaeensis, a hardy species from Europe, evergreen tufted perennial with 50-80 cm long stems, reddish and almost erect, up to 2.0-7.5 cm long leaves lanceolate, bright glaucous and with blunt tips, and raylet leaves greenish-yellow.

E obesa from Cape Province (S. Africa), a very curious-looking grey-green, leafless, spineless, striped and brown-purple banded (chequered looking) succulent (stem) growing up to 15 cm in spherical to cylindrical form with eight broad ribs, and cyathia dioecious, bell-shaped, bearing minute and sweet-scented green inflorescence.

E. palustris from Europe suitable for wet soils, wide clump-forming hardy herbaceous perennial, stem erect and with a height of 1 m or more, leaves 2-6 cm long, grey-green but may change to purple in autumn, lanceolate to narrowly oblong, pseudumbels large, and raylet leaves yellow.

E. pseudocactus from Natal (S.A.) is a succulent and cactoid shrub growing 1-2 m high with erect or inclined branched stems which are constricted into ovoid segments with 3-5 angles, each angle horny with toothed edge and bearing paired spines some 1.2 cm long, and arching V-shaped green patterning between the ribs.

E. pulcherrima from Mexico, a deciduous species suitable for cutting, growing to 0.6-4.0 m or even more and branched, spread 30-90 cm, bright green leaves elliptic, shallowly lobed or entire and from 7.5-15 cm long, and flowers 30 cm or more across are surrounded by many 15 cm or more long elliptic and deep crimson, scarlet, pink or white bracts which appear during winter.

E. robbiae from NW Turkey, shooting adventitiously through roots to form wide colonies soon, hardy evergreen perennial, leaves lustrous green and obovate, and raylet leaves bright yellow-green.

E. seguierana from Europe and SW Asia, an hardy and tufted perennial with up to 60 cm tall stems, leaves up to 3.5 cm long, linear to elliptic-oblong, pointed and glaucous and raylet leaves sulphur-green.

E. seguierana niciciana (syn. *E. niciciana*) from Balkan Peninsula, more robust than *E. seguierana* and with larger flowering clusters.

E. sikkimensis from E. Himalaya, very similar to *E. griffithii* but young plants glossy-red, mature leaves white-veined, and raylet leaves greenish-yellow.

E. tirucallii from S. Asia and suitable for warm regions, a shrubby leafless and small succulent tree having a dense crown of slender and cylindrical whorled branches first curving outward then erect, joints some 10 cm long, twigs 6-13 mm thick, and leaves narrow and about 2.5 cm long but fall off immediately.

E. valida, a succulent species from Cape Province grows some 1 m in spherical form though plant broadens and flattens with age advancement, stems dark green with 8-10 ribs with lateral markings, flowers green, small, sweet-scented, flowering from May to September, and floral stalk woody persisting for years, *etc*.

In a nutshell the most important species with their ornamental values for which these are grown are *Euphorbia heterophylla* and *E. marginata* as annual species where former one produces beautiful bract and flowers though in latter case bracts are insignificant but its variegated foliage is used in floral arrangements; *E. lathyrus* is biennial and suitable only for wild gardens; *E. fulgens* bears beautiful scarlet bracts and is most suitable for greenhouse growing, *E. pulcherrima* again a greenhouse species with beautiful bracts suitable even as cut flowers, and *E. milii* again a free-flowering greenhouse species with beautiful flowers; other perennial species are *E. biglandulosa* bearing beautiful bright green-yellow bracts, *E. characias* bearing beautiful sulphur-yellow inflorescences, *E. cyparissias* most suitable as ground cover, *E. griffithii* is attractive whether in bloom or without, when foliage becomes very attractive, *E. myrsinites* is a beautiful trailing plant whether in flower or in foliage and most suitable in front row of the border, *E. palustris* bears beautiful sulphur-yellow bracts, *E. polychroma* bears beautiful bright yellow flowers, *E. robbiae* is a shade-loving plant bearing leathery leaves and yellow-green bracts, *E. sikkimensis* has bright red new growths and leaf veins and beautiful chrome-yellow bracts, *E. wulfenii* (syn. *E. characias wulfernii*) bearing spectacular rounded bright yellow-green bracts; and the succulent species are *E. bubalina* with green and some 19 mm thick stem having markings of leaf scars and bearing bell-shaped green flowers, *E. echinus* is a leafless bisexual species with 6-7 angled green stems having wavy ridges and small green bell-shaped flowers, *E. obesa* is a spherical, leafless and spineless 8-broad ribbed, striped and banded stem 15 cm high which bears small green inflorescence with fragrant flowers where male and female appearing on separate plants, *E. valida* is a spherical 7.5 cm high and 8-10 ribs with lateral markings, male and female flowers appearing on separate plants, flowers small, green and sweetly scented, *etc*.

Poinsettia varieties can be divided into two groups, (i) free-branching, and (ii) restricted branching (Dole and Wilkins, 1991a, 1994). When young plants are pinched, free-branching types produce innumerable axillary buds at almost every node though restricted branching type only 2-4 axillary shoots due to apical dominance being established thereupon over lower shoots as in case of 'Eckespoint Celebrate'. This free-branching characteristic is due to a phytoplasma closely related to 'peach X' and 'spirea stunt' (Lee *et al.*, 1997) which is graft-transmissible and therefore when such genotypes are grafted over restricted-branching type, the latter also acquires the same characteristics (Dole and Wilkins, 1992; Dole *et al.*, 1993; Ruiz-Sifre *et al.*, 1997) though through seeds it reverts back therefore again seedlings require to be grafted with free-branching types (Dole and Wilkins, 1991). According to Gopalaswamienger (1991) the present day cultivars have special features such as dwarf habits, broader and brighter bracts, more number of bracts clustering together and with large clusters, long flowering season, and suitability for planting in the grounds and in pots. Some of the cultivars are 'Alba' (bracts creamy-white but some with green edges near apex, contrasting with crimson or scarlet flowers), 'Annette Hegg' (a compact bush, leaves ovate, bracts amall and broad red), 'Annette Hegg Marble' (bracts apricot-yellow, inside rosy-pink along the mid-vein, terminal inflorescences densely clustered with bicoloured bracts in a circular arrangement), 'Annette Hegg Supreme' (a sport of 'Dark Red Hegg', compact with bright red bracts), 'Dark Hegg Red' (bracts dark red), 'Jingle Bells' (gorgeous and sensational plant with broadly ovate-elliptic spirally arranged red bracts splashed with pink and rosy dots, flecks and streaks), 'Lutea' (Compact, broad ovate leaves, petiole yellowish and bracts attractive yellow), 'Plenissima' (branches droopy, flowers crimson with crinkled bracts), 'Plenissima Starlite' (bracts red, long narrow elliptic to lanceolate with 2-4 rows of clusters arising over shoots), 'Rosea' (plants dwarf bushy bearing heads of ovate rose bracts with darker veins), 'Rosy Pink' (tall with single row of rosy-pink elliptic bracts), 'Yellow Star' (tall and hardy with yellow bracts), *etc.* Some other noteworthy varieties are 'Dueimco', 'Duepre', 'Duespot', 'Eckadire', 'Fislemon', 'Fismarble Silver', 'Fismille', 'Freedom White', 'Ice Punch', 'Kamp Burgundy', 'Lemon Drop', 'Pink Peppermint', 'Polar Bear', 'Prestige Red', *etc.* Under *E. leucosephala* there are two common varieties, (i) 'Snow Flake' with snow-white bracts which glisten brightly in the sun, and (ii) 'Pink Finale' where bracts appear as white but sooner turn to pink.

Red is the most important colour in poinsettias though other colours are pink, white, pink margined broad white *e.g.* 'Marble'; pink flecked red *e.g.* 'Pink Peppermint', 'Monet', *etc.*; red flecked pink *e.g.* 'Jingle Bells'; white flecked red *e.g.* 'Candy Cane'; yellow *e.g.* 'Lemon Drop'; with variegated foliage *e.g.* 'Silverstar', etc. The cultivars of poinsettia are series-specific, each series including red, pink, and white and a few as bicolours such as 'Marble' and 'Jingle Bells'patterned; and these series are 'Freedom' series such as 'Freedom Pink', 'Freedom Red', 'Freedom White', 'Freedom Marble', and 'Freedom Jingle Bells' and so the other cultivar series are 'Angelika', 'Cortez', 'Eckespoint Freedom', 'Gross Supjibi', 'Gutbier V-14 Glory', ' Gutbier V-17', 'Nutcracker', 'Peter Jacobsen's, 'Peter star' and 'Sonora'. In each series, it is the red colour which is most important and the first colour and other colours have developed through mutation. There are also many individual varieties which have nothing to do with series. The cv. 'Pepride' is unusually dwarf, most suitable to 10-13 cm pot. Besides branching and colour, poinsettias also vary in plant height, foliage colour, leaf retention, blooming date & flowering duration and post harvest life (Dole and Wilkins, 1999).

Propagation

Cuttings of ***E. pulcherrima*** are planted in a sterile and porous medium, however, cutting equipments and cutting areas should be clean and sterile to prevent diseases. For commercial cutting propagation, blocks of foam or rockwool are commonly used though these may also be planted directly into the final pot but the media should be well-drained with a pH of 5.6 to 6.2 (Dole and Wilkins, 1999). Cuttings propagated in rockwool or foam blocks should be applied with 100-150 ppm N from ammonium nitrate, calcium nitrate or potassium nitrate, first 10 days after insertion and then weekly and these cuttings will be ready for transplanting in 4-5 weeks (Dole and Wilkins, 1999). When the cuttings should be taken, depends on two things ~ the cultivar, and the number of weeks from the date of last pinching which is normally six weeks (Anonymous, 1994). Immature cuttings are slow and very irregular in rooting. Grueber (1985) stated that for maximum cutting production, the original rooted cuttings should be allowed to develop 7-12 nodes before the first pinch. Pinching below 7 nodes forces axillary shoots from axillary buds to develop which are slow in development, and leaving more than 12 nodes on original plant causes growth of too much axillary shoots making the cuttings weak due to competition so only 9-11 nodes should be left on main stem for optimum cutting production. For enough mature plants, the first pinch may be used for raising another crop of stock plants but such cuttings may form premature flower buds. The distance for stock plants may be maintained from 20 × 20 cm to 45 × 45 cm (Ecke *et al.*, 1990), depending on the weather conditions. Those planted in dry season should be closer as compared to those when weather is humid and improved. However, pinching once will produce 8-9 cuttings per plant over a 3-week period, pinching twice over 4-5 week period will give 15-20 plants though closely spaced plants will be unpredictable. A media temperature range between 18-21 °C, and air temperature range at night 20-22 °C and at day 24-27 °C have been defined as optimum (Williams, 1993). If medium temperature reaches below 24 °C, these require to be bottom-heated (Dole and Wilkins, 1999). Temperature above 30 °C to the stock plants will weaken branching, *vis-à-vis* cutting diameter (Faust and Heins, 1996). For rooting, the temperature should never drop below 21-22 °C, however, the optimum requirement is of 24-25 °C. Reduced light levels and increased air circulation will make poinsettia stock plants capable of tolerating warm temperatures. If required, night temperatures may be reduced to 18 °C, however, cooler temperatures may cause increase in premature flower initiation (splitting), leaf mottling and reduced cutting production (Williams, 1993). Temperatures below 10 °C increase the chances of root rot infections (Berghage *et al.*, 1987). Light levels above 2,000 fc will cause leaf burning and increase in wilting though below 1,000 fc will slow down rooting and will result in stem elongation, however, the light intensity should be increased as soon as cuttings root (Dole and Wilkins, 1999). Relative humidity should be high, *i.e.* 70-80 per cent when cuttings begin rooting but after rooting is accomplished the humidity is lowered. The size of terminal tip cutting (commercial propagation method) should be 7.5 cm and thickness 6 mm with at least one fully mature leaf intact. Direct 'sticking' into final pot saves lobour, production costs, keeps the plants healthy as foliar infection is reduced due to large spacing, and decreases production time by 1 week though requires more propagation space than using foam or rockwool blocks. In this case, medium should be light, porous, non-waterlogged, fertile but should have good water holding capacity. Cuttings will ooze a milky sap which can be stopped by placing the cut end in water before potting. Terminal cuttings are dipped in IBA at 0.1 to 0.3 per cent for 3 minutes and planted in sand medium under mist for good results where these root within 10-14 days and these accomplish full rooting within 21 days.

Small cuttings stripped of their basal foliage in case of ***E. leucocephala*** should be taken from mature new growth and rooted during summer. These cuttings should be inserted into well-drained moist soil suitable for potting mixture and kept under filtered situation having high humidity or if possible, in mist chamber. When these root well in 4-6 weeks, these are transplanted into individual pots or in already prepared pits. Though not so urgent but IBA at 500 ppm will certainly improve rooting percentage.

Slow growth rate of plantlets, few micro-shoots per explant, and slow root growth rate are restrictions of *in vitro* propagation of poinsettia (*Euphorbia pulcherrima*), that is why growers resort to producing their own cuttings from stock plants. Dinum Perera and Brian Trader (2010) with poinsettia 'Prestige Red' developed an efficient *in vitro* proliferation. The explants (apical and axillary buds) placed on MS basal medium containing 6-benzylaminopurine (BA) and combinations of BA and IAA, mostly produced red callus. Addition of IAA into the rooting medium increased rooting efficiency. Plantlets grown on half-strength MS salts and vitamins with 28.5 mM IAA initiated rooting 11 days earlier than the plantlets grown with no PGRs.

Cultural Practices

It is widely grown in tropical and subtropical climates. Forenoon sun, and shading in hotter part of the day is necessary for proper colour development. It is very comfortable in the temperature range of 12-30 °C and the persisting 10 °C temperature is not congenial for its growth and development, albeit it incites infection of root rot disease in case of cutting propagation (Berghage *et al.*, 1987). For pot planting, one cutting per 10-15 pot and 1 to 3 cuttings per 20 to 25 cm pots should be planted. It can also be planted in hanging baskets and there only dwarfest varieties should be used. Thiride is recommended at one week after planting to reduce chances of disease infection during early phases of production. **Medium** or **soil** for pot culture should be well-drained, porous, light and with a pH of 5.6 to 6.2. Sandy loam **soil** is preferable. Dolomite lime is preferred for media pH control. Poinsettias prefer a more acidic medium than many other flowering plants. Care should be taken to keep pH within the preferred range. Though the crop is highly tolerant of high media EC even then sometimes problems of yellowing, abscission and marginal necrosis of lower leaves, stunting and increased susceptibility to root and stem diseases, and in exceptional case upper foliage may show interveinal chlorosis. Therefore the level of EC should be below 3 dS/m in root medium (Yelanich and Biernbaum, 1990) when using saturated medium extract or 1 dS/m when using 1:2 medium:water from mid October until flowering.

In commercial production, many growers use black cloth to either produce earlier crops or to make the entire crop more uniform. Higher **light** decreases plant height and increases bract colour, and light less than 3,500 foot candles is detrimental to growth. In fact, 3,500-4,500 fc should be used with dark foliage cultivars and 5,000-6,000 fc for other cultivars provided greenhouse temperatures are maintained below 32 °C (Williams, 1993). Before rooting, 20-30 per cent shading may be provided but soon after rooting the shading is removed. Light levels can be brought down to 2,000 fc after bracts are fully mature otherwise bract fading and burning may occur (Dole and Wilkins, 1999). Day **temperatures** can vary between 24-27° C and night temperatures should remain above 15° C throughout production period. Poinsettia crops are delayed when night temperatures drop below 10 °C. Once bracts are fully formed, many growers lower temperatures to 10 °C at night to intensify bract colouration. Rooted cuttings should be planted in July-August for better establishment. *Euphorbia pulcherrima* cvs 'Lilo' and 'Starlight' were grown under five different day/night temperature (DT/NT) regimes from the start of short days (SD) to flowering for two different starting periods. The average daily temperature (ADT) was kept at 18–20 °C. Plants were given supplementary lighting with high pressure sodium lamps (HPS) for 10 h day^{-1} at three photosynthetic photon flux densities (PPFD), *i.e.* 12, 37 and 73 mol m^{-2} s^{-1}. Both plant height and plant diameter were lower when the plants were grown at DT < NT (negative DIF) compared with plants grown at DT > NT (positive DIF) or with constant temperature (zero DIF) (Moe *et al.*, 1992). DIF is monitoring of day and night temperatures for height control. Reduction of temperature to 13-16 °C enhances bract colour.

Poinsettias may be grown without **pinching** for production of large inflorescences but this way only one cutting will have to be taken per 15 cm or larger pot though pinching induces 4-7 inflorescences per pot. Pinching to 6-7 nodes should be carried out at approximately 2-3 weeks after planting. First pinch is removal of 1.25 to 2.50 cm of the top growth. If the plants are allowed to grow larger before pinching, crop delays can result. If a grower is harvesting cuttings from plants while pinching, the pinch should be delayed until 3-4 weeks after planting. Pinching the plant during growing season will result in a compact plant at flowering time. Berghage *et al.* (1987) described three types of pinching ~ soft, hard and very hard. With the soft pinch only apex is removed above a young immature leaf and it is done only when propagation and plant establishment are delayed, a hard pinch is the most common pinch and it is done above the first fully mature leaf and this pinch only produces uniform and vigorous axillary shoots, and a very hard pinch is made in older stem tissue which results in erratic branching and poor quality plants (Dole and Wilkins, 1999). More gap between pinching and start of the SD encourages good vegetative growth and node numbers, therefore, tall varieties require to be propagated later in the season than the short varieties.

Fertilizer dose of 225-300 ppm N in over-head irrigated plants, and 100-225 ppm N for subirrigated plants are recommended (Berghage *et al.*, 1987; Ecke *et al.*, 1990; Yelanich and Biernbaum, 1990; Dole *et al.*, 1994). The requirement of fertilizer is about 25 per cent less in dark-leaf cultivars than light-green leaf ones. Fertilizers including superphosphate, dolomitic limestone and micronutrients should be incorporated into the medium before planting but only after medium analysis (Dole and Wilkins, 1999). Controlled-release fertrilizers (CRF) are effective with poinsettias and can best be combined with constant liquid fertilization, 50-75 per cent CRF and rest with constant liquid fertilization. Whipker and Hammer (1997) stated that plant nutrtient uptake is lowest at potting and highest at flowering. **Nitrogen** deficiency causes yellowing of leaves starting from older ones. Smith and Dole (1993) stated that in case when CRF is used, application of a drench of 300 ppm N from ammonium nitrate within one week of planting will improve plant quality. Ku and Hershey (1997) advocated application of nitrogen by combining both ammonium and nitrate forms, 60 per cent nitrogen through ammonium or urea-based fertilizers in the summer while only 30 per cent in winter and rest through nitrate form during both the seasons, however, some one month before marketing the plants, application of ammonium form should be stopped. Ammonium form in warm conditions stimulates lush growth with dark green leaf colour but in cool conditions as its conversion into nitrate is quite slow therefore it accumulates and causes yellow chlorotic bands on dull green base and downward curling of young leaves *cum* falling, stunting and poor root growth. Application of any complete fertilizer will prevent **phosphorus** deficiency as its deficiency causes plant stunting, and yellowing *cum* necrosis of older leaves. Obvious **calcium** deficiency symptom is though rare but may be expressed in terms of weak stems, *vis-à-vis* leaf and bract necrosis so the proper level of calcium in the medium should be maintained and medium pH should be kept at 5.5 or higher and then calcium chloride as foliar spray at 200 to 400 ppm alongwith some spreader should be applied weekly starting from bract colouration to anthesis, however, calcium nitrate is used on stock plants as in flowering plants it causes phytotoxicity of bracts (Dole and Wilkins, 1999). Further they have mentioned the deficiency and toxicity symptoms of various other elements. Poinsettias have high demand for **magnesium** so the plants should be regularly sprayed at the rate of 40 ppm otherwise its deficiency may result yellow mottling of just mature leaves starting from the margin and between veins, and in severe cases the leaves are hardened and become highly chlorotic. Poinsettias are highly sensitive to **boron** deficiency which is expressed in terms of cessation of terminal growth and distortion of stem and leaves whereas when in excess its toxicity is expressed as necrosis or yellowing of older leaf margins. The deficiency symptoms of **molybdenum** is expressed in terms of recently mature leaves showing marginal necrosis, edge burn and upward rolling. A complete micronutrient application is recommended at 1-2 weeks after planting.

Poinsettias need less frequent **watering** than other ornamentals. Soil is allowed to dry deep before irrigating though wilting should not occur otherwise lower leaves will start falling. Controlled watering is not effective in case of height control albeit plants become poor. Due to heavier canopy late in the growth period, sprinkler irrigation will not be effective as water will hardly reach the medium. Regular leaching will have to be required when water quality is not good and EC is high. Overwatering is in no case permitted as this will cause breaking of stems due to heavy growth. Mulches help prevent water loss during hot, windy or sunny weather.

Euphorbia leucocephala plants grow in full to partial sun in any fertile, porous and well-drained soil. If these are to be planted in holes the holes are prepared by digging some 45 cm deep and 90 cm in diameter, the holes are filled with dug up soil after mixing equal volume of FYM and half the quantity of coarse sand. If the soil is heavy, it may be amended through addition of organic matter further. In these pits the plants are planted with intact root balls and from all around the soil is pressed and then watered. Plants become quite dense and compact in habit with exposure to proper illumination. They are pruned back hard in early spring to throw out new and compact shoots with neat and clean foliage mass. After the growths have picked up fully, before mid-summer light trimming may again be carried out for proper shaping but stem tips should be retained intact as these produce flowers later. The plants being phototropic, require a long period of darkness to initiate blooms. After flowering they are pruned to about 1/3rd of the plants to avoid the plants becoming leggy. The plants may be mulched to conserve moisture and to control the weeds and 2-4 times per year they are fertilized with a balanced fertilizer followed by irrigation every time and further as per requirement of the plants and prevailing weather conditions. These are fertilized with a balanced NPK fertilizer first when the plants are picking up growth in summer and then next in the fall. Per pit some 60 grammes of nitrogen, 90 grammes of phosphorus and if the soil is deficient in potash then 40 grammes of potash through some potassic fertilizer should be mixed in the soil before filling in the pit. This dose may be repeated 2 to 4 times at 30 to 45 days intervals.

Growth and Development

To accommodate a wide range of cultivars, flowering dates, container sizes, climates and growing styles, *viz.* single stem, pinched, tubs, hanging baskets, various schedules have been developed, and the most commonly grown type of poinsettia is a single pinch plant in 15 cm pot. As a general rule, about 4 weeks are required for propagation, 1.5 to 3 weeks from planting a rooted cutting until pinching so that plants before pinching are properly established, and 1.5 to 3 from pinching to start of SD so that sufficient growth before flower initiation has occurred (Dole and Wilkins, 1999).

Plants are to be **pruned** after blooming in early spring. They should be cut back to within 30 to 45 cm of the ground level. After four weeks or when it is 30 cm long, new growth should be cut back leaving four leaves on each shoot and it is to be repeated up to September. New growth after the last pinch will usually grow to a length of 20 to 25 cm and, in the first week of October, will initiate flower buds. However, pruning after September does not allow enough time for side shoots to grow and develop before bud initiation in early October. As a result, the bracts will be much smaller than those on a plant where the last pinch was made before September.

Poinsettia is a **short day** plant as it flowers when the nights are longer (more than $11^3/_4$ hours) than days. Cloudy weather before flower initiation time, i.e even longer night lengths than that required for initiation as is also accomplished naturally due to lengthening of nights during autumn, causes earlier flower initiation and flower development (Dole and Wilkins, 1999). A 14 h night treatment by covering the crop with black cloth induces earlier flowering than those provided natural photoperiods. Berghage *et al.* (1987) stated that primary cyathium forms only after 4-8 short days treatment which can be seen within 15 days under microscope but the treatment should continue further up to anthesis stage (presence of pollen on cyathia) so that proper bract is developed. LD treatment to the plants after flower initiation but before anthesis may cause leaf-like and greening of immature bracts (Dole and Wilkins, 1999). Grueber (1985) also stated that night temperatures above 21 °C may delay flower initiation.

The gap between pinching and start of SD treatment, day-night temperature manipulation (DIF), light intensity, plant spacing and use of chemicals may suppress the **plant height** of poinsettia in pots. **Pinching** of tall or otherwise varieties late in the season and then subjecting them to SD treatment controls the height. Berghage and Heins (1991) stated that difference between night and day temperatures control height of the plants. Irrespective of average daily temperature, the plants subjected to the same level of **DIF** (16 and 21 °C) will result into similar plant height, which may be used in altering the plant height by altering day and night temperature treatments to the plants, and thus proper DIF may be worked out. Increase of DIF above zero will greatly influence the plant height than a similar reduction below zero DIF. Temperature being lowered in the morning starting at sunrise (negative DIF) for 2-6 hours (the longer the better) will effectively reduce the plant height as compared to maintaining a low DIF all the day (Cockshull *et al.*, 1995; Moe *et al.*, 1995). After pinching, the growth of poinsettia is divided into three phases ~ (i) axillary shoots developing slowly for 1 to 2 weeks after the pinch, (ii) stems elongate rapidly, and (iii) stem elongation slows as cyathia and bracts develop therefore uncontrolled temperature and inadequate growth retardant application during second phase will result in taller plants (Dole and Wilkins, 1999). Growth regulators [cycocel, B-Nine, A-Rest (ancymidol), Bonzi (paclobutrazol) and Sumagic (uniconazole)] also control plant height in poinsettia. Except A-Rest which is applied as drench, others are applied as sprays. **CCC** is the most commonly used one at 1,000 to 1,500 ppm concentration, but in case its application causes foliage yellowing or reduction of bract size, its lower concentrations should be used though mild yellowing turns back to green colour soon. **B-Nine** (750-2,500 ppm) is usually not used alone but is combined with CCC (1,000-1,500 ppm) though in case of phytotoxicity the concentration in multiple applications may be lowered. In warm climates 1,500 ppm CCC and 2,500 ppm B-Nine and in cooler conditions 800 ppm B-Nine and 1,000 ppm CCC but the mix should not be used after start of the SD, and should be used only when by any other chemical or means the height is unable to be controlled. Barrett (1996) recommended spray applications of **Bonzi** at 5 to 20 ppm and **Sumagic** at 2 to 5 ppm in cool regions though in warmer regions the concentrations may be increased from 30 to 45 ppm for Bonzi and 5 to 10 ppm for Sumagic, however, after start of SD none of these chemicals should be applied. The Bonzi drench is more effective than spray applications, and it should be used from 0.5 to 3 ppm, and on var. 'Eckespoint Freedom' it was 1 ppm in southern US (warmer part) while 0.5 to 0.75 ppm in northern areas (cooler part), and depending on the vigorosity of the variety *cum* regional climate, *i.e.* warmer region the concentration may be raised up to 3 ppm (Barrett, 1996). This has also been recorded that spray or drench applications of Bonzi can reduce bract size when applied from flower initiation to the end of October though early November drench application is not as injurious, and spray application any time after flower initiation is rather more injurious than drenching as it delays flowering and reduces bract size (Barrett, 1996). James *et al.* (2001) reported that Bonzi (paclobutrazol) drenches (0.118 mg/container) controlled late-season stem elongation in poinsettia cv. `Freedom Red' grown under natural photoperiods. About **A-Rest**, Barrett (1996) mentions that as drench it can be applied at 1 to 4 ppm, 1 to 2 ppm in cool areas and 2 to 4 ppm in warmer climates, however, if the medium has pine bark, Bonzi and A-Rest will be less effective and then the rates will have to be increased by up to 25 per cent. PGRs should be applied only after the pinch when axillary shoots are 4 to 5 cm long, and in warm climates their application to the stock plants should be made before propagation, to cuttings during propagation, and to young plants 3 to 7 days before the pinch (Dole and Wilkins, 1999). In America, the peak demand is in the first week of December but for earlier flowering as occurs by certain cultivars by mid November in North America, the growers will have to give artificial SD during late August and September (Dole and Wilkins, 1999), however, in producing tree poinsettias, multiple GA_3 sprays at 250 to 750 ppm to the well established cuttings in the pots should be given to promote rapid stem elongation prior to pinching (Mynett and Wilkońska, 1989).

In case of *Euphorbia leucocephala*, ancymidol and chlormequat are effective in reducing plant height. Ancymidol increases five times more number of cyathia through application of 0.25 mg per pot though highest quality plants are produced through 0.125 mg ancymidol per pot.

Post Harvest

In poinsettia, reduction of temperature, nutrition and light is required 2-3 weeks before marketing. Flowers are harvested when bracts are fully developed and expanded and pollen is visible in 1-2 cyathia. In most of the varieties, bracts develop full colour before cyathia mature. Before maturation of bracts if plants are transported the young bracts may turn green and may not develop proper colour. Over-mature plants are not suitable for transportation as they are susceptible to bract necrosis, bract discolouration, leaf drop and *Botrytis* infection. After maturation of bracts the light intensity is reduced to 2,000 fc to save the bracts from sun-burning and fading. Sleeving is not desirous due to build up of ethylene. During transit the temperatures should be 10 to 16 °C, preferably 13 °C, however, below 10 °C temperatures will cause falling of leaves. A vase life of 20.6 days was recorded after cut poinsettia cv. 'Renaissance Red' was placed directly into vases with 22 °C de-ionized water + 200 ppm 8-HQS and up to 1 per cent sucrose without floral foam. Maturity of stems at harvest, ranging from 0 to 4 weeks after anthesis, had no effect on vase life or days to first abscised leaf. Holding stems in the standard floral solution increased vase life and delayed leaf abscission compared with deionized or tap water only, with further improvement when stem bases were re-cut every three days. A 1 per cent to 2 per cent sucrose concentration in the vase solution produces longest postharvest life for stems placed in foam but has little effect on stems not placed in foam (Dole *et al.,* 2004). Thompson and Kofranek (1978) reported that sleeved poinsettia cvs. 'Annette Hegg Supreme' and 'Annette Hegg Dark Red' stored best at l0 °C. They found lower temperatures (2-7 °C) inducing chilling damages as manifested mainly by bract bluing while higher temperatures (up to 16 °C) resulted into increased leaf petiole epinasty and bract drooping. The bract blueing and leaf petiole epinasty disorders were highly pronounced as storage duration increased from 2 to l0 days though bract drooping in the same period was observed decreased. Plants sleeved and stored in paper were generally of higher quality upon removal than those sleeved and stored in plastic. Under relatively static conditions (l5m/minute air speed), poinsettias freeze at about -4 °C. Sleeving poinsettias delayed low-temperature damage. The injury of sleeved poinsettias was related to temperature, air speed and exposure time. However, the sleeves from poinsettia should be removed immediately after reaching the assigned destination otherwise petiole epinasty (twisting and curling of upper foliage and bracts) may occur, and they are also sensitive to exogenous ethylene. After opening the consignments, poinsettias are put in water and cooled to 16 °C whereas cut flowers can be stored at 15 °C, *vis-à-vis* exposed to 75-225 fc light.

Insect-pests, Diseases and Physiological Disorders

The adult **whitefly** looks like a small fly covered with white powder and can be found either on the plant or soil surface. The adults are more of a nuisance than a real problem, but the larvae do feed on leaf tissues and their severe infestation distorts the leaves and make their surface sticky. These excrete honey dew so invite black sooty mould. Greenhouse whitefly (*Trialeurodes vaporariorum*) and the silverleaf whitefly (*Bemisia argentifolia*) are of special importance, where the latter has been found to be more attracted to fertilized vegetative poinsettias due to more nitrogen being present there (Benz *et al.,* 1995). In both the cases the whiteflies are covered with a waxy powder. Their bright white eggs soon change to dark gray and their pupae remain perpendicular to the leaf surface. In its infestation the plants should be sprayed with 0.2 per cent Pyrethrum or methyl parathion. **Fungus gnat** (*Bradysiai* spp.) is smaller than shorefly, the antennae longer than its head, and wings have a Y-shaped vein starting half the way from the end of the wing, and the larvae are whitish with black heads. Fungus gnat larvae feed on roots, stems and leaf tissues, and their population is favoured by peat moss and abundance of moisture. Fungus gnat adults transmit a variety of diseases including *Pythium* and *Thielaviopsis*. **Shore fly** (*Scatella stagnalis*) is larger than fungus gnat, has shorter antennae than head, wings almost parallel-veined, and larvae light tan with no clear heads. Its larvae do not damage the crop but the adults are a nuisance and also transmit disease. Controlling growth of algae through hydrated lime, copper sulphate or bromine, maintaining proper sanitation, aeration and plant spacing, *vis-à-vis* following sub-irrigation system will keep both the pests under check. Predatory microorganisms such as *Bacillus thuringiensis* var. *israelensis*, mites such as *Hypoaspis miles* and nematodes such as *Steinernema bibionis, S. carpocapsae* and *S. feltiae* predate over the larvae of these pests. Drenching the pots with insecticides or through granules will control larvae and aerial spray of insecticides will kill adults. **Mealy bugs** having a waxy powdery covering are very difficult to control. The young tend to move around until they find a suitable feeding spot where these hang out in colonies and feed. Mealy bugs through sucking the plant sap distort and yellow the foliage and weaken the plants, and in their serious attack leaves start falling. They also excrete a sweet substance called honeydew which invites ants and this can lead to an unattractive black fungal growth called sooty mould. Ladybird predator (*Cryptolaemus montrouzieri*) predates over this pest and controls this partially. Some of the suggested insecticides include Monocrotophos at1.7 ml/l of water or Imidacloprid at 0.5ml/l of water or Carboryl at 3 g/l of water which may be sprayed as and when these pests are noticed. **Aphids** (*Aphis gossypii, Myzus persicae* and many others) suck plant sap from their tender parts, especially floral organs and tender leaves. Their infestation causes plant stunting, and deformed leaves and buds, apart from transmitting harmful plant viruses. Aphids also excrete a sweet substance called honeydew which attracts ants and can lead to a black fungus growth known as sooty mould. They are often massed at the tips of branches feeding on succulent tissues. Parasitic wasp (*Aphidius colemani*), aphid midge (*Aphidoletes aphidomyza*), lacewing predators (*Chrysoperla carnea, C. rufilabris*), Asian beetle (*Harmonia axyiidus*), and parasitic fungus (*Beauveria bassiana*) predate over aphids in the garden. While controlling the mealy bugs, aphids will also be controlled. **Lewis mite** (*Eotetranychus lewis*), a serious pest of poinsettia thrives in high humid conditions (80 per cent or more) and cool temperatures of 16 °C, especially in heated houses causing plants to appear yellow and stippled though its heavy infestation causes even leaf drop and ultimate death of the plants, and also this webs the affected part. It ranges in colour from red to green to yellow. To manage the mites, infested plants are burnt and field is kept weed-free, dryness in the polyhouse is maintained and glasshouse temperature is maintained higher than 16 °C. Use of predator (*Neoseiulus fallacis*), predatory beetles (*Metaseiulus occidentalis* or *Stethorus punctum*), predatory midge (*Feltiella acarisuga*) and predatory mites (*Neoseiulus californicus*, *Phytoseiulus persimilis* and *P. longipes*) will predate over this pest. Thorough spraying of Dicofol at 1.7 ml/l or any other systemic insecticide will control this pest.

Poinsettias are susceptible to many diseases from propagation to flowering. **Gray mould** (*Botrytis cinerea*) during production period can cause tan to brown water-soaked lesions on poinsettia stems, flower bracts and leaves, during propagation the cuttings are destroyed and during postharvest this affects the nectaries and cyathia clusters as symptomized by covering of gray fuzzy fungus under high humid conditions. The fungus readily invades already damaged tissues, including those caused by bract burn. After the true flowers have fully developed, *Botrytis* attacks and covers them with the typical smoky and gray spores. Poinsettias are particularly susceptible to gray mold late in the season when it can cause significant damage to the aesthetic quality of plants. To manage this disease the field or polyhouse will have to be properly cleaned, aerated, and the humidity should be maintained below 80 per cent. Fungicides including Chlorothalonil @ 0.2 per cent and copper + mancozeb can be used to protect foliage. Kutek and Floryszak-Wieczorek (2002) reported benzothiadiazole (BTH) at 0.3 mM highly effective in increasing resistance of two poinsettia cvs 'Coco White' and 'Malibu Red' which are moderately susceptible to *Botrytis cinerea*. It was also observed that the applied inducer at 0.03 and 0.3 mM had a favourable influence on the increase of poinsettia systemic resistance of SAR type (systemic acquired resistance). **Powdery mildew** (*Oidium* spp.) disease is very common on poinsettias as its infection spreads a powdery white covering on foliage, and this is favoured by a mild day temperature below 29 °C, cool nights, reduced light intensities and no-free moisture on the plant. The white fungal colonies mar the beauty of leaves as well as of coloured bracts. Lowering the humidity in the polyhouse will keep this disease under check. The most effective sprays for control are Triadimefon, Triflumizole and Myclobutanil at the rate of 0.3 per cent. ***Pythium* root rot** is the serious disease of poinsettia affecting rooted cuttings. In its infection the base of the cutting becomes brown and water-soaked. Surviving infected plants are stunted and these plants often wilt during hot day-time though in the night appear to be recovering. The outer layer of root tissues of infected roots of established plants strip off leaving a bare strand of inner vascular tissue exposed. Contact fungicide drench of the planting at 1-2 months intervals is quite effective. For the management of *Pythium* root rot, biological agents like *Trichoderma*, *Bacillus*, *Gliocladium*, and *Streptomyces* are also applied to the potting mix. In general, it is recommended that fungicides should not be applied to the potting mix 10 days before and 10 days after application of biological control agents. Biological control agents and fungicides (Metalaxyl MZ @ 0.2 per cent) may have to be applied more than once in order to maintain adequate protection throughout the season, particularly on stock plants. *Rhizoctonia solani*, *Thelaviopsis basicola*, *Phytophthora parasitica* and *Fusarium* spp. also cause root and stem rot which can also be controlled like to that of *Pythium*. Ecke *et al.* (1990) reported *Rhizopus stolonifera* blight, *Choanephora cucurbitarum* wet rot, *Alternaria euphorbicola* blight, *Corynespora cassiicola* bract and leaf spot and *Sphaceloma poinsettiae* scab infecting poinsettias.

Bacterial stem and leaf spot (*Erwinia carotovora*) is very common during propagation in hot sunny weather when cuttings receive heavy misting, therefore, foliage will have to be kept dry and the field should be kept clean. *E. chrysanthemi* also affects this crop causing stem rot. *Pseudomonas viridiflava* causes greasy canker and *Corynebacterium flaccumfaciens* bacterial canker. During production, ***Xanthomonas campestris*** causes leaf spot of pinhead type, dull gray to brown and slightly water-soaked. As the lesions mature they turn yellow to tan and are scattered across whole leaf surface. Sometimes they enlarge and become angular in shape due to limited movement across leaf veins. Severe infections can cause distortion of new leaves as well as complete chlorosis and finally abscission of older leaves. Use of copper bactericides such as COC at 0.3 per cent may be partially effective in controlling this disease but is rarely effective in stopping an outbreak once infection has occurred. All the bacterial diseases are favoured by warm humid conditions and free water so avoiding such conditions will prevent infection of these pathogens in poinsettia plantings.

Silver marbling of leaves on *Euphorbia pulcherrima* occurs due to infection of 'silver leaf virus', and occasional mild mottling of poinsettia due to infection of 'poinsettia mosaic virus' (PnMV) and 'poinsettia cryptic virus'

(PnCV) that occur during cool weather and low light conditions of mild winter with no obvious symptoms. All the virus infected plants should immediately be uprooted and burnt.

Many poinsettias are highly light sensitive so one should be careful that dark cycle is never broken. The glare of a street light or light entering beneath a door is often sufficient to prevent buds from forming. **Bract edge burn** (bract necrosis) is due to high salt concentration in the medium. Small brown dots appear along outer edges of poinsettia bracts in early- to mid-December. This may be due to over-fertilization during flowering when tissue expansion has stopped. It may also indicate a calcium deficiency due to inadequate calcium level in the nutrient supply, localized calcium deficiencies in the bracts, or calcium being tied-up due to high salts. **Marginal bract necrosis** in poinsettia is because of poor calcium uptake due to low calcium levels, low medium temperature, high ammonium medium levels, high humidity and root rot (Woltz and Harbaugh, 1986; Hartley, 1992), high concentrations of other competitive cations such as K and Mg (Woltz and Harbaugh, 1985; Strømme *et al.,* 1994) in planting media, and also due to low medium pH which allows rapid leaching of Ca from the medium. Supplementary light in an early phase of bract development (SD weeks 6 and 7) decreases or prevents marginal bract necrosis while a high light supply later on (SD weeks 9 and 10) increases the occurrence of the disorder exponentially. Weekly calcium chloride sprays at 200 and 400 ppm starting from bract colouring until anthesis prevents bract necrosis but in case of stock plants calcium nitrate is sprayed instead of calcium chloride as the latter causes more phytotoxicity on bracts.Ca spraying under high light supply during a late phase of bract development (Wissemeier *et al.,* 2000) has been found quite effective. Use of 4 mM sodium silica sprays before anthesis or by one 100 ppm benzyladenine spray at anthesis reduces bract necrosis (McAvory and Bible, 1995a,b). **Leaf edge burn** starts as downward curling and marginal yellowing to tan colouration of young leaves due to low Ca levels in the leaf margins which can be corrected by reduction of humidity levels and ammonium-nitrate ratio in the field, *vis-à-vis* Ca spraying (Bierman *et al.,*1990). **Cycocel burn** appears as large irregular spots on the upper leaf surface or near the margin shortly after application of CCC though later it becomes less apparent as the plant matures. In severe burn the leaves may drop and die whereas in slight damage yellow areas may regreen. While applying CCC, one should be careful to use only lower concentrations with repeat applications. **Splitting** (premature initiation and development of terminal flower buds on vegetative or young reproductive plants) is usually surrounded by a whorl of three shoots, each subtended by a leaf so such inflorescences should be removed. Such problems can be prevented by frequent pinching of the stock plants and by application of GA_3 at 10 to 25 ppm (Evans *et al.,* 1992a,b). **Leaf crippling** or distortion is leaf puckering, narrowing, thickening and curving which occurs either on cuttings just after rooting or on axillary shoots just after pinching. Its occurrence on newly rooted cuttings is due to excessive phosphorus absorption which ties up minor elements causing a transitory minor element deficiency (Dole and Wilkins, 1999). Therefore, after fertilization the plants should be rinsed with water. **Centre bud drop** in poinsettia is premature abscission of cyathia due to insufficient carbohydrate accumulation in plants when they are grown under excessively low light conditions, low plant spacing, high night temperatures and water stress (Miller, 1984). **Rabbit tracks** (bilateral bract spots) is the development of silvery spots on either side of the main vein of bracts occurring late in the crop cycle primarily due to high nutritional levels and warm night temperatures (Dole and Wilkins, 1999). Other disorders are yellowing of leaves, their curling and ultimate falling due to too dry and hot weathers and darkness which may be corrected through regular irrigation, feeding and water spraying, and by keeping the plants at a place having sufficient light; leaf shrivelling and drying due to gas fumes so such plants should be shifted to fume-free space; and drooping of whole plant due to draughts so such plants should be moved to a protected place.

References

Anonymous, 1994. Cutting quality. *The Poinsettia*, **9**: 2.

Bailey, L.H. 1960. *Euphorbia*. In: *The Standard Cyclopedia of Horticulture*, vol. I, pp. 1167-1175.The Macmillan Company, New York, USA.

Barrett, J. 1996. Poinsettia height control. *Greenhouse Product News*, **7**(8): 12-14.

Beckett, K.A. 1985. *The Concise Encyclopedia of Garden plants*, pp. 150-151. Orbis Publishing Limited, London.

Beckett, K.A. 1987. *Euphorbia*. In: *The RHS Encyclopaedia of House Plants Including Greenhouse Plants*, pp. 246-249. Salem House Publishers, Massachusetts, USA.

Bempong M.A. and K.C. Sink,1968. Meiotic chromosome analysis in the 42-chromosome poinsettia, *Euphorbia pulcherrima*. *Canadian J. Gen. Cytol.*, **10**(1): 198-199.

Benz, J., J. Reeves III, P. Barbosa and B. Francis, 1995. Within plant variation in nitrogen and sugar content of poinsettia and its effects on the oviposition pattern, survival and development of *Bemisia argentifolii* (Homoptera: Aleyrodoideae). *Population Ecol.*, **24**: 271-277.

Berghage, R.D. and R.D. Heins, 1991. Quantification of temperature effects on stem elongation in poinsettia. *J. Amer. Soc. hort. Sci.*, **116**: 14-18.

Berghage, R.D., R.D. Heins, W.H. Carlson and J. Biernbaum, 1987. *Poinsettia Production*. Michigan State University Extension Bulletin E-1382. East Lansing, Michigan, USA.

Bierman, P.M., C.J. Rosen and H.F. Wilkins, 1990. Leaf edge burn and axillary shoot growth of vegetative poinsettia plants. Influence of calcium, nitrogen form, and molybdenum. *J. Amer. Soc. hort. Sci.*, **115**: 73-78.

Cockshull, K.E., F.A. Langton and C.R.J. Cave, 1995. Differential effects of different DIF treatments on chrysanthemum and poinsettia. *Acta Hort.*, No. 378, pp. 15-25.

Dinum Perera and Brian W. Trader, 2010. Poinsettia 'Prestige Red' (*Euphorbia pulcherrima) in vitro propagation. HortSci.*, **45**(7): 1126–1128.

Dole, J.M. and H.F. Wilkins, 1991. Vegetative and reproductive characteristics of poinsettia altered by a graft-transmissible agent. *J. amer. Soc. hort. Sci.*, **116**: 307-311.

Dole, J.M. and H.F. Wilkins, 1992. *In vitro* characterization of the graft-transmissible agent in poinsettia. *J. Amer. Soc. hort. Sci.*, **117**: 972-975.

Dole, J.M. and H.F. Wilkins, 1994. Graft-transmissible free-branching agent. In: *The Scientific Basis of Poinsettia Production* (ed. Strømme, E.), pp. 45-48. Agricultural University of Norway, Aas, Norway.

Dole, J.M. and H.F. Wilkins, 1999. *Euphorbia*. In: *Floriculture Principles and Species*, pp. 331-347. Prentice Hall, New Jersey, USA.

Dole, J.M., H.F. Wilkins and S.L. Desborough, 1993. Investigations on the nature of a graft-transmissible agent in poinsettia. *Canad. J. Bot.*, **71**: 1097-1101.

Dole, J.M., J.C. Cole and S.L.von Broembsen, 1994. Effect of irrigation methods on water use efficiency, nutrient leaching and growth of poinsettias. *HortSci.*, **29**: 858-864.

Dole, J.M., Paul Fisher and Geoffrey Njue, 2004. Optimizing postharvest life of cut 'Renaissance Red' poinsettias. *HortSci.*, **39** (6): 1366-1370.

Ecke, P. Jr., O.A. Matkin and D.E. Hartley, 1990. *The Poinsettia Manual* (3rd ed.). Paul Ecke Poinsettias, Encinitas, California, USA.

Evans, M.R., H.F. Wilkins and W.P. Hackett, 1992a. Meristem ontogenetic age as the controlling factor in long-day floral initiation in poinsettia. *J. Amer. Soc. hort. Sci.*, **117**: 961-965.

Evans, M.R., H.F. Wilkins and W.P. Hackett, 1992b. Gibberellins and temperature influence long- day floral initiation in poinsettia. *J. Amer. Soc. hort. Sci.*, **117**: 966-971.

Ewart, L. C. and Walker, D. E. 1960. Chromosome numbers of poinsettia, *Euphorbia pulcherrima* Klotzsch. *J. Hered.*, **51**: 203-208.

Faust, J.E. and R.D. Heins, 1996. Axillary bud development of poinsettia 'Eckespoint Lilo' and 'Eckespoint Red Sails' (*Euphorbia pulcherrima* Willd.) is inhibited by high temperatures. *J. Amer. Soc. hort. Sci.*, **121**: 920-926.

Hartley, D. 1992. Bract edge burn. *The Poinsettia*, **4**: 2-7.

Hay, R. and K.A. Beckett, 1971. *Reader's Digest Encyclopaedia of Garden Plants and Flowers*, pp. 267-271. The Reader's Digest Association Limited, London.

Gopalaswamienger, K.S. 1991. *Complete Gardening in India*, pp. 443-445. Published by Gopalaswamy Parthasarthy, Bangalore, India.

James, E. Faust, Pamela C. Korczynski and Robert Klein, 2001. Effects of paclobutrazol drench application data on *Poinsettia* height and flowering. *Hort Techn.*, **11**(4): 557-560.

Kutek, B. and J. Floryszak-Wieczorek, 2002. Local and systemic protection of poinsettia (*Euphorbia pulcherrima* Willd.) against *Botrytis cinerea* Pers. induced by benzothiadiazole. *Acta Physiologiae Plant.*, **24**(3): 273-278.

Ku, S.M.C. and D.R. Hershey, 1997. Growth response, nutrient leaching, and mass balance for potted poinsettia. I. Nitrogen. *J. Amer. Soc. hort. Sci.*, **122**: 452-458.

Lee, I.M., M. Klopmeyer, I.M. Bariszyk, D.E. Gundersen-Rindal, T.-S. Tau, K.I. Thomson and R. Eisenreich, 1997. Phytoplasma induced free-branching in commercial poinsettia cultivars. *Nature Biotechn.*, **15**: 178-182.

McAvory, R.J. and B.B. Bible, 1995a. Benzyladenine and daminozide sprays applied after initial anthesis affect bract necrosis in poinsettia (abstr.). *HortSci.*, **31**:584.

McAvory, R.J. and B.B. Bible, 1995b. Silica sprays reduce the incidence and severity of bract necrosis in poinsettia. *HortSci.*, **31**: 1146-1149.

Miller, S.H. 1984. Environmental and physiological factors influencing premature cyathia abscission in *Euphorbia pulcherrima* Willd. M.S. Thesis, Michigan St. University, East Lansing, Michigan, USA.

Moe, R., K. Willumsen, I.H. Ihlebekk, A.I. Stupa, N.M. Glomsrud and L.M. Mortensen, 1995. DIF and temperature drop response in SDP and LDP, a comparison. *Acta Hort.*, No. 378, pp. 27-33.

Moe, R., T. Fjeld and Leiv M. Mortensen, 1992. Stem elongation and keeping quality in poinsettia (*Euphorbia pulcherrima* Willd.) as affected by temperature and supplementary lighting, *Scient. Hort.*, **50**(1-2): 127-136.

Mynett, K. and A. Wilkońska, 1989. Growth regulators application in the shape forming of some pot plants. *Acta Hort.*, No. **251**, pp. 311-314.

Ruiz-Sifre, G., J.M. Dole, B.A. Kahn, P.E. Richardson and J. Ledford, 1997. Correlation of poinsettia graft union development with transmission of the free-branching characteristic. *Sci. Hort.*, **69**: 135-143

Smith, R.M. and J.M. Dole, 1993. Effects of supplemental NH_4NO_3 drenches and no-leach production on poinsettias grown with controlled-release fertilizers (abstr.). *HortSci.*, **28**: 549.

Strømme, E., A.R. Sewlmer-Olsen, H.R. Gislerød and R. Moe, 1994. Cultivar differences in nutrient absorption and susceptibility to bract necrosis in poinsettia (*Euphorbia pulcherrima* Willd. Ex Klotzsch.). *Gartenbauwissenschaft*, **59**: 6-12 (in German).

Thompson, J. F and A.M. Kofranek, 1978. Postharvest characteristics of poinsettias as influenced by handling and storage procedures. *J. Amer. Soc. hort. Sci.,* **103**(6): 712-715.

Whipker, B.F. and P.A. Hammer, 1997. Nutrition uptake in poinsettia during different stages of physiological development. *J. Amer. Soc. hort. Sci.*, **122**: 565-573.

Williams, J.E. 1993.Stock plant production from the ground up. *The Poinsettia*, **5**: 2-15.

Wissemeier, A.H., A.-K. Puschel., F. Weinhold and W.J. Horst, 2000. Effect of light regime on marginal bract necrosis in poinsettia. *Proc. ISHS* (*Acta Hort.*), No.519. *XXV int. hort. Congr.*, Part 9: Computers and Automation, Electronic Information in Horticulture, Belgium.

Woltz, S.S. and B.K. Harbaugh, 1985. Effect of nutritional balance on bract and foliar necroses of poinsettia. *Proc. Fla St. Hort. Soc.*, **98**: 122-123.

Woltz, S.S. and B.K. Harbaugh, 1986. Calcium deficiency as the basic cause of marginal bract necrosis of 'Gutbier V-14 Glory' poinsettia. *HortSci.*, **21**: 1403-1404.

Yelanich, M.V. and J.A. Biernbaum, 1990. Effect of fertilizer concentration and method of application on media nutrient content, nitrogen runoff and growth of *Euphorbia pulcherrima* V-14 Glory. *Acta Hort.*, No. 272, pp. 185-189.

17

Ferns and Allied Plants

R.K. Dubey, Sanyat Misra, Parminder Singh, R.L. Misra and Simrat Singh

(**Group**: Pteridophyta; **Order**: Filicales; and **Families**: Actiniopteridaceae, Adiantaceae, Aspidiaceae, Aspleniaceae, Athyriaceae, Blechnaceae, Ceratopteridaceae, Cyatheaceae, Davalliaceae, Dennstaedtiaceae, Dicksoniaceae, Gleicheniaceae, Hemionitidaceae, Hymenophyllaceae, Lomariopsidaceae, Marattiaceae, Marsiliaceae, Oleandraceae, Onocleaceae, Ophioglossaceae, Osmundaceae, Parkeriaceae, Polypodiaceae, Pteridaceae, Salviniaceae, Schizaeaceae, Selaginellaceae, Sinopteridaceae, Vittarriaceae; Salviniaceae for *Salvinia* and Selaginellaceae for *Selaginella*).

[**Common names for ferns**: Adder's tongue fern (*Ophioglossum vulgatum*), American wall fern (*Polypodium virginianum*), Australian maidenhair/Rose maidenhair (*Adiantum hispidulum,* syn. *A. pubescens*), Australian tree fern/Tasmanian tree fern (*Dicksonia antarctica*), Banana leaf fern (*Microsorium musifolium*), Bear's foot fern (*Humata tyermannii*), Bear's paw fern (*Phlebodium aureum*), Beech fern (*Phegopteris*), Bird's nest fern (*Asplenium nidus-avis*), Black tree fern/Mamaku (*Cyathea medullaris*), Bladder fern (*Cystopteris*), Boston fern (*Nephrolepis exaltata* var. *bostoniensis*), Brake fern (*Pteridium, Pteris*), Brittle bladder fern (*Cystopteris alpina, C. regia*), Bristle fern (*Trichomanes*), Buckler fern/Wood fern (*Dryopteris*), Button fern (*Pellea rotundifolia*), Californian gold fern/Gold fern (*Ceropteris triangularis*), Chain fern (*Woodwardia virginica*), Christmas fern (*Polystichum acrostichoides*), Cinnamon fern/Fiddleheads (*Osmunda cinnamomea*), Cliff brake (*Pellaea*), Climbing bird's nest fern (*Microsorium polycarpon*), Climbing fern/Japanese climbing fern (*Lygodium japonicum*), Common polypody (*Polypodium vulgare*), Creeping fern (*Microgramma*), Crested buckler fern (*Dryopteris cristata*), Crown stag's horn (*Platycerium coronarium,* syn. *P. biforme*), Cup fern (*Denstaedtia*), Deer fern (*Lomaria*), East Indian holly fern (*Polystichum aristatum,* syn. *Arachnoides aristata*), Ebony spleenwort (*Asplenium platyneuron*), Elephant's ear fern (*Elaphoglossum crinitum,* syn. *Acrostichum crinitum; Platycerium angolense*), Elk's horn fern (*Platycerium alcicorne, P. hillii*), Female fern/Lady fern (*Asplenium filix-faemina*), Filmy fern (*Hymenophyllum*), Floating fern/Water fern (*Ceratopteris pteridoides*), Florida ribbon fern (*Vittaria lineata*), Flowering fern (*Anemia, Osmunda regalis*), Giant holly fern/Western sword fern (*Polystichum munitum*), Golden-scaled male fern (*Dryopteris borreri,* syn. *D. pseudomas*), Green brake fern (*Pellaea viridis,* syn. *P. adiantoides, P. hastata*), Grape fern (*Botrychium*), Hard fern (*Blechnum spicant*), Hard shield fern/Prickly shield fern (*Polystichum aculeatum*), Hare's foot fern (*Davallia canariensis, Polypodium aureum,* syn. *Phlebodium aureum*), Hart's tongue fern (*Phyllitis scolopendrium,* syn. *Asplenium scolopendrium*), Hartford fern (*Lygodium palmatum*), Hay-scented buckler fern (*Dryopteris aemula*), Hay-scented fern (*Dennstaedtia punctilobula*), Holly fern (*Polystichum lonchitis*), Interrupted fern (*Osmunda claytoniana*), Japanese buckler fern/Shield fern (*Dryopteris erythrosora*), Japanese holly fern (*Polystichum falcatum,* syn. *Aspidium falcatum, Cyrtomium falcatum*), Japanese painted fern (*Athyrium goeringianum*), Jointed pine fern (*Phlebodium subauriculatum*), Lace fern (*Cheilanthes gracillima, Dryopteris intermedia, Nephrolepis exaltata* var. *whitmannii*), Lacy ground fern (*Dennstaedtia davallioides*), Lacy pine fern (*Polypodium *subauriculatum*), Ladder fern (*Nephrolepis exaltata, Pteris vittata*), Lady fern (*Athyrium filix-femina*), Leather fern/ Leather-leaf fern (*Acrostichum aureum, Rumohra adiantiformis*), Lip fern (*Cheilanthes*), Liquorice fern (*Polypodium glyrrhiza*), Maidenhair fern (*Adiantum capillus-veneris, A. pedatum* and other *Adiantum* spp.), Maidenhair spleenwort/ Spleenwort (*Asplenium trichomanes*), Male fern (*Dryopteris filix-mas*), Marsh fern (*Dryopteris thelypteris*), Mother fern (*Asplenium daucifolium,* syn. *A. viviparum*), Mother spleenwort (*Asplenium bulbiferum*), Narrow-leaved strap fern/Narrow-leaved ribbon fern (*Polypodium angustifolium*), Necklace fern (*Asplenium flabellifolium*), Oak fern (*Phegopteris dryopteris*), Oak leaf fern (*Drynaria quercifolia*), Ostrich fern/Osrich feather fern/Shuttlecock fern (*Matteuccia struthiopteris*), Parsley fern (*Cryptogramma crispa*), Pod fern (*Ceratopteris thalictroides*), Purple-stemmed cliff brake (*Pellaea atropurpurea*), Rabbit's foot fern (*Davallia fijiensis,* syn. *Davallia solida fijiensis*),Rattlesnake fern (*Botrychium virginianum*), Regal elkhorn fern (*Platycerium grande*), Resurrection fern (*Phlebodium polypodioides*), Ribbon fern/Table fern/Cretan brake (*Pteris cretica*), Rigid buckler fern (*Dryopteris villarsii*), Royal fern (*Osmunda regalis*), Rusty-back fern (*Ceterach officinarum*), Sago fern/Black tree fern (*Cyathea medullaris*), Sensitive fern (*Onoclea sensibilis*), Shield fern/Soft shield fern (*Dryopteris erythrosora, Polystichum setiferum*), Silver lace fern (*Pteris ensiformis*), Silver tree fern/Ponga (*Cyathea dealbata,* syn. *Alsophila tricolor*), Snow brake fern (*Pteris ensiformis*), Soft shield fern (*Polystichum setiferum*), Soft tree fern (*Cyathea smithii,* syn. *Alsophila smithii, Hemitelis smithii; Dicksonia antarctica*), Spider fern (*Pteris multifida,* syn. *P. serrulata*), Squirrel's foot fern {*Davallia mariesii* (Ball fern), *D. pyxidata, D. trichomanoides,* syn. *D. dissecta*}, Stag-horn fern (*Platycerium bifurcatum,* syn. *P. alcicorne*), Strap/Ribbon fern (*Campyloneurum phyllitidis,* syn. *Polypodium phyllitidis*), Strap water fern (*Blechnum patersonii*), Sun fern (*Phegopteris*), Sweet fern (*Nephrolepis exaltata*), Sword brake (*Pteris ensiformis*), Sword fern/Ladder fern (*Nephrolepis cordifolia, N. exaltata, Polystichum munitum*), Table fern (*Pteris cretica*), Tasmanian tree fern (*Dicksonia*), Tree fern (*Cyathea*), Tsusina holly fern (*Polystichum tsus-simense*), Vegetable fern/Paco (*Diplazium esculentum,* syn. *Athyrium esculentum*), Venus maidenhair fern (*Adiantum capillus-veneris*), Walking fern (*Camptosorus rhizophyllus*), Walking maidenhair fern (*Adiantum caudatum*), Wall fern (*Polypodium vulgare*), Wall-rue (*Asplenium ruta-muraria*), Washington fern (*Nephrolepis exaltata* var. *washingtoniensis*), Water fern (*Ceratopteris pteridoides, C. thalictroides*), West Indian tree fern (*Cyathea arborea*), *etc.*]

[**Common names for *Salvinia* and *Silaginella***: Climbing clubmoss (*Selaginella willdenowii*), Floating moss (*Salvinia auriculata*), Plume clubmoss (*Selaginella clubmoss*), Rainbow fern/Peacock moss (*Selaginella uncinata*), Resurrection plant/Rose of Jericho (*Selaginella lepidophylla*), Spike moss/Club moss (*Selaginella* species), Spreading clubmoss (*Selaginella kraussiana*), Sweat plant (*Selaginella pallescens*)].

Introduction and Origin

Ferns are among the primitive but higher plants that have existed for nearly 400 million years, long before flowering plants evolved, *i.e.* the earliest periods of earth's history. During carboniferous era when gigantic forests of tree-sized ferns covered the planet, the ferns were in their most abundant population and contributed significantly in the formation of our present-day coalfields as is evident from the fossilized ferns being found constantly. Though these are non-flowering vascular plants possessing true roots, stems and complex leaves usually with branching-vein system but differ from the higher plants most conspicuously in their mode of reproduction, *i.e.* start of life in the form of microscopic dust-like specks (spores). These in size alone range from only 2 to 3 mm tall to huge tree ferns 10 to 25 metres in height and occur all over the world, most abundantly in the tropical rainforests (more than 900 species in Costa Rica alone, twice as many as North America or North Mexico) with steady temperatures, shade and high humidity which are the typical climatic conditions required by tropical ferns and which also constitute most of the indoor ferns belonging to a number of families which are exclusively tropical such as Blechnaceae, Cyatheaceae, Davalliaceae, Gleicheniaceae, Marattiaceae and Schizaeaceae. In the tropics, two-third of the ferns grows as epiphytes on the shaded lower trunks and branches or on the crowns of trees. Terrestrial ferns, growing on the ground, may also possess such modifications, especially those that grow in salt-marshes (*Agrostichum aureum, Rumohra adiantiformis*) and in fully open and exposed sites (bracken and *Pteris*). Such tropical ferns include many of the important ones such as *Adiantum, Davallia* and *Platycerium* species which like even day temperatures in between 18-25 °C and night temperature of as low as 16 °C. Most of the other families occur in both the tropics and the temperate zones and only certain genera are primarily temperate and Arctic. Though less in number than tropical rainforests, but various species are distributed as far as sub-arctic regions, such as *Adiantum pedatum, Asplenium trichomanes, Athyrium, Cystopteris, Dryopteris* and *Polystichum*; certain others in exposed positions on rocks such as *Ceterach officinarum*; certain others even in regions with hot and dry seasons such as *Davallia denticulata* which during adversities though dry up but leave behind live rhizomes; while certain others like *Salvinia* or *Azolla* (both fern-allied) float on the surface of the water. Ground-dwellers (terrestrial), usually occur in the rainforests, along the springs, on the banks of streams or rivers in soil and in the rocky habitats where they occur in cracks and crevices of boulders and rocks, as all these require high humidity and filtered light; while there are ferns such as *Asplenium nidus* and *Platycerium* which comparatively require more light therefore colonize tree trunks, on the fork and crown of the branches, especially in areas with high rains so that their requirement of water is met regularly, and mostly such ferns have evolved the special plant structure in the form of funnel to collect falling leaves and other plant wastes to convert them into humus over a period of time. However, the ferns such as *Nephrolepis* may be found growing as terrestrial as well as epiphytic. Some fern species have been introduced into tropical or subtropical areas, *e.g.* southern Florida and Hawaii, and in some cases, these have become naturalized and have spread into the native forests.

Ferns, by and large, are liked for their graceful foliage, unique growth phases and for their low light requirement which make them most suitable plants for landscaping any home, office or garden. Though they are neuter foliage plants (bearing no flower and no seed) but are attractive in their own right and serve as foils for flowering plants. They are suitable as specimen plants, for pots and containers, in hanging and wooden baskets such as epiphytic ferns including *Platycerium*, as fern pillars where all the epiphytic ferns can be used, as epiphyte trunks such as *Davallia*, *Dryopteris pedata*, *Polypodium* or *Pteris*, in the terrarium such as *Actinopteris australis*, *Hemionitis arifolia*, *Pellea rotundifolia*, *Polystichum tsus-simense* and small *Pteris* along with fern-allied *Selaginella*, in the window sills, in mixed displays, and as a filler to give mass effect in atrium, greenhouses and conservatories. The conservation of various species can be done in ferneries.

Botany

Ferns are herbaceous perennials but neuter plants, producing neither flowers nor fruits, nor seeds. These are mostly evergreen but only a few are deciduous. Leaves of the fern may be undivided as in *Asplenium nidus*, *Hemionitis arifolia* or *Polypodium*; hand-shaped, divided fronds as in *Dryopteris pedata* and *Phlebodium aureum* or coarsely-leaved as in *Platycerium*, pinnatifid (single-feathered) with undivided single feathery leaves growing from the main leaf rib as in *Blechnum gibbum*, *Cyrtomium falcatum* and *Nephrolepis exaltata;* bipinnate (double, triple or multi-feathered fronds) with the feathers divided two or more times as in *Adiantum raddianum* and *Polystichum tsus-simense* (double) and *Davallia trichomanoides* (3-4 divisions). The smaller segments that make up the entire **frond** (a very finely divided fern leaves consisting of small feathery sections which grow from one central leaf stem, rib or rhachis) are called pinnae, an individual section is known as the pinnule.

Their life cycle demonstrates a clear **alternation of generations**. A mature fern plant is a spore-forming (**sporophytic**) asexual generation where capsulated minute spores on the underside or on the margins of the **fronds** (very finely divided fern leaves) or in a few cases are grouped in spikes or panicles, or in rare cases over entire under-surface of the leaves are formed through division as they have only half a set of chromosomes (haploidy) and when the spores are mature they are catapulted out of the capsules and wander through wind and settle in an ideal conditions, they begin to germinate and then these become **gametophytes**, the sexual phase. A germinating gametophyte gives rise to green, small, flat, heart-shaped and lobed **prothallium** which has male and female organs with only half a set of chromosomes (haploidy). The prothalli produce **rhizoids** (hair-shaped root-like growths) to anchor them firmly in the ground, and the sexual organs from the underside of prothallium, and this is the second stage of sexual generation. The **archegonia** (female cells) form in the multicellular region beneath the heart-shaped notch. Their lower section containing the ova is sunk into the tissue of the prothallium and the upper part forms a neck. **Antheridia** (male cells) are formed in the growths of the rhizoids which later release mobile sperm (spermatozoa) which swim across the surface moisture towards the female cells of the prothallus to fertilize the female organ's egg cells (**ova**). Since both the male and female cells are haploid, so on fusion these become diploid with full set of chromosomes, and then the zygote begins to grow into a mass of cells called an embryo which develops into a new fern directly. When this plant is large enough to survive on its own, the prothallium withers and dies. Artificial crossing in ferns is generally not possible but they hybridize frequently in nature at prothallium stage.

Usually the ferns have microscopic **spores** (single-celled propagation unit) contained in the **sporangia** (spore capsule with stalks) which are further contained in the **sori**, on the underside of the fronds, though not always. These sporangia may differ from species to species. The sporangia are grouped together in characteristic clusters called sori (singular sorus, which is groups of spores forming different shapes on the underside of leaf fronds), which during their development are usually protected by growths on leaves

of narrow and a little serrate segments; ***A. lunnulata*** bearing smooth rachises, herbaceous and thick fronds, and 12.5-15.0 cm long 20-30 pairs of segments with finely serrate margins; ***A. oligocarpa*** from Colombia bearing smooth and greyish-straw-coloured rachises, some 45-60 cm long pinnules where segments are deeply pinnatifid and ligulate with blunt lobes, and sori are median and 4-6 on the lower lobes; ***A. rebeccae*** from Australia bearing fibrillose rachises, about 20 cm long fronds, and 30-38 cm long pinnae which bear 5.0-7.5 cm long, serrate or crenate and 20-30 pinnules on each side of the fronds; ***Alsophila smithii*** (*Cyathia smithii, Hemitelia smithii*) from New Zealand grows up to 8 m, bearing up to 1.5 m long and bi- to tri-pinnate fronds which are bright green above and paler beneath, and the secondary pinnules are 3-6 cm long; ***A. tricolor*** (*Cyathia dealbata*; Ponga/Silver tree fern) from Lord Howe Island (New Zealand), Tasmani and Victoria bears 9 m tall trunks though in containers only 2-3 m, woolly-brown scaling on rachis and ribs, and 1.8 m long and tripinnatifid fronds where pinnules are some 3.75 cm long and 1.5 mm wide with green above and white-powdery beneath.

Anemia (also Aneimia; family - Schizaeaceae)

A genus of 90 small to medium evergreen fern species widespread in tropics and sub-tropics of South Africa, Florida to South America and Mexico. Its name derives from the Greek *aneimon* meaning 'naked' and *heima* meaning 'clothing', in reference to its unprotected sporangia. Plants are tufted or creeping in habit bearing simply bi- or tri-pinnate fronds, fertile ones with lowest pair of pinnae converted into an elongated stalk which bears a panicle-like sporangia which seem tiny floral buds, hence vernacularly it is known as flowering fern. These are propagated through spores and division. ***A. adiantifolia*** from tropical South America to Florida has creeping rootstock which bears bi- or tripinnate and 45-90 cm tall fronds and where the pinnules are oblong-obovate to narrowly wedge-shaped; ***A. dregeana*** from South Africa bears 20-30 cm long fronds with ovate-deltoid pinnae which individually are 2.5-4.0 cm long; ***A. phyllitidis*** from Mexico to Brazil is tufted to clump-forming semi-xerophytic fern with 30-60 cm tall fronds which are simply pinnate and its individual leaflets are little lustrous, firm-textured, ovate-oblong and some 5-15 cm long; ***A. rotundifolia*** from Brazil is clump-forming in wild but when grown in conservatories, the frond-tips root and form the plantlets for creating the colonies. Its fronds are simply pinnate, arching and 25-45 cm long though leaflets are roughly rounded and generally rhombic to fan-shaped.

Arachnoides (Polypodiaceae)

A genus with one noteworthy species, ***A. aristata*** (*Polystichum aristatum*) from southern Asia to Australia. It has creeping and branching rhizomes where rich green fronds in the conservatory form colonies, each frond growing up to 60 cm in length, being bi- to tripinnatifid and triangular-ovate. The pinnae of its var. 'Variegatum' are banded palest-green. Though this fern is rarely offered as a pot plant but is one of the rarities in the greens in the cut flower arrangements. It is a cooler fern liking 10-18 °C night/day temperatures for its luxurious growth. Its spore-producing specimens cause irritation in the skin.

Asplenium (Aspleniaceae)

A genus of 650 tufted to shortly rhizomatous fern species found worldwide from tropical to cool temperate zones (east tropical Africa, temperate and tropical Asia and Australasia). It derives its name from the Greek *a* meaning 'not' and *spleen* meaning 'the spleen', in allusion to its supposed medicinal properties to cure spleen ailments. These are excellent as specimen plants. Its leaves are soft green, soft, quite elegant, simple, entire to tripinnatifid and arranged in V-shape. Its sori on pinnae are produced along the veins underside in linear to oval form. It is propagated through division and offsets, *vis-à-vis* spores. ***A. bulbiferum*** from Australasia and Malaysia is a temperate fern bearing 90 cm long, arching and bi- or tripinnate fronds with oblong-triangular blade and lanceolate-triangular pinnules. Its sori are oblong. The upper side of the mature fronds produce bulbils and afterwards small plantlets. ***A. daucifolium*** (*A. viviparum*) from Reunion and Mauritius, a tropical species, is though similar to *A. bulbiferum* but smaller having less cuts and with more graceful fronds, and the segments of the deeply lobed pinnae are smaller and forked. Fronds bear plantlets. ***A. nidus*** (*A. nidus-avis*), an epiphytic tropical species from tropical Asia and Australia, bears shining green, semi-erect and lanceolate fronds growing more than 90 cm in length in the form of broad spreading funnel where midrib is black and the sori are linear. It can tolerate dry condition for shorter duration. ***A. platyneuron*** is a tufted evergreen species for cool regions with fronds growing more than 30 cm in length with lustrous brown stalks, and narrowly oblong pinnae some 2.5 cm long with one basal lobe uppermost, *etc.*

Athyrium (Athyriaceae)

A genus of 180 tufted or clump-forming and a few even slender rhizomatous fern species in cosmopolitan distribution, bearing pinnate to tri-pinnate fronds, and underneath the sori are borne at the end of smaller branched veins. It derives its name from the Greek *thyrion* meaning 'a small door', referring to the indusium (a shield-like sorus cover) which appears to be persistent. Most species are hardy though many are tender to sub-hardy. These grow best in rich soil and partial shade and are propagated through division of clumps or by spores. ***A. felix-femina***, a hardy and clump-forming species from northern temperate zone, tropical mountains to southern S. America grows some 45-90 cm in height and spread, bearing fresh green, dainty, arching and bi- to tri-pinnate fronds, and oblong or oblong-lanceolate and 0.3-2.0 cm long pinnules, fertile ones with two rows of 1 mm wide sori beneath and the indusium is persistent. Various forms differing in shape and size of pinnae and pinnules, *vis-à-vis* crested or tessellated versions are found; and ***A. goeringianum*** (*A. iseanum, A. nipponicum*) from Japan and Taiwan is tufted species with deciduous, arching, bipinnate and 20-30 cm long fronds, and deltoid-lanceolate and asymmetrically lobed some 1.0-1.7 cm long pinnules. Sori small and are covered with an indusium. *A. g.* 'Pictum' is the commonest and most decorative form in cultivation and bears silvery-grey fronds with purple shading along the midrib.

Blechnum [Blechnaceae (Polypodiaceae)]

A genus of 220 usually evergreen and highly decorative fern species of cosmopolitan distribution (a few are deciduous and tolerate drier conditions). *Blechnum* derives from the classical Greek name for an unknown species of fern which was adopted by Linnaeus for this genus. Not all but a few of the species have fast-growing rhizomes and form mats. Majority of them show upright trunks which become distinct with age, and bear a rosette of simple-feathered and slightly overhanging fronds which spread about 90 cm at the top similar to those of palms. Usually the pinnate or pinnately lobed fronds are stalkless but the central ribs are quite strong. Normally the sterile fronds produce broader feathery leaflets though fertile ones only narrow having spore capsules in long strips along the central ribs. For its growing the well-drained soil enriched with peat or leaf mould at a sheltered and filtered situation is best. These are propagated through division or spores. ***B. auriculatum*** (*B. hastatum*) from temperate South America bears lanceolate, pinnate and up to 60 cm long fronds whereas the pinnae are lanceolate-falcate with bases of upper ones expanding as ear-like lobe where spore-bearing ones are narrower having bands of sori midway to the margins and the ribs; ***B. brasiliense*** from Brazil and Peru grows up to 1 m in height (growing much higher if given a free hand as it is a tree fern) with oblong-lanceolate and pinnatipartite fronds when youngs are reddish and unfurl from the centre of the leaf rosettes that are on a brown, scaly and trunk-like rhizome, and there is no difference in shape and size between fertile and sterile fronds, however, narrow pinnae are finely toothed, and its form 'Crispum' possesses smaller and wavy-margined fronds which are coppery-red when young; ***B. gibbum*** is a graceful miniature tree fern excellent as a specimen plant, does well at 13-35 °C, the slender and black-scaly trunk of which attains up to 90 cm height and on top of which the wide-spreading rosettes are formed, however, in pots it remains smaller; and its lanceolate to narrowly ovate and pinnatipartite fronds are 60-90 cm long with very slender and pointedly tapering pinnae, and its form 'Platyptera' is larger in size and grows fast; ***B. moorei*** (*Lamaria ciliata*) from New Caledonia is a neat and spreading rosetted fern, bearing 20-30 cm long, oblong-ovate and pinnate fronds, *vis-à-vis* oblong, wavy and lobed or toothed pinnae. Fertile fronds are distinct, narrowly linear and the underside is fully covered with sori; ***B. occidentale*** from tropical South America is though highly leafy fern forming rosettes but also spreads through underground creeping stolons. Its similar sterile and fertile fronds are ovate, pinnatipartite and 35-70 cm long, and the pinnae closely set, slightly wavy and narrowly oblong; ***B. patersonii*** from Queensland to Tasmania in Australia bears slowly creeping rhizomes which produce dense tufts of fronds which are pink when young but on maturity become deep green. Fronds are variable in form, being completely entire to deeply lobed, 30-60 cm long, strap-shaped with pointed tips and the fertile fronds are narrower; ***B. pennamarina*** (*Lomaria alpina*) from southern temperate zone is a hardy mat-forming species, bearing spreading, lanceolate, pinnate and 5-20 cm long fronds, and where the pinnae are linear-oblong, the fertile ones being narrower and widely spaced; ***B. spicant*** from northern temperate zones is though larger and more erect otherwise similar to *B. pennamarina*, and there are its forms with larger and crested pinnae; *etc.*

Bolbitis [Lomariopsidaceae (Aspidiaceae)]

Bolbitis derives from the Greek *bulbo* 'a bulb', referring to the curiously swollen leaf veinlets. The genus comprises 85 species mostly of epiphytic ferns but also a few found on the mossy rocks by streams. Their spore-bearing fronds are much reduced and their undersides are coated with small sporangia. Many of the species are rhizomatous and prefer to be grown in loose peaty compost. They are propagated by division or spores. Only two species are in common cultivation and both have rhizomatous rootstocks. They grow well under humid and shady places in pots. ***B. cladorrhizans*** from Central America and West Indies has short rhizomes and is clump-forming. Its fronds are green, ovate to triangular-ovate, 30-60 cm or more long, arching, basally pinnate, above pinnatifid and the basal pinnae bearing large rounded teeth. ***B. heteroclita*** from tropical Asia bears slender rhizomes and forms colonies. Its fronds are bright green, usually pinnate, occasionally entire, erect, 15-30 cm long, and are divided into 3-7 broadly thin and broadly lanceolate pinnae.

Campyloneurum (Polypodiaceae)

These, strap-shaped, tufted and colonizing ferns growing through underground runners are native to American tropics with narrow, straight-sided and both end-pointed leaves. ***C. angustifolium*** bears erect, narrow, strap-shaped, thin and papery fronds some 60 cm long and 15.5 cm wide. The vars 'Corkscrew' has undulating leaf edges, 'Narrow' bears leaves like a light green grass with length being almost double of the width and 'Villa Tenaril' throws out thin and bluish-green runners on the ground bearing small rectangular leaves with rounded tips. ***C. phyllitidis*** sends out small runners and bears crowded and strap-shaped leaves some 90 cm long and 10 cm wide. ***C. punctatum*** (*Microsorium polycarpon*; Indian strapleaf) bears erect, thick and leathery leaves some 90 cm long and 7.5 cm wide, and its var. 'Cristatum' bears fronds which are widened towards the tips with crested forkings. ***C. sphenodes*** produces crowded leaves which are shining yellowish-green and some 75 cm long.

Ceratopteris (Parkeriaceae)

A genus of four tufted aquatic fern species for warm water from tropics and sub-tropics which bears bi- to tri-pinnate fronds which produce plantlets freely when above the water. The name derives from the Greek *keras* meaning 'a horn' and *pteris* 'a fern', referring to the shape of the pinnae of the sterile fronds. They can grow completely submerged in water after getting rooted in compost or sand kept in the bottom of the tank. Both the types of fronds (barren and fertile) are edible. They can be propagated through division, aerial plantlets and spores. ***C. pteridoides*** from Florida to Brazil produces sterile fronds widely triangular, free-floating, more than 20 cm long and simply pinnate with triangular pale-green pinnae while fertile fronds are usually bi- or tri-pinnate and up to 40 cm long with narrowly oblong to linear pinnules; ***C. cornuta*** (*C. thalictroides cornuta*) from tropical Africa, Malagasy to Indonesis and northern Australia, bearing pale-green lanceolate sterile fronds some 30 cm long, and the pinnae irregularly lobed, submerged or floating but rooting into the medium. Fertile fronds are larger than sterile ones, bi-, tri- or quadri-pinnate and the pinnules are linear; ***C. thalictroides*** (*C. siliquosa*) from southern Japan, SE Asia and Pacific islands is similar to *C. cornuta* except that sterile fronds are oblong, best for aquarium decoration and is most widely used as salad; *etc.*

Ceratopteris (Parkeriaceae)

Ceratopteris derives from the Greek *keras* 'a horn' and *pteris* 'a fern', referring to the shape of the pinnae of the non-spore bearing (sterile) fronds. The genus comprises four species of fast-growing aquatic ferns (even completely submerged) from tropics and sub-tropics, suitable for warm water aquaria and pools. These are tufted ferns with bi- to tri-pinnate fronds that produce plantlets freely when above the water. Those planted on wet compost with aerial parts in the air produce thicker fronds. Whether in water or above, both the frond types are edible. ***C. pteridoides*** (Floating fern/Water fern) from Florida to Brazil bears fertile and sterile fronds separately, fertile ones some 40 cm long and usually bi- or tri-pinnate with pinnules being narrowly oblong to linear whereas sterile fronds are widely triangular, 20 cm or more long, pinnate with triangular pale-green pinnae and usually free-floating. ***C. cornuta*** (*C. thalictroides cornuta*; water fern) from tropical Africa, Malagasy to Indonesia and N. Australia is similar to *C. thalictroides* but sterile fronds are lanceolate and the pinnae are erratically lobed. ***C. thalictroides*** (*C. siliquosa*; Water fern) from S. Japan, Pacific Islands and SE Asia is widely grown ornamental for use in aquaria and for salad and vegetables. This produces bi-, tri- or quadric-pinnate fertile fronds larger than sterile ones and with linear pinnules, whereas sterile fronds are submerged or floating with their roots into the substratum, some 30 cm long, oblong and pale-green.

Ceterach (Polypodiaceae)

A genus of evergreen or semi-evergreen hardy ferns useful for planting in shaded crevices and does quite well if soil is chalky. Faded fronds are removed regularly to make them in shape and neat. They are propagated through division or by spores. ***C. officinarum*** grows 15 cm in length and breadth and bears dark green, lanceolate and leathery fronds divided into alternate and bluntly rounded lobes. While young, the back of the fronds are covered with silvery scales which mature to reddish-brown.

Cibotium (Dicksoniaceae)

A genus of 10 species of large ferns (some being tree-like) from tropical America, suitable for warm regions. It takes its name from the Greek *kibotos* meaning 'a small box', referring to the shape of leathery indusial covering of sori. It is extremely graceful with arching fronds so make a graceful specimen. These bear large, gracefully arching and usually tripinnate fronds where the base of the long stalks are closely long-hairy. These are propagated by spores. ***C. chamissoi*** from Hawaii is a tree fern but becomes quite slow when planted in pots, its fronds are tripinnate and grow up to 2 m with stipes (stalks) having yellow-brown and blackish hairs, lower pinnae broadly lanceolate and some 45 cm long, pinnatisect in the lower part and the ultimate pinnules are oblong and blunt; ***C. glaucum*** from Hawaii is quite similar to *C. chamissoi* though the fronds are glabrous, have more glaucous tint, and pinnules are linear, pointed and slender; ***C. schiedei*** from Mexico and Guatemala is a tree fern with tripinnate fronds growing 1.2 m or more in length and the bases of its stipes are covered densely with long, lustrous and brownish-yellow hairs, while the pinnae are pale-green above, glaucous beneath, 30-45 cm long and oblong-lanceolate and the pinnules toothed and lanceolate; *etc.*

Cryptogramma (Polypodiaceae)

A genus of deciduous or semi-evergreen and fully hardy ferns. It requires partial shade and moist but well-drained neutral or acidic soil. Fading fronds are removed and it is propagated through spores. ***C. crispa*** grows 15-23 cm high and 15-30 cm across with bright pale-green, broadly oval to triangular and finely divided fronds resembling parsley which turn bright rusty-brown in autumn and remain so for whole winter.

Cyathea (*Alsophila, Sphaeropteris*; Tree fern; Cyathaceae)

A genus of 600 species of tree ferns distributed world over in the mountainous tropics and temperate areas of southern hemisphere. *Cyathea* derives from the Greek *kyatheion* meaning 'a little cup', referring to the shape of sori that cover the spores. These are frost-tender evergreen ferns grown for their beautiful fronds and overall appearance. Most species can not bear persisting temperatures below 10 °C though throughout the growing period these need humid atmosphere, sun or partial shade, humus-rich, moisture-retentive and well-drained soil. When small these make good pot plants but later these form palm-like trunks closely woven with brown aerial roots and topped with the rosette of bi- or tri-pinnatifid fronds. These are propagated through spores. ***C. arborea*** from Puerto Rico, Cuba and Jamaica is evergreen upright tree fern with a slender trunk which grows 7-10 m high with 2-3 m wide canopy, producing bright yellow-green, 2-3 m long, arching and tripinnate fronds delicately divided into small, oblong and serrated pinnae; ***C. australis*** from Tasmania and Australia is an evergreen and upright-growing tree fern with a robust and nearly black trunk, bearing straw-coloured rachises, 2-4 m long and finely divided fronds with primary pinnae being light green above, bluish beneath, 45 cm long and 15-25 cm wide with deeply pinnatifid, ovate-oblong, sharply serrate but segments at the base broadest; ***A. cooperi*** from Queensland bears rachises with pale-brown scales, pinnae spear-shaped, and linear pinnules some 10-13 cm long; ***C. medullaris*** is an evergreen upright tree fern bearing slender and black trunk with height of 7-16 m and canopy spread of 6-12 m, topped with 7 m long arching fronds which are dark green above and paler beneath, and the pinnae are glossy, small and oblong; *etc.*

Cyrtomium (syn. *Phnerophlebia*; Aspidiaceae)

A genus of about 10 species with densely scaly rhizomes from SE Asia and Polynesia. It takes its name from the Greek *kyrtos* meaning 'arched', due to arching fronds. Cyrtomium can withstand draughts and dry air more than other ferns. The plants are easily adaptable and grow some 60 cm in height where fronds are dark glossy green, leathery, pinnate with broad and informal leaflets which may have entire or toothed edges similar to holly leaves. Frond undersides bear abundance of brown capsules of spores with centrally located indusia. It tolerates dry air, drafts, low light levels and freezing temperatures and is most suitable as indoor pot plant. It is propagated through division and spores. ***C. falcatum*** (syn. *Aspidium falcatum, Polystichum falcatum*) from South Africa, Hawaii and Asia (from India to Korea and Japan) is a tufted fern with 35-65 cm long, arching and pinnate fronds bearing shining deep green above and velvety brown-hairy with a lot of sori on the undersides, oblong-ovate, up to 23 cm long, wavy- but smooth-edged and tail-ending pinnae. Its form *C. f. rochfordianum* is strong grower with broader fronds having wavy and holly-like distinctly toothed margins. ***C. fortunei*** (*Aspidium falcatum fortunei*) from China, Japan and Korea is though smaller in all its parts than *C. falcatum* otherwise quite similar and hardier. Fronds have almost double number of less taper-pointed leaflets which are less glossy.

Cystopteris (Athyriaceae)

A fully hardy genus of 18 small and tufted to rhizomatous deciduous fern species from arctic, temperate and sub-tropical regions, mainly from mountains in the warmer areas. Its name derives from the Greek *kystis* meaning 'a bladder' and *pteris* 'a fern', alluding to the globular sori. The fronds are bi- to quadri-pinnate where fertile pinnae bear two rows of globular sori with hooded indusia either side of the midribs. They like well-drained but water-retentive humus-rich soil in partial shade and are most suitable for planting in the rock gardens. Their propagation is through division and spores, *vis-à-vis* bulbils as a few species form these. ***C. bulbifera*** from North America is deciduous and tufted with fronds being pinnate and up to 50 cm long, the pinnules being oblong, sterile ones being lanceolate and shorter than fertile ones which are quite long and slender with tapering tips where below rounded bulbils are formed which soon fall off and start new growth. ***C. dickieana*** (*C. fragilis dickieana*) from arctic Asia and Europe is highly variable, some forms being very similar to *C. fragilis*, however, the pinnules are shallowly lobed or crenate-toothed and overlapping. ***C. fragilis*** has same origin as *C. dickieana*, is deciduous and tufted with thick and short rhizomes, grows to 15 cm with spread of 23 cm, and bears 5-35 cm tall, delicate, pale-green and broadly lanceolate fronds which are much divided (bi- to tri-pinnate) into oblong, indented and pointed pinnae. *C. alpina* and *C. regia* which are normally included in *C. fragilis*, have more dissected fronds.

Davallia (Davalliaceae)

A genus of 40 small to medium-sized epiphytic ferns with creeping and scaly rhizomes. It is native to Korea, Japan, Malaysia, New Guinea, Indonesia, SW Europe, Canary islands, Malagasy and tropical and sub-tropical Pacific Islands. It is named so to honour Edmond Davall (1763-98), a Swiss botanist and friend of James Edward Smith (1759-1828), founder of the Linnean society of London. Its dark green fronds are 3-4 pinnate, triangular to ovate and a little leathery. It grows 30-45 cm tall and is ideal for placing along north facing window, on tree bark, moss sticks, in pans and in the hanging baskets indoors or in the conservatory but during hot weather this requires additional humidity. Its plants become dormant during winters for a short time though *D. canariensis* which comes from dry localities, remains dormant during summer for 4-6 weeks.Its rhizomes are thick, hairy and grow over the edge of the pot. It is propagated through division of clumps, by rhizome cuttings and through spores. ***D. canariensis*** from Canary Islands, Spain and North Africa bears stout rhizomes with dense covering of fur and narrow chaffy-brown scales. The fronds are quadripinnatifid, triangular, 25-45 cm long and carrot-like. ***D. fijiensis*** (*D. solida fijiensis*) from Fiji bears stout rhizomes with scales and long marginal hairs, fronds being bright green, long-stalked, quadripinnatifid, broadly triangular and the blades 30-60 cm long. A very charming species. ***D. amriesii*** from Japan, Korea, Taiwan and China bears light brown scales over its rhizomes, the fronds are quadripinnatifid, ovate to triangular, 15-20 cm long, and the segments are 1.0-2.5 mm long, sharp-toothed and oblong-lanceolate. ***D. pyxidata*** from Australia bears slender rhizomes densely covered with hairy-margined brownish scales, and bright glossy green, 25-45 cm long, tri- to quadripinnatifid and triangular fronds. ***D. solida*** from Malaya to NE Australia is similar to *D. fijiensis* but little coarser and tripinnatifid. ***D. trichomanoides*** (*D. dissecta*) from Malaysia is similar to *D. amriesii* even in hardiness but the scales on the rhizomes are whitish to pale-yellowish-brown, and the fronds 30-45 cm long.

Dennstaedtia [Dennstaedtiaceae (Polypodiaceae)]

A fully hardy pantropical fern genus of 70 deciduous or semi-evergreen, creeping and hairy rhizomatous species from temperate South America, North America, Australasia and Japan. *Dennstaedtia* commemorates August Wilhelm Dennstedt (1776-1826), a German botanist. These make good ground cover and graceful specimen foliage plant in the shaded locations. Multipinnate fronds bear sori protected by cup-shaped indusia. These are propagated through division or spores. ***D. bipinnata*** from Mexico to South America and West Indies bears up to 2.5 m long fronds with stalks, which are bi- to tri-pinnate with rhomboidal lobes. The lobes are glossy green and toothed. It makes a graceful foliage plant in a tub. ***D. davallioides*** from Queensland to Victoria in Australia has long creeping rhizomes which makes extensive colonies. The fronds with stalks are 90-120 cm long, tripinnate, segments oblong and deeply pinnatifid. It makes a proper ground cover in temperate areas though is invasive. ***D. obtusifolia*** from central America, West Indies, Trinidad and south to Paraguay is a tropical fern bearing 90-180 cm long and elegant fronds which are triangular, quadri- to even more pinnate and slightly hairy. ***D. punctilobuta*** grows some 30 cm in height and spread where the fronds are mid-green, delicate, oval to triangular, multi-pinnate and lace-like which die in winter.

Dicksonia (Dicksoniaceae)

A genus of 30 true tree ferns from Malaysia to Australasia, St. Helena and tropical America. It was named so to honour James Dickson (1738-1832), a British nurseryman and botanist. It may grow up to 13.5 m tall, depending upon the space available. Its trunk takes many years to develop and is covered with brown fibrous roots. The fronds are light green, large, arching, very finely divided, *i.e.* bi- or tri- pinnate, elegant and lacy, and is a beautiful specimen plant at shady locations. It is propagated by spores at temperature around 15 °C. ***D. antarctica*** from Queensland to Tasmania in eastern Australia which grows up to 5 m tall, is covered with a rusty brown surface of matted roots. At younger stage the fronds start from ground level but later in terminal rosettes, each being 1.3-3.0 m long and tripinnate. The sori are formed at the end of the veins close to the margins of leaflets which curl to protect these. *D. antarctica* and

D. squarrosa can be enjoyed as a pot plant for quite sometime, but when well grown should be taken out.

Didymochlaena (Aspidiaceae)

A genus of one dwarf evergreen tree fern species in circum-tropical distribution, especially the Natal. It derives from the Greek *didymos* meaning 'twin' and *chlaina* 'a cloak', referring to the indusium which covers and protects the sporangia. It is a graceful specimen growing in a quite dense shade under tropical conditions where humidity is high and the plants can be raised from division and spores. Its fronds are leathery, brownish-green and bear shiny leaflets in a double herringbone pattern. ***D. trunculata*** (*D. lunulata, D. sinuosa*) grows to 1.2 m high with erect stem, crowned with dense, rosetted, bipinnate and some 1.2 m long fronds, and the pinnules almost oblong, 2.0-2.5 cm long and wavy-margined. In the wild the trunk height and frond length roughly double. The texture of the fronds is almost similar to *Asplenium*.

Diplazium [Athyriaceae (Aspidiaceae)]

A widespread genus of about 400 evergreen to deciduous and medium to large size fern species distributed to the tropics and northern hemisphere. *Diplazium* derives its name from Greek *diplasios* meaning 'double', referring to the sporangia occurring in pairs to either side of a leaf veinlet. It is highly variable genus, closely related to *Athyrium* and *Asplenium*. They may be tufted, rhizomatous or tree type. The fronds are simple to bi- or tri-pinnate, and evergreen species make excellent foliage specimens. ***D. esculentum*** (*Athyrium esculentum*) from India to Polynesia has trunk-like erect rhizome some 60 cm in height, bearing up to 1.2 m or more long bright green fronds in rosettes on the top. When young the fronds are simply pinnate but afterwards bipinnate. Does well in full light and partial shade. ***D. proliferum***

Rumohra adiantiformis

Nephrolepis exaltata

Adiantum tenerum

Asplenium nidus

Cyrtomium falcatum

Davallia trichomanoides

from tropical Africa, Asia and Polynesia is tufted with 60-120 cm or more long simply pinnate and arching fronds where pinnae are sessile, bright green, leathery and narrowly triangular. Main leaf stalk of the mature fronds bears plantlets in the axils of the pinnae.

Drynaria (Polypodiaceae)

A rhizomatous genus of some 20 fern species, mainly epiphytic from Old World tropics and Australia, mostly making very effective hanging baskets. *Drynaria* derives its name from the Greek *drys* meaning 'oak' as the sterile fronds are somewhat similar to certain oaks. Similar to *Platycerium*, these also produce two distinct types of fronds, sterile ones which are stalkless but cling securely to the host even by sending out fern roots and collect plant and other organic debris in them to convert these into humus, whereas the fertile fronds are long-stalked and simply pinnate to pinnatifid. These can be grown on section of tree fern stem, on a slab of rough bark and in a basket where these make compact mounds of overlapping sterile and pleasantly arching fertile fronds. These are propagated through division and spores. ***D. quercifolia*** from India, Malaysia and Australia bears short, woody and thick rhizomes, bearing obovate, lobed and 20-30 cm long sterile fronds, and deeply pinnatifid and 60-90 cm long (including the leaf stalk), by 30 cm or more wide fertile fronds. Sporangia in two rows are formed between the lateral veins of the pinnae. ***D. sparsisora*** is quite similar to *D. quercifolia* but sporangia are scattered irregularly.

Dryopteris (Aspidiaceae)

A genus of 150 evergreen or deciduous fern species distributed worldwide from tropical to cool temperate regions, having short and thick rhizomes, along with bi- or tri-pinnate arching rosettes of fronds. It derives its botanical name from the Greek *drys* meaning 'oak' and *pteris* 'a fern'. Most of its species are hardy and suitable for cold conservatory, unheated porch or houses. They are propagated through division of multi-crowned plants or by spores. ***D. erythrosora*** from China, Japan and islands of Philippines is an evergreen fern with coppery-red fronds when young. The fronds are broadly ovate, bipinnate and 30-60 cm or more long, the pinnules are coarsely toothed, and the sori which are bright red initially become dark with maturity of spores. ***D. filix-mas*** from north temperate zones and mountains in the southern hemisphere south to Peru, Malagasy, Java and Hawaii is deciduous to semi-evergreen, bearing oblong to elliptic-lanceolate, 30-120 cm long and pinnate fronds. Pinnae are very deeply lobed and toothed appearing almost bipinnate. Various frond shapes are available such as 'Cristata' (crested pinnae tips), 'Decompositum' (more finely cut fronds), 'Linearis' (slender and more elegant fronds), 'Polydactyla' (frond and pinnae tips crested), *etc.* ***D. pseudomas*** (*D. borreri*) from Europe and SW Asia bears usually evergreen, thick-textured, almost erect and narrower fronds with stalks and midribs having shaggy and brown scales, and the pinnules with only a few blunt teeth on their apices otherwise similar to *P. filix-mas.*

Elaphoglossum [Lomariopsidaceae (Aspidiaceae)]

A genus of probably 400 evergreen and almost epiphytic fern species from tropics and sub-treopics, mainly South America, though appearance of these ferns is quite un-fernlike. *Elaphoglossum* derives from the Greek *elaphos* meaning 'a stag' and *glossa* 'a tongue', alluding to the shape of the fronds of certain species. The fronds are little feathery, where the fertile ones beneath are entirely covered with sporangia. They are cultivated similar to a bromeliad, and propagated through division or spores. ***E. crinitum*** (*Acrostichum crinitum*) from West Indies, Mexico and central Ameica is initially for a few years forming solitary rosette due to being slow-growing but later on clump-forming. The fronds are thick-textured, ciliate but with scattered hairs, a little wavy and 25-45 cm or more long, and their stalks being long and scaly-hairy. ***E. hirtum***, a rhizomatous species from tropics that forms small colonies with fronds being lanceolate, 15-30 cm long, gradually tapering to base and tip, thickly scaly-hairy to both the surfaces and being reddish on the margins, and the spore-forming fronds are narrower. ***E. longifolium*** from West Indies and tropical America bears slow-growing and thick rhizomes which form clumps, and the fronds are glabrous, narrowly lanceolate, slightly wavy and 30-45 cm long.

Hemionitis [Hemionitidaceae (Polypodiaceae)]

A genus of seven tufted to small clump-forming fern species from northern tropical America and tropical Asia bearing dark- and long slender-stalked, rounded to cordate, a little leathery, neat and small fronds bearing spores but when young these bear shorter stalks with compact plants, giving the plant a two-tier effect, and so being most effective for growing indoors and in conservatories. It derives from the Greek *hemionos* meaning 'a mule', as in the ancient times the native women wore these ferns to prevent pregnancy. These are propagated through division and spores. ***H. arifolia*** from tropical Asia grows about 30 cm in height with smooth rich green, sharply sagittate and 5-13 cm long fronds, whose stalks are brown-scaled and black. ***H. palmata*** (*Asplenium hemionitis*) from tropical America grows 20-30 cm in height with 15 cm wide, palmate and brown-stalked fronds which are deeply cut into five pubescent lobes, each being edged with small, shallow and rounded lobes.

Humata (Bear's-foot fern; Polypodiaceae)

Humata in Latin is for 'earth', which refers to creeping habits of its rhizomes. The genus comprises some 50 species of small ferns related to *Davallia*, having small but wide-creeping rhizomatous runners covered with linear white scales that turn light brown. Only one very handsome species, *H. tyermannii* is grown as basket plant indoors. Its fronds are 30 cm long, thick, deltoid and triangular with tough, suborbicular or reniform indusium which have free sides and apex but with broad base. ***H. tyermannii*** from central China bears 10-15 cm long, deltoid and 3-4-pinnatifid leaves where lower pinnae are largest, the lowest pinnules are cuneate-oblong or deltoid and sori at the base of the ultimate lobes. This species appears as *Davallia trichomanoides*, tolerating an wide range of temperatures and light.

Lygodium (Schizaeaceae)

A genus of 40 species of climbing ferns from tropics to warm temperate regions (tropical Africa, eastern Australia, Cuba to Brazil and tropical to eastern temperate Asia including Japan, Korea and India). It derives its name from the Greek *lygodes* meaning 'like a willow', referring to the flexuous and twisting leaf stalks. The rhizomes are formed underground in clumps from where slender fronds of unlimited growth (some 1.0 m) arise with attractive scalloped edges and their stem-like twining midribs to climb up the supports.The pinnae (lateral leaflets) may be pinnate or bipinnate and look like individual fronds. The sporangia are formed on greatly shortened pinnules which appear spiky. These elegant ferns are ideal for growing in bright light in pots, in hanging baskets, on trellises and in the border of conservatory. These may go dormant during late winter. ***L. japonicum*** from Japan and eastern temperate Asia produces more than 2 m length of fronds with pinnate leaflets, each with 5-11 narrowly ovate and lobed and/or asymmetrically toothed pinnules. ***L. microphyllum*** (*L. scandens*) from tropical Africa, Asia and Australia is similar to *L. japonicum* though pinnules are broader with absent lobes or a few small basal lobes with a bluish-green cast. ***L. volubile*** from Cuba to Brazil possesses 2-3 m long fronds which bear pinnate leaflets being composed of 5-13 lanceolate pinnules, each some 5-15 cm long.

Matteuccia [Onocleaceae {Aspidiaceae or Polypodiaceae}]

Matteuccia is named so for Carlo Matteuci (1800-1868), an Italian scientist. The genus contains only three species from northern temperate regions, one of which is generally available. It is a fully hardy rhizomatous fern, growing from 80 to 130 cm in height with fronds up to 50 cm long. Though deciduous but it is an attractive garden fern, preferring semi-shade, fertile and moist soil and is propagated through offsets by division of crowded plants and spores. If the substrate becomes too dry, its leaves turn yellowing early. ***M. struthiopteris*** (*Onoclea struthiopteris, Struthiopteris germanica*) from Europe and Asia is deciduous and far-creeping rhizomatous fern growing up to 1 m in height and 45 cm in breadth, with lanceolate, erect and divided fronds arranged in rosettes like a shuttlecock. Outermost fresh green sterile fronds surround densely produced dark brown fertile fronds, are arching, slenderly elliptic to oblanceolate, deeply bipinnatifid and 60-120 cm long, while fertile fronds are quite smaller, being formed in the centre of the rosettes, and pinnate, and the the pinnae are rolled around the sporangia until they ripen.

Microgramma (Polypodiaceae)

This creeping genus is native to tropical America and Africa. Its fronds are short and oval to elliptic, snd suitable for growing on long-fibre sphagnum moss in shallow but wide pots, in terrarium and on tree fern slabs without addition of any fertilizer. These grow comfortably between 13-27 °C temperature and at a minimum of 50 per cent relative humidity. These are propagated through separation of runners and spores. ***M. heterophyllum*** grows through its creeping and branching runners, bearing upright, elliptical to spoon-shaped and some 1.5 cm long fronds. It can be grown from shade to filtered sun with a minimum of 150 fc light for 10 h daily. ***M. lycopodioides*** produces creeping and dangling runners similar to hairy strings, and fronds being lime-green with darker veining, elliptical and some 5 cm long. ***M. palmerii*** produces large and flat runners with silvery furry, and 20 x 7.5 cm spoon-shaped fronds. It is excellent as a small basket plant. ***M. piloseloides*** has branching runners with mat-forming nature, having closely set, elliptical and 5 cm long fronds. ***M. squamulosum*** produces thin and white runners with 2.5 cm long fronds. ***M. vaccinifolium*** produces white runners, is fast-growing and bears 5 cm long lanceolate fronds.

Microlepia [Dennstaedtiaceae (Pteridaceae)]

An evergreen, deciduous or semi-evergreen frost-tender (minimum 5 °C) genus of 45 species with creeping rhizomes or thick, slow-growing and clump-forming rhizomes from Old World tropics, Japan and New Zealand. It derives its name from the Greek *micros* meaning 'small' and *lepis* 'a scale', referring to the small scale-like indusia. These make elegant foliage plants for the home and shady conservatories, in pans and hanging baskets. The fronds are mostly tripinnate, and the pinnules of fertile fronds bear quite thin (membranous) crescent-shaped indusia. These are propagated through spores or division. ***M. platyphylla*** (*Davallia lonchitidea*) from India to Eastern Asia grows in the wild up to 2.1 m high with fronds about 90 cm tall and is clump-forming. Fronds are mellow-green, slightly glaucous, tripinnatifid, white-hairy when young but almost hairless when mature. It can overwinter at 15-17 °C temperature. It requires the substrate of leafmould,farmyard manure and sand in the proportion of 3:2:1. ***M. pyramidata*** (some authorities claim it as *M. hirta*) from tropical Asia is clump-forming and appear similar to *M. platyphylla* in size and shape. Fronds are triangular, usually hairy beneath on the veins and on rachis (midrib) and often quadripinnate. ***M. speluncae*** from tropics and sub-tropics is strong-grower and an adult plant bears 1-2 m long, pale-green, soft, tri- or quadripinnatifid and white-hairy fronds. Its forms *M.s. hirta* (*M. hirta*) bear rich green and up to tripinnatifid fronds with purple-brown stalks; and 'Cristata' have broader-tipped fronds. ***M. strigosa*** from SE Asia and Polynesia is a widely clump-forming species, growing to 90 cm high with 60 cm spread, having creeping rootstock with evergreen, pale-green, broad, irregularly lanceolate to narrowly ovate, slightly hairy, bipinnatifid, deeply cut and divided fronds which grow 45-120 cm in length. It can tolerate to a minimum temperature of 5 °C.

Microsorium (Polypodiaceae)

Microsorium takes its name from the Greek *micros* meaning 'small' and *soros* 'a heap', obliquely referring to the small clusters of sporangia. It is an evergreen epiphytic genus with far-creeping rhizomes of about 60 species from Old World tropics (Asia and western Pacific) and range from medium to large. This genus is related to *Polypodium* under which it was formerly included, and most species blend well with bromeliads and orchids, either on bark of trees or in hanging baskets. Their fronds are simple, strap- to lance-shaped and pinnately lobed. They are propagated through division or spores. ***M. diversifolium*** (*Phymatodes diversifolium*) from Australasia produces branching rhizomes running to several metres and covered with slightly glaucous dark brown and about 1 cm long scales. Stalked fronds are 5-20 cm long, dark glossy green, leathery and producing blades of three forms, *i.e.* simple, entire and lanceolate to elliptic-oblong. These all are almost irregularly more or less lobed or regularly densely-pinnately lobed. These blade-forms vary greatly, *i.e.* 10-40 cm. It withstands cooler climates and is most suitable as ground cover. ***M. elegans*** is erect growing from 30.0-37.5 cm high where frond-tips are multi-forked. ***M. integrifolium*** 'Cristatum' fronds are strap-shaped, to 75 cm long, forked and overlapping-cristate while 'Serrulatum' possesses cresting to the full length of the fronds. ***M. musifolium*** produces black runners, 75-90 cm long and 10 cm wide fronds, which are creamy-green with darker spreading veins. It survives a minimum temperature of 18.5 °C and relative humidity of 65 per cent. ***M. polycarpon*** is a climbing fern with 90 cm long and 7.5 cm wide fronds, which are shiny and leathery with rounded tips. In its form 'Grandiceps' which is a terrestrial fern the fronds are erect and heavily crested, while in 'Grandiceps Compactum' the fronds are 30 cm long and crested. ***M. scolopendria*** (*Phymatodes scolopendria*) produces green and highly branched rhizomes running to several meters, fleshy at first but later becoming woody, bearing simple and oblong-lanceolate, or deeply pinnately lobed, deep lustrous green, stalked fronds some 4-20 cm long though blades up to 60 cm, leathery and with 10 or more pairs of deeply cut segments, or simple leaves or those with undulating edges.

Nephrolepis [Oleandraceae (Davalliaceae)]

A runner-forming genus of about 30 rosette-forming fern species widely distributed world over to the tropics and sub-tropics, extending to temperate Japan and New Zealand. It derives from the Greek *nephros* meaning 'a kidney' and *lepis* meaning 'a scale', alluding to the membranous indusium which covers the spores. The fronds are evergreen, oblong-lanceolate and pinnate bearing rounded sori near the leaf margins at the tips of the vein. These ferns are ideal for growing indoors and in conservatories, and those with arching habits are suitable for hanging baskets. They are propagated by removal of the offsets, through division and by spores. ***N. cordifolia*** from New Zealand to Japan, is a tufted, evergreen and often tuberous fern with short and scaly stems, 30-75 cm long, light green and almost erect fronds which are narrowly pinnate on short stalks, the pinnae are closely set and the individual pinnae are 4 cm long and sharply toothed. *N. c. duffii* bears much narrower fronds which have often forked tips and rounded pinnae some 1.5 cm wide. *N.c. plumosa* bears almost pinnate leaflets. ***N. exaltata*** from tropical areas around the world is tuberless tufted fern with erect initially but arching later and 60-120 cm long fronds which produce many narrow pinnae some 8 cm long. The growing of the species is surpassed by its innumerable varieties such as 'Bostoniensis' (being economically important foliage plant since the 1920's; Smith and Scarborough, 1981; McConnell *et al.*, 1989) which has more strongly drooping fronds which are some 15-20 cm wide and grows faster than the type; 'Bostoniensis Compacta' is similar to the former but smaller in stature; 'Elegantissima' bears bright green, compact and bi- to quadri-pinnate fronds with feathery effect, the pinnae being closely set and overlapping, and occasionally it reverts back to the pinnate form; 'Hillii' bears light green, crinkled and bipinnate fronds; 'Marshallii' bears pale-green, broad and densely crested fronds; 'Rooseveltii' bears dark green fronds having wavy leaflets; 'Teddy Junior' is a compact sport of 'Rooseveltii'; 'Todeoides' bears pale-green, feathery and finely divided fronds; 'Whitmannii' bears light green and bi- to tri-pinnate fronds some 45 cm long, having small pinnules which cause it to appear very lacy; *etc.*

Onoclea [Aspidiaceae (Polypodiaceae or Athyriaceae)]

In Greek *onos* means 'a vessel' and *kleio* for 'to close', alluding to closely-rolled fertile fronds. A genus of one hardy species from temperate regions in both hemispheres, which grows well in wet soil and therefore is grown along the banks of streams as ground cover. ***O. sensibilis*** is rhizomatous fern growing 30-60 cm in height. The fronds are long-stemmed, pale-glaucous-green, pinnate and triangular. The sterile fronds which appear first are solitary, to 60 cm or more, long-stalked, the blade triangular-ovate, deeply pinnatipartite and the pinnae wavy-toothed, though fertile fronds are much smaller, bipinnate, pinnae erect and pinnules rolled around the sori so resemble masses of small beads, and dark brown to blackish on maturity. Propagated by division and spores.

Osmunda (Osmundaceae)

A genus of 10 clump-forming species of large ferns distributed worldwide but not in Australasia. The exact derivation of the name of *Osmunda* is not certain, however, it may have originated from the Latin *os* meaning 'mouth' and *mundare* 'to purify', or for Asmund-Osmundus – an 11th century Sandinavian writer instrumental in preparing the way for Christianity to be accepted in Sweden. They grow up to 3 m in height in its native haunts though not so when cultivated. All these are deciduous and hardy with bipinnate fronds, the fertile pinnules of which are quite reduced and bear marginal sporangia. For their growing these require shade though *O. regalis* is sun-tolerable and does well in very wet conditions. They are propagated through division and spores. ***O. cinnamomea*** from North and South America, and East Asia is a deciduous fern with 1 m height and 45 cm spread, bearing pale-green, lanceolate and divided sterile fronds with deeply cut pinnae in the outer side, which surround the brown fertile fronds, all arising from a fibrous rootstock. ***O. claytoniana*** from North America is deciduous, some 60 cm high and 30 cm wide, bearing pale-green, lanceolate and divided into oblong and blunt pinnae. Outer larger fronds are sterile while smaller central ones are fertile. ***O. regalis*** from Europe, West Asia, India, South Africa and North and South America grows to 1.2-2.0 m with long fronds which are bright green, elegant, woolly, broadly oval to oblong or lanceolate and bipinnate though pinkish-brown when young. Outer ones are sterile, the inner ones have barren lower pinnae, and fertile upper pinnae with no pinnules, while their veins are covered with spore capsules which on ripening turn brown and look like dead flowers of an astilbe. In fact, on maturity the plants bear tassel-like and rust-brown fertile rachis at ends of taller fronds. These plants gradually build up a mass of crowns and a matting of black roots some 60-90 cm above the ground, and these roots are used as orchid substrate (compost) under the name of osmunda fibre. *O. r.* 'Cristata' growing to 90-120 cm in height and spread with crested pinnae. *O. r.* 'Purpurascens' is a fine waterside plant which grows 1.2-1.5 m tall with young fronds being deep copper-pink, maturing to copper-green with purple midribs.

Pellaea [Sinopteridaceae (Polypodiaceae)]

A genus of 80 evergreen, tufted or rhizomatous species from temperate and sub-tropical regions including New Zealand and Africa, majority being in the Americas. *Pellaea* derives from the Greek *pellaios* meaning 'dark' due to the dark (purple to black) stalks of the fronds in many species. The plants of the genus bear small leathey and pinnate to tripinnate fronds having small linear to orbicular and evenly arranged pinnae along slender and arching stems, where close to the margins, globular to oblongoid sori are formed so to cover them the pinnae edges become curled. These can tolerate more dryness than many of such other ferns, admire for filtered sun and survive well at the temperature range of 7-27 °C. These are propagated by spores, and by division at the time of repotting during spring. These are eye-catching ferns for any brightly lit room. ***P. atropurpurea*** from North America and Guatemala is a clump-forming fern with short rhizomes. Fronds are erect-type, below bipinnate and at the ends pinnate, 30 cm long and 15 cm wide, and the pinnae are leathery, lanceolate to linear, 1-5 cm long and with shining purple stalks. ***P. falcata*** from S. Asia to New Zealand bears pendant, up to 60 cm long and 7.5 cm wide fronds with short stalks while its pinnae are wing-shaped, all roughly equal in size and some 3.75 cm long. ***P. rotundifolia*** from New Zealand, Australia and Norfolk Islands grows and spreads through prostrate and wiry runners (creeping rhizomes). Fronds some 30-90 cm long are arching to prostrate on wiry and black stalks which are covered with red-brown scales and bristly hairs, and the pinnae are 20-60 to a frond, quite dark green, alternate, round or broadly oval and some 1 cm in diameter. ***P. sagittata*** (*P. cordata sagittata*) from Texas to southern Mexico is clump-forming, bearing 45 cm or longer fronds which are firm-textured, bipinnate and narrowly triangular. Pinnules are rich green, sagittate to triangular-lanceolate and 1-2 cm long. ***P. viridis*** (*P. adiantoides, P. hastata*) from South Africa and West Indies is a clump-forming fern with more than 60 cm long, erect, ovate and pipinnate fronds where the leaf stalk is purple-black.Ovate to broadly lanceolate pinnules may be 2 cm or more long and those near the midrib are sometimes auricled. Its var. 'Macrophylla' bears paler and much larger leaflets.

Phlebodium (Polypodiaceae)

A genus of 10 species from tropical to sub-tropical America can tolerate to a minimum of 16 °C winter temperature whether in sun or under room condition and grow up to 1 m in height. Its most cultivated species is ***P. aureum*** (*Polypodium aureum*) which grows from 90 to150 cm high and about 60 cm wide, bearing creeping golden-scaled aboveground rhizomes with evergreen, bluish or glaucous, long-stalked, arching and deeply indented fronds, where beneath are covered with round and orange-yellow clusters of sporangia. Var. 'Mandaianum' grows to about 75 cm with glaucous, evergreen, arching and deeply lobed fronds having wavy leaf edges, and bearing attractive orange-yellow sporangia beneath. Pinnae are deeply cut and wavy. It can tolerate as low as 5 °C temperature. Var. 'Cristatum' has curly leaf tips.

Phyllitis (Aspleniaceae)

A genus of eight fern species from Europe, northwest Africa and as far east as Iran, Korea, Japan and southeastern Russia. This grows to 15-30 cm long and possesses straight or curved strap-shaped leathery leaves. ***P. scolopendrium*** (*Asplenium scolopendrium*), the hardy species from Europe bears rich glossy green, 10-60 cm long, entire and strap-shaped fronds which are basally heart-shaped and with highly wavy margins. It is a shade-tolerant beautiful fern with many of the cultivars having lobed, waved or crested fronds. The fronds are ruffled-edged in 'Antiquorum', strongly crisped margins in 'Crispum' and 'Undulatum', both being the most graceful cultivars in the containers

Platycerium (Polypodiaceae)

A genus of 17 species of large and epiphytic ferns from Africa, Australia, New Guinea, southeast Asia, Malaysia and South America. It derives its name from the Greek *platys* meaning 'broad' and *keros* meaning 'a horn', due to shape of its fronds. They have two types of leathery fronds, (i) sterile at the base which are simple ensheathing each other and set firmly against their support and these collect the plant debris, decay them and convert into humus, and (ii) spore-bearing fertile fronds above are deer/elk-horned though not necessarily, erect, arching or drooping. The discrete areas of the undersides of the fertile frond lobes are completely covered with brown sporangia. All the species are tropical in nature and can easily be grown indoors or in the warm conservatory as hanging basket plants and by clinging to the slabs of tree bark outside. These are propagated through division and by spores. It relishes bright indirect or filtered sunlight. Staghorns need very high humidity that is hard to achieve in most houses. ***P. angolense*** from tropical Africa bears rounded to broadly oblong, wavy-edged and 30-60 cm long sterile fronds, while undivided, ovate, cuneate and up to 45 cm long fertile fronds. ***P. bifurcatum*** (*P. alcicorne*) from eastern Australia to Polynesia is popular because of its evergreen, bright green, spreading to drooping fertile fronds which are long-cuneate at the base, and with several dichotomous narrow lobes above and at the tips these bear spores, while sterile fronds are downy when young, rounded to kidney-shaped, somewhat lobed or undulated and 15-30 cm wide. It is a highly variable species which forms huge colonies in its natural habitat. Its some of the various cultivars are 'Majus' which is a little larger with brighter green and erect fertile fronds; 'Netherlands' (*P. alcicorne* 'Regina Wilhelmina') has grey-green shorter but broader fronds radiating to all the directions;' Ziesenherne' is slow growing and smaller; *etc.* ***P. coronarium*** (*P. biforme*) from SE Asia bears deeply lobed, erect and up to 60 cm long sterile fronds while fertile fronds are bright green, 2-3 m long, pendulous and divided into thong-like forked lobes. ***P. grande*** from Malaya, Java and N. Australia is an evergreen species growing 60 cm or more in length and breadth. Though it is similar to *P. bifurcatum* but is more vigorous and larger, the fan-shaped fertile fronds are 45 to 150 cm long, arching to pendulous and have deeper lobes, while upper set of barren fronds are glossy bright green, cuneate-ovate, some 60 cm or more wide, erect with boldly lobed top and fan-like. ***P. hillii*** from Australia is similar to *P. bifurcatum* but the lobes of the fronds are broader, erect and blunt-tipped. ***P. madagascariense*** from Madagascar forms an interesting net-like pattern on its sterile fronds which have protruding veins. It is most successful in cooler temperatures.

Pityrigramma (Hemionitidaceae)

A most attractive rare genus mostly from tropical America, South Africa and West African islands. It tends to colonize dry lands. It likes bright locations with a cool atmosphere (15 °C). The fronds in all the species of this genus are characterized by a white, light yellow or golden-yellow floury film on the undersides, and the spores are formed along the veins.

Polypodium (Polypodiaceae)

A creeping rhizomatous genus which once included some 1,000 fern species but now reduced by Copeland and Ching only to 75. These bear feathered or undivided fronds. They are cosmopolitan in distribution, some being indigenous to Europe such as *Polypodium vulgare* whose roots were years ago often used as a planting medium for orchids and epiphytes, and many others are generally found in the tropics or sub-tropics. *Polypodium* derives its name from the Greek *polys* meaning 'many' and *podium* (*pous*) 'a foot', referring to its branched rhizomes. Most of the species are accustomed to warm climates and will only flourish indoors in a temperate climate. They are mostly evergreen but some others are semi-evergreen or deciduous and are grown for their sculptured fronds, requiring minimum winter temperature of not less than 10 °C. These grow as terrestrial or rock ferns and as epiphytic plants. Their rhizomes are fleshy and branched and usually with pinnate or pinnately lobed fronds though a few species possess almost entire fronds. The spores being formed on the underside of fronds are without coverings (indusium). These are propagated by rhizome divisions, clump division and spores. ***P. angustifolium*** from tropical America is a dense clump-forming epiphytic un-fern-like fern coming up from rhizomes, and is most suitable for hanging baskets. Fronds are dark green, densely borne, leathery, linear, entire, 30-60 cm long and arching. ***P. aureum*** fronds are some 30-60 cm long, deeply cut and on thin stalk. Its thick rhizomes creep along the surface and does well in the dry air of the living room. Its beautiful variety is 'Mandaianum' bearing blue-green and wavy-edged leaflets. ***P. glycyrrhiza*** from the tropics and sub-tropics is a deciduous fern growing some 45 cm in length and breadth, bearing mid-green, oblong-triangular to narrowly oval and divided fronds arising from liquorice-scented rootstock, and the pinnae are lanceolate to oblong. ***P. longifolium*** from tropics and sub-tropics bears very narrow (2 cm wide) and long (60 cm) leaves. ***P. lucidum*** from tropics and sub-tropics bears feathered leaves. ***P. musifolium*** (*Microsorium musifolium, Phymatodes musaefolia*) from Malaysia and New Guinea bears semi-woody rhizomes, the plants are robust and epiphytic, fronds are stalkless or so, dark green but with an attractive dark vein patterning, 45-90 cm long, narrowly oblanceolate and widely undulate. It is most suitable for hanging baskets and pot planting. ***P. phyllitidis*** (*Campyloneurum phyllitidis*) from tropical America bears fleshy, short and creeping rhizome. An epiphytic fern with lustrous brightish-green and stalkless fronds that are entire, narrowly lanceolate to strap-shaped and 30-90 cm long. It is suitable for planting on tree bark, in pots and baskets. ***P. piloselloides*** from tropical America is an epiphytic fern with fairly slender and far-creeping rhizomes. Fronds are variable, 2.5-8.0 cm long and entire, the sterile ones glossy and leathery, and ovate to obovate, and fertile spore-bearing ones a little narrower and longer. To contain its wandering rhizomes, it can be grown as tree fern stem, on tree branches or in the hanging baskets. ***P. polypodioides*** from the tropics and sub-tropics comes up from scaly rhizome, is a frost-tender, semi-evergreen and creeping fern with 10 cm height and 15 cm spread. Its fronds are mid-green, lanceolate with widely spaced pinnae. ***P. punctatum*** (*P. integrifolium, P. polycarpon, Phymatodes irioides*) from tropical Africa, Asia, Australia and Polynesia is an epiphytic fern with dark brown, thick and scaly rhizomes. The fronds are light to yellow-green, somewhat glassy, leathery, entire, narrowly lanceolate to strap-shaped and 30-90 cm long which in some varieties may be dentate or forked at the tip as in 'Grandiceps'. ***P. scandens*** from tropics and sub-tropics bears feathered leaves. ***P. scouleri*** from tropics and sub-tropics bears spreading rootstocks, from where arises an evergreen and creeping fern with leathery, triangular to oval and divided fronds. ***P. subauriculatum*** from tropical Asia to Australia with creeping rhizomes is usually epiphytic most suitable for hanging baskets, bearing 90 cm or more lengthy fronds that are bright glassy green, spreading to pendulous, lanceolate and pinnate, and the pinnae are wavy, entire or toothed. Its form 'Knightiae' appears to a *Nephrolepis* fern, is slow-growing, slightly smaller and the pinnae are cut into various narrow and pointed lobes. ***P. tsus-simense*** is a small plant most suitable for growing indoors as it does not mind dry air, and grows hardly 30 cm in height with dark green fronds. ***P. virginianum*** from temperate America is a frost-hardy, semi-evergreen and creeping fern growing 30 cm high and 23 cm wide and its fronds are mid-green, lanceolate and divided. ***P. vulgare***, most renowned species in cultivation is from northern hemisphere and South Africa. It is a mat- or clump-forming epiphytic and terrestrial evergreen with densely brown scaly rhizomes usually creeping on the surface. It grows 15-38 cm in height and breadth. Its fronds are solitary, mid-green, ovate to oblong or linear-lanceolate, 10-30 cm or more long, elegantly drooping and pinnatisect or pinnatipartite in the form of herring-bone-like where beneath the fertile pinnae bear large, rounded, yellow to brown sori in single rows between midrib and margin. It is most suitable for rock gardens for cooler regions. It is a highly variable species with various forms, *viz.*, 'Cambricum' which bears some 30-50 cm long with quite wide fronds and pinnae cut to many narrow lobes; 'Cornubiense' is usually infertile and bears variously tri-, quadri-pinnate and pinnatifid fronds (sometimes normal pinnate so such fronds should be removed) on the same plant; 'Cristatum' is a fully hardy creeping fern with rhizomes having copper-brown scales, height and breadth being 25-30 cm, and fronds being mid-green and narrowly lanceolate with semi-pendulous terminal crests; 'Longicaudatum' being quite similar to original species but its frond-tips are much elongated; 'Macrostachyon' is similar to the type species but is more robust with broader fronds that taper to a long point; 'Pulcherrimum' bears tripinnatipartite fronds; *etc.* These can effectively be grown in shade beneath trees or in dry walls in well-drained humus-rich soil, and once established, these do not require to be disturbed.

Polystichum [(Polypodiaceae (Aspidiaceae)]

This genus is cosmopolitan (worldwide in temperate to subtropical regions often in mountainous forests and along the banks of rivers) with 120-135 terrestrial species having thick and erect rhizomes, and with clusters or rosettes of pinnate or bipinnate fronds, where fertile pinnae have two rows of rounded sporangia beneath. It derives its name from the Greek *polys* meaning 'many' and *stichos* meaning 'a row' alluding to the rows of sori on the fronds. These are known mainly as outdoor ferns except the small species *P. tsus-simense*. The fronds of *P. munitum* are prized by florists for use as greenery with flower arrangements and bouquets. ***P. acrostichoides*** (*Asplenium acrostichoides*) from EN America is semi-cold-hardy evergreen species with 60 cm long, lustrous green and pinnate fronds, pinnae being lobed at base, lanceolate, fertile ones smaller and towards the tips, its mutants with forked, twisted or crested pinnae are also available; ***P. aculeatum*** from Asia, Europe and South America is evergreen with 1 m long and a little lustrous green fronds which are a little leathery, rigid, lanceolate and bipinnate, and the pinnules are toothed and bristle-tipped; ***P. falcatum*** (*P. lonchitis, Cyrtomium falcatum*) from Arctic and northern temperate zone, *vis-à-vis* farther south mountains, so requires shaded cold greenhouse, is evergreen with somewhat lustrous green fronds which are linear or narrowly lanceolate, pinnate and 15-60 cm long, pinnae unequal at base, often overlapping, ovate-lanceolate, 1-3 cm long and sharply bristle-toothed; ***P. munitum*** from Alaska to Montana and California is similar to *P. acrostichoides* even in cold-hardiness but the fronds are more straight and the length is up to 105 cm; ***P. setiferum*** (*P. angulare*) from temperate climates of north and south hemisphere, though is similar to *P. aculeatum* but the fronds are arching or drooping, rather soft, and the pinnules have fewer and larger teeth, and its var. 'Acutilobum' (Proliferum) is quite showy and very hardy, bearing numerous tiny plantlets especially along the midrib and the pinnules are dense and narrower, and its another var. 'Divisilobum Densum' bears tri- to quadric-pinnate and plumose fronds; ***P. tsus-simense*** from temperate to subtropical regions of Asia, mainly Korea, Japan (Tsushima islands), China and Taiwan is a terretrial fern growing up to 30 cm in height and 15-20 cm across, suitable for cooler locations especially in bottle gardens, bearing long, dark green, leathery, oval and bipinnate fronds where the lower part of the stalk is dark-brown-scaly while upper section hairy, the round clusters of spores have shield-shaped veils forming two regular rows between the edge of the frond and its central rib; *etc.*

Pteris [Pteridaceae (Polypodiaceae)]

A tufted to clump-forming genus of 250 deciduous, semi-evergreen or evergreen fern species of cosmopolitan distribution (tropics and sub-tropics with some species even in temperate regions). They have long-stalked pinnate to bipinnate fronds having marginal spores covered by their rolled edges to the reverse side. The leaf stalks are thin, smooth, yellow, green or brown. They are frost-tender, prefer bright to semi-shady situation, green frond types also prefer shady locations with no sun and cool temperatures (15-18 °C), and variegated ones warmer (18-20 °C) though can tolerate as minimum as 3 °C winter temperature. Most of the species prefer high humidity. They are suitable for growing indoors and in cool greenhouses. These are propagated by division of clumps or spores. ***P. argyraea*** (*P. biaurita quadriaurita*) produces grey-green pinnate fronds with broad and deeply lobed pinnae, each pinna in the centre having a broad and pale-green-white band which extends up to the base. ***P. cretica*** from SW Europe, Mediterranean islands, Iran, India and Japan is an evergreen or semi-evergreen fern growing to 30-45 cm in height and 22.5-38.0 cm in spread, bearing pale-green, pinnate and triangular to broadly oval fronds, and 5-13 narrowly lanceolate to strap or finger-shaped pinnae some 7-15 cm long, and the basal one being often divided into 2-3 lobes. Its spores can be observed germinating in the pots of other plants in a cool greenhouse. Its forms 'Albolineata', an evergreen or semi-evergreen fern growing up to 45 cm in height and 30 cm in spread, stems wiry, bearing pale-green, triangular to broadly oval and divided fronds, having a white central band on each finger-shaped pinna; 'Alexandrae' bears strap-shaped crested fronds with forked and/or lobed tips; 'Childsii' bears light green, waved and frilled, and lobed pinnae; 'Major' is a green-leaved fern more tolerant to shade, wiry leaf stalk, each with whorled pinnae which are wavy and/or shallowly crenate, and the lower pinnae are often deeply cut; 'Major Cristata' ('Wimsetti') is trade name of the group having heavily crested varieties, fronds are irregularly lobed, and the toothed ends of pinnae are often crested; 'Mayi' is similar to the dwarf 'Albolineata', growing to 30 cm with crested pinnae tips; 'Parkeri' bears broad and rough fronds on long stalks, pinnae lanceolate and pointed; *etc.* ***P. ensiformis*** from Himalayas and Sri Lanka to Australia, is a deciduous or semi-evergreen fern growing 30-50 cm in height and 23-30 cm in spread, bearing dark green bipinnate fronds often with grayish-white around the midribs and are coarsely divided into lanceolate to linear or finger-shaped pinnae, where sterile fronds are smaller with slight broader lobes. This fern is a slender and deep green version of *P. cretica* with more elongated pinnae. Its forms 'Arguta' grows to 45 cm in height bearing deeper green fronds with silver-white marks in the centre; 'Evergemiensis' is small and very elegant with green and white variegation on fronds; 'Victoriae' (*P. cretica* 'Victoriae') bears dark green, small and pinnate to bipinnate, slender, upright, irregularly white-variegated along the pinnae-centres, and fertile fronds, while barren fronds are smaller and more pendent, on long wiry stalks; *etc.* ***P. multifida*** (*P. serrulata*) from East Asia is a deciduous or semi-evergreen fern with 30 cm height and 23 cm spread, and bears mid-green, narrow and much-divided wispy fronds some 45-60 cm in length and 20-30 cm in breadth, pinnate, upper pinnae entire and linear, the lower ones twice or thrice forked, and the midrib is winged between pairs of the pinnae. Its robust cultivar is 'Rochfordii' which has crested fronds. ***P. quadriaurita*** from tropics is a graceful pinnate fern with 30-60 cm height and 25-75 cm spread with straw-coloured leaf stalks. The fronds are bi- to tri-pinnate with ovate to lanceolate pinnae which are finely toothed. Its cv. 'Tricolor' bears greenish-red fronds. ***P. tremula*** is a graceful fern from New Zealand, Australia and Fiji growing comfortably and seeding freely in a cool greenhouse. It may reach to 90-150 cm in height and spread, bearing bipinnate upper half of the soft bright green frond though lower half tripinnate. ***P. vittata*** from tropics and sub-tropics of Asia, Africa and Australia is fast-growing, bearing simply pinnate fronds 90 cm or more in height, each pinna being dark green, oblong to strap-shaped, 10-15 cm long, slender-pointed and toothed.

Pyrrosia (Polypodiaceae)

A genus of about 100 species of mostly epiphytic ferns from tropical and temperate regions of Asia and Africa. They are closely related to *Polypodium* and *Microsorium*. Their rhizomes are scaly which wander along mossy and branches bearing alternate and largely simple fronds. The sori are large and close together and are formed on the interveinal rows of fertile fronds below. The fronds are dissected and many of them are most suitable for growing in hanging baskets. These are propagated through rhizome cuttings and spores. ***P. lingua*** from China, Japan and Taiwan produces slender and branching rhizomes up to 1.2 m or more in length, and covered with rusty scales. Fronds are brown-felted when young but with age become rich green above, 10-25 cm long, lanceolate and wavy. The leaf tips of the form 'Cristata' ('Obake') are broader and frequently deeply forked. The leaf margins in 'Montrifera' ('Hagoromo', 'Lacerata') are irregularly but deeply fringed. 'Nankin-shishi' though similar to 'Cristata' but with more finely and densely crested leaf tips. 'Variegata' and 'Nakogriri-ba' bear fronds with rounded teeth patterning of yellowish streaks and bands. ***P. longifolia*** from Australia, Malaysia and Polynesia produces shorter but thicker rhizomes than *P. lingua*, and the fronds are 20-60 cm long, strap-shaped, thick, arching to pendant, sub-glossy above and grey-hairy beneath.

Rumohra [Davalliaceae (Polypodiaceae)]

A genus of one evergreen rhizomatous species, ***R. adiantiformis***, formerly included in *Polystichum* though resembling a *Davallia*, from southern hemisphere in tropical to warm temperate regions. It is named in honour of Dr. Carolus de Rumohr Holstein. The species bears quite showy and long-lasting fronds with a very high post-harvest life, hence is highly valued in the international flower market as cut green for floral arrangements (D'Souza *et al.*, 2006). Most of the world's production of *R. adiantiformis* is in Florida where it is grown in controlled environments (Stamps, 2004; Stamps *et al.*, 1994). It grows through brown crawling runner and its leaves are triangular and some 90-120 cm long though indoors only up to 30 cm. Frond blades are light green, ovate to triangular, bi- or tri-pinnate and leathery while pinnules are oblong with large sori beneath where each one is covered with an umbrella-like crenate-margined indusium. It is an ideal fern for pot planting and hanging baskets, and tolerates minimum temperature of 10 °C, low humidity and light, and fluctuating temperatures. It is propagated by division and spores.

Salvinia (Salviniaceae)

A genus of about 10 aquatic plant species closely related to the ferns. These are distributed in the tropics and sub-tropics worldwide (Europe, North Africa, Asia, and tropical America including Mexico and Cuba to Argentina). It was named for Antonio Maria Salvini (1633-1729), Professor of Greek at Florence University. These plants provide shelter to fish in the well-lit aquaria, and are ideal for a well-illuminated conservatory pool in the tropics where minimum temperature rarely falls below 15 °C, however, these can survive to a minimum of 10 °C non-persisting temperature. These are rootless, perennial and free-floating water plants, deciduous in temperate regions and evergreen under tropical conditions, where slender stems bear whorls of three leaves, two being flat and floating but the third one below the water is deeply pinnately dissected into slender segments to function as non-existent roots. Fading foliage from aquaria/pools are regularly removed and when it becomes overcrowded the thin and weak plants are also removed. These are propagated by separating young plants. ***S. auriculata*** from Mexico and Cuba to Argentina grows to 25 cm across in spreading colonies, the pale to mid-green boat-shaped and rounded surface leaves appearing in pairs on branching stems, are sometimes suffused purplish-brown, 2.5 cm long and 4.5 cm wide, and have clothing of erect and stiff hairs in fours from minute papillae on the upper surface; ***S. natans*** from Asia, N. Africa and central and SE Europe is slightly cold-tolerant species, growing to 15 cm across with mid-green and oval surface leaves on branched stems, and which are 1.0-1.5 cm long and 0.5-0.9 cm wide where papillae bear 3-4 hairs; ***S. rotundifolia*** from tropical America grows to 7 cm across with almost circular surface-leaves having 1.5 cm length and 2.0 cm width, with many papillae on the upper surface, each bearing four hairs.

Selaginella (Selaginellaceae)

A genus of 700 species of spore-bearing plants closely allied to ferns. It is mainly circum-tropical with a few species extending into temperate and arctic zones. *Selaginella* named so by Linnaeus, is diminutive of the Latin *selago*, once used for *Lycopodium selago*, the common or fir club moss or spike moss of the northern temperate zone. These are lycophytes that are members of an ancient vascular plant lineage that first appeared in the fossil record some 400 million years ago. These lycopsids lack true leaves and roots. Under dry conditions, some species of *Selaginella* roll into brown balls. In this state, they may be uprooted. Under moist conditions the brown balls become green. Other species are tropical forest plants that appear at first glance as the ferns though genuinely not. This class developed its wide range of shapes and forms millions of years ago. The scale and seal trees of this class, that dominated the forests of the carboniferous age, grows up to 40 m tall, though most of the species are leafy. The shoots of club moss have usually forked branching and are covered in four rows of scale-like tiny leaves. The spore bearing leaflets form 4-sided terminal spikes, each spore leaf carrying only one spore capsule which is formed where the leaf emerges. Male sporangia with many minute spores and female ones with four large spores are produced and grow together on one spike. Development of male and female cells start before spores leave their capsules. Apart from trees, there are climbers, prostrate- and low-growing and upright ones, but the ones being grown are small fern-like, terrarium plants requiring minimum winter temperature of 10 °C. The species may differ in leaf colouration from yellow to bright green and blue-green, *vis-à-vis* variegated white-green and yellow-green types. These are propagated through spores, through pinning down the branch tips in the compost and through division, *vis-à-vis* plantlets being formed on the leaves of a few species. The most popular species are ***S. apoda***, a low-growing species from North America which looks like a moss; ***S. denticulata***, a prostrate mat-forming cool-location species from Macronesia and Mediterranean region to Syria, which grows 30 cm or more wide with pale-green leaves, and when mature these redden in bright light, and the largest leaves being broadly ovate and about 2 mm long; ***S. erythropus***, an erect-growing 30-cm high and bronzy-red plant from tropical America, has much divided, triangular and fern-like fronds; ***S. gracilis***, a species though looks daintier but grows to 90 cm tall and is native to Polynesia; ***S. kraussiana***, a dense carpet-forming prostrate-growing species from tropical and South Africa which bears up to 30 cm long branches, *S. kraussiana* 'Aurea' bears golden-green foliage, 'Brownii' forms mossy hummocks some 15 cm across, 'Variegata' is cream-variegated, *etc.*; ***S. lepidophylla***, from Texas and Arizona to El Salvador, is a rosette-forming cool-location plant growing up to 10 cm with stiff branches radiating from a short central stem, and when dry these branches roll inwards and make a ball but expand again when moistened; ***S. martensii***, a 30 cm tall upright growing tufted to clump-forming temperate species from Mexico which has support of very strong roots, *S. m.* 'Watsoniana' bears silvery tips; ***S. pallescens*** (*S. cuspidata, S. emmeliana/S. emiliana*), a tufted to clump-forming temperate species from Mexico to Colombia and Venezuela, where frond-like 20-30 cm high branching appears on the erect central stems, and its var. 'Aurea' bears bright yellow-green foliage; ***S. plana*** from tropical America, grows to 30 cm tall with erect stems and spaced leaves, and is excellent as terrarium plant; ***S. tamaracina*** from Japan, a mat-forming, low-growing and preferring cooler regions, is a variable species with a number of varieties, and are presentable in various colours of leaves, *viz.* various shades of green, and variegated ones in yellow or golden shades; ***S. uncinata***, a native of China, bears densely tufted stems which radiate outwards first in ascending order and then somewhat horizontal, growing up to 60 cm long while sending out aerial roots, leaves

Dicksonia antarctica

Lygodium japonicum

Pellea rotundifolia

Phyllitis scolopendrium

Platycerium bifurcatum

***Selaginella* species**

are metallic blue and is most suitable for hanging basket at cool locations; ***S. victoriae*** from Europe has strong and erect growing stem topped by broad leaves, and is suitable in underpotted terrarium; ***S. vogellii***, an erect growing, 30-60 cm tall and branching plant from West Africa, where above branches are bright green and frond-like, which in light show bronzy tints; ***S. willdenowii***, a climbing species bearing blue-green fronds from tropical Asia and Africa which grows several metres with the help of its protruding frond shoots; *etc.*

Thelypteris [Thelypteridaceae (Polypodiaceae)]

A genus of four rhizomatous and fully hardy fern species with cosmopolitan distribution. It takes its name from the Greek *thelys* meaning 'female' and *pteris* 'a fern'. The fronds are tufted, rosette-forming or solitary bearing small rounded sori near the margins of pinnae. They prefer shade or partial shade and are propagated by spores or division. ***Thelypteris hexagonoptera*** (*Dryopteris hexagonoptera, Phegopteris hexagonoptera*) from northern hemisphere is a deciduous fern growing to 45 cm high and with 30 cm breadth from the creeping rootstock, bearing mid-green, lanceolate and much-divided fronds with pinnae being oblong to triangular and indented. It requires shaded position. ***T. oreopteris*** (*T. limbosperma, Oreopteris limbosperma*), a thick, short and ascending rhizomatous fern is scattered in northern hemisphere south to Madeira. Yellow-green but deciduous fronds appear in rosettes, which are lanceolate, bipinnate, 30-90 cm long, firm-textured, and lemon-fragrant when bruised, and the sori are <0.5 mm wide. ***T. palustris*** from northern hemisphere is a deciduous fern, growing up to 75 cm tall and 30 cm across with creeping, wiry and blackish rhizomes, bearing pale-green, strong, erect and lance-shaped fronds which have widely separated and deeply cut pinnae. ***T. phegopteris*** (*Phegopteris connectilis*) is scattered in northern hemisphere south to Virginia (USA) and bears underground slender and creeping rhizomes, bearing solitary, triangular-ovate, long-staked but blades at almost right angles, pinnate, up to 20 cm long and pinnae deeply lobed, and the sori are <0.5 mm wide.

every 2-3 days for about a month so that plants may not desiccate due to harsh weather, already existing roots settle properly and new root emergence is not hampered. Around each plant some tree guard should be used to protect them from being damaged by animals and human browsing. The establishment of the plant is ensured when these are planted during rainy season. Afterwards, the plants are cared well for any outgrowing or broken branch, for termite infestation and for any other damage. The plants are kept always in shape so that when full grown the plants may not look odd and all the branches have grown properly to all the directions. Proper shaping at earlier stage will save the plants against breaking of their branches during strong winds.

Soil in the pot is allowed to dry a little between **waterings**. During dry weather the plants are watered once each week to keep the soil consistently moist throughout the growing season but during rainy season and in winters frequency of watering is drastically reduced. In case of potted plants, these are watered twice in a week but very lightly. Always it should be kept in mind that these plants do not like water-logged conditions at all. Regular supply of **fertilizers** during active growth should be made only when the plants either in pots or in their infancy when grown as specimen trees or along the roads. Liquid fertilizer of slurry in case of potted plants will be very useful once in a fortnight. During low light intensities, less feeding is recommended. Generally 14:14:14 ratio NPK at the rate of 5-10 g per young plant is applied at regular intervals, but to the grown up plants the quantity is increased to five times or even more, depending on age and growth of the plant. Feed using a high phosphorous 5-10-5 NPK fertilizer during the first two months of growth, and then switching to a high nitrogen, *i.e.* 10-5-5- NPK fertilizer for the rest of the life of the plant. Fertilizers should be applied once every 3-4 weeks during spring and summer, and once every 5-6 weeks during fall. Only half the recommended dose as prescribed by the manufacturer should be given for best results.

General deficiency of primary nutrients in the plants is expressed as general light greening or paling of older leaves due to chlorosis. **Potassium deficiency** is expressed in the lower leaves with marginal chlorosis and necrosis. Leaf tissue analysis is the best way to confirm the deficiency of a particular nutrient.In case of potassium deficiency, it should be supplemented by application of potassium chloride, potassium nitrate or the most common muriate of potash. **Mg-deficiency** shows chlorosis in the lower leaves at the most distant margins first. Plants grown with adequate soil-incorporated dolomite should not develop Mg-deficiency symptoms. Foliar or soil application of magnesium sulphate at 3g/l of water will correct its deficiency. **Mn-deficiency** occurs as the interveinal chlorosis in the terminal leaves. Prevention is accomplished through incorporation of a microelement blend at the rate of 3-6 g per square metre of potting mix. Mn-deficient plants can be sprayed with manganese sulphate at the rate of 4 g/l of water.

Regular **weeding** in the nursery beds, pots, borders and beds wherever these plants are being nurtured or grown should be done at regular intervals so that these may not get any chance to rob the main plants of their nutrients available in the soil or in pot. Moreover, organic mulches spread over the beds and in pots will conserve the moisture and check the weeds to grow.

Growth and Development

Since majority of the cultivated members in the genus *Ficus* are trees or shrubs which on growing indoors become unfit in a few years to be enjoyed indoors due to their growth so their growths are required to be arrested for their enjoyment for much longer time. Growth retardants inhibit cell elongation and interfere with gibberellin-synthesis in the plants so an early spray will retard growth of main developing trunk *vis-à-vis* together other plant parts. Watson (1987) states that these move upward in the xylem to active growing points though these retardants are immobile in the phloem. Bonzi applied as a foliar spray (35, 40, 75 ppm) or as a drench (0.125, 0.25, 0.5, 0.75 or 1.0 mg/10-cm pot) reduced plant height, leaf production, internodal length and leaf size, the results being more pronounced when these were drenched under higher concentrations up to 0.5 mg *a.i.* per 10-cm pot, and drenching of A-Rest at 1 ppm per pot improved growth of plants (Jim Green, 1987).

Post Harvest

Growing, buying and selling light-acclimatized rubber plants are a must because plants grown under reduced light can be subsequently stored for up to 20 days in the dark and still perform very well under interior environments. Such light-acclimatized plants had 80 per cent less leaf drop than those grown under higher light. Often the plants will drop foliage when moved to lower light levels and/or when placed under water stress or watered too much. Floralife® Leafshine is a ready to use aerosol spray that can be applied on leaves of *Ficus elastica* to enhance their overall appearance. This look entices more buyers for such plants at the store level, keeps interiorscaping more attractive and charming, and gives designed arrangements more life. This product can be used on hard leaf foliage plants and fresh cut foliage greens. The spray removes dust and residue deposits on leaves and gives a clean appearance and restores the natural shine so its application should be repeated as and when required to maintain healthy and shiny appearance of leaves (Ranwala, 2010).

Insect-Pests, Diseases and Physiological Disorders

Rubber trees known for their glossy evergreen foliage are weakened by attack of certain insect-pests and diseases though are not of major concern. Occasionally scales, mealy bugs, thrips and spider mites are some important pests.

Mealy bugs are white, waxy insects that sometimes colonize the undersides of rubber tree leaves. Early signs of infestation include drooping or dry-looking leaves and the appearance of cotton-like masses along leaf attachment sites and on the undersides of the leaves. Manual removal of mealy bugs is difficult and seldom effective, although spraying the affected area with an alcohol and water solution can lessen the severity of the infestation. Honeydew and sooty mold are often present and infested plants become stunted and, with severe infestations, plants parts die. They can be prevented and controlled by spraying with Imidacloprid @ 0.5 ml/l of water.

Scale infested plants become weakened or stunted and die. Scales can be found feeding on leaves, petioles or stems. Their shapes, sizes and colours are variable and many are hard to distinguish from the plant material on which they are feeding. **Thrips** are tiny winged insects. Although some cultivars of rubber tree are resistant to thrip infestation, most commonly available houseplants are susceptible to them. Early symptoms include silvery-gray scarring under the leaves where feeding has occurred, often causing infested leaves to curl and drop. They can be controlled by spraying Metasystox @ 1.5 ml/l of water.

Rubber plants are infested with **foliar nematodes** which enter initially the lower leaves from nearby vegetation infested by them. Rectangular leaf spots start near the mid-vein on lower leaves and extend to the margin. Proper sanitation of the area of rubber plantation will keep this problem under check. **Lesion nematode** greatly reduces the vigour of the root system which appear rotted in many cases. Since the symptoms caused by fungal root pathogens are so similar, accurate diagnosis of the problem is critical to disease control. Use of sterile soil and growing of plants off the ground, if possible, are the only alternatives to get rid of this problem. **Root knot nematodes** (*Meloidogyne incognita*) cause galling of roots which ultimately hampers the growth of roots, thereby leaves are cupped, plant growth is slowed down and plants appear sickly. Sterile soil may give rid of this pest. Use of Furadan at 1 g/m^2 followed by watering will also minimize its population drastically.

Anthracnose is characterized initially by yellow but later dark brown spots anywhere on the leaf. Yellowish masses of spores form in zones along leaf veins or in concentric rings in the spot, and finally abscission of leaves occur. *Ficus elastica* cultivars are commonly infected with this pathogen during summer months and appear especially susceptible when they are being rooted under mist conditions. To escape infection of this disease the plants are kept away from water and heat stresses. Any cutting having spots of this type should not be taken for propagation from mother stock. Overhead irrigation on rooted plants should also be avoided. ***Cercospora* leaf spot** causes tiny and slightly raised red or dark green spots on lower surface of *Ficus elastica* leaves. ***Corynespora* leaf spot** causes small to large reddish leaf spots on the youngest mature leaves, with leaf abscission common in severe infections when leaf spots expand interveinally. This disease occurs on both green and variegated forms of *Ficus benjamina* and *Ficus nitida* but is more severe on the variegated cultivars. Overhead watering should be restricted and fertilizer applications should be only at recommended levels. ***Botrytis* blight** occurs primarily on *F. elastica* and especially on cuttings during chilly weather and causes large tan to brown leaf spots with concentric rings, usually found between the leaf and sheath or on leaf tips. Sanitation alone is not sufficient for minimizing *Botrytis* blight. Prevention should be the main focus in management programme. An integrated strategy combining environmental management, cultural practices, and fungicides will most effectively manage this disease. Chemical control is by use of benzimidazole and a less pervasive dicarboximide@. **Southern blight** is caused by the fungus *Sclerotium rolfsii* and is a moisture-related infection during summer months. The symptoms include reddish or yellow spotting on the lower leaves, particularly on plants that are regularly misted or overwatered. As the infection progresses, masses of cottony fungi with hard brown lumps appear around the base of the stems and eventually spread up the stems and onto the leaves. The sclerotia usually form on the basal portion of stems of infected plants but may also be found on infected leaves. Eventually the entire cutting or plant may be covered with the fungus. Management strategies include use of sterilized potting medium for repotting and water should be given sparingly during the growing season, taking care to keep the leaves dry. Fungicides used for controlling *Botrytis* blight will, to some extent, control even this disease.

Rubber plants suffer most due to certain **physiological disorders**. Too much water or too little light causes **yellowing of leaves** and their drop. Potting mixture should be watered only when required and between two watering it should be allowed to dry out a little. Even if the symptoms persist further the plants are placed at a more illuminated spot. **Droopy leaves** are the results of underwatering, therefore, at this time till recovery of the plants there should be sufficient watering. Small reddish spots 1-3 mm across on the undersides of leaves appear due to **moisture stress**. The symptom is usually observed on stock plants grown in full sun and not in shade, which have been so frequently air-layered during driest months of December through June. Leaf spotting can be prevented by timely irrigation of stock and use of the vertical slit technique of air layering versus the girdling procedure. Permanent petiole wilt and stem shrinking can be avoided by providing adequate soil moisture and high humidity during propagation. Prevention of excessive root development outside the container eliminates most of the shock that occurs when large segments of roots are severed from plants because roots extend into soil below the pot. Use of plastic ground covers, drip irrigation and root pruning during production usually eliminates root pruning shock when plants are removed. **Excessive light intensity** causes the leaves turning medium to light green and the sides of leaf blades curving upward and/or margins becoming wavy. Angles of the branch in some species of rubber plant tend to become narrow and branches stiff. Such plants are not light- acclimatized and usually defoliate excessively when moved to interior conditions. Plants must be exposed to shade levels of 70 per cent or more for a minimum of 2-6 months depending on plant size (2 months for up to 20 cm pot, 3 to 4 months for 40-50 cm pot, and 4 to 6 months for larger containers). **Excessive soluble salts** are highly damaging to rubber plants. Plant damage from high salinity can be placed in two categories based on stage of development. Plants in production with excessive salinity in the root zone become stunted and in severe cases defoliate starting with the oldest leaves, and eventually die if the condition is not corrected. Foliage in early stages of stress from excessive fertilizer in the soil appears dark green, but later new growth appears chlorotic and wilted if soil conditions are not corrected. Root tips of plants exposed to excessive salinity shrivel and eventually die. Plants placed indoors under less than 200 fc, with fertility levels above that recommended for production, usually defoliate excessively and die in some cases. Over-application of fertilizer and soil mix components with high salinity should be avoided. The plants should be irrigated with palatable water quite low in salts. Salt accumulation can be corrected in most cases through thorough leaching of soil. Soil fertility at the end of the production cycle should be reduced as a part of the acclimatization process.

References

Bailey, L.H. 1929. *Ficus*. In: *The Standard Cyclopedia of Horticulture*, pp. 1229-1238. The Macmillan Company, New York, USA.

Balakrishna M. and S.K. Bhattacharjee, 1991. Studies on propagation of ornamental trees through stem cuttings. *Indian J. Hort.*, **48**(1): 87-94.

Beel, E and H.B. Herregods, 1970. Vegetative propagation of *Ficus elastica* 'Decora'. *Meded. BedrVoorlDienst Oosf-Vlaanderen*, **55**: 5.

Brickell, C. 1994. *The Royal Horticultural Society Gardeners' Encyclopedia of Plants and Flowers*, pp. 496. Dorling Kindersley Limited, London, Great Britain.

Hassan, A.A. and M. Rahmani, 2003. Investigation on the chemical constituents of the leaves of *Ficus elastica* Roxb and their antimicrobial activity. *Pertanika J. Sci. & Technol.*, **11**(1): 57–63.

Hay, R. and K.A. Beckett, 1971. *The Reader's Digeswt Encyclopaedia of Garden Plants and Flowers*, pp. 276-277. The Reader's Digest association Limited, London, England.

Hellyer, A. 1982. *The Collingridge Illustrated Encyclopedia of Gardening*, pp. 94-95. Collingridge Books, Middlesex, England.

Herwig, R. 1987. *2850 House and Garden Plants*, pp. 65-67, 205. Crescent Books, New York, USA.

Jim Green, 1987. Chemical growth regulators, *Ornamentals. Northwest Archives*, **11**(3): 3-6.

Ranwala, A. 2010. Effectiveness of Floralife® Leafshine spray. *Floralife*, **12**(5): 5-7.

Siddiqui, M. I. and S. A. Hussain, 2007. Effect of indole butyric acid and types of cuttings on root initiation of *Ficus* 'Hawaii'. *Sarhad J. Agric.*, **23**(4): 919-924.

Sultan, S.M., T.M. Al-Chalabi and A.O. Al-Atrakchi, 1990. Effect of planting media and indole butyric acid on the rooting of *Ficus elastica* var. 'Schryveriana'. *Mesopotamia J. Agril.*, **22**(4): 45-51.

19

Flowering Indoor Plants

Sanyat Misra and R.L. Misra

Introduction

Anyone can grow house plants and make the interior looking attractive. Whether one goes to restaurants, hotels, banquet halls, shops & shopping malls, educational institutions, offices, airports, harbours, railway stations, bus terminus, palatial houses, private residences large or small and even in the cluster-houses, everywhere one finds the ornamental plants outside, in the backyards, in corridors, on terraces, in halls, on podiums, in living rooms and on the tables as well as dish and miniature gardens in the living rooms, in patios and balconies, in kitchen, even in the bathrooms, in window boxes & hanging baskets from the eaves of roofs, on the roofs, along the staircases, on the entrances, and even on the walls. This is not only the tradition now but apart from keeping the environment clean these give seclusion for concentration and enchantment to the people if kept properly maintained. Ornamental plants are available everywhere in plenty now. To look them beautiful, they require proper maintenance specific to their requirements. Requirement of all the house plants may not be similar, some may require ample light, some survive for years indoors without any direct sunlight, some require dry environment while many others ample relative humidity, some hot tropical conditions while certain others cold conditions. House plants come from such places as the rain forests of India, Malaysia, South America, Africa and the temperate regions of Himalayas and northern China. A vast account of such plants is furnished by Brickell (1994), Anderson (1998) and Rosenfeld and Strong (1998). Their uses; propagation; potting mixture, potting and repotting; and insect-pests, diseases and physiological disorders are the same as described in the next 'Chapter on **Foliage Plants**'.

Genera, Species and Varieties

Abutilon (Chinese bellflower/Flowering maple/ Parlour maple; Malvaceae)

It is an Arabic name for a type of mallow which comprises about 100 species of tender to half-hardy deciduous to evergreen free-flowering annuals, herbaceous perennial, small to large shrubs or trailing plants from the tropics and subtropics, especially South America where they are found growing wild in the light woodland and scrub at a sunny situation. They are effective pot plants and can also be used for beddings. It is a well-branched shrub bearing long-stalked and 3-5-lobed, palmate and mostly hairy leaves and bell-shaped pendulous flowers some 4 cm long on some 2.5 cm stalks. Leaf colour is mostly medium green but those of indoor types have variegated leaves, and the flowers may either be single-coloured or in two colours or in two shades of a single colour and appear singly or in pairs from the leaf axils throughout the summer and autumn. They are propagated either by seeds or by soft cuttings. Most of the species and varieties are sun-loving (Pizzetti and Cocker, 1975) but those suitable as house plants are described here with. ***A.*** × ***hybridum*** (syn. *A. globosom*, *A. darwinii* × *A. striatum* and other species). Plants resemble *A. darwinii*, starts flowering while still young and usually single pendent flower appears in shades of white, yellow, orange or red. Its varieties for cold regions are 'Ashford Red' (crimson), 'Boule de Neige' (white), 'Canary Bird' (yellow, turning red with age), 'Fireball' (orange-red), 'Golden Fleece' (golden-yellow), 'Master Hugh' (rose-pink), 'Nabob' (crimson), 'Orange Glow' (orange-yellow), 'Savitzii' (leaves with white splashing), 'Souvenir de Bon' (pale orange-yellow with veining of light red-purple and the leaves broadly margined creamy-white), *etc.* ***A. megapotamicum*** (trailing abutilon) from Brazil is a deciduous shrub growing to a height and spread of 2.5 m but in pots up to 1.2 m. It is a half-hardy slender-stemmed branching shrub, bearing bright green, 6-10 cm long, ovate-cordate but sometimes 3-lobed, acuminate and serrate leaves. The calyces are red and about 2.5 cm long and covering the corolla roughly to its half; corolla yellow and about 4 cm long; and anthers dark brown, cluster in the centre and protude some 1.2 cm beyond the corolla. For outdoor growing, it requires sheltered situation. Its var. 'Variegatum' has yellow-mottled leaves. ***A. pictum*** (*A. striatum*) from Brazil though in the wild is about 1.8 m tall but in pots only 1.2 m. It is a tender shrub with slender stems, bearing mid- to deep green, palmate with 3-5 lobes and serrate leaves; and about 4 cm long orange flowers with crimson veins. Var. 'Thompsonii' bears smaller but heavily yellow-mottled leaves.

Acacia (Leguminosae)

It takes its name from the Greek *akis* 'a point' alluding to the spines as some species possess. A genus of 800 evergreen or deciduous shrubs or trees from tropics and subtropics of whole world. Acacia is a useful shrub where space is not a problem. Flowers appear during early spring which are used for floral arrangements. After flowering, the straggly and unwanted branches are removed. These do not like overwatering but require well-lit spots. It can tolerate a minimum winter temperature of 4.4 °C. It should be repotted every 2-3 years but after flowering and is propagated through stem cuttings. ***A. armata*** (kangaroo thorn) is the best known which grows up to 0.9-1.2 m tall covered with modified leaf stalks in the form of leaves (phylloclades). Its deep yellow flower heads are fluffy and fragrant. ***A. dealbata*** (silver wattle) is more attractive than the *A. armata* but is taller.

Acalypha hispida (Copperleaf; Euphorbiaceae)

The name derives from the Greek *akalephes* which means 'nettle'. The genus comprises of 450 evergreen perennial species, mostly shrubby from the tropics and subtropics. Some species have attractive foliage while others have, apart from the attractive foliage, long catkin-like floral spikes. They are frost tender and survive a minimum temperature of 10-13 °C, but best at 16 °C night temperature. Though these can easily grow under open sun but performance is better when it is partial shade. These all are highly responsive to nutrients and for better performance these require to be irrigated regularly, however, waterlogging should be avoided. To make bushy growth the stem tips are removed to encourage lateral shoots. They are propagated through softwood, greenwood and semi-ripe cuttings. They are most suitable for growing in the borders, in living rooms, in verandahs and in the greenhouses but wherever these are planted, constantly the atmosphere should have moist air because dry air will cause leaf fall and infestation of red spider mites. Mealy bugs, whitefly and red spider mites may sometimes trouble these plants. ***A. godseffiana*** (lance copperleaf) from New Guinea is smaller and more bushy than *A. wilkesiana*. Its leaves are gracefully green and ovate-lanceolate with a broad serrated creamy-white margins. Flower spike is tufted catkin-like with inconspicuous tiny greenish-yellow flowers. The leaves in *A.g. heterophylla* are narrowly lanceolate. ***A. hispida*** (Chenille plant, Philippine medusa, red-hot cat's tail) from New Guinea is a soft-stemmed erect-growing shrub to a height of 2 m and with a spread of 1-2 m. The leaves are gracious deep green, oval, toothed and up to 15 cm long. The spike up to 50 cm long is drooping and catkin-like compact with tassels of tiny crimson flowers which continue appearing throughout the year intermittently but the peak being during the rains. 'Alba' is its unusual white form. ***A. wilkesiana*** (beefsteak plant, copperleaf, fire dragon plant, Jacob's coat, match-me-if-you-can) from Pacific Islands is an erect-growing shrub growing up to 2 m high, bearing decorative, coppery-green, radiating, ovate, serrated and 10 cm or more long leaves which may be mottled or splashed with shades of red, rather copper-green. Flower spikes are small and insignificant. Its forms 'Macafeeana', 'Macrophylla' and 'Musaica' have variously toned leaves with amount of colouration differing. 'Obovata' leaves are obovate and pink-margined.

Achimenes (Hot water plants, cupid's bower, magic flower, nut orchid; Gesneriaceae)

It derives probably from the Greek *achaemaenes* meaning 'a magic plant'. A genus of 50 herbaceous perennial species from tropical America, mostly Mexico. The common name 'hot water plant' probably is the instruction not to water it with cold water. They are excellent hanging basket plants, especially the spreading ones. These produce a spectacular combination of foliage and flowers during spring and summer. It is a rhizomatous plant, rhizomes about 2.5 cm long and 0.6 cm thick, and a single plant from 7.5 to 8.0 cm high with several stems, emerging from a rhizome, bearing dark green, short-stalked, velvety, cordate and toothed leaves in opposite pairs. Narrow-tubed flowers flaring out into five broad lobes appear on the short stalks from the leaf axils one by one for a long duration though single flower is short-lived. The taller species have tendency to sprawl and are therefore excellent for planting in the hanging baskets. After the flowering is over, the leaves start shrivelling and dry out, and then the stems just above the surface are sheared off to make them to begin their life the next season. ***A. erecta*** (*A. coccinea*) produces 45 cm long, hairy, green to reddish and trailing stems suitable for hanging baskets. Leaves are dark green above and pale-green to red on the underside, and 2.5-5.0 cm long & 1.2 to 2.5 cm wide. Bright red flowers appear with about 2 cm long tubes and mouth apart. ***A. grandiflora*** grows upright with up to 45 cm tall, hairy and green or red stems, bearing up to 15 cm long and 7.5 cm wide, rough and hairy leaves with upper side dark green but lower surface pale-green or red. Flowers are deep reddish-purple with white throat, and about 3.8 cm long and wide. It is one of the parents for the popular hybrid 'Purple King'. ***A. longiflora*** produces some 60 cm long trailing stems which bear toothed leaves about 8.75 cm long and 3.75 cm wide. Blue flowers with white throat appear with about 5.0 cm long and 7.5 cm across the mouth. Its white form is known as 'Alba' which is one of the parents involved in developing 'Ambroise Verschaffelt' (white flowers with purple lines in the throat). There are many more hybrid forms such as 'Fritz Michelssen' (blue), Minuet' (deep pink), 'Tarantella' (salmon-pink), 'Valse Bleu' (blue), *etc.* Achimenes grow best at 16-27 °C, though higher temperatures become unbearable. They are propagated through stored or detached rhizomes as well as 7.5 cm long tip cuttings taken during summer.

Adenium obesum (Apocynaceae)

See chapter on '**Succulents Other Than cacti**'.

Aechmea (Bromeliaceae)

See chapter on '**Bromeliads**'.

Aeonium (Crassulaceae)

See chapter on '**Succulents Other Than cacti**'.

Aeschynanthus lobbianus (Basket vine/Lipstick vine; Gesneriaceae)

See chapter on '**Foliage Plants**'. *A. lobbianus* is an epiphytic climbing or trailing plant running up to 60 cm with elliptic leaves. Flowers are crimson with yellow throat and about 4 cm long. Calyx is silky and dusky-red. It is most suitable for hanging baskets. Many species produce beautiful flowers such as *A. boschianus, A. bracteatus, A. ellipticus, A. evrardii, A. hidebrandii, A. javanicus, A. longiflorus, A. marmoratus, A. micranthus, A. nummularius, A. obconicus, A. parasiticus, A. pulcher, A. radicans, A. speciosus, A. tricolor, etc.* Out of these, apart from *A. lobbianus, A. hildebrandii, A. marmoratus, A. pulcher* and *A. speciosus* are highly promising as indoor ornamentals. ***A. hildebrandii*** is a small creeping plant with red flowers. ***A. marmoratus*** is worth for its 10 cm long and 3.75 cm wide foliage which is mottled above and red below, though greenish-yellow flowers having dark brown splashing in the throat are also attractive. This is one of the parents in case of 'Black Pagoda' hybrid. ***A. pucher*** is almost similar to *A. lobbianus* but the flowers are less hairy, calyx is green with a purple tinge and corolla is 6.25 cm long. ***A. speciosus*** is the most spectacular of all the species where stems grow up to 1 m tall, bearing 10 cm long and 3.75 cm wide leaves in pairs or in whorls of 3 though at the tips 4-8 leaves surround the 6-20 flowers. Calyx and corolla are slightly hairy. The erect flowers are orange, inside orange-yellow with a dark red bar having scarlet borders across the lower lobes These plants like bright light but not the direct light. They are propagated through 10-15 cm long tip cuttings which root in 3-4 weeks.

Agapanthus (African lily, lily-of-the-Nile; Alliaceae)

Agapanthus derives from the Greek *agape* meaning 'love' and *anthos* 'a flower'. A clump-forming hardy to semi-hardy bulbous genus of 12 species from South Africa bears fleshy tuberous roots, strap-shaped mid-green and deciduous or evergreen leaves and a compact & rounded umbel comprising from 30 to 200 of 6-petalled widely to narrowly funnel-shaped flowers 2.5-7.5 cm long which are produced from June to September in temperate regions as in the plains these do not produce flowers. These grow 45 cm to 1.2 m tall and their flowers are useful for cutting as these last for 3-4 weeks. Their beauty is superb if grown in an irregular terrain giving a sense of movement or in the background of rocks. Indoors, these should be kept at an illuminated corner otherwise these may be brought in when flowering starts. These are always preferred to be planted near the water among groups of bog plants. The plants with their fleshy roots are divided and planted during April or May for propagation. These can also be propagated through seeds sown in April at 13-15 °C temperatures in a seed compost where seedlings take 2-3 years to attain flowering size. These require filtered full light during growing period. Important species are *A. africanus* (*A. umbellatus*), *A. campanulatus, A. caulescens, A. inapertus, A. orientalis* and *A. praecox*. ***A. africanus*** from Cape is a semi-hardy evergreen species growing up to 75 cm tall, producing about 18, erect, tough and leathery leaves 10-15 cm long and 0.8 cm wide, and crowded umbels of sky blue flowers. Its var. *atro-caeruleus* bears dark violet flowers. ***A. campanulatus*** from Natal is a hardy deciduous species growing up to 75 cm tall, bearing crowded umbel of pale-blue flowers in late summer. Its var. 'Isis' bears large heads of clear lavender-blue flowers. ***A. caulescens*** from Swaziland is hardy deciduous species growing up to 75 cm tall bearing crowded umbels of bright to deep blue flowers. Several of its sub-species have varying shades of blue flowers, and one is having white form. ***A. inapertus*** from South Africa is a hardy deciduous species growing up to 1.2 m tall and the few-flowered loose umbels of deep blue to violet-blue flowers. ***A. orientalis*** from Cape is a semi-hardy evergreen species growing up to 75 cm tall, bearing densely packed blue flower heads. It has now been classified as a sub-species of *A. praecox*. Also white and double blue forms are available. ***A. praecox*** from East Cape Province and Natal is a half-hardy evergreen species growing up to 75 cm tall, bearing dense umbels of bright to pale-blue flowers 5-7.5 cm long from July onwards. It also has one white-flowered form.

Aichryson (Crassulaceae)

See chapter on '**Succulents Other Than Cacti**'.

Allamanda cathartica (Golden trumpet; Apocynaceae)

Allamanda was named for Dr. Frederick Allamand (1735-1776?), a Swiss botanist who collected it from Surinam. It is a genus of 15 species of evergreen climbers and sprawling shrubs distributed over tropical areas of northern South America and the West Indies. *A. cathartica* from Brazil and Guyana is a climber with glossy dark green and lanceolate leaves 10-15 cm long and 3.75-6.25 cm wide which appear generally in whorls of four. Its flowers are yellow with 2.5-3.75 cm long tubes which flare into five petals spanning to 12.5 cm wide from summer to autumn. This is best suited for planting on the windows. It requires 3-4 hours of bright full sunlight every day and cold temperature. It is propagated through 7.5-10.0 cm long tip cuttings of early spring growth.

Aloe (Liliaceae)

See chapter on '**Succulents Other Than Cacti**'. Some of the species producing attractive flowers are *A. arborescens, A. aristata, A. barbadensis, A. brevifolia, A. ferox, A. variegata, etc.*

Alpinia purpurata (Zingiberaceae)

See chapter on '**Ornamental Gingers**'.

Amaryllis belladonna (Belladonna lily; Amaryllidaceae)

The name *Amaryllis* originates from the Greek feminine name Amaryllis derived from the verb *amaryssein* 'to shine'. Also there is a legend that it was named so due to a beautiful Greek shepherdess of the same name.It is a South African genus of one single bulb species, *A. belladonna*. Its bulbs are very large and pear-shaped with several skins.Its deciduous leaves are medium green, glabrous, strap-shaped, 45-75 cm long, up to 4 cm wide and develop one above the other after the flowers fade. Leafless stem appearing directly from the centre of the bulb is stout, solid, erect, up to 90 cm long and bears 1-4 sweetly scented, pale-pink and trumpet-shaped flowers 10-15 cm across from late summer to autumn which are useful for cutting. Its varieties produce 6-10 pale-pink lily-like flowers per stem. Once established, it is left undisturbed for many years. It is extremely poisonous plant. Its popular varieties are 'Cape Town' (dark reddish-pink), 'Durban' (deep carmine-red with white centre), 'Hathor' (white), 'Kewensis' (dark pink shading to yellow, late flowering), 'Parkeri' (deep red with 10-12 flowers per stem, of Australian origin, believed to be a cross of *A. belladonna* × *Brunsvigia josephinae*), *etc*. These are propagated through division of the clump when leaves have started yellowing, and through seeds sown at 16 °C in the greenhouse. Though these are most suitable for growing in the conservatory but are brought home when in bloom and are kept at an illuminated corner.

Ananas (Bromeliaceae)

See chapter on '**Bromeliads**'.

Anguloa (Tulip orchid; Orchidaceae)

Anguloa was named for Don Francisco de Angulo, a Spanish botanist of the 18th century. A native to tropical South America, especially in the Andean forests, the genus consists of 10 epiphytic or terrestrial clump-forming species, bearing ovoid to oblong pseudobulbs. Its leaves are lanceolate and prominently veined. It produces solitary erect flowers with overlapping tepals and appear as flowers of tulips. Propagation is through division. Though it is a greenhouse plant but can be brought inside the house for flowering. ***A. clowesii*** from Colombia and Venezuela produces 11-15 cm tall pseudobulbs, broadly oblanceolate leaves up to 60 cm long, pale to golden-yellow and waxy flowers about 8 cm long but hinged, boat-shaped and lobed labellum is white to orange-yellow. ***A. ruckeri*** from Colombia though smaller but is similar to *A. clowesii*. It produces up to 9 cm long fragrant flowers which in the outside are greenish-brown and inside yellow with dense red, blood-red or white spots. ***A. uniflora*** from Colombia to Peru produces a little angular pseudobulb 10-18 cm tall with 45-60 cm long, folded and broadly lanceolate leaves, and the cup-shaped and unpleasantly fragrant cream to white and waxy-textured flowers, which are sometimes brown-spotted externally and pink-spotted internally.

Angraecum eburneum (Orchidaceae)

Angraecum derives from the Malaysian name *angurek*. This genus of 220 epiphytic orchid species native to tropical and sub-tropical areas of Africa, Malagasy, Indian Ocean islands and the Philippines is without pseudobulbs. These orchids grow erect with stems bearing two rows of fleshy and narrowly ovate to strap-shaped leaves. Starry, usually quite large and white & green or rarely yellow flowers appear in racemes or sometimes even solitary, having a lip with slender spur which elongates mostly to a great length. These plants requiring more humidity are most suitable for growing in conservatories but are brought home when to flower. These can be propagated through cuttings of side shoots. ***A. eborneum*** (*A. superbum*) from Mascarene Islands grows up to 2 m high with single stem, bearing up to 90 cm long ovate-lanceolate leaves which are bi-lobed on the tips. Green to greenish-white, fragrant and up to 7 cm long flowers with a white labellum having some 10 cm long spur appear during autumn to winter in two ranks on a stem growing up to or more than 90 cm long. ***A. eichlerianum*** from tropical West Africa grows 1.2 m or more in length with pendulous or climbing stems, bearing leathery and oblong-elliptic leaves which are notched at the tips. Strongly fragrant and 8-9 cm across yellow-green flowers appear 1-3 during autumn to winter having a green-marked white labellum with a spur about 4.5 cm long. ***A. infundibulare*** from tropical West Africa has more than 1.2 m long climbing to pendulous stems, bearing up to 10 cm long oblong-elliptic and leathery leaves. Fragrant yellow-green and up to 9 cm across solitary flowers appear during autumn to winter with a white labellum. ***A. rhodostictum*** (*Aerangis rhodostictum*) from Cameroons, Ethiopia, Kenya and Tanganyika generally grows pendent with short stems, bearing bright green and strap-shaped leaves up to 15 cm long and bi-lobed at the tip. Arching inflorescences grow to about 35 cm long bearing two parallel ranks of up to 20 white to cream or palest-yellow flowers during winter to Spring, each flower about 2.5-3.0 cm across with a red column and green-tipped spurs. ***A. sesquipedale*** from Malagasy is a robust plant with more than 60 cm long and erect stems, bearing blue-green leaves some 30 cm long which are unequally bi-lobed at the tips. In winter its 18 cm across, fragrant and fleshy white flowers appear with a cordate lip and up to 25 cm long spur.

Anthurium (Araceae)

Anthos in Greek means 'a flower', and *aura* means 'a tail', referring to the spadix. Anthuriums are evergreen tropical ornamentals. There are some 550 species from the humid rain forests of tropical America and the West Indies. These are propagated by division or through seeds. These are not prized solely for their foliage though these are very difficult to grow indoors, there are some more easily cultivated kinds that provide an additional bonus ~ striking and long-lasting (up to 8 weeks) inflorescences, each comprising a large flat spathe surrounding a thin and twisted spadix. They have a fleshy rootstock with a small root system and their long-stalked leaves are oval to cordate with pointed end. Important flowering species worth cultivating indoors are *A. andraeanum* (oilcloth flower, painter's palette,flamingo flower or flamingo lily) and *A. scherzerianum* (flamingo flower, pigtail plant or tail flower). ***A. andraeanum*** from Colombia grows up to 45 cm tall, bearing handsome dark green and cordate foliage 20-25 cm long, 10-15 cm wide, *vis-à-vis* long-lasting and very attractive, cordate, pointed, shining waxy red or white floral spathes about 10 cm long, 7.5 cm wide and with a cylindrical 7.5 cm long spadix. Flowering occurs from May to September on the hills. Now the species has been replaced by a number of varieties in various colours and sizes. ***A. scherzerianum*** from Guatemala grows up to 25 cm high with spread of up to 45 cm, bearing dark green lanceolate and leathery leaves up to 18 cm long and about 7.5 cm wide on a 15-20 cm long stalks. From April to October it bears palette-shaped, waxy and brilliant scarlet spathes some 7.5-10.0 cm long and a spirally twisted orange-red spadix 5.0-7.5 cm long. In some forms the spathe is darker-red spotted with white. This is most suitable as house plant. These survive well at medium light, *i.e.* at a slightly shaded window, and a constant temperatures of 18-21 °C. In spring, their clump for propagation is divided, each section having some fleshy roots and a growing point and planted.

Aphelandra (Acanthaceae)

Aphelandra derives from the Greek *apheles* meaning 'simple' and *aner* for 'male' as the anthers have only one cell. It is a genus of 200 evergreen shrubs from tropical and sub-tropical America, but only two are commonly grown indoors and these are *A. chamissoniana* and *A. squarrosa* (saffron spike or zebra plant) which can be grown indoors for any window having limited space. Both the species grow 30-45 cm high with stout stems, bearing dark glossy green, in opposite pairs, leathery, pointed, elliptic to ovate leaves, often veined or mottled grey-silver to white. The cone-shaped dense floral spikes of tubular and 2-lipped bracts from the top of the plants and sometimes even additional spikes between the upper leaves are produced in spring with yellow or orange-yellow flowers in a formal pattern where flower bearing bracts are 2.5-3.75 cm long. The small flower may last only a few days but the spike of bracts remains fresh for several weeks. These plants can also be grown in terrarium. They are propagated by cuttings of young shoots in spring. ***A. chamissoniana*** from Brazil is a slender species up to 1.2 m tall, bearing 7-10 cm long close-set and elliptic leaves, having a broad silver-white pattern. Bright yellow narrow-pointed flower bracts appear during late autumn and winter. ***A. squarrosa*** from Brazil grows more than 1.2 m high with deep glossy green and pointed ovate leaves 15-25 cm long contrasting with white veins which are borne in opposite alternate pairs. Tubular bright yellow overlapping bracted flowers 3.75 cm long, sometimes red-edged, appear on angular and cone-shaped terminal spike up to 10 cm long, from late summer to winter.

Aporocactus (Cactaceae)

See chapter on '**Cacti**'.

Ardisia (Coral berry/marlberry/spiceberry/spear flower; Myrsinaceae)

Ardis in Greek is for 'a point', owing to its anthers being spear-shaped. It is a genus of some 400 species of evergreen trees and shrubs from tropics and sub-tropics of Asia, America and Australia. ***Ardisia crispa*** (*A. crenata*), a shrub from SE Asia is a popular house plant from temperate regions growing up to 1.5 m high, bearing shining dark green, elliptic-lanceolate, undulate, leathery and up to 15 cm long and 5 cm wide leaves. Star-shaped white or rose-pink fragrant and 5-petalled flowers about 6 mm long in thick clusters appear from leaf axils in early summer, and these flowers are followed by 6 mm across shining and persistent red berries that are held on horizontal stalks. These are most suitable for window gardens. Though these prefer open sun but do well under filtered light and a maximum temperature of 16 °C. These are propagated through seeds sown in spring or through heel cuttings from lateral shoots during late spring or early summer which root in 6-8 weeks when planted in peat moss + sand mixture.

Ascocentrum (Orchidaceae)

It is a genus of nine orchid species allied to *Ascoglossum* and *Vanda*, native in the Himalayas to SE Asia to Philippines, Borneo, Java, S. China and Taiwan. Once the genus was included in *Saccolabium*. These dwarf, compact, erect and leafy-stemmed monopodial epiphytes (mostly) to lithophytes (rarely) bear stem – ensheathing basally, alternate, carinate, fleshy, stiff, linear, apex with two teeth and up to 30 cm long leaves, many-flowered erect to spreading inflorescence which may be sometimes more than one with closely set flowers and as long as the leaves, and small (1-3 cm), closely set, erect, flat, often showy, long-lasting and wide-opening flowers coming up in succession with similar petals and sepals, which appear in lateral upright racemes. The tepals are five, lower two sometimes larger than others, the top one is hooded and the labellum is small with long narrow nectar spur. Their colour is yellow, orange-red, scarlet or purple with a spur. They are best all timer for indoor cultivation. They are propagated in spring after flowering is over, through stem tips having aerial roots. Important species are *A. ampullaceum*, *A. curvifolium* and *A. miniatum*. ***A. ampullaceum*** (*Aerides ampullaceum, Gastrochilus ampullaceum, Saccolobium ampullaceum*) from Himalayas to Myanmar produces up to 25 cm long stems, bearing leathery, linear, irregularly toothed at the tips and about 13 cm long leaves. Rose-red to rose-purple flowers appear from spring to summer with labellum sometimes toned paler and column white. ***A. curvifolium*** from NE India and Indo-China Peninsula also bears erect spikes bearing large scarlet florets with yellow blotched-lip. ***A. miniatum*** (*Gastrochilus miniatum, Accolobium miniatum*) from Himalayas to Borneo and Malaysia produces thick, woody and 10 cm tall stems, bearing firm-textured, fleshy, linear and 8-20 cm long leaves. Bright orange-red, sometimes orange-yellow to vermillion, and 2 cm across flowers appear from spring to early summer.

Astrophytum asterias (Sand dollar cactus/sea urchin cactus/star cactus; Cactaceae)

See chapter on '**Cacti**'.

Begonia (Begoniaceae)

Begonia was named for Michel Bégon (1638-1710), patron of botany, and one time Governor of French Canada. A genus of about 2,000 tender perennial and evergreen or deciduous species including sub-shrubs and climbers grown for their flowers and foliage, widely distributed in tropical to warm temperate climates, most frequent being in Mexico & Central and South America (Argentina, Bolivia, Brazil, Colombia, Ecuador, Guatemala, Panama, Peru, West Indies, Venezuela, *etc.*), but almost absent from Australasia. These are also distributed over Asia (Bhutan, China, India, Japan, Java, Nepal, Pakistan, Philippines, Sri Lanka, Sumatra, Taiwan, *etc.*) and South Africa (Republic of South Africa ~ Cape of Good Hope and Natal, Congo, Guinea, Jamaica, Madagascar, Tanzania, Zambia) and one species in South Pacific Ocean (Solomon Islands), *etc.* As their number suggests, they are highly varied in appearance and habit, in size ranging from tiny ground-hugging creepers to stout-stemmed bamboo-like shrubs up to

3 m in height, even though these all share a number of similar characteristics. The almost ear-shaped leaves in all the cases are almost asymmetrical and appearing always alternately along the stems and new ones emerging from the leaflike sheaths known as stipules. The flower size is characteristic of all the three groups (tuberous, rhizomatous and fibrous-rooted), but varying only in size.

They are unisexual, male and female flowers being borne on the same plant, male being more showy as their petals are of different shapes and sizes, and the females having the petals more nearly alike and are distinguished by the prominent winged ovaries looking like a three-lobed appendage immediately behind the petals, followed by triangular and often winged seed capsules. Female flowers though may appear a little faded but usually last for weeks or even months though male flowers drop off within 2-3 days of opening. Generally, the male and female flowers each have four ovate-oblong petals, in males two being larger though in females they are similar, almost equal in size but sometimes up to five petals. Rhizomatous begonias are evergreen so majority is being grown for their decorative foliage borne on a creeping rootstock. Tuberous types are generally deciduous and bear single female and double male flowers. Fibrous-rooted ones are also evergreen having tall, erect and rarely shrubby stems but are grown mainly for their flowers.

The showiest blooms are generally borne by the tuberous begonias (*B.* × *tuberhybrida*, syn. *B.* × *tuberosa* hort. and not *B. tuberosa* Lam.; Tuberhybrida hybrids or hybrid tuberous begonias), the pendulous basket begonias flowering during summer and autumn. *B.* × *tuberhybrida* was first named so in 1896, and this by crossing with *B. boliviensis, B. clarkei, B. davisii, B. froebeli, B. gracilis, B. pearcei, B. rosaeflora* and *B. veitchii*, all of South American origin, and *B. socotrana* native to South Africa, has produced many beautiful cultivars, with or without stems, and erect to pendulous, about 60 cm high, flowers single or double in axillary clusters of various colours, *viz.*, white, yellow, orange, pink to red, bicoloured and fragrant but neither blue nor mauve. The second group of Lorraine or Cheimantha hybrids under *B.* × *tuberhybrida* ~ the old favourites as they bloom around Christmas, and the third group the Hiemalis hybrids which perpetually bloom in any season and their blooms last for months; and some fibrous-rooted ones, especially the hybrids of *Begonia semperflorens* which become dormant after a long flowering period with vividly coloured blooms, *vis-à-vis* often possessing attractive foliage, and are among the most popular and effective of all present-day annuals for summer bedding and terrace cultivation. Most of these bear clustered flowers on short stalks arising from or near leaf axils, each cluster being normally composed of either all male or all female flowers. The species under tuberous groups are *B. biserrata* (flowers in dense axillary racemes with white tepals), *B. bogneri* (4-tepalled male and 6-tepalled female flowers, outer pair pale-pink and inner white), *B.* × *cheimantha* (*B. dregei* × *B. socotrana*; usually large pink flowers), *B. cinnabarina* (fragrant orange-red), *B. davisii* (bright red), *B. dregei* (syn. *B. parviflora*; white; *B. suffruticosa*, probably a variety of it with white to pale-pink flowers), *B. froebelii* (crimson to scarlet), *B. gracilis* (syn. *B. bicolor, B. diversifolia*; pink and its var. *maritiana* bears fragrant flowers), *B. grandis* ssp. *evansiana* (syn. *B. evansiana* (fragrant pink or white), *B. homonyma* (syn. *B. caffra*, very close to *B. dregei*, flowers white); *B. x intermedia* (*B. boliviensis x B. veitchii*, its var. 'Bertinii' evolved by crossing *B. boliviensis* with *B. x intermedia*, vermillion); *B. josephi* (pink); *B. micranthera* (white or pale-pink); *B. octopetala* (stemless, ivory-white); *B. partita* (flowers small white); *B. picta* (fragrant pink); *B. sikkimensis* (bright red, and its forms are 'Gigantea' and 'Variegata'); *B. socotrana* (rose-pink, rarely in cultivation but superseded by the hybrid strain known as 'Lorraine begonias', a parent of many important winter-flowering hybrids, all with pink flowers); *B. sutherlandii* (orange to orange-red); *B. x tuberhybrida*; *B. veitchii* (syn. *B. rosiflora*; fragrant bright scarlet); *B.*× *weltoniensis* (*B. dregei* × *B. sutherlandii*; white or pink, its var. 'Alba' is white); *B. wollnyi* (green-white), *etc.*

Fibrous-rooted (Cultorum hybrids, the wax begonia) are *B. albo-picta* (flowers green-white), *B. coccinea* (bright coral-red, a hybrid variety 'President Carnot' is more vigorous with paler flowers), *B. corallina* (bright coral-pink), *B. fuchsioides* (bright red or pink flowers), *B. haageana* (pink tinged white), *B.* × *hiemalis* [syn. *B.* × *elatior*, parents being *B. socotrana* × (*B.* × *tuberhybrida*)], commonly known as 'winter flowering begonias' where tubers are absent but tending to die back to swollen bases, flowers single or double, white to pink, yellow, orange or red and flowering in winter or year-round], *B. ludwigii* (flowers many, cream-white with green and pink marks externally), *B. metallica* (flushed-pink white flowers), *B. semperflorens* (red, pink or white flowers), *etc.* Many begonias do not require continuous direct sunlight and by this fact they are particularly suitable for indoor use.

Beloperone guttata (Shrimp plant; Acanthaceae)

Beloperone derives from the Greek *belos* 'an arrow' and *perone* 'a buckle or rivet', referring to the way the anther lobes are connected. It is a genus of some 60 species of evergreen shrubs now often included in *Justicia* but only one species of which is in general cultivation, usually as house plant and this is *B. guttata*. A native of Mexico, it grows up to 90 cm tall with the spread of up to 45 cm, bearing long-stalked (2.5-3.75 cm), shining, soft green, slightly hairy, overlapping, ovate to elliptic and paired leaves 2.5-8.0 cm long. Though its 2-lipped white flowers being produced in abundance from April to December are inconspicuous but these are protected by decorative, cordate, overlapping, reddish-brown or brown-pink terminal bracts up to 2.5 cm long. Entire inflorescence some 15 cm long forms a pendent body resembling the body of a shrimp. To keep it in proper shape, this plant is pruned regularly every year. The tips of the young plants are pinched several times to make it bushy otherwise these become leggy and then there is no any other option except to replace such plants. Direct sunlight helps in developing properly coloured bracts otherwise filtered light is alright. Warm room temperature is conducive for its growth and development but in winter the recommended temperature is 18 °C. They are propagated through 5.0-7.5 tip-cuttings taken in spring.

Billbergia pyramidalis (Queen's tears/Summer torch; Bromeliaceae)

See chapter on '**Bromeliads**'.

Bouvardia ×domestica (Rubiaceae)

Bouvardia was named for Dr. Charles Bouvard (1572-1658), who at one time was in charge of the Jardin du Roi, Paris. It is a temperate but frost-tender genus of 50 species of deciduous, semi-evergreen or evergreen greenhouse shrubs from tropical and subtropical America with showy and fragrant flowers being produced for a long period. Undivided and usually ovate to lanceolate leaves appear in opposite pairs or whorls of three or sometimes even more. The fragrant flowers which start appearing from mid-summer to early winter are very showy, tubular, 4-lobed and in large flattish terminal or axillary corymbs or clusters above the pointed leaves. These start flowering even when small. They are propagated through 7.5 cm long softwood cuttings taken in spring or by greenwood or semi-ripe cuttings in summer, and the cuttings root within three weeks at a temperature of 21 °C. These tolerate a minimum winter temperature of 7-10 °C though for winter-flowering species it should be 13-15 °C. During autumn and winter, the maximum light should be provided for their growth so the plants should be kept at a well illuminated place in the house, however, during summer these may be given filtered light. Established plants are stopped twice, between April and the end of May to obtain flowers in July and August, and the autumn-flowering plants are pinched out several times until the end of August. These require ample of water and nutrition. They should be fed with weak liquid manure weekly from May to September. After flowering up to six weeks they are kept just moist and then lateral growths in the beginning of February should be drastically pruned leaving only 2.5 cm at the base, followed by frequent watering to encourage fresh growth. After flowering the plants are kept barely moist until growth retards in Februaary or March and then they are annually potted in March to produce vigorous growth. Plants in pots deteriorate after two years so they are raised annually from cuttings. *Bouvardia ×domestica* and *B. longiflora* are worth for indoor cultivation. ***Bouvardia × domestica*** is a hybrid species developed by crossing three species, *viz. B. leiantha, B. longiflora* and *B. ternifolia*. This can be grown to a minimum height of 60 cm with a spread of 45 cm. Its leaves are mid-green, ovate and arranged in opposite pairs. This produces 15 cm across terminal clusters of white, pink or red flowers freely from June to November. The flowers are tubular with four spreading petals. It has produced many varieties such as 'Bridesmaid' (double pink), 'Mary' (white and pink), 'President Cleveland' (bright crimson-scarlet), 'Rosea' (rose-pink), *etc.* ***B. longiflora*** from Mexico can be trained up to 90 cm tall with a spread of 60 cm. Its leaves are glossy mid-green and ovate. Flowering appears from October to December in 10 cm across loose clusters bearing fragrant, white and tubular flowers.

Brassavola nodosa (Lady of the night; Orchidaceae)

Brassavola was named for Antonio Musa Brassavola (1500-1555, an Italian doctor *cum* botanist & professor at Ferrara). This genus comprises of 15 evergreen, epiphytic and clump-forming species producing stem-like cylindrical and slender pseudobulbs which terminate into a single and fleshy leaf. The showy flowers are single or racemose with similar sepals and petals (tepals) with a larger base-rounded labellum which has a slender tail-like tip. These can easily be grown as basket plant, on tree ferns or on the bark of the trunks. They are propagated through division of densely-formed clumps. ***B. cucullata*** (*B. appendiculata, B. odoratissima, Epidendrum cucullatum*) from Mexico to South America bears up to 13 cm long pseudobulbs, arching to pendulous 60 cm long slender leaves, and 1-3 fragrant and white to cream or greenish-white flowers which are drooping with up to 9 cm long-tailed and fringed tepals. This flowers during winter. ***B. digbyana*** (*Rhyncholaelia digbyana, Laelia digbyana*) from Mexico to Guatemala produces up to 15 cm or more long, a little flattened and club-shaped pseudobulbs which bear grey-green, up to 20 cm long, tough, fleshy and linear to narrowly elliptic leaves, and night fragrant creamy-white solitary flowers up to 18 cm across having greenish throat. ***B. glauca*** (*Rhyncholaelia glauca, Laelia glauca*) from Mexico to Panama produces spindle-shaped pseudobulbs about 10 cm long, bearing grey-green, leathery, stiff, up to 12 cm long and oblong-elliptic leaves and long-lasting, white to lavender or green fragrant flowers 12 cm across with pink spots or lines in the lips. ***B. nodosa*** (*B. grandiflora, Epidendrum nodosum*) from Mexico to Peru bears up to 15 cm tall stem-like pseudobulbs. Leaves are very fleshy, erect, linear and up to 30 cm long. White to yellowish-green and long-lasting night-fragrant flowers 9 cm across appear in racemes of 1-6 though continue appearing intermittently throughout the year, especially during autumn and winter.

Brassia (Spider orchid; Orchidaceae)

Brassia was named for William Brass (*d.* 1783), who in West Africa collected many plants for Sir Joseph Banks. This genus contains some 50 evergreen and epiphytic species from tropical America, extending from Mexico to Brazil, and the West Indies thriving in intermediate temperatures, and possessing large flattened pseudobulbs. The long-tailed sepals and petals are the characteristic feature of *Brassia*. They are borne by creeping rhizomes and large, oblong (egg-shaped), flattened and crowded pseudobulbs where each one is topped with dark green, 1-3, thick, leathery and narrowly elliptic leaves. Long, stiff and arching floral stalks in the form of racemes arise from the base of the pseudobulbs and bear closely spaced florets on the upper part where each flower has three outer tepals much larger than the inner. Quite long sepals and petals with strange shape emerge from a narrow base, *vis-a-vis* taper to a narrow point in such a manner that both appear as a tail and the short lip often hangs down between the lower sepals like a tongue. They are best grown in baskets or on slabs of bark or tree fern, as well as in pots. They are propagated by division. ***B. brachiata*** from Guatemala produces oblong (egg-shaped), compressed and 7.5-12.5 cm long pseudobulbs, some 30 cm long leaves and 6-12 flowered racemes. The sepals and petals are light yellowish-green with a few purple spots towards the base and the sepals are longer than the petals. Centrally constricted light yellow lip has orbicular base, upper part ovate-triangular and acuminate, and bears dark green warts. ***B. caudata*** is native to West Indies. Its pseudobulbs are yellowish-green, cylindrical, 7.5-15.0 cm high and some 2.5 cm wide, bearing 2-3 leaves some 17.5-23.0 cm long and 6.3 cm wide and up to 46 cm long floral stalk which may bear up to 12 fragrant flowers some 12.5 cm long and 7.5 cm across. Sepals and petals are light greenish-yellow with brown spots and bars near the base and the lip is triangular, light yellow and with reddish-brown spots. ***B. gireoudeana*** is from Costa Rica. Its pseudobulb is much compressed, 7.5-10.0 cm long and 3.75-5.0 cm wide, 1-leaved and leaf some 30 cm long, racemes 6-12-flowered, sepals and petals yellowish-green but sepal base brown-spotted and petal base brown, lateral sepals 12.5 cm long and little longer than dorsal sepals, petals half as long as dorsal sepals, and the clawed lip is yellow with brown spots where upper part is almost orbicular and acute. ***B. lanceana*** from Guiana is very similar to *B. lawrenceana* with regard to shape, size and colouration of pseudobulb, leaves, sepals and petals except that the sepals may be slightly smaller and petals its half, lip oblong, wavy, acute, yellow and without or with a few basal brown spots. ***B. lawrenceana*** is from Brazil with much compressed, ribbed and 7.5-12.5 cm long pseudobulbs, bearing two leaves which measure some 30 cm long. Its racemes are 7-12-flowered with light yellow sepals and petals which are basally brown-spotted, sepals some 7.5 cm long, petals 3.75 cm long and the lip 3.75 cm long, elliptic wavy and acute. ***B. longissima*** (*B. lawrenceana* var. *longissima*) is from Costa Rica. Its pseudobulb is compressed, oblong-elliptic, 7.5-15.0 cm long and 1-leaf or sometimes paired with about 20-60 cm length and elliptic. The raceme comprises of numerous deep orange-yellow to greenish-yellow flowers with a few large basal blotches, lateral sepals some 17.5-30.0 cm long, *i.e.* almost double the length of dorsal sepals, 6 mm wide at base, petals 5.0-15.0 cm long and the lip cream to pale-yellow with purple spots at the base, elliptic, acuminate and some 7.5-15.0 cm long. ***B. maculata*** is from Jamaica. Its 7.5-10.0 cm long pseudobulb bears solitary leaf some 23 cm long and 5-10-flowered raceme. Tepals (sepals and petals) yellowish-green with lower part brown-spotted. The length of petals is $^{2}/_{3}{}^{rd}$ of sepals and sepals are some 7.5 cm long. Lip is creamy-white with purple dots, where upper part is broadly ovate and acute and the claw broad. Var. *Guttata* from Guatemala bears small and green flowers. ***B. verrucosa*** is from Guatemala. Its much compressed pseudobulbs are ovoid (egg-shaped), little furrowed and some 10 cm long with two elliptic leaves some 35 cm long. Its up to 60 cm long racemes have 8-15-waxy flowers with light yellowish-green tepals (sepals and petals) which are spotted with darker green or red basally. The sepals are up to 10 cm long with petals being half to their length, and the lip is usually smaller but sometimes up to 20 cm long, white with many dark green warts at the base, claw broad and dilated, upper part almost orbicular and acute. Some of the intergeneric hybrids are *Brapasia* (*Brassia* × *Apasia*), *Brassidium* (*Brassia* × *Oncidium*), *Miltasia* (*Brassia* × *Miltonia*), *Odontobrassia* (*Brassia* × *Odontoglossum*), *Rodrassia* (*Brassia* × *Rodriguezia*), etc.

Browallia speciosa (Methyst or bush violet; Solanaceae)

Browallia was named for John Browall (1707-1755), a Swedish botanist, Bishop of Abo and a champion of Linnaeus. It is a genus of six half-hardy annual and shrubby species from northern South America and the West Indies. They are suitable for planting in the hanging baskets where their trailing stems will make attractive specimens, and in pots but there they will require thin support to aid their slim and branching stems. They are bushy plants with opposite or alternate, ovate and pointed leaves. Their loose and racemose or solitary, tubular, violet-shaped and blue, violet or white flowers appear with five broad petal-like lobes where upper one is broader. They make very colourful plants for fall and winter-flowering. Only two species are recommended for indoor cultivation and these are *B. speciosa* and *B. viscosa*. ***B. speciosa*** from Columbia grows up to 1.2 m tall but in pot cultivation it can be managed up to 60 cm through pruning and pinching as even otherwise these activities are necessary to make the plant compact and bushy, however, for cut flowers the unpinched plants are better. Its leaves are bright green, ovate and pointed and violet-blue flowers 5.0 cm across which appear during autumn. Its form *B.s. major* bears deep blue flowers and is winter-flowering though 'Silver Bells' is white. ***B. viscosa*** from Colombia is a more compact species growing to a height of up to 30 cm only. This is most suitable, apart from bedding, for pot-growing Its leaves are short-stalked, mid-green, little sticky-hairy, ovate, blunt and 2.5-3.75 cm long. The flowers 2.5 cm across appearing from July to September are bright blue with highly notched petals and white throat. Its forms *B.v.* 'Alba' is white while *B.v.* 'Sapphire' is deep blue. These require at least 4 h of sunlight every day and a temperature range of 12-18 °C. These are propagated through seeds.

Brunfelsia (Yesterday-today-and-tomorrow; Solanaceae)

Brunfelsia was named for Otto Brunfels (1489-1534), a German physician, botanist and Carthusian monk who produced some of the earliest good plant drawings. It is a genus of 30 species of evergreen shrubs up to 60 cm tall from tropical America and West Indies, bearing alternate, glossy, leathery, 7.5-15.0 cm long, elliptic and ovate to lanceolate leaves, and showy, long-tubed, and often fragrant flowers which at the mouth are spreading and 5-lobed. They are plants for filtered light and humid conditions most suitable for large pots and containers. These require pinching as well as pruning to keep the plants in shape and for good display of flowers. They do well at room conditions but during winters the plants are moved for about six weeks at 10-12 °C temperature to complete their resting period so that better flowering may occur. They are propagated through tip cuttings during spring. There is only one species grown indoors and that is ***B. pauciflora calycina*** native to Brazil and Peru growing up to 60 cm tall. Its leaves are shining mid-green, oblong-lanceolate and about 10 cm long. Fragrant flowers opening singly about 5 cm across appear in clusters of up to 10 on the ends of long stems from April to August but at temperatures 13-16 °C this continues flowering throughout the year. The flowers are 5-lobed, with a small white puckered eye. Their flowers open violet-purple, fade to pale-lavender-blue and become almost white and dies by fourth day. The flowers of *B.c. macrantha* are 7.5 cm across.

Bryophyllum (Crassulaceae)

See *B. daigremontianum* and *B. tubiflorum* (*Kalanchoe tubiflora*) under the chapter on '**Succulents Other Than Cacti**'.

Calanthe (Orchidaceae)

Calanthe derives its name from the Greek *kalos* for 'beautiful' and *anthos* for 'a flower'. This evergreen to deciduous genus of 120 species of mainly terrestrial orchids, is native to the tropics and sub-tropics from South Africa eastwards to SE Asia and northern Australia, and one species in the Americas. Evergreen species which have large conical to ovoid pseudobulbs, rarely constricted close to the middle are suitable for cool houses while deciduous ones which have smaller and more globular pseudobulbs nearly hidden by the leaf bases are suitable for growing at warmer situations, especially in the home, and both the types have pleated and elliptic to lanceolate leaves. Flowers without any distinction between sepals and petals are spurred with large and lobed labellum. These are propagated through division at potting time. ***C. brevicornu***, an evergreen temperate species from Nepal to Assam bears up to 30 cm long oblong to elliptic leaves. It bears purple-brown flowers 2-3 cm across on up to 60 cm long racemes during spring and these flowers shade to yellowish-buff in the centre, and the labellum is red-purple with a white edge and a yellow keel. ***C. masuca*** (*Bletia masuca*) from Himalayas, an evergreen temperate species which blooms during summer and autumn, bears strongly pleated, up to 60 cm long and narrowly elliptic leaves. Deep violet-purple flowers up to 4 cm across with labellum darker and having golden-yellow patch, appear on up to 75 cm long racemes. ***C. vestica*** from SE Asia is a deciduous tropical species producing silvery-green, ovoid-conical, centrally constricted and15-20 cm tall pseudobulbs which bear up to 90 cm long lanceolate leaves. Flowers appear during winter before expansion of the leaves, they are white though in some cases the labellum is flushed pink,and moré than 4 cm across flowers appear on up to 90 cm long arching inflorescences.

Calceolaria (Scrophulariaceae)

See chapter on '**Annuals**'.

Calliandra (Leguminosae)

In Greek *kallos* is 'beautiful' and *aner* is for 'male', referring to the beatuful stamens it posesses. It is a genus of beautiful shrubs with about 120 species native of India and tropical America, the important ones being *C. brevipes, C. haematocephala, C. houstonii, C. inaequilatera, C. speciosa* and *C.tweedyi*. These are beautiful shrubs with handsome bipinnate, feathery and shining leaves composed of a large number of leaflets, and powder-puff-like flowers where petals are usually small and it is ball-shaped bristle-like stamens which look very charming, and when colour of the flowers is referred to, it is in fact, entirely the colour of the stamens. These flower from winter to spring, one plant lasting for 6-8 weeks. They are excellent for shrubbery, along the paths, as a specimen plant, and a few as indoor plants, such as *C. inaequilatera* and *C. tweedyi*. Calliandras are though very popular in USA but in England these are rarely grown, however, in India these may be found planted as hedge and in the centre of wide roads, especially at shady locations. To keep these in shape and to check its growth, after flowering is over it is pruned in every spring to a height of 60-90 cm. A winter temperature below 15 °C is not congenial for its survival. These are propagated after flowering is over in spring through stem cuttings. ***C. inaequilatera*** (red powder puff) from South America bears dark green foliage dissected into large leaflets, and a large, globose and dense with silky bright red stamens. ***C. tweedyi*** from Brazil is an unarmed and lightly pubescent shrub with 20-30-yoked shining, linear, leathery and obtuse leaflets and 3-4-yoked pinnae, peduncles axillary and 2.5-5.0 cm long which arise from large scaly buds, calyx and corolla are silky with erect lobes and the stamens are long, numerous and purple (Bailey, 1929).

Callistemon citrinus (Bottlebrush; Myrtaceae)

In Greek *kallos* is for 'beauty' and *stemon* for 'a stamen'. It is an evergreen genus of 25 shrubs and small trees. The leaves are mid-green, dense, alternate, leathery, linear to lanceolate and also clustered at the top of the inflorescence. Showy yellow, red or purplish-flowered terminal spikes appearing in summer, each flower is structured with long protruding stamens similar to that of a bottlebrush. The species suitable for indoor cultivation is ***Callistemon citrinus*** from Australia, growing usually up to 1.8 m tall, bearing grey-green, stiff and linear to lanceolate leaves 4-9 cm long & some 2.0 cm wide which along with the young stems are covered with silky hairs, and in July 5-10 cm long spikes appearing near the tips of the branches with yellow-tipped hair-like red hundreds of stamens but no petals. It is the colour of stamens which is referred to as colour of the flowers. The inflorescence stalk continues growing as stem even after the falling off the flowers but cylindrical capsules persist over there up to two years or even more. These flower freely and to keep them in proper size and shape, they are pruned in spring regularly, but in summer when flowering is over the pots are kept outdoors. In large pots or tubs these make effective specimens about 90 cm tall. They are propagated through seeds or through 7.5-10.0 cm long heel cuttings in summer. A persisting winter temperature of 7 °C may prove dangerous.

Camellia japonica (Theaceae)

Camellia commemorates George Josef Kamel (1661-1706), a Czeck Jesuit pharmacist who wrote about the flora of the Philippines. This genus is from India to Indonesia, China and Japan comprising of 80 species of evergreen trees and shrubs, some with attractive foliage and large spectacular blooms. Many of these grow best in containers in cool conservatories and may be brought indoors when in flower in the late winter. Many of the species are frost-hardy, preferring a neutral to acidic planting medium and continuous moist and humid conditions. They are propagated through stem or leaf bud cuttings in late summer, through layering in autumn and by fresh seeds. Most indoor camellias are varieties of one species, ***Camellia japonica*** from Japan and Korea growing up to 3 m in containers, bearing woody trunks & stems, and rich glossy green, alternate, leathery, broadly ovate & tapered-pointed, 6-12 cm long and 5 cm wide leaves. These flower solitary to clustered from winter to spring at the ends of the branchlets, flowers varying in shape (single or double) and colours (white, pink, red, purple and bicolour) and are 6-15 cm across, bearing five red petals in the true species. This has numerous varieties to choose from. These can be grown in the bright filtered light and at 7-16 °C temperatures but not above 18 °C.

Campanula isophylla (Campanulaceae)

See chapter on '**Annuals**'.

Capsicum annum (Solanaceae)

It has derivation from the Greek word *kapto* meaning 'to bite' as the fruits are hot-tasting. It is a genus of 50 species of annuals, sub-shrubs and shrubs from the warm temperate to sub-tropical areas of the Americas, mostly grown as annuals and become attractive when their fruits mature to yellow and red colours and persist longer on the plants. They have alternate and ovate to lanceolate leaves, 5-lobed erect to pendulous flowers, each with a cone-shaped cluster of five stamens in the centre. Their fruits are slender-cylindrical to oblong. They are propagated through seeds. The most popular forms are bushy growing ***Capsicum annum*** (Christmas pepper) probably from Peru which grows only up to 40 cm tall and across. Their stems are woody with thin dark green branches, bearing green and up to 2.5 cm long leaf-stalks, the leaves are 3.75-10.0 cm long and 1.25-3.75 cm wide, and the insignificant white flowers are produced from the leaf axils in early summer, followed by tapering-long to round fruits which after maturity persist on the plants for up to 8-12 weeks and it is only because of the fruit beauty these are grown though the fruits are pungent and edible in spices. The cherry, cone and cluster peppers are familiar indoor potted plants. Cherry peppers are berry-like, bright yellow or purplish-white or red and about 2.5 cm in diameter. Cone peppers are cylindrical or cone-shaped up to or more than 5.0 cm long, tapered, green or ivory-white, yellow, orange, red or purple but are liable to change the colours after ripening. Cluster peppers are clustered in 2-3, slender-pointed and about 7.5 cm long. These require forenoon bright light and do well at normal room temperature and when the temperature is 13-16 °C their life on the plant becomes more. These are propagated through seeds.

Catharanthus roseus (Apocynaceae)

See chapter on '*Catharanthus*'.

Cattleya (Orchidaceae)

Cattleya was named for William Cattley (*d.* 1832), a collector and grower of rare plants.This epiphytic mostly tropical or sometimes temperate orchid, loving partially shaded conditions, has sympodial growth and originates in South America. It is related to *Brassavola, Encyclia, Laelia, Sophronitis,* etc. It is the most popular and widely grown in the entire Orchidaceae. It has some 65 species, several thousand hybrids and numerous varieties. These are native to tropical Americas and at one time these were very popular for corsages. New hybrids are available in a variety of sizes and colours from white to purple and make attractive potted plants. Grow best in greenhouses. Hager (1957) stated that by proper selection of hybrids and cultivars the flowers can be harvested daily. Through photoperiodic adjustments some cultivars may be flowered twice a year, and once the flower buds have been formed their development can be retarded or hastened through temperature adjustments. Recommended cattleyas for indoor growing are *C. intermedia, C. labiata, C. loddigesii,* and *C. trianaei*. ***C. intermedia*** bears 25-40 cm high and 1.25 cm in diameter cylindrical pseudobulbs, each bulb topped by two 20 cm long and 5 cm wide dark green leaves. The flowering stalk that appears during early summer is about 10 cm long which bears up to six rose-pink flowers about 10 cm across with dark purple lip and these flowers last for about five weeks. ***C. labiata*** produces pseudobulbs some 25 cm long, 2.5 cm wide and about 2 cm thick, each pseudobulb being topped by one medium green, thick, leathery and up to 25 cm long & about 7.5 cm wide leaf. Up to 10 cm long floral stalk bears up to five fragrant flowers about 12.5 cm across which appear during fall to winter and last for about five weeks. Flowers are wavy-edged, rose-pink with a dark crimson-magenta frilled lip which is streaked bright yellow in the throat. ***C. loddigesii*** produces 20-30 cm high and about 1.25 cm thick cylindrical pseudobulbs, each one is topped by two gray-green leaves 10-15 cm long and 2.5 cm wide. During summer, up to six pale-lilac flowers with deep purple lip splashed with yellow throat and about 10 cm across are produced on a 15 cm long stalk, each flower lasting for about four weeks. ***C. trianaei*** bears 20-25 cm high, about 2 cm wide and 2.5 cm thick club-shaped pseudobulbs, each topped with a single deep green leaf 15-25 cm long and 5-7.5 cm wide. Light to deep lilac 2-5 flowers about 18 cm across, having deep crimson-purple frilled lip which has yellow marking in the throat, appear on a 10 cm long stalk during late winter and a flower lasts for about three weeks. Cattleyas require bright but indirect sunlight and 13-16 °C temperature. There should be regular water misting on the plants in case temperature rises to 21 °C. They are propagated by dividing the rhizomes into halves.

Celosia argentea (Cockscomb; Amaranthaceae)

In Greek, *kelos* stands for 'burning', alluding to the flame-like shape and colour of some of the species. A compact-growing perennial though mostly grown as an annual native to Tropical Asia, growing up to 1 m tall, bearing 5-7 cm long, pale-green, alternate and lanceolate to ovate leaves, and erect or feathery plume-like, sometimes somewhat drooping flowers which including the bracts are silvery-white 7.5 cm across and appear from July to September though from next year the same plant continues blooming throughout the year even through the winter over tropical conditions. *C. a.* var. *cristata* is a dwarf compact plant which bears mid-green and oval leaves and feathery *cum* fan-like and cristate flowerheads in yellow, orange, pink, apricot and red shades. *C.a. pyramidalis* syn. *plumosa* again from Tropical Asia growing to 60 cm tall and bearing mid-green and ovate leaves produces feathery floral plume 7.5-15.0 cm high from July to August in a wide colour range and its flowers are suitable even for drying. These all are propagated through seeds sown in February-March at 18 °C temperature. This species is grown as pot plant suitable for conservatory and for bedding and before flowering these can br brought in for indoor display and kept at an illuminated corner.

Chamaecereus silvestrii (Gherkin cactus/ Peanut cactus; Cactaceae)

See chapter on '**Cacti**'.

Chionodoxa (Glory of the snow; Liliaceae)

In Greek *chion* is for 'snow' and *doxa* for 'glory of', as in the wild these flower as the snow melts. It is an temperate genus of 6-7 small bulbous plants found wild in alpine areas of Cyprus, Crete and Turkey. This is closely related to *Scilla* but differs in having shortly tubular base and broad flattened stamen filaments. It is a bulbous plant producing small rounded bulbs with 2-4 mid-green, linear and blunt-tipped leaves often with bronze margins when young. Racemes are short, loose and bear starry, 6-petalled, and blue or blue & white flowers in early spring. These are planted under semi-shaded situation during autumn. Their propagation is through offset bulbs when dormant or through seeds in autumn or spring. These plants are suitable for planting in the temperate areas in the rock gardens, at the front of borders and for naturalizing effect. These are grown on window sills inside the house or grown outside but during flowering are brought inside. The indoor species are *C. giganteam, C. luciliae* and *C. sardensis*. ***C. grandiflora*** from Asia Minor grows up to 20 cm high with the spread of about 8 cm. The spikes comprise of several violet-blue flowers 3.75 cm across each having small white centre and flowers appear from late February to April. ***C. luciliae*** from Crete and Asia Minor grows up to 15 cm with spread of about 5 cm. The flowers are light blue with white centre and 2.5 cm across which appear from February to March. 'Alba' (white) and 'Rosea' & 'Pink Giant' are pink forms. ***C. sardensis*** from Asia Minor grows 10-15 cm high with spread of about 5 cm. The bulbs are quite small and produce 2-folded leaves, and the stems are slender carrying nodding sky-blue flowers about 2 cm across with a tiny white centre from March to early May.

Citrus (Rutaceae)

Citrus is an ancient name of a fragrant African wood, which was afterwards given to the citron. The common names for indoor ornamental types are citron (*Citrus medica*), lemon (*C. limon*, lemon) and orange (*C. limonia*, Otaheite orange; *C. mitis*, calamondin orange; *C. sinensis*, sweet orange). There in the genus are 15 evergreen species of spiny trees and shrubs originating in eastern Asia. These are the shrubby species which are grown for decoration and their fruits mature at a temperature range of 18-24 °C. These shrubs bear spiny stems and branches; short-stalked, lustrous dark green, ovate to elliptic (roughly oval) and glandular leaves; and solitary to clustered up to five fragrant flowers about 2.5 cm across which are usually white with five oblong and blunt-ending petals curving outwards and bold stamens. These usually flower during late spring to summer but sporadic flowering continues occurring at any time. However, under favourable conditions, the lemons continue flowering whole of the year. Initially the fruit is green up to the full development, and then it slowly ripens to yellow, yellowish green or orange as per species and this process takes about three months or even more and these ripened fruits keep on hanging on the branches for several months. Only young and small plants are grown in the living room because of their lustrous dark green and glandular foliage, plentiful of fragrant white flowers and overall for their bright-coloured fruits. The grown up ones are taken in the conservatory or greenhouses or in the open field under Indian conditions. All these require at least four hours of direct sunlight daily and normal room temperature but winter temperature should be maintained not below 10-13 °C. They are propagated through 7.5-15.0 cm long stem cuttings dipped in some rooting hormone and then planted. They are also multiplied through seeds. ***C. limon*** (lemon) from East Asia is an ornamental species, especially the *C.l.* 'Meyer' (Meyer's lemon) and *C.l.* 'Ponderosa' (American wonder lemon) which grows to a height of 1.2 m, bearing highly narrowed wings to stalkless, elliptic-ovate, toothed and 5-10 cm long leaves, and fragrant white red-flushed flowers 2-4 cm across which are borne from April to June from the pink to purplish buds, followed by thin-skinned oval-round fruits 5-10 cm long with acidic juice which take about a year to ripen. ***C. × limonia*** (*C. taitensis, C. otaitensis*; Otaheite orange) is considered to be a hybrid between *C. limon* and *C. nossoste* (mandarin orange). It is a thornless small bush producing flowers with purple-tinged petals, followed by rounded, deep yellow to orange fruits 5 cm in diameter. ***Citrus medica*** (citron) from SW Asia is an ornamental shrub growing up to 3 m high bearing stout spines though in containers its height can be managed up to 1.2 m. Its stalks are wingless and the leaves are 5-10 cm long, toothed and ovate-oblong. Flowers are white though at bud these are pink and 3.5 cm or more across. Rough and thick-skinned fruits are yellow when ripe, oval, 5-10 cm long and possesses little and sour juice. ***C. mitis*** (*Citrofortunella mitis*, a hybrid, calamondin orange, *hazara*) is the most popular indoor citrus from Philippines which is almost thornless and grows up to 45 cm high in pots and is therefore best for pot cultivation. It flowers and fruits freely even when small. Its leaves are deep green, 5-10 cm long, and lanceolate, and 3-4 clustered fragrant white flowers 1.25 cm across are produced, followed by rounded orange-yellow fruits 2.5-3.75 cm in diameter. ***C. sinensis*** (*C. aurantium sinensis*; sweet orange) from China grows up to 1.2 m in height as a house plant with stout stems having sharp spines. It is a round-headed plant with dark green and ovate-oblong leaves up to 10 cm long. From April to June, fragrant white flowers appear 2.5 cm across. It bears smooth-skinned bright orange solitary fruits about 7 cm in diameter where juice is sweet.

Clerodendrum thomsoniae (Bleeding heart vine, glory bower; Verbenaceae)

In Greek, *cleron* is for 'chance' and *nossos* for 'a tree', being of no significance. It is a genus of 400 species of shrubs, climbers and small trees from the tropics and sub-tropics of the Old World. The leaves are in opposite pairs, usually undivided and ovate. The flowers are tubular and appear in axillary or terminal clusters, some species being showy by their colourful calyx while in some cases the berry-like fruits are also attractive. They are suitable for potting and for displaying in the conservatory and indoor cultivation. *Clerodendrum thomsoniae* can also be trained as basket plant. The only species ***C. thomsoniae*** from tropical W. Africa has turned out to be a popular house plant (Huxley and Gilbert, 1979). This is an evergreen, vigorous, woody-stemmed and twining climber growing up to 4 m in length, bearing shining rich green, coarse, broadly ovate to cordate, 7-15 cm long and 5 cm wide leaves which have a quilted look and paler vein markings. Its terminal or axillary pendulous panicles up to 15 cm long appear during spring, summer and autumn in clusters of 10-30 where flowers have crimson petals and bell-shaped pure white calyces about 2.5 cm long but with stigma and anthers protruding. Individual flower consists of a white lantern-shaped calyx and a crimson starry corolla. For room display its tips are pinched out in winter and some support is provided for twining or trailing around. These should be rested during winter by subjecting them to 13.0-15.6 °C temperature. They are propagated through stem cuttings 10-15 cm long in spring.

Clianthus (Lobster's claw; Leguminosae)

Clianthus derives from the Greek *kleos* meaning 'glory' and *anthos* for 'a flower', on account of the bright coloured flowers they produce. It is a genus of three species of evergreen herbaceous plants from warm temperate areas of New Zealand and Australia and one species in SE Asia. Only two species are reported growing indoors and these are *C. dampieri* (*C. formosus*) and *C. puniceus*. They are scrambling climbers having alternate and pinnate leaves. The flowers are pea-like with pointed keel where standard petals resemble a parrot's beak. ***C. dampieri*** (Dampier's pea, glory pea) from Australia grows prostrate up to 60 cm long with white-woolly stems, bearing buff-green and woolly leaves with 11-21 leaflets which are 1.0-2.5 cm long and oval to obovate in shape. The brilliantly red coloured flowers 3-6 in an umbellate cluster with a large projected black blotch at the base of the standard petals appear during late winter to early spring and in the temperate regions during summer. Though it is a perennial plant but behaves as an annual and is grown every year through seeds. 'Alba' is its white form. ***C. puniceus*** (kaka beak, lobster's claw, parrot's bill) from New Zealand is a climber growing up to 3 m long, having slender stems and is a best indoor plant. The leaves are hairy with closely preseed silky hairs beneath, and are composed of up to 31 leaflets which are narrowly oblong and 1.0-2.5 cm long. Red to scarlet flowers 10-12 cm long are borne in pendent racemes during late spring. Its forms 'Albus' (creamy-white with a touch of green) and 'Roseus' (scarlet and pink) are quite outstanding. These grow during winter at a minimum winter temperature range of 10-13 °C, though *C. puniceus* prefers even lower than 10 °C temperature. *C. dampieri* is propagated through seeds sown in February on the hills but in September in the plains and kept at a temperature of 13-16 °C but *C. puniceus* through stem heel cuttings 7.5 cm long of lateral shoots taken during summer (June or July) and kept at 16-18 °C temperature under temperate conditions. *C. puniceus* can be propagated even through seeds similar to that of *C. dampieri*.

Clivia miniata (Amaryllidaceae)

Clivia was named for the Duchess of Northumberland, née Charlotte Florentina Clive (granddaughter of Robert Clive), in whose garden first it flowered in Great Britain. It is an evergreen leek-like genus of three perennial species from the warm dry forests of South Africa but only ***C. miniata*** (kaffir lily) from Natal is a familiar house plant growing up to 45 cm high with orange to red flowers and yellow throat. It is a clump-forming bulbous plant with very thick and fleshy roots. Itss leaves are glossy dark green, arching, leathery, strap-shaped, up to 60 cm long, and narrow to over 7.5 cm wide fan out from a leek-like thickly layered leaf bases, new ones appearing during each summer but almost equal number dieing the each autumn to winter. Being a subject of the temperate regions, the funnel-shaped and 6-tepalled 10-60 flowers 5.0-7.5 cm long appear in terminal umbels on stout, cylindrical but one-side inflated scape during early spring on the hills. The flower colour is a combination of orange-red or yellow and bright orange. They are excellent house plants tolerating a lot of shade. During winter these remain comfortable at 4.5-10 °C temperatures for giving it a brief rest. Its cultivation becomes easy if these are kept in cool house or onservatory well-spaced, no fertilizer and just enough water, no moving of the pots or repotting when in bud or flower and no repotting until the plants become pot-bound. In spring when flowering is over and the plant is pot-bound, the crown is divided and single crown is planted in a 12.5 cm pot at a temperature of 16 °C. These plants are watered properly and regularly. When these have sufficiently grown, these are shifted to 20 cm pot without dividing and in the next spring these are shifted to a 25-30 cm pot without damaging the roots at any time.

Coelogyne (Orchidaceae)

Coelogyne derives from the Greek *koilos* meaning 'a hollow' and *gyne* for 'female', due to the somewhat hollow stigma. It is an epiphytic genus of about 200 species from India to Malaysia, W. China and the Pacific Islands, bearing rounded to flask-shaped pseudobulbs, each bulb usually with two leathery and linear to lanceolate leaves. Flowers either appear singly or in racemes, having spreading tepals with a prominent labellum which may be lobed and deeply keeled. These can be grown anywhere in the home where sufficient humidity can be provided and those with pendulous flowers are suitable for growing in the hanging baskets. These are propagated through division. Recommended species for home growing are *C. barbata, C. corymbosa, C. cristata, C. dayana, C. nossos, C. ochracea, C. pandurata, etc.* ***C. barbata*** from Himalayas bears up to 10 cm high, ovoid and clustered pseudobulbs, bearing 45 cm long and oblong-lanceolate leaves. White and fragrant flowers 5-8 cm wide having crested and fringed labellum with brown sepia, appear during autumn to winter on 45 cm long on erect or arching racemes. ***C. corymbosa*** from Himalayas produces somewhat angled, 5 cm long and ovoid to oblong pseudobulbs. Leaves are 15 cm long and oblong-lanceolate. Creamy-white 3-5 flowers 3-5 cm across appear during summer and autumn on a 20 cm long raceme, flowers having yellow-brown throat streaking, and cream with bright yellow labellum with brown-ringed spots. ***C. cristata*** from Himalayas produces about 6 cm long, clustered and oval to globular pseudobulbs which with age become wrinkled. Each pseudobulb is topped by two 30 cm long linear-lanceolate leaves. Its fragrant, white with highly waving tepals and some 8-10 cm across flowers with labellum having five deep yellow keels, appear during winter and spring on 15-30 cm long arching to pendulous racemes. ***C. dayana*** from Malaysia bears strongly ribbed, 13-25 cm tall and narrowly conical pseudobulbs. Leaves are 60 cm long, pleated and erect. Musk-scented pale yellow-brown flowers 6-7 cm across appear in spring and summer on 45-90 cm long pendent racemes with labellum's keel and margins being white but veined chocolate-brown. ***C. nossos*** from Himalayas produces up to 8 cm high and spindle-shaped pseudobulbs, and glassy-green and narrowly lanceolate leaves about 30 cm long. Waxy-white fragrant and 7-12 flowers some 4 cm across with golden-yellow centred labellum appear on a 25 cm long arching to pendulous raceme. C. ***ochracea*** from Himalayas is quite similar to *C. corymbosa* though leaves and pseudobulbs are larger. Flowers are white and its labellum is yellow with red-flushed keels, and these are borne during spring and summer on erect racemes.***C. pandurata*** (black orchid) from Borneo and Malaysia produces quite flattened, oblong and up to 13 cm long pseudobulbs which continue appearing at intervals on a strong creeping rhizome. Lustrous leaves are up to 45 cm long, and narrowly elliptic to lanceolate. Bright pale-green and strongly fragrant flowers 8-10 cm across and where labellum is violin-shaped with black markings and wrinkled margins, appear during summer to autumn.

Columnea microphylla (Gesneriaceae)

Columnea was named for Fabio Colonna (1567-1640) through Latinizing his name to Columna, an Italian botanist and author of first work to use copperplate illustrations. A genus of 200 species of evergreen perennials and sub-shrubs, most being epiphytic. They are of two types, entirely trailing with long and thin stems, and partially erect but deeply arching stems but both are prolific bloomer and can bloom any time of the year. The leaves are short-stalked, opposite but in unequal overlapping pairs, somewhat elliptic, and pointed to both the ends. These flower solitary or in clusters from the leaf axils, bearing calyx of striking shape and colour and tubular corolla flaring into five differently shaped lobes, two upper lobes joining together to form a hood and stamens and stigma protruding. One large plant may contain up to 100 flowers at a time. Recommended columneas for indoor gardening are *C. × banksii, C. gloriosa, C. linearis* and *C. microphylla.* These require bright but indirect sunlight and for short-day plants even fluorescent light is sufficient during winter. They have luxurious growth and flowering at 18-35 °C year round temperature. They are propagated through 7.5-10.0 cm long tip cuttings taken at the time of repotting and these cuttings root within 3-4 weeks. ***C. × banksii*** (*C. oerstediana × C. shiediana*) is a sub-shrubby hybrid with pendulous stems 60-90 cm long bearing glossy dark green leaves, fleshy, 2.5-4.5 cm long and 1.25-2.0 cm wide. The orange-red flowers being produced from

November to April have orange markings in the throat. ***C. gloriosa*** from Costa Rica produces basally branched trailing stem about 90 cm long, bearing dark green ovate-oblong leaves in unequal pairs, 1.25-3.0 cm long, 1.25-1.5 cm wide with dense purplish-red hairs on upper surface and reddish on the lower. Very showy about 8 cm long fiery-red flowers with yellow throat and hooded upper petal are freely produced from late autumn through the whole winter to the early spring. It is an excellent plant for hanging basket. ***C. linearis*** from Costa Rica is an erect and shrubby species growing to 45 cm. Its leaves are above glossy deep green, 9 cm long and linear-lanceolate. Flowers 4.5 cm long appearing during spring are rose-pink with silky-white hairs. ***C. microphylla*** from Costa Rica bears pendulous stems up to 1.8 m long, bearing light green and broadly ovate leaves covered with purple hairs. Between November to April, the bright orange-scarlet flowers are produced some 3.75-5.0 cm long.

Crassula nossos (Crassulaceae)

See chapter on '**Succulents Other Than Cacti**'.

Crinum (Spider lilies; Amaryllidaceae)

Crinum derives from the Greek *krinon* which means a lily. It is a long-necked bulbous genus of more than 100 species spread over tropics and sub-tropics worldwide. The bulbs are large up to 15 cm across, up to 30 cm long and stalkless, the leaves are large sword or strap-shaped and 5-8 clustered flowers 7.5-13.0 cm long with mouths about 15 cm across are lily-like funnel-shaped which are carried in umbels emerging from the spathe-like bracts on stalks 60-90 cm long and about 2.4-4.0 cm thick, and the flowers last for about one month. They are propagated through removal of offsets, by division and by fresh seeds. When planting the bulbs, upper third should be kept exposed. After planting, it takes about four years for flowering. It is only the ***Crinum bulbispermum*** (*C. capense, C. longifolium*) from South Africa which is recommended for cultivation indoors. Its bulb is 7.5-10.0 cm thick, oval and flask-shaped with a long neck and from which some 10 sword-shaped, spreading, 60-90 cm long, 5.0-7.5 cm wide and slender-pointed leaves emerge. It produces white flowers deeply tinged with pink on the outer surface of petals. Its var. 'Album' produces pure white blooms. The var. 'Powellii' produces 10 cm thick globular bulb with about 20 sword-shaped leaves, each leaf measuring 90-120 cm long and 7.5-10.0 cm wide at base. Its flowers are pinkish-red, shaded green at the petal-bases. Var. 'Ellen Bosanquet' bears shorter but broader leaves and flowers deep wine red. Indoors, these require a corner with sufficient light. At mid-winter, these require three month's resting at 10 °C temperature.

Crocus (Iridaceae)

Crocus derives from the Semitic *karkom* for 'a yellow dye' obtained from the stigmas of *C. sativus*, and in Greek *krokos* is for 'saffron'. It is a hardy cormous genus with 75 species from Europe, Western Mediterranean to Central Asia and Pakistan. Its corm is small, rounded, sometimes flattened and covered with tunic as remnant of the leaf bases of the previous season growth. These grow maximally to the height of 13 cm. Their foliage is slender-narrow, grooved, grassy with a central silvery-white line, and 6-tepalled flowers opening almost flat in the sun. Crocus flower onsists of a slender corolla tube arising directly from the corm so nearly half being underground and expanding into six ovate petals where three stamens are also situated. Corolla tube down towards the corm is seated at the apex of the corm in the form of ovary. The cut-shaped flowers may be mauve, purple, bronze, yellow, white or with bicoloured-striping. Winter to spring-flowering species flower along the emergence of the leaves while the autumn-flowering ones before emergence of the leaves, and the leaves continue growing even after the flowering. In fact, these survive filtered light but for the sake of proper growth and development, these are brought inside only when flowering starts. They are propagated through seeds or by dormant offsets. Instead of the true species, the Dutch hybrids which have twice or thrice the size of the flowers than the true species are ideal house plants. These require cold conditions for their growth and development so these can be grown only in temperate areas.

Crossandra infundibuliformis (Firecracker flower; Acanthaceae)

Crossandra takes its name from the Greek *knossos* meaning 'fringe' and *aner* for 'male', owing to its fringed anthers. It is an evergreen tropical perennial genus of 50 species of sub-shrubs, bearing lanceolate leaves either in opposite pairs or in whorls. Terminal or axillary floral stalk is 4-angled and cone-like, bearing overlapping bracts and five broad-petalled lobes in shades of white, yellow, orange or orange-red which are long-lasting, and suitable for greenhouse, conservatory and for indoor cultivation. The most popular crossandra recommended for indoor cultivation is ***Crossandra infundibuliformis*** (*C. undulifolia*) from India and Sri Lanka, is a slow-growing plant growing up to 60 cm in height and starts flowering when seedlings are only a few months old. This bears 1.25-2.5 cm long leaf stalks, glossy dark green, lanceolate, undulate, 5.0-12.5 cm long and up to 5 cm wide leaves. Tubular and orange-red to salmon-pink flowers always with a yellow eye are produced on a 15 cm long spike from spring to autumn, mostly from the terminal leaf axils, partly hidden by small and triangular bracts, each bloom flaring out into flattened 5-lobed disc 3.75 cm across. Its dwarf form 'Mona Walhed' grows hardly up to 30 cm tall bearing dark leaves and salmon-pink flowers. In winter these require direct sunlight otherwise only medium light, and the plants do not tolerate temperature below 18 °C. This is propagated through 5.0-7.5 cm long tip cuttings taken during spring where rooting occurs in 4-6 weeks.

Cryptanthus **(Bromeliaceae)**

See chapter on '**Bromeliads**'.

Cuphea **(Lythraceae)**

Cuphea derives from the Greek *kyphos* meaning 'curved', due to curved seed capsules. There are about 250 annual, perennial and sub-shrubby species under the genus native to tropics and subtropics of Americas. Out of all the species, only two are worth growing indoors, *viz. C. hyssopifolia* and *C. ignea*. ***C. hyssopifolia*** (firecracker plant) from Guatemala and Mexico is a branched shrubby species growing up to 60 cm high with wiry stems and its width is almost higher than the length. Leaves are dark green, leathery, linear, heather-like, 0.6-1.9 cm long, 6 mm wide and crowded together on the stem. Purple, pink or white, bell-shaped, 6-petalled and green-tubed flowers some 1.0-1.25 cm long appear on the raceme-like clusters during summer and autumn. ***C. ignea*** (*C. platycentra*; cigar flower) from Mexico grows up to 30 cm, bearing mid-green, lanceolate and 2.5-5.0 cm long, and 1.25 cm wide leaves. From spring to summer, solitary tubular dark red flowers 2-3 cm long appear from the leaf axils with a dark band and white ring at the end in the form of glowing cigar. They are propagated through seeds and cuttings.

Cyclamen **(Primulaceae)**

Cyclamen derives from the Greek *kyklos* meaning 'circular', the way the stems spiral down. A native to Europe and the Mediterranean region to Iran, the genus comprises 20 species of perennials with rounded and nearly woody cormous tubers. The ovate-cordate to rounded leaves which are mostly marked with silver, and are long-stalked arise directly from the tuber. The flowers are solitary, 5-petalled, pendent and so reflexed as to appear a characteristic shuttlecock. These appear from September to December and one plant remains decorative for 2-3 months. The flowers may be pure white, pink, salmon, red, mauve and purple. After fertilization the stems coil (except *C. persicum*) pulling the rounded capsule to ground level. These are temperate ornamentals suitable only for a cool house. Their tubers are dried during spring in the pots and are repotted when growth begins and only then these are watered. These require bright but indirect sunlight and for their growing 13-18 °C temperature. Only ***C. persicum*** is treated as house plant and that too when it is to flower is brought inside. It is from Mediterranean region (Algeria to Lebanon) growing up to 23 cm tall. Its dark green rounded-cordate to broadly ovate and toothed leaves are marbled with silver. Fragrant rose to pale-pink and white with elegantly twisted flowers 2.5-3.75 cm long are borne during winter and spring.

Cymbidium **(Orchidaceae)**

Cymbidium derives from the Greek *kymbe* for 'a boat', referring to the boat-like hollow shape in the labellum. It is a genus of 40 epiphytic and terrestrial clump-forming orchids from Tropical Asia and Australia bearing usually short, erect and oval to conical pseudobulbs that arise from a woody rhizome, and these bulbs are sheathed by bases of the leathery, arching and ribbon-shaped linear leaves. The ones suitable for home are exclusively epiphytic. Flowers appear on either erect or drooping stems. The best modern cymbidiums are those miniature forms that have less than 5 cm tall pseudobulbs and where leaves rarely grow longer than 35 cm. These can produce as many as 30 scented or non-fragrant flowers, each about 7.5 cm across and in the colour range of mahogany red to pink, yellow, green or white. The individual flowers open along the stems during a period of several weeks during late spring and early summer, and a flower lasts for about six weeks. An established miniature *Cymbidium* in a 15- 25 cm pot can produce up to six flower stems in one season. Recommended *Cymbidium* species and varieties for home growing are the miniature *C. devonianum*, and the hybrids such as 'Munuet', 'Peter Pan', *etc.* ***C. devonianum*** from Himalayan and Khasi Hills is a miniature clump-forming species where about 5 cm tall pseudobulbs are hidden by the bases of 3-5 pale-green, leathery, 20-35 cm long and 3.5-7.5 cm wide leaves that are broadly oblanceolate and tapering to a long petiole. Pendulous floral stalks are about 30 cm long with 12-18 yellowish-green waxy-textured flowers having deep purple markings, the lip being purplish-red and the flower size 3.75 cm across. The flowers appear during spring and summer. 'Minuet' produces 25-40 cm tall spikes with about 20 green, brown or yellow flowers about 2.5-3.75 cm across where lips are spotted darker. 'Peter Pan' floral stems grow from 25-35 cm long, bearing 10-15 greenish-yellow flowers about 2.5-3.75 cm across, and the lips are spotted red. These are grown under bright but indirect light and at 16 °C temperature. These require regular water misting inside. These are propagated through division of rhizomes, each rhizome piece having at least two pseudobulbs and a few roots.

Cytisus (Leguminosae)

Cytisus has derived from the Greek *kytisos*, a name for several woody plants of the pea family. It is a genus of 30 species of mainly deciduous but a few evergreen shrubs from Europe, Mediterreanean region, Asia Minor and North Atlantic Islands and are related to *Genista*. They are prostrate to bushes up to 3 m long. These have simple to trifoliate alternate leaves which are usually short-lived to perform their functions by the green stems. These have typical pea-family flowers, often fragrant. There are only two species, *Cytisus canariensis* and *C. racemosus* recommended for indoor cultivation. As these do neither flourish at normal room temperature at any time nor these flower the following year if not overwintered below 15.6 °C, *vis-à-vis* before next flowering they take more than 11 months gestation period, so instead of retaining these after flowering for the next season these are discarded and replaced with the new ones in waiting. Their requirement of light is medium except during spring and flowering time when bright light with at least three hours per day is necessary. For flowers to last more these are kept at a temperature below 15.6 °C throughout the flowering period in the room. Its commercial propagation is through seeds but it takes 2-3 years to produce a flowering plant. They are also propagated through 7.5-10.0 cm long stem cuttings taken in spring when temperature is 13-15.6 °C and these cuttings root in 4-6 weeks. ***C. canariensis*** (*Genista canariensis, Teline canariensis*; florist's genista) from Canary Islands grows up to 1.8 m high with a spread of up to 1.5 m in the wild but as a pot plant its height is about 45 cm and spread 30 cm. It is a green-stemmed well-branched evergreen shrub having short-stalked or stalkless, hoary and trifoliate leaves with leaflets being 6-12 mm long and obovate to elliptic. It produces several short terminal spikes bearing 5 cm long fragrant yellow flowers from winter to spring. ***C.* × *racemosus*** (*C. fragrans, C.* × *spachianus, Teline* × *spachianus*; a natural garden hybrid between *C. stenopetalus* × *C. canariensis*, though often in the garden is confused with *C. canariensis* as in most aspects the hybrid resembles the latter) grows up to 2.4 m high and up to 1.8 m in spread though as a pot plant it grows up to 45 cm only. This is a twiggy evergreen shrub with 6 mm leaf-stalks, grey-green, and trifoliate leaves with 1-2 cm long obovate leaflets which are green above and silky-hairy beneath, and fragrant bright yellow flowers up to 1.2 cm long which appear on slender terminal racemes 5-10 cm long during winter and spring. The flowers in this hybrid last longer than *C. canariensis*.

Dactylorrhiza (Orchidaceae)

It is a moderately hardy terrestrial orchid genus for sunny or partially shaded sites with popular species of *D. elata, D. fuchsia, D. maderensis,* and *D. purpurea* (Boyd, 1994). They produce ensheathing lanceolate leaves, which are sometimes speckled with brown. The flowers are lobed and in shades of purple, magenta and red. Their floral stalk is leafy, stout and tall where hundreds of flowers are closely set in pyramidal shape, opening from below towards the top and look very charming. These make a durable cut flower. ***D. elata*** grows up to 60 cm tall, bearing dull green leaves and densely clustered deep lilac-purple flowers. ***D. fuchsia*** grows up to 50 cm tall, bearing faintly mottled leaves and the densely set spikes of reddish-brown flowers spotted with lilac-pink. ***D. maderensis*** (lady orchid) grows up to 45 cm high, bearing shining green leaves and red-purple flowers. ***D. purpurea*** (lady orchid) grows up to 45 cm high, bearing broader leaves and closely set clusters of fragrant purple-pink flowers. These are planted during autumn or spring in humus-rich soil and kept in sun or at a semi-shaded situation. The soil should remain moist all the time. They are propagated through rhizomes divided in spring.

Datura (Solanaceae)

See chapter on '**Annuals**'.

Dendranthema (Asteraceae)

Dendron in Greek is for 'a tree' and *anthos* 'a flower'. It is a genus of about 200 species of annuals and perennials from northern temperate zone. These are the temporary house plants which are taken out after the flowers are gone. These all produce usually showy daisy-like flowers almost freely produced for a long period and are brought inside the house only when in bloom. They are propagated through cuttings and suckers, and also through seeds if they produce. The two species most commonly grown indoors are ***Dendranthema frutescens*** (*Argyranthemum frutescens*; Paris daisy, white marguerite) from Canary Islands and ***D. morifolium*** (*Dendranthema vestitum, D. hortorum*; florist's chrysanthemum). Both are bushy, soft-woody-stemmed plants with terminal clusters of daisy-like flowers, *D. frutescens* flowering only in summer while *D. morifolium* which is a short-day plant flowering from late autumn to early winter. ***D. frutescens*** grows up to 90 cm tall but in pots these are pinched out to restrict their growth up to 45 cm. This bears pale-green, alternate, 5-10 cm long and up to 7.5 cm wide deeply dissected leaves into lobed leaflets on short stalks. Numerous clusters of terminal flowers appear so hugely that most of the foliage is covered, each flower being 5.0-7.5 cm across where a single dense circle of white petals surrounds a raised yellow disc. ***Dendranthema morifolium*** in a pot grows up to 30 cm with dark green leaves similar to *D. frutescens*. The flowers 2.5-7.5 cm across are also similar to *D. frutescens* but each flower has densely overlapping circles of petals so central disc is hidden and the colour is white, cream, yellow, orange, pink, bronze or purplish. Both like filtered sunlight and 13-18 °C growing temperatures.

Dendrobium (Orchidaceae)

Dendrobium derives from the Greek *dendron* meaning 'a tree' and *bios* for 'life', referring to these plants being epiphytic. The genus from Asia to Australasia and the Pacific Islands, comprises of 1,400 epiphytic deciduous or evergreen orchids, most of which having stem-like or club-like pseudobulbs, that bear unpleated but rolled under or flat lanceolate leaves and the racemose flowers. The tepals are spreading, outer three often narrower and the labellum is entire or lobed and sometimes fringed. The species suitable for home cultivation as mentioned below bloom from late spring to early summer, each flower lasting for 4-6 weeks. Deciduous types should be given cool and dry spell after flowering though evergreen types require only a short resting period. These are propagated through division of the rhizomes into segments, each segment with at least four pseudobulbs out of which at least one should not have flowered. Also as some dendrobiums develop new growths at the top of old psseudobulbs and when such growths have developed roots about 2.5 cm long, these are cut away and treated as a mature plant. They are best at temperature range of 16-21 °C during active growth but during rest period in winter these should be provided 15.6-18.3 °C temperature during daytime and 10-13 °C during night time. The species suitable for home culture are *D. infundibulum, D. kingianum* and *D. nobile*. ***D. infundibulum*** from Myanmar produces medium green pseudobulbs 25-50 cm tall, topped by several dark green, strap-shaped, 7.5 cm long and about 2 cm wide leaves. Leaf sheaths encasing the pseudobulbs as well as floral buds have short black hairs. White flowers numbering 2-6 appear on one floral stalk, each flower measuring 10 cm across with wavy tepal edges, and these have a deep yellow mark in the throat of the tubular lip. ***D. kingianum*** from Australia is a temperate evergreen species producing clustered, reddish-green, club-shaped with mostly thicker one below and thinner one above, often branched and about 7.5-45.0 cm tall pseudobulbs, each pseudobulb being ensheathed by bases of 3-6 grayish-green, lanceolate, 7.5-15.0 cm long and about 2.5 cm wide leaves. Pale to purplish-pink, fragrant and cup-shaped flowers 1.25- 2.5 cm across, with darker marks on the lips (rarely all white flowers), appear during winter to spring in a terminal clusters of 2-12. ***D. nobile*** from Himalayas to Taiwan (SE Asia) is a temperate semi-deciduous species which produces yellowish-green, stem-like to narrowly club-shaped and erect pseudobulbs from 60 cm to 1.2 m tall, topped with several narrow and oblong leaves up to 10 cm long and 2.5 cm wide with notched tips that appear in early fall though in the late spring the leaves of a just to flower pseudobulb fade and are replaced with branched floral stalks, each stalk carrying 2-4 flowers about 7.5 cm across. Mostly fragrant flowers are white, lavender to deep purple, tepal margins wavy, the tips pink, labellum white, the lips are large and rounded having tubular base and with a deep maroon blotch in the centre. Its pure white form is *D.n. virginale*.

Dicentra (Dielytra; Fumariaceae, formerly Papaveraceae)

In Greek, *dis* is for 'twice' and *kentron* for 'a spur', as its two nectaries in some species are shaped spur-like. It is a clump-forming or rhizomatous genus of 20 highly temperate ornamentals found wild in Western Himalaya to Siberia and North America. Its curving stems bear highly decorative leaves which are glabrous and fern-like deeply dissected. Its heart-shaped curious flowers attract more attention and the whole plant is an all-time favourite perennial. The flowers are pendent, 4-petalled, outer two petals pouched and spurred while inner ones are smaller. Recommended species of indoor cultivation for temperate regions is ***Dicentra spectabilis*** (bleeding heart, Dutchman's beeches, lady in the bath) from Japan, Korea and China which is a clump-forming, growing up to 50 cm with branched stems, bearing long-stalked, decorative fern-like and biternate foliage where leaflets are 3-5-lobed and glaucous. Flowers appear on arching racemes in spring and are cordate, outer petals rose-crimson, protruding inner ones white. The plants are grown outside but brought indoors when start flowering but are again shifted to coldframe when foliage starts yellowing and these are repotted annually. These are so hardy that can survive even -18^0 C temperature and like to be grown best in the filtered sunlight. These are propagated through division of clumps during autumn and through seeds sown during spring or autumn.

Dipladenia splendens (Apocynaceae)

See ***Mandevilla*** in this chapter.

Duchesnea (Rosaceae)

See chapter on '**Lawn**'.

Dyckia (Bromeliaceae)

See chapter on '**Bromeliads**'.

Echeveria (Crassulaceae)

See chapter on '**Succulents Other Than Cacti**'.

Echinocereus baileyi (Cactaceae)

See chapter on '**Cacti**'.

Echinopsis (Cactaceae)

See chapter on '**Cacti**'.

Epidendrum (Orchidaceae)

Epidendrum derives from the Greek *epi* meaning 'upon' and *dendron* standing for 'a tree', named so because of its epiphytic habit. It is an epiphytic genus of 400 species native to Tropical America, but one species to West Africa. Some of the species bear tall and reed-like stems though others may have rounded, ovoid, cylindrical or conical pseudobulbs up to 30 cm high and 2.5 cm thick, topped with two medium green, 10-12.5 cm long, 2.5 cm wide, elliptic to strap-shaped and cylindrical leaves. Flowers comprise of three sepals and two petals, both being similar and a simple to 3-lobed lip. These require daytime temperature of 21 °C and a minimum nighttime temperature of 12 °C with sufficient humidity. These are propagated during spring when the new pseudobulbs are about 2.5 cm tall. Rhizomes are divided into segments, each segment with at least three pseudobulbs where one should carry the leaves. ***E. cochleatum*** (cockle-shell orchid) from Mexico to Brazil produces clustered, ovoid and 10-20 cm high pseudobulbs. Oblong to linear-lanceolate leaves are about 30 cm or more long. Greenish-yellow-tepalled flowers that are upside down are 7-10 cm across, lip cockle-shell-like, overlaid with black-purple below, centre white with pale and radiating lines. Seven flowers appear on 30-45 cm long racemes. ***E. pentotis*** produces light green and little cylindrical pseudobulbs that are about 30 cm high and 2.5 cm thick, and are topped with two medium green, 10-12.5 cm long and 2.5 cm wide leaves. The floral stalks are about 10 cm long that carry two yellowish or creamy-white fragrant flowers each about 7.5 cm across and the flowers are borne during late spring and summer. The pointed lips are oval and white with purple stripes. ***E. vitellinum*** from Mexico and Guatemala produces bluish-green pseudobulbs about 7.5 cm high and 2 cm thick, and topped with 2-3 bluish-green, about 10 cm long and 1.25 cm wide leaves. Flower appears during summer with stalks about 30 cm long, carrying up to 18 flowers each, flowers being 3.75 cm across of bright orange-red colour and the lip is yellow and tubular.

Epiphyllum (Cactaceae)

See chapter on '**Cacti**'.

Episcia (Gesneriaceae)

See chapter on '**Foliage Plants**'. Only four species, *viz. E. cupreata, E. dianthiflora. E. lilacina, E. reptans* and varieties 'Acajou' (flowers bright red-orange), 'Cleopatra' (flowers red-orange) and 'Cygnet' (flowers white with prominent purple spots), are recommended for indoor cultivation. ***E. cupreata*** (flame violet) from Colombia and Venezuela is a creeping plant with more than 40 cm long stems, bearing 6-8 cm long, elliptic, toothed, puckered or quilted and coppery flushed green leaves with white or pink veins. The flowers are 2.0-2.5 cm across and bright red with yellow spots internally. ***E. dianthiflora*** (lace flower) from Mexico is a tufted plant with creeping and hairy runners with small plantlets. Leaves are dull green with purple midrib, thickish. 4 cm long, oval, crenate and softly downy above. Flowers are milky white with deeply fringed petals and about 2.5 cm long. ***E. lilacina*** from Central America is a tufted species producing runners with small plantlets. Its leaves are green and hairy, suffused often with bronze to reddish, 6-8 cm long, ovate-lanceolate and the central veins paler above and pink below. White flowers are slender-tubed with yellow patch in the throat. ***E. reptans*** from Colombia to Brazil has trailing stems more than 60 cm long, bearing ovate, 5-12 cm long and quilted dark green with red flushing and veins silver. Deep fiery-red flowers are 4 cm long with fringed apex. These are light-loving plants so fluorescent lighting above 10-15 cm should be provided. These require high humidity and temperature range of 15-29 °C.

Erica (Heaths or heathers; Ericaceae)

Erica derives from the Greek word *ereike* or the Latin *erice*, the ancient name for the tree heath, *Erica arborea*, which is common in the Mediterranean region. It is a genus of 600 species of wiry-stemmed evergreen shrubs mainly from South Africa though a few even from North Atlantic Islands. These bear narrow and linear to oblong leaves that are usually rolled under along the margins. Being free-flowering, the flowers are bell- or urn-shaped. These mostly like acidic peaty compost to facilitate the growth of symbiotic endotrophic (in roots) mycorrhiza which helps in providing essential minerals. These are propagated through heel cuttings during spring with exception of *E. carnea* which is propagated during summer, and the species also by seeds. Since these come from the cool temperate areas so require cool conditions for their growth when grown indoors, therefore, it is hard to grow them satisfactorily even for a brief period so they are treated as temporary indoor plants. The recommended ones for indoor cultivation are *E. canaliculata, E. gracilis, E.* × *hyemalis* and *E. ventricosa*. ***Erica canaliculata*** (Christmas heather) from South Africa grows erect 1 m or more in the pots though in the wild up to 5m tall having white-hairy stems. Leaves are carried in whorls of three, are recurved and 4-6 mm long. White to pale-pink and bell-shaped flowers about 3 mm long having black centres appear from spring to early summer in leafy panicles. ***Erica gracilis*** (Cape heath) from South Africa grows from 30-45 cm tall with the spread of about 30 cm having numerous side shoots. It bears tightly packed hairless, pale-green, linear, 3-6 mm long leaves in whorls of four or more, and the shoots tipped with clusters of 3-4 rose-purple, globe-shaped flowers about 3-6 mm across and 1.25 cm long appearing from October to January. Its white form is listed as *E. nivalis*. ***Erica* × *hyemalis*** (French heather) is an erect hybrid from South Africa growing up to 60 cm tall with a spread of up to 40 cm. The tightly packed leaves are mid-green, finely hairy, thread-like linear and 0.5-2.0 cm long. White, pendent and tubular flowers 1.5-2.0 cm long, about 6 mm across with a flush of pink are produced on long terminal racemes from November to January. ***Erica ventricosa*** from South Africa is an erect bushy shrub growing up to 2 m in large containers. Crowded leaves are in whorls of four, grey-green, linear to awl-shaped, ciliate and 2.0-2.5 cm long. Red, pink or white and ovoid flowers 1.2-2.0 cm long & 6 mm across, inflated at the base but constricted at the mouth appear in dense terminal umbels during summer and autumn. These for growing require bright filtered light, a temperature range of 7-10 °C and high humidity.

Eucharis grandiflora (*E. amazonica*; Amazon lily; Amaryllidaceae)

Eucharis in Greek means 'charming or pleasing'. It is a tropical South American genus of about 10 species of bulbous plants allied to *Hymenocallis*. These are hothouse bulbous plants with pleasantly fragrant flowers blooming in winter and spring. Their bulb is tunicated and 2.5-5.0 cm in diameter, leaves are radical, broad-ovate, narrowed into distinct petioles and parallel-ribbed, the highly showy flowers are white, umbellate and stand on long stout scapes, perianth tube erect or curved, segments six, dilated and spreading, and the central cup or corona is formed of outgrowths from the bases of the stamen filaments, and the throat is dilated. Only ***E. grandiflora*** from Colombia and Peru is described as the most popular indoor species. Its leaves are evergreen, lustrous deep green, long-stalked, oblong to elliptic-ovate and blades up to 20 cm long. Floral stalk is erect, up to 60 cm long and is topped with 3-6 narcissus-like fragrant white flowers 8 cm across having a slender curving tube, appearing during summer.

Euphorbia (Euphorbiaceae)

See *Euphorbia leucocephala, E. milii and E. pulcherrima* in the chapter on '***Euphorbia***'.

Exacum affine (Gentianaceae)

See chapter on '**Annuals**'.

Faucaria (Aizoaceae)

See chapter on '**Succulents Other Than Cacti**'.

Felicia (Asteraceae)

See chapter on '**Annuals**'.

Ferocactus (Cactaceae)

See chapter on '**Cacti**'.

Fortunella (Kumquats; Rutaceae)

Fortunella was named to honour Robert Fortune (1812-1880), a Scottish horticulturist, plant collector, traveller and author responsible for establishment of the tea industry in India (from China) and who introduced many garden plants to Britain. It is a genus of 4-6 species of evergreen shrubs and small trees from eastern Asia and the Malay Peninsula. Though outside these grow to 4.5 m tall but indoor in pots these can be grown up to 1.2 m in bush-shape and some of these have even thorns. The leaves are dark green above, paler-green below, alternate, thick, leathery, slightly pitted and with short leafstalks. Usually solitary flowers appear from the leaf axils with short stalks during spring and summer which are white, heavily fragrant, 5-petalled and 1.25 cm across. Fruits ripen slowly and turn orange on maturity hanging on the bush for many weeks, an additional beauty. These require direct light as much as possible and do well at normal room temperature, however, during winter rest these should be subjected to 13-15.6 °C temperature. They are propagated through fresh seeds. Recommended indoor species are *F. japonica* and *F. margarita* (Nagami kumquat). ***F. japonica*** from Japan is spiny bearing 7.5 cm long, 5.0 cm wide and elliptic leaves, fragrant white flowers and about 3.0 cm across orange-yellow round fruits. ***F. margarita*** from southern China is an erect, thornless and bushy shrub, growing up to 3 m tall in the wild but in pots only 1.2 m. This bears lustrous deep green, broadly lanceolate to ovate, up to 10 cm long and 6.25 cm wide leaves. Flowers are white and fragrant. Fruits are broadly oblongoid, rich orange-yellow, some 4 cm long and 2 cm wide where juice is flavoured but sour though rind is sweet.

Freesia (Iridaceae)

Freesia was named for Frederick Freese (*d.* 1876), German doctor and friend of the botanist who named the genus. A South African genus with 20 species of cormous plants with narrow leaves mostly in the basal fans. The spike about 30-45 cm long is slender where to one side the fragrant florets facing upward are borne. Florets are tubular funnel-shaped with segments flaring out in six lobes, three stamens and one style. Florets in shades of white, yellow, orange, pink, red, lilac or blue appear during winter to spring. They are long-lasting cut flowers. These are propagated through corms and cormlets and also through seeds. ***Freesia* × *hybrida*** (*F.* × *kewensis*, *F. armstrongii* × *F. refracta*) is suitable for growing indoors. This has mid-green, narrowly lanceolate & sword-shaped and up to 30 cm long leaves. The spike is up to 60 cm in length and bears funnel-shaped fragrant flowers 2.5-5.0 cm long. This has plenty of varieties to choose from. The freesia pots should be brought inside when to flower and kept at a bright light condition. After the flowers fade, these pots are taken out for corm development.

Fuchsia (Lady's eardrops; Onagraceae)

See chapter on '***Fuchsia* × *tuberhybrida***'.

Gardenia jasminoides (Cape jasmine; Rubiaceae)

Gardenia was named for Dr. Alexander Garden (1730-1791), a physician and naturalist of Scottish origin who lived in Charleston, S. Carolina (USA), and who had correspondence with Linnaeus. It is a genus of 250 species of evergreen shrubs and trees from warmer parts of Asia and Africa. Their leaves are entire, in pairs or in whorls of three or more. The flowers usually white or cream comprise of 5-11 waxy petals, and their pots are brought indoors at the time of flowering and kept at a well illuminated place at a temperature of 16-17 °C temperature with sufficient humidity. They are propagated by taking 7.5 cm long tip cuttings in early spring. The indoor species is ***Gardenia jasminoides*** (*G. grandiflora, G. florida*) from S. China growing in pots up to 45 cm in height and spread, bearing shining green, leathery, lanceolate to obovate, 10 cm long and in opposite pairs or in whorls of three leaves. The semi- to fully-double white flowers 5-10 cm across appear solitary from the leaf axils near the end of the shoots. This species has many forms such as 'Bedmont', 'Fortuniana', 'Veitchii', *etc.* all with pure white flowers.

Gasteria (Liliaceae)

See chapter on '**Succulents Other Than Cacti**'.

Gesneria (Gesneriaceae)

Gesneria honours Conrad von Gesner (1516-1565), a famous German naturalist of his time, author of the comprehensive *Historiae Animalium*, the originator of bibliography and much more. This evergreen perennial genus of 50 species of shrubs from tropical America and West Indies produces stolon-like underground runners and new rosette-shaped plants wherever the tips of the runners break the surface of the potting mixture, and from the aerial parts the attractive foliage and showy flowers. They are slow-growing, short-stemmed and normally rosette-forming. They have stalkless, alternate, simple, paired, prominently veined, sometimes elliptic, and often hairy leaves. Tubular to funnel-shaped, solitary or in clusters of up to four, the flowers continue appearing on long stalks from the upper axils throughout the year if congenial atmosphere is provided, and these flowers slightly widen at the mouth. Their humidity requirement is high therefore these make a spectacular terrarium specimen. These require 3-4 hours of bright but filtered light every day and a minimum of 18 °C temperature at all the times. They are propagated by separating the plantlets growing on the underground runners. Recommended kind for indoor cultivation is ***G. cuneifolia*** from Puerto Rico, Cuba and Hispaniola which is tufted to clump-forming and with age producing short erect stems up to 30 cm tall. It produces cuneate, up to 15 cm long, oblanceolate, smooth and with upper leaf surface dark green. Solitary red or yellow flowers about 3 cm long appear from the leaf axils which is tubular and slightly curved.

Gloriosa (Glory lily; Liliaceae)

Gloriosa is the Latin name for glorious. It is a genus from Africa and Asia of five species of tender tuberous-rooted perennial climbers, related to the genus *Lilium*. The tubers are brittle and narrow. Stems are wiry and slender so require support to grow and bear oblong-lanceolate leaves where tip is often elongated to form tendrils. Upper leaf axils produce solitary long-stalked showy 6-tepalled blooms in red or yellow or both the colours combined. The tepals are quite reflexed and usually wavy and beneath them there are six stamens spreading widely. Tuber is potted in October in the tropical conditions and in spring in the temperate areas, where these produce flowers generally after 70 days of tuber planting. They are brought in the house only at the time of flowering or are planted at sunny window sills. After yellowing of the leaves the tubers are collected and kept dry in cold storage. These should be planted in groups in large pots. They are propagated through separating the offsets or tubers. Only two species are recommended for indoor cultivation and these are *G. rothschildiana* and *G. superba*. ***G. rothschildiana*** from tropical Africa grows up to 2 m long, bearing up to 18 cm long, 5 cm wide and broadly ovate-lanceolate leaves. This produces, in succession, 5-8 cm long and 7.5-10.0 cm across solitary red flowers on long stalks with yellow margins, from the axils of the upper leaves, and with age the yellow at margins changes to red, tepals are strongly reflexed and edges are wavy. ***G. superba*** from tropical

Africa and Asia grows up to 1.8 m long, bearing 10-15 cm long leaves. Flowers initially emerge as yellow but change to orange and then red, petals are slightly reflexed, up to 7 cm long and are strongly undulated and crimped at the margins. These require brightly lit spots and a normal room temperature but not less than 15.6 °C during growth period.

Guzmania (Bromeliaceae)

See chapter on '**Bromeliads**'.

Gymnocalycium mihanovichi (Cactaceae)

See chapter on '**Cacti**'.

Gynura (Asteraceae)

See chapter on '**Foliage Plants**'.

Haemanthus (Blood flower, blood lily, football lily; Amaryllidaceae)

Haemanthus takes its name from the Greek *haima* for 'blood' and *anthos* for 'a flower', as most of the species have red flowers. It is a bulbous genus of 50 species from tropical southern Africa, Arabia and Socotra with evergreen or deciduous broadly strap-shaped leaves and a many-flowered umbels of small colourful flowers, all making together a football-like shape. Its floral stalk is stout. When leaves start yellowing, the watering should be highly restricted and when planting the neck of the bulbs should peep out the soil. These can be propagated through offsets at the time of potting, and these are repotted after every 4-5 years. Though it is highly tolerant to shade but in winter the minimum temperature should not go below 10 °C. Recommended species for indoor cultivation are *H. albiflos, H. coccineus, H. katherinae, H. magnificus, H. multiflorus, H. natalensis* and *H. pole-evansii*. ***H. albiflos*** (paintbrush) from South Africa bears dark green, tongue-shaped, fringed with hairs, 20 cm long and 10 cm wide leaves. Scape is stout and about 20 cm long bearing white flowers in 5 cm across umbels with yellow protruding anthers, followed by bright red fruits. ***H. coccineus*** from South Africa bears dark green, broadly strap-shaped, deciduous, 30-60 cm long and 15 cm wide leaves. Red flowers within two scarlet bracts which overtop the umbels, have white anthers tipped with yellow and these appear during autumn on a 30 cm long and stout scape before the emergence of leaves. ***H. katherinae*** (blood flower) from South Africa (Natal and Transvaal) and Zimbabwe produces dark green, oblong and 20-30 cm long leaves. Stout scape is 30-60 cm long and appears during summer bearing bright salmon-red flowers 15-25 cm across with drooping or spreading bracts and bold stamens. ***H. magnificus*** from South Africa (Natal and Transvaal) bears leafy stems up to 60 cm tall with dark green, 6-8, ovate to oblong-ovate, undulate, 30-40 cm long leaves which develop after the flowering, narrowing to a clasping base. The peduncle is stout and about 30 cm long and bears a globose umbel up to 15 cm across with bright scarlet flowers. ***H. multiflorus*** (blood flower) from tropical central Africa bears lanceolate and 30 cm long leaves. Stout scape 30-60 cm tall appears during spring bearing umbels up to 15 cm across with deep red flowers which are markedly stalked. ***H. natalensis*** (*H. puniceus, Scadoxus puniceus*; royal paintbrush) from South Africa (Natal) grows up to 40 cm high and 45 cm spread, bearing shining green, semi-erect and elliptic leaves 30 cm long and 10 cm wide in basal cluster and leaf bases are joined to form a false stem. Flower stem appears during autumn bearing conical umbel containing up to 100 tubular and scarlet flowers, subtended by a whorl of 7-8 red bracts. ***H. pole-evansii*** from mountain forests of Zimbabwe produces fresh green, deciduous, oblanceolate and 30-60 cm long leaves.Starry deep salmon-pink flowers 4.5 cm across are produced from winter to spring on 90-120 cm long and stout floral stalks.

Hamatocactus (Cactaceae)

See chapter on '**Cacti**'.

Hedychium gardnerianum (Kahili ginger; Zingiberaceae)

In Greek, *Hedychium* is from *hedys* meaning 'sweet' and *chion* is for 'snow', referring to the first described species having pure white fragrant flowers. A genus of 50 clump-forming rhizomnatous species from India, SW China, Malaysia and Malagasy, bearing unbranched, erect and reed-like stems which are topped with spikes of flowers which look alike orchids and are often fragrant. The leaves are entire, lanceolate and are borne in two parallel ranks. There are many popular species such as *H. coccineum, H. coranarium, H. densiflorum, H. flavum, H. garderianum, H. greenei,* and *H. spicatum*. ***H. garderianum*** from North India is, in fact, an epiphytic hedychium which can be seed growing on tree trunks on the hills of Sikkim and Darjeeling though it can easily grow on ground as well. This grows up to 2 m tall, bearing mid-green and lanceolate leaves 40 cm long and 13 cm wide. Its spikes may grow up to 50 cm long bearing architecturally arranged numerous densely-packed pale-yellow flowers about 5 cm across with brilliant red stamens, opening from July to September. These can be grown in open situation but indoors these can be grown on window sills and in balconies. These tolerate a minimum winter temperature of 7 °C. These are propagated through division of rhizomes when repotting. Also see the chapter on '**Ornamental Gingers**'.

Heliocereus (Cactaceae)

See chapter on '**Cacti**'.

Heliconia angustifolia [Lobster claws; Heliconiaceae (Musaceae)]

Heliconia derives from Mt. Helicon (Greece), the seat of the mythological Muses. It is a tropical American genus with 80 perennial species, and the genus hangs between *Strelitzia* and *Musa*, requiring climate and cultivation practices as to those of *Alocasia, Anthurium, Calathea, Musa* or *Strelitzia*. They are almost clump-forming with mostly long-stalked paddle-shaped leaves. The recommended kind for indoor cultivation is ***Heliconia angustifolia*** (*H. bicolor*) from South America. Plant is dwarfer and grows only up to 1.2 m tall, bearing erect-peduncled, glabrous, green, 45-75 cm long, and 7.5-15.0 cm wide leaves. Flowers are yellowish-green and 6-10 in each glabrous red bract which appear one by one. These like ample atmospheric humidity, soil moisture, filtered light and temperature not less than 18 °C. Also see chapter on '**Foliage Plants**'.

Heliotropium (Boraginaceae)

In Greek, *helios* stands for 'sun' and *trope* for 'turning', an allusion to the fallacy that flowers turn towards the sun. It is a genus of 250 tender to half-hardy species of annuals, sub-shrubs and shrubs from tropics to warm temperate regions. It is only the evergreen ***H. × hybridum*** (*H. corymbosum × H. peruvianum*) garden race which is used as a pot plant for indoor cultivation, which grows to a height of 30-45 cm in pots though outside in beds more than 60 cm, and with a spread of 30-40 cm, bearing mid- to dark-green, finely puckered and oblong-lanceolate leaves. The corymbose flowers more than 7.5 cm across, fragrant and 'forget-me-not'-like appear usually from May to October, varying in colours in different cultivars as dark violet through lavender to white, but flowering time can easily be manipulated through cultural practices to produce flowers at any time of the year. A few of the outstanding named cultivars are 'Florence Nightingale' (pale-mauve), 'Lemoine's Giant' (a true to type seed strain, growing vigorously with large purple flowers), 'Marguerite' (a true to type seed strain with large heads of dark blue-purple flowers having white eye), 'Marina' (violet-blue), *etc.* These are overwintered at a temperature of 7-10 °C and for early flowering a 16 °C winter temperature is provided. A well-lit corner should be provided for its growing inside. These are propagated either through seeds sown in February or through 7.5-10.0 cm long cuttings taken in July. They also require drastic pruning (half to two-third) in February and March.

Heterocentron elegans (*Schizocentron elegans, Heeria elegans*; Spanish shawl; Melastomataceae)

In Greek, *heteros* is for 'variable or diverse' and *kentron* for 'a spur or point', on account of the dimorphic stamens, the larger ones having bristle-like appendages or spurs. This genus from Mexico and Central America comprises 12-27 species of shrubs, sub-shrubs and perennials, but the species in general cultivation is ***H. elegans*** from Mexico, Guatemala and Honduras which is a sub-shrubby perennial suitable for conservatory borders and as indoor plants but being with trailing stems, therefore is most suitable for hanging baskets. The plant is mat-forming with more than 60 cm long, slightly hairy and reddish stems where base becomes woody when plants are aged. Leaves are usually rich green, smooth, in opposite pairs, pointed-oval, 1.2-2.5 cm long and 1.25 cm wide with 6 mm long leaf stalks. Rounded bright purple-rose 4-petalled flowers 2.5 cm across with spreading petals and a cluster of prominent purple stamens in the centre appear in summer in profusion. These should be kept indoors at a well illuminated place otherwise they will produce nil to a few flowers only. These do well at normal room temperature but in winter the temperature going below 13 °C may prove injurious. These are propagated through 7.5 cm long tip cuttings with 3-4 pairs of leaves taken during autumn.

Hibiscus (Malvaceae)

Hibiscus is the Greek name for mallow, applied by Linnaeus to this closely related genus. This genus comprises up to 300 hardy and tender species of evergreen or deciduous annuals, perennials, shrubs and trees from tropical and subtropical regions. Only three species are generally recommended for indoor cultivation and these are *Hibiscus mutabilis, H. rosa-sinensis* and *H. scizopetalus*. The leaves have 2.5-5.0 cm long leaf stalks, and are dark green, alternate, roughly pointed-oval, 5.0-7.5 cm long, 2.5-3.75 cm wide and tooth-edged. Flowers are widely funnel-shaped, solitary, showy, short-lived and appear from spring to summer from the leaf axils situated at the tips of stems and shoots. Every day these require bright light and normal room temperature, however, for winter rest these should be subjected to 13 °C temperature for 2-3 months. These are propagated in spring or summer through 7.5-10.0 cm long tip- or heel-cuttings. For vigorous growth and floriferousness these should be cut back within 15 cm of the base. ***H. mutabilis*** from China grows in the wild up to 3 m height but in pots it is kept low up to 90 cm only through pruning and other cultural practices. Its leaves are mid-green, a little downy, cordate and angular. The flowers last for one day, may be double or single, solitary or in clusters, 5-cm stalked and appear from the upper leaf axils which on opening are white or pale-pink and change to deep red as the day advances. ***H. rosa-sinensis*** (blacking plant, Chinese hibiscus, Chinese rose, rose mallow or rose of China) from tropical Asia, especially China is an evergreen shrub up to 2 m in a pot. Its leaves are glossy green, broadly ovate, up to 15 cm long and pointed, coarsely toothed and sometimes shallowly lobed. Red flowers 10-15 cm across are borne in the upper leaf axils from June to September in the type species but varieties are available in white, yellow, orange, apricot or pink colours, and even singles or doubles, *vis-a-vis* larger flowers. The stamens in the 5-petalled forms are united in a tubular column up to 5 cm long whereas in many-petalled forms these grow in loose clusters. The flowers are short-lived. ***H. schizopetalus*** (Japanese hibiscus or Japanese lantern) from E. Africa is an evergreen shrub growing up to 3 m tall with slender and little arching branches which need support or best grown against a wall. Its leaves are mid-green, elliptic to ovate, 2-13 cm long and toothed. Pendent pink flower about 6 cm across on about 6 cm long stalks appear where petals are deeply fringed, reflexed and backward curving along with about 7.5 cm long stamen column that superimposes the beauty of the flower.

Hippeastrum hybrida (Amaryllis, hippeastrum; Amaryllidaceae)

Hippeastrum derives from the Greek *equus* 'a horse' and *hippeus* 'rider', as the floral buds within their spathes are formed so to suggest a horse and rider. From Central to South Amerca, this genus comprises of 75 species of bulbous plants, each bulb being about 19 cm in diameter. These produce mid-green, strap-shaped and up to 45 cm long leaves from the neck of the bulbs in early spring or sometimes in late winter before emergence of the floral stalk, and arch over alternately on opposite sides. Hollow but thick floral stalk about 45 cm tall appear from one side of the bulb generally in early spring, bearing 2-4 blooms some 15-18 cm across, each bloom lasting for 2-3 weeks. In case of larger bulbs, even two floral stems appear. In hippeastrum bulbs, the floral bud is already initiated at the time of planting. Hippeastrums need bright light throughout the active growth period and normal room temperature but at flowering time, if possible, these should be provided a temperature not higher than 18 °C. These are propagated through bulbs and smaller bulbs. The recommended species for indoor cultivation are ***Hippeastrum* × *hybrida*** (*H.* × *ackermannii*) evolved through involvement of *H. aulicum*, *H. elegans*, *H. reginae*, *H. reticulatum* and *H. striatum*. These hybrids are vigorous and thick-textured with dark green and up to 60 cm long leaves. The floral stems are robust, 60 cm or more tall and bear 2-4-clustered and large lily-like flowers in shades of white, orange, pink, red and striped. The flowers either open just before or with the young leaves in late winter or spring.

Hoya australis (Porcelain flower, wax plant, wax vine; Asclepiadaceae)

See chapter on '**Succulents Other Than Cacti**'. *H. australis*, *H. bella* and *H. carnosa* are recommended kinds for house growing. These require at least three hours of indirect bright light every day and do well at room temperatures. They are propagated through 7.5-10.0 cm long stem cuttings taken in spring.

Hyacinthus orientalis (Dutch hyacinth; Liliaceae)

Hyancnthus derives from an early pre-Greek language in which *hyakinthos* denoted blue. It is a central and eastern Mediterranean genus now considered to contain only one bulbous species, ***Hyacinthus orientalis*** but they are of three types. The most popular one is Dutch hyacinth which produces single flower spike 10-15 cm long crowded with bell-shaped flowers 2.5-5.0 cm long, on a 5.0-7.5 cm long stalk.Their petals are arching and in shades of white, yellow, pink, red or blue. Second type is the Roman hyacinth which produces 2-3 thinner floral stalks only 15 cm high with fewer and more widely spaced white, pink or blue blooms. Third type is the multiflora or cynthella which per bulb produces several 15 cm long stalks with loosely carried blooms in the basic colour range. Leaves in all the types are basal, variable in length and width and surround the central flower spike(s). The Dutch type for flower production is in the range of 16-19 cm bulb-circumference and the larger ones produce taller spikes so require support in the bowls. Roman and multiflora hybrids do not require any support. The bulbs have a paper thin tunic over them. They are planted before beginning of the winter, only half buried in the moist mixture in the pots, bowls or pans and kept in dark for about six weeks at 10 °C temperature to ensure good root production, and then these are gradually acclimatized to the light with ample watering and then these are kept cool until flower bud emergence. Now these are provided with indirect and direct lights at leaf and bud-producing stages. After the flowering is over, the bulbs are planted outdoors as these can not be used again as house plants.

Hydrangea macrophylla [*H. hortensis*; Common French hydrangea; Hydrangeaceae (Saxifragaceae)]

In Greek, *hydro* is for 'water' and *aggos* 'a jar', referring to the tiny cup-shaped seed capsules. It is a genus of 80 species of deciduous and evergreen flowering shrubs and climbers from northern temperate zone and central and South America. They all have opposite pairs of entire leaves toothed minutely or deeply. The flowers are normal fertile ones and those much larger are sterile ones, all with 4-5 petals, appearing at the end of branches in corymbs or panicles. Only one species, ***H. macrophylla***, a deciduous shrub from Japan is common house plant but it is difficult to carry over from one year to another indoors as it requires constantly cool and aerated conditions in order to bloom. Therefore only potted hydrangeas are brought inside when bud formation has started in early spring, retained inside up to flowering and then planted outdoors. Its leaves are fresh green, broadly ovate to obovate and 10-15 cm long. It produces trusses of white, pink or blue flowers. Flowers produced are in two combinations ~ all sterile or mixed as in the wild type. Hortensias (mop-headed type) are in fact *H. m. macrophylla* cultivars having slmost all sterile flowers, *H. m. serrata* cultivars bear flat corymbs of small fertile flowers encircled by a ring of large-petalled sterile ones and these are known as 'lacecaps'. Hydrangea flowers last for up to eight weeks when kept at or below 21 °C temperature. They are propagated through cuttings from spring to late summer.

Hypocyrta glabra (Clog plant; Gesneriaceae)

In Greek, *hypo* is for 'under' and *kyrtos* for 'pouch-like, arched or curved', referring to the lower side of corolla being pouch-like. This genus from tropical America comprises 10 shrubby, erect, creeping or climbing species, but only two species, *H. glabra*, a small shrubby species and *H. nummularia*, a trailing plant are worth growing indoors. When not in flower, ***H. glabra*** is an attractive quick-growing foliage plant with shining dark green, small, oval-elliptic and succulent leaves but when it produces a large number of tiny goldfish-shaped waxy-orange flowers the beauty is superimposed. Its branches are erect or gracefully arching. ***H. nummularia*** is recorded among the moss on trunks in Costa Rica and Guatemala (Everard and Morley, 1970). It has peculiarly formed corolla with its pouch-like lower side and small mouth reminiscent of certain deep-sea fishes. These are most suitable for planting in mixed arrangements, bottle gardens and hanging baskets. These require moist air so careful frequent water-misting daily is advocated to raise the humidity in the atmosphere. In winter the plants are placed in a cool and sunlit spot and the stems are cut back to encourage flowering canes for next spring. In winter the temperature should not go below 10 °C. These are repotted every two years in spring. These are propagated through stem cuttings taken in spring or summer.

Impatiens walleriana (Busy Lizzie, patience plant, patient lucy, snap weed, sultana, touch me not; Balsaminaceae)

See chapter on '**Impatiens**'.

Isoplexis (Scrophulariaceae)

In Greek, *isos* is for 'equal' and *pleko* for 'to contrive or to plait', due to the equally contrived lower lobes at the mouth of the corolla tube which in *Digitalis* are not equal. It is a genus of four shrubby evergreen species from Canary Islands and Madeira, closely allied to and initially belonged to *Digitalis* having racemes of showy tubular flowers, best for growing in conservatory or brought indoors at the flowering time. They are propagated through seeds or cuttings taken in spring. Only two species are worth trying indoors, and these are *I. canariensis and I. sceptrum.* ***Isoplexis canariensis*** (*Digitalis canariensis*) from Tenerife though naturally grows up to 1.5 m high but in containers much less and sparingly branched, bearing basally clustered, dark glossy green, fairly thick, narrowly ovate to lanceolate, toothed and up to 15 cm long leaves. Columnar terminal spikes some 30 cm long appear in summer bearing bright orange-red to brownish-orange tubular, closely set and drooping flowers about 3 cm long. ***I. sceptrum*** (*Digitalis sceptrum*) from Madeira bears oblong-oval to obovate leaves up to 25 cm long, and terminal spikes some 15 cm long bearing closely set racemes with up to 4 cm long, orange and drooping flowers.

Ixia speciosa (African corn lily, corn lily, grass lily; Iridaceae)

Ixia is the Greek word for bird lime, as these plants have a very sticky sap. The genus comprises about 40 semi-hardy species of cormous plants from South Africa. The leaves are mid-green, narrowly sword-shaped, about 30 cm long, all making the form of a fan and the tall wiry but erect spikes comprise of star-shaped, somewhat crocus-like and 6-petalled dark-centred gaily-coloured flowers which are worth growing in pots for flowering indoors and for cutting. Only two Cape species are in general cultivation. They are propagated through seeds in spring or by separating the corms when dormant. ***I. maculata*** produces linear to lanceolate and 15-30 cm long leaves, and yellow-orange to white and 4 cm wide flowers with a brown-purple eye on a 45 cm high spike. ***I. viridifolia*** produces 45 cm long and linear leaves in tufts of 5-7. Metallic blue-green flowers 4 cm across with narrower petals having purplish black throat appear during sping. ***I. × hybrida*** is the garden hybrid derived by crossing *I. maculata* (orange to white), *I. patens* (pink to light red), *I. speciosa* (crimson) and certain other species and these range in colours from white to yellow, orange, pink, red and purple and in various shapes such as starry, bowl-shaped, *etc.*

Ixora coccinea (Flame of the woods/Indian jasmine; Rubiaceae)

See chapter on '**Ixora**'.

Jacobinia (Acanthaceae)

Jacobinia is probably a personal name. This genus from tropical, subtropical and temperate North America comprises of 300 species of glabrous perennial sub-shrubs and shrubs, bearing entire and ovate to elliptic leaves in opposite pairs, and showy racemes or spikes of red & purple to yellow & white, slender, tubular and 2-lipped (1 lip 2-lobed and the other 3-lobed) solitary to clustered flowers usually borne in profusion in small clusters from the leaf axils often with conspicuous bracts though in few species they are carried in dense terminal cone-like spikes. These are cultivated for their narrow-tubular red, orange or yellow flowers. They make good pot plants and only two species are worth cultivating indoors. ***Jacobinia carnea*** (*Justicia carnea, J. pohliana, J. velutina*; Brazilian plume, king's crown, pink acanthus) from N. South America, especially Brazil is a shrub growing up to 2 m tall with about 1 m of spread though in pots their height and spread are reduced to 60 cm. These bear dark green, hairless to shortly velvety-hairy, oblong to ovate, acutely pointed, up to 25 cm long, 6.25 cm wide and deeply veined leaves in opposite pairs on 5 cm long leafstalks. Pink to rose or purple-pink flowers 5 cm long, supported with closely packed green bracts are produced in dense cone-shaped floral heads 10-15 cm long during summer and autumn. ***J. pauciflora*** (*J. rizzinii, J. floribunda, Libonia floribunda*) from Brazil is a soft shrub growing up to 60 cm high producing soft-downy and free-branching stems. The leaves are mid-green, opposite, produced in unequal pairs with short stalks, are oblong to broadly obovate, 2 cm long and 1.25 cm wide. The nodding scarlet flowers 2.5 cm long, tipped yellow, are produced from autumn to late spring in small axillary clusters from leaf axils. These require bright filtered to direct sunlight for at least four hours a day throughout the year. They do well at normal room temperature from early spring to late autumn though *J. carnea* requires a 13 °C winter temperature for winter rest, and *J. pauciflora* from end of the winter when its flowering stops till new growth begins is also given the same temperature treatment. Cuttings 5-10 cm long taken from the stems or tips in spring are used for its propagation.

Jasminum (Jasmine; Oleaceae)

Arabic name Yasmin has given the generic name *Jasminum*. The genus comprises some 300 species of deciduous to evergreen shrubs and climbers from Asia. Africa and Australia, mostly from tropical regions. These have leaves arranged along the stems in opposite pairs, each having three or more leaflets up to 7.5 cm long. Only three species are recommended for growing indoors and these are *J. mesnyi, J. officinale* and *J. polyanthum*. Out of these the former one is a fast- growing non-fragrant rambling shrub while the latter two are strong-growing climbers which cling to any available support with heavily fragrant blooms. ***Jasminum mesnyi*** (*J. primulinum*; yellow or primrose jasmine) from W. China has 4-angled rambling stems with leaves growing along having three leaflets. It produces solitary, short-tubed, bright yellow non-scented flowers with darker centres and usually more petals per flower as compared to other indoor species. Plants may require slight support of thin sticks. ***J. officinale*** (common white jasmine, summer jasmine, poet's jasmine) from Caucasus to Western China is a squarish-stemmed climber which along bears almost stalkless leaves with 5-7 leaflets. From mid-summer to mid-autumn, strongly fragrant long-tubed white flowers are produced at the end of the stems. Its some forms may have even pale-pink flowers. ***J. polyanthus*** (pink jasmine, Chinese jasmine) from W. China is a multi-branched climber where leaves bear 5-7 leavlets. Its small plants even up to six-month old have tendency to flower indoors. Long-tubed fragrant flowers appear from mid-winter to mid-spring in large clusters from leaf axils near the stem ends. At bud stage these flowers are rosy-pink but open pure white. These all require bright light including the direct sunlight for a few hours per day for proper flowering. Though these require relatively cool conditions for their proper growing, however, will do best when grown at 15.5 °C temperature throughout. However, below 7 °C during winter may prove injurious to the plants. They are propagated through short tip cuttings or heel cuttings taken from the side-shoots during spring or in autumn.

Jatropha podagrica (Euphorbiaceae)

See chapter on '**Succulents Other Than Cacti**'.

Kalanchoe blossfeldiana (Flaming caty, Tom thumb; Crassulaceae)

Apart from this species there are many other species producing brilliantly coloured flowers. Therefore for all these see chapter on '**Succulents Other Than Cacti**'.

Kohleria amabilis (Tree gloxinia; Gesneriaceae)

It was named for Michael Kohler, a 19th century Swiss teacher of natural history. It is a genus of 50 tender and evergreen species of shrubs and perennials from Mexico, central America and northern South America, growing from scaly rhizomes. Underground scaly rhizomes produced by the plant in a year are many and 6 mm thick. There are kohlerias which can flower indoors throughout the year. Each rhizome from the tip produces a thickly hairy stem which bears lanceolate to ovate leaves either in opposite pairs or in whorls of 3-4, which at the edges are either scalloped or toothed. Bell-shaped flowers in shades of either red to pink or lavender or white appear with 5-lobed mouth which is often speckled, and these are carried on pendulous stalks from the upper leaf axils either singly or in small clusters where green calyx is hairy with lobes sometimes curving outward and upward. These during winter are kept barely moist, repotted annually in spring where after discarding the whole plant the scaly rhizomes are separated. They are propagated through seeds, scaly rhizomes and 7.5 cm long tip cuttings in spring. During dormancy these do not require any light otherwise bright sunlight and normal room temperatures up to 27 °C with little day-night differences are best. The plants die at or below 10 °C temperature though rhizomes in the pots can be stored at 7 °C. The kinds suitable for indoor cultivation are *K. eriantha, K. lindeniana* and the var. 'Rongo', the latter being the best one. ***K. eriantha*** (*K. hirsuta*) from Colombia is a shrubby species growing up to 1.2 m tall with red or white hairs on the stems, producing up to 15 cm long rhizomes, bearing 2.5 cm long stalks, mid-green and ovate to elliptic leaves some 7-13 cm long and 6.25 cm wide which are densely red-hairy at the margins and woolly beneath. Orange-red solitary to 3-4-clustered flowers up to 5 cm long appear during summer with about 2 cm wide mouth where lower three lobes are spotted yellow. This has hybrids also with various markings on the lobes. ***K. lindeniana*** bears up to 7.5 cm long rhizomes and slender, reddish and 30 cm tall white-haired stems. The leaves are 3.75 cm long-stalked, cordate, 3.0-7.5 cm long, 2-5 cm wide, upper surface being velvety deep green with prominent silvery pale-green veins and lower surface being hairy and pale-green flushed with red. Solitary or paired white flowers 1.25 cm long and about 3 cm wide at the flared end on 6.25 cm long-stalks appear in late summer through early fall with a yellow throat but having a large spot of dark lavender. '**Rongo**' produces 5 cm long rhizomes and about 30 cm tall stems with white hairs. Leaves are 5-cm-long-stalked, mid-green, hairy, 10 cm long and 5 cm wide. Usually solitary bright magenta flowers 5 cm long and 2.5 cm wide at the mouth with white mouth-veining appear throughout the year on 6.25 cm long stalks. This is probably the best kohleria hybrid for indoor use.

Laelia (Orchidaceae)

Laelia was named after one of the Vestal Virgins. It is native of Mexico to tropical South America. It is a genus of 30 epiphytic orchid species, closely allied to *Cattleya* but separated by number of pollinia as in *Cattleya* it is four though in *Laelia* it is eight. Its pseudobulbs bear 1-2 slightly channelled fleshy leaves, channelling being into a 'V'-shape, with a distinct midrib.Each pseudobulb produces one floral stem from the top which may carry one to several blooms, each lasting for up to six weeks. The flowers are more or less starry with the lip being tubular or trumpet-shaped. These have winter rest period. These like light but not the mid-day direct sunlight. Daytime temperature should be around 16 °C while night time it should be 10 °C throughout the year but never less than 9 °C. These should be propagated through division of the rhizomes when active growth has initiated. At this time the laelia rhizome should have produced eight or more pseudobulbs. The rhizomes are cut having at least four pseudobulbs per division in the frontal region. Recommended laelias for home cultivation are *L. anceps, L. cinnabarina* and *L. purpurata*. ***L. anceps*** from Mexico to Belize produces medium green often flushed with purple, flattened, 4-angled, 7.5-15 cm tall and ovoid-oblong pseudobulbs, clustering or spaced apart on stout creeping rhizomes. Each pseudobulb produces single (sometimes two) oblong-lanceolate leaf 15-23 cm long and 3.75 cm wide. Arching floral stalk 45-90 cm long carries 4-8 winter-flowering blooms, each bloom 10 cm across with tepal colour being lilac-pink while labellum yellow-striped basally in the throat and margins purple. ***L. cinnabarina*** (*Cattleya cinnabarina*) from Brazil produces reddish-green, cylindrical, erect, clustered and 13-25 cm tall pseudobulbs, bearing mostly single but sometimes two leathery, narrowly oblong, often purple-flushed, 30 cm long and 1.25-2.5 cm wide leaves. Floral stalk is wiry, erect to arching, 25-60 cm long, racemose and bears 5-15 winter-blooming flowers. Flowers are 6 cm across with orange-red, narrow and spreading tepals and the lip small, tubular and darker orange. ***L. purpurata*** from Brazil produces medium green, narrowly club-shaped and 30-60 cm tall pseudobulbs, each bearing single oblong leaf 50-75 cm long and 5.0-7.5 cm wide. On short (25-75 cm), thick and upright floral stalk some 3-7 fragrant and white or pale-purple flowers 15-20 cm across with wavy-edged tepals are borne during early to late summer. Sometimes the flowers are flushed and streaked with darker purple, the labellum is yellow with purple streaks inside and the lip is tubular and rich velvety purple. ***Laeliocattleya*** orchid has been developed by crossing *Laelia* and *Cattleya* and their fragrant hybrids such as 'Anna Ingham', 'Derna', 'Dorset Gold', *etc.* are worth growing indoors.

Lantana camara (Common lantana, shrub verbena, yellow sage; Verbenaceae)

Lantana derives from an old Latin name for *Viburnum*, an unrelated genus though with somewhat similar flowers. It is a genus of 150 evergreen shrubs and perennials from tropical America and West Africa. Lantanas are prized for their clusters of small fragrant flowers. Only one species, ***L. camara*** flourishes indoors which in its natural habitat grows up to 1.5 m tall but in pots only up to 40 cm high maintained mainly through pruning starting from very younger age. However, in late winter to early spring the plants should be cut back to within 15 cm of the base. They bear 1.25 cm-stalked, in whorls of three or in opposite pairs, mid-green, elliptic, 7.5 cm long, 3.75 cm wide, toothed and quite rough leaves. They produce round flower heads about 5-cm across with densely packed tubular flowers on about 5-cm long stalks produced from leaf axils from late spring to mid-fall. Flowers open successively starting from outer rows and the flowers darken with age, and if it is yellow, it first changes to orange and finally to red. These have named varieties in shades ranging from white to red via yellow, orange and pink. They require bright light, atleast three hours a day and normal room temperature from early spring to the end of the flowering period but the plants are given a short winter rest at 10 °C. These are propagated through stem cuttings taken in mid-summer or through seeds.

Lapageria rosea [Chilean bellflower, copihue; Philesiaceae (Liliaceae)]

Lapageria was named for Napoleon's Empress, Josephine de la Pagerie, who was a keen gardener. It is a half-hardy evergreen genus of one species found in thickets and woods of central Chile. A semi-woody slender-stemmed climber growing up to 4.5 m long with handsome foliage and uniquely attractive flowers which in Chile is a national flower, locally called as '*copihue*'. For its foliage as well as flowers, its plants are trained up on some support. Its twining stems bear dark green, leathery, ovate-pointed and 8-14 cm long alternate leaves. Its often solitary or sometimes 2-3-clustered, waxy rose-crimson and bell-shaped flowers some 7.5 cm long appear from July to October with six petals, followed by edible yellow-green oblong fruits. Its var. 'Albiflora' produces white flowers. These should either be kept on window sills or in lighted patio. It tolerates a minimum winter temperature up to 7 °C. These are propagated through seeds sown in spring at 16-19 °C temperature. Also these are propagated through layering but it takes longer time.

Lilium (Liliaceae)

Lilium is the form of Latin vernacular name for the Madonna lily (*L. candidum*). It is an usually hardy genus of about 80 bulbous species from northern hemisphere, the bulbs being formed of a number of separate overlapping scales. They can be massed together in a bed or border, placed among the shrubs, on the large rock gardens and planted in the pots, and over all for cutting because of the shapely blooms and erect stems. The *Lilium* worth for growing indoors are *L. auratum* (golden-rayed lily), *L. longiflorum* (Easter lily), *L. pumilum, L. regale* (Regal lily), *L. speciosum* (Japanese lily), 'Citronella', 'Empress of China', 'Fiesta Hybrids', 'Grand Commander', 'Golden Splendor', Mid-Century hybrids ('Brandywine', 'Chinook', 'Cinnabar', 'Connecticut King', 'Destiny', 'Enchantment', 'Paprika', 'Prosperity', 'Sterling Silver', 'Tobasco', *etc.*), 'Mount Everest', 'Royal Gold', *etc.* These for their cultivation require indirect bright light, cool temperature, *i.e.* not above 10 °C night temperature during growth period and in no case less than 1.7 °C and ample relative humidity. They are propagated through bulbils produced on the aerial stems as in a few cases, bulblets produced underground and through bulbs. ***L. auratum*** from Japan grows from 1.5-2.4 m tall but in pots much shorter, is a stem-rooting lily so can be propagated even through stem cuttings. Leaves are green, scattered along the stem, ovate to lanceolate and up to 23 cm long. White, highly fragrant, bowl-shaped and up to 10 flowers 25-30 cm across are borne on each stem in late summer, having widely spreading tepal with red and yellow spots, *vis-à-vis* a golden-yellow band along the centre. ***L. longiflorum*** from Japan grows up to 90 cm tall, bearing glossy green, lanceolate, pointed and up to 18 cm long leaves. Fragrant white flowers 13-18 cm long appear during summer. ***L. pumilum*** (*L. tenuifolium*) from Siberia and China is stem-rooting, growing up to 1.2 m tall and are also propagated through stem cuttings. Turk cap's, bright red and nodding flowers 3.75-5.0 cm across appear in June. ***L. regale*** from China is a hardy stem-rooting species growing up to 1.2 m tall, bearing dark green, linear, scattered along the stem and up to 13 cm long leaves. Funnel-shaped fragrant white flowers up to 13 cm long are loosely set on the stem and appear during July. ***L. speciosum*** (*L. lancifolium*) from Japan, China and Taiwan is a half-hardy stem-rooting species growing up to 1.8 m tall, bearing broadly lanceolate leaves along the stem up to 18 cm long. Fragrant white, bowl-shaped and nodding flowers 7.5-13.0 cm long with recurved and wavy tepals appear during August heavily shaded with crimson.

Liriope (Lily turfs, liriope; Liliaceae)

See chapter on '**Foliage Plants**'.

Lithops (Living stones/pebble plant/stone face; Aizoaceae)

See chapter on '**Succulents Other Than Cacti**'.

Lobivia aurea (Golden lily cactus; Cactaceae)

See chapter on '**Cacti**'.

Lycaste (Cinnamon orchid; Orchidaceae)

In Greek mythology, *Lycaste* is the name for one of the daughters of King Priam of Troy. The genus native to mountains of tropical America and Cuba, comprises 40 deciduous epiphytic species, producing conical to ovoid (egg-shaped) pseudobulbs that are furrowed and wrinkled with age and produce 1-3 dark green, oblong-lanceolate and prominently veined or pleated leaves from the tops and bases, narrowing at the bases and at the top portion but centrally some four times wider. Erect flower stalks appear from the bases of the newer pseudobulbs, each stem bearing long-lasting single fragrant flowers with waxy texture, with or before the emergence of young foliage. The flowers are either flat with outspread or bowl-shaped sepals. In the flowers, three outer tepals (sepals) are broader and three inner tepals (petals) are smaller, and these arch over the small 3-lobed labellum. These require a fairly long winter rest period though a few bloom even in winter. Their light requirement all the year is medium, the daytime temperature closer to 18 °C though during night 13 °C and with adequate humidity. They are propagated by division when repotting. Rhizomes are divided at the joints so that each piece has a growing point and at least two pseudobulbs. Recoomended kinds are *L. aromatica, L. cruenta, L. deppei* and *L. virginalis*. ***L. aromatica*** from Mexico to Honduras produces 5.0-8.5 cm tall and 2.5-5.0 cm wide pseudobulbs with usually three leaves 17.5-45.0 cm long and 7.5 cm across at their widest.Each pseudobulb produces several floral stalks 7.5-15.0 cm long, each with 7.5 cm across flat, waxy and lemony fragrant flowers with reddish-yellow sepals, golden-yellow petals and orange-yellow lips spotted with red and the midlle lobe of the labellum is wavy-edged. ***L. cruenta*** from Mexico, Guatemala and El Salvador produces 8-10 cm long, 5 cm wide, flattened and long-ovoid pseudobulbs, each with three pointed-elliptic leaves 38 cm long and 15 cm wide. One pseudobulb produces about 18 cm long several floral stalks in spring, each with a bowl-shaped flowers up to 7.5 cm across. Sepals are yellow to yellowish-green, petals orange-yellow but basally spotted red and the labellum is orange-yellow but basally blotched red. ***L. deppei*** from Mexico and Guatemala produces long-ovoid, almost flattened 10 cm high and 5 cm wide pseudobulbs with 3-4 leaves that are 23-45 cm long and 10 cm wide. Each pseudobulb produces several 15 cm long floral stalks in spring and autumn, each with a 7.5-11 cm across fragrant and waxy-textured flowers, the sepals are large, outspreading and pale-green with red or purple spots, the petals white to creamy-white and the labellum bright yellow with red markings. ***L. virginalis*** (*L. skinneri*) from Mexico to Honduras produces oblong-ovate, furrowed and 8-13 tall pseudobulbs. Leaves about 67 cm long and 15 cm wide come out from the pseudobulbs. Pure white fragrant flowers but sometimes with pink to a purplish flush, and red-spotted lip appear during autumn and winter.

Mammillaria zeilmanniana (Rose pin cushion; Cactaceae)

See chapter on '**Cacti**'.

Mandevilla (Dipladenia; Apocynaceae)

Mandevilla was named for Henry John Mandeville, British minister in Argentina and the introducer of its first species into cultivation. It is a genus of 100 species of woody climbers from Central and South America. Their leaves are entire in whorls of opposite pairs, and racemes with twisted buds which open into funnel-shaped blooms with five spreading lobes. For proper flowering, these are planted in large pots. They are propagated through seeds sown in spring or by 7.5 cm long tip cuttings taken in summer from the short new lateral growths produced in spring. They prefer filtered light and grow well at normal room temperature but to force these for resting period during winter these are kept at 13 °C. The only species suitable for indoor cultivation is ***M. sanderi*** (*Dipladenia sanderi*) from Guinea which is a woody-stemmed climber with winding stems bearing 1.25 cm long leafstalks, and glossy green, leathery, pointed-oval, up to 5 cm long, 1.25 cm wide leaves growing in opposite pairs, and the rose-pink solitary flowers with orange throat are trumpet-shaped, 5-petalled and some 7.5 cm across. These start flowering at an early age when they have yet not attained even 30 cm height. For their support the plants are staked by pushing the sticks into the soil mixture and are pruned regularly to maintain these in proper shape and growth.

Manettia bicolor (*M. inflata, M. luteo-rubra*; Rubiaceae; Candy corn plant, firecracker plant, firecracker vine)

Manettia commemorates Saverio Manetti (1723-1785), prefect of the Florence Botanic Garden. A genus from central and tropical South America, this comprises 100-130 species of evergreen perennial and woody-stemmed climbers. These have usually simple leaves in opposite pairs and tubular flowers appearing from the upper axils. The species are decorative, flower when young and make a good pot plant. These are propagated through cuttings in spring and summer. All the year these require indirect but bright sunlight. During active growth these are comfortable at normal room temperature but during winter rest period the ideal temperature should be 13-16 °C. ***Manetttia bicolor*** (firecracker vine) from Paraguay and Uruguay is the only house plant species with branched, thin and twining stems some 3 m long, bearing stalkless or up to 2.5 cm stalked, bright green, lanceolate to narrowly ovate & pointed, 3-6 cm long and 2.5 cm wide leaves. Slightly base-inflated, tubular, densely short-hairy scarlet with yellow tips and about 2 cm long flowers appear from early spring to late fall from the leaf axils on 2.5-3.75 cm long stalks with green, erect or spreading, recurved to reflexed sepals which are divided into five arching sections.

Masdevallia (Orchidaceae)

Masdevallia was named for José Masdevall (*d.* 1801), a Spanish botanist. A genus of 375 epiphytic orchid species without pseudobulbs from central and tropical South America and West Indies. The leaves are linear to obovate and fleshy. Flowers solitary but are often carried in abundance. The three outer tepals (sepals) are larger, fused at base to form a tube or cup and often end into a tail-like tips, and the two inner tepals and the labellum are smaller. Under indoor environment these orchids are most suitable provided humid atmosphere is maintained. They are propagated through division at the time of potting or repotting during early autumn or in spring. The recommended kinds are *M. amabilis, M. bella* and *M. coccinea*. ***M. amabilis*** from Peru produces glossy, leathery, narrowly oblanceolate and 13-18 cm long leaves. Floral stalk is erect and 30 cm long only with one flower about 2.5 cm across which appears in winter to spring. In the flower the upper sepal is orange with shading of red and large up to 4 cm long and the lateral sepals are red, all with tails. ***M. bella*** from Colombia produces oblong to lanceolate and 13-18 cm long leaves. Flowers are solitary, pale-yellow with red-brown spots, up to 23 cm long including their long tails, and a white labellum, and these are borne on pendent scapes during winter to spring. ***M. coccinea*** from Colombia is an evergreen epiphytic orchid for a cool greenhouse, growing erect up to 30 cm tall, bearing 15-23 cm long narrowly oval leaves and solitary crimson, scarlet, orange or yellow or rarely even white or magenta flowers 7-9 cm long during spring and summer. In the flower the upper sepal is shorter than two laterals.

Maxillaria (Orchidaceae)

Maxillaria derives from the Latin *maxilla* for 'a jaw' as the flowers appear as insect jaws. This genus comprises about 300 species of mainly epiphytic orchids (a few being terrestrial) from Americas (Florida to Argentina). Those suitable for indoor cultivation are exclusively epiphytic with egg-shaped pseudobulbs that are spread evenly on a creeping or climbing rhizome, or clustered in tufts. The leaves are mid to dark green, leathery, generally strap-shaped, arranged in basal tufts or on the pseudobulbs. Erect floral stalks arise from the base of the newest pseudobulb, each producing several stalks at a time, and each stalk with only one flower. Tepals are usually long, narrow and pointed so flowers become starry and appear insect-like. In a few species the flowers are inconspicuous. It requires filtered sunlight, daytime temperature of 21 °C and night time of 16 °C along with sufficient humidity. They are propagated through division in spring, each rhizome division having a growing point and at least two unwithered pseudobulbs. For indoor cultivation the recommended kinds are *M. picta, M. praestans* and *M. tenuifolia*. ***M. picta*** from Brazil produces ovoid and 5-8 cm tall pseudobulbs clustered or spaced along the rhizomes, each bearing two (rarely 1) glossy dark green and narrowly oblong leaves 23-40 cm long and 1.25 cm wide. Solitary creamy-white flowers 5-6 cm across on a 10-20 cm long stalk, flowers heavily spotted and dotted with maroon, and the white labellum also has maroon markings, are borne in winter and spring which, in fact, is their rest period. Sepals arch inward and the petals are small. ***M. praestans*** produces 2.5 cm long pseudobulbs, each bulb bearing one 15-18 cm long and 2.5-3.75 cm wide strap-shaped leaf. Almost an erect floral stalk 10-12.5 cm tall carries a reddish-brown flower 2.5-3.75 cm across where petals which are small point-upward while the dark red lip is surrounded by much larger sepals though lip encloses the bright yellow column. Tepals are lengthwise marked with yellow streaks. The blooming time may vary but generally it is in the early summer. ***M. tenuifolia*** from central America to Mexico produces ovoid and 2.5-5.0 cm high pseudobulbs set at intervals along the erect and vertically climbing rhizomes, and these pseudobulbs bear linear strap-shaped, dark green and channelled leaves about 35 cm long. Flower stalks 3.75-5.0 cm long appear one each from newer pseudobulbs during spring with narrow, pointed and yellow sepals. The flowers are 3.75-5.0 cm across heavily barred with crimson. Labellum is oblong, dark red at the base and shaded to yellow with red spots.

Medinilla magnifica [Love plant, rose grape; Melastomataceae (Melastomaceae)]

Medinilla was named for Jose de Medinilla y Pineda, governor of the Ladrones (Marianas Islands) in *c.* 1820. This evergreen genus comprises of 150 tropical shrubs and climbers, sometimes epiphytes growing on the forks of tree branches, from tropical Africa, SE Asia and Pacidic Islands. These produce pairs or whorls of undivided leaves and pink or white flowers in panicles or clusters, rarely with showy bracts. The species ***M. magnifica*** from Philippine Islands is most suitable for growing in pots in warm conservatory or indoors but its requirement for humidity is high. It is propagated through cuttings taken in spring and summer. ***Medinilla magnifica*** is a robust shrub growing up to 2.4 m though indoors hardly up to 1.2 m high with similar spread, having woody, and winged or 4-angled stem with numerous branches carrying stalkless, shining rich green, coarse-textured, strongly undulate, leathery, ovate to obovate and pointed-oval in shape, in opposite pairs, up to 30 cm long, 12.5 cm wide leaves and with veins prominently marked. This plant is very difficult to grow even under the best possible controlled conditions. However, this plant becomes bulky and therefore takes a large space. The flower stalks are drooping, up to 45 cm long and appear from the branch-tips during late spring. The flower is composed of a number of pinkish papery bracts, each 5.0-10 cm long, 5.0 cm wide, bracts arranged in 2-3 tiers along the pyramidal-shaped floral panicle, and between each tier hangs a cluster of about 20 cherry-red flowers with yellow stamens and purple anthers, though tip of the stalk contains a large cluster of about 40 blooms. Each flower is short-stalked and has up to 1.25-2.5 cm long and wide flowers. The flowers in 'Rubra' is darker and in 'Superba' larger and darker.

Miltonia (Pansy orchid; Orchidaceae)

Miltonia was named for Viscount Milton (1786-1851), a patron of horticulture. It is an evergreen epiphytic genus of 25 clump-forming rhizomatous orchid species from central and tropical South America, producing greyish-green, shiny & smooth and flattened-ovoid pseudobulbs about 5-10 cm high, 3.75 cm wide and 1.25 cm thick, bearing normally 1-3 pale-green, long, narrow, gracefully arching and almost transluscent leaves at the tips, and 1-10 short-stalked, large, flat, richly coloured and often fragrant flowers 5-10 cm across, resembling pansies that emerge on erect stalks 15-45 cm long and these stalks rise from the base of pseudobulbs. Flower texture may be, especially of the hybrids, velvety. Tepals are oblong so giving overlapping appearance and their colour contrasts with lip colour as in one of the two larger lobes in the upper part there is a contrasting blotch. Flowering is usually during late spring through summer but sometimes there is repeat in the fall. Each bloom lasts 4-5 weeks. They may be given direct morning or late afternoon light only for 2-3 hours daily during winter but for rest of the year only medium light. For their effective cultivation, the daytime tempearature should be around 21 °C and nighttime 18 °C with high humidity. In no case the temperature should rise above 24 °C and fall below 16.7 °C. They are propagated through division of the rhizomes, each segment having at least two pseudobulbs. Apart from growing in the home, these can also be grown in the terrarium, and the species recommended for the purpose are *M. roezlii, M. spectabilis, M. vexillaria* and *M. warscewiczii*. ***M. roezlii*** (*Odontoglossum roezlii*) from Colombia is closely allied to *M. vexillaria* and produces 2.5-5.0 cm high and narrowly ovate pseudobulbs, bearing numerous slender, nerved and narrowly linear-lanceolate leaves 20-30 cm long. Flower stalks about half the length of the leaves, each stalk with 2-3 white, large and flat flowers 7.5-8.25 cm across and having a purple band at the base of the petals, and large and broadly obcordate labellum at its base is more or less marked reddish-brown. Tepals are ovate-oblong and acute. Labellum in the sinus has a tooth and a spur-like horn projecting backward on each side of the column. It flowers twice, in winter and spring. ***M. spectabilis*** from Brazil is a robust plant, producing 7.5-10.0 cm tall pseudobulbs, each with two strap-shaped and up to 15 cm long leaves. Floral stalks are up to 20 cm tall, topped with solitary flowers 7.5-10.0 cm across from July to September. The tepals are white to cream and tinged pink towards the bases. The labellum is bright rose-purple marked with a central blotch, margins pale-rose & wavy and veins deeper. Its form 'Moreliana' bears purple tepals, wine-purple lip and more heavily veined. ***M. vexillaria*** (*Odontoglossum vexillarium*) from western slopes of Andes, Colombia produces 3.75-6.25 cm high pseudobulbs, each with 3-5 strap-shaped and 20-30 cm long leaves. Several floral stalks 45-50 cm tall appear at one time from each pseudobulb, each bearing 3-10 pale-rose-mauve (rarely white), flattened and fragrant flowers 8.75 cm across from May to June. Labellum is large (up to 6.25 cm across) and veined bright yellow. ***M. warscewiczii*** (*M. weltonii, Odontoglossum weltonii, Oncidium fuscatum, O. weltonii*) from Peru produces much flattened and 7.5-12.5 cm long pseudobulbs. The leaves are linear-oblong, obtuse and 12.5-15.0 cm long. Panicles appearing in March are nodding, branched and bear numerous pale-reddish-brown flowers 5.0 cm across with whitish tips. Tepals are cuneate-obovate, waving and crisped and white labellum is oblong, fan-shaped, bifid having a large rose-purple disc where in the centre there is a large brownish-yellow blotch.

Mimosa pudica (Humbleplant/Sensitive plant; Leguminosae)

See chapter on '**Foliage Plants**'.

Musa (Musaceae)

See chapter on '**Ornamental Banana**'.

Myrtus communis (Myrtaceae)

See chapter on '**Foliage Plants**'.

Narcissus tazetta (Amaryllidaceae)

Narcissus was named after the Greek youth of mythology who was turned by the gods into a flower after being entranced by his own reflection in a pool. This genus contains 60 species of bulbous plants from Europe, N. Africa and Asia, mainly SW Europe. The leaves are strap-shaped, linear, filiform or flat to cylindrical, glossy deep green and erect stems bearing solitary or clustered flowers during late winter or early spring where in many cases the clustered flowers are pleasantly scented. Only clustered polyanthus types deriving only from ***Narcissus tazetta*** (Division 8) from Canary Islands, S. Europe. N. Africa, Iran, Kashmir (India), China and Japan growing up to 45 cm tall, bearing strap-shaped leaves are suitable for growing indoors. These produce 2.5-3.75 cm across flowers with white petals and cup-shaped but short lemon-yellow coronas, flowering earliest from January to February. 'Cheerfulness', 'Cragford', 'Geranium', 'Paper White', 'Soleil d'Or', *etc.* from *N. tazetta* group, and large-flowered ones such as, 'Carlton', 'Celebrity', 'Jack Snipe', 'King Alfred', 'Mount Hood', 'Valiant Spark', *etc.* are all suitable for growing indoors but only under temperate conditions which all after once forcing indoors are discarded. These require growing temperature of less than 21 °C. These are propagated through bulbs and bulblets.

Neoregelia (Bromeliaceae)

See chapter on '**Bromeliads**'.

Nerine bowdenii (Amaryllidaceae)

Nerine derives from *Nereis* 'a Greek water nymph', name given so as the first species was brought ashore from a ship stranded in Guernsey. It is a genus of 30 bulbous species from South Africa bearing arching and narrowly strap-shaped leaves and umbels of short-tubed funnel-shaped flowers having six tepals which are narrow and waved. These can be grown in the conservatory but can be brought indoors when to flower during late summer or in autumn. These are propagated through removal of dormant offsets or fresh seeds. The most preferred kinds for indoor cultivation are ***Nerine flexuosa*** which produces about 30 cm long leaves and 60 cm tall floral stalks in autumn bearing up to 10-20 flowered clusters of pink or white wavy petalled bell-shaped flowers with 5-8 cm long tepals; and ***N. sarniensis*** (Guernsey lily) which bears strap-shaped green or glaucous-green leaves up to 30 cm long and 2 cm wide, and floral stems some 25-50 cm long, bearing 4-8 recurved crimson, scarlet or pink flowers with 3-4 cm long tepals, sometimes wavy.

Nerium oleander (Oleander, rose bay; Apocynaceae)

Nerium is the ancient Greek name for oleander. It is a genus of 2-3 species of evergreen shrubs from Mediterranean region across Asia to Japan, growing up to 6 m tall in the wild, bearing short-stalked shining deep green pairs or whorls of 3-4 leathery and narrowly lanceolate leaves 25 cm long and 2.5 cm wide with a prominent central vein, and large showy flowers in terminal cymose clusters. The most widely cultivated species is ***Nerium oleander*** (*N. indicum, N. odorum*) which if grown indoors, its height rarely exceeds 1.8 m in a pot. Its stems are woody, indoor up to 1.8 m tall, erect, smooth, branching into two or more stems almost at an interval of 45 to 60 cm length and bearing whorls of three deep green leaves at almost equidistance around the stems, and 2.5-5.0 cm across single or double and in shades of rose-pink, red, purple, yellow, orange or white flowers in a cluster of six to several at the ends of the branched stems during summer, rains and autumn. Single ones have 5-lobed flowers flared out flat from tubular base while double ones have many petals and resemble a rose. These all require bright light and normal indoor temperature all the year round but these should once in winter should be rested at or less than 21 °C temperature. They are propagated through 7.5-10.0 cm long tip cuttings taken in early summer when the plant is not in flowering.

Nertera (Rubiaceae)

See chapter on '**Foliage Plants**'. Though flowers are inconspicuous in this genus but they are very decorative when the plant is laden fully with pea-size orange-coloured berries.

Notocactus leninghausii (Cactaceae)

See chapter on '**Cacti**'.

Odontoglossum (Crown orchid/tiger orchid; Orchidaceae)

Odontoglossum derives from the Greek *odontos* meaning 'a tooth' and *glossa* for 'a tongue', as the labellum is toothed. A genus of 250 temperate epiphytic orchid species from cool upland forests of Andes with usually clustered and flattened, ovoid or rounded pseudobulbs appearing on short rhizomes, each pseudobub in the side and apex bearing 1-2 (rarely 3) evergreen, mid-green, strap-shaped to lanceolate leaves which have 'V' shape pleating from the midrib. From the bases of the pseudobulbs, arching stalks appear carrying up to 30 long-lasting but short-stalked and flat flowers arranged alternately at upper two-third portion, and in some cases these flowers are fragrant. Tepals may or may not overlap partially. Flowering all the year is continuous. Direct sunlight is beneficial during winter short-days but in summer only bright filtered light should be provided. These do well at 15-16 °C continuous temperatures and a high degree of humidity. They are propagated through division of overcrowded clumps at repotting. Recommended kinds are *O. bictoniense, O. citrosmum* (*O. pendulum*), *O. crispum, O. grande* (tiger, baby or crown orchid), *O. pescatorei* (*O. nobile*) and *O. pulchellum* (lily-of-the-valley orchid). ***O. bictoniense*** from Mexico to Central America produces ovate, compressed and 10-18 cm long pseudobulbs, each with 2-3 elliptic-oblong to lanceolate leaves more than 30 cm long. Generally unbranched and erect floral spike about 75 cm long bears about 12 flowers 3.75 cm across from autumn to early spring. The tepals are yellow-green with heavy brown spots. The labellum is cordate and soft pink or nearly white with yellow markings. It has many forms such as 'Album' (white lip and red-brown tepals), 'Sulphureum' (yellow tepals and white lip), 'Superbum' (chocolate-brown tepals and purple lip), *etc.* ***O. citrosmum*** from Mexico and Guatemala produces smooth, compressed and subrotund pseudobulbs, each with one thick, oblong and obtuse leaf shorter than raceme. Stalk is pendent, 8-12-flowered, flowers 7.5 cm across and white to rose with a violet, reniform and unguiculate labellum, and tepals oblong and obtuse. Flowers appear during May and June. ***O. crispum*** from Colombia bears sturdy, ovate and 2-leaved pseudobulbs about 10 cm high. The leaves are strap-shaped and 23-38 cm long. Floral stem is arching, about 75 cm long and per stem with 20-30 flowers 7.0-12.5 cm across. Though variable but generally white flowers start appearing from late winter to early spring. Wavy petals are broader than sepals. Labellum is spotted red and marked yellow. There are innumerable varieties and forms under this species. ***O. grande*** from Mexico to Guatemala produces ovoid and clustered 2-leaved pseudobulbs 6-10 cm high, leaves are glaucous green, ovate to broadly lanceolate and up to 35 cm long, stalk 30-40 cm long but few flowered (7-8 blooms) and flowers 10-15 cm across appear in autumn and spring, sepals lanceolate though laterals keeled with yellow colouration and banded with deep reddish-brown spots, petals are broader, obtuse, oblong, subundulate and apex yellow, and labellum yellow with rusty-blotched bands and spots and with a large-lobed tubercle on the claw. ***O. pescatorei*** from Colombia produces ovate 2-leaved pseudobulbs about 10 cm high. Leaves are strap-shaped. Stalks 60-90 cm long, diffuse and bears numerous (up to 100) white membranaceous flowers tinged slightly with rose. Flowers 7.5 cm across appear from March to May. Tepals are ovate-oblong and little undulate but petals wider than sepals. Labellum is pandurate, cordate-oblong and with a yellow fimbriate crest and a few rosy spots. ***O. pulchellum*** from Mexico to Costa Rica produces 5-8 cm high oblong and flattened pseudobulbs, each with paired but rarely three and 30-40 cm long and linear leaves. Pendent racemes are moré than 30 cm long, bearing up to 10 upside down fragrant white flowers 2.5-3.0 cm across, opening in autumn and spring. Labellum has a yellow crest.

Oncidium (Dancing lady orchid; Orchidaceae)

Oncidium derives from the Greek *onkos* for 'a tumour', referring to the swellings on the labellum. A native from Florida (USA) to Argentina, *i.e.* American sub-tropics, this genus comprises about 400 evergrenn epiphytic orchid species closely allied to *Miltonia* and *Odontoglossum.* They have showy flowers similar to *Odontoglossum* but in this case the labellum is larger and quite prominent. Among the species, one can find great variation in their structure as some have pseudobulbs with one or two large leaves coming from the apex while others have stems rising from a rhizome and carry large fleshy leaves, and there are some others where leaves are shaped fan-like similar to the small irises. Long erect, arching or pendent and several feet long stalks arise from the base of the pseudobulbs or from the leaf axils and may bear numerous small and showy flowers on short stalk-branches. The flowers also vary in shape, size and colour (brown, red, pink, green or white). They have one thing in common that there are petal-like wings on the column, a bump below the stigma and a large and many-toothed crest at the base of the lip. The flowers may be small but they are bright and showy and generally appear in yellow or brown. They all require a definite resting period, 3-4 weeks either in winter or for winter-flowering species immediately after the flowering is over. Those suitable for home growing have generally egg-shaped pseudobulbs with two large and medium green leaves growing from apex. Except the midday, these like direct sunlight and during short days these will benefit from supplementary lighting. During active growth period, the optimum temperature is 18.3 °C, however, above this and up to 21 °C the humidity level is increased though during rest period it should be 13 °C when humidity is decreased. They are propagated in spring at the time of repotting by dividing the large plants and by separating individual pseudobulbs. They are suitable indoors and to display on bark slabs or in hanging baskets. The recommended kinds are *O. altissimum, O. cheirophorum, O. crispum, O. flexuosum,*

O. ornithorhynchum, O. papilio, O. sphacelatum, O. tigrinum, O. varicosum and *O. wentworthianum*. ***O. altissimum*** from West Indies produces dark green, smooth and flattened pseudobulbs 10 cm high, each bearing one mid-green strap-shaped leaf. In August and September, up to 2.1 m long stalks with branching near the tips appear carrying numerous flowers 3.75 cm across. The pale-yellow, narrow but spreading tepals with wavy margins and large brown blotches are borne out on the spike along with a sulphur-yellow labellum with a broad band across it and around the crest. ***O. cheirophorum*** from Colombia and Costa Rica is a small-growing species with 2.5 cm high pseudobulbs and mid-green lanceolate leaves. The 15 cm long branching spikes appear between October and December, bearing numerous fragrant bright yellow flowers about 2.0 cm across. White-crested labellum is larger. ***O. crispum*** from Brazil produces dark brown, rough, oblong and sulcate pseudobulbs 10-12.5 cm high, topped with a pair of about 15-23 cm long, 5.0 cm wide, leathery and oblong-lanceolate leaves. Arching spike is 30-120 cm long, bearing 20-50 large and shining brown flowers 3.75-7.5 cm across with a few yellow and red marks on the segment bases, appearing at various seasons, mostly during autumn. Sepals are half in length than the petals, the tepals are obovate, obtuse, wavy, recurved and crisped, and middle lobe of the labellum is large, cordate, wavy and crisped while lateral lobes horn-like. ***O. flexuosum*** (dancing doll orchid) from Brazil, Paraguay and Uruguay produces flattened, ovate and 4-7 cm long pseudobulbs. Leaves are paired or solitary, linear-oblong and about 15-20 cm long. Arching and quite slender flower stalks 60-90 cm long appear during late summer and winter, bearing 2-3 cm across bright yellow flowers with brown streaking, especially near the base and yellow labellum is much larger patterned with red spots. ***O. ornithorhynchum*** (dove orchid) from Mexico to Costa Rica *via* Guatemala produces glaucous green, ovoid and clustered pseudobulbs up to 5-13 cm high, bearing strap-shaped to narrowly lanceolate, paired and up to 30 cm long leaves. Pendent or arching flower stalks up to 60 cm long appear during October to December, bearing numerous sweetly scented lilac-pink flowers about 2.5 cm across during autumn and winter. Labellum is darker rose-lilac with yellow crest. In its variety 'Album' the crest is yellow. ***O. papilio*** (butterfly orchid) from West Indies (Venezuela to Brazil and Trinidad) produces rounded pseudobulbs 5 cm high, topped with erect, solitary and elliptic-oblong leaf 15-22 cm long with thick dark green or red-purple marks. The 60-150 cm long inflorescence which is jointed and winged at the upper portion, bears crimson-red but sometimes yellow-barred flowers, with outer tepals quite narrow and upward-pointing, and appear throughout the year in succession. ***O. sphacelatum*** from Mexico to West Indies *via* Honduras produces clustered, ovoid to oblong or spaced along the stout rhizomes. Leaves are 2-3, narrowly strap-shaped and more than 60 cm long. Flowering appears during spring and summer on 1-2 m long drooping stalk, with yellow flowers being 3 cm long spotted with purplish-brown. ***O. tigrinum*** from Mexico produces rounded, clustered and 7-10 cm high pseudobulbs, bearing 2-3 narrowly oblong, blunt-tipped and 30 cm long leaves. Usually erect but loosely branched up to 90 cm tall stalks bearing bright yellow flowers 7 cm long with brown-blotched tepals and yellow labellum, and the flowers appear in autumn and winter. ***O. varicosum*** (golden butterfly orchid) from Brazil produces flattened, oblong to ovoid and 7.5-13 cm high clustered pseudobulbs, each with 2-3 mid-green, lanceolate to strap-shaped and 25 cm long leaves. Nodding and branched floral stalks 90-150 cm long bearing up to 90 fragrant flowers about 3 cm across appear between September and November. Tepals are yellow-green barred with pale red-brown and the reniform lip is large and bright golden-yellow with a dark red blotch in front of the crest. ***O. wentworthianum*** produces a little flattened and 7.5 cm high pseudobulbs, each topped with a pair of strap-shaped leaves about 25 cm long and 13 mm wide. Floral stalks up to 1.2 m long have many small branches, each with many 2.5 cm across flowers and these appear in summer. There are up to 50 blooms per stalk, the blooms being of yellow colour but blotched red-brown and the labellum is lobed at the base.

Pachystachys lutea (Golden hops, golden shrimp plant, lollipop plant; Acanthaceae)

In Greek *pachys* is for 'thick' and *stachys* is for 'a spike', owing to its shape of inflorescence. A tropical American genus of 5-6 evergreen species of perennial shrubs closely related to *Jacobinia*. These have opposite pairs of entire leaves and tubular flowers with quite unequal lobes being borne in twos or threes subtended by a large bract, and many such bracts making a spike-like head or cluster. They are excellent pot plants. They are propagated by cuttings taken in summer. They like bright filtered light throughout as direct full sunlight becomes intolerable but insufficient lighting fail them to flower. They can be grown at normal room temperature but in winter the minimum persisting temperature should not be below 15.6 °C. Only one species, ***P. lutea*** from Peru is suitable shrub with erect stems for indoor cultivation which grows up to 50 cm high in 18-20 cm pot, though outside it can grow up to 1 m tall or even more but then it looses its lower leaves. With age its stems become woody. Its leaves are glossy dark green at the upper surface, in opposite pairs, lanceolate to narrowly ovate and 10-18 cm long with puckered surface and strongly marked veining, *vis-à-vis* slightly undulate margins. On each stem, erect white flowers 4-5 cm long within golden-yellow bracts are being borne from late spring to autumn on spikes of 10-15 cm length. The bracts are erect, cordate and tipped green. Though each bloom lasts only a few days but the bracts retain their colours up to 8-12 weeks and flowering continues up to full summer.

Paphiopedilum (Lady's slipper orchid/slipper orchid/Venus' slipper; Orchidaceae)

Paphiopedilum derives from the Latin *paphos* meaning 'sandal', and also as Paphos is city of Cyprus ~ sacred to Venus. As its lip is pouched fancifully like a helmet or the toe of a slipper or moccasin hence the vernacular name is 'slipper orchids'. The genus with 50-60 epiphytic and terrestrial species (mostly terrestrial) is native to southern Asia from Himalaya east to New Guinea. For a quite long time these were under *Cypripedium*, a genus which is now used only for the frost-hardy terrestrial orchids. They are without pseudobulbs. They have tufted habit with thick, fleshy, pointed evergreen basal leaves with prominent midrib, usually being strap-shaped, which arise directly from a short rhizome. Flowers vary considerably in size, shape and colour and generally appear solitary (sometimes in racemes on long stalks) on a stalk that appears from the centre of a group of leaves. The flowers are fleshy, polished and waxy and each has an erect standard tepal usually in the contrasting colour to other tepals (especially the petals) and labellum, two fused one appearing as one and pointing down, two smaller and narrower laterals usually curving downwards and a pouched slipper-like labellum is formed. The pouch-like labellum hides the lower sepals (tepals) which are joined together. These flower generally from fall to spring and each flower lasts for 8-12 weeks. These should never be exposed to direct sunlight but supplementary fluorescent lighting can be given from late fall to early spring. If humidity is not a limiting factor, these can be grown at ordinary room temperature in between 16-21 °C but plants require regular water misting if temperature reaches 21 °C. They are propagated by dividing those plants which have formed six or more clumps of leaves. The kinds recommended for home cultivation are *P. appletonianum*, *P. barbatum*, *P. bellatulum*, *P.callosum*, *P. fairieanum*, *P. hirsutissimum*, *P. insigne*, *P.* × *maudiae*, *P. rothschildianum*, *P. spiceranum*, *P. sukhakulii*, *P. tonsum* and *P. venustum*. ***P. appletonianum*** from Thailand is mostly erroneously named as *P. siamense* or *P. sublaeve*. Stem and leaves attain up to 45 cm length. Tessellated leaves are blue-green. Stalks about 30 cm long bears solitary flower 7.5 cm across from January to June. Standard is pale-green, petals mauve-tipped and labellum is brown. ***P. barbatum*** from SE Asia produces pale-green and tessellated leaves, spotted dark green. From January to June, about 30 cm long spikes emerge with 1-2 flowers 6.25 cm across. In the flowers the standard (dorsal sepal) and petals are white to pale-green with vertical purple stripes on roundish standard, the petals bear shiny crimson hairy warts along the edges and the labellum is deep purple-brown. ***P. bellatulum*** from Myanmar and Thailand is a tropical species producing dark green with mottling on the upper surface or paler spots and red-purple below, about 15 cm long and narrowly elliptic to tongue-shaped. Flower stalk is small, flowering is in spring, flowers about 6 cm across, colour is white with red-purple spotting, tepals are broad and overlapping and forming a rounded flower, labellum is ovoid and small. ***P. callosum***, a tropical species from Thailand and Cambodia produces only few darker green leaves which are mottled with bright green and are about 30 cm long and 5 cm wide. Flower stalk is 45 cm long which carries 1-2 flowers 7.5-10 cm across during mid-spring, tepals being white with purple and green stripes, down sweeping wings are purplish-green with hairy, black and raised spots along the top edge, and the labellum is glossy brownish-purple. ***P. fairieanum*** from Himalayas is a temperate species bearing pale- green, up to 20 cm long and 2.5 cm wide leaves. Flower stalk is 25-30 cm long, carrying 1-2 flowers 3.75-6.0 cm across with wavy and white tepals that are veined with thin and purple lines and flowers appear during autumn. Petals curve downward but turning upward at the end and their top edges have tiny black hairs, and the labellum is green flushed with reddish-purple. ***P. hirsutissimum*** from Himalayas is a temperate species producing dark green, 30 cm long and 2.5 cm wide leaves with slight mottling of light green. Floral stalk is about 25 cm long with tiny black hairs, bearing solitary flower 12.5 cm across which appears in spring. The tepals are pale-green, the standard is wavy-edged, with purple-brown spots in the centre of the bottom, petals are green, spotted with reddish-brown though tip is pale-pink, and green labellum is tinted brown with darker spots. ***P. insigne***, a temperate species from Himalayas of North India produces stems up to 25-30 cm long, bearing pale to mid-green leaves about 30 cm long, The waxy solitary (rarely in pairs) flowers 12.5 cm across appear between September and February, with standard being oval, light green and patterned with numerous purple-brown longitudinal veins and the lip is yellow-green flushed with brown. ***P.* × *maudiae***, a hybrid between *P. callosum* and *P. lawrenceanum*, introduced in 1900, is the first green orchid. Its leaves are pale-green but sometimes spotted black. The stem is 20-30 cm long which bears a few flowers 10 cm across from January to June. Tepals are white and striped with green and the labellum is large pale-green though sometimes suffused with light brown colouration. ***P. rothschildianum*** from Borneo is a robust growing species some 90 cm high. The leaves are deep green and strap-shaped. Spike is up to 60 cm long bearing up to five flowers 15 cm across during summer. Their sepals are pale-white patterned with brown-purple longitudinal stripes, the petals are yellow-green with a brown-yellow lip, and the staminode is like a large spider's leg about 6.25 cm long, bent and hairy. ***P. spiceranum*** produces 15 cm long and 2.5 cm wide bright green leaves whose underside is purple. Floral stalk is about 20 cm long carrying 1-2 flowers 6.25 cm across in the fall. Standard (dorsal sepal) is white with centrally long stripe and the petals are yellowish-green with a cross-banding of crimson in the centre and the labellum is crimson. ***P. sukhakulii*** from Thailand is a tropical species producing up to 20 cm long green leaves which have dark mottling at the upper surface. Its stalk is about 20-30 cm long, appearing during winter with solitary flowers about 10 cm across, the standard white with regular green lining, and wings are horizontally flat and greenish-purple spotted with brown-purple. ***P. tonsum*** from Sumatra is a tropical tessellated deep green-leaved species bearing 45 cm long stalks in September and October, bearing solitary waxy flowers 7.5 cm across at the top of the stalk. In the flower, the standard sepal and the petals are green-white, but petals have dark brown spots and purple flushing though labellum is brown-green. ***P. venustum*** from Himalayas is a temperate species producing tessellated mid-green leaves about 20 cm long with darker mottling above and blotched-purple beneath. Purple stems are 15-23 cm long, appearing in autumn and spring, bearing solitary 7 cm wide flowers where standard is white with regular green streaks, wings are narrow, wavy, ciliate, greenish, upper half overlaid with purplish lines and the lower with small blackish warts, and the labellum is yellow-green with darker veining and purple flushing.

Parodia sanguiniflora (Prairie fire cactus/Tom thumb cactus; Cactaceae)

See chapter on '**Cacti**'.

Passiflora caerulea (Blue passion flower, common passion flower, passion vine; Passifloraceae)

Passiflora derives from the Latin *passio* for 'passion' and *flos* for 'a flower', from the interpretation of the flower by Jesuit missionaries to South America. It is a genus of about 400 species of climbers chiefly from Americas, but a few from Asia and Australia. The stems are thin and sometimes grow even more than 6 m long producing tendrils, and shining deep green, alternate, 5-9-lobed, about 10 cm across and fan-shaped leaves which are stalked 2.5-3.75 cm long. The flowers are short-stalked, usually solitary, petals and sepals, each being white of equal lengths encircling a wheel-shape collection of fine and purple-base colourful filaments which in the centre are white and blue at the tip, with five prominent golden anthers and three brown stigmas in the centre, flowers some 7.5 cm across, from the upper leaf axils, followed by rounded to ovoid berries, sometimes edible as in case of *P. edulis*. The showy flowers are large, tubular at base but 10-tepalled which spread out flat or overlapped bowl-shaped. ***Passiflora caerulea*** from Central America and western South America is the only species commonly grown as the house plant. It should be kept at a well illuminated place and at a cool place though during winter rest period it should be provided about 10 °C temperature. It is propagated through 7.5-10.0 cm tip cuttings taken during summer.

Pelargonium (Geranium; Geraniaceae)

See chapter on '***Geranium* & *Pelargonium***'.

Pentas lanceolata (*P. carnea*; Egyptian star cluster, Egyptian star flower; Rubiaceae)

In Greek, *pentos* is for 'five', as this genus contains its floral parts in fives, and this is its distinct character from other genera of this family. A genus of perennial shrubs comprising of 35-50 species from Africa, Arabia and Malagasy. ***P. lanceolata*** is the only species grown as house plant. This bushy shrub is soft-wooded which in pots grows up to 45 cm in height producing 20 or more extremely attractive short-stalked small starry flowers per cluster in the shades of white, pink, lavender or magenta on flat heads some 10 cm across from the end of the branches and stems starting from autumn to winter. The stem is highly branched bearing short-stalked leaves in opposite pairs or in whorls of three or more which are bright green, lanceolate, hairy, 7.5-10.0 cm long and up to 2.5 cm wide. The flower base is tube-like some 2.5 cm long but flaring out into five petals some 1.25 cm across. For up to four hours it requires bright light per day and a temperature of 18 °C. During winter,below 10 °C temperature is intolerable. Tip cuttings 5.0-7.5 cm long from vegetative shoots at any time of the year are used for its multiplication.

Pfeiffera (Cactaceae)

See chapter on '**Cacti**'.

Phalaenopsis (Moth orchid; Orchidaceae)

Phalaenopsis derives from the Greek *phalaina* 'a moth' and *opsis* 'like'. The genus with 35-55 evergreen epiphytic orchid species is distributed from India and

Indonesia to the Philippines, New Guinea and northern Australia. These single-stemmed orchids do not produce pseudobulbs but instead have a short rhizome with two rows of alternate, mid-green, large, wide, leathery, thick, limp, oblong-lanceolate to shortly strap-shaped and densely arranged succulent leaves. Most of these have short and thick stems up to 7.5 cm high. Stem among the lower leaves produces some long aerial roots that cling tightly to the sides of the container including baskets. The floral stalks arising from the leaf axils are arching, up to 90 cm long with up to 25 cm long branches. These may be even shorter, erect or pendulous. Stalk bears up to 30 pansy-like flowers resembling a flight of moths and one flower lasting for up to three weeks. These like usually tropical hot and humid climate and in the winter a definite resting period. For its successful growing, a minimum temperature of 20 °C throughout the year should be maintained. These can be propagated through keikis and the shoots emerging from the joints of the branches of floral stalks after flowering is over. For indoor cultivation the recommended species are *P. amabilis, P. leuddemanniana, P. rosea, P. sanderiana, P. schilleriana* and *P. stuartiana*. ***P. amabilis*** from Malaysia produces dark green leaves 30 cm long, 10 cm wide with reddish undersides, which are arranged on 2.5 cm high stems. One or more arching floral stalks 75 cm long, which is sometimes branched, emerges during winter with up to 15-30 white and waxy flowers, each 10 cm across. Petals are broad and spread flat, and the labellum is white, spotted red, throat flushed yellow with large side lobes which at the base are spotted red while the central lobe is spear-shaped and bifurcates into two long backward-curving tendrils. ***P. leuddemanniana*** from Philippines produces 15 cm high stems, and oblong-elliptic leaves up to 30 cm long. Spike is pendulous, 60 cm long and bears a few highly fragrant white flowers 5 cm across in May and June and these are lined with brown and purple while 3-lobed white labellum is flushed rose-purple. ***P. rosea*** (*P. equestris*) from Philippines bears bright green leaves up to 20 cm long. Branched floral stalk is up to 60 cm long with many brightly coloured flowers 3.75 cm across and these appear from spring to autumn. White tepals are flushed with rose, the side lobes of the labellum are light rose-purple with darker lining and the mid-lobe is bright rose-purple but the base is brown. ***P. sanderiana*** from Philippines bears elliptic and up to 30 cm long leaves. Floral stalk is long, arching, sometimes with branched racemes and emerge in spring with pink flowers up to 8 cm across and the labellum is spotted and streaked with red and purple. ***P. schilleriana*** from Philippines bears dark green and ovate-elliptic leaves mottled silver-grey but reddish-purple beneath about 45 cm long and 7.5 cm wide. Arching or erect flower stalks about 90 cm long emerge during spring carrying up to 30 rosy-lilac to white blooms 6.25-8.0 cm across, the two lower sepals with reddish-brown spots, and the labellum is pink with red spots and yellow throat. ***P. stuartiana*** from Philippines bears grey-green, elliptic-oblong, up to 40 cm long, 7.5 cm wide and silver-mottled leaves though underneath mottled with red-purple. Arching to erect and sometimes branched floral stalk up to 90 cm long appears from autumn to spring, carrying up to 20 greenish-white flowers 5 cm across. Tepals are speckled heavily with reddish-purple on the bottom half of the lower sepals, and the yellow labellum is spotted purplish-red and the throat is flushed yellow.

Pittosporum (Australian laurel, houseblooming mock orange; Pittosporaceae)

In Greek *pitta* means 'pitch' and *spermum* is for 'a seed', on account of the sticky coating of the seeds. A tender to half-hardy genus from eastern Asia, Australasia, the Pacific Islands and Africa, this comprises some 150 species of evergreen shrubs and small trees bearing alternate, entire, ovate or obovate to oblong leaves, and solitary to clustered, shortly tubular with spreading or reflexed petals and 5-petalled flowers (fragrant in some species) that appear from spring onwards. The seeds are embedded in a sticky mucilage of the rounded to ovoid capsule. These require atleast three hours of bright sunlight every day. Though these do well at normal room temperature, but during winter it would be better if these are rested at 10 °C temperature. These are propagated in late spring by taking 5.0-7.5 cm long tip cuttings from new growth and from seeds. In spring the plants should be pruned, if required, of their thin, overlong, twiggy growths just above the whorls of leaves to keep them in shape. Only one species, ***Pittosporum tobira*** the much-branched rounded shrub only suitable indoor plant from China and Japan grows hardly up to 1.5 m tall in a large pot and is very slow growing though outside it can grow even more than 6 m high. Its leaves have 6-13 mm long stalks, but they are shining deep green, elliptic to obovate, thick-textured, tip-rounded, pointed towards the side of stalk, 5-10 cm long, 2.5 cm wide with edges slightly rolled beneath and are produced in whorls or in loose rosettes on the woody stems. Its white to pale-yellow and up to 5.0 cm wide flat-headed clusters of flowers about 1.0 cm across are fragrant, tubular and up to 1.3 cm long. Its form 'Variegata' bears leaves marked with white to cream along the edges.

Pleione (Orchidaceae)

Pleione derives from the Greek goddess of that name, mother of the seven Pleiades. It is native to SE Asia, from India to China, closely allied to and once included in *Coelogyne*. It is a genus of 10 temperate deciduous terrestrial or semi-epiphytic orchid species with pseudobulbs, and in the wild these are often found growing on mossy rocks, tree trunks and branches. These are suitable for growing in a cool greenhouse or in an alpine house or for growing as house plants except the *P. bulbocodioides* and its varieties which survive outdoors in sheltered rock gardens. Their pseudobulbs are small, each producing mid-green, 1-2, a little pleated, elliptic to lanceolate and prominently ribbed leaves which appear after the flowering rising from the bases of the short and square pseudobulbs and survive one season only. The flowers often large for the size of the plant have spreading similar tepals and a 3-lobed trumpet-shaped to tubular labellum fringed at the mouth are borne singly (rarely 2) on the stems arising from the pseudobulbs. They are available in shades of white, yellow, pink or mauve often heavily marked with other contrasting colours and are most suitable for growing in shallow containers. The medium is kept dry since the leaves start yellowing till new foliage start appearing in spring. They are propagated by dividing the clumps or by potting individual pseudobulbs separately which have been formed on top of the clump. These require filtered but bright sunlight, however, during short days supplementary fluorescent lighting may be helpful, alongwith a temperature range of 10-16 °C. Important species are *P. bulbocodioides, P. forrestii, P. humilis,* and *P. praecox*. ***P. bulbocodioides*** is native to Tibet to Formosa *via* China and Taiwan. This is the most common species in which *P. formosana* (blush of dawn), *P. delavayi, P. limprichtii, P. pogonoides* and *P. pricei* are now included after the revision. This is suitable for growing in a cold frame, on a sheltered rock garden and in a living room. It produces dull green (sometimes partly blackish-purple) and nearly rounded pseudobulbs nearly 2.5 cm high. Flowering stalks nearly 15 cm long and the leaves about 20 cm long on maturity and 3-6 cm wide appear together from January to May. Pure white to deep mauve-pink flowers 7-10 cm across are borne with white to paler labellum which are fringed, crested and spotted with red, purple or yellow. This has produced many cultivars such as 'Alba' (graceful pure white), 'Blush of Dawn' (*P. formosana;* pale lilac-pink tepals and white labellum tinged with pale-mauve), 'Limprichtii' (tepals magenta-purple, labellum paler with dense red spots and streaks), 'Oriental Splendour' (*P. pricei*; pseudobulbs dark purple and more flask-shaped, flowers pale-violet, labellum white but finely lined with orange), 'Polar Sun' (flowers pure white), *etc*. ***P. forrestii*** from SW China and Myanmar bears pseudobulbs gradually narrowing towards the top, blooms opening after the death of the leaves in summer, flowers are yellow to orange-yellow, and marked with reddish-brown on the labellum. ***P. humilis*** from North India and Nepal is a half-hardy dwarf species with 2.5-3.75 cm high pseudobulbs. Solitary white to pale-mauve flowers about 7.5 cm across appear on a 10 cm long stalk from January to May where lip is frilled, white-keeled and heavily spotted with lines of purple, crimson or yellow-brown spots. ***P. praecox***, a native of Nepal, China and India, is a tender species which produces about 3.75 cm high, deep green and barrel-shaped pseudobulbs spotted with maroon. Its floral stems are 15 cm long, carrying a single 7.5 cm across flower between November and January. Tepals are deep rose-pink, and labellum is pale-pink with a pale-yellow disc and five keels.

Plumbago (Leadworts; Plumbaginaceae)

Plumbago derives from the Latin *plumbum* for 'lead', as one species has ability to cure lead poisoning. There are10-20 evergreen species of annuals and perennials (sub-shrubs, shrubs and climbers) widespread in the warmer regions of the world except Australia. The only species grown indoors is ***P. auriculata*** (formerly *P. capensis*; Cape leadwort) from South Africa. The species produces thin straggly and slender stems growing 4.5 m long but with the help of supports as it is a climber, however, in pot it grows to 1.2-1.8 m long. Its leaves are short-stalked, mid-green, evergreen, alternate, entire, elliptic to obovate or oblong, spathulate, and 5.0-7.5 cm long which tend to curve downward. Terminal clusters of about 20 sky-blue and primrose-like slender-tubed flowers 2.5-4.0 cm long with 5 petals are produced having 3.75 cm long tube from April to November in the temperate regions or in cool places. Each petal bears a thin dark blue line in the centre. Its varity 'Alba' bears white flowers. The species responds well to hard pruning in winter when flowering is over. It is propagated through cuttings of the non-flowering shoots during summer or by seed in spring. It requires maximum light in the home and in winter the temperature range of 7-10 °C.

Primula (Primulaceae)

Primula derives from the *primus* meaing 'first', referring to the early spring primrose. Though they are found worldwise but mainly in the northern hemisphere and particularly Asia. There are some 400 species but four species and two of garden origin are recommended for indoor cultivation and these are *P.× kewensis, P. malacoides, P. obconica, P. sinensis, P. × tommasinii* and *P. vulgaris*. Though they are perennials which like cold or temperate climate for their growing but most are treated as annuals to be enjoyed while in bloom and then disposed off. All the tender ones bloom principally during late winter and early spring as in the plains though there are many others which flower during other seasons especially in temperate climate. They are tufted to clump-forming evergreen or deciduous perennials, bearing oblanceolate or orbicular leaves in basal rosettes and tubular flowers with five lobes on leafless scapes. This scape may be sometimes so small that stalks of the flowers which make the umbels seem to have originated separately from the base of the plant as in case of *P. vulgaris*. The tip-notched petals are usually obovate. These may have pin-eyed (stigma longer than the stamens) or thrum-eyed (stamens longer than the stigma). The hardy ones such as *P. vulgaris* and *P.* × *tommasinii* are potted from late autumn onwards which start flowering in about a month from being brought into warmth (Beckett, 1987). After flowering they are moved to open condition. The tender ones are full-time house plants, they are short-lived perennials though are treated as annuals and so are sown from spring to late summer which flower 9-15 months late, *i.e.* winter and spring. For winter flowering, these require temperate conditions. They are propagated through seeds or by division of hardy species. They require maximum possible sunshine at home and a temperature range of 10-13 °C, however, at 15.6 °C the floral life is shortened. ***P.*× *kewensis*** (*P. floribunda* × *P. verticillata*, originated in Kew), a perennial hybrid can grow to a height of 40 cm, bearing light green, spathulate, wavy-edged and toothed leaves with whole plant including the flowers covered with waxy-white farina. From December to April, 2-5 fragrant yellow flowers some 2.0 cm across appear in whorls one above the other on the 25-30 cm long erect stalks. This is the only yellow-flowered species. Its var. 'Red Gold' is free of farina. ***P. malacoides*** (baby primrose, fairy primrose) from western China is a perennial species (usually grown as an annual) growing up to 45 cm tall, bearing pale-green, ovate and hairy leaves with rounded teeth. From December to April it produces slender stalks bearing whorls of 20-30 star-like pale-lilac-pink through red to white flowers 1.25 cm across. This has several varieties in mixed and separate colours all with fragrant flowers such as 'Fire Chief' (brick-red), 'Jubilee' (cherry-red), 'Lilac Queen' (soft lilac-purple), 'Rose Bouquet' (carmine-red), 'Snow Queen' (white), 'Snow Storm' (double white), *etc.* ***P. obconica*** from northern China is a perennial species though grown as an annual which grows up to 40 cm high, bearing long-stalked, light green, ovate to almost rounded and slightly glandular-hairy, coarse and slightly indented, and 6-10 cm long leaves with wrinkled edges, which may cause painful allergic reactions on sensitive skins among some people. From December to May, clusters of numerous 2.5 cm across and long-lasting lilac-purple to pink flowers appear though its varieties are available in white, pink, red, salmon, lilac or blue-purple colours, rarely with wavy edges, each with a large apple-green central eye on a 15-30 cm tall stalks. Seveal varieties, most of them being giant or large-flowered ones up to 5.0 cm across such as 'Apricot Brandy' ('Appleblossom', apricot but cream at bud stage), 'Caerulea' (purple-blue), 'Fasbender Red' (deep red), 'Giant White' (white), 'Salmon King' (salmon-red), 'Snowstorm' (white), 'Wyaston Wonder' (giant, deep crimson), *etc.* ***P. sinensis*** (Chinese primrose) from China is a perennial species though grown usually as an annual, growing up to 25 cm high with an erect rosette of long-stalked, bright mid-green, 7-13 cm long, broadly ovate to rounded, softly hairy, toothed and lobed leaves. From December to March this bears thick and 15-25 cm long scapes with 2-3 whorls of white, pink, red or purple flowers 2.5-3.75 cm across having a yellow central eye and with notched petals. It has numerous varieties, and most of the present varieties have delicately fringed and undulate petals. Its varieties are 'Dazzler' (vivid orange-scarlet), 'Empress Mixed' (colour range outstanding), 'Royal Blue' (deep purple-blue), 'Stellata' (starry flowers in 2-tiered umbels), *etc.* ***P.* × *tommasinii*** (*P. veris* × coloured forms of *P. vulgaris*; polyanthus) is though basically outdoor growing one but there are varieties for indoor cultivation. It bears obovate, wrinkled, often downy beneath and up to 25 cm long leaves. Large trussed flowers 2.5-3.75 cm across in shades of white, yellow, pink, red, purple and blue appear in umbels on stout scapes up to 20 cm long. Some of the indoor varieties are 'Biedermeier Strain', 'Mother's Day', 'Pacific Strain', *etc.* ***P. vulgaris*** (syn. *P. acaulis*; primrose) from Europe including Great Britain and Asia is a compact species growing up to 15 cm high and spread of about 20-25 cm, bearing a rosette of bright green, oblanceolate to obovate, and corrugated leaves which are usually downy beneath and up to 25 cm long. In March and April, about 2-3 cm across yellow or in shades of pink, red and purple flowers are produced with deep yellow centre. This species with *P. judiae* has produced many of the excellent dwarf and compact pot plants such as 'Colour Magic' (15 cm high, mixed colours), 'Julion Bicolor' (10 cm high, mixed), 'Panda' (10-12 cm high, mixed), *etc.* In this case the varieties when are to flower are brought in and kept at the sunny site inside. These are most suitable for border planting.

Punica granatum nana (Dwarf pomegranate; Punicaceae)

Punica derives from the Latin *malum punicum* meaning 'apple of Carthage'. A native to SE Europe to Himalaya, and Socotra, this is tender genus of two small deciduous tree (*P. granatum*) and shrub (*P. venusta* syn.

P. ignea) species, often with spine-tipped branches. The leaves are entire and opposite, coppery when young but afterwards becoming shining green. The bright orange 5-8 crinkle-petalled flowers appear from tubular and leathery calyces with the same number of lobes as the petals, followed by green but on maturity bright orange fruits though in case of mini 'Nana' the fruits never ripen. They are propagated by seed in spring or summer or by softwood cuttings in summer. Only the miniature form of *P. granatum* 'Nana' is suitable for pot cultivation indoors (Huxley and Gilbert, 1979) or for home window sill (Beckett, 1985; Hessayon, 1987). It forms a bushy and compact growth up to 1.2 m tall though indoors only 90 cm. In spring the unwanted growth is cut back while in summer the plants may be stood outdoors but in winter a cool and frost-free spot should be selected where these become dormant and drop their leaves. It does not bear persisting 4 °C temperature in winter. It can grow at normal room temperature but is shifted to 13 °C in late autumn when it has dropped its most of the leaves. It bears short-stalked, in opposite pairs or in whorls of 3-4, mid-green, leathery, glossy, about 2.5 cm long and 1.25 cm wide, and lanceolate leaves; and from spring to summer solitary to 2 or more bright scarlet (orange-red) and bell-shaped flowers about 2.5 cm long with crinkled petals, which emerge from the purplish-red calyces, followed by 5.0 cm across, rounded, yellow or yellow-orange fruit with tufted calyx attached at the apex. Flowering occurs when it has attained 30 cm height and the flowers appear at the tips of side branches.

Rebutia krainziana (Crown cactus; Cactaceae)

See chapter on '**Cacti**'.

Rechsteineria cardinalis (Cardinal flower; Gesneriaceae)

Though the genus comprises of 75 tender, herbaceous and tuberous-rooted perennials, mostly from tropical South America, grown for their attractive tubular flowers, *vis-a-vis* handsome foliage. Only one species, ***R. cardinalis*** from Brazil is recommended for indoor cultivation. It may grow up to 45 cm tall with a spread of about 30 cm. The whole plant of the species is covered with light purple hairs. Its leaves are shining green and broadly ovate. The bright scarlet flowers with a paler throat and some 5.0 cm long appear usually in flat terminal clusters from June to August though through staggered sowing it can be flowered throughout the year. They are propagated through seeds and by this method these flower within six months of seed sowing. Even their dormant tubers can be started into growth by placing them in moist peat at 21 °C (Hay and Beckett, 1971). When the growth is up to 5.0 cm high, individual tubers are moved to 15 cm pots filled with JIC No. 2 or in a proprietary peat compost. They are grown at a minimum temperature of 16-18 °C. From May to September the plants should be fed with liquid manure fortnightly. These require copious watering during the growing season. Its tubers are stored with their pots in a dry place at a minimum temperature of 12 °C.

Rhipsalidopsis (Cactaceae)

See chapter on '**Cacti**'.

Rhododendron (Ericaceae)

Rhododendron derives from *rhodon* for 'a rose' and *dendron* stands for 'a tree', and in Greek it is the name of pink flowered oleander which was later chosen by Linnaeus as the scientific name for this genus. The genus comprising of about 800 evergreen or deciduous shrub and tree species spread to the temperate regions of the northern hemisphere, especially the Himalayas to China, the tropics in the mountains of SE Asia, from Malaysia to New Guinea and with one species in NE Australia. They have linear to ovate, entire and leathery leaves. Their terminal flowers are tubular, bell- to funnel-shaped with 5-8 lobes and usually in umbellate clusters though sometimes solitary. Half-hardy and tender species, both are suitable for the cool houses in the temperate conditions. Most of the tender species are epiphytes in the wild in the rain forests in SE Asia. When in bloom all the bushy types can be brought indoors for their beautiful blooms. The recommended house plant kinds are *R. obtusum* hybrids and *R. simsii* hybrids. These both grow to a height and spread of up to 50 cm, having leathery, generally oval and 2.5 cm long leaves. The flowers appear at the end of the stems during mid-spring though it is possible to take succession of blooms from early winter to spring by manipulating cultural practices. ***R. obtusum*** (*R. indicum* var. *obtusum, R. edgeworthii, Azalea obtusa*) from the lower slopes of Himalaya to Japan, grows up to 1.2 m in containers though double of this when in the wild, bearing handsome foliage and flowers, having deep glossy green leaves, with upper surface being boldly veined and finely puckered, and lower surface densely buff-felted. Fragrant, waxy-textured, funnel-shaped and white solitary or 2-3 flowers 2.5 cm across are produced which at bud stage are pink-tinted. ***R. simsii*** (*R. indicum* var. *simsii, R. indicum* var. *ignescens, Azalea indica*) from W&S China is a semi-deciduous to almost fully evergreen bushy species growing up to 90 cm in the containers though in the wild up to 1.5 m tall, bearing bristly-haired margins of the leaves which are of two types ~ spring ones elliptic-oblong and up to 5 cm long though summer ones obovate to oblanceolate and 4 cm long. Red and broadly funnel-shaped flowers usually with darker markings are borne some 5 cm across in small clusters of 2-5, each single or double flower 3.75-5.0 cm across where petals are sometimes ruffled. It is the main parent of all the Indian azaleas with colour range of white, pink and purple. Both of these have produced many hybrids which are easily grown indoors. During floral initiation stage these require indirect bright winter light though other times at illuminated but sunless spot. They require cool position where temperature can be maintained 7-10 °C.

Rivina humilis (*R. laevis*; baby pepper, bloodberry, rouge plant; Phytolaccaceae)

Rivina is named for August Quirinus Rivinus (1691-1725), professor of botany at Leipzig. It is a genus of 1-3 bushy perennial species, one of which *i.e.* ***R. humilis*** from southern USA, central America and West Indies is grown as indoor pot plant for its chains of red or yellow berries. It grows erect and bushy about 90 cm tall though less in pots. It is a perennial but grown as an annual from seeds. It bears slender-petioled, alternate, ovate to lanceolate or cordate-ovate and 5-10 cm long leaves with small and caduceus stipules. Inflorescence is racemose and longer than the leaves, axillary or rarely terminal, flowers small, perfect, perianth greenish or pinkish and 4-parted, pea-sized red fruits appear in autumn profusely in chains in the form of red berries somewhat similar to currants and these berries persist through the winter. The plant has its beauty in its fruits.

Rochea (Crassulaceae)

See chapter on '**Succulents Other Than Cacti**'.

Ruellia (Acanthaceae)

Ruellia was named for Jean Ruel (1475-1537), a French physician and herbalist serving king Francois I. A genus of five species of evergreen perennials and sub-shrubs from tropical South America, especially Brazil, out of which only one species, ***Ruellia makoyana*** (monkey or trailing velvet plant) is commonly grown as a house and hanging basket plant which is prized for its winter-blooms and satiny leaves. It is a trailing plant growing up to 45 cm long, bearing violet-tinged deep green opposite pairs of leaves which at upper surface is patterned with white veins, purple beneath, velvety-hairy, narrowly elliptic, up to 8 cm long, 2.5-3.75 cm wide and on 2.5 cm stalk. Bright carmine to purple-red trumpet-shaped flowers 4-5 cm long and about 6.25 cm across up to the flared end appear singly in leaf axils near the growing tips during winter. It likes illuminated place. For this plant minimum tolerable temperature is 13 °C. During summer, it is propagated through 7.5-10.0 cm long tip cuttings having 3-4 pairs of leaves.

Saintpaulia (African violet; Gesneriaceae)

See chapter on '**Saintpaulia**'.

Sanchezia (Acanthaceae)

See chapter on '**Foliage Plants**'.

Schlumbergera (Cactaceae)

See chapter on '**Cacti**'.

Scilla (Bluebell, squills, wood hyacinths; Liliaceae)

Scilla is the Greek vernacular for sea squill (*Urginia maritima*), included in this genus by Linnaeus. It is a genus of 80 species of bulbous plants from tropical and southern Africa, and temperate Europe and Asia. The small South African species are now classified as *Ledebouria*. They have bulbs of about 1.25 cm diameter covered with tunic. These form rich shiny green basal tufts of stalkless linear to strap-shaped leaves 5-23 cm long, 2.0-2.5 cm wide, and produce 3-4 leafless scapes bearing several pendent, brightly coloured and bell- to star-shaped flowers. They are attractive pot plants for indoor display. The hardy European and Asian species require treatment similar to *Crocus*. They are propagated through seeds or through dormant offsets. Hardy bulbs are poor multiplier though bulbs of the tender species produce the offsets profusely. Indoor scillas are hardy as well as tender, and the hardy ones will not flower again if kept indoor for more than one season so they are brought inside when in flower and taken out when flowering is over. Tender types can be held indoors for several years though they have winter rest. Hardy bulbs loose their top growth after the flowering whereas the leaves of tender ones remain decorative throughout the year. The flowers of hardy ones are comparatively more brighter. ***S. adlamii*** from the regions of cool and temperate South Africa is a tender species producing green bulbs with transparent tunic, bearing stolons. The leaves are fleshy, greenish-brown with darker striping, up to 25 cm long, linear-lanceolate and with longitudinal fine brown lines. The floral scapes 15 cm long appear in spring bearing 3 mm long densely packed bell-shaped and purplish-mauve flowers at the end of scape. ***S. ovalifolia*** from South Africa is a tender species producing pale-green bulbs with transparent tunic, bearing only a fewer pale-green leaves than most other scillas and sometimes only 2-3 per bulb with the length of 6.25-7.5 cm where margins are undulated and the upper surface is spotted darker green. Quite small and greenish flowers appear in spring at 5.0-7.5 cm long spike formed on a 10-15 cm long scape. ***S. sibirica*** (Siberian squill) from Turkey, Iran and Caucasus is a hardy species, bearing white bulbs with a deep violet-purple tunic, bright glossy green, narrow strap-shaped, channelled and up to 15 cm long 2-5 leaves together, and deep blue nodding flowers some 2 cm across in early spring, arranged generally in threes at the top of 7.5-13.0 cm long floral stalk. Its white form is 'Alba' and deeper blue form is 'Atrocoerulea'. ***S. tubergeniana*** (*S. mischtschenkoana*) is a hardy species from Iran and Caucasus. It is though similar to *S. sibirica* in many aspects but starts flowering in early spring from the ground itself even before development of the leaves, and escape elongates with the development of the blooms. Its bulbs are yellowish with a transparent tunic. Its pale-blue flowers arranged in threes with some 2 cm wide mouth are starry having dark central vein on each tepal, and are arranged at the top of 7.5-13.0 cm long stalks, some being nodding, some others erect while certain others grow horizontally. Its leaves are pale-green and 10 cm long.

Senecio cruentus (*Cineraria cruentus, C. hybridus*; Asteraceae)

See *Senecio* under the chapter on '**Annuals**'. These, in fact, are hybrids of complex origin, so the plants vary in height, flower colour and shape from variety to variety, flowering from late winter to spring in various shades such as pastel shade of red, blue, mauve or purple surrounding a circle of white and which ultimately in turn surrounds a typical daisy centre though in some forms it is all single colour throughout. The flowers 2.5-7.5 cm across are carried in a dome-shaped cluster of up to 25 cm across. The leaves are cordate to triangular, softly hairy and toothed and the underside of the leaves are mostly tinged purple. For their short stay indoors as flowering plants, these require indirect bright light, a coolest place and high humidity to prolong the flowering period. Plants continue flowering for several weeks.

Siderasis (Commelinaceae)

In Greek, *Siderasis* derives from *sideros* for 'iron', referring to the rust-coloured hairs which cover all but the petals. A genus of one species, ***Siderasis fuscata*** (*Pyrrheimia fuscata, Tradescantia fuscata*) from Brazil, this is propagated by seeds or through division in spring. These never require bright light so only medium light is provided. These plants are extremely sensitive to dry air so these should get ample humidity along with 21-24 °C temperature throughout with only little difference in day/night temperatures. It is a tufted to small clump-forming rosette, herbaceous plant best for terrarium culture. Its leaves are dark green above with a longitudinal band of silvery-white, red-purple flushing beneath, elliptic, 10-20 cm long and 7.5 cm wide which rise from a short underground stem. Entire plant is covered with very fine, short and rust-coloured thickly set hairs. In late summer hairy floral stalks emerge from the middle of the rosette bearing a beautiful 2.5 cm across violet to rosy-purple 3-petalled flowers.

Sinningia (Gesneriaceae)

Sinningia was named for Wilhelm Sinning (1794-1874), a German gardener. A genus of 75 species recorded from Mexico to Argentina which have erect to ascending stems, ovate leaves and 5-lobed tubular flowers arising either solitary or in cymes from the upper leaf axils. Now *Sinningia* is one entity by combining *Corytholoma, Gesneria, Gloxinia* and *Rechsteineria*.

Broadly these can be classified into three categories, *viz.* (i) ***Rechsteineria***, (ii) **florist's gloxinias** or **Fifiana** (the original *Gloxinia speciosa* which has now become *Sinningia speciosa*) and **velvet slipper sinningias**, and (iii) the special group of **miniature species such as *S. pusilla* and its hybrids** with special characteristics.

In ***Rechsteineria*** group (*S. cardinalis, S. leucotricha*), the stems are erect and hairy and produce short-stalked handsomely hairy leaves, clustered tubular flowers from upper leaf axils, where each flower is subtended by 5-lobed calyx along with a narrow tubed corolla which at the mouth is 5-lobed. These flower from seeds in 5-6 months. Short dormancy follows after the flowering where top growth dies down. ***S. cardinalis*** (*Gesneria cardinalis*; cardinal or helmet flower) from Brazil grows with erect stems some 15-25 cm tall, bearing 10-15 cm long ovate-cordate leaves densely covered with short hairs and with deeper green veins. Bright scarlet, tubular, 5 cm long and 2-lipped flowers are produced in summer. ***S. leucotricha*** (*Rechsteineria leucotricha*; Brazilian edelweiss) from Brazil has erect or reclining stems up to 25 cm tall. Broadly ovate to obovate leaves up to 15 cm long are thickly covered with silvery hairs. Three- to five-clustered, tubular and salmon-red flowers 3-4 cm long are borne in summer.

The ***Gloxinia*** group includes the *Sinningia gymnostoma* (*Achimenes gymnostoma*), *S. eumorpha, S. perennis* (*S. maculata*), *S. regina, S. speciosa* (gloxinia of the horticulturist), velvet slipper gloxinias, *etc.* These tender and tuberous-rooted flowering plants are native to central and tropical South America. These are similar to *Achimenes* to which these are allied too, having scaly rhizomes. These all make handsome pot plants, taken indoors when in flowering and when given a night temperature of 18-21 °C these are brought to flowering from seed in five months. Its flowers are bell-shaped, velvety, and foxglobe-like with short rounded lobes at the mouth. In **velvet slipper** types the plants have short stems, velvety leaves and trumpet-shaped flowers from white to red to purple with a spreading 5-lobed calyx and a 5-lobed mouth. These plants have 5-6 month's period of complete dormancy. ***S. gymnostoma*** (botanical gloxinia) from Argentina produces erect stems up to 60 cm tall, where leaves are long-stalked, ovate, toothed, hairy, conspicuously veined and about 8 cm long. Funnel-shaped solitary rose-pink flowers with red spots and about 3.5 cm long appear during summer from the upper leaf axils, having 5-petalled lobes with lower three larger. ***S. eumorpha*** (botanical gloxinia) from Brazil grows up to 20 cm tall with about 25 cm spread. It is a short-stemmed species with glossy to downy bronze-green ovate leaves. Milky-white pendent flowers about 5 cm long with lilac and yellow tinges in the throat are produced freely from May-June to October. ***S. perennis*** from Colombia to Peru produces erect and up to 60 cm long stems. Its leaves are long-stalked, lustrous deep green but sometimes lower surface is red-tinted, broadly ovate, 13-18 cm long and coarsely toothed. Lavender-blue solitary and racemose flowers 2.5-4.0 cm long having deep purple throat, emerge from the axils of the smaller upper leaves in summer, each flower having five corolla lobes with central lower one toothed. ***S. regina*** from Brazil grows up to 23 cm tall with spread of up to 30 cm. Its leaves are short-stemmed, velvety, above bronze-green, below deep red, ovate-elliptic and white-veined. The violet-purple pendent flowers about 5 cm long are produced from May to July. ***S. speciosa*** (gloxinia) from Brazil is the most popular species among the horticulturists, which is almost stemless and grows up to 25 cm tall with a spread of up to 30 cm. Its large leaves are dark green, velvety, fleshy and oblong-ovate. It is a profuse-bloomer with violet or purple flowers some 5-10 cm long which are produced usually from May to August though through successive sowings it can be flowered throughout the year. In cultivation, its large-flowered 'Fifiana' group is most prevalent as florist's gloxinias ranging in colour from shades of red, pink and purple to white. Its some of the varieties are 'Emperor Fredrick' (scarlet with white edges), 'Emperor William' (blue-purple, bordered white), 'Mont Blanc' (white), 'Tiger Red' (deep red, petals crimped), *etc.* Seed strains including the double types are also available.

Miniature sinnigias (species and hybrids) are mostly stemless little plants though a few produce up to 7.5 cm tall stems, with profuse blooming, having rossetted leaves and if the conditions are ideal these continue blooming throughout the year. These are most suitable for a terrarium. The most generally indoor-grown species is ***Sinningia pusilla*** which in most cases is a parent of most of the popular hybrids. Everard and Morley (1970) mention that the diminutive species *S. pusilla* can be grown in a thimble and is probably the smallest member of the Gesneriaceae and which if in groups are rooted on old tree fern trunks look most attractive. Its stems are hardly 1.25 cm high, and 3 mm-stalked leaves are mid-green with darker veins above, pale-green with reddish veins beneath, almost round, scallop-edged, tiny-hairy and 1.25 cm across. Trumpet-shaped solitary and violet or lavender flowers sometimes with darker lines in the white throat are borne on 2.5 cm long stalks with 2 cm long and flared to 1.25 cm wide corolla mouth having its two upper lobes smaller than the three lower ones, and are supported with 3 mm long green calyces. In case of *S. pusilla* and its varieties as these are everblooming, if one flower bearing stem starts to die back, new ones rise from the top of the pea-size tuber. In the process, the plant flowers and sheds the seeds so new plants also continue arising nearby. Its attractive white-flowered form is 'White Sprite'. Some other outstanding varieties are 'Bright Eyes' (flowers light purple with darker lobes), 'Dollbaby' (lilac to light blue flowers suffused with white), 'Little Imp' (flowers lavender, lobes magenta), 'Pink Petite' (pink), 'Wood Nymph' (tiny, flowers reddish-purple, throat with white spots), *etc.*

All of these categories have one thing in common that these produce brightly coloured flowers, their stems rising from fibrous-rooted woody tubers, several stems in most species from one tuber, and roots grow from upper surface of the tubers. Actively growing sinningias should be provided with bright filtered light though it is of no consideration when dormant. Miniatures perform well under 12 h fluorescent lighting when kept 25 cm below. During growth these do well at 18-24 °C provided humidity is not a limiting factor. During dormancy, the tubers should be kept at 7-15.5 °C. Young stem cuttings 2.5-10.0 cm long are used for propagating them during summer in mist chamber where within six weeks tubers and roots are formed and new growths initiate. They are also propagated through leaf cuttings. The miniatures and others can also be propagated through seeds.

Smithiantha (Temple bells; Gesneriaceae)

Smithiantha was named for Matilda Smith (1854-1926), a botanical artist at Kew. A genus of 4-8 tender rhizomatous species of perennials from Mexico whose former name was *Naegelia*. These are normally cultivated in the pots in greenhouses and brought indoors only at the time of flowering when these look very handsome. Its fleshy rhizome is tuber-like, stem is erect and the leaves are velvety and broadly ovate. Tubular and nodding flowers are 5-lobed foxgloves-like which appear on long stalks in summer and autumn. These are propagated at the time of potting from late winter to early spring by separating their rhizomes. Successive fortnightly potting will extend the flowering period. The species are

S. cinnabarina (temple bells) growing up to 60 cm tall, bearing glossy deep green and purple leaves 8-15 cm long and up to 13 cm wide with red-hairy leaf-stalks and red-brown-hairy leaves which give it a velvet sheen. Brick-red to scarlet flowers some 4 cm long often have creamy-yellow on the lower side and in the throat, spotted with pale-red. ***S. gulgida*** is similar to *S. cinnabarina* in many aspects but differs as it has simple green leaves and scarlet flowers where lobes and yellow throat are red-spotted. ***S. multiflora*** grows up to 90 cm tall bearing velvety, dark green above, pale-green below, cordate and up to 15 cm long leaves which are covered with long glandular hairs. It bears creamy-yellow to white pendent flowers borne freely from June to October. The flowers are about 4 cm long with yellow throat but without any mark over them. ***S. zebrina*** grows up to 90 cm tall with spread of up to 45 cm, bearing dark green cordate leaves some 17.5 cm long and wide with brown to purple colouration around the veins and covered with soft silky hairs. Yellow-banded red nodding flowers about 4 cm long appear from June to October with red-spotted yellow throat, and have two upper smaller lobes coloured with orange-yellow while other three yellow. These all the species along with the one orange-red *S. fulgida* have produced numerous small (20-30 cm tall) and free-flowering cultivars with pyramidal-shaped panicles having white, yellow, orange, pink, carmine and scarlet flower colours, most suitable for indoor display. These all grow in medium light provided by shading and above 18 °C temperature with high humidity.

Solanum (Nightshade; Solanaceae)

Solanum is an old Latin name probably for *S. nigrum*, the black nightshade. It is a cosmopolitan genus of about 1,700 species of half-hardy and tender annuals, herbaceous perennials, and climbing evergreen shrubs and subshrubs, bearing dark green and alternate leaves. Only two species are popular house plants, *viz. S. capsicastrum* and *S. pseudocapsicum*. Both are bushy shrubs growing up to 45 cm tall and spread with twiggy branches which carry small dark green leaves and insignificant star-shaped white flowers. The flowers appear in summer, followed by long-lasting highly decorative but non-edible orange to red berries. These are taken indoors when berries have started changing their colours from green to yellow. The yellow colour changes to orange and then finally to orange-red. When these berries after a few months start shrivelling and falling, the plants are discarded though if kept well these will again flower and set berries in the following year if these after falling off the berries are kept out. These require bright light from early autumn to early spring when these are brought indoors which is the berry period. During fruiting period these should be provided 15.5 °C and below temperatures and high humidity. The lowest temperature these can tolerate is 10 °C. These are propagated through seeds sown in early spring. ***S. capsicastrum*** (false Jerusalem cherry, winter cherry) is a half-hardy evergreen subshrub from Brazil growing up to 60 cm tall or more but when grown as an annual it hardly exceeds 30 cm. The leaves are 6 mm long-stalked, densely arranged along the short branches, dark green, obovate to oblong-lanceolate, 4.0-7.5 cm long, 3.75 cm wide, margin undulated, each large leaf with a small one at its base and covered with short branching hairs. White, star-shaped and solitary to paired or in threes flowers appear 1.25 cm across on 2.5 cm long flower-stalks from the leaf axils in June and July, followed by oval, pointed and 1.25-2.0 cm scarlet berries during winter. ***C. pseudocapsicum*** (Jerusalem cherry) from E. South America grows up to 1 m tall, bearing glossy green, in pairs of unequal size, oblong to lanceolate, smooth and 5-10 cm long leaves. White flowers from solitary to threes and 1.5 cm across appear during summer, followed by 1.0-1.5 cm across, globose, red and poisonous berries. There are two of its dwarf forms ~ *S.p.* 'Nanum' and *S.p.* 'Tom Thunb' which hardly exceed 30 cm in height.

Sonerila (Melastomataceae)

Sonerila is the Latinized version of the native Malabar name *soneri-la*. This genus comprises of 100-175 species of evergreen perennials and small shrubs from tropical Asia. Only one species, ***S. margaritacea*** from Java to Myanmar is a low-growing bushy plant widely cultivated for its beautifully patterned leaves and attractive flowers suitable for warm humid conservatory, indoors and in closed containers or terrarium. Its stems are tufted, erect to decumbent, branched and red growing up to 20 cm in height. The leaves are in opposite pairs, rich coppery-green but beneath purple, with silvery-white spots, ovate-lanceolate and 5-10 cm long. Racemose rosy-mauve and 3-petalled flowers 1.2 cm across are protruded on 7.5 cm long panicles from May to September. These require bright filtered light, 18 °C temperature and a very high humidity throughout the year.

Sparmannia (Tiliaceae)

See chapter on '**Foliage Plants**'.

Spathiphyllum (Peace lily/Spathe flower/White flag/White Sails; Araceae)

See chapter on '**Spathiphyllum**'.

Sprekelia formosissima (Aztec lily, Formosa lily, Jacobean lily; Amaryllidaceae)

Sprekelia was named for Johann Heinrich von Sprekelsen (1691-1764), a German lawyer, gardener and an amateur botanist. This is a half-hardy bulb monotypic genus from Mexico which is closely related to *Amaryllis* and with which it has produced an intergeneric hybrid also. It is a handsome lily for temperate climate suitable for indoor cultivation as well as for conservatory. It can be planted on the sunny window sills. When planting the neck should peep out the soil. In winter when leaves start yellowing in October, they are kept dry around 7-10 °C. They are propagated through offsets or seeds. Its leaves are mid-green, strap-shaped, up to 30 cm long and emerge as the flowers die. It produces a 30 cm robust and erect scape usually in April, on which solitary 6-tepalled bright crimson flowers 10 cm long appear in June, where upper one is broader and erect or tip-curving at the back, the two arching laterals are narrower, spreading and tip-curving, and the three bottom falls overlap and form labellum-like structure.

Stanhopea (Orchidaceae)

Stanhopea was named for Philip Henry Stanhope (1781-1855), 4th Earl of Stanhope and president of the Medico-Botanical Society of London. This genus is distributed from Mexico to Brazil with about 25-45 epiphytic orchid species bearing ovoid and clustered pseudobulbs 5-7 cm high, each carrying one pleated leaf 30-60 cm long and 2-6 fragrant flowers usually on pendent stalks. The tepals are reflexed, outer ones (sepals) wider than inner ones (petals), the labellum is round to shoe-shape, terminating into a tongued-tip or two horn-like projections near the base, and the long column arches over the tip of the labellum. They are most suitable for growing on bark slabs and in hanging baskets indoors. They grow best at 16-21 °C temperatures and high humidity. These are rested after completion of the growth, i.e in October-November and then temperature is also lowered. They are propagated by dividing the psudobulbs in July or August. Important species are *S. costaricensis* (*S. graveolens*), *S. devoniensis*, *S. eburnea* (*S. grandiflora*), *S. ecornuta*, *S. oculata*, *S. tigrina* (*S. hernandezii*) and *S. wardii*. ***S. costaricensis*** (*S. graveolens*) from Mexico to Brazil bears ovoid and clustered pseudobulbs about 3.75 cm high. In late summer to early autumn, each pseudobulb produces 12.5 cm long sheathed floral stalks with two flowers up to 12.5 cm across from spring to early summer. The flower 10-13 cm long comprises of buff-yellow, concave and 7.5 cm long sepals patterned with light red ring-spots, narrow and buff-yellow petals with wavy margins and small but bold red spots, and a pair of dark red patches at the base of the yellow-white lip. ***S. devoniensis*** from Mexico produces pale-yellow flowers 10 cm across with red-brown to deep red-purple spots, and the cream labellum blotched and flushed with purple, along with two horns. The flowers appear during summer. ***S. eburnea*** (*S. grandiflora*) from Brazil bears 3.75 cm high pseudobulbs, and the floral stalks which appear from August to October are 15-25 cm long with 1-2 flowers 10 cm across having long and narrow white tepals, and the labellum is narrow and white with a purple base but lacks the usual long curving horns. ***S. ecornuta*** from Central America produces 10-12 cm across flowers in summer bearing creamy-white tepals having purple spots basally, and the yellow labellum is without horns and shade to orange at the base with age. ***S. oculata*** from Mexico bears 5 cm high pseudobulbs, and the inflorescence which appears during July to October is more than 30 cm long with 3-10 fragrant and 10 cm across flowers. Light yellow tepals have red spots, and the labellum is narrow, base orange-yellow with 2-4 large black spots and the tip of the labellum is white with red to purple spots. ***S. tigrina*** (*S. hernandezii*) from Mexico produces in summer the orange-yellow flowers 20 cm across bearing purple-brown blotches and horned labellum. ***S. wardii*** from Mexico to Panama *via* Guatemala and Venezuela bears 5 cm high clustered pseudobulbs, producing slender floral stalks 23-30 cm long at any time from July to November. Stalks bear 6-10 strongly fragrant flowers 7.5-13.0 cm across with tepal colour being golden-yellow but spotted purple, and the deep yellow labellum is horned and patterned with two circular and velvet-purple blotches.

Stapelia (Asclepiadaceae)

See chapter on '**Succulents Other Than Cacti**'.

Stephanotis floribunda (Madagascar jasmine, wax flower; Asclepiadaceae)

In Greek *stephanus* is for 'crown' and *otis* for 'ear', referring to the crown of stamens which have outgrowths looking as ears. It is a genus of 15 evergreen woody climbers from Malagasy, east of southern Asia to Malaysia and Peru. Only one species, ***Stephanotis floribunda*** from Malagasy is a well known climbing house plant growing up to 6 m long so requires canes, wires, strings or trellis to support the long climbing shoots. In pots its height can be managed up to 60 cm but even then it will require a low support for twining. The leaves are glossy dark green, in opposite pairs, undivided, broadly elliptic, tip mucronate, leathery, with prominent central rib, 5-10 cm long, up to 5 cm wide and on 1.25 cm stalks, while fragrant white flowers 3.75 cm long are tubular jasmine-like flaring to five pointed lobes, waxy and borne in axillary cymes from spring to autumn in loose clusters of 10 or more. These require to be properly pruned after the flowering in winter before the new growths start. These are propagated through 7.5-10.0 cm long semi-hardwood tip cuttings from nonflowering shoots, layering and through seeds.

Indirect bright light and 18-21 °C constant temperature with sufficient humidity is required for its growing.

Strelitzia reginae (Bird of paradise flower, crane flower, crane lily; Strelitziaceae)

Strelitzia was named for Charlotte of Mecklenburg-Strelitz (1744-1818), who had been wife of George III of England. It is a genus of 4-5 evergreen perennial species from sub-tropical to southern Africa. These are clump-forming plants having long-stalked large and dark green leaves in a fan-like arrangement, and strangely shaped unique flower heads in the form of a bird's head, bearing strikingly colourful flowers. Also there are tree-like species growing up to 7.5 m tall in the wild, but only ***Strelitzia reginae*** from South Africa is a clump-forming species which can be managed growing in a pot up to 1 m tall instead of up to 2 m in the wild. It forms a 2-ranked fan of large, glaucous mid-green, ovate, leathery and undivided leaves 25-45 cm long & 7.5-15.0 cm wide borne on cylindrical stalks 30-75 cm long. When the plants are about six years old, it throws out a stem about 1 m long which at the top bears 15-20 cm long inflorescence in April and May. The inflorescence bears a purple-flushed green, boat-shaped and beaked horizontally-borne bract from which comes out a succession of long, keeled, orange & blue flowers standing erect to make a crest-like appearance. The flowers are asymmetrical with three narrow orange or yellow sepals and three petals, two blue ones joined together to form a tongue-like structure and the third smaller one is orange. The flowers continue emerging during the course of several weeks. These require filtered light and normal room temperature for successful growing. To give winter rest, these should be subjected to 13 °C temperature. They are propagated by division of the clumps or by separating a section with 2-3 leaves and some roots.

Streptocarpus (Cape primrose/Cape cowslip; Gesneriaceae)

See chapter on '**Streptocarpus**'.

Strobilanthes (Acanthaceae)

See chapter on '**Foliage Plants**'.

Thunbergia alata (Black-eyed Susan vine/Clock vine; Acanthaceae)

See chapter on '**Annuals**'.

Tibouchina (Melastomataceae)

It is a Latinized form of a Guyanan name. It is a genus from central and South America with 350 species of perennial sub-shrubs, shrubs and climbers. These bear undivided opposite leaves and showy flowers usually with five sepals and petals. The only species suitable as house plant is ***Tibouchina urvilleana*** (*T. semidecandra*; glory bush, princess flower, purple glory tree) from Brazil which is a shrub up to 5 m tall though in the pots it is managed up to 1.2 m in height. Under tropical conditions it becomes evergreen and flowers perpetually but under temperate conditions it becomes deciduous with restricted flowering only during summer and autumn. Its 4-angled stems are covered with red hairs and bear rich green, velvety hairy, paired, ovate to oblong or pointed-oval with 3-7 depressed longitudinal veins, toothed, 7-15 cm long and 2.5-3.75 m wide leaves. The attractive saucer-shaped purple flowers 7-10 cm across with cluster of curiously hooked and darker stamens appear at branch tips from midsummer to early winter. These require bright filtered light and normal room temperature for growth but in winter for rest these should be given a temperature of 10 °C. These are propagated through 7.5-10.0 cm long stem or tip cuttings taken during spring.

Tillandsia cyanea (Pink quill; Bromeliaceae)

See chapter on '**Bromeliads**'.

Trichopilia (Corkscrew orchid; Orchidaceae)

Trichopilia in Greek stands for 'hair and cap', as the anther is concealed under a cap surrounded by three tufts of hairs. This epiphytic orchid comprises of 20 species ranging ftrom Mexico to South America.This is very handsome genus grown in pots and most suitable for indoor cultivation. Its short rhizomes are crowded with flattened and elongated pseudobulbs, each with one leaf and a tuft of dry scales at the base. Leaves are solitary, large, erect, fleshy and keeled. Scapes are short, decumbent or nodding and bear abundance of flowers. Tepals are narrow but spreading and twisted, labellum is large & most conspicuous and united with the column below. Its flowers are long-lasting. For its growing it requires an intermediate temperature. Propagation is through division of rhizomes. Important species are *T. crispa, T. fragrans, T. galeottiana, T. marginata, T. nobilis, T. suavis* and *T. tortilis*. ***T. crispa*** from Costa Rica is closely related to *T. marginata*. It produces dark green, ovate, flattened, 5.0-7.5 cm long and 1-leaved pseudobulbs. Leaves are fleshy, acute-pointed, keeled, 15 cm long and 5 cm wide. Drooping but short floral stalks emerge from the base in May or June bearing three flowers 5 cm long, pedicel included. Tepals are brownish-yellow, spreading, twisted & wavy-edged, 6.25 cm long and 1.25 cm wide, and labellum is deep crimson with a white margin, 3.75 cm across, folded over the column but spreading in the front. ***T. fragrans*** (*Pilumna fragrans, T. bachhousiana*) from Colombia bears 7.5-12.5 cm high, flattened and clustered 1-leaved pseudobulbs. Leaves are oblong-lanceolate, acute and 15-20 cm long. Scape some 30 cm long is pendent with six flowers having 7.5 cm long pedicels and appear in summer. Greenish-white tepals are linear-lanceolate, twisted & undulated, spreading and 6.25-7.5 cm long, and a little lobed labellum is white with yellow blotch over throat, folded over the column and spreading in the front. ***T. galeottiana*** from Costa Rica and Mexico bears flattened and narrow pseudobulbs about 12.5 cm long. Leaves are about 15 cm long, oblong and acute. Scapes are short, usually 1-flowered, flowers yellowish-green with sometimes banding below the middle, tepals cuneate-lanceolate, and the labellum is cup-shaped and whitish patterned with purple blotches (dots and streaks) in the centre. ***T. marginata*** (*T. coccinea, T. crispa* var. *marginata*) from Central America bears oblong, compressed and clustered pseudobulbs. Leaves are subauriculate at the base, broadly lanceolate and suddenly acuminate. Scape is generally small and with three large flowers appearing during May and June and these are whitish externally and reddish-purple internally, tepals are linear-lanceolate having white margins with slightly twisted petals, and the labellum is large, wavy, rounded and a cup-shaped 4-lobed blade. ***T. nobilis*** (*Pilumna nobilis, T. candida*) from Venezuela produces large pseudobulbs with broadly oblong-acute leaves. Flowers are white, tepals are little twisted, linear-oblong, acute and 5 cm long, and the labellum is large and white having a yellow spot in the throat. ***T. suavis*** from Central America produces thin and compressed pseudobulbs 5 cm high, bearing 20 cm long and broadly oblong leaves. Stalks are long, arching or pendent and with three creamy large flowers appearing in May and June where lanceolate-acuminate tepals are straight but wavy and about 5 cm long, and the labellum is also cream to white with pale-purple or yellow spots in the throat and the limb is large-lobed, wavy and crenate. ***T. tortilis*** from Mexico produces little curved, compressed and oblong pseudobulbs 5-10 cm high. Leaves are oblong, acute, solitary and 15 cm long. Stalk is decumbent, shorter than the leaves and 1-flowered and flowers about 4.5 cm across being borne in profusion during summer and again in winter. Flowers are brown with yellowish margins, tepal is 5 cm long and spirally twisted, and the labellum where upper expanded portion is 4-lobed, white with crimson spots outside & within entirely crimson, and this forms a tube around the column.

Vallota speciosa (Guernsey or Scarborough lily; Amaryllidaceae)

Vallota was named for Pierre Vallot (1594-1671), a French doctor and botanist. It is a genus of only one bulbous species, the bulbs having brown skins, and these maturing for flowering only when the bulbs have attained 3.75 cm of diameter. ***V. speciosa*** (*V. purpurea*) is closely allied to *Cyrtanthus* and when in flower the whole plant looks like a slender *Hippeastrum*. It is an excellent pot plant for a sunny window or for conservatory. It does not like disturbance of its roots so the plants are repotted when have become pot-bound. Its evergreen leaves are dark green often basally tinged with bronzy-red, basal, broadly linear, 30-50 cm long and 1.2-1.9 cm wide. It flowers from late summer to autumn, being borne in terminal clusters (umbels) from 3-8 on a 30-60 cm tall stems, bearing funnel-shaped and 7-10 cm long bright red tepals with a flower diameter of 7.5-10.0 cm. Its form 'Alba' is white and 'Delicata' has pale-salmon-pink flowers. These should be given some direct sunshine. At room temperature these grow well and flower but during rest in winter these are provided with 10-13 °C temperature for a better performance the next season. The bulbs in the clumps are separated and are planted for further propagation though it can also be multiplied through seeds, albeit the process is slow.

Vanda (Orchidaceae)

Vanda is the Hindi Indian name for this orchid. This tropical orchid is related to *Aerides* and *Rhyncostylis* genera. *Vanda coerulea* has been recorded growing above 1,500 m height though *V. teres* and *V. tesellata* up to 700 m. This sun-loving monopodial evergreen epiphytic orchid, in habit being similar to *Arachnis*, has highly attractive large flowers in all the species, has some 70 species spread in East India and the Malay Islands with outlying species in China and New Guinea. The species may be short-stemmed and erect (compact) or tall and branched (sometimes climbing to some height) and without pseudobulbs. The inflorescence is erect or pendulous and arises from the leaf axils. Based on leaf shape, it may be terete (cylindrical-leaved), semi-terete and strap-leaved. There are two types: (i) with strap-shaped leaves, and (ii) with cylindrical leaves. The species for indoor cultivation are ***V. cristata*** from Himalaya producing up to 60 cm tall stems carrying nearly opposite, strap-shaped, arching, blunt, deeply channelled, 12.5-18.0 cm long and 1.2-2.0 cm wide leaves which are slightly toothed. Floral stalks 10-15 cm long appear from the leaf axils during spring and summer, bearing up to seven flowers 5 cm across. Tepals are yellowish-green to creamy-yellow, and the short and oblong labellum is green-and-yellow, basally with deep purple-red linings. ***V. sanderiana*** from Philippines produces up to 60 cm tall stems with closely packed opposite pairs of strap-shaped lightly arching leaves 30-40 cm long and 2.5 cm wide. Stalks 25-30 cm long emerge out from the leaf axils during late summer and autumn bearing up to 10 flat and disclike flowers 12.5 cm across.

Upper tepals are rosy-pink suffused with white, lower sepals are reddish-yellow with darker markings, and the quite small and round labellum is reddish-yellow. ***V. teres*** from Thailand to Myanmar and East Himalaya throws out branched stems up to 4 m tall in the wild though under cultivation up to 2 m, so requires support. Leaves are dark green, 10-20 cm long, 1.25 cm thick, alternate and nearly cylindrical. In cultivation the flower stem up to 30 cm long arise from points opposite the leaves, appearing from June to August, bearing 3-6 flowers 7.5-10.0 cm across with white sepals and rose-magenta petals, tepals being broadly diamond-shaped and with wavy margins, and in the labellum the side lobes are yellow-orange marked with red while the mid-lobe is deep red-purple. ***V.t. suavis*** is the form of *V. tricolor* which is most commonly grown. Its sweetly scented flowers are large, more freely borne and are white with less heavy magenta to purplish spotting than the type species. Vandas prefer bright but indirect sunlight, and the supplementary lighting during winter shortdays will benefit them. A difference of 8 °C day-night temperature is beneficial. They are propagated by removing the plant tops and through shoots from the cut back plant as cuttings when they have produced aerial roots.

Veltheimia (Forest lily; Liliaceae)

Veltheimia was named for August Ferdinand von Veltheim (1741-1801), a German patron of botany. A genus of two bulbous plant species from South Africa, the plants have a very attractive basal rosettes of seasonal, bright green, fleshy and oblong-lanceolate leaves and thickly packed racemes of tubular flowers on stout erect scapes similar to those of *Kniphofia*. They are most suitable for pot planting. They may be either evergreen (*V. viridiflora*) requiring moist condition throughout the year or deciduous (*V. capensis*) requiring a resting period in summer once the leaves begin to yellow. They are propagated through offsets when repotting or through leaf cuttings taken when these attain full size. These prefer direct sunlight of 3-4 hours every day even during rest period so these are required to be kept at a sunny situation of the house or when swelling of the scape starts these can be taken in and after the flowering these may be taken out in the open. It would be better if there remains a difference of 8 °C between day and night temperatures. ***V. bracteata*** (*V. undulata, V. viridiflora*) has broadly ovoid to rounded bulbs. The leaves up to 12 per bulb are evergreen, glossy, broadly strap-shaped, up to 40 cm long and 10 cm wide with wavy margins. Pinkish-purple densely clustered flowers, faintly speckled yellow, and about 3-4 cm long appear in spring on a glaucous green and purplish-mottled stem growing 30-60 cm tall. Its form 'Rose-alba' developed in Holland bears cream and rose flowers. ***V. capensis*** (*V. glauca*) bears narrowly ovoid bulbs. The leaves are deciduous, glaucous blue-green, lanceolate, up to 30 cm long and 2.5-5.0 cm wide with strongly wavy margins as compared to *V. bracteata*. Pale-pink pendent flowers tipped green and 2.5 cm long are borne on 30 cm long and stout terminal stalk which is spotted purple.

Vriesea splendens (Flaming sword; Bromeliaceae)

See chapter on '**Bromeliads**'.

Zantedeschia rehmannii (Calla lily/pink arum/pink calla/trumpet lily; Araceae)

Zantedeschia (formerly known as *Richardia*) was named for Francesco Zantedeschi (1797-1873), an Italian botanist. It is a genus of 8-9 tropical to southern African species of fleshy rhizomatous perennials with no stem, bearing long-stalked, large and arrow-shaped to lanceolate leaves. The flowers are small, petalless and typical arum-like, arranged thickly on a thick and cylindrical spadix which is surrounded by a large colourful spathe. They are good pot plants to be kept indoors as like filtered situation but require dry rest in summer when leaves start yellowing after flowering, however, *Z. aethiopica* does not require any rest. Their requirement of light is ample when in leafy condition. From early autumn onwards when these start into growth, they are kept cool at a temperature of 10-13 °C for about three months but thereafter they are kept at 15.6 °C for *Z. aethiopica* and 18 °C for other species and varieties until these have started flowering when they are given room temperature but never above 21 °C. They are propagated through detachment of the offsets or by dividing the rhizomes at the time of repotting. The species suitable for indoor cultivation are *Z. aethiopica* (arum lily), *Z. albomaculata* (spotted calla), *Z. elliottiana* (golden calla) and *Z. rehmannii* (pink calla). ***Z. aethiopica***, a deciduous species from South Africa grows up to 90 cm tall, bearing mid- to deep green, little glossy and arrow-shaped leaves up to 45 cm long and 25 cm wide which stand on a leaf stalk up to 90 cm long. From late winter to early spring, any time the spadix (floral stalk) appears bearing golden-yellow spadix surrounded by a milky-white cordate and pointed spathe 12.5-25.0 cm long with slightly outward curving edges. ***Z. albomaculata*** (*Z. melanoleuca, Richardia albomaculata*) from South Africa to Zambia grows up to 60 cm tall, bearing 20-45 cm long on a stalk of up to 90 cm long, 5.0-7.5 cm wide at base, narrowly triangular and shining dark green leaves with silvery-white spots. Its more trumpet-shaped creamy-white to greenish-yellow tubiform spathes 10.0-12.5 cm long are patterned with a dark red-purple blotch at the base inside while the spadix is white. ***Z. elliottiana*** from Transvaal grows up to 90 cm with a spread of up to 60 cm and is a deciduous species bearing up to 28 cm long and 23 cm wide with heavily laid white-spots on elongated heart-shaped mid-green leaves. The floral stalk may be up to 60 cm tall with 15 cm long and showy greenish-yellow spathe, bright yellow inside in the form of an open trumpet around the yellow spadix. ***Z. rehmannii*** from Tansvaal growing up to 60 cm high with a spread of about 30 cm, is a deciduous species, bearing mid-green lanceolate leaves up to 20 cm long, which sometimes is spotted with silvery-white flecks. The pale-pink to wine-red and 7.0-13.0 cm long spathe appear from April to June.

Zephyranthes candida (Fairy lily/flowers of the western wind/rain lily/swamp lily/zephyr lily/zephyr flower; Amaryllidaceae)

Zephyranthes derives from the Greek *zephryos* meaning 'the west wind' and *anthos* for 'a flower', in reference to their origins in the western hemisphere. This is a bulbous genus from warmer parts of Americas, with about 40 species all having evergreen and basal leaves. These are clump-forming plants with an erect crocus-like flower on each stem where the six tepals open widely. They are good pot plants for indoor display or for conservatory and can be flowered easily at an unlighted corner which is not dark. These are not repotted until quite necessary as the congested clumps flower more freely. Their pots are kept dry during winter. Either they are propagated through their black seeds which they produce amply or by dividing the clumps at repotting. The species most suitable for indoor cultivation are *Z. candida, Z. citrina* and *Z. grandiflora*. ***Z. candida*** from Uruguay bears rush-like slightly fleshy leaves 20-30 cm long. White though rarely pinkish flowers but rarely with greenish shade near the base and 3-5 cm long open widely during summer and autumn on 10-20 cm long stalks. ***Z. citrina*** from South America bears narrowly linear, channelled, blunt and 30 cm long leaves. Bright yellow-gold but basally green flowers 4.0-4.5 cm long are carried on 15-25 cm long stalks in summer and autumn. ***Z. grandiflora*** from Mexico, Guatemala and West Indies bears narrowly linear leaves some 25-40 cm long. Pink to rosy-red with 2 cm long petals, the flowers 6-10 cm long appear during late summer and autumn on 20 cm tall stalks.

Zygopetalum (Orchidaceae)

Zygopetalum derives from the Greek *zycos* meaning 'a yoke' and *petalon* for 'a petal', a swelling at the base of the labellum seems to join or yoke the petals together. The evergreen genus from central and tropical South America (Brazil and Guianas) comprises 20 terrestrial and epiphytic orchid species, producing clump-forming ovoid pseudobulbs, each bearing two or more mid to deep green, distichous, narrow lanceolate and terminal leaves sheathing a short stem, and often racemes (sometimes even solitary) of large blooms which are often very attractive. Usually purple-spotted green flowers bear similar five radiating tepals where sepals and petals are often united to each other at the base, the sepals forming a short chin with the foot of the column and a large fan-shaped labellum of contrasting pattern. After flowering is over, these are propagated by division in spring. The ones worth trying indoors are *Z. crinitum, Z. intermedium, Z. mackayi* (*Z. mackaii*), *Z. maxillare* and *Z. perrenoudii*, all from Brazil. ***Z. crinitum*** is similar in habit to *Z. intermedium*. Its leaves are broadly linear-lanceolate. Floral scape is stout and long which bears green with sparsely brown-blotched tepals that are 5.0 cm long and oblong-lanceolate. Labellum is white with purple veins which radiate from the thick crest, scarsely emarginate, 5 cm across, spreading & wavy, and disc hairy. For flowering, it is not particular about the time. *Z. mackaii* var. *crinitum* has comparatively fewer brown blotches on the tepals than *Z. intermedium*. The veins are dark blue in the var. *caeruleum*. This species has several varieties with vein colour of the labellum being pink, blue or almost colourless. ***Z. intermedium*** produces bright green pseudobulbs about 9 cm tall and with a basal sheath. The leaves are bright green, lanceolate-elliptic to nearly strap-shaped, 3-5-clustered or fan-shaped, arching and 30-45 cm. Fragrant, waxy, long-lasting, 8 cm across with almost similar size of tepals, each tepal being usually yellow-green but blotched red-brown while labellum is white patterned with branched purple-red radiating veins where margins are wavy or crimped, are borne during autumn and winter. ***Z. mackayi*** is the most popular evergreen epiphytic orchid species in the genus, growing up to 30 cm in height with broad, ovoid and 5.0-7.5 cm high pseudobulbs.Its leaves are ribbed, narrowly oval and about 30 cm long. During autumn (November to February) on a 30-60 cm long flowering stalk, the plants produce fragrant and purple-brown-blotched yellow-green flowers 5-8 cm across with spreading tepals where lips are white veined with violet-purple and the crest is ridged. ***Z. maxillare*** produces 5 cm high pseudobulbs with 30 cm long lanceolate leaves. Floral stalk (scape) is 20-25 cm long with 6-8 green but small flowers 1.25 cm across which are borne out during winter. Tepals are ovate-oblong and acute with brown transverse bands. Purple labellum is horizontal with a quite large glossy purple notched horse-shoe-shaped crest where central lobe is waved and roundish. ***Z. perrenoudii*** is an evergreen epiphytic orchid for a cool house. This grows up to 30 cm tall, bearing ribbed, 30 cm long and narrowly oval leaves. The spikes bear violet-purple-lipped fragrant dark brown flowers 8 cm across in winter.

References

Anderson, M. 1998. *The World Encyclopedia of Cacti & Succulents*, 256 p. Anness Publishing Ltd., Hermes House, London, Great Britain.

Bailey, L.H. 1929. *The Standard Cyclopedia of Horticulture* (3 vols), 3639 p. The Macmillan Co., New York, USA.

Beckett, K.A. 1985. *The Concise Encyclopedia of Garden Plants*, 438 p. Orbis Publishing Ltd., London, Great Britain.

Beckett, K.A. 1987. *The RHS Encyclopedia of House Plants Including Greenhouse Plants*, 492 p. Salem House Publishers Massachusetts, USA.

Brickell, C. 1994. *Gardeners' Encyclopedia of Plants and Flowers*, 640 p. Dorling Kindersley Ltd., London, Great Britain.

Boyd, L. 1994. *Successful Gardening A – Z of Perennials*, 176 p. The Readers' Digest Association Inc. of Pleasantville, New York, USA.

Everard, B. and B.D. Morley, 1970. *Wild Flowers of the World*, plate 185. Octopus Books Ltd., London, U.K.

Hager, H. 1957. Control of flowering in *Cattleya*. *Proc. of the 2nd World Orchid Conf.*, pp. 130-132., Honolulu Orchid Society, Honolulu, Hawaii.

Hay, R. and K.A. Beckett, 1971. *Reader's Digest Encyclopaedia of Garden Plants and Flowers*, 798 p. The Reader's Digest Association Ltd., London, Great Britain.

Hessayon, D.G. 1987. *The Gold Plated House Plant Expert*, 256 p. Century Hutchinson 'Ltd.,London, Great Britain.

Huxley, A. and R. Gilbert, 1979. *Reader's Digest Success with House Plants*, 480 p. The Reader's Digest Association, Inc., Pleasantville, New York, USA.

Pizzetti, I. and H. Cocker, 1975. *Flowers A Guide for Your Garden* (two vols), 1384 p. Harry N. Abrams, Inc., Publishers, New York, USA.

Rosenfeld, R. and G. Strong, 1998. *A Gardener's Guide to Perennials*, 112 p. Mererhurst Limited, London, Great Britain.

20

Foliage Plants

Sanyat Misra

Introduction

Foliage plants are centuries-old ornamentals grown usually in pots or other containers for their attractive foliage, which may be plain; green, blue-green, pale-green, red, variegated or variously coloured; and are valued for their beautiful leaves than the flowers, though there are many such ornamentals where flowers are also equally beautiful and attractive which add the bonus point. Most of the foliage plants are meant for interior landscaping, and there are a few which beautify our indoor environment continuously for years in absence of sunlight, such as *Aspidistra* and *Ruscus*, however, most require periodical filtered sunlight for good growth and develoment. Such plants are used in the public gardens, terraces, inside the houses and bungalows (veranda, patios, balconies, porches, corridors, living rooms, rooms, dish, window boxes and kitchens), airports, harbours, railway stations, bus terminals, conservatories, public and institutional buildings, offices, malls, showrooms, hotels and restaurants. Most of these are grown in pots for instant landscaping so are worth decorating on occasional ceremonial gathering, in pandals and banquet halls, in front of gates and entry points and along side the paths and streets in the garden or where certain special gathering (social, religious or otherwise) is to be held. There are many which provide shade on roads and streets, some work as specimen plants in the garden, some are most suitable for mass effects, some are used as hedges and/or edges, some others in hanging baskets while there are certain foliage plants used in floral arrangements. Foliage plants in nature are innumerable but approximately 1,000 are in trade worldwide.

Most of such plants have their native haunts in tropical and sub-tropical regions of the world having varying degrees of rainfall and relative humidity, temperature fluctuations from one area to the other, sunlight availability, soil types and differing elevations. Maximum species of foliage plants are comfortable between 13.0 °C and 18.0 °C temperatures with sufficient relative humidity in the atmosphere throughout the year but temperatures above 29 °C and below 10 °C may prove injurious to many of such plants so these should be protected during hot summers and winters. During summer, direct sunlight may also affect the plants adversely and their leaves and even the growing tips are scorched or burnt.

Foliage Plants for Various Purposes

Plants Suitable for Aquariums (Wardian Cases)

Aquariums are an interesting addition to the window garden where oxygenating plants on permanent basis are planted in adequate number but commensurate to the number of fishes so that these may be useful to the fish and may require little care after once these are planted. However, the water for the benefit of the fishes are changed regularly. The oxygenating plants are *Aponogeton crispus*, *Azolla caroliniana*, *A. filicoloides*,

Cabomba caroliniana, Callitriche autumnale, C. stagnalis, C. verna, Cardamine lyrata, Ceratophyllum demersum, C. submersum, Chara aspera, C. fragilis, C. hispida, C. vulgaris, Cryptocoryne becketti, C. ciliata, C. cordata, C. griffithii, C. wrightii, Egeria densa (*Elodea densa*), *Elodea canadensis* (*Anacharis canadensis*), *Fontinalis antipyretica, Hottonia palustris, Hydrocharis morsus-rannae, Hydrotrida caroliniana* (*Herpestris amplexicaulis, Bacopa amplexicaulis*), *Lagarosiphon major* (*Elodea crispa*), *Lemna* (invasive), *Ludwigia mulertii, L. palustris, Lysimachia, Myriophyllum alterniflorum, M. spicatum, M. verticillatum, Najas flexillis, Potamogeton cristus, P. densus, P. natans* (invasive), *P. perfoliatus, Proserpinaca palustris, Ranunculus aquatilis, Rorippa nasturtium aquaticum* (*Nasturtium officinale*), *Sagittaria, Stratiotes aloides, Tillaea recurva, Utricularia vulgaris* (insectivorous), *Vallisneria americana, V. spiralis, etc.* Whatever the size of the aquarium may be, but roughly per gallon of water some 5.0 cm of goldfish is sufficient.

Plants for Dish Gardens

Here only small plants are planted in a dish or dish-like structures by putting soil mixtures at the base of the dishes. Usually this is made for the living room table decoration. Suitable plants for the purpose are *Acorus, Agave parviflora, Aloe, Begonia, Bertolonia, Calathea, Cryptanthus, Cyanotis, Deuterocohnia, Dyckia, Ethretia, Episcia, Fittonia, Gasteria, Haworthia, Iresine, Juniperus compressa, Kalanchoe, Maranta, Ophiopogon, Pellionia, Pilea, Polyscias, Rosmarinus, Sansevieria, Soleirolia, etc.*

Plants Suitable for Basket Planting

Plants in baskets give enchantment inside the house, and these may be kept on the corner tables in the living room and bed-rooms, hanged with the walls as well as with the hooks of the roofs. The hanging baskets should have plants all around the wall. The foliage plants suitable for this purpose are *Aeschynanthus* (even for flowers), *Asparagus, Begonia, Calathea, Callisia, Chlorophytum, Cissus, Coleus, Cyanotis, Epipremnum, Episcia, Excoecharia, Fittonia, Gynura, Hemigraphia, Hoya, Kalanchoe, Oplismenus, Pellionia, Peperomia, Pilea, Plectranthus, Rhoicissus, Sageretia, Scindapsus, Setcreasia, Siderasis, Soleirolia, Stenotaphrum, Tetrastigma, etc.*

Plants as Ground Covers

These foliage plants grow quite fast to cover the land. The suitable plants for the purpose are *Acanthus mollis latifolius, Alchemilla mollis, Alternathera, Blechnum spicant*, bromeliads, *Calathea, Callisia, Chamaeranthemum, Chlorophytum, Codiaeum, Coleus, Dennstaedtia bipinnata, D. davallioides* and *D. obtusifolia* (cup ferns), *Epipremnum, Euonymus fortunei radicans, E. fortunei* f. *carrierei, E. radicans* 'Silver Queen' with variegated foliage, *Euphorbia robbiae, Ficus, Glechoma hederacea, Hedera canariensis* 'Azorica', *H. helix, Hosta crispula, H. decorata, Iresine, Juniperus communis* ssp. *depressa, J. c. jackii* and the varieties 'Depressa Aurea', 'Effusa', 'Repanda', *etc., J. conferta*, its varieties 'Harbor' and 'Glauca', *J. horizontalis, J. sabina tamariscifolia, J.* × *media* 'Pfitzeriana', *Lamium galeobdolon variegatum, L. maculatum, Liriope muscari, Luzula maxima, Mahonia aquifolium, Maranta, Muehlenbeckia axillaris, Obscura marginata, Ophiopogon japonicus* (stemless), *Pellionia, Peperomia, Pilea, Plectranthus, Polygonum campanulatum, Pseuderanthemum, Sarcococca humilis, Scindapsus, Sedum spathulifolium, S. spurium, Senecio* 'Sunshine', *Setcreasia, Strobilanthes, Tradescantia, Vinca major, V. minor* 'Bowles', *Waldsteinia ternata, Wedelia oblonga, Zebrina, etc.*

Plants Suitable for Growing in Soilless Water

There are certain plants whose cuttings, without soil, can be put in a container of water where these will grow and last for years and such plants are *Aglaonema simplex, Catharanthus minor, Cissus rhombifolia, Dieffenbachia, Euonymus fortunei* (*E. radicans*) and its varieties, *Hedera helix, Nephthytis picturata* (*Rhektophyllum mirabile*), *Pachysandra terminalis, Pedilanthus, Peperomia crassifolia, Philodendron, Scindapsus, Tradescantia, etc.* However, in its water 20 g/l of 4-12-4 complete NPK fertilizer at six monthly interval will be useful.

Plants Tolertating Low Light (less than 400 fc)

Such plants are *Aeschynanthus, Aglaonema, Alocasia, Amorphophalus, Anthurium, Aspidistra, Beaucarnea, Begonia, Caladium, Calathea, Chlorophytum, Cordyline, Cyanastrum, Dieffenbachia, Dracaena, Ehretia, Epipremnum,* ×*Fatshedera, Ficus, Fittonia, Haemaria, Haworthia, Homalocladium, Homalomena, Kaempferia, Kalanchoe tubiflora, Leea, Macodes, Maranta, Monstera, Myrtus, Pellonia, Philodendron, Pilea, Plectranthus, Pleomelee, Polyscias, Pothos, Raphidophora, Rhoeo, Rhoicissus, Ruscus, Schismattoglottis, Scindapsus, Siderasis, Spathiphyllum, Tradescantia, Zantedeschia, Zanthosoma, Zebrina, etc.*

Plants Requiring Warm Temperature (16°C and above)

Acalypha, Acanthus, Aeschynanthus, Aglaonema, Aloe, Alpinia, Alternanthera, Anthurium, Aralia, Bertolonia, Blakea, Brassaia, Caladium, Calathea, Callisia, Carludovica, Carissa, Coleus, Cordyline, Dianella, Dieffenbachia, Dracaena, Ehretia, Epipremnum, Episcia,

Eranthemum, Excoecaria, Ficus elastica, Ficus pumila, Hemigraphis, Hoffmannia, Homalomena, Hoya, Iresine, Kaempferia, Kalanchoe, Lavandula, Leea, Macodes, Maranta, Monolena, Monstera, Muehlenbeckia, Myrciaria, Nautilocalyx, Necodemia, Neoregelia, Pellionia, Peperomia, Pereskia, Peristrophe, Phyllanthus, Philodendron, Pilea, Piper, Pleomelee, Polyscias, Portulacaria, Pseudoranthemum, Raphidophora, Reineckia, Rhoeo, Sansevieria, Schismattoglottis, Scindapsus, Spathiphyllum, Stromanthe, Syngonium, Tetrastigma, Tradescantia, Trevesia, Triolena, Tupidanthus, Vriesea, Xanthosoma, Zamioculcas, Zenophia, Zebrina, etc.

Plants Tolerating a Temperature Below 13°C

Such plants are *Acanthus, Acorus, Aechmea, Aeonium, Aloe, Aspidistra, Ballota, Bergenia, Buxus, Carex, Cissus, Dianella, Dioscorea, Dyckia,* ×*Fatshedera, Geogenanthus, Haworthia, Hosta, Ligularia, Mimosa, Myrtus, Ophiopogon, Parthenocissus, Pelargonium, Reineckia, Rhoicissus, Rosmarinus, Ruscus, Sageretia, Sagittaria, Senecio, Zantedeschia, Zamia, etc.*

Plants Requiring More Than 50 per cent Humidity

Such plants are *Alocasia, Alpinia, Amorphophalus, Anoectochilus, Anthurium, Bertolonia, Campelia, Chirita, Cissus, Colocasia, Ctenanthe, Cyrtosperma, Dizygotheca, Episcia, Geogenanthus, Guzmania, Haemaria, Hoffmannia, Homalomena, Macodes, Monolena, Monstera, Nautilocalyx, Neoregelia, Philodendron, Pilea, Pseuderanthemum, Rhektophyllum, Rhodospatha, Schismattoglottis, Soleirolia, Spathiphyllum, Strobilanthes, Trevesia, Xantheranthemum, Xanthosoma, etc.*

Propagation

All the ferns produce powder-like **spores** and not the seeds, so their propagation through spores is challenging but highly promoing, especially at home as its germination is not like other seeds, *vis-a-vis* quite time consuming. These spores from ripe spore-sacs, usually located on the underside of the fronds are collected in paper bags and for about three weeks are kept as such to dry its external moisture and then are spread thinly over the steam-sterilized moist seed compost which is filled in a flat plastic bag and then the pot is covered with a sheet of glass and kept at a filtered but lighted place where these germinate and when grown sufficiently these are transplanted in phases in pots of different sizes. As these grow it is not immediately necessary to water them. Since gametophytes change into sporophytes, it will be necessary to separate the small plants in order to give them room to grow. Transplanted sporelings take several months to grow and eventually they can be removed from the containers. In asexual propagation, new fern plants are produced from rhizomes, stolons, tubers, stipules, roots, buds, cuttings and attached aerial stems (layering). Spreading types are easily propagated by **division** in early spring just when new growth begins. Many of the *Nephrolepis* and *Adiantum* groups can be divided easily but certain others not. Another way of obtaining plants is through **bulbils** or **babies** on the mother plant itself. There are groups of ferns producing **offshoots** on the plants such as *Asplenium* and *Diplazium*. Commercial propagation of *R. adiantiformis* is done by **rhizome division**, but frequent replanting is necessary. Asexual fern propagation also includes apospory and apogamy (Kottackal *et al.*, 2006). **Apospory** is the development of a gametophyte from an epidermal cell or cells of a sporophyte (Ambrozic-Dolinsek *et al.*, 2002), while **apogamy** is the development of a sporophyte directly from a gametophyte without sexual fusion (Kottackal *et al.*, 2006). Ramage and William (2002) stated the influence of absolute and relative amounts of nitrate and ammonium on induction and differentiation of plant cell cultures for a number of *in vitro* systems. Gonzalez *et al.* (2006) reported 32 per cent spore germination in *Nephrolepis exaltata* within 28-30 days after culture initiation on MS media while Chang *et al.* (2007) in *Drynaria fortunei* reported spore germination rate of 15.3 per cent after 7 days on MS medium under light which is one of the most important factors affecting events in the life cycle of a fern. Weinberg and Bruce (2007) mentioned that the spore culture of *Anemia phyllitidis* needed light to induce its germination.

Seed sowing is not the popular method of propagating foliage plants as this requires time, skill for raising the seedlings and conditions for germination of seeds. Sometimes when no option is left, one is bound to go for such propagation. However, in this system, very little cost is involved as this way one gets large number of plants. *Acanthus, Coleus, Mimusa, Pelargonium, etc.* can easily be grown through seeds. In this method the pots or pans are filled with seed compost, mixture is pressed firmly with finger-tips and watered lightly to settle the soil properly. Seed compost can be formulated by taking 1 part coarse and fresh sand, ½ part properly sieved compost or farmyard manure and ½ part sandy-loam soil. After sowing of the seeds thinly, minute ones are just covered with soil while large ones slighatly deep. Sometimes, after sowing the seeds the mouth of the container is covered with perforated polythene bags. Such pots are kept at a place with filtered sun or in a frame and at a temperature of 16-21 °C temperature. As soon as seeds germinate, the pots are exposed to a bright spot but away from bright sunlight after removing the plastic bag. Pots are rotated daily to avoid the lop-sidedness and mixture is kept moist. As soon as the seedlings are large enough to handle, these are pricked out individually in small pots filled with potting compost. From time to time, the pots are changed looking into the size of the plants.

Foliage plants are mostly **propagated vegetatively** through their underground organs, *viz.* bulbs, corms, rhizomes, tubers, *etc.*; division of clumps; separation of offshoots (offsets); removal of aerial plants as formed on the inflorescence in certain species such as *Agave, Chlorophytum, Saxifraga sarmentosa, Tolmiea, etc.* or on mature leaves such as *Bryophyllum* and *Asplenium bulbiferum*; cuttings (stem or shoot tip cuttings, cane cuttings, leaf cuttings and root cuttings); and layering (ground or air). With every plant described here, method of propagation is given there. The **underground organs** as formed in several plants such as many ferns, *Alocasia, Amorphophalus, Asparagus, Caladium, Colocasia, Dioscorea, etc.* are removed on maturity and planted in pots filled with moistened potting mixture or directly in the field with sufficient moisture as such or by dividing them as per size and demand of the species and then irrigated only once, but after sprouting these are cared well. Certain species also produce **runners** or **stolons** such as *Fittonia* which are taken out with nodes and planted in moist soil. Many species are **clump-forming** with daughter rosettes such as many ferns, *Aspidistra, Calathea, Carex, Chlorophytum, Cyperus, Fittonia, Helxine, Maranta, Ophiopogon, Oplismenus, Pellionia, Rohdea, Sansevieria, Scirpus, Siderasis, Stenotaphrum, etc.* which are lifted, their clumps divided with roots and individual shoots or segments are planted separately. **Offsets** are produced with many plant species such as bromeliads, *Aloe, Beaucarnea, Bowiea, Dieffenbachia, Pandanus, Rhoeo, Sansevieria, Yucca, etc.* which should be cut off as near the main stem as possible without destroying the roots in any way, and planted gently in the prepared field or in pots filled with potting mixture and then watered. Offsets taken out from the plants should better be ¼th to $^1/_3$ rd the size of the mother plant after flowering. If possible, the part of the plant producing the **aerial plantlets** should be pegged down in the moist seed compost directly in the soil or in pots where these will root and establish themselves as separate plants. Plantlets producing roots on the plants themselves should be separated and planted up as a rooted cutting. Many of the plants are propagated through **cuttings** whether it be shoot **tip-cuttings** such as *Coleus, Dieffenbachia, Dracaena, Glechoma, Hedera, Monstera, Pleomele, etc.*; **leaf-cuttings** such as *Begonia masoniana, Begonia rex, Bryophyllum, Crassula, Echeveria, Peperomia, Saintpaulia, Sansevieria, Streptocarpus, etc.*; **stem-cuttings** such as *Abutilon, Acalypha, Aglaonema, Aucuba, Breynia, Buxus, Cissus, Cleyera, Coccoloba, Coffea* with rooting hormones, *Codiaeum* with rooting hormones, *Coleus, Cupressus, Dichorisandra, Dizygotheca* with rooting hormones, *Euonymus, Ficus* with rooting hormones, *Fittonia, Geogenanthus* with rooting hormones, *Gynura, Fatshedera, Fatsia, Hemigraphis, Heptapleurum, Hypoestes, Iresine, Laurus, Monstera, Mimosa, Mikania, Nicodemia, Oplismenus, Osmanthus, Pandanus* with rooting hormones, *Pelargonium, Pellionia, Peperomia, Philodendron, Pilea, Piper* with rooting hormones, *Plectranthus, Podocarpus* with rooting hormones, *Polyscias* with rooting hormones, *Radermachera, Rhoicissus, Schefflera* with rooting hormones, *Scindapsus, Selaginella, Senecio, Sonerila* with rooting hormones, *Strobilanthes, Syngonium, Tradescantia, etc.*; and **cane-cuttings** such as *Cordyline, Dieffenbachia, Dracaena, Pleomele, Yucca, etc.* Only a few species are propagated through **air layering** such as *Aglaonema, Dieffenbachia, Ficus elastica, Dracaena, Monstera, Philodendron, etc.* Many climbers and trailers with long and flexible stems are propagated through **layering** where stems are pinned down into a pot containing rooting compost mixture after making a longitudinal cut with a sharp knife below the stem and this part is buried in the mixture with pressure of weight and watered regularly. When roots have appeared these should be detached and planted separately. This way at a time many layers can be made out of one plant and these layers may be cut and buried at several places if the cane is quite long.

Genera and Species

While describing the foliage plant genera and their species, the author took help from various literature sources (Bailey, 1929; Laurie and Ries, 1950; Hay and Beckett, 1971; Huxley and Gilbert, 1979; Hellyer, 1982; Beckett, 1985; Randhawa and Mukhopadhyay, 1986; Beckett, 1987; Hessayon, 1987; Virginie and Elbert, 1989; Brickell, 1994).

Acalypha (Euphorbiaceae)

Acalypha derives from the Greek *akalephes* for 'nettle'. There are 450 evergreen, semi-evergreen or deciduous and almost shrubby species from tropics and sub-tropics. They are suitable for conservatory, as house plants, for hedging and a few even for edging. They are propagated through cuttings. The two species described below are foliage decorative plants. ***A. godseffiana*** (Lance copperleaf; Euphorbiaceae) from New Guinea is a small-growing bushy species bearing lustrous green and ovate-lanceolate leaves with broad creamy-white serrated margin and insignificant greenish-yellow flowers appearing in catkin-like spikes. Leaves in *A. g. heterophylla* are narrowly lanceolate. ***A. wilkesiana*** (Copperleaf/Jacob's coat) from Pacific Islands is an evergreen species bearing highly decorative, ovate, twisting and serrated leaves some 10 cm or more long which are coppery green but usually splashed with red shades, and the flowers are insignificant. This has many cultivars in various shades and shapes such as 'Macafeeana', 'Macrophylla', 'Mudaica', 'Obovata', *etc.*

Acorus (Araceae)

Acorus takes its name from the Greek word *akoron* because of the yellow flag iris though later it was taken for sweet flag (*Acorus gramineus*). The genus comprises two hardy evergreen perennials from north temperate and sub-tropical zones growing through their rhizomes. They are marginal and submerged water plants. Their small greenish flowers carried on a spadix being insignificant, but the plants are grown for their often aromatic and narrow iris-like foliage emerging directly from the underground creeping rhizomes. These withstand well to poor conditions as draught, waterlogging or cold does not affect its growth adversely but prefer an open and sunny position. Multiplied through division, they are excellent pot plants for porch or veranda in the cool regions. ***A. gramineus*** (Sweet flag) from temperate India to Japan grows with grassy leaves, having fan-shaped tufts, up to 50 cm long with a 5-10 cm long spadix. *A. g.* 'Variegatus' (Myrtle flag/Sweet flag) is marginal water plant, having mid-green, sword-shaped and tangerine-scented leaves which are vertically cream-striped and in spring is flushed rose-pink. This is most suitable for terrarium planting. *A. g.* 'Pusillus' is a more tufted, and smaller plant suitable for pot culture.

Adansonia (Bombacaceae)

Adansonia is named after M. Adanson, a French botanist. A genus of 10 species of tropical shrubs and trees from Africa, Madagascar and Australia, closely related to *Bombax*. ***A. digitata*** (Baobab tree) from South Africa grows up to 18 m high with enormous swollen trunk, *i.e.* up to 9 m, and in the full-grown tree in its native haunts the girth of the trunk sometimes becomes more than its height but indoors it can be grown at a sunny window up to 3 m with enlarged girth. Thus it is a most unusual tree. Its bark yields a fibre. Stalks are up to 30 cm long and the leaves are palmate with three leaflets in the young plants but 5-7 in older ones, each being elliptic with a length of up to 13 cm. Solitary drooping flowers some 15 cm across bear 5 obovate petals curling up which are creamy with numerous purplish anthers and 7-10-rayed stigma, having mild aroma and are borne on long axillary peduncles during summer though in winter it may go dormant. Its ovary is 5-10-celled, fruit oblong, woody, indehiscent and with a mealy pulp containing many seeds. Seeds are used for its further propagation and sometimes it also produces offshoots. It is said that this plant never dies as even if blown by winds, it comes up from the side shoots again.

Adiantum (Adiantaceae)

See under the chapter '**Fern and Allied Plants**'.

Aechmea (Bromeliaceae)

See chapter on '**Bromeliads**'.

Aeonium (Crassulaceae)

See chapter on '**Succulents Other Than Cacti**'.

Aeschynanthus longicaulis (*A. marmoratus, A. zebrinus*; Gesneriaceae)

Aeschynanthus derives from the Greek *aischune* 'shame' and *anthos* for 'a flower', referring to the beautiful flowers as the sweet and shy blushing of a charming girl. Really, in flowering these seem so. The evergreen genus *Aeschynanthus* from India, China and Malaysia comprises 80-100 species of climbing and trailing sub-shrubs, many being epiphytic in the wild, and all being beautifully colourful indoor plants except one, *A. longicaulis*, an epiphyte from Myanmar and Thailand which is grown for its foliage. Its whip-like stem trail up to 60 cm long, bearing short-stalked, opposite, 7-10 cm long and elliptic leaves some 4 cm long which are handsomely netted yellow-green above and purple beneath. Its pale-green 3 cm long flowers with chocolate mottling continue appearing from summer to autumn as a bonus. It is a good subject for a shade-tolerant hanging basket and is easily propagated through cuttings.

Agathis (Araucariaceae)

In Greek, *Agathis* is for 'glome', due to the flowers in clusters. These are tender dioecious conifers, allied to *Araucaria*, from Australia, New Zealand and the Philippines, tolerating a minimum temperature of 10 °C. There are 30 species but only one, *A. robusta* is worth growing indoors. Leaves are coriaceous, and not needle-like as other conifers but usually broad, parallel-veined, petioled or sessile and opposite or alternate, cones are axillary, ovate or globular and consist of persistent bractless scales. ***A. robusta*** (*Dammara robusta, D. brownie*; Queensland kauri) from Australia produces a little verticillate, spiralled, closely alternate and horizontal branches or erect stems, leaves are broad, stiff, elliptic, 10 cm long, obtuse and stalkless, and the cones are oval, 7.5-10.0 × 5.0-7.5 cm size. It is a very tall tree so indoors it can be grown as a juvenile pot plant only. Two other important tall trees, ***A. australis*** (*Dammara australis*; Kauri pine) from New Zealand and ***A. orientalis*** (*Dammar orientalis*) from E. Indies are also leafy but are not amenable for indoor culture.

Agave (Agavaceae)

See chapter on '**Succulents Other Than Cacti**'.

Aglaomorpha meyeniana (Polypodiaceae)

See chapter on '**Ferns and Allied Plants**'.

Aglaonema (Araceae)

See chapter on '***Aglaonema***'.

Alloplectus (Gesneriaceae)

In Latin, *Alloplectus* means 'diversely plaited', owing to appearance of the calyx. These are tender, evergreen and tropical shrubby plants from South America, grown like *Gesneria*. These bear opposite leaves, out of which one in each pair remains smaller than the other and the leaves beneath are usually reddish or purplish. Flowers are tubular, yellowish and axillary. ***A. repens*** from E. Indies, while rooting from the nodes between pairs of leaves, trails out on the ground bearing hairy or smooth, ovate and coarsely serrate leaves, pale-green calyx blotched with purple, and corolla yellow, red-tinged with tube swollen at base. The limb is with 4 spreading segments where uppermost is doubly cut. ***A. schlimii*** from tropical S. America is erect growing shrub, leaves bearing green above, purple-violet below, oblong but base rounded or subcordate and acuminate, flowers axillary either in pairs or sometimes many, calyx with green spots, and corolla ~ below yellow-scarlet but above with violet shading. ***A. sparsiflorus*** from Brazil is an erect growing species, bearing entire, ovate-oblong and acute leaves, in which lower side of petiole and nerves are usually red. Calyx is composed of 5 dark blood-red or purple, heart-shaped or triangular sepals, and 5-parted with equal segments of thickly hairy and club-shaped corolla of yellow colour. Other important species are *A. capitatus* and *A. lynchii*.

Alocasia (Araceae)

In Greek, *Alocasia* evolves from *a* 'without' and *Colocasia*, an allied genus from which *Alocasia* was separated. The genus comprises of 70 evergreen perennial species from tropical Asia but only about 20 are grown indoors. Its erect and stem-like rhizomes (sometimes tuberous) are thick and fleshy. Leaves are highly ornamental, deep green, peltate and sagittate to cordate. Undecorative spadix bearing petalless sterile flowers emerge from a spathe. These are propagated through suckers and division of rhizomes. ***A. × amazonica*** (*A. sanderiana* × *A. lowii*) grows with its leaves up to 30-40 cm long. The leaves are deep glossy green contrasting with wide white veins and narrower white margin, sagittate-peltate, margins undulating, and pointed. In ***A. × argyrea*** (*A. longiloba* × *A. pucciana*), leaves are leathery and firm with a silvery patina, 30-50 cm long and peltate with a cordate base. ***A. × chantrierana*** (*A. × chantrieri*) was raised in France by Chantrier Bros, Mortefontaine by crossing *A. cuprea* × *A. sanderiana*. It produces underground tubers. The leaves are 30 cm or more long, wavy, cordate-ovate with short basal lobes, purple below, main veins curving horizontally, midribs and lobe-ribs white while side veins greyish and the leaves between the veins are quilted. ***A. × chelsonii*** (*A. cuprea* × *A. longiloba*) is very slow-growing, bearing rich metallic green leaves above and purple reverse, oval with no basal lobes, narrowly and shortly pointed, convex, some 40 cm long stalk, leaf size being 35-40 × 15-20 cm, midrib raised and greenish-white while side veins silvery. ***A. cucullata*** (Chinese taro) from West Bengal (India) and Myanmar is bushy, slow-growing, bearing stiff and spreading stalks (30 cm or more in length) in spiral form with pale-green, cordate, horizontal, little cupped, long-lasting leaves some 25-30 × 15-20 cm size. ***A. cuprea*** (*A. metallica*) from Borneo is a compact plant, bearing 30 cm long, 15 cm wide, pointed, ovate and peltate-cordate leaves with metallic green above and purplish beneath where midrib and lateral veins are dark green and margins white and veins depressed. Var. 'Fantasy' comprises of both the types, large and dwarf ones, the latter being more popular, and the leaf stalks are long and bright pink. Leaves are 20 × 10-12.5 cm, thick-textured, deep pinkish-green above and purple below, cordate, lobes round-tipped, and little cupped. Midrib and lateral veins broadly green with silver overlay. ***A. guttata*** var. *imperialis* is a rare plant, producing purple-spotted leaf stalks, leaves some 25 cm long, thick and leathery-textured, cordate, convex, and deeply cleft with wide lobes, blue-grey above, blue-black near the veins and reverse purple, while midrib and lateral veins deeply embedded. It is resistant to mites. Its cv. 'Hilo Beauty' leaves are irregularly cordate with blunt tips, and upper green surface is marked with shapeless brownish-yellow island formations. It is resistant to mite. ***A. korthalsii*** from Malaya is similar to *A. lowii* or *A. × chelsonii*, producing very long leaves some 60 cm long having slightly cupped blades, lower surface purple, upper surface deep olive-green with a touch of grey, midribs, veins and margins greyish to grey-white and leaf-blades leathery. ***A. lindenii*** from New Guinea emits a strong odour when leaves are bruised. Its petioles are almost white, the leaves are bright green, long-pointed, cordate-ovate and some 20-30 cm long, having yellowish veins which curve from the midrib and vanish near the margins. ***A. longiloba*** (*A. gigantea*) from Malaysia,

Borneo and Java bears greenish-white mottled purple and some 60 cm long petioles. Leaf blades are 30-65 cm long, sagittate, narrowly triangular, upper surface deep lustrous green with silvery to grey bands along veins and midrib, leaf under-surface light purple, and basal lobes erect and very long, some more than half as long as the main lobe. Its var. 'Magnifica' has larger leaves. ***A. lowii*** from Borneo bears broadly sagittate leaves to about 60 cm long or even more when mature and being different from juvenile phase, deep green overlaid with metallic sheen and silvery veins above, purple below, and the two basal lobes are jointed for about one-quarter of their length. Outline of the blades being usually curved so making these cordate with a distinct cleft. ***A. macrorhiza*** (Giant elephant's ear) from Sri Lanka and Malaya bears trunk-like rhizomes some 4.5 m tall in the wild. The sagittate leaves are more than 60 cm long, spathe yellow-green and some 20 cm long with 9-12 pairs of veins. Its var. 'Variegata' leaves are blotched with cream and grey-green. *A. m. rubra* (*A. plumbea*) looks as the smaller version of the type species but with larger leaves which are variably purple-flushed. ***A. micholitziana*** (Green velvet) from Philippines bears 30-45 cm long stalks, 30-40 × 15-16.5 cm size and narrowly sagittate leaves, *i.e.* from arrow-shaped to quite oval in outline, rich velvety-green and slightly convex, lobes rounded, margins wavy and turning downward, main veins few, thick and crystalline-white. ***A. odora*** from tropical Asia is similar to *A. macrorhiza* but with peltate leaves where main lobe bears only 6-10 pairs of veins. ***A. portei*** from Philippines produces some 2 m long and very robust leaf-stalks which are green marbled with red-purple. Ovate-sagittate deep green leaves with a metallic luster are about 2 m long. It is suitable for large pots for keeping in larger conservatory. In juveniles the leaves are deeply sinuous only though in mature plants deeply lobed. ***A. sanderiana*** (Kris plant) from Philippines is similar to *A. × amazonica*, bearing metallic silver-green leaves with grey veins and white margins, 30-40 cm long, narrow, sagittate and deeper lobed. ***A. × sedenii*** (*A. lowii × A. cuprea*) appears, in general, similar to its first parent but the basal lobes are jointed $2/3^{rd}$ of their length. ***A. veitchii*** (*A. lowii veitchii*) from Malaya has green-striped petioles, and the plant resembles *A. lowii* in shape and general colour but midrib and main veins have grey-green border but secondary veins are whitish. Deep green leaves with purple beneath and light grey margins are sagittate, narrowly triangular and some 45 cm long. ***A. watsoniana*** from Sumatra produces up to 90 cm long ovate-sagittate leaves, upper surface tinted glaucous with veins being silvery. Basal lobes of the leaves join together for about half their length. ***A. wentii*** (*A. whinckii*) from New Guinea bears deep glossy green leaves with somewhat wavy margins, ovate, peltate or sagittate, tip-pointed and the basal lobes rounded and join together for half to two-third of their length. ***A. zebrina*** from Philippines bears triangular-sagittate, leathery-textured, some 35 cm long leaves, green on both the surfaces, however, 50-90 cm long and green leaf stalks have brownish-black zebra stripes, hence, its specific name is *zebrina*. Its basal lobes are obtuse.

Aloe (Liliaceae)

See chapter on '**Succulents Other than Cacti**'.

Alpinia sanderae (Zingiberaceae)

See chapter on '**Ornamental Gingers**'.

Alsophila (Cyathia; Cyathaceae)

See chapter on '**Ferns and Allied Plants**'.

Alternanthera (Amaranthaceae)

Alternanthera derives from the Latin *alternans* 'alternate' and *anthera* 'anthers', due to the alternate barren anthers. The genus comprises 200 species from worldwide tropics and sub-tropics, most of them are perennials with attractive foliage. These may be erect, bushy, mat-forming or tufted, bearing narrow leaves in opposite pairs, often patterned colourfully, and the flowers are small and insignificant. These are propagated through division or cuttings. They make good pot plants for home and conservatory and used for carpeting, bedding and edging. ***A. amoena*** (*R. ficoidea* var. *amoena*) from Brazil is a mat- or hummock-forming species, bearing elliptic to lanceolate leaves some 2.5-7.5 cm long which are veined and blotched with red, orange or yellow. ***A. bettzickiana*** (Calico plant) from Brazil is a nearly sub-shrubby species growing to 30 cm or more in height, bearing narrowly spathulate, green and red leaves. The leaves are bright yellow in 'Aurea Nana' ('Nana'). ***A. dentata*** (*A. ramisissima*) from northern South America and West Indies grows erect, bushy and 30-60 cm tall, bearing up to 9 cm long, ovate and tip-pointed leaves, and some 2.5 cm long white to greenish-white heads though of no use. The leaves in 'Rubiginosa' are red-purple. ***A. ficoidea*** (Parrot leaf) from Brazil is a mat-forming species and bears green, broadly lanceolate, wavy-margined and variously marked with red and purple. ***A. versicolor*** (*A. ficoidea* 'Versicolor') from Brazil is a mat-forming species growing erect with bushy nature and attaining up to 30 cm height, where leaves are obovate, and bronze-green shaded and margined with red pink and also sometimes yellow.

Amicia (Leguminosae)

Amicia was named for Giovanni Battista Amici (1786-1863), astonomy and microscopy professor at Florence. The genus comprises eight species from warm-temperate areas of Mexico to South America. These are shrubs with soft stems, bearing alternate and pinnate leaves, and the pea-shaped flowers appear from the leaf axils on short racemes but of no use as the plants are admired only for its unusual foliage. These are propagated from heel cuttings taken from the new growths. ***A. zygomeris*** from Mexico is an erect-growing deciduous shrub to about 2 m tall, bearing pinnate leaves with 4-6 obcordate leaflets some 5 cm long, with two inflated and large leaf-like stipules at the base. Yellow and purple flowers some 2.5 cm long appear during September-October.

Amomum (False cardamom; Zingiberaceae)

In Greek, *Amomum* is for an Indian spice, closely allied to *Alpinia* and *Elettaria* and yields cardamom (aromatic tonic seeds) similar to *Elettaria.* A genus of 150 evergreen and hard-rhizomatous perennial species from old world tropics (Asia, Africa and Pacific Islands), one (*A. compactum*, syn. *A. cardamomum*) of which is cultivated indoors for its durable and decorative foliage and as a substitute for cardamom, forming dense masses of handsome erect or spreading annual stems and linear, lanceolate or elliptic leaves. Flowers appear in dense cone-like spikes or racemes where half are hidden in the floral bracts. Calyx is funnel-shaped, splitting down to one side and slightly toothed. Corolla tube cylindrical and longer than calyx, the upper lobe curved and the two lower narrow and spreading, lip (staminode) large and petaloid, filament slender and fruits ovoid. It is propagated by division. ***A compactum*** from Java is a clump-forming grassy species with leafy cane-like stems growing up to 90 cm tall, bearing stalkless, thin, alternate, linear-lanceolate and amicably aromatic leaves some 25 cm long. Flowers arising from the rhizomes are pale-yellow, 1.5 cm long, tubular and are hidden in the leaves.

Ananas (Bromeliaceae)

See chapter on '**Bromeliads**'.

Anthurium (Araceae)

Anthos in Greek means 'a flower', and *aura* means 'a tail', referring to the spadix. Anthuriums are evergreen tropical ornamentals, and the species *A. crystallinum* (crystal anthurium) from Peru and Colombia and *A. warocqueanum* (queen anthurium) from Colombia are highly valued for their delicate look and unusually attractive foliage. There are some 550 species from the humid rain forests of tropical America and the West Indies. These are propagated by division or through seeds. In ***A. crystallinum*** the leaves are lustrous deep green patterned with silver-white veins, cordate, and 25-50 cm long and some 30 cm wide. Non-decorative green spathe is some 9.0 cm long and linear-oblong. It is the most beautiful among all the foliage anthuriums. ***A. warocqueanum*** bears 90 cm × 30 cm heart-shaped emerald-green leaves which are smooth with ivory-white veins. Spathe is green or yellowish, linear-lanceolate and some 10 cm long while spadix up to 30 cm.

Aphelandra (Acanthaceae)

Aphelandra takes its name from the Greek *Apheles* meaning 'simple' and *aner* for 'male', owing to its anthers having only one cell. It is a genus of 20 evergreen shrubs from tropical and sub-tropical America, many being nice house plants. The leaves are leathery, elliptic to ovate often veined or mottled white or grey-silver and appear in opposite pairs. The 2-lipped tubular flowers appear in dense terminal spikes or in the axils of the coloured bracts. Either these are propagated through seeds or by cuttings of young shoots. ***A. aurantiaca*** (*A. fascinator*) from Mexico grows up to 90 cm, bearing dark green leaves veined silver, 10-15 cm long and broadly ovate. Flowers are bright orange-scarlet. *A. a. roezlii* produces compact plant with twisted leaves suffused silver-grey. ***A. bahiensis*** from Brazil bears some 10 cm long elliptical leaves, green above and purple beneath, and is a beautiful foliage plant but is grown rarely. Bracts and tubular flowers, both are yellow with flowers some 2.5 cm long. ***A. chamissoniana*** from Brazil is a slender plant growing up to 1.2 m in height, bearing 7-10 cm long elliptic leaves having broad silver-white veining. Bright yellow flowers and bracts appear during autumn. ***A. fuscopunctata*** from northern S. America grows more than 60 cm tall, bearing green stems, and the leaves are hairy dark green above, paler beneath, ovate and more than 10 cm long. Floral bracts are red and glandular from where coffee –coloured flowers appear. ***A. ignea*** is a spreading plant growing only up to 10 cm high with some 10 cm long leaves which have basal cleft, and this is sometimes confused with *Xantheranthemum igneum.* Yellow flowers are some 3.8 cm long. ***A. sinclairiana*** from Central America is up to 1 m tall, bearing thin, glossy green, boldly veined and more than 10 cm long leaves. Clustered floral spikes appear at the stem tips bearing orange bracts and rose-pink flowers. ***A. squarrosa*** (Zebra plant) from Brazil grows up to 1.25 m tall, bearing deep glossy green with dense white veins, pointed lanceolate, opposite alternate pairs and 15-25 cm long leaves. Tubular flowers some 3.75 cm long and bracts are bright yellow, the latter sometimes is with red margins, and both appear in angular cone-shaped floral spikes borne terminally. *A. s.* 'Louisae' is a compact form with smaller leaves. ***A. tetragona*** from northern S. America and West Indies is of free-growing nature growing up to 2 m tall, bearing 15 cm long, slender-pointed and broadly ovate leaves. Bright red flowers in dense terminal and axillary spikes appear some 5-8 cm long.

Aporocactus (Rat's tail cactus; Cactaceae)

See chapter on '**Cacti**'.

Aralia (Araliaceae)

Aralia is the Latinized name of an old French-Canadian name *aralie*. It is a genus of some 35 species of trees, shrubs, climbers and herbaceous perennials from N. America, E. Asia and Indo-Malaysia. Many species are suitable for growing in pots indoors or in conservatories. These bear compound pinnate to bipinnate leaves and small, 5-petalled green to purplish or white flowers in rounded umbels on branched racemes or panicles. Many of its species such as *A. japonica, A. moseri* and *A. sieboldii* have now become species of *Fatsia*. *Aralia elegantissima* is now known as *Dizygotheca elegantissima*, *A. balfouriana* as *Polyscias balfouriana* (*Panax balfourii*), *A. filicifolia* as *Polyscias filicifolia* (*Panax filicifolius* or *P. filicifolium*), *A. guilfoylei* as *Polyscias guilfoylei* (*Nothopanax guilfoylei*), *A. veitchii* as *Dizygotheca veitchii* and so on. The fruit is a berry. ***A. chinensis*** (Angelica tree) from NE Asia is an erect, deciduous and suckering shrub or small tree with thick and spiny stems, some 1.25 cm long tripinnate leaves, and white flowers on large branched racemes which appear from summer to autumn. ***A. elata*** (Angelica tree) from Japan is a sparsely-branched stout, erect, prickly, deciduous and suckering shrub bearing rosette of leaves some 1 m long towards the stem tips and the flowers are similar to *A. chinensis*. *A. e.* 'Aureomarginata' has leaflets irregularly bordered with silvery-white, *A.e.* 'Variegata' (*A. e.* 'Albovariegata') bears leaflets margined and splashed with creamy-white, aging to silvery-white.

Araucaria (Araucariaceae)

See chapter on '**Araucaria**'.

Archontophoenix (*Ptychosperma*; King palms; Palmae)

Archontophoenix derives from the Greek *archontos* 'a chieftain' and *phoenix* 'the date palm', referring to the size of the trees sometimes going up to 30 m in height. A genus of two species from Australia which at younger stage and also being slow-growing, are excellent for growing in containers. Leaves are arching, pinnate and composed of may narrow leaflets. These produce white or lilac flowers and red fruits in trusses but not on containerized plants. These are propagated through seeds sown above 18 °C temperature. Once germinated, the seedling growth is quite fast. ***A. alexandrae*** (Alexandra palm/Northern bungalow palm) from Queensland may grow up to 3 m in large containers, bearing some 90 cm long and green leaves which are pale-greenish beneath. ***A. cunninghamiana*** (Illawarra palm/Piccasbeen bungalow palm/Piccabeen palm) from Queensland and New South Wales also in large containers grows some 3 m tall, bearing green and some 90 cm long leaves.

Ardisia (Myrsinaceae)

Ardisia derives from the Greek *ardis* 'a point' as the anthers are spear-pointed. It is a woody genus of 400 evergreen species of shrubs and trees from tropical and sub-tropical regions, mostly from Asia and America and some in Australia though only 3-4 are worth as architectural columnar foliage plants for growing indoors. Though their small flowers in terminal or axillary clusters are not that much significant but leaves are deep green, alternate and elliptic to lanceolate. These have special and unparallel beauty of the clustered red berries which persist for long. These like filtered to partial sun and require cool temperatures during summer. They are propagated by seeds and heel cuttings. ***A. crenata*** (Coralberry) from India to Japan is a most popular indoor and greenhouse plant growing up to 1.8 m high outdoors though indoors only up to 90 cm. Its leaves are shining dark green, leathery, elliptic, 10 cm long and closely waving. Clusters of small white flowers appear in summer, followed by showy red berries which persist until the next season of flowering arrives. It is a cold-liking shrub. ***A. crispa*** (Coralberry/Marlberry/Spiceberry) from East Indies and Japan is a shrub growing to 90-150 cm in height and 30-45 cm in spread with upright habit. Its leaves are shining alternate, dark green, oblong-lanceolate and 5-10 cm long with wavy margins. Star-shaped, sweet-scented, 5-petalled, 1.25 cm across and cream-white flowers are tinged red which appear in June in axillary umbels, followed by round scarlet berries some 6-8 cm in diameter. It differs from *A. crenata* principally in having smooth-edged leaves. ***A. japonica*** from Japan is a very slow-growing semi-hardy small shrub growing to 45 cm tall and this is used frequently for training as a bonsai. Its elliptic leaves are some 10 cm long. Its various cultivars having different pattern of variegation are 'Hakubotan', 'Hinotsukasas', 'Maculata', 'Marginata', 'Matsu-Shima', *etc.* which are worth growing at the cool windowsills.

Areca [Arecae (Palmae)]

Areca is the Latinized form of Malayan name to this genus. A genus wth 14 species of graceful and

(entire leaf nearly yellow), 'Lance Leaf' (green, male), 'Longifolia' (glossy green, female), 'Maculata' ('Variegata' with variegated leaves), 'Speckles' (male), 'Variegata Maculata' (female), *etc.*

Ballota (Labiatae)

Ballota is the ancient Greek name for black horehound, *B. nigra* an original species in the genus named by Linnaeus. It is a genus of 35 perennial or sub-shrubby and almost hardy species from Europe, W. Asia and Mediterranean. They have opposite pairs of hairy, generally ovate to rounded leaves and 2-lipped tubular flowers where upper lip is hooded. These are cultivated for their attractive foliage. They can be grown from sunny to partial shade. They are propagated through seeds and division in case of perennial species, *vis-à-vis* cuttings in case of sub-shrubs. ***B. acetabulosa*** from Greece, Crete and Aegean is a spreading evergreen sub-shrub growing erect up to 60 cm in height with woolly stems, bearing 1.25 cm long leaf stalks, grey-green, highly felted-white, cordate to arrow-shaped, crenate, 2.5 cm long and 3-4 cm wide leaves, and rose-purple, white & 1.5-1.8 cm long woolly flowers appearing from the joints, and 1.0-1.5 cm long and salver-shaped calyx. ***B. pseudo-dictamnus*** from S. Aegean (eastern Mediterranean) is a white-woolly, much-branched and comparatively smaller than *B. acetabulosa*, growing from 30-60 cm in height with almost equal spread. Its whole plant is yellowish-woolly, leaves are opposite, woolly-grey, cordate and 1.5-2.0 cm wide, whorled flowers are 2-lipped, tubular, 1.25 cm long, white with purple spots and appear in July, and the calyx are funnel-shaped and 7-8 mm wide.

Bambusa (Bambusoideae)

Bambusa is one of the 23 genera of bamboos, and one of the most important four genera, *viz. Arundinaria, Bambusa, Dendrocalamus* and *Phyllostachys. Arundinaria* has persistent sheath and cylindrical stems and these are the only major distinction with *Phyllostachys* as here sheaths are early deciduous and the internodes above the base are flattened to one side. The size of *Dendrocalamus* is huge and this is the distinction with others. *Arundinaria* and *Bambusa* are very similar to each other horticulturally. Under bamboos, there are more than 200 species, out of which more than 160 are from Asia, about 70 in America and five in Africa (Bailey, 1942). They are found from sea level to 3,000 m in the Himalayas and 4,500 m in the andes, and growing up to 32 m with a diameter of culm 20-30 cm (Bailey, 1942). Only those are suitable for planting as ornamentals which are tufted or clump-forming. These are propagated by dividing the clumps when new growth has started. Still there are many species under the genus *Bambusa* though most of these are also, at occasions, placed under various other genera such as *Arundinaria, Phyllostachys, Pleioblastus, Sasa, Shibataea* and others. ***B. eutuldoides*** from China forms large clump and the stiff, erect and hollow canes grow up to 9 m tall but in containers 2-3 m only and diameter of the culm some 2.0-2.5 cm, bearing olive-green leaves patterned with longitudinal handsome greenish-white or ivory stripes. ***B. gracillima*** from China is up to 2 m tall but arching with culm diameter 1 cm, young ones being green but older ones becoming orange. ***B. multiplex*** var. 'Alphonse Karr' (syn. *B. nana* 'Suochiku') from Japan is a tall plant (up to 12 m high) but in containers grows up to 2.5 m, and forms rosette of bright yellow stems as well as culms, striped green irregularly, and the culm sheaths bear yellow stripes. ***B. nana*** (*B. disticha*) is a beautiful species growing in zigzag fashion up to only 1 m high having green colour or sometimes purple-tinged. Green leaves are produced in two ranks. ***B. pygmaea*** (Japanese bamboo) from Japan is a very hardy species growing hardly 30 cm in height with slender, much-branched and purple stem with prominent nodes. It is used as undergrowth where it outgrows quickly and forms a carpet. ***B. ventricosa*** (Buddha's belly bamboo) is from China with swollen internodes. ***B. vulgaris*** is from India, West Indies, S. & C. America, Africa and Java is a very tall bamboo.

Beaucarnea (Agavaceae)

See chapter on '**Succulents Other Than Cacti**'.

Begonia (Begoniaceae)

See chapter on '**Begonia**'.

Berberidopsis corallina (Coral plant; Flacourtiaceae)

Berberidopsis derives from *Berberis*, the genus and the Greek *opsis* 'like', owing to the general appearance of the plant. The genus contains only one species, *B. corallina* native to warm-temperate rain forests. It can be trained to the required shape against the roof space, the walls and in conservatories at a shady position. It is propagated by seeds, cuttings and branch-layering. It is an evergreen, woody-stemmed and twining climber for temperate areas growing up to 6 m in length, bearing dark green leaves at upper surface, glaucous below, oval to cordate, leathery, some 8 cm long and edged with small spines. Clustered deep red flowers, rounded or bowl-shaped and some 1 cm across appear on long red stalks in pendent racemes from late summer to autumn.

Bergenia (Saxifragaceae)

It was named for Carl August von Bergen (1704-1760), a German physicist and botanist. It is a genus of six hardy, evergreen, herbaceous and perennial species bearing semi-woody rhizomes, and large, mid-green, leathery, glossy and paddle-shaped leaves which sometimes change the colour to copper during autumn. They are native to central and east Asia. In spring to summer these produce showy panicles of white, pink or red bell-shaped flowers with five or more petals. These are suitable as ground cover and on slopes for binding the soil. They are propagated through division or seeds. ***B. ciliata*** (*B. ligulata*) is an evergreen clump-forming half-hardy perennial growing up to 30 cm high and with some 50 cm spread, bearing green, decorative, large, leathery, rounded and hairy leaves. Clustered white flowers which appear from late winter to summer, age to pink. ***B. cordifolia*** from Siberia is fully hardy, evergreen and clump-forming species growing to a height of about 45 cm with 60 cm spread, bearing rounded with heart-shaped base, puckered and crinkle-edged leaves, and light pink, open cup-shaped racemose and drooping flowers some 2.5 cm long in spring. *B. c. purpurea* bears purple-tinged leaves and pink-purple flowers. ***B. crassifolia*** from Siberia is a clump-forming fully hardy species growing up to 30 cm tall, bearing evergreen ovate to spoon-shaped, flat and fleshy leaves, and in panicles bell-shaped pale-pink flowers some 2.5 cm long from late winter to spring. ***B. purpurascens*** (*B. beesiana, B. delavayi*) is a fully hardy evergreen and clump-forming perennial growing up to 45 cm high and which is similar to *B. crassifolia* but leaves are narrowly elliptic often slightly convex and dark green but turning beetroot-red during autumn. Purple-red to pink petalled flowers with purple-brown calyces and cup-shaped flowers some 2.0-2.5 cm long appear in loose panicles during spring. ***B. × schmidtii*** is a fully hardy clump-forming plant growing up to 30 cm high and 60 cm spread, bearing flat and oval leaves with toothed edges, and sprays of cup-shaped soft pink flowers are produced on small stalks in spring. ***B. stracheyi*** from Himalayas is a fully hardy, evergreen and clump-forming perennial growing from 23 to 30 cm tall, bearing clear rosettes of small, rounded and flat leaves, and branched heads of cup-shaped white or pink flowers.

Bertolonia (Melastomataceae)

Bertolonia was named for Antonio Bertolini (1775-1869), professor of botany at Bologna. A genus of 10 frost-tender tiny evergreen perennial species with highly ornamental foliage from the tropical rain forests of Brazil and Venezuela, which tolerates minimum temperature of 15 °C. A small plant bearing a little fleshy, ovate to cordate or elliptic, oftenly patterend with white, metallic white or silvery with crystalline luster. Small flowers are borne in terminal clusters. These are the terrarium plants. These require fairly shaded position and high humidity but not the waterlogged soils for their proper growth. They are propagated by tip or leaf cuttings. ***B. hirsuta*** (Jewel plant) bears hairy and vivid green leaves but veins are centrally-zoned-reddish, and the flowers are small and white. ***B. × houtteana*** bears 10-18 cm long, ovate-elliptic and dull green leaves with bold marks of silvery pink on cross-veins, and small pink flowers. ***B. maculata*** from Brazil has tufted rosette of leaves, rich green with an irregular silvery stripe in the centre, ovate-cordate and 10-20 cm long, whereas spike is short and clustered with rose-pink to violet-purple flowers. ***B. marmorata*** (*Eriocnema marmoratum*) from Ecuador is rosette-forming and grows about 15 cm in height when in bloom and 45 cm spread, bearing leaves in vivid velvety green above, reddish-purple below, a little fleshy and broadly oval but with cordate-base, patterned with parallel white veins on the upper surface, and saucer-shaped pinkish-purple flowers off and on. *B. m. aenea* has lustrous coppery or bronze leaves without silvery markings. *B. m.* 'Mosaica' bears leaves with broader silvery vein marked with pinkish tints. ***B. sanguinea*** bears leaves with bronze sheen, centrally silvery-banded and below reddish.

Beschorneria (Agavaceae)

Beschorneria was named to honour Friedrich W.C. Beschorner, a German amateur botanist. A Mexican genus of 10 evergreen perennial species with a strong tuberous rootstock. Their leaves are basal, thick, narrowly lanceolate and rosette-forming. The plants are suitable for container growing in larger conservatory as well as outside. These are propagated by seeds or through division of the sucker-like offsets. ***B. bracteata*** is stemless and bears grey-green some 20-30 leaves having length of 45-60 cm and width of 5 cm. Its greenish-yellow flowers some 4 cm long appear on a branched flowering stalk some 2 m long. The flowers become red with age. ***B. yuccoides*** is stemless with 15-20 grey-green leaves which are 30-45 cm long. Unbranched flowering stalk about 1.2 m tall bears bright green, tubular, drooping and some 5 cm long flowers with red bracts.

Billbergia (Bromeliaceae)

See chapter on '**Bromeliads**'.

Biophytum (Oxalidaceae)

In Greek, *bios* is for 'life' and the *phytum* is for 'a plant', referring to the sensitive leaves. The genus being pan-tropical (Tropical Asia, Africa and America) in nature, comprising of 70 perennial species which are oxalis-like and often woody at the base. Its two species are grown as curiosity because leaflets when touched, fold back from their midrib similar to that of *Mimosa pudica*, and are suitable for terrarium planting. Their leaves are abruptly pinnate with many leaflets, and capsules split explosively by opening the valves up to the base. They are propagated from seeds. ***B. foxii*** from Peru has 5 cm high stems where at the apex some 8 leaves are formed in a whorl, each with 3-6 pairs of leaflets and the flowers are white. ***B. sensitivum*** (Sensitive plant) is the most popular species with 10 cm long woody stems, and bearing rosette of 5-8 cm long pinnate leaves with 6-15 pairs of small and oblong leaflets. Flowers are small, yellow and funnel-shaped, followed by splitting capsules.

Blakea (Melastomataceae)

Blakea was named to honour Stephen Blake, gardener of the island of Antigua (West Indies) who wrote the book "**Complete Gardener's Practice**" in 1664. There are some 30 species from West Indies and South America. They are diffuse, woody to non-woody shrubs, with large, opposite, leathery, 3-7-nerved, petioled, entire or nearly so, often rusty-pubescent beneath, vertical and broadly elliptical or lanceolate leaves which have a set of three deeply set principal veins from leaf base to the apex. Large showy, rose-purple or white, solitary or fascicled flowers appear in the axils, calyx with 4 or more scales or bracts at base, oblong or obovate petals 6, stames 12 with fleshy filaments, calyx-jointed ovary which is 4-6-celled and the fruit is a fleshy berry. It is not an easy plant for indoor cultivation as soon this succumbs to fungus so is suitable only in the glasshouse environment. Though difficult but these are propagated by cuttings. ***B. gracilis*** from Costa Rica is a spreading shrub with shiny, lanceolate and about 15 cm long leaves and 6-petalled some 5 cm long flowers though seldom appear indoors. ***B. trinervia*** from Mountain woods of Jamaica produces round and green stems and its branches which support themselves on neighbouring bushes, bearing cupped, elliptical, basally notched, some 20 cm long though stalk length some 5 cm, with 3 deep veins and numerous parallel transverse veining. Flowers purplish but appear rarely.

Blechnum [Blechnaceae (Polypodiaceae)]

See chapter on '**Ferns and Allied Plants**'.

Boehmeria (Urticaceae)

Boehmeria was named in commemoration of George Rudolf Boehmer (1723-1803), a German professor of botany. A genus of about 100 species of evergreen perennials, shrubs and trees, widespread in the tropics and sub-tropics, a few of which are grown for their large and attractive foliage. Insignificant tiny, petalless and greenish flowers are produced in the leaf axils. These make pleasant site in conservatory, in the house and outside. They are propagated by division, cuttings and suckers, the most usual being cuttings. ***B. argentea*** from Mexico grows in the wild up to 3 m in height but in containers indoors up to 2 m where these are replaced every 1-2 years to avoid it becoming leggy. It is propagated through cuttings. Its leaves are glaucous green and some 30 cm long, having alternate depressions on upper surface with silvery-white margins. ***B. nivea*** (China grass/Ramie) from tropical Asia is clump to colony-forming erect perennial some 1.0-1.5 m tall, bearing medium to deep green leaves on the upper surface and white-felted below, some 15 cm long, broadly ovate and slender-pointed. The plant remains elegant if each year is cut from the ground to encourage precocious sucker formation. They are propagated through division, suckers or cuttings. Its stems yield a very fine silky fibre better than cotton but is very difficult to process.

Bolbitis [Lomariopsidaceae (Aspidiaceae)]

See chapter on '**Ferns and Allied Plants**'.

Bowiea (Liliaceae)

See chapter on '**Succulents Other Than Cacti**'.

Brachychilus horsefieldii (*Brachychilum horsefieldii*; Zingiberaceae)

Brachychilus derives from the Greek *brachys* meaning 'short' and *chilos* meaning 'lip', as the petals (labellum) are short and lip-like. In the genus there are only two herbaceous epiphytic species forming clumps of hairless leafy stems, native to islands of SE Asia. These are closely allied to *Hedychium* or *Alpinia*. Successfully these can be grown in pots, going dormant during the winter though requiring shade and high humidity in summer. These are propagated through seeds and by division. ***B. horsfieldii*** (*Hedychium horsfieldii*) from Java is sometimes confused with and has been cultivated as *Hedychium calcarata*. It is clump-forming with stems growing up to or more than 90 cm high, bearing dark green, leathery, lanceolate, up to 30 cm long and slender-pointed leaves. White and tubular flowers appear in terminal spikes which may be up to 8 cm long, bearing 6 perianth-segments and a lip-like appendage formed by two aborted stamens, followed by very attractive and long-lasting oblong capsules which on opening show orange colouration and reveal three rows of red seeds. Another species is ***B. tenellum*** from Moluccas though not under common cultivation.

Brachychiton (Australian bottle tree; Sterculiaceae)

Brachychiton derives from the Greek *brachys* 'short' and *chiton* for 'imbricating hairs', referring to the short imbricating hairs these plants possess. The genus comprises one dozen of species of semi-succulent shrubs and trees with swollen trunks from Australia. Three species are suitable for growing indoors as pot plants. They are propagated from seeds or cuttings. ***B. acerifolium*** (Flame tree) is a very attractive and sculptural slow-growing tree growing to 12 m in the wild but indoors only up to 1 m. Stems are quite thick at base with erect branches, bearing 3- to 5-lobed green and leathery leaves, each lobe narrowly lanceolate and with a prominent midrib, 3.0-15.0 cm long, basal smallest and central the longest. ***B. paradoxus*** (*B. ramiflorus*) is a woody shrub, bearing nearly round, felted and lobed leaves, lobes being up to 15 cm in diameter. ***B. populneus*** is a woody shrub with a thick base, bearing stiff, leathery and axe-shaped leaves some 12.5 cm long patterned with a complex thin white veining. At younger stage it looks like a bonsai specimen. Fourth species, ***B. rupestris*** from Australia, is though not an indoor plant but can be trained so. It is a succulent caudiciform tree growing about 8 m tall with bottle-shaped trunk having about 1.5 m diameter and is the most drought tolerant in the genus though persisting drought defoliates the plant, and tolerates intense sun and a -7 °C temperature if is only for a brief period. In pots it may be kept up to 1.5 m tall. For indoor display, these should be kept at a sunny side. Mature plants bear single leaves some 7-10 cm long and 0.75 cm wide but young plants bear 4-5 lobed leaves. It is propagated through seeds.

Brassaia (Araliaceae)

Brassaia was named for Samuel von Brassai (*d.* 1897), a Hungarian botanist. The genus originates from India to Malaysia, Philippines and NE Australia with about 40 species of trees and shrubs having digitate, alternate and with mostly handsome large leaves. Its small 5-petalled flowers are borne in racemes or panicles, several together forming terminal clusters, followed by small berries. It is propagated through seeds or by air layering. ***B. actinophylla*** (syn. *Schefflera actinophylla*, Octopus tree/Queensland umbrella) from Queensland (Australia) is a popular indoor foliage plant when young, growing up to 3.0 m when planted in pots or tubs though it is a tree growing up to 12.0 m in nature. Its long-stalked leathery and shining rich green leaves are composed of 5-16 ovate-oblong and 15-30 cm long leaflets. Dark reddish flowers appearing in 30-60 cm long racemes from the mature stem tips seldom appear in potted plants. ***Heptapleurum arboricola*** (*Schefflera arboricola*; Parasol plant) from SE Asia is a fast-growing tree-like plant with about 10 leaflets radiating from each leaf-stalk. Indoor in pots, it grows some 1.0 m in height if growing-tips of the main stem are regularly removed. Its stalked and digitate leaves are 10-15 cm long with 7-16 stalked and arching leaflets which may be some 8-15 cm long with curved margins and pointed tips. It is less popular than *Schefflera actinophylla* as being of recent introduction. Its popular varieties are 'Hayata' (leaves greyish), 'Hong Kong' (dwarf grower) and 'Variegata' (leaves splashed yellow). ***Schefflera*** from tropical and warm temperate areas of Asia, Australasia and the Pacific Islands with some 150-200 species of evergreen shrubs and trees was named for J.C. Scheffler, a 19th century German botanist. The long-stalked and digitate leaves are alternate. Small 5-petalled flowers have umbellate grouping, and together make up racemes or panicles. They are propagated through seeds, cuttings or air layering. ***Schefflera digitata*** from New Zealand can be grown to about 2 m in pots indoors at a sunny position though in nature it attains some 8 m height. Shining dark green leaves have 7-10 unstalked, obovate and sharply toothed leaflets some 18 cm long. This species is meant only for sub-temperate to temperate regions.

Breynia disticha (*B. nivosa, Phyllanthus nivosa*; Snow bush; Euphorbiaceae)

Breynia was named for German merchant Jacob Breyne (1637-1697) and his doctor son Philip Breyne (1680-1764), both authors of works on lesser known plants. The genus comprises of 25 species of shrubs and small trees from SE Asia to Australia and the Pacific Islands though only one species is in cultivation for its foliage. Alternate leaves appear on branches in two ranks and normally the flowers are insignificant. It is basically a greenhouse shrub with slender branches densely clothed with highly decorative colourful leaves. Since it requires higher level of humidity therefore it may be kept indoors only for a short duration otherwise in the greenhose where temperature should not be less than 13 °C. It requires bright light but not direct sun, quite moist mixture, should be repotted every two years and leaves should be misted frequently. Its propagation is through mature stem cuttings, each cutting having a heel at the base. ***B. nivosa*** (*Phyllanthus nivosa*; Snow bush) from Pacific Islands grows to a height of about 1 m with slender habit having green stems, short-stalked green leaves with conspicuous white marbling, which are broadly ovate and 2.5-5.0 cm long, and greenish or reddish petalless flowers arise from the leaf axils. Variety 'Atropurpurea' bears purplish leaves and 'Roseopicta' is the natural choice having red, pink, white and green leaf variegation which appear as colourful flower.

Bryophyllum (Crassulaceae)

See chapter on '**Succulents Other Than Cacti**'.

Bucida buceras (Black olive; Combretaceae)

The genus *Bucida* comprises of six species of trees

and shrubs from Central America and West Indies. These are very slow-growing, tough and long-lasting. Generally one is worth indoor cultivation and that is ***B. buceras*** (Oxhorn bucida) whose up to 18 m trunks are leafy, irregularly thorny and branched horizontally but this can be kept up to 3 m by trimming to a required size in neat round form or in any other formal shapes in artistic pots. Its leaves are shiny dark green, alternate, and spoon-shaped with the size of 8.75 × 2.5 cm where leaf stalk length is roughly 2.5 cm. Its other most common species is ***B. spinosa*** whose trunks and branches are regularly thorny.

Butia (Pindo palm; Palmae)

This genus comprises of 12 palm species from Brazil and Argentina, having single stems. These can be grown in partial to full sun and propagated through seeds. ***B. capitata*** grows up to 6 m in height with thick trunk, bearing bluish-grey leaves where leaflets are angled upward-sided making a 'V' shape and the tips are curved downward. Its seeds take 3-4 months to germinate. ***B. mitis*** is a clustered fan-palm growing up to 9 m high though indoors the specimens reach to a height of up to 2.4 m in pots, bearing 90-120 cm long fronds where leaflets some 12 pairs and sub-leaflets are about 30 cm and 10 cm long, respectively. The plants are propagated through division. ***B. urens*** is single-stemmed and grows up to 12 m in height, bearing 6 m long leaves. Height indoor may be restricted at 3.6 m through potting and feeding and the leaf size to 1.8 m in length. Leaves and leaflets are similar to *B. mitis* but are less pleated, narrowly wedge-shaped and more irregular.

Buxus (Box; Buxaceae)

The genus is a Latin name taken up by Linnaeus. A genus of 70 evergreen shrubby or tree species from Europe, Africa, America (north and south) and Asia. They have opposite or alternate, small, oval to oblong leaves and clustered, petalless, inconspicuous and monoecious flowers where males have four stamens. Fruit is a horned capsule which opens explosively. For its growing, it requires well-drained soil in sun or shade. Propagation is through cuttings or seeds. ***B. microphylla*** is highly slow-growing attaining the height up to 20 cm with bare trunk having spreading and twisting branches, bearing closely set shining, oval and 1.25 cm long leaves. Its varieties are 'Baby Boxwood', 'Compacta' (Kingsville box) a favourite of bonsai makers, 'Green Beauty', 'Morris Dwarf', *etc.* with slight differences in growth habits ***B. sempervirens*** (common box) from Europe, W. Asia and N. Africa is the most popular bushy species which is ultra slow-growing and at the most attaining 3 m height though in pots it hardy reaches 1.5 m with a spread of about 80 cm. Stem is stout and square. The leaves are opposite, glossy dark green, ovate to oblong and notched at the apex. Honey-scented inconspicuous pale-green flowers with yellow anthers are borne in April. This is subject of the temperate regions. Varieties include 'Elegantissima' which is dense and compact with grey-green leaves edged silver, 'Handworthensis' with erect growth suitable for hedges and screens, 'Latifolia Maculata' with compact growth having broad leaves variegated yellow growing hardly up to 1.8 m, 'Pyramidalis' growing erect as a clipped bush, 'Suffruticosa' is a dwarf form good for edging, *etc.*

Caladium (Araceae)

The name *Caladium* has origin from a vernacular Malayan name *kaladi*. A genus of 15 perennial species coming up from underground tubers, is native to Tropical America and West Indies. These are stemless plants with long-stalked, memberanous and cordate to lanceolate decorative leaves liking filtered situation (never the direct sun) and most suitable for indoor gardening which may dry up during winter if not protected but leaves start emerging when summer arrives. Leaves may grow more than 30 cm across. Leaf colours may be green, pink, red, deep crimson, purple, bronze or metallic white, having variously striped, dotted, blotched, splashed and marked. Inconspicuous flowers are small in the form of arum-like spathes though appear seldom. They are propagated through offsets and division of tubers. They show their best performance during rains. Persisting temperature at 18.5 °C or below is dangerous to these plants. Though its all the species are only of botanical interest except ***C. humboldtii*** from Brazil where leaves are long-stalked, bright green with white patterning between the veins. ***C. × hortulanum*** (Angel's wings) are the hybrid varieties derived mainly from *C. bicolor* and *C. picturatum*. Leaves are white to pink and red with variable portion being green, however, white and pink ones are more susceptible to sun scorch. ***C. bicolor*** from Tropical South America grows up to 40 cm in height, bearing arrowhead-shaped leaves of varying sizes on long leaf stalks, colour being in shades of white and red along with green and in combination of these and the veins may be white, cream, red or crimson. Caladiums have many named varieties such as 'Candidum' (white leaves with green veins), 'John Peel' (densely veined with metallic red), 'Macahyba' (scarlet veins and lilac blotches), 'Pink Beauty' (pink mottled green leaves with darker pink veins), 'Pink Cloud' (mottled pink), 'Seagull' (deep green with broad white veins), 'Stoplight' (suffused crimson with margins narrow green), *etc.*

Calathea (Peacock plant; Marantaceae)

Calathea derives from the Greek *kalathos* meaning 'a basket' as its flower clusters fancifully resemble a basket of flowers, flowers keeping intact within their bracts. A tufted or clump-forming perennial genus of 150 evergreen species is from tropical America and West Indies. A few species even form tubers. Their leaves are ovate to lanceolate and attractively patterned which give its common name. They are excellent house and conservatory plants propagated by division. Its asymmetrical flowers may appear in clusters though only seldom. Its about 30 species are generally used for indoor cultivation, such as *C. aemula, C. argyraea, C. bella, C. carolina, C. crocata* (much preferred), *C. cylindrica* (quite long leaves up to 90 cm), *C. eximia, C. kegeliana, C. lancifolia* (syn. *C. insignis*), *C. leopardina, C. lietzei, C. lindeniana, C. louisiae, C. makoyana* (syn. *Maranta makoyana*), *C. medio-picta, C. metallica, C. micans, C. musaica, C. ornata, C. picturata, C. princeps, C. roseo-picta, C. rotundifolia, C. rufibarba, C. stromata, C. veitchiana, C. vittata, C. warscewiczii, C. wiotii* and *C. zebrina*. The leaf patterning can broadly be divided into narrow stripes, vertical striping and feathering, as curved brush strokes, and distinct patterning. Leaves this way are really very handsome for keeping the plants indoors.

Calibanus (Agavaceae)

See chapter on '**Succulents Other Than Cacti**'.

Callisia (*Spironema*; Commelinaceae)

Callisia derives from the Greek *kallos* for 'beauty', owing to these being the beautiful foliage plants. These are the prostrate evergreen perennials with 12 species from Tropical America, bearing ovate to oblong-lanceolate and clasping leaf bases. Stalkless flowers being inconspicuous arise in clusters usually from leaf axils or similar bracts. These are suitable for hanging baskets as well as in pots where these grow from tufts of roots which colonize. These being rampant straggly grower are propagated every year through cuttings otherwise the plants will last only one or two years. The leaves of ***C. congesta*** are broadly lanceolate and some 5 cm long and when young these are purplish. Its var. 'Variegata' bears violet leaves with white vertical stripes. ***C. elegans*** (*Setcreasia striata*; Wandering Jew/Pinstriped inch plant/Striped inch plant) is from Mexico. The whole plant is clothed with velvety-hairs. Highly branched stems can trail more than 60 cm in length, with triangular and clasping leaves which are greyish to olive-green with striped paler above and purple below and about 5 cm long. White 3-petalled flowers some 2 cm across are borne during autumn and winter. ***B. fragrans*** (*Spironema fragrans, Tradescantia dracaenoides*) from Mexico forms rosette initially but with age elongates and forms long runners. Leaves are glossy green, little fleshy, narrow lanceolate and some 20-25 cm long and 4 cm wide. White fragrant flowers some 1.5-2.0 cm across appear in attractive terminal panicles. The leaves in the var. 'Melnickoff' are striped cream and/or brownish-purple in varying widths, and stems run up to 90 cm. ***C. navicularis*** (*Tradescantia navicularis*) is a creeping and rooting perennial growing more than 50 cm long though in height only hardly up to 10 cm. The oval leaves appearing in two rows are keeled and clasping, some 2.5 cm long, and 3-petalled pinkish-purple flowers appear in leaf axils during summer to autumn. ***C. repens*** from Mexico, Central America and South America as south as Brazil, is a dark-stemmed trailer growing up to 1 m long with rooting wherever it comes in the contact of soil. Overlapping and bright green leaves which are some 2.5 cm long, ovate and ciliate, crowd when given full light conditions, and small, inconspicuous and white flowers appear from the upper leaf axils. ***C. tehuantepecana*** stem is pendent and grows up to 90 cm in length, bearing closely packed spiralling bright green leaves some 2.5 cm long.

Campelia zanonia (Mexican flag; Commelinaceae)

The derivation of *Campelia* is not certain. The genus native of Mexico to Brazil and West Indies, comprises an erect clump-forming perennial species, *C. zanonia is* allied to *Dichorisandra, Geogenanthus* and *Tradescantia*. It is the most striking of the cultivated Commelinaceae. It is a tall fleshy herb growing up to 1.8 m tall with fleshy stems and requires partial sun. The leaves are rich green, lustrous, stem-clasping, spiralling, broadly oblanceolate to elliptic, size being 18-30 × 7.5 cm, tip-pointed and narrowing to the broad but short leaf stalks. It is propagated through cuttings. In cultivation, it is represented by the var. 'Mexican Flag' ('Albolineata', *Dichorisandra* 'Albomarginata') where leaves are strongly white-striped and narrowly red-edged.

Campyloneurum (Polypodiaceae)

See chapter on '**Ferns and Allied Plants**'.

Canistrum (Bromeliaceae)

See chapter on '**Bromeliads**'.

Carex (Sedge; Cyperaceae)

Carex derives from the Greek *keiren* meaning 'to cut', referring to the sharp edges of the leaves of some species. It is a very large genus with some 1,500-2,000 species of grassy-leaved perennials, out of which only a few provide durable ornamental pot plants. The ornamental species are tufted to clump-forming, evergreen and with arching leaves though their catkin-like flowers are undecorative. These are propagated through division of clumps. These are very slow-growing and require cool condition, plenty of sun, continuous moistening of media with a good drainage, and proper ventilation. ***C. brunnea*** native to China, Japan, Taiwan and Philippines is clump-forming where leaves are bright- to yellowish-green and 25-45 cm long. The leaves of its var. 'Variegata' are striped gold and bronze. ***C. buchanani*** (Leatherleaf sedge) is a fully hardy reddish-brown tufted plant growing up to 60 cm high and with 20 cm spread, where stems are solid and triangular and looks always as if dried and the insignificant spikelets are brown which appear during summer. ***C. conica*** from Japan and Korea is clump-forming dark green stiff and some 20 cm long leaves. The leaves of its var. 'Variegata' are cream-striped. ***C. elata*** (*C. stricta*; Tufted sedge) is a fully hardy tufted evergreen perennial sedge growing up to 1 m high and with about 15 cm spread. The stems are solid and triangular and bear blackish-brown spikelets during summer. Its leaves are little glaucous, very narrow and coppery but bases red. The var. 'Aurea' (Bowle's golden sedge) bears golden-yellow leaves. ***C. foliosissima*** 'Albo-mediana' is a tufted plant growing up to 13 cm in height, bearing bright green leaves edged with a white stripe. ***C. grayi*** (Mace sedge) is a fully hardy tufted evergreen perennial sedge growing up to 60 cm high with spread of 20 cm. Its leaves are bright green and the large female spikelets appearing during summer mature to greenish-brown, pointed and knobby fruits. ***C. morrowii*** (*C. oshimensis*; Japanese sedge grass) from Japan is a fully hardy tufted and clump-forming evergreen perennial sedge growing up to 50 cm high with a spread of some 25 cm, having solid and triangular stems and insignificant spikelets.It has semi-lustrous mid-green and narrow leaves some 20-45 cm long. It has produced certain variegated varieties. 'Evergold' ('Aurea') bears yellow-striped leaves some 20 cm long. *C. m. variegata* (*C. m. albomarginata*, *C. m. expallida*) bears arching leaves some 10 cm long, and almost entirely white with a narrow green stripe. There are many other varieties where leaves may be of differing size and patterned even on the reverse. ***C. ornithopoda*** from Europe to Turkey is a tufted to clump-forming plant, bearing light to mid-green and soft leaves some 20 cm long. *C. o.* 'Variegata' leaves are striped creamy-white. ***C. pendula*** (Pendulous sedge) is a graceful tuft-forming evergreen perennial sedge growing up to 1 m in height and 30 cm in spread, bearing solid triangular stems, narrow green leaves some 45 cm long, and pendent greenish-brown floral spikes in summer. ***C. riparia*** (Greater pond sedge) has a very popular and fully hardy var. 'Variegata' which is vigorous, evergreen and grows up to 60 cm in height, bearing solid and triangular stems, broad mid-green white-striped leaves, and dark brown spikelets that are narrow and bristle-tipped.

Carludovica (Cyclanthaceae)

Carludovica was named so for Charles IV and his Queen Louisa of Spain. A native of Ecuador, this is stemless (sometimes with a lax creeping stem) palm-like plants with soft leaf texture, often united with Pandanaceae. It survives well under filtered situation but full day continuous lighting, at the least 60 per cent humidity and at least 18.5 °C minimum temperature. The leaves are usually stalked (sometimes sessile) and flabellate. The flowers are monoecious, the two sexes being on the same spadix inclosed in a 4-leaved spathe, staminate flowers having many lobed-calyx and many stamens out of which 4 surround the pistillate flowers which has a 4-sided ovary, 4 barren stamens and 4-lobed calyx, and the fruit is a 4-sided many-seeded berry. *C. palmata* and certain others are the source of Panama hats and fibres. They are suitable for bedding. Their ripe fruits are ornamental until their bursting. Though difficult, but the propagation is through seeds and division. Seeds are very small which should thoroughly be washed free from the pulp and sown on the surface of finely chopped sphagnum moss where these will germinate within two weeks if the medium is moist with proper drainage and briskly hot conditions. ***C. atrovirens*** from Colombia bears deep green and glabrous leaves cut very deeply into two lobes. ***C. × elegans*** leaves are 4-5-lobed, and very deeply cut into strap-like straight divisions. ***C. humilis*** from Colombia is one of the best dwearf species where blades are angular and 2-lobed at the summit though not divided completely but with high protruded lobes which may be up to one foot wide. ***C. imperialis*** from Ecuador bears short and prostrate caudex, tumid-based, canaliculated and purplish petioles, and shining green 2-lobed blades with ovate-lanceolate and entire segments having quite prominent veins and the lobes being about 12.5 cm wide. ***C. palmata*** from Peru is the most common trunkless semi-aquatic species growing up to 2.4 m high with petiole length of 0.9-1.8 m in the wild but indoors these require winter protection. The plants are terete, glabrous and unarmed, leaves are dark green, gracefully spreading up to 90 cm in diameter, 3-5-lobed, each lobe cut into narrow segments and drooping at the margin. ***C. plumerii*** (*C. palmaefolia*) from Martinique bears erect caudex where blades are divided into two lanceolate but fan-like folded divisions, upper surface being bright green, below paler and spadices pendent. ***C. rotundifolia*** from Costa Rica produces shorter but distinctly pubescent petioles and much larger orbicular blades with 3-4 lobes.

Caryota (Fishtail palm; Palmae)

Caryota derives from the Greek *karyon* 'a nut', referring to the large seeds. A genus of 12 large and noble palm species from Sri Lanka, Indo-Malaysia, Solomon Island and NE Australia, having distinctive appearance. Leaves are fern-like bipinnate with lop-sided leaflets. These flower after attaining the maturity and die after the fruits ripe. Young plants are very attractive and are kept indoors for enjoyment. The large ones are impressive in lobby and other public places. These are propagated by seeds. ***C. mitis*** from Myanmar, Malaysia and Philippines grows up to 12 m with erect stems when planted out but in pots only up to 3 m. With age, these sucker and form the clumps with some 1.5 m long leaves which bear lopsidedly fish-fin or tail-shaped leaflets. ***C. urens*** (Sago palm/Toddy palm/Wine palm) from India, Sri lanka and Malaysia is a fast-growing, erect and solitary-stemmed plant up to 24 m in height under natural conditions but in containers only up to 5 m. Leaves are up to 3 m long (nearly half in container-grown ones) and the opposite or alternate leaflets are broad, oblique, unevenly lobed and pleated with or without toothed tips.

Cephalocereus (Old man cactus; Cactaceae)

See chapter on '**Cacti**'.

Ceratonia siliqua (Algaroba/Carob/Locust/St. John's bread; Leguminosae)

Ceratonia derives from the Greek *keras* 'a horn', as the seed pods are as hard as the horns. The genus comprises one evergreen tree species from the Mediterranean region growing up to 12 m in height in the wild, having biblical associations, bearing red flowers and carob fruits though it is grown in containers mainly for its attractive foliage where it is tamed to grow not more than 2 m. Its leaves are dark green, leathery, paripinnate with 4-10 leaflets, each measuring some 2.5-8.0 cm long. Flowers are small and petalless, which are followed by tough leathery pods some 10-20 cm long, filled with a sweet edible pulp and shiny, oval and hard seeds. Only old plants produce the flowers. It is propagated by seeds.

Ceratopteris (Parkeriaceae)

See chapter on '**Ferns and Allied Plants**'.

Ceratozamia (Cycadaceae)

In Greek language, *Ceratozamia* means 'horned zamia', owing to the horned scales of the cones which is the only its distinguishing feature when compared with the genus *Zamia*. These are very slow-growing but handsome Mexican foliage plants with six species, having cycas-like leaves, though in less cultivation. These have erect trunk which are crowned with whorls of cycas-like pinnate, petiolate and unarmed leaves. Flowers in the form of stalked-cones, appear from among the leaves, which rarely contain the seeds. These are propagated mainly through offsets and sometimes by seeds. By burning out centre, the plant is forced to produce a number of offsets from the crown as well as from the trunk which can be separated and grown as separate plants. ***C. kuesteriana*** from Mexico bears a very slow-growing, short, thick and rounded trunk though probably branches from the base. Leaves are limited in number and some 1.2 m in length, leaflets are leathery, alternate, lanceolate, some 15-25 cm long and the edges turning downward. ***C. mexicana*** from Mexico is an excellent decorative plant having short but thick trunk which have scars of fallen leaf stalks. Prickly petioles are some 12-15 cm long and at younger stage are quite untidy. Leaves are dark green, pinnatifid with very numerous lanceolate leaflets which grow from 15 to 30 cm or more in length. Male and female cones are borne on separate plants annually, females being 23-30 cm long and 10-15 cm thick with 2-horned scales (the only difference with cycas), and the longer male cones are narrower and on hairy stalk though scales are represented with two small teeth. If closely observed by planting together, only minor differences in leaf shape and size are noticed among the varieties such as *C. m. latifolia*, *C. m. longifolia* and *C. m. miqueliana*.

Ceropegia (Hearts entangled/Hearts-on-a-string/Rosary vine; Asclepiadaceae)

See chapter on '**Succulents Other Than Cacti**'.

Chamaecereus (Peanut cactus; cactaceae)

See chapter on '**Cacti**'.

Chamaecyparis [Cupressaceae (Pinaceae); False cypress]

Chamaecyparis has derived from the Greek *chamai* meaning 'dwarf' and *kuparissos* for 'cypress', referring to its affinity with *Cypress*. The genus comprises six species native to North America and East Asia and is allied to *Cupressus*. These yield valuable timbers but the horticultureal species are often shrubby. They are evergreen with opposite scale-like leaves in four rows which are very closely set on the compressed branchlets. Flowers are small and monoecious, pistillate globose and inconspicuous but staminodes are oblong, conspicuous by their abundance and are red or yellow. Globular cones are small, having 6-11 bracts, each with 2-5 winged seeds which ripen the first season though in case of *Cupressus* ripening takes two seasons and the cones are larger with 4 or more seeds. These grow best under moist sandy-loam soil which is not waterlogged and under filtered situation. They should be multiplied through seeds and cuttings. No species as such are worth growing for ornamental use because of their size, however, the dwarf-growing varieties are suitable for the purpose such as ***C. lawsoniana*** 'Ellwoodii' growing erect up to 3 m with incurved blue-grey leaves, *C. l.* 'Gnome' growing only 50 cm tall with greenish-blue foliage, *C. l.* 'Minima' growing globular up to 1 m with light green foliage, ***C. obtusa*** 'Coralliformis' growing up to 50 cm high with thread-like shoots, *C. o.* 'Intermedia', a globular plant some 30 cm high with downward spread and light green foliage, *C. o.* 'Kosteri', a sprawling bush growing up to 2 m with lustrous and twisted foliage, *C. o.* 'Nana' growing up to 1 m with flat topped shape, *C. o.* 'Nana Aurea' growing upto 2 m with golden-yellow leaves, *C. o.* 'Nana Gracilis' growing up to 2 m with glossy foliage, *C. o.* 'Nana Pyramidalis', a conical bush growing up to 60 cm high with horizontal and cut-shaped leaves, ***C. pisifera*** 'Filifera Nana' growing up to 60 cm high with whip-like branches, *C. p.* 'Nana' growing up to 50 cm high with dark bluish-green foliage, *C. p.* 'Plumosa Rogersii' growing up to 2 m high with yellow foliage, *C. p.* 'Andelyensis', a conical bush growing up to 3 m high with aromatic blue-green leaves, *etc.*

Chamaedorea (Palmaceae)

Chamaedorea derives from the Greek *chamai* for 'dwarf' and *dorea* meaning 'a gift', as bright fleshy fruits of some species are coriaceous. A genus of 100 species of small palms from Mexico to northern South America. Only a few species grow with solitary stems but others sucker from the base, producing bamboo-like stems and pinnatifid (rarely entire) leaves. Pendent or erect spike-like flowers are dioecious, followed by single-seeded berry like fruits. If atmosphere is not too dry, these can be enjoyed indoors in containers and in conservatory. These are multiplied through detaching the suckers or by seeds which can be germinated by giving 20-24 °C temperature. ***C. costaricana*** from Costa Rica is a clump-forming, suckering and a handsome foliage plant, bearing dark green stems and growing up to 5 m in the wild but indoors these can be grown up to 2 m in pots. Its leaves are rich green, some 60-90 cm long and pinnatifid with about 20 or more pairs of linear-lanceolate leaflets. ***C. elegans*** (*Neanthe elegans*; Dwarf mountain palm/Parlour palm) from Mexico and Guatemala is a suckering plant with slender stems growing up to 3 m high, bearing dark green, leathery and some 60-120 cm long pinnatifid leaves with broadly lanceolate leaflets. *C. e.* 'Bella' (*Neanthe bella*) is similar to its parent but is slow-growing attaining the height of hardly 1 m. It has an added beauty as in pots indoors it flowers with erect clusters of yellow flowers, followed by small globular fruits. ***C. erumpens*** (Bamboo palm) from Honduras is a clump-forming slender palm growing up to 3 m high but indoors less and tolerates dry atmosphere. It bears some 60 cm long leaves with 5-15 pairs of ovate and recurved leaflets. ***C. graminifolia*** from Guatemala is a clump-forming, reed-like, and slender palm growing up to or more than 2 m, bearing dark green, pinnatifid and some 60 cm long leaves having some 30 cm long and about 50 slender leaflets which arch elegantly. Flowering stems grow erect. ***C. metallica*** (Miniature fish tail palm) from Mexico is a handsome single-stemmed palm some 60 cm high, bearing dark bluish-green undivided leaves some 30 cm long and 15 cm wide. The leaves are cleft at apex similar to fish-tail. ***C. microspadix*** from eastern Mexico is similar to *C. erumpens* but the flowers are large and white and the fruits are red. ***C. oblongata*** from Mexico to Nicaragua grows to 3 m height with solitary stem, bearing deep green, leathery, 1.2 m or more long and pinnatifid leaves with lanceolate and 12-18 leaflets. Its flowers are green and white and the fruits black. ***C. seifrizii*** (Reed palm) from Mexico, a highly decorative room plant is clump-forming with slender canes, bearing some 60 cm long and pinnatifid leaves with 26-30 well-spaced narrow and lacy leaflets.

Chamaeranthemum (Acanthaceae)

In Greek, it is formed of two words, *chamai* 'dwarf or on the ground' and *Eranthemum*, the related taller genus. The genus comprises 4-8 small, sub-shrubby, perennial and evergreen species from Tropical America. These are allied to *Eranthemum* but are quite smaller. These are grown for their decoratively patterned leaves though its small, tubular and white to mauve or yellow flowers are insignificant. They are suitable for pot-planting to be enjoyed indoors and in warm conservatory. These are multiplied by division or through cuttings. Only three species are cultivated. ***C. gaudichaudii*** from Brazil produces mat-forming prostrate stems growing about 45 cm across, bearing deep green leaves with a bold silvery-grey patterning, elliptic to oblong-ovate and some 5-10 cm long, and white and lavender flowers. It is most suitable for growing as hanging basket plant, *vis-a-vis* ground cover. ***C. igneum*** (*Aphelandra igneum*, *Stenandrum igneum*, *Xantheranthemum igneum*) from Peru produces mat-forming prostrate stems growing about 25 cm across, bearing deep velvety bronzy-green leaves patterned with clear yellow to reddish veins, oblanceolate to oblong-elliptic and 5-8 cm long. This species produces comparatively showy yellow flowers. ***C. venosum*** from Brazil produces mat-forming prostrate to decumbent stems growing about 40 cm across, bearing grayish-green leaves overlaid with silvery veins, downy, leathery, broadly ovate but base often heart-shaped and up to 8 cm long, and the flowers are lavender and white. Var. 'India Plant' bears doubly large leaves.

Chamaerops humilis (Dwarf fan palm/European fan palm; Palmaceae)

Chamaerops derives from the Greek *chamai* meaning 'dwarf' and *rhops* 'a bush', owing to its short growing stature. The genus comprises only two species of fan palm from western Mediterranean but only one is cultivated. Though it is not as decorative as *Chamaedorea* but withstands cooler conditions. It is propagated through suckers and seeds. ***C. humilis*** is a clump-forming and suckering palm, growing up to 1.5 m high though indoors in pots even low. Its leaf stalks are spiny-margined and some 1 m long, bearing fan-shaped, some 60-90 cm across and deeply cut into slender and linear lobes. Small flowers appear in dense panicles from a large spathe with six yellowish perianth lobes, and are followed by small date-like yellow to brown fruits.

Chirita (Gesneriaceae)

See chapter on '**Succulents Other Than Cacti**'.

Chlorophytum (Liliaceae)

Chlorophytum derives from the Greek *chlorus* meaning 'green' and *phytun* 'a plant'. It is a genus of 215 stemless perennial species native to tropical and southern Africa, but has become naturalized in other parts of the world, including western Australia and San Francisco (California). Their roots are shortly rhizomatous, fleshy and some 5-10 cm long. *Chlorophytum comosum* grows to about 60 cm high. Variegated forms in particular are used as house plants. The long, narrow and linear to lanceolate-ovate leaves with parallel veins reach a length of 20–45 cm and a width of 6–25 mm. Starry flowers are 6-tepalled in loose racemes on slender and branched stems. The inflorescences carry plantlets at the tips of their branches, which eventually droop and touch the soil, developing adventitious roots. The stems (scapes) of the inflorescence are sometimes loosely called as 'stolons'. These are propagated through plantlets and by division of clumps. Spider plants are easy to grow, being able to thrive in a wide range of conditions. They will tolerate temperatures down to 2 °C, but grow best at 18-32 °C. Plants can be damaged by high fluoride or boron levels. Spider plants have also been shown to reduce indoor formaldehyde air pollution, and approximately 15 plants would neutralize formaldehyde production in a representative energy-efficient house. Nordal and Thulin (1993) described nine new species of *Chlorophytum* such as *C. aplanatum,. filifolium, C. hiranense, C. littorale, C. nervosum, C. pendulum, C. petraeum, C. pterocarpum* and *C. ramosissium*, while further one more species, *C. zingiberastrum* was described from SE Africa (Nordal and Poulsen,1998). **C. *capense*** (*C. elatum, Anthericum elatum*; Spider plant) from South Africa has branched stems some 1.2 m long, bearing pale-green leaves some 25-60 cm long but no plantlets. Its white flowers are 1.5-2.0 cm across. Its var. 'Variegatum' leaves are longitudinally striped with pale creamy-yellow. ***C. comosum*** (*Anthericum comosum*; Airplane plant/Hen-and-chickens/Spider plant) is though similar to *C. capense* but arching leaves some 20-45 cm long in rosettes are borne on branched stems. Inflorescence bears about 2 cm across flowers along with plantlets so stems become pendent and only because of this habit it is used in hanging baskets. Its var. 'Mandaianum' has a beautiful pale-yellow central variegation, 'Variegatum' has marginal yellow-striping, and 'Vittatum' (St. Bernard's lily/Spider plant) leaves are 23 cm long and green with creamy-white longitudinal band in the centre. ***C. laxum*** from Tropical Africa is tufted to clump-forming with arching, linear to narrowly oblanceolate and 13-20 cm long leaves. Flowers are white some 2 cm across in slender panicles. *C. l.* 'Variegatum' (*C. bichetii*) is a colonizing plant with solid mat of foliage. Its leaves are 20 cm long and cream-striped and margined.

Chrysalidocarpus (Palmaceae)

The generic name derives from the Greek *chryos* 'gold' and *karpos* 'a fruit' or perhaps from the *chrysalid*, the gold-splashed pupa of the butterfly, owing to appearance of seeds of certain species. A genus of 20 species from Madagascar, Comoro and Pemba Islands, areca palm grows to 6-12 m tall (in pots indoors up to 2 m) with branched basal stems and prefers bright filtered sun as direct bright light burns the fronds. Its leaves are 2–3 m long, arched and pinnate, with 40-60 pairs of leaflets. It bears panicles of yellow flowers in summer. Arecas are propagated from seeds by soaking them for 10 minutes in a solution of hot sulphuric acid, which may germinate in about 6 weeks. Fresh yellow to ripe seeds should be planted with the top of the seed barely visible and germination temperature maintained between 26.6-29.4 °C. Lower temperatures may double the germination time. Seed storage at low humidity and low temperature is detrimental to germination. Cleaning seed is not essential if they are planted immediately. If seeds are to be stored, clean the yellow to fully-ripened red seeds, air-dry them properly, treat them with a seed protectant, and store at 24 °C. These can be planted in pots for keeping indoors where these can be retained till these are young or up to 2m in height. ***C. lutescens*** (*Areca lutescens*; Areca palm/Bamboo palm/Butterfly palm/Golden cane palm/Golden feather palm) from Malagasy produces clustered stems

2-3 m tall in pots or containers, bearing pinnate, arching and 0.9-1.8 m long leaves with many linear leaflets and yellowish stalks.

Cibotium schiedei (Dicksoniaceae)

See chapter on '**Ferns and allied Plants**'.

Cissus (Vitaceae)

See chapter on '**Succulents Other Than Cacti**'.

Cleistocactus (Silver torch; Cactaceae)

See chapter on '**Cacti**'.

Cleyera (*Eurya*; *Theaceae*, formerly Ternstroemeaceae)

Cleyera was named for Andrew Cleyer, Dutch physician of the 17th century. The genus comprises of 17 evergreen species of shrubs and trees from Asia including Japan, and these are distinguishable by free or scarcely coalesced petals, hairy anthers, numerous ovules and scarcely bracted flowers, and flowers comprise of 5 sepals with 2 bractlets, 5 petals and 2-3 stigmas, and 2-3-celled berries. The genus is a rarity, liked by their foliage and flowers, and flowering occurs during summer. These are half hardy plants preferring semi-shady position, and moist acidic soil. These are propagated through semi-ripe cuttings. ***C. ochnacea*** (*C. japonica* Sieb. & Zucc., *Eurya ochnacea*) attains the height of up to 1.8 m and bears glossy, oval-oblong and entire leaves which are acute at both the ends and are veined above. It produces numerous fragrant white flowers, followed by red berries. However, the most cultivated species is ***C. japonica*** from Japan which is also a shrub growing indoors up to 60 cm in pots and bears oblong-lanceolate, veinless, and minutely serrate leaves but only at the apex. Flowers are small, white, saucer-shaped and fragrant and the fruits are small spherical and black. *C. j. tricolor* (*C. j. variegata*) has glossy dark green leaves with greyish variegation, more prominent being in younger ones, and the leaf margins are white and rose. This plant is most frequently grown. It is a slow-growing shrub which can be kept in size and shape by occasional removal of shoot tips.

Clusia (Guttiferae)

Clusia is named for Charles de l'Ecluse (Carolus Clusius of Artois) (1526-1609), the famous Belgian botanist and author of *Rariorum Plantarum Historia* and many other works. An evergreen genus of 145 small trees, shrubs and climbers, often living epiphytically on forest trees and mossy rocks of tropical and sub-tropical America, Malagasy and New Caledonia. Though some species are grown only for their handsome foliage but others for both leaves and flowers. The leaves are little lustrous, rich green with lighter zones, leathery, oval and obovate. Small or large flowers, sometimes beautiful have 4-6 sepals, 6-9 petals and appear in terminal clusters. They are propagated by cuttings or layering. Only two species, ***C. grandiflora*** (Scotch attorney) from Guyana is a wide-spreading robust shrub growing epiphytically up to 6 m tall on large trees but indoors hardly 90 cm. Its leaves are clustered towards the tip, dark green above and paler beneath with dark lines, elliptic to obovate and 15-30 cm long. Pink and white flowers some 5 cm across appear only on outdoor planted shrubs; and ***C. rosea*** (Autograph tree/Balsam apple/ Fat pork tree) from Florida and W. Indies to Venezuela is worth growing indoors for its handsome foliage. It is a strangling epiphytic shrub some 6 m in height but indoors it is grown not more than 90 cm so that its foliage beauty is enjoyed, therefore, overgrown ones should be replaced, bearing short (2.5 cm) leaf stalks, and the leaves in opposite pairs, very thick, lustrous deep green, spathulate, 20 cm long, 10 cm wide with tips square and horizontally oriented. Flowers some 5.0 cm in diameter are pink or white, followed by 8 cm wide, globose and greenish fruits but flowers appear only when planted outdoors. 'Aureo-Variegata' has thick irregular stripes of yellow and cream on green background. 'Marginata' leaves are edged with yellow.

Coccoloba uvifera (Polygonaceae)

Coccoloba derives from the Greek *kokkolobis* for a kind of grape, on account of appearance of its fruits. A genus of 150 species of shrubs, trees and woody climbers from tropical and sub-tropical America. These are unusual foliage plants especially for semi-tropical regions, propagated through seeds, cuttings and layerings. Only 2-3 species are grown in containers indoors. ***C. diversifolia*** (*C. floridana*, *C. laurifolia*; Pigeon plum) from Florida grows to 6 m high in the wild but indoors only up to 1 m, bearing short and thick-stalked leaves some 10 cm long and broadly oval to oblong. ***C. uvifera*** (Seaside grape) from sandy sea shores of Florida to Brazil is a tender species growing in the wild up to 6 m high but indoors only up to 2 m in pots. Its branches are thick, spreading and nearly prostrate. The leaves are lustrous green, stiff and leathery, with or without a notch at the base, nearly round to kidney-shaped and some 20 cm across, red-veined but with age changing to pale-cream, and somewhat wavy-margined. Fragrant white flowers some 25 cm long appear in dense racemes, followed by long purple grape-like berries some 2 cm long though usually it does not flower indoors. This is most suitable for the conservatory or greenhouse, however, indoors these should be kept at a most sunny site with protection in winter. Its varieties such as 'Aurea' and 'Variegata' are heavily and irregularly variegated with creamy patches.

Cocos nucifera (Coconut palm; Palmae)

Cocos derives from the *coco*, which in Portuguese is for 'monkey', as the nuts resemble with the head of certain monkeys. The genus comprises one species of elegant palm whose place of origin is not certain as some authorities state this as a native of Melanesia though some others tropical western South America. The plants attain a natural height of up to 24 m though in containers only 2 m. Leaves are pinnate, some 2 m long and arching. The plants yield the popular coconut fruit. This is sometimes used as container-grown house plant though its requirements are akin to those of sun-loving plants with high humidity, a minimum night temperature of 18 °C, and a well-drained sandy soil rich in lime and potash. It is propagated through seeds (nuts) at 26-30 °C.

Codiaeum (Croton; Euphorbiaceae)

See chapter on '**Codiaeum**'.

Codonanthe (Gesneriaceae)

Codonanthe derives from the Greek *kodon* 'a bell' and *anthos* 'a flower', as in some species the tubular base of the flowers is bell-shaped. There are 15 tropical American epiphytic evergreen species of perennials, sub-shrubs and climbers, some of which are suitable as house plants and for hanging baskets. Leaves appear in pairs and are ovate to elliptic, and in their axils these bear small tubular flowers with five corolla lobes. The fruits are berry-like and attractive brightly coloured. These are propagated through seeds and cuttings. ***C. carnosa*** from Brazil is a sub-shrubby, compact and spreading species with neat growth, having about 40 cm height and width, bearing elliptic to obovate leaves some 8 cm long and which are sometimes in whorls of 3-4. Its white or pink-flushed flowers with yellow throat are about 2 cm long and appear from spring to summer. ***C. crassifolia*** from Central America to Peru bears some 30 cm long, node-rooting, prostrate or hanging stems. Leaves are green, little fleshy, red-glandular below, elliptic to ovate and some 5 cm long. White or pink-flushed flowers are about 2.5 cm long with yellow throat, followed by showy red or pink and egg-shaped berries about 1.2 cm long. ***C. gracilis*** from central America bears 30-60 cm long, prostrate or hanging stems with slender-pointed, elliptic to lanceolate leaves which are 2.5-4.0 cm long. Flowers are white and some 2 cm long. ***C. macradenia*** from central America is almost similar to *C. crassifolia* but leaves are broader and only 3 cm long, white flowers are red-spotted or with throat mottling and some 2.5-3.0 cm long with curved tube and the red-purple fruits are egg-shaped.

Coffea arabica (Arabian coffee; Rubiaceae)

Coffea derives from the Arabic word *kahwa*. A native of the Old World, mainly Africa, the genus comprises 40 evergreen shrubs with opposite pairs of lanceolate to ovate leaves, and starry, tubular and white flowers in axillary clusters, followed by fleshy and red berry-like fruits containing two seeds each. They are elegant house plants grown for their attractive foliage but require high humidity. They are propagated through seeds and tip-cuttings. ***Coffea arabica*** is the most common one evergreen plant growing in the wild up to 5 m tall but indoors in containers only up to 1 m, bearing dark glossy green, ovate to elliptic, wavy-edged, some 5-15 cm long and conspicuously veined. Fragrant white flowers some 1 cm across with narrow petals appear solitary to six in a cluster in summer, followed by 1.5 cm long fruits. *C. a.* 'Nana' grows quite compact and is most suitable for growing indoors.

Coleus (Coleus/Painted nettle; Laminaceae)

Coleus derives from the Greek *koleos* meaning 'a sheath', as filaments unite to enclose the style. An annual to perennial genus of 150 herbaceous species is native to tropical Africa and SE Asia. Its important cultivated species are *C. blumei* (*Solenostemon scutellarioides*), *C. frederichii*, *C. pumilus* (*C. rehneltianus*) and *C. thyrsoideus*. Coleus is used for its brightly coloured leaves that come usually in multicolours. In addition, leaf margins may be entire, serrated, or lobed and fimbriated. It is a tender perennial highly susceptible to frost. Different coleus cultivars are adapted to different amounts of light from shade to full sun. Coleus is propagated by seeds and herbaceous cuttings. Coleus thrives well above 21 °C though is susceptible below 10°C temperatures.

Colocasia (Taro; Araceae)

Derivation from the Greek name *kolocasia*, originally used for the rhizomes of sacred lotus. A native of SE Asia to Polynesia, it is a perennial genus of eight species grown ornamentally for their long-stalked leaf clusters originating from the ground level though their leaf-stalks, the leaves and rhizomes or tubers are also edible as cooked vegetables. These can be grown in warm conservatories and in the house with ample humidity. These are propagated through division of their mature rhizomes or tubers. ***C. esculenta*** (Cocoyam/Dasheen/Elephant's ear/Taro) is a stemless simi-aquatic plant from SE Asia where leaves grow up to 1 m including the leaf-stalks some 20-60 cm long, from starchy rhizomes directly. Leaves are mid- to dark-green, basally cordate with prominently raised white veins. Flowering occurs on large plants only if left undisturbed for many years with plenty of water and organic matter. The spathes are in cylindrical shoots, closely overlapping in plume-like branchlet-sprays. Juvenile leaves are slender and pointed. It is similar to *Chamaecyparis* but differs in having their cylindrical shoots arranged in plume-like branchlets and in their larger cones. Male strobili are small, cylindrical and rounded to ellipsoid, female strobili are woody cones having angular and spiny scales with a central bump, and many winged seeds. They are propagated through seeds or cuttings with heel. ***C. arizonica*** (Rough barked Arizona cypress) from Arizona grows up to 10.5 m tall with an ovoid head but in container the height can be reduced to its one-fourth. Its bark is greenish-brown, finely fissured and stringy, the leaves are green-grey some 2 mm long, sharply pointed, and the cones are often clustered, globose and 1.5-2.0 cm wide. ***C. glabra*** from Arizona is of the same size as *C. arizonica*. Its bark is smooth, purple and blistered until peeling of large chunks and leaving the pale-yellow patches. Leaves are small, blue-grey with many having a small white spot, blunt, closely pressed and fully cover the twigs. Pale-yellow male flowers are clustered on the tips of the shoots during autumn, these become bright yellow when shedding the pollen in April, and the cones which are 2.5 cm across are purple-brown. ***C. macrocarpa*** (Monterey cypress) from California is taller than both the above species, younger trees columnar with conic tops but mature ones are flat-topped. Its red-brown bark is criss-crossed with shallow ridges, and densely clustered shoots are clothed with rich green leaves which give an aroma when crushed. Golden male flowers appear in terminal clusters, and the shiny brown cones are 2.5-3.75 cm across clustering along the shoots. ***C. sempervirens*** (Mediterranean or Italian cypress) from France to Iran is 7.5 m tall which forms a square-topped column when mature. The foliage is dull dark green, and the cones about 2.5 cm across are scattered all over the crown which with age turn dull grey.

Cussonia (Araliaceae)

Cussonia commemorates Dr. Pierre Cusson (1727-1785), a French doctor and professor of botany at Montpelier. It is a genus of 25 species of evergreen shrubs and trees native to Tropical and South Africa. Most of the species are grown for their handsome digitate foliage when they are juvenile but when old enough, they add a bonus of flowers. Leaves in mature plants become simple. They are propagated by seeds, cuttings and layering. Only 2-3 species are worth growing indoors. ***C. paniculata*** (Cabbage tree) from South Africa grows in containers up to 2 m tall though up to 5 m in the wild. Leaves some 30 cm in diameter are palmate and composed of 7-12 radiating leaflets, each leaflet pinnately lobed and some 15-30 cm long, each lobe spine-tipped and sometimes toothed. Bottlebrush-shaped spikes are produced on some 30 cm long panicles, bearing white or yellow flowers some 8 cm long. ***C. spicata*** from Zambia to South Africa grows up to 2 m when containerized, bearing greyish-green leaves composed of 6-8 radiating, narrow and obovate leaflets up to 13 cm long, each leaflet being pinnatifid from upper one to three pinnules but the lower ones are fused to appear as a fish-tail. Flowers appear in spikes which are borne on panicles some 25 cm long. ***C. thyrsiflora*** leaves are palmate, the segments spoon-shaped and up to 7.5 cm long.

Cyanotis (Commelinaceae)

See chapter on '**Succulents Other Than cacti**'.

Cyathea (Tree ferns; Cyatheaceae)

See chapter on '**Ferns and Allied Plants**'.

Cycas (Bread palm; Cycadaceae)

It is an ancient Greek name for a palm, adopted by Linnaeus for this genus. A genus of 20 species of primitive seed-bearing palm-like plants from Malagasy, E & SE Asia, Indo-Malaysia, Australasia and Polynesia but only two are worth growing indoors. Their trunks are stout and woody and sucker at the base. Leaves are leathery, hard-textured, pinnate and in terminal rosette. Male flowers are cone-like and the females appear as much reduced woolly leaves bearing scattered globules. These are propagated through seeds, and well developed suckers. They are long-lasting and drought resistant Important cultivated species are *C. circinalis* and *C. revoluta* (Sago palm). ***C. circinalis*** (Fern palm/Sago fern palm) from Old World tropics grows up to 2 m in containers and is utterly slow in growth, bearing more than 1.5 m long leaves with crowded leaflets which are glossy rich green, linear-lanceolate and 1.25-2.5 cm wide. It tolerates minimum persisting temperature of 16 °C. ***C. revoluta*** (Japanese sago palm) from Ryukyu Islands (SW Japan) is also utterly slow in growth and grows up to 1.5 m or more when planted in container though more than 6 m high in the open. Leaves are bright rich green and some 75 cm long, crowded with leaflets some 15 cm long and 6.25-12.5 cm wide with revolute margins and spiny tips. Seeds are bright red, ovoid, flat and some 4 cm long but are produced when the plants of both the sexes are planted together. It tolerates the minimum persisting temperature of 10 °C.

Cyclanthus bipartitus (Cyclanthaceae)

Cyclanthus has derived from two words, *cycle* meaning 'circle' and *anthus* for 'a flower', *i.e.* flowers in a circle. It is a tropical American genus of four species, closely allied to *Carludovica* and *Dicranopygium* though only one, *C. bipartitus* from Guiana is in cultivation. These are acaulescent plam-like herbaceous plants producing milky latex, the leaves being long-stalked, entire or bifurcate with 1-nerved and lanceolate segments, and the flowers are fragrant. ***C. bipartitus*** leaves are long-petioled (90-180 cm) and sometimes divided into two long narrow lobes, the spadix being erect, 60 cm long, cylindrical and appear from a 4-leaved yellow spathe. ***C. cristatus*** from Colombia bears short-stalked and bilobed leaves, the lobes being curved and tapered to a point as well as connivent, and the spadix is some 20 cm long. ***C. discolor*** (from North America) young leaves have brown-orange streakings. The leaves are bilobed and the lobes are lanceolate and acuminate with frilling on the margins. ***C. godsffianus*** leaves are rich green, oblong to obovate and taper to a sheathing stalk.

Cyperus (Umbrella sedge; Cyperaceae)

Cyperus is a Greek word for a sedge. An evergreen genus of 550 species, mostly perennial but sometimes annual, native to tropical, sub-tropical and warm temperate regions throughout the world. All cultivated ones are perennials, mostly rhizomatous or stoloniferous, and some with tubers. They all have grass-like tapering leaves. Flowers are umbellate, petalless, each composed of a single ovary and 1-3 stamens arranged in flattened and grass-like spikelets. Though these are excellent aquatic plants but leaves along with the portion of stems are used as fillers in floral arrangements. These are propagated through division and seeds. Important cultivated species are *C. alternifolius, C. diffusus and C. esculentus.*

Cyphostemma (Vitidaceae)

See chapter on '**Succulents Other Than Cacti**'.

Cyrtomium falcatum

See chapter on '**Ferns and Allied Plants**'.

Cyrtosperma johnstonii (*Alocasia johnstonii*; Araceae)

In Greek, *Cyrtosperma* derives from *kyrtos* 'curved or arched' and *sperma* 'seed', owing to their curved seeds. The genus comprises some 10-15 rhizomatous or tuberous herbaceous perennial species from the tropics but only one is under cultivation. Their leaf- and floral-stalks are often warty to spiny, leaf-petioles long and sheathing at base. ***C. johnstonii*** (*Alocasia johnstonii*; Cobra plant) from Solomon Islands is a tuberous plant, growing up to 75 cm high indoors. Leaf stalks are spiny, some 60 cm long and brown-spotted. Its leaf-blades are olive-green, arrow-shaped (sagittate), centrally depressed, deeply cleft, having upright held tips and with a size of 45 × 20 cm though basal lobes up tp 12.5 cm long. The leaf surface is bright oily green with a blood-red vein network.

Cyrtostachys renda (*C. lakka*; Palmae)

Cyrtostachys derives from the Greek *kyrtos* 'curved' and *stachys* 'a spike', referring to the flowering spike. A genus of feather palms from Malaysia, especially New Guinea and Solomon Islands with 10 species which are clump-forming and suitable for conservatory and houses, though only one species is under common cultivation. These are propagated through seeds grown at 27 °C and above temperature. ***C. lakka*** (Sealing wax palm) from Borneo is an erect but slow-growing and a robust bamboo-like slender palm which though grows up to 8 m in height with orange stems but for many years it can be enjoyed in containers where it can be kept only 2-3 m high. Its erect to slightly arching pinnate leaves with orange (sealing-wax-red) leaf sheaths, along with leaf-stalks which are 60-120 cm long with 20-50 slender, linear and grey-green showy leaflets which glow even in poor light.

Dasylirion (Agavaceae, formerly Liliaceae)

It has derived from the Greek *dasys* meaning 'thick' and *lirion* for 'a lily', as the plants are dense (thick) and lily-like. It is a xerophytic genus of about 15 evergreen perennial species, usually with palm-like erect caudex (woody base) and agave or yucca-like narrow leaf appearance from Mexico and SE USA. Numerous long, rigid and linear leaves with sharply toothed margins in rosette-form crowd near the top of the trunk. Small dioecious flowers in dense racemes crowd into an elevated narrow compound panicle going up to 3 m long bearing campanulate perianth where segments are white- to whitish-green, toothed, nearly equal and obtuse, the six stamens exserted, style short, stigmas three and the fruit is indehiscent, dry, 3-winged and 1-seeded. These are highly ornamental plants most suitable for rockeries, in the lawn, on staircases and terraces. These can be propagated through seeds and by cuttings of the branches as these do not sucker. ***D. acrotriche*** (*D. acrotrichum, D. gracile, Barbacenia gracilis, Bonapartea gracilis, Littaea gracilis, Roulinia gracilis, Yucca acrotricha, Y. gracilis*) from Mexico bears woody trunk some 60-90 cm high when old, greenish to dull-pale leaves some 60-90 cm long and 0.9 mm wide (quite narrow), prickles pale-yellow but tipped brown and with brush-like fibres, inflorescence 2.7-4.5 m tall and the fruits round-cordate and some 4.5 × 6.0 mm in size with shallow notching. ***D. glaucophyllum*** (*D. glaucum, Bonapartea glauca*) from Mexico is similar to *D. acrotriche* in general appeearnace and bears 90-120 cm long and 1.25 cm wide leaves with yellowish-white prickles but leaf tips lack brush-like would be expected from the number of stamens. A native to Australasia, the genus comprises some 17 species of shrubs and small trees but generally only three are in cultivation such as *D. elegantissima*, *D. kerchoveana* and *D. veitchii*. Stems are slender canes, bearing long-stalked spirally-arranged leaves which are alternate and digitate with 7-11 linear-lanceolate leaflets mostly with wavy lobe-margins, lower leaves not long-lasting which fall off soon under greenhouse forcing. These are propagated through seeds, by air layering and by cuttings (tip cuttings and stem section cuttings). ***D. elegantissima*** (*Aralia elegantissima*) from Vanuatu (New Hebrides) is usually non-branching shrub growing up to 5 m high, bearing palmately compound and dark reddish-brown leaves when juvenile though turning green on maturity, marbled white on 10-15 cm long brown stalks, leaflets 6-10 with circular arrangement, triangular, saw-toothed, 12.5-25 cm long and 1.25-2.50 cm wide. ***D. kerchoveana*** (*Aralia kerchoveana*) from Polynesia is similar to *D. elegantissima* except the leaves which are lustrous deep green with a pale midrib, spoon-shaped, broader, more toothed and upcurving with usually 9-11 leaflets some 1.5 cm wide. ***D. veitchii*** (*Aralia veitchii*) from New Caledonia is also similar to above with slenderous erect stems up to 2 m high, leaves are digitate, narrowly elliptic, leaflets of juvenile plants are coppery-green above, deep red beneath but adult leaves are green above with margins only undulating.

Dodonaea viscosa (Akeake/Hop bush; Sapindaceae)

Dodonaea was named for Rembert Dodoens (1518-1585), professor of medicine at Leiden, Royal physician, herbalist, and author of works on plants. There are some 60 evergreen species of trees and shrubs in the genus from tropics and sub-tropics, especially Australia. These bear alternate and shining green leaves and insignificant flowers. Out of all these, only one is grown for its foliage and as an effective hedge plant. These are propagated through seeds and cuttings. ***D. viscosa*** grows up to 6 m high but in containers only up to 2 m, bearing alternate, shining green, firm, narrowly ovate to elliptic, 6-10 cm long and about 2.5 cm wide leaves. Greenish-yellow petalless and insignificant flowers some 4 mm long appear in small panicles, followed by 2-3-winged, reddish to purple and 1.5 cm across seed capsules. A form with coppery or purple foliage is 'Purpurea' which is in great demand.

Dolichothele (Cactaceae)

See chapter on '**Cacti**'.

Dorstenia (Moraceae)

Dorstenia was named so for Theodore Dorsten (1492-1552), a German botanist and professor of medicine at Marburg. A native of tropical America and Africa with at least one species in India, it is a genus of 170 species of perennials and sub-shrubs, some being evergreen, some others deciduous and remaining semi-succulents adapted to arid climates. The tips of each flowering stem expand and form the little fleshy and hollow or flat platforms wherein numerous tiny green florets are embedded. The fruit contains only one tiny white seed. ***D. argentata*** (*D. argentea*) from Brazil is an erect sub-shrub growing up to 30 cm high, bearing broadly lanceolate, 13 cm long and deep green leaves with a broad irregular silvery zone along the centre or covering most part of the leaf. Concave inflorescence is margined with purplish tubercles. ***D. barteri*** from West Africa is a shrubby species growing some 60 cm high with sparingly-branched erect stems.Leaves are elliptic, usually hairy and 13-18 cm long. Inflorescence is almost circular, flat, 2.5-5.0 cm wide and bears clustered greyish flowers, inclosed by a green memberanous margin from where slender ray-like projections radiate. ***D. contrajerva*** (Torus herb) from northern S. America to Mexico and West Indies grows up to 30 cm tall, bearing very slow growing, thick and fleshy stem with creeping rhizomes. Leaves are glossy dark green mottled silver-green, thin and hairy, roughly triangular, 10-20 cm long on long stalks (15 cm), and divided into seven deeply cut jagged segments. Flower receptacle comprises of an oblong surface with a crimped rim bearing numerous minute flowers which come up continuously for months. Inflorescence is flattish, 2.5-5.0 cm wide, a little 4-angled and irregularly lobed. Surface-visible one tiny white seed is produced by each fruit. ***D. foetida*** from southern Arabia in the wild forms low mounds of thick-contorted stems arising from the flattened caudex, but in cultivation the stems grow more taller up to 15 cm, erect and more succulent, bearing leaves only at the stem tips. Its leaves are narrowly ovate to lanceolate, patterned with pale vein and finely crimped marginally. Inflorescence circular and some 2 cm wide, and flattened, with many slender rays on the margins which give it starry look. Bright light conditions colour the inflorescences in a reddish tinge. ***D. gigas*** from southern Arabia and Socotra is a shrub-like, bearing highly slow growing, sparingly branched and swollen stems growing up to 1.2 m high with glossy green oblanceolate and 10-15 cm long leaves which are finely puckered with recurved margins. Inflorescence is similar to *D. foetida* but with fewer and shorter rays. ***D. turnerafolia***, an excellent foliage house plant bears dark green, lanceolate and oily-surfaced leaves some 25 cm long on a long maroon stalk.

Dracaena (Female dragon; Liliaceae)

Dracaena derives from the Greek *drakaina* meaning 'a dragon' because of the *D. draco* (the dragon tree) of Canary Islands. It comprises of 150 species of trees and shrubs from tropical and sub-tropical Africa and Asia. Leaves are lanceolate and are formed either in tufts on the tips or along the length of the stem and make them attractive foliage plants most suitable indoors. They are propagated through basal, stem or tip cuttings. Important cultivated species are *D. angustifolia, D. deremensis, D. draco, D. goldieana, D. hookeriana, D. marginata, D. phrynioides, D. sanderiana, D. surculosa, D. thalioides* and *D. umbraculifera.*

Drynaria (Polypodiaceae)

See chapter on '**Ferns and Allied Plants**'.

Dryopteris (Wood fern; Polypodiaceae)

See chapter on '**Ferns and Allied Plants**'.

Duranta (Verbenaceae)

Duranta commemorates Castore Durantes (1529-1590), a Papal doctor and botanist in Rome. It is a genus of 36 species of trees and shrubs from Mexico, Central and South America, Florida and West Indies. These are suitable for pot-planting, in conservatory and as a hedge. These are propagated through seeds, cuttings and layering. ***D. repens*** (*D. plumieri*; Golden dewdrop/ Pigeon berry/Skyflower) from Florida to Brazil is a shrub growing up to 3 m or more in height but its height is drastically reduced when grown in pots. Moreover, its growth is utterly slow therefore it is also grown as hedge plant where it takes many years to attain the maximum size though through continuous maintenance these are never able to attain. The free-branching stems are erect to spreading and sometimes spiny. Leaves are green, ovate to obovate, toothed, 1.2-8.0 cm long and appear in opposite pairs.Lilac-blue flowers about 1 cm across, shortly tubular with five rounded petals appear intermittently during rainy season on simple to branched raceme some 15 cm long, followed by yellow berry-fruits about 1 cm across in trusses and singly. Its forms 'Alba' bears white flowers, 'Variegata' with white-margined leaves and 'Goldeana' with light yellow foliage.

Dyckia (Bromeliaceae)

See chapter on '**Bromeliads**'.

Echeveria (Crassulaceae)

See chapter on '**Succulents Other Than Cacti**'.

Echinocactus (Golden ball; Cactaceae)

See chapter on '**Cacti**'.

Echinocereus (*Hedgehog cactus*; Cactaceae)

See chapter on '**Cacti**'.

Echinopsis (Sea urchin cactus; Cactaceae)

See chapter on '**Cacti**'.

Elaphoglossum (Polypodiaceae)

See chapter on '**Ferns and Allied Plants**'.

Elettaria (Cardamom; Zingiberaceae)

It is an East Indian name taken for this genus. Its native haunts are India. It differs from *Amomum* in having slender tube of the perianth, the presence of the internal lobes within and the filaments are not prolonged beyond the anther. Probably the genus comprises one species, ***E. cardamomum*** (*Cadamomum officinale, Amomum cardamomum*) which produces true small cardamom capsules of the commerce. Its plants are horizontally rhizomatous, herbaceous, shade-loving and grow to 0.75-3.0 m tall, some 2.0 cm thick, bearing closely sheathed, curving and jointed stems with short-stalked (2 cm long), some 30-60 cm long, 7.5 cm wide, green, entire, in opposite pairs, oblong-lanceolate and acuminate leaves. Flowers are insignificant, striped purple, capsules are oblong to globular and indehiscent with many vertical ribs and the seeds are small and angled. After 3-4 crops the plants exhaust and require to be planted afresh. It is propagated through seeds and division.

Encephalartos [Kaffir bread; Zamiaceae (Cycadaceae)]

Encephalartos derives from the Greek *en* 'within', *kephale* 'a head' and *artos* 'bread', as seeds of certain species yield a starchy flour which African natives used as food or the bread like interior of the trunk. A genus of 20-30 dioecious sycads from tropical and South Africa which are grown for their attractive evergreen foliage. This is allied to *Dioon* and *Macrozamia*. They are primitive palm-like plants with stout cylindrical often fleshy trunks which on the crown have long, rigid, leathery and often spiny pinnate leaves in rosettes. Flowers appear in cones, staminate cones being oblong, ovoid or cylindrical containing thick and often rough, broadly or elongate-cuneate and imbricate scales in many series having anthers on the under surface, pistillate cones are thick and ellipsoid or oblong with many scales which are imbricate and peltate and in many series below which are ovules. These plants are most suitable for large conservatories and in the open as well and prefer a sunny and tropical conditions. Many of the species are so slow-growers that the trunks grow only a few inches in many years. Their cones are also very decorative. The woolliness of the stem and leaf segments varies with the age of the plants and of the leaves. They are propagated by seeds. ***E. altrensteinii*** (bread tree cycad) has a 15 m tall and stout trunk without wool. Leaf stalk bases are swollen, the leaves are some 60-180 cm and leaflets 15 cm long and 2.5 cm wide, leaflets are mostly opposite, paler beneath, narrowly oblong to oblong-acuminate, the edges and apex spiny. Yellow-green cones some 30-45 cm long are formed only on fully mature specimens. ***E. borridos*** grows with short and stout trunks which may or may not be woolly. Leaves are some 1.8 m long with top-reflexed, and the leaflets are opposite or alternate, lanceolate, entire or toothed and with a sharp spine at the apex. Its forms are 'Glauca' and 'Trispinosus', former having more glaucous leaflets. ***E. caffer*** bears the trunk with height up to 5.4 m high, diameter about 30 cm, leaves some 1.2 m and leaflets about 15 cm long. Petioles are 3-angled, rachis glabrous, leaves rigid and recurved, leaflets narrower at base, alternate, twisted and the new ones with 1-2 teeth. *E. c.* var. *brachyphyllus* (*E. brachyphyllus*) bears pubescent rachis and blades of the lower leaflets, sessile male cones, and erect, longer and narrower pinnae. ***E. cycadifolius*** (*E. cycadifolius* var. *friderici-guilielmi*, *E. friderici-guilielmi*) bears globular trunk with several inch in diameter and which are initially woolly. Petioles and rachis are ashy-pubescent, and the leaflets are linear, opposite as well as alternate and margins revolute. ***E. ferox*** grows to about 60 cm in height though in pots it takes many years to form the trunk. Leaves are glaucous and 1-2 m long, lower leaflets are modified into spines though ovate-oblong upper ones some 15 cm long are toothed and spine-tipped. Flowering cones are some 25-30 cm long, where female cones are quite thick and shaded with pink to red. ***E. lehmannii*** (*Cycas lehmannii*) produces some 2 m tall trunks without wools, petiole and rachis are obtusely 4-angled, richly glaucous green leaves are 90-150 cm long and linear leaflets some 13-20 cm long, opposite, lanceolate, with a few and small teeth and brown-spiny apex. Flowering cones are 25-40 cm long and the female cones are plumpy and reddish-brown. ***E. longifolius*** bears tall trunk without any wool over it as this soon vanishes, petioles and rachis 4-angled but flattish above, and lowest leaflets often 1-3-toothed with little revolute margin. Its botanical varieties are *E. l. angustifolius* (narrower and flat leaflets), *E. l. revolutus* (distinctly revolute margins) and var. 'Hookeri' (rachis not woolly, leaves intense green but not glaucous and leaflets narrowly lanceolate). ***E. villosus*** produces short, thick, woolly and scaly trunks some 1.8 m tall. Leaves are up to 1.8 m long with a quite many leaflets which are opposite or alternate, linear-lanceolate, spiny-toothed and pointed.

Epipremnum (Araceae)

Epipremnum derives from the Greek *epi* meaning 'upon' and *premnon* 'a trunk', as in the wild it tends to climb up the tree trunks. This genus with about 10 root-climbing species is native to tropical SE Asia and Eastern Pacific, out of which three are cultivated indoors for their handsome foliage including *Pothos*. The stems are climbing and the leaves are alternate with thin stalks. These require filtered light for their growing. They are frost tender. Propagation is through stem cuttings. ***E. aureum*** (*Scindapsus aureus*) is a rampant vine with some 90 cm long cordate leaves, split horizontally in several narrow segments on either side of the midrib though juvenile leaves are only up to 10 cm long and entire. The leaves are shining dark green or splotched with yellow and white. 'Golden Queen' has shining light yellow foliage blotched with green. 'Marble Queen' is fairly fast-growing woody-stemmed root-climber with shining white leaves streaked and marbled white though sometimes reverting back to entire green, and the stems may run up to 10 m. 'Tricolor' is profusely marbled with yellow, cream and light green. ***E. falcifolium*** is initially a trailer with rapid growth. Leaves are elliptic with attenuated tips, 30 × 1.25 cm size initially but much bigger (60 × 12.5 cm) in mature plants. Veins in the leaves are dark green with lighter silvery in between. ***E. pinnatum*** (*Monstera mechodonnii*; *Raphidophora pinnata*) mature plants produce leaves up to 1 m in length though juveniles produce alternate segmented leaves on each side of the midrib, each broad-based segment of 15 × 1.25 cm size is sword-shaped and with 1-3 vertical veins. It is a quite handsome plant.

Episcia (Gesneriaceae)

Episcia derives from the Greek *episkios* meaning 'shaded', referring to their natural habitat. A genus of 40 species of prostrate to trailing evergreen perennials from tropical America and West Indies grown for their handsome foliage, *vis-a-vis* beautiful flowers in bluish, white, yellow, orange or red colour. Though not so popular as its close relative African violet, but in no case these are less charming. They are most suitable as ground cover and for the hanging baskets. These require high humidity and a fairly shaded position in humus-rich well-drained soil. These are propagated through division, rooted runners and stem cuttings. These bear pairs of ovate to elliptic leaves, often patterned with white and gold, and mostly with metallic sheen. Tubular flowers open to five flared and rounded lobes. ***E. cupreata*** (Flame violet) from Colombia and Venezuela is an evergreen trailing to creeping plant, growing up to 10 cm high though stems running up to 40 cm or more in length, bearing green with a coppery flush, small (6-8 cm long), slightly toothed, downy and wrinkled, puckered or quilted purple leaves, generally having inconspicuous silver, white or pink veins and/or banding and hairy scarlet flowers marked yellow within. The varieties 'Metallica' bears oval, downy and wrinkled leaves, having broad silvery bands along midribs and tinged pink to copper, and the orange-red flowers being funnel-shaped are marked yellow within; and 'Tropical Topaz' bears yellow flowers. ***E. dianthiflora*** (*Alsobia dianthiflora*; Lace flower) from Mexico is initially a tufted plant, creeping with runners which produce small plantlets, and growing up to 10 cm high, bearing dull green, thick, velvety, crenate and some 4 cm long oval leaves with brown-purple midribs and pure white flowers some 2.5 cm across having highly frilled petals. ***E. lilacina*** from Central America is a first tufted plant and then producing creeping runners growing up to 10 cm high bearing plantlets. Leaves are pale-green, hairy, ovate-lanceolate and 6-8 cm long often with bronze or reddish suffusion and the central vein paler-green above and pink beneath. Flowers are white with yellow eyes and tinged mauve, slender-tubed and appear in small clusters from autumn to spring. Its var. 'Cuprea' bears bronze-tinged leaves; whereas 'Quilted Beauty' and 'Shaw's Garfens' display heavy toweling effect similar to *E. lilacina*. ***E. reptans*** (*E. fulgida*) from Colombia to Brazil produces stems which trail up to 60 cm length, bearing dark green, ovate, quilted and 5-12 cm long leaves flushed with mahogany and veins silver. Deep flame-red flowers some 4 cm long with no interior spot are toothed or fringed at the mouth.

Episcias have produced a number of named attractive cultivars and hybrids with variously coloured and patterned leaves and flower colours including cream to yellow. 'Chocolate Soldier' leaves are large, brown and silver-veined. 'Cleopatra' is a beautiful terrarium plant with sharper colour pattern. 'Ember Lace' is a trailer with many suckers, and the leaves are brown splashed with bright pink. 'Moss Agate' bears large, quilted green leaves with silver veins. 'Shimmer' and 'Silver Sheen' bear silver-centred leaves with brown margins.

Eranthemum (Acanthaceae)

Eranthemum in Greek means 'a lovely flower', *errano's* meaning 'lovely' and *anthemon* meaning 'flower'. It is an ancient name used for this genus by Linnaeus. The genus comprises 30 evergreen species of shrubby perennials, suitable for indoor display, in borders and in the conservatories. Their colourful foliage is very handsome but in humid and warm climates these also produce attractive loose flowers. Propagation is through softwoot cuttings. These grow in shaded and semi-shaded locations. Usually the leaves are ovate-elliptic, entire (rarely coarsely toothed) and green with suffusion of various shades, especially purple, reddish-mauve and silvery-white. Flowers appear in white to lilac, rosy or red with narrow and small bracts and bractlets, long, slender and cylindrical corolla tubes and then 5-parted, stamens 2, ovules 2 in each cell and seeds 4 or less. ***E. andersonii*** from Trinidad bears simple, lanceolate or elliptic leaves narrowed into a short stalk. Spike grows some 15 cm long with lower middle lobe of corolla larger and speckled purple. ***E. albomarginatum*** (*Pseuderanthemum albomarginatum*) from Polynesia is a shrubby foliage plant growing up to 1.5 m high. Lower surface of leaf is green and upper surface is broadly margined with white and irregularly suffused with grey. ***E. bicolor*** (*Pseuderanthemum aspersum*) from Solomon Islands grows to a height of 1.5-1.75 m, bearing green leaves in the centre, suffused with greenish-yellow patches, and a part of leaf lamina and margins yellow. The flowers are white suffused with pink. ***E. laxiflorum*** (*Pseuderanthemum laxiflorum*) from Fiji grows up to 0.6-1.2 m high with varying shape and size of leaves. Those near the flowers are 5.0-8.75 cm long, ovate-oblong, obtuse but narrowed at the base and widest below. It does well under semi-shaded conditions though full shade proves lethal especially during rains. It produces purplish flowers in cymes profusely during monsoon though flowering is continued for a long duration constantly. ***E. nigrum*** (*E. nigrescens*, *Pseuderanthemum atropurpureum*) from Pacific Islands grows to a height of 1.5-1.8 m. Upper leaf surface is blackish-purple, lower purplish with darker veins, and white flowers spotted rose at the base appear in erect terminal spikes. ***E. pulchellum*** (*E. nervosum*, *Daedalacanthus nervosum*; Blue sage) from India is an erect growing shrub some 1.0-1.2 m high, producing 8-15 cm long, elliptic to ovoid deep green leaves in opposite pairs, veined prominently. Flowers are blue, tube about 3 cm long with 5 petals, petal lobes rounded and these appear during spring in terminal and axillary spikes up to 8 cm long. ***E. tricolor*** (*Pseuderanthemum tricolor*) from Polynesia is a straggling shrub growing up to 1.5 m tall, bearing leaves with upper surface greyish-purple on a dark green base, lower reddish with violet spots, and the flowers are red spotted-white. ***E. tuberculatum*** from unknown origin has many small roundish and rough swellings on the stem branches. Leaves are small, almost sessile and some 9 mm-1.5 cm wide, broadly elliptic, obtuse or notched. Flowers are borne in plenty but singly in the axils, corolla tube very long and slender and 3.75 cm long, limb 2.5 cm across and the stamens are scarcely exserted.

Espostoa (Cotton ball cactus; Cactaceae)

See chapter on '**Cacti**'.

Ethretia (Carmona) microphylla (Philippine tea; Boraginaceae)

Ehretia was named for G.D. Ehret (1708 or 1710-1770), a botanical painter who was born in Germany and died in England. The genus comprises 40-50 tender trees and shrubs from tropical world, mostly from the Old World tropics but only one is worth indoor cultivation and that is ***E. microphylla*** from Himalayas and China. This highly branched woody little shrub growing hardly up to 30 cm in height forms a pseudo-bonsai in a pot. Its joints are corky, the leaves are shiny dark green, bristly above, soft-hairy beneath, stalkless, spathulate to broadly elliptic, acuminate, serrate, finely blistered, shallowly lobed near the tip and some 15-20 cm long. Insignificant small white flowers appear almost throughout the year on terminal panicles, followed by globose red berries which drop down soon after ripening. It is suitable for growing on windowsills. It is highly amenable to trimming and is propagated through seeds and cuttings.

Eucalyptus (Blue gum; Myrtaceae)

In Greek, *eu* is for 'well' and *kalypto* for 'to cover', owing to the united sepals and petals forming a cap over the numerous showy stamens which is shed when flower opens. *Eucalyptus* is an evergreen genus of about 500 species of trees and shrubs from Australia and Indo-Malaysia. They have aromatic, waxy and leathery leaves where shape of the juveniles (whether in young plants or leaves emerging after annual cutting back) may differ and appear in pairs in most species to those of senile or adult plants where the leaves are often willow-like, and it is only the juvenile foliage which has demand in the trade. *E. globulus* (blue gum) has brightly hued leaves, the juvenile leaves are ovate-cordate and blue-white while the adult ones are lanceolate and more green than grey. To retain the juvenile leaves, they are required to be cut back annually or every two years at the end of harvest season or in spring when new growth is to begin. The flowers are petalless and multi-stamened which may be white, cream, yellow, pink or red. The species grown for cut foliage are *Eucalyptus gunnii, E. parvula, E. perriniana, E. pulverulenta, E. mourei* and *E. rubida.* These are propagated by seeds. *E. cinerea* 'Silver Dollar' and *E. gunnii* bear rounded twin leaves, *E.populus* and *E. stuartina* oval leaves, while *E. parvifolia* and *E. nicolli* bear thin and longer leaves. For indoor display, the suitable species are *E. globulus* though it grows fast and taller so is trimmed frequently, *E. gunnii* (cedar gum) and *E. citriodora* (lemon scented gum), latter both growing dwarf and slow.

Euonymus (Celastraceae)

In Greek, *euonymon* means 'of good name' as many species are poisonous to livestock. A genus of about 170 deciduous or evergreen species of green shrubs, trees and climbers of cosmopolitan distribution, mainly Asia. They have simple leaves in opposite pairs and insignificant greenish, whitish or purplish flowers in small axillary cymes. They are propagated through seeds and cuttings. The species most commonly used as fillers are *E. fortunei, E. japonica* and *E. lucidus* (*E. fimbriatus*). However, for the purpose of indoor display, the suitable species is *E. japonica* with various varieties, especially those with variegated leaves such as 'Mediopictus', and *E. microphyllus*, the dwarf species.

Eupatorium (Asteraceae)

Eupatorium honours Mithridates Eupator, King of Pontus, an ancient district of modern Turkey, who discovered that species which acts as an antidote to poison. The genus comprises some 1,200 species of shrubs and perennials mainly from America, though a few from Europe, Africa and Asia. Most of the species are grown for their abundantly borne clusters of tine groundsel-like flower heads though a few for their foliage in pots or as edges. They are propagated from seeds and cuttings. ***P. atrorubens*** is from Mexico. ***E. cannabinum*** from Europe, N. Africa and Asia grows from 60-120 cm tall, bearing mid-green, opposite and broadly lanceolate leaves. Flower heads are terminal, rounded, some 10-13 cm across and the flowers are small and red-purple which appear during rainy season. *P. c.* 'Plenum' bears more attractive purple-pink flowers. ***E. micranthum*** (*E. ligustrinum, E. weinmannianum*) from Mexico is a bushy species growing more than 2 m tall, bearing light green, smooth, elliptic to lanceolate and some 4-9 cm long leaves in opposite pairs. White fragrant flowers, sometimes flushed pink, appear in flattish corymbs some 10-20 cm wide in autumn. It does well even in temperate regions. ***E. purpureum*** from N. America grows up to 15 cm tall, bearing mid-green, slender and pointed leaves in whorls. Flower heads some 10-13 cm wide are borne on branched stems with rose-purple flowers from August to September. ***E. rugosum*** (*E. ageratoides, E. fasteri*) from N. America grows some 10 cm tall with oval and opposite mid-green leaves. Slender branching stems produce flat and fluffy floral heads which are 7.5-10.0 cm across, opening with white flowers during July to August. ***E. sordidum*** (*E. ianthinum*) from Mexico is a temperate species growing up to 1 m tall, bearing robust stems, and long-stalked, deep green, ovate to broadly oblong and toothed leaves some 10 cm long in opposite pairs. Violet-purple fragrant flowers appear during winter in dense corymbose clusters some 9 cm wide.

Euphorbia **(Spurge; Euphorbiaceae). See chapter on 'Euphorbia'.**

Eurya japonica **(Theaceae)**

Eurya derives from the Greek *euru* 'broad' but in what sense is not known. A genus of 130 species of evergreen shrubs and trees is from Eastern Asia to the Pacific region. These are allied to *Camellia* but the flowers are small and the fruits are berry-like. These are propagated through seeds or cuttings. ***E. japonica*** is a handsome foliage plant native to China, Korea, Japan, Taiwan and Malaysia. In containers it can be tamed to grow up to 1.2 m tall. The plant has dense habit and bears alternate, lustrous deep green, leathery, elliptic to oblong-lanceolate, toothed and 4-8 cm long leaves. Solitary or small clusters of white to greenish-yellow flowers some 6 mm wide appear from the leaf axils, followed by glossy black berries to the size of the flower.

Evodia hortensis **(Rutaceae)**

Evodia in Greek is for 'pleasant odour', due to odour of the leaves which these emit after bruishing. The genus comprises 50 species of deciduous or evergreen trees or shrubs from Madagascar to Polynesia but only one is grown indoors as furnished below. These are woody ornamental plants cultivated for their handsome foliage. It grows well under filtered sun and is propagated through seeds and cuttings. ***E. hortensis*** (*E. lepta*; Lacy lady) from SE Asia is an unusual, little and quite charming shrub some 60 cm in height when planted in pots indoors, having 5 cm long leaf stalks, trifoliate (trident) leaves with length of leaflets being some 12.5 cm and width 1.25 cm and the midrib sides unevenly wide and wavy. Insignificant tiny white flowers appear during winter followed with clustered berries.

Excoecaria **(Euphorbiaceae)**

It derives from the Latin *excoecares*, referring to its effect on eyes. The genus contains tropical trees or shrubs with poisonous milky juice. The genus comprises some 25 species which are glabrous with leaves opposite or alternate, entire to crenate or serrate, the inflorescences are usually axillary, the flowers are dioecious or monoecious, the calyx is imbricate with 2-3 sepals which may be free or connate at base, no corolla but stamens 2-3, filaments free, ovary 3-celled and the seeds not caniculate. In ***E. bicolor*** (*Croton bicolor*) shrub, the leaves are opposite and red beneath and this is worth growing indoors at a sunny site. Its tree species, ***E. agallocha*** (Agallocha/Blinding tree) which has alternate leaves is used as river poison for fishing in South Asian coasts.

×Fatshedera lizei **(*Fatsia japonica* 'Moseri'× *Hedera helix hibernica*; Ivy tree/Tree ivy; Araliaceae)**

This hybrid, inheriting slight climbing habit of *Hedera* was developed in 1910 by the French nursery firm Lize Frères of Nantes. A good indoor pot plant growing up to 3 m high is highly tolerant to heat, cold and poor light though remains more comfortable around 21 °C tremperature and in reasonably bright light but requires regular spring-pinching to keep it bushy and in shape. Its ivy-shaped leaves are palmately-lobed with 5-9 divisions, and 20-40 cm wide having wavy margins. Rounded and umbellate pale-green flowers on terminal panicles are individually 5 mm across. Variety 'Variegata' is not so strong-grower but the leaves are informally cream-margined.

Fatsia japonica **(Paper plant; Araliaceae)**

A native of Japan, Korea and Taiwan, this evergreen shrub comprises only one species. *Fatsia* is Latinized old Japanese name *fatsi*. Its leaves are dark green with prominently white-veined, alternate and palmate. Loose umbellate small whitish flowers appear in panicles. Though it is a good specimen and house plant but not so strong as ×*Fatshedera* and prefers slightly cool conditions. It should be pinched each spring. It is propagated through seeds, air layerings and cuttings. Its varieties, 'Variegata' has cream-edged foliage, and 'Moseri' compact growth habit.

Faucaria **(Aizoaceae)**

See chapter on '**Succulents Other Than Cacti**'.

Ferocactus **(Devil's tongue; Cactaceae)**

See chapter on '**Cacti**'.

Ficus **(Moraceae)**

See chapter on '**Ficus**'.

Fittonia **(Acanthaceae)**

Named for Elizabeth and Sarah Mary Fitton who wrote '**Conservations on Botany**' in 1817. *Fittonia* comprises of two evergreen creeping perennials from Peru, bearing handsome foliage where in colour contrasts with rest of the foliage colour though flowers are insignificant. Apart from growing indoors with plenty of humidity, these are excellent terrarium as well as hanging basket plants. These are multiplied through careful divisions, as well as through cuttings taken in summer. ***F. argyroneura*** (*F. verschaffeltii argyroneura*; Silver net leaf) is a very decorative plant with clear white veining. *F. argyroneura* 'Minima' is similar to above but is too small, not exceeding more than 10 cm in height bearing

2.0-3.5 cm long leaves. *F. a. nana* (snake skin plant) is quite dwarf. ***F. gigantea*** either creeps or grows erect up to 30 cm in height, bearing shining green and 8-12 cm long leaves having slightly depressed red veins. Yellowish flowers are borne on some 15 cm long spikes though are not attractive but not unsightly. ***F. verschaffeltii*** (Nerve plant/Painted net leaf) is creeping to tufted and some 15 cm tall, bearing dark green oval leaves some 7-10 cm long with networks of red veins on the leaf surface, and small reddish to yellowish flowers appear on short spikes. In *F. v. pearcei*, the foliage veins are pink. They are very difficult to grow, especially the large-leaf types as these require constant warmth, however, these make best terrariums or bottle gardens.

Furcraea (Agavaceae)

See chapter on '**Succulents Other Than Cacti**'.

Gasteria (Liliaceae)

See chapter on '**Succulents Other Than cacti**'.

×*Gastrolea* (*Gasteria* × *Aloe*; Liliaceae)

See chapter on '**Succulents Other Than cacti**'.

Geogenanthus undatus (Commelinaceae)

In Greek, *ge* is for 'earth', *genea* 'birthplace' and *anthos* 'a flower', on account of its flowering habit. It is a tropical South American genus comprising of 3-4 evergreen creeping perennial species, one of which (*G. undatus*) is a decorative indoor plant. Their flowers appear from the leafless nodes near the base and often quite close to the ground. ***G. undatus*** (*Dichorisandra musaica undata*; Seersucker plant) from Brazil and Peru is a thickened-rhizomatous plant from where erect and unbranched stems some 15-25 cm tall appear in small groups bearing leaves in terminal rosettes of 3-4. With age these stems decline. Leaves are highly short-stalked, appear in pairs some 10 cm long and 5 cm wide, and are puckered with silky surface, dark green with silvery-green vein stripes above, purple-red beneath, broad, lanceolate, leathery and fleshy. Short-lived purple flowers some 2 cm wide with three fringed petals lasting only for a day are produced on quite mature plants of several years old. It is a most suitable plant for terrarium gardening. It is propagated through cuttings.

Glechoma hederacea (*Nepeta hederacea, N. glechoma*; Labiatae)

In Greek, *glechon* is for a kind of 'mint'. The genus comprises 10-12 evergreen perennial creeping species from temperate Europe and Asia. Most of the species are creeping weeds. The leaves are green and kidney-shaped and flowers small, 2-lipped and mauve. Even ***G. hederacea*** (Ground ivy), a hardy species is also considered a weed but its variegated form is decorative and worth growing for its foliage beauty indoors, especially in hanging baskets and as ground cover. *G. hederacea* has prostrate-growing stems which root at nodes forming extensive mats and is invasive. Leaves in opposite pairs are softly hairy, 1-3 cm wide, foetid, rounded to broadly ovate-cordate, and softly hairy with crenate margins. Its tubular and violet flowers some 1.5-2.0 cm long are 2-lipped and appear in whorls from the upper axils of the opposite leaf-pairs in spring. It tolerates sun and shade both. It is propagated by division or soft-wood cuttings. *G. h.* 'Variegata' (Variegated ground ivy) is the only type grown as a cool house plant but is comparatively less hardy, evergreen, pale-green and carpeting perennial with trailing stems which bear small heart-shaped toothed leaves marbled and/or edged white, especially on the apex portion though flowers are insignificant.

Graptopetalum (Crassulaceae)

See chapter on '**Succulents Other Than Cacti**'.

Graptophyllum pictum (Caricature plant; Acanthaceae)

Graptophyllum derives from the Greek *graptos* meaning 'to write or paint' and *phylum* 'a leaf', named so from the patterning of the leaves. An evergreen tropical genus of only two shrub species from Australia and Pacific Islands, bearing opposite, mostly entire and colourful leaves, and red and purple tubular flowers with wide mouth. These are mainly grown for their foliage. These are propagated through cutting. Only one species, ***G. pictum*** from New Guinea is in common cultivation for conservatory and house, as well as for hedge making, growing 2 m in the wild but in pots less than 1 m in height. Its stems are pink, leaves glossy green, elliptic-ovate, pointed and 12-15 cm long with marbling of cream, yellow, purple and pink, more clearly along the main vein and stalk. Its tubular red to purple flowers some 4 cm across appear on short terminal spikes in spring and autumn. 'Aureo-marginatus' bears green leaves with irregular yellow edges. 'Tricolor Aureo Medium' bears green leaves with deep yellow and cream zoning in the centre.

Greenovia (Crassulaceae)

See chapter on '**Succulents Other Than Cacti**'.

Griselinia [Griseliniaceae (Cornaceae)]

Griselinia was named for Francesco Griselini (1717-83), an Italian naturalist. The genus comprises six species of evergreen shrubs from New Zealand, Brazil and Chile.

The species being described below have decorative foliage but insignificant flowers which are not produced on small plants. Being wind and salt resistant, these are normally planted along the cool coastal areas as hedges or windbreaks as well as in cool homes and conservatories. These are propagated through semi-ripe cuttings. ***G. littoralis*** from New Zealand though grows up to 10 m in the wild but indoors these can be tamed up to 1.5 m in containers. Leaves are glossy yellowish-green, leathery, broadly ovate to oblong and 3-10 cm long. Insignificant unisexual flowers are yellow-green. 'Dixon's Cream' has centrally creamy-white leaf variegation. 'Variegata' bears white-variegated foliage. ***G. lucida*** from New Zealand has similar growth as to *G. littoralis* but with broadly ovate and larger leaves (10-18 cm long) that are glossy mid-green. It is suitable only for the mildest areas and conservatory.

Guaiacum officinale (Lignum-vitae; Zygophyllaceae)

Guaiacum is a W. Indian vernacular name. The genus comprises four species of shrubs and trees from tropical America. Leaves are opposite, leathery, entire and abruptly pinnate with 2-14 leaflets. Peduncles appear in pairs with 1 flower each of unshowy blue or purple colour, sepals are 4-5 and unequal, petals broadly obovate and 4-5 and stamens 8-10 and inserted. Propagation is through seeds and cuttings. ***A. officinale*** from Florida to Venezuela bears evergreen, 2-3 pairs of leaflets which are oval or obovate, blunt and 6-13 mm long. This tree is source of hard wood and a commercial gum. ***G. sanctum*** from West Indies is similar to above but the leaves are obliquely lanceolate-elliptic and the smooth sepals about half as long as the petals. This also yields a valuable wood.

Guzmania (Bromeliaceae)

See chapter on '**bromeliads**'.

Gymnocalycium (Rose plaited cactus; Cactaceae)

See chapter on '**Cacti**'.

Gynura (Asteraceae)

Gynura derives from the Greek *gyne* meaning 'female' and *oura* 'a tail', referring to the long stigma. It is a genus of some 100 evergreen perennial species from Africa to E. Asia and Malaysia, having shrubs and climbers with entire, alternate and little fleshy leaves which are covered with shiny purple hairs, so it requires sunny position so that hairs could maintain their shine. Only four species are worth indoor culture, *viz. G. aurantiaca, G. bicolor, G. procumbens* and *G. sarmentosa.* Its flowers have a heavy pungent smells which may be unpleasant to some persons so the flowers should be nipped in the bud. It is propagated through cuttings. ***G. aurantiaca*** from Java grows up to 90 cm with the spread of some 45 cm where leaves are bright green and the stems are covered thickly with bright violet-purple hairs. Broadly eliliptic leaves are almost triangular, entire or little lobed and some 15 cm long. Small cluster of orange flowers some 2.5 cm long appear on erect stems in the form of flower heads. ***G. bicolor*** from Moluccas is a branching sub-shrub growing up to 60 cm tall with ovate-lanceolate leaves of the size of 15 × 3.75 cm, which are downy, having 7 angular coarse teeth or lobes, pale greenish-purple above and purple below and are radially clustered at the tips. Orange-yellow flowers appear during winter and are 1.5-2.0 cm wide. ***G. procumbens*** is a trailing species, producing branched stems and green but overlaid purple leaves, measuring about 12.5 × 5.0 cm. It makes good basket plant for a sunny location. Its flowers are burnt yellow. ***G. sarmentosa*** (Purple passion vine) from India grows about 1.5 m tall and is trailing species so requires support and is most suitable for planting in baskets. Its leaves are narrow and rich purple. It produces pale-orange flower heads some 1.25 cm long in terminal clusters. In 'Aurea-Variegata' the leaves are splashed with cream and yellow.

Hadrodemas warszewiczianum (*Tradescantia warszewicziana*; Spironema/Tripogandra; Commelinaceae)

In Greek, *hadros* is for 'well developed or of good size', and *demas* is for a 'living body', referring probably to the large size of the plant. It is a genus of only one species native to Guatemala. An evergreen perernnial, it is most suitable for growing indoors and in conservatories as a handsome specimen plant. Its stem is robust, erect, unbranched but sometimes forking and grows up to 90 cm tall and then may sprawl above. Leaves are stem-crowding and clasping to the stem but arching outwards and so making dracaena-like appearance, bright green, little fleshy, narrowly ovate and some 30 cm long. Small flowers are numerous on a wide panicle, having 3 persistent purplish to lilac sepals and 3 rose-purple and rounded petals lasting for a day. Its floral stalks form the plantlets so it is propagated through these and by cuttings.

Haemaria discolor (*Goodyera discolor, Ludisia discolor*; Orchidaceae)

Haemaria is a Greek name owing to the blood-red under-surface of the leaves. It is a dwarf terrestrial orchid with four species from China and Malaya, and is valued for its foliage. In trade, it is normally called as *Goodyera* though in *Goodyera* the lip-blade is small and without claw whereas in *Haemaria*, the lower lip swells above the base making a wide claw adorned with a pouch-like basal sac. This has creeping rhizomes. ***H. discolor*** from China and Brazil bears leaves ~ upper green and red underneath, blades 7.5 cm long and 1.9 cm wide, with some plants displaying white longitudinal lines. About one dozen of small white flowers on the densely hairy scape appear. Leaf colouration in ***H. dawsoniana*** (*Goodyera dawsonii, Anoectochalus dawsonianus*) from Myanmar and Philippines has foliar colouration similar to *H. discolor* but is brighter and attractive because of beautiful netting above with red or yellow. Leaf blades are elliptic, 7.5 cm long and some 3.0 cm wide, and the veins in two leaves blood-red though in others yellow.

Hamatocactus (Strawberry cactus; Cactaceae)

See chapter on '**Cacti**'.

Haworthia (Liliaceae)

See chapter on '**Succulents Other Than Cacti**'.

Hechtia (Bromeliaceae)

See chapter on '**Bromeliads**'.

Hedera (Araliaceae)

A genus of about 15 evergreen climbing species from Europe to Caucasus, Himalaya and Japan, Canary Islands, and Madeira. Juvenile stems (runner growth) have short clinging roots which help these plants to climb. They have usually palmately-lobed ovate leaves though mature plants (arborescent growth) have unlobed leaves but with wavy margins. The arborescent growth is produced from the summit of the runner growth when it reaches the top of its support. Insignificant small greenish-yellow and 5-petalled flowers are produced in terminal umbels. They make an ideal pot and hanging basket plants. They are propagated through cuttings, however, those taken from non-climbing stems produce bushy, i.e. so-called 'tree ivies'. Important cultivated species are *H. canariensis, H. colchica* and *H. helix*.

Helichrysum (Asteraceae)

In Greek, *helios* is for 'sun' and *chrysos* for 'golden', due to the flowerheads of some species. This is native of Old World subtropics with 400 species but only two are worth growing indoors. Both are lovely shrubs with up-facing leaves which are clustered on the joints on very short stalks and are propagated through seeds or cuttings. ***H. angustifolium*** (curry plant) is from Mediterranean having smell of leaves as to that of Indian curry though is not used for culinary purposes. It is a handsome little shrubby plant amenable to trimming to a round shape, growing more than 30 cm tall, bearing grey-woolly and narrowly oblong leaves some 3.75 cm long. ***H. petiolatum*** (licorice plant) from South Africa is a shrub growing up to 1.2 m tall, with a spreading-trailing habit suitable for baskets, bearing 2.5 cm long, broadly lanceolate and white-woolly leaves.

Heliconia (False plantain, lobster claws; Heliconiaceae)

Heliconia is named so to honour Mt. Helicon (in Greece), the seat of the mythological Muses, or derives from the Greek *helios* 'sun and chrysos or golden', referring to the flower heads of some species. The genus is close to *Musa* in many aspects but differs chiefly in having a dry, mostly dehiscing 3-locular, capsular blue fruit containing only three seeds, the lamina anatomy and as Musaceae blades have an irregular apex (Triplett and Kirchoff, 1991). It is also close to *Strelitzia* where perianth segments are fused and there are many ovules in each chamber while *Heliconia* perianth segments are only partly fused and each chamber contains a single ovule (Everard and Morley, 1970). *Heliconia* consists of over 250 species and some 350 varieties, majority of these native to Tropical America (Mexico to Central America and northern South America, the Caribbean Islands and some of the South Pacific islands), South East Asia and Polynesia (Everard and Morley, 1970). Most of the species are from moist or wet regions though some are also from seasonally dry areas. Although most heliconias are reported to thrive in the humid lowland tropics at elevations below 500 m, the greatest number of species is found in middle elevations rain- and cloud-forest habitats. Heliconias have gained importance because of being most attractive of all exotic tropical flowering plants with diverse flower forms and colours, *vis-à-vis* attractive and evergreen foliage. It is also used in creating boundaries and screens for privacy and for special mass effects in the gardens. Its planting also checks erosion. In India, it is West Godavari district of A.P. which produces half of the total heliconias in the country, followed by Karnataka and Kerala. In general, there are three types of leaf arrangements in heliconias, *viz.* musoid (growing vertically with long petioled leaves), cannoid (angular growing with short to medium petioles) or zingiberoid (horizontal growing with short petioles) with banana-like foliage and have long-lasting (several days to several months) erect to pendulous terminal inflorescences (Anderson, 1985) composed of 3-30 boat-shaped and colourful bracts (mostly brilliantly coloured in hues of yellow, orange, pink and red or combination of these with variously splashed colours and green markings) arising from a central floral axis, bracts distichous or spiralling, usually large, carinate, waxy or leathery, each enclosing few to many small and 6-petalled clustered flowers protruding from the bracts in most species, on short or long pedicels. In appearance and relationship, its flowers are in between *Musa* and *Strelitzia*. Heliconias are normally large or giant evergreen and herbaceous perennials coming up from rhizomes, clump-forming and with often long stalked, paddle-shaped leaves. Important species are *H. bihai* (*H. distans*), *H. latispatha* and *H. shiedeana*.

Heliocereus (Sun cactus; Cactaceae)

See chapter on '**Cacti**'.

Helxine soleirolii (*Soleirolia soleirolii*; Urticaceae)

See '***Soleirolia***'.

Hemigraphis (Acanthaceae)

It takes its name from the Greek *hemi* for 'half' and *graphis* 'a brush', referring to the way the hair covers the stamen filaments. The genus comprises of some 100 annual, perennial and sub-shrub species closely allied to *Ruellia*. Their origin is tropical SE Asia and Australia. Generally there are two cultivated species which are spreading in nature and become woody at base when old. Their leaves appear coloured in opposite pairs and the flowers are small, tubular and white. Both have attractive foliage and are excellent as indoor, conservatory and hanging basket plants, *vis-à-vis* ground covers especially in conservatory borders. They are propagated through cuttings. ***H. colorata*** (*H. alternata*; Red/Flame ivy) probably from Malaysia has some 30 cm long prostrate stems which root while creeping, leaves metallic blue-purple above and red-purple beneath, little puckered, 6-10 cm long and broadly ovate-cordate with rounded teeth, and inconspicuous white flowers with purple lines are some 1.2-2.0 cm long.Its form 'Exotica' from New Guinea is more bushy with smaller flowers. ***H. repanda*** from Malaysia produces 25 cm long, rooting and prostrate to decumbent stems. Leaves are lustrous purple-green above and deep purple beneath, up to 6.0 cm long, broadly toothed and lanceolate to linear and the flowers are similar to *H. colorata*.

Hemionitis (Polypodiaceae)

See chapter on '**Ferns and allied Plants**'.

Heptapleurum arboricola (Araliaceae)

See '***Brassaia***'.

Hoffmannia (Rubiaceae; Taffeta plant)

Hoffmannia was named for George Franz Hoffmann (1761-1826), a Dutch professor of botany at two institutuions, Gottengen and Moscow. A native to Mexico and northern Argentina, the genus contains some 100 species of perennial shrubs, but only three are suitable for growing indoors for their handsome foliage. Leaves in opposite pairs with conspicuous veins and the clustered axillary with 4- or sometimes 5-petalled flowers are insignificant. These are propagated through cuttings. ***H. ghiesbreghtii*** from southern Mexico and Guatemala though in nature grows up to 1.2 m in height but in pots up to 60 cm only. Its 4-angled or winged stems bear 30 cm long oblong-lanceolate leaves which are deep velvety bronze-green above with silvery or pink veins and red below. Yellowish flowers appear hidden by the leaves. Its var. 'Variegata' has foliage marked with pink and cream. ***H. refulgens*** from Mexico and Central America grows to about 30 cm with often hairy and cylindrical stems. Some 15 cm long and obovate leaves are wrinkled in between the veins, metallic purplish or green above and purple-red beneath. Flowers bright red but insignificant. The veins and margins on the upper surface of leaves in the var 'Vittata' are grey, var. 'Fantasia' bears satiny greenish-copper surface, and 'Nana' is a dwarf shrub with corrugated, coppery-green and much smaller leaves. ***H. roezlii*** has leaf length of about 20 cm in oval to round shape in attractive dull coppery tone with prominently green veins. A beautiful form.

Homalocladium platyclados (Ribbon/Tapeworm plant; Polygonaceae)

See '***Muehlenbeckia***'.

Homalomena (Araceae)

In Greek, *homalos* is 'flat' and the *mene* is for 'moon', having no sense as originally its vernacular Malayan name was translated so in Greek. A native to tropical Asia and South America, the genus comprises about 130 species of evergreen perennials, most of these being with handsome foliage. They are clump-forming, mostly with arrow to cordate leaves. Petalless tiny flowers appear normally in a greenish spathe hidden among the leaves. These are propagated through division of clumps or by seeds. ***H. lindenii*** from New Guinea grows to about 75 cm high, bearing bright green and triangular-ovate leaves with a long slender point and where all the veins are yellow. ***H. picturata*** (*Curmeria picturata*, *Rhektophyllum kewense*, *R. picturata*) from Colombia bears pilose petioles and midribs, petiole some 10 cm long, blades 25-30 cm long and 20 cm wide and blotched silvery-white only near the midrib. ***H. rubescens*** (*H. rubrum*) petioles are slender and red and the leaves are cordate-sagittate, reddish-green with brownish edge which is tinged red and below it is purplish. ***H. sulcata*** from Borneo grows up to 45 cm tall with leaves being green above, coppery below, nearly triangular ovate and 13-20 cm long. ***H. wallisii*** (*Curmeria wallisii*) from Venezuela and Colombia is quite small and grows only 15 cm in height though spread is more. Leaf-stalks are short (3.75 cm long) with obtuse or acute base, the blades are 13-20 cm long and 5.0-6.75 cm wide, glabrous throughout, arching or recurving, ovate to triangular-cordate, reflexed, and bright green with pale-yellowish-green blotches and marbling.

Hosta (*Funkia*; Day lily/Plantain lily; Liliaceae)

Hosta was named after Nicholas Tomas Host (1761-1834), an Austrian botanist and doctor. It is a genus of about 40 species of hardy herbaceous perennials from East Asia, mainly Japan. Due to the luxuriance of their foliage and charming nature, these are sought after every garden large or small, as ground cover where once established these grow for many years, along the walks and drives, in the angles against the buildings, for balconies *cum* patio gardening, on the borders, in containers, on the water side, damp and shady situations giving an exotic touch, also as they are shade and moisture-loving plants so they are also liable to be infested with slugs and snails. Their foliage dies during frost but comes up in the spring. These vary in size from plants a few centimeters high to vigorous forms making a clump up to 1.5 m across in a few years. They are elegant for their leaves as these are diverse in size, shape, texture and colouration with variously shaded and variegated. Many produce even beautiful spikes of flowers as a bonus to the foliage elegance, rising above the foliage during summer in the temperate and sub-temperate regions. The root-leaves are clump-forming and long-stalked, ovate to ovate-lanceolate and prominently ribbed while those on the floral stalks are smaller and bract-like. Flowers are white or blue which appear in terminal racemes or spikes with flowers opening well above the leaves, perianth is funnel-shaped, almost irregular and 6-parted, the stamens 6 with filiform filaments and oblong *cum* versatile anthers, pods are angled, oblong, many seeded and split into three valves with flat, winged and black seeds. They are propagated through division of clumps and a few species even through seeds collected from the ripe pods and sown immediately. The cultivated species are ***H. albomarginata*** (*H. sieboldii*) from Japan which is similar to *H. lancifolia* in habit but with slightly smaller leaves which are margined-white. It grows up to 45 cm tall with green but narrowly bordered with white, ovate-elliptic to narrowly lanceolate and some 15 cm long leaves. Pale-violet funnel-shaped flowers with deeper stripes and some 5 cm long appear on 60 cm tall racemes during late summer. ***H. crispula*** from Japan grows up to 60 cm tall bearing dark green with clear white margins, broadly lanceolate to elliptic, 20 cm long and long-pointed leaves with short stalks. Lilac-purple flowers some 5 cm long are produced on racemes some 50-80 cm long from late summer to autumn. ***H. elata*** (*H. fortunei gigantea*) from Japan is a robust species growing up to 90 cm in height, bearing a little glossy, 30 cm long, dark green and ovate-elliptic with wavy-margined foilage. White to pale-violet flowers some 5.0-6.25 cm long are produced on stiff and slender stalk some 90 cm long well above the foliage during June and July.***H. fortunei*** from Japan grows to a height of up to 90 cm bearing long-stalked grey-green leaves which are boldly veined, ovate but cordate at base and some 13 cm long. Lilac to violet flowers some 3.75 cm long are produced in July densely arranged on a 90 cm tall raceme. ***H. lancifolia*** from China and Japan grows up to 60 cm in height, bearing glossy mid-green and narrowly lanceolate leaves, and pale-lilac some 3.75 cm long flowers appear from July to September. ***H. plantaginea*** from China and Japan grows up to 60 cm tall, producing lustrous pale-green, some 25 cm long and ovate-cordate leaves, and some 10.0-12.5 cm long fragrant white flowers on 60-75 cm tall spikes during August and September. In *H. p. grandiflora*, the perianth lobes are longer than the width and the leaves are more elongated than the type. ***H. rectifolia*** (*Funkia longipes*) from Japan grows up to 105 cm tall, bearing some 15 cm long, dark green, erect and broadly lanceolate leaves, and some 5 cm long violet-mauve flowers with darker lines appearing freely on slender spikes some 60-90 cm long in July. ***H. sieboldiana*** (*H. glauca*, *Funkia sieboldii*) from Japan is the most robust species, growing up to 60 cm high, bearing glossy mid-green, thick textured, strongly veined, 25-38 cm long and broadly lanceolate to ovate leaves. Off-white to pale-lilac flowers tinged purple and some 3.75 cm long appear barely above the foliage in August on about 60 cm long racemes. Var. 'Elegance' bears glaucous leaves and pale-lilac flowers which appear during July and August. ***H. tardiflora*** (*H. lancifolia tardiflora*, *H. sparsa*) from Japan bears deep glossy-green, lanceolate, long-tipped and some 15 cm long leaves. Pale-lavender flowers some 4-5 cm long appear during autumn on 60 cm tall racemes. ***H. undulata*** (*H. lancifolia undulata*) from Japan grows up to 60 cm tall, bearing elliptic to ovate and strongly waving leaves some 15 cm long which are mid-green with white or silvery markings. Pale-violet to lavender and funnel-shaped flowers some 5 cm long appear during late summer on up to 90 cm tall raceme. *H. u.* 'Albomarginata' leaves are narrowly white-margined. *H. u. erromena* bears simple mid-green, larger and slightly waving foliage than the type and little darker flowers. The leaves in the var. 'Medio-variegata' are green variegated with yellow in the centre and the mauve flowers are carried on 60 cm long spike. ***H. ventricosa*** from East Asia grows up to 90 cm in height bearing glaucous green and broadly ovate-cordate leaves up to 24 cm long. Funnel-shaped violet-mauve flowers some 3.75-5.0 cm long are produced freely during July-August on about 90 cm tall racemes. Var. 'Variegata' is shorter than the type with dark green but yellow-margined foliage.

Howea (*Kentia*; Palmae)

Howea was named after Lord Howe Island where they are found. A genus of two species of palm from Lord Howe Island and SE Pacific, growing with erect single stem up to 20 m in height, bearing arching and pinnate leaves. At younger stage these make attractive pot plants. These are propagated by seeds. ***H. belmoreana*** (*Kentia belmoreana*; Curly palm/Sentry palm) is slow-growing, slender and elegant, growing 2-3 m in pots, bearing short-stalked and strongly arching leaves some 1 m long with crowded, 2.5 cm wide, linear-lanceolate and drooping leaflets from the rachis and pointing outwards. ***H. fosterana*** (*Kentia fosterana*; Kentia palm/Paradise palm/Sentry palm/Thatchleaf palm) grows 2-3 m in containers, bearing long-stalked and strongly arching leaves about 2.7 m long with fewer, quite long and linear-lanceolate leaflets pointing downwards.

Hoya carnosa (Asclepiadaceae)

See chapter on '**Succulents Other Than Cacti**'.

Humata tyermannii (Polypodiaceae)

See chapter on '**Ferns and Allied Plants**'.

Hypoestes (Acanthaceae)

Hypoestes derives from the Greek *hypo* meaning 'under' and *estia* 'a house', a little obscure reference to the way the calyces are covered by bracts. A genus with 150 species of evergreen and perennial sub-shrubs and shrubs from tropical Africa and Asia, especially Malagasy, but only two are grown as foliage pot plants. The leaves are in opposite pairs, simple and sometimes colourfully variegated. The spikes are terminal with clusters of 2-lipped tubular flowers. They like filtered sun. These are propagated through cuttings and seeds. ***H. aristata*** from South Africa is a herbaceous plant growing up to 90 cm high though indoor its height is only up to 60 cm. Leaves are soft green, elliptic-ovate and 5-8 cm long. Rose-purple flowers some 2.5 cm long appear on long and tiered spikes and at flowering this looks very spectacular as it produces the best flowers among all the flowering *Hypoestes*. ***H. phyllostachya*** (*H. sanguinolenta*; Freckle lace/Polka-dot plant) from Malagasy is a branching herbaceous plant growing up to 50 cm in height, bearing dark green, thin, elliptical to ovate and 5-6 cm long leaves freckled with irregular dense pink spots though in most cases the dots revert back to green when grown indoors. The plants are fast-growing so require pruning to keep it in shape and it would be better if every year these plants are replaced with new ones. Lavender flowers some 2 cm long are borne singly from the leaf axils. *H. p.* 'Splash' has larger and brighter leaf spots. There are many other varieties differing in their colourations.

Ilex (Holly; Aquifoliaceae)

Ilex is the ancient Latin name of *Quercus*. The genus comprises 400 ornamental woody plants from North and South America, tropical and temperate Asia, Africa, Australia and Europe. *Ilex* has innumerable number of cultivars. They are grown for their handsome foliage and attractive berries which are mostly red. They are evergreen (mostly trees) or deciduous (mostly shrubs) plants, with alternate, petioled, simple and medium-sized leaves having small and caducous stipules which may be spiny, flowers are insignificant, whitish, dioecious and generally in a few-flowered axillary cymes or sometimes solitary, calyx lobes, petals and stamens usually 4, style very short, ovary superiuor, and fruit a berry-like black, red or yellow drupe which contains one stony seed, and these berries hang on the plants until next spring. During summer, these should be protected if temperature rises above 27 °C. They are propagated through seeds though these take a very long time to germinate hence are sown after stratification, however, the evergreens, especially the shrubby ones may also be propagated through cuttings of ripened woods, grafting or budding on seedlings of *I. aquifolium* or *I. opaca*. Only a few, coming from the warmer habitats are suitable for growing indoors. ***I. cinerea*** from China is a large shrub, bearing stalkless leaves which are lanceolate, serrate with black tips and some 12.5 cm long and 3.75 cm wide, and the berry is red. ***I. dimorphophylla*** from Okinawa is a shrub most suitable as a house plant, bearing shiny light green, 2.5 cm long and very spiny leaves and red berries. ***I. hanceana*** from Hong Kong is a small shrub, bearing 2.5 cm long, spathulate, unspiny and leathery leaves and red berries. ***I. paraguariensis*** from Yerba Mate is a tree growing up to 6 m tall in the wild but indoors these are tamed in pots up to 1.5 m. Its leaves are dark green, elliptic, 12.5 cm long and 6.25 cm wide and with scalloped margins, and the berries are red. ***I. pubescens*** from Taiwan is a shrub growing up to 3 m in the wild but indoors only up to 1 m, producing 4-angled branches. The leaves are olive-green, spathulate, 5.0 cm long and partly toothed and berries are red.

Iresine (*Achyranthes*; Bloodleaf; Amaranthaceae)

Iresine derives from the Greek *eiros* meaning 'wool', as the flowers are woolly. There are some 80 herbaceous species, a few being climbers from tropical and warm temperate regions but only two are grown indoors. Stems are weak, fleshy, herbaceous and coloured. Leaves are ovate to lanceolate and colourful but flowers are quite insignificant. These are suitable for edging, conservatory, hanging baskets and window sills, doing well in full or partial sun. These are propagated through cuttings. ***I. herbertii*** (Beefsteak plant) from South America grows up to 60 cm or more in height with red stems, deep purplish-red leaves with paler to yellowish veins, sometimes puckered, notched at apex, ovate to rounded and about 13 cm long. *I. h. aureo-reticulata* has the same vein pattern as to the parent species but leaves are bright green. The leaves in 'Brilliantissima' is deep purplish-red with scarlet veins. ***I. lindenii*** (Blood leaf) from Ecuador grows up to 60 cm tall, bearing narrowly lanceolate, pointed, 5-8 cm long and deep red leaves. *I. l. formosa* leaves are green and broader with clear yellow veining.

Jubaea (Palmae)

Jubaea honours king Juba of Numidia, Africa (*d.* 46 B.C.). It is a large palm with one species which when young or small makes a good display as pot plant. It is propagated by seeds. ***J. chilensis*** (*J. spectabilis*; Chilian wine/Coquito/Honey palm) from coastal region of Chile is though quite slow-growing but grows very tall, up to 30 m in height with a thick and grey-black trunk. Usually the plants in containers do not form the trunk. In the wild the leaves may attain even up to 3 m length but in pots these hardly grow 1.2 m. Leaves are usually greyish-green and long but short-stalked and pinnate, and the lanceolate to linear leaflets are some 30-45 cm long. Only large specimens produce panicles some 90 cm long, bearing small, maroon and yellow flowers, followed by 2.5-4.0 cm long and nearly spherical fruits which are almost yellow when ripening but turning brown with a very hard shell having coconut-like flesh used in confectionary. The tree trunks also yield a sweet sap which is boiled and converted to molasses and sugar or even wine.

Kaempferia (Zingiberaceae)

Kaempferia honours Engelbert Kaempfer (1651-1716), a German doctor and botanist who travelled widely in the east, particularly Japan where he lived for two years, later writing books on the plants and history of that country. A genus of 70 species of rhizomatous and tuberous perennials from tropical Africa and SE Asia, grown for both foliage and flowers. Their aromatic rhizomes produce clumps of simple, aromatic and arching leaves and flowers with three true petals and an equally large to larger staminodal labellum (lip) similar to orchids. These are suitable for growing in warm conservatories where humidity is not a problem, however, in the home these can be kept only for a short periods. They are propagated by division. ***K. pulchra*** from Thailand and Malaysia is a horizontally spreading species, producing shoots with 1-2 broadly ovate, quilted, 13 cm long and dark bronze-green leaves having a broken grey zone between the margins and the midribs. Light purple flowers with basal lip colour yellow having white spots appear during summer. ***K. raskoana*** (Dwarf ginger lily/Peacock plant) from Myanmar is in many respect similar to *K. pulchra* but the leaves are purple-bronze and more iridescent having a pale-green zone. ***K. rotunda*** (Tropical crocus) from SE Asia produces shoots with two erect and narrowly elliptic to oblong-lanceolate leaves some 30 cm long which are dark with light green variegation above and purple beneath. The white flowers some 5 cm wide with a pale-purple lip appear in clusters of 5-10 in summer.

Kalanchoe (Crassulaceae)

See chapter on '**Succulents Other Than Cacti**'.

Kleinia (Hot dog plant; Asteraceae)

See *Senecio* under the chapter on '**Succulents Other Than Cacti**'.

Kochia (Chenopodiaceae)

See chapter on '**Annuals**'.

Latania (Latan palms; Palmae)

Latania is a Latinized form of *latanier*, the local name used in Mauritius where in Mascarene Islands, it is found wild. It is a genus of four erect, unbranched and elegant fan palm species topped by a crown of greyish leaves, with deeply cut, slender, long and sharply pointed lobes that are deeply pleated. The trees are dioecious in nature and produce clusters of tiny flowers borne in the leaf axils which are insignificant and remain hidden in the leaves but usually pot specimens do not produce flowers. They are decorative house plants for many years. They are propagated through seeds. ***L. loddigesii*** (*L. glaucophylla*; Silver latan) from Mauritius though grows to a height of 10 m bearing blue green, rigid and about 1 m wide leaves with veins and long stalks being orange to red-brown but in young ones, *vis-a-vis* in pots, the size is drastically reduced, and the growth is also slowed down. ***L. lontaroides*** (*L. borbonica*, *L. commersonii*; Red latan) from Mauritius is most attractive species growing in the open up to 15 m tall, bearing long, orange and thorny stalks and some 2 m wide leaves with finely toothed grey-green segments which at younger stage is reddish- to purplish-flushed while irrespective of the stage the veins and margins remain reddish to purplish throughout. The height, *vis-à-vis* leaf size are highly reduced in potted specimens coupled with quite slow growth. ***L. verschaffeltii*** (*L. aurea*; Yellow latan) from Rodriguez Islands attains up to 15 m plant height, bearing long, orange to yellow and glabrous stalks and with 1 m wide leaves in its native habitat but in pots the height and leaf size are drastically reduced. Leaves are green above, white and soft-woolly beneath.

Laurus (Laurels; Lauraceae)

Laurus is an old Latin name for bay laurel. It is a genus of two evergreen tree species from Mediterranean region and Macronesia (North atlantic Islands), of which one is in common cultivation as a decorative foliage plant and for its leaves used to flavour the cooked food. Through regular pruning and pot culturing, these can be tamed for home gardening. They are propagated by seeds and cuttings. ***L. nobilis*** (Bay/Sweet bay) from Mediterranean region, in the wild grows up to 20 m tall but in pots only up to 1.2 m high, bearing glossy deep green, leathery, narrowly oval, some 5.0-8.5 cm long leaves which emit a strong pleasing aroma when crushed. Flowers are unisexual and appear in axillary clusters in spring, followed by 1.0-1.5 cm long, black and berry-like ovoid drupes. *L. n.* 'Aurea' leaves are golden-tinted.

Lavandula dentata (Lavender; Labiatae)

Lavandula derives from the Latin *lavo* 'to wash', on account of its use in perfuming soaps, shampoo, scents and other toiletries. It is a hardy genus of evergreen and strongly aromatic shrubs of 28 herbaceous species scattered over Mediterranean region, south and west of Somalia, Macronesia and India. Out of all the species, only ***L. dentata*** from Spain and Balearic Isles is worth growing indoors for its aromatic and culinary leaves and fragrant flowers. It grows up to 60 cm in height with highly branched stems having thin and sweeping branches which while growing incurve upward to make the form of a globose-cone. Plants when aged become leggy so require replacement after 4-6 years. Leaves are strongly aromatic, linear to oblong, and evenly pinnately cut into many bluish-green small lobes in the form of small teeth, each leaf measuring some 2.5-4.0 cm long and 6 mm wide. It is propagated through cuttings or seeds. Its dark purple flowers some 8 mm long appear in spikes which are tipped by a cluster of purple bracts. It has many varieties to choose from.

Ledenbergia sequieriodes (Phytolaccaceae)

Ledenbergia was named for von Ladenberg. The genus has only one shrub species from tropical America (Antilles), allied to *Rivina* with decorative foliage. Its leaves are green, slender-stalked, simple, alternate, elliptic or elliptic-ovate and acuminate, and the small greenish flowers which are slender-pedicellate, appearing on long, peduncled and drooping racemes from lateral axils, bearing 4-parted perianth with linear-oblong or linear-obovate and obtuse segments that extend to form the fruit, stamens 12 and fruit is a small achene.

Leea (Leeaceae/Vitaceae)

Leea is named for James Lee (1715-1795), a Scotch nurseryman. The genus originates in tropical Africa, Asia, Australia and Pacific with 70 species of shrubs and trees, out of which some eight species are suitable for growing indoors in tropical area for their coloured foliage and stately habit. Leaves are alternate, simple, 1-3 times pinnate and the petioles are dilated at the base, leaflets are entire or serrate, flowers large or small, red, yellow or green in cymes opposite the leaves, calyx 5-toothed, petals 5, connate at the base and jointed with stamen tube, ovary 3-6-celled, cells 1-ovuled and fruit a globose berry flattened at top. It differs from *Vitis* as the latter has climbing habit, 2-celled ovary and each cell with two ovules. These are propagated by cuttings. ***L. aculeata*** from Malaysia is a shrub with trident leaves on long stalks. Leaflets are elliptic 7.5-15.0 cm long and toothed. ***L. amabilis*** from Borneo bears handsome silvery vine-like velvet bronze-green foliage of compound leaves some 60 cm long with 5-9 elliptical leaflets some 10 cm long, pairing opposite. Leaflet midrib is bordered with a broad white stripe. The var. 'Splendens' is more reddish. It is most suitable against the pillar-posts or in pots but necessarily requires winter dormancy of partially drying out. ***L. coccinea*** (West Indian holly) from Myanmar and Borneo is a shrub growing up to 1.4 m tall, bearing elliptical and toothed leaflets in opposite pairs (single or with 3 subleaflets, each 7.5-10.0 cm long). Flowers scarlet in bud but opening pink and some 60 in a cluster, appearing some 7.5 cm across (each flower some 1.25 cm across) in a trichomous flat-topped cluster, and the stamens are yellow and exserted. ***L. micholitzii*** from New Guinea produces slender stems, large to gigantic, pinnatifid and arching leaves, and rich green leaflets having white veins in the young leaves and marked bright red. ***L. rubra*** is similar to *L. coccinea* but bears dark red leaves. ***L. sambucina*** (*L. roehrsiana*) from India, Malaya, Philippines and Tropical Australia is a quite handsome shrub with long stalked, pinnate or 3-pinnate compound leaves some 1.2 m long, and paired leaflets are oblong but cordate at base, acuminate and coarsely crenate, and the size being 16.3 cm long and 6.2 cm wide.

Licuala (Small fan palm; Palmae)

Licuala is the Latinized version of the Moluccan vernacular name *leko wala*. This genus comprises 100 species of small palms from SE Asia, Malaysia, Bismark and Solomon Islands but only 3-4 are grown indoors.

These are solitary or cluster-stemmed plants with palmate to costapalmate (a mid way to palmate and pinnate) and entire, lobed or long-segmented leaves. Red or orange tubular or urn-shaped 3-lobed flowers appear on panicles. Certain species make attractive specimen pot plants. These are propagated through seeds or offsets. ***L. grandis*** (*Pritchardia grandis*) from New Hebrides Island is a solitary-stemmed, erect and slender palm growing up to 2 m in pots though natural height is up to 6 m. Leaf stalks and leaves are of almost equal length, and the leaves are lustrous bright green, palmate, boldly toothed or shortly lobed, shallowly pleated by the numerous radiating veins and some 60-90 cm in diameter. ***L. pumila*** (*L. elegans, L. gracilis*) from Java and Sumatra may have solitary or clustered stems growing up to 1.5 m high, bearing rich green, palmate leaves about 45 cm wide and deeply segmented into 6-8 or even up to 24 narrow 2-ribbed divisions. It produces the offsets. ***L. spinosa*** from SE Asia and W. Pacific Islands produces clustered stems up to 3.5 m high. Leaves are almost round and 90 cm in diameter which are segmented into as many as 15 narrowly wedge-shaped leaflets measuring up to 45 cm long. It likes quite moist environment.

Ligularia tussilaginea (*Senecio kaempferi*; Asteraceae)

Ligularia derives from the Latin *ligula* meaning 'a strap', alluding to the shape of the ray florets. These are propagated through seeds or division. The genus comprises 80-150 perennial species where most are deciduous though ***L. tussilaginea*** (*Farfugium grande, F. japonicum, Senecio kaempferi*) from China, Korea, Japan and Taiwan is a stemless evergreen and bold foliage plant suitable for planting in large containers indoors. It is clump-forming with wide spreading habit, only a few kidney-shaped leaves coming up from the base which are glossy green with shallowly lobed margins, up to 16 × 30 cm length and width having some 30 cm long stalks, and are little leathery. Yellow corymbose-clustered flowers in the form of daisy in 4-6 cm across flowerheads appear during autumn. *L. t.* 'Argentea' ('Albovariegata', 'Gingetsu' or 'Variegata') leaves are edged irregularly in creamy-white and grey. *L. t.* 'Aureo-maculata' (Leopard plant) leaves are covered with large yellow spots and this is most commonly grown plant. *L. t.* 'Crispata' leaves are green, crisp and frilled at edges. Var. 'Glasshouse Gold' leaves are splashed with yellow and grey. These prefer sunny and cool site, and do well between 13-27 °C temperatures.

Liriope (Lily-turfs; Liliaceae)

Liriope was named after the wood nymph of Greek mythology who was the mother of *Narcissus*. A genus of 5-6 evergreen perennial species with swollen fleshy rhizomes, it is native of China, Japan and Vietnam, bearing narrowly linear and leathery leaves and 65-tepalled mauve racemose flowers and berry-like black fruits. They are suitable as ground covers, for edging and as attractive pot plants. They can be multiplied through seeds or by division. ***L. exiliflora*** (*L. muscari exiliflora*) from China and Japan is a tufted, colonizer and rhizomatous species, bearing 30-40 cm long strap-shaped dark green leaves arching at the top. Purple-tinted floral stems arise above the leaves bearing violet-purple some 8 mm long expanded flowers in loose racemes. ***L. hyacinthiflora*** (*Reineckea carnea*) from Japan and China is a dense clump-forming rhizomatous perennial, bearing 10-40 cm long basal and very dark green linear to lanceolate leaves. Spikes some 8 cm long bear 0.8-1.2 cm long, bell-shaped and pale-pink flowers with 6 reflexed lobes. Fruit is a berry. 'Variegata' has longitudinal creamy-white stripes. This species is better known as *Reineckea carnea* with only one species under the genus, and named so for Johann Heinrich Julius Reinecke (1799-1871), a German gardener. ***L. muscari*** (*L. platyphylla, L. graminifolia densiflora*) from China and Japan is an evergreen perennial growing about 30 cm high, with narrow, glossy and dark green leaves, and in autumn, carries rounded, bell-shaped, clustered and lavender or purple-blue flowers. 'Majestic' is a frost-hardy rhizomatous perennial with 30 cm height and 45 cm spread, bearing linear and glossy bright green leaves and spikes of thickly clustered, rounded and bell-shaped flowers in autumn. Its varieties are 'Big Blue', 'Blue Spire', 'Gold Banded', 'Majestic', 'Munroe White', 'Variegata', *etc.* ***L. spicata*** (*L. graminifolia, Dracaena graminifolia*) from China and Vietnam is a fully hardy evergreen perennial growing to 30 cm high with a spread of 30-40 cm, bearing grass-like and glossy dark green leaves and pale-lavender rounded bell-shaped flowers in spikes in summer.

Lithops (Aizoaceae)

See chapter on '**Succulents Other Than Cacti**'.

Livistonia (Chinese fan palm, fountain palm; Palmae)

Livistonia commemorates Patrick Murray, Baton of Livingstone, whose fine plant collections prior to 1680 proved the nucleus of the Royal Botanic Garden, Edinburgh. A genus of about 30 species mainly of large fan palms from Indo-Malaysia and Australia.They have single straight trunks and terminal cluster of leaves that are halfway between pinnate and palmate (fan-shaped) and grow up to 1.8 m long including sharply spined petioles though in containers these are quite small. The divided leaves have long, tapering, ribbon-like segments which gracefully sway beneath the leaves, creating an overall fountain-like effect. The inconspicuous flowers are hidden among the leaves and are followed by small, blue-black and olive-like fruits. Although Chinese fan palm has long been used as a container plant, its neat leaf habit and interesting form make it ideal for further landscape uses, such as for staggered groupings, as free standing specimens or as street trees. They form a closed canopy when planted about 3 m apart along a walk or street. They grow well in confined soil spaces. The palm is self-cleaning of old leaves and will require little or no pruning. Tolerant of full sun, young specimens should be partially shaded. Any reasonably fertile and well-drained soil including the alkaline ones is suitable for its cultivation. They are propagated through seeds. Since these are the tall plants so these can be enjoyed indoors in containers only when young. ***L. australis*** (Australian fountain palm/Gippsland fountain palm) from eastern Australia is fairly slow-growing in containers. Leaves are rich glossy green, orbicular up to 1.2 m long in containers with divisions to halfway to the midrib into several narrow lobes and long-stalked which are edged with spines. ***L. chinensis*** (*Latania borbonica*; Chinese fan or fountain palm) from southern Japan and China is fairly slow-growing in containers with 90 cm long glossy rich green leaves which are rounded and divided nearly halfway to the midrib into several narrow pendulous lobes.

Lobivia (Cob cactus; Cactaceae)

See chapter on '**Cacti**'.

Lycopodium (Club moss; Lycopodiaceae)

It takes its name from the Greek *lykos* meaning 'a wolf' and *podion* 'a foot', as a part of the plant looks like wolf's foot. These epiphytic and terrestrial plants closely related to *Selaginella* and so allied to ferns are found worldwide in the tropics and subtropics with about 450 species. Some species may be prostrate and others erect, bearing branched stems with small, crowded and overlapping leaves. Leafy spikes bear sporangia terminally. They make excellent foliage plants suitable for conservatory, and if brought indoors, they should be provided with plenty of humidity. These can be propagated through division and by cuttings. ***L. cernuum*** is a clump-forming to colonizing terrestrial species growing up to 30 cm in height in the form of a tiny coniferous tree which is highly branched and bears awl-shaped small leaves. ***L. phlegmaria*** (Queensland tassel fern) is tufted to clump-forming terrestrial species producing erect stems which with age branch and droop. The leaves are glossy rich-green, sometimes slightly overlapping, awl-shaped and some 2.5 cm long. ***L. squarrosum*** from tropical Asia is a tufted to clump-forming epiphytic species with initially erect stems but with age drooping, bearing narrow, bright green and more crowded leaves compared to *L. phlegmaria*. ***L. taxifolium*** from Jamaica is an epiphytic plant quite similar to *L. phlegmaria* but with little thickened stem tips and some 1.2 cm long leaves are brighter green.

Lygodium (Schizaeaceae)

See chapter on '**Ferns and Allied Plants**'.

Macropiper excelsum (Piperaceae)

Macropiper derives from the Greek *macros* 'large' and *Piper*, the allied climbing genus which included it initially. There are probably six species of evergreen shrubs in the genus found from Polynesia to New Guinea and New Zealand, though only one, ***M. excelsum*** from New Zealand and adjacent islands is grown as a highly decorative foliage house plant. It is propagated through seeds and cuttings. It is a robust large shrub which can be tamed to grow low through container-culture and regular pruning, *vis-à-vis* regular replacement with newly propagated plants. Its aromatic leaves are deep lustrous green, 6-12 cm long, broadly ovate to sub-orbicular, basally cordate and tip-pointed, and the insignificant tiny flowers which are unisexual appear on 2-8 cm long spikes. *M. e. majus* (*M. e. psittacorum*) recorded from an Island north of New Zealand is superior to the type species with up to 20 cm long leaves.

Macrozamia (Zamiaceae)

Macrozamia derives from the Greek *macros* 'large' and the allied genus *Zamia* in which are included the smaller growing species. It is an Australian member from the group of cycads, less allied to *Cycas* than to *Dioon*, *Encephalartos* and *Zamia*, with dioecious palm-like structure, containing 12-14 species in the genus. Similar to cycads, it is a living fossil of the primitive seed-bearing plants which flourished at the end of the Triassic and beginning of the Jurassic periods. All the members form a dense rosette of stiff and pinnatifid leaves arising from the ground level or from the top of the caudex, if present. These aquire both, the trunk and leaves of *Cycas* but without midrib in the pinnae though is distinctly striate on the underside with several parallel equal veins throughout the leaf. Mature plants from the centre of the rosette produce cone-like stalked floral spikes. Scales of the female cones which are peltate with thickened shield, are produced usually into an erect and acuminate blade. They make excellent specimen plants in the garden soil or in pots for indoor decoration. ***M. miquelii*** (*M. mackenzei*) bears 60-120 cm long leaves with straight or falcate pinnae which are 20-30 cm with finer and less prominent longitudinal vein. Petiole base is woolly, rachis of the leaves broad and flat with thick but variable cone scales, cones being cylindrical and 15-20 cm long but female thicker and scales with a long point. ***M. moorei*** from Queensland and New South Wales remains stemless for many years though in the wild have been recorded up to 6 m probably on 300 years old specimen. Its deep green fronds grow 1-2 m having 20-30 cm long pinnae which are linear with basal ones reduced to spines and the upper ones spine-tipped. ***M. paulo-guilielmi*** (*M. plumosa*) grows with short caudex, sometimes only protruding above the ground with woolly bases of old petioles, the fronds some 30-90 cm long with narrow rachis are often flat on top and the pinnae which are many are slenderly very narrow, cylindrical and smooth. Male cones are about 7.5 cm long and the females about half in thickness but 10 cm long. ***M. peroffskyana*** (*Lepidozamia denisonii, L. peroffskyana*) from Queensland and New South Wales grows taller and attains 5.4-6.0 m height with some 30 cm diameter of the trunk. Leaves are 2.1-3.6 m long with angular petioles, the linear pinnae 30-60 cm long and 1.25 cm wide with finely marked parallel veins. Cones are stalkless with hairy scale tips, male cones are ovoid, 10-15 cm long and 7.5-10 cm in diameter but females are 20-25 cm long and some 15 cm in diameter. ***M. riedlei*** (*M. fraseri*) from W. Australia is usually stemless though very old specimens exert protruding trunks from the ground. Leaves are glaucous green, 1-2 m long with linear leaflets some 25 cm long and spine-tipped. ***M. spiralis*** from New South Wales bears invisible to short trunk, having leaves 60-120 cm long with spiralling midrib and 10-20 cm long leaflets. Male cones measuring some 15-25 cm long with much flattened scales but female cones are shorter and thicker with scales having incurved short point.

Mammillaria (Rose pincushion; Cactaceae)

See chapter on '**Cacti**'.

Manihot (Euphorbiaceae)

Manihot is the Latin version of the Brazilian vernacular name *manioc*. It is a tuberous-rooted genus of 160 perennial species of trees and shrubs of which one *M. esculenta* is widely cultivated in the tropics as food. The raw roots are poisonous, containing hydrocyanic acid but on heating these become harmless. It is propagated through young tip cuttings or some 15-20 cm long cuttings of stemless stems. These yield milky latex and the plant is usually glaucous and glabrous, the leaves are alternate, entire to lobed or palmate. Flowers are insignificant, apetalous and monoecious, and appear in terminal or axillary racemes or panicles, with sepals united basally, imbricate, often petaloid, stamens 10, ovary 3-locular with one ovule in each and the seeds are carunculate. ***M. esculenta*** (*M. utilissima*; Bitter cassava/Cassava/Manioca/Sweet potato tree/Tapioca) from Mexico, Guatemala and Brazil is a herbaceous shrub growing up to 2.5 m tall with fleshy and clustered roots, leaves are deep green, pubescent and 3-7 deeply parted, lobes entire, lanceolate, acuminate and 7.5-20 cm long, stipules entire and small, flowers in panicles with pale-yellow or red-tinged calyx which is pubescent inside and up to 1.2 cm long, and the wing-angled fruit is a capsule. There is one of its variegated form, *M. e.* 'Variegata' where leaf centre is irregularly patterned yellow, and this is most admired as ornamental foliage plant.

Maranta (Prayer plant; Marantaceae)

Maranta was named for Bartolommeo Maranti, an Italian physian and botanist living in Venice in 1559. There are some 20 rhizomatous and clump-forming species grown indoors for their ornamental foliage. Their leaves have sheathing stalks and are entire. These clump-forming herbs are indigenous to tropical Americas, primarily South America and that too Brazil. These are fleshy and low-foliage house plants. The leaves are alternate, almost oval in shape, some 15 cm long and their stalks clasp the stems. ***M. arundinacea*** (Arrowroot) from tropical America produces slender stems 1-2 m tall, bearing long-stalked and lanceolate leaves some 30 cm long. Mature plants produce small white flowers. Its var. 'Variegata' has leaf patterning in three shades of green. ***M. bicolor*** from Brazil and Guyana is tuberous, tufted to clump-forming with a very short and fleshy stem, bearing elliptic to oval, wavy, up to 15 cm long bluish-green leaves with an irregular pale zone near the centre, bordered by two rows of dark to brownish-green spots though beneath the leaves are violet. Its small white flowers are adorned with red-purple lines. ***M. leuconeura*** (Prayer plant) from Brazil is non-tuberous and bears creeping, spreading and pendant, branched but short stems growing up to 30 cm tall with green having grey or red lateral veins and blotched-purple beneath, elliptic to oblong, 30 cm long leaves which are almost blunt. *M. l. erythrophylla* (syn. *M. tricolor*, 'Fascinator'; Herringbone plant) bears on the leaves prominent red veins with metallic central zone and crimson veins curving to the leaf margins. *M. l.* var. *leuconeura* (*M. l. massangeana*) has blackish-green leaves, light green mid-vein and metallic central zone. *M. l. kerchoveana* (Rabbit foot/Rabbit tracks) bears light green leaves with a row of brown blotches between the veins which turn green with age. Marantas are versatile plants indoors because they can be used as small specimen plants, in hanging baskets which cascade, as ground covers in interiorscapes, and in dish gardens and other combination plantings. Marantas are usually started from cuttings selected from stock beds. Marantas are best grown under a light level range of 1,000 to 2,500 fc in greenhouses where moisture and temperature can be controlled. Temperatures of 21-27 °C are ideal for maranta rooting and growth. Good growth occurs up to 32.2 °C temperature but beyond this the growth is hampered. Maranta grows well in a potting medium with good aeration, high water holding capacity and a pH of 5.5 to 6.0. Peat-based mixes generally require addition of dolomite to raise the initial pH and addition of a micronutrient blend product, such as MicroMaxR (500 g/m^3) is very effective.

Marsilea (Aquatic fern/Pepperworts/Water clover; Marsileaceae)

Named so to honour Count Luigi Ferdinando Marsigli (1658-1730), an Italian botanist. It is a rhizomatous genus of about 60 evergreen species of wet-ground to aquatic fern allies from tropical to temperate regions grown for their oranamental foliage, however, a few may dry up in dry weathers but again coming up from their rhizomes under congenial conditions. These rhizomes help them to creep and spread nearby. The barren fronds which are long-stalked and quadrifoliate remain erect in wet soil as well as in shallow water though float in deep water. The greatly modified stalkless leaves having small capsular-shape and known as sporocarps bear sporangia. These are propagated by division. These are suitable plants for aquaria, for conservatory-pool and for pots stood permanently in water saucers. ***M. drummondi*** (Water clover) from tropical to temperate Australia has free-branching rhizomes from which crowded tufts of foliage appear with broadly obovate-culineate, silky-bloomed and wavy-edged leaflets some 2.5-4.0 cm long. ***M. quadrifolia*** from Europe and Asia bears quite thick rhizomes with only a few branches. Its leaflets are glossy rich green, entire obovate-deltoid and some 2.0 cm long. Though it survives in cold areas but when very cold it dies back, especially in winter. It is ideal for conservatories which remain cool during summer.

Matteuccia [Onocleaceae {Aspidiaceae or Polypodiaceae}]

See chapter on '**Ferns and Allied Plants**'.

Melianthus (Melianthaceae)

Melianthus derives from the Greek *meli* 'honey' and *anthos*' a flower', referring to the nectar which the plant freely produces. A genus of six shrubby perennials or shrubs from South Africa and India, grown by their two species in pots or tubs for their ornamental foliage. These are suitable for cool conservatory and the short one in cool rooms. The leaves are pinnate and coarsely toothed. They are propagated annually through division or suckers and cuttings. ***M. major*** (Honeybush) from South Africa which has restricted branching, grows up to 3 m tall, bearing glaucous to grey-green leaves some 25-45 cm long and pinnatifid into 7-13 ovate leaflet-divisions, each 5-13 cm long. Terminal racemes some 30 cm long produce brownish-red and some 2.5 cm long flowers in summer. ***M. minor*** from South Africa is similar to *M. major* but is smaller (nearly half) in all its parts and is most suitable for growing indoors.

Miconia (Melastomataceae)

It was named so to honour Spanish physician, *Miconia.* This comprises 518 species of trees and shrubs from tropical America with handsome foliage. The foliage are opposite or verticillate and strongly veined. Small white, yellow, rose or purple flowers appear in corymbs or panicles with 4-8 apex-rounded and spreading or reflexed petals, stamens 8-16 with variable shape and polymorphous anthers and a 2-5-loculed dry or leathery berry-fruit having a few or many seeds. These are warm-house subjects. These are propagated through seeds and cuttings. ***M. magnifica*** (*Meconia velutina, Cyanophyllum magnificum, Tamonea magnifica*) from Mexico is a robust tree but indoors it can be tamed to a height of 2-3 m, bearing 60-75 cm long most striking leaves where upper surface is lustrous green, lower red, broadly ovate, wavy, arched, rugose and prominently veined lightly or with white. Flowers are white and appear in panicles. ***M. spectanda*** (*Cyanophyllum spectandum*) from Brazil bears some 45 cm long and ovate leaves which are 15-18 cm wide in the centre with prominently grey midribs, upper surface dark lustrous green and lower greenish-red. ***M. denticulata*** (*Melastoma denticulatum*) from Ecuador produces obtusely 4-angled branches which when young bear furfuraceous leaves and calyx, while in usual case the leaves are narrowly ovate with rounded base, little obtuse, 3-nerved and minutely serrulate, pedicellate flowers aggregate in the panicles, calyx hemispherical and petals are small and subrotund.

Microcachrys (Podocarpaceae)

In Greek, *micros* is for 'small' and *kachrys* 'a cone'. It is an alpine evergreen genus of small coniferous shrub comprising one species, ***M. tetragona*** from Tasmania. It grows prostrate with sinuous stems having scale-like 4-ranked overlapping leaves giving the stems a 4-angled appearance. Flower clusters are ovoid, small, cone-shaped, dioecious, female cones being 6-8 mm long with mature seeds partially embedded in the scales which are fleshy and red. It is suitable for growing in the alpine house in a pot and is propagated by seeds or cuttings.

Microcoelum (Weddel palm; Palmae)

This genus is a compact palm comprising just two species, and only one, ***M. weddellianum*** is used as house plant. Indoor it does not grow taller than 1.2 m with about 60 cm spread. The fronds of the full grown plants are shiny dark green, on a 7.5-15.0 cm long stalks, and the leaves some 90 cm long and 23 cm wide with 20-30 pairs of opposite to alternate pinnae (leaflets) in herringbone fashion, attached with central rib which is covered with black scales. No flowers are produced on indoor plants. These are propagated through seeds and offsets.

Microlepia (Polypodiaceae)

See chapter on '**Ferns and Allied Plants**'.

Microsorium (Polypodiaceae)

See chapter on '**Ferns and Allied Plants**'.

Mikania ternata (Asteraceae)

Mikania is named in honour of Joseph Gottfried Mikan (1783-1814), professor of botany at Prague or probably his son Johann Christian Mikan who collected plants in Brazil and followed his father as a professor at Prague. The genus comprises some 250 perennial species of climbers and shrubs from tropical America, West Indies and South Africa. ***M. ternata*** (*M. apiifolia*), a climbing species from Brazil, which is propagated through cuttings or layering, grows up to 5 m with scrambling and purple-woolly stems, bearing purple-green leaves above and deep purple beneath in opposite pairs, and these leaves are hairy and digitate with 1.5-4.0 cm long and lobed leaflets which are 5 (in the wild 3-7), stalked, thromboidal to broadly obovate and undulated. Its *Senecio*-like small yellowish flowers appear in loose corymbs.

Milium (Millet; Gramineae)

Milium is an ancient Latin name for millet. Genus comprises 6 species of annual and perennial grasses from northern temperate zone. ***M. effusum*** 'Aureum' (Golden wood millet) from Europe, Asia and north-eastern region of North America is the yellow-leaved tufted perennial which is in common cultivation. Its stems are 0.6-1.5 m tall and bear some 10-30 cm long and 0.5-1.5 cm wide, linear and arching leaves. It bears ovate, nodding and loose panicles with 3-4 mm long spikelets which appear during autumn. It likes shaded locations and is propagated through division or seeds.

Mimosa pudica (Leguminosae)

In Greek, *Mimosa* is for 'a mimic', referring to the sensitivity of leaves in some species. The genus comprises 450-500 species of annuals, perennials, climbers, shrubs and trees from tropical and sub-tropical America, Africa and Asia, having bipinnate leaves and tiny tubular flowers in globular heads with protruding stamens. ***M. pudica*** (Humble plant/Sensitive plant) from tropical America is a popular house or conservatory plant grown for amusement as leaflets close and the leaves droop by slight air movement or through mild touch but after sometime again become as usual. It is propagated through seeds. It is a sub-shrubby perennial growing up to 90 cm high, though normally grown as an annual. Its stems have hooked prickles, leaves bipinnate where four pinnae are divided into 15-25 pairs of small and narrowly oblong leaflets. Its tiny pale-pink flowers appear from the leaf axils in globular heads and then plants look quite attractive.

Miscanthus (Gramineae)

Miskos in Greek is for 'a stem' and *anthos* 'a flower', probably due to the stalked spikelets. The genus comprises 20 species of tall perennial grasses from Asia. Those being grown in the gardens are hardy and form bold clumps with wide grassy leaves topped by feathery panicles of paired and awned spikelets. These are propagated through division and grown in sun or partial shade. ***M. sacchariflorus*** (*Imperata sacchariflorus*; Amur silver grass) produces thick rhizomes which make spreading colonies with 3 m or more tall stems and 2 cm wide leaves. Almost awnless and grey-brown spikelets are covered with long silky hairs. ***M. sinensis*** (*Eulalia japonica*) bears thick, dense and short rhizomes which form the clump and produce up to 3 m tall stems. The leaves are 1 cm wide with whitish midrib. Some 20 cm long panicles bear awned spikelets covered with silky hairs. *M. s.* 'Gracillimus' is a dwarf variety with 6 mm wide leaves, *M. s.* 'Variegatus' has cream stripes in the centre of leaves, *M. s.* 'Zebrinus' leaves have creamy-yellow horizontal banding, *etc.*

Molinia (Gramineae)

Molinia was named for Juan Ignacio Molina (1740-1829), natural historian of Chile. It is a clump-forming genus of 5 perennial species of grasses from Europe and Asia requiring full sun to partial shade for its growth. Their leaves are narrow, grassy, long and arching and the panicles are slender with small and 2-5-flowered spikelets. They are propagated through division and by seeds. ***M. arundinacea altissima*** (*M. altissima, M. litoralis*) grows taller with 1.4-1.8 m tall stems and panicles 30-50 cm long. ***M. caerulea*** (Purple moor grass) from Europe and Asia is a deciduous clump-forming grass, stems growing up to 75 cm with 20 cm long and 4-7 mm wide leaves, and up to 20 cm long and purplish panicles. A. most popular form *M. c.* 'Variegata' bears creamy-white striped leaves, while young ones often have pink-tinting.

Monolena primulaeflora **(syn.** ***Bertolonia primulaeflora*****; Melastomaceae)**

Monolena is a Greek, meaning single spur-like appendage on the anterior side of the anther-connective. The genus comprises five stemless and little fleshy herbaceous species from Central America, Colombia and Peru. The rootstocks are composed of clusters of short and thick rhizomes which have prominent scars by the falling of the leaves. Leaves are long-petioled, oblong to orbicular and entire or dentate. Pink flowers that appear primrose-like on slender scapes, are 5-petalled and some 2.5 cm across. One species, ***M. primulaeflora*** from Colombia is a small and glabrous foliage plant suitable for indoor cultivation and for terrarium. Leaf stalks are about 25 cm long, green or reddish. Its 10-15 cm long, 7.5-10.0 cm wide, metallic-green leaves, rosy-purple beneath, are leathery, heavily quilted and broadly elliptical to lanceolate having 3-5 parallel veins and with toothed margins. These arise directly from the partly underground stem (rhizome) some 8 cm or more across. The calyx lobes are broadly ovate-rounded and flowers attractive pink. It is propagated through division of the thickened stems. Several varieties of this species are available.

Monstera **(Araceae)**

See chapter on '***Monstera***'.

Muehlenbeckia **(Polygonaceae)**

Muehlenbeckia was named for Henry Gustave Muehlenbeck (1798-1845), a physician and botanist in Alsace (France). It is a genus of 15-20 climbing to prostrate shrubby species with stiff wiry stems and alternate leaves, sometimes only in the form of scales. Its tiny greenish to whitish insignificant flowers appear in clusters in the leaf axils. These are propagated either through seeds or cuttings. ***M. complexa*** (Maidenhair vine/Wire-vine) from New Zealand is climbing to sprawling shrub bearing dark purple-brown wiry stems and rounded to cordate (sometimes lobed) dull green leaves some 5-10 mm long oftenly having purple-brown margins. It is an excellent house plant and most suitable for hanging baskets but requires cooler conditions. ***M. platyclada*** (*Homalocladium platyclados*) from Solomon Islands is the most common tropical house plant growing up to 2 m high with sprawling habit, bearing leaves reduced to scale-like. The plants are green with arching and flattened ribbon-like stems.

Murdannia scutifolia variegata **(Commelinaceae)**

Murdannia was named after Murdan Aly, a plant collector and keeper of the herbarium at Saharanpur (India). The genus comprises 40-45 annual and perennial herbaceous species native to Old World tropics. It is a herbaceous plant with handsome foliage, doing well in shade to filtered sun. It has weak but fleshy stems which grow very fast to 75 cm high with reddish joints, bearing green, opposite and clasping leaves some 20 cm long and 1.25 cm wide, having thin white stripes. It is a most suitable indoor plant. It is propagated through cuttings which root easily in water. Apart from this, there are two more important species, *e.g. M. nudiflora* from Asia and *M. simplex* from tropical Asia and Africa.

Murraya **(Rutaceae)**

Murraya was named to honour Johann Andreas Murray (1740-1791), a Swedish professor of botany and medicine at Gottingham University, and formerly a student of Linnaeus. It is a genus of 4-12 evergreen trees and shrubs from SE Asia, Indo-Malaysia and Pacific Islands. The genus is allied to *Citrus*. Shining leaves are pinnatifid and leaflets alternate or sub-opposite, oval and acute to acuminate. Flowers are 5-petalled and appear in terminal or axillary panicles. Its unedible fruits are globular and colourful in some species. These make nice pot plants and are worth growing in the conservatory. These may also be grown as an intermittent indoor plant. They are propagated through seeds or cuttings. ***Murraya exotica*** (*M. paniculata*; Orange jasmine) from SE Asia and Malaysia is a densely branched tall shrub growing up to 4 m in height. Rich glossy green leaves are composed of 3-9 obovate leaflets which are 5-7 cm long.White fragrant corymbose flowers some 1.2-2.0 cm long appear during summer with pointed petals. Fruits are ovoid, bright red and some 1.2 cm long. It is most suitable for hedging. ***M. koenigii*** (Curry leaf) from India and Sri Lanka is a tall shrub growing up to 4 m though in pots up to 1.5 m only. Leaves are deep green, having a strong pleasing aroma and with 11-21 ovate to lanceolate leaflets some 2.5-5.0 cm long. Clustered white to cream flowers some 1.0 cm long appear in terminal cymes during rainy season, followed by trusses of black ovoid berries which have quick germination ability if sown after plucking. Its leaves, bark and roots are used as tonic in India, apart from the extensive use of its leaves in curry, vegetables and in various other preparations due to its pleasant aroma and taste.

Musa **(Musaceae)**

See chapter on '**Ornamental Banana**'.

Myrciaria myriophylla (Myrtaceae)

Myrciaria in Greek probably means *Myrtus*-like. As their fruits are edible and similar to *Eugenia*, these are sometimes clubbed with the latter. These are Brazilian trees and shrubs comprising of about 40 species but only one is worth growing indoors. Their leaves are entire and opposite, flowers are almost sessile, axillary, clustered, sometimes solitary or in panicles, calyx 4-lobed, petals are 4 and perigynous, stamens many, filaments free and filiform, anthers oval to oblong and dehiscing longitudinally, style filiform, stigma simple but rarely capitate, ovary inferior and bilocular with 2 ovules in each, fruit is a berry and seeds 1-4. These are propagated through cuttings and seeds. ***M. myriophylla*** is a slow-growing and much-branched little shrub bearing 3.5-5.0 cm long and 0.6 cm wide leaves which are strap-shaped, pointed and in opposite pairs.

Myriophyllum (Water milfoil; Haloragidaceae)

In Greek, *myrios* is for 'many' and *phyllon* for 'a leaf', referring to the many and often dissected leaves. A cosmopolitan genus with 45 species of marsh and submerged aquatics, having slender and flexible stems, bearing narrow, alternate, opposite or whorled leaves and petalless or 4-petalled minute flowers in terminal spikes or from upper leaf axils appearing above the water. These are suitable for growing in aquarium and at sunny pools, some for the cold water pools and others for warm water up to a minimum of 13 °C water temperature. These are propagated by cuttings. ***M. aquaticum*** (*M. brasiliense, M. proserpinacoides*; Water feather/Parrot's feather) from Brazil, Argentina and Chile is a half-hardy species growing up to 2 m with stem-tips above the water. Leaves are glistening blue-green, 2-3 cm long, pinnatifid and 4-6 per whorl on stem tips. ***M. hippuroides*** (Western milfoil/Red water milfoil) from California to Washington is almost a hardy species bearing up to 60 cm long stems with tips above the water, leaves pinnatifid, 1.5-3.0 cm long, pectinate and 4-6 per whorl. ***M. spicatum*** (Spiked water milfoil) from Europe, Asia, N. America and N. Africa is a hardy species sending out stems up to 2.5 m long with 1.5-3.0 cm long, pinnate and 4 leaves per whorl. ***M. verticillatum*** (Whorled water milfoil/Myriad leaf milfoil) from northern temperate zone, N. Africa and S. America is a hardy species with stems growing up to 3 m long, bearing 2.5-4.5 cm long, pinnate and 5 leaves per whorl. It produces detached over-wintering resting buds (turions) which is a further means of propagation.

Myrsine (Myrsinaceae)

Myrsine is the Greek vernacular for myrtle, taken up by Linnaeus for this genus. The genus comprises 5-7 evergreen shrub and tree species from Azores and Africa to China. These require partial sun to grow and are propagated through cuttings and seeds. They have alternate, leathery, often entire, elliptic to obovate and coriaceous leaves, and floral parts in 4-5's where the lobes are imbricated in the bud, tiny, whitish or yellowish, sessile or peduncled and polygamo-dioecious flowers in axillary or lateral clusters, anthers are short and normally blunt, and the fruit is a berry-like, dry or fleshy 1-stoned drupe with globose seed. Two species are worth growing indoors. ***M. africana*** (African boxwood/Cape myrtle) from Africa is a woody shrub with reddish-brown branches, most suitable as pseudo-bonsai, growing up to 1.8 m tall but in pots indoors only 80 cm high, bearing alternate, 0.6-2.5 cm long and tear-drop-shaped aromatic leaves, pale-brown flowers which are 4-lobed and dioecious, and the fruits are pale to deep blue-purple, 0.6 cm wide and globular. ***M. nummularia*** (*Rapanea nummularia, Sutttronia nummularia*; Creeping matipo) from New Zealand is a prostrate-growing sub-alpine hardy shrub with red-brown and joint-rooting wiry stems growing up to 50 cm long, bearing obovate to oblong or nearly orbicular and 0.4-1.8 cm long but stalkless leaves, 4-petalled flowers, and blue-purple and some 0.5-0.6 cm wide fruits.

Myrtus communis (Myrtaceae)

Myrtus is the Greek vernacular name for this genus. The genus comprises some 100 evergreen trees and shrubs from tropics and the sub-tropics. These bear entire leaves in opposite pairs which are lanceolate to obovate. The flowers are 4 to 5-petalled and the berries fleshy which may be edible. They are handsome pot plants propagated through seeds or heel cuttings. It is only ***M. communis*** from West Asia which can be grown indoors. In the wild, it is highly branched and grows up to 4.5 m in height though indoors up to 2 m only in pots. Its various varieties which are low-growing, are most suitable for indoor cultivation. Its leaves are glossy rich green, hard, lanceolate to ovate, 2.5-5.0 cm long, aromatic when crushed and in alternate and opposite pairs. White solitary flowers some 2 cm across are borne, followed by 0.9-1.2 cm long purple-black berries. It is amenable to trimming to keep it into shape. 'Buxifolia' with small buxus-like leaves, 'Compacta' with compact plant growth, 'Microphylla' is a low-growing with 2 cm long leaves, 'Minima' having quite short stature with quite small leaves, 'Nana' with very small leaves and 'Variegata' tiny leaves are margined creamy-white and very suitable for bonsai making.

Nandina domestica [Heavenly bamboo/Sacred bamboo; Berberidaceae (Nandinaceae)]

Nandina is the Latinized form of the Japanese

vernacular *nanten*. This genus from India to E. Asia (China) likes cooler climate and comprises only one evergreen species liked for its handsome foliage though flowers are also graceful and the fruits are too colourful. ***N. domestica*** is an erect, slender and sparsely-branched shrub with many stems together from the ground, growing up to 2 m in height, bearing 30-45 cm long, numerous, alternate and bi- to tri-pinnate leaves with odd-numbered, opposite, 3-7 cm long, well-spaced and lanceolate to elliptical leaflets some 5 cm long and 1.25 cm wide, which have coppery-red flushing when young but tinted-red to purple in autumn. Flowers some 6 mm long are formed of several whorls of tepals (petals about 6 but numerous sepals gradually turn into petals) on a 20-35 cm long terminal panicle where upper tepals are petal-like, outer ones being small, green and leathery, the inner larger and whiter, followed by the 2-seeded globular red berries some 6-8 mm wide. It is propagated by seeds and heel cuttings. It has several cultivars such as 'Compacta' growing up to 90 cm high, 'Nana Harbour Dwarf' up to 60 cm high, 'Nana Purpurea' with young leaves purple, 'Wood's Dwarf' growing up to 40 cm tall, *etc.*

Nautilocalyx (Gesneriaceae)

Nautilocalyx derives from the Latin *nautilus* 'the marine shell' and *calyx* 'the ring of sepals' which protects the floral bud or shell-like calyx. A frost-tender genus of 15 erect to decumbent and evergreen perennial species, native to South America which like high humidity and partial shade for their growth. The leaves are bullate or quilted, quite handsome and in opposite pairs. Tubular flowers are borne with five rounded petal lobes in axillary clusters though in some species these emerge from coloured calyces. They are propagated by stem or leaf cuttings and seeds. Six species are grown indoors but these can not be called as foliage plants. ***N. bullatus*** (*N. tessellatus*) from Peru is a tender bushy plant grown for its handsome foliage and graceful flowers. In height and spread its thick and hairy stems attain up to 60 cm height, bearing dark green paired, opposite-alternate leaves up to 23 cm long, narrowly oval, wrinkled, heavily veined, short-toothed and with a bronze sheen above, reddish-green beneath. White-haired pale-yellow small tubular and clustered flowers appear in the leaf axils in summer. ***N. forgettii*** from Peru grows with erect stems up to 60 cm high, bearing bright green leaves, reddish-brown veining above and purple beneath, elliptic, quilted and some 15 cm long. White flowers having yellow throat and some 4 cm long appear from large red calyces. ***N. lynchii*** from Colombia grows robust with erect stems up to 60 cm long, bearing winged short stalks, and the leaves are lustrous greenish-maroon above, purple beneath, broadly lanceolate, slightly wrinkled and slender-pointed. Pale-yellow tubular and red-haired flowers some 4 cm long appear from red calyces.

Neoregelia (Bromeliaceae)

See chapter on '**Bromeliads**'.

Nepenthes (Nepenthaceae)

See chapter on '**Carnivorous Plants**'.

Nephrolepis (Polypodiaceae)

These ferns are excellent for hanging baskets and as specimen plants. ***N. cordifolia*** and ***N. exaltata*** (sword fern) have erect & stiff fronds and plain-edged leaflets with basic leaf pattern in herringbone style, having long leaflets arranged on either side of the midrib but when about one-and-a-quarter century ago a drooping mutant in Boston was discovered bearing gracefully arching fronds which was named ***N. exaltata bostoniensis*** (Boston fern), this revolutionized the *Nephrolepis* cultivation. This now has scores of types such as *N. e. b. maassii* (compact plant with undulating leaflets), *N. e. b. rooseveltii* (large with undulating leaflets), *N.e.b. scottii* (compact plant with rolled leaflets), *etc.* Also, the varieties with double herringbone pattern (leaflets being divided into herringbone-type or sometimes even further) to make a feathery or lacy effect, such as 'Fluffy Ruffles' (feathery leaflets in a double herringbone-pattern), 'Smithii' (fine lacy leaflets in a quadruple herringbone pattern), and 'Whitmanii' (lacy leaflets in a triple herringbone pattern).

Nephthytis (Araceae)

Nephthytis derives from the goddess of the same name in Egyptian mythology, wife of Typhon and mother of Anubis. Native to tropical Africa, it is a genus of four species of tufted evergreen perennials similar to *Syngonium*. These bear horizontally creeping rhizomes from where long-stalked, leathery and arrow- or spear-shaped leaves arise directly. Typical arum-like flowers arise with green spathe and spadix, followed by orange fruits. These are most suitable for conservatory and by managing humidity can also be grown indoors as their requirement for humidity is high. These are propagated through division. ***N. afzelii*** from Sierra Leone and Liberia is an evergreen creeping rhizomatous perennial growing to 75-100 cm high, bearing tufts of dark green, sagittate and lobed leaves some 25-45 cm long carried on a stalk to 45 cm long. The spathe some 7 cm long and 3 cm wide, which are greenish and hooded continue appearing intermittently enclosing a green spadix, followed by orange spherical berries some 1 cm long. ***N. poissonii*** from Cameroon is somewhat similar to *N. afzelii* but leaves are 5 cm long and rich green, spathes green with fine brown spots and berries are ellipsoid and some 3 cm long.

Nertera granadensis (*N. depressa*; Rubiaceae; Bead plant)

In Greek, *nerteros* is for 'lowly', referring to the mode of growth. It is a hummock- or mat-forming evergreen genus of 12 small creeping perennial species from SE Asia, Australia, New Zealand and islands of the Pacific and western Central and South America, having a very slender thread-like stems. Leaves are small and appear in opposite pairs. Minute inconspicuous flowers appear in mass, followed by spherical to bead-like, orange to red and 2-seeded drupes. They are propagated by seed, division or tip-cuttings. ***N. granadensis*** from Mexico is a prostrate plant forming a dense cushioned-mat of tiny bright green, oblong to broadly ovate leaves which are 2-4 mm long, and in summer bears minute greenish-white flowers, followed by mass of shiny orange berries (drupes) some 6 mm wide.

Nicodemia (Loganiaceae)

Nicodemia was named in honour of Gaetano Nicodemo, who commited suicide in 1803. He was Italian botanist and curator of the Lyons Botanic Garden during 1799-1803. A genus of six species of shrubs from Sri Lanka, Mauritius and Mascarene Islands, separated from *Buddleia* since the members of this genus have fleshy berry-like fruits though *Buddleia* has dry capsules. These look most attractive in the borders or in pots in the conservatory or indoors. These are propagated through seeds or cuttings. ***N. diversifolia*** from Africa is a fast-growing branching shrub growing up to 90 cm high, bearing leaves in opposite pairs, which are oak-like, some 10 cm long with a metallic bluish tinge. It is a handsome indoor shrub. ***N. madagascariensis*** (*Buddleia madagascariensis*) from Malagasy grows about 3 m in height with lanceolate and 13 cm long leaves which are dark green above and white-felted beneath. It bears slender, pyramidal and terminal panicles from winter to spring with small bright orange flowers, followed by purple fruits.

Nidularium (Bromeliaceae)

See chapter on '**Bromeliads**'.

Notocactus (Cactaceae)

See chapter on '**Cacti**'.

Onoclea [Aspidiaceae (Polypodiaceae)]

See chapter on '**Ferns and Allied Plants**'.

Ophiopogon (Lily turf/Mondo grass; Liliaceae)

In Greek, *ophis* means 'snake' and *pogon* means 'beard'. A genus of more than 10 species of evergreen perennials from Himalayas to Japan and Philippines. They are tufted, rhizomatous or stoloniferous, bearing leathery and grassy leaves, and racemes of small, 6-tepalled, bell-shaped, nodding and normally white flowers. Though it may often be confused with *Liriope* but the latter bears individual flowers facing outwards or upwards on straight stalks. They can be grown in humus-rich well-drained soil in shade. They are propagated through plant offsets and seeds. Important cultivated species are *O. jaburan* (variegated), *O. japonicus,* and *O. nigrescens.* ***O. arabicus*** (*O. planiscapus*) from Japan is tufted and mat-forming, bearing deep green and 30-50 cm long numerous strap-shaped leaves. Scape some 20-30 cm long appears with 5-7 cm long white or pale-purple flowers. Its var. 'Nigrescens' is clump-forming, growing up to 23 cm high and with 30 cm spread, the leaves are black and grass-like, racemes are lilac-flowered and the fruits are black. ***O. jaburan*** from Japan is though similar to *O. japonicus* but more robust having tight tufts of leaves and is clump-forming. It grows to 15 cm high with a spread of 30 cm and is frost-hardy. Leaves are dark green, 45-90 cm long and some 6 mm wide, multi-nerved, scape 15-60 cm long, raceme 7.5-15.0 cm long, flowers bell-shaped, white to liac and appear in groups of 6-9, followed by deep blue berries. Its var. 'Caeruleus' is blue-flowered, 'Aureus variegatus' has leaf-striping of golden-yellow, 'Argenteus Variegatus' has white-striped foliage and 'Variegatus' has longitudinal white or yellow stripes. ***O. japonicus*** (*Mondo japonicum*), a clump or mat-forming species from Japan, Korea and China is a compact stemless glabrous plant with stoloniferous rhizome that produces numerous narrowly linear, erect but curving at tips, 15-30 cm long, 3 mm wide and 5-7 nerved grassy and glossy dark green leaves. Scape is 10-15 cm long, the raceme 5.0-7.5 cm long with a few violet-purple to lilac to whitish and drooping flowers though lower flowers appear in a group of 2-3. It grows to 30 cm high with indefinite spread. The leaves in the var. 'Atakai' is striped white and some 15 cm long, in 'Kyoto Dwarf' black-green, some 5.0 cm high and longitudinally ribbed, in 'Old Gold' the leaves are ivory with green stripes and up to 25 cm long while var. 'Variegatus' bears variegated leaves. ***O. regnieri*** from China is stemless. Leaves marked pale-green and yellow, lanceolate, erect or reflexed, some 30 cm long with slender petioles and form a rosette. Scape is 2-edged, some 20 cm long and bears a many-flowered raceme. Violet-white flowers have greenish segment tips, and are some 1.9 cm across.

Oplismenus hirtellus (Graminae)

Oplismenus derives from the Greek *hoplismus* 'armed for war', with reference to the bristle-tipped awns on the spikelets. The genus from tropical and sub-tropical areas, comprises 15-20 species of delicate branching grasses with slender trailing stems, one with handsome foliage worth growing indoors. The leaves are narrowly ovate to lanceolate and the panicles are narrow with spreading 1-sided raceme. Spikelets are 1-flowered. ***O. hirtellus*** (Basket grass) from southern USA to Argentina is a creeping and node-rooting, charming house and conservatory plant, particularly when its leaves are variegated, *vis-a-vis* a good hanging basket plant. It is propagated by cuttings. Its stems are wiry, spreading, about 1 m long and bear 10 cm long and 2 cm wide leaves which are green and wavy, narrowly ovate and acuminate. Flowers are insignificant. Form 'Variegatus' (*Panicum variegatum*) leaves are white-striped, sometimes tinted pink.

Opuntia (Prickly pear; Cactaceae)

See chapter on '**Cacti**'.

Osmanthus heterophyllus (Oleaceae)

Osmanthus derives from the Greek *osme* 'fragrance' and *anthos* 'a flower'. A genus of 30-40 species of evergreen trees and shrubs native to East Asia, eastern USA, Mexico and Pacific Islands. Leaves are green, opposite, often leathery, sparsely and irregularly serrated. Fragrant white to rarely creamy-yellow, tubular and 4-lobed flowers appear usually in axillary clusters. A few species as described below can be enjoyed indoors but for a while, however these may be kept in conservatory and on window sills when young. They are useful foliage and flowering plants. They are propagated by heel cuttings. ***O. fragrans*** from China though grows up to 12 m in height but in container it can easily be maintained at 2 m. Leaves are dark green, 8-11 cm long, broadly lanceolate, acuminate and small but widely spaced toothing on the edges. Strongly fragrant white or pale-yellow, and bell-shaped flowers are borne in clusters. ***O. heterophyllus*** (*O. ilicifolius*) from Japan though in the wild grows up to 8 m in height but in containers it can easily be tamed up to 2 m, bearing dark green, quite holly-like, 4-6 cm long, ovate to elliptic and sparsely but spiny-toothed leaves. Highly fragrant, whte and small flowers appear in summer in clusters. Form 'Aureo-marginatus' has yellow margined leaves, 'Purpureus' has purplish-black young leaves which age to purplish-green, *etc*.

Oxalis (Wood sorrels; Oxalidaceae)

Oxalis originates from the Greek *oxys* 'sharp' and *als* referring to the 'acid taste'. In the wild there are 850 species distributed on every continent but most are concentrated along coastal South Africa and from Mexico in the North America to the southern part of South America. It is also recorded from Europe, SE Asia, Australia and New Zealand. These are half-hardy to hardy bulbous, tuberous rooted, rhizomatous or stoloniferous annual to perennial herbaceous to woody tree-like shrubs with clover-like leaves which are alternate, mostly digitately compound (3-12 leaflets, mostly 3-foliate), one to many-flowered with sepals, petals and styles 5, stamens 10 (5 longer and 5 shorter), ovary 5-celled each with many ovules, fruit a capsule with loculicidal dehiscence. Flower colour ranges from white to yellow and orange, to pink and rose, to red, to lilac and purple in various shades and these remain only when there is sunshine. Leaves also close in the night. These are suitable for a conservatory, sunny window sill in the home and in hanging baskets. These are propagated by division, and seeds, *vis-à-vis* cuttings of the succulent species. Only a few species are grown for foliage as described below. ***O. corymbosa*** (*O. martiana*) ***areo-reticulata*** is a rhizomatous plant growing into a lovely mound of 15 cm long foliage. Its leaf segments are broadly cordate, widest at the top and have unparallel gold-veining though pink flowers do not match with leaf colouration. ***O. hedysaroides*** 'Rubra' (Fire fern) from Ecuador is fibrous-rooted and the branched stems are some 40 cm long indoors. Leaflets are deep red, some 2.5 cm long, little lobed and turn downward at night. ***O. ortgiesii*** (Tree oxalis) from Peru and Andes grows up to 45 cm high with persistent and generally unbranched stems, bearing long-stalked handsome, trifoliate and hairy leaves which are olive-green above and red-purple beneath, with 5 cm long leaflets which are divided at the tips into two large triangular lobes. Its about 1 cm wide flowers are yellow with darker veins and appear well above the leaves in long-stalked cymes. It prefers cool climatic conditions. To bring this into shape and size, this is amenable to trimming. ***O. regnellii*** (*O. cathariensis*; Real shamrock) from Argentina, Bolivia, Brazil, Paraguay and Peru bears leaf stalk some 15 cm long, leaves with strictly triangular leaflets having 2.5 cm long segments. Its flowers are white and are held above the foliage. This is very easy to maintain indoors. ***O. siliquosa*** (*O. vulcanicola*) from Costa Rica produces a little succulent and persistent stems growing more than 20 cm high with good spread and is mat-forming. Leaves are trifoliate, leaflets obcordate, 1 cm or more long, red-flushed above and magenta beneath. Yellow flowers with purple-red lines inside appear in umbel-like cymes above the foliage in spring. It has winter dormancy.

Othonna capensis (Asteraceae)

See chapter on '**Succulents Other Than Cacti** '.

Pachyphytum (Crassulaceae)

See chapter on '**Succulents Other Than Cacti**'.

Pachypodium (Apocynaceae)

See chapter on '**Succulents Other Than Cacti**'.

Pandanus (Screw pine; Bromeliaceae)

See chapter on '**Bromeliads**'.

Parochetus communis (Leguminosae)

In Greek, *para* is for 'near or close to' and *ochetos* 'a brook', as the plants grew in such situations in the wild. A native to mountains of Central Africa and Himalayas, the genus comprises one or perhaps two species of prostrate perennials, suitable for hanging basket or in pots, best in the alpine house. They are propagated through division, cuttings or by seed. ***P. communis*** (Blue shamrock pea) from mountains of East Africa, is a mat-forming prostrate perennial having 2.5-4.5 cm across and clover-like trifoliate leaves and obcordate leaflets. The flowers are pea-like, brilliant blue, usually solitary from the leaf axils and 1.5-2.0 cm long, and these continue appearing throughout the year, but optimum season being from autumn to spring when day length is short.

Parodia (Cactaceae)

See chapter on '**Cacti**'.

Pedilanthus tithymaloides (Euphorbiaceae)

See chapter on '**Succulents Other Than Cacti**'.

Pelargonium (Geraniaceae)

See chapter on '**Succulents Other Than Cacti**'.

Pellaea (Polypodiaceae)

See chapter on '**Ferns and Allied Plants**'.

Pellionia daveauana (Watermelon pellionia; Urticaceae)

Pellionia was named for Adolphe Odet Pellion (1796-1868), a French Admiral who accompanied Luis Freycinet on a world voyage. It is a low-growing genus of 50 perennial species from tropical Asia and Polynesia, bearing alternate leaves but only at one side of the stem, and insignificant small flowers. Their leaves are quite handsome and suitable for conservatory and indoor planting, as well as for hanging baskets. They are propagated through division or by cuttings. Only two species are worth growing indoors. ***P. daveauana*** (*P. repens*; Trailing watermelon begonia) from Myanmar, Vietnam and Malaysia produces succulent and pinkish stems up to 60 cm long, bearing short stalks, and bronze to olive-green above with purple-suffused edges and a broad paler-green band along the centre of each leaf, pink below. The leaves are little fleshy, basally asymmetrical, 2.5-5.0 cm long and oval to lanceolate. The leaves in var. *viridis* (*P. argentea*) are greyish-green with crisped edges, suitable for a terrarium or small basket; 'Frosted' (Jack frost vine) is a trailing plant with closely set puckered leaves having frosty-white veins extending up to crimped edges; 'Satin Creeper' bears grey and olive-green leaves with brownish-umber markings, *etc.* ***P. pulchra*** (Watermelon patch) from Vietnam produces nearly succulent stems some 60 cm long, bearing green above with a dark brown vein network, purplish beneath, oblong-elliptic, basally asymmetrical and 5 cm long leaves.

Peperomia (Piperaceae)

See chapter on '**Succulents Other Than Cacti**'.

Pereskia aculeata (Cactaceae)

See chapter on '**Cacti**'.

Peristrophe (Acanthaceae)

In Greek, *peri* is 'around' and *strophe* 'to twist', referring to the twisted corolla lobes. The genus comprises of 15 fleshy, herbaceous, and sub-shrubby perennial species from tropical Africa and Asia but only one is worth growing indoors for handsome foliage. These bear entire and opposite leaves, and tubular flowers in clusters. These are propagated by cuttings and division. ***P. hyssopifolia*** (*P. angustifolia*) from Java is wide spreading species with angled and thin stems growing up to 90 cm tall. Leaves are dark green, 5-8 cm long, and narrowly elliptic to lanceolate. Red-purple small flowers emerge from within two ovate bracts where one is larger. Its variety 'Aureo-Variegata' bears leaves having veins and central zones yellow.

Persea americana (Lauraceae)

Persea is the Greek vernacular for *Cordia myxa*, an Egyptian tree. A genus of 150 evergreen tree species from tropics and sub-tropics, mainly South America and SE Asia (Taiwan, Korea, Japan), where one (*P. americana*) is grown indoors. They have entire and alternate leaves, small 6-tepalled greenish flowers, and berry-like fruits as in case of avocado. They are propagated by seeds. ***P. americana*** (*P. gratissima*; Avocado) from Central America is a fast-growing tree up to 20 m in the wild but in tubs only up to 4 m. Its leaves are dull green with lighter veins, glaucous beneath, alternate, lanceolate or elliptic to oblong and 5-20 cm long. Insignificant green and clustered flowers appear, followed by dark green or purple-tinted pear-shaped fruits. Through seeds when seedlings appear and attain the 4-true-leaf stage, upper two should be removed to induce branching and the branches are also trimmed to make the plant to manageable size and to last as a small plant for a few years.

Pfeiffera (Cactaceae)

See chapter on '**Cacti**'.

Philodendron (Araceae)

See chapter on '***Philodendron***'.

Phoenix (Date palms; Palmae)

Phoenix is the Greek vernacular name for the date palm. The genus originates in tropical and sub-tropical Africa and Asia with 17 palm species, bearing terminal rosettes of narrowly pinnate leaves where lower leaflets are often spine-like. Small dioecious flowers arise from among the foliage in dense panicles, followed by fleshy fruits where each one contains a single and grooved seed. They are handsome pot plants only when young. These are propagated through seeds grown at 21 °C or above temperature, *vis-à-vis* suckers as we grow cuttings. ***P. canariensis*** (Canary Islands date palm) from Canary Islands is a solitary-stemmed palm growing up to 16 m high, bearing feather-like, finely divided and some 5 m long (containerized specimens to less than half its length) and these form a dense crown in a course of time. ***P. dactylifera*** (Date palm) either from North Africa or SW Asia is a base-suckering species growing up to 30 m tall, bearing 5 m long leaves (less than half of its length in containers) with grey-green, stiff and linear leaflets. Not very attractive but worth growing. ***P. roebelenii*** (Pygmy date palm) from Assam (India) to Vietnam is quite slow-growing and may sometimes produce suckers. It grows to 2-4 m high, bearing 1.0-1.2 m long, arching, feathery and leathery leaves, and many dark glossy-green, 25 cm long and 9 mm long leaflets. Flowers appear on some 45 cm long panicles, followed by 1.2 cm long and ovoid black fruits.

Phormium [New Zealand flax; Agavaceae (Liliaceae)]

The generic name derives from the Greek *phormion* 'a mat', as in New Zealand the leaf fibres of the plant is used for making cloth, mats, baskets and many other things. A native to New Zealand and other adjoining islands, the genus comprises two clump-forming and evergreen perennial species bearing sword-shaped arching and leathery leaves. Their stout panicles bear tubular dark red or bronze flowers, followed by cylindrical capsules containing shining black, papery and flattened seeds. The plants are suitable for growing indoors and in conservatory. Propagation is through seeds or by division of clumps. ***P. cookianum*** (*P. colensoi*; Mountain flax) bears glossy green, up to 1.5 m long and erect leaves which arch when grown and then droop when fully mature. Floral stalk is up to 2 m high bearing greenish to orange flowers some 2.5-4.0 cm long, followed by twisted, pendent and some 10 cm long capsules. Some of the cultivars are 'Bronze Baby' (more than 60 cm long and bronze foliage which are glaucous beneath), 'Cream Delight' (growing up to 90 cm with green leaves having a cream band in the centre), *etc.* ***P. tenax*** (New Zealand flax) produces up to 2 m long or more and little arching leaves which are glaucous beneath, dull to brownish-green above, usually with orange border. Flowering stems grow up to 5 m though indoors in pots to its half, bearing dull red and 2.5-5.0 cm long flowers, followed by 5-10 cm long and erect capsules. Its cultivars are 'Maori Surprise' (up to 1 m tall bearing pinkish leaves which are striped bronze), 'Thumbelina' (up to 30 cm tall with bronze-purple leaves), 'Yellow Wave' (up to 80 cm tall bearing green leaves banded with yellow), *etc.*

Phyllitis scolopendrium (Hart's tongue; Polypodiaceae)

A fern from North America, especially USA, it takes its name from the Greek *phyllon* meaning 'a leaf', referring to its simple foliage form. Its leaves in tufts of up to 50 or more and on dark brown stalks are simple, strap-shaped, some 25-40 cm long which may be initially erect but afterwards arching with waving frond-edges, usually cordate at base and some 2.5-5.0 cm wide, and the sori almost at right angles to the midrib. Its vars 'Crispum' and 'Undulatum' have frilled margins. It has numerous varieties, mostly developed in Great Britain.

Phyllostachys aurea (Bambucaceae)

Phyllostachys derives from the *phyllon* 'a leaf' and *stachys* 'a spike'. The genus with 30-40 rhizomatous and clump-forming bamboo species, is a native of temperate eastern Asia and the Himalayas. These bear lightly zigzag stems having alternately flattened and grooved internodes and two lateral branches at each node with lanceolate leaves. Occasionally, on maturity these produce 2-4-flowered spikelets with greenish grass-like flowers.These are propagated through division or cuttings of rhizomes. ***P. aurea*** (Fishpole bamboo/Golden bamboo) from China grows up to 8 m tall and yellow stem with about 30 cm or more girth. Just below each node there is a brown swollen ring, and the branches bear mid-green, linear-lanceolate and 5-11 cm long leaves.

Pilea (Urticaceae)

See chapter on '**Succulents Other Than Cacti**'.

Pimenta dioica (Myrtaceae)

Pimenta derives from the Spanish *pimento* 'allspice'.

It is allied to *Eugenia* and *Myrtus*, but differs in having circular to spiral embryo and 2-celled ovary with 1-6 ovules which are pendulous from the apex of each cell. The genus comprises 5-6 aromatic tree species native to tropical America. Their leaves are long-stalked, large, leathery, feather-veined and with black dots on the reverse. Numerous small white flowers appear in terminal or axillary trichotomous cymes with 4-5 petals, numerous stamens and 1-2-seeded drupes. They are propagated by seeds or cuttings. ***P. dioica*** (Allspice) trees grow to 12 m outside but in pots only 2 m where its height is controlled through pot culture and trimming. It is a much-brnached plant, bearing shining leaves some 15 cm long, 5 cm wide, elliptical, leathery and deliciously aromatic. 'Allspice' which has flavour of clove, cinnamon and nutmeg, is the unripe berry which is collected and sun-dried. ***P. officinalis*** (*Eugenia pimento*; Allspice/ Pimento) from Cuba, Jamaica, Mexico and Central America which grows up to 12 m in height outside but can be grown indoors in pots to a height of 2 m only through pot-culture and trimming. It yields the real 'allspice'. Its petiole is about 1.25 cm long, the leaves are oblong and 5-15 cm long, the calyx 4-lobed and drupes globose. ***P. racemosa*** (Bay rum tree) leaves are highly aromatic and transfer the lemon aroma in cooking. Its leaves are similar to *P. dioica* but veins are comparatively more netted.

Piper (Pepper; Piperaceae)

Piper is from Greek *peperi*, which was taken from Indian name pepper *i.e. Piper nigrum*. The genus comprises of 1,000-2,000 species of climbers, shrubs and trees of pantropical origin. These bear deep green, simple, alternate, ovate or cordate leaves which in some species are marbled with pink, purple, red or white. Its cylindrical spikes bear insignificant tiny and petalless flowers, followed by fleshy and clustered berry-like drupe-fruits. They are propagated through seeds or cuttings. ***P. crocatum*** (*P. ornatum crocatum*; Ornamental pepper) from Peru is a climber with 2-3 m long wiry vines, bearing deep purple beneath and lustrous green peltate and cordate leaves above, some 8-13 cm long, dotted with bands of silver-pink along the veins. ***P. nigrum*** (Black pepper) from Western Ghats (India) is widely grown in India and Sri Lanka for spices. It is a charming climber growing up to 5 m by clinging to its support by aerial roots. Cordate leaves are dark green, more than 10 cm long, thick, prominently veined and tip-pointed.Flowers are green and insignificant, followed by 4-6 mm wide fruits which while ripening become red and then black which are dried and then ground to obtain black pepper but to obtain white pepper the fruits are kept in running water to remove the fleshy coats and then the seeds are dried. This is a good house plant. ***P. ornatum*** from Celebes is a climber similar to *P. crocatum* but leaves are broader like to those of betel vines with more even silver-pink marbling, which are more prominent on ribs, veins and veinlets though the marbling ages to white.

Pisonia umbellifera (*Heimerliodendron brunonianum*; Bird-catching tree/Parapara; Nyctaginaceae)

Pisonia is named after Willem Piso (*d.* 1648), a Dutch physician and naturalist at Amsterdam. It is a genus of about 50 species of shrubs and trees, one of which makes a decorative foliage plant for keeping in the conservatory or indoors under filtered to partial sun. It is propagated through seeds, cuttings and layering. In the wild, it is a small tree but indoors through regular pruning and container-planting, it is amenable to maintain at or below 2 m in height. Leaves are glossy green with firm texture when mature, are ovate-oblong and usually appear in opposite pairs or in alternate whorls of three up to 25 cm long on some 10 cm long stalks. Tiny funnel-shaped 5-lobed flowers are greenish, yellowish or pinkish which appear in terminal panicles. Fruits are very sticky, 5-ribbed, narrowly oblongoid and 2.5 cm long. The leaves of 'Variegata' are marbled green with a broad and irregular cream margin often splashed pink which makes it a most handsome foliage plant.

Pistia stratiotes (Corn lettuce; Araceae)

Pistia derives from the Greek *pistos* meaning 'watery'. The genus comprises one species, ***Pistia stratiotes*** (Water lettuce) which is native to worldwide tropics to sub-tropics and is a free-floating evergreen perennial. It roots while floating in a very shallow water. It is a suitable plant for a pool in a conservatory or in a large tub or tank in a home garden at a water temperature around 21 °C. It is propagated through division as it produces offsets quickly that are easy to separate. Its plants are stoloniferous and rosette-forming some 60 cm in diameter which float with their roots, freely in the water. Stalkless leaves are glowing grey to yellow-green, fuzzy, broadly obovate, truncate, velvety, vertically ribbed and 5-13 cm long. Minute petalless flowers in the form of typical aroid-spathes some 2 cm long appear with one female and several males together.

Pithecolobium flexicaule (*P. texense*, *Acacia flexicaulis*; Texas Ebony tree; Leguminosae)

Pithecolobium in Greek means monkey and ear-ring, owing to its ring-like coiled fruits. The genus comprises some 100 species from Texas to Lower California, but only the slow-growing one (*P. flexicaule*) is suitable for growing indoors as pseudo-bonsai to 10-25 cm in height topped with flat branching. It bears short stout stipular spines. Its leaves are long-petioled and compound with 4-6 pairs of oblong leaflets with rounded tips, each being 0.6-1.25 cm long but lowest pair the shortest and these remain green if winter temperature does not fall below 13 °C. Its flowers are insignificant. It is raised from seeds easily.

Pittosporum tobira (Ausralian laurel; Pittosporaceae)

Pittosporum derives from the Greek *pitta* meaning 'pitch' and *spermum* 'a seed', owing to the sticky coating of the seed. An evergreen genus of trees and shrubs with 150 species from E. Asia, Australasia, Pacific Islands and Africa. Leaves are alternate, entire, and normally ovate to oblong or linear to obovate. Flowers are solitary or clustered, 5-petalled and often with shortly tubular bases. They are excellent tub plants and quite amenable to geometrical forms. They prefer moderately fertile soil which is well-drained and at sheltered sunny sites. They are propagated through seeds at 13-15 °C temperatures or through cuttings given with bottom heat of 18 °C. ***P. crassifolium*** (Karo) from New Zealand grows to about 3 m high in a container with white-downy leaves and new shoots. Leaves are dark glossy green at upper surface and downy below, leathery obovate to lanceolate with thick margins and some 5-7 cm long. Dull red flowers some 1 cm across appear in terminal umbels, followed by dark green to black small fruits. ***P. dallii*** from New Zealand grows up to 1.2 m in containers, bearing mid-green leathery and ovate-lanceolate leaves. Young shoots are glossy red. Fragrant white flowers some 1.25 cm across are borne in 3.75 cm across terminal clusters. ***P. eugenioides*** from New Zealand grows spreading to erect some 3 m tall in a container, bearing foetid and dark glossy green leaves which are oblong-elliptic with undulating margins and some 5-10 cm long. Honey-scented yellow flowers in umbellate clusters appear some 5-7 mm across. Var. 'Variegatum' leaf margins are creamy-white. ***P. tenuifolium*** (*P. mayi*; Kohuhu) from New Zealand grows up to 3 m in a container having dark purplish-black shoots. Leaves are rich glossy green, oblong to ovate and undulate, and 2.5-6.0 cm long. Fragrant, dark purple to black and some 1 cm across flowers appear from the leaf axils in small clusters. The leaves in 'Atropurpureum' is dark purple, flushed pink with white variegation in 'Garnetti', golden suffusion,especially during winter in 'Golden King' and silver suffusion markings in 'Silver Queen'. ***P. tobira*** (Japanese pittosporum/Tobira) from China and Japan is an oval slow-growing shrub, in a container up to 3 m high, bearing shining rich green, thick, obovate and some 5-10 cm long leaves. Creamy-white fragrant flowers some 1 cm long appear in attractive umbels. Its var. 'Variegatum' leaves are marked silvery-grey with cream margin. ***P. undulatum*** (Victorian box) from SE Australia grows to 4 m in a container with lanceolate to oblong, undulate and 6-13 cm long leaves. Fragrant white flowers some 1 cm across appear in umbel-like terminal clusters, followed by orange-brown fruits. ***T. viridiflorum*** from South Africa grows up to 3 m in containers with aromatic, thick-textured, obovate, wavy and 2.5-10.0 cm long leaves. Yellow-green and some 6 mm across flowers appear in crowded terminal clusters. Among all these, *P. tobira* is quite popular.

Platycerium (Polypodiaceae)

See chapter on '**Ferns and Allied Plants**'.

Plectranthus (Swedish ivy; Labiatae)

For details of the genus see chapter on '**Succulents Other Than Cacti**'. However, the species suitable as house plants are furnished below. ***P. australis*** from Australia is a hairy shrub growing erect up to 60 cm, bearing broadly lanceolate, some 3.75 cm long and coarsely toothed leaves with stalks being 2.5 cm long. ***P. coleoides*** (Candle plant) from S. India is a compact, closely-clustered basal branching, with purple stems growing up to 90 cm in height. Leaves are velvety with depressed veins, opposite to alternate, overlapping, and lanceolate to 10 × 6.25 cm. *P. c.* 'Marginatus' bears green leaves mottled lighter and edged white. *P. c.* 'Marginatus Minimus' with trailing branches so most popular as a basket plant and the leaves are 2.5-3.75 cm long with white margins. *P. c.* 'Variegatus' is similar to above but with more depressed veins and white border. ***P. fruticosus*** (*P. fosteri*) from S. Africa grows erect up to 90 cm high with 4-angled stems which bear some 10 cm long, broadly lanceolate and coarsely scalloped leaves. Its var. 'Marginata' has broad white margins. ***P. madagascariensis aliciae*** (Busy mint) from Africa is a very small trailer with tiny crimped leaves having quite depressed veins and toothed margins. ***P. nummularius*** from S. Africa is a trailer with succulent and branched-stems growing up to 30 cm in length. Leaves are silvery green above and grey-green with purple veins beneath, fleshy, broadly ovate to orbicular (nearly round) with 6.25 cm diameter and similar to *P. oertendahlii* but quite asymmetrical and coarsely toothed. White to pale-lavender flowers some 1 cm long are borne in about 30 cm long racemes. *P. n.* 'Freckles' leaves are spotted yellow. *P. n.* 'Variegatus' is blotched irregularly with white. ***P. oertendahlii*** (Brazilian coleus/Swedish ivy) from S. Africa is a fine-hairy branched-trailer, stems growing up to 45 cm high and is an excellent basket plant. Leaf stalks 2.5 cm long, leaves nearly round with 6.25 cm diameter, broadly scalloped and vein network white. ***P. prostratus*** (Pillow plant) from Tanzania is a mat-forming plant rooting at stem joints. Stems are finely hairy with branches growing some 5.0 cm long. Leaves are some 2 cm thick and succulent. ***P. tomentosus*** (Woolly Swedish ivy) from S. Africa is a small shrub covered with white wools, and the leaves are succulent, nearly round with about 8.75 cm diameter having scalloped margins.

Pleomele reflexa (Liliaceae)

A native of Africa and India, it is an excellent foliage house plant. Its stem is very narrow and branched with densely clustered leaves which are glossy green, short, narrow undulating and reflexed. When lower leaves fall off, it may require stakes for support. It does not require direct sunlight though bright light is essential to maintain lustrous look of the foliage. Temperature below 13 °C is not conducive for its growth. Its compost should be kept always moist but only seldom it should be watered during winter months, however, leaves should be misted regularly. The plants should be repotted every two years. ***P. r. angustifolia*** produces strap-shaped leathery leaves which are crowded around the stem. ***P. r. variegata*** (Song of India) from India and Sri Lanka bears variegated leaves where on the margins there are two broad bands of yellow or cream which make this plant quite handsome. ***P. thalioides*** is a green-leaved plant.

Podocarpus macrophyllus (Buddhist pine/ Southern yew; Taxaceae/Podocarpaceae)

In Greek, *podos* is for 'a foot' and *karpos* 'a fruit', on account of the way the fruit in some species is carried on the end of a swollen stalk. It is a genus of 75-100 species of evergreen trees and shrubs from the S. Hemisphere (north to Japan and tropics of West Indies) but only a few (about one dozen) shrubby ones are worth growing indoors, some for cooler areas and others for warm tropical areas. The stalkless, straightsided, narrow strap-like, elliptical, firm-textured, shiny and some 7.5 cm long leaves are borne on upright stems. These can be shaped and retained compact by regular pruning. They are propagated by seeds or cuttings. ***P. elongatus*** (weeping podocarpus) from Tropical Africa is a tall tree so should be grown indoors only when young and also as it is slow-growing. It bears pendent branches. ***P. gracilior*** (fern pine) from Tropical Africa is similar to *P. elongatus* but the branches are more erect and the leaves are numerous, closely set and often black-tipped. ***P. macrophyllus*** (Buddhist pine) from China and Japan is quite slow-growing with horizontal branching where leaves are lustrous deep green but yellowish below and 10-18 cm long. Its fruit is dark red with exposed 1 cm long greenish seed. Its house plant variety is 'Maki' (southern yew) with small leaves and more compact growth habit. ***P. nagi*** (broadleaf podocarpus) from Japan is though similar to *P. macrophyllus* but leaves are wider (7.5 × 2.5 cm) and growth is columnar so suitable as a specimen plant. It prefers cool climate. ***P. nivalis*** (false yew) from alpine New Zealand is much-branched and spreading (about 1 m) shrub, bearing bright to yellow-green leaves which are 0.6-1.5 cm long, and bright red fruit with exposed greenish seed.

Polypodium (Polypodiaceae)

See chapter on '**Ferns and Allied Plants**'.

Polyscias (Araliaceae)

Polyscias derives from the Greek *polys* for 'many' and *skias* 'a sunshade or canopy', on account of the flowering umbels. The genus contains 80 species from tropical Asia and Western Pacific. They are special shrubby foliage plants having twisted and branched stems, alternate, palmate or pinnate and leathery leaves with margins of leaflets cut into ferny-segments and often edged white or differently, and where rachis and stalks may be spotted. These are propagated through cuttings by using rooting hormone. These are not easy to grow under room conditions until these are facilitated with good light, moist air and media and good warmth during winter otherwise these will immediately drop their leaves. The leaves should be misted frequently, plants should be watered moderately from spring to autumn and sparingly in winter, should be provided with bright light but not direct sun and the temperature should never drop below 13 °C. ***P. balfouriana*** (dinner plate aralia) from New Caledonia though grows in the wild up to 6 m but indoor in pots hardly 1.5 m, bearing dark green, cordate and 10 cm wide leaves, having leaflets speckled with grey or pale-green. The foliage in 'Pennockii' is yellow-veined and in 'Marginata' white-bordered. ***P. fruticosa*** (Ming aralia/ parsley aralia) is quite distinct as its stems are twisted and the leaves are quite long, parsley-crisped and divided into many irregular and saw-edged ferny-leaflets. Its varieties 'Elegans' and 'Plumata' are very compact forms good for training as pseudo-bonsai. ***P. filicifolia*** (fern-leaf aralia) from Polynesia is quite spreading with purplish branches, and leaves up to 30 cm long with leaflets widely spaced and in 4-5 pairs with one at the tip. Its varieties 'Marginata' and 'Variegata' have white-edged leaflets. ***P. guilfoylei*** (geranium leaf aralia/lace aralia/wild coffee) from Polynesia has compound leaves up to 40 cm long with 3-4 pairs of thin, cupped, elliptical and variously toothed leaflets with one at the tip. ***P. g. quinquefolia*** 'Elegans' is very compact with green leaflets and does well indoors. 'Elegans Variegata' with lemon-yellow leaflets which age to green and where the stems are woody and the bark corky. 'Laciniata' bears deeply lobed leaflets edged white. 'Monstrosa' leaflets are irregularly toothed, grey-blotched and edged white. 'Victoriae' (lace aralia) is slow-growing compact plant bearing bipinnate and feathery greyish-green leaves edged white. ***P. paniculata*** from Mauritius may grow up to 2.0 m in a pot with pinnatifid leaves, terminal pinna the largest, leaflets broadly elliptic to oblong and toothed. Its var. 'Variegata' is blotched pale-green and cream.

Polystichum (Polypodiaceae)

See chapter on '**Ferns and Allied Plants**'.

Portulaca (Portulacaceae)

See chapter on '**Succulents Other Than Cacti**'.

Portulacaria afra (Portulacaceae)

See chapter on '**Succulents Other Than Cacti**'.

Pothos (Araceae)

The Latinized version of the Sinhalese name *potha* is *Pothos*. A native of Indo-Malaya and Malagasy, the genus comprises of 70-75 species of evergreen climbers, having aerial roots and ivy-like mode of growth. The leaves are alternate, cordate, acuminate and long-stalked, stalks winged, and grooved above. The leaves may be glossy green or variously variegated in cream to yellow on green base. Leaves are usually around 10 cm in length and slightly less in width but when given free access to grow up a tree with rich organic nutrition, these grow sometimes quite large, *i.e.* up to 45 cm in length and breadth. These seldom produce floral spathes which are greenish and anthurium-like. These are propagated through cuttings by inserting in some medium or in water. ***P. scandens*** from Malaysia produces highly branched slender stems running several metres, the younger branches being highly succulent. ***P. seemannii*** from S. China and Formosa is similar to *P. scandens* in many aspects but here winged stalk is only about 2.5 cm long and the blades are ovate to obovate-oblong to lanceolate and about 8.0 cm long.

Pritchardia thurstonii (Coryphae (Palmae)]

Pritchardia was named in honour of W.T. Pritchard, a Bristish consul at Fiji in 1860. There are 36 species of slow-growing and spineless fan palms from islands of the South Pacific, especially Fiji Islands. Only one species is grown indoors which is native to Fiji. It is a thin-stemmed 4.5 m tall-growing and some 2.4 m spreading palm though indoors its height is less than 1.5 m. The umbrella-shaped blades some 75 cm in diameter bear about 70 small segments (quite smaller than the blade) as leaflets. It is a magnificent palm with most graceful appearance. Other species are very tall. These are propagated by seeds.

Pseuderanthemum (Acanthaceae)

In Greek, *Pseuderanthemum* derives from *pseudo* 'false' and the allied genus *Eranthemum*. It is a genus of 50-120 (varying number being dependent as per various botanical authorities) species of shrubby perennials widely spread in the Tropics. These are closely allied to *Eranthemum*. Most of these can be grown in the conservatories and in the house, mostly for their handsome foliage though some even for their flowers. In habit, these are erect to spreading, having opposite pairs of simple leaves and slender-tubed flowers with five clear petal lobes. These are propagated through cuttings. ***P. atropurpureum*** (*P. kewense*, *Eranthemum atropurpureum*) probably from Polynesia grows erect up to 1.2 m high and is a colourful but straggly shrub, bearing usually prominently purple-flushed leaves though sometimes green or patterned with white, green, yellow and pink, ovate to elliptic and 10-15 cm long. White or purple-flushed flowers some 2 cm across with a red-purple eye appear on short terminal or axillary spikes in summer and last up to autumn. Its varieties 'Tricolor' and 'Variegatum' are splashed with white, cream and pink and red-spotted white flowers appearing on mature plants. ***P. reticulatum*** (*Eranthemum reticulatum*, *E. schomburgkii*) from Australia is a shrubby species growing to a height of 0.9-1.5 m, bearing rich green leaves, netted or reticulated with golden-yellow, being quite conspicuous in young leaves, however, in the older leaves the netting is almost absent but veins are prominently yellow. Upper leaves are 5.0-17.5 cm long, ovate-lanceolate and netted with yellow, lower 15-25 cm long and unnetted. Flowers white and some 2.5 cm across in racemose terminal panicles, corolla mouth speckled with blood-red in throat and spotted on the lower petal while anthers are reddish-brown and exserted. Its var. 'Eldorado' with ovate leaves is a very handsome foliage plant. ***P. sinuatum*** (*Eranthemum cooperi*) probably from New Caledonia is a small-growing (30 cm high) shrub, bearing linear to lanceolate leaves with irregularly shallow lobing, mottled green above and purplish below. White flowers some 4 cm across with lower petals spotted-purple appear in short terminal racemes.

Psilotum nudum (Whisk fern; Psilotaceae)

The genus of a very primitive plant is from S. USA with two semi-aquatic herbaceous species, where one is foliage plant for indoor growing up to 60 cm height. It requires shade to filtered sun, so suitable for growing indoors. Its propagation is through spores and by division. It is shallowly rooted with highly flexible and shining stems, having triangular branches and sub-branches in cross-section. For pot culture, this is a very curious and charming plant where these last long by following repotting every year. Its hardly noticeable tiny leaves are spiky. It forms yellow balls of spores on the joints throughout the year.

Pteris cretica (Polypodiaceae)

See chapter on '**Ferns and Allied Plants**'.

Ptychosperma (Palmae)

In Greek, *ptyche* is for 'a fold' and *sperma* 'a seed', alluding to the grooved seeds of some species. It is a genus of 38 species of palms from N. Australia, New Guinea and Solomon Islands, one of which making a decorative pot plant though 3-4 more species are under cultivation. It is propagated by seeds. ***P. caryotoides*** is a small palm producing slender stem with dense white woolly scales on green leaves which form stem-like structure (crownshaft), arching leaves with wedge-shaped pinnae, green to yellowish flowers and bright red fruits. ***P. elegans*** (*Seaforthia elegans*; Alexander palm/Solitare palm) from Queensland is an elegant, slender-stemmed and slow-growing palm which attains up to 2 m in a few years in a pot. Its leaves are 60-120 cm long and pinnatifid with many linear leaflets. Though young, *vis-à-vis* containerized plants do not produce flowers, otherwise they bear small green and white flowers and red berry-like clustered fruits. ***P. macarthurii*** (MacArthur palm) is a most popular palm as a clump-forming neat hedge in the tropical gardens. It may grow up to 7 m tall but in pots up to 2 m only, with stem diameter being up to 5 cm. Leaves are dark green, arching and pinnatifid, and the fruits ellipsoid and red. ***P. microcarpum*** is a clump-forming handsome indoor palm growing up to 8 m high in the wild, but tamed one up to 2 m, bearing dark green, arching, pinnatifid leaves with clustered leaflets, and clustered fruits which are red.

Putranjiva roxburghii (Indian Amulet plant/ Wild olive; Euphorbiaceae)

The genus is an evergreen tree from tropical Asia with two or more species, but only one is commonly cultivated either for avenue planting or in the hedge. Its dark green and undulating leaves are alternate and simple, and the insignificant and apetalous flowers are either solitary or in small clusters with imbricate calyx, 1-4 stamens, broad and spreading styles, 2-3-celled ovary with two ovules in each cell and fruit a 2-celled drupe. ***P. roxburghii*** from India is almost a glabrous tree with slender branches. Its leaves are short-petioled, shining deep green, elliptic, smooth, serrulate and 5.0-7.5 cm long. Insignificant flowers are light yellow, and white-tomentose fruits are ovoid and some 1.25 cm thick.

Pyrrosia (Polypodiaceae)

See chapter on '**Ferns and Allied Plants**'.

Radermachera danielle (*Radermachia Danielle, R. sinica, Stereospermum chelonoides, S. sinicum*; Bignoniaceae)

It is a house plant of eighties of the 20th century, introduced to Europe from Taiwan. *Radermachera* (*Radermachia*) commemorates J.C.M. Radermacher (*d.* 1783), a Dutch amateur botanist resident in Java. A genus of 40 evergreen tree and shrub species from India to China, Philippines and Java, bearing opposite pairs of pinnatifid to tripinnatifid leaves and panicles of funnel-shaped flowers with five rounded lobes. The species ***R. danielle*** from China when young is admired most for its decorative foliage, so it is enjoyed in conservatory and at home where it can be tamed up to the height of 1.2 m by planting in containers and through pruning. When young its leaves may be bipinnatifid but when well grown are mostly tripinnatifid with leaflets being lustrous rich green especially at younger stage, deeply veined, ovate, suddenly slender-pointed and 3-6 cm long. Though young containerized plants do not produce the flowers but its flowers are yellow. It propagates from seeds and stem cuttings.

Raphidophora (also *Rhaphidophora*; Araceae)

The generic name derives from the Greek *rhaphis* meaning 'a needle' and *phoros* 'to bear', referring to the sharp crystals of calcium oxalate particularly in those of the fruits. The genus originates in Indo-Malaysia and New Caledonia and comprises some 60-100 creeping to climbing species, closely related to *Epipremnum*, *Monstera* and *Scindapsus*, though only half a dozen of them grow indoors. Their stems are quite thick and run several meters climbing on tall trees with the help of their aerial roots. Leaves are alternate, ovate but in some cases are pinnately cut to the both sides of the midrib or perforated similar to *Monstera*. Seldom these produce flowers but sometimes those planted in humid areas up a tree with good supply of organic material, these produce petalless flowers in the form of spadix, subtended by a boat-shaped spathe. These are used similar to *Philodendron* in a conservatory or at home. These are propagated through leaf-buds or stem-tip cuttings. This genus like a few others in Araceae such as *Epipremnum, Pothos* and *Scindapsus* has been invented to create great confusion as mostly the species of one genus is synonymous to other genera and even as there flowering being rare the plants at different stages of growth (seedling to maturity) and at different agro-climatic conditions show differently shaped and sized leaf blades, so these all together merely confound the problem. Therefore, all these four genera should be merged together to avoid any such confusion. However, the important house plant species under *Raphidophora* are ***R. cavalieri*** (*Epipremnum falcifolia*) which is a low-light climber bearing grey-streaked green leaves of 20 × 3.75 cm size, lanceolate, leathery and strongly held out on highly channeled stalks some 15 cm long. ***R. celatocaulis*** (Shingle plant) is a vigorous climber with alternate and leathery leaves which are initially cordate having size of 20 × 10 cm, then elongated to 30-37.5 × 12.5 cm with wavy edges, and then finally oblong and some 60 cm or even more longer with top portion from both sides of the midrib cut into rectangular lobes halfway similar to *Monstera*. Likewise, initially the stalks are also short then as long or longer than the blades. ***R. decursiva*** from Southern Asia is a quite vigorous climber, having leaf stalks almost equal to the size of the leaves and leaves are compound about 60 cm long, with 8-12 pairs of star-shaped lobes to both sides of the midrib and a single lobe at the tip. ***R. peepla*** is similar to above except that the leaves are pendent and longer with depressed veins. ***R. pinnata*** (see *Epipremnum pinnatum*). ***R. silvestris*** is very tall-growing climber having short and channelled-stalked lanceolate leaves some 12.5 × 2.5 cm size.

Ravenala madagascariensis [Traveller's tree; Strelitziaceae (Musaceae)]

Its generic name is the vernacular name to this plant in Madagascar. A genus of one species from Malagasy which is allied to *Musa* and *Strelitzia*. When young it makes an impressive foliage plant for conservatory and large rooms. It may attain its natural height up to 10 m with a palm-like trunk and fan-shaped head of lustrous green, alternate, oblong and long-stalked banana-like leaves some 3 m long which are packed so closely one by one that stems look absolutely flat. The leaves in their ovarlapping bases form a goove to collect rain water for use in emergency by travelling human beings. White strelitzia-like insignificant flowers appear in small clusters in leaf axils but when grown directly in the soil and then the beauty of the plant is marred a little. In containers its growth is very slow and takes many years to attain the optimum size where its leaves attain only up to 1.5 m in length. It is propagated through seeds.

Rebutia (Crown cactus; Cactaceae)

See chapter on '**Cacti**'.

Reineckia carnea (*Liriope hyacinthiflora*; Liliaceae)

Reineckia was named for Johann Heinrich Julius Reinecke (1799-1871), a German gardener. A tender genus of one rhizomatous grassy perennial species from Japan and China bears decorative tufted foliage and is suitable as ground cover and house plants, for window sill and for the conservatory. It grows from 30-45 cm in height and is propagated through seeds or division. It is a dense clump-forming plant with dark green, radical, channelled, linear to lanceolate, 10-40 cm long and 5-7 mm wide leaves. Its bell-shaped sessile pale-pink flowers some 0.8-1.2 cm long with 6 reflexed lobes appear during summer on 8 cm long loose spikes. Its perianth tube is cylindrical, ovary is 3-loculed each with a few seeds and the fruit is a globular red fruit. Its var. 'Variegata' has vertical creamy-white stripes.

Reinhardtia (Palmae)

This comprises of five species of dwarf clustering palms from Central America, of which one is grown indoors. These are propagated by division by separating the rooted pups. ***R. gracilis*** is a spectacular little palm growing very slow up to 2.4 m in height and in containers its height is less than its half. It bears strong stems, leaf stalk is short and the fronds are segmented into four short and broad lobes, measuring 17.5 cm long, 15 cm wide, one pair facing upward and the other downward and each segment having 10 bold and parallel veins while the upper ends are saw-toothed coarsely. This palm requires constant warmth, moisture and high humidity but adapts well to low light conditions.

Rhapis (Palmae)

Rhapis derives from the Greek *rhapis* 'a needle', referring to the narrow leaflets or their pointed tips. It is a genus of 15 small clump-forming palm species from southern China to Java with erect, bamboo-like and unbranched stems suckering freely at base and bearing neat foliage. These palms look quite elegant, leaf-stalks are fibrous and ensheath the stem and leaves are palmate or digitate. Insignificant yellowish and dioecious flowers appear in small clusters among the leaves. They are propagated through suckers, division and seed. ***R. excelsa*** (Bamboo palm/Ground rattan/Little lady palm/Miniature fan palm) from southern China grows some 1.5 m high in containers though twice this in the wild, stems covered with coarse leaf sheath fibre. Its leaves are deep lustrous green, little puckered and composed of 3-10 lanceolate leaflets 20-30 cm long. *R. e.* 'Variegata' bears ivory-white stripes on the leaflets. ***R. humilis*** from southern China is similar to *R. excelsa* but smaller and thinner having reed-like stems and the leaves are composed of 9-20 glabrous, narrower and smaller leaflets.

Rhektophyllum mirabile (Araceae)

See ***Homalomena picturata***.

Rhipsalidopsis (Easter cactus; Cactaceae)

See chapter on '**Cacti**'.

Rhipsalis (Cactaceae)

See chapter on '**Cacti**'.

Rhodospatha (Araceae)

In Greek, *rhodon* is for 'rose' and *spatha* for 'spathe', referring to the colour of the spathe in some species. A herbaceous climbing genus with six species from tropical America, out of which some three are grown indoors as foliage plants. The genus is quite similar to *Epipremnum*. These are excellent indoor plants and for warm conservatory. Its branches often root while climbing. The leaves are distichous, elliptic-oblong and acuminate, spathe is cymbiform, beaked, deciduous and longer than spadix, spadix is cylindrical and densely set usually with petalless perfect flowers or sometimes pistillate ones below on the spadix, stamens are 4 and the fruit is a small 2-celled berry which is oblong and truncate. ***R. forgetii*** from Costa Rica is a stem-climbing species, bearing spreading oblong-lanceolate leaves some 25-50 cm long with the petiole length of about 25 cm. Some 15 cm long peduncle is topped by about 15 cm long broadly elliptic spathe coloured with dirty rose-white exterior and dirty rose interior. ***R. hastata*** does not have complete hastate leaves as basal lobes are spreading at least at maturity and the blades are strap-shaped and pointed. Leaf stalk is short and winged while the leaves are 30-40 cm long and about 5 cm wide. ***R. latifolia*** from Brazil bears leaf stalks as long as the blades, leaves are light green, some 45 cm long, 25 cm wide, elliptic, surface corrugated with plenty of horizontal veins and edges wavy. ***R. pictum*** (*Spathiphyllum pictum*) from South America (Brazil) bears stems running up to 75 cm long, bearing short and clasping leaf stalks, glossy dark green, little fleshy and broadly ovate-elliptic leaves that are up to 60 cm long and 20 cm wide, and mottled crosswise along the transverse veins and blotched golden-green.

Rhoeo spathacea (Boat lily; Commelinaceae)

See ***Tradescantia spathacea*** in chapter '**Succulents Other Than Cacti**'.

Rhoicissus (Vitaceae)

Generic name has probably originated from Latin *rhoicus* meaning 'similar to *Rhus* (sumach)' together with the Greek *kissos* 'ivy'. It is a genus of 10-12 evergreen climbing species where stem tips are modified into tendrils and the leaves are alternate. Tiny greenish 5-petalled flowers are followed by small grape-like edible fruits. It is propagated through seeds or cuttings. ***R. capensis*** (*Cissus capensis*, *Vitis capensis*; Evergreen grape vine/Cape grape) climber from South Africa can grow up to 5 m though in pots it can be confined to 2 m of length through pruning and other cultural means. Its leaves are glossy green, simple, cordate, deeply crenate and some 20 cm across. ***R. rhomboidea*** (Grape ivy/ Natal vine) climber from South Africa grows up to 6 m long with trifoliate leaves and rhomboid leaflets where lateral ones are asymmetrical though this species is not in cultivation and the one with this name is, in fact, *Cissus rhombifolia* which has forked tendrils though in *Rhoicissus rhomboidea* the tendril is unbranched. *R. r.* 'Ellen Danica' (see '***Cissus***' under '**Succulents Other Than Cacti**').

Rohdea japonica (Lily of China; Liliaceae)

Rohdea commemorates Michael Rohde, a physician and botanist at Bremen. It is a monotypic genus of tender foliage plants from Japan and China. It bears cylindrical rootstocks with fleshy fibres. Leaves are erect, green, numerous (9-12), rosette-forming, radical, thick, leathery, oblanceolate, arching downwards, margins little wavy and 15-45 cm long. Thick spikes some 30 cm long appear in spring from among the leaves with densely packed pale and tiny flowers which have bell-shaped globular perianth, sessile anthers, peltate stigma, nearly wanting style and nearly 1-seeded globular berry. This is an excellent plant for cool living rooms. This is the most favourite plant of the Japanese, as well as Chinese. In Japan they have developed many varieties and Duch Bulb Growers offer numerous varieties for sale. 'Marginata'(Sacred Manchu lily) leaves are near-black edged with white while 'Variegata' bears pale-yellow bands. All these are propagated through division when potting.

Rosmarinus (Rosemary; Labiatae)

Rosmarinus is the old Latin name for rosemary, derived from *ros* 'dew' and *maritimus* 'maritime'. It is a genus of three evergreen aromatic woody shrub species from the Mediterranean but only two are in cultivation. They have opposite pairs of sweetly aromatic linear-needle-shaped leaves with rolled margins, and tubular 2-lipped flowers in purple shades from the upper leaf axils. Though they are foliage plants throughout the year, adding a bonus of leaves for culinary purposes but during blooming flowers add an additional bonus as then the plants add to an extra beauty. *R. officinalis* is the most suitable plant for growing indoors even as bonsai specimen where this develops gnarled and barked trunks with age but these require frequent exposure to sun, therefore should be placed at a sunny corner. Their proper shape and sizes are maintained through frequent trimming, sometimes even the roots. They are propagated through semi-ripe cuttings in cold frame. ***R. lavandulaceum*** (*R. officinalis prostratus*) from Europe is a half-hardy, prostrate and mat-forming plant growing only up to 15 cm high and spread almost its 10 times more, with leaves being dark green above, grey-white-hairy beneath and 1.0-1.5 cm long. It produces pale-blue flowers some 1.2-2.0 cm long from April to September. Its var. 'Seven Sea' is less hardy and compact bearing blue flowers on low-growing arching branches. ***R. officinalis*** (Rosemary/Old man) from S. Europe and Asia Minor grows up to 2.1 m high with 1.5-1.8 m spread. An erect to spreading handsome shrub, this bears numerous mid- to dark-green leaves above and white beneath. Mauve flowers some 1.25-2.0 cm long appear in March and April in short racemes and continue producing sporadically up to September. 'Albiflorus' ('Albus') bears white or very pale-blue flowers, 'Erectus' ('Fastigiatus') as well as 'Pyramidalis' have erect and pyramidal habit, 'Jessop's Upright' growing vigorous and erect with lighter mauve flowers, 'Severn sea' bears bright blue flowers, *etc.*

Roystonea (Royal palm; Palmae)

Roystonea commemorates General Roy Stone (1836-1905), an American army engineer who served in Puerto Rico during Spanish-American war. It is a genus of 12 large and fast-growing palm species from central and tropical South America, Carribean region, West Indies, Mexico and Florida (USA). Only three of its species are grown indoors when young and these are *R. borinquena* (Puerto Rican royal palm) from Puerto Rico, *R. oleracea* from Barbados, Colombia, Trinidad and Venezuela and *R. regia* (*Oreodoxa regia*; Cuban royal palm) from Cuba and adjacent islands. Adult specimens have grey and smooth trunks, sometimes swollen at the base or in the middle. Leaf blades are so tightly rolled around eath other on the stem apex that these form a green stem like structure which is known as crownshaft. The arching leaves are pinnatifid and grow up to 6 m in length. Large panicles emerge from below the crownshaft bearing tiny flowers. They are propagated through seeds sown in spring. ***R. borinquena*** though is similar to *R. regia* but the trunk is bulged above the middle instead of the base, grows 15 m high with about 40 cm trunk diameter, and the leaves are dark green and feathery with multi-ranked leaflets. As these grow very high so are suitable for growing as avenue trees and in the parks, however, indoors these can be kept only for a few years when these are young and up to 2 m high. ***R. oleracea*** is tallest among the royal palms, growing up to 30 m high but indoors these are tamed to grow up to 2 m high with smooth stems having the same width throughout and with shiny green crownshaft, the leaves are spreading 2 m long with rich green and linear leaflets on one plane. ***R. regia*** is a national tree of Cuba having characteristic bulging at the base of the trunk. The tree grows up to 25 m tall and 60 cm in diameter, bearing dark green feathery and arching leaves on the large crown with linear leaflets arranged in four ranks.

Rumohra adiantiformis (Leather fern; Davalliaceae/Polypodiaceae)

See chapter on '**Ferns and Allied Plants**'.

Ruscus hypoglossum (Butcher's broom; Liliaceae)

Ruscus is the Latin name of bucher's broom (*Ruscus aculeatus*). An evergreen, tufted to clump-forming genus of 3-7 shrubby species with rhizomatous rootstock from Atlantic Islands, through Mediterranean Europe to the Caucasus and Iran. They have green stems which broaden to form leaf-like cladodes, and from the axils of which tiny 6-tepalled starry flowers emerge. The plants are dioecious. They grow in humus-rich soil in shade to partial shade and are propagated through seeds and division of clumps. Important cultivated species are ***R. aculeatus*** with dark green, erect, stiff and branched stems, bearing dark green, rigid, 2-4 cm long, ovate and spine-tipped cladodes; and ***R. hypoglossum*** with flexible, arching and branched stems, bearing 5-11 cm long, glossy and elliptic to oblanceolate cladodes.

Sageretia (Rhamnaceae)

Sageretia was named to honour Augustin Sageret (1763-1851), a French botanist. A genus of about 15 armed to unarmed shrub species, is native to tropical Asia and America (North Carolina to Mexico) but only one species, *S. thea* is in general cultivation as a house plant. These shrubs are often scandent, bearing small, deciduous or persistent, opposite to nearly opposite, and serrulate or entire leaves. Tiny whitish, perfect, sometimes fragrant and pentamerous flowers with hooded petals appear in terminal or axillary spikes or panicles. These are propagated through seeds and cuttings. ***S. thea*** is a much-branched woody shrub growing up to 1.2 m high with a spread of about 60 cm, bearing long, wiry and drooping branches. The leaves are short-stalked, in pairs, dark shiny green and some 2.5 cm long with a small notch at the base. Flowers are quite small, whitish and pentamerous and the fruit is a purple berry.

Sanchezia speciosa (Acanthaceae)

Sanchezia was named to honour José Sanchez, a 19th century professor of botany at Cadiz in Spain. The genus comprises 60 species of shrubs and climbers but only one species, ***S. speciosa*** (*S. glaucophylla*, *S. nobilis*) probably from Ecuador or Peru, is in general cultivation for its striking foliage and attractive flowers. It is suitable for growing indoors and in conservatory but only for a short duration as these require partial sun exposure. It is propagated by cuttings. Though this shrub grows with long canes up to 1.8 m high but in containers it can be tamed to grow up to 75 cm tall. Leaves are rich glossy green with main veins clearly yellow-banded, in opposite pairs, oblong-ovate to elliptic, 15-30 cm long and pointed on stalks of the same length. Yellow tubular flowers some 4-5 cm long with 5 short and reflexed lobes at the mouth, and subtended by red bracts appear densely on short terminal panicles.

Sansevieria (Agavaceae)

See chapter on '**Succulents Other Than Cacti**'.

Saxifraga (Saxifragaceae)

Saxifraga derives from the Latin *saxum* 'a rock' and *frago* 'I break', as some species are found growing in the rock crevices as if these have split the rock. It is a genus of about 350 mostly tufted species of a few annuals and many perennials from the mountains of northern temperate zone and South America. Mostly these form rosettes of linear to orbicular leaves and producing 5-petalled flowers. Many species are suitable for growing in cool room, conservatory or porch. They are propagated through ripened seeds, by division of clumps or by offsets. ***S. cuscutiformis*** probably from China looks as *S. sarmentosa* reduced in all its parts including flowers. Its dodder-like branched red runners are produced in profusion. The leaves are medium green with whitish veins, elliptic and 2-5 cm long. It is a shy-bloomer and is suitable for hanging basket. ***S. sarmentosa*** (*S. stolonifera*; Mother of thousands/Strawberry geranium) from Eastern Asia is a mat-forming species spreading through slender and branched red stolons which produce plantlets. Its leaves are silvery-hairy above, reddish lower, rounded, up to 10 cm across and coarsely toothed. White flowers some 2 cm wide with 2 or sometimes only 1 petal appear during summer. *S. s.* 'Tricolor' (Magic carpet/ Mother of thousands) leaves are smaller and pink-flushed with white markings. ***S. umbrosa*** from Pyrenees is a hardy hammock to mat-forming plant, bearing ciliate stalks, the leaves being leathery and deep green, obovate to oblong, broadly crenate and up to 6 cm long. White flowers dotted-red and some 8 mm across appear in panicles some 30-45cm tall in summer. ***S. × urbium*** (*S. spathularis* × *S. umbrosa*; London pride) is hardy and more vigorous than *S. umbrosa* and thrives well under total shade, providing contrasts with ferns. Var. 'Variegata' leaves appear in rosettes on long stalks with clavate-ovate leaves which are patterned irregularly with beautiful cream-yellow patches and dots of dissimilar sizes. *Saxifraga* has many named varieties to choose from such as 'Bob Hawkins', 'Brookside', 'Carmen' ('Elizabethae'), 'Cloth of Gold', 'Cranbourne', 'Crenata', 'Gloria', 'Gregor Mendel', 'Hindhead Seedling', 'Irvingii', 'Jenkinsiae', 'Minor', 'Plena', 'Rubrifolia', 'Ruth Draper', 'Southside Seedling', 'Tumbling Waters', 'Valerie Finnis', 'Water Irving', 'Wisley Variety', *etc.*

Schefflera (Araliaceae)

See '***Brassaia***'.

Schinus (Anacardiaceae)

Schinus derives from the Greek name *schinos* used for *Pisracia lentiscus* as some species yield a similar gum or resin. A genus of about 30 evergreen species of dioecious shrubs or trees, a few of which can successfully be grown as foliage plants in temperate or tropical conditions in tubs for use in the conservatories and houses. They have alternate, simple or pinnately compound leaves and panicles of tiny 5-petalled flowers, followed by berry-like fruits. They are propagated from seeds or cuttings. ***S. molle*** (Mastic tree/Peruvian pepper) from Andes region of Peru is though a small to medium-sized tree but easily tamed up to 2 m in height in a container. Leaves are rich green, alternate, 10-22 cm long and pinnate with 15-41 narrowly lanceolate leaflets. Flowers are tiny and whitish and if trees of both sexes are planted together, the flowers are followed by tiny pea-like rose-red fruits. ***S. terebinthifolius*** from Brazil as also although a large shrub or small tree but is liable to be tamed up to 2 m in containers and if required the plant is also shaped through pruning. Its alternate leaves are pinnatifid and 8-18 cm long with 3-13 broadly lanceolate to obovate leaflets. Tiny white flowers are borne in panicles, and if the plants of both the sexes are planted together, the flowers are followed by red, globular and some 5 mm across fruits.

Schismatoglottis (Araceae)

In Greek, *Schismatoglottis* means 'falling tongue', referring to the limb of the spathe which soon falls off. Native to Malay Archipelago, the genus comprises about 75 herbaceous species with stoloniferous rhizomes and the caudex above ground. These are closely allied to *Aglaonema* and *Alocasia*. These can be cultivated in the warm greenhouses and indoors where day temperature of 21 °C during winter can be maintained throughout. Its leaf petiole is sheathing at base or nearly to half its length, and the leaves are oblong or ovate-cordate, sometimes hastate or lanceolate, and often marbled, maculate or striped. Peduncle is singly or dascicled, spathe cylindrical, spadix sessile and included in the spathe, upper part of spadix cylindrical or clavate with male flowers, lower female, petalless flowers are monoecious, stamens 2-3 and ovary oblong. The plants, in appearance, resemble closely to *Dieffenbachia*. ***S. asperata*** from Borneo bears a very short caudex with petioles and blades of almost equal length. The leaves are green with very minute white dots above and pale-black dots below, ovate or obovate, base cordate or emarginate, and apex acuminate. The leaves in the var. 'Albo-maculata' (*S. crispata*) are silvery but the midrib, the veins and margins are green. ***S. calyptrata*** produces creeping runners. Petioles are erect and up to 50 cm long with pinkish sheaths. The blades are cordate, 15-25 cm long and half as wide. ***S. concinna*** (*S. lavallei*) from Malaya bears 15-20 cm long petioles with reddish sheath, the blades 12.5-18.0 cm long, 3.3-5.6 cm wide, lanceolate or oblong-lanceolate, rounded or narrowed at the base, and blotched variously-sized with silvery-white. Its var. 'Immaculata' (*Lavallei* var. 'Lansbergiana') bears purple sheaths and leaf stalks, upper surface of foliage green and purple below. Var. 'Purpurea' is a Sumatran form having foliage blotched-grey above and dark purple below. ***S. latifolia*** (*S. rupestris*) from Java, Celebes and the Philippines bears thick aboveground caudex. Petioles longer than the blades. Leaf surface is dull green, lower paler, leaves are ovate, deeply cordate, acute and some 15-45 cm long and 15-30 cm wide with posterior lobes semi-ovate and sinus acute. ***S. longispatha*** from Borneo produces petioles one-and-a-half times longer than the blade. Upper surface of the leaves is dull white-striped centrally and rest green, shape is ovate with little cordate base and apex acuminate. ***S. neoguineensis*** (*S. novoguinensis*) from New Guinea bears ovate-cordate, 20-23 cm long, 12.5-17.0 cm wide and bright green leaves, where lesser part is blotched informally with pale-yellowish-green. The petiole is 22.5-30.0 cm long. ***S. ornata*** petioles are about 15 cm long and the blades some 15 cm long and 3.75 cm wide. Leaves are dark green and velvety with grey-green midrib. ***S. picta*** (Painted tongue) from Java produces petioles longer (20-25 cm) than leaf-blades (15.0-17.5 cm), bearing ovate-cordate leaves with short basal lobes, upper surface dark green and marked with glaucous spots in the middle on each side of midvein and between the nerves. ***S. pulchra*** (*S. decora*) from Borneo produces 7.5-11.3 cm long petiole for a 10-13 cm long and 4.2-6.5 cm wide blades which are ovate to obliquely cordate, and irregularly variegated in the larger part of the area with silvery-white blotches so only little part remains green. ***S. rutteri*** leaf-stalks are around 30 cm long, and blades are bright green, 30 cm long and 15 cm wide, and landeolate to triangular. ***S. tecturata*** (*S. variegata*, *Colobogynium tecturatum*) from Borneo, bears dark green leaves above, marked white along the midrib, oblong-lanceolate, obtuse or rounded at the base and long cuspidate at the tip and the petioles are 7.5-10.0 cm long or half the length of the blade.

Schlumbergera (*Claw cactus*; Cactaceae)

See chapter on '**Cacti**'.

Scindapsus (*Epipremnum*; Araceae)

Scindapsus is an ancient Greek name for a climbing plant akin to ivy. The genus originates in SE Asia from China to Malaysia with about 20 evergreen climbing species closely allied to *Philodendron*. Stems run long for many metres producing aerial roots, stalked, cordate, entire and attenuated leaves with stalk bases sheathing the stem. These are good house plants. These are propagated through leaf bud or stem-tip cuttings. *S. aureum* (see *Epipremnum aureus*). ***S. pictus*** from Malaysia and Indonesia is a climber but growing up to 2 m in pots indoors with 10-15 cm long, dark green patterned with lighter green, ovate-cordate and attenuated leaves. Var. 'Argyraeus' (*Pothos argyraeus*; silver vine) has slender stems and smaller ovate-cordate and tip-pointed foliage which are dull green blotched with silver and has thin white line around the edge. This is the most common house plant.

Scrophularia auriculata (Water figwort; Scrophulariaceae)

Scrophularia derives from the Latin *scrofula* 'a disease' that some of its plants were throught to cure. It is a genus of about 200 species of biennials, perennials and sub-shrubs native to Europe, Asia and tropical Africa, however, the species ***S. auriculata*** native to Europe and North Africa is grown only for its handsome foliage. It is a clump-forming perennial growing up to 1 m tall with 4-winged stout stems. Its leaves are opposite, ovate, crenate and some 6-12 cm long. Purple-brown and nearly globular and 5-lobed flowers some 1 cm long are borne on terminal cymose panicles from summer to autumn. *S. a* 'Variegata' leaves have bold and irregular broad creamy margins, new leaves mostly entirely cream but afterwards intensity of the green colouration inceases in few leaves. It is propagated through basal shoot cuttings or division. It likes full sun to partial shade but direct sun burns its leaves.

Scirpoides holoschoenus (*Scirpus holoschoenus*; Rounde-headed clubrush Cyperaceae)

Scirpoides means *Scirpus*-like. Its origin is in Europe and Asia and is a wetland plant though can even be grown in dry lands. It is a hardy evergreen tuft-forming perennial rush growing up to 1.2 m in height. Its leafless stems are green, stiff, rush-like and cylindrical and produce oval, awned and quite numerous brown spikelets on stalked and closely packed 1 to many spherical heads during summer. Its stems arise from the stout rootstocks. There are 1-2 basal leaves which are erect, stiff, narrow and furrowed, the larger one appears in the form of continuing the stem. Its var. 'Variegatus' bears horizontally banded stems with green and yellowish-white.

Scirpus cernuus (*Isolepis gracilis*, Bulrush/ Miniature bulrush/Sedge; Cyperaceae)

Scirpus is a Latin name for a rush. The genus comprises some 300 Cosmopolitan species found in wet places. They are tufted and often colony-forming rushes with grassy leaves or reduced only to basal sheath but in latter case the leaf functions are performed by the triangular or cylindrical green stems. Petalless tiny green to chocolate flowers cluster into compact spikelets. Their need is persisting moist soil so either these should be grown as marginal water plants or the pots in which these are planted should stand in a saucer of water. These are propagated by division. ***S. cernuus*** is a densely tufted evergreen rush growing up to 15 cm high with initially erect stems which arch afterwards, bearing filiform leaves. Its greenish or brownish flowers some 5 mm long appear in solitary spikelets in summer.

Sedum (Crassulaceae)

See chapter on '**Succulents Other Than Cacti**'.

Selaginella (Selaginellaceae)

See chapter on '**Ferns and Allied Plants**'.

Semele androgyna [Cimbing butcher's broom; Ruscaceae (Liliacaeae)]

It is named for Semele, the mother of Bacchus. The genus contains only one species, ***Semele androgyna*** from Canary Islands and Madeira. It is an evergreen climber related to *Asparagus* and *Ruscus*. It is suitable for covering the back wall of the conservatory where it makes a lustrous green curtain. It is propagated by seeds or division. It is though a high climbing twiner (climbing up to 18 m) but in pots it can be maintained around 3 m in height. Its stems are highly branched and bear leaf-like flattened cladodia which are glossy bright green but changing to darker, alternate, leathery, 5-10 cm long, solitary at the axils of dark greyish-brown minute memberaceous scales, ovate-lanceolate and acuminate. Small, yellow and insignificant flowers with six spreading tepals and 6 anthers appear on the margins of cladodes in the clusters of 2-6, followed by globular, black and pulpy berry-fruits some 1 cm across.

Senecio macroglossus (Asteraceae)

See chapter on '**Succulents Other Than Cacti**'.

Serissa foetida (*S. japonica*; Rubiaceae)

Serissa is a Latin version of an East Indian name. It is a genus with one species of small evergreen shrub which makes a decorative pot plant for its foliage beauty, *vis-a-vis* the flowers. These are easily propagated through cuttings. ***S. foetida*** from SE Asia (Japan) is a bushy and much-branched bonsai shrub growing 30-60 cm tall and more than 1.0 m in spread, with highly-branched wiry stems which may be glabrous or puberulent and foetid when bruised, bearing deep green, leathery elliptical to ovate and highly short-stalked leaves in opposite pairs some 0.6-1.8 cm long where the sides of the blades are raised over the midrib. Sparingly, solitary or fascicled, and white and funnel-shaped flowers some 1.2 cm wide appear from the leaf axils or terminally with 4-6 spreading, ovate and little fringed petals. Var. 'Pink' bears pink flowers, the form 'Plena' has double flowers similar to tiny white roses, 'Snow Rose' with double white flowers and the var. 'Variegata' leaves are beautifully cream-margined.

Setcreasea (Commelinaceae)

See chapter on '**Succulents Other Than Cacti**'.

Siderasis fuscata (*Tradescantia fuscata*, *Pyrrhemia fuscata*; brown spiderwort/ spiderwort; Commelinaceae)

Siderasis derives from the Greek *sideros* 'iron', on account of rust-covered hairs which covers the whole plant except the petals. It is a genus with one species ***S. fuscata*** native to Brazil. It is a stemless, tufted to small clump-forming evergreen perennial herb, bearing 10-20 cm long and 7.5-10.0 cm wide, closely packed, horizontally carried, elliptic and decorative foliage which are dark green above with a silvery-white longitudinal band, and flushed with red-purple beneath. Rose-purple 3-petalled flowers some 2.5 cm across are borne in summer. It is propagated through offsets and seeds. It is an aristocratic plant for terrarium conditions.

Soleirolia soleirolii (*Helxine soleirolii*; Baby's tears/Mind your own business; Urticaceae)

It is a creeping ground cover from the Mediterranean (Corsica and Sardinia), named so for Joseph Francois Soleirol (*d.* 1863) who collected plants in Corsica. The genus contains only one species which is creeping and mat-forming evergreen perennial valued for its dense foliage. It is propagated through division. Its stems are slender, filiform, rise up to 2.5 cm high and make a neat and interlacing mat. Its alternate and densely borne leaves on stems are almost rounded, little hairy and short-stalked. Minute unisexual flowers appear singly in the leaf axils. 'Argentea' bears silvery-green and 'Aurea' yellow-green leaves.

Sonerila margaritacea (Frosted sonerila; Melastomataceae)

Sonerila is the Latinized version of the native Malabar name *soneri-ila*. The genus comprises 175 species of mostly small evergreen perennials and a few small shrubs native to tropical Asia. Popular for their beautifully patterned leaves, which are memberanaceous, entire or serrulate and 3-5-nerved. Normally rosy large and small flowers appear in scorpioid racemes or spikes where calyx is 3-lobed, glabrous or setose and tube turbinate, petals are 3, ovate, obovate or oblong, stamens 3 (rarely 6 with 3 smaller), ovary is 3-celled and capsule is 3-valved. They grow well in a close and moisture-laden atmosphere having temperature of around 24 °C. These are best suited in Wardian cases and terrariums. They are propagated through cuttings of the mature growth. ***S. laeta*** from China is closely related to *S. maculata*. This grows erect up to 15 cm high having terete, glandular and hairy stems. The leaves are petioled, simple, green but white-spotted above, purple- and green-spotted beneath, ovate or elliptic-ovate, and the larger leaves being 10 × 5 cm in size. Flowers 7 in number appear in terminal cyme, calyx is purple and petals oblong. ***S. maculata*** from India bears leaves in unequal sizes especially at the base, in each pair the larger one being some 7.5-12.5 cm long and the smaller ones its half or one-third. Leaves are ovate or oblong, minutely toothed and 9-11-nerved, and the flowers are violet. ***S. margaritacea*** from Java is the most important species growing up to 15 cm high with reddish creeping stems, named so due to the regular rows of pearly (margaritacea = pearly) spots between the nerves and parallel with them, while purplish beneath. Leaves are dark green, up to 10 cm long, ovate-lanceolate, serrate, glabrous, acute at base and 7-9-nerved, and the flowers which appear on some 7.5 cm long panicles are 3-petalled and rosy. *S. m. hendersonii* is probably the main parent involved in the development of numerous hybrids with blotched foliage. From the main species, it differs in having a broader leaf with rounded base and being covered with irregularly placed same-sized blotches not crossing the nerves. *S. m. h.* var. *argentea* (*S. argentea*) has silvery foliage similar to certain begonias with dark green nerves and this character has come in the varieties when this was used as a parent in breeding programme. ***S. picta*** from Sumatra grows erect having scaly or hairy branches. Leaves are short-petioled, lanceolate but wedge-shaped at the base, minutely dentate, 7-nerved and white lining along the primary nerves. ***S. speciosa*** from India is the only species grown for its flowers. It grows up to 30 cm high, with leaves green above and sometimes crimson beneath, cordate-ovate, opposite and having 7-9-nerves, and the flowers some 2.5 cm across are borne with 4-14 in a cluster of purple or rose colour.

Sparmannia africana (Tiliaceae)

Sparmannia was named for Dr. Andreas Sparman (1748-1820), a Swedish botanist who collected plants in the Pacific region during second voyage of Capt. Cook, and in South Africa with Thunberg. The genus is native to Africa (including Malagasy) with 3-7 species of trees and shrubs, usually with large palmately-lobed leaves and umbels of white flowers, both being attractive. The broad leaves appear in a dense mass, are cordate, dentate and/or lobed. White flowers appear in terminal umbelliform little cymes with 4 sepals, 4 petals, numerous free stamens, 4-celled ovary and spiny capsules which are globose. These are quite suitable container plants to be displayed in the house or conservatory where these will even flower, and whose height can be tamed through pruning. These can also be grown as annuals through tender cuttings each year, rooting and then planting in spring. These also require pinching of the tips of the new plants to make the plants more compact and bushiness. ***S. africana*** (African hemp) from South Africa is a 3-6 m high shrub though as an annual plant it can be maintained up to 60 cm in height, bearing long-stalked, 5-7-angled, alternate, cordate-acuminate and 12.5-15.0 cm long leaves in a dense moss which are unevenly toothed and under-surface 7-9-ribbed. White flowers some 3-4 cm wide with yellow and red-purple stamen filaments appear in plenty on peduncles and sometimes the load of the blooms is so heavy that branches arch and touch the ground so proper and regular pruning after flowering is required. While in bloom, it surpasses the beauty of many other shrubs such as *Hydrangea* and *Viburnum* while when not in flowering its leaves are enjoyable so throughout the year this gives an enchantment. Its capsules are 5-celled. Var. 'Flore Plena' bears double flowers. ***S. palmata*** from South Africa is a small slender shrub, much smaller in all its parts than *S. africana*. Its branches are half-herbaceous bearing long petioles. The lobes of the leaves are long acuminate, margins indented with unequal tooths and wavy with 5-7 prominent nerves below. The white or purple densely arranged flowers appear on the subterminal peduncles during summer, and the capsule is 4-celled.

Spartina (Graminae)

Spartina has derived from the Greek word *spartine* which means 'a cord', on account of its tough leaves. This comprises of about 10 perennial species, mostly found in the saline marshes of seacoast, mostly in USA while certain others in other parts of the world. Its jointed hollow stems (culms) are rigid and reed-like. Leaves are basal, stem-sheathing, coarse and rough, linear, long and the fully grown up ones bending down roughly from the middle. Spikes are two to several in a raceme, and the spikelets are 1-flowered, sessile, strongly flattened and closely imbricated in two rows on one side of a narrow rachis. ***S. pectinata*** 'Aureomarginata' ('Aureovariegata') growing up to 2 m high, is a spreading rhizomatous grass with long, arching and yellow-striped leaves which turn orange-brown in late autumn to winter.

Spathiphyllum (Araceae)

See chapter on '***Spathiphyllum***'.

Stapelia (Toad plant; Asclepiadaceae)

See chapter on '**Succulents Other Than Cacti**'.

Stenandrium lindenii (Acanthaceae)

Stenandrium derives from the Greek *stenos* 'narrow' and *aner* 'a man', referring to the very slender stamens. It is a genus of 30-60 evergreen perennials from tropical and sub-tropical America, of which only one from Peru is under common cultivation. It is tufted and hummock-forming plant spreading to more than 20 cm across, bearing coppery-green, puckered, elliptic to broadly elliptic leaves in opposite pairs some 2.5-8.0 cm long with a clear yellow pinnate-vein-pattern. Yellow tubular 5-lobed and 2-lipped flowers some 2 cm long appear on spikes some 5 cm long. This requires warm and humid conditions for its growth, hence are most suitable for growing as terrarium plant. It is propagated by division, seeds and cuttings.

Stenocarpus (Firewheel tree/Wheel of fire; Proteaceae)

In Greek, *stenos* is for 'narrow' and *carpus* is for 'a fruit', referring to the follicles which are long but narrow. There are some 18 species, three endemic to Australia and the others to New Caledonia but only ***S. sinuatus*** (*S. cunninghamii*) is worth growing indoors. These are shrubs to trees bearing alternate or scattered, entire or deeply pinnatifid leaves with few lobes. Racemes which are sometimes short are either terminal or appear from the upper axils and sometimes several in an umbel, each with an umbel of pedicellate red or yellow flowers. Bracts falling off early or none, the flowers are hermaphrodite with little irregular perianth, tube opening along the lower side, limb almost globular and recurved, anthers broad and sessile within the concave laminae, ovary stipitate, ovules several and imbricate down in two rows and the follicles are coriaceous and narrow. ***S. sinuatum*** from Australia is a tree growing to 18-30 m tall in the wild but indoors it grows up to 1.6 m only. This has a glabrous or with minutely tomentose inflorescence, leaves are glossy pale-green but tinged red beneath and midrib pale-green, alternate, 2.5-7.5 cm petioled and either entire, elliptic, 15-45 cm long and 23 cm wide, or pinnatifid, glabrous, reddish beneath, mostly obtuse, penninerved and minutely reticulate and some 45 cm long with 1-4 oblong lobes either side with a single segment at the tip end. Its some of the forms have smaller, lanceolate, entire or only slightly lobed leaves but the type species is more attractive. Peduncles terminal, 2 or more together or several at different ranks which form a short but broad raceme each with a peduncle length of 5-10 cm long, each umbel with 15-20 wheel-shaped bright red flowers having 2.5 cm long perianth. As this usually does not branch out so when planting indoors, the growing point of the stem should be nipped out to make it branched and bushy. Indoors these do not flower so are planted just to enjoy its handsome deeply lobed foliage.

Stenotaphrum secundatum (Graminae)

See chapter on '**Lawn**' under St. Augustin grass.

Strobilanthes (Acanthaceae)

It takes its name from the greek *Stobilus* 'a cone' and *anthos* 'a flower', due to bracted floral spikes of some species. It is a genus of 250 perennial shrubby species from tropical Asia and Malagasy though only 2-3 species are grown indoors such as *S. dyeranus, S. isophyllus* and *S. lactatus* (*S. maculatus*). An erect herbaceous shrub, it is a lovely foliage plant when young with dark green long pointed leaves having silvery purple sheen which diminishes with age. This prefers semi-shaded and humid atmosphere for its growth and is propagated through seeds and cuttings. ***S. dyeranus*** (Persian shield) from Myanmar is an erect herbaceous shrub growing more than 60 cm high with sparingly branched stems, bearing finely toothed leaves in opposite pairs, leaves with stalks some 20 cm long, narrowly ovate to lanceolate, upper surface green blotched purple and silver which extends up to the margins, and purple-colouration below. Pale-blue or lavender narrowly funnel-shaped flowers with an oblique mouth and five rounded lobes, some 2.5-4.0 cm long, appear in dense terminal spikes. ***S. isophyllus*** (Willow-leaf Persian shield) from India is a much-branched low-growing herbaceous plant swollen at joints and growing to 60-90 cm high. Leaves are short-petioled, opposite, narrowly lanceolate, and entire or distantly notched. Axillary peduncles shorter than the leaves appear having several blue and white flowers with 2.5 cm long funnel-shaped corolla with emarginated lobes. **S.**

lactatus grows some 20 cm high bearing tightly spiralled leaves which are blue-green with silver plumes on both sides of the midrib.

Stromanthe (Marantaceae)

Stromanthe derives from the Greek *stroma* meaning 'a bed or couch' and *anthos* 'a flower', on account of the shape of the inflorescence. A native of tropical America, this evergreen genus comprises about a dozen species of rhizomatous perennials related to *Maranta* and *Calathea*. Stems arising from the thick horizontal rhizome, are erect, leafy, little branched and base are covered with long leaf-sheaths. Their leaves are simple and quite handsome so suitable for growing indoors for a short duration. These prefer high humidity and partial shade. They are propagated through division and cuttings. The flowers are small, insignificant, asymmetrical, 3-petalled and are borne in the axils of red bracts. ***S. amabilis*** (*Calathea amabilis, Maranta amabilis*) from Brazil is a clump-forming species growing up to 90 cm high, bearing about 30 cm long elliptic-oblong leaves with grey zonings in between the veins while lower surface is red. ***S. porteana*** (*Maranta porteana*) from Brazil is almost clump-forming with stems growing up to 2 m in height, bearing dark glossy green and some 45 cm long elliptic or ovate-lanceolate and acuminate to nearly obtuse leaves which are slightly puckered with silvery-white vein patterning. ***S. sanguinea*** (*Thalia sanguinea*) from Brazil is prominently clump-forming strong perennial with stems rising up to 1.5 m in height, bearing lustrous rich green oblong-acuminate to lanceolate leaves up to 40 cm long where upper surface has paler midribs though under surface is red or striped green.

Syagrus weddelliana (Palmae)

Syagrus derives from the Greek *syagros*, the name of a type of date palm. It is a genus of 30-50 pinnate-leaved palms from South America, closely allied to *Cocos* (coconut). They are low-growing and sometimes even without a false trunk as in some species so are quite suitable for pot-planting. These are propagated through seeds grown at or above 21 °C temperature. The cultivated species is ***S. weddelliana*** (*Cocos weddelliana, Microcoelium weddellianum*) from Organ Mountains of Brazil attaining the ultimate height of up to 3 m though remains smaller for dozen of years. In pots, its arching fronds grow up to 60 cm long forming a terminal rosette and then after so many years the slender stems are formed. It does not flower or fruit in pots.

Synadenium grantii (Euphorbiaceae)

It is a Greek name which indicates the united involucral bracts. This almost succulent genus is from tropical Africa with 13 species out of which 3-4 are worth growing indoors. These are thick-branched tropical shrubs having the generic characters of *Euphorbia* except that the glands of the involucres are united so as to form a ring around the lobes. ***S. amabilis*** is a herbaceous stem succulent having oval and blunt leaves with yellowish midrib having wide deep green patterning alternate with almost more wider yellowish-green throughout the leaves starting from the side of the midrib and running upward making an inward curvature and merging at the margins. ***S. arborescens*** from South Africa grows erect and unbranched up to 1.2 m high, bearing obovate-cuneate and inconspicuously crenulate leaves where midrib is keeled and mostly denticulate, bracts puberulent and involucres yellow. *A.* ***cupulare*** is an erect-growing high stem-succulent with light coppery splashing throughout the plant including the leaves beneath though upper surface is green, however, the young ones remain coppery. The leaves are almost stalkless, thick, leathery, ovate-acuminate with ample vein networking, and are formed on the stem all around and sometimes even two together, the old lower ones falling one by one leaving prominent scars on the solitary stem while new ones continue emerging at the apex. Inflorescence is light coppery and terminal bearing insignificant creamy flowers. ***S. grantii*** (African milkbush), a tropical African spectacular, sturdy and branched shrub growing up to 3 m at its natural habitat but only 90 cm indoors, bearing very short-stalked obovate-spathulate and obtuse leaves some 10-15 cm long with rounded midrib. Cymes are dichotomous with tomentose bracts and red involucres. Its var. 'Rubra' bears dull red leaves.

Syngonium (Araceae)

Syngonium is derived from the Greek *syn* meaning 'together' and *gone* 'womb', an allusion to the united ovaries. A native to the tropical areas of Americas including West Indies, there are 20 evergreen epiphytic climbing species though now grown terrestrially. Leaves are variable in adult form than those at juvenile stage, initially being entire to sagittate but later changing to trifoliate to pedate. Only mature adult plants produce the flowers typical to aroids with white, green or purple spathes. Though these tolerate dry air but prefer humid environment in the house. These require moss sticks to climb on otherwise these will spread here and there. ***S. angustatum*** produces palmate leaves with 3-5 elliptic segments, central one being larger and up to 15 cm on a 25 cm stalk, and the veins are yellowish. Mostly its 3-leaf stage is known as *S. podophyllum*. Its var. 'Ruth Fraser' produces rich green leaflets with silvery veins. ***S. auritum*** (*Philodendron auritum*; Five fingers) from Jamaica produces leaves some 20 cm long which are trifoliate or pedately divided into five leaflets, with central one being the longest whereas th side lobes are small ear-like at the bases. Spathe is greenish-yellow and up to 28 cm long. Var. 'Fantasy' leaves are variegated in cream and white. ***S. erythrophyllum*** (Copper nephthytis) is a very slow-growing climbing plant with slender woody stems growing more than 1 m in length. Juveniles have arrow-shaped leaves flushed purple beneath, but on maturity these turn trifoliate with long and thick leaf stalks. ***S. hoffmannii*** (*Nephthytis hoffmannii*) from Costa Rica produces 2.0-3.5 m long stems, bearing grey-green with silvery-white veining and in the middle area, initially sagittate when young but at adulthood trifoliate, having largest (15 cm or more) central lobe. About 12 cm long spathe is white with a purplish throat. ***S. podophyllum*** (Arrowhead vine/Goosefoot plant) from Mexico to Panama remains compact, small with sagittate leaves often having silvery-white suffusion or veining initially when young, but when attain maturity the stems become climbing and grow about 2 m in length bearing first with trifoliate leaves and then pedate with 5-9 leaflets, central one being the largest. Spathe is white and some 30 cm long. *S. p.* 'Albovirens'produces leaves with creamy-silver bands along the veins, 'Atrovirens' produces dark green leaves with silvery suffusion and veins, 'Emerald Gem' produces roughly trilabiate and green leaves with midrib and veins prominently creamy-white, 'Gold Butterfly' blades are irregularly marked with a chartreuse centre, 'Green Gold' bears green trifoliate leaves with suffusion of greenish-pale on and along the midrib, *vis-à-vis* veins to more than its half the length though irregularly, 'Imperial White' leaves are green and trifoliate with white spots and suffusion to most part, 'Mottled Arrowhead' blades are deep green with comparatively wider tip-sgments and are mottled yellow-ochre, 'White Butterfly' is really a dwarf variety having white leaves with green edges, 'Trileaf Wonder' at juvenile stage produces green and trifoliate leaves with a silvery-grey patterning, 'Variegatum' (*Nephthytis liberica*) produces sagittate leaves with asymmetrical cream-white splashing, *etc.* ***S. standleyanum*** leaves are green and attain senile stage soon with trifoliage form, central lobe is elliptic with 25 × 20 cm in size, and basal segments are spreading and 22.5 × 7.5 cm in size with curved bottom edge. ***S. wendlandii*** (*Nephthytis wendlandii*) from Costa Rica bears sagittate to hastate juvenile leaves that are trifoliate but rarely entire, velvety deep green with white veins, while senile leaves are trifoliate with central being larger up to 18 cm × 7.5 cm and dark green but without white patterning. Green spathes are purplish within and some 13 cm long.

Tacitus bellus (Crassulaceae)

See *Graptopetalum* in the chapter on '**Succulents Other Than Cacti**'.

Talinum (Portulacaceae)

See chapter on '**Succculents Other Than Cacti**'.

Tapeinocheilos (*Tapeinochilus*) *ananassae* [Costaceae (Zingiberaceae)]

Tapeinocheilos derives from the Greek *tapeinos* 'low' and *cheilos* 'lip', referring to the short labellum-like petals. A native of Moluccas, and New Guinea to Queensland, there are some 20 large perennial species in the genus, one having handsome foliage for use in the warm conservatory and indoors. It is propagated by division. ***T. ananassae*** from Moluccas is a clump-forming species which bears stems of two types ~ (i) leafy and 1.5-2.5 m tall but non-flowering, and (ii) leafless and smaller (0.45-1.2 m tall) but flowering. Leafy one stem is unbranched, erect with tips being in half spiral form and bears deep green and some 15 cm long narrowly obovate leaves which are dense above and scattered below. Leafless flowering stem appears directly from the ground and terminate in a cone-shaped 15-20 cm long spike bearing recurved crimson to orange-red bracts where at the base the small yellow flowers are concealed.

Tetrapanax papyriferus (Rice paper plant; Araliaceae)

Tetrapanax derives from the Greek *tetra* 'four' and the allied genus *Panax*, owing to the floral parts being present in fours rather than fives as in *Panax*. A native to China and Taiwan, the genus is related to *Fatsia* and comprises one species, ***T. papyriferus*** (*Aralia papyrifera*, *Fatsia papyrifera*). It is a single-stemmed shrub with ornamental foliage which is propagated by suckers. Stems are erect, robust, sparingly branched and 2-3 m high. Leaf stalk is stiff and about 60 cm long. Leaves are deep green, white felted beneath, long-stalked, alternate, orbicular and 30-60 cm across with 5-16 broad, pointed, partly jagged and finely toothed lobes, each with a prominent pale-green mid-vein. Whitish to yellowish small flowers appear in spherical umbels, followed by small berry-fruits. Stem pith which is white is the source of Chinese rice-paper. Its var. 'Album' is white while 'Variegatus' is patterned with irregular white splotches.

Tetrastigma (*T. voinierianum*, *Cissus voinieriana*; Chestnut vine/Lizard plant; Vitaceae)

In Greek *tetra* is for 'four' and *stigma* is the pollen-receptive part of the ovary, on account of its stigma being 4-parted. The genus comprises 90 species of woody-stemmed climbers from SE Asia, Indo-Malaysia and Australia. These are allied to *Cissus* and *Vitis*. These should be propagated through cuttings and layering. Normally 3-4 species are grown as ornamental foliage plants indoors. *T. **harmandii*** from Philippines is a large climbing plant having support of tendrils. Leaf stalks are 30 cm long. Leaves are irregularly palmate with 3-7 lanceolate leaflets which may be up to 15 cm long having finely but sparingly toothed margins. ***T. obovatum*** from China is a climbing plant bearing brown-felted palmate leaves with 5 spathulate leaflets some 20 cm long having wavy margins and depressed veins. ***T. voinierianum*** (*Cissus voinierianum*, *Vitis voinierianum*; Chestnut vine) from Laos (Vietnam) is a large, coarse vine with brown-hairy and drooping branches, most suitable as basket plant. Leaves are lustrous pale-green, fuzzy beneath, palmate with 3-5 leaflets which are individually broadly oval, 10-20 cm long, conspicuously toothed, wavy-margined and drooping. Shoot tips and young leaves are covered with hairs similar to mature leaves. Tiny and greenish flowers appear in axillary clusters with four petals or sepals, and the fruit is acidic berry similar to grapes.

Thalia (Marantaceae)

The genus commemorates Johannes Thal (1542-1583), a German doctor who authored a flora of the Hartz Mountains (*Sylvia Hercynia*), published posthumously in 1588. This genus from tropical America and Africa comprises 7-11 aquatic perennial species, out of which *T. dealbata* provides an elegant accent on the edges of the conservatory pools. It is a hardy species propagated through seeds or division. ***T. dealbata*** from SE USA is a clump-forming perennial, bearing long-stalked, obovate to narrowly elliptic and some 20-40 cm long leaves covered with a white waxy powder. Flowering stems are slender, some 2 m long and branching with dense panicle at the top, and with a few basal leaves. Purple flowers have three small petals and a large petaloid staminode about 1.2-1.5 cm long.

Thrinax (Thatch palms; Palmae)

Thrinax is the Greek name for a trident, referring to the forked tips of the leaflets or lobes. A genus of 4-12 species of smooth-trunked palms from Mexico to Belize, Florida and West Indies. They are small to medium size with rounded heads of palmate to digitate leaves. These are propagated by seeds. ***T. excelsa*** (Jamaican thatch plant) produces single stem up to 10 m high and 20 cm in diameter. The leaves are mid-green above, silvery-white beneath and have dense scale-covering, nearly circular, and up to 3 m across with acute segments, paniculate flowers are pink to purple and the fruits are 1 cm across, round and white. ***T. microcarpa*** (*T. morrisii*; Key palm) from S. Florida and West Indies may grow in the wild up to 10 m in height but is decorative only when it is young and then it can be grown in pots where its growth is very slow. Its leaves are 60-100 cm wide, orbicular and divided into 20-30 lanceolate lobes (leaflets) whose lower surface is silvery-white. ***T. parviflora*** (mountain thatch palm, palmetto thatch palm) from Jamaica is similar to *T. microcarpa* though undersurface of the leaves is not silvery-white. Its stems are single, 15 cm high and 15 cm in diameter, having leaves on sparse open crown, flowers are paniculate and cream to yellowish. ***T. radiata*** (Florida thatch palm) is a small growing plant up to 4 m high with 15 cm trunk diameter, and the leaves are deep green above, white beneath and palmate. It makes an attractive pot plant.

Tolmeia menziesii (Mother of thousands/Pick-a-back/Piggyback plant/Thousand mothers/Youth-on-age; Saxifragaceae)

Tolmiea was named for Dr. William Fraser Tolmie (1812-1886), surgeon at the Hudsom Bay Company depot at Fort Vancouver. This hardy genus from western North America comprises one species of low evergreen perennial excellent for pot planting or for hanging baskets. It is propagated by division or through plantlets formed on the leaves. ***T. menziesii*** is a clump-forming up to 60 cm tall plant when in bloom, otherwise remains only 15 cm high. This plant has short creeping rhizome. Leaves are mid-green, long-stalked, broadly ovate-cordate, maple-like and lobed, hairy and toothed and some 4-10 cm long. From the junction of leaf stalk and the leaf blades, small plantlets arise which root when touched to the soil. Tubular greenish-purple five petaloid calyx lobes and four thread-like deep brown petals in the flowers which are about 1.2-2.0 cm long appear on slender racemes during summer. 'Variegata' has leaf variegation marbled white with green midrib and veins.

Trachycarpus (Palmae)

In Greek *trachys* is 'rough' and *karpos* is 'fruit'. A native of Himalaya to SE Asia, the genus is hardy and comprises 6-8 species of small to medium-sized palms. These bear fan-shaped leaves divided to more than half-way into about 40 or more narrow and linear segments. They are suitable for a cool room or conservatory. These are propagated through seeds. ***T. fortunei*** (*T. excelsa*, *Chamaerops excelsa*; Chusan palm, windmill palm, Chinese windmill palm) from Borneo and East China grows up to 10 m tall and 20 cm diameter in the wild but only 2 m in containers. Stem is solitary, slender and covered with brown and fibrous leaf bases. Leaf stalks are some 90 cm long and sharply toothed. Leaves are glaucous to mid-green below, dark green above, rounded, palmate (fan-shaped), some 90 cm across, deeply divided into many narrow, linear and pleated segments. When young or in pots this normally does not produce flowers. However, small yellow flowers in dense panicles up to 60 cm long appear during summer, sometimes followed by blue-black globose fruits some 1.25 cm wide. ***T. takil*** when young produces creeping stems though grows to a height of up to 12 m. The stem is covered with brown fibres, the leaves are shiny green above, glaucous beneath and 1 m across while the fruits are bean-shaped.

Tradescantia (Commelinaceae)

See chapter on '**Succulents Other Than Cacti**'.

Trevesia (Araliaceae)

The genus *Trevesia* honours the family of Trèves the Bonfigli of Padua, patrons of botany in the 18th century. A. genus native to Indo-Malaysia and Pacific, includes 4-10 species of shrubs and small trees, propagated through heel-cuttings under mist, and seeds. The species ***T. palmata*** (Snowflake aralia) native to N. India to SW China has a very graceful foliage, most suitable for container-planting indoors. In the wild it is a small tree but can be tamed up to 2 m high indoors in pots. Its stems are erect, scarcely branched and sometimes prickly. Leaves are deep glossy green and lightly corrugated, 25-60 cm across, long-stalked, alternate, circular, digitate, divided into 7-11 leaflets, of which the midribs are attached to the edge of a central web (duck's-foot-like) of leaf tissue. At younger stage the leaflets are ovate to oblong-ovate and sparingly toothed but on maturity these are deeply cut into bluntly triangular lobes. Small and yellowish flowers appear on erect panicles in rounded umbels only on mature plants. Its var. *micholitzii* bears nearly white young leaves. Apart from *T. palmata*, ***T. burckii*** (*T. sanderi*) which is a shrub growing up to 1.5 m tall is highly leafy and shaped as *T. palmata* but leaves are undulating, crinkled and about 60 cm across. ***T. sundaica*** from S. Asia is a small tree of up to 7.5 m high though can be tamed up to 2 m in pots, bearing leaves with some 90 cm stalk length and leaf diameter of 75 cm, leaves divided into 9 leaflets which are further lobed, elliptic or rarely fiddle-shaped and with tip-drooping.

Trichocereus (Golden column; Cactaceae)

See chapter on '**Cacti**'.

Tristania (Myrtaceae)

Tristania was named to honour Jules M.C. Tristan (1776-1861), a French botanist. The genus contains some 23 evergreen trees or shrubs from Malaya, New Caledonia and Australia but 1-3 species are worth growing indoors as foliage plants. These require filtered partial sun and can tolerate up to a minimum temperature of 10 °C. They are propagated through seeds and half-ripe cuttings. Leaves are alternate or little whorled, rarely opposite and congregate at the ends of the branches. Small yellow or white flowers appear in axillary peduncled cymes with turbinate-campanulate calyx tube and limb with 5 short segments, petals 5, stamens numerous and ovary is either inferior or semi-superior. ***T. conferta*** (*Lophostemon arborescens*; Brisbane box) from Queensland is a very tall evergreen shade tree attaining up to 45 m in height, though indoors it can be managed up to 2.4 m tall, tolerating -6.7 °C low temperature and is valuable for boulevard avenue planting in hot dry regions and as a durable timber tree. Its young shoots and calyx are hairy, leaf stalks are some 5 cm long, the leaves are elliptic, glabrous, verticillate and some 7.5-15.0 cm long and 5 cm wide with narrow tips, flowers on branches appear well below the leaves with white petals about 6 mm long, spotted and fringed. Its var. 'Aureo-Variegata' leaves have yellow variegation in jagged splotches similar to *Graptophyllum*. ***T. lactiflua*** (small tree) and ***A. laurina*** (a shrub) both from Australia can also be used as house plants.

Tupidanthus calyptratus (Araliaceae)

In Greek, *Tupidanthus* derives from the Greek which means 'mallet' and 'flower', owing to the shape of its floral buds. The genus comprises one species, ***T. calyptratus*** from India which initially is a glabrous shrub but with age becoming a tall climber, bearing large digitately compound leaves arranged in a circle with 45 cm long stalks, leaflets are drooping, 7-9, up to 30 cm long, 5 cm wide and tips narrow with petiolules 5 cm long, entire, narrow-oblong, short-acuminate, narrowed at base and coriaceous, stipules connate and the main umbel about 3-rayed, branches 7.5 cm long with large bracts at the bases, umbellules 5-7-cleft and are arranged in a short panicle or compound umbel. Green flowers are large where petals are connate in a leathery or fleshy hood, stamens 50-70, disc convex and the fruits are subglobose fleshy to leathery. For its growing it likes filtered to partial sun and is propagated from cuttings.

Veitchia (Adonidia; Palmae)

Veitchia honours James Veitch (1815-1869) and his son John Gould Veitch (1839-1870), famous nurserymen of Chelsea and Exeter. A genus of 18 species of palms found in the Philippines, New Caledonia, New Hebrids and Fiji. These require partial to full sun. These are propagated by seeds and through the growths occurring on stem. Only one species, ***V. merrillii*** (Christmas palm, Manila palm) is suitable as an indoor plant with filtered light and that too only when it is juvenile but when full sun is available its beauty declines. In the wild it grows up to 5 m tall though is fairly slow-growing. The leaves in young plants are flatter. Leaf bases tightly overlap one upon the other to form a smooth cylindrical shape of the stem. Leaf stalks are strong, and the leaves are pinnate, up to 2 m long, ascending and arching with numerous deep green and linear leaflets which are also ascending and arching. From just below the crown shaft large panicles of small flowers emerge, followed by bright red, 2.5-4.0 cm long and oval fruits but only on well-grown plants. ***V. joannis*** growing up to 30 m high with 25 cm stem diameter is an elegant and slender palm, bearing shiny dark green, arching, pinnatifid and up to 3 m long with drooping leaflets. Its 5-6 cm long and red fruits appear in clusters.

Vittaria lineata [(Polypodiaceae (Vittariaceae)]

See chapter on '**Ferns and Allied Plants**'.

Vriesea (Bromeliaceae)

See chapter on '**Bromeliads**'.

Washingtonia (Palmae)

Washingtonia commemorates George Washington (1732-1799), the first President of the USA. A genus of two species of large fan-shaped plams from California, Arizona and Mexico. When young, these palms make handsome foliage plants in containers and there these seldom form the trunks. In the wild, their height may go up to 32 m clothed with base-remnants of the dead leaves. Their leaves are costapalmate with spiny petioles. Large axillary panicles which arch out among the leaves extending beyond the leaf length, bear whitish flowers, followed by berry-like black drooping fruits. They are propagated through seeds. ***W. filifera*** (*Pritchardia filifera*; Desert fan plam/Petticoat palm) from California and Arizona is a fast-growing palm at younger stage. Its stalks are green and fronds grey-green and some 1-2 m in diameter with many deeply cut, narrow and linear segments having filamentous margins. ***W. robusta*** (*W. gracilis, Pritchardia robusta*) from Baja California bears bright green leaf stalks, *vis-a-vis* leaves which are 1-2 m in diameter, segments linear, firm, longer and drooping with less or no filaments on the margins. At maturity, this palm is more slender than the other species.

Woodwardia (Blechnaceae)

See chapter on '**Ferns and Allied Plants**'.

×*Xantheranthemum* (*Chamaeranthemum* × *Eranthemum igneum*; Acanthaceae)

In Greek, *xanthos* is for 'yellow' and *Eranthemum* a genus closely allied to *Xantheranthemum*, developed in Peru with only one hybrid species, ×*Xantheranthemum igneus*. It is a low colonizing herb with closely set upward-facing velvety and oval to spoon-shaped brownish-green leaves some 10 × 3.74 cm in size but on a short stalk, where veins and midribs are red to burnt orange similar to fish bones and the margins are curved down. It is propagated by offsets and cuttings.

Xanthorrhoea (Xanthorrhoeaceae, formerly Liliaceae; Black boys/Grass gums/Grass trees)

In Greek, *xanthos* is 'yellow' and *rheo* 'to flow', on account of yellow resin being exuded from the caudex. Native to Australia, the genus comprises some 14 species of woody-based perennials, bearing dense and large tufts of coarse, brittle, long-linear grass-like recurved or spreading leaves atop a very short and blackish palm-like caudex which may be zero to 4.5 m raised from the ground in case of aged specimens as in *X. preissii*, well adapted to greenhouse culture or to create a picturesque spot in desert or in arid gardens. Caudex is densely covered by the bases of old dead leaves cemented with oozing yellow resinous gum which turn black. Scapes are terminal often a few metres long which terminate into a dense terminal spike, bearing numerous sessile and greenish flowers where perianth is persistent with 6 segments, 3 outer glume-like, erect and concave or nearly hooded at the top, 3 inner much thinner, erect along the outer ones but protruded beyond them, stamens 6, ovary sessile and 3-celled, and shining, hard, brown, ovoid or acuminate capsules protrude from the perianth. The species are long-lasting if planted in dry and rocky places. They are propagated by offsets. ***X. arborea*** (Botany bay gum) caudex is some 3 m high in a well mature plant and attaining of this takes many years. Densely set leaves are linear, flattened, some 60-120 cm long and 4-6 mm wide, first erect and then arching and hanging, and the flower spikes are up to 1.2 m tall. ***X. australis*** (Australian grass tree/Blackboy) is to some extent similar to *X. arborea* but the trunk is shorter and the leaves are narrower and only up to 3 mm wide. Flower spikes are wider and most prominent than *X. arborea*. ***X. hastilis*** caudex is very short, the leaves are 0.9-1.2 m long and linear, scape often up to 3.5 m long, spikes many and 5-60 cm long and is distinguished by the dense and rusty tomentum covering the end of the bracts and outer perianth segments. ***X. minor*** caudex is very short, the leaves are 30-60 cm long and scapes even longer, spikes 7.5-20.0 cm long and less than 2 cm wide, inner perianth segments having a white blade clearly spreading over the outer ones. ***X. preissii*** (Common blackboy) caudex generally grows up to 1.5-1.8 m high and in quite aged ones sometimes even up to 4.5 m which is sometimes even branched, leaves are linear, very brittle and fragile when young and grow 60-120 cm long, scapes inclusive of spike some 60-180 cm long where spike may be its one-half or nearly all its length, and the fruit is long. ***X. undulatifolia*** also has the same length of the trunk as *X. preissii* with about 30 cm diameter. This bears a large crown of fragile, sword-shaped, reflexed and rhomboidal leaves some 1.5 m long, scapes some 3.6 m long, cylindrical and erect with a dense spike of golden-yellow flowers.

Xanthosoma (Araceae)

In Greek, *xanthos* is for 'yellow' and *some* for 'a body', on account of the inner tissues of the tubers and rhizomes of some species. A native of Mexico, and tropical America including West Indies, the genus comprises some 40-45 tuberous and rhizomatous perennial species looking like *Alocasia* or *Colocasia*, where the tubers or rhizomes and even the leaves of certain species are edible in the tropical areas. These are suitable as pot plant for keeping in the house. When repotting, these are divided and planted for further multiplication. These require shaded moist conditions but with good drainage. Their long-stalked leaves appearing directly from the rhizomes from below the surface are arrow- to spearhead-shaped and sometimes variegated. The spathes are arum-like, having separate columns for tiny petalless male and female flowers. ***X. atrovirens*** leaf blades are about 60 cm long, blades are green above, greyish below, cordate and deeply but narrowly notched, and the basal lobes are broad and some 90 cm long but much less indoors. Its var. 'Albomarginatum' and 'Albomarginatum Monstrosum' are curious-looking plants, 'Albomarginatum' blades are contorted with white blotches along the edges, though 'Albomarginatum Monstrosum', in addition to the above, bears hardly 15-20 cm long blades and has fusion of the edges at the tips to form a triangular sack with an elongated narrow point. ***X. daguense*** leaf stalks are some 60 cm long, and the blades are narrowly arrow-shaped some 30 cm long with a deep notch. ***X. lindenii*** (*Phyllotaenium lindenii*; Indian kale) from Colombia is tuberous-rooted, bearing large (45 × 12.5 cm in size) arrow-shaped shining deep green, thin-leathery with down-curving leaf edges, basal lobes short and broad, central and main veins brightly silver-white and depressed, and stalks as long as the leaves. Spathe is green and white and about 13 cm long. Its varieties 'Alboscens' bears narrower leaves with basal lobes somewhat longer and 'Magnificum' produces larger leaves with more prominent cream to white veins. ***X. sagittifolium*** is similar to *X. atrovirens* but with broader notch and some 90 cm long leaves. Its rhizomes are also edible. X. ***violaceum*** (*Alocasia violacea*; Blue taro/Yautia) from West Indies is similar to *X. sagittifolium*, bearing 1.5 m long leaf stalks, 45 cm long, deep green and triangular-ovate to arrow-shaped blades, and the edges, veins and stalks have violet tinges. Spathes are glaucous-purple and some 30 cm long. It is cultivated for its high quality starchy tubers which are edible.

Xenophya lauterbachiana (*Schizocasia lauterbachiana*; Araceae)

It is a superb but rarely grown decorative plant with two species but only one is sometimes found in cultivation. This leafy herb is native to Moluccas and New Guinea. It can be grown in filtered to partial sun with minimum humidity level of 60 per cent and at a minimum temperature of 13 °C. Its propagation is by division of clumps as its pups arise at some distance from the mother plant. ***X. lauterbachiana*** bears tight spirals of erect but lobed leaves up to 1.2 m long where stalk is usually shorter (50-60 cm) than the blade and is stiff and channelled. Blades are satiny blue-green above, silvery-purple below, generally 60 × 10 cm in size, older ones becoming sword- to sickle-shaped, each vein tipped by a shallow lobe with an upward curving hook. Each blade may have slightly irregularly placed 13-17 lobes, one slender acuminate at the tip, and is somewhat similar to certain *Philodendron*.

Yucca (Adam's needle/Spanish bayonet; Agavaceae)

Yucca is the Carib Indian name for manihot (*Cassava*), erroneously chosen for this genus by Gerarde. The genus comprises of 40 hardy and tender evergreen shrub and tree species from the deserts of southern USA, Mexico and West Indies, out of which many can be grown indoors. These differ from *Agave* in having perianth segments and stamens which are inserted at the base instead of the top of the ovary, as in *Agave*. They are long-lived and thrive well in seaside gardens in poor sandy soil. These may be stemless or nearly so or with erect and robust stems which may sometimes be branched nominally. Long, strap- to sword-shaped, fibrous and pointed (usually spine-tipped) leaves form a rosette on top of a stout trunk. White to cream and bell-shaped, usually pendent flowers which open at night with unusual fragrance, appear on stout stems in large and usually erect panicles, generally overtopping the leaves. Pollination rarely occurs and that too through the small white moth, *Pronuba yuccasella* as its larvae feed on the maturing seeds of these plants. Their fruits are erect capsule. These are propagated from stem cuttings, rhizome cuttings, offsets and seeds. ***Y. aloifolia*** (*Y. crenulata, Y. serrulata*) from S. USA, Mexico and West Indies becomes a small tree in the wild though is slow-growing and can easily be maintained up to 2 m in height indoors. Stems are erect, robust and unbranched or sparingly branched, bearing 60-75 cm long, sword-shaped, rigid and spine-tipped leaves with finely toothed margins. Though containerized plants flower rarely, otherwise up to 60 cm long panicles bear white or purple-flushed flowers some 8-10 cm wide. The leaves in the varieties such as 'Marginata' are yellow-margined while in 'Tricolor' ('Quadricolor') the stripe is broadly cream or white centrally but at younger stage the variegation is suffused with pink. *Y. a. draconis* produces free-branching stems and curved but pliable leaves. ***Y. australis*** (*Y. baccata australis, Y. filifera*) from EC Mexico is a large tree where leaves are smooth, some 2.5 cm wide and almost flat with short coarse threads. Inflorescence is up to 1.8 m long and hanging with small flowers and fruits. ***Y. baccata*** from S. Colorado to New Mexico and Nevada is a nearly acaulescent plant with rough, concave and some 5 cm wide leaves having coarse threads on the margins. Erect inflorescence grows some 90 cm tall bearing very large (7.5 cm long) flowers, followed by sometimes up to 20 cm long fruits. ***Y. brevifolia*** (Joshua tree) from California to Utah (W. USA) is a tree in the wild bearing many crooked branches which terminate in bunches of dagger-like green leaves. Leaves are dark green, some 35 cm long, stiff, narrow, spine-tipped and minutely toothed. When in the wild though not in containers, it bears some 50 cm long panicles bearing about 7 cm long flowers which are green, greenish-yellow or cream. *Y. b. brevifolia* is a slow-growing which branches out when has attained 2-3 m height. *Y. b. herbertii* branches at ground level and grows up to 5 m high. ***Y. desmetiana*** probably from Mexico is caulescent and some 1.8 m high, bearing purplish, short, stiffly recurved, sometimes may be basally rough-edged and some 2.5-3.75 cm wide leaves. It is a quite less known species. ***Y. elata*** (*Y. angustifolia elata, Y. angustifolia radiosa, Y. radiosa*; Palmella/Soap tree) from S. Arizona, W. Texas and Mexico grows up to 6 m tall but indoors it can be kept up to 2 m in a containers. Its usually unbranched stems are erect and strong, bearing pale-grey-green, linear, some 90 cm long and 0.6-1.3 cm wide leaves with thread-bearing narrow white margins. In the wild this bears large panicles some 2-3 m tall with white to green and often pink-tinted flowers some 5 cm long. ***Y. elephantipes*** (*Y. ghiesbreghtii, Y. gigantea, Y. guatemalensis, Y. lenneana, Y. mooreana, Y. roezlii, Dracaena ehrenbergii, D. fentelmanni, Y. lenneana, D. lennei, D. yuccoides*; Spineless yucca) from Mexico grows up to 10 m in the wild but its height can be tamed up to 2 m for quite some time indoors in containers for displaying its graceful foliage. Its normally unbranched stems especially when young are erect and robust with spineless but rough-edged, linear and up to 1.2 m long and some 10 cm wide leaves which form a compact rosette. Non-containerized plants display large panicles of white flowers. *Y. e.* 'Variegata' leaves are margined white. ***Y. filamentosa*** from SE USA is almost stemless and clump-forming foliage plant growing up to 75 cm high, bearing mid-green, stiffly erect, narrow and somewhat glaucous leaves up to 75 cm long (nearly half the length when grown in pots) where the margins are decorated with many curly threads. Plume-like panicles are 1.0-3.0 m tall, and bear white to cream bell-shaped flowers 5.0-7.5 cm long in summer but only when plants have become 2-3 years old. *Y. karlsruhensis* was developed by crossing *Y. filamentosa* × *Y. glauca* and *Y. rekowskiana* by *Y. filamentosa* × *Y. gloriosa*. ***Y. flaccida*** from SE USA is similar to *Y. filamentosa* but the leaves are tapered, more flaccid, quite flexible and recurving with straight thread on the margins. ***Y. glauca*** from S. Dakota to New Mexico (USA) is a stemless and clump-forming plant with linear leaves some 60 cm long and 1.0-1.5 cm wide and having only a few threads along the margins. Its inflorescence rises to about 1 m tall during summer bearing about 6 cm long greenish-cream flowers with a reddish-brown tinge. *Y. draco* was developed by crossing *Y. flaccida* × *Y. aloifolia*. ***Y. gloriosa*** (Spanish dagger) from SE USA (coastal areas from North Carolina to Florida) is a some 2 m tall trunk-forming species bearing dense rosette of deep green, thick and stiffly erect leaves up to 75 cm long at the top of a slow-growing woody trunk. Floral stems some 2 m tall appear during autumn in dense and erect panicles, bearing 6-10 cm long cream-white bell-shaped flowers some 7.5 cm long often with a red tinge outside. The species does not flower until it is about five years old. ***Y. recurvifolia*** (*Y. pendula, Y. recurva*) from SE USA (coastal regions of Georgia, Alabama and Mississippi) is a short-stemmed plant growing up to 1.8 m tall and is very closely allied to *Y. gloriosa* though comparatively the leaves are less stiff and narrower and more arching than *Y. gloriosa*. This hardy species was introduced into Britain in 1794. Its gracefully recurved leaves (curving being increased with age) are green, quite thin, about 5 cm wide and flat on the upper surface for most of their length. Abundantly branched pyramidal panicles up to 1.8 m high on the inflorescence are loosely set with bell-shaped cream-white flowers some 5.0-7.5 cm long during autumn and the fruit is erect. It seldom flowers until the woody trunk attains 90 cm height. Its forms *variegata* has median yellow leaf variegation though *elegans* has a reddish stripe. ***Y. treculeana*** (*Y. agavoides, Y. argospatha, Y. contoria, Y. cornuta, Y. dispera, Y. longifolia, Y. vandervinniana*) from Texas to E. Mexico is a small tree which bears rough, very concave and 2.5-5.0 cm wide leaves, at first entire or little denticulate but afterwards becoming slightly filiferous. Inflorescence bears large bracts below. Variety *canaliculata* (*Y. canaliculata, Y. revoluta, Y. undulata*) is a form having broader leaves. ***Y. whipplei*** (*Hesperoyucca* *whipplei*; Our Lord's candle) from lower California and Mexico is a spectacular plant having short stem to only 30 cm high, topped with a thick rosette of numerous glaucous dark green and linear leaves that are stiff, some 45-60 cm long with brownish and yellowish finely-toothed margins and spiny tips. Floral stem is about 4 m in height, half above being the tapered cylindrical panicle that bears fragrant, pale-cream (occasionally flushed purple) and some 6 cm long flowers. The plant dies after the flowering. However, in case of *Y. w. caespitosa* the plant does not die after flowering and it continues producing offsets even from an early age. *Y. w. intermedia* requires several years to form a good rosette worth flowering though growth of inflorescence is quite rapid. This produces only a few offsets after flowering.

Zamia [Zamiaceae (Cycladaceae)]

Zamia derives from *zamie* a mis-rendering of '*azaniae*' 'pure cones' in the original texts of Pliny the Elder (A.D. 23-79), the Roman soldier, observer and writer. The genus comprises some 40 woody-based evergreen perennial species native to tropical and sub-tropical America and West Indies. They originate from the primitive cycad group, one of the earliest families of Gymnosperms. Zamias are either trunkless or bear only the short ones (caudex), not like most other cycad genera. They resemble, for some respects to the ferns while for some other respects palms. Plants are utterly slow-growing with caudex usually a low trunk aboveground or below, simply lobed or branched. Leaves developing one after the other, are evergreen, stiff, leathery and long pinnatifid, parallel-veined but without midrib, entire or serrate, and broad or narrow pinnae which are articulate at base, and the petiole is smooth or spinulose. Cones are small, glabrous (sometimes scurfy), flowers are dioecious with peltate and not-horned floral scales, anthers are numerous and ovoid *cum* sessile ovules in pairs and pendulous, cones oblong-cylindric but males smaller and thinner than females. These normally make the good house plants. They are propagated by seeds and offsets, *vis-à-vis* division when there is more than one crown. ***Z. angustifolia*** from Bahamas and Cuba bears short and almost round stems with up to 90 cm long leaves that are glabrous at maturity, elongated leaflets narrowly linear and usually alternate though base not narrow, 12-30 pairs, 12.5-25.0 cm long, 1.25 cm wide, 6-8-nerved, and apex obtuse and serrulate or entire. Pistillate cones are obtuse but cuspidate. ***Z. debilis*** from West Indies bears cylindrical trunk some 15 cm high, with 75 cm long leaves where narrowly lanceolate leaflets are in 12-25 pairs. ***Z. fischeri*** from Mexico is a very graceful compact plant, bearing up to 20 cm high and cylindrical trunk, 50 cm long leaves with short stalks, and leaflets alternate, lanceolate and up to 20 pairs with 12.5 × 2.5 cm in size. ***Z. floridana*** (Comptie/Coontie/Seminole bread) from Florida bears tuber-like trunk, often partially or wholly underground, bearing 60 cm long leaf including 10 cm long stalk, triangular petiole in outline, sericeo-tomentose at base and scarcely hairy at upper surface, and the leaves are upright, glossy deep green and ovate or ovate-lanceolate with dense and mostly opposite and very stiff leaflets in 14-24 pairs that have linear, 9-15 cm long and about 6 mm wide, falcate, reflexed, glabrous above, scarcely hairy below, apex obtuse, narrowing at base, and margins revolute with few obscure teeth, the male flowering cones are 4-6 cm long and the pistillate cones at maturity are oblong, 12-17 cm long with projected scales and densely tomentose. Local Indian in USA use its macerated roots as soap. *Z. f. portoricensis* is taller with widely spaced leaflets. ***Z. furfuracea*** (Florida arrowroot) from Florida, West Indies and Mexico bears cylindrical trunks either underground or partially exposed, growing up to 60 cm high with dilated, basally concave and small-prickly petioles. Prickly stalk and leaf together are some 60-120 cm long. Leaves comprise of 10-13 pairs of very close, opposite or alternate leaflets, glossy green above, lower brown-scurfy, oblanceolate, acute or obtuse, and upper half serrate or jagged. Cones are pale-yellowish-brown, pedunculate, downy, oval-conical, males 10 cm long and often in small clusters and pistillate cones broader and somewhat longer. ***Z. integrifolia*** from Florida and West Indies bears erect, globular or oblong trunks some 30-45 cm high, stalks angled, leaves glabrous with 7-18 pairs of alternate, oblong to linear-lanceolate to lanceolate, mostly obtuse, entire or dentate towards the apex, and cones short-peduncled, oblong and obtuse. ***Z. kickxii*** from Cuba is probably a variety of *Z. pygmaea* with a very short trunk, leaves up to 38 cm long with 10-15 pairs of short spathulate to nearly wedge-shaped leaflets some 5 cm long and 2.5 cm wide. ***Z. attonis*** from Cuba is similar to *Z. kickxii* with under-surface of leaves being brownish-fuzzy. ***Z. lindenii*** from Ecuador bears cylindrical trunk some 30-60 cm high with dilated, basally concacve, prickly and little woolly petioles. Opposite leaflet pairs 20 or more, sessile, glabrous or puberulous, long lanceolate and acuminate and are toothed towards the top. ***Z. pseudoparasitica*** (*Z. roezlii*) from Panama bears cylindrical trunk with leaflets that are entire, glabrous, lanceolate, sinuose-falcate, basally acute and cuspidate at apex and with 18 prominent nerves which bifurcate twice. ***Z. pumila*** from Florida, Mexico and West Indies bears 15 cm high caudex, finely toothed stalks, 1.2 m long leaves and 15 pairs of short (15 cm long), variable, lanceolate and broad leaflets that are little reflexed and toothed towards the tips. The cones are non-umbonate and the seed-bearing scales are thin and more flattened at the outer end. ***Z. pygmaea*** from Cuba is smallest of the zamias with 10-25 cm long leaves and 8 pairs of lanceolate leaflets some 5 cm long and 2.5 cm wide.

Zamioculcas zamifolia (*Z. loddigesii, Caladium zamaefolium*; Cast iron plant; Araceae)

Zamioculcas is made of two words, *Zamia* and *Culcas*, adapted by Linnaeus for the genus. A native of tropical Africa, the genus comprises only one species which has pinnatifid leaves, an exception to other genera of the family Araceae. It can survive at a minimum temperature of 10 °C, relative humidity above 40 per cent and prefers partial sun to shaded conditions. Propagation is by division, tubers and from leaf cuttings . ***Z. zamifolia*** is an evergreen perennial herb with strong creeping rootstock having glabrous and radical leaves which arise directly from the runners and stand erect to 60-90 cm in length. Leaf stalk is stout, much thickened towards the base and cylindrical. Leaves bear about 12 usually opposite but sometimes even alternate, very leathery, elliptical and acute leaflets some 15 cm long that are jointed to the petioles, and which after falling form a small tuber at the base. Green and glabrous spathe is convolute at the base either expanded or with reflexed blade, and the clavate spadix which is 2.5-3.75 cm long and 1.25 cm in diameter bears white flowers with female flowers in the lower part while the male above in the larger part. Also a small form named as *Z. lanceolata* is described but, in fact, it is a form of above with smaller leaves and shorter stalks. These are very care-free indoor plants

Zebrina pendula (Silvery inch plant/Wandering Jew; Commelinaceae)

Zebrina is named so due to animal zebra's stripes similar to this plant. The genus originates from Mexico and Guatemala with 2-5 decumbent or trailing perennial species, though only one is generally grown. *Zebrina* is closely related to *Tradescantia*. The leaves are generally glistening and multicoloured above though purple below, clasping the stem, alternate, and ovate to oblong. The flowers are small, 3-petalled and purple, fused at the base to form a tube and subtended by a pair of leaf-like bracts. It is excellent shade plant as ground cover and for hanging basket. Easily multiplied by cuttings. ***Z. pendula*** (Silvery inch plant/Wandering Jew) from Mexico is prostrate or trailing species growing to 90 cm long with node-rooting, bearing stem-clasping green leaves, longitudinally striped with two broad silvery bands above, and red-purple beneath, 2.5-7.5 cm long and narrow to broadly lanceolate. Small rose-purple flowers some 1 cm across appear from two dissimilar bracts. 'Daniel's Hybrid' bears thick-textured purple leaves. 'Discolor' (Tricolor inch plant) bears crystalline coppery-green leaves suffused purplish and overlaid by two narrow silvery bands and edged purple. 'Discolor Atropurpurea' bears purple leaves with delicate silver strips. 'Discolor Multicolor' leaves below are purple, above green striped lighter pink and irregularly placed rusty-red spots. 'Minima' leaves to its half the length are same as *Z. pendula*. 'Purpusii' (bronze inch plant) bears green leaves flushed purplish and lavender-pink flowers. 'Quadricolor' leaves are irregularly banded with pink, red and white.

Cultural Practices

Foliage plants come from diverse habitats and are of diverse nature, *viz.* trees, shrubs, sub-shrubs, climbers, annuals, biennials, perennials, herbaceous, cacti, other succulents, aroids, bromeliads, gesneriads, bamboos, palms, grasses. ferns, selaginellas, bonsais, *etc.* and that too requiring full sun, filtered situations, shady places, cold or warm conditions and so on so that way their cultural requirements also differ to a great extent. An excellent specimen brought from anywhere if is not given required growing condition, becomes undecorative and sickly within a season or two. Therefore it is necessary to grow them in proper medium and conditions to enjoy maximum decorative value for years together. For this, the choice of pots, growing medium, nutrition, top dressing, watering, foliage sprinkling, repotting, trimming, *etc.* should be properly chosen and executed. If any plant shows sickly signs, immediately that should be cured. Moreover, larger plants or trees which are grown inside the house require to be kept at manageable height through potting, pruning and regular trimming. Also such plants should all along be grown in the nursery so that if growing beyond the manageable height, these are easily replaced.

Potting Mixture

Basically the plants are of two types, epiphytic and terrestrial. Epiphytes are perched on spongy masses of decomposed vegetation though terrestrials which are grown directly in the soil have advantage as every year the soil is fortified with additional nutrients. To grow epiphytes directly into the soil, we will have to create those natural conditions by addition of very fibrous material such as **peat** combined with perlite and vermiculite in different proportion to satisfy all sorts of plants. Such soils are light, aerated and moisture-retentive. These soils can also be mixed with coarse sand, good garden loam, and well rotten compost or farmyard manure. For making artificial soil, peat moss is the first ingredient. Sphagnum peat is formed through decomposition of sphagnum moss found in the marshy places. Properly decomposed peat moss is brown, fibrous and has capacity to absorb water multifold. Its pH value is quite acidic and the nutritive value is very low. Sedge-peat is formed of the decomposition of sedge plants of the family Cyperaceae but it is still poor in nutrition.

Vermiculite is a transformed form of mica (a silica crystal) which is extremely thin and transparent sheets which when subjected to high temperatures, the books exfoliate and the high grade horticultural mica is formed which is chemically neutral and looks like small, silvery, crisp and brittle cubes or chips consisting of many small sheets. Though it does not absorb water but its many surfaces hold the water due to surface tension.

In commercial mixes, the **bark** is also found, especially as replacement to perlite, as this is light and less water absorbing so most suitable for succulents. **Perlite** is formed by subjecting obsidian (a volcanic stone) to intense heat which causes it to become granular, gritty, firm, light and pure white which has various grades on the basis of its size. It is a substitute to sand but is light in weight, permits more aeration and becomes more water -aborbing. **Pumice** is a crushed lava stone which is very light though slightly heavier than perlite, porous, firm and gritty and much better over perlite so is used as a better substitute to perlite but is little costly. **Sand** of horticultural grade is clean, gritty, sharp and coarse but heavy, and evaporates moisture too quickly. It is a substitute for perlite and for succulents it is a necessary ingredient.

Sterilized sandy loam garden **top-soil** is another ingredient to be added in the peat mix, less succulent plants requiring more peat and less top soil. A standard mix contains 1 part peat and 4 parts top soil.

Where rooting is required, *i.e.* for soft cuttings **live sphagnum moss** should be collected and used. The fresh one survives for about six months if kept properly moist in a container. It is quite good in a mix in case of aroids. **Long-fibre sphagnum** is a partly cleaned and dried moss containing long strands of the plant and is used as to that of live sphagnum moss. This moss permits excellent aeration with a high water absorption, to some extent resists the invasion of fungi and is excellent for many of the aroids. **Milled sphagnum moss** is a dried and finely dust-ground moss and has same resistance to fungi as to that of long-fibre sphagnum moss, therefore, it is an excellent medium for seeding and for rooting of cuttings as there is no root damage while lifting the rooted cuttings as it clings to the roots and root hairs. Before use, it should be kept in a container with water and covered for whole night to absorb maximum moisture. It is a substitute for peat in the mixes. Use of moss in the medium makes the mix acidic so use of **dolomite lime** in the form of horticultural grade chips and powder becomes necessary. Crushed egg shells are also excellent source of lime. In one litre of acidic medium, the lime should be added @ 15-30 g.

Potting mixture can be made as per requirement of the crop. Sphagnum peat moss, vermiculite and gritty material (sand/perlite/pumice) should be mixed together in a large container as per the specified ratios. This is now kept in a smaller container, poured with water to ¼th of the volume of the mixture and then it is allowed to stand overnight so that moisture permeates the mixture properly. Now in different ratios (by volume) for different types of plants this mixture is prepared and designated by numbers (Virginie and Elbert, 1989), such as:

House Plant Mix # 1

It is 1-1-1, that is 1 part peat moss, 1 part vermiculite and 1 part perlite. This mixture is suitable for nurseries for small and juvenile tropical plants, and is an excellent medium for first plantings of seedlings or cuttings, for succulents and for epiphytes.

House Plant Mix # 2

It is 2-1-1, that is 2 parts peat moss, 1 part vermiculite and 1 part perlite, which is best for non-woody foliage plants.

House Plant Mix # 3

It is 3-2-1, that is 3 parts sphagnum moss, 2 parts vermiculite and 1 part perlite, which is best for woody plants such as trees and shrubs and most vines.

Cacti & Succulent Mix # 1

It is 1-1-2, that is 1 part sphagnum moss, 1 part vermiculite, 2 parts perlite/pumice/gritty sand and when mixture is having pH of 6 or less, lime is also added.

Cacti & Succulent Mix # 2. It consists of 2 parts sterilized garden soil, 1 part peat or milled sphagnum moss and 1 part gritty sand. If pH of the soil is 6 or less, lime should be added.

Pots, Potting and Repotting

Earthen (clay) pots with bottom holes are most common and best for plants as these permit aeration through, damaging salts are leached down from the compost, waterlogging is less likely to occur due to these being porous in nature, adjust temperature variation during summer or winter, and these pots being heavy normally do not topple over during windy conditions if plant is proportionally not too tall. Plastic pots do not permit aeration through the body and become very hot during summer and very cold during winter which is highly injurious to the plants, though these may be colourful in many shades to choose from, do not require crocking as have many bottom holes, water loss is less as there is no evaporation in the sides and are light weight. For very heavy plants, large earthen tubs or cemented pots are available which can be used. There are self-watering plastic pots also especially for windowsills, or for floor-standing tub to house a large indoor garden. These pots work on two-pot principle, one which contains mixture or compost along with the plant and the other below is filled with water and nutrients and topped up through a filler tube from which the water or liquid feed is added and this reaches to the plant above through capillary action. During winter the compost becomes too wet in this system.

While **potting** either in September-October or in February-March, the pots should be absolutely clean or fresh. If old pots are used, these should be cleaned properly with hot water and potassium permanganate solution then sun-dried and then filling should be done. A few clean crocks at the bottom should be kept, one above the other, so that it permits the leaching of the excess water properly. Above this, 3-5 cm layer of gravel or coarse sand is filled and then above this potting compost but leaving some 2.5 cm space at the rim. This soil should be firmly pressed with fingers and in the centre the plants are installed by spreading the roots and then the soil around the plant is gently and firmly pressed and watered immediately. The planted pots are stacked at a filtered place for a few days and when these plants have become quite normal, should be taken to their desired places. **Repotting** is the process of planting the pot-bound plants after dividing and pruning their roots and this process is also followed when there is transition in weather, *i.e.* September-October or February-March. For successful repotting, Hessayon (1987) has given seven steps:

1. Already used pots should be thoroughly scrubbed out, and if it is the new clay pot it should be soaked in water overnight
2. In case of clay pots, drainage holes are covered with crocks, *i.e.* broken pieces of pots and bricks and then over it a shallow layer of potting compost
3. The plant should be watered and one hour later the plant should be removed from the pot by spreading the left hand fingers over the soil surface invert and then rim of the pot should be gently knocked on the edge of a table and then removing the pot with right hand
4. From the ball the old crocks are removed along with root mats outside and the rotten roots but not all the roots, however, nearly $^{2}/_{3}{}^{rd}$ of the roots are removed and the earthball is loosened
5. Now plant is placed on the top of the compost layer in the new pot and then moist mixture is filled around the soil ball
6. Compost is now firmed up with thumb and fingers up to the base of the stem
7. Now the pots should be watered carefully and placed at a shaded place for about a week where the leaves should be misted daily to avoid wilting and then take to its growing quarters

Light

Unlike the human beings, the plants manufacture their own food in presence of light and CO_2. A few plants manufacture their food in very little or almost no light no darkness, some others in diffused light while many others under bright light. Some require only momentary light, some others for a few hours daily, some others 6-8 hours while many others more than eight hours daily. **Direct sunlight** means 100 per cent as in case of south-facing windows throughout most of the day but east-, south-east, south-west or west-facing windows receive the light only for a few hours. **Bright filtered light** means 60-70 per cent light falling on the object, being filtererd through a translucent blind, plastic venetial blind, curtain or through a leafy tree. Lower latitude sunlight is as very strong, so should be filtered through thicker curtains otherwise plants may be burnt or scorched. **Bright light** means 20-25 per cent light which is available in a quite sunny room. **Medium light** means 9-10 per cent of the total light as close to north-facing window, or obstructed lighting to east or west-facing windows where building or trees deflect the sun's rays. **Poor light** means 3-5 per cent light as in case of spots not facing the windows. Moreover, when compared the too little light is less dangerous than too much of light, as in the former case the results appear after certain period of time but in latter the harmful effect is seen at once. High light intensities are not required by most potted plants, apart from cacti and those plants grown for their flowers. Plants greatly differ in their light requirement. Foliage plants are not exception so keeping them in the house does not mean that for whole of the year these will continue giving enchantment there. In 3-4 days or at the most fortnightly these plants should be taken out gradually under filtered to well-lit situations, and in turn, others kept outside should be brought in and this cycle will go on throughout. Foliage plants may require bright light though not direct sunlight. Plants with variegated or purple leaves require more bright light to retain their striking colourations than entirely green ones but in case they do not get it for a few weeks or are kept in medium light they will not fade irrevocably. During active growth and flowering the right type of light should essentially be provided. Flowering plants can tolerate direct sunlight to a great extent. Generally, the plants living at ground level in tropical rain forests flourish well in shade while cacti in open desert are fully exposed to sun hence require full sunshine. Likewise, the shade-loving ferns will not survive when kept under open sun or when sun-loving pelargonium is kept in a dark corner. Shade loving plants such as *Aglaonema, Anthurium,* bromeliads, *Caladium, Calathea, Cotyledon, Dieffenbachia,* ferns, *Fittonia, Gynura, Hedera, Helxine, Hemigraphis, Haworthia, Maranta, Peperomia, Philodendron, Plectranthus, Saxifraga, Sedum, Sonerila, Tolmiea, etc.* are less tolerant of the wrong kind of light albeit they can not survive in the dark or the bright to direct sunlight. Plant kept in less lighted area as in a room, their leaves will turn towards the more lighted area, *i.e.* towards the window so such plants should be turned up regularly but there are also exceptions such as stiff-leaved plants such as *Sansevieria* and rosette-forming bromeliads whose leaves do not turn towards the light source. No turning of plants is required during flowering. During winter the natural light duration and intensity inside the room can be supplemented through artificial lighting not by the bulbs but through fluorescent tubes mounted under a reflector. Generally, it is 20 watts per 30 cm^2 area. Response of plants with compact growth and colourful leaves such as begonia, bromeliads, cineraria, gloxinia, orchids, peperomia and saintpaulia is quite encouraging.

Temperature

Foliage plants, especially the ones growing indoors require fairly constant and moderate temperature during growing period and a lower temperature during the resting season. At quite humid location plants can tolerate constant temperature as high as 30 °C but under ordinary humidity conditions 24 °C. Tender house plants such as *Aglaonema, Caladium, Dieffenbachia, Dizygotheca* and *Syngonium* tolerate a minimum constant temperature of 15 °C, non-hardy ones such as *Aphelandra, Araucaria, Asparagus, Begonia, Beloperone,* bromeliads, *Coleus, Dracaena,* ferns, *Ficus, Gynura, Hoya, Kalanchoe, Maranta, Monstera,* orchids, palms, *Pandanus, Peperomia, Philodendron, Pilea, Rhoeo, Sansevieria, Schefflera, Scindapsus, Spathiphyllum, etc.* tolerate a constant minimum temperature of 10-13 °C and the hardy plants such as *Aspidistra, Chlorophytum, Fatshedera, Fatsia, Grevillea, Hedera, Helxine, Laurus, Pelargonium, Saxifraga, Tradescantia, Yucca,* and various succulents and vines tolerate 4.5-7.2 °C constant minimum temperatures. Most house plants tolerate 18-24 °C temperatures in the temperate zones with a 5-10 °C lower in the night, however, the constant great variation is harmful. During summer, 27-32 °C will not be harmful to most of the house plants if humidity level is high. During highly cold nights, certain plants kept near the windows may chill out so during nights such windows should be closed with heavy curtains to insulate the room. Likewise, the plants kept near the heating ducts or radiators are liable to be leaf-scorched due to hot and dry air.

Humidity

Humidity is the relative amount of water vapour contained in the air hence it is called relative humidity which is 0-100 per cent (0 means absolutely dry air and 100 means full saturation of air with humidity). Air humidity has no relation with the moisture present in the potting mixture. Fog indicates the air is fully saturated. This is measured through hygrometer. The leaves have innumerable tiny stomatal pores through which plants absorb vital gases from the atmosphere and with opening of the stomata there is moisture loss which is known as transpiration and this water loss affects the plant. In case the air is sufficiently humid, it can not accommodate more of moisture further and thus transpiration rate diminishes. The imbalance of external humidity will cause leaf shrivelling and drying, falling of the buds and premature withering of flowers. A 40 per cent relative humidity is the absolute requirement for most plants including cacti. Species introduced from tropical rain forests require more humid atmosphere and that is 60 per cent. Thick-leaved plants stand better in dry air than those plants having papery leaves. The amount of moisture that the air should contain to maintain a given level of humidity increases with the rise in temperature and rise of temperature from 10 °C to 21 °C will have to double the moisture level in the atmosphere to attain the same percentage. That is why, during hot weathers, growers spray a fine mist of water daily, preferably in the morning to their indoor plants though this has only a short-lived effect. Alternatively, the most effective way to ensure the build up of air humidity around the plant is to fill a waterproof tray or dish with 5-15 cm deep gravel or small pebbles along with water slightly below the level of the grave, over which the pot with plant should be stood. The tray should be as wide as spread of the plant so that whatever water vapour rises above should benefit the leaves of the plant. Care should be taken that dish water should not dry up. Grouping of plants ensures increase of moisture in the air due to vaporization of the moisture present in the potting mixture, *vis-à-vis* transpiration by the foliage of the surrounding plants and so a humid micro-atmosphere around the plants is formed though this situation normally does not occur in case of isolated plants.

Watering

Water used for watering the plants should not be cold as it checks the growth and spots the foliage. It should be lukewarm or at room temperature and soft, *i.e.* lime-free. If convenient, better the collection of rain water should be made and used for watering the plants or the distilled or demineralized water. Lightweight watering can having long and thin spout is best to water the plants. It would be advisable to give water near the stem so that danger of *Botrytis* infection during winter is gone. Also there is a system of self-watering where a simple water reservoir is placed below the potted plant so as per the need, through capillary action, the potting mixture sucks the needed water. Another simple method of watering the plant is putting of the pots of the plants for a while in water-filled tubs slightly below the rim and this way the necessary water will be sucked by the potting mixture and when the job is over the water bubbles are seen on the top surface of the mixture. Desert plants, cacti, other succulents, *etc.* require less watering than others. Marsh-dwelling rushes and many other plants require more watering. Many plants such as *Calathea, Chlorophytum, Cordyline, Dracaena, Ficus elastica, Fittonia, Maranta, Pilea, Pleomele, etc.* require moderate watering.

Nutrition

During rest, plants do not require any nutrient but when in growth or flowering these require to be fed. Nitrogen (N), phosphorus (P) and potash (K) are the major nutrients where nitrogen keeps the leaves performing properly, phosphorus the roots and potash the flowers. Nitrogen in the form of nitrate helps in building up the stems, branches, leaves and its energy-making green colour, *i.e.* chlorophyll. It also facilitates metabolism of other elements. Its excess causes plants becoming lanky with weak growth, sappy and prone to attack of various diseases and insects. Phosphorus is given to the plants as either phosphoric acid or phosphate and encourages root formation and is more helpful in bud formation and flowering. Potassium in the form of potash helps plants becoming sturdy, helps the plants for the production of flowers and fruits and brings resistance against diseases but its metabolism is helped when other elements are also present amply. Calcium (Ca), magnesium (Mg), sulphur (S), iron (Fe), manganese (Mn), zinc (Zn), copper (Cu), boron (B), molybdenum (Mo), chlorine (Cl) and nickel (Ni) are recognized universally as essential elements for diverse group of organisms, *viz.* algae, bacteria, fungi and green plants. Cobalt (Co) has been established to be essential for leguminous plants only. Those required in large or macro amounts are known as **macronutrients** such as C, H, O, N, P, K, Ca, Mg and S, and out of these C, H and O in the plants are absorbed from air and water to a tune of 90-95 per cent; N, P and K are the primary nutrients and Ca, Mg and S secondary nutrients. Those required by the plants in small or micro-amount (<500 mg/kg dry matter except Fe and Mn) are termed as **micronutrients** such as Fe, Mn, Zn, Cu, B, Mo, Cl and Ni. These are further divided into **micronutrient cations** (Fe, Mn, Zn, Cu and Ni) and **micronutrient anions** (B, Mo, and Cl). Calcium is instrumental in cell wall formation and is present in nearly neutral soil. Its addition as lime reduces acidity. Foliage plants for greenness of the foliage require ample nitrogen though its excess is bad. Farmyard manure (FYM) or compost has almost all the nutrients required by the plants so for field crops it is incorporated in the soil at the rate of 500 quintals per hectare. This would be sufficient for one year to the crop and then this will not require any other nutrient. However, in most of the cases for pot-filling 1 part rich garden soil, 2 parts FYM and 1 part each of vermiculite and perlite (all by volume) will be sufficient for luxurious growth of the plant for one year, and then every subsequent year 5.0-7.5 cm thick layer of FYM may further be incorporated in the potting mixture for better results. In case repotting is done, the whole mixture will have to be substituted afresh. In case of heavy-feeder plants, slow-release fertilizer such as Osmocote can be used which will be effective up to six months. Weekly, when plants are taken out, these may be fed with slurry during their growth period only.

Ventilation

Unlike the human beings, plants do not have to be provided with air to enable them to breathe. Some plants grow quite happily in sealed glass containers or in wardian cases but still majority of the plants require fresh air though green leaves themselves manufacture food. A change of air lowers the temperature during hot weather, since moist conditions encourage *Botrytis* infection so the fresh air lowers the humidity levels especially in case where these are overcrowded, strengthens the stems and increases disease and pest resistance, and removes the traces of toxic vapours. Stagnating air creates congenial atmosphere for development and spread of pathogens and therefore sometimes stored plant materials rot so rapidly. However, fresh air draws the excess humidity out, and helps in drying the plant stacked in the room so disease development is minimized. Fresh air can be provided by opening the door and/or windows especially during summer to most of the house plants such as *Araucaria,* cacti, *Fatsia, Pelargonium,* succulents, *Tolmiea, etc.*

Grooming

Dust spoils the appearance of the foliage, blocks the leaf pores so hindrance in breathing of the plant is caused, forms the screen to block the light which weaken the plants, and in industrial areas this may contain harmful chemicals which may kill the plants. Therefore, the leaves should be cleaned by immersing the plants in a bucket of water, syringing and sponging. The plants are when taken out weekly or fortnightly for providing gradual sunlight, these should be jet-sprayed in the morning to clean the pores. Young leaves should be cleaned neither by jet-spraying nor with sponging but with syringing otherwise these may be damaged. However, in case of cacti and other succulents, *vis-à-vis* those having hairy leaves should be cleaned with soft brushes. With age, the leaves in some cases become dull and the new ones loose their glossy sheen so to get rid of these problems diluted vinegar, milk or beer can be used for getting back the glossy sheen. Olive oil may be used though this afterwards collects dust which causes damage. Sometimes aerosol sprays may be used though its frequent use is dangerous. Anything which is used for shining the leaves, that should be gently applied for wiping the leaves through a piece of cotton wool impregnated with liquid. New leaves should never be polished. Non-trailing foliage plants usually do not require **support** but there are a few such as *Dizygotheca* and *Fatshedera* where canes droop if not held up with some support. Support is usually a thin stake, generally of split bamboo, inserted in the centre of the potting mixture if it is one though in the sides when it is more than one, with which the stems or canes are loosely tied with garden twines or raffia to keep the plants erect and in shape. All the branches should not be tied together but individually one by one. For climbing plants, especially the ivies, various patterns can be woven around the stakes. Since *Philodendron* produces nodal roots as haustorium so a moss stick will be quite useful and this stick is kept moist throughout. *Cissus, Parthenocissus* and *Rhoicissus* are trained into an open rectangle. Various other types of supports, as per the convenience and to maintain proper ornamental value of the plants, can be used and these are wire frames with feet or stand in the form of trellises, hoops, coils, globes, pyramids and obelisks. To train a plant for giving it a definite shape, sometimes the plant parts are even sheared off and this is known as **pruning**. Pruning includes trimming, pinching & disbudding, and cutting back of excessive growth. **Trimming** is removal of dead leaves, damaged parts, faded flowers and clipping of some growth to bring the plant into proper shape. Many of the plants grow out in informal way if not cared well. From the very beginning the plants are to be shaped by regular pruning and trimming. Those plants giving abnormal growth with stems having abnormally small and pale leaves as in case of *Hedera* and *Philodendron* climbers when kept too warm in winter so such branches should be sheared away when winter ends. Diseased plant parts should also be removed and cuts should be treated with Bordeaux paste. No green shoots on variegated plants should be allowed to compete with others. Crowded stems should also be removed. Old leafless twigs and overgrown branches should also be removed. Dead-heading of the faded flowers will enhance flowering duration. While pruning conscience should be applied as some species produce flowers on old shoots as in case of *Hoya*, some others only on new shoots as *Pelargonium*. **Stopping** (pinching) is the practice to induce branching in many bushy and trailing plants such as *Buxus, Codiaeum, Coleus, Euonymus, Fatshedera, Ficus benjamina, Ficus pumila, Ficus radicans, Glechoma, Gynura, Hedera, Hemigraphis, Heptapleurum, Osmanthus, Peperomia, Pilea, Pisonia, Pseuderanthemum, Podocarpus, Polyscias, Radermachera, Tradescantia, Tetrastigma, Zebrina, etc.* This way plants become compact and bushy. In climbers the weak side shoots are removed and one or two main stems are allowed to grow. **Disbudding** is practiced in case of those plants where large flowers are desired.

Insect-Pests, Diseases and Physiological Disorders

Insect-Pests

Insect-pests infesting foliage plants are in plenty though those plants kept indoors are infested usually with tiny insects, mostly being sucking type and some chewing type. Sucking type insects such as mites and scale insects since suck the plant sap so yellowing and distortion of the leaves occur. Chewing insects bite off pieces of leaf edges, make holes in the leaves and feed on the stems and growing tips. Leaf-mining grubs tunnel the leaves eating into the plant tissues as they move. There are some insects whose larvae may infest on the underground plant parts which may result into plant wilting. Therefore, at the time of repotting the underground parts should be carefully examined for such infestation. The insects feeding on the foliage plants are therefore dealt here with.

Whole plant of certain indoor plants, shrubs, begonia and many bulbous plants are infested by **vine weevil larvae** (*Otiorhynchus sulcatus*) which measure some 1 cm long with little curve and brown head but no legs. In its infestation, the growth of plants becomes slow, and such plants afterwards collapse and die. **Fungus gnats**, sciarid flies or mushroom flies (*Bradysia* spp.) infest the plants being grown under cover, *vis-à-vis* seedlings, and cuttings fail to grow. Here greyish-brown flies may be seen running or flying over the surface or among the plants. Its white larvae (maggots) some 6 mm long with black heads may commonly be seen. Though this feeds on dead roots and leaves but may attack young roots. This is not a problem on established plants.

Root-knot nematodes, mainly *Meloidogyne* spp. exhibit swelling or galling on **roots and tubers** of many foliage plants, thereby plant growth is arrested, leaves curl and yellow and the plants are killed. **Chafer larvae** (summer chafer, *Amphimaleon solstitialis*; cock chafers, *Melolontha melolontha*; and garden chafers, *Phyllopertha horticola*) are 'C'-shaped, some 2 cm long with 3 pairs of legs and a brown head. These feed on the underground parts eating their bases so the young plnts die soon. They also infest on leaves, flowers and fruits of various ornamental trees and shrubs. Though damage by adult beetle is seldom but the larvae inhabiting in the soil can cause serious injury to roots, tubers, bulbs and stems of many ornamentals including lawn. **Cut worms** (*Agrotis* and *Noctua*) feed on underground parts of low-growing perennials. Their infestation causes extensive root damage, tuber cavities and ultimate death of the plants. The feeding caterpillars are usually creamy-brown and some 4.5 cm long. **Leatherjackets** (*Nephrotoma* spp. and *Tipula* spp.) are the legless larvae of crane flies (daddy-long legs) some 3.5 cm long with tubular and greyish-brown bodies which feed on roots and stems, and in the process the stems are severed from the roots so causes ultimate plant's death.

Underground structures such as **bulb, corm and rhizome** are infested by **bulb flies** (*Eumerus* spp. and *Merodon equestris*). The *Merodon* looks like humblebees and lays eggs singly on sound bulbs, larvae are up to 2 cm long and brownish-white and these feed inside the bulbs and fill the bulb centres with muddy excrement. *Eumerus* infests on injured bulbs, thereby either such bulbs do not grow or produce only a few grass-like leaves. **Bulb mite** (*Rhizogyphus echinopus*) is small pearly-white soil mite which often invades rotting tissues of bulbs as secondary pest. **Bulb scale mite** (*Steneotarsonemus laticeps*) feeds and breeds inside certain bulbs, causing rusted streaks, distortion and stunting on leaves and flowers. The **nematode** (*Ditylenchus dipsaci*) infests certain bulbs inside as well as foliage, causing bulb rotting, growth distortion and stunting. *Meloidogyne* nematode is very harmful to many of the foliage plants as their juveniles feed and colonize the roots in the form of knots.

Stem, branch and leaf-bud are infested with **scale insects** (*Aonidiella aurantii, Aspidiotus dictyospermi* var. *arecae, A. orientalis, A. transparens, Chionaspis* spp., *Iceria aegyptiaca, Iceria formicarum, Iceria purchase, Lindingaspis rossi*, and *Saissetia hemisphaericum*) which feed on the stems of indoor and greenhouse plants, cacti and other succulents, trees, shrubs, *etc*. Their infestation makes the stems sticky and excrete honeydew which entices sooty mould. Plants become stunted, show sick growth and sometimes die. **Froghoppers** or spittlebugs (*Philaenus spumarius*) is a creamy sap-sucking insect which secretes a protective froth over themselves. Adults are up to 4 mm long and darker in colour. Its infestation causes little distortion.

Leaf and floral problems are innumerable in ornamentals. **Slugs** (*Agriolimax reticulatus, Arion, Deroceras, Milax* spp., *etc.*) and **snails** (*Helix aspera, etc.*) feed either after rains or in the night on seedlings, herbaceous perennials, climbers, *etc.* by either completely battering the leaves or making holes in the leaves and stripping in the stems, leaving the silvery slime-trail on leaves, stems and on the soil surface. Stems are also damaged underground, leaving the large cavities inside. **Earwigs** (*Forficula auricularia*) are yellowish-brown and some 15 mm long insects with a pair of curved pincers. They are nocturnal and during daytime hide themselves. Both adult and young earwigs feed on young leaves and blooms of many ornamentals during summer. **Woodlice**, slaters or pill bugs (*Armadillium vulgare, Oniscus, Porcellio, etc.*) feed during night on seedlings and other soft growths including flowers and fruits. They are grey to brownish-grey, sometimes with white or yellow markings, hard with segmented bodies and up to 1 cm long. They normally eat on the decaying plant materials or on already damaged plants or their parts. Generally they invade near the shoot tips and cause holes. **Bugs** [Capsid bugs, tarnished plant bug or bishop bug (*Lygus rugulipennis*), the common green capsid (*Lygocoris pabulinus*), and the apple capsid (*Plesiocoris rugicollis*)] are green or brown, 6 mm long and suck plant sap from leaves, shoot tips and flowers of many ornamentals by making numerous small holes, *vis-a-vis* their toxic saliva kills plant tissues so the leaves are torn. Regarding their damage on the flowers, Capsid bug infestation causes uneven flower development with under-sized ray petals while in *Fuchsia*, flowers abort completely. Both young and adult rhododendron bug (*Stephanitis rhododendri*) punctures the lower surface of rhododendron leaves and feeds on the sap so fine light mottling on the upper surface, *vis-à-vis* rusty-brown or chocolate spotting on the lower surface may be seen. **Mealybugs** (*Pseudococcus* spp., *Planococcus* spp.) are soft-bodied, wingless, greyish-white and some 5 mm long, often with white and waxy filaments trailing from their bodies. These infest on indoor and greenhouse plants, cacti and other succulents and many other ornamentals. Their infestation causes appearance of a fluffy white substance in leaf and stem axils. They also excrete honeydew to entice sooty mould. Roots are also infested, and there are even root mealybugs (*Rhizoecus* spp.) which feed on roots, especially cacti and other succulents, causing discolouration and wilting. **Aphids** [bean aphid (*Aphis fabae*), beech aphid (*Phyllaphis fagi*), cabbage aphid (*Brevicoryne brassicae*), cherry blackfly (*Myzus cerasi*), cypress aphid (*Cinara cupressi*), *Aphis gossypii,* leaf curling plum aphid (*Brachycaudus helichrysi*), mottled arum aphid (*Aulacorthum circumflexum*), orchid aphid (*Cerataphis orchidearum*), peach-potato aphid (*Myzus persicae*), *Microsiphoniella sanborni,* spruce aphid (*Elatobium abietinum*), tulip bulb aphid (*Dysaphis tulipae*), waterlily aphid (*Rhopalosiphum nymphaeae*), willow stem aphid (*Tuberolachnus saligna*), woolly aphid (*Eriosoma lanigerum*), *etc.*] including those known as blackfly, greenfly or plant lice are up to 5 mm long, winged or wingless, and grey, green, yellow, pink, brown or black and the woolly beech aphid is covered in a fluffy-white wax. While feeding on tender parts of shoots, leaves, buds and flowers, the plants lose vigour as their attack is in large number, leaving sticky excretion to entice sooty mould. Some aphids also infest on roots and the underground parts, *etc.* They also transmit viral diseases. **Whiteflies** (*Aleyrodes proletella, Trialeurodes vaporariorum*) are 2 mm long white-winged insects which invade indoor and greenhouse plants. Their scale-like nymphs are immobile, flat or oval and whitish-green though adult flies are quite active and suck the cell sap so leaving sticky honeydew to attract sooty mould. These may also transmit viruses. **Thrips** (*Frankliniella occidentalis, Kakothrips pisivorus, Taeniothrips simplex* syn. *Thrips simplex, Thrips fuscipennis, Thrips tabaci, etc.*) nymphs and adults rasp the plant fluid leaving a silver-white discolouration on leaves with tiny black dots on the upper surface of leaves. They also feed the floral buds by sucking the sap and by entering inside during dry summers. They are brownish-black insects up to 2 mm long with narrow bodies which are sometimes banded pale, and survive well in hot and dry conditions. Nymphs resemble the adults except that they are pale-yellow-orange. These infest the indoor and greenhouse plants apart from many other plants. **Leaf miners** [celery leaf miner (*Euleia heracleii*), chrysanthemum leaf miner (*Phytomyza syngenesiae*), beech leaf mining weevil (*Rhynchaenus fagi*), the birch leaf mining sawfly (*Fenusa pusilla*), lilac leaf miner (*Caloptilia syringella*, a moth caterpillar which is highly abnoxious and damages by burrowing into syringa leaves, *vis-à-vis* leaf-mine blotches on ligustrum and remain active from June onwards), *etc.*] are larvae of various flies, moths and beetles which infest various foliage plants, annuals and perennials, chrysanthemum, trees and shrubs including *Ilex* and *Syringa* by mining their leaves and on herbaceous plants even their stems which can be witnessed by the visibility of the white or brown areas (linear, irregular or circular characteristic shape for the particular leaf miner). **Suckers** or psyllids [bay sucker (*Triozaalacris*), the box sucker (*Psylla buxi*)] infest perennials, trees including *Laurus nobilis* (bay), shrubs including *Buxus* (box) and climbers and cause distortion and stunting of the leaves throughout the summer. Infestation on new shoots of *Buxus* produces cabbage-like clusters of malformed leaves. *Laurus* leaf margins become yellow, thick and curled, though the box leaves are stunted. The nymphs are usually grey-green, small (2 mm long), winged and aphid-like. **Earwigs** (*Forficula auricularia*) is a 1.5 cm long yellowish-brown insect which eats away the petals. **Leafhoppers** [glasshouse leafhopper (*Hauptidia maroccana*), and the rhododendron leafhopper (*Graphocephala fennahi, G. coccinea*), *etc.*] infest on various ornamentals including foliage plants, and their infestation causes pale spotting on upper leaf surface. During spring, these are covered with black fungal bristles. Glasshousse leafhoppers are green or yellow, 2-3 mm long, wider-bodied behind the head and then tapering. **Leaf-cutting bees** (*Megachile* spp.) feed on many shrubs and trees during night by taking out circular to lozenge-shaped uniform cuts from the leaf margins to build their nests. These bees are 1 cm long with ginger-hairs on their abdomen beneath. **Bumble bees** (*Bombus* spp.) sometimes bite the corolla tubes or spurs of tropaeolum and make small holes. **Millipedes** (*Blaniulus, Brachydesmus, Cylindroiulus* spp., *etc.*) are black, brown, grey or creamy-white and feed at soil level or underground. The body is hard and segmented with two pairs of legs per segment though **centipedes** which are beneficial predators bear only one pair of legs. Spotted snake millipede (*Blaniulus guttulatus*) is slender bodied, creamy-white and some 2 cm long, bearing a row of red dots along each side. Millipedes have high population in the soil which has high organic matter. They infest on the seedlings and other soft growths though the damage is never alarming. Among the **caterpillars** [brown webber tail moth (*Euproctis chrysorrhoea*), lackey webber moth (*Malacosoma neustria*), small ermine webber moth (*Yponomeuta* spp.), juniper webber moth (*Dichomeris marginella*), the hawthorn webber moth (*Scythropia crataegella*), *etc.*], the first three have hairy and some 5 cm long bodies though the caterpillars of the last two species are up to 2 cm long. They feed on many shrubs including *Crataegus* and *Cotoneaster*, trees such as *Euonymus, Juniperus, Salix,* and many other garden plants. The caterpillars defoliate the shoots by covering the feeding area with dense greyish-white and silk webbing. Tortrix moth caterpillars (*Cacoecimorpha pronubana*) are dark green with brown heads measuring up to 2 cm long, and feed on leaves either by binding two leaves together, or one leaf folded and then bound in itself or the leaf attached in similar fashion with fruit while dwelling inside and feeding on the inner surface of leaves. When disturbed, these caterpillars wriggle backwards rapidly. Winter moth caterpillars (*Operophtera brumata*) are pale-green and some 2.5 cm long, and feed from bud burst to late spring on leaves, buds, flowers and fruitlets of many other deciduous trees. Cabbage caterpillars [large cabbage white butterfly (*Pieris brassicae*), small cabbage white butterfly (*P. rapae*) and cabbage moth (*Mamestra brassicae*)], the first one being yellow and black with distinct hairs, the second one is pale-green with velvety hairs and the caterpillar of the last one is yellowish-green to brown adorned with a few hairs. These feed on nasturtium, the members of Brassicaceae, certain annuals and perennials, *etc.* These feed on plants making holes in the foliage from late spring to early autumn and can be easily seen on leaves and flowers. **Beetles** [asparagus beetle (*Creoceris asparagi*), flea beetle (*Phyllotreta* spp.), pollen beetle (*Meligethes* spp.), poplar beetle (*Melosoma populi*), red lily beetle (*Lilioceris lilii*), viburnum beetle (*Pyrrhalta viburni*), waterlily beetle (*Galerucella nymphaeae*), willow beetle (*Phyllodecta vitellinae*), *etc.*] attack various ornamentals. Adult asparagus beetle is 7 mm long, having reddish thorax and yellow & black wing cases, its larvae are greyish-yellow, and both feed on asparagus stems and leaves from late spring to early autumn and from the feeding place the upper portion dries. Flea beetle is small (2 mm long) with enlarged hind legs for leaping, black to metallic blue and sometimes with a yellow stripe running down each wing case. Its feeding causes small pitting and holes on upper leaf surface which may cause plant death. It overwinters in plant debris. Pollen beetles are small (2 mm) and black, and feed on the pollen of many garden plants though other parts of the plant are not damaged. Poplar beetle larvae resemble to those of ladybird and feed on the leaf tissues of poplar and skeletonize the foliage leaving only the veins, and the adult beetle eats on leaves making irregular holes. Greyish-brown viburnum beetle larvae are creamy-white 7 mm long with black markings which feed by making holes on viburnum leaves, in early summer by larvae and in late summer by adults. Waterlily brown adult beetle and its fat black larvae feed on the upper surface of waterlily by making holes in leaves and such leaves disintegrate and rot. Adults and larvae of willow beetles feed on salix leaves, and their severe infestation causes extensive defoliation. **Vine weevil** (*Otiorhynchus sulcatus*) adult is greyish-black and about 9 mm long with a pair of elbowed antennae and a short snout. It is nocturnal in habit and keeps itself hidden during daytime. It feeds by making notches on the leaf margins, generally close to ground, from mid-spring to mid-autumn on shrubs, especially azaleas, camellias, *Euonymus*, hydrangeas and many other herbaceous plants. **Sawfly larvae** (*Diprion, Nematus, Pristiphora, etc.*) are generally green, caterpillar-like, some 3 cm long and sometimes with black spots. The larvae of the Solomon's seal sawfly (*Phymatocera aterrima* spp.) are usually greyish-white. All these feed on *Aquilegia, Aruncus dioicus,* conifers, *Geum, Polygonatum* (Solomon's seal), *Salix* (willow) and various other perennials, bulbous plants, trees and shrubs and defoliate the branches or shoots by feeding the leaves bit by bit. **Gall midges** [violet gall midge (*Dasineura affinis*), the gleditsia gall midge (*D. gleditchiae*), *etc.*] are the larvae of midges (flies). The flies are tiny and greyish-brown. They are whitish-orange fly-maggots some 2 mm long that feed within the galled tissues and complete 3-4 generations in one summer. Their infestation on violet (*Viola*) causes thickening of leaves which fail to unfurl though in *Gleditsia* the leaflets swell and fold over to form pod-like galls. **Scale insects** [soft scale (*Coccus hesperidum*), mussel scale (*Lepidosaphes ulmi*), horse chestnut scale (*Pulvinaria regalis*), *etc.*] infest the indoor and greenhouse plants, cacti and other succulents, trees and shrubs. They are yellow, brown, dark grey or white, flat or raised, circular, pear-shaped or oval and up to 6 mm long. They feed on lower leaf surface and stems and excrete sticky honeydew which attracts sooty mould. Their infestation slows down the plant growth. **Fasciation** in the form of peculiar flattening of stems occurs on certain ornamentals, especially *Celosia, Daphne, Delphinium, Forsythia, etc.*due to early damage to the growing point caused by mis-handling, frost injury, slug or insect infestation, bacterial infection or genetic malfunction.

Red spider mites (*Tetranychus urticae*) are 8-legged 2-spotted black-marked creatures, yellowish-green though may change to orange-red during autumn, and less than 1 mm long. There are several other species. Their infestation causes pale mottling on the upper leaf surface, leaf dullness, yellowish and ultimate premature falling. They also develop a fine silk-webbing on the entire plant. **Bryobia mites** (*Bryobia* spp.) are closely related to red spider mites and are troublesome on hedera. They feed on leaves which first develop a light freckling on upper side of leaves which afterwards becomes bronze and wither.

Leaf and bud nematodes, *Aphelenchoides* spp. and others infest inside the leaves and buds of annuals and most herbaceous perennials and cause brownish-black patches between the larger veins of leaves. *Ditylenchus dipsaci* infests on many of the ornamentals by feeding on the plant tissues, causing spotting of leaves with rough, small and yellow patches (spickels).

Diseases

Major **diseases** infecting various parts of foliage plants are sooty mould, *Alternaria* and other leaf spots, anthracnose, powdery and downy mildews, grey mould, corky scab (Oedema), damping off, root or tuber rot, wilt, black leg, crown and stem rot, rusts, bacteria and viruses. **Sooty mould** is a non-parasitic black or grayish-green fungus growing on foliage and other surfaces which have had an accumulation of honeydew. Honeydew is an excretion deposited on the upper surface of leaves by aphids, whiteflies, mealybugs and some scale insects. Though as such this is not a problem but such spots on plants appear unsightly.

Leaf spots are of various kinds caused by various pathogens in various crop plants. *Arbutus* leaf spot (*Septoria unidonis*) appears on *Arbutus* leaves as numerous small spots with dry and grey-white central area bordered by a dark purple band but its infection does not affect the vigour of the plants, however, proper sanitation should be maintained. Leaf and stem spotting (various species of *Colletotrichum*) which may turn into sunken lesions and ringed stems, causes death of full plant or a part thereof. Some of them is believed to be seed-borne and their spread is facilitated through water. The seeds of such plants should not be collected and as soon as the disease is noticed the part or whole plant should be removed and destroyed. Ivy leaf spot (*Colletotrichum trichellum, Mycosphaerella hedericola, Phyllosticta hedericola*) appears on green leaves of ivies as brown or grey and almost circular spots though on variegated or pale-leaved forms the spots are more severe and surrounded by a reddish-purple narrow band. Though this problem does not lower the overall vigour but plants become unsightly and to limit this problem such leaves should promptly be removed. *Yucca* leaf spot is caused by *Cercospora concentrica* and *Coniothyrium concentricum*. Affected leaves display raised, circular, brown with black fruiting bodies and pin-head size spots which may enlarge in due course and coalesce. Its spores spread by water. Severely affected leaves should be removed and the plant should be sprayed with mancozeb. ***Anthracnose*** (*Marsonina salicicola*) infection in willow causes leaf curling, yellowing and falling off immaturely, especially during mid-summer. Leaves develop pin-head-sized dark brown spots which produce spores and the stems develop raised and rough lens-shaped canker which when enlarges and girdles the stem, dieback may occur. The fungus is encouraged by moist conditions. The fungus overwinters on fallen leaves, bud scales and on stem cankers. The infected stems should be cut off and proper sanitation should be maintained in the vicinity. **Powdery mildew** (*Erysiphe, Microsphaera, Oidium, Phyllactinea, Podosphaera, Uncinula*) disease is found on many ornamentals, each usually only infects a single genus or closely related plants. The spread of the disease is favoured by plants growing in dry soils but with damp or humid air around the top growth. First the disease spreads on the upper surface producing white powdery fungal growth which may extend even to lower surface, *vis-à-vis* on other aerial plant parts depending on the fungal species and the host plant. New leaves become rapidly distorted. On begonia, *Microsphaera begoniae* develops greyish-white powdery patches and spots on upper leaf surface and these leaves either dry and turn brown or become yellow and soggy if fleshy. The disease is very serious on indoor begonias where stems and flowers are also affected. Keeping the root zone moist, proper aeration to reduce air humidity, no overhead watering, prompt removal of infected leaves, and use of fungicides such as carbendazim, mancozeb, triforine with bupirimate, penconazole or sulphur will be quite effective against this disease. **Downy mildew** (*Bremia* spp., *Peronospora destructor, P. parasitica, Plasmopara* spp.) is generally common on young plants and those growing in moist environment. In its infection, yellow or otherwise discoloured areas develop on upper leaf surface, each corresponding to a slightly fuzzy and grayish-white or sometimes purplish fungal growth beneath which afterwards may cover entire leaves and their eventual death. Infected leaves should be removed immediately, air circulation in the cropping area should be improved, overhead watering should be avoided and such plants should be sprayed with mancozeb fungicide. ***Pestalotiopsis*** spp., especially *P. guipini* infection is favoured by high humidity and it infects shrubs and trees producing grey-brown necrotic patches on leaves and shoots and thus whole plant becomes scorched with premature leaf drop and extensive dieback.

Grey mould (*Botrytis cinerea*) spores spread through air currents, rain or water splash. Its sclerotia present on the plant debris and in the soil spread the disease from year to year. Its infection causes grey to off-white or grey-brown fuzzy fungal growth. It enters the host through wounds or when a part of the plant has been damaged or weakened due to frost injury. It also infects the fruits via its infection in the flowers. Its infection causes discolouration or browning of tissues and softness, *vis-à-vis* yellowing and wilting of leaves, flowers and fruits. The fungus produces numerous small black sclerotia (resting bodies) which stand a wide range of growing conditions and then infecting the plants when conditions become favourable. Once infection has appeared its control becomes difficult. All dead or injured plant parts should be removed, proper sanitation should be maintained and carbendazim should be sprayed.

Damping off (*Phytophthora, Pythium, Rhizoctonia*) is a soil- and water-borne fungal disease of seedlings where seedlings show discolouration and collapsing after rotting at the base. At initial infection, at the base the water-soaked lesions appear, soon resulting into mass death of seedlings. Such seedlings show white and fluffy fungal growth on the surface. The spread of the disease is favoured by over-wet compost and prolonged high temperatures. *Phytophthora cinnamomi* infection to the woody plants causes blackish-brown discoloured root rotting, hence growth becomes poor, the foliage sparse and discoloured, stems show decline and whole plant dies. Infected plants should be removed with soil and destroyed. To avoid its infection, full hygiene should be maintained about the soil, water, trays or pots, mixture and implements, *vis-à-vis* sowing of seeds thinly with improved drainage. The compost or soil should be drenched with a copper-based fungicide at sowing and afterwards the seedlings should be treated with the same fungicide at regular intervals till the development of the seedlings. ***Sclerotinia* rot** (*Sclerotinia sclerotiorum*) occurs in many foliage and other plants and its spread is favoured by cool and damp conditions and its sclerotia fall into the soil where these remain dormant until spring and then these produce spore-yielding cup-shaped fungal growths (apothesia) which infect the crop in the following season. Its infection causes tissues becoming discoloured, brown and wet and finally rotting. Cotton-wool-like fungal growth in large number develops with scattered large black sclerotia (resting bodies) on all the aerial parts of the plants, more pronounced being fruit and stem bases. In store, it causes rotting of corms, rhizomes and tubers. Thorough sanitation of the field and its proper disposal, crop rotation, prompt removal and burning of affected parts are few of the measures to keep this problem under check. **Dieback** is a very serious problem in case of many plants. This is caused due to infection of various fungi where some actively invade the healthy plants though some enter through wounds. This may also be caused due to poor management practices such as waterlogging, drought and poor establishment of young plants. However, the dieback occurring due to fungal infection is symptomatically from stem tips to downwards or sometimes from the base or part-way up the stem. At the starting point, dark blotches or sunken patches develop, leaves yellow, wilt and die and then symptoms may spread down into the base of the plant or its crown which may cause death of the whole plant. Affected parts should be cut back and cultural practices should be improved. **Blackleg**, black rot or black stem rot is a form of grey mould caused by various soil- or water-borne fungi and those responsible for causing damping off. It is caused due to overwatering or if potting mixture is too water-retentive. Stems of afflicted plants become black and rot at the base. As soon as the cuttings start forming roots, the base of affected cuttings darkens and entire cuttings discolour and die. Its control is similar to damping off, however, at planting the cuttings should also be dusted with rooting hormone and a fungicide. **Foot and root rots or wilt** are caused by various fungi such as *Fusarium* and *Verticillium*, *vis-a-vis* those responsible for damping-off disease of seedlings. It is infection of the tissues around the stem base which causes wilting of the upper part of the plant, discolouration and then die back. Its infection causes atrophy, darkening, discolouration, and sometimes tissue softening of the stem base, *vis-a-vis* root rotting and ultimate collapse of the plant. Once disease appears, there is no cure, however, if any such symptom appears, promptly the affected plants along with the adjoining soil or mixture should be removed and destroyed and proper sanitation with respect to soil, fertilizer, water, and implements should be done. ***Fusarium* wilt** (various types of *Fusarium*, the major one being *F. oxysporum* with various forms where these are fairly host-specific infecting only one genus or only a few related ones) where this blocks the vascular tissues in stems and roots partially or wholly by forming gum-like substance. The fungus remains viable in the soil for years even without a proper host plant. Its infection causes wilting of leaves in woody plants and wilting of whole plant in case of soft-stemmed plants. There is no effective chemical control once disease appears. Its infection can be minimized by following proper crop rotation, use of healthy planting material, proper sanitation and through prompt removal of infected plants and adjoining soil or mixture. Spraying of healthy crop with thiram 0.2 per cent alternate with Bavistin 0.1 per cent fortnightly will save the crop from infection. **Crown rot** is caused to various herbaceous and a few woody plants due to presence of various soil- and water-borne fungi and bacteria. In their infection the crown (the junction of the stems and roots) of the plant deteriorates which may cause ultimate death. As and when such malady appears the plant parts should immediately be sheared off and destroyed.

***Pelargonium* rust** (*Puccinia pelargonii-zonalis*) is a serious problem when atmosphere is humid during winter and the plants are overcrowded. In its infection, oftenly the concentric rings of dark brown pustules develop on lower leaf surface along with yellow blotches on upper surface so leaves become discoloured, wither and die and in its severe infection the plants may also die. Plants should be spaced properly, air circulation should be improved, infected leaves should be promptly removed and the plants should be sprayed with mancozeb, triforine with bupirimate or penconazole. **Periwinkle rust** (*Puccinia vincae*) produces a fungal mycelium that invades all parts of the plant and its roots. The dark brown spore pustules develop on reverse side of leaves so leaves become pitted and the whole plant distorted with no flowering. Promptly such plants should be uprooted and destroyed.

Bacterial diseases are caused by *Erwinia*, *Pseodomonas* and *Xanthomonas*. The symptoms of their infections differ from species to species and bacterium involved. In their infection the leaves usually become spotted often with a yellow 'halo' around the edge of the leaf spot though fleshy organs such as bulbs, corms, rhizomes, tubers, roots and stems discolour and disintegrate. Their texture becomes slimy and emits an unpleasant odour. Main cause is infection of *Erwinia*. Crown gall (*Agrobacterium temifaciens*) is the variable (diameter being from some mm to 30 cm) but rounded swellings, generally in small groups due to rapid proliferation of cells, which are off-white when young but harden and turn brown with age. This enters the plants initially through wounds and affects crowns of stems and roots. When this develops on woody plants, it persists for long though is short-lived and softer on herbaceous plants. These galls disintegrate soon, followed with entry of secondary organisms. If the galls have not ringed the stem, the vigour of the plant will not be affected. However, splitting of root or stem due to its infection may cause entry of secondary organisms which may cause dieback.

Mycoplasma (virus-like), similar in structure and activity to both bacteria and viruses, though produce host-specific but virus-like symptoms such as yellow markings on leaves often combined with distortion and stunting, and greening of flowers. Once infected, there is no cure of the disease so affected plants will have to be uprooted promptly and destroyed. **Pelargonium viruses** develop yellow flecks, streaks and ring spots on leaves, and sometimes with leaf distortion. Its infection also causes colour-breaking on petals. Though plant vigour is rarely affected but plant appearance is spoiled. Such plants should be uprooted and burnt as there is no cure for viruses. Since there may be many viruses involved in the complex so insects, especially aphids and nematodes should be controlled as these transmit the diseases.

Physiological Disorders

Chlorosis is the yellowing (loss or deterioration of green pigment, *i.e.* chlorophyll) of plant tissues, most commonly in the foliage, due to deficiencies of Fe or Mn (lime-induced), N and Mg, or due to infection of virus, effect of weedkiller, low temperature or waterlogging. Slightly yellow or blue-green discolouration, especially on peaty soils, which sometimes follows by death of plants is due to copper deficiency. Supply of copper nutrition will correct this malady. To reclaim this, the cause should be diagnosed and then appropriate treatment should be given. **Corky cactus scab** is common in many of the cacti and other succulents, especially *Epiphyllum* and *Opuntia* cacti which occurs due to unsuitable growing conditions, especially excessive levels of humidity and light. In this case, buff to tan corky-brown markings develop on the skin which may or may not be sunken. To control this problem, the plants should be shifted to a cooler situation with improved aeration. **Oedema** is the raised wart-like outgrowth, usually on lower leaf surfaces which occurs due to plant's taking more water than its consumption so small groups of leaf cells swell up storing water inside. Colour of this abnormal part is initially the same but afterwards may turn pale-green and warty, and if the growing conditions do not improve these cells rupture, die off, turn brown and become corky. Since this is a physiological disorder and removal of leaves will further aggravate the situation as water loss will be further reduced. However, drainage should be improved, watering should be reduced and air circulation should be increased. **Fasciation** in certain ornamentals is very common. Shoots or floral stems become enlarged, flattened and ribbon-like though leaf buds and flowers develop normally on such stems. It occurs due to early mechanical or frost injury or by early insect or slug feeding of the growing point. Some other factors may also be involved. Though it normally does not affect beauty of the plant but if need be these should be cut back. **Frost injury** of the foliage and tender stems, usually towards the shoot tips or to the exposed parts of the plant causes browning, blackening or scorching-like symptoms so plants become unsightly and sometimes even death of the plants occur. Such plants should be given winter protection. **Scorching** is caused due to hot or bright sun. Normally, soft and hairy-leaved plants are often most severely affected. Injury appears as pale-brown, bleached or scorched patches due to direct rays of hot sun falling on the foliage or other aerial parts. Such symptoms may also appear due to toxicity of certain pesticides or due to use of certain contact weedkillers. Use of unharmful chemicals is only recommended. During summer the plants should not be persistently exposed to burning sun, particularly from 10 a.m. to 3 p.m.

References

Ambrozic-Dolinsek, J., M. Camloh, B. Borhancec and J. Zel, 2002. Apospory in leaf culture of staghorn fern *Platycerium bifurcatum*. *Plant Cell Report*, **20**: 791-796.

Anderson, L. 1985. Musaceae. In: *Flora of Ecuador*, No. 22 (eds Harling,G. and B. Sparre), pp. 1-87. Swedish Research Council Publishing House, Stockholm, Sweden.

Bailey, L.H. 1942. *The Standard Encyclopedia of Horticulture*. The Macmillan Co., New York.

Beckett, K.A. 1985. *The Concise Encyclopedia of Garden plants*. Orbis Pub. Ltd., London, U.K.

Beckett, K.A. 1987. *The RHS Encyclopaedia of House Plants Including Greenhouse Plants*. Salem House Publishers, Massachusetts, USA.

Brickell, C. 1994. *The Royal Horticultural Society Gardeners' Encyclopedia of Plants and Flowers*. Dorling Kindersley, London, U.K.

Chang, H. C., D. C. Agrawal, C. L. Kuo, J. L. Wen, C. C. Chen and H. S. Tsay, 2007. *In vitro* culture of *Drynaria fortunei*, a fern species source of Chinese medicine "Gu-Sui-Bu". *In vitro Cell Dev. Biol. Plant*, **43**: 133-139.

Everard, B. and Brian D. Morley, 1970. *Wild Flowers of the World*, pp. 189. Peerage Books, London.

Gonzalez, R. H., J. A. Herrera and A. C. Ramos, 2006. Multiplicacion *in vitro* de *Nephrolepis exaltata* (L.) Schott, a partir de esporas. *Revista Chapingo* (Serie Horticultura), **12**: 141-146.

Hay, R. and K.A. Beckett, 1971. *Reader's Digest Encyclopaedia of Garden Plants and Flowers*. The Reader's Digest Association Limited, London, Great Britain.

Hellyer, A. 1982. *The Collingridge Illustrated Encyclopedia of Gardening*. Collingridge Books, England.

Hessayon, D.G. 1987. *The Gold Plated House Plant Expert*. Century Hutchinson Ltd., London.

Huxley, A. and R. Gilbert, 1979. *Reader's Digest Success with House Plants*. The Reader's Digest Association, Inc, Pleasantville, New York, USA.

Kottackal, P. M., S. Sini, C. L. Zhang, A. Slater and P. V. Madhusoodanan, 2006. Efficient induction of apospory and apogamy in vitro in silver fern (*Pityrogramma calomelanos* L.). *Plant Cell Report*, **25**:1300-1307.

Laurie, A. and V.H. Ries, 1950. *Floriculture Fundamentals and Practices*, pp. 426-442. McGraw-Hill Book Company, Inc., New York, USA.

Nordal, I. and A.D. Poulsen, 1998. *Chlorophytum zingiberastrum*, a new species from south-eastern Africa. *Kew Bull.*, No. 53, pp. 937-942.

Nordal, I. and M. Thulin, 1993. Synopsis of *Anthericum* and *Chlorophytum* (Anthericaceae) in the Horn of Africa, including the description of nine new species. *Nordic J. Bot.*, **13**:257-280.

Ramage, C. M. and R. R. William, 2002. Mineral nutrition and plant morphogenesis. *In Vitro Cell. Dev. Biol.-Plant*, **38**: 116-124.

Randhawa, G.S. and A. Mukhopadhyay, 1986. *Floriculture in India*, pp. 250-273. Allied Publishers Pvt. Ltd., 13/14 Asaf Ali Road, New Delhi-110 002.

Triplett, J.K. and B.K. Kirchoff, 1991. Lamina architecture and anatomy in the Heliconiaceae and Musaceae (Zingiberales). *Canad. J. Bot.*, **69**(4): 887-900.

Virgine, F. and G.A. Elbert, 1989. *Foliage Plants for Decorating Indoors*. Timber Press, Portland, Oregon, USA.

21

Fuchsia × *hybrida* (Family: Onagraceae)

Sapna Panwar, Namita, Poonam Kumari and R.L. Misra

[**Common names:** Dancing lady, Earring flower, Fuchsia, Fuchsia angel earrings, Hardy fuchsia, Ladies' eardrops, Tree fuchsia (F. arborescens, syn. F. syringaeflora)]

Introduction

Fuchsia is a distinctive genus in its family, comprising nearly 110 species, mainly shrubs mostly confined to cool and moist habitats, with entire leaves in opposite pairs or whorls, placed in 12 sections that have been distinguished based on geographical, morphological, molecular and cytogenetic characters (Berry *et al.*, 2004). *Fuchsia* was named for Leonhart Fuchs (1501-1565), a German doctor and herbalist. It originates from Central, *i.e.* Mexico to Pacific Islands, and South America, *i.e.* Northern Argentina to Colombia and Venezuela, four in New Zealand (*F. colensoi*, a shrub; *F. excorticata*, a tree; *F. perscandens*, a woody climber; and *F. procumbens*, a prostrate creeper), two in Hispaniola in the Carribean, and one in Tahiti. Its three-quarters of the species are concentrated in the tropical Andes alone. It is the only genus in the family having bird-pollinated flowers with largely biporate pollen and fleshy berries. Like today, they were once very popular during the late Victorian and Edwardian periods. The history of the garden fuchsias shows that they are of hybrid origin. The genus *Fuchsia* has rich genetic diversity, though breeders have utilized only a small number of species (Hoshino and Berry, 1989; Godley and Berry, 1995). Hundreds of named cultivars have been derived primarily from hybridizing members of **section Quelusia** during 19th and 20th centuries (Reiter, 1944).

The first *Fuchsia* was discovered in 1623 by the missionary Father Charles Plumier, who was also botanist to Louis XIV, then working in the Santo Domingo. He made three journeys to West Indies and after the third he published details of this discovery in '**Plantarum Americanum Genera**' in 1703 and called the new plant *Fuchsia triphylla flore coccineo* after Leonhart Fuchs, author of **Historia Stirpum**, a most beautiful and richly illustrated 16th century herbaria. According to Solander in '**Aiton's Hortus Kewensis**' and the '**Botanical Magazine**' of 1789, *Fuchsia coccinea* was the first to reach the Kew, England in 1788. It was the end of the 18th century that other *Fuchsia* species were introduced such as *F. magellanica* and *F. arborescens*, following *F. fulgens* in the early 19th century. Although the results of early hybridization are not known, it seems likely that these species were the ancestors of most of our modern *Fuchsia* hybrids. The *Fuchsia* species known today all come from Central and South America, New Zealand and Tahiti. These species ranged from trees such as *F. excorticata* to creeping plant such as *F. procumbens*. Other species, such as *F. tunariensis* and *F. tuberosa* are epiphytic thriving in the vegetative detritus that collects in the forks of trees or on rocks. All the modern hybrids have inherited the dislike of a hot dry atmosphere. In addition to hybrids, chance seedlings were also developed and 'Mieke Meursing' was the most famous seedling.

Uses

Fuchsias are rich source of phenolic compounds. Flavone glycosides were found in only three species: luteolin 7-glucoside in *F. splendens* and luteolin and apigenin 7-glucuronides and 7-glucuronidesulphates, tricin 7-glucuronidesulphate and diosmetin 7-glucuronide from *F. procumbens* and *F. excorticata.* Luteolin 7- glucuronidesulphate is reported for the first time in *Fuchsia* (Williams *et al.,* 1983). Helsper *et al.* (2003) reported that the levels of various antioxidants (glutathione, ascorbate, carotenoid, and flavonoids) were increased significantly in petals, leaves and sepals when fuchsia plants were exposed to UV-A (320- 400 nm) radiations. 3- Glucosides and 3, 5- diglucosides of pelargonidin, cyanidin, peonidin, delphinidin, petunidin and malvidin have been identified as flower pigments in *Fuchsia* sp. These pigments solely or in mixtures appear to be responsible for different flower colours in this genus. Their production and inheritance seems to be under a complex system of genetic control.

Fuchsias are used for various ornamental purposes such as colourful house and conservatory plants, containers (upright or spreading) and hanging baskets, trellises, standard and specimen plants, bedding (spring and summer), informal hedging, *etc. Fuchsia* petals are edible (Rop *et al.,* 2012) and used in salad or as garnish but have no distinct flavour. According to Wilson (2013), *Fuchsia* flowers are not delicious as they have a slightly acidic flavour but if used sparingly, they make wonderful garnish. Some recipes of *Fuchsia* flowers include fuchsia and potato mash, cold chicken and fuchsia salad (Roberts, 2000), *etc.* The fruits of all the species and cultivars of *Fuchsia* are edible, however, quality of the fruit varies from species to species. Some of the species and cultivars are tasteless, whereas others have unpleasant taste. The fairly large fruits of *Fuchsia splendens* have citrusy and peppery tang and are used for jam making. The wood of *Fuchsia magellanica* is utilized for making black dye. The flower of *Fuchsia arborescens* is even eaten and being used on bites, scratches and grazes and its juice has a relieving effect on itching and taking away the redness. They are also used to relieve inflamed blisters and sunburn. *Fuchsia* flowers and berries are used to make a superb jelly that includes lemon juice, apple juice and a dash of brandy which is used as a remedy for sore throat, tonsillitis and to strengthen the voice (Roberts, 2000).

Botany

Fuchsia leaves are opposite, verticillate or occasionally alternate, or in whorls of 3–5, simple, 1-25 cm long, deciduous or evergreen depending upon the species, lanceolate or sometimes dentate though entire in some of the species. The flowers are very decorative and have a pendulous 'teardrop' shape, axillary, solitary or clustered in a racemose inflorescence. The floral tube elongates and terminates in a calyx with four sepals (sometimes 3) which are usually curved and spreading. These are displayed in profusion throughout the summer and autumn and all the year in tropical species. Fuchsias bear slender sepals and four shorter, broader petals. In many species the sepals are bright red and the petals are purple (colours that attract the humming birds that pollinate them), but the colours may vary from white to dark red, purple-blue, pink and orange. Very often the sepal's colour contrasts the petal colour. A few species have yellowish tones. In majority of the cases there are eight stamens though species with 3-4 stamens are also recorded such as *Fuchsia triphylla* and *F. tetraphylla.* The ovary is inferior and the fruit is a small (5–25 mm) dark reddish green, deep red, or deep purple cherry-like berry, containing numerous but very small seeds.

Gametic chromosome number of *Fuchsia* is n = 11 (*Fuchsia*, RHS) with diploid (2n=22) and tetraploid (2n=44) species existing in nature (Berry, 2004). Diploid species (2n=22) are *Fuchsia arborescens* Sims, *F. encliandra* Steudel, *F. excorticata, F. fulgens* DC, *F. microphylla, F. minutiflora* Hemsl., *F. procumbens* R.Cunn ex a.Cunn, *F. splendens* Zucc. and *F. triphylla;* whereas tetraploid (2n=44) species include *F. boliviana* Carrière, *F. glazioviana* Taub., *F. hatschbachii* P.E. Berry and *F. magellanica* Lam (Talluri and Murray, 2009).

Propagation

Fuchsias can be propagated from cuttings and seeds. Plants produced from seed will not necessarily be the same as the parent plant from which the seeds were taken. The commercial and best method of propagation is through cuttings.

Seeds are removed from the ripe pods by squeezing onto a piece of absorbent paper. Seeds are sown in trays or pots. The trays are kept at 21-24 °C temperature where germination commences in a fortnight. Pricking is done when seedlings are large enough (7.5 cm) to handle and then are placed at 20 °C.

Fuchsia **cuttings** can be taken anytime from spring through fall, with spring being the most ideal time. A young growing tip is severed about 5-10 cm in length just above the second or third pair of leaves but below a leaf joint (node) with a sharp knife in one attempt to avoid possibility of any jagged edge so that any infection is discouraged. The advantage of taking cuttings just below a joint is because of maximum hormonal concentration at this point. In the process, bottom pair of leaves are carefully removed. If the stem is left too long on the cutting it encourages the possibility of *Botrytis* and then rotting eventually starts from the point of cutting. The cutting is inserted in the medium about half the length of the stem below the second pair of leaves. The tip cutting should be inserted as much as possible. Three or four cuttings can be inserted in a 7-8 cm diameter pot or numerous cuttings in a planting tray, into a moist growing medium such as sand, perlite, vermiculite, peat moss or sterilized soil. Cuttings are covered with ventilated plastic to retain moisture and humidity and to speed up the rooting process. Before inserting any cutting into the striking medium, it may be essential to dip the cuttings in a fungicidal solution, especially 0.2 per cent thiram or indofil M-45 which prevents *Botrytis* infection which is encouraged by insufficient air circulation within the propagating chamber. If any flower bud is present, it is advisable to remove it first because the cuttings need all the energy for rooting. In fact, the cuttings should be taken when the shoot is vegetative. Regular watering is required. Time required for cuttings to root is 3-4 weeks. During the ideal period and with favourable conditions of April or May, the rooting time could be shorter. The cuttings when rooted should be a slightly darker green and take on a 'perky' look. At this stage, it is necessary to introduce them to the ambient air temperature gradually. When they are acclimatized, these are transplanted into individual pots, preferably 6-cm size. As soon as the roots begin to show around outside of the compost, it is right time to transplant them to a pot that is one size larger (http://www.solentfuchsia.co.uk/index.html).

The **callus proliferation and regeneration** of *Fuchsia × hybrida* was successfully carried out by Chow *et al.* (1990), and being more rapid in callus derived from ovary tissues (Dabin and Vaerman, 1985). Rapid development of axillary buds from shoot-tips and nodes of 18 cultivars of *Fuchsia × hybrida* has been obtained on solid MS medium with BAP (6- benzylaminopurine) and an auxin (Kevers *et al.*, 2003). Tissue culture and rapid propagation of *Fuchsia alba-coccinea* Hort. was also carried out by using explants of stem fragments with nodes on MS medium containing 0.8 mg/l 6-BA, 0.10 mg/l NAA and 1.0 g/l PVP (Fugen *et al.*, 2006).

Classification, Species and Varieties

Fuchsias were arranged into 12 sections by taxonomist, Dr. Paul Berry of the University of Michigan at Ann Arbor. Some of the species grow in the northern part of South America, Central America and Mexico and like warmth. They have long tubes, up to 10 cm and often are in the shade of orange. Others, with red sepals longer than their tubes and purple corollas are winter hardy on the West Coast of the Pacific Northwest and have parented many hardy hybrids. One section from Mexico includes species with wee blossoms, the smallest being only 0.75 cm long and they are often hardy as well (*Fuchsia* species: http://www.nwfuchsiasociety.com/).

Section Skinnera

It is from New Zealand and Tahiti. The species in this section have a floral tube with a swelling above the ovary. The sepals curve back on themselves and the petals are small or nearly absent. The species include *Fuchsia antiqua, F. cyrtandroides, F. excorticata, F. perscandens, F. × colensoi, etc. F. excorticata* is a 12 m tall tree in New Zealand and the source of our 'aubergine' hybrids. At maturity it is very dark in the tube and sepals. It is winter hardy but likes to bloom in spring when it is too cold in most inland areas.

Section Schufia

Mexico to Central America (3 taxa). Species in this section bear flowers in an erect, corymb-like panicle. *F. paniculata* subsp. *paniculata* looks like a lilac, likes warm climate and has serrated leaves. This section has two more species such as *F. arborescens* with entire leaves and *F. paniculata* subsp. *mixensis*, both from Mexico, which look similar to *F. paniculata* ssp. *paniculata* for many aspects.

Section Hemsleyella

Central America to northern South America (15 species). These species are characterised by a nectary that is fused with the base of the flower tube and petals that are partly or completely lacking. They have tuberous roots, and like warm climate. Some of these species are *F. apetala, F. tillettiana* (native to Venezuela), *F. pilaloensis* (Ecuador), *F. juntasensis* (Bolivia and requires warmth for its growing, blooming mostly after the leaves drop), *F. inflata* (recognized as a fuchsia only in Peru), *etc.*

Section Procumbentes

New Zealand. *F. procumbens* is a common ground cover in the PNW (The Pacific Northwest). It is hardy, perky and easy to grow. It is the only yellow species in the genus and yellow colour only in the tube. The variegated form is less vigorous but very attractive. *F. procumbens* is a trioecious species in which the female is heterogametic. Both male sterility and female fertility is controlled by single dominant gene and they are linked. Male sterility is constant in expression but female fertility is invariable (Godley, 1963).

Section Encliandra

Mexico and Central America (6 species; 14 taxa). Flowers on the six species in this section have flat petals and short stamens and are reflexed into the tube. Fruits contain few seeds. Fuchsias from this section cross so easily that it is very difficult to find true species today. This section includes *Fuchsia* × *bacillaris*, *F. cylindracea* (cylindrical blossoms), *F. encliandra, F. e.* ssp. *encliandra, F. e.* ssp. *microphyloides, F. e.* ssp. *tetradactyla, Fuchsia microphylla, F. m.* ssp. *aprica, F. m.* ssp. *chiapensis, F. m.* ssp. *hemsleyana, F. m.* ssp. *hidalgensis, F. m.* ssp. *microphylla, F. m.* ssp. *quercertorum, F. obconica, F. ravenii, F. thymifolia, F. t.* ssp. *minimiflora, F. thymifolia* ssp. *thymiflora, etc.*

Section Ellobium

Mexico to Central America (4 taxa). This section contains three species, *viz. Fuchsia decidua, F. fulgens* and *F. splendens* (*F. splendens* var. *cordifolia*, and *F. splendens* var. *splendens*). *Fuchsia splendens* is fairly winter hardy, it likes to bloom in early spring in the PNW when it is too cold. There is the pinch in the tube. Heart-shaped leaves stand out. Also, there is a longer-tubed version with quite obvious pinch. *F. splendens* var. *cordifolia* bears lighter but larger leaves and orangish blossoms with a less obvious pinch. Finding a true *F. fulgens* is very difficult, but there are a lot of hybrids with it in their backgrounds.

Section Kierschlegeria

This section consists of a single species (*F. lycioides*) from Chile with pendulous axillary pedicels. The leaves are sparse. The sepals are reflexed and slightly shorter than the tube.

Section Quelusia

Brazil and Chile/western Argentina (9 species; 11 taxa). This is the best known section in the PNW because the species are winter hardy to varying degrees and some of the plants have historical significance. All the species are in shades of red in the tube and sepals, and with shades of purple in the petals. The sepals are longer than the tube. The nine species in this section have the nectary fused to the base of the tube or hypanthium. The hypanthium is cylindrical and is generally no longer than the sepals. The stamens are long and are exserted beyond the corolla. *Fuchsia magellanica* is believed to be one of the most significant species whose blood is found in many winter hardy hybrids. *F. magellanica* and *F. coccinea* are believed to be the first fuchsias which entered Europe for cultivation around 1733. Their similarities led to much confusion. These species can be differentiated from each other through leaves. *F. magellanica* grows in Chile and western Argentina. It mutates easily to find all shades of red and pink over all shades of purple to pink and different leaf sizes and shapes. *F. magellanica* 'Alba' ('Molinae') is a light pink natural mutant of red over purple species. It is also very hardy and grows to about 1.8 m tall in the PNW where it is very chilly. *F. alpestris* grows in Brazil. *F. bracelinea* is a very hardy species blooming in June, and as it has somewhat lax stem so is good in a rockery. It has 3-5 lush green leaves per node with short internodes and clusters of blossoms. *F. brevilobis* bears soft leaves and has a very attractive blossom. *F. campos-portoi* is characterised by its small plants, leaves and blossoms and loves cool conditions. *F. glazioviana* has shiny, glazed leaves. *F. hatschbachii* is very hardy and has long, thin leaves. *F. regia radicans* is another form. *F. regia* ssp. *regia* grows very fast and can be used as climber, with the largest blossoms in this section. *F. regia* ssp. *reitzii* is the hardiest species. *F. regia* ssp. *serrae* has lax stems with large shiny leaves.

Section Fuchsia

It is the largest section having 65 species, 67 taxa and shares the same name as the genus itself. These species grow in the warmer areas, *viz.* Central and northern South America. They have long tubes and shorter sepals. The flowers are perfect, normally orange and with convolute petals. The stamens are erect, opposite the petals are shorter and may or may not be exserted from the corolla, and the fruit has many seeds. *F. andrei* and *F. ayavacensis* grow in Ecuador and Peru. *F. hartwegii, F. crassistipula, F. magdalenae* and *F. venusta* are from Colombia. *F. cinerea* and *F. vulcanica* are from Colombia and Ecuador. One of the best known species in this section in the Pacific Northwest is *F. boliviana* var. *luxurians*. It has a 10 cm long tube, large leaves and tall growth. Its shorter-tubed version is *F. boliviana* var. *boliviana* which grows from Mexico through northern South America. *F. boliviana* (*F. boliviana* var. *luxurians* 'Alba') produced a spontaneous mutant with white tube. *F. furfuraceas* and *F. santae-rosae* are from Bolivia. *F. macrophylla* is from Peru. *F. mathewsii* (and/or *F. denticulata*) from Peru is a winter bloomer. *F. campii, F. dependens, F. lehmannii, F. loxensis, F. orientalis, F. scabriuscula, F. scherffiana, F. steyermarkii* and *F. summa* are from Ecuador. *F. pringsheimii* and *F. triphylla* are native to Dominican Republic.

The Types

Their varieties may be hardy and non-hardy in singles (having 4 petals), semi-double (5-8 petals), doubles (layers of 8 or more petals) and Triphylla (having clustered flowers) (http://homeguides.sfgate.com/types-

Single

Semi-Double

Double

Triphylla

fuchsias-26272.html; http://weidners.com/fuchsia-varieties-2/). The hybrids are derived chiefly from crosses between *F. fulgens* and forms of *F. magellanica*, *vis-à-vis* certain other species.

Trailing Fuchsias for Hanging Baskets

The trailing habit of some fuchsias makes them ideal choices for hanging baskets or trellises, where their tendrils help them attach to the support. This group is very diverse, with plants that produce flowers in various colours, including some that are bicolour, with sepals and corolla having contrasting shades. Some of these cultivars also produce exceptionally large flowers, with curly or double-petalled corollas, making them especially showy. Examples of trailing fuchsias are 'Ann's Delight', 'Bicentennia', 'Blueberry Fizz', 'Cascade', 'Ceil Peller', 'Cindy Robins', 'Dark Eyes', 'Deep Purple', 'Falling Stars', 'First Love', 'Flamenco Dancer', 'Fluffy Ruffles', 'Golden Anniversary', 'Golden Marinka', 'Hula Girl', 'Infierno', 'Kaleidoscope', 'Miss California', 'Royal Mosaic', 'Starry Trail', 'Swingtime', *etc.*

Semi-Upright for Pot Culture

These varieties can be planted in a basket and pots or trained as an upright tree or a bush. Their flowers are allowed to cascade downwards in mid-air and these are 'Abbé Fargées', 'Annabel' (double), 'Autumnale', 'Beacon Rosa', 'Bella Forbes' (double), 'Blue Satin' (double), 'Border Queen', 'Brutus', 'Cascade', 'Charming', 'Coachman', 'Display', 'Cygnet', 'Display', 'Dollar Princess', 'Fascination' (double), 'Feather Duster', 'Flying Cloud' (double), 'Frosted Flame', 'Grand Les Roi', 'Harry Gray', 'Jack French', 'Jack Shahan', 'Jingle Bells', 'La Campanella', 'Leonora', 'Marinka', 'Miss California' (semi-double), 'Mrs. Churchill', 'Mrs. Lovell Swisher', 'Mrs. Marshall', 'Pink Galore' (double), 'Pink Lena', 'Pinto de Blue', 'Red Spider', 'Rose of Denmark' (semi-double), 'Swingtime' (large flowered and double), 'White Spider', 'Winston Churchill' (double), *etc.*

Upright for Bedding

Some fuchsias have an upright shrub-like growth habit, making them useful as accent plants for the outdoor garden. Containers are normally in the form of pots, tubs and troughs and the planted containers are normally displayed on pedestals in patios or elsewhere so that when in bloom their beauty can be viewed from below. *Fuchsia magellanica* grows up to 3 m in height in frost-free areas. In regions that have some winter frost, the plant may regrow in spring after being frozen back to the ground. Generally, their hybrids reach to a height of 1.2-1.5 m. Most suitable cultivars are 'Aintree', 'Beacon', 'Black Prince', 'Blush Fondant', 'Citation', 'Constellation' (double), 'Display', Fandango' (semi-double), 'Gartenmeister Bonstedt', 'Gray Fallbrook', 'Lyes Unique', 'Marin Glow', 'Nancy Lou', 'National Velvet' (double), 'Orange Crush', 'Pink Dessert', 'Pink Rain', 'Royal Purple', 'Royal Velvet' (double, very large flowers), 'Rufus the Red', 'Snowcap' (semi-double), 'Thalia', 'Ting-a-Ling' (pure white), 'Violetta', 'White Spider' (pure white), 'Winston Churchill', *etc.*

Spreading for Hedging and Borders

The varieties in this group are of arching habit and therefore are ideal for planting around the edges of the containers so that flowers cascade down covering the containers. The varieties in this group are 'Annabel' (double), 'Beacon', 'Brutus', 'Dancing Flame' (double), 'Dark Eyes' (double), 'Enchanted' (double), 'Eva Boerg' (semi-double), 'Flying Cloud' (double), 'Gartenmeister Bonstedt' (Triphylla type), 'Gay Paree' (trailing, double), 'Leverhulme' (Triphylla type), 'Muriel', 'Quaser' (trailing, double), 'Sunny Smiles' (a little trailing), 'Texas Longhorn' (large double), 'Vanessa Jackson' (trailing), *etc.*

Standards as Specimen Plants

In larger containers the upright growing varieties are planted in the form of half-standard which produces the two-tiered effect. Such plants give an exciting display when kept beside a garden seat. Suitable varieties are

'Annabel', 'Border Queen', 'Cella Smedley', 'Cloverdale Pearl', 'Dawn', 'Devonshire Dumpling' (double), 'Dollar Princess' (double), 'Estelle Marie', 'Fascination' (double), 'Garden News' (semi-double), 'Gypsy Queen' (double), 'Jack Acland', 'Joy Potmore', 'Lilian Lampard', 'Lye's Unique', 'Marin Glow', 'Mieke Meursing', 'Miss California' (semi-double), 'Morning Glow' (semi-double), 'Prodigy' (semi-double), 'Royal Velvet' (double), 'Rufus the Red', 'Snowcap' (semi-double), 'Swanley Gem', 'Tennessee Waltz' (arching, double), 'Thalia' (Triphylla type), *etc.*

Hardy Types for Outdoor Growing

There are frost-hardy fuchsias which can be planted directly in the soil in the garden outside where these continue producing flowers from the late summer until autumn or early winter. During severe winter a few varieties may die down to ground level but new shoots appear from below the ground later in the spring. Ideal varieties are 'Army Nurse', 'Bambini', 'Brutus', 'Checkerboard', 'Chillerton Beauty', 'Firefly', 'Flash', *Fuchsia magellanica* 'Alba', *F. m.* 'Aurea', *F. m. gracilis* 'Variegata' (leaves variegated), 'Genii' (stem red), 'H.E. Brown', 'Lena' (double), 'Margaret' (1.5 m tall), 'Margaret Brown', 'Mrs. Popple', 'Pee Wee Rose' (dwarf), 'Rose of Castle Improved', 'Rufus', 'The Tarns', 'Tom Thumb' (dwarf), 'Whitenights Blush', *etc.*

The Species

Most important species are:

F. arborescens (F. syringaeflora)

A native to Guatemala, Mexico and Costa Rica, this species grows up to 7.5 m high where main trunk and larger branches are leafless and the branches when young are reddish-pink. Leaves are smooth, verticillate, in threes, elliptically oval, 10-15 cm long and 2.5-5.0 cm wide, entire and acuminate at both the ends. Lilac scented small pale reddish-pink flowers appear during winter (Oct.-Feb.) in erect terminal racemes. It likes a cool temperataure of 13-16 °C.

F. austromontana (F. serratipetala)

It is a bushy shrub from Peru which may grow more than 3 m high, bearing paired or whorled, narrowly oblong, 3-7 cm long and toothed leaves. Its flowers are carmine-red, some 5 cm long and appear during summer.

F. boliviana

It originates from Guatemala, Ecuador, Bolivia, Peru and Argentina and is a small tree though in pots it can be restricted up to 2 m in height. Leaf stalk is red, and the leaves are pubescent, elliptic-ovate, some 10-18 cm long and about 6 cm wide, with mid-vein, acuminate and dentate. Crimson flowers appear in branched drooping racemes or narrow panicles from summer to autumn, dark red sepal lobes are slender, 1.9 cm long, spreading then recurved, light to dark red petals are a little shorter and drooping and the flowers have a slender trumpet-shaped dark red tube some 4-6 cm long. *F. b. luxurians* from El Salvador to Ecuador is quite similar to *F. corymbiflora*, bears 6 cm long floral tube and black-red edible fruits having fig-like flavour.

F. corymbiflora

It is a lax scrambling shrub from Ecuador and Peru, growing up to or more than 3 m in height, bearing hairy, oblanceolate, opposite, green tending to bronze and 7-13 cm long leaves where midribs and leaf stalks are reddish. Deep red drooping flowers appear in terminal clusters during summer with slender tubes some 6 cm long, and the petals are longer than non-reflexing sepals.

E. excorticata

A most hardy and a true arborescent species from New Zealand growing up to 11 m high (Everard and Morley, 1970) in the lowland and foothill forests of both the islands of New Zealand, bearing dark green leaves suffused with purple and whitish-pale underneath, alternate, oval-lanceolate, pointed, 3.0-8.8 cm long and 0.8-2.8 cm wide. Axillary flowers appearing from May to August are solitary, sepals first yellowish-green then turning dark red, petals are dark purple and stamens blue. Flowers are sometimes borne on naked branches. Its 6 mm long fruits are blackish-purple berries which mature by June to September. Its var. *purpurascens* bears crimson-purple foliage.

F. fulgens

It is some 2 m tall shrub with reddish stems and branches, native to Mexico, bearing ovate-cordate to oval-lanceolate, opposite, entire or inconspicuously denticulate, 10-20 cm long and 5.0-12.0 cm wide leaves. Scarlet terminal cluster of flowers appear in pendulous leafy racemes during summer with slender tubes some 10 cm long where tepals are tipped green. Its stems are slightly succulent and the upper part of the roots is enlarged to form a tuber.

F. macrantha

A native of Peru, this is a prostrate-growing trailing species found as an epiphyte at its native habitat. Its leaves are 3.7-9.0 cm long and 2.8-6.5 cm wide, the petalless flowers appearing in April to May are largest, *i.e.* 8.0-11.3 cm long, pinkish-red, coral-red or scarlet in colour where sepals are 0.5-1.9 cm wide, pink and with greenish-yellow tips.

F. magellanica (*F. macrostemma* and erroneously *F. coccinea*)

It is a hairless erect shrub some 4 m tall native to Chile along the coastal belt near the Strait of Magellan and Argentina, bearing purple-red stems, branches and veins of the leaves. Leaves are in opposite pairs or tri-verticillate (whorls of 3) or sometimes even in fours, lanceolate to oblanceolate, pointed and 1.8-5.0 cm long with a very short stalk. Axillary, short-stalked, hairy and reddish flowers are some 4-5 cm long which appear solitary or in fours from summer to late autumn. Here sepals are red, petals purple and end curling, and the funnel-shaped tube is pendulous, reddish-scarlet and short. Its form *F. m. macrostemma* (*F. gracilis*) is slender in habit and bears narrow leaves and flowers. *F. m. molinae* (*F. alba*) bears quite pale-pink flowers. The varieties are 'Aurea' (foliage golden), 'Variegata' (foliage flushed cream and leaf margins pink), 'Versicolor' (leaves grayish, pink-tinted at younger stage but becoming white with maturity), 'Pumila' (smaller in all its parts) and 'Riccartonii' (hardier, petals broader and sepals darker). It has produced numerous varieties where many are hardy.

F. microphylla

A native of Mexico, this species is a densely branched erect shrub from Mexico, growing up to 1.8 m high, bearing shining dark green, leathery, 0.6-2.0 cm long and 1.5-4.5 mm wide, oblanceolate or elliptic and dentate leaves. Quite small flowers with cherry-red tube and sepals, pink petals and red stamens appear during August to October.

F. procumbens

A native of New Zealand, this species grows prostrate with trailing habit among the rocks or sand above the high-water mark in coastal areas of North Island, first discovered by Richard Cunningham in 1834 near the village of Matauri opposite the Cavallos Islands and was introduced into European gardens before 1870. It has much branched stems that easily root on the ground. The leaves are alternate, rounded to ovate, upper surface pale-green and whitish below which may be 0.6-2.0 cm long with long slender stalks. Narrow-tubed purple and red corolla-less axillary flowers some 1.2-2.0 cm long are solitary and erect which appear in summer with green and reflexed sepals which are tipped with purple and the attractive fruits are some 2 cm long, pinkish-purple to dull red.

F. splendens

A native of Mexico, Guatemala and Costa Rica, this species is quite similar to *F.fulgens*. This erect shrub grows to about 2.5 m tall, bearing almost triangular stems, covered with hairs along with the leaf stalks which are long and pale-green. Leaves are opposite or tri-verticillate, cordate, highly dentate and pointed. Scarlet-red stalked pendulous flowers which appear during December to February, are axillary, solitary, 3.0-3.8 cm long, tubes are flattened on the sides, swollen at base and pink to bright red, sepals green-tipped, petals shorter than sepals and yellowish-green, and the stamens are yellow. This species is suitable only for greenhouse growing.

E. triphylla

A native of Haiti and San Domingo, this is a densely hairy shrub some 1.2 m tall, bearing opposite or tri-verticillate but sometimes even in fours, some 4-10 cm long, rarely dentate, pointed, and lanceolate to oblanceolate leaves that are reddish-green but brownish beneath. Orange-scarlet, slender and narrow flowers some 4 cm long continue appearing throughout the year in terminal leafy corymbs with a long and slender tapering tube. Its cv. 'Thalia' is very similar to the parent species but is free-flowering and its leaves are red. It flowers all the year round if minimum winter temperature does not touch below 15 °C. Its popular varieties are 'Coralie', 'Fulgens', 'Gartenmeister Bondstedt', 'Heinrich Heinkle', 'Leverhulme', 'Mary', 'Traudchen Bondstedt', *etc.*

Cultural Practices

Fuchsia thrives in a warm-cool subtropical climatic regime. It grows well in fertile, sandy loam moist but well-drained **soil**, with shelter from drying winds in the ground or container. Soil pH should be 6.0-6.5. *Fuchsia* is intolerant to drought or water-logged conditions. This prefers shade for hottest part of the day and even otherwise grows best in filtered sunlight though too much shade shows leggy growth and produces fewer flowers. Fuchsias have a growth habit proportional to the amount of light they receive. Plants in bright conditions will be tight bushes as opposed to the long flowering plants that are grown in the shade. A temperature of 2-3 °C should be sufficient to keep plants alive over winter. An ideal temperature in summer would be around 16-21 °C. At temperatures above 24 °C, the plants stop growing. Further, the relative humidity should be maintained below 70 per cent with good air circulation so that *Botrytis* problems may be avoided.

Fuchsia can be **planted** directly in the ground or in pot. For planting directly in the soil, a hole double the size of pot is dug up, soil removed and water crystals and Osmocote are added. Poor quality soil is removed and well rotten compost or manure is added to half and half. Outermost roots are clipped and planting is done about 5-7 cm deep in the soil and then the soil is filled around the plant and pressed firmly. For a bush fuchsia, the leaves of the plant should not be in contact with the ground. In fact, leaves should be well clear as the flowers may be damaged because of the weight of flowers which further pull branches much lower. Upright cultivars can be planted directly in the ground, in well-drained soil, as bushes or trees some 1.5 m tall, grouped as hedges, and trained on trellises or in rings.

Fuchsia can be grown in **containers**. Ideal containers are 30 cm wooden baskets. Other suitable containers are moss baskets or clay or plastic pots. Any fuchsia can be grown in a basket, but the trailing types generally make the most attractive ones. Soil for baskets should be a light, fast-draining type. A recommended soil is 2 parts nitrolized redwood compost, 2 parts potting soil, and 1 part sponge rock (perlite). Pizzetti and Cocker (1975) advocated 2 parts ordinary garden soil or fibrous loam, 1 part fairly coarse sand, 1 part well-decomposed leafmould, ½ part completely decomposed or dried cow manure, plus little of bone meal.

Pinching (stopping) is an important operation in the successful production of quality blooms in fuchsia. Once the cutting is established, the next thing to consider is to produce a plant with an abundance of flowers. For a bush or pyramid-trained specimen, after rooting of the cuttings and when some growth has taken place, the tips are pinched out at about 10-15 cm. Leaving the new shoots which are strong growers, the rest of the canes, especially the weaker ones are removed at 4-5 leaf pairs and when the leading shoot has made another 8-10 cm, the stopping practice is repeated until the desired height is achieved. Standards are raised only from a strongly growing cutting and here all the side growths are pinched out until the main stem is 3-4 leaf pairs taller than the required height. The tips or any subsequent shoot arising from the top buds are pinched out. The hanging basket plants, similar to bush specimen, require two early pinchings to facilitate forming of sufficient stems. Pinching involves removing the tips of side shoots at one or two pairs of leaves from the origin of their growth, in case of already established old plants. However, in early spring these should be potted annually, cutting back the previous year's canes to 2-3 cm. Pinching will encourage more side-shoots to grow from the leaf joints on the remaining stems, thus achieving a plant with a potential of producing a greater number of blooms. For a bush, pinching is repeated when the new lateral growths are about 10 cm and a second time, if required, to make it really a well-balanced plant. Repetition of the stopping more the times will induce the plant to produce an even greater number of side shoots that will eventually become the branches which in turn will yield more blooms. Flowering period varies among the singles and doubles after pinching. It is recommended to allow 60 days after the final stop for the single varieties, 70 days for semi-doubles and at least 80 days for doubles for getting blooms.

Pruning is done to control the size and shape of the plant and to ensure lots of new wood as fuchsias only bloom on new wood. To ensure full growth and flowering, they need to be pruned lightly every year after flowering but before stacking for protection against frost. Mid-November is the best time to prune off entire foliage, leaving the main stems about 15 cm long or to the edge of the basket for a basket plant but those directly planted in the soil should be pruned off to $^{1}/_{3}{}^{rd}$ of the plant. Pruning may also be carried out during March or April except where the climate is sufficiently mild to grow fuchsias as deciduous shrubs. Together, all the dead or diseased parts should also be removed. Greenhouse species and varieties are trimmed back lightly in February if not overgrown, however, overgrown ones are cut hard back. Bush fuchsias are pruned in spring by heading back the stems to a permanent low framework. 'The British Fuchsia Society' recommends the stem lengths after pruning 15-25 cm for miniature standards, 25-45 cm for quarter standards, 45-75 cm for half standards and 75-105 cm for full standards. This measurement is taken from top of the soil to the first break (branch). In fact, fuchsia standards have a main stem topped with a dense head of foliage created through pinching and pruning, for making these specimen plants. The head of a finished 'standard' should be approximately one-third of the total height, and the width approximately two-thirds of the height. Trailing or basket fuchsias require hard pruning to encourage growths from the base of the plant, oldest stems are pruned in the spring when the fresh buds begin to break, and remaining stems are reduced to restrict their vigorous growth to the available space. At the time of pruning, no plant should be fertilized, and very little watering should be carried out. New growth will start emerging after 4-5 weeks of pruning and then pinching should start. As a branch develops, it will form a pair of leaves.

Young fuchsia stems are allowed to grow upright and all the side shoots are removed as they develop. Main stem is tied into a cane to provide the **support** as it grows. Stem tips are pinched out when the plant reaches 20 cm height. New side shoots are produced at the top of the plant and these will form the head of the standard. Pinching of the tips of each side shoot is done when it reaches 2-4 sets of leaves. Pinching and pruning will be continued until a rounded head has formed. The leaves on the main stem will be shed naturally in time, or can be carefully removed.

Water requirement is directly related to the soil radiation received by the plant. Fuchsias do not like water-logged conditions though should be watered regularly to maintain proper soil moisture. In containers, these need frequent watering depending on the size of the container and weather conditions. Hanging baskets should be watered at least once a day during hot summer. Fuchsias that are planted directly into borders become more self-sufficient once established. Established plants should not be allowed to dry excessively which otherwise results in leaf yellowing and subsequent falling off.

Proper **nutrition** is very essential for obtaining optimum plant growth and high quality blooms as fuchsias are heavy feeder. Large plants in active growth or during flowering will require weekly or even twice-weekly feeds. It is always better to give frequent weekly feeds rather than infrequent heavy doses. Fertilization requirements vary by method of propagation and stage of plant development. Seeds are sown in a substrate with a recommended pH of 6.0-6.5. Once cotyledons appear, 14-0-14 NPK at a concentration of 50 to 75 $mg{\cdot}l^{-1}$ nitrogen (N) be applied and changed to 20-10-20 at 100-150 $mg{\cdot}l^{-1}$ N when true leaves appear (Hamrick, 2003). In case of cutting-propagated plants, the recommended pH ranges vary from 5.0-5.5 (Hamrick, 2003) to 6.0-6.6 (Heins *et al.*, 1990). There are no reports of low or high pH problems; however, Hamrick (2003) recommended the application of 20-10-20 which is an acidic fertilizer. Cuttings should be fertilized with 50-75 $mg{\cdot}l^{-1}$ N at sticking stage but increased to 100-200 $mg{\cdot}l^{-1}$ N once a week of 20-10-20 NPK when approximately 50 per cent of the cuttings have initiated roots. It is also recommended to alternate between acidic (20-10-20) and basic (15-0-15) fertilizers. Fuchsia should be potted in a substrate that is well-drained, disease-free, has a high initial starter charge, and a pH of 5.0-5.5. Hamrick (2003) recommended to continue alternating between 20-10-20 and 15-0-15 when fuchsia are potted, increased to 150-200 $mg{\cdot}l^{-1}$ N while the plants are becoming established, and once well rooted there should be further increase to 200-300 $mg{\cdot}l^{-1}$ N. Nutritional problems reported pertain to over-fertilization or under-fertilization. Over-fertilization can lead to excessive vegetative growth (Hamrick, 2003) or plants that will not flower or have delayed flowering (Dole and Wilkins, 2005). Poor branching will result if plants are under-fertilized (Hamrick, 2003).

Growth and Flowering

Fuchsia is a long day plant. Under warm or low-light conditions, B-Nine (1,250 to 2,500 ppm) and Cycocel (800 to 3,000 ppm) are effective in controlling stretch. In warm temperature regions, products like Sumagic and Topflor are more effective in controlling height. Vince (1977) found increased internodal length in fuchsia with increase in the daily duration of light when light was given throughout an otherwise dark period of 16 h, and with increase in illuminance to a saturation value of 200 lx from tungsten lamps. Elongation increased as a linear function of decrease in photostationary state of phytochrome down to P_{fr}/PC"0.3; however, internodes were shorter in far-red light than in 25 per cent red/red+far-red. It was concluded that stem length is a net response to two modes of phytochrome action. An inductive effect of P_{fr} inhibits a late stage in internode expansion, and a phytochrome reaction which operates only in light (and may involve pigment cycling) promotes an early stage of internode development. Stem elongation is thus a function both of the daily duration of light and its red/red+far-red content. The outgrowth of axillary buds was controlled by the first type of phytochrome action only. Kim (1995) observed that reduction in the stem length, leaf length and width of *Fuchsia × hybrida* 'Corallina' was proportional to the concentration of uniconazole under both type of treatments, *i.e.* foliar spray (5–200 ppm) and soil drench (0.025–1.0 mg a.i.per plant). But the effect was more prominent in the soil drench than in the foliar spray. Uniconazole also reduced the flower size as a result of reducing the peduncle, sepal and petal length. Flowering was accelerated by uniconazole treatment and the number of flowers increased. Moe *et al.* (2002) grew *Fuchsia* at a PPF of 200 $\mu mol\ m^{-2}s^{-1}$ in growth cabinets and recorded reduced plant height of about 20 per cent when the R/FR ratio in the light spectrum was changed from 1.1 to about 8, while a combination of high R/FR ratio and additional blue light reduced plant height about 30 per cent. Maas and Hattum (1997) observed that in absence of applied gibberellins, paclobutrazol (> 0.32 $\mu mol\ plant^{-1}$) strongly retarded shoot elongation in fuchsia.

Insect-Pests, Diseases and Physiological Disorders

Various *Aphis* species damage fuchsias with their feeding, especially at the tips of branches, or encourage the growth of diseases developing on their sugary secretions. They are small wingless insects, which may be green or bluish in colour. They suck the sap from young growing parts of the plant and excrete honeydew which leads to the sooty mould. A number of low-key or organic methods, including lady bugs or even simple sprays of water, are often very effective. Additionally, insecticidal soaps or chemical controls might be used. Insecticides like dimethoate, malathion and diazinon are effective against aphids. Whitefly (*Trialeurodes vaporariorum*) is a very small flying insect usually found feeding on the undersides of leaves. Numbers of these insects can be seen characteristically flying off erratically in a cloud when heavily infested plants are disturbed. The scale-like nymphs also do damage with their feeding and the resulting sugary secretions can encourage the formation of sooty moulds. Synthetic analogues of pyrethrum (a naturally occurring insecticide) are very effective against whitefly and are relatively non-poisonous to human beings. Permethrin when applied at 10 days interval is the best control measure for this troublesome pest. Alternatively, the introduction of parasitic wasp, *Encarsia* will provide good biological control. Vine weevil (*Otiorhynchus* spp.) feeds during nights on the fuchsia leaves and makes small semicircular notches while feeding. The creamy-white grubs eat the stems, roots and leaves and often cause the whole plant to wilt or die. The insect can be deterred by stirring a little gamma HCH dust onto surface of the soil. Caspid bug (*Lygocoris pabulinus*) is small insect, causing considerable amount of disfiguring damage to growing fuchsias by sucking cell sap, especially from the tips of shoots which causes the surrounding plant tissue to die. Consequently small holes and characteristic 'tears' appear and disfigure the leaves as they grow. Significantly the flowering is delayed as new growth has to develop in compensation. Systemic insecticides containing gamma HCH and insecticidal soaps are both effective when applied at the first sign of its damage. Caterpillars (*Deilephila elpenor*) are polyphagus and feed on every growing point, leaves and flowers. They are controlled by hand removal and by spraying either Dipel or dusting the carbaryl. Sciarids adults are small black flies, but larvae live in soil and cause damage only to seedlings or cuttings. They can be controlled by stirring a little gamma HCH powder into the surface of the soil around the infected plant. Leaf hoppers and thrips sometimes also feed on fuchsias which can be controlled by spraying the infested plants with insecticide containing gamma HCH or diazion.

Cyclamen mites (*Phytonemus pallidus*, syn. *Steneotarsonemus pallidus*) have also been recorded causing damage to fuchsias. Spraying such plants with insecticide containing gamma HCH or diazion will control this pest. Fuchsia gall mite (eriophyid mite; *Aculops fuchsiae*) infests the growing points, young leaves, and blossoms of fuchsia. As a result of hormonal-like substances that the mites inject into plant tissue as they feed, infested growth becomes twisted, distorted, stunted, grotesquely swollen and blistered, and often reddened. Pruning off mite-galled tissues whenever and wherever these are seen will keep plants of low to moderate susceptibility. Plants of higher susceptibility require pruning first to remove all galled growths, then thorough spraying with carbaryl. This process should be repeated two to three weeks later. If carbaryl is used, a miticide such as dicofol should be added to the spray tank to prevent spider mite outbreaks, but this may not be necessary if only two applications of carbaryl are made. After two applications of carbaryl, several months of mite-free fuchsia growth can be expected. Carbaryl is a more appropriate insecticide for the home gardener because of its much lower toxicity. Growing fuchsias are highly resistant to gall mite. Also there are a few cultivars which are resistant to gall mite. Some of the gall mite resistant species are *Fuchsia arborescens, F. asplundii, F. brevilobus, F. colensoi, F. denticulata, F. excorticata, F. glazioviana, F. loxensis, F. microphylla* var. *hidalgensis, F. microphylla var. quercetorum, F. paniculata, F. parviflora, F. procumbens* and *F. regia*. Spider mite (*Tetranychus urticae*) is very common and its infestation causes speckling and bronzing of the leaves, and sometimes premature leaf fall. The tiny mites spin fine web, especially on the underside of the leaves. Hot and dry conditions being congenial for mite build up, therefore such conditions should be avoided in the growing environment. Malathion and dimethoate are effective against mites. The introduction of predatory mite *Phytoseilus persimilis* will provide good biological control against this mite.

Fuchsia rust (*Pucciniastrum epilobii*) is the most serious disease of fuchsia. This disease is characterized by the raised, round and orange spots on the underside of leaves. All infected parts must be removed and burnt. The plants should be sprayed with fungicides such as thiram, maneb or zineb at fortnightly intervals until the problem disappears. This disease is also harboured by willow herbs and fir and to prevent the reinfection, these plants should be removed from vicinity of the greenhouse and garden. Grey mould (*Botrytis cinerea*) infection covers the entire leaves and stems with a downy mass of grey spores. Cold, damp and stagnant air favours the growth of the fungus. Plenty of ventilation, proper plant density, avoiding overhead irrigation and use of heat will help to control the disease. Benomyl spraying is also helpful.

Scorching of leaves is caused by excessive light levels and high temperature. It can be cured by increasing relative humidity. Parts of the leaves remain green at first, but afterwards leaf discolouration occurs in the form of yellow, brown and purple spots. When the days become shorter, certain natural changes take place within the plant itself so older leaves discolour though this is not alarming. Sunburn can also leave brown marks with purple edge on affected leaves. A purple tinge of the leaves in summer, especially on purple or deep red-flowered sorts is often due to incipient nitrogen starvation, and then a high nitrogen feed should be applied to correct this problem. Reddish foliage spots are caused due to phosphorus deficiency so in such cases sufficient amount of phosphorus as nutrition should be provided. As the situation becomes alarming at cold temperatures so immediately the growing temperature should be raised.

References

Berry, P.E. 1982. The systematics and evolution of *Fuchsia* Sect. *Fuchsia* (Onagraceae). *Ann. Missouri bot. Gard.*, **69**(1): 1–99.

Berry, P.E., W.J. Hahn, K.J. Sytsma, J.C. Hall and A. Mast, 2004. Phylogenetic relationship and biogeography of *Fuchsia* (Onagraceae) based on monocoding nuclear and chloroplast DNA data. *Amer. J. Bot.*, **91**(4): 601-614.

Chow, Y.N., B.M.R. Harvey and C. Selby, 1990. An improved method for callus proliferation and regeneration of *Fuchsia hybrida. Pl. Cell Tis. Org. Cult.*, **22**: 17-20.

Dabin, P. and A.M. Vaerman, 1985. Callogenese et organigenese chez deux cultivars de fuchsia (Constance et Swingtime). *Bull. Soc. Royal Bot. Belgium*, No. 118, pp. 172-178.

Dole, J.M. and H.F. Wilkins, 2005. *Bedding Plants*. In: *Floriculture Principles and Species* (2nd ed.), p. 934. Pearson-Prentice Hall, Upper Saddle River, New Jersey, USA.

Everard, B. and B.D. Morley, 1970. *Wild Flowers of the World*, plate 133. Peerage Books, London.

Fugen, G., C. Ruiqing, W. Zhigang and S. Bingya, 2006. Tissue culture and rapid multiplication of *Fuchsia albacoccinea* Hort. *J. Pl. Resources and Environ.*, **15**(3): 55-59.

Godley, E.J. 1955. Breeding systems in New Zealand plants, *Fuchsia. Ann. Bot.*, **19**: 549-559.

Godley, E.J. 1963. Breeding systems in New Zealand plants. 2. Genetics of the sex forms in *Fuchsia procumbens. N.Z. J. Bot.*, **1**: 48-52.

Godley, E.J. and P.E. Berry, 1995. The biology and systematics of *Fuchsia* in the South Pacific. *Ann. Missouri bot. Gdn*, **82**: 473-516.

Hamrick, D. 2003. *Fuchsia* (Ball Redbook, vol. 2, 17th ed.), pp. 393-396. Ball Publishing, Batavia, Illinois, USA.

Heins, R., R. Moe, T. Dudek and G. Draheim, 1990. Tips on growing quality fuchsia. PPGA News, p. 8.

Helsper, J.P.F.G., C.H. Ric de Vos, F.M. Mass, H.H. Jonker, H.C. van der Broeck, C.S. Jordi W. Pot, L.C.P. Kleizer and A.H.C.M. Schapendonk, 2003. Response of selected antioxidants and pigments in tissues of *Rosa hybrida* and *Fuchsia hybrida* to supplemental UV-A exposure. *Physiologia Plant.*, **117**: 171-178.

Hoshino, T. and P.E. Berry, 1989. Observations on polyploidy in *Fuchsia* sects. Quelusia and Kierschlegeria (Onagraceae). *Ann. Missouri bot Gdn*, **76**: 585-592.

Kevers, C.L., M.F. Coumans-Gilles, M. Coumans and T.H. Gaspar, 2003. *In vitro* vegetative multiplication of *Fuchsia hybrida. Sci. Hort.*, **21**(1): 67-71.

Kim, H.Y. 1995. Effects of uniconazole on the growth and flowering of *Fuchsia × hybrida* 'Corallina'. *Acta Hort.*, No. 394, pp. 331-335.

Maas, F.M. and J. Hattum, 1997. The role of gibberellins in the thermo- and photocontrol of stem elongation in *Fuchsia*. *Acta Hort.*, No. 435, pp. 93-104.

Moe, R., L. Morgan and G. Grindal, 2002. Growth and plant morphology of *Cucumis sativus* and *Fuchsia × hybrida* are influenced by light quality during the photoperiod and by diurnal temperature alterations. *Acta Hort.*, No. 580, pp. 229-234.

Pizzetti, I. and H. Cocker, 1975. *Flowers A Guide for Your Garden*, pp. 473-491. Harry N. Abrams Inc., Publishers, New York.

Reiter, V. 1944. Notes on the history of *Fuchsia* breeding. *J. Calif. hort. Soc.*, **5**: 144-192.

Roberts, M.J. 2000. Edible and Medicinal Flowers, 160 p. New Africa Publishers, Claremont.

Rop, O., J. Mlcek, T. Jurikova, J. Neugebauerova and J. Vabkova, 2012. Edible flower – a new promising source of mineral elements in human nutrition. *Molecules*, **17**: 6672-6683.

Talluri, R.S. and B.G. Murray, 2009. DNA C-values and chromosome numbers in *Fuchsia* L. (Onagraceae) species and artificial hybrids. *N.Z. J. Bot.*, **47**(1): 33-37.

Vince, D. 1977. Photocontrol of stem elongation in light-grown plants of *Fuchsia hybrida. Planta*, **133**(2): 149-156.

Williams, C.A., J.H. Fronczyk and J.B. Harborne, 1983. Leaf flavonoid and other phenolic glycosides as indicators of parentage in six ornamental *Fuchsia* species and their hybrids. *Phytochem.*, **22**(9): 1953-1957.

Wilson, H. 2013. Chelsea Flower Show : Guide to Edible Flowers.

22

Geranium & Pelargonium (Family: Geraniaceae)

Sapna Panwar, Sanyat Misra and R.L. Misra

[**Common names for true *Geranium***: Alumroot/Chocolate flower (*Geranium maculatum*), Bloody crane's bill/ Streaked crane's bill (*G. sanguineum*), Crane's bill (*Geranium*), Dusky cranesbill/Mourning widow (*G. phaeum*), Fox geranium/Herb Robert/Red Robin/Red Shanks (*G. robertianum*), Geranium, Gray crane's bill (*G. cinereum*), Iberian crane's bill (*G. ibericum*), Meadow crane's bill (*G. pratense*), Silver-leaved crane's bill (*G. argeneum*), Siberian crane's bill (*G. sibiricum*), Spotted/Wild crane's bill (*G. maculatum*), *etc.*]

[**Common names for true *Pelargonium***: Apple geranium (*Pelargonium odoratissimum*), Cactus geranium/Prickly geranium/Sweetheart geranium (*P. echinatum*), Fern-leaved geranium (*P. fruticosum*), Geranium, Ivy-leaved geranium/ Hanging basket geranium (*P. peltatum*), Lemon geranium/Prince Rupert geranium (*P. crispum*), Nutmeg geranium (*P.* × *fragrans*), Oak-leaved geranium (*P. quercifolium, P. terebinthaceum*), Pelargonium/Regal pelargonium/Butterfly pelargonium/Florist's pelargonium/Leopold pelargonium (*P. macranthum*, syn. *P. grandiflorum, P. regale*), Peppermint geranium/Peppermint-scented geranium (*P. tomentosum*), Regal geranium/Regal pelargonium/Royal geranium/ Fancy geranium/Lady Washington geranium/Martha Washington geranium/Queen of the garden geranium/Show geranium (*P.* × *domesticum*), Rose/Scented geranium (*P. graveolens*), Rose-scented geranium (*P. capitatum*), Scented geranium (various species or varieties with distinct leaf aroma), Stork's bill, Zonal/Fish/Horseshoe/Common house geranium (*P. zonale*), Zonal/Common house geranium (*P.* × *hortorum*), Tuberous-rooted pelargonium (*P. triste*), *etc.*]

Geranium

The genus *Geranium* derives from the Greek *geranus* 'a crane', which refers to the beaked fruit and so its common name 'crane's bill'. It is a genus of 300-400 evergreen (though some of which are semi-evergreen) herbaceous annual to perennial species of cosmopolitan distribution, *i.e.* temperate zones of the Northern Hemisphere. Though *Geranium* is different than *Pelargonium* but both are closely related. These are fast- and vigorous-growing, many preferring partial sun but there are species and varieties liking full sun, usually hardy plants with long flowering habit, and are excellent as ground cover, for herbaceous borders, for naturalizing effect under trees and the large outcroppings of rocks, against dry walls, in rock gardens, in alpine house, in the paving stones of a large terrace, in the side of stone steps, and in hanging baskets. These are most suitable for temperate gardens. The trade 'geranium oil' is misnomer, as commercial geranium oils are derived from *Pelargonium* hybrids and cultivars. The only true *Geranium* species which yields an essential oil in small quantities is *G. macrorrhizum* which is native of Bulgaria and its oil is used in adulterating the rose oil (Guenther, 1950). Its cultivated species are described below.

Geranium argenteum

A native to Europe, this biennial to perennial herbaceous species has prostrate growing habit growing only up to 10 cm high, therefore is most suitable for planting in the rock gardens. It is highly tolerant to sun.

The leaves are very attractive, silvery and multi-lobed with no regular number of lobes per leaf but it is usually 5 and more with spreading lobes. Pink flowers some 2.5 cm wide are violet-veined and very attractive. Old plants develop a thick and woody rhizomatous rootstock.

Geranium cinereum

A native of Balkans and Pyrenees, this tufted alpine species grows up to 10-15 cm high with spread of about 25-30 cm, bearing round to kidney-shaped grey-green leaves some 3 cm wide with 5-7 wedge-shaped lobes, each lobe having 3 teeth or segments some 2 mm long. Flowers are 2.5-3.5 cm across, crimson to magenta with almost black centre which appear freely from May to October. Its form *G. c. subcaulescens* is mot popular and highly compact with large flowers. 'Album' (white), 'Ballerina' (white, feathered crimson-purple), 'Claridge Druce' (purplish-pink; *G. endressii* × *G. versicolor*), *etc.* are its very popular varieties.

Geranium dalmaticum

A native of Dalmatia, SW Yugoslavia and N. Albania, this species grows up to 15 cm high with spread of 25-30 cm, and it forms a dense and low cushions with glossy mid-green and palmately 5-lobed some 3 cm or more wide leaves which get tinted red and orange during autumn. Light pink clustered flowers, each flower some 2.5 cm across appear from June to August. Its form 'Album' is pure white.

Geranium endressii

A native of West Pyrenees and Southern Europe (SW France and Spain), this is a clump-forming hardy rhizomatous species making excellent ground cover and naturalizes well at lightly shaded position where it assumes a more compact growth, and grows up to 30-45 cm tall with spread of 45 cm and leaf breadth of 5-8 cm. Its palmate (ranunculus-type) and deeply 4-5-lobed and toothed leaves are mid-green. Lightly red-veined pale-pink flowers some 2.5-3.0 cm across appear from May to August having long flowering period. 'Claridge Druce' is its named hybrid growing up to 45 cm high bearing some 5.0 cm across lilac-pink flowers. Its var. 'A.T. Johnson' grows up to 30 cm high with silver-pink flowers, 'Rose Clair' 45 cm tall with salmon-rose flowers having purple veining, and 'Wargrave Pink' grows up to 45 cm high with pink flowers.

Geranium grandiflorum (G. himalayense, G. meeboldii)

A native of Sikkim (India), this hardy, rhizomatous and colonizing species grows to 30-45 cm in height and some 45-60 cm spread with leaves some 7-10 cm wide. Its long-stalked and 5-7-lobed leaves are round and mid-green, where each lobe is further 3-lobed, deeply toothed and prominently veined. Blue-purple but red-veined flowers some 3.75-5.50 cm across appear from June to July. It is suitable for growing under full sun or partial shade. Its popular variety is 'Alpinum' ('Gravetye') which is very compact and where flowers are reddish-stained in centres.

Geranium ibericum (G. ibericum × G. platypetalum, G. × magnificum, G. platypetalum)

A native of Caucasus, this is a hardy, clump-forming and vigorous species growing 30-60 cm in height with spread of 45 cm though with this name now it is available in nurseries only. In fact, it is a sterile hybrid. Leaves are mid-green, erect, hairy, 10 cm or more wide and deeply 5-7 pinnatifid-lobed. In July and August, this produces 2.5-3.25 cm across glossy violet-blue flowers veined reddish. 'Album' is its white form.

Geranium incisum

A native of western North America (British Columbia to California), it is a hardy perennial growing up to 45 cm high, bearing 3-5-segmented dentate leaves. From June to August, purple-pink flowers some 2.5 cm in diameter appear.

Geranium macrorrhizum

A native of Southern Europe (SE Alps, the Balkan Peninsula, Carpathians and Apennines), this rhizomatous and colonizing species grows up to 40 cm tall with a spread of up to 60 cm and the leaf breadth of 4-10 cm. It is a very handsome plant for semi-shady borders. Its old specimens develop a thick and woody rootstock. Its leaves are palmate, mostly basal, dentate, mid-green but slightly suffused red in autumn, aromatic when rubbed and deeply 5-7-lobed where each lobe is pinnatifid. Pale-magenta-pink flowers appear in trusses, each some 2.5 cm across from May to July. Its var. 'Album' is white but calyces reddish,'Variegatum' with leaves splashed creamy-white, 'Walter Ingwersen' bears pink flowers, *etc.* Out of all the *Geranium* species, only this species yields essential oil which is used for adulterating the rose oil.

Geranium maculatum

A native to the eastern United States, this hardy perennial herb has been recorded growing to a height of 45 cm in stony soils of the light woodland areas giving a naturalizing effect. Its green leaves are deeply 3-5-lobed. Its pinkish-purple flowers with about 2.5 cm diameter appear from April to May.

Geranium napuligerum (G. farreri)

A native of Yunnan (China), it is a tufted, decumbent and slow-growing species growing only up to 15 cm high with spread of 30 cm, suitable for screes. Its close mat-forming mid-green leaves are about 3 cm wide, palmate and deeply 3-5-lobed. Its flowers are pale-rose, some 2 cm across and appear from May to August. Its var. 'Album' is white.

Geranium nodosum

A native from Pyrenees to C. Italy and C. Yugoslavia, it is a rhizomatous species growing up to 20-50 cm high with erect or reclining stems. Its leaves are dentate, ovate and deeply 3-5-lobed. Pink, lilac or reddish flowers some 2-3 cm across appear during summer.

Geranium phaeum

A native of Europe and W. USSR, it is a clump-forming species growing to 60 cm high with 6-18 cm wide 5-7 leaves. Leaves are boldly toothed and 7-lobed, where central one is smallest. Lilac- to brownish-purple flowers some 2.5-3.9 cm wide with or without paler eye appear during summer with irregularly and minutely notched petal margins. Its varieties are 'Album' (white), 'Lividum' (brownish-purple), *etc.*

Geranium pretense

A native of North Europe, especially Great Britain, it is a clump-forming species growing up to 50-80 cm tall with spread of 60 cm and leaf breadth of 6-12 cm. Its mid-green leaves are long-stalked and deeply divided up to base into 5-7 lobes, each lobe being pinnatisect. Blue to violet-blue flowers veined crimson and some 3-4 cm across appear from July to September. Its vars 'Mrs Kendall Clarke' bears single blue flowers, 'Flore Pleno' double blue, 'Album' white, 'Plenum Album' double white, 'Plenum Caeruleum' double lavender-blue, 'Plenum Violaceum' double deep violet-blue, 'Roseum' pink veined darker, 'Striatum' syn. 'Bicolor' petals striped or sectioned with white and purple flecks, 'Johnson's Blue' with large lavender-blue and darker veined flowers, and 'Rectum Album' from Kashmir having large white flowers with purple veins. Its hybrid 'Johnson's Blue' grows up to 38 cm in height with deeply lobed, dark green and palmate leaves, and light blue flowers some 5 cm wide which appear from July. The chromosome number in *Geranium pretense* is 2n=28.

Geranium psilostemon (G. armenum)

A native of Armenia (USSR), it is a clump-forming and robust species growing up to 75-120 cm tall with spread of 60 cm and leaf breadth of 10-20 cm, which is suitable for planting in the herbaceous border. Its leaves are mid-green, palmate, 5-lobed and deeply toothed and turn red in autumn. Flowers which appear during June and July are 3.5-4.0 cm across and vivid magenta with a black central spot.

Geranium pylzowianum

A native of Kansu (China), this rhizomatous ground cover species grows to 10-30 cm in height with a spread of 15-23 cm. It is a fast-growing species, especially suited for rock gardens. Its mid-green leaves are 5 cm wide, rounded and deeply 5-lobed, each lobe being trifid. Pink-purple flowers some 3 cm across appear sparsely during June and July.

Geranium renardii

A native of Caucasus, this clump-forming rock garden species grows to 20-30 cm in height and with 30 cm spread, bearing rounded grey-green, upper surface wrinkled, lower surface felted and 5-6 cm wide leaves having broadly 5-7 lobes with attractive veining. It makes a neat mound. Pale-lavender flowers some 2.5-4.5 cm across with purple-veined petals appear from May to July.

Geranium robertianum (Robertiella robertianum)

A native to Europe, Asia and North America, this hardy annual to biennial species with erect branching is always found under shady positions such as light and moist woodlands, and in the crevices of damp rocks and walls, and makes an excellent ground cover and hanging baskets. It bears red stems, grows to about 25 cm tall, is very effective in informal plantings of the garden and continues appearing each year from the self-sown seeds. Its dentate leaves have pleasant aroma when crushed, are palmate and multi-lobed. Small pinkish-violet flowers appear in profusion from June to October.

Geranium sanguineum

A native of Europe including Great Britain and W. Asia, this hardy, rhizomatous and decumbent species grows to a height of up to 30 cm and to a spread of 45 cm, and is an excellent ground cover forming wide mats, bearing 3.5 cm wide, mid-green and gracefully deeply divided leaves into 3-7 pinnatisect lobes. Crimson-magenta flowers 2.0-3.8 cm across appear for a long time from June to September. It thrives well in sun and partial to complete shade and even in the poorest soils providing there is no dearth of water. Var. 'Album' is smaller than parent having pure white flowers. The form *G. s. lancastrense* is a mat-forming quite dwarf strain growing up to 7.5-10.0 cm tall with a spread of up to 38 cm. Its leaves are soft dark green and pale-pink flowers are 2.5 cm wide with darker veins. 'Prostratum' grows nearly prostrate, leaves smaller than parent and flowers pale-pink veined red.

Geranium sessiliflorum

A native of Australia and New Zealand, this is a tufted species forming low hummocks. Its leaves are shining green, 3-4 cm wide and 5-7-lobed. White flowers some 1.0-1.5 cm wide appear during summer, often hidden in the leaves. The leaves of *G. s. nigricans* are suffused purple-brown.

Geranium staphianum roseum

It is akin to *G. pylzowianum.* It bears beaded rhizomes and makes the spreading mats, stems grow up to 15 cm high, leaves 4-6 cm wide, 5-7 narrow lobes cut up to the base and each lobe again deeply cleft and often tipped/margined dull purple-red, bright lilac-pink flowers some 3 cm wide appear during summer.

Geranium subcaulescens

A native of SW Balkan Peninsula and CS Italy, this grows to a height of 15 cm and spread of 30 cm, and is most suitable for planting in the rock gardens. Its lobed leaves are grey-green and orbicular and the lobe-segments some 5 mm long. Similar to *G. cinereum*, its flowers are bright crimson-magenta with black centres but lack the dark veining, some 2.5 cm wide and appear in masses during May to October. Its var. 'Splendens' bears large and paler flowers. Hybrid 'Russsell Prichard' grows up to 20 cm tall with a spread of 45 cm, bears palmate pale-grey leaves, prostrate floral stalk and masses of magenta flowers some 3.75-5.0 cm across. This is most suitable as ground cover.

Geranium sylvaticum

A native of Europe to Siberia, this clump-forming species grows to a height of 60 cm having branched stems above and the spread of 45 cm. Its 5-12 cm wide leaves are silver-green, rounded and deeply 5-7-lobed, each lobe being toothed and pinnetisect. Purple flowers some 3 cm wide appear in loose terminal clusters some 2.5-3.75 cm across in May and June. Variety 'Album' is pure white, 'Mayflower' is lavender-blue, and 'Roseum' & 'Wanneri' are pink.

Geranium tuberosum

A native of S. Europe, this nearly hardy tufted species has rounded-tubererous rootstocks and grows to 30-60 cm tall. It prefers sunny, hot and dry position, unlike to other species. It is ideal for rock gardens, dry walls and for pot culture. Leaves are divided up to the base into 5-7 pinnetisect lobes. Pale-pinkish-purple flowers some 2.5-3.0 cm wide appear in May-June.

Geranium traversii

An almost hardy species native of Chatham Islands and New Zealand, this tufted species bears silvery grey-green, 2-5 cm wide and 5-7-lobed leaves which form a 15 cm mound. Pink or rarely white flowers some 2.5 cm wide continue appearing from summer to autumn. Variety 'Elegans' bears darker-veined pink petals.

Geranium wallichianum

A native of Himalayas, it is a hardy and tufted species with prostrate or decumbent silky and hairy stems, growing to a height and spread of up to 60 cm and is suitable for planting in the front row of the border. It prefers damp and shady positions. Its light green, wedge-shaped, 3-5-lobed leaves are deeply toothed and covered with silky hairs. Beautiful purple-blue flowers some 2.5-4.0 cm across appear from July to September. The flowers in the var. 'Buxton's Blue' is blue with white eyes.

Geranium wlassovianum

A native of Siberia, Manchuria and NE China, this is a clump-forming species growing to 38-60 cm tall with a spread of 60 cm. Its stems are loose and branching-type, and bear velvety dark green, bean-shaped and 3-5-lobed leaves. Deep lavender-blue with darker veined flowers some 3-4 cm across appear from June to September.

Pelargonium

The genus *Pelargonium* derives from the Greek *pelargos* 'a stork', in allusion to its long-beaked fruits (seed case) (Pizzetti and Cocker, 1968; Beckett, 1983; Bown, 2001; Miller, 2002). It is reported that the name *Pelargonium* was introduced by Johannes Burman in 1738. It is widespread throughout warm temperate areas of the new world, centred in South Africa, apart from North Africa, adjacent Atlantic Islands and eastwards to Arabia and southern India, as well as Australasia. The genus comprises of nearly 280 species which is mostly native to South Africa (Kellen, 1983) though a few species are also from Australia, Eastern Africa, New Zealand, the Middle East and the islands of Madagascar, St. Helena, and Tristan de Cuhna (www.herbsociety.org). The first pelargonium to reach Europe was probably *Pelargonium zonale* as the plants were sent in 1609 by the Governor of Cape Colony (South Africa) to Holland. The first evidence of pelargonium in cultivation dates back to 1633. Dutch East India Company in 1652 established a trading post at Table Bay which with time developed into a colony so from here various pelargonium species were imported into Europe. While revising Gerard's Herbal, Thomas Johnson observed geranium in blooms in the garden of John Tradescant which he referred to that as *Geranium indicum noctu odoratum* and the species was enlisted in the catalogue of the Botanic Garden in University of Leyden during 1668. A large number of pelargonium species were imported into Europe between the end of the 18th century and the beginning of the 19th century so in 1802 the first monograph on Geraniaceae, and Derek Clifford (Blandford, London) in 1958 published a complete book 'Pelargoniums, including the Popular Geraniums' (Pizzetti and Cocker, 1968). The genus *Pelargonium* comprises of diversified group of plants with a varying growth habit, and a few of *Pelargonium* species are known for their commercial traits. Hence, commercial pelargoniums are mostly suited as bedding plant, pot plant, container-grown plants especially in window boxes and hanging baskets. Pelargoniums are also well known for their medicinal use and more than 120 different chemical compounds have been identified in its oil (Williams and Jeffrey, 2002). Major components of pelargonium essential oils include citronellol, geraniol, linalool, menthone, isomenthone, limonene, pinene and methyl eugenol. The species especially *Pelargonium graveolens* is known to cure many ailments such as bleeding, wound healing, ulcers and skin disorders, diarrhoea, dysentery, *etc.* The species are also reported of having antibacterial and insecticidal properties.

Botany

Geraniaceae comprises of three genera, *viz. Geranium*, *Erodium*, and *Pelargonium*, differing only in floral morphology. The genus *Geranium* and *Erodium* bear regular flowers, while in *Pelargonium* the flower symmetry is zygomorphic. *Geranium* consists of 10 fertile stamens with anthers while in *Erodium*, some of the stamens do not possess anthers.

The members of the genus *Geranium* cultivated in gardens are generally perennials, some even with tuberous rootstocks, certain being prostrate-growing but there are even annual species. They are tufted, clump-forming or wide-spreading through rhizomes, bearing long-stalked alternate or opposite, and simple rounded or palmately-lobed leaves. The flowers are often attractive and symmetrical but lacking spurred calyx and are in shades of pink through purple to blue, 5-petalled, flattened or reflexed and bowl- or saucer-shaped, followed by splitting-type long-beaked fruits containing 1-5 seeds.

Pelargonium exhibits wide variations in respect of plant habit and flower forms and generally bears beak-like fruit by a nectar tube. The most characteristic feature of the *Pelargonium* is a tube that runs from the uppermost sepal along the flower pedicel. *Pelargonium* is more commonly known as geranium but differs from *Geranium* in being semi-woody, frost-tender and is meant only for bedding and pot culture worldwide. Pelargoniums prefer warm and sub-tropical climate such as of South Africa from where most species are recorded but a few are native even to Asia Minor, New Zealand and Australia with warmer climate. Its flowers are generally irregular with two upper petals different than others in shape, size and often in colour, and here the calyx spur is fused to the flower stalk though *Geranium* proper has no spurred calyx but symmetrical flowers (Everard and Morley, 1970). Certain morphological characters pertaining to *Pelargonium* are tabulated below.

Habit	Perennial shrubs, shrublets, acaulescent geophytes, scramblers or annuals
Stem	Erect or decumbent, often woody at base, soft-wooded or sub succulent, sometimes succulent, often aromatic, variously hairy, often glandular, sometimes with persistent spine-like stipules or petioles
Leaves	Usually petiolate, stipulate, alternate or opposite, entire to much dissected (sometimes on the same plant) or compound, variously hairy, often glandular and aromatic
Inflorescence	Two to many-flowered pseudo-umbel, bracts present
Flowers	Rarely solitary, zygomorphic, pentamerous
Sepals	Imbricate, connate at base, receptacle forming a hypanthium with a nectariferous spur opening (hypanthium) at base of posterior sepal, lower end of spur thickened and with a nectariferous gland
Petals	Usually 5, rarely 4, seldom 2, in 2 groups of 2 upper and 0-3 lower, imbricate, unguiculate (clawed) or sessile, rarely lacerate, usually variously coloured
Stamens	10, connate at base, 2-7 filaments bearing anthers (fertile), remaining ones often vestigial (staminodes), anthers dorsifixed
Ovary	5-lobed, 5-locular, beaked (rostrate), with 2 ovules in each locule, hirsute; style of varying length; stigmas 5, usually filiform and reflexed
Fruit	rostrate schizocarp; mericarps rostrate, 1-seeded, tapering from the apex to the base, long-haired
Seeds	More or less oblong-obovoid; endosperm absent; embryo curved.

Source: Van der Walt (1977).

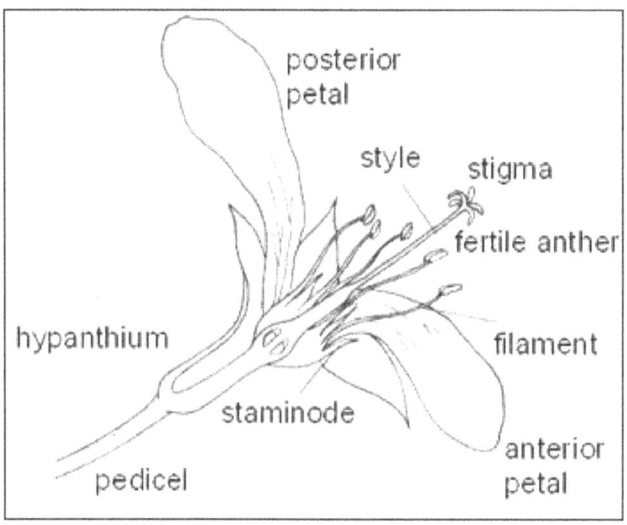

***Source*: www.pelargonium.si**

The chromosomal studies among several *Pelargonium* species and cultivars was done by Daker (1969) who stated that most of aromatic pelargoniums are derived from species *Pelargonium crispum* and *Pelargonium graveolens*, having basic chromosome number x = 11. The zonal and the ivy-leaved pelargoinums are reported with basic chromosome no. x = 9 while *P.* × *fragrans* x = 8. Diploid numbers recorded in 37 species are 16, 18, 20, 22, 36, 45, 60, 72, 81, 88 and 90 with x = 8, 9, 10 and 11 (Federov, 1974), while Demame (1989) reported 2n = 8x = 88 for *P. radens, P. graveolens, P. vitifolium* and 2n = 6 x = 66 for *P. capitatum*. The chromosome numbers of some of the species are tabulated below.

Pelargonium Species	Chromosome No (2n)	Pelargonium Species	Chromosome No (2n)
Pelargonium ovale	20	P. denticulatum	44
P. exhibens	22	P.capitatum	66
P. acetosum	18	P. radens	88
P.frutetorum	18	P. gibbosum	22
P. inquinans	18	P. exstipulatum	16
P. peltatum	18	P. odoratissimum	16
P. zonale	18	P. endlicherianum	34
P. echinatum	22	P. cucullatum	22
P. grandiflorum	22	P. carnosum	22
P. oblongatum	22	P. fulgidum	22
P. trifidum	22		

Source: www.geraniumsonline.com

Classification, Species and Varieties of *Pelargonium*

Important *Pelargonium* species have been categorized into six groups as following:

Zonal

The large and diverse group of cultivars of a hybrid race are included in Zonal group, as also known as *Pelargonium* × *hortorum* (Zonale hybrid with chromosome number of 2n=18) which are largely derived from *P. zonale*. The true ancestry of zonal pelargonium is though not well known but the species such as *P. inquinans, P. zonale, P. hybridum* and *P. frutetorum* have been considered to be involved in its ancestry. These are propagated by cuttings though many of its hybrids are propagated through hybrid seeds. These are succulent-stemmed and branched perennials with upright and bushy appearance, and single or double flowers having attractive foliage and most commonly used for bedding displays. They have rounded, pale to mid-green leaves with a conspicuous zone of bronze or maroon. Flowers are 1.25-2.5 cm across, in shades of white, pink, orange and red, appearing in dense rounded umbels from the upper leaf axils during summer and autumn.

Out of several seed strains available, the most prominent one is 'Carefree' which includes 'White', 'Bright Pink', 'Deep Salmon' and 'Scarlet' and their mixture; other one is 'Irene seed strain' which includes 'Electra' (crimson, semi-double)', 'King of Denmark' (salmon-pink, semi-double), 'Maxim Kovaleski' (orange), *etc.* F_1 hybrid seed strain includes 'Deep Salmon', 'Light Pink' and 'Scarlet'.

Named outstanding varieties are 'Always' (double, creamy-white, centrally flushed pink), 'A.M. Mayne' (double, magenta, centre flushed scarlet), 'Baron de Layres' (double white), 'Barrone A. de Rothschild' (large semi-double, pale-pink), 'Beauty of El Segundo' (double creamy-pink), 'Belvedere' (mauve-pink), 'Bob Legge' (leaf bronze-zoned, flowers double pink), 'Brenda Kitson' (semi-double, rosy-mauve), 'Brocade' (double red, centre shaded white), 'Burgenland Girl' (semi-double red-pink), 'Caroline Schmidt' (leaves edged cream, double red flowers), 'Clarona' (single, scarlet-red), 'Cleopatra' (soft pink), 'Coronation' (rich pink), 'Countess Mariza' (large semi-double, coral-pink), 'Countess of Jersey' (salmon-pink, large but single), 'Cover Girl' (semi-double, pale magenta-pink), 'Decorator' (scarlet semi-double), 'Delight' (vermilion-red), 'Double Jacoby' (double red), 'Du Barry' (salmon-pink), 'Evesham Wonder' (large semi-double, salmon-pink), 'Favourite' (double, pure white), 'Fiat Queen' 'salmon semi-double', 'Fleurette' (height 20 cm, leaves dark and flowers double salmon-red), 'Glenn Barker' (pink double), 'Golden Lion' (single pale-orange flowers), 'Gustav Emich' (vermilion-red, semi-double), 'Hans Rigler' (red semi-double), 'Hermione' (double white), 'Hildegarde' (semi-double, orange-red), 'Improved Ricard' (semi-double, orange-scarlet), 'Irene Cal' (double pale-salmon-pink), 'Jewel' (deep pink, double), 'King of Denmark' (semi-double, rose with marking and veining of geranium lake), 'Lady Ilchester' (satin rose, double), 'Lief' (large semi-double, orange-pink), 'Lioness' (single, violet), 'Lorelei' (double, attractive salmon-pink), 'Lucia' (single, rose), 'Marktbeherracher' (semi-double, rose-carmine marked darker), 'Meteor' (double, scarlet-red), 'Millie' (semi-double, brick-red), 'Modesty' (large semi-double, white and free-flowering), 'Mrs Henry Cox' (leaves bright tricolour), 'Mrs Lawrence' (double soft pink), 'Mrs Parker' (leaves edged white, flowers double rose-pink), 'Noele Gardon' (double pale-pink, prolific), 'Olympia' (semi-double pink), 'Orange Ricard' (large semi-double orange), 'Paul Crampel' (oldest and best, single, scarlet-red), 'Paul Humphries' (double purple with crimson tinge), 'Queen of Denmark' (semi-double salmon-pink),

'Queen Sofia' (double, salmon-pink), 'Radiance' (double coral red, eye white), 'Rainbow' (single, pink), 'Red Black Vesuvius' (height 20 cm, foliage black-green, flowers single bright red), 'Red Denmark' (bright scarlet double), 'Royal Purple' (double purple, eye small white), 'Santa Maria' (large semi-double, salmon), 'Shimmer' (semi-double, apricot with white centre), 'Silverlachs' (semi-double, pale-pink), 'Sultan' (semi-double, salmon-pink), 'Venus' (single white flowers), 'Volcano' (single, red), *etc.* Miniature varieties growing from 15-30 cm in height are 'Caligula' (height 15 cm, leaves dark green, flowers double and scarlet), 'Carolyne' (flowers clear pale-pink, foliage dark green), 'Claudius' (20 cm, blackish-green foliage, flowers large white suffused pink), 'Dick's White' (white double), 'Nero' (15 cm, stems red, blackish-green with red marks in the centre, flowers red), 'Red Black Vesuvius' (leaves flushed heavily with black-purple and flowers bright scarlet), 'Silver Kewense' (10 cm, foliage cream-white marked green, flowers crimson), 'Tiberius' (20 cm, leaves greenish-black, flowers red), 'Timothy Clifford' (20 cm, foliage dark green with darker mark, flowers double and pink), *etc.*

Varieties for foliage beauty are 'Black Cox' (foliage vivid dark green, widely zoned black, flowers single, pale-pink, centre darker),'Caroline Schmidt' (leaves green and pale-yellow), 'Chelsea Gem' (leaves green variegated silver-cream, flowers double red), 'Crystal Palace Gem' (leaves golden with butterfly-type marking, flowers single and pink), 'Distinction' (foliage dark green with darker ring, flowers single, red and pink), 'Flower of Spring' (foliage margins white), 'Freak of Nature' (stems pure white, foliage apple-green with butterfly markings, flowers single pink), 'Happy Thought' (foliage green with gold butterfly mark, flowers single crimson), 'Henry Cox' (leaves splashed and zoned with cream, green, red and maroon, flowers single and pink), 'Lass O'Gaurie' ('Carse O'Gowrie; as to Henry Cox but basic colour silver instead of gold'), 'Maréchal MacMahon', 'Miss Burdett-Coutts' (leaves silvery-green suffused pinkish-carmine, flowers red), 'Mme Stelleron' (silvery-white flushing on foliage), 'Princess Alexandra' (foliage silver-margined light green, flowers double mauve), 'Skies of Italy' (leaves maple-like, splashed with orange and crimson and edged creamy-white, flowers single vermilion), 'Snowbody' (double white variegated leaves), 'Sophie Dumaresque' (height 25 cm, leaves similar to 'Henry Cox', flowers single red and vermilion), 'Verona' (foliage pale-gold, flowers single pink but centrally veined darker), *etc.*

Regal

Regal pelargonium is, in fact *P.* × *domesticum*, mainly derived from *P. capitatum, P. cucullatum* (*P. angulosum*), *P. fulgidum* and *P. grandiflorum* and is the most beautiful of the flowering geraniums propagated through cuttings. This consists of a large group of plants that were known by varied names such as 'Show', 'Grandiflora', 'Large-flowered', 'Decorative' and 'Fancy'. They have also been named as 'Martha' or 'Lady Washington' pelargoniums in the United States. It includes erect, evergreen, perennial and branching shrubs with bushy appearance, growing to a height of 38-60 cm. The leaves are mid-green, roundish to palmate, sometimes lobed and/or partially toothed or deeply serrated, slightly hairy, odourless or with faint pleasant smell. Flowers open flat or slightly bell- to trumpet-shaped in umbels from the upper leaf axils, 3.75-5.0 cm across, single type (rarely double) in shades of mauve, pink, purple or white, usually veined or blotched with darker shades. These are grown for outdoor or indoor display. These flower only when temperatures are about 10 °C. Most of the ancestral regal pelargoniums flower only once in spring, however, the modern types are repetitive in flowering and continue to bloom from May to October or first frost and some varieties even during winter.

Varieties under this group are 'Alberts Choice' (orange-salmon, white throat), 'All My Love' (mauve, base creamy-white), 'Applause' (pink, centre white, edged frilled and pearly-white), 'Ashley Stephenson' (orange-salmon, throat white), 'Aztec' (bright red, margined white, veined red-purple), 'Beauty of Bath' (pale-mauve and white ruffled), 'Black Knight' (black-purple, petal edges white-picotee), 'Black Magic' (blackest), 'Bredon' (large wine-red), 'Caprice' (deep cherry-red), 'Carisbrooke' (large frilled pink flowers, blotched maroon), 'Cezanne' (upper petals purple, lower pale-lavender), 'Cherie' (white, petals marked maroon), 'Chew Magna' (pinky-white, blotched and veined red), 'Chorus Girl' (bright salmon with lavender edges), 'Country Girl' (blotched pink), 'Doris Frith' (creamy-white veinred red), 'Dubonnet' (red), 'Dunkery Beacon' (orange), 'Elgar' (mauve and purple), 'Elsie Hickman' (pink, throat white, overlaid black and maroon), 'Fascination' (pale-pink, marked red and maroon), 'Flame' (fiery-red), 'Geoffrey Horseman' (purple, deep markings), 'Georgia Peach' (pink), 'Golden Princess' (green-gold variegation, flowers white), 'Grand Slam' (rose-red with violet shading), 'Harlequin' (pink and red), 'Hazel' (large violet-purple), 'Horace Parsons' (white, petals marked crimson), 'House and Gardens' (red, petal top maroon, throat pale mauve-pink), 'Jim Field' (purple-black), 'Joan

Fairman' (white, flushed pink, blotched maroon), 'Joy' (frilled salmon-pink, throat white), 'Julie Smith' (pretty lavender and purple), 'Lavender Grand Slam' (silvery-mauve with a light maroon blotch), 'Lowood' (purple-mauve, marked deeper), 'Madrilène' (intense lilac), 'Marie Rober' (large violet, petals marked darker), 'May Magic' (large salmon-orange, edges and throat white), 'Mendip' (large salmon-pink, white reverse, top petals flushed vermilion and feathered black), 'Morocco' (rich violet), 'Muriel Hawkins' (large pink), 'Noche' (maroon-red, petal edges red), 'Nomad' (flowers large white, upper petals blotched), 'Pansy' (purple-red), 'Pinocchio' (petals white, margined red), 'Pompeii' (nearly black, petals edged pinkish-white), 'Purple Emperor' (large lavender-purple), 'Quakeress' (mauve with purple markings), 'Quantock' (salmon-pink, flushed orange, upper petals marked purple), 'Rita Coughlin' (pale-lavender), 'Robbie Hare' (salmon), 'Rogue' (large crimson), 'Rosy Dawn' (pinkish-white), 'Sandringham' (rose), 'Smile' (bright pink), 'Sunrise' (large orange-salmon, throat white), 'Sybil Bradshaw' (large violet-lavender), 'Ted Dutton' (orange-salmon, throat white), 'Thea' (salmon-orange), 'Victoria Regina' (white, splashed deep purple), 'Violetta' (purple, centre pink), 'Wedding Gown' (large white), 'White Glory' (white), *etc.*

Ivy-leaved

The ivy-leaved pelargoniums (*P. peltatum*) derive their names from ivy (*Hedera helix*) because of close resemblance of leaves in both the genera. It has been reported that some cultivars of the group such as 'Lilac Gem', bear strong smell of ivy. The group includes evergreen perennials with trailing nature, ideal for hanging baskets, propagated primarily from stem cuttings though a few cultivars are seed-propagated. Rounded and lobed leaves are stiff, fleshy, often shining and some cultivars possess dark zone in the centre. It bears single or double flowers similar to those of Zonal pelargoniums. The varieties usually grown for the pot plant trade are 'Galilee' (large rose-pink), 'Madame Crousse' (pale-pink veined maroon), 'Scarlet Crouse' (scarlet sport of Madame Crouse), 'Souvenir de Charles Turner' (deep pink veined maroon), *etc.*

Scented-leaved

The characteristic feature of the group is that the members (species/hybrids) have scented foliage though flowers are small, and often irregularly star-shaped. These are shrubby evergreen perennials, some cultivars are cutting-propagated while others seed-propagated. Leaves are scented, often distinctly lobed, toothed and/or incised or variegated. It is stated that *P. crispum* and *P. graveolens* have contributed more to the development of many attractive plants whereas *P. tomentosum* and *P. odoratissimum* have contributed only to a lesser extent. The flowers of this group are often considered less important. All these species are from South Africa such as *Pelargonium odoratissimum* (apple fragrance, flowers white), *P. crispum* (lemon scented, flowers pale-lavender, var. 'Variegatum' bears variegated leaves), *P. abrotanifolium* (artemesia scented), *P. limoneum* (strongly lemon scented, flowers pink), *P. radula* (rose fragrance, flowers small, pale-purple), *P. graveolens* (strong rose scent, flowers small mauve-pink), *P. capitatum* (rose-scented, flowers small pink), *P. tomentosum* (peppermint-scented, flowers small pink), *etc.* Scented varieties are 'Attar of Roses' (rose-scented, flowers mauve-pink), 'Citriodorum' (citrus scent, pale-mauve flowers), 'Citrodifolium' (syn. 'Queen of Lemons', leaves lemon scented and pale-pink flowers), 'Clorinda' (rose-scented, large purple flowers), 'Crispum Variegatum' (lemon scent, mauve flowers), 'Endsleigh (aroma pepper-like, flowers lavender), 'Fair Helen' (pungent aroma, flowers purple), 'Filicifolium' (pungent aroma, flowers small purple), 'Fragrans' (nut-meg aroma, flowers small white), 'Graveolens' (orange-scented, pink-mauve flowers), 'Joy Lucille' (peppermint smell, white flower), 'Lady Plymouth' (rose-scented and pale-pink flowers), 'Mabel Grey' (strong citrus scent, purple flowers), 'Odoratissimum' (apple scented, white flowers), 'Prince of Orange' (orange-aroma and pale-pink flowers), 'Quercifolium' (spicy smell, purple flowers), 'Radula' (rose-lemon scent, mauve flowers), 'Tomentosum' (strong peppermint smell, white flowers), 'Variegated Fragrans' (pine scent, white flowers), *etc.*

Angel

The angel pelargoniums are almost similar to regals and are classified as miniature regal pelargoniums with a difference that they are more compact and bushy in appearance. Angel name is considered to be derived from a dwarf pelargonium 'Angeline' which is not taller in cultivation and group is mostly derived from *P. crispum*.

Unique

This group includes large shrubby evergreen perennials with Regal-like single brightly-coloured flowers that are borne continuously through the season. The leaves are often deeply lobed with sometimes having pleasant smell. It produces sparse flowers in shades of white, pink, mauve and red. The ancestry is uncertain but it is thought that *P. fulgidum* played an important role in its ancestry.

Also there is an important group 'Irenes' introduced since World War II in America with semi-double flowers, all raised from the original variety 'Irene'. Their florets have no clear forms that old semi-doubles have but other way they are superior. They are easy to grow and propagate, provide plenty of cuttings per plant and produce the saleable plants in much less time. The varieties under this group are 'Apache' (darkest red), 'Better Times' (short-jointed plant, flowers soft salmon-pink), 'Cal' (salmon), 'Corsair' (light red), 'Dark Red Irene' (very large dark red bloom), 'Electra' (rose tinged purple), 'Genie' (red), 'Irene' (crimson, parent of all Irenes), 'Jeweltone' (short jointed plant, deep crimson), 'La Jolla' (large crimson), 'Lollipop' (orange-scarlet), 'Modesty' (white, free flowering), 'Party Dress' (pale rose-pink), 'Penny' (pink with large white eye), 'Radiant' (short jointed plant, light brick-red flowers), 'Rose Irene' (rose-pink, eye white), 'Toyon' (crimson), 'Treasure Chest' (bright scarlet), 'Warrior' (red shaded salmon), *etc. P. floribunda* with no botanical standing is a new hybrid group which resulted by crossing *P.* × *hortorum* and *P. peltatum*. This group is commonly known as 'cascade geranium' and these are cutting-propagated, most suitable for hanging baskets. These are insensitive to high temperature as plants do not lose vigour as to that of *P. peltatum*.

The Species

Pelargonium abrotanifolium

A native of South Africa, this shrubby species grows up to 1 m tall, bearing grey-downy, aromatic, 2.5 cm across, 3-lobed and each lobe finely incised. White or pink flowers veined red are some 2 cm wide with two upper petals widest and these appear during summer.

Pelargonium acetosum

A native of South Africa, this almost glaucous species is evergreen with grey-green foliage, and grows erect with strong stems to a height of up to 60 cm and with a spread of some 25 cm. Its stems are succulent with fleshy grey-green leaves, often margined red. Its graceful single salmon-pink flowers are freely produced from May to July.

Pelargonium affrapaceum

Widely distributed along the coast of W. South Africa, this species is tuberous, hence so its specific name which in Latin means turnip. It is a geophytic succulent with a caudex up to 20 cm in diameter, producing a tuft of pinnate, highly divided leaves during the winter season. In summer, flower stems bear a cluster of white to yellow flowers.

Pelargonium angulosum

A native of South Africa, this shrubby to sub-shrubby species is well-branched and grows to a height of up to 90 cm, bearing firm, rounded to broader, some 6 cm wide, 5-angled and toothed leaves. Carmine-purple flowers with darker veins and some 3.0 cm wide appear during summer in 3-7 stalked umbels.

Pelargonium betulinum

A native of South Africa, this shrubby species grows erect with slender stems up to 90 cm in height, bearing 1-2 cm long, broadly ovate, leathery, toothed and glaucous-tinted leaves. White or pink flowers having carmine vein patterning at the base of two upper petals appear during summer in 2-3 clustered umbels.

Pelargonium bowkeri

A native to the Eastern Cape (South Africa), it is a succulent and tuberous species with a large underground tuber, named after Henry Bowker, soldier and naturalist of 19th Century who collected the plant. It bears pinnate, feathery leaves with inflorescence of light yellow-green flowers.

Pelargonium capitatum

A native of South Africa, this straggling shrubby species is a decumbent plant bearing white-hairy stems growing to 30-60 cm tall and the leaves are softly hairy, 3-5-lobed and toothed, and up to 5 cm across. Pink flowers with red-purple veins appear during summer months in a cluster of 8-24 per umbel on slender stalks, individual flowers being 2 cm long. This species is quite similar to *P. karooense* which is more erect and bears deeply lobed leaves.

Pelargonium carnosum (P. crithmifolium, P. dasycaule)

A native from Namibia to NW Cape (South Africa), it is a shrubby succulent plant with thick, waxy yellow-green and knobby stems up to 1.2 m tall bearing thick roots, and glaucous, 5-10 cm long, succulent and bipinnatifid leaves which are produced in tufts from growing points along the stem. The inflorescence comprises of 3-7-flowered cluster bearing white flowers some 2.5 cm wide with pink markings on the base of the upper pair of petals. *P. crithmifolium* and *P. daisycaule* are exactly similar to *P. carnosum* except that flowers in former case are lilac-pink while in latter case these are white.

Pelargonium × citrosum (P. crispum hybrida)

Where it originated is unknown but in general appearance it is very similar to *P. crispum*, bearing larger leaves in two ranks and when bruised these strongly emit a lemon scent. Though flowers are also similar to *P. crispum* but are lilac-purple having darker veining and spotting on the upper petals.

Pelargonium cotyledonis

Its specific name *cotyledonis* is named so as leaves of this species resemble to cotyledons. A native to the island of St. Helena, it is a slow-growing shrubby species with a few thickened and succulent branches, bearing simple, summer-deciduous and heavily veined leaves. Inflorescence bears clustered large white flowers.

Pelargonium crispum

A native of South Africa, this erect, slender and much-branched shrubby species grows to a height of 60-90 cm, bearing mid-green to greyish, small, fan or wedge-shaped, densely arranged, finely crisped and toothed two-ranked leaves up the stems which are 2-3 cm wide, shallowly 3-lobed and have lemon aroma (balm-like) when crushed. White to pink 1-3 narrow-petalled flowers some 2.5 cm long and wide with darker reddish veining appear in threes from May to October in the upper leaf axils. The leaves of its var. 'Variegatum' are yellow-blotched and margined cream-white.

Pelargonium cucullatum

A native of South Africa, this freely branched shrubby species is softly hairy and grows up to 2 m tall, bearing bean-shaped, cupped and 6 cm across leaves with finely toothed or scalloped margins. Purplish-red flowers with darker veins and 4-5 cm across appear during summer. It also has a purple sort.

Pelargonium denticulatum (P. filicifolium)

A native to South Africa, this shrubby species grows up to 1 m in height, bearing ferny, bipinnate-partite, toothed, 4-8 cm long and balsam-scented leaves. Lilac, rose-pink to pale-purple flowers having darker markings and 1.5-2.0 cm width are produced during summer. The leaves in its form 'Filicifolium' (*P. filicifolium*) are narrower and more finely dissected.

Pelargonium desertorum

A native to the N. Cape (South Africa), it is a deciduous shrubby plant with smooth stems, bearing small 5-lobed leaves which have aroma. The flowers are white with a purple tinge.

Pelargonium × domesticum (P. grandiflorum, P. × macranthum, P. regale)

A group of hybrid pelargonium evolved by involving *P. angulosum, P. cucullatum, P. fulgidum, P. grandiflorum* and a few others. As a great number of its hybrids were being grown at the Royal Gardens at Sandringham (England), these were named as Regal or Royal pelargoniums. They are woody, evergreen and erect perennials mostly compact but some are even loose-growing, up to 40 cm to 1.5 m tall, and flowering from May to September. During summer these are grown outdoors. Their leaves are stiff, glabrous, shallowly 3-lobed, irregularly triangular, dentate, some 5-10 cm wide and the edges wavy or toothed. Inflorescence is an umbel whose size is highly variable from one to other variety, and the individual flowers some 4-5 cm wide are butterfly- or azalea-shaped of one colour to bicoloured in white, pink, red and purple, *vis-a-vis* blotched with darker shades and/or picotee edges. Also see Regal type under 'Classification, Species and Varieties' above.

Pelargonium echinatum

A tuberous-rooted basally-woody shrublet native of rocky soils and cliffs of Richtersveld and Namaqualand (South Africa), this is a deciduous perennial adorned with persistent soft curved spine-like stipules on its thick, fleshy, upper-succulent and sprawling stems, up to 40 cm in height. It bears heart-shaped, broadly ovate and shallowly 5-7-lobed summer-deciduous leaves some 3.8 cm wide where upper surface is green and covered with whitish felt, and crenate margins. The globular inflorescence is comprised of a cluster of white to reddish-pink to purple flowers with a heart-shaped red spot occurring on one of the upper white petals and reddish-purple blotches of variable sizes near the base of the lower white petals. The species flowers from May to September.

Pelargonium × fragrans

It is a hybrid of *P. extipulatum* and *P. odoratissimum*, morphologically very similar to the latter parent but leaves are almost 3-lobed with aroma of nutmeg, lemon or pine. Its flowers are small white where upper petal has red veins.

Pelargonium fruticosum

A native to the S. Cape (South Africa), it is a shrubby species with smooth brown stems. The leaves are slightly succulent, highly-divided, trifoliate and are bright-green in colour. It bears pale to dark pink flowers with darker pink central markings.

Pelargonium fulgidum

A native of South Africa, this is a sub-shrubby species growing to more than 60 cm tall, bearing cordate, pinnately lobed, silvery-hairy and 7cm long leaves. Bright scarlet flowers with darker veining appear during summer some 4 cm across.

Pelargonium gibbosum

A native to the coastal W. Cape (South Africa), it is a sprawling sub-shrub with thickened stems producing a mass of annual growth with glaucous semi-succulent pinnate leaves. The inflorescence is a panicle of night-scented yellow flowers.

Pelargonium grandicalcaratum

A native to Namibia and the NW Cape (South Africa), it is a woody shrub with stems bearing succulent pinnate leaves. Persistent petioles along the stems give appearance of spines. When crushed, its leaves emit thyme-aroma. The small flowers are white to pale- yellow with fine red markings on the upper petals.

Pelargonium graveolens

A native of dry rocky slopes of Cape province in South Africa, this is a branching-type spreading shrub, growing to a height of up to 1 m, bearing hoary-grey-green, aromatic, palmate, and deeply 5-lobed toothed leaf segments. Three lobes are cut deeply up to the base. From June to October, the rose-pink flowers with a dark purple spot on each of the upper two petals appear, each flower some 3-4 cm across, in a 5-10-flowered terminal umbel. This species has a number of forms involving *P. radens* as other parent. This is grown for rose scented essential oil extracted from its leaves and tender shoots.

Pelargonium × hortorum

It is a complex hybrid group involving *P. inquinans, P. zonale* and certain other species, and is popular with the name of zonale pelargoniums. They are shrubby, more than 1.5 m tall, bearing mid-green leaves with a darker brown or bronze horse shoe-shaped mark in the centre, rounded to bean-shaped, 6-13 cm across and with waved margins. White, orange, pink, red, purple or bicoloured single to double flowers where some may have quilled petals or even picottee edges, and some 2-5 cm across are produced from May to October. There are cultivars where foliage is variegated with various colours, *viz.* green, dark green, white, yellow, crimson, purplish and brown. For certain other details and cultivars, see Zonal pelargoniums under Classification, Species and Varieties above.

Pelargonium laevigatum

Its specific name refers to smoothness. A native to the S. Cape (South Africa), it is a succulent shrubby species with smooth reddish-brown stems and narrow lanceolate to trifoliate leaves which are slightly incurving. The flowers have white, cream or pink petals with central reddish-purple markings and pink stamens.

Pelargonium odoratissimum

A native to South Africa, this is a shrubby species with sprawling stems growing to 45 cm tall, bearing velvety-hairy, ovate-cordate, 2-3 cm across and wavy-margined leaves emitting sweet apple aroma when these are bruised. Flowers white spotted and veined red appear during summer some 2 cm across.

Pelargonium paniculatum

A native to S. Namibia, it is an infrequently-branching succulent shrub with a thick 90 cm trunk patterned with persistent petioles, bearing a head of summer-deciduous pinnate succulent leaves. The inflorescence which appears during summer is a panicle of small white flowers with pink markings on the base of the upper pair of petals.

Pelargonium papilionaceum

A native of South Africa, this is a shrubby species with well-branched stems growing up to 1 m in height, bearing rounded to cordate and toothed leaves some 10 cm across. Flowering stem is multi-branched and carries several umbels of 5-10 flowers, each flower composed of two large and obovate upper pink petals with white base which are some 2 cm long with a carmine blotch and veined pattern though three lower petals are almost rudimentary. With this name many of its varieties developed by using either *P. capitatum* or *P. vitifolium* are available with the nurseries.

Pelargonium peltatum (Geranium peltatum)

A native to the coastal tip of the Cape (South Africa), this evergreen species is a scrambling succulent with slender stems, growing up to 0.9-1.8 m, bearing mid-green, small (5-7 cm wide), glabrous, fleshy and 5-lobed peltate (ivy-like) leaves which may exhibit some zonation. This is one of the most decorative of all popular geraniums, and flowers for a very long duration, *i.e.* from March to November in frost-free zones. This was introduced into European gardens in 1701 and is most suitable for hanging baskets, wall brackets, window baskets, terraces, balconies, against trellis or nettings, and in large receptacles. Inflorescence is pinkish and appears at the end of a long stalk in the form of a loose umbel up to 10 cm in diameter. Flower buds are quite hairy. Carmine to pink flowers with darker markings on the upper petals and some 3-4 cm across appear from May to October in 5-7-flowered umbels. This plant is ancestral to many horticultural varieties. Its popular varieties are 'Abel Carrière' (semi-double, soft rose-purple), 'Amethyst' (amethyst, double), 'Blue Peter' (mauve), 'Galilee' (double, vivid pink, free-flowering), 'Gardenia' (large double, white), 'Gloire d'Orléans', (double, pink) 'La France' (double, mauve with upper surface of petals flecked maroon), 'L'Elegante' (flowers white, leaves cream-edged but turning purple in autumn), 'Lord

Dickinson' (double, scarlet-red), 'Madame Crousse' (double, bright pink), 'Mrs W.A.R. Clifton' (flowers double scarlet), 'Sir Percy Blakeney' (single, crimson), 'Souv. Mmme. Amelia' (double, deep violet), 'The Pearl' (semi-double, pinkish-white), 'Zazy' (double, violet), *etc.*

Pelargonium quercifolium (P. terebinthaceum)

A native of South Africa, this semi-woody, evergreen and well-branched shrubby species grows erect up to 1.2 m, bearing short-stalked, mid-green with occasional brown to dark purple markings, 5-10 cm long, oblong-triangular to triangular (heart-shaped at the base), hairy, deeply pinnately lobed and toothed aromatic leaves having undulating margins similar to those of oak leaves. Flowers some 2.5-3.0 cm across are pink to pinkish-purple, veined with deep purple where upper petals each bears a central deep purple blotch, appearing in 3-7-flowered umbels from the upper leaf axils from May to August.

Pelargonium quinquelobatum

A native of South Africa, this evergreen, rare but very beautiful species grows to a height of 40 cm, bearing leaves divided into five lobes, and curious-looking greenish flowers suffused red appear during May-June. It can easily be cultivated outdoors, making an excellent specimen in pots.

Pelargonium radens (P. radula)

A native to South Africa, this shrubby species grows to about 1 m tall, bearing roughly hairy bipalmatipartite leaves some 4-7 cm long with slender and toothed segments which roll inward at the margins. Pink to rose-pink flowers with red-purple markings and some 3 cm wide appear during summer.

Pelargonium saxifragoides

This was discovered in 1888 growing at Royal Horticultural Society's garden, Chiswick, having no any record, hence is thought to be of garden origin, probably a hybrid of *P. peltatum*. Its plant being dwarf resembles a saxifrage, hence, its specific name *saxifragoides*. It is a succulent plant, bearing small dark green fleshy leaves having slightly thickened edges and light pink flowers during summer.

Pelargonium scandens

A native of South Africa, this scrambling and sparingly branched shrubby species grows to more than 1.5 m high, bearing greyish-green leaves with a dark horse shoe-shaped zone, and these are crenate and almost orbicular. Pink to light red flowers some 2.5-3.0 cm wide appear during summer in umbels of 5-12. This species is suitable for hanging baskets, on walls and around a window embrasure.

Pelargonium stipulatum

A native to South Africa, this is an evergreen perennial growing only up to 15 cm height, bearing clear mimosa-yellow flowers during May-June.

Pelargonium tetragonum

A native to the Cape region of South Africa, this scrambling-shrubby species grows some 1 m tall, bearing succulent, jointed and 4-angled stems, hence its specific name *tetragonum*. It produces non-succulent, almost pubescent, deciduous and 5-lobed leaves with rounded teeth at the stem joints having zonal markings. The inflorescence which appears during summer is a cluster of cream to pink flowers, flowers 2.5-3.5 cm across with four pink petals, upper two larger and deep-coloured than the lower and feathered with purple veining.

Pelargonium tomentosum

A native of South Africa, this species is a hummock-forming semi-prostrate which grows to a length of 1.5 m if trained as a low climber. Leaves emit a strong peppermint-type aroma when crushed. Leaves are pale-green, soft and dense-haired, palmate (triangularly cordate), 13 cm across and shallowly 5-lobed and the lobes are rounded. White flowers with red markings are insignificant, narrow-petalled, 1.2-2.0 cm across and appear in umbels from the tip of the shoots from June to September. The foliage of its variety 'Variegatum' bears creamy-white margins.

Pelargonium triste

A native to the W&SW Cape (South Africa), it is a tuberous-rooted evergreen perennial growing up to 40 cm in height with hairy stems, bearing large, pinnate, hairy and quite handsome foliage similar to those of carrot leaves. Its leaves hide a substantial portion of caudex and spreading tuberous root, from which the seasonal foliage arises. Originally it was described by Linnaeus as *Geranium triste*. It is one of the first South African pelargoniums to be brought to England by John Tradescant in 1632. Its pleasantly night-scented, dark maroon and spotted flowers appear during May-June.

Pelargonium violareum (P. splendidum violareum, P. tricolor)

A native of South Africa, this shrubby and well-branched species is a highly fascinating evergreen perennial bearing tricoloured blooms with white, vivid red and black petals, flowering from May to July. This grows up to 45 cm tall and almost the same of spread, bearing grey-pubescent, about 4 cm long, usually ovate to lanceolate, irregularly scalloped leaves, sometimes with a few narrow lobes. Inflorescence is branched, each branch with two or more umbels, each umbel with 2-6 blooms, and each bloom some 2.5 cm wide, composed of five rounded petals ~ upper two red to red-purple with darker base and the lower ones white to palest-pink.

Pelargonium zonale (Geranium zonale)

It was introduced into England in 1710. A native of South Africa, this woody-based shrubby species has succulent and somewhat hispid branches at younger stage. Leaves are glabrous or pubescent, long-stalked, round-cordate, shallowly multi-lobed, upper surface with deeper zoning or horseshoe marking and the margins are crenate-dentate. Stipules are cordate-oblong and broad, and the peduncles long. Inflorescence has many but almost sessile flowers in shades of white, orange, pink, red, scarlet or crimson and these have spathulate segments. Also see *Pelargonium × hortorum* above.

Source: Succulent-plant.com, Bailey (1960), Pizzetti and Cocker (1968), Everard and Morley (1970), Hay and beckett (1971), Beckett (1983, 1987).

Propagation

Geraniums are propagated by division in spring or any time from September to March *in situ*. **Stem cuttings** from the vegetative shoots are taken, preferably with three nodes and planted under light in sand, perlite or vermiculite where these root within a fortnight and then these are transferred to prepared pots for ornamentation. Pelargonium stocks are easily raised through tip-cuttings during active vegetative growth, *i.e.* during spring or in autumn. If air temperature is 10 °C, the cuttings can be taken any time of the year, preferably from February to October. However, it is suggested that temperature of rooting media should be 21-24 °C for rapid callusing and rooting for all *Pelargonium* types (Dole and Wilkins, 1999). The most favourable time for induction of rooting is active growth period, *i.e.* in early spring and late summer. Even with good bottom heat having ample air temperature, cuttings root indifferently during November, December and January due to short days and poor light conditions (Key, 1985). The use of rooting hormones by dusting the cut-bases is recommended for quick initiation of rooting, especially during winter months when temperature is quite low. If cuttings in the temperate regions are taken during August or September there is no need of bottom heating. The cuttings should have three nodes (5-8 cm long) and the growing tips, and their leaves, *vis-a-vis* stipules should be trimmed leaving only the immature ones at the tips. These cuttings are inserted some 1.25 cm deep in the medium with just lower node buried but next node should not touch the medium. Up to one fortnight of cutting insertion, no watering should be given. Equal portions of damp peat + silver sand are the best for rooting though others like perlite or vermiculite can also be used either singly or as a mixture. In a 12-cm pot some six cuttings should be planted. Cutting-planted pots are taken to the greenhouse bench with bottom heat of 16 °C for the first fortnight. It would be better if there is reduction of temperature every week by 5-6 °C but when this temperature matches to that of air temperature no further reduction is required. In no case these cuttings should be covered with polythene. The cuttings will root in 12-21 days (Dole and Wilkins, 1999), Zonal pelargoniums root in about 10-14 days at a temperature of 18 °C whereas Regal pelargoniums in 20-40 days. While watering, the water should dribble from the spout of the can between cuttings or such pots or boxes should stand in a larger pan containing water in a manner that only lower three-fourth of the outside surface should come in contact of water and this way when the medium becomes wet the pot or box should be removed. When watering again, the rooting compost should be allowed to nearly dry out and the wetting again. The cuttings should be protected from excessive sunshine so as to prevent excessive wilting.

The **seed-propagation** method, nowadays, is becoming commercially popular in geraniums and pelargoniums. Seed raised strains of Zonal pelargoniums are quite popular among commercial growers. The commercial seeds are mostly F_1 hybrids and these are quite expensive. Approximately, the seed number per gramme is 28-30. Seeds are sown normally in February. The temperature of the chamber should be maintained at 16-18 °C to ensure proper and uniform germination. Key (1985) stated that seeds require a minimum temperature of 13 °C to germinate, however, the ideal is 19 °C. Vermiculite, however, it is used as medium. Due to hard seed coat in geranium or pelargoniums, it would be better to go for seed abrasion or chemical scarification for improving germination rates. Germination is enhanced when chamber is quite illuminated. Seeds germinate in 1-2 weeks time. Seedlings are pricked out into boxes, later transferring to 10 cm pots in John Innes Potting Compost No. 2 (loose bulk loam 7 parts: peat 3 parts: sand 2 parts, all by volume; and then addition of some 6 g of the fertilizer mixture containing 2 parts bone meal: 2 parts lime superphosphate: 1 part potassium sulphate, all by weight, and 1.2 g of ground chalk or limestone to every litre of potting compost). **Tissue culture** can effectively be employed to propagate quickly a large number of plants of the elite cultivars with additional advantage of producing disease-free plants.

Cultural Practices

Pelargoniums and geraniums are sun-loving plants as most of the species flower profusely in full **sunlight**, however, some species like *P. odoratissimum, P. graveolens, P. grossularioides* and *P. tomentosum* prefer shady conditions (Amidon and Brobst, 2005). They thrive well if they avail two-thirds of the daylight (at least six hours) where double and semi-double types may also be grown. Such flowers last longer, are larger, have proper standing, and shattering is to the minimum even during inclement weather conditions. Some of the species perform well in partially shaded conditions but the flowering is not prolific. However, singles prefer sheltered or filtered situations. In windy areas, short-jointed cultivars are preferred in all colours, singles, semi-doubles or doubles. Full shade is always disappointing. Shading against high temperature (>32 °C) and high light levels during summer months will be required to prevent scorching of plants. Various groups of pelargoniums differ slightly in their light requirements, such as ivy-leaf pelargoniums require high quantum of light whereas Regal pelargoniums require shading from strong sunlight, especially during flowering. The flowering of many Zonal pelargoniums will also last longer if lightly shaded. Floral initiation is dependent on total cumulative light energy, especially for *P.* × *hortorum*. During low light periods of winter, supplemental lighting hastens flowering if provided during early stages of seedling growth (Carpenter and Rodriguez, 1971; Armitage and Tsujita, 1979). Bethke and Carlson (1985) proved that supplemental lighting of at least 350 fc is most important for the first 4-6 weeks after germination as high pressure sodium lamps from germination to transplanting accelerated floral initiation, however, Quatchak *et al.* (1986) reported hastened flowering when supplemental lighting was given after transplanting, though Kaczperski and Heins (1995) stated that response of supplemental lighting was quite pronounced in case of 28-35 day old seedlings than those of 7-21 day old seedlings. Aimone (1985) stated that light intensity of 3,500–5,000 fc and 2,500-3000 fc are optimum for Zonal geraniums and ivy-leaved geraniums, respectively, and he further mentioned that for every 1 per cent increase in irradiance, 1 per cent growth increase should be expected if optimum temperature is available. Growing plants under 60 per cent shade causes development of 22-24 nodes prior to flower initiation and so requiring 110 days for flowering but in case of full ambient light only 16-18 nodes were found developed so plants flowered only in 37 days, the difference being only of 8 days from floral initiation to visible floral buds between shaded and full sun treatments which strongly suggests that floral initiation is a light-driven process though floral development is a temperature-driven process (Armitage and Wetzstein, 1984), *i.e.* from visible bud stage to opening of the first floret of the inflorescence is less dependent on light but more on temperature (Armitage *et al.*, 1981).

Pelargoniums grow best at night/day **temperatures** of 10.0-15.5/18.3-23.8 °C (Becker and Brawner, 1996). The gray-green-leaved pelargoniums are tolerant to high heat levels. *P. grossularioides* is adaptable to extreme conditions and survives temperatures of more than 48.8 °C in greenhouse when grown in gravel with no water (Amidon and Brobst, 2005), though in most of the cases, the growth and development of most pelargoniums are hampered above 32.0 °C. Trellinger (1997) stated that an average daily temperature of 20 °C is best for ivy-geraniums as this makes compact growth while day/night temperatures of 24/16 °C cause greater stem elongation. Zonal pelargoniums are tolerant to a wide range of temperatures with respect to floral initiation and development and if growth has occurred the flowering will also occur, however, 21 °C night temperature proves better for promotion of flowering over 15 °C or 10 °C in cutting-grown plants (Carpenter and Carlson, 1970) but in seed-geraniums the increase of temperatures from 10 °C to 24 °C progressively increased the rate of growth and flowering, optimum being 20/17 °C day/night temperatures though before floral initiation there had been 17, 16 and 14 nodes at 10, 16 or 21 °C, respectively (Carpenter and Rodriguez, 1971; Erwin and Heins, 1992).

They make luxurious growth in heavily manured **soil** with handsome foliage but that is on the cost of flowering as such plants do not flower well. Slightly under-manured soils are best for their cultivation. Well established plants flower well even in a dry situation. It requires well-drained soil having good air circulation. However, the species have wide variations in their requirements as *P. graveolens* prefers moist soil and *P. scabrum* is well suited to dry and sandy areas (Amidon and Brobst, 2005). The pH of soil should be 5.6-6.0 though for *P. peltatum* the pH should be slightly lower, *i.e.* 5.3 (Tellinger, 1997). For sale, majority of *P.* × *hortorum* plants are produced in 10- to 15-cm pots and these should be grown pot-to-pot until leaves start overlapping. For a 10-cm pot the spacing should be 15-18-cm whereas spacing for 15-cm pots should be 21-23-cm, however, for *P.* × *domesticum* production final spacing requirement is 30.5 × 30.5 cm (Tayama and Klesa, 1991). For ground planting, raised beds are preferred so as to provide good drainage. Ventilation is also of greatest importance since these plants dislike a stuffy, stagnant atmosphere, and even during very hot days the greenhouse temperature should never exceed 32.2 °C (Pizzetti and Cocker, 1968). Both, low and high levels of humidity adversely affect plant growth and development. The excessive dry air during summers is highly detrimental to both, vegetative and reproductive phases so misting of greenhouse is recommended to increase humidity. High humidity on wintery cold days causes rotting of flowers and flower buds, so it is necessary to lower atmospheric humidity by heating the air to prevent the occurrence of fungal diseases. Ivy-leaved pelargonium is more prone to a physiological disorder called oedema under high humidity conditions. CO_2 concentration in the greenhouse should be 800-1,500 ppm for stock plants, propagation through cuttings and for seed geranium production and so the vents should be kept closed (Shaw and Rogers, 1964).

Re-potting is done in case of old plants in either the same size pot or slightly larger in case when the plant has become pot-bound or potting mixture has turned completely sour or exhausted of nutrients. In this case old soil is removed from the roots of a plant before being potted on into new pot. Generally older plant needs repotting if they have been growing in the same pot for a longer duration while in younger plant it is done with those having a root problem. It is best to re-pot one year old plants at the initiation of growth, however; the exact timing will depend on the environmental conditions. It is advisable to avoid any root injury at re-potting time. It is a good time to re-pot the plants with fresh compost after the plants have been pruned.

It is advisable to shift the cutting or plant as soon as they are well-rooted as delay in shifting will damage the root system, and ultimately result into stunted growth. After rooting the cuttings are individually **potted** on to 7.5 cm clay or plastic pots using John Innes No. 2 or a soilless compost. It is advisable to avoid over-firming of potting mixture as it minimizes the air space thus leading to water-logging and soury conditions, and at the same time of the potting the growing tips are **pinched** out so that their apical dominance is broken and the plants make their first break low down to ensure that plants grow compact and well-shaped. At this time the plants may have 5-8 cm height with 2 to 3 pairs of leaves. Morval and Deacon series pelargoniums will naturally produce neat and beautifully dome-shaped plants without being pinched and if pinching is practiced it will result in an ugly flattened appearance. Most modern Regal pelargoniums and miniature and dwarf Zonal pelargoniums should not be pinched frequently, however, occasional pinch during the later growth may benefit some of them. Pinching is practiced when the plant is in active growth, *i.e.* in spring or early summer. The potting of cuttings during November, December and January should be avoided due to poor light conditions, but from Early February when atmospheric temperature is above 5 °C, these can be potted on but in case the temperature is below these can be retained in their pots or boxes until April but they should regularly be given liquid feed (slurry) every week starting from January when growing tips start to grow. For bedding out, these can be retained in their 7.5 cm pots but for pot-planting these require 10-cm pots. When their early breaks (shoots) have 2-3 breaks with two nodes, their growing tips should be removed. This practice should be repeated further as frequent pinching will make the plants producing more shoots, and compact & bushy plant growth full of flowers. Any flower bud appearing should be removed until a three fortnights before scheduled date of flowering and then stopping is stopped so that flowering appears in time.

Normally, pinching is carried out till May end, and in the end of June the plants will require another potting in a 15-cm pot where it remains till the end of the season. Spent flowers are continuously removed to save the energy being spent towards seed formation so that more flowers are formed. In October beginning, these plants are **headed-back** to almost 10 cm stump with all the leaves removed if still any, and the pots are kept closely on the benches of the greenhouse where these are given occasional watering for their just growth. Alternative to heading-back, **pruning** is done to produce healthy, attractive, sturdy, much-branched, bushy (by preventing upright leggy growth) and well-shaped plants. Amount of pruning required depends on the growth rate of the plant. Smaller-leaved cultivars need more conservative pruning whereas rose- and oak-leaved types require aggressive pruning. Young plants of trailing ivy-leaved cultivars are best pruned back so as to promote branching (www.rhs.org.uk). It is advisable to remove any dead or diseased branch. Pruning should always be done with sharp secateurs so that crushing and bruising of the stems are avoided. It is advisable not to hard-prune the plants during late autumn, winter or early spring so that stem rot is prevented. Vigorously tall-growing pelargonium cultivars are to be trained into standards or tree forms. In January beginning, these plants are potted on into John Innes Compost No. 2 or potting compost having 2 parts fibrous loam +1 part each of coarse sand, peat and leaf soil (all by volume) and a little of bone meal, using 20-cm pots and are put in full light. Whether planted directly in the beds or in pots, the medium should have pH range of 6.0-6.8. Container geraniums require sterilized potting mixture (Pizzetti and Cocker, 1968). When new shoots appear these are pinched out in the manner as defined above. At this time, 4-5 split canes some 30 cm long should also be inserted around the plants, leaning outward and tied with the main shoots as the plants grow and these canes will be hidden by the plant when fully grown and such plants will give enchantment through their flowers throughout summer if high potash liquid feed is provided weekly with initiation of the floral buds.

Most commonly, the pelargoniums are grown as seasonal plants, however, following some procedures the plants can be carried throughout the winter (www.rhs.org.uk.) and this is known as **overwintering**. Softwood cuttings taken in late summer are allowed to root and then further shifted to compost-filled trays and finally kept under well-lit indoor environment. It is advisable to lightly fertigate in late winters. Pinching is recommended so as to get bushy appearance. Potting should be done in spring when temperature starts warming up and after the hardening process is over, and then these are shifted outdoors. Even in temperate regions, plants are easily saved during winters if there is no greenhouse or glassed enclosure. Plants are saved by keeping them on a bench in the glassed veranda or on the windowsills. Arranging five 12-cm pots, each with six cuttings taken in August from around the lower part of the plants and planted around the outside rim of each pot and standing the pots in a shady part of the garden until mid-September when these root, and then the pots are brought inside the house on the windowsills (5 pots per windowsill) to overwinter and where these are watered sparingly, *i.e.* fortnightly. From early January, these are fed weekly with general liquid feed. In John Innes Compost No. 1, these are potted singly in 6-cm pots in February when these are also pinched of their growing tips as defined earlier, and the pots are kept again on windowsills up to March where weekly liquid feed is continued as earlier. In April, these cuttings are potted into 8-cm pots in John Innes Compost No. 2 and then shifted outside into the cold frame or in a sheltered corner of the garden. These plants will be ready to be planted outside in early May.

Pelargoniums are grown under well-drained soils as mostly they adapt well to dry conditions since most species are native to low-rainfall areas (Miller, 1996), although some are found naturally growing near streams or in areas with winter rainfall (Amidon and Brobst, 2005). Geraniums require thorough **watering** during hot summers but the soil should be allowed to dry out between waterings. Generally, container-grown plants require more frequent watering than field-planted ones, however, waterlogged conditions are not conducive for their growth. The automated irrigation system, *i.e.* pot-drip irrigation method is more economical to commercial growers. Regal pelargoniums have more water needs than other groups. Ivy-leaved pelargoniums such as *P. tetragonum, P. echinatum* and *P. gibbosum* are succulent in nature so should be allowed to dry almost completely between each watering otherwise they will develop soft growth and also become more prone to diseases. The **nutritional requirements** of pelargonium groups varies as Regal and Unique pelargoniums require more potassium, whereas remaining groups including Zonal and ivy-leaved show high requirement for nitrogen. Their requirement of Ca and Mg is high, therefore their nutrient regimes are mostly based on KNO_3 and $Ca(NO_3)_2$ combinations if phosphorus is added to the medium as superphosphate or in the water as phosphoric acid so that proper pH is maintained, apart from micronutrients. However, stock plants should regularly be monitored for their nutrient requirements as they are frequently in production. The container-grown plants require regular fertilization for better presentation but plug trays especially of *P. × hortorum* are sensitive to high soluble salts due to restricted drainage in the plug-flats. Cutting-propagated Zonal pelargoniums have requirement similar to *P. × hortorum*. Seedlings should be feritilized with 100-150 ppm N but after proper rooting of the transplants the dose should be increased to 250 ppm N, at inflorescence stage the dose should be decreased to 150 ppm N, and it is further reduced to 50 ppm prior to transportation provided the plants are grown at 13 °C (Dole and Wilkins, 1999). Craig (1986) reported that *P. peltatum* and *P × domesticum* should not be fertilized with more than 200 ppm of nitrogen. Iron and manganese toxicity is reported in seed geraniums due to low pH and then plants become stunted, show necrotic spots on leaves and leaf edges, and lower leaf necrosis, hence it is recommended to raise pH of the medium above 6 so that they are less available to the plants. Foliar tissue levels of Fe and Mn should not exceed 500 ppm (Erwin, 1997). A fertilizer dose containing 1:0:2 or 1:1:2 NPK is found most suitable during winter and summers for Regal pelargoniums. In case of low potassium levels, a dose of 1:0:3 NPK can occasionally be used as frequent use will lead to magnesium deficiency, though it can be overcome by spraying with $MgSO_4$ @25g/25 l water (Clark, 1988). Zonal and ivy-leaved pelargoniums require more nitrogen and less potassium than Regal pelargoniums. A fertilizer dose of 1:1:1 NPK is most suitable with an occasional NPK dose of 2:1:1. For having brightly coloured pelargoniums, the cultivars with gold or tricoloured leaves should be fertilized with high potassium fertilizers similar to that of Regal pelargonium (Clark, 1988).

Nutrient Concentrations at Different Production Stages (Dole and Wilkins,1999)

Sl.No.	*Stage*	*N (ppm)*
1.	Seedling stage	100-150
2.	After transplanting	250
3.	Inflorescence begins to colour	150
4.	During the final weeks prior to shipping when plants are grown at 13 ºC.	50

Growth and Flowering

The geraniums are more commonly grown as the pot plants, hence, to control the plant height dwarfing chemicals are used. In fact, all commercial *Pelargonium* cultivars are considered day-neutral. The rate of floral initiation and subsequent development is dependent on total light energy received (intensity × duration) at the appropriate temperatures (Langton and Runger, 1985). In some seed cultivars of geraniums juvenility occurs (Bethke and Carlson, 1985) but when this phase is over the light intensity and duration become critical. Chlormequat (CCC) hastens floral initiation in some seed cultivars (Quatchak *et al.,* 1986), critical time for spray-application being a fortnight after commencement of germination to hasten floral initiation, though spraying after 40 days no response could be obtained (White and Warrington, 1984). Spraying the crop with A-Rest @26 to 66 ppm, Bonzi @5 to 20 ppm, Cycocel @800 to 1,500 ppm or Sumagic 3 to 6 ppm for cutting-geraniums and 2 to 4 ppm for seed-geraniums are recommended (Bailey and Whipker, 1914). Spraying of Cycocel @1,500 ppm will be useful for inducing early flowering in seed geraniums. Florel @300 to 500 ppm spraying increases the number of laterals, providing bushier appearance to the plants. Armitage (1994) stated that prior to flowering at least 15 nodes should be formed, and upon flowering, as per Armitage and Tsujita (1979) the photosynthetic energy of the plant is directed to axillary shoots and secondary flowers.

Key (1985) mentions that for winter flowering, geraniums require scheduling even when one has varieties for this purpose such as 'Alberta' (pink and white), 'A.M. Mayne' (large double deep crimson), 'Beatrix Little' (vermilion), 'Belvedere Glory' (pink), 'Caledonian Maiden' (rose), 'Cramden Red' (scarlet), 'Dark Red Irene' (large semi-double crimson), 'Double Jacoby' (double red), 'Dove' (mauve), 'Genie' (semi-double red), 'Gustav Emich' (semi-double scarlet), 'Hanchen Anders' (semi-double pink), 'Heidi' (red and white), 'Irene' (semi-double crimson), 'Madame Dubarry' (red), 'Paul Gotz' (scarlet), 'Penny' (semi-double pink, eye large white), 'Pink Raspail' (double pink), 'Prince of Wales' (crimson, overlaid darker), 'Rachel Fisher' (mauve-pink), 'Rosamunda' (semi-double pink, dwarf), 'Royal Purple' (double purple, eye small white), 'Snowstorm' (white), 'Toyon' (semi-double crimson), 'Vera Dillon' (light cerise, centre scarlet), 'Willingdon Beauty' (large trusses of rosy-salmon blooms), *etc.* In this case one must have heated house with a minimum of 7 ºC temperature. Cuttings are rooted in April, potted on into 7-cm pots, stopped and potted on as stated under '**Cultural Practices**' and so by September these should be in 12-cm pot having JIC No. 2 and during this time the flower buds if emerging any should be removed but from September onwards these are allowed to develop in natural course and the plants are fed weekly throughout the winter with a high potash liquid feed. Low temperatures are critical with *P.* × *domesticum* since in this case flower initiation takes place in lighted coolers (fluorescent/incandescent lighting greater than 1,950 fc, preferably for 3-6 weeks at 7-10 ºC; Erwin and Engelen, 1992) when temperature is less than 17 ºC (Erwin, 1991), though this temperature response is variable among various cultivars, *vis-a-vis* dependent on total light energy (Post, 1949), however, usually initiation requires four weeks. The requirement of cold treatment may be longer under low irradiance. Increase of the average daily temperature for cooling from 2 to 14 ºC delays floral induction but increase in the night temperature from 2 to 6 ºC increases the number of flowers (Erwin and Engelen, 1992). Temperature higher than 17 ºC during forcing may abort the flower buds (Dole and Wilkins, 1999).

Post Harvest

Seedling plugs can be stored up to four weeks at 2 ºC (Heins *et al.,* 1994) and their performance under fluorescent light (incandescent lamp is next preference) is better than dark-storage, however, plugs tolerate better storage or shipping when have been irrigated previously with 150 ppm nitrogen for two weeks (Kaczperski *et al.,* 1996a). Areca *et al.* (1996) reported poor post harvest life of pelargonium cuttings when transported at warm temperatures due to build up of ethylene because all the pelargoniums are sensitive to ethylene where apical meristem can totally abort, leaves may become thickened and cupped, veins exaggerated, stems thickened and swollen and flower buds abnormally developed or aborted (Dole and Wilkins, 1999). Even the seed geranium plants shatter the petals during transportation due to build up of ethylene so STS sprays at 175 ppm 14-21 days prior to transportation will be quite effective in case of Zonal and *P.* × *domesticum,* and then these should be transported at 5 ºC (Deneke *et al.,* 1990). Unrooted cuttings can be stored for 4-6 weeks at -0.5 ºC (Eisenberg *et al.,* 1978) while rooted ones for up to 8 weeks at 5 ºC and 250 fc illumination for 9 h per day (Kaczperski *et al.,* 1996b).

Essential oils are extracted from the tiny glands distributed over surface of leaves and stem. For good recovery, harvesting is usually done manually in the morning during sunny weather, first harvest being taken normally 4-6 months of transplanting depending upon the growth of the foliage. It is the proper time for harvesting when sparse flowering has started, basal leaves start turning yellow, when lemon-like aroma changes to rose-like and on pressing, the leaves emit a delicate rosy aroma. Best portion is terminal cuttings with 10-12 leaves. Leaf blades contain maximum oil, followed by petioles. Mani and Sampath (1981) reported that at Kodaikanal conditions, harvesting four times at 90 days interval during January, April, July and October gave better oil recovery of good quality. Guenther (1950) describes that yield of geranium oil is high if the herbage is air-dried in shade and allowed to wither for few hours prior to distillation by spreading in thin layers to avoid fermentation. Mani *et al.* (1981) recorded maximum oil recovery of 0.125 per cent (geraniol out of this being 61.6 per cent) with the herbage dried for 12 hours prior to distillation at Kodaikanal conditions though drying beyond 12 hours reduced the oil recovery and after 72 hours of drying it reached to 0.05 per cent. The distillation equipment consists of a boiler, distillation stills, condensers and receivers and through hydro-steam distillation at atmospheric pressure the oil is extracted though oil obtained through steam distillation is of better quality. The oil obtained through distillation should be completely moistuare-free so even traces of water should be removed by sprinkling anhydrous sodium sulphate at 20-30g/l, stirring for about 15 minutes, separating the water layer and then filtering through a good filter paper. This oil should be kept in a light-proof air-tight container and stored at a cool place. This oil contains large amounts, *i.e.* 60-70 per cent of alcohol (primary citronellol and geraniol), esters, *i.e.* 20-30 per cent (geranyl tiglate, geranil acetate, *etc.*), aldehydes and ketones (citronellol, citral, *etc.*) (Petrovski, 1971). Over 100 compounds have been detected in the oil. Khan and Dimri (1969) reported oil yield of more than 25 kg/ha though normally it comes to 7-12 kg/ha. A plant population of 25,000/ha can produce 15,000-18,000 kg of herbage/year which yielded about 15 kg of oil at Kodaikanal conditions (Arumugam and Kumar, 1979). Mani and Sampath (1981) at upper Palani hills with proper fertilizer application obtained 24,680 kg/ha of herbage and 21 kg/ha of oil. However, through Egyptian culture in Hyderabad the herbage yield obtained was 60 tonnes and of the oil 60 kg/ha though under Bengaluru conditions the herbage yield obtained was 40 tonnes/ha and of the good quality essential oil 40 kg (Rajeswara Rao *et al.,* 1989).

Pests, Diseases and Physiological Disorders

Pests

Geranium **sawfly** (*Protoemphytus carpini*) larvae infest only the geranium leaves (not the pelargoniums). In its infestation, some 1.2 cm long greyish-green larvae may be seen feeding on leaf undersides and making rounded holes. Though leaves become perforated but plant health is not affected so control is also not necessary. However, to save the leaves being holed the pest may be controlled with bifenthrin, derris, permethrin, pyrethrum or primiphos methyl. **Aphids** (greenflies, plant lice) suck cell sap and transmit the virus diseases. They most commonly feed on the young, tender parts such as newly growing stems, leaves, flowers, *etc.* They also excrete the excess sugar obtained from plant sap and this sticky material provides a food source for fungi which lead to development of sooty mould. Infested plants should be sprayed with a diluted soap solution, growing marigold side by side in the greenhouse will attract hoverflies whose larvae feed on aphids, and malathion spraying will kill aphids. **Caterpillars** are larvae of various types of moths and butterflies, infesting badly on outdoor Zonal pelargoniums and others. Budworms feed on the floral buds as well as young shoots. An introduction of *Bacillus thuringiensis* will kill the caterpillars. Permethrin insecticide spraying is also very effective. **Whitefly** (*Trialeurodes vaporariorum*) infests Zonal and ivy-leaved pelargoniums in general and Regal pelargoniums in particular and excrete a sugary substance which entices sooty mould. Yellow sticky traps are very effective, *Encarsia* wasps should be introduced and spraying of aldicarb can be effective. **Mealybugs** are sucking insects which infest young soft plant tissues. These can be killed by wiping the affected part with a cotton swab dipped in alcohol. Thrips, woodlice and slugs are also sometimes found feeding on this crop but these are of minor importance.

Spider mite (*Tetranychus urticae*) is found mostly infesting on the undersides of the leaves, spinning a fine web there. Their heavy infestation may kill the plant. Red spider mite is now known to be resistant to most organophosphorus insecticides such as diazinon, dimethoate and malathion, however, smoke of azobenzene can be effective. Introduction of the predatory mite, *Phytoseiulus persimils* may prove very effective.

Diseases

Downy mildew (*Peronospora* spp.) of geranium causes pale-green or light brown discolouration throughout the year, mostly in angular patches on leaves, and off-white slightly fuzzy fungal patches develop beneath. In severe infection, the leaves wither and die. Its spread is favoured by moist or humid conditions and poor air circulation. **Powdery mildew** of geranium causes white powdery fungal growth on the upper leaf surface and on leaf stalks. To control the spread of powdery and downy mildews, infected leaves should immediately be removed and destroyed and the leaves along with leaf stalks should be sprayed thoroughly with a fungicide containing mancozeb. **Leaf spot** (*Phyllosticta geranii*) causes small circular and brownish spots on leaves which make these turning yellow and falling. Mancozeb spraying will control this problem also. Leaf spots caused by *Cercospora brunkii* produces light-brown or pale brick-red circular to oval spots on the leaves with the prominent borders and these spots coalesce to make these larger in size. Leaves in severe cases turn yellow and fall. Spraying with 0.1 per cent bavistin, 0.2 per cent dithane Z-78 or 0.2 per cent copper oxychloride will control this disease effectively. **Alternaria leaf spots** (*Alternaria tenuis*) appear in the form of water-soaked spots on the old leaves below but later on new ones are also engulfed. These spots turn brown to dark brown forming concentric rings. Irregular necrotic spots are formed later on when these coalesce together. In its infection diseased leaves should be collected and destroyed, proper sanitation should be maintained, cuttings should be collected only from the healthy plants and 0.2 per cent dithane M-45 should be sprayed.

Pelargonium rust (*Puccinia pelargonii-zonalis*) forms concentric rings of dark brown fungal pustules on lower leaf surface with corresponding yellow blotches on the upper surface. Sometimes fungal pustules engulf even upper surface. Such leaves wither and die and if not checked in time even the whole plant may be killed. Its spores spread through air and water and are favoured by humid atmosphere and poor ventilation. Infected leaves should be removed, air circulation should be improved and spraying with fungicides such as mancozeb, penconazole or triforine with bupirimate should be done. **Black leg** (*Pythium* spp.) affects mainly cuttings but sometimes also the mature plants. The first symptoms appear on the basal portion of the stem as black spots, rapidly engulfing even the top portion, and then death of entire plant. Sometimes gross overfeeding of the plants also produces similar symptoms as of black leg without any sign of fungus. Against this disease, prevention is better than cure hence full sanitation should be maintained and the pots and media used should be sterilized. In case of black mould, the fungus grows on the excretion of the aphids and whiteflies and then black sooty mould appears. It is not a real problem except that it disfigures the plant so such spots should be washed out along with control of aphids and whiteflies. **Grey mould** (*Botrytis cinerea*) primarily affects dead and dying tissues. The spores appear as furry-grey coatings on the plant, especially on old flowers and that too in severe form during rains since the spores germinate only under cool and damp conditions. Situation is alarming among poorly spaced plants due to poor ventilation. Maintaining proper sanitation of the field and pots, dead-heading of affected flowers and the floral stalks when weather is dry will keep down this problem to the minimum. Spraying the crop with benomyl + Activex 2 will control this problem. **Bacterial leaf spots** (*Xanthomonas campestris, X. pelargonii, Pseudomonas cichorii, P. syringae* and *Acidovorax* spp.), especially that which occurs with infection of *Xanthomonas campestris*, is the most serious disease of pelargoniums as it spreads through propagation and irrigation, for many years its infection remains mute and with chemicals it is difficult to control (Roberts, 1997). **Vascular bacterial blights** (*Pseudomonas solanacearum*), **verticillium wilt** (*Verticillium alboatrum*), **root and stem rots** (*Pythium, Rhizoctonia, Fusarium* and *Thielavapsis*) and bacterial **stem fasciation** (*Clavibacter fascians*) also occur on geraniums (Horst, 1990; Daughtrey and Chase, 1992). **Galls** (fasciation) appear on the main stems in the form of knobby growths at ground level of the plants, and are more common with miniature Zonal pelargoniums. These galls should be removed as soon as these appear otherwise these will hamper growth of the plants when grow larger. Pelargonium **viruses** cause yellow markings in the form of flecks, ring spots, streaks and distortion of leaves and flowers may also have pale streaks on the petals. Some of the viruses are transmitted by aphids, soil nematodes or mechanically. Such plants should be uprooted and destroyed.

Physiological Disorder

Oedema physiological disorder more commonly affects the ivy-leaved pelargoniums but to a lesser extent the Zonal pelargonium. It is characterized by appearance of watery bumps on the leaves. There is appearance of raised, light brown, irregularly shaped regions on the undersurface of the leaves which are produced by the rupturing of the plant cells. The main reason is sudden increase in sap pressure which is caused by watering a very dry plant. Highly damaged leaves should be removed and no over-watering should be done.

References

Aimone, T. 1985. New geranium technology. *Growertalks*, **48**(11): 128, 130, 132, 134.

Amidon, C and J. Brobst, 2005. To grow pelargoniums is to know them. *The Herbarist*,**71**: 4-10.

Areca, R.N., J.M. Arteca, T.W. Wang and C.D. Schlagnhaufer, 1996. Physiological, biochemical, and molecular changes in *Pelargonium* cuttings subjected to short-term storage conditions. *J. Amer. Soc. hort. Sci.*, **121**: 1063-1068.

Armitage, A.M. 1994. *Growing on*. In: *Ornamental Bedding Plants*, pp. 43-94. CABInternational, Oxon, United Kingdom.

Armitage, A.M. and H.Y. Wetzstein, 1984. Influence of light intensity on flower initiation and differentiation in hybrid geranium. *HortSci.*, **19**: 114-116.

Armitage, A.M. and M.J. Tsujita, 1979. The effect of supplemental light source, illumination and quantum flux density on the flowering of seed-propagated geranium. *J. hort. Sci.*, **54**: 194-198.

Armitage, A.M., W.H. Carlson and J.A. Flore, 1981. The effect of temperature and quantum flux density on the morphology, physiology, and flowering of hybrid geraniums. *J. Amer. Soc. hort. Sci.*, **106**: 643-647.

Arumugam, R. and N. Kumar, 1979. Geranium cultivation in Kodaikanal hills. *Indian Perfumer*, **23**(2): 128-130.

Bailey, D. and B. Whipker, 1914. Best management practices for plant growth regulators used in floriculture. *Horticulture Information Leaflet*, No. 529, Revised 10/98, published by North Carolina Cooperative Extension Service, College of Agriculture and Life Sciences, North Carolina State University, p 16.

Bailey, L.H. 1960. *The Standard Cyclopedia of Horticulture* (vols II, III), pp. 1330-1332, 2525- 2534. The Macmillan Company, New York.

Becker, J. and F. Brawner, 1996. *Scented Geraniums: Knowing, Growing and Enjoying Scented Pelargoniums*. Loveland, CO: Interweave Press.

Beckett, K.A. 1983. *The Concise Encyclopedia of Garden Plants*, pp. 163-165. Orbis Publishing Limited, London.

Beckett, K.A. 1987. *The RHS Encyclopedia of House Plants Including Greenhouse Plants*, pp. 380- 383. Salem House Publishers, Massachusetts, USA.

Bethke, C.L. and W.H. Carlson, 1985. Seed geraniums- 18 years of research. *GrowerTalks*, **49**(6): 58, 60, 62, 64, 66.

Bown, D. 2001. *The Herb Society of America New Encyclopaedia of Herbs & Their Uses*. D.K. Publishing, New York.

Carpenter, W.J. and R.C. Rodriguez, 1971. Earlier flowering of geranium cv. Carefree Scarlet by high intensity and supplemental light treatment. *HortSci.*, **6**: 206-207.

Carpenter, W.J. and W.H. Carlson, 1970. The influence of growth regulators and temperature on flowering of seed propagated geraniums. *HortSci.*, **5**: 183-184.

Clark, D. 1988. *Kew Gardening Guides: Pelargoniums*, p. 124. Published by The Royal Botanic Gardens, Kew, in association with Collingridge.

Craig, R.1986. Regal and ivy leaved geraniums. *Bedding Plant News*, **17**(6): 6-10.

Daker, M. G. 1969. Chromosome numbers of *Pelargonium* species and cultivars. *J. roy. hort. Soc.*, **94**: 346.

Daughtrey, M. and A.R. Chase, 1992. *Geranium*. In: *Ball Field Guide to Diseases of Greenhouse Ornamentals*, pp. 97-113. Ball Publishing, Geneva, Illinois, USA.

Demame, F.E. 1989. Genetic Improvement of geranium roast (*Pelargonium* sp.): *Systematical, karyological and biochemical contribution*. D.Sc. Thesis, University of Paris.

Deneke, C.F., K.B. Enensen and R. Craig, 1990. Regulation of petal abscission in *Pelargonium* × *domesticum*. *HortSci.*, **25**: 937-940.

Dole, J. M. and H. F. Wilkins, 1999. *Floriculture Principles and Practices*, pp. 451-460, 613. Published by Prentice Hall, Inc., USA.

Eisenberg, B.A., G.L. Staby and T.A. Eretz, 1978. Low pressure and refrigerated storage of rooted and unrooted ornamental cuttings. *J. Amer. Soc. hort. Sci.*, **103**: 732-737.

Erwin, J. 1991. Cool temperatures are still critical on regals. *Minnesota Com. Fl. Grs Asscn Bull.*, **40**(3): 3-4.

Erwin, J. 1997. Irrigation water considerations. *Minnesota Com. Fl. Grs Assoc. Bull.*, **45**(6) & **46**(1): 1-10.

Erwin, J. and G. Engelen, 1992. Regal geranium production. *Minnesota Com. Fl. Grs Asscn Bull.*, **41**(6): 1-9.

Erwin, J.E. and R.D. Heins, 1992. Environmental effects on geranium development. *Minnesota Com. Fl. Gr. Assoc. Bull.*, **41**(1): 1-9.

Everard, B. and B.D. Morley, 1970. *Wild Flowers of the World*, pp. 14, 32, 74. Peerage Books, London.

Federov, 1974. *Chromosome Numbers of Flowering Plants*. Otto Koeltz Science Publishers, Germany.

Guenther, E. 1950. *The Essential Oils*, vol. IV, pp. 672-673. D. Von Nostrand & Co. Inc., New York.

Hay, R. and K.A. Beckett, 1971. *Reader's Digest Encyclopaedia of Garden Plants and Flowers*, pp. 299-301, 507-510. The Reader's Digest Association Limited, London.

Heins, R., N. Lange, T.F. Wallace, Jr. and W. Carlson, 1994. *Plug Storage, Cold Storage of Plug Seedlings*. Meister Publishing, Willoughby, Ohio, USA.

Horst, R.K. 1990. *Geranium*. In: *Westcott's Plant Disease Handbook* (5th ed.), pp. 653-654. Nostrand Reinhold, New York, USA.

Kaczperski, M,.P., A.M. Armitage and P.M. Lewis, 1996a. Performance of plug-grown geranium seedling preconditioned with nitrogen fertilizer or low-temperature storage. *HortSci.*, **31**: 361-363.

Kaczperski, M,.P., A.M. Armitage and P.M. Lewis, 1996b. Using temperature, light and fungicides to prolong the storage life of rooted geranium cuttings *HortSci.*, **31**: 656.

Kaczperski, M.P. and R.D. Heins, 1995. The effect of timing and duration of supplemental irradiance on flower initiation of plug grown geraniums. *HortSci.*, **30**: 760.

Kellen, V. 1983. *Pelargonium × hortorum* or a geranium by any other name. *Florists' Review*, **173**: 19–24.

Key, H. 1985. *Pelargoniums*, 63 p. Published by Cassell Educational Limited, London for Royal Horticultural Society, London.

Langton, F.A. and W. Runger, 1985. *Pelargonium*. In: *Handbook of Flowering*, vole IV (ed. Halevy, A.H.), pp. 9-21. CRC Press, Boca Raton, Florida, USA.

Mani, A.K., S. Mohan Dass, N. Kumar and V. Sampath, 1981. Effect of storage time of herbage prior to distillation on oil recovery and its quality in geranium (*Pelargonium graveolens* L. Herit). *Indian Perfumer*, **24**(3-4): 35-36.

Mani, A.K. and V. Sampath, 1981. Seasonal influence on the oil content and quality of oil in geranium. *Indian Perfumer*, **25**(3-4): 41-43.

Miller, D. 1996. *Pelargoniums: A Gardener's Guide to the Species and Their Cultivars and Hybrids*. Timber Press, Portland, Oregon, USA.

Miller, D. M. 2002. The taxonomy of *Pelargonium* species and cultivars, their origins and growth in the wild. In: *The Genera Geranium and Pelargonium* (ed. Lis-Balchin, M). Taylor & Francis, New York.

Petrovski, S.N. 1971. Effect of some essential oils on cholera and para-cholera vibroios. *Sauvrem Med.*, **26**(6): 51-56.

Pizzetti, I. and H. Cocker, 1968. *Flowers A Guide for Your Garden*, pp. 519-524. Harry N. Abrams, Inc., Publishers, New York.

Post, K. 1949. *Pelargonium*. In: *Florist Crop Production and Marketing*, pp. 730-737. Orange Judd Publishing, New York, USA.

Quatchak, D.J., J.W. White and E.J. Holcomb, 1986. Temperature, supplemental lighting and chlormequat chloride enhances on flowering of geranium cuttings. *J. Amer. Soc. hort. Sci.*, **111**: 376-379.

Rajeswara Rao, B.R., E.V.S. Prakasa Rao and M.R. Narayana, 1989. Rose geranium, an economical crop in the South Indian plains. *Indian Hort.*, **34**(2):14-17.

Roberts, D.L. 1997. Major geranium diseases. *Professional Plat Growers Association News*, **28**(1): 21-22.

Shaw, R.J. and M.N. Rogers, 1964. Interactions between elevated carbon dioxide levels and greenhouse temperatures on the growth of roses, chrysanthemums, carnations, geraniums and African violets. Part 5. Various flowers. *Flor. Rev.*, **135**(3491): 19, 37-39.

Tayama, H.K. and J.M. Klesa, 1991. *Scheduling*. In: *Tips on Growing Zonal Geraniums* (2nd ed.; eds Tayama, H.K. and T.J. Roll), pp. 34-36. Ohio Cooperative Extension Service, Ohio state University, Columbus, Ohio, USA.

Trellinger, K. 1997. Top 10 tips for perfect ivy geraniums. *GrowerTalks*, **61**(8): 73, 75.

Van der Walt, J. J. A. 1977. "*Pelargoniums of Southern Africa*" (vol. 1). Purnell, Cape Town, S. A.

White, J.W. and I.J. Warrington, 1984. Effect of split night temperatures, light and chlormequat on growth and carbohydrate status of *Pelargonium × hortorum*. *J. Amer. Soc. hort. Sci.*, **109**: 458-463.

Williams, C. A. and B. H. Jeffrey, 2002. *Phytochemistry of the genus Pelargonium*. In: *Geranium and Pelargonium: The genera Geranium and Pelargonium* (ed. Lis-Balchin, Maria), pp. 99-115. Taylor and Francis, New York.

23

Hedera (Family: Araliaceae)

Sapna Panwar, Namita and Sanyat Misra

[**Common Names**: Algerian ivy/Canary island ivy (*Hedera canariensis*), British common ivy/Common ivy/ English ivy (*Hedera helix*), Irish ivy (*Hedera hibernica*), Ivy, Japanese ivy (*H. rhombea*), Nepal ivy (*H. nepalensis*), Persian ivy (*H. colchica*), Russian ivy (*H. pastuchovii*)]

Introduction

Hedera is the Latin name for ivy. It originates from Europe to Caucasus, Western Asia (Himalaya and Japan), and also Canary Islands (coast of NW Africa) and Madeira Islands (EN Atlantic Ocean), and comprises 5-11 species. However, some reports recognize 15 *Hedera* species (12 true species and 3 subspecies; Ackerfield and Wen, 2002). The genus *Hedera* is one amongst 55 genera belonging to family Araliaceae (Evans, 2009). Existence of a close relationship between family Araliaceae and Apiaceae has been reported by Zomlefer (1994) and both these families are found to be the major group members belonging to order Apiales (Stevens, 2001). The genus *Hedera* belongs to the subfamily Aralioideae like to that of other popular genera such as *Aralia*, *Oplopanax* and *Panax* (Anon., 2007). *Hedera* is known for its ornamental value and is commonly used in the outdoor as well as indoor landscaping. It was used as a decorative and celebratory symbol 3,000 years ago by the Egyptians and later by the Romans. Its popularity peaked in Victorian times when it was used extensively inside the home and outdoors in the gardens. The genus is categorised mostly in the group of climbers and twiners. Depending on the species and cultivar, it can be utilized in interiors as an excellent foliage plant for hanging planters and baskets due to its cascading or trailing habit. However, it can also be grown in pots, dish gardens, terrariums, for topiaries, as ground covers, against walls and trellis, on tree trunks, *etc*. In exteriors, it can be utilized as a softscape element to cover the drabness of concrete walls or as climbers over tree trunks to provide natural and aesthetic look. Nowadays, cut vines varieties of *Hedera* are used in many floral arrangements, decorations, wreaths, *etc*.

Ivy is found to be rich in various bioactive compounds like flavonoids (rutin, kaempferol-3-rutinoside), saponins (hederin with its sub-units, hederacosides A and C, hederagenin, glucopyranosyl-hederagenin, cauloside F), polyphenolic acids (caffeic and chlorogenic acids) along with other active principles like traces of alkaloids (emetine), polyacetylenes (falcarinone; falcarinol), phytosterols (sitosterol, stigmasterol, spinasterol, campesterol), sesquiterpenes (germacranene, β-elemene), iodine, *etc*. (www.centrechem.com). The saponins in *Hedera* are reported of having *in vitro* anti-elastase and anti-hyaluronidase activities that are mostly used in the treatment of vein insufficiency disorders. The flavonoids are reported to lower the permeability and increase the resistance of blood capillaries (Bruneton, 2001), thereby, are mainly used in the treatment of blood vessel-related disorders. The *Hedera* extract also finds applications in preparation of many cosmetic products because of its stimulating action on blood circulation. The cosmetic application of *Hedera* is presented in Table 1. The presence of certain compounds in leaves and fruits have also been reported which can restrict breathing process or may induce coma-like conditions. The fruits which are actually berries have been utilized for tanning of leather and also in textile industry for dyeing purpose (Foote and Jones, 1989).

Table 1: Cosmetic Applications of *Hedera*

Active Compounds	*Action*	*Cosmetic Application*
Saponins	Blood circulation	Anti-edema
Flavonoids	Stimulation	Anti-cellulite,
		Vessel-protector and venotonic
Saponins	Anti-inflammatory	Anti-irritant
Saponins	Antimicrobial	Purifying

Source: www.centrechem.com.

Botany

The most characteristic feature of the genus *Hedera* is its evergreen ornamental foliage. *Hedera* is one amongst several genera that express the dimorphism, *i.e.* first form being juvenile and the second adult which is called arborescence when these finish growth, form the flowers and produce seeds. Its flowering stems produce leaves of a different shape to those lower on the same stem. Irrespective of the shape of leaves of the juvenile form, the new leaves are oval and form bracts to the flowering heads. The growth habit of both the forms is also different as juveniles are mostly climbing in nature and produce roots on every node, whereas adult-forms lack adventitious roots in stems and their growth is erect, rigid and shrubby or tree-like. There are reports of existence of ivy trees of 12 m tall in Europe. Ivy shrubs with more branching habit can be utilized in form of hedges (Coon, 2001; Chen *et al.,* 2014). The juvenile foliage is 3-5 lobed, 5-10 cm wide at base with palmate venation whereas adult foliage is larger and usually unlobed than juvenile forms (www. public.asu. edu). Rutherford (1997) reported variation in foliage colouration and stated that foliage colour of *H. helix* is usually a blackish-green and that of *H. hibernica* yellow-green. The leaf description is presented in Figure 1. The comparison of juvenile and adult form is presented in Table 2.

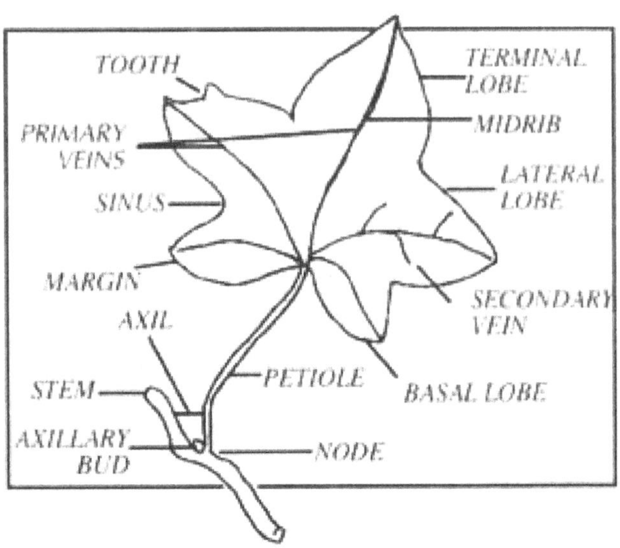

Figure 1: Leaf Description of *Hedera* sp.

Table 2: Comparison of Juvenile and Adult Growth Phase of English Ivy

Characteristic	*Juvenile*	*Adult*
Growth habit	Plagiotrophic	Orthotrophic
Flowers	Absent	Present
Leaf arrangement	Alternate	2/5th spiral
Leaf production	1 leaf/week	2 leaves/week
Shoot growth	Vigorous	Slight
Leaf shape	Lobed	Entire
Rooting ability	Good	Weak
Aerial roots	Present	Absent

Source Stein *et al.,* 1969.

The leaf characteristics vary widely in genus *Hedera.* According to the American Ivy Society (Naples, FL), leaf shapes can be classified into nine categories in reference to the system of Pierot (1974) and these are arborescents or senile (only a few in cultivation, grow stiffly upright and frequently produce flowers), bird's foot (varieties such as 'Brokamp', 'Green Feather', 'Irish Lace', 'Needlepoint', 'Perfection', *etc.*; these cultivars bear leaves with narrow lobes), curlies (these have leaves with ruffles, ripples or pleats such as 'Big Dewal', 'Ivalace', 'Manda's Crested', *etc.*), fans (leaves with lobes of equal length such as 'California Fan', 'Fan', *etc.*), heart-shapes (leaves shaped as a valentine such as 'My Heart', 'Sweertheart', *etc.*), ivy-ivies (though leaves are typical of species but with pronounced terminal, lateral, and basal lobes such as 'Hahn', 'Pittsburgh', *etc.*), miniatures (leaves less than 1 cm such as 'Jubilee', *etc.*), oddities (unusual forms as fasciated stems or distorted leaves though only few in cultivation), and variegation (where leaves are multicoloured such as 'Glacier', 'Gold Dust', 'Gold Heart', 'Hahn', 'Kolibri', *etc.*). *Hedera* is valued only for its foliage and not the inflorescence which occurs in arborescents but its inflorescence is usually a globose umbel mostly produced in the compound panicles, rarely solitary, each umbel usually comprising of 10-15 or sometimes even up to 60 light green flowers and these help in enticing the bees, butterflies, wasps and various other insects. Its berry-fruits are ornamental and may be in colours of black to dark purple or dark blue, and even yellow ('Sand Hill') and orange (*Hedera nepalensis* and *H. helix* 'Poetica'). *Hedera helix* inflorescence is a terminal racemose umbel (Radford *et al.,* 1968), flower colour yellowish-green and the size 5-7 mm across (Hitchcock and Cronquist, 1973), peduncle size 2-8 cm, pedicel size 5-10 mm, sepals and petals 5 and the size of sepal 2-4 mm, petal 3 mm (Radford *et al.,* 1969), stamens 5 (Gleason, 1963), carpels 3-5 (Radford *et al.,* 1968), fruits nearly black and spherical berries, seeds per berry 3-5 (Gleason, 1963) and seed size 6-9 mm (Hitchcock and Cronquist, 1973).

Hedera hibernica and its cv. 'Hibernica' are tetraploid (4x = 96), while *H. helix* is diploid (2x = 48) (McAllister, 1982, 1984). Man-made hybrids between *Hedera helix* and *Hedera hibernica* are not reported (Rutherford, 1984). In nature also there is no record of mating these two species together as there is difference in their flowering time (Rutherford, 1984).

Propagation

Hedera sp. is easily propagated by taking 7.5-12.5 cm long tip and node cuttings in July or August. In case a bush plant is desired, the cuttings are to be taken from arborescent growth but in case of climbers the cuttings should be taken from runner growth. The node cuttings are most commonly used for propagation as yield of cuttings per plant is more as compared to tip cuttings. Single node cutting refers to cuttings with single leaf and a bud while double node cutting comprises of two leaves and two buds. Fast rooting and growth are observed in case of two-leaf cuttings as compared to single leaf cuttings. Cuttings are directly placed into the pot containing moist medium having combination of peat, pine bark and perlite/vermiculite. Some media composition such as peat moss (50 per cent), + pine bark (50 per cent) or peat moss (80 per cent)+ perlite (20 per cent) are used for propagation. Hay and Beckett (1971) advocated equal parts of peat and sand by volume in a closed frame or under mist propagation, and then potting of rooted cuttings individually in a 7.5 cm pot and then growing them on in a cold frame or greenhouse until needed for planting in their permanent positions. They further suggested an alternative that the 15 cm cuttings from the ripe shoots should be taken in October or November, their soft tips removed, and then these inserted in the prepared beds of sandy soils in a sheltered position outdoors directly. Huxley (1979) stated that 7.5-10.0 cm long cuttings root easily when placed in a glass of water at warm room temperature in bright indirect light and when roots become 2.5-3.75 cm long, these cuttings are planted in pots containing standard potting mix recommended for adult plants. He further advocated inserting 3-4 tip cuttings in a 7.5 cm pot containing a moistened equal parts mixture of peat moss or coarse sand or perlite and kept at a heated propagating chamber having bright indirect light where these cuttings will root in 2-3 weeks. The medium should have adequate water-retention capacity and aeration, soluble salts should range between 1-3 dS/m and a pH of 5.5 to 6.5. It was also reported that rooting of cuttings is best under mist in a shadehouse with light intensity of 1,000–1,500 fc and temperature range of 21.1 to 30 °C (Chen *et al.,* 2014). Ivies can also be propagated through layering.

Hedera is never propagated from seeds for commercial purposes as it is time-consuming and the seedlings are not genetically true to type. However, some novel variants can be isolated from seedlings which are further evaluated and released as the new cultivars like *H. helix* 'Rotunda' (Coon, 1997) and *H. helix* 'Goldfinch' (Coon, 2000).

Species and Varieties

Genus *Hedera* comprises 4-15 species, however, the *Hedera helix* is the one amongst several species which is commercially exploited as an ornamental plant. The description of the various species is as follows.

Hedera azorica

A native of Azores, it is a fast-growing very hardy climber. Its two most popular varieties are 'Typica' bearing 5-7-lobed leaves covered with soft pale hairs, and 'Variegata' with 5-7 lobed cream-yellow leaves that are splashed, spotted, striped and stippled with green.

Hedera canariensis (H. algeriensis)

It is native of the Canary Islands, the Azores (N. Atlantic Ocean), Madeira and NW Africa and is an ideal plant for trellises. It is a hardy creeper growing up to 6 m in length with reddish twigs and petioles; leaves some 15 cm wide are dark glossy green, leathery, broadly ovate with cordate base, bearing 3-5 shallow lobes, and covered with grayish- white scales. *H. c. arborescens* 'Variegata' leaves are ovate, hard with creamy variegation on light green or grey base. *H. c. azorica* bears light green and shallowly lobed leaves. *H. c.* 'Variegata' ('Gloire de Marengo') leaves are 6-10 cm long, thin, leathery, centre green, joined by a zone of grey-green and creamy-white in the margin. Other noteworthy varieties are 'Algerian', 'Margina Maculata' and 'Ravensholst'. The latter is quite vigorous with large leaves and makes good ground cover.

Hedera colchica (H. amurensis)

A native of Caucasus to N. Iran, it is a strong-growing species some 9 m long, bearing dark glossy green, triangularly cordate but unlobed leaves that are leathery, 7.5-25 cm long and 12-15 cm wide, and emit an aroma when crushed. The vine is best suited for hanging basket in indoor environment. Its var.'Dentata' has some 20 cm long and distantly toothed leaves suitable for pot culture; 'Dentata Variegata' has dark green leaves shaded with grey-green and variegated with greyish streaks and white margins; Gloire de Marengo' is an excellent clone similar to 'Dentata Variegata' and is easy to maintain, and 'Sulphur Heart' has broad yellow centre and pale veins.

Hedera helix

A native from Europe (Great Britain) to Caucasus, it is an incredibly varied group with a large diversity of foliage colours and shapes. It grows up to 30 m long bearing glossy dark green leaves, often with silver markings along the veins. Post (1949) stated that greater variation exists among the various ivy cultivars than among the cultivars of any other ornamental. More than 300 cultivars of *H. helix* are reported which vary in leaf size, shape, colour, and variegation patterns and all these cultivars were selected from sports. Many of the cultivars are commonly seen growing as house plants while still many can survive a few degrees of frost. Its leaves are cordate, diamond-shaped or 3-5-triangular lobed, and narrow to broader, tiny to large, plain or frilled and tight to loose on stems. Its popular varieties under large green leaves are 'Asterisk', 'Atropurpurea', 'Bowles Ox Heart', 'Brokamp', 'Buttercup' ('Golden Cloud' or 'Russell's Gold'), 'California', 'Conglomerata', 'Deltoidea', 'Digitata', 'Dragon Claw', 'Erecta', 'Flamenco', 'Fleur de lis', 'Fluffy Ruffles', 'Garland', 'Glymii', 'Gracilis', 'Greenheart', 'Green Ripple', 'Helvetica', 'Hibernica Hamilton', 'Irish Lace', 'Ivalace' ('Lace Ivy'), 'Koniger's Auslese', 'Lobata Major', 'Merion Beauty', 'Parsley Crested', 'Pedata', 'Pin Oak Improved', 'Pittsburgh', 'Plume d'Or', 'Poetica', 'Professor Friedrich Tobler', 'Ralf', 'Ritterkreuz', 'Sagittifolia', 'Shamrock', 'Spear Point', 'Succinata', 'Telecurl', 'Wilson', 'Woerner', *etc.*; under miniature green leaves are 'Baltica', 'Direktor Badke', 'Duckfoot', 'Filigran', 'Gnome', 'Green Finger', 'Helena', 'Irish Lace', 'La Plata', 'Merion Beauty', 'Needle Point', 'Spetchley' (a real gem among miniatures suitable for rock crevices and between paving stones as the leaves are smaller than fingernail), 'Spinosa', 'Sulphurea', 'Sweetheart', 'Tobler', *etc.*; and under variegated leaves are 'Adam', 'Angularis Aurea', 'Anna Marie' ('Harold'), 'Aureo-variegata' ('Chrysopylla'), 'Bruder Ingobert', 'Caenwoodiana Aurea', 'Cavendish', 'Ceridwen', 'Chester', 'Conglomerata Erecta', 'Discolor', 'English', 'Eva', 'Glacier', 'Gold Child', 'Gold Dust', 'Golden Esther' ('Ceridwen'), 'Golden Ingot', 'Gold Heart', 'Heise Denmark # 1 & # 2', 'Hahn's Self-Branching', 'Hahn's Variegated', 'Harold', 'Hibernica Variegata', 'Ingelise', 'Ingrid Liz', 'Itsy-Bitsy', 'Jubilee', 'Kolibri', 'Lalla Rookh', 'Little Diamond', 'Maculata', 'Manda's Crested' ('Curly Locks'), 'Marginata Major', 'Melanie', 'Midas Touch', 'Mrs Pollock', 'Nigra', 'Plume d'Or Glasshouse', 'Sagittifolia Variegata', 'Silver Queen' ('Marginata'), 'Tess', 'Tricolor' ('Marginata-rubra'), *etc.*

Hedera hibernica (H. Helix var. *hybernica)*

It is hardy in nature hence mainly grown as an evergreen outdoor vigorous climber growing up to 5 m. Its large leaves are mid-green. It is suitable for covering a large area, either on the ground or against a wall. Its popular indoor cultivars are 'Sweetheart Ivy', 'Deltoidea', *etc.* 'Deltoidea' is slow-growing, but dense with dark leathery leaves having prominent veins.

Hedera nepalensis

It is half-hardy species from Nepal which grows up to 4 m long. It produces prominently grey-veined, soft grey-green, toothed and oval to triangular leaves which are basally notched and about 6.25 cm long on slender stems, giving a very lacy effect, especially in hanging baskets and against a sheltered wall. Its popular varieties are 'Marbled Dragon' (leaves 7.5 cm long, veins light grey), 'Sinensis' (deeper notch, basal lobe larger), 'Suzanne' (a trailing plant with finely hairy leaves), 'Typica' (drooping serrated leaves), *etc.*

Hedera pastuchovii

A fully hardy species native of Russia, it is very distinct ivy, growing up to 2.5 m long, having shield-shaped glossy black-green and elongated leaves. It is suitable for growing against a wall. Its form 'Typica' bears blue-green, long, unlobed and hanging leaves.

Hedera rhombea

It is a frost-hardy species from Japan and Korea. It grows 1.2 m long, producing mid-green, fairly thick, diamond-shaped (triangularly cordate) and unlobed leaves which are attached to long, trailing and purple-flushed stems. It is suitable for growing against a low wall. It is most tolerant to higher temperatures. Its most common form is *H. r.* 'Variegata'.

Cultural Practices

Ivy is well suited to be grown under diverse agro-climatic conditions. **Light** plays an important role in production of quality plant material. It has been observed that ivies are able to tolerate moderate to deep shade levels, however, in some of the variegated types the cultivars grown under very low light levels may lose their variegation. Ivy grows well under light intensities of 1,500- 2,500 fc. High quality plants are produced under protection (Henley *et al.*, 1991; Pierot, 1974). These do not prefer direct sunlight for whole part of the day as this causes leaf scorching, however, some solid green cultivars can tolerate full light for short duration of the day. Generally, ivies thrive in an environment of bright filtered to low light under indoor environment. Indoors, under low light regime, the supplemental light through fluorescent tubes can be provided to compensate the existing low light (Pennisi *et al.*, 2003). They also tolerate a wide range of temperature regimes, therefore these are being grown indoors as well as outdoors. Some species and varieties are frost-hardy under outdoor environment while others perform well under indoor environment. It. can be best grown at a temperature range of 18.3-29.4 °C. However, the day/night temperatures of 18.3°C/10.0°C are optimum for their excellent growth. The temperatures below 4.4 °C may result into differential shade patterns in their leaves. Change in pigment patterns, especially in green ivies will occur as the leaves turn red to purple and there is development of maroon streaks over yellow-leaved cultivars. Sometimes, marginally variegated ivies turn pink at the edges (www.flowershopnetwork.com). At higher temperature, *i.e.* above 32.2 °C, root growth is affected poorly. Growth inhibition at temperature above 29-32 °C has been observed in some cultivars. The optimum temperature for propagation medium is 21 to 22 °C and a temperature of 18-24 °C is required afterwards for luxurious growth. Henley *et al.* (1991) reported that some landscaping cultivars of *Hedera helix* can survive as low as -23 °C, however, many cultivars can survive -12 to -7 °C if proper acclimatization is accomplished. It is advised to avoid the fluctuating temperatures as these can badly affect the performance. Ivy prefers an evenly moist environment and a relative humidity above 50 per cent is required in order to grow ivy successfully under indoor environment. Misting is the best way to achieve this and should preferably be given twice a day. Nowadays, humidifiers are also used to enhance humidity level. An excellent method to provide extra moisture is by allowing the plants to stand on shallow trays filled with pebbles and water.

Garden **soil** with a pH of 5.5-6.5 is one of the most common and easily available potting mixture. A good potting mixture should be able to hold the plant and should retain optimum water and nutrients for proper growth. One of the recommended **potting mixture** is loam, sphagnum peat moss and coarse sand/horticultural perlite (equal parts) or soil mix (one kilogramme of good garden soil + 3 g of dolomitic lime + 6 g of bonemeal). These both the mixes are soil-based but where soil is substituted with peat moss that is known as peat-based soilless mixture. Ivies are comfortable in soil-based medium, *i.e.* 1 part sterilized fibrous soil, 1 part medium grade peat moss/ground tree bark/leaf mould, 1 part coarse sand/fine perlite, and 1 part dehydrated cow manure. It is advisable that all the ingredients of potting mixture should be thoroughly mixed. Nowadays, many synthetic soilless mixes are also commercially available and are very popular among growers. Considering both, the cost and convenience, commercial growers prefer plastic pots and synthetic soil mixes, however, clay pots are the best choice. The overgrown plants are **repotted** from one container to the other. The symptom shown by overgrown plants is when the roots of the plant begin to grow out of the bottom of the container or when the plants become top-heavy or root-bound or dry out too rapidly. It is the appropriate time to transplant the plant into the next larger pot. The soil mixture should contain sterilized houseplant potting soil (1 part), peat moss (1 part) and sand/perlite (1 part) (www.nybg.org). It is recommended to repot every 2-3 years. The plant should be repotted in immediate next size pot and the new pot should not be more than 2.5 cm larger in diameter than the earlier one (www.clemson.edu).

Like other plants, ivies also have requirement of **fertilizers** for N, P and K for their proper vegetative growth and reproductive development. Chen *et al.* (2014) recommended N: P_2O_5: K_2O at 3:1:2 or 2:1:2 at the rate of 12.2 kg N/100 m^2 annually. Henley *et al.* (1991) advocated 12.2 to 14.6 g/m^2 of actual N per month from spring through fall to the stock plants for taking cuttings. It is also advisable to use water soluble mineral fertilizer or 20:20:20 balanced liquid fertilizer at bi-weekly or monthly intervals during active period of growth for enhanced plant performance. Suggested levels for continuous application are 150 ppm N, 50 ppm P_2O_5, and 100 ppm K_2O per month. Growers report good results using liquid feed with 9:3:6 NPK along with micronutrients (Pennisi *et al.,* 2003). It is suggested that for better production of ivy in both cool and warm seasons is to reduce recommended fertilizer levels by 25 per cent during December to February and raise these by 25 per cent from the recommended fertilizer levels from June to September. Slow release fertilizers like Osmocote with 2 to 12 months release period also hold good promise. Ivies prefer an evenly moist environment so **watering** with soft water at weekly interval will help them maintaining their turgidity and in reducing the infestation of the plants from spider mites. The scarcity of water will result in stunted growth of the plant whereas excess water will cause blocking of air spaces in the medium which may result in suffocation and subsequent death of roots. Between two waterings, the mixture should be allowed to appear dry. After every application of fertilizer in granule form, the plants should be watered.

Ivy produces long vines mostly of trailing nature, which require some structure for **support**. As a potted specimen, it is recommended to initially train the vine in a manner that further it can grow naturally over the containers. However, supporting structures are provided for training of plants into wreaths, hearts, topiary, trees and other shapes (www.guide-to-houseplants.com). Ivy tends to produce long vines with few leaves during winter season. The plants present woody look so it is recommended to prune the woody growth along with a few leaves. A few strong branches should be retained and weak, diseased or pest-infested shoots as well as spindly growths that interfere with plant growth should be removed. Light pruning can be done any time of the year (www.guide-to-houseplants.com).

Growth and Flowering

The growth regulators like A-Rest (ancymidol), CCC (chlormequat) and B-nine (daminozide) are efficient in retarding the growth and vigour of the vines (Adriansen, 1985). However, gibberellic acid can be used for internodal length elongation. Christenden (1973) when propagated *H. canariensis* in early September, the production time in a 9-cm pot required 290 days but when grew in January or February, these required only 140 days, so was the important role of the radiant energy.

Post Harvest

Ivy is reported to be moderately sensitive to even low levels of ethylene, and causes epinasty of immature growth of the vines (Nowak and Rudnicki, 1990). *H. canariensis* is transported at 15 °C with relative humidity of 90 per cent (Sterling and Molenaar, 1986). Dole and Wilkins (1995) advocate shipping of ivy plants at 4-13 °C.

Insect-Pests, Diseases and Physiological Disorders

Insect-Pests

Aphids (blackfly, greenfly; *Myzus persicae* and certain others) build up their large population in a little time in hot weather, their nymphs and adults cluster around the tender parts of the plants such as shoots, leaves and flowers and suck their sap which results into distortion of new growth and withering of flowers, and their excretion becomes sticky and invites sooty mould. To prevent them, healthy planting material should be used, yellow sticky traps should be laid up, and to kill them tar-oil or DNOC/petroleum wash should be sprayed during winter. Use of dimethoate, heptenophos with bifenthrin, permethrin, primicarb or primiphosmethyl, or malathion at 0.1 per cent spraying will kill the aphids of any kind. Primicarb being a selective insecticide kills only the aphids and not the bees, ladybirds and other beneficial insects. Insecticiadal soaps, pyrethrum and derris are the effective organic insecticides. **Whiteflies** (*Trialeurodes vaporariorsum*) breed continuosly throughout the year (except winter) on indoor plants and their aduls as well as immobile, oval and scale-like nymphs suck cell sap on underside of leaves whereas upper surface becomes sticky with honeydew excretion enticing sooty mould. Their infestation causes small yellow spots on the leaves, followed by yellowing and sooty mould. There is a pesticide resistance problem with this pest. A tiny parasitic wasp (*Encarsia formosa*) and *Beaveria basiana* are the best remedy in case of indoor plants from spring to autumn and temperatures then are congenial for the parasites. Hanging of yellow sticky trap is quite effective. Insecticidal soaps do not have lasting effect on *Encarsia* though reduces whitefly population considerably. Bifenthrin, permethrin, primiphos methyl and pyrethrum are effective if resistant strains of whitefly are not present, however, these insecticides prove dangerous to the parasite. Nymphs and female adults of **mealybugs** (*Planococcus* spp. and *Pseudococcus* spp.) covered with white coating may be noticed clustered on the shoots in the leaf axils and between twining stems, at the tips, peduncles and pedicels, and on tender leaves, remaining there and sucking their sap and through excreted honeydew attracting sooty mould. Mealybugs are soft-bodied some 4 mm long greyish-white or pink insects. Their infested plants become sticky due to excretion of honeydew, new growth distorted and flowers aborted. During summer, **ladybird predator** (*Cryptolaemus montrouzieri*) will predate over them but during other times thorough spray of insecticidal soap or malathion should be done. Nymphs and adults of the **western flower thrips** (*Frankliniella occidentalis*) and banded greenhouse thrips (*Parthenothrips dracaenae*) feed on tender plant parts by which leaves curl and distort and mark silver-grey scars at the place of infestation, *vis-a-vis* act as a vector of impatiens necrotic spot virus. These are controlled through yellow sticky traps. Three fortnightly sprayings of bifenthrin, permethrin or pyrethrum will control banded greenhouse thrips. Three fortnightly sprayings with malathion-based insecticides and/or **predatory mite** (*Amblysetus degenerans*) is effective against western flower thrips. Caterpillars of various moths, butterflies, sawflies and leaf miners bore the flowers and chew the leaves and make holes there. Some also feed on the roots. Spraying with bifenthrin, permethrin or pyrethrum is effective. Use of *Bacillus thuringiensis* kills these larvae. Grubs and adults of **vine weevil** (*Otiorhynchus sulcatus*) are the serious pests of pot-grown plants which eat away the roots, and plants start wilting. Introduction of **pathogenic nematode**, *Heterorhabitis megidis* or *Steinernema* through water into the potting mixture in late summer will protect the plants against vine weevil grubs and for the adult weevils, plants can be sprayed at dusk with bifenthrin or primiphos methyl or dusting with lindane.

Scales (*Lichtensia viburni*) are the sap-feeding insects, feeding on the upper leaf surface making the surface sticky with the honeydew excreted by them but live on the lower leaf surface next to a leaf vein. These are more troublesome on plants growing against a wall or under sheltered places. In mid-summer, the underside of leaves should be sprayed with malathion to control this pest.

Bryobia mites (*Bryobia* sp.) are up to 1 mm long and black-bodied mites having four pairs of pinkish legs. These suck the sap on the upper leaf surface from late winter to early summer and so a fine pale mottling on upper surface of leaves is formed. These are controlled by three fortnightly spraying with bifenthrin, malathion or primiphos methyl. **Two-spotted spider mite** (*Tetranychus urticae*) infestation causes pale-mottling, speckling and webbing on infested leaves but later the leaves change to yellowish-white. The yellowish-green mites are less than 1 mm long and have two large dark markings towards the head end of the body. Infested leaves dry up and fall prematurely. Such plants should be treated with several fortnightly sprays of bifenthrin, malathion, primiphos methyl or an insecticidal soap, or only *Phytoseiulus persimilis* **predatory mite** should be introduced. **Cyclamen mite** (*Phytonemus pallidus*) is minute whitish-brown that infests the flower buds causing their distortion, and developing leaves at the shoot tips causing 'rat-tail' symptoms. Due to its infestation, new foliage fails to expand properly, stunts and distorts. To prevent mites, pest-free planting material should be used and infected plants should be destroyed. Through chemicals, it cannot be controlled.

The damaging pattern of **slugs** is similar to those of caterpillars. These feed during night on all tender parts of plants leaving a slimy trail behind and hiding them on the under-surface of leaves and in the plant debris found below the plant. Their attack batters the plants completely. Treatment with methiocarb and metaldehyde baits can be very effective.

Diseases

***Xanthomonas* leaf spot** is caused by bacterium *Xanthomonas campestris* pv. *hederae* where brown to black, and circular to irregular spots often with yellow to red halo and the water-soaked margin are formed first on the older leaves. To prevent this problem, use of healthy planting material, proper sanitation, avoiding overhead irrigation, pruning of the diseased part and use of copper hydroxide, copper sulphate pentahydrate or fosetyl aluminium are suggested.

Botrytis blight (*Botrytis cinerea*) occurs during cold and humid conditions and its infection causes brown to tan lesions on leaves, stems and petioles in the form of fuzzy fungal growth. Since this fungus lives in most living and dead plant material, its spores are always present in the air. To prevent this disease, proper plant density should be adopted to permit proper ventilation so that relative humidity is also properly maintained. Beds and the surroundings should be clear of all the plant debris. Use of copper hydrochloride, iprodione or carbendazim is suggested. **Anthracnose** (*Amerosporium trichellum* or *Colletotrichum trichellum*) is characterized by brown and sunken lesions on foliage and sometimes with black specks in the centre of the lesions. Use of healthy planting material, no-overhead irrigation, pruning of the infected portion, proper sanitation of the field, maintaining of proper plant density and use of copper hydroxide, manzate or iprodine will keep this disease at bay. **Black leaf spot** (*Phyllosticta concentrica*, *P. hedericola*) is also sometimes noticed on ivies producing sunken and brown longitudinal marks on stems with colour change of the leaves which later on die down and if the weather is wet, a pink-coloured slimy fungal growth may occur. To control this, infected parts should be removed immediately. Nomrod-T is effective in controlling this disease. **Root rot** (*Phytophthora* spp.) causes browning and downward curling of lower leaves, leaf chlorosis, wilting, root die-back, *etc.* Apart from cultural operations adopted, the plants may also be treated with etridiazole + thiophanate methyl or mefenoxam. ***Rhizoctonia* root rot and aerial blight** is caused by *Rhizoctonia solani*. Its infection causes brown irregular lesions, blighting and wilting of plants. To prevent the occurrence of this pathogen, potting media should be sterilized, planting material should be healthy, over- and overhead watering and over-fertilization should be avoided, affected portion should be removed and burnt, field should be clean and the plants should be sprayed with etridiazole + thiophanate methyl, oriprodione or PCNB (terrachlor). *Glomerella cingulata* and *Sphaceloma hederae* also cause leaf spot diseases in ivies (Horst, 1990). **Powdery mildew** (*Oidium* spp.) infection occurs on the leaves, leaf petioles, stems and buds in the form of white powdery coating, especially during spring and summer. Usually, upper leaf surfaces are affected first most severely and then spreading to lower surfaces and other parts. Its infection is encouraged by humid atmospheres. To prevent this disease, proper ventilation should be maintained and plants should be sprayed with carbendazim, copper sulphate pentahydrate, mancozeb, propiconazole, triadimefon, triforine with bupirimate or sulphur.

Physiological Disorders

Slow growth of vines is due to high temperatures which may also cause leaf scorching or edge browning. Ivies should never be exposed to direct sunlight during summer or higher than 29 °C temperature. Proper ventilation and higher humidity should also be maintained. **Leaf edge browning** also occurs due to spider mite infestation so such insects should not be allowed to crop up. Thin, lanky and spindly growth of vines is due to excess of watering and fertilization, *vis-a-vis* low light levels so such plants should be moved to indirect but bright light conditions and adequate supply of fertilizer and water should be made. **Leaf burn** is due to high concentration of soluble salts in the growing medium. Appropriate salt levels (1.0 to 3.0 dS/m) in the medium should be maintained. Appearance of small **yellow leaf spots** is due to more intake of water by the roots than can be given off by the leaves, and sometimes leaf cells become inflated with water and swell. In such cases the temperature and ventilation around the plant should be increased to facilitate moisture loss through the leaves and over-watering should be avoided. **Loss of variegation** from young leaves is due to very low light levels or because of advancing age so in such cases the plants should be moved to brighter conditions at regular intervals.

References

Ackerfield, J. and J. Wen. 2002. A morphometric analysis of *Hedera* L. (the ivy genus, Araliaceae) and its taxanomic implications. *Adansonia*, **24**(2):197-212.

Adriansen, E. 1985. *Kemisk Vaekstregulering*. In : *Potteplanter I- Produktion, Metoder, Milder* (eds Christensen, O.V., A. Klougart, I.S. Pedersen and K. Wikesjo) (Danish). Gartner INFO, Kobenhavn, Denmark, pp. 142-162.

Anonymous, 2007. The Plants. Database (http://plants.usda.gov, 12 November 2007). National Plant Data Center (USDA, NRCS), Baton Rouge, LA 70874-4490 USA.

Bruneton, J. 2001. *Farmacognosia. Zaragoza*: ed. Acribia, pp.305-341.

Chen, J., Dennis B. McConnell and Kelly C. Everitt, 2014. *Cultural Guidelines for Commercial Production of Interiorscape Hedera*. Published by IFAS Extension, University of Florida, pp: 1-4.

Christenden, O.V. 1973. Seasonal variation in production time of *Hedera canariensis* Willd. 'Gloire de Marengo'. *Tidsskrift for Planteavl*, **77**: 224-231.

Coon, C. L. 2000. [2000 New Registrations. III.] *Hedera helix* 'Goldfinch'. *Ivy Journal*, **26**: 27-30.

Coon, C. L. 2001. All about adult ivies: An overview. *Ivy Journal*, **27**: 24-35.

Dole, J.M. and H.F. Wilkins, 1995. *Floriculture Principles and Species*, pp. 361-363. Prentice Hall, Upper Saddle River, New Jersey, USA.

Evans, W. C. 2009. *Trease and Evans Pharmacognosy* (16th ed.). Elsevier Science limited, UK, p. 48.

Foote, L.E. and S.B. Jones, Jr. 1989. *Native Shrubs and Woody Vines of the Southeast: Landscaping Uses and Identification*. Timber Press, Portland, Oregon, USA.

Gleason, H.A. 1963. *The New Britton and Brown Illustrated Flora of the Northeastern United States and Adjacent Canada*. Hafner Publishing Company, Inc., New York, USA.

Hay, R. and K.A. Beckett, 1971. *Reader's Digest Encyclopaedia of Garden Plants and Flowers*, pp. 325-326. The Reader's Digest Association Limited, London, U.K.

Henley, R.W., A.R. Chase, and L.S. Osborne, 1991. English Ivy. Central Florida Research and Education Center, Foliage research Note RH-91-15, Apopka, Florida.

Hitchcock, C.L. and A. Cronquist, 1973. *Flora of the Pacific Northwest: An Illustrated Manual*. University of Washington Press, Seattle, Washington, USA.

Horst, R.K. 1990. *Ivy, English*. In: *Westcott's Plant Disease Handbook* (5th ed.), p. 693. Van Nostrand Reinhold, New York.

Huxley, A. (ed.), 1979. *Reader's Digest Success with House Plants*, pp. 234-235. The Reader's Digest Association, Inc., Pleasantville, Canada Ltd., Canada.

McAllister, H. A. 1982. New work on ivies. *International Dendrology Society Yearbook 1981*, pp. 107-108.

McAllister, H.A. 1984. [McAllister raises var. *hibernica* to species status.], pp. 33-34. "The great ivy debate: the status of *Hibernica*. *Ivy Journal*, **10**(1): 33-46.

Nowak, J.M. and R.M. Rudnicki. 1990. *Postharvest Handling and Storage of Cut Flowers, Florists Greens and Potted Plants*. Timber Press, Portland, Oregon, USA.

Pennisi Bodie V., D. Oetting Ronald, E. Stegelin Forrest, A. Thomas Paul and L. Woodward Jean, 2003. Commercial Production of English Ivy (*Hedera helix* L.). *Ivy Journal*, **29**: 16-26.

Post, K. 1949. *Hedera helix* (English Ivy) In: *Florist Crop Production and Marketing*, pp. 558-559. Orange Judd publishing, New York,

Pierot, S.W. 1974. *The Ivy Book-The Growing and Care of Ivy and Ivy Topiary*. Macmillan Publishing, New York.

Radford, A.E., H.E. Ahles and C.R. Bell, 1968. *Manual of the Vascular Flora of the Carolinas*. The University of North Carolina Press, Chapel Hill, North Carolina, USA.

Rutherford, A. 1984. Rutherford's commentary on *Hedera hibernica* and *Hedera helix*. The Status of *Hibernica*, The Great Ivy Debate. *Ivy Journal*, **10**(1): 33-46.

Rutherford, A. 1997. Plant Crib. Published by Botanical Society of the British Isles in association with National Museums & Galleries of Wales.

Stein, Otto L. and B. Fosket, Elizabeth, 1969. Comparative developmental anatomy of shoots of juvenile and adult *Hedera helix*. *Amer. J. Bot.*, **56**(5): 546-551.

Sterling E.P. and P. Molenaar, 1986. The influence of time and temperature during simulated shipment in the quality of pot plants. *Acta Hort.*, No. 181, pp. 429-434.

Stevens, P. F. 2001. Angiosperm Phylogeny Website. Version 8, June 2007 [accessed 13 November 2007]. http://www.mobot.org/MOBOT/research/APweb/.

Zomlefer, W.B. 1994. *Guide to Flowering Plant Families*. University of North Carolina Press, Chapel Hill, North Carolina, USA.

24

Holmskioldia sanguinea (Family: Verbenaceae)

B. Sathyanarayana Reddy, S.Y. Chandrashekar, S.K. Nataraj and S. Latha

[**Common names**: Chinese hat plant/Cup and Saucer plant/Mandarin's hat plant (*Holmkioldia sanguinea*), Northern Chinese hat/Purple Chinese hat/Tahitian hat plant (*H. tattensis*)]

Introduction and Origin

Holmkioldia commemorates the Danish botanist Theodor Holmskiold (1732-1794). There are 11 species, all being native of the Old World (Everard and Morley, 1970), especially Madagascar in East Africa, and India to Malaysia in Asia. However, *H. sanguinea* is from Indian Himalaya which displays brick red calyces and deep orange to darker red corollas, all are commonly cultivated in the New World tropics. *H. tattensis* from Africa grows to the height of 6 m and is a very conspicuous plant displaying dull orange calyces and deep mauve corollas. *Holmskioldia* is a straggling shrub grown for its beautiful blossoms and the whole plant when in flower, from a distance, looks like *Bougainvillea*. The most noted feature of the plant is the bright orange or yellow flowers which resemble the Chinese hat. The flowers have scooped corollas with a pronounced lower lip surrounded by a coloured papery calyx shaped like a shallow dish. Its common name 'cup and saucer plant' is because of its hat-like calyx and tubular corolla (Graf, 1985).

It is grown both in tropical and subtropical countries. The shrub is a popular plant in gardens and parks as a specimen plant, in the form of standard, as hedge, against the walls and boundaries, and as border plants as it produces attractive flowers throughout the year. It would be better to place these towards the back

Single Flower (Close up)

Single Twig Bearing Flowers

H.S. 'Citrina'

Holmskioldia sanguinea

of shrubbery border to hide the bottom of the plants and together displaying beautifully coloured flowers on the top of the border. The juice of the leaves of *Holmskioldia sanguinea* Retz is taken orally against hypertension, bodyache and fever. The leaves and flowers are used as ingredient for preparation of indigenous hair lotion (Jamir, 2008).

Botany

Holmkioldias, the evergreen straggling shrubs grow at a moderate rate and attain the height of 3-9 m and the spread of 1.5 to 3.0 m depending upon the species and their heights. Its leaves are simple, opposite or sub-opposite, ovate, acuminate, serrated or entire and some 10 cm long. Flowers appearing in axillary racemose or cymose clusters along the branches throughout the year are orange to red or yellow, calyx membranous, bell-shaped, spreading and about 2.5 cm in diameter, the central corolla are tubular and curved, oblique tubes with 5 short lobes and some 2.5 cm long, stamens 4, didynamous, and the anthers exserted, and the fruit is a obovoid drupe, 4-lobed and included in the enlarged calyx (Bailey, 1929). The chromosome number of *Holmskioldia* is 2n= 32, 36 with basic chromosome number being x=5-7, 9 (Bhattacharya and Johri, 1998).

Species and Cultivars

Out of the 11 reported *Holmskioldia* species, only *Holmskioldia sanguinea* is most commonly grown one which is a shrub or scrambling climber. However, other cultivated species are *H. gigas, H. speciosa* and *H. tattensis. Holmskioldia sanguinea* is an evergreen species native of the tropical and sub-tropical foothills of India. The flowers of these are having tubular corolla 5 cm long and a spreading bell-shaped pink calyx. It is a frost-tender and can tolerate a minimum of 16 °C temperature (Brickell, 1994). A variant of the species *Holmskioldia sanguinea* 'Citrina' recorded from sub-tropical Indian Himalayas producing attractive cup and saucer flowers with yellow tubular corolla is most commonly available type in tropical climates, flowering from summer to rainy season (Fred and Chittenden, 1951). *H. tattensis* is an African relative which bears disc-shaped flowers in summer on an upright or vase-shaped plant. This charming species has mauve lavender cup and saucer-shaped flowers (Horace and James, 1977).

Propagation

It is commercially propagated through softwood or semi-ripe stem **cuttings** in frames during spring or outside during rains but only when plants are not in flower. Cuttings strike better roots when taken at the start of the rains in summer to full rainy season and there these strike roots in 4-5 weeks and will be ready for planting in 4-5 months. Stem cuttings some 20-30 cm long from semi-hardwood root better than those from hardwood cuttings. Treatment with rooting hormones, particularly with IBA will enhance the rooting with more success. The media used for rooting should be highly porous, well-drained but should have capacity to retain sufficient moisture and the cuttings should be inserted at a site where there is sufficient sunlight. They can also be propagated through **seeds**. For this either nursery is prepared or these are sown in containers containing coarse sand, good garden loam and sterilized compost all in equal volume and thoroughly mixed. These germinate within 10 days but take some two month's time to attain the appropriate size for potting. From seeds it takes some two years for planting at its permanent place.

Cultural Practices

Holmkioldia thrives best in **soils** with normal pH, but it can tolerate acidic to slightly alkaline soils, *i.e.*5.5-7.5. It comes up well in light to medium **soils**, particularly the sandy loam is preferable. It performs very well in moderate to high humid and warm conditions, and is also moderately tolerant to drought. Though it likes quite exposed position but performance in partial **sunlight** is not unsatisfactory. A **temperature** range of 25-35 °C is most suitable for its good growth and development but its growth is checked at a minimum temperature of 16 °C, and 10 °C becomes dangerous (Graf, 1985, 1992).

For its nursery raising, raised or flat beds are prepared looking into the rains, the flat ones for moderate rains and raised beds for high rainfall areas. Beds are prepared by thorough soil digging and mixing of coarse sand and quite decomposed farmyard manure at 3-5 kg/m^2. This mixture will ensure proper drainage, aeration and fertility in the soil. Cuttings measuring 20-30 cm length are prepared from the middle portion of the shoots and planted in the beds at close spacing. Rooting is improved with use of rooting hormones. The cuttings will strike roots in 4-5 weeks and will be ready for planting in containers or in the garden in 4-5 months. The recommended spacing for planting either way in the garden is 1.0-1.5 m. For planting at the required spacing, the pits of the dimension of 30 × 30 × 30 cm will have to be prepared and filled with the excavated soil by mixing sufficient quantity of FYM and coarse sand in order to ensure proper establishment and stand of the plants. This composition will also ensure that neither the soil may be water-logged nor its water-holding capacity, *vis-à-vis* aeration is jeopardized. Plants are removed gently from the container along with ball of earth to keep the root and soil intact. While planting, the top of ball of earth must be in level with the surface of the soil. After planting the pits are sufficiently watered to settle the roots. Plants require staking in case when they are quite large (Bose *et al*, 2001). Plants have to be **pruned** once or twice a year to keep them in proper shape and size and to remove the dead and diseased parts. Once they are pruned before the end of winter and the other when the plants are outgrown or when flowering is over. Since **weeds** compete with the plants for light, moisture and nutrients, whenever, these are seen are taken out either manually or through chemical weedicides if the area is large.

Nutrition and Irrigation

Plants need plenty of organic matter for its growth and development. Plants are supplied with FYM at the rate of 3-5 kg/plant every year. Chemical fertilizers having nitrogen, phosphorus and potassium may be supplied in the rainy season at the rate of 50-100 g/plant. They are irrigated deeply to facilitate roots penetrating quite deep, but not so frequently. If plants are irrigated shallowly and frequently, plants tend to develop shallow root system. Even under dry climatic conditions, they need to be watered less frequently. Mulching helps in conserving the moisture and prevents development of weeds.

Insect-Pests and Diseases

The crop is easy in cultivation as no serious insect-pest or disease has been encountered. However, soil borne nematodes may present the biggest challenge to this plant. The other minor insects include **whiteflies, red spider mites and mealy bugs.** The adult **whitefly** is quite small and covered with white powder. Fungus gnats are small, dark-coloured flies about 3 mm long. They can be found either on the plant or soil surface. The adults are more of a nuisance than a real problem, but the larvae do feed on plant tissue. **Mealy bugs** are wingless, dull-white and soft-bodied insects covered with a waxy-powder mass over whole body. They are slow-moving and adult females and nymphs through their piercing and sucking type of mouth parts suck sap from the tender parts of the plant by congregating in large number at one place. Adult male is winged. Mealy bugs can weaken the plants leading to yellowing of foliage and leaf drop. They also produce a sweet substance called honeydew which entices ants, and afterwards unattractive black sooty mould. Mealy bugs are managed by isolating infested plants from the healthy ones. Natural enemies such as lady bird beetles are encouraged in the garden. Some of the suggested insecticides include Monocrotophos@1.7 ml/l of water or Imidacloprid @ 0.5ml/l of water or Carbaryl @ 3 g/l of water sprayings. **Spider mites** are small, 8-legged and spider-like creatures which thrive in hot and dry conditions, and multiply quickly, female laying up to 200 eggs in a life span of 30 days. Spider mites feed with piercing mouth parts and their infestation causes plants to appear yellow and stippled. Heavy infestation causes leaf drop and ultimate plant death. They also produce a web which can cover infested leaves and flowers. To manage the mites, infested plants should be removed and the field and surroundings will have to be kept weed-free. Since dry temperatures seem to worsen the problem, so it should be made sure that plants are regularly watered. Natural enemies such as ladybugs will have to be encouraged. Spraying of miticide such as Dicofol @1.70ml/l of water will manage this pest. The leaves of *Holmskioldia sanguinea* showing necrotic spots during dry seasons of summer (May to mid-August) can be controlled by spraying with Captan at 0.2 per cent.

References

Graf, A.B. 1985. *Exotica International Pictorial Cyclopedia of Exotic Plants from Tropical and Near Tropic Regions*, p. 2113. Roehrs Company Publishers, U.S.A.

Graf, A.B. 1992. *Hortica Colour Cyclopedia of Garden Flora in All Climates Worldwide and Exotic Plants Indoors*, p. 899. Roehrs Company Publishers, U.S.A.

Bailey, L.H. 1929. *The Standard Cyclopedia of Horticulture* (vol. II), p. 1498. The Macmillan Company, New York.

Bhattacharya, B. and B.M. Johri, 1998. Flowering Plants ~ Taxonomy and Phylogeny, p. 753. Narosa Publishing House, New Delhi.

Bose, T.K., B. Chowdhary and S.P. Sharma, 2001. *Tropical Garden Plants in Colour*, p. 194. Horticulture and Allied Publishers, Kolkatta, India.

Brickell, C. 1994. *The Royal Horticultural Society Gardeners' Encyclopedia of Plants and Flowers*, p. 513. Dorling Kindersley Limited, London.

Everard, B. and B.D. Morley, 1970. *Wild Flowers of the World*, plate 80. Peerage Books, London.

Fred J. and O.B.E. Chittenden, 1951. *Dictionary of Gardening ~ A Practical and Scientific Encyclopedia of Horticulture*, p. 1006. Oxford at the Clarendon Press, U.K.

Horace, F. C. and C.H. James, 1977. *The Hawaii Garden Tropical Shrubs*, p. 65. The University press of Hawaii, U.S.A.

Jamir, N.S. 2008. Traditional knowledge of medicinal plants used by Ao-Naga tribes of Mokokchung district, Nagaland (India). *Proceedings of International Seminar on 'Multidisciplinary Approaches in Angiosperm Systematics'*, pp. 602-607, held at Nadia (W.B.).

25

Impatiens (Family: Balsaminaceae)

M.K. Singh, Sanjay Kumar, T. Janakiram and Sanyat Misra

[**Common names**: Balsam/Rose balsam/Touch-me-not (*Impatiens balsamina, I. noli-tangere*), Bedding impatiens/ Busy lizzie/Patience plant/Patient Lucy/Snapweed (*Impatiens walleriana,* syn. *I. holstii*), Busy lizzie/Sultana impatiens (*I. sultanii*), Himalayan touch-me-not/Indian balsam/Policeman's helmet (*I. glandulifera*), New Guinea impatiens (*I. hawkeri*)]

Introduction and Origin

Impatiens derives from the Latin 'impatient', alluding to the seed capsules that explode explosively when touched. It is a genus of about 8,500 species (Dole and Wilkins, 1999) but many of the taxonomists put it somewhere in between 400 to 700 species of hardy, half-hardy and tender annuals and evergreen sub-shrubs, though for garden purposes the genus has been little exploited with only 15 species in general cultivation (Pizzetti and Cocker, 1968). Impatiens in Kew Gardens reached in 1884 though *Impatiens balsamina* was introduced into Europe in 1596. The tender species require greenhouse cultivation and are also popular as house plants. They are particularly suitable for outdoor bedding, especially the busy lizzies (*Impatiens holstii* × *I. sultani*). Their area of distribution is quite extensive, majority being indigenous to tropical and subtropical Asia and tropical Africa. Only eight species are native to temperate Asia, Europe and America. The genus comprises of annuals, evergreen perennials and sub-shrubs. It has been found through phylogenetic analysis that cultivated *Impatiens* species actually originated in SE Asia from where these got dispersed to India, Africa and Madagascar (Yuan *et al.*, 2004). *I. balsamina,* which is most commonly grown in India as a summer annual, is native to tropical areas of China, India, and the Malay Peninsula. *I. walleriana* is a succulent perennial herb indigenous from Mozambique to Tanzania on the east coast of equatorial Africa. It is the most important bedding plant in North America, along with geraniums (Voigt, 1994). *I. hawkeri*, the progenitor of the New Guinea *Impatiens* and all other species involved in developing modern hybrids are all native to New Guinea, Java and the Celebres (Bailey and Bailey, 1976) The Popularity of impatiens has been increased due to rapid introduction of a wide range of new colours and plant forms. Genetic improvement has resulted in improved self-branching habit combined with production of white, yellow, orange, pink, deep red, violet and bicoloured flowers. Introduction of variegated dark green to bronze coloured leaves with some red and yellow shades in the foliage of New Guineas (*I. hawkeri*) makes it an excellent bedding as well as hanging basket plant. Recently, it has become popular as a garden flower and also as a potted flowering plant for decorating balconies and terraces. It can also be planted in the interspaces of the shrubbery, on roof garden and in a shaded place between two buildings. Bushy and upright type *I. petersiana* is also used as a house plant because of its attractive foliage which remains on the plant all the year round in addition to its attractive flowers. *I. hawkeri* apart from being a bedding and potted flowering plant, can effectively be used in hanging baskets.

In India, the juices extracted from *Impatiens balsamina* and *I. biflora* are used for dyeing silk and wool in yellow and red. Ethiopian women through the red dye extracted from *I. balsamina* strain their hands and feet in red or black colour, and Japanese women paint their nails with the red dye extracted from *I. balsamina.*

Botany, Cytology and Breeding

The members of this genus are hardy, half-hardy and tender semi-succulent herbaceous annuals and evergreen perernnial sub-shrubs, many being recorded from lower mountainous regions and subalpine meadows. Stems in many of the species are fleshy and succulent. Leaves are usually alternate but sometimes opposite, ovate to lanceolate, undivided, upper ones often in whorls, and often toothed. Spurred and cup-shaped or flat flowers appear from the upper leaf axils either solitary or sometimes in racemes or panicles, and these flowers are distinctively asymmetrical, with 3-5 small sepals out of which one is petal-like (petaloid) and spurred, and five petals, the upper (standard) often hooded or helmet-shaped while remaining four united into deeply lobed pairs or wings. There are 5 stamens with appendaged filament, fruit a 5-valved pod which on maturity bursts scattering the seeds far and wide.

The chromosome numbers in eight species of *Impatiens* have been worked out by Gill and Chinnappa (1977), and they recorded n = 5 in *I. assurgens,* n = 7 in *I. epiphytica* and *I. hamata*, n = 8 in *I. ulugurensis*, n = 10 in *I. digitata* and n = 13 in *I. kilimanjari*, and Rao (1978) reported new basic chromosome number, *i.e.* n = 3 in case of *I. leschenaultii* and *I. latifolia*. Earlier reported basic chromosome numbers were n = 6, 7, 8, 9,10,12,13 and 14. This was a smallest and a new basic chromosome number ever recorded in the Balsaminaceae. Govindarajan and Subramaniam (1985) analysed the karyotypes of four species and determined that somatic chromosome number in *I. balsamina* var. *areuata* is 2n = 44, in *I. orchioides* 2n = 18, in *I. hensloviana* and *I. phoenicea* 2n = 16. Govindarjan and Subramaniam (1986) studied the karyotaxonomy of 14 taxa of *Impatiens*. The chromosome number ranged from 2n = 6 to 2n = 44.Each taxon had a distinct karyotypes of chromosome indicating that the karyotype alterations of chromosomes might play important role in speciation.

The main **breeding** objectives in New Guinea impatiens are to produce superior new cultivars with dwarf ideotypes suitable for various climatic conditions, stunning colours, large flower size, foliage variegation, high or low irradiance and drought tolerance, *vis-à-vis* resistance to various insect-pests and diseases (Strefeler, 1995; Strefeler and Quene, 1995; Strope and strefeler, 1997; Whealy, 1995). Swarup *et al.* (1975) studied the heterosis and inheritance pattern in *Impatiens balsamina* through five quantitative characters, *viz*, plant height, number of branches per plant, number of flowers per plant, number of petals per flower and size of the flower. Heterosis was observed for all the characters and the performance of F_1 hybrids was better than the parental lines. Heterosis for flowers per plant was maximum (145 per cent) followed by number of branches per plant (62 per cent). Heterosis in other characters ranged from 40 per cent in plant height to 10 per cent in number of petals per flower. The best performing F_1 hybrid was also the best heterotic hybrid (having the highest heterosis percentage) for flowers per plant and flower size, but it was not true for the remaining three characters. Regarding the inheritance of characters, the gene action was found to be additive in nature without any epistasis for all the characters, except flowers per plant, where epistasis was of complementary nature. The association of excess of recessive genes with high values was observed for all the characters. Autoploidy of *Impatiens balsamina* (n=7) was studied by Jalil *et al.* (1977) through selection for four generations. It was noted that tetraploid plants increased in height, increased the size of leaves, flowers and petals, and also in days to flower and flowering duration.

Species and Varieties

There may be more than 700 annual and perennial *Impatiens* species, but for ornamental purposes some 20-30 species have been exploited which are being described below.

Impatiens acaulis

This perennial species is native of South India in Kerala, which grows about 70 cm tall with tuberous roots, having single leaf with 6-9 cm long blades, and leafless and umbellate inflorescence up to 10 cm tall, bearing some 3.0 cm across pink flowers with wide petal lobes. Seeds of this species should be kept slightly moist until planting otherwise its viability may be lost.

Impatiens arguta

A dwarf and perennial species growing up to 40 cm tall, forming roots at the nodes, is native of SW China. It grows in damp shady places preferring moist but well-drained soil. It bears blue to pink or sometimes white flowers some 2-3 cm across.

Impatiens assurgens

This perennially creeping species is a native of central Africa in wet but exposed places, growing up to 60 cm high with fleshy roots, bearing opposite leaves which are usually sessile, glabrous to pubescent, suborbicular to ovate or lanceolate, and the flowers are white, light pink or mauve, often spotted yellow near the base, and with 1.5-2.8 cm long petals.

Impatiens balfourii

It is though grown as an annual but is a beautiful non-hardy herbaceous perennial growing erect up to 1 m in height and is native of western Himalayas. Its stems are reddish, leaves dark green, oval, acuminate and dentate, and the flowers which appear during June to September are pink marked with white and yellow. Its seeds can be germinated at a temperature above 18 °C, at the place of planting directly.

Impatiens balsamina

A native of warm and moist tropical East Asia (India, China and Malaysia), *I. balsamina* is fast-growing, erect, compact, sparingly branched and bushy annual species, growing up to 75 cm tall and some 45 cm in spread with lance-shaped and 7-15 cm long leaves. The flowers are cup-shaped, spurred, pink or white and some 4 cm across, the hybrids being mostly white, yellow, pink, red, scarlet, lilac, mauve and blue, and emerge from the leaf axils close to the stem in summer and early autumn. The flowers are often double. In some cultivars flowers are spotted, blotched or variegated. Seed pods are hairy and greenish-white. Its variety 'Blackberry Ice' is half-hardy and bears large double white flowers splashed with purple.

Impatiens biflora

It is a hardy annual suitable for moist shade, growing to a height of 90 cm and is native to North America. Leaves are mid-green, ovate and toothed and the flowers are orange spotted red-brown, some 2.5 cm long and appear from July to October.

Impatiens flaccida (syn. *I. latifolia* Hook. not Linn.)

This perennial species is a slender, little branched and glabrous, growing about 70 cm high, and is native to Southern India and Sri lanka. The species is glabrous, leaves are stalked, alternate, ovate or lanceolate, 5.0-12.5 cm long, crenate and the petioles are with or without glands. Mostly solitary flowers or in the groups of 2-3 are borne directly from the leaf axils. These are flat, rose-purple or rarely white, 3.75 cm across, petals broad and both (standard and wings) 2-lobed but standards spurred, lip boat-shaped with a long and slender spur which is curved and about 3.75 cm long.

Impatiens glandulifera (syn. *I. roylei*)

It is a vigorously growing hardy annual species native to Himalayas but naturalized in Britain, grows up to 1.5 m in height and flowers from August to September. Stems are succulently branched, leaves are light green, ovate and deeply serrated while flowers of reddish-purple to dark pink or yellow, spotted crimson are borne in loose clusters in terminal and axillary panicles with short spurs having length of each flower about 5 cm long. Its var. *candida* (syn. *Impatiens candida*) bears white flowers marked with crimson.

Impatiens hawkeri

A native of New Guinea and South Sea Isles (Java and Celebres) is a bushy perennial species with stems growing up to 60 cm tall. Leaves are ovate-oblong in opposite pairs in whorls of three. The flowers are deep to brick-red with white eye in the centre and 5-7 cm across and the spur is about 7 cm long. Flowers appear from June to September. This species reached to Kew Gardens in 1884, is originally a spring bedding species but together suitable for hanging baskets and pot culture. It is a progenitor of New Guinea *Impatiens*. Its propagation is through asexual means, *i.e.* cuttings though plug seedlings raised through seeds as well as cuttings are also available with specialists in Sweden, Britain, Germany and USA. It has hundreds of varieties and some 90 are listed by Whealy (1995) alone.

Impatiens holstii

It is a herbaceous perennial though is grown as an annual and is native to tropical Africa. It grows up to 90 cm and flowers from June to October. Its stems are slender, leaves are fleshy, oval, acuminate and some 7.5 cm long, and the flowers are flat, scarlet and 2.5 cm in diameter with spurs some 3.75 cm long.

Impatiens hookeriana (syn. *I. biglandulosa, I. sultani alba*)

A perennial species growing up to 90 cm tall with branched and fleshy stems, is native of Sri Lanka. Leaves are ovate-lanceolate, serrated and long-petioled, and the flowers which appear from July to September are large white with purple markings on the lower petals, and the spurs horn-shaped and longer than the flowers. It is propagated by cuttings.

Impatiens marianae

This sprawling and creeping perennial species with 20 cm long stems, suitable for humid and shady locations, is native of Assam in India. It is an attractive foliage plant where leaves are alternate, broadly oblong-ovate, serrated with round tooths, 2.5-6.0 cm long, blue-green painted with silvery-white between the prominent white veins, leaf blades about 6 cm long, and the small (2-3 cm long) pale-purple flowers appear either solitary or in pairs from the upper leaf axils.

Impatiens niamniamensis (syn. *I. bicolor*)

This upright flowering perennial species with fleshy stems, growing up to 60 cm tall is native to Western and Central Tropical Africa though uncommon in cultivation. It is either unbranched or sparingly branched and the stems are reddish-green. Alternate leaves are long-stalked and are found mainly on the upper part of the stems, blade ovate, elliptic or lanceolate, crenate and 5-22 cm long. Some 3 cm long greenish-yellow to pure red flowers *via* red & white, red & pink and red & yellow, each with a long orange, red, purple or crimson pouch-like spur, are borne from summer to autumn on short stalks in the leaf axils. Its special cultivar 'Congo Cockatoo' is very showy with red, green and yellow flowers and the flowers are borne from the leaf axils from summer to autumn on short stalks in clusters. It can be easily grown in warm humid conditions.

Impatiens noli-tangere

Specific name *noli-tangere*, in Latin means 'touch-me-not'. It is a most hardy species among all the annual types though collapsing with the first autumn frost in the temperate regions, and is native to Central and Northern Europe eastwards to Siberia, growing up to 1 m, and flowers from June to September. A charming green plant most useful for shady locations of the garden, with sappy stems and swollen nodes, leaves are oval-lanceolate and dentate, and the flowers are brilliant yellow though throat spotted-red.

Impatiens parasitica

A small annual species with about 20 cm tall slender stems, preferring cool-shady locations is native of South India in the Cardamom hills in Kerala. It bears red flowers some 4-5 cm across with curved spurs. Another similar species close to *I. parasitica* is *I. jerdoniae* Wight, which bears red and green flowers.

Impatiens platypetala

This perennial species from Java is grown as an annual. This shrubby plant grows about 45 cm in height with almost succulent branches. Leaves are whorled, ovate to lanceolate, prominently veined and about 8 cm long. Flowers which appear during summer are white, pink, red or purple, flat, about 4 cm across and with a slender spur. *I. p. aurantiaca* from Celebes bears orange-yellow flowers with red eye.

Impatieans pseudoviola

From Kenya to northern Tanzania, this perennial trailing species may sometimes grow about 30 cm tall with decumbent stems, and is most suitable for planting in the hanging baskets. Leaves are ovate, crenate and up to 5 cm long. It is a free-flowering species, flowering from spring to autumn with slightly cupped or flattish flowers about 2 cm across and in purple to pinkish-purple or in palest-pink colour with a violet stripe in the centre of the four lower petals, where two lower petals are much narrower.

Impatiens quisqualis

This tuberous rooted perennial and up to 1 m upright growing species but with thin stems is native of Malawi in the cool moist conditions. Stalked leaves are alternate with blades being pubescent on lower surface and some 4-14 cm long. Flowers in clusters of 2-3 are generally pale-yellow, white or light pink, lower petals 2.0-3.5 cm long with yellow or orange spots near the base and the lower sepal with 5-8 mm long spur. Another shrubby species found in the same region is ***I. shirensis*** bearing up to 3 m tall stems with glabrous leaves.

Impatiens recurvicornis

This species found in the moist places of West China is a creeping perennial with up to 20 cm long ascending stems, bearing about 7 cm long leaves and about 2 cm across flowers with spurs extending horizontally.

Impatiens repens

A native to India and Sri lanka the species is a short-lived perennial or a procumbent annual most suitable for planting in the hanging baskets. The leaves are broadly ovate-cordate and up to 4 cm long. Yellow and hairy flowers appear during summer which are up to 3 cm long with a quite broad spur.

Impatiens schlechteri

A native of New Guinea, it is a shrubby perennial growing up to 60 cm high, bearing whorls of 3-7 ovate to oblanceolate leaves which are glossy deep green, oftenly having red flushing, are 5-10 cm long and serrated. Bright red, pink or orange flowers which are flat, some 4.5 cm across with pale and slender spurs, appear from spring to autumn. Though the species is not common in cultivation but is one of the parents for developing New Guinea hybrids. These hybrids may have green, red or bronze leaves, mostly variegated with yellow mid-vein stripe, and the flowers range from red to orange, pink, purple and white.

Impatiens sodenii (syn. *I. magnifica*, *I. oliveri*)

A strong-growing shrubby evergreen perennial species growing in the well exposed rainforests with succulent stems up to 3 m tall, is native of southern Kenya to northern mountainous Tanzania. The sessile and narrowly oblanceolate and toothed leaves some 5-18 cm long are produced in dense whorls of 6-10. Flowers usually solitary, 4-6 cm across, white or pale-pink to mauve, lower petals 3.0-3.7 cm long often with red veins towards the base, and the lower sepal with 0.6-1.0 cm long spur, are produced during autumn-winter.

Impatiens subecalcarata

A delicate annual with 20 cm long stems is native of cool-moist areas of West Yunnan province of China. Its leaves are long-stalked, crenate and flowers are about 2.5 cm across.

Impatiens tinctoria subsp. *tinctoria*

A huge clump-forming perennial species with tuberous roots and up to 2 m tall succulent stems is native of Ethiopia, Uganda, Zaire and Sudan. The short-stalked leaves are alternate, narrowly ovate, elliptic or lanceolate with 9.5-19.5 cm long blades. White scented flowers some 4-6 cm across appear in clusters of 2-8, with lower petals 3.0-5.5 cm long and oftenly streaked red near the base, and lower sepal with 0.8-1.2 cm long spur. For its growing it requires shade and shelter. Roots need lifting in winter to protect from frost injury. C. Grey-Wilson (1980) identified 5 subspecies such as *I. tinctoria* subsp. *latifolia* Grey-Wilson, subsp. *elegantissima*(Gilg) Grey-Wilson, subsp. *abyssinica* (Hook. fil.) Grey-Wilson, and subsp. *songeana* Grey-Wilson.

Impatiens verticillata

A low bushy perennial species with 70 cm or more long stems, preferring warm and shady locations with moist but well-drained soil is native of South India in the Cardamom hills of Kerala. Leaves are narrowly lanceolate, blade some 7 cm long, and the flowers are orange-scarlet and 2.5 cm across having 4.5 cm long spur.

Impatiens walleriana (syn. *I. holstii*, *I. petersiana*, *I. sultani*)

It was named so to honour Horace Waller, a British missionary in Africa. This perennial species well-suited to semi-shady conditions is a native in the high rainfall areas of East Africa in Kenya, Tanzania and Mozambique with self-branched thick stems and compact growth, producing numerous flowers. Plants are pinched for bushy growth. It is propagated through seeds though double-flowered and leaf-variegated types through cuttings. It is a half-hardy, fast-growing, evergreen and bushy perennial, growing up to 60 cm in height and spread and is usually grown as an annual. Leaves are oval and fresh green, and the solitary to three flowers in a group are 5-petalled, flattish, spurred and bright red which appear from spring to autumn. ***I. sultani***, a native of Zanzibar having succulent stems up to 60 cm tall, bearing alternate, smooth and toothed leaves, and pink flowers with 5 petals, was first named so by Sir William Jackson Hooker to honour the Sultan of Zanzibar to whom East Africa is greatly indebted for his enlightened and philanthropic government, though ironically the Sultan made his fortune in the slave trade. This non-hardy succulent herbaceous perennial species bears a robust main stem with strong branches but when full grown requires support. Leaves are elliptic to lanceolate, alternate at the base but whorled upward. Flowering stalk is fragile and bears 1 to 3 flowers per stalk. Flowers in original species are vivid scarlet with about 3.75 cm diameter having upward curving spurs. Its var. *episcopi* bears carmine flowers with purple markings; the var. *holstii* bears intensely vivid scarlet flowers and is best for pot-culture; though var. *petersiana* in its all parts is red and bronze and the stems are little hairy. This has replaced the most popular species of *I. balsamina*, and in Germany, England and the United Stastes the *I. sultani* has become most favourite as with little care this can be bloomed through out the year. *I. sultani* being exactly similar to *I. walleriana*, it was afterwards shifted to *I. walleriana*. Its most popular varieties are 'Accent', 'Blitz', 'Carousel', 'Carpet', 'Dazzler Stars', 'Dazzler and Super Elfin series', 'Expo', 'Impulse', 'Lipstick', 'Nana', 'Novette Stars', 'Red Ripple', 'Swirl (Tempo Frosts)', 'Showstopper', 'Tempo', *etc.* Busy Lizzie '**Duet Series**' are fast-growing evergreen, bushy and half-hardy perennials with 30 cm height and spread, usually grown as annuals, having fresh green oval leaves, and flattish, spurred, semi-double and double flowers in shades of red and orange suffused white. Busy Lizzie '**Confection Series**' cultivars are mostly fast-growing evergreen bushy perennials with height and spread of 20-30 cm, usually grown as annuals, having fresh green leaves and small, flat, spurred, double to semi-double flowers in reds and pinks, appearing from spring to autumn. Busy Lizzie '**Novette Series**' cultivars are mostly fast-growing evergreen bushy perennials growing with height and spread of 15 cm, usually grown as annuals, having fresh green pointed-oval leaves and small, flat and spurred flowers which appear from spring to autumn. Some varieties have red flowers others of various other colours. Its var. 'Red Star' bears star-shaped red flowers with white markings. Its salmon type bears salmon-pink flowers from spring to autumn. Busy Lizzie '**Super Elfin Series**' are half-hardy, fast-growing, evergreen and bushy perennials, growing up to 20 cm in height and spread but are grown as annuals. Leaves are oval fresh green, and the flowers are flattish, 5-petalled, spurred and in mixed colours which appear during spring to autumn. Its var. 'Lipstick' grows to 20 cm in height and spread and bears pointed-oval, fresh green leaves and small, flat, spurred rose-red flowers, appearing from spring to autumn. Busy Lizzie '**Tom Thumb Series**' is a half-hardy fast-growing evergreen perennial, growing up to 20-30 cm in height and spread but are grown as annuals. The leaves are oval fresh green, and the 5-petalled flowers are rounded, some 8 cm across, spurred and double in shades of red, purple, pink or white.

Impatiens zombensis

A bushy perennial species with soft and fleshy stems, growing up to 75 cm tall and preferring cool-moist conditions is native of Malawi and Mozambique. Stalked leaves are alternate, glabrous or with scattered hairs, blades 3-9 cm long, ovate, lanceolate or elliptic, cuneate at the base and acuminate. Flowers are usually solitary, light pink, 1.5-2.0 cm across, lower petals longest, and the lower sepals having 1.6-3.2 cm long spur.

New Guinea *Impatiens* hybrids

It is a cross between *I. hawkeri* having large brick-red flowers with *I. linearifolia* in which leaves have usually white centre. Leaves of hybrids are 8-15 cm long, and the flowers 3-7 cm across. These modern hybrids are suitable for summer bedding in warm climates having cool shady positions. These are excellent pot plants for the greenhouse, conservatory and balconies. Whealy (1995) divided the various cultivars of New Guinea *Impatiens* according to colours, *viz.* (i) fuchsia, (ii) lavender, (iii) orange, bicolour orange, dark orange, (iv) pink, bicolour pink, hot pink, light pink, medium pink, (v) purple, bicolour-purple, blue-toned purple, red-toned purple, (vi) bicolour red, cherry red, scarlet red, (vii) salmon, dark coral, light salmon, and (viii) white, blush white. Double flowering cultivars are also available.

Some of the most popular varieties of New Guinea hybrids are 'Anaca', 'Argus', 'Aurora', 'Celeria', 'Celsia', 'Cethosia', 'Dewas', 'Duniya', 'Eclipse', 'Fieste Cerise', 'Flambee', 'Isis', 'Isopa', 'Jasius', 'Mirage', 'Phoebis', 'Sesia', 'Thecla', 'Veronica', 'Vulcane', *etc.*

Colourwise Reference Varieties

- ☆ **White**: 'Ballet', 'Innocence', 'Moorea'
- ☆ **Bluish white**: 'Samoa'
- ☆ **Orange**: 'Antigua', 'Bingo','Escapade', 'Nebulous', 'Tanna'
- ☆ **Dark orange**: 'Ambrosia', 'Marpesia', 'Susanne', 'Timor'
- ☆ **Light pink**: 'Blues', 'Equinox', 'Tahiti', 'Tina'
- ☆ **Medium pink**: 'Rosetta', 'Barbara', 'Gemini', 'Kallima'
- ☆ **Dark pink**: 'Dandin', 'Impulse'
- ☆ **Hot pink**: 'Aglia', 'Bonaire', 'Doerte', 'Lambada'
- ☆ **Pink bicolour**: 'Pago Pago'
- ☆ **Lavender**: 'Alice', 'Flamenco', 'Serenade', 'Tonga',
- ☆ **Purple** (blue shade): 'Antares', 'Aruba', 'Bora-Bora', 'Debka', 'Elvira', 'Swing'
- ☆ **Purple** (Red shade): 'Apollon', 'Danserra', 'Papete', 'Samba', 'Syrtaki', 'Wanda'
- ☆ **Cherry red**: 'Anguilla', 'Anna', 'Martinique', 'Pulsar'
- ☆ **Red scarlet**: 'Lanai', 'Karina', 'Mirach', 'Prepona'
- ☆ **Red bicolour**: 'Ambience', 'Danlight'

Propagation

Impatiens is grown from seeds as well as through vegetative means. 'New Guinea *Impatiens*' is mostly vegetatively-propagated but recently its **seed**-propagated types were also introduced to the market. The species of *Impatiens walleriana* contains approximately 1,760 seeds per gramme, and its seeds are stored at 4 °C and 25-30 per cent humidity (Corr, 1997; Dole and Wilkins, 1999). The New Guinea *Impatiens* contains approximately 400-600 seeds per gramme though cultivars may differ slightly. Seeds are sown at around 18 °C in pots or pans filled with seed compost containing 2 parts good quality loam which is sterilized preferably at 93 °C for 20 minutes, 1 part fibrous or granular peat free from fine dust and 1 part quite coarse *cum* sharp sand having particle size of 3 mm, all by volume. One litre of this mixture should be fortified with 1 g of superphosphate of lime and ½ g of either finely ground chalk or finely ground limestone. This whole mixture should now be sieved through a 12 mm mesh. Now the seeds should be sown thinly in this mixture when in spring the danger of frost is over. Seeds germinate within 21 days and young seedlings can be shifted to planting beds after 4-6 weeks. Seedlings are pricked out in boxes when large enough to handle and later into 8-10 cm pots or containers filled with planting mixture, *i.e.* 7 parts loose bulk loam, 3 parts peat and 2 parts sand. Each litre of this mixture should have been fortified with 3 g of 2:1 superphosphate of lime and sulphate of potash + ½ g of finely ground chalk or ground limestone.

Cuttings of New Guinea *Impatiens* are taken from the vigorous shoots at any time from April to September and rooted in a medium containing equal volume of peat and sand, at a rooting temperature of 16-21°C, preferably 20 °C (Dole and Wilkins, 1999). Leaf turgidity should be maintained by providing adequate moisture. Cuttings of the size of 7.5-10.0 cm are taken containing at least one pair of fully expanded leaves, one pair of expanding leaves and the growing point. Rooting hormones can be used for proper rooting. The pH of rooting medium is preferred in the range of 5.5-6.5 (Mikkelsen, 1995). Cuttings in plug trays develop callus by day 7, roots are visible by day 10-12 and the cuttings are ready for transplanting by 3 to 4 weeks (Dole and Wilkins, 1999). Rooted cuttings are transplanted individually in 7.5-10.0 cm pots or containers. The growing points are pinched out to encourage bushy growth.

Cultural Practices

Outside, the *Impatiens* grows well under ordinary conditions when soil is fertile and well-drained, in full sun or partial shade, except the *I. arguata, I. roylei* and *I. recurvicornis* which are suitable only at the damp sites. However, greenhouse perennials such as *Impatiens acaulis, I. arguta, I. assurgens, I. flaccida, I. hawkeri, I. holstii, I. hookeriana, I. marianae, I. niamniamensis, I. petersiana, I. platypetala, I. quisqualis, I. recurvicornis, I. schlechteri, I. sodenii, I. sultanii, I. tinctoria, I. verticillata, I. walleriana* and *I. zombensis* are grown with minimum winter temperature of 13 °C though can survive at 7 °C if kept just moist. High humidity and partial shade is conducive for growth for most of the species. Annual types such as *I. balsamina, I. biflora, I. noli-tangere, I. parasitica, I. repens, I. roylei, I. subecalcarata, etc.* can be grown in full sun or partial shade.

Impatiens is a day-neutral plant. The quantum of flowering depends on total irradiance received at an appropriate temperature. It takes 4 weeks to flower after transplanting in warmer places and about 7 weeks in colder places depending upon the individual management of the crop and the choice of the cultivar. In balsam, barring the perpetual types or the ones growing in the glasshouse, the blooming period lasts for about 30-45 days. New Guinea *Impatiens* takes 10 to 14 weeks from transplanting to flowering, differing on the type of cultivars grown. Large pots require 2 or more additional weeks to flower. Sufficient **light** is important for growth but in excess it reduces the growth. However, internodal elongation occurs if too much shade is provided (Kaczperski and Carlson, 1982) but plants become lanky and weak. *Impatiens* can survive in depleted light or partial shade but excessive shade can deteriorate the quality and quantity of flowers. Armitage and Vines (1982) found that *Impatiens* is more efficient in CO_2 fixation at low irradiance (30 $\mu mol.s^{-1}\ m^{-2}$) than at the higher level. New Guinea *Impatiens* in winter when glasshouse vents are closed, a concentration of 1,000 ppm CO_2 results in superior plants (Christensson, 1985). Carlson and Klire (1985) suggested 1,500-2,000 ppm CO_2 as most appropriate for New Guinea hybrids but he recommended that at this time 3-6 °C temperature should be increased when CO_2 is being used. For germination, seed should be put under high pressure sodium, fluorescent or incandescent lights. An intensity × duration interaction exists. It has been reported that 86 per cent germination occurred when seeds are irradiated at 2 $\mu mol.s^{-1}\ m^{2}$ for 3 continuous days; at 20 $\mu mol.s^{-1}\ m^{-2}$ for 2 days and at 100$\mu mol.s^{-1}\ m^{-2}$ only 1 day is required. No continuous lighting is required and it should be stopped when the roots begin to elongate (Ball, 1991, 1995; Carpenter, 1994; Carpenter *et al.*, 1994; Corr, 1998). In New Guinea *Impatiens*, light intensity below 600 $\mu mol.s^{-1}\ m^{-2}$ increases internode elongation, *vis-a-vis* days to flowering. Light intensity above 5,000 fc (1,000 $\mu mol.s^{-1}\ m^{-2}$) can result in reduced growth (Erwin *et al.*, 1992). **Temperature** plays an important role for germination of *Impatiens* seeds. At 21 to 26 °C, germination occurs within a fortnight, though for production of plants, 21-24/16-18 °C day/night temperatures are quite favourable. Leaf damages will occur if temperature along with light is too high (Armitage *et al.*, 1982; Kaczperski *et al.*, 1988). New Guinea *Impatiens* has a very narrow temperature range for its proper growing, *i.e.* 25 to 27°C (Erwin *et al.*, 1992; Erwin, 1995). Inhibition of growth occurs when temperature is above 27 °C and below 17 °C. Studies on the effects of differences (DIF) between day and night temperatures, and constant temperature regimes (15, 20 or 25 °C) on the elongation and development of *Impatiens walleriana* seedlings revealed that 25/15 °C (day/night, *i.e.* positive DIF) compared with constant 20 °C (zero DIF) regime treatment promoted plant height, internode and petiole lengths, whereas 15/25 °C (negative DIF) suppressed growth compared with zero DIF, however, leaf length and leaf unfolding rate responded significantly different to different constant temperatures, while not being significantly affected by DIF treatments, therefore, it is inferred that DIF effects may be advantageous to facilitate production of compact seedlings but that short-term control of elongation may be more easily achieved by altering average temperature than by changing DIF (Ito *et al.*, 1997).

The **pH** of the germination medium should be 6.0-6.5 though for New Guinea *Impatiens*, the pH of the medium should be 5.5 to 6.5. A well-drained medium, preferably sandy-loam with adequate water holding capacity is required. Most seeds are germinated in plug containers in germination rooms where high humidity is maintained. Seeds should be covered lightly with vermiculite or any other mulching material such as leaf mould to maintain moisture but should permit light to filter through. For preparation of land, with the first ploughing, farmyard manure or compost should be mixed in the soil at the rate of 250 quintals per hectare and the soil should be ploughed to a depth of 30 cm to permit the roots to spread fully. Any foreign material including the rootstocks of all the perennial weeds should be taken out and then the soil should be planked then again ploughed and left out rootstocks should again be taken out and planked. At third ploughing, there should be sufficient moisture in the soil otherwise before ploughing the field will have to be irrigated and then at proper stage it will have to be ploughed the third time to make the soil fully pulverized, planked and levelled and then beds of convenient sizes should be made after making provision for irrigation channels and bunds. Irrigation channels can work as walking paths, *vis-à-vis* service lines so these should be at least 60 cm wide having bunds to both the sides to demarcate the beds though main irrigation channels should be from 90-100 cm wide. It would be convenient if the width of the beds are 1.6 m or little less for proper facilitation of the cultural operations. Now the beds are ready for planting. After planting, the beds will have to be properly irrigated to settle the roots properly and to overcome the lifting shock in the seedlings. The seeds are **sown** in January-February for summer flowering and in May-June for the rainy season crop. In the hills, seeds are sown in the month of March-April, but under most areas of Karnataka, Kerala, Tamil Nadu, Goa and Andaman conditions, the flowers can be obtained throughout the year. Seeds are germinated within 21 days and the young seedlings can be shifted to planting beds after 4-5 weeks when these attain 6-leaf stage. These can be planted in the beds at a **spacing** of 20 × 30 cm. Sometimes seeds are sown *in situ* but it requires thinning by maintaining the spacing of 20-30 cm between the plants and at least 30 cm between the rows. For vigorous varieties, a wider spacing of 30-50 cm may be maintained. However, Arent and Voigt (1994) recommended 20 to 25 cm apart spacing.

Overcrowding of plants does not produce quality flowers so timely **thinning** out the unwanted plants is necessary. Though no pinching or disbudding is required in balsam but overgrown plants can be **pruned** to shape them and to help them standing erect. In many of the varieties of *I. walleriana* and in certain other species also which are tall and vigorous growing, there is requirement of pinching at earlier stages to make the plants bushy and to regulate flowering and to encourage profusion in flowering, especially in case of potted plants. To obtain large flowers, only three shoots should be allowed by pruning all side shoots. Chemical like ethephon promotes development of axillary shoots without damaging the apical meristem or growing points. Ethephon also inhibits flower initiation, aborts young flower buds already initiated, and encourages more vegetative growth. Old perennial plants in the greenhouse are also cut back in March leaving 7.5-10.0 cm of the base so that new vigorous shoots are initiated.

Internodal elongation is maximized when plants are kept continuously moist. Moderate **water stress** results in short plants (Kaczperski and Carlson, 1982). New Guinea *Impatiens* transpires large quantity of water which can be related to their large leaf-canopy mass and warm production temperatures. Over-watering of newly transplanted seedlings should be avoided. Root formation and its development is better in the moist than wet medium. Excessive use of nutrition can encourage overall plant growth and stem elongation. Bedding impatiens, *I. walleriana* responds well to limiting nutrition in reducing plant height. Reducing nutrition, especially nitrogen level, will slow down the plant growth. Application of 100 ppm N is adequate for every second or third irrigation. Excess nitrogen results in long internodes, delayed flowering and numerous leaves (Kaczperski and Carlson, 1982). The NO_3 to NH_4 ratio should be 75:25 to 50:50 for maximum growth (Nelson, 1994). Application of fertilizer affects the plant height apart from overall growth of the plant.

Nutrient Values found in Healthy *Impatiens walleriana* Leaves (Nelson, 1994)

Dry Weight (per cent)					*ppm*		
N	*P*	*K*	*Ca*	*Mg*	*F*	*Mn*	*B*
3.90	0.71	3.47	2.86	0.71	495	56-282	103

New Guinea *Impatiens* is highly sensitive to excessive soluble salts. A range of 100 to 200 ppm N from a balanced **fertilizer** is adequate. Controlled-release fertilizers can be used if the medium moisture level remains high (Haver and Schuch, 1996). New Guinea *Impatiens* is also susceptible to iron and manganese toxicity, which occurs often due to low to medium pH. Symptoms of the toxicity include stunting and twisting or malformations of the upper leaves. Foliar levels of these elements should not exceed 500 ppm (Erwin, 1997). If toxicity occurs, the pH is raised from medium to above 6.0 to make iron and manganese less available to the plants.

Removal of **weeds** is necessary to save the nutrients being given to the main crop, and to control the population of virus-vector thrips and certain other insects, *vis-à-vis* many of the pathogens which harbour on these weeds. Post-emergence herbicides like difenopenten and diclofop when applied to container-grown balsams did not injure the nursery crop (Grewe and Williams, 1981). Sethoxydim, cycloxydim, fluazifop-butyl and AXF 1294 provided total control of grasses without any injury to *Impatiens* (Kühns *et.al.*, 1986). Besides chemical control, the crop can be freed of weeds even manually and through various organic and inorganic mulches.

Growth and Flowering

PGRs are generally not required if plants are not overwatered or overfertilized. However, spray of paclobutrazol at 5 to 18 ppm or uniconazole at 5 ppm can be used for **height control**. In *Impatiens balsamina*, which is a qualitative short-day plant, GA_3 extends the growth (height) till 56 days under 8 and 24-hour photoperiods. Cycloheximide which decreases height slightly under inductive conditions, at a later stage does not affect the GA_3- promoted growth extensions. Both GA_3 and cycloheximide cause enhancement of the rate of differentiation, although this effect is temporary in case of GA_3. Cycloheximide does not affect photoperiodic induction, whereas it hastens and increases the magnitude of GA_3-induced flowering (Nanda *et al.*, 1973). Most cultivars of New Guinea *Impatiens* are compact and well branching. All commonly used plant growth regulators are effective as sprays including anycmidol (A-Rest) at 67 ppm, daminozide (B-Nine) at 1,500 ppm, paclobutrazol (Bonzi) at 5 ppm and uniconazole (Sumagic) at 2 ppm (Corr, 1995; Pasutti and Weigle, 1980). **Flower regulation** can be obtained due to treatment with various hormones. In an experiment on *Impatiens balsamina*, Nanda *et al.* (1969) observed that gibberellins A_3, A_{4+7}, A_{13} and (-) kaurene or TIBA initiated floral buds under strictly non-inductive conditions. In another experiment with the same plant on the effect of decapitation and growth retardants (Phosphon-D and cycocel), Nanda *et al.* (1969) noticed inhibition of flowering by auxin. Agnihotri and Srivastava (1964) found that pre-sowing of seeds of *Impatiens balsamina* in 5 ppm IBA solution produced more flowers than the plants obtained from untreated seeds. Jayanthi *et al.* (1987) observed that when *Impatiens balsamina* plants were pinched after 27 days of planting, it produced more compact and bushy plants with increased number of branches (7.5 per plant) and more number of flowers (44.1). Kanwar and Nanda (1985) recorded that flower bud initiation was advanced considerably and the number of flower buds was higher in GA_3 than in tannic acid treated plants. The combined treatment had the greatest effect regardless of day length.

Post Harvest

Doi *et al.* (1992) reported that with the spray of silver thiosulphate (STS) under low light conditions flower retention was increased by 60 to 80 per cent over a period of 30 days. It has been found that ethylene induces flower abscission and STS increases the postharvest quality of flowers when sprayed at 0.3 to 0.5 mM concentration (Nowak and Rudnicki, 1990). Flowers of *Impatiens hawkeri* and *I. walleriana* are found to be very sensitive to ethylene (Redman, 1990; Sacalis 1993). New Guinea hybrids are susceptible to low temperature below 6 °C. STS pretreatment sprays (1.0 mM) reduce ethylene-induced flower abscission and prolong postharvest life (Dostal *et al.*, 1991). Heins *et al.* (1994) suggested storing of the seedlings up to 42 days at 7-10 °C but plants require to be exposed at 5 fc. During transit, the flowering plants of *Impatiens* should be shipped at 13-16 °C temperature.

Insect-pests, Diseases and Physiological Disorders

The insects infesting balsam are **aphids** (*Macrosiphum impatientis*) which usually feed on young leaves and flower buds. These suck the sap of foliage and buds and cause retarded growth of the plant with poor quality flowers. These also act as a vector for viral diseases. Two spray of Malathion or Metasystox @ 1.0-1.5 ml per liter of water at 15-20 days interval will control these. The body of **spotted cucumber beetle** (*Diabrotica undecimpunctata howardii*) is black and the wings are yellowish-green having six large spots on them. While feeding on flowers and petioles, these make holes there. It can be controlled by regular spraying with Rogor or Methoxychlor or Diazinon. **Tarnished plant bug** (*Lygus linearis*) infests on the new shoots of balsam and blacken them. It can be controlled by regular spraying with Rogor or Methoxychlor. **Cyclamen mite** (*Steneotarsonemus pallidus*) is one of the smallest insects that mostly infests on the young parts of the plant which afterwards are unable to form normal buds. Blooms get distorted and usually drop off early. Infested leaves become deformed, curl from the outside and become wrinkled. It can be controlled by spraying with Dicofol @ 1.0-1.5 ml or Propargite @ 0.3-0.5 ml per liter of water. The young **caterpillars** of *Pericallia ricini* feed extensively on leaves, defoliating the plants completely. These larvae can be controlled by spraying with Malathion or Monocil @ 1.0-1.5 ml per liter of water. Sometimes **semiloopers** (*Sylepta textalis*) have been found feeding on the leaves, particulary when the weather is warm. While feeding on the foliage and the floral buds, it rolls up the leaves and the buds. It can be controlled by spraying with Malathion or Metasystox @ 1.0-1.5 ml per liter of water. **Whiteflies** multiply very fast and feed mostly on young foliage and underside of the leaves by sucking their sap, which in turn causes plant retardation and poor quality flowers. They also act as vector for transmission of viruses. Regular spraying of Acetamiprid @ 3 g per 10 litres of water at 'run off' stage on the plant will control these flies. **Grasshoppers** are active particularly in summer and rainy seasons. They generally feed on leaves and retard the plants. Regular spraying with Monocil or Metasystox @ 1.0-1.5 ml per litre will control these pests.

Root-knot nematodes (*Meloidogyne incognita*) have been reported infesting balsam (*I. amphorata, I. balsamina, Impatiens* sp.) in India (Khan *et al.,* 1964) which can be controlled by spreading Furadan or Carbofuran granules in the infested field. Another nematode, *Pratylenchus penetrans* also infests on balsam. Control measures include sterilization of soil by steam or drenching by formaldehyde @ 2 per cent before planting or incorporation of Furadan in the soil before planting. Application of neem-based products and crop rotation with marigold can also minimize the nematode infestation.

Flower or **corolla rot** (*Aspergillus* sp.) on *Impatiens* sp. is recorded by Thom and Church (1926). **Foot rot** in *I. balsamina* is caused by *Rhizoctonia solani* (Siradhana *et al.,* 1964). Young balsam plants are mostly affected by **'damping-off'** caused by *Pythium ultimum* and *Pellicularia filamentosa*. These can be controlled by using fresh soil or moist sphagnum moss for seed sowing. **Stem rot** (*Pellicularia rolfsii*) and **wilt** (*Verticillium albo-atrum*) are two most disastrous fungal diseases which affect balsams. Use of fresh soil or treating the infected soil with heat or formaldehyde before planting will give control of these problems. **Powdery mildews** in *Impatiens balsamina* are caused by *Erysiphe cichoracearum* (syn. *Oidiopsis balsaminae, Oidium balsaminae*) (Petch, 1912, 1919; Patel, 1949, Narayanaswamy and Ramakrishnan, 1967-1968); by *Laveillula taurica* (syn. *Oidiopsis taurica*) (Butler and Bisby, 1931; Mundkur, 1938; Patel *et al.,* 1949); by *Oidium erysiphoides* (Rajderkar, 1966; Roy, 1965, 1968); and by *Sphaerotheca fuligenea* (Patil, 1965). These fungi infect all the aboveground parts of the whole plants by forming grayish-white powdery patched structures, especially on the leaves during high humid conditions and such plants do not flower properly. These can be controlled by providing adequate ventilation and by keeping the crops dry during wet seasons. Regular spraying with Karathane at 0.3 ml to per litre of water at 10-15 day's interval will control all these pathogens. **Downy mildew** in *Impatiens* sp., *I. Amphorata* and *I. balsamina* is caused by *Plasmopara obduscens* (Sydow *et al.,* 1907) which can be controlled through the same treatments as to that of powdery mildew. **Leaf rust** in *Impatiens* sp. and *I. amphorata* is caused by *Puccinia argentata* (Arthur and Cummins, 1933; Sydow and Sydow, 1904-1920) and *Puccinia komorovi* (Barclay, 1890). **Leaf spots** in *Impatiens balsamina* are caused by *Alternaria alternata* (syn. *A.tenuis*) (Rao, 1963), *Cercoseptoria balsaminae* (Sydow and Mitter, 1935; Sydow *et al.,* 1937; Agarwal and Hasija, 1961), *Colletotrichum capsici* (Roy, 1965; Sahni, 1966), *Phyllosticta impatiens* (Rao, 1962), *Septoria balsaminae* (Sohi *et al.,* 1972); and **leaf spot** in *I. acaulis, I. chinensis* and *I. lawii* is caused by *Cercospora fukushiana* (Patel *et al.,* 1949; Ramakrishnan and Ramakrishnan, 1949; Chiddarwar, 1959). *Septoria noli-tangeris* and *Stemphyllium botryosum* also cause more or less circular small brown leaf spots and in their infection the affected leaves should be removed and burnt and to check further spread Dithane M-45 or Dithane Z-78 should be sprayed at 0.2 per cent at 10-12 day's intervals.

Stem gall in *Impatiens dalzelli* is caused by *Synchytrium balsaminii* (Patil and Mahabale, 1964) and in *I. chinensis* by *S. travancoricum* (Karling, 1966). Generally **bacterial wilt** (*Pseudomonas solanacearum*) which causes balsam wilting, also affects many Solanaceous plants. It can be controlled by removing and destroying affected plants. Infected soil should be steam-sterilized or chemically treated before planting and the balsam should not be planted close to solanaceous crops. Seed inoculation with *Pseudomonas fluorescens* strain E6 frequently restricted growth of balsam by 7-13 per cent (Yuen and Schtoth, 1986).

Symptoms of '***Inpatiens* necrotic spot virus**' (INSV) on infected seed type *Impatiens* plants are small black to brown necrotic flecks and ring spots on leaves and white rings on flower petals. Necrosis may be found along the midribs of diseased leaves which show yellowing, malformation and stunting. Infected leaves drop prematurely. Symptoms of chlorosis, necrosis and ringspots also occur on the stems of infected plants. In New Guinea hybrids though symptoms are the same but diagnosis is difficult due to high anthocyanin pigmentation in the leaves and also as some leaves do not show any symptom. Temperature above 27 °C and high light above 1,500 foot candles generally suppress the symptoms of INSV in naturally infected seed-produced and New Guinea *Impatiens* plants. Symptoms produced by **'tomato spotted wilt virus'** (TSWV) on infected *Impatiens* plants are similar to those produced by INSV and both these are most severe diseases of balsam and are transmitted by 'western flower thrips' (*Frankliniella occidentalis*) and these carry the viruses when its younger stage feeds on the infected plants

(German *et al.*, 1992). '**Tobacco streak ilarvirus**' (TSV) was isolated from *Impatiens holstii* showing stunting, leaf twisting and deformation and such plants produce poor quality flowers with poor yield though there was no floral deformity observed (Lockhart and Betzold, 1980b). '**Tobacco ringspot nepovirus**' (TobRSV) was isolated from infected *Impatiens holstii* showing leaf mottling, stunting and leaf abscission (Lockhart and Pfleger, 1979) with obvious symptoms of mosaic and chlorotic or necrotic ringspots on leaves, sometimes coupled with leaf abscission, and which at 28-30 °C temperature than in plants grown at 20-24°C shows more visible TRSV symptoms. A rod-shaped virus named as '***Impatiens* latent virus**' (Helenium S Carlavirus), measuring 520 µm long was isolated from *Impatiens holstii* showing local lesions and systemic mottling (Lockhart and Betzold, 1980a). The same virus from the same species of balsam once more was isolated and was identified as 'Helenium S carlavirus' (Pleše *et al.*, 1988). '**Cucumber mosaic cucumovirus**' (CMV) was found to infect *Impatiens sultanii* which produced stunting, mosaic and deformation of leaves (Herold, 1964). Identification of CMV was based on test plant symptoms and serology. Sastry (1980) recorded two viruses, *viz.* '**leaf curl disease**' and '**phyllody disease**' on *Impatiens balsamina.* From Markazi (Iran), Farzadfar *et al.* (2005), for the first time in 2003, recorded TuMV causing chlorotic spots, leaf yellows and black stem streaks on *Impatiens balsamina.* To check the infection or further spread of the viruses, the measures applied are sanitation with appropriate weed control in the field, removal of weeds from the field as alternate hosts which harbour the vectors, rouging-off and destruction of all the infected plants from the field during crop season and when flowering is over, control of the insect-vectors or thrips through some effective insecticides by the alternative use of pyrethroids, carbamates, chlorinated hydrocarbons, organophosphates and soaps, procurement of plants raised from certified virus-free mother stocks, covering of the entire field with 400 mesh screens to prevent entry of thrips, and monitoring of thrips in the field with the use of yellow sticky strips.

There are certain **physiological problems** in balsam. Yellowing of leaves and premature leaf falling are due to water-logging. Seed germination can be a problem in *Impatiens* due to imperfect pH, EC, temperature and light which can be corrected by lowering the pH of the medium at 6.2, lowering of water EC, maintaining the temperature at 20-22 °C and providing the proper lighting up to radical emergence (Khademi Karlovich, 1995). Presence of ethylene can abort meristem tips (Daughtrey and Chase, 1992). In New Guinea *Impatiens*, cold damage leading to plant death can occur with extended temperatures of 6 to 8°C.

References

Arent, G.L. and M.L. Voigt, 1994. *Landscaping.* In: *Bedding Plants IV* (ed. Holcomb, E.J.), pp. 75-91. Ball Publishing, Batavia, Illinois, USA.

Agarwal, G.P. and S.K. Hasija, 1961. Fungi causing plant diseases at Jabalpur (M.P.)-V. *Proc. Nat. Acad. Sci., India,* **31**: 99-108.

Agnihotri, B.N and J.C. Srivastava, 1964. Contribution to our knowledge on the effect of indole butyric acid on the germination, flowering and fruiting of *Impatiens balsamina* L. *Indian Agric.,* **8**(2): 183-184.

Armitage, A.M. and H.M. Vines, 1982. Net photosynthesis, diffusive resistance and chlorophyll content of shade and sun tolerant plants grown under different light regimes. *HortSci.,* **17**: 342-343.

Arthur, J.C. and G.B. Cummins, 1933. Rusts of the North West Himalayas. *Mycologia,* **25**: 397-406.

Bailey, L.H. and E.Z. Bailey, 1976. *Impatiens.* In: *Hortus Third ~ A Concise Dictionary of Plants Cultivated in the United States and Canada.* Macmillan Publishing, New York, USA.

Ball. V. 1991. Plugs-the way of the 1990s. In: *Ball RedBook* (15th edn; ed. Ball, V.), pp. 137-153. J. Ball Publishing, West Chicago, Illinois, USA.

Ball. V. 1995. *Impatiens* germination. *Grower Talks,* **58**(11): 8.

Barclay, A. 1890. Descriptive list of the Uredineae occurring in the neighbourhood of Simla (Western Himalayas). I. *J. Asiatic Soc. Bengal,* **56**: 223-226.

Butler, E.J. and G.R. Bisby, 1931. *The Fungi of India. Imperial Council of Agricultural Research Indian Science Monograph* 1, XVIII, Calcutta.

Carlson, W. and M. Klire, 1985. Try New Guinea impatiens. *Greenhouse Grower,* **3**(8): 12-13.

Carpenter, W.J., 1994. Germination. In: *Tips on Growing Bedding Plants* (3rd edn; eds Tayama, H. K., T.J. Roll and M.L. Gaston), pp.10-14. Ohio Florists Association, Columbus, Ohio, USA.

Carpenter, W.J., E.R. Osmark and J.D. Cornell, 1994. Light governs the germination of *Impatiens walleriana* Hook.f. seed. *HortSci.,* **29**: 854-857.

Chiddawar, P.P. 1959. Contribution to our knowledge of the Cercosporae of Bombay state I. *Sydowia,* **13**: 152-163.

Christensson, H. 1985. *Odling av Nya Guinea-Impatiens.* Swedish University of Agricultural sciences, Trätgård 288, Alnarp, Sweden (Swedish).

Corr, B. 1995. *Seed New Guinea Impatiens: Seed to Plug to Finish.* In: *New Guinea Impatiens* (eds Banner, W. and M. Klopmeyer), pp.105-111. Ball Publishing, Batavia, Illinois, USA.

Corr, B. 1997. The ABC's of plug production. *Greenhouse Product News*, **7**(2): 21-23.

Corr, B. 1998. *Impatiens* (bedding plant). In: *Ball RedBook* (16th edn; ed. V. Ball), pp. 567-575. Ball Publishing, Batavia, Illinois, USA.

Daughtrey, M. and A.R. Chase, 1992. *Impatiens*. In: *Ball Field Guide to Diseases of Greenhouse Ornamentals*, pp.125-131. Ball Publishing, Geneva, Illinois, USA.

Doi, M., T. Mizuno and H. Imanishi, 1992. Postharvest quality of potted *Impatiens walleriana* Hook. f. ex. D. Oliver as influenced by silver thiosulfate applications and light conditions. *J. Japanese Soc. hort. Sci.*, **61**: 643-649.

Dole, J.M. and H.F. Wilkins, 1999. *Floriculture Principles and Species*, pp. 388-396. Prentice Hall, Upper Saddle River, New Jersey, USA.

Dostal, D.L., N.H. Agnew, R.J. Gladon and J.L Weigle, 1991. Ethylene, simulated shipping, STS, and AOA affect corolla abscission of New Guinea impatiens. *HortSci.*, **26**: 47-49.

Erwin, J. 1995. Light and temperature. In: *New Guinea Impatiens* (eds Banner, W. and M. Klopmeyer), pp.41-45. Ball Publishing, Batavia, Illinois, USA.

Erwin, J., M. Ascerno, F. Pfleger and R. Heins, 1992. New Guinea impatiens production. *Minnesota Commercial Flower Growers Association Bulletin*, No. 41(3), pp. 1-15.

Farzadfar, S., K. Ohshima, R. Pourrahim, A.R. Golnaraghi, S. Jalali and A. Ahoonmanesh, 2005. Occurrence of turnip mosaic virus on ornamental crops in Iran. *Pl. Path.*, **54**(2): 261.

German, T.L. R., D.E. Ullman and J.W. Moyer, 1992. Tospoviruses: diagnosis, molecular biology, phylogeny and vector relationships. *Ann. Rev. Phytopath.*, **30**: 315-348.

Gill, L.S. and C.C. Chinnapa, 1977. Chromosome number from herbarium sheets in some Tanzanian *Impatiens* L. (Balsaminaceae). *Caryologia*, **30**(4): 375-379.

Govindarajan, T. and D. Subramaniam, 1985. Karyomorphological studies in South Indian Acanthaceae. *Cytologia*, **50**: 473-482.

Govindarajan, T. and D. Subramaniam, 1986. Karyotaxonomy of South Indian Balsaminaceae. *Cytologia*, **51**: 107-116.

Grewe, L. R. and D.J. Williams, 1981. The feasibility of using post-emergence grass herbicides in nursery crops. *Proc. North Central Weed Control Conf.*, **36**: 65. Illinois University, USA.

Grey-Wilson, C. 1980. *Impatiens of Africa*. A. A. Balkema, Rotterdam, Holland.

Haver, D.L. and U.K. Schuch, 1996. Production and postproduction performance of two New Guinea *Impatiens* grown with controlled-release fertilizer and no leaching. *J. Amer. Soc. hort. Sci.*, **121**: 820-825.

Herold, F. 1964. Natural infection of *Impatiens sultanii* with cucumber mosaic virus. *Pl. Dis. Rep.*, **48**: 603-605.

Ito, A., T. Hisamatsu, N. Soichi, M. Nonaka, M. Amano and M. Koshioka, 1997. Effect of altering diurnal fluctuations of day and night temperatures at the seedling stage on the subsequent growth of flowering annual. *J. Japanese Soc. hort. Sci.*, **65**(4): 817-823.

Jalil, R., S.N. Zadoo and T.N. Khoshoo, 1977. Colchiploid balsams. *Nucleus*, **17**: 118-124.

Jayanthi, R., B. Shiva Raj, G.K. Ramchandrappa and K.N. Chandrashakarappa, 1987. Pinching induces branching in balsam. *Current Research* (Univ. of Agric. Sci., Bangalore), **16**(6): 82-83.

Kaczperski, M.P. and W.H. Carlson, 1982. Production pointers for *Impatiens*-1. *Bedding Plant Crop, Bedding plants Incorporated*, **19**(1): 6-7.

Kanwar, K and K.K. Nanda, 1985. Effects of gibberellic acid and tannic acid on flowering of *Impatiens balsamina* L. in relation to the number of inductive and non-inductive photoperiodic cycles. *Indian J. exp. Biol.*, **23**(7): 404-405.

Karling, J.S. 1966. The chytrids of India with a supplement of other zoosporic fungi. *Sydowia*, **4**: 1-100.

Khademi, J. and P. Karlovich, 1995. How you can avoid *Impatiens* tip abortion. *Greenhouse Management and Production*, **14**(2): 32-37.

Khan, A.M., M.R. Siddiqi, E. Khan, S.I. Hussain and S.K. Sexena, 1964. List of Stylet-bearing Nematodes Reported from India-1. Nematology Publication No.1, Aligarh Muslim University, Aligarh.

Kühns, L.J., C. Haramaki, J. Becker and G. Lyman,1986. Pre-emergence and post-emergence grass herbicides on annual flowers. *Proc. 40th ann. meet. Northeastern Weed Sci. Soc.* Pennsylvania St. Univ., USA, pp. 260-262.

Lockhart, B.E.L and J.A. Betzold, 1980a. Latent infection of impatiens by a potexvirus. *Acta Hort.*, No., 110, pp. 81-83.

Lockhart, B.E.L and J.A. Betzold, 1980b. Leaf-curl of impatiens caused by tobacco streak virus infection. *Plant Dis.*, **64**: 289-290.

Lockhart, B.E.L and F.L. Pfleger, 1979. Identification of a strain of tobacco ringspot virus causing a disease of impatiens in commercial greenhouses. *Plant Dis. Rep.*, **63**: 258-261.

Mikkelsen, E.P. 1995. *Rooting*. In: *New Guinea Impatiens* (eds Banner, W. and M. Klopmeyer), pp. 81-86. Ball Publishing, Batavia, Illinois, USA.

Mundkur, B.B. 1938. *Fungi of India* (Supplement-I). *I.C.A.R. Science Monograph*, **12**: 1-54.

Nanda, K.K., H.N. Krishnamoorthy, K.L. Toky and K. Lata, 1969. Effects of gibberellins A_3, A_{4+7}, A_{13} and of (-) kaurene on flowering and extension growth of *Impatiens balsamina* L. under different photoperiods. *Planta*, **69**: 249-257.

Nanda, K.K., M. Kumar, S. Sawhney and N. Sawhney, 1973. Effect of cycloheximide and GA_3 response on photoperiodic reactions on *Impatiens balsamina* L. *Ann. Bot.*, **37**: 107-111.

Narayanaswamy, P. and K. Ramakrishnan, 1967-68. Powdery mildews of Coimbatore, Madras State. *The Madras Univ. J.*, **37-38 (B)**: 84-99.

Nelson, P.V. 1994. Fertilization.In: *Bedding Plants IV* (ed. Holcomb, E.J.), pp. 151-175. Ball Publishing, Batavia, Illinois,USA.

Nowak, J. and R.M. Rudnicki.1990. *Postharvest Handling and Storage of Cut Flowers, Florist Greens and Potted Plants.* Timber Press, Portland, Oregon, USA.

Pasutti, D.W. and J.L. Weigle, 1980. Growth-regulated effect on New Guinea *Impatiens* hybrids. *Scient. Hort.*, **12**: 293-298.

Patel, M.K., M.N. Kamat and V.P. Bhide, 1949. *Fungi of Bombay* (Supplement I). *Indian Phytopath.*, **2**: 142-155.

Patil, S.D. and T.S. Mahabale, 1964. The genus *Synchytrium* in Maharashtra-II. *J. Univ. Pune*, **26**: 67-79.

Patil, S.D. 1965. Contribution to the fungi of Maharashtra-II. *Journal of University Pune*, **30**:2-37.

Petch, T. 1912. Ustilagineae and Uredineae of Ceylon. *Ann. roy. bot. Gard., Peradeniya*, **5**: 223.256.

Petch, T. 1919. Gasteromycetae Zeylanicae. *Ann roy. bot. Gard., Peradeniya*, **7**: 57-78.

Pizzetti, I. and H. Cocker, 1968. *Flowers ~ A Guide for Your Garden*, pp. 643-651. Harry N. Abrams, Inc., Publishers, New York, USA.

Pleše, N., Ž. Eriæ and M. Krajaèiæ, 1988. Further information of infection of *Impatiens holstii* with Helenium virus S. *Acta Hort.*, No. 234, pp. 477-484.

Rajderkar, N.R. 1966. Occurrence of two powdery mildews on balsam, *Impatiens balsamina* L. *Mycopathol. et Mycol. Appl.*, **28**(1): 149-152.

Ramakrishnan, T.S. and K. Ramakrishnan, 1949. Additions to fungi of Madras-VI. *Proc. Indian Acad. Sci.*, **29B**: 48-58.

Rao, V.G. 1962. The genus *Phyllosticta* in Bombay, Maharashtra. *Sydowia*, **16**: 275-283.

Rao, V.G. 1963. Some new host records for *Alternaria* species from India. *Mycopath. et Mycol. Appl.*, **19**: 181-183.

Rao, R.S.V. 1978. New basic chromosome number of three in the genus *Impatiens*. *J. Indian bot. Soc.*, **57**: 77-78.

Redman, P.B. 1990. Vase life determination and postharvest evaluation of specialty cut flowers. M.S. Thesis, Oklahoma State University, Stillwater, Oklahoma, USA.

Roy, A.K. 1965. Additions to the fungus flora of Assam-I. *Indian Phytopath.*, **18**: 327-354.

Roy, A.K. 1968. Additions to the fungus flora of Assam-II. *Indian Phytopath.*, **21**: 182-189.

Sacalis, J.N. 1993. *Cut flowers, Prolonging Freshness* (2nd ed.; ed. Seals, J.L.). Ball Publishing, Batavia, Illinois, USA.

Sahni, V.P. 1966. Deuteromycetes from Jabalpur-II. *Mycopath. et Mycol. Appl.*, **29**: 226-244.

Sastry, K.S. 1980. *Plant Virus and Mycoplasamal Diseases in India: A Bibliography*. Bharti Publication, Delhi, pp.1-270.

Siradhana, B.S., J.P. Agnihotri and R.L. Mathur, 1964. Foot rot in balsam. *Sci. & Cult.*, **30**: 400-401.

Sohi, H.S., S.L. Sharma, S.K. Nayar and K. Shyam, 1972. New records of fungi from Himachal Pradesh. *Indian Phytopath.*, **25**: 439-441.

Strefeler, M.S. 1995. *Genetics*. In: *New Guinea Impatiens* (eds Banner, W. and M. Klopmeyer), pp. 227-247. Ball Publishing, Batavia, Illinois, USA.

Strefeler, M.S. and R.J.W. Quené, 1995. Variability in water loss patterns of New Guinea *Impatiens* cultivars and breeding selections. *J. Amer. Soc. hort. Sci.*, **120**: 527-531.

Strope, K.M. and M.S. Strefeler, 1997. Analysis of heat tolerance in New Guinea impatiens (*Impatiens hawkeri*) utilizing diallel analysis. *HortSci.*, **32**: 499.

Swarup,V., S.P.S. Raghava and K.A. Balakrishnan, 1975. Heterosis in balsam. *Indian J. Genet. Pl. Br.*, **35**: 69-75.

Sydow, H. and P. Sydow, 1904-20. Novae fungorum species 1904. No. I, *Annales Mycologici*, **2**: 162-174. 1911. No. VI, *Annales Mycologici*, **9**: 142-146. 1912. No. VIII, *Annales Mycologici*, **10**: 405-410. 1914. No.XII, *Annales Mycologici*, **12**: 195-204. 1915. No. XIII, *Annales Mycologici*, **13**: 35-43. 1916. No. XIV, *Annales Mycologici*, **14**: 256-261. 1917. No. XV, *Annales Mycologici*, **15**: 143-148. 1920. No.XVI, *Annales Mycologici*, **18**: 154-160.

Sydow, H. and J.H. Mitter, 1935. Fungi Indici-II. *Annales Mycologici*, **33**: 46-71.

Sydow, H., J.H. Mitter and R.N. Tandon, 1937. Fungi Indici-III. *Annales Mycologici*, **35**: 222-243.

Sydow, H., P. Sydow and E.J. Butler, 1907. Fungi Indiae Orientalis.Part II. *Annales Mycologici*, **5**: 485-515.

Thom, C. and M.B. Church, 1926. *The Aspergilli*. The Williams & Wilkins Co., Baltimore, USA.

Voigt, A.O. 1994. *Marketing Trends*. In: *Bedding Plants IV* (ed. Holcomb, E.J.), pp. 17-33. Ball Publishing, Batavia, Illinois, USA.

Whealy, C.A., 1995. Commercial varieties. In: *New Guinea Impatiens* (eds Banner, W. and M. Klopmeyer), pp. 213-226. Ball Publishing, Batavia, Illinois, USA.

Yuan, Y.M., Y. Song, K. Geuten, E. Rahelivololona, S. Wohlhauser, E. Fischer, E. Smets and P. Kupfer, 2004. Phylogeny and biogeography of Balsaminaceae inferred from ITS sequences. *Taxon*, **53**: 391-392.

Yuen, G.Y. and M.N. Schroth, 1986. Interactions of *Pseudomonas fluorescens* strain E6 with ornamental plants and its effect on the composition of root-colonizing microflora. *Phytopath.*, **76**(2): 176-180.

26

Ixora (Family: Rubiaceae)

Jayoti Mazumder, Sellam Perinban, R.L. Misra and Sanyat Misra

(Common names: Flame of the woods (*Ixora coccinea*), Ixora, Jungle geranium (*I. javanica*), Cambodia-Kam rontea, Bengali-Rangan, Hindi-Rukmini, Malayalam-Chethi, Tamil-Vedchi, Telugu-Nooru, Vietnam-Dun do)

Introduction

Ixora derives from the Portuguese rendering of the Sanskrit 'Ikvana' or 'Israra' meaning 'Ishwara' (the Hindu god Shiva), to whom the flowers are offered and *coccinea* means scarlet. *Ixora* is a tongue-twister name as it is pronounced *icks-sore-ah*. This genus has some 500 species of shrubs and trees from Pan-tropical regions, *i.e.* India, China, Java, Malayan Archipelago, Moluccas, Madagascar, Guinea, East Indies, *etc.* The ixoras were used as house plant in England around 1870 and numerous varieties were developed during that period and at the turn of the century. All ixoras produce clusters of star-shaped flowers on stem tip or terminal corymbs varying in colour from red to orange to yellow, pink and white and are used for offering to gods, especially the Shiva and for making garlands apart from its effective use in shrubberies, hedging and for garden landscaping. Often the leaves and stems are used as an ablution for infantile and is sedative. The flowers are used externally to sores, chronic ulcers, scabies and in certain types of dermatitis and the roots in case of cholera, diarrhoea, dysentery, gonorrhoea and fever.

Ixora coccinea Linn, a national flower of Myanmar, is a medium size beautiful landscape shrub of Indian origin which is grown throughout the world in the tropics and sub-tropics for its exquisite flowers. In India though it is found throughout but is very common in west peninsula in scrub jungles. Its loose flowers are used for religious offerings to gods and goddesses, especially the Shiva.

Botany and Classification

The *Ixora* L. is the third largest genus of coffee family. Ixoras are popular ornamental plants grown in the tropics for their beautiful inflorescences which are in different shades of white, yellow, orange, pink and red. In general, the classification and identification of the species and cultivars in this genus has been very difficult. The existence of natural hybrids among the species cannot be ruled out. Taxonomists have tried to distinguish the taxa morphologically, especially using the nature of leaves and floral parts. Alternative approaches for intra-and inter-specific classification have been reported based on geographic origin, morphology, karyotype or isozyme analysis (Kaser and Steiner, 1983). These criteria are either influenced by environmental factors and stage of plant development or reveal only limited variation. It's a finest shrub or tree ensuring for itself a perennial position in the garden. The leaves are simple in opposite pairs and terminal corymbs of tubular, mostly 4-petalled flowers which in a few species are fragrant. There are 15-30 or even more magnificent flowers arranged across the corymb. In *Ixora coccinea*, the large terminal flowers of orange scarlet colours are born in profusion during summers and rains but in places such as Bangalore the flowering continues throughout the year. The shrub is generally 2.0-3.0 m tall with 5.0-7.5 cm long leaves which are bronzy when young, later turning to a glistening dark green. Although most *Ixora* plants of shrubby nature are much-branched and compact so are ideal for hedges, borders, screens, or as a specimen planting. Ixora fruit is a subglobose berry which ripens to red and black.

Ixora displays colourful, nocturnally fragrant, nectariferous, tubular flowers. The flowers are dichogamous. Therefore, it adopts self pollination (Latha *et al.*, 2012). Ixora flowers are strongly protandrous and it has 'ixorid' secondary pollen presentation mechanism (Nilsson, 1998). In Ixora, the natural fruit set decreases with increasing pollination population (Nayak *et al.*, 2010). Both the fertile and sterile appendages are born on the receptacle (Pandey, 2006). The cultivated shrub such as *Ixora bandhuca* (67.0 per cent) and *Ixora barbata* show low pollen fertility (Bedi *et al.*, 1981). Principal pollination adoption probably occurs by a noctuid moth (Nilsson, 1998). SEM studies of *I. chinensis* in China revealed that if once the terminal apex reaches the inflorescence-bud-stage, it would flower without abortion. The sepals, petals, stamens, and pistil are well developed thereafter and anthesis achieves in January through March in the following year. The haploid chromosome number of *Ixora* is n=11 (Bedi *et al.*, 1981) and *Ixora rosea* is a triploid with 2n=33 (Sharma and Chatterjee, 1960).

Species and Cultivars

Ixora comprises of about 500 species (Mouly *et al.*, 2009) of shrubs and small trees distributed in tropical and subtropical regions of the world and valued primarily for their colourful flowers though one shrubby species ***I. borbonica*** from Reunion Island growing some 2.0 m with off-white small flowers and bearing 15-25 cm long, lanceolate, leathery-textured and lustrous mossy-green leaves veined carmine-red is valued for its handsome foliage looking like *Codiaeum*. Phylogenetic analyses of Ixoreae reveal that *Ixora* tribe can be subdivided into three major clades: the Mascarene/neotropical/Malagasy/African clade, the Pacific clade, and the Asian clade. Other conflicting positions for the cultivated species are most likely due to anthropogenic hybridization. Most of cultivated species are native to southern Asia and China but only a few of them are grown as landscape plants (Rajasegar *et al.*, 1997).

In India the genus is represented by 46 species (Husain and Paul, 1991; Sivadasan and Mohanan, 1991; Deb and Rout, 1992; Pradeep, 1997). Among them the popular species grown are ***Ixora acuminata***, a glabrous shrub having various types of elliptic to linear-oblong leaves but floral leaves rounded or obovate and sessile and others petioled, cyme 5-10 cm across and corymbose comprising of densely packed florets which are pure white and fragrant; ***C. amboinica***, a shrub bearing glabrous, ovate-oblong, undulate, acuminate, large but short-petioled leaves, cyme trichotomously divaricately compound and flowers long-lasting showy orange-yellow; ***I. barbata*** from India, a large glabrous shrub bearing thin, stalked, elliptic, and a little acute leaves though upper pair sessile, small and cordate, cyme short-stalked and quite wider (maximally 30 cm) than length, and corolla white being woolly at mouth; ***I. chinensis*** native to Malaysia and China, growing up to 1.5 m tall, very similar to *I. coccinea* but smaller in all its parts and petals with blunt tips, a glabrous shrub growing 60-90 cm bearing obovate or obovate-oblong and sessile to subsessile, glossy and deep green leaves, slender-tubed orange flowers in dense corymbs which appear in summer. This species offers many cultivars, many of which produce yellow, yellow-orange, orange, red, pink and white flowers such as 'Alba' (white), 'Rutilans' (crimson), and 'Singapore Yellow' of 1.5-2.0 m in height with bright to dark green leaves and yellow-orange flowers which continue coming up throughout the year; ***I. coccinea*** L. (syn. *I. bandhuca* Roxb.) from India and East Indies grows to 2.5-3.0 m with new growth reddish, bearing elliptic to oblong with rounded or cordate base, glossy and deep green and up to 19 cm long leaves, and 13 cm wide corymbs with 4 cm long orange-scarlet flowers (cultivars in various shades of red, pink, orange-yellow and yellow) having broadly pointed petal tips coming up in summer. The hybrids being the natural crosses among the species account for majority of ixoras for landscaping. The hybrids 'Nora Grant' grows to 1.5 m with large and light to bright green leaves and clusters of pink flowers which continue appearing throughout the year; 'Mauli' grows to 1.5 m with compact growth bearing smaller leaves and orange-red flowers, and 'Rosea' (pink); *I. congesta* from India, an evergreen glabrous tree (only cyme pubescent), leaves stalked, acute or acuminate, elliptic or elliptic-oblong and 15-30 cm long, cyme sessile or so bearing orange-yellow flowers which change to reddish; ***I. duffi***, a larger growing glabrous and tall species native of East Indies with large green (30 cm), linear-oblong to oblong-lanceolate leaves and deep red tinged crimson flowers in large corymb (20 cm across) and its most common cultivar is 'Superking' reaching to 3.0 m in height and bearing orange-red and showy flowers in large clusters which continue coming up almost throughout the year; ***I. fulgens*** from India is an erect shrub with polished branches, leaves linear-oblong to obovate-oblong, acute or acuminate with sunken nerves and petioled, cymes corymbose, large, sessile or short peduncled and orange to scarlet flowers having too short pedicel; ***I. javanica*** from South East Asia grows from a large shrub to a small tree (7.5 m) in its homeland, bears 20 cm long, ovate-oblong, leathery-textured with waxy shine rich green leaves and 6 cm or more across cluster with 4 cm long salmon-red flowers appearing during summer; ***I. laxiflora*** from Guinea, a 90-120 cm slender shrub, leaves oblong-lanceolate, acuminate and short-petioled, cymes large, open and trichotomous, and flowers very fragrant and white tinged pink; ***I. lutea*** (syn. *I. coccinea* var. *lutea*) of garden origin, which differs from *I. coccinea* by laxer inflorescence and pale-yellow flowers though corolla lobes are larger ovate-rhomboid; ***I. odorata***, a small fragrant shrub from Madagascar bears large (up to 30 cm), thick, ovate to obovate-lanceolate, acute or acuminate leaves, cyme 30 cm or more across with purplish branches, and corolla 10.0-12.5 cm long and white but changing to yellowish-brown and highly fragrant; ***I. parviflora***, an evergreen tree from India, bearing subsessile oblong or elliptic-obtuse leaves 7.5-15.0 cm long, cymes sessile bearing flowers in subglobose clusters, and corolla glabrous and white; ***I. taiwanensis*** 'Petites' seldom reaches over 90 cm with quite compact growth, bears tiny leaves and red, yellow or pink flowers. Among hybrids with unknown parentage are 'Angela Busman' (flowers shrimp-pink), 'Frances Perry' (large trusses of deep yellow flowers), 'Gillettes Yellow' (pale-yellow), 'Helen Dunaway' (deep orange), 'Henry Morat' (fragrant pink), 'Herrera's White' (beautiful white), 'Superking' (deep red and floriferous), *etc. Ixora chinensis* is a spontaneous seedling selection of unknown origin (Zaundam, 1997). 'Frankie Hipp' is a variety derived from 'Nora Grant' through spontaneous mutation in Florida and is being grown there (Fulghum Leigh, 2008). Some other hybrids, varieties or species in **whites** are *Ixora colei* (*I coccinea* × *I. chinensis* 'Alba'), an English hybrid developed around 1870 by E. Cole of Manchester (England) but when once lost there, it was again introduced from Calcutta in 1964 by Swedroe's Nursery, is more cold-hardy and prefers filtered situation having compact and low spreading growth bearing massive pure white flowers blooming intermittently during early summer and late fall and remains in flower for a considerable time; 'Herrera's White', a sport of 'Herrera's Pink', originating at Swedroe's Nursery in Fort Lauderadale (Florida), is a cultivar liking partial shade and grows upright and compact with constant blooming habit, producing pure white flowers of medium size; in **yellows** are 'Frances Perry' ('Gillete's Yellow' × 'Singapore No. 1'), developed by Swedroe in 1963, is compact with broader and dark green leaves and bears best pure yellow flowers. *Ixora javanica flava* was introduced by Swedroe in January 1964 from Bangkok (Thailand), has bushy growth and bears apricot-gold rounded flowers as in *I. chinensis* with which it hybridizes easily and has produced numerous hybrids. 'Singapore No. 1', a chance seedling of *I. aureo-rosea* recorded in 1952 from Singapore with similar flowers to *I. macrothyrsa*, likes partial shade and is prolific bloomer bearing orangish-yellow to pinkish flowers as large as 12.5 cm across; in **orange and salmon** are'Angela Busman' ('Superking' × 'Singapore No. 1') developed by Swedroe in 1963, bearing large and dark green leaves, and the flower colour is shrimp-pink to orange when first open but turns old rose as they age. 'Lois Shore' (a seedling of 'Singapore No. 1') originated at Swedroe Nursery in 1965, prefers partial shade and flower colour is soft orange similar to 'Singapore No. 1'; in **pink and rose** are *I. fraseri* (probably a cross between *I. ambonensis* and *I. chinensis*), a very beautiful shade-loving shrub but highly susceptible to root-knot nematode, first exhibited in England about 1874 and was introduced in Fort Lauderdale around 1952, bearing large and dark green leaves, flowering in very large clusters with open buff-orange flowers fading to salmon; in **pinks** are 'Henry Morat' (*I. acuminata* × *I. coccinea*) developed in Miami (Florida) in 1943, having compact growth with upright branches bearing large and dark bluish-green leaves and fragrant light pink flowers being produced in great profusion over a long period; 'Herrera's Pink', imported from Trinidad in 1952 and introduced in 1960 to the horticultural trade is an upright shrub loving shade, bearing quite large and light green leaves and the flowers are dark rose similar to *I. rosea; I. hydrangeaeformis* (syn. 'Trinidad Pink'), a fast growing, shade-loving, cold-tolerant and tall but compact shrub most suitable for hedging was imported in 1952 from Trinidad and has since been growing in south Florida for its pink flowers similar to 'Henry Morat'; 'Kelly Gent' (*I. acuminata* × 'Singapore No. 1') was developed by Swedroe in about 1964, is profuse seasonal bloomer with beautiful orange flowers at opening but turning salmon with age; 'Pinksing' (*I. acuminata* × 'Singapore No. 1') was developed by Swedroe around 1960 with slightly fragrant rich pink flowers appearing from spring to summer; *I. rosea* was introduced by Edwin A. Menninger through seeds brought into Florida from Madras (India), likes partial shade, is susceptible to nematode and chlorosis and has compact growth with free branching habit, leaves small and dark green, and the flowers are vivid deep rose appearing throughout the growing season; *I. westii* is a natural hybrid recorded in 1882 and was introduced to Fort Lauderadable by Swedroe in 1964 from Calcutta, flowers opening to pale-rose but turning to bright rose afterwards; in **reds** are *I. chinensis* discovered in South China around 1822, has dwarf and compact growth with good cold hardiness, very variable plant with white, yellow, pink and orange forms, so much used by English hybridizers, mostly free-flowering with orange-red flowers; *I. williamsii* (syn. 'Trinidad Red'), a free-flowering plant is highly amenable to severe trimming in late winter, bears very large and dark green leaves and the large flowers are the deepest of the reds; 'Superking' (a *I. macrothyrsa* cultivar), a large bushy plant introduced to the South Florida trade at about 1950 by Swedroe though was already available there since 1932, probably most spectacular of all the ixoras, bearing large brilliant red flowers being produced throughout the year in Florida, leaves very large, up to 25 cm long, new ones usually pale-green but turn dark green as they mature; and with **variegated leaves** are 'Variegated', appearing to be a mutant cultivar of *I. coccinea* in which the majority of the leaves show patches of non-green (white) sectors (Rajaseger *et al.*, 1999). Dwarf cultivars are also available like *I. compacta 'Sunkist'* growing to a height of only 60 cm with orange flowers; *I. taiwanensis,* a dwarf and compact flowering shrub. A new variety named 'Diora' which is of dwarf hybrid class of milk white colour is under patent.

Propagation

Whistler (2000) in Puerto Rico when grew its **seeds** in commercial potting mixture, he obtained 70 per cent germination. Seeds germinate well in acidic soil (Holtum & Enoch 1991) if sown during autumn (Randhawa and Mukhopadhyay, 1994). Through seeds also these normally come true to the type but it takes long time in attaining the full growth.

It's generally accepted that perennial ornamental plants are multiplied and propagated through vegetative means such as cuttings, layering or grafting (Elgimabi, 2008). Traditionally *Ixora* is propagated from stem cuttings of mature shrubs and is raised in polybags though rooting in such cuttings is very difficult. The **cuttings** should be taken from young and fast growing tips from matured shrubs (Khan *et al.*, 2004) in summer or spring. Since rooting is a problem, therefore IAA and IBA rooting hormones and bottom heating should be employed for efficient rooting. Annual and semi-mature lignified twigs of middle and lower portion of *Ixora chinensis* when were subjected to mixture of various hormonal concentrations and rooting media, it was recorded that sand and clay in half medium was best among the various media used and the mixture of IBA 75 mg^{-1}+ NAA 75 mg^{-1} recorded highest root rate and root number (Sheng-Jian and Jun-Bang, 2011). When cuttings of *I. chinensis* were treated with various concentrations of NAA, IBA and NAA + IBA, the latter at 1,000 ppm recorded best rooting as compared to NAA or IBA treatment alone (Jinsui Lin, 2009). The 2 cm bases of *Ixora chinensis* cuttings taken from October to February were when dipped for 15 seconds in solutions of IAA, IBA or NAA, each at 1,000-6,000 ppm, highest rooting (80-86 per cent) was obtained with cuttings taken in October and treated with IBA at 2,000-6,000 ppm though those taken in January showed poorest response (Singh, 1980). The soilless media cocopeat and teak sawdust produce better root growth and leaf formation (Adetimiria *et al.*, 2008). Taking plant ash as basic culture material and 800 ppm IBA treatment,the *Ixora duffii* cv.'Super King' cuttings presented the best rooting with highest root number. Summer propagation gives the best performance in number and length of both roots and leaves of the cuttings. Plastic tunnels with water mist also give the best rooting and vegetative growth followed by cuttings planted under plastic tunnels without mist, irrespective of season. The control treatment without plastic cover has been found giving the poorest results (Elgimabi, 2008).

There are some difficulties in growing a good quality flowering plant in various agroclimatic conditions. Foliage chlorosis caused by mineral deficiencies, nematodes and other factors often limit the use of many at the showy varieties in the landscape. In effort to minimize the damage caused by these factors, trials on various rootstocks with **grafting** were conducted. It was discovered as early as 1946 that *Ixora parviflora* rootstock gives resistance to nematode infestation, chlorosis club rot and cold damage, *vis-à-vis* provides great vigour to the scion species and varieties. Cleft grafting can be successfully performed with unrooted or rooted cuttings of *I. parviflora* (Mouly *et al.*, 2009).

Propagation through cuttings is not economical since the collection of stem cuttings leads to arrest of growth and development of the mother plant. Moreover, the market demand for propagules is hardly met with such cuttings. Mass propagation for commercial cultivation requires a simple, economical, rapidly multiplying and highly reproducible protocol employing tissue culture technology. Since sexual propagation is time consuming and labour oriented (Lakshmanan *et al.*, 1997) and cuttings being difficult to root, therefore a more economical, rapid and an elaborate means of propagation in *Ixora* is desired in order to meet the ever-rising market demand of this beautiful plant (George and Sherrington, 1993), and the only alternative left with us is micropropagation. The callus culture was initiated from leaf (explants) of *Ixora* taken from expanding young leaves in MS salt medium containing various concentrations of 2, 4-dichlorophenoxy acetic acid (2, 4-D) where it was found that callus initiation was significantly higher in MS medium having 2, 4-D at 3 mg/l. Direct root hair formation occurred from callii induced on 2, 4-D at 1 and 2 mg/l. The best combination which showed excellent multiplication was both, *i.e.* woody plant medium (WPM) and MS supplemented with gibberellic acid at 2.5 µM (Noreen *et al.,* 2001). Nodal segments were found to be best explants for axillary shoot formation on MS medium with 0.5 mg/l BA + 0.1 mg/l NAA. There was rooting on half strength MS medium having 0.1-0.5 mg/l of IAA, NAA or IBA (Amin *et al.,* 2002). *Ixora singaporensis* when cultured in MS medium with $AdSO_4$ (10 mg/l) gave maximum shoots per explant, and when sub-cultured in B_5 medium with IBA 2.0 mg/l, it gave highest number of roots. The shoot tip explants were cultured in WPM supplemented with 0.5 mg/l BAP and incubated for 10 weeks initially and then sub-cultured onto fresh medium after every 4 weeks. The proliferating clusters were cultured in WPM basal medium fortified with peptone (40 mg/l) and 3 per cent sucrose for elongation of shoots, formation of root initials and further proliferation of axillary shoots. The most suitable medium for rooting was half-WPM enriched with IBA at a concentration of 0.05 mg/l (Khan *et al.,* 2004).

Cultural Practices

Basically *Ixora* prefers tropical to subtropical conditions. It grows best in the warm climate and flowers continuously throughout the year though in cooler climates of sub-tropical and sub-temperate regions it will flower during spring to summer and summer months, respectively. Alamu (2012) stated that minimum temperature for its active growth should be 16 °C and maximum 27 °C. In ixora, 33 °C induced the highest leaf carbon assimilation (LCA) (0.40 mg CO_2/m^2 per second) in the greenhouse at 1,200 h, but there were no apparent differences in midday LCA between plants with root zone temperature (RZT) of 28, 33 and 38 °C in the growth room. Effects of RZT and environment on the daily fluctuations of gaseous exchange processes raise questions about using measurements at only one time during the day to separate treatment effects. When root temperature of *Ixora coccinea* was increased from 34 to 40 °C, shoot- root weight ratio increased and shoot sugar -starch ratio decreased. In general, 34 °C appeared to be the optimum root temperature (Ingram *et al.,* 1986). The requirement of humidity by this plant is also high, and the preferable one is generally above 70 per cent. Plants have only moderate drought-tolerance capacity (Ingram *et al.,* 1986). *Ixora* being a tropical plant, prefers full sunlight for its growth and development though at flowering certain species prefer filtered situation to avoid blossom burn. Exposure to greater amount of light could result in compact growth and more flower bud formation. Since almost all the ixoras grow and flower in partially-shaded spots, so it can be grown as a good pot plant for terrace and verandas. These plants like a well-drained, moist and acidic soil that is high in organic matter. While preparing its pits for planting it would be better to fill the pits with $^2/_5$ th part of good garden soil, $^1/_5$ th part of coarse sand and $^2/_5$ th part of well decomposed organic matter in the form of compost, cow dung manure or FYM or peat moss. The preferred pH of the soil should be 5.0-6.5. In fact, the ixora is intolerant to alkaline soil that makes the leaves chlorotic or a sticky yellow-coloured. Use of mulch is required to conserve the moisture. The run-off from concrete may cause the soil to become alkaline. To avoid this problem, *Ixora* should be planted at least a few metres away from concrete structures.

It can be planted in every season except during heavy rains. The land should be fertile with proper provision of watering. The land should be ploughed twice to get good tilth and aeration. Since the plant is maintained at least for 10 years, enough quantity of FYM should be applied. Ridges are formed at 2.5-3.0 m spacing and the plant to plant distance is kept 0.9-1.5 m. However, in pots also almost the same mixture is used but the size should be 30 cm and above looking into the plant size at final growth. Grafts are used for planting. The polythene sheet should be removed before planting. However, at initial period of growth some intercrops such as onion, marigold, coriander, bean, cowpea, black gram, *etc.* can be taken but the weeds will have to be removed regularly and the irrigation should be followed taking into account the main crop, the type of intercrops grown, climatic conditions, soil types and natural rains. In case of drip irrigation, it should be followed on alternate days. Per plant of main crop, 50 g of cake (groundnut, castor, *Pongam*, *neem* or *mahua*) should be applied once in 60 days.

In *Ixora*, suckers and sprouts may grow up to 1 m the first year after which the growth slows down. Individual stems live at least for 15 years but by coppicing and suckering a plant may live indefinitely. Maintenance of ixora in hedges requires frequent **pruning** (Whistler, 2000). Taller growing varieties should be pruned sufficiently after flowering, cutting back by about one-third of the plant. Dwarf varieties don't need anything more than light tip pruning when young to promote bushiness (Ellis *et al.,* 2003).

During the most active growth period, from spring through summer, plants should be **fertilized** once a week though during autumn and winter only once a month. Recommended fertilizer formulation is NPK at 4-8-8 (Keeler and Schoellhornm, 1999). Another option is to alternate an all purpose fertilizer, such as 20-20-20 which can specifically be geared to support bloom production (Gilman, 2003). These are moderate feeders, especially when grown under high light, fertilize once a week using ¼ tsp of fertilizer per gallon of water. Using a balanced fertilizer like a 15-15-15 or a blooming fertilizer like a 7-9-5 of NPK can prove effective (Gilman, 1988). During the winter or when growth has stopped the feeding should be stopped because new growth needs more nutrients that will not be available at that time. Thus more red spots and/or chlorosis, an yellowing of the leaves due to iron or manganese deficiency (another common ailment with ixora) will appear. Use of slow release nitrogen based fertilizer is a better solution for this. The same is the situation even with K as it is not retained in the root zone of sandy soil. Due to P and K deficiencies, purplish leaf spots occur as is also enhanced due to cold snap (Rahuj, 1999). Chlorosis (yellowing of leaves) indicates a high pH in the root zone (alkaline soil). Treatment with iron and manganese reduces this problem. Slow release fertilizer with a ratio of 8N-4P-12K-4Mg + other micronutrients or in the range of 9-11-11 + iron and micronutrients should satisfy ixora requirements (Szilard Paula, 2003). In older plants, 6 grammes of nitrogen per m^2 and in younger plants 12 g of actual nitrogen per m^2 of root zone for the first two years should be applied. Organic manure like cassava peel, organic compost and tithonia manure increases number of branches, internodal length and plant spread (Alamu, 2012). For regular feeding, a 9-11-11 NPK fertilizer first in February, second in May and the last in August should be applied each time with 112 g per plant (Dickey, 1977). Mulching in this will eventually contribute to the build-up of organic matter in the root zone and improves nutrient availability (Thomas and Gilman, 1991). More organic matter will also reduce nematode population that can also plague older varieties of ixora. Often, the plants are grown in containers too long which leads to a compromised root system causing their girdling and thus reducing the nutrient absorption rate (Gilman, 1988). *Ixora* does not like excess **watering** though moist condition is conducive for its proper growth and waterlogged conditions are injurious as root rot occurs.

Growth and Development

Application of GA_3 at 100 ppm on shoot cuttings produced higher number of leaves, leaf area and dry weight per plant in ixora (Fagge and Manga 2011). Trimtect® (8 per cent paclobutrazol as active ingredient) when was applied to trimmed ixora, it reduced the regrowth by 30-70 per cent over a season, produced very compact plants and the leaves produced were darker green (Idun *et al.*, 2011).

Post Harvest

Ixora flowers remain fresh for a week on the plant and up to even three days it does not dry up when severed fresh from the plant. The flowers can be plucked even two days later of the opening. For a few initial years the yield may be less but in subsequent years it goes on increasing up to 5-6 years. Usually it starts blooming after six months of planting and then flower plucking may start on weekly basis or up to two days of opening and this way one average plant can yield up to 5 kg of flowers per plucking which is sold in the market at the rate of Rs. 50-80/kg. Those having red flowers have good demand in the market. In its flower harvesting the ant is a nuisance as this collects honey from the flowers. *I. chinensis* has a longer vase life which makes it a potential cut flower (Chen and Huang, 1996). A new tool, 1-methylcyclopropene (1-MCP) extends the shelf life and quality of cut flowers. Its cut flowers treated with various concentrations of sucrose, KH_2PO_4, 8-HQ and citric acid increased the vase life, the best being 2 per cent sucrose + 150 mg^{-1} KH_2PO_4 + 120 mg^{-1} 8-HQ + 30 mg^{-1} citric acid where the vase life was recorded of six days (Luan-Mei *et al.*, 2010). STS solution spraying at 0.3 to 0.4 mmol/l 10 days before storage prevents or reduces flower falling (Luan-Mei et al., 2010) caused by ethylene. For the antioxidant property it can be harvested and stored in air-tight desiccator at 30 °C (Ratnasooriya *et al.*, 2005).

Insect-pests and Diseases

Anthracnose (*Colletotrichum* spp.) attacks various evergreen and deciduous shrubs and trees, including dwarf ixora. This disease occurs in humid weather and causes black, tan or red spots, yellowing, falling and wilting of leaves, and blackish canker on the stems. Plants should be watered only when needed and that too through sub-irrigation system and overhead irrigation will help dispersal of disease through wet leaves. Infected branches and leaves during autumn and winter should be removed. Once infection has spread, no fungicide will be effective. **Leaf spot** is caused by *Pseudocercospora ixoricola* and many other pathogens. This shows discoloured spots, raised areas or brownish yellow patches on the leaves of infected plants. In severe cases, leaf fall may occur but the long-term damage is rare. The symptoms of leaf spot disease are sometimes confused with signs of a bacterial pathogen or insect infestation. If the ixora is healthy and vigorous, the plant can tolerate leaf spot disease with no lasting damage. The disease can be kept under check by over-watering and sanitation. This disease can be prevented by spraying mancozeb 750 WB at 2-4 kg a.i./ha during spring. **Mushroom root rot** (*Armillaria* spp.) infects vascular tissues and once the infection is noticed, the cambium is already damaged and therefore roots also die. Aerial symptoms include leaf dropping, branch dieback, discoloured stems and cankers. As the fungal disease progresses, mushroom clusters will begin to grow at the base of the plant. Mushroom root rot is fatal problem for plants and occurs in soggy soils with poor drainage. The best way to avoid infection of this fungus is to plant the specimens in well-drained soils. Dead or dying plants should be removed with as much root system as possible and the soil should be replaced or sterilized before replanting. Dithane M-45 at 300 mg/l and Ridomyl MZ 500 mg/l drenching may prevent this disease. The black mess is a fungus called **sooty mould** (*Acetomycetes* fungi) which spreads through sap sucking insects such as scales, mealy bugs and aphids. They prefer the shady north side of the house. After feeding, the insects excrete their waste that drips down onto parts of the plants which attracts the black mess fungus. Products containing Imidacloprid will give long-lasting control. Horticultural oil, usually a 2 per cent mixture of a paraffinic mineral oil may also be effective. This fungus can also be washed away by using light insecticidal oil. Most often, however, unaffected new leaves replace the sooty mould affected leaves but even then to check further spread the insects will have to be controlled. A systemic soil drench product called Bayer Advanced Garden™ Tree and Shrub Insect Control (contains 1.47 per cent Imidacloprid) is reported to give 4 to 12 months residual effects, depending on the organic matter content of the soil.

Ixoras growing in calcareous sandy soils show an intense reddish leaf spot which is a **physiological disorder**. Symptoms appear as irregular diffuse brownish-red blotches on slightly chlorotic but oldest leaves. The symptoms of necrotic spotting appear sharply and prominently on the old leaves of potassium deficient ixoras and infected shoots are able to retain only a few leaves, however, phosphorus-deficient plants though show no spotting but exhibit uniformly brownish-red older leaves and olive-green younger foliage. Plants deficient in both elements display symptoms similar to those observed on landscape plants (Broschat, 1989). Iron and manganese deficiency causes light green to yellow leaf chlorosis with veins remaining dark green. A slow release fertilizer in the range of 9-11-11 + iron and micronutrients should therefore be applied. These problems can be overcome through application of nitrogenous fertilizer as defined under **cultural practices**, organic matter in the root zone to reduce nematode population plaguing older plants, and soil applications of Sequestrene 138 or foliar applications of Sequestrene 330 to alleviate iron and manganese chlorosis.

There have been found only sucking type of insects on ixoras. **Scale insects** will not be dislodged even when dead, **mealy bugs** are usually very white in colour and **aphids** are often green and pear-shaped. Horticultural oil spray, weekly for five weeks in the afternoon can control these insects. **Nematodes** can be controlled with Nemacur or Furadan at 1 g/m^2 followed by irrigation.

References

Adetimiria, V.O., S.K. Kim and M. Szezeeh, 2008. Factors associated with emergence of ornamental flowers. *Sri Lanka J. agric. Sci.*, **144**: 68.

Alamu, L.O. 2012. Enhancing environmental management through a luxuriant vegetative improvement of *Ixora coccinea* by means of organic manuring. *Intern. J. Acad. Res.in Business & Social Sci.*, **2**(6): 252-259.

Amin, M.N., A. Shahrear, S. Sultana, M.R. Alam and M.A.K. Azad, 2002. *In vitro* rapid clonal propagation of an ornamental plant ~ *Ixora fulgens* Roxb. *Online J. boil. Sci.*, **2**(7): 485-488.

Bedi, Y.S., S.S. Bir and B.S. Gill, 1981.Cytopanology of woody taxa of family Rubiaceae from north and central India. *Proc. Natl. Sci. Acad.*, **6**: 708-715.

Broschat, T.K. 1989. Potassium deficiency in south Florida ornamentals. *Proc. Fla St. hort. Soc.*, **102**: 106-108.

Chen, L.Y. and M.C. Huang, 1996. Effect of shading and winter heating on growth and flowering of Chinese ixora (*Ixora chinensis*). *J. agric. Asscn. China* (New Series), **174**: 72-82.

Deb, D.B. and R.C. Rout, 1992. Two new species of *Ixora* (Rubiaceae subfam. Ixoroideae) from India and Burma. *Kew Bull.*, No. 47, pp. 295-300.

Dickey, R.D. 1977. Nutritional deficiencies of woody ornamental plants used in Florida landscapes. *Co-op. Extn. Bull.*, p. 791. University of Florida, USA.

Elgimabi, M.E.N.E. 2008. Effect of season of cutting and humidity on propagation of *Ixora chinensis*. *Adv. Boil. Res.*, **2**(5-6): 108-110.

Ellis, W.O., L. Atuah, P. Kumar and J.A. Bakang, 2003. An overview of the floriculture industry in Ghana. *Ghana J. Hort.*, **3**: 47-56.

Fagge, A.A. and A.A. Manga, 2011. Effect of sowing media and gibberellic acid on the growth and seedling establishment of *Bougainvillea, Ixora coccinea* and *Rosa chinensis*. 2. Root characters. *Bayero J. Pure & Appl. Sci.*, **4**(2): 155-159.

Fulghum Leigh, 2008. Flowering shrubs in Florida ~ 'Frankie' and 'Nora' *Ixora*. *Daily Post in Floridaplants.com*. United States Patent PP09200.

George, E.F. and P.D. Sherrington, 1993. Plant propagation by tissue culture, 709 p. Exegetics Ltd. Press.

Gilman, E.F. 1988. Tree root spread in relation to branch drip line and harvestable root ball. *HortSci.*, **23**(2): 351-353.

Gilman, E.F. 2003. *Ixora coccinea*. University of Florida Cooperative Extension Fact Sheet FPS-291.

Holtum, R.E. and I. Enoch, 1991. Gardening in the Tropics. *Times Publication* (Singapore), **6**:134-137.

Husain, T. and S.R. Paul, 1991. *Ixora manantoddii*, a new species of *Ixora* L. (Rubiaceae - Pavetteae) from India. *Bull. Du Jardin bot. Belgium*, No. 61, pp. 15-19.

Idun, I.A., P.A. Kuah and H.V. Adzraku, 2011. Rooting and vegetative growth response of difficult to root *Ixora coccinea* and *Ficus benjamina* cv. 'Straight' to different stem cutting types and soilless media. *African J. Pl. Sci.*, **5**(3): 773-780.

Ingram, D.L., C. Ramcharan and T.A. Nell, 1986. Response of container-grown banana, ixora, citrus and dracaena to elevated root temperatures. *HortSci.*, **21**: 254-255.

Jinsui Lin, 2009. Effect of plant growth regulator on rooting and cutting of *Ixora chinensis*. *Chinese Agr. Sci. Bull.*, No. 685, pp. 105-109.

Kaser, H.R. and A.M. Steiner, 1983. Subspecific classification of *Vicia faba* L. by protein and isozyme patterns. *Fabis Newletters*, **7**: 19-20.

Keeler, G. and R. Schoellhornm, 1999. *Ixora* for South Florida. University of Florida IFAS Extension (ENH 955).

Khan, S., M. Iftikhar and B. Saeed, 2004. An economical and efficient method for mass propagation of *Ixora coccinea*. *Pakistan J. Bot.*, **36**(4): 751-756.

Lakshmanan, P., C.L. Lee and C.J. Goh, 1997. An efficient *in vitro* method for mass propagation of a woody ornamental *Ixora coccinea*. *Pl. Cell. Rep.*, **16**: 572-577.

Latha, L.Y., I. Darah and S. sreenivasan, 2012. Pharmacological screening of methanolic extract of *Ixora* species. *Asian Pacific J. Trop. Biomed.*, **2**(2): 149-153.

Luan-Mei, L.U.L., J. Lin and M.X. Zhi, 2010. Effects of different preservative solution on cut flowers of *Ixora chinensis*. *Acta Hort. Sinica*, **37**: 1351-1356.

Mouly, A., S.G. Razafimandimbison, A. Khodabandeh and B. Bremer, 2009. Phylogeny and classification of the species-rich pantropical showy genus *Ixora* (Rubiaceae-Ixoreae) with indications of geographical monophyletic units and hybrids. *Amer. J. Bot.*, **96**(3): 686-706.

Nayak, R., K. Geetha and Priya Davida, 2010. Pollination limitation and effective breeding system on plant reproduction in forest fragment. *Acta Oecologia*, **36**: 191-196.

Nilsson, L.A. 1998. Deep flowers for long tongues, *Trends in Eco. and Evol.*, **13**: 259-260.

Noreen, R., M.A. Khan, M.J. Jaskani and N. Hussain, 2001. Callogenesis and embryogenesis from leaf discs of *Ixora chinensis*. *Int. J. Agri. Biol.*, **3**(1): 65-67.

Pandey, A.K. 2006. Strructure development and reproduction in flowering plants. Ph.D. thesis. T.M. Bhagalpur University (Bihar).

Pradeep, A.K. 1997. *Ixora sivarajiana*, a new species of Rubiaceae from India. *Nordic J. Bot.*, **17**: 315-317.

Rahuj, B. 1999. *Ixora. J. hort. Sci. & Ornam. Plants*, **2**(2): 78.

Rajasegar, G., H.T.W. Tan, I.M. Turner and P.P. Kumar, 1997. Analysis of genetic diversity among *Ixora* cultivars (Rubiaceae) using random amplified polymorphic DNA. *Ann. Bot.*, **84**: 253-257.

Rajasegar, G., H.T.W. Tan, I.M. Turner, L.G. Saw and P.P. Kumar, 1999. Random amplified polymorphic DNA variation among and within selected *Ixora* (Rubiaceae) populations and mutants. *Ann. Bot.*, **84**:253-257.

Randhawa, G.S. and A. Mukhopadhyay, 1994. Floriculture in India. Allied Publishers Pvt. Ltd., New Delhi.

Ratnasooriya, W.D., G. Galhena, S.S.P. Liyanage and J.A.C. Jayakody, 2005. Antinociceptive action of aqueous extract of the leaves of *Ixora coccinea*. *Acta Biologica Hungarica*, **56**: 1556.

Sharma, A.K. and T. Chatterjee, 1960. Chromososme studies in ixora. *Genetics*, **31**: 421-447.

Sheng-Jian, M.A. and Y. Jun-Bang, 2011. Effect of hormone mixture and matrix on the rooting of *Ixora chinensis*. *Heiloingjian Agric. Bull.*, No. 685, pp. 99-105.

Singh, S.P. 1980. Response of varying concentrations of auxins to rooting of *Ixora bandhuca* cuttings during winter under intermittent mist. IV. *Progressive Hort.*, **12**: 21-25.

Sivadasan, M. and N. Mohanan, 1991. *Ixora agasthyamalayana*, a new species of Rubiaceae from India. *Bot. Bull. Academia Sinica*, **32**: 313-316.

Szilard Paula, 2003. Ixoras for year round enjoyment. University of Florida, Cooperative Extension Fact Sheet 5/2/2003.

Thomas, H.Y. and E.F. Gilman, 1991. Fertilization recommendations for trees and shrubs in home and commercial landscapes. University of Florida, Cooperative Extension Circular No. 948.

Whistler, W.A. 2000. Tropical Ornamentals ~ A Guide, 542 p. Timber Press, Portland, Oregon, USA.

Zaundam, D. 1997. Ixora Plant Diora. United States Patent PP09814.

27

Lathyrus (Family: Fabaceae)

R.K. Dubey, Simrat Singh and Sanyat Misra

[**Common names**: Beach pea/Sea or Seaside pea (*Lathyrus maritimus*), Black pea/Black bitter vetch (*L. niger*), Everlasting pea (*L. grandiflorus, L. latifolius*), Everlasting pea/Sweet pea (*L. odoratus*), Lord Anson's blue pea (*L. magellanicus*), Marsh pea/Wing-stemmed wild pea (*L. palustris*), Perennial pea/Wild pea (*L. latifolius*), Persian everlasting pea (*L. rotundifolius*), Prairie vetchling (*L. polymorphus*), Pride of California/Campo pea (*L. splendens*), Spring vetch/Spring bitter vetch (*L. vernus*), Tangier scarlet pea (*L. tingitanus*), Tuberous pea (*L. tuberosus*), Tow-flowered pea (*L. grandiflorus*)]

Introduction and Origin

Lathyrus, named so by Theophrastus, derives from the Greek *lathyros* meaning 'a pea'. It is a genus of 100-150 species of hardy flowering annuals, sub-shrubs and herbaceous perennials, most being the climbing plants supporting themselves by means of tendrils these produce. *Lathyrus* is native to northern temperate zone, mountains of Africa and South America. Its area of distribution extends throughout Northern Europe (excluding the coldest zones), Asia, the Americas and the mountains of tropical Africa but always in the temperate regions. *L. odoratus*, a best known annual species native to the eastern Mediterranean region from Sicily east to Crete (Rice, 2002), has numerous garden varieties. It was first described in 1696. When Queen Anne reigned the Great Britain, in 1699 the seeds of *Lathyrus odoratus* were sent from Palermo (Sicily) by Father Francisco Cupani to his friend Dr. Uvedale, a surgeon living near London and who was a famous enthusiastic gardener there at that time. About 20 years later in London flower market when one day this climbing plant reappeared bearing strongly fragrant red and purple flowers, it was named sweet pea, and in the market its seeds were available by 1730. In England in the year 1800, five fragrant sweet pea colours appeared, by 1837 there had been only six colours, but after 1870 Henry Eckford began to select seedlings and succeeded in enlarging the size of their flowers and the range of colours, and so giant-flowered types started appearing, out of which one var. 'Primodonna' gave rise to a separate class with frilled standard petals as mutation in 1901 which paved the basis for development of the modern strain of 'Spencer sweet peas'. Eckford produced a number of excellent varieties with larger and more impressive flowers than the original species *Lathyrus odoratus* which are named as 'grandifloras' or 'grandiflora sweet peas' though when compared with the modern standards the size of these grandiflora varieties is quite small so they are now frequently listed as 'heirloom', 'antique' or 'old fashioned' varieties. However, these varieties are famous for their strong spicy fragrance and come in a huge range of colours and patterns. Blue, lavender, mauve, white, orange, salmon and pink colours were by now developed in the varieties. Father Gregor Mendel, born in 1822 of Austro-Silesian parents, who early in his life entered the Augustinian monastery at Brünn (Bruno), experimented this plant and published the results in 1865 which founded the basis of moderrn science of heredity though his work got recognition only in 1900, 16 years after his death. Sweet pea cultivation is popular in the United Kingdom, United States, Russia, Egypt, New Zealand, Australia, Japan (Hambidge, 1996) as well as India. There are some reports of commercial sweet pea cut flower production, however, the quantity of cut flower production worldwide is unknown due to lack of statistical data available (Hammett, 2006). The majority of sweet pea seed is produced in California (USA) and New Zealand (Hambidge, 1996; Rice, 2002). *Lathyrus* contains the alkaloid lathyrin which if absorbed over a long period, especially under cold and damp climates, causes symptoms of paralysis in the lower limbs (Pizzetti and Cocker, 1968). The root exudates have been reported to contain ñ-hydroxybenzoic acid (Asao *et al.*, 2007). The edible tubers of *Lathyrus tuberosus* have high sugar content and a flavour similar to that of chestnuts, and the flowers were used for extraction of essential oil during sixteenth century but was later superseded by *L. odoratus*.

These are planted in pots, in containers, as dividers of the beds, in the herbaceous borders, in the lawn, in the beds, along the paths in front of the house and for cutting but in the open or in the semi-shade conditions and these are kept erect through stakes and strings, and wherever these are sown after flowering emit a sweet fragrance in the surroundings.

Botany and Breeding

Lathyrus odoratus grows to a height of 1–2 m, with the help of their tendrils. Its roots have nodules containing colonies of the bacterium *Rhizobium*. The stems are angular or winged and not circular in cross-section. The leaves are alternate, mainly pinnate with two leaflets where a terminal leaflet of the climbing types is modified to a tendril, which twines around supporting plants and structures helping the sweet pea to climb. The axillary racemes of pea flowers in various colours (white, orange, pink, mauve, lavender, violet and purple but not yellow), especially purple in the original species, 2.0-3.5 cm broad, appear in the wild plant but larger and very variable in colour in many of the cultivars. Sepals are narrower not leaf-like and the style is flattened and not reflaexed at the margins. Tendrils, adapted for clasping and gripping nearby objects, thus help to support the plant growing erect. Usually a sweet pea leaf has one pair of leaflets and several pairs of tendrils, but if the plant is growing vigorously there may be two pairs of leaflets and fewer tendrils. Not all sweet peas have tendrils; the 'acacia leaf' varieties have pinnate leaves with several pairs of leaflets and no tendrils. *Lathyrus odoratus* seeds vary considerably in colour, size and shape. This species is diploid with $2n = 14$. The yellow flowered species such as *L. annus, L. chloranthus* and *L. hirsutum* (all annuals) are also diploid though perennial species, *L. pratensis* has diploid as well as tetraploid cytotypes. Even attempts to cross *Lathyrus odoratus* with *Lathyrus belinensis* (yellow) failed to produce hybrids so yellow flower in sweet peas still remains a dream. Chemical analysis of *L. belinensis* confirms three pigment types, *i.e.* carotenoid, flavonoid and anthocyanin. A hybrid was established by means of embyo rescue which showed very similar karyotypes to its parents, and the meiotic pairing in the hybrid indicated considerable homology. The hybrid between *L. belinensis* and the cream-flowered 'Mrs. Collier' (a cultivar of *L. odoratus*) showed novel flower colour but with reduced vigour. Crossing of *L. odoratus* with *L. annus* and *L. pratensis* failed to produce seeds though *L. hirsutum* × *L. odoratus* produced hybrids, however, the hybrids produced by crossing *L. odoratus* with *L. chloranthus* was quite distinct from its parents with regard to its vegetative and floral characters.

Classification, Species and Varieties

Sweet peas have been cultivated since 17th century. As defined by various literature sources (Bailey, 1929; Pizzetti and Cocker, 1968; Everard and Morley, 1970; Hay and Beckett, 1971; Beckett, 1983; Brickell, 1994), the noteworthy *Lathyrus* species are being defined hereunder. ***Lathyrus odoratus***, an annual climber is a native of southern Italy and Sicily and the most cultivated species growing up to 3 m in height. Its stems are rough, hairy and winged, leaflets in single pairs and are 2-6 cm long, smooth, mid-green, ovate-oblong to elliptic, mucronulate and ending in a tendril, stipules lanceolate, peduncles much longer than the leaves and with 2-4 intensely fragrant flowers each about 2.5 cm long, and in the shades of dark red and purple though its various varieties are available in blue, purple, red, pink and white. Most varieties produce flowers suitable for cutting. Its 'Galaxy strain' continues flowering for several months with more number of flowers per stalk and longer stalk length as compared to quite short flowering period of the original species. 'Cuthbertson Floribunda varieties' include 'Jenny' (white) and 'Robert' (mid-blue) which grow up to 2.4 m in height bearing 5-7 flowers per stem. 'Multiflora Gigantea' includes 'Ramona' (orange) and 'Colorama' (with a mixture of colours) which grow up to 2.4 m in height with each stem carrying 6-10 flowers some 5 cm across. The 'Spencer varieties', the most popular of all, grow up to 3 m tall bearing 4-5 blooms on each long stalk. Its popular varieties are 'Air Warden' (scarlet), 'Carlotta' (carmine), 'Gertrude Tingay' (lavender), 'Lady Diana' (pale-violet-blue), 'Leamington' (lilac), 'Noel Sutton' (blue-purple), 'Princess Elizabeth' (cream flushed salmon-pink), 'Red Ensign' (rich scarlet), 'Royal Flush' (cream-pink), 'Selana' (large white flowers flushed pink), 'Sonata' (salmon-pink), 'Spotlight' (ivory flushed pink), 'Stylish' (blue), 'Swan Lake' (white), 'Tell Tale' (white edged pink), 'Xenia Field' (pink and cream), *etc.* 'Bush type dwarf varieties' are 'Americana' which bears long-stemmed flowers and grows about 60 cm tall; 'Bijou' in mixture of pink, red or blue or in pure colours grows up to 45 cm tall with 5 cm across flowers which continue blooming longer as it is weather-resistant; 'Colour Carpet' grows up to 25 cm tall bearing three flowers per stalk though shy bloomer in wet weather and is most suitable for growing in rock gardens and edgings; and 'Knee-Hi' (pink, red, blue or white) grows up to 1.2 m tall, bearing 5-10 blooms per stalk, and may require support. ***L. tingitanus*** from Mediterranean region which was introduced in 1680, is a half-hardy annual pea growing up to 1.8 m in height and is most suitable for garden cultivation in temperate areas. It is a beautiful species but could not become popular in the northern Europe due to its doubtful hardiness though in the southern regions it is quite choicest plant. Leaves with only 1 pair of leaflets are mid-green, narrow, lanceolate and pointed and these terminate into a 3-4-fid tendrils, each branch being quite long. The flowers are bicoloured (deep purple and bright red), up to 3.75 cm across and are borne in a 2-flowered raceme. Its var. *roseus* bears pink flowers.

Following are the **perennial** species. ***L. cirrhosus*** from Pyrenees is a very vigorous perennial species growing up to 1.8 m in height, bearing dark green leaves, each with 2-3 pairs of leaflets. Floral racemes are some 30 cm long and bear about 1.9 cm across rose-purple flowers. ***L. grandiflorus***, an introduction to England in 1814 from Southern Europe, a hardy perennial climber growing up to 2 m tall, is a spectacular species with large-size unscented flowers which are carried 2-3 per stalk over a long season. The flowers are a rich magenta with a deep red keel. Stipules are narrow and the leaves are typical of the annual sweet pea with only 1 pair of leaflets and normally terminate into a trifid tendril with short branches. Seed is rarely produced in cultivation and is not commercially available. It is, however, readily propagated from its spreading tuberous roots. ***L. latifolius***, an European vigorous perennial climber is hardiest, most commonly cultivated and widely naturalized in USA. It is most suitable for trellis work, for training over walls, porches and for growing at rough places. It grows up to 3 m in height bearing broad stipules and two dull to dark green, ovate to lanceolate and mucronate leaflets some 4-10 cm long with multi-branched tendrils, each branch being quite long. Peduncles are 25-30 cm long, many flowererd (5-15) and longer than the leaves, and the flowers are non-fragrant, some 2.5 cm wide and rose-purple or white. It bears flat and 10-13 cm long pods. Its varieties *albus* (white), *roseus* (deep pink), *splendens* (dark purple and red), 'Snow Queen' (white), 'White Pearl' (white) are very popular. ***L. luteus*** (syn. *L. montanus, Orobus luteus*) from Europe is a hardy non-climbing perennial of low spreading habit with angled, smooth and simple stems, growing up to 45 cm tall with large orange-yellow flowers, a rare colour found in perennials. Its leaves are dark green, 3.75-5.0 cm long and divided into 5-8 pairs of small, elliptic-lanceolate and pointed leaflets but without tendrils. It is most suitable in mixed borders and rockeries, and produces an abundance of small pea-like orange-yellow flowers appearing in the erect and spike-like clusters. Its var. *aureus* bears yellow and brown flowers. ***L. magellanicus*** (syn. *L. nervosus*) from Straits of Magellon is woody, strong-growing and almost evergreen species covered with a bluish bloom. Its growth is accelerated due to maritime salts as it grows comfortably under such situations. Its stems are smooth, angled, slightly branched and up to 1.5 m long. Grey-green leaves each have a pair of leaflets which are ovate to oblong-linear and tendrils are 3-branched. Stipules are cordate-sagittate and wide while peduncles are long with 3-4 fragrant and deep purple-blue flowers. ***L. ornatus*** is an unusual non-climbing perennial species with glabrous, often glaucous and occasionally branched stems where paired leaflets are linear-hairy, rigid and strongly veined, stipules are entire and peduncles are longer than the leaves. Racemes are usually 4-flowered and the flowers are very showy, large and purple. ***L. pubescens***, a native of Tropical South America from Valparaiso to Chiloe in Chile is a perennial branching shrubby species with 4-angled stems. Stipules are broad and the leaves rarely with two pairs of leaflets, and terminate into a stout trifid tendril. ***L. rotundifolius*** from West Asia, Crimea and Russia is a fully hardy low-growing (only up to 80 cm tall) perennial and winged species whose dentate stipules are narrow and the leaflets are two (1 pair), ovate-orbicular to elliptic which terminate into 3-branched tendril, 2.5-6.0 cm long, and peduncles are longer than the leaves with many (3-8) large rose-pink to purplish racemose flowers, each about 2 cm across. It can be grown in a cool, shady, sheltered and stony positions. ***L. splendens***, a native of southern California is an attractive semi-shrubby and slender perennial growing up to 2.1 m in height. Stems are quite strong, the leaves with up to 5 pairs of leaflets, oval-oblong to linear, acute and 2.5 cm in length, stipules are narrow and the flowers appear in the clusters of 6-12 which may vary in colour from pink to violet or reddish-purple, each about 2.5 cm across. Pods are beaked at the end, smooth and some 7.5 cm long. ***L. sylvestris*** is fully hardy, has winged stems and grows up to 2 m tall. Leaves have narrow stipules, a pair of leaflets and a terminal branched tendril. Racemes are small with 4-10 rose-pink flowers. ***L. tuberosus*** from Europe grows up to 90 cm in height and has invasive but tuberous roots. Sometimes the roots are used for culinary purposes which taste like those of Jerusalem artichoke. It is a perennial species bearing paired light green and elliptic leaves and brilliant pink fragrant flowers. Once established, it is very difficult to eradicate. ***L. vernus*** (syn. *Orobus vernus*) from Central and South Europe is a fully hardy, quick-growing and clump-forming erect perennial some 30-60 cm tall with simple and somewhat pubescent stems, bearing entire stipules, 2-4 pairs of light green, ovate to lanceolate and acuminate leaflets (sometimes only 1 pair), peduncles shorter than the leaves and racemes of 3-10 nodding red-purple and blue flowers some 1.3-2.0 cm long where keels are shaded with green. When in flower the species is quite spectacular with proper contrast of green leaves and purple-blue flowers which appear in plenty. It is highly difficult to transplant. *L. v.* 'Albo-roseus' (syn. *L. cyaneus, L. digitatus*) bears fern-like soft and much divided leaves, and pink and white flowers. Many other species of *Lathyrus* are distributed in both the old and new worlds. Several are widely cultivated, and others are limited for their use as a garden plants as they become more widely available.

The **acacia-leaf sweet peas** are the reversion to the primitive ancestral form where the leaflets are linear and are not mutated into tendrils. These have much denser foliage on compact bushy plants as to that of *Lathyrus ornatus*.

Between 1880 and 1910, it was the **grandifloras** which first gave the sweet pea its popularity as a garden flower. Though their growth is informal but are adapted to a wide range of growing conditions and generally produce highly scented flowers. **Spencer** or **English sweet peas** are the modern large-flowered strains. These are available in a very wide range of colours, many of which are very sweetly scented. The Spencers are the standard exhibition sweet peas, and are also the most popular for providing cut flowers from the garden in temperate areas. **Early-flowering sweet peas** were originally developed through chance mutation as a winter cut flower for the Christmas market but these are less vigorous than the later-flowering sorts, and require a good winter light levels for best results. Over the time, they have differentiated into the winter- and spring-flowering strains of cut flower sweet peas. **Intermediate sweet peas** grow only 90-120 cm tall, requiring less support than the taller ones and are most suitable in the mixed border. Contrary to popular belief, these sweet peas are not the result of a cross between tall and dwarf varieties. The original dwarf sweet pea found in 1893 by C. C. Morse & Co. of San Francisco (California) grew only about 15 cm high with spreading habit and had white flowers similar to its parent, and so was named as **'Cupid'**, which subsequently became the common name for all the dwarf sweet peas. In 1895, another dwarf sport was found in the variety 'Blanche Ferry' and this was the same pink and white **'Pink Cupid'** which is popular even today. Though most of the cupids developed through spontaneous mutation some 100 years ago but crossing the cupids with existing grandiflora varieties yielded cupids in a wide range of patterns and colours. Subsequently the cupids were crossed with the new 'Spencer' sweet peas' to produce a strain of large-flowered cupids. Cupids were also crossed with the early or winter-flowering strains of sweet peas to give a race of early flowering cupids. These came into flower about a month earlier than normal, and were considered to have great potential for use in areas with warm winters. Dwarfs require plenty of sunshine, well-drained soil and make excellent specimens for tubs and hanging baskets.

Various types and varieties of sweet peas can be classified as '**Fancy**' which are of various patterns such as '**picotees**', '**striped**', '**flakes**', '**marbles**' and the '**bicolours**'. Marbles have patches of different colours on the background of ground petal colours in the form of marbling. Picotees may be normal or wire rim type while the bicolours include the recently evolved reverse bicolours. '**Picotee**' is the simplest form of '**fancy sweet peas**' having a thin line (very narrow clearly defined line in the form of wire rim) or band of colour (a broader more diffused band) round the edge of each petal. The '**stripe**' pattern is quite distinctive in that the pigmentation is predominantly on one face of each petal. Spots and stripes of pigment radiate from a central blotch, and each petal exhibits a strong picotee edge. '**Flakes**' have pigments equally distributed on both surfaces of the petals contrasting with white or cream base colour. **Clear true blues** are rare in sweet pea flowers though tinge of pink is most common, however, true blues are also susceptible to weather damage and the *Botrytis* infection also leaves clear white spots on petals. The grandiflora variety, 'Flora Norton' is one of the clearest blues available. **Cream sweet peas** are most valuable as it blends well with most other colours. The Spencer variety 'Jilly' has long been the mainstay of gardeners and exhibitors as to that of old grandiflora 'Mrs Collier'. **Lavender sweet peas** are the most traditional ones with a wide choice and are more adaptable to cool climatic conditions. **Rose-pink sweet peas** are quite popular, generally vigorous and have a wide range of shades on either a white or a cream background. 'Mrs Bernard Jones' is one of the choicest variety. Good **red sweet peas** are only a few. Crimsons are usually over-vigorous with coarse spikes and badly positioned flowers. Some scarlets develop a blue edge to their petals in hot weather, although this tendency has now largely been eliminated by careful breeding. A few red sweet peas such as 'Restormel' can be relied upon to produce top quality exhibition racemes. **White sweet peas** have always been popular, and there have been many excellent varieties such as grandiflora 'Dorothy Eckford' bred in 1901. It makes a fascinating contrast to modern varieties like 'White Frills'. Older varieties tended to be white-seeded but the most recent sorts are black-seeded. Black seeded varieties are considered easier to grow, but have the disadvantage that flowers may develop a pink tinge in hot weather if the plants are under stress.

'**Mimi**' sweet pea (*Lathyrus odoratus* L.) is a new cultivar in Japan with pink flowers, released by the Miyazaki Agricultural Research Institute (Kaoru *et al.*, 2008) and is also suitable as a cut flower. 'Mimi' produces 145 per cent more marketable flowers than the main pink cv. 'Super Rose'.

Propagation

Annual *Lathyrus* is propagated through seeds by direct sowing at the permanent site or containers in the early spring (March) in the temperate areas of India while before onset of winter (September-October) in the tropical and sub-tropical conditions in the open but under semi-shade conditions, however, after germination the seedling pots are shifted to full sun. Sweet pea seeds have a hard protective coating. Seeds are soaked in lukewarm water for 18 to 24 hours to speed up the germination process. One *Lathyrus* seed is generally sown on the surface of the compost in the middle of the plug pot or 7.5 cm pots or in trays. Care should be taken that the seeds are not bunched up in the middle or to the side of the pots or trays. Sow them evenly over the surface of the compost. After sowing, the seeds are covered with a very light dusting of sieved compost. Seeds are sown in the containers filled with seed compost at a temperature of around 16 °C and when the seedlings attain some 10 cm height these are pinched for precocious branching. Perennial species can also be propagated through seeds the same way as the annual sweet peas. Perennial species such as *Lathyrus latifolius* and *L. vernus* can also be propagated through division of rootstock in March though is not quite successful in other perennial species except the two mentioned above. Sweet peas can also be raised *in-vitro via* axillary shoots (Ochatt *et al.*, 2010).

Generally, each gramme of sweet pea seeds contains some 10 seeds. Pink and picotee spencer varieties tend to have larger seeds, some 8 seeds per gramme, while many lavender and blue varieties have very shrivelled seeds, with 12 or 14 seeds per gramme. Sweet peas are classified as being 'black' or 'white' seeded. 'Black' seeds can be grey or brown, while 'white' seeds are cream or fawn (yelloweish-brown) in colour. White seeds seem to be more prone to rotting under unfavourable conditions, but the flowers of white seeded sweet peas often display clearer colours. Many blue and lavender varieties have mottled seed with darker patches which are slightly sunken. This is entirely normal, and all attempts to isolate a pathogen from such seed have failed. Like the edible garden pea, sweet peas can have wrinkled or round seeds. Sweet peas with striped flowers such as 'Mars' and 'Lilac Ripple' tend to have large wrinkled seeds. The difference is of no consequence to the gardener as both perform equally well. Sweet pea seeds have a 'scar' or hilum which marks the point of attachment of the funiculus, the short stalk which connects the seed to the inside of the pod, or legume. Although the dried funiculus is normally shed by the ripe sweet pea seed, in some varieties it is persistant and remains as a pale crescent-shaped fragment of tissue covering the hilum. This may cause concern to gardeners who think that the seed has started to germinate and that the funiculus is a dried root.

Cultural Practices

Sweet peas prefer cooler temperatures and bloom in the late spring through the summer but in very hot climates flowers will fade in intense heat. Good quality commercially available peat-free compost is ideal for raising sweet peas. Sieved compost devoid of any rubble or concrete is most suitable for its growing provided the soil is rich in humus with **pH** reaction of slightly acidic to neutral, *i.e.* 6.0-7.5, and is well-drained. Hot and dry climate is not suitable for its growing though it feels quite comfortable at a sunny situation in cool and well aerated garden. In rows, the trenches are made well in advance of sowing some 30-40 cm deep, incorporated with some 10-15 cm deep well-decomposed manure and then covering it with good garden soil up to the level, and then only the seeds are sown directly in the trench some 2.5 cm deep and about 5 cm apart from plant to plant and row to row but when the atmospheric temperature is 20-16 °C (day-night). In no case the temperature should fall down 13 °C. When the seedlings have attained 3-true-leaf-size, their tips should be pinched to encourage strong basal shoots. In the pots when basal shoots start appearing, these should be planted gently at their permanent sites or in pots of 7.5-9.0 cm size singly. The plants are adequately supported with **stakes** to climb upon or these may be supported with nylon netting or chicken wire. When flowering begins, the spent flowers with their stalk should be removed immediately to avoid forming the pods, if the seed production is not the aim. During growth but before flowering the plants should be fed fortnightly with **liquid manure** prepared through fresh cow dung, and mixed with little bone meal. In fact, 120 g bonemeal per square metre area is quite encouraging.

Watering should be done with great care. Watering is applied either with sprinkling can or mist sprayer filled with tepid water. Lightly spray the surface of the compost taking care not to disturb the seed. Avoid over-watering as the saturated compost may cause rotting of seed resulting in poor germination. Plants should be gradually hardened off by placing them outside, in a sheltered position, during the day. A cold frame with the lid open can be used for hardening.

Nowak and Rudnicki (1975) when treated sweet pea cv. 'Cuthberson Jimmy' flowers with Proflovit-70 (0.3 g 8-HQS+0.05 g CCC+50 g sucrose/l) and compared with $AgNO_3$ at 20-50 ppm, $KMnO_4$ at 4 ppm, streptomycin at 200 ppm, 8-HQS at 200 ppm, sucrose 5 per cent or CCC 50 ppm, they recorded Proflovit-72 giving best results with regard to **vase life** and flower quality. The effects of STS and 1-MCP were studied by Dole *et al.* (2005) on vase life of cv. 'Winter Elegance' by treating the unpacked cut stems and placing them either in deionized water (DI) and subjected to 1-MCP (740 nl l^{-1}) or ambient air for 4 h or DI + STS at 0.2 mM for 4 h. After treatment, stems were removed, placed in polyethylene sleeves and stored either wet in DI water or dry in plastic-lined floral boxes at 5 °C in the dark for 4 days. After storage, bunches were placed in DI water under 12 h light (76 to 100 μmol $m^{-2}s^{-1}$) per day where it was found that both STS and 1-MCP increased the vase life, more being under STS.

Insect-pests. Diseases and Physiological Disorders

Sweet peas are less susceptible to pests and diseases than many other garden plants. They are most vulnerable as seedlings, before the waxy surface has had time to develop, or when the plants are old or under stress and the natural immune system is breaking down. The most serious problem is **greenfly** (aphids) which can transmit serious virus diseases to sweet peas. The waxy leaf coating, to some eextent, even hinders the absorption of chemical sprays so that contact sprays are necessary. Natural predators like ladybirds, lacewings, *Aphidius* and *Aphidoletes* control the aphid populations. Only under unavoidable circumstances the contact insecticides should be used. **Leaf hoppers** though are minor pests but can transmit virus diseases from one plant to the other, especially from clovers so the patches of clover plants should be eradicated from nearby sweet pea planting or a suitable insecticide should be sprayed. Attack of **leaf miner** is not a serious problem on sweet peas, and most often its infestation is seen on late-planted crops. The larvae enter the epidermis of the leaves and bore the serpentine mines clearly visible to the naked eye. Common species are easily controlled with pyrethroid insecticides. A small **black beetle** feeds on the pollen. They crawl inside the keel of sweet pea causing distortion of flowers. They are most prevalent in areas where oilseed rape is grown, and are attracted to the yellow colour, so yellow sticky traps are useful to ward off the pollen beetle. They are also attracted to light, so placing affected flowers in a dark shed with a bright light source may lure them out. Minute **thrips** (thunderflies), especially the western flower thrips are the pests of many of the glasshouse crops but in the open in the temperate areas when temperature during summer is warm, these sometimes become a nuisance, though is not a serious problem on sweet peas. Contact insecticide will control these.

Slugs have little liking for the tough waxy leaves of mature sweet peas. They do, however, relish the tender growing point of young plants, and can do great harm to plants before they have started to climb. Methiocarbamate-based baits are more effective than metaldehyde-based ones. Nematodes (*Phasmarhabditis* sp.) offer an organic alternative for controlling this pest. **Mice** can devastate a newly sown crop of sweet peas by burrowing in the soil and eating the germinating seeds and sometimes severing the top growth from its roots. Baits and traps are effective means of control. **Sparrows** feed on the flowers and **blackbirds** uproot young plants when searching for food, while hungry **pigeons** will occasionally decimate young seedlings. Bird scarers should be used to scare them in the earlier stages of seedling growth.

Sweet pea leaves have a thick waxy coating which gives considerable protection against **pathogens**. **Anthracnose**, a seed-borne fungal disease has been reported from the southern USA which is characterized by small white spots on the leaves, flowers and shoots. It spreads rapidly, causing affected parts to wilt and die. Initially, it attacks the younger growth but afterwards spreads downwards. Young shoots become pale and brittle. Damp and dull conditions encourage spread of ***Ascochyta* leaf spot** which is very difficult to control. Systemic fungicides with added wetting agent may give some control. This disease is seed-borne, so seed should not be collected from infected plants. **Cottony rot** is caused by *Sclerotinia*. Its spores can lodge in the junction of the two leaflets on a sweet pea leaf, usually in late spring, causing an innocuous looking rot which will rapidly spread back up the petiole and into the main stem, killing the plant. Cutting off the infected leaves solves the problem if carried out promptly before the disease starts to spread. **Black root rot**, a quite widespread soil-borne fungus (*Rhizoctonia*) attacks a wide range of plants. Symptoms vary, but in sweet peas it shows a black discolouration of the roots. Although development of the disease is most rapid in the temperature range of 20 °C - 25 °C, it seems to establish itself when a plant is under stress and is favoured by cold wet soils and poor hygiene. Excessively high air temperatures can also render a plant liable to infection. Maintaining the acidic soil pH (below 5.6) controls this disease as the fungus cannot survive in acidic soils. ***Botrytis* (grey mould)** occurs when weather is dark and in its infection the flowers develop white spots, especially the blue varieties as well as the white varieties which develop brown spots, though other colours are almost resistant. These form grey fluffy mould which may kill the whole plant. Ambient soil conditions should be maintained to avoid the fungus. The disease likes mild damp conditions with stagnant air, and consequently is discouraged by air movement, and high or low temperatures. **Downy mildew** at early stage is difficult to diagnose though it is less common than powdery mildew. Seedlings showing weak thin growth with small leaves, tending to curl upwards and inwards, and showing a blistered surface on plants with good roots and a clean neck, should be suspected of carrying this infection. The faint greyish deposit on the underside of the leaves is the prominent symptom as the disease progresses. Fungus prefers cold damp conditions and proliferates in poor air circulation. **Powdery mildew** appears as irregular powdery white spots on the upper surface of the lower leaves. It can best be controlled by fungicides containing *Pyrifenox*. Cutting off old lower leaves of sweet peas may help by improving air circulation, and reducing the source of further infection. To control *Pythium splendens* on sweet peas seedlings, when Niebisch and Kelling (1986) used Previcur N (propamocarb) and Ridomil (metalaxyl) + zineb, they recorded Ridomil superior over Previcur N, and application of Captan 80 or Thiram FW on seeds and seedlings, they recorded increased yield of cut flowers. **Reddish streaks** and **tomato spotted wilt virus** also affect sweat peas so in such cases such plants should be rogued out and burnt.

Sometimes the growing points of cordon-trained sweet peas may go **blind** in the middle of the season due to the plants being suffered a very cold spell over winter, though degree of resistance may differ from variety to variety. To overcome this problem, removal of too many side shoots should be avoided, otherwise the vine may appear barren devoid of flowers. **Bud drop** is most severe on cordon-grown Spencer varieties which can be caused due to drastic temperature and/or watering fluctuations when the plants are growing vigorously. Bud drop can also be due to ethylene pollution. Sweet peas are among the most sensitive of plants to traces of ethylene.

Chimera is a minor genetic disorder, though easily not detectable in sweet pea flowers, petioles and/or stems, but makes the flowers more attractive. It variegates the flowers with clearly defined edges so differing from other disease symptoms.

References

Asao, T., H.U.K. Kitazawa, Y. Sueda, T. Ban and M.H.R. Pramanik, 2007. Autotoxicity in some ornamentals with the means to overcome it. *HortSci.*, **42**(6): 1346-1350.

Bailey, L.H. 1929. *The Standard Cyclopedia of Horticulture* (vol. II), pp. 1824-1827. The Macmillan Company, New York.

Beckett, K.A. 1983. *The Concise Encyclopedia of Garden Plants*, pp. 206-207. Orbis Publishing Ltd., London.

Brickell, C. 1994. *The Royal Horticultural Society Gardener's Encyclopedia of Plant and Flowers*, pp. 168, 169, 170, 171, 174, 234, 272, 276, 530. Dorling Kindersley Ltd., London.

Dole, J.M., W.C. Fonteno and S.L. Blankenship, 2005. Comparison of silver thiosulfate with 1-methylcyclopropene on 19 cut flower taxa. *Acta Hort.* (*Proceedings of the Fifth International Postharvest Symposium*, eds Mencarelli, F. and P. Tonutti), at Verona, Italy, on 6-11 June, 2004; No. 682 (vol. 2), pp. 949-956.

Everard, B. and B.D. Morley, 1970. *Wild Flowers of the World*, pp. 10, 29. Peerage Books, London.

Hambidge, C. 1996. *The Unwins Book of Sweet Pea*. Silent Books, Cambridge, UK

Hammett, K. 2006. Pilgrimages. *Natl. Sweet Pea Soc. Ann.*, pp. 87–89.

Hay, R. and K.A. Beckett, 1971. *Reader's Digest Encyclopaedia of Garden Plants and Flowers*, pp. 388-390. The Reader's Digest Association Ltd., London.

Kaoru, N., H. Hino, S. Gunji, N. Hattanda, T. Murata, H. Tominaga and K. Fukumoto, 2008. 'Mimi' sweet pea for forcing culture. *HortSci.*, **43**(7): 2238-2239.

Niebisch, R.M. and K. Kelling, 1986. Results of chemical control of fungal diseases in ornamental plant production (German). *Gartenbau*, **33**(7): 215-218.

Nowak, J. and R.M. Rudnicki, 1975. The effect of "Proflovit-72" on the extension of vase-life of cut flowers. *Prace Instytutu Sadownictea w Skierniewicach, B*, **1**: 173-179.

Ochatt, S.J., C. Conreux and L. Jacas, 2010. *In vitro* production of sweet peas (*Lathyrus odoratus* L.) *via* axillary shoots. *Methods Mol. Biol.*, No. 589, pp. 293-301.

Pizzetti, I. and H. Cocker, 1968. *Flowers ~ A Guide for Your Garden*, pp. 723-730. Harry N. Abrams., Inc., Publishers, New York.

Rice, G. 2002. *The Sweet Pea Book*. B T Batsford, London, UK.

28

Lawn

R.L. Misra and Sanyat Misra

Introduction

Lawn is an open area of closely-mown grass space, often part of a garden or an area of turf kept closely mown. **Turf** means a surface covering of grass, clover or some other meadow-forming plants. Lawn is the most important and striking feature in a garden without which a garden is seldom complete. Grass when well tended and trimmed remains perennially green and gives us highly pleasing effect during winter when the garden looks at its drabbest. A well-kept lawn is a show-piece in the garden, visually appealing and pleasant to walk on, softens hard surfaces, complements planting, acts as a canvas for colourful beds and borders, is a relaxing haven, a playing surface for children, separates and defines distinctive features, unifies the garden as a whole, protects our valuable non-renewable soil surfaces, controls water erosion and wind, entraps gaseous pollutants, water and airborne particulate matters, helps in biodegradation of organic chemicals, conversion of CO_2 emissions, *vis-a-vis* as a catchment of the run off water as it absorbs rainfall six times more than a grain field so automatically enhances ground water recharge, and enhances heat dissipation and temperature moderation.

Lawns may also be created for sports and games in orchards and wild gardens, on wide paths and on gently sloping banks, *vis-à-vis* on roof-tops (the so-called terraces).

Establishing the Lawn through Seeds

Preparing the Top-Soil

Making of lawn takes considerable time. There are many instant methods of establishement such as transplanting of blocks of turf from neighbouring garden or field but a perfect lawn of uniform colour and even surface free of weeds can only be established by thorough and careful preparation of soil and by sowing carefully selected grass seed mixture. Seeded lawn is uniform in evenness, colour, growth and texture though it takes considerable time to bring it to working condition such as those laid down with good turves where spot events can be organized even in the first year itself. The soil for the purpose should be at a sunny location with a soil pH of 5.5-7.5, fairly moist, well-drained and not remaining spongy and soft for long after rains. To avoid this, a 10-15 cm layer of ashes should be maintained 20-30 cm below the top-soil and the top-soil should be a good fibrous-loam rich in humus or any other good garden soil. Good top-soil will assist the drainage in the lawn and will save the lawn from being scorched during hot and dry weather. However, the top-soil borrowed from elsewhere or from the same plot, may have various weed-seeds which after lawn-grass seeding may prove disastrous for lawn. Therefore, the land should be worked out and watered frequently before seeding or sodding so that all the weed-seeds have germinated and knocked down while preparing the land. The site should be prepared by thorough digging in the spring at a depth of 30 cm by incorporating some 3-5 kg/m^2 of Okhla sludge, well-decomposed cattle dung manure, farmyard manure or compost. Introduced top-soil if is heavy, it should be incorporated with plenty of finely sifted ashes or coarse sand to make it porous for proper root penetration and function, and if the soil is too light, well decayed manure 1 part to 3 parts of soil (v/v) should be incorporated thoroughly in the top 25 cm layer, levelled through spirit leveller with the help of pegs, straight edge and water filling in the small prepared beds, and then left for considerable time for weathering. Watering the field while preparing it will cause weed seeds present in the manure to germinate which should also be knocked down and certain other perennial weed's underground perennating organs as of *Cyperus rotundus* (rhizomes), *Oxalis* (bulbs), *Trifolium* (rootstock), *etc.* should be destroyed by collecting through rakes and hand picking. In lime-deficient soils, ground lime @ 175 g/m^2 should be incorporated and then the land should be left fallow for 3-4 months and then again the soil should be worked into.

Levelling

Whether it is tennis-lawn or the general garden lawn, it is necessary to level it properly, barring a few exceptions where lawn-mounds for better effects, especially in the informal gardens are to be made. After selection of the site, it is essential to dig up and level the land for lawn making. However, for tennis-court the lawn should be quite level. It should start months before so that soil is properly settled before seeding or turfing the field. The soil should be levelled only after making the soil highly porous. For levelling, spirit leveller should be used after making small beds throughout the plot with pegs at appropriate places.

Seeding

For new soils, Dawson (1963) recommends 15 grammes each of ammonium sulphate and potassium sulphate and 30 grammes each of superphosphate and bone meal per square metre of soil before 7-10 days of seed sowing. Seeds of the cool season grasses such as *Poa compressa* and *P. pratensis* should be sown in March or in late summer months when rains commence in temperate regions and where temperature ranges from 15-24 °C, but in the plains and tropical conditions with temperature range of 26-35 °C, February is the best time for seed sowing. Seeds in the plains can also be sown during rains, *i.e.* July to September. Lawn through seeds is quite uncommon in India. Sowing of cool-season grasses in autumn will provide cool fall temperatures for establishment. It takes 2-3 weeks for seed germination. Since the seeds are very light so sowing during windy days will hamper even distribution so it would be better to sow it on non-windy days. At the time of seed sowing the soil should be friable and not wet otherwise the distribution will not be even and the soil will also stick to the feet. Immediately after seed sowing, there should be some 6-8 mm layer of finely sifted soil over the seeds and then the ground rolled over lengthwise and across, so that winds after sowing may not carry the seeds, *vis-à-vis* birds and ants may also not be able to collect and carry these in their respective nests or burrows. Light water spraying with fine rose-can should be made after rolling to make the medium moist and to facilitate early and proper germination. After germination when the seedlings are 2.5-3.0 cm high, this should be rolled with a light wooden roller in dry weather. When the grasses have attained 5.0-8.0 cm of height, carefully the weeds should be removed and the grass should be cut through scythe, and this practice will have to be followed repeatedly and regularly for months together until the grass is strong enough to bear the mower as this pulls up the young grasses by their roots.

Seeds should be sown at 10-25 g/m^2. Seed rate per metre of ground is 3-5 g for *Agrostis* (bents), 5-8 g for *Cynodon dactylon* (Bermudagrass), 5-10 g for *Zoysia* grasses, 8-12 g for *Buchloe* and *Axonopus* (carpetgrasses), 8-15 g for *Poa pratensis* (Kentucky bluegrass), 15-20 g for fineleaf *Festuca*, 15-25 g for *Eremochloa ophiuroides* (centipedegrass), 30-40 g for *Paspalum notatum* (bahiagrass), 10-15 g for fineleaf *Festuca* + *Agrostis*, 20-25 g for *Lolium perenne* (perennial ryegrass) + other species, 25-30 g for *Festuca arundinacea* (tall fescue) + *Poa pratensis* (Kentucky bluegrass), 25-35 g for *Lolium perenne* (perennial ryegrass) + *Festuca arundinacea* (tall fescue), *etc.* Shiels (1984) recommended 35 to 100 g of seeds per square metre of land, depending upon the genera and species of the grass, the soil type, the region, and the prevailing weather conditions. For finest lawn, the seed requirement for *Festuca rubra* ssp. *commutata* and *Agrostis tenuis* (80:20) is 35 g/m^2 though for thick-seeded lawn the seed quantitity may go up to 50-100 g/m^2. Thick sowing does not allow weed-seeds to germinate among them even if these are there. Thick sowing is not liable to be damaged by scorching sun or winter frost but over-crowding encourages diseases such as 'damping off', *Fusarium*, *etc.* However, if any seedling of a weed is coming up, it should immediately be removed with its roots and the lawn area where lawn seeds have been sown, should be kept thoroughly weed-free throughout otherwise afterwards their removal becomes very difficult.

Early cutting by mower would cause uprooting of young plants so first 3-4 clippings should be carried out only by a sharp scythe. After about three months of sowing, *i.e.* in June in temperate regions and in the plains during October-November, the mower can be used by raising its blades about 2.5 cm above the normal level for the first three cuttings so that instead of usual 1.2-1.8 cm length as necessary for well-established and mature grass, it is 2.5-5.0 cm in length. First mowing to a height of 5.0 cm is carried out when grasses have grown to a height of 7.5 cm, though lawn height and its mowing frequency are governed on the type of grass species grown, the purpose for which the lawn is established and the season in which mowing is to be done. While mowing any time, it should not be mown more than one-third of the leaf growth. For cool season lawn, it should be frequently mown during flush of growth, occasionally during winter and no mowing during summer in India, *i.e.* May-June. High quality lawn may be mown as low as 1.25 cm every 2-3 days during active growth, *i.e.* during rains and when there is transition in weather. However, utility lawns are mown at 5-8 cm, *i.e.* longer than high quality lawn. Fineleaf fescues if grown in shade are mown even at more longer height, *i.e.* 10-12 cm. Shady lawns are mown every 2-3 weeks in summer.

Grass Seeds for Different Situations

Warm or Tropical Area Grasses

Bentgrass [(*Agrostis elegans*, *Agrostis nebulosa*, *Agrostis pulchella*), carpet grass (*Axonopus affinis*, *A. compressus*), buffalo grass (*Buchloe dactyloides*), Bermuda grass (*Cynodon dactylon*, common Bermuda grass; *Cynodon incompletus*, Bradley Bermuda grass; *Cynodon × magennisii*, Magennis Bermuda grass; *Cynodon trasvaalensis*, African Bermuda grass), centipede grass (*Eremochloa ophiuroides*), Italian ryegrass (*Lolium multiflorum*), Bahia grass (*Paspalum distichum*, bahia or joint-grass; *Paspalum notatum*, Argentine bahia; *Paspalum vaginatum*, seashore paspalum), *Polytrias diversiflora*, Javagrass; St. Augustine grass (*Stenotaphrum secundatum*); zoysia (*Zoysia japonica*, Japanese lawn grass/Korean grass; *Zoysia matrella*, Manila grass; and *Zoysia tenuifolia*, Korean velvet grass)].

Warm or Tropical Area's Hard-Wearing Grasses

Agrostis elegans, *Agrostis nebulosa*, *Agrostis pulchella*, *Axonopus affinis*, *Axonopus compressus*, *Buchloe dactyloides*, *Cynodon dactylon*, *Eremochloa ophiuroides*, *Lolium multiflorum*, *Paspalum distichum*, *Paspalum notatum*, *Paspalum vaginatum*, *Polytrias diversiflora*, *Stenotaphrum secundatum*, *Zoysia japonica*, *Zoysia matrella* and *Zoysia tenuifolia*.

Temperate Area Grasses

Bentgrass [(*Agrostis canina*, velvet bentgrass; *Agrostis palustris* {*A. stolonifera*}, creeping bentgrass; *Agrostis tenuis*, colonial bentgrass); crested dog's tail grass (*Cynosurus cristatus*), orchard grass (*Dactylis glomerata*); fescue (*Festuca elatior*, tall or meadow fescue; *Festuca rubra*, red fescue; *Festuca tenuifolia*, fine-leaved sheep's fescue), perennial ryegrass (*Lolium perenne*); Bluegrass (*Poa compressa*, Canada bluegrass; and *Poa pratensis*, Kentucky bluegrass)].

Temperate Area's Hard-wearing Grasses

Agrostis canina, *Agrostis stolonifera*, *Agrostis tenuis*, *Cynosurus cristatus*, *Festuca elatior*, *Festuca rubra*, *Festuca tenuifolia*, *Lolium perenne* and *Poa pratensis*.

For general use: *Cynosurus cristatus*, *Festuca tenuifolia*, *Poa nemoralis* var. *sempervirens*, *Poa pratensis* and *Poa trivialis*.

For annual growing: *Agrostis elegans*, *Agrostis nebulosa* and *Agrostis pulchella*.

For very fast growing: *Axonopus compressus* and *Lolium multifolium*.

For low-pH heavy soils: *Eremochloa ophiuroides*.

For high-pH heavy soils: *Polytrias diversiflora*.

For light soil: *Cynosurus cristatus*, *Festuca longifolia* (*F. duriuscula*), *Festuca tenuifolia*, *Festuca rubra* and *Festuca pratensis*.

For sandy soil: *Axonopus affinis* and *Axonopus compressus*.

For poor and shallow soil: *Festuca longifolia* (*F. duriuscula*).

For dry soil: *Buchloe dactyloides*, *Eremochloa ophiuroides*, *Festuca rubra*, *Stenotaphrum secundatum*.

For conditions of moist, shady and under trees: *Paspalum notatum*, *Paspalum vaginatum* and *Poa trivialis*.

For shady, sandy, heavy and dry areas: *Stenotaphrum secundatum*.

Near sea: *Buchloe dactyloides*, *Cynodon dactylon*, *Festuca rubra*, *Paspalum notatum*, *Paspalum vaginatum* and *Stenotaphrum secundatum*.

For sport's grounds: *Agrostis canina*, *Agrostis stolonifera*, *Agrostis tenuis*, *Cynodon dactylon*, *Festuca elatior*, *Festuca rubra* and *Festuca tenuifolia*.

For town gardens: *Poa annua* and *Poa nemoralis*.

Lawn from Turf

The ground for turfing should be well-dug to a depth of 25 cm and well-rotten farmyard manure should sufficiently be mixed to this layer. This requires preparation of the land at least three months before the turfing by cultivating it thrice followed by planking every time, all the perennial weeds along with their rootstocks taken out along with other foreign materials such as wooden, bricks, stone, glass and polyethylene pieces, with the help of forks, for making a smooth lawn. Turfing is carried out either in spring but never during April to June, or during autumn but never after October. After preparing the land, the ground should be applied with a basal fertilizer of 15 grammes of potassium sulphate and 30 grammes each of superphosphate and bone meal per square metre 7-10 days before turfing and thoroughly mixed (Dawson, 1963), then rolled and levelled, leaving only some 1.5 cm of the top soil properly raked so that the roots of the grass may be facilitated to work there properly and together bind the soil. The turves to be laid down should be of fine texture and absolutely weed-free otherwise the maintenance will be very difficult afterwards. The turves (**sods**) are usually cut some 5.0 cm thick and 30 cm^2 (Coutts *et al.*, 1963), 90 × 30 cm^2 rectangles or any other required size and laid alongside a straight edge in the prepared field and a plank is placed on it to kneel on or rolled to tear them easily, and this will help laying them out properly in the ground so close as the bricks in a wall, and then finely sifted soil should be worked in to fill the crevices, followed by watering and then pounded thoroughly and evenly with turf-beater, and then rolled. For some time, it should be watered daily by the evening. It should frequently be rolled with a light roller but only when the lawn is not too wet. For first two months of planting, the grass should be cut twice a week with scythe and then afterwards once a week with mower, followed by rolling after each cutting in the first year. Though sodding is the quickest method of developing a lawn but is very expensive, therefore, it is generally done only for improving patchy lawn for quick results.

Production of lawn through **sods** is traditionally very costly and the sods also become excessively heavy in weight making its carriage very difficult. Through traditional system, the sod is produced directly in the soil after thorough ploughing, harrowing, grading and seeding, *vis-à-vis* continuous mowing, irrigation, feeding and pest control up to two years until the sod becomes knitted, *i.e.* when it becomes capable of being rolled by holding together in a sheet when harvesting. One more drawback concerning the producer is that machine used for harvesting the sods cuts off the bulk of the root system leaving behind in the soil, apart from losing the nutrient-rich topsoil. After its placing at a new place it requires several weeks for regenerating the new roots for knitting and water requirement becomes very high to keep it moist. However, Decker (2001) suggested production of sods over plastic in soilless media for rapid establishment, easy portability and for saving the topsoil. In this case the topsoil is replaced by composted yard mulch, wellrotten manure, dry sewage sludge, riverbed flume sands, fly ash mixed with manure and so on. Here the sods are produced in the prepared growing media over any root-impervious barrier, especially plastic sheets (preferably 1.2 m wide) where through seeds which germinate within 2-3 weeks, for rooting 4-5 weeks further are required, and the primary fibrous root system is found well established within four months through binding and knitting the growing medium into a sod as plastic is impervious to roots that compels these to spread horizontally and making a perfect knit, easily rollable within a few weeks (5-10), however, this way the medium is given frequent nutrient-rich mix evenly across the surface of the plastic. While rolling such sods, plastic may be left as such for taking another sod, and the bare-rooted sods may be laid at a new site where it will bind and root rapidly, however, in case of temporary laying for instant effect these may be carried with plastic base which will help its smooth rolling and carriage and after the programme is over it may be taken back. The **bases** used for such programmes are cement, plastic films (perforated and unperforated, 0.5-6.0 mm thick white to black), various plastic foams, the biodegradable matting materials (bases) such as wood-chips and bark, tissue paper, crepe-paper, papier-mâché, paper wastes (cardboards, mould-proof papers, brown craft papers, or any other type of waste paper), cotton, flax, pulp fibre mulch, ground corn cobs and bagasse, seed hulls, spun plastic, *etc.* with 1-3 layers (at least 3 cm of depth) of fibrous materials as **growing media** such as vermiculite, sewage sludges (the popular Okhla sludge), packed clay, cocopeat, sawdust, and other cellulose fibres (compressed cellulose), peat, peat moss (sphagnum), manure, composted leaves, decomposed garbage, various kind of sand, rockwool, many of the above-mentioned bases may also act as growing media, *etc.*, and the **binding materials** such as burlap or jute netting, cotton fabrics and waddings, plastic foams and films with cellulose wadding attached, polyester and polypropylene fibre netting, hydrogels, and their various combinations, *etc.* which should be joined to the base sheet by an adhesive binder, *i.e.* rubber cement or gel particles. Pregerminated seeds on burlap blankets can also be rolled up and sent to the site of installation for further establishment. However, in case of liquid mulch sod planting system (LMSP) over plastic, the requirement of growing medium is barely 1.3 cm.

To quicken up the process of establishment in comparatively quite less time, the **sod-strips** of the quick-growing grasses such as *Zoysia* are laid in long rows at 15-25 cm apart. This is more convenient and easy but is very expensive due to requirement of more planting material. Immediately after planting, this requires thorough irrigation. In **plugging**, small pieces of spreading type of grass-sods (turf-plugs measuring 5-10 cm are removed from sod-strips to be planted in 15-40 cm centres) such as *Stenotaphrum secundatum* and *Zoysia* that are planted 15-30 cm apart, as a small plant in the soil, by making small holes with a small spade or the bulb planter at regular intervals similar to that of sod-stripping, where base areas are filled in by the spreading stolons and rhizomes, followed by thorough watering. Closer planting makes earlier establishment. When compared with sodding, the process is quite slow, *vis-à-vis* creates more problem for lawn by encouraging growth of various weeds. Plugging may take 1-3 seasons for complete establishement, depending upon the type of grass used, initial planting distances and environmental conditions.

Planting of the grass is also carried out through **dibbling, sprigging or stolonization** and for this purpose the grass selected should be rooted or unrooted but well-matured, sun-grown and short in length with close joints (nodes) having at least five nodes. Those collected from shady areas, *vis-à-vis* after rains are sappy, long and with high intermodal length, therefore, such grasses have poor stamina for good establishment. For **dibbling** in the prepared soil, the holes are made at 7-10 cm apart when the weather is moist, and then the right sides of the turves are pressed in these holes with only a little part peeping out, followed by immediate irrigation. **Sprigging** is placing of the stolons in the furrows spaced 15-20 cm apart and this way some 150-200 litre volume of grass will have to be required for planting 100 m^2 of area. Immediately after planting the sprigs, the field should be irrigated. Sprigging requires less quantum of grass and establishes well in four months. **Stolonizing** is the process under which the stolon-bits are spread on the surface of the prepared field and then covered with a thin layer of garden soil, followed by irrigation.

Lawn may also be established through **turf plastering**. In this method, the thoroughly prepared land having sufficient moisture is plastered with a paste of chopped up fresh roots, stolons, stems or rhizomes of desired grasses, mixed with fresh cow dung and water which is plastered at the site which is sufficiently wet. Immediately afterwards, over the plastering the 2.0 cm layer of garden soil is spread over it, followed by quite slow watering. Watering is carried out every day for about a week. This method is now out of practice as the job is not so clean and it is not quite successful in dry and variable climate.

Genera, Species and Varieties of Grasses and Other Herbage

Details pertaining to different lawn genera and species were scanned from different literature sources (Bailey, 1942; Laurie and Ries, 1950; Dawson, 1963; Hay and Beckett, 1971; Beckett, 1983; Hellyer, 1983; Shiels, 1984; Randhawa and Mukhopadhyay, 1986;Brickell, 1994; Misra and Sanyat Misra, 2012).

A.Warm Area Hard-Wearing Lawn Grasses (Tropical and sub-tropical regions thriving at 26-35 °C temperature ranges)

Bermudagrass (Family: Gramineae; common 'Bargusto' Bermuda- or wire-grass, *Cynodon dactylon*, which contains some 400 seeds per gramme). *Cynodon* derives its name from the Greek *kuon* 'a dog' and *odons* 'a tooth'. Some four species are reported, out of which only *C. dactylon* (*Capriola dactylon*) from India is economic for making lawn and for pasture, though *C. incompletus* (blue couch-grass) from East and South Africa and New South Wales only as pasture grass but it is reported causing poisoning at certain stages of its growth producing hydrocyanic (prussic) acid. *Cynodon dactylon*, a low-growing perennial grass bearing flattened, slender, creeping and node-rooting stems, and so producing numerous white, slender and stout creeping rootstocks having leafy stolons with narrow, straight, base-hairy leaves and inflorescence a slender digitate spike some 2.5-4.0 cm long, spikelets compressed, 1-flowered, in 2 rows and along one side of the rachis. It is resistant to drought and high temperatures, most suitable at sunny locations on clays and loams, for sandy beaches, school grounds, public parks, tennis court, polo fields and golf greens in a soil having neutral pH, though this grass is getting replacement in *Zoysia matrella*; 'Bradley' Bermuda grass, *C. incompletus* var. *hirsutus*; 'Magennis' Bermuda grass, *C.* × *magennisii*; and 'African' Bermuda grass, *C. transvaalensis*. A native of India, *Cynodon* has stoloniferous and trailing habit, and gives excellent results if it is planted at a sunny location in any soil except that which is extremely wet, is kept well fed with nutrients, and frequently mown, however, it turns brown with first approach of frost and presentation remains poor during the winter. During winter also it can be kept properly displayable by mowing it close to the ground, followed by raking or harrowing and then applying with ample amount of bonemeal then again raking and rolling. Promising varieties of Bermuda grass propagated vegetatively are 'Bayshore' ('Gene Tift'), 'Everglades', 'Midway', 'Ormond', 'Sl-1', 'Sl-2, 'Sl-3', 'Sunturf', 'Texturf IF', 'Texturf 10', 'Tiffine', 'Tifgreen', 'Tiflawn', 'Tifway', 'Tufcote', 'U-3', 'Uganda', *etc.*

Bahia grass or joint-grass (Family: Gramineae; *Paspalum distichum*) from the moist and low-grounds of South USA is suitable for making a lawn which is too wet for planting Bermuda grass though it is low-creeping and is not of quite dense nature. *Paspalum* derives from the Greek word *paspalos*, which is an ancient name for millet. *Paspalum* comprises of 150 species from warmer parts of both the hemispheres, more abundantly in America. It provides a reasonably durable and attractive surface. *Paspalum notatum* (bahia grass/Argentine bahia) and *P. vaginatum* (seaspray seashore paspalum) are coarse-textured and have better salt-tolerance so are suitable for shady areas. *P. compressum* (carpet-grass) from USA grows from 15 to 60 cm in height with wiry and creeping stems and broad leaves and makes an excellent lawn in Gulf region of USA. Its floral culms are 15-30 cm long bearing a pair of divergent spikes. *P. membranaceum* is a slender perennial.

Buffalo grass (*Buchloe dactyloides*; Gramineae) is a native of USA, propagated through seeds or stolons and is a best substitute for *Poa pratensis* in hot dry regions as is quite drought-resistant though during autumn and spring it appears dry with dull-attraction. Its seeds do not germinate so quickly and also as are expensive so it is normally propagated through sodding with small pieces during rains, planted 30 to 60 cm apart where this will make a complete covering in one season. Its requirement for mowing and watering is quite little due to it being drought-resistant.

Javagrass (Family: Gramineae; *Polytrias diversiflora*) is a very soft-leaved grass, most suitable at sunny locations and in very heavy loam and alkaline soils.

St. Augustine- or buffalo creeping- or Charleston- or goose-grass (Family: Gramineae; *Stenotaphrum secundatum*) is tender perennial native of the southern states of USA. In Greek, *stenos* means 'narrow' and *taphros* for 'a trench', referring to the spikelets being partially embedded in the rachis. It is an early-growing coarse and creeping perennial bearing compressed culms, flat divergent and channelled leaves, and flat spikes embedded in the surface of a broad rachis-forming terminal spikes. One species is native along the Gulf coast, especially in Florida and is similar to Bermuda grass adapting well to sandy soils or sandy-coast lands or salt-water soils and even in swamps and ponds where it binds such soils through its rhizomes and creeping habits. It is suitable for shady, sandy, heavy and drier areas, and endures salty sprays so is most suitable at the seashore. Its varieties propagated vegetatively are 'Bitter Blue', 'Floratine', *etc.* One of its forms is the most valuable lawn grass growing almost in every soil type, especially the heavy ones and thrives even in shade, bearing rather broad and at the most 15 cm high and requires little mowing. Neither it ever becomes coarse nor holds rain or dew, is excellent for home-lawns and is grown through cuttings with rootlets by planting in rich soils. Its water requirement is quite less as to that of *Cynodon*. *S. secundatum* (*S. americanum*) bears stolons on the surface of the ground with erect flowering branches some 15-30 cm high and requires frequent mowing. It is propagated through rootlets. Its var. 'Variegatum' has white stripes on its leaves and is excellent for basket planting.

Zoysiagrasses [Family: Gramineae; *Zoysia japonica*, *Z. matrella* and *Z. tenuifolia* (*Osterdamia matrella*)]. Zoysias are very suitable for the plains of India with a very warm summers and chilling winters but all these grasses become dormant and loose their colour at temperatures below 10 °C. *Zoysia* was named after Karl von Zoys, an Austrian botanist. *Zoysia* is a creeping maritime perennial which provides a reasonably durable and attractive surface in the lawn. These are propagated by cuttings of the rhizomes. Flowering is in close spike-like panicles with closely appressed spikelets, having one flower. The genus is native to SE Asia and Australasia. Zoysias require two years to form a closely-set lawn, but once it is established, requires little attention and does not allow the weeds to grow due to its dense turfs. In the first year of its establishment, it should be properly cared through weeding, watering and fertilization at regular intervals, and in the second year when it is well-set, becomes reasonably drought resistant. Since leaves of *Zoysia* are very low, hence, require flat-blade rotary mowing, especially in the first year of its establishment, but afterwards when the growth is set, it requires hardly any mowing. This grass requires plenty of sunshine. Sometimes these warm-season lawn grasses may be used even in temperate conditions but during winter when temperature goes down to 10 °C or below, these become dormant and turn brown. Usually, warm-season grasses are sown as single species since these have vigorous and good creeping habit and if mixed with other species, the appearance becomes patchy, unsightly and of differing colours and texture.

Zoysia japonica (Japanese/Korean lawn grass/palm-beech grass) is little coarser than *Z. matrella* (Korean grass) and is native to Japan and China, and its name Korean grass comes due to its introduction into USA from Korea. Its blades are some 3 mm wide and the panicles about 2.5 cm long and often purplish. It is most suitable for making lawn in the sandy-soils. The varieties

'Meyer' and 'Midwest', both growing vegetatively are very fast-growing, have intensive root system, tolerate even the acidic soils, *vis-à-vis* more shades than even Manila grass but thrive best in full sunshine. Its hybrid 'Meyer Z-52' is hardier and is drought resistant.

Z. matrella, var. 'Matrella', (syn. *Z. pungens, Osterdamia matrella*; Manila/Japanese carpet-grass; grown vegetatively) should be planted in the well-drained soil otherwise the pH of the soil, due to improper drainage, will go down and the pH below 5.5 is detrimental to its roots. *Zoysia matrella*, a native of SE Asia and East Indies, is a slow-growing grass with creeping stems from which emerge numerous short leafy shoots and flowering stems. It bears firm, some 2.5-7.5 cm long, and crowded leaves terminating into a sharp hard point, spikes some 2.5-5.0 cm long and spikelets some 3 mm, smooth and hard.

Z. tenuifolia (bluegrass, Mascarene grass, velvet grass) from Mascarene Islands looks like dark green velvet as the growth is quite thick, and this smothers other vegetations including the devil- or Bermuda-grass. Even in high snow-fall areas when the plants are frozen, it comes up from the roots with improvement of temperature. Its water requirement is very little. *Z. tenuifolia* var. 'Emerald', propagated vegetatively, has 0.5-1.0 mm wide blades (very thin needle-like) as compared to *Z. matrella* (1-3 mm) and *Z. japonica* (3-4 mm). The velvet grass when well established makes dark green mounds and is most suitable in rock gardens. This normally does not need mowing but to overcome wavy surface, frequent mowing and rolling is necessary. This grass provides springing texture which character most other grasses do not possess.

B.Temperate or Cool-Area Lawn Grasses (15-24 °C temperature range)

Bents (Family: Gramineae; *Agostis* spp.; cloud grass). The Greek *agros* is for 'a field', the ancient name for a forage grass similar to those of *Festuca*. Commonly known as browntop and highland bent, these bear fine leaves so making fine lawns. *Agostis* is an annual or perennial grass with erect or creeping stems having open panicles comprising small flowers. Spikelet is one-flowered, glumes almost equal and acute, the lemma is shorter and more delicate than the glumes. Some 100 species are available with worldwide distribution, especially in the north temperate zone, some making good lawn, some others as ornamental grasses while many others as forage grasses, and the panicles are used for making bouquets. The species suitable for lawn making are *A. alba*, *A. canina*, *A. gigantea*, *A. setacea*, A. *stolonifera* and *A. tenuis*. These all perform well in poor soils, vegetative reproduction is rapid through rhizomes and/or stolons, and also these set copious seeds, and mixing with bristle-leaved fescues these make very good lawn. These are low-growing hence can tolerate a very low mowing. At young stage the bents have rolled leaves. For tennis courts or sports, a mixture of bents and fescues (especially fine-leaved) is used for high quality lawn for tolerating some wear. *Agrostis elegans, A. nebulosa* and *A. pulchella* are good annual species either for borders or for pot culture and are effectively grown in the warm tropical areas.

Agrostis alba (herd's grass, red-top) from Europe is a perennial lawn grass. It contains some 1,000-1,250 seeds per gramme. It has usually decumbent base with underground creeping stems, having short rootstocks, produces erect culms some 60-90 cm tall with smooth sheaths. Ligules are membraneous, blades are flat, some 6 mm wide, strongly nerved, scabrous and pointed, panicles are pyramidal or oblong from small up to 30 cm in length and reddish, the glumes are scabrous on the keels, lemma is awnless and palea$^1/_2$-$^2/_3{}^{rd}$ of the size of lemma. It is successful in wide range of conditions, wet or dry and is suitable for holding banks to prevent erosion. Var. 'Aristata' is though similar to 'Vulgaris' but the lemma bears an exserted awn from near the base. 'Maritima' (*A. alba* var. *stolonifera*; creeping bent-grass) bears long stolons and the panicles are narrow and contracted. This makes a good lawn. Var. 'Vulgaris' (fine bent-grass, red-top) is rather a more delicate grass growing some 30 cm tall, producing 2.5-7.5 cm long panicles and is most suitable for lawn making.

A. canina (brown bent, Rhode island bent, velvet bentgrass) from Europe is unaggressive, caespitose, creeping, slender and erect grass, growing about 15 to 60 cm high, bearing soft, velvety and much finer in leaf than *A. stolonifera*, with lemma having about the middle an exserted bent awn and the seeds are deficient in palea. It likes a soil pH of 5.0-6.0, makes a good lawn, especially in the damp or shady soils such as pond margins but areas with a lot of wear and tear are not suitable for this grass. The areas experiencing a drought are at all not suitable for its growing where it quickly dies. It spreads by stolons but is not amenable to mower if drawn quite low. Its promising varieties are 'Kernwood', 'Kingstown', 'Raritan', *etc.*

A. gigantea (redtop) and many others including the natural hybrids as the species hybridizes easily in nature. It creeps by underground stems but being coarse and open, is not so preferred.

A. setacea (brittle-leaved bent) gows naturally on sandy and peaty heaths or moors. It is a tussock-former and does not blend intimately with itself or others.

A. stolonifera (*A. palustris*; creeping bentgrass) is aggressive in growth, produces fine sward for making a good lawn, is hard-wearing and creeps by overground stolons though is less common than *A. tenuis*. Neither it relishes ill-drained soils, especially during winters nor the summer drought at any time and prefers a soil pH of 5.4-6.5. Though it is coarser than *A. tenuis* but can produce very fine lawns by producing vigorous stolons so if left unmown for sometime, it makes a spongy lawn. It should be raked regularly for better results. Its maintenance part is little expensive. Its varieties propagated through seeds are 'Penncross', 'Penneagle', 'Seaside', *etc.* and through stolons are 'Arlington', 'Cohansey', 'Collins', 'Congressional', 'Dahlgren', 'Evansville', 'Metropolitan', 'Pennlu', 'Toronto', 'Washington', *etc.*

A. tenuis (browntop bent, colonial bentgrass) from New Zealand is most important of all bents as it adapts well under a wide range of conditions, much better in medium (loam) and heavy clay soils which are rather dry and acidic (pH 5.5-6.5) but not vigorous in light soils, however, its shallow rooting quality makes it undesirable for drought conditions. It is a tufted perennial creeping underground through its stems, rhizomes or short stolons. It is slow in establishment but through regular mowing and frequent top-dressing of compost because of being shallow-rooted, it produces a very dense turf of high quality with fine sward. Its leaves are pleasing in colour, fine and tapering to a fine point, tough with lightly ribbed surface, hard-wearing and makes a close and uniform sward. It blends well with fescues and makes a good lawn. Per gramme seeds in this species is from 12,000-15,000. Its promising varieties are 'Astoria', 'Bardot', 'Exeter', 'Highland', 'Holfior', 'Tracenta', *etc.*

Crested dog's tailgrass (Family: Gramineae; *Cynosurus cristatus*, perennial; *C. elegans*, annual) is most suitable for general use and for light soils. *Cynosurus* derives from Greek word *kuon* for 'a dog' and *oura* for 'a tail', hence its common name dog's tail grass. The genus comprises 6 species in the north temperate regions of the Old World but only two are used as ornamental grasses and for making lawn. This is caespitose grass with flat blades and spike-like panicles. Spikelets are of two forms, the terminal being perfect but lower sterile with several empty plumes. *C. cristatus* (crested dog's tail) is 30-60 cm tall perennial used sometimes in lawn mixtures as by itself it does not make a good turf and bears 2.5-7.5 cm long spike. In both, light and heavy soils it grows as a bottom grass. It is a tough and resistant to wear and tear. *C. elegans* is an annual growing 15 to 45 cm high from Europe which bears one-sided loose panicle some 2.5 cm long with silky awns and longer than the lemmas. It is most suitable for dry arrangements.

Fescues (Gramineae; *Festuca* spp.). *Festuca* is ancient Greek name for a kind of grass. There are about 300 species in the genus from the cooler parts of the world. It contains some 1,000-1,250 seeds per gramme. These are of two types, the 'red fescues' such as *F. arundinacea*, *F. rubra* ssp. *commutata* (*F. rubra* ssp. *fallax*), *F. rubra* ssp. *littoralis*, and *F. rubra* ssp. *rubra*; and the 'sheep's fescues' such as *F. cinerea*, *F. elatior*, *F. longifolia*, *F. ovina*, and *F. tenuifolia*.

F. arundinacea Schreb. ('Barvado' red tall fescue) with promising varieties such as 'Alta', 'Fawn', 'Goar', 'Kentucky 31', 'Kenwell', *etc.*

F. cinerea (*F. glauca*; blue fescue), a coarser-textured hardy perennial with round and handsome tufts, grows 20-40 cm, tolerates more summer heat and drought better than most cool-season types and is most suitable for pavings, as ground covers, in the rock gardens and in heather gardens with its blue-grey and very thin leaves, and its vars 'Blaufuchs' is bluish-green, 'Harz' dark bluish-green, both growing to 30 cm.

F. elatior (tall or meadow fescue). It is from Europe which grows 30-60 cm tall, and is frequently grown as meadow or pasture grass. Its blades are narrow, inflorescence paniculate and few-flowered, spikelets awnless, 5-8 flowered and some 1.2 cm long. Its varieties are 'Ensign', 'Mimer', 'Trader', *etc.*

F. longifolia (*F. duriuscula*; hard sheep's fescue) from Europe is tussock-forming and suitable for light, poor and dry or shallow soil. Its blades are firm, thick and rough.

F. ovina 'Silberreiher' (true sheep's fescue) from Europe is a tussock-forming hardy perennial growing up to 25 cm, most suitable to well-drained alkaline soil, excellent as ground cover, in the rock gardens and pavings and is also drought resistant. Its 5.0-120.0 cm long panicles contract after flowering.

F. rubra (red fescue) from Europe grows 15-60 cm in height with red stem bases. Its blades are narrow and involute. It is suitable for general use, for dry soil, near the sea or in sport's grounds with a soil pH of 4.3-5.5. Its promising varieties are 'Dawson', 'Duraturf', 'Illahee', 'Merlin', 'Olds', 'Pennlawn', 'Rainier', 'Jamestown', *etc.* ***Festuca rubra*** var. ***commutata*** (*F. rubra fallax*; Chewing's fescue) from New Zealand is suitable for planting on sandy and gravelly soils, and in shady locations. It creeps by underground stems, bearing light green, stiff and sharp leaves, and produces a fine-textured turf. Due to poor germination of seed and utterly slow growth rate of the germinated seedlings, its seed requirement is quite high. Its one gramme contains some 1,000-1,200 seeds. Its varieties are 'Atlanta', 'Frida', 'Highlight', 'Lobi', *etc.*

Shiels (1984) has referred various types of seed mixtures for making various types of lawn, as given here under. For making fine ornamental lawn, its 80 per cent seeds should be mixed with 20 per cent of *Agrostis tenuis* by weight and the lawn should be mown at 5-6 mm from the ground. To maintain a fine lawn cut at 10 mm from the ground, the seed mixture by weight should contain 40 per cent *F. rubra* var. *commutata* + 40 per cent *F. rubra* var. *rubra* + 20 per cent *Agrostis tenuis*. Making a lawn for the purpose of hard-wearing, the seed mixture by weight should contain 40 per cent *F. rubra var. commutata* + 30 per cent *Lolium perenne* + 20 per cent *F. rubra* var. *rubra* + 10 per cent *Agrostis tenuis*. ***F. rubra*** **ssp.** ***littoralis*** (creeping red fescue) survives drought much better and is more wear-tolerant than Chewing's fescue but does not tolerate close mowing. Its promising varieties are 'Dawson', 'Merlin', *etc.* For making lawn under light shaded areas with taller growth, the seed mixture by weight should consist of 35 per cent *F. rubra* ssp. *littoralis* + 30 per cent *F. rubra* var. *commutata* + 25 per cent *Poa pratensis* + 10 per cent *Agrostis tenuis*. Making of lawn along the banks and steep slopes, the seed mixture by weight should contain 45 per cent *F. rubra* var. *rubra* + 30 per cent *F. rubra* var. *commutata* + 20 per cent *Agrostis tenuis* + 5 per cent *Phleum bertolonii*.

F. tenuifolia (fine-leaved sheep's fescue) is suitable for heavy shade and poor soils or general use. It naturally occurs on soils of a peaty or acid nature and as it forms tussock, does not blend easily with other grasses.

Meadowgrass (Gramineae; *Poa* spp.). *Poa* is a ancient Greek name for grass or fodder. Though there are a few annuals but mostly they are perennial grasses of some 100 species from temperate and cold regions of low growth, having glossy, folded, ribless leaves and spreading paniculate flower-heads of 2-8 spikelets. Glumes are shorter than the lemmas and awnless, lemmas are back-keeled, membraceous, 5-nerved, awnless, scarious-margined and often cobwebby at base.

Poa annua (annual meadowgrass, in America it is called annual bluegrass) is the commonest meadowgrass in lawn turf which is a short-lived annual and it quickly runs to seed making patches among fine-leaved grasses, hence is considered as weed but is suitable for town gardens. It persists in lawn as it forms a large number of low-growing panicles despite frequent mowings so through seeds it continues coming up in the lawn year after year even invading the fescues though does not like dry conditions.

P. compressa (Canada bluegrass) from Europe is relatively tolerant of droughty locations, especially on steeps and is often used in mixtures with *P. pratensis* due to its leaf colour which is blue-green similar to *Festuca pratensis* hence is also sometimes spelt as English bluegrass which, in fact, is a common name for *F. pratensis*. *P. compressa* differs from *Poa pratensis* with its distinctly flattened culms, *vis-à-vis* its short and much contracted panicles. It reproduces itself by rhizomes and flourishes well in sterile soils. It is amenable to close mowing and makes a handsome lawn even at those dry conditions where *P. pratensis* does not survive.

P. nemoralis (wood meadowgrass) from Europe is for shady woodland conditions and town gardens and is seldom found growing where the turf is short mown in the open. Its panicles are short and narrow with short branches, culms are 30-90 cm high, and glumes are 3-nerved and acuminate. Sometimes its seeds are mixed with other grasses for use in shady areas. *P. nemoralis* var. *sempervirens* is for general use.

P. pratensis ('Barrister' Kentucky bluegrass, Junegrass, Kentucky bluegrass or smooth-stalked meadowgrass) from cooler regions of northern hemisphere was brought to Kentucky (USA) by early settlers. It is a very attractive lawn grass, and in wear-tolerance it is second to *Lolium perenne*. It makes a nice turf on light to medium soils with a 6.0-7.0 soil pH. Its underground creeping stems (runners) bear a tuft of leaves at the tips, each leaf bearing two very distinct lines running down the centre of each leaf and the young leaves are crinkled. Its leaves are green (not the blue so one should not confuse by its common name bluegrass), V-shaped in cross-section with tips being in the shape of bow of a boat. Its shoot is flat as at young stage each leaf is folded. Its 7.5-10.0 cm long, open and pyramidal panicles bear 3-6-flowered and some 4 mm long spikelets, first glume 1-nerved, second 3-nerved, lemma cobwebby at base and culm normally 30-60 cm high. It grows well on light to medium well-drained and rich soils which may be alkaline to slightly acidic. It is resistant to wear and tear, grows through rhizomes that fill the bare patches, is amenable to mowing as close as 2.5 cm and is quite suitable for light soil and general use. Its one gramme may contain some 3,000-6,000 seeds. Promising varieties are 'Baron', 'Bensun', 'Cambridge', 'Cougar', 'Delta', 'Kenblue', 'Merion', 'Newport', 'Parade', 'Park', 'Pennstar', 'Prato', 'Troy', 'Windsor', *etc.*

P. trivialis (rough-stalked meadowgrass) from Europe resembles *P. pratensis* but is lighter in colour and has rough stalks. It creeps slowly at the surface of the ground through short stems (runners). It is most suitable for general use, in moist or under trees and shady situations better than *P. perennis,* but is not amenable to close mowing, heavy wear and tear and in hot summer even in the shade. It does not make a good turf by itself but by mixing with others. It contains some 1,000 seeds per gramme of sample. Its variegated form is var. *follies albo-vittatis*.

Orchard grass (*Dactylis glomerata*; Graminae). In Greek *dactulos* is for 'a finger'. It is native of north temperate regions of the Old World. It is a coarse perennial tufted hardy grass growing to 60-90 cm tall where blades are flat with thin but prominent ligules and the sheaths are closed nearly to the throat. The panicles are glomerate, spikelets 2-5 flowered, nearly sessile in dense 1-sided fascicles, lemmas hispid-ciliate on the keels, awn-tipped and compressed. Though it is a temperate pasture and meadow-grass but makes an effective shady lawn under the trees. It is of easy cultivation and is propagated through division. Its var. *variegata* is a dwarf form with compact growth where foliage is variegated in silver and green and is most suitable for planting on the borders.

Ryegrass (*Lolium* spp.), a perennial, tufted and rapid-growing grass requiring highly fertile soil with heavy nutrition, is extremely hard-wearing (durable) so most suitable in rugby and soccer pitches though more prone to summer diseases. However, being tufted its spread is limited, therefore to thicken the sward it is normally mixed wth fine grasses, though appearance of perennial ryegrass is even easily distinguishable by the brightness of the leaf-undersurface, dull-ribbed leaf-uppersurface and the red shoot bases, *vis-à-vis* having ligules at each leaf junction which only a few grasses possess. This prefers higher soil pH. *Lolium multifolium* (Italian ryegrass) though smothers as a lawn very quickly but on mowing dies down within two years, so is suitable only as ground cover. It is most suitable for growing in warm tropical regions. However, in Bermuda grass lawns, the sowing of Italian ryegrass seeds during autumn will make the lawn lush-green during winter. *Lolium perenne* (perennial ryegrass) with fast establishment and growth, does well even in heavy clay soils with a pH of 6.0-7.0 and is the most common grass in hard-wearing areas but is coarse-textured so not amenable to close mowing though requirement of mowing is quite frequent because of its quick growth habit. During autumn or under drought conditions there is leaf tip-browning in *L. perenne*. Similar to other coarse grasses, *L. perenne* demands heavy feeding and watering than very fine grasses. One gramme seed packet of *L. perenne* contains some 600 seeds. Its varieties are 'Derby', 'Majestic', 'Manhattan', 'Norlea NK100', 'Pelo', 'Sprinter', *etc*. For a very durable lawn with high wear and tear, a mixture of *Poa pratensis* and *Lolium perenne* or *Festuca elatior* is quite suitable.

C. Other Miscellaneus Grasses (Gramineae)

For cold areas are recorded which though do not make a good turf themselves alone but are widely distributed in the lawn plantings as weeds having potentialities as a constituent of turf and together make a compact turf and these grasses are defined here under.

Agropyron (wheatgrass) originates from the Greek words *agros* for 'field' and *pyros* for 'wheat', hence the common name is wheatgrass.The genus comprises 30-40 perennial species from temperate regions of both the hemispheres, producing creeping rootstocks. The spikelets are sessile, 3-to many-flowered, glumes are equal, firm, many-nerved, acute or awned, lemmas 5-7-nerved and more or less awned and the palea is ciliate on the keels. Only *A. cristatum* and *A. smithii* are the two worth for lawn making.

Aira praecox (early hairgrass). *Aira* is ancient Greek name for Darnel, and is native to Mediterranean region with 6 species. This is an annual grass with delicate culms and open capillary panicles, spikelets 2-flowered where one or both are awned, palea and lemma both almost equal in size. It is very similar to annual meadow-grass.

Anthoxanthum odoratum (sweet vernal). In Greek, *anthos* is for 'a flower' and *xanthos* is for 'yellow'. The genus comprises four species from Europe. Panicles of the perennial *A. odoratum* are spike-like, spikelets with one awnless floret and two 2-lobed sterile lemmas. It confers the sweet smell to new mown hay.

Avena flavescens (golden oatgrass). *Avena* is the old Latin name. The genus comprises of some 50 annual species from the temperate and cooler regions of the world. The panicles are open with large and 2-6-flowered spikelets, and the rachilla are bearded below the florets. Nearly equal glumes of the length of spikelets are membranaceous and many-nerved. Lemmas are bidentate at the tips with stout twisted awns on the back. It is a leafy plant and is sometimes noticed growing in the lawn.

Axonopus compressus (carpet grass) from the West Indies is the low-creeping but the fastest-growing grass out of all the lawn grasses, covering the ground within 60 days and makes a closely-set thick turf bearing broad and wavy-margined leaves some 1-2 cm wide with blunt leaf tips. It has compressed and 2-edged creeping stems with rooting at each node. It becomes very comfortable in sandy soils if the soil contains sufficient moisture. It requires seldom flooding. It is most satisfactory for sunny as well as shady lawns. It tolerates a variety of soils, even acidic ones. This and *A. affinis* are low-maintenance grasses having coarse-textured leaves and both are effectively grown in the warm tropical conditions.

Buchloe dactyloides (buffalograss) is a native of USA, propagated through seeds or stolons and is a best substitute for *Poa pratensis* in hot dry regions as is quite drought-resistant though during autumn and spring it appears dry with dull-attraction. It is low-maintenance and drought-resistant grass with little mowing and is suitable for shady, sandy or heavy conditions producing rhizomes and enduring salty sprays so is suitable even at the seashore. Its seeds do not germinate so quickly and also as are expensive so it is normally propagated through sodding during rains. Its requirement for mowing and watering is quite little.

Deschampsia caespitosa (hassock grass, tufted hair-grass). The genus was named for Deslongehamps (1774-1849), a French botanist. The genus comprises some 20 species from the cooler regions of the northern hemisphere. *Deschampsia* is a tufted perennial with narrow or loose panicles bearing shining 2-flowered spikelets where rachilla has a hairy elongation, the glumes are almost equal to florets and lemmas are toothed with a dorsal awn. *D. caespitosa* is a troublesome grass due to its coarse tussocks, grows 30-90 cm tall with firm and narrow blades, open panicles and slender branches. *D. flexuosa* (wavy hairgrass) is very similar to fescue and grows up to 60 cm tall with flexuous branches having spikelets at the ends, culms are slender and the blades are numerous and capillary. It is suitable for shades.

Eremochloa ophiuroides (centipede grass) from China is a coarse-textured grass so is low-maintenance and requires less frequent mowing. It is deep-rooted, drought-resistant, stolon-creeping, sun-loving, excellent for heavy clay soils having low pH and makes a dense mat of yellowish- to bluish-green foliage. Its coarse-textured leaves do not provide smooth and fine finishes. It can effectively be grown in the warm tropical regions.

Holcus lanatus (syn. *Notholcus lanatus*; common names: velvet grass, Yorkshire fog, Yorkshire fog proper). In Latin, *Holcus* has derived from *holcos* 'attractive', and applies to a kind of grass. It is also found as weeds in the lawn in circular patches during autumn, with light colour and hairy leaves and persists in the lawn even after mowing. This grass is from Europe, suitable for sterile soil and grows from 60 to 90 cm tall with velvety blades and panicles greenish or purple-tinged. *H. mollis* is another soft-grass species for lawn making with creeping habit.

Phleum bertolonii (Timothygrass). *Phleum* derives from *phleos*, an old Greek name for a kind of reed. *Phleum bertolonii* from Europe does well on heavy and wet soils with 6.0-7.0 pH but, at all, this is unsuitable on dry and shallow soils. Shoot bases have a quite distinct bulge similar to spring onion. Its foliage is very attractive and displays a pleasant colour contrast with other grasses. Its per gramme of seeds contain some 4,000 seeds. Its varieties are 'Nobis', 'Ramona', *etc. P. pratense* (Timothy) is also found in turf which is light green. Though its meadow-types are not in much use in turf production but its var. 'S. 50' which originated at Aberystwyth is though expensive but is capable as a constituent of turf.

All are sometimes found in acid soils in lawn and have potentialities of making a constituent of turf. *Agrostis canina, A. stolonifera, A. tenuis* and *Festuca ovina* give pleasing natural appearance even without mowing; *F. rubra* ssp. *commutata* is highly suitable to light soils and quite tolerant to dry conditions, and *F. rubra* ssp. *littoralis* is more wear-tolerant and survives drought much better than and *F. rubra* ssp. *commutata*. *Agrostis* and *Festuca* are the two lawn genera that form the basis for most fine lawns, *vis-à-vis* used to thicken up the sward in more hard wearing areas, whereas the lawn grasses most suitable for coarse or hard-wearing lawns are *Lolium perenne, Phleum bertolonii* and *Poa pratensis*. In fact, cool-seaon lawn is mostly a mixture of more than one species, though in certain specific areas either of the two can be grown on its own. The mixture for making a good cool-season lawn consists of rough-stalked bluegrass (*Poa trivialis*) with Kentucky bluegrass (*P. pratensis*), and fine-leaved fescue (*Festuca tenuifolia*) that stands well to close mowing as well as wearing. In dry shades, fine-leaved fescue (*F. tenuifolia*) with some Kentucky bluegrass will make a good lawn. However, in the lawn of heavy wear or that which is under dense shades, the seeds of perennial ryegrass (*Lolium perenne*) are sown for better effects. In excessive dry regions, western wheatgrass (*Agropyron smithii*) or buffalograss (*Buchloe dactyloides*), or fairway-crested wheatgrass (*Agropyron cristatum*) amenable for mowing up to 5 cm height, are quite effective. To minimize disease problems and for overall colour impact in single species lawn, several of its varieties should be sown/planted together.

D. Other Herbage as Substitute to Lawn Grasses

Sometimes, instead of grasses certain other herbage is used such as *Arenaria balerica, A. montana, A. verna, Chamaemelum nobile, Cotula muelleri, Dichondra micrantha, Duchesnea indica, Fragaria chiloensis, Hedera colchica, Mazus reptans, Ophiopogon japonicus, Phyla nodiflora, Potentilla cinerea, Prunella vulgaris, Sagina subulata, Thymus serphyllum, Trifolium repens, Veronica repens, etc.*

Arenaria balearica (Caryophyllaceae; Corsican sandwort). The generic name *Arenaria* derives from the Latin *arena* 'for sand' as many of the species are native to sandy soil. It is native to Balearic Islands and Corsica. It makes a dense creeping mat of interlacing stems up to 7.5 cm high bearing 2-4 mm long, needle-like ovate, glossy, thick and ciliated leaves and 5-6 mm wide white flowers on long peduncles from spring to summer. It has cold hardiness to the extent of -15 to -21 °C. This likes shade and moist soil for lawn making.

Arenaria montana (Caryophyllaceae; mountain sandwort). It is a loosely mat-forming species growing up to 2.5 cm high, and is native to SW Europe, bearing grass-like, grey-green, hairy, 1-2 cm long and oblong-lanceolate leaves and solitary cymose flowers that are white and some 2 cm wide but quite long-stalked from spring to summer. It has cold-hardiness to the extent of -23 to -29 °C. It makes a good lawn even in poor soils.

Arenaria verna (syn. *Alsine verna*; Caryophyllaceae; moss sandwort). A native of rocky mountains of Europe, it is a mat-forming species developing lumps unless occasionally thinned out. It grows 2.5-7.5 cm high, bearing bright green, erect, flat, strongly 3-nerved and linear-subulate needle-like leaves and the flowers appear on filiform peduncles with strongly 3-nerved sepals. Its var. *caespitosa* is a leafy form, compact and makes dense moss-like masses all summer. It has cold-hardiness to the extent of -29 to -37 °C.

Chamaemelum nobile (Asteraceae; chamomile) with a non-flowering clone 'Treneague' is most suitable for making lawn as it is low-growing though tolerates drought better than grass lawns, and its leaves when crushed release a sweet apple-like odour. It is frost-tolerant and survives from -23 to -29 °C temperatures, and withstands drought conditions and traffic tolerance excessively.

Cotula (Asteraceae) in Greek means a 'small cup', alluding to the clasping leaves forming a basin. There are about 50-60 annual to perennial species, largely in the southern hemisphere, especially New Zealand, are strong smelling and yellow-flowered. The leaves are fern-like, alternate, toothed, lobed or pinnatifid, heads many-flowered, discoid, pedunculate and hemispherical or bell-shaped, outer florets almost apetalous, and the disc-florets are 4-toothed. It is more hard-wearing as it forms a thick carpet of creeping stems, and flourishes well under moist conditions. *C. dioica*, a dwarf and compact carpeter produces some 30 cm long and creeping stems which may be glabrous to slightly hairy and which bears solitary to tufted and stalked leaves up to 5.0 cm long, linear-obovate to spathulate, obtuse, and serrate to pinnatifid. *C. muelleri* grows with long, stout, creeping and rooting stems bearing ascending branches. Its leaves are linear-obovate, deeply pinnatifid, glandular-dotted and 5.0-12.5 cm long.

Dichondra micrantha (Convolvulaceae) from New Zealand and Australia is distributed throughout the world in the tropical and sub-tropical regions and has a flat and mat-like creeping growth habit with cordate leaves. In Greek, *di* means 'two' and *chondros* stands for the 'grain' due to its 2-lobed capsule. Though there are 10 species under this genus but *D. micrantha* is the most important. It is most suitable for the tropics of India but a -4°C winter temperature is highly injurious though is tolerant of the very hot summer temperatures as is found in the North Indian plains. Its lawn should be established either in spring or during summer when soil temperatures are sufficient for seed germination and rapid growth. In large areas planting with *Dichondra* is highly expensive so a small area should only be planted with it only through plug system by cutting flats of 5.0 cm^2 or larger and transplanted 15-30 cm apart. It is a high-maintenance plant which requires regular watering and frequent fertilization, *i.e.* every 2-3 weeks at 500 g of high nitrogenous fertilizer per 100 m^2 area for making the lawn attractive (Rice and Rice, 1986). In shade with abundant moisture it requires monthly mowing but when grown in full sun the mowing requirement becomes nil.

Duchesnea indica (syn. *Fragaria indica*; Rosaceae; mock or yellow strawberry). *Duchesnea* was named for A.R. Duchesne, monographer of *Fragaria* in 1766. It is native to the waste grounds of New York, Afghanistan and Japan, and grows up to 5.0 cm high. It is a neat ground-trailing and pubescent plant with leafy runners. Leaves are strawberry-like; leaflets petioled, coarsely crenate, rhombic-ovate and obtuse. Yellow flowers which appear during spring are pedunculate and some 1.9 cm across, and the fruits are red and some 1.2 cm in diameter. It can tolerate cold-hardiness to the extent of -21 to -23 °C.

Fragaria chiloensis (Rosaceae; beach strawberry). It takes its generic name from the Latin 'fragrance', owing to the fragrance of the fruit. It is native to the Pacific coastal region from Peru to Patagonia. It is densely tufted and grows low but is stout in all its parts and throws out runners only when fruiting is over. Leaves are thick, deep glossy-green above and densely downy beneath, the leaflets are broadly obovate, 2-5 cm long and blunt-toothed. White flowers are clustered, long-rayed, 2-3 cm across with short peduncles so lopping on the ground. Berries are darker, 1.5-2.0 cm wide, firm and with musk flavour. Its cold hardiness is to the extent of -23 to -29 °C. Its lawn is good for sandy soil and spreads through runners.

Hedera colchica (syn. *H. roegneriana, H. coriacea*; Araliaceae; colchis or Persian ivy). *Hedera* is the ancient Latin name of the ivy. It is native to Asia Minor, Caucasus and Iran. It is robust, vigorous and high-climbing plant, bearing bright green leaves that are leathery, cordate, almost entire but rarely somewhat 3-lobed, some 25 cm long and 18 cm wide, and of the flowering branches normally oblong-ovate. Leaves overlap the ground tightly. Its cold-hardiness is to the extent of -21 to -23 °C. It is shade-tolerant.

Mazus reptans (Scrophulariaceae; mazus). *Mazus* takes its name from the *teats* or tubercles in the mouth of the corolla. *M. reptans* is native to Himalayas. It is a 2.5-5.0 cm high tufted perennial with slender, prostrate and node-rooting stems. The plant looks like the small lobelias. Leaves are opposite, lanceolate to elliptic, sparsely-toothed and deciduous. Racemes are erect, 2-5-flowered, and the flowers are purplish-blue but lower lips are blotched red-purple, yellow and white. It prefers moist soil and partial shade. Its cold-hardiness is to the extent of -29 to -37 °C.

Ophiopogon japonicus (Liliaceae; dwarf lily turf). In Greek, *aphis* is for 'a snake' and *pogon* for 'beard'. This perennial, stemless, glabrous and evergreen species is native to Japan, Korea and China. It is tufted and mat-forming plant with underground stoloniferous rhizome, which bears long, slender, fibrous and nodulose roots. Numerous leaves appear from the rootstocks which are dark green, grass-like, 10-20 cm long with 5-7-nerved and its fast growth covers an area rapidly. Racemes are 7-12 cm tall bearing 4-5 mm long lilac to white flowers in summer. Fruits are pea-sized and deep blue. It grows comfortably in sandy soils. It can tolerate as low as -12 to -15 °C temperatures. Its var. 'Variegatus' bears variegated leaves.

Phyla nodiflora (Family: Verbenaceae; frog plant or garden lippia) is a drought-tolerant groundcover on sunny banks. In Greek, *phyle* is for 'clan or tribe', an illusion to the many florets in a tight head. The genus comprises some 15 species from warmer regions of Americas. It is mat-forming perennial creeping or procumbent and stem-rooting herb, sometimes with woody base, bearing grey-green leaves which withstands treading and mowing though occasional to maintain even-surface. It prefers warm conditions and sunny situations and is highly drought and traffic tolerant. It tolerates up to -18 °C temperature. Its other important species are *P. canescens, P. lanceolata* and *P. dulcis* (*Lippia dulcis*).

Potentilla cinerea (syn. *P. tommasiniana*; Rosaceae; rusty cinquefoil). In Latin *protens* stands for powerful, owing to the medicinal properties formerly accorded to certain species. It is native to south-east and central Europe. It is a mat-forming with prostrate, slender, ascending, rooting and somewhat woody stem growing 5.0-10.0 cm high. Leaves are gray-green, with starry and simple hairs and basal ones palmately 3-5-foliate. Leaflets thick, strawberry-like, 0.5-2.0 cm long, sessile or almost so, narrowly to nearly oblong to broadly obovate, toothed and apex-rounded. Pale-yellow flowers in a cluster of 3-5 and some 1-2 cm wide appear in summer. Its cold hardiness is to the extent of -29 to -37 °C.

Prunella vulgaris (Family: Labiatae; common names: Heal all, Self-heal). *Prunella* has arisen from *Brunella* which was its pre-Linnean name, however, its origin is obscure but may stem from *Die Breaune*, German for quinsy ~ a throat infection which these plants allegedly cured. The genus comprises seven species native to temperate Northern Hemisphere. Its semi-evergreen leaves lie flat on the ground. Floral spikes bear purple flowers which appear during summer. In damp soils its growth is quite rapid and rampant. With regard to its cold-hardiness, it can tolerate -29 to -37 °C temperatures. Its other noteworthy species are *P. grandiflora* with several colour variants such as 'Alba' (white), 'Rosea' (pink), and 'Rubra' (red), *P. hyssopifolia, P. laciniata* and *P. vulgaris.*

Sagina subulata (syn. *S. pilifera, Spergula pilifera*; Caryophyllaceae; heath pearlwort). *Sagina* in Latin is for fodder, as a species *Spergula arvensis* formerly included under the genus *Sagina* was used for sheep-feeding. It is native to Corsica. It is an evergreen, densely tufted, much-branched, creeping and mat-forming plant covering the ground like a sheet of moss, growing up to 10 cm tall, and bearing needle-like, stiff, marginally aristate, linear and 6-15 cm long leaves. White flowers are 5-petalled, some 0.5 cm long and appear during summer on 2-4 cm high thread-like stalks. It prefers shady locations and is useful in carpet-bedding. It can bear -23 to -37 °C temperatures. Its var. 'Aurea' has leaves marked with yellow.

Thymus spp. (Labiatae; creeping thyme). *Thymus* is ancient Greek name used by Theophrastus either for thyme or for savory. It is a genus of 300-400 aromatic species of hardy perennial herbaceous plants and sub-shrubs, some growing prostrate and others up to 30 cm tall. *T. caespititius* (*T. azoricus, T. micans*) from Spain, Portugal and Azores is low hummock- or mat-forming with base-ciliate leaves which are some 8 mm long and narrowly spathulate, and the flowers purplish-pink to almost white and are some 6 mm long. *T. praecox* from Europe is mat-forming and some 5-10 cm tall, bearing 5-14 mm long leaves which are obovate to almost orbicular, base-ciliate and the flowers in great profusion are 8 mm long and purple. This species is highly variable and is divided into 5 sub-species, out of which only one is in general cultivation and that is *T. praecox arcticus* (*T. drucei*; mother of thyme) where leaves are obovate, 5-8 mm long and flowers variable in colour and about 6 mm long. *T. praecox* var. *arcticus* 'Coccineus' (*T. coccineus*) is dark-leaved with crimson flowers, *T. praecox* var. *arcticus* 'Doone Valley' is another form worth lawn making, *etc.* *T. serphyllum* is evergreen which grows less than 2.5 cm in height so is best for lawn making, and produces small purple flowers during summer. It withstands poor soils, shade and drought and tolerates temperatures as low as -29 to -37 °C. *T. lanuginosus* and *T. lanicaulis* are since taller so are less tolerant of foot traffic.

Trifolium repens (Leguminosae; white clover, trefoil). In Latin, *tri* is for 'three' and *folium* for 'a leaf.' There are some 300 species but *T. repens* from Europe, NW Asia and North Africa is only important species for the purpose of lawn making. It is mat-forming with stems creeping some 50 cm long with node-rooting, leaves digitately 3- but rarely 3-7-foliate, leaflets some 1-3 cm long, obovate to obcordate and fragrant flower-heads some 1.5-2.0 cm long are globular and white which age to pink or entirely pinkish or purplish. It is much used with lawn seed mixtures. It contains some 1,500 seeds per gramme of sample.

Veronica repens (Scrophulariaceae; creeping speedwell). The genus *Veronica* was erected for St. Veronika. A native of Corsica, this plant is slender, prostrate and grows up to 10 cm tall in compact and dense masses. Its leaves are shining green, moss-like, ovate, a little crenate and 0.6-1.2 cm long. Racemes which appear during spring are slender, few-flowered, flowers rose to bluish-white, and the capsules are more in diameter than their length and deeply notched. Its var. 'Alba' is a white-flowered form. Its lawns are successful in the open situation with moderately dry soil though preferring moist corners, and where grasses can not be grown, this can be grown for making a rapid sod. It can tolerate the temperatures as minimum as -21 to -23 °C.

Various other species of moss though not standing well up to heavy and continuing wear, these may primarily be used for ornamental areas to creep over the patios and courtyards, and along the path-edges, however, are not the best choice for the main lawn. These all the plants are mat-forming, most suitable for creating 'tapestry lawn' with a patchwork effect though these should not be used so frequently.

Division of Floriculture and Landscaping, I.A.R.I., New Delhi is maintaining *Agrostis palustris* (for cool regions), *Cynodon dactylon* var. *bargusto*, 'Palna', 'Panam', 'Panama' and 'Selection 1' (all for warm regions), *Dichondra repens* (a lawn-alternative in tropical areas though not a true lawn grass), *Eragrostis curvula* (annual lovegrass for tropical areas), *Lolium perenne* (hard-wearing ryegrass for cool areas), *Paspalum notatum* (tropical region bahiagrass or Argentine bahia) and *Poa pratensis*, *i.e.* hard-wearing Kentucky blue meadow grass for cool regions (Namita and Janakiram, 2012). *Eragrostis* (Gramineae) derives from Greek *er* for 'spring' and *agrostis* 'a grass', and is closely allied to *Briza* and *Poa*. *E. curvula* with wide distribution is a hardy annual with more or less diffuse but with feathery and graceful panicles of many-flowered small compressed spikelets. It is propagated through seeds. Its fully developed inflorescences should be harvested and dried for dry-flower arrangement in case grown for ornamentation.

Lawns for Sports

Construction and maintenance for lawns used for tennis, croquet, putting and other sports, are a bit different than usual lawns as in these cases it should be liable to withstand maximum wear and tear whereas usual ornamental lawns which usually are banned for public may have soft and succulent cover of grass. **Hard-wearing grasses suitable for warm and tropical areas** are *Agrostis elegans*, *A. nebulosa*, *A. pulchella*, *Axonopus affinis*, *A. compressus*, *Buchloe dactyloides*, *Cynodon dactylon*, *Eremochloa ophiuroides*, *Lolium multiflorum*, *Paspalum distichum*, *P. notatum*, *P. vaginatum*, *Polytrias diversiflora*, *Stenotaphrum secundatum*, *Zoysia japonica*, *Z. matrella* and *Z. tenuifolia*; whereas **hard-wearing grasses suitable for temperate and cold regions** are *Agrostis canina*, *A. stolonifera*, *A. tenuis*, *Cynosurus cristatus*, *Festuca elatior*, *F. rubra*, *F. tenuifolia*, *Lolium perenne* and *Poa pratensis*.

For **tennis courts**, the lawn should follow the principles of trueness, firmness, uniformity and durability so that it is possible to get a good accurate bound to the ball. As much as possible, the orientation of courts should be north to south with full of natural light, however, with suitable background of shrubs and trees. There should be sufficient room with plenty of run-back (6.0-6.3 m) and side run (3.6-4.2 m) though double courts should have a gross area of 23.4 × 10.8 m [(11.7 × 2 m both sides net length and 8.1 m net width) while side space 2.7 m (1.35 m each side to the court's length)]. The site should be graded uniformly and carefully throughout to help water run-off besides giving a layer of ashes 10-15 cm thick below at least 10 cm of the ground but after consolidation. Though seeding takes more time but it makes the best turf for playing whereas turfing requires less time. The court should be mown regularly and rolled heavily so that true hard surface is formed and at the end of each playing season the turf should be spiked and every three years a tubular forking should be carried out. Good top dressing and scarification should also be carried out. No earthworm should be allowed to invade the court and weeds but if any should immediately be taken out or controlled through selective weedicide, lawn raked and then dressing with ammonium sulphate is recommended to cover the gaps created by removal of weeds. Regular watering should be given especially during dry weather.

Croquet lawns are just like the golf greens but without undulation. The turf used should be fine and fast-growing, *vis-à-vis* quite amenable to wear and tear, and the surface should be quite flat. Regular and systematic spiking and forking are necessary to avoid soil compaction. Other treatments are just like the ornamental lawns. **Putting lawns**, mostly having gentle undulations with true surfaces so that golf ball holds to the line along which it is struck, are meant for playing 2-, 3-, or 4-ball competitions or for practicing putting holes.

Maintenance

In October, the lawn should be well swept and raked to take out the dead grass, weeds and mosses, followed by rolling with spiked roller having spikes 7.5-10.0 cm apart and to a depth of up to 15 cm so that it may facilitate proper aeration in the lawn. In January-February, the lawn should be dressed with clean sharp sand, and equal volume of clean fibrous loam & leaf mould together, along with finely-powdered bonemeal. The maintenance part requires various operations as mentioned here with.

Mowing

Initially for up to 2-3 months of turf-planting, the grasses are kept in height with scythe for fear of uprooting the turf but afterwards the height is controlled through mower. Before cutting or mowing it would be better to sweep the whole lot to keep out pieces of bricks, rocks and stones, any other hard material or worm-casts so that blades of the scythe or mower are not damaged. Scythe should be used early in the morning when the dew is still on the grass and the stems are stiffer while the mower is used from March until mid- November at the time of the day when the grass is dry, *i.e.* after morning hours. During spring and autumn, the lawn should be mown every 10 days though in summer at weekly intervals and after commencement of rains every five days. During mild winter also when the grass has grown above 5.0 cm in height, it should be mown otherwise mowing is not required throughout winter. Lawn edges should also be kept neat and in proper shape through regular shearing with the edging shears.

Rolling

For correcting minor variations on lawn surface caused by walking, worm castings, and burrowings, *etc.*, rolling is carried out. While rolling, one should be careful that worm casts should not be rolled over but rolling should be done only after sweeping these. In fact, earthworms are beneficial to the plants and their burrowing causes aeration in the region of root but they do more harm than good as the surface becomes soft and muddy for pedestrian traffic, grass becomes weak and wears easily and provides the passage for weed growth. Watering the earthworm-menaced lawn with potassium permanganate at 3 g/l of water is very effective to control them. Rolling helps the lawn grass to anchor itself securely by spreading the roots, grows horizontally on the surface, by making compaction of the soil by removing air-filled porosity, and over all, makes the soil surface levelled. Rolling is carried out more frequently on a newly laid lawn, a few in the early autumn and another 3-4 in the spring, just before the commencement of mowing. During frost or just after, the rolling is prohibited. Rolling is recommended after every weeding so that surface remains levelled. Heavy soils require light rolling while light soils heavy rolling. Light rolling is always beneficial with a light roller than one with a heavy roller. The ideal stage of rolling the lawn is when soil density is 1.6 g/cm^3. Rolling of the lawn should be done throughout the year when soil is not too wet and sodden, but from June to August normally this operation is not required.

Scarifying, Scraping, Raking, Spiking and Slicing

Scarification is the vigorous use of rake or similar equipment to remove weeds and thatch from the surface of the lawn where they form an impenetrable barrier to air, water and fertilizer. In spring, the scarification is light than during autumn. Regular mowing, rolling and treading results into the hard crust formation, *vis-à-vis* lower stratum of the lawn becomes matted and hard, therefore to loosen the soil **scraping** with a small antiquated tool (*khurpi*), a small pick-axe or the sharpened metal hooks is carried out during March-April and also in September-October. There is much to be gained from a light raking during even growing season, apart from during spring and autumn. Strong multi-prong **rakes** may also be used to drag out all the weeds, dead grass, the moss and to remove the surface-thatch formed by the grass itself, made up of its roots, stems and leaves, whenever the need be. Scraping, raking, spiking and slicing practices are much less severe than scarifying though all these operations are almost one and the same. Spongy lawns mostly suffer due to thatch layer. Thatch occurs where the rate of production of organic matter in a lawn exceeds the rate of decomposition. A 6 mm layer of thatch works as a mulch, reduces the wear of the grass and water loss but thickness more than this encourages shallow and surface-rooting and becomes a barrier for water penetration. This apart from loosening the soil also makes provision for aeration in the zone of stolons and roots. Raking is carried out length- and cross-wise in a well-set lawn and then the mower is set as close as possible to the ground. After raking, the lawn should be dressed with **spiked roller** or the garden forks having spikes some 15 cm long and 7.5-10.0 cm apart for **aerating** the root zone. In this process, the hard and compacted layer of lawn is broken for almost every month, especially during autumn for water penetration, free drainage and proper aeration up to the pores close to the growing roots. For such works in the just moist soils the tines are allowed to enter. Tines are of three types, *viz.* knife-shaped quite wide at the end as to that of a *khurpi* which is helpful for stimulating root growth and to break through the layer of thatch so that its decomposition is hastened, the solid tine is almost round and slender for penetrating the quite hard and compacted lawn where penetration is very difficult, and the hollow tine is meant especially in poorly drained soil for removing the core of the soil to facilitate aeration and for working proper top-dressing of soil, sand and peat into the lawn. After coring in poorly drained soils, the cores may be filled with sandy materials. The net length of each tine is 10 cm with a provision of handle to the other end. However, in February, the lawn should be dressed with clean sharp sand, and equal parts of fibrous loam and leaf-mould, along with sprinkling of finely powdered bonemeal. **Slicing** in lawn is a though summer activity wherein the V-shaped knives mounted on discs are pressed in the soil 7.5-10.0 cm deep similar to spiking in heavily trafficked turfs that have high surface compaction but this can be repeated as often in any season to loosen the compact soil. This stimulates root and shoot growth around the holes. Sometimes, instead of these activities, the soil is cored which causes soil coming out of the holes along with grass roots and stolons so this is not that much desirable.

Fertilizing and Irrigation

The lawn should be **fed** once during winter, *i.e.* October-November with wellrotten manure @ 1-2 kg/m^2, in March with 30 g/m^2 potassium sulphate + 15 g/m^2 sodium nitrate, and in October in light soils with 50 g bone meal + 50 g superphosphate of lime/m^2. In late February, 20 g/m^2 of ammonium sulphate should be applied. After every feeding, as a general rule, the lawn should be watered. Slurry (liquid manure) may be applied in October and March. Slurry is prepared by adding three parts of water to one part of fresh cow dung and 8 per cent urea to total volume, and then fermented in an earthen pot covered with an earthen lid under sun for 7-15 days depending upon the atmospheric temperature and then strained through fine mesh or cloth and then diluted to its four times, and this is applied on the lawn. In light soils when grass is poor, in March application of 30 g/m^2 potassium sulphate will be very useful; poor and weak-growing grass in damp and sour soils can be corrected by applying pulverized chalk at 300 g/m^2 in November or December; in light sandy or gravelly soils if the grass is scorched, it should be applied with 1.25 cm thick layer with equal parts of well-sieved fibrous-loam and leaf mould with a sprinkling of soot in February; in case grass is poor and weak in heavy soils, in October or February 60 g/m^2 of bone meal/equal parts of bone meal + superphosphate of lime should be applied; when grass is coarse in heavy soils, application of 1.25 cm thick layer of clean and sharp river-sand in early February will rectify the problem; in heavy soils if the grass is showing scorching, 15 g/m^2 of nitrate of soda should be applied in March which encourages grass-growth in cold weather and discourages clover; and the soils which are mossy, exposed soot should be applied to the lawn sufficient to blacken the lawn during October-November just before the rains (Coutts *et al.*, 1963). Shiels (1984) recommends ammonium sulphate 1.5 kg, superphosphate 2.0 kg, bone meal 0.5 kg, potassium sulphate 0.5 kg, calcinated iron sulphate 0.5 kg and if possible dried blood 0.5 kg to each 100 m^2 area when in spring the grass has started taking growth and/or during autumn only if grass cover appears thin or weak otherwise not, however, ammonium sulphate at 170 g/100 m^2 should be applied twice during summer, each at 45 days interval. Before fertilizer application the lawn should be spiked so that nutrients reach to the roots, and after every fertilizer application, it should be followed with irrigation. Alkaline soils (pH >7.0) not being suitable to turf grass due to poor availability of nutrients so in such cases elemental sulphur should be applied in the established lawn @ 2.2 kg/10m^2 during spring or fall and this practice is known as **sulphuring**.

A well-cared-for lawn is the one which is adequately fed, regularly scarified and aerated, and such lawns survive long periods of dry weather as the grass has a deep and extensive root system, however, only a few signs of suffering may be seen but in such cases immediate recovery is made when **water** is immediately supplied. When grass colour becomes dull with a distinct blue tint, footprints persist longer since the grass is becoming limp and then subsequently it starts drying up, it means that there has been a long gap of watering. Rice and Rice (1986) also advocated that just by walking on the lawn if the grass stands upright quickly it has sufficient moisture in the soil but if it takes long time or does not recover the turf is likely to wilt soon for want of moisture so immediately the lawn is to be watered. When limping is noticed, immediately it should be watered. Lawn should be watered twice in a week during summer and at 7-10 days during winter. Lawns maintained at roof-tops should be watered every 2-days during summer and every 6-days during winter. Watering of the lawn though depends much on the prevailing environmental conditions and the soil type. Clay or heavy soils show sign of water-stress later than the sandy soils as latter has poor water reserves. In dry and hot weather the lawn may lose 25 mm of water in 7-10 days for which the lawn requires about 27 litres of water for every one metre square area (Shiels, 1984). In dry weathers the lawn should be saturated with water as light watering will simply skim the surface which will force the grass to encourage only shallow rooting, and as a result the lawn devastates. Water should be given at least to a depth of 15 cm which can be ascertained by digging test holes before and after every watering.

Lawn Renovation

This applies only to those lawns which are either patchy or in even poor condition but not so poor as to require complete re-establisment. The poor condition may be due to a very poor maintenance, a very poor soil quality, low fertility levels, weed invasion, thatch accumulation, soil compaction or a combination of a few of these or other causes. The renovation can be taken up in any month but the late-summer is best for the cool-season grasses while late-spring for warm-season ones. Before starting the renovation work one should be sure about the cause of the decline and the possible reclamation measures. Lack of proper maintenance results into weedy growth, paling of lawn and footpath formation. Nutritional and/or pH imbalance also cause poor show of lawn. The cause for patchy lawn is due to poor drainage, excess of shade, underground insect infestation or infection of pathogens. These all the problems are discussed one by one in the following paragraphs with their remedial measures.

Uneven Surface

In the process of lawn settlement and compaction the hollows and bumps are caused, especially in all the new lawns, though good preparation of the ground minimizes it maximally. Addition of organic manure also causes such problems due to their further decomposition. In such cases, during autumn as winter gives it ample time for settlement, the turf should be turned back to remove or add the soil below as per requirement and then after firming the soil the turf is replaced to its original position.

Cracking in the Soil and between New Turves

The soil cracks normally occur in spring-sown lawn when soil dries out. In such cases the soil should be given thorough irrigation followed by top-dressing of some 2.5-5.0 cm layer of equal parts of wood ash, pulverized fibrous loam and properly decomposed manure to fill in the cracks, and if need be a little more seed is sown to thicken up the lawn.

When after turf-laying there had been long gap in watering, it is usual that cracks will occur, and in case when after laying lawn the cracks were not given sandy top-dressing immediately, this problem is more likely to occur. To mitigate this problem, there should be regular watering, especially with no any gap during dry weather and sandy top-dressing in the affected portion.

Waterlogging

Long-standing of water on the surface due to poor ground management during soil preparation on or afterwards due to soil compaction when water can not escape through this to the subsoil or drains is the cause of this problem. Correction of soil compaction through aerating it by hollow tine fork may solve this problem if while preparing the land for planting the provision of drainage has been made by mixing sufficient sand in the soil and by keeping the provision of piped drainage system below.

Spongy Surface

If surface of the lawn is soft holding water for long, it may be due to peat-like material formed thereon due to thatching. Thatching occurs faster than its disintegration so thatch-thickness is gradually increased considerably, and when it is more than 6 mm in thickness it becomes quite harmful, and neither permits top-dressing nor fertilizers down to the plant roots. Thatch traps the water in its thickness though soil remains dry which causes shallow-rooting and so the grass becomes highly susceptible to drought. Thatch formation is mostly due to soil compaction and often with shallow watering. Therefore, it is advised to aerate the lawn every year and water the lawn deeply every time so that thatch build up is avoided. Thorough aeration and scarification will solve this problem.

Thin or Sparse Grass

Thin sward or sparse grass is caused on a new lawn due to poor seed-bed preparation. Apart from this, this may also be caused due to persistent shade, inadequate feeding, severe soil compaction and/or poor mowing. This can be prevented through careful management of these factors.

Coarse Grasses

Weeds of the coarse grasses invade the poorly maintained or neglected lawn. The situation is sometimes aggravated due to excess of fertilization or watering, *vis-à-vis* infrequent cutting or mowing when the lawn grass has grown exceesively higher. To avoid these problems, one should be very careful in carrying out all these operations correctly and timely so that lawn is not neglected and remains clean and beautiful. The lawn should also be regularly scarified or the clumps should be slashed with knives, and during summer between two waterings the lawn should be allowed to dry.

Yellow or Pale-Green Areas

Usually yellowing of the lawn is associated with inadequate fertilization, though other factors also contribute to this problem. Thatch problem does not allow water and nutrients reaching below to roots so roots start forming shallowly which aggravates the situation and the soil becomes compact and waterlogged. Certain pests also feed on the roots. Removal of thatch, aeration in the soil, timely and correct feeding, *vis-à-vis* watering and/or incorporation of insecticidal granules will solve this problem.

Damaged Edges

When the lawn edges are damaged or broken, correction may be done by trimming the affected edges and filling the gaps with trimmed turf pieces or compost and seed so that the look may be neat and nice.

Bare Patches

Bare patches are caused by hard wear and tear such as on tennis court. This is caused due to persistent disturbance to the planting and wear and tear, *vis-a-vis* severe soil compaction. This is also caused due to disease infection or lawn mower scalping. The patchy soils should be water-filled, soil-scarified and planted with similar turves by inserting thereon but at the height of some 1.25 cm than the original level so that it settles to the level after settlement. While inserting the turves, more than the actually worn portion should be stripped away so that the point of camouflage of the new and old turf may be vigorous and healthy, and together the empty spaces should be levelled with fine soil and seeded. These reclaimed patchy lawns should be reused only after the grass has fully recovered.

Black Surface-Slime

The sign of black or black-green slime in the lawn is due to presence of algae which causes a thin turf and waterlogging. This requires all together stopping of lawn rolling and intensive aeration of the soil. Incorporation of sand to the upper surface of soil and the each 100 m^2 of lawn should be treated with 34 grammes of copper sulphate dissolved in 136 litres of water.

Diseases and Insect-Pests

Brown Patches, Fairy Ring and Other Diseases

The straw-coloured or brown patches with whitish or pink tinge or mould covering are due to *Fusarium* infection, however, fescues may suffer even from dollar spots. Sometimes brown patches appear when its roots are eaten by some insect-pests such as leatherjackets. Bitch urination in the lawn during dry season also causes brown patches.

Brown patch disease (*Fusarium nivale, Rhizoctonia solani*) occurs at the end of the growing season which becomes more serious when the turf is with fine swards, *vis-à-vis* overfed with nitrogen though death of the grass is casual. The first sign appears in the form of small, brown and circular patches about 10 cm in diameter, afterwards pink or white mould covers the grass which makes the grass slimy with eventual death. A soil with high pH is comparatively more susceptible so lime should not be used in high pH soils. Bavistin or Quintozene is very effective. Chlorothalonil, Flutolanil, Propiconazole, mancozeb, thiram, Triademefon or Triconazole is also effective. ***Corticium*** **patch disease** (*Corticium fuciforme*) usually appears during summer and may continue until autumn, creating seriously damaged and discoloured grasses though these are not killed. The grasses give a bleached look with small pink-branched needles of the fungus developing on the sheaths and blades, most serious being on fescues. Ordinarily, nitrogenous fertilizer application improves such lawns but in severe cases Bordeaux mixture should be applied. ***Ophiobolus*** **patch** (*Gaeumanomyces graminis* syn. *Ophiobolus graminis* var. *avenae*) is a ring disease observed mainly on the sport's turf planted with fine grasses such as *Agrostis* though fine grass *Festuca* is resistant. This disease is quite prevalent on chalky or limy soils or on lawns which have been limed. In its infection, bleached patches develop and spread and the fungal growth is apparently visible among the dead, brown roots and stolons of the grasses. These patches are soon filled up with weeds. Though small outbreaks can be corrected by re-turfing but larger ones are best encountered with an organo-mercury fungicide + fertilizer and renovation with seed. Removal of affected areas and several inches of soil below the affected grasses, then filling with fresh topsoil but no liming and then re-seeding is the best remedy.

Thatch fungi disease in lawn is caused by various

fungi whose infection produces bleached, yellowed or sometimes reddish patches of grass in the lawn and such grasses are severely stressed or killed and fungal growth occurs around the base of the grasses. Though these fungi are often not directly pathogenic but live on organic debris, mowings and on thatch, *i.e.* remains of old grasses around the grass bases and form a strong water-repellent layer so grasses die due to drought stress. New turves not maintained properly are the worst affected ones. One can get rid of this problem by proper lawn maintenance, its scarification, spiking, and weed & moss control. No organic debris should be allowed to accumulate in the lawn. Badly affected areas should be removed and re-seeded or re-turfed. Common grasses are affected with **slime moulds** which tend to thrive during periods of heavy rain. Clustered beige, orange or white fruiting bodies smother individual blades of grass and then the spores are released which provide it a grey-appearance. Though this is not harmful but makes the appearance of lawn unsightly during late spring and early autumn. Affected areas should be jet-sprayed to clean these off.

Fairy ring disease (*Agrocybe* spp., *Lepiota morgani*, *Lycoperdon* spp., *Marasmius oreades*, *Psalliota compestris*) of lawn which appears in late summer to autumn is of three types ~ **one** of quite little importance is that which forms only a ring of toadstools in random arrangements or in straight lines but normally this does not damage the lawn though for slowing down the growth of the fruiting bodies, sometimes magnesium sulphate or iron sulphate is used; the **second type** causes a dark green ring of stimulated grass, probably due to fruiting bodies being formed at the edges, and this can be overcome with application of nitrogenous fertilizer; and the **third type** of ring is serious and is caused due to infection of *Marasmius oreades* which feeds on the organic matter in the soil and releases nitrate as a waste product which is taken by nearby growing grasses where growth is accelerated and the grass becomes darker and so the green rings are formed, and the rings are two, one inner and the other outer whereas in between there is dead grass. Death of the grass occurs due to soil being full of fungal mycelium which hinders water penetration so grasses die due to creation of drought conditions. The rotting of the mycelium in the soil releases more nitrate to be absorbed by nearby grasses hence second ring of darker colour is formed. Ringmaster, the fungicide based on oxycarboxin is effective to kill the fairy ring fungus. Bordeax mixture (4 kg copper sulphate: 4 kg hydrated lime: 50 litres of water) drenching will control these problems. Spray of azoxystrobin, flutolanil or triademefon is also effective.

Red thread disease is caused generally after heavy rain by infection with *Corticium fuciforme*, especially on fine grasses such as fescues though perennial rye grass and annual meadow grass are rarely affected. It is also problematic when soil is nitrogen-deficient or poorly aerated. Its initial symptoms are formation of 7.5-8.0 cm diameter of patches of the reddish-brown to bleached grasses coupled with dark pink horn-like gelatinous growth out of the leaves which turn pale-pink and slightly fluffy. However, when the needles are broken off, the fungus becomes dormant till occurrence of congenial conditions and then these germinate and start a new infection. The pathogen rarely kills the grass but weakens it and spoils its appearance. This disease is common in light and sandy soils deficient in nitrogen and its infection is favoured by mild and damp weather. Its infection is visible from June to December but not afterwards. Good feeding schedule especially with sulphate of ammonia, improved drainage, proper aeration and use of carbendazim will control its infection.

Turf snow mould (*Monographella nivalis* syn. *Fusarium nivale* and certain others) is a disease of temperate lawn with poor aeration and damp conditions and at this time but after the fall of the snow if the lawn is walked on, the disease spreads there too. Patches of dying grass develop and spread, and the grass which after infection becomes yellow enlarges soon, turns brown and die. In damp weather the dying patches become covered with a white to pale-pink fluffy fungal growth which causes the grass blades to stick together. Routine regular spiking or any other means of improving aeration, use of carbendazim fungicide and no walking in the lawn during winter will keep this problem at bay.

Dollar spot (*Sclerotinia homeocarpa*) is common in fine-leaved bents and creeping fescues, *vis-a-vis* in compact lawn with heavy, compact and wet soils with high pH or alkaline soils and those treated with lime and its spread is favoured by poor turf growth and high humidity in summer though it is usually over by the winter. The grass first shows straw-yellow patches which turn brown and dies, the patch starting from 1.0 cm in diameter and then finally coalesce and reach to some 7.5-10.0 cm, especially during autumn. Middle of the leaf blades is also girdled due to its infection though leaf tips may remain green. It can be prevented by improving drainage *cum* soil aeration, spiking in autumn and through proper feeding and its control can be accomplished by applying quintozene, benomyl, carbendazim or thiophanate methyl.

Damping off occurs in the seedlings due to infection with a complex of various fungi, where the grasses rot at ground level. Thick sowing, poor drainage and high soil moisture are the main causes for this problem. Seeds dressed with fungicides should be sown to discourage these pathogens.

Leaf spot (*Bipolaris sorokiniana, Drechslera* spp., *Exserohilum* spp.) forms elongated lesions throughout the year with tan centre and purple to black border on blades which blights entire leaves and sheaths in due course of time. Proper height of mowing, sufficient watering at regular intervals, avoiding nitrogenous feeding, removal of thatch and spraying Chlorothalonil, liquid copper concentrate fungicide, Iprodione or Mancozeb will minimize the incidence and control the disease. ***Curvularia* blight** causes leaves to become yellow, brown or black in fading lawns during hot and humid weathers. Too close mowing, thatching and non-aeration of lawn aggravate this problem. Spraying the lawn with maneb (Dithane M-45) 0.2 per cent and carbendazim (Bavistin) at 0.1-0.15 per cent controls this disease (Parminder Singh and Premjit Singh, 2014).

Dog lichens (*Peltigera canina*) are most troublesome in badly drained, poorly aerated, compacted or shaded lawns. In its attack, the groups of greenish-black, curled or leafy growths with pale-creamy-white undersurface appear in the lawn. To cure the problem drainage and aeration in the lawn should be improved, the lichens should be raked and the soil and the lawn should be fed and top-dressed regularly. In compacted sites with poorly drained and poorly aerated lawn, green or greenish-black slippery patches on the surface are formed due to **growth of gelatinous lichens and algae**. Any plant or tree shading the lawn should be severely pruned and proprietary lawn-care products that contain cichlorophen should be applied to kill lichens and algae.

Smut (*Ustilago cynodontis*) appears when flowerheads are formed though foliage remains free. However, where the seed production is the aim, pre-sowing seed treatment with Captan or Thiram 0.3 per cent is effective.

Insect and Other Pests

Earthworms (*Allolobophora* and others) though themselves are not injurious to the plants but their castings make the lawn unsightly, produce the slippery surface and brings about surface-unevenness which blunts the mower edges and hinders in its smooth working, *vis-a-vis* these often smear to leave a patch of bare soil which entices weed growth. Increase in the earthworm population also entices moles to feed on them by further burrowing the soil which makes the lawn a terrible site. Use of acid fertilizers to maintain a low pH, avoiding lime and as the casts appear exclusively during rainy to autumn season so this time Carbaryl should be used to control these. **Moles** (*Talpa europaea*) damage the lawn by burrowing in the soil and feeding on earthworms and other invertebrates. These can be trapped or killed through zinc phosphide poison baits. **Dogs** to clean their stomach, lick the grasses especially in the morning when dew is there, and then they vomit which burns the grasses in the vomited region and where even afterwards certain insects and microorganisms start working which causes even stolons to rot so dogs should not be permitted in the lawn. The fouling of the grassed area with the droppings of dogs and cats damage the lawn. Their urine scorches the lawn leaves. Their entry in the grassed area should be restricted.

Leatherjackets as the grayish-brown, tubular and legless larvae up to 4.5 cm long and without obvious heads of some eight craneflies or daddy-long-legs (*Tipula paludosa, Nephroloma* spp. and others) feed voraciously on the roots of the grasses during winter and spring though eggs are laid in the turf during the previous summer. Their infestation causes yellowish-brown patches. Leatherjacket infested lawn attracts certain birds, especially the starlings. Proper aeration in the turves and application of HCH watering will minimize and control the pest population. Pathogenic nematode, *Steinernema carpocapsiae* will control the older larvae when the soil is moist and warm above 14 °C. **Cockchafer grubs** (*Hoplia philanthus, Melolontha melolontha, Phyllopertha horticola* and others) are the curved larvae of the beetles feeding on the roots of the grass during summer and spring which can be controlled by applying Carbaryl. **Termites** cause massive damage to lawn under dry soil conditions, feeding mainly on the roots but spreading further up to the leaf blades and convert the lawn into the organic heap. Soil drenching with 0.05 per cent chlorpyriphos or 0.1 per cent malathion before lawn planting will be quite useful. The larvae of **feverfly** (*Dilophus febrilis*) and **St. Mark's fly** (*Bibio johannia*) which resemble small leatherjackets though with shining brown heads sometimes feed in turf during spring and autumn below 1.8-2.5 cm of the surface. As such these are not harmful to the turves as these feed on the decaying organic matter but birds devouring on them damage the lawn. Control is the same as for leatherjackets. **Burrowing bees** (*Andrena fulva* and many others) form small conical heaps of fine soil in lawns. There may be cluster of heaps together. Though they are at all not damaging to the lawn and instead are beneficial in pollination but their presence in the lawn causes a fear among the human beings when working in the lawn. Various **ants** (black ant, *Lasius niger*; yellow meadow ant, *Lasius flavus*; red ant, *Myrmica* spp. and many others) feed on the honeydew excreted by aphids and other sap-feeding insects and make fine heap of soil in the lawn. Though directly these are not harmful to lawn but their ever increasing population is a matter of great concern. Proper and regular watering may damage their colonies.

Weeds

Common weeds occurring in turfgrasses in India are moss-grasses, annual bluegrass (*Poa annua*), barnyard grass (*Echinochloa crusgalli*), congress grass (*Parthenium hysterophorus*), globe amaranth (*Gomphrena celosioides*), goose grass (*Eleusine indica*), horse purslane (*Trianthema portulacastrum*), Indian crabgrass (*Digitaria longiflora*), red-leaved clover (*Alysicarpus vaginalis*), milkweed or dudhi (*Euphorbia microphylla*), red spurge (*E. hirta*), roadside itsit (*Boerhaavia diffusa*), *tinpatia* (*Desmodium triflorum*), tropical crabgrass (*D. bicornis*), witchgrass (*Panicum capillare*), *etc.* as **annual weeds**; black medic (*Medicago lupulina*), carpet weed (*Mollugo verticillata*), prickly lettuce (*Lactuca scariola*), prostrate pigweed (*Amaranthus blitoides*), slender amaranth (*A. viridis*), prostrate spurge (*Euphorbia maculata*), *etc.* as **broad-leaved annual weeds during summer**; common chickweed (*Stellaria media*), redstem filaree (*Erodium cicutarium*), southern plantain (*Plantago virginica*), Virginia pepperweed (*Lepidium virginicum*), *etc.* as **broad-leaved annual weeds during winter**; dallisgrass (*Paspalum dilatatum*), Johanson grass (*Sorghum halepense*), orchard grass (*Dactylis glomerata*), oxalis (*Oxalis corniculata*), quackgrass (*Agropyron repens*), smooth brome (*Bromus inermis*), Timothy (*Phleum pratense*), *etc.* as **perennial weeds**; blackseed plantain (*Plantago rugelii*), buckhorn (*Plantago lanceolata*), bushy aster (*Aster dumosus*), chicory (*Chicorium intybus*), curly dock (*Rumex crispus*), red sorrel (*R. acetosella*), dichondra (*Dichondra carolinensis*), mugwart (*Artemisia vulgaris*), southern fleabane (*Erigeron quercifolius*), spiny sowthistle (*Sonchus asper*), white clover (*Trifolium repens*), wild carrot (*Daucus carota*), wild violet (*Viola papilionacea*), yarrow (*Achillea millefolium*), *etc.* as **perennial broad-leaved weeds**; **sedges** [annual sedge (*Cyperus compressus*), hurricane grass (*C. spathacea*), purple nutsedge (*C. rotundus*), yellow nutsedge (*C. esculentus*), *etc.*], *etc.*

Poor drainage, soil compaction, shade, excessive thatch and very high or low pH cause invasion of **moss** (tufted type *Ceratodon purpureus* which is highly problematic in acid soils, trailing type *Hypnum cupressiforme* and *Eurynchium* which are quite common in soft and spongy turf, and upright type *Polytrichum commune* which rarely poses a problem though is common on very impoverished soil) commonly during spring and autumn in a weak lawn. The cause of their invasion should be worked out and corrected and then the mosses can be controlled by spreading iron sulphate, ammonium sulphate and sand in the ratio of 1-3-20 (by weight) at a rate of 140 g/m^2 (Seals, 1984) in a morning when there is a heavy dew on the leaves of the mosses so that chemicals stick to the leaves of the moss. After two days, if there is no rain, it should be watered. This will cause blackening of moss and other weeds which should be raked in. The grass will recover soon and cover the space occupied by the weeds. However, Dichlorophen also kills the mosses.

MCPA controls **creeping buttercup** (*Ranunculus repens*), **plantain** (*Plantago* spp.), **chamomile** (*Chamaemelum nobile*) and **dandelion** (*Taraxacum officinale*); 2, 4-D is effective against creeping buttercup, common **chickweed** (*Stellaria media*), plantain, **daisy** (*Bellis perennis* with a repeat spray), chamomile, **yarrow** (*Achillea millefolium* with a repeat spray), **cat's ear** (*Hypochoeris radicata* with a repeat spray) and dandelion; Mecoprop + 2, 4-D for common chickweed and yarrow (with a repeat spray); and Mecoprop against **pearlwort** (*Sagina procumbens*), **parsley piert** (*Aphanes arvensis* with a repeat spray), **white clover** (*Trifolium repens* with a repeat spray) and cat's ear (with a repeat spray).

In the lawn, the weedicides used for control of sedge (*Cyperus* spp.) are sulfonylureas (halosulfuron-methyl, Ethosulfuron) and benzothiadiazole (Bentazon); for **non-selective control** of weeds the herbicides are bipyridiliums (Diquat) and organophosphorous (Glufosinate-AM); the **pre-emergence weedicides** are acetamide (Metolachlor, Napropamide), amide/benzamide (Isoxaben, Bensulide), dinitroaniline (Benefin, Pendimethalin, Oryzalin) and pyridine/substituted pyridine (Dithiopyr, Oxadiazon); **post-emergence annual grass killers** are organic arsenicals (DSMA, MSMA), PPO inhibitors (Quinclorac) and cyclohexanediones (Sethoxydim); **selective perennial grassy weed-killers** are sulfonylureas (Chlorsulfuron) and triazines (Atrazine, Simazine); and the **post-emergence broad-leaved herbicides** are phenoxyalkanoic acids (2, 4-D, 2, 4-DP, MCPA), benzoic acid (Dicamba), benzonitrile (Bromoxynil) and pyridines (Triclopyr). Dacthal (DCPA) is the safest pre-emergence seed-grown herbicide for most of the turf grasses when used at greening stage. Siduron (Tuperson) can be used in the newly seeded or just established cool season turves for controlling broad-leaved seedlings and warm season grass weeds but should not be used in warm season turves. When the lawn grass has tillered out and has been mown 2-3 times, Dicamba and Turflon or 2, 4-D can be used for controlling broad-leaved weeds (Das, 2013).

References

Bailey, L.H. 1942. *The Standard Cyclopedia of Horticulture* (three vols.). The Macmillan Co., N.Y.

Beckett, K.A. 1983. *The Concise Encyclopedia of Garden Plants*. Orbis Publishing Ltd., London.

Brickell, C. 1994. *The Royal Horticultural Society Gardeners' Encyclopedia of Plants & Flowers*. Dorling Kindersley Ltd., London.

Coutts, J., A. Osborn, A. Edwards and G.H. Preston, 1963. The Lawn: Its Construction and Upkeep. In: *The Complete Book of Gardening*, pp. 73-85. Ward, Lock & Co., Limited, London.

Das, T.K. 2013. Weed Management in Turf Grasses. In: *One Day Dialogue on Turf Grass Research and Management* (eds Namita, T. Janakiram, Prabhat Kumar, Sapna Panwar, Ritu Jain, M.K. Singh, M. Lakshmipathy. Palmsey Sangama and P. Pavan Kumar), pp. 18-26. Held in the Division of Floriculture and Landscaping, IARI, New Delhi on December 13.

Dawson, R.B. 1963. *Lawns for Garden and Playing Field*. Penguin Books Private Ltd., England.

Decker, H.F. 2001. Producing sods over plastic in soilless media. In: *Horticultural Reviews*, **27**: 317-351.

Hay, R. and K.A. Beckett, 1971. *Reader's Digest Encyclopaedia of Garden Plants and Flowers*. The Reasder's Digest Association Limited, Great Britain.

Hellyer, A. 1983. *The Collingridge Illustrated Encyclopedia of Gardening*. Collingridge Books, England.

Laurie, A. and V.H. Ries, 1950. *Floriculture Fundamentals and Practices*. McGraw-Hill Book Company, New York.

Misra, R.L. and Sanyat Misra, 2012. Lawn, pp. 61-62. In: *Landscape Gardening Design Elements, Garden Planning and Pollution Monitoring*. Westville Publishing House, New Delhi.

Namita and T. Janakiram, 2012. Turf grasses for enhancing aesthetic beauty. *Indian Hort.*, **57**(3): 36, inside cover page and inside of the next cover page.

Parminder Singh and Premjit Singh, 2014. *Lawn Development and Maintenance*. Centre for Communication and International Linkages, Punjab Agricultural university, Ludhiana.

Randhawa, G.S. and A. Mukhopadhyay, 1986. *Floriculture in India*. Allied Publishers Private Ltd., New Delhi.

Rice, L.W. and R.P. Rice, Jr. 1980. Lawn Establishment and Care. In: *Practical Horticulture*, pp. 267-285. Prentice Hall, Engelwood Cliffs, New Jersey, USA.

Shiels, G.R. 1984. *The Lawn*. World Book Ltd., in conjunction with City and Guilds of London Institute, Surrey, England.

29

Monstera (Family: Araceae)

T. Janakiram, Namita and Sapna Panwar

[(**Common Names**: Anana japonez (Japanese pineapple in Brazil), Arum du pays/Arum troud (French Guinea), Caroal/Liane percee/Liane franche (Guadeloupe), Ceriman/Fruit salad plant/Split-leaf philodendron/Swiss-cheese plant/Window leaf (*Monstera deliciosa*), Harpon or Arpon comun (Guatemala), Hojadillo (Colombia), Hurricane plant, Lace-leaf monstera (*M. friedrichsthalii*), Locust and wild honey, Mexican bread-fruit, Ojul or Huracan (Venezuela), Pinnanona or Pina anona (Latin America), Shingle plant, Siguine couleurre (Martinique)]

Introduction and Origin

The genus *Monstera* takes its name from the Latin *monstrum* which means monster or marvel, possibly referring to the size and shape of the leaves. Some refer to the peculiar perforations (fenestrae) of the giant leaves of many species, the perforations occurring due to self-protection mechanism of the plant against tropical storm or hurricane. The species name *deliciosa* refers to the edible fruits these plants produce in the wild. *Monstera* is native to the rain forests of Central America, South America, and the tropical climates of North America, such as Mexico. It was introduced into cultivation into England in 1752, reached Singapore in 1877 and in India it was introduced into NBRI, Lucknow in 1878. During the 19th Century, *Monstera deliciosa* was grown for both its ornamental foliage and its fruit. Under glasshouse conditions the arum-like flowers develop into white, banana shaped fruits called cerimans, which taste like a cross between pineapples, bananas and mangoes. Massachusetts Horticultural Society exhibited its fruit specimens in 1874 and 1881. It has become familiar as an indoor ornamental, most suitable for pot-growing in most of the warm countries—especially in conservatories and greenhouses—though it does not bloom and fruit in confinement. The fruits are marketed to some extent in Queensland and, in the past, were sometimes transported from Florida to gourmet grocers in New York and Philadelphia. In Guatemala, it is raised in containers in patios to prevent too rampant growth. Under glasshouse conditions the arum-like flowers develop into white banana shaped fruits. The common potted foliage plant species include the *Monstera deliciosa* which is widely traded as cut floral green (Will, 1985) for various floral arrangements. In India, *Monstera* is little known though it is a beautiful foliage plant for pot-growing, for planting along an old tree to climb on in filtered shade and humid conditions, where in rich organic medium if not disturbed for long, it also produces delicious sour fruit to relish on. It will be found occasionally in a conservatory where it is grown for its foliage.

It is a striking house plant and does well in containers with good drainage. It may also be used on a trellis, on fences and tree stumps, to cover walls and steep rock-cuts and at all the shaded locations.The climbing habit and large and deeply incised leaves of *Monstera deliciosa* make it a popular choice in large displays or as a stand-alone specimen. The flesh of fruit can be cut away from the core and eaten. Its fruits are low in calories, high in potassium and vitamin C. It contains 77.88 per cent moisture, 19.19 per cent sugar, 1.81 per cent protein, 0.2

per cent fat, 0.57 per cent fibre and 0.85 per cent ash, apart from 737/kg of calories. Eating five daily servings of fruits and vegetables lowers the chances of cancer. A recent study has found that eating nine or ten daily servings of fruits and vegetables, combined with three servings of low-fat dairy products, is effective in lowering blood pressure. The fruit takes a little longer than a year (roughly 15 months) to mature to an edible stage. It may be served as dessert with a little light cream, or may be added to fruit cups, salads or icecream. The aerial roots have been used as ropes in Peru, and to make baskets in Mexico. In Mexico, a leaf or root infusion is drunk daily to relieve arthritis. In Martinique, the root is used to make a remedy for snake-bite.

Botany

Monsteras are evergreen vines, though woody but herbaceous in nature, growing to 21 m on trees and supports. Stem is cylindrical and quite thick, covered with leaf scars and from it develop numerous, long, cord-like aerial roots. Aerial roots act as hooks to climb over trees and other supports and also for sucking nutrients from soil and bark through their 4-6 root branchlets, and those entering the ground may grow over 3.6 meters in length. Heart-shaped leaves are green (sometimes variegated) and heteroblastic, *i.e.* entire or deeply incised (sometimes finger-like deeply lobed) but perforated (young plants bear entire and non-perforated leaves) by conspicuous holes and in some species a series of large, regularly spaced perforations extend from the midrib to the leaf margin and may break through the leaf edge, forming pinnately dissected or pinnatifid leaves. Leaf perforations may serve as heat trtansfer as other compound leaves do and may reduce effective leaf size (Madison, 1977), therefore, its 60-90 cm long petiole is quite strong to support the weight of the leaf. The leaves are alternate, leathery, dark green, very large (15-75 cm wide and 25-90 cm long, up to 130 cm long in *M. dubia*), often with holes in the leaf blade. A single inflorescence (spadix) some 5-45 cm long, with innumerable flowers, emerges from leaf axil. The inflorescence is a dense spike (spadix) of small, bisexual flowers without a perianth (perigone) and is protogynous as is normal in Araceae (Mayo *et al.*, 1997). Except for a few sterile ones near the base of the spike, the flowers are perfect (containing both male and female parts). Chouteau *et al.* (2006) reported pollen-ovule ratios in some neotropical Araceae and their putative significance. The flower spike is surrounded by a white boat-like bract. The stigmas are wet at anthesis and by the time the anthers ripen and shed their pollen, the stigmas have shrivelled and are probably no longer receptive. According to Gibernau (2003), pollination in *Monstera* has been reported by only three authors (Madison, 1977; Ramirez and Gomez, 1978; and Ramirez, 1980), the pollinators being either trigonid bees or scarabaeid beetles. In *Monstera deliciosa*, diurnal *Trigona* bees are common visitors which are attracted to the gums which its flowers produce (Ramirez and Gomez, 1978). The fruit looks like an elongated closed pine cone and the scales like to those of the pineapple. The fruit is up to 25 cm long and about 5.6 to 6.5 cm in diameter at the widest part and slightly tapering to a rounded apex. The fruit is a cluster of white berries. The fruit on ripening, which occurs gradually bit by bit (some 3 cm) daily, turns paler or yellowish green and emits a powerful odour. The ripened portion of the fruit breaks away from the stalk if touched. The mature fruit has a yellow-green, violet-spotted rind of hexagonal plates covering a creamy-white soft pulp. The actual pulp is grey-white in colour and very strongly scented. This fruit has to be fully ripe before eating as sharp calcium oxalate crystals of unripened fruit can irritate the membranes of the mouth, tongue and throat. The oxalic acid in the unripe portions of the fruit will cause an itching of the throat and it is recommended that only those portions from which the scales have fallen should be eaten. The chromosome number of *Monstera adansonii* var. *klotzschiana* is reported as 2n= 60 by Ramalho (1994).

Classification, Species and Varieties

The genus is divided into four sections, *viz. Tornelia* and *Echinospadix*, each with single species, and *Monstera* and *Marcgraviopsis*, the former having free juvenile leaves, while latter tightly appressed (Madison, 1977; Mayo *et al.*, 1997). This plant, before it was finally placed as *M. deliciosa* by Liebmann, had been allotted to various other genera such as *Arum hederaceum, Calla dracontium, Calla pertusa, Dracontium pertusa, Heteropsis pertusa, Philodendron pertusum, Monstera linnea, Pothos cannaefolia, Serangium* sp., *Tornelia fragrans, etc.* 'The Plant List' includes 148 scientific plant names of species rank for the genus *Monstera*. Of these 37 are accepted species names. Hay and Beckett (1971) mention that *Monstera* consists of 50 species of evergreen climbers. 'The Plant List' includes a further 16 scientific plant names of intra-specific rank for the genus *Monstera. Monstera florescanoana*, a new species in section *Monstera*, endemic to central Veracruz (Mexico) is identified. This species appears to be most closely related to *Monstera siltepecana* Matuda and *Monstera dubia* (Kunth) Engl. et K. Krause (Croat *et al.*, 2010). Other species are *Monstera acacoyaguensis* Matuda, *M. acuminata* K. Koch, *M. adansonii* Schott, *M. amargalensis* Croat & M.M. Mora, *M. aureopinnata* Croat, *M. barrieri* Croat, Moonen & Poncy, *M. buseyi* Croat & Grayum, *M. costaricensis* (Engl. & K.Krause) Croat & Grayum,

M. dissecta (Schott) Croat & Grayum, *M. epipremnoides* Engl., *M. dubia* (Kunth) Engl. & K.Krause, *M. deliciosa* Liebm., *M. filamentosa* Croat & Grayum, *M. cenepensis* Croat, *M. obliqua* Miq., *M. glaucescens* Croat & Grayum, *M. gracilis* Engl., *M. molinae* Croat & Grayum, *M. kessleri* Croat, *M. minima* Madison, *M. lechleriana* Schott, *M. membranacea* Madison, *M. lentii* Croat & Grayum, *M. luteynii* Madison, *M. oreophila* Madison, *M. pinnatipartita* Schott, *M. pittieri* Engl., *M. praetermissa* E.G.Gonç. & Temponi, *M. punctulata* (Schott) Schott ex Engl., *M. siltepecana* Matuda, *M. spruceana* (Schott) Engl., *M. standleyana* G.S.Bunting, *M. subpinnata* (Schott) Engl., *M. tenuis* K.Koch, *M. tuberculata* Lundell, *M. vasquezii* Croat and *M. xanthospatha* Madison. A brief description of some of the species are being furnished here with.

Monstera acuminata uncut young leaves overlap and cling close to the support, perforate at maturity; ***Monstera deliciosa*** (ceriman), a native of Mexico, commercially known as *Philodendron pertusum*, is the popular florist plant that sinks its roots into the bark slab and holds its foot-long perforated or split leaves well out from the climbing stem, young leaves 30-60 cm long or even larger, leathery, pinnately cut, perforated, more or less distichous, from lanceolate to oblong and broader, entire, petiole prominent and sheathing, peduncles terminal, solitary or fascicled, bearing an ovate or oblong boat-shaped spathe that opens widely after flowering and finally becomes deciduous, spadix shorter than spathe, cylindrical or nearly so, dense-flowered, having hermaphrodite or perfect flowers above and sterile flowers below, fertile or perfect flowers with no perianth, 4 stamens and a 2-celled ovary with 2 ovules in each cell, fruit very small berries, crowded or formed into a multiple fruit or cone-like structure, *M. deliciosa* 'Albo-Variegata' bearing smaller leaves than that of the species with sections of green and creamy white, *M. deliciosa* 'Marmorata' bears irregularly variegated, blotched or intermingled leaves with cream to greenish-yellow and different shades of green in an attractive pattern, *M. deliciosa* 'Borsigiana', a vining type with smaller glossy leaves having pinnate lobes widely and evenly separated, leaf-stalk wrinkled where it joins the leaf, *M. deliciosa* 'Variegata' with an irregular variegation where parts of the leaf may be entirely green and other sections marbled-cream to greenish-yellow, or entirely cream, new growth may revert back to green form; ***M. friedrichsthalii***, bearing the medium-sized leaves with more regular, egg-shaped perforations, wavy on the edge; ***M. guttiferyum***, looks more like the philodendrons, leaves of varying shapes, with indented veins; ***M. obliqua***, a tall foliage climber having large elliptic leaves with tapering apex and large irregular holes on older leaves, *M. obliqua* 'Compacta', having ovate leaves, smaller in size, arranged closer than the species with perforation, is a compact and attractive foliage climber which was recorded from Bangkok; and ***M. schleirama*** (*Philodendron leichtlini*), with many lacy perforations has very little leaf surface.

Propagation

Monstera deliciosa is easy to propagate and can be multiplied throughout the year. Monsteras can be propagated through seed, suckers or offshoots, tip cuttings, stem cuttings, air layering, tissue culture, *etc.* **Suckers** or offshoots (lateral shoots) with or without roots can be separated from parent plants when these are 15 cm long and transplanted successfully. Monsteras seldom produce lateral shoots. Mulching is desirable as well as watering until new roots have become well-established. In case of **tip cuttings**, monsteras can be propagated by removing the tips of the growing stem with one mature leaf and inserted during monsoon in sterilized medium (equal parts of peat and sand, both by volume) filled in a 10-cm container and kept at 24-27 °C temperature. **Stem cuttings** are the best and easiest means of propagation. In this case the top of the stems can be cut into several pieces in lengths, each consisting of three to four joints (7.5 cm long) with one leaf per piece. Each piece is inserted in a propagating case heated to 24-27 °C. When well rooted, these are potted separately in 10-cm pots and returned to the glass case until established and then are shifted into 15- or 18-cm pots. Plants generated from cuttings may come into bearing in 4 to 6 years, whereas suckers begin fruit production in 2 to 4 years. Though not very popular but it can also be propagated through **air layering** similar to other plants. **Seeds** are occasionally produced but seedlings take a long time to develop fully. In some European nurseries, it is raised from imported seeds. **Micropropagation** with stem cuttings as explants has been successfully achieved in Denmark. The optimum medium for *Monstera deliciosa* induction was MS+BA 5.0 mg/l +ZT 0.5 mg/l+NAA 0.1 mg/l. The appropriate medium for multiplication was found to be MS+BA 3.0 mg/l+ZT 0.3 mg/l+NAA 0.1 mg/l whereas for rooting it was ½ MS+IBA 0.5 mg/l+NAA 0.2 mg/l (Han De-wei, 2010).

Cultural Practices

Monstera is grown in tropical and warm subtropical areas of the world and in protected culture in temperate areas. It can withstand cold conditions provided it is sheltered from frost and cold winds. It thrives best in semi-shaded conditions having high humidity. Therefore, in nature rain forests are best suited. Average warm indoor **temperatures** are appropriate. It grows best between the temperatures of 20 °C and 30 °C. It can not tolerate persisting temperatures of 10 °C and below. Hay and Beckett (1971) mentions that *Monstera* plants do not feel uneasy at 10 °C winter temperature but will come to little harm at 7 °C for short periods provided they are kept fairly dry, and growth starts as soon as the temperature reaches 18 °C. Freezing temperatures are always dangerous to *Monstera* because leaves at 0 °C and stems at -2 °C are damaged due to cold injury. Since *Monstera* has its origin in tropical humid areas so it requires proper management of **humidity** levels. Misting is done to increase the humidity level. This can be done once daily using a spray bottle filled with tepid water. A humidifier can also be utilized to increase the humidity at all times. Excessive deficiency of moisture leads to wilting and ultimately dropping of foliage. Dry air tends to make the leaf edges go brown and papery. In case of excess moisture the plant may develop fungal infection. It prefers lightly shaded conditions, so the plants are placed in those places which receive filtered **sunlight** throughout the day. The growth and quality parameters are affected by both high and low light conditions. Under low light conditions plants become leggy, stalks spindly and leaves small and with less divisions. The overall growth of plant can stop under too dark conditions. Average indoor light is appropriate. Direct exposure to sun is also avoided. Excessive exposure to direct sunlight causes leaf scorching and ultimately withering of whole plant. *Monstera* that receives high light will have much larger leaves with more slits. It is suggested that while growing plants outdoors, only filtered light should be provided. It tolerates a wide range of **soil** types including limestone though saline soils are at all not suitable for its cultivation. However, it prefers and grows more rapidly in a well-drained soil rich in organic matter. In greenhouses an ideal mix contains two parts garden loam, two parts peat moss and one part sand or perlite. The planting medium though should be porous but should have water retaining capacity. Alternatively, compost should consist of three parts of turfy loam, two of peat moss or leaf mould, one of coarse sand and one of well-decayed manure.

Monstera is a great favourite as a pot plant for indoor décoration as well as outdoors when grown in conservatories and rockeries where it grows on posts or tree trunks. For planting, the first step is to choose a healthy nursery plant. Commonly, nursery *Monstera* vines having no constrictions and free of insects or diseases and those which have attained 60 to 120 cm length are grown in a container having 12-15 litres capacity and when grown sufficiently, which may take about 3-4 months, these are repotted to a larger permanent container having some 45-50 litres capacity. **Planting** is done in February or March. *Monstera* vines should be planted as an understorey plant, *i.e.* in partial shade of large overhanging trees but some 2-3 metres away from the trees or posts. When the plants have outgrown the pots or have become pot-bound these are transplanted. *Monstera deliciosa* requires repotting every two years.

Pruning is limited to the removal of an unwanted stem, leaf, *etc.* to shape the plant, or when it is desired to reduce the length of a vigorous shoot. *Monstera deliciosa* is cut back with pruning shears to control spread. Leaves should be cut close to the stem and the basal end will fall off naturally, the fruit stalk should also be cut off after collecting the fruit. However, removal of roots is not advised. It is advisable to periodically cut back the vines, to maintain them in their appropriate space, otherwise they will take over a large part of the landscape.

Mulching around the base of the plant should be done with an organic material such as shredded bark, leaf mould, rice husk or cocopeat to hold moisture in the soil and to make the plant looking tidy. Replenish the mulch each year as it degrades into the soil to keep an even layer in place. *Monstera* produces cord-like roots that soak up moisture from the air. These roots are provided with **support** on which to climb, such as a moss-covered pole, which can be made by securing sphagnum moss to a wooden pole with fishing wire. Moss is kept on the pole moist at all the times by misting it. *Monstera* having no clinging roots to hold it to its support, is apt to bend away from this, especially with the pull of the aerial roots when they take a grip of the soil so it should be given spiral support with a thick wire in a fashion that stems are not damaged. Maintain a **weed-free** cultivated area at least 60 cm in diameter around the stem of the plant to reduce competition for nutrients in the soil.

Nutritional and Water Requirement

Monsteras when grown in containers have a limited volume of soil from which these extract mineral nutrients (fertilizer). The supply of **nutrients** rapidly becomes exhausted when the plant is actively growing. Replenish nutrients regularly, better with soluble fertilizer which is easily available with the dealers. Since they vary in strength (per cent of fertilizer nutrients), dilute or dissolve them according to the labelled directions. Mix only enough solution to water your plants. Nutrients will evaporate or settle out if the fertilizer solution is kept too long. Soluble fertilizers applied over the whole plant during the warmer months of the year will encourage establishment. Established plants generally require no fertilizing. Feed every two weeks with a balanced house-plant fertilizer in spring and summer otherwise once every month. *Monstera* fertilizer requirements do not appear to be high. After planting when new growth begins, application of 113 g of a complete dry fertilizer mix with 20 to 30 per cent of the nitrogen from organic sources should be provided. A complete mix includes nitrogen, phosphate, potash, and magnesium. Repeat this every eight weeks for the first year, then gradually increase the amount of fertilizer to 227 g, 341 g and 454 g but decrease the frequency to 2-3 times per year as the vines grow. Application of magnesium and micronutrients such as zinc and manganese may be made in ground applications to vines growing in sandy soil with a low pH (4-7). However, foliar applications of zinc, manganese, and magnesium is more effective for vines growing in highly calcareous soil having high pH (7.0-8.5). Micronutrient applications should be made 2-3 times per year, generally during the growing season. Iron should be applied in a chelated formulation (Crane and Balerdi, 2005). It has been reported that when grown as pot plant, application of Multicote 6 which is a controlled release fertilizer @ 4-5 kg/m ^{3} (www.paton.com.au) is very effective.

Monstera is drought tolerant, but the growing medium should never be allowed to dry out completely. After planting, *vis-à-vis* pruning, it is **watered** immediately. This will lessen the stress on the plant and will support roots to throw out new growths. The container should have drainage hole at its bottom so that excess water drains out and water-logging is avoided. Over-watering is highly dangerous to plant health as it produces stem rot and yellowing of foliage, especially in winter. If grown in soil, keep the soil barely moist from November to March. For the remaining part of the year water thoroughly but allow the surface of the compost to dry between watering. For watering, hard water is avoided and instead distilled or reverse-osmosis water may be used. Watering should be done about once a week during spring, summer and fall and once in every 10 days during winter though frequency depends upon prevailing weather conditions and soil types.

Post Harvest Management

In general, the average vase life of *Monstera* is about 5 to 7 days. Shanan and Shalaby (2011) significantly revealed that glycerol at 2 or 4 per cent extended vase life of *M. deliciosa* cut leaves by 7-fold of the control (7 days) and better than other treatments. Also, glycerol treatment at the mentioned concentrations showed lowest leaf weight reduction, as well as water loss rate, which obviously reflected on extending leaf vase life. The response of glycerol on prolonging leaf vase life was accompanied by a decrease in the degradation of pigments and protein as well as decrease in the percentage of defense enzymes (superoxide dismutase and catalase) and this correlated with decreasing leaf water loss. Farahat and Gaber (2010) investigated the effect of preservative chemicals on vase life of cut leaves of *Monstera deliciosa*. Freshly cut leaves were put in vases containing different treatments like 8-hydroxyquinoline sulfate (8-HQS) at 200 and 400 ppm with sucrose 30 g/l, calcium chloride ($CaCl_2$) at 500 and 1000 ppm, and GA_3 at 25 and 50 ppm, in addition to control (distilled water). The holding solution GA_3 at the rate of 50 ppm was the most efficient treatment in increasing the vase life period (59 days), followed by GA_3 at 25 ppm (51 days). The maximum amount of total soluble sugars resulted from GA_3 at the rate of 50 ppm after 6 and 12 days. Also, GA_3 at 50 and 25 ppm gave the highest values of chlorophyll after 6 and 12 days, respectively. Calcium chloride at 1000 ppm and GA_3 at 50 ppm were the most suitable preservative solutions for improving water uptake percentage. It was finally concluded that GA_3 could increase the vase life of cut leaves of *Monstera deliciosa*.

Insect-Pests, Diseases and Physiological Disorders

When grown outdoors, they are usually pest-free but under indoor conditions the plants are subject to infestation by many pests such as aphids, grasshoppers, mealybugs, scales, spider mites and root-knot nematodes. In India, wire cages are placed around developing fruits to protect them from rats, squirrels, monkeys and other creatures. **Aphids** are soft-bodied insects that multiply rapidly. In indoor situations, all aphids are females. Aphid infestations often are evident by the white cast skins that are shed and left behind by the aphids when moulting. Through targeted jet spraying or through tobacco decoction spraying these can be controlled. **Lubber grasshopper** (*Romalea microptera*) has been recorded from Florida. During dry seasons, it rapidly consumes entire leaves leaving only the base of the midrib and the petiole. These are controlled through use of any contact insecticide. **Mealybugs** are the most common and difficult to control. All foliar-feeding mealy bug species have sucking mouthparts which remove plant fluids. A sticky honeydew is excreted, which coats foliage below the infested area. Mealy bugs, can be removed with tweezers or a cotton swab dipped in alcohol. These can also be finger-trapped and killed. *Cryptolaemun montrouzieri* is a ladybug beetle predator of mealy bugs. **Scales** are sap sucking insects and also excreting honeydew. Females produce up to 1,000 eggs underneath their protective scale. The eggs hatch into tiny crawlers, which spread about the plant. After dispersing, crawlers settle and feed in one location for the remainder of their life. The length of the life cycle varies with each species, ranging from 1-8 or more generations per year. Cotton swab soaked in Rogor when rubbed to the affected part with the help of the forceps will kill these. **Spider mites** (*Tetranychus urticae*) have a wide host range, and very few plants are immune to attack. Adult spider mites are quite minute creatures and are usually found on lower leaf surfaces. Feeding injury on many plant species usually involves lighter coloured stippled areas on leaves along with webbings. Severe spider mite infestations cause leaves to dry and fall off from the plants. Spider mites can be sponged off with soapy water. Regular leaf cleaning and an adequately humid environment help prevent these problems. Its natural effective predator is another mite (*Phytoseiulus persimilis*) which normally develops twice as fast as spider mites, so it is able to reduce the population rapidly. **Root-knot nematode** (*Meloidogyne incognita*) causes gall formation (nodules) on roots, impairing root function, which results in stunting of the plant. Growing medium pasteurization prior to planting will kill adults as well as eggs of root-knot nematodes. Fumigants are as effective as steam for this purpose. Soil mixing of Furadan granules in the vicinity of the plant, followed by watering will control this pest.

Leaf spot or blight is caused by *Leptosphaeria* sp., *Macrophoma philodendri, Phytophthora* sp. and *Pseudomonas cichorri*. Its symptoms are dark green water-soaked spots that may turn tan, dark brown or black with a yellow border. The spots can enlarge until the entire leaf blade is involved. Sometimes these lesions spread into the petioles. Control of these diseases generally involves prompt removal of infected plant parts. After handling the plants, hands should be washed with soap and the concerned tools should be disinfected in 70 per cent alcohol. Fixed copper sprays can be used, though at present no fixed copper spray is registered for use on indoor plants so the plants will have to be taken outdoors for treatment. **Root rot** is caused by *Pythium splendens* and *Rhizoctonia solani* which live in the growing medium. Control of these fungi involves medium pasteurization and sanitation programmes. **Anthracnose** is caused by *Glomerella cingulata* which can be controlled through spraying with 0.1 per cent Bavistin. **Bacterial soft rot** is caused by *Erwinia carotovora* which can be controlled through spraying with streptocyclin or any other bactericide. **Brown-tipped leaves** occur due to under-watering or too-dry conditions. **Yellowed leaves** occurs due to **o**ver-watering if many leaves are involved and if signs of wilting and rotting are present. If it is not due to over-watering, yellowed leaves may be the sign of nutrient deficiency.

References

Chouteau, M., D. Barabe and M. Gibernau, 2006. Pollen–ovule ratios in some Neotropical Araceae and their putative significance. *Plant Systematics and Evolution*, **257**: 147–157.

Crane, J.H. and C.F. Balerdi, 2005. Monstera growing in the Florida home landscape (Extension Folder No. HS1071), 5 p. Florida Coop. Extn Service, Institute of Food and Agricultural Sciences, University of Florida, USA.

Croat, T.B., Thorsten Krömer and Amparo Acebey, 2010. *Monstera florescanoana* (Araceae), a new species from central Veracruz, Mexico. *Revista Mexicana de Biodiversidad*, **81**: 225- 228.

Farahat, M.M. and A. Gaber, 2010. Influence of preservative materials on postharvest performance of cut window leaf foliage (*Monstera deliciosa*). *Acta Horticulturae*, No.877, pp. 1715-1720.

Gibernau, M. 2003. Pollinators and visitors of aroid inflorescences. *Aroideana*, **26**: 66–83.

Hay, R. and K.A. Beckett (eds), 1971. *Reader's Digest Encyclopaedia of Garden Plants and Flowers*, pp. 447-448. Reader's Digest Association, London.

Madison, M. 1977. A Revision of *Monstera* (Araceae). *Contributions from the Gray Herbarium of Harvard University*, No. 207, pp. 3-100.

Mayo, S.J., J. Bogner and P.C. Boyce, 1997. *The Genera of Araceae*, 370 p. Royal Botanic Gardens, Kew, London.

Paton Fertilizers, 126 Andrews Rd, Penrith, NSW, 2035 www.paton.com.au

Ramalho F.C. 1994. *Taxonomia e nu´mero cromosso^mico de representantes da famý´lia Araceae em Pernambuco.* M.Sc. Dissertation, Universidade Federal Rural de Pernambuco, Recife.

Ramirez, W.B. and L.P.D. Gomez, 1978. Production of nectar and gums by flowers of *Monstera deliciosa* (Araceae) and of some species of *Clusia* (Guttiferae) collected by New World Trigona bees. *Brenesia*, **14–15**: 407–412.

Ramirez, W.B. 1980. Informe cientifico. *Brenesia*, **18**: 367–368.

Shanan, N.T.and E.A. Shalaby, 2011. Influence of some chemical compounds as antitranspirant agents on vase life of *Monstera deliciosa* leaves. *African J. agric. Res.*, **6**(1): 132-139.

Will, A.A. 1985. Foliage for flower arrangements from your Florida garden. *Proc. Fla State Hortic. Soc.*, **98**: 349-350.

30

Mussaenda (Family: Rubiaceae, sub-family: Ixoroideae)

T. Janakiram and Ritu Jain

[**Common names**: Dhobi tree (*Mussaenda frondosa*), Mussaenda, Prophet's tears (*M. erythrophylla*), Red flag bush, Red mussaenda, Tropical dogwood (*Mussaenda glabra*), Virgin tree/Kahoy dalaga/Tropical dogwood (*M. phillipica*); Hindi: *bebina*, *bedina*; Kannada: *billoothi*, *hastygida*; Konkani: *mithai phool*; Malayalam: *parathole*, *vellila*; Manipuri: *hanu-rei*; Marathi: *bhutkes*, *burthkasi*, *lavasat*; Tamil: *vellaiyilai*]

Introduction and Botany

Mussaenda is a native name in Ceylon which has been adapted as generic name to this plant. Though it is an evergreen shrub where certain species grow as high as 9 meters, and is native to Asia such as Thailand and India, West Indies and parts of tropical West Africa. It produces many stems and sometimes resembles to a rambler. In tropical climate, it grows as an evergreen shrub but it becomes deciduous in colder climates. Leaves are opposite, bright to dark green, and rounded to elliptic, pubescent (with fine silky hair), and prominently veined (ribbed). The flowers are small tubular and often yellow, white or orange in colour. The calyx is five lobed, with one lobe apparently enlarged, leaf-like and usually brightly coloured. Sometimes this enlarged sepal is termed a calycophyll. In many of the cultivars all the five sepals are enlarged, and range in colour from white to various shades of pink to carmine red. Fruit is a small up to 18 mm long, fleshy, somewhat elongated berry containing many seeds. The blossom colour comes from bracts and not from the flowers which are small, and located at the centre of each bract. Bracts may be seen in several colours including rose, white, red, pale-pink and some mixtures. The principal ornamental feature of these plants is the inflorescence. The flowers are small and tubular. The corolla is five-lobed, spreading and bright yellow to white. They are borne in terminal clusters (cymes or panicles). The basic chromosome number in *Mussaenda* is x=11. Mussaendas are valued for their colourful bracts and are most suitable for landscaping. In tropical zones, these make a beautiful specimen plant and are most suitable for gardens. White Mussaenda can be grown as flowering ornamental in parks and public gardens or along roadsides, bylanes and highways.

The development of the mussaenda hybrids gained momentum when a white mutant of the species *Mussaenda philippica* was discovered in 1915 and this variety was named as 'Aurorae'. This does not produce seed, the pollen is fertile and has been used in the breeding of a number of outstanding cultivars in crosses with *M. erythrophylla* and *M. frondosa.* Variety 'Queen Sirkit' was developed by backcrossing the F_1 hybrid between *M. erythrophylla* to *M. philippica* 'Aurorae'. It is one of the most spectacular mussaendas with all five calyx lobes enlarged up to 9 cm in shades of ivory to pale pink.

Species and Varieties

There are some 40 species of *Mussaenda* but only a few are in cultivation. *Mussaenda erythrophylla* is native to West Africa where it can grow to 9 metres as a scandent shrub. It is usually found in woodlands where it often grows into surrounding trees. When cultivated, it grows less than 3 meters. Its enlarged sepals are blood red in colour. The remainder of the flower is composed of a five-lobed, tubular, white to cream corolla with a red felt-like centre. *M. erythrophylla* has an open sprawling habit and requires careful pruning to maintain it as a low spreading shrub. Alternatively, it can be allowed to grow and should be given with support such as a trellis or an adjacent tree. Because of the conspicuous red bracts, *M. erythrophylla* has been suggested as a year-round landscape plant which provides colour most of the times in a year. Its varieties 'Ginang Imelda' and 'Rosea' are quite popular in India. *Mussaenda philippica* is native to the Philippines, and is known commonly as virgin tree or, less often tropical dogwood, and forms a shrub or small tree 3-5 metres tall. The dark green leaves are similar to those of *M. erythrophylla*, though less ovate and not as prominently veined. The flowers which are borne in terminal cymes, consist of a yellow, tubular corolla with one lobe of the calyx greatly enlarged white, showy and leaf-like. The var. 'Aurorae' is more showier than the wild type, most suitable as a specimen plant, having all five calyx lobes greatly enlarged, white and pendant. The corolla is a deep golden-yellow and is often hidden by the profusion of enlarged bracts. Although it does not produce seeds, the pollen is fertile and this variety has been used in the breeding of a number of outstanding cultivars in crosses with *M. erythrophylla* and *M. frondosa*. *Mussaenda frondosa* is found from Indo-China to Malaysia. It is somewhat smaller and more upright than the above two species, 2-3 metres tall, with an equal spread. The foliage is lighter green, and the terminal flower clusters have orange to yellow and tubular corollas with a single white enlarged calyx lobe. This species is often grown in clumps. *Mussaenda incana*, a native from India to Malaysia, is much smaller than the above mussaendas, growing to not more than one metre in height. It has flat-topped flower clusters (corymbs), with bright yellow corollas and a single enlarged calyx lobe that is yellow to cream. In the landscape it is most effective in mass plantings. Var. 'White Wings' is excellent for mass plantings. *Mussaenda flava* is a small, sub-shrub with delicate pale-yellow blooms and has naturalized in India. Golden star-shaped blooms are surrounded by creamy-yellow bracts. The unusual flowers of dwarf *Mussaenda* feature large sepals in palest-yellow-white and small yellow corollas. The leaf on the end of the stem is white to indicate that a future bloom will appear and thereafter this leaf falls off. It is an ever-blooming small evergreen shrub excellent for containers. It blooms well during warm months. *M. luteola*, a native of Tropical Africa grows to 1.5 m or more, leaves narrowly ovate to lanceolate and up to 5.0 cm long, flowers about 2.5 cm long, yellow with an orange-red eye, enlarged sepals white, 2-6 cm long and broadly ovate and flowering from April to September. *M. sanderiana*, a very showy species is a native of Indo-China which grows to 1.8 metres tall or sometimes prostrate but is compact in habit, leaves are silky hairy, nearly sessile, cordate and lanceolate, tubular flowers are small and yellow in numerous terminal cymes, and silky-hairy calyx-lobes are petaloid (white) and more than 7.5 cm long. *Mussaenda* species and varieties are separated from one another on the basis of their colours but more authentic way is to classify them on the basis of their style and stamen lengths, *i.e.* 'pin' and 'thrum' types. The 'pin' type cultivars are 'Dona Eva', 'Dona Esperanza', 'Dona Helaria', 'Maria Makiling' and 'Queen Sirikit' while 'thrum' type cultivars are 'Baby Aurora' (*M. flava*), 'Diwata', 'Dona Luz', 'Dona Trining' (*M. erythrophylla*), 'Mutya', *etc.* though 'Lakambini' contains stamen and style equal in length, an intermediate case. Certain other popular varieties are 'Aurorae' (*M. philippica*), 'Dona Aurora', 'Dona Pacencia', 'Paraluman', 'Rosea' (*M. erythrophylla*), *etc.* Mukherjee *et al.* (2012) through RAPD (random amplifaied polymorphic DNA) studies of *M. erythrophylla* cvs 'Ginang Imelda' & 'Rosea', *M. luteola*, *M. mermelada* and *M. philippica* cv. 'Aurorae' reported *M. luteola* as quite distinct from others at molecular level while *M. erythrophylla* cv. 'Rosea' had maximum similarity (84 per cent) with *M. philippica* cv. 'Aurorae' though the two varieties of *M. erythrophylla* were found related only at 56 per cent which suggests that *M. pilippica* may itself be a hybrid of *M. erythrophylla* cultivars.

Propagation

Propagation of mussaenda is accomplished primarily with **cuttings**. The soft wood cuttings about 5-6 cm long taken from the tip of strong healthy shoots can be rooted after treating them with rooting hormone and grown under a mist propagation system. In Philippines, it was observed that auxin application is not necessary for certain varieties like 'Dona Luz', 'Queen Sirikit', 'Paraluman', 'Dona Hilaria' and 'Dona Aurora', while for 'Mutya', 'Ginang Imelda', 'Dona Pacencia', 'Lakambini', 'Diwata' and 'Dona Eva', auxin is needed to enhance and accelerate rooting. Best rooting is observed from cuttings taken from current growth, which are partially mature and green. These cuttings of mussaenda cultivars responded well to a 15-30 minute dip in 100 ppm IBA before planting into the rooting bed. IBA combined with dopamine or paclobutrazol improved rooting in cuttings of mussaenda var. 'Lakambini' (http://redb.uplb.edu.ph/index.php/technology-database/article/7-propagation-of-mussaenda-by stem-cuttings). For multiplying potted mussaendas, the healthy shoot tip slant cuttings having 3-5 nodes and some 10-12 cm long are taken, leaves are removed leaving 2-3 pairs of nodal leaves, cut ends dipped in rooting hormone and 2-3 such cuttings are planted in 10-cm pots filled with one part coir dust, ½ part sharp sand (http://blog.agriculture.ph/tag/dona-mussaenda) and pots are placed in mist chamber at least for four weeks and then these pots are transferred from mist chamber to partial shade for hardening (Junelyn, 2003). In case of non-availability of mist chambers the chance of success using soft wood shoots is very much reduced. Therefore, propagating mussaendas in the absence of a mist unit is advised by hard-wood cuttings. In order to take hardwood cuttings, after flowering the plants are cut-back hard till the right shape of bush is obtained. Hard-wood cuttings of 15-20 cm length which are straight and 1.0 cm in diameter are taken from pruned branches. To these cuttings always a slanting cut is given just below a node, the flat top cut is given just above the node which helps to distinguish the top of the cutting from the bottom. A sloping cut at the bottom exposes more of cambium layer which eventually produces new roots. Time gap should be minimum between preparations and planting of the cutting. Cutting taken should be prepared immediately, the bases are dipped in rooting powder, rubbing off any excess powder and then the cuttings are inserted into clean sand at 1/3rd of its length and watered. Further regular watering is required as it may take several weeks to root these cuttings. Success is indicated when the cuttings start sending out tiny green shoots, but it is important to remember that at that stage there will be no root formation. Wait a couple of weeks to lift and pot off the cuttings. Red coloured mussaendas have always been difficult to propagate. In such cases where stem cuttings are extremely difficult to root, **cleft grafting** becomes most effective method of propagation. Best time for cleft grafting is from January to April when the plant is dormant. In case of *M. formosa,* **air layering** is used to generate new plants. Mussaenda can be ***in vitro*** propagated from **axillary bud explants** by culturing them on MS medium supplemented with 40 mg/l adenine sulphate and 2.25 mg/litre 6-benzylaminopurine without callus intervention. The shoots were rooted on half strength MS basal medium supplemented with 0.5 mg/l indole butyric acid and having 800 mg/l thiamin-HCl (Maity *et al.,* 2001).

For multiplying potted mussaendas, the healthy shoot tip slant cuttings having 3-5 nodes and some 10-12 cm length are taken, leaves are removed leaving 2-3 pairs of nodal leaves only, cut ends dipped in rooting hormone and 2-3 such cuttings are planted in 10-cm pots filled with one part coir dust, ½ part sharp sand (http://blog.agriculture.ph/tag/dona-mussaenda) and pots are placed in mist chamber at least for four weeks and then these pots are transferred from mist chamber to partial shade for hardening (Junelyn, 2003).

Cultural Practices

Mussaenda plants grow well in tropical **climate** with ample sunlight, good rainfall and high humidity. Dwarf and white mussaendas can be grown in partial shade. The plant responds well to humus-rich **soil** conditions with adequate drainage. In sandy soil, watering and proper addition of manure are required for the proper growth of the plant. Planting beds are enriched with organic matter such as well-rotten compost, sphagnum peat, or coir. Do not use black topsoil, since it is too heavy. The soil should be friable and moist but not wet. Mussaenda requires a **sunny location** for best bract colour development as bracts start appearing in spring and last through summer though in perfect tropical climate it flowers throughout the year, however, white varieties require partial shade. In places where **temperatures** rarely fall below 16 °C mussaendas flower year-round. Below 12-15°C plant growth slows down, and below 4°C mussaenda plants are damaged with loss of most leaves and flowers. Temperatures below 4°C can damage the stems, especially if the cold period is prolonged, and if it drops persistently below 4 °C, plants will die. Therefore, if temperature goes in winter months low, mulching is required. *M. erythrophylla* mostly requires humid conditions. As plants flower during the warm season, faded flower clusters should be removed, especially during extended periods of wet weather which encourages further flower production, and this also minimizes chances of further fungal infection.

For **transplanting** of mussaendas, a pit is dug twice the diameter of the plant container and some 30 cm depth, pit is filled with water to saturate it fully, now after soaking of the water the pit is filled with good garden loam soil + FYM in equal parts+coarse sand about 1/3rd of the total volume of mixture and little BHC. The plants from the containers are now taken out, its roots loosened and akin to its previous depth it is planted in the pit and then the mixture is pressed around firmly, followed by thorough watering (http://www.ehow.com/how_8111247_plant-mussaenda-bush.html).

Mussaendas are not naturally straight plants and have a tendency to straggle. This can be corrected quite easily if the plants are **pinched** back from early stage in order to make them in proper shape and bushy. In fact, quite old plants can also be cut back and shaped very nicely if they have been neglected in their early years. It is good to **prune** mussaenda heavily in late fall to late winter. This will contribute in shaping the plants and will produce many more new branches which will produce a lot of colourful bracts and flowers. If plants are grown as single specimen, these should be kept at 1.5 meter or more.

Dwarf mussaenda performs well in **pots** and where temperature goes below 4°C, planting is done in pots so that these may be kept indoors in such situations. The potting media should be composed of 1 part garden soil, 1 part coarse sand or perlite, 1 part well-rotten organic matter or leaf mould and a light dusting of lime.

Regular hoeing and weeding are carried out so that nutrients supplied to the plants are fully utilized by them and the medium also remains porous with proper aeration.

Irrigation and Fertilization

Mussaendas are not drought tolerant plants and will need to be **watered** necessarily during periods of hot and dry weathers. The plant prefers regular watering, however, it should never be over-watered, or else the roots will rot. The plant is also sensitive to prolonged flooding. After planting mussaenda, watering should be every other day for two weeks, twice a week for two weeks and once a week thereafter. Dwarf mussaendas should be watered freely when in full growth.

Mussaendas are not very demanding, only organic manure can be applied once in a year. **Fertilize** lightly four weeks after planting and every month thereafter, with a blooming plant fertilizer. For large trees, application of about 200 grammes of complete fertilizer before and after the rainy season is sufficient (Bautista, 2009). Organic fertilizers can also be an option to synthetic fertilizers. In case of potted mussaendas one tablespoonful of Osmocote can be given at three –four months intervals or fed monthly during summer with a dilute liquid fertilizer. Potted mussaendas should be fed through a balanced fertilizer like a 15-15-15 or lower at 2.5 ml/3.5 l of water at regular intervals during active growing season. Fertilization is discontinued in fall and winter and resumes when new growths begin in the spring.

Growth, Development and Flowering

Plant height reduced with increase in the concentration of growth retardants. The most attractive potted plants can be produced with two spray applications of daminozide at 5000 mgl^{-1} or two drench applications of ancymidol at 0.5 mg/pot (Cramer and Bridgen, 1998). Spray and drench application of paclobutrazol at 5-15 per cent concentration showed dwarfing effect on mussaenda seedlings. However, spray method of paclobutrazol application increases the number of petaloids and number of true flowers (Neri *et al.*, 2007).

Insect-Pests and Diseases

Mussaendas are relatively insect-pest and disease-free. Avoid soil from being water-logged to prevent most fungal and bacterial diseases in the roots. Mussaenda is prone to mealy bugs (*Maconellicoccus hirsutus*) and red spider mite (*Tetranychus urticae*) which can be controlled through regular misting or spraying of water or spraying of Malathion at 10-15 days intervals. White flies (*Bemisia tabaci*) are controlled by spraying with a dilute soap solution on the whole plant as a insect repellant while scale **insects** (*Paratachardina lobata*) are controlled by treating the plants with rape seed oil or pyrethrum containing chemicals.

References

Bautista Norberto R. 2009. Landscaping with mussaenda hybrids. *The Urban Gardener*, **2**(11): 2-14.

Cramer, C. S. and M. P. Bridgen, 1998. Growth regulator effects on plant height of potted *Mussaenda* Queen Sirikit. *Hort Science*, **33**(1):78-81.

http://www.ehow.com/how_8111247_plant-mussaenda-bush.html.http://blog.agriculture.ph/tag/dona-mussaenda. http://redb.uplb.edu.ph/index.php/technology-database/article/7-propagation-of-mussaenda-by- stem-cuttings (Office of the Vice-Chancellor for Research and Extension - University of the Philippines, Los Banos).

Junelyn, S. de la Rosa, 2003. Mussaenda blooms year-round. *Bar Digest*, **5**(4): 43.

Maity, S.K., K.K. De and A.K. Kundu, 2001. *In vitro* propagation of *Mussaenda erythrophylla* Schum and Thom cv. Scarlet through multiple shoot regeneration. *Indian J. exp. Biol.*, **39**(11):1188-1190.

Mukherjee, A.K., L.K. Acharya, T. Mohapatra and P. Das, 2012. Genetic variability studies on *Mussaenda* species, variation among five *Mussaenda* species detected by random amplified polymorphic DNA. *Indian J. Hort.*, **69**(2): 226-230.

Neri, F.R, E.B Abalos, J. Valmores and F.T. Orpilla, 2007. Mussaenda collections: their propagations and dwarfing technology. http://www.neda.gov.ph/knowledge-emporium/details1.asp?DataID = 284.

31

Ornamental Banana (Family: Musaceae)

Ritu Jain and T. Janakiram

[**Common names**: Abaca/Manila hemp (*M. textilis, syn. M. formosana*), Abyssinian banana/Bruce's banana/Ethiopian black banana/Plantain/Wild banana (*E. ventricosum*), Balbis banana/Devil banana/Mealy banana/Seeded apple banana/Seedy banana/Starchy banana (*M. balbisiana*), Blood banana/Maroon-variegated banana plant/Seeded red banana/Sumatra ornamental banana (*Musa acuminata* ssp. *zebrina*), Cheesman banana (*M. cheesmani*), Chinese banana (*M. splendida*), Chinese dwarf banana/Golden lotus banana (*M. lasiocarpa*), Coccinea/Flowering Red Thai banana/Okinawan banana flower/Okinawa torch/Red ornamental banana/Red torch banana/Scarlet banana/Thai red banana (*M. coccinea*), Elephant banana/Seeded sweet banana/Snow banana/Virgin banana/Wild banana (*Ensete glaucum*), Fiji fruit/Fei banana (*Musa troglogytaquorum*), Japanese fibre banana (*Musa basjoo*), Pink velvet banana (*Musa velutina*)]

Introduction and Origin

Musa is named so after Antonio Musa, physician to Octavius Augustus, first emperor of Rome. *Musa beccarii* was named in honour of Odoardo Beccari, an Italian naturalist who did splendid work on banana in Borneo. The scientific name for *Musa* is a Latinization of the Arabic mauz (ãæÒ) meaning the fruit. Mauz is also the Turkish and Persian name for the fruit. The word banana came to English from Spanish and Portuguese, which in turn apparently got it from a West African language (possibly wolof). The domestication of bananas took place in its native haunts, *i.e.* south-eastern Asia and nearby tropical and sub-tropical countries as still many species of wild bananas occur in New Guinea (considered to be the primary centre of diversification and where the earliest domestication had occurred; Simmonds, 1962), Malaysia, Indonesia, and the Philippines. Africa is considered as a secondary centre of diversity.

Ornamental banana with its varied characteristics is suitable for a wide range of landscapes and interiorscapes in tropical areas. Some of the ornamental bananas are with colourful bracts and flowers which are used in flower arrangements. Some of the banana species can be used to combat noise pollution due to their dense, leathery, waxy and giant leaves. *Musa balbisiana* can be grown as indoor pot plant. *Musa coccinea*, a crimson flowered plant can be used as living fence (Elevitch, 2006). However, such plants have beauty that is long-lasting and exotic.

Botany

Banana plant is monocotyledonous, tree-like and largest herbaceous perennial. There are three genera of banana, *i.e. Musa, Musella* and *Ensete*. The genus *Musa* contains 30-40 species, with all wild species being diploids (2n=2x= 14, 18, 20, 22) and native to South East Asia (Stover and Simmonds, 1987). Based on the basic chromosome numbers, orientation and arrangement of flowers in the inflorescence, *Musa* is grouped into 5 sections. Two of the sections contain species *Callimusa* and *Australimusa* with a basic chromosome number of n = 10 (2n=20), two other sections have species *Eumusa* and *Rhodochlamys* with a basic chromosome number of 11 (2n=22). The species in the sections *Callimusa* and *Rhodochlamys* are only of ornamental interest and do not produce edible fruits. The banana tree trunk is called pseudostem because it does not lignify or undergo secondary growth like woody plants. The pseudostem is a cylinder of tightly bound leaf petioles that arise directly from an underground stem, *i.e.* rhizome. The main pseudostem is monocarpic, *i.e.* dieing after flowering and then its place is taken over by the next oldest sucker. Due to their suckering ability, these tend to increase their clump all around in area though *Ensete* does not sucker, and dies after flowering. The succulent and juicy pseudostem is not very strong but can support banana plant over 7.5 m tall. The colour of the pseudostem may be green, red or purple and/or black and can contribute to the ornamental quality of the plant. The pseudostem of *Musella* is swollen at the base whereas the pseudostems of *Musa* and *Ensete* tend to have the same width over the entire length. Each pseudostem produces a single terminal inflorescence which hangs down beneath the leaf canopy on a long flower stalk in *Musa* and *Ensete*, but faces upward on a short flower stalk in *Musella.* The leaves are the main ornamental feature of the banana plant that are generally dark green in colour, sometimes with variegation or sectors of white, red, maroon or purple colouration on the leaf blade and impart a majestic look to the garden. These are smooth, waxy, generally quite large, reaching up to or more than 60 cm length and 15 cm width on dwarf plants, and up to 2.7 m length and 60 cm width on large ones. The leaf may have a contrasting red colour midrib while reverse side contrasts with the front side and on windy days these colours look majestic. New leaves may open up as one colour but gradually as they turn older the colour changes. The leaves which emerge tightly curled, are arranged in a spiral pattern around the top of the pseudostem. A single pseudostem may have as few as four to several dozen at a time. Banana flowers have very remarkable look but the pseudostem of each species has a definite growing period to attain flowering stage and normally by this time these have 9-12 leaves. *Musa velutina* is the only species that can die to the ground in winter and flower and fruit in the following season, requiring only 20 weeks for completing its life cycle. *Musella lasiocarpa* takes several seasons to produce a pseudostem large enough to flower and its inflorescence lasts for several months. Other bananas must retain a pseudostem for more than one growing season in order to flower and produce fruit. The pseudostem of some species such as *Musa basjoo* remains viable even at -9.5 °C temperature and subsequently flowers in the following season. Technically, the flowers are inflorescences and a single inflorescence forms on a spike at the top of the plant. *Musa* and *Ensete* flower stalks are long and hang down beneath the leaf canopy, but *Musella* inflorescences are borne on short stalks and face upward. The individual florets are slim and tubular and are subtended by very large, brightly-coloured bracts that may be red, purple, orange, or yellow. The inflorescence starts as a large purple tapered bud which elongates during opening and reveals bracts which surround whorls of florets. Banana plants are monoecious meaning separate male and female flowers are produced on the same inflorescence. The female florets are grouped together in 5 to 15 rows at the basal end of the inflorescence, followed by a region of hermaphrodite or neutral flowers. Finally, there is a zone of male flowers near the tip of the inflorescence. The flowers open sequentially from base to tip and the male flowers shed few days after opening while female flowers grow into bunches of bananas. The banana fruits are technically a type of berry (a soft multi-seeded fruit developed from a single compound ovary). The young green fruits resemble green fingers and dangle down in clusters from the top of the plant. All of the bananas on a single stalk are called a bunch. Each cluster of young bananas (which form at a node on the stalk) is called a hand and each banana (fruit) is known as finger. As the fruit matures, it changes colour from green to familiar yellow or to less familiar shades of red or white. The fruit may even be striped with multiple colours. Bananas range from 5-30 cm in length and from 18 mm to 5 cm in width. The flesh of the ripe fruit ranges from pure white to various shades of yellow. Wild type bananas are lax in flesh but are filled with 3 mm to15 mm hard black seeds and upon ripening these bananas peel themselves to expose the flesh and seeds.

Classification, Species and Varieties

Genera and species having ornamental importance in family Musaceae are genus~***Ensete*** (2n=18, x=9) which contains as many as nine species, most being annuals and all monocarpic unbranched herbs those sucker rarely and are used for food, fibre and as ornamentals, resembling banana plants but their wide spreading and immensely long, paddle-shaped leaves with usually crimson midribs make them fully distinguishable, and the fruits similar in appearance to those of banana but they are dry, seedy, and inedible. These are propagated through seeds. Among the species in this genus are *Ensete glaucum, E. superbum, E. ventricosum* and its sub-species *E. ventricosum* 'Maurrelli', *vis-a-vis* other dwarf types suitable for home gardening and growing along the paths and roads. Genus ~***Musella*** (2n=18, x=9), the most fascinating and unique member of the family *Musaceae* is a genus comprising two species (*Musella lasiocarpa* and *M. splendida*) native to SE Asia, including southwest China (Yunnan and Guizhou),

Vietnam, Laos and Myanmar. Genus ~ ***Musa*** (2n=14, 20, 22, 33, 44; x=7, 10, 11) where a great number of important species are available bearing edible fruits and floral parts apart from being ornamentals and plantains, used for various medicines, beverages, fibres, dyes, fuel, cordage, wrapping materials, *etc.*, and the *Musa* has four sections such as *Australimusa, Callimusa, Musa* (formerly known as *Eumusa*), and *Rhodochlamys* where a majority of species in the sections have significant ornamental values. ***Section- Callimusa*** where bracts are plain, firm, shiny on the outer surface, rarely glaucous and strongly imbricate when closed. These plants are most important as ornamental, mostly bear upright flower stalks of variously coloured buds and flowers and small seeded fruit. ***Section-Rhodochlamys*** contains a total of nine species of which seven (*Musa ornata, M. laterita, M. velutina, M. rosea, M. mannii, M. rubra, M. aurantiaca*) are well recognized as ornamentals (Wu Delin and Kress, 2000) and the remaining two, *viz. M. siamensis* and *M. sanguinea* are less known and are of somewhat less definite status. *Rhodochlamys* consists of the only *Musa* species adapted to withstand seasonal droughts, which is common in the monsoon areas to which they are native. The natural habitat of *Rhodochlamys* species is NE India, Bangladesh, Myanmar and Northern Thailand (exception *Musa rosea* being native to Cambodia and S. Vietnam). *Musa sanguinea* is also known to occur in Yunnan (China). Species with ornamental potential in *Rhodochlamys* are suitable for indoor and greenhouse culture and can also be grown outdoors during growing season. As in their natural habitats, these species are seasonal plants with flowering, fruiting and dormant periods, they flower easily every year. Grown as indoor potted plants, they normally go into dormancy during the darkest winter months. During this period when growth ceases, soil should be kept slightly moist so that plants do not lose their leaves even without extra lighting. In the greenhouse, plants normally go into semi-dormancy in winter with very slow growth even under good growing lights. When grown outdoors in temperate climates, plants should be cut before they freeze and rhizomes should be stored dry without soil. However, rhizomes can be potted and stored as indoor plants.

Ornamental bananas under genus ***Ensete*** (chromosome numbers 2n =2x =18) are ***E. glaucum*** which grows some 3 m high having thick bluish trunk and giant bluish-green leaves; ***E. perrieri***, a robust plant with beautiful bluish-waxy short pseudostem distinctly swollen at the base, leaves with yellowish midribs are crowned as shuttlecock, spike short and the flowers large and maroon, and suitable for many temperate and tropical areas; and ***E. ventricosum*** growing some 3 m in temperate climates and 6 m in tropical with 3 m long leaves flushed wth burgundy-red at emergence and the colour being more intense under bright sun, and is most suitable for summer border. Under genus ***Musella*** (2n = 2x = 18) are ***M. lasiocarpa*** which is cold-hardy, dwarf (60 cm), flowering starts normally the next year bearing 20-25 long, erect and densely packed flowers with yellow bracts, each subtending 4-5 erect yellow flowers lasting for a few months and these flowers before opening resemble the lotus, and the round and dark brown fruits contain 6 seeds, and in its var. *rubribracteata* the leaf-midrib and petiole at abaxial surface are reddish to purple-red, inflorescence deltoid while the bracts tightly attached, markedly imbricated and orange-red to red (Ma Hang *et al.*, 2011); and ***M. splendida*** grows about 1.0-1.2 m tall, leaves elliptical, inflorescence bud ovate but apex open as the long pointed tips of individual bracts spreading out, basal flowers hermaphrodite or female in some other types and the fruits are seedless. Under genus ***Musa*** (x = 7, 10, 11; 2n = 14, 20, 22, 33, 44) are ***Callimusa*** which possesses x = 10 where ***Musa beccarii*** (deriving this name in honour of Odoardo Beccari, an Italian naturalist who did much work on banana in Borneo), growing some 3.0-3.5 m tall, is a beautiful banana from Borneo which starts flowering when 1.0-1.5 m high, sheaths with a brown-purple rim, petiole 30-35 cm long, leaves bright green, oblong, lanceolate, up to 1 m long and 30 cm wide, and hairy inflorescence bearing narrow, erect, elliptical, bright scarlet bud with green tipped bracts and flowering for a long time, closely overlapping bracts are shiny green and red tipped yellow, fruits 1-3, 5-15 cm long and 2 cm in diameter, upright and cylindrical but narrowing towards the tip; ***M. campestris*** emits many stolons, pseudostem slender, 1.5-2.0 m high, green and blotched brackish-purple, leaves 2 m long and 40 cm wide are almost erect and narrowing at base, petioles 50-70 cm long with scarious margins, inflorescence erect, peduncle pubescent, bract sterile and usually 1, usually persistent at the opening of the male flowers, 5-7 basal hands female, upper hands male, bracts bright purple, flowers orange-yellow, fruit very pale and whitish-green irregularly mottled with purple, fruit bunch compact, its peduncles and rachis pubescent, the fingers inflexed to stand almost parallel to the rachis, individual fruit 8-13 cm long and about 2 cm in diameter, cylindrical with 4-5 angles, apical part bottle-neck-shaped with truncate apex, pericarp thin, powdery green and usually blotched reddish-purple, seeds cylindrical or pyriform, tuberculate, 4-5 mm long and 2.5-3.0 mm in diameter, and its varieties are *M. c. campestris, M. c. lawasensis, M. c. limbangensis, M. c. miriensis, M. c. sabahensis, M. c. sarawakensis, etc.*; ***Musa coccinea*** where plant stooling freely, pseudostem up to 1.5 m high, 5 cm in diameter at base, green, devoid of wax, leaf blades up to 1 m long, 25 cm wide, narrowed gradually to a rounded apex, rounded at base at right side, usually longer than the other, shining dark green above, paler beneath but not glaucous, midribs green like the lamina above, paler beneath, petioles up to 35 cm with narrow erect margins clasping the pseudostem, not becoming scarious, inflorescence quite erect, its peduncle scarcely emerging from the sheath of the subtending leaf, the rachis glabrous, last foliage leaf reddish-petiole, sterile bracts usually 2, bright scarlet with green leaf like tips, first fertile bract about 15 cm long, flowers in the basal bracts female, clusters 1-4 and sometimes more, upper flowers male, fruit oblong, 4-5 cm long and 2.0-2.5 cm wide, somewhat laterally compressed dorso-ventrally, rounded at the sessile base, narrowed to a truncate apex, crowned by persistent withered perianth, pericarp about 1.5 mm thick, orange-yellow at full maturity with a waxy bloom, pulp white, seed almost cylindrical, black about 6 mm long, a little wider at the top than at the base, with a distinct waist marking the base of the perisperm chamber within, 4 mm in diameter at the waist, and the surface marked with longitudinal warty ridges; other species in this category being *M. aurantiaca, M. exotica, M. gracilis, M. lawitiensis* (syn. *M. suratii*), *M. mannii, M. rubra, M. siamens, M. violascens;* ***M. sanguina*** having slender plant with pseudostem as thick as a stout cane, reddish and growing to about 1.0-1.5 m high, leaf midribs red on both sides of young leaves, later becoming green only above, inflorescence horizontal, bracts dark pink or pale-crimson and the whole bud usually aborts before the fruit ripes, the staminate flowers orange-yellow, fruit stalk is red and velvety, and fruit greenish-yellow when ripe; ***Musa troglogytaquorum*** has robust plant bearing erect bunches of brilliant orange-gold fruit triangular in shape, plant is ornamental and fruits nutritious containing high level of beta carotene; ***M. peekelii*** ssp. ***peekelii*** with pseudostem up to 10 m high and 80 cm girth, predominantly rich brown colour in the lower part, brown or non-waxy green above, leaf lamina bright green, non-waxy above, paler and very waxy below, inflorescence attractive with yellowish-green bracts, peduncle glabrous-green, the bunch hanging geotropically vertical, moderately to extremely lax, basal bracts long, ligulate, brown or green, shiny, deciduous or often at least partly persistent, basal flowers functionally female but often with staminodes, fruit tipped with a narrow green bud which is pendent, mature fruit rich orange in colour, extremely sweet when ripe, indehiscent, cylindrical, about three times as long as broad, blunt or bottle-nosed and seeds very irregular, 6-7 mm in diameter, characteristically broader than deep and hilum depressed; ***M. textilis*** (syn. *M. formosana* Hayata) having pseudostem 2.5-4.0 m high, 15-20 cm in diameter at base, green or more or less purplish or even almost black towards the base, leaf sheaths and petioles devoid of wax, leaf blades oblong, narrowing towards the apex, 1.5-2.0 m long, 40-50 cm wide, inflorescence at first sub-horizontal, and its peduncle and rachis glabrous or minutely pubescent. Under section ***Eumusa*** (x = 11) are ***M. balbisiana*** which is extremely robust, fast-growing, and drought resistant, one of the parents of many edible seedless bananas and is native to Southeast Asia from Sri Lanka to the Philippines; ***Musa acuminata*** ssp. *zebrina* is characterized by very slender pseudostems and small, slender and predominantly beaked fruits full of grape-like seeds, plant thriving under heavy shade and bears striking dark green leaves blotched with variably sized maroon patches; ***M. flaviflora*** (syn. *M. thomsonii*) is a highly spectacular banana with a dark purple-red pseudostem, highly waxy around the shoulders, trunk 6 m or more and suckering, leaf's abaxial surface with dark red midrib and lamina so waxy as to appear almost white, leaves 3 m long, glaucous beneath, inflorescence drooping, bracts oblong-lanceolate, lower 20-25 cm long, 10 cm wide, uppermost 15-18 cm long, nearly 7.5 cm wide, red outside, bright shining orange inside, each enclosing 18-20 flowers in two rows, perianth orange, 6.25 cm long, 1.25 cm wide, 4-lobed, free petal ovate-lanceolate, over 2.5 cm long and acute, fruit angled, 12.5-15.0 cm long, stalked and not recurved, and seeds nearly 1.25 cm long and 8 mm wide; ***Musa basjoo***, the world's cold-hardiest banana, planted in the ground at -18.3 to -31.3 °C but with protective mulching otherwise leaves and pseudostem may freeze, even though the rhizomes survive underground, growing up to 2.4-2.7 m in containers and up to 5.4 m in the ground with long, slender, bright green leaves, and the inflorescence one of the most beautiful of all bananas; ***Musa sikkimensis*** is relatively cold hardy, leaves wide, deep reddish underside in emerging new leaves but changing to green when leaves reach maturity, new leaves with deep maroon variegation on the leaf surface but on maturity these turn green retaining only a red midrib, though all the plants do not exhibit this variegation and colouring on the underside of the leaves, inflorescence growing out at a stiff angle to the upright stem, but at developing and fruiting stage it becomes almost horizontal, fruits about 12.5-15.0 cm long and angled and when ripe though pulp is lax but sweet, and the seeds numerous large, black and angled; ***M. cheesmani*** with bushy, dense and deep reddish-brown pseudostem, leaf green upper and grey underside with purplish-brown midrib, flowers white, fuit bunch very lax, up to 10 hands of fruit, fruits biseriate in the centre of most hands (uniseriate in small bunches), curved-spreading and angular, tapering to the long (4 cm) pedicel, acuminate at the apex, and the seeds flattened-subglobose, 8-10 × 5-7 mm deep and intensely rough-warty; *etc.*

Krewer *et al.* (2008) evaluated some commercial banana cultivars (from section *Musa* and plantain group) in southern Georgia for ornamental purpose and divided these cultivars into three types based on height and their ornamental values and reported tall-growing cultivars as 'Belle', 'Ice Cream', 'Kandarian', 'Manzano', 'Saba', medium growing cultivars as 'Dwarf Namwah', 'Dwarf Orinoco', 'Gold Finger', 'Rajapuri', and 'Super Plantain', and the dwarf-growing ones as 'Dwarf Nino', 'Gran Nain', 'Kru' and 'Sum X Cross'.

Propagation

Ornamental banana can be propagated through 2 to 4 month old suckers or rhizomes and seeds. Genus like *Ensete* is propagated through **seeds** which germinate readily after harvesting and show no signs of dormancy. **Sword-suckers** conical in shape with actively growing central bud having lanceolate leaves and well developed rhizome weighing some 500 to 750 g are generally good taking up as the new plants. **Cut rhizomes** (bits and peepers) are also used to propagate the plants but this takes more time to develop into flowering size plants. Propagation through **shoot-tip** culture is cost-effective for production of disease-free plants. The incorporation of growth retardants such as ancymidol (ANC) or paclobutrazol (PBZ) in liquid culture media during multiplication stage of bananas decreases the excessive growth of stems and leaves (Albany *et al.*, 2005).

Cultural Practices

Hill banana (*Musa sikkimensis, Musella lasiocarpa*) is well suited for cultivation in humid subtropical to semi-arid subtropics up to 2,500 m above mean sea level with a **temperature** of 15-35 °C and an average rainfall of 500-2000 mm per year. Most of the species are susceptible to low temperature and frost, except the members belonging to genera *Musella* and some of the *Eumusa* species like *Musa sikkimensis* and *Musa basjoo* (hardiest bananas tolerating -19 to -29 °C temperatures) though require protective mulching at -19 °C or below temperatures. Frost kills the aerial leaves and the plants up to the ground but on appearance of congenial temperatures these re-grow from the underground buds present in the rhizomes. In colder areas, new planting is carried out in each spring. In semi-temperate condition also most of the species can be grown as indoor plants by maintaining the required temperatures artificially. Mean temperature of 20-30 °C is optimum for growing most of the species. Temperatures above 36-38°C cause scorching effect with increased rate of transpiration while below 16°C results in restricted growth. High temperature along with water stress causes loss in growth, however, water stagnation in poorly drained soils also leads to slow growth. Apart from temperature and water, wind is also a major constraint in banana cultivation as high winds result in uprooting and collapsing of plants.

It can be grown in all kinds of **soils** having good drainage. Sandy loam soil having a pH range of 6.5-7.5 is optimum (Nelson *et al.*, 2006) though can tolerate pH upto 8.5 with suitable amendments. The planting site should be free from wind and frost. Pit method of planting is generally followed. Pits of 60 × 60 × 60cm size are prepared and filled with mixture of soil, coarse sand and farmyard manure in 1-1-1 ratio. Suckers or seedlings are planted in the centre of pit and soil around compacted. Dwarf ornamental bananas may be planted as close as 2.5 meters and tall plants 4.0 m apart. *Musella* and other dwarfer species can be grown in containers. Most of the bananas attain large size quickly, hence, require a 15 gallon or even larger container. Quantum of watering is dependent on size of the container, the plant and prevailing weather conditions. Young plants in large containers may need to be watered every 2-4 days but when it becomes pot-bound it requires a daily drenching. A pot-bound banana rhizome can easily split a plastic pot wide open so before splitting these should be repotted. Banana should be re-potted at least every 3 years and the old soil must be replaced with a high quality potting mix. During winters the container can be kept inside to protect the plants from frost damage. Care should be taken to provide plenty of light, humidity and the required amount of water. Avoid exposing the container to extreme temperatures.

Every time when the flowers have lost their beauty the plants are cut back just above the rhizome to give room for other suckers developing around the rhizome. The same plant does not repeat its flowering as in fact the aerial plant is composed of only ensheathing leaves and no stem, though inflorescence comes directly from the rhizomes. **Pruning** is normally practiced only to provide suckers for propagation, as most banana plantings are allowed to grow freely in mats of several plants of varying ages and sizes. Suckers can be quickly separated from its mother plant when these attain some 20-45 cm length and their brown leaves are trimmed as and when observed. Tattered older leaves can be removed after they break and hang down along the trunk. Most banana plants are susceptible to freezing temperatures, therefore, cut the trunk to the ground level before onset of cold temperatures and provide protection to the underground stem from cold through mulching and thatching. The main purpose for **winter care** is to protect the pseudostem which allows for large plants that may flower and fruit, along with increasing survivability of marginal plants. The pseudostem of plants during winter is either protected by lifting and repotting in the fall and keeping indoors at a warm location or stored at a cool location above freezing temperature at a place with proper aeration.

Regular **weeding** is important during the early stage of plant growth. Either hand weeding or pre-emergence application of Diuron (1kg a.i./ha) or Glyphosate (2 kg a.i./ha) is effective in controlling grasses and broad-leaved weeds.

Control of Growth and Flowering

Dormancy can be removed by giving pre-sowing treatments like scarification of seeds with sulphuric acid, chipping of testa, softening testa by soaking and use of alternating temperatures to promote germination. Application of growth retardants significantly reduces the plant height. Application of PP333 at 0.25 mg per plant at the end of hardening stage reduces the plant height two months after application. Soil application of paclobutrazol with 1-2 g a.i./plant was found reducing plant height up to 25 per cent but did not affect the number of days from planting to flowering in two cvs 'Prata Ana' and 'FHIA-01' (Maia *et al.*, 2009). Liquid pulse treatment of combined cytokinins and auxins for 60 minutes (BA:kinetin, 1:1 at 50 mg l^{-1} for multiple shoot proliferation, and NAA:IBA, 1:1 at 100 mg l^{-1} for rooting) was found quite effective, however, high concentration of growth regulators caused necrosis and reduction in shoot and root formation during *in vitro* propagation (Madhulatha *et al.*, 2004). Soil application of paclobutrazol with 1-2 g a.i./plant was found reducing plant height up to 25 per cent but did not affect the number of days from planting to flowering in two cvs 'Prata Ana' and 'FHIA-01' (Maia *et al.*, 2009). Liquid pulse treatment of combined cytokinins and auxins for 60 minutes (BA:kinetin, 1:1 at 50 mg l^{-1} for multiple shoot proliferation, and NAA:IBA, 1:1 at 100 mg l^{-1} for rooting) was found quite effective, however, high concentration of growth regulators caused necrosis and reduction in shoot and root formation during *in vitro* propagation (Madhulatha *et al.*, 2004).

Insect-Pests and Diseases

The banana **weevil borer** (*Cosmopolites sordidus*) causes wilt-like yellowing of leaves and stunting of plant growth. The presence of tunnels in the rhizome and lower pseudostem confirms its infestation as its larvae feed tunnelling these parts. Its infestation reduces bunch weight and causes toppling or snapping of banana plants. Therefore, only healthy planting stock should be used, however, its infestation can be checked by hot water treatment before planting for 15-27 minutes at 52-55 °C of cleaned and trimmed suckers. Application of 60-100 g of *neem* seed powder or *neem* cake at planting and again at every 4-month interval will diminish the damage. However, field sanitation is a must.

Burrowing nematode (*Radophilus similis*) and **root-knot nematode** (*Meloidogyne* spp.) cause severe damage to banana crop. The root-knot nematode (*Meloidogyne* spp.) causes wilt-like symptoms of leaves and both the primary and secondary roots of infested bananas show galls and swellings. When the roots are cut open, necrotic lesions caused by these nematodes are observed. Root-knot nematode infestation can also result in narrow and yellowing of leaves, stunting, small bunches and reduced plant growth. Burrowing nematode (*Radophilus similis*) causes more severe damage on banana though symptoms are almost similar to *Meloidogyne* infestation. Management practices include soil fumigation, removing of host plants and crop rotation with non-host crops as well as planting trap crops such as *Crotalaria* or *Tagetes* which reduces nematode population in the soil before planting,. Furthermore, before planting cut the rhizomes in order to remove dark spots (infestation) and then soak them in hot water (55° C) for 20 minutes.

Panama disease or *Fusarium* wilt is caused by soil-borne fungus, *Fusarium oxysporum* f. sp. *cubense* and is one of the most destructive diseases as the fungus survives in infected soil for many years and the only viable method is to replace susceptible varieties with resistant ones. The symptoms show premature yellowing of old leaf-margins and progressing towards the younger ones. When leaves eventually collapse at the petiole, they hang down around the pseudostem. Longitudinal splits may also develop in the pseudostem which spread locally through the use of infected rhizomes or suckers and soil attached to planting material, implements or vehicles. The pathogen spreads slowly from plant to plant, but if the spores are carried on surface through run-off water or contaminated irrigation reservoir the disease will spread very rapidly. Quarantine and exclusion procedures are the only effective measures to control the disease. **Speckle disease** or *Mycosphaerella* speckle is caused by *Mycosphaerella musae* which is not a very serious problem in the tropical areas. However, it can become a problem on commercially grown cultivars in the subtropics. Since 2000, *Mycosphaerella* speckle leaf disease of banana dominated the scene. The first noticeable symptoms are light brown or tan-coloured irregular blotches on the lower leaf surface that may show as smoky patches on the upper surface. The blotches darken in colour and eventually become dark, irregularly-shaped and with speckled areas. Leaf tissues in and around the speckled areas become yellow and later necrotic. Extensive death of leaf tissue is seldom seen above the eighth leaf of an actively growing plant. However, when leaf production ceases with bunch emergence, extensive damage and defoliation can occur before fruit is ready for harvest. Banana speckle may be damaging in subtropical areas where it can cause leaf death, as fungus primarily affects the older leaves of the plant. Fungicides like mancozeb or propiconazole give adequate control of *Mycosphaerella* speckle. Removing leaves damaged or killed by this fungus can further reduce inoculum levels.

Ralstonia solanacearum is an aerobic soil-borne and motile, having a polar flagellar tuft, non-sporing, gram-negative plant pathogenic **bacterium**. It colonises the xylem, causing wilt in a very wide range of potential host plants. It causes typical yellowing and wilting of older leaves, blackening and shrivelling of flowers and discolouration of the vascular tissues. If an area becomes infected, it would be better to eliminate all of the infected plants and follow strong sanitation practices to reduce the spread of disease. Integrated disease management (IDM) is the best strategy to reduce any impact of the pathogen. Using pathogen-free planting materials is a necessity and planting resistant cultivars will also aid in minimising the ill-effects of the pathogen. Finally, a good rotation system that replaces susceptible crops with resistant or immune crops can assist in diminishing the incidence.

Banana bunchy top virus (BBTV) is transmitted by an aphid vector (*Pentalonia nigronervosa*) and is considered to be the most economically destructive of all the viral diseases affecting bananas worldwide. BBTV is one of the most serious diseases of banana. Once established, it is extremely difficult to eradicate or manage. The virus is spread from plant to plant by aphids and from place to place by people transporting planting materials obtained from infected plants. The initial symptoms of BBTV consist of dark green streaks in the veins of lower portions and the leaf-midrib. Also, dark green, hook-like extensions of the leaf lamina veins can be seen in the narrow, light-green zone between the midrib and the lamina. Mature plants infected with BBTV, hardly produce new leaves and those produced are narrower than normal, are wavy rather than flat, and have yellow (chlorotic) leaf margins. They appear to be bunched at the top of the plant, the symptom for which this disease is named. There is no direct cure for BBTV, therefore the only solution left is to control the aphids with pesticides which in turn will check further spread of the virus, *vis-a-vis* complete sequencing of the genomes will allow researchers to try to develop BBTV resistant banana varieties. The **banana bract mosaic disease** is a viral problem transmitted in a non-persistent manner by several aphid species including *Aphis gossypii, Pentalonia nigronervosa* and *Rhopaiosiphum* sp. This disease is also transmitted in vegetative planting material including bits, suckers and tissue-cultured plantlets. The symptoms of banana bract mosaic disease are usually very distinct. The characteristic dark reddish-brown mosaic pattern on the bracts of the inflorescence distinguishes this disease from all other known viral diseases of banana. Initial symptoms include green or reddish-brown (depending on cultivar) streaks or spindle shaped lesions on the petioles and a tendency towards a congested leaf arrangement. Leaf lamina symptoms may or may not occur and are most prominent on the younger leaves. The symptoms consist of spindle-shaped chlorotic streaks running parallel to the veins when the dead leaf sheaths are pulled away from the pseudostem and then the distinctive dark coloured mosaic patterns, stripes or spindle-shaped streaks are visible. Chlorotic streaks may occur on the bunch stalks and high disease incidence is associated with increased levels of fruit rejection on commercial plantations. Management of aphid vectors and use of disease-free planting material help to reduce the disease.

References

Albany, N. R., J.A. Vilchez, L. Garcia and E. Jiménez, 2005. Comparative study of morphological parameters of Grand Nain banana (*Musa* AAA) after *in vitro* multiplication with growth retardants. *Plant Cell, Tissue and Organ Cult.*, **83** (3): 357-361.

Elevitch, C.R. 2006. *Traditional Ttrees of Pacific Islands.* Permanent Agricultural Resources. Hawaii, USA, 701 pp.

Krewer, G., *E.G.* Fonsah, M. Reiger, R. Wallace, D. Linvill and B. Mullinix, 2008. Evaluation of commercial banana cultivars in southern Georgia for ornamental and nursery production, *Hort. Techn.*, **18**(3): 529-535.spp. AAA). http://www.traditionaltree.org

Madhulatha, P., N. Anbalagan, S. Jayachandran and N. Sakthivel, 2004. Influence of liquid pulse treatment with growth regulators on *in vitro* propagation of banana (*Musa* spp. AAA). *Plant Cell Tissue & Organ Cult.*, **76**: 189-191.

Maia Emanuel, Dalmo L. Siqueira, Luiz C.C. Salomão, Luiz A. Peternelli, Marilla C. Ventrella and Rithiely P.Q. Cavatte, 2009. Development of the banana plants 'Prata Anã' and 'FHIA-01' under the effect of paclobutrazol applied on the soil. *Ann. Brazilian Acad. Sci.*, **81**(2): 257-263.

Nelson, S.C., R.C. Ploetz and A.K. Kepler, 2006. *Musa* species (bananas and plantains, ver. 2.2). In: *Species Profiles for Pacific Island Agroforestry* (ed. Elevitch, C.R.).

Permanent Agriculture Resources, Holualoa, Hawaii (http://www.traditionaltree.org).

Simmonds, N.W. 1962. The evolution of the bananas. Longmans, Green and Co., Ltd. London. 170 pp.

Stover, R.H. and N.W. Simmonds, 1987. Bananas (3rd ed.). Longman Scientific and Technical, London, UK, 468 pp.

Wu Delin and W.J. Kress, 2000. *Musaceae*. In: *Flora of China*, **24**: 297-313. http://www.fna.org/china/mss/volume24/MUSACEAE.published.pdf

32

Ornamental Gingers (Family: Zingiberaceae)

V.L. Sheela and A. Sheena

[**Common names**: Beehive ginger (*Zingiber spectabile*), Black ginger/Midnight ginger (*Zingiber malaysianum*), Bronze peacock/Peacock ginger (*Kaempferia elegans*), Butterfly ginger/Butterfly gingerlily/Garland flower/White gingerlily (*Hedychium coronarium*), Dancing ladies/Globe ginger (*Globba winitii*), Indian ginger (*Alpinia calcarata*), Patumma/Siam tulip/Summer tulip/Thai tulip (*Curcuma alismatifolia*), Pinecone ginger/Pink porcelain lily/Shampoo ginger (*Zingiber zerumbet*), Porcelain flower/Torch ginger/Wax flower (*Etlingera elatior*), Red ginger/Fire ginger plant (*Alpinia purpurata),* Shell ginger *(Alpinia zerumbet*), Variegated ginger (*A. sanderae*)]

Introduction and Origin

Ornamental gingers are recent introductions to the landscape as well as cut flower industry. They enjoy a special position in the botanical kingdom with their elegance in form and texture, sparkling colour and amazing symmetry and some are even known as queen of the flowers in the plant world. The long lasting inflorescences and ability to perform well in partial shade make them ideal choices for cut flowers and for growing in specific situations. Gingers are classified as herbaceous perennials and once planted they continue growing and flowering for many years. The flower characteristics in Zingiberaceae are highly specialized as what look like petals are often sterile and modified stamens, called staminodes. About 1,200 species of gingers are distributed worldwide and nearly 250 species are cultivated as ornamental plants. Around 200 species of gingers are present in India, of which about 60 species are ornamental (Sabu *et al.*, 2013). The genera of flowering gingers include many but the important ones of high ornamental values are *Alpinia, Brachychilum, Burbidgea, Curcuma, Elettaria, Etlingera, Globba, Hedychium, Kaempferia, Monocostus, Tapeinochilos* (*Tapeinocheilos*) and *Zingiber.*

Alpinia, found in wild in Polynesia, Japan and East India, is the largest genus with more than 250 species, comprising a very large and diverse group of everblooming gingers valued both for its tropical foliage and large colourful flower heads, apart from bearing different forms and varying heights. The name *Alpinia* commemorates the Italian Prospero Alpino (1553-1616), Professor of Botany at Padua. ***Alpinia purpurata*** is native to the Pacific Islands (Moluccas, Yap to New Caledonia) and is widely cultivated in the tropics and subtropics. The inflorescence has bright red floral bracts and inconspicuous white flowers. It is quite popular as an ornamental and cut flower, both for the home and for commercial sale. Essential oils and aqueous extracts obtained by hydro-distillation of flowers of red and pink variants of *Alpinia purpurata* can be used as larval insecticides against *Aedes aegypti*, the dengue mosquito. The floral oil is antibacterial in action with 42 essential oil components with pinene and caryophyllene being the major constituents (Santos *et al.,* 2012). ***Alpinia zerumbet*** is a fabulous and cold hardy ginger. It is distributed widely in the tropical and sub-tropical regions of the world including south-east Asia and India. This herbaceous plant produces pendant sprays of delicate flowers resembling 'white sea shells', tipped pink with a red and yellow throat. The blooms are excellent for arrangements. Even without flowers, the foliage is very attractive, and it makes a good screening plant. Dwarf cultivars are suitable for pot planting. It has been used widely in folk medicine. In Japan, leaves of *Alpinia zerumbet* are sold as herbal tea, and are commonly used to flavour noodles and to wrap rice cakes. Decoction of leaves has been used during bathing to alleviate fevers. *Alpinia* essential oil is obtained by steam distillation of leaves and is used in cosmetics, perfumes, and soaps. More than 20 products are commercialized from the stem fibre such as Japanese paper, *kariyushi*-wear, and cloth. The rhizomes can be used as spices and in beverages. The plants possess pharmacological activities, used as an antioxidant and in the treatment of intestinal disorders, hypertension and inflammation (Chompoo *et al.*, 2012). The main components in leaf oils are 1, 8-cineol, camphor and methyl cinnamate, whereas the major oils in the rhizome are 5, 6-dehydrokawain (DDK) and methyl cinnamate. During essential oil production and fibre isolation, large volumes of solid and squeezed wastes are produced and subsequently discarded. Disposed wastes may be utilized in foodstuffs as a cheap source of natural antioxidants (Tawata *et al.*, 2008). Ferulic acid, p-hydroxybenzoic acid and syringic acid are the predominant phenolics in flowers (Elzaawely *et al.*, 2007). From the leaves, flavonoids, kava pyrones, and phenolic acids have been isolated. Leaves have higher inhibition of b-carotene oxidation and radical–scavenging activity than rhizomes (Chan *et al.*, 2009). It could be used as a source of bioactive compounds against HIV-1 integrase and neuraminidase and that DDK and dihydro-5, 6-dehydrokawain may have possibilities in the design of drugs against the viral diseases, *viz.* AIDS and influenza (Upadhyay *et al.*, 2011).

The genus ***Curcuma*** comprises approximately 120 species (Skornickova aand Sabu, 2002; Skornickova *et al.*, 2004) of rhizomatous herbs originating in the Indo-Malayan region and the genus is widely distributed in the tropics of Asia to Africa and Australia. *Curcuma* is the Latinized version of an Arabic name (*kurkum* = yellow) for the yellow dye plant turmeric. *Curcuma alismatifolia*, a species native to Indo-China has gained popularity recently in the international market as a new ornamental plant. In India, the genus is represented with 40 species (Roxburgh, 1820; Baker, 1890-1892; Velayudhan *et al.*, 1996; Sabu, 2006), having maximum diversity of species in peninsular region. Curcumas are valuable ornamentals for their beautiful spikes and foliage. The spikes are produced basally or from the central part of the plant. They have broad and veined leaves, and small but brightly coloured flowers which emerge from large bracts of the inflorescence rising about 90 to 100 cm above the foliage. The flowers range in colour from white to pink, orange and shades of violet and the shape of the bracts resembling to that of tulip. The flower spikes last for several weeks. *Curcuma* species exhibit inter- and intra-specific variation for the biologically active principles coupled with morphological variations. The species name alludes to the distinctive foliage, which resembles leaves of the water plantain family, Alismataceae. It is used as a bedding plant in tropical countries, as a pot plant throughout the world, and as a cut flower.

The genus ***Etlingera*** includes approximately 60 species of perennial evergreen rhizomatous herbs from Sri Lanka to New Guinea via Indomalaysia and naturalized throughout tropical Asia. The genus *Etlingera* is named after the German botanist 'Andreas Ernst Etlinger'. *Etlingera elatior*, a native of Malay Peninsula and Indonesia (Java) is the only widely cultivated species and it is naturalized throughout tropical Asia. The specific name *elatior*, in Latin means 'taller'. The varying colours of the bracts and flowers make this species a very attractive plant for ornamental purposes. The petals are actually bracts, which later reveal small yellow true flowers inside. The bracts are red, pink or white. The pink torch ginger is the most widely grown, but the most attractive one is the red one. It's all the parts are aromatic. It is grown commercially for cut flowers and is also used as a landscape plant. The rhizome is extensive and etlingeras are generally not suitable for containers. *Etlingera* plant has extensive traditional uses. The young shoots, flower buds or fruits are consumed raw by indigenous communities, used as a condiment, as a spice for food flavouring or cooked. The young flowering shoot (often called a flower bud) of the torch ginger is an indispensable ingredient used to flavour both *rojak* and *laksa*, which are popular dishes in Malaysia and Singapore. Chan *et al.* (2009) isolated three caffeoylquinic acids including chlorogenic acid (CGA), and three flavonoids quercitrin, isoquecitrin and catechin from the leaves of torch ginger and reported the antioxidant activities of leaves and rhizomes of this plant. A total of 55 compounds were identified in the oil and extract. The predominant chemical classes of the essential oil were alcohol (44.25 per cent) followed by acids (24.42 per cent), aldehydes (19.54 per cent), esters (10.51 per cent), and sesquiterpenes (0.99 per cent). The chemical compound 1-dodecanol (25.2 per cent) is the most abundant alcohol in the inflorescence. The oil and extracts derived from the inflorescence of torch ginger has rich antibacterial activity and possess great potential to be used as a preservative in the food and pharmaceutical industries (Wijekoon *et al.*, 2013).

Genus ***Globba*** was first discovered in 1924 by a Siamese (Forestry Officer) in NE Thailand, *i.e.* Lampung Province. In Indonesia (Amboina), its native name is *galeba* from where it has taken its botanical name. It is one of the largest genera in the family but after *Alpinia*. It consists of about 100 species mainly distributed in tropical Asia with maximum diversity in Indonesia, Malaysia, Thailand, Myanmar to southern China. In India it is represented by 13 species and is widely distributed in Western India, South India and Andaman & Nicobar Islands. *Globba winitii* is a tender perennial native to Thailand and Vietnam. Dancing ladies, it's common name is derived from the way the flowers seem to dance in the air suspended from the bracts. It is small enough to be grown in pots, and will tolerate low light, making it highly suitable as an indoor plant, and an ideal container plant for patios and verandahs. Besides potted plant use, they can be marketed for use in year round interior landscapes and as either perennials or summer annuals in exterior landscapes, depending on the climate. They are also excellent for cutting and arranging.

Hedychium in Greek is *hedys* meaning sweet (for fragrance) and *chion* meaning snow (for white colour of flowers). *Hedychium* is one of the most fragrant types of gingers for which India is the centre of origin with maximum diversity of species being in the NE India. It is native to Tropical Asia, Himalayas and Madagascar. *H. coronarium* is distributed in Bhutan, China, India including Sikkim, Indonesia, Malaysia, Myanmar, Nepal, Sri Lanka, Thailand, Vietnam and Australia (Sharma *et al.*, 2011), however, Morley (1970) mentions that there are some 50 species in the genus spreading in abundance in Madagascar, Indonesia, Malaysia and SW China. *Hedychium coronarium*, is widely cultivated although *H. coccineum, H. elwesii, H. flavescens, H. flavum, H. garderianum, H. greenii, H. rubrum, H. spicatum, H. wardii*, etc. are some of the other beautiful butterfly gingers. *H. coronarium* is the *gandasuli* of India, a Sanskrit word meaning 'the fragrance of the princess' or 'the queen's perfume'. The rhizomes have a fragrance somewhat reminiscent of ginger but not suitable for flavouring food. Individual flowers resemble butterflies, hence the common name as butterfly ginger for *H. coronarium*. The flowers are used for making garlands, and worn in the hair by women. Their fragrant flowers can be effectively utilized in specialized gardens like the moonlight gardens, shade gardens and gardens for visually impaired. The flowers are also edible. It is industrially recognized for its use in manufacturing paper and perfumes as well as medicinal and antimicrobial properties (Joy *et al.*, 2007). The seed and rhizomes are also aromatic, carminative and stomachic. The root is antirheumatic, excitant and tonic. The ground rhizome is used as a febrifuge. An essential oil from the roots is carminative and has antihelmintic indications. *H. spicatum* is a native of Nepal and *H. flavescens* of Madras and Bengal.

Genus ***Kaempferia*** is named so in honour of Englebert Kaempfer (1651-1716), German Physician. It includes about 70 herbaceous perennial species, two-third of which are found in Asia and the rest in Africa though now all of its African species fall under the genus *Siphonochilus* so number of species has also reduced only to 50. The *Kaempferia* species are grown primarily for their beautiful foliage. Most of the *Kaempferia* species have a silver feather pattern in the middle of the upper side of the leaf, radiating outwards with various shades of green. Kaempferias produce small white, pink, purple or orange flowers and bloom sporadically throughout the growing season. *Kaempferia elegans* is a small ornamental with wide popularity among floriculturists. The beautifully zoned patterns on leaves with silvers, blues, blacks, and shades of green make this plant very effective even without flowers. Iridescent purple flowers are produced over the top of the plants. The plant is widely adapted to tropical climate and grown all over India, Myanmar and Malaysia. It is cultivated as ornamental ground cover or indoor plant or as a potted plant for shady areas. The plant produces leaves and flowers during May to October and becomes dormant in summer months, *i.e.* January to April in South India (Sabu *et al.*, 2013).

The name ***Zingiber*** was used by Dioscorides which means ginger. There are some 100 species under the genus. *Zingiber* is seasonally dormant rhizomatous herb from Southeast Asia and Indomalaysia to northeastern Australia. ***Zingiber spectabile*** is a medium to tall species that has a large yellow and red inflorescence. The cones open to produce brown and yellow spotted flowers - the resulting combination looks like bees in a hive, giving the plant its name. It is used frequently in the cut flower industry and has a very long lasting inflorescence. The inflorescence is prized as cut flower and is also most beautiful when left on plant lasting for months. All parts of the plant exude a pleasant aroma of ginger. Flowering occurs during spring/summer and lasts for about 4 months. In the tropics, the plant grows year round, but naturally goes dormant in winter in cooler areas. *Zingiber zerumbet* is a native of tropical Asia. It is known as 'shampoo ginger' for the creamy liquid substance in the cones and it is in fact used as a shampoo in Asia and Hawaii, and as an ingredient in several commercial shampoos. Rhizome is useful in colic, headache, haemorrhoids, respiratory disorders and cough, asthma, leprosy and skin diseases. It makes an excellent fast-growing landscape plant for tropical effect, and the flowers are long lasting and useful for cut flower arrangements. It grows to about 2.1 m tall with long narrow leaves arranged opposite along the stem. In mid to late summer, separate stalks grow out of the ground with green cone-shaped bracts that resemble pine-cones. The green cone turns red over a couple of weeks and then small creamy yellow flowers appear on the cone.

Botany, Genetics and Breeding

Alpinia bears ginger-like rhizomes which throw out many reed-like stems bearing many lanceolate leaves in 2 ranks at right angles. Inflorescence is usually showy and terminal, and sometimes in certain species racemose or paniculate, often in axils of bracts, bracteoles tubular, open to base or absent, calyx 3-lobed, wide tubular or almost bell-shaped but tubes short, petals 3, posterior petal largest, lip 2-lobed, larger than petals, staminodia 1 or more and lateral staminodes small or absent. Stamens are reduced to 1 pollen-bearing organ. One of the staminodia is toothed and showy, and longer than corolla. Fruit is an ovoid or orbicular capsule. Plants of *Alpinia purpurata* are tall, normally 1.5-2.1 m high, upright and herbaceous perennials. Leaves are oblong with 45 to 56 cm long and 10-15 cm wide. This evergreen clumping plant attains a height of 3.0 m when grown in sun or light shade in a rich and moist soil with protection from wind. The inflorescence is a compact spike, 15-30 cm long with a cluster of bracts overlapping to form the shape of a cone or funnel. This cone-shaped bloom appears as the central attraction of the plant. Inflorescences are normally erect but droop if large. Bracts subtend small, tubular, white 6 mm flowers having a narrow lip. The flowers open a few at a time. Rhizomes and stalks are aromatic. Chromosome number is 2n = 48 and it is a tetraploid. Intergeneric hybridization was attempted in red ginger. Plants of *Alpinia purpurata* 'Eilen McDonald' were cross-pollinated with a pink selection of *Etlingera elatior* both with same chromosome number. About 20 per cent of the pollinations yielded viable hybrid seeds and the hybrid seedlings were fertile and produced more flowers compared to *A. purpurata*. Fruits were formed from these flowers without manual pollination. Subsequently, these intergeneric hybrid plants of *Alpingera martinica* were back-crossed to *A. purpurata* 'Eilen McDonald'. A population of 300 seeds from this back-cross was treated with gamma radiation at a dose of 30 Gy. Within the treated seedlings new variations in colour and shape of the inflorescence were observed. This led to the first *A. purpurata* variety with pure white flowers (Fereol *et al.*, 2010). ***Alpinia zerumbet*** grows up to 1.8 to 3.0 m or more with similar spread in ideal conditions, but is usually less, particularly outside the tropics and sub-tropics. Leaves are broadly oblong to lanceolate up to 60 cm long.

Curcuma alismatifolia bears thick rhizome with swollen roots. It is a herbaceous perennial, whose erect shoot is a pseudo-stem comprising of an axis covered by overlapping and sheathing leaf blades. Leaves are narrowly spear-shaped, lanceolate to oblong and with purple midribs. At the base of a pseudo-stem is the true stem, an underground structure known as a tuberous rhizome. Inflorescences are terminal on short leafy shoots or on a distinct space, lower bracts floriferous, fused for one-third to half length forming pockets where each may carry 2 to 7 flowers, upper bracts sterile, erect, quite large up to 7.5 cm long, petal-like and brightly & distinctly coloured from the lower ones, *vis-à-vis* cup-like called coma, bracteoles open to base and thickened at centre. Colour varies from white to pink to dark pink with a purple lip. Petals 3, thin-textured, posterior erect, hooded, lateral ones decurved, lateral staminodes petaloid, folded under posterior petal, lip broad, base tubular, fused with corolla tube, anthers versatile, usually spurred, fruit ellipsoid and a thin-walled capsule. The plant has two underground storage organs, a rhizome and a spherical tuberous root designated as 't-roots' or milk sacs, attached to the rhizome by a connecting root. The 't-roots' are swollen and egg-shaped root-ends that are thought to act as storage organs for plant growth during dormancy and emergence thus playing an important role in growth and development. This plant flowers approximately 60 days after emergence and every 30 days thereafter during the summer months. Malvidin 3-rutinoside is responsible for the pink colour of the bracts. The concentration of this pigment increases with increase in bract colour (Nakayama *et al.*, 2000). Chromosome number is 2n = 32.

Velayudhan (2015) collected a total of 105 accessions devoid of sessile tubers where 40 accessions belonged to 13 identified species, two varieties and three unidentified entities which were then assessed in *in situ* and *ex situ* mainatenance plots to assess their ornamental values. He found *Curcuma mutabilis* and *C. inodora* securing 9 marks each followed by *C. albiflora, C. oligantha* var. *oligantha, C.o.* var *lutea* and *Curcuma* species 1 and 2 with a score of 8 each, *C. karnatakensis* and *C. thalakaveriensis* with a score of 7 each and being ahead of species such as *C. decipiens, C. coriacea, C. pseudomontana, C. neilgherrensis, C. kudagensis* and *Curcuma* sp. 3 with a score of 6 and *C. vamana* of Sect. *Stolonifera* with the lowest score of 5.

Etlingera is a large growing herbaceous rhizomatous plant. The leafy cane-like shoots of a mature specimen can reach a height of about 3 meters with a spread of about 4 cm. The lanceolate leaves that line alternately on the leafy shoots can grow up to a length of about 80 cm. They have lovely porcelain inflorescences from 0.9-1.2 m tall plants. The terminal inflorescences are produced separately from the leafy shoots on long stalks usually about 0.9 to 1.2 m high (although these may grow up to 1.8 m high). The inflorescences are crowned with involucres of overlapping and sterile bracts, each bract subtending 1 flower. The bracteoles are tubular. The flower is as large as 25cm long and 15cm wide. Flowers are tubular with white or gold margins, bracts waxy, red, pink or purple with white margins, in whorls around a sub-globular head. Sterile outer bracts enclose the inner fertile bracts until head is fully expanded, then reflex downward on a leafless scape. Flowers have 3 unequal petals with erect to spreading lips which roll inwards when withering, lips ovate, emarginate, margin frilled, lateral staminode lacking or highly reduced as appendages, fertile stamen 1, filament short, anther emarginated, and smooth or sometimes hairy, fleshy and indehiscent fruits are clustered into globose head. The plant requires two years of growth in humid warm climate for its profuse flowering. According to Eksomtramage *et al.* (2002), the chromosome number of *E. elatior* (pink) and *E. elatior* (white) are 2n = 48 suggesting that the two forms of *Etlingera* may differ in a colour coding gene. Torch ginger requires continuous improvement in certain characters like vase life, flower colour, morphology, size, fragrance and decreased time to flower formation. Crop improvement of this plant *via* traditional breeding is handicapped by incompatibility problem, poor fruit set and also low seed production.

Globba bears slender rhizomes with fibrous roots and reed-like leafy stems some 1 m in height. The foliage of dancing ladies grows to about 60 cm high with long, lance-shaped leaves on short stems which has a heart-shaped base and is slightly hairy underneath. Leaves are alternate, 2-ranked, lanceolate or oblong and sub-sessile. Each stem produces a terminal pendant racemose inflorescence with reflexed bracts that are showy white to purple with a slender, curved and yellow corolla. Flowers are borne in cincinni on slender branchlets appearing from axils of bracts and where lower flowers are replaced with bulbils. Bracteoles open to base, calyx is funnel-form, 3-lobed and from where corolla with 3 small and subequal petals protrudes with long bent tube so exceeding the calyx, posterior petal is spurred and then corolla tube is topped by 3 greenish yellow lobes, and 2 waxy and yellowish staminodes, where another arrow-shaped staminode points upward to the spurred stamen whose filament encloses the style as in case of *Hedychium*, staminodes are petaloid, equal or exceeding petals with 1 fertile stamen, filament long and curved, and anther with 2 small triangular projections on each side. The flowers are characterized by relatively small labellum, attached with the filament forming on androecial tube above the attachment of corolla tubes, and long arching filament like a fish-hook and the anther at the distal end. Ovary is 1-celled with numerous ovules and the fruit is a capsule with arillate seeds. Important species are *G. atrosanguinea*, *G. marantina* and *G. winitii*. *Globba winitii* has a pseudostem which is 20 to 30 cm tall, terminated with an inflorescence. It produces flowers at the end of emerging shoots and will continue to do so during the growing season. The production cycle of this ginger is 90 to 120 days.

Hedychium bears stout rhizomes with numerous reed-like erect stems with terminal inflorescences. From the rhizomes emerge tall leafy shoots which bear spikes of flowers at their tips. These leafy shoots may be 0.6 to 4.0 m tall depending on *Hedychium* species or cultivar. Leaves are large, sessile or on short petioles, oblong to lanceolate and arranged in 2 neat ranks that run the length of the stem. The inflorescence is cylindrical or cone-shaped with fragrant flowers on dense spikes where floral heads are sheathed by inflorescence bracts. The flowers have a long tube and 2 staminodes fused to form a 2-lobed lip while another 2 staminodes look like petals on either side of it. Bracts spaced or imbricate, bracteoles tubular, calyx tubular and irregularly toothed, corolla tube slender, petals 3 and are relatively inconspicuous narrow appendages, lip long with 2 entire lobes, lateral staminode petaloid, stamen filament long and slender, anther without spur at base, having 1 functional stamen, and in the groove along the upper surface of the filament there fits the style with its stigma projecting beyond this fertile stamen, ovary inferior and 3-celled, and fruit a capsule which on ripening bursts and 3 aril covered seeds come out being attached to the central axis of the ovary. *H. coronarium* flowers consist almost exclusively of volatile terpenoid (monoterpenes and sesquiterpenes) and benzenoid compounds due to which these emit fragrance. *Hedychium* oil contains more than 97 monoterpene compounds. The predominant components are 1, 8-cineole, linalool, *dinna*-pinene, *dippa*-pinene, and (*E*)-nerolidol. Essential oil extracted from *Hedychium* rhizomes was extremely inhibitory to growth of *Aspergillus flavus* and *Fusarium verticillioides*, two of the most important fungal species responsible for mycotoxin contamination of several food crops under pre-harvest or storage conditions (Rajasekaran *et al.*, 2012). The oils show promise as an adult mosquito repellent, but they would make rather poor larvicides or adulticides for mosquito control. *Hedychium* oils act either as a fire ant repellent or attractant, depending on plant genotype and oil concentration (Sakhanokho *et al.*, 2013). Flowers are hermaphrodite, showy and white or pale-yellow in colour. The floral tubes are long and slender and frequently extend well beyond the bracts. The perianth is 3-lobed. There are 2 distinctive linear pendent staminodes. The single fertile stamen is held perpendicular to the staminodes. The outer edge of the lip is often lobed and the base is clawed. The 3-valved fruit splits open revealing bright orange arils and dark red seeds which are attractive to birds. Pollens are either binucleate or trinucleate. Chromosome number is 2n = 34. Self-sterility is common in *Hedychium*, so species of this genus hybridize readily (Sakhanokho and Rajasekaran, 2010). Tremendous diversity exists within this genus, providing breeders with an opportunity to develop new cultivars or varieties through hybridization. However, differences in ploidy levels can constitute serious barriers to hybridization. Polyploidy was successfully induced in several *Hedychium* hybrids and species, including *H. ousigonianum*, leading to the creation of dwarf forms of this species. Embryogenic calli from selected *Hedychium* species and hybrids were subjected to various concentrations of sodium azide (NaN_3) and ethyl methane-sulphonate (EMS) to induce mutations. *H. muluense* × 'White Starburst' calli treated with 20 mM NaN_3 for 6 hours produced variegated plants. These *in vitro* assisted breeding techniques are accelerating the development of new and improved *Hedychium* cultivars (Sakhanokho *et al.*, 2010).

Kaempferia hardly attains 45 cm height, mostly the height being below 10 cm, though spreads 30 cm and more. During growing season this spreads quickly to form a good ground cover. The species sheds off its leaves at the onset of winter in the plains of India and produces new sprouts at the onset of spring-summer. It possesses thick aromatic tubers and irregularly thickened roots. The plant produces flowers profusely and sets seeds. It is a rhizomatous herb, often stemless or so and with leaves clumping at the base or sometimes distichous in 2 ranks on short stems, leaves mostly broader than lanceolate, flowers precocious in spikes on terminal leafy stems appearing before or with the leaves or on radical, scaly, terminal scapes, in a bracted tuft (1 per bract) or small cluster in the centre of leaf-clump or in a peduncled raceme, bracteoles open to base, very thin and deeply split, bracts often large and showy in white, yellow, violet or purple, calyx cylindrical or funnelform, split one side and toothed, corolla tubular, exserted with narrow lobes, petals 3, narrow, lip large and deeply 2-lobed, staminodia petal-like and are the showy parts, one of them being a broad lip, staminodes large and petaloid, the fertile stamen 1, and anther with prominent reflexed crest.

Zingiber is a perennial herb with thick aromatic rhizomes which have precocious branching habit, and which produce leafy reed-like stems. Leaves in 2 ranks, usually lanceolate and ligulate. The inflorescences are borne at ground level on short stalk, each on separate lateral and erect scape (shoot) from the leaves. Flowers appear in pairs, from the bottom of the inflorescence upwards. Each flower blooms only one day. Flowers are single in axils of colourful (cream with buff and purple reticulate markings, and when aging under sunlight the colour passes to pale-yellow with gingery-red and purple-black reticulations), waxy and imbricate bracts, bracteoles open to base, calyx thin and tubular, corolla tube slender, petals 3 where posterior petal is broader than 2 lateral petals, lip 3-lobed, lateral staminodes absent or very small, anther with elongated crest, bracts persistent where 3-valved fleshy fruits are held and seeds arillate. *Zingiber zerumbet* contains Zerumbone (ZER), a sesquiterpene which is a potent inhibitor of tumor promoter-induced Epstein-Barr virus activation and has been implicated as a promising chemo preventive agent (Ohnishi *et al.*, 2013). The species is a potential resistant donor against soft rot disease caused by *Pythium aphanidermatum*. The ZzR1 gene characterized from the resistant accession represents a valuable genomic resource for ginger improvement programmes due to the economic importance of soft rot disease and the environmental impact associated with chemical control measures (Nair and Thomas, 2013).

Species and Varieties

Several cultivars of ***A. purpurata*** are available, displaying light pink to dark pink and double-red inflorescence. New dwarf cultivars are increasingly popular as container plants and for indoor use. Some important varieties are 'Jungle King' (large, rounded, globe-shaped inflorescence, not producing aerial offshoots, stems sturdier than common red and slow growing); 'Red Dwarf' (excellent potted plant with compact lush foliage); 'Tahitian Ginger' (tight, multiple head, torch-shaped, producing numerous aerial offshoots, leaves large 90 cm by 20 cm wide, flower head large, often 15 cm in diameter, made up of a series of red bract clumps, dense, multi-branched inflorescences and red bracts, good for large containers); 'Madikera White' (first pure white cultivar, produced by mutation breeding), *etc.* ***Alpinia zerumbet*** 'Variegata'(Reaches up to 1.2 m, the leaves are quite attractive, striped in yellow and green and the plants produce the same white and pink, shell-like flowers, widely demanded for decorative purposes). *Alpinia zerumbet* 'Variegata Dwarf'(plants striped green and yellow, looks exactly like to its parent, *A. zerumbet* 'Variegata', but grows much smaller, reaching 30 cm in height at maturity). *Alpinia zerumbet* 'Nana' grows 30 to 60 cm in height and spread. Other important ornamental *Alpinia* species are *A. calcarata, A. rafflesiana* and *A. sanderae.* Major ***Curcuma*** varieties are 'Chiang Mai Pink' (commercial cut flower, 70 per cent shade required), 'Tropical Snow' (white flower, 50cm tall), 'Thai Beauty' (flowers dark pink, 30cm tall), 'Siam Tulip' (flowers pink/mauve, 40cm tall), 'Precious Petuma', 'Shalom', 'Pink Pearl', 'Red Shalom', *etc.*

The inflorescence of ***Etlingera*** is available in three main colours, *viz.* pink, red and white. When not in flower, plants from the three varieties cannot be easily differentiated via their aerial parts as all plants will be similarly green in appearance. However, there is a torch ginger variety that produces red inflorescence having purplish-red leaves underside, and the leafy shoots also take on a similar reddish tinge. In general, the pink variety is the one that is the most floriferous which is followed by the red and white varieties. Important cultivars are 'Thai White' (early to mid-flowering variety with pink flowers, 100-150 stems per plant, vase life 6-10 days), 'Hintze Red' (mid-flowering tulip-type, red, less than 100 flowers, vase life 6-10 days), 'Helani Tulip' (dark pink to red, with yellow highlights in the centre, flowers do not open and look more tulip- like).

Cornukaempferia aurantiflora has a silver feather pattern on the outer edge of the leaf with a deep maroon underside. ***Kaempferia gilbertii*** '3D' has a white margin on a deep green leaf. Other cultivars are *Kaempferia* spp. 'Grande', *K. elegans 'Shazam'* and *K. elegans* 'Mae Ping'.

Commonly, ***Globba winitti*** has mauve-purple bracts, but there are several popular cultivars in other colours too, important ones being 'Mt. Everest' (magnificent selection of the white colour form of *Globba winitii* propagated through tissue culture, excellent pot plant, flowering through summer; the flowers last for more than six weeks, extremely good for landscaping and for cut flower use), 'White Dragon' (pure white bracts with yellow flowers), 'Red Leaf' (pinkish-purple bracts and a reddish tint on the undersides of the leaves), *etc.* Other important cultivars are 'Ruby Queen', 'Pristina Pink', 'Purest Angel' (white) and 'Blushing Maiden' (pink and white), *etc.*

Hedychium cv. 'Ramata' (dwarf, about 60 cm) is the first ornamental ginger cultivar that combines both dwarfism and stable variegation, two highly desired but rare traits in hedychiums. It is well suited to a variety of landscapes as a specimen plant, contrasts in mixed plantings, and in smaller gardens. Flowers are mildly fragrant and produce little to no pollen (Sakhanokho *et al.,* 2012). 'Yellow Spot' (grows to 2.0 m and flower throat has golden spots and is suitable for full to partial sun). 'Yellow Butterfly' (*Hedychium flavescens*) produces pale-yellow flowers with a deep yellow centre and sweet, lemony fragrance. It grows to 2.0-2.5 m, is late bloomer and grows well in partial sun. 'White Pincushion' (*Hedychium thrysiforme*) bears lush, arching, waxy and broad foliage with a rippled texture. It produces masses of small white flowers with long stamens in clusters like snowballs. In the early stages of opening, the flower resembles those of spider mums. They are very attractive in the landscape. 'Dr. Moy' is a gorgeous variegated foliage ginger. The great magnolia scent of its deep yellow blooms is an added attraction. It blooms from mid-summer to November and grows to 2 m. It can be grown in partial to full sun. 'Gold Flame' is a ginger which smells like jasmine. Beautiful pale double-toned flowers are yellow with a pink centre. Blooms in mid-summer and grows to 2 m. It can be grown in partial to full sun. 'Multiflora White' bears extremely fragrant large white blooms and is an excellent choice for shade or moon gardens. It blooms from July through October and grows to 1.5-2.0 m in light shade to partial sun. 'Peach' produces peachy-pink flowers with a rich fruity fragrance and blooms in mid-summer. It grows to 2 m and can be grown in partial to full sun. 'Orange Brush', 'Anne Bishop', 'Filigree', 'Daniel Weeks', 'Andromeda', 'Ness Botanical Garden', 'Gold Spot', *etc.* are other popular cultivars.

Zingiber spectabile has many beautiful cultivars out of which the important ones are 'Giant Red Pinecone Ginger' (*Z. spectabile,* foliage strappy to 1.2 m and odd flower cones that are bright red with small yellow flowers, cone inside contains a liquid which is used to make shampoo), 'Variegated Pinecone Ginger' (*Z. spectabile,* growing to 0.9 m, preferring partially shaded locations, and bears beautiful variegated white and green foliage), 'Golden Sceptor' (*Z. spectabile,* golden beehive ginger, 1.8 to 2.4 m tall, bracts golden-copper but becoming reddish with maturity under full sun), 'Pink Maraca' (*Z. spectabile,* cut flower, cones open to produce brown and yellow spotted flowers which are around 10-15 cm long, long lasting and early blooming), *Z. malaysianum*, a native of Johor (Malaysia) is more of a foliage plant with shiny, almost black or purple-bronze leaves, inflorescence with short stalks arising from a short distance of the rhizome, bracts sulphur-yellow but turning bubble-gum pink with age, and the flowers appearing from the bracts are creamy, *etc.*

Propagation

Alpinias produce seeds rarely, however, when produced these are sown shallowly in a bit-acidic and well-drained but moist organic medium. Seeds germinate in 2-3 weeks. The seedlings may be transplanted into larger pots as soon as they are large enough to handle. With heavy fertilizer application, some flowers will be produced in 2-3 years. Inflorescences develop aerial offshoots (small plantlets) from the sides of the bracts and these can be used as the source of new plants. For propagation with offshoots, the whole flower head can be bent into a pot and covered with soil. These offshoots can initially be separated and planted individually in the pots or after formation of roots in the offshoots several weeks later, the mass of rooted offshoots can be cut off from the mother plant and planted in the pots. Though these can be rooted without the aid of rooting hormone but the rooting is improved with 500 ppm auxin (IBA or NAA) treatment. Red ginger propagated from offshoots makes a very attractive foliage plant in 15-cm or larger pots. In two years the plants through these offshoots produce flowers of marketable size and quality. The cultivars not developing offshoots are propagated by rhizome divisions. The horizontally grown rhizomatous mat is divided into small clumps of one to four stems. If the roots are not well developed on the rhizome, the upright stem should be cut back to reduce water loss. The individual pieces are dusted with a fungicide and planted 5 cm below the surface in a well-drained medium. They should be kept in a warm place but not in full sun. The rhizome with growing buds or new shoots can be used for propagation. The rhizome should be planted with the top (leaf stem/new shoots/buds pointing upwards) no more than 3-4 cm under the soil. New shoots or larger buds should be above the soil. It is very important not to plant the rhizomes too deep as roots may rot. Thorough watering is recommended after planting but no watering should be given until soil gets dry and afterwards only periodical watering is given until the plants are fully established. Soil should be evenly moist but not wet when shoots grow and leaves start to unfold. If planted in pots, the pots should be of sufficient size, *i.e.* at least 10 to 15 cm wider than rhizome size. The pots are kept in a warm and sunny place. When leaves start to unfold, the plants should be planted out. Rhizome-propagated plants when planted in beds typically produce marketable flowers within a year. Illg and Faria (1995) reported multiple shoot formation in *Alpinia purpurata* from inflorescence buds inoculated on MS medium containing 10 μM 6-benzyladenine and 5 μM NAA with a mean increase of 15 to 20 new shoots every 4 weeks. ***Curcuma alismatifolia*** is usually propagated from geophytic units which comprise of a rhizome and several 't-roots'. The rhizome has buds that will produce next season's leaves and inflorescence. The numbers of 't-roots' that are attached to a rhizome affect time to flower, inflorescence stem length, and number of stems per rhizome. The more the number of 't-roots', the earlier will be the shoot emergence and flowering. Breaking the tuberous roots from the rhizome may decrease flowering and delay time to emergence. Thus, care should be taken to keep the entire tuberous root intact. Micropropagation is also employed in *Curcuma*. Young inflorescence segment and lateral buds from rhizome can be used as explants. The bud from dried rhizome is better than one from fresh rhizome. A modified micropropagation system in twin-flasks temporary immersion bioreactors (TIB) was developed from inflorescences cultured in modified MS solid medium supplemented with 10 mg/l BA for 8 weeks (Topoonyanont *et al.*, 2011).

In ***Etlingera***, the seeds are soaked overnight in water to ensure quick germination. Seeds should be sown in pots and kept in warm and sunny spot for germination. Seedlings grow slowly and take two or three years from germination to flowering. Rhizomes with growing buds or new shoots as well as the leaf stem can be planted with the buds pointing upwards not more than 3- 4 cm under the soil. It is better to use a clump of several leafy shoots rather than one that just consists of one leafy shoot. New shoots or larger buds should be above the soil. Deep planting is avoided as this will invite fungi and cause root rot. Freshly planted rhizomes need oxygen to grow new roots and will die if the planting medium is too heavy or too wet. When planting, the leafy shoots can be cut away to reduce the loss of water from the leaves and potted up in a well-drained medium. They are kept in a sheltered position until new shoots sprout. For micropropagation, MS medium supplemented with 13.32 μM BAP was found most ideal for both induction and multiplication of shoots of torch ginger from axillary bud explants. MS medium without PGR was recommended for profuse and strong rooting while the combination of soil:sand:peat moss (1:1:1) was chosen as the most suitable potting medium for acclimatization (Yunus *et al.*, 2012).

Globba winitii is propagated by division of clumps during growing season, by cutting pieces of dormant rhizome or through micropropagation. Propagation by seed is possible in the spring. In colder climates, the rhizomes may be lifted in winter and stored in a cool place in slightly damp peat. A hastening in emergence and flowering time was achieved when rhizome's storage duration was prolonged up to 3 weeks for cold and hot storage. Short rhizome storage durations of only 1 week at any temperature prolonged emergence up to 30 days.

The cold storage condition that hastened emergence and flowering was 15 °C and the hot storage temperature was 30 or 35 °C, indicating that due to its tropical nature the species prefers higher storage temperatures. If rhizomes are not stored or stored for only 2 weeks at 25 °C, shoots would have emerged in approximately 56 days (Paz *et al.*, 2004). Jala *et al.* (2013) reported the highest average number of new shoots in young *Globba* embryos cultured on MS medium supplemented with 5 mg/l BA.

Hedychium is easy to grow from seed and will flower after 2 to 3 years. Seeds are best sown as soon as these are ripe. Seedlings are transplanted into individual pots when they are large enough to handle. **Rhizome** clumps are dug out and divided with a sharp knife, making sure that each division has a growing shoot. Larger clumps can be planted out direct into their permanent positions. Some cultivars like 'Assam Orange' and 'Stephen' are clones and should be propagated only by division of the rhizomes. Axillary bud explants were cultured on MS medium supplemented with 3 mg/l benzylaminopurine, 3 mg/l kinetin (KIN), and 0.2 mg/l thidiazuron, yielding a maximum of 13.2 ± 0.3 shoots. Shoot clusters containing 3 to 5 shoots were successfully rooted in KIN (3 mg/l) and IAA (0.5 mg/l), yielding a maximum of 6.3 ± 0.5 roots (Parida *et al.*, 2013). Micropropagated plants sometimes take a little longer to come into flower than usual.

Kaempferia can be propagated through seeds and rhizomes. Both mother rhizome and split-rhizome can be used for planting. The longer the rhizomes are stored, the less time it takes for emergence. Breaking the tuberous roots from the rhizome may decrease flowering and delay time to emergence. Thus, care should be taken to keep the entire tuberous root intact.

Zingiber spectabile and other zingibers are propagated through seeds, cuttings and division of rhizomes. *In vitro* propagation of ***Z. zerumbet*** has been reported via induction of multiple shoots from *in vitro* shoot explants using different culture systems such as the agar-gelled medium cultures, shake flask system and temporary immersion system (Stanley *et al.*, 2010).

Cultural Practices

A. purpurata may grow up to approximately 480 m elevation though ordinarily it is grown in the plains as basically it is a tropical plant. An emerging stalk flowers in 135 to 150 days when minimum temperature is above 21 °C, however, a little yellowing may occur at high temperatures. It grows well in rich soil and in wet habitats but with well-drained soils having a pH range of 5.5 to 6.8 under full sunlight or partial shade, better under 30 per cent shade (Misra *et al.*, 2002) at temperatures above 15.6 °C, though it can also be grown in the dry areas as well. When the temperature drops below 10 °C, its growth becomes quite slow, turns yellowish green, and produces small, tight cone-like inflorescences that do not open normally. In Hawaii, plants are grown on volcanic cinder. Under optimum temperature, adequate moisture and proper nutrition, flowering could be obtained year-round, with greater production during summer. Flower yield and rate of development depends on the amount of sunlight received by the plant. A spacing of 1.2-1.9 m is recommended within rows. Only healthy shoots should be retained and the weak ones should be thinned out. Closer spacing though increases yield per unit area but the quality and per plant yield is reduced. After flowering, the spent shoots should be removed from the ground to save the energy and to initiate more shoot formation. Yellowed and unsightly leaves are also removed. ***Alpinia zerumbet*** prefers growing in well-drained soils rich in organic matters. Though it tolerates full sun but prefers up to 50 per cent partial shade with protection from strong winds. It has moderate salt tolerance. It may go dormant in winter in very cold areas.

In case of ***Curcuma***, well-drained fertile soil having alkaline pH reaction is required. It grows under partial sun to bright filtered light. For production of brightly coloured bracts and deep green leaves, it should be grown in full sun. If the species are grown under shaded conditions, the flower stems and petioles tend to elongate and topple. The bracts of the inflorescence tend to fade and post-production longevity is shortened. Short day photoperiods promote rhizome formation as compared to long day photoperiods. The flower stalks need staking to avoid toppling. Due to its rhizomatous root system, it requires dividing every 2-3 years to achieve best results. Rhizomes are dug and after removing the adhering soil, these are cut into 5 cm sections with intact roots. Dried blooms are continuously removed as low to the foliage as possible to produce quality flowers. As the leaves die back, to promote new growth and to prevent pathogenic infections, they should also be removed. When plants go dormant all the old foliage are removed down to the soil surface. For cut flower purpose, flower stems should be cut as low as possible to the foliage just as the bloom begins to open.

Etlingera requires two years of growth in semi-shaded condition and humid warm climate for its profuse flowering. It can, however, be acclimatized to grow under higher light levels. When exposed to more sunlight, the plant grows shorter in stature. It should be planted in rich well-drained soil which does not have waterlogging. Organic materials incorporated into clayey soils will help to improve aeration and soil structure, will improve drainage as well as retain moisture. Plants should be protected from winds as the leafy shoots may be damaged and constant winds may shed the leaves. *Etlingera* plant is generally not quickly invasive as the clump of leafy shoots is quite tight and advances at quite a manageable pace. However, rhizomes which spread to unwanted areas in the garden or advancing in a wrong direction can be stopped at its tracks by breaking it with a shovel. These rhizomes can be dug up and used as material for propagation.

Globba winitii needs a well-drained, slightly acidic potting mixture containing fertile organic soil. It needs to be watered abundantly while it is growing actively, but soil should be well-drained. *Globba* needs day temperatures of 18.3 °C or higher. It grows best and flowers under 30 per cent shade. *Globba* requires morning sun or filtered sun as under direct sun the leaves will burn. Growth is controlled by division of plant and repotting.

Hedychium can be though cultivated in light, medium and heavy soils but loamy soils are best for its growth. They are tolerable to acid, neutral and alkaline soils. They can be grown in full sun with sufficient moisture though flowers remain fresh for long in somewhat filtered light. It prefers moist or wet soil. It also succeeds in shallow water and can also be grown in a sunny border as a summer sub-tropical bedding plant. For pot culture, only large-size pots are preferred, vis-à-vis through splitting are repotted every year as the rhizomes are very strong and can easily break the pots. If the plant is split into relatively large pieces the flowering may be obtained the same year.

Kaempferias prefer full to partial shade for their growing. If this plant receives too much sun, the leaves will curl up in defense from the hot rays. The plant grows well at a soil pH of 7 to 7.5. It can be planted in the ground or pots. The container should provide the best drainage and room for placing the tuberous roots attached to the rhizome towards the bottom of the pot. The top of the rhizome should be covered with approximately 2.5 cm of media. The media should have excellent drainage and water holding capacity.

Feeding and Irrigation

The best flower quality is achieved with generous irrigation in gingers. ***R. purpurata*** requires 2.5 cm or more of water per week from irrigation during dry periods. Flower production increases with increasing levels of nitrogen fertilizer, however, complete fertilizer is required to be applied every month. High levels of nitrogen fertilizer do not adversely affect the postharvest life of the flowers. ***A. zerumbet*** if once established requires only a moderate amount of water, and tolerates fairly dry conditions. A slow release fertilizer is ideal in the beginning. Larger and already established plants require more fertilizing, and of course, more water. In ***Curcuma***, boron accumulators may lead to marginal necrosis. Therefore, plants should be fertilized with a water-soluble fertilizer low in boron or no boron. During growing season fertilizer application is carried out about once per month. Fertilizer application is withheld when plants have lost their foliage and have gone dormant. Once growth resumes regular fertilizer schedule can be resumed. A high level of N supply to the curcuma plant increases new rhizome formation because of increased flower numbers, but depressed new storage root formation because of reduced starch accumulation (Ohtake *et al.*, 2006). Endophytic bacteria carrying out nitrogen fixation and IAA synthesis such as *Sphingomonas pseudosanguinis*, *Bacillus drentensis* and *B. methylotrophicus* were identified in *C. alismatifolia*. These isolates act as growth promoters and could stimulate the rapid growth of host plant (Thepsukhon *et al.*, 2013).

Thorough watering is needed in *Etlingera* after planting. The soil should be kept evenly moist but not wet when shoots grow and leaves start to unfold. Organic mulch consisting of dried leaves or compost around the root zone is ideal and mulching helps to maintain a cool constant temperature and reduces water loss from the roots during hot and dry weather, *vis-à-vis* provides nutrients for the plants when they break down. Additional feeding will promote growth and should be done using slow release organic fertilizer but it is usually not required.

Globba winitii is more drought-resistant than most other gingers but it prefers regular moisture. Watering should be reduced during its winter dormancy, and light fertilization is sufficient during the growth period. When night temperatures fall below 18.3 °C, the plant goes dormant and then watering is restricted to a minimum of once a week.

Kaempferia requires consistently moist soil and should not let it dry out between watering. Soil should be well-drained during winter so that the dormant rhizomes may not rot. After the initial irrigation, plants should be given a preventive fungicide treatment. Some gingers are boron accumulators which may lead to marginal necrosis. Therefore, plants should be fertilized with a water-soluble fertilizer low in boron or without boron, such as a tropical foliage fertilizer.

Post Harvest

Alpinia purpurata floral spikes are harvested about 4-5 months after stem emergence. Inflorescences are harvested in the early morning while still turgid. They should be cut when the bracts are about 2/3rd to 3/4th open as an immature flower has a longer shelf life than a mature one. The entire shoot is harvested at the ground to have longer post harvest life. To extend shelf life, trimming of all or all but the top 1 to 3 leaves from the stem in the field or at the packing shed prior to cleaning is recommended. Stem bases are kept in water during transport from the field to the packing area. After harvest the flowers are to be placed in a bath containing a commercial preservative and thoroughly washed. The major postharvest problems are negative geotropic response and insect infestation. Soaps can be used to clean the flowers and kill the insects. The end of vase life of *Alpinia purpurata* is partially determined by browning in the middle of the bracts. A positive correlation is observed between the vase life and the sugar level of its stem. Postharvest life is increased by use of floral preservatives containing 2 per cent sucrose + 200 ppm 8-HQC, antitranspirants, or simply recutting the stems. Sucrose 2 per cent + 200 ppm 8-HQC as holding solution increased the vase life of *A. purpurata* from 10 to 15 days (Broschat and Donselman, 1988). Hot water treatment of red ginger at 49-50 °C for 12-15 minutes extends postharvest life, kills most of the pests that infest red ginger, and reduces the geotropic response. Postharvest vase life varies from 5 days in young flowers (stem diameter <1 cm) to 25.5 days for standard size flowers. Sugar will extend their postharvest life by at least a week. A 200 mg/l BA spray extends the vase life of red ginger inflorescence and attached leaves (Kobayashi *et al.*, 2007). Flower stems are packed flat, singly or bunched, in standard or insulated fibreboard boxes. Single stems are layered in rows in the box. Bunches may be wrapped in a polyethylene film, or moistened newspaper (shredded) may be packed in between bunches, with non-shredded newspaper separating the layers. Bunches are fastened to the box to minimize mechanical damage due to shifting. To prevent geotropic bending during shipping, it is preferable that the boxes be kept upright, so that the stems are in a vertical orientation. **Curcuma** have excellent post-production longevity of up to 40 days. Vase life is limited by browning at the bract tips. This browning may relate to ethylene production as it is hastened with exogenous application of ethylene. The flowers are chilling-sensitive, and cannot be stored dry but they can be stored in water at 7 °C for about 6 days. Since vase life is rather long, it is also possible to store the flowers in water for a few days at ambient temperatures (Bunya-Atichart *et al.*, 2004). Treatment with 1-MCP has been found to extend the vase life and post-harvest quality (Chutichudet *et al.*, 2011).

Etlingera is identified as a minor commercial ginger due to the negative aspect of poor vase-life. To overcome this problem in *Etlingera*, stems are generally harvested as closed buds with long stem known as 'candlesticks' for distant markets. Being a chilling sensitive flower they must be held at warmer temperatures. Ideal temperature is 12.5-15°C. Gingers are packed flat in standard or insulated fiberboard boxes. Their large size makes them difficult to manage. It is seen that 1-MCP at a concentration of 1.5 g m^{-3} is efficient against the actions caused by ethrel. The association between the 1-MCP and certain commercial floral preservatives has promoted greater longevity and greater quality in the postharvest conservation of torch ginger for additional 3 days when compared to those kept in water only (Unemoto *et al.*, 2011).

Insect-Pests, Diseases and Physiological Disorders

Alpinia purpurata is infested with ants, aphids, soft scales and mealy bugs in the field. If left unchecked, pest build up can make postharvest disinfestation time consuming and difficult. Field sanitation is part of good pest management for red ginger. All mature flowers are to be removed from the field regardless of marketability, so that they do not serve as hosts where pests can multiply. Wide spacing should be followed when planting, and plants trimmed back to avoid over-grown fields, that are difficult to spray. Wide spacing helps preventing easy spread of pests from plant to plant in the landscape. Pink cultivars suffer from a tip burn disorder that is lessened with 30 per cent shade. Chlorosis due to high pH is a common problem in calcareous soils. ***A. zerumbet*** usually has a few pest problems, but the leaves will brown on the edges if the soil is not kept moist or if touched by frost. Curcumas are hardy and moderately resistant to disease and pests. The most common disease problems are rhizome rot, wilt, anthracnose caused by *Colletotrichum musae* and other leaf spots. A new fungal species, *Plectosporium delsorboi* was found causing leaf and stem lesions in *Curcuma alismatifolia* and a few other Zingiberaceae in greenhouses in Southern Italy (Vincenzo *et al.*, 2001). The most widespread pests tend to be nematodes, sucking or leaf eating insects and shoot borers.

Etlingera is a relatively pest and disease-free plant. The most common pest is the grasshopper that chews along the leaf margins. Sucking insects such as spider mites congregate on the underside of the leaves while aphids on the young shoot and leaf, though these are also minor pests. In general, all these pests rarely do great damage to an established plant but attention must be paid to newly established young plants.

Hedychium is relatively free of pest and diseases. Slugs and snails will eat the shoots and flowers. In the greenhouse or conservatory, spider mite is the biggest problem. Keeping high humidity by misting deters spider mites.

References

Baker, J.G. 1890-1892. Scitamineae. In: *Flora of British India*, **6**: 198-264.

Broschat, T.K. and H. Donselman, 1988. Production and post harvest culture of red ginger in south Florida. *Proc. Annual Meeting Florida St. hort. Soc.*, **101**: 326-327.

Bunya-Atichart, K., S. Ketsa and W.G. van Doorn, 2004. Postharvest physiology of *Curcuma alismatifolia* flowers. *Postharvest Biol. and Techn.*, **34**(2): 219-226.

Chan, E.W.C., Y.Y. Lim, S.K. Wong, K.K. Lim, S.P. Tan, F.S. Lianto and M.Y. Yong, 2009. Effects of different drying methods on the antioxidant properties of leaves and tea of ginger species. *Food Chemistry*, **113** (1): 166–172.

Chompoo, J., A. Upadhyay, M. Fukuta and S. Tawata, 2012. Effect of *Alpinia zerumbet* components on antioxidant and skin diseases-related enzymes. *BMC Complementary and Alternative Medicine*, **12**: 106.

Chutichudet, P., B. Chutichudet and K. Boontiang, 2011. Influence of 1- MCP fumigation on flowering weight loss, water uptake, longevity, anthocyanin content and colour of Patumma (*Curcuma alismatifolia*) cv. Chiang Mai Pink. *International Journal of Agricultural Research*, **6**(1): 29-39.

Eksomtramage, L., P. Sirirugsa, P. Jivanit and C. Maknoi, 2002. Chromosome counts of some Zingiberaceous species, *Songklanakarin J. Sci. Tech.*, **24**(2): 311-319.

Elzaawely, A.A., T.D. Xuan and S. Tawata, 2007. Essential oils, kava pyrones and phenolic compounds from leaves and rhizomes of *Alpinia zerumbet* (Pers.) and their antioxidant activity. *Food Chemistry*, **103**(2): 486-494.

Fereol, L.F., F. Luc-Cayol and M. Guitteaud, 2010. Use of intergeneric hybridization and mutagenesis to go ahead to new colours of *Alpinia purpurata* (ginger lily). *Acta Hort.*, No. 855, pp. 131-136.

Illg, R.D and R.T. Faria, 1995. Micropropagation of *Alpinia purpurata* from inflorescence buds. *Plant Cell, Tissue and Org. Cult.*, **40**(2): 183-185.

Jala, A., N. Chanchula and T. Taychasinpitak, 2013. Multiplication of new shoots from embryo culture on *Globba* spp. *Intern. Transaction J. Engineering, Management, & Applied Sci. & Techn.*, **4** (3):207-214.

Joy, B., A. Rajan and E. Abraham. 2007. Antimicrobial activity and chemical composition of essential oil from *Hedychium coronarium. Phytother. Res.*, **21**:439–443.

Kobayashi, K.D., J. McEwen and A.J. Kaufman, 2007, Ornamental ginger, red and pink. Cooperative Extension Service, College of Tropical Agriculture and Resources, University of Hawaii at Manoa. Ornamentals and Flowers, OF-37, 8 p.

Misra, R.L., Vinod Kumar and Sanyat Misra, 2002. Greenhouse Management of Ornamental Plants ~ A Review. In: *Floriculture Research Trend in India*, pp. 13-18 (eds Misra, R.L. and Sanyat Misra). Indian Society of Ornamental Horticulture, IARI, New Delhi.

Morley, Brian D. 1970. Wild Flowers of the World, pp. 278-279. Peerage Books, London.

Nair, R.A. and G. Thomas, 2013. Molecular characterization of ZzR1 resistance gene from *Zingiber zerumbet* with potential for imparting *Pythium aphanidermatum* resistance in ginger. *Gene*, **516**: 58–65.

Nakayama, M., M.S. Roh, K. Uchida, Y. Yamaguchi, K. Takano and M. Koshioka, 2000. Malvidin 3-rutinoside as the pigment responsible for bract color in *Curcuma alismatifolia. Biosci., Biotech., and Biochem.*, **64**(5): 1093-1095.

Ohnishi, K., E. Nakahata, K. Irie and A. Murakami, 2013. Zerumbone, an electrophilic sesquiterpene, induces cellular proteo-stress leading to activation of ubiquitin–proteasome system and autophagy. *Biochem. and Biophys. Res. Commun.*, **430**: 616–622.

Ohtake, N., S. Ruamrungsri, K. Sueyoshi, T. Ohyama and P. Apavatjrut, 2006. Effect of nitrogen supply on nitrogen and carbohydrate constituent accumulation in rhizomes and storage roots of *Curcuma alismatifolia* Gagnep. *Soil Sci.& Pl. Nut.*, **52**(6):711-716.

Parida, R., S. Mohanty and S. Nayak, 2013. *In vitro* propagation of *Hedychium coronarium* Koen. through axillary bud proliferation. *Plant Biosystems.* doi-0.1080/11263504.2012.748102. (IF: 1.4).

Paz, M.P., J.S. Kuehny, G.B. McClure, R. Criley and C.J. Graham, 2004. Storage temperature and duration effects on growth and development of ornamental gingers. *HortSci.*, **39** (4): 834-835.

Rajasekaran, K., H.F. Sakhanokho and N. Tabanca, 2012. Antifungal activities of *Hedychium* essential oils and plant extracts against mycotoxigenic fungi, *J. Crop Imp.*, **26**:3: 389-396.

Roxburgh, W. 1810. Description of several monandrous plants of India. *Asiat. Res.*, **11**: 318-362.

Sabu, M. 2006. *Zingiberaceae and Costaceae of South India.* Indian Association for Angiosperm Taxonomy. Department of Botany, Calicut University, India, 282 p.

Sabu, M., P.K.M. Kumar, V.P. Thomas and K.V. Mohanan, 2013. Variability studies in 'Peacock Ginger', *Kaempferia elegans* Wall. (Zingiberaceae). *Ann. Pl. Sci.*, **2** (5): 138-140.

Sakhanokho, H.F and K. Rajasekaran, 2010. Pollen biology of ornamental ginger (*Hedychium* spp. J. Koenig). *Scientia Hort.*, **125** (2010) 129–135.

Sakhanokho, H.F., K. Rajasekaran, N. Tabanca, B.J. Sampson, L.M. Nyochembeng, C.T. Pounders, D.E. Wedge, N. Islam-Faridi and J.M. Spiers, 2010. Induced polyploidy and mutagenesis of embryogenic cultures of ornamental ginger (*Hedychium* J. Koenig). *Acta Hort.*, **35**:121-128.

Sakhanokho, H.F., B.J. Sampson, N. Tabanca, D.E. Wedge, B. Demirci, K.H. Baser, U.R. Bernier, M. Tsikolia, N.M. Agramonte, J.J. Becnel, J. Chen, K. Rajasekaran and J. M. Spiers, 2013. Chemical composition, antifungal and insecticidal activities of *Hedychium* essential oils. *Molecules,* **18**(4):4308-27.

Sakhanokho, H.F., A.L. Witcher, C.T. Pounders Jr. and J. Spiers, 2012. 'Ramata': a new dwarf and variegated *Hedychium* J. Koenig cultivar. *HortSci.*, **47**(6):803-805.

Santos,G.K.N., K. A. Dutra, R. A. Barros, C.A.G. da Camara, D.D. Lira, N.B. Gusmao and M.A.F. Daniela, 2012. Essential oils from *Alpinia purpurata* (Zingiberaceae): Chemical composition, oviposition deterrence, larvicidal and antibacterial activity. *Industrial Crops and Products*, **40**: 254–260.

Sharma,G.J., P. Pukhrambam Chirangini and R. Kishor, 2011, Gingers of Manipur: diversity and potentials as bioresources. *Genet. Resource Crop Evol*, **58**:753–767.

Skornickova, J. and M. Sabu, 2002. The genus *Curcuma* L. in India: resume and future prospects. In: *Prospectives of Biology* (ed. Das, A.P.), pp. 45-51. Bishen Singh Mahendrapal Singh, Dehradun.

Skornickova, J., M. Sabu and Prasanth Kumar, 2004. *Curcuma mutabilis* (Zingiberaceae): a new species from South India. *Garden's Bull. Singapore*, **56**: 43-54.

Stanly, C., A. Bhatt and C.L. Keng, 2010. A comparative study of *Curcuma zedoaria* and *Zingiber zerumbet* plantlet production using different micropropagation systems. *African J. Biotech.*, **9**(28): 4326-4333.

Tawata, S., M. Fukuta, T.D. Xuan and F. Deba, 2008. Total utilization of tropical plants, *Leucaena leucocephala* and *Alpinia zerumbet*. *J. Pestic. Sci.*, **33**(1): 40–43.

Thepsukhon, A., S. Choonluchanon, S. Tajima, M. Nomura and S. Ruamrungsri, 2013. Identification of endophytic bacteria associated with N_2 fixation and indole acetic acid synthesis as growth promoters in *Curcuma alismatifolia* Gagnep. *J. Pl. Nut.*, **36**(9): 1424-1438.

Topoonyanont, N., S. Jaikanta and P. Boonmanee, 2011. *Curcuma alismatifolia* Gagnep., micropropagation in twin-flasks temporary immersion bioreactor. *Acta Hort.*, No. 886, pp. 267-271.

Unemoto, L.K., R.T. Faria, L.S.A. Takahashi, A.M. Assis and A.B. Lone, 2011. Longevity of torch ginger inflorescences with 1- methylcyclopropene and preservative solutions. *Acta Scientiarum Agron.*, **33**(4): 649-653.

Upadhyay, A., J. Chompoo, W. Kishimoto, T. Makise and S. Tawata, 2011. HIV-1 integrase and neuraminidase inhibitors from *Alpinia Zerumbet*. *J. Agric.and Food Chem.*, **59**(7): 2857-2862.

Velayudhan, K.C. 2015. Ornamental *Curcuma* species in Western Ghats of India. *Indian J. Plant Genet. Resour.*, **28**(3): 269-277.

Velayudhan, K.C., V.A. Amalraj and V.K. Muralidharan, 1996. The conspectus of the genus *Curcuma* in India. *J. Econ. Taxon. Bot.*, **20**: 375-382.

Vincenzo, A., G. Walter and M. Fabrizio, 2001. *Plectosporium delsorboi* nov. sp., a pathogen of *Curcuma*, Zingiberaceae. *J. Econ. Ent.*, **94**(5): 209-214.

Wijekoon, M.M.J.O., R. Bhat, A.A. Karim and A. Fazilah, 2013. Chemical composition and antimicrobial activity of essential oil and solvent extracts of torch ginger inflorescence (*Etlingera elatior* Jack.). *Int. J. Food Properties*, **16**(6):1200-1210.

Yunus, M.F., M.A. Aziza, M.A. Kadira and A.A. Azmi Abdul Rashida, 2012. *In vitro* propagation of *Etlingera elatior* (Jack) (torch ginger), *Scient. Hort.*, **135**:145–150.

33

Petunia × *hybrida* (Family: Solanaceae)

Namita, Meenu Kumari, Sapna Panwar and Sanyat Misra

Introduction

In Latin, *petun* in Brazil is for 'tobacco', and the genus name was coined by Antoine Laurent Jussieu due to similarity of the plant to tobacco. *Petunia* is a S. USA to S. American genus of about 40 species of half-hardy tufted perennials but usually grown as half-hardy annuals which flowers the same season from seeds. The common types of petunia are weedy in habit, but their profusion of blooms under all the conditions makes them useful and popular garden plant. These are small herbs grown for their showy flowers. Most of the petunias now grown are garden hybrids. They are grown in beds & borders, edgings, rock gardens, window boxes, hanging baskets, tubs and pots due to longer flowering period. Petunia is considered to be the first cultivated bedding plant and has remained as a commercially important ornamental crop since the early days of horticulture. Apart from its significance as an ornamental crop, petunia has proved to be one of the most excellent model crops for studies on gene regulation, genome structure, *etc.*

Origin and Distribution

The geographic distribution of petunia includes temperate and subtropical regions of Argentina, Uruguay, Paraguay, Bolivia and Brazil with a centre of diversity in southern Brazil. *Petunia* is endemic to South America, mostly with subtropical distribution ranging from 22° to 39° S. The major species are found in Brazil followed by Argentina, Uruguay, Paraguay and Bolivia. All the species occur in Brazil, except for *P. occidentalis* which has a distribution restricted to the Sub-Andean mountains in north western Argentina and southern Bolivia. The Serra do Sudeste and neighbouring places in southern Rio Grande do Sul have a low altitude mountain range and a set of diverse edaphic conditions. *Petunia integrifolia* (syn. *Petunia violacea, P. phoenicea*) with violet-purple flowers and *P. axillaris* (syn. *P. nyctaginiflora, Nicotiana axillaris*), the parental species of garden petunia are sympatric in these areas. Sink (1984) stated that it was *P. axillaris* subsp. *parodii* which, in fact, is one of the parents of garden petunias. Five species of *Petunia* grow in Serra do Sudeste, out of which three ~ *P. bajeensis, P. exserta* and *P. secreta* are strict endemics.

Botany

Petunias are small soft plants of straggling or decumbent habit, pubescent and usually more or less sticky glandular hairy with solitary large showy flowers. The leaves are simple and entire, alternate or opposite, spathulate to lanceolate or ovate and viscid-pubescent with fine hairs which are borne along entire length of the trailing slender stems. Funnel- or salver-shaped flowers are borne on solitary, terminal and axillary peduncles. The calyx is 5-lobed and corolla funnel-form. The flowers have 5 stamens (2+2+1), attached in the tube, one of them sometimes sterile, ovary small and 2-celled, style slender with dilated stigma and fruits are 2-celled capsules. Varieties of the present day are modification of two stem types. Inflorescence is monochasial with opposite leaf-like bracts, aestivation is imbricate and symmetry is actinomorphic or zygomorphic.

Certain petunia types have deeply fringed fully-double flowers whereas others have star-like markings radiating from the throat and extending nearly or quite to the margin of the limb. The first double forms were introduced in 1849. They are available in wide variety of colours such as yellow, cream, white, pink, red, pale-blue, violet, mauve, purple or bicoloured and in combinations of various colours. Its flowers are usually insect-pollinated with exception of *P. exserta* which is a rare hummingbird-pollinated species. The *Pseudonicotiana* species are pollinated by nocturnally active hawkmoths (*Manduca contracta* and *M. diffusa*), while the *Eupetunia* species are pollinated by a diurnally active bee (*Hexantheda* sp.). Petunia have homozygous double flowers that are female sterile but have capability to produce viable pollen since double character is dominant. Double Grandiflora petunia hybrids are produced from the crosses of single and double. Pollen germination and pollen tube growth of *P. alba* are sensitive to sulphur dioxide. Moist pollen grains are most sensitive to sulphur dioxide than dry pollen grains, and pollen tube growth is more sensitive than pollen germination (Varshney and Varshney, 1981). The ethylene associated with the pollination has no effect on pollen tube growth in style, but other pollination-induced factors may lead to an acceleration of growth (Folkert *et al.,* 1998). In P. *inflata*, better pollen tube growth was obtained on heterozygous than on homozygous styles (Takats, 1992). Compatible and incompatible pollen growth in diploid *P. inflata* showed pollen tube losses at different points in growth from stigma to the ovary (Takats, 1993). Pollen of *P. hybrida* can tolerate temperature as high as 60 °C and this does not affect the pollen viability, vigour or ability to set fruits and seeds (Rao *et al.,*1995). Geitmann and Cresti (1998) proposed a hypothetical model for the mechanism that controls pulsating growth in pollen tubes. In self-compatible and self-incompatible petunia clones, ethylene acts as a mediator of the pollen-tube growth in the sporophyte pistil tissues. The pistil tissues are the primary target of the pollination-induced ethylene (Holden *et al.,* 2003). Characterisation of pollen growth transition in self-incompatible *P. inflata* suggests that pollen competition is most intense during the pre-PGT (pollen growth transition) (Lubliner *et al.,* 2003). Pollen fertility is regulated mainly by genetic factors. Rapid pollen tube growth places unique demands on energy production and biosynthetic capacity.

Pollen stored for 2-3 days at -18 °C reduces infection by microorganisms for up to 12 h and improves germination and pollen tube growth on a 10 per cent sucrose solution (Andreichrenko, 1983). Polyester and nylon powders are used as pollen diluents that preserve pollen germination and tube growth in controlled pollinations (Yi *et al.,* 2003). Petunia seeds germinate in 3 to 10 days.

Genetics and Breeding

The first monograph published on petunia in 1984 carried an illustration of the 7 chromosomes with a total of 60 assigned genes (Cornu, 1984). The 1984 report increased the number of roughly mapped loci to 74, with the introduction of results from classic two point test crosses, including a number that relied on analysis of isozyme variants (de Vlaming *et al.,*1984). The list of 122 genes published just a few years later by the Dutch-French collaborators (Gerats *et al.,*1987) with 88 markers assigned to chromosomes was the final version lacking assignments based on molecular data. The updated CSH map released six years later (Gerats *et al.,* 1993), in which 134 of 165 listed nuclear genes had chromosomal assignments, included for the first time locii mapped by restriction fragment length polymorphisms (RFLPs). Most petunias are diploid with 14 chromosomes and are inter-fertile with other petunia species (Ando *et al.,*2001). However, Watanabe *et al.* (1996) stated interspecific cross incompatibility among the *Petunia* spp. with differeing chromosome numbers. Using relative length and arm ratios, it is possible to distinguish all the chromosomes except 4 and 5 (Benzer *et al.,* 1971; Maizonnier and Cornu, 1971). These two chromosomes can be distinguished by their quinacrine fluorescence pattern (Smith *et al.,*1973) and DNA content (White and Rees, 1987). Chromosome 2 has a secondary constriction on its short arm due to nucleolar organizer. Pollen diameter (Ferguson and Coolidge, 1932), chloroplast number in guard cells (Mitchell *et al.,* 1980), stomata length (Santos and Handro, 1983) and microfluorimetry (Galbraith *et al.,* 1980) have been used to simplify the determination of ploidy level. Under certain instances, these techniques may not be reliable. For example, chloroplasts counts only measure the polyploidy level of the epidermis and will not be able to determine the ploidy of the gametes (Kamo and Griesbach, 1989). In petunia, the chromosome number is 2n = 14 for which Wijsman and Hendriks (1986) suggested that all those with this chromosome number should be placed under a separate genus *Stymoryne*. Gerats *et al.* (1982) stated that *Inl* gene controls the rate of anthocyanin synthesis and mutation frequency of the gene *Anl* in *P. hybrida*.

The major **breeding** objectives of the petunia are to create new and novel colours, reduction in time between sowing and flowering, prolonging the duration of flowering under field conditions, maintenance of a compact plant habit throughout the growing season, reduction in plant lodging, *etc.* Petunia flowers bear stamens on corolla where four stamens are fertile and one is smaller or rudimentary. *Petunia* species are primarily self pollinating. Both emasculation and pollination are done by hand which enables the breeder to have wider choice of parents for F_1 hybrid seed production. In petunia, single pollination produces numerous small seeds in a capsule. To produce 100 g petunia seeds, nearly 700 flowers are required. Since numerous seeds per pollination are obtained, the cost of F_1 hybrid seed production is adequately compensated. Petunia fruit is two-celled capsule, with many seeds. The generalized breeding system in petunia is out-breeding. In an outbred population, all plants become F_1 hybrids. Two genetically distinct parent plants are crossed irrespective of their state of homozygosity. The self incompatibility system operating in petunia is gametophytic, which can be utilized in practical application of cross pollination. Under open field condition, during the first fortnight of the harvesting season the seed capsules are initially handpicked due to irregular flowering. At the peak season of flowering, however, the flowers mature uniformly, hence to collect the ripe seeds the plants are shaken over the metal trays. In this way seeds can be collected quickly with less manpower. Harvesting should be done when the seed capsule turns brown and show minute apical splitting. Delay in picking will cause shattering of seeds beneath the plant and due to minute size of seeds these cannot be collected after shattering. The development of the commercial petunia hybrids typically requires the development of homozygous inbred lines, the crossing of these lines and the evaluation of the crosses. Pedigree breeding and recurrent selection breeding methods are used to develop inbred lines from breeding population.

Classification, Species and Cultivars

Petunia × hybrida that includes all the present day hybrid garden petunias, has been developed by crossing *P. axillaris* (syn. *P. nyctaginiflora, Nicotiana axillaris*) with dull-white flowers and *P. violacea* (syn. *P. integrifolia, P. phoenicea*) with violet-purple flowers. These compact hybrids grow from 15 to 60 cm in height in spreading or trailing form with a vast range of colours and floral shapes & forms, bearing funnel-shaped flowers where the tube is broader than *P. axillaris* and the mouth is wider than *P. violacea*. These are groups of various types such as:

- ☆ **Large-Flowered Grandiflora Hybrids** (catalogue name *Petunia hybrida grandiflora*) which are highly free-flowering, growing up to 60 cm in height and width and having the largest flowers up to 10 cm in diameter, widest variety of forms and colours and well-suited to growing in hanging baskets, window boxes and elsewhere. The single grandifloras are the most common or popular group on market, however, new cultivar releases have dramatically increased interest in both the single multiflora and single floribunda classes (Nau, 1991). These are not suitable for exposed sites as plants can be blown over in winds. Its F_1 variety is 'Cascade' which is available in a mixture of white, pink, red and blue colours. Its promising varieties are 'Blue Frost' (flowers violet-blue edged white), 'Cascade Series' (single flowers in wide range of colours), 'Colour Parade' (ruffled single flowers in wide colour range), 'Flash Series' (single flowers in a range of bright colours), 'Fluffy Ruffles' bears large ruffled flowers in shades of white, pink, crimson, lavender and purple with darker veins, 'Happiness' (rose-pink), 'Magic Cherry' (single cherry-red compact flowers), 'Miss Blanche' (white), 'Mariner' (deep blue), 'Picotee series' (mixed colour range of brightly coloured single flowers with contrasting margins), 'Polynesia' (coral salmon), 'Razzle Dazzle' (single, white-striped flowers in a wide range of colours), 'Recoverer series' (single flowers in single or mixed colours), 'Star Series' (wide colour range, single but white-striped flowers), 'Superbissima Mixed' has largest flowers (up to 13 cm across) of all the petunias with highly ruffled petals and veined throat markings, 'Victorious Series' (ruffled, flared, trumpet-shaped double flowers in a mixture of colours), *etc.* 'Pan American All Double Mixed' also bears large ruffled flowers.
- ☆ **Large-Flowered Dwarf Compact** (catalogue name *Petunia nana compacta grandiflora*) grows 25 cm high bearing 5 cm flowers. These are highly branched but the plants are neat, bushy and compact. Flowers are large, entire, ruffled and appear above the foliage. Under this group the mixtures are popular. Separate colours include 'Alderman' (indigo-violet), 'Blue Bedder' (blue), 'Dwarf Giants of California' (large ruffled blooms in mixed colours), 'Dwarf Resisto' (blue, red, pink), 'Fire Chief' (scarlet), 'Rose of Heaven' (pink), *etc.*
- ☆ **Large-Flowered Dwarf Fringed** (catalogue name *Petunia grandiflora fimbriata nana*) is a basally well-branched plant growing from 20 to 30 cm in height, is free-flowering and most suitable for pot cultivation.
- ☆ **Large-Flowered Dwarf** (catalogue name *Petunia grandiflora superbissima nana*) grows to a

height from 25 to 30 cm, most suitable for pot cultivation but is little free-flowering than the tall giant-flowered class though with same cultural practices except the staking.

- **Small-Flowered Dwarf Compact** or **Multifloras** (catalogue name *Petunia hybrida nana compacta*) type which grow up to 25 cm tall, are quite free-flowering but with no rain damage or high sun damage and are excellent for massing. Their flower size is roughly half of the Grandiflora Hybrids, *i.e.* 5 cm across. 'Single Mixed' is the most widely grown strain. Its F_1 varieties are 'Plum Crazy' with a mixture of yellow, pink, lavender and purple, and variously veined with darker shade in the centre. Single coloured F_1 varieties are 'Apple Blossom' (pale-pink), 'Bonanza Series' (frilled, flared, trumpet-shaped double flowers in a mixture of colours), 'Brass Band' (deep cream), 'Dream Girl' (pink to deep rose), 'Gypsy' (salmon-red), 'Jamboree Series' (stems pendulous, flowers single in a range of colours), 'Mirage Velvet' (large, flared, trumpet-shaped, rich-red single flowers with black centre), 'Pearl Series' (small and single flowers in a wide range of colours), 'Picotee Ruffled Series' (ruffled single flowers, edged white in a range of colours), 'Plum Crazy' (single flowers with contrasting veins and throats, colours various), 'Polaris' (white starred deep blue), 'Red Satin' (scarlet), 'Sugar Plum' (pink, veined mauve), *etc.* Double type F_1 is 'Cherry Tart' (carnation type, pink, and mixture). Certain other promising varieties are 'Plum Blue' (soft blue), 'Snowdrift' (pure white), 'Summer Sun' (yellow), *etc.*
- **Giant-Flowered** (catalogue name *Petunia grandiflora superbissima*) petunias grow up to 75 cm or more in height, require staking to remain in position, and are suitable for culturing in pots of 20 cm or larger size. These are comparatively heavy feeder. The flowers are heavily ruffled and large but fewer in number than the small-flowered ones. Their corolla is striped, mottled and veined in contrasting colours.
- **Spreading** types of petunia suitable for hanging and for balconies (catalogue name *Petunia hybrida pendula*) are characterized by their low height (usually about 15 cm height and 24-36 cm to 1 m spread), 5 cm across flowers with long slender growth. These make a very effective ground cover in a landscape of any area with pendulous or trailing habit. These are suitable for growing in pots, in window boxes, in hanging baskets or for planting on the top of the walls. The blooms are of medium size and the plant is free-flowering. It's a special good strain is *Petunia longissima pendula*. The colour of the flowers is the same as in Multiflora (*Compacta*) group except the red. 'Avalanche' (mixed colours) and 'Balcony Blended Mixed' are its very popular strains.
- **Double-Flowered Petunias** (catalogue name *Petunia flora plena*) include the both, the large- and small-flowered types of double-flowered petunias, varying in height from 25 to 45 cm. These have a very long flowering period, are excellent for pot cultivation but not toerant to bad weather, especially the rains that spoil the flowers.
- Recently, two new types of petunias have been released ~ **Millifloras** with numerous small flowers only about 2-3 cm and **Supertunias,** with vigorous and vining growth (Weidner, 1994) because of the blood of *P. axillaris*.

Basically there are only four types ~ the Multiflora, the Grandiflora, the Dwarf *Nana Compacta*, and Pendula varieties.

The first *Petunia* species was collected by P. Commerson in Uruguay, but described by Lamarck (1793) as *Nicotiana axillaris* due to its similarity to *Nicotiana* though leaves were quite small. It was soon discovered that *N. axillaris* and *Petunia nyctaginiflora* were very closely related and then *Petunia nyctaginiflora* was transferred to *Nicotiana* as *N. nyctaginiflora*. By 1825, it was well recognised that these two plants were the same taxon and were not a *Nicotiana* spp. The taxon was commonly known as *Petunia nyctaginiflora*. The genus *Petunia* comprises of many species, most of which are native to South America. The species was first sent from South America to Paris in 1823. Petunia species with 2n=14 should be grouped in a separate genus. The common garden petunia (*P.*× *hybrida* Vilm.) is a cultigen likely resulting from a series of hybrids between *Petunia axillaris* and *Petunia violacea* and perhaps *Petunia inflata* (Bailey, 1951; 1976). Ando and Hashimoto (1993, 1994,1995, 1996) recognised six *P. integrifolia* taxa as distinct species such as *P. altiplana*, *P. bajeensis*, *P. bonjardinensis*, *P. guarapuavensis*, *P. interior* and *P. riograndensis*. Tsukamoto *et al.* (1998) resurrected *P. occidentalis* as a separate species. Other species in the genus are *Petunia alpicola, P. axillaris, P. exserta, P. helianthemoides, P. humifusa, P. inflata, P. integrifolia, P. ledifolia, P. littoralis, P. mantiqueirensis, P. parviflora, P. patagonica, P. pubescens, P. reitzii, P. saxicola,*

P. scheideana, P. variabilis, P. villadiana, etc. The modern day *Petunia* hybrids, for convenience named as *Petunia × hybrida*, have derived from crosses between *P. axillaris* and *P. integrifolia*. Numerous cultivars are available in a wide range of colours and colour combinations under all the classes. Below are the cultivated species.

Petunia axillaris (syn. *P. nyctaginiflora, Nicotiana axillaris*)

A native of South Brazil to Argentina, this grows erect up to 60 cm in height and produces highly fragrant flowers, fragrance being very strong during nights. It is one of the parents of modern garden petunia hybrids. Its leaves are 5-11 cm long, and ovate to lanceolate with sticky hairs. Flowers are off-white, funnel-shaped 3.8-5.0 cm in diameter and 6.5 cm long.

Petunia inflata

A native to Argentina and Paraguay, this grows 35-45 cm tall, bearing small and linear foliage, and inflated *cum* swollen corolla tube otherwise it is quite similar to *Petunia violacea*. It is one of the parents of garden petunia hybrids.

Petunia parviflora

A native to Central and South America, this elegant plant grows to a height of up to 45 cm having graceful slender stems. Leaves are very small and covered with soft silky hairs. A mass of light purple flowers of the size of about 1.2 cm in diameter continue appearing from December to May. These provide excellent mass effect.

Petunia violacea (syn. *P. integrifolia*)

A native to Argentina, this grows up to 30 cm in height and is another important parent for evolving modern hybrid petunias. Its prostrate and slender stems are sticky and bear short-stalked, oval and sticky leaves. Pinkish-red to reddish-violet flowers about 3.8 cm long appear from December to May. Its mass effect is very attractive.

Propagation

Propagation is primarily done through seeds. Each gramme contains some 8,600-10,000 seeds. Petunia seeds may be primed to enhance germination as priming controls the hydration of seeds and allows pre-germination metabolic activities to proceed though prevents actual radical emergence (Heydecker and Coolbear, 1977; Bradford, 1986; Koranski, 1988). In this case the seeds are actually soaked at 15-20 °C in aerated solutions of high osmotic potential for 5-21 days with 20-30 per cent polyethylene glycol (Mechel and Kauffmann, 1973) and so water enters the seed coat slowly though not sufficient so germination process is interrupted and when these seeds are removed from the gel the germination is arrested. These seeds are air-dried, packaged and marked with sowing date. Such seeds provide uniform germination under various environmental range than standard seeds. Generally, seeds of petunia are sown in controlled space for better handling owing to their small size. In general, 10-15 cm raised nursery beds of 1 m width and 2-3 m length are prepared. The soil should be well prepared by mixing 5-10 kg/m^2 well rotten farmyard manure. To check the soil-borne diseases in nursery, soil should be drenched with 0.2 per cent brassicol or captan or soil can be sterilized with 2 per cent formaldehyde, followed by covering with polythene. For better handling of small seeds, these are mixed with bulk material like sand or ash. Seed compost should comprise of 1 part leafsoil, 1 part peat, ¼ part sand and ¼ part ordinary garden soil or fibrous loam. For pricking out, the compost should comprise of ½ part ordinary garden soil or fibrous loam, 1 part leafsoil, ¼ part peat, ¼ part sand, and little part bone meal. Seeds are sown by hand in line 5-6 cm apart and 0.5 cm deep. After sowing, seeds are covered with well-sieved mixture of farmyard manure and soil. Watering should be done twice a day with rose can over the straw spread on the nursery bed. When the seeds start germinating, immediately the covering is removed to prevent the seedlings becoming lanky and crooked. Seedlings are pricked off into the seed boxes. It takes about 2-3 weeks to one month after sowing for sufficient growth of the seedlings for transplanting with 2-4 true leaf stage. For plug production, the plants remain in plugs for about 5-6 weeks prior to transplanting. At 3rd or 4th stages of plug production, temperatures are lowered to 17-18 °C and after transplanting a temperature of 15-17 °C should be maintained. Flowering time, plant height and lateral branching are directly correlated to average temperature (10-25 °C). Warmer temperatures make the plants taller with fewer branches but faster flowering. For raising one hectare of crop some 400 g seeds are required. Seeds start germinating in 2-3 days and the germination is complete within 10 days at 24-26 °C. After visibility of the seedlings the temperature should be lowered to 22 °C. Uncovered seeds germinate within 10-12 days at 15.5 °C. Light is not necessary for seed germination but once germinated in the dark, these are immediately moved to the light. However, growth room sowing requires lighting. A few cultivars with variegated leaves as well as Supertunias™ are propagated by cuttings.

Cultivation

Petunias are not very selective for **soil** as these can be grown from rich sandy loam to clay loam having pH range of 6.5-7.5, however, rich loam soils are quite suitable for successful petunia production provided these are not waterlogged and the drainage system is proper. The growing of petunia at a pH of 5.5 to 5.8 will prevent iron deficiency.

The **soil preparation** starts by mixing some 20 tonnes of farmyard manure per hectare thoroughly in the soil before planting. The soil should be ploughed thrice, followed by planking each time. All the rootstocks of the perennial weeds, plant debris, wood, bricks and stone pieces, polythene pieces, *etc.* should be taken out before planting and soil should be made fully pulverized. Bunds, channels and beds are prepared as per requirement after levelling the soil. Planting is usually done in the month of November on the hills (temperate regions) and during October-November in the plains. These flower from December to May in the plains and from June to September in the temperate regions. In general, 50 kg urea, 250 kg of single superphosphate and 60 kg of muriate of potash should be incorporated in the soil by thorough mixing at the time of planting. The application of remaining 100 kg urea should be applied in two equal split doses, one and two months of transplanting of seedlings. Lower levels of ammonium fertilizer should be used to prevent toxicity if cold temperatures are used to slow down the growth or to store the plugs. Phosphorous is normally incorporated into the medium prior to transplanting of the seedlings. Nelson (1985) has reviewed the nutrition requirement on petunia. In case of plug production, at stage 2 of growth 50-75 ppm nitrogen of KNO_3 may be applied but in the plugs nitrogen level may be raised to 150 ppm. After transplanting when active growth starts, plants can be given 150-200 ppm of N from a complete fertilizer at each irrigation or 300-400 ppm once a week. Immediately after transplanting, which should be accomplished when brightness of sun is gone, light irrigation should be applied which helps in better establishment of seedlings. The **distances** for petunia should be 45 × 45 cm in the field, and for container, hanging basket, tubs or garden should be 25.5 cm across (Nau, 1991). For potting, the compost should comprise of 2 parts fibrous loam or ordinary garden soil, 1 part leafsoil, 1 part well-rotted dry manure, ¼ part sand, ¼ part peat, and a little of bonemeal.

Petunias are quite tender and require proper attention throughout their life. Regular **watering** is essential for successful raising which should be done according to the requirement of the crop. After use of fertilizer, immediately the irrigation should be carried out. Sufficient moisture is required from transplanting till harvesting of seed. Depending upon season and soil type, frequency of irrigation is decided. In rainy season, generally irrigation is not required except during dry spell. During winter season, irrigation is required at 10-12 days whereas during summer season it should be done at every 4-5 days. Regular weeding and hoeing are essential for proper development of seedling into healthy plants.

Seed producer should have detective eyes, so any **off type** plant appearing in the field should immediately be uprooted. Constant vigil for this should be kept from the beginning of the crop to the maturity. Moreover, if any plant exhibits superior trait in terms of vigour, earliness, colour, or size of flower, *etc.*, such plants should be protected up to seed maturity.

Pinching is not required generally in petunia crop. However, in case if plants become too tall, these require to be pinched. **Disbudding** is neither practiced nor is practical. Spent flowers should be removed to aid in *Botrytis* control in the large flower types. Lanky plants in the garden can be sheared back to allow for regrowth and renewed flowering later in the season.

Growth and Flowering

Petunias become reproductive at the sixth leaf stage. The rate of vegetative and reproductive development is dependent on light duration, intensity and quality interacting with temperature. Flowering will occur under LD or SD, however, plants will flower earlier with longer photoperiods and higher light intensities (Karlsson, 1996). Consequently, petunias are facultative LD (quantitative long day, *i.e.* flowering at any photoperiod but rapid flowering under long days) and thermo-photoperiodic (at temperatures less than 20 °C plants are well-branched irrespective of photoperiod but faster flowering at long days) plants. At temperatures less than 20 °C, plants become well-branched irrespective of photoperiod though under long days the flowering is significantly faster. Short days coupled with greater than 20 °C temperature the branching is more than the long days and flowering is delayed. Under cool temperatures (13 °C) petunias are day neutral, between 13-22 °C they are long day plants while temperatures above 22 °C they flower soon under both the photoperiods though rate of growth remains faster with long day conditions (Piringer and Cathey, 1960; Wolnick and Mastalerz, 1969). The critical night length for flowering is between 13 and 10 h (Wilkins and Pemberton, 1981). Piringer and Cathey (1960) observed that the plants grown with light durations of over 12 h were in flower after nine weeks, while those receiving 9-10 h light duration the flower buds were visible and with 8 h of duration flower buds were only microscopic. Plants grown under LD, however, have strong apical dominance and are poorly branched. Light intensity interacts with duration, and photosynthesis is enhanced, which hastens the transition from vegetative juvenile stage to reproductive stage. At any photoperiod, higher light intensity will result in more rapid flowering. Supplemental HID lighting speeds the transition from vegetative to reproductive, even if HID lighting is for only 15 days commencing immediately after germination (Carpenter and Carlson, 1973). Light quality also influences rate of flowering. Far red light sources promote flowering when compared with red light sources in LD species (Lane *et al.*, 1965; Piringer and Cathey, 1960).

Temperature influences rate of flowering and the number of axillary branches and interacts with photoperiod. Flowering is rapid at 21-26 °C and slow at 10-15 °C. Branching is enhanced by cooler temperatures even under LD. Rate of flower development is most rapid at warmer temperatures even under SD (Piringer and Cathey, 1960; Wilkins and Pemberton, 1981). In fact, petunias are high light plants, where low light intensity results in taller plants but delayed flowering as in case of high temperatures.

Restricted watering slows down the growth process and thus prevents rapid flowering, but may reduce plant quality. In greenhouse crops, supplemental **carbon dioxide** (1,000 ppm) for only 2 weeks during the early juvenile phase of growth hastens flowering (Krizek *et al.*, 1968). This fact again illustrates that optimal growth is essential for rapid flowering.

Plants grown at cool temperatures generally have no height contol problem but high temperatures and long photoperiods necessitate **controlling of the height** and for this DIF of 2-3 °C higher during nights than the days is quite effective at plug stage. Daminozide at 2,500-5,000 ppm initial application is effective when plants are 5 cm and then 7-10 days apart. Similarly paclobutrazol and uniconizole are also effective.

Insect-Pests, Diseases and Physiological Disorders

The petunias are commonly attacked by *Helicoverpa* sp, aphids, leaf miner, *etc.* Protective control measures should be adopted well in advance to avoid any loss. Metasystox @ 250 ml/200 l of water is effective to control aphids. Root-knot nematodes (*Meloidogyne* spp.) may cause problems outdoors (Horst, 1990).

Rhizoctonia and *Sclerotinia* infection results in basal stem canker and lower leaf yellowing, *vis-a-vis* collapse (Daughtrey and Chase, 1992). *Botrytis* is a common problem on open flowers and can quickely eliminate summer floral displays after a rain or period of high humidity. Bavistin (1 g/kg) or captan (3 g/kg) seed dressing or drenching the infected nursery beds with 0.2 per cent brassicol or bavistan (0.1 per cent) is quite effective. 'Cucumber mosaic virus' and 'tomato spotted wilt virus' infections cause severe stunting, leaf distortion, brown rings or either mottling or browning of the veins so such plants should be uprooted and burnt.

Chlorosis can occur with iron deficiency or low temperatures. Boron deficiency is also often observed with petunia where foliage becomes hard, distorted and mottled, terminal buds abort and side shoots proliferate so in this case the pH should be maintained at 5.5-5.8 and boron should be supplemented. Plants are also sensitive to ozone.

Post Harvest

Plugs can be stored at 0-10 °C in the dark or under 1μmols^{-1}m^{-2} for six weeks, the optimum temperature being 2.2 °C though there *Botrytis* can be a major problem (Heins *et al.*, 1994). The plants for sale should be stored or transported at 10-13 °C. In case of long storage, 3,000 f.c. lighting should be used (Armitage and Kowalski, 1983; Armitage, 1985). Exposure of flowers to **ethylene** causes quick wilting (Nowak and Rudnicki, 1990) but spraying the flowers with 0.2 to 0.5 mM STS proves a boon against ethylene damage (Moe and Fjeld, 1987).

Seed harvesting is done individually before capsules start splitting. The harvested seeds are spread over the tarpauline under shade or in ventilated room for a week but reshuffled daily. The seeds are cleaned and sieved with different types of seed machine, and then finally cleaned by hand winnowing or using table fan to separate light seeds. These seeds are dried to 8 per cent level of moisture and packed in moisture-proof envelops.

References

Ando, T. and G. Hashimoto, 1993. Two new species of *Petunia* (Solanaceae) from southern *Brazil. Bot. J. Linn. Soc.*, **111**: 265-280.

Ando, T. and G. Hashimoto, 1994. A new Brazilian species of *Petunia* (Solanaceae) from the Serra da Mantiueria. *Brittonia*, **46**: 340-343.

Ando, T. and G. Hashimoto, 1995. *Petunia guarapuavensis* (Solanaceae): a new species from Planalto of Parana and Santa Catarina, Brazil. *Brittonia*, **47**: 328-334.

Ando, T. and G. Hashimoto, 1996. A new Brazilian species of *Petunia* (Solanaceae) from interior Santa Catarina and Rio do Sul, Brazil. *Brittonia*, **48:** 217-223.

Ando, T., M. Kurata, J. Tsukhara, H. Watanabe, H. Kokubin, T. Tsukamoto, G. Hashimoto, E. Marchesi and I. Kitching, 2001. Reproductive isolation in a native population of *Petunia. Ann. Bot.*, **88**: 403-413.

Andreichenko, S. V. 1983. *In vitro* production of a sterile culture of *Petunia hybrida* pollen grains. *Fiziologyiai-Biokhimiya-Kul Turnykh-Rastenii*, **15** (2): 144-147.

Armitage, A. M. 1985. *Petunia*. In: *Handbook of Flowering* (vol. IV, ed. Halevy, A.H.), pp. 41-46, CRC Press, Boca Raton, Florida, USA.

Armitage, A. M. and T. Kowalski, 1983. Effects of light intensity and air temperature in simulated post-production environment on *Petunia hybrida* Vilm. *J. Amer. Soc. hort. Sci.*, **108**: 115-118.

Bailey, L.H. 1951. *Manual of Cultivated Plants*. Macmillian, New York.

Bailey, L.H. 1976. *Hortus Third: A Concise Dictionary of Plants in the United States and Canada*. Macmillian, New York.

Benzer, B., R. Bothmer, L. Engstrand, M. Gustafsson and S. Snogerup, 1971. Some sources of error in the determination of arms ratios of chromosomes. *Bot. Notiser.*, **124**: 65-74.

Bradford, K.J. 1986. Manipulation of seed water relations via osmotic priming to improve germination under stress conditions. *HortSci.*, **21**:1103-1112.

Carpenter, W.J. and W.H. Carlson, 1973. Comparison of photoperiodic and high intensity lighting on the growth and flowering of *Petunia hybrida* Vilm. *Flor. Rev.*, **154**(3998): 27-28, 68-71.

Cornu, A. 1984. Genetics. In: *Petunia Monographs on Theoretical and Applied Genetics 9* (ed. Sink, K.C.), pp. 34-48. Springer-Verlag, Berlin.

Daughtrey, M. and A. R. Chase, 1992. *Petunia*. In: *Ball Field Guide to Diseases of Greenhouse Ornamentals*, pp. 153-154. Ball Publishing, Geneva, Illinois, USA.

De Vlaming, P., A.G. Geras, H. Wiering, H.J.W. Wijsman, A. Cornu, E. Farcy and D. Maizonnier, 1984. *Petunia hybrida*: A short description of the action of 91 genes, their origin and their map location. *Plant Mol. Biol. Rptr.*, **2**: 21-42.

Ferguson, M.C. and E.B. Coolidge, 1932. A cytological and genetical study of *Petunia*. IV. Pollen grains and the method of studying them. *Amer. J.Bot.*, **19**: 644-659.

Folkert, A., P. Hoekstra and R. van Tineke, 1998. Effect of previous pollination and stylar ethylene on pollen tube growth in *Petunia hybrida* styles. *Plant Physiol.*, **86**(1): 4-6.

Galbraith, D.W., T.J. Mauch and B.A. Shields, 1980. Analysis of the initial stages of plant development using 33258 Hoechstz: reactivation of the cell cycle. *Physiol. Plant.*, **51**: 439-447.

Geitmann, A and M. Cresti, 1998. Ca^{2+} channels control the rapid expansions in pulsating growth of *Petunia hybrida* pollen tubes. *J. Pl. Physiol.*, **152**: 439-447.

Gerats, A. G. M., E. Souer, J. Kroon, M. McLean, E. Farcy and D. Maizonnier, 1993. *Petunia hybrida*. In: *Genetic Maps: Locus Maps of Complex Genomes* (6th edn, ed. O'Brien, S.J.), pp. 613-623. Cold Spring Harbor Laboratory Press, New York.

Gerats, A. G. M., W. Veerman, I. de Flamming, H. Wiering, A. Cornu, E. Farcy and D. Maizonnier, 1987. *Lineage map of Petunia hybrida* (2n=14). In: *Genetic Maps: Locus Maps of Complex Genomes* (6th edn, ed. O'Brien, S.J.), pp. 746-751. Cold Spring Harbor Laboratory Press, New York, USA.

Gerats, A.G.M., P. Devlming, M. Doodeman, B. Al and A.W. Schram, 1982. Genetic control of the conservation of dihydroflavonols into flavonols and anthocyanins in flowers of *Petunia hybidra*. *Planta*, **155**: 364-368.

Heins, R., N. Lange, T. F. Wallace, Jr. and W. Carlson, 1994. Plug storage, cold storage of plug seedlings. Meister Publishing, Willoughby, Ohio, USA.

Heydecker, W. and P. Coolbear, 1977. Seed treatments for an improved performance – survey and attempted prognosis. *Seed Sci. & Tech.*, **5**: 353-425.

Holden, M. J., J. A. Marty, and C. A. Singh, 2003. Pollination induced ethylene promotes the early phase of pollen tube growth in *Petunia inflata*. *Journal of Plant Physiology*, **160**: 261-69.

Horst, R. K. 1990. Petunia. In: *Westcott's Plant Disease Hand book* (5th edn), pp. 768-769. Van Nostrand Reinhold, New York.

Kamo, K.K. and R.J. Griesbach, 1989. Determination of polidy level in Mitchell *Petunia*. *Plant Sci.*, **65**: 119-124.

Karlsson, M. 1996. Control of flowering in *Petunia* by photoperiod and irradiance. *HortSci.*, **31**: 681.

Koranski, D.S. 1988. Primed seed, a step beyond refined seed. *Grower Talks*, **51**(9): 24, 26-27, 29.

Krizek, D. T., W. A. Bailey, H. H. Klueter and H. M. Cathey, 1968. Control environments for seedling production. *Proceedings of International Plant Propagator's Society*, **18**: 273-280.

Lamarck, J.B. 1793. Tableau encyclopedique et methodique. *Botanique*, **2**: 7.

Lane, H. C., H. M. Cathey and L.T. Evans, 1965. The dependency of flowering in several long day plants on the spectral composition of light extending the photoperiod. *Amer. J. Bot.*, **52**: 1006-1014.

Lubliner, N., D. T. Singh-Cudy, and A. Singh- Cudy, 2003. Characterization of the pollen growth transition in self-incompatible *Petunia inflata*. *Sexual Plant Reproduction*, **15** (5): 243-53.

Maizonnier, D. and A. Cornu, 1971. A telocentric translocation responsible for variegation in *Petunia*. *Genetica*, **42**: 422-436.

Mechel, B.E. and A. Kauffmann, 1973. The osmotic potential of polyethylene glycol 6000. *Pl. Physiol.*, **51**: 914-916.

Michell, A.Z., M.R. Hanson, R.C. Skvirsky and F.M. Ausbel, 1980. Another culture of petunia genotypes with high frequency of callus, roots, or plantlet. Z. *Pfanzenphysiol*, **100**: 131-146.

Moe, R. and T. Fjeld, 1987. Keeping quality of potted plants as influenced by ethylene. *Gartner Tidende*, **101**: 1580-1583.

Nau, J. 1991. *Petunia*. In: *Ball Red Book*, (15th edn, ed. Ball, V.), pp. 690-692. George J. Ball publishing, West Chicago, Illinois, USA.

Nelson, P. 1985. Fertilization. In: *Bedding Plants III* (eds Mastalerz, J.W. and E. J. Holcomb), pp. 182-211. Pennsylvania Flower Growers, University Park, Pennsylvania, USA.

Nowak, J. and R. M. Rudinicki, 1990. *Postharvest Handling and Storage of Cut Flowers, Florist Greens and Potted Plants.* Timber Press, Portland, Oregon.

Piringer, A. A. and H. M. Cathey, 1960. Effect of photoperiod, kind of supplemental light and temperature on the growth and flowering of petunia plants. *Proc. Amer. Soc. hort. Sci.*, **76**: 649-660.

Rao, G.U., K.R. Shivanna and V.K. Sawhney, 1995. High-temperature tolerance of petunia and *Nicotiana* pollen. *Current Sci.,* **69**(4): 351-355.

Santos, R.F. and W. Handro, 1983. Morphological patterns in *Petunia hybrida* plants regenerated from tissue cultures and differing by their ploidy *Theor. Appl. Genet.*, **66**: 55-60.

Sink, K.C. 1984. Protoplast fusion. In: *Petunia* (ed. Sink, K.C.), pp. 133-138. Springer-Verlag, Berlin.

Smith, F.J., J.L. Oud and J.H. de Jong, 1973. A standard karyogam of *Petunia hybrida. Genetica*, **44**: 474-84.

Takats, S.T. 1992. Effect of female heterozygosity on incompatible pollen growth in styles of *Petunia hybrida. Plant cell Incompatibility News*, No., 24, pp. 54-57.

Takats, S. T. 1993. Compatible and incompatible pollen tubes growth in diploid *Petunia inflata*. Plant Cell Incompatibility Newsletter, 25:55-58.

Tsukamoto, T., T. Ando, M. Kurata, H. Watanabe, H. Kokubun, G. Hashimoto and E. Marchesi, 1998. Resurrection of *Petunia occidentalis* inferred from a cross compatibility study. *Jap. J. Bot.*, **73**: 15-21.

Varshney, S.R.K. and C.K. Varshney, 1981. Effect of sulphur dioxide on pollen germination and pollen tube growth. *Environmental pollution*, **24**(2): 87-92.

Watanabe, H., T. Ando, S. Iida, A. Suzuki, K. Buto, T. Tuskamoto, G. Hasimoto and E. Marchesi, 1996. Cross-compatibility of petunia cultivars and *P. axillaris* with native taxa of petunia in relation to their chromosome number. *J. Japanese. Soc. hort. Sci.*, **65**: 625-634.

Weidner, E. 1994. Supertunias TM - more than just a petunia. *Ohio Florists' Association Bulletin*, No. 777, pp. 4-5.

White, J. and H. Rees, 1987. Chromosome weights and measures in *Petunia. Hered.*, **58**: 139-143.

Wijsman, H.J.W. and T. Hendriks, 1986. On the interrelationships of certain speciecs of *Petunia*. V. Inheritance of flower morphology. *Acta Bot. Neerlandica*, **35**(1): 35-37.

Wilkins, H. F. and H. R. Pemberton, 1981. Interaction of growth regulators and light. *Proc. 14th int. Bed. Pl. Conf.*, **14**: 182-188, held in Seattle, Washington.

Wolnick, D.J. and J.W. Mastalerz, 1969. Response of petunia cultivars to selected combinations of electric light, photoperiod, temperature and B-Nine. *Pennsylvania Fl. Gr. Bull.*, No. 216, pp. 1-7.

Yi, W. G., S. E. Law and H. Y. Wetzstein, 2003. Polyester and nylon powders used as pollen diluents preserve pollen germination and tube growth in controlled pollinations. *Sexual Plant Reproduction*. **15** (5): 265-69.

34

Philodendron (Family: Araceae)

T. Janakiram, Namita and Usha Sonkhle

(**Common names**: Elephant's ear (*Philodendron domesticum*), Philodendron, Heart-leaf philodendron, Sweetheart vine, *etc.*)

Introduction and Botany

The genus *Philodendron* is the Greek word referring to the compound for 'tree-loving' owing to the fact that most of the species attach themselves to trees either by climbing up the trunk or nestling in the crotches of branches and in rough bark. It is the second largest genus in the Araceae with 700 or more species (Croat, 1969, 1979, 1985, 1988, 1990) though Bailey (1930) mentions some 222 and Hay and Beckett (1971) some 275. Philodendrons are evergreen climbing, shrubby or seldom arboreous plants, most suitable for growing in the greenhouse or as house plants. Its natural haunts are fresh-water swamps, stream banks, regrowth forest, rocky outcrops and road banks since most of the species occur in virgin humid forests. The genus provides a wide variety of choicest ornamental plants for horticulture but it is still very poorly known taxonomically, especially in India. *Philodendron* is apparently mentioned in pre-Colombian arts and medicines during the 16th century and from there herbarium material was collected by George Marcgraf as early as 1644 (Mayo, 1990). However, Charles Plumier made the first effective introduction of the genus to Europe. He collected five or six species inhabiting Martinique, St. Thomas, and Hispaniola and gave phrase names beginning with 'Arum' or 'Dracunculus'. Resolution of many of the remaining generic problems with the Araceae awaited Austrian botanist Heinrich Wilhelm Schott who was the first to devote himself almost exclusively to the taxonomy of the Araceae (Nicolson, 1960). In 1829, Schott described the genus *Philodendron* (published as *Philodendrum*) in one of his first publications after returning from Brazil in 1821. The first species placed in the genus was *Philodendron ligulatum,* originally described as *Arum ligulatum* L., a member of *Philodendron* subgenus *Pteromischum.*

Philodendron is mostly vigorous evergreen climber suitable for greenhouse or for indoor gardening. These produce aerial roots at every node, (i) clinging roots which fix the plants to its support, and (ii) hanging roots which also ultimately grow down into the soil. Internodes are more or less elongated; cauline leaves distinctly alternate, radical, simple, entire to frequently bipinnately or palmately divided, often broad, long-petioled and net-veined; inflorescence in the form of spadix subtended by a large spathe; flowers sessile, numerous and closely packed on the spadices, monoecious (trimerous or dimerous) and without perianth, *i.e.* naked, female flowers at the base; stamens 2-6 and united into a sessile body pyramidal in shape, anthers 2-celled and dehiscing by terminal pore; and the pistillate flowers with 2-10 loculed ssuperior ovary often with parietal placentation, and some staminodia; and the fruit is a berry with or without endosperm. Only 10 per cent of the species have been studied cytologically for their chromosome counts, with a predominance of 2n=32, 34 and 36, and the isolated counts of 2n=30 and 33 (Contia-de-Oliveira *et al.*, 1999).

Origin, Distribution and Uses

Holdridge (1967) mentions that the genus is found ranging from Tropical moist forest to Premontane rain forest, especially in Central and South America (Brazil, Colombia, Costa Rica, Ecuador, Guiana, Panama, Peru, Venezuela, *etc.*) and only a few in North America (Mexico). *Philodendron* has 119 Central American species including 128 taxa, distributed in two subgenera of *Philodendron*. This Central America revision includes only members of *Philodendron* subgenus *Philodendron* and includes 103 taxa, including 95 species and 8 varieties or subspecies. A total of 65 taxa are new to science. These include 59 species, 6 subspecies or varieties, and 2 combinations. Alternatively, *Philodendron* subgenus *Pteromischum*, revised separately by Grayum (1996) of the Missouri Botanical Garden, contains 24 species (including 25 taxa) from Central America. Philodendron species can be found in many diverse habitats in the tropical Americas and the West Indies (*Gonçalves* and Mayo, 2000). It is a most important genus in the neotropics, inhabiting in a wide range of mesic habitats from sea level to over 2,000 m and in life zones (Croat and Yu, 2006). Species of this genus are often found clambering over other plants, or climbing the trunks of trees with the aid of aerial roots. Philodendrons can also be found in Australia, some in Pacific islands, Africa and Asia.

Philodendrons look quite elegant in pots when these have enveloped the moss-sticks fully from all the sides. When taking new growth, usually these produce purple or red coloured foliage and shoots which also look very beautiful. Their leathery and shining leaves which persist quite long on the plant are quite charming. Since most species are beautiful climbers for indoor use so these may be used anywhere where other climbers are used.

The resin produced during the flowering of *Philodendron* is known to be used by Trigona bees in the construction of their nests, from where native Indians from South America take the resin from the bees' nests and use it to make their blowguns air- and water-tight (Schott, 1832; Sakuragui, 2001). Though philodendron contains calcium oxalate crystals, even the white and sweet berries of some species (*Philodendron bipinnatifidum*) are eaten and the aerial roots are also used for preparation of ropes. Venezuelan red howler monkeys eat on the leaves of *Philodendron* (Schott, 1856). The leaves and stems of an unknown species are used in preparing a particular recipe for curare by the Amazonian Taiwanos. The leaves and stems are mixed with the bark of *Vochysia ferruginea* and with some parts of some species in the genera *Chondodendron* and *Strychnos*. A Colombian Amazon tribe uses *Philodendron craspedodromum* to add poison to the water for temporary stunning of the fish. Through its use in the fishing water, fishes rise up to the water surface where these can be easily scooped up. To add the poison to the water, the leaves are cut into pieces and tied together to form bundles and allowed to ferment for a few days. The bundles are crushed and added to the water into which the poison will dissipate. Although the toxicity of *Philodendron craspedodromum* is not fully known, it is possible that the active ingredients in the poisoning of the fish are coumarins which are formed during the fermentation process (Schott, 1860). Some philodendrons are also used for ceremonial purposes (Sellers *et al.*, 1978). Among the Kubeo tribe of Colombia, *Philodendron insigne* is used by witch doctors for treating patients. They use the juice of the spathe to stain their hands red since many such tribes view the colour red as a sign of power.

Classification, Species and Varieties

The first intrageneric system of classification for *Philodendron* was published in 1856 by Schott which contained only three groups that are still recognized today. Schott's final classification of *Philodendron* was published four years later in 1860 in the '**Prodromus Systematis Aroidearum**', a more rigorous work that came to be his last comprehensive self- published work since he died at the age of 71 in 1865. This revision included 110 genera, almost all of which are still recognized as genera or subgenera. Some of the species included by Schott (1861) in his publication '**Prodromus Systematis Aroidearum**' for Central America are *Philodendron acrocardium* Schott, *P. advena, P. anisotomum* Schott, *P. aurantiacum* Schott, *P. aurantiifolium* Schott, *P. brevispathum* Schott, *P. daemonum* Liebm. (syn *P. sagittifolium* Liebm.), *P. dagilla* Schott [syn. *P. tripartitum* (Jacq.) Schott], *P. fragrantissimum* (Hook.) Kunth, *P. gracile* Schott, *P. hederaceum* Schott (*P. jacquinii* Schott), *P. hoffmannii* Schott, *P. impolitum* Schott (*P. radiatum* Schott), *P. inaequilaterum* Liebm., *P. ligulatum, P. micans* Klotzsch, *P. oxycardium* Schott, *P. polytomum* Schott (syn *P. radiatum* Schott), *P. pterotum* K. Koch, *P. sagittifolium* Schott, *P. scandens* K. Koch & Sello (syn. *P. hederaceum*), *P. seguine* Schott, *P. subincisum* Schott, *P. tanyphyllum* Schott (syn. *P. sagittifolium* Liebm.), *P. tenue* K. Koch, *P. tripartitum* (Jacq.) Schott, *P. verrucosum* Mathieu, *P. warczeewiczii* and *P. wendlandii* Schott.

Very little work was done with *Philodendron* after Schott's death in 1865 until Adolf Engler, working at the Munich Botanical Garden (later on at the Berlin

Botanical Garden) began his revisionary work on the Araceae. The species included in Engler's 1899 revision in each section of *Philodendron* subgenus *Euphilodendron* for Central America are *Pteromischum* Schott [*P. aurantiifolium* (*P. guttiferum* Kunth], *P. guatemalense* Engl., *P. inaequilaterum*, *P. seguine*, *P. talamancae* Engl.), under section *Baursia* Reichb (*P. wendlandii* Schott), 'gruppe' *Solenosterigma* Klotzsch (*P. oxycardium* = *P. hederaceum* ssp. *oxycardium*), *P. purpureoviride* (called as *P. purpureoviridis* from South America), *P. micans* (Klotzsch) K. Koch (*P. hederaceum* ssp. *hederaceum* forma *micans*), *P. scandens* (*P. hederaceum* var. *hederaceum*), 'gruppe' *Platypodium* Schott (*P. pterotum*), `gruppe' *Cardiobelium* Schott (*P. brevispathum*, *P. gracile* = *P. tenue*, *P. schottianum*), `gruppe' *Achyropodium* Schott (*P. verrucosum*), 'gruppe' *Macrobelium* Schott (*P. sagittifolium*, *P. daemonum*, *P. sanguineum* Regel = *P. sagittifolium*, *P. mexicanum* Liebm.), 'gruppe' *Belocardium* Schott (*P. ligulatum* Schott, *P. immixtum* Schott, *P. advena*, *P. subovatum* Schott = *P. advena*, *P. smithii* Engl.), `gruppe' *Oligocarpidium* Engl. (*P. pittieri* Engl. = *P. hederaceum*), section *Tritomophyllum* Schott (*P. anisotomum*, *P. tripartitum*, *P. fenzlii* Engl. = *P. tripartitum*), section *Schizophyllum* Schott (no species represented), section *Polytomium* Schott (*P. augustinum* K. Koch = *P. radiatum*, *P. radiatum* Schott, *P. warszewiczii* K. Koch & Bouché), section *Macrolonchium* Schott (*P. fragrantissimum*), section *Macrogynium* Engl. (*P. hoffmannii* Schott sensu Engl. = *P. jacquinii* Schott).

Engler made no changes in his revision but went on to publish 26 additional species (Engler, 1905). In addition, seven species were described by Alfred Barton Rendle, Ignatz Urban, Ambroise Gentil and N. E. Brown between 1901 and 1908. Kurt Krause, who began working with Engler at the Berlin Botanical Garden on January 1, 1905, described two additional species before preparing his revision of *Philodendron* for *Das Pflanzenreich* (Krause, 1913). Krause's revision is a slightly reworked version of Engler's 1899 revision but did include the description of a new section, *P.* sect. *Camptogynium,* with a single species in *P.* subg. *Euphilodendron* and included 55 more species. There were 32 other new species published in *P.* subg. *Philodendron.* Six of these were in *P.* sect. *Pteromischum* (*P.* subg. *Pteromischum*) while one was in *P.* subg. *Meconostigma.* The remaining 25 were in *P.* subg. *Euphilodendron* (now *Philodendron*). Most of them were members of *P.* sect. *Baursia* and *P.* sect. *Polyspermium* (*Philodendron*) with a single species each in the following sections: *Oligospermium* (*Calostigma*), *Schizoplacium*, *Macrolonchium*; and three species in *P.* sect. *Polytomium.* Only two species, *P. grandipes* K. Krause and *P. panamense* K. Krause (both in current *P.* sect. *Philodendron*) were from Central America.

One hundred forty two (142) new species of *Philodendron* have been introduced todate since the time of Krause's revision. *Philodendron* spp. described between the time of Krause's revision and the completion of modern work are *Philodendron apocarpum* Matuda = *P. jacquinii* Schott, *Philodendron glanduliferum* Matuda, *Philodendron auriculatum* Standl. & L.O. Williams, *Philodendron harlowii* I.M. Johnst. = *P. hederaceum*, *Philodendron basii* Matuda, *Philodendron jamapanum* G.S. Bunting = *P. sagittifolium* Liebm., *Philodendron brenesii* Standl., *Philodendron jodavisianum* G.S. Bunting, *Philodendron davidsonii* Croat, *Philodendron lancigerum* Standl. & L.O. Williams = *P. sagittifolium*, *Philodendron dressleri* G.S. Bunting, *Philodendron latisagittatum* Matuda = *P. mexicanum*, *Philodendron erlansonii* I.M. Johnst. = *P. jacquinii*, *Philodendron lundellii* Bartlett ex Lundell = *P. jacquinii*, *Philodendron davidsonii* Croat, *Philodendron jamapanum* G.S. Bunting = *P. sagittifolium*, *Philodendron dressleri* G.S. Bunting, *Philodendron erlansonii* I.M. Johnst., *Philodendron lundellii* Bartlett ex Lundell, *Philodendron lancigerum* Standl. & L.O. Williams = *P. sagittifolium*, *Philodendron latisagittatum* Matuda = *P. mexicanum*, *Philodendron jodavisianum* G.S. Bunting, *Philodendron glanduliferum* Matuda, *Philodendron harlowii* I.M. Johnst. = *P. hederaceum*, *Philodendron miduhoi* Matuda = *P. hederaceum*, *Philodendron microstictum* Standl. & L.O. Williams, *Philodendron monticola* Matuda = *P. advena* Schott, *Philodendron mirificum* Standl. & L.O. Williams = *P. pterotum* K. Koch & Augustin, *Philodendron platypetiolatum* Madison, *Philodendron pleistoneurum* Standl. & L.O. Williams = *P. grandipes* K. Krause, *Philodendron microstictum* Standl. & L.O. Williams, *Philodendron brenesii* Standl., *Philodendron miduhoi* Matuda = *P. hederaceum*, *Philodendron basii* Matuda, *Philodendron mirificum* Standl. & L.O. Williams = *P. pterotum* K. Koch & Augustin, *Philodendron auriculatum* Standl. & L.O. Williams, *Philodendron monticola* Matuda = *P. advena*, *Philodendron apocarpum* Matuda = *P. jacquinii*, *Philodendron platypetiolatum* Madison, *Philodendron trisectum* Standl.=*P. anisotomum*, *Philodendron pleistoneurum* Standl. & L.O. Williams = *P. grandipes* K. Krause, *Philodendron pseudoradiatum* Matuda = *P. radiatum* var. *pseudoradiatum* (Matuda) Croat, *Philodendron pseudoradiatum* Matuda = *P. radiatum* var. *pseudoradiatum* (Matuda) Croat, *Philodendron trisectum* Standl. = *P. anisotomum* Schott.

The Flora of Guatemala (Standley & Steyermark, 1958) was much more accurate and complete in the percentage of the total taxa of *Philodendron* subg. *Philodendron* which were included. *The Flora of Guatemala* treated 11 species of *Philodendron*, only 8 of them being members of P. subg. *Philodendron*. These were *P. anisotomum*, *P. hederaceum*, *P. hoffmannii* = *P. jacquinii*, *P. radiatum*, *P. sagittifolium*, *P. smithii*, *P. tripartitum*, and *P. warszewiczii*. Properly named Costa Rican species recognized by Standley were *P. brenesii*, *P. ligulatum*, *P. pterotum*, *P. radiatum*, *P. schottianum*, *P. tripartitum*, *P. verrucosum*, and *P. wendlandii*. Species that are now synonymized include *P. gracile* = *P. tenue*, *P. hoffmannii* = *P. jacquinii*, *P. pittieri* = *P. hederaceum*, *P. trisectum* Standl. = *P. anisotomum*.

Phylogenetic and phenetic analyses by Mayo (1986, 1989) have shown *Philodendron* to have three distinct subgenera which are distinct in vegetative and floral morphology, floral anatomy and to some extent by distribution. The three subgenera of *Philodendron* in general can be most easily separated by the characters as presented below (modified after Mayo, 1991).

Subgenus (i) *Pteromischum*: Stem of mature flowering plants with a succession of short sympodial segments each bearing a cataphyll and a single leaf with the inflorescence(s) 1-10 and appearing to be borne in the leaf axils; petioles of adult plants with short, usually inconspicuous petiole sheath which is borne on the side of the stem, not encircling it at the base, usually only winged near the base on adult plants (sometimes fully winged on juvenile plants); inflorescences produced with each new leaf though frequently aborted. Stems often aborescent with conspicuous leaf scars and frequently interpetiolar scales persisting around at least the above margins of the petioler scars; male flowers conspicuously elongated, up to 10 times longer than wide; staminodal zone between staminate and pistillate zones of the spadix subequal or longer than fertile zone.

Subgenus (ii) *Meconostigma*: Stems rarely arborescent, often scandent, stout or slender and with interpetiolar scales lacking; male flowers only 2-3 times longer than wide; staminodal zone between staminate and pistillate zones of spadix much shorter than the fertile staminate zone.

Subgenus (iii) *Philodendron*: There are also a number of anatomical characteristics separating this subgenus. Vegetative buds of *Philodendron* subg. *Philodendron* are always located below the point of overlap in the sheath margins of the cataphyll whereas they are lacking in *P.* subg. *Pteromischum* (Ray, 1987). *Philodendron* subg. *Pteromischum* is also distinct in having a style with a shallow compitum with a subepidermal concentration of raphide crystals (Mayo, 1986,1989) and a total lack of tannin cells in the stamens (Mayo, 1986). In addition, while hypophyllous stem segments are typical for *P.* subg. *Philodendron*, they are ambiphyllous, hyperphyllous or peraphyllous in *P.* subg. *Pteromischum*. In addition, *P.* subg. *Philodendron* is characterized by having continuous parenchyma from the cortex to the centre of the stem. In contrast *P.* subg. *Pteromischum* has a central cylinder with a solid ring of fibres around the central cylinder.

In terms of growth habit *Philodendron* is clearly one of the most variable genera in the Araceae (Blanc, 1977a, 1977b, 1978, 1980). On the basis of habit, *Philodendron* is terrestrial to epiphytic or hemi-epiphytic. Very few *Philodendron* species are terrestrial and these are *P. glanduliferum*, *P. grandipes*, *P. malesevichiae* and *P. hammelii*. Some species are hemi-epiphytic such as *P. basii*, *P. roseospathum* var. *roseospathum* and *P. warszewiczii*. *Philodendron knappiae* is about equally terrestrial or hemiepiphytic, depending on the situation.

Philodendron eichleri is known as 'the king of the philodendrons'. It has magnificent leaves measuring some 2.1 m × 0.9 m in well grown specimens. There is a distinctly metallic cast to the foliage which gives them an extra air of permanence. *Philodendron squamiferum* is hardy and easy to grow, which is steadily gaining popularity. It can be trained to a totem pole of moss or the stem can be made to grow horizontally and twined around a large container, making a sort of hanging basket effect. The striking dagger-shape leaf is effective against the background of a modern living room. *Philodendron cordatum*, a Brazilian species is the commonest among all philodendrons having heart-shaped leaves, and twining over trees or wire. This species persists under the most adverse growing conditions and will bring living green to many dark nooks. It is believed to be the most fool-hardy vine in existence. Sprigs of it are often used in dish gardens, terrariums and novelty containers, where these either hang gracefully, trail along the soil or climb upon trellises. When this vine is allowed to grow freely as in conservatories, it will produce very large leaves and make the plants looking gorgeous. It flowers each year (unless the vine is pruned heavily) and the spathes are always of a creamish white colour on the outside with spathes of green showing on the lower half. The spathe is creamish yellow for two-third of the way down and brilliant red in the lower third. *Philodendron micans*, a native of Panama has small leaves with trailing habit. Sometimes it is listed under the name *P. scandens*. It has same general habit as *P. cordatum* except that the upper side of the leaf is velvety-green and iridescent while the underside is reddish. *Philodendron verrucosum*, a native of Costa Rica is one of the most beautiful trailers with delicate satiny-green leaves that are a combination of salmon and violet underneath. The petioles are conspicuously covered with bright red bristles and green hairs. The only drawback is its habit of loosing the lower leaves, maintaining only a few at upper end. This species likes an extremely hot and humid atmosphere in order to keep its leaves persisting. *Philodendron sodiroi* is a neat vining type of slow growing habit. The leaves are broadly heart-shaped with large irregular mottling pattern giving a silvery appearance. *P. bipinnatifidum*, a Brazilian species is vigorous growing climber with large, waxy green, stiff leaves up to 1 m long, having 10-12 segments on each side of midrib. *P.* 'Black Cardinal' has compact growth, broadly ovate leaves, cordate base, leathery, shining deep green, young leaves dark coppery red, and petiole and stem red. *P.* 'Black Cardinal Sport' is a more handsome and vigorous plant than 'Black Cardinal'. Leaves are larger, narrow base, thick glossy deep green, new leaves are maroon in colour, petiole reddish brown. *P.* 'Blue Mist' is slow growing but unusually attractive plant, leaves narrow shaped, 15-20 cm in length, deeply lobed base and tapering apex.

Philodendron cannifolium, a Brazilian species is slow-growing with long, leathery, almost erect, shining green leaves on stout petioles. *P.* 'Ceylon Gold' is quick growing attractive climber, leaves oblong, base cordate, 15-25 cm long, yellow, gradually yellowish-green with age. *P.* 'Charm' has smaller and thinner leaves than 'Venus Pluto', up to 25 cm long, lower part of the lamina deeply lobed, and new leaves reddish brown. *Philodendron cruentum* has waxy green leaves, oblong pointed with cordate base and depressed veins, wine-red underneath and petioles winged. *P. deflexum* is a creeper with robust, thin, leathery, waxy dark green leaves and purplish blotches on the lower side. *P. domesticum* (elephant's ear) having leaves arrow-shaped, bright green, undulate, veins pale and raised, showy and spathe red. *P. domesticum* 'Variegatum' has light to dark green, irregularly variegated, splashed or blotched green, yellow and cream-white leaves. *P. domesticum* 'Dubonet' is a robust but slow growing plant having large (20-25 cm), leathery, glossy-green leaves with serrated margins, and prominent midrib and veins. *P. elegans* possesses large, thin, leathery, dark geen and deeply pinnatifid leaves with narrow segments. *P.* 'Emerald Duke' is a hardy vigorous and spreading type with large hastate leaves which are shining green, leathery with yellowish midrib. *P.* 'Emerald King' is a large and hardy philodendron where leaves are spade-shaped, 30 cm or even longer and mid-vein light or yellowish green. *P. erubescens*, a native of Colombia is a beautiful plant having roots at every node and the leaves are arrow-shaped, waxy bronzy-green with red edges, wine-red beneath and petioles green with red. *P. erubescens* 'Gold' has large, broadly arrow shaped leaves, golden-yellow when young, changing to yellow and finally yellowish-green with age. P. 'Florida' is a slender attractive climber, having soft, deep green leaves cut into almost symmetrical (on both sides) five-pointed main lobes with a pale midrib. *P.* 'Florida Compacta' is a slow growing creeper, developed from *P. quercifolium* × *P. squamiferum*. Leaves are lobed, thick and leathery, and deep waxy-green on long petioles. *Philodendron fragrans* is also a slow growing creeper having large, broad cordate leaves, glossy green, with primary veins sunken and flowers fragrant. *P. lacerum* has ovate, crenately lobed leaves with undulate margins when young, and mature leaves are deeply incised, up to 65 cm long, and glossy-green with light green veins. *P. lacinatum* is large, hardy climber with oddly shaped, lobed and serrated leaves, and young leaves bright yellowish-green which turn deep green at maturity.

Propagation

Depending on the type of *Philodendron* grown, propagation can be accomplished by **stem cuttings**, by **air layering**, or by **offsets** removed from the parent plant. Stem pieces that contain at least two joints can be used as cuttings and inserted in pots containing sand or mixture of sand and peat moss. The pots should be kept at 21-24 °C temperature and shaded from direct sunlight until they are well rooted. Trailing varieties of philodendron will often root at any point where the stem comes in contact with the soil. An offset which emerges from the base of the plant or from the roots, and has sufficient root system to support itself, can be removed gently from the parent plant by cutting it off with a sharp and clean knife but with intact roots. Philodendrons can also be propagated from **seeds** but this is a long and slow process to get a healthy specimen-sized plant. The seeds must be kept moist and at a temperature of 24-27 °C for germination, which take 15 to 30 days.

Cultural Practices

Philodendrons are adaptable to many types of soils. The soil which has a pH of 6.0-6.5 is suitable for their cultivation. Philodendrons require indirect light. They can tolerate low light but if there is too little light, the new leaves will develop smaller in size and farther apart on the stem. However, direct sunlight burns the foliage, and plants become stunted. Philodendrons, in general, should be grown in shade-house with temperatures between 21 °C and 32 °C and a relative humidity of 60 - 100 per cent. When day temperature exceeds 35 °C or night temperature drops down to 18 °C, plant quality deteriorates and growth rate begins to decline. Swapna (1996) stated that at 25 and 50 per cent shade levels the plant height, leaf number & area, and number of side shoots were found increased when *Philodendron* var. 'Wendlandii' was grown in combination of peat, mud pot and soluble (controlled release) fertilizer at its higher concentration, and even better being under 50 per cent shade level, however, the height was significantly superior under 75 per cent shade level but over all performance was better only under 50 per cent shade level. For its cultivation, the soil should be highly porous which may permit air circulation freely and which may allow draining down of the excess water immediately but it should have sufficient organic matter for luxurious growth of the plant. It would be better if the planting medium is made of good garden soil+sharp sand+well decayed organic matter, all in equal proportion by volume. This substrate should also be fortified with Osmocote so that slowly it releases necessary nutrients required by the plant. The container should be kept weed-free throughout and any damaged part of the plant should immediately be removed. When any leaf is showing yellowing that should also be cut back. During winter the temperature should be at or above 13 °C though for a short period this can tolerate a temperature of as low as 5 °C. These require to be kept moist throughout the year, and in no part of the year these should be allowed to dry out completely. The climbing species and varieties require support of trellis-work, strings, wires or twiggy sticks, all padded with moss and kept moist at all the times. These are repotted every second year in February-March when the weather is still cool.

Water and Nutritional Requirements

Since philodendrons are native to tropical American rain forests, they enjoy humidity and moisture. They require only moderate watering but care should be given not to let the roots dry out completely. As they prefer humid conditions, the humidity level found in home is sufficient but outside these require high humidity for healthy growth.

There is relatively little information on the optimization of philodendron nutrition. Controlled-release or water-soluble fertilizers or a combination of both can be used for *Philodendron* production. Use of a fertilizer with a 3:1:2 or 3:1:3 NPK @ 60 g/plant twice, once when repotting in February-March and the second after four months will be sufficient in case no Osmocote has been used. Nutrient concentrations in *Philodendron* leaves given by Dennis (2003) for commercial production are 2.6-4.5 per cent N, 0.2-0.5 per cent P_2O_5, 2.0-3.5 per cent K_2O, 1.0-2.5 per cent Ca, 0.25-0.5 per cent Mg, 0.2-0.5 per cent S, 60-200 ppm Fe, 40-200 ppm Mn, 25-100 ppm Zn, 10-100 Cu, and 20-50 ppm boron.

Pot Culture

Plastic, non-porous ceramic pots and clay pots with proper drainage holes, all are ideal for plants which flourish in moist compost. Watering is needed less frequently in plastic and ceramic or glazed pots than in clay pots. Clay pots should be soaked in water for several hours before use. Philodendrons grow best when their roots are slightly cramped, so these should be planted in medium to large sized pots. When plants appear to be pot-bound, *i.e.* roots are growing through drainage hole it should be repotted during late spring and early rains. Good drainage is essential for proper growth and flowering, therefore, clean stones or broken crocks are placed over the drainage hole in a fashion which may permit proper draining out of the excess water but should not allow draining out of the potting compost. *Philodendron* requires strong support and for this wooden or plastic rods covered with sphagnum moss are best to induce rooting from the nodes and for support. Growing media play an important role in the successful growing of *Philodendron* plants. The potting media may consist of 100 per cent organic matter to 50 per cent inorganic matter and 50 per cent organic matter. The growing media should be selected on the basis of aeration, moisture retention, nutrient status, availability, weight, cost, *etc.* The pH of media should be in between 5.5 and 6.5. A soil mixture containing one part each of loam, leaf mould and sand is best for its cultivation.

Insect-Pests, Diseases and Physiological Disorders

Insects are not a major problem for philodendron, but spider mites, aphids, mealy bugs and scales may sometimes become a nuisance. **Spider mites** (*Tetranychus urticae*) are among the most serious house plant pests. These multiply rapidly and cause injury, defoliation and plant death. These mites are oval in shape and yellowish or greenish in colour. Spider mites thrive in dry and warm conditions. These mites first feed on the undersides of leaves, then expand their territory as populations increase, moving from stem to stem and onto nearby plants by means of fine webbing. They damage plants by piercing leaf tissues with needle-like mouthparts and feeding on sap. Usually, the first symptom is a mottling or pin-prick yellow discolouration on the undersides of leaves. Washing with bifenthrin, insecticidal soap or plant oil extracts should be carried out to control spider mites. **Scale**s are usually found on plant stems and the undersides of leaves especially along mid-veins. They use needle-like mouthparts to feed on plant sap, secreting sticky honeydew. Heavy feeding causes leaves to yellow and drop, slows the growth and stunts the plants. These can be controlled by washing, physical removal, bifenthrin, permethrin, resmethrin, insecticidal soap, pyrethrins, disulfoton, imidacloprid or use of plant oil extracts. Since their waxy covers are very impervious to insecticides, addition of a few drops of liquid soap or detergent will help the material slide under the edges of the shells. **Mealybugs** are soft-bodied insects and are most common along veins on the undersides of leaves and at axils where leaves join stems. They pierce plant tissues with sharp mouth parts and suck the sap which causes yellowing of the leaves, leaf drop and poor growth. These can be controlled by washing, through physical removal, and with the use of bifenthrin, permethrin, imidacloprid, resmethrin, pyrethrins (at least 2-3 applications sprayed once every 10 - 14 days).

Xanthomonas leaf spot (*Xanthomonas campestris* pv. *dieffenbachiae*) causes 'red edging' on the heart-leaved philodendron (*Philodendron oxycardium*). Under warm and moist conditions, large areas of the leaf blade can become infected, but lesions are generally confined to interveinal areas. The *Xanthomonas* bacterium enters the leaf through hydathodes on the leaf margins, stomates on the lower surface, or wounds. Symptoms appear within 7 -18 days after infection. The newly developing leaves on an infected plant do not necessarily become infected, since the bacterium is rather slow-growing and its infection is favoured by warm (21 ° to 32 °C), moist conditions. The organisms are easily spread from leaf to leaf and plant to plant by splashing water, contaminated tools, insects, handling infected plants, and by propagating infected plants. In case of infection of ***Pseudomonas* leaf spots and blights**, lesions are first small and water-soaked but rapidly enlarge (up to 3 cm) and turn dark brown or black, often with concentric light and dark rings and sometimes surrounded by a bright yellow halo. Most of the lesions are roughly circular to irregular and are rarely confined to interveinal areas. The centre of lesions is tan and brittle if allowed to dry. Lesions develop anywhere on the leaf blade but are more common at the leaf margins. Under very damp conditions infected leaves may drop off. The *Pseudomonas* bacterium enters a leaf primarily through hydathodes and wounds. Severe symptoms appear in a short time, *i.e.* 3 days after infection, particularly if plants are misted before infection occurs. The number of lesions increases linearly as the misting period increases. Disease severity is enhanced at 28 ° to 29 °C with little disease developing above 32°C. The organism is easily spread from leaf to leaf and plant to plant by splashing water, contaminated tools, insects, and by handling infected plants. **Erwinia leaf blight** (*Erwinia carotovora* subsp. *carotovora*) is characterized by leaf infections as minute spots that are water-soaked and yellow to pale-brown develop. The lesions are sometimes surrounded by a diffused yellow halo. When the humidity is high and temperatures are warm to hot, the spots expand rapidly, becoming slimy, irregular and sunken with light tan centres, darker brown borders, diffused yellow margins, and may involve the entire leaf in a few days. Large watery blisters may develop on the lower leaf surface when conditions are humid. When the air is dry and temperatures are low, the leaf lesions continue to enlarge, become papery, brittle, yellow to tan, and commonly tear away. Disease development is favoured by the presence of moisture and temperatures of 22 ° to 34 °C. The bacterium is transmitted from an infected to a healthy plant by splashing water, insects, and by contaminated knives, tools, gloves, and tray carriers. Use of culture-indexed, pathogen-free plants from reputable commercial propagators is always preferred. Overhead irrigation and splashing water on the leaves should be avoided. Crowding plants, heavy shade, poor air circulation, over-watering, mechanical injuries to plants, and high humidity should be avoided. Sterilization of the contaminated pruning knives, tools, carrying trays, and plants by dipping or swabbing in a solution of 70 per cent alcohol, and liquid household bleach, *etc.* are recommended. If possible, lower the temperature at which plants are grown to 21°C. On the appearance of symptoms of bacterial leaf spots and blights, application of several sprays of streptomycin formulation (Agrimycin) at 200 ppm of active ingredient at 4 to 7 days interval during damp weather is recommended.

Pale colouration of leaves occurs due to winter chilling. Symptoms firstly appear on lower leaves and then spread to upper or older leaves. Plants should be avoided to be exposed to chilling temperatures. Light colour can also be attributed to excessively high light levels or low nutrition. Therefore, only recommended light levels should be used with proper fertilizer application. When petioles become long and leaves become lobed due to low light intensity, the plants should be exposed to higher light intensity. **Mg deficiency** causes V-shaped chlorosis on older leaves, which spreads from the petiole attachment to the leaf margins but midrib remains green. Application of magnesium sulphate at 3-5 g/litre of water will correct this problem. New leaves become slightly twisted and distorted due to **Ca deficiency**. Increase of calcium levels in liquid fertilizer or application of chelated calcium will prevent this disorder.

References

Bailey, L.H. 1930. *The Standard Cyclopedia of Horticulture* (vol. III), pp. 2583-2585. The Macmillan Co., New York.

Barreto, 1999. Chromosome numbers for *Anthurium* and *Philodendron* spp. (Araceae) occurring in Bahia, Brazil. *Genet. Mol. Biol.*, **22**(2): 237-242.

Blanc, P. 1977a. Contribution a l'etude Aracées. I. Remarques sur l'eroissance monopodiale. *Rev. Gén. Bot.*, **84**: 115-126.

Blanc, P. 1977b. Contribution à létude Aracées. II. Remarques sur l'eroissance sympodiale chez l'*Anthurium scandens* Engl., le *Philodendron fenzlii* Engl., et de *Philodendron speciosum* Schott. *Rev. Gén. Bot.*, **84**: 319-331.

Blanc, P. 1978. Aspects de la ramification chez des Aracées tropicales. Thése du Diplôme de Docteur éme Cycle, 83 p. Université Pierre & Marie Curie.

Blanc, P. 1980. Observations sur les flagelles des Araceae. *Adansonia*, **20**(2): 325-388.

Bose, T. K., B. Chowdhury, S. P. Sharma and P. Pal, 2003. *House Plants*. In: *Garden Plants in Colour,* 188 p. Chakraberia Lane, Calcutta.

Contia-de-Oliveira, Anna Lucia Pares; Maria Lenise Silva Guedes and Ervene CerqueiraBarreto, 1999. Chromosome numbers for *Anthurim* and *Philodendron* spp. (Araceae) occurring in Bahia, Brazil. *Genet. Mol. Biol.*, **22**(2): 237-242.

Croat, T.B. 1979. The Distribution of Araceae. In: *Tropical Botany* (eds Larsen, K. and L.B. Holm-Nielsen), pp. 291-308. Academic Press, London.

Croat, T. B. 1990. A comparison of aroid classification systems. *Androideana*, **13**(1-4): 44-64.

Croat, T.B. and G. Yu, 2006. Four New Species of *Philodendron* (Araceae) from South America. *Willdenowia* (Botanischer Garten und Botanisches Museum, Berlin-Dahlem), **36** (2): 885–894.

Engler, A. 1905. Araceae-Pothoideae. In: *Das Pflanzenreich IV* (ed. Engler, A.), pp. 1-330. Publsihed by 23B (Heft 21), W. Engelmann, Leipzig and Berlin.

Grayum, M.H. 1996. Revision of *Philodendron* subgenus *Pteromischum* (Araceae) for Pacific and Caribbean Tropical America. *Monograph Systematic Bot. Missouri bot. Garden*, **47**: 1-233.

Gonçalves, E.G. and S. J. Mayo, 2000. *Philodendron venustifoliatum* (Araceae): a new species from Brazil. *Kew Bulletin* (Springer), **55** (2): 483–486.

Hay, R. and K.A. Beckett (eds), 1971. Philodendron. In: *Reader's Digest Encyclopaedia of Garden Plants and Flowers*, pp. 518-519. The Reader's Digest Association Ltd, London.

Holdriche, L.R. 1967. *Life Zone Ecology*, 206 p. Tropical Science Center, San Jose, Costa Rica.

Krause, K. 1913. Araceae:Philodendron-Philodendreae-Philodendrinae. In: *Das Pflanzenreich, IV* (eds Engler, A. and K. Krause). Published by 23Db (Heft 60), W. Engelman, Leipzig and Berlin.

Mayo, S.J. 1986. Systematics of *Philodendron* Schott (Araceae) with special reference to inflorescence characters. Ph. D. Thesis, University of Reading, U.K.

Mayo, S.J. 1989. Observations on gynoecial structure in *Philodendron* (Araceae). *J. Linn. Soc. Bot.*, **100**: 139-172.

Mayo, S.J. 1990. History and intrageneric nomenclature of the genus *Philodendron* Schott (Araceae). *Kew Bull.*, **45**(1): 37-71.

Mayo, S.J. 1991. A revision of *Philodendron* subgenus *Meconostigma* (Araceae). *Kew Bull.*,**46**: 601-681.

Nicolson, D.H. 1960. A brief review of classifications in the Araceae. *Baileya*, **8**: 62-67.

Ray, T.S. 1987. Leaf types in the Araceae. *Amer. J. Bot.*, **74**: 1359-1372.

Schott, H.W. 1832. Araceae. In: *Meletemata Botanica* (eds Schott, H.W. and S. Endlicherf), pp. 16-22. C. Gerold, Vienna.

Schott, H.W. 1861. Aroideologisches. *Bonplandia*, **9**: 367-369.

Sakuragui, C.M. 2001. Two new species of *Philodendron* (Araceae) from Brazil. *Novon* (Missouri Botanical Garden Press), **11**(1): 102-104.

Sellers, S.J., K. Mairaqlee, C.E. Aronson and A.H. Der Marderosian, 1978. Toxicologic assessment of *Philodendron oxycardium* Schott (Araceae) in domestic cats. *Veterinary and Human Toxicology*, **20**(2): 92-96.

Standley, P.C. and J.A. Steyermark, 1958. Araceae. In: *Studies of Central American plants* III. *Publ. Field Mus. Nat. Hist., Bot.*, **23**: 1-28.

Swapna, S. 1996. Environmental effects on the growth of *Philodendron* 'Wendlandii' (M.Sc. Thesis). Kerala Agricultural University, Vellanikkara, Kerala.

35

Primula (Family: Primulaceae)

Sapna Panwar, R.L. Misra, Thaneshwari and Poonam Kumari

[**Common names:** Auricula (*Primula auricula*), Baby primrose (*P. forbesii*), Bird's eye primrose (*P. farinosa*), Chinese primrose (*P. sinensis*), Common/English primrose (*P. vulgaris*, syn. *P. acaulis*), Cowslip (*P. veris*, syn. *P. officinalis*), Drumstick primrose (*P. denticulata*), Fairy primrose/Baby primrose (*P. malacoides*),German primrose (*P. obconica*), Giant cowslip (*P. florindae*), Himalayan cowslip (*P. sikkimensis*), Japanese primrose (*P. japonica*), Oxlip (*P. elatior*), Primrose (*P. vulgaris*, syn. *P. acaulis*) and Polyantha/Polyanthus (hybrids of *P. vulgaris* but called as *P. polyantha* or *P.* × *tommasinii*, and these have evolved with *P. veris* × *P. elatior* × *P. vulgaris*)]

Introduction

Primula is derived from the Latin *primus* which means 'first or early', referring to the early spring primrose. It is found worldwide, mainly in the northern hemisphere and particularly Asia. With approximately 430-500 species in the genus, *Primula* is the largest genus under the family primulaceae (Hu and Kelso, 1996). The first account on *Primula* was compiled by Linnaeus during 1753 after fairly investigating its taxonomy. Systematic treatments of *Primula* have been based on multiple aspects of anatomy, morphology, cytology and more recently molecular studies, which provide a foundation for division of the genus into well-defined sections. The genus *Primula* includes 38 sections, out of which the Sino-Himalayan region represents 26 sections of the genus with more than 300 species. So the Sino-Himalayan region is regarded as the geographical origin and centre of diversity of this genus (Richards, 2003). It is mainly distributed in the temperate and arctic area of the Northern Hemisphere, with a few in the mountainous region of Africa (Ethiopia), tropical Asia (Java and Sumatra) and South America. Some primroses are found almost at sea level, such as the Italian *Primula palinuri,* found at Salerno, but they are indigenous to hills and mountains, even up to altitudes of 3,000 metres (Pizzetti and Cocker, 1975). The Indian Himalayan region is also considered as one of the centres for rich diversity of *Primula* species with 106 species distributed from western to eastern Himalaya (Basak *et al.*, 2014). Ghosh (1978) after critically examining different herbaria of the country for the genus *Primula* and various literature sources concluded that Eastern Himalaya (excluding Nepal) is the chief abode of the genus. Hooker (1882) reported 43 species of *Primula* from different parts of the Indian Himalayan region, whereas about 40 species of *Primula* have been reported only from Arunachal Pradesh (Boardman *et al.*, 2010). Furthermore, the occurrence of about 50 species from Sikkim Himalaya (Grierson and Long, 1999) and 16 from Himachal Pradesh (Chowdhery, 1984) has been reported. A total of 25 species has been reported in western Arunachal Pradesh out of which five species, *viz. Primula ioessa* W.W.Sm., *Primula munroi Lindley, Primula obliqua* W.W.Sm., *Primula prolifera* Wall. and *Primula jigmediana* Hook. f. and Thomson ex Watt, are new to Arunachal Pradesh (Bawri *et al.*, 2015).

Most of the *Primula* species bear beautiful attractive flowers, and some species such as *P. malacoides*, *P.*

obconica and *P.× polyantha* are now ranked among the most economically important ornamental plants. *P. malacoides, P. obconica, P. sinensis* and *P. veris* are grown as popular flowering potted plants in Europe; *P. × polyantha* and *P. vulgaris* are popular flowering plants for display of their vivid colours so are used as bedding plants, potted indoor plants and for baskets; *P. × polyantha* and *P. vulgaris* being quite hardy so are most suitable growing outdoors (Dole and Wilkins, 1999). In Japan, *P. sieboldii* which is classified into sect. Cortusoides, subsect. Cortusoides (Richards, 1993), is an important ornamental species, with approximately 300 cultivars having been produced since Edo Era (about 300 years ago) (Torii, 1985). Globally the commercial cultivation of *Primula* is mainly confined to the U.K., CN Europe, the USA, China and Japan. In India, commercial cultivation is being undertaken mainly in hilly areas comprising North East (Sikkim), Jammu and Kashmir (Srinagar Valley), Himachal Pradesh (Shimla Hills), Uttarakhand (Mussoorie and Almora Hills) and Tamil Nadu (Ooty Hills).

Impotance and Uses

These are excellent winter- and spring-flowering hardy herbaceous plants with usually compact cum dwarf growth stature and free-flowering habit which make these most desirable ornamental plant. English Primrose (*Primula acaulis*), the cowslip (*Primula veris*), the oxlip (*Primula elatior*) and the polyanthus (*Primula polyantha*) are excellent subject for massing or naturalizing in open woodland, sheltered banks or any position where they are not too shaded. They are almost indispensable in gardens where a spring display of flowers in formal and informal way is wanted. Several *Primula* species are used as bedding plants or flowering potted plants. There are primroses suitable for almost every type of site, *i.e.* in pots, beds and borders, screen garden, rock garden, bog garden, along the bank of the water sources, greenhouse and alpine house. In addition to flowering potted plants, *Primula* is used together with other plants in pot-arrangements, colour bowls and hanging baskets or as a potted plant for outdoor planting in areas with moderate summer temperatures. The flower clusters of some species such as *Primula malacoides* and *P. obconica* are used in artistic flower arrangements.

This genus is well known for its medicinal values as for many years it was used in traditional herbal medicine. The flower and leaves of primroses, cowslips and polyanthus are edible. Young leaves of *Primula* are cooked in soups. The dried or fresh leaves are used as a tea substitute. They create a decorative addition to the salad bowl. The magenta and dark red flowers of *Primula polyantha* were found to contain an anthocyanin pigment named 'primulin' which consisted of 3- monoglucosidylmalvidin chloride (Scott Moncrieff, 1930). The major carotenoids detected in the yellow and pale-green petals of *Primula × polyantha* and yellow-petalled *P. helodoxa* were (9Z)-violaxanthin, (all-E)-violaxanthin, lutein and antheraxanthin (Yamamizo *et al.*, 2011). This species is now almost at the brink of extinction due to the destruction of the habitats, farming practices and over-collection in the past 100 years. When it was available in abundance, the flowers were harvested in high quantity in the spring and were used for preparation of a tasty wine with sedative and nerving properties. *Primula* plants have been used mainly in treating conditions involving cramps, spasms, paralysis and rheumatic pains. The plant contains saponins, which has an expectorant effect, and salicylates which are the main component of aspirin and have anodyne, anti-inflammatory and febrifuge effects. Flowers are anodyne, diaphoretic, diuretic and expectorant (Jager *et al.*, 2006). *Primula* is also recommended for treating over-activity and insomnia, especially in children. It is potentially precious in the treatment of asthma and other allergic conditions. The species may also contain allergens and some species are used traditionally to treat epilepsy and convulsions (Paulsen *et al.*, 2006; Tokalov *et al.*, 2004). A *Primula* species had flavonoids that possessed strong cytostatic properties against HL 60 cells even at low concentrations (Adebayo and Ishola, 2009). Leaf and root extracts of *P. vulgaris* contain a broad range of antimicrobial activity against microorganisms (Majid *et al.*, 2014). In Tibetan traditional therapy system, Amchis use whole plant of *Primula macrophylla* for food poisoning, fever, indigestion, dysentery, ulcer, *etc.* and flower of *Primula sikkimensis* for blood vein disorders in case of children (Pandey, 2006). An ethnobotanical investigation of western Indian Himalaya revealed that *P. denticulata* is a religious plant (Shreekar and Samant, 2010). Hassan and Mohammad (2010) reported that infusion from young stem of plant base is used to improve eye sight and control of ophthalmia. *Primula macrophylla* possesses antileishmanial activity, cytotoxic activity and antifungal activity which is due to presence of flavone compound, *i.e.* 2-phenylchromone (Saqib *et al.*, 2009).

Botany

Primulas are evergreen to deciduous, mainly herbaceous perennials (rarely annual), glabrous or pubescent, often farinose (evergreen and semiwoody in *P. suffrutescens*), tufted to clump-forming, rarely mat-forming (*P. suffrutescens*; sometimes stoloniferous as in *P. nutans*), and slightly to moderately succulent. In some primroses, the flowers, stems, leaves, sepals and occasionally section of petals are covered with waxy powder known as farina. Primulas are mainly rhizomatous, though some have poorly developed rhizomes and are short-lived (*Primula malacoides*), and the roots are fibrous. Stems (scapes) are ascending and simple. Leaf petiole is absent or obscure, winged or not; leaves in single basal rosette (multiple rosettes in *P. suffrutescens*), simple with margins entire, dentate or denticulate, apex toothed, acute, obtuse, rounded or spatulate, surfaces glabrous to rarely with simple hairs (*P. veris*); blade linear, broadly lanceolate, oblanceolate, oblong-obovate, rhombic or elliptic to cuneate or spatulate, base tapered or rounded and abruptly narrowed. Inflorescences are involucrate umbels (racemes or spikes); flowers 2-25 (sometimes solitary), tubular, and the bracts are 1-5; pedicels are erect, spreading, arching, nodding, arcuate or slightly reflexed. Flowers often heterostylous (sometimes homostylous); calyx broadly campanulate to cylindric or urceolate, ± 5-angled, weakly keeled or not keeled, glabrous, pilose, or puberulent, sepals 5, green, lobes not reflexed, length 0.5-1.0 times the tube; corolla campanulate, lobes not reflexed, length 1-2 times the tube, apex rounded; petals 5, lavender, magenta, pink, rose, violet, white, yellow or red; stamens included; filaments distinct; anthers connivent; capsules globose, cylindric or ellipsoid, valvate, dehiscent to 1/3rd length; and the seeds 10-1,000, brown, ovoid or oblong, somewhat 4-angled, reticulate or vesiculate. Basic chromosome number of genus *Primula* is 8, 9, 10, 11 or 12.

Around 90 per cent of the species are characterized by heterostyly (Richards, 2003), a condition in which populations consist of two floral morphs: 'pins', with anthers in the lower and stigmas in the upper portion of the corolla tube, and 'thrums' with a reverse arrangement of the sexual organs. This morphological differentiation is usually coupled with an incompatibility mechanism that hampers fertilization within the same morph (Barrett, 2002). Heterostyly is a genetically controlled breeding system that likely evolved to promote outcrossing (Barrett and Shore, 2008). Pollination is by bees, mainly bumblebees, but other long-tongued pollinators including syrphids, bee-flies and even butterflies may be locally important. *P. vulgaris* is an obligate outbreeder, with two genetically determined self-incompatible morphs ('pin' and 'thrum'). A third morph ('homostyle' or 'long homostyle') with a stigma like that of 'pin' but anthers like that of 'thrum', has been found in 'Somerset' and 'North Dorset'. In seasons when pollinators are scarce, homostyles have higher reproductive success than 'pins' and 'thrums', suggesting that reproductive assurance could have had a profound effect on the evolution of homostyly in *P. vulgaris*.

Propagation

Traditionally, primulas can be propagated from seeds, division and root cuttings. Plants produced from seeds will not necessarily be the same as the parent from which the seed were taken. To increase stock of named varieties it is best to propagate these vegetatively.

Primulas are commercially propagated by their very small **seeds** which require light for germination though light intensity is not critical and even natural and fluorescent light is acceptable (Thompson, 1969a, b). Chinese, English and polyanthus primroses have similar sized seeds, weighing 990 seeds per gramme. German primrose has smaller with 6,170 seeds/g and the fairy primrose the smallest with 12,700 seeds/g. They should be shown in tray, pans or small shallow flats in early spring on a seed bed of moist peat moss which has been layered over sterile potting soil having a soil pH of 5.0-5.7 for *P. obconica* and *P. vulgaris* with low soluble salts (Straver, 1970). They can also be sown in a fine mixture of loam, leaf mould and sand of about equal proportion. The tray or pans must be chilled in the refrigerator for 3-4 weeks, after which it should be kept at 18-20 °C (never above 21 °C; Hammer, 1992) and left uncovered where these germinate (Cathey, 1969a, b) in 3-6 weeks, though in case of *P. sinensis*, Rohde and Albert (1972) reported seed germination equally well at temperatures ranging from 19-27 °C irrespective of the light or dark regime. Chavagnat and Jeudy (1981) reported most favourable temperature range of 15-25 °C for seed germination in *P. obconica* where 75 per cent seeds germinated in one fortnight in the dark though germination percentage was found increased, *i.e.* 81 per cent when the germination was effected in light. Seed stratification from 0 to 9 °C for 1-12 weeks or seed soak in a 2 per cent KNO_3 solution + 500-1,000 ppm GA_3, Wikesjö (1975) though recorded good germination in seeds of *P.* × *polyantha* and *P. vulgaris* but suggested that key to successful germination was moisture as has also been reported by Cathey (1969a, b). Wikesjö (1975) further stated that these seedlings can be transplanted after 6-7 weeks. Goldsberry (1980) got germinated the seeds of *P.* × *polyantha* 'Pacific Giants' and F_1 seeds of *P. vulgaris* when subjected them to 21 °C under mist, shifting them to 16-18 °C and then transplanting when seedlings attain two sets of leaves. A sheet of clear plastic over the tray will help to retain the moisture in the tray until the seed germination and then the cover sheet should be removed. Individual seedlings are transplanted in pots when they are some 5 cm tall. By the middle of May, they would have attained substantial growth and as they do not flower the first season, they may be planted out in lines at a sheltered part of the garden till September, and then they are lifted and planted where they will flower in spring. Also those intended for flowering in the greenhouse should be potted at this time. Seeds can also be sown in a cold frame in April or May.

Propagation by **division** is practised preferably in September when the plants become rather large, or to perpetuate some important varieties when they are dormant. It consists simply of dividing the plants or clumps into two or more parts and replanting them in highly porous medium having peat and leaf mould. Auricula primroses should be propagated by gently separating the **offsets** in early spring or early autumn. *Primula denticulata* and its varieties can also be propagated by means of **root cutting** at any time except winter.

As many varieties are F_1 hybrids, clonal propagation of parental lines may enable the production of uniform progenies in such cross-fertilizing species. **Micropropagation** of *Primula obconica* was described by Coumans *et al.* (1979) who used inflorescence tips as explants. *In vitro* propagation of *P. vulgaris via* the proliferation of axillary buds was achieved by Merkle and Gotz (1990). A simple regeneration method is reported for *Primula heterochroma* by Hamidoghli *et al.* (2011) where seedling-derived leaf explants were cultured on MS basal medium enriched with different concentrations of 6- benzylaminopurine (BA) or thidiazuron (TDZ) in combination with different concentrations of 1-NAA and results showed that the highest frequency of shoot regeneration was 93 per cent and 94 per cent on the medium containing 2 $mg.l^{-1}$ BA + 2 $mg.l^{-1}$ NAA and 2 $mg.l^{-1}$ TDZ + 1 $mg.l^{-1}$ NAA, respectively, along with largest number of shoots per leaf. Jia *et al.* (2014) developed a protocol for successful callus induction and plant regeneration from *Primula forbesii* anthers. The highest induction and proliferation of indefinite buds (55.2 per cent) was produced on MS + 0.2 mg/l BAP + 0.01 mg/l NAA. Plant regulator-free MS was also found to be the best rooting medium. Among a total of 516 anther-derived plantlets, flow cytometry and cytological analysis identified 2 per cent to be haploid, 65 per cent diploid, 9 per cent triploid, 5 per cent tetraploid, 2 per cent hexaploid, and 17 per cent mixoploid. This protocol provides a useful foundation for further research towards the development of homozygous *P. forbesii* and other species.

Classification, Species and Varieties

Classification (Hu and Kelso, 1996 who placed 300 Chinese natives into 24 sections)

1. Sect. **Monocarpicae**. *Primula cavaleriei, P. duclouxii, P. effusa, P. forbesii, P. interjacens, P. malacoides, P. pellucida, P. petrocallis.*
2. Sect. **Obconicolisteri**. *Primula ambita, P. densa, P. obconica, P. rubifolia, P. sinolisteri.*
3. Sect. **Cortusoides**. *Primula latisecta, P. mollis, P. sinomollis, P. vaginata, P. palmata, P. pauliana.*
4. Sect. **Malvacea**. *Primula aromatica, P. bathangensis, P. celsiiformis, P. malvacea.*
5. Sect. **Auganthus**. *Primula filchnerae, P. rupestris, P. sinensis.*
6. Sect. **Pycnoloba**. *Primula pycnoloba.*
7. Sect. **Ranunculoides**. *Primula cicutariifolia, P. merrilliana.*
8. Sect. **Primula**. *Primula veris.*
9. Sect. **Carolinella**. *Primula chapaensis, P. henryi, P. kweichouensis, P. kwangtungensis, P. levicalyx, P. rugosa, P. wangii.*
10. Sect. **Bullatae**. *Primula bracteata, P. bullata, P. forrestii, P. rockii.*
11. Sect. **Petiolares**. *Primula bracteosa, P. davidii, P. hookeri, P. odontocalyx, P. hilaris, P. sinuata, P. strumosa, P. veitchiana, P. whitei.*
12. Sect. **Proliferae**. *Primula aurantiaca, P. mallophylla, P. melanodonta, P. poissonii, P. secundiflora, P. serratifolia, P. smithiana, P. wilsonii.*
13. Sect. **Amethystina**. *Primula faberi, P. kingie, P. valentiniana, P. virginis.*
14. Sect. **Sikkimensis**. *Primula alpicola, P. reticulata, sikkimensis, P. waltonii.*
15. Sect. **Crystallophlomis**. *Primula calliantha, P. diantha, P. elizabethiae, P. elongata, P. macrophylla, P. megalocarpa, P. minor, P. ninguida, P. russeola, P. woodwardii, P. youngeriana.*
16. Sect. **Cordifoliae**. *Primula baileyana, P. caveana, P. littledalei, P. rotundifolia.*
17. Sect. **Aleuritia**. *Primula caldaria, P. concinna, P. glabra, P. meiotera.*
18. Sect. **Minutissimae**. *Primula annulata, P. barbatula, P. bella, P. primulina, P. walshii.*
19. Sect. **Souliei**. *Primula aliciae, P. blinii, P. homogama, P. humilis, P. rupicola, P. souliei.*
20. Sect. **Dryadifolia**. *Primula dryadifolia, P. triloba, P. tsongpenii.*

21. Sect. **Denticulata**. *Primula atrodentata, P. denticulata, P. monticola, P. pseudodenticulata.*
22. Sect. **Capitatae**. *Primula capitata, P. glomerata.*
23. Sect. **Muscarioides**. *Primula gracilenta, P. inopinata, P. vialii, P. violacea, P. watsonii.*
24. Sect. **Soldanelloides**. *Primula buryana, P. cawdoriana, P. eburnea, P. flaccida, P. klattii, P. sandemaniana, P. sapphirina, P. spicata, P. wollastonii.*

Botanists have grouped the species with characters in common into 30 named sections such as *Amethystina, Auricula, Bullatae, Candelabra, Capitatae, Carolinella, Cortusoides, Cuneifolia, Denticulata, Dryadifolia, Farinosae, Floribundae, Grandis, Malacoides, Malvaceae, Minutissimae, Muscarioides, Nivales, Obconica, Parryi, Petiolares, Pinnatae, Pycnoloba, Reinii, Rotundifolia, Sikkimensis, Sinensis, Soldanelloideae, Souliei* and *Vernales*. However, the species are broadly grouped into seven sections, *viz. Auriculata, Candelabra, Farinosae, Sikkimensis, Vernalis,* Hardy primulas, and Non-hardy primulas (Pizzetti and Cocker, 1975) though florists normally put them into four major categories such as Alpine, Border, Indoor and Polyanthus.

Auricula Section

This includes hardy herbaceous perennial primroses mostly suitable for rock gardens, narrow borders and for pot-growing. These are usually evergreen, farinose, leaves fleshy and mostly have a preference for calcareous gritty soils, such as *Primula allionii, P. auricula, P. auriculata, P. glaucescens, P. marginata, P. palinuri, P. pedemontana, P. pubescens, P. viscosa.*

Candelabra Section

This includes hardy herbaceous perennial primroses that remain fully dormant during winter so are not evergreen but have a long flowering season as blooms appear in tiers one above the other arranged in circles around a tall stout scape in the form of candelabrum, flowering from the lower circles to the above in succession. These all are moisture-lovers and most suitable for woodlands where many of these naturalize from their self-sown seeds, and for their growing these require semi- or complete shade in humus-rich but normally non-calcareous soils, and these species are *Primula anisodora, P. aurantiaca, P. beesiana, P. bulleyana, P. cockburniana, P. helodoxa, P. japonica, P. pulverulenta.*

Farinosae Section

This includes hardy dwarf herbaceous perennial primroses which grow prostrate and possess very small leaves, all requiring cool and moist position in semi- or complete shade. Here inflorescences are usually umbellate with grouped flowers such as *Primula farinosa, P. frondosa, P. involucrata, P. rosea.*

Sikkimensis Section

This includes moisture-loving tall *cum* hardy herbaceous perennial primroses that bear top-clustered bell-shaped pendulous flowers on erect scapes such as *Primula alpicola, P. florindae, P. sikkimensis.*

Vernalis Section

This includes the common yellow-flowered hardy primroses, its various forms and varieties which for their cultivation require cool, moist and moderately fertile soil rich in organic matter, such as *Primula juliae, P. vulgaris.*

Miscellaneous Hardy Primroses of Minor Sections

This includes *Primula capitata, P. cortusoides, P. denticulata, P. elatior, P. nutans, P. sieboldii, P. veris, P. winteri.*

Non-hardy Section

This includes three important species cultivated in pots especially for indoor or greenhouse gardening, *vis-à-vis* summer bedding since these do not require great heat for their cultivation, such as *Primula malacoides, P. obconica, P. sinensis.*

Cultivated Species and their Popular Cultivars

Primula allionii

A native to Maritime Alps, this outdoor **alpine** species is suitable for planting in alpine house or rock gardens. It grows 5.0 cm tall with 15 cm spread forming a hummock. The leaves are mid-green, sticky and spathulate. When in flower during March and April, whole plant is hidden by its purple and rose-red to white flowers some 2.5 cm across which are borne singly or in twos on a stem.

Primula amoena (P. altaica)

A native to Europe (probably of garden origin), this **alpine** species grows to 20 cm tall with 10 cm spread. Its spathulate leaves are irregularly crenate and pale to mid-green. In April and May, its pink flowers some 2.5 cm across appear in umbels. Many named varieties in various pink shades are available.

Primula aurantiaca

A native to the Yunnan Province (China), it is an **alpine border** perennial species of moisture-loving

Candelabra group growing up to 30 cm with basal rosette-forming, large (up to 20 cm long), oblong and dentate foliage where stalk is winged. Flowers which appear during summer are small, 1.25 cm long, handsomely golden-red and in terminal whorls.

Primula auricula

A native to mountainous areas of central and southern Europe, it is a dwarf **alpine** species growing 15-20 cm tall and preferring calcareous rocky and heavy soils having good drainage. Leaves are evergreen, pale to grey-green, 5.0-10.0 cm long, fleshy, ovate or nearly rounded and slightly dentate, sometimes covered with farina. Inflorescence is an umbel bearing dark yellow or purple flowers about 2 cm across which appear during March to May. There are numerous beautiful hybrids such as 'Blue Fire' (blue), 'Blue Velvet' (10 cm high bearing dark velvet-blue flowers with white eye), 'Dusty Miller' (15 cm tall with strongly scented yellow flowers while foliage almost white with farinose surface), 'Gold of Ophir' (yellow), 'Jean Walker' (growing 15-20 cm high bearing fragrant mauve flowers with creamy eyes), 'Old Yellow Dusty Miller' (yellow), 'Red Dusty Miller' (15 cm high with red flowers), 'The Mikado' (dark red), 'Willowbrook' (yellow), *etc.*

Primula auriculata

A native to Asia Minor and Iran, it is an evergreen species with 12.5-15.0 cm long, glabrous and lanceolate leaves. Inflorescence is umbellate bearing numerous reddish-purple flowers during May-June.

Primula beesiana

A native to China, it is a very vigorous and large growing **border** species of moisture-loving Candelabra group with basally rosette-forming, 15 cm long and obovate foliage. Purple-red flowers of velvety texture appear in whorls one above the other on tall rigid stems some 60 cm high in May-June. This species requires a damp soil and is ideal for planting near a stream or pond.

Primula × bileckii

This **alpine** garden hybrid grows to 5 cm tall with spread of some 15 cm and makes a neat tuft of dark green dentate leaves. Deep rose flowers about 2.5 cm wide appear just above the foliage in May. This requires a soil mixture of peat or leaf mould with plenty of stone chippings.

Primula bulleyana

A native to China, it is a large and vigorous **alpine border** species of moisture-loving Candelabra group with large basal leaves forming a dense mass of vegetation. The leaves are ovate, dentate and thick-textured. This is one of the best species for naturalising. Inflorescence appears in May-June and is composed of whorls of orange and apricot-coloured fragrant flowers on tall erect stems. This species has been crossed with *Primula beesiana* to produce a strain of variously coloured hybrid.

Primula burmanica

A native to Myanmar, this is a taller species suitable for **border** planting, growing to the height of 60-90 cm and with spread of 22-30 cm. It is a moisture-loving species of Candelabra group. Its leaves are mid-green and oblanceolate, and the flowers which appear during June to July are red-purple and some 2 cm across.

Primula capitata

A native to the Himalayas, this **alpine** perennial species is of stout habit, bearing widely lanceolate and dentate basal leaves up to 12.5 cm long with greyish under-surface. Inflorescence appears during April to May in the form of a globose head composed of minute violet-coloured flowers with silver-white calyces.

Primula chionantha

A native to China, this **alpine** perennial species bears winged leaf-stalk, the leaves some 25 cm long, oblong-lanceolate and tapering towards the base where undersurface is covered with a yellowish powder. Inflorescence is an umbel bearing white scented flowers some 2.5 cm across on a 25 cm high stem which appear during April to May.

Primula clarkei

A native to Kashmir, this **alpine** species grows only 5 cm high with about 15 cm spread. Its leaves are pale-green and round to oblong. This requires semi-shaded position and well-drained soil for its cultivation. Bright rose-pink flowers about 2 cm across appear in spring when still the leaves are unfolding.

Primula clusiana

A native to the Austrian Alpines, this dwarf **alpine** perennial species bears shiny elongated-oval pointed basal rosette-forming leaves some 5.0-7.5 cm long. Flowers are lilac-pink with white eye which appear during April in umbellate clusters of 5.0-15 cm.

Primula cortusoides

A native to western Siberia, this perennial **alpine** species bears dark green basal rosette of soft-textured, ovate or cordate lobed leaves with ruffled edges where lobes are up to 10 cm long. Inflorescence is a terminal cyme of semi-pendulous wine-red flowers carried on slender stems 25-30 cm high, which appear during April.

Primula denticulata

A native to the Himalayas and China, this perennial vigorous species is the most beautiful of all the **alpine border** species of moisture-loving Candelabra group which is usually grown as an annual or biennial attaining about 30 cm height. It is most suitable for growing on the waterside, in rock gardens and for spring bedding. Its leaves are compact rosette-forming, pale-green, oblong, dentate, slightly pointed, 12.5 cm long and farinose. Its dense and globular umbellate inflorescence bearing lavender to purple or purple-crimson flowers appear from March to May. The floal heads measure 5.0-7.5 cm across. Some of its best varieties are 'Alba' (white), 'Bengal Rose' (ruby red), 'Cashmiriana' (purple), 'Prichard's Ruby' (ruby red), 'Ruby' (rose-purple), 'Rubra' (magenta-red), *etc.*

Primula edgeworthii (P. winteri)

A native to W. Himalayas, it is an **alpine** species growing to 10 cm in height and 20-25 cm in spread and flowers earlier from January onwards. Its grey-green leaves are spathulate to triangular-ovate with wavy and dentate margins, and covered with farina. Its pale-lavender flowers some 2 cm across have a yellow eye and appear in dense umbels.

Primula elatior

A native to Europe, the Caucasus and Iran, this is an **alpine** common European perennial similar to *Primula veris* but with larger flowers. Leaves are 5.0- 10.0 cm long. Umbellate inflorescence composed of numerous pendulous yellow or orange- red flowers in globose form at the end on 15-25 cm long stems appear during April and May. This species together with *Primula veris* and *Primula vulgaris* has produced the modern race of superb, large and vividly coloured (white, cream, yellow, pink, crimson and various shades of blue) polyanthus-type garden primulas. Its var. 'Garryarde Guinevere' is a novelty with purple foliage and pink flowers.

Primula farinosa

A native to Europe, including Great Britain and N. Asia, this **alpine** perennial species grows up to 15 cm high with spread of 15 cm, bearing some 15 cm long, basal and broadly lanceolate leaves, dusted with a farinose powder where undersurface is white. It is most suitable for planting in the rock or scree gardens. Umbellate inflorescence appears in March to April, bearing many pinkish-purple flowers with white or pale-yellow eye, each flower with about 8-13 mm in diameter.

Primula florindae

A native to SE Tibet (now under Chinese occupation) and China, this **alpine border** species of moisture-loving Candelabra group was discovered by the famous botanical explorer, Captain Kingdon Ward, who named it on the name of his wife. This is one of the largest species of the genus with a height of 0.9-1.2 m when in bloom. A true novelty, this plant bears long reddish leaf-stalks, broadly ovate-cordate leaves up to 20 cm long, and tall (1 m or more) erect floral stems bearing pendulous clusters of fragrant sulphur-yellow and powdery flowers some 2 cm across which appear during June-July. It prefers planting in boggy soil or waterside.

Primula frondosa

A native to Balkans and Bulgaria, this rosette-forming **alpine** perennial species grows up to 12.5 cm tall with a spread of 15 cm and is excellent for naturalising. Its thin-textured obovate leaves are basal and dentate with undersurface white-mealy. Minute lilac-pink flowers with yellow eyes, each measuring about 1.25 cm across, are produced in an umbel of 10-30 which appear during April-May on miniature branching stems. It grows properly in a lightly shaded moist position.

Primula gracilipes

A native to Sikkim (India), it is an **alpine** species growing to a height of 7.5 cm and the spread of 23-30 cm, forming a compact tuft of rosettes with mid-green, finely toothed and wavy leaves. The orange-yellow eyed lavender-pink flowers about 2.5 cm across appear in April.

Primula helodoxa

A native to Myanmar and the Yunnan province (China), it is a vigorous **alpine border** species of moisture-loving Candelabra group, requiring very wet soil for its successful cultivation. It grows from 60-90 cm in height. Its large basal leaves are shiny light green, narrow oblanceolate and dentate, and are retained on the plants throughout the winter. Pointed whorls of golden-yellow semi-pendulous flowers, each measuring some 2 cm across, appear during June-July in tiers on rigid stems.

Primula japonica

A native to Japan, it is an **alpine border** species of moisture-loving Candelabra group, being one of the most beautiful of the Candelabra group of primulas, with rigid tall erect floral stalk some 75 cm long, bearing a candelabrum-like inflorescence in tiers of large flowers in whorls during May-July. It is a short-lived perennial found along the waterside. Its rosette-forming leaves are pale-green and oblong-obovate. Its scapes are quite stout and bear several whorls of flowers, each measuring about 2 cm across. Flower colour is white to pink and reddish-purple. Some of the best hybrids under this species are 'Miller's Crimson' (crimson-purple), 'Postford White' (ivory-white), 'Rosea' (deep pink), 'Splendens' (crimson-red), *etc.*

Primula juliae (P. × garryarde)

A native to Caucasus, this perennial **alpine** hybrid grows only up to 8 cm in height with a spread of 25 cm, forming a flat mat of creeping stems with mid-green rounded leaves having cordate base. This flowers quite early from December to March on short scape, producing some 2 cm wide red-purple flowers with yellow eye. This has importance being grown in pots either in the cold frames or in an unheated greenhouse, where it begins to bloom in December-January. Its some of the best hybrids are 'Alba' (white), 'Betty Green' (crimson), 'E.R. Janes' (cherry-red), 'Guinevere' (soft pink), 'Jewel' (crimson-purple), 'Kinlough Beauty' (lilac-mauve), 'Pam' (red), 'Victory' (purple), *P.* × *pruhoniciana* (*P.* × *juliana*) var. 'Wanda' (purple-red), *etc.*

Primula × kewensis

A perennial hybrid originated at Kew between *P. floribunda* and *P. verticillata* of which the cultivated form comes true from seed. It is an **indoor** hybrid. Finely white-farinated leaves are up to 20 cm long, obovate, dentate, spathulate and with undulated margins. Whorled 2-5 fragrant yellow flowers some 2 cm across appear from December to April on 30 cm tall scapes. Its var. 'Red Gold' bears farinaless bright green foliage.

Primula malacoides

A native to China, this non-hardy **indoor** species though is a perennial but is grown as an annual. Its graceful appearance of foliage and flowers has given it the common name 'Fairy Primrose'. The plant forms a compact basal mass of long-stalked, pale-green, thin-textured, ovate to oblong-elliptic, cordate, 7 cm long, pointed, hairy beneath and deeply rounded-dentate foliage with ruffled edges. Umbellate inflorescence appears from December to April on a graceful slender scape about 30-50 cm tall, bearing star-like, 1.3 cm across, scented pale-lavender through red to white flowers in 3-6 whorls. In cultivation, it is represented with its forms (even dwarf forms with 15-20 cm tall) bearing single or double flowers in pink, red and white colours. These are available mostly in mixed forms though single colours are also sometimes offered. Its notable hybrids are 'Carmine Pearl' (dwarf, carmine-red), 'Cherry Glow' (cherry red), 'Fimbriata' (mauve-pink), 'Fire Chief' (brick red), 'Jubilee' (cherry red), 'Lilac Queen' (double, soft lilac-purple), 'Mars' (deep lavender), 'Rosea' (pink), 'Rose Bouquet' (carmine-red), 'Royal Rose' (crmine-pink), 'Snow Queen' (pure white), 'Snow storm' (pure white, double), 'Tyrian Rose' (deep pink), 'White Pearl' (dwarf, white), *etc.* Most of these varieties are fragrant.

Primula marginata

A native to Alpines and other mountainous districts of Europe, this is an **alpine** evergreen perennial, bearing rosette-forming, oblong, 10 cm long, very attractive leaves with silver-white margins, and these are often covered with a white mealy powder. Umbellate inflorescence which appears in May bears several slightly scented lavender-blue flowers 2.5 cm in diameter. Var. 'Linder Pope' bears wider foliage and large violet blooms.

Primula minima

A native of mountains of Europe, it is an **alpine** rosette-forming species growing only 5 cm high with the spread of 15 cm, bearing mid-green, serrated and blunt-tipped leaves and is suitable for planting in the rock or scree gardens or in the crevices. The rose-pink flowers some 2.5 cm across appear singly or in twos on the scape in April and May having deep petal-indentations.

Primula nutans

A native of Yunnan and Szechwan (China), it is an attractive **alpine** species growing up to 30 cm high with the spread of 22-30 cm, bearing grey-green, lanceolate and hairy leaves. Its bell-shaped soft lavender-violet flowers some 2.0 cm across appear in June.

Primula obconica

A native to China, it is an **indoor** perennial species though growing to 22-38 cm tall as an annual, having 6-15 cm long, light green, ovate-cordate and glandular-hairy leaves with petioles 5-10 cm long. Loose umbellate or clustered flowers some 2.5 cm wide and in colours of pink, red, lilac or blue-purple appear from December to May on 15-18 cm long scapes. Sometimes the petals are wavy-edged. Cultivars are available with flowers even in much larger size and in many pleasing soft pastel shades from white to lilac, purple, pink and orange. Expected production time for German primrose varies from 4 to 6 months. The leaves of some German primrose cultivars produce the primin allergen that may cause skin dermatitis. The German primrose series **Juno** (S&G Seeds, Downers Grove, Ill.) has been grown extensively in the United States, and this grows to a height of 30 cm with relatively small leaves and abundant flowering. The recent development and introduction of cultivars that do not produce primin has reduced the skin rash problem and renewed the interest in German primroses. The first cultivars stated to lack primin were introduced in

1990 under the names 'Freedom' and 'Beauty' (Richards, 1993). In 1995, the **Libre** series (Goldsmith Seeds, Inc.) was released as the first true primin-free selection of the German primrose. The Libre series members grow 20-25 cm tall, shorter than the Juno series and are suitable for 10-15 cm pots. These flower in several colours such as white, pink, salmon, red and blue. Schoneveld (Twello, The Netherlands) recently released the primin-free **Twilly Touch Me** series (James and Beytes, 2001). Most promising hybrids under this species are 'Alba' (white), 'Apple Blossom' (syn. 'Apricot Brandy'; pale-pink changing to darker tone), 'Atrosanguinea' (deep red), 'Caerulea' (blue), 'Fasbender Red' (deep red), 'Giant White' (pure white), 'Peach Blossom' (orange-salmon), 'Salmon King' (salmon-red), 'Salmon Pink' (pink), 'Snowstorm' (white), 'True Blue' (lavender-blue), 'Wyaston Wonder' (flowers extra large, bright crimson), *etc.*

Schnoveld b.v. of the Netherlands has developed primin-free *Primula obconica* F_1 hybrids and some of the exciting series are as (i) Twilly series 'Touch Me F_1' (primin-free non-itching type) in colour range of red-white, lilac red-white, blue-white, wine red, blue, white, orange, violet, pink, light pink and salmon; (ii) Twilly series 'Touch Me Midi F_1' (primin-free non-itching type) such as Midi red-white, Midi lialc red-white, Midi blue-white, Midi dark blue, Midi white, and Midi violet; and (iii) Twilly series 'Original F_1' (primin-containing itching type) such as red-white, lilac red-white, blue-white, red, dark blue, white, orange, violet, pink and light pink.

'S&G Seeds' have developed Juno F_1 hybrid series. Junos are earlier and their improved plant habit with small leaves result in a nicely rounded plant. They are quite free-flowering, vigorous and produce plants in all the nine colours, such as 'Juno Blue' (shining deep blue), 'Juno Blue & White' (flowers open white but turn blue *via* light blue), 'Juno Light Blue' (pale pastel-blue), 'Juno Orange' (deeper orange apple blossom), 'Juno Pink' (soft pastel-pink), 'Juno Red & White' (initially white but turning light rose then dark rose), 'Juno Red Picotee' (intense rose-red with small white edge on the petals), 'Juno Rose' (brilliant carmine-rose), 'Juno White' (pure white), 'Juno Formula Mix' (blend of all the colours), *etc.*

Some of the varieties under Ariane Series of *Primula obconica* are 'Ariane Deep Blue' (deep blue), 'Ariane White' (white), 'Ariane Orange' (orange), *etc.*

Primula palinuri

A native to southern Italy, it is an **alpine** evergreen perennial of spreading habit closely related to *P. auricula* but with ovate and large dentate leaves up to 15 cm long and 7.5 cm wide, which is found growing among the coastal rocks and cliffs at an altitude of about 180 m above sea coast. Inflorescence is an umbel bearing a cluster of semi-pendulous yellow flowers on a rigid stem appearing in March-April.

Primula parryi

A native to the rocky mountains of W. USA, it is an **alpine** perennial bearing winged leaf stalks, and elongated-oblong 20 cm long leaves which taper at the base. This species is free-flowering and the inflorescence is an umbel bearing bloom 2.5 cm in diameter, purple with yellow centre that appear from June to August.

Primula poissonii

A native to China, this **border** species of moisture-loving Candelabra group grows to a height of 30-45 cm with the spread of 15-23 cm, bearing shiny mid-green oblong-obovate leaves. Bright purple-red flowers some 2 cm across appear on thin scape in June and July.

Primula polyantha (P. × tommasinii)

The polyanthus group are the hybrids, primarily among cowslip (*P. veris*), oxlip (*P. elatior*), English (coloured forms of *P. vulgaris*) and Julian (*P. juliae*) primroses, and are most suitable **alpines** for growing outdoors and also for pot culture. Leaves are downy beneath, wrinkled, up to 25 cm long and obovate. The flowers of the size of 2.5-4.0 cm across develop in umbellate clusters on a 10-15 cm extended main flower scape which is quite stout. The colour is dominated by yellows and reds but cultivars with white, pink, red, maroon, purple, bronze or gold with or without a yellow or white eye are available. For years, the most important polyanthus cultivar has been the 'Pacific Giant' series. Sakata Seed Corporation has been maintaining and developing the Pacific Giant series since 1968 (Ward, 1997). This vigorous growing cultivar series with long peduncles has large clusters of flowers. The newer 'Concorde' (Daehnfeldt, Inc.), 'Hercules' (Royal Sluis Ornamentals) and 'Rumba' (Goldsmiths Seeds, Inc.) series have more compact but stronger peduncles than the older Pacific Giant series. The Hercules series is also known for its postharvest ability to ship well. 'Gold Laced' bears small dark red flowers neatly edged with gold patterning and 'Pacific has 5 cm across flowers. Flowering period for this species is April to May.

Primula × pubescens

This excellent **alpine** hybrid originated between *P. auricula* and *P. rubra* which grows up to 10 cm in height with the spread of 15 cm, bearing mid-green obovate leaves and umbellate flowers that appear during April or May. Its varieties 'Argus' produces compact purple flowers with white eye, 'Faldonside' crimson and 'Mrs. J. Wilson' a profusion of rich lilac-purple.

Primula pulverulenta

A native to China, this very beautiful **alpine border** species of moisture-loving Candelabra group, is one of the best Chinese species for garden cultivation, growing from 60-90 cm in height. The basal rosette of vegetation is composed of narrow 30 cm long pale-green obovate or oblanceolate leaves with dentate margins. Flower stems up to 90 cm high, covered with white powder, bears inflorescence in the form of successive whorls of semi-pendulous bell-shaped flowers of the size of about 2 cm across, wine red in colour with a darker eye and these flowers appear in May-July. Some of the hybrids under this species are 'Aileen Aroon' (vivid red), 'Bartley Strain' (pink shades), 'Invereve' (vivid orange-scarlet and propagated only through division), 'Red Hugh' (crimson), 'Lady Thursby' (pink, eye yellow), *etc.*

Primula reidii

A native to NW Himalayas, this **alpine** species grows to a height of 10 cm with the spread of 15 cm, bearing characteristically primrose-like pale-green leaves which are softly hairy. Its semi-pendent bell-shaped fragrant ivory-white flowers some 2 cm across appear during May in flower heads which comprise of 3-10 flowers per head. Its var. 'Williamsii' bears more fragrant soft blue flowers.

Primula rosea

A native to NW Himalayas, this **alpine** species forms a compact tuft and grows to a height of 15 cm with the spread of 15-20 cm. It bears mid-green, obovate and toothed leaves which are not fully developed at the time of flowering. It is the most vividly coloured, smallest **alpine** perennial species blooming in March and April providing splashes of intense colouration with its bright pink flowers which are borne out several together in loosely formed umbels, each umbel with 4-14 flowers and each flower some 2 cm across. Every year it is grown from seeds and this likes boggy soil. Its var. 'Delight' ('Visser de Geer') bears brilliant pink flowers and the plants are more robust, 'Grandiflora' produces deeper pink flowers of larger size.

Primula rubra

A native to Central Alps and Pyrenees, this rosette-forming **alpine** species grows to a height of 5-10 cm with spread of about 15 cm, bearing mid-green, hairy and ovate to obovate leaves and numerous umbels of pink flowers some 2.0-2.5 cm across in March and April.

Primula secundiflora

A native to Yunnan, it is an evergreen **alpine border** species of moisture-loving Candelabra group which grows to a height of 45 cm. Its oblong-obovate leaves are mid-green, and the wine-red flowers are pendent bell-shaped, 1.25-2.0 cm across and appear during July.

Primula sieboldii

A native to Japan, this **alpine** perennial grows to 15-23 cm high with 30 cm spread and is most suitable for planting in the woodland. The leaves are tufted, pale-green, 7.5 cm in diameter, soft-textured, heart-shaped (ovate) with clearly rounded teeth, and completely covered with minute hairs. Umbellate inflorescence composed of small lilac-pink fragrant flowers some 2.5-4.0 cm across appear in May and June with dentate (fringed) petals. Some of the hybrids under this species are 'Alba' (white), 'Croix de Malte' (blue), 'Elsiew Berger' (dark pink), 'Queen Victoria' (white), 'Rosea-alba' (pinkish-white), *etc.*

Primula sikkimensis

A native to Himalayas [Myanmar, Bhutan, Nepal, Sikkim (India), Tibet (under Chinese occupation) and Yunnan (China)], it is an evergreen **alpine border** species of moisture-loving Candelabra group which grows up to 45 cm high with the spread of about 25 cm. Its pale-green leaves are oblanceolate and finely toothed. Its fragrant flowers are funnel-shaped, pendent, pale-yellow and 2.0-2.5 cm across which appear during June and July. Its forms *P. s. hopeana* and *P. s. pudibunda* bear deeper yellow and larger flowers.

Primula sinensis (syn. P. chinensis, P. stellata)

A native to China, it is a non-hardy **indoor** species, generally grown in pot as an annual. Dark green thick-textured large rounded to broadly ovate leaves are long-stalked, softly hairy, 7-13 cm long, lobed and dentate or scalloped, and form an erect and thick basal rosette where underneath is red. Clustered large flowers (2.5-4.0 cm wide) in 2-3 whorls and in colours of white, pink, red and purple appear on 15-25 cm long scapes from November to March. The 'Fanfare' series (Daehnfeldt, Inc.) produce flowers in 5- 6 months with exceptionally good shelf life. Most promising varieties of the species are 'Alba' (white), 'Dazzler' (orange-scarlet), 'Empress Mixed' (many colours), 'Pink Beauty' (pink), 'Royal Blue' (deep blue), 'Ruby Queen' (deep red), 'Stellata' (2-tiered umbels with starry flowers), *etc.*

Primula veitchii (syn. Primula polyneura)

A native to China, it is an **alpine** perennial, bearing long-stalked rounded foliage some 10 cm in diameter having prominent silver-white undersurface. Umbellate inflorescence bearing pinkish-red flowers are freely produced in May on slender stem about 22-25 cm long

Primula veris (syn. P. officinalis)

A native to Western Europe and Western Asia, it is an **alpine** perennial used in rock, alpine or other such gardens. Its extended peduncle produces fragrant bright yellow flowers in April-May, though now its varieties in orange, apricot, crimson, light purple and white are also available.

Primula vialii

A native to NW Yunnan and SW Szechwan, this **alpine** perennial grows to a height of 30 cm with spread of some 22 cm, making the tufts of pale-green, farinose and lanceolate leaves. The lavender-blue flowers with scarlet floral buds and calyces appear in June and July on a dense spike some 7.5-12.5 cm long.

Primula viscosa (P. latifolia)

A native to mountainous regions of Europe, this is prostrate-growing and mat-forming **alpine** perennial species growing up to 20 cm in height with throughout slightly sticky surface. This bears small palish-green oval foliage which when crushed emits an unpleasant aroma. Its umbellate inflorescence bears slightly fragrant, violet-red flowers some 1.25 cm across in May-June.

Primula vulgaris (syn. P. acaulis)

A native to Europe and Asia, this **border** species of moisture-loving Candelabra group bears compact rosette of oblanceolate to obovate and wrinkled leaves some 5-25 cm long with usually downy beneath. The flowers some 2-3 cm across and in shades of yellow (ecotype found in west Europe), pink, red and purple (farther east) usually with a yellow eye develop on individual 5-20 cm long pedicels from the centre of the plant since scape is very short or nil. Even double forms are also available. This with *P. juliae* has produced a number of dwarf and compact primroses often with a ring of red around the yellow eye. Through breeding efforts and the development of F_1 hybrids, cultivars of English primrose are now available with flowers in many colours from white to red, blue and purple. The flowers may have a yellow or white eye and are sometimes fragrant. The 'Danova' series was introduced in 1989 by Daehnfeldt, Inc. (Odense, Denmark) in more than 20 colours, with many cultivars having two-toned flowers. The Danova series is intended for early season marketing in November and December in the northern hemisphere from the June seeding. Other early season cultivars include the 'Lovely' and the 'Pageant' series bred by the Sakata Seed Corporation (Yokohama, Japan) and the 'Quantum' series from Goldsmith Seeds, Inc. (Gilroy, Calif.). The Lovely series has smaller flowers than the Pageant series although they both have a compact growth habit suitable for 10 cm or smaller pots. Several red and pink bicolours are included in the Pageant series. The Quantum series is also compact with high and uniform germination rates and good postharvest quality. The 'Dania' (Daehnfeldt, Inc.), 'Finesse' (Ernst Benary Seed Growers Ltd., Hann. Muenden, Germany) and 'Gemini' (Goldsmith Seeds, Inc.) series are suitable for midseason marketing (January and February). Compared to the Danova series, the Dania series has somewhat larger flowers but a more limited colour range. A unique feature of the Finesse series is the narrow silver or gold petal borders. The limited leaf growth of the Gemini series makes it suitable for 10-cm pots or in combination with other plants in patio planters or colour bowls. Cultivars for late season marketing (February and March) include the 'Daniella' (Daehnfeldt, Inc.), 'Joker' (Ernst Benary Seed Growers Ltd.) and 'Paloma' (Royal Sluis Ornamentals, Leyland, UK) series. The Daniella series was introduced in 1995 as a complement to the Danova series to extend the marketing period. The Joker series consists of both bicoloured and clear flowers and the Paloma series is a leading cultivar in Europe. 'Colour Magic' comes up in mixed colours and grows to 15 cm high, 'Julian Bicolor' produces mixed colours and grows up to 10 cm high, while 'Panda' grows to 10-12 cm high and comes up in mixed colours.

'S&G Seeds' has developed the following range of series, each with its specific features and benefits. **'Peso F_1 hybrid' series** is a very early strain for autumn flowering. Ideal for warmer areas as it needs less cold temperatures for flower induction. Its flowers are very large on compact plants and is available in eight colours and a formula mix such as 'Peso Blue', 'Peso Carmine', 'Peso Deep Rose', 'Peso Orange', 'Peso Pink', 'Peso Scarlet', 'Peso White', 'Peso Yellow', 'Peso Formula Mix'. **'Lira F_1 hybrid' series** is an improvement of the Ducat Series and is specially bred for the forced production of top quality large-flowered pot primulas in the early winter period. Lira is characterized by high uniformity in both plant habit and flowering time. It is well suited to forcing and has strong tolerance against yellowing. Lira is available in nine different colours and a formula mix such as 'Lira Blue', 'Lira Carmine', 'Lira Golden Yellow', 'Lira Bronze', 'Lira Deep Rose', 'Lira Orange', 'Lira Pink', 'Lira Scarlet', 'Lira White', 'Lira Yellow', 'Lira Formula Mix'. **'Corona F_1 hybrid' series** is a strong improvement on the original Crown Series in uniformity of plant habit and earliness. Corona shows large flowers on compact, round and well-shaped plants. The Corona series being much more tolerant to cold and unfavourable weather conditions than the very early varieties which are ideal for flowering in mid/late winter. Corona is available in 11 different colours and a formula mix such as 'Corona Blue', 'Corona Deep Rose', 'Corona Golden Yellow', 'Corona Orange', 'Corona Pink', 'Corona Pink with Eye', 'Corona Rose with Eye', 'Corona Scarlet', 'Corona Flame', 'Corona White', 'Corona Yellow', 'Corona Formula Mix'. **'Twilight F_1 hybrid'** has large soft rose flowers with a yellow centre surrounded by a dark rose ring. **'Sterling F_1 hybrid' series** is a late blooming primrose. It is frost-tolerant and should be grown unheated. It comes into bloom naturally in late winter. Sterling is available in seven different colours and a formula mix such as 'Sterling Blue', 'Sterling Dark Blue with Edge', 'Sterling Rose', 'Sterling Rose with Eye', 'Sterling Scarlet', 'Sterling White', 'Sterling Yellow', 'Sterling Formula Mix'.

Benary of Germany has evolved the primrose series such as **'Ernst Benary series F_1'** which are known for their abundance of beautiful flowers, and these include the varieties such as 'E.B. Blue', 'E.B. Golden Orange', 'E.B. Scarlet', 'E.B. Show Musture' and 'E.B. White'. **'Joker series F_1'** is a mixture of 50 per cent unique bicolours and 50 per cent beautiful clear colours such as 'Carmine with Eye', 'Cherry', 'Red and Gold' and 'Violet with Eye'. **'Finesse series F_1'** is a blend of special 'Picotee' colours. Now Finesse is available in separate colours, each ringed with a silver border or a delicate tinge of gold such as 'Blue Shades', 'Carmine Rose', 'Peach Shades', 'Purple', 'Red and Gold', 'Rose Shades' and 'Violet'. Under **'Lucent F_1' series,** 'Lucento' mix is a mid season variety and is a formula mix of clear pink, salmon- pink, amaranth, apricot and bright red. It is slightly later than the Joker Series and Finesse Series and earlier than the Finale Series. **'Finale series F_1'** was developed specially for late season sales. The flowers are large, round and arranged on sturdy compact plants. It includes the varieties such as 'Carmine Rose', 'Deep Blue', 'Golden Orange', 'Rose', 'Scarlet', 'White' and 'Yellow'.

Primula warshenewskiana

A native to Afghanistan, this early flowering **alpine** primula grows to 7.5 cm high with spread of 15 cm and is similar to *P. rosea* though a little smaller, and grows to perfection in a moist position. Its leaves are mid-green and oblanceolate, and the flowers brilliant rose-red which appear in April on short stems, measure about 2 cm wide.

Out of these, four kinds (*Primula kewensis, P. malacoides, P. obconica* and *P. sinensis*) are grown in large quantity as pot plants, where *P. obconica* is most admired species because it is grown throughout the year in Europe though *P. malacoides* and *P. sinnesis* are marketed there only during December to early March. *P. sinensis* has one more drawback as its plants are brittle and therefore some of the growers prefer *P. kewensis* over *P. sinensis* albeit being poor multiplier.

Cultural Practices

Indoor species are grown through fresh seeds at 16 °C, then seedlings transplanted in a 10-cm pot in JIC 2 but with start of flowering weekly liquid feed is given. In September or early October these should be shifted to a 15-cm pot and the pots are kept moist all the times. For spring-flowering, they are given 10-13 °C temperature. For *P. obconica*, larger pots are required where it can be grown even the second year and then discarded. **Alpine** and other outdoor species can also be raised through fresh seeds by growing them in JIC 1 from May to September as stated earlier but the varieties and many species such as *P. clarkei, P. denticulata, P. juliae, P. rosea, P. florindae* and *P. japonica* are propagated through division. Dwarf, tufted or mat-forming species such as *P. auricula, P. minima* and *P. marginata* are propagated by taking 2.5-5.0 cm long cuttings or small rooted shoots. Alpine species do well in a well-drained gritty soil containing plenty of humus and these are usually planted from September to March. **Border** species do well in any fertile soil which does not dry out during summer or spring and these are planted from October to March.

Primulas prefer cool climate with partial shade. Mid and high hill areas are very suitable for its growing. **Temperature** plays an important role to control the floral induction and further development. Efficiency and quality could improve if proper temperature is applied at both the stages, *i.e.* floral induction and floral development. Holcomb (1983) recommended 4-7 °C as optimum, while Hammer (1992) 5-7 °C from transplanting until visibility of 2-3 flower buds, then forcing at 13 °C until flowering, though Smith (1969) and Goldsberry (1980) recommended warm temperatures up to 20 °C for faster development. During production no shade should be used except for temperature control in the late spring and summer. The plant likes a damp and humid atmosphere. In general, primrose prefers a neutral or slightly acidic **soil**, i.e pH 5.6-7.5, however, a slightly acidic pH (5.5-5.7) is said to be optimum. For pot production, primulas can be grown in varieties of media but media should be well drained. Perry (1981) advocated a well-drained medium which is high in peat-moss or a peat-lite and produced high quality plants in 1:1:1 of sand/perlite:peat:loam with a pH of 5.5-6.5 though Straver (1970) suggested a medium of 50 per cent loam compost and 50 per cent moist peat having pH of

5.5-5.7. A well-drained medium high in peat moss or peat-like medium is good for its production. By reducing the amount of **solar radiation** in the greenhouse during summer months and by increasing during winter months, high quality flowers or plants can be produced, however, during production usually no shade is required except for temperature control and in case of *P.* × *polyantha* and *P. vulgaris* when production temperature is low, high light is beneficial (Karlsson, 1996). Increase in temperature reduces optimum light levels. Shading can be done by using liquid shading compounds and movable screens. Shading up to 30 per cent is beneficial for primulas.

The germination process requires 10 to 14 days. After 6 to 8 weeks, when the seedlings have 2-3 true leaves these are suitable for **transplanting**. A single seedling is planted in a 7.5-10.0 cm pot. For larger pots, two or more seedlings are used. Similar to the germination medium, the growing medium should be well-drained and rich in organic matter with a pH of 5.5 to 6.0. The planting depth should be the same as in the seedling flat to avoid crown rot and other diseases. **Repotting** is an important cultural operation for keeping plants healthy. The best time of year to repot is in spring, before the new flush of summer growth. Pot-grown primulas should be repotted annually. The plants are **spaced** when the leaves reach the edge of the pot. Suitable final spacing for 10-cm pots is 42 to 44 pots/m^2. Finaly the pots are spaced 15 × 15 cm to 18 × 18 cm.

Regular hand **weeding** in early stages should be done. In large scale it may be expensive, so pre-emergence herbicide application can be used. Weeds can also be effectively controlled by solarization of planting media.

Fertilizing should start as soon as the cotyledons begin to develop, about 2 weeks from seeding. The initial fertilizer rate should be low at levels of 60 ppm of nitrogen and potassium. The rate can increase up to 200 ppm nitrogen immediately prior to transplanting. During the production phase, fertilizer rates of 90 to 100 ppm nitrogen and potassium from a complete fertilizer with micronutrients are suitable. Excessive nitrogen and fertilizer easily result in plants with too much leaf growth. *Primula* is sensitive to high soluble salt levels that may result in necrotic leaf margin burns. Regular monitoring of pH, soluble salts and the nutrient balance in the medium through soil tests is highly recommended. Holcomb (1983) and Mortensen and Holcomb (1984) described deficiency symptoms of nitrogen, phosphorus, potassium, calcium, magnesium and iron in polyanthus primulas. Nitrogen deficiency appears as chlorosis in newly developing leaves and in the form of chlorosis and necrosis of older plant tissues. Premature flowering may also occur under nitrogen deficiency. Mostly due to cool growing conditions of primroses, the conversion of ammonium to nitrate nitrogen by soil microorganisms is reduced and release of other nutrients of the medium is slowed down, therefore ammonium-based fertilizers should not be used to prevent excessive ammonium accumulation. The initial symptoms of a phosphorous deficiency are bronzing of older leaves, inward curling of younger leaves and leaf tip necrosis. Similar to nitrogen deficiency, plants may prematurely flower. Signs of low potassium are chlorotic lower leaves with slow necrotic and curling of leaf margins. Some plants may also die following unexpected and sudden wilting with potassium deficiency. Polyanthus grown with low or no calcium has poor root growth and pale-green foliage. Magnesium deficiency appears as interveinal chlorosis with tip and marginal necrosis of older leaves (Mortensen and Holcomb, 1984). Styer and Koranski (1997) list primula as an indicator crop for insufficient iron. Early symptoms of iron deficiency are chlorotic new growth followed by completely bleached white tissue due to the lack of chlorophyll. A boron deficiency initially appears as a light green colouration, later turning into chlorosis of recently matured leaves. The leaves may get cupped or crinkled with leaf edges turning downward. The veins become excessively prominent especially on the lower side of the leaf. If a boron deficiency is not corrected, the apical growing point dies and the stem becomes hollow. The German primrose requires higher fertilizer levels than the other primula species. On the other hand, German primrose and the fairy primrose are highly sensitive to elevated soluble salts, which manifests as leaf edge necrosis. Leaching at regular intervals is recommended to avoid salt build up. Primula is a cool temperature crop and greenhouse venting is often limited during long periods of the production. Under conditions with restricted venting, **carbon dioxide** enrichment at 900 to 1,000 ppm has been reported beneficial for primula production (Karlsson, 1997).

During germination and early seedling development, the medium should never be allowed to dry out. The seedlings are, however, sensitive to **over-watering** and **waterlogged** conditions. The fairy and German primroses are more sensitive to moisture stress and the media should be kept continuously moist. Low-temperature (12 °C and below) grown plants require low water, *vis-à-vis* low fertilizer requirements as in case of *Primula sinensis* (Dipner, 1979). Plants allowed to dry out or grown at uneven moisture, readily develop brown and dried leaf edges. High salt levels in the medium result in similar symptoms as water stress in fairy and German primroses.

Growth and Flowering

Final plant height is normally not a problem in primula production. B-Nine (daminozide) spray at 1,000 to 2,000 ppm can be used to **control height** in Chinese (*P. sinensis*), English (*P. vulgaris*), fairy (*P. malacoides*) and polyanthus (*P.* × *polyantha*) primroses but is ineffective in German (*P. obconica*) primrose (Karlsson, 1997). Proper spacing, temperature and irrigation are better methods for controlling stem elongation and height than plant growth regulators in primroses. Flower initiation in *Primula vulgaris* occurs in several weeks at 4-10 °C. The **temperature** is lowered from 16 or 18 °C when plants have developed 6–10 leaves. When the flower buds are visible, the temperature can be increased to 10–13 °C or below 10 °C (Hartnett, 1993). Cooling *P. vulgaris* plants for up to 10 weeks is expected to increase plant quality through reduced pedicel length and leaf size, and increased flower number (Lo¨fenberg, 1988). However, flower initiation has been observed in plants grown continuously at 15–20 °C (Karlsson, 1997). Photoperiod, daily light integral (DLI) and other production conditions interact with temperature to determine flower and plant development.

P. malacoides is either a SD plant or day neutral depending upon temperature; as plants flowered under SD or neutral day when temperatures were 10-16 °C though these plants became obligate SD when temperatures were increased from 16 to 24 °C, however, plants were when grown at LD at 16 to 21 °C, these never flowered and even growing at 10-16 °C under LD the flowering was found delayed (Post, 1949; Zimmer, 1985). Similarly, Rünger and Wehr (1971) with *P. malacoides* recorded flowering under SD or LD at 5-17 °C temperatures, however, 8-11 °C was found optimum in case of LD, though, irrespective of SD or LD, a 10 °C temperature induced earliest flowering but after 40 days of flower induction subsequent flower development was found stopped at 25 °C; at 20 °C some flowers opened and some aborted; at 15 °C floral initiation was more rapid than when plants were continuously kept at 10 °C; whereas, plants died at 30 °C four weeks after being induced to flower at 10 °C. Zimmer (1969) and Hammer (1992) reported that temperatures above 13 °C caused weak and elongated scapes in case of *P. malacoides*.

P. obconica is usually produced at 18-20 °C though before marketing this is subjected to 15-18 °C to improve flower colour and plant quality, however, there is no any specific photoperiod or cold treatment for flowering (Karlsson, 1997).

Armitage and Billingsley (1983) reported seeds of *P.* × *polyantha* 'Pacific Giant Dwarf Jewell Station' germinating at a constant temperature of 20 °C under 18 h LD and after about 60 days when these seedlings were transplanted and moved to natural SD in a greenhouse at either 10 or 20 °C, at both the temperature treatments the day temperatures varied from 18-25 °C, where plants in both the temperature treatments produced excellent quality of flowers at 98 days after transplanting (158 days from germination), which proves that despite a common belief that *P.* × *polyantha* and *P. vulgaris* are traditionally cool crops requiring a 6-week cold period at 4 to 10 °C for flower initiation is not correct as have also been reported by Welander and Selander (1981), Armitage and Billingsley (1983) and Karlsson (1996). Welander and Selander (1981) found *P. vulgaris* taking longer time for flower initiation at 9 °C than at 13 °C and Karlsson (1996) reported that in this species flower initiation was faster at 12 or 16 °C than at 8 or 20 °C and advocated that temperatures below 10 °C are unnecessary for floral initiation in *P. vulgaris*.

Primula is considered a low light crop with maximum photosynthetic photon flux (PPF) at 600 µmol m^{-2} s^{-1}. Since *P. vulgaris* does not grow well at temperatures above 20 °C, high natural light conditions may warrant shading to prevent sunscald and to improve temperature control. Perry (1981) stated that *P. vulgaris* is a short day plant. However, light intensity and duration are effective only when optimum temperature is made available in both, *i.e. P.* × *polyantha* and *P. vulgaris* for floral initiation as well as development. Increase in temperature decreases optimum light intensity, and cool temperatures of >9-16 °C cause fastest floral initiation at high irradiance and LD, however, 9 °C or below temperatures cause delayed initiation, development and production compared with warmer temperatures irrespective of photoperiod, though high light intensity and LD decrease the duration for floral initiation (Armitage and Billingsley, 1983; Karlsson, 1996).

Post Harvest

Primula is marketed when the first 5-7 flowers have opened. Proper temperatures for shipping and holding are 2 to 6 °C, and maintaining the plants well-watered is vital for longevity. *Primula* is highly sensitive to ethylene (Dole and Wilkins, 1999). A silver thiosulphate spray application at 65 to 165 ppm (0.2 to 0.5 mM) has successfully improved the keeping quality of English primrose (Nowak and Rudnicki, 1990). The home environment is often at low relative humidity and higher temperatures than the preferred 15.5-18.3 °C. Although the keeping quality is expected to be limited under these conditions, a high quality *Primula* should flower and remain attractive for 10-12 days (Wandås, 1991).

With proper care, the German primrose is expected to continue flowering for 2-4 weeks in an appropriate postharvest environment (Hartnett, 1993). For cut flowers as in case of *P. cortusoides, P. denticulata, P. edgeworthii, P. elatior, P. farinosa, P. florindae, P. japonica, P. malacoides, P. obconica, P. palinuri, P. prolifera, P. pulverulenta, P. sinensis, P. veris, P. vialii* (*P. littoniana*), *etc.* may be severed from the plants only when one-half of the florets are open in the inflorescence.

Insect-Pests, Diseases and Physiological Disorders

Insect-Pests

Cowpea and green peach **aphids** (*Myzus persicae*) are serious pests of primrose. These small, soft-bodied insects feed by inserting special mouthparts into leaf, stem and floral tissues. The shiny black cowpea aphid also injects toxins into plant materials that can kill tissues or whole plants. Green to pink-green peach aphids are less damaging, but can exhaust plants if their numbers are high. Insecticidal soap and horticultural oils are very effective against aphids, but one should be sure primroses are not sensitive to these chemicals. **Black vine weevil** (*Otiorhynchus sulcatus*) larvae often injure plants in nurseries and ornamental plantings by feeding on the roots. The tops of affected plants first turn yellow, then brown, and so the severely injured plants die. Leaf notching by adults can also be unsightly. The 1.25 cm long adult weevil is black, with a beaded appearance to the thorax and scattered spots of yellow hairs on the wing covers. Adults are flightless and nocturnal in habit. The legless grub is white with a brown head and is curved like to those of other weevil grubs. Adults and large larvae overwinter, emerging from May-July. The adults have to feed for 3-4 weeks before being able to lay eggs. Treating the soil with insect pathogenic nematodes may control the larvae and should be the first line of defence for landscape plantings. Acephate and fluvalinate are among the compounds used for control of this pest and may be applied when there is adult feeding and before the egg-laying starts. The usual timing for these foliar sprays is during May, June and July at three week intervals. **Greenhouse whitefly** (*Trialeurodes vaporariorum*) is another sap-sucking insect found on the underside of leaves in the greenhouses. Whitefly is easily identified by its ability to take flight when disturbed. The narrow-bodied, white to yellow winged greenhouse whitefly is sometimes a problem of primrose when their natural enemies have been destroyed by pesticides or environmental pressures. Whiteflies can be controlled with insecticidal soap or ultra-fine horticultural oil. Biological control is possible with *Encarsia formosa*, a species of parasitic wasp. Season-long control can be achieved with imidacloprid applied as a systemic insecticides taken up by the roots. Both the long-tailed and obscure **mealybugs** are problems on primrose, though distinguishing between the two at a glance can be difficult. They are both greyish with white waxy coatings, but long-tailed mealybug has two protruding filaments that are at least as long as its body. They feed on the sap of primrose and form white cottony masses on affected plants. Spraying with insecticidal soap will give sufficient control. When large areas of leaves have been eaten from the margin inward, these are **foliage-feeding caterpillars** similar to black cutworm or the beet armyworm. Black cutworms are brown to gray and night-feeding caterpillars that hide in the soil near their chosen plants. Beet armyworms appear in a variety of greens with distinctive white stripes and can be seen feeding in the daytime. Both caterpillars can be controlled with applications of *Bacillus thuringiensis* or spinosad. The larval stage of the serpentine **leaf miner** tunnels its way through the leaves of primrose as it feeds. The mines are clearly visible, marring the foliage of affected plants. While these mines are largely cosmetic problems, leaf miner larvae can occasionally introduce diseases. Treating these insects is difficult, since they are protected by plant tissues - often the best control is to remove affected foliage and burn it. If infestation is extensive, feed and water the plant to support it while targeting adults with sprays of azadirachtin or pyrethrin. Both the long-tailed and obscure **mealybugs** are problems on primrose, though distinguishing between the two at a glance can be difficult. They are both greyish with white waxy coatings, but long-tailed mealybug has two protruding filaments that are at least as long as its body. They feed on the sap of primrose and form white cottony masses on affected plants. Spraying with insecticidal soap will give sufficient control. When large areas of leaves have been eaten from the margin inward, these are **foliage-feeding caterpillars** similar to black cutworm or the beet armyworm. Black cutworms are brown to gray and night-feeding caterpillars that hide in the soil near their chosen plants. Beet armyworms appear in a variety of greens with distinctive white stripes and can be seen feeding in the daytime. Both caterpillars can be controlled with applications of *Bacillus thuringiensis* or spinosad. **Glasshouse leafhopper** (*Hauptidia maroccana*) is a sap-feeding insect from the upper leaf surface though hiding at the undersurface and breeds throughout the year in the greenhouses, causing a coarse and pale mottling. These are controlled by spraying with bifenthrin, dimethoate, heptenophos with permethrin, permethrin, or pyrethrum. The larval stage of the serpentine **leaf miner** tunnels its way through the leaves of primrose as it feeds. The mines are clearly visible, marring the foliage beuty of affected plants. While these mines are largely cosmetic problems, leaf miner larvae can occasionally introduce diseases. Treating these insects is difficult, since they are protected by plant tissues - often the best control is to remove affected foliage and burn it. If infestation is extensive, feed and water the plant to support it while targeting adults with sprays of azadirachtin or pyrethrin.

Mites- two-spotted red spider, *Tetranychus urticae*; bryobia mite, *Bryobia* spp., former infesting the undersides and latter the uppersides of the leaves, and their infestation make the plants becoming light yellow and unhealthy. Sometimes the mites form webs, which more or less enclose the upper as well the lower leaf surface. Insecticidal soap and ultrafine horticultural oil are widely used to control this pest. Spraying with insecticidal soap will give sufficient control if applied at least twice at 7-10 day intervals. The predatory mite, *Neoseiulus fallacis*, is most commonly found feeding where there are mite infestations. A single application of ultrafine horticultural oil (0.5-1 per cent) can be effective if predatory mites are present. Special care should be taken with soap or oil to obtain thorough spray coverage as they only work on contact. Commercial growers may use hexythiazox or abamectin to control spidermites. Application of carbaryl or pyrethroids should be avoided as these tend to be much more toxic to the predators than to the pest spider mites.

In the garden, primroses are often eaten by **slugs** (*Arios ater, A.s. hortensis, Deroceras reticulatus,* and *Milax* spp.). Moist and shaded conditions are favoured by both primroses and slugs. Slugs feed mostly at night, rasping holes in tender leaves or along leaf margins. Slug damage is easily recognized from the slimy, iridescent trail left by their crawling. At maturity they can be up to 10 cm long. They lay clusters of eggs, encased in slime, in the fall. Slugs can be controlled by lightly cultivating the ground in the spring to destroy dormant slugs and their eggs. A band of diatomaceous earth put around newly set plants will control slugs by rupturing their epidermis. During the day, slugs hide in dark and moist places, so they can be trapped by placing a small board in the flowerbed for them to hide. Scrape the slugs off the board and into soapy water to kill them. A bowl of beer, sunk into the ground with a roof of some sort, will act as a bait and the slugs will drown. Metaldehyde bait is best to control this pest.

Diseases

Flowers turn a papery brown and become covered with gray, fuzzy masses due to infection of **gray mould** (*Botrytis cinerea*). Senescing flowers are particularly highly susceptible. Tan to brown spots with a target-like appearance can also develop on the leaves during dry seasons. These patches are often associated with flowers which have dropped onto the leaf surface. This disease is particularly troublesome during periods of extended cloudy, humid and cold weather. The most effective method of preventing this disease is ventilation and temperature control. Gray mould cannot spread in conditions of low humidity and temperatures under 12 °C. Also, greenhouse films that block near-ultra-violet rays are effective in controlling *Botrytis* infection. Iprodione, mancozeb, copper sulphate pentahydrate, thiophanate methyl and polyoxins are effective in controlling gray mould. **Leaf spot** (*Ramularia primulae, Alternaria* sp.) appears on the leaves in conditions of high humidity, over-fertilization of nitrogen or lack of fertilizer and insufficient ventilation during all stages of primula growth, especially when producing outdoors. This disease can be prevented by soil sterilization prior to planting, by providing good ventilation and applying fertilizer moderately. Thiophanate methyl, zineb, polycarbamate, chlorothalonil and captan are effective against leaf spot. **Root rots** (*Pythium* sp., *Phytophthora primulae*) occur due to over-watering, poorly drained media or standing water. As a result of infection plants wilt and die. *Pythium* and *Phytopthora* are soil-borne fungi, hence soil sterilization and crop rotation are best ways to control the disease. Disease-free planting material should be used. Use of a potting medium with good drainage characteristics maintains aeration to the roots. If possible, the plants should be grown on raised benches to limit splashing spores from the soil in the vicinity. Fertilizers should not be applied in excessive amount. Etridiazole, fosetyl-al, mefenoxam and propamocarb are effective against these pathogens. **Rhizoctonia crown rot and bud blight** (*Rhizoctonia solani*) infection causes root and stem browning, wilting of the plants and ultimate death. Older plants may fail to flower or produce new leaves. Growing media should be sterilized before planting. Controlling the environment, crop rotation, using resistant varieties, and minimizing soil compaction are effective against this fungus.

Bacterial soft rot (*Erwinia carotovora*) infection is favoured by high temperature, high humidity and highly moist soil conditions. This disease infects the basal part of plant and is characterized by a dark green rot, resulting in the basal part becoming thinner and finally wilting of the whole plant. This disease can be controlled by soil sterilization with chloropicrin or dazomet prior to planting. **Bacterial Leaf Spot** (*Pseudomonas primulae*) causes irregular to circular spots with distinct yellow halos developing on the leaves. When spots are numerous, they coalesce and may kill entire foliage. This disease can be minimized by improving air circulation through thinning the plants and by avoiding overhead irrigation since these bacteria are easily spread in splashing water. Picking and destroying infected leaves and cleaning up all plant debris in the fall are also very helpful. Any equipment or tools that come in contact with diseased plants should be disinfected with 10 per cent household bleach, 70 per cent alcohol, or any of the commercially available compounds.

Aster yellows (*Phytoplasma)* infection causes stunting with excessive branching, strap-shaped leaves and yellowing. Affected plants should be removed from the roots and burnt. Good insect and weed control should be maintained in the greenhouse so that no vector or alternate host may be present. **Cucumber mosaic virus** (CMV) is a common viral disease and is carried by aphids and through plant contamination. Primary symptoms of CMV are stripes on flower petals and leaves. Infected plants should be removed and destroyed to prevent the spread of the disease. It is required to weed in and around the greenhouse and to use a preventative insecticide spray programme. Nicotine-sulphate and benfuracarb are effective in controlling CMV.

Cultural problems include unattractive plants with **long and large leaves**. This problem can usually be traced back to high nitrogen and fertilizer levels (Karlsson, 1997). English primrose plants sometimes fail to develop flowers. Exposure to cool temperatures before the plant has developed a good root system or temperatures higher than 20 °C during the initial stages of floral induction are possible causes for blindness (Hartnett, 1993). Stress at any time during plant development has been suggested to cause elongated flower stems in English primrose similar to the inflorescence of polyanthus. Bleaching of upper foliage due to cool soil conditions which inhibits iron uptake is also a problem in primula. It is most often seen in white flowering varieties.

References

Adebayo, E. A. and O.R. Ishola, 2009. Phytochemical and antimicrobial screening of crude extracts of *Terminalia gaucescens*. *African J. Pharm. and Pharmac.*, **3**:217-221.

Armitage, A.M. and J.W. Billingsley, 1983. Influence of warm night temperatures on growth and flowering of *Primula × polyanthus*, *HortSci.*, **18**: 882-883.

Barrett, S. C. H. 2002. Evolution of sex: the evolution of plant sexual diversity. *Nat. Rev. Genet.*, **3**: 274–284.

Barrett, S. C. H. and J.S. Shore, 2008. *New Insights on Heterostyly: Comparative Biology, Ecology and Genetics.* In: *Self-Incompatibility in Flowering Plants* (ed. Franklin-Tong, V.E.), pp. 3-32. *Springer*, Berlin/Heidelberg, Germany.

Basak, S. K., G.G. Maity and P.K. Hajra, 2014. The Genus *Primula* L. in India: A Taxonomic Revision. Bishen Singh Mohendra Pal Singh, Dehra Dun (India).

Bawri, A., P.R. Gajurel, A. Paul and M.L. Khan, 2015. Diversity and distribution of *Primula* species in western Arunachal Pradesh, eastern Himalayan region, India. *Journal of Threatened Taxa*, **7**(1): 6788–6795.

Boardman, P., A. Chambers, P. Eveleigh and J. Richards, 2010. Discovering primulas of subsection Agleniana. *Plantsman*, **9**(2): 88–92.

Chavagnat, A. and B. Jeudy, 1981. Study of the germination in the laboratory of *Primula obconica. Seed Sci. & Tech.*, **9**: 577-586.

Chowdhery, H. J. 1984. Primulaceae. In: Flora of Himachal Pradesh, vol. 2 (eds Chowdhery, H. J. and B.M. Wadhawa). Botanical Survey of India, Kolkata (India).

Cathey, H.M. 1969a. I. Guidelines for the germination of annual pot plant and ornamental herb seed. *Flor. Rev.*, **144**(3742): 21-23, 58-60.

Cathey, H.M. 1969b. II. Guidelines for the germination of annual pot plants and ornamental herb seed. *Flor. Rev.*, **144**(3743): 18-20, 52-53.

Coumans M., M.F. Coumans-Gillès, J. Delhez and T. Gaspar, 1979. Mass propagation of *Primula obconica*. *Acta Hort.*, No. 91, pp. 287– 293.

Dipner, H. 1979. Sind die Chinesen-Primeln vergessen? *Deutscher Gartenbau*, **33**: 774-775.

Dole, J.M. and Wilkins, H.F. 1999. *Floriculture: Principles and Species*. Prentice-Hall, Inc., Upper Saddle River, N.J.

Ghosh, R.B. 1978. An analysis on the distribution of the Indian taxa of the genus *Primula* Linn., in the Eastern Himalaya with remarks on the species of Assam. Central National Herbarium, Calcutta (India).

Goldsberry, K. L. 1980. Primula: A flowering mini pot plant. *Colorado Greenhouse Growers Assn. Bul.*, No. 359, pp. 1–2.

Grierson, A. J. C. and D.G. Long, 1999. Primulaceae. In: Flora of Bhutan: Including A Record of Plants from Sikkim and Darjeeling, vol. 2, part II (ed. Long, D.G.), pp. 516-554. Royal Botanic Garden, Edinburgh (UK).

Hamidoghli, Y., A.R.N. Sharaf and H. Zakizadeh, 2011. Organogenesis from seedling-derived leaf explants of primrose (*Primula heterochroma* Stapf.) as a scarce Iranian plant. *Australian J. Crop Sci.*, **5**(4):391-395.

Hammer, P. A. 1992. *Other Flowering Pot Plants*, I. *Primula*, In: *Introduction to Floriculture*, 2nd ed. (ed. Larson, R. A.), pp. 492-493. Academic Press, New York.

Hartnett, G. 1993. Focus on primula production. *Professional Plant Growers Assn. News*, **24** (7), 4–7.

Hassan, S. and A. Mohammad, 2010. Economically and ecologically important plant communities in high altitude coniferous forest of Malam Jabba, Swat, Pakistan. *Saudi J. boil. Sci.*, **18**: 53-61.

Holcomb, E. J. 1983. The primrose path to nutrition. *Greenhouse Grower*, **1**(10):224–225.

Hooker, J. D. 1882. Flora of British India (vol. 3, Indian Reprint 1983), pp. 482-495. Bishen Singh Mohendra Pal Singh, Dehra Dun (India).

Hu, C. M. and S. Kelso, 1996. Primulaceae. In: *Flora of China*, vol. 15 (eds Wu, Z.Y. & P.H. Raven). Science Press, Beijing, & Missouri Botanical Garden Press, St. Louis.

Jager, A.K., B. Gauquin, A. Adsersen and L. Gudiksen, 2006. Screening of plants used in Danish folk medicine to treat epilepsy and convulsions. *J. Ethnopharmac.*, **105**: 294-300.

James, L. and C. Beytes. 2001. International events. Four days at the HortiFair. *Grower Talks*, **64**(9):76–78,80.

Jia, Y., Q.X. Zhanga, H.T. Pana, S.Q. Wanga, Q.L. Liub and L.X. Sun, 2014. Callus induction and haploid plant regeneration from baby primrose (*Primula forbesii* Franch.) anther culture. *Sci Hort.*, **176**: 273–281.

Karlsson, M.G. 1996. Temperature, light and daylength for flowering primula. *Professional Pl. Gr. Assoc. News*, **26**(5): 10-11.

Karlsson, M.G. 1997. *Primula*. In: *Tips on Growing Specialty Potted Crops* (eds Gastron, M.L., S.A. Carver, C.A. Irwin and R.A. Larson), pp. 107–111. Ohio Florists' Assn., Columbus.

Lo¨fenberg, E. 1988. Kylperiodens och drivings—temperaturens effekt pautveckling och kvalitet hos *Primula vulgaris*, Tradgard 341. Swedish University of Agricultural Sciences Research Information Centre, Alnarp.

Majid, A., S. Hassan, W. Hussain, A. Khan, A. Hassan, A. Khan, T. Khan, T. Ahmad and M.U. Rehman, 2014. *In vitro* approaches of *Primula vulgaris* leaves and root's extraction against human pathogenic bacterial strains. *World Applied Sciences J.*, **30**(5): 575-580.

Merkle, S. and E.M. Götz, 1990. Micropropagation of inbred lines of *Primula acaulis* for breeding purposes. In: *7th int. Congr. Plant Tissue Cell Cult.* (abstr.), p. 116, held on 24-29 June, 1990 at Amsterdam, the Netherlands.

Mortensen, W. and E.J. Holcomb. 1984. Nutrient deficiency symptoms on primula. *Pennsylvania Flower Growers*, No. 352, pp. 1, 6, 7.

Nowak, J. and Rudnicki, R.M. 1990. *Postharvest Handling and Storage of Cut Flowers, Florist Greens and Potted Plants.* Timber Press, Portland, Oregon, USA.

Pandey, M. R. 2006. Use of medicinal plants in traditional Tibetan therapy system in Upper Mustang, Nepal. *Our Nature*, **4**: 69-82.

Paulsen, E., L.P. Christensen and K.E. Andersen, 2006. Miconidin and miconidin methyl ether from *Primula obconica* Hance: new allergens in an old sensitizer. *Contact Dermatitis*, **55**: 203-209.

Perry, L.P. 1981. Primulas make a comeback as a 5-month cool crop. *Flor. Rev.*, **168**(4344): 22, 55.

Post, K. 1949. *Primula* (Primrose). In: *Florists Crop Productionn and Marketing*, pp. 746-748. Orange Judd Publishing, New York, USA.

Pizzeti, I. and H. Cocker, 1975. *Primula*. In: *Flowers- A Guide for Your Garden* (vol. II), pp. 1038-1054. Henry N. Abrams, Inc., New York, USA.

Richards, J. 1993. *Primula*. Timber Press, Portland, Oregon, USA.

Richards, J. 2003. Primula. *Timber press (rev. ed.), London, UK, 346p.*

Rohde, J. and G. Albert, 1972. Keimung bei *Primula sinensis fimbriata* 'Helle Zukunft'. *Gartenwelt*, **72**: 146.

Rünger, W. and B. Wehr, 1971. Einfluss von Tageslänge und Temperatur auf die Blütenbildung und - entwicklung von *Primula malacoides*. *Gartenbauwiss.*, **36**: 51-62.

Saqib, N., F. Alam and M. Ahmad, 2009. Antimicrobial and cytotoxicity activities of the medicinal plant *Primula macrophylla*. *J. Enzyme Inhib Med Chem.*, **24**: 697-701.

Scott Moncrieff, R. 1930. Natural anthocyanin pigments: The magenta flower pigment of *Primula polyanthus*. *Biochem J.*, **24**(3):767-78.

Shreekar, P. and S.S. Samant, 2010. Ethnobotanical observations in the Mornaula reserve forest of Kumoun west, Himalaya, India. *Ethnobotanical Leaflets*, **14**: 193-217.

Smith, D.R. 1969. Controlled flowering of *primula malacoides*. *Experim. Hort.*, **20**: 22-34.

Straver, N. 1970. Een proef met verschillended zaaigronden bij primula. *Proefstation Vakblad Bloemist.*, B25B: 1758.

Styer, R. C. and D.S. Koranski, 1997. *Plug and Transplant Production: A Grower's Guide*. Ball Publ. Batavia, Illinois, USA.

Thompson, P.A. 1969a. Germinating primula seed. *J. roy. hort. Soc.*, **93**: 133-138.

Thompson, P.A. 1969b. Comparative effects of gibberellins A_3 and A_4 on the germination of seeds of several different species. *Hort. Res.*, **9**: 130-138.

Tokalov, S.V., B. Kind, E. Wollenweber and H.O. Gutzeit, 2004. Biological effects of epicuticular flavonoids from *Primula denticulata* on human leukemia cells. *J. agric. Food Chem.*, **52**: 239-245.

Torii, T. 1985. Sakurasou (Japanese). Nihon-terebi, AkatukiInsatu press, Tokyo.

Wandås, F. 1991. *Primula vulgaris (acaulis)*, en odlingsbeskrivning. Trädgård 361. Swedish Univ. Agr. Sci., Alnarp.

Ward, P. 1997. *Primroses and Polyanthus:A Guide to the Species and Hybrids*. Batsford, London.

Welander, T. and C. Selander, 1981. Flower formation in *Primula vulgaris* (Swedish). *Swedish J. agr. Res.*, **11**: 41-47.

Wikesjö, K. 1975. Production programs for pot plants and cut flower in Sweden (Swedish). *Hort. Advisors Bull.* No. 89. Agriculture College, Alnarp, Sweden.

Yamamizo, C., M. Hirashima, S. Kishimoto and A. Ohmiya, 2011. Carotenoid composition in the yellow and pale green petals of *Primula* species. *Bull. Natl. Inst. Flor. Sci.*, No. 11, pp. 67 -72.

Zimmer, K. 1969. Zur Blütenbildung bei *Primula malacoides*. *Gartenwelt*, **69**: 137-138.

Zimmer, K. 1985. Primula. In: *Handbook of Flowering*, vol. IV (ed. Halevy, A.H.), pp. 137-138. CRC Press, Boca Raton, Florida, USA.

36

Proteaceous Ornamentals (Family: Proteaceae)

Kalkame Ch. Momin, Y.C. Gupta, S.R. Dhiman, R.L. Misra and Sanyat Misra

Introduction

Scientific research furthers our understanding of plant growth, delving into biochemical pathways to reveal patterns in growth and flowering regulation. Fresh flowers play a unique role in our lives, both in celebration and in sorrow. Unlike most other horticultural crops, demand for fresh flowers is often related to fashion trends. Our growing multicultural population has seen an increased use of flowers outside of the traditional peaks of Valentine's Day, Mother's Day and Christmas. For other festive occasions like Chinese New Year and Orthodox Easter, fresh flowers are used extensively. New flower lines and cultivars are being introduced all the time, either in new colours and shapes of traditional flowers like roses and chrysanthemums, or as new Australian native crops. Over the last decades, utilization of the members of the Proteaceae for cut flowers has intensified and changed from wild harvesting to cultivation for improved quality of the floral product.

The family name Proteaceae is given due to genus *Protea*. Plant life research indicates that Proteaceae probably originated in South Africa along the southern coastal mountain ranges. Proteaceae occurs in region where rainfall varies from as low as 180 mm to as high as 2,500 mm per annum. The family Proteaceae includes over 1,400-1,500 species in over 60 genera of trees, shrubs and herbs from the southern hemisphere, especially Australia (30°0' S, 135°0' E) and South Africa (33°55' S, 18°22' E). Out of these species, over 800 in 45 genera are from Australia; Africa is home for about 400 species, including 330 in 14 genera from the Western Cape; about 90 species including *Embothrium* occur in CS America; 80 on the islands of New Guinea; and 45 in New Caledonia; though Madagascar, New Guinea, New Zealand (*Knightia, Persoonia, etc.*) and SE Asia host only small numbers of species (Rebelo, 1995). Today, the members of Proteaceae are cultivated for their flowers or colourful involucral bracts largely in South Africa and also in other countries like Australia, USA, Chile, Israel, New Zealand, Portugal, Spain and Zimbabwe.

The export of Proteaceae cut flowers provides an important source of foreign exchange and livelihood for these countries. Total area of Proteaceae under cultivation is now more than 6,000 ha, out of which 3,000 ha is in South Africa and 2,000 ha in Australia. The area under cultivation of Proteaceae in South America is approximately 8 ha, with 0.5 ha in Chile and 7.5 ha in El-Salvador. Spain and Portugal have approximately 30 ha of cultivated Proteaceae, located mainly on the islands of Madeira and Tenerife. The *Protea* cut flower industry started by using indigenous genetic material from both South Africa and Australia in 1940s (Parvin *et al.*, 2003). The products were initially harvested only from wild growing plants in their natural habitat, and then after when the markets were fascinated by the range and novelty of protea products, demand exceeded the supply, it was only then that its commercial cultivation started.

Growers shifted from collection in the wild to cultivation, from seed-growing where plants do not come true to the type to clonal propagation to perpetuate desired types, assessment of the product for use in the cut flower industry and to develop speciality crop novelties for pot culture, keeping in view the growing floral industry based on Proteaceous products. This also caused improvement in its packaging, in its shipment through sea routes and air, *vis-à-vis* development of new markets so that higher income is generated.

Banksia ericifolia, B. robur, B. spinulosa, Grevillea thelemanniana, Isopogon formosus, Leucadendron adscendens, L. argenteum, L. comosum, L. eucalyptifolium, Leucospermum catherinae, L. conocarpodendron, L. cordifolium, L. cuneiforme, Protea cynaroides, P. eximia, P. grandiceps, P. longiflora, P. obtusifolia, P. repens, etc. are quite suitable for use in **landscaping**. *Leucospermum catherinae, L. conocarpodendron, L. cordifolium, L. reflexum, Protea barbigera, P. compacta, P. cynaroides, P. eximia, P. lacticolor* hybrid, *P. pulchra, P. repens, Serruria florida, Telopea speciosissima, etc.* are suitable as **cut flowers**.

Typical Features of Proteaceae

Proteaceous plants tend to be sclerophylls, and they have leathery leaves which implies that these do not require a high atmospheric moisture leavel, *vis-à-vis* can tolerate moisture stress to a great extent. Their leaf buds are not protected by scale leaves which make them susceptible to frost-damage. Mild climates are quite congenial for their growth throughout if the soil remains sufficiently moist, and lack of water coupled with warm conditions causes these plants to go dormant for a while in summer. Many genera have proteoid roots (*Coprosma* or *Coriaria*) which enable them to draw nutrients from the soil even when nutrient level there is quite low (Thomas, 1979; Matthews and Carter, 1983). However, in nature, Proteaceous plants grow generally on infertile soils, and some even in sandy soil. Certain members of the family produce lignotubers, a short of swelling at trunk base which throws out new shoots, especially when main trunk is damaged. One more feature is quite prevalent in such plants that within the same species, very local forms have developed differing in flower colour, flowering time and growth habits. In *Grevillea* and *Protea*, bracts and floral parts both are attractive but in *Leucospermum* the bracts are small though styles and stigma are centre of attraction whereas in *Leucadendron* and *Telopea* the leaves surrounding the flowers are point of attraction (Salinger, 1985). Except for some *Grevillea*, the flowers grow mostly in clustered heads as in Asterceae.

Researchable Issues

Protea has become an established horticultural crop, with a world sale of approximately 8 million flowering stems. In South Africa 3.01 million stems are exported, 1.14 million sold through the formal market, and 1.01 million sold by the informal sector. Other producing countries do not have figures for sales of *Protea*, but total sales are estimated at 3.0 million. Less than 1.5 million stems are sold through the Dutch auction system annually. Proteas create great interest wherever they are seen and offer not only a new exotic appeal, but a good value as well. Protea industry has come a long way- from harvesting wild flowers in South Africa and Australia to cultivated plantations of superior cultivars in many different parts of the world. The continued international interest in Proteaceous cut flower crops provides the challenge for the plant breeders to supply new cultivars. This goal will be facilitated by a number of developments which require action over the next decade. One of the most important factors is the international co-operation in all aspects of cultivar development. In the last decade, research efforts enabling floriculture of Proteaceae have been concentrated on species selection, breeding, gene bank maintenance, propagation techniques, plant protection and post harvest techniques. The basics of plant nutrition and root growth aided the establishment of protea plantations. Knowledge of pests and diseases found in natural stands gave an idea of what to expect in cultivation, often enabling early control. To support the change from wild harvesting to cultivation, problems associated with the seed germination and plant establishment were studied. Research topics then began to explore ways to intensively cultivate proteas to achieve higher yields of acceptable quality at times of peak market demand. To enable greater control of plant growth and to provide uniformity, protea plantations were established using clonal material, rather than the seedlings. Propagation research provided protocols detailing propagation media, climatic conditions and treatments essential to successful root formation. The use of rootstocks was investigated to improve plant adaptability, but the higher costs associated with these techniques precluded widespread adoption of these results at this time. Plantations established using clonal material formed a monoculture in which pest and disease problems were enhanced. Studies on the susceptibility of species and cultivars to diseases allowed matching of cultivars to climatic areas to assist disease control. To offset the increased costs associated with the commercial production in plantations high productivity became essential. Despite the fact that proteas characteristically grow in nutrient-poor soils in the wild, the increased demands made on plants in cultivation depleted nutrient reserves. Research on nutrition management contributed to maintaining plant vigour. Investigations into flowering control aimed at matching supply of product to periods of high market demand. Global concerns about the impact of wild picking on conservation of natural flora put pressure on the industry to harvest fewer species from the wild. Consequently new products became economically viable to grow commercially and current management regimes, including propagation, were tested to validate their applicability. Cultivar registration of new products facilitated differentiation in the market between cultivated and wild-picked products, protecting market share for superior products. As prices increased research focused on aspects of quality to help industry attain a higher income per unit area. Post harvest problems such as leaf blackening were investigated. Cultivation of disease resistant or tolerant cultivars reduces production costs while providing better quality product. In addition to improved quality, higher income was attained by preferential market share captured by novelty products. Development of new products was supported by breeding and selection trials. As the production of cut flowers and plant material expanded, worldwide distribution costs came under close scrutiny. While growers increased packing efficiency, research sought reliable methods to use alternative to air transport. The high costs of air freight also forced attention onto reducing wastage in consignments. Improved fumigation techniques and processes have contributed to reduction in shipment rejection due to failure to comply with phytosanitary measures.

Crop Improvement

The main advantage of proteaceous blooms is their conspicuous and showy nature, which renders them ideally suited as standard blooms which form the focus or centerpiece of large floral arrangements. This prominence also means that the quality of the bloom is paramount and achieving this quality is one of the main challenges facing the plant breeders. The cut flower industry is subjected to changes in consumer preferences and plant breeding is by nature a long term undertaking. Sensitivity to and appreciation of potential future trends in the industry must be taken into account in the development of the selection criteria. Novelty is an important factor in success and the breeder must exploit available germplasm to maintain the flow of new and interesting cultivars. For most Proteaceous crops, the important selection criteria should be number of blooms per bush or per unit area; flower quality should qualify the international standard so there should be long straight stems with large, attractive and durable blooms; attractive leaf shape, texture and colour; resistance or tolerance to devastating root rot fungus *Phytophthora cinnamomi*; *etc.*

The Species

Protea is the most widely known genus of the Proteaceae, and other genera in this family that are widely used in floriculture are *Leucospermum* (Criley, 2001), *Banksia* (Sedgley, 1995) and *Leucadendron*. Other important genera, in addition to four mentioned above, are *Aulax*, *Dryandra*, *Grevillea*, *Hakea*, *Isopogon*, *Mimetes*, *Serruria* and *Telopea* which are being described here under. Collectively, all these genera would henceforth be called as Proteaceous due to the family Proteaceae or under the umbrella of *Protea* due to this being the most dominating genus.

Aulax

This is a South African genus of small- to medium-sized shrubs. This genus and *Leucadendron* are the only dioecious genera (separate male and female plants) of Proteaceae. Seeds of all three species, *viz. Aulax cancellata*, *A. pallasia* and *A. umbellata* are available but only *A. cancellata* is commonly planted. It grows to 1.5-2.0 m × 1 m and has fine needle-like leaves. In spring, female plants produce red-edged yellow flowers that develop into red seed cones. The catkin-like male flowers are yellow, as are those of *A. pallasia* and *A. umbellata*, the female flowers of which are not very showy. *A. pallasia* grows to about 3 m and *A. umbellata* some 1.5m high. All are hardy to about -5 °C and are usually raised from seeds.

Banksia

An evergreen genus with 60 species, having handsome foliage, is found throughout Australia and into New Guinea, but only a few species are suitable as cut flowers either due to by-passing the flowerheads or suitable stem lengths are not available. The genus *Banksia* is named so in honour of the famous English naturalist and patron of science Sir Joseph Banks (1743-1820) who sailed with Captain Cook on his first voyage to the Pacific. Banksias come true through seeds with only a little variation in growth habits or flower colours. Its species range widely from the ground cover (low prostrate) to medium and tall trees, and so different species prefer different climates and soil conditions but planting these in the garden will attract birds of different kinds. Their foliage is very attractive, ranging from fine needle-like to wide serrated leathery, and the colourful blooms or brushes range in size. The flowering season is primarily from late winter to late spring and most species have cylindrical cone-like flowerheads composed of densely packed filamentous styles radiating from a

Banksia

Dryandra

Hackea

Leucadendron

Aulax

Grevillea

Isopogon

Protea

Leucospermum

Mimetes

Serruria

Telopea

central core. Creamy-yellow to light golden-yellow is the predominant colour range, although a few species, such as *B. ericifolia* (heath banksia) and *B. praemorsa* have golden-orange flowers and those of *B. coccinea* are red. Flowers of *Banksia* are suitable for drying of their flowerheads. The large leaves of *B. grandis* or *B. speciosa* are used by florists and the mature fruiting heads which are known as cobs or cones are used as dry flowers. Most species have narrow serrated leaves that are mid to deep green above and silvery-grey below but *B. ericifolia* has fine needle-like dark leaves. *B. ericifolia* is a dense shrub growing about 3.0 m, bearing long russet-coloured flowers and is most suitable for planting in the coastal gardens. In *Banksia*, leaf size varies from very small up to 50 cm long as in *B. grandis*. Hardiness varies with the species, some are quite frost-tender but some others can bear even -10 °C. From New South Wales, there is a species *B. serrata* which was discovered by Banks and Solander in 1770 from Botany Bay, as well as from Victoria and Tasmania. It grows up to 6.0 m high with tomentose young branches, leaves are 7.5-15.0 cm long, oblong-lanceolate, coriaceous and with deeply cut regular serration. Exterior of the broader tips of the perianth lobes are often hairy and bluish-grey but with opening of the flowers further when style predominates, it shows yellow colouration. *B. baueri* grows to about 2.0 m tall and is a dense-growing shrub. The flowers are large furry, yellowish inside and grey outside and appear during winter and spring. *B. candolleana* is a low shrub with yellow flowers. The flowers of *B. coccinea* are striped-red. *B. collina* (syn. *B. spinulosa*) is an open shrub growing up to 3.0 m tall, bearing narrow and notched foliage and honey-coloured brushes with black pins, and it grows well under filtered light positions. *B. grandis* grows up to 7.0 m high with yellowish flowers. *B. littoralis* from swampy areas of Western Australia is a tree some 18 m tall. Its inflorescence is charming yellow. *B. integrifolia*, a hardy species grows 3.0-4.0 m or sometimes even up to 12.0 m high, bearing 15 cm long and 2.5-4.0 cm wide, entire or occasionally slightly dentate leaves whose upper side is dark green but lower silvery-white, and inflorescence 7.5-15.0 cm long are produced during autumn, winter and spring, with about 2.5 cm long and greenish-yellow perianths. *B. menziesii* is a beautiful small tree growing up to 8.0 m tall and bears red flowers. *B. occidentalis* grows up to 3.0 m tall with red flowers. *B. paludosa* grows only up to 3.0 m tall and bears yellow flowers. *B. prionotes* is an attractive tree growing up to 8.0 m tall with long deeply toothed foliage, and the large orange flowers are produced during autumn and winter. *B. marginata* reaches to a height of up to 5.0 m, bearing narrow foliage which is white underside, and the small yellow flowers are produced during summer, autumn and winter. *B. robur* which grows up to 3.0 m tall, bears large foliage and the flowers appear as if these have developed along the limbs and up the trunk, initially appearing as bottle-green but turning yellow, bronze and brown as they mature. *B. speciosa* grows up to 4.0 m tall and bears greenish flowers.

Dryandra

An evergreen Australian genus of around 60 species of shrubs ranging in height from about 1.0-4.0 m in height and are found only in the south-western regions of Western Australia. The genus commemorates Jonas Carlsson Dryander (1748-1810), a Swedish botanist and one time Librarian to Sir Joseph Banks. The leaves are narrow, mid to deep green, that are often very long and narrow with sharply toothed edges. Its flowers are similar to those of the related genus *Banksia* though differs in having less woody seed capsules and more dense heads of flowers surrounded by basal bracts. The rounded flowerheads, which appear from mid-winter, are usually light to bright yellow. The flowerheads of *Dryandra* are suitable for drying. *D. drummondii* which is similar to *D. calophylla*, both are found wild in the Swan river area of SW Australia, bearing pale-yellow flowers inside while ouside has covering of gingery perianth hairs. In the flowerhead, outer flowers open first, the gingery perianth lobes curl back to expose the yellow style and stigma and then the process proceeds to the inner side of the head up to the centre. Its young leaves are almost covered with hairs that are lost with maturity. It differs from *D. calopylla* as it does not have lateral subterranean shoots and small flowering heads. *D. formosa*, which grows to about 3.0 m and is hardy to around -5 °C once established, though most other species are less hardy. Dryandras are superb long-lasting cut flowers and some even dry well. They will grow on extremely poor soil and generally react badly to most fertilisers.

Grevillea

With some 250 species, this is the largest of the Australian Proteaceous genera, with a few found in New Guinea, New Hebrides and New Caledonia. Most of the common garden species and cultivars are ground covers to medium-sized shrubs (up to 3.0 m) with needle-like foliage. However, some species are far larger and attain the height of a complete tree. Its flowers resemble *Hakea* though differs in fruits. *G. alpina* (syn. *G. alpestris*) is found in Tasmania, Victoria and New South Wales and was first grown in 1857 in the Australian House in Kew. Its flowers have combination of yellow and red or pink and red. *G. australis*, a frost-hardy species grows prostrate and bears fragrant white flowers. *G. anethifolia* grows up to 2.0 m, is hardy and bears fine-pointed foliage, and creamy and greenish flowers. *G. banksii* has pinnate leaves similar to *G. robusta* but grows only to about 3.5 × 3.0 m in height. *G. baueri* is a prickly plant which grows up to 2.0 m tall, doing well in damp soils and bears deep red spider flowers twice in the year. *G. gaudichaudi* bears lobed foliage and burgundy toothbrush flowers during spring or summer. *G. hookerana* is a frost-resistant, grows up to 3.0 m high and bears red toothbrush flowers in spring. Some of the variants of *G. juniperina* are nearly prostrate to semi-prostrate, ideal for embankments though others may be spreading tall shrubs bearing flowers in the colour range of cream, yellow, apricot or red. *G. juniperina* 'Molonglo' bears spreading arching branches with numerous apricot-coloured flowers. *G. juniperina* f. *sulphurea* (syn. *G. sulphurea*) is a rounded bushy shrub with almost needle-like leaves, above dark green and recurved though silky-haired below. It produces clusters of small spider-like pale-yellow flowers during spring-summer. *G. lanigera* produces a massed display of cream and pinkish-red flowers. *G. laurifolia* is frost-hardy and spreads above 4.0 m across. *G. punicea* (red spider flower) was first introduced into western Europe around 1857 from its native place of New South Wales. The species bears both large- and small-leaved plants and apricot or red flowers. *G. repens* (rock grevillea) grows prostrate and does well in demi-shaded conditions but in well-drained soil. Its leaves are holly-like and the flowers toothbrush-shaped red and green. *G. robusta* (silky oak), which is often seen in mild areas, can grow to 20 m and is common with most of the larger species. It has large pinnate leaves. The more densely-foliaged plants, especially *G. juniperina* and *G. rosmarinifolia* are often used as hedging plants. The plants of *G. rosmarinifolia* are frost-tolerant, grow 1.5-3.0 m high and bear green needle-like leaves and small red and cream flowers appearing from spring. For brightness and profuse flowering the forms of *G. speciosa* are highly suitable, and its ssp. *dimorpha* bears brilliant red flowers during winter. *G. sericea* (pink spider flower) grows about 2.0 m tall, is hardy and bears narrow pointed foliage and pink flowers most of the year. *G. thelemanniana* (*G. preissii*) is a spreading shrub having soft-tomentose young growths, and the leaves are pale or glaucous, 2.5-5.0 cm long, pinnatifid, lower pinnae divided with linear segments. Racemes are up to 4.0 cm long, dense and terminal, and bear pink flowers with green tips. *G. triloba* is hardy, grows up to 2.5 m high and bears grey spiky foliage and fragrant white flowers. There are many attractive *Grevillea* hybrids introduced over the years, especially in Australia, and some of these are 'Robin Gordon' with large reddish heads at the ends of the branches throughout the year, which grows to 2.0 m with spread of 3.0-4.0 m and responds very well to pruning; 'Misty Pink' grows up to 3.0 m, bearing heavily divided silvery foliage and large pink spidery flowers for most of the year; 'Pink Surprise' is much taller and its soft and pink bushes are produced mainly during autumn and summer; 'Poorinda Tranquility' is 1.0 m tall with serrated foliage and mauve flowers in spring; *etc. Grevillea* flowers are often described as 'spider flowers'. This refers to the styles of some species, which tend to radiate from the centre like a spider's legs. Some species have 'toothbrush' flowers; the styles are all on one side like the bristles of a toothbrush. The best known example of this type of flower is the common red-flowered cultivar 'Robin Hood'.

Hakea

Hakea was named so to honour Baron von Hake, a German botanist. This evergreen Australian genus includes about 130 species, few of which are widely cultivated even indoors. They are drought-resistant. The most common is probably *H. laurina* (*H. eucalyptoides*), the 'pincushion or sea urchin hakea', growing up to 9.0 m high with reddish-brown bark, bearing up to 15.0 cm long and some 2.5 cm wide, bluish-green, elliptic or lanceolate and sickle-shaped leaves tapering to a petiole, and producing strikingly handsome flowers of crimson colour in a globular involucre heads, from which numerous golden-yellow styles protrude in every direction. When not in flower, this species could easily be mistaken for a small eucalyptus. Its mature trees have a slightly weeping habit. The flowers appear in late autumn and early winter, opening cream and turning to orange and red as they age. This shrub is hardy to about -5 °C once well established and is easily grown in most well-drained soils. *H. elliptica* is a shrub which bears 5.0-9.0 cm long, spiny, oval or elliptic with undulated margins, almost sessile and attractive bronze foliage when young, and its flowers appear in white globose cluster. *H. saligna* is a glabrous and pale shrub (younger shoots hairy), growing up to 2.5 m high. Its leaves are 7.5-15.0 cm long, oblong or lanceolate, tapering to a short petiole and pinnately veined. Numerous small white and dense flowers are borne in clusters with recurved corolla. *H. ulicina* bears erect branches, with 2.5-20.0 cm long and hardly 3 mm wide, dense, narrowly linear, entire and acute foliage in appearance to ulex, and the flowers are very small. *H. suaveolens* (*H. pectinata*) is a 4.5 m high rounded shrub bearing 5.0-10.0 cm long, 1-5 irregularly terete-lobed leaves with spiny tips, and the flowers are fragrant white. Of the other species, the most common ones are *H. salicifolia*, *H. prostrata* and *H. sericea* (syn. *H. lissosperma*). They are hardy to about -8 °C or slightly lower and are easily grown in most soils. *Hakea salicifolia* has narrow, willow-like leaves; and spidery, white flowers that are produced in spring. It grows up to 5.0 m high and will tolerate poor drainage. *H. prostrata* and *H. sericea* have fine needle-like leaves and white or pale-pink flowers in winter and early spring. It grows to about 3 × 2 m high. All the members of this genus are usually raised from seeds, some can be grown from even cuttings, and a few, such as *H. franciscana*, are weak growers that often perform better when grafted onto more vigorous stocks, such as *H. salicifolia*.

Isopogon

The genus is commonly referred to as 'drumsticks' due to shape of the floral stems and unopened buds, though the name is often used for *Isopogon anemonifolius*. It is an Australian genus having 34 species of small to medium sized shrubs, most of which grow 1.0 to 2.0 m high and wide. It has a preference for poor but well-drained soil and collapses quickly if over-watered or overfed. Most species have about 7.5 cm long and narrow lanceolate leaves, though some such as common *I. anemonifolius* which is hardy, grows up to 2.0 m high, and bears finely cut foliage reminiscent of marguerite daisy or anemone leaves. The floral cones open in spring mainly to white, yellow or pink, predominantly yellow. The two most widely grown species, *I. anemonifolius* and *I. anethifolius* are hardy and can tolerate up to -5 °C, but many other species, such as *I. cuneatus* and the temptingly beautiful pink and yellow-flowered *I. latifolius* are damaged at temperatures below -2 °C. *Isopogon* species are usually raised from seeds. *I. dawsonii* (conebush) grows only up to 1.0 m high, does well in moist soils and bears yellow-cone-like flowers in spring.

Leucadendron

In Greek, *Leucadendron* means white tree. *Leucadendron* and *Aulax* are the only dioecious genera (separate male and female plants) of Proteaceae. Species of this genus are most widely grown of the South African Proteaceae and many are valued for their long-lasting qualities of flower bracts once cut. Most are medium-sized shrubs around 1.0-2.5 m high. However, one of the best known species, the silver tree (*Leucadendron argenteum*), can grow up to 10.0 m high though is less widely grown. Many species and cultivars are grown, but probably the most widely planted is 'Safari Sunset'. It is a hybrid between *L. laureolum* and *L. salignum* and is fairly typical of the genus. It has narrow, lanceolate leaves some 10.0 cm long. Some species, such as *L. argenteum* have tomentose foliage but 'Safari Sunset' does not. The upward-facing foliage densely covers the narrow, upright branches and develops deep red tints at the flowering tips. As the insignificant flowers near maturity, the bracts become intensely coloured. 'Safari Sunset' has red bracts but others appear in cream, yellow, pink or orange tones. The species and hybrids vary considerably in hardiness but most will tolerate frosts of at least -3 °C provided they have good drainage with proper humidity as it does not like excess humidity. 'Safari Sunset' is hardy to about -8 °C as to those of *L. salignum* and *L. laureolum*. North Island leucadendrons, generally thrive in all but the coldest central areas and they can be grown with varying degrees of success in all coastal areas of the South Island. Both, *Leucadendron* and *Protea* have similar requirements for their cultivation but nutritional requirement in case of *Leucadendron* is quite limited.

Leucospermum

A South African genus of about 50 species, most of which are medium to large shrubs that grow to about 1.5-3.0 m high. *Leucospermum* has two classes based on the type of flowerhead. The first group includes those species which have single flowerheads such as *L. cordifolium*, *L. lineare* and *L. tottum*. The second group has multiple flowersheads as seen in *L. cordifolium*, *L. muiri* and *L. mundii*. Some, such as *L. reflexum* have strongly upright growth habits but most including the commonly cultivated species *L. cordifolium*, are dense and bushy. Both of these species have tomentose greyish-green leaves that are usually broadly oval-shaped, often with small red-tipped lobes. The flowers are variously described as 'Catherine wheels', 'pincushions' and 'sky rockets', all of which refer to the numerous radiating styles. These are often incurved, creating a cupped effect. The flowers usually appear in late spring and continue for about two months. They are attractive when fresh, but often become unsightly once they die off. Most garden leucospermums are cultivars of *L. cordifolium* and are hardy to occasional frosts of about -5 °C, but they resent wet or humid winter conditions, which can often lead to tip die back. *L. reflexum* is an erect shrub with ascending branches. Leaves are dense, small and grey- to blue-green. Tubular, slender and crimson flowers with long styles appear in tight rounded heads during spring and summer.

Mimetes

This South African genus includes 16 species with faintly coloured leaf-like bracts, and all the species are more or less similar to *M. hottentoica*. *Mimetes* is easily recognized by the compact clusters of often brightly coloured flowers embedded on a sessile flowerhead around which the subtending bracts are also occasionally coloured. This is a bizarre genus bearing softly hairy silver leaves, withered and reflexed perianth segments of the yellow hue, and red and yellow styles that contrast with almost black stigma. Anthers come out from the spoon-shaped tips of the perianth lobes. *M. stokoei* bears tinged-red bright yellow styles, almost black stigmas, pinkish bracts subtending the flower clusters and satiny leaves. *M. cucullatus* is widely grown species, growing to about 1.5 × 1.5 m and is hardy to around -3 °C, having 4.0 cm long oblong leaves with small lobes at the tips, that densely cover the branches like upward facing scales, and the plant prefers moist, well-drained soil and is not very drought resistant. The small white flowers are enclosed within leaf bracts that change colour to a bright red as the flower buds mature. This species is usually raised from seeds. *Mimetes* may flower throughout the year but is usually at its best in late-spring when the new growth appears, as this is also red.

Protea

Protea was named so by Linnaeus in 1753, referring to the Greek mythical god, Proteus, who could change his shape at will, and this is an apt name due to the wide diversity of the genus. It is a large evergreen genus with 136 species of which 70 are distributed in the temperate zones of the southern hemisphere and the balance distributed in the sub-tropical to tropical zones of southern hemisphere. The natural habitat of *Protea* ranges at elevations from sea level to over 2,000 m and inhabits the well-drained, moderately acid and low fertility areas but not the humid ones. Once planted, they are low-maintenance plants. *Protea* are suitable for drying of their flowerheads. The most widely recognized species in the genus is *Protea cynaroides* (meaning 'like a globe artichoke'), the 'king protea', the national flower of South Africa. It has flowerheads up to 30.0 cm across with widely spaced bracts arranged around a peak of flowers that vary in colour from near white to soft silvery-pink to deep rose-pink to crimson. *Protea magnifica*, the 'queen protea' is grown for its large 15-20 cm flowerheads of white to rose-pink to salmon colours. *Protea compacta* has lanky flower stems on a stiffly upright, sparsely branched shrub that grows to 3.5 m tall. The prominent flowerheads, unobscured by foliage, make fine winter cut flowers. *P. obtusifolia* is found growing in the white sandy soils in the coastal region of southern Cape Province with one of its white bracted variety occurring in the Albertinia district of South Africa. It bears smooth, shiny and brittle chaffy-textured floral bracts. *P. amplexicaulis* from southern Cape Province being found at altitudes between 300-610 m, is named so as the leaves clasp the stem from which they arise. Its remarkably coloured red-brown inflorescences are placed near the ground on one of the aerial branches, bracts are velvety exterior and paler interior. *P. grandiceps* is recorded from Table Mountain, Devil's Peak (near Cape division), Langeberg (near Swellendam), the Jonkershoek Mountains (near Stellenbosch) and the Cockscomb Mountains (near Port Elizabeth). It is a bush growing up to 1.5 m high. Its leaves only near the inflorescence are edged red though leaves below are entirely green. It is slow-growing and flowers only when very old. Its red bracts never open fully but even look very beautiful because of the pure white, grey or brownish beards on the top of the terminal inflorescence bracts. Other important species include *P. eximia, P. lacticolor, P. longiflora, P. neriifolia, P. repens* and *P. susannae*. Proteas prefer acidic soil having a pH range of 5.0-5.5, require plenty of nutrition for best growth and flowering but at planting no fertilizer should be used but soil should have sufficient humus, and afterwards only compost gives better results and not the animal manures. Its requirement for potash is very limited. After planting in spring or autumn, these do not require frequent disturbance so only little cultivation should be done only when necessary. Since its plants grow haphazard, so when needed these may be pruned to keep in shape. Apart from cut flowers when these last for a few weeks, its flowers can also be used after drying. Seeds can be sown 4-6 months after the flowering is over and the flowersheads have been allowed to mature on the plants. Seeds are sown deeper to their width and soils are kept moist until germination, and then seedlings should be grown in the 15-cm pots before planting these. Through seeds these take 3-4 years to bloom.

Serruria

Serruria florida (blushing bride) is very popular with florists because its nigella-like papery white bracts are very delicate and last well as cut flowers. The bracts, which are surrounded with finely cut lacy leaves, are produced freely in winter and spring. Blushing bride can be difficult to grow, because not only it is frost-tender (it tolerates only occasional exposure to -2 °C), it must also have full sun and absolutely perfect drainage. *Serruria* is a genus of 44 species from South Africa, of which the only other species commonly grown is *S. rosea*. It is a densely foliaged 70 × 90 cm bush with small pink bracts and is slightly hardier and definitely easier to grow than *S. florida*. *Serruria* species should be raised from seeds.

Telopea

Commonly known as waratah, native to Australia, this evergreen genus includes just four species. The New South Wales waratah (*Telopea speciosissima*) is most commonly grown species, having oblong, finely serrated leaves that are up to 12.5 cm long with small notches or lobes at the tips. It develops into a large shrub or small tree up to 5 × 5 m. The flowers, which are produced in spring and carried at the tips of the branches, are impressively large, bright red, and composed of numerous incurving styles surrounded by red foliage bracts. Several cultivars, such as the semi-dwarf 'Forest Fire' (2 × 2 m) are commonly available. The 'Victorian Waratah' (*Telopea oreades*) is a similar plant with slightly lighter coloured leaves and flowers. Both of these species and the cultivars are hardy to around -8 °C. *T. truncata* (Tasmanian waratah) is a upright-growing shrub which becomes bushy with age. The leaves are densely formed and deep green. The flowerheads are rounded bearing small, tubular and crimson flowers in late spring and summer.

Propagation

Seed is still widely used for the establishment of varieties not represented by cultivars and increasingly for the establishment of foliage plantations. The decline in research reports on seed germination techniques for commercial Proteaceous varieties probably reflects the increased availability of cultivar material and/or that agriculturally satisfactory seed dormancy breaking techniques had been established. Dormancy seems to be imposed by a low temperature requirement and by the action of the pericarp, which prevents simultaneous germination of all achenes. Scarification, stratification, and incubation in pure oxygen improved the germination of *P. compacta.* Imbibition of *P. eximia* and *P. nerifolia* seeds to 100 ppm GA_3 for 24 h improved significantly the percentage and rate of germination (Rodríguez Pérez, 1995). *Leucadendron* species germinated optimally when temperature fluctuations between 20 pC (daytime) and 10pC (night) were used (Sedgley *et al.,* 2001). Many species only required a low temperature to germinate optimally (11 pC). *Leucadendron* can be divided into two sections based primarily upon characteristics of the seed. Section A 'Leucadendron' has rounded nut-like flattened seeds, while section B has seeds that are flattened and have small wing. Flattened seeds from section B are reported to be easier and faster to germinate than the round nut-like seeds of section A, requiring no pretreatment prior to being sown in a 1:1 sand-perlite mix.

Most commercial Proteaceous species are propagated by using approximately 20 cm long terminal **semi-hardwood cuttings**. In general, a 5 second basal dip in 1,000 to 4,000 ppm IAA is followed by setting the cuttings in well aerated medium with intermittent mist and bottom heat at 22 to 25 °C (Malan, 1995). Rooting generally occurs within 6-16 weeks. Auxin concentration, auxin carrier, and hormone mixtures all influence rooting success. Specific requirements have to be adapted for each cultivar for optimum results. The time of taking the cuttings is important in *Protea,* where growth flushes are not always well synchronized (Malan 1995), because the physiological status of the new growth flushes may not be consistent. Scarring of the base of the cutting is effective in promoting rooting of some *Protea* cultivars. The propagation of *Protea obtusifolia* by stem cuttings following the standard technique, can be improved upon, when prior to hormone treatment, four longitudinal cuts, 2cm long, equally spaced are made in the bark of the cutting-base. Control of diseases while plants are rooting is important to ensure success and includes proper sanitation in the mother plants. Rodriquez *et al.* (2001) recommended the terminal cuttings for the vegetative propagation of *Leucospermum* spp., although some commercial nurseries also use basal cuttings. In some plant species, the use of basal wounding technique alone or combined with hormonal treatments (IBA) has stimulated root formation in stem cuttings. Use of this technique has improved the rooting in *Protea obtusifolia*, *Leucadendron* 'Safari Sunset' propagated in spring when rooting is difficult and in *Leucospermum* 'Sunrise'. Wounding followed by treatment with 4,000 ppm of IBA is quite effective for propagation of *Leucospermum cordifolium* 'California Sunshine' by terminal stem cuttings in order to shorten rooting process, although unwounded cuttings treated with 2,000 ppm of IBA were not successful.

Tissue culture techniques for propagation of *Protea* (Rugge, 1995) have been developed. The major factors that limit the tissue culture of the Proteaceae apparently relates to obtaining sterile explants from the field grown plants, phenolic browning of the medium and explants as well as difficulties in getting axillary buds on explants to sprout. However, shoot proliferation has been obtained in *P. cynaroides, P. obtusifolia* and *P. repens*. Successful transplanting of rooted shoots to soil has not been achieved. Rugge (1995) studied the micropropagation of *Protea repens* cv. 'Embers' and concluded that treatment of the actively growing axillary shoots on field grown mother plants with 200 mg/l BA significantly reduced browning and promoted bud sprouting *in vitro*. Through the use of GA_3 *in vitro* on the sprouting of multinodal explants of *Protea repens* cv. 'Embers', Rugge (1995) reported maximum axillary bud sprouting with 3-6 mg/l though increase in the GA_3 concentration to 9 mg/l did not result in the number of sprouted shoots. Croxford *et al.* (2006) obtained multiplication of *Leucadendron* hybrids on MS medium containing 20 g/l sucrose and 3 g/l Phytogel. They also studied the effect of rooting substrate and environment on the root strike and survival of *in vitro* grown shoots of *Leucadendron* genotype 007.

Cultural Practices

Small-leaved Proteaceous plants such as *Grevillea* are not so specific in their requirements though larger-leaved plants such as *Leucospermum* are very specific. Normally, the Proteaceous plants require acidic medium with a **pH** of 5.0 to 5.5 (Eliovson, 1965) for growing and flowering. The plants in case of most of the genera and species are raised in pots first and then planted at their permanent positions when these are 30-60 cm long. **Soils** for their planting should be free-draining, especially during winter months when evaporation and transpiration are very low. In water-logged soil, these are infected with *Phytophthora cinnamomi*, the root rot fungus, collapsing entire plant. However, some genera such as *Hakea* and some *Banksia* species feel comfortable in clay soils. **Planting** is carried out from late summer to early autumn or late spring to early summer in the sub-tropical and sub-temperate areas as during both the seasons plants experience warmer days when roots are encouraged to appear, however, planting in mid-summer is avoided as high temperatures irrespective of the soil moisture, cause stress to the plants thereby resulting into leaf scorching. Planting in Proteaceae is effected closely, and var. 'Safari Sunset' is planted in double-row system in a triangular placing at one metre distance from plant to plant, and so is the case with *Leucadendron xanthoconus* with one metre distance though the plants grow to a two metre high. Close spacing encourages upright growth.

However, large plants such as *Banksia* and those being planted in high humidity areas especially hairy-leaf types which have tendency to hold moisture should be given more spacing. Planting distances vary as per genera and species as some are prostrate, certain grow only 30-60 cm tall, many others are shrubs growing 1-7 metres tall, some being small trees while others are tall trees. Those growing below one metre in height or are prostrate-growing, are planted very close, *i.e.* 30-45 cm distance, while shrubs growing taller than one metre and for trees, distances range from one metre to five metres. The planting may be carried out in lines or in the pots. Planting in pots may be done at any time and during inclement weathers may be brought inside the greenhouse while for liners which are planted outside, should be planted only during proper season. The liners are planted in deep furrows drawn by tractors. Planting should be done at a place (coastal areas; sandy or sandy-loam, heavy or light, moist or dry soils; at temperate, sub-tropical or tropical conditions; and in open sun, semi-shaded or shaded locations) as per requirement of a particular species to be planted. After planting the plants are looked after properly till these are properly established. Through seeds it may take many years to bloom though cutting-raised and even the tissue-cultured plants may start blooming the second year.

In Proteaceous plantings the **weeds** require to be controlled properly and timely. Spreading of black polythene of 900 mm width as mulch will control almost all the weeds and the film will last at least for two years. Alternatively, organic wastes, bark splitting, dry grasses and straw, dry leaves, husks of rice and groundnut, sawdust, *etc.* can effectively be used as mulches. Glyphosate herbicide is also effective in controlling weeds if the herbicide does not touch the leaves.

Flowers of the Proteaceae constitute a considerable proportion of the market, both locally and overseas. Proteas grow normally on leached and acidic soils which are poor in available minerals. Soil texture plays an important role in protea development. Surprisingly, the **nutritional** requirements of Australian and South African natives used in cut flower production are very poorly understood, particularly with respect to phosphorus. For example, advice from growers ranges from advocating the use of superphosphate to the use of no-phosphorus fertiliser. Because of the lack of information on nutritional requirements, fertiliser management is not scientifically based, for example, red/purplish coloured new leaves are commonly cited by growers as a sign of good health. In fact it is well known that such symptoms are a sign of phosphorus deficiency. Observations of plantations and descriptions from growers indicate that many plants in commercial plantations are growing sub-optimally due to either nutrient deficiency or toxicity. It is clear that growers need guidelines, based on hard facts, for nutrition management of their crops. Many Proteaceous plants grow poorly and develop necrotic and chlorotic leaf symptoms when given fertiliser at rates considered normal for other cultivated plants. This response has been attributed to phosphorus toxicity. The growth of Proteaceous plants in their native habitat gives a good indication of their cultural requirements. Nutritional status of *Protea, Leucadendron* and *Leucospermum* using soil test and plant analysis (Anon., 2001) is presented in Table 1.

Table 1: Soil and Leaf Analysis

Analyte	*Unit*	*Range*
	Soil	
pH	pH	5.0–5.5
Olsen phosphorus	mg/l	5.0–20
Potassium	me/100	0.40–0.80
Calcium	me/100	4.0–10
Magnesium	me/100	0.70–3.0
Sodium	me/100	0.0–0.50
CEC	me/100	12–25
	Leaf	
Nitrogen	per cent	0.75–1.5
Phosphorus	per cent	0.050–0.14
Potassium	per cent	0.35–0.60
Calcium	per cent	0.50–1.0
Magnesium	per cent	0.10–0.25
Sulphur	per cent	0.10–0.25
Sodium	per cent	0.0–0.40
Iron	mg/kg	20–70
Manganese	mg/kg	50–400
Boron	mg/kg	8.0–25
Zinc	mg/kg	10–50
Copper	mg/kg	3.0–8.0

Proteaceous species have both a lower requirement, *vis-a-vis* lower tolerance to fertilizer compared with plant species from other families. However, all the species grow better with applied fertilizer than without fertiliser, and large differences are found in responsiveness and sensitivity even within the same genus. It is also clear that these plants cannot be grown effectively without use of fertilisers. The lower requirement for fertiliser is based on the ability of these Proteaceous species to utilise nutrients efficiently. These features of the group are important for survival on the low fertility soils in which they have evolved. Specific knowledge on optimal nutrition will

Table 2: The Range of Fertiliser Rates and Mature Leaf Macronutrient Contents Associated with Optimum Growth

Species	NPK (kg/m³)	N (per cent)	P (per cent)	K (per cent)	Ca (per cent)	Mg (per cent)
Banksia hookeriana	1.75–10.3	2.60–3.55	0.09–0.32	0.80–1.20	0.58–1.00	0.31–0.32
Protea cv. Masquerade	1.50–6.75	0.90–2.30	0.06–0.17	0.75–1.40	0.50–0.70	0.13–0.30
Protea cv. Clarks Red	1.75–4.25	1.24–1.93	0.09–0.14	1.23–1.28	0.40–0.60	0.10–0.13
Leucodendron cv. Sundance	0.50–4.00	1.00–2.20	0.05–0.13	0.60–1.00	0.70–0.75	0.20–0.22

Table 3: The Range of Fertiliser Rates and Mature Leaf Micronutrient Contents Associated with Optimum Growth

Species	Al (ppm)	Cl (per cent)	Cu (ppm)	Fe (ppm)	Mn (ppm)
Banksia hookeriana	62–71	0.52–0.80	7 – 8	68–111	345–500
Protea cv. Masquerade	30–40	0.35–0.37	<10–<10	20–25	170–280
Protea cv. Clarks Red	15–20	–	<10–<10	20–22	180–280
Leucodendron cv. Sundance	30–30	0.55–0.55	<10–<10	30–35	350–380

aid the consistent production of quality plants and cut flowers (Tables 2 and 3).

Parvin *et al.* (1973) stated that desired nutrient levels in *Protea neriifolia* leaves should be 0.5-1.0 per cent Ca, 0.03-0.06 per cent phosphorus, 0.3-0.7 per cent potassium and 0.1-0.3 per cent magnesium. Compost is preferred over animal manure and that too in autumn or spring when plants have started taking new growth. However, at the time of planting also the soil should be mixed with well rotten compost or any other manure. If the soil is properly mixed with compost, further there is no any need of fertilizer application, and even otherwise fertilizers should be discouraged, especially the phosphatic types. In the first year of planting, the plants require utmost care in **watering**. From mid-June to September, no watering is required because of frequent rainfall during the period, from November to February there is only nominal evaporation when also watering is generally not required, in October and from March to mid-June the weather is dry so during this period light weekly irrigation in the first year should be given so that plants may not die due to desiccation. After establishment, there is, generally no requirement of watering. Trickle irrigation is most economical and convenient.

Most members of Proteaceae produce flowers on new year's growth so after flowering, if seed is not required, these should be **pruned**. Pruning should be carried out to bring the plants into shape and also for better flowering the next season so the stems on which the flowers have appeared should be cut back to allow new fowering canes to appear.

Post Harvest

Flowering stems are harvested at any stage between soft-bud, or anthesis of the outer ring of florets. The stems are best placed immediately in water, with cooling to 2 to 5 °C within 60 minutes after harvest. Thereafter the cool chain should be maintained until the stems are sold to the florist or consumer. In exporting countries, the cold chain is of necessity though broken during air transport. The stem length categories for export standards from South Africa starts at a minimum of 40 cm, with an increase in length of 10 cm for the next category. Maximum allowable blemishes, either physical or due to disease, on the involucral bracts and leaves are also defined, but each importing country sets its own phytosanitary restrictions. Malan (1995) specified the harvesting stages for different genera of Proteaceae. *Leucospermum* flowers can be harvested at 33-50 per cent styles reflexed stage, resulting in effective style opening and optimum travel ability and shelf life. *Protea* is harvested at soft-tip stage, except for *P. aurea*, *P. lacticolor* and *P. mundii* which are harvested at the emergence of the styles. Single stem of *Leucadendron* is harvested only during the period when it satisfies the consumer. Poorly coloured *Leucadendron* 'Safari Sunset', for instance, will harm the product's market position. *Leucadendron* foliage is harvested at any time subsequent to leaf maturity. *Serruria* flower is harvested when anthesis of second inflorescence occurs.

Stephens *et al.* (2003) pulsed the *Leucospermum* stems in either 0 or 2 per cent glucose solution for 20 hours at 18±1°C before being transferred to individual vases with tap water (Table 4). Vase life was assessed in a temperature controlled room subjected to natural light, and the number of leaves with at least 10 per cent blackened leaf area was determined daily.

From the Table 4 it can be seen that there was a significant increase in vase life of 'Cordi', 'Gold Dust', 'High Gold' and 'Succession' stored at 1°C for 3 days, followed by 2 per cent glucose pulse. No significant improvement in 'Scarlet Ribbon' or 'Tango' was found with 2 per cent glucose pulsing. Further, Stephens *et al.* (2003) reported that the glucose pulsing solutions significantly extended the vase life of 'Brenda', 'Carnival', 'Pink Ice', 'Susara' and 'Sylvia' cultivars of *Protea*.

Table 4: Effect of Glucose Pulsing Solution on Vase Life of *Leucospermum* Cultivars

Cultivar	*Control (water)*	*2 per cent Glucose Pulse*
	Vase Life (Days)	
Cordi	12	17
Gold Dust	15	24
High Gold	12	14
Scarlet Ribbon	16	16
Succession	10	15
Tango	20	18

Table 5: Effect of Glucose Pulsing Solution on Vaselife of Cut Flowers of Protea Cultivars

Cultivar	*Glucose Concentration (per cent)*						
	0	*1*	*2*	*3*	*4*	*5*	*10*
	Vase Life in Days						
Brenda	5	6	5	7	8	9	1
Cardinal	4	4	4	3	4	3	1
Carnival	6	6	7	7	8	10	1
King Protea	12	12	13	12	14	14	1
Pink Ice	5	6	8	10	1	1	1
Susara	6	6	7	8	9	12	14
Sylvia	5	5	6	8	9	14	12

From Table 5, it can be concluded that significant differences exist in the response of various *Protea* cultivars to glucose supplementation. Five of the seven cultivars responded positively to glucose supplementation. Increasing glucose concentration was associated with a significant improvement in the vase life of *Protea* cultivars. Phytotoxicity was observed in all proteas pulsed with 10 per cent glucose solution.

Gibberellic acid spray is usually known to cause elongation of the stem internodes. There are rare cases where GA spray enlarges the flower size. Sahebat and Zieslin (1995) reported an increase in weight of the detached rose petals or petal discs when imbibed in GA_3 containing sugar. Four weekly sprays of GA_3 at 1,000 ppm when the flowering buds of *Protea* cv. 'Pink Ice' were 4.0 cm long, caused an elongation of the involucres of bracts and thus increased the inflorescence size of this cultivar (Ben-Jaacov, 2006). The treated inflorescences were not only larger but had better appearance, *vis-a-vis* longer shelf life.

Large flowers such as *Hakea* & king and queen *Protea* are packed individually with rolled paper or cellophane though in many of the other cases where flowers are smaller are packed in fives in dry conditions in cartons lined with clean paper, however, for long storage the lining of the cartons should be done with polythene films to limit water loss. *Leucospermum* and *Protea*, after being sprayed with some botrycide can be stored dry for 2-3 weeks at 2 °C with 90-95 per cent relative humidity. Storage avoids glut in the market because there are certain species such as *Leucospermum cordifolium* that tend to flower at one time. For cool storage of the Proteaceous flowers in water for more than a few days, it is desirable to have a low light levels of about 400 lux during the daytime to break the continuous darkness. Blackening of leaves after harvest is also most prevalent in Proteaceous plants, especially *Protea* (*P. eximia, P. magnifica* and *P. neriifolia*) and Leucospermum. Post harvest life of the flower is dependent upon the leaf and stem carbohydrates for complete opening and quality development, though it is more complex morphologically and physiologically. As fresh flowers, Protea is limited by leaf blackening and loss of fresh flower appearance. Often, leaf blackening occurs within a week of harvest and occasionally upon removal from the shipping carton less than three days from harvest. Leaf blackening leads to loss of decorative value, loss of market and possible rejection of the consignment. Blackening after cutting occurs due to enzyme activity, the production of phenolic compounds and the breakdown of the cell tissues and this occurs particularly during transport when surface moisture of cut flowers was not dried before packing (Paull *et al.*, 1980).

Insect-Pests, Diseases and Physiological Disorders

These are not the subject to many pests and diseases. **Root rot** is most serious disease of Proteaceous crops caused by *Phytophthora cinnamomi* which can be avoided by managing proper drainage and plant-soil water relations. Since rooting in Proteaceous crops is almost very slow so use of chemicals will not prove effective. Other diseases are leaf spots as well as floral rots which can be controlled by spraying with 0.2 per cent Dithane M-45 or Dithane Z-78. ***Botrytis cinerea*** is the most common pathogen which affects flowers at low temperatures when humidity is high for which Dithane is very effective. Shoot blackening occurs when water is lodged on large hairy leaves of many of the Proteaceous plants, and in case of *Hakea* and *Telopea* which have flowerheads full of florets and nectors, so water lodges there in the heads causing development of *Botrytis*, ***Rhizopus*** and a few other pathogens including moulds that make rotting of the flowers. Silver leaf disease (*Chondrostereum purpureum*) seriously infects

Leucadendron 'Safari Sunset' and greys the normal red-purple shoots. Though there is no any effective control but Difolatan sprayings reduce its occurrence. After cutting when surface moisture is not dried, the leaves in *Protea, Leucospermum* and in certain other Proteaceous leaves turn black within a week of harvest or in three days when taking out from the cartons which is due to enzyme activity, the production of phenolic compounds and the breakdown of the cell tissues. One can get rid of leaf blackening by drying the surface-moisture of cut material before packing, removing field heat and storage at cool temperatures (Salinger, 1985). Insect-pests are the same as the other plants such as aphids, thrips, bugs, grasshoppers, humble bees and caterpillars though not of serious nature. **Leaf roller caterpillars** feed on the tips of the shoots during active growth and tie the leaves together. This may be controlled by applying systemic organophosphates or pyrethrins.

Future Prospects of Protea Cultivation in India

India has an ancient heritage when it comes to floriculture. It has grown flowers for various purposes ranging from aesthetic to social and religious purpose. A consistent increase in demand for cut and potted flowers has made floriculture one of the most important commercial trades in Indian agriculture. A mild winter, abundant sunlight, suitable agro-climatic conditions, low labour costs, availability of skilled manpower are factors that are beneficial for the growth and development of this sector into a potential earner of foreign exchange. Moreover, the thinking that only rose is the best exchange earner will have to be changed and one should go for various other options through product diversification. Various Proteaceous crops can be introduced into India for commercial cultivation. Places like Bengaluru, Pune, Nilgiri hills, lower Darjeeling hills, Koraput region of Odisha are some of the best places for introducing various Proteaceous crops. After introduction, the germplasm can be assessed for their performance and then the cultivation on commercial scale of the most promising ones should be taken. Afterwards, the improvements in the agro-technology may be undertaken along with the genetic improvement for developing novel hybrids and varieties to match with international competitors. A private company called 'Casablanca' in Bengaluru is evaluating certain Proteaceous plants for its suitability. Likewise, places having a mild climate, low rainfall with good amount of humidity need to be identified for popularizing their cultivation.

Efforts within the individual producer countries to develop quality standards based on the requirements of the consumer countries, should be coordinated internationally. These standards can then be used to set the selection criteria or individual breeding programmes. A programme on gene mapping of *Banksia* is already in progress so other Proteaceous crops should also be assessed for varietal identification. Detailed research into breeding systems of Proteaceous genera will assist in development of hybridization methods to produce a wider range of cultivars. Genus *Leucospermum* has so far proved to be the most flexible in terms of cultivar production *via* interspecific hybridization. Many of the other commercial genera are more difficult to manipulate, but are equally spectacular in terms of the cut flower market. Through genetic engineering, transfer of useful genes across taxonomic barriers can be accomplished. While research into gene isolation and control is quite advanced in a number of organisms, there is a major bottleneck with many plants as regeneration *in vitro* is generally required for transformation. Thus, an important goal of Proteaceous research should be *in vitro* regeneration of whole plants. When this has been achieved, then the path will be open for genetic manipulation and transformation of these genera. Refinement of cultivation practices, such as pruning, fertilization and irrigation are required to maintain the economic return of *Protea* as a crop and to ensure the delivery of quality blooms to a very competitive international market. The challenges of cultivating *Protea* differ from region to region, but the basic plant physiology controlling the plant's reaction to environmental stresses remains the same. The international flower markets are always searching for new and exciting products. *Protea* can fulfil this demand. A larger variety of cultivars with different forms and colours, longer vase life, exceptional quality, and extended availability during the year are needed to maintain and increase the market share. These goals will only be achieved by continued research.

References

Anonymous, 2001. http://www.hilllaboratories.com. Crop Guide to Proteas.

Ben-Jaacov J. 2006. Gibberellic acid spray increased size and quality of *Protea* 'Pink Ice' flowers- A preliminary experiment. *Acta Hort.*, No. 716, pp. 135-140.

Criley R A. 2001. Proteaceae ~ beyond the big three. *Acta Hort.*, No. 545, pp. 79-85.

Croxford, B., G. Yan and R. Sedgley, 2006. Micropropagation of *Leucadendron. Acta Hort.*, No. 716, pp. 25-33.

Eliovson, S. 1965. *Proteas for Pleasure.* Howard Timmins, Capetown, South Africa.

Malan, D. G. 1995. Crop Science of Proteaceae in Southern Africa ~ Progress and Challenges. *Acta Hort.* No. 387, pp. 55-72.

Matthews, L.J.,and Z. Carter, 1983. South African Proteaceae in New Zealand. Matthews Publishing, Manakau, New Zealand.

Parvin, P. E., R.A. Criley and R.M. Bullock, 1973. Proteas ~ developmental research for a new cut flower crop. *HortScience*, **8**(4): 290-303.

Parvin, P. E., R.A. Criley and J.H. Coetzee, 2003. Proteas ~ A dynamic industry. *Acta Hort.*, No. 602, pp. 123-126.

Paull, R., T. Goo, R. Criley and P.E. Parvin, 1980. Leaf blackening in *Protea eximia*: Importance of water relations. *Acta Hort.*, No. 113, pp. 159-166.

Rebelo T. 1995. SASOL Proteas. In: *A Field Guide to the Proteas of Southern Africa*. Fernwood Press, Vlaeberg, South Africa.

Rodriguez-Prez, J. A. 1995. Effects of treatment with gibberellic acid on germination of *Protea cynaroides, P. eximia, P. nerifolia* and *P. repens*. *Acta Hort.*, No. 387, pp. 85-89.

Rodriguez-Prez, J. A., M.C. Vera-Batista, A. M. de Leon-Hernandez and P.C. Armas, 2001. Influence of cutting position, wounding and IBA on the rooting of *Leucospermum cordifolium* 'California Sunshine' cuttings. *Acta Hort.*, No. 545, pp. 171-175.

Rugge, B. A. 1995. Micropropagation of *Protea repens*. *Acta Hort.*, No. 387, pp. 121-125.

Sahebat, A. and N. Zieslin, 1995. Promotion of postharvest increase in weight of rose (*Rosa* × *hybrida* cv. Mercedes) petals by gibberellins. *J. Plant Physiol.*, **145**: 296-298.

Salinger, J. P. 1985. Proteaceae. In: *Commercial Flower Growing*. Butterworths of New Zealand, pp. 216-239.

Sedgley, M. 1995. Cultivar development of ornamental members of the Proteaceae. *Acta Hort.*, No. 387, pp. 163-169.

Sedgley, R., B. Croxford and G. Yan, 2001. Breeding new *Leucadendron* varieties through interspecific hybridization. *Acta Hort.*, No. 545, pp. 67-75.

Stephens, A. I., D.M. Holcroft and G. Jacobs, 2003. Post harvest treatments to extend the vase life of selected Proteaceae cut flowers. *Acta Hort.*, No. 602, pp. 155-159.

Thomas, M.B. 1979. Nutrition of container-grown plants with emphasis on the Proteaceae. Ph.D. Thesis, Canterbury Univ., New Zealand.

37

Saintpaulia (Family: Gesneriaceae)

Sapna Panwar and Sanyat Misra

(Common names: African Violet)

Introduction

Saintpaulia was named for Baron Walter von Saint Paul-Illaire (1860-1910), district commissioner of Tanga province, who discovered the first species in 1892 from Tanganyika (now Tanzania) and sent its seeds back to his father, an amateur botanist in Germany.Two British plant enthusiasts, Sir John Kirk and Rev. W.E. Taylor, had earlier collected and submitted specimens to the Royal Botanic Gardens, Kew, in 1884 and 1887, respectively, but the quality of specimens was insufficient to permit scientific description at that time. This belongs to the family of the popular tropical plants, *Achimenes, Episcia,* and *Gloxinia*. It originates in Eastern Africa mainly Tanzania (mainly Usambara Mountains) and Kenya (Bailey and Bailey, 1976) and comprises of 21 rosette-forming or tufted species of evergreen herbaceous perennials (Burtt, 1958). The African violets have long been associated with mothers and motherhood and for this reason they have been a traditional gift to mothers in many cultures around the world. The species *Saintpaulia ionantha* Wendl. is a popular ornamental pot-plant grown worldwide. The species name *ionantha* translates for 'resembling a violet', in reference to the flower colour (www.gardeninghelp.org). A New York florist introduced this to United States in 1894 and when in 1938 fluorescent lights were introduced, this became the most important indoor-flowering pot plant worldwide. Until the early 1970s most of the cultivars were from *S. ionantha* having rosetted stemless plants with single or double flowers but afterwards this species was hybridized with certain other species which gave us long-stemmed character with pendulous and trailing cultivars though with use of small-growing *S. pusilla* which resulted into a race of mini African violets (Beckett, 1987). African violets are shade-tolerant, their propagation being very easy, having a high aesthetic appeal and remaining in flowering almost throughout the year under artificial light conditions (Grout,1990) which have made these a favourite decorative plant worldwide for indoor cultivation. *Saintpaulia ionantha* comprises of many cultivars with varying plant size, leaf shapes & patterns, and flower forms & colours. As it is highly adaptable to indoor environment, it can be utilized as a specimen plant, for bottle gardening,and for planting in the dish & terrarium. The African Violet Society of America is the international cultivar registration authority for the genus *Saintpaulia* and its cultivars.

Botany

African violets are rosette-forming or tufted, tender, evergreen and herbaceous perennials with long-stemmed or stemless plants. The leaves are thick, fleshy, obovate to almost orbicular, often cordate and with scalloped margins, and the leaves as well as petioles are covered with thin hairs. The inflorescence arises from the leaf axils and the terminal meristem remains vegetative in nature. The short-peduncled flowers which arise singly or in loose clusters, are shortly tubular having five rounded lobes as petals in the shades of white, pink, red, blue and violet. Bailey and Bailey (1976) reported that peduncles of *S. confusa* are shorter than leaves while it is longer in case of *S. ionantha*. African violet is a day-neutral plant (Hildrum and Kristoffersen, 1969). Some species are known to be epiphytic in nature while others grow among stones at the edges of streams.

The basic chromosome number in *Saintpaulia* is reported to be n=15. A number of hybridization attempts between many of the species revealed hardly any genetic barrier among them as most species can be freely crossed with any other and produce fertile off springs. Tschermak-Seysenegg (1953) reported sterility in most saintpaulia flowers either due to failure of the anthers to burst or their particular manner of pollen discharge. Arisumi (1964) when attempting interspecific crosses using 15 *Saintpaulia* species, observed that 13 out of the 15 species had close affinities and crossed readily among themselves with no loss or increase in vigour among the selfed or hybrid seedlings and all the selfed or back-crossed seedlings were fertile. However, *S. nitida* and *S. shumensis* though are fertile with each other but do not cross freely with other species. Sparrow *et al.* (1960) when irradiated the petioles of a diploid *Saintpaulia ionantha* variety with X-ray at 2,000 and 3,000 r, they recorded 10.06 and 21.38 per cent, respectively, more non-chimaeric mutants than untreated controls.

Classification

Most of the modern day African violets are results of hybridization between the two species, *viz. Saintpaulia ionantha* and *S. confusa* and afterwards a few others and thus the hybrids are available in a wide array of growth habits, foliage types and flower types as described below.

I. Growing Habit

It is based on size of single crown of a fully grown plant and is grouped into five categories as below.

- ✰ Micro-Miniature. Less than 7.5 cm size of single crown of a fully grown plant.
- ✰ Miniature. 7.5-15 cm size of single crown of a fully grown plant.
- ✰ Semi-miniatures. 15-20 cm size of single crown of a fully grown plant.
- ✰ Standard. 20-40 cm size of single crown of a fully grown plant.
- ✰ Large. More than 40 cm size of single crown of a fully grown plant

II. Foliage Types (www.gardeninghelp.org)

- ✰ Boy. Standard plain green leaves, named for 'Blue Boy', the most famous of the earlier varieties.
- ✰ Girl. Green leaves with wavy edges and a white spot at the base of the leaf blade, named for the 'Blue Girl'.
- ✰ Variegated. Green leaves blotched, edged or spotted with yellow, cream or white.
- ✰ Red reverse. Leaf reddening that is especially visible from the back and very dark on top, often nearly black.
- ✰ Oak leaf. Cultivars of this variety tend to grow very large and have indented leaf margins.
- ✰ Quilted. Leaves of this large cultivar exhibit marked raised areas between the veins.
- ✰ Fringed. Leaves overall are extremely wavy and serrated giving the foliage a lacy appearance. These plants are difficult to grow so that the foliage forms a perfect wagon wheel; the leaves tend to twist and fold.
- ✰ Trailing. Plants produce variable foliage; crowns cover the top and hang down the sides of the pot in a trailing manner.

III. Flower Types

The flowers are grouped into four categories on the basis of petal number.

- ✰ Singles. It includes the flowers having 5 petals.
- ✰ Semi-doubles. Over 5 petals but less than 10 with yellow stamens visible.
- ✰ Double. Over 10 petals but less than 20, completely covering the central anthers.
- ✰ Double plus. Over 20 petals, *i.e.* twice that of a double type.

IV. Flower Forms

The flowers are grouped into various categories (Anon., 2014).

- ✰ Pansy Single. Five petals, two on the top being smaller.
- ✰ Star. Five petals of same size and spaced evenly.
- ✰ Double Star. Ten petals or more on a star-shaped bloom.
- ✰ Bell. 5 to 6 petals which are connected to form a narrow bell.
- ✰ Cup. 5 to 7 petals that are joined to form a shallow cupped star.
- ✰ Wasp. Small twisted petals, bunny ears, tubular narrow or curved lobes.
- ✰ Fluted petal. Petals have a bowed-shape at the tip.
- ✰ Ruffled. Plants with slightly wavy petal edges.

- ✰ Frilled. Plants with very wavy petal edges.
- ✰ Flat petals. Lower petals forming an almost flat plane.

V. Flower Colours

The flowers are grouped into various categories. The flower colour in African violet is available in wide spectrum of colours, *viz.* white, blue, violet or purple, mauve, lavender, red, maroon, burgundy, pink, *etc.*

- ✰ Bicolour. The flower colours with two values of one hue, *i.e.* two shades of the same colour.
- ✰ Multi-coloured. The flower colours with two or more hues, *i.e.* petals consisting of two or more different colours.
- ✰ Picotee. The petals are in white or pastel shades having contrasting dark edges. In some varieties the edges are slightly frilled.

Species and Varieties

Early introduction of *Saintpaulia*, especially of *S. ionantha* and *S. confusa*, and their intercrossing resulted in generating considerable variation among hybrids. Wide cross-fertility within the genus *Saintpaulia* is reported by Arisumi (1964). Later on, several other species were used in breeding programme. In 1958, B. L. Burtt recognized 19 species, and stated *S. amaniensis* and *S. magungensis* as synonymous (Burtt, 1964), and reported 20 species by recognition of *S. brevipilosa* and *S. rupicola* as separate species (Burtt, 1964). Beckett (1987) reportsd that there are 21 species under the genus *Saintpaulia*.

Sainpaulia amanuensis (*S. magungensis*)

A native to Usambara Mountains of Tanzania, this is a tufted species with prostrate stems running more than 15 cm long, bearing shortly stalked, broadly ovate to orbicular, crenate and up to 7 cm long leaves. Purple flowers about 2 cm across appear in a cluster of 2-4 from the leaf axils. *S. a. minima* grows to slightly more than half of its type species and the flowers 2/3rd the size.

Saintpaulia confusa

A native to Usambara Mountains of Tanzania, this is a rosette and clump-forming species which develops a short thick stem with age. In the past, this was being confused with *S. diplotricha* and *S. kewensis* (syn. *S. ionantha*). Its leaves are long-stalked, appressed hairy, elliptic to ovate with rounded tip and some 4 cm long. Blue-violet flowers appear 2-3 cm across.

Saintpaulia dipotricha

A native to Usambara Mountains of Tanzania, this species is quite similar to *S. confusa* except that the leaf blades are purple-flushed at upper surface and bears erect hairs and not appressed as in *S. confusa*.

Saintpaulia grotei

A native to eastern Usambara Mountains of Tanzania, this is a tufted species with prostrate stems more than 10 cm long. Quite long-stalked leaves are orbicular with coarsely rounded teeth, up to 9 cm long and conspicuously veined. Pale-mauve flowers with a darker eye appear 3 cm across.

Saintpaulia intermedia

A native to eastern Usambara Mountains of Tanzania, this is a tufted species with more than 15 cm long prostrate branched-stems with rooting at nodes. Long-stalked and dark green leaves are broadly ovate, crenate, white-hairy above and purple- or red-flushed underneath and up to 5 cm long.

Saintpaulia ionantha (*S. kewensis*)

A near-threatened species is native of coastal Tanzania and adjacent SE Kenya in eastern tropical Africa. Most of the species are concentrated in the Nguru Mountains of Tanzania. This is a very popular rosette-forming dwarf species often in clumps, bearing somewhat fleshy-textured coppery-green leaves which are long-stalked, broadly ovate to rounded, 4-6 cm long and with silky hairs. Violet, blue, red, pink and white flowers about 2.5 cm across are held above the leaves on 5-10 cm long scapes and these come up in succession throughout the year. It has numerous cultivars. Its forms 'Alba' originated from blue-flowered parent and bears white or rose-pink flowers whereas 'Diana hybrid' bears attractive single pink flowers. Most of the varieties and hybrids have appeared by crossing *S. ionantha* with *S. confusa*. Three of the main U.S. breeders/propagators are Holtcamp Greenhouses (offering the 'Optimara', 'Ballet', and 'Rhapsody' series), Arnold Fisher Greenhouses, and Nortex Nursery Industries (www.ag.auburn.edu). Its important varieties in various flower colours are:

Blue. Blue Caprice', 'Blue Fairy Tale', 'Blue Queen', 'Bright Eyes', 'Colorado' (flowers frilled magenta), 'Delft' (semi-double), 'Diana Blue', 'Double Delight', 'Elfriede', 'Porcelain' (semi-double, flowers white, edged purple-blue), 'Rhapsody' (semi-double), *etc.*

Red and pink. 'Blushing Bride', 'Diana Double Pink' (semi-double), 'Diana Pink', 'Diana Red', 'Flash', 'Fuchsia Red', 'Grandiflora Pink', 'Kristi Marie' (semi-double dusky red edged white), 'Pink Miracle', 'Pip Sqeak', 'Rococo Pink' (double), 'Ruffled Queen', 'Star Pink', *etc.*

White. 'Garden News' (double), 'Miss Pretty' (single, frilled, flushed-pink), 'Snow Prince', *etc.*

Saintpaulia orbicularis

A native to Usambara Mountains of Tanzania, this is a clump-forming rosetted species which sometimes develops a short stem with age. Long-stalked leaves are pale-green, broadly ovate to orbicular, coarsely toothed and 3-6 cm long. Blue eyed light blue, violet to white 8-10 clustered flowers appear from the leaf axils, each flower measuring 2.0-2.5 cm across.

Saintpaulia pendula

A native to Usambara Mountains of Tanzania, this species is quite similar to *S. intermedia* but differs that stems of the species are very fast-growing and bears solitary blue-purple flowers having darker throats.

Saintpaulia pusilla

A native to Tanzania, this is a rosetted and clump-forming species making neat small hummocks. This is the smallest known species which by crossing with *S. ionantha* has produced mini African violets suitable for terrarium, bottle gardens and for hanging baskets. Upper surface of leaves is pale-green, lower surface flushed purple and the length of leaves is 2.0-3.5 cm. From the leaf axils, some 2.5 cm across flowers appear in small clusters where upper petals in the flowers are blue while lower white.

Saintpaulia rupicola

The species presented well shaped pretty appearance. The leaves are 7-8 cm long and glossy in appearance with pale silvery green beneath. The flowers are light blue with dark centre and about 3 cm across.

Saintpaulia tongwensis

A native to Usambara Mountains of Tanzania, This is a rosetted and clump-forming species, sometimes developing a thick and short stem with age. Leaf stalk equal to the length of the leaves, and leaves are dark green with pale midrib, ovate to elliptic with pointed tips, coarsely crenate and about 8 cm long. Four to six clustered flowers of lavender shade and about 3 cm across, appear from the leaf axils.

Propagation

African violets are propagated sexually by seeds, especially the species, and vegetatively by leaf stalk cuttings and divisions in case of species and varieties. However, the most common method of propagation is through leaf cuttings. Nowadays, micropropagation is also being utilized for production of elite planting material.

African violets are mostly propagated from **seeds**, especially the species but not the varieties because of the limitation that they rarely come true to type. However, the method is very effective for cultivar development as the seedlings obtained from natural or artificial crosses will generate wide variability with regard to colour, shape and size of flower, leaf and entire plant. Depending on growing conditions, it takes about 6 to 9 months to produce flowering plants from seeds. Seeds are very small, *i.e.* approximately 3,000-4,000 seeds/g. Usually the seeds are sown thinly in March or April in the temperate regions while during autumn in the sub-tropical region of India when atmospheric temperatures are 18.3-21 °C. The medium in which the seeds are sown must be quite fine. At commercial level, a mixture of horticultural grade vermiculite (topmost layer of fine grade) + fine sphagnum peat is recommended. After sowing, the trays are covered with glass and dark paper. Light irrigation with a fine mist is given in the early days of sowing. A 18-21 °C temperature is optimum for seed germination and at these temperatures these germinate within 2-3 weeks. Immediately after germination, the glass and papers are removed. When all the seeds have germinated, the temperature should be lowered to 15.5-18.3 °C. When large enough to handle, the seedlings are pricked out in pans, boxes or trays using JIP I, and when well grown singly into their final 9-10 cm pots having growing compost. The seed germination requires light conditions but avoid placing them under direct sunlight. However, good germination percentage is observed under fluorescent light.

Vegetative propagation method is useful for production of identical plants similar in all aspects to the parent plant. Propagation through **leaf stalk cuttings** is the most popular and commercial method of multiplying the identical plants. The production cycle is about 6 to 9 months, *i.e.* duration from leaf cutting to full blooming stage. Leaf cuttings are taken prior to initiation of their reproductive stage. The medium-sized leaves which are midway from the edge of the plant to the centre of the crown are preferred. It is advisable to remove the leaf with a sharp knife after making clean cut at the main stalk of the mother plant. The leaf cuttings should be taken from two month old vegetative plants from fully grown leaves having 2.5-3.75 cm of stalk attached, between June to September in the temperate regions and during autumn in the tropical and sub-tropical regions of India. Pringale (1957) planted saintpaulia leaf cuttings in John Innes Seed Compost in sealed polythene bags where these rooted in 3-4 weeks, produced plantlets in 6-8 weeks, and were ready for separation from the leaf a month later but transferring the leaf cuttings to the soil when the roots had formed the plantlet development was much better. However, Von Hentig (1976) suggested that the length of petioles should be about 1-5 cm with about 1.25 cm deep being inserted in container having rooting media, normally without rooting hormone but with high humidity in the atmosphere. It is advisable to firmly press the rooting medium around the petiole. Different rooting media include vermiculite, sphagnum moss, perlite, sand or their combinations. Equal volumes of vermiculite and sand make an ideal medium. If medium temperature is around 24 °C and of the air 18 °C with relatively high humidity, it proves highly conducive for rapid induction of rooting. Plummer and Leopold (1957) initiated buds in the basal callus of leaf cuttings of *Saintpaulia ionantha* with application of 25 ppm of kinetin or adenine sulphate. In general, the leaves are inserted 5.0 cm apart and some 2.5 cm deep with erect stalks in peat-sand propagating compost contained in a 5.0 cm seed trays and these containers are staged in a propagating house at 18-21 °C temperature but shaded from direct sunlight, there these plantlets form on the stalks and push through the surface of the propagating medium, and there these are ready for potting-off in about six weeks. Leaf blade should be in contact with the rooting medium. Leaf cuttings start rooting in about 2-3 weeks at 24 °C and new harvestable plantlets emerge from the base of the petiole in around 6-8 weeks following root development, and these plantlets in 8-12 weeks become 2.5 cm long for harvesting. For uniformity of the propagating material, only largest plantlets should be harvested for replanting the leaves and then the cycle is repeated two weeks later for a total of three harvests. Young plants can be placed in plugs, then grown on for 2-6 weeks and then transplanted or sold into 10-cm pots (Dole and Wilkins, 1999). The total duration for obtaining transplantable plant having 3 to 5 leaves per crown from this method is approximately 14 to 16 weeks. **Division** is preferred for multiplication of old multiple-crown plants. In this method, crown portion of plant having root system is individually removed from the plant and potted in soil mixture. Sometimes plants show wilting symptoms till the new root development takes place. To overcome this problem, it is advisable to cover the plant with plastic or setting in a glass jar.

Micropropagation is largely exploited for large-scale production of quality planting material of uniform size, introduction of genetic variability for new cultivar development, or to maintain and multiply chimeras which cannot be maintained by means of leaf cuttings (Bilkey *et al.,* 1978; Peary *et al.,* 1988). A successful *in vitro* propagation protocol for rapid micropropagation of the *Saintpaulia ionantha* has been developed by Zaid *et al.* (2008). Micropropagation of African violet from various explants, *viz.* leaf discs, petioles, petals, and anthers has been reported and also regeneration has been obtained through direct as well as an indirect mode of organogenesis (Shukla *et al.,* 2013).

Cultural Practices

A **potting mixture** should fulfill four basic needs, *viz.* nutrition, moisture, aeration, and physical support required for overall development of plant. African violets grow best in well-drained, loose and porous soil. The soil mixture should be slightly acidic with a pH of about 6.0 to 6.5 (www.ext.vt.edu). A poorly-drained soil is mostly responsible for poor root development. Quality plants can be grown in mixtures containing equal parts (by volume) of soil, peat and horticultural-grade perlite. Soilless mixtures are also nowadays available in market. They are light and inert in nature which permit proper air exchange around the roots. Soilless mixes containing mixture of sphagnum peat moss and sand, or horticultural vermiculite or perlite can be recommended for commercial production. Potting mixture, commonly known as 'soil-mix' may be used as such or modified to meet specific cultural requirements and composition of various constituents (www.gardeninghelp.org) such as 1 part peat moss, humus, or leaf mould, 1 part garden soil and 1 part perlite, vermiculite or sand. Nagy (1968) grew saintpaulias in acid peat, 70 per cent acid peat+30 per cent river sand, perlite or perlite+ sand, to each of which 2 per cent NPK+Ca+Mg was added. He obtained best results with peat+sand medium. Ryser (1969) reported that saintpaulias require a well aerated substrate lightly fertilized during preparation. He obtained best results with a substrate comprising of equal parts of peat and good garden soil with the addition of Plantomaag at 1.5 kg/m^3, and the results were highly encouraging when 0.1 per cent solution of 18:12:24 fertilizer was provided with irrigation water.

Choice of **container** is an important consideration in case of African violets. Clay or plastic containers are mostly preferred and they have an important effect on the frequency of watering. Because of porous nature, clay pots tend to lose water due to excessive evaporation whereas plastic pots tend to retain more moisture. Hence, plant grown in clay pots require more frequent watering than plastic pots. Use of decorative glazed pots without drainage holes should be avoided as plant growth is affected due to lack of drainage. Pot size must be related to plant size and age and *Saintpaulia* is commonly potted and finished in 10-15 cm pot sizes.

Light plays a critical role in the growth and development, *vis-a-vis* ultimate quality production of

Saintpaulia. When other environmental conditions are favourable, the failure of flowering in *Sainpaulia ionantha* is attributed to insufficient light for flower initiation and its development (Stinson and Laurie, 1954). The light requirements may vary from cultivar and growing season. The duration and intensity are the major factors that affect growth and flowering. Generally, the light requirement for flowering of African violet is less than many other flowering plants but proper light levels are very critical as both high and low light levels affect the plant growth. Low light levels will result in deeper leaf colour, longer petioles and thin leaves, and in extreme cases the plant will produce scarce or no flowers or sometimes will fail to bloom. In case of high light intensity, the leaves exhibit paler appearance, slow growth though plants remain compact. African violets should not be directly exposed to sunlight during the summer months, however, early morning exposure of direct sunlight in winters will not be detrimental for plant growth. A northern or eastern exposure during summer is considered as the best option. A good foliage colour can be obtained at light intensity of 100 fc for 12 hours/day. Hanchey (1955) through 600 fc of fluorescent lighting for 18 hours a day obtained earliest flowering with greatest number of flowers whereas 100 fc light for 6, 12 or 18 h daily was not found sufficient for satisfactory flower initiation or development though vegetative growth was recorded better than the natural lighting. However, high light intensity of around 1,000 foot-candles for 6 to 8 hours/day is needed for blooming of African violets (Culbert and Hickman, 1966). The irradiance of 1,000 to 1,500 fc for adult plants and 500 to 800 fc for young vegetative plants was found optimum by Laurie *et al.* (1969) and Post (1949). African violets can be successfully grown under artificial light conditions. Supplemental light is necessary for those locations where there is lack of natural light, and in that case the supplemental light serves a boon in promoting flowering. The common artificial light sources used are fluorescent and incandescent lamps. However, the best results are obtained with the use of fluorescent lamps as they are cheap to operate and also produce less heat in comparison to incandescent lamps. Generally the plants produce maximum flowers per plant with fluorescent lighting of 800 to 1,200 fc. For entire plant production under fluorescent light regime, an approximate light intensity of 600 fc for about 15 hours per day is required.

Saintpaulia requires night **temperatures** of 12 -15 °C and day temperatures of 21-23 °C (Bose *et al.*, 2013). However, Fisher (1991) reported the optimum temperature (day and night) in the greenhouse as 21-26 °C. Extreme high and low temperature affects plant performance badly. Growth and flowering process is slowed down under the prolonged high temperature regimes. Similarly, if the temperature drops to 15 °C for long duration, both the phases (vegetative and flowering) are affected. Brittleness, discolouration and downward curling of leaves, *vis-a-vis* deformed flowers may be borne due to lower temperatures. Further lowering of temperature below 10 °C will result in stoppage of growth and ultimate plant death. Under extreme cold conditions, the plants suffer from cold injury symptoms, *i.e.* they develop water-soaked lesions and shrivelling (Culbert and Hickman, 1966). The day temperature above 29 °C also induces poor growth and premature flowering. Normally the raised temperatures speed up the flowering while reducing the levels delay it. African violets can be divided into nine stages with regard to flower development (www.ag.auburn.edu) such as (i) visible bud stage in the leaf axil (2 mm long), (ii) flower stalk beginning to elongate, (iii) flower stalk beginning to bend, (iv) flower stalk curves over to protect primary bud, (v) flower stalk completely curves, (vi) inflorescence pokes through leaf canopy and flower starts to straighten, (vii) flower stalk straightens out, (viii) primary flower opens, and (ix) five flowers open per plant.

Optimum **humidity** level is dependent mostly on two factors, *viz.* light and temperature. African violets flourish well at 50 -70 per cent relative humidity (Kessler and Pennisi, 2004). High light and temperature regimes during summer months result into low humidity conditions. Low humidity desiccates the flower and produces petal burn injury. It may be raised by use of certain measures such as evaporative cooling or simply by wetting the walks and side walls of greenhouse. High humidity >90 per cent during winter, especially at night will lead to development of *Botrytis* blight.

African violets require balanced **fertilizer** in small amounts right from when the plants are half grown and becoming pot-bound. In a sand culture trial, Kohl *et al.* (1956) found decreased growth when a complete nutrient solution containing more than 13 m.e./l total salts was applied, and the plants showed high ability of accumulating Na, especially in the older leaves though moderate excess of Na, Cl, SO_4 and Ca ions did no particular harm whereas NH_4 ion in moderate excess was found very harmful. Nagy (1964) when grew *S. ionantha* var. 'Blaues Märchen' in peat, perlite and mixtures of peat+sand and perlite+sand with nutrient solution containing N, P, K, Ca and Mg at 2 per cent during planting and 0.4 per cent every 10 days thereafter, he obtained largest plants in peat and the earliest and greatest number of flowers in perlite. Feeding with a dilute fertilizer dose (1:1:1 NPK) at each watering or the liquid fertilizer is highly appreciable. Newly potted plants should not be fertilized until the proper development of root system takes place. Balanced fertilizer dose of 1N:1P:1K is preferred for most of growing conditions (www.gardeninghelp.org). However, fertilizer dose which is higher in phosphorus such as 12-36-14 or 15-30-15 is preferred to produce more and larger blooms (www.avsc.ca/species.htm). Use of 2.25-2.7 kg/m^3 of 13-6-11 or 14-6-12 NPK formulation during winter months and 3.7-3.6 kg/m^3 during summer months are recommended though some differences in the performance exist (Payne and Adam, 1980; Poole *et al.,* 1986). Fertilizer dose for African violets grown under artificial lighting should be 1.25 g/litre of water while under natural light conditions it should be some 600 mg/litre of irrigation water. Micronutrients should also be included in the fertilizer schedule. Electrical conductivity (EC) of the substrate should be 1.0 - 1.7 dS/m during the active phase of crop growth (Chen and Henny, 2009). Plants are more efficient users of fertilizer during active growth phase, *i.e.* late spring, summer, to early fall and will require less water and fertilizer during rest period. Slow-release fertilizers have not been successful for African violets and growers mostly opt liquid fertilization programme. Symptoms of over-fertilization appear as small, pale and deformed flowers, and leaves develop black spots along the margins of the leaf-underside.

Watering is one of the most important factors influencing *Saintpaulia* production. The frequency and amount of watering will greatly vary with many factors such as soil mixture used, drainage, and the conditions of light, temperature and humidity under which the plants are grown. African violets require thorough watering and then allowed to dry moderately between each watering. Plants may be watered through various ways such as subirrigation through capillary mats (Payne and Adam, 1980; Poole *et al.,* 1986) or ebb-and-flow (Brown-Faust and Heins, 1991), surface irrigation and overhead watering. Brown-Faust and Heins (1991) and Raabe (1957) recommended heating of irrigation water up to 26 °C to prevent foliar spotting. Dole and Wilkins (1999) stated that if irrigation water is 5-8 °C cooler than the leaves, the leaves may develop cream-coloured blotches, spots or streaks. Overhead watering early in the morning is preferred as the foliage dries up during the day time. African violets are highly susceptible to disorder called 'leaf-spotting' which is caused by extreme cold water splashs on the leaves during irrigation. Both over and underwatering hampers the production cycle. Overwatering will result in yellowing of the young leaves and growing point due to saturated medium which deprives the roots of the oxygen. It further aggravates the development of crown rot disease. Underwatering will result in stunted growth. Foliar damage is known to occur both at very high or very low temperature of irrigation water. It is suggested to maintain 18 to 23 °C water temperature for irrigation (Kessler and Pennisi, 2004).) Sub-irrigation methods, such as capillary mats or drip irrigation are generally used for large scale production of *Saintpaulia.*

Quality African violet plants can be produced by **carbon dioxide** enrichment in greenhouses. Additional carbon dioxide concentration of 800 to 1,000 ppm produces quality plants even at lower light intensity. Carbon dioxide enrichment is required during the winter months when light levels are lower (www.ag.auburn.edu). Münch and Leinfelder (1967) when used CO_2 at 0.12-0.15 volume percentage in the controlled greenhouse from December end to mid-April (winter months) in Germany, they noted vigorous plants with dark green leaves and improved bud formation and flower set.

Leaf pruning from outside is the common practice to stimulate blooming by providing chance to the inside rows to produce flowers. Old yellow and bleached leaves should be regularly sheared. However, some leaves may need to be occasionally removed to maintain the plant symmetry. **Desuckering** is essential to remove excess of suckers from the base of the plant and to maintain plant shape. After flowering is over, it is advisable to **de-head** the spent out flowers and dead stems so as to encourage them to flower again and also to make the plants more aesthetic in appearance.

Growth and Flowering

They are day-neutral plants where flowering is controlled by the total accumulated photosynthetic irradiance (Post, 1942; Hanchey, 1955; Hildrum and Kristoffersen, 1969). Plant growth modelling techniques is utilized in **scheduling** a crop for specific day as recommended by Faust and Heins (1994). The schedule shows production stages and timing for African violets in 10-cm pots. Individual stages may require more or less time depending on the cultivar, time of the year, and geographic location (Kessler. and Pennisi, 2004). Brown-Faust and Heins (1991) stated that plantlet production takes 12 weeks, plantlets developing into plugs 30-40 days, and for flowering into 10-cm pots 10 weeks, *i.e.* in total 29-36 weeks from leaf cutting to finished plant. Root zone heating with 23.5-25 °C temperature when air temperature was 18-21 °C, the time required from potting of plug to flowering decreased from 64 to 59 days (Vogelezand, 1988). However, in case of too advanced production schedule, these can be subjected to 13-15 °C temperature as plants can survive up to three weeks at 13 °C temperature though lower temperature can be very dangerous (Brown-Faust and Heins, 1991). Dole and Wilkins (1999) mention that plants acclimatize to the light by developing new leaves as when plants were grown at 13 klx they stopped flowering but when were subjected to 2.0 klx flowering started again after three months. Herklotz (1964) grew two clones of *S. ionantha* at 15°, 20°, 25° and 30° C and their combinations and with photoperiods of 8, 9 and 16 hours, and obtained optimum growth, development and flowering at a constant temperature range of 20-25° C though 15° and 30° proved detrimental so this effect could be corrected with use of prolonged photoperiods. The flowering in African violet is in acropetal succession.

Saintpaulia should never be allowed to become pot-bound as it will retard growth and development and such plants require more frequent watering and fertilization. When the plant becomes root-bound and begins to develop an extended crown, *vis-a-vis* overgrown roots or heavy root mass start coming out of drainage holes, it is the time to **repot** the plants into the next large size containers. African violets need repotting so as to keep them growing well and in good proportion to their pot size. The plant along with root-ball is taken outside and shifted to a larger container without disturbing the roots. Repotting is done every year or once in every two years. Fertile *cum* well-drained soil mixture containing equal parts sterilized houseplant pot-soil and peat moss should be used. However, these require to be divided during repotting and suckers removed to re-establish a single crown plant.

Post Harvest

Plants are typically harvested (marketed) when at least 5-6 floral spikes have appeared bearing 10-12 fully opened flowers in a pot so that these may provide good display of the blooms from day 1. Packaging of African violets is done by placing individual plants in polythene sleeves and further in appropriate boxes for transporting at a temperature of 13 to 16 °C (Dole and Wilkins, 1999). Chilling injury may occur if shipping temperature drops below 10 °C for more than 12 hours. African violets are also reported to be highly sensitive to ethylene. Conover and Poole (1981) recorded no influence of nutrient levels on postharvest life.

Insect-Pests, Diseases and Physiological Disorder

African violets are infested with both, *i.e.* foliar and soil **mealybugs**. Foliar mealy bugs appear as white cottony mass on aerial plant parts, *viz.* underside of leaves, stem and leaf axils. Two major species, *i.e.* Comstock mealybug (*Planococcus citri*) and the citrus mealybug (*Pseudococcus comstocki*) infest the African violets. Mealybugs feed by sucking the sap from plant resulting in their distortion and stunting. They also excrete honeydew which attracts ants and sooty mould. There are reports of existence of another form of mealy bug, *i.e.* Prichard mealy bug (*Rhizoecus pritchardi*) which lives in the soil and suck the juice from the roots. They destroy the root hairs and symptoms of infestation include yellowing of plant leaves, appearance of wilting, stunting and bloom reduction. Little infestation can be prevented by wiping these out through a cotton swab dipped in alcohol though for soil mealy bugs, infested soil will have to be drenched with a contact insecticide. During mid-summer when temperatures are high, introduction of ladybird predator, *i.e. Cryptolaemus montrouzieri* in the greenhouse can reduce the population. In case of heavy infestation in the aerial parts, plants are required to be sprayed with an hoirticultural oil and/or an insecticide containing malathion. Snetsinger (1966) controlled this mealybug through routine greenhouse hygiene coupled with drenches of aldrin or endrin. Thin-winged and tiny **thrips** (*Frankliniella occidentalis*) feed on the tender plant parts such as flowers and leaves. Damage appears as silvery streaks on leaves with blotches and small black excrement-dots on the infected surfaces. The damage on flowers appear as stunting, distortion, discolouration and short life. Infested plants may be treated with diazinon or the whole plant may be immersed in a malathion solution. Introduction of predatory mite, *Amblysetus degenerans* will also reduce the population of thrips drastically. The small and soft-bodied **aphids** damage the plants by sucking plant sap from the tender parts while clustering around the crown and making the infested parts sticky. They secrete honeydew to invite sooty mould and ants. These can be controlled through jet water sprays, sprayijng with nicotine sulphate or tobacco decoction or malathion.

Cyclamen mites [*Tarsonemus* (*Phytonemus*) *pallidus*] are spider-like pests which suck the plant sap. They prefer high humidity (80 to 90 per cent) and temperatures around 15.7 °C. The damage is first seen on new growth which leads to stunting of new leaves, leaf curl, and a grayish appearance. Flower buds may also fail to develop properly, becomes misshapen and finally wither off. The evidence of **red spiders** (*Tetranychus urticae*) appears as minute webs over the aerial plant parts like leaves, flowers and flower stalks. They suck the sap by piercing the plant tissues and damage appears as bleached or yellowish spots on the leaves. In severe cases, infested plants show stunted appearance. To control the problem, the plant along with the pot should be discarded in case of heavy infestation but when infestation is just beginning the affected parts should be removed and burnt and some effective miticide should be used. For outdoor African violets, *Phytoseiulus* spp.can be introduced during summer.

Root-knot nematode (*Mrloidogyne* spp.) is most destructive to African violets. Goidànich and Garavini (1959) in Italy recorded *Meloidogyne arenaria thamesii* invading the roots and stem base and gradually killing saintpaulia plants by destroying parts of the vascular tissue. This nematode penetrates the root system and feeds on the cell contents. Root-knot nematode infestations form gall-like root swellings which interferes with the functioning of the roots. Ultimately leaves become dull yellow-green, cupped and fragile. Its severe infestation and stem invasion may cause eventual death of the infested plant. The seedlings for planting should be free of nematodes as can be observed by examining their roots while planting. At planting, soil, soil-mix or media, *vis-à-vis* earthen pots should be steam sterilized. Infested plants should be lifted gently along with adjoining soil and destroyed. *Saintpaulia* **foliage and stem nematode** (*Aphelenchoides ritzemabosi*) which readily transmits to chrysanthemum and *vice versa* can be controlled by immersing the cuttings (rooted ones surviving HWT better) for 1 h at 43 °C or for 15 minutes at 46 °C HWT without any damage to the cuttings.

Botrytis blight or grey mould (*Botrytis cinerea*) develops under high humidity and poorly aerated conditions. The disease appears as small water-soaked lesions on the petiole which enlarges rapidly and covers the entire leaf mass. The flowers also appear as water-soaked. In its infection, the aerial parts first exhibit the blighted appearance, then afterwards turning dark brown to grey and ultimately makes a cover of fuzzy coating. McDonough and McGray (1957) attributed the cause of *Botrytis cinerea* infection to *Saintpaulia* probably due to injury of the plants by *Tarsonemus pallidus* mites. When they tried 25 per cent di-(p-chlorophenyl) methyl carbinol at 1-2 ml/gal, the plants were found free from mites as well as from *Botrytis*. Proper hygiene should be maintained by collecting all the plant debris and destroying it, by providing proper air circulation to the plants and spraying of zineb at 0.2 per cent every week of infection. **Root and crown rot** (*Phytophthora nicotinae parasitica* for root, stem and leaf rot and *Rhizoctonia solani* for root rot; and *Pythium ultimum* for root and crown rot both) is one of the serious fungal diseases. Varieties have different reactions to infection of *Phytophthora* but the infection could be eliminated by use of 0.02-0.04 per cent Dexon suspension immediately before inoculation and after four weeks (Kröber and Plate, 1973). Out of various fungi and bacteria isolated from root and crown rot infected plants, *Pythium ultimum* was found to be most potent and infected readily the leaf cuttings, rooted cuttings, and the petioles and leaves of plants in contact with moist infested soil, and its inoculum was when placed on the crown of a healthy plant, that also showed slow rotting (Thompson, 1958). The disease is favoured by conditions like excessive watering, poor drainage and deep planting, and infected plants turn soft and mushy. Fungi invade the stems at or below the soil line, may appear as completely girdled and destroyed which finally leads to complete wilting of the entire plant. The young leaves of the plant become stunted, black and die though older leaves droop down. While planting, the containers and potting mix should be thoroughly sterilized and the deep planting should be avoided. Highly infected plants should be discarded and destroyed but when infection is lighter the dead roots or soft stems should be trimmed off, *vis-à-vis* ferbam should be applied in the soil. **Basal stem rot** is caused by *Rhizoctonia* spp. which is favoured by poor drainage, overwatering and deep planting. Copper fungicide drench in the soil may prevent infection of this disease. If any plant is showing symptoms of its infection, entire plant along with adjoining soil should be taken out and burnt. **Powdery mildew** symptoms appear as white or gray powdery patches on aerial plant parts, soon covering entire aerial plant parts. Among the plants, there should be proper air circulation and overhead irrigation should be avoided. Humidity levels should also be controlled in the polyhouse. In case of infection, infected parts should be removed and destroyed and the plants should be dusted with finely ground sulphur at weekly intervals. Karathane or mildex spray may also be effective.

Hadorn (1942) observed **mottling** in *Saintpaulia* and other Gesneriaceae due to too intensive radiation as well as by variations in the greenhouse temperature, for prevention of which he suggested to adopt proper cultural practices. Chen and Henny (2009) reported certain **physiological disorders** associated with *Saintpaulia*. Circular **light yellow or green spots** are formed on the upper leaf surface due to overhead irrigation with cold water having temperature lower than the leaf surface. Hollings (1955) attributed the ring and line patterns on the leaves due to sudden chilling and that too at a minimum temperature of 18 °C under sunlight conditions and the plants at 2,000 fc were highly susceptible though below 30 fc no such symptoms could be produced. Durbin (1957) reported chloroplast breakdown in the palisade layer of *Saintpaulia ionantha*, resulting in the physiological ring spot disease which appears in two stages, first one appearing as darkened areas in the leaf tissue within a few seconds when leaf temperature is lowered by contact with cold water or air irrespective of the illumination though light may induce injury by increasing leaf temperature above the ambient, and in the second stage the chlorophyll in the dark areas is degraded producing yellowish ring patterns and blotches and this stage is light-dependent as in the dark the darkish areas may disappear and the plastids appear to return to normal, without affecting the respiration rate. Production of light green leaves with **chlorotic edges** is due to high light levels or some nutrienat deficiency or both. To avoid this problem, African violets should be grown at decreased light levels and only that much light should be made available which is necessary for blooming. Macro- and micro-nutrient levels should also be slightly increased. **Long petioles** occur due to low light and high fertilizer dose so in such cases nutrient level should be lowered and photoperiod should be increased. Immature **flower bud-drop** occurs from the plant after turning brown and shrivelling. Sometimes fully open flowers also drop prematurely. It is due to varied reasons like gas injury, dry air, and extreme fluctuations in soil moisture, temperature, light intensity and humidity so such situations should be avoided. **Bloom failure** is due to plants being produced under low light conditions which cause unopening of the blooms. In such conditions, light intensity should be increased by placing the plants closer to light source to make them available at least 1,000 fc irradiance. **Multiple-crown plants** soon become the problem due to overcrowding in the pots and this occurs due to too deep planting of the young plantlets which results in the development of too many adventitious shoots.

References

Anonymous, 2014. *Introductory Guide to the Wild World of African Violet.* Optimara Group, Hermann Holtkamp Greenhouses, Inc. Nashville Tennessee.

Arisumi, T. 1964. Interspecifi¢ hybridization in African violets. *J. Hered.*, **55**: 181-183.

Bailey, L.H. and E.Z. Bailey, 1976. Saintpaulia. In: *Hortus Third: A concise Dictionary of Plants Cultivated in the United States and Canada*, p. 994. Macmillan Publishing, New York, USA.

Beckett, K.A. 1987. *The RHS Encyclopedia of House Plants Including Greenhouse Plants*, pp. 423-424. Salem House Publisherss, Massachusetts, USA.

Bilkey, P.C., B.H. McCown and A.C. Hildebrandt, 197. Micropropagation of African violet from petiole cross-section. *HortSci.*, **13**: 37-38.

Bose.T. K., B. Chaudhary, S.P.Sharma and P. Pal, 2013. *Garden Plants in Colour: House Plants*, p. 189. Naya Udyog, Kolkata, India.

Brown-Faust, J. and R. heins, 1991. Cultural notes on African violets. *Greenh. Gr.*, **9**(2): 74, 76-77.

Burtt, B. L. 1958. Studies on the Gesneriaceae of the Old World XV: the Genus *Saintpaulia. Notes from the Royal Botanic Garden Edinburgh*, **22**: 547–568.

Burtt, B. L. 1964. Studies on the Gesneriaceae of the Old World XXV: additional notes on Saintpaulia. *Notes Roy. Bot. Gdn Edinb.*, **25**: 191-195.

Chen, J. and R. J. Henny, 2009. *Cultural Guidelines for Commercial Production of African Violets* (*Saintpaulia ionantha*). Published by Univeristy of Florida, Institute of Food and Agricultural Sciences Extension Bull., pp 1-4.

Conover, C.A. and R.T. Poole, 1981. Light acclimatization of African violet. *HortSci.*, **16**: 92-93.

Culbert, J. R. and D. Hickman,1966. African Violets. Cooperative Extension Work, University of Illinois, College of Agriculture and U.S. Department of Agriculture, pp 1-31.

Dole, J. M. and H. F. Wilkins.1999. *Floriculture Principles and Species*, pp. 508-513. Published by Prentice Hall, Inc., USA.

Durbin, R.D. 1957. Physiology of chloroplast breakdown in saintpaulia (abstr.). *Phytopath.*, **47**: 519.

Faust, J. E. and R. D. Heins, 1994. Modelling inflorescence development of African violet (*Saintpaulia ionantha* Wendl.). *J. Amer. Soc. hort. Sci.*, **119**: 727-734.

Fischer, A. W. 1991. Saintpaulia. In : *Ball Redbook*, 15th ed. (ed. Ball, V.), pp. 759-762. George J. Ball, West Chicago, Illinois, USA.

Goidànich, G. and C. Garavini, 1959. Dying of *Saintpaulia ionantha* due to infestation by *Meloidogyne arenaria thamesii. Riv. Ortoflorofruttic. Ital.*, **43**: 381-385.

Grout, B. W. W. 1990. *African Violet*. In: *Handbook of Plant Cell Culture*, vol.5(eds Ammirato, P.V., D.A. Evans, W.R. Sharp and Y.P.S. Bajaj,), pp. 181-205. McGraw-Hill, Inc., USA.

Hadorn, C. 1942. Mottling or yellow leaf spot of *Saintpaulia* and related species (German). *Geb. Gartenb.*, H. 1, pp. 13-15.

Hanchey, R.H. 1955. Effects of fluorescent and natural light on vegetative and reproductive growth on *Saintpaulia. Proc. Amer. Soc. hort. Sci.*, **66**: 378-382.

Herklotz, A. 1964. The effect of constant and daily alternating temperatures on the growth and development of *Saintpaulia ionantha. Gartenbauwiss.*, **29**: 425-438.

Hildrum, H. and T. Kristoffersen, 1969. The effect of temperature and light intensity on flowering in *Saintpaulia ionantha* Wendl. *Acta Hort.*, No. 14(vol. 1), pp. 249-255.

Hollings, M. 1955. Physiological ring pattern in some Gesneriaceae. *Plant Path.*, **4**: 123-128.

Kessler, R. and Bodie Pennisi, 2004. *Greenhouse Production of African Violets*, pp 1-8. Published by The Alabama

Cooperative Extension System (Alabama A&M University and Auburn University), USA.

Kröber, H. and H.P. Plate, 1973. *Phytophthora* rot of saintpaulias [causal agent: *Phytophthora nicotinae* var. *parasitica* (Dast.) Waterh.] (German). *Phytopathologische Zeitschrift*, **76**(4): 348-355.

Kohl, H.C., A.M. Kofranek and O.R. Lunt, 1956. Effect of various ions and total salt concentrations on *Saintpaulia*. *Proc. Amer. Soc. hort. Sci.*, **68**: 545-550.

Laurie, A., D.C. Kiplinger and K.S. Nelson, 1969. African violets (*Saintpaulia ionantha* – Gesneriaceae). In: *Commercial Flower Forcing*, pp. 377-381. McGraw-Hill, New York.

McDonough, E.S. and R.J. McGray, 1957. *Botrytis* on saintpaulia and its relation to mite control. *Phytopath.*, **47**: 109-110.

Münch, J. and J. Leinfelder, 1967. The result of a supplementary CO_2 experiment with saintpaulias (German). *Gartenwelt*, **67**: 250-1, 253.

Nagy, B. 1964. Raising saintpaulias with nutrient solution in peat and perlite media. *Kert. Szõl. Föisk. Közlem.*, **28**(1, part 1): 205-218.

Nagy, B. 1968. Experimental evaluation of media for growing ornamental plants in pots. *Kert. Szöl. Föisk. Közlem.*, **32**(5, part I): 43-52.

Payne, R.N. and S.M. Adam, 1980. Influence of rate and placement of slow release fertilizer on pot plants of African violet grown with capillary mat watering. *HortSci.*, **15**: 607-609.

Peary, J.S., R.D. Lineberger, T.J. Malinich and M.K. Wertz, 1988. Stability of leaf variegation in *Saintpaulia ionantha* during *in vitro* propagation and during chimera separation of a pin-wheel flower form. *Amer. J. Bot.*, **75**: 603-608.

Plummer, T.H. and A.C. Leopold, 1957. Chemical treatment for bud formation in *Saintpaulia*. *Proc. Amer. Soc. hort. Sci.*, **70**: 442-444.

Poole, R.T., C.A. Conner and Y. Ozeri, 1986. Response of African violets to fertilizer source and rate. *HortSci.*, **21**: 454-455.

Post, K. 1942. Effects of day length and temperature on growth and flowering of some florist crops. *Cornell Univ. Agric. Exp. St. Bull.*, No. 787, pp. 1-10.

Post, K. 1949. Saintpaulia. In: *Florist Crop Production and Marketing*, pp. 804-805. Orange Judd Publishing, New York.

Pringle, C.O. 1957. Rooting cuttings in polythene bags. *Gdnrs' Chron.*, **141**: 275.

Raabe, R.D. 1957. Physiological breakdown of chlorophyll in the Gesneriaceae. *Phytopath.*, **47**: 28.

Ryser, J.P. 1969. The importance of the substrate and of fertilizers in flower growing. A case in particular: systematic trials on saintpaulia. *Rev. hort. Suisse*, **42**: 240-244.

Shukla, M., J. A. Sullivan, M. Jain, S. J. Murch and P.K. Saxena, 2013. *Micropropagation of African violet* (*Saintpaulia ionantha* Wendl.) In: *Protocols for Micropropagation of Selected Economically Important Horticultural Plants. Methods in Molecular Biology* (eds Maurizio Lambardi *et al.*). Springer Sciences and Business Media, New York, USA.

Snetsinger, R. 1966. Biology and control of a root-feeding mealybug on *Saintpaulia*. *J. econ. Ent.*, **59**: 1077-1078.

Sparrow, A.H., R.C. Sparrow and L.A. Shairer, 1960. The use of X-rays to induce somatic mutations in saintpaulia. *African Violet Mag.*, **13**(4): 32-37.

Stinson, R.F. and A. Laurie, 1954. The effect of light intensity on the initiation and development of flower buds in *Saintpaulia ionantha*. *Proc. Amer. Soc. hort. Sci.*, **64**: 459-467.

Thomas, P.R. 1968. *Aphelenchoides ritzemabosi* on saintpaulia. *Plant Path.*, **17**: 94-95.

Thompson, H.S. 1958. *Pythium* rot of *Saintpaulia*, the African violet. *Canad. J. Bot.*, **36**: 843-863.

Tschermak-Seysenegg, E. 1953. Flowering and pollination in sainpaulias (German). *Bodenkultur*, **7**: 109-111.

Vogelezand, J.V.M. 1988. Effect of root-zone heating on growth, flowering and keeping quality of *Saintpaulia*. *Scint. Hort.*, **34**: 101-113.

von Hentig, W.U. 1976. Results of propagation with leaf cuttings of *Saintpaulia ionatha*. *Acta Hort.*, No. 64, pp. 55-63.

Zaid, S., Emad Al-Tiawni and Ahmad Abdul-Kader, 2008. Effect of some components of nutrient media on micropropagation of African violet (*Saintpaulia ionantha* L.).*Tishreen Univ. J. Res. & Scientific St. - Biological Sciences Ser.*, **30** (3): 131-134.

38

Spathiphyllum (Family : Araceae)

M. Ganga, Sanyat Misra, V. Jegadeeswari, K. Padmadevi, P. Mekala and R.L. Misra

(**Common names**: Peace lily, Spathe flower, Spathiphyllum, White anthurium)

Introduction

Commonly it is known as peace lily due to its creamy white spathes. *Spathiphyllum* is a Greek word referring to the leaf-like spathes, *spathe* meaning bract and *phyllon* meaning leaf. This is the tropical plant which originated in Central America and the northern part of South America, and some of its species are native to SE Asia. Bunting (1960) reported *Spathiphyllum* to be native of Central America, northern South America, and the eastern Malay archipelago. These are probably among the house plants the Victorians domesticated in the mid to late 1800s. There are some 35 species of this evergreen perennial with long-stalked ovate to lanceolate leaves forming a dense clump. These are ideal pot plants for the home gardens and conservatories requiring humid atmosphere for their proper growing.

Botany and Breeding

Spathiphyllum is a terrestrial herb with a rhizomatous stem which is short, erect or creeping (seemingly acaulescent). The plant forms clumps with many stems emerging from the soil. Leaves arise in a loose rosette and are simple, long-petioled, and the petiole being geniculate at apex, with a long sheath. The nodes often produce roots, leaves, and the axils. The interesting flower known as spathe is supported on a stalk known as peduncle. A node is the point where the plant produces its roots and holds buds which may also grow into shoots of various forms. The stem's roots then anchor the plant either to the ground, a tree or to a rock depending on the species and genus. In *Spathiphyllum*, a stem may even spread as a repent rhizome creeping across the soil but is often just beneath the surface. Stems may either grow aboveground, underground or partially above the soil. Different species and varieties may vary in size of the leaves and overall size of the plant. Some species and hybrids have often small and narrow lanceolate leaves while others have much larger and broader leaf blades.

The white spathes produced by *Spathiphyllum* plants are referred to as their flowers. Botanically, a spathe is a bract (modified leaf) enclosing a flower cluster or spadix (fleshy spike). Some species produce small inflorescences while others a spathe and spadix those are quite large. Inflorescence is solitary and erect, peduncle shorter than the petiole, spathe herbaceous, not enclosing the spadix, ascending, marcescent, white, with distinct mid-vein and pinnate laterals, spadix sessile or short-stipitate, cylindrical and flowering from base to apex. Flowers bisexual, the perianth segments 4-6, free or connate into a short cup, stamens 4-6, free, ovary 2-4 locular, with 2-8 axile ovules per locule, the style long, the stigma 2 or 3 lobed, sub-capitate or punctiform. Berry 1-8 seeded, green and seeds oblong, ellipsoid or ovoid. The spathe unfurls and exposes the spadix several days before flowers are receptive for pollination. The unfurling of the spathe reveals numerous uniformly sized bisexual flowers located along the entire spadix. All flowers on a spadix mature simultaneously. The stigmatic surfaces become dry and brown before pollen is dehisced and hence floral emasculation is not required. A *Spathiphyllum* inflorescence will produce pollen over a period of 3-4 days. The pollen of *Spathiphyllum* is lighter and may be dispersed by air and hence a soft brush should be used for transferring the pollen. It is always best to use fresh pollen. However, if pollen is in short supply, it can be stored in a container in a high humidity environment in a refrigerator. *Spathiphyllum* pollen may be stored for several days or weeks if necessary (Henny, 2000).

Breeding new varieties is a continuous process to dominate the market. However, hybridization within the aroid family can be difficult to achieve because of sporadic flower production and specific requirements for successful pollination. Research by the 'Foliage Breeding Program' at the University of Florida [Mid-Florida Research and Education Center (MREC), Apopka] has resulted in standardization of new techniques for plant breeders to manipulate flowering and pollination in ornamental aroids for hybridization purposes. These techniques have made breeding easier and practically possible for *Spathiphyllum* and other aroid genera those were previously difficult to hybridize. Henny (1981) reported that potential barrier to breeding caused by flower unavailability in *Spathiphyllum* can be overcome through single foliar spray of 250 ppm GA_3 to stimulate flowering throughout the year, and the plants start flowering within 3-5 months of spraying though with seasonal variation.

Differences in the natural flowering cycles, even within the same genus makes breeding difficult. Flowering times are affected by plant size, maturity, environmental conditions and cultural conditions such as light, irrigation and nutrition. Different *Spathiphyllum* species and cultivars may not flower simultaneously. Natural flowering of spathiphyllum involves three parameters: (i) photoperiod, (ii) temperature, and (iii) plant genetic factors (including age/maturity and cultivar genetics) as are being elaborated under **cultural practices**. Alternative tool to conventional system is mutation breeding and polyploidization which can create a lot of variation even in a given variety. Feckhaut *et al.* (2004) obtained polyploid *Spathiphyllum wallisii* Regel 'Speedy' plants from somatic embryos induced on anther filaments exposed to mitosis inhibitors. Primary embryos yielded less polyploid plantlets than secondary embryos. Colchicine could be efficiently replaced by oryzalin or trifluralin (10 μM), resulting in an average yield of 5 per cent polyploids. The mitosis inhibitors were directly added to the induction medium. This way tetraploids can be developed with an altered morphology, or triploids with a reduced fertility. The chromosome number in spathiphyllum is 2n=30, 60 or more.

Species and Varieties

Spathiphyllum is a genus of about 41 species (Bunting, 1960). The most likely used species to create the recent day hybrids include *Spathiphyllum wallisii, S. floribundum, S. friedrichsthalii* and *S. cannifolium* which are among the most widespread species in Colombia and certain other parts of South America. The spathe of *S. floribundum* and *S. cannifolium* are close matches to the spathe observed in many of the hybrid house plants. *S. cannifolium* appears restricted to the Amazon basin where it is known from practically all major tributaries of the Amazon basin, at least in Colombia. Within Colombia, *S. cannifolium* occurs at 200 to 1,000 meters elevation in tropical wet forest zones and produces an inflorescence almost any time of the year. *S. minor, S. perezii* and *S. lancaefolium* are found in tropical wet rain forests at 200 to 300 metres, while *S. silvicola* is found in both rain forests and tropical wet rain forests. The species accepted by the botanists are 51, *viz. Spathiphyllum atrovirens* Schott (published in 1858), *S. barbourii* Groat (2005), *S. bariense* G.S. Bunting (1988), *S. blandum* Schott (1857), *S. brent-berlinii* Groat (2005), *S. brevirostre* (Liebm.) Schott (1853), *S. buntingianum* Groat (2005), *S. cannifolium* (Dryand. Ex Sims) Schott (1853), *S. cochlearispathum* (Liebm.) Engl (1879), *S. commutatum* Schott (1857), *S. cuspidatum* Schott (1857), *S. diazii* Croat (2005), *S. dressleri* Croat & F. Cardona (2004), *S. floribundum* (Linden & André) N.E.Br. (1878), *S. friedrichsthalii* Schott (1853), *S. fulvovirens* Schott (1858), *S. gardneri* Schott (1853), *S. gracile* G.S.Bunting (1960), *S. grandifolium* Engl (1905), *S. grazielae* L.B.Sm. (1968), *S. humboldtii* Schott (1853), *S. jejunum* G.S.Bunting (1960), *S. juninense* K.Krause (1932), *S. kalbreyeri* G.S.Bunting (1960), *S. kochii* Engl. & K.Krause (1908), *S. laeve* Engl. (1905), *S. lanceifolium* (Jacq.) Schott (1832), *S. lechlerianum* Schott (1860), *S. maguirei* G.S.Bunting (1960), *S. matudae* G.S.Bunting (1960), *S. mawarinumae* G.S.Bunting (1988), *S. minor* G.S. Bunting (1960), *S. monachinoi* G.S. Bunting (1960), *S. monachinoi* var. *monachinoi* G.S. Bunting, *S. monachinoi* var. *perangustum* G.S.Bunting (1988), *S. montanum* (R.A.Baker) Grayum (1997), *S. neblinae* G.S. Bunting (1960), *S. ortgiesii* Regel (1870), *S. patinii* (R.Hogg) N.E.Br. (1878), *S. patulinervum* G.S.Bunting (1960), *S. perezii* G.S.Bunting (1975), *S. phryniifolium* Schott (1857), *S. quindiuense* Engl. (1905), *S. schlechteri* (Engl. & K.Krause) Nicolson (1968), *S. schomburgkii* Schott (1857), *S. silvicola* R.A.Baker (1976), *S. solomonense* Nicolson (1967), *S.*

Power Petite

Emerald Star

High Five

Milkyway

Sensation

Supreme

Sweet Chico

Sweet Dario

Sweet Lauretta

Sweet Pablo

tenerum Engl. (1905), *S. uspanapaensis* Matuda (1976), *S. wallisii* Regel (1877), and *S. wendlandii* Schott (1858). Various popular varieties of spathiphyllums are 'Emerald Star' growing to 45 cm high with thick and dark green foliage and pure white 10 cm long flowers, 'High Five' bearing dark green foliage with 10-13 cm long pure white flowers, 'Milkyway' bearing broad and variegated (white and light green over green base) foliage but with normal milky flowers, 'Power Petite' bearing tough, thick, dark and shiny green leaves with usual white flowers, 'Sensation' bearing dark green, tough and attractively ribbed foliage and showy white flowers, 'Supreme', a disease resistant cultivar bearing deep green and wide leaves, 18 cm long snow-white blooms and growing well even under adverse conditions, 'Sweet Chico', a highly preferred cultivar for its branching and blooming habit and its suitability to small pot-planting, 'Sweet Dario' bearing dark green symmetrical foliage, maintaining its greenness even during hottest days of summer, and large and pure white flowers, 'Sweet Lauretta' bearing a large structured plant with glossy foliage and large snow-white spathes, 'Sweet Pablo' which is a quick growing cultivar having white flowers and with consistent flowering habit, *etc.* Certain other promising cultivars are 'Alfa (Alpha)', 'Chris', 'Claudia', 'Cervin', 'Codys Color', 'Daniel', 'Double Take', 'Londonii', 'Lyskamm', 'Mascha', 'Petite', 'Queen Amazonica', 'Rica', 'S1007', 'S1008', 'S18', 'S4002', 'Showpiece', 'Sierra', 'Sonya', 'Tasson', 'Textura', 'Vanessa', 'Vickylynn', *etc.*

Propagation

To ensure **seed set** for *Spathiphyllum*, no special environmental manipulation such as controlling relative humidity is needed after pollen transfer. Pollinated flowers develop mature fruits within 4-6 months. As seeds mature, the spadix begins to turn yellowish and soften. Breeding for ornamental tropical foliage plants can be a lengthy process due to the limited number of seeds obtainable and prolonged germination time. Several years may be needed to grow the hybrids to reach sufficient size for evaluation. Once a hybrid is selected and tested, tissue-culture methods may be applied so that a new aroid hybrid cultivar can be increased rapidly enough to reach commercial production levels within 2-3 years. The production of new hybrids in crops like *Spathiphyllum* leads to new plant introduction, and these foundations keep the foliage market fresh and profitable.

Propagation of spathiphyllum is not difficult and can be made by **division**. It can be done any time of the year, preferably in February. A sharp and clean knife is used

Clustered Offshoots in Spathiphyllum.

Division of Plant for Offshoot Propagation.

to cut away a crown from the parent plant. Rhizomes are pulled apart gently making sure that each piece has at least two or three leaves attached. This crown should be properly rooted. Individual pieces are planted in 7 cm pots containing a moistened mixture of equal parts peat moss and perlite or coarse sand. Newly potted rhizomes are not fertilized for three months.

In order to maximize yield, *Spathiphyllum* is cultured ***in vitro*** in novel culture vessels termed '*Vitron*'. The best growth is obtained by culturing plantlets on sugar-free liquid medium under CO_2 enrichment (3 000 μmol $^{-1}$ 24 h^{-1} d^{-1}) at a low photon flux density (PPFD of 45 μmol m^{-2} s^{-1}), suggesting that the novel '*Vitron*' culture system is suitable for the photoautotrophic micropropagation of *Spathiphyllum*. Four types of basic media and 12 combinations of plant growth regulators, *viz.* 6-BA and NAA at different concentrations were tested for *in vitro* culture of shoot tips of *Spathiphyllum* 'Xiangshui'. MS medium containing 2 mg/litre of 6-BA and 0.1 mg/litre NAA was the best medium for shoot tip culture and subculture, though for rooting, ½ MS containing 0.5 mg/litre IBA + 0.5 g/litre active carbon was the best medium. A medium of ash of rice husk and bran of coconut shell (3:1) was best for transplanting micropropagated plantlets (Chen BiHua, 2000). Kaçar *et al.* (2005) subcultured the shoots of spathiphyllum cv. 'Sweet Pablo' in a MS medium supplemented with either of cytokinins ~ BA, PBA and MS medium and on 1 mg l^{-1} BA media, the multiplication rate began to increase starting from the 1st subculture and reached its maximum at the 4th subculture with 8.49 shoots per plant. They recorded PBA as the more favourable cytokinin source than BA. Dewir *et al.* (2005) recorded greater fresh and dry weights of roots and shoots, number of leaves and roots per plantlet, and shoot length in mixtures consisting of peat moss:carbonized rice hull:vermiculite (1:1:1) out of the five media mixtures cultured in microponic system in case of micropropagated *S. cannifolium* shoots. In the past, when a new plant was found to be desirable, if it did not lend itself to asexual production or if it was sterile or a poor seed producer, it would never reach the market place in substantial numbers. Even if such a plant could be produced in limited number from seeds, the off springs often varied in appearance from seedling to seedling, due to genetic variations. In the past, when a sport or a new hybrid was developed, it took four to five years before sufficient stock could be grown to produce enough cuttings for the plant to be released to the trade. Now, through tissue culture, this time has been cut to one to two years.

Cultural Practices

The ideal pH for spathiphyllum cultivation is 5.8-6.5. This requires a **potting mix** with good drainage and water holding capacity. Generally a 1:1:1 ratio of peat, perlite and bark is used as a common potting mix which provides excellent results. Pogroszewska and Hetman (2004) recorded hotbed soil mixed with highmoor peat and hotbed soil mixed with pine bark as the best substrates in Poland out of homogenic organic substrate or a mixture of highmoor peat, composted pine bark, clay, sand, hotbed soil and lowmoor peat on cvs 'Londonii' and 'Queen Amazonica'. The characteristics of the good potting soil are springy texture with no or little odour and the dark colour. Heemers (1999) in Belgium when planted three *Spathiphyllum* cultivars, *viz.* 'Alfa', 'Lyskamm' and 'Cervin' in the greenhouse at the plant density of 16, 20 and 24 plants/m^2 at 12, 24, 37 and 48 weeks of sowing, high plant density deteriorated vegetative development in the cv. 'Cervin'. Summer cultivation, *i.e.* at 12 weeks and autumn planating, *i.e.* at 24 week of planting enhanced vegetative development first and then generative development occurred while planting at 37 week (winter) and at 48 week (spring) first caused generative development and then followed by vegetative development. When the plants become crowded, these are **repotted** in the same container and the containers of the same size or one size larger. The best sign for repotting is that when plants have constant need of water. The pot is inverted and pushed through the bottom hole or the gentle jerks are given to all sides of the container turn by turn so that the plants come out of the container with ball of earth. The earth ball is removed along with the excess roots and then divided plants are placed in the same sanitized or in new pots akin to their requirements and properly potted. The bottom should have enough soil above the crocks so that when the plant is placed on it the top of the existing soil on the plant is the level required it to be in the new pot. No need to cover the top of the existing soil with new soil. Just fill in around the edges with new soil, pat down and water. Add a little more soil around the edges after the watering as it causes some settling.

Light intensities for production of spathiphyllums are somewhat cultivar dependent although a range of 9,000–27,000 lux is commonly used. Plants grown in the lower light intensities tend to have longer petioles, reduced branching, a softer appearance and darker green colour. Under higher light intensities, the plants tend to be more compact, exhibit more branching and are lighter in colour. Plants grown under excessive light intensities exhibit curled, pale or chlorotic leaves. Plants grown at the extremes of the light intensity range may produce fewer flowers than those grown in the central range. Due to **photoperiods**, three types of flowering responses have been reported in *Spathiphyllum*, *viz.* (i) the cultivars start flowering after attaining a specific age or maturity, (ii) the cultivars start flowering after attaining a specific age but flower only during the particular seasons (mostly during springs and summers), and (iii) the cultivars which flower more than once in a year.

The optimum **temperature** range for *Spathiphyllum* is 20 °C at night and 32 °C at day, but will tolerate lows of 7 °C and highs of 35 °C under special circumstances. Frost or even short term freezing temperatures would lead to foliar damage and crop loss. Plants grown at temperatures above 32 °C for extended periods can exhibit narrow leaves (strap leaf), loss of colour, inhibited root development and reduced flower quality and quantity. The optimal air temperature for flowering of spathiphyllum is 22 °C whereas root-zone temperatures in the range of 20–26 °C were of less importance. Growth, development, plant height and leaf size of *Spathiphyllum* were enhanced by higher air temperatures up to 23 °C. Elevated root-zone temperatures increased the shoot/root ratio and decreased the number of side shoots, but the dry weight per shoot was increased. Plant shape of *Spathiphyllum* can therefore largely be influenced by the root-zone and air temperatures (**Vogelezang**, 2003). However, exposure to a low temperature of 12 °C can induce flowering 3-4 weeks earlier. The high initial investment necessary for the cost of treatment chambers and the increased energy and labour costs to operate chambers make chilling treatment an impractical alternative to chemical treatment for the average grower.

Feeding and Watering

N-P-K ratio of 3:1:2 as a slow release or liquid feed produces high quality plants. Slow release dry fertilizers, constant feed liquid fertilization or combinations of both are equally effective methods of applying nutrients. Magnesium deficiency, a problem with some cultivars, appears as golden-yellow margins on lower leaves. Prevention of Mg deficiency via supplemental Mg is essential. Iron and manganese deficiencies exhibit reduced growth rates and chlorotic leaves can occur during winter months when the soil temperature is below 18 °C. Sulphur deficiency, exhibits as overall chlorosis of foliage and is sometimes seen when using highly refined, low sulphur fertilizers. Boron deficiency may be a cause of longitudinal ribbing of the leaves, often seen on new growth. Yeh *et al.* (2000) studied spathiphyllum cv. 'Sensation' by growing it hydroponically and recorded that N deficiency causes reduction in leaf number, leaf area and chlorophyll content; P deficiency causes slow plant growth; K deficiency causes many small yellow specks of 2 mm or less in diameter on adaxial surface of lower leaves; Ca deficiency causes suppression of leaves and necrosis on the young leaves at the middle and basal leaf blade; Mg deficiency causes distortion of the fully expanded leaves and lower chlorophyll content; Fe deficiency causes interveinal chlorosis in young leaves; and B deficiency causes marginal necrosis at leaf apex, distortion and crinkling of petioles that break at the leaf blade.

Irrigation frequency should be designed to keep the soil medium evenly moist during all phases of the crop cycle. Spathiphyllums easily tolerate overhead irrigation and do exceptionally well with drip. Spathiphyllums do not tolerate saturated soil conditions for extended period of time. Various diseases can easily infect over-watered spathiphyllums causing wilted or collapsed leaves, necrosis along leaf margins and extensive root damage.

Growth and Development

Spathiphyllum plants will flower continuously once reaching maturity, similar to their close relative *Anthurium*. McConnell *et al.* (2003) reported that growing **temperature** above 30 °C results into narrow leaves and delayed flowering as had been shown through their studies on two cvs 'Petite' and 'Tasson' which were exposed to 29 °C, 35 °C and 41 °C for 12 h daily during day time and 21 °C in the night up to 12 weeks, which caused narrowing of leaves above 29 °C temperature and the growth rate was found considerably decreased with 6 °C rise in temperature, hence emphasized the need to develop heat tolerant cultivars. Heemers and Oyaert (2001) in Belgium pot-planted the spathiphylum cvs 'Alfa', 'Lyskamm' and 'Cervin' on February 14, 2000 at 20/21 °C (day/night) and placed the light covers over these on April 27 and May 29 and left as such on the same place up to July 3 which was the expected initiation date. They recorded cv. 'Lyskamm' initiating much earlier after light reduction, *i.e.* 47 days in case light covers placing on April 27 while 71 days in case when light covers were placed on May 29, though in case of cv. 'Cervin' the initiation was found more pronounced by biomass than external environmental factors. Two growth regulators, *viz.* benzyladenine and GA_3 are commonly used on spathiphyllums. **BA** is quite effective in enhancing the branching and fullness of the plant and is generally used at the young plant (liner) stage. In addition to young plant treatments, some growers also apply BA shortly after young plants have been planted into larger pots. Enhancing branching qualities and fullness are especially important to the small pot growers since shorter production time limits the impact of natural branching. BA can be applied as a spray or drench at 250 to 1000 ppm. BA treatments can inhibit root development if applied before roots are well established. The overall effect of BA application is dependent on its concentration, the cultivar, stage of growth, application method and season.

GA_3 is used extensively to force early or year round flowering (timed flowering) of spathiphyllum. With maturity, spathiphyllum will naturally flower consistently in the spring and sporadically during the rest of the year. Since the market demand for plants with flowers is always higher, growers use GA_3 to gain year round sale advantage while also allowing programming crops for holiday's promotions or weekly orders. With GA_3, growers can also force early flowering to allow the production of smaller pot sizes. A standard treatment is a single foliar spray of 150-250 ppm 8 to 15 weeks prior to sale. The spray concentration and time between treatment and flowering depend on cultivar and season of the year. Henny *et al.* (1999) when grew 39 spathiphyllum cultivars in greenhouse and sprayed the plants with 0.25 per cent GA_3, the plants in all the varieties flowered barring control. After 16 weeks, 'Vickilynn' produced 14.1 flowers per plant, 'Piccolino' with 12.8, 'Mascha' 12.6, 'Chris' and 'Alpha' 11.7, 'Daniel' 11.0, 'S4002' 5.6, 'S18' 5.5, 'Vanessa' 5.1,'Sonya' 4.3, 'Rica' 3.4, 'S1008' 3.2 and 'Sierra' 2.5 though 'Alpha', 'Textura', 'Daniel', 'Mascha', 'S1007' and 'Showpiece' produced best quality flowers while 'S1008', 'Codys Color' and 'Petite' produced poorest flowers. The potential barrier to breeding caused by flower unavailability can be overcome by the use of GA_3 sprays to stimulate flowering. Treatment of *Spathiphyllum* with GA_3 sprays induces plants to flower throughout the year (Henny, 1981). Treatment generally consists of a single foliar spray of 250 ppm GA_3. Following treatment, plants flower within 3-5 months depending on the time of year. Also GA_3 induces more number of flowers.

Some cultivars produce good quality flowers after treatment while others do not. Treated plants may exhibit narrowing of new leaves, stretching of the petioles and distorted flowers. Prior to any large scale growth regulator treatment, growers should test small samples of each cultivar for phytotoxicity and growth regulator response. All growth regulators should be applied carefully and uniformly over the entire crop to ensure consistent results. Never apply growth regulators when plants are under stressful conditions. Application of the growth regulators may not be proper in all growing areas. Many growers have concerns that currently labeled growth regulators will eventually be eliminated or their application severely restricted. Spathiphyllum breeders are aware of these concerns and have recently introduced new cultivars that naturally exhibit improved branching and flowering habits.

Post Harvest

The flowers of *Spathiphyllum* could be cut through the peduncle with or without the leaf sheath. *Spathiphyllum* spp. are severely injured when exposed to very low temperatures. The ideal temperatures for shipping of *Spathiphyllum* foliages and whole plants are 10-16 °C.

Insect-Pests, Diseases and Physiological Disorders

Henny *et al.* (2006) have given a detailed account of insect-pests and the physiological disorders while Simon *et al.* (1994) of diseases being observed in spathiphyllum cultivation.

The arthropod pests of *Spathiphyllum* are of relatively minor importance, but include caterpillars, mealy bugs, scales, and thrips. Mealybug and scale infestations often begin by bringing infested plant material into the greenhouse. Moths and thrips invade the greenhouse by flying in from weeds and other infested plants outside. **Aphids** are pear-shaped, soft-bodied insects which vary in colour from light green to dark brown. Infestations may go undetected until honeydew or sooty mould is observed. Aphids can cause distortion of new growth or in extreme cases the infested plants can be stunted. Root aphids have been controlled with soil drenches with any of the insecticides and those infecting the aerial parts can be controlled through forced water spraying or through spraying with tobacco decoction or any one of the insecticides at 0.1-0.2 per cent dilution. **Caterpillars** (worms) feed on all the aerial parts of the plants, making holes in the leaves and the buds similar to the damage caused by the slugs, and leave the excreta on the plants while crawing from one part to the other. **Fungus gnats** are small black flies some 3 mm long. Its larvae spin webs similar to spider webs, on the soil surface and feed on roots and root hairs, lower stem tissues and on the leaves touching the surface of the soil. Spraying the soil surface or soil drenching with 0.2 per cent Metasystox will control this pest. No algal growth should be allowed to occur and watering should be done as less as possible. Certain nematodes which feed on such insects may also be used to control these. **Mealybugs** appear as white cottony masses in leaf axils, on the lower surface of leaves and on the roots. Honeydew and sooty moulds are often present and infested plants become stunted, and with severe infestations plant parts begin to die. Because of the white powder covering on their body it is not easy to control them, however, before planting the soil should be sterilized to kill their eggs if the soil is already infested. Metasystox at 0.3 per cent will also control this pest. **Scales** feed on leaves, petioles or stems by which plants stunt, weaken and die. Rogor at 0.2 per cent is effective in controlling the scales. **Thrips** rasp the juice of the plants by which leaves curl and distort and help in spreading 'tomato spotted wilt virus'. These are controlled by spraying with 0.2 per cent Metasystox. **Whiteflies** including the sweet potato whitefly feed on the underside of leaves making small yellow spots. This may also be controlled while controlling the thrips.

Two most common pathogens affecting *Spathiphyllum* production are *Cylindrocladium spathiphylli* causing **root** or **brown rot** (chlorotic lower leaves and wilting) and *Phytophthora parasitica* causing **wilting**, root die-back, leaf chlorosis and discolouration. *C. spathiphylli* is most prevalent during warm summer months, particularly between 20-26 °C temperature range, pH 5-9 and around 90 per cent RH, especially for spore germination though temperatures above 34 °C and below 15 °C are quite unfavourable for this fungus as recorded by Wu *et al.* (2004) on *S. pallas*. Its infection initially exhibits reddish-brown lesions on roots which spreads rapidly and causes total root collapse. Since spores move from one plant to another via water so care will have to be taken that roots and soils of healthy plants do not come in contact with contaminated water. Affected plants should be removed while the remaining ones should be treated with trifumizole. In Japan, this fungus was recorded for the first time from Gifu city in 1998 in greenhouse grown spathiphyllums causing root and petiole rot (Horiuchi *et al.*, 2000). *P. parasitica* colonizes soil and its zoospores swim across the wet leaf surfaces causing black lesions on leaves and finally die back. In this case also the plants should be kept on raised benches so that soil or water of one plant may not come in contact with healthy plants, leaf surfaces should be kept dry, infected plants should be removed and the rest should be treated with either fosetyl aluminium or metalaxyl. Sometimes spathiphyllum plants are infected with *Pythium* spp.which can also be controlled similar to *Phytophthora* infection.

There are four additional pathogens which are encountered to a lesser extent, but under appropriate conditions can also cause extensive damage. These four additional pathogens are *Myrothecium roridum* (**wilt**), *Rhizoctonia* spp. (**root rot**), *Sclerotium rolfsii* (**Southern blight**) and *Xanthomonas campestris* pv. *Dieffenbachiae* (**leaf necrosis**). *Myrothecium* causes extensive damage to young leaves and stems, especially tissue-cultured plantlets damaged by excessive handling. It develops mounds of dark black spores on infected tissues which release thousands of spores that spread between plants by splashing water. Infected plants should be removed and the rest treated with triflumizole or thiophanate methyl + mancozeb. *Rhizoctonia* and *Sclerotium* both colonize the soil and move to other plants through soil medium. *R. solani* produces small hyphal mats (sclerotia) which splashes between the plants and *S. rolfsii* forms many small hard round sclerotia which vary in colour from yellow to brown. Such infected plants should be discarded and remaining ones should be treated with PCNB. *Xanthomonas campestris* pv. *dieffenbachiae* bacterium though causes necrotic water-soaked lesions on leaf margins but is not a problem as the crop is grown usually through tissue culture and also if it is grown away from other aroids, especially *Anthurium, Dieffenbachia* or *Syngonium*. This bacterium moves through splashing water so care should be taken not to wet the leaf surface when watering, and by removing the infected leaves. However, application of fosetyl aluminium lowers its outbreak.

Saturated soil medium causes necrosis along the leaf margins and its collapse, sparse root system and that too with black root tips. These problems aggravate during winter months (cold weathers) when plant's water requirement is less but the soil is saturated which causes poor soil aeration. Reducing the frequency of watering and use of potting medium with high porosity will reduce this problem significantly. Due to deficiencies of Fe and Mn in the soil, plants develop reduced growth with chlorotic leaves, which aggravates the situation during cold weathers. Increase of temperature (18.3 °C and above) and aeration in the soil will improve this situation as incorporation of Fe and Mn nutrients in the soil will be of little benefit to the plants if weather is cold. Due to **excessive light** or **temperature,** leaves may become pale, chlorotic to necrotic (burnt), curled and with burnt tips and margins, so the growing temperatures should range from 18.3-31.0 °C and level of light should be less than 25,000 lux, for optimum production. However, use of nitrogenous fertilizer will improve the plant greening. Young plants take about 9-15 months (at higher growing temperatures plants flower earlier than grown under lower temperatures) for flowering but sometimes **lack of flowering** occurs mainly due to prevailing winter growing temperatures coupled with poor growth due to improper nutritional application. Flowering may be induced earlier through spraying the plants with 250 ppm GA_3 and this induces the flowering after 12 to 16 weeks of its treatment.

References

Bunting, G.S. 1960. A revision of *Spathiphyllum* (Araceae). *Mem. New York bot. Gard.*, **10**:1-53.

Chen BiHua, 2000. Studies on tissue culture and rapid propagation of *Spathiphyllum* 'Xiangshui' (Chinese). *J. Fujian Coll. Forestry*, **20**(3): 273-275.

Dewir, Y.H., Hahn EunJoo, Jeong Hyuncheol, Hong SoonDal and KeeYoeup Paek, 2003. Rooting and growth of micropropagated *Spathiphyllum* shoots in various mixtures of growing media. *J. Korean Soc. hort. Sci.*, **46**(4): 269-274.

Eeckhaut, T.G.R., S.P.O. Werbrouck, L.W.H. Leus, E.J. van Bockstaele and P.C. Debergh, 2004. Chemically induced polyploidization in *Spathiphyllum wallisii* Regel through somatic embryogenesis. *Plant Cell, Tissue & Organ Cult.*, **78**(3): 241-246.

Heemers, L. 1999. Effect of season and plant density on the vegetative and generative development of *Spathiphyllum* (Belgian). *Verbonsnieuws*, **43**(14): 30-32.

Heemers, L. and E. Oyaert, 2001. Light reduction during summer cultivation of *Spathiphyllum* (Belgian). *Verbondsniwuws*, **45**(15): 28-29.

Henny, R.J. 1981. Promotion of flowering in *Spathiphyllum* 'Mauna Loa' with gibberellic acid. *HortScience*, **16** (4): 554-555.

Henry, R.J. 2000. Breeding ornamental aroids. In: *Breeding Ornamental Plants* (eds Callaway, D.J. and M.B. Callaway), pp. 121-132. Timber Press, Portland, Oregon (USA).

Henny, R.J., A.R. Chase and L.S. Osborne, 2006. A Foliage Plant Research Note. *CFREC-RH*, pp. 91-32.

Henny, R.J., D.J. Norman and T.A. Mellich, 1999. *Spathiphyllum* cultivars vary in flowering response after treatment with gibberellic acid. *HortTech.*, **9**(2): 177-178.

Horiuchi, H., H. Hagiwara, S. Izutsu and Y. Taguchi, 2000. Occurrence of *Cylindracladium* root and petiole rot of *Spathiphyllum* (Japanese). *Proc. Kansai Pl. Protect. Soc.*, No. 42, pp. 15-22.

Kaçar, Y.A., M. Mazmanoðlu, Y.Y. Mendi, S. Serçe and S. Çentiner, 2005. The effect of cytokinin type and concentration on multiplication rate of *Spathiphyllum* (Fam. Araceae). *Asian J. Pl. Sci.*, **4**(4): 401-404.

McConnell, D.B., J. Chen, R.J. Henny, S.V. Pennisi and M.C. Kane, 2003. Growth responses of *Spathiphyllum* cultivars to elevated production temperatures. In: *Asian Plants with Unique Horticultural Potential ~ Genetic Resources, Cultural Practices and Utilization* (eds Lee JungMyung and Zhang Donglin). *Proc. XXVI intern. hort. congr.*, held at Toronto (Canada) on August 11-17, 2002. *Acta Hort.*, No. 620, pp. 273-279.

Pogroszewska, E. and J. Hetman, 2004. Effect of substrate in the growth and blooming of *Spathiphyllum* Schott. (Polish). *Folia Universitatis, Agriculturae Stetinensis, Agricultura*, No. 94, pp. 153-157.

Simon G, M. Elliott and R. Mullin, 1994. *Florida Plant Disease Control Guide*, Vol. 1. Plant Pathology Dept., University of Florida, Gainesville FL. 362 pp

Vogelezang J.V.M. 2003. Effect of root-zone and air temperature on flowering and growth of *Spathiphyllum* and *Guzmania minor* 'Empire'. *Scientia Horticulturae*, **49**(3-4):311-322.

Wu HuiXiong, Jian Wang, Jun Wang, BiungZhan Cen, QianZhu Yu and JingDu Yao, 2004. Study on biological and pathogenic characteristics of the brown rot of *Spathiphyllum pallas* (Chinese). *J. South China agric. Univ.*, **25**(4): 30-34.

Yeh, D.M., L. Lin and C.J. Wright, 2000. Effects of mineral nutrient deficiencies on leaf development, visual symptoms and shoot-root ratio of *Spathiphyllum*. *Sci. Hort.*, **86**(3): 223-233.

39

Streptocarpus (Family: Gesneriaceae)

Sapna Panwar, Namita, Sanyat Misra and R.L. Misra

[**Common names:** Cape primrose (*Streptocarpus rexii*), False African violet (*S. saxorum*), Red nodding bells (*S. dunnii*)]

Introduction

Streptocarpus is very popular amongst flowering houseplants with wide array of flower colours. It is native to southern Africa and Madagascar islands (http://web.extension.illinois.edu.). *Streptocarpus rexii* and *Streptocarpus johannis* are perennials and native to the humid tropical areas of South Africa (Kimmins, 1992). In its natural habitat, the species is found to be growing on shady areas of rocky cliffs, rock crevices, *etc.* The name *Streptocarpus* is derived from two Greek words, *streptos* meaning 'twisted' and *carpus* meaning 'a fruit', as the slender seed pods are longitudinally spirally twisted. There are about 132 recognized evergreen perennial species under *Streptocarpus.* The first species reported is *S. rexii*, which is involved in development of many cultivars suitable for indoor and outdoor cultivation. In 1826, *Streptocarpus rexii* was introduced to the Royal Botanic Gardens, Kew (England) by James Bowie. Most of the modern *S.* × *hybridus* is the result of hybridization of various species in which *S. rexii* and *S. johannis* are probably the major species involved. *Streptocarpus* is used in landscape garden to provide an aesthetic beauty as bedding, border, in hanging baskets, as pot plants, *etc.* The pendant types of *Streptocarpus* are best suited for hanging baskets while upright and tall types for pot planting. They are highly valued when in flower for their large beautiful flowers which bloom for longer duration. It also makes excellent cut flower, especially the long-stemmed varieties. It is suitable for planting in the borders, as bedding plants, in the conservatory or as house plants, in the hanging baskets, especially the spreading types and as pot plant but these require additional humidity during hot weathers. Hutchings (1989) also suggested use of *Streptocarpus* sp. in gynaecological problems and leaf infusions of some species of *Streptocarpus* are given in Zulu medicine to ease birth pains (www.plantzafrica.com).

Marston (1964) reported that the ornamental value of the genus *Streptocarpus* was earlier not exploited fully. However, the genus has been recognised as a potential pot plant with commercial values in seventies of 20th century in the United States (White, 1975; Widmer and Platteter, 1975) and Europe (Davies, 1974; Krause, 1974) as now various scientific reports are available on *Streptocarpus* (Carne and Lawrence, 1956; Lawrence, 1957; Lawrence and Sturgess, 1957; Appelgren and Heide, 1972). In 1947 in England, cultivar 'Merton Blue' was crossed with *Streptocarpus johannis* resulting in production of cv. 'Constant Nymph'. In 1952, Weismoore hybrids were introduced in Germany. During 1960s, the radiation treatment applied resulted into the var. 'Nymphs'.

Botany

These are tufted or clump-forming, compact and rosetted with basal leaves, or shrub-like with pairs of small leaves on branched stems, or unifoliate with a quite enlarged cotyledon (30-35 cm long in the wild but 15-20 cm long when cultivated) but seldom with a few small basal leaves, and the plant is usually monocarpic, dieing after the flowering. New leaves usually emerge out from the upper bases of older leaves (Marston, 1964), and in this case the leaves are oblong and hairy, never stop the growth and continue to grow throughout the life. Its inflorescence is cymose with up to six flowers, the flowers are tubular to funnel-shaped opening with five spreading and rounded lobes, and corolla in various shades of white, violet, blue and bicolour (Bailey and Bailey, 1976), followed by slender pod-like and spirally twisted capsules. Flowers develop from the leaf, originating from the upper portion of the midrib of a leaf in young plants. Its diploid chromosome number is 2n=32 (Lawrence, 1957).

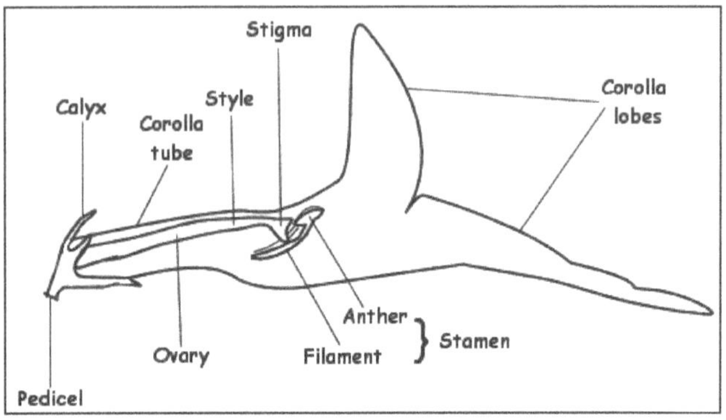

Flower Structure.
***Source*: www.streptocarpus-info.com.**

Genus *Streptocarpus* has two main subgenus:

I. **Subgenus *Streptocarpus*:** It produces plants with strap-like leaves, grows either with numerous leaves emerging directly from the ground in the form of irregular rosette or as a single leaf from the plant.

II. **Subgenus *Streptocarpella*:** It produces plants with stems and leaves, and flowers emerge directly from the leaf axils.

Propagation

Streptocarpus is easily raised through **seeds** and this process is adopted even in case of hybrids apart from the F_1 hybrids (Royle, 1979). Its seeds are very minute, counting 32,000-35000 seeds per gramme. Seeds are sown thinly but without being covered in a seed tray containing mixture of loam, coarse sand and peat moss (2:2:1½). The optimum temperature for seed germination should be 18-24 °C. After sowing, germination takes about 10-14 days under interrupted mist, in light or dark conditions. When seedlings develop a true leaf, they need to be pricked out into boxes and transplanted later into 8-10 cm pots singly. Fertilization with 100 ppm 20-20-20 fertilizer can be started when seedlings are 30 days old (Aimone, 1985).

Vegetative propagation is done commonly through **leaf cuttings** and such plants require quite less time to develop, *vis-à-vis* remain true to the type. Leaf cuttings from the mid portion of the plant are taken from young healthy leaves in spring or early summer. The leaf is given a cut either half way along the midribs or they are cut across the leaf at 5cm intervals to produce numerous sections. Basal end of the leaf is inserted in the seed tray containing mixture of equal parts peat-free compost and perlite as the medium (www.rhs.org.uk) and the seed-trays are shifted to a high moisture greenhouse at 21 °C. The roots develop within three weeks and this way some 20-30 plantlets are obtained within 2-3 month period (Aimone, 1985; Kimmins, 1992). Heide (1964) when had cut along the leaves at each side of the main vein, removed and placed in a rooting medium with the cut-edge downwards at 20-22 °C, between 15-20 days the rooting occurred and an average of 26 plants were found formed per half-leaf. After proper root system in these plants have developed, these are shifted to pots. Kathan (1989) produced small plants by growing 2.5-3.5 cm long cuttings of *S. saxorum* in pots under mist at 18/16 °C day/night temperatures and these plants were transferred to 5.5 cm pots after 30 days, where soon these plants were ready for sale. Stock plants grown under 9 h short days than 15 h long days produce comparatively more plantlets (Yelanich, 1997). *Streptocarpus* can also be propagated through **division** of clumps of older plants.

Possibility of *in vitro* propagation in *Streptocarpus* was also reported by various workers (Handro, 1983; Raman, 1977; Simmonds, 1982, 1985b). *In vitro* cultures were performed using leaf sections of 1.5-2 cm square pieces in MS medium supplemented with different auxins and cytokinins.

Species and Varieties

The genus comprised of approximately 132 species, of which the most important species which contributed to development of most of modern cultivars are *S. candidus, S. caulescens, S. dunnii, S. fasciatus, S. floribundus, S. formosus, S. gardenii, S. holstii, S. johannis, S. primulifolius, S. rexii, S. roseoalbus, S. saxorum, S. silvaticus, S. wendlandii, etc.* The details pertaining to the most cultivated species are furnished herewith.

Streptocarpus caulescens

An erect perennial native to East Africa, this is a shrubby species growing up to 75 cm tall, bearing dark green, fleshy, velvety-hairy, elliptic to ovate and up to 6 cm long leaves. Flowers violet or white patterned with violet stripes, some 2 cm long having slender tubes appearing in cymes from the leaf axils intermittently.

Streptocarpus dunnii

A native to South Africa, it produces grey-green, oblong-ovate, velvety-hairy and single crenate leaf 30-45 cm long with undulated margins which may have downward curving. Inflorescence arises from the base of the plants, growing up to 25 cm panicle, bearing stalked and clustered rose to brick-red, drooping and foxglobe-like flowers up to 5 cm long with narrow funnel-shaped tube, appearing in May and June.

Streptocarpus holstii

A native to East Africa, this is a erect-growing species similar in appearance and height (45 cm high) to the shrub-like species *S. caulescens* with succulent branching stems though possesses comparatively more slender stems that are swollen at the nodes. Leaves are deep green, opposite, ovate, lower surface being prominently veined and comparatively less hairy. Violet-like blue-purple flowers some 2.5 cm across with white throat appear freely from June to September.

Streptocarpus × *hybridus*

This is of garden origin with a height of 22-30 cm, started evolving about a century ago, and has evolved through contribution of about a dozen of species, mainly *S. rexii* and therefore the plants are also similar in appearance to *S. rexii* but the tufted leaves are mid-green, shortly strap-shaped, corrugated, hairy and larger, and small-clustered *cum* variably coloured (shades of white, red and purple) flowers are also foxglove-like (funnel-shaped), 4-8 cm long and appear from May to October. In the international markets their F_1 hybrids 'Concord' and 'Melody' and hybrid strains such as 'Mixed Hybrids', 'Prize Strain', and 'Triumph Hybrids' are available as seed mixtures. Its varieties 'Constant Nymph' (blue-purple, throat darker veined) and 'Merton Blue' (purple-blue with white throat) are quite popular.

Streptocarpus nobilis

A native to Nigeria, this species was first grown through seeds at John Innes Institution in England (Lawrence, 1943) where during summer it grew to 75 cm high producing only one leaf and scores of deep purple and cleistogamous flowers. This requires at least 11 h of daylength for full development.

Streptocarpus primulifolius

A native to South Africa, this tufted to clump-forming species bears 30-45 cm long, arching and tongue- to strap-shaped leaves. Floral peduncles grow up to 25 cm in length, bearing 1-4 narrowly funnel-shaped blue-purple flowers some 6-11 cm long.

Streptocarpus rexii

A native to South Africa, *i.e.* Knysna district in estates of George Rex, therefore it is named *S. rexii*. It is rhizomatous tufted to clump-forming stemless perennial growing up to 30 cm high and is sensitive to frost. Its deep green, rosette, shortly strap-shaped, narrowly oblong, hairy, puckered and crenate leaves are from 10 to 30 cm long. The inflorescence some 10-15 cm long is a cymose cluster consisting of 1-6 funnel-shaped violet to blue-purple flowers which are tinged white and some 4-7 cm long (http://hortuscamden.com). The flowers appear in May and June. In cultivation, this is frequently represented by its modern hybrids. Wóycicki (1947) obtained fertile and prolific F_1 hybrids by crossing it with *S. polyanthus* and the resultant hybrid dominated the leaf characters of *S. rexii* while the floral structure of *S. polyanthus*, though inflorescence, flower size and fruit characters were intermediate.

Streptocarpus saxorum

A native to Tanzania (East Africa), this woody-based, semi-succulent, semi-prostrate, sub-shrubby, rounded and evergreen species forms the low mounds of hairy and slender stems 15-30 cm long, bearing dark to mid-green, fleshy, softly hairy, some 3 cm long and elliptic to ovate leaves which appear either in pairs or in a whorl of three. Solitary to paired violet-like blue to lilac flowers some 2.5-4.0 cm across with white tube and mouth appear on long and slender peduncles from the upper leaf axils between April and October where three lower lobes are larger than the upper two. This species is suitable for hanging baskets. Kathan (1989) reported this species flowering more abundantly in summer than during winter with flowers taking about two weeks to open and lasting for 16-22 days.

Streptocarpus wendlandii

A most spectacular species native to South Africa, it grows from 45 to 75 cm tall with 30-45 cm spread producing solitary downward-curving hairy leaf which is deep green above and red-purple beneath. Its 22-30 cm long, rounded and branched panicles appear in May and June from the base of the leaf, bearing violet-like blue-purple and white flowers.

Most of the *Streptocarpus* cultivars available to growers and consumers have been developed through hybridization involving different species, and were reported to be first cultivated in the early 1800s in South Africa (Marston, 1964). The use of radiation has also contributed for the development of certain cultivars (Broertjes, 1969; Davies and Hedley, 1975). Broertjes *et al.* (1969) irradiated the leaves of *Streptocarpus* cv. 'Constant Nymph' with X-rays at various dosages where 4.5 krad and above dosages killed the leaves but 3 krad dose was found optimum, however, when they treated 41 leaves with fast neutrons, a total of 2,800 plantlets were obtained with 850 clear mutants, out of which four outstanding ones were named as 'Blue Nymph', 'Mini Nymph', 'Nelta Nymph' and 'Purple Nymph'. Their treatment of leaves with colchicine in combination with irradiation produced several tetraploids, including 'Cobalt Nymph'. The breeding objectives in *Streptocarpus* had been to produce symmetrical and balanced plants resistant to certain biotic and abiotic stresses, and to reduce the leaf length and number. The most popular cultivars are 'Albatross' (tetraploid, white with yellow throat), 'Concord Blue' (blue), 'Crystal Ice' (flowering all the year round with white flowers, veined blue), 'Falling Stars' (light blue flowers with prominent veining and white throat), 'Fiona' (pink), 'Harlequin Blue' (plant compact with masses of flat bicolour flowers with lower petals yellow while upper blue), 'Heidi' (blue), 'Joanna' (frilled, velvet red), 'Karen' (magenta pink), 'Kim' (purple, throat white), 'Lisa' (pink with white throat), 'Maassens Wit' (white), 'Margaret' (deep purple), 'Marie' (dusty purple), 'Mighty Mouse' (blue with white throat), 'Nicola' (semi-double, pink), 'Olga' (cerise), 'Paula' (blue), 'Rosa Nymph' (pink), 'Snow White' (white), 'Susie' (red, throat yellow), 'Tina' (pink), *etc.*

Cultural Practices

Temperature is one of most critical factors in *Streptocarpus* culture. Temperature of 18 to 21 °C for 2-3 weeks at initial stage is optimum for its growth. A night temperature of 16-18 °C is recommended to be optimum for the best performance, though day temperature exceeding 27 °C becomes detrimental (Widmer and Platterer, 1975). At SD and 15 °C the plants remain vegetative, and even when moved to SD at 25 °C, these remain vegetative (Simmonds, 1985a). Modern *Streptocarpus* hybrids are reported to be day neutral (Widmer and Platterer, 1975) though there are reports that long days may improve flowering, however, *S. nobilis* is a short day plant as Simmonds (1982) and Handro (1983) recorded quicker flowering under SD at 25 °C though plants were eventually found flowering under LD at 25 °C. *S.* × *hybridus* plants fail to produce flowers during winter when light level is low which could not be improved even when LD as night breaks was provided from 2200 to 0200 h though at high irradiance for 15 h plants flowered quickly with more number of flowers (Widmer and Platterer, 1975). The optimum **light** requirements for flowering in *Streptocarpus* is stated to be 1,000- 3,000 fc by Yelanich (1997) though Bodnaruk and Tjia (1991) reported it only as 1,000 fc. *Streptocarpus* prefers bright but indirect light exposure as direct exposure causes scorching so in a greenhouse or conservatory proper shade arrangement during summer should be done. Light levels below optimum results in the large-sized leaves with a few or no flowers so it may be supplemented through fluorescent lights indoors. Kimmins (1992) reported that *Streptocarpus* plants require about 12.9 klx light levels for quality flower production which is slightly higher as required for African violets. However, the minimum light levels for *Streptocarpus* should be 500 fc (Dole and Wilkins, 1999). When *Streptocarpus* is grown under 15 h of daylength, it results in production of more than 50 per cent more flowers compared to those provided with 9 h daylengths. They can also be grown in east and west facing windows. It requires the humidity levels of 60-70 per cent as optimum. In many advanced countries, growers inject supplemental **CO_2** at 500-1,000 for commercial production of *Streptocarpus* in the greenhouses.

Light and well-drained **medium** is found to be best for overall growth. Widmer and Platteter (1975) reported use of media combinations (1:1:1) of peat:moss:perlite/sand/vermiculite with a media pH of 5.5 to 6.0 (Kimmins, 1992; Maria *et al.*, 2004). Many growers prefer the media mixture of 4:1:1 potting soil:perlite:vermiculite. The crop requires 4-7 months from propagation to becoming over.

Generally, young seedlings and plantlets are grown in pots either in 10-cm one or in a 13-cm pot, hence, pot to pot **spacing** is very important. A 15 × 15 cm spacing is recommended for 10 cm pots and 20 × 20 cm for 13 cm pots. Generally, one plant per pot should be adequate for pot sizes upto 15 cm, however, for a large basket of 20-25 cm in diameter, 2-3 plants may be planted. Spacing is also essential to avoid spread of *Botrytis* and other diseases. Hence, it is recommended that transplanted plants should be kept at a proper spacing where leaves of the neighbouring plants do not touch each other.

Streptocarpus plants have been characterised with fine root system and there are many chances of being easily overwatered. Therefore, the soil should be allowed to dry in between the **waterings** (Aimone, 1985; Kimmins, 1992). During hot summers, water can be spread on the plants twice a day but in winter water spraying is normally not required and watering should be adjusted in such a way so as to maintain moist conditions. *Streptocarpus* is susceptible to root rot and the biggest danger of overwatering is that it kills the roots and causes the plant to wilt. Surface watering in this crop is the best option but once irrigated the water should not stagnate in the field. Balanced **fertilization** is required as it will increase both quality and quantity of flowers. The fertilizer which is rich in potash, this at ¼th strength along with irrigation is recommended, however, high dose is avoided as it damages the leaves and the root system (www.plantzafrica.com). The nitrogen and potassium should be used regularly at the rate of 75-100 ppm. Nitrogen deficient conditions will result in reduction of leaf size. Balanced fertilizer in 20-20-20 at $^1/_4$ or $^1/_8$th the strength is recommended for the container-grown plants. Burghardt (1967) recommended 4.5-6.0 g of a chloride-free complete fertilizer (12:12:20) per litre of substrate to sustain the plants for a period of 7.5 months with some 30 flowers per plant, 2/3rd being applied as basal dose and remaining 1/3rd as top-dressing.

Repotting should preferably be done during spring which is the beginning of the new growth, and to accommodate in the proper size of the pot this process may be repeated every six months so that these may not become woody and remain in active growth. *Streptocarpus* has a shallow root system, so it does not require potting in deeper pots. Avoid destroying too many fine roots of the plants near the soil surface. During repotting avoid use of heavy soils which retains more water. Only fast-draining soils are better for *Streptocarpus* cultivation. Old, yellow, brown, blemished or diseased leaves are regularly removed for good appearance and to check further spread of diseases if any. Regularly plants should be groomed by removing the spent flowers so that look is better, growth becomes active and coming flowers have the same vigour.

Growth and Development

The use of growth regulators in *Streptocarpus* influences the vegetative and flowering phase. Gibberellic acid$_{4+7}$ sprayed at 100 ppm is reported to accelerate the flower development and to increase the number of flowers if applied 30 days of transplanting when the buds are less than 3 mm long (Lyons *et al.*, 1985; Orvos *et al.*, 1989). GA_3 treatment is responsible for production of more elongated, upright and to less brittle leaves. Hess (1959) reported that GA is not a substitute for cold treatment. Heide (1967) when halved the *Streptocarpus* leaves after removing the midrib, he recorded 26 plantlets per half leaf and when these plantlets were grown at three temperatures and three light intensities given continuously and in two other daylengths from December 8, first plants to flower on January 22 were those grown with 6,000 lux at 21 °C, and those flowered after March 13 were grown in normal daylight at the same temperature but the daylength had no any effect on the time of flowering. Under all the light conditions, flowering was much slower at lower temperatures of 15 °C and 18 °C, irrespective of the light intensities. Where increased light intensities for 8, 16 or 24 h per day was given at a constant temperature of 21 °C, rate of flowering was not affected due to daylength though vegetative growth was best with 6,000 lux given for 16 h/day, however, the number of flowers per stem was also greatest irrespective of the daylength. In an effort to convert trailing habit of *S. saxorum* to upright form, Kathan (1989) treated the plants from November 22, 1987 to March 1, 1988 with 1.5 per cent CCC or 0.125 per cent daminozide which caused the plants becoming compact and upright.

Post Harvest

Streptocarpus plants can be transported at 10-16 °C (not below 10 °C or above 16 °C) when 1-2 flowers are open, and after reaching to the destination the retailers should place these at a well illuminated place after unpacking. *Streptocarpus* is sensitive to ethylene gas as it results in wilting of blossoms, so Nowak and Rudnicki (1990) recommended STS sprays at 0.3 to 0.6 mM one week prior to dispatch so that flower drop is slowed down and ethylene damage is prevented.

Insect-Pests and Diseases

Mealy bugs (*Pseudococcus, Planococcus*), root mealybug (*Rhizoecus* sp.), larvae and adults of vine weevil (*Ottorhynchus sulcatus*), thrips (*Frankniliella occidentalis*), aphids, cyclamen mites (*Phytonemus pallidus*), Tarsonemid mite (*Polyphagotarsonemus laetus*), whiteflies and leafhopper (*Hauptidia maroccana*) infest *Streptocarpus* in various ways (Aimone, 1985; Yelanich, 1997). The mealybugs can be controlled by wiping off with a cotton swab dipped in rubbing alcohol. During summer, introduction of a ladybird predator, *Cryptolaemus montrouzieri* in the greenhouse will feed on the mealybugs. Spraying malathion and insecticidal soap will also control these pests. Root mealybugs are though difficult to control but drenching with malathion will reduce their population. Vine weevil grubs can be controlled by introducing one of the pathogenic nematodes, *Heterorbabditis megidis* or *Steinernema carpocapsae* through watering in the medium during summer. Pirimiphos methyl or bifenthrin or lindane dust is also effective. Adult weevils can be collected at dusk and killed. Thrips can be controlled by introducing a predatory mite, *Amblyseius degenerans*. Malathion-based insecticides are also effective. Aphids are controlled through use of tar oil wash in winter and by the use of dimethoate, pirimiphos methyl, permethrin, bifenthrin or pirimicarb. Frequent misting of plants during hot weather is recommended for reduction in mite infestation, however, considerably infested plants should be uprooted and burnt as there is no any effective control of mite. Pyrethrum or insecticidal saop spray to the underside of the leaves will control the whiteflies. Permethrin, bifenthrin or pirimiphos methyl are also quite effective. Leafhoppers are controlled through the insecticides such as dimethoate, fenitrothion, heptenophos, malathion, permethrin, pirimiphos methyl and pyrethrum.

Crown rot (*Pythium* spp., *Phytophthora* spp. and various other soil- and water-borne fungi and bacteria) appears as water-soaked lesions at the crown portion (the junction of the stems and roots) of the plants and then collapse of the whole plant. It should be dealt with very promptly to save the infected plant. The affected areas should be cut well back into healthy tissue. To prevent further spread of the disease, diseased plants are discarded and proper sanitation is maintained. It is recommended to always use sterile propagating and potting mixtures. Grey mould (*Botrytis cinerea*) occurs when there is ample humidity and poor aeration as occurs when plants are crowded. Its control is difficult as the fungus is so widespread, so while planting proper spacing should be provided so that plants neither overcrowd nor air may stagnate, and overhead watering should also be avoided. All the injured and dead plant parts should be removed and proper sanitation should be maintained. Carbendazim should be used to control the disease. **Powdery mildew** affects generally the upper leaf surface and then spreading even to entire foliage and other plant parts and distort them, especially the young leaves and such leaves may fall prematurely. Infected parts should be removed promptly and the plant should be sprayed with bupirimate, carbendazim, mancozeb, triforine or sulphur.

References

Aimone, T. 1985. Cultural notes *Streptocarpus* (Cape primrose) family: Gesneriaceae, Genus, species: *Streptocarpus hybridus*. *GrowerTalks*, **48**(11): 22.

Appelgren, M. and Ola M. Heide, 1972. Regeneration in *Streptocarpus* discs and its regulation by temperature and growth substances. *Physiol. Plant*., **27**(3): 417-423.

Bailey, L.H. and E.Z. Bailey, 1976. *Streptocarpus*. In: *Hortus Third: A concise Dictionary of Plants Cultivated in the United States and Canada*, pp.1079-1088. Macmillan, New York.

Bodnaruk W. H. and B. Tjia, 1991. Gesneriads for the Florida Gardener. *Circular* No. 465, pp. 1-4. Florida Cooperative Extension Service, Institute of Food and Agricultural Sciences, University of Florida.

Broertjes, C. 1969. Mutation breeding of *Streptocarpus*. *Euphytica*, **18**: 333-339.

Broertjes, C., L. Leffring and B. Leuning, 1969. New varieties of *Streptocarpus* produced by irradiation. *Vakblad Bloemist*., **24**: 806-807.

Burghardt, H. 1967. Nutrient supply for *Streptocarpus* (German). *Dtsche Gartnerbörse*, **67**: 609-611.

Crane, M.B. and W.J.C. Lawrence, 1956. *The Genetics of Garden Plants*, pp. 79-80. Macmillan and Co. Ltd., London.

Davies, D.R. 1974. New AYR Cape primrose able to cope with low light and power economies. *The Grower*, **81** (12): 563.

Davies, D. R and C.L. Hedley, 1975. The induction by mutation of all year round flowering in *Streptocarpus*. *Euphytica*, **24**: 269-275.

Dole, J.M. and H.F. Wilkins, 1999. *Floriculture Principles and Species*, pp. 530-533. Prentice Hall, New Jersey, USA.

Handro, W. 1983. Effects of some growth regulators on *in vitro* flowering of *Streptocarpus nobilis*. *Plant Cell Reports*, **2**: 133-136.

Heide, O.M. 1964. *Streptocarpus* (Norwegian). *Garteneryket*, **54**: 762-763.

Heide, O.M. 1967. *Streptocarpus* – root formation and flowering (Norwegian). *Gartneryket*, **57**: 8-11.

Hess, D. 1959. The effect of gibberellic acid on development and the nitrogen and nucleic acid contents of *Streptocarpus wendlandii* (German). *Naturwissenschaften*, **46**: 408-409.

Hutchings, A. 1989. A survey and analysis of traditional medicinal plants as used by the Zulu, Xhosa and Sotho. *Bothalia*, **19**(1): 111-123.

Kathan, J.G. 1989. *Streptocarpus saxorum* (German). *Deutscher Gartenbau*, **43**(12): 770-772.

Kimmins, R.K. 1992. *Stemless Streptocarpus Hybrids*. In: *Introduction to Floriculture* (ed. Larson, Roy A.), pp. 298-300. Academic Press, San Diego, CA, USA.

Krause, W. 1974. Constant Nymph may have opened door for the easy launching of other varieties. *The Grower*, **81**(22): 1049.

Lawrence, W.J.C. 1943. Photoperiodism in *Streptocarpus*. *Gdnrs' Chron*., **113**: 156.

Lawrence, W.J.C. 1957. Studies on *Streptocarpus*. IV. Genetics of flower colour patterns. *Heredity*, **11**(13): 337-357.

Lawrence, W.J.C. and V.C. Sturgess, 1957. Studies on *Streptocarpus*. III. Genetics and chemistry of flower colour in the garden forms, species and hybrids. *Heredity*, **11**(3): 303-336.

Lyons, R.E., R.E. Veilleux and J.N. Booze-Daniels, 1985. Relationship between GA application and phyllomorph length in *Streptocarpus*. *J. Amer. Soc. hort. Sci*., **110**: 647-650.

Maria, C., D. Stana and I. Pop, 2004. *Streptocarpus – Flowering Pot Plant -Propagation and Culture*. *Nat. Bot. Hort. Agrobot. Cluj*, XXXII, pp.15-19.

Marston, M.E. 1964. The morphology of *Streptocarpus* hybrid and its regeneration from leaf cuttings. *Scientific Hort.*, **17**: 114-120.

Nowak, J. and R.M. Rudnicki, 1990. *Postharvest Handling and Storage of Cut Flowers, Florist Greens and Potted Plants.* Timber Press, Portland, Oregon, USA.

Orvos, A.R., R.E. Lyons and R.L. Grayson, 1989. Effect of GA_{4+7} on flower initiation and development and vegetative growth of *Streptocarpus* × *hybridus* Voss. 'Hybrid Delta'. *Scient. Hort.*, **41**: 131-140.

Raman, K. 1977. Rapid multiplication of *Streptocarpus* and gloxinia from *in vitro*-cultured pedicel segments. *Zeitschrift fur Pflanzenphysiologie*, **83**: 411-418.

Royle, D. 1979. *Streptocarpus* strategy, *The Grower*, **92**(14), 28, 30, 35.

Simmonds, J. 1982. *In vitro* flowering of leaf explants of *Streptocarpus nobilis*. The influence of culture medium components on vegetative and reproductive development. *Canad. J. Bot.*, **60**: 1461-1468.

Simmonds, J. 1985a. Effect of temperature on flower induction of *Streptocarpus nobilis* (abstr.). *Pl. Physiol.*, **77**(7): 110.

Simmonds, J. 1985b. *In vitro* propagation of leaf tissue of *Streptocarpus nobilis*. *Biologia Plantarum* (Praha), **27**(4-5): 318-324.

White, J.W. 1975. New and renewed pot plants-*Streptocarpus*. *Pa Fl. Grow. Bull.*, No. 279, p. 3.

Widmer, R.E. and R.J. Platteter, 1975. *Streptocarpus* (Cape primrose) culture. *Minnesota St. Florist's Bull.*, pp.3-5.

Wóycicki, S. 1947. Hybrids of *Streptocarpus rexii* × *S. polyanthus* (Polish). *Reports Warsaw Scientific Soc.*, Sect. IV (Biol. Sci.), **39-40**: 40-56.

Yelanich, M. 1997. *Streptocarpus*, pp 122-124. In: *Tips on Growing Specialty Potted Crops* (eds Gaston, M.L., S.A. Carver, C.A. Irwin and R. A. Larson). Ohio Florists' Association, Columbus, Ohio, USA.

40

Succulents Other than Cacti

Sanyat Misra and R.L. Misra

Introduction and Origin

The word succulent has derived from the Latin *succus* or *succulentus* which means 'juice', referring to the plants being excessive juicy (thick and fleshy) in certain of their parts. These generally show one of two growth forms such as (i) swollen stems and/or its branches where leaves are reduced or absent, and the (ii) swollen leaves where these act as water-storage organs and these are normally small but plump. Sometimes third type is also found and this is root-type stems, *i.e.* caudex. All these organs adapt to store water. Some succulents store water in their stems in the same way as cacti and may have leaves, while others are leaf succulents having fleshy leaves with arrangement of water storing in their tissues. It is believed that all the succulents evolved from other related plants growing in a normal environment. Succulents may not be limited to one or a few species under a genus, one or a few genera in a family or only one or a few of the families in the whole plant kingdom. It is widespread and may occur in widely varied plant families. The classical habitat of the succulents, especially of the cacti is hot semi-desert (xeromorphic) though species extend north and south of Central America to British Columbia and Patagonia, and ascend to over 3,050 m in the Andes. Where no rain falls, and the region is quite arid, no cacti can be found, though deserts having infrequent rainfalls are home for many of cacti and other succulents. They are usually recorded from the harsh environments where water is scarse. Certain other succulents, such as *Agave, Echeveria, Sedum* and *Yucca* are native to America, and are found growing together with cacti along with other xerophilous non-succulent plants. Original homes for majority of the succulent plants other than cacti are African deserts, notably the Karroo; some species from parts of southern Asia such as Arabia, India, Iran and Turkey; and also from the countries of the Near East. It is the frequent drastic adversities in climatic conditions, especially the rainfall as well as habitat that these plants have learnt to store water in their tissues to survive the prolonged drought conditions by making most economical use of water, *vis-à-vis* making their firm settlement in the unfavourable habitats. Their distinctive appearance is due to fleshy water-storing tissue which expands when moisture is plentiful and contracts during droughts. Flowers of these succulents vary greatly in shape and size from genus to genus. Since this publication has a separate chapter dealing exclusively with cacti, hence here only non-cacti succulents are being dealt with.

Cytogenetics

Panda and Das (1995) stated that since succulents comprise of various diverse groups therefore their chromosome numbers also differ. Basic chromosome number as present in the pollen grains and egg cells is 11 in Stapelieae, 7 in Aloineae, 9 in Aizoaceae, 10 in succulent euphorbias, 30 in *Agave* and *Yucca* with diploids being very common as these are the ancestral forms, along with polyploidy and aneuploidy with odd chromosome numbers as the later derivation. In *Haworthia tessellata* the numbers vary in multiple of

14, *i.e.* 14, 28, 42 and 56 in different populations. Even a single plant in a few members under Aizoaceae in its large watery cells of succulent tissue may contain up to 16 times the basic numbers but these cells not being involved in reproduction process, produce only normal haploid pollen and egg cells. Even in the forms also the chromosome numbers may differ as in Aloineae with basic chromosome number n = 7, each set is made up of four long and three short chromosomes, whereas in *Agave* and *Yucca*, 5 large and 25 minute chromosomes have been observed in each set.

Propagation

Other succulents can be successfully grown from seeds and cuttings of leaves or stems, and through division of clumps or offsets, as well as through viviparous bulbils or plantlets as are formed in certain genera or species. It takes one season to several years for the seedlings to attain proper size specimen because of slow growth rate in certain species. Small seedlings require special care for their nourishment. As certain succulents cannot be propagated from cuttings or offsets because of their growth habits, raising them from **seeds** becomes necessary. Moreover, it is compulsory to raise the hybrids resulting from inter-generic or inter-specific crosses from seeds. Raising plants from seeds is, however, not difficult but cumbersome. Seeds are preferably germinated in greenhouses in a medium commonly known as seed compost. A good seed-compost is composed of equal parts of well decomposed leaf mould and fine sand. A little methyl parathion powder may be mixed with the compost to keep off the ants which may take away the tiny seeds after sowing. The compost should be sterilized before sowing. Seed is sown in a moist growing medium and then kept in a covered environment, until 7–10 days after germination, to avoid drying out. A very wet growing medium causes to rot the seeds and seedlings. A temperature range of 18–30 °C is considered ideal for germination, soil temperature of around 22 °C promotes best root growth. Low light levels are adequate during germination, but afterwards semi-desert succulents need higher light levels to produce strong growth. Succulents grow well in cool, humid and partially shaded situations. A light intensity of 1,500 to 3,500 fc may be optimum for most of the species.

Reproduction by **cuttings** makes use of parts of a plant that can grow roots. Some succulents produce 'joints' that can be detached or cleanly cut off while others produce offsets that can be detached and planted. Stem cuttings can be made ideally from relatively new growth. It is recommended that any cut surface be allowed to dry for several days to several weeks until a callus is formed over the cut surface. Rooting can then take place in an appropriate growing medium at a temperature of around 22 °C. This method has some limitations also. The most suitable time to take cuttings is from March to September. The plant from which cuttings are to be taken should not be watered for few days and thus the plant is forced to utilize its excess water reserve if any. This operation helps the cutting to be ready for use quickly and chances of rotting are minimized. Cuttings are taken from the mother plants with a sharp knife so that the cut is clean and horizontal. Immediately after cutting, the cut ends should be put with fine dry sterilized sand and kept in shade for 4-5 days depending upon the size of the cutting. Then the cutting should be planted in a compost as suggested earlier under glass or polythene cover. August and September are the most ideal months for Delhi growers. Generally, roots strike within 12-15 days of planting. In some cases more time may be required.

Some succulents other than cacti which are difficult to root, slow-growing or some special clones where through seeds these may not appear true, are propagated through **grafting** by using fast growing stocks. *Euphorbia nerifolia* is used as understock in case of *Euphorbia obesa, E. picidermis, E. valida* and certain others. *Pachypodium lamerei* is used for *Pachypodium geayi* and *P. lamerei* and the *Pereskiopsis velutina* rootstock has been found compatible for *Didiera madagascariensis*. Certain members of the Didieraceae on the vigorous *Allaudia procera*, Apocynaceae on oleanders, Asclepiadaceae on the tubers of *Ceropegia woodii* and a few other species on *Stapelia*, the elongated-stemmed species of Crassulaceae are grafted on certain *Kalanchoe* species and so on (Das and Panda, 1995). Grafting should be carried out during dry and warm weathers in a clean environment to avoid infection.

Genera, Species, Hybrids and Varieties

There are more than 50 plant families of succulents spreading to some 10,000 species, the family Cactaceae being the largest one as there are some 134 genera with about 1,650 species out of which all are succulents except *Pereskia*. Cacti have been taken up as a separate chapter, hence in this chapter these are excluded. The plant families with members of succulents are Acanthaceae, Agavaceae (Nolinaceae), Aizoaceae, Aloeaceae, Amaryllidaceae, Anacardiaceae, Apocynaceae, Araceae, Araliaceae, Asclepiadaceae, Asphodeliaceae, Asteraceae, Basellaceae, Bombacaceae, Bromeliaceae, Burseraceae, Cactaceae, Caricaceae, Chenopodiaceae, Cochlospermaceae, Commelinaceae, Convolvulaceae, Crassulaceae, Cucurbitaceae, Didiereacae, Dioscoreaceae, Eraniaceae,

Euphorbiaceae, Fouquieriaceae, Geraniaceae, Gesneriaceae, Icacinaceae, Labiatae, Liliaceae, Malvaceae, Menispermaceae, Moraceae, Moringaceae, Orchidaceae, Passifloraceae, Pedaliaceae, Peperomiaceae (Piperaceae), Phyllanthaceae, Phytolaccaceae, Portulacaceae, Rubiaceae, Sterculiaceae, Tamaricaceae, Urticaceae, Vitaceae, Xanthorrhocaceae, Zygophyllaceae, *etc.* which all may be quite diverse. The common ornamental succulents with their families are being described below as per description given by Bailey (1930), Šubik and Kaplická (1968), Everard and Morley (1970), Hay and Beckett (1971), Nicholson *et al.* (1980), Hellyer (1982), Herwig (1986), Randhawa and Mukhopadhyay (1986), Beckett (1985, 1987), Virginie and Elbert (1989), Huxley *et al.* (1992), Brickell (1994) and Anderson (2003). Certain families may have succulent members but of little value in their aesthetic taste so these are not being described here. Cactaceae, Crassulaceae, Liliaceae, Euphorbiaceae, Aizoaceae, Asclepiadaceae and Agavaceae are the dominant succulent plant families. The most popular genera of succulents are being described here below.

Abromeitiella

See ***Deuterocohnia*** under the chapter on **Bromeliads**.

Adansonia (Bombacaceae)

Adansonia is named after M. Adanson, French botanist. It is a genus of 10 deciduous or semi-evergreen species of shrubs and trees from South Africa, closely related to *Bombax* (*Salmalia*). They are grown for their curiously swollen trunks, their shining and palmate foliage and as its foliage give protection against fierce sun. They are propagated through seeds. ***Adansonia digitata*** (African calabash tree, baobab tree, cream of tartar tree, monkey-bread tree; Hindi – Gorakh imli) inhabits arid, sandy environments throughout tropical and southern Africa. It is a giant deciduous tree growing the heights of 14-21 m and the circumference of the massive bole being from 25-43 m, mostly the width in the old mature trees being much more than the height, appearing disproportionately large in comparison to the small crown. In diameter, it is the most thickened tree in the world. Though the trees are generally indestructible except the ravages by the elephants, and when young due to strong winds and cyclones. However, due to excessive loss of water from the wood, the tree collapses rapidly into a pile of bleached fibrous material as it is basically a succulent tree. Botanists through dating of the wood through radiocarbon method have shown its age approximately 1,000 years. Under the South African Forestry Act 1941, this is a protected tree. The leaves are palmate with 3 lustrous green leaflets in young plants though 5-7 in the old specimens. The flowers are long-stalked, solitary, pendent, fragrant, 15 cm across, almost white and with 5 reflexed petals, purplish anthers and 7-10-rayed stigma at full anthesis, appearing after the leaves and these open by midnight and fade by afternoon of the first day. The flowers open wide, something like a spreading hibiscus. The fruits are sausage-shaped, some 30 cm long, woody when ripe, brown and edible. White fruit pulp in which various black seeds are embedded, after extraction of the seeds when dried to a powder and mixed with water, makes a drink rich in citric acid and tartaric acid. Fresh pulp can not be used by human beings due to its rapid discolouration and unpleasant smell but provides food for the baboons. Wood after long pounding and processing yields spongy fibre which is difficult to chop.

Adenia (*Modecca*; Passifloraceae)

Aden is the Arabic vernacular name for *Adenia venenata*, the type species of *Adenia* collected in Yemen. *Adenia* is a herbaceous to undershrubby and highly succulent genus of about 20 species of vines and shrubs from East Africa, Madasgascar, Australia and Myanmar and which may have grossly thickened stems, thorns and stems (branches) terminating into tendrils. Leaves are alternate, glandular at leaf bases, entire or 3-5 palmately lobed. Flower stalks are axillary and terminate into a tendril, greenish or whitish flowers are unisexual, male flower with 5 very narrow petals are included within the floral tube with 4-5 stamens and staminodes or disc glands, female flowers with 10 staminodes or disc glands in 2 series and a stipiate 1-celled ovary, and both types of flowers with a specially designed corona, and the fruit is a coriaceous or a fleshy 3-valved or indehiscent capsules. These are propagated easily through cuttings. ***Adenia senensis*** (*Clemanthus senensis*, *Modecca senensis*) is a tall climber from Senna, Zambesi and other parts of tropical Africa. The plant is glabrous and glaucous and bears pale-green leaves that are glaucous beneath, about 2.5 cm long and 5-parted with narrow and obtuse segments, petals rudimentary and within the calyx, calyx narrow campanulate, 2.5 cm long, pale-yellow and corolla-like. ***Adenia spinosa*** from Transvaal forms thickened stems ranging from bottleshaped to those shaped like formless boulders up to 1 m high and 2.4 m in diameter, with smooth or bumpy surfaces. Quite short to sufficiently long branches, oftenly with thick thorns and sometimes even with tendrils, cover the entire surface. Juvenile plants may differ drastically from those of the adult ones as the latter may develop 1.8 m across tubers with smooth green rind though juveniles normally bear bottleshaped tubers up to 20 cm long with a rubbery brownish-green smooth skin. Branches are generally many and hghly spiny. The short-lived leaves are small, quite soft textured, 3-lobed, about 3.75 cm long and oval. Certain other commonly grown species are *A. fruticosa, A. glauca, A. keramanthus* and *A. volkensii*.

Adenium (Apocynaceae)

The name *Adenium* derives from the Arabian country Aden (now a part of the Yemeni Arab Republic), the country from which the first species was described. It is a genus of 15 species of semi-succulent shrubs, most of which have the unusually thickened main trunk at the base, sometimes with formation of a tuber. Its native haunts are tropical and sub-tropical Africa and arid regions of Arabia. These bear somewhat thickened and lustrous deep green leaves which are short-lived and fall off during the dry season in their natural habitats though in cultivation these remain on the plants and provide an exact contrast to their tubular, colourful and 5-petalled flowers. They are propagated through cuttings or by seeds. ***Adenium obesum*** (desert rose) from drier parts of northern tropical Africa, right across from the west to the east and in Arabia, is very slow-growing erect shrub, attaining the maximum height of 3 m, and its thick succulent tissues are filled with poisonous milky sap. The stems and branches are succulent and its base is curiously thickened (caudex) which in form appears a miniature 'baobab' though the poison it contains interferes with the nervous action of the heart, similar to that of digitalin. In cultivation, flowers and foliage are seen together due to upsetting of its flowering rhythm by providing all the congenial atmosphere throughout the year but in the wild the flowers appear on bare branches. Glossy deep green and ovate leaves, tapering towards the base, slightly fleshy and 4-10 cm long appear at the end of the stems. Bright pink flowers some 5 cm long are carried in terminal umbel-like cymes, first appearing in summer and continue till autumn. Now it has developed varieties in various shades, starting from white to red via cream, yellow, orange, pink and scarlet. Its form ***A. o. multiflorum*** grows taller than its parent and bears white flowers with pink margins. ***A. somalense*** from East Africa though resembles *A. obesum* in form but has glaucous green leaves with wavy margins, and pink or white flowers are smaller.

Adromischus (Crassulaceae)

Adromischus is an evergreen genus of 50 species of perennial succulents from South Africa, closely related to *Cotyledon*. It takes its name from the Greek *hadros* meaning 'thick' and *mischos* meaning 'a stalk', referring to the short and thick stems of the plants. These are small plants whose ovate to lanceolate, rounded, thin or fleshy and clustered leaves often marbled to darker hues, look attractive all the year round. They bear small reddish or white 5-petalled flowers in summer. They are multiplied through seeds or cutting of stems or leaves. ***Adromischus alveolatus*** from Cape and Namaqualand is a tuberous rooted and clump-forming succulent with erect stems. Leaves are whitish-green, rough-textured to tuberculate, globular to obovoid or narrower and up to 4 cm long. Several green and maroon flowers some 1.5 cm long are borne on a stem about 15 cm tall. ***A. cooperi*** bears thick, grey-green and wavy-margined leaves splashed with purple. Reddish hairs seen at the base are, in fact, aerial roots. ***A. cristatus*** from South Africa produces short-stalked, 2.5-4.0 cm long and paddle-shaped green leaves with wavy margins. Whitish-red flowers appear on 15-20 cm long and erect stem. ***A. festivus*** (plover's eggs) from South Africa is a slow-growing and clump-forming succulent growing to 10 cm high and with 15 cm across spread. The leaves are grey-green but purple-blotched, very fleshy, 2.5 cm long and are held more or less erect, and egg-shaped rounded to spathulate, each one with wavy edging especially near the compressed tips. Small (6 mm long), tubular and pink flowers appear in summer on a 30 cm long floral stalk. ***A. maculatus*** (syn. *Cotyledon maculata, Crassula maculata*; common name 'calico hearts') from South Africa is a clump-forming perennial succulent shrublet, growing 6-10 cm in height and with 10-15 cm spread, with glossy green blotched brown, rounded and tip-waving leaves with horny edges. Floral stalk is about 30 cm long and bears purplish-white tubular flowers during summer. ***A. marianiae*** from Cape is a clump-forming shrublet with semi-erect and with age the bare stems, bearing fleshy green leaves mottled or spotted darker though in bright sunlight becoming flushed with pinkish or reddish-brown, lanceolate, sharply edged, 5 cm long, and a little convex and channelled above. Red and green or white flowers are borne in clusters above the leaves. ***A. tricolor*** is widespread in Cape. It is a clump-forming succulent, bearing fleshy grey-green leaves marked purplish, cylindrical to narrowly oblong-ovoid and about 6 cm long. White or with purple-tinted and 1.5 cm long flowers appear on panicles some 25 cm long. ***A. trigynus*** from Cape is a low-growing, clump-forming leaf succulent with compact habit. Leaves are fairly crowded, greyish-green with dense brown spots on either side, erect and some 4 cm long, oval to orbicular, very thick and flat. It is propagated by breaking off the leaves and rooted ia a sandy mixture. Reddish to purple-brown flowers about 1 cm long in clusters appear on a 25 cm long spike. ***A. umbraticola*** from Cape and Transvaal is a shrublet forming clumps having glaucous green leaves with or without a light marbling, obovate-cuneate to elongate-oblong, waving or rounded at the tip and some 5 cm long.

Small purplish flowers appear well above the leaves on simple or branched stems.

Aeonium (Crassulaceae)

Aeonium is the Latin name for what is now *A. arboreum*. It is a genus of 40 species of succulents from Arabia, Ethiopia, the Mediterranean region and the North Atlantic Islands. Even otherwise they are decorative but in blooms these become a spectacular sight. These have rosetted and fleshy leaves, some forming solitary rosette while certain others grow in colonies forming the clumps. Many species are in the form of shrublet, the rosettes standing on a woody stem. The rosettes may be flat-saucer shape with densely-packed leaves as in case of *A. tabulaeforme*, or a looser arrangement on top of the branched stems as in *A. arboreum*. The colour of the leaves ranges from yellow to almost black. The starry flowers with 6-12 petals are borne in terminal, rounded or pyramidal panicles. Solitary growing species die completely after flowering while a few from the colonizing ones also die though certain other clump-forming ones remain alive even after flowering. They are propagated through seeds, and cuttings of leaves and stems. ***Aeonium arboreum*** from Portugal, Spain, Morocco, Sicily and Sardinia is a thinly branched shrublet growing up to 1 m high, bearing glossy green to brown, 5-9 cm long and fleshy leaves with oldies falling off leaving scars on the stem, and yellow-flowered panicles in which flowers appear at the end of the branches during winter and spring. Its form 'Atropurpureum' differs from the type species due to its dark and glossy purple leaves. Variety 'Achwarzkopt' bears near-black foliage. ***A. decorum*** from Canary Islands is a much-branched shrublet growing to 60 cm tall, each branch terminating into 10.0-12.5 cm across rosette with bluish and spathulate leaves some 2.5-5.0 cm long and with red margins. Stems are encircled with rows of hard and whitish scales. ***A. canariense*** from Canary Isles (Tenerife) produces short, thick and thinly branched stems, bearing downy, light green, ovate to spathulate and cup-shaped leaves in rosettes with finely hairy margins. Bright yellow flowers appear during spring on up to a 60 cm tall panicled stalk. ***A. gomerense*** from Canary Isles (Gomera) is thinly branched with lax and slender habit, and grows up to 90 cm high, bearing glaucous green leaves which are red-margined, fleshy and some 6-10 cm long. It bears white flowers during spring. ***A. haworthii*** from Canary Isles (Tenerife) is though a small shrublet growing up to 60 cm tall but with free-branched stems, bearing loose rosettes of glaucous green leaves with red-edges, and yellow to rose-pink flowers appearing in spring. ***A. holochrysum*** from Canary Islands (Gomera, Hierro, La Palma and Tenerife) produces up to 1 m tall stout and fleshy shrublet, bearing to 20 cm across dense rosettes with shining green narrow leaves about 20 cm long, and conical inflorescence with yellow flowers. ***A. lindleyi*** from Canary Islands (Tenerife) produces about 30 cm long shrublets with copiously thin-branched stem, bearing bright glossy green leaves oftenly with a reddish tinge, thickly-fleshy and sticky-hairy, and golden-yellow flowers opening from summer to autumn. ***A. manriqueorum*** from Gran Canaria produces up to 1 m high branched shrublet with shining green and red-striped leaves which are spathulate and up to 9 cm long in about 20 cm across rosettes which are initially conical but later becoming flat, and pyramidal inflorescence with yellow flowers. ***A. simsii*** (syn. *A. caespitosum*) from Gran Canaria is cushion-forming succulent shrublet, bearing almost stalkless rosettes comprising of green and linear-lanceolate leaves some 3-6 cm long where margins are conspicuously armed with soft white hairs, and free-blooming golden-yellow flowers which appear during spring. When in flower, this is most attractive species. ***A. tabuliforme*** from Canary Islands (Tenerife) forms sessile and solitary rosette which is flat and plate-like and up to 50 cm across, bearing green and spathulate leaves armed with fine hairs along the margins. After attaining of 2-3 years of age, the rosette elongates into thickly branched inflorescence bearing yellow flowers and then dies. It is propagated through seeds or leaf cuttings. ***A. undulatum*** (syn. *A. pseudo-tabuliforme*) from Gran Canaria produces up to 30 cm across rosetted succulents, bearing thinly-branched silvery-grey stems growing up to 1 m tall with red-brown leaf scars. End-frilling wide leaves are to 9-15 cm long. Bright yellow flowers are borne on pyramidal inflorescence. ***A. urbicum*** monocarpic shrublet from Canary Islands (Gomera and Tenerife) grows up to 2 m high when in full bloom and bears erect, unbranched and robust stems which terminate into a solitary and about 25 cm wide rosette. Leaves are spathulate, ciliate and pale-green to glaucous with red margins. The plants attain maturity in 5-7 years and only then develop huge terminal panicle some 60 cm long with white to pink-tinted flowers, followed by seed formation and death.

Agave (Agavaceae)

Its common name is 'century plant'. It takes its name from the Greek word *agavos* meaning 'admirable' as plants appear quite spectacular when in full bloom. It comprises some 300 succulent species which are evergreen perennials occurring from southern USA into South America. These very short-stemmed ornamental plants form huge rosettes of stiff and thickly-fleshy leaves with spiny tips and are most suitable for planting in parks where these bear abundance of beautiful, large, tubular or bell-shaped flowers of greenish or brownish colour on quite stout and long inflorescence arising during summer from the centre of the rosette supplanting the growing point, florets opening in acropetal succession but normally after flowering the main plant dies throwing out several offshoots around the base, *vis-à-vis* aerial stem bulbils. However, many species are peculiar in having long life cycles, only flowering after 60-100 years spent in the vegetative condition. With the exception of *A. americana*, agaves seldom flower in Great Britain. The younger ones when small can be enjoyed indoors. These were introduced into Europe around 1521-1525 after the conquest of Mexico. ***A. albicans,*** probably from Mexico is a stemless succulent with grayish-blue, glaucous, stiff and fleshy leaves some 35 cm long and about 10 cm wide at base, and with needle-shaped narrowly grooved spines on the leaves where margines are minutely spiny. The inflorescence some 90 cm long bears nearly sessile and reddish-green paired flowers. The plant dies after the flowering. ***A. a.*** 'Albopicta' has lighter central stripe on the leaves. ***A. americana*** (American aloe, century plant, maguey) from Mexico grows up to 1.2 m with the basal rosettes of narrow, grey-green and sharply pointed toothed leaves which in the wild may go up to 90 cm in length. It bears bell-shaped yellow-green flowers about 9 cm long on 4.5-8.0 m long and branched tapering flowering stems in late summer. Rosettes die after the flowering. ***A. a.*** 'Marginata' (syn. Variegata) has yellow leaf margins, and flowers only when attains 10-30 years of age. ***A. a.*** 'Stricta' grows very tall and vigorous with narrow to wide yellow stripes in the centre of leaves or elsewhere, and where the length of leafy spike goes up to 6 m. ***A. angustifolia*** with unknown habitat is a clump-forming succulent with short-stemmed dense rosettes, stems from a smaller one to 40 cm long. Light green to grey-green sword-shaped leathery leaves are 50-90 cm long and spine tipped. Greenish flowers are borne on a narrowly pyramidal panicle up to 2.7 m in length. *A. a.* 'Marginata' on its leaf margins has wide creamy lining. ***A. atrovirens*** from Mexico is similar to *A. americana* in shape and habit though plant is little smaller and the leaves are quite dark green. ***A. attenuata*** (syn. *A. glaucescens*) from Mexico is though initially short-stemmed or stemless but after attaining the full growth, the stem growth extends 90 to 120 cm with leaves measuring about 90 cm. Stems are thick and crowned by a rosette of sword-shaped and spineless leaves. Leaves are pale-green but sometimes waxy-white beneath, ovate and tapering to a point. Inflorescence stalk is a cylindrical dense raceme more than 2 m long, strongly arching and bears larger greenish-yellow flowers held at right angles to the inflorescence axis, followed by fruits at the basal part and the viviparous plantlets towards the apex. ***A. bracteosa*** from Mexico produces clump-forming rosettes with grey to pale-green, arching, slender but unspiny tips, 30-35 cm long and margins sharply and finely toothed. It bears quite appealing cylindrical inflorescence about 2 m long bearing dense creamy-yellow flowers. ***A. desertii*** from Colorado desert of California is a wide and dense clump-forming succulent with crowded rosettes. Grey-green and channelled leaves are up to 30 cm long, triangular-lanceolate, spine-tipped and toothed. Chrome-yellow flowers are borne on 2-3 m tall panicles. ***A. falcata*** from Mexico produces dense rosettes, bearing grey-green leaves which are variously flushed reddish-brown at sunny sites, 40-50 cm long, stiff, sharp pointed and very slender. Panicles are bracted, about 2 m long and with yellowish flowers. ***A. ferdinandi-regis*** (syn. *A. victoriae-regina laxior, A. v. nickelsii*) from New Mexico is a small agave quite similar to *A. victoriae-reginae* though each rosette comprises of fewer leaves which taper gradually to a dark terminal spine. ***A. filifera*** (thread agave) from Mexico grows about 25 cm high with basal spherical rosette, bearing numerous shiny green, tapering, linear to narrowly lanceolate and stiff leaves some 2.5 cm wide and tip-spined, edges comprising of white-horny tissue which breaks up into threads, and the upper surface is marked with 2-3 white lines. The length of the flowering stem is up to 2.5 m long and the flowers are borne in dense cylindrical raceme. ***A. franzosinii*** from Mexico is similar to *A. americana* but differs in the leaf characters. Leaves are blue-white to blue-grey, longer and slimmer, as well as elegantly arching. ***A. horrida*** from Mexico does not produce offsets so it is propagated only through seeds. It bears glaucous dark green, stiff, about 40 cm long and hefty marginal spines. Flower stalk about 3-4 m in the wild bears yellow flowers. ***A. palmeri*** from SW USA forms dense rosettes bearing dark to grey-green leaves up to 1.5 m long. Yellowish-green flowers are borne on the top of a 2.5-6.5 m long stalk. ***A. parrasana*** from Mexico (Sierra de Parras) produces dense, compact and almost spherical rosette, with grey-blue, 35 cm long, obovate, fleshy and 2.5 cm long, hard and spiny-tipped leaves, and black and large hooked teeth on the margins of the upper surface. Yellow flowers are borne on a stiff and tufted panicle some 3 m tall. ***A. parryi*** from Arizona and New Mexico forms the basal rosettes of broad grey-green leaves some 30 cm long and edged spiny. Pale-yellow flowers are borne on a 2.5 m tall stem in summer. ***A. parviflora*** USA (Arizona) and Mexico grows up to 10 cm high with dense basal rosettes of dark green leaves having 10 cm length and 1.25 cm width, marked above with a few white lines though lower margins are toothed. White threads split off from the horny upper margins of the leaf. Tubular flowers are borne on a 1.5 m long simple flowering axis. ***A. polyacantha*** from Mexico produces rarely clump-forming plant but normally after the flowering, with dense rosettes which bear broadly lanceolate dark green or a little grey-green, 50-70 cm long, tapering to a slender and sharp terminal spine, and minute and dark teeth along the margins. Greenish flowers appear on a dense raceme about 2 m tall. ***A. schidigera*** from Mexico is a quite decorative plant bearing rosette of narrowly oblong and spreading dark green and fine hairy leaves about 30 cm long and 1.25 cm wide with flattened spines. ***A. striata*** from Mexico produces large rounded rosettes, bearing grey-green but striped darker and narrow leaves about 45 cm long. Green flowers are borne on a 3.5 m long stalk. ***A. s.*** 'Nana' is very slow-growing dwarf form with about 30 cm long leaves. ***A. stricta*** from Mexico grows up to 35 cm high with dense rosettes having closely packed narrow dark green leaves, about 40 cm long and 1.25 cm wide, stiff and slender, and ending in a terminal spine about 2.5 cm long. The trunk may branch out at maturity forming many rosettes. In this species, rosettes do not die after the flowering is over. About 2 m long inflorescence axis bears greenish flowers on a 75 cm spike-stretch. The plants do not die after the flowering. ***A. utahensis*** from Utah and California produces stemless basal rosettes of 30 cm long grey-green and rigid leaves which taper to a point with hooked and spiny margins. Floral stalk some 2.5 m high is produced in a cylindrical raceme with yellowish flowers. ***A. victoriae-reginae*** from Mexico is named so in honour of Queen Victoria and is one of the beautiful smallest agaves (15 cm high) which is propagated through seeds. It is a very slow-growing, basal rosetted, usually stemless succulent bearing broadly spherical rosettes of dark green and spineless leaves marked with white stripes on the sides and edges, erect, angular, stiff, thick, and centrally curving, 15 cm long and at base about 7 cm wide and tapering leaves spreading some 50-70 cm across. The leaves terminate into three black-brown spines, two small and one strong some 2 cm long and after many years it bears flowers on 4 m long stalk.

Aichryson (Crassulaceae)

Aichryson is the classical Greek name for *Aeonium arboreum*, which the botanists Webb and Berthelot assigned to this related genus. It is a genus of 10 annual and perennial succulent species from Islands of Macronesia, grown for their spoon-shaped to rounded, fleshy and hairy leaves being crowned in the form of rosettes at the top of the spreading stems or on its branches, if any. Most species are short-lived and die after flowering. The starry yellow 6 to 12-petalled flowers apear on terminal panicles. They are propagated through seeds and by single-rosetted stem cuttings. ***Aichryson* × *domesticum*** (syn. *Aeonium* × *domesticum*) is thought to be a hybrid of true *A. tortuosum*. A mound-forming loose shrublet growing about 10 cm high with more spread when aged. Rounded-spathulate leaves form a rosette some 2-5 cm wide. Yellow flowers appear freely with 7-8 petals each, during summer. *A.* × *d.* 'Variegatum' (cloud grass) is a prostrate perennial succulent, growing about 15 cm high and with 40 cm spread, and produces stems crowned by rosettes of white to cream-bordered green leaves or sometimes even the entire rosette is pure cream. Star-shaped yellow flowers are borne during spring. ***A. laxum*** from Canary Islands is an erect annual or biennial hardly growing 30 cm tall usually with repeatedly forked stems, bearing densely hairy rounded to spathulate leaves some 1.5 cm long in loose elongated rosettes. Pale-yellow 9-12-petalled flowers are borne on weak-airy panicles in summer. ***A. l. foliis-purpureis*** bears suffused purple leaves and through seeds it maintains its characters. ***A. tortuosum*** from Canary Islands (Fuerteventura, Lanzarote) is a dense hummock-forming succulent growing up to 10 cm tall, bearing sticky-hairy obovate leaves 1.2-1.5 cm long, and 8-petalled golden-yellow flowers which appear in summer. *A. villosum* from Azores (archipelago in the N. Atlantic Ocean and west of Portugal) and Madeira (the island resorts in the E. North Atlantic Ocean) is a bushy and spreading annual or biennial growing up to 20 cm tall, bearing broadly rhomboidal leaves about 2.5 cm long, and 6-9-petalled deep yellow flowers in early summer.

Albuca [Asparagaceae(Asphodeliaceae)]

From South Africa, the bulbous genus has only a few species and is gaining popularity still. Some species are so reduced that in whole of the growing season only one leaf, little more than 1 cm across appears. This genus is easy to cultivate provided it is kept fairly dry when dormant, and each protected from frost. *A. spiralis* has its leaves tightly spiraled when grown in bright light. Its bulbs are fleshy and about 5 cm in diameter and with time these cluster together. Its unbranched 15 cm tall spike appears during late summer bearing small pendulous flowers with greenish shade.

Alluadia (Didiereaceae)

This genus from Madagascar is a single-trunked shrub to tree branching well above the ground, and is often confused with members of the genus *Fouquieria* because the *Alluadia* is closely related to the Cactaceae. Some members can grow up to 15 m tall. Excepting one, all other species produce deciduous leaves emerging from the same nodes every year. In sexual orientation, they are dioecious, *i.e.* male and female flowers being on separate plants of the same species. Mature plants produce small whitish flowers. These make excellent patio plants under full sun until these become too large. Seed is rarely formed so these are propagated only through cuttings. *A. dumosa* is a shrub till its juvenility but becomes 10 m tall tree when matures. Its leaves are highly reduced, inconspicuous and look like the scales. This prefers tropical climates for rapid growth and can tolerate only mild frost. *A. procera* is a typical species in the genus as has basal branching and sends up the main single trunk though at maturity it attains 15 m height and becomes a tree. It can not tolerate even the mild frost.

Alluadiopsis (Didiereaceae)

An endemic genus with two species native to the sub-arid zones of Madagascar, this succulent is quite rare, spiny and grows in thickets. *A. marnieriana* is a shrubby species growing 2.0 m high with branching from the caudex as many of the side shoots are formed though usually only the main one is thick with further aerial branching. It produces needle-like small leaves and spines throughout the branches. The plant is not straight-grower but bears greenish-yellow flowers. It is easy to grow and is propagated through cuttings.

Aloe (Aloeaceae)

Aloe derives from the Arabic name *Alloeh.* They are found growing wild in warm and tropical parts of Africa (south of Equator, Malagasy and Arabia) in arid and semi-desert areas. It is a genus of some 330 species of evergreen leaf succulents. These are liked for their ornamental leaves and showy orange, red or yellow flowers. These are perennials with great variation in height, some being large and erect tree-like plants, some other smallest ones are clump- or mound-forming, while there are intermediates with shrubby, sprawling or semi-climbing habits with greatly elongated stems and well-spaced leaves. The tufted or rosetted leaves are generally tapering lance-shaped, mostly mottled or patterned and fleshy, and are arranged in rosettes at the end of shoots. The long-stemmed inflorescence from the leaf axils is in simple or branched racemes and well above the leaves and the flowers are tubular or bell-shaped, mostly showy with 6 narrow tepals. They are propagated by seeds, stem cuttings and through offsets. ***Aloe acutissima*** from Malagasy produces 1 m tall but reclining stems, bearing grey-green leaves tinged red and some 30 cm long which taper to a long slender point. Scarlet flowers some 3 cm long are produced during autumn on branched racemes. ***A. aethiopica*** from Ethiopia is a stemless succulent with basal rosettes, bearing glaucous green, 60-70 cm long and sword-shaped leaves which are broadly grooved and edges red-horny with sharp triangular teeth. Dark red flowers some 2.5 cm long appear on branched raceme up to 1.35 m tall. ***A. africana*** from Transvaal which has been recorded growing in the wild up to 2-4 m tall with unbranched stems and covered with old dry foliage but in cultivation its height may be checked to 30 cm. It has rosettes of 25-40 leaves up to 65 cm long and 12.5 cm wide, which are dark to bluish-green, tapering and grooved, with a row of reddish spines on the back and along the margins. Yellow to orange flowers about 3.5 cm long are borne normally on branched racemes. ***A. arborescens*** (tree aloe) is a shrubby species from South Africa growing about 2 m in height and breadth, each stem being crowned by rosettes of widely spreading, dull blue-green, up to 60 cm long, slender and curved leaves with conspicuously jagged edges. Simple and long scarlet racemes appear with masses of tubular to bell-shaped red flowers from 4 to 4.5 cm long in loose clusters in late winter and spring. Leaf sap has a soothing effect on burns. ***A. a. variegatum*** is a bushy succulent shrub, growing up to 2 m in length and breadth, with stems crowned by rosettes of blue-green, long and slender leaves with cream stripes and toothed edges. It produces numerous inflorescences of red flowers in late winter and spring. ***A. aristata*** (lace aloe, torch plant) from Cape (South Africa) is a clump-forming species, growing up to 15 cm high and with 30 cm spread, forming offsets freely and bearing stemless basal rosette of pointed grey-green, densely clustered and incurving leaves some 8-10 cm long and 1.25 cm wide, covered with white tubercles, horny white edges and long hairs from the tips. Orange-red flowers about 4 cm long appear in May-June on 30 cm high racemes. ***A. ausana variegata*** from South Africa bears 12.5 cm long and 5 cm wide and broadly triangular leaves, distinctly dotted in rows. ***A. bainesii*** from South Africa is a slow-growing tree-like succulent growing up to 18 m high with thick trunk which bears tight rosettes of 1 m long, recurved and sword-shaped leaves. Salmon-pink, red or yellow flowers about 3-4 cm long are borne during summer. ***A. barbadensis*** (syn. *A. vera*, *A. vulgaris*; common name 'medicine plant' or 'burn plant') probably from NE Africa and Arabia is a free-offset and clump-forming succulent with short stems, growing up to 60 cm in height with indefinite spread, and with compact basal rosettes of mottled-green but later becoming grey-green, tapering, pointed, to 50-75 cm long and fleshy leaves having white spots and serrated margins. Under dry and cool winters, the leaves may become red-tinted. The bell-shaped some 2.5 cm long yellow or red flowers appear on up to 1 m tall, simple or branched racemose spike during summer. Though it is a less attractive aloe, but is valued for its medicinal and cosmetic virtues. ***A. barbertoniae*** from South Africa has dense rosette of fleshy leaves, spotted white in horizontal bands, about 40 cm long and 10 cm wide with a long hair at tip. ***A. bellatula*** from Malagasy is stemless and tufted with linear, minutely warted and spotted pale-green leaves. Coral-red flowers about 1.2 cm long are borne in simple racemes during autumn. ***A. brevifolia*** from South Africa is a clump-forming succulent, growing to about 15 cm tall and with 30 cm spread, having basal rosettes with many offsets, bearing blue-green, fleshy, broadly sword-shaped, and from 7-19 cm long and 2.5 cm wide leaves with a few teeth on the edges, where underside is boat-shaped. Narrowly bell-shaped bright red flowers about 1.5 cm long on about 50 cm long stems appear in spring. ***A. bulbilifera*** from Malagasy is usually stemless and bears lanceolate and tapered leaves about 60 cm long. Scarlet flowers about 2.5 cm long are formed on well-branched panicle-like racemes, and later on main flowering stalk also produces bulbils. ***A. cameronii*** from Malawi and Zimbabwe is an erect to spreading shrubby succulent growing up to 1 m in height. Usually green leaves but often flushed coppery-red during excessive dry-cold conditions, are about 40-50 cm long with prominently toothed margins. Rich bright scarlet and slightly incurved flowers about 4.5 cm long are produced on branched racemes. ***A. camperi*** (*A. abyssinica, A. eru*) from Ethiopia and Kenya is a shrubby succulent with almost erect stems, each stem with a dense rosette of leaves which are glossy dark green with oblong white spots, highly fleshy, sword-shaped, 40-60 cm long, grooved above, large reddish teeth on margins and tips spiny and keeled. Orange-yellow flowers about 1.5 cm long are borne on branched racemes. ***A. candelabrum*** from Natal grows with normally unbranched erect stem about 2-4 m tall. Glaucous green persisting leaves about 60-90 cm long have spines on the margins and a row down the middle of the back. Orange, pink or scarlet flowers, about 3 cm long are produced on branched racemes. ***A. capitata*** from Madagascar forms stemless rosette, bearing red-tipped, fleshy, up to 70 cm long and 5cm wide leaves. This has many forms and varieties. ***A. castanea*** from South Africa bears 1-5 m tall and branched stems and 40 cm long leaves. Red-brown flowers appearing on panicle-like racemes during summer are some 2 cm long. ***A. chabaudii*** from Transvaal to Rhodesia is a stemless and clump-forming succulent, bearing grey to blue-green, lanceolate and up to 50 cm long leaves. Dull red some 3 cm long flowers are borne in panicle-like racemes. ***A. ciliaris*** (climbing aloe) from South Africa is a climbing succulent with slender, flexible, scrambling or prostrate and branched stems growing up to 5 m long with 30 cm spread, crowned with a rosette of green, narrow and linear-lanceolate leaves about 15 cm long, which on the base where joining the stem bear white teeth. Bell-shaped, 3 cm long and scarlet flowers with yellow-green mouths are borne in spring on usually simple racemes. ***A. comosa*** from South Africa grows up to 1.8 m high, bearing bluish and 35-50 cm long leaves with toothed edges. ***A. comptonii*** from Cape is a rosette, clump-forming and stemless or shortly stemmed succulent, which bears glaucous green leaves which are sometimes red-flushed. Leaves are 9 cm boader at the base with a length of up to 30 cm, and pale-brown teeth along the margins and a few even on the back. Scarlet flowers some 3.5-4.0 cm long are borne on branched raceme. ***A. concinna*** from Zanzibar has 10-15 cm long and up to 2.5 cm thick leaves arranged in rosette, outer surface convex, pale-green and covered with white elongate spots. Leaf edges are toothed and the tips little curved. Inflorescence up to 25 cm bearing 2.5 cm long and orange flowers with green edge. *Concinna* means gentle or lovely. ***A. cooperi*** from Cape has a distichous rosette, either solitary or in clumps, and growing stemless, bearing slender, tapering and channelled bright green leaves spotted-white at the base, 30-50 cm long and white-margined teeth. Pale-orange to pinkish-green flowers appear in simple conical racemes. ***A. cryptopoda*** from Mozambique, Zimbabwe and South Africa (Transvaal) is a solitary or small clump-forming stemless succulent, bearing glossy deep green leaves some 60 cm long. Reddish-orange flowers some 3-4 cm long are borne on simple raceme. ***A. dawei*** from Uganda is a branched short-stemmed succulent, bearing bluish, toothed, up to 50 cm long and 7.5 cm wide leaves. ***A. dichotoma*** from South Africa grows like a tree up to 9 m tall with thick trunk which bears leaves in spiral rosettes some 35 cm long and 5 cm wide, having small and hard teeth. ***A. distans*** from Cape produces prostrate stems which branch at the base and creep around, bearing broadly ovate and glaucous leaves which are about 9 cm long, with a few off-white tuberculate spots and large whitish marginal teeth. Scarlet flowers some 2.5-4.5 cm long appear on branched racemes. ***A. esculenta*** from Zambia and Botswana is a stemless succulent, bearing grey-green spotted white leaves though rarely flushed reddish, and measuring about 40 cm long. Flowers are reddish-orange. ***A. ferox*** (syn. *A. supralaevis*) from South Africa is a slow-growing succulent tree-species having unbranched woody stem, growing to about 4.5 m tall with 1.5-2.0 m spread, crowned by a dense rosette of sword-shaped and lanceolate, some 90 cm long and 15 cm wide and glaucous bronzy-green but pimple-surfaced leaves which are covered with brown spines on the edges and on the back. It is a very handsome species when young. It produces an erect spike of bell-shaped orange-scarlet flowers some 3 cm long which are borne densely on branched racemes in spring. Efficient purgative drug the 'cape aloes' is prepared from the exuded juice of *A. ferox* cut leaves. The juice is collected and allowed to evaporate and when the residue is solidified, it is ready for use. The racemes up to 90 cm long of red flowers appear in terminal clusters in March. ***A. fosteri*** from Transvaal is normally a stemless succulent. Its leaves are dark grey-green, linear-lanceolate and 40-50 cm long with vertical pale striping. Flowers are orange-red and yellow and some 3-4 cm long. ***A. harlana*** from South Africa is one of the most spectacularly marked aloes, especially when young. Leaves are up to 50 cm long and 15 cm wide but in juvenile plants they are light green with darker vertical markings, crowded in rosettes with short, broad, triangular, having toothed edges. ***A. haworthioides*** from Madagascar is the gem among the aloes which produces densely packed little rosettes with leaves some 2.5 cm long, 6.25 cm wide and 9-10 mm thick with white hairs lined on the thin edges and white beads on the surface. The shape and arrangement of the leaves make watering a difficult task in this case. ***A. humilis*** from Cape is an offset-forming rosette succulent growing up to 10 cm with 30 cm spread. Its blue-green, 10-15 cm long, fleshy, often erect but with incurving tips and narrowly sword-shaped leaves edged transparent white-spiny are borne in dense basal rosette. Floral stalk some 25-40 cm long bears a simple racemose spike of narrowly bell-shaped orange-red flowers with green markings in spring. A hybrid, ***A. × spinosissima*** (*A. humilis echinata × A. arborescens pachythyrsa*) is though initially stemless but later forming stems up to 1 m high, which bear blue-green, 30 cm long, tapering and serrated leaves in basal rosettes. Clustered orange-red flowers appear on some 60 cm long non-branched inflorescence. ***A. jucunda*** from Somalia is a very popular and attractive stemless or very shortly stemmed dwarf succulent forming dense clumps. Leaves are broadly ovate, spreading to recurved, dark green above with densely pale-green to white transparent dots, the edges are red-brown and sharply toothed, and the length and width of the leaves are 4.0-12.5 cm and up to 2.5 cm, respectively. It produces about 30 cm tall simple racemes with about 2 cm long pink flowers. ***A. krapohliana*** from South Africa bears stemless rosettes with 5-10 cm long and 2.5 cm wide and incurved, 6-10 cm long, and grey-blue leaves covered with bluish powder and the edges are toothed and reddish. Flowers are scarlet and tipped green. ***A. laeta*** from South Africa is stemmed to stemless with separated leaf joints. Leaves are narrow and tapering, some 15.0-22.5 cm long, surface pimpled and edges toothed. ***A. latifolia*** (syn. *A. saponaria latifolia*) from Cape is much like *A. saponaria* but more robust succulent with larger and greener leaves having less number of spots but with more elongation. ***A. longystylis*** from South Africa is stemless, bearing vertically striped leaves with pimples more on the upper surface, and the length and breadth of the leaves are 10-15 cm and 2.5 cm, respectively. ***A. marlothii*** from South Africa (Natal) to Botswana grows up to 3.6 m high though in the wild but much less in cultivation, bearing sharply toothed and showy pale- or bluish-green rosettes of broadly lanceolate leaves with 50-60 cm length and 10 cm width. Leaves are studded with small reddish spines. Orange flowers appear on branched racemes. ***A. mitriformis*** from South Africa produces stems with sprawling habit or ascending at the tips, bearing dark blue-green, ovate-lanceolate (triangular), cupped or incurved and pointed, some 15-20 cm long and 6 cm wide leaves with yellow incurving-spines on the edges. Older plants bear dense groups of prickly rosettes. During summer the dense clusters of scarlet-red flowers about 4-5 cm long appear in loose panicles on 38-60 cm long stems. ***A. ortholopha*** from Zimbabwe is stemless and forms densely leafy rosettes. The green leaves glaucous below are more or less erect, up to 55 cm long, lanceolate to narrowly ovate and with a horny-toothed margin. Yellowish-red flowers some 4 cm long are borne in a panicle to a 90 cm tall stalk. ***A. parvula*** from Malagasy is a stemless succulent. Its grey-green leaves are spreading, lanceolate, tapering to a slender point, 7-12 cm long with soft whitish teeth and the upper surface with white tubercles. Red flowers some 2 cm long are borne on 20-30 cm tall simple loose raceme. ***A. peglerae*** from South Africa is stemless and bears up to 35 cm long and 5 cm wide leaves which are boat-shaped beneath. ***A. plicatilis*** from Cape grows up to 4.5 m high with branched stems by repeated forking (a dichotomous case), every branch ending in a fan of narrow, flat and rigid leaves arranged in two parallel ranks (a distichous case). The leaves are smooth, closely layered bluish-green, opposite, about 30 cm long, strap-shaped with 30 cm length and 5 cm width and with rounded tips. The red flowers some 2.5 cm long are densely borne in simple racemose inflorescence. ***A. polyphylla*** from Lesotho produces only up to 10 cm tall stem with solitary or clustered rosettes. Leaves are grey-green with red-brown and hard tips, broadly lanceolate and tapering, upper surface with a cartilaginous margin which bears a few 5-8 mm long pale and triangular teeth. Flowers are green with purple tips and some 4 cm long. ***A. rauhii*** from Madagascar bears separated, spiralled, and white-spotted deep blue-green leaves of some 7.5 cm length and 2.5 cm breadth with horizontal growth. A very showy species. ***A. saponaria*** from South Afrca grows in dense rosettes with thick, 15-20 cm long and 5.0-7.5 cm wide, and blue-green leaves covered with lighter or white oblong spots in rows. Sometimes the leaves are tinged red. Orange-red drooping flowers are borne in a dense flower-head. ***A. squarrosa*** from Socotra and Zanzibar is short-stemmed, bearing 10-20 cm long and 18.6 cm wide leaves which are lightly spotted in parallel lines. ***A. striata*** (syn. *A. albo-cincta*; common name 'coral aloe') from South Africa grows quite large, bearing smooth grayish-blue leaves, some 50 cm long and 5 cm wide with a thin white or red lining on their margins. The flower stalks are branched with clusters of coral-red flowers. ***A. succotrina*** from South Africa is almost stemless. Blue to grey-green leaves having paler stripes and spots are about 50 cm long. Red flowers tipped-green and some 3.5 cm long are borne in simple raceme. ***A. suprafoliata*** from South Africa is a curiously sculptured small succulent bearing deep green, closely set, overlapping and concave leaves which are opposite, horizontal, large toothed, up to 32.5 cm long and 7.5 cm wide. This has many forms and varieties. ***A. suzannae*** from Malagasy produces a thick stem which bears up to 1 m long and thick leaves. White flowers tinged pink and some 3 cm long are borne on a long cylindrical raceme. ***A. thraskii*** from Natal (South Africa) grows in its native haunts in sand at the very edge of shore vegetation. It has deep chanelled, very concave and outward curving, up to 1.2 m long and 32.5 cm wide leaves which have small teeth on the margins and with the habit of *A. recurvifolia* but when in flower, the former bears multiple inflorescence while the latter only one. The flowers are composed of dense racemes bearing yellowish petals and exserted orange stamens. The stem which grows up to 3 m tall is clothed with persistent dead leaves. It is a very handsome plant when young. ***A. transvalensis*** from Transvaal is a stemless succulent forming small clumps. Pale-green and narrowly lanceolate leaves bearing tranverse bands of oval white spots, are 20-25 cm long. Pink to coral-red flowers some 1.5 cm long are borne on branched racemes. ***A. variegata*** (falcon feather, partridge-breasted aloe, tiger aloe), a very handsome clump-forming humped succulent from Cape (S.A.) with 30 cm height of the stem and 10 cm width of a rosette, is named so due to coloured dark grey-green leaves strongly marked with irregular bands of white spots. The leaves are triangular-ovate, up to 11 cm or more long and up to 3.5 cm wide in ascending order, arranged spirally in three closely overlapping ranks (rows) in rosette form, and are keeled beneath towards the base. Leaves are triangular and very concave, indented on the inner side, edges bordered with fine, white and gristly teeth. Floral stalk is a racemose spike of about 30 cm length comprising of tubular and some 20-30 flesh-red to orange flowers edged green, in loose raceme in March and April. ***A. zebrina*** from SW Africa is a dense clump-forming succulent, bearing linear-lanceolate, 15-30 cm long, patinated glaucous deep green leaves banded transversely with irregularly arranged whitish dots, and brown-tipped sharp horny teeth on the margins. Well-spaced, base inflated red flowers are borne in spring and summer in branched racemes.

Aloinopsis (Aizoaceae)

Aloinopsis derives from *Aloe*, a separate genus and *opsis* meaning 'like', though in no case it has any resemblance to *Aloe*. It is a genus of 16 species from South Africa with tuberous rootstocks, of dwarf, tufted, and perennial succulents, bearing thickly-fleshy and keel-tipped leaves shaped and marked similar to their rocky habitats. These produce daisy-like flowers opening either in the afternoon or by dusk during late summer to early spring, just above the leaf tips. These are propagated through seeds or by division. ***A. malherbei*** (syn. *Nananthus malherbei*) from Cape produces leaves which are quite fleshy, grey-green with prominently white tubercles on the margins, some 2 cm long and wide, erect and broadly spathulate. It likes limy soil. Flesh- to pinkish-brown flowers some 2.5 cm across are borne in autumn. ***A. peersii*** (syn. *Cheiridopsis noctiflora, Nananthus soehlemannii*) from Karroo region of Cape produces leaves which are glaucous, about 2 cm long, narrowly ovate-spathulate and recurved, and the yellow flowers about 2.5 cm across open after sunset in autumn. ***A. rosulata*** from Willowmore Division of Cape produces deep glaucous green, about 3 cm long, spathulate and the tip margins with whitish tubercles. Yellow flowers about 3.0-3.5 cm across are borne in autumn. ***A. schooneesii*** from Cape is a quite dwarf succulent with tuberous roots, growing up to 3 cm high with about 7 cm spread in mounds, bearing very fleshy blue-green and almost spherical leaves arranged tightly in tufts. It produces flattish yellow flowers during winter-spring. ***A. setifera*** (syn. *Titanopsis setifera*) from Namaqualand region of Cape produces glaucous to red-tinted leaves some 2 cm long, narrowly spathulate, uniformly covered with closely set tubercles and with bristle-toothed tips. Yellow flowers about 2.5 cm across appear in autumn. ***A. spathulata*** (syn. *A. crassipes, Titanopsis crassipes, T. spathulata*) from Sutherland Division of Cape produces gray-green and very broadly spathulate leaves, often with red-tinted margin, which are about 2 cm long and are covered with minute tubercles at the terminal margins. Deep pink flowers which are paler in the bud, appear about 3 cm across in autumn.

Anacampseros (Portulacaceae; common names 'love plant')

Anacampseros derives from the Greek *anakampseros* probably for a kind of stonecrop which on touching brings back love. The genus comprises of 71 xerophytic and succulent herbs from dry areas of South Africa, especially from Cape of Good Hope but one species from Australia. These are slow-growing small plants with mostly tuberous roots. The leaves are fleshy, ovate and mostly with long and hair-like or large membranous stipules coming out from their axils, the larger ones hiding the smaller leaves in some species. Short-lived 5-petalled flowers which are showy in most cases, open only in full sunshine, and sometimes only for a few hours. They are propagated through seeds or cuttings. ***Anacampseros albissima*** from Cape to Namibia is some 10 cm across mat-forming prostrate-growing species with decumbent stems, bearing closely packed minute leaves covered with white membranous stipules. It produces more than 6 mm across cream to white flowers. ***A. arachnoides*** from Cape is a cobwebbed and green-leaved succulent with simple racemes of white flowers. ***A. karasmontana*** from South Africa grows only 2 cm tall with 1.2 cm long and wedge-shaped leaves which are sometimes reddish, along with curled stipular hairs. Flowers are pink and some 2 cm across. ***A. namaquensis*** from South Africa grows up to 13 cm tall bearing white-hairy, pear- or wedge-shaped and 4-5 mm long leaves, and white stipular hairs. Flowers are white and 8-10 mm across. ***A. papyracea*** from Cape to Namibia is clustered or small clump-forming succulent with prostrate or decumbent growing stems some 5 cm long. Leaves are small and crowded and covered with bright white membranous stipules. Insignificant flowers are very small and greenish-white. ***A. rufescens*** from South Africa grows with up to 5-8 cm long erect or prostrate stems, bearing some 2 cm long obovate to ovate-lanceolate leaves which in good sunlight are red or bronze-flushed, and the stipulaar hairs are white, about 2 cm long and often wavy. Floral stalks are normally 10 cm long bearing 3-4 cm across purple or pink flowers. ***A. telephiastrum*** is widespread in cape region of South Africa. Though it is a slow-growing but with age forms mats comprising of a number of rosettes. Leaves are green, smooth, some 2 cm long, ovate and fleshy with bristly hairs. Rose-carmine flowers some 3-4 cm across appear solitary or 2-4 per floral-stalk well above the leaves.

Ananas (Bromeliaceae)

It is a genus of five perennial evergreen species from tropics of Central and South America. *Ananas* is the South American name for these plants. These are clump-forming with erect rosettes of sword-shaped, thick and spiny-edged leaves. Blue or red 3-petalled flowers are borne in spikes, followed by fleshy fruits which fuse to form a syncarp. They are propagated through suckers or by leafy shoots which are borne on the crown of the fruits. ***Ananas bracteatus*** (wild or red pineapple) from Brazil and Paraguay bears dark green leaves which in the wild measure up to 1.2 m and the width 3.75 cm but under cultivation the length is quite less, the juvenile leaves being 15-25 cm long in rosettes and the margins are sharply serrated (spiny). Flowers are lavender with pink-red bracts and the fruits are brownish-red. Its form *A. b.* 'Striatus' ('Tricolor') has leaves margined with yellow and a bronze-green centre with red spines. Its fruits are fleshy and take long time to grow and mature. This is the ornamental plant.

Antimima (Aizoaceae)

A native from Namibia to South Africa (Cape Province), *Antimima* is a symmetrically dense cushion-forming leaf succulent. Leaves are green, highly fleshy, broadly strap-shaped, alternate and form several individual rosettes to make the cushion. Pink to purple flowers continue appearing throughout the year either solitary or in clusters from the centre of the mature rosettes. These are easily propagated through seeds or by stem cuttings. Large species are most suitable for rockery though small ones for pot-planting. *Antimima granitica* is the most popular species suitable for rockeries.

Apicra (Liliaceae)

Apicra in Greek means 'not bitter'. The genus is native to Cape region of south Africa with 14 species of *Aloe*-like succulents, most suitable for planting in cactus-house or rockeries. These succulents have spirally arranged or crowded with small green leaves in such an amicable manner that whole stem is systematically covered from bottom to top. These leaves may be a few to many depending on the species. Flowers are greenish and often striped white, straight, tubular or prismatic with short, flat or spreading white limb longer than stamens. ***Apicra aspera*** (syn. *Aloe aspera, Haworthia aspera*) grows up to 15 cm high with erect stems having 3.75 cm diameter while ensheathed with leaves, leaves green, somewhat globose, acuminate and little keeled, 1.5 cm long, smooth on the convex upper surface while warty-green on the back and margins granular. Unbranched inflorescence up to 25 cm high bearing smooth rosy flowers some 1cm long, *i.e.* double the length of pedicel. ***A. bicarinata*** (syn. *Aloe bicarinata*) bears up to 30 cm high erect stems. Leaves green, smooth on concave upper surface while irregularly white-warty on the back, short-ovate, acute, keeled with rough texture, margins roughened and 1.5 cm long and 1.9 cm wide. ***A. bullulata*** (syn. *A. pentagona bullulata, Aloe bullulata*) has erect stems and the leaves are pale, 1.5 cm long and some 3 cm wide, lanceolate, acute, obliquely keeled, back irregularly white-warty and the margin and keel finely denticulate, and the flowers are yellowish. ***A. congesta*** (syn. *Aloe* *congesta*) has solitary and erect stems, some 30 cm high and 10 cm in diameter when sheathed with leaves. Leaves glossy green with granular margins, 2.5-3.0 cm long and 3.75 cm wide, broadly ovate, acuminate, flat above and keeled. Inflorescence simple and 30-40 cm high with greenish some 1.5 cm long, thrice longer than the pedicels, and the segment tips are large and spreading. ***A. deltoidea*** (syn. *Aloe deltoidea*) produces clustered stems some 15 cm high and 5.6 cm in diameter including leaf sheaths. Leaves glossy green, margin and keel serrulate, 2.5 cm long and 3.0-3.75 cm wide, distinctly 5-ranked, ovate, somewhat concave, sub-acute and low-keeled. Inflorescence is simple, some 30 cm in height, flowers yellow-green, nearly sessile, about 1 cm long with large spreading white segments but rosy tips. ***A. foliolosa*** (syn. *Aloe foliolosa, Haworthia foliolosa*) produces clustered and erect stems about 30 cm in height and some 2.8 cm in diameter including those of leaves when ensheathed. Leaves glossy green, some 1 cm long and 1.5 cm wide, broadly ovate-acuminate, flat above, acutely keeled and margin granular. Inflorescence simple and 30 cm or more in length, bearing greenish and some 1 cm long flowers which is twice the length of the pedicel. ***A. pentagona*** (syn. *Aloe pentagona, Haworthia pentagona*) produces usually solitary and erect stems up to 30 cm high and about 10 cm in diameter alongwith leaves. Leaves are green with white dots, 1.9 cm long and 3.75-5.0 cm wide, distinctly 5-ranked, broadly triangular-lanceolate, acute, and the margins little granular. Inflorescence up to 45 cm long and occasionally forked, bearing greenish flowers some 1 cm long, being twice the length of the pedicel. ***A. spiralis*** (syn. *A. imbricata, Aloe cylindrica, A. imbricata, A. spiralis, Haworthia imbricata*) produces somewhat clustered stems up to 30 cm high and 5.6 cm in diameter with those of leaves. Leaves glaucescent, 1.25 cm long and 3.0-3.75 cm wide, broadly triangular-acuminate and margins slightly granular. Inflorescence simple and about 30 cm long and bears greenish, 1.25 cm long flowers (twice the length of the pedicel).

Aptenia (Aizoaceae)

In Greek *opten* means 'wingless' as the valves of the capsules are without wings. The genus comprises only two species, *A. cordifolia* from Cape and *A. lancifolia* from South Africa. The members of this genus are fast-growing perennial succulents with free-branching trailing stems which make a good ground cover. These are propagated by seeds and stem cuttings. ***Aptenia cordifolia*** grows 5 cm high with indefinite spread due to trailing nature of its stems, and bears glossy bright green, fleshy and ovate-triangular leaves. Small daisy-like bright pink flowers appear in summer. Its 'Variegata' form bears creamy-white margined leaves.

Argyroderma (Aizoaceae)

Argyroderma derives from the Greek *argyros* meaning 'silver' and *derma* meaning 'skin', referring to the silvery-green appearance of the plants in some species. It is a genus of solitary or growing in small tufts or groups of perennial succulents of 50 species from South Africa (mainly Cape and Namaqualand), grown for their grey-green egg-shaped prostrate form, composed of one or two-pairs of extremely swollen leaves. In some of the species, the two-leaved organs are so closely pressed together to appear as almost one entity with a deep slit at the top. From the central fissure of the united leaves, daisy-like large sessile flowers emerge out. They are propagated through seeds or by careful division of the tufted kinds. ***A. aureum*** bears 2-leaved solitary shoot some 4 cm long, being connate for 2/3rd of their length then gaping apart. Bright yellow flowers some 3-4 cm across appear in late summer from the centre of the slit. ***A. blandum*** (syn. *A. delaetii*) grows prostrate and is egg-shaped some 3 cm high and 5 cm wide with two united, fleshy and silvery-green leaved shoots being such of the forms one, two or rarely more, from the fissure of such paired-leaves shoots pink-purple daisy-like 5 cm across flowers appear in late summer. ***A. brevipes*** (syn. *A. fissum*) is a clump-forming perennial succulent, growing up to 15 cm high and 10 cm in spread, with green, finger-like, 5-10 cm long, and under sun often reddish-tipped leaves. From the centre of the paired leaves, light red flowers appear in summer. ***A. framesii*** has tufted shoots with two fleshy jointed-leaves some 2 cm high where roughly half the length below is connate but opening out above to form a 6-8 mm wide gap. Rosy-purple flowers some 2 cm across appear during summer. ***A. f. minus*** is smaller in size. Various other forms are also available for commerce. ***A. octophyllum*** (silver skin) grows usually with solitary shoot which bears 2-4 leaves some 3 cm long and as wide, and jointed in pairs. The leaves are connate below, lower leaf surface strongly convex but upper flat with a wide gap in the centre, from where yellow flowers some 4 cm across appear during summer oftenly with spirally arranged petals. ***A. patens*** is a wide clump-forming succulent, each shoot comprising of paired leaves some 2.5-3.0 cm long, connate for 1/3rd below, enclosed in a sheath, and the top with wide slit and bluntly keeled. Yellow flowers some 3 cm across arise in summer. ***A. pearsonii*** (syn. *A. schlechteri*) is a prostrate-growing egg-shaped perennial succulent with 3 cm height and 5 cm width, bearing longitudinally united pair of silvery-gray and very fleshy leaves having deep slit in between the paired leaves from which some 3 cm across red flowers appear in summer. ***A. roseum*** (syn. *A. delaetii roseum*) produces 1-2 or sometimes even more shoots with 2 or 4 leaves some 3.5 cm large and a bit wider, connate for half of their length, upper surfaces flat though lower prominently convex. Red-purple flowers about 8 cm across appear during summer with lax and drooping petals over the whole plant. ***A. testiculare*** is a clump-forming succulent bearing one to several shoots, each with 2 cm long paired leaves which individually are hemispherical. White or cream flowers some 4.5 cm across are borne in summer.

Avonia (Portulacaceae)

This genus was erected in 1994 by splitting *Anacamperos.* All the species have their origin in South Africa (north-east, east and southern Africa). The species may be either fibrous-rooted or tuberous and the stems appear in clusters. These bear thin papery scale covering the much-reduced leaves. These tolerate fairly hot and bright light but no frost and wet soil. *A. papyracea* has strong taproot system on which several stems of the thickness of about 1.2 cm are formed. It prefers warm climate at the growth stage, and the frost at any stage is detrimental. Emergence of short-lived small white flowers during summer in the hot afternoon on stem tips having the same colouration as the leaf scales, normally goes unnoticed until small fruits are formed.

Beaucarnea [syn. Nolina; family – Nolinaceae (Agavaceae)]

Its common names are bottle palm, elephant foot tree and 'ponytail plant'. The derivation of *Beaucarnea* is not certain though it has only six species, carved out from *Nolina*. *Nolina* is named for C.P. Nolin (*d* 1755) of France, joint author of an essay on agriculture. The genus is native to southern USA and Mexico with 30 evergreen species, though *Beaucarnea* is described only with six species such as *B. bigelowii* (syn. *Dasylirion bigelowii, Nolina bigelowii*), *B. gracilis* (syn. *B. Oedipus, Nolina histrix*), *B. guatemalensis, B. recurvata* (syn. *B. tuberculata, Nolina recurvata, N. tuberculata, Princenectitia tuberculata*), *B. r.* var. *intermedia* and *B. stricta* (syn. *B. glauca, B. princenectitia glauca, B. purpusi, B. recurvata stricta*). It was placed under the tribe Nolineae which apart from the genera *Beaucarnea*, *Calibanus* and *Dasylirion*, also contains *Nolina*, all being so closely related that at one time these all were placed under one genus *Nolina*, all with unarmed leaves. However, *Nolina* has panicled small polygamo-dioecious flowers and wingless 3-lobed, 1-3 seeded and often inflated fruits; *Calibanus* differs from it having neither lobed nor inflated fruits; though *Beaucarnea* similar to *Dasylirion* bears 3-winged fruits which are neither lobed nor inflated, and the latter differs in having more swollen trunk-base as compared to *Nolina* (Bailey, 1930). It is a very slow-growing genus of evergreen shrubs and trees with dracaena-like leaves, grown mainly for their intriguing overall appearance. They are propagated by seeds, through stem-tip cuttings and from suckers. ***Nolina longifolia*** (syn. *Beaucarnea longifolia, Dasylirion longifolium, Roulinia karwinskiana, Yucca barrancaseca, Y. longifolia*) is a highly rough-barked 1.5-3.0 m tall shrub with a little swollen base and scantly top-branched, bearing green, thin, rough-edged and more or less gracefully pendant leaves some 1 m long and 2.5 cm wide. Inflorescence sessile or so bearing 1.6 mm long flowers, 6 mm long and 12 mm wide inflated fruits and roundish seeds with 3 mm diameter. ***Beaucarnea bigelowii*** from USA (Arizona, California) and Mexico (Baja California) though in the wild grows up to 3 m tall but in cultivation up to 1.5 m in a large container. It is an erect but slow growing woody-stemmed and usually unbranched evergreen perennial, however, the old specimens may have scarse branching or when the top is damaged. Leaves are formed in a dense radiating tuft or in globular rosette, linear, first straight then arching and up to 1 m long. ***B. gracilis*** from C. Mexico is a small tree growing up to 9 m high with highly swollen base and variously branched trunk having crowded leaves on the top of the trunk. Leaves are straight, very glaucous, up to 50 cm long and 6 mm wide and rough-edged. Inflorescence is short-stalked bearing 1-5 mm long flowers and 9 mm long and wide fruits. ***B. guatemalensis*** from Guatemala is a much-branched small tree growing up to 6 m in height with thin, green, up to 90 cm long and 2.5 cm wide, recurving and smooth-edged leaves. Inflorescence is short-stalked and bears 3 mm long flowers and 1.5 cm long and 1.25 cm wide fruits. ***B. recurvata*** (elephant's foot and pony tail) from SE Mexico grows up to 9 m in the wild though much smaller (about 2 m) when cultivated, is a slow-growing and with age or in case of broken top it is sparsely branched shrub or tree with several tufts of about 40-70 tightly top-crowded, 1 m long and some 10-12 mm wide and recurving leaves which persist on the plants even after turning yellow or brown. The trunk is markedly swollen at the base (caudex) and looks like elephant's foot in shape, colour and texture, the ratio of swelling of the base to trunk-diameter being roughly 7:1, and in 5-6 years old plant the base being 14-16 cm whereas the trunk diameter 1.0-1.2 cm. Leaves are mid-green, gradually tapering towards the free side where it becomes thread-type narrow, smooth but margins finely serrated and with up-curving to both the sides from the midrib. ***B. stricta*** from south Mexico is a little-branched small tree growing up to 9 m high, bearing top-crowded, straight, pale-green, 60-90 cm long and 6-13 mm wide, and slightly rough-edged leaves. Inflorescence is short-stalked which bears 1.5 mm long flowers and 13 mm long and 6 mm wide fruits. ***B. recurvata*** var. *intermedia* grows half to its parent.

Bergeranthus (Aizoaceae)

A native to the spring and autumn rainfall area of the eastern Cape (South Africa), this has thick rootstock with growth in summer. The plant with projected rootstock gives the impression of a bonsai. These rosette-forming plants bear mid-green, smooth, triangular and succulently thick leaves, and clustered yellow flowers appearing from the centre of the rosette. Flowers may appear at any time of the year, preferably during autumn. It requires to be watered regularly. It is propagated through seeds. The species *B. multiceps* is most common.

Beschorneria (Agavaceae)

It was named for Friedrich W.C. Beschorner (1806-1873), an amateur German botanist. A genus of 10 almost stemless evergreen perennial succulents from Mexico bearing strongly tuberous rootstocks, and narrowly lanceolate, thick-textured and erect leaves at the base in rosette form. This makes excellent tub plant. It is propagated though division of offsets or by seeds. ***B. bracteata*** is stemless clump-forming succulent with 20 to 30 rosetted leaves some 45-60 cm long and 5 cm wide. Flowering stems some 2 m tall, branched and with greenish-yellow flowers some 4 cm long which fade to red. ***B. yuccoides*** is stemless clump-forming succulent growing up to 1 m tall and about 3 m in spread, bearing basal rosettes of 15 to 20 rough and grayish-green leaves from 30 cm to 1 m long and 5 cm across. It bears bright green, tubular, pendant, 6-tepalled and 5 cm long flowers with large rose-red bracts in summer on a 1.2 m tall and unbranched red spike.

Bijlia cana (Aizoaceae)

It is endemic to the small Karoo region (western Cape of South Africa). It bears bluish-green, highly fleshy, irregularly triangular to quadrangular rosetted leaves where both the leaves in the pair are similar. To sustain them compact, these should be grown under bright sunlight. These require regular watering during active growth period in the autumn when several large yellow flower heads appear from the centre of the rosette. They are propagated through seeds.

Boophane (Amaryllidaceae)

This is native to both, summer and winter rainfall areas of South Africa. These are spectacular even when not in bloom as the bulbs produce a neat fan of undulating grey leaves. These have extensive root system so wherever it is planted the soil should be highly porous and well-drained. The genus comprises of three species; *B. disticha, B. ernestii-ruschii* and *B. haemanthoides* all bearing quite large bulbs which are toxic to cattle. One should be careful as during flowering its pollen irritate the eyes. After flowering the umbel enlarges in the process to ripen the seeds. It is easily propagated through seeds.

Boswellia (Burseraceae)

There are some 16 species of shrubs and trees from North Africa, Arabia and India, main one being *B. sacra* which is the main source of frankincense, an aromatic resin. All the species yield resin but only four yield aromatic resins though resins obtained from other species are not so aromatic. This highly basally- and aerially-branched tree grows up to 6 m high in driest conditions on rocky slopes with the trunk having papery peeling bark and the deciduous leaves are compound.

Bowiea (Liliaceae)

Bowiea is named so to honour James Bowie (1789-1869), a gardener sent out from Kew to collect plants, working chiefly in Brazil and South Africa. A genus of three species native of South and East Africa with silvery-green, globose and somewhat flattened bulbs. Though it is a bulbous plant as described by Misra and Misra (2013) but its green bulbs are fully set and exposed above the ground, hence is being described here as a bulbous succulent bearing scrambling, cylindrical and much-branched green stems about 2 m long which are herbaceous, slender and climbing and function as leaves for photosynthetic works along the green scales of the bulbs in the absence of leaves as improper small linear leaves fall off soon. Being native of seasonally dry areas, it is tolerant of irregular watering. Largely, though its value is as a curiosity but it is quite decorative in form, as its new shoots grow quite fast. The flowers are small, star-shaped and green which are produced at tips of the stems in summer. Mosly it is propagated through seeds or sometimes with offsets being produced by the bulbs. The genus consists of three species, *viz. Bowiea griepensis, B. kilimandscharica* (syn. *Schizobasopsis*), *B. volubilis* but out of this it is only *B. volubilis* which is common in cultivation hence this is being described here under. ***Bowiea volubilis*** (climbing onion) from South Africa has flattened, rounded and silvery green bulbs being formed above the ground with only base-rooting and attains up to 20 cm diameter when fully grown. Stems are cylindrical, fleshy, green and twining, bearing 6-tepalled starry greenish-white flowers some 1 cm across at the upper part of the vine from summer to autumn.

Brachychiton (Sterculiaceae)

See chapter on **Foliage Plants.**

Brachystelma (Asclepiadaceae)

In Greek, *brachys* means 'short' and *stelma* means 'garland'. The genus is native to Tropical Africa, Ethiopia, South Africa and Namibia with 100 species. ***Brachystelma barberiae*** is recorded from the arid environment of eastern Cape and Transvaal to Rhodesia, discovered by James Henry Bowker (1822-1900) from Transkei and named by him in honour of his sister, Mrs. Mary Barber. This is a small plant (12-15 cm height) with no spines though leaves are hairy. It has a hard tuber some 5.0 cm high and 7.5 cm wide with fibrous roots, short stalk, i.e 2.0 cm long and 0.45 cm thick and the zone of leaves and flowers some 8.0 cm long and 13.5 cm wide. Tuber is ridged from bottom to top as corms and dull yellow-brown, stem also dull yellow-brown, the leaves some 7.5 cm long and centrally 2.3 cm wide as to those of *Calotropis gigantea*. Though the leaves and flowers are only for a brief period but their presence make it a remarkable sight since the tips of the petals continue to adhere to one another even after opening of the blooms, making these the curious structures resembling bird cages. Though petals are white but the edges including the originating bases are deep purple.

Bryophyllum (Crassulaceae)

In Greek, the generic name means 'sprouting leaf' due to plantlets being formed on the edges of the leaves, and only this character distinguishes this genus from other related genera. The planting of a mature leaf or a piece of it in a warm moist place, the notches on the leaf margins will soon produce the young plants. It is a native of the tropical areas of both the hemispheres including Madagascar and South Africa. It is small genus famous for its foliage novelty though some produce good flowers. It produces innumerable fibrous roots, stems are usually erect and simple (sometimes base-branching), leaves are petioled, simple or pinnately compound, opposite and fleshy and the inflorescence is cymose or paniculate. Flowers are drooping with 4 petals which are united into an inflated calyx, covering nearly half of the corolla, the corolla is cylindrical ending with 4 petal tips, 4 ovaries and 8 stamens. ***B. crenatum*** *from Madagascar* grows 60-90 cm high bearing 2.5-7.5 cm long, simple and very fleshy leaves which become smaller as on the upper stems. Leaves when young are bluish-glaucous but afterwards becoming bright green with purplish margin, are ovate with cordate or rounded base, and the margins are coarsely crenate to dentate. Upper side of petioles and rachis are not grooved. Flowers are in corymbose cymes, calyx is pink and membranous and the corolla is red and ends in 4 rounded segments. ***B. pinnatum*** from the tropics of both the hemispheres grows up to 1.2 m high with ovate leaves where base is cordate or rounded, and then upward pinnatifid and then further up 3-5 short-stalked leaflets. Leaf margins crenately double-serrate, petioles and rachis narrowly grooved above. Young stems, petioles and leaf margins along the veins purplish. Flowers drooping in terminal clusters, calyx purplish-green with lighter dots, corolla greenish-white with purple shades and with spreading tips. ***B. daigremontianum*** (*Kalanchoe daigremontianum*) is an erect, simple, 60-90 cm long plant with opposite and triangular leaves which are held stiffly to the stem at an angle. These have serrated edges which curl inward and from which tiny plantlets arise. ***B. proliferum*** from South Africa is very robust and grows up to 3.6 m high. New growing stem is 4-angled but turning cylindrical, leaves pinnate to pinnatifid with quite thickened base of the pinnae, leaf margins crenate, and the petioles and rachis are conspicuously grooved. ***B. tubiflorum*** (*Kalanchoe tubiflora*) grows about 1 m high, bearing series of fleshy-tubular leaves in circle all around the stem with purplish bands and dots, and on the tips a group of plantlets are formed.

Bulbine (Asphodelaceae)

A native to Tropical and South Africa and Australia, this is named so due to bulb-like small dome-shaped tubers being produced by many of the species. These may be shrubby and weedy perennials, short geophytes or soft annuals. Most of the *Bulbine* species have tuberous roots or caudex. These resemble *Aloe* and *Haworthia*, with soft, succulent and needle-like mid-green leaves. Many of the *Bulbine* species are grown in the gardens, and *B. frutescens* is the major one. Flowers, mostly yellow (sometimes white, orange or pink) with bearded stamens, are borne either on the lax or compound racemes. *Bulbine* remains dormant from late-spring to mid-autumn, though there may be differences from species to species. In the process of being dormant, leaves die and fall down whereas roots contract into the caudex with no visible sign of life.

Bursera (Burseraceae)

This is a New World (Mexico to Arizona and California) genus which makes an excellent succulent bonsai. Though several species are typical shrubs or small trees, many bear thickened trunks with colourful or peeling bark, and numerous species are deciduous in nature. The leaves may be solitary to compound with thin leaflets. Most of the species contain terpenes in their sap so on pruning, different species give a distinctive aroma, however, many of the species produce resinous sap for burning. Flowers are inconspicuously small, each flower producing one seed. They are propagated through seeds and cuttings. These require to be protected from frost. When they are dormant, their mixture should not be wet. ***B. fagaroides*** from Mexico grows 5 m tall with thick trunks though certain attaining hardly 20 cm height and forming a natural bonsai as mature trees. The bark on the mature stems is tan and peels away in layers. Leaves are glossy green, compound and deciduous, and sometimes turning bright yellow in autumn. It requires frost-protection. It is propagated through seeds. ***B. hindsiana*** from Baja California (Mexico) attains 3 m height. Its bark is red, the leaves pubescent, light green and deciduous though a few are left on the plant for the winter, and the branch-ends bear the clustered white flowers. It is highly susceptible to frost. It is propagated through seeds or cuttings. ***B. microphylla*** from Mexico to Arizona and California grows into a small tree attaining up to 10 m height. Small plants develop thick roots long before the trunk thickens. At maturity, the trunk becomes white with peeling bark. In extreme climates the plants become natural bonsai. Its sap is aromatic and the resin is used as incense. Leaves are dark glossy green and reduced to thin leaflets, compound and deciduous. It requires frost-protection. It is propagated through seeds and cuttings. ***B. schlectendahlii*** from Mexico has reddish peeling bark when mature enough. Leaves are deciduous and single or compound with three leaflets often with a waxy coating. They can not tolerate frost. These can be propagated through seeds or cuttings.

Calandrinia (Portulacaceae)

The genus *Calandrinia* was named in honour of Jean Louis Calandrini (1703-1758), professor of mathematics and philosophy at Geneva. A genus of 150 species from WN & WS America and Australia. They are mainly low-growing, sub-shrubby and often with fleshy, linear to obovate leaves, and flat, bowl-shaped 5-7 petalled flowers from pink to red-purple which are borne in racemes or umbel-like clusters. ***Calandrinia balonensis*** (parakeelya) from Australia and in the New World from Canada to Chile is named so after the Balonne river in Queensland where it was first collected. It is found growing in the arid areas of southern Australia and the northern Territory. It grows up to 38 cm in height and has broad and fleshy leaves and large long-lasting flowers. ***C. discolor*** from Chile bears about 7 cm long, fleshy and obovate leaves which are pale-veined above and purple beneath and the bright purple-red and bowl-shaped flowers some 4-5 cm across are borne during summer. ***C. crassifolia*** from Chile is a succulent sub-shrub growing to 25 cm in height. Leaves are grey-green, thick-fleshy, oblong-ovate to spathulate and about 3 cm long. Bowl-shaped red-purple flowers some 3-4 cm across are borne in summer. ***C. umbellata*** from Peru is a trailing perennial with ascending floral stalks to about 10 cm bearing linear and fleshy leaves, and crimson-purple flowers some 2 cm across appear in umbel-like clusters during summer.

Calibanus [Nolinaceae (Agavaceae)]

Only one species, ***Calibanus hookeri*** is under cultivation. It may well be described as a dwarf relative of *Beaucarnea recurvata* with its hemispherical thickened base some 40 cm across and 25 cm high with thick bark which splits when mature, bearing 10-20 whorled, linear and sprawling leaves on the top. These leaves are dark green, tough, up to 75 cm long and 1 cm wide. It is highly slow-growing plant. With age, the caudex (the thickened base) produces offsets above the soil level. It is much of curiosity among the growers.

Caralluma (Asclepiadaceae)

Caralluma is thought to have been derived from the mistaken belief that the vernacular name 'car-allum' was used in India. A genus of some 100 succulent species allied to *Stapelia*, this has originated in the dry regions of Africa, the Mediterranean and through the arid regions of India and Myanmar. These are low-growing, bearing 4-6 angled cylindrical, tufted or clump-forming fleshy stems with rounded tubercles. Leaves are scale-like and falling soon. Flowers are small to large, showy, 5-lobed, starfish- or bell-like and sometimes with smell of rotting flesh. These are propagated through base-pulled up stems, dried for a few days (2-4) before planting singly in small pots or through seeds which germinate within a few days around 25 °C. ***Caralluma europaea*** is from Island of Lampedusa (Mediterranean), from east and west coast of West Africa to southern coast of Spain, It grows to 10 cm high with stems bearing 1.0-1.5 cm thick, quadrilateral, blunt-toothed, having greenish-brown flecks, and 5-pointed 1.0-1.5 cm across flowers appearing in 10 or more in clusters at the tip of the shoot and are pale-brown striped brownish-red. ***C. joannis*** from Morocco produces green, 6-10 cm long, 4-angled and serrated stems, bearing 2-10 clustered and velvety hairy flowers with bell-shaped base, yellow centre with red-purple merged spots and 1.2-2.5 cm across. ***C. laterita*** from Botswana, Namibia and Zimbabwe produces gray-green stems some 20 cm long with conical teeth. Brick-red flowers with velvety hairs and some 5-8 cm across are borne in cluster but smell like rotten meat. ***C. retrospiciens*** from Kenya to certain other northern African countries was discovered in 1820 from Dahlak Island in the Red Sea by Christian Gottfried Ehrenberg (1795-1876) which grows to about 2.2 m tall with hairy flowers.

Carpobrotus (Aizoaceae)

In Greek *karos* means 'a fruit' aand *brotos* means 'edible'. The genus comprises some 25-30 prostrate succulent species from South Africa with a few from Australia, America and Pacific islands. This genus was formerly included in *Mesembryanthemum*. They are mat-forming and are excellent for binding sandy soils. The genus has dark green, cylindrical to angular, fleshy and paired leaves. The members of this genus can effectively be grown in the hanging baskets. They mostly have daisy-like and colourful showy blooms. Their propagation is through cuttings and seeds. ***C. acinaciformis*** (hottentot fig) from Cape is a prostrate-growing succulent with more than 90 cm long articulated stems, bearing pale gray-green, opposite, curved and thicker at one edge and some 9 cm long leaves with keeled edges. Red-purple flowers largest in the genus some 9-12 cm across with 14 stigmas are borne in summer, followed by gooseberry-sized edible fruits. ***C. chilensis*** (sea fig) from sea coasts of Oregon to Chile is prostrate growing succulent with stems running up to 60 cm or more, bearing 3-5 cm long, thickly-fleshy and roughly triangular leaves when cross-sectioned. Red-purple flowers some 8-9 cm across are borne in summer. ***C. edulis*** (hottentot fig, kaffir fig) from Cape is a carpeting succulent growing 15 cm high, 90 cm in length and with indefinite spread, bearing prostrate and rooting branches, with bright green, 8-12 cm long and 1.5 cm thick, opposite, curved, triquetrous leaves when cross-sectioned, lower edge keeled and serrated. It produces showy yellow or purple flowers ageing pink with 8 stamens, and flowers about 12 cm across by mid-sunny-day during spring-summer, followed by fig-like brownish fruits which are edible.

Carruanthus (Aizoaceae)

Carruanthus takes its name from *carru* meaning 'Karroo', a place in Cape Province of South Africa and *anthos* in Greek means 'flower'. It is native to South Africa and the genus comprises of two species. It is short-stemmed mat-forming highly branched leaf succulent where roots are also fleshy. Leaves are green to grey-green, crowded at shoot-tips, opposite and decussate, oblanceolate to clavate, erect to spreading, narrower towards the tip, flat above and carinate beneath with the keel angle pulled forward, and the margins are dentate. Yellow flowers are mostly solitary where pedicel bears 2 bracts. ***C. peersii*** from Cape bears entire or 1-2-toothed leaves on each margin, expanded above, narrowing towards apex, some 5 cm long, 1 cm wide at base, 1.5 cm at apex and 1.4-1.6 cm thick. Yellow flowers tipped pink are some 4 cm in diameter. ***C. ringens*** from Cape bears densely crowded grey-green leaves that are oblanceolate to clavate, triquetrous, keel angle expanded, pulled forward with the dentate margins towards the tips, 5-6 cm long and 1.6-1.8 cm wide, and narrower at the base. Yellow flowers stained with red below and are 4-5 cm in diameter.

Cavanillesia (Malvaceae)

It is a massive *Adansonia*-like columnar and deciduous tree with erect, tall, smooth, cylindrical and bottle-shaped swelled solitary trunk, having umbrella-like branching only on the top. It is native to Brazil, Bolivia and Panama and grows very quickly in hot regions where after being established, this has ability to grow in dry soil. Though when in flower the tree is very attractive, but even otherwise the tree looks handsome. The tree is fully covered with a huge mass of delicate pink and white flowers, *vis-à-vis* star-shaped fruits before leaves appear.

Cephalopentandra ecirrhosa (Cucurbitaceae)

It is a monotypic genus from Ethiopia, Kenya and Uganda which forms large caudex but only a few vines. Seedlings of this plant produce light green rosette leaves on the top of the small caudex. In summer when there is sufficient warmth and provided with plenty of water, the seedlings develop vines bearing long grey-green leaves and now the caudex turns conical and more than 30 cm across. Its creamy dioecious flowers some 4 cm across appear, followed by fruits which when ripe are bright orange. It does not tolerate frost and is propagated through seeds.

Cephalophyllum (Aizoaceae)

Cephalophyllum derives from the Greek *kephale* meaning 'a head' and *phyllon* meaning 'a leaf', probably referring to dense tufts of long leaves in some species. The genus comprises about 70 species of South African origin. They are mat-forming or tufted evergreen perennials of clustered or paired green, narrow and semi-cylindrical or cylindrical fleshy leaves, bearing showy daisy-like flowers during summer but when plants have attained maturity, *i.e.* after one or two years. They are propagated by seeds or cuttings. ***C. alstonii*** (red spike) from Cape is a prostrate growing succulent some 45 cm long and with 1 m spread. Gray-green and semi-cylindrical leaves up to 7 cm long are borne in erect tufts. Very showy dark red flowers with purple anthers are 5-8 cm across and appear in summer. ***C. cupreum*** from Cape and Namaqualand grows to 45 cm or more long with prostrate stems, bearing pale-green, semi-cylindrical, keeled, upper half clavately thickened and up to 6 cm long leaves in erect tufts with close pairs. Yellow and coppery-red flowers some 8 cm across being the largest are borne in summer.

C. pillansii from Cape and Namaqualand is a large tufted clump-forming succulent growing up to 8 cm long and 60 cm spread, bearing decumbent stems with dark green dotted-darker, slender pointed, cylindrical and 8 cm long leaves, though the length may range from 5 to 15 cm, in pairs. Daisy-like yellow flowers with red centre and some 6 cm across are produced on short stalks from spring to autumn. ***C. regale*** from Namaqualand is mat-forming succulent some 30 cm across with prostrate stems. Leaves are mid-green marked with fine darker dots, tapering to a truncate tip and 5-9 cm long. Bright pinkish-purple flowers some 5 cm across with about 4 cm long pedicels are borne during summer months. ***C. surrulatum*** from Cape is a mat-forming succulent growing 30 cm or more across with prostrate stems, beaing bright green paired leaves up to 7 cm long which are triangular in cross-section, and the angles are finely toothed. Rose-purple flowers are borne up to 5 cm across. ***C. subulatoides*** from Cape is a thick and large tufted succulent bearing decumbent to prostrate, short and much branched stems. Leaves are closely paired, gray-green, semi-triangular in cross section and 5-7 cm long. Red-purple and 4 cm or more across flowers are freely borne during summer.

Ceraria (Portulacaceae)

A small genus from South Africa, it is a small succulent shrub growing from 20-150 cm tall with thickened trunk, small leaves and 5 mm across pink or white dioecious flowers that appear near the ends of the branches. All the species are excellent as succulent bonsai. In winter, they should be kept fairly dry. They do not tolerate frost. ***C. pygmeae*** is a dwarf species growing up to 20 cm high but little wider in spread. As it forms a large thickened trunk and roots, disproportionate to its size, and growing only 10 cm in a year it makes an excellent succulent bonsai. It is propagated through seeds or cuttings. Though it can bear hot and bright light and a well-drained mixture but can not tolerate frost. ***C. fruticulosa*** is a small shrub growing hardly 30 cm in height though each shoot may grow up to 20 cm long in a season, bearing some 1.5 cm long and 5 mm wide leaves in the wild while in cultivation only to its $1/4^{th}$ size. Its stems and the thickened roots have red bark. Pink 5-petalled flowers some 3 mm wide open during summer at the end of the branches. It tolerates hot and bright sunlight but not the frost. It is propagated through seeds or cuttings.

Ceropegia (Asclepiadaceae)

Ceropegia takes its name from *keros* meaning 'wax', and *pege* meaning 'a fountain', due to some species producing fountain-like flowers as if formed of wax. *Ceropegia*, a native of Natal (S.A.), Asia, Malagasy, Canary Isles and Australia has some 160 drooping, creeping or twining semi-shrubby species with fleshy branches and lantern-shaped flowers. Leaves are linear to rounded, appearing in opposite pairs, however, in some species falling soon. Flowers are tubular often with inflated base, five corolla tubes may be reflexed or tip- joined forming a cage or with expanded lobe-tips forming umbrella-like shape, and the petals are often spotted darker. These are propagated either through seeds or by cuttings. ***Ceropegia ampliata*** from SW Africa and E. South Africa is a shortly creeping succulent with thick stems bearing small and short-lived leaves which fall off before blooming. White flowers with green veins, 5-6 cm long, inflated like a small balloon at the base, tubular up to the mouth, the mouth is capped with a cage of narrow red petals, and these appear in clusters of 2-4 but at a time only one opens. ***C. caffrorum*** (lamp flower) from SE Africa grows more than 1 m in length with twining habit, bearing ovate-lanceolate to linear, 1-3 cm long and fleshy leaves. Three to five green clustered flowers with purple streaks appear which are 5 cm long and hairy with slender lobes. ***C. debilis*** from Malawi, Zambia and Zimbabwe grows more than 60 cm in length with pendant or twining stems, bearing 2-3 cm long, linear and fleshy leaves. Stems at the nodes form aerial tubers which root easily and can be used for its further multiplication. Base-inflated flowers which are expanded at the mouth with narrow lobes, are hairy, 2.5 cm long with green tinged-reddish exterior and purple-red interior. ***C. dichotoma*** from Tenerife (Canary Islands) grows up to 90 cm tall with erect and tip-branched stems which are clothed with waxy-white or purple-tinted patina. Leaves are 4 cm long, linear and deciduous. Clustered yellow flowers which are sessile, tubular and some 2 cm long appear at the nodes. ***C. fusca*** from Tenerife and Gran Canaria (Canary Islands) is very similar to *C. dichotoma* in appearance though the stems are more gray-white and flowers dark brownish-red. The plant is clump-forming, each clump producing 1-2 dozens of jointed, leafless, erect, thickly fleshy and unbranched but rarely only a very few stems bearing 3-5 forked branches at the tips, especially when the stem is somehow broken or damaged. Leaves are linear and short-lived, corolla dull reddish-brown and corona light yellow. ***C. haygarthii*** from South Africa is a semi-evergreen, climbing and succulent sub-shrub growing up to 2 m or more in length, bearing semi-succulent, oval or rounded dark green and 2-5 cm long leaves. Small, solitary but rarely clustered white or pinkish-white flowers some 4 cm long with a pitcher-shaped tube which is bent at the base appear during summer, the flowers flaring widely at the mouth but the five narrow, star-like, cream or pale-pink lobes spotted-maroon unite at the tips making about 8 mm across hairy and hollow knob. ***C. hians*** from La Palma (Canary Islands) has greater similarity to *C. dichotoma* but the stems are reddish and the flowers borne on them are yellow. ***C. sandersonii*** (parachute plant, fountain flower) from Natal (South Africa) has green and strong but entangled creeping branches often several metres long and 0.5 cm thick. The leaves are 3-5 cm long, thick, fleshy, heart-shaped, in opposite pairs, and 10-20 cm apart on short stalks. Flowers which emerge on the top of the plants, are solitary, some 7 cm long and up to 4.5 cm across with swollen base, pale-green lined and mottled with darker shades, pentagonal at the top and looking like a parachute some 5 cm across when open, short-stalked and the aperture-edges are bordered with fine teeth and radiating hairs. ***C. stapelliformis*** from Cape (S.A.) is either climbing, creeping on the ground or burrowing into the ground, the plant having dull green spotted grey-brown and dotted white, short and gnarled stems, growing up to 1.5 m long and 1.5-2.0 cm thick. Greenish-brown leaves are small and pointed. About 6 cm long and erect flowers having slender but slightly swollen tube and with funnel-shaped mouth, appear in a cluster of 2-4 on short stalks. Flower exterior white with brown spots, and interior white and hairy. The blooms are fragile and last for several days on the plant. ***C. woodii*** (*C. linearis* subsp. *woodii*; common names 'rosary vine', 'string of hearts') from Natal (South Africa) produces tuberous-rooted, slender about 60 cm long creeping or pendant succulent sub-shrub, oftenly with aerial tubers being formed at the nodes, and these tubers root easily. Leaves with short leaf stalks are purple, upper surface with strong silver marbling, 1.5-2.0 cm long, cordate and fleshy but under sun these show intense colouration. Red to reddish-brown narrow-lobed, slightly curved and hairy flowers some 2 cm long are borne from spring to autumn with base-inflated but expanded at the mouth.

Cheiridopsis (Aizoaceae)

A large genus of South African species, this produces two differently-shaped pairs of leaves, the smaller one sheathing the larger inner pair. This is a clump-forming genus most suitable for rock garden planting where only light frost occurs. During spring some 5 cm across yellow flowers appear. ***C. peculiaris*** from Cape is an exception in its growth habit, quite apart from other species of the genus. The first pair of leaves lies flat on the ground while second is upright. In winter it requires long rest at a dry place. It is propagated through seeds. ***C. vanzijlii*** resembles as the pebbles in its natural haunts and forms a cushion of rounded stems. This handsome species lacks typical growth form and bears all the leaves of equal size. It produces 6 cm across yellow flowers during spring. It is propagated through seeds and can tolerate up to non-persisting -4 °C temperature.

Chirita (Gesneriaceae)

Chirita derives from the Nepalese vernacular name for a gentian as its some species have gentian-blue flowers. The genus with 80 species is native to Indonesia, Malaysia, SE Asia and southern China. The members of this genus are evergreen perennials, sub-shrubs or annuals of varied appearance, grown mostly for their flowers. They are generally put in three groups, *viz.* (i) rosette-forming, (ii) with leafy stems, and (iii) with short erect stem topped with one very large leaf and a cluster of smaller ones. All the species bear tubular flowers with usually rounded five petal lobes. They are propagated through tip cuttings or seeds. ***Chirita lavandulacea*** from Asia (Indo-China region) is an erect annual or evergreen perennial with fleshy, erect and leafy stems, growing up to 60 cm in height and almost with the same spread having downy pale-green, elliptic-oblong and cordate leaves up to 20 cm long. Clusters of lavender-blue flowers some 3-4 cm long with white tubes appear from the leaf axils during summer and autumn.This species is highly ornamental due to its translucent and hairy leaves which set off the trusses of pale-levender flowers shaped like to those of dog flowers. Its anthers are bearded and the corolla tube is quite broad. ***C. micromusa*** from Thailand is an annual or evergreen perennial growing to 30 cm high with erect and fleshy stems topped by one large ovate-cordate leaf some 25 cm long together with several others having similar shape but quite smaller. Orange-yellow flowers 1.2-1.8 cm long with pale-yellow tube appear from the leaf axils during summer and autumn. ***C. sinensis*** from China and Hong Kong is an evergreen, stemless and rosetted perennial, growing up to 15 cm in height and some 25 cm in spread. Its leaves are rich green, stalked, 13-20 cm long, oval, thick-textured, fleshy and corrugated hairy surface but with silvery-white patterns. Clusters of lavender flowers some 3-4 cm long with yellow-marked white tubes, held well above the leaves, are produced during spring-summer.

Cibirhiza echirrhosa (Apocynaceae)

These prostrate to scrambling perennial herbs or shrublets are found in Arab Peninsula to Malagasy and dry western Africa with irregularly turnip-shaped tuber about 30 cm in diameter. Deciduous leaves are long-stalked, stalks half to as long as the leaves, opposite and broadly ovate to obovate. Umbelliform, each umbel with 5-20 flowers appear in axillary clusters where floral bracts are small, ovate and ciliate.

Cissus (Vitaceae)

Cissus is derived from the Greek *kissos* meaning 'ivy'. The genus with 34 species of woody straggling shrubs or climbers, along with 10 stem succulent species, native to tropical, sub-tropical and even a few to temperate regions of the world. They have woody or herbaceous stems bearing alternate leaves, and most with tendrils. Though in cultivation, succulent species seldom flower but others may flower with insignificant minute yellow or green-tinted and 4-petalled flowers in starry axillary and branched clusters, followed by berry-like fruits. They are propagated through stem cuttings. Out of 10 succulent species, some have been placed under *Cyphostemma*. ***Cissus antarctica*** (kangaroo vine) from Australia is a rampant woody climber growing more than 5 m long though much less in containers. Leaves are leathery and glossy, dark green above and paler beneath, ovate to oblong and some 7-15 cm with slightly toothed margins though the opposite of each leaf has a forked tendril. The greenish flowers are insignificant and appear in small axillary clusters. It is a subject of the temperate regions. ***C. cactiformis*** from East Africa grows up to 3 m in length and 0.9-1.25 m in spread against a wall bearing green and branched stems which are 4-angled and winged climbing succulents having brown and crenate horny edges. The stems are fleshy, up to 5 cm thick but constricted at the nodes, and from here aerial roots also develop. Leaves, which are mid-green, small and ovate and appear in autumn with the new growths, occur only on new growth but soon fall off in two months. It flowers and fruits rarely. It is a temperate growing species. ***C. discolor*** (Rex begonia vine) from Cambodia and Java is a woody climbing species growing to a length of some 2 m with dark red stems and tendrils. Leaves maroon beneath, little quilted and velvey-green above with silvery-white patterning between the veins, oblong-ovate to cordate, some 10-15 cm long and marginally toothed. It is not suitable for temperate areas as it needs warmth and humidity and is most suitable for a hanging basket. ***C. quadrangularis*** from southern and tropical Africa grows up to 2.4 m in length and is quite similar to *C. cactiformis* but with less stout clambering stems and non-crenate on their rough and horny edges, though these are thick (less than 5 cm), fleshy, 4-angled and winged, and constricted at nodes. It's vines bear 12.5 cm apart single leaf at each node, the leaf being mid-green, palmate and some 2.5 cm broad and long but fall off in two months duration. Aerial roots develop on the nodes. It is a very ornamental species which hangs in several strands from the hanging basket. ***C. rhombifolia*** (*Rhoicissus rhomboidea*, grape ivy, Natal vine) from Mexico to Brazil and West Indies is an evergreen woody climber with forked tendrils though in *Rhoicissus rhomboidea* the tendril is unbranched, growing up to 6 m in length. The trifoliate (irregularly diamond-shaped) leaves are initially silvery covered with pale-brown hairs but on maturity become glossy dark green and boldly toothed. It is temperate in nature. *R. r.* 'Ellen Danica' (Mermaid vine) with 3-5 leaflets, each some 6.25 cm long and deeply lobed, 'Fionia' is slow-growing, more compact, trifoliate with bright green, each leaflet some 5 cm long, deeply lobed and little broader than other varieties while 'Jubilee' leaflets are large and dark green. ***C. rotundifolia*** from East Africa and Arabia bears 4-angled velvety stems and almost round leaves which are slightly toothed, thick and about 7.5 cm in diameter. It is a good basket plant. ***C. sicyoides*** (Princess vine) from South America (especially Brazil) survives only at 65 per cent and above RH and continuous moisture. It grows up to 3 m in pots but in the wild up to 6 m. Its leaves are dark green with stalk being bright crimson, cordate, broad, some 10 cm long and pentafoliate with elliptic leaflets some 5 cm long and 2.5 cm wide.

Coccinia (Ivy gourd, Scarlet gourd; Cucurbitaceae)

A native to sub-Saharan Africa with one species in S&SE Asia, the genus comprises of 28-30 species, the best known one, *Coccinia grandis* is being cultivated as vegetable. Their succulent form is caudex in many of the species. These are easily propagated through stem cuttings.

Commiphora (Burseraceae)

A native of Africa and Madagascar, this genus comprises many of the species making fine succulent bonsai though size and growth habit of the plants and the size and shape of the leaves vary greatly from species to species, some making natural bonsais with thickened trunks, others being naturally dwarf in windswept styles. Most species are deciduous in nature and like fairly dry conditions during winter. These produce small flowers, followed by 1-seeded fruit about 5 mm across. If watering is not the limiting factor, they will tolerate intense sun but not the frost. They are propagated through seeds or cuttings. ***C. glandulosa*** from Africa is easy to cultivate and a fast-growing species which in a year or two develops thickened trunk with peeling bark, and the leaves are light green, deciduous and 3-lobed. Though uncommon in the genus, this species develops spiny short branches. It can tolerate enormous sun-heat but not the frost.

Conophytum (Aizoaceae)

Commonly known as 'living pebbles' or 'cone plant', *Conophytum* is a genus of 270 species from southern Africa, taking its name from the Greek *konos* meaning 'cone', and *phyton* meaning 'a plant', referring to the shape of their bodies as the leaves are almost completely fused, in contrst to *Lithops*, and resemble small inverted cone. The plants belonging to *Conophytum* and *Lithops* mimic their surrroundings. These have tiny spherical bodies growing among stones, so closely appear to each other that these are distinguishable only when flowers push their way out of the apices. These grow solitary or in clumps, each shoot consisting of two completely fused fleshy leaves except for a slit at the top. These grow from August to March at 10-12 °C temperature, and flowering freely during autumn, producing daisy-like flowers which normally hide the entire plant below. These are propagated by separating the congested plants in the clumps, or through seeds which may take 2-3 years to flower. ***C. albescens*** is a clump-forming succulent growing up to 3.75 cm high and 2 cm in spread with grey-green plant body which is divided into two lobes at the top marked with fine red lines on the crown. Yellow flowers 1.25 cm across appear in September. ***C. bilobum*** grows in clumps of succulent plant bodies up to 5 cm in height and 2.5 cm spread with prominent flattened lobes. The 2-lobed grey-green plant body gradually turns red and with the age several plant bodies are formed all joined to one stem. Yellow flowers 2.5-3.0 cm across appear in September. ***C. calculus*** grows to about 5 cm in height and 7.5 cm in spread with spherical and flattened pale-green body up to 2 cm in diameter having a shallow some 0.6 cm long fissure across the crown. Tipped-brown daisy-like deep yellow flowers some 2.5-3.0 cm across are borne in October-November. ***C. concavum*** is a clump-forming succulent growing 5 cm in height and some 2.5 cm in spread with green-purple globular body, bearing two jointed fleshy leaves which on the crown are a little concave with a shallow fissure in the centre from where white flowers some 1.25 cm across are borne during October. ***C. corculum*** is clump-forming with roughly triangular paired leaves, which are fused into a spherical and top-flattened body with a split in the centre from which the small yellow daisy-like flowers emerge during August to September. Leaves are light green, very thick growing up to 5 cm in height. ***C. cornatum*** bears dark green body of two fused leaves into a spherically-shaped body ~ the corpusculum, having a fissure and depression in the centre of the crown from where orange flowers appear during August to September. ***C. ernianum*** bears grey-green plant body of jointed leaves with numerous dark green dots having a fissure at the top from where pink flowers some 1.25 cm across appear during August and September. It grows some 2.5 cm high and with some 2 cm spread. ***C. framesii*** grows 1.2 cm high and ovoid with galucous grey-green plant bodies bearing a few darker marks. Flowers are cream and some 1 cm across. ***C. frutescens*** is a clump-forming succulent growing as a shrublet up to 10 cm high with about 5 cm spread where stems are tipped by a dark green plant body shaped like *C. bilobum* but only 3 cm long. Plant body divided in two lobes is dark green. Rich orange-yellow flowers some 2.5 cm across appear from July to September. ***C. globosum*** bears smooth glossy-green and grey-tinged body some 2 cm in diameter, 4 cm in height and with 7.5 cm spread, topped with small fissure from the centre of which some 1.25 cm across and pink flowers appear during September. ***C. gratum*** has loosely clustered a few plant bodies which are broadly inverted pear-shaped, glaucous with darker dots and about 2.5 cm high. Red-purple flowers are some 1.2 cm across. ***C. halenbergense*** is clump-forming, blue-green speckled darker with almost cordate body up to 2 cm in diameter, growing to 1.25-2.0 cm high and with 7.5 cm spread, and topped with a fissure from the centre of which yellow flowers some 1.25 cm across are borne during October and November. ***C. meyerae*** (*C. meyeri*) is a clump-forming succulent with prostrate and woody branches growing slowly and reaching to 6-10 cm high with about 10 cm spread. The dark grey-green, egg-shaped and apex-notched plant body some 2.0-2.5 cm tall is divided into two egg-shaped flattened and lobed leaves which are faintly marked with darker lines and dots and a little hairy. Daisy-like yellow flowers some 1.5-2.0 cm across are borne from the centre of the fissure in autumn. ***C. minusculum*** was discovered by N.S. Pillans from its native place Clanwilliam district and is an excellent ornamental succulent worth growing. Its pink to scarlet flowers are proportionately larger in length and breadth to the leaves (plants) and last for about two weeks. ***C. minutum*** was introduced into cultivation by Francis Masson in 1795 from its native of Van Rhynsdorp area near Bakhuis but its flowers though of the same colour as to *C. minusculum* but are smaller and less frequently produced. During resting period the plants dry up to the extent that these are at all at that time not recognizable. To recognize them in the wild these should be allowed to fatten and flower. Its plants have vertical branching. ***C. multipunctatum*** is a clump-forming with broadly inverted pear-shaped pale grey-green plant bodies bearing darker dots. White flowers some 1.2 cm across appear by dusk. ***C. mundum*** grows some 1.2 cm tall, is clump-forming, bearing grey-green obconical plant bodies with transparent dots, top-flattened and centre-sunken. Yellow flowers appear some 1.2 cm across. **C.**

muscosipapillatum is a small branched shrublet growing some 8 cm high, each stem ensheathed with brown sheaths of older leaves. Terminal grey-green to nearly whitish plant bodies some 4 cm high with conspicuous flattened lobes bear fine velvety hairs. Rich yellow flowers are borne some 3-4 cm across. ***C. nelianum*** forms some 2-4 cm high colonies of the plant bodies that bear grey-green dotted darker with minute protuberances on lobes with conspicuous flattening and keeling. Yellow flowers appear some 1.5 cm across. ***C. notabile*** is a 3-cm high with indefinite spread, spherically-borne and slow-growing perennial succulent forming clumps of light blue-green edged red or oftenly with a red spot on fissure edge between the two lobes, and the lobes being somewhat flattened and very fleshy. Copper-orange flowers some 1 cm across appear in late spring to summer. ***C. obcordellum*** is a clump-forming succulent in various shades of green, dotted darker green or purplish, and often suffused pink or red, with broadly obconical plant bodies some 0.6-4.0 cm high and with almost the same width, bearing usually flattened top which is sunken in the centre. White to cream flowers some 1.5 cm across appear in October. ***C. ornatum*** is a clump-forming some 2.5 cm long blue-green succulent with sparingly darker dots, bearing inverted conical bodies which are flattened at the top and sunken in the centre. Yellow flowers some 2.5 cm across are borne with red stamens and stigmas. ***C. pearsonii*** grows in clumps some 2.0 cm high and wide in quite compact and broadly obconical form with flattened to slightly convex top, and is smooth and glaucous bluish-green. The well-established ones become cushion-like. The pale-violet, glossy and some 2.0 cm across flowers appear on the top from about 0.3 cm long fissure (slit). ***C. pictum*** grows some 1.5 cm high and 1 cm wide, bearing deep green obconical plant bodies with sides being usually reddish and purplish though sides dotted above with brownish veins. Cream flowers appear some 1 cm across. ***C. pillansii*** grows about 4 cm high and 2.5 cm in spread bearing solitary or in groups of 2-3, bright yellow-green globular plant bodies with short broad lobes bearing translucent spots and about 6 mm deep cleft on the top. Purple-red flowers about 2.5 cm across are produced during September to October. ***C. praegratum*** has dark blue-green pear-shaped body dotted with olive-green and possesses a shallow oval fissure on the top. It grows some 2.5 cm high and 2 cm in spread. Pink flowers some 1.25 cm across are borne in October. ***C. quaesitum*** has pear-shaped blue-green plant bodies marked with darker dots and is equipped with a minute fissure at the top. White flowers about 2 cm across are borne during October and November. ***C. scitulum*** bears grey-green, cone-shaped plant body which is red-tinged and marked with broken red-brown lines, and bears a narrow fissure at the top. White flowers about 2 cm across appear during October and November. ***C. sitzlerianum*** bears blue-green plant bodies having two blunt lobes with red tips and the plant grows some 4 cm in height and 1.25 cm in spread. Golden-yellow flowers some 2 cm across are borne in September and October. ***C. subrisum*** forms obconical and smoothly white glaucous plant bodies some 1.5 cm tall in small groups with centrally depressed flattened top. Rich yellow flowers about 2 cm across appear with red petal-tips. ***C. taylorianum*** grows with the height and spread of about 2.0-2.5 cm. The dark grey-green body speckled with red is cordate and has a shallow fissure in each lobe from the centre of which some 1.25 cm across pink flowers appear during October. ***C. tischeri*** produces short woody stems when plants become old. The plant is loosely clump-forming, bearing grey-green dotted darker and broadly obcordate plant bodies some 1.2 cm tall and almost as wide having slightly laterally flattened with a central cleft some 4 mm deep. Pale-lilac flowers are borne some 1.5 cm across. ***C. truncatum*** (syn. *C. truncatellum*) is a 1.5-2.0 cm high with about 15 cm spread, slow-growing and clump-forming perennial succulent with blue-green, dark-spotted and obconical pea-shaped leaves each with a shallow sunken groove at the crown. Cream flowers some 1.5 cm across are produced during autumn. ***C. uvaeforme*** (*C. uviforme*) is a very compact, clump-forming and makes small hummocks when old, bearing globular and yellowish to grey-green body with darker green speckles or sometimes flushed with red or purple, and grows some 2 cm high and 15 cm across in spread though individual clump some 1.2 cm wide. The top has a cleft. Creamy-white flowers some 1.25 cm across appear during October. ***C. wettsteinii*** is a broadly matted, clump-forming and broadly obconical succulent growing some 1.5 cm high and by 2-3 cm wide, composing of 2-jointed-leaved bodies of grey-green, minutely white-dotted with a quite large corpuscular on flattened or slightly convex top often bulging over the sides. Reddish-mauve flowers are borne some 3 cm high and across. ***C. wiggettae*** is a clump-forming succulent with some 2 cm height and 15 cm spread, each flat globular body is grey-green, dotted red-brown, and bears a fissure at the top. White flowers some 1.25 cm across appear in October.

Corallocarpus bainesii (Cucubitaceae)

A climbing to rarely prostrate monoecious herb with raised caudex. Its leaves and tendrils are simple and the flowers are yellowish. Male flowers are subsessile to pedunculate, appearing in thick axillary clusters with small corolla lobes which unite at the base and the stamens are 5. Female flowers appear from leaf axils as solitary to clustered, ovary smooth with a cup-shaped base, corolla small and alike the male flowers, stigma 2- 3-lobed, fruits ofen beaked with persistent base and on maturity turning red.

Cotyledon (Crassulaceae)

Cotyledon comprises 41 species, all from South Africa except one *C. barbeyi* which is from East Africa, Eritrea and Arabia. *Cotyledon* derives from the Greek *kotyledon* meaning 'hollow' or 'a cup', referring to cupped leaves of 'navelwort' then under this genus but now it is *Umbilicus rupestris*. These are valued for their leaves which are of two rankings, first those having opposite pairs of fleshy leaves on shrubby and evergreen species and in winter these require warmer and less dry conditions for their cultivation, and the next those where leaves are alternate on the short, fleshy, thickened and deciduous stems requiring cool and almost dry winter rest. Evergreen species are comparatively more attractive than the deciduous ones. Leaves are either flattened or cylindrical, and in most species these are poisonous such as *C. pearsonii, C. ventricosa, etc.* which form the deciduous group. The tubular, 5-lobed and terminal flowers are formed on the umbel-like clusters. These are propagated through leaf or stem cuttings, and also through seeds. ***Cotyledon barbeyi*** from E. Africa, Eritrea and Arabia is a branched and shrubby succulent growing with the height and spread of 60 cm. The leaves are glossy light grey-green, obovate, 6-14 cm long and fleshy. Finely hairy orange-red tubular flowers some 2.5 cm long appear on long panicles during autumn to winter. It is a temperate ornamental. ***C. jacobseniana*** is a sub-shrubby free-branching and semi-prostrate evergreen succulent growing 30 cm wide, bearing densely set green or waxy-white patinated leaves which are narrowly ovoid and tapering to both the ends, and are 2-3 cm wide. Greenish-red pendant flowers in 3-10 umbel-like clusters appear on a 5-13 cm stalk above the leaves in summer. This is a temperate species. ***C. ladismithensis*** is a low-growing compact shrubby evergreen succulent growing to 15-20 cm tall with the spread of 30 cm and the stems becoming woody with age. Leaves are dark green, obovate, top-rounded but base tapering, coarsely white-hairy and sticky with toothed-apex bearing 2-3 teeth, 3-8 cm long and thickly fleshy. Though rarely flowers but the clustered brownish-red flowers some 1-2 cm long appear on a 15 cm stalk during summer. It is a temperate species. ***C. orbiculata*** may grow up to 1.2 m in height and 60 cm in spread. A branched evergreen shrubby succulent bears grey-green with waxy-white covering, round and blunt, 4-14 cm long, lanceolate to ovate, a little convex below and fleshy leaves with thin and red margins. Orange-yellow tubular flowers some 1.2-2.0 cm long appear in panicles on long stalks during July-August. This has given certain very popular cultivars. ***C. paniculata*** (syn. *Tylecodon paniculata*) is an extremely slow-growing deciduous succulent with brown and fleshy stems greatly thickened at base, and attains the height of up to 20-25 cm under cultivation though in the wild it grows up to 2 m, and the spread is 30-38 cm. Branches bear the leaves clustered only at the tips and these are grey-green, lanceolate to ovate and 5-10 cm long. The leaves produced during late autumn are light green, about 2.5 cm long and narrowly spoon-shaped. Though not a free-flowerer, dark red tubular flowers streaked green and some 2.5 cm long appear during July-August. It likes cool climate. ***C. papillaris*** is a prostrate-growing shrubby evergreen succulent growing 30 cm wide with branches mostly in opposite pairs having soft hairs. Mostly rosetted, dark green, red-tipped or margined leaves only at the tips are obovate to wedge-shaped, 1.5-4.5 cm long and 0.3-0.6 cm thick. Together merged shades of red, green and yellow flowers some 0.8 cm long appear in 3-15 umbel-like clusters on a 5-25 cm stalk length above the leaves during summer. This is a temperate species. ***C. reticulata*** (syn. *Tylecodon reticulata*) is very similar to *Cotyledon paniculata* but smaller, branched near the base, deciduous and attaining the height of 15-30 cm and the spread of 30-38 cm. Leaves which are deciduous are produced in October and are almost cylindrical, fleshy and 1.5-2.0 cm long. The green-yellow erect tubular flowers some 0.8-1.25 cm long are borne from June to August. It likes cool temperatures. ***C. undulata*** grows as an evergreen bush some 50 cm high with thickly-fleshy, oval, up to 12 cm long and 6 cm wide leaves having wavy edges and tips, and heavy chalky bloom. Floral stalks are up to 40 cm long and bear 2.5 cm long golden-orange flowers. ***C. wallichii*** (syn. *Tylecodon wallichii*) is a deciduous shrubby succulent with fleshy stems growing up to 30 cm tall. Leaves are grey-green, semi-cylindrical, clustered around the apices and fall off after the short growing season. Green flowers spotted red on the tips are borne on a 30 cm long stalk. Likes cool situations.

Crassula (Crassulaceae)

Crassula in Latin means thickish, referring to the thick leaves and stems. Though main centre of its origin is arid South African deserts but species are found throughout the world. There are 300 *Crassula* species, mostly succulents with annuals, evergreen perennials and shrubs. These are difficult to cultivate indoors due to watering frequency it demands, the shortage of sunlight and the problem of keeping them cool during summer. Stems and leaves are rubbery and the latter being often extraordinarily thick and waxy due to patination of bluish or grayish powder, rounded to linear and opposite. Flowers being small are more or less starry with 3-9 petals, mostly 5 which appear in terminal panicles. These are propagated through division of plants, stem cuttings and seeds. ***Crassula arborescens*** (syn. *Cotyledon arborescens, Crassula cotyledon*) from Cape (Namaqualand) to Natal under cultivation grows up to 1 m high and almost the same across with robust stems. It is a well-branched shrubby succulent with strong stems, bearing grey-green but differently red-edged, broadly ovate and 4-7 cm long leaves, and white flowers fading to pink are borne on terminal panicles some 5.0-7.5 cm across during May and June. It requires good light for proper foliage colour development.***C. barbata*** from South Africa is a clump-forming succulent bearing glossy-green with long white hair-margined, crowded in angular rosettes, broadly ovate and up to 4 cm long leaves. Small white flowers are borne on an erect spike some 30 cm long. Flowering rosettes, after flowering and seed-setting, die down producing offsets for further generation. ***C. barklyi*** (syn. *C. teres*; common name 'rattlesnake tail') from Namaqualand and Namibia produces normally solitary, erect and up to 8 cm high stems, bearing green with ciliate margins, alternate-opposite pairs which securely overlap each other creating a column some 2 cm wide, cupped and orbicular. White or yellow flowers appear in terminal clusters during autumn. ***C. columnaris*** from Cape (Namaqualand, Karroo) is usually single-stemmed with occasional branching, growing 10-15 cm high with stems completely hidden by the short, broad, thickly fleshy and overlapping leaves which are so properly set as to imitate some masonry work to make square or round columns some 10 cm high and 4 cm thick. Young actively growing leaves are dark green but the dormant ones brownish-green. Fragrant flowers are white or orange-yellow and appear during autumn. ***C. conjuncta*** from South Africa is multi-stemmed plant with strong but arching stems bearing patinated opposite pairs of silvery green, fleshy and ovate *cum* boat-shaped leaves with pink edges and with rosette at the tips. ***C. cooperi*** from Cape and Transvaal with about 7.5 cm height and 30 cm spread is a low tufted or mat-forming succulent with mid-green but flushed and dimpled with red and narrowly lanceolate leaves. From May to July the pale-pink flowers appear in sparsely flowered clusters some 1.2-1.8 cm across. ***C. corymbulosa*** from coastal regions of Cape is usually solitary to tufted, erect, sparingly with base-branching and the entire stems are covered with pale grey-green dotted darker, alternate pairs of narrowly triangular and some 4 cm long leaves. Small white flowers appear in clusters at the top of the stalk some 30 cm tall. ***C. deceptor*** (syn. *C. deceptrix*) from Cape is an erect and small succulent up to 5 cm tall with initially solitary stems but later base-branching making a small clump. White-grey leaves are thickly fleshy and compact, papillose, keeled and triangularly rounded. Small white flowers appear just above the foliage during autumn. This is winter-growing plant. ***C. deltoidea*** (syn. *C. rhoboidea*; common name 'silver beads') from Cape (Namaqualand) is a quite small sub-shrub growing up to 10 cm tall, bearing thickly fleshy paired leaves which perfoliate the stem, are rhombic-ovate and grey-farinated, bluntly keeled, up to 1.5 cm long and upper surface ridged. Off-white to pinkish small flowers are borne in autumn in terminal cymes. ***C. falcata*** (syn. *Rochea falcata*; common name 'airplane plant') from Cape and Natal is a very attractive erect-growing shubby succulent with about 1 m height and some 30-45 cm spread. Its thickly-fleshy and sparingly branched stems bear numerous blue-green, 7-10 cm long, very fleshy, in two rows pointing alternately to left and right, laterally flattened, and somewhat sickle-shaped curving leaves, being horizontal near the stem but tip-twisted. Bright orange-red paint-brush shaped crowded flowers in flattish clusters at the top of the grey-white stem are borne in a 7.5 cm wide heads from June to August. It requires regular pruning to keep it in shape. ***C. lactea*** from Natal (Transvaal) is prostrate to semi-erect shrubby succulent growing about 30 cm tall and with about 1 m spread, bearing white-dotted narrowly triangular-oval and glossy dark green leaves. Small white and fragrant flowers appear on long stalks in dense terminal panicles in winter. ***C. longifolia*** (syn. *C. perfoliata*) from Cape and Natal is a 60 cm tall sparingly branched erect shrubby succulent with grey-green lanceolate leaves which are 10-15 cm long, tapering to a point and somewhat channelled above. Small red or white flowers in a dense terminal corymbose cluster some 10 cm wide are borne during summer. ***C. lycopodioides*** (rat tail plant) from Cape (Namaqualand) to Namibia is a 30 cm long with some 40 cm spread, twiggy, clump-forming, sub-shrubby and woody-based perennial succulent with erect, slender and branched stems covered with tiny, triangular-ovate, scaly and mid-green leaves densely and neatly overlapping in four rows so as to completely concealing the stems and fitting together like roof-tiles, and the tiny white-yellowish flowers which arise in the leaf axils are quite insignificant. ***C. marnieriana*** from Cape is a sparingly branched, sprawling and semi-erect sub-shrub growing to about 15 cm wide. It bears red-edged glaucous, almost orbicular, about 0.8 cm across and thickly-fleshy leaves. Small white flowers in compact terminal cymes are borne in spring. ***C. mesembryanthemopsis*** from Cape (Namaqualand) is a low hummock-forming rosetted perennial succulent with some 10 cm width. Leaves are whitish-green but in sun red-tinted, about 3 cm long, oblong-cuneate with swollen tips and thickly fleshy. Tiny white flowers are borne in small sessile cymes in autumn. ***C. milfordae*** (syn. *C. sedifolia*) from Basutoland is a dense cushion-forming evergreen succulent with tiny grey-green rosettes, growing about 3 cm tall but with 30-40 cm spread. It is a quite hardy species but seldom flowers. White flowers appear from crimson buds of the rosettes on 2.5 cm heads in June and July. Though the true species seems to be extinct but is replaced with almost similar one, *i.e. C. sedifolia*. ***C. multicava*** from Natal (Transvaal) is a bushy, erect to spreading perennial succulent growing up to 15-30 cm high and with about 1 m spread, bearing 2.5-8.0 cm across brightish grey-green but in sun red-flushed, obovate to oval and subcordate leaves with superficially pitted surface. Star-shaped white flowers in numerous clusters from pink-splashed buds appear on elongated stems in terminal panicles from late winter to spring, followed by small plantlets. ***C. nealeana*** (syn. *C. perfossa minor, C. rupestris minor*) from Cape is a tiny procumbent perennial shrubby succulent growing to about 15 cm width, bearing glaucous or purple-flushed and red-edged, perfoliated and crowded, broadly oval but shortly pointed and about 2 cm long leaf pairs. Small buff-yellowish clustered flowers on elongated stem tips in paniculate inflorescence some 4-5 cm high appear in spring. ***C. orbicularis*** from Cape and Natal is 15 cm wide, mat-forming perennial succulent, composed of loosely-set flat rosettes, bearing glossy green, white ciliate, spathulate-obovate and 4-5 cm long leaves. Small white flowers in some 20 cm tall spike-like inflorescence appear in spring. ***C. ovata*** (syn. *C. argentea, C. portulacea*; common name 'friendship tree', 'jade plant', 'money tree') from Namaqualand to Transvaal grows up to 4 m high and some 2 m in spread, is a perennial succulent with a quite swollen, robust, woody and freely branched stem which is crowned with some 4 cm long, thickly fleshy, opposite, obovate to spathulate, glossy green leaves, sometimes edged-red. Star-shaped small and pinkish-white flowers appear in clusters during winter or spring. ***C. perforata*** (syn. *C. perfossa*) from Cape (S.A.) is a prostrate and fleshy plant suitable for hanging baskets, later becoming woody. Joined pairs of leaves on stems appear as beads on strings. Leaves are 1.5-2.0 cm long and 0.9-1.3 cm wide, broadly ovate with short pointed tips, at right angle to the stem and grey-green dotted on the margins with small whitish dots. Flowers are red. ***C. rupestris*** (bead vine, rosary vine) from Cape (Namaqualand and Karroo) is an attractive leaf succulent most suitable for hanging baskets, with decumbent to erect and repeatedly forking stems with about 25 cm spread, bearing perfoliate pairs of thickly-fleshy triangular leaves, 1.0-2.5 cm long and wide, grey-green and edged reddish-brown, and small yellowish umbellate flowers appear in terminal clusters in spring and summer. ***C. sarcocaulis*** from mountainous Cape of South Africa growing more than 20 cm tall with spread of some 15 cm, is a small growing knotty perennial shrublet most suitable for alpine house. Stems are woody with small pointed green and lanceolate leaves some 1 cm long which is mostly flushed red during summer when kept dry. Clustered minute pink to red flowers appear from the attractive crimson buds on the 1.25 cm across flower-heads during July to September. ***C. schmidtii*** from Namibia, and Transvaal & Natal of South Africa is mostly confused with *C. gracilis, C. hookeri, C. rubicunda* and *C. schmidtiana* when grown in the gardens. It is a low growing tufted to mat-forming (carpeting) perennial succulent growing up to 10 cm high with 30 cm spread. The rosettes are dense with dark green, pitted and marked, linear and 3-4 cm long leaves. Clustered bright pinkish-red star-shaped flowers are borne in winter. ***C. socialis*** from Cape is a winter-growing, free base-branching and spreading perennial succulent growing some 5 cm high and 5-8 cm spread forming mats of rosettes. Its rosettes are short and 1 cm across, bearing bright green, fleshy, sparingly toothed margins and triangular *cum* horny leaves. Clustered star-shaped and bell-like white flowers on 3 cm tall stems and some 1.25 cm across panicles appear during spring from the centre of the rosettes. ***C. tetragona*** from Cape is sparingly branched and erect growing shrubby succulent some 60 cm in length. Leaf-pairs are light green, united at their bases, fairly close to well-spaced on the stems, almost decussate cylindrical and awl-shaped, and some 2-4 cm long. Small white flowers appear on terminal and dense panicles during spring to summer.

Cussonia (Araliaceae)

A genus with two succulent species, *Cussonia* originated in South Africa and Madagascar. *C. spicata* (cabbage tree) is a fast growing and long-lived tree attaining the height of up to 15 m. *C. paniculata* is a sparsely branched, slow-growing thick tree attaining a height of up to 5 m. The bark of this tree is grey, thick, corky and with longitudinal fissures. It has swollen roots. It is a pachycaul (fibrous-wooded) succulent as the stem bases have swelling like to those of tubers and if grown through seeds these swellings appear early.The cabbage blue leaves that appear at the end of a long stalk with spring flush, when fully grown are up to 60 cm, mostly with thick waxy layer, digitately compound, composed of 7-13 leaflets which may measure up to 30 cm in length, in some forms leaflets may be deeply lobed.

Cyanotis (Commelinaceae)

It derives from Greek *kyanos* meaning 'blue' and *anthos* meaning 'flower'. A genus of 50 tropical and sub-tropical species from both the hemispheres of creeping or tufted, ascending or weak-branching, thin-stemmed, often woolly perennial and semi-succulent herbs, mostly evergreen, and very much alike to *Tradescantia*. Most suitable for hanging baskets though is properly maintained through trimming back of lanky stems and by watering sparingly. Leaves are several, small, alternate, linear to ovate and the bases ensheath the stems. Flowers are 3-petalled and usually purple-blue. Propagation is through cuttings. ***Cyanotis kewensis*** (teddy bear plant) from India grows up to 10 cm high with main stem which is reddish-hairy, and bears opposite-alternate, crowded, 2.5-3.75 cm long, hairy green but red-tipped and sword-shaped leaves which overlap the base at the stems. Flowering stems are pendant with small reddish flowers. ***C. somaliensis*** (pussy ears) from East Africa grows with stiffy trailing stems some 20 cm long, bearing shiny green clothed with long white hairs, and linear-lanceolate to ovate-lanceolate triangular leaves some 4 cm long but edging hairs in this species are longer than the above. Flowers are bright violet-blue which appear from winter to spring.

Cynanchum (Dog strangling vine; Apocynaceae)

About 300 species of this genus are distributed worldwide in the tropics and subtropics, as well as in the temperate regions as its species are perennial herbs or subshrubs, mostly forming rhizomes though most of these are non-succulent climbers and twiners. Usually the leaves are opposite, sometimes with petioles, and the inflorescences and flowers appear in various shapes on various species. Flowers in case of *C. macrolobum* are pedicellate, white, wide open, 5-angled, flat, wavy, like to those of Convolvulaceae and these are seated on quite thick calyx which is 5-lobed, each lobe much larger to the flower. The flowers are borne at the end of the branched shoot.

Cyphostemma (Vitaceae, syn. Vitidaceae)

Cyphostemma derives from the Greek *kyphos* meaning a 'tumour' or 'swelling' and *stemma* meaning 'a crown' as the main stem or crown of the plant has swelling due to water storage tissues. It is a genus of at least four succulent shrubs formerly included under *Cissus*. The caudices are very thick, woody and fleshy throwing out many woody branches with soft stems having yellowish papery bark. Leaves are deciduous, fleshy and its underside exudes small droplets of resin. It is propagated through seeds though is very difficult to grow. ***Cyphostemma bainesii*** (syn. *Cissus bainesii, Vitis bainesii*) from West Africa is a perennial succulent shrub growing some 60 cm tall and wide when cultivated though in the wild it grows higher. It possesses thick, swollen and bottle-shaped trunk often unbranched and is covered with yellow, papery and peeling bark. Deciduous leaves are shortly stalked, fleshy, silvery-green, often trifoliated into three oval leaflets, each leaflet some 11 cm long with deeply serrated margins. When young the leaves have silvery hairs. The flowers are small yellow-green, cup-shaped and appear in small corymbs during summer, followed by grape-like red fruits. ***C. juttae*** (syn. *Cissus juttae*) from Namibia (SW Africa) is quite similar to *C. banesii* but with less serrated glaucous leaves and the stems are often lobed and covoluted. It grows up to 6 m in the wild with barrel-shaped trunks though under cultivation the height is much less, yellowish-grey papery bark peeling off in strips as the plant ages, and bears scandent but deciduous branches bearing broad pointed-ovate, glaucous-green and some 15 cm long leaves. Inconspicuous yellow-green flowers appear in summer followed by yellow or red fruits. ***C. macropus*** from Namibia is though quite similar to *C. bainesii* but the trunk is more freely branched and leaves are smaller with no leaflet. It matures to a shrubby appearance. ***C. montagnacii*** from Madagascar throws out thin vines from large underground roots that are oftenly branched, some 30 cm long and 10 cm thick, vines bearing 6 cm long and 4 cm wide compound leaves. The roots should be protected from direct intense sunlight though plant will tolerate heat but not the frost. ***C. glabra*** from South Africa is still listed as *Cissus* as is quite similar to *C. saundersii* but the leaves are larger and not pubescent. It produces several metres long vines during summer which all drop down during winter. Its main stem is quite thick and some 50 cm high which is produced by quite large underground roots which require to be saved from intense sunlight as well as from frost and the plant should be kept fairly dry when dormant during the winter.

Dasylirion [Nolinaceae (Agavaceae)]. A native of Mexico, they are xerophytes where some species develop thickened trunk but most are not the succulents. All the species bear thin, grassy and very long ball-shaped rosetted leaves and hundreds of tan to yellow flowers on a long spike. ***D. longissimum*** from Mexico bears roughly 4-angled quite slender leaves more than 1.2 m long and about 6 mm wide and thick with slender tapered point which top the stem some 2 m long but only old specimens develop the woody trunk. It does not tolerate below -10 °C temperature. Though not usual but can be grown in containers and is propagated through seeds.

Decabilone (Asclepiadaceae)

Decabilone grandiflora (syn. *Tavaresis grandiflora*) was discovered about 1886 in South-West Africa by Hans Schinz and is also found extending into Botswana,

Cape Province and Transvaal (S.A.) and Rhodesia, the easterly specimens having straight-tubed corollas while those from the west with slightly curved tubes. It is a clump-forming plant growing up to 16-20 cm high and some 10 cm wide, individual stem being 1.2 cm thick bearing 6 ribs with closely set tubercles, each tubercle bearing one stout and two thin yellowish-brown spines some 2.3 to 3.5 cm long and the stem-tips with numerous soft emerging spines. The vegetative organ of this plant greatly resemble those of columnar cacti. Brownish-yellow flowers are bell-shaped with spreading mouth having 5 lobes, each lobe acute. Whole flower is densely pin-head spots of brown colour from bottom to top and even on the margins.

Delosperma (Aizoaceae)

Delosperma takes its name from the Greek *delos* meaning evident and *sperma* meaning a seed due to capsules not being fully closed and so seeds can be seen at all the stages of development. This genus originates in the South Africa with about 150 species of mainly woody-based dwarf shrubby perennials, though some are prostrate-growing, bearing opposite pairs of succulent leaves, and often showy daisy-like flowers. Prostrate-growing species are most suitable for window sills and for hanging baskets. Either they are propagated through seeds or through cuttings.***D. echinatum*** (syn. *D. pruinosum*) from Cape is a densely growing bushy species having up to 60 cm height and width. Highly decorative light green leaves are covered with bristle-tipped tubercles, ovoid to hemispherical in shape and crowded. White or yellow flowers some 1.5 cm across continue coming up throughout the year. ***D. lehmannii*** from Cape is a hummock-forming succulent bearing stem branches some 30 cm in length. Leaves of the side shoots.are quite smaller and crowded though of the main shoot grey-green, up to 2.5 cm long, base connate, 3-angled and with convex sides. Pale yellow flowers some 4 cm across open in summer during mid-day. ***D. lydenburgense*** from Transvaal is hummock-forming producing about 20 cm long stems, bearing grey-green, linear and pointed, and 4.5-5.5 cm long leaves, Purple flowers appear in loose clusters some 2.5 cm across in summer. ***D. nubigenum*** from Cape is hummock-forming succulent having well-branched some 15 cm long stems. Leaves are pale-green, often red-tinted, thick, minutely tubercled, ovate to elliptic or almost linear and 0.6-1.2 cm long. Orange-red to bright yellow flowers some 2 cm across appear during spring and summer. ***D. tradescantioidesi*** from South Africa is spreading succulent growing to 10 cm high but with indefinite spread. It produces trailing stems. The broad light green, fleshy and some 3 cm long leaves are cylindrical and 3-angled. Daisy-like white flowers are produced during summr.

Dendrosicyos socotranus (Cucumber tree; Cucurbitaceae)

This is the only tree-form species among all the Cucurbitaceous plants, endemic to island of Socotra in Yemen. It has a bottle-shaped pachycaul base, the trunk thickening to a diameter of up to 1 m, and which throws out numerous small branches. Leaves form on the main trunk and its branches, singly or in a cluster of up to 6, are deep green, ovate-rounded, irregularly and shallowly lobed and sparsely toothed, and covered with bristled-hairs. The species is bisexual, and yellow male or female flowers of 3 cm diameter appear on the same plant. Fruits measure 3 cm long and 5 cm across, initially green but maturing to brick red, and this fruit produces many seeds which are used for further propagation.

Didierea (Didiereacae)

Didierea is named so for E. Didier (1811-1889), student of the flora of Savoy. This genus is native of Madagascar (southern and south-western coastal regions) only with two succulent species, *viz. Didierea madagascariensis* (syn. *D. mirabilis*) and *D. trollii*. *Didierea madagascariensis* is found in the southern part of western coastal area and *D. trollii* in the southern end of the island. The mature stems of *D. madagascariensis* are divided internally by transverse diaphragms of pith into a series of horizontal chambers which make them strong in construction and together light-weighted though in case of *D. trollii* which bears smaller leaves and stems, latter bending horizontally. Entire stem is covered with cluster of spines, each cluster being the modified shoot. From the apex of each spine cluster, a series of elongated leaves and afterwards orangish-yellow male or female flowers are borne, both on separate plants, during rainy season, hiding the plants completely. The flowers have red throat. With age these spines are shortened into stouter ones. Male flowers have 2 small sepals and 4 larger petals surrounding the 8 stamens though in case of female flowers every thing is the same but with an ovary terminated by a stigma and 3 to 4 irregular lobes on the place of stamens.

Dieffenbachia (Araceae)

Dieffenbachia was named so to honour J.P. Dieffenbach (1790-1863), the Administrator of the Royal Palace Gardens at Schonbrunn in Vienna. The genus with 30 species originated in tropical America and West Indies and is valued for its foliage. It is a shrubby evergreen genus with erect and robust stems which are sometimes very fleshy, with large oblong to ovate terminal leaves.

Tiny flowers like to those of arum are produced on a spadix within a narrow arum-like spathe which may be creamy to light orange. Its sap being poisonous causes dermatitis and is also harmful if swallowed or touched to the eyes. Its common name being 'dumb cane' due to its causing of speechlessness if swallowed. It is propagated through cuttings of the stem tips or by stem sections. ***Dieffenbachia amoena*** (giant dumb cane) from Tropical America is quite robust and grows about 2 m high with 7.5 cm diameter, bearing elliptical to oblong leaves some 50 cm long and 20 cm wide on 30 cm leaf stalk, glossy dark green and with irregular splashing, lines and spots of white or cream along the lateral veins. Insignificant greenish-white flowers are clustered on th spadix, surrounded by a narrow leaf-like spathe. ***D. maculata*** (syn. *D. picta*; common name 'common dumb cane') from Brazil is a stem succulent growing some 1 m in height, bearing green marked creamy-white, and oblong-elliptic to ovate leaves some 25 cm long. It has various forms and varieties. ***D. seguine*** (mother-in-law plant) from Tropical America is one of the largest species and as big as *D. amoena*, with stems growing about 2 m high, bearing glossy dark green, broadly lance-shaped, heavy-textured and thick-stalked leaves to 50 cm long and 13 cm wide. Insignificant minute greenish-white flowers are clustered on the spadix, surrounded by a narrow leaf-like spathe.

Dinteranthus (Aizoaceae)

Dinteranthus honours Prof. Kurt Dinter (1868-1945), a German botanist who studied the flora of South-West Africa. This genus is native of southern Africa with six perennial succulent species, closely resembling and allied to *Lithops*, but each shoot with 1-3 pairs of leaves. Flowers appear in late summer to autumn, except in *D. pole-evansii* where it is spring. These are propagated through seeds. ***D. inexpectatus*** from Namaqualand bears a single pair of leaves either solitary or in clusters. The leaf pair is grey with translucent green spots, broadly ovoid with a transverse slit, and each leaf top about 2 cm long from the central slit. Yellow flowers some 2.5 cm across open in autumn. ***D. microspermus*** from Namaqualand is quite similar but larger than *D. inexpectatus*, except of the deeper slit between each leaf pair and yellow petals tipped-reddish. ***D. pole-evansii***.from Cape bears grey but oftenly tinted red or yellow leaves in single pairs, generally solitary and oblongoid with a deep and narrow transverse slit, and each leaf top from the central slit is 1.4 cm long. Rich silky-yellow flowers some 4 cm across appear in late spring. ***D. punctatus*** (syn. *D. puberulus*) from Cape has small clustered plants. Leaves usually in two pairs bearing coppery grey-green with darker dots and minute velvety hairs, each leaf pair uniting up to half their length, are obovoid and slightly angular and some 2.5-3.0 cm long. Yellow flowers some 2.5 cm across open during autumn. ***D. wilmotianus*** from Cape produces grey leaves in single pair having pink tints and deep purple dots, emerge in solitary pair in the shape of a wide boat. Each leaf is keeled at the tip and is some 3.5 cm long from the central suture. Deep yellow flowers some 3 cm across open during autumn.

Dioscorea (Dioscoreaceae)

Dioscorea honours Pedanios Dioscorides, the first century Greek doctor, herbalist and author of the original *Materia Medica*. This genus comprises of about 500 species, widespread to the tropics and sub-tropics in every continent. There are many species which produce edible or medicinal tubers or rhizomes, but there are many other species of ornamental foliage value. They are deciduous herbaceous perennials with twining stems and yellow flowers which appear during autumn. Only a few species are succulent such as *Dioscorea elephantipes* (syn. *Testudinaria elephantipes*). They are propagated by division of the rootstock, by cuttings of young basal shoots or by seeds. ***Dioscorea elephantipes*** (elephant's foot, hottentot bread) from South Africa is an unusual and very slow-growing tuberous succulent with a domed woody trunk, mostly aboveground, attaining a diameter of up to 90 cm and is humped like the carapace of a tortoise to a maximum height of 20 cm. The tuber is brown-corky and packed in thin layers. Its body is covered with 5 to 7 sided protuberances some 2.5 cm high. The tubers are deeply furrowed and cracked with very rough skin. A thick and strong stem emerges from near the centre and grows up to 9 m as a climber with hanging branches. Leaves are light green, kidney-shaped and some 5 cm wide, and the flowers yellow-green on a little tassel but insignificant. It is propagated through seeds and cuttings. ***D. macrostachya*** from Mexico is quite similar to above but the tubers are not so geometrically plated and the leaves are heart-shaped and some 12.5 cm long.

Dischidia (Apocynaceae)

Dischidia derives from the Greek *dischides* 'twice parted or cleft', referring to the corona segments. It is a genus of 80 species of epiphytic climbing plants allied to *Hoya* even the flowers resembling the same, originating in China, Indo-Malaysia to Polynesia and Australia. It clings to its support by climbing stems and/or aerial roots. The leaves have two types of modifications ~ in some species these are modified into pitchers to store probably rain water and organic matter, as frequently even roots for taking water and food matertials grow into the pitchers such as *D. complex*, *D. major* and *D. vidalii*, while in many other species the leaves are imbricate which hold tightly to the growing surface, where the underside of the leaves has a space which is filled with roots such as *D. astephana, D. imbricata* and *D. platyphylla* and such ones are called shingle plants because the formation of such structures. Outer segments of the flowers curve down to form an urn or bell-shaped structure though these are insignificant. While growing them, these can be trained in conservatory or at home through moss-sticks or any other support. They are multiplied through seeds or through cuttings. Both the species described below form pitchers even on some of their roots. ***D. bengalensis*** from India is roughly 2 m long but height is drastically reduced indoors when trained on moss poles. Fleshy leaves are in pairs, glaucous, 2.5-4.0 cm long and obovate to elliptic, and the flowers some 5 mm long are formed in axillary umbels. ***D. rafflesiana*** (Malayan urn vines) from Malaysia and New Guinea grows to several metres, bearing grey-green, ovate to circular, fleshy and paired leaves some 2.5 cm long, some of the leaves forming pear-shaped 5-13 cm long pitchers which are purplish within. Flowers are green with a purple throat and some 6 mm long.

Dorotheanthus (Aizoaceae)

Dorotheanthus comprises of two words, Dorothea, mother of Prof. C. Schwantes, a *Mesembryanthemum* specialist, and the Greek *anthos* meaning a flower. *Dorotheanthus* was formerly classified in *Mesebryanthemum*, and this comprises 10 species from South Africa. Leaves are linear or spathulate to almost cylindrical, and the flowers are stalked and daisy-like which open only in the sun and remain closed during cloudy weather. These are attractive succulent annuals for beds, in edgings and on banks, in the rock gardens, as hanging basket plants and for pot growing for instant landscaping. These are propagated through seeds. ***Dorotheanthus bellidiformis*** (syn. *Mesembryanthemum criniflorum*) from Cape is commonly known as 'livingstone daisy'. It is a prostrate mat-forming succulent annual growing more than 30 cm in width and about 15 cm in height. This bears light green, obovate to almost cylindrical and some 8 cm long fleshy leaves covered with tiny pearly-glistening spots, clearly visible at an angle of sun. This may produce solitary flowers but in masses during winter to spring in the Indian plains and during June to August in temperate regions, in shades of red, crimson to pink, orange to orange-gold or white to buff but with darker centre and sometimes even bicoloured, stamens being blackish-purple, and measuring 2.5-4.0 cm across. ***D. tricolor*** (syn. *Mesembryanthemum gramineus, M. tricolor*) from Cape is a spreading and mat-forming annul succulent growing 30 cm across and 8 cm high. Dark green, narrow, linear to spathulate to cylindrical leaves are 5-8 cm long. Flowers in shades of red, pink to deep rose or white with contrasting centre and black-purple stamens are borne during winter and spring in the Indian plains while during June to August on Indian hills, some 4 cm across.

Dorstenia (Moraceae)

Dorstenia honours Theodore Dorsten (1492-1552), a German botanist and professor of Medicine at Marburg. The genus comprises 170 species from tropical parts of Africa, America and Asia, including one species in India, of which a large number are African succulents. The genus may have deciduous or evergreen perennials and sub-shrubs, some being succulents or adapted to desert conditions. It is a very curious genus though not so showy, widely distributed in all tropical areas, and is easily recognizable for its intriguing, unique, flat or hollowed and fleshy inflorescence arising from the tip of the stem. The inflorescence bears minute and unisexual perianthless green flowers, followed by minute one-seeded capsule which on maturity is ejected by force to a distance. Few of the succulent species are described here with. ***Dorstenia crispa*** from tropical Africa is though normally found in the moist rain forest habitats but adapts well to life in arid conditions. It grows wild in the dry places from the mountains of Somaliland, south into Kenya in the Mombasa region and develops a fat and succulent stem about 40 cm tall crowned with pale-green, crinkled, vein-pitted and pedicellate leaves some 30 cm long having fimbriated margins and acute tips. Stems have protective tubercular leaf-scars. Inflorescences with 35-40 cm long hollow peduncles arise singly from leaf axils, bearing flat disc-shaped receptacle with many unisexiual flowers spread over the surface, female ones being sunk into little pits, and these after ripening are thrown out due to natural process of the contraction of the surrounding tissues of the receptacle. ***D. foetida*** from southern Arabia forms the low mounds of thickly contorted, erect and succulent stems up to 15 cm tall, arising from the caudex which is a flattened and swollen stem and is partially embedded in the soil. Lanceolate to narrowly ovate and green leaves some 12-20 cm long are patterned with paling veins and with finely crimped margins and are clustered on the crown of the stem in a dozen or so. Inflorescence is small, some 2 cm wide, flattened and circular with some eight slender and marginal starry rays, and continue appearing throughout the summer.

Drosanthemum (Aizoaceae)

Drosanthemum derives from the Greek *drosos* meaning dew and *anthos* meaning a flower, referring to dew-like leaf papillae. The members of this genus were formerly included under *Mesembryanthemum* but structurally it is related to *Lampranthus* and is found only in South Africa with 70-95 species. This dwarf shrubby plant grows with pairs of cylindrical to angled fleshy leaves bearing translucent papillae as to that of *Drosera*. The members of this genus are free-flowering, with large and showy daisy-like flowers opening only in the afternoon during summer to autumn in the temperate India and during winter to spring in the plains of India. These are excellent ornamentals for pots and hanging baskets. These are propagated through seeds and cuttings. ***Drosanthemum floribundum*** from Cape is a freely-branched creeping succulent forming 30 cm or more across hummocks, bearing light green, narrowly obovoid and cylindrical leaves some 1.0-1.5 cm long. Pale glowing pink flowers some 2 cm across appear in plenty. ***D. hispidum*** from Namibia to Namaqualand is a well-branched shrubby succulent growing in the wild some 60 cm tall and 90 cm or more in breadth though in cultivation it can be tamed to less than half of its dimension. Its glossy light green leaves turning reddish tinted in bright light, are some 1.5-2.5 cm long and cylindrical in well-spaced pairs. Glossy deep purple flowers some 3 cm across appear in plenty from spring to autumn on the hills. ***D. speciosum*** from Cape, commonly known as 'dewflower' is a quite spreading succulent shrub, growing more or less erect up to 60 cm tall and wide and bearing slender and branched stems. Bright green leaves with well-spaced pairs are up to 1.5 cm long. Bright orange-red flowers with green-tinted centre and up to 5 cm across appear any time of the year from spring to autumn. ***D. splendens*** was discovered in 1932 by R. Pickard from the Montague district of South Africa. This produces some 15 cm long pink and branched flowering stem with angled to cylindrical, green and fleshy opposite leaves on each node some 2.5-3.0 cm long, and 3.2 cm across orange flowers with black centre.

Dudleya (Crassulaceae)

Dudleya is named so to honour William Russel Dudley (1849-1911), an early professor of Botany at Stamford University, California. The native haunts of this genus are WN USA (Oregon) to Mexico, comprising of some 40-60 succulent species closely allied to *Echeveria*, and are mostly clump-forming basal-rosetted evergreen perennials though many species produce stems and then become semi-shrubby. Many species hybridise freely and therefore are highly variable. The leaves are usually fleshy, grey to white-waxy farinose but sometimes grey-green, and are borne in terminal tufts or rosettes. Five-petalled flowers are borne in cymose clusters on stems well above the leaves, and the broadness with which the flowers expand determines their shape whether these are starry or bell-shaped. White-leaved sorts are most suitable for container growing. These are propagated through seeds and offsets. ***Dudleya attenuata*** (syn. *Echeveria attenuata, E. edulis*) from Baja California sea coasts is a semi-shrubby succulent growing to 30 cm high, bearing erect, glaucous green, narroiwly obovoid to cylindrical and up to 10 cm long leaves. Flowering stems grow up to 25 cm long with many branches bearing yellow starry flowers lined red in summer with up to 8 mm long petals. *D. a. orcuttii* bears white flowers flushed rose. ***D. brittonii*** from California is basal-rosetted and about 60 cm in height when in flower and with about 50 cm spread. Leaves are silvery-white, narrowly lance-shaped, tapering, and fleshy. Clustered pale-yellow star-shaped flowers appear in summer. ***D. densiflora*** (syn. *Echeveria nudicaulis*) from California is thick, semi-shrubby, erect and sparingly branched stems some 10 cm tall. Leaves are waxy-white, 6-15 cm long, cylindrical and tip-pointed. Branched flowering stalk grows 15-30 cm high bearing white or pink-tinted starry flowers with 1 cm long petals, which appear during summer. ***D. farinosa*** (syn. *Echeveria compacta, E. eastwoodiae, E. septentrionalis*) from coastal region of California is a well-branched semi-shrubby succulent growing up to 30 cm high, bearing grey-white or green but sometimes in strong sunlight leaves are red-flushed, 3-6 cm long, ovate-oblong and pointed. Flowering stalk is 3- to 5-branched, some being forked above and about 30 cm high, bearing lemon-yellow bell-shaped flowers in summer with some 1 cm long petals. ***D. greenei*** (syn. *D. regalis, Cotyledon greenei*) from California (Santa Catalina, Santa Cruz and Santa Rosa, and San Miguel Islands) is base-branching forming a clump of 3.7 m wide with stem thickness of 5 cm. Leaves are rosette forming, green and fleshy. Brownish-pale inflorescence up to 20 cm tall with nodding top appearing from the side of the plant, bearing clustered 5-petalled greenish-yellow flowers some 1.7 cm across with green petal-midrib. This species is closely related to *D. caespitosa*. ***D. hassei*** (syn. *Echeveria hassei*) from Santa Catalina and Guadalupe Islands grows up to 30 cm high with semi-shrubby and much-branched stems, bearing white starch-yielding, linear-lanceolate, obtuse and 4-10 cm long leaves. Flowering stalk some 30 cm tall comprise 2-4 branches, each again forked or simple and bears starry flowers in summer, where petals are white and some 1 cm long. ***D. pulverulenta*** from California is basal-rosetted succulent growing up to 60 cm in height and about 30 cm in breadth when in flower with silvery-grey strap-shaped and pointed leaves.The clustered star-shaped red flowers emerge during spring-summer.

Duvalia **(Asclepiadaceae)**

It was named so to honour Duval, an early botanist. They comprise of some 20 dwarf, herbaceous and leafless succulent species from South Africa, closely related to *Stapelia*. They are clump-forming or carpeting, growing erect or decumbent and sometimes even subterranean with the tips appearing above the surface. Stems are short, thick, leafless, 4-6 angled bearing teeth, each tooth with a rudimentary leaf which are deciduous. Flowers are solitary or in small cymes appearing usually in the middle of young stems and star-shaped with thick and fleshy petals which are recurved at tips. Corolla lobes are linear-lanceolate to ovate and folded longitudinally. They are propagated through seeds and cuttings. The cultivated species, ***Duvalia corderoyi*** is clump-forming and grows prostrate with leafless stem bearing 6 indistinct ribs which are often purple. The plant height is 5 cm and spread 60 cm. Flowers are star-shaped, dull green with purple hairs and appear during summer to autumn.

Dyckia **(Bromeliaceae)**

See ***Dyckia*** under the Chapter **Bromeliads.**

Echeveria **(Crassulaceae)**

It is native of S.W. USA to Mexico and Argentina. *Echeveria* was named for the Spanish botanical artist Athabasio Echeverria Godoy who accompanied an expedition to Mexico in 1787-1797. It is a genus of 150-200 species of evergreen succulent perennial, usually stemless but with rosettes of fleshy leaves which are often attractive mostly with waxy sheen. These spread to form usually a thick carpet. The bell-shaped and tubular flowers are produced in loose sprays on arching stems with five fleshy petals. These are propagated through division, detachment of rosettes, leaf cuttings and through seeds. ***Echeveria affinis*** (black echeveria) bears stemless dense rosettes some 10 cm wide, solitary or in small groups. The specimens at water-starved and sunny sites bear brownish- to greenish-black leaves though under poor light and in well irrigated plants these are deep to bright green. Leaves are oblanceolate with a short pointed tip. Inflorescence is 20-35 cm long, composed of several forked and simple cymes, bearing scarlet flowers some 1.2 cm long during late summer to early autumn. ***E. agavoides*** (syn. *Urbinia agavoides*) is an agave-like, slow-growing with dense and largest rosetted species, growing solitary or in small clumps forming stemless rosettes 10-25 cm wide and 15 cm tall. Leaves are pale-green, tapering with tips deep red to brownish and spiny, thick, ovate and often red-margined. Inflorescence is composed of 1, 2 or more simple cymes and of 20-45 cm tall, bearing some 1.2 cm long, cup-shaped and red-petalled flowers with yellow-tipped petals in May-June. *E. a.* 'Cristata' is a comb-like green-leaved cultivar turning red under sunny situation. ***E. atropurpurea*** (syn. *E. sanguinea*) grows some 20 cm tall robust and unbranched stems bearing 20-30 cm across rosettes on the crown. Leaves are coppery-red to purple with a glaucous patina, spathulate to obovate and more or less boat-shaped. Inflorescence racemose and 30-60 cm tall, bearing 1.2 cm long and red flowers in summer. ***E. crenulata*** is a rosette short-stemmed at younger stage but with age extends to 60 cm, sometimes with branching. Stems are crowned with rounded green leaves and usually with wavy red margins, some 15-25 cm long and obovate to oblong-spathulate. Cup-shaped light red flowers some 2.5 cm long are borne in autumn and winter on paniculate inflorescence about 60 cm tall. ***E. derenbergii*** (painted lady) is a clump- or mat-forming, stemless or shortly stemmed rosetted succulent growing 5.0-7.5 cm high and 30 cm wide though the breadth of one rosette is up to 7.5 cm, and produces a short-stemmed rosette of tightly arranged rounded and glaucous grey-green leaves some 6 cm long and covered with a white bloom, bearing reddish margins and tips, and are broadly spathulate and concave. It produces offsets freely. Cup-shaped yellow-and-red or orange flowers some 1.0-1.5 cm long are borne in 8 cm long arching racemes in summer. This has been used extensively in breeding programme. ***E. elegans*** (syn. *E. harmsii, E. perelegans, Oliveranthus elegans*) is a bushy, branched and sub-shrubby succulent growing up to 20 cm high and 30 cm spread with erect stems which are crowned with a 15 cm across rosette of pale-green, up to 4 cm long and narrowly lance-shaped leaves keeled beneath and covered with short hairs. Red tipped-orange and cup-shaped elongated flowers some 2-3 cm long which are yellow within and 1-3 per stem are produced in profusion on 10-20 cm long stalks during late spring. ***E. gibbiflora*** is quite similar to *E. crenulata* but leaves are light grey-green tinged purple-red., *e.g.* 'Carunculata' which is a shrubby succulent with large rosette, bearing fleshy, broadly spathulate, little concave, margins highly wavy and beautiful mauvish-green leaves up to 30 cm long on a plant of 60 cm height and where older leaves at the bottom tend to die off. Upper surface of the leaves have wart-like protuberances, most conspicuous during autumn, and due to this reason the leaves are wavy. Red flowers on a 45 cm long inflorescence are borne in autumn. ***E. gigantea*** when old produces new shoots quite often which can be removed, dried and used as cuttings. It bears large, fleshy, ovate, stem-clustered and purplish-green leaves with blunt tips. ***E. glauca*** (syn. *E. secunda glauca*; 'blue echeveria') is almost stemless, clump-forming, and rosettes some 10 cm across, bearing glaucous blue-grey patinated leaves sometimes with narrow reddish margins, being broadly obovate and some

2.5-8.0 cm long. Bright red racemose flowers appear in spring and summer on erect and 20-30 cm tall raceme that archs at the tips. ***E. leucotricha*** is commonly known as 'Chenille plant', producing 30 cm or more high branching stems with about 8 cm long and fleshy leaves arranged in 15 cm wide rosette and covered with white hairs below and brownish at the tip, and some 2 cm long red flowers appear on 30 cm or more longer inflorescence forming a thyrsoid spike. ***E. multicaulis*** (copper roses) is a shrubby succulent with stems growing up to 90 cm high and tipped with a rosette some 10 cm across. Leaves are lustrous green, obovate and edged red. Racemose inflorescence up to 30 cm long bears 1.2 cm long flowers that are yellow within and red outside which appear during winter. ***E. nodulosa*** bears sparingly branched stems some 45 cm tall, each stem towards the tips leafy and the tips crowned by rosettes some 8-15 cm across. Leaves are little keeled and flushed red along with the margins, obovate and covered with minute and whitish papillae. Inflorescence is racemose, 40-60 cm tall bearing tan-yellow red-splashed flowers some 1.2 cm long which appear during summer. ***E. peacockii*** (syn. *E. desmetiana, E. subsessilis*) produces usually solitary and almost stemless rosettes up to 13 cm across. Leaves are blue-green patinated densely waxy-white and are oblong. Inflorescence is a simple cyme and up to 35 cm tall bearing bright red flowers some 1.2 cm long which appear in early summer. ***E. pulvinata*** is commonly known as 'plush plant'. Its semi-woody stem attains up to 25 cm in height and has a few branches topped with loose rosettes up to 10 cm wide. Top of the stem is covered with white hairs though base with brown hairs. Obovate leaves are dark green, 4-5 cm long, 2-4 cm wide and up to 1 cm thick with short pointed tip, and are covered with dense coat of soft-white hairs. Inflorescence is 20-30 cm long, terminating in a thyrsoid raceme, and emerges between the leaves bearing several bright red or yellow and red flowers some 2 cm long with fleshy petals. Its noteworthy varieties are 'Frosty' (extra white-hairy leaves), 'Ruby' (leaves narrower with red hairs at the tip and along the margins), 'Doris Taylor' (*E. pulvinata* × *E. setosa*) with long orange blooms and has acquired characters from both the parents, 'Pulv-Oliver' (*E. harmsii* × *E. pulvinata*) with reddish-tipped and densely hairy leaves and coral-red flowers, *etc.* ***E. purpusorum*** (formerly *Urbinia purpusii* as christened by Dr. Rose) is though little cultivated but is the prettiest member of the whole genus due to thickness, shape and colouration (triangular, oval with short pointed tips, closely set, grey-green and finely dotted) and arrangement in a compact rosette some 10 cm across of its leaves. Yellow-tipped red flowers some 1.0-1.2 cm long appear in racemose inflorescence some 30 cm long. ***E. rosea*** is commonly known as 'desert rose', growing 30 cm high with branching stems where tips are rosetted 12-18 cm wide. Leaf bases semi-cylindrical, and leaves oblanceolate to spathulate, pale to grey-green though in winter these may be reddish. Racemose inflorescence 15-35 cm long, bearing 1.2 cm long yellow flowers, appears in the axils of red bracts (sepals). ***E. runyonii*** grows up to 8 cm high or may be stemless, bearing some 8 cm long blue-green, spathulate, upcurving and waxy leaves arranged in a loose and flattened rosette some 10 cm across. It forms offsets freely. Pink or white flowers some 2 cm long are borne during autumn on 15-20 cm long forking racemes. ***E. secunda*** is a clump-forming, usually stemless and some 10 cm across rosette forming. Broadly obovate, glaucous and light green leaves are up to 8 cm long with reddish tips. Red flowers are 0.8-1.0 cm long on 20-30 cm tall racemose inflorescence which arches at the tips. ***E. setosa*** (firecracker plant or Mexican firecracker) was discovered by Dr. J.A. Purpus (1860-1932) in 1907 from southern Mexico during his explorations of 1907-1908 in the mountains of Puebla State. It is distinguished from other fellow species by its succulent and hairy leaves arranged in a hundred or even more leaves in a dense but loose basal rosette. Its stemless thick rosettes are some 10 cm high, 10-15 cm across and some 30 cm in spread of long and mid-green and 5 cm long leaves with short and thick white hairs, and small red flowers tipped yellow which are produced on 20-30 cm tall inflorescences in racemose form during spring. ***E. shaviana*** produces stemless and usually solitary rosettes spreading 10-20 cm across with glaucous and spathulate leaves having crimping and pink tinging on the margins. Deep pink flowers some 1.6 cm long appearing in summer are borne on some 30 cm tall inflorescence with 1-2 simple cymes. ***E. subrigida*** was discovered by C.G. Pringale in 1892 in the Tultenango Canyon, the only known habitat in Mexico, initially named as *Cotyledon* species, first grown in the Missouri Botanic Garden and then introduced to Kew in 1905 where it flowered for the first time in 1911. It is a splendid flowering plant but also with foliage beauty. Its leaves are oval-pointed, grey-green, fleshy and purple-margined. Strong and erect paniculate inflorescence some 40-50 cm tall arise from near the centre of the rosette, bearing numerous cup-shaped, red, spreading and stalked flowers some 2.5 cm across.

Echidnopsis (Asclepiadaceae)

A native to South Arabia, Tropical and South Africa and Socotra, the genus comprises a number of perennial, succulent and prostrate herbs having sparse branching and where the stems are 8-10 angled and the ribs are divided into tubercles. Young tubercles may bear small and thin leaves which fall off soon. Wine red flowers some 1 cm across and pubescent in the centre appear from the furrows present between the ribs. These plants can grow under sun or filtered sun but hot summers are not congenial. *S. cereiformis* is a very handsome species.

Edithcolea **(Asclepiadaceae)**

A small genus from Africa which comprises odd-looking stapeliads with greenish-brown and spiny stems, in the wild forming the mounds that die from the centre outwards as occurs in case of 'fairy ring disease' caused by a fungus. Its flowers have varying amounts of red-brown spots on a brownish-yellow background. ***E. grandis*** (Persiasn carpet flower) from NE Africa bears 1.5 cm thick and spiny stems and is most suitable for hanging basket. Solitary flowers being more than 8 cm across appear on the new growth with dark reddish spots though some clones turn up almost entirely black. After producing many cuttings the old plants often rot and the cuttings root at a minimum of 27 °C temperature. It is allergic to prolonged cold and wetness in the root zone.

Eulophia **(Orchidaceae)**

Eulophia in Greek means 'handsome crest'. It is a terrestrial genus of small plant with 50-60 species from the tropics of both the hemispheres and is cultured similar to *Calanthe*. It develops from the pseudobulbs with membranous leaves. Scapes form on the base with several flowers where sepals and petals are spreading type, labellum is 3-lobed and pollinia 2. ***E. maculata*** from Brazil bears ompressed and ovate pseudobulbs, spotted or blotched leaves and small flowers where upper sepal is hooded while laterals acuminate and reddish-brown, petals wide, white or pale-rose, labellum cordate with two crimson spots near the base otherwise white. ***E. quartiniana*** from Central and East Africa, Sudan and Ethiopia was introduced into Europe around 1895. It was discovered by French botanist and explorer Richard Quartin-Dillon who died in Ethiopia in 1841. It is an erect growing species with long, wide and green overlapping leaves, the inflorescence sralk growing to more than 50 cm in length with about 20 cm rachis, bearing mauve drooping flowers. Greenish-brown sepals are back-reflexing. The species is very attractive and most suitable as cut flowers. ***E. scripta*** from Madagascar produces linear and subdistichous leaves, purple and yellow flowers, sepals and petals are linear-oblong, labellum is 3-lobed with lateral lobes rotund at the apices.

Euphorbia **(Euphorbiaceae)**

Euphorbia is named so, for Euphorbos, doctor to the king of Mauritania. Euphorbias are cosmopolitan in origin with about 2,000 species of annuals, herbaceous biennials, perennials, sub-shrubs and shrubs of both the nature, *viz.* deciduous and evergreen, many of them being succulent, and with respect to form and structure of flowers they are almost alike though vegetatively quite diverse in almost every respect. The herbaceous and sub-shrubby species in cultivation are normally hardy though shrubby and succulent species are generally not. They all have a small cup-shaped whorl of bracts fused together on the stem tips in dichasial cymes, called cyathium which consists of several male flowers as reduced to single stamen each and one female flower in the form of only a 3-lobed ovary. The stalks of the cyathia cymes are known as rays and are borne in umbel-like clusters (pseudoumbels) with separate pairs of bracts, known as raylet leaves, at cyathia bases, and a ring of larger bracts called pseudumbel leaves at each pseudumbel base, and these can be brightly coloured as in *Poinsettia.* The bracts may sometimes have petal-like structures though have often crescent-shaped nectar glands. The fruits are 3-lobed loud-dehiscing capsules. There are species requiring short days for flowering such as *E. fulgens* and *E. pulcherrima* (syn. *Poinsettia pulcherrima*). These require 3-month's 9-10 hours of darkness just before requirement of flowers so that these can initiate flowering. The milky latex of most *Euphorbia* species is highly irritant to eyes, mouth and skin. Succulent species are propagated through cuttings by dipping the cut ends in powdered charcoal and then drying them for several days before planting, whereas others can be propagated through seeds, cuttings or divisions, whatsoever be possible in any case. The hardy herbaceous *E. characias*, *E. wulfenii* and tender shrubby *E. pulcherrima* are suitable for cutting. Most of these have spines. Important succulent species are being described here with. ***E. antiquorum*** from India is a 3 m tall shrub with erect and succulent branches which are jointed and 2.5-5.0 cm thick with wavy and toothed angles bearing paired spines 2.5 cm apart. Leaves are roundish and very small. ***E. aphylla*** from Canary Islands is a mound-forming bushy succulent growing 45 cm or even more in height, bearing bright greyish to yellowish-green, leafless, erect and fleshy stems. ***E. atropurpurea*** from Canary Islands grows 1-2 m tall, rarely branched shrub with somewhat succulent stems. Oblanceolate and pale glaucous green leaves, 5-10 cm long are formed in deep rosettes or in tufts at the stem tips. Umbel is 5-10-rayed and the cyathia are surrounded by large purple and broadly ovate bracts. ***E. bubalina*** from South Africa is palmoid with green stem and grows up to 30 cm in height and some 1.9 cm thick, some 10 cm long lanceolate green leaves are borne on its top portion which are deciduous and falling off in winter leaving marked leaf scars. Flowers are minute, green and bell-shaped which appear during rains and set seeds. ***E. bupleurifolia*** is highly prized among the succulent growers as to those of *E. fasciculata, E. horrida, E. meloformis, E. obesa* and *E. valida. E. bupleurifolia*, a South African (Natal) species grows erect up to 20 cm in height and up to 8 cm thickness with usually unbranched stem, is ovoid to shortly cylindrical and is boldly tubercled due to protruding remains of leaf stalk bases. Light green, lanceolate and deciduous leaves are 10-15 cm long and appear on the top of the plant. Cyathia are solitary on 5 cm stalks bearing two green to red via bright yellow raylet leaves. ***E. canariensis*** from Canary Islands is the most common and very attractive but slow-growing succulent euphorbia forming wide mound-like masses with stems branching at the base though sometimes even above. It is a shrub to small tree growing up to 6 m tall with several 4-6-angled, green, suberect and jointed branches some 7.5 cm thick with waxy-white patina, the branches are almost leafless and the angles have horny margins bearing paired, 1.5 cm long black spines. Cyathia are reddish-green, solitary and short stalked at stem crown though are rarely formed in container-grown plants. ***E. caput-medusae*** from South Africa is commonly known as 'Medusa's head'. It is a succulent with thick stem which hardly grows above 15 cm height. Its branches are 45 cm or more long, greyish-green, prostrate to decumbent, fleshy, crowded and radiating from the top as to that of multi-armed octopus. Linear-lanceolate leaves are short-lived, 1.5-2.5 cm long and only on the tips of the branches. ***E. cereiformis*** from South Africa is 90 cm high succulent shrub with dark green and base-branched stems. Stems after a height is furrowed into 9-15 ribs with tubercled and spiny ridges. Leaves are short-lived and minute. Cyathia are solitary at the stem tips with purple raylet leaves. ***E. echinus*** from Morocco is a bisexual, green, highly branched but leafless succulent, growing up to 1 m in height with 6-7-sided wavy ridges that bear paired grey spines. The tiny flowers are green and bell-shaped and appear during rains. ***E. gorgonis*** from Cape (S.A.) is commonly known as 'Gorgon's head'. It is more or less globose bearing crowded, fleshy, succulent and tubercled stem with persistent swollen leaf bases, and grows some 10 cm high and wide. Main stem is much ribbed and is crowned by stiffly radiating 3-5 rows of green (rarely red-tinted) and prostrate branches, some 5 cm long. Leaves are minute and falling off soon. Cyathia bear brownish to bright crimson bands, are solitary and appear only from the main stem in the form of rounded heads of small fragrant flowers with cup-shaped yellow bracts. ***E. grandicornis***, commonly known as 'cow's horn' from Natal (S.A.), Tanganyika and Kenya is a thick succulent bush growing up to 2 m tall bearing pale-green and triangular branches which are constricted into rounded or ovoid segments with three wing-like angles, each angle wavy, horny-margined and with wide-spreading pairs of pale-brown turning grey, 2-5 cm long, sharp and dangerous spines. The flowers are small and yellow and the fruit red. Its latex is poisonous to the eye and wounds. ***E. grandidens*** from Cape (S.A.) is though a succulent growing up to 15 m in the wild but in containers it is tamed up to 2 m. Stems are erect but with whorled branching, and the branches have 3-4 often spirally twisted angles, each angle bearing horny and toothed margin which bears 4-6 mm long and paired spines. The whole plant looks like a statue bearing short-lived minute leaves. ***E. heteracantha*** is a leafless rusty-green succulent with several closely set dark green lateral branches thoughout the main body and up to the top, the lateral stems being rarely branched, the stems being usually 4-angled (rectangular), angle notches with sharp and small bristles. It is propagated through cuttings. Its milky latex is very poisonous. ***E. horrida*** from Cape to Karroo (S.A.) is spherical when young but turning columnar with age, base-branching to form an erect-growing clump as well as top-branching, is cactus-like shrubby, growing up to 1 m in height and up to 15 cm across, each stem with up to 14 furrows where ridges are toothed and bear up to 1-4 cm long 1-3 or clustered spines set 1-2 cm apart and come out from the hardened flowering peduncles. Stem is dark green with grey tinge, and corky brown at the base. Cyathia solitary and green, the flowers dioecious and nondescript. ***E. ingens*** from southern Africa, north to Zimbabwe and Zambia is a succulent tree growing to a height of 10 m in the wild. At younger stage and the branches of older specimens on maturity are dark green and somewhat constricted at the end of each growing season. The body is leafless and bears 3-5 wing-like ribbed and wavy angles, sometimes with minute spines. Cyathia are formed only on the currently grown joints on the top of the plant bearing yellowish-green bracts. ***E. lactea*** (syn. to perennial and succulent *E. havanensis* Hort. and not Willd.), a most common succulent from East Indies is quite similar to *E. antiquorum* but with a white-marbled area running through the middle of the branches. It is also confused with *E. hermentiana*. It also has monstrous forms. ***E. mammillaris*** from Cape (S.A.) is a small cactoid and suckering succulent growing some 20 cm high and the width being roughly double of it. Cylindrical stems are 5 cm or even more thick, semi-erect or prostrate and bearing 7-17 tuberculate angles with flat ribs. Flowering peduncles are woody, persistent and become scattering of 1.2 cm long spines. Leaves are scale-like and short-lived. Cyathia are solitary with yellowish or purple glands. ***E. meloformis*** from South Africa is a globose-cactoid succulent growing 10-12 cm high with 8-10 prominent ribs having grey-green skin or reddish with a paler ribbed pattern. This closely resembles the Mexican

Echinocactus ornatus. Leaves are linear, minute and short-lived, whereas inflorescences bear yellow-green flowers and their 2 cm or more long lignified-woody peduncles form scattered blunt-tip spines. Propagation is easily through seeds. ***E. milii*** (syn. *E. splendens*) from Malagasy (Madagascar) is commonly known as 'crown of thorns' and is very attractive flowering pot plant. It is a grooved and highly thorny semi-succulent and semi-prostrate evergreen shrub with woody base, spreading to 60-100 cm with height of 30-60 cm, bearing 1.5-10.0 cm long, mid-green and lanceolate-obovate to almost rounded leaves. Cyathia are borne in small and branched cymose umbels some 5.0-7.5 cm wide which continue emerging all the year round though optimally during the winter months from the upper parts of stems bearing two kidney-shaped crimson bracts extending beneath the flowers. In the form *lutea*, the bracts are lemon-yellow. ***E. obesa*** from Cape (S.A.) is commonly known as 'Gingham golf ball'. It is a pale to dark green to brownish-green curiously-looking spherical (ball-like), later turning slightly columnar (pear-shaped) small succulent growing up to 15 cm in height and 8-10 cm in diameter with slightly depressed crown. The body is spineless and leafless bearing reddish-brown horizontal stripes when kept in sun. The body bears 5-8 broad and flat longitudinal ribs with many small tubercles on the edge.The plant bears nondescript bell-shaped dioecious flowers (all staminate or all pistillate) with greenish-yellow small cyathia, the inflorescence being minute, green and sweetly-scented. ***E. pseudocactus*** from Natal (S.A.) is commonly known as 'cactus spurge'. It is a 3-5-ribbed (angled), articulated, shrubby cactoid succulent growing up to 2 m in height with about 5 cm wide stems which have medium green to deep green with a contrasting yellowish V-shaped arching markings between the ribs. Each rib is horny and toothed bearing paired spines some 1.2 cm long. Stems are erect or inclined, branched and constricted into ovoid segments. Upper part of the plant produces minute yellow flowers from the notches on the ribs. ***E. pugniformis*** is a prostrate-growing succulent with several branches from the ground which all bear 5-bracted large yellow flowers in profusion. It is an excellent pot plant. While taking its cuttings, these are plunged into water to prevent bleeding otherwise the sap will seal off the cut surface and will impede rooting. ***E. stolonifera*** from Laingsburg district of Cape Province (S.A.) was discovered by Rudolph Marloth in 1920. It has an underground much reduced main stem from which develop some 60 cm long 8-15 or sometimes even more spineless succulent branches above and long spreading rhizomes below to produce a diffuse plant. These main branches may or may not be jointed and branched. Each stem of the main body is green, cylindrical, thickened and tapering at both the ends, and produces a central, short-stalked male cyathium surrounded by 5-8 branches, each terminating in a bisexual cyathium. The bracts are small and bright yellow. Only on new growth some 5 mm long juvenile leaves appear which soon fall off. ***E. tirucalli*** from Tropical and South Africa and South Asia is commonly known as 'rubber spurge', 'Indian tree spurge', 'finger tree' or 'milk bush'. It is a green, smooth, round, jointed (joints about 10 cm long), precocious and whorled branching succulent growing 2 m or more in height with finger-like, smooth and 0.6-1.3 cm thick stems. Sometimes it attains the dimension of a small tree. Stems and branches bear very small and linear leaves that drop off in the autumn. ***E. trigona*** is a beautiful succulent with green, erect and slender stems having 3-4-sided branching. The central zone between the two winged-angles is pale-green which radiates sidewards between the two notches. Spines are minute but the oval leaves one from each notch are quite large and deciduous. ***E. valida***, an unisexual plant from Cape (S.A.) is spherical at younger stage but broadens and flattens becoming cylindrical with age and may grow 7-10 cm high and slightly less in width. The skin is dark green and the body bears 8-10 ribs with lateral markings. Green and sweetly scented flowers appear on male and female plants during summer to autumn and the floral stalks become woody and persist for years as spines.

Faucaria (Aizoaceae)

The genus takes its name from the Latin *faux* meaning a throat, the paired leaves resembling a gaping mouth like that of a tiger. The common name for the plants of this genus is 'tiger jaws' due to prominent upstanding whitish teeth on the edges of flat upper surface. It is a dwarf, succulent and perennial genus of 37 species from Cape (S.A.). The plant is almost stemless and consists of several pairs of tufted highly fleshy bright green leaves arranged crosswise one above the other, broad at the base, keeled underneath and toothed along the margins. Large and showy daisy-like, chiefly yellow though only one or two species producing white flowers which emerge by the evening in the end of summer and continue throughout the autumn. Buds and spent flowers may appear orange or red. There is very little difference among the species. They are propagated through division of the clumps and seeds. Seedlings bear flowers in 2-3 years. Most important species are briefly described below. ***F. felina*** is commonly known as 'cat's jaws'. Leaves are bright green dotted white but rarely flushed reddish after maturity, 4.5 cm long, boat-shaped with a prominently keeled tip and 3-5 teeth on each of the upper edges. Yellow flowers some 5 cm in diameter and commensurately larger than the size of the plant emerge in the late afternoon from late summer to late autumn. ***F. lupina*** is commonly known as 'wolf's teeth' on account of its edge-teeth structure as also its specific name *lupina* meaning wolf's. When young it grows singly but later forms some 15 cm across clump with a short stem. The triangular leaves are intense green, some 4 cm long and 1.5 cm wide. Upper part of the leaf is flat with long and fine teeth curving towards the plant, and the lower part keel-shaped with blunt tip. Yellow flowers some 3.5 cm in diameter emerge in the centre of the plant in the late afternoon from late summer to late autumn. It is propagated through division of clumps and from seeds. ***F. tigrina*** was introduced into Europe in 1790 by Fracis Masson and is commonly known as 'tiger's jaws' due to the structure of its 8-10 teeth present on the edges of upper part of the leaves. This species is more or less the longer version of *F. felina* and is possibly the most spectacular species in the genus. Leaves are broad greyish-green covered with white dots. Ray-shaped rich yellow flowers emerge during autumn some 5 cm across. ***F. tuberculosa*** is an almost stemless plant with 3-4 pairs of deep green somewhat rhomboidal leaves some 2 cm long with a bluntly rounded keeled tip, 3-5 large and a few smaller teeth on each edge and blunt to wart-like tubercles on upper surface. Gradually the plant forms a clump some 15 cm across. Yellow flowers some 4 cm across emerge from the centre of the plant from late summer to autumn.

Fenestraria (Aizoaceae)

The genus *Fenestraria* was erected by N.E. Brown. *Fenestraria* derives from the Latin word *fenestre* which means a window due to its leaf character as shall be described below. Its common name is 'window plant' as in case of *Frithia, Haworthia maughanii, H. truncata, Lithops, Ophthalmophyllum* and certain others. The genus *Fenestraria* from South Africa has two species of flowering succulents whose leaves are buried up to the tips in the desert sand but their transparent windowed tips make contact of their chlorophyll with the sunlight for photosynthetic activities. Flowers are daisy-like and disproportionately large and showy. These are propagated through seeds and division of stock. *F. aurantiaca* from Cape has tufted growth forming small hummocks. Rosette of long, thin and cylindrical leaves are narrowly obovoid with little flattened tip, greyish on the sides and 2-3 cm long. Bright yellow flowers some 3-5 cm across emerge fron late summer to autumn from the side of the plants. *F. rhopalophylla* from Namaqualand and Namibia is a rosette of cylindrical leaves which form the clump some 10 cm across. It is very similar to the former with respect to foliage but individual leaves are a little shorter and thicker. Leaves are pale-green, some 5 cm long and 0.5-1.0 cm thick, club-shaped at the tip, and terminate in a convex, nearly circular and green transparent window. Flowers are white some 1.8-3.0 cm across and emerge during autumn.

Ficus palmeri (Moraceae)

A native of Mexico, this species is closely related to *F. petiolaris*. It is a caudiciform where caudex grows faster and forms a thick squat plant. Roots of its squat caudex grip the rocks tightly and mimic as an ideal succulent bonsai. Though it tolerates deep shade to full sun but when too dry their leaves fall down. It is propagated through seeds and does not tolerate frost. Apart from *F. palmeri*, other caudiciform species are *F. brandegei, F. glumosa, F. petiolaris,F. obtusifolia*, and *F. soldanella*.

Fockea (Asclepiadaceae)

This genus of 10 species belongs to southern Africa (South Africa and Angola) and was named so to honour Gustav Woldemar Focke, an early 19th century German physician and plant physiologist. These are caudiciform perennial succulents with napiform or tuberous stems bearing thin, twining or erect branches with flat and oblong leaves having wavy margins. Flowers are starfish-like, dioecious, some 4 cm in diameter and appear solitary or several in dense clusters from axils of leaves, bearing narrow petals. ***F. angustifolia*** from Cape has large caudex, producing several minutely hairy stems, up to 70 cm long, erect or climbing, bearing 1.5-10.0 cm long and 0.2-0.6 cm wide leaves. Flowers appear in clusters of 2-6 with green lobes. ***F. crispa*** from Karroo (South Africa) has velvety hairs throughout the plant body. Its large napiform caudex is almost entirely subterranean with thin, twining or prostrate stems. Leaves are oval-pointed, 2.3 cm long and 1-2 cm wide. Flowers are borne 2-3 together and are green-grey blotched minutely brown. ***F. dammarana*** from Namibia bears thickened caudex, with branches bearing thinly spread wooly hairs when young. Leaves are linear-lanceolate, acuminate, 1.2 -2.3 cm long and 3-6 cm wide, and only a few flowers are borne on the stem. ***F. edulis*** from South Africa produces large white and warty caudex with oblong to elliptical leaves. Lime-green flowers are solitary or in a group of 2-3, bearing white corona with long tubes and curved petals. ***F. multiflora*** from Angola has large caudex bearing stout but misshapen branches, small and flat leaves with wooly-white below and flowers are numerous in clusters.

Fouquieria (Fouquieriaceae)

Some authors consider its appropriate family as Tamaricaceae. *Fouquieria* is named so to honour Pierre Ed. Fouquir, professor of medicine at Paris in the first part of 19th century. The common name to the plants of this genus is 'candlewood'. There are 10 species mosly in the deserts of Mexico, but one, *F. splendens* extends to the United States and sometimes is found planted in rockeries of California, and this species is of more importance. *Fouquieria* species are small trees or shrubs which become leafless during dry weather. Leaves are fleshy, obovate, and clustered in the axils of thorns. The plants produce tubular flowers in terminal racemes or panicles bearing 5 sepals, spreading 5-lobed corolla, stamens 8 to many, styles 3 individually or united and seeds either winged or hairy. ***F. splendens,*** commonly known as 'ocotillo', 'coach whip', 'Jacob's staff' and 'vine cactus', is a shrub growing to a height of 1.8-7.5 m with erect rod-like stiff branching at the base which are furrowed and grey bearing 1.3-2.5 cm long fleshy leaves where apex is rounded and the base is wedge-shaped, and in masses 2.5 cm long showy scarlet or brick-red flowers appear in elongated clusters on racemose or thyrsoid inflorescenses with 8-12 exserted stamens. The stout and sharp thorns are almost as long as the leaves. Capsules are 12-17 mm long with white seeds which have thick hairs. The plants of this species make effective natural fences and their stems yield a sort of wax, gum and resin.

Frithia (Aizoaceae)

Frithia commemorates Fredrick Frith, a South African succulent enthusiast. A rosette-forming flowering leaf-succulent from Transvaal (S.A.) with one species, *F. pulchra* which is very similar to *Fenestraria.* Here too, to prevent the water loss from its body and to protect the plants from intense desert-sunlight as well as from the pests, the entire plant is covered with desert-sand except the windowed leaf-tips which absorb the sunlight and export it to the side tissues where the chlorophyll is present though when growing it in the home or conservatory the plants should be grown with their leaves exposed to our less intense sunlight. It is propagated through seeds or by careful division. ***F. pulchra*** is of tufted habit and forms hummocks. Nine or more rosetted dark green leaves are obovoid with a shallow longitudinal groove and rounded tip with flat top, 2.0-3.5 cm long and grey-green on the sides. The tips are formed by an asymmetrically oval, pale-green and transluscent window some 5-8 mm across. Carmine flowers with white centre or entirely white flowers emerge in the centre of the plant, being 2 cm across.

Furcraea (Agavaceae)

Furcraea (also pronounced as *Fourcroya*) is named so to honour Antoine Francois Fourcroy (1755-1809), a French naturalist and chemist much involved in the French Revolution. A semi-desert genus of 20 species from Tropical America, superficially resembling *Agave* in foliage but differs in its short-tubed flowers. The plants may bear bulbils on lower stems. When young such as *F. foetida* and *F. selloa* or those which are stemless may be containerized for keeping indoors though flowering can be obtained only when kept outside. Certain species are stemless but others may have short to tall trunk-like stems with a dense rosette of sword-shaped fleshy leaves at the top. After many years of growth, only the large rosettes bloom with very tall and highly branched flowering stems. The entire rosette dies after setting of seeds leaving only small offsets to grow further or even none at all. Flowers are cream to white, starry and 6-tepalled. These are propagated through bulbils, offsets and seeds. ***F. foetida*** (syn. *F. gigantea*) from northern S. America is commonly known as 'Mauritius hemp' as in Mauritius and St. Helena it is grown for its fibres, otherwise it is highly ornamental elsewhere. It is usually trunkless though with age it is shortly stemmed, bearing 1.0-2.5 m long deep green leaves where margins are slightly wavy having a few curved, spiny but small teeth. Spread of the plant may be up to 5 m and the flowering stems may grow up to 8 m in height bearing many 4 cm wide bell-shaped green flowers with white interiors. The leaves in var. 'Mediopicta' bears a quite wide longitudinal band of silvery-white or mixed white, green and whitish-green and sometimes in the broken stripes, in the centre. ***F. selloa*** from Colombia grows with trunk some 90 cm tall after many years of age. Leaves are well-spaced glossy deep green and 90 cm or more in length with hooked brown teeth. Flowering stems grow to the height of 6 m with pendent green and white flowers. Var. 'Marginata' bears margined white leaves which age to creamy-yellow.

Gasteria [Asphodeliaceae (Liliaceae)]

An almost stemless genus of some 70 clump-forming succulent species from southern Africa, especially Cape region of South Africa, having clustered leaves arranged in compacted two ranks, and often with white marbled markings. The species defined below are only from Cape. Henrik Bernard Oldenland and Carl Peter Thunberg, at the end of the 17th century were probably the first to collect *Gasteria* species though the species were described between 1819 and 1826 by Adrian Hardy Haworth (1768-1833) only when he received these specimens from James Bowie. *Gasteria* derives from the Greek *gasta* meaning a belly referring to the swollen floral tube. Their leaves are decorative throughout. The racemes are taller, some 12 cm long and wiry or arching, bearing usually 2.5 cm long drooping red tubular flowers with swollen base and have petals oftenly fused except for a small portion at the tip of each, forming a gently curving perianth tube. These are propagated through seeds, leaf cuttings and clump division. Some of the most common species are described herewith. ***Gasteria armstrongii*** was discovered by W. Armstrong after whom the species is named. He in 1912 sent its seeds to the botanical garden in Vienna from where the plants spread to other botanical gardens. It is a small plant generally not exceeding 10 cm in size, producing distichously arranged tongue-shaped inclined leaves when young but afterwards become prostrate and 5 cm long. The leaves are minute white-warty on deep green base. Old plants bear sowewhat longer triangularly-ovate leaves in rosettes. The flowers are alternately arranged, fairly small, pink and loosely arranged on 30 cm long wiry raceme. ***G. batesiana*** (syn. *G. carinata* var. *verrucosa, G. subverrucosa marginata*) is a clump-forming succulent with only a few offsets. Leaves are stiff, dark green, triangular-lanceolate, keeled, tapering, some 10-15 cm long, rosettes spirally arranged, bearing tubercled white dots and incurved edges. It grows 15 cm in height with 30 cm spread bearing bell-shaped orange-green flowers some 10 cm long. ***G. bicolor*** (syn. *G. liliputana*) is the smallest species growsing up to 7 cm in height with 10 cm spread, forming spirally arranged rosettes of some 5 cm long and dark green leaves keeled underneath and blotched white. Racemes are 15 cm long bearing orange-green bell-shaped flowers. ***G. brevifolia*** bears 8-15 cm long distichously arranged leaves which are tongue-shaped, rounded tips with a small abrupt point, dark green with tubercled and confluent white spots in transverse bands and toothed margins. It produces 60 cm long racemes bearing 2 cm long orange-red flowers. ***G. caespitosa*** is a clump-forming succulent with fan-shaped leaf arrangement and grows up to 15 cm in height with 30 cm spread having leaf length of 15 cm. Leaves are triangular, dark green and quite fleshy with toothed margins. Upper leaf surface bears numerous white to pale-green spots in diagonal rows. Flowers are orange-green and bell-shaped on long stalks. ***G. candicans*** grows about 15 cm in height and some 25 cm in spread, bearing dark green spirally arranged leaves having a few blunt teeth towards the tip and with clustering of intermittent white to pale-yellow dots throughout upper surface. Upper surface of leaves is slightly cupped. Flower stalk is more than 50 cm long and bears pink flowers with green lobes at mouth. ***G. humilis*** bears trinangular, deep green and 8-10 cm long leaves in spiralling rosettes with upper surface concave, tubercled and spotted white, spots either scattered or irregularly arranged into transverse lines, sometimes confluent, while lower deeply keeled. Floral stalk is about 30 cm long with orange flowers. ***G. liliputana*** bears dark green patterned with white marbles and spots, upper surface hollowed, lower keeled and rounded, 5-10 cm long, lanceolate and spiraling rosettes of leaves. Red flowers 1.5 cm long are carried in 10 cm long racemes. ***G. lutzii*** at early stage bears distichous rosettes then spiralling with up to 25 cm long, green, sometimes with reddish tinge or suffusion and with a few large paler spots, and strap-shaped leaves which are acuminate but rounded at tips. Floral stalks are up to 90 cm or even more bearing up to 2.5 cm long flowers which are green flushed red. ***G. maculata*** produces distichously-rosette or slightly spiraling 16-20 cm long and 5 cm wide stiff leaves which are dark green with bands of merging or otherwise white spots which become pinkish under sun, tongue-shaped, upper surface glossy, flattened or slightly convex and lower hollow. Racemes appear frequently in the season, are 0.3-1.2 m long bearing 1.5-2.0 cm long flowers, swollen at base and red with green border, It is normally propagated through leaf cutting or division of offsets but least through seeds though these germinate fairly quickly. ***G. marmorata*** bears rosette of distichous leaves which are dark green with merging pale-green spots, lanceolate and up to 15 cm long. Floral stalk is up to 75 cm long, and the flowers are reddish and 2 cm long. ***G. obtusa*** bears spiralling rosettes with younger leaves erect but older ones spreading. Leaves are 10-15 cm long, triangular-lanceolate, tips are blunt, glossy green above with transverse bands made of many small and indistinct spots, and either 3-angled beneath or keeled. ***G. obtusifolia*** is composed of distichous rosettes which on maturity may contain up to 14 tongue-shaped green leaves measuring 15-18 cm each. Tranverse bands with whitish-green spots occur on the leaves. Racemes are some 75 cm in length with 2.5 cm long red flowers. Var. 'Variegata' bears creamy-white marks on the leaves. ***G. picta*** is a clump-forming rosette succulent. Leaves are mostly glossy, and dark green with irregularly arranged white dots on cross-bands, strap-shaped or lanceolate, and distichous or arranged spirally. Racemes are about 45 cm with densely formed reddish-pink flowers. ***G. pseudonigricans*** is formed of distichous rosette of leaves which are triangular-lanceolate to nearly strap-shaped. Young leaves are first erect but later spreading or curving, 15-20 cm long and shining green with a white spot. ***G. pulchra*** has a 15-30 cm high stem on which some 20-30 cm long glossy green leaves are arranged more or less in distichous form. The leaves are narrowly sword-shaped and triangularly arranged with white spots in transverse bands. Panicles are up to 90 cm with 2 cm long reddish flowers. ***G. verrucosa*** (Ox tongue) has distichous rosettes having dull green lanceolate leaves dotted with small white warts with rectangular apices, upper surface concave but rounded beneath and some 23 cm long. Some 2.0-2.5 cm long flowers emerge on 60 cm long racemes.

×*Gastrolea* (Liliaceae)

It is a hybrd genus comprising of some 15 such hybrids between *Gasteria* and *Aloe*. In appearance these are more towards *Gasteria* with tubular flowers appearing in racemes above the leaves with usually reddish ones having green tips. As defined under, only two such hybrids are notable. ×***Gastrolea beguinii*** (*Aloe aristata* × *Gasteria verrucosa*) bears 15 cm long, deep green, erect, triangular-ovate, acuminate and in light the tips are reddish, and banded transversely with white spots. ×***G. smaragdina*** (*Aloe variegata* × *Gasteria candicans*) bears up to 20 cm long, stiff, succulent and spreading leaves. They are light green, ornamented with greenish-white spots in transverse pattern but in irregular form, chanelled and keeled towards the distal part and narrowly triangular to triangular-lanceolate in shape.

Gerrardanthus (Cucurbitaceae)

A native to Tropical and South Africa, it is a flowering perennial climber with height up to 5 m. The plants appear from a swollen tuberous base having a thickness of up to 1.5 m. Its stems are herbaceous but turn grey barked and woody with age. Leaves are green and palmate as other cucurbits have though in certain other members these may be needle-shape or ferny. *Gerrardanthus lobatus* and *G. microrhizus* are the best examples of this.

Gibbaeum (Aizoaceae)

Gibbaeum derives from the Latin *gibba* meaning swollen or humped on account of very fleshy leaves in certain species. It is a flowering succulent genus with some 30, mostly variable species, especially from Little Karoo (S.A.). These are clump-forming having loose growth with more elongated leaves, and the flowers are daisy-like, appearing in spring, summer or during autumn. These are propagated through seeds or division of clumps. ***G. album***, a succulent with grayish-white clumps of firmly adpressed leaves, is restricted to whitish quartz outcrops of Karoo covering many acres there. Each shoot comprises of two fused, highly minute white-downy and informally hemispherical leaves which together measure some 2.5 cm wide. The flowers are pure white and some 2.5 cm across. ***G. fissoides*** (syn. *G. nelii, Antegibbaeum fissoides*) is a tufted to small clump-forming where each shoot is composed of one or two pairs of grayish-green leaves fusing at base, each leaf being semi-erect, nearly cylindrical but flat to the inner side, tip rounded and measuring some 5 cm in length. Flowers are rose-purple and 2.5-5.0 cm across. ***G. gibbosum*** (syn. *G. perviride*) is the first species to be described in the genus. It has tufted compact habit with woody, branched and tubercled rootstock some 6 mm thick and of chocolate colour with about 10-15 cm in height, each shoot of one very unequal pair of leaves, the largest being 6 cm long, i.e triple the size of the smallest one, semi-erect, semi-cylindrical and with flattened inner surface. Flowers are pink to magenta-purple with yellow centre and up to 3 cm across. ***G. petrense*** is a very small mat-forming species growing 3 cm in height but 30 cm or more in spread. Each branch (shoot) comprises of 1-2 pairs of pale-grey-green, connate for 1/3rd of their length and lie parallel, fleshy and some 5 mm thick and wide and 1 cm long leaves which are flattened to the inner side and keeled towards tips on the back. Flowers are daisy-like, pink-red and 1.5 cm across. ***G. schwantesii*** (syn. *G. muirii*) comprises of one very unequal pair of leaves on each shoot but a little larger, broad-based, keeled and hooded tips when compared to *G. gibbosum*. Leaves are dark green, sometimes overtoned greyish or brownish and very minutely velvety pubescent. Smaller leaves are half the size of larger ones. Flowers are pure white and 5 cm across. ***G. velutinum*** is a clump-forming and prostrate-growing, growing up to 8 cm in height and 30 cm in spread, each shoot 1-2 paired, pairs unequal, bluish grey-green, velvety- pubescent, finger-like but triangular in outline and some 6 cm long leaves. With the age the plants are though quite small but become woody covered with dry remains of older leaves. Longer leaves are 6 cm long and 3 cm basally wide though smaller ones 4 cm long, both fused basally. Flowers are daisy-like, white, pink or lilac and 5 cm across.

Glottiphyllum (Aizoaceae)

A genus of flowering succulents from South Africa, especially Cape region with 50 species. The generic name derives from the Greek *glotta* meaning 'a tongue' and *phyllon* meaning 'a leaf', referring to the texture and shape of the leaves, therefore, commonly these are known as 'tongue leaf'. It is a tufted clump-forming plant with long, semi-cylindrical, soft-textured, thick and very juicy leaves, often with broadened tips. The flowers are daisy-like and emerge from autumn to spring. They are propagated by seeds and division. ***G. fragrans*** is a clump-forming succulent bearing yellow-green, some 8 cm long and 2.5 cm wide, and slightly obliquely tongue-shaped leaves which resemble a cluster of tongues. Fragrant yellow flowers are 8-10 cm across but self-sterile so these may set seeds only when an individual plant is raised nearby. ***G. linguiforme*** is a clump-forming succulent with distichously arranged, glossy green and some 6 cm long leaves. Flowers are intense yellow and some 5-7 cm wide. ***G. nelii*** is a clump-forming succulent growing up to 5 cm in height and 30 cm in spread with green, semi-cylindrical and fleshy leaves some 5 cm long in two rows. Short-stemmed flowers emerge during autumn to spring with daisy-like, golden-yellow and 4 cm across.***G. regium*** has restricted growth as compared to other species. Tufted plants appear with shoots composing of 1-2 pairs of smooth, thick and light green leaves. Leaves are semi-erect, narrow, 4-10 cm long and 10-15 cm wide. Flowers are yellow and some 4 cm across. ***G. semicylindricum*** is a clump-forming succulent growing 8 cm in height and 30 cm in spread bearing bright green, some 6 cm long, and as the name applies the semi-cylindrical leaves which are toothed half-way along the margins. Flowers are daisy-like, 4 cm across and golden-yellow which emerge during spring-summer.

Grahamia (Portulacaceae)

A new genus from New Mexico & western Texas, Mexico, Bolivia & Argentina and Australia, recently separated from *Anacampseros*, range from small herbaceous plants with tuberous roots to small shrubs. It produces pink or white flowers and abundance of seeds for further propagation though is also propagated through cuttings. ***G. coahuilense*** from Mexico produces limping stems about 10 cm long that emerge from the clustered thickened roots, stems bearing succulent leaves. Flowers some 2 cm wide of pink shade appear throughout the whole summer. Water scarcity forces this to become limp and wrinkled but recover soon when watered. This requires protection from harsh sun and frost.

Graptopetalum (Crassulaceae)

It takes its name from the Greek *graptos* meaning 'to write or paint', referring to the lines on the petals of some species. This rosette succulent genus with 10-12 species from SW USA and Mexico is similar to *Echeveria* with which it hybridizes. This has entire and fleshy leaves with umbel-like trusses of flowers which are borne in the leaf axils. It is propagated through seeds and stem or leaf cuttings. The species being described below have their native haunts in Mexico. ***Graptopetalum amethystinum*** (syn. *Pachyphytum amethystinum*) is a clump-forming and prostrate-growing succulent some 40 cm high and 1 m wide. It produces blue-grey to red, fleshy and oblong-laceolate to rounded leaves some 3-6 cm long in terminal rosettes. Flowers are 1-2 cm across, yellow-and-red and appear in spring-summer. ***G. bellum*** (syn. ***Tacitus bellus***) from Mexico is basal-rosetted and grows some 3 cm high and with 15 cm spread. *Tacitus* is described only with one species, and is Latin for 'silent' as the corolla tube is very close-mouthed. It is a low-growing succulent combining the features of the allied *Echeveria* and *Sedum*. It's rosettes are compact, flattened, clustering, 4-8 cm wide and hummock-forming. Its foliage remains attractive throughout the year and when in flower during summer it becomes very spectacular. It bears dark grey-green with mild marginal reddening, triangular, thickly fleshy and 3-5 cm long leaves. Clustered bright carmine-pink or also a form with yellow flowers, 2-4 cm across are starry, 5-petalled and appear in 3-10 cm wide-spreading panicles just above the leaves. ***G. pachyphyllum*** grows prostrate to decumbent, base-branching and some 20 cm long. Leaves are blue-grey but red-tipped under sunlighat, club-shaped and some 2 cm long. White spotted-red flowers some 1.2 cm across emerge on small terminal panicles during early summer. ***G. paraguayense*** (syn. *Echeveria paraguayense, Sedum weinbergii*) is commonly known as 'mother-of-pearl plant'. It is basal-rosetted clump-forming succulent, 10 cm tall and 1 m in width. Basal rosettes are some 15 cm across bearing 5-7 cm long, obovate-spathulate, flat above and keeled beneath, and glaucous blue-grey leaves oftenly tinged pink. Yellow-and-red or white spotted red star-shaped flowers some 1.5-2.0 cm across emerge during summer on small panicles.

×*Graptoveria* (Crassulaceae)

It has been evolved by crossing *Graptopetalum* with *Echeveria*, former being tolerant to hot sun so this hybrid is beautiful and hardy. They are propagated from cuttings and leaves. **×G. 'Silver Star'** is a hybrid of *G. filiferum* and forms a low and dense rosettes some 12 cm wide of offsets bearing pale-green leaves. It requires only a few hours of sun in the morning and tolerates up to -4 °C temperature only for a brief period. It is propagated through offsets. ×***G. amethorum*** is small form which forms a 5 cm wide rosette of green and thick leaves where tips in the bright light turn purple. It also behaves similar to that of var. 'Silver Star'.

Greenovia (Crassulaceae)

Greenovia is named so in the honour of George Bellas Greenough (1778-1855), the English geologist. This genus originates in Canary Islands with four species of succulents, closely allied to *Sempervivum*. These normally grow during autumn to spring and rest during summer. These are almost decorative, doing well in containers and produce usually starry yellow flowers. Their multiplication is through seeds, and offsets. ***Greenovia aizoon*** is a clump- or hummock-forming succulent having several 5-8 cm wide rosettes, bearing light green, broadly spathulate, white-waxy and softly hairy leaves, Floral stalk is 10-15 cm long with panicles bearing 1-2 cm across yellow flowers composed of about 20 linear petals. ***G. aurea*** is a small clump-forming and clustering but in the wild, while in cultivation it is generally solitary with 13-30 cm wide rosette. Leaves are bluish-green, cuneate to spathulate and thickly overlapping. Floral stalk is up to 45 cm long bearing 2.5 cm wide yellow flowers with 30-35 petals.

Haworthia [Asdphodeliaceae (Liliaceae)]

Once *Aloe, Gasteria* and *Haworthia* belonged to one genus. *i.e. Aloe* as they have many features in common though *Haworthia* species are smaller and frequently marked white with pink or green transverse stripes on thick, stiff and tapering leaves. *Haworthia* is named so for Adrian Hardy Haworth (1768-1833), an English entomologist and an authority on succulents. The native haunts of this genus are southern Africa, especially Cape Province of South Africa. It is a genus of 150 species mostly of clump-forming succulents with triangular-shaped and fleshy leaves which appear either in rosettes or crowded on short erect stems. The long and wiry inflorescence is racemose with tubular and often curved greenish-white flowers which are though composed of six narrow lobes but appear to be 2-lobed. These are multiplied through division or by offsets. A few of the important species are described here with. ***Haworthia arachnoidea*** (syn. *H. setata*) is a slow-growing and some 5 cm tall, 10 cm wide, and clump-forming succulent bearing rosetted triangular leaves, softly serrated white along the margins. Flowers are white and appear from spring to autumn. ***H. armstrongii*** has erect stem with 10 cm or more in height bearing crowded and upward pointing some 3-4 cm long leaves which are deep green, lanceolate and keeled below with slightly waxy-white patina and white pointed tip. ***H. attenuata*** is rosette-forming clumpy succulent growing 7 cm in height and some 25 cm in breadth and produces a basal rosette of dark green, triangular-oblong and tapering to a point, fleshy and 6-8 cm long leaves ornamented with bands of tubercled white dots which are rather more dense on the backside. White tubular and bell-shaped flowers appear during spring to autumn on long slender stems. ***H. a.*** var. ***clariperla*** is clump-forming with pronounced bands of white dots on the back of leaves bearing white spreading flowers. ***H. batesiana*** is hummock-forming succulent with rosetted leaves, each leaf pale-green, broadly lanceolate to oblong with tessellated darker veins and incurving bristle-tips. ***H. coarctata*** grows erect with stem height of about 20 cm. Leaves are deep green, 6 cm long, triangular-lanceolate, tapering to a point, and bear small greenish-white tubercled lines. ***H. cooperi*** appears as the larger form of *H. obtusa* and is rosette-forming. ***H. cuspidata*** is a clump-forming succulent growing 5 cm in height and 25 cm in breadth, producing a basal rosette of light green leaves some 2.5 cm long with translucent marks, smooth, fleshy, ovate-triangular though with rounded back, and narrowing to an abrupt pale-green point with transparent tip. White tubular to bell-shaped flowers emerge on long slender stems from spring to autumn. ***H. cymbiformis*** forms large clumps with 7-10 cm across rosettes of pale-green and very thick leaves. Leaves are boat-shaped, 4 cm long and 2.5 cm wide with pale and transparent faintly striped windows on the upper side, while lower surface strongly concave and keeled towards tip. Tip is shortly pointed and transparent. White and small flowers emerge frequently on 30 cm stalk. ***H. fasciata*** is commonly known as 'zebra haworthia', grows some 15 cm high and 30 cm across and is clump-forming with leaf-rosettes some 8 cm across though individual leaf some 4-6 cm long and 1.3 cm wide. Leaves are glossy green, triangular and flat at the upper surface, and convex with large white horizontal warts underside. Racemose inflorescence is erect, some 30-40 cm long and bears tubular to bell-shaped whitish small flowers thoughout the season, *i.e.*spring to autumn. ***H. greenii*** bears more or less erect stems some 10-20 cm long which become procumbent with age. Leaves are reddish to copper green with horizontal rows of whitish tubercles, broadly lanceolate, some 4 cm long and with a keeled and incurving tip. ***H. limifolia*** is a slow-growing rosette-forming short-stemmed succulent growing up to 5 cm in height. Leaves are dark green marked with glossy whitish transverse lines (ridges) of tubercles, stiff, broadly ovate-triangular, gradually tapering, clearly convex underneath and roundish keeled. Offsets are formed in plenty. Flowers are produced from mid-summer to autumn. ***H. margaritifera*** (syn. *H. pumila*) is commonly known as 'pearl plant', bearing broad rosette of dark green, some 8 cm long, upcurving, lanceolate (triangular-ovate) and tip-keeled leaves and grows some 8 cm in height. Leaves are initially erect but spreading with age. Leaves are densely marked in bands of white-pearly tubercles. This forms offsets freely. Flowers are produced from June onwards. Its var. 'Maxima' is larger and 'Minima' smaller in all its parts. ***H. maughanii*** is a rarest and choicest one growing very slowly and only 2.5 cm high. Leaves are mid-green, almost vertical and arranged in spiral rosettes with flattened and almost transparent top similar to *H. truncata* but in this case the leaves are more or less cylindrical. The transparent tops work as windows to transport the light to the inner tissues of the plant where chlorophyll is present so that assimilation activity may go on. In the wild these plants are completely covered with desert-sand with only exposed windows. This rarely forms offsets under cultivation. White flowers on a 12-15 cm long raceme are formed in August. ***H. obtusa*** (syn. *H. cymbiformis obtusa*) is a thick rosette-forming succulent with densely arranged leaves. Leaves are erect, 2.5-3.0 cm long, almost cylindrical, cuspidate, ovoid to obovoid and upper half to one-third translucent but with dark vertical veining.

H. planifolia is rosette-forming bearing light greyish-green leaves which are 3.5-5.0 cm long, shortly ovate but tapering abruptly to a short point, upper surface flattened and lower convex. ***H. pygmaea*** is rosette-forming succulent bearing 2.5 cm or longer leaves which are glossy dark green, narrowly oval and the tips are abruptly truncated and translucent. ***H. reinwardtii*** is a widespread but variable succulent with respect to density of tubercles and size of leaves and stems. It bears 10-15 cm tall more or less erect stems where 5 cm long incurving leaves closely overlap each other. They are deep green with clear whitish tubercles. ***H. reticulata*** is a rosette-forming succulent bearing pale-green, 3-5 cm long and long bristle-tipped leaves which are oblong-lanceolate with toothed margins and where upper one-third is nearly transparent but with darker veins. ***H. retusa*** is a rosette-forming bearing up to 5 cm long, highly fleshy and almost erect leaves where upper half bends almost at right angle and its triangular surface almost flat and translucent with vertical pale veins. ***H. setata*** is a highly different and variable species worth for collection. It is a rosette-forming rounded succulent comprising of deep green and tip-keeled leaves being 2.5 cm or more in length, slender and lanceolate bearing numerous pale bristles on the margins. *H.s. gigas* bears quite large rosettes of 6 cm long leaves with very prominent bristles on the edges. ***H. subfasciata*** is though quite similar to *H. fasciata* but proportionately very large, rosette-forming and the leaves are up to 13 cm long. ***H. tesselata*** is commonly known as 'star window plant'. It is spirally rosette-forming some 6-10 cm across and is composed of shortly tapering, stiff, fleshy, 4-5 cm long and 2-5 cm wide triangular and green leaves, where upper surface becoming dark brownish-green in the sun, is often flat, glossy, translucent, with 5-7 branched longitudinal lines to form a network. Small whitish-green flowers are produced continuously throughout the season on up to 30 cm long loose racemes. ***H. truncata*** is a slow-growing and clump-forming succulent growing 2.5 cm in height and 10 cm in width, having 6-8 fan-shaped but in opposite rank leaf arrangement with incurved and some 2-4 cm long, broadly cylindrical, roughly tuberculate and blue-grey to brownish transparent leaves bearing pale-grey bands and flat truncated ends as if these have been cut off. Offsets are formed very rarely. Flowers are central, small, white and tubular with spreading petals which emerge during rains on 20-30 cm long raceme. ***H. venosa*** is though rosette-forming but elongate with age forming a short stem. Leaves are triangular-lanceolate, 5-7 cm long, spreading to recurved, upper surface green but sometimes with purple splashing, mostly with tessellated vein pattern, tips keeled and margins serrated. Other important haworthias are *Haworthia bilineata, H. lepida, H. ryderiana, H. turgida, etc.*

Hechtia (Bromeliaceae)

See chapter on '**Bromeliads**'.

Hesperaloe (Agavaceae)

In Latin, *Hesperaloe* means western *Aloe*. These are acaulescent plants with basal-rosetted, dark green, very narrow, outcurved, grooved and filiferous leaves like many of the agaves and yuccas, which are soft-pointed and often with white coarse fibres at the margins. These produce offsets freely at the base. Propagated through seeds or by divison.These bear short-lived greenish or red flowers opening during daytime. Inflorescence is lax-branched, flowers are oblong, filaments slender and attached to the base of perianth, pistil with ovoid ovary, style slender and stigma small. There are two species, *H. parviflora* from USA and *H. funifera* from Mexico, ***H. parviflora*** (syn. *Yucca parviflora*) is basal rosetted succulent growing up to 1 m or more in height and some 2 m in spread, often with peeling white fibres at leaf margins. Floral stalk bears a raceme of bell-shaped pink to red flowers during summer to autumn. ***H. funifera*** (syn. *H. davyi, Agave funifera, Yucca funifera*) is a source of Zamandoque fibre, one of the kinds of Tampico hemp. This succulent is similar to former but with double the size producing purplish green 2.5 cm long flowers.

Hoodia (Asclepiadaceae)

Hoodia is named so to honour Mr. Hood, a succulent collector around mid 19th century. It is a succulent genus with 100 species from Tropical and South Africa (South Africa, Namibia and Angola) of clump-forming, more or less erect, cylindrical, fleshy and green stems, generally branching at the base. These are allied to *Huernia* and *Stapelia*. Stems bear thickly set with rows of spine-tipped cactus-like tubercles on low ribs. Solitary or small clustered bowl- or bell-shaped flowers of yellowish, purplish or orange-browny tinged gold appear from the stem grooves on the top. The most cultivated species are described below. ***Hoodia bainii*** from Cape and Namibia is clump-forming, growing 20-30 cm high and with 15 cm spread, individual stem up to 4 cm thick, bearing dull green stem with 12-15 spiral rows of spine-tipped tubercles which terminate to sharp thorn each. Flowers are bell-shaped, pale-yellow or buff-coloured often tinged pale-pink or purple, 5-merging lobes and 7 cm across which emerge during summer to autumn. ***H. gordonii*** from Cape, Namaqualand and Namibia is a variable clump-forming and erect species, growing 30- 80 cm high and 30 cm spread but individual stems only 5 cm thick, often branching into clumps. Stems are green equipped with up to 14 rows of spine-tipped tubercles. Flesh-coloured to brownish with yellow stripes, 5-lobed, saucer-shaped flowers some 8-10 cm across emerge during late summer. ***H. macrantha*** from Namibia grows to about 90 cm in height in the wild but when cultivated often only half of this. The stems bear some 15 rows of spine-tipped tubercles. The plant height and the flower size are the largest in this species in whole of the genus. The bowl-shaped light purple flowers with yellowish vein emerge during summer up to 20 cm across where the lobes are darker purple and hairy.

Hoya (Asclepiadaceae)

Hoya was named for Thomas Hoy (*d.* 1809), head gardener to the Duke of Northumberland at Syon House, Isleworth. The genus with 200 species originated in SE Asia and the Pacific Isles. These are evergreen shrubby-climbers, some even epiphytic, with opposite pairs of fleshy and entire leaves, and large waxy flowers with 5 petals in large pendent umbels. These are propagated through cuttings. ***Hoya australis*** from Australia and New Guinea is a succulent climber growing more than 5 m long but in container less than half its length. Leaves are rich green, quite fleshy, rounded to broadly obovate and some 8 cm long. Fragrant white flowers with five red-purple spots in the centre, up to 50 in each umbel and some 1.2 cm wide are borne in summer. ***H. bella*** is commonly known as 'miniature wax plant'. It is epiphytic branching shrub from India, growing up to 45 cm tall with arching to pendulous stems. Leaves are ovate-lanceolate and 3 cm long. Flowers of 1 cm width are white with red centre, in umbels of 8-10 that open in summer. For better performance, *H. bella* is grafted on *H. carnosa*. ***H. carnosa*** (syn. *Asclepias carnosa*), commonly known as 'wax plant', is from S. China to Northern Australia, and is temperate in nature. It grows more than 6 m in length with ovate to obovate, 5-8 cm long leaves, and highly night-fragrant white to light pink flowers ageing to pink with a pink centre which are borne in large umbels from late spring to autumn. *H. c.* 'Exotica' is a mutant of *H. carnosa* with white flowers and centrally cream-variegated leaves but full open sun is injurious for its growing. *H. c.* 'Variegata' has cream-edged leaves. ***H. cinnamomifolia*** from Java is a twiner growing more than 6 m in the wild though in containers up to 3 m. Leaves are fleshy-thick with quite thick stalk, ovate, slender-pointed and 8-13 cm long. Bright yellow-green flowers with a deep magenta centre and some 1.6 cm wide are borne densely in globular umbels in summer. ***H. coriacea*** was first discovered by Carl Ludwig Blume (1796-1862) from Java having leaf texture in between leathery and fleshy. It has yellow flowers with small purple eye having about two dozen per inflorescence. ***H. coronaria*** from Java is a slow-growing thick-stemmed. climber growing 2-3 m in containers. Leaves are oblong to oval, leathery-textured, hairy beneath and up to 15 cm long. Flowers are largest among the species of the genus, bell-shaped, yellow to white with penta red-dots in the centre and 2.5-3.5 cm across are produced in axillary umbels during summer. ***H. longifolia*** from Central Himalaya is a climbing succulent growing more than 3 m in length but when cultivated the length is reduced to its half in containers. The leaves are thickly fleshy, up to 20 cm long and oblanceolate. Flowers are white or pink with deeper centre, 2.4 cm across and open in summer. *H. l. shepherdii* is thinner-stemmed and smaller growing with smaller leaves and 1.2-2.0 cm wide flowers. ***H. multiflora*** from Malacca (Malaysia) is a climber and in cultivation it makes a loose bush some 30-90 cm high and wide. Leaves are light green, up to 10 cm long and elliptic. Flowers in loose umbels are white to straw-yellow with white or brownish centre, about 2.5 cm long and with sweeping back petals which emerge during summer. ***H. purpureo-fusca*** is a woodland climber first collected by Thomas Lobb from Panarang area of Java. Its leaves resemble cinnamon but with a raised silvery-pink mottle.There are some two dozen of flowers per inflorescence with mauve-coloured petals though with purple centre and the petals with white hairs. ***H. sussuela*** (syn. *H. imperialis*) was introduced into Europe by its discoverer Sir Hugh Low from Borneo and the Moluccas around mid 19th century. It is composed of some 14 longer lasting fragrant flowers in each inflorescence, each flower measuring 7.5 cm across. The flower colour at upper surface of the petal is purple and the under surface yellow. ***H. kerrii*** from Thailand and Laos though grows up to 3 m in the wild but quite less in the container. Its leaves are dark green, up to or more than 10 cm in length, hairy, obovate to orbicular-cordate and with notched tip. White flowers with rose-purple centre and 1.2 cm wide are borne during summer in many-flowered umbels.

Huernia (Asclepiadaceae)

The 30 species of *Huernia* extend from South and tropical Africa into southern Arabia and are related to *Hoodia* and *Stapelia* with stems much like *Hoodia* though with fewer and more prominent ribs. It is clump-forming succulent producing 4-angled green stems. The stem angles are tubercled mostly with spines and notched. Leaves are minute and short-lived. Though these produce flowers similar to those of *Hoodia* and *Stapelia* but are with bell-shaped at bases, arising from the base or below the middle of the young stems and have a unpleasant odour. These are propagated by seeds and stem cuttings. ***Huernia barbata*** from South Africa bears beards of conspicuous long purple hairs at the mouth of the yellowish red-spotted flowers. ***H. brevirostris*** has sinuous serrated stems, Flowers are star-shaped, attractively marked and have an unpleasant smell. ***H. confusa*** from South Africa has a fleshy shiny ring of yellow-crimson colour around the throat, the flower contrasting bizzarely with the greenish and red-patched petals. ***H. hystrix*** (porcupine huernia) was introduced into cultivation at Kew in 1869 from material sent by M.J. McKen of Durban. It was collected from eastern region of South Africa and north into Rhodesia and Mozambique. It bears basally branched grey-green stems which are 5-ridged and notched with pointed tubercles, and grow up to 12 cm tall, erect to ascending or decumbent. Flowers dull yellow, fleshy, bell-shaped, 4 cm across and petals densely hairy throughout and even on the margins, with thick but broken chocolate-banding throughout the petals, and a large darker round zone in the centre of the flowers and an yellow eye. ***H. macrocarpa*** from Ethiopia produces some 10 cm long finger-like grey-green stems with five ribs bearing large pointed tubercles. Flowers are bell-shaped with recurved tips, their exterior are greenish-yellow, interior yellow, and are ornamented with narrow purple-brown transverse stripes. Form 'Arabica' bears white hairs on its dark purple flowers, 'Cerasina' produces flowers with red interior, 'Flavicoronata' with maroon and yellow, and 'Penzigii' entirely dark maroon with larger blooms. ***H. oculata*** has constrasting white mouths (throat) and purplish black petals giving the impression of eyes peering from the base of the stem. ***H. pillansii*** from Cape is clump-forming deciduous succulent growing some 5 cm high, some 2 cm thick and with 10 cm spread. Its stems are light green, finger-like and have 15 or more ribs, thickly edged with short tubercles having hair-like tips. Leaves are short-lived. Flowers are bell-shaped with five long-pointed lobes in starfish-shape, 3 cm across, pale-yellow to creamy-red with red spots (papillae) and appear during summer to autumn from the base of new growth. ***H. primulina*** (syn. *H. thuretti* var. *primulina*) from Cape is clump-forming, growing up to 8-10 cm tall and 15 cm spread, individual stem being 1.2 cm thick, having light grey-green but sometimes red-spots, and stems having 4-5 ribs bearing teeth-like tubercles with deciduous pointed tips. Bell-shaped dull yellow flowers some 2 cm across having reflexed petals and blackish tip appear during summer to autumn at the base of new growth. ***H. zebrina*** (owl eyes) from South Africa (Transvaal, Natal) and Namibia is deciduous and clump-forming succulent growing up to 10 cm high with 15 cm spread, individual stem being 2.5 cm thick, and is very similar to *H. primulina* but bears pale-yellow-green and some 4 cm wide flowers banded red-brown. The stem is 5-angled, each angle with spreading teeth-like tubercles.

Hydnophytum (Antplants; Rubiaceae)

A native to SE Asia, the Pacific region and northern Australia (Queenland), this is an epiphytic myrmecophyte (ant plant) comprising of 55 species, out of which 44 are recorded from the vicinity of New Guinea Islands. In Greek, *hydnon* is for 'a tuber' and *phyton* for 'a plant', alluding to its swollen and succulent stems. The species grow on the trunks and the tree branches making a symbiotic relationship with ants similar to its related genus *Myrmecodia*. Its caudex is structured with smooth walled tunneling within, along with opening to the outside to facilitate ants for colonizing it. *H. ferrugineum* with spiny swollen trunk and *H. moseleyanum* with smooth trunk are sometimes found in cultivation.

Ibervillea (Cucurbitaceae)

This small genus of caudiciforms occurs in Texas, south into Mexico and west into Baja California. Its thick and white-skinned caudices are either subterranean or partially exposed which produce annual vines. Its leaves are lobed to highly dissected. When dormant, its caudex is highly prone to rotting. This is propagated only through seeds. ***I. tennuisecta*** from Texas and Mexico bears subterranean caudex of the size of 20 × 20 cm (depth and width). The leaves are round and heavily dissected into thin leaflets. Its round fruits ripen red during late summer to autumn having only a few seeds. They are not tolerant to frost and in winter generally the caudex rots so new plants appear from seeds or sometimes through cutings.

Idria (Fouquieriaceae)

A monotypic genus from small area of mainland and Baja California (Mexico), its succulent species ***I. columnaris*** (boojum tree) is sometimes included in the genus *Fouquieria*. This is a single-stemmed plant growing slenderously erect about 20 m tall with chalky trunk, tolerating intense heat and mild frost. Main trunk bears spiny and thin lateral branches bearing grey-green small leaves in winter though most falling in summer. Seedlings are lodged in leaves and grow for first 2-3 years without entering dormancy. For maturity, seedlings take many years and when more than two metres tall, straw-coloured flowers are produced during autumn. It tolerates intense sun and up to -4 °C temperature only for a brief period. It is propagated through seeds.

Ipomoea (Convolvulaceae)

The members of this genus with about 500 species are found worldwide in the tropical to warm temperate regions. *Ipomoea* derives from the Greek *ips* meaning worm and *homoios* meaning 'similar to', may be due to its twining stem tips or the curling roots of some weedy windweeds. These are annuals, perennials and shrubs, many being climbers. They all have tubular to funnel-shaped flowers. They are propagated through seeds. There are only a few succulent species with fleshy leaves such as *Ipomoea pes-caprae* (syn. *I. maritima*) and *I. ternata* (syn. *I. horsfalliae* var. *alba, I. horsfalliae* var. *thomsoniana, I. thomsoniana*). ***I. pes-caprae*** from Tropical coasts of both hemisphere bears creeping stems, seldom twining, and growing up to 18 m in length with roots up to 3.6 m long and 5 cm thick. Leaves are fleshy, roundish and normally broader than long and clearly pinnately veined. Peduncles few-flowered, corolla bell-shaped, 5 cm long and scarsely lobed and flowering from August to October. ***I. ternata*** from West Indies bears stems of woody base, leaves 3-lobed with fleshy segments which are elliptic or elliptic-oblong and smooth, and flowers are trumpet-shaped and 5 cm across.

Jacaratia (Caricaceae)

A native to Central and South America, this is a genus of five species closely related to *Carica papaya*. The most common species is *Jacaratia corumbensis* (syn. *J. hassleriana, J. heptaphylla*) which is dioecious and is well adapted to semi-arid sandy soils. Its succulent storage roots develop into sizeable caudex.

Jatropha (Euphorbiaceae)

Jatropha derives its name from the Greek *jatros* meaning a physician and *trophe* meaning food. The genus comprises 175 species of perennials, shrubs and trees, some of them have their native haunts in arid regions having water storage tissues, and are mainly from tropics and sub-tropics with a few species in temperate North and South America. Most of the species are shrubs with alternate leaves and terminal corymbs of colourful 5-petalled flowers. They are propagated by seed or by cuttings. Only one species, ***Jatropha podagrica***, commonly known as 'gout plant' or 'tartogo' from Central America has base of the main stem club-shaped, swollen with water storage tissues. It is a deciduous shrub growing more than 90 cm in height, at younger stage almost nil to 2-3 branches though even after maturity there in no much progress as then also it is sparingly branched. Stems are rough with horny stipules. Leaves are smooth, a little waxy-glaucous, long-stalked, with peltate blades, 3-lobed, 10-20 cm broad and clustered at the top of the stem. Scarlet to coral-red flowers are small and in clusters up to 5 cm wide which continue appearing throughout the year. Becomes deciduous during winter in the sub-tropical regions though it can not survive in the open under temperate regions.

Jovibarba (Crassulaceae)

The generic name derives from the Latin *iovis* meaning 'Jupitor' and *barba* meaning 'beard'. This evergreen herbaceous genus with about 5 species from Europe is fleshy mat-forming, monocarpic and rosetted perennial spreading through short stolons and growing to 30 cm. They are grown for their symmetrical rosettes of oval to strap-shaped, pointed and fleshy leaves. They are most suitable for growing in rock gardens, alpine houses, screes, walls and banks. Though they are very hardy but take several years to attain flowering size. After every flowering the plants die but leaving numerous offsets around for regeneration. ***Jovibarba allionii*** (syn. *Sempervivum allionii*) from S. Alps produces 2-3 cm wide rosettes which are yellow-green, rarely with red-flushed tips, globular, downy and ciliate. Greenish-white flowers are produced in summer. ***J. arenaria*** (syn. *Sempervivum arenarium*) from E. Alps bears globular, ciliate and 0.5-2.0 cm across rosettes which produce greenish-white flowers on 7-12 cm tall stems. ***S. heuffelii*** (syn. *Jovibarba heauffelii*) from mountains of SE Europe grows up to 5 cm high, rosettes up to 10 cm across though variable in size and spread up to 30 cm. The leaves are mid-green to glaucous, occasionally tipped brown though other forms have red-flushing. Bell-shaped yellow flowers some 1.3 cm across appear on 15-20 cm long stems from July to September. ***J. hirta*** (syn. *Sempervivum hirtum*) from E. Alps, and Carpathians to N. Balkan Peninsula is rosette-forming some 2.5-7.0 cm across, ciliate, green to yellow-green though in some other forms red-flushed at sunny and dry sites. Pale-yellow to greenish-white flowers are borne by 10-20 cm tall stems. ***J. sobolifera*** (syn. *Sempervivum soboliferum*; common name 'hen and chickens houseleek') from N. Europe to C. USSR, forms 2-3 cm wide, vigorous, evergreen, rounded, and greyish- or olive-green rosettes often tinged red, which spread with short stolons up to 20 cm, offsets are produced quite freely on top of the rosettes, but flowering seldom though flowers are small greenish-yellow, cup-shaped, 6-petalled (rarely 5 or 7), and are produced in terminal clusters during summer on 10-20 cm long stems.

Kalanchoe (Crassulaceae)

Kalanchoe (syn. *Bryophyllum*) is a Latinized form of the Chinese name for one species. The genus comprises of some 200 species of succulent shrubs and perennials distributed throughout tropical Africa into South Africa and Madagascar and east into Java and China (southern Asia), with one species in tropical South America. Some of its species have been included in *Bryophyllum* and *Kitchingia*. Leaves are fleshy, cylindrical, oval or linear in opposite pairs and the flowers are tubular or bell-shaped, 4-petalled and in terminal panicles. These are excellent house plants, for sunny window sills and certain for hanging baskets. They are propagated through seeds, cuttings of leaf and stems, and offsets. Certain species generate clustered leaf-edge-plantlets emerging from the indented margins in plenty, especially on older leaves which when rooting can be separated and planted individually. ***Kalanchoe beharensis*** is named so after the Behara place in southern Madagascar where from the species originated. Its thick short-haired coloured leaves are borne in a cabbage-like cluster at the top of a woody leafless stem some 2.7 m tall with a maximum length of 30 cm. Lower part of stem is leafless as the plant tends to lose its lower leaves and become leggy. Under cultivation in containers its height is around 60 cm. Leaves are wavy, triangular-lanceolate (roughly heart-shaped), some 12-20 cm long, olive-green, finely dark brown hairy above and pale beneath, giving a velvety feel. The length of inflorescence is about 60 cm with bell-shaped yellow flowers about 1.2 cm long appearing in late winter in the wild and usually not in cultivation as it flowers only after attaining the plant height of at least 2.0 metres. ***K. blossfeldiana***, commonly known as 'flaming katy' from Madagascar is a erect bushy 30 cm tall well-branched succulent having glossy dark green leaves which are broadly ovate, some 7 cm long, with a narrow red edge, and toothed edges. Clusters of scarlet flowers some 1.25-5.0 cm long and 5 mm across appear in spring on short leafless stem arising from the upper leaf axils. A number of varieties using this species with others are developed with flowers in shades of red, pink, orange, yellow and white such as 'Kuiper's Orange' (flowers orange), 'Solfereno Purper' (flowers lilac-orange), *etc.* ***K. diagremontiana*** (syn. *Bryophyllum daigremontianum*), commonly known as 'devil's backbone' or 'Mexican hat plant' due to the shape of its leaves, from Malagasy is nearly shrubby and up to 90 cm tall. Leaves are long and narrowly boat-shaped triangular, marbled red-brown beneath, 10-20 cm long and with tiny plantlets with red-margined tiny leaves along the red-lined margins from each notch. Flowers are umbellate, pendent, glaucous-purple and some 2 cm long on stem tops which appear in winter. ***K. farinacea*** from Madagascar is an erect growing beautiful architectural plant with almost oppsosite greyish-green, oval, symmetrical and slightly cupped leaves covered with silvery-white powder which spreads even up to inflorescence though lower stem is reddish along with the basal portion of leaves, however, the larger part of the leaf is still greyish though upper leaves only greyish thoughout. Clustered bell-shaped reddish-orange flowers similar to kaffir lily appear on the top in spring. ***K. fedtschenkoana*** or *K. fedtschenkoi* (South American air plant) is a bushy succulent growing some 1 m in height and spread. The leaves are blue-grey, oval and indented with tiny plantlets in each notch. Flowers are bell-shaped, 2 cm long, brownish-pink and appear in late winter. *K. f.* forma *variegata* bears blue-green edged-cream leaves which may have even splashing of red colouration. Other characters are just the same as its parent. ***K. flammea*** from Somali Republic is a scarcely branched erect succulent bearing light grey-green, some 8 cm long, obovate and entire or sinuately toothed leaves. Orange-red to pink flowers some 2 cm wide with yellow tube in corymbose clusters appear during winter and spring. ***K. gastonis-bonnieri*** from Malagasy is basally branched and erect growing succulent some 60 cm tall. Leaves are narrowly ovate to lanceolate or spathulate, crenate, base of each leaf pair amplexicaule, clothed with whitish powder and some 13-20 cm long. The pink, reddish or yellowish petalled pendulous flowers in a 15 cm wide terminal corymb are borne with each bloom having an inflated and long ovoid calyx some 2-3 cm long and a 4 cm long corolla with recurved flowers. ***K. grandiflora*** from India and East Africa is a base-branching erect succulent growing some 60 cm in height. Leaves are waxy blue-purple to patinated reddish, up to 8 cm long, obovate and crenate. Yellow flowers some 2.0-2.5 cm wide in dense corymbose cyme appear shyly on the plant. It is valued for its leaves as well as the flowers. ***K. laxiflora*** 'Fedschenko' (syn. *K. crenata, Bryophyllum crenatum*) from Madagascar is an erect but loosely growing plant with fleshy stem and leaves. Leaves are small grey-green, slightly covered with silvery powder, fleshy, 2.5-4.5 cm long and with indented margins. Blooms are bluish and cover the leaves when in full bloom. Higher temperatures make the plants straggly. ***K. longiflora*** from Natal (S.A.) is a shrubby and erect succulent growing to 60 cm in height bearing 4-angled stems. Leaves are pale grey-green, tinged orange to yellow under strong sun or low temperatures, 4-9 cm long, obovate to ovate-oblong and edges of upper half clearly toothed. Flowers are yellow and are borne in terminal corymbose cymes. ***K. manginii*** from Malagasy is semi-erect to decumbent with wiry stems branched mainly at the base. Leaves are lanceolate or spatulate, fleshy and some 2.5-4.5 cm long. Flowers are bright red, drooping in loose terminal clusters of 3-9, narrowly bell-shaped, about 2.5 cm long and appear when the day is shorter than 12 hours in spring. ***K. marmorata*** from Eritrea, Ethiopia and Somali Republic is a shrubby succulent up to 90 cm tall, though in cultivation only up to 30 cm, with erect to decumbent stems which are mainly base-branching. The leaves are blue-green, mottled with chocolate-brown, stalkless, some 10 cm long, devoid of hairs and obovate with toothed margins. White tubular flowers in terminal cymes are some 6-8 cm long appearing during late winter and spring, and sometimes even continuing from March to May. ***K. marnieriana*** from Malagasy bears woody-based semi-erect to decumbent stems often with aerial roots. Leaves are blue-green spotted purple on the upper surface, quite fleshy, overlapping, broadly oval, some 3 cm long and bear plantlets on the margins. Flowers of pink colour some 2-3 cm long appear from winter to spring in terminal clusters. ***K. millotii*** from Malagasy is a well-branched shrub growing some 40 cm high bearing ovate and hairy leaves some 3-4 cm long. Flowers are yellow and orange and some 1 cm long. ***K. orgyalis*** from Madagascar bears ovate leaves some 7 cm long and 4 cm wide, clothed with short brown hairs on the upper surface and pale beneath. ***K. pinnata*** is with uncertain origin because of its wide spread by man in the tropical and sub-tropical areas worldwide. It is a slightly branched shrubby succulent growing to 40-180 cm tall. Leaf margins are wavy and bear tiny plantlets, pinnately divided into 3-5 leaflets, each with 7-20 cm long, oblong to rounded and at the margins bear tiny plantlets. Greenish-white flushed-red flowers some 3.5 cm long appear from within the bell-shaped calyces. ***K. pumila*** from Madagascar is a semi-prostrate succulent growing 25 cm high and is most suitable in hanging baskets with highly decorative pale-pink clustered flowers some 1.25 cm long which appear during winter, *i.e.* mid-January to mid-March. The leaves are some 2.5 cm long, lanceolate and coarsely toothed. Stems and leaves both are completely covered with waxy-white patina and tinged with pink. ***K. thyrsiflora*** from Cape to Transvaal (S.A.) is an erect and glaucous whitish succulent from 60 to 120 cm tall, bearing some 15 cm long obovate to spathulate leaves with waxy-white surface. Numerous branching clusters of individual yellow flowers arise from late winter to spring from a single main stem in terminal panicles in urn-shaped and 1.5 cm long individual blooms. ***K. tomentosa***, commonly known as 'panda plant' or 'pussyears' from the semi-desert scrub of south-west Madagascar grows up to the height of some 75 cm. Whole plant (stem, leaves and inflorescence) is covered with short but dense silvery felting, including the outer surface of the petals. The leaves are fleshy, narrowly oblong-obovate with blunt tip, decussate in arrangement, and 4-8 cm long and 2 cm wide. The young leaves have toothed tips and margins stained brown. Yellowish to purplish clustered flowers are some 1 cm long. ***K. tubiflora*** (syn. *Bryophyllum tubiflorum*), commonly known as 'Chandelier plant' from Malagasy is an erect and less-branched shrubby succulent growing up to 1 m in height, bearing solitary or in threes the grey-green spotted reddish-brown cylindrical leaves, 3-12 cm long, grooved and with plantlets at or near the tips which are flattened and notched. Flowers are umbellate, pendent, tubular some 2.5 cm long and salmon-red to scarlet which appear in late winter. ***K. uniflora*** (syn. *Kitchingia uniflora*) from Malagasy is a creeping and epiphytic succulent growing up to 6 cm tall with indefinite spread and is excellent hanging basket plant. Leaves are rounded to ovate, mid-green and 5 mm to 3 cm long. Flowers are reddish-purple flushed yellow, urn- or bell-shaped and about 2 cm long which appear during winter. ***K. velutina*** from Angola through the Congo, Malawi and Tanzania, to Zanzibar comprises of creeping and rooting robust fleshy stems which are densely hairy with velvety feel. Leaves are up to 10 cm long and narrowly ovate. Flowering stems are erect and bear narrower leaves and with terminal rounded clusters of yellow or sometimes pink blooms some 1.2 cm long. *Kalanchoe* 'Tessa' is prostrate to pendent succulent growing up to 30 cm in length and bearing green and narrowly oval leaves some 3 cm long. This bears orange-red drooping tubular flowers some 2 cm long in late winter. *Kalanchoe* 'Wendy' is semi-erect succulent growing up to 30 cm in height, bearing glossy green, 7 cm long and narrowly oval leaves, and bell-shaped 2 cm long, drooping and pinkish-red flowers with yellow tip-margins. It is an excellent succulent for hanging basket.

Kedrostis (Cucurbitaceae)

This genus comprises of 23 species of climbing or prostrate herbs which are usually monoecious and often with sufficiently swollen caudiciform bases. Leaves are palmate or pinnate, entire and lobed with usually simple tendrils. Small green, white or yellow flowers appear, the male being racemose or corymbose with campanulate calyx which is 5-lobed and the lobes are ovate to linear-lanceolate, corolla being rotate and 5-lobed, stamens 3-5, free or coherent and uni- or bilocular with short filaments, though in solitary or clustered female flowers the calyx and corolla are similar to its male counterpart but staminodes are either 3 or absent, ovary is ovoid and rostrate, placentas and stigmas each 2-3, and the ovules are a few to many. Fruit is a sessile or subsessile and indehiscent berry which is fleshy, few to several seeded, rostrate and ovoid to subglobose. Only one species, ***K. africana*** from Africa is a succulent with caudiciform swollen base, bearing slender and climbing stems growing up to 6 m. Leaves are usually glabrous with 1.2 cm long petioles and 10 cm diameter, orbicular to cordate but deeply lobed in palmate- or pinnate-form with segments being filiform or elliptic. Minute flowers are white to yellow-green, male inflorescence up to 8 cm bearing a few to 12 flowers, calyx lobes narrowly triangular to linear and some 2 mm long, and corolla lobes finely papillose and up to 2 cm long, though female inflorescence is up to 6 mm with glabrous ovary. Fruit is subglobose, glabrous, red on ripening, few-seeded and 1.5 cm in diameter.

Kensitia (Aizoaceae)

A succulent genus named after Dr. Louise Bolus (née Kensit), who first described its species in the genus *Mesembryanthemum*, the genus *Kensitia* consists of the single species, ***K. pillansii*** from its native haunts of South Africa. This species is entirely different from the stone flower as it grows up to 46 cm high and where the flowers do not close at night. The stem is brown, branched and straggling-erect but strong enough to properly support the flowers, and bears green, top splashed pink, quite fleshy, some 4 cm long and 1.5 cm broad and almost opposite leaves in pairs but sometimes lower leaves are from 2-5 on each node. The pinkish-mauve flowers some 5 cm across appear terminally from each stem and branches bearing curious-looking spoon-shaped petals where a series of staminodes enclose the inward bent stamens around the 8-10 stigmas, but the staminodes later part to allow pollination.

Kleinia articulata (*Senecio articulatus*, candle plant; Asteraceae)

See *Senecio.*

Lampranthus (Aizoaceae)

Lampranthus derives from the Greek *lampros* meaning shining and *anthos* meaning a flower, referring to the satiny sheen to the petals. It is a South African (almost Cape region) genus of 160 dwarf, creeping, sub-shrubby and perennial succulents allied to *Mesembryanthemum* to which it initially belonged but afterwards got established as an independent genus. These are valued because of the ease in their cultivation, the growing speed and their silky, brilliantly coloured daisy-like flowers which continue appearing over a long period from June to October in the temperate climatic conditions as these are semi-hardy. These are densely branched bearing numerous flowers borne more than one at a time or singly at the branch-ends, usually opening after sunrise and each lasting for about a week. The older plants after several years, if properly fed and cared, become woody. These are propagated through seeds and stem cuttings. The important cultivated species are being described here under. ***L. amoenus*** from Cape grows to a height of 25 cm and spread 30 cm across with tapering semi-cylindrical bright green leaves which turn red in bright sunshine. The flowers are brilliant purple and some 7.5 cm across. ***L. aurantiacus*** from Cape grows initially erect to a height of about 50 cm with a spread of 70 cm, and then prostrate with sparse branching bearing tapering glaucous grey-green, short and semi-cylindrical leaves and daisy-like masses of bright orange flowers some 3.5-5.0 cm across which appear during summer. ***L. aureus*** from Cape is a bushy and erect succulent with up to 40 cm height. Leaves are glaucous green with tiny transparent dots, almost triangular and some 5 cm long. Bright orange flowers some 6 cm across are borne in summer. ***L. blandus*** from Cape is more or less erect and bushy succulent with red stems, growing up to 45 cm in height. Leaves are pale grey-green ornamented with quite small transparent dots, equi-triangular and some 3-5 cm long. Pale to pink flowers some 6 cm across are borne in summer. ***L. brownii*** from Cape grows about 30 cm in height with tapering and triangular leaves and 1.8 cm wide bright orange flowers maturing to deep red. ***L. coccineus*** from Cape grows up to 40 cm in height and 30 cm in spread bearing dull grey-green and triangular leaves and brilliantly carmine flowers some 3.75 cm in diameter. ***L. conspicuus*** from Cape grows up to 45 cm in height and 30 cm in spread bearing incurved bright green tipped red leaves which are either triangular or semi-cylindrical, and some 5 cm across purple-red flowers. ***L. elegans*** from Cape grows to a height of 40 cm and width of 30 cm bearing semi-cylindrical grey-green leaves, and some 3.75 cm across purple-red flowers. ***L. howorthii*** from Cape is shrubby and more or less erect growing up to 60 cm tall bearing grey-green semi-cylindrical and 2.5-4.0 cm long leaves, and pale-purple 5-7 cm wide flowers in summer. ***L. multiradiatus*** (syn. *L. roseus*) from Cape is a shrubby, and erect to spreading some 60 cm tall succulent bearing glaucous blue-green with translucent dots, triangular and 2-3 cm long leaves, and light pink to pale rose-purple flowers some 4 cm wide in summer. ***L. spectabilis*** is a mat-forming succulent from Cape growing up to 30 cm in height with semi-erect to prostrate habit, bearing incurved and glaucous green, some 5-7 cm long, and somewhat triangular leaves, and purple flowers some 7 cm across in late spring to late summer. ***L. stipulaceus*** from Cape is a 40 cm tall shrubby succulent with many short shoots appearing from leaf axils. Leaves are light green with translucent dots, triangular to semi-cylindrical and 4-5 cm long, and the purple flowers which appear in summer are 4 cm across. ***L. zeyberi*** from Cape grows to a height of 38 cm and spread of 30 cm bearing cylindrical bright green leaves and some 7.5 cm across purple-violet flowers.

Lapidaria (Aizoaceae)

A monotypic genus where leaves mimic as have been carved from stones. It is grown as *Lithops*. ***L. margarettae*** from Namibia is a very handsome miniature forming several clumps some 8 cm in diameter. Its previous year's leaves are retained on the plant to make each head with two pairs of leaves. Each head produces 4 cm wide bright yellow flowers in the afternoon during autumn which persist for several days. From mid-winter to early spring new leaves start emerging and the oldest ones shriveling. Sporadic light watering in winter but during summer it should be watered when medium looks dry. It tolerates intense heat, indirect bright light and for a brief period up to -4 °C temperature. It is propagated through seeds.

Larryleachia (Apocynaceae)

A native of Namibia, Angola, Botswana and the northern Cape Province of South Africa, this smooth stemmed genus was formerly included under *Trichocaulon*. This has cactus-like appearance. The most popular species being *Larryleachia cactiformis* which bears soft tuberculate stems about 10 cm high and 5 cm across, with dotted yellow and banded maroon flowers. This species is not easy to cultivate due to sudden rot of the plant starting underground.

Lewisia (Portulacaceae)

Lewisia commemorates Meriwether Lewis of the famous 'Lewis and Clark Expedition' across North America during 1804-1806. Its family Portulacaceae is related to the Cactaceae. It is a western American perennial genus with some 20 semi-succulent to succulent species, some even with attractive flowers. They are suitable for rock gardens and alpine houses. Some members of this genus are evergreen with rosettes of succulent leaves and long tap roots. Herbaceous species which are deciduous are propagated through seeds while evergreens through seeds and offsets both. The species described below are evergreen bearing profusion of widely funnel-shaped flowers except *L. rediviva* which is deciduous. ***Lewisia brachycalyx*** from New York grows up to 7.5 cm high and 22.5 cm spread, bearing narrow, glaucous and fleshy leaves from a flat rosette, short-stemmed, solitary and silky white flowers sometimes tinged pink opening flat under bright sun in May. *L.* × 'Physellia' (*I. brachycalyx* × *I. cotyledon*) is dwarf as to its former parent and with flower brightness as to its latter parent, bearing mid-green oblong-spathulate leaves from fleshy rosettes and white to pink flowers with a deeper pink veining. ***L. columbiana*** from Cascade Mountians of British Columbia grows some 30 cm tall and up to 8 cm in spread. Wiry inflorescences bearing 3.75 cm wide heads with magenta or white flowers veined red in May-June appear from rosettes of linear-spathulate leaves. *L. c. beckneri* from California produces mid-green, ovate-oblong to spathulate leaves and branched clusters of some 5 cm across rose-pink flowers from April to July. *I. c. bowellii* from Oregon bears oblong-ovate mid-green leaves in flat rosettes with finely crested margins, and clustered racemose 5 cm across pink flowers striped carmine from April to June. There are many of its hybrids in colours ranging from pale-pink to apricot. 'George Henley' has brick-red flowers and 'Rosea' with purple-red. ***L. nevadensis*** (syn. *I. pygmaea nevadensis*) from Nevada is closely related to *L. rediviva* bearing linear fleshy leaves in tufted rosettes which die down at flowering. Stemless solitary white flowers up to 3 cm across appear from June to August from the leaf axils. ***L. rediviva*** (bitter root) from USA grows up to 8 cm high and 15 cm in spread bearing narrow red-green rosette leaves which die down when stemless solitary flowers appear. The flower is white or rose-pink and some 5 cm across resembling a small waterlily. ***L. tweedyi*** is the alpine succulent found in the Wenatchee Mountains of Washington State between 1,830-2,130 m a.s.l. and is one of the largest flowered lewisias. It has large tuberous rootstock and succulent leaves. It grows up to 15 cm high with a spread of 23 cm and is the best among all the evergreen lewisias. The leaves in loose rosettes are wide, mid-green and more erect than other lewisias. Palest-pink to soft apricot flowers with silky sheen are 5 cm across. The var. 'Rosea' bears rose-pink flowers with darker petal tips.

Lithops (Living stone/Stoneface; Aizoaceae)

In Greek *lithos* means 'stone' and *ops* means 'like' as the plants under this genus closely resemble stones. *Conophytum* and *Lithops* species, the tiny spherical objects growing among stones, are often indistinguishable from the surrounding pebbles until a flower pushes its way out of its apex. The plants of this genus are commonly known as 'living stones' which are native to southern Africa, especially Namibia and South Africa with about 75 species of solitary or clump-forming succulents, where each stem comprises of two highy fleshy, thick and fused leaves making an obconical plant body slit at the flattened or sometimes rounded crown. As the pair of new leaves develop there is corresponding shrinkage in the old pair and the new growth almost sucks dry the older one relegating the pair as a protective sheath around the bare leaves. The tissues in the upper surface of the two leaves as are devoid of chlorophyll and store water in their transparent tissues so work as window to transmit light to the chlorophyll-containing tissues in the sides of the plant which only carry out photosynthetic activities. In nature the plant sides, *i.e.* undersurface of the leaves are covered by desert sand. Such plants survive up to 18-19 months of drought. Many petalled large flowers emerge from the slit during autumn which completely hide the plant body beneath. They are propagated through seeds and by separating the plant bodies. ***Lithops alpina*** from Namaqualand grows up to 2.5 cm thick with light brown fleshy leaves which are mottled darker and produce yellow blooms easily in June. ***L. aucampiae*** from Transvaal bears 2 cm height and 3 cm breadth with slightly convex and reddish-brown upper part which is divided by a shallow slit into 2 unequal halves having red network markings. Adjacent to slit are dark green and translucent 'windows'. This bears yellow flowers in the early autumn some 2.5 cm across. ***L. bromfieldii*** from South Africa where the green body is composed of 2 semi-circular jointed leaves having flattened top with red markings. From slits daisy-like yellow flowers appear in September. ***L. dorotheae*** from South Africa bears the pale pink-yellow to steel-green body of 2 semi-circular unequal jointed leaves with good structured darker blots on upper surfaces. Body thickness being up to 3 cm and the breadth 5 cm. The daisy-like yellow flowers appear from the slits in summer or autumn. ***L. erniana*** from Namaqualand grows up to 2.5 cm thick with grey plant body having a network.of red-brown lines. The white flowers appear in September. ***L. fulleri*** from South Africa has egg-shaped, dove-grey to brown-yellow body with 2 cm height and width. Upper surface is convex and sunken with darker marks. The white flowers are borne in late summer or early autumn. ***L. insularis*** from South Africa is egg-shaped, brown and some 3 cm high and 2 cm wide. Upper surface is slightly convex with dark green 'windows' patterned with red lines and dots. Yellow flowers are produced in late summer or early autumn. ***L. julii*** from South Africa has egg-shaped jointed pairs with pearl- to pink-grey leaves which have slightly convex top marked darker. White flowers appear in late summer or autumn. ***L. karasmontana*** from Namibia is egg-shaped succulent made through 2 conjunct unequal fleshy leaves which are of grey stone-merging colour though upper surface is sunken with darker pink markings. It grows 4 cm high, 5 cm wide and produces white flowers in late summer or early autumn. *L. k.* subsp. *bella* (syn. ***L. bella***) from Namibia is a clump-forming with plant bodies up to 3 cm thick and 2 cm wide with convex tops.Body is marked brownish-yellow with branched darker lines which produces white flowers in summer.***L. lesliei*** was discovered by T.N. Leslie in Transvaal, apart from occurring in Griqualand, the Orange Free State and Cape Province. It is similar to *L. aucampiae* with solitary or small clump-forming, egg-shaped flattened top and light to reddish-brown body up to 4.5 cm high and 6 cm wide and marked with dark greenish-brown spots and pits, and its non-fragrant golden-yellow flowers emerge in the late afternoon of summer. Its var. *albinica* growing up to 3 cm high and 5 cm in width, with its body in 2-unequal size, convex, pale-green but upper surface dark green, having clustered yellow dots and white flowers in late summer to early autumn. ***L. marmorata*** from South Africa is glaucous green growing up to 3 cm in height and 5 cm in breadth. Its body is egg-shaped jointed with 2 unequal leaves with dark grey markings on convex upper surface. It bears pure white flowers in late summer or early autumn. ***L. mundtii*** from Namaqualand grows up to 2.5 cm with grey-brown plant body patterned with brown lines and green dots. Yellow flowers are produced in September and October. ***L. olivacea*** from South Africa has dark green and egg-shaped paired body having darker 'windows' on upper convex surfaces. The length and breadth of the body is 2 cm each. Yellow flowers appear in late summer or early autumn. ***L. optica*** from South Africa is found in the coastal desert of Prince of Wales Bay and Namaqualand. It grows on quartz hills and screes in its native haunts. This grows up to 2 cm high with grey-green plant body making the easy clumps. It produces white flowers in December. A form *rubra* with reddish leaves grows scattered amongst the true species in its native haunts.Not like other *Lithops*, it has a deep slit at the joint of the 2 leaves having translucent tips from where light penetrates the interior tissues of the plants. ***L. otzeniana*** from South Africa grows some 3 cm high and 2 cm wide and has paired grey-violet bodies having convex upper surface on both which have light bordered large semi-translucent 'windows'. Yellow flowers appear in late summer or early autumn. ***L. pseudotruncatella*** from Namibia is clump-forming, egg-shaped broadly obconical succulent with light brownish-grey body some 3 cm high, and patterned with reddish-brown lines and dots, having prominent fissures across the wide and flattish tops which extends from side to side but only on mature plants. With respect to colouring and patterning of the body, the plant is highly variable. Rich yellow flowers some 3.5 cm wide appear in late summer or early autumn. Var. *pulmonuncula* is though similar to original species but has grey body with dark green and red marks on upper surfaces. ***L. rushchiorum*** from South Africa with pale grey-green body having jointed two leaves so closely that a small groove is left. A new corpusculum (body) is formed under the old and shrivelled skin in May each year. Flowers are of yellow colour appearing during summer. ***L. salicola*** from Orange Free State grows solitary or in small clumps growing up to 2.5 cm in height in obconical shape. The body is pale-grey with transparent areas and shallowly convex above with a dark grey-green mottling. White flowers some 2.5 cm across appear during summer. ***L. schwantesii*** from Namibia is a clump-forming succulent with 2 jointed parts of greyish to yellow-red colour marked with reddish-brown dots and lines, often bordered rusty-yellow. The body is up to 4 cm in thickness having flattened to convex top. The yellow flowers appear from slits during summer. *L. s.* var. *kuibisensis* has egg-shaped body of some 3 cm thickness and breadth, divided into 2 unequal-sized leaves, and having blue or red marks on the upper surface. Slits produce yellow flowers during late summer or autumn. ***L. terricolor*** is the most widely distributed species in the Cape Province where it is found in the clumps of 6-10 plants growing in areas with as little as 5 cm rain in a year. It bears leaves of greenish-yellow to greenish-pink with upper surface bearing brownish-green spots. The flowers are of buff-white colour which appear during summer. ***L. turbiniformis*** (syn. *L. hookeri*) from Cape grows up to 4 cm high and 2 cm wide, singly or in small clumps. The body is obconical with tan and nearly flattened top, networked with sunken brown lines on paired brown leaves. Bright yellow flowers some 4 cm across appear in late summer or early autumn. ***L. volkii*** from SW Africa has caespitose body with 2-4 stems per clump. The grey-blue body is up to 4 cm high, sometimes with reddish tinge. Leaves though with flattened to slightly convex upper surface are 3 cm across tinged white and divided by 2 unequal halves which are covered with red lines formed through merging of dots though some plants may be found without markings but having a number of pale and nearly transparent spots. Dark yellow flowers some 2.5 cm across appear in early July.

Malephora (Aizoaceae)

Malephora is made up of two Greek words, *male* means 'armhole', and *phorein* means 'to bear'. Malephora from South Africa and Namibia, is a genus of some 15 species of erect or spreading shrubby succulents having long semi-cylindrical and soft-fleshy leaves which are shortly united at the base, prismy-triangular in cross-section and unspotted but with slightly blue- powdery coating. Short-pedicellate and axillary or terminal flowers which are golden-yellow, yellow or pink, 5 cm in diameter with usually 4 sepals and up to 8 feathery stigmas, are borne in late summer to winter. These are propagated through seeds and stem cuttings. ***M. crocea*** from Cape grows 20 cm in height and up to 1 m in spread with stout and more or less gnarled stems, bearing 2.5 to 4.5 cm long and 0.6 cm wide, pale- to blue-green, pruinose and semi-cylindrical to indistinctly triquetrous leaves crowding together on small shoots, and shortly uniting at the base. It produces daisy-like orange-yellow, solitary, 3 cm across flowers with golden-yellow interior and reddish exterior in spring-summer. ***M. engleriana*** from Namaqualand and Namibia is a dense and soft shrub growing to 30 cm tall and 50 cm across with green, waxy, very soft and 2-4 cm long leaves that are obtuse-triquetrous and curved above. Flowers glossy orange-yellow inside and orange-red outside and 2 cm across appear in spring-summer. ***M. herrei*** from South Africa is almost prostrate-growing with rooting stems. Leaves are green, triquetrous, rounded carinate, 5 cm long and 0.5 cm wide. Flowers are golden-yellow, orange below, axillary and some 55 cm in diameter which appear during summer. ***M. lutea*** from Cape is an erect growing shrubby succulent producing short shoots, bearing yellow-green, white-wooly, spreading, compressed triquetrous and 2.5-4.5 cm long and 0.4 cm wide leaves. Yellow flowers which are orange below and some 3-4 cm in diameter appear during spring. ***M. thuribergii*** from Cape grows with rigid, nodose but prostrate stems, bearing crowded fresh green and spotted, semi-cylindrical leaves which are bluntly triangular above, about 5 cm long and some 6 cm across. Yellow flowers appear during spring to summer some 3.5-4.0 cm in diameter.

Matelea (Asclepiadaceae)

A Mexican genus of small shrubs and vines is rare in cultivation, excepting ***M.* (*Gonolobus*) *cyclophylla***. Its conical trunk (caudex) can grow more than 30 cm across and resembles as a tan version of *Dioscorea* but it has soft corky bark. During late spring or in summer when night temperatures exceed 18 °C, one to several vines emerge from the apex bearing cordate leaves up to 10 cm wide. Every to several nodes, the foul-smelling velvety dark green or purple flowers some 2.5 cm across appear which on pollination, each flower produces single seed horn. Though vines die back to the caudex but the fruits remain on the dry vine which ripen and open in the spring. Its dormant period lasts for about five months. If water is not a limiting factor, it will tolerate intense heat and bright light but neither over-watering nor frost. It is propagated through seeds and its growth is very fast.

Mesembryanthemum (Aizoaceae)

See '***Dorotheanthus***'.

Mestoklema (Aizoaceae)

This small genus from South Africa has only a few shrubby succulent species of interest. *M. arboriforme* and *M. tuberosa* both have tuberous roots and are excellent as succulent bonsai. ***M. tuberosa*** produces thick roots which further thicken and lengthen with age, developing a dark red bark. Stems grow with thin and straggling branches which through proper pruning and training can be brought into desired bonsai shape to a height of 20 cm. Stem tips produce 7 mm wide gold to orange clustered flowers during whole summer. It tolerates intense heat, half day bright sun, and for a brief period the -4 °C temperature. It is multiplied through seeds and cuttings.

Momordica rostrata (Cucurbitaceae)

The tropical African and Asian genus *Momordica* comprises some 50 species of vining plants, less than a dozen of species are of interest to the landscaper. They require cold-protection though adjust well to hot and bright light when watered well. ***M. rostrata*** from Africa is a dioecious species which forms a fluted caudex in few seasons. It produces plenty of vines from the top of the caudex, bearing 3-lobed dark leaves. Flowers are yellow, in females a few and in males numerous with 2 cm across, having black spots at the bases of three petals. Fruit ripens to dark orange, 6 cm long and 2 cm wide resembling the small peppers. It is propagated through seeds.

Monadenium (Euphorbiaceae)

An African (Ethiopia, Kenya *etc.*) genus with 50 thin annuals coming up from underground tubers, to thick and up to 1 m long perennials forming small shrubs. They are closely related to *Euphorbia* but in their inflorescence there is a single ring-like gland surrounding the cyathia and the cyathia are borne laterally on the floral stalks, above the cyathia a pair of hood-like bracts appear together which are sometimes brightly coloured, and mostly a second series of cyathia emerge from the bracts forming a branched inflorescence, and those species where inflorescences are brightly coloured these series of cyathia are quite attractive and last for more than 20 days. However, *Euphorbia* contains five petal-like glands. *Monadenium* bears fleshy and annual leaves which vary considerably in size and colouration often with patterned markings. Though these like bright indirect light but many species will tolerate heavy shade and not the frost so during winter these should be kept fairly dry and warm. ***M. coccineum*** from Africa is mostly grown for its inflorescences than its form. These have almost erect stems which grow to about 1 m tall, bearing light green leaves about 3 cm long and 2 cm wide with undulating margins. During autumn, several brightly red branching inflorescences appear on each stem whose brightness is more charming when the specimen is under the sun and where these last for months together. It is propagated through cuttings or seeds. ***M. ellenbeckii*** from Ethiopia is a very handsome plant which forms a shrub up to 1 m tall having only a very few branches. Its 1 mm long and 8 mm wide leaves are simple green which are short-lived and appear during the growing season. Short-stalked small yellowish-green inflorescences cluster at the tip of the stem. It is propagated through cuttings or seeds. ***M. reflexum*** from Ethiopia is a very handsome species but with slow growth where plants are branched with 3-6 cm thick stems, bearing long and below-pointing tubercles. At seedling stage the plants bear leaves often with attractive purple veins but on maturity these disappear. It is propagated through seeds. ***M. ritchiei*** from Kenya is the most commonly cultivated species which is quite short-lived as the plants rot mostly due to over-watering, especially during winter. It is a cluster-forming species attaining the height and spread of up to 20 cm though individual stems are only 3 cm thick. The stems are green and tuberculate, bearing roughly-rounded leaves which appear sparingly and are 3 cm across. Sometimes no leaf is borne if the conditions are not congenial for growth. In autumn and during spring, small light pink inflorescences appear freely from stem apices. It is propagated through cuttings or seeds. ***M. rubellum*** (*M. montanum* var. *rubellum*) from Kenya is known for its very attractive inflorescence, *vis-à-vis* tuberous root. Its white and branching roots which become irregularly tuberous in a few seasons, produce thin and long stems where leaves are narrow, green and mottled red. Especially during autumn and also throughout the year many light pink inflorescences are borne on the stems. After pruning and training its stems, *vis-a-vis* raised tubers, it makes a nice display in bonsai pots. It is quickly propagated through cuttings, and also from seeds. All these species tolerate indirect bright light and intense heat but not the frost, and over-watering especially during winters.

Monilaria (Aizoaceae)

An odd genus from South Africa having two different pairs of leaves each season as in the case of *Mitrophyllum* and *Conophyllum*. In *Monilaria*, the pair of 'resting leaves' form a rounded body to protect second pair of leaves. During its growing season which is winter, second pair bursts through the dead remains of the first. Its leaves are long and green often with glassy cells which thickly cover the surface giving a crystalline sheen. The plants should be watered generously when in active growth and perfect dryness when dormant. This produces white to pinklish and yellow flowers. ***M. pisiforme*** bears jointed stems like a necklace similar to many other members of the genus, and produces white flowers with reddish centre. It tolerates moderate heat but not the frost. Propagation is through seeds.

Monsonia (Geraniaceae)

This is an old genus from Namibia to South Africa and with disbanding of the genus *Sarcocaulon* all of its members have been combined with *Monsonia* now. These differ from *Pelargonium* having actinopmorphic flowers though *Pelargonium* has zygomorphic (bilaterally symmetrical) flowers. Whether these are summer growers or winter but wherever these are planted, should be watered throughout the year to keep them awaken even if not in active growth, and the leaf growth should always be there. Leaves from one species to other vary greatly from glabrous to pubescent and entire to deeply cut and the flowers from white to pink and yellow.

Mostly they are propagated through seeds but can be grown even from cuttings. ***M. penniculinum*** from those areas of Namibia which receives no measurable periodic rainfall for years together so under such dry periods the species remains dormant under its waxy bark so the next time when it receives the rainfall, it produces a flush of incised and short-woolly hair-like leaves directly from the tubercle of the main thick trunk. It is a small shrub about 7 cm tall and more than 15 cm spread with thick stems. It produces light pink flowers about 2 cm across. It tolerates intense heat and half-day sun but no frost. It is propagated through seeds. ***M. vanderietiae*** from South Africa is easier and faster to grow though its stems lack that sort of succulence and thickness which other species have. The plant forms a flat shrublet up to 30 cm wide and this flattened growth makes it an excellent succulent bonsai. Its dark green and elliptic leaves are end-notched. It produces pale-pink flowers some 3 cm wide. It tolerates intense heat and half day sun. It is propagated from seeds or cuttings.

Moringa droughardtii (Moringaceae)

This small genus includes several caudiciform trees from Africa and Madagascar. Its leaves are deciduous, quite large and compound that give its foliage a lacy appearance. Small white flowers produce long pods with large seeds. ***M. droughardtii*** from Madagascar is a fast-growing species with a free root-run if copiously watered and attains the height of some 8 m with a thickness of 1 m in its natural haunts. Lacy and compound leaves are about 30 cm long and 20 cm wide. It is propagated from seeds, can tolerate intense heat and full sun but no frost.

Myrmecodia (Rubiaceae)

A native to SE Asia and Australia (Queensland), this is a myrmecophyte epiphytic plant allied to *Hydnophytum* which grow on trunks and branches of the trees. *Myrmecodia* may be found growing on the trees as a symbiont with its tubers hanging downward on bare branches with insufficient substrate. Its greyish-brown swollen caudex which develops spines over it with age, stores water and food, and the thick and unbranched stems are covered in clypeoli and alveoli which also grow spines and are densely filled with dry bracts. Alveoli produces unattractive white flowers, followed by bright orange fleshy berry on maturity, each berry containing six small seeds which are carried by birds and through their droppings on the trunks and branches of the trees new plants come up.

Nananthus (Aizoaceae)

These low-growing plants from South Africa are often included under *Aloinopsis* and often produce large tuberous roots which are raised for effect. Yellow-petalled flowers some 2 cm wide appear during fall, sometimes with reddish stripe in the centre. These tolerate intense heat, bright indirect sunlight and for a brief period up to -4 °C temperature. They are propagated from seeds or cuttings. ***N. schooneesii*** is probably the first species collected in the genus.Its roots are tuberous, tapering and some 6 cm wide at the crown as well as in depth. Its leaves are 1 cm wide and form a dense clump of more than 7 cm wide. Its golden flowers about 2 cm across are borne in the autumn. Requires protection from direct summer sun though tolerates intense heat and light frost. It is propagated through seeds.

Oophytum (Aizoaceae)

In Greek, the *oon* of *Oophytum* means 'egg' and *phyton* means 'a plant', referring to the egg-shaped plant bodies. The genus comprises of 2 species from Cape region of South Africa. It is not easy to cultivate. It is a mat- or clump-forming highly succulent, dense and compact perennial, bearing egg-shaped soft and ovoid bodies with two highly fleshy and fused leaves which when resting are covered in dry papery sheaths but in spring these sheaths split open to reveal a new pair of leaves. It differs from *Conophytum* by the fleshy and non-membranous calyx tube. It produces mostly white flowers on the crown from a slight central fissure. It is propagated through seeds and stem cuttings. ***O. nanum*** from Cape is a clump-forming succulent growing only 2 cm in height and 1 cm in spread with some 5-7 cm across spherical and minutely papillose body bearing bright green, fleshy and paired leaves fused together and ensheathed in a dry papery covering which splits open in spring producing a new pair of fused leaves. Red-tipped white flowers are daisy-like some 1 cm across which appear during autumn. ***O. oviforme*** from Cape grows 1.2-2.0 cm high bearing glossy papillose, olive-green and often flame-coloured bodies having fissures 1.0-1.2 cm across, small and only slightly gaping. Flowers are white below, purple-pink above and with some 2.2 cm diameter.

Operculicarya (Anacardiaceae)

A native of SW Madagascar, the genus comprises of three species. The members of this genus grow up to 10 m high forming a thick trunk which tolerates intense heat but not the frost as even 4 °C temperature proves injurious. ***O. decaryi*** is excellent as succulent bonsai which has thickened roots. Its bright green leaves continue appearing through whole summer though during winter these drop for a while. Bright sunlight turns the leaves reddish. In old specimens the trunks become thickened and warty. Seeds have poor germination therefore it is propagated only through cuttings.

Ophthalmophyllum (Aizoaceae, common name 'eye leaves')

Its derivation is not certain. The genus comprises some 19 species from South Africa. These are clump-forming succulent very allied to *Lithops*. Each plant bears two erect, very fleshy, cylindrical and fused leaves for most of their length with a fissure in between at the top. Each leaf has a rounded upper surface with a translucent 'window'. These are propagated through seeds and stem cuttings. ***O. longum*** (syn. *O. herrei*) is a clump-forming, 3 cm high and 1.5 cm wide succulent bearing two grey-green to brown united leaves which are erect, very fleshy and cylindrical. Daisy-like white to pale-pink flowers some 2 cm across appear in late summer. ***O. maughanii*** (syn. *Conophytum maughanii*) is a clump-forming, 4 cm high and 10 cm wide succulent bearing 2 yellowish-green and almost united leaves which are erect, very fleshy and cylindrical with 0.5-1.0 cm deep fissure on the upper surface. Daisy-like white flowers some 1.5 cm across appear in late summer. ***O. triebneri*** (syn. *Conophytum friedrichiae*) from Namibia is almost solitary having 0.8-4.0 cm high, 0.6-1.5 cm wide and 0.4-1.0 cm thick body with pale-brown, membranous and not persistent tunic. Body is cylindrical, bilobed at the truncate or rounded apex, soft and rubbery in texture, top translucent, brown- to red-green or grey-green, very glossy when turgid otherwise dull, very shortly papillate, often with raised idioblasts. Fissure extending across the entire top, often postulate at base. Flowers white to mauve-pink, fragrant or nonfragrant, 1.2-2.0 cm across and open in the late morning or in the afternoon. Bracts prominent, highly succulent to leaf-like. Capsules 4-8-locular, to 0.7 cm across and light brown. Seeds pale-brown, almost smooth, numerous, 0.3-0.45 mm in size. ***O. villetii*** is a clump-forming, 2.5 cm high and 1 cm broad succulent bearing two grey-green, erect, broad and fleshy leaves which are fused for most of their length though clearly bi-lobed at the top. Flowers are pale-pink and appear in late summer.

Orbea (Asclepiadaceae)

It has derivation of its generic name from the *orbis* which means 'a disc', referring to the annulus of the mouth of the corolla tube. Its native haunts are in South and East Africa. These are some 20 leafless perennial species under this genus, and the genus is closely related to and often included in *Stapelia*. It is clump-forming succulent growing some 10 cm in height with basal stem branching, bearing 4-angled and erect to decumbent stems with acutely and prominently indented edges from where deciduous small rudimentary and caducous leaves appear which last only for a few weeks. There is solitary or a few-flowered cyme. These are propagated through seeds and stem cuttings. ***Orbea variegata*** (syn. *Stapelia variegata*) is commonly known as 'star flower' and grows some 10 cm high with indefinite spread. It is a clump-forming and base branching succulent bearing 4-angled indented stems. Flowers are star-shaped, variable in colour and blotched yellow, purple- or red-brown and appear during summer to autumn.

Orbeopsis (Asclepiadaceae)

This African genus of 10 species came into existence by splitting *Caralluma*. They are easy to grow, their stems are light green and prostrate producing roots on the ground, and yellow to dark purple flowers are foul-smelling. They require bright but indirect light and warm conditions though cold and wet roots rot them. ***O. melanantha*** bears 3 cm wide dark purple but foul-smelling flowers, rarely with light centre and dark hairs on the petal-margins. Its propagation is through cuttings or seeds.

Orostachys (Crassulaceae)

In Greek *oros* means 'a mountain' and *stachys* means 'a spike'. There are some 10 species of biennial leaf-succulent herbs from North Asia to Europe. These grow 5-30 cm high. A genus of basal-rosetted and short-lived perennial succulents in a dense hemispherical to globose form with spiny tips bearing sword-shaped very fleshy leaves. Inflorescences are produced only in the second or third year with leaves of the floral stem being alternate. Flowers in yellow-green or white to red are subsessile and its sepals are fleshy. Through seeds it takes three years to attain flowering stage but it dies after flowering leaving offsets around the body and seeds in the floral stalk. It is propagated through seeds and division of clumps. ***O. aggregata*** from Japan is monocarpic perennial growing to 45 cm in height with individual stems from 10-25 cm. Leaves are glaucous green, spathulate-oblong with blunt and rounded tips, numerous, densely packed, 2-4 cm long and 1-2 cm wide. White and sub-sessile numerous dense flowers appear during autumn on a 5-20 cm racemose inflorescence. ***O. chanetii*** from China is crowded basal-rosetted succulent growing up to 15 cm high and 8-10 cm wide with 10-15 cm long inflorescence, bearing grey-green rosetted and linear leaves with a small cartilaginous spine and these leaves are shorter in the centre of the rosette. Star-shaped white-pink flowers some 1-2 cm across emerge during autumn on dense pyramidal-shaped tapering spikes some 20 cm long. ***O. erubescens*** from Japan, China and Korea is a monocarpic perennial growing to 25 cm long. Densely leafy stems are 6-15 cm long. Leaves are spathulate, spine tipped, sparsely toothed, 1.5-3.0 cm long, 0.4-0.7 cm wide, fleshy and cartilaginous but in summer these lack cartilaginous margins. Racemes have many sub-sessile red turning purple flowers, raceme dimension some 4-10 cm in length and 1.5-2.0 cm in breadth and floral length 6-8 mm, appearing during autumn. ***O. fimbriata*** from Tibet, China, Mongolia and Japan is similar to *O. chanetii* and grows up to 15 cm in height. Leaves are cartilaginous with long spines, oblong tip and some 2.5 cm long. Leaves on the floral stems are linear-lanceolate and 1-3 cm long, racemes dense and branched, floral bracts spiny and pedicels long. ***O. furusei*** from Japan is a perennial succulent, producing erect or decumbent, thin, short and sterile stems, bearing flat, obovate and fleshy leaves 1-2 cm long and 0.5-1.0 cm wide. Many sessile flowers on a 5-10 cm long raceme are borne in autumn. ***O. iwarenge*** from China is a monocarpic perennial growing to 45 cm tall, bearing 10-25 cm long stems. Leaves are numerous and densely arranged, glaucous, spathulate-oblong, blunt and the length and width of the lower leaves are 3-7 cm and 0.7-2.8 cm, respectively. Many-flowered dense racemes some 5-10 cm long with white sub-sessile flowers are borne during autumn. ***O. malacophylla*** from Mongolia, China and Japan forms crowded rosettes, bearing blunt-tips with no spines, lanceolate-oblong or elliptic leaves though floral stalk leaves are alternate and some 7 mm long. Elongated racemes appear during late summer with sometimes branched ones, many flowered, flowers pale green-yellow and covered with bracts. ***O. spinosa*** from east USSR to North and Central Asia forms saucer-like crowded rosettes 2-7 cm across, growing to 35 cm in height with 10-30 cm long floral stems. Leaves are oblong with apical spines and margins white, 1.5-2.5 cm long and 2-4 mm wide. Leaves on the flower stalk are sessile, spine-tipped, 1.0-2.5 cm long and 0.2-0.5 cm wide. Racemes are compact and many-flowered and the flowers yellow-green.

Oscularia (Aizoaceae)

It is a genus of five sub-shrubby flowering succulent from South Africa, which has been named so due to Latin word *osculum* which means 'a little mouth' due to mouth-like appearance of its leaf-pairs or due to circular hole at the top of the tight cone-like mass of stamens. Its two widely grown species which produce masses of flowers are being described below. ***O. caulescens*** (*Mesembryanthemum caulescens*) from Cape and certain other parts of South Africa, is a thick mat-forming or low-mound succulent with reddish branching stems growing about 25 cm long, and bearing glaucous grey-green paired leaves with reddish edges. The leaves are top-flattened, some 2 cm long, and expanded and keeled below in the upper half. This produces fragrant mauve to lilac-pink flowers some 1.5 cm across. ***O. deltoides*** (syn. *Mesembryanthemum deltoides*) from Cape and some other parts of South Africa is quite similar to *O. caulescens* except the leaves which are smaller, more glaucous, thickly squarish, with a few teeth on the upper margins and a distinct keel, and the flowers which are pink. ***O. d. majus*** bears a little larger flowers.

Othonna (syn. *Othonnopsis*, family- Asteraceae)

Othonna though has been used by Linnaeus to describe this genus but it was a Greek name once used to describe a group of unspecified group of succulents. This genus is native to tropical and South Africa with about 140 perennial shrubby species where most of these have fleshy leaves and/or roots. Only one species, *Othonna capensis* is commonly grown as an effective flowering succulent in hanging baskets or in shallow-wide pans. It is propagated through cuttings. ***Othonna capensis*** from Cape is a beautiful mat-forming evergreen succulent whose branched and slender stems grow more than 60 cm in length, bearing greyish-green cylindrical leaves some 2.5-5.0 cm long. Sometimes the leaves are purple-tipped. Yellow solitary flowers some 1.2-2.5 cm across appear on slender stems above the foliage during spring to autumn. ***O. cheirifolia*** (syn. *O. cheirifolia, Othonnopsis cheirifolia, Herita cheirifolia*) from Algeria is an evergreen and decumbent shrub growing up to 30 cm in height and spread with fleshy, narrow, glaucous grey-green, about 5 cm long and broadly paddle-shaped leaves in 2 alternate facing ranks. Solitary daisy-like yellow flowers some 3-4 cm across on erect stems appear in early summer.

Pachycormis (Elephant tree; Anacardiaceae)

A monotypic genus from Baja California (Mexico), in natural form away from the seasides it grows up to 8 m tall. They are, in habitat, winter-growers which sheds its leaves during late spring and bears sprays of minute pink flowers. When small the plants can be damaged by frost. ***P. discolor*** is quite suitable as succulent bonsai. It should be watered sparingly when leaves start yellowing otherwise the roots will rot. Through seeds it grows quickly, forming a thick taproot and then a thickened aboveground stem.

Pachycymbium (Asclepiadaceae)

Pachycymbium derives from the Greek *pachys* meaning thick and *kymbion* meaning a small cup, referring to its thick, fleshy and campanulate flowers. It is a genus of 32 leafless succulent perennial herbs with native haunts in Africa and Arabia. Its stems are rhizomatous, bluntly quadrangular, dark mottled, tuberculate, teeth flattened-deltoid or conical, subulate, apices soft, about 20 cm high and are formed by indistinct leaves. Inflorescence is the 1- to few-flowered cyme where lobed flowers are campanulate to flat. ***P. baldratii*** with its native haunts being Eritrea, Kenya and Tanzania grows with about 10 cm tall, branched and quadrangular stems often tinged purple. Flowers are sessile, pale-brown or cream with minute red spots, and have 2-8 cm diameter. ***P. carnosum*** from Transvaal bears grey-green spotted red quadrangular stems some 6-15 cm long, 4.5 cm wide where hard and pointed teeth are some 1.2 cm long. Campanulate 1-3 flowers of about 1 cm diameter are coloured grey-mauve spotted red outside, deep cream spotted dark red and dense tuberculate hairs inside. ***P. gemugofanum*** from South Ethiopia produces about 20-25 cm tall, erect or procumbent stems with 2.2 cm long tuberculate teeth. Fleshy, flat to shallowly campanulate and deeply lobed 1-3 flowers with 0.5 cm long pedicels are borne having bright or green-yellow to pale-brown flower colour. ***P. rogersii*** from Transvaal bears green, glabrous, quadrangular, ascending, 10 cm long and 0.8 cm wide stems with 1.5 cm long teeth. Pale-yellow 3-4 clustered flowers with slender lobes are borne along the stem. ***P. sprengeri*** from Ethiopia, Somalia and Sudan produces free-branching quadrangular, erect to procumbent and pale-green stems some 15 cm tall and 1.5 cm wide with minute grey spots or stripes and some 1.5 cm long teeth. Flat and deeply lobed flowers in clusters of 5-6 with stout pedicels are borne having corolla of 2-3 cm diameter and yellow to pink-brown corona.

Pachyphytum (Crassulaceae)

Pachyphytum derives from the Greek *pachys* meaning thick and *phyton* meaning a plant, commonly known as 'moonstones'. This Mexican genus comprises some 12 species of rosetted succulents allied to *Echeveria* with which this even hybridises. The leaves are very thick, fleshy and attractively shaded with grey to blue-green, bell-shaped, cymose and 5-petalled flowers appear above the leaves. These are propagated through seeds, and stems and leaf cuttings. ***Pachyphytum bracteosum*** is the most attractive of the 12 species in its genus. The stems are fleshy, rise from the rosettes of smooth and thick leaves, and the old plants attain up to 30 cm height. Flowers with red petals appear in April and May at its native place. ***P. brevifolium*** from Mexico is more or less erect and base-branching with more than 25 cm of stem. Leaves are blue-white, sometimes tinged red, crowded at the tips and are obovate. Dark or carmine red flowers some 1.2 cm long are borne during summer. ***P. compactum*** from Hidalgo State of Mexico grows up to 15 cm high having indefinite spread having basal rosette of green leaves with angular and paler edges, narrowing to a blunt point, and bears smaller inflorescence bracts than *P. bracteosum*, each inflorescence with 3-10 flowers bearing green to pink calyces and orange petals. ***P. funifera*** from Mexico is rosette-forming succulent with thick, roughly cylindrical and oval leaves, and red flowers. It is propagated through leaf cuttings. ***P. heterosepalum*** from Puebla State of Mexico bears spreading sepals and red flowers. ***P. maneriana*** from Mexico is rosette-forming succulent with small, very fleshy and ovate leaves. It produces red flowers. Its propagation is through rooted leaves or through tip cuttings. ***P. oviferum*** from Mexico is commonly known as 'moonstones' or 'sugar almond plant'. It is a clump-forming succulent. Stems are erect when young but later becoming decumbent or prostrate, sparingly branched, up to 10 cm in length and 30 cm across. Leaves are in basal rosette, obovate, oval in cross-section, patinated blue-white, sometimes tinted lavender or reddish, 2-4 cm long and crowded at stem tips. Rich red bell-shaped flowers some 10-15 per stem and about 1.2 cm long on 5-13 cm tall cymes appear in late winter and spring having powder-blue calyces and orange-red petals. ×***Pachyveria*** is a cross between *Echeveria* and *Pachyphytum*, probably *P. bracteosum*. These hybrids are vigorous than either of the parents, though in general these are halfway to their parents in habits, foliage and flower characters, and are sometimes almost stemless. The best one among many of the hybrids is ×***P. scheideckeri*** (*Echeveria secundum* × *Pachyphytum bracteosum*) bearing grey-white rosette of leaves and some 13 cm across orange flowers tipped yellow which appear during spring. ×***Pachyveria clavifolium*** bears blue-green, clavate and fleshy leaves margined reddish in loose rosette. Flowers have white exterior and red interior. ×***P. glauca*** is a clump-forming succulent growing with the height and spread of 30 cm with a dense basal-rosette of silvery-blue patinated grey, angular to oval, spatular and fleshy leaves some 6 cm long. Flowers are star-shaped and yellow tipped-red and appear during spring.

Pachypodium (Apocynaceae)

In Greek *pachys* means thick and *pous* or *podium* for foot, referring to the thick roots of the genus. This comprises some 17 spiny and deciduous xerophytic stem-succulent miniatures to shrubs and trees up to 8 m in height from Madagascar, and Tropical and South Africa, out of which only a few are cited in horticultural literature. These are normally propagated through seeds or cuttings. It is closely related to *Adenium* though most species are spiny and difficult to grow. It differs from *Adenium* by the armature and pappus found at the one end of the seed. Mostly the trunk is much swollen, the deciduous leaves are simple, entire, leathery and the bases elevated on tubercles arranged spirally round the stems, stipules are stout spines in twos or threes (in some species even third or fourth additional spine is formed above the leaf bud) spreading irregulary on the swollen part, the cymes are terminal with a few to many sessile or peduncled tubular flowers in pink, yellow or white colours, sepals 5, corolla salver-form and constricted at base or funnelform to campanulate, 5-lobed, anthers forming a cone. ***P. baronii*** from Madagascar has a massive shapeless caudex and crown of prickly branches up to 3 m in length. Flowers are salverform, brilliant red and some 5 cm long and wide. ***P. brevicaule*** from Madagascar is a quite dwarf species growing only a few inches above the soil level. Initially it bears potato-like caudex which afterwards broadens into an irregularly lobed flat cake some 30 cm or more in diameter. Leaves are dark green, subsessile when young, come out from a cluster of 0.3 cm spines and are 3-4 cm long and 1.2-1.6 cm wide. Salverform bright yellow and narrow-tubed flowers appear in the condensed cyme of 5 cm long and 2-6 cm wide with corolla having 1.5 cm length and 2.5 cm width. ***P. decaryi*** from Madagascar has quite reduced armature, smooth branches and very small 0.5 cm spines. White flowers are highly perfumed, 5-8 cm long and 12 cm wide. ***P. densiflorum*** from Madagascar is quite similar to *P. rosulatum* but the deep yellow flowers are differently formed with almost flat-open upper part of the tube and with a cone of the filaments exposed in a shallow depression in the centre. ***P. geayi*** from Madagascar is a succulent tree growing about 9 m in height with cactus-like spiny trunk which branches at the top. Leaves are long and narrow and are borne in terminal tufts. ***P. lamerei*** from South Africa resembles a cactus with cylindrical body growing up to 6 m tall and 2 m in spread, crowned with linear to lanceolate leaves. Stems usually branch out after each flowering. The flowers are fragrant, trumpet-shaped and creamy-white which appear during summer on the plants which have attained 1.5 m height. ***P. l*** var. ***cristata*** is the monstrous form of the species. It bears hairless spines. Small lanceolate leaves crowd in abundance at the top of the crest which fall off during autumn and reappear during spring. ***P. lealii*** from Angola and Namibia produces a massive caudex being more broader than its height or elongated conically to 6 m, is irregularly branched and tapering into thick tuberculate branches, and the spines are three, spreading with laterals 2-4 cm long. Leaves are deciduous, elliptic, obovate or oblanceolate with slight wavy margins, 2.5-8.0 cm long and 1.2-4.0 cm wide, and are scattered along the new shoots with crowding at tips. White, salverform and showy flowers some 2.5-4.0 cm long are produced with asymmetrical lobes which are 1.5-2.5 cm long with curled margins. ***P. namaquanum*** from South Africa grows up to 1.8 m tall and about 38 cm in diameter with fleshy, tubercled, spiny and tapering upward growth. The leaves are obovate-oblong to oblong and crowded at little crown on the top of the trunk. Flowers are reddish tinged yellow and green. ***P. rosulatum*** from Madagascar produces pyriform stems when young with massive wide caudex more than its height, a crown of thick and forked branches up to 3.5 m high, smooth below but tip-armed having pairs of conical prickle-like 0.6-1.1 cm long spines. Leaves are sessile, elliptic and 3-8 cm long. Yellow and tubular some 4-10 flowers having 2-7 cm length and 0.4-2.3 cm breadth are held high on a 7-40 cm long forking peduncle bearing 5 rounded lobes some 1.0—1.5 cm long. ***P. succulentum*** from South Africa has turnip-like basally-tuberous underground caudex some 15 cm in diameter, which produces twiggy shrublet measuring 20-60 cm long and 0.5-1.2 cm wide, with 30 cm spread, and produces several fleshy and green to grey-brown stems which may be somewhat branched. Leaves scattered along the body and young shoots, with clustering near the tips are tomentose above, oblong-lanceolate to oblanceolate or linear, some 3.5-6.0 cm long and 1 cm wide and with a pair of 2.5 cm long stipular spines, rarely with 1-2 additional smaller spines above the leaf bud. Cymes are terminal and short-stalked, and bear a few white, pink or partially coloured flowers some 1-2 cm long with narrow tubes.

×*Pachyveria* (Crassulaceae)

It is a hybrid of *Pachyphytum* and *Echeveria* where succulent leaves of the former have been combined with the many-leaved rosettes of the latter. There are many such hybrids that tolerate more heat and light exposure than *Echeveria* as well as cold hardiness up to -4 °C for a brief period. These are mostly cluster-forming perennials.

Pedilanthus (Euphorbiaceae)

Pedilanthus comes from Greek *pedilon* meaning a sandal or shoe and *anthos* meaning a flower, and is commonly known as 'bird cactus', 'Jew bush', 'redbird cactus', 'slipper plant' and 'slipper spurge'. This stem succulent genus of some 14 species, closely related to *Euphorbia*, has its native haunts in the Florida and Mexico to tropical South America and West Indies. It differs from *Euphorbia* in its cyathia, which face to the side and make a point like a bird's beak. It is propagated through cuttings and seeds. It is a bushy genus with highly branched stems which are cylindrical, thick and fleshy, and its juice is milky. Leaves are alternate, boat-shaped and oftenly keeled below, with thickened midrib. Inflorescence is terminal or cymose axillary and bears small bird's head-shaped bracts which are yellowish-green, pink, red or brown. ***Pedilanthus macrocarpus*** from Baja California (Mexico) forms clumps of whitish erect stems up to 1.5 m tall and 1.5 cm thick and red cyathia appear at the tips of the branches in the form of a bird's head. It tolerates low light but as compared to sun-grown plants the shade-grown ones do not grow erect. It survives fairly light frost for a brief period. It is propagated through seeds or cuttings. ***P. tithymaloides*** from northern South America including West Indies to Mexico and Florida may grow up to 3 m in height and up to 30 cm in spread though in cultivation, it usually grows up to 90 cm. Stems are mid-green, thin and erect with ascending branches which zigzag at every node. Leaves are mid-green, ovate to broadly lanceolate, pointed, 5-10 cm long, prominently ribbed beneath and with keeled midrib. Cyathia (floral heads) appear in terminal clusters with red to yellowish-green bracts which appear as a slipper upside down or a bird's head. *P. t.* 'Nana Compacta' is dwarfer and more compact than its parent, *P. t. smallii* (Jacob's ladder) bears prominently zigzag stems, *P. t.* 'Variegata' (redbird flower) possesses angled stems at each node, has white-margined leaves with sometimes irregular variegation, oftenly flushed pink, small greenish flowers at stem tips in red to yellowish-green bracts appear in summer.

Pelargonium (Geraniaceae)

Pelargonium derives from the Greek *pelargos* meaning a stork, an illusion to the long beaked fruit. The genus consists of 250-280 species of perennial shrubs and sub-shrubs, though only a few are annuals, and also there are some succulent species. The genus is widespread throughout warm temperate areas of the new world, majority being in South Africa, though also in North Africa and the adjacent Atlantic Islands and eastwards to Arabia and southern India, as well as in Australasia. These are adapted to areas of low rainfall. Though commonly all the pelargoniums are called geraniums, but in real sense, the *Pelargonium* has a calyx spur fused to the flower stalk and has asymmetrical flowers though the *Geranium* has no spurred calyx and has rounded symmetrical flowers. The common zonal pelargoniums are frequently known as geraniums but should not be confused with the genus *Geranium*. *Pelargonium* is deciduous or evergreen, 5-petalled, upper two being often larger and strongly veined or suffused with a darker colour. In *P.* × *hortorum* and certain other species, the petals are of even size, forming a saucer-shaped flower. ***P. alternans*** from South Africa is a small and highly branched shrub up to 60 cm tall with 2-3 cm thick stems. Its flowers are small and inconspicuous and appear during autumn. ***Pelargonium acetosum*** from Cape is rare plant growing along the rivers up to 60 cm in height and up to 25 cm in spread. It has fleshy stems and its branches with grey-green but often margined red and rarely divided leaves. Flowers are single, starry and salmon-pink where upper petals are smaller than others and two stamens are shorter than remaining five. ***P. carnosum*** (syn. *P. crithmifolium, P. succulentum*) from South Africa is a deciduous pelargonium growing 30 cm in height and spread with thick and succulent stems and a woody, swollen, tuber-like rootstock. Leaves are long, grey-green and deeply lobed triangular leaflets. Flower head is branched, umbel-like and bears white or greenish-yellow flowers where upper petals are streaked-red and shorter than the green sepals. ***P. echinatum*** (cactus geranium) from Namaqualand is tuberous-rooted with stems woody below and fleshy above, bearing persistent spiny stipules on the aged stems though at younger stage these are herbaceous. Leaves are kidney-shaped, palmate or rarely finely divided having long stalks with hairy under-surface. The 5-7 lobes are shallow-toothed. Flowers are reddish-pink to purple along with its 6-7 white through pink to purple stamens though in some cases the upper petals in the centre have heart-shaped deeper red blotches. ***P. laxum*** from South Africa has thickened roots and grows with branched stems some 30 cm high. Its grey-green leaves are fleshy, 8 cm long, 3 cm wide and heavily dissected. Lemon-scented small pink flowers continue appearing throughout the winter. ***P. klinghardtense*** from Namibia grows about 40 cm tall, bearing light blue-green, smooth, thick and fleshy stems, and 4 × 2 cm leaves with entire but undulating margins. During mid-winter to spring small white flowers appear. ***P. peltatum*** from South Africa, commonly known as 'ivy-leaved geranium', is a trailing, evergreen and brittle-jointed succulent growing 1.5 m in height and spread, with leaves similar to ivy in shape. Leaves are mid-green, fleshy and with pointed lobes. Carmine-pink flowers some 2.5 cm across are produced in 5-7 flowered umbels from summer to autumn. Its popular varieties are 'Blue Peter' (mauve), 'La France' (double, mauve with maroon flecking above), 'Lachskönigin' (semi-double, mauve-purple), 'L'Elegante' (white flowers and cream-edged leaves), 'Madame Crousse' (double, bright pink), 'Sir Percy Blakeney' (crimson), 'Tavira' (crimson), *etc.* ***P. reniforme*** from South Africa is a tuberous-rooted small shrublet with shallowly lobed and bluish-green leaves and 2 cm wide bright magenta flowers. ***P. tetragonum*** from South Africa is a succulent species growing to about 1 m tall with jointed, 4-angled, greenish-yellow and fleshy stems. Leaves are greenish-yellow, hairy, circular and inconspicuously 5-lobed with rounded teeth. Out of the four pink petals in the flower which is 2.5-3.5 cm across, the upper two are larger and deeper in colour with more deeper veining. Everard and Morley (1970) described that bizarre, fleshy and woody habit of some pelargoniums such as *P. ferulaceum* and *P. gibbosum* found in north-western districts of South Africa exemplify adaptation to arid conditions in nature. Other important species and their specific hybrids are *P. abrotanifolium, P. angulosum, P. betulinum, P. capitatum* (rose geranium), *P.* × *citrosum, P. crispum* (lemon geranium), *P. crithmifolium, P. cucullatum, P. denticulatum, P.* × *domesticum, P.* × *fragrans* (nutmeg geranium), *P. fulgidum, P. graveolens* (rose geranium), *P.* × *hortorum, P. odoratissimum* (apple geranium), *P. papilionaceum, P. quercifolium* (*P. terebinthaceum*), *P. radens* (*P. radula*), *P. scandens, P. tomentosum* (mint geranium), and *P. violareum* (*P. tricolor, P. splendidum violareum*).

Peperomia (Peperomiaceae, syn. Piperaceae)

Peperomia derives from the Greek *peperi* meaning pepper and *homoios* meaning resembling or alike because of the resemblance of some species to pepper, *Piper*. Its native haunts are tropics and sub-tropics worldwide. This genus has about 1,000 species of tufted small annuals and perennials, many of which are epiphytic. Leaves are lanceolate to orbicular, often with decorative markings and veins. Spikes are usually of insignificant flowers which are borne in short spikes of small white or yellow colour, excepting in a few species where the flowers are also ornamental and appear as mice tails. They are excellent pot plants for indoor as well as for bottle gardening. They are propagated through stem and leaf cuttings and also through seeds. ***Peperomia clusiifolia*** (syn. *P. obtusifolia* var. *clusiifolia*; common names 'baby rubber plant', 'desert privet') from tropical South America is one of the hardiest species growing to about 20 cm high with the spread of 30 cm. Sometimes its stems are prostrate growing. It is a much-branched succulent with purple stems which bear dark green edged purple, thickly fleshy, broadly ovate or rounded leaves some 10-15 cm long. A number of floral spikes bearing 5 cm long stretch of white flowers are produced from June to September. *P.c.* 'Variegata' bears red bordered leaves. ***P. dolabriformis*** from Peru grows some 10 cm in height with erect and stiff stems which are sparingly branched. Leaves are very fleshy, tightly folded and fused, broadly sickle-shaped and 10-15 cm long. Flowering spike is paniculate, long and narrow. ***P. fraseri*** (syn. *P. resediflora*, common name 'flowering mignonette') from Ecuador is a tufted species growing erect with fleshy branches. Leaves are deep dull-green, broadly ovate to rounded cordate and some 4.5 cm long. Hundreds of white flowers on some 5 cm length on the top of some 30 cm long, branched and fluffy spike open in acropetal succession. Though the species is stem-fleshy but is the only species grown for its attractive flowers. ***P. galioides*** from NW South America is quite succulent with red, erect and branched stems growing up to 30 cm or sometimes even more in height. Leaves appear in whorls of 3-6, are narrowly elliptic to oblanceolate and 2.0-2.5 cm long. ***P. glabella*** (wax privet) from tropical America is evergreen, erect to spreading with reddish and spreading stems some 15 cm long. Leaves are glossy bright green, waxy, alternate, ovate to broadly elliptic, fleshy and some 3-5 cm long. Flowers insignificant. ***P. nivalis*** from Peru is a tuft-forming or clump-forming succulent growing up to 10 cm or more in height with semi-erect and branched stems. Leaves are thick and fleshy, boat-shaped, tightly folded and fused with transparent folded edge which forms a linear window and are about 1.25 cm long. ***P. obtusifolia*** (syn. *P. magnoliifolia*; common name 'desert privet') is an evergreen species from Panama, West Indies and northern South America, growing up to 25 cm or longer and wide with erect, freely-branched and bending downward stems when old. Leaves are glossy dark green, firm, thick and fleshy in texture, alternate, obovate to rounded and some 5-10 cm long. Flowers are insignificant. *P. o.* 'Green Gold' bears larger leaves with wide cream margins. *P. o.* 'Variegata' bears oval fleshy leaves some 20 cm long on short stalks which have beautifully silvery yellowish-green to creamy-white margins and other parts of leaves with little portion of irregular greening along the midrib though with age leaves become more greener. Certain other important peperomias are *P. argyreia* (*P. sandersii*; watermelon peperomia), *P. caperata* 'Emerald Ripple', *P. c.* 'Variegata', *P. hederaefolia* (*P. griseoargentea*; ivy peperomia), *P. prostrata* (creeping peperomia), *P. scandens* 'Variegata' (cupid peperomia), *P. rotundifolia* and *P. verticillata* (whorled peperomia).

Petopentia (Asclepiadaceae)

A native to the subtropical wet forests of South Africa, this monotypic genus is characterized by a single species, *Petopentia natalensis* which is a vine growing up to 15 m with large underground caudex. The buds on the vines are glabrous and reddish with corky bark. The leaves are broadly oblong but basally rounded to cordate, glabrous, 7-11 cm long and 3.5 cm wide with purple to brown reverse.

Phyllanthus mirabilis (Phyllanthaceae)

This is the only caudiciform succulent in the family from Thailand and Myanmar where it is found naturally growing under filtered sun in peaty soil with a lot of water during rainy season. Its stem can grow up to 30 cm in diameter and with a height of up to 8 m. Its leaves fold together during the night and are deciduous during winter.

Phyllobolus (Aizoaceae)

A native to the Namibia and South Africa, this caudiciform genus grows during winter but during summer this sheds its leaves as well as stems so during this time they are kept completely dry. Its fleshy leaves are covered with tiny pearly-glistening spots in the form of water cells similar to that of *mesembryanthemum*, and these spots are clearly visible at an angle of sun. It is propagated by seeds or division.

Phytolacca dioica (Elephant tree; Phytolaccaceae)

A native to South America, this is the only succulent tree species in the genus which may grow up to 10 m high, having thick and fleshy roots, and brown and succulent stems which are wider at base. Thick branches when young are green but later turning brownish and the lenticels are large and white.Green leaves are spirally arranged at the apex with extended tips and the veins pinkish-red. Petioles are green and wide. Flowers are inconspicuous and whitish. Propagation is through seeds.

Piaranthus (Asclepiadaceae)

This small stapeliad from South Africa is similar to *Davalia* having 4-angled stems which form a low cluster. Its yellow flowers some 3 cm across open flat and are densely covered with red dots or lines. These are propagated from seeds or cuttings. ***P. foetidus*** produces up to 4 cm long and 1 cm wide stems, bearing 2.5 cm wide yellow flowers with red lines during summer.

Pilea (Urticaceae)

Pilea derives from the Latin *pileus* meaning 'a cap'. This genus has its native haunts worldwide in the tropics and sub-tropics. There are 200-400 species of annuals and perennials with obovate to ovate leaves in opposite pairs, sometimes with decorative patterning. Flowers are inconspicuous and petalless. These are excellent container plants. These are propagated through cuttings. ***Pilea depressa*** from Puerto Rico is 30 cm wide mat-forming species with prostrate and rooting stems, and is most suitable for hanging baskets. Leaves are succulent, bright green, obovate to rounded and some 6 mm wide. ***P. microphylla*** (syn. *P. muscosa*; common name 'artillery plant') from tropical America northwards to Florida grows to 30 cm in height with freely branched and somewhat fleshy stems. Leaves are bright green, dense, 3-9 mm long and covering entire plant so that the plants appear to those of ferns. ***P. peperomioides*** from Yunnan (Hunnan) of China grows to 30 cm in height and spread and is clump- or mound-forming, producing numerous suckers direct from the roots. Each stem is unbranched and becomes woody when old. A tuft of foliage is formed at the top of the stem. The leaves are bright green, fleshy, ovate to rounded, peltate and 6-11 cm long. Tiny greenish-yellowish flowers are insignificant. Certain other important ones most suitable as foliage plants are *P. cadierei* (aluminium plant), *P. crassifolia*, *P. involucrata*, *P. nummulariifolia* (creeping Charlie), *P. repens* (black leaf Panamiga) and *P. spruceana*.

Plectranthus (Labiatae)

Plectranthus derives from the Greek *plectron* meaning 'spur' and *anthos* meaning 'a flower'. *Plectranthus* from Africa to eastern Asia, south to Australasia and the Pacific Islands, comprises 250 species of evergreen perennials and sub-shrubs, bearing 4-angled stems, entire leaves in opposite pairs, and whorls of 2-lipped tubular flowers appearing in panicles or in racemes. These are propagated through cuttings or by division. ***P. nummularius*** from S. Africa is a trailer with succulent and branched-stems growing up to 30 cm in length. Leaves are silvery green above and grey-green with purple veins beneath, fleshy, broadly ovate to orbicular (nearly round) with 6.25 cm diameter and similar to *P. oertendahlii* but quite asymmetrical and coarsely toothed. White to pale-lavender flowers some 1 cm long are borne in about 30 cm long racemes. *P. n.* 'Freckles' leaves are spotted yellow. *P. n.* 'Variegatus' is blotched irregularly with white. ***P. prostratus*** (Pillow plant) from Tanzania is a mat-forming plant rooting at stem joints. Stems are finely hairy with branches growing some 5.0 cm long. Leaves are some 2 cm thick and succulent. ***P. tomentosus*** (Woolly Swedish ivy) from S. Africa is a small shrub covered with white wools, and the leaves are succulent, nearly round with about 8.75 cm diameter having scalloped margins.

Pleiospilos (Aizoaceae)

Pleiospilos derives its name from the Greek *pleios* meaning many and *spilos* meaning a speck or spot, as most species have prominently dotted leaves. Its native haunts are in South Africa with about 35 succulents which are in almost stemless rosettes and mimic stones similar to *Lithops*. They emerge either solitary or in clumps with highly short stems, each such stem bearing 2-4 or sometimes even more pairs of very thickly fleshy leaves similar to those of granite pieces, each with a flat upper surface and each pair united at the base. The daisy-like showy flowers larger to the size of the plant appear mostly during summer. These are propagated by seeds or through division. ***P. bolusii*** (living rock, mimicry plant) from Cape grows up to 10 cm in height with 20 cm spread, having one or sometimes two pairs of pitted coarse leaves very much like bits of stone. Each grey-green leaf with numerous dark dots is roughly half ovoid but thickened towards the apex. The leaves are often wider than the length and become narrow at recurved tips. Under bright light the leaves are flushed reddish. Bright yellow flowers some 6-8 cm across are borne in the afternoon in summer to autumn. ***P. compactus*** (syn. *P. simulans*) from Cape is clump-forming succulent growing to 10 cm high and 30 cm in spread with 1-2 pairs of grey and thick leaves some 8 cm long. Coconut-scented yellow flowers appear during early autumn. ***P. nellii*** from Cape is though very similar to *P. balusii* but differs in having two pairs of a little smaller and almost hemispherical leaves, and turmeric-yellow flowers with a white centre.

Plumeria (Apocynaceae)

Plumeria from tropical America is a genus of seven species of shrubs and trees with thick and succulent branches and large entire leaves. *Plumeria* was named so to honour Charles Plumier (1646-1704), a French Franciscan monk who travelled widely and made accurate drawings of the plants he saw. It is a genus of mainly deciduous fleshy-branched shrubs and trees grown for their highly fragrant and showy flowers but its sap is poisonous. The flowers are tubular with 5 overlapping lobes of corolla, appearing in terminal panicle-like clusters with 20-60 blooms per cluster. These are propagated through stem cuttings. ***P. rubra*** (syn. *P. acuminata, P. acutifolia, P. rubra* f. *acutifolia*; common name 'frangipani') from Central America is a deciduous, spreading and sparsely branched shrub or tree growing up to 8 m in height and in canopy spread. In container it can be tamed up to 3 m in height. Leaves are alternate, lance-shaped to oval, pointed at both the ends and up to 30 cm long. Flowers some 5 cm or more across are fragrant rose-pink to red with yellow eye, having 5 spreading petals and about 2.5 cm long perianth tube. Seed pods are 15-25 cm long with plumed seeds. *P. r. acutifolia* (pagoda tree) bears large white flowers with prominent yellow eyes. *P. r. lutea* bears yellow flowers sometimes tinged pink.

Portulaca (Portulacaceae)

The native haunts of *Portulaca* are tropics and warm temperate regions, especially the Americas. *Portulaca* is from the Latin vernacular name for purslane. It is a genus of up to 200 species of fleshy and herbaceous annuals and perennials with flowers that open in sun with 5 spreading petals and close in shade. Their leaves are alternate or almost opposite, ovate and linear and often fleshy. These are propagated through seeds or cuttings. ***Portulaca grandiflora*** (rose moss, sun plant) from Argentina, Brazil and Uruguay is a tufted annual growing about 15-20 cm in height and 15 cm in spread with spreading-prostrate to ascending stems. Leaves are bright green, lance-shaped cylindrical or oval, fleshy and about 2.5 cm long, Flowers are white, yellow, orange, pink, red or purple with yellow stamens, each flower sitting over a leafy collar. There are many varieties of this available worldwide in various colours and variegations in single, semi-double and double forms. 'Cloudbeater' is double-flowered. 'Sunnyside Series' has rose-like double flowers. ***P. molokaiensis*** from Isle of Molokai (Hawaii) is now rarely found in the wild but is in common cultivation. It is a small shrublet bearing thick stems and round fleshy leaves. It is propagated through seeds or cuttings.

Portulacaria [Elephant bush, Jade plant; Didiereaceae (Portulacaceae)]

It is a glabrous succulent shrub with two species, originally recorded from South Africa, but only one, *i.e. Portulacaria afra* is in cultivation. The leaves are dark jade-green, smooth, fleshy, waxy, opposite, obovate, dissimilar, maximally 1.25 cm wide and are borne in clusters on short twigs attached to the thicker branches. Flowers are small, rose, fascicled in the upper axils with leafy panicles, having 2 short sepals, 4-5 longer petals, 3-cornered, 1-ovuled and free ovary and indehiscent 3-winged capsules. These are propagated through cuttings. ***Portulacaria afra*** in cultivation grows up to 2 m long but generally it is 35 cm in height, 60 cm in stem length and some 1.2 m in spread with multi-branched hanging stems, branching opposite, leaves obovate-roundish, and flowers pink and small. Their growth is kept under check through frequent trimming so that these may retain only 30 cm of height.

Pseudolithos (Asclepiadaceae)

The genus from Africa comprises only a few species having an unusual rock-like succulent stems. These require bright light, moderate heat and a well-drained potting soil. Small flowers some 1 cm in size appear in various colours depending on the species planted. ***P. migurtinus*** from Somalia is often short-lived with spherical and tuberculate stems which offsets with time. In autumn, small brown flowers appear in clusters laterally. It likes bright sunlight and moderate heat. Grafting may prolong the life of this species. It is propagated through seeds.

Pterodiscus (Pedaliaceae)

Pterodiscus derives from the Greek *pteron* meaning 'wing', and *diskos* meaning 'a disc', referring to the flat disc-shaped winged seeds. This genus with some 18 species has its native haunts in Troical East Africa, South Africa, Angola and Namibia. These are small perennial herbs to shrubs, often semi-succulent and grow up to 30 cm in height. These have swollen caudex and tuberous roots. Stems are simple or branched and solitary or several. Leaves are variable in shape with entire but undulating margins, dentate or laciniate. Variously coloured solitary flowers emerge from the leaf axils with small calyx and funnel-sahped corolla tube, often a little basally tuberculate, 2-lipped, and with spreading limb, lobes dissimilar, ovate, circular or elliptic. These are propagated through seeds. ***P. angustifolius*** from Tanzania is basally branched. The branches are spreading, globose, purple, 9-20 cm long and fleshy. Leaves are dense, dark green, oblong-lanceolate, a little glandular at initial stage of growth, 2.5-13.0 cm long and 6-12 cm wide, tips obtuse or pointed, margins entire and sometimes undulating but dentate towards apex and petioles 2.5 cm long. Yellow to orange flowers appear with blotched purple in its tube, lobes are ciliate and hairy inside. ***P. aurantiacus*** from Angola, Namibia (Great Namaqualand) and South Africa (Kalahari region of Cape) produces bottle-shaped caudex some 30 cm high with several branches at the top. Leaves are smooth, tinged blue, sinuate and oblong-lanceolate to ovate-spathulate. The flowers are brilliant red. ***P. coeruleus*** from Somalia and Kenya produces 5-20 cm high, simple or branched stems bearing 0.4-2.5 cm long petiole, leaves some 1.3-4.0 cm long and 0.4-1.5 cm wide which are basally cuneate with undulating margins. White flowers are suffused mauve in the throat with lobes having red veins. ***P. kelleranus*** from Somalia has edible and fleshy caudex. Bassl leaves are elliptic with undulated margins while other narrowly lanceolate with usually entire or somewhat incised margins, and sometimes are distinctly pinnatisect. ***P. luridus*** from South Africa (Kalahari) and Namibia produces obconical and fleshy caudex which may be 4-8 cm high and 0.5-2.0 cm wide, bearing some 4-20 cm long 20 stems. Leaves are numerous, dark green above and tinged white or blue below, 7-8 cm long and 2.5 cm wide, oblong, spathulate at base though laciniate apically with linear lobes and pruinose all along. Flowers are yellow with externally dotted red. ***P. ruspolii*** from Kenya, Ethiopia, Sudan and Somalia produces basally thick and fleshy caudex which grows some 4-8 cm high and 0.5-2.0 cm wide. This produces some 4-20 cm long around 20 stems. Leaves are obovate to elliptic, apex rounded, margins entire or undulating, glandular beneath, 1.5-6.5 cm long and 0.8-3.5 cm wide, and petiole 0.5-3.5 cm long. Light yellow to orange flowers may or may not bear red or purple blotches in the centre and the lobes are often ciliate. ***P. speciosus*** from South Africa (Cape and Transvaal) produces conical to cylindrical caudex some 15-50 cm high and 6 cm wide and bears few branches at the crown growing up to 15 cm. Leaves are numerous, linear to linear-oblong, irregularly dentate or slightly incised, and 3-6 cm long and 0.5-1.0 cm wide. Light purple-red, almost regular in shape and flat and some 2.5-3.0 cm across flowers with 3 cm long corolla tube and 5-lobed limb are borne during spring to summer.

Puya raimondii (Bromeliaceae)

See chapter on '**Bromeliads**'.

Pyrenacantha (Icacinaceae)

A native to the eastern Africa, this is the only genus in the family with caudiciform succulence having two species ~ *P. kaurabassana* with subterranean caudex and vining stems, and *P. malvifolia* with boulder-sized aboveground caudex and almost deciduous climbing or vining stems. Minute flowers emerge from the leaf axils. These are propagated through seeds.

Quaqua (Apocynaceae)

An endemic genus to the SW Africa, especially in Namaqualand, this genus bears a tough and 4- to 5-sided stems on which conical tubercles are borne with tapering spikes at their ends though a few other species either have smoothly rounded tubercles or no spike. The genus produces numerous inflorescences from each stem at or near its tops, and usually the attractive red and papillate flowers are small (maximally 2.7 cm across) and have pleasing sweet smell. Most popular species is *Quaqua mammillaris.*

Raphionacme (Asclepiadaceae)

It is a small African genus which is preferred for its large underground caudices from where annual vines some 1 m long emerge which produce green to purple and smaller than 5 mm flowers. ***R. flanaganii*** forms a brown-skinned caudex more than 10 cm long and wide, producing 1 m long vines in summer with 4 cm long, 1 cm wide dark green leaves. It is propagated through seeds. In case of ***R. zeyheri*** the caudex is egg-shaped.

Rhombophyllum (Aizoaceae)

Rhombophyllum derives from the Greek *rhombos* meaning 'lozenge' and *phyllon* means 'a leaf', referring to the leaf shape which is obliquely diamond-shape. A genus from Cape with three species of compact mat-forming perennial succulents having napiform roots, bearing dense basal rosettes of opposite and decussate, united at base, linear or semi-cylindrical leaves, keeled above, underside pulled forward and chin-like, sometimes producing a secondary 'tip', each expanding towards tip or the middle with reflexed or incurved leaf tips, upper surface almost linear, margins entire or with 1-2 short teeth, smooth, almost glossy, deep green with pale translucent dots. Golden-yellow 3-7 flowers on a pedicel are formed in early summer or early autumn. It is propagated through seeds and cuttings. ***R. delabriforme*** from east Karroo (Cape) is much-branched mat-forming but later on becoming shrubby, growing up to 30 cm high, bearing 2.5-3.0 cm long green leaves with translucent dots, tapered on the upper surface and semi-cylindrical on the lower, extended above up to 1.1-1.5 cm with a hatchet-shaped keel which is pulled forward making a tooth-like tip. It produces golden-yellow flowers some 4 cm in diameter. ***R. nelii*** from Karroo (Cape) is similar to *R. delabriforme* but pale blue-green or grey-green leaves with darker dots. Leaves are bifid, 1.5 cm long and 0.4-0.6 cm wide, and upper surface being 0.6-0.8 cm wide. Yellow flowers some 4 cm in diameter are borne out during summer. ***R. rhomboideum*** from Cape is a stemless clump-forming succulent growing up to 5 cm in height and 15 cm in spread bearing glossy grey-green with translucent dots and linear, 4-5 pairs of unequal leaves resting on the soil surface, leaves being 2.5-5.0 cm long and 1.2 cm wide, rhombic from above and rounded below, and then thickened and keeled towards the tip and pulled forward and chin-like, and margins are white to pale and rarely 1-2 toothed. Yellow flowers tinged-red externally, 3-7 in number and some 3-4 cm across are borne in summer.

Rhytidocaulon (Apocynaceae)

A native to Ethiopia, Kenya, Somalia, Oman, Saudi Arabia and Yemen, this little-branched perennial genus of stem succulents is closely related to *Caralluma*, though differs in having wrinkled, rough and irregular stems covered with a waxy layer. Its inflorescences are sunken into the stems. It requires semi-shaded situation for its growing and as it is difficult to grow on its own feet, it is mostly grafted. *R. fuller* is the most popular species.

Rochea (Crassulaceae)

Rochea is a genus from South Africa with 3-4 evergreen succulent sub-shrubs formerly included and closely related to *Crassula*. It is named so in honour of Daniel de la Roche (1743-1813), a Swiss doctor and botanist. These have pairs of entire leaves, 5-petalled flowers, generally petal edges outcurving. After every flowering the stems need cutting back some 5 cm top during late winter. These are propagated by seeds and stem cuttings. ***Rochea coccinea*** (syn. *Crassula coccinea*) from South Africa grows erect up to 60 cm tall and 30 cm spread with about a dozen of green and stiff stems growing together, bearing leathery, oblong to obovate, hairy-margined and 2.5-4.0 cm long numerous mid-green leaves in four close ranks (in alternate pairs, each uniting at the base) at right angles around the stem in rows from bottom to top and even when not in flowering the plants in leaves look beautiful. Carmine-red, tubular (tube 2.5 cm long), fragrant and umbellate flowers some 4 cm across are borne during July to September in dense terminal clusters of 7.5-13.0 cm, each flower with five spreading lobes.

Ruschia (Aizoaceae)

It is a large heterogeneous genus of the plants where some 200 species are listed but this requires immediate revision.

Sansevieria [Dracenaceae (Agavaceae)]

Sansevieria was named for Prince Raimond de Sanagrio de Sanseviero (1710-1771). The native haunts of this genus are Africa and southern Asia with about 60 ground-dwelling or epiphytic species of succulent perennials which are usually leaf-like. They have rosettes or clumps of stiff cylindrical or sword-shaped leaves and almost tubular flowers in panicles or axillary racemes which afterwards bear red berries. These are fast-spreading so require repotting every 2-3 years to avoid cracking of the pots. These are propagated by division and cutting of leaf sections. ***S. aethiopica*** from South Africa grows up to 40 cm height, bearing glaucous with darker cross-banding having linear to lanceolate and with white awl-shaped tip. Raceme some 30 cm long bears white flowers. ***S. cylindrica*** from tropical Africa though grows up to 1 m tall but under cultivation it is up to 45 cm. Its leaves are cylindrical, tapering, grooved along the length and cross-banded grey-green at early stage of growth. White or pink-tinted flowers some 4 cm long are sometimes produced on 30-90 cm long racemes. ***S. grandicuspis*** (star sansevieria) from tropical Africa bears up to 50 cm long and arching, green, grey cross-banded and narrowly lanceolate leaves which are sharply pointed with brown horny margin. ***S. grandis*** from East Africa is epiphytic which produces long stout runners with offsets. Leaves are about 20 cm long, ovate to lanceolate, dark green, red-edged and with metallic cross-banding. Narrow panicles appear with white flowers well above the leaves. ***S. hahnii*** from tropical West Africa growing up to 15 cm high is a rosette-forming species with triangular-ovate leaves some 10 cm long and 6.25 cm wide. The leaves are dark green with grey and yellow cross-bands. Its variegated form bears two longitudinal yellow stripes near the edge. ***S. hyacinthoides*** (syn. *S. spicata, S. thyrsiflora*) from South Africa grows up to 45 cm in height. Leaves are dark green with pale cross-bands and yellow margins, lanceolate and taper to a channelled petiole. Flowers are fragrant greenish-white on up to 75 cm tall racemes. ***S. kirkii*** from East Africa bears greyish-green leaves up to 1.8 m long mottled with pale-green bands. ***S. senegambica*** (syn. *S. cornui*) from West Africa bearing leaves that are green sometimes marked darker, oblanceolate up to 60 cm long and tapered pointed. Flowers white, racemose and either above the leaves or equal to. ***S. trifasciata*** (syn. *S. guineensis*; common name 'mother-in-law's tongue', 'snake plant') from South Africa is a clump-forming stemless and rhizomatous succulent bearing five leaves per plant, with leaves growing up to 90 cm long though in containers almost its half, stiff, erect, deep green but cross-banded paler greens and linear-lanceolate. Sometimes racemes of greenish-cream, highly tubular and 6-lobed flowers appear in abundance which emit sweet fragrance during night. *S. t.* 'Golden Hahnii' (Golden bird's nest) is quite dwarf but with same width of leaves, no cross-banding and has quite wide yellow margins so looking very attractive. Sometimes bears small and pale-green flowers. *S. t.* 'Hahnii' is a dwarfer form of *S. trifasciata* but with same width of leaves. *S. t.* 'Laurentii' is a mutant with yellow margins. To maintain this character, it is grown through division as it reverts back when grown through leaf cuttings. ***S. zeylanica*** (devil's tongue) from Sri Lanka grows up to 30 cm tall bearing grey-green lanceolate leaves with a soft spine-shaped tip, dark green cross-banding, keeled below and slightly channelled above and some 30 cm long.

Sarcocaulon (Geraniaceae)

A native to W. South Africa and Namibia, the genus is a semi-erect or erect spiny succulent shrublet with short fleshy stems which have branching at the ground level that are covered with waxy and translucent bark. Leaves are long and dimorphic (two forms), and long- to short-petioled. Short petioles appear singly or in a groups of 2-7 and the long ones appear singly which are either blunt are as sharp as spines, both in the axils of long spines. The tapered-base leaves may or may not be segmented, entire or toothed, elliptic to ovate to obovate, often pleated and the tips are notched. Solitary and nearly white flowers are subtended by two bracts and appear from the axils of the leaves. *S. crassicaule* is the most common species.

Schwantesia (Aizoaceae)

This genus was named for Dr. M.H.G. Schwantes (1881-1960), professor at the University of Kiel. The stemless genus is native of Cape (South Africa) and Namibia with 10 species having many-headed cushion-forming succulents with rosettes of unequal pairs of keeled leaves. Young leaves or those of the seedlings show its close relationship with the 'bodied' genera *Lithops* and *Conophytum* but the adult ones show similarity to genera such as *Carruanthus* and *Faucaria*. Stems are 1 cm thick and its branches up to 0.7 cm thick and are densely clothed with dry remains of the yesteryear's leaves. Leaves are in unequal pairs, entire or dentate or lobulate, 2.5-5.0 cm long and 0.5-1.7 cm wide, keeled beneath and oftenly reddened along the margins, upper surface flat, acute or somewhat rounded, glabrous or velvety, tinged blue or pale blue-green and oftenly marbled. Solitary daisy-like velvety yellow flowers some 2.8-5.7 cm in diameter appear in summer. These are propagated through seeds and stem cuttings. ***S. acutipetala*** from Little Namaqualand (Cape) bears leaves which are blue-grey, flat above, semi-circular below, shortly tapered, acutely angled and measuring 5.2 cm long, 1.5 cm wide and 0.5 cm thick. Flowers are yellow and 4 cm in diameter. ***S. herrei*** from Namibia is compact cushion-forming bearing pale blue-green, smooth and 2-3 pairs of leaves per stem, leaves measuring 2.5-3.5 cm long and 1.6 cm wide, entire, acute, keel pulled forward, chin-like towards the obtuse tip, and often with several teeth. Yellow flowers are 3.2-4.2 cm in diameter. ***S. pillansii*** from Cape bears blue-green and obliquely acute to tapered 2 leaves per stem measuring 5.3 cm long, 5 cm wide and 0.8 cm thick, obliquely keeled below along with a red line. Flowers are yellow and 3.2 cm in diameter. ***S. ruedebuschii*** from Namibia is a dense mat-forming succulent growing to a height of 5 cm with spread of 20 cm, bearing 4-6 leaves per shoot of bluish-green, cylindrical, erect, 3-5 cm long, 1.0-1.2 cm wide, and 1 cm thick basally, with expanded and obtuse tips, navicular margins rounded, suffused blue-green with white mottling, and each of the leaf edges produce 3-7 minute but broad teeth with blue-brown tips, tooth being 0.4 cm long. It produces daisy-like light yellow flowers some 3.5-4.0 cm in diameter in summer. ***S. triebneri*** from Cape bears white-green to blue-green or yellow-green 3 pairs of leaves per stem with red dots and angles, leaf size being of 4-6 cm long, 1 cm wide and 0.5 cm thick, quite expanded and mucronate above, round below and keeled towards tip. Flowers are yellow and 4-5 cm in diameter.

Sedum (Crassulaceae)

This genus with 500-600 species, mostly succulent, originates in the Northern Hemisphere. These are mostly perennials and sub-shrubs though with a few annuals and biennials. The genus has derived from the Latin *sedo* meaning 'I sit' referring to those species which appear to sit upon rocks and walls. Sedums are erect to prostrate in habit with entire succulent leaves that are cylindrical, ovoid or rounded. Flowers are 5-petalled in terminal cymes. These are propagated by seeds and by leaf and stem cuttings. *Sedum* and *sempervivum* species are quite similar to each other except that former has leaves arranged on long stems and flowers with less than 5 sepals though the latter has basal rosette of leaves and flowers with more than 8 sepals. ***Sedum acre*** (biting stonecrop) from Europe, North Africa and West Asia grows some 5 cm in height and about 25 cm in spread and is mat-forming alpine evergreen succulent with pale-green overlapping leaves which are bluntly conical. Yellow flowers some 2.5-4.0 cm across are produced freely on flattened heads during June and July. ***S. adolphii*** (golden sedum) from Mexico is a low semi-shrubby with fleshy or semi-woody branches more or less erect, bearing 3.5 cm long and 1.5 cm wide, ovate and dull yellow-tinted leaves with distinctly keeled margins, and those at the end of the branch are quite compact. White flowers are borne loosely in the paniculate cyme. ***S. aizoon*** from Japan and Siberia is a perennial succulent growing up to 30 cm in height. It is compact plant with mid-green, shiny and oblong-lanceolate leaves toothed coarsely and irregularly. Golden-yellow flowers some 5.0-7.5 cm across are borne on flattened heads in July. ***S. album*** from Europe, North Africa and West Asia is mat-forming evergreen alpine succulent growing up to 15 cm in height and up to 40 cm in spread. Leaves are oblong and cylindrical and are borne on pink stems. Profuse white flowers some 7.5 cm across appear in loose clusters in July. ***S. allantoides*** from Oaxaca (Mexico) is a sub-shrubby succulent growing up to 30 cm in height with erect and robust stems which branch out mainly at the ground level. Leaves are whitish-green, clavate, incurving, some 2.5 cm long and with waxy-white covering. Flowers are white with greenish tinge, some 1.6 cm across, and are borne in summer in short terminal panicles. ***S. anglicum*** (English stonecrop) from England grows only a few inches high with spreading stem, bearing fleshy and hairless leaves, and almost sessile flowers often tinged pink. ***S. brevifolium*** from SW Europe and Morocco is a mat-forming prostrate succulent 1.25 cm high and some 30 cm in spread with white-green, tiny and ovate leaves closely arranged along the stems. White flowers some 1.25-2.5 cm across are borne in July in flattened clusters. ***S. burrito*** was first collected from eastern Mexico in 1972 and described in 1977 is quite similar to *S. morganianum* as it has pendent growth habit but with stiffer stems, bluish and more crowded leaves which are some 1.5 cm long, ellipsoid and with blunt tips. Pink and bell-shaped flowers some 8 mm long and with darker lines on the petals appear in terminal clusters of 10-30. ***S. caeruleum*** from Morocco and Algeria, Corsica, Sardinia and Sicily is an annual succulent growing with smooth and small stems up to 15 cm in height. Leaves are ovate pale-green. Stems and leaves both turn red slowly when flowering commences. Pale-blue flowers some 2.5 cm across with white centres appear in July. ***S. dasyphyllum*** from South Europe and North Africa is a hardy succulent forming a dense mat some 1.25 cm high and about 30 cm in spread densely borne out with tiny blue-green ovate leaves. Minute white flowers some 1.25 cm across appear in flat clusters in June. ***S. dendroideum*** from Sierra Madre del Sur and Guatemala mountains in Mexico is though semi-shrubby succulent but with trunk-like main branched stem growing up to 30 cm in height bearing leaves only at the stem tips. Leaves are glossy bright green, some 3 cm long, obovate to spathulate, flat but thickly fleshy and at the margins minute reddish glands. Many bright yellow 5-petalled flowers some 1.2 cm across appear in panicle-like terminal cymes during spring. ***S. griseum*** from Mexico is a compact semi-shrubby and erect-growing succulent bearing green and cylindrical leaves and white flowers. ***S. hintonii*** from Michoacan and Pinzan of Mexico is tufted and small hummock-forming succulent with some 10 cm long prostrate to decumbent stems. Leaves are pale grey-green, rosetted and crowded, oblongoid to obovoid and slightly incurving, highly fleshy, some 2.5 cm long and are densely covered with short, white and bristly hairs. Several white 5-petalled flowers some 1 cm across in about 10 cm long dense terminal leafy cyme appear in spring. ***S. morganianum*** (burro's tail, donkey's tail) was discovered by Eric Walther in 1935 in Vera Cruz State of Mexico and named so to honour Dr. Meredith Morgan of Richmond of California who was the person to get it flowered first in 1938. It has tail-like hanging branches, sometimes reaching to 90 cm long and are clothed with overlapping, pale-green, some 2.5 cm long and sharply pointed cylindrically-fleshy leaves which look quite spectacular and elegant in hanging baskets. The star-shaped flowers about 1 cm long are pink-red in small clusters which appear during spring. ***S. nussbaumerianum*** from Vera Cruz and Zacuapan (Mexico) is a sub-shrubby hummock-forming succulent growing up to 15 cm in height with yellowish-green and red-streaked stems, branching at the base and above. Leaves are yellowish-green bordered red, very fleshy, elliptic to oblong-elliptic or lanceolate, 3 cm long, flattish above and rounded beneath, and are formed mainly at the stem tips. Yellow 5-petalled flowers some 1.6 cm across appear in dense lateral cymes during winter and spring. ***S. oaxacanum*** from high mountains of Oaxaca (Mexico) is a mat-forming succulent interlaced with decumbent stems growing 10 cm high and about 30 cm across, with waxy-white above and keeled beneath, crowded, obovate and 5-8 mm long leaves. Bright yellow flowers some 1.5 cm across are borne in small terminal clusters in late spring and summer. ***S. pachyphyllum*** (jelly beans) from Oaxaca mountains, San Luis and Sierra Madre del Sur of mexico is a sub-shrubby succulent with main stems arising decumbent from the base then getting erect and spreading, bearing fresh green red-tipped clavate leaves which become abruptly mucronate, some 2.5 cm long and cylindrical in cross-section. Bright yellow 5-petalled flowers some 2 cm across are borne in dense lateral cymes in spring. ***S. palmeri*** from Sierra Madre del Sur (Mexico) grows up to 15 cm in height having semi-shrubby, erect to spreading and base-woody stems, Leaves are formed mainly at the stem tips, and they are flat and thickly fleshy, spathulate, glaucous and about 2.5 cm long. Orange 5-petalled flowers some 1 cm across are borne in lateral branched cymes from early spring to autumn. ***S. prealtum*** from Mexico and Guatemala mountains grows some 60 cm in height with branches arching down to the soil and rooting. Flowering occurs from late spring to early summer with bright yellow flowers. ***S. rubrotinctum*** (*S. guatemalense*, Christmas cheer) from Mexico grows some 20 cm high and 30 cm in spread and is a base-branched sub-shrubby succulent with thin shoots and bright green, cylindrical and fleshy leaves having club-shaped reddish-brown tips. Whole plant turns red at sunny positions. Its var. 'Aurora' bears grey-green leaves tinged rose-red. ***S. sieboldii*** from Japan is tufted succulent with carrot-shaped tuberous rootstock. ***S. stahlii*** from mountainous regions of Mexico is loosely mat-forming succulent with slender, prostrate or decumbent branches growing to 20 cm in length. Leaves are flushed brownish-red, minutely pubescent, ovoid but circular in cross-section, crowded along the stem and some 1 cm long. Bright yellow 5-petalled flowers some 1.2 cm across are borne in late summer in 2-forked terminal cymes. ***S. suaveolens*** from Mexico has *Sempervivum*-type rosettes of bluish splashed purple especially towards upper side, broadly ovate and quite fleshy and pointed leaves arranged alternately in the rosettes. ***S. treleazii*** from Sierra Madre del Sur of Mexico grows some 30 cm in height and 15 cm in spread. It is a semi-shrubby succulent producing only 1-2 stems from the base though these branch out afterwards at a low level. Leaves crowd along the stem, are glaucous blue-green, 2-3 cm long, ovate, incurving and very fleshy. The rounded clusters of bright yellow flowers some 1.25 cm across appear terminally in March and April in cymose panicles. Everard and Morley (1970) mention two species, ***S. hispanicum*** from Switzerland and Italy to the Caucasus and northern Iran as an annual to bienneial species reaching to the height of 5-15 cm; and ***S. nevii*** from Illinois to Alabama in USA as an unusual one among the sedums due to its liking for moist habitats. Its specific name commemorates its discoverer Reverend Dr. Nevius.

Sempervivum (Crassulaceae)

Sempervivum (commonly known as houseleek) in Latin means 'living forever', *i.e. semper* meaning 'always', and *vivus* meaning 'alive', referring to the nature of plants as they are evergreen and very tenacious of life. This genus with about 40 species of perennial succulents, has its native haunts in the mountains of Europe, North Africa, and West Asia (Turkey and eastward to the Caucasus) and is closely allied to *Sedum* but differs in having mats of dense basal rosettes (each rosette monocarpic) of leaves and flowers with 8-17 spreading petals that open out flat though *Sedum* species have leaves arranged on long stems and the flowers with 5 or less number of petals. The plants of this genus are most suitable for planting in the rock gardens, screes, alpine house, on dry walls, as edgings and in containers. Their leaves are oblong to obovate to elliptic, often flushed or tinted red or purple and pointed, curving around each other to form close symmetrical rosettes, fleshy and green, tipped red or purple or in some species appearing leaves are green but already appeared ones are either fully red or red tipped-green. The flowers appear in sprays and star-like though in *S. beuffelii* and *S. sobolifera* the petals being fused appear bell-shaped, however, such species by some botanists are placed under section II, *i.e.* genus *Jovibarba* having only 6-7 petals that are fringed and stay erect, and this genus is described above separately. The plants take several years to attain flowering size and after flowering the rosettes die producing several offsets each year up to flowering. These are propagated through their offsets. The members of *Sempervivum* are valued for their leaves, however, *S. arachnoideum* and *S. tectorum* bear even attractive flowers. The important species are described below. ***S. andraeanum*** from N. Spain produces rosettes 2-3 cm wide, blue-green tipped darker, flowers pale-pink and deeper at the base and are borne on 10-12 cm long stems. ***S. arachnoideum*** (cobweb houseleek) from Pyrenees, Alps, Apennines and Carpathians has its globulaer rosettes growing up to 2.5 cm high and 2.5-4.0 cm across, but spreading up to 30 cm. The leaves are green, sometimes flushed red, and the tips spun together with a white cobweb-like mat of white hairs. Stems are 8-15 cm tall bearing rose-red flowers some 1.9 cm across which appear in June and July. This is the probably most popular one among whole of the species. Its several forms and hybrids are available. ***S. ballsii*** from NW Greece produces rosettes some 3 cm across with a few or no offset, outer leaves bronze to red-tinted, and flowers pink. ***S. calcareum*** (syn. *S. tectorum calcareum*) from French Alps is quite similar to *S. tectorum* but produces flowers seldom and the rosettes are grey-green with large purple-brown tips. ***S. ciliosum*** from Bulgaria, Yugoslavia and NW Greece produces 3.5-5.0 cm across rosettes which are ciliate, hairy and a little hoary with outer leaves being red-flushed. Greenish-yellow flowers are produced on a 10 cm tall stem. ***S. dolomiticum*** from eastern Alps grows up to 5 cm high, rosettes up to 5 cm across and spread 15-23 cm across. Its rosettes are bright green, and the flower stems up to 15 cm long bearing deep rose-red flowers some 1.8 cm across in June and July. ***S. erythraceum*** from Bulgaria grows up to 5 cm high with up to 23 cm spread and the flat rosettes up to 5 cm across which are composed of grey-green leaves, overlaid with a faint purple sheen. The deep rose flower some 2.5 cm across are borne on 15 cm long stems in July. ***S. giuseppii*** (syn. *S.* × *giuseppii*) from NW Spain is vigorous, evergreen and prostrate perennial succulent, forming rosettes up to 4 cm across, spreading up to 10 cm. Leaves are pale green tipped-brown, densely downy - necessarily during spring, stiffly ciliate and with dark spotted-tips. Star-shaped terminal clusters of deep pink or red flowers appear in summer on 12-15 cm tall stems. ***S. grandiflorum*** (syn. *S. globiferum*) from S. Switzerland and N. Italy produces rosettes variable in size, up to 20 cm across, red-tinted dull green with unpleasing aroma, sticky and dense glandular downy. Large greenish-yellow flowers are borne in loose terminal clusters on 15-30 cm tall stem in summer. ***S. kosaninii*** from NW Yugoslavia bears rosettes 4-8 cm wide, dark green tipped-red, ciliate and glandular hairy. Green and white flowers with red anthers are produced on 15-20 cm tall stems. ***S. marmoreum*** (syn. *S. schlehanii*) from E. Europe and the Balkans is a very variable species but similar to *S. tectorum* though young leaves are hairy. ***S. montanum*** from Pyrenees, Corsica, Alps and Carpathians forms 2.0-4.5 cm across rosetted mat with 10 cm spread and growing 10-15 cm tall, bearing dark green, glandular hairy and fleshy leaves. It is a very variable species and a parent of many hybrids. Star-shaped violet-purple flowers are borne on terminal clusters on a 10-15 cm tall stem in summer. ***S. nevadense*** from Sierra Nevada and Spain bears rosettes 2.5-3.5 cm wide, ciliate, pinkish in summer and scarlet in winter. Deep pink flowers appear on 8-13 cm tall stems. ***S. octopodes*** from SW Macedonia (formerly a constituent republic of Yugoslavia) is closely allied to *S. ciliosum*. It grows up to 2.5 cm high, 1.0-2.5 cm wide and with about 20-25 cm spread. Hairy rosettes are pale to mid-green and rarely tipped-maroon, densely glandular downy and ciliate. New rosettes are formed at the end of quite long (7 cm) slender thread-like stolons. Yellow flowers flushed red at the base and some 1.9 cm across appear in June and July on a 15 cm tall stem. ***S. tectorum*** ('common or roof houseleek') ranges from the Pyrenees across the Alpine system to the Apennines and the northern Balkan region, as far north as Ireland and Scandinavia and as far east as Iran. It grows from 5.0-7.5 cm in height and some 30 cm in spread with width of rosette 5.0-18.0 cm. The leaves are bright to mid-green, often tipped-maroon, and white ciliate. Floral stalk is from 30 to 50 cm bearing rose-purple flowers with greenish centre and orange-brown anthers, having 2.5-6.0 cm across dimension and emerge during rains. Everard and Morley (1970) from Alps recorded *S. arvernense* and *S. rupestre* apart from certain other species.

Senecio (Asteraceae)

It takes its name from the Latin *senex* meaning 'an old man', obliquely referring to the hoary pappus. *Senecio* with about 3,000 species distributed throughout most part of the world, is the largest genus of flowering plants including half-hardy and tender annuals and biennials, hardy herbaceous and tender succulent perennials as well as evergreen trees or shrubs. The evergreen shrubby species which include climbers are hardy, half-hardy or tender. Some of the shrubby species are now referred to the genus *Brachyglottis*. They have alternate leaves in a wide variety of forms and normally terminal loose clusters of daisy-like flower-heads often in shades of yellow, at occasions without any ray floret. Certain most common succulent species are being described here under. ***Senecio articulatus*** (syn. *Kleinia articulatus*; common name 'candle plant') from South Africa is a succulent shrub up to 60 cm tall with grey-green, thickened and cylindrical bloom-coated stems divided with marked joints and 1.5-2.0 cm across. Leaves that appear on the crown in winter only on young stems are short-lived, deeply 3- to 5-lobed and up to 5 cm long. Yellowish-white flower-heads appear some 1.5 cm long.These are propagated through removal of the stem sections which sometimes even fall down. ***S. citriformis*** from Cape (South Africa) is a dwarf shrubby succulent perennial forming the radiating tufts of procumbent and fairly fleshy stems up to 10 cm long and up to 30 cm spread which normally branches out from a central rootstock. The stems are covered with almost sessile, erect, spindle-shaped and succulent blue-grey leaves some 1.5-2.0 cm long which are spirally arranged on the stems and are shaped like lemons. Leaves have a waxy farina and numerous vertical translucent markings. Yellow-white flowers are produced on the 15 cm long flowering stems during December and January. ***S. fulgens*** (syn. *Kleinia fulgens*) from Natal (South Africa) is a sub-shrubby succulent with tuberous base and grows up to 60 cm in height. Leaves are pale glaucous-green, obovate to spathulate and 5-10 cm long. Orange to red terminal flowers are solitary or in pairs and some 2.5 cm long which are borne on 20-30 cm long stalks. ***S. haworthii*** (syn. *Kleinia haworthii, K. tomentosa*) from South Africa grows erect to a height and then spreads to about 60 cm. Robust stems at first emerge solitary, later on branching sparingly but from the base. The blue-grey succulent leaves are almost erect, cylindrical but tapering slightly at both the ends, 2-4 cm long, little arching inward and highly crowded. Leaves and stems both are densely and completely covered with silver-white woolly hairs. Orange-yellow solitary flower-heads are borne in July. ***S. herreianus*** (syn. *Kleinia gomphophylla, K.*

herreiana; common name 'lady's necklace') from Buchu mountains of Namibia is a beautiful prostrate-growing leaf succulent most suitable for hanging basket as its stems are not strong enough to grow erect. Similar to it is better known *S. rowleyanus* where leaves are spherical while in this case these are pointed-globular and marked with dark green and fine longitudinal lines, and are borne at intervals along the stems. At the most it grows straight from 5-8 cm, then its stems are hanging down to a length of more than 60 cm. It also spreads up to 60 cm when left in the ground to grow. Its stems and leaves are glaucous-green. Its flowers are insignificant though it rarely flowers. ***S. kleinia*** from Canary Islands is an erect growing, thick, segmented, poorly branched and succulent shrub which in containers attains something more than 1 m in height though in the wild it grows up to 3 m. Leaves are grey-green, lanceolate, fleshy, 6-15 cm in length, appearing in autumn and falling in the beginning of summer. Cream flower-heads in terminal corymbs of 30 or more and some 2 cm long are borne in winter or spring. ***S. macroglossus*** (Cape ivy, Natal ivy, wax vine) from South Africa is a semi-succulent evergreen twiner growing more than 3 m in length. Initially its stems are succulent but with age become woody. Leaves are dark glossy-green, triangular-hastate, up to 6 cm long, 3-5 pointed lobes and fleshy. Solitary and terminal flower-heads bear loose daisy-like florets in winter where ray florets are white, disc yellow, and measure about 5-6 cm across. *S. m.* 'Variegatus' (Cape ivy/Wax vine) bears waxy succulent leaves blotched yellow with irregularly wide-yellow margins. ***S. mikanioides*** (syn. *Delairea odorata*; 'German ivy/Parlour ivy) from South Africa is semi-woody evergreen climber with twining stems, growing up to 3 m bearing bright green, 6 cm long, rounded with widely cordate base and fleshy leaves which have 5-7 radiating, broad and pointed lobes. Large clusters of small and yellow flower-heads appear in axillary and terminal corymbs during autumn-winter. ***S. radicans*** (syn. *Kleinia radicans*) from South Africa is prostrate-growing leafy succulent which grows up to 10 cm high with 60 cm spread in the form of a mat. The stems are slender and much-branched, bearing numerous glaucous-green, almost erect and spherical with pointed tips, some 2-3 cm long and small grape-like leaves, and each leaf is marked with darker, translucent and longitudinal stripe down the middle. White, tèrminal and solitary or paired flower-heads emerge on short stalks in December though it rarely flowers in cultivation. ***S. rowleyanus*** (string-of-beads) from SW Africa is a mat-forming leaf succulent with slender creeping stems rooting freely and growing up to 90 cm long. The leaves are glaucous-green, 0.5-1.2 cm across, globose or grape-shaped, each leaf with a vertical translucent band and minutely pointed tip. Solitary and sweet-scented flower-heads up to 5 cm long comprise of long white disc florets with purple stamen tubes and stigmas, in shape similar to those of tiny shaving brushes, are borne from September to November. These are most suitable as carpeting or basket plant. ***S. serpens*** (syn. *S. repens, S. succulenta, Kleinia repens*) from Cape (South Africa) is a mat-forming and base-branched shrubby succulent with trailing and then erect fleshy stems up to 30 cm high and with about 60 cm spread, most suitable for hanging baskets. Glaucous-blue, linear to lanceolate and almost cylindrical leaves some 4 cm in length with bluish waxy patina, spirally arranged on the stem, tapering to each end, chanelled above and form loose rosettes at the stem tips. White to cream flower-heads in small cymes are borne in July and August. ***S. stapeliiformis*** (syn. *S. stapeliaeformis, Kleinia gregorii, K. stapeliiformis*) from Cape (South Africa) is a succulent shrub growing about 25 cm in height, bearing grayish-green, base-branching, cylindrical and 5-7-ribbed stems, each rib with a row of dark green cushion-like leaf base. At younger stage these leaf bases terminate into the minute lance-shaped leaves which wither soon. Solitary, red and brush-like flower-heads some 2.5-4.0 cm across are borne on long slender stalks.

Sesamothamnus (Pedaliaceae)

This is a caudiciform shrub from Africa with only a few species, bearing small deciduous leaves. White to yellow flowers, sometimes with a blush of pink appear when the plants are leafless. Its seeds are sometimes winged. ***S. lugardae*** from SW Africa has fleshy stems bearing small tan spines. In the wild it has been recorded forming 2 m thick caudex. This bears felted grey-green leaves that fall up in autumn. It is propagated through seeds and cuttings.

Sesuvium (Aizoaceae)

It is of unknown derivation. There are some 13 species under the genus, widespread in the temperate and tropical regions of the world. The plants are herbaceous or sub-shrubby, branched, erect or prostrate,and succulent. Leaves are opposite, fleshy and linear or oblong. Flowers are solitary, pedunculate or sessile, bractless or with two bracts, clustered or rarely subcoymbose, pink to purple, calyx 5-lobed, oblong and obtuse with coloured inside, petalless and stamens 5. ***S. portulacastrum*** from the tropical regions of both hemispheres is a diffuse procumbent or prostrate herb with entire plant colour being blood-red or purple. Leaves are succulent, linear to oblanceolate or obovate-oblong and much narrowed at the base. Axillary purplish or rose flowers with calyx deeply 5-lobed appear from the plant.

Setcreasea (Commelinaceae)

Setcreasea is a name of unknown derivation. It is a genus of six species of tender perennials allied to *Tradescantia*, that is mainly grown for their decorative foliage. It is a native of USA (Texas) to Mexico. These are tufted to clump-forming and little fleshy with alternate ovate to oblong leaves and 3-petalled flowers borne in the boat-shaped bracts. They are grown mainly as indoor decorative plants. They are propagated by division of clumps or by cuttings. Only one species, *i.e.* ***Setcreasea pallida*** from Mexico can be considered as a succulent as its leaves are sufficiently fleshy. It is a erect to trailing plant with slender stems some 40 cm long, bearing fleshy, purple and oblong leaves up to 15 cm long, and mauve to lavender-pink clustered flowers some 2 cm across at stem ends

Seyrigia (Cucurbitaceae)

A genus from Madagascar has four species with tuberous roots, succulent and leafless vining stems coming out from the top of the tuber in clustered form though radiating all around. Stems may be round or square and smooth or covered with fine white wool. The flowers about 2-3 mm wide in female plants appear singly from the nodes and in the males in trusses from a short spur. This tolerates bright light and moderate heat but no frost. ***S. humbertii*** is very handsome and stout in the genus and forms potato-like tubers in masses which are attached with the plant with thin stolons. White felt on the square stem is about 1 cm thick. Stems grow about 30 cm every season. It tolerates intense heat and indirect bright sunlight but not the hot sun. It is propagated through stem cuttings and rarely through seeds.

Stapelia (Asclepiadaceae)

Stapelia honours Johannes Bodaeus van Stapel (*d. circa* 1636), a Dutch physician. This genus consists of some 90 species of perennial succulents found throughout tropical and southern Africa. The plant branches out freely from the base forming a dense cluster which spreads outwards, bearing thickened stems appearing as cacti and are coarsely toothed, though sometimes these have spine-tipped scale leaves. Star-shaped, hairy and 5-lobed, and brilliantly coloured flowers originate from the base of the plant during late summer and early autumn but have a dull unpleasant smell to attract the pollinating flies including blowflies. Centre of the flower is comprised of a 2-whorled corona, the outer five segments flat and spreading and the inner five horn-like erect. They are propagated through division when repotting, cuttings and seeds. The cuttings before insertion in the media are dried for several days, and the seeds germinate within 24 hours. ***Stapelia flavirostris*** from South Africa is a clump-forming succulent bearing green, rectangular, indented and hairy stems some 20 cm high with indefinite spread. Purple-brown flat flowers are some 10 cm across, having lighter transverse banding and ridging decorated with white or purple hairs, being more conspicuous on the margins. Each lobe is almost rectangular and the lobed portion is less than half of the upper part of the flower. ***S. gigantea*** from Cape and Natal (South Africa) and Zimbabwe bears the largest flower in the family, *i.e.* corolla diameter being from 15-46 cm of pale-yellow colour, formed with fine transverse crimson wrinkles, and divided into large ovate lobes with tail-like tips. The upper surface of the flower is clothed with dense and shaggy purple hairs and margins with white hairs, making the surface velvety but flowers are foul-smelling as carrion. Stems are ascending, pale-green, angles wing-like and toothed, and some 15-20 cm long. ***S. grandiflora*** from Cape (South Africa) is clump-forming succulent growing up to 30 cm high. Stems are green, quadrangular, 7-10 cm thick, convex, and covered having fine hairs with edges being compressed and toothed. Many baloon-shaped swollen pinkish-green and short-stalked pointed buds appear from the base of new shoots, one from each. Purple corolla lobes are ovate and tail-like pointed, with transverse stripes, and thickly covered with soft grey hairs. ***S. hirsuta*** (hairy toad plant) from South Africa is rapid grower and produces 20-25 cm long, slender and bronze-green stems covered with short and soft hairs. Flowers are borne in August, are 1-3, yellowish-red with yellow base, marked with red transverse lines and ornamented with brown-red hairs, and 10-20 cm across, with slender and pointed lobes. ***S. leendertziae*** from South Africa is a fast-growing species where stems are 4-angled, 10 cm tall, 1.5 cm thick, base-branched and where a small piece can grow up to 30 cm in a season but is easily contained in a shallow hanging basket. Its newest growth produces flesh-coloured tubular flowers with 4 cm wide tubes and petals more than 10 cm long, petals unite to their half length. It also has a cristate form growing easily but does not flower. It tolerates intense heat but prefers partial shade and frost-protection. It is propagated through cuttings or seeds. ***S. nobilis*** from South Africa and Mozambique resembles *S. gigantea* but bears darker yellow smaller flowers (20-25 cm across), a deeper throat, more purple hairs and more compact stems. ***S. pillansii*** from South Africa is a 15 cm tall vigorous growing succulent with dark green velvety-haired stems. Deep purple-black flowers some 15 cm across and covered with dark hairs are produced in August. ***S. plantii*** from Cape (South Africa) produces minutely pubescent erect stems some 15-20 cm long where angles are compressed with setting of small well-spaced teeth. Brownish-purple flowers some 10-13 cm across, with narrow transverse and yellow wrinkles, ornamented with long purple hairs on the margins, lobed halfway, each lobe a little reflexed, ovate-lanceolate and taper-tipped. ***S. variegata*** (syn. *Orbea variegata*; common name 'carrion flower') from South Africa is dense clump-forming and branching succulent bearing grey-green to coloured-green, rectangular, ascending and indented stems with erect teeth, some 5-10 cm long that are mottled reddish-purple under sun, and is most commonly grown species. Freely produced flowers 1-5 together, smooth-surfaced, 5-8 cm across, lobes broadly triangular-ovate, wrinkled transversely, yellow, marked with reddish-brown spots and banding though highly variable, and foul-smelling are borne in August.

Stapelianthus (Asclepiadaceae)

The members of this small genus from Madagascar are creeping stapeliads which have difficulty in their growing. They are slow-growing and produce urn-shaped flowers with a small opening. Their requirement for light and temperature is moderate. They are propagated through seeds or cuttings. ***S. neronis*** forms a low compact clump with soft and purple stems which are 4 cm long and 1.5 cm wide. During autumn, urn-shaped flowers appear along the outer edges of the cluster where a small pore is present at the tip. Petal interior is white and exterior is velvety purple. For its cultivation, it requires moderate warmth and indirect bright sunlight.

Stephania (Menispermaceae)

A native to Asia, Africa and Australia, the genus comprises of 26 species, most of them being climbers. Mild climate with semi-shaded condition is suitable for its growing. It is not more demanding and does well in coarse soil with low organic matter and moderate watering that too during active growth period. It is propagated through seeds and stem cuttings. ***Stephania perrieri*** possesses quite thick caudex and palmate leaves though ***S. rotunda*** has largest caudex up to 2.8 m in diameter which may weigh more than 215 kg. Its climbing stems are slender, pale-green and several metres long, the leaves are green, thin with wavy edges, peltate, more or less round and petioles long, and yellow minute flowers are dioecious and clustered.

Streptocarpus (Gesneriaceae; common name 'Cape primrose')

Streptocarpus derives from the Greek *streptos* meaning 'twisted' and *karpos* 'a fruit'. A genus of 132 species of evergreen, tufted or sub-shrubby tender plants, often with foxglove-like flowers, from tropical to southern Africa and Malagasy, has three types of growth forms, such as tufted or clump-forming with basal leaves which can be propagated by leaf or leaf section cuttings, shrub-like with pairs of small leaves on branched stems which can be propagated through stem cuttings, or unifoliate ~ a curious development with each plant producing one quite enlarged cotyledon and occasionally a few small basal leaves throughout their lives and the plant usually dies after the flowering (monocarpic). Unlike majority of dicotyledons whose young shoots end in 2 seed-leaves, followed by others that gradually come to resemble adult leaves, a number of streptocarps, though strictly 2-leaved, develop one at the expense of the other, and this single enlarged seed-leaf when matures sometimes exceeds 76 cm in length. All the types can be propagated through seeds. The flowers are tubular and funnel-shaped with five rounded spreading lobes, followed by pod-like, spirally-twisted and slender capsules. Only a few species are succulent as described here with. ***Streptocarpus holstii*** from East Africa is allied and similar to *S. caulescens* in general habit including height. It is shrub-like erect branching plant growing to 75 cm tall and 45 cm spread with more slender succulent stems though swollen at nodes. Leaves are dark green, up to 6 cm long, opposite, hairy with prominent veins on the undersurfaces. Flowers that appear from June to September in axillary cymes are 2.0-2.5 cm long with slender tube, violet or white with violet sriping and white throat. ***G. saxorum*** (false African violet) from East Africa (Tanzania) bears succulent stems and was first described in 1893 from material found on rocks by C. Holst in the Usambara Mountains near Lutindi. It grows up to 30 cm in height and 45 cm in spread. It is a semi-prostrate sub-shrubby succulent with dark to mid-green, elliptic to ovate, some 3 cm long, softly hairy and fleshy leaves that appear in pairs or whorls of three. Lilac flowers some 3-4 cm across with white mouth and tube are borne singly or in pairs on slender stalks from the upper leaf axils between April and October. Three lower lobes of the flower are larger than the upper two.

Talinum (Portulacaceae)

Though its derivation is exactly not known but *Talinum* is probably derived from an African (Senegal) vernacular name. This summer-flowering genus has been recorded mainly from Mexico and North America, and also from South America, Africa and Asia and are useful for planting in the rock gardens, troughs and alpine houses and as pot plants. It is a genus of about 50 species of succulent annuals, perennials and sub-shrubs, sometimes becoming woody at the base with age. Leaves are simple, alternate to almost opposite, cylindrical or flat but in some species aggregated into terminal clusters or rosettes. Flowers are small, short-lived but appearing in succession, in terminal cymes, racemes or panicles and rarely solitary, axillary or lateral, sepals 2, petals 5 and ephemeral, stamens 5 or more, ovary many-ovuled and capsules globose or ovoid and 3-valved. They are propagated by seeds or by cuttings. The cultivated succulent species are described here with. ***Talinum aurantiacum*** from New Mexico and Texas and south Mexico forms 10 cm long solitary or branched underground tubers and the thick annual stems about 15 cm tall which may be sparingly branched. Yellow self-fertile flowers 1.5 cm across are produced in the hot afternoon from mid- to late-summer and remain open for several hours. It is propagated from seeds or cuttings. ***T. guadalupense*** from Guadalupe Island (Mexico) has fleshy globose to cylindrical roots; thickened and knotted stems growing to 60 cm in height; and blue-green margined-red, spathulate, fleshy and some 5 cm long leaves in terminal clusters or rosettes. Pink flowers some 2.0-2.5 cm across appear in panicles well above the leaves with 3-4 days field life. ***T. okanoganense*** is low-growing (4 cm high), cushion- or mat-forming prostrate perennial spreading up to 10 cm. Stems are succulent and bear tufts of greysih-green, cylindrical and succulent leaves and small cup-shaped white flowers. ***T. paniculatum*** (syn. *T. patens*) from southern USA to Central America bears tuberous roots and one to several green and succulent stems growing up to 60 cm in height, and bearing 5-10 cm long, mostly opposite, oval and abrupt basal-tapering succulent leaves clustered at the stem tips. Rose-red or sometimes yellow flowers some 1.2 cm across are borne on leafless panicles some 25 cm tall. ***T. triangulare*** from West Indies, Brazil and Peru bears alternate and obovate-lanceolate leaves. Red or white flowers in corymbose cymes are borne with triangular pedicels.

Tavaresia (Asclepiadaceae)

A native of south and SW Africa, the stapeliads belonging to this genus have erect stems patterned with many ribs which are covered with small brittle spines. Morphologically, these appear as cactus. Cream coloured funnel-shaped flowers open widely and have maroon markings. They tolerate fairly high temperatures at a shaded place and are propagated through cuttings or seeds. ***T. barklyi*** from South Africa now includes *T. grandiflora*. This clump-forming species with a spread of about 15 cm in a few seasons produces light green stems growing to 10 cm high and 2 cm wide. In autumn, flowers of varying sizes, preferably 10 cm long and 7 cm wide appear. It tolerates intense heat and indirect bright sunlight but are susceptible to frost.

Tillandsia (Bromeliaceae)

See chapter on '**Bromeliads**'.

Titanopsis (Aizoaceae)

Its derivation is from *titan* 'the Greek sungod' and *opsis* meaning 'like', due to its yellow flowers. This genus originated in South Africa, mainly Cape and comprises 6-8 succulent species which mimic limestone rock of its homeland. It is basal-rosetted forming small but dense clumps and when mature become hummock-forming. Fleshy, erect, 2-3 cm long and triangular leaves are borne in 6-8 opposite pairs and are narrow at stems. Each leaf is spathulate with triangular tip densely covered with white to grey-white tubercles of irregular size. Usually large and attractive yellow flowers are borne in autumn. Its two of the species such as *T. crassipes* and *T. setifera* are described under *Aloinopsis* as *A. spathulata* and *A. setifera*, respectively. ***T. calcarea*** from Cape is a clump-forming succulent growing not more than 3 cm in height with a basal rosette of leaves. Leaves are crusty and white blotched similar to the limestone substrates on which the plants grow, blue-grey, 2.5 cm long, tips bluntly and broadly triangular and the wart-like grey-white tubercles are sometimes tinged faintly bluish. Golden yellow to orange flowers some 2 cm across are borne from autumn to spring. ***T. fulleri*** from Cape is quite similar to *T. calcarea* but the size of the leaves is about 2 cm in length with brownish-grey tubercles on the purplish background. Here flowers are dark yellow and some 1.5 cm across. ***T. shwantesii*** from Cape is a clump-forming perennial succulent growing 3 cm high and with 10 cm spread. Leaves are in a basal rosette, grey-blue, triangular with rounded corners and are covered with small, wart-like and yellow-brown tubercles. Light yellow daisy-like flowers some 2 cm across are borne in summer-autumn.

Tomatriche (Asclepiadaceae)

A small genus from South Africa, produces sparsely branched long stems that bear brown to purple flowers having reflexed petals. They require frost-protection but can tolerate intense heat and indirect bright sunlight. These are propagated through seeds and cuttings. ***T. revoluta bicolor*** produces more than 25 cm long and 1.5 cm thick stems which from near end and laterally produce 2.5 cm wide dark purple and mildly carrion-scented flowers with apple-green centre having back-recurved petals where edges have purple hairs. It tolerates intense heat and bright sunlight but not the frost.

Trachyandra (Asphodelaceae)

A native to East Africa (southern and tropical, and western Cape) and Madagascar, they are geophytic succulent perennials with rhizomes or tubers, as well as shrublets. Their green leaves are mostly long and twisted as in case of *T. tortilis*, fleshy and sometimes hairy. The flowers are short-lived, white and star-shaped with 6 petals and stamens.

Tradescantia **(Commelinaceae; common name 'spiderworts')**

Tradescantia was named for John Tradescent (1608-1662), gardener to King Charles I. It is an evergreen or deciduous perennial, erect and bushy to trailing genus, and rooting at the nodes, from North and South America (Manitoba to Argentina) with 20-60 species (as per classification followed). Its stems are simple or diffusely branched, bearing entire, alternate and somewhat fleshy leaves, and 3-petalled flowers basically triangular in outline emerge out from the leaf- or spathe-like bracts with bearded stamens in terminal clusters. They are grown for their beautiful foliage or flowers. These are propagated through division and tip-cuttings. ***Tradescantia albiflora*** (wandering jew) from South America is a mat-forming foliage species with trailing, about 60 cm long and succulent stems which is quite suitable for hanging baskets. Leaves are stemless, glossy light green, broadly ovate, and some 3-5 cm long. Flowers are white and some 0.8 cm across. There are several forms and varieties of this species with variegated leaves, especially striped with cream such as 'Albovittata', 'Variegata', *etc.* ***T. blossfeldiana*** (syn. *T. cerinthoides*) from Argentina is an evergreen semi-erect to creeping perennial with 5 cm height and indefinite spread where stems are fleshy, purple and grow some 20 cm long, Leaves are narrowly dark glossy green above and purple with long white hairs below, oblong-elliptic, stem-clasping and some 10 cm long. Compact tiny pink flowers some 1.2-1.8 cm across with white centres surrounded by two leaf-like bracts are borne out intermittently from March to July on mature plants in terminal clusters. Its form 'Variegata' (Flowering inch plant) is striped with cream. ***T. fluminensis*** (syn. *T. fluviatilis*; common name 'Wandering Jew') from South America is similar to *T. albiflora* but differs in having trailing stems which grow about 60 cm in length and bearing almost glossy bluish-green leaves above and purple-flushed below, elliptic-ovate and longer on short leaf stalks. Under continuous bright light the leaf's underside turn pale-purple. Sometimes it is confused with *T. zebrina*. Its form 'Variegata' (Inch plant/Wandering Jew) bears regular striping of leaves in white and cream. ***T. geniculata*** (*Gibasis geniculata*) from tropical America grows up to 45 cm long with prostrate to decumbent stems, bearing softly hairy leaves some 10 cm long which are lanceolate to narrowly ovate. Inflorescence some 22 cm long in the form of erect leafy stems appear in summer with small white flowers in loose terminal panicles. *T.* ***navicularis*** (syn. *Callisia navicularis*) is from Mexico or Peru. *Callisia* derives from Greek *kallos* meaning 'beauty'. As a genus, *Callisia* comprises 12 species but except *C. navicularis*, others are not so fleshy. Christened with separate ranking after getting separated from *Tradescantia*, the *Callisia navicularis* was retained with *Tradescantia* itself. It is a tufted to mat-forming prostrate evergreen perennial with 15 cm long stems, bearing grey-green, ciliate, thickly fleshy, ovate to oblong-lanceolate (boat-shaped), distichous, and crowded leaves with clasping bases, most suitable for hanging baskets apart from pot culture. Clustered but stalkless flowers some 1.2-2.0 cm across arise in small cymes from the axils of leaves or the similar bracts during summer though sometimes even at other times. ***T. sillamontana*** (*T. pexata, T. velutina, Cyanotis sillamontana, C. veldthoutiana*) from Mexico grows with semi-erect and branched stems some 30 cm tall, and covered with dense white bloom. Leaves are crowded on the stem, elliptic to ovate and some 5-7 cm long. Bright magenta-rose flowers some 2 cm across are borne during summer. ***T. spathacea*** (syn. *Rhoeo discolor, R. spathacea*; common names 'boat lily', 'moses-in-a-boat') is from Mexico, Guatemala and the West Indies. As *Rhoeo discolor*, it is only one species under the genus *Rhoeo* which is of unknown derivation. Stems are solitary or base-branching, erect but often becoming decumbent, and 20 cm long. Leaves are dark glossy green above and purplish below, linear to strap-shaped, 30 cm long, succulent and overlapping at the base. White 3-petalled flowers some 1.5 cm across appear from deeply boat-shaped bracts intermittently all the year round. Its form 'Variegata' (syn. Vittata, common name 'moses-in-the-cradle') bears longitudinally pale-yellow striped leaves with purple underside.

Trichocaulon **(Asclepiadaceae)**

Though not sure but it is thought that this is defunct genus as those having stiff raised ribs or stiff tubercles tipped with short stiff spines are now considered the members of *Hoodia*, and the group with soft, spineless and tuberculate stems have been shifted to couple of genera including the genus *Lavrania* but yet no consensus on the issue. These are spineless species which cluster rarely and throws 15 cm long and 5 cm thick stems, bearing numerous 1 cm across flowers on tips of the stem. The flowers are normally yellow with maroon dots or stripes, with almost darker centre. They can be propagated through seeds and then grafted. These tolerate indirect bright sunlight and well-drained potting compost but not the intense heat or frost.***T. cactiforme*** produces soft tuberculate stems some 10 cm high and 5 cm wide which rarely clusters at the base. Its flowers are yellow, dotted and banded maroon.

Trichodiadema (Aizoaceae)

Trichodiadema derives from the Greek *trichos* meaning 'a hair' and *diadema* meaning 'a diadem' or 'crown', referring to the hairy crown, *i.e.* the curious tuft of bristles found on each leaf tip of this genus. A genus of 30 species from Cape (South Africa) of succulents formerly put under *Mesembryanthemum*. They are tufted and mat- or hummock-forming. Its rootstock is woody with tuberous roots, and the stems branched, tubercled and warted, and of chocolate colour with about 10-15 cm length. Narrow and spindle-shaped cylindrical to semi-cylindrical succulent green leaves in opposite pairs are formed in the upper part of the body. Daisy-like and mostly showy flowers open from autumn to spring only when the weather is sunny. They are propagated through seeds or cuttings. ***T. bulbosum*** is a low-growing, rounded and heavily branched shrublet making natural succulent bonsai, producing some 5 mm long light green leaves crowned with soft white spines. Cuttings or seedlings take several years to develop several thickened roots. It flowers almost freely during summer producing 2 cm across flowers even on small cuttings. It tolerates intense heat as well as up to -4 °C temperature for a brief period and likes morning sun. ***T. densum*** has thickly-fleshy roots, and short, green and prostrate stems form the caudex. Stems are up to 10 cm high and 20 cm in spread forming rounded tufts, and greysish-green cylindrical and paired leaves with each being 1.2-2.0 cm long and tipped with clusters of white bristles. Carmine to purple-red and daisy-like flowers some 3-5 cm across are borne on stem tips during summer. ***T. intonsum*** is loosely tufted with well separated semi-cylindrical, outcurving and densely papillose leaf pairs tipped with short, erect and dark brown bristles, and each leaf measuring about 1.2 cm long. Pink or white flowers some 2 cm across are borne on stem tips during summer. ***T. mirabile*** is a bushy, prostrate and hummock-forming succulent growing some 15 cm high with 30 cm spread. Stems are about 8 cm in height bearing white bristles. Pairs of the dark green leaves some 1.2-2.5 cm long, are papillose and unite at their bases, a little flattened above and tipped with dark brown erect bristles. Pure white flowers some 4 cm long appear on stem tips from spring to summer. ***T. stellatum*** is quite similar to *Drosanthemum* except that former has bristly leaves and plumed stigmas. Sometimes it confuses with a cactus but cactus stems are succulent and bear spiny leaves though in this case the leaves are succulent. Its roots are fleshy and the woody rootstock is hummock-forming, branched, tubercled and warted, of chocolate colour with about 5-15 cm in height and 6 mm thick. Light green spindle-shaped succulent leaves some 1.0-1.3 cm long in trusses of 3-10 appear from a point of stem in the upper part of the body but tapering to both the ends and bearing about a dozen of soft hairs on the tips. Leaves are covered with dense white spots all around. Red flowers some 2.8-3.0 cm across with yellow eye emerge from the plant during summer.

Tylecodon (Crassulaceae)

A native of Namibia to South Africa, this genus was carved out of *Cotyledon* having caudiciform or tuberous stems, succulent deciduous leaves and flowers being borne upright than being pendulous, and many of these species are worth making handsome succulent bonsai specimens. Being winter-growers, these should be provided bright light and warmth during cold periods as for a brief period these can tolerate 0 °C night but the day temperatures should be around 27 °C. ***T. paniculata*** from South Africa is the fastest growing and tallest (2 m tall and 50 cm thick) species in the genus with main trunk being branched and rebranched, and this is quite easy to cultivate. During autumn it produces handsome succulent leaves measuring 6 cm long and 3 cm wide. It tolerates intense heat, bright indirect sunlight and light frost. It is propagated through cuttings and rarely from seeds.

Uncarina (Pedaliaceae)

The genus mostly endemic to Madagascar comprises of several species which in the wild form much-branched trees and shrubs growing up to 7 m tall, bearing large, green, deciduous and short felted leaves and a thickened base. In a season the stems can grow up to 2 m in length. Though these require frost-protection and well-drained soil but frequent watering, and can tolerate intense heat and sun. During summer yellow to orange flowers are borne, which is followed by flattened pointed pod armed with many barbed hooks. *U. decaryi* becomes a sizeable shrub in a few seasons with white-barked tapering trunk some 1.5 cm thick which is thickened at base and topped with light green and felted leaves. These prefer winter-dryness though are susceptible to 0 °C temperature. It is propagated through seeds or cuttings.

Veltheimia (Asparagaceae)

A native to South Africa, this genus comprises of two species ~ *V. bracteata* (eastern part) and *V. capensis* (drier western part). This is a bulbous plant with basal rosette of succulent, bright green and oblanceolate leaves and a dense raceme of tubular flowers.

Xanthorrhoea (Blackboy, Grass trees; Xanthorrhocaceae)

See chapter on '**Foliage Plants**'.

Xerosicyos (Cucurbitaceae)

Its derivation is unknown. There are some 4 climbing perennial species in this genus from Madagascar. Stems may be smooth to little hairy bearing usually bifid tendrils. Leaves are short-petioled, entire, alternate, suborbicular to oval and thickly fleshy. Flowers are unisexual. Male flowers small green-yellow in axillary umbelliform clusters, calyx short cup-shaped, 4-lobed and lanceolate, petals 4, glabrous, oval-acuminate and alternate to the 4 stamens, stamens equal and free and anthers reniform. Female flowers in loose panicles or in umbelliform clusters, calyx long-obconic, staminodes 4, ovary unilocular, ovules 4, styles 2, stigma bi-lobed, fruit obconic and seeds oblong, compressed and winged. ***X. dangui*** bears glabrous and striate stems. Leaf petiole 1.5 cm long, leaves glabrous, glaucous above, sub-orbicular, 3.5-5.5 cm long, 2.5-5.0 cm wide and thickly fleshy. Male flowers yellow-green, numerous, slightly shorter than female, calyx lobes 1-2 mm, petals 3 mm long, 1 mm wide, glabrous and recurved. ***X. decaryi*** is similar to *X. dangui* but leaf petioles are 1-2 mm long and leaves light green, glabrous, oblong-elliptic, rounded at apex, 2.5 cm long and 1 cm wide. ***X. perrieri*** is a tall climber with stems being slightly woody at base. Leaf petioles are 1.0-12.5 cm long, and leaves light green, glabrous, sub-orbicular, thick, 1.8-2.0 cm long and 1.6-1.8 cm wide. Floral pedicels are 2-3 cm long, male flowers slightly smaller than female, 3-20 and pale green-yellow, calyx lobes 3 mm long and 1 mm wide, and petals glabrous, 1.0 cm long and 6-7 mm wide. Fruit is yellow-brown, 2.5 cm long, 2 cm wide and the seeds 1.0 cm long and 7 mm wide.

Yucca (Agavaceae)

A large xerophytic genus with 40 species from North America where most species are not succulent. These range from more or less stemless rosettes of thin and long leaves to tree-like species with sizeable trunks. They make clustered rosettes, with a few exceptions where solitary rosette is formed, when reach to flowering size. After flowering the rosettes remain alive and not like *Agave* where these die, and continue flowering year after year or as in some species every two years. White flowers which are usually waxy appear on an upright stalk though sometimes the stalk is pendulous. Most species tolerate intense heat and sun but there are species which tolerate even substantial cold. They are propagated by dividing the clusters or by seeds. ***Y. carnerosana*** from Texas and Mexico grows up to 10 m tall and is most appealing plant for landscaping. It flowers only after attaining 1 m height, the flower spike is roughly 1.5 m long which appears every two years. It can tolerate -12 °C temperature for a brief period. It is propagated through seeds. ***Y. rigida*** from Texas and Mexico forms about 2 m tall trunk topped with handsome rosettes of buff-blue leaves which are about 1 m long. It is most suitable as a landscape or container plant. Its cultural requirements are the same as in *Y. carnerosana*.

Zamiacaulcas (Araceae)

A native of Africa, these odd succulents are the ideal houseplant thriving in low or bright light and tolerate moisture and drought both. They form a large underground tuber which from the apex produces the long green leaves which mimic *Zamia* and from top the roots. They are propagated from leaves as cutting with one leaflet produces plantlets. ***Z. zamiafolia*** is one of the most common species in cultivation. Its tubers attain 15 cm across and throws out roots from the top and 40 cm long and 10 cm wide leaves from the apex. It survives well in low light and cooler temperatures and tolerates intense heat and bright indirect sunlight. It is propagated from leaves.

Zygophyllum (Zygophyllaceae)

Its common name is 'caper bean'. In Greek *zygo* means 'yoke' and *phyllon* means 'leaf', referring to the paired leaflets. This genus comprises some 80 species from Mditerranean to Central Asia, South Africa and Australia. The members of this genus are herbs or shrubs bearing fleshy branches which may sometimes be jointed. The stems bear opposite leaves which may sometimes be pinnatifid and stipulate. Leaflets and stipules are usually 2 each. Stipules are often spiny and membranous. Axillary flowers are borne solitary or rarely in pairs with 4-5 somewhat fleshy, and rarely basal-connate sepals. Overlapping petals, sometimes clawed, are 4-5. Floral disc 8-10 angled is fleshy with 8-10 disc-base-inserted stamens, which normally exceed the petals. Ovary is superior, 4-5 angled and taper into style, bearing minute stigma. Fruit is a 4-5 locular angular or winged capsule with generally 1 seed in each locule and the seeds are crustaceous. ***Zygophyllum album*** from Arabia and N. & C. Africa produces procumbent, papillose or hoary stems growing up to 40 cm long with irregular branching. Leaf petioles are 4-10 mm long and fleshy, the leaflets 1 pair, cobwebby, cylindrical to ellipsoid and fleshy. White flowers are solitary with 3-6 mm long pedicel, petals are rounded and clawed and the fruits are 5-angled. ***Z. fabago*** (Syrian bean caper) from Syria to Afghanistan is a deep fleshy-rooted perennial succulent, growing to about 1 m. Leaflets 1 pair, flat, smooth and obovate-orbicular to elliptic. Rachis exceeds the leaflets. Flowers with erect pedicels, solitary and yellow having coppery brick-red at base appear during summer. Fruits are oblong-cylindrical and 2-3 cm long.

Cultural Practices

Soil mixes containing greater amount of organic material, *i.e.* peat moss, bark or wood chips absorb the nutrients quickly at pH 5.5-6.0. Sterilized garden soil 2 parts, peat or milled sphagnum moss 1 part and gritty sand 1 part is most suitable mix for most of the succulents, and if pH is less than 6.0, it requires addition of lime. Larger collections of succulents are mostly housed under **protective structures** of polythene and glass cover where these remain safe and get optimum conditions throughout the year. Many succulent lovers who own smaller collections, they house these in glassed veranda, to both sides of staircases and by stacking them around their houses. Glasshouses are though expensive structures but provide favourable growing conditions in areas where maximum and minimum temperatures vary widely in summer and winter months. The conditions in a polyhouse or glasshouse also favour better germination of seeds and growth of young seedling. Those very sensitive succulents which can be grown outside, can effectively be germinated and grown inside. Plants are commonly grown in pots arranged on shelves or benches. Some growers prefer to raise plants in groups by planting them directly on raised beds or in the soil on the glasshouse/polyhouse floor. A succulent grown under semi-shade, if is suddenly removed to a more sunny situation, develops burns due to scorching, therefore, such plants should be exposed to sunny site gradually. Many a succulents love **sunshine**, exposure to the morning sun should be preferred. In regions where it becomes very hot during summer, it is advisable to avoid exposure to mid-day sun for a few hours. Once grown under shade causes the plants to become lanky, their foliage become dull and often fail to flower, or flower sparsely. Excess of water is harmful for most of the species. For production of flowers in large numbers and for development of healthy plants, abundant circulation of fresh air is essential. Hence, a succulent house should have only the roof made of glass, fibre-glass or polythene, supported on pillars, while the sides should remain open. Such houses should run in the north-south direction (Das and Panda, 1995). Under outdoor conditions most succulents prefer a sunny situation. Many genera are well adapted to thrive in conditions of great **light intensity** and summer heat. An excessive buildup of heat in glass or polythene covered areas can be avoided by shading with arrangements for adequate **ventilation**.

Succulents being of diverse nature as come from different climatic conditions therefore are grown under varied **temperature** regimes in different parts of India. In Delhi, Kalimpong and Darjeeling where winter temperature goes down to 2 °C, many of the succulents may enjoy a winter rest and to a large extent suspend growth. However, best growth has been recorded in most species when the temperature is between 15-30 °C. The period from March to April under our condition, when average temperature variation is from 25-35°C, is the best time for most species to flower. Therefore, under glass or polythene covers most succulents can be grown throughout the year in different climatic zones of India.

Succulents grown on porous soil mixture require superficial **watering** at 3-4 days intervals during summer months. However, during winter the frequency of watering is drastically reduced and it is normally once in a week or in a fortnight depending on the species grown. Soil should be given time to dry out between watering, rather than being kept wet. Freshly potted plants should not be watered until rooting, usually a week or two after planting. Dripping water is fatal, therefore, any crack or hole, if detected in the roof of the succulent house need to be repaired immediately. Water should ideally have a slightly acidic reaction, be close to air temperature and have an abundance of dissolved oxygen. Succulents like *Adenium, Agave, Beaucarnea, Euphorbia, Furcraea, Pachypodium* and *Yucca* are well adapted to humid tropics and can be grown outdoors in direct sunlight though *Haworthia* and *Gasteria* dislike intense sunlight. *Fenestraria, Haworthia, Lithops, Opthalmophyllum*, and many of such other genera have 'windows', the non-chlorophyllous tissues which are usually situated at the tips of the leaves, in order to make access of light to the assimilating tissue.

For balanced growth and flowering of succulents, in general, a good porous **compost** is quite helpful. One part of sandy-loam soil, one part of well-decomposed cowdung manure, two parts of screened leaf mould, one part of fine sand, small quantity of charcoal powder and a little of bone meal make a good compost for succulent growing. With sufficient precautions while handling the plants, **potting** or **planting** of succulents in pot or ground can be done either during September-October or during February-March, depending upon the requirement of the species. It is best to transfer the plants when they are dormant, *i.e.* during winter months. New growth starts from March and continues up to October. However, potting and planting should be avoided during rainy season. Porous earthen **containers** are always superior to glazed plastic or metallic types. Exchange of air takes place through pores which facilitates healthy root growth. The pots should be clean and dry while potting. Size of the pots should be just large enough to accommodate the plants, too big pots with enough compost keep extra water which may damage the plants. In small pots plants become pot-bound quickly and growth ceases. Two to three layers of clean crocks should be put with concave side towards the bottom so as to facilitate easy drainage. Careful examination of root of the plants to be repotted or planted is essential. The dead or diseased roots and roots with tubercles caused by nematodes should be cut off. The wounds may be dressed with sulphur dust. Root mealy bugs which cling to the roots very often, if present, may be washed with methylated spirit in decoction of nicotine. Crystals of para-dichloro-benzene (PDB) may be put on the bottom layer of the compost above the crocks to keep mealy bugs away from roots. The roots of the treated plants should not be moist at the time of planting.

Insect-Pests and Diseases

When the media are excessively porous and the compost is sufficiently dry, the ants (*Formica* spp.) may colonize by tunnelling the plants below and exposing the roots which results into drying of the old fibrous roots and the new ones not coming into the contact of the soil. It is advisable to treat the medium with methyl parathion dust while potting. However, the affected plants should be repotted by soil treatment. **Mealy bugs** (*Pseudococcus* spp. on stem and *Rhizoecus falcifer* on roots) are slow-moving creatures with thick powdery covering and can be found clustering at a point whether on the roots or on the aerial parts, sucking the plant sap. **Stem mealy bugs** (*Pseudococcus*) are though not common but sometimes its infestation is noticed on certain plants which may be controlled by swabbing the stems with strongly soapy water, plain rubbing alcohol, Malathion or pirimiphos methyl sprayings fortnightly. The thick body coating protects the mealy bugs from insecticides. Unless the plants are examined of their roots for poor or sickly appearance, it is difficult to detect the incidence of root mealy bugs (*Rhizoecus falcifer*). Roots of plants affected by mealy bugs should be washed in decoction of nicotine containing methylated spirit. The excess water on roots should be allowed to dry off before planting. If the old pots and crocks are to be used, those should be cleaned in boiling water for 15 minutes before use. Crystals of para-dichloro-benzene (PDB) should be spread sparingly over the lowermost of compost on crocks and then the pots should be filled and planted with the treated plants. The fumes emitted by this chemical will keep off mealy bugs from roots. Introduction of ladybird (*Cryptolaemus montrouzieri*) will also control mealy bugs. **Root mites** similar to red spiders are almost unknown to the growers but they cause a lot of damage. The infected plant may appear to be bleached and become spotted. Plants growing under shade and in drier parts are observed to be more susceptible to mites. Decoction of nicotine may be drenched at fortnightly intervals to get rid of this pest. With infestation by **scale insects** (*Cactoblastics* sp., *Dactylopius* sp.), the plant body becomes sticky with excrement (honeydew) and blacken with sooty moulds. Plant growth is also adversely affected. They are yellow, brown, dark grey or white and up to 6 mm long and are found sticking to the shady portion of the body or on the joints. Tar oil wash in winter or spraying with Malathion or pirimiphos methyl during summer and autumn proves very effective in controlling these when newly hatched nymphs are present. **Red spider mites** (*Tetranychus urticae*) appear to be tightly fixed to the plants and are difficult to eradicate. Only way to save the plant is to use the healthy part of affected plant for propagation. The infected parts should be destroyed. The female mite overwinters in the crevices of dead parts or in the soil. Malathion, dimethoate or formothion will protect the plants from its attack. Regular wetting of plant body with water will help checking the pest. Fumigants in the greenhouse will be very effective. **Woodlice** also known as slaters or pill bugs are grey or brownish-grey, sometimes with white or yellow markings, and up to 1 cm long with hard segmented bodies feeding at night though hide away in dark shelters during daytime. In fact, these feed on the decaying materials and are often found on plants already damaged by other pests and diseases. These may make holes in the buds and flowers and seedling plants, and sometimes on the tender parts of the plant body, especially the growing tips, though in fact, they are not real pests. These are controlled by cleaning the plant debris and the dead plant parts. Scattering the methiocarb slug pellets will also control these pests. Lepidopterous **caterpillars** (*Agrotis segetum* and other species, *Helicoverpa armigera, Mamestra brassicae, Pieris brassicae, P. rapae, Plusia orichalcea*) feed on tender growing shoots, buds and flowers except the *Agrotis* which feed normally underground during daytime though at night above soil level. They can be picked and killed. Spraying pirimiphos methyl or fenitrothion in the soil for cutworms and spraying on aerial parts for other caterpillars will be effective. Alternatively, the soil treatment with chlorpyriphos + diazinon as soil application for cutworms will be effective. *Bacillus thuringiensis* bacterium can also be introduced which will parasitize these. **Grasshoppers** (*Hieroglyphus* spp.) feed on the plants by chewing the tender parts. All the chemicals used for controlling caterpillars will also control grasshoppers. **Whiteflies** (*Trialeurodes vaporariorum, Aleyrodes proletella*) are very active tiny creatures some 2 mm long with white wings. These rest beneath the young leaves. The immobile nymphs are scale-like whitish-green. These feed on the tender leaves and flowers covering the affected areas with sticky white excrement (honedew) and a form of sooty mould. These can be controlled through spraying with permethrin, pyrethrum, pirimiphos methyl or insecticidal soaps thrice or four times at 5-days intervals. Parasitic wasp (*Encarsia formosa*) is an effective biological control in the polyhouse. **Slugs** and **snails** feed on seedlings, stem parts and flowers when other succulents are grown at humid and darker places. While feeding, these batter the plants completely leaving a trail of silvery slime on the surface as well as on the earth surface. They are attracted to organic fertilizers and mulches. Soil should be cultivated regularly to expose their eggs and slug pellets prepared of metaldehyde or methiocarb should be scattered around the affected plants. Regular sanitation and polyhouse fumigation for control of any insect will together control all these pests also.

Nematodes (*Melodogyne incognita, M. hapla, M. javanica*), the microscopic creatures are highly devastating to many of the succulents as these interfere with the growth of plants by reducing the size of root system, its rotting and galling. These restrict supply of water and nutrients to the plants from the soil. No satisfactory permanent remedy to nematode infestation is known as these invade the plant tissues and feed inside the plant body. Temporary relief can be obtained by cutting away the affected roots from the plants and planting them again after dressing the cut ends of roots with sulphur dust against possible fungus attack through the wounds. Treating the potting compost before potting or repotting with Furadan or Thimet G-3 @ 2.5 g per cubic feet will effectively control root-knot nematodes.

Fungi, bacteria and viruses attack the succulents, the first two particularly when plants are over-watered. **Root and stem rot** are caused by *Pythium* spp. and *Phytophthora* spp. among many other fungi. Stem rot is almost sudden and occurring at the soil level where stems turn black and/or mushy, and afterwards this extends to the upward position. Once the symptoms have been noticed there is no cure but further precaution should be taken so that other plants may not be infected. Soil sterilants may be used. **Powdery mildew** appears as a grayish powdery layer completely covering the green leaves and sometimes even the whole plant. Here many fungi may be involved and the disease occurs in excessive humid and shady conditions. **Downy mildew** (*Peronospora* spp.) infection is encountered when the growing environment is too wet and warm. **Gray mould** (*Botrytis cinerea*) occurs when the weather is chilly and humid. These can be checked through proper ventilation of the chamber or field and through weekly sprayings with 0.2 per cent Dithane M-45 and/or Dithane Z-78. **Anthracnose** (*Mycosphaerella opuntiae*), **leaf scorch** (*Hendersonia opuntiae*), **leaf spot** (*Septoria cacticola*) and **stem rot** (*Aspergillus alliaceus, Helminthosporium* spp.) are also sometimes observed. Infected parts should immediately be removed and burnt and use of zineb or ziram (0.2 per cent) will control all these diseases. '**Damping off**' or young seedling collapse is caused by *Pythium, Rhizoctonia* and *Botrytis cinerea* due to rotting of the base in saline medium or through wounds inflicted on plant body. It becomes too late to control this when the symtoms become clearly visible. Therefore low pH value peat-mix and crushed bricks in the medium will prevent this problem. The drenching of seedling pots or pans with benomyl, Captan or Bavistin will help preventing the mould. The plants or plant organs should be planted only after getting dried the wounded parts under sun for a few days and by treating with sulphur dust. The main fungi which attack other succulents are *Phytophthora, Rhizoctina* and *Fusarium*. *Fusarium* can gain entry through a wound and causes rotting accompanied by red-violet mould. ***Helminthosporium*** and ***Phytophthora* rots** are also commonly encountered. Fungicides may be of limited value in combating these diseases. Unless the attack is detected at an early stage it is difficult to save the affected plants and in most of the cases they are lost. Isolation of the diseased plant should be done immediately and if required, the plant has to be destroyed in order to save other plants from infection. Prevention is better than cure holds good in tackling disease problem in succulents. Correct watering and sterilized compost helps in bringing down disease incidence by fungus considerably. Excess watering is harmful because aeration of roots is impaired. Drenching the soil and spraying the plants with Thiride or Blitox once every month may prove useful.

Symptoms of several **mycoplasma** and **viruses**, such as chlorotic (pale green) spots and mosaic effects (streaks and patches of paler colour), bunchy top, little leaf, phyllody and broom-type structures appear in certain plants of certain species. However, in an *Agave* species, 'cactus virus X' has been shown to reduce growth, particularly when the roots are dry. There is no treatment for these diseases except destroying such plants to check further infection.

Off-colouration and **deformed** winter growth can be prevented by providing proper exposure of sunshine and temperature akin to their requirement, *vis-à-vis* reduction in the frequency of watering during winter. Excessive heat and strong sunlight may cause **brown scarring** of plants so placing the plants at correct positions with proper ventilation will prevent this problem.

References

Anderson, M. 2003. *The World Encyclopedia of Cacti & Succulents*, 256 p. Anness Publishing Ltd., Hermes House, London.

Bailey, L.H. 1930. *The Standard Cyclopedia of Horticulture* (vol. I-III). The Macmillan Co., New York, USA.

Beckett, K.A. 1985. *The Concise Encyclopedia of Garden Plants.* Orbis Publishing Ltd., London.

Beckett, K.A. 1987. *The RHS Encyclopedia of House Plants Including Greenhouse Plants.* Salem House Publishers, Topsfield, Massachusetts, USA.

Brickell, C. 1994. *The Royal Gardener's Encyclopedia of Plants and Flowers.* Dorling Kindersley Limited, London, Great Britain.

Das, P. and P.C. Panda, 1995. Protected cultivation of cacti and other succulents. *Advances in Horticulture,* Vol – 12. *Ornamental Plants*, Part II. (eds Chadha, K.L. and S.K. Bhattacharjee), pp. 819-851. Malhotra Publishing House, New Delhi.

Everard, B. and B.D. Morley, 1970. *Wild Flowers of the World.* Peerage Books, London, Great Britain.

Hay, R. and K.A. Beckett, 1971. *Reader's Digest Encyclopaedia of Garden Plants and* Flowers. The Reader's Digest Association Limited, London, Great Britain.

Hellyer, A. 1982. *The Collingridge Illustrated Encyclopedia of Gardening.* Collingridge Books, England.

Herwig, R. 1986. *2850 House & Garden Plants.* Crescent Books, New York, USA.

Huxley, A., M. Griffithsand and M. Levy, 1992. *The New Royal Horticultural Society Dictionary of Gardening* (4 volumes). The Macmillan Press Ltd., USA.

Misra, S. and R.L. Misra, 2013. *Bowiea.* In: *Commercial Ornamental Bulb Science,* pp. 17-18. Westville, Paschim Vihar, New Delhi.

Nicholson, B.E., S. Ary and M. Gegory, 1980. *The Illustrated Book of Wild Flowers.* Peerage Books, London.

Panda, P.C. and P. Das, 1995. Genetic Resources of Cacti and Other Succulents. In: *Advances in Horticulture ~ Ornamental Plants* (vol. 12, Part I; eds Chadha, K.L. and S.K. Bhattacharjee), pp. 243-266. Malhotra Publishing House, New Delhi.

Randhawa, G.S. and A. Mukhopadhyay, 1986. *Floriculture in India.* Allied Publishers Private Ltd., New Delhi.

Šubik, R. and J. Kaplická, 1968. *A Concise Guide in Colour ~ Cacti and Succulents.* The Hamlyn Publishing Group Ltd., London.

Virginie, F. and G. A. Elbert, 1989. *Foliage Plants for Decorating Indoors.* Timber Press, Portland, Oregon. USA.

41

Viola (Family: Violaceae)

K.K. Dhatt, Sanyat Misra, R.L. Misra, D.V.S. Raju and M.K. Singh

[**Common names**: Bird's foot viola or violet (*Viola pedata*),Common dog violet (*V. riviniana*), Common violet (*V. calcarata*), English violet/Florist's violet/Garden violet/Sweet violet (*V. odorata*), European wild pansy/Field pansy/ Heartsease/Johnny-jump-up/Miniature pansy/Viola (*V. tricolor*), Garden pansy/Pansy (*V. × wittrockiana,* syn. *V. × hortensis*), Horned violet/Bedding pansy/Viola/Violet (*V. cornuta*), Ivy-leaved violet/Australian violet (*V. hederacea*, syn. *Erpetion hederaceum, E. reniforme, E. petiolare, E. spathulatum*), Violet (*V. cornuta, V. odorata*), *etc.*]

Introduction

The herbaceous genus *Viola* comprises of about 500 evergreen to deciduous (Beckett, 1983) hardy annual, biennial, perennial and sub-shrubby species of cosmopolitan distribution, scattered in the northern as well as southern hemispheres, extending to Hawaii, Andes in South America, Australasia, New Zealand and tropical and subtropical Africa. Fourteen species are indigenous to Europe including the one most beautiful species, *V. calcarata*, the common violet. *Viola* is the Latin name for a violet, perhaps originally from the Greek (Beckett, 1987). In French, the word *pansee* is for 'thought'. Though *Viola tricolor* was commonly found throughout Europe but its garden cultivation was taken up after the first quarter of 16th century so up to 18th century all the varieties still belonged to *V. tricolor* var. *hortensis*. Lord Gambier and William Thompson attempted crosses among *V. tricolor, V. lutea* and *V. altacia* which resulted in the evolution of 'face' type of pansies with huge block of colours on the lower petal, and the first pansy hybrids with dark central blotches appeared in 1814, and by 1835 there were 400 named varieties available. 'Medora Series' was discovered in 1839 and by 1850 many new strains of pansies became popular among pansy growers and breeders throughout Europe. About 1810, Lady Bennett took interest in her father's (Lord Tankerville) small garden at Walton where in a heart-shaped bed she put all the varieties of pansies she got from anywhere so by 1813 there were 14 excellent forms and soon the number of collections rose to 20 (Pizzetti and Cocker, 1968). Wild pansies have a strong tendency to form natural hybrids resulting in a lot of variation among the species. This was start of the building up of the varieties. By about 1850, Benary, Heinemann and certain others in Germany, a few in England and Vilmorin in France started specializing on this flower and small multiflora types were developed which proved a starting point for giving way to other strains. All the British strains raised by 1861 from the imported continental types were classified as 'Fancy pansies' which were perfectly circular in outline, measuring 7 cm across where petals were flat, smooth and velvety and three lower petals were with large blotches.By the turn of the 20th century, Dr. Charles Stewart, the Scottish grower grew the clear pansies without a face. After some time the Britishers took the lead and after 1945 the Netherlands developed Giant Aalsmeer strain. Next important viola, probably more important than pansy, is *Viola odorata* which Theophrastus called *V. oscura* whereas Dioscorides & Pliny *V. purpurea*. People in Athens so admired this flower that there was no house or garden without violets. Garden varieties are referred to as *Viola × wittrockiana* as these have originated due to crossing among *V. tricolor, V. lutea* and *V. altaica*. From the second half of the 20th century, pansy breeding is being concentrated in Germany, the United States and Japan to improve the quality such as plant vigour, free-flowering nature and the heat tolerance. All the cultivated forms are perennials though are grown as annuals from seeds. These are suitable as ground cover, in beddings and borders, as edgings, in pots and in the hanging baskets. The giant-flowered pansies such as 'Swiss Giant', 'Masterpiece' or 'Butterfly Hybrids' with ruffled petals are suitable for edging, borders and pot cultivation. For mass effects in hanging baskets, window baskets and rockeries the most effective pansy varieties are from *V. tricolor* and *V. cornuta*. For bedding. The most suitable ones are small-flowered types which flower profusely.

All parts of the plant contain a bitter alkaloid, violine and a glucoside, and flowers a blue pigment. Because of these substances, they are used as antiseptics and expectorants, in blood disorders and in catarrhal affections. The glucosidal principles found in the leaves of *V. odorata* has antiseptic properties and gives pain relief in case of cancerous growths and even in the treatment of cancer.

Botany, Genetics and Breeding

Pansies are small perennial, trailing and clump-forming and sometimes stoloniferous herbs in the temperate regions though behave as an annual in the plains and biennials on the hills. The leaves are linear to orbicular, ovate to lanceolate, cordate, scalloped or palmate with stipules that are often leaf-like, the margins are crenate and texture pubescent. Andean species mimic houseleeks. The horizontally borne or nodding flowers are usually pedicellate, flat, bracteate, hermaphrodite, hypogynous, zygomorphic and pentamerous: five green and leafy sepals; five petals, 2 overlapping large upper petals, 3 lower comprising the lower half of the flowers ~ a large central which is spurred and 2 side petals, each petal is covered with tiny hairs, giving the flowers the velvety look. In the three lower petals, the lowermost is folded to hold the nector and the internal 2 lower petals normally have blotch of contrasting colour. Female part of the flower is in the pistil where petals meet, the pistil in the shape of a vase comprising of ovary, style and stigma; and the male parts of the flower surrounding the stigma are 5 stamens, each comprising of a filament, topped by an anther which produces yellow pollen. The flowers have saprophytic self-incompatibility therefore bud pollination is successful, the fruit is a ovoid capsule bearing obovate and smooth seeds and these capsules on maturity split into three boat-shaped valves having thick rigid keels. Many species produce tiny cleistogamic flowers that never open and produce seeds by self-fetilization. The group of species and hybrids with flat-faced flowers and prominent leafy and lobed stipules are known as pansies.

Basic chromosome number in *Viola × wittrockiana* (a high fertile but self-incompatible) is n=6 (octoploid 6x=48), and this garden species was developed by crossing *V. tricolor* (n=13), *V. lutea* (n=24, 2n=48) and *V. altaica*. *V. cornuta* has basic chromsosome number as n=11. Polyploidy is a potent tool in developing novel and vigorous cultivars with desirable phenotypic traits such as compactness, novel colour combinations, floral shape & floral size, time of flowering, flower duration & floriferousness, and tolerance to biotic & abiotic stresses. Pansies also express cleistogamy where self-pollination at bud stage occurs. Also there exists sporophytic self-incompatibility which is quite helpful in F_1 hybrid seed production. Certain breeders and companies have developed many inbred lines as varieties which are being used to produce F_1 hybrids of desired traits. The first commercial F_1 hybrid was bred in the 1960s, and since then a dozen of F_1 hybrids are now available worldwide.

Propagation

Violets (*Viola cornuta, V. odorata*) are propagated **vegetatively**. Established plantings can be divided during winter and planted (Post, 1949) in pots or the prepared field and these come out easily as separate plants. *V. odorata* forms trailing stems so from the terminal regions 2-leaf-stem cuttings are planted in the rooting medium after removing the lower leaves on the stem. These root out easily. All the species can be propagated from 2.5-5.0 cm long cuttings taken from the basal non-flowering shoots (Hay and Beckett, 1071). Pansies (*Viola × wittrockiana* syn. *V. × hortensis*) and violas (*Viola tricolor*) where per gramme it may contain some 600-800 seeds, are usually grown through **seeds** only. Even otherwise, all these can be propagated only through seeds. The seeds can be sown in the dark at temperatures ranging from 13 °C to 26 °C, lower the temperature the percentage gertmination is generally poor and higher the temperature the germination percentage is normally higher. There was 55 per cent germination at 13 °C, 48 per cent at 16 °C, 60 per cent at 18 °C, 46 per cent at 21 °C and 64 per cent at 24 °C (Cathey, 1969; Holcomb and Mastalerz, 1985). Corr (1997) stated that primed seeds can be subjected to quite high temperatures for germination as to high temperatures these are less sensitive. However, violas are more temperature-sensitive than the pansies and when day temperatures are above 21 °C, the night temperature should be kept around 16 °C for seed germination (Dole and Wilkins, 1999). For seed propagation, the nurseries should be raised in pans, pots, boxes or in the nursery beds. The mixture should comprise of good garden soil, farmyard manure and sand, all equal by volume. The mixture should be quite porous but moisture-retentive and free from germs so it would be better to treat these with formaldehyde. The sowing is done in rows spaced 4-6 cm apart and thinly in the rows in the month of September-October. These germinate within 10 days. After one month of sowing these are ready for transplanting as by this time the seedlings would have attained 4-leaf stage.

Classification, Species and Varieties

Pansies are classified under Section Melanium of the genus *Viola*. The flowers may be either of pure or clear colours or mixed multicolours blotched or faced with very dark blue to black centres. Also there are blotched pansies with distinct colour blotches than the usual dark blue to black face. Other multicolour pansies have white or light-coloured edges or petals of differing colours. Based on the size of the flowers, the pansies are classified as large (9-11.5 cm), medium (6-9 cm) and multiflora (4-6 cm).

Under **large** blooms are 'Accord Series' introduced in 1992 which have abundance of blooms on short but sturdy stems, and this has eight single colours ~ as 'clear', 'faced' and 'mixed'. Free-flowering 'Lyric Series' introduced in 1985 are weather tolerant and available in five colours either all 'faced' or 'mixed' ones. Free-flowering 'Majestic Giant Series' are quite tolerant to heat or cold, all with six colours and a mixed one, all with 'faces'. Majestic Giant Mix and Majestic Giant White Face, under the Majestic Giant Series had won All America Selections Award 1966. 'Medallion Series' pansies have extra large flowers in six colours and the pansy face. 'Swiss Giant Series' are the old-fashioned pansies bred in Switzerland which contributed in developing many of the today's hybrids. Some of these large-flowered ones having long stalks are good for cut flower use.

Under **medium** blooms are 'Crown Series' pansies with nine bright 'clear' colours and a 'mix' which are early spring flowering. 'Imperial Series' pansies have vigorous growth with non-fading colours and mostly with 'faced' blooms, out of which the 'Imperial Blues' that were All America Selections Award Winners' in 1975 and a long favourite of gardeners. 'Joker Series', the AAS Award Winner in 1990, has compact plants, spreading up to 20 cm and the flowers are velvety purple with orange face. 'Roc Series' were introduced in 1985 with nine colours with or without a face. 'Springtime Series' having a wide rainbow of 17 colours, has heat and cold tolerance and has a reliable performance.

Under **multiflora** pansies are 'Crystal Bowl Series' which are compact with no sprawling habit, all having different 11 clear colours but with no face; 'Maxim Series' having compact plant habit with 14 clear colours *cum* faced flowers, prolific bloomer, and its one of the series 'Maxim Marina', an A.A.S. 1991 Award winner, having unusual light blue colouration with dark blue face but outline white, though 'Padparadja', a brilliant orange-sapphire colouration and 1991 A.A.S. Award winner whose flowers do not fade under heat stress; and 'Universal Series', a cold- or heat-tolerant with faced or 13 clear colours and a mixture, where flowers appear early.

In the country, P.A.U., Ludhiana has developed two pansy varieties in 2013 with the name of 'Punjab Purple-Wave' and 'Punjab Choco-Gold'. 'Punjab Purple-Wave' produces 30 cm tall plants with dark green leaves which are tinged purplish, blooming in 64 days with a flower duration of 76 days and flower number of about 70 per plant, and the flower colour is purple with darker blotch. 'Punjab Choco-Gold' grows up to 31 cm tall, leaves are light green, blooming in 66 days with 83 days flowering duration and 83 flowers per plant, and the flowers are yellow with chocolate blotch. These are the first pansy varieties developed in the country.

Important perennial cultivated *Viola* species are described as under:

Viola aetolica (syn. *V. saxatilis aetolica*)

A native to East Europe, especially Balkan Peninsula, this tufted species is quite similar to *V. tricolor* but more prostrate and grows to 10 cm tall, bearing about 2 cm long, greyish-green, ciliate, and ovate to lanceolate leaves with wedge-shaped base, and bright yellow pansy-like flowers some 1.8-2.5 cm across.

Viola biflora

A native of N. Hemisphere, this species grows hardly 8 cm high but spreads up to 30 cm wide. The leaves are bright fresh green and reniform. The flowers are vivid yellow, violet-shaped and 1.2-2.0 cm across.

Viola bosniaca (syn. *V. elegantula*)

A semi-trailing species native to SW Europe grows up to 13 cm in height and bears pointed oval leaves with undulated margins. Flowers are small and pinkish-purple with a yellow mark at the lower petal's base.

Viola canina

A native to Europe, a shade-loving hardy species with all the characteristics of *V. odorata* but without scent.

It grows up to 23 cm tall, bearing dark green, dentate and pointed leaves that are 7.5 cm across. Flowers are large, violet-blue or white. It is propagated through division.

Viola cornuta

A native to Pyrenees, this rhizomatous species is clump-forming and grows to 10-30 cm tall, bearing ascending to decumbent stems, mid-green, ovate to oval and rounded-toothed leaves that are 2-5 cm long and hairy beneath, flowers are long-stalked, prolific for several months, violet-purple to lilac, fragrant, 2-3 cm or more across with spur some 1.0-1.5 cm long. It is one of the chief parents of the popular modern garden violets. Its varieties are 'Alba' (white), 'Campanula Blue' (sky-blue), 'Chantreyland' (suffused apricot-orange), 'Jersey Gem' (rich blue-purple), *V.c. minor* (plants smaller, flowers lavender), 'Purple Perfection' (red), 'White Lady' (white), 'White Perfection' (cream-white), 'Yellow Bedder' (golden-yellow), *etc.*

Viola cucullata (syn. *V. papilionacea, V. sororia*)

A deciduous and thick rhizomatous species from eastern N. America, grows up to 15 cm tall, bearing broadly ovate-cordate to kidney-shaped, crenate and 5-9 cm wide leaves, and flowers are white to violet-purple with a greenish-white eye and about 2 cm across. *V.c. albiflora* is white and is similar to *V. sororia*.

Viola elatior

A deciduous, tufted to clump-forming species native to Europe and W. Asia, it grows up to 50 cm tall with erect stems and bearing shortly hairy, lanceolate to narrowly ovate and 4.0-7.5 cm long leaves, and the flowers are pale-blue and some 2.0-2.5 cm across.

Viola gracilis

A mat-forming species from Balkan Peninsula and Asia Minor, growing up to 15 cm tall with decumbent stems, the leaves are mid-green, broadly ovate to oblong with rounded teeth and 2-3 cm long, and the pansy-shaped prolific flowers are yellow or violet-blue, velvety and 2-3 cm across. Its vigorous varieties are 'Alba' (white), 'Black Knight' (purple-black), 'Lutea' (golden-yellow), 'Major' (deep purple-blue), 'Moonlight' (pale-yellow), *etc.*

Viola hederacea

A stoloniferous and mat-forming tender species native to SE Australia grows to 10 cm tall and spreads from 30 to 60 cm, bearing pale-green, kidney-shaped, undulated and 2.5-4.0 cm across leaves, and the flowers are violet-purple with white tips, about 2 cm across and appearing well above the leaves throughout the year though sparsely. It makes an excellent pot plant.

Viola labradorica

A mat-forming species native of the northern part of N. America to Greenland, this grows up to 13 cm tall, bearing mid-green, broadly ovate to orbicular and cordate leaves about 1.5-3.0 cm long, and the flowers are violet-like light purple-blue (mauve) and 1.5-2.0 cm across. Its var. 'Purpurea' bears suffused deep purple leaves.

Viola lyallii

A tufted species native to New Zealand which grows up to 10 cm or more in height, bearing ovate to rounded and cordate leaves some 1-3 cm long, and white, streaked lilac and yellow flowers some 1-2 cm across.

Viola odorata

A tufted mat-forming species with thick rhizomes native to Europe including Great Britain, N. Africa and Asia, grows to 15 cm in height, bearing mid to dark green, orbicular to ovate-cordate leaves about 4-7 cm long. The fragrant violet-purple or white flowers are 1.5-2.0 cm across. It has produced many cultivars in shades of white, cream-yellow, pink, lavender, violet, purple and lilac and also larger in size and doubles. 'Alba' (white), 'Christmas' (small white blooms with green eye), 'Coeur d'Alsace' (rich pink), 'Czar' (deep violet-purple), 'Marie Louise' (double mauve), 'Parma Violet' (pale-lavender double and intensely fragrant), 'Princess of Wales' (rich violet flowers with long stalks, worthy of cutting), 'Sulphurea' (interior apricot-yellow, exterior purple-tinted), 'White Czar' (white, 12.5 cm across), are some of the prominent varieties.

Viola papilionacea (syn. *V. sororia*)

A deciduous species is native of eastern N. America. It comes up from the thick fleshy rhizomes and grows up to 10 cm high. Leaves are broader than length, ovate, toothed with underside softly hairy. White-eyed flowers are rich violet, light blue or reddish and about 3 cm across. Its forms 'Albiflora' is white with purple veining and is confused with *V. cucullata albiflora;* and *V.s. papilionacea* (*V. papilionacea*) bears hairless leaves.

Viola pedata

A tufted to small clump-forming and hardy species from eastern and midwestern USA grows to 15 cm tall, bearing only basal palmate leaves some 5-10 cm across with 3-5 lobes like a bird's foot, each lobe having further divisions, and the lilac-purple flowers are flat and about 3 cm across. Three lower petals are lovely delicate lilac, and the two upper petals are dark violet.

Viola pedunculata

A native of USA (California), growing up to 20 cm high, bearing broadly oval and dentate leaves and the flowers are rich yellow. It flowers during late summer and prefers semi-shaded situation. As it likes drier condition for growing, it is most suitable for rock gardening.

Viola riviniana

A tufted and colonizing species from Europe including Iceland, N. Africa and Madeira which grows hardly 15 cm tall, bearing ovate-orbicular and cordate leaves from 2 to 8 cm long, and the flowers are blue-violet and 1.5-2.0 cm across. Its flowers appear to *V. odorata* but is scentless and the plant is non-stoloniferous.

Viola rupestris

A native to N. Hemisphere, it grows up to only 4 cm tall, bearing 5-10 mm long leaves and pale blue-purple flowers 1.0-1.5 cm across. Its white, pink and reddish-purple forms are also available. It appears as the small version of *V. riviniana.*

Viola septentrionalis

A native to NE North America, it is though similar to *V. papilionacea* but smaller. Its leaves are narrower but toothed and the flowers are deep violet-purple to lilac or white and the petals are downy.

Viola tricolor

A short-lived tufted true hardy pansy species which is native to Europe including Great Britain and Asia with trailing square stems running up to 40 cm in length. It grows more than 15 cm tall and spreads from 15 to 40 cm with ascending to decumbent stems. Leaves are mid-green, ovate to lanceolate or cordate with rounded teeth and up to 5 cm long. Flowers are cream and yellow to blue-violet and often tricoloured, solitary, pansy-like 1.2-3.8 cm or even more across. With *Viola altaica* and *V. lutea*, it has produced the modern *V.* × *hortensis* (*V.* × *wittrockiana*). The some of the most interesting pansies under this form are 'Swiss Giants', 'Dutch Giants', 'Majestic Giants', 'Gay Jester', 'Jumbo Giants', 'Oxford Blue', 'Raspberry Rose', 'Snow White', 'Ullswater', *etc.*

Viola × *williamsii*

A garden species covering several sorts of tufted pansies or violas derived by crossing *V.* × *wittrockiana* and *V. cornuta*, having character of *V. cornuta* but flowers are pansy-like round and usually in self-colours. This is propagated through division or cutting. Its cultivars include 'Admiration' (deep blue-purple), 'Irish Molly' (copper-yellow), 'Maggie Mott' (silvery-mauve), 'Norah Leigh' (lavender-blue), 'Primrose Dame' (light yellow), 'White Swan' (pure white), *etc.*

Viola × *wittrockiana* (syn. *V. tricolor hortensis*, *V. tricolor maxima*, *V.* × *hortensis*)

It is a common garden pansy derived by crossing *V. tricolor*, *V. lutea* and *V. altaica*. This is a short-lived perennial much like the *V. tricolor* but the plants are more robust, erect to decumbent and some 20 cm tall, the leaves are ovate, crenate and up to 8 cm long, and the flowers quite large (4-12 cm across) and in wide range of colours (white, buffs, yellow, orange, red, purple and blue), usually with a mask-like black blotch in the centre which fancifully make it a 'funny faced'. It has varieties blooming during different seasons such as spring, the summer, autumn and the winter. Spring to early summer blooming varieties are 'Azure Blue' (bright blue), 'Majestic Giants' (white, yellow, red, blue and certain other colours), 'Sunny Boy' (black-blotched golden-yellow), *etc.* Summer-blooming varieties are 'Arkwright Ruby' (fragrant, bright crimson), 'Blue Heaven' (blue with yellow eye), 'Clear Crystals' (mixed self colours), 'Engelman's Giant' (mixed colours), 'Jackanapes' (dark blue-purple and yellow bicolour), 'Roggli Swiss Giants' [single-coloured named shades such as 'Berna' (deep violet), 'Brunig' (red), 'Jungfrau' (creamy-white), 'Monch' (yellow), *etc.*].

Viola yakusimana (syn. *V. verecunda yakusimana*)

A tufted species native to Japan (Yakushima) is the tiniest violet, growing hardly 3 cm tall. Its leaves are reniform, cordate, shallowly toothed and up to 6 mm long. Flowers are white with lower lobes veined purple and some 7-8 mm across.

Cultural Practices

These prefer growing under a semi-wild woodland conditions under partial shade and thrive best at a cool place in a humus-rich soil having pH of 5.5-6.0 (Post, 1949), however, if pH rises above 6.5, foliar chlorosis may occur. Sometimes, however, these are recorded growing even in highly alkaline (calcareous) soils. Any soil or medium which does not have water-logging problem but is light and moisture-retentive is suitable for its cultivation. Peaty soil is most suitable when growing these in greenhouses. At the time of soil preparation, the field should be thoroughly ploughed 2-3 times after incorporating 20 tonnes/ha of farmyard manure in the soil, every time the ploughing should be followed by planking. The rootstocks of the perennial weeds and any other hard material and waste polythene sheets should be removed. Now the soil properly levelled, lay out is prepared, channels and bunds are drawn and the beds of convenient sizes are prepared and then again the beds are levelled. At this time the soil should not dry out. Planting of one month old seedlings is carried out, followed immediately with watering to establish the roots properly. For planting in the beds, the small flowered pansies are spaced 15-20 cm apart though larger ones 20-25 cm to 30 cm apart. The large spreading plants are planted 40 × 30 cm apart in the beds. Planting time is generally October-November whether it is tropical plains or in the temperate regions.

During winter season they are **watered** at 12-15 days intervals, however, during summer season these require to be watered every week. If the soil is sandy, light and porous the frequency of watering is more but in heavy soils the frequency is quite less. Pansies never tolerate water-stress at any level. Pansies are highly responsive to **nutrition**. Since this is a cold-growing crop, so nitrogen should neither be used as urea nor ammonium, however, a regime of 100 ppm N is sufficient. During excessive cold when leaves develop blue-purple pigmentation, this may be due to phosphorus deficiency. When new foliage show cupping and distortion, *vis-a-vis* abortion of the growing tips, it may be due to boron deficiency, however, the cupping and foliage distortion are also associated to calcium deficiency. Styer (1997) stated that boron deficiency is expressed due to insufficient boron, high medium pH or high calcium levels in the soils. If the pH of the medium is low, *i.e.* less than 5.5, Ca deficiency, as well as Mn & Fe toxicity in the form of rusty mottling of foliage are expressed.

These crops require to be weed-free to maintain the vigorosity of the crop otherwise weeds compete for light, nutrients, water and harbour the vectors and sometimes work as hosts for certain insect-pests and pathogens. Timely weeding keeps the crop healthy. To conserve moisture in the field, sometimes these are mulched with organic mulches which decompose after the crop is over. Pinching, disbudding, staking, *vis-à-vis* CO_2 enrichment in this crop are not required. In case of plug production, however, elevated CO_2 may be beneficial. Violets do not require to be treated with dwarfing chemicals as these sprawl and do not grow erect.

Growth and Flowering

Viola odorata and *V. cornuta* plants remain vigorous during summer months due to high temperatures and **long days** though during autumn to winter (October to March) these continue producing chasmogamous (open-petalled) flowers throughout autumn, winter and early spring along with only a few runners (Post, 1949) as during this time plants get short days only, though when there are long days during summer these produce cleistogamous flowers (Allard and Garner, 1940; Cooper and Watson, 1952; Mayers, 1986). Quality flowers start appearing 10 weeks after start of the 8-h SD when the plants are grown at 10 °C night temperature, though spring-propagated stock starts producing the flowers in the following autumn (Cooper and Watson, 1952). Most of the species tolerate full sun in the garden but these grow best under shaded situation, however, heavy shading is required during summer in the polyhouses. In case of pansies and violas for plug production, during winter these may not require any shading but in summer, especially during later days when summer is very hot, the shading is required to reduce the temperatures (Post, 1949; Seeley, 1985; Ball, 1991; Healy, 1998; Nau, 1998). For an usual crop growing, application of long days through high density discharge (HID) supplementary lighting during 2-6 weeks after germination is very effective and this decreases the time for first flowering (Erwin *et al.*, 1997), and increases the flower number, *vis-a-vis* intermodal lengths, though Seeley (1985) recorded short and sturdy stems under SD. If temperatures remain cool, the pansies can continue flowering throughout the year, however, high summer temperatures limit the vigorosity of the plants.

Post (1949) stated that violets are cold-house crops so grow well during cool days of the year and the preference for **temperature** during growth period is 4-10 °C, however, Cooper snd Watson (1952) recorded greater flower size at 10 °C night tempersture than those grown at 4 °C. Pansy and viola established plants can be grown best at 4-13 °C temperatures (Post, 1949) as above this temperature range the growth becomes weak and internodal lengths increase (Holcomb and Mastalerz, 1985).

If plants of pansy and viola are grown at appropriate cool temperatures (4-13 °C) and high light, these grow proportionately with no extended height so height control is not required but growing at 16 °C and above temperatures increases the intermodal length so plants become taller when 2-15 ppm of paclobutrazol (bonzi), 3-6 ppm uniconazole (sumagic) or 2,500 ppm daminozide (B-Nine) is required to be sprayed to control the height. While applying dwarfing chemicals, one should be careful as pansies are highly sensitive to over-application. In case of pansy plugs, 10 ppm spraying of ancymidol (A-Rest) twice is quite effective (Wieland *et al.*, 1997). Excessive growth can be arrested through sufficient lighting, and controlled fertilizing, especially restricted use of phosphorus and reduced rate of ammonium in the medium.

Insect-Pests

Aphids, thrips and the Hymenoterous sawfly larvae infest on these plants, especially on emerging shoots, the new leaves, and flowers. Sawfly larvae are voracious feeder of leaves and their severe infestation defoliates the plants completely. Infestation of these insects make the leaves distorted and buds fail to open. White ants feed on the roots and stems of the plants and so entire plant dies. Aphids can be controlled by spraying bifenthrin, dimethoate, heptenophos with permethrin, malathion, permethrin, pirimicarb or pirimiphos methyl. Thrips and sawfly larvae can be controlled effectively by application of dimethoate, malathion, permethrin, pirimiphos methyl or pyrethrum. For controlling ants, their nests should be dusted with bendiocarb, lindane, permethrin or pirimiphos methyl, or sprayed with chlorpyriphos, cypermethrin, deltamethrin or permethrin.

Diseases

Leaf spot diseases such as *Alternaria violae, Cercospora granuliformis, C. violae, Marsonina violae, Phyllosticta violae, Septoria violae* (Horst, 1990) and *Ramularia* spp. (Greenwood and Halstead, 1997) have been found infecting outdoor pansies, violas and violets. *Alternaria violae* causes grey or brown spots on leaves forming concentric rings of fruiting bodies and these spots may coalesce and kill the leaves. *Cercospora, Phyllosticta* and *Ramularia* cause brown, purple-brown or very pale buff-coloured spots on the foliage, starting from the old foliage and making shot holes. Likewise, *Septoria* also causes shot holes. **Stem and leaf lesion anthracnose** or **scab**, an important disease, caused by *Sphaceloma violae* (Forsberg, 1979) and *Colletotrichum violae-tricoloris*. The spots are typically grey to pale with thick and black margins, sometimes patterned with concentric rings on leaves, petioles, peduncles and stems. Eventually the lesions may girdle the stem and kill the plants. Sometimes the brown spot anthracnose symptoms also appear on the flowers. The disease is favoured by warm and wet conditions as occur during spring. Generally the leaf spot diseases are caused by weak parasites, so mostly these are not so serious. Affected leaves are removed and destroyed, proper sanitation is maintained and affected plants are sprayed with mancozeb.

Rusts such as *Puccinia violae, P. ellisiana* and *Uromyces andropogonis* (Horst, 1990) infect outdoor growing pansies, violas and violets on their leaves and stems. Bright orange to dark brown gelatinous spore masses or pustules are formed and its spread is favoured by moist to damp environment. Infected leaves are removed, air circulation among the plants are improved and the plants are sprayed with bupirimate along with triforine, mancozeb, myclobutanil or penconazole.

Viruses (Horst, 1990) have also been recorded infecting outdoor pansy, viola and violets by causing streaking, distortion, mosaic and certain other symptoms. Such plants should be uprooted and burnt.

Under **greenhouse environment** the diseases occurring on pansies are **leaf spots** caused by *Mycocentrospora* sp., **grey mould** caused by *Botrytis cinerea*, **petiole and stem rot** caused by *Myrothecium* sp., **damping off** caused by *Pythium* spp. and *Rhizoctonia solanii*, and **root or crown rot** diseases caused by *Fusarium oxysporum, Phytophthora* spp., *Pythium* spp., *Rhizoctonia* sp. or *Thelaviopsis basicola* (Horst, 1990; Daughtrey and Chase, 1992; Styer, 1997). Leaf spot diseases and grey mould can be controlled by spraying the plants with mancozeb or zineb.Damping off usually occurs on the seedlings where these show discolouration and deterioration around the stem base and finally toppling over. Initially the basal lesions appear water-soaked and soon other seedlings are also affected. Such seedlings collapse immediately. The disease is favoured by over-wet compost and prolonged high temperatures. Strict hygiene should be observed, clean containers and implements should be used and the compost should also be sterilized. The medium should be soaked or drenched with copper-based fungicide at sowing and afterwards to the developing seedlings at intervals. Spraying the plants with captan alternate with bavistin at fortnightly intervals will save them from petiole and stem rot. *Thelaviopsis* is a widespread and destructive fungus causing black root rot of pansies. Infected plants become stunted, their leaves turn yellowing and plants show wilting due to dark spots or bands being formed on the roots.

Post Harvest

The pansies with long stalks can be used as cut flowers where these remain fresh up to 4-5 days in ordinary tap water. The flowers in case of *Viola odorata* and *V.* × *wittrockiana* are harvested when these are almost fully open and these can be stored for a week at 1-5 °C (Nowak and Rudnicki, 1990). Rogers (1985) stated that pansies are sensitive to ethylene gas. Brightly coloured flowers can be press-dried for various uses. Heins *et al.* (1994) stated that pansy plugs can be stored up to 16 weeks at 5 °C with 5 fc lighting or at 0-2 °C under darkness.

References

Allard, H.A. and W.W. Garner, 1940. Further observastions on the response of various species of plant to the length of the day. *USDA Tech. Bull.*, No. 727, pp. 1-64.

Ball, V. 1991. Plugs – The way of the 1990s. In: *Ball RedBook* (15th edn), pp. 137-153. Geo. J. Ball Publishing, Illinois, USA.

Beckett, K.A. 1983. *Viola*. In: *The Concise Encyclopedia of Garden Plants*, 425-428. Orbis Publishing Ltd., London, Great Britain.

Beckett, K.A. 1987. *Viola*. In: *The RHS Encyclopedia of House Plants Including Greenhouse Plants*, p. 473. Salem House Publishers, Massachusetts, USA.

Cathey, H.M. 1969. Guidelines for the germination of annual pot plant and ornamental seeds. *Flor. Rev.*, **144**(3742): 21-23, 58-60; **144**(3743): 18-20, 52-53; **144**(3744): 26-28, 75-77.

Cooper, C.C. and D.P. Watson, 1952. Influence of daylength and temperature on growth of greenhouse violets. *Proc. Amer. Soc. hort. Sci.*, **59**: 549-553.

Corr, B. 1997. The ABC's of plug production. *Greenh. Product News*, **7**(2): 21-23.

Daughtrey, M. and A.R. Chase, 1992. *Pansy*. In: *Ball Field Guide of Diseases of Greenhouse Ornamentals*, p. 148. Ball Publishing, Geneva, Illinois, USA.

Dole, J.M. and H.F. Wilkins, 1999. *Viola*. In: *Floriculture Principles and Species*, pp. 545-550. Prentice Hall, Inc., Upper Saddle River, New Jersey, USA.

Erwin, J.E., R. Warner, T. Smith and R. Wagner, 1997. Photoperiod and temperature interact to Affect *Viola × wittrockiana* Gams. development (sbstr.). *HortSci.*, **32**: 466.

Forsberg, J.L. 1979. Pansy, Violets. In: *Diseases of Ornamental Plants*, pp. 126-129. Special Publication No. 3, College of Agriculture, University of Illinois, Urbana, USA.

Greenwood, P. and A. Halstead, 1997. *The Royal Horticultural Society Pests and Diseases : The Complete Guide to Preventing, Identifying & Treating Plant Problems*, pp. 99, 124, 126, 128, 132, 143, 154, 165, 173, 176, 186. Dorley Kindersley Ltd., London.

Hay, R. and K.A. Beckett, 1071. *Viola*. In: *Reader's Digest Encyclopaedia of Garden Plants and Flowers*, pp. 733-735. The Reader's Digest association Ltd., London, Great Britain.

Healy, W. 1998. *Viola × wittrockiana* (Pansy). In: *Ball RedBook* (16th edn; ed. Ball, V.), pp. 777-782. Ball Publishing, Batavia, Illinois, USA.

Heins, R., N. Lange, T.F. Wallace, Jr., and W. Carlson, 1994. *Plug Storage, Cold Storage of Plug Seedlings*. Meister Publishing, Willoughby, Ohio, USA.

Holcomb, E.J. and J.W. Mastalerz, 1985. Seedling and seedling production. In: *Bedding Plants III* (3rd edn., eds Mastalerz, J.W. and E.J. Holcomb), pp. 87-125. Pennsylvania Flower Growers, University Park, Pennsylvania, USA.

Horst, R.K. 1990. Pansy. In: *Westcott's Plant Disease Handbook* (5th edn), p.753. Violet. In: *Ibid.*, pp. 859-860. Van Nostrand Reinhold, New York, USA.

Mayers, A.M. 1986. *Viola odorata*. In: *Handbook of flowering* (vol. V, ed. Halevy, A.H.), pp. 372-384. CRC Press, Boca Raton, Florida, USA.

Nau, J. 1998. *Viola* (Johnny-jump-up, violet). In: *Ball RedBook* (16th edn, ed. Ball, V.), pp. 775-776. Ball Publishing, Batavia, Illinois, USA.

Nowak, J. and R.M. Rudnicki, 1990. *Postharvest Handling and Storage of Cut Flowers, Florist Greens and Potted Plants*. Timber Press, Portland, Oregon, USA.

Pizzetti, I. and H. Cocker, 1968. *Viola*. In: *Flowers A Guide for Your Garden* (vol. II), pp. 1365-1375. Harry N. Abrams, Inc., Publishers, New York, USA.

Post, K. 1949. *Viola tricolor* and *Viola cornuta*; *Viola odorata*. In: *Florist Crop Production and Marketing*, pp. 839-845. Orange Judd Publishing, New York, USA.

Rogers, M. 1985. Air pollution. In: *Bedding Plants III* (3rd edn, eds Mastalerz, J.W. and E.J. Holcomb), pp. 274-314. Pennsylvania Flower Growers, University Park, Pennsylvania, USA.

Seeley, J.G. 1985. Finishing bedding plants – Effects of environmental factors : Temperature, light, carbon dioxide, growth retardants. In: *Bedding Plants III* (3rd edn, eds Mastalerz, J.W. and E.J. Holcomb). Pennsylvania Flower Growers, University Park, Pennsylvania, USA.

Styer, R.C. 1997. Diagnosing fall pansy problems. *Greenh. Product News*, **7**(8): 38-40.

Wieland, C.E., J.E. Barrett, C.A. Bartucka, D.G. Clark and T.A. Nell, 1997. Growth regulator effects on development of three bedding plant plugs (abstr.). *HortSci.*, **32**: 508-509.

COLOUR PLATES

Aglaonema 'Abidjan'

Aglaonema 'Amelia'

Aglaonema 'B.J. Freeman'
(syn.'Cecelia' or 'Gabrielle')

Aglaonema 'Black Lance'

Aglaonema 'Brilliant'

Aglaonema 'Cory'

Page: 2

Aglaonema 'Deborah'

Aglaonema 'Emerald Star'

Aglaonema 'Emerald Bay'

Page: 3

Aglaonema "Golden Bay"

Aglaonema "Green Lady"

Aglaonema "Illumination"

Page: 4

Aglaonema 'Jewel of India'

Aglaonema 'Silver Queen'

Aglaonema 'Silver Frost'

Page: 4

Aglaonema 'White Rain'

Aglaonema 'Stripes'

Aglaonema 'White Lance'

Page: 4

Various Variegated and Non-variegated Bougainvillea Varieties. (Page 118)

Page: 138

Page: 139

Page: 140

Aechmea fasciata
Photo: Blossom World Bromeliads

Billbergia nutans
Photo: Michael Andreas

Billbergia pyramidalis
Photo: Dorothy Berg

Page: 146

Page: 157

Page: 158

Page: 160

Page: 163

Page: 163

Codiaeum variegatum

***C.v.* var. 'Blume'**

Red Croton

Croton Planting

Page: 251

Euphorbia leucocephala

E. pulcherrima

Poinsettia (Red)

Potted Euphorbia

Page: 269

Rumohra adiantiformis

Nephrolepis exaltata

Adiantum tenerum

Asplenium nidus

Cyrtomium falcatum

Davallia trichomanoides

Page: 295

Dicksonia antarctica

Lygodium japonicum

Pellea rotundifolia

Phyllitis scolopendrium

Platycerium bifurcatum

Selaginella species

Page: 305

Ficus elastica

F. elastica 'Doescheri'

F. elastica 'Rubra'

F. elastica 'Variegata'

Page: 313

Single

Semi-Double

Double

Triphylla

Page: 467

Single Flower (Close up)

Single Twig Bearing Flowers

H.S. 'Citrina'

Holmskioldia sanguinea

Banksia

Dryandra

Hackea

Leucadendron

Page: 507

Aulax

Grevillea

Isopogon

Protea

Leucospermum

Mimetes

Serruria

Telopea

Page: 636

Power Petite

Emerald Star

High Five

Milkyway

Sensation

Supreme

Sweet Chico

Sweet Dario

Page: 661

Sweet Lauretta Sweet Pablo

Page: 661

Clustered Offshoots in Spathiphyllum. Division of Plant for Offshoot Propagation.

Page: 661

Glossary

Abaxial. Of the surface or part of a lateral organ facing away from the axis and towards the plant base.

Acaulescent. Stem of a plant being absent or appears to be absent, being very short or subterranean.

Achene. A one-seeded dry and indehiscent small fruit (nut) with a tight thin pericarp as in certain members of Ranunculaceae.

Acropetalous. Leaves or flowers that grow in order from the base of a plant or stem towards the apex.

Actinomorphic. Regular flowers with radial symmetry, developing equally on all sides.

Acute. Having a short sharp point.

Acuminate. Where the tips or the base of leaves and perianth segments taper gradually to a point.

Adaxial. Surface or a part of a lateral organ facing towards the axis and apex.

Adnate. Being attached by its whole length to the face of an organ.

Adpressed (appressed). Hairs (indumentum), leaves, *etc.* lying flat or close to the stem or leaves to which these are attached.

Adventitious. Originating from an uncommon place as viviparously produced plantlets, the roots from the stem or leaf axil than the radicles, or the buds originating along the stem than at leaf axils.

Alternate. Arrangement of branches on a plant or leaves on the stem alternatively to each other at different heights.

Amplexicaul. Leaf sessile at the base as in case of *Tradescantia.*

Androecium. The stamen or stamens (the male component) of a flower as a whole.

Angular. The laterally projecting angles as in case of longitudinal ridging and angular stems.

Annual. A plant completing its entire life cycle (seed to seed *i.e.* germination, flowering and death) within one season or in a year.

Anther. The pollen-bearing organ of the stamen.

Appendix. The long and narrowed development of the spadix in Araceae.

Apiculate. Of leaf tips which terminate abruptly in a short, often firm and sometimes sharp point.

Aquatic. A plant growing naturally in water entirely or submerged.

Arborescent. Tree-like.

Arboretum. A man-made garden of planted trees as a museum or collection.

Areole. A small pin-cushion like sunken area on the body of cacti functioning as a point for throwing out spines, producing flowers and working as a generating organ.

Aril. Fleshy layer covering the seed of a plant.

Arillate. Fleshy appendage of the hylum or funiculus enveloping the testa of seed entirely or partially.

Aristate. Bearing bristles on awns, often appearing as if bearded.

Articulate. Jointed.

Asymmetrical. Informal (irregular) or lop-sided, usually of leaves which have one half larger than the other.

Attenuate. Gradually long and tapering, as in case of a leaf.

Auriculate. Eared as in case of leaves of some plant species in which leaf has two basal lobes.

Awl-shaped. Appearing an awl, a short and broad needle-like tool usually flattened on one face.

Axillary. Arising from an axil (a junction point of stem and leaf).

Baccate. Berry-like.

Back bulb. A pseudobulb in case of an orchid, produced at the back of a flowering stem and one which has borne a flower in the previous season.

Basipetalous. In case of inflorescence where flowers open from apex to base

Beard. Referring to a tuft or zone of hairs on or within a flower as in bearded irises, the tuft or line of hairs on the falls.

Bending. A method used for increasing the yield of flowers and/or fruits through arching or bending of prospective young branches.

Berry. A fleshy and pulpy many seeded indehiscent fruit with a soft covering.

Bidentate. The apex or margins with two teeth, the teeth may also be toothed.

Biennial. The plants that grow and die (seed to seed) within two seasons, producing leaves only in the first season or year, flowers and seed the next.

Bilabiate. Two-lipped.

Bipinnate. Compound leaves with both the primary and secondary divisions pinnated.

Bipinnatisect. Of leaves, bracts or stipules which are twice or doubly pinnatisect, *i.e.* with leaf lobes which are again lobed.

Bisexual. Flowers having both the sexes, *i.e.* stamens and pistils.

Blade. The extended portion of a leaf.

Bloom. A flower or a thin layer of white waxy powder on certain plants on their stems, leaves, flowers and fruits.

Bract. A modified leaf usually associated with an inflorescence, in the axils of which flowers arise. Some bracts are scale-like and insignificant, others are large and coloured as in *Bougainvillea* and *Euphorbia pulcherrima.*

Bracteole. A little bract.

Break bud. A flower bud which eventually appears at the end of the solitary main stem if left to grow in natural manner, but before sending out side growths. Normally such buds shrivel and do not develop into flowers.

Bud. A shoot in embryonic form containing within it a miniature stem, leaves and sometimes flowers. Also it is applied to a flower before it opens.

Bulb. A modified usually subterranean bud (growing point) consisting of a short thickened and solid stem surrounded by one or many tightly overlapping fleshy scales (modified leaves) attached to the basal plate (stem) enclosing the growing point.

Bulbil. Diminutive bulbs being formed in the axils of the leaves or in the inflorescences.

Bulblet. Small bulbs being formed by parent bulb underground directly around it.

Burr. Fruits or seeds with hooked or barbed hairs or bristles which cling to animal fur or feathers and are thus dispersed well away from the parent plant.

Caducous. Abscising or falling very early.

Callus (pl. calli). In orchid a ridge-, bump- or club-shape thickening in the labellum.

Calyx. The external whorl of floral organs composed of sepals, mostly protect the petals but sometimes become petaloid.

Campanulate. A broad tube of corolla extending in a flared limb or lobes forming bell-shaped structure.

Candelabra. Flowers of an inflorescence arranged in tiers up the stem.

Capillary. Hair-like.

Capitate. A flower head composed of small flowers arranged in dense cluster as in Asteraceae, or terminating in a knob or in somewhat spherical tip.

Capitulum. A flower head that looks like a large single flower but consists of numerous sessile or subsessile small flowers clustered together on a disc.

Capsule. A fruit becoming dry after ripening and opening through one or more vertical valves.

Carnivorous (insectivorous). A group of plants which catch insects and other small animals, such as *Darlingtonia, Dionaea, Nepenthes, Pinguicula* and *Sarracenia.*

Carpel. The floral organ (female sporophyll or a simple pistil) bearing ovules.

Cataphyll. A reduced or little developed leaf, bract-like and often sheathing emerging shoots.

Catkin. Slender inflorescence of small and unisexual, usually wind-pollinated flowers.

Caudate. The apex of a leaf or a perianth-segment tapering gradually to form a long tail-like appendage.

Caudex. A trunk of a tree that bears leaves only at its apex as in palm and tree ferns, or the swollen stem base of certain perennial plants from where new growths appear.

Cauliflorus. Flowers or inflorescences being produced directly from the trunks and larger branches of the trees.

Cauline. Arising from the stem.

Cephalium. The area of stem in cacti, often at or near the apex from which the flowers emerge year after year from a dense mass of woolly hairs or bristles as in case of *Espostoa* and Melocactus.

Channelled (candiculate). A narrow leaf or leaflet with up-turned margins and forming a channel or gutter-shape.

Ciliate. Fringed with hairs.

Cladode. A green stem or branch that has taken over the major part of a plant's photosynthetic function.

Clavate. Thickening to the apex from a tapered base. Club or basketball bat shape.

Cleistogamous. Self-pollination occurring in the closed flower.

Climber. A plant not having strong stem to grow erect and requires support to grow straight in terms of twining stems, tendrils, hooked thorns, clinging roots and the leaves with reflexed stalks.

Column. The central part of an orchid flower which combines ovary, stigma and pollinia.

Compound. A division into two or more subsidiary orders.

Connate. Joined together as in case of leaf bases or sometimes flowers.

Conservatory. A structure composed partly or entirely of glass attached to the house and within which a large number of plants are grown.

Contractile. These are roots contracting the plants deeper into the soil akin to their natural habitat.

Cordate. The heart-shaped leaves and leafy stipules which have rounded base and a deep basal sinus or notch where the petiole is inserted.

Coriaceous. Leathery and tough but smooth in texture.

Corm. An annual solid and swollen part of a stem which is bulb-like underground structure having growing point at the apex and the basal plate below which forms the roots, and all around the body it has rings as scars of the previous season's leaves. Every ring has a prominent node just opposite the buds of its adjoining rings and many invisible buds. Every part of the corm has regenerating capacity. Entire corm is covered with shredded tunic as dry leaf bases.

Cormel (Cormlets). A small corm developing from and around the new developing (daughter) corm which normally has a very tough and intact tunic. It is as senile as the corm and can develop the flowers similar to a corm under optimum conditions.

Cormule. Similar and smaller than corm but normally larger than the cormels, with shredded tunic, developing 1-4 laterally or terminally at the first nodes but on the cost of spike. No further spike development in case it is terminal.

Corolla. The collective name for petals and if there are two whorls the inner one.

Corona. An extra cup-like appendage or crown as a development of the perianth (*Narcissus*), of the staminal circle (*Asclepias*), or located between corolla and the stamens (*Passiflora*).

Corymb. An indeterminate flat-topped or convex inflorescence where outer flowers open first.

Costa-palmate. The leaves being halfway between palmate and pinnate as in case of palms.

Cotyledon. The primary or seed leaf appearing immediately after germination.

Creeper. A plant not having strong trunk for growing erect so spreads horizontally prostrate over the ground or on a wall or any other object, usually throwing out clinging aerial roots or adhesive pads.

Crenate. Margins of leaf blades and other flattened organs with blunt and rounded teeth.

Crest. A ridge formed on the falls as in *Iris*, which is most commonly orange or yellow in colour.

Crisped. Of hairs, leaves or petal margins being wavy or waved.

Cristate. Crested or less commonly 'crest-like' or ruffled.

Crown. A corona, or the branching pattern, the foliage and the shape (spreading, conical, dome, *etc.*) of a tree or a shrub, or a part of herbaceous plant where the roots (or underground structures) and the aerial parts meet, or a plane (*i.e.* the point on rhizome or root tubers) having strong terminal bud for propagation.

Crownshaft. The sheathing frond bases wrapped around one another to form a smooth, usually green cylindrical shape with an apparent continuation of the trunk as in case of *Roystonia regia*.

Cultivar. A cultivated variety that has originated and persisted under cultivation.

Cuneate. Wedge-shaped or inversely triangular.

Cuspidate. Apices terminating abruptly in a sharp inflexible point.

Cutting. An unrooted part of stem, leaf or root of a plant taken for producing similar plant.

Cyme. More or less flat-topped and determinate inflorescence being formed of main and secondary stem where the central or terminal flower opens first.

Dealbate. Covered with a white powder.

Deccusate. At right angles to one another.

Deciduous. Of leaves, not persistent or evergreen, and generally fall in the autumn.

Decumbent. Having prostrate stems with tips erect.

Deliquescent. Appearing to melt or become fluid as in petals of *Commelina*.

Dehiscent. The fruit, pollen or spore which spontaneously open when ripe.

Dentate. Margins of leaf blades and other flattened organs having shallow and equilateral triangular teeth.

Denticulate. Having many very small teeth.

Determinate. Inflorescences as in cymes that terminate in a bud where terminal or central flower opens first.

Dichotomous. A method of branching in which each axis bifurcates at the tip.

Dicotyledon. A plant having two primary seed-leaves when germination takes place.

Didynamous. Containing two pairs of stamens, one pair shorter than the other.

Digitate. A compound leaf palmately arranged in which the leaflets arise from the same point at the apex of the petiole

Dioecious. Indicating male and female flowers on separate plants.

Diploid. Plants having two matching sets of chromosomes.

Diplont. An organism whose cells (other than reproductive) have a diploid number of chromosomes in their nuclei.

Diplontic selection (extra somatic selection). In case of mutation in plants when once the exhibited mutation does not appear again in the successive generation(s) as mutated (abnormal) tissues are not allowed to develop further due to suppression by the surrounding normal tissues, and which results in the loss of the mutation, it is known as *diplontic selection*.

Disc (Disk). The central part of the flower head (capitulum) of a composite plant made up of many tubular flowers, or a fleshy (or raised) development of the torus within the calyx or within the corolla and stamens or composed of coalesced nectarines or staminodes and surrounding the pistil, or the basal plate (condensed stem) of a bulb or a corm.

Discoid. Leaves with round fleshy blades and thickened margins, or the capitula of Asteraceae formed entirely of disc florets.

Dissected. Divided into numerous segments.

Distichous. Arrangement of leaves on stems or its branches in two opposite ranks.

Dormancy. The state of cessation of growth of a plant during adverse environmental conditions, usually during winter.

Dorsal. The back of an organ or the surface turned away from the axis *i.e.* abaxial.

Drupe. A fruit consisting of a more or less succulent pericarp which encloses a single many-celled stone.

Ellipsoid. Midway between oblong and ovate with equally narrowed or rounded ends and widest across the middle but 3-dimensional.

Elliptic. Midway between oblong-ovate but with equally narrowed or rounded ends and widest across the middle, i.e oval but narrowing towards each of the rounded ends.

Emarginate. Apex shallowly notched with acute indentation, *i.e.* petals or leaves having a notched tip.

Embryo. It is part of the seed after fertilization but before germination and this develops into a new plant, *i.e.* the rudimentary plant within the seed.

Endemic. The natural haunts being confined only to a particular region and not elsewhere.

Endosperm. The tissue surrounding the embryo in a plant seed to nourish the embryo while germinating.

Ensiform. Sword-shaped and with an acute point, especially the foliage.

Entire. The leaf margins not toothed.

Epicalyx. A ring of bracts just below the calyx or sepals and resembling them.

Epigynous. The sepals, petals and stamens growing on top of the ovary or appearing so.

Epipetalous. Borne upon the petals of a flower.

Epiphyte. Plants that grow habitually in the wild on the tree trunk or its branches, poles, telephone wires, or rock, *etc.* but are not parasitic.

Equitant. The conduplicate leaves overlapping inside each other in two ranks in a strongly compressed fan as in *Iris* or *Gladiolus* and in other members of Iridaceae.

Evergreen. A plant that remains in leafy conditions through all the weathers.

Exotic. An introduced plant not native to that area.

Exserted. Projecting or extending beyond the organs surrounding it as stamens coming out of the corolla.

Eye. The centre of a flower, particularly when of a contrasting colour.

F_1, F_2. Symbols used to indicate first or second generations from a deliberate cross.

Falcate. Curved and tapering to a point like a sickle.

Falls. Outer parts of the flowers which seem to fall downwards or outwards, often bearded, as in *Iris*.

Family. A plant or a group of plants having certain common botanical characteristics, especially in the flowers, and contain a number of genera.

Farina. The mealy or powdery coating on stems, leaves and sometimes on flowers of certain plants.

Farinaceous (Farinose). Having a mealy or granular texture.

Fasciated. The malformation with flattened stems looking as if many stems have joined together as in *Amaranthus* and *Celosia.*

Fascicle. A cluster of roots, stems, leaves, racemes or flowers though arising always independent but appear as if coming out from one common point.

Fastigiate. Of erect habit with all or most stems steeply ascending, creating an overall shape taller than wide.

Filament. The thread-like stalk of a stamen.

Filiform. Thread-like long and very slender leaves, branches, *etc.* and rounded in cross-section.

Fimbriate. Bordered with a fringe of slender processes usually derived from the lamina rather than attached as hairs.

First crown bud. The first bud which appears at the end of a lateral growth from the main stem of the plant.

Flabellate. A fanlike structure.

Flore-pleno. Double flowered, usually with stamens and sometimes with carpels transformed to petals or petaloids.

Floret. A small individual flower, generally one of a group or a cluster.

Flower. A part of the plant containing reproductive organs. It may either be unisexual or bisexual.

Flowerhead (capitulum). A dense cluster of flowers of more or less regular size.

Foliaceous. Leaf-like in appearance or structure.

Form. A variant of a wild species not so distinctive to be called a subspecies or a variety.

Frond. An usually large, much divided or compound leaf which appears consisting of many leaves such as in case of ferns and palms.

Fruit. The seed-bearing organ of a plant.

Frutescent. Shrubby or becoming so.

Fruticose. Though without a single main trunk, but bearing stems or branches similar to shrubs.

Fugacious. Withering soon.

Fusiform. Spindle-shaped.

Funnel-form. Flowers shaped like the inverted cone or funnel where petals may either be fused or overlapping and usually joined into a short tube at the base.

Fusiform. Swollen in the middle and tapering at both the ends *i.e.* spindle-shaped.

Gamopetalous. Petals united by their margins to form a tubelike corolla.

Gamosepalous. Having sepals united by their margins.

Genotype. The underlying genetic make up of an individual plant or clone.

Genus. The next upper rank above species in botanical classification.

Geophyte. A plant growing with stem or bulb (bulbs, corms, rhizomes and tubers) below ground.

Gesneriad. A plant belonging to African violet (Gesneriaceae) family. The most popular gesneriads are *Columnea, Gloxinia, Saintpaulia* and *Streptocarpus.*

Glabrous. Smooth, without hairs.

Gland. Usually the base of tepals secrete normally a sweet substance to attract insects for pollination.

Glaucous. Coated with a fine bloom mostly grey or blue-green in colour being formed by the waxy coating of the foliage, which may easily be rubbed off.

Globose (Globular). Spherical or near spherical in shape.

Glochid. Tiny barbed spine or bristle growing from areole in cactus.

Grex. A name applied to a group of individual plants, sometimes difficult to separate individually, as they differ slightly or much from each other but with common origin. Usually applied to a batch of hybrid plants of the same parentage, particularly orchids.

Gynoecium. Female organ of a flower, the carpels collectively.

Gynophore. Stalk of a pistil which raises it above the receptacle.

Gynostegium. A covering of any kind of gynoecium, a perianth.

Habitat. The district or geographical area where a plant is found in the wild.

Haft. The narrow basal portion of the 'falls' as in *Iris*.

Halophyte. A plant tolerant or adapted to saline soil.

Half-hardy. A plant requiring a minimum temperature of 10-13 °C for healthy growth.

Hardy. A plant withstanding prolong exposure to temperature at or below 10 °C.

Hastate (Sagittate). Arrow-shaped, in triangular fashion, both downward lateral points and one upward.

Heel. A strip of bark and wood remaining at the base of side shoot cutting pulled of a main shoot.

Helophyte. A plant growing seasonally or permanently in mud.

Herbaceous. A plant with no persistent woody stem above ground though it may be evergreen or deciduous.

Hermaphrodite. Flowers having in them both types of sex organs, *viz.* stamens and pistils.

Heterogamous. In Asteraceae where two types of flowers (ray and disc) are borne, or where flowers bear abnormally arranged sexual organs, or the functions being transferred from one sex to another.

Heteromorphic. Where plants assume different forms at different stages of the life cycle, which is opposite to polymorphic where different forms occur at the same stage of plant development.

Heterophyllous. Where a plant bears two or more forms of leaves on the same plant at the same or at different times.

Heterostylous. Species where flowers differ in the presence or number of styles, some being unisexual while some others hermaphrodite.

Hirsute. Covered with long, coarse, spreading hairs.

Hispid. Covered with bristle-like hairs.

Hoary. Leaves and stems being covered with very short, usually branched or star-shaped white hairs.

Homogamous. Either the flowers are hermaphrodite or of the same sex.

Hyaline. The margins of leaves or bracts being transparent or translucent.

Hybrid. The offspring of crosses involving two plants of the same variety or race, or two different varieties, species or even genera.

Hydrophyte. The plants growing in water.

Hypogynous. All or any of the floral organs situated below the ovary.

Hysteranthous. Where leaves develop after the flowers.

Imbricate. Leaves and bracts overlapping in a regular pattern, sometimes even encircling the axis.

Imparipinnate. Pinnate leaf with an odd terminal leaflet.

Imperfect. When flowers become unisexual due to certain parts being though present but are undeveloped.

Incised. Having the leaves or petals deeply cut at the margins, generally in an irregular manner.

Incomplete. Where a flower lacks one or more of the four organs.

Incurved. Inwards or upwards curving nature of leaves and petals.

Indehiscent. Fruit or seed pod and spores splitting self for dispersal of seeds or spores.

Indeterminate. An inflorescence not terminating by a single flower but bears many flowers where opening starts from the bottom.

Indumentum. A covering of hairs, scales or scurfs, used in the sense of hairs as in *Anigozanthos*.

Indusium. The membranous covering over the sporangia of ferns which gives each sorus its particular shape.

Inflated. Bladdery, frequently used to describe a type of fruit or seed vessel.

Inflorescence. It refers to the spike of individual flowers on a stem or a branch. This may be a corymb, cyme, panicle, raceme or spike.

Internode. The part of a stem between two nodes.

Introrse. Facing inward or towards the centre.

Involucre. A single, a pair or a ring (whorl or whorls) of small bracts or leaves subtending a flower or inflorescence.

Involute. Rolling inward towards the uppermost side.

Irregular. Asymmetrical or zygomorphic.

Juvenile. Sexually immature phase of a plant's life.

Keel. A prominent ridge running longitudinally down the centre of the under-surface of a leaf, petiole, bract, petal or sepal as a longitudinal ridge on the foliage of *Kniphofia*.

Keiki. An offshoot or offset of certain plants as in orchids.

Labellum. The lowest of three petals in the orchid flower commonly known as lip. It is modified in a variety of ways to aid insect-pollination.

Labiate. The calyx or corolla possessing a lip or lips.

Labium. One of the lip-like divisions of the calyx or caorolla.

Laciniate. An irregular but finely cut as if slashed.

Laevigate. Appearing smoothly polished.

Lamina. Leaf blade, the extended part of a leaf.

Lanate. Covered with thick woolly hairs.

Lanceolate. Lance-shaped tapering towards the apex, three to six times as long as broad, broadest point before the middle.

Lateral. Shoots arising on the sides of main or leading stems.

Leggy. Abnormally tall and spindly growth.

Ligule. The strap-shaped petals or leaves.

Linear. Slender and elongated with margins parallel or nearly so.

Lip. A staminode or petal differentiated from the others, or in Labiatae, one of the two distinct corolla divisions, the upper often hooded and the lower often forming a flattened platform for landing of pollinators. In orchids, it is labellum.

Lithophyte. A plant growing on rocks or stony soil by deriving nourishments from the atmosphere.

Littoral. Growing on the sea-shore.

Lobe. Divided usually into rounded segments and separated by adjacent lobes with sinuses as in petals and leaves.

Loculicidal. Longitudinal and dorsal splitting through a capsule wall into the capsule.

Long day plant. A plant which requires light for a longer period than it would normally receive from daylight in order to induce flowering, such as *Saintpaulia*.

Lorate. Strap-shaped.

Lyrate. Pinnatifid with a large rounded terminal lobe.

Membranous. Thin, soft, more or less translucent.

Mericarp. A one-seeded carpel where one of a pair splits apart at maturity from a syncarpous or schizocarpous ovary.

Meristem. Tissues yet not differenatiated but capable of developing into organs.

Mesophyte. The vast majority of plants which are midway to hydrophytes and xerophytes.

Midrib. The thicker central vein of a leaf.

Monadelphous. Stamens fused by their filaments into a single bundle.

Moniliform. A cylindrical or terete organ with regular constrictions to appear as string of beads.

Monocarpic. Fruiting or flowering only once after several years of growth and then dying as in case of bromeliads and sempervivums. Technically annuals and biennials are also monocarpic, but horticulturally the term is used for plants which live more than two years before flowering as in case of some *Agave* species.

Monocotyledon. Having only one cotyledon or seed leaf at germination.

Monoecious. Male and female flowers being borne on the same plant, with one flower of one sex.

Monopetalous. Only with one petal.

Monopodial. A stem or rhizome growing continuously year after year from the apical or terminal bud, with little or no secondary branching.

Monotypic. It describes a taxon, in particular a genus, family, order, class or division that has only one member at its next lower level.

Mucronate. An apex terminating suddenly with an abrupt spine or spur developed from the midrib.

Mutation. A sudden change in the genetic make-up of a plant, leading to a new feature which may be inherited.

Natural break. Sending out of side growths from the main stem after appearance of the break bud at the end of the main stem which has yet not been removed.

Natural 1st crown. It is, in fact, 1st crown bud but is termed so when the plant has made a natural break

– as distinct from a 1st crown bud which appears on a plant that has been stopped.

Nectary. Often a projected or depressed gland which secretes or sometimes absorbs the nectar.

Netted. Net-veined or reticulate.

Node. The part of stem where the leaf is joined and a lateral shoot grows out.

Obconical. Of a leaf, roughly conical but with the stalk at the narrow end.

Obcordate. Like to that of cordate but with the sinus at the apex rather than at the base.

Oblanceolate. Similar to lanceolate but with the broadest part above the middle and tapering to the base rather than toward apex.

Oblong. Linear and 2-3 times as long as broad with almost parallel sides terminating obtusely at both ends.

Obovate. Ovate but broadest at half past so narrow at the base.

Obovoid. As ovoid but broadest before the half.

Obtuse. Apex or base terminating gradually in a blunt or rounded end.

Odd-pinnate. A pinnate leaf with odd number of leaflets, though also implying that there is a single terminal leaflet.

Offset. A lateral growth at base of the plant used for propagation.

Oleiferous. Those plants which have leaves and seeds producing oil.

Opposite. Of leaves and flowers being borne on different sides of axis with bases on the same level.

Orbicular. Circular or nearly so.

Ovary. Usually the swollen base of a pistil containing one or more ovules, which after fertilization become the seeds.

Ovate. Egg-shaped, 1.5-2.0 times as long as broad, broadest before half past and rounded at both ends.

Over-potting. Repotting a plant into a pot which is too large to allow successful establishment.

Ovoid. Egg-shaped ovate but 3-dimentional.

Ovule. One or more minute roundish structures occurring within the chamber of the ovary, and each such structure contains an egg cell which after fertilization develops into a seed.

Palmate. A leaf with several lobes attached to the stalk like the fingers of a hand.

Palmitifid. Palmate with the sinuses reaching about half-way down, *i.e.* rounded leaves lobed in a palmate fashion to half their length.

Palmatisect. Rounded leaves palmately lobed almost to the base.

Panicle. Inflorescence in the form of a branched or compound raceme.

Papillae (papillose). Small, usually soft nipple-like protubrarances on a leaf, stem or fruit.

Pappus. A tuft or whorls of delicate bristles or scales on the place of calyx as in certain Asteraceous plants.

Paripinnate. Pinnate leaf with an even number of leaflets.

Parthenocarpic. Fruiting without fertilization.

Pathenogenesis. Development of seed without fertilization.

Peat. Partially decomposed sphagnum moss or sedge used in making composts. It is valuable for its pronounced air and water-holding capacity and its freedom from weeds and disease organisms.

Pedate. A palmatisect leaf with atleast the two basal lobes again lobed.

Pedicel. The stalk supporting an individual flower or fruit.

Peduncle. The stalk of entire inflorescence.

Peltate. A leaf whose stalk instead of being attached at the edge is attached inside its margin normally at the centre below as in *Nelumbium*.

Pendent. Hanging downwards.

Pentamerous. Organs in a group of five or multiples thereof.

Perennate. The survival from year to year by overwintering.

Perennial. Any plant that lives from several to many years, but in the strict sense applied to non- woody species which produce new stems annually from at or near ground level.

Perfoliate. A leaf whose base clasps the stem as in *Papaver somniferum* or paired leaves which fuse around the stem.

Perianth. The calyx and corolla together.

Perigynous. The flowers in which the perianth and the stamens are basally united and borne on the margins of a cup-shaped rim, itself being borne on the receptacle of a superior ovary, neither above nor below it.

Persistent. An organ neither withering nor falling.

Petal. A part of flower which forms the corolla.

Petaloid. Various organs of a flower adapted to give the appearance of a petal.

Petiole. The stalk of a leaf.

Photosynthesis. The food-making process which occurs usually in the leaf. This process requires water, air and adequate light.

Phylloclad. A flattened stem or branch to function as a leaf.

Phylloclade. Another name for cladode.

Phyllode. An expanded petiole functioning as a leaf blade.

Picotee. Of petals with a narrow margin having a contrasting colour.

Pilose. Covered with long ascending hairs.

Pinching. The act of removing the tip of a growth, either of the main stem or/and of a side growth.

Pinna (pl. pinnae).One of the primary divisions of a pinnate leaf (leaflet).

Pinnate. In compound leaves the arrangement of pinnae (leaflets) in two rows along the rachis. In bipinnate leaves the pinnae are divided into rachillae and petiolules bearing pinnules (leaflets). Applied also to feather-like arranged veins.

Pinnatifid. Deeply lobed to about half-way down or more with the lobes pinnately arranged.

Pinnatisect. A leaf divided pinnately to the midrib or almost so, but not rounded off into leaflets.

Pinnule. The secondary or more complex division of a fern or palm frond.

Pips. Corms, rhizomes or tubers obtained by ripening-off seedling plants a few months after flowering as in case of anemone and convallaria.

Pistil. Female seed-bearing organ made up of ovary, style and stigma, *i.e.* gynoecium.

Pitcher plants. Species of carnivorous plants in which the trapping mechanism is of pitcher-like form, with a pool of digestive liquid in the bottom, such as *Cephalotus, Heliamphora, Darlingtonia, Napenthes* and *Sarracenia.*

Plicate. Pleated or folded lengthwise, mainly of leaves.

Plumose. Feather-like with long fine hairs which further have fine secondary hairs.

Pollen. The male spores which are developed in the pollen sacs or loculi of anthers.

Pollinia. A group of waxy pollen grains found in orchids and in the members of Asclepiadaceae.

Polygamous. A species bearing both unisexual and bisexual flowers on the same or on different plants.

Pot-bound. A plant growing in a pot which is too small to allow proper leaf and stem growth.

Potting on. The repotting of a plant into a proper-sized larger container which will allow continued root development.

Pricking out. The moving of the seedlings from nursery, tray or pot in which they were sown to other receptacles where they can be spaced out individually.

Prickle. A sharp-pointed broad-based outgrowth of a stem, leaf or other organ as in roses, as opposed to a thorn or spine.

Priming: Prior to sowing, it is the preparation of seeds to enhance the partial water inbibitions to help germination and establishment as in case of petunia. Polyethyline glycol 600 is quite effective substance used to activate processes leading to germination and then seeds are stored dry. Such seeds germinate quite even and more quickly after sowing.

Procumbent. Loosely trailing on ground without rooting.

Prophyll. The bract-like leaves subtending the entire plant or inflorescence in monocots as in *Gladiolus.*

Prostrate. Lying flat on the ground.

Protandrous. In a flower where mature anthers release the pollen before stigma receptivity, but when it continues even after the stigma receptivity it is known as incomplete protandy which enables self-pollination.

Prothallus. A tiny leaf-like body which develops from the spore of a fern, on which are borne the sex cells or gametes which upon fusion give rise to a new fern plant.

Pseudobulb. The fleshy bulb-like thickened aerial stem found in many epiphytic orchids.

Pseudostem. The aerial stem composed of only overlapping sheaths and stalks of basal leaves.

Pubescent. Hairy. When an organ is covered with silky or downy hairs.

Punctate. Dotted with minute, translucent impressions, pits or dark spots.

Pyramidal. Conical in shape but with more angular sides.

Raceme. An indeterminate and unbranched inflorescence composed of pedicellate flowers. Here each flower is borne on an individual stalk along a central stem.

Rachis. The axis of a compound leaf or of an inflorescence.

Radical. The leaves rising directly from the roots.

Ray floret. A small flower with a tubular corolla having expanded and flattened limb in a strap- like ligule (blade), normally occupying the peripheral rings in a radiate capitulum as in Asteraceae.

Receptacle. The elongated and either flat, concave or convex end of the stem from which the perigynous zone (floral parts) originates.

Reniform. Kidney-shaped.

Repent (repens)**.** Growing flat on the ground, usually of stems that root at intervals.

Reticulate. Netted like a fish-net. Network of veins, ribs and colouring. Reticulate tunic is composed of a lattice of fibres in certain bulbs and corms.

Retrorse. A minor organ attached to a larger part such as prickle on a stem, a barb on a leaf or a callus on a lip, and so on, and turned, curved or bent backwards or downwards, away from the apex.

Revolute. Margins rolling towards the dorsal surface.

Rhizome. A specialized slender or swollen, simple or branched subterranean (sometimes close to soil surface, but exposed in epiphytes) and normally creeping stem which produces roots, stems, leaves and inflorescences along its length and at its apex.

Rhomboid. In case of leaves and tepals being angularly oval, the base and apex with acute angles and both the sides having obtuse angles.

Rootstock. The base of a stem from which the roots emerge but in graftage the rootstock refers to the locally adapted plants on whose rooted stems the desired variety is grafted or budded.

Rosette. Leaves clustered together in circular pattern.

Rosulate. Bearing a rosette.

Rugose. Wrinkled, puckered or corrugated appearance of leaves.

Runner. A slender, prostrate stem bearing plantlets at the well-spaced nodes.

Sagittate. Spear- or arrow-shaped leaves where triangular basal lobes point towards the stalk or downward.

Salverform. Where corolla have long and slim tube with an abruptly expanded flattened limb.

Scabrous (scabrid)**.** Rough, usually with short and hard hairs.

Scale leaves. Specialized leaves which are fleshy and scale-like and cover the buds or make the bulbs.

Scandent. Climbing or scrambling stems.

Scape. A leafless erect stalk bearing a terminal inflorescence or flower.

Schizocarp. A dry fruit splitting at maturity into two or more one-seeded carpels which remain closed.

Second crown bud. Removal (stopping) of the tips of the lateral growths (first breaks) from the main stem, forces appearance of further side growths (second breaks) in the leaf axils of these laterals, and these laterals grow longer terminating into a bud at the end, and this is termed as second crown bud.

Secund. Where all the parts of the leaves or flowers are borne along one side of the axis or appear so due to twisting of the stalk.

Segment. One of the parts of a compound leaf or the flowers where sepals and petals are similar.

Sepal. One of the separate parts of the calyx.

Sericeus. Silky-hairy.

Serrate. Leaf blades and other flattened organs having sharp saw-toothed margins curving forward.

Sessile. Without a stalk as in case of leaves amd flowers.

Short day plant. A plant which requires light for a shorter period than it would normally receive from daylight in order to induce flowering, *e.g. Dendranthema* and *Poinsettia.*

Shrub. A woody plant with a framework of branches and little or no central system.

Simple. One piece leaf, neither composed nor divided.

Sinuate. With undulated or wavy margins.

Sinus. The space between two lobes or divisions.

Solitary. Used in case of flowers when they are borne one to a stem of leaf axil.

Sorus. A group of sporangia usually of rounded or oval shape on the back of a fern frond. Each sorus may or may not be covered with an indusium.

Spadix. A spike of flowers on a thick and fleshy axis, often subtended by spathe as in Araceae.

Spathe. A modified, often papery bract or leaf enclosing the inflorescence.

Spathulate (Spatulate). An oblong and spatula-shaped organ which is narrow and tapering at base but rounded at apex.

Spicate. Borne in a spike-like inflorescence.

Spike. An indeterminate and unbranched inflorescence bearing sessile flowers.

Spikelet. A very small, usually compact spike, generally used for the basic unit of a grass inflorescence.

Spine. A hard and sharp pointed structure formed by a modified twig.

Sporangium. An asexually formed spore produced by ferns and certain fungi.

Spore. Minute reproductive bodies formed of one or a few cells together, which give rise to new individuals, either directly as in fungi or indirectly as in ferns.

Sport. A plant which shows a marked and inheritable change from its parent; a mutation.

Stalk. The petiole, peduncle, filament, stipe (stalk of the pistil), *etc.* which provide stem-like support to their respective organs.

Stamen. Anther and filament constituting the male pollen-bearing part of the flower.

Staminate. A male flower bearing stamens and no functional pistils.

Staminode. A stamen which is sterile and often enlarged and petal-like.

Standard. The perianth segments of inner whorl, often erect as in *Iris*.

Stellate. Star-shaped.

Stigma. The part of the pistil which is receptive when pollen over it is deposited.

Stipe. An alternative name for the leaf stalks of ferns. Also used for the stems of mushrooms and toadstools.

Stipule. A bract-like or leafy appendage, often in pairs, at the base of a petiole, shedding soon.

Stolon. An underground trailing stem giving rise to plantlets at its nodes and apex or cormlets and bulblets at its simple or branched apices.

Stool. It is the root of an old plant with a portion of the old stem and its surrounding young shoots.

Stopping. The act of removing the tip of a growth, either of the main stem and/or the side one(s).

Stove plant. A plant which requires warm greenhouse conditions in winter.

Strain. A selection of a variety, cultivar or species which is raised from seed.

Striate. Fine parallel lines or stripes on the surface.

Style. The area of ovary between stigma and ovary.

Sub-shrub. A plant which at the adult stage has woody stems near the base and a framework of soft green growth.

Succulent. Fleshy, juicy as in cacti and many other succulents.

Sucker. A rooted shoot developing from an underground stem or the plant's root.

Sympodial. In growth of a plant where terminal bud terminates into an inflorescence or dies, and successive growth starts in the ensuing season through secondary axes from lateral buds.

Synanthous. Where leaves appear with the flowers.

Syncarpous. Where two or more carpels fuse in a compound pistil.

Synflorescence. A compound inflorescence which possesses a terminal inflorescence and lateral inflorescences.

Syngenecious. Stamens united into a cylindrical form by the anthers.

Taking a bud (securing a bud). The act of removing all leaf axil shoots and unwanted buds except the bud which is to develop into a flower.

Tender. A plant which requires a minimum temperature of 15.6 °C. Occasional short exposure to temperatures below this level may be tolerated.

Tendril. A slender twining organ that enables a plant to climb.

Tepal. An unit of undifferentiated perianth which can not be distinguished as a petal or a sepal.

Terete. Slender and cylindrical, circular in cross-section.

Terminal bud. It is a bud which develops surrounded by other flower buds, and this marks end of further vegetative growth.

Ternate. Usually of leaves, in whorls or clusters of three.

Terrarium. A partly or entirely closed glass container used to house a collection of indoor plants.

Terrestrial. Growing on the ground.

Tessellated. Chequered or marked with a grid of small squares.

Tetramerous. With parts in fours or group in fours.

Throat. The opening between the tube and the limb of the tubular part of calyx, corolla or the perianth.

Thyrse. A panicle which is broadest in the middle and tapers to base and apex.

To break. The branching or sending out of side growths.

Tomentose. A downy covering of tiny, wooly and stiff hairs on leaves and other plant parts.

Topiary. Art of clipping and training woody plants to form geometric shapes or intricate patterns.

Toothed. Dentated. Bearing small teeth on the edges.

Torus. A receptacle of a flower, the region in which floral parts are inserted. An elongated torus is a gynophore, the stalk of a pistil which raises it above the pistil.

Transpiration. The loss of water through the pores of the leaf.

Tree. A woody plant with a distinct woody trunk.

Trifoliate. With three leaves or leaflets.

Trimerous. Having parts in threes.

Trumpet. When the tepals (perianth segments) are fused or appearing to be fused into the flared bell of a trumpet as in *Hymenocallis, Narcissus,* etc.

Truncate. Ending abruptly or bluntly.

Tuber. A generally subterranean fleshy and swollen stem with eyes over the surface as in tuberous *Begonia,* or fleshy and swollen roots having buds (eyes) only on the crown, the part which is attached with the aerial stem as in *Dahlia.*

Tubercle. A small wart or knob-like projection on a stem, leaf, fruit, *etc.*

Tuberous. Producing tubers or resembling a tuber.

Tunic. The papery or netted covering of corms derived from widely expanded leaf bases.

Tunicate. Enclosure of a tunic, the papery covering of a bulb or a corm.

Umbel. Umbrella-type inflorescence, normally flat-topped and corymb-like where all the flowered pedicels arise from the same point at the apex of the main axis.

Uncinate. Hooked as in the spines of certain cacti.

Undulate. The wavy margins of leaves, bracts, petals, *etc.*

Unilateral. Arranged only at one side.

Unisexual. Having single-sexed flowers, though both may be on the same plant.

Urceolate. Of flowers, urn-shaped calyx or corolla with an inflated tube narrowed in the mouth but slightly broadening again at the tip.

Vaginate. Enclosed by a sheath.

Variegated. Mottled, streaked, edged or striped with colours of leaves mostly white to yellow other than the normal green.

Velamen. A multilayered corky covering of dead cells on some aerial roots as in some orchids, which enable them to absorb moisture from the atmosphere.

Ventral. The inner (adaxial) of the upper side of a leaf or other surface that faces towards the stem.

Virescens. Appearance of green pigmentation on the tissues not ordinarily green.

Vernal. Appearing in spring.

Verrucose. Covered with small wart-like outgrowths, usually on stems.

Verticil. An alternative name for whorl.

Verticulate. Arranged in a whorl or in a series of whorls.

Vesicle. A small bladder or blister-like organ.

Villous. Covered somewhat shaggily with long soft hairs.

Viscid. Sticky, glutinous.

Viviparous. The buds or bulbs which sprout and become plantlets or the seeds which germinate, while still attached to the parent plant.

Whorl. A circular arrangement of foliage or flowers round an axis, at a node, as in case of leaves of some lilies.

Woolly. For leaves and stems, covered with long soft hairs that are usually white or grey.

Xerophyte. A plant which is able to live under very dry conditions.

Zygomorphic. Capable of being divided into two equal parts in one plane (only along a vertical axis), *i.e.* bilaterally symmetrical.

Index

C

M

N

O

P

Q

R

S

T

U

V

W

X

Y

Z

9 789359 191041

Printed at Repro India Ltd.